Manufacturing Engineering and Technology

FIFTH EDITION

Serope Kalpakjian
Illinois Institute of Technology

Steven R. Schmid
The University of Notre Dame

PEARSON
Prentice Hall

Upper Saddle River, NJ 07458

Library of Congress Cataloging-in-Publication Data on File

Vice President and Editorial Director, ECS: *Marcia J. Horton*
Executive Editor: *Eric Svendsen*
Associate Editor: *Dee Bernhard*
Executive Managing Editor: *Vince O'Brien*
Managing Editor: *David A. George*
Production Editor: *Rose Kernan*
Director of Creative Services: *Paul Belfanti*
Creative Director: *Heather Scott*
Cover Designer: *John Christiana*
Art Editor: *Xiaohong Zhu*
Manufacturing Manager: *Alexis Heydt-Long*
Manufacturing Buyer: *Lisa McDowell*
Senior Marketing Manager: *Holly Stark*

About the Cover: Car assembly line image courtesy of Corbis; machining image courtesy of Kennametal (Kennametal Inc., 1600 Technology Way, Latrobe, PA USA 15650); Boeing assembly plant image courtesy of Boeing (Boeing, 100 North Riverside, Chicago, Illinois 60606); spine image courtesy of Intel (Intel Corporation, 2200 Mission College Blvd., Santa Clara, CA 95052)

© 2006 by Pearson Education, Inc.
Pearson Prentice Hall
Pearson Education, Inc.
Upper Saddle River, NJ 07458

Printed in the United States of America

10 9 8 7 6 5 4 3 2

ISBN 0-13-148965-8

Pearson Education Ltd., *London*
Pearson Education Australia Pty., *Sydney*
Pearson Education Singapore, Pte. Ltd.
Pearson Education North Asia Ltd., *Hong Kong*
Pearson Education Canada, Inc., *Toronto*
Pearson Educación de Mexico, S.A. de C.V.
Pearson Education—Japan, *Tokyo*
Pearson Education Malaysia, Pte. Ltd.
Pearson Education Inc., *Upper Saddle River, New Jersey*

To

a special sister

Mari Yegiyayan

in gratitude

and to

a beautiful daughter

Carly Petronis Schmid

CONTENTS

5   Ferrous Metals and Alloys: Production, General Properties, and Applications   149

6   Nonferrous Metals and Alloys: Production, General Properties, and Applications   169

7   Polymers: Structure, General Properties, and Applications   191

8   Ceramics, Graphite, and Diamond: Structure, General Properties, and Applications 219

9   Composite Materials: Structure, General Properties, and Applications 238

Part II: Metal-Casting Processes and Equipment 259

10   Fundamentals of Metal Casting 261

17 Processing of Metal Powders   483

18 Processing of Ceramics, Glass, and Superconductors   513

19 Forming and Shaping Plastics and Composite Materials   534

22 Cutting-Tool Materials and Cutting Fluids   647

23 Machining Processes Used to Produce Round Shapes: Turning and Hole Making   674

24 Machining Processes Used to Produce Various Shapes: Milling, Broaching, Sawing, and Filing; Gear Manufacturing   723

Part V: Fabrication of Microelectronic Devices and Micromanufacturing   865

28 Fabrication of Microelectronic Devices   868

29 Fabrication of Microelectromechanical Devices and Systems (MEMS)   908

36   Quality Assurance, Testing, and Inspection   1110

Part IX: Manufacturing in a Competitive Environment   1142

37   Automation of Manufacturing Processes   1144

38 Computer-Aided Manufacturing   1191

39 Computer-Integrated Manufacturing Systems   1218

40 Product Design and Process Selection in a Competitive Environment   1238

Index   1278

Case Studies

PREFACE

The science, engineering, and technology of manufacturing processes and systems continue to advance rapidly on a global scale and with major impact on the economies of all nations. In preparing this fifth edition, our goal throughout has been to provide a comprehensive and state-of-the-art manufacturing engineering and technology textbook, with the additional aims of motivating and challenging students to study this important discipline.

As in the previous four editions, the text presents topics with a balanced coverage of relevant fundamentals and real-world practices to help the students develop an understanding of the often complex interrelationships among the many technical and economic factors involved in manufacturing.

While this new edition basically follows the same introductory nature, format, and organization as the fourth edition, it now includes increased emphasis on (a) the influence of materials and processing parameters in understanding individual processes and operations; (b) design considerations, product quality, and manufacturing costs; and (c) the global competitive context of each manufacturing process and operation, highlighted with numerous illustrative examples and case studies.

What is New in this Edition

A page-by-page comparison with the fourth edition will indicate that literally thousands of changes have been made to improve the clarity and thoroughness of the numerous topics covered.

- As a general guide for the student, each chapter now begins with a brief, highlighted outline of the chapter objectives, the topics to be described, and their relevance. Wherever appropriate, a list has been included regarding typical parts made by the processes described in the chapter and the alternative methods of producing the same parts.

- Most of the illustrations in the book have been revised thoroughly for improved graphic impact and clarity, and numerous new photographs have been added.

- There are now two chapters on the topics of microelectronic and microelectromechanical device and system manufacturing, including MEMS.

- There are about 120 examples and case studies, all highlighted.

- The questions and problems in each chapter have been updated, about 20% of which are new to this edition. Also, the last section of the questions and problems is now labeled "Synthesis, Design, and Projects" to better reflect the increased emphasis on these topics throughout the book.

- The text has more cross-references throughout to other relevant sections, chapters, tables, and figures in the book.

- The bibliographies at the end of each chapter have been thoroughly updated.

Study Aids

- Each topic is presented within a larger context of manufacturing engineering and technology, using various flowcharts and schematic diagrams whenever appropriate.

- There is a continued emphasis on the practical uses of the concepts described and information presented.
- Attempts have been made to provide analogies, discussions, and problems designed to stimulate the student's interest and curiosity about consumer and industrial products and how they are manufactured while minimizing production costs.
- Extensive data and reference materials, including numerous tables, illustrations, graphs, and bibliographies are presented.
- Several new examples and case studies have been included to highlight important concepts and techniques in manufacturing.
- Numerous tables compare advantages as well as limitations of major competitive manufacturing processes.
- A chapter summary and a list of key terms is included to help and remind students of the topics covered in the chapter.

Audience

As in the previous editions, this fifth edition has been written for students in mechanical, manufacturing, industrial, biomedical, aerospace, and metallurgical and materials engineering programs. It is hoped that by reading and studying this book, students will come to appreciate the vital nature of manufacturing engineering and technology as an academic subject that is as exciting, challenging, and important as any other discipline.

We would be grateful for any comments from instructors and students regarding any suggestions about the numerous topics presented or any errors that may have escaped our attention during the preparation of this text.

Acknowledgments

This book, together with its previous editions, represents a total of about 20 years of effort. It could not have been written and produced without the help of numerous colleagues and former students. It gives us great pleasure to acknowledge the assistance of the following in the preparation and publication of this fifth edition: K. E. McKee Illinois Institute of Technology; K. J. Weinmann, Michigan Technological University, P. J. Guichelaar, Western Michigan University, Z. Liang, Indiana University-Purdue University, Fort Wayne, and R. Abella, University of Toledo. We also acknowledge Kent M. Kalpakjian as the original author of the chapter on Fabrication of Microelectronic Devices.

We would like to thank our editors Dorothy Marrero and Eric Svendsen at Prentice Hall for their enthusiastic support and guidance, to Rose Kernan for her meticulous editorial and production supervision and the interior design of this book, and to Xiaohong Zhu for the development of all the new illustrations in the book.

We are happy to present the following cumulative list of all those individuals who, in one way or another, made various contributions to this and the previous editions of this book:

B. J. Aaronson	M. Dugger	J. Kotowski	M. Savic
S. Arellano	D. R. Durham	S. Krishnamachari	W. J. Schoech
R. A. Arlt	D. Duvall	K. M. Kulkarni	S. A. Schwartz
V. Aronov	S. A. Dynan	T. Lach	M. T. Siniawski
A. Bagchi	J. El Gomayel	L. Langseth	J. E. Smallwood
E. D. Baker	M. G. Elliott	M. Levine	J. P. Sobczak
J. Barak	E. C. Feldy	B. S. Levy	L. Soisson
J. Ben-Ari	J. Field	X. Z. Li	J. Stocker
G. F. Benedict	G. W. Fischer	B. W. Lilly	L. Strom
S. Bhattacharyya	D. A. Fowley	D. A. Lucca	A. B. Strong
J. T. Black	R. L. French	L. Mapa	K. Subramanian
C. Blathras	B. R. Fruchter	A. Marsan	T. Sweeney
G. Boothroyd	D. Furrer	R. J. Mattice	W. G. Switalski
D. Bourell	R. Giese	C. Maziar	T. Taglialavore
B. Bozak	E. Goode	T. McClelland	M. Tarabishy
N. N. Breyer	K. Graham	L. McGuire	K. S. Taraman
C. A. Brown	P. Grigg	K. E. McKee	R. Taylor
R. G. Bruce	B. Harriger	K. P. Meade	B. S. Thakkar
J. Cesarone	D. Harry	R. Miller	A. Trager
T.-C. Chang	M. Hawkins	T. S. Milo	C. Tszang
R. L. Cheaney	R. J. Hocken	S. Mostovoy	S. Vaze
A. Cheda	E. M. Honig, Jr.	C. Nair	J. Vigneau
S. Chelikani	S. Imam	P. G. Nash	G. A. Volk
S.-W. Choi	R. Jaeger	J. Nazemetz	G. Wallace
A. Cinar	C. Johnson	E. M. Odom	K. J. Weinmann
R. O. Colantonio	K. Jones	S. J. Parelukar	R. Wertheim
P. Cotnoir	D. Kalisz	J. Penaluna	K. West
P. Courtney	J. Kamman	C. Petronis	J. Widmoyer
P. Demers	S. G. Kapoor	M. Philpott	K. Williams
D. Descoteaux	R. Kassing	J. M. Prince	G. Williamson
M. F. DeVries	R. L. Kegg	W. J. Riffe	B. Wiltjer
R. C. Dix	W. J. Kennedy	R. J. Rogalla	J. Wingfield
M. Dollar	B. D. King	A. A. Runyan	P. K. Wright
D. A. Dornfeld	J. E. Kopf	G. S. Saletta	
H. I. Douglas	R. J. Koronkowski	M. Salimian	

We are grateful to numerous organizations that supplied us with many illustrations and case studies. These contributions have been specifically acknowledged throughout the text.

Finally, our many thanks to Margaret Jean Kalpakjian for her assistance during the editing of this book.

SEROPE KALPAKJIAN

STEVEN R. SCHMID

About the Authors

Serope Kalpakjian is a professor emeritus of mechanical and materials engineering at the Illinois Institute of Technology, Chicago. He is the author of *Mechanical Processing of Materials* (Van Nostrand, 1967) and a co-author of *Lubricants and Lubrication in Metalworking Operations* (Dekker, 1985). Both of the first editions of his books *Manufacturing Processes for Engineering Materials* (1984) and *Manufacturing Engineering and Technology* (1989) have received the M. Eugene Merchant Manufacturing Textbook Award. He is the author of numerous technical papers and articles in handbooks and encyclopedias, and has edited several conference proceedings. He has been editor and co-editor of various technical journals and is on the editorial board of *Encyclopedia Americana*.

Among other awards, Professor Kalpakjian has received the Forging Industry Educational and Research Foundation Best Paper Award (1966), an Excellence in Teaching Award from IIT (1970), a Centennial Medallion from the ASME (1980), the International Education Award from SME (1989), a Person of the Millennium Award from IIT (1999), and the Albert Easton White Outstanding Teacher Award from ASM International (2000). The SME Outstanding Young Manufacturing Engineer Award for 2002 was named after him. He is a Life Fellow of ASME, Fellow of SME, Fellow and Life Member of ASM International, an emeritus full member of CIRP (International Institution for Production Engineering Research), and is a founding member and past president of NAMRI/SME. He is a high-honor graduate of Robert College (Istanbul), Harvard University, and the Massachusetts Institute of Technology.

Steven R. Schmid is an Associate Professor in the Department of Aerospace and Mechanical Engineering at the University of Notre Dame, where he teaches and conducts research in the general areas of manufacturing, machine design, and tribology. He received his Bachelor's degree in Mechanical Engineering from the Illinois Institute of Technology (with Honors) and Master's and Ph.D. degrees, both in Mechanical Engineering, from Northwestern University. He has received numerous awards, including the John T. Parsons Award from the Society of Manufacturing Engineers (2000), the Newkirk Award from the American Society of Mechanical Engineers (2000), the Kaneb Center Teaching Award (2000 and 2003), and the Ruth and Joel Spira Award for Excellence in Teaching (2005). He is the recipient of a National Science Foundation CAREERS Award (1996) and an ALCOA Foundation Award (1994).

Dr. Schmid is the author of over 80 technical papers, has co-authored the texts *Fundamentals of Machine Elements*, *Fundamentals of Fluid Film Lubrication*, and *Manufacturing Processes for Engineering Materials*, and has contributed two chapters to the *CRC Handbook of Modern Tribology*. He serves on the Tribology Division Executive Committee of the American Society of Mechanical Engineers, is an Associate Editor of the *Journal of Manufacturing Science and Engineering*, and is a registered Professional Engineer and Certified Manufacturing Engineer.

General Introduction

The objectives of this chapter are to explain:

- What manufacturing is and, with examples, show its role in our daily lives.
- The product design process and the importance of selection of materials and processes.
- The role of computers in all aspects of manufacturing.
- Manufacturing costs and their role in a global economy.
- General trends in manufacturing.

I.1 | What is Manufacturing?

As you begin to read this Introduction, take a few moments and inspect various objects around you: your clock, cell phone, chair, beverage can, light switches, coffee cup, and your computer. You will soon realize that all these objects and their individual components had a different shape at one time; you would not find them in nature as they appear in your room. They have been transformed from raw materials into various shapes and assembled into the products that you now see.

You will readily note that some objects are made of a single part, such as a nail, bolt, fork, plastic coat hanger, or bicycle tire. However, most objects, such as automobile engines, shown in Fig. I.1 (invented in 1876), washing machines (1910), toasters (1926), air conditioners (1928), refrigerators (1931), ballpoint pens (1938), photocopiers (1949), and thousands of other products are constructed by assembling a number of parts (Table I.1) and components made from a variety of materials. All of the products just mentioned are made by various processes called *manufacturing*.

Manufacturing, in its comprehensive sense, is the process of converting raw materials into products. Manufacturing also involves activities in which the manufactured product, itself, is used to make other products. Examples could include large presses to shape sheet metal for appliances and car bodies, machinery to make fasteners, such as bolts and nuts, and sewing machines to make clothing. A nation's level of manufacturing activity is related directly to its economic health; generally, the higher the level of manufacturing activity in a country, the higher the standard of living of its people.

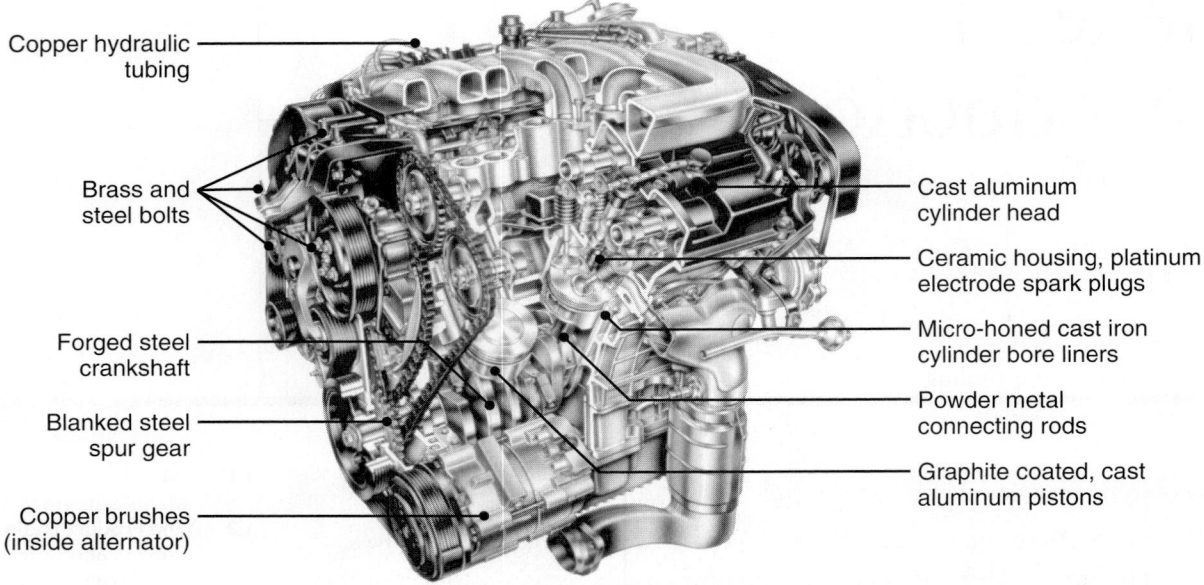

Copper hydraulic tubing

Brass and steel bolts

Forged steel crankshaft

Blanked steel spur gear

Copper brushes (inside alternator)

Cast aluminum cylinder head

Ceramic housing, platinum electrode spark plugs

Micro-honed cast iron cylinder bore liners

Powder metal connecting rods

Graphite coated, cast aluminum pistons

Polymer manifolds removed for clarity

FIGURE I.1   Illustration of an automotive engine—the Duratec V-6—showing various components and the materials used in making them. *Source:* Courtesy of Ford Motor Company. Illustration by David Kimball.

TABLE I.1

Number of Parts in Some Products	
Product	Number of parts
Rotary lawn mower	300
Grand piano	12,000
Automobile	15,000
C-5A transport plane	>4,000,000
Boeing 747–400	>6,000,000

The word *manufacturing* is derived from the Latin *manu factus*, meaning made by hand. The word *manufacture* first appeared in 1567, and the word "manufacturing" appeared in 1683. The word *product* means something that is produced, and the words "product" and *production* first appeared sometime during the 15th century. The words "manufacturing" and "production" often are used interchangeably.

Because a manufactured item typically has undergone a number of processes in which pieces of raw material have been turned into a useful product, it has a **value**—defined as monetary worth or marketable price. For example, as the raw material for ceramics, clay has some small value as mined. When the clay is made into the ceramic part of a spark plug, a vase, a cutting tool, or an electrical insulator, value is added to the clay (**value added**). Similarly, a wire coat hanger or a nail has a value over and above the cost of the piece of wire from which it is made; thus manufacturing has the important function of *adding value*. The term

high value-added is used to identify such products. Examples include: computer chips, engine blocks, gears, and gym shoes.

Manufacturing may produce **discrete products**, meaning individual parts, or **continuous products**. Nails, gears, balls for bearings, beverage cans, and engine blocks are examples of discrete parts, even though they are mass produced at high volumes and production rates. On the other hand, wire, sheet metal, and plastic pipe and tubing are continuous products, which are then cut into individual pieces and thus become discrete products.

Manufacturing is generally a complex activity involving a wide variety of resources and activities, such as the following:

- product design
- machinery and tooling
- process planning
- materials
- purchasing
- manufacturing
- production control
- support services
- marketing
- sales
- shipping
- customer service

It is essential that manufacturing activities be responsive to several demands and trends:

1. A product must fully meet **design requirements, product specifications**, and **standards**.

2. A product must be manufactured by the most **economical and environmentally friendly** methods.

3. **Quality** must be *built into the product* at each stage, from design to assembly, rather than relying on quality testing after the product is manufactured.

4. In the highly competitive environment of today, production methods must be sufficiently **flexible** to respond to changing market demands, types of products, production rates, production quantities, and to provide on-time delivery to the customer.

5. Continuous developments in **materials, production methods**, and **computer integration** of both technological and managerial activities in a manufacturing organization must be evaluated constantly with a view to their appropriate, timely, and economical implementation.

6. Manufacturing activities must be viewed as a large **system**, all parts of which are interrelated to varying degrees. Such systems can now be modeled in order to study the effect of factors such as changes in market demands, product design, materials, and production methods on product quality and cost.

7. The manufacturer must work with the customer for timely feedback for **continuous product improvement**.

8. A manufacturing organization constantly must strive for higher levels of **productivity**, defined as the optimum use of all its resources such as materials, machines, energy, capital, labor, and technology; output per employee per hour in all phases must be maximized.

I.1.1 A brief history of manufacturing

Manufacturing dates back to about 5000–4000 B.C. (Table I.2). It is older than recorded history. Primitive cave or rock markings and drawings were dependent on some form of a marker or brush using a "paint" or some means of notching the rock. Appropriate tools had to be made for these applications. Manufacturing of products for various specific uses began with the production of articles made of wood, ceramic, stone, and metal. The materials and processes first used to shape products by casting and hammering have been developed gradually over the centuries, using new materials and more complex operations, at increasing rates of production and at higher and higher levels of quality.

The first materials used for making household utensils and ornamental objects included metals such as gold, copper, and iron, followed by silver, lead, tin, bronze, and brass. The production of steel in about 600–800 A.D. was a major milestone; since then, a very wide variety of ferrous and nonferrous metals have been developed. Today, the materials used in advanced products such as computers and supersonic aircraft include engineered materials (tailored materials) with unique properties, such as advanced ceramics, reinforced plastics, composite materials, and nanomaterials.

Until the *Industrial Revolution*, which began in England in the 1750s, goods had been produced in batches, requiring much reliance on manual labor in all phases of production. (The Industrial Revolution is also called the First Industrial Revolution, the second one beginning in the mid-1900s with the developments of solid-state electronic devices and computers.) Modern *mechanization* began in England and Europe with the development of textile machinery and of machine tools for cutting metals. This technology soon moved to the United States where it was developed further, including the important advance of designing, making, and using *interchangeable parts*, first conceived by Eli Whitney in the early 1800s. Prior to the introduction of interchangeable parts, a great deal of hand-fitting was necessary because no two parts could be made exactly alike. By contrast, it is now taken for granted that we can replace a broken bolt of a certain size with an identical one purchased years later from a local hardware store. Further developments soon followed, resulting in countless products that we cannot now imagine being without, because they are so common.

Beginning in the early 1940s, major milestones have been reached in all aspects of manufacturing. For example, note from Table I.2 the progress made during the past 100 years, and especially during the last three decades with the advent of the computer age, as compared to that during the long period from 4000 B.C. to 1 B.C. Although the Romans had factories for mass producing glassware, production methods were at first very primitive and generally very slow, with much manpower involved in handling parts and running the machinery. Today, with the help of *computer-integrated manufacturing systems*, production methods have been advanced to such an extent that, for example, aluminum beverage cans are manufactured at rates of 500 per minute, holes in sheet metal are punched at rates of 800 per minute, and light bulbs at more than 2000 per minute.

TABLE I.2

Historical Development of Materials and Manufacturing Processes

Period	Dates	Metals and casting	Various materials and composites	Forming and shaping	Joining	Tools, machining, and manufacturing systems
Egypt: ~3100 B.C. to ~300 B.C. Greece: ~1100 B.C. to ~146 B.C. Roman Empire: ~500 B.C. to 476 A.D. Middle Ages: ~476 to 1492 Renaissance: 14th to 16th centuries	Before 4000 B.C.	Gold, copper, meteoric iron	Earthenware, glazing, natural fibers	Hammering		Tools of stone, flint, wood, bone, ivory, composite tools
	4000–3000 B.C.	Copper casting, stone and metal molds, lost-wax process, silver, lead, tin, bronze		Stamping, jewelry	Soldering (Cu-Au, Cu-Pb, Pb-Sn)	Corundum (alumina, emery)
	3000–2000 B.C.	Bronze casting and drawing, gold leaf	Glass beads, potter's wheel, glass vessels	Wire by slitting sheet metal	Riveting, brazing	Hoe making, hammered axes, tools for ironmaking and carpentry
	2000–1000 B.C.	Wrought iron, brass				
	1000–1 B.C.	Cast iron, cast steel	Glass pressing and blowing	Stamping of coins	Forge welding of iron and steel, gluing	Improved chisels, saws, files, woodworking lathes
	1–1000 A.D.	Zinc, steel	Venetian glass	Armor, coining, forging, steel swords		Etching of armor
	1000–1500	Blast furnace, type metals, casting of bells, pewter	Crystal glass	Wire drawing, gold- and silversmith work		Sandpaper, windmill-driven saw
	1500–1600	Cast-iron cannon, tinplate	Cast plate glass, flint glass	Water power for metalworking, rolling mill for coinage strips		Hand lathe for wood
	1600–1700	Permanent-mold casting, brass from copper and metallic zinc	Porcelain	Rolling (lead, gold, silver), shape rolling (lead)		Boring, turning, screw-cutting lathe, drill press

(continues on next page)

TABLE I.2

Historical Development of Materials and Manufacturing Processes (cont.)

Period	Dates	Metals and casting	Various materials and composites	Forming and shaping	Joining	Tools, machining, and manufacturing systems
Industrial Revolution: ~1750 to 1850	1700–1800	Malleable cast iron, crucible steel (iron bars and rods)		Extrusion (lead pipe), deep drawing, rolling		
	1800–1900	Centrifugal casting, Bessemer process, electrolytic aluminum, nickel steels, babbitt, galvanized steel, powder metallurgy, open-hearth steel	Window glass from slit cylinder, light bulb, vulcanization, rubber processing, polyester, styrene, celluloid, rubber extrusion, molding	Steam hammer, steel rolling, seamless tube, steel-rail rolling, continuous rolling, electroplating		Shaping, milling, copying lathe for gunstocks, turret lathe, universal milling machine, vitrified grinding wheel
	1900–1920		Automatic bottle making, bakelite, borosilicate glass	Tube rolling, hot extrusion	Oxyacetylene; arc, electrical-resistance, and thermit welding	Geared lathe, automatic screw machine, hobbing, high-speed-steel tools, aluminum oxide and silicon carbide (synthetic)
	1920–1940	Die casting	Development of plastics, casting, molding, polyvinyl chloride, cellulose acetate, polyethylene, glass fibers	Tungsten wire from metal powder	Coated electrodes	Tungsten carbide, mass production, transfer machines
WWI	1940–1950	Lost-wax process for engineering parts	Acrylics, synthetic rubber, epoxies, photosensitive glass	Extrusion (steel), swaging, powder metals for engineering parts	Submerged arc welding	Phosphate conversion coatings, total quality control
WWII	1950–1960	Ceramic mold, nodular iron, semiconductors, continuous casting	Acrylonitrile-butadiene-styrene, silicones, fluorocarbons, polyurethane, float glass, tempered glass, glass ceramics	Cold extrusion (steel), explosive forming, thermomechanical processing	Gas metal arc, gas tungsten arc, and electroslag welding; explosion welding	Electrical and chemical machining, automatic control

TABLE I.2

Historical Development of Materials and Manufacturing Processes (cont.)

Period	Dates	Metals and casting	Various materials and composites	Forming and shaping	Joining	Tools, machining, and manufacturing systems
	1960–1970	Squeeze casting, single-crystal turbine blades	Acetals, polycarbonate, cold forming of plastics, reinforced plastics, filament winding	Hydroforming, hydrostatic extrusion, electroforming	Plasma-arc and electron-beam welding, adhesive bonding	Titanium carbide, synthetic diamond, numerical control, integrated circuit chip
Space Age	1970–1990	Compacted graphite, vacuum casting, organically bonded sand, automation of molding and pouring, rapid solidification, metal-matrix composites, semisolid metalworking, amorphous metals, shape-memory alloys (smart materials), computer simulation	Adhesives, composite materials, semiconductors, optical fibers, structural ceramics, ceramic-matrix composites, biodegradable plastics, electrically conducting polymers	Precision forging, isothermal forging, superplastic forming, dies made by computer-aided design and manufacturing, net-shape forging and forming, computer simulation	Laser beam, diffusion bonding (also combined with superplastic forming), surface-mount soldering	Cubic boron nitride, coated tools, diamond turning, ultraprecision machining, computer-integrated manufacturing, industrial robots, machining and turning centers, flexible-manufacturing systems, sensor technology, automated inspection, expert systems, artificial intelligence, computer simulation and optimization
Information Age	1990–2000s	Rheocasting, computer-aided design of molds and dies, rapid tooling	Nanophase materials, metal foams, advanced coatings, high-temperature superconductors. machinable ceramics, diamondlike carbon	Rapid prototyping, rapid tooling, environmentally friendly metalworking fluids	Friction stir welding, lead-free solders, laser butt-welded (tailored) sheet-metal blanks, electrically conducting adhesives	Micro- and nano-fabrication, LIGA (a German acronym for a process involving lithography, electroplating, and molding), dry etching, linear motor drives, artificial neural networks, six sigma

Source: J.A. Schey, C.S. Smith, R.F. Tylecote, T.K. Derry, T.I. Williams, S.R. Schmid, and S. Kalpakjian.

7

EXAMPLE I.1 Paper clips

The paper clip (Fig. I.2) as we know it today was developed by a Norwegian, Johan Vaaler, who received a U.S. patent for it in 1901. In this example, we will identify the important factors involved in the design and manufacturing of paper clips.

Assume that you are asked to design and produce paper clips. What type of material would you choose to make this very simple product? Should it be metallic or can it be nonmetallic, such as plastic? If you choose a metal, what type of metal and in what condition? If the material that you start with is in the shape of wire, what should be its diameter? Should it be round or have some other cross-section? Is the wire's surface finish and appearance important, and, if so, what should be its surface finish and appearance? Furthermore, how would you take a piece of wire and shape it into a paper clip? Would you do it by hand on a simple fixture, or, if not, what kind of special machine would you design and make, or purchase, to make paper clips? If, as the owner of a company, you were given an order for 10,000 paper clips versus an order for millions of clips, would your approach to manufacturing them be different?

It should be obvious that the paper clip must meet its basic functional requirement: to hold pieces of paper together with sufficient clamping force so that the papers do not slip away from each other. Consequently, it must be designed properly, especially in shape, size, feel, and appearance. The material selected must have a certain stiffness and strength. For example, if the stiffness (a measure of how much it deflects under a given force) is too high, a level of force uncomfortable to or inconvenient for users may be required to use the clip, just as a stiff spring requires a greater force to stretch or compress it than does a softer one. If the clip stiffness is, on the other hand, too low, it will not exert sufficient clamping force on

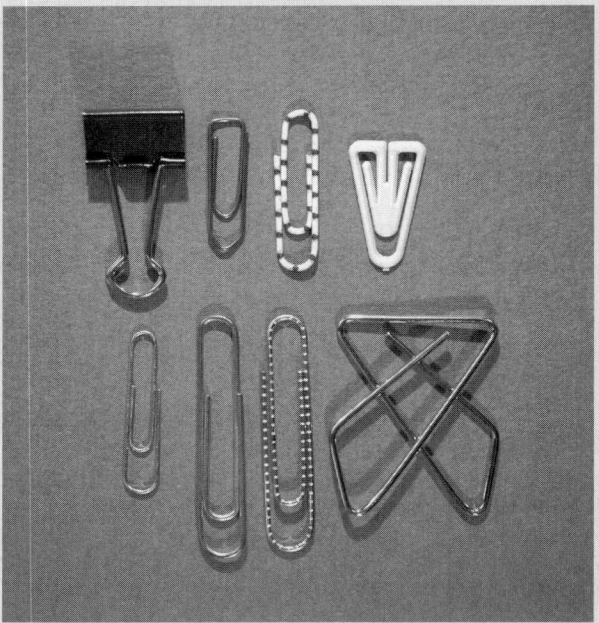

FIGURE I.2   Examples of the wide variety of materials and geometries for paper clips.

the set of papers. Furthermore, if the yield stress of the wire material (the stress required to cause permanent deformation of a material) is too low, the clip will bend permanently during normal usage and thus will be difficult to reuse, as we all have experienced. Note that the stiffness and strength of the paper clip also depend on the wire diameter and the dimensions and the design of the clip.

After finalizing the clip design, a suitable material has to be selected. Material selection requires a knowledge of the *function* and *service requirements* of the product, and it leads to choosing materials that, preferably, are commercially available. The selection of the material also involves consideration of its corrosion resistance, because the clip is handled often and is subjected to moisture and other environmental attacks. Note, for example, the rust marks left by paper clips on documents stored in files for a long period of time.

Several other questions concerning the production of the paper clips must be asked. Will the material selected be able to undergo bending during manufacturing without cracking or breaking? Can the wire be easily cut from a long piece without causing excessive wear on the tooling? Will the cutting process (shearing) produce a smooth edge on the end of the wire, or will it leave a burr (a sharp edge) that will interfere with its intended use? Finally, what is the most economical method of manufacturing this part at the desired production rate, so that it can be competitive in the marketplace. A suitable manufacturing method and tools, machinery, and related equipment then must be selected.

EXAMPLE I.2 Incandescent light bulbs

The first incandescent lamp was made by T.A. Edison (1847–1931) and was lit in 1879. However, a typical bulb had a life of only about 13.5 hours. Since then, many improvements have been made in the materials and manufacturing methods for making bulbs. In this example, we will describe the sequence of methods used in manufacturing these bulbs, which are produced on highly automated machines at rates of 2000 bulbs a minute.

The components of a typical light bulb are shown in Fig. I.3a. The light-emitting part is the filament, which, by the passage of current, due to its electrical resistance, is heated to incandescence; that is, to temperatures between 2200° and 3000°C (4000° and 5400°F). Edison's first successful lamp had a carbon filament, although he and others had also experimented with various materials such as carbonized paper and metals like osmium, iridium, and tantalum. However, none of these materials has the strength, high-temperature resistance, and long life of tungsten (Section 6.8), which is now the most commonly used filament material.

The first step in manufacturing a light bulb is making the glass stem that supports the lead-in wires and the filament and connects them to the base of the bulb (Fig. I.3b). These components are positioned, assembled, and sealed while the glass is heated by gas flames. The filament is then attached to the lead-in wires.

The completed stem assembly (the mount) then is transferred to a machine that lowers a glass bulb over it, and gas flames are used to seal the rim of the mount to the neck of the bulb. The air in the bulb is exhausted through the exhaust tube (an integral part of the glass stem), and then the bulb is either evacuated or filled with inert gas. For bulbs of 40 W and higher, the gas is typically a mixture of nitrogen and argon. The exhaust tube is then sealed off. The next production step consists of attaching the base to the bulb, using a

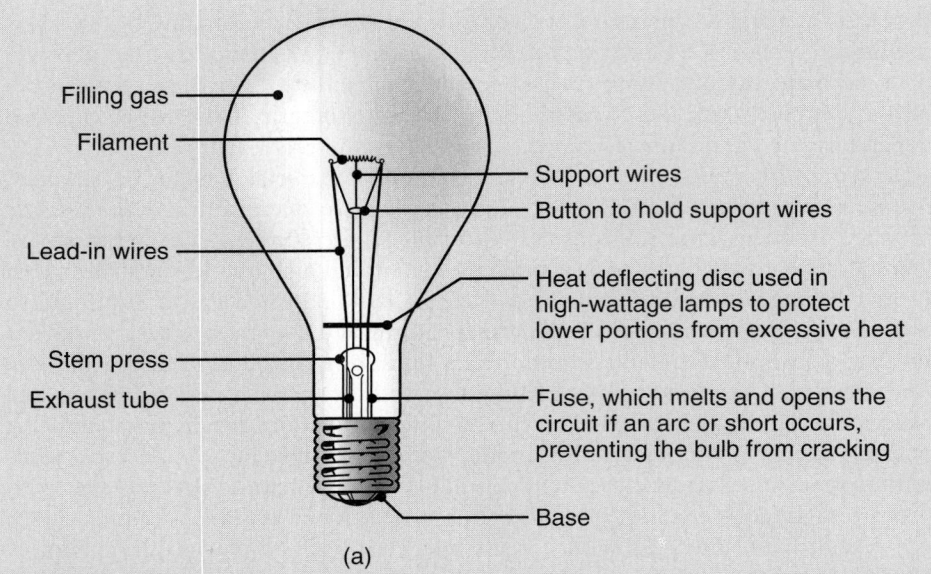

Filling gas

Filament

Support wires

Button to hold support wires

Lead-in wires

Heat deflecting disc used in high-wattage lamps to protect lower portions from excessive heat

Stem press

Exhaust tube

Fuse, which melts and opens the circuit if an arc or short occurs, preventing the bulb from cracking

Base

(a)

FIGURE I.3a   Components of a common incandescent light bulb. *Source:* Courtesy of General Electric Company.

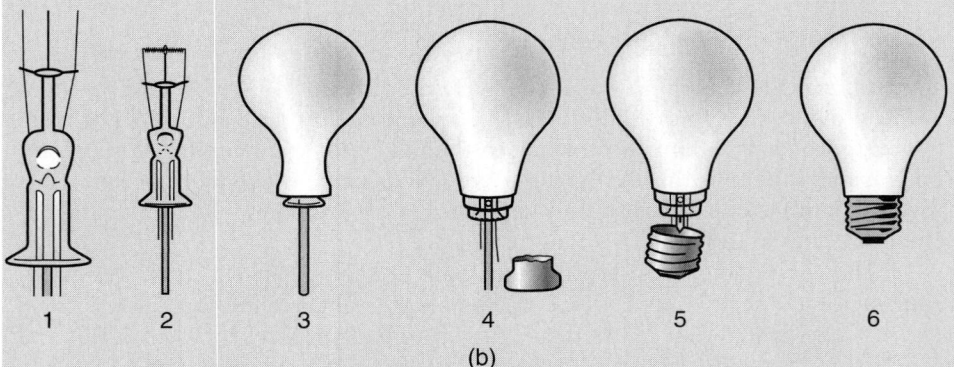

1   2   3   4   5   6

(b)

FIGURE I.3b   Manufacturing steps in making an incandescent light bulb. *Source:* Courtesy of General Electric Company.

special cement. The machine that performs the attaching operation also solders or welds (Part VI) the lead-in wires to the metal base to provide the electrical connection.

The filament is made by first pressing tungsten powder into ingots and sintering it (heating without melting; Section 17.4). The ingot is then shaped into round rods by rotary swaging (Section 14.4), and the rods are drawn through a die, in a number of steps, into thin wire (Section 15.7). The wire is coiled to increase the light-producing capacity of the filament. The wire diameter for a 60-W, 120-V lamp is 0.045 mm (0.0018 in.), and it must be controlled very accurately because if the wire diameter is only 1% less than that specified, the life of the lamp may be reduced by as much as 25%. The spacing between coils also must be very accurate in order to prevent localized heat concentration, and thus possible shorting.

Lead-in wires are usually made of nickel, copper, or molybdenum, and the support wires are made of molybdenum (Section 6.8). The portion of the lead-in wire embedded in the stem is made of an iron-nickel alloy, which is coated with copper. The wire has essentially the same coefficient of thermal expansion as does the glass (Chapters 3 and 8); thus, thermal stresses that otherwise might cause the stem to crack do not develop. The bulb base is generally made from aluminum (replacing, for lower cost, the brass that was used earlier), specially coated to permit easy insertion into the socket.

The glass for the bulbs is commonly made by blowing molten glass into a mold (Section 18.3.3). Several types of glasses are used, depending on the kind of bulb desired. The inside of the bulb is either frosted (translucent) to reduce glare and to better diffuse the light, or is plain (transparent). The filling gas must be pure, as otherwise the inside surfaces of the bulb will blacken. For example, just one drop of water in the gas used for half a million lamps will cause blackening in all of them.

I.2   The Product Design Process and Concurrent Engineering

Product design is a critical activity because it has been estimated that 70 to 80% of the cost of product development and manufacture is determined by the decisions made in the *initial* design stages. The design process begins with the development of an original product concept. An innovative approach to design is highly desirable at this stage, and even essential, for the product to be successful in the marketplace. Innovative approaches can also lead to major savings in material and production costs.

The design of a product first requires a thorough understanding of the *functions* and the *performance* expected of that product. The market for a product and its anticipated uses must be clearly defined, with the assistance of market analysts and sales personnel who bring in valuable and timely input from the marketplace to the company. The product may be new, or it may be a modified and newer version of an existing product; note, for example, how the design and style of cell phones, calculators, appliances, automobiles, and aircraft have changed over the years.

Traditionally, design and manufacturing activities have taken place sequentially (Fig. I.4a), a methodology that at first may seem to be logical and straightforward, but in practice is extremely wasteful of resources. In theory, a product can flow from one department in an organization to another, produced, and then placed directly into the marketplace, but there are usually difficulties encountered. For example, a manufacturing engineer may wish to taper the flange on a part to improve its castability, or may decide that a different alloy is more desirable. Such changes necessitate a repeat of the design analysis stage in order to ensure that the product will still function satisfactorily. These iterations, also shown in Fig. I.4a, waste resources and, more importantly, they waste time.

Driven by the consumer electronics industry a great need developed for bringing products to the market as quickly as possible. The rationale was that products introduced early enjoy a greater percentage of the market and, hence, higher profits. They also have a longer life before obsolescence. For these reasons, **concurrent engineering**, also called **simultaneous engineering**, came to the fore, which led to the

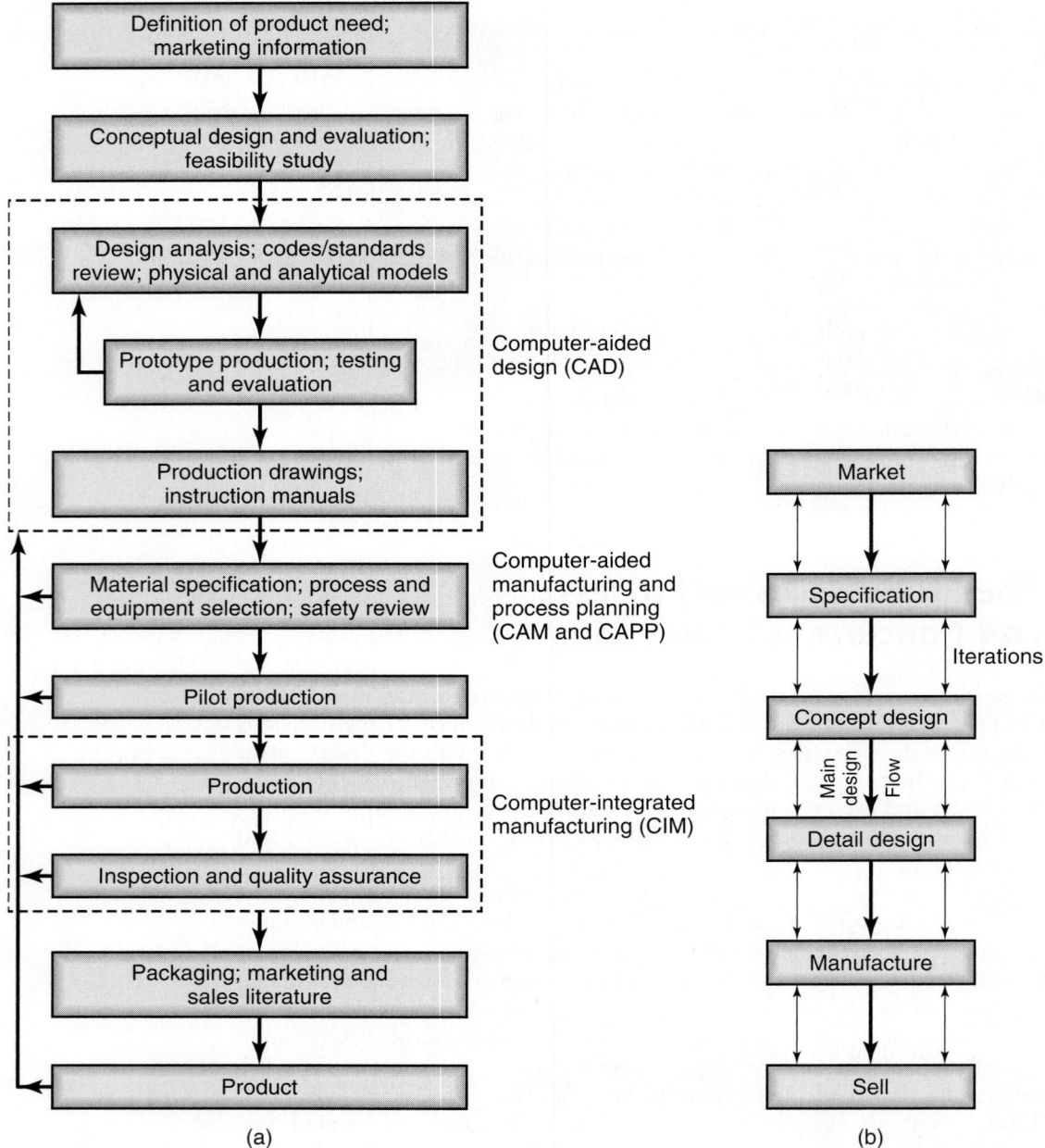

FIGURE I.4   (a) Chart showing various steps involved in design and manufacturing a product. Depending on the complexity of the product and the type of materials used, the time span between the original concept and the marketing of a product may range from a few months to many years. (b) Chart showing general product flow in concurrent engineering, from market analysis to selling the product. *Source:* After S. Pugh, *Total Design.* Addison-Wesley, 1991.

product development approach shown in Fig. I.4b. While it still has a general product flow from market analysis to design and manufacturing, it contains several deliberate iterations. The main difference from the older approach is that all disciplines now are involved in the early design stages so that the iterations, which by nature occur, result in less wasted effort and time. A key to this approach is the now well-recognized importance of *communication* among and within various disciplines. While there must be communication between engineering, marketing, and service functions, so too must there be avenues of interaction between activities, such as design for manufacture, design for recyclability, and design for safety.

Concurrent engineering integrates the design and manufacture of products with a view toward optimizing all elements involved in the **life cycle** of the product. This approach reduces (a) the changes in a product's design and engineering, and (b) the time and costs involved in taking the product from its design concept to its production and its introduction into the marketplace. The life cycle of a new product typically consists of the following stages: (a) Product startup, (b) rapid growth of the product in the marketplace, (c) product maturity, and (d) decline.

The concept of **life-cycle engineering** requires that the entire life of a product be considered in the design stage: The design, production, distribution, use, and recycling/disposal must be considered simultaneously. Thus, a well-designed product is:

- functional (design)
- well manufactured (production)
- well packaged (arriving safely at the end user or customer)
- durable (functions effectively for its intended purpose)
- maintainable (has components which can be easily replaced, maintained, or repaired)
- resource efficient (can be disassembled so that components can be recycled)

While this textbook mainly emphasizes the *production* aspects of a product's life cycle, the need for integration of multiple disciplines in product development pervades the life-cycle; for example, recycling is best addressed during product development, by selecting materials that are easily recyclable. Although the concept of concurrent engineering appears to be logical and efficient, its implementation requires considerable time and effort when those using it either do not work as a team or they fail to appreciate its real benefits.

There are numerous examples of the benefits of concurrent engineering. One automotive company, for example, reduced the number of components in one of its engines by 30% and, as a result, decreased the engine weight by 25% and cut its manufacturing time by 50%. The concurrent engineering concept can be implemented in companies, both large and small, particularly in view of the fact that 98% of U.S. manufacturing establishments have fewer than 500 employees.

Product design often involves preparing *analytical and physical models* of the product as an aid to studying factors such as forces, stresses, deflections, and optimal part shape. The necessity for such models depends on product complexity. Today, constructing and studying analytical models is highly simplified through the use of **computer-aided modeling and design** (CAD), **computer-aided engineering** (CAE), and **computer-aided manufacturing** (CAM) techniques. CAD systems are capable of rapidly and completely analyzing designs, from a simple bracket or a shaft, to large and complex structures. The two-engine Boeing 777 passenger airplane, for example, was designed completely by computer (**paperless design**), with 2000 workstations linked to eight design servers. Unlike previous models, no prototypes or mockups are built, and the airplane is constructed directly from the CAD/CAM software developed.

Using computer-aided engineering, the performance of structures subjected, for example, to static or fluctuating loads and to temperature gradients can be simulated, analyzed, and tested efficiently, accurately, and rapidly. The information developed can be stored, retrieved, displayed, printed, and transferred anywhere in the organization. Designs can be optimized, and modifications can be made, directly and easily, at any time.

Computer-aided manufacturing (Section 39.5) involves all phases of manufacturing, by utilizing and processing further the large amounts of information on materials and processes gathered and stored in the organization's database. Computers now assist manufacturing engineers and their associates in organizing such tasks as programming numerical control of machines, programming robots for material handling and assembly, designing tools, dies, and fixtures, and maintaining quality control.

On the basis of the models developed using the foregoing techniques, the product designer selects and specifies the final shape and dimensions of the product, its dimensional accuracy and surface finish, and its component materials. The selection of materials is made with the advice and cooperation of materials engineers, unless the design engineer is also experienced and qualified in this area. An important design consideration is how a particular component is to be *assembled* into the final product; for example, lift the hood of your car and observe how hundreds of components are fitted into a limited space, or do the same for a light switch or a telephone.

The next step in the production process is often to make and test a **prototype**; that is, an original working model of the product. An important technology is **rapid prototyping** (Chapter 20), which relies on CAD/CAM and on various manufacturing techniques (using metallic or nonmetallic materials as workpieces) to produce prototypes rapidly and at low cost, in the form of a solid physical model of a part. For example, prototyping new automotive components by traditional methods, such as shaping, forming, and machining, may cost hundreds of millions of dollars a year, and some components may require a year or so to produce. Rapid prototyping can cut these costs and the associated development times, significantly. These techniques are now being advanced further so that they can be used for low-volume economical production of actual parts to go into products.

Tests for prototypes must be designed to simulate as closely as possible the conditions under which the product is to be used. These conditions include environmental factors (such as temperature and humidity) and the effects of vibration and of repeated use, as well as misuse, of the product. During testing of prototypes, modifications in the original design, materials, or production methods may be necessary. After this phase has been completed, appropriate process plans, manufacturing methods, equipment, and tooling are selected, with the cooperation of manufacturing engineers, process planners, and others involved in production.

Virtual prototyping is a totally software form of prototyping, and uses advanced graphics and virtual-reality environments to allow designers to examine a part. In a way, this technology is used by CAD packages to render a part so that designers can observe and evaluate the part as it is drawn. However, virtual prototyping systems should be recognized as highly demanding cases of rendering part details.

1.3   Design for Manufacture, Assembly, Disassembly, and Service

It is apparent from the foregoing discussions that design and manufacturing should never be viewed as separate activities. Each part or component of a product must be designed so that it not only meets design requirements and specifications, but also can be manufactured economically. **Design for manufacture** (DFM) is a comprehensive approach to production of goods, and it integrates the design process with materials, manufacturing methods, process planning, assembly, testing, and quality assurance. This methodology requires that designers have a fundamental understanding of the characteristics, capabilities, and limitations of materials, manufacturing processes and related operations, machinery, and equipment. This knowledge includes such characteristics as variability in machine performance, dimensional accuracy, and surface finish of the workpiece, processing time, and the effect of processing method on part quality.

Designers and product engineers must be capable of assessing the impact of design modifications on process selection and on assembly, inspection, tools and dies, and product cost. Establishing quantitative relationships is essential in order to optimize the design for ease of manufacturing and assembly at minimum product cost. **Expert systems** (ES), which have optimization capabilities and thus can expedite the traditional iterative process in design optimization, are powerful tools in such analysis.

The individual components manufactured have to be assembled into a product. **Assembly** is an important phase of the overall manufacturing operation and requires consideration of the ease, speed, and cost of putting parts together (Fig. I.5). Furthermore, products must be designed so that **disassembly** is also possible, in order to enable the product to be taken apart with relative ease for maintenance, servicing, and recycling of its components. Because assembly operations can contribute significantly to product cost, **design for assembly** (DFA), as well as **design for disassembly,** are important aspects of manufacturing. Typically, a product that is easy to assemble is also easy to disassemble. Further important developments include **design for service**, the goal of which is that individual parts or sub-assemblies in a product be easy to reach and service.

Methodologies and computer software are available for DFA, utilizing 3-D conceptual designs and solid models. Subassembly and assembly times and costs are thus minimized while product integrity and performance are maintained; the system also improves the product's ease of disassembly. A natural outcome of these developments is **design for manufacture and assembly** (DFMA), which recognizes

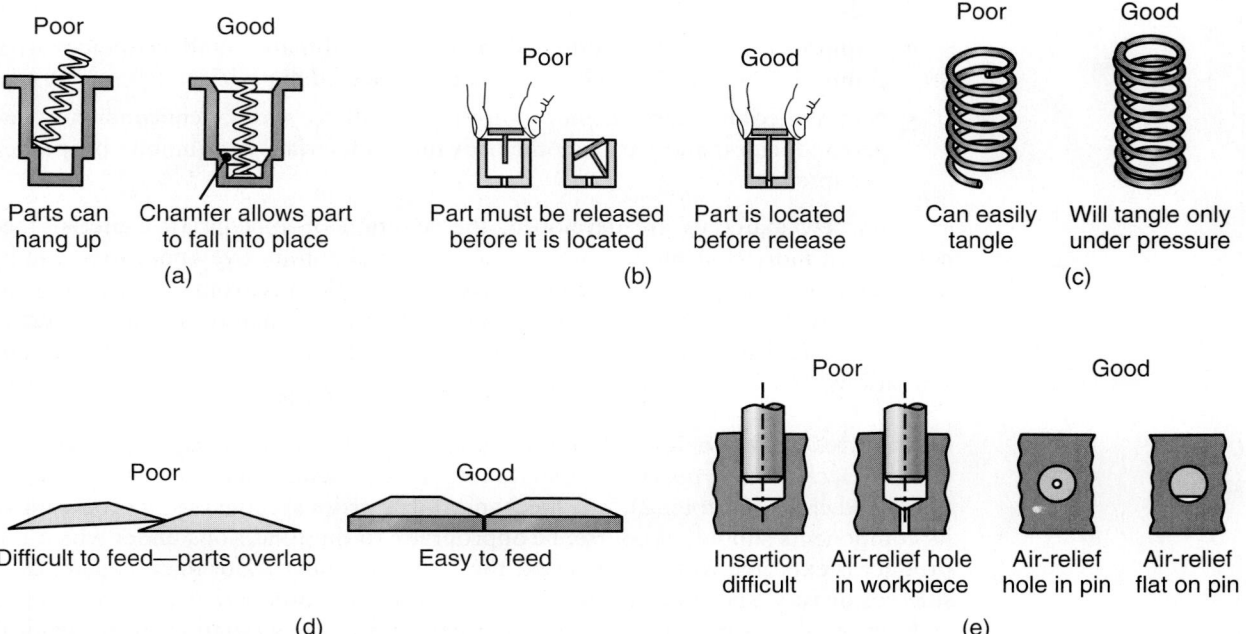

FIGURE I.5   Redesign of parts to facilitate assembly. *Source:* Reprinted from G. Boothroyd and P. Dewhurst, *Product Design for Assembly*, 1989. Courtesy of Marcel Dekker, Inc.

the inherent interrelationship among the manufacturing of components and their assembly into a final product.

There are several methods of assembly (e.g., use of fasteners or adhesives, or by welding, soldering, and brazing), each with its own characteristics and each requiring different operations. The use of a bolt and nut, for example, requires preparation of holes that must match in location and size. Hole generation, in turn, requires such operations as drilling or punching, which require additional time, require separate operations, and also produce scrap. On the other hand, products assembled with bolts and nuts can be taken apart and reassembled with relative ease.

Individual parts may be assembled either by hand or by automatic equipment and robots. The choice depends on factors such as the complexity of the product, the number of parts to be assembled, the care and protection required to prevent damage or scratching of finished surfaces of the parts, and the relative cost of labor compared to that of the machinery required for automated assembly.

I.4 | Selecting Materials

An ever-increasing variety of materials is now available, each having its own characteristics, applications, advantages, limitations, and costs (Part I). The following are the general types of materials used in manufacturing, either individually or in combination with other materials:

- **Ferrous metals:** carbon, alloy, stainless, and tool and die steels (Chapter 5).
- **Nonferrous metals:** aluminum, magnesium, copper, nickel, titanium, superalloys, refractory metals, beryllium, zirconium, low-melting alloys, and precious metals (Chapter 6).
- **Plastics (polymers):** thermoplastics, thermosets, and elastomers (Chapter 7).
- **Ceramics, glasses, glass ceramics, graphite, diamond,** and **diamond-like materials** (Chapter 8).
- **Composite materials:** reinforced plastics, metal-matrix and ceramic-matrix composites (Chapter 9), also known as **engineered materials.**
- **Nanomaterials, shape-memory alloys, amorphous alloys, semiconductor, superconductors,** and various other **advanced materials** with unique properties (Chapter 6).

As new materials are developed, the selection of appropriate materials becomes even more challenging. Aerospace structures, automotive applications, military equipment, and sporting goods have been at the forefront of new-material usage. There are continually shifting trends in the use of materials in all products, trends that are driven principally by economic considerations, but also by other considerations, as will be described below.

Properties of materials. When selecting materials for products, we first consider their *mechanical properties*: strength, toughness, ductility, hardness, elasticity, fatigue, and creep (Chapter 2). The mechanical properties specified for a product and its components should, of course, be appropriate to the conditions under which the product is expected to function. Next the *physical properties* of materials are considered: density, specific heat, thermal expansion and conductivity, melting point, and electrical and magnetic properties (Chapter 3). A combination of mechanical and physical properties is the strength-to-weight and stiffness-to-weight ratios of materials, particularly important for aerospace and automotive applications, as well

TABLE I.3

General Manufacturing Characteristics of Various Alloys			
Alloy	Castability	Weldability	Machinability
Aluminum	E	F	E–G
Copper	G–F	F	G–F
Gray cast iron	E	D	G
White cast iron	G	VP	VP
Nickel	F	F	F
Steels	F	E	F
Zinc	E	D	E

Note: E, excellent; G, good; F, fair; D, difficult; VP, very poor.

as for sports equipment. Aluminum, titanium, and reinforced plastics, for example, generally have higher such ratios than steels and cast irons.

Chemical properties also play a significant role, both in hostile and in normal environments. Oxidation, corrosion, general degradation of properties, toxicity, and flammability of materials are among the important factors to be considered. In some commercial airline accidents, for example, many deaths have been caused by toxic fumes from burning nonmetallic materials in the aircraft cabin.

The *manufacturing properties* of materials determine whether they can be cast, formed, machined, joined, and heat-treated with relative ease (Table I.3). The method or methods used to process materials to the desired final shapes can affect the product's performance, service life, and cost.

Cost and availability.   Cost and availability of raw as well as processed materials and of manufactured components are major concerns in manufacturing. The economic aspects of material selection are as important as technological considerations of the properties and characteristics of materials (Chapter 40). If raw or processed materials or manufactured components are not available in the desired shapes, dimensions, and quantities, substitutes and/or additional processing will be required, and these can contribute significantly to product cost. For example, if we need a round bar of a certain diameter and it is not available in standard form, then we have to purchase a larger rod and reduce its diameter by some means (such as machining, drawing through a die, or grinding). However, a product design can often be modified to take advantage of standard dimensions of raw materials and, thus, avoid extra manufacturing costs.

Reliability of supply, as well as demand for the material, affects cost. Most countries import numerous raw materials that are essential for manufacturing. For manufacturing purposes, the United States, for example, imports most of the following materials: natural rubber, diamond, cobalt, titanium, chromium, aluminum, and nickel. The broad geopolitical implications of such reliance on other countries are self-evident. Reliability of supply is also important in industries, especially automotive, when it is crucial for materials and components to arrive at then plant at the proper time intervals.

Different costs are involved in processing materials by different methods. Some methods require expensive machinery, others require extensive labor, and still others require personnel with special skills, a high level of education, or specialized training.

Appearance, service life, and recycling.   The appearance of materials, after they have been manufactured into products, influences their appeal to the consumer. Color, feel, and surface texture are characteristics that we all consider when making a decision about purchasing a particular product.

Time- and service-dependent phenomena, such as wear, fatigue, creep, and dimensional stability, also are important. These phenomena can significantly affect a product's performance and, if not controlled, can lead to malfunction or failure of the product. Similarly, compatibility of materials used in a product is important. Friction and wear, corrosion (including galvanic corrosion between mating parts made of dissimilar metals), and other phenomena can shorten a product's life or cause it to fail prematurely.

As we have become increasingly conscious of the need for conserving resources and for maintaining a clean and healthy environment, recycling or proper disposal of component materials at the end of a product's useful service life has become a major consideration. Note, for example, the use of biodegradable packaging materials and of recyclable plastic bottles and aluminum beverage cans. The proper treatment and disposal of toxic wastes and materials is also a major consideration.

EXAMPLE I.3   Material selection for U.S. pennies

Billions of pennies are minted each year. The materials used by the U.S. Mint to make pennies have undergone significant changes throughout history, mainly because of material shortages and the fluctuating cost of raw materials. The following table shows a chronological development of material substitutions (see Chapters 5 and 6) in pennies. These materials, or their combinations, need to impart appropriate properties to the pennies during their circulation and use.

1793–1837	100% copper
1837–1857	95% copper, 5% tin and zinc (bronze)
1857–1863	88% copper, 12% nickel (nickel-bronze)
1864–1962	95% copper, 5% tin and zinc (bronze)
1943 (WW II years)	Steel (plated with zinc)
1962–1982	95% copper, 5% zinc (bronze)
1982–present	97.5% zinc, plated with 2.5% copper

EXAMPLE I.4   Material selection for baseball bats

Baseball bats for the major leagues are generally made of northern white ash because of its high dimensional stability, elastic modulus, strength-to-weight ratio, and shock resistance. The bats are made on semiautomatic lathes (Section 23.3.4) and then subjected to finishing operations (Section 26.7). The straight uniform grain required for such bats has become increasingly difficult to find, particularly when the best wood comes from ash trees that are at least 45 years old.

For the amateur market, aluminum bats (top portion of Fig. I.6) have been made for some years by various metal forming techniques (Part III). Although at first their performance was not as good as that of wooden bats, the technology has been advanced such that aluminum bats are now made from high-strength aluminum tubing and possess desirable performance characteristics such as weight distribution, center of percussion, and sound and impact dynamics. They are usually filled with polyurethane or cork for sound damping, and for controlling the balance of the bat. However, aluminum bats can be sensitive to surface defects that develop during their normal use, and they usually fail from fatigue due to repeated use (cyclic loading, Section 2.7).

FIGURE I.6   Cross-sections of baseball bats made of aluminum (top portion) and composite material (bottom portion).

Developments in bat materials include composite materials consisting of high-strength graphite and glass fibers in an epoxy resin matrix (Chapter 9). The inner woven sleeve (lower portion of Fig. I.6) is made of Kevlar fibers, which add strength to and dampen vibrations in the bat. The price of these bats is about $125, and they perform and sound much like wooden bats. *Source*: Mizuno Sports, Inc.

I.5   Selecting Manufacturing Processes

An extensive and continuously expanding variety of manufacturing processes are used to produce parts and there is usually more than one method of manufacturing a part from a given material. The broad categories of the processing methods for materials are as follows, referenced to the relevant part in this text and illustrated with examples for each:

a. **Casting:** Expendable mold and permanent mold (Part II); Fig. I.7a.

b. **Forming and shaping:** Rolling, forging, extrusion, drawing, sheet forming, powder metallurgy, and molding (Part III); Figs. I.7b through 1.7d.

c. **Machining:** Turning, boring, drilling, milling, planing, shaping, broaching, and grinding, ultrasonic machining; chemical, electrical, and electrochemical machining; and high-energy-beam machining (Part IV); Fig. I.7e; this category also includes **micromachining** for producing ultraprecision parts (Part V).

    d. Joining: Welding, brazing, soldering, diffusion bonding, adhesive bonding, and mechanical joining (Part VI); Fig. I.7f.

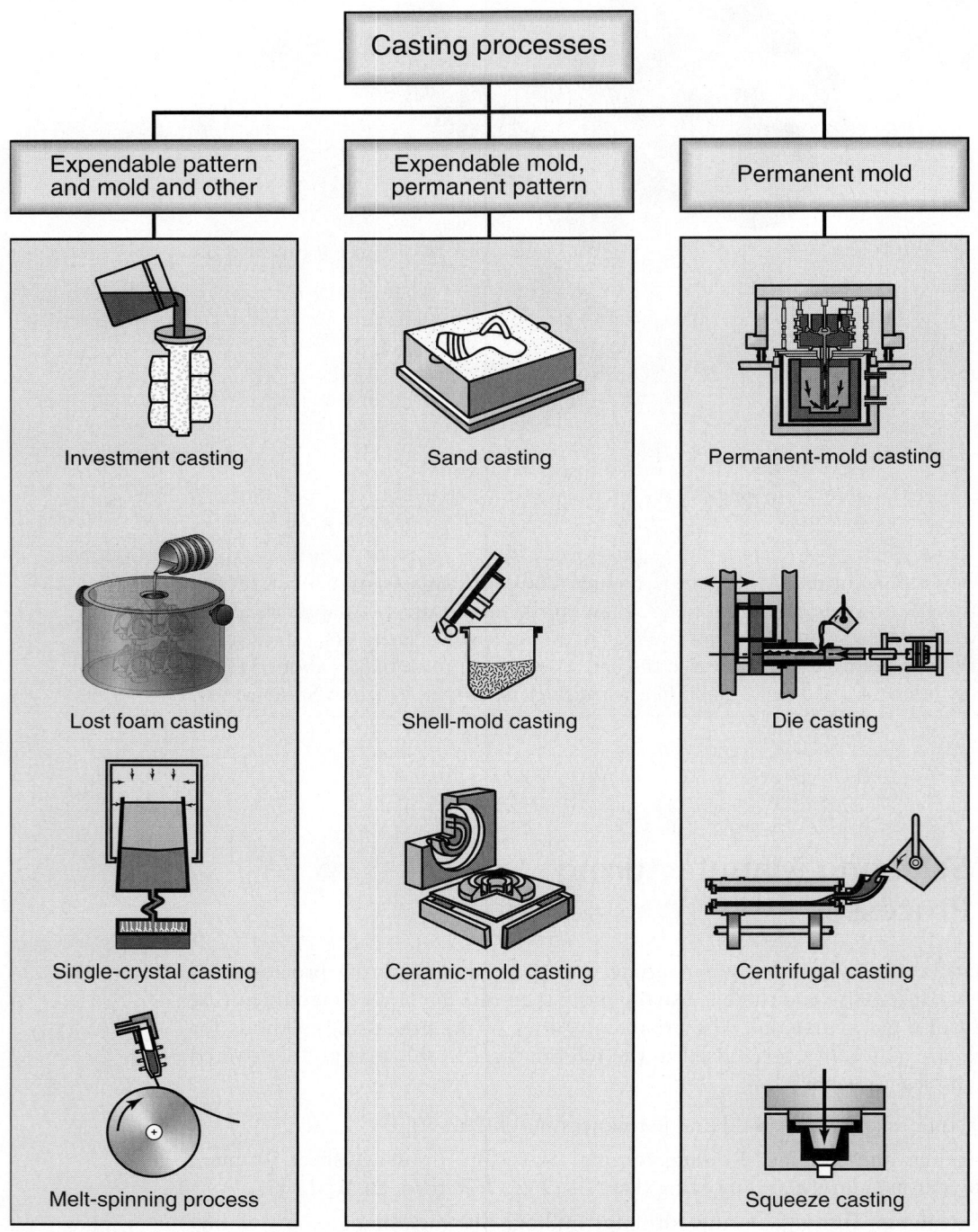

FIGURE I.7a   Schematic illustrations of various casting processes.

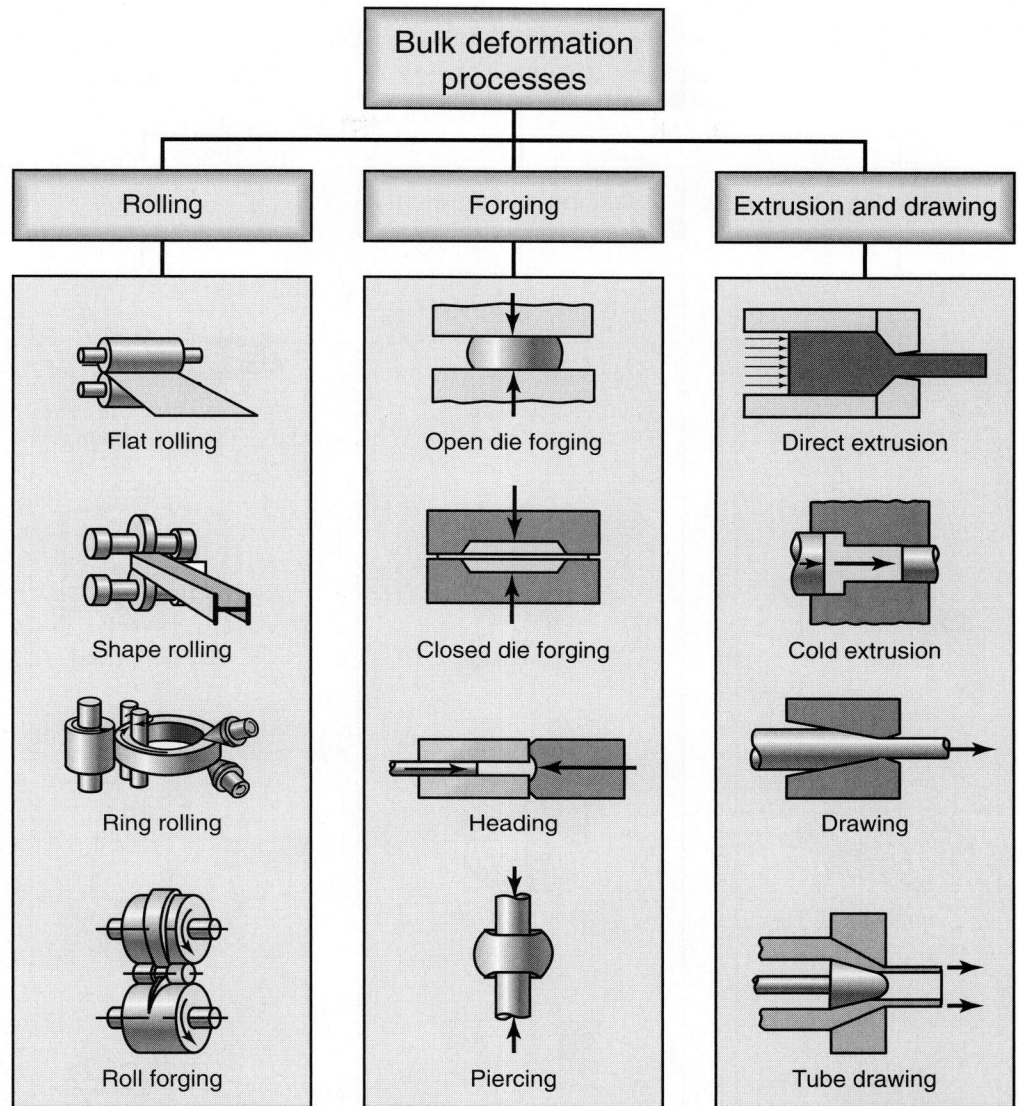

FIGURE I.7b   Schematic illustrations of various bulk deformation processes.

e. **Finishing:** Honing, lapping, polishing, burnishing, deburring, surface treating, coating, and plating (Chapters 26 and 35).

f. **Nanofabrication:** It is the most advanced technology and is capable of producing parts with dimensions at the nano level (one billionth); it typically involves processes such as etching techniques, electron-beams, and laser-beams. Present applications are in the fabrication of microelectromechanical systems (MEMS) and extending to nanoelectromechanical systems (NEMS), which operate on the same scale as biological molecules.

Selection of a particular manufacturing process, or a sequence of processes, depends not only on the shape to be produced but also on other factors pertaining to material properties. As examples, brittle and hard materials cannot be shaped or

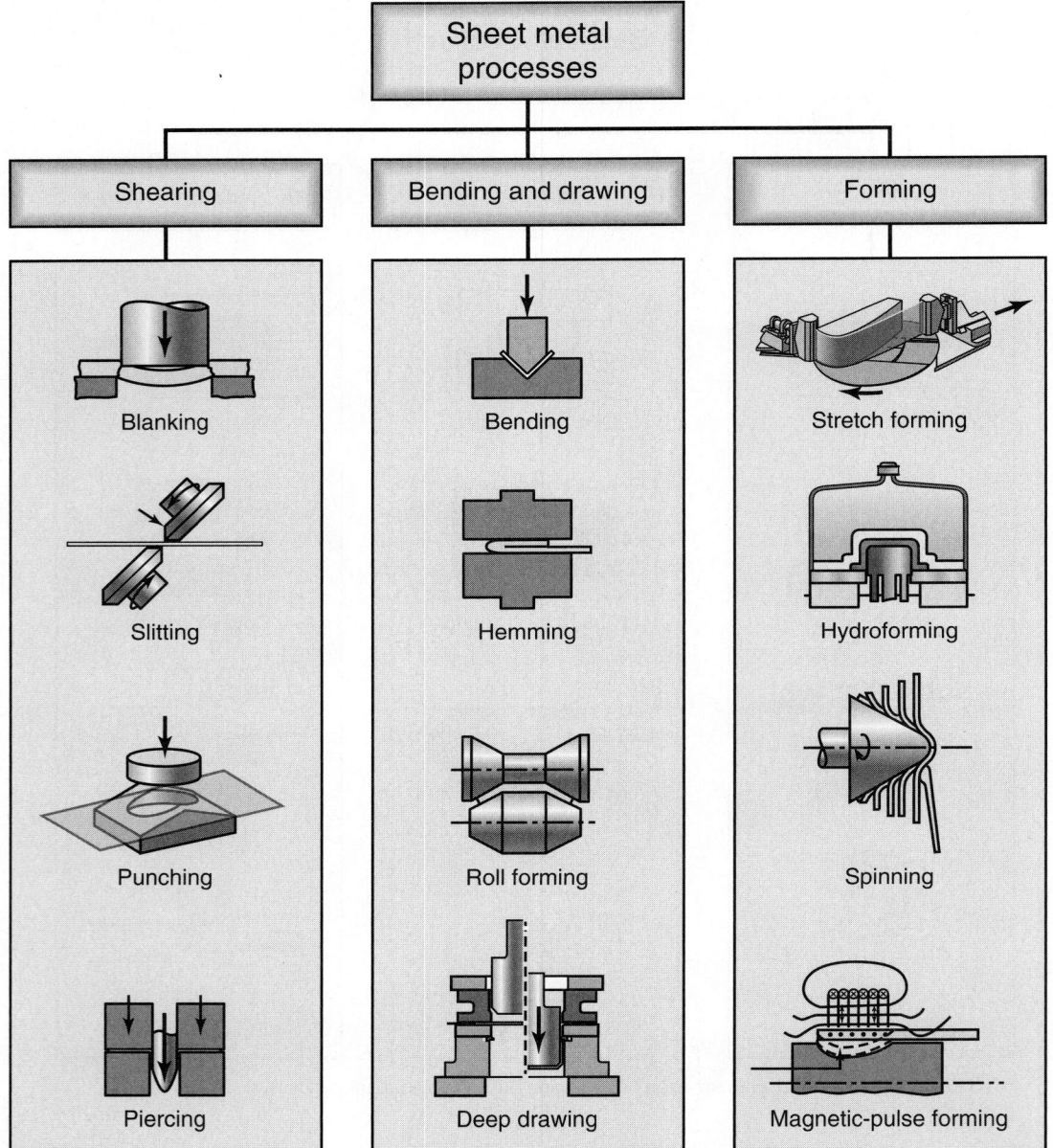

FIGURE I.7c   Schematic illustrations of various sheet-metal forming processes.

formed easily, whereas they can be cast, machined, or ground. Metals that are previously formed at room temperature become stronger and less ductile than they were before processing them, and thus will require higher forces and be less formable during subsequent (secondary) processing.

As described throughout this text, each manufacturing process has its own advantages and limitations, as well as production rates and product cost. Manufacturing engineers constantly are being challenged to find new solutions to manufacturing problems and cost reduction. For example, sheet metal parts traditionally have been

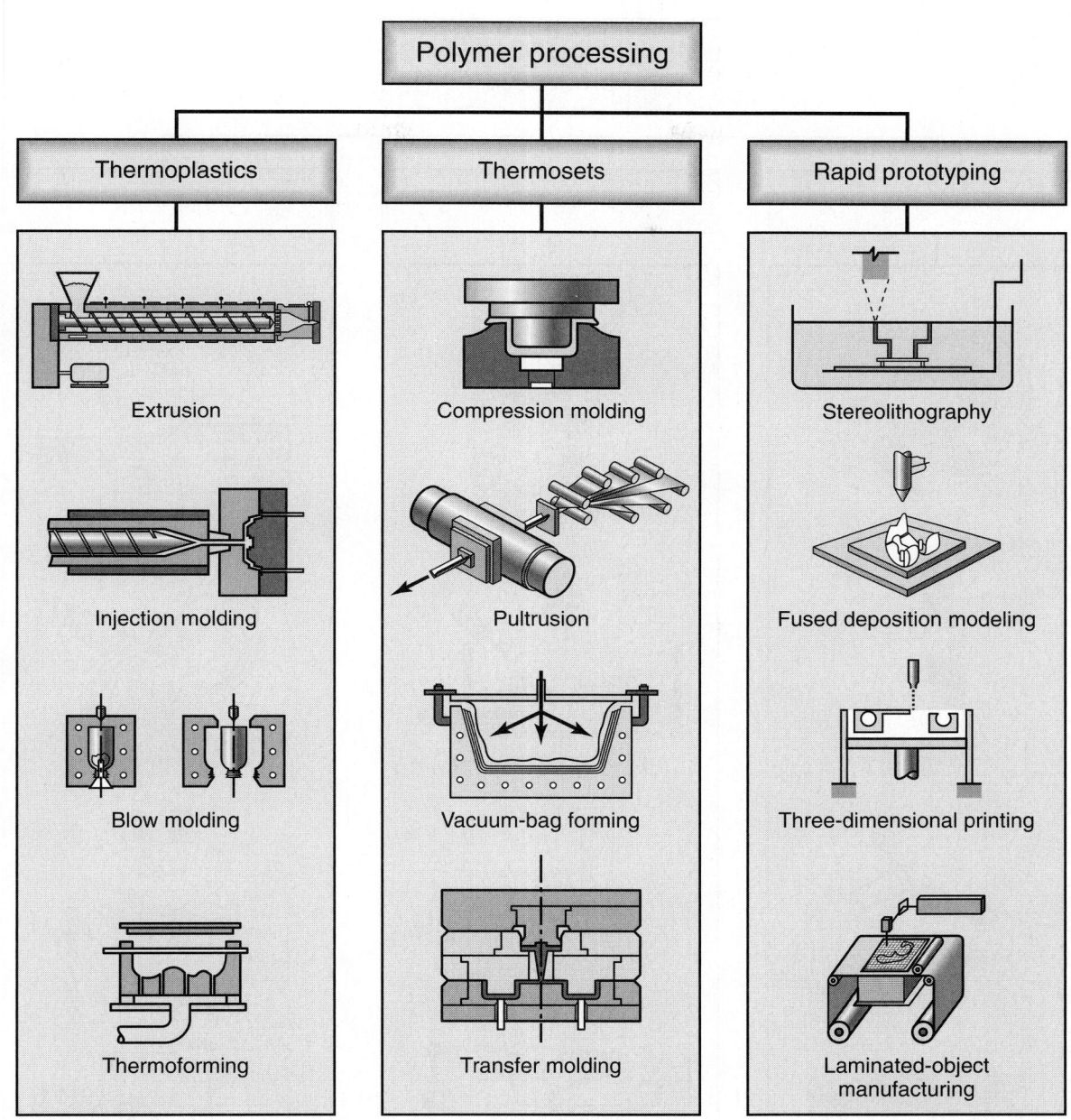

FIGURE I.7d   Schematic illustrations of various polymer processing methods.

cut and fabricated using common mechanical tools, punches, and dies. Although still widely used, some of these operations are being replaced by laser-cutting techniques (Fig. I.8) in which the path of the laser can be controlled, thus increasing the capability for producing a wide variety of shapes accurately, repeatedly, and economically, as well as eliminating the need for punches and dies. However, as expected, the surface produced by punching has different characteristics than that produced by laser cutting.

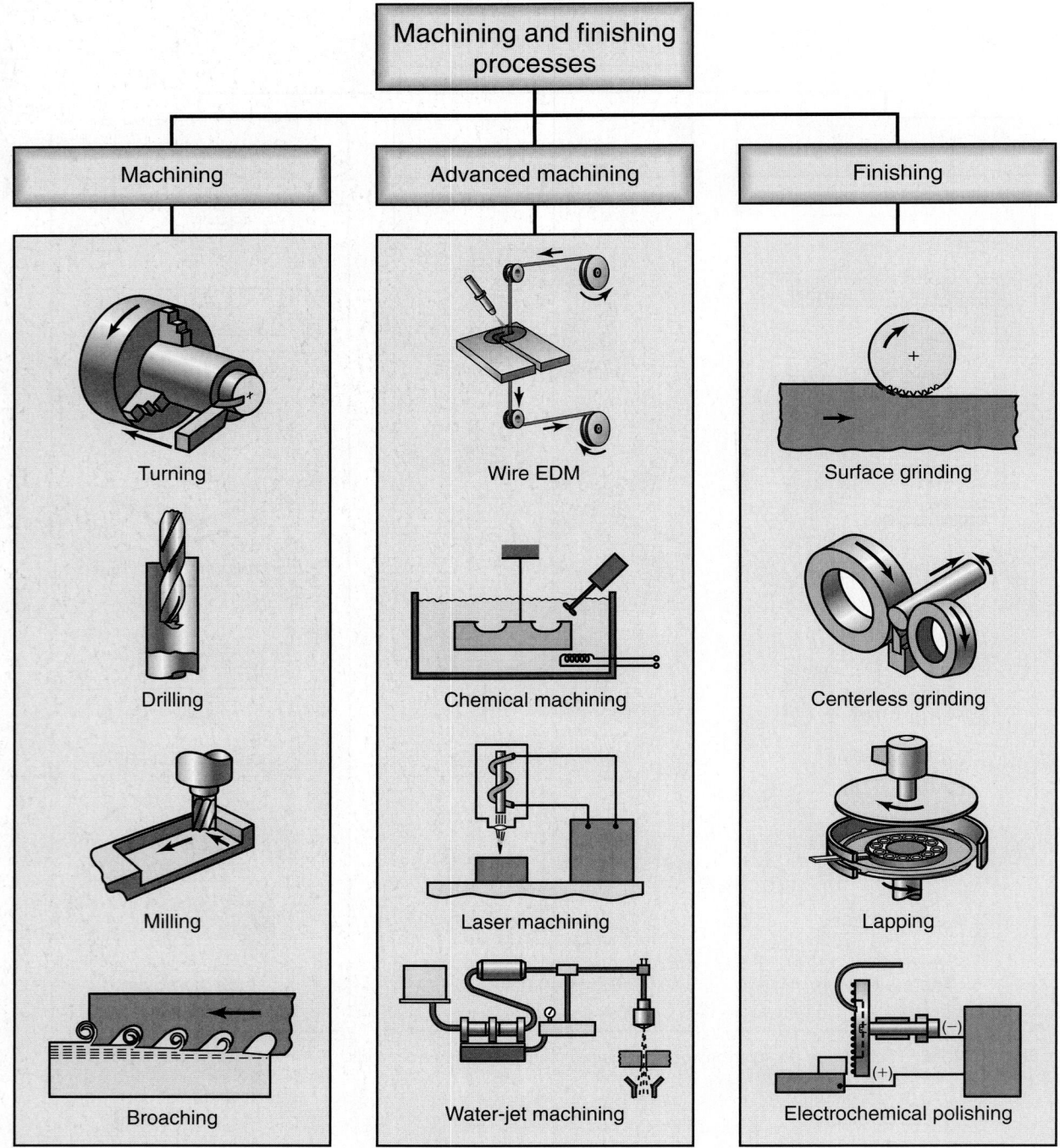

FIGURE I.7e Schematic illustrations of various machining and finishing processes.

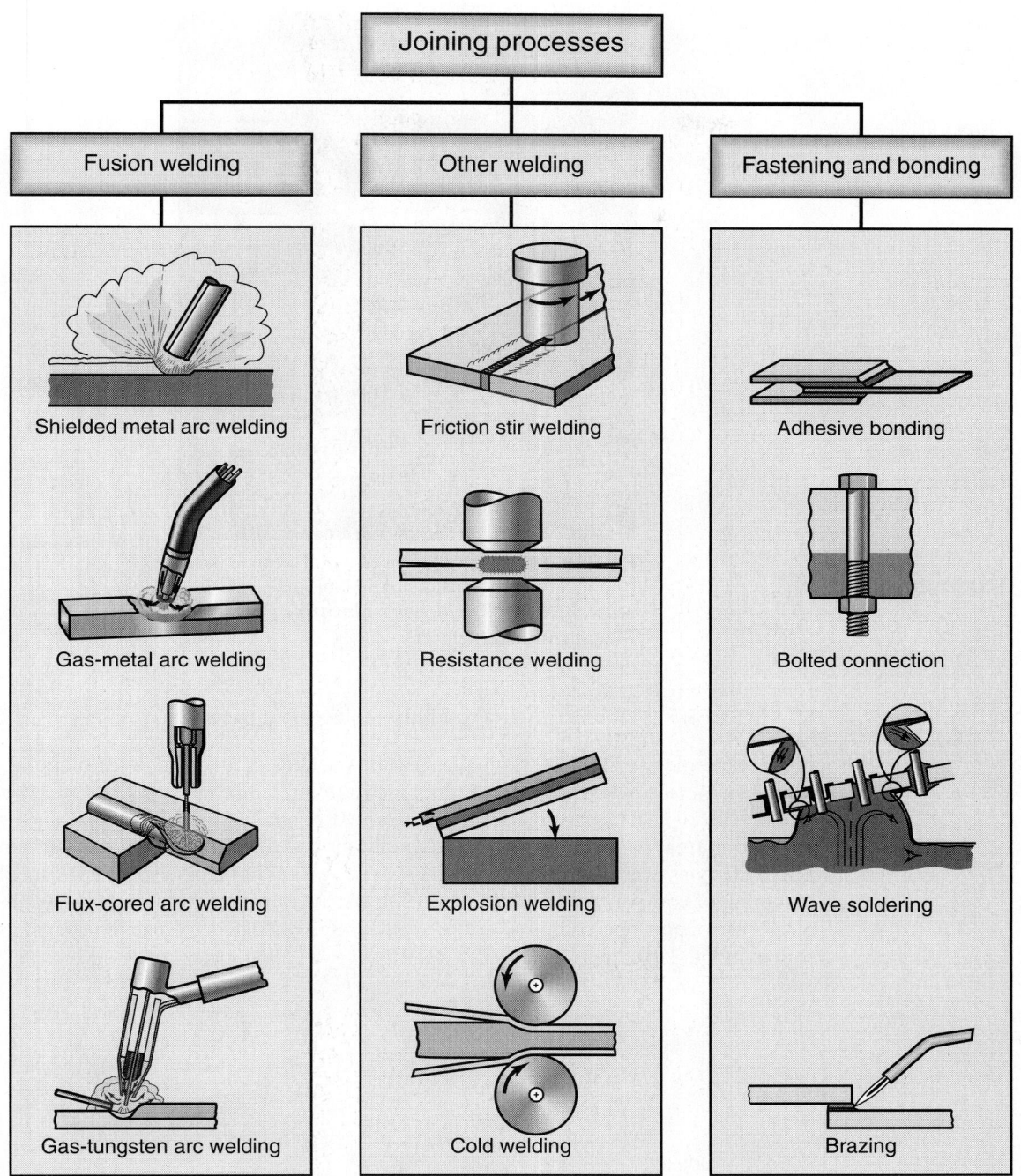

FIGURE I.7f   Schematic illustrations of various joining processes.

FIGURE I.8   Cutting sheet metal with a laser beam. *Source:* Courtesy of Rofin-Sinar, Inc. and *Manufacturing Engineering Magazine*, Society of Manufacturing Engineers.

EXAMPLE I.5 Manufacturing a total hip replacement stem

In this example, we describe the challenges and the choices involved in the manufacture of an increasingly important product such as a hip replacement insert. Over two million hip replacement surgeries are performed annually. Total hip replacements are used to treat painful, often arthritic conditions and allow their users to maintain active lifestyles even in later stages of their lives. A total hip replacement consists of a number of components (Fig. I.9), including a stem, ball, cup and liner. This example focuses upon one component, the stem, and is restricted to metallic stems, as shown in Fig. I.10.

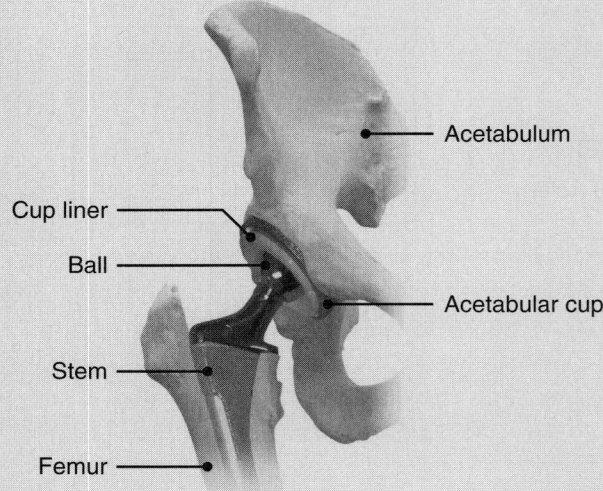

FIGURE I.9   Components of a total hip replacement. *Source:* Courtesy of Zimmer, Inc.

The hip stem is subjected to a very demanding environment. The normal load encountered by the stem is approximately 3 to 4 times the user's body weight during each step, and about two million steps are taken each year, thus the implant is subjected to fatigue loading (Section 2.7). Furthermore, when stumbling or tripping, the load can be 10 to 15 times the body weight. Once implanted, the stem is surrounded by tissue and fluids, and the implant's interaction with these tissues could lead to its corrosion.

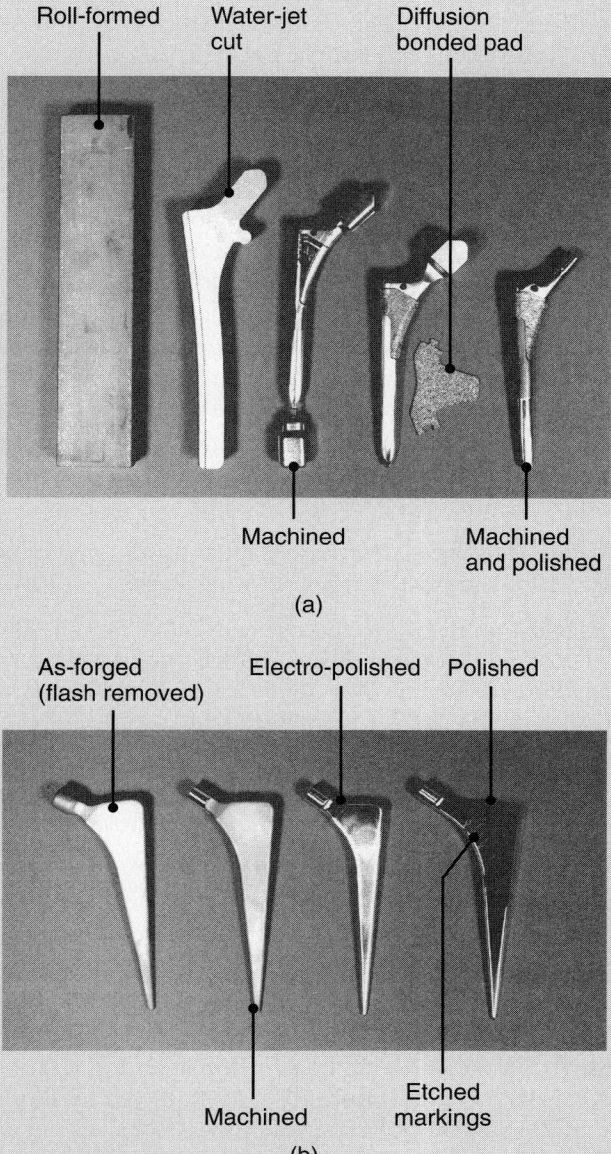

Roll-formed   Water-jet cut   Diffusion bonded pad

Machined   Machined and polished

(a)

As-forged (flash removed)   Electro-polished   Polished

Machined   Etched markings

(b)

FIGURE I.10   (a) Manufacturing steps in the production of a roll-formed and machined total hip replacement stem; (b) Manufacturing steps in the production of a forged stem. Hip stems also can be produced by investment casting, metal injection molding, insert injection molding, and assorted other processes. *Source:* Courtesy of Zimmer, Inc.

A number of material and process combinations have been used successfully for total hip replacements. Materials are restricted to options that are approved by governmental agencies for use inside the human body. Most hip stems are constructed from titanium alloy (Ti-6Al-4V; Section 6.7) or cobalt-chrome alloy (Section 6.6), although tantalum and stainless steels are also used.

There are several manufacturing strategies for producing total hip stems. Four approaches are outlined here.

Rolled and Machined Stems. One method of producing hip stems is shown in Fig. I.10a and involves the following steps:

- Titanium in the form of shape-rolled bar stock is used to produce blanks for the stem. Rolled stock has excellent material properties and is readily available. The blank is produced through water jet cutting (Section 27.8).
- The general shape of the stem is produced on a CNC milling machine (Section 24.2.8). Note that the blank at this stage still has some material for fixturing in the machine; these regions subsequently are removed.
- A fiber metal pad is then attached to a portion of the blank through diffusion bonding (Section 31.7). The pad is intended to promote bone ingrowth and provide good bonding between the implant and bone.
- The blank is finish machined and polished (Section 26.7). The machining produces the tapered head, which has tight tolerances so that it can mate properly with the ball. Note that some exterior surfaces have been highly polished in order to improve fatigue and corrosion resistance.
- Before it is ready for insertion into a patient, there are extensive additional processes performed, such as cleaning, passivation (Section 3.8), sterilization and packaging operations.

Forged and Machined Stems. Another method for producing hip stems is shown in Fig. I.10b, and involves the following steps:

- A blank is first produced by hot forging a cobalt-chrome alloy blank (Section 14.3). The blank shown in Fig. I.10b has had the flash trimmed from the blank, and is oversized to allow for subsequent machining to tight tolerances.
- The blank is then machined on a CNC milling machine, the remaining flash is removed, and the required contours and tolerances are obtained. Note that the taper has been machined into the stem at this step, since no further machining operations or fixturing of the blank into a workholder will be needed.
- Hot forging results in a significant oxide layer on the part, which is then removed by electropolishing (Section 26.7).
- The stem is then polished to obtain the smooth surface shown in the figure. A close examination indicates some additional markings that have been laser etched into the implant to aid the surgeon in gauging the degree of insertion into the femur.

Cast Stems. A common method for hip stem manufacture is to produce the stem from a cast cobalt-chrome alloy using the method of investment casting (Section 11.2.6). In this process, wax models of the stem are produced and assembled into a "tree," then coated with a ceramic coating. The mold is inverted and placed into a furnace, where the wax melts away, and the ceramic particles fuse to increase the mold strength. The heated mold is then placed into a casting machine, where the mold is inverted and filled with molten metal. The mold is allowed to cool, solidifying the metal. The solid metal is then removed from the tree and machined, as described above, to achieve the desired dimensional tolerances and surface finishes.

Powder Metallurgy. Metal injection molding (MIM) also can be used to produce hip stems. In this process (Section 17.3), a heated mold with a cavity matching the hip stem geometry is filled with a viscous mixture of metal powder and binders. Inside the heated mold, the binder fuses, so when the part is ejected it has sufficient strength for handling purposes. The stem is then transferred to a furnace, where the binder burns off and the metal particles sinter, or bond, to provide sufficient strength for the implant.

Clearly, a variety of manufacturing process and material combinations are available and have been commercially successful for making these implants. Many products have more than one viable manufacturing path. It is apparent that selecting a manufacturing path is among the creative tasks and challenges faced by manufacturing engineers.

Source: Courtesy of M. Hawkins, Zimmer, Inc.

Dimensional accuracy and surface finish. The dimensions and shape complexity of the part have a major bearing on the manufacturing process selected to produce it. For example, (a) flat parts with thin cross-sections cannot be cast properly; (b) complex parts cannot be formed easily and economically, whereas they may be cast or else fabricated from individual pieces; (c) dimensional tolerances and surface finish obtained in hot-working operations cannot be as fine as those obtained in cold-working (forming at room temperature) operations because dimensional changes, warping, and surface oxidation occur during processing of materials at elevated temperatures; and (d) some casting processes produce a better surface finish than others because of the different types of mold materials used.

The sizes and shapes of manufactured products vary widely. For example, the main landing gear for the twin-engine, 400-passenger Boeing 777 jetliner is 4.3 m (14 ft) high, with three axles and six wheels; it is made by a combination of forging and machining processes. The diameter of a runner (buckets and shroud) for a hydraulic turbine is 4.6 m (180 in.) in diameter, and weighs 50,000 kg (110,000 lb); it is made by sand casting. The rotor for a large steam turbine is made by hot forging and machining, and weighs 300,000 kg (700,000 lb).

At the other dimensional extreme are the production of microscopic gears (Fig. I.11a) and the product shown in Fig. I.11b. This is a mechanism in which a mirror is suspended on a torsional beam and can be inclined through electrostatic attraction by applying a voltage on either side of the mirror (note the 100 μm scale at the bottom of the figure). These manufacturing operations are called *nanofabrication.*

Ultraprecision manufacturing techniques and machinery are being developed rapidly and are coming into more common use. For machining mirror-like metal surfaces, for example, the cutting tool is a very sharp diamond and the equipment has very high stiffness; it is operated in a room where the temperature is controlled to within 1°C. Highly sophisticated techniques such as molecular-beam epitaxy and scanning-tunneling engineering are being implemented to obtain accuracies on the order of the atomic lattice (0.1 nm; 10^{-8} in.).

Operational and manufacturing costs. The design and cost of tooling, the lead time required to start production, and the effect of workpiece material on tool and die life are among other major considerations. Depending on product size and shape the cost of tooling can be substantial. For example, a set of steel dies for stamping sheet-metal fenders for automobiles may cost about $2 million. For parts made from expensive materials, the lower the scrap rate, the more economical the production process will be; thus, every attempt has to be made to *minimize waste.* Because it generates chips, machining may not be economical compared to forming operations, all other factors being the same.

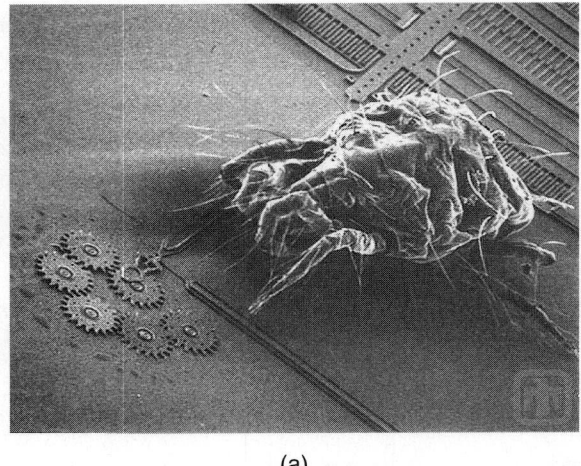

(a)

(b)

FIGURE I.11   (a) Microscopic gears with dust mite *Source:* Courtesy Sandia National Laboratory; (b) A movable micromirror component of a light sensor. *Source:* Courtesy of Richard Mueller, University of California at Berkeley.

Availability of machines and equipment within the manufacturing facility, and the experience of operating personnel, are also important cost factors. If these capabilities are not available, some parts have to be manufactured by outside companies (*outsourcing*). Automakers and appliance manufacturers, for example, purchase many parts from outside vendors, or they have them made by outside firms to their specifications.

The *quantity* of parts to be made and the *production rate* (pieces per hour) are important in determining the processes to be used and the economics of production. Beverage cans, gears, and computers, for example, are produced in numbers and at rates much higher than are jet engines, tractors, or ships. Small quantities (small lot size) can be manufactured in **job shops** on general-purpose machines. **Small-batch production**, for annual quantities from about 10 to 100, is done on similar machines as well as on machines with various computer controls. **Batch production** usually involves lot sizes between 100 and 5000, and **mass production** in lot sizes often over 100,000, using special-purpose machinery (*dedicated machines*) and various automated equipment for transferring materials and parts.

The particular manufacturing process and the operation of machinery can have significant environmental and safety implications. Some processes adversely affect the environment, such as the use of oil-based lubricants in hot metalworking operations, or the use of various chemicals in heat treating, coating, and cleaning operations. Unless properly controlled, these and other activities cause air, water, and noise pollution. The safe use of machinery is an additional important consideration, requiring precautions to eliminate hazards in the workplace.

Consequence of improper selection of materials and processes. Numerous examples of product failure can be traced to improper selection of materials or of manufacturing processes, or to insufficient control of process variables. A component or a product generally is considered to have failed when:

- It stops functioning (broken shaft, gear, bolt, cable, or turbine blade).
- It does not function properly or perform within required specifications (worn bearings, gears, tools, and dies).
- It becomes unreliable or unsafe for further use (crack in a shaft, poor connection in a printed-circuit board, or delamination of a reinforced plastic component).

Throughout this text, we describe the types of failure of a component or product that may have resulted from design deficiencies, improper material selection, material defects, manufacturing-induced defects, improper component assembly, and improper product use.

Net-shape manufacturing. A particular manufacturing process may not produce a finished part, and thus additional operations may be necessary. For example, a forged part may not have the desired dimensions or surface finish; as a result, additional operations such as machining or grinding may be necessary. Likewise, it may be difficult, impossible, or uneconomical to produce a part with holes in it by using a single manufacturing process, and thus a subsequent process may be required, such as drilling or producing the hole using various advanced methods, such as chemical or electrical means. Furthermore, the holes produced by a particular manufacturing process may not have the proper roundness, dimensional accuracy, or surface finish, and thus they may require an additional operation, such as honing.

Additional operations on a part can contribute significantly to the cost of a product. Consequently, the concept of *net-shape* or *near-net-shape manufacturing* has become very important, in which by the first operation the part is made as close to the final desired dimensions, tolerances, surface finish, and specifications as possible. Typical examples of such manufacturing methods are near-net-shape forging (Chapter 13) and casting (Chapter 11), stamped sheet metal (Chapter 16), injection molding of plastics (Chapter 19), and components made by powder-metallurgy techniques (Chapter 17).

EXAMPLE I.6 Manufacturing a salt shaker and pepper mill

A commonly used household product is a salt shaker and pepper mill set. The one shown in Fig. I.12 contains metallic as well as nonmetallic components. The main components are made by injection molding of a thermoplastic, such as acrylic, which has both transparency and other desirable characteristics and is easy to mold. The top of the salt shaker is made of sheet metal and is plated for appearance. The knob on the top of the pepper mill is made by machining and is threaded on the inside to allow for screwing and unscrewing of the knob. A square rod, connecting the top portion of the pepper mill to the two pieces shown at the bottom of the figure, is made by rolling. The two grinder components are made of stainless steel by powder-metallurgy techniques. An analysis indicated that casting or machining them would have been too costly.

FIGURE I.12   A salt shaker and pepper mill set. The two metal pieces (at the bottom) for the pepper mill are made by powder-metallurgy techniques. *Source:* Reproduced with permission from *Success Stories on P/M Parts*, Metal Powder Industries Federation, Princeton, NJ, 1998.

I.6 Environmentally Conscious Design and Manufacturing

In the United States alone, nine million passenger cars and about 300 million tires are discarded each year; about 100 million of those tires are reused in various ways. More than five billion kilograms of plastic products are discarded each year, and every three months, industries and consumers discard enough aluminum to rebuild the country's commercial air fleet. Furthermore, (a) lubricants and coolants are often used in most manufacturing operations; (b) various fluids and solvents are used in cleaning manufactured products, some of these fluids pollute the air and waters during their use; (c) many by-products from manufacturing plants have been discarded for years (i.e., sand containing additives used in metal-casting processes; water, oil, and other fluids from heat-treating facilities and from plating operations; slag from foundries and from welding operations); and (d) a wide variety of metallic and nonmetallic scrap, produced in operations such as sheet forming, casting, and molding.

The adverse effects of these activities, their damage to our environment and to the earth's ecosystem, and, ultimately, their effect on the quality of human life are by well recognized. Major concerns are water and air pollution, acid rain, ozone depletion, the greenhouse effect, hazardous wastes, landfill seepage, and global warming. In response, numerous laws and regulations have been promulgated in the United States and in other industrialized countries.

Much can be gained by detailed and careful analysis of products, their design, the types of materials used in them, the manufacturing processes and practices employed in making them, and the waste produced. Certain guidelines can be followed in such an analysis:

- Reducing waste of materials, by refinements in product design and reducing the amount of materials used.
- Reducing the use of hazardous materials in products and processes.
- Conducting research and development into environmentally safe products and into manufacturing technologies.
- Ensuring proper handling and disposal of all waste.
- Making improvements in recycling, waste treatment, and reuse of materials.

Major developments continually are taking place regarding these matters and the term **environmentally conscious design and manufacturing** is now in common usage in manufacturing industries. A major emphasis is on **design for the environment** (DFE) or **green design**. This approach anticipates the possible adverse environmental impact of materials, products, and processes, so that it can be taken into account at the earliest stages of design and production. The main objectives are to prevent pollution at the source and to strongly promote recycling and reuse instead of disposal. These goals have led to the concept of **design for recycling** (DFR). Green design has far-reaching implications for many manufacturing processes. For example, in the automotive industry there is a strong desire to improve fuel economy without compromising performance, safety, or luxury. One way of simultaneously achieving these goals is to use materials with high strength-to-weight ratios, which has encouraged consideration of aluminum alloys, metal-matrix and fiber-reinforced polymer composites, and optimizing designs throughout automobiles.

In the U.S. automotive industry, for example, about 75% of car parts, mostly metals, are now recycled, and there are continued plans to recycle the rest as well, including plastics, glass, rubber, and foam. The benefits of recycling also are evident from a study showing that producing aluminum from scrap, instead of from bauxite ore, costs only one-third as much, and reduces energy consumption and pollution by more than 90%. Cartridges for copiers and printers are now returnable by the customer to the manufacturer, which repairs and replaces some of the parts and resells the cartridge. This means that cartridges must be designed for ease of disassembly, such as by the use of snap fits instead of screws which require more time for their removal.

The ISO 14000 standards concern environmental management. These standards address what a company can do to minimize harmful environmental effects cause by its activities and achieve continual improvement of its environmental performance.

I.7 | Computer-Integrated Manufacturing

Beginning with computer graphics and computer-aided modeling, design, and manufacturing, the use of computers has been extended to **computer-integrated manufacturing** (CIM) in which software and hardware are integrated from product concept through product distribution in the marketplace. Computer-integrated manufacturing is particularly effective because of its capability for making possible:

- Responsiveness to rapid changes in market demand and product modifications.
- Better use of materials, machinery, and personnel, and reduction in inventory.
- Better control of production and management of the total manufacturing operation.
- The manufacture of high-quality products at low cost.

The following is an outline of the major applications of computers in manufacturing, described in detail in Chapters 38 and 39:

a. **Computer numerical control** (CNC). First implemented in the early 1950s, this is a method of controlling the movements of machine components by direct insertion of coded instructions in the form of numerical data.

b. **Adaptive control** (AC). The parameters in a manufacturing process are adjusted automatically to optimize production rate and product quality, and to minimize cost. Parameters such as forces, temperatures, surface finish, and dimensions of the part are monitored constantly; if they move outside the acceptable range, the system adjusts the process variables until the parameters again fall within the specified range.

c. **Industrial robots.** Introduced in the early 1960s, industrial robots (Fig. I.13) have been replacing humans in operations that are repetitive, dangerous, and boring, thus reducing the possibility of human error, decreasing variability in product quality, and improving productivity. Robots with sensory-perception capabilities have been developed (*intelligent robots*), with movements that simulate those of humans.

d. **Automated handling of materials.** Computers have made possible highly efficient handling of materials and components in various stages of completion (*work in progress*), such as when being moved from storage to machines, from machine to machine, and at points of inspection, inventory, and shipment.

FIGURE I.13   Automated spot welding of automobile bodies in a mass production line. *Source:* Courtesy of Ford Motor Company.

e. **Automated** and **robotic assembly systems.** These systems mainly have replaced costly assembly by human operators, although humans still have to perform some of these operations. Products are now designed or redesigned so that they can be assembled more easily and faster by machines.

f. **Computer-aided process planning** (CAPP). This system is capable of improving productivity by optimizing process plans, reducing planning costs, and improving the consistency of product quality and reliability. Functions such as cost estimating and monitoring of work standards (time required to perform a certain operation) are also incorporated into the system.

g. **Group technology** (GT). The concept of group technology is that parts can be grouped and produced by classifying them into families, according to similarities in design and similarities in the manufacturing processes employed to produce the part. In this way, part designs and process plans can be standardized and families of similar parts can be produced efficiently and economically.

h. **Just-in-time production** (JIT). The principal of JIT is that supplies of raw materials, parts, and components are delivered to the manufacturer just in time to be used, parts and components are produced just in time to be made into subassemblies and assemblies, and products are finished just in time to be delivered to the customer. As a result, inventory-carrying costs are low, part defects are detected right away, productivity is increased, and high-quality products are made at low cost.

i. **Cellular manufacturing** (CM). This system utilizes workstations (*manufacturing cells*) that usually contain several production machines controlled by a central robot, each machine performing a different operation on the part.

j. **Flexible manufacturing systems** (FMS). These systems integrate manufacturing cells into a large unit, all interfaced with a central computer. Although very costly, FMS is capable of efficiently producing parts in small runs and of changing manufacturing sequences on different parts quickly; this flexibility enables them to meet rapid changes in market demand for all types of products.

k. **Expert systems** (ES). These systems basically are complex computer programs; they have the capability to perform various tasks and solve difficult real-life problems much as human experts would.

l. **Artificial intelligence** (AI). This important field involves the use of machines and computers to replace human intelligence. Computer-controlled systems are now capable of learning from experience and of making decisions that optimize operations and minimize costs. **Artificial neural networks** (ANN), which are designed to simulate the thought processes of the human brain, have the capability of modeling and simulating production facilities, monitoring and controlling manufacturing processes, diagnosing problems in machine performance, conducting financial planning, and managing a company's manufacturing strategy.

In view of these continuing advances and their potential, some experts have envisioned the **factory of the future.** Although highly controversial, and viewed as unrealistic by some, this is a system in which production will take place with little or no direct human intervention. Ultimately, the human role is expected to be confined to the supervision, maintenance, and upgrading of machines, computers, and software.

EXAMPLE I.7 Application of CAD/CAM to make a mold for sunglasses

The metal mold used for injection molding of plastic sunglasses is made on a CNC milling machine, by using a cutter (ball-nosed end mill), as illustrated in Fig. I.14. First, a model of the sunglasses is made in a computer-aided design software package, from which a model of the mold is automatically generated. The geometric information is sent to a computer numerical control milling machine and the machining steps are planned.

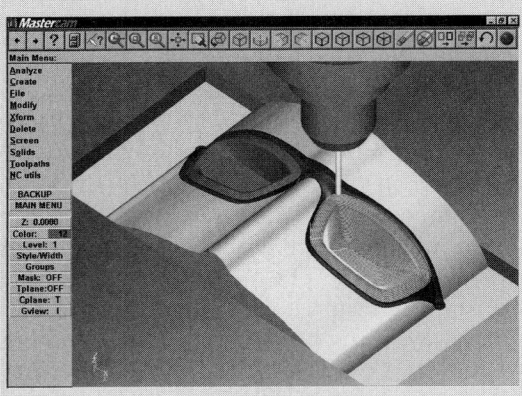

(a)

(b)

FIGURE I.14 Machining a mold cavity for making sunglasses. (a) Computer model of the sunglasses as designed and viewed on the monitor. (b) Machining of the die cavity using a computer numerical-control milling machine. (c) Final product. *Source:* Courtesy of Mastercam/CNC Software, Inc.

(c)

FIGURE I.14   (*continued*)

Next, an offset is added to each surface to account for the nose radius of the end mill, thus determining the cutter path (the path of the center of the machine spindle). The NC programming software then executes this cutting program on the CNC milling machine, producing the die cavity with appropriate dimensions and accuracy. Electrical-discharge machining (Section 27.5) can also be used to make this mold; however, it was found that EDM was about twice as expensive as CNC machining the mold and it had less dimensional accuracy. *Source:* Courtesy of Mold Threads Inc.

I.8   Lean Production and Agile Manufacturing

The trends described thus far have led to the concept of lean production or lean manufacturing. Although not a novel concept, *lean production* is a methodology that involves a thorough assessment of each of the activities of a company in order to minimize waste at all levels. These include the efficiency and effectiveness of all its operations, the efficiency of the machinery and equipment, the number of personnel involved in each operation, and the possible dispensing of some of its operations and managers. This approach continues with a comprehensive analysis of the costs of each activity, including those due to productive and to nonproductive labor.

This strategy requires a fundamental change in corporate culture, as well as an understanding of the importance of *cooperation and teamwork* among management and the work force. Its results do not necessarily require cutting back on resources; rather, it aims at continually improving the efficiency and the profitability of the company by removing all waste and by dealing with problems as soon as they appear.

Agile manufacturing is a term indicating the implementation of the principles of lean production on a broad scale. The principle behind agile manufacturing is ensuring agility (hence *flexibility*) in the manufacturing enterprise, so that it can respond rapidly to changes in product demand and in customer needs. This flexibility is to be achieved through people, equipment, computer hardware and software, and advanced communications systems. As an example, it has been predicted that the automotive industry could configure and build a car in three days, and that, eventually, the traditional assembly line will be replaced by a system in which a nearly custom-made car will be produced by connecting individual modules.

These approaches require that a manufacturer *benchmark* its operations, that is, the competitive position of other manufacturers with respect to its own, and realistic goals for the future. Thus, benchmarking becomes a reference from which various measurements can be made and to which they can be compared.

I.9 Quality Assurance and Total Quality Management

Product quality is one of the most important aspects of manufacturing because it directly influences the marketability of a product and customer satisfaction. The practice of inspecting products after they are made largely has been replaced by the broader view that *quality must be built into a product*, from the design stage through all subsequent stages of manufacture and assembly. Because products typically are made by using several manufacturing processes, each of which can have significant variations in its performance even within a short period of time, the control of processes is a critical factor in product quality. Thus, the objective should be *to control processes, not products.*

Producing defective products can be very costly to the manufacturer, creating difficulties in assembly operations, necessitating repairs in the field, and resulting in customer dissatisfaction. **Product integrity** is a term that is generally used to define the degree to which a product: (a) is suitable for its intended purpose; (b) responds to a real market demand; (c) functions reliably during its life expectancy; and (d) can be maintained with relative ease.

Total quality management (TQM) and **quality assurance** must be the responsibility of everyone involved in designing and manufacturing a product. Pioneers in quality control, primarily Deming, Taguchi, and Juran, have emphasized the importance of management's commitment to product quality, pride of workmanship at all levels of production, and the use of **statistical process control** (SPC) and **control charts** (Chapter 36) for on-line monitoring of part production and for rapidly identifying sources of quality problems. The major goal is to prevent defects from occurring, rather than to detect and reject defective products after they are made.

We now have the capability of producing computer chips such that only a few parts out of a million may be defective. The levels of defects are identified as, for example, three sigma and four sigma, with the highest being **six sigma**. Three sigma in manufacturing would result in 2700 defective parts per million, which is an unacceptable rate in modern manufacturing. In fact, it has been estimated that at this level no modern computer would function reliably. At six sigma, defects are reduced to only 3.4 defects per million.

Important developments in quality assurance include the implementation of **experimental design**, a technique in which the factors involved in a manufacturing process and their interactions are studied simultaneously. For example, variables affecting dimensional accuracy or surface finish in a machining operation can be identified readily, and thus making it possible for appropriate preventive actions to be taken.

Global manufacturing and competitiveness has created an obvious need for international conformity in the use of, and for consensus, regarding the establishment of quality control methods. This need has resulted in the ISO 9000 standards series on Quality Management and Quality Assurance Standards, as well as QS 9000 (Chapter 37). A company's registration for this standard, which is a *quality process certification* and not a product certification, means that the company conforms to consistent practices as specified by its own quality system. ISO 9000 and QS 9000 have permanently influenced the manner in which companies conduct business in world trade, and they have become the world standard for quality.

Product liability. We are all familiar with the consequences of a product's malfunction or failure, possibly causing bodily injury (or even death) and financial loss to a person or an organization. This important topic is referred to as *product liability*. Laws generally vary from state to state, and from country to country. Designing and manufacturing safe products is an essential and integral aspect of a manufacturer's responsibilities. All those involved with product design, manufacture, and marketing must fully recognize the consequences of product failure, including failures occurring during possible misuse of the product. Numerous examples of products that could involve liability can be cited, such as: (a) a grinding wheel that shatters and blinds a worker; (b) a cable that snaps, allowing a platform to drop; (c) brakes that become inoperative because of the failure of a component; (d) machines with no guards or inappropriate guards; and (e) electric and pneumatic tools without proper warnings.

Human-factors engineering and **ergonomics** deal with human versus machine interactions and thus are important aspects of the design and manufacture of safe products. Examples include: (a) an uncomfortable or unstable workbench or chair; (b) a mechanism that is difficult to operate manually, causing back injury; and (c) a poorly designed keyboard that causes pain to the user's hands and arms after repetitive use.

I.10   Global Competitiveness and Manufacturing Costs

The economics of manufacturing have always been a major consideration, and it has become even more so as **global competitiveness** for high-quality products (**world-class manufacturing**) and *low prices* have become a necessity in worldwide markets. Beginning with the 1960s, the following trends developed that have had a major impact on manufacturing:

- Global competition increased rapidly, and the markets became multinational and dynamic.
- Market conditions fluctuated widely.
- Customers demanded high-quality, low-cost products and on-time delivery.
- Product variety increased substantially and products became complex, and product life cycles became shorter.

In the late 1990s, a further important trend has been the wide disparity in manufacturing labor costs (by an order of magnitude) among various countries. Table I.4 shows the estimated relative hourly compensation for production workers in manufacturing, based on a scale of 100 for the United States.

It is not surprising that many of the products one purchases today are either made in China or assembled in Mexico, where labor costs, thus far, are the least, but

TABLE I.4

Approximate Relative Hourly Compensation for Production Workers, for 2003. United States = 100. Compensation Costs Vary Depending on Benefits and Allowance.			
Denmark	147	Ireland, Italy	85
Norway	144	Spain	67
Germany	136	Israel	53
Belgium, Switzerland	127	New Zealand, Korea	48
Finland, Netherlands	123	Singapore	33
Austria, Sweden	116	Portugal, Taiwan	27
United States	100	Czech Republic	20
France	96	Brazil, Mexico	11
United Kingdom	93	China, India	10
Australia, Canada, Japan	90		
European countries	111		
Asian countries	33		

Source: Courtesy of U.S. Department of Labor, November 2004.

are bound to increase as the living standards in those countries rise. Likewise, software development and information technology are now far more economical to implement in India than they are in Western countries. Keeping costs at a minimum is a constant challenge to manufacturing companies and an issue crucial to their very survival. The cost of a product is often the overriding consideration in its marketability and in general customer satisfaction. Typically, manufacturing costs represent about 40% of a product's selling price.

The concept of design for manufacture and assembly and of concurrent engineering, described in Section I.4, also includes design principles for economic production:

- The design should make the product as simple as possible to manufacture, assemble, disassemble, and recycle.
- Materials should be chosen for their appropriate manufacturing characteristics.
- Dimensional accuracy and surface finish should be specified as broadly as is permissible in order to minimize manufacturing costs.
- Net-shape manufacturing of parts should be emphasized, and secondary and finishing operations should be avoided or minimized.

The **total cost** of manufacturing a product consists of the costs of materials, tooling, and labor, the fixed costs, and capital costs. Several factors are involved in each cost category. Manufacturing costs can be minimized by analyzing the product design to determine whether the part size and shape are optimal, and whether the materials selected are the least costly ones that also possess the desired properties and manufacturing characteristics. The possibility of substituting one material for another is an important consideration in minimizing costs.

Fixed costs and capital costs depend on the particular manufacturer and plant facilities. Although computer-controlled machinery (a capital cost) can be very expensive, economic analyses have continually indicated that, more often than not, such an expenditure is warranted in view of its long-range benefits. Direct labor costs are usually only a very small portion of the total cost, typically ranging from 10 to 15%. Increased automation, computer control of all aspects of manufacturing, implementation of modern technologies, and the increased efficiency of operations have greatly helped reduce the direct labor component of manufacturing costs. Consequently, companies in more industrialized countries are beginning to be more competitive.

I.11 | General Trends in Manufacturing

With advances in all aspects of materials, processes, and production control, there have been certain important trends in manufacturing, as briefly outlined below.

Materials and processes. The trend is for better control of material compositions, purity, and defects (impurities, inclusions, flaws) in order to enhance their overall properties, manufacturing characteristics, reliability, service life, and recycling, while keeping material costs low. Developments are continuing on superconductors, semiconductors, nanomaterials, nanopowders, amorphous alloys, shape-memory alloys (smart materials), and coatings. Testing methods and equipment are being improved, including use of advanced computers and software, particularly for materials such as ceramics, carbides, and composites.

Concerns over material and energy savings have lead to better recyclability, as well as weight savings by improving design and engineering considerations, such as higher strength-to-weight and stiffness-to-weight ratios. Thermal treatment of materials are being conducted under better control of relevant variables for more predictable and reliable results, and surface treatment methods are being advanced rapidly. Included in these developments are advances in tool, die, and mold materials to improve their performance. Challenging recent developments in processing involve ultraprecision, micro, and nanomanufacturing, approaching atomic levels.

Computer simulation and modeling continue to be used widely in design and manufacturing, resulting in the optimization of processes and production systems, and better prediction of the effects of relevant variables on product integrity. As a result of such efforts, the speed and efficiency of product design and manufacturing is improving greatly, also affecting the overall economics of production and reducing product cost in an increasingly competitive and global marketplace.

Manufacturing systems. Continuing developments in control systems, industrial robots, automated inspection, handling and assembly, and sensor technology are having a major impact on the efficiency and reliability of all manufacturing processes and equipment. Advances in computer hardware and software, communications systems, adaptive control, expert systems, and artificial intelligence and neural networks have all helped enable the effective implementation of concepts such as group technology, cellular manufacturing, and flexible manufacturing systems, as well as modern practices in the efficient management of manufacturing enterprises.

Organizational trends. There have been important trends in the operational philosophy of manufacturing enterprises. Traditionally, the emphasis was on top-down communication in the organization and on strong control by management, with priorities for quick financial return (*profits first*) and growth and size (*economy of scale*). The major trend is now toward broad-based communication *across* the organization.

With **global competition** and the requirements for **world-class manufacturing**, corporate strategies are continually undergoing major changes. Manufacturing has become an integral part of long-range business planning for companies that must maintain their competitive positions and increase their market share. These are complex issues because they involve a broad range of considerations such as product type, company size, changing markets, laws and business practices in different countries, tariffs and import restrictions, geopolitics, and especially the major trends in rapidly increasing manufacturing activities in countries where labor costs are about a tenth of those in traditionally more industrialized countries. The rapidly growing field of **information technology** (IT) can provide the tools to help meet these major global challenges.

It has been increasingly recognized that for a manufacturing enterprise to be successful, it must respond to the following:

- View the *people* in the organization as important assets, and emphasize the importance and need for *teamwork* and involvement in problem solving and in decision-making processes in all aspects of operations.
- Encourage *product innovation* and improvements in *productivity*.
- Relate product innovation and manufacturing to the *customer* and the *market*, seeing the product as *meeting a need*.
- Increase *flexibility* of operation for rapid response to product demands, in both the domestic and the global marketplace.
- Encourage efforts for *continuous improvement in quality*.
- Ultimately and, most importantly, focus on *customer satisfaction* on a global scale.

Fundamentals of Materials: Behavior and Manufacturing Properties

Part I of this text begins by presenting the behavior and engineering properties of materials, their characteristics, applications, advantages, and limitations that influence the choice of materials in the design and manufacture of products.

In order to emphasize the importance of the topics that are to be discussed, let's look at an automobile as an example of a common product that uses a wide variety of materials (Fig. I.1). These materials were selected primarily because, among the materials that possess the desired properties and characteristics for the intended functions of specific parts of the automobile, these were the ones that can be manufactured at the lowest cost.

Steel was chosen for much of the body because it is strong, easy to form, and inexpensive. Plastics were used in many components because of characteristics such as a wide choice of colors, light weight, and ease of manufacturing into various shapes at low cost. Glass was chosen for the windows not only because it is transparent, but also because it is hard, easy to shape and clean, and resistant to scratching. Similar observations can be made for each component of an automobile, which typically is an assemblage of some 15,000 parts, ranging from thin wire to bumpers.

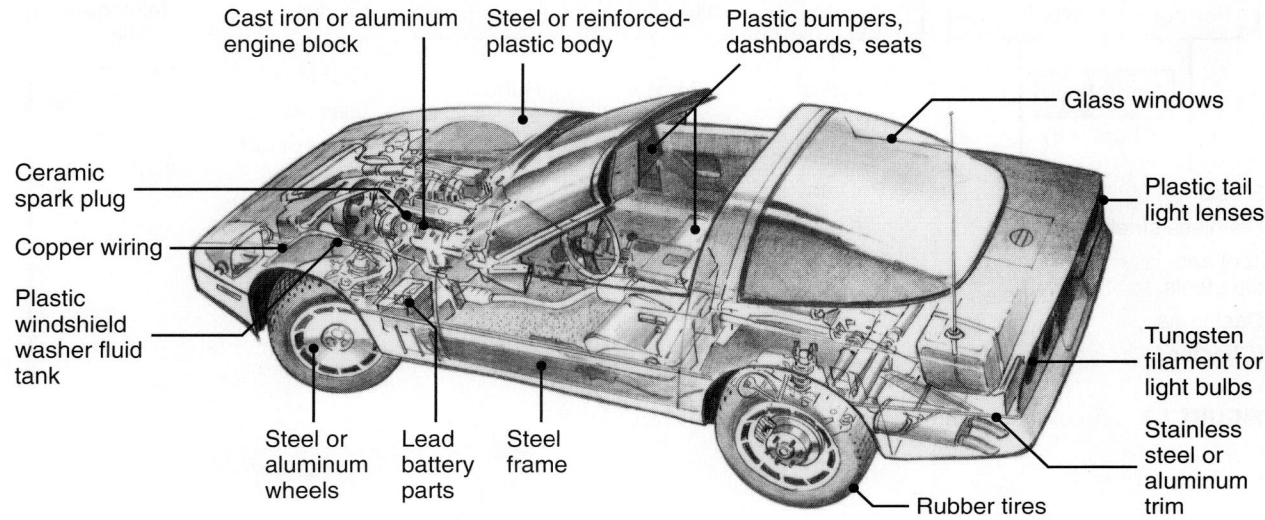

FIGURE I.1 Some of the metallic and nonmetallic materials used in a typical automobile.

As stated in the General Introduction, selection of materials for individual components in a product requires a thorough understanding of their properties, functions, and costs. Note that by saving just one cent on the average cost per part, such as by selecting a different material or manufacturing technique, the manufacturer can reduce the cost of an automobile by $150. The task of manufacturing engineers thus becomes truly challenging, especially with the ever increasing variety of materials now available (Fig. I.2).

A general outline of the various topics presented in Part I is given in Fig. I.3. The fundamental knowledge presented on the behavior, properties, and characteristics of materials will help the reader understand their relevance to the manufacturing processes described in Parts II through VI of this book. This knowledge will also make it possible to analyze the often complex relationships among materials, manufacturing processes, machinery and tooling, and the economics of manufacturing operations.

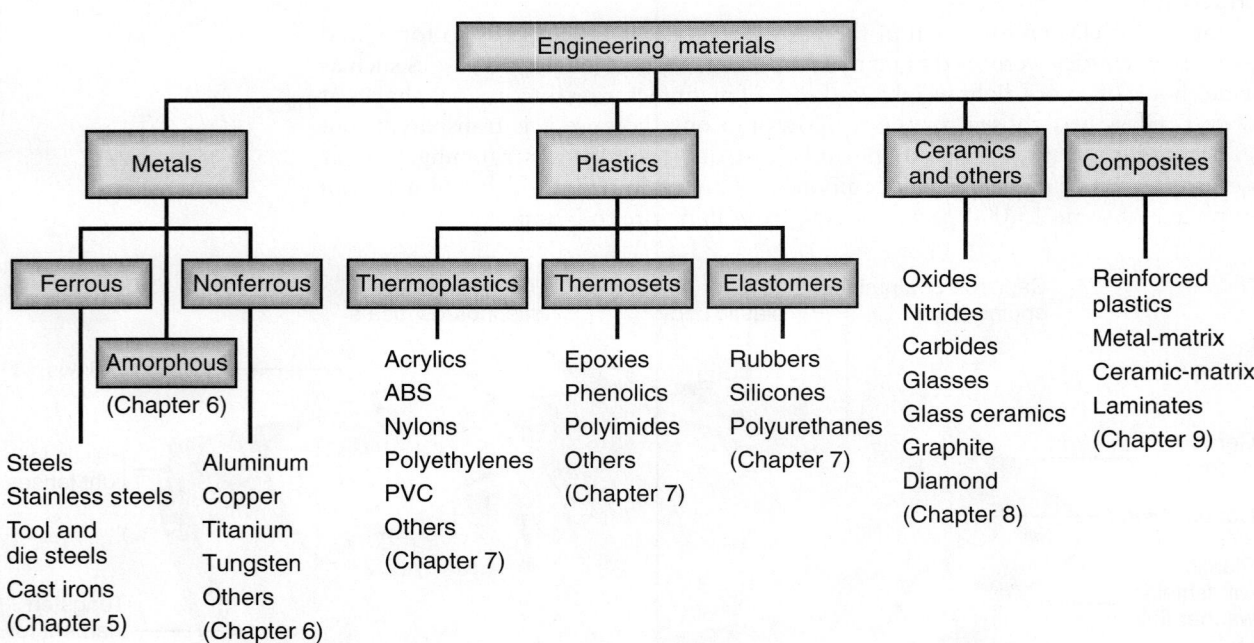

FIGURE I.2   An outline of the engineering materials described in Part I.

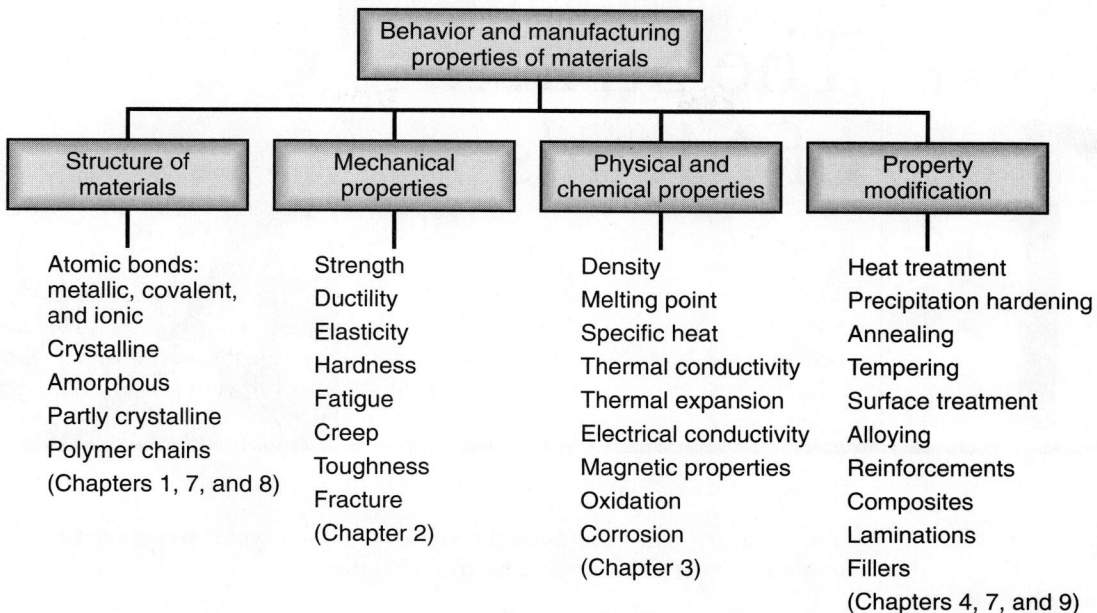

FIGURE I.3   An outline of the behavior and the manufacturing properties of materials described in Part I.

The Structure of Metals

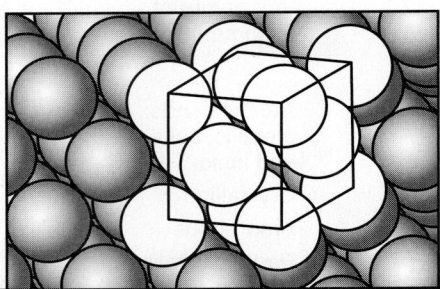

This chapter will introduce the subject of the crystal structure of metals and explain why it is important to study them:

- How atoms are arranged in a metal structure.
- Types of imperfections that exist in crystal structures and their effects.
- How grains and grain boundaries are developed.
- Mechanisms of strain hardening and anisotropy.
- What effects these atomic arrangements, imperfections, grain boundaries, and temperature have on metal properties relevant to manufacturing.

1.1 | Introduction

Why are some metals hard and others soft? Why are some metals brittle, while others are ductile and can be shaped easily without fracture? Why is it that some metals can withstand high temperatures, while others cannot? These and numerous similar questions can be answered by studying the **atomic structure** of metals—that is, the arrangement of the atoms within metals—because they greatly influence their properties and behavior. This knowledge then guides us in controlling and predicting the behavior and performance of metals in various manufacturing processes. Understanding the structure of metals also allows us to predict and evaluate their **properties**; this helps us make appropriate selections for specific applications under particular external physical and environmental conditions. In addition to atomic structure, various other factors also influence the properties and behavior of metals. Among these are the composition of the metal, impurities and vacancies in the atomic structure, grain size, grain boundaries, environment, size and surface condition of the metal, and the methods by which metals and alloys are made into useful products.

The topics covered in this chapter, and their sequence, are outlined in Fig. 1.1. The structure and general properties of materials other than metals are described in

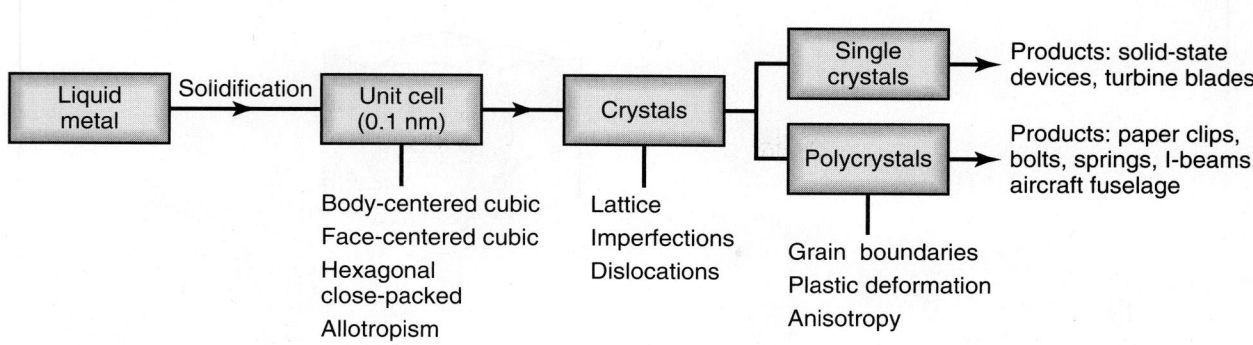

FIGURE 1.1 An outline of the topics described in Chapter 1.

Chapter 7 (polymers), Chapter 8 (ceramics and glasses), and Chapter 9 (composite materials). The structure of metal alloys, the control of their structure, and heat-treatment processes are described in Chapter 4.

1.2 | The Crystal Structure of Metals

When metals solidify from a molten state, the atoms arrange themselves into various orderly configurations, called **crystals**, and this arrangement of the atoms in the crystal is called **crystalline structure**. The smallest group of atoms showing the characteristic **lattice structure** of a particular metal is known as a **unit cell**. It is the building block of a crystal, and a single crystal can have many unit cells.

The three basic atomic arrangements in metals are:

1. **Body-centered cubic** (bcc)—alpha iron, chromium, molybdenum, tantalum, tungsten, and vanadium.

2. **Face-centered cubic** (fcc)—gamma iron, aluminum, copper, nickel, lead, silver, gold, and platinum.

3. **Hexagonal close-packed** (hcp)—beryllium, cadmium, cobalt, magnesium, alpha titanium, zinc, and zirconium.

These structures are represented by the illustrations given in Figs. 1.2 through 1.4; each sphere in these illustrations represents an atom. The order of magnitude of the distance between the atoms in these crystal structures is 0.1 nm (10^{-8} in.) These models are known as **hard-ball** or **hard-sphere** models; they can be likened to tennis balls arranged in various configurations in a box.

As shown in Fig. 1.2, each atom in the bcc structure has eight neighboring atoms. Of the three structures illustrated, the fcc and hcp crystals have the most densely packed configurations. In the hcp structure, the top and bottom planes are called **basal planes**. The manner in which these atoms are arranged determines the properties of a particular metal. These arrangements can be modified by adding atoms of some other metal or metals, known as **alloying**; it is often done to improve various properties of the metal.

The reason that metals form different crystal structures is to *minimize* the energy required to fit together in a regular pattern. Tungsten, for example, forms a bcc

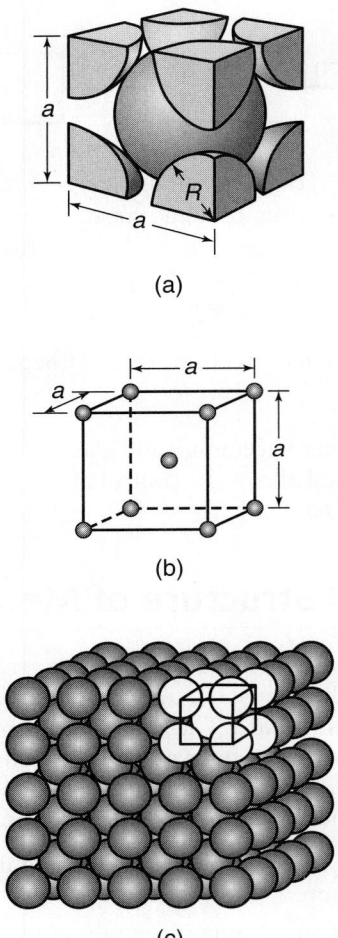

(a)

(b)

(c)

FIGURE 1.2   The body-centered cubic (bcc) crystal structure: (a) hard-ball model; (b) unit cell; and (c) single crystal with many unit cells.

structure because that structure involves less energy than do other structures; likewise, aluminum forms an fcc structure. At different temperatures, however, the same metal may form different structures, because of a lower energy requirement at that temperature. For example, iron forms a bcc structure (alpha iron) below 912°C (1674°F) and above 1394°C (2541°F), but it forms an fcc structure (gamma iron) between 912° and 1394°C.

The appearance of more than one type of crystal structure is known as **allotropism** or **polymorphism** (meaning "many shapes"). Because the properties and behavior of a metal depend greatly on its crystal structure, allotropism is an important factor in the heat treatment of metals as well as in metalworking and welding operations (Parts III and V). Single crystals of metals are now produced as ingots in sizes on the order of 1 m (40 in.) long and up to 300 mm (12 in.) in diameter, with applications such as turbine blades and semiconductors (see Sections 11.4 and 28.3). However, as described in Section 1.4, most metals used in manufacturing are polycrystalline.

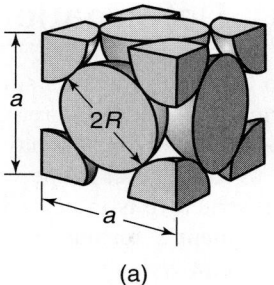

(a)

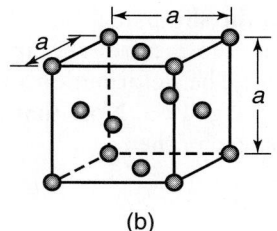

(b)

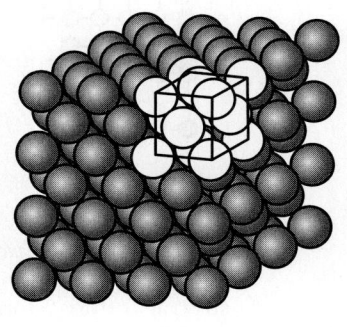

FIGURE 1.3   The face-centered cubic (fcc) crystal structure: (a) hard-ball model; (b) unit cell; and (c) single crystal with many unit cells.

(c)

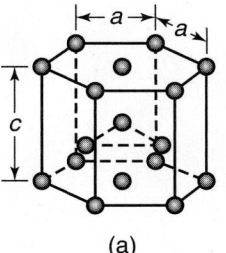

(a)

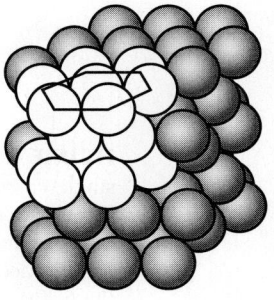

FIGURE 1.4   The hexagonal close-packed (hcp) crystal structure: (a) unit cell; and (b) single crystal with many unit cells.

(b)

1.3 | Deformation and Strength of Single Crystals

When a single crystal is subjected to an external force, it first undergoes **elastic deformation**—that is, it returns to its original shape when the force is removed. An analogy to this type of behavior is a helical spring that stretches when loaded and returns to its original shape when the load is removed. However, if the force on the crystal structure is increased sufficiently, the crystal undergoes **plastic deformation** or **permanent deformation**—that is, it does not return to its original shape when the force is removed.

There are two basic mechanisms by which plastic deformation takes place in crystal structures. One is the slipping of one plane of atoms over an adjacent plane (called the **slip plane**, Fig. 1.5a) under a **shear stress** (Fig. 1.5b). Shear stress is defined as the ratio of the applied shearing force to the cross-sectional area being sheared. The deformation of a single-crystal specimen by slip is shown schematically in Fig. 1.6a. Note that this behavior is much like the sliding of playing cards against each other.

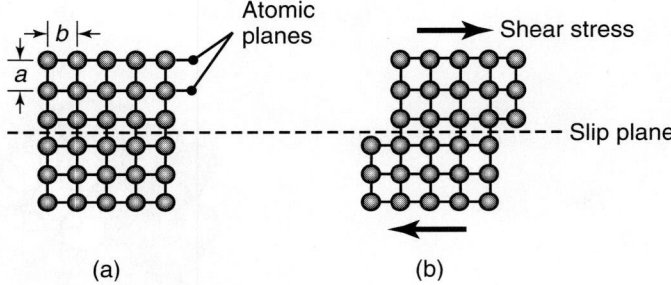

(a)   (b)

FIGURE 1.5 Permanent deformation (also called plastic deformation) of a single crystal subjected to a shear stress: (a) structure before deformation; and (b) permanent deformation by slip. The *b/a* ratio influences the magnitude of the shear stress required to cause slip.

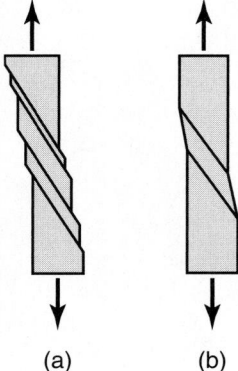

(a)   (b)

FIGURE 1.6 (a) Permanent deformation of a single crystal under a tensile load. Note that the slip planes tend to align themselves in the direction of the pulling force. This behavior can be simulated using a deck of cards with a rubber band around them. (b) Twinning in a single crystal in tension.

Just as it takes a certain amount of force to slide playing cards against each other, a crystal requires a certain amount of shear stress (**critical shear stress**) to undergo permanent deformation. Thus, there must be a shear stress of sufficient magnitude within a crystal for plastic deformation to occur, otherwise the deformation is elastic only.

The shear stress required to cause slip in single crystals is directly proportional to the ratio b/a in Fig. 1.5a, where a is the spacing of the atomic planes and b is inversely proportional to the atomic density in the atomic plane. As b/a decreases, the shear stress required to cause slip decreases. Therefore, it can be stated that slip in a single crystal takes place along planes of *maximum atomic density* or, in other words, that slip takes place in closely packed planes and in closely packed directions.

Because the b/a ratio varies for different directions within the crystal, a single crystal has different properties when tested in different directions; this behavior is called **anisotropic**. A simple example of anisotropy is the behavior of woven cloth, which stretches differently when we pull it in different directions; another common example is plywood, which is much stronger in the planar direction than along its thickness direction (note, for example, how plywood splits easily).

The second and less common mechanism of plastic deformation in crystals is **twinning**, in which a portion of the crystal forms a mirror image of itself across the *plane of twinning* (Fig. 1.6b). Twins form abruptly and are the cause of the creaking sound ("tin cry") that occurs when a tin or zinc rod is bent at room temperature. Twinning usually occurs in hcp metals.

Slip systems.   The combination of a slip plane and its direction of slip is known as a *slip system*. In general, metals with 5 or more slip systems are ductile, whereas those with fewer than 5 slip systems are not.

1. In **body-centered cubic** crystals, there are 48 possible slip systems. The probability is therefore high that an externally applied shear stress will operate on one of these systems and cause slip. Because of the relatively high b/a ratio in the crystal, however, the required shear stress is high. Metals with bcc structures generally have good strength and moderate ductility.

2. In **face-centered cubic** crystals, there are 12 slip systems. The probability of slip is moderate, and the required shear is low because of the relatively low b/a ratio. These metals generally have moderate strength and good ductility.

3. The **hexagonal close-packed** crystal has 3 slip systems, and therefore has a low probability of slip; however, more slip systems become active at elevated temperatures. Metals with hcp structures are generally brittle at room temperature.

Note in Fig. 1.6a that the portions of the single crystal that have slipped have rotated from their original angular position toward the direction of the tensile force; note also that slip has taken place only along certain planes. With the use of electron microscopy, it can be shown that what appears to be a single slip plane is actually a **slip band**, consisting of a number of slip planes (Fig. 1.7).

1.3.1 Imperfections in the crystal structure of metals

The actual strength of metals is approximately one to two orders of magnitude lower than the strength levels obtained from theoretical calculations; this discrepancy has been explained in terms of **defects** and **imperfections** in the crystal structure.

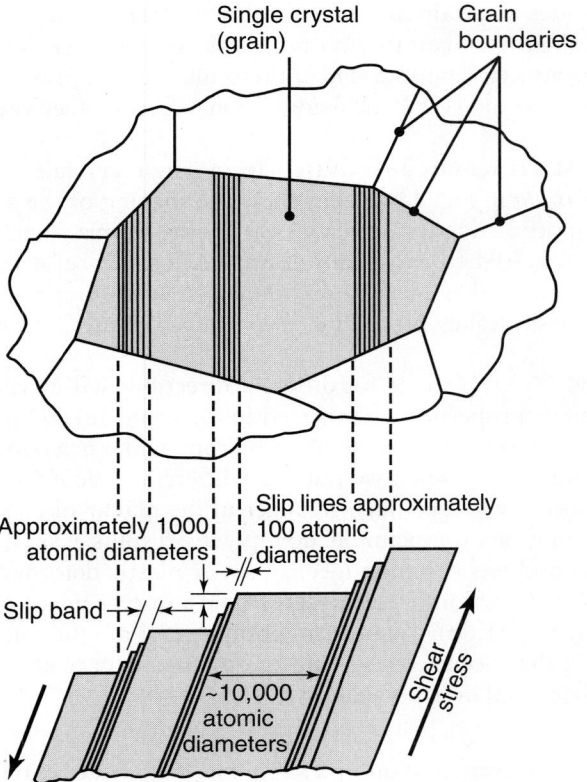

FIGURE 1.7   Schematic illustration of slip lines and slip bands in a single crystal (grain) subjected to a shear stress. A slip band consists of a number of slip planes. The crystal at the center of the upper illustration is an individual grain surrounded by several other grains.

Unlike the idealized models described above, actual metal crystals contain a large number of defects and imperfections, which are categorized as follows:

1. *Point defects*, such as a **vacancy** (missing atom), an **interstitial atom** (extra atom in the lattice), or an **impurity** (foreign atom that has replaced the atom of

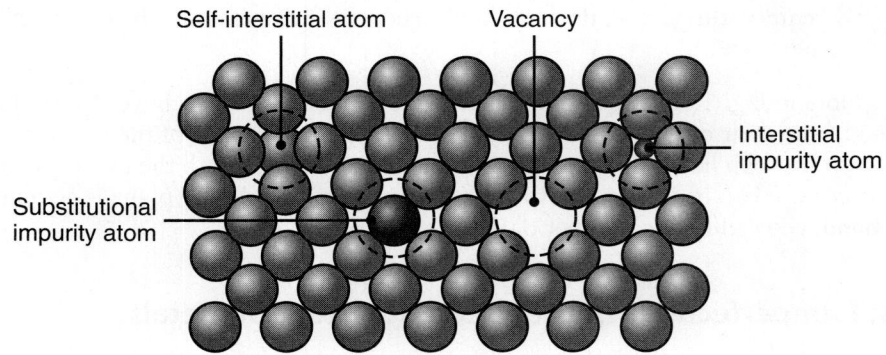

FIGURE 1.8   Schematic illustration of types of defects in a single-crystal lattice: self-interstitial, vacancy, interstitial, and substitutional.

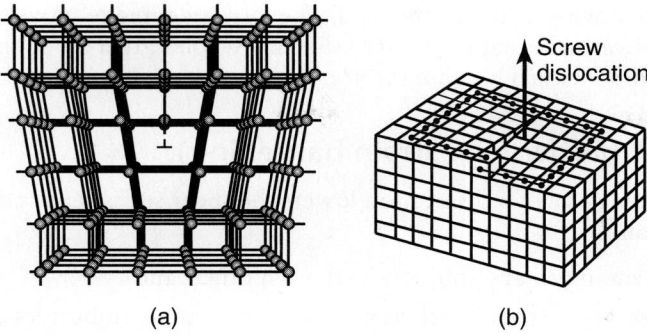

(a)   (b)

FIGURE 1.9   Types of dislocations in a single crystal: (a) edge dislocation; and (b) screw dislocation.

the pure metal) (Fig. 1.8).

2. *Linear*, or *one-dimensional, defects*, called **dislocations** (Fig. 1.9).

3. *Planar*, or *two-dimensional, imperfections*, such as **grain boundaries** and **phase boundaries** (see Section 1.4).

4. *Volume*, or *bulk, imperfections*, such as **voids, inclusions** (nonmetallic elements such as oxides, sulfides, and silicates), other **phases**, or **cracks**.

Mechanical and electrical properties of metals, such as yield stress, fracture strength, and electrical conductivity, are adversely affected by these defects; these are known as **structure-sensitive** properties. On the other hand, physical and chemical properties such as melting point, specific heat, coefficient of thermal expansion, and elastic constants (e.g., modulus of elasticity and modulus of rigidity) are not sensitive to these defects; these are known as **structure-insensitive** properties.

Dislocations.   First observed in the 1930s, *dislocations* are defects in the orderly arrangement of a metal's atomic structure. A slip plane containing a dislocation (Fig. 1.10) requires less shear stress to allow slip than does a plane in a perfect lattice. Dislocations are the most significant defects and help explain the discrepancy between the actual and theoretical strengths of metals.

There are two types of dislocations: **edge** and **screw** (Fig. 1.9). One analogy used to describe the movement of an edge dislocation is the earthworm, which moves forward by means of a hump that starts at the tail and moves toward the head. Another analogy for an edge dislocation is moving a large carpet by forming a hump

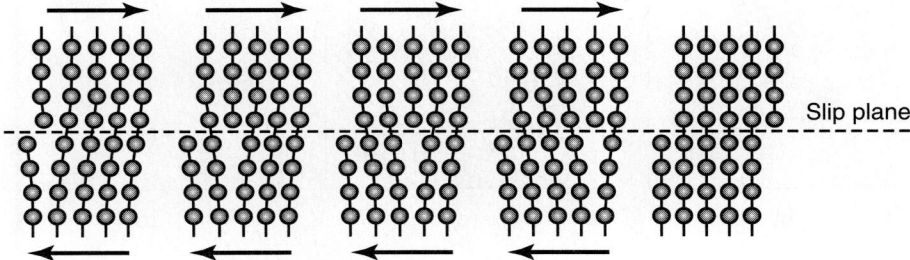

Slip plane

FIGURE 1.10   Movement of an edge dislocation across the crystal lattice under a shear stress. Dislocations help explain why the actual strength of metals is much lower than that predicted by theory.

at one end and moving it to the other end. The force required to move a carpet in this way is much lower than that required to slide the whole carpet along the floor. Screw dislocations are so named because the atomic planes form a spiral ramp.

1.3.2 Work hardening (Strain hardening)

Although the presence of a dislocation lowers the shear stress required to cause slip, dislocations can:

1. Become entangled and interfere with each other; and
2. Be impeded by barriers, such as grain boundaries and impurities and inclusions in the material.

Entanglements and impediments increase the shear stress required for slip. Entanglement is like moving two humps at different angles across a carpet: where they cross, the two humps interfere with each other's movement, and their combined effect is to make it more difficult to move the carpet.

The effect of an increase in shear stress, thus causing an increase in the overall strength and the hardness of the metal, is known as **work hardening** or **strain hardening**. The greater the deformation, the greater the number of entanglements, hence an increase in the metal's strength. Work hardening is used extensively for strengthening metals in metalworking processes at ambient temperature. Typical examples are producing sheet metal for automobile bodies and aircraft fuselages by cold rolling, making the head of a bolt by forging, and strengthening wire by reducing its cross-section by drawing it through a die.

1.4 | Grains and Grain Boundaries

Metals commonly used in manufacturing various products consist of many individual, randomly oriented crystals (**grains**); thus, metal structures typically are not single crystals but **polycrystals** ("many crystals"). When a mass of molten metal begins to solidify, crystals begin to form independently of each other at various locations within the liquid mass; they have random and unrelated orientations (Fig. 1.11). Each of these crystals then grows into a crystalline structure or grain.

The number and size of the grains developed in a unit volume of the metal depends on the *rate* at which **nucleation** (the initial stage of formation of crystals) takes place. The number of different sites at which individual crystals begin to form (seven in Fig. 1.11a) and the rate at which these crystals grow both influence the median

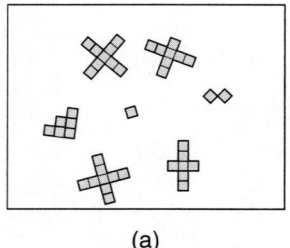

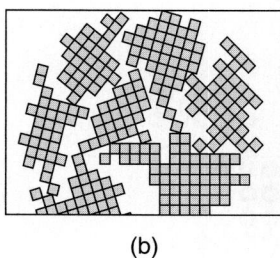

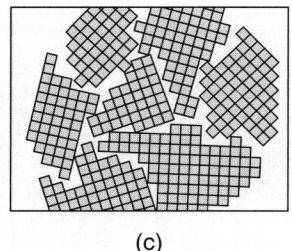

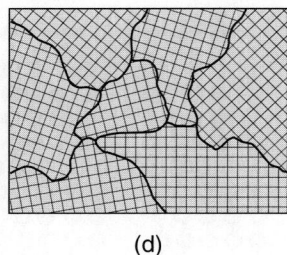

(a)   (b)   (c)   (d)

FIGURE 1.11   Schematic illustration of the stages during solidification of molten metal; each small square represents a unit cell. (a) Nucleation of crystals at random sites in the molten metal; note that the crystallographic orientation of each site is different. (b) and (c) Growth of crystals as solidification continues. (d) Solidified metal, showing individual grains and grain boundaries; note the different angles at which neighboring grains meet each other.

size of the grains developed. If the nucleation rate is high, the number of grains in a unit volume of metal will be large and thus grain size will be small. Conversely, if the rate of growth of the crystals is high (as compared to their nucleation rate), there will be fewer grains per unit volume, and their size will be larger. Generally, rapid cooling produces smaller grains, whereas slow cooling produces larger grains.

Note, in Fig. 1.11d, how the growing grains eventually interfere with and impinge upon one another; the surfaces that separate these individual grains are called **grain boundaries**. Each grain consists of either a single crystal (for pure metals) or a polycrystalline aggregate (for alloys). Also note that the crystallographic orientation changes abruptly from one grain to the next across the grain boundaries. (Recall from Section 1.3 that the behavior of a single crystal or a single grain is anisotropic.) Thus, because its grains have random crystallographic orientations, the ideal behavior of a piece of polycrystalline metal is **isotropic**: its properties do not vary with the direction of testing.

1.4.1 Grain size

Grain size significantly influences the mechanical properties of metals. At room temperature, for example, a large grain size generally is associated with low strength, low hardness, and low ductility. Large grains, particularly in sheet metals, also cause a rough surface appearance after the material has been stretched (see **orange peel**, Section 1.6). Grains can be so large as to be visible with the naked eye, such as those of zinc on the surface of galvanized sheet steels.

Grain size usually is measured by counting the number of grains in a given area, or by counting the number of grains that intersect a length of a line randomly drawn on an enlarged photograph of the grains (taken under a microscope on a polished and etched specimen). Grain size may also be determined by comparing it to a standard chart. The ASTM (The American Society for Testing and Materials) grain size number, n, is related to the number of grains, N, per square inch at a magnification of $100\times$ (equal to 0.0645 mm^2 of actual area) by the formula

$$N = 2^{n-1} \tag{1.1}$$

Because grains are typically extremely small, many grains can occupy a piece of metal

TABLE 1.1

Grain Sizes		
ASTM No.	Grains/mm^2	Grains/mm^3
−3	1	0.7
−2	2	2
−1	4	5.6
0	8	16
1	16	45
2	32	128
3	64	360
4	128	1,020
5	256	2,900
6	512	8,200
7	1,024	23,000
8	2,048	65,000
9	4,096	185,000
10	8,200	520,000
11	16,400	1,500,000
12	32,800	4,200,000

(Table 1.1). Grain sizes between 5 and 8 are generally considered fine grains. A grain size of 7 is generally acceptable for sheet metals for making car bodies, appliances, and kitchen utensils.

1.4.2 Influence of grain boundaries

Grain boundaries have an important influence on the strength and ductility of metals, and because they interfere with the movement of dislocations, grain boundaries also influence strain hardening. These effects depend on temperature, deformation rate, and the type and amount of impurities present *along* the grain boundaries.

Grain boundaries are more reactive than the grains themselves, because the atoms along the grain boundaries are packed less efficiently and are more disordered; as a result, they have lower energy than the atoms in the orderly lattice within the grains, and can be more easily removed or chemically bonded to another atom. Consequently, a surface becomes rougher when etched or subjected to corrosive and hostile environments.

At elevated temperatures and in materials whose properties depend on the rate at which they are deformed (worked), plastic deformation also takes place by means of **grain-boundary sliding**. The **creep** mechanism (that is, elongation under stress over a period of time, usually at elevated temperatures) results from grain-boundary sliding (Section 2.8).

Grain-boundary embrittlement. When brought into close atomic contact with certain low-melting-point metals, a normally ductile and strong metal can crack when subjected to very low external stresses. Examples of such behavior are (a) aluminum wetted with a mercury-zinc amalgam or liquid gallium and (b) copper (at elevated temperature) wetted with lead or bismuth. These embrittling elements weaken the grain boundaries of the metal by *embrittlement*. The term **liquid-metal embrittlement** is used to describe such phenomena, because the embrittling element is in a liquid state. However, embrittlement can also occur at temperatures well below the melting point of the embrittling element, a phenomenon known as **solid-metal embrittlement**.

Hot shortness is caused by local melting of a constituent or an impurity in the grain boundary at a temperature below the melting point of the metal itself. When subjected to plastic deformation at elevated temperatures (*hot working*), the piece of metal crumbles and disintegrates along its grain boundaries; examples are antimony in copper, leaded steels, and leaded brass. To avoid hot shortness, the metal usually is worked at a lower temperature in order to prevent softening and melting along the grain boundaries. Another form of embrittlement is **temper embrittlement** in alloy steels, which is caused by segregation (movement) of impurities to the grain boundaries.

EXAMPLE 1.1 Number of grains in the ball of a ballpoint pen

Assume that the ball of a ballpoint pen is 1 mm in diameter and has an ASTM grain size of 10. Calculate the number of grains in the ball.

Solution A metal with an ASTM grain size of 10 has 520,000 grains per cubic millimeter (see Table 1.1). The volume of the 1-mm diameter ball is given by

$$V = \frac{4\pi r^3}{3} = \frac{4\pi (0.5 \text{ mm})^3}{3} = 0.5236 \text{ mm}^3$$

To obtain the total number of grains, the volume is multiplied by the grains per cubic millimeter or

No. grains $= (0.5236 \text{ mm}^3)(520,000 \text{ grains/mm}^3) = 272,300$ grains

1.5 | Plastic Deformation of Polycrystalline Metals

If a polycrystalline metal with uniform *equiaxed grains* (having equal dimensions in all directions, as shown in the model in Fig. 1.12a) is subjected to plastic deformation at room temperature (*cold working*), the grains become deformed and elongated. The deformation process may be carried out either by compressing the metal, as is done in forging to make a turbine disk, or by subjecting it to tension, as is done in stretching sheet metal to make a car body. The deformation within each grain takes place by the mechanisms described in Section 1.3 for a single crystal.

During plastic deformation, the grain boundaries remain intact and mass continuity is maintained. The deformed metal exhibits higher strength because of the entanglement of dislocations with grain boundaries. The increase in strength depends on the degree of deformation (*strain*) to which the metal is subjected; the higher the deformation, the stronger the metal becomes. The increase in strength is higher for metals with smaller grains, because they have a larger grain-boundary surface area per unit volume of metal and, hence, more entanglement of dislocations.

Anisotropy (Texture).   Note in Fig. 1.12b that, as a result of plastic deformation, the grains have elongated in one direction and contracted in the other. Consequently, this piece of metal has become *anisotropic*, and thus, its properties in the vertical direction are different from those in the horizontal direction. Many products develop anisotropy of mechanical properties after they have been processed by metalworking techniques (Fig. 1.13). The degree of anisotropy depends on how uniformly the metal is deformed; note from the direction of the crack in Fig. 1.13a, for example, that the ductility of the cold-rolled sheet in the transverse (vertical) direction is lower than that in its rolling (longitudinal) direction.

Anisotropy influences both mechanical and physical properties of metals. For example, sheet steel for electrical transformers is rolled in such a way that the resulting deformation imparts anisotropic magnetic properties to the sheet. This arrangement reduces magnetic-hysteresis losses and improves the efficiency of transformers. (See also *amorphous alloys*, Section 6.14.) There are two general types of anisotropy in metals: *preferred orientation* and *mechanical fibering*.

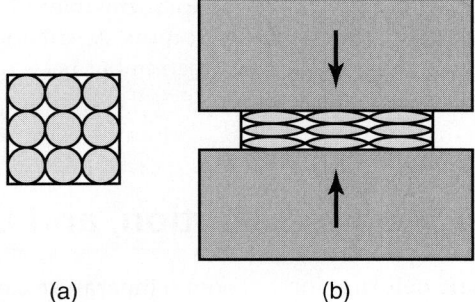

(a)   (b)

FIGURE 1.12   Plastic deformation of idealized (equiaxed) grains in a specimen subjected to compression (such as occurs in the forging or rolling of metals): (a) before deformation; and (b) after deformation. Note the alignment of grain boundaries along a horizontal direction; this effect is known as preferred orientation.

Top view

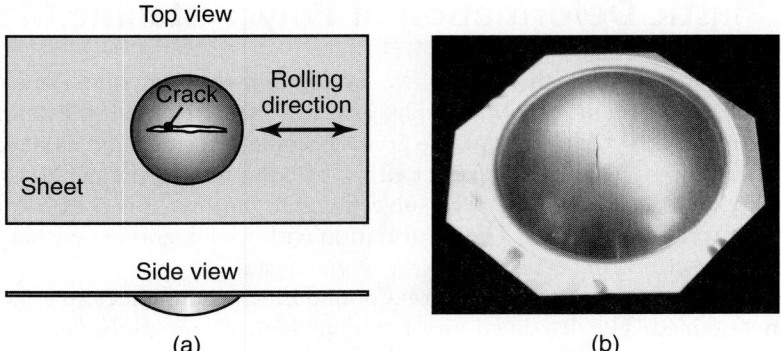

(a)   (b)

FIGURE 1.13   (a) Schematic illustration of a crack in sheet metal that has been subjected to bulging (caused by, for example, pushing a steel ball against the sheet). Note the orientation of the crack with respect to the rolling direction of the sheet; this sheet is anisotropic. (b) Aluminum sheet with a crack (vertical dark line at the center) developed in a bulge test; the rolling direction of the sheet was vertical. *Source*: Courtesy of J.S. Kallend, Illinois Institute of Technology.

Preferred orientation.   Also called **crystallographic anisotropy**, *preferred orientation* best can be described by referring to Fig. 1.6a. When a metal crystal is subjected to tension, the sliding blocks rotate toward the direction of the pulling force; as a result, slip planes and slip bands tend to align themselves with the direction of deformation. Similarly, for a polycrystalline aggregate with grains in random orientations, all slip directions tend to align themselves with the direction of pulling force. Conversely, slip planes under compression tend to align themselves in a direction perpendicular to the direction of the compressing force.

Mechanical fibering.   This type of anistropy results from the alignment of inclusions (**stringers**), impurities, and voids in the metal during deformation. Note that if the spherical grains in Fig. 1.12a were coated with impurities, these impurities would align themselves generally in a horizontal direction after deformation. Because impurities weaken the grain boundaries, this piece of metal will be weak and less ductile when tested in the vertical direction. An analogy to this is plywood, which is strong in tension along its planar direction but peels off easily (splits) when pulled in tension in its thickness direction.

1.6 | Recovery, Recrystallization, and Grain Growth

It was noted that plastic deformation at room temperature causes the deformation of grains and grain boundaries, a general increase in strength, and a decrease in ductility; it also causes anisotropic behavior. These effects can be reversed (and the properties of the metal can be brought back to their original levels) by heating the metal to a specific temperature range for a period of time; this process is called **annealing** (Section 4.11). The temperature range and the amount of time required depend on the material and on other factors. Three events take place consecutively during the heating process:

1. Recovery. During *recovery*, which occurs at a certain temperature range below the **recrystallization temperature** of the metal, the stresses in the highly deformed regions are relieved. Subgrain boundaries begin to form (a process called **polygonization**) with no significant change in mechanical properties such as hardness and strength (Fig. 1.14).

2. Recrystallization. The process in which (within a certain temperature range) new equiaxed and strain-free grains are formed (replacing the older grains) is called *recrystallization*. The temperature for recrystallization ranges approximately between 0.3 and $0.5T_m$, where T_m is the melting point of the metal on the absolute scale.

The recrystallization temperature generally is defined as the temperature at which complete recrystallization occurs within approximately one hour. Recrystallization decreases the density of dislocations, lowers the strength, and raises the ductility of the metal (Fig. 1.14). Lead, tin, cadmium, and zinc recrystallize at about room temperature; as a result, they do not work harden when cold worked.

Recrystallization depends on the degree of prior cold work (work hardening): the more cold work, the lower the temperature required for recrystallization to occur. The reason is that, as the amount of cold work increases, the number of dislocations and the amount of energy stored in dislocations (**stored energy**) also increase. This energy supplies some of the work required for recrystallization. Recrystallization is

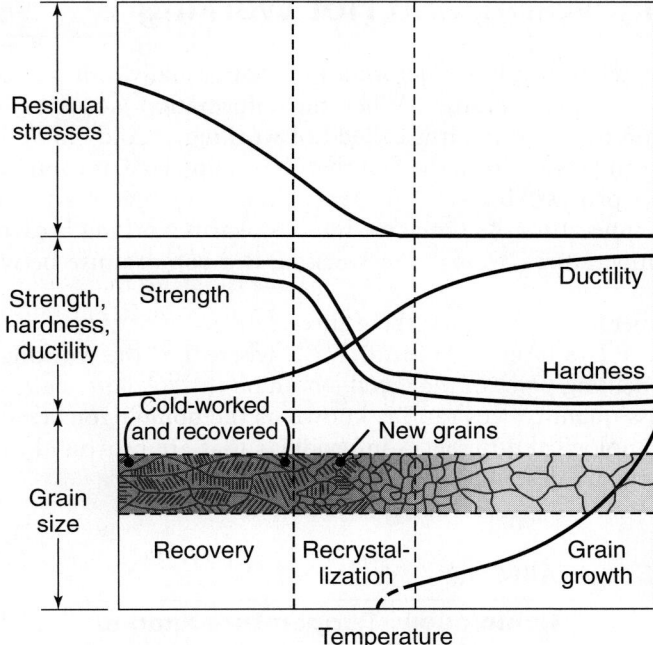

FIGURE 1.14   Schematic illustration of the effects of recovery, recrystallization, and grain growth on mechanical properties and on the shape and size of grains. Note the formation of small new grains during recrystallization. *Source*: After G. Sachs.

also a function of time, because it involves **diffusion**—the movement and exchange of atoms across grain boundaries.

The effects on recrystallization of temperature, time, and plastic deformation by cold working are as follows:

- For a constant amount of deformation by cold working, the time required for recrystallization to occur decreases with increasing temperature.
- The more the prior cold work, the lower the temperature required for recrystallization.
- The greater the degree of deformation, the smaller the grain size becomes during recrystallization. This effect is a common method of converting a coarse-grained structure to one having a finer grain and, hence, improved properties.
- Anisotropy due to preferred orientation usually persists after recrystallization. To restore isotropy, a temperature higher than that required for recrystallization may be necessary.

3. Grain growth. If we continue to raise the temperature of the metal, the grains begin to grow, and their size may eventually exceed the original grain size; this phenomenon is called *grain growth*, and it affects mechanical properties (Fig. 1.14). Large grains produce a rough surface appearance on sheet metals, called **orange peel**, when they are stretched to form a part or when a piece of metal is subjected to compression (such as in forging).

1.7 | Cold, Warm, and Hot Working

Cold working refers to plastic deformation that (usually, but not necessarily) is carried out at room temperature. When the deformation is carried out above the recrystallization temperature, it is called **hot working**. "Cold" and "hot" are relative terms, as can be seen from the fact that deforming lead at room temperature is a hot-working process, because the recrystallization temperature of lead is at about room temperature. As the name implies, **warm working** is carried out at intermediate temperatures, thus warm working is a compromise between cold and hot working.

The temperature ranges for these three categories of plastic deformation are given in Table 1.2 in terms of a ratio, T/T_m, where T is the working temperature and T_m is the melting point of the metal—both on the absolute scale. Although it is a dimensionless quantity, this ratio is known as the **homologous temperature**. The important technological differences in products that are processed by cold, warm, and hot working are described in Part III.

TABLE 1.2

Homologous Temperature Ranges for Various Processes	
Process	T/T_m
Cold working	<0.3
Warm working	0.3 to 0.5
Hot working	>0.6

SUMMARY

- There are three basic crystal structures in metals: body-centered cubic (bcc), face-centered cubic (fcc), and hexagonal close-packed (hcp). Grains made of these crystals are not perfect, because they contain various defects and imperfections, such as dislocations, vacancies, impurities, inclusions, and grain boundaries. Polycrystalline metals are composed of many crystals or grains in random orientations.

- Plastic deformation in metals takes place by a slip mechanism. Although the theoretical shear stress required to cause slip is very high, actual stresses are much lower because of the presence of dislocations (edge or screw type). Dislocations become entangled with one another or are impeded by barriers such as grain boundaries, impurities, and inclusions. As a result, the shear stress required to cause further slip is increased. Consequently, the overall strength and hardness of the metal is also increased (work hardening or strain hardening).

- Grain size has a significant effect on the strength of metals; the smaller the size, the stronger is the metal.

- Grain boundaries have a major influence on metal behavior. They can undergo embrittlement, which severely reduces ductility at elevated temperatures (hot shortness); they are also responsible for creep, which is due to grain-boundary sliding.

- Metals can be deformed plastically (worked) at room, warm, or high temperatures. Their behavior and workability depend largely on whether deformation takes place below or above the recrystallization temperature. Deformation at room temperature (cold working) results in higher strength but reduced ductility of the metal; it also generally causes anisotropy (preferred orientation or mechanical fibering)—a state in which the properties are different in different directions.

- The effects of cold working can be reversed by annealing the metal: heating it in a certain temperature range for a period of time and, thereby, allowing successive processes of recovery, recrystallization, and grain growth to take place.

KEY TERMS

Allotropism	Grain size	Recrystallization
Anisotropy	Hexagonal close-packed	Shear stress
Basal plane	Homologous temperature	Slip band
Body-centered cubic	Hot shortness	Slip plane
Cold working	Hot working	Slip system
Creep	Lattice structure	Strain hardening
Crystals	Mechanical fibering	Structure-insensitive
Dislocations	Orange peel	Structure-sensitive
Elastic deformation	Plastic deformation	Twinning
Embrittlement	Polycrystals	Unit cell
Face-centered cubic	Polygonization	Vacancy
Grains	Polymorphism	Warm working
Grain boundaries	Preferred orientation	Work hardening
Grain growth	Recovery	

BIBLIOGRAPHY

Ashby, M.F., and Jones, D.R.H. *Engineering Materials*, Vol. 1, *An Introduction to Their Properties and Applications*, 2nd ed., 1996; Vol. 2, *An Introduction to Microstructures, Processing and Design*, 1986. Pergamon Press.

Callister, W.D., Jr., *Materials Science and Engineering*, 6th ed., Wiley, 2002.

Shackelford, J.F., *Introduction to Materials Science for Engineers*, 5th ed., Prentice Hall, 2000.

Shackelford, J.F., and Alexander, W., *CRC Materials Science and Engineering Handbook*, 3rd ed., Lewis Pub., Inc., 2000.

REVIEW QUESTIONS

1.1   Explain the difference between a unit cell and a single crystal.

1.2   What effects does recrystallization have on properties of metals?

1.3   What is strain hardening, and what effects does it have on the properties of metals?

1.4   Explain what is meant by structure-sensitive and structure-insensitive properties of metals.

1.5   What influence does grain size have on the mechanical properties of metals?

1.6   What is the relationship between the nucleation rate and the number of grains per unit volume of a metal?

1.7   What is a slip system, and what is its significance?

1.8   Explain the difference between recovery and recrystallization.

1.9   What is hot shortness, and what is its significance?

1.10   Explain the differences between cold, warm, and hot working of metals.

1.11   Describe the orange peel effect.

1.12   Why can't some metals, such as lead, become stronger when cold worked?

1.13   Describe the difference between preferred orientation and mechanical fibering.

1.14   What is twinning? How does it differ from slip?

QUALITATIVE PROBLEMS

1.15   What is the significance of the fact that some metals undergo allotropism?

1.16   Is it possible for two pieces of the same metal to have different recrystallization temperatures? Is it possible for recrystallization to take place in some regions of a part before it does other regions of the same part? Explain.

1.17   A cold-worked piece of metal has been recrystallized. When tested, it is found to be anisotropic. Explain the probable reason.

1.18   Explain the advantages and limitations of cold, warm, and hot working, respectively.

1.19   Do you think that it might be important to know whether a raw material for a manufacturing process has anisotropic properties? What about anisotropy in the finished product? Explain.

1.20   Explain why the strength of a polycrystalline metal at room temperature decreases as its grain size increases.

1.21   What is the significance of the fact that such metals as lead and tin have recrystallization temperatures at about room temperature?

1.22   It has been noted that the more a metal has been cold worked, the less it strain hardens. Explain why.

QUANTITATIVE PROBLEMS

1.23 Plot the data given in Table 1.1 in terms of grains/mm^2 versus grains/mm^3, and state your observations.

1.24 By cold working, a strip of metal is reduced from 40 mm in thickness to 20 mm. A similar strip is reduced in a similar way from 40 mm to 30 mm. Which one of these cold-worked strips will recrystallize at a lower temperature? Why?

1.25 A paper clip is made of wire that is 5 in. long and 1/32 in. in diameter. If the ASTM grain size is 9, how many grains are there in the paper clip?

1.26 How many atoms are on the surface of the head of a pin? Assume the head of a pin is spherical with a 1 mm diameter and has an ASTM grain size of 2.

1.27 A technician determines that the grain size of a certain etched specimen is 8. Upon further checking, it is found that the magnification used was 180× instead of the 100× that is required by the ASTM standards. Determine the correct grain size.

1.28 If the diameter of the aluminum atom is 0.5 nm, how many atoms are there in a grain with an ASTM grain size of 5?

SYNTHESIS, DESIGN, AND PROJECTS

1.29 By stretching a thin strip of polished metal (as in a tension-testing machine), demonstrate and comment on what happens to its reflectivity as it is stretched.

1.30 Draw some analogies to mechanical fibering (for example, layers of thin dough sprinkled with flour or butter between each layer).

1.31 Draw some analogies to the phenomenon of hot shortness.

1.32 Obtain a number of small balls made of plastic, wood, marble, and so forth, arrange them with your hands, or glue them together to represent the crystal structures shown in Figs. 1.2 through 1.4. Comment on your observations.

1.33 Take a deck of playing cards, put a rubber band around them, and slip them against each other to represent Figs. 1.6a and 1.7. If you repeat the same experiment with more and more rubber bands around the same deck, what are you accomplishing as far as the behavior of the material is concerned?

1.34 Give examples where anisotropy is scale dependent. For example, a wire rope can contain annealed wires that are isotropic on a microscopic scale, but the rope as a whole is anisotropic.

1.35 In Section 1.3, the movement of an edge dislocation was described by means of an analogy involving a hump in a carpet and how the whole carpet can be moved by moving the hump forward. The entanglement of dislocations was described in terms of two humps at different angles. Demonstrate these phenomena using a piece of cloth placed on a flat surface.

CHAPTER

2

Mechanical Behavior, Testing, and Manufacturing Properties of Materials

Metals can be processed into various shapes by deforming them plastically under the application of external forces. The effects of these forces on material behavior are described in this chapter, including:

• Types of tests for determining the mechanical behavior of materials.

• Elastic and plastic features of stress-strain curves and their significance.

• Relationships between stress and strain and their significance, as influenced by temperature and deformation rate.

• Characteristics of hardness, fatigue, creep, impact, and residual stresses, and their role in materials processing.

• Effects of inclusions and defects in the brittle and ductile behavior of metals.

• Why and how materials fail when subjected to external forces.

2.1 | Introduction

In manufacturing operations, numerous parts and components are *formed* into various shapes by applying external forces to the workpiece, typically by means of various tools and dies. Common examples are forging turbine disks, extruding various components of aluminum ladders, drawing wire for making nails, and rolling metal to make sheets for kitchen utensils and car bodies. Forming operations may be carried out at room temperature or at elevated temperatures and at a low or a high rate of deformation. These operations also are used in forming and shaping nonmetallic materials, such as plastics and ceramics.

As was noted in Fig. I.2, an extensive variety of metallic and nonmetallic materials are now widely available, and they have an equally wide range of properties, as shown qualitatively in Table 2.1. This chapter covers those aspects of mechanical properties and the behavior of metals that are relevant to the design and manufacture of parts and includes commonly used test methods employed in assessing the various properties of materials.

TABLE 2.1

Relative Mechanical Properties of Various Materials at Room Temperature, in Decreasing Order. Metals are in their Alloy Form.				
Strength	Hardness	Toughness	Stiffness	Strength/Density
Glass fibers	Diamond	Ductile metals	Diamond	Reinforced plastics
Graphite fibers	Cubic boron nitride	Reinforced plastics	Carbides	Titanium
Kevlar fibers	Carbides	Thermoplastics	Tungsten	Steel
Carbides	Hardened steels	Wood	Steel	Aluminum
Molybdenum	Titanium	Thermosets	Copper	Magnesium
Steels	Cast irons	Ceramics	Titanium	Beryllium
Tantalum	Copper	Glass	Aluminum	Copper
Titanium	Thermosets		Ceramics	Tantalum
Copper	Magnesium		Reinforced plastics	
Reinforced thermosets	Thermoplastics		Wood	
Reinforced thermoplastics	Tin		Thermosets	
Thermoplastics	Lead		Thermoplastics	
Lead			Rubbers	

2.2 | Tension

The **tension test** is the most common test for determining such *mechanical properties* of materials as strength, ductility, toughness, elastic modulus, and strain-hardening capability. The test first requires the preparation of a **test specimen**, typically as shown in Fig. 2.1a. In the United States, the specimen is prepared according to ASTM specifications. Otherwise, it is prepared to the specifications of the appropriate corresponding organization in other countries. Although most tension-test specimens are solid and round, they also can be flat or tubular.

Typically, the specimen has an **original gage length**, l_o, generally 50 mm (2 in.), and a cross-sectional area, A_o, usually with a diameter of 12.5 mm (0.5 in.). The specimen is mounted between the jaws of a tension-testing machine. These machines are equipped with various accessories and controls so that the specimen can be tested at different temperatures and rates of deformation.

2.2.1 Stress–strain curves

A typical sequence of deformation of the tension-test specimen is shown in Figs. 2.1a and 2.2. When the load is first applied, the specimen elongates in proportion to the load; this behavior is called **linear elastic**. If the load is removed, the specimen returns to its original length and shape in an elastic manner similar to stretching a rubber band and releasing it.

The **engineering stress** (**nominal stress**) is defined as the ratio of the applied load, *P,* to the original cross-sectional area, A_o, of the specimen:

$$\text{Engineering stress, } \sigma = \frac{P}{A_o} \tag{2.1}$$

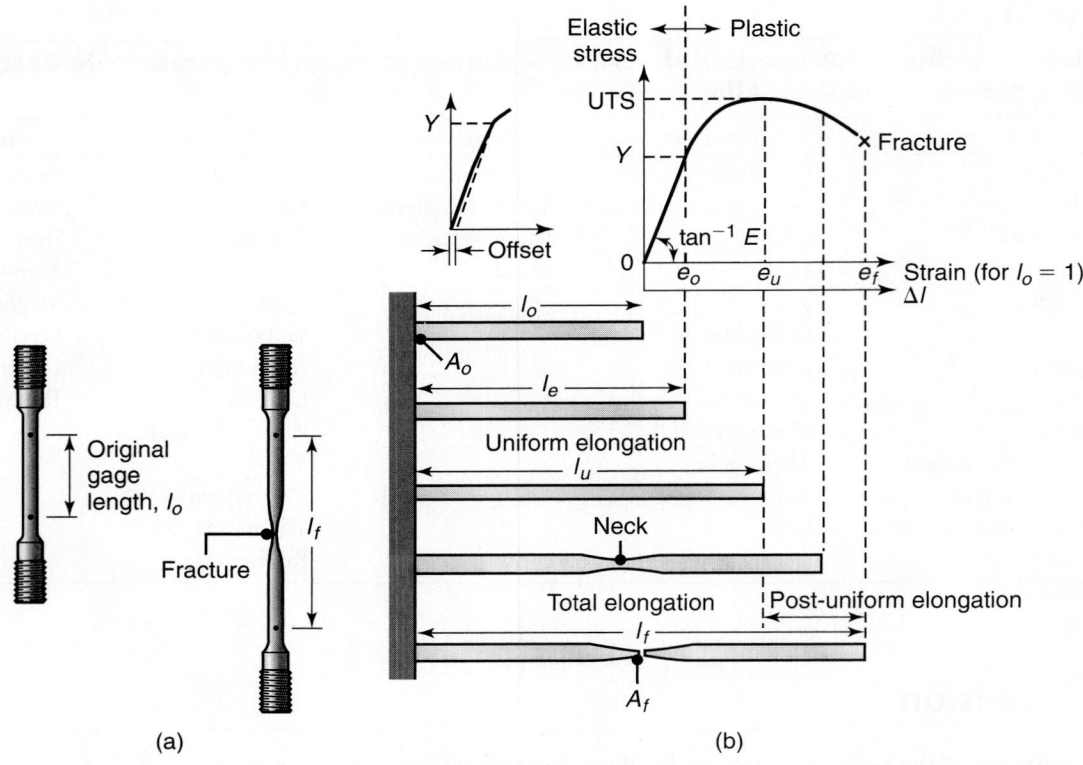

FIGURE 2.1 (a) A standard tensile-test specimen before and after pulling, showing original and final gage lengths. (b) A tensile-test sequence showing different stages in the elongation of the specimen.

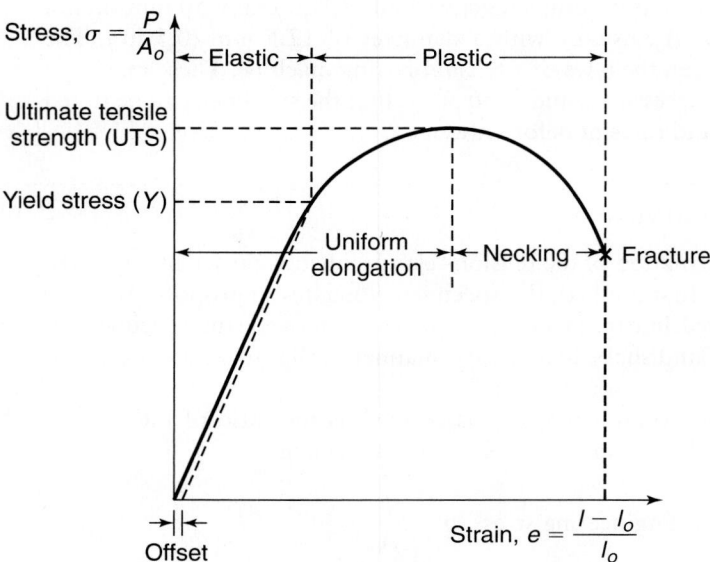

FIGURE 2.2 A typical stress–strain curve obtained from a tension test, showing various features.

The **engineering strain** is defined as

$$\text{Engineering strain, } e = \frac{(l - l_o)}{l_o} \tag{2.2}$$

where l is the instantaneous length of the specimen.

As the load is increased, the specimen begins (at some level of stress) to undergo **permanent (plastic) deformation**. Beyond that level, the stress and strain are no longer proportional, as they were in the elastic region. The stress at which this phenomenon occurs is known as the **yield stress, Y,** of the material. The yield stress and other properties for various metallic and nonmetallic materials are given in Table 2.2.

For soft and ductile materials, it may not be easy to determine the exact location on the stress–strain curve at which yielding occurs, because the slope of the straight (elastic) portion of the curve begins to decrease slowly. Therefore, Y usually is defined as the point on the stress–strain curve that is **offset** by a strain of 0.002, or 0.2% elongation. This simple procedure is shown on the left side in Fig. 2.2.

As the specimen (under a continuously increasing load) begins to elongate, its cross-sectional area decreases **permanently** and **uniformly** throughout its gage length.

TABLE 2.2

Mechanical Properties of Various Materials at Room Temperature				
Metals (Wrought)	E (GPa)	Y (MPa)	UTS (MPa)	Elongation in 50 mm (%)	Poisson's ratio (ν)
Aluminum and its alloys	69–79	35–550	90–600	45–4	0.31–0.34
Copper and its alloys	105–150	76–1100	140–1310	65–3	0.33–0.35
Lead and its alloys	14	14	20–55	50–9	0.43
Magnesium and its alloys	41–45	130–305	240–380	21–5	0.29–0.35
Molybdenum and its alloys	330–360	80–2070	90–2340	40–30	0.32
Nickel and its alloys	180–214	105–1200	345–1450	60–5	0.31
Steels	190–200	205–1725	415–1750	65–2	0.28–0.33
Titanium and its alloys	80–130	344–1380	415–1450	25–7	0.31–0.34
Tungsten and its alloys	350–400	550–690	620–760	0	0.27
Zinc and its alloys	50	25–180	240–550	65–5	0.27
Nonmetallic Materials					
Ceramics	70–1000	—	140–2600	0	0.2
Diamond	820–1050	—	—	—	—
Glass and porcelain	70–80	—	140	0	0.24
Silicon carbide (SiC)	200–500	—	310–400	—	0.19
Silicon nitride (Si_2N_4)	280–310	—	160–580	—	0.26
Rubbers	0.01–0.1	—	—	—	0.5
Thermoplastics	1.4–3.4	—	7–80	1000–5	0.32–0.40
Thermoplastics, reinforced	2–50	—	20–120	10–1	—
Thermosets	3.5–17	—	35–170	0	0.34
Boron fibers	380	—	3500	0	—
Carbon fibers	275–415	—	2000–3000	0	—
Glass fibers	73–85	—	3500–4600	0	—
Kevlar fibers	62–117	—	2800	0	—
Spectra Fibers	73–100	—	2400–2800	3	—

Note: In the upper table, the lowest values for E, Y, and UTS and the highest values for elongation are for pure metals. Multiply gigapascals (GPa) by 145,000 to obtain pounds per square in. (psi) and megapascals (MPa) by 145 to obtain psi.

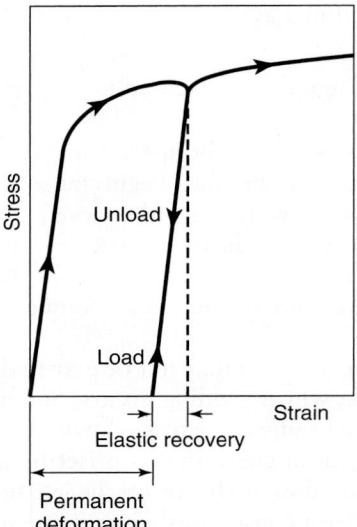

FIGURE 2.3   Schematic illustration of the loading and the unloading of a tensile-test specimen. Note that, during unloading, the curve follows a path parallel to the original elastic slope.

If the specimen is unloaded from a stress level higher than the yield stress, the curve follows a straight line downward and parallel to the original slope of the curve (Fig. 2.3). As the load is further increased, the engineering stress eventually reaches a maximum and then begins to decrease (Fig. 2.2). The maximum engineering stress is called the **tensile strength** or **ultimate tensile strength** (UTS) of the material. Values for the UTS for various materials are given in Table 2.2.

If the specimen is loaded beyond its ultimate tensile strength, it begins to **neck** or *neck down*. The cross-sectional area of the specimen is no longer uniform along the gage length and is smaller in the necked region. As the test progresses, the engineering stress drops further and the specimen finally fractures at the necked region (Fig. 2.1a). The engineering stress at fracture is known as the **breaking** or **fracture stress**.

The ratio of stress to strain in the elastic region is the **modulus of elasticity**, *E*, or **Young's modulus** (after T. Young, 1773–1829):

$$\text{Modulus of elasticity, } E = \frac{\sigma}{e} \tag{2.3}$$

This linear relationship is known as **Hooke's law** (after R. Hooke, 1635–1703).

Note in Eq. (2.3) that, because engineering strain is dimensionless, *E* has the same units as stress. The modulus of elasticity is essentially a measure of the slope of the elastic portion of the curve and, hence, the **stiffness** of the material. The higher the *E* value, the higher the load required to stretch the specimen to the same extent and, thus, the stiffer the material. Compare, for example, the stiffness of metal wire with that of a rubber band or plastic sheet when you try to stretch them to the same extent.

The elongation of the specimen under tension is accompanied by lateral contraction; this effect can be observed easily by stretching a rubber band. The absolute value of the ratio of the lateral strain to the longitudinal strain is known as **Poisson's ratio** (after S. D. Poisson, 1781–1840) and is denoted by the symbol ν.

2.2.2 Ductility

An important behavior observed during a tension test is **ductility**: the extent of plastic deformation that the material undergoes before fracture. There are two common measures of ductility. The first is the **total elongation** of the specimen:

$$\text{Elongation} = \frac{(l_f - l_o)}{l_o} \times 100 \tag{2.4}$$

where l_f and l_o are measured, as shown in Fig. 2.1a. Note that the elongation is based on the *original* gage length of the specimen and that it is calculated as a percentage.

The second measure of ductility is the **reduction of area**:

$$\text{Reduction of area} = \frac{(A_o - A_f)}{A_o} \times 100 \tag{2.5}$$

where A_o and A_f are the original and final (fracture) cross-sectional areas, respectively, of the test specimen. Reduction of area and elongation generally are interrelated, as shown in Fig. 2.4 for some typical metals. Thus, the ductility of a piece of chalk is zero, because it does not stretch at all or reduce in cross-section. By contrast, a ductile specimen, such as putty or chewing gum, stretches and necks considerably before it fails.

2.2.3 True stress and true strain

The engineering stress is based on the original cross-sectional area A_o of the specimen. However, the instantaneous cross-sectional area of the specimen becomes smaller as it elongates, just as the area of a rubber band does; thus, engineering stress does not represent the *actual* stress to which the specimen is subjected.

True stress is defined as the ratio of the load, P, to the actual (instantaneous, hence true) cross-sectional area, A, of the specimen:

$$\text{True stress, } \sigma = \frac{P}{A} \tag{2.6}$$

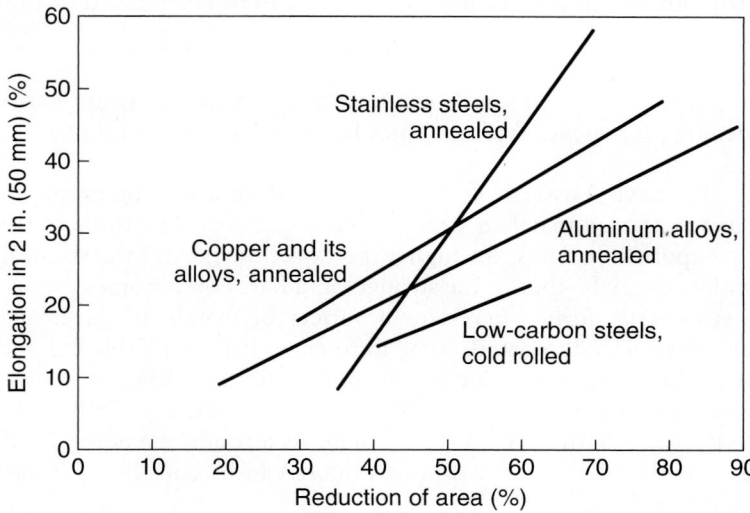

FIGURE 2.4 Approximate relationship between elongation and tensile reduction of area for various groups of metals.

For true strain, we first consider the elongation of the specimen to consist of increments of instantaneous change in length. Then, using calculus, it can be shown that the **true strain** (*natural* or *logarithmic strain*) is calculated as

$$\text{True strain, } \varepsilon = \ln\left(\frac{l}{l_o}\right). \tag{2.7}$$

Note from Eqs. (2.2) and (2.7) that, for small values of strain, the engineering and true strains are approximately equal. However, they diverge rapidly as the load increases. Thus, for example, when $e = 0.1$, $\varepsilon = 0.095$, and when $e = 1$, $\varepsilon = 0.69$.

Unlike engineering strains, true strains are consistent with actual, physical phenomena in the deformation of materials. Let's assume, for example, a hypothetical situation. A specimen 50 mm (2 in.) in height is compressed between flat platens to a final height of zero; in other words, we have deformed the specimen infinitely. According to their definitions, the engineering strain that the specimen undergoes is $(0 - 50)/50 = -1$, but the true strain is $-\infty$. Note that the answer will be the same regardless of the original height of the specimen. Clearly, true strain describes the extent of deformation correctly, since the deformation is indeed infinite.

2.2.4 Construction of stress–strain curves

The procedure for constructing an engineering stress-engineering strain curve is to take the load-elongation curve (Figs. 2.2 and 2.5a) and then to divide the load (vertical axis) by the original cross-sectional area, A_o, and the elongation (horizontal axis) by the original gage length, l_o. Because these two quantities are divided by constants, the engineering stress–strain curve obtained (shown in Fig. 2.5b) has the same shape as the load-elongation curve shown in Fig. 2.5a. (In this example, $A_o = 0.056 \text{ in}^2$, and $A_f = 0.016 \text{ in}^2$.)

True stress–true strain curves are obtained similarly, by dividing the load by the instantaneous cross-sectional area, with the true strain being obtained from Eq. (2.7). The result is shown in Fig. 2.5c. Note the *correction* to the curve; this reflects the fact that the specimen's necked region is subjected to three-dimensional tensile stresses, as described in more advanced texts. This state gives higher stress values than the actual true stress; hence, to compensate, the curve must be corrected downward.

The true stress–true strain curve in Fig. 2.5c can be represented by the equation

$$\sigma = K\varepsilon^n \tag{2.8}$$

where K is known as the **strength coefficient** and n as the **strain-hardening** (or **work-hardening**) **exponent**. Typical values for K and n for several metals are given in Table 2.3.

When the curve shown in Fig. 2.5c is plotted on a log–log graph, it is found that the curve is approximately a straight line (Fig. 2.5d). The slope of the curve is equal to the exponent n. Thus, the higher the slope, the greater the strain-hardening capacity of the material—that is, the stronger and harder it becomes as it is strained.

True stress–true strain curves for a variety of metals are given in Fig. 2.6. (When they are reviewed in detail, some differences between Table 2.3 and Fig. 2.6 will be noted; these discrepancies result from the fact that different sources of data and different specimens are involved.) Note that the elastic regions have been deleted, because the slope in this region is very high. As a result, the point of intersection of each curve with the vertical axis in this figure can be considered to be the yield stress, Y, of the material.

The area under the true stress–true strain curve at a particular strain is the *energy-per-unit volume of the material deformed* (**specific energy**) and indicates the

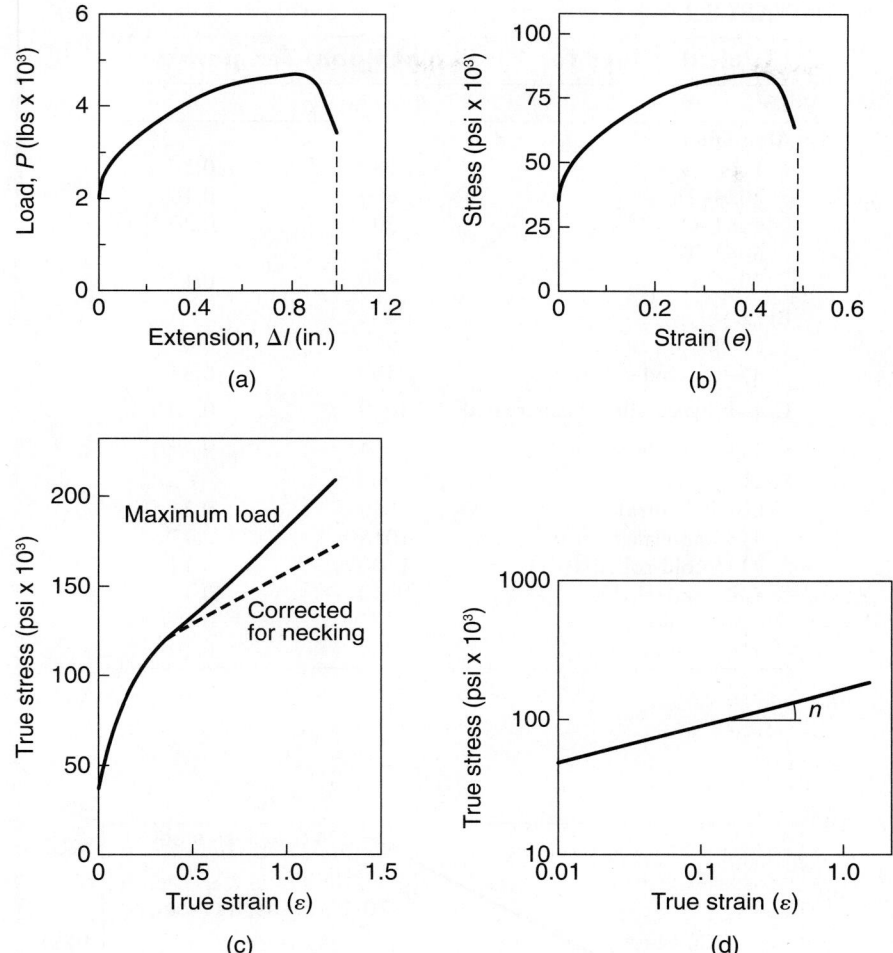

FIGURE 2.5   (a) Load-elongation curve in tension testing of a stainless steel specimen. (b) Engineering stress–engineering strain curve, drawn from the data in Fig. 2.5a. (c) True stress–true strain curve, drawn from the data in Fig. 2.5b. Note that this curve has a positive slope, indicating that the material is becoming stronger as it is strained. (d) True stress–true strain curve plotted on log–log paper and based on the corrected curve in Fig. 2.5c. The correction is due to the triaxial state of stress that exists in the necked region of the specimen.

work required to plastically deform a unit volume of the material to that strain. The area under the true stress–true strain curve up to fracture is known as the material's **toughness** (that is, the amount of energy per unit volume that the material dissipates prior to fracture). Note that toughness involves both the height and width of the stress–strain curve of the material; whereas strength is related only to the *height* of the curve, and ductility is related only to the *width* of the curve.

2.2.5 Strain at necking in a tension test

As noted previously, the onset of necking in a tension-test specimen corresponds to the ultimate tensile strength of the material. Note that the slope of the load-elongation curve at this point is zero, and it is at this point that the specimen begins to neck.

TABLE 2.3

Typical Values for *K* and *n* at Room Temperature		
	K (MPa)	*n*
Aluminum		
1100–O	180	0.20
2024–T4	690	0.16
6061–O	205	0.20
6061–T6	410	0.05
7075–O	400	0.17
Brass		
70–30, annealed	900	0.49
85–15, cold-rolled	580	0.34
Cobalt-based alloy, heat-treated	2070	0.50
Copper, annealed	315	0.54
Steel		
Low-C annealed	530	0.26
4135 annealed	1015	0.17
4135 cold-rolled	1100	0.14
4340 annealed	640	0.15
304 stainless, annealed	1275	0.45
410 stainless, annealed	960	0.10

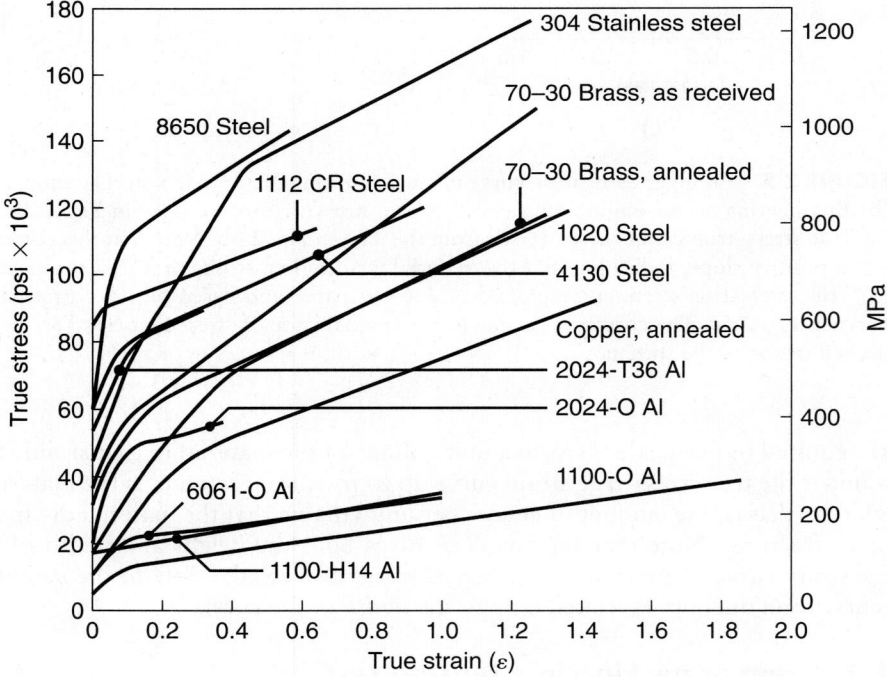

FIGURE 2.6   True stress–true strain curves in tension at room temperature for various metals. The curves start at a finite level of stress: The elastic regions have too steep a slope to be shown in this figure, and thus each curve starts at the yield stress, *Y*, of the material.

The specimen cannot support the load because the cross-sectional area of the neck is becoming smaller at a rate that is higher than the rate at which the material becomes stronger (strain hardens).

The true strain at the onset of necking is equal numerically to the strain-hardening exponent, n, of the material. Thus, the higher the value of n, the higher the strain which a piece of material can experience *uniformly* throughout before it begins to neck. This observation is important, particularly in regard to sheet-metal forming operations that involve the stretching of the workpiece material (Chapter 16). It can be seen in Table 2.3 that annealed copper, brass, and stainless steel have high n values; this means that they can be stretched uniformly to a greater extent than can the other metals listed.

EXAMPLE 2.1 Calculation of ultimate tensile strength

This example will show that the UTS of a material can be calculated from its K and n values. Assume that a material has a true stress–true strain curve given by

$$\sigma = 100,000\varepsilon^{0.5} \text{ psi}$$

Calculate the true ultimate tensile strength and the engineering UTS of this material.

Solution Because the necking strain corresponds to the maximum load and the necking strain for this material is

$$\varepsilon = n = 0.5$$

the *true* ultimate tensile strength is

$$\sigma = Kn^n = 100,000(0.5)^{0.5} = 70,710 \text{ psi}$$

The true area at the onset of necking is obtained from

$$\ln\left(\frac{A_o}{A_{neck}}\right) = 0.5$$

Thus,

$$A_{neck} = A_o\varepsilon^{-0.5}$$

and the maximum load, P, is

$$P = \sigma A_{neck} = \sigma A_o e^{-0.5}$$

where σ is the true ultimate tensile strength. Hence,

$$P = (70,710)(0.606)(A_o) = 42,850A_o \text{ lb}$$

Since UTS $= P/A_o$,

$$\text{UTS} = 42,850 \text{ psi}$$

2.2.6 Temperature effects

Increasing the temperature generally has the following effects on stress–strain curves (Fig. 2.7):

a. It raises the ductility and toughness.
b. It lowers the yield stress and the modulus of elasticity.

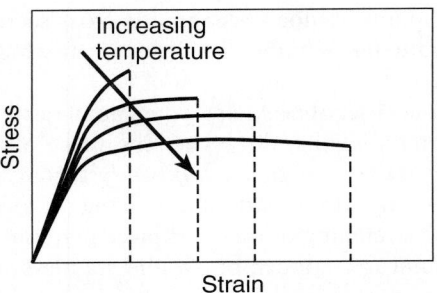

FIGURE 2.7   Typical effects of temperature on stress–strain curves. Note that temperature affects the modulus of elasticity, the yield stress, the ultimate tensile strength, and the toughness (area under the curve) of materials.

Temperature also affects the strain-hardening exponent of most metals, in that n decreases with increasing temperature. The influence of temperature, however, is described best in conjunction with the rate of deformation.

2.2.7 Rate-of-Deformation effects

Just as we can blow up a balloon or can stretch a rubber band at different rates, we also can shape a piece of material in a manufacturing process at different speeds. Some machines, such as hydraulic presses, form materials at low speeds; others, such as mechanical presses, form at high speeds. To simulate such differences, the test specimen can be strained at a rate corresponding to that which the metal will experience in the actual manufacturing process.

Deformation rate is defined as the speed at which a tension test is being carried out, in units of, say, m/s or ft/min. The **strain rate**, on the other hand, is a function of the specimen length. A short specimen elongates proportionately more during the same time period than does a long specimen. For example, let's take two rubber bands, one of 20 mm and the other of 100 mm in length, respectively, and elongate them both by 10 mm within a period of 1 second. The engineering strain in the shorter specimen is $\frac{10}{20} = 0.5$ and in the longer is $\frac{10}{100} = 0.1$. Thus, the strain rates are 0.5 s^{-1} and 0.1 s^{-1}, respectively, with the short band being subjected to a strain rate five times higher than that for the long band (although they are both being stretched at the same deformation rate)

Deformation rates typically employed in various testing and metalworking processes and the true strains involved are given in Table 2.4. Note the considerable difference in magnitudes. Because of this wide range, strain rates usually are stated in terms of orders of magnitude, such as 10^2 s^{-1}, 10^4 s^{-1}, and so on.

The typical effects that temperature and strain rate jointly have on the strength of metals are shown in Fig. 2.8. Note that increasing the strain rate increases the strength of the material (**strain-rate hardening**). The slope of these curves is called the **strain-rate sensitivity exponent**, m. The value of m is obtained from log–log plots, provided that the vertical and horizontal scales are the same (unlike those in Fig. 2.8). A slope of 45° would, therefore, indicate a value of $m = 1$. The relationship is given by the equation

$$\sigma = C\dot{\varepsilon}^{m} \tag{2.9}$$

where C is the **strength coefficient,** similar to (but not to be confused with) the strength coefficient K in Eq. (2.8). The constant C has the units of stress, while $\dot{\varepsilon}$ is the true strain rate, defined as the true strain that the material undergoes per unit time.

Note, in Fig. 2.8, that the sensitivity of strength-to-strain rate increases with temperature; in other words, m increases with increasing temperature. However,

TABLE 2.4

Typical Ranges of Strain and Deformation Rate in Manufacturing Processes		
Process	True strain	Deformation rate (m/s)
Cold working		
Forging, rolling	0.1–0.5	0.1–100
Wire and tube drawing	0.05–0.5	0.1–100
Explosive forming	0.05–0.2	10–100
Hot working and warm working		
Forging, rolling	0.1–0.5	0.1–30
Extrusion	2–5	0.1–1
Machining	1–10	0.1–100
Sheet-metal forming	0.1–0.5	0.05–2
Superplastic forming	0.2–3	10^{-4}–10^{-2}

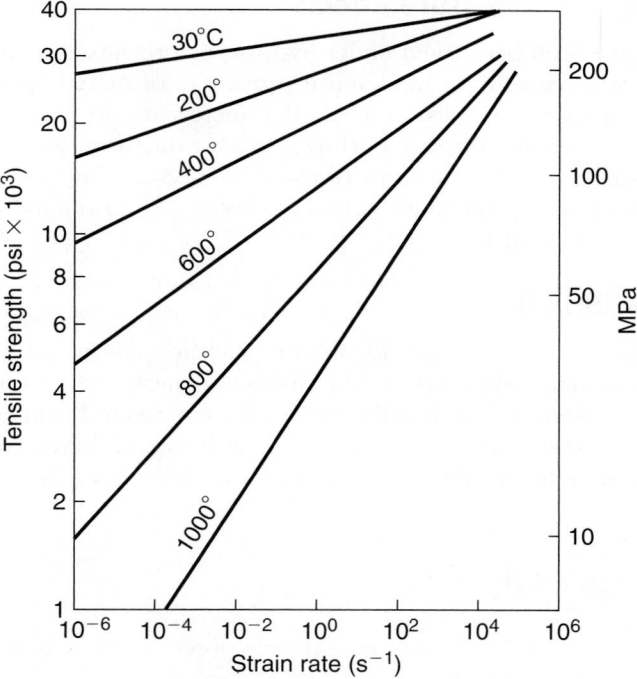

FIGURE 2.8 The effect of strain rate on the ultimate tensile strength for aluminum. Note that, as the temperature increases, the slopes of the curves increase; thus, strength becomes more and more sensitive to strain rate as temperature increases. *Source*: After J. H. Hollomon.

note that the slope is relatively flat at room temperature—that is, m is very low there. This condition is true for most metals but not for those that recrystallize at room temperature, such as lead and tin. Some typical ranges of m for metals are as follows:

> **Cold working:** up to 0.05
> **Hot working:** 0.05 to 0.4
> **Superplastic materials:** 0.3 to 0.85

The magnitude of the strain-rate sensitivity exponent significantly influences necking in a tension test. With increasing m, the material stretches farther before it fails; thus, increasing m delays necking. Ductility enhancement caused by the high strain-rate sensitivity of some materials has been exploited in the **superplastic forming** of sheet metal (Section 16.10).

Superplasticity. The term *superplasticity* refers to the capability of some materials to undergo large, uniform elongation prior to necking and fracture in tension. The elongation may be on the order from a few hundred percent to as much as 2000%. Common non-metallic materials exhibiting superplastic behavior are bubble gum and glass (at elevated temperatures) and thermoplastics. As a result, glass and thermoplastics can be formed successfully into complex shapes, such as beverage bottles and lighted advertising signs. Among metals exhibiting superplastic behavior are very fine-grain (10 to 15 μm) titanium alloys and alloys of zinc-aluminum; when heated, they can elongate to many times their original length.

2.2.8 Hydrostatic pressure effects

Various tests have been performed under hydrostatic pressure to determine the effect of hydrostatic pressure on mechanical properties of materials. Test results at pressures up to 3.5 GPa (500 ksi) indicate that increasing the hydrostatic pressure substantially increases the strain at fracture, both for ductile and for brittle materials. This beneficial effect of hydrostatic pressure has been exploited in metalworking processes, particularly in hydrostatic extrusion (Section 15.4) and in the compaction of metal powders (Section 17.3).

2.2.9 Radiation effects

In view of the use of many metals and alloys in nuclear applications, extensive studies have been conducted on the effects of radiation on mechanical properties. Typical changes in the properties of steels and other metals exposed to high-energy radiation are increased yield stress, tensile strength, and hardness and decreased ductility and toughness. Radiation has similar detrimental effects on the behavior of plastics.

2.3 | Compression

Many operations in manufacturing, particularly processes such as forging, rolling, and extrusion (Part III), are performed with the workpiece subjected to compressive forces. The **compression test,** in which the specimen is subjected to a compressive load, gives information useful for estimating forces and power requirements in these processes. This test usually is carried out by compressing a solid cylindrical specimen between two well-lubricated flat dies (platens). Because of friction between the specimen and the platens, the specimen's cylindrical surface bulges; this effect is called **barreling** (see Fig. 14.3). Friction prevents the top and bottom surfaces from expanding freely. Note that slender specimens can *buckle* during this test; thus, the height-to-diameter ratio of the solid cylindrical specimen typically should be less than 3:1.

Because the cross-sectional area of the specimen now changes along its height (being maximum in the middle), obtaining stress–strain curves in compression can be difficult. Furthermore, friction dissipates energy, so the compressive force is higher than it otherwise would be in order to supply the work required to overcome friction. With effective lubrication, friction can be minimized, and a reasonably constant cross-sectional area can be maintained during the test.

When the results of compression tests and tension tests on *ductile* metals are compared, the true stress–true strain curves for the two tests coincide. This comparability does not hold true for *brittle* materials, which generally are stronger and more ductile in compression than in tension (see Table 8.2).

When a metal with a certain tensile yield stress is subjected to tension into the plastic range, and then the load is released and applied in compression, the yield stress in compression is found to be lower than that in tension. This phenomenon is known as the **Bauschinger effect** (after J. Bauschinger, who reported it in 1881), and it is exhibited in varying degrees by all metals and alloys. Because of the lowered yield stress in the direction opposite the original load application, this phenomenon is also called **strain softening** or **work softening**.

Disk test.   For brittle materials, such as ceramics and glasses, a **disk test** has been developed in which the disk is subjected to compression between two hardened flat platens (Fig. 2.9). When the material is loaded as shown, tensile stresses develop perpendicular to the vertical centerline along the disk, fracture begins, and the disk splits in half vertically. The *tensile stress, σ,* in the disk is uniform along the centerline and can be calculated from the formula

$$\text{Tensile stress, } \sigma = \frac{2P}{\pi dt} \tag{2.10}$$

where P is the load at fracture, d is the diameter of the disk, and t is its thickness. In order to avoid premature failure at the contact points, thin strips of soft metal are placed between the disk and the platens. These strips also protect the platens from being damaged during the test. The fracture at the center of the specimen has been utilized in the manufacture of *seamless tubing* (Section 13.5).

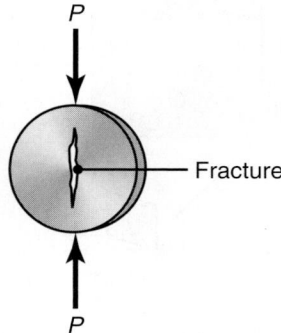

FIGURE 2.9   Disk test on a brittle material, showing the direction of loading and the fracture path.

2.4 | Torsion

In addition to tension and compression, a workpiece may be subjected to **shear strains** (Fig. 2.10), such as in the punching of holes in sheet metals and in metal cutting (Section 21.2). The test method generally used for determination of properties of materials in shear is the **torsion test.** In order to obtain an approximately uniform stress and strain distribution along the cross-section, this test usually is performed on a thin tubular specimen.

The torsion specimen usually has a reduced cross-section in order to confine the deformation to a narrow zone. The **shear stress** can be calculated from the formula

$$\text{Shear stress, } \tau = \frac{T}{2\pi r^2 t} \tag{2.11}$$

where T is the torque, r is the average radius of the tube, and t is the thickness of the tube at its narrow section.

The **shear strain** can be calculated from the formula

$$\text{Shear strain, } \gamma = \frac{r\phi}{l} \tag{2.12}$$

where l is the length of tube subjected to torsion and ϕ is the **angle of twist** in radians.

The ratio of the shear stress to the shear strain in the elastic range is known as the **shear modulus**, or **modulus of rigidity**, G (a quantity related to the modulus of elasticity, E). The angle of twist, ϕ, to fracture in the torsion of solid round bars at elevated temperatures also is useful in estimating the forgeability of metals. The greater the number of twists prior to failure, the better the forgeability (Section 14.5).

2.5 | Bending

Preparing specimens from brittle materials, such as ceramics and carbides, is difficult because of the problems involved in shaping and machining them to proper dimensions. Furthermore, because of their sensitivity to surface defects and notches,

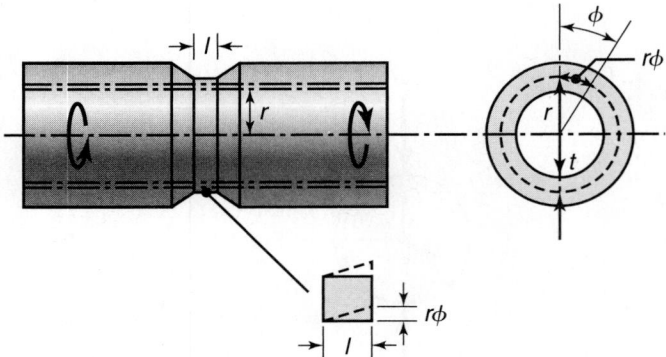

FIGURE 2.10   A typical torsion-test specimen; it is mounted between the two heads of a testing machine and twisted. Note the shear deformation of an element in the reduced section of the specimen.

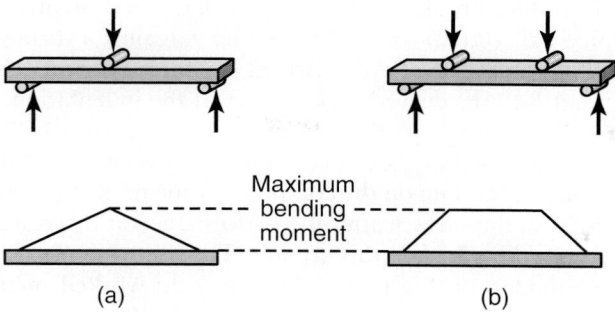

FIGURE 2.11 Two bend-test methods for brittle materials: (a) three-point bending; (b) four-point bending. The areas on the beams represent the bending-moment diagrams, described in texts on mechanics of solids. Note the region of constant maximum bending moment in (b); by contrast, the maximum bending moment occurs only at the center of the specimen in (a).

clamping brittle test specimens for testing is difficult. Also, improper alignment of the test specimen may result in nonuniform stress distribution along the cross-section of the specimen.

A commonly used test method for brittle materials is the **bend** or **flexure test**. It usually involves a specimen that has a rectangular cross-section and is supported at its ends (Fig. 2.11). The load is applied vertically at either one point or two points; as a result, these tests are referred to as **three-point** and **four-point bending**, respectively. The longitudinal stresses in these specimens are tensile at their lower surfaces and compressive at their upper surfaces.

These stresses can be calculated using the simple beam equations described in texts on mechanics of solids. The stress at fracture in bending is known as the **modulus of rupture, flexural strength**, or **transverse rupture strength** (see Table 8.1). Note that, because of the larger volume of material subjected to the same bending moment in Fig. 2.11b, there is a higher probability that defects exist in this volume than in that in Fig. 2.11a. Consequently, the four-point test gives a lower modulus of rupture than the three-point test.

2.6 | Hardness

Hardness is a commonly used property; it gives a general indication of the strength of the material and of its resistance to scratching and to wear. More specifically, **hardness** usually is defined as *resistance to permanent indentation*; thus, for example, steel is harder than aluminum, and aluminum is harder than lead. However, hardness is *not* a fundamental property, because the resistance to indentation depends on the shape of the indenter and on the load applied.

2.6.1 Hardness tests

Several test methods have been developed to measure the hardness of materials, using different *indenter* materials and shapes. Commonly used hardness tests are described next.

Brinell test.   Introduced by J. A. Brinell in 1900, this test involves pressing a steel or tungsten-carbide ball 10 mm (0.4 in.) in diameter against a surface with a load of 500, 1500, or 3000 kg (Fig. 2.12). The Brinell hardness number (HB) is defined as the ratio of the load P to the curved surface area of the indentation. The harder the material to be tested, the smaller will be the impression; hence, a 1500-kg or 3000-kg load usually is recommended in order to obtain impressions sufficiently large for accurate measurement. Depending on the condition of the material, one of two types of impression develops on the surface after the performance of this test (Fig. 2.13) or of any of the other tests described in this section. The impressions in annealed metals generally have a rounded profile (Fig. 2.13a); in cold-worked metals they usually have a sharp profile (Fig. 2.13b). The correct methods of measuring the indentation diameter, d, are shown in Fig. 2.13a and b.

The indenter, which has a finite elastic modulus, also undergoes *elastic* deformation under the applied load; as a result, hardness measurements may not be as accurate as expected. One method for minimizing this effect is to use tungsten-carbide balls. Because of their high modulus of elasticity, they distort less than steel balls do. Carbide balls usually are recommended for Brinell hardness numbers greater than 500.

Rockwell test.   Developed by S.P. Rockwell in 1922, this test measures the *depth* of penetration instead of the diameter of the indentation. The indenter is pressed

Test	Indenter	Shape of indentation Side view	Shape of indentation Top view	Load, P	Hardness number
Brinell	10-mm steel or tungsten carbide ball	D / d	d	500 kg / 1500 kg / 3000 kg	$HB = \dfrac{2P}{(\pi D)\left(D - \sqrt{D^2 - d^2}\right)}$
Vickers	Diamond pyramid	$136°$	L	1–120 kg	$HV = \dfrac{1.854P}{L^2}$
Knoop	Diamond pyramid	$L/b = 7.11$ / $b/t = 4.00$ / t	b / L	25 g–5 kg	$HK = \dfrac{14.2P}{L^2}$
Rockwell A C D	Diamond cone	$120°$ / $t = $ mm		60 kg / 150 kg / 100 kg	HRA / HRC / HRD $= 100 - 500t$
B F G	$\frac{1}{16}$- in. diameter steel ball	$t = $ mm		100 kg / 60 kg / 150 kg	HRB / HRF / HRG $= 130 - 500t$
E	$\frac{1}{8}$- in. diameter steel ball			100 kg	HRE

FIGURE 2.12   General characteristics of hardness-testing methods and formulas for calculating hardness.

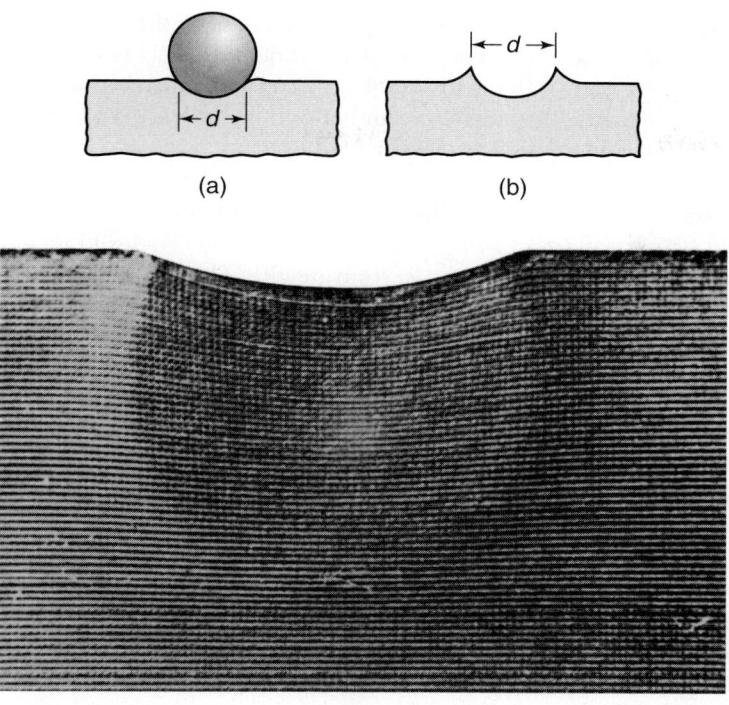

(a)

(b)

(c)

FIGURE 2.13   Indentation geometry in Brinell hardness testing: (a) annealed metal; (b) work-hardened metal; (c) deformation of mild steel under a spherical indenter. Note that the depth of the permanently-deformed zone is about one order of magnitude larger than the depth of indentation. For a hardness test to be valid, this zone should be developed fully in the material. *Source:* Courtesy of M. C. Shaw and C. T. Yang.

onto the surface first with a minor load and then with a major load; the difference in the depths of penetration is a measure of the hardness of the material. Some of the more common Rockwell hardness scales and indenters used are shown in Fig. 2.12. Rockwell **superficial hardness** tests also have been developed using the same type of indenters but at lighter loads.

Vickers test.   This test, developed in 1922 and formerly known as the *diamond pyramid hardness* test, uses a pyramid-shaped diamond indenter (Fig. 2.12) and a load that ranges from 1 kg to 120 kg. The Vickers hardness number is indicated by HV. The impressions obtained typically are less than 0.5 mm (0.020 in.) on the diagonal. The Vickers test gives essentially the same hardness number, regardless of the load, and is suitable for testing materials with a wide range of hardness, including heat-treated steels.

Knoop test.   This test, developed by F. Knoop in 1939, uses a diamond indenter in the shape of an elongated pyramid (Fig. 2.12) with applied loads ranging generally from 25 g to 5 kg. The Knoop hardness number is indicated by HK. Because of the light loads that are applied, it is a **microhardness** test; therefore, it is suitable for very small or very thin specimens and for brittle materials, such as carbides, ceramics, and glasses.

This test also is used for measuring the hardness of the individual grains and components in a metal alloy. The size of the indentation is generally in the range from 0.01 to 0.10 mm (0.0004 to 0.004 in.); consequently, surface preparation is very important. Because the hardness number obtained depends on the applied load, Knoop test results should always cite the load used.

Scleroscope.   The *scleroscope* (from the Greek *skleros*, meaning hard) is an instrument in which a diamond-tipped indenter (hammer) enclosed in a glass tube is dropped onto the specimen from a certain height. The hardness is related to the *rebound* of the indenter: the higher the rebound, the harder the material. The impression made by a scleroscope is very small. The instrument is portable, and it simply is placed on the surface of the part. Thus, it is useful for measuring the hardness of large objects that otherwise would not fit into the limited space of conventional hardness testers.

Mohs hardness.   Developed in 1822 by F. Mohs, this test is based on the capability of one material to *scratch* another. The Mohs hardness is based on a scale from 1 to 10, with 1 being the measure for talc and 10 for diamond (the hardest substance known). A material with a higher Mohs hardness number always scratches one with a lower number. Soft metals have a Mohs hardness of 2 to 3, hardened steels about 6, and aluminum oxide (used for cutting tools and as an abrasive in grinding wheels) of 9. Although the Mohs scale is qualitative and is used mainly by mineralogists, it correlates well with Knoop hardness.

Durometer.   The hardness of rubbers, plastics, and similar soft and elastic nonmetallic materials generally is measured with an instrument called a *durometer* (from the Latin *durus*, meaning hard). An indenter is pressed against the surface, and then a constant load is applied rapidly. The *depth* of penetration is measured after one second; the hardness is related inversely to the penetration. There are two different scales for this test. Type A has a blunt indenter and a load of 1 kg; it is used for softer materials. Type D has a sharper indenter and a load of 5 kg; it is used for harder materials. The hardness numbers in these tests range from 0 to 100.

Hot hardness.   The hardness of materials at elevated temperatures (see Fig. 22.1) is important in applications in which higher temperatures are involved, such as the use of cutting tools in machining and of dies in hot-working and casting operations. Hardness tests can be performed at elevated temperatures using conventional testers with some modifications, such as enclosing the specimen and indenter in a small electric furnace.

2.6.2 Hardness and Strength

Because hardness is the resistance to *permanent* indentation, it can be likened to performing a compression test on a small volume on the surface of a block of material (Fig. 2.13c). Studies have shown that (in the same units) the hardness of a cold-worked metal is about three times its yield stress, Y; for annealed metals, it is about five times Y.

A relationship has been established between the ultimate tensile strength (UTS) and the Brinell hardness (HB) for steels as measured for a load of 3000 kg. In SI units, the relationship is

$$UTS = 3.5(HB) \tag{2.13}$$

where UTS is in MPa. In traditional (English) units,

$$UTS = 500(HB) \qquad (2.14)$$

where UTS is in psi.

2.6.3 Hardness-testing procedures

For a hardness test to be meaningful and reliable, the **zone of deformation** under the indenter (see Fig. 2.13c) must be allowed to develop freely. Consequently, the *location* of the indenter (with respect to the *edges* of the specimen to be tested) and the *thickness* of the specimen are important considerations. Generally, the location of the indenter should be at least two diameters from the edge of the specimen, and the thickness of the specimen should be at least 10 times the depth of penetration of the indenter. Successive indentations on the same surface of the workpiece should be far enough apart so as not to interfere with each other.

Moreover, the indentation should be sufficiently large to give a representative hardness value for the bulk material. If hardness variations need to be detected in a small area or if the hardness of individual constituents in a matrix or an alloy is to be determined, the indentations should be very small, such as those obtained in Knoop or Vickers tests using light loads. While *surface preparation* is not critical for the Brinell test, it is important for the Rockwell test and even more important for the other hardness tests because of the small sizes of the indentations. Surfaces may have to be polished to allow correct measurement of the impression's dimensions.

The values obtained from different hardness tests (on different scales) can be interrelated and can be converted using Fig. 2.14. Care should be exercised in using these charts because of the many variables in material characteristics and in the shape of indentation.

2.7 | Fatigue

Various components in manufacturing equipment, such as tools, dies, gears, cams, shafts, and springs, are subjected to rapidly fluctuating (cyclic or periodic) loads in addition to static loads. **Cyclic stresses** may be caused by fluctuating mechanical loads (such as on gear teeth) or by thermal stresses (such as on a cool die coming into repeated contact with hot workpieces). Under these conditions, the part fails at a stress level below that at which failure would occur under static loading. Upon inspection, failure is found to be associated with cracks that grow with every stress cycle and propagate through the material until a critical crack length is reached and the material fractures. Known as **fatigue failure**, this phenomenon is responsible for the majority of failures in mechanical components.

Fatigue *test methods* involve testing specimens under various states of stress, usually in a combination of tension and bending. The test is carried out at various *stress amplitudes* (S), and the number of cycles (N) it takes to cause total failure of the specimen or part is recorded. Stress amplitude is defined as the maximum stress (in tension and compression) to which the specimen is subjected. Typical plots of **S–N curves** are shown in Fig. 2.15. These curves are based on complete reversal of the stress—that is, maximum tension, then maximum compression, then maximum tension, and so on—such as that imposed by bending a rectangular eraser or a piece of wire alternately in one direction and then the other. The test can also be performed on a rotating shaft with a constant downward load. With some materials, the S–N curve becomes horizontal at low stresses, indicating that the material will

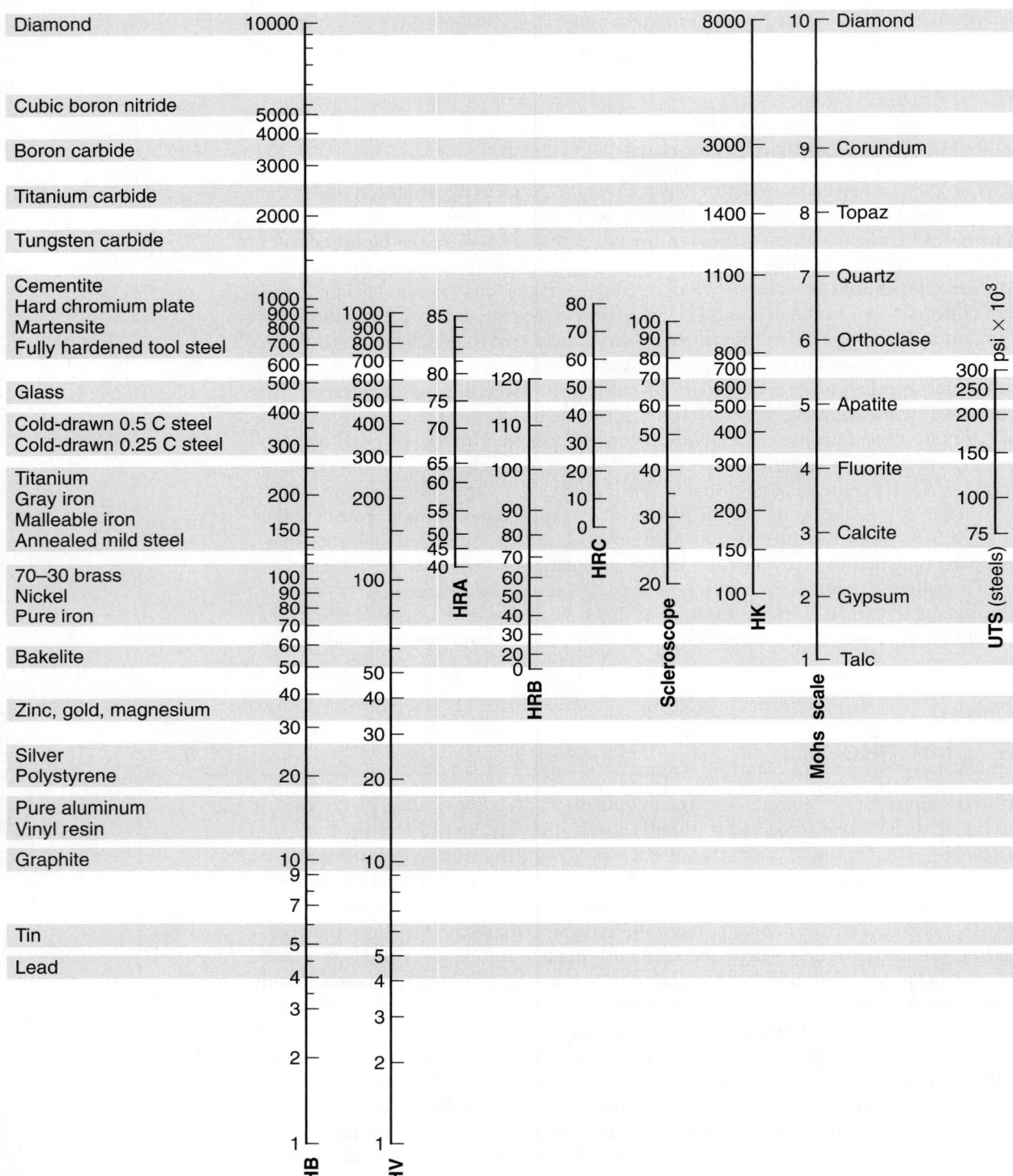

FIGURE 2.14 Chart for converting various hardness scales. Because of the many factors involved, these conversions are approximate.

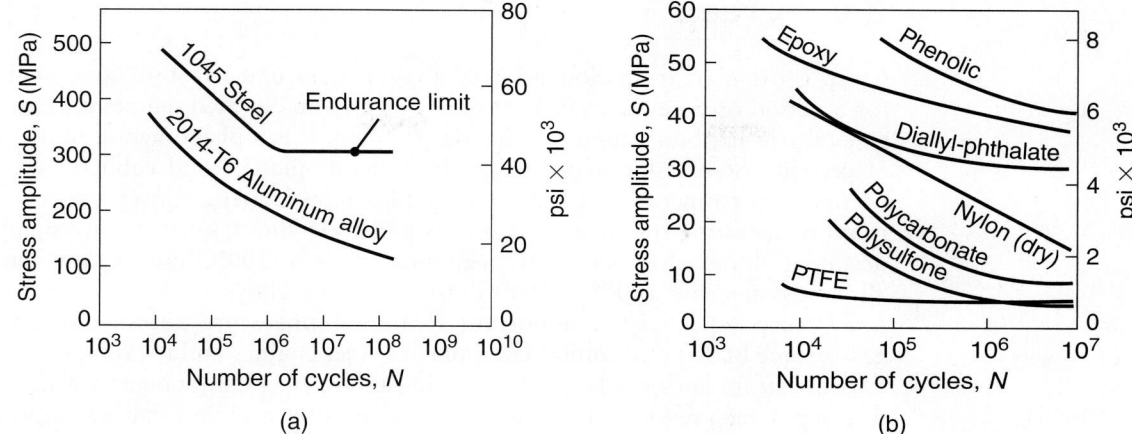

FIGURE 2.15   (a) Typical S–N curves for two metals. Note that, unlike steel, aluminum does not have an endurance limit. (b) S–N curves for common polymers.

not fail at stresses below this limit. The maximum stress to which the material can be subjected without fatigue failure (regardless of the number of cycles) is known as the **endurance** or **fatigue limit**.

Although many materials (especially steels) have a definite endurance limit, aluminum alloys do not have any, and the S–N curve continues its downward trend. For metals exhibiting such behavior, the fatigue strength is specified at a certain number of cycles, such as 10^7. In this way, the useful service life of the component can be specified. The endurance limit for metals can be related approximately to their ultimate tensile strength (Fig. 2.16). For carbon steels, the endurance limit is usually 0.4 to 0.5 times the tensile strength, although particular values can vary.

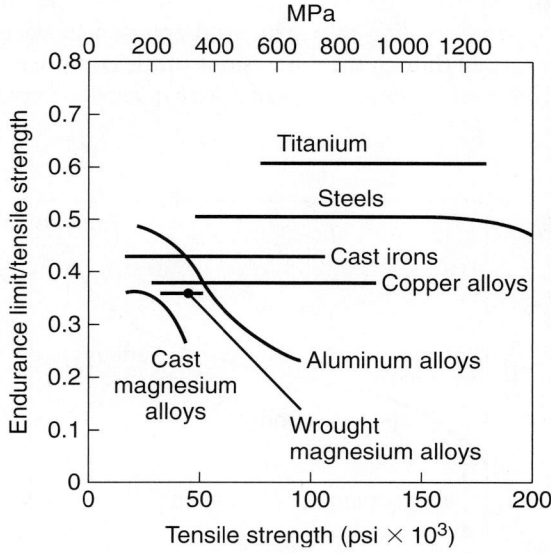

FIGURE 2.16   Ratio of endurance limit to tensile strength for various metals, as a function of tensile strength. Because aluminum does not have an endurance limit, the correlations for aluminum are based on a specific number of cycles.

2.8 | Creep

Creep is the permanent elongation of a component under a static load maintained for a period of time. The mechanism of creep at an elevated temperature in metals generally is attributed to **grain-boundary sliding**. It is a phenomenon of metals and of certain non-metallic materials, such as thermoplastics and rubbers, and it can occur at any temperature. Lead, for example, creeps under a constant tensile load at room temperature. However, for metals and their alloys, creep of any significance occurs at elevated temperatures, beginning at about 200°C (400°F) for aluminum alloys and at about 1500°C (2800°F) for refractory alloys.

Creep especially is important in high-temperature applications, such as gas-turbine blades and similar components in jet engines and rocket motors. High-pressure steam lines, nuclear-fuel elements, and furnace components are also subject to creep. Creep deformation also can occur in tools and dies that are subjected to high stresses at elevated temperatures during hot-working operations, such as forging and extrusion.

The *creep test* typically consists of subjecting a specimen to a constant tensile load (hence, constant engineering stress) at a certain temperature and measuring the changes in length at various time increments. A typical creep curve usually consists of *primary, secondary*, and *tertiary* stages (Fig. 2.17). The specimen eventually fails by necking and fracture, called **rupture** or **creep rupture**. As expected, the creep rate increases with specimen temperature and the applied load.

Design against creep usually involves a knowledge of the secondary (linear) range and its slope, because the creep rate can be determined reliably only when the curve has a constant slope. Generally, resistance to creep increases with the melting temperature of a material; that fact serves as a general guideline for design purposes. Thus, stainless steels, superalloys, and refractory metals and alloys are used commonly in applications where resistance to creep is required.

Stress relaxation.   Stress relaxation is related closely to creep. In *stress relaxation*, the stresses resulting from loading of a structural component decrease in magnitude over a period of time, even though the dimensions of the component remain constant. One example is the decrease in tensile stress of a wire in tension between two fixed ends

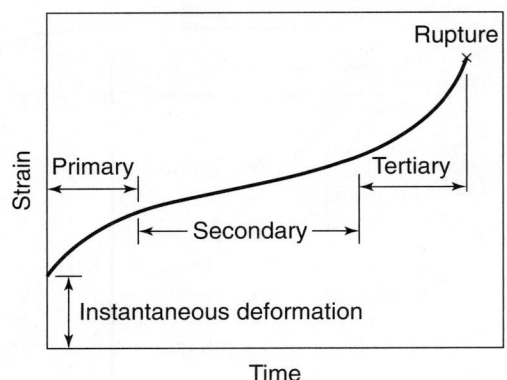

FIGURE 2.17   Schematic illustration of a typical creep curve. The linear segment of the curve (secondary) is used in designing components for a specific creep life.

(such as a piano), while other examples include rivets, bolts, guy wires, and similar parts under either tension, compression, or flexure. Stress relaxation is common and important in thermoplastics.

2.9 | Impact

In many manufacturing operations, as well as during the service life of components, materials are subjected to **impact** or **dynamic loading** (for example, in high-speed metalworking operations, such as heading to make bolt heads and in drop forging; Section 14.8). A typical *impact test* consists of placing a notched specimen in an impact tester and breaking it with a swinging pendulum.

In the **Charpy** test, the specimen is supported at both ends (Fig. 2.18a); in the **Izod** test, it is supported at one end like a cantilever beam (Fig. 2.18b). From the amount of swing of the pendulum, the energy dissipated in breaking the specimen can be obtained; this energy is the **impact toughness** of the material. Unlike hardness-test conversions (Fig. 2.14), no quantitative relationships have yet been established between Charpy and the Izod tests. Impact tests are useful particularly in determining the ductile–brittle *transition temperature* of materials (Section 2.10.1). Materials that have high impact resistance generally are those that also have high strength and high ductility and, hence, high toughness. Sensitivity to surface defects (**notch sensitivity**) is important, as it significantly lowers impact toughness, particularly in heat-treated metals and in ceramics and glasses.

2.10   Failure and Fracture of Materials in Manufacturing and Service

Failure is one of the most important aspects of material behavior, because it directly influences the selection of a material for a particular application, the methods of manufacturing, and the service life of the component. Because of the many factors involved, the failure and fracture of materials are a complex area of study; this section focuses only on those aspects of failure that are of particular significance in selecting and processing materials. There are two general types of failure:

1. **Fracture:** through either internal or external cracking. Fracture is further subclassified into two general categories as *ductile* or *brittle* (Figs. 2.19 and 2.20).

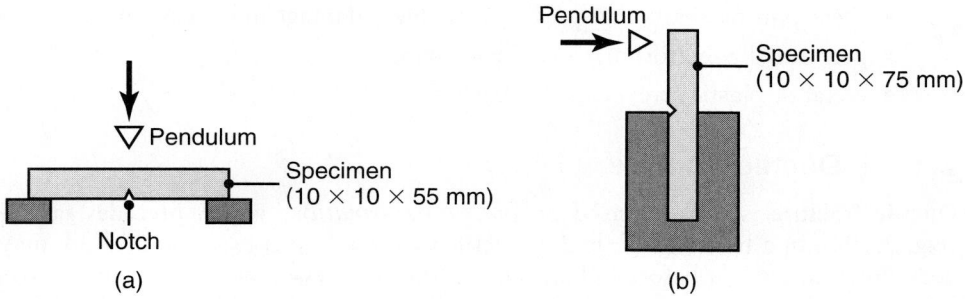

FIGURE 2.18   Impact test specimens: (a) Charpy; (b) Izod.

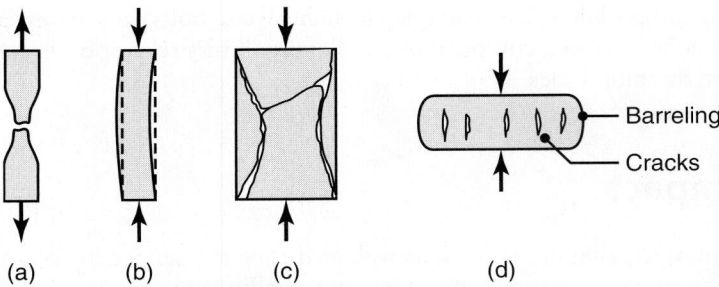

FIGURE 2.19 Schematic illustration of types of failures in materials: (a) necking and fracture of ductile materials; (b) buckling of ductile materials under a compressive load; (c) fracture of brittle materials in compression; (d) cracking on the barreled surface of ductile materials in compression.

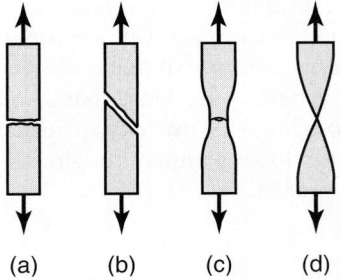

FIGURE 2.20 Schematic illustration of the types of fracture in tension: (a) brittle fracture in polycrystalline metals; (b) shear fracture in ductile single crystals—see also Fig. 1.6a; (c) ductile cup-and-cone fracture in polycrystalline metals; (d) complete ductile fracture in polycrystalline metals, with 100% reduction of area.

2. **Buckling** (Fig. 2.19b): Although failure of materials generally is regarded as undesirable, some products are designed in such a way that failure is essential for their function. Typical examples are:

   * food and beverage containers with tabs (or entire tops) which are removed by tearing the sheet metal along a prescribed path
   * shear pins on shafts that prevent machinery damage in the case of overloads
   * perforated paper or metal as in packaging
   * metal or plastic screw caps for bottles

2.10.1 Ductile fracture

Ductile fracture is characterized by *plastic deformation*, which precedes failure (Fig. 2.19a). In a tension test, highly ductile materials, such as gold and lead, may neck down to a point before failing (Fig. 2.20d). However, most metals and alloys neck down to a finite area and then fail. Ductile fracture generally takes place along planes on which the *shear stress is a maximum*. Thus, in torsion, a ductile metal fractures along a plane perpendicular to the axis of twist—that is, the plane on which the shear stress is a maximum. Fracture in simple shear, by contrast, is a result of extensive slip along slip planes within the grains (see Fig. 1.7).

Close examination of the surface of ductile fracture (Fig. 2.21) shows a *fibrous* pattern with *dimples*, as if a number of very small tension tests have been carried out over the fracture surface. Failure is initiated with the formation of tiny *voids*, usually around small inclusions or preexisting voids, which then *grow* and *coalesce*, developing into microcracks which grow in size and eventually lead to fracture.

In a tension-test specimen, fracture begins at the center of the necked region as a result of the growth and coalescence of cavities (Fig. 2.22). The central region becomes one large crack, as can be seen in the mid-section of the tension-test specimen in Fig. 2.22d. This crack then propagates to the periphery of the necked region. Because of its appearance, the fracture surface of a tension-test specimen is called a **cup-and-cone fracture**.

Effects of inclusions. Because they are nucleation sites for voids, *inclusions* have an important influence on ductile fracture and, consequently, on the workability of materials. Inclusions may consist of impurities of various kinds and of second-phase

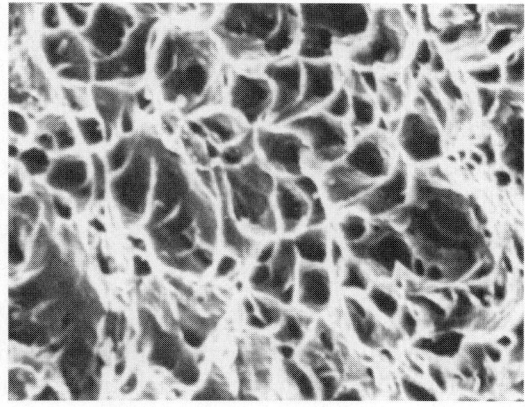

FIGURE 2.21 Surface of ductile fracture in low-carbon steel, showing dimples. Fracture usually is initiated at impurities, inclusions, or preexisting voids (microporosity) in the metal. *Source:* Courtesy of K.-H. Habig and D. Klaffke.

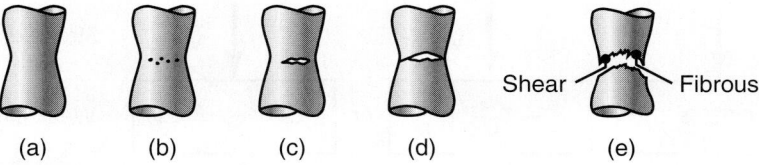

Shear   Fibrous

(a)   (b)   (c)   (d)   (e)

FIGURE 2.22 Sequence of events in the necking and fracture of a tensile-test specimen: (a) early stage of necking; (b) small voids begin to form within the necked region; (c) voids coalesce, producing an internal crack; (d) the rest of the cross-section begins to fail at the periphery, by shearing; (e) the final fracture surfaces, known as cup- (top fracture surface) and cone- (bottom surface) fracture.

particles, such as oxides, carbides, and sulfides. The extent of their influence depends on such factors as their shape, hardness, distribution, and fraction of total volume: the greater the volume fraction of inclusions, the lower the ductility of the material. Voids and porosity also can develop during processing of metals, such as ones resulting from casting and metalworking processes such as drawing and extrusion.

Two factors affect void formation:

a. The strength of the bond at the interface between an inclusion and the matrix: If the bond is strong, there is less tendency for void formation during plastic deformation.

b. The hardness of the inclusion: If it is soft, such as one of manganese sulfide, it will conform to the overall shape change of the workpiece during plastic deformation. If the inclusion is hard (for example, carbides and oxides), it could lead to void formation (Fig. 2.23). Hard inclusions, because of their brittle nature, may also break up into smaller particles during deformation.

The alignment of inclusions during plastic deformation leads to **mechanical fibering** (Section 1.5). Therefore, subsequent processing of such a material must involve considerations of the proper direction of working for maximum ductility and strength.

Transition temperature.   Many metals undergo a sharp change in ductility and toughness across a narrow temperature range called the *transition temperature* (Fig. 2.24). This phenomenon occurs mostly in body-centered cubic and in some hexagonal close-packed metals; it is rarely exhibited by face-centered cubic metals. The transition temperature depends on such factors as the composition, microstructure, grain size, surface finish and shape of the specimen, and deformation rate. High rates, abrupt changes in workpiece shape, and the presence of surface notches raise the transition temperature.

Strain aging.   *Strain aging* is a phenomenon in which carbon atoms in steels segregate to dislocations, thereby pinning them and, in this way, increasing the resistance to dislocation movement; the result is increased strength and reduced ductility. Instead of taking place over several days at room temperature, this phenomenon can occur in just a few hours at a higher temperature; it is then called **accelerated strain aging**. An example of accelerated strain aging in steels is **blue brittleness**, so named because it occurs in the blue-heat range where the steel develops a bluish oxide film. This phenomenon causes a marked decrease in ductility and toughness and an increase in the strength of plain-carbon and of some alloy steels.

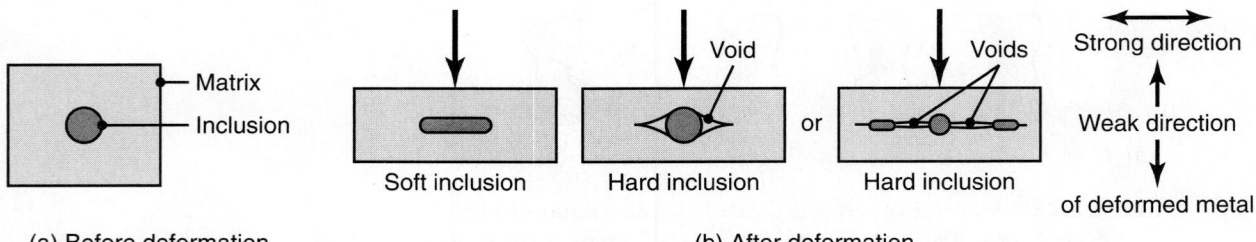

FIGURE 2.23   Schematic illustration of the deformation of soft and hard inclusions and of their effect on void formation in plastic deformation. Note that, because they do not conform to the overall deformation of the ductile matrix, hard inclusions can cause internal voids.

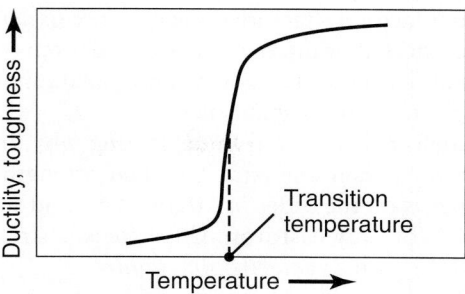

FIGURE 2.24   Schematic illustration of transition temperature in metals.

2.10.2 Brittle fracture

Brittle fracture occurs with little or no gross plastic deformation; in tension, fracture takes place along the crystallographic plane (**cleavage plane**) on which the normal tensile stress is a maximum. Face-centered cubic metals usually do not fail by brittle fracture, whereas body-centered cubic and some hexagonal close-packed metals fail by cleavage. In general, low temperature and a high rate of deformation promote brittle fracture. In a polycrystalline metal under tension, the fracture surface has a bright *granular* appearance, because of the changes in the direction of the cleavage planes as the crack propagates from one grain to another (Fig. 2.25). Brittle fracture of a specimen in compression is more complex and fracture may even follow a path that is theoretically at an angle of 45° to the direction of the applied force.

Examples of fracture along a cleavage plane are the splitting of rock salt and the peeling of layers of mica. Tensile stresses normal to the cleavage plane, caused by pulling, initiate and control the propagation of fracture. Another example is the behavior of brittle materials, such as chalk, gray cast iron, and concrete. In tension, they fail in the manner shown in Fig. 2.20a. In torsion, they fail along a plane at an angle of 45° to the axis of twist (Fig. 2.10)—that is, along a plane on which the tensile stress is a maximum.

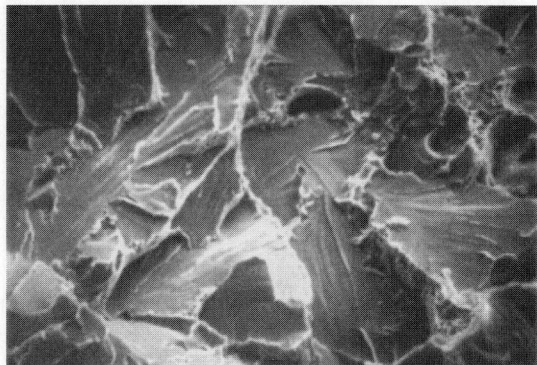

FIGURE 2.25   Fracture surface of steel that has failed in a brittle manner. The fracture path is transgranular (through the grains). Magnification: 200x. *Source*: Courtesy of B. J. Schulze and S. L. Meiley and Packer Engineering Associates, Inc.

Defects.   An important factor in fracture is the presence of *defects*, such as scratches, flaws, and pre-existing external or internal cracks. Under tension, the sharp tip of the crack is subjected to high tensile stresses, which propagate the crack rapidly because the material has little capacity to dissipate energy.

The presence of defects is essential in explaining why brittle materials exhibit such weakness in tension (when compared to their strength in compression); see Table 8.2. The difference is on the order of 10 for rocks and similar materials, about 5 for glass, and about 3 for gray cast iron. Under tensile stresses, cracks propagate rapidly, causing what is known as *catastrophic failure*.

In polycrystalline metals, the fracture paths most commonly observed are **transgranular** (*transcrystalline* or *intragranular*); that is, the crack propagates *through the grain*. In **intergranular** fracture, the crack propagates *along the grain boundaries* (Fig. 2.26); it generally occurs when the grain boundaries are soft, contain a brittle phase, or have been weakened by liquid- or solid-metal embrittlement (Section 1.4).

Fatigue fracture.   Fatigue fracture typically occurs in a basically brittle nature. Minute external or internal cracks develop at pre-existing flaws or defects in the material; these cracks then propagate over a period of time and, eventually, they lead to total and sudden failure of the part. The fracture surface in fatigue generally is characterized by the term **beach marks**, because of its appearance (Fig. 2.27). Under high magnification (typically more than 1000 ×), a series of **striations** can be seen on fracture surfaces, each beach mark consisting of several striations.

Improving fatigue strength.   Fatigue life is influenced greatly by the method of preparation of the surfaces of the part or specimen (Fig. 2.28). The fatigue strength of manufactured products can be improved overall by the following methods:

a. Inducing compressive residual stresses on surfaces—for example, by shot peening or by roller burnishing.

b. Case hardening (surface hardening) by various means.

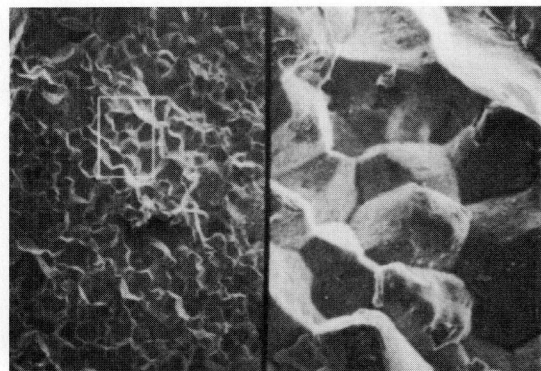

FIGURE 2.26   Intergranular fracture, at two different magnifications. Grains and grain boundaries clearly are visible in this micrograph. The fracture path is along the grain boundaries. Magnification: left, 100x; right, 500x. *Source*: Courtesy of B. J. Schulze and S. L. Meiley and Packer Engineering Associates, Inc.

c. Providing a fine surface finish and thereby reducing the effects of notches and other surface imperfections.

d. Selecting appropriate materials and ensuring that they are free from significant amounts of inclusions, voids, and impurities.

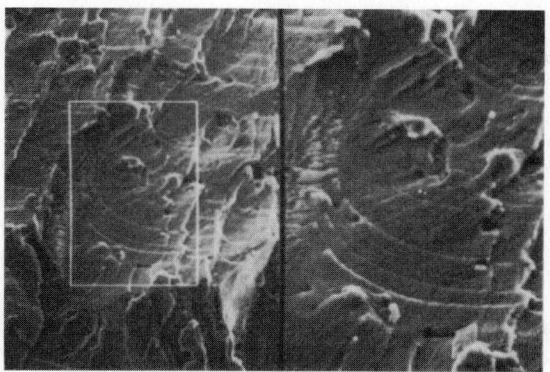

FIGURE 2.27   Typical fatigue-fracture surface on metals, showing beach marks. Magnification: left, 500X; right, 1000X. *Source:* Courtesy of B. J. Schulze and S. L. Meiley and Packer Engineering Associates, Inc.

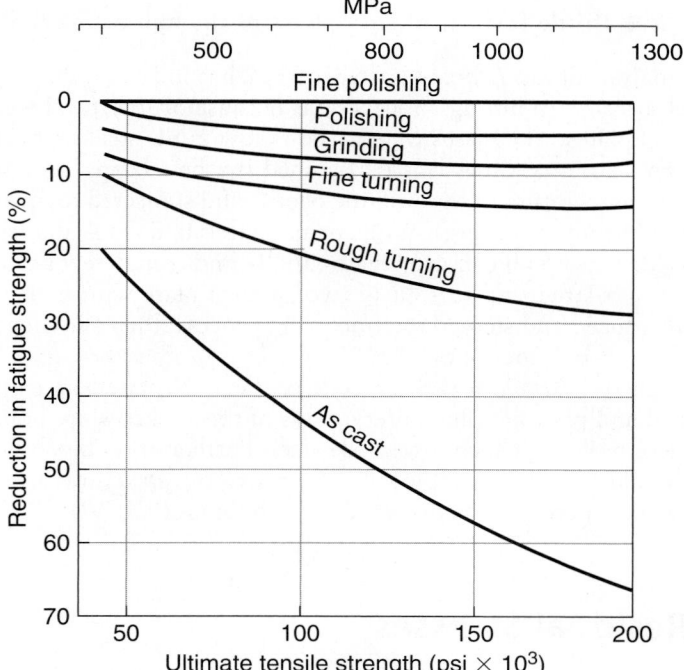

FIGURE 2.28   Reductions in the fatigue strength of cast steels subjected to various surface-finishing operations. Note that the reduction becomes greater as the surface roughness and the strength of the steel increase. *Source:* After M. R. Mitchell.

Conversely, the following factors and processes can reduce fatigue strength: decarburization, surface pits (due to corrosion) that act as stress raisers, hydrogen embrittlement, galvanizing, and electroplating.

Stress–corrosion cracking.   An otherwise ductile metal can fail in a brittle manner by *stress–corrosion cracking* (also called **stress cracking** or **season cracking**). Parts free from defects may develop cracks, either over a period of time or soon after being manufactured into a product. Crack propagation may be either intergranular or transgranular. The susceptibility of metals to stress–corrosion cracking depends mainly on the material, on the presence and magnitude of *tensile residual stresses*, and on the environment. Brass and austenitic stainless steels are among metals that are highly susceptible to stress cracking; the environment includes corrosive media such as salt water or other chemicals. The usual procedure to avoid stress–corrosion cracking is to *stress relieve* the part just after it is formed. Full annealing also may be done, but this treatment reduces the strength of cold-worked parts.

Hydrogen embrittlement.   The presence of hydrogen can reduce ductility and can cause severe embrittlement and premature failure in many metals, alloys, and nonmetallic materials. Known as *hydrogen embrittlement*, this phenomenon is especially severe in high-strength steels. Possible sources of hydrogen occur during melting of the metal, pickling (removal of surface oxides by chemical or electro-chemical reaction), and electrolysis in electroplating; other sources are water vapor in the atmosphere and moist electrodes and fluxes used during welding. Oxygen also can cause embrittlement, particularly in copper alloys.

EXAMPLE 2.2 Brittle fracture of steel plates of the hull of the *R.M.S. Titanic*

A detailed analysis of the *Titanic* disaster in 1912 has indicated that the ship sank not so much because of hitting an iceberg as because of structural weaknesses in the ship's steel plates. The plates were made of low-grade steel with a high sulfur content; it had little toughness (as determined by the Charpy test, Section 2.9) when chilled (temperature of the Atlantic ocean) and subjected to an external impact loading (hitting an iceberg). With such a material, a crack that starts in one part of a welded steel hull can propagate rapidly and completely around the hull and cause even a large ship to split in two. A steel plate with higher toughness would have reduced and slowed the fracturing process, thus allowing the ship to stay afloat longer and flood more slowly.

Although the *Titanic* was built with brittle steel plates, as we now know from physical and photographic observations of the sunken ship, not all ships of that time were built with such low-grade steel. Furthermore, better construction techniques could have been employed to improve the structural strength of the hull, among them practicing better welding techniques (Part VI).

2.11 | Residual Stresses

When workpieces are subjected to plastic deformation that is not uniform throughout the part, they develop **residual stresses**. These are stresses that remain within a part after it has been formed and all the external forces (as applied through tools and dies) are removed; a typical example is the bending of a metal bar (Fig. 2.29). The bending moment first produces a linear elastic stress distribution (Fig. 2.29a).

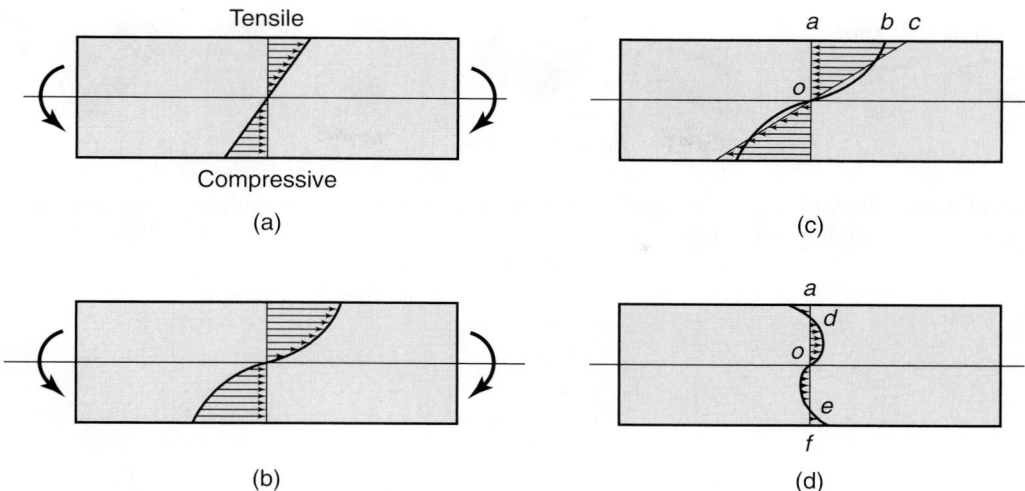

FIGURE 2.29   Residual stresses developed in bending a beam having a rectangular cross-section. Note that the horizontal forces and moments caused by residual stresses in the beam must be balanced internally. Because of nonuniform deformation, especially during cold-metalworking operations, most parts develop residual stresses.

As the external moment is increased, the outer fibers in the bar reach a stress level high enough to cause yielding. For a typical strain-hardening material, the stress distribution shown in Fig. 2.29b is eventually reached, and the bar has undergone permanent bending.

Let's now remove the external bending moment on the bar. Note that this operation is equivalent to applying an equal but opposite moment to the bar. Consequently, the moments of the areas *oab* and *oac* in Fig. 2.29c must be equal. Line *oc* (which represents the opposite bending moment) is linear, because all unloading and recovery is *elastic* (see Fig. 2.3). The difference between the two stress distributions gives the residual stress pattern within the bar, as is shown in Fig. 2.29d. Note the presence of compressive residual stresses in layers *ad* and *oe*, and the tensile residual stresses in layers *do* and *ef*. Because there are no external forces applied, the internal forces resulting from these residual stresses must be in static equilibrium. Although this example involves only residual stresses in the longitudinal direction of the bar, in most cases these stresses are three dimensional.

The equilibrium of residual stresses in Fig. 2.29d may be disturbed by the removal of a layer of material from the part, such as by machining or grinding. The bar then will acquire a new radius of curvature in order to balance the internal forces. Such disturbances of residual stresses lead to the **warping** of parts (Fig. 2.30). The equilibrium of residual stresses also may be disturbed by *relaxation* of these stresses over a period of time (see below). Residual stresses also can be caused by *temperature gradients* within a body, such as those occurring during cooling of a casting or a forging. The local expansion and contractions caused by temperature gradients within the material produce a nonuniform deformation, such as is seen in the permanent bending of a beam.

Tensile residual stresses on the surface of a part generally are undesirable because they lower the fatigue life and fracture strength of the part. This is due to the fact that a surface with tensile residual stresses cannot sustain additional tensile stresses from external forces as high as those of a surface free from residual stresses. This reduction in strength is characteristic particularly of brittle or less ductile materials, in

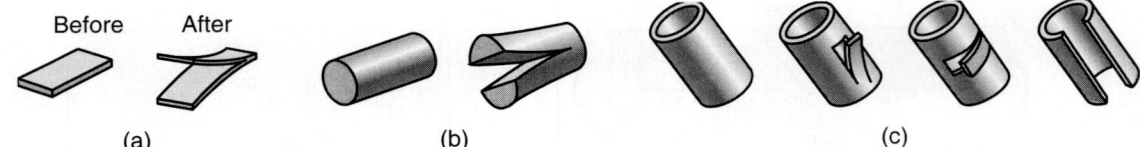

Before   After

(a)   (b)   (c)

FIGURE 2.30 Distortion of parts, with residual stresses, after cutting or slitting: (a) flat sheet or plate; (b) solid round rod; (c) thin-walled tubing or pipe.

which fracture takes place with little or no plastic deformation preceding fracture. Tensile residual stresses also can lead, over a period of time, to *stress cracking* or to *stress–corrosion cracking* of manufactured products. Compressive residual stresses on a surface, on the other hand, generally are desirable. In fact, in order to increase the fatigue life of components, compressive residual stresses can be imparted to surfaces by techniques such as shot peening and surface rolling.

Reduction and elimination of residual stresses. Residual stresses can be reduced or eliminated either by *stress-relief annealing* or by a further *deformation* of the part, such as stretching it. Given sufficient time, residual stresses also may diminish at room temperature (by *relaxation* of residual stresses). The time required for relaxation can be reduced greatly by raising the temperature of the workpiece.

2.12 | Work, Heat, and Temperature

Almost all of the mechanical work of deformation in plastic working is converted into **heat**. This conversion is not complete, because a portion of this work is stored within the deformed material as elastic energy. Known as **stored energy** (Section 1.6), it is generally 5 to 10% of the total energy input; in some alloys, however, it may be as high as 30%.

In a simple frictionless deformation process and assuming that work is converted completely into heat, the theoretical (adiabatic) *temperature rise*, ΔT, is given by

$$\Delta T = \frac{u}{\rho c} \qquad (2.15)$$

where u is the **specific energy** (work of deformation per unit volume), ρ is the density, and c is the specific heat of the material. It can be seen that higher temperatures are associated with large areas under the stress–strain curve and with smaller values of specific heat. However, such physical properties as specific heat and thermal conductivity also depend on temperature, and thus, they must be taken into account in the calculations.

The temperature rise for a true strain of 1 (such as occurs in a 27 mm-high specimen when it is compressed down to 10 mm) can be calculated to be as follows: aluminum, 75°C (165°F); copper, 140°C (285°F); low-carbon steel, 280°C (535°F); titanium 570°C (1060°F). In actual operations, heat is lost to the environment, to tools and dies, and to lubricants or coolants used, if any. If the deformation process is performed rapidly, the heat losses will be relatively small over that brief period. If the process is carried out slowly, the actual temperature rise will be only a fraction of the calculated value.

SUMMARY

- Many manufacturing processes involve shaping materials by plastic deformation. Consequently, such mechanical properties as strength (yield strength (Y) and ultimate tensile strength (UTS)), modulus of elasticity (E), ductility (total elongation and reduction of area), hardness, and the energy required for plastic deformation are important factors. These properties, in turn, depend, to various extents, on the particular material and on its condition, temperature, deformation rate, surface condition, and environment.

- The tensile test is the most commonly used test to determine mechanical properties. From these, true stress–true strain curves are constructed that are needed to determine the strength coefficient (K), the strain-hardening exponent (n), the strain-rate sensitivity exponent (m), and the toughness of materials.

- Compression tests are subject to inaccuracy due to the presence of friction and to resultant barreling of the specimen. Torsion tests are conducted on tubular specimens subjected to twisting. Bending or flexure tests commonly are used for brittle materials to determine their modulus of rupture or the transverse rupture strength.

- Several hardness tests are used to determine the resistance of a material to permanent indentation or scratching. Hardness is related to strength and wear resistance of a material, but it is, itself, not a fundamental property.

- Fatigue tests indicate the endurance limit or fatigue limit of materials—that is, the maximum stress to which a material can be subjected without fatigue failure, regardless of the number of cycles.

- Creep is the permanent elongation of a component under a static load maintained for a period of time. The specimen eventually fails by rupture (necking and fracturing).

- Impact tests determine the energy required to completely break the specimen, called the impact toughness of the material. Impact tests also are useful for determining the transition temperature of materials.

- Failure and fracture are an important aspects of a material's behavior when it is subjected to deformation in manufacturing operations. Ductile fracture is characterized by plastic deformation preceding fracture, and it requires a considerable amount of energy. Brittle fracture can be catastrophic, because it is not preceded by plastic deformation. It requires much less energy than ductile fracture. Impurities, inclusions, and voids play major roles in the fracture of metals and alloys.

- Residual stresses are those that remain in a workpiece after it has been deformed plastically and then has had all external forces removed. Surface tensile residual stresses generally are undesirable; they may be reduced or eliminated by stress-relief annealing, further plastic deformation, or by relaxation over a period of time.

KEY TERMS

Bauschinger effect

Blue brittleness

Buckling

Compression

Creep

Deformation rate

Disk test

Ductility

Elongation

Engineering strain

Engineering stress

Fatigue

Fatigue failure

Fracture

Hardness

Impact loading

Inclusions

Microhardness

Modulus of elasticity

Modulus of rigidity

Modulus of rupture

Poisson's ratio

Reduction of area

Residual stresses

Rupture

Shear

Shear modulus

Strain aging

Strain-hardening exponent

Strain rate

Strain-rate sensitivity exponent

Strength coefficient

Stress–corrosion cracking

Superplasticity

Tension

Torsion test

Toughness

Transition temperature

True strain

True stress

Ultimate tensile strength

Yield stress

BIBLIOGRAPHY

Ashby, M.F., *Materials Selection in Mechanical Design* 3rd ed, Elsevier, 2005.

Ashby, M.F., and Jones D.R.H., *Engineering Materials*, Vol. 1, *An Introduction to Their Properties and Applications* (2d ed.), 1996; Vol. 2, *An Introduction to Microstructures, Processing and Design* (2d ed.), 1998; Vol. 3, *Materials Failure Analysis: Case Studies and Design Implications*, 1993. Pergamon.

ASM Handbook, Vol. 8: *Mechanical Testing and Evaluation*, ASM International, 2000.

Atlas of Stress–strain curves, 2nd ed., ASM International, 2002.

Budinski, K.G., *Engineering Materials: Properties and Selection*, 7th ed., Prentice Hall, 2001.

Chandler, H., *Hardness Testing*, 2nd ed, ASM International, 1999.

Chawla, K.K., and Majers M.A., *Mechanical Behavior of Materials*, Prentice Hall, 1998.

Courtney, T.H., *Mechanical Behavior of Materials*, 2nd ed., McGraw-Hill, 1999.

Davies, J.R., (ed.), *Tensile Testing*, 2nd ed., ASM International, 2004.

Dowling, N.E., *Mechanical Behavior of Materials: Engineering Methods for Deformation, Fracture, and Fatigue.* 2nd ed., Prentice Hall, 1999.

Herzberg, R.W., *Deformation and Fracture Mechanics of Engineering Materials*, 4th ed, Wiley, 1996.

Wulpi, D.J., *Understanding How Components Fail*, 2nd ed., ASM International, 1999.

REVIEW QUESTIONS

2.1 Distinguish between engineering stress and true stress.

2.2 Describe the events that occur when a specimen undergoes a tension test. Sketch a plausible stress–strain curve and identify all significant regions and points between them. Assume that loading continues up to fracture.

2.3 What is ductility, and how is it measured?

2.4 In the equation $\sigma = K\varepsilon^n$, which represents the stress-strain curve for a material, what is the significance of the exponent n?

2.5 What is strain-rate sensitivity, and how is it measured?

2.6 What testing procedures can be used to measure the properties of brittle materials, such as ceramics and carbides?

2.7 Describe the differences between brittle and ductile fracture.

2.8 Explain the difference between stress relaxation and creep.

2.9 Describe the difference between elastic and plastic behavior.

2.10 Explain what uniform elongation is in tension testing.

2.11 Describe the difference between deformation rate and strain rate. What unit does each one have?

2.12 Describe the difficulties involved in making a compression test.

2.13 What is Hooke's Law? Young's Modulus? Poisson's ratio?

2.14 What is the reason that a yield strength is defined as a 0.2% offset strength?

2.15 Explain the difference between transgranular and intergranular fracture.

QUALITATIVE PROBLEMS

2.16 Using the same scale for stress, the tensile true stress–true strain curve is higher than the engineering stress–engineering strain curve. Explain whether this condition also holds for a compression test.

2.17 With the aid of a simple sketch, explain whether it is necessary to use the offset method to determine the yield stress, Y, of a material that has been highly cold worked.

2.18 Explain why the difference between engineering strain and true strain becomes larger as strain increases. Does this difference occur for both tensile and compressive strains? Explain.

2.19 If a material does not have an endurance limit (for example, aluminum), how would you estimate its fatigue life?

2.20 Which hardness tests and scales would you use for very thin strips of metal, such as aluminum foil? Why?

2.21 Which of the two tests, tension or compression, requires higher capacity of testing machine, and why?

2.22 List and explain briefly the conditions that induce brittle fracture in an otherwise ductile metal.

2.23 List the factors that you would consider in selecting a hardness test and then interpreting the results from this test.

2.24 Using Fig. 2.6 only, explain why you cannot calculate the percent elongation of the materials listed.

2.25 If you pull and break a tension-test specimen rapidly, where would the temperature be highest, and why?

2.26 Will the disk test be applicable to a ductile material? Why or why not?

2.27 What hardness test is suitable for determining the hardness of a thin ceramic coating?

2.28 In a Brinell hardness test, the resulting impression is found to be an ellipse. Give possible explanations for this result.

2.29 Some coatings are extremely thin—some, as thin as a few nanometers. Explain why even the Knoop test is not able to obtain reasonable results for such coatings. Recent research has attempted to use highly polished diamonds (tip radius around 5 nanometers) to indent such coatings in atomic force microscopes. What concerns would you have regarding the appropriateness of the results?

2.30 Select an appropriate hardness test for each of the following materials. Justify your answer.

   a. Cubic boron nitride
   b. Lead
   c. Cold-drawn 0.5% C steel
   d. Diamond
   e. Caramel (candy)
   f. Granite

QUANTITATIVE PROBLEMS

2.31 A paper clip is made of wire 1 mm in diameter. If the original material from which the wire is made is a rod 18 mm in diameter, calculate the longitudinal engineering and true strains that the wire has undergone during processing.

2.32 A strip of metal is 250 mm long. It is stretched in two steps, first to 300 mm and then to 400 mm. Show that the total true strain is the sum of the true strains in each step—that is, that true strains are additive. Show that, in the case of engineering strains, the strains cannot be added to obtain the total strain.

2.33 Identify the two materials in Fig. 2.6 that have the lowest and the highest uniform elongations. Calculate these quantities as percentages of the original gage lengths.

2.34 Plot the ultimate strength versus stiffness for the materials in Table 2.2, and prepare a three-dimensional plot for the metals in Table 2.2 where the third axis is maximum elongation in 50 mm.

2.35 If you remove the layer of material *ad* from the part shown in Fig. 2.29d—for instance, by machining or grinding—which way will the specimen curve? (*Hint*: Assume that the part shown in sketch *d* in the figure is composed of four horizontal springs held at the ends. Thus, from the top down, you have compression, tension, compression, and tension springs.)

2.36 Percent elongation always is described in terms of the original gage length, such as 50 mm or 2 in. Explain how percent elongation varies as the gage length of the tensile specimen increases. (*Hint*: Recall that necking is a local phenomenon, and think of what happens to the elongation as the gage length becomes very small.)

2.37 Make a sketch showing the nature and distribution of residual stresses in Fig. 2.30a and b, before they were cut. (*Hint*: Assume that the split parts are free from any stresses, then force these parts back to the shape they had before they were cut.)

2.38 You are given the K and n values of two different metals. Is this information sufficient to determine which metal is tougher? If not, what additional information do you need?

2.39 A cable is made of two strands of different materials, A and B, and cross-sections as follows:

For material A: $K = 70,000$ psi,

$n = 0.5, A_o = 0.6$ in^2

For material B: $K = 25,000$ psi,

$n = 0.5, A_o = 0.3$ in^2

Calculate the maximum tensile force that this cable can withstand prior to necking.

2.40 On the basis of the information given in Fig. 2.6, calculate the ultimate tensile strength (engineering) of annealed copper.

2.41 In a disk test performed on a specimen 1.25 in. in diameter and 1/2 in. thick, the specimen fractures at a stress of 30,000 psi. What was the load on it?

2.42 A piece of steel has a hardness of 300 HB. Calculate the tensile strength in MPa and in psi.

2.43 A material has the following properties: UTS= 50,000 psi and $n = 0.25$. Calculate its strength coefficient, K.

2.44 A material has a strength coefficient $K= 100,000$ psi and $n = 0.2$. Assuming that a tensile-test specimen made from this material begins to neck at a true strain of 0.2, show that the ultimate tensile strength of this material is 59,340 psi.

2.45 *Modulus of resilience* is defined as the area under the elastic region of the stress–strain curve of the material; it has the units of energy-per-unit volume. Derive an expression for the modulus of resilience in terms of the yield stress and modulus of elasticity of the material.

2.46 What is the modulus of resilience for a highly cold-worked piece of steel having a hardness of 275 HB? For a piece of highly cold-worked copper with a hardness of 100 HRB?

2.47 Using only Fig. 2.6, calculate the maximum load in tension testing of a 304 stainless steel specimen with an original diameter of 5 mm.

2.48 Calculate the major and minor pyramid angles for a Knoop indenter and compare to Vickers and Rockwell A indenters.

2.49 If a material has a target hardness of 350 HB, what is the expected indentation diameter?

2.50 A Rockwell A test was conducted on a material and a penetration depth of 0.1 mm was recorded. What is the hardness of the material? What material typically will give such hardness values? If a Brinell hardness test were to be conducted on this material, give an estimate of the indentation diameter if the load used is 1500 kg.

2.51 A material is tested in tension. Over a 1-in. gage length, the engineering strain measurements are 0.01, 0.02, 0.03, 0.04, 0.05, 0.1, 0.15, 0.2, 0.5, and 1.0. Plot the true strain versus engineering strain for these readings.

SYNTHESIS, DESIGN, AND PROJECTS

2.52 List and explain the desirable mechanical properties for (a) an elevator cable, (b) a paper clip, (c) a leaf spring for a truck, (d) a bracket for a bookshelf, (e) piano wire, (f) a wire coat hanger, (g) a gas-turbine blade, and (h) a staple.

2.53 When making a hamburger, have you observed the type of cracks shown in Fig. 2.19d? What can you do to avoid such cracks? *Note*: Test hamburger patties by compressing them at different temperatures and observe the crack path (i.e., through the fat particles, the meat particles, or their interface).

2.54 An inexpensive claylike material called Silly Putty™ is often available in stores that sell toys and games. Obtain some and do the following experiments. (a) Shape it into a ball and drop it onto a flat surface. (b) Put the re-rounded ball on a table and place a heavy book on it for five minutes. (c) Shape it into a long cylindrical piece and pull it apart—first slowly, then quickly. Describe your observations and refer to the specific sections in this chapter where each particular observation is relevant.

2.55 Make individual sketches of the mechanisms of testing machines that, in your opinion, would be appropriate for tension, for torsion, and for compression testing of specimens at different rates of deformation. What modifications would you make to include the effects of temperature on material properties?

2.56 In Section 2.6.1, we described the Mohs hardness test. Obtain small pieces of several different metallic and nonmetallic materials, including stones. Rub them against each other, observe the scratches made, and order them in a manner similar to the Mohs numbering system.

2.57 Demonstrate stress relaxation by tightly stretching thin plastic strings between two nails on a long piece of wood. Pluck the strings frequently, to test the tension as a function of time and of temperature. (*Note:* Alter the temperature by placing the fixture in an oven set on "low.")

2.58 Demonstrate the impact toughness of a round piece of chalk by first using a triangular file to produce a V-notch, as shown in Fig. 2.18a, and then bending the chalk to break it.

2.59 Using a large rubber band and a set of weights, obtain the force–displacement curve for the rubber band. How is this different from the stress–strain curves shown in Fig. 2.5?

2.60 Find or prepare some solid circular pieces of brittle materials (such as chalk, ceramics, etc.) and subject them to the type of test shown in Fig. 2.9 by using the jaws of a simple vise. Describe your observations as to how the materials fracture. Repeat the tests using ductile materials (such as clay, soft metals, etc.) and describe your observations.

2.61 Devise a simple fixture for the bend tests shown in Fig. 2.11 and test sticks of brittle materials by loading them with dead weights until they break. Verify the statement that the specimens on the right in the figure will break sooner than the ones on the left.

2.62 By pushing a ball bearing against the top surfaces of various materials (such as clay, dough, etc.) observe with a magnifier the shape of the indentation, similar to those shown in Fig. 2.13a and b.

2.63 Embed a small steel ball in a soft block of material and compress it as shown in Fig. 2.23a. Then cut it carefully along the center plane and observe the deformation of the material. Repeat the same experiment by embedding a small round jelly bean in the material and deforming it.

2.64 Devise a simple experiment and perform tests on materials commonly found in a kitchen by bending them for a qualitative assessment of their transition temperature, as shown in Fig. 2.24.

2.65 Locate some solid and some tubular metal pieces and cut them as shown in Fig. 2.30 to determine if there are any residual stresses in the parts prior to cutting them.

Physical Properties of Materials

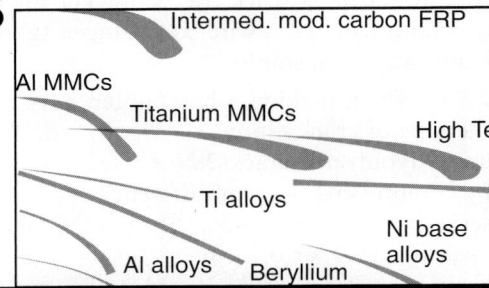

Physical properties can have a wide variety of roles in the processing as well as the use of materials. In this chapter, we specifically describe:

• Thermal, electrical, magnetic, and optical properties, with examples of their roles.
• Corrosion and its importance in the service life of components.
• How a combination of physical properties are often important in the processing and use of materials.

3.1 | Introduction

Why is electrical wiring generally made of copper? Why are aluminum, stainless steel, and copper so commonly used in cookware? Why are their handles usually made of wood or plastic, while other types of handles are made of metal? What kind of material should be chosen for the heating elements in toasters? Why are the metallic components in some machines being replaced with ceramics? Why are commercial airplane bodies generally made of aluminum, and why are some airplane components being replaced gradually with those made of various composite materials, including reinforced plastics?

It is apparent from these questions that one of the important considerations in material selection is their **physical properties** (that is, density, melting point, specific heat, thermal conductivity, thermal expansion, electrical and magnetic properties, and resistance to oxidation and corrosion). Combined with mechanical properties, the strength-to-weight and stiffness-to-weight ratios of materials are equally important, particularly for aircraft and aerospace structures. Also, for example, high-speed equipment such as textile and printing machinery, and forming and cutting machines for high-speed operations, require lightweight components to reduce inertial forces and, thus, keep machines from excessive vibration. Several other examples of the importance of physical properties are discussed in this chapter, with each physical property being presented from the viewpoint of material selection and manufacturing and its relevance to the service life of the component.

3.2 | Density

The **density** of a material is its mass per unit volume. Another term is **specific gravity**, which expresses a material's density in relation to that of water, and thus, it has no units. The range of densities for a variety of materials at room temperature, along with other properties, is given in Tables 3.1 and 3.2.

Weight saving is important particularly for aircraft and aerospace structures, for automotive bodies and components, and for other products where energy consumption and power limitations are major concerns. Substitution of

TABLE 3.1

Physical Properties of Various Materials at Room Temperature				
Material	Density (kg/m³)	Melting point (°C)	Specific heat (J/kg K)	Thermal conductivity (W/m K)	Coefficient of thermal expansion (μm/m°C)
Metallic					
Aluminum	2700	660	900	222	23.6
Aluminum alloys	2630–2820	476–654	880–920	121–239	23.0–23.6
Berylium	1854	1278	1884	146	8.5
Columbium (niobium)	8580	2468	272	52	7.1
Copper	8970	1082	385	393	16.5
Copper alloys	7470–8940	885–1260	337–435	29–234	16.5–20
Gold	19300	1063	129	317	19.3
Iron	7860	1537	460	74	11.5
Steels	6920–9130	1371–1532	448–502	15–52	11.7–17.3
Lead	11350	327	130	35	29.4
Lead alloys	8850–11350	182–326	126–188	24–46	27.1–31.1
Magnesium	1745	650	1025	154	26.0
Magnesium alloys	1770–1780	610–621	1046	75–138	26.0
Molybdenum alloys	10210	2610	276	142	5.1
Nickel	8910	1453	440	92	13.3
Nickel alloys	7750–8850	1110–1454	381–544	12–63	12.7–18.4
Silicon	2330	1423	712	148	7.63
Silver	10500	961	235	429	19.3
Tantalum alloys	16600	2996	142	54	6.5
Titanium	4510	1668	519	17	8.35
Titanium alloys	4430–4700	1549–1649	502–544	8–12	8.1–9.5
Tungsten	19290	3410	138	166	4.5
Nonmetallic					
Ceramics	2300–5500	—	750–950	10–17	5.5–13.5
Glasses	2400–2700	580–1540	500–850	0.6–1.7	4.6–70
Graphite	1900–2200	—	840	5–10	7.86
Plastics	900–2000	110–330	1000–2000	0.1–0.4	72–200
Wood	400–700	—	2400–2800	0.1–0.4	2–60

TABLE 3.2

Physical Properties of Materials, in Descending Order					
Density	Melting point	Specific heat	Thermal conductivity	Thermal expansion	Electrical conductivity
Platinum	Tungsten	Wood	Silver	Plastics	Silver
Gold	Tantalum	Beryllium	Copper	Lead	Copper
Tungsten	Molybdenum	Porcelain	Gold	Tin	Gold
Tantalum	Columbium	Aluminum	Aluminum	Magnesium	Aluminum
Lead	Titanium	Graphite	Magnesium	Aluminum	Magnesium
Silver	Iron	Glass	Graphite	Copper	Tungsten
Molybdenum	Beryllium	Titanium	Tungsten	Steel	Beryllium
Copper	Copper	Iron	Beryllium	Gold	Steel
Steel	Gold	Copper	Zinc	Ceramics	Tin
Titanium	Silver	Molybdenum	Steel	Glass	Graphite
Aluminum	Aluminum	Tungsten	Tantalum	Tungsten	Ceramics
Beryllium	Magnesium	Lead	Ceramics		Glass
Glass	Lead		Titanium		Plastics
Magnesium	Tin		Glass		Quartz
Plastics	Plastics		Plastics		

materials for the sake of weight saving and economy is a major factor in the design both of advanced equipment and machinery and of consumer products, such as automobiles.

A significant role that density plays is in the **strength-to-weight ratio (specific strength)** and **stiffness-to-weight ratio (specific stiffness)** of materials and structures. Table 3.3 shows the ratio of maximum yield stress to density for a variety of metal alloys. Note that titanium and aluminum are at the top of the list; consequently, and as described in Chapter 6, they are among the most commonly used metals for aircraft and aerospace applications.

TABLE 3.3

Ratio of Maximum Yield Stress to Density for Assorted Metals	
Alloy	Maximum yield stress/ density (in. $\times 10^3$)
Titanium	1250
Aluminum	800
Steels	750
Magnesium	675
Nickel	550
Copper	500
Tantalum	375
Molybdenum	215
Lead	5

The range for specific strength and specific stiffness at room temperature for a variety of metallic and nonmetallic materials is given in Fig. 3.1. Note the positions of composite materials, as compared to those of metals, with respect to these properties; these advantages have made composites become one of the most important materials (see Chapter 9). At elevated temperatures, specific strength and specific stiffness likewise are important considerations, because of the temperatures at which certain components and systems operate, such as automotive and jet engines and gas turbines. Typical ranges for a variety of materials are given in Fig. 3.2.

Density is an important factor in the selection of materials for high-speed equipment, such as the use of magnesium in printing and textile machinery, many components of which usually operate at very high speeds. To obtain exposure times of 1/4000 s in cameras without sacrificing accuracy, the shutters of some high-quality 35-mm cameras are made of titanium. The low resulting mass of the components in these high-speed operations reduces inertial forces that otherwise could lead to vibrations, to inaccuracies, and even (over a period of time) to part failure. Because of their low density, ceramics (Chapter 8) are used for components in high-speed automated machinery and in machine tools. On the other hand, there are applications where weight is desirable. Examples are counterweights for various mechanisms (using lead or steel), for flywheels, and for components of self-winding watches that utilize the inertia effects of a mass when the wrist is in motion (using high-density materials such as tungsten).

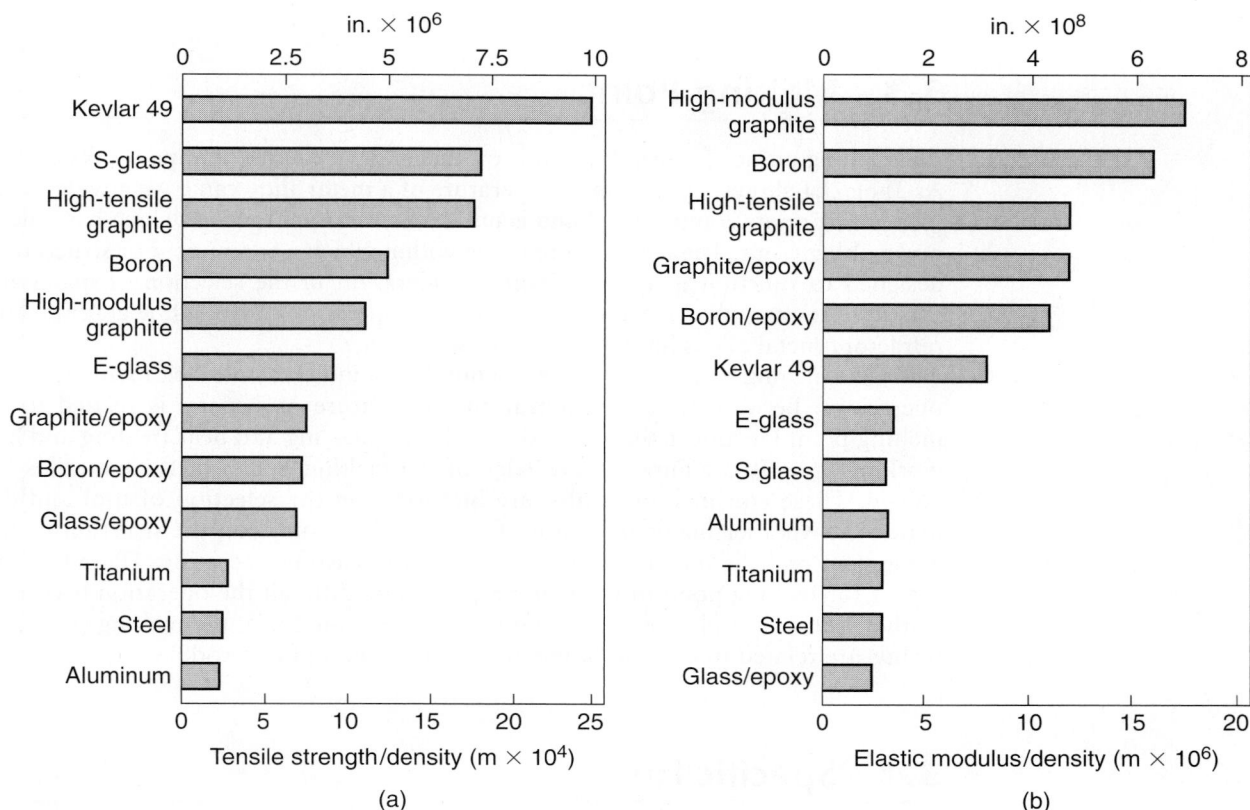

FIGURE 3.1   Specific strength (tensile strength/density) and specific stiffness (elastic modulus/density) for various materials at room temperature.

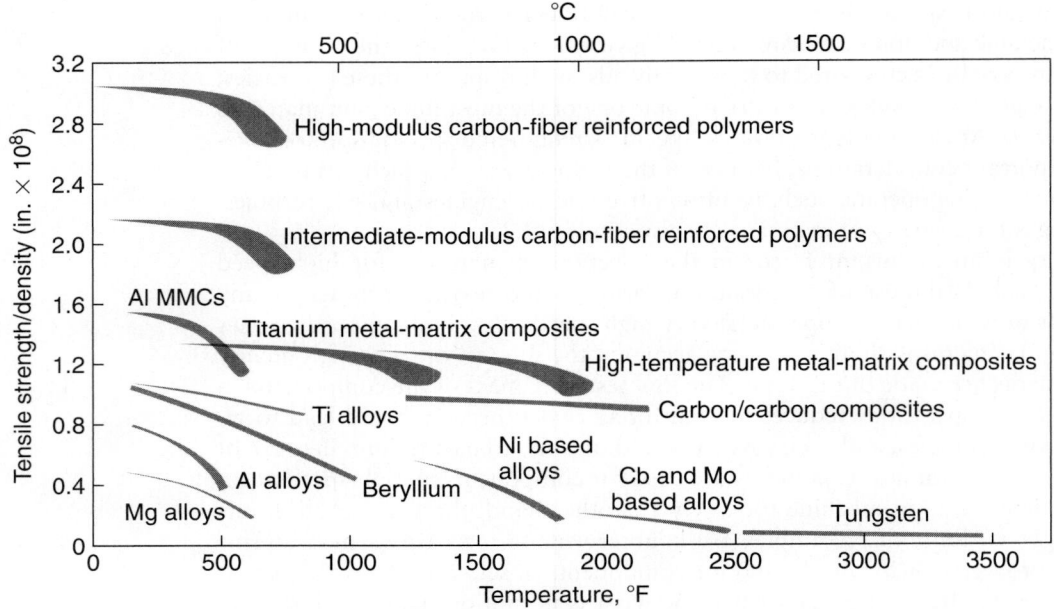

FIGURE 3.2 Specific strength (tensile strength/density) for a variety of materials as a function of temperature. Note the useful temperature range for these materials and the high values for composite materials.

3.3 | Melting Point

The **melting point** of a metal depends on the energy required to separate its atoms. As Table 3.1 shows, the melting temperature of a metal alloy can have a wide range (depending on its composition) and is unlike that of a pure metal, which has a definite melting point. The temperature range within which a component or structure is designed to function is an important consideration in the selection of materials. Plastics, for example, have the lowest useful temperature range, while graphite and refractory-metal alloys have the highest useful range.

The melting point of a metal has a number of indirect effects on manufacturing operations. Because the recrystallization temperature of a metal is related to its melting point (Section 1.6), operations such as annealing and heat treating and hot working (Part III) require a knowledge of the melting points of the materials involved. These considerations also are important in the selection of tool and die materials. Another major influence of the melting point is on the selection of the equipment and the melting practice employed in casting operations (Part II). The higher the melting point of the material, the more difficult the operation becomes. In the electrical-discharge machining process (Section 27.5), the melting points of metals are related to the rate of material removal and of electrode wear.

3.4 | Specific Heat

A material's **specific heat** is the energy required to raise the temperature of a unit mass by one degree. Alloying elements have a relatively minor effect on the specific heat of metals. The temperature rise in a workpiece resulting from forming or machining operations (Parts III and IV, respectively) is a function of the work done and of the specific heat of the workpiece material (Section 2.12). Temperature rise in a workpiece, if excessive, can decrease product quality by adversely affecting its surface finish and dimensional accuracy, can cause excessive tool and die wear, and can result in undesirable metallurgical changes in the material.

3.5 | Thermal Conductivity

Thermal conductivity indicates the rate at which heat flows within and through a material. Metallically bonded materials (metals) generally have high thermal conductivity, while ionically or covalently bonded materials (ceramics and plastics) have poor conductivity (Table 3.2). Because of the large difference in their individual thermal conductivities, alloying elements can have a significant effect on the thermal conductivity of alloys, as can be seen by comparing the metals with their alloys in Table 3.1.

When heat is generated by plastic deformation or due to friction, the heat should be conducted away at a rate high enough to prevent a severe rise in temperature. The main difficulty experienced in machining titanium, for example, is caused by its very low thermal conductivity. Low thermal conductivity can also result in high thermal gradients and, in this way, cause inhomogeneous deformation of workpieces in metalworking processes.

3.6 | Thermal Expansion

The **thermal expansion** of materials can have several significant effects, particularly the relative expansion or contraction of different materials in assemblies such as electronic and computer components, glass-to-metal seals, struts on jet engines, and moving parts in machinery that require certain clearances for proper functioning. The use of ceramic components in cast-iron engines, for example, also requires consideration of their relative expansions. Typical coefficients of thermal expansion are on the order of 10×10^{-6} per °C for steels, 24×10^{-6} per °C for aluminum, and up to 200×10^{-6} per °C for thermoplastics. (See also *Invar* later in this section.) Generally, the coefficient of thermal expansion is inversely proportional to the melting point of the material. Alloying elements have a relatively minor effect on the thermal expansion of metals.

Shrink fits utilize thermal expansion and contraction. For example, a part, such as a flange or a lever arm with a hole in it, is to be installed over a shaft. It first is heated and then slipped over a shaft or spindle which is at room temperature. When allowed to cool, the part shrinks and the assembly effectively becomes an integral component. Thermal expansion in conjunction with thermal conductivity plays the most significant role in causing **thermal stresses** (due to *temperature gradients*) in manufactured components, in tools and dies, and in molds for casting operations. This consideration is particularly important in, for example, a forging operation during which hot workpieces repeatedly are placed over a relatively cool die, thus subjecting the die surfaces to thermal cycling. To reduce thermal stresses, a combination of high thermal conductivity and low thermal expansion is desirable. Thermal stresses also can be caused by **anisotropy of thermal expansion** (that is, the material expands differently in different directions) generally observed in hexagonal close-packed metals, ceramics and composite materials.

Thermal expansion and contraction can lead to the cracking, warping, or loosening of components in the structure during their service life, and they can lead to the cracking of ceramic parts and in tools and dies made of relatively brittle materials. **Thermal fatigue** results from thermal cycling and causes a number of surface cracks, especially in tools and dies for casting and metalworking operations (*heat checking*). **Thermal shock** is the term generally used to describe the development of cracks after a single thermal cycle.

To alleviate some of the problems caused by thermal expansion, a family of iron-nickel alloys with very low thermal-expansion coefficients has been developed, called **low-expansion alloys**. The low thermal-expansion characteristic of these alloys often is referred to as the **Invar effect** (after the metal *Invar*). The thermal coefficient of expansion is typically in the range of 2×10^{-6} to 9×10^{-6} per °C. Typical compositions are 64% Fe-36% Ni for Invar and 54% Fe-28% Ni-18% Co for *Kovar*. These alloys also have good thermal-fatigue resistance and good ductility. As a result, they easily can be formed into various shapes. Applications include (a) bimetallic strips consisting of a low-expansion alloy metallurgically bonded to a high-expansion alloy (the strip bends when subjected to temperature changes), and (b) high-quality glass-to-metal seals in which the thermal expansions are matched.

3.7 | Electrical, Magnetic, and Optical Properties

Electrical conductivity and the *dielectric* properties of materials are important not only in electrical equipment and machinery but also in such manufacturing processes as the magnetic-pulse forming of sheet metals (Section 16.11) and the electrical-discharge machining and electrochemical grinding of hard and brittle materials (Chapter 27). The units of electrical conductivity are mho/m or mho/ft, where *mho* is the reverse of *ohm*, the unit for electrical resistance. The influence of the type of atomic bonding on the electrical conductivity of materials is the same as that for thermal conductivity. Alloying elements have a major effect on the electrical conductivity of metals. The higher the conductivity of the alloying element, the higher the electrical conductivity of the alloy.

Dielectric strength.   A material's *dielectric strength* is its resistivity to direct electric current. This property is defined as the voltage required per unit distance for electrical breakdown and has the units of V/m or V/ft.

Conductors.   Materials with high electrical conductivity, such as metals, generally are referred to as *conductors*. **Electrical resistivity** is the inverse of electrical conductivity. Materials with high electrical resistivity are referred to as **dielectrics** or **insulators**.

Superconductors.   *Superconductivity* is the phenomenon of near-zero electrical resistivity that occurs in some metals and alloys below a critical temperature. The temperatures involved often are near absolute zero (0 K, −273°C, or −460°F). The highest temperature at which superconductivity has been exhibited to date is 150 K (−123°C, −190°F), but advances in high temperature superconductivity continue to be made. These developments indicate that the efficiency of such electrical components as large high-power magnets, high-voltage power lines, and electronic and computer components can be improved markedly.

Semiconductors.   The electrical properties of *semiconductors* (such as single-crystal silicon, germanium, and gallium arsenide) are extremely sensitive to temperature and to the presence and type of minute impurities. Thus, by controlling the concentration and type of impurities (**dopants**), such as phosphorus and boron in silicon, electrical conductivity can be controlled. This property is utilized in the semiconductor (solid state) devices, used extensively in miniaturized electronic circuitry (Chapter 28). They are very compact, very efficient, and relatively inexpensive; furthermore, they consume little power and require no warm-up time for operation.

Ferromagnetism and Ferrimagnetism.   *Ferromagnetism* is a phenomenon characterized by high permeability and permanent magnetization that are due to the alignment of iron, nickel, and cobalt atoms into domains. It is important in such applications as electric motors, electric generators, electric transformers, and microwave devices. *Ferrimagnetism* is a permanent and large magnetization exhibited by some ceramic materials, such as cubic ferrites.

Piezoelectric effect.   The *piezoelectric effect* (*piezo* from Greek, meaning "to press") is exhibited by some materials, such as quartz crystals and some ceramics, in which there is a reversible interaction between an elastic strain and an electric field. This property is utilized in making transducers, which are devices that convert the strain from an external force into electrical energy. Typical applications are force or pressure transducers, strain gages, sensors, sonar detectors, and microphones.

Magnetostriction.   The phenomenon of expansion or contraction of a material when it is subjected to a magnetic field is called *magnetostriction*. Materials such as pure nickel and some iron-nickel alloys exhibit this behavior. Magnetostriction is the principle behind ultrasonic machining equipment (Section 26.6).

Optical properties.   Among various other properties, color and opacity are relevant, particularly to polymers and glasses. These two properties are described in Sections 7.2.2 and 8.4.3, respectively.

3.8 | Corrosion Resistance

Metals, ceramics, and plastics all are subject to forms of corrosion. The word **corrosion** itself usually refers to the deterioration of metals and ceramics, while similar phenomena in plastics generally are called **degradation**. Corrosion leads not only to surface deterioration of components and structures (bridges and ships) but also reduces their strength and structural integrity. The direct cost of corrosion to the U.S. economy alone has been estimated to be $275 billion per year, which is approximately 3% of the gross domestic product. Indirect costs of corrosion are estimated at twice this amount.

   Corrosion resistance is an important aspect of material selection for applications in the chemical, food, and petroleum industries, as well as in manufacturing operations. In addition to various possible chemical reactions from the elements and compounds present, environmental oxidation and corrosion of components and structures is a major concern, particularly at elevated temperatures and in automobiles and other transportation vehicles.

   Resistance to corrosion depends on the composition of the material and on the particular environment. Corrosive media may be chemicals (acids, alkalis, and salts), the environment (oxygen, moisture, pollution, and acid rain), and water (fresh or salt water). Nonferrous metals, stainless steels, and nonmetallic materials generally have high corrosion resistance. Steels and cast irons generally have poor resistance and must be protected by various coatings and surface treatments.

Corrosion can occur over an entire surface, or it can be *localized*, called **pitting**. (Pitting is a term that also is used for fatigue wear or failure of gears and in forging; see Section 31.5) Corrosion also can occur along grain boundaries of metals as **intergranular corrosion** and at the interface of bolted or riveted joints as **crevice corrosion**.

Two dissimilar metals may form a **galvanic cell** (after L. Galvani, 1737–1798)— that is, two electrodes in an electrolyte in a corrosive environment including moisture—and cause **galvanic corrosion**. Two-phase alloys (Chapter 4) are more susceptible to galvanic corrosion (because of the physical separation of the two different metals involved) than are single-phase alloys or pure metals; as a result, heat treatment can have a significant influence on corrosion resistance.

Stress–corrosion cracking (Section 2.10.2) is an example of the effect of a corrosive environment on the integrity of a product that, as manufactured, has residual stresses in it. Likewise, cold-worked metals are likely to have residual stresses, hence they are more susceptible to corrosion than are hot-worked or annealed metals.

Tool and die materials also can be susceptible to chemical attack by lubricants and by coolants; the chemical reaction then alters their surface finish and adversely influences the metalworking operation. One example is that of carbide tools and dies having cobalt as a binder (Section 22.4) in which the cobalt is attacked by elements in the metalworking fluid (**selective leaching**). Thus, compatibility of the tool, die, and workpiece materials with the metalworking fluid under actual operating conditions is an important consideration in the selection of materials.

Chemical reactions should not be regarded as having adverse effects only. Advanced machining processes, such as chemical and electrochemical machining, are indeed based on controlled reactions. These processes remove material by chemical action in a manner similar to the etching of metallurgical specimens. The usefulness of some level of **oxidation** is exhibited in the corrosion resistance of aluminum, titanium, and stainless steel. Aluminum develops a thin (a few atomic layers), strong, and adherent hard-oxide film (Al_2O_3) that better protects the surface from further environmental corrosion. Titanium develops a film of titanium oxide (TiO_2). A similar phenomenon occurs in stainless steels, which (because of the chromium present in the alloy) develop a protective film on their surfaces. These processes are known as **passivation**. When the protective film is scratched and exposes the metal underneath, a new oxide film begins to form.

EXAMPLE 3.1 Selection of materials for coins

There are five general criteria in the selection of materials for coins.

1. The *subjective factors* include the *appearance* of the coin: its color, weight, and ring (the sound made when striking it). Also included in this criterion is the *feel* of the coin. This term is difficult to describe, because it combines many human factors. It is similar in effect to the feel of a fine piece of wood, polished stone, or fine leather.

2. The intended *life of the coin* is another consideration; its duration will reflect resistance to corrosion and to wear while the coin is in circulation. These two factors basically determine the span over which the surface imprint of the coin will remain identifiable and the ability of the coin to retain its original luster.

3. The *manufacturing of the coin* includes factors such as the formability of the candidate coin materials, the life of the dies used in the coining operation (Section 14.4), and the capability of the materials and processes to resist counterfeiting.

4. Another consideration is the *suitability of the coin for use* in coin-operated devices, such as vending machines, turnstiles, and pay telephones. These machines generally are equipped with detection devices that test the coins: first, for proper diameter, thickness, and surface condition; and second, for electrical conductivity and density. The coin is rejected if it fails these tests.

5. Final considerations are the *cost* of raw materials and processing and whether there is a sufficient *supply* of the coin materials.

SUMMARY

- Physical and chemical properties can have several important influences on materials selection, on manufacturing, and on the service life of components. These properties and other material characteristics also should be considered because of their effects on product design, service requirements, and compatibility with other materials, including tools, dies, and workpieces.

- Thermal conductivity and expansion are major factors in the development of thermal stresses and thermal fatigue and shock, effects which are important in tool and die life in manufacturing operations. Low-expansion alloys (such as Invar and Kovar) have unique applications.

- Electrical and chemical properties are important in many advanced machining processes, such as electrical-discharge, chemical, and electrochemical machining.

- Chemical reactions, including oxidation and corrosion, are considerations in material selection, design, and manufacturing, as well as on the service life of components. Passivation and stress–corrosion cracking are two significant phenomena.

- Certain physical properties are utilized in manufacturing processes and their control, such as the magnetostriction effect (for ultrasonic machining of metals and nonmetallic materials) and the piezoelectric effect (for force transducers and various other sensors).

KEY TERMS

Corrosion	Heat checking	Specific stiffness
Degradation	Magnetostriction	Specific strength
Density	Melting point	Stress–corrosion cracking
Dielectrics	Oxidation	Superconductivity
Electrical conductivity	Passivation	Thermal conductivity
Electrical resistivity	Piezoelectric effect	Thermal expansion
Ferromagnetism	Selective leaching	Thermal stresses
Galvanic corrosion	Semiconductors	
Invar effect	Specific heat	

BIBLIOGRAPHY

Callister, D.C., Jr., *Materials Science and Engineering*, 5th ed., Wiley, 2000.

Mark, J.E., *et al.*, *Physical Properties of Polymers*, 3rd ed., Cambridge Univ. Press, 2004.

Pollock, D.D., *Physical Properties of Materials for Engineers*, 2nd ed., CRC Press, 1993.

Schweitzer, P.A., *Encyclopedia of Corrosion Technology*, Marcel Dekker, 1998.

Talbot, D., and Talbot, J., *Corrosion Science and Technology*, CRC Press, 1997.

REVIEW QUESTIONS

3.1 List reasons why density is an important material property in manufacturing.

3.2 Why is the melting point of a material an important factor in manufacturing processes?

3.3 What adverse effects can be caused by the thermal expansion of materials?

3.4 What is the piezoelectric effect?

3.5 What factors lead to the corrosion of a metal?

3.6 What is passivation? What is its significance?

3.7 What is the difference between thermal conductivity and specific heat?

3.8 What is stress–corrosion cracking? Why is it also called season cracking?

3.9 What is the difference between a superconductor and a semiconductor?

QUALITATIVE PROBLEMS

3.10 Describe the significance of structures and machine components made of two materials with different coefficients of thermal expansion.

3.11 Which of the properties described in this chapter are important for (a) mechanical pencils, (b) cookie sheets for baking, (c) rulers, (d) paper clips, (e) door hinges, and (f) beverage cans? Explain your answers.

3.12 You will note in Table 3.1 that the properties of the alloys of metals have a wide range as compared to the properties of the pure metals. What factors are responsible for this?

3.13 Does thermal conductivity play a role in the development of residual stresses in metals? Explain.

3.14 What material properties are desirable for heat shields, such as those placed on the space shuttle?

QUANTITATIVE PROBLEMS

3.15 If we assume that all of the work done in plastic deformation is converted into heat, the temperature rise in a workpiece is (a) directly proportional to the work done per unit volume and (b) inversely proportional to the product of the specific heat and the density of the workpiece. Using Fig. 2.6 and letting the areas under the curves be the unit work done, calculate the temperature rise for (i) 8650 steel, (ii) 304 stainless steel, and (iii) 1100-H14 aluminum.

3.16 The natural frequency, f, of a cantilever beam is given by $f = 0.56EIg/wL^4$, where E is the modulus of elasticity, I is the moment of inertia, g is the gravitational constant, w is the weight of the beam per unit length, and L is the length of the beam. How does the natural frequency of the beam change, if at all, as its temperature is increased? Assume that the material is steel.

3.17 It can be shown that thermal distortion in precision devices is low for high values of thermal conductivity divided by the thermal expansion coefficient. Rank the materials in Table 3.1 according to their suitability to resist thermal distortion.

SYNTHESIS, DESIGN, AND PROJECTS

3.18 From your own experience, make a list of parts, components, and products that have corroded and have had to be replaced or discarded.

3.19 List applications where the following properties would be desirable: (a) high density, (b) low density, (c) high melting point, (d) low melting point, (e) high thermal conductivity, and (f) low thermal conductivity.

3.20 Give several applications in which specific strength and specific stiffness are important.

3.21 Design several mechanisms or instruments based on utilizing the differences in thermal expansion of materials, such as bimetallic strips that develop a curvature when heated.

3.22 For the materials listed in Table 3.1, determine their specific strength and specific stiffness. Describe your observations.

3.23 The maximum compressive force that a lightweight column can withstand before buckling depends on the ratio of the square root of the stiffness to the density for the material. For the materials listed in Table 2.1, determine (a) the ratio of tensile strength to density and (b) the ratio of elastic modulus to density. Comment on the suitability of each for being made into lightweight columns.

3.24 Describe possible applications and designs using alloys exhibiting the Invar effect of low thermal expansion.

4

Metal Alloys: Structure and Strengthening by Heat Treatment

This chapter explains how the properties of metals and alloys can be altered by subjecting them to various heat treatments. In this chapter, we describe the following:

• Microstructure of ferrous alloys and characteristics of phase diagrams.
• Various phases in alloys and their significance in manufacturing.
• Mechanisms by which properties are altered.
• Procedures used for heat-treating ferrous metals to enhance their properties.
• Microstructures of nonferrous alloys and their significance.
• Procedures used for heat-treating nonferrous alloys.
• Characteristics of heat-treating furnaces.
• Design considerations for successful heat treating.

4.1 | Introduction

The properties and behavior of metals and alloys during manufacturing and their performance during their service life depend on their composition, structure, and processing history, as well as on the heat treatment to which they have been subjected. Important properties (such as strength, hardness, ductility, toughness, and resistance to wear) are influenced greatly by alloying elements and heat-treatment processes. The properties of non-heat-treatable alloys are improved by mechanical working operations, such as rolling, forging, and extrusion (Part III).

The most common example of property improvement is *heat treatment* (Sections 4.7 through 4.10). Heat treatment modifies microstructures and, thereby, produces a variety of mechanical properties that are important in manufacturing, such as improved formability and machinability. These properties also enhance the service performance of the metals when used in machine components (such as gears, cams, and shafts; see Fig. 4.1) and in tools, dies, and molds.

This chapter follows the outline shown in Fig. 4.2, beginning with the effects of various alloying elements, and followed by the solubility of one element in another, phases, equilibrium phase diagrams, and the influences of composition, temperature,

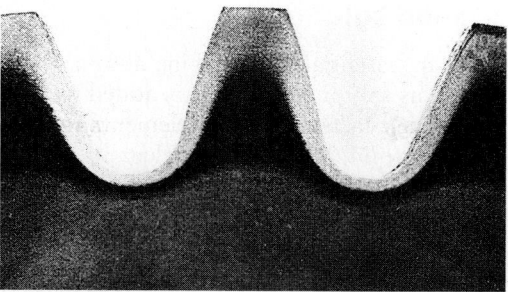

FIGURE 4.1 Cross-section of gear teeth showing induction-hardened surfaces. *Source*: Courtesy of TOCCO Div., Park-Ohio Industries, Inc.

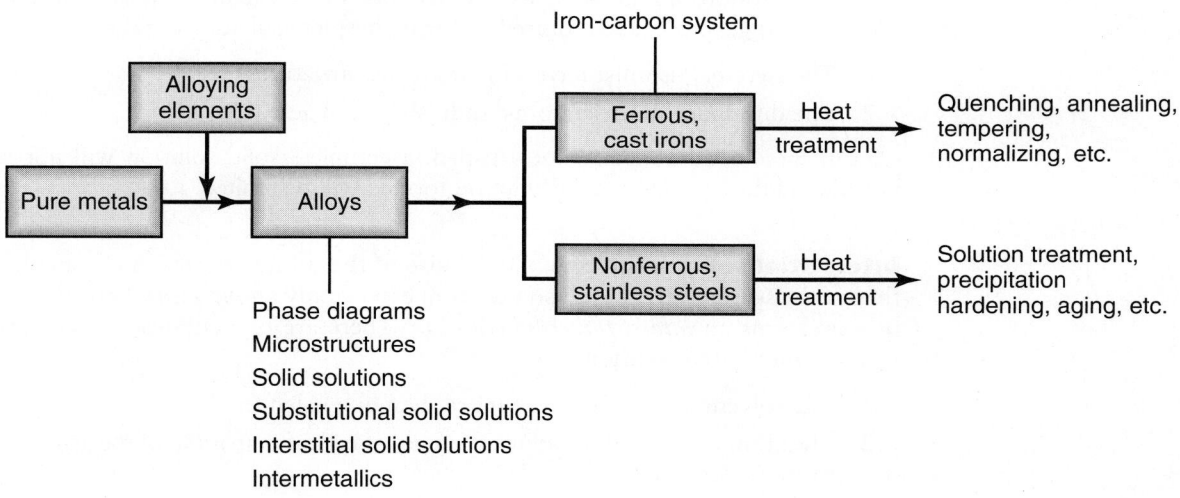

FIGURE 4.2 Outline of topics described in Chapter 4.

and time. It also discusses methods and techniques of heating, quenching, tempering, and annealing and describes the characteristics of the equipment involved.

4.2 | Structure of Alloys

When describing the basic crystal structure of metals in Chapter 1, it was noted that the atoms are all of the *same* type, except for the presence of rare impurity atoms. These metals are known as **pure metals**, even though they may not be completely pure. *Commercially pure* metals are used for various purposes: aluminum for foil, copper for electrical conductors, nickel or chromium for plating, and gold for electrical contacts. However, pure metals have somewhat limited properties; these properties can be enhanced and modified by **alloying**. An **alloy** is composed of two or more chemical elements, at least one of which is a metal. The majority of metals used in engineering applications are some form of alloy. Alloying consists of two basic forms: *solid solutions* and *intermetallic compounds*.

4.2.1 Solid solutions

Two terms are essential in describing alloys: **solute** and **solvent**. Solute is the *minor* element (such as salt or sugar) that is added to the solvent, which is the *major* element (such as water). In terms of the elements involved in a crystal structure, the solute (composed of *solute atoms*) is the element that is added to the solvent (composed of *host atoms*). When the particular crystal structure of the solvent is maintained during alloying, the alloy is called a **solid solution**.

Substitutional solid solutions.   If the size of the solute atom is similar to that of the solvent atom, the solute atoms can replace solvent atoms and form a *substitutional solid solution* (see Fig. 1.8). An example is brass (Section 6.4), an alloy of zinc and copper, in which zinc (solute atom) is introduced into the lattice of copper (solvent atoms). The properties of brasses thus can be altered over a range by controlling the amount of zinc in copper.

Two conditions (known as *Hume-Rothery rules*, after W. Hume-Rothery, 1899–1968) generally are required to form complete substitutional solid solutions:

1. The two metals must have similar crystal structures.
2. The difference in their atomic radii should be less than 15%.

If these conditions are not satisfied, a complete solid solution will not be obtained, and the amount of solid solution formed will be limited.

Interstitial solid solutions.   If the size of the solute atom is much smaller than that of the solvent atom, each solute atom can occupy an *interstitial* position; such a process forms an *interstitial solid solution*. There are two conditions necessary for forming interstitial solutions:

1. The solvent atom must have more than one valence.
2. The atomic radius of the solute atom must be less than 59% of the atomic radius for the solvent atom.

If these conditions are not met, limited interstitial solubility may take place—or even not at all.

An important family of interstitial solid solutions is **steel** (Chapter 5), an alloy of iron and carbon, where carbon atoms are present in interstitial positions between iron atoms. The atomic radius of carbon is 0.071 nm, which is less than 59% of the 0.124 nm radius of the iron atom. The properties of carbon steels can be varied over a wide range by controlling the proportion of carbon in the iron. This controllability is one reason that steel is such a versatile and useful material with a wide variety of properties and applications, in addition to being inexpensive.

4.2.2 Intermetallic compounds

Intermetallic compounds are complex structures consisting of two metals in which solute atoms are present among solvent atoms in certain proportions. Some intermetallic compounds have solid solubility, and the type of atomic bond may range from metallic to ionic. Intermetallic compounds are strong, hard, and brittle. Because of their high-melting points, their strength at elevated temperatures, their good oxidation resistance, and their relatively low density, they are candidate materials for advanced gas-turbine engines. Typical examples are the aluminides of titanium (Ti_3Al), nickel (Ni_3Al), and iron (Fe_3Al).

4.2.3 Two-phase systems

It has been noted that a solid solution is one in which two or more elements in a solid state form a single homogeneous solid phase in which the elements are distributed uniformly throughout the solid mass. Such a system is limited by some maximum concentration of solute atoms in the solvent-atom lattice, just as there is a solubility limit for sugar in water. Most alloys consist of two or more solid phases and may be regarded as mechanical mixtures; such a system with two solid phases is known as a **two-phase system**.

A **phase** is defined as a physically distinct and homogeneous portion in a material; each phase is a homogeneous part of the total mass and has its own characteristics and properties. Let's consider a mixture of sand and water as an example of a two-phase system. These two different components have their own distinct structures, characteristics, and properties. There is a clear boundary in this mixture between the water (one phase) and the sand particles (a second phase). Another example is ice in water; in this case, the two phases are the same chemical compound of exactly the same chemical elements (hydrogen and oxygen), even though their properties are very different.

A typical example of a two-phase system in metals occurs when lead is added to copper in the molten state. After the mixture solidifies, the structure consists of two phases: one having a small amount of lead in solid solution in copper, the other having lead particles (roughly spherical in shape) *dispersed* throughout the structure (Fig. 4.3a). The lead particles are analogous to the sand particles in water described earlier. This copper-lead alloy has properties that are different from those of either copper or lead alone. Lead also is added to steels to obtain leaded steels, which have greatly improved machinability.

Alloying with finely dispersed particles (**second-phase particles**) is an important method of strengthening alloys and controlling their properties. In two-phase alloys, the second-phase particles present obstacles to dislocation movement and, thus, increase strength. Another example of a two-phase alloy is the aggregate structure shown in Fig. 4.3b, where there are two sets of grains, each with its own composition and properties. The darker grains may have a different structure from the lighter grains. For example, the darker grains may be brittle, while the lighter grains are ductile. Defects may appear during metalworking operations, such as forging or extrusion; such flaws may be due to the lack of ductility of one of the phases in the alloy. In general, two-phase alloys are stronger and less ductile than solid solutions.

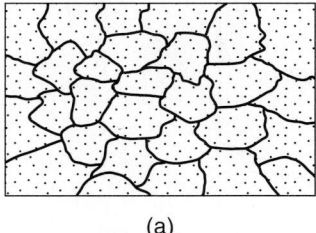   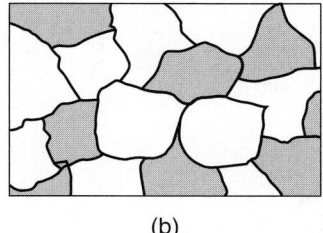

(a)   (b)

FIGURE 4.3   (a) Schematic illustration of grains, grain boundaries, and particles dispersed throughout the structure of a two-phase system, such as a lead-copper alloy. The grains represent lead in solid solution in copper, and the particles are lead as a second phase. (b) Schematic illustration of a two-phase system consisting of two sets of grains: dark and light. The dark and the light grains have separate compositions and properties.

4.3 | Phase Diagrams

Pure metals have clearly defined melting or freezing points, and solidification takes place at a *constant temperature* (Fig. 4.4). When the temperature of the molten metal is reduced to the freezing point, the energy of the *latent heat of solidification* is given off while the temperature remains constant. Eventually, solidification is complete, and the solid metal continues cooling to room temperature.

Unlike pure metals, alloys solidify over a *range* of temperatures (Fig. 4.5). Solidification begins when the temperature of the molten metal drops below the **liquidus**; it is completed when the temperature reaches the **solidus**. Within this temperature range, the alloy is in a *mushy* or *pasty state*; its composition and state are described by the particular alloy's phase diagram.

A **phase diagram**, also called an **equilibrium** or **constitutional diagram**, shows the relationships among the temperature, the composition, and the phases present in a particular alloy system under equilibrium conditions. *Equilibrium* means that the state of a system remains constant over an indefinite period of time. The word *constitutional* indicates the relationships among the structure, the composition, and the physical make-up of the alloy. As described in detail next, types of phase diagrams include: (1) complete solid solutions; (2) eutectics, such as cast irons; and (3) eutectoids, such as steels.

One example of a phase diagram is shown in Fig. 4.5 for the copper-nickel alloy; it is called a **binary phase diagram**, because there are two elements (copper and nickel) present in the system. The left-hand boundary of this phase diagram (100% Ni) indicates the melting point of nickel; the right-hand boundary (100% Cu) indicates the melting point of copper. (All percentages in this discussion are by weight, not by number of atoms.)

Note that for a composition of, say, 50% Cu-50% Ni, the alloy begins to solidify at a temperature of 1313°C (2395°F), and solidification is complete at 1249°C

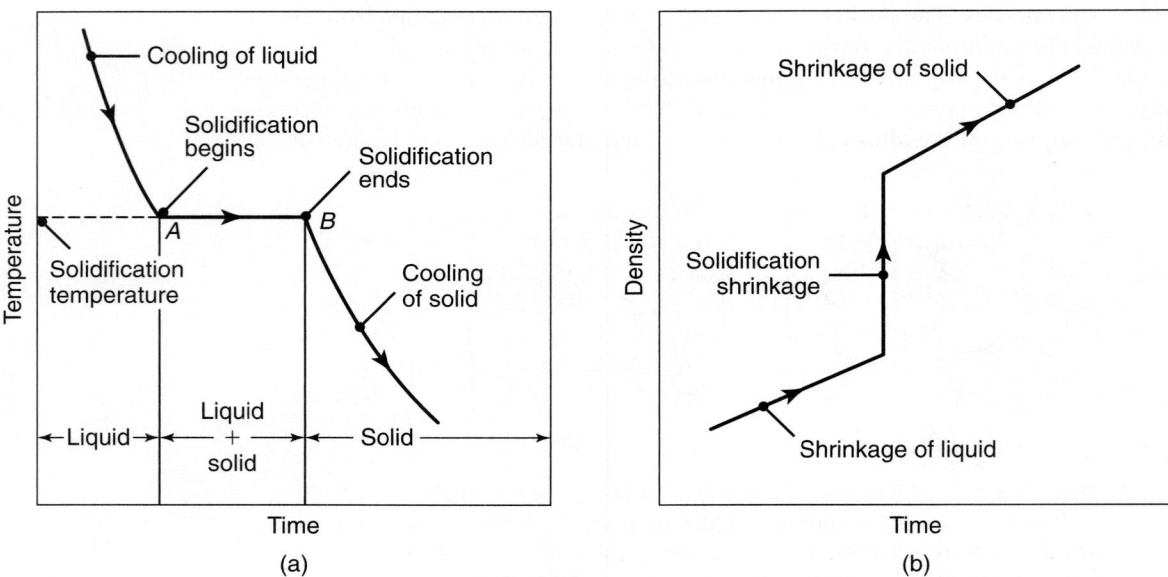

FIGURE 4.4   (a) Cooling curve for the solidification of pure metals. Note that freezing takes place at a constant temperature; during freezing, the latent heat of solidification is given off. (b) Change in density during the cooling of pure metals.

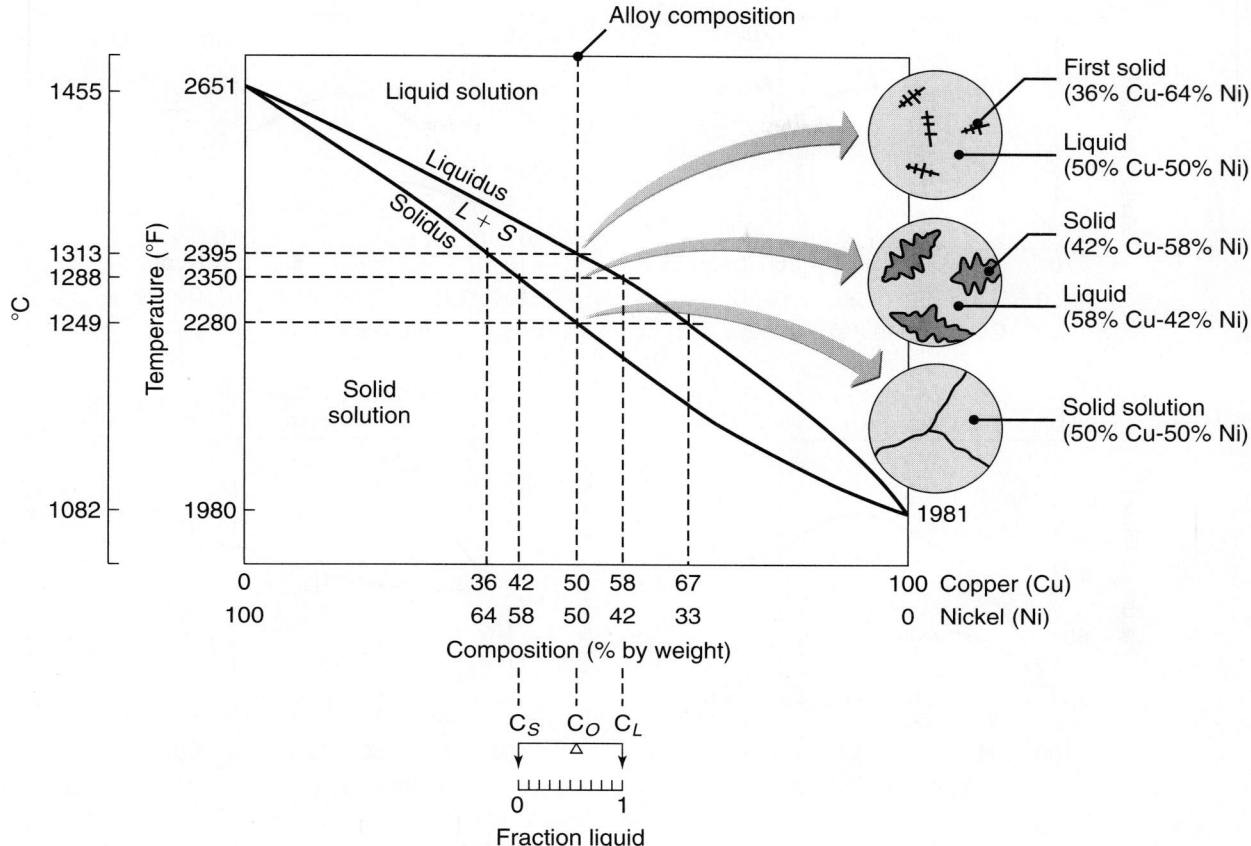

FIGURE 4.5 Phase diagram for nickel-copper alloy system obtained at a slow rate of solidification. Note that pure nickel and pure copper each has one freezing or melting temperature. The top circle on the right depicts the nucleation of crystals. The second circle shows the formation of dendrites (see Section 10.2). The bottom circle shows the solidified alloy with grain boundaries.

(2280°F). Above 1313°C, a homogeneous liquid of 50% Cu-50% Ni exists. When cooled slowly to 1249°C, a homogeneous solid solution of 50% Cu-50% Ni results. However, between the liquidus and solidus curves and at a temperature of 1288°C (2350°F), there is a two-phase region: a **solid phase** composed of 42% Cu-58% Ni, and a **liquid phase** of 58% Cu-42% Ni. To determine the *solid composition*, we go left horizontally to the solidus curve and read down, obtaining 42% Cu. The *liquid composition* (58%) is obtained similarly by going to the right to the liquidus curve. The procedure for determining the compositions of various phases in phase diagrams (called the **lever rule**) is described in detail in texts on materials science and metallurgy.

The *completely solidified* alloy in the phase diagram shown in Fig. 4.5 is a **solid solution,** because the alloying element Cu (solute atom) is dissolved completely in the host metal Ni (solvent atom), and each grain has the same composition. The atomic radius of copper is 0.128 nm and that of nickel is 0.125 nm, and both elements are of face-centered cubic structure. Thus the Hume-Rothery rules are obeyed.

The mechanical properties of solid solutions of Cu-Ni depend on their composition (Fig. 4.6). Up to a point, the properties of pure copper are improved upon by

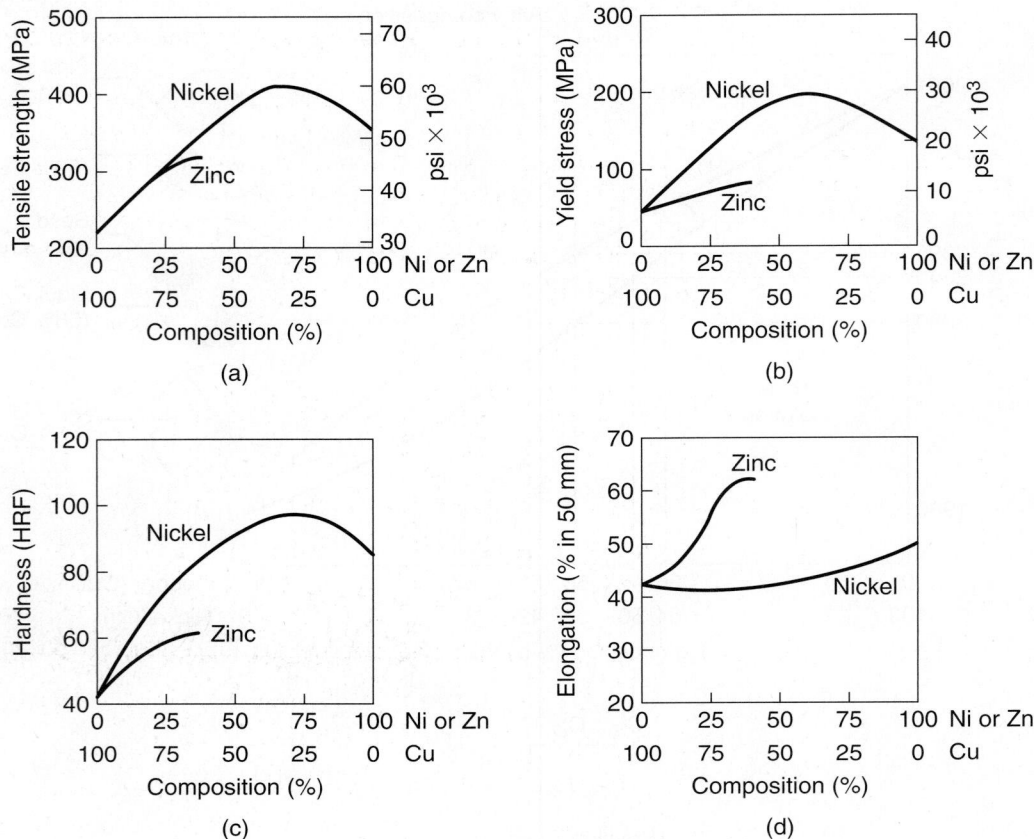

FIGURE 4.6   Mechanical properties of copper-nickel and copper-zinc alloys as a function of their composition. The curves for zinc are short, because zinc has a maximum solid solubility of 40% in copper.

increasing the nickel content; there is an optimal percentage of nickel that gives the highest strength and hardness to the Cu-Ni alloy. Figure 4.6 also shows how zinc, as an alloying element in copper, affects the mechanical properties of the alloy. Note a maximum of 40% solid solubility for zinc (solute) in copper (solvent), whereas copper and nickel are completely soluble in each other. The improvements in properties are due to *pinning* (blocking) of dislocations at substitutional nickel or zinc atoms, which also may be regarded as impurity atoms. As a result, dislocations cannot move as freely, and the strength of the alloy increases.

Another example of a two-phase diagram is shown in Fig. 4.7 for the lead-tin system. The single phases, alpha and beta, are solid solutions. Note that the single-phase regions are separated from the liquid phase by two two-phase regions: *alpha + liquid* and *beta + liquid*. Figure 4.7 shows the composition of the alloy (61.9% Sn-38.1% Pb) that has the *lowest* temperature at which the alloy is still completely liquid, namely, 183°C (361°F).

This point is known as the **eutectic point**, and it is the point at which the liquid solution decomposes into the components alpha and beta. The word *eutectic* is from the Greek *eutektos*, meaning easily melted.

Eutectic points are important in applications, such as soldering, where low temperatures are desirable to prevent thermal damage to parts during joining. Although there are various types of solders, tin-lead solders commonly are used for

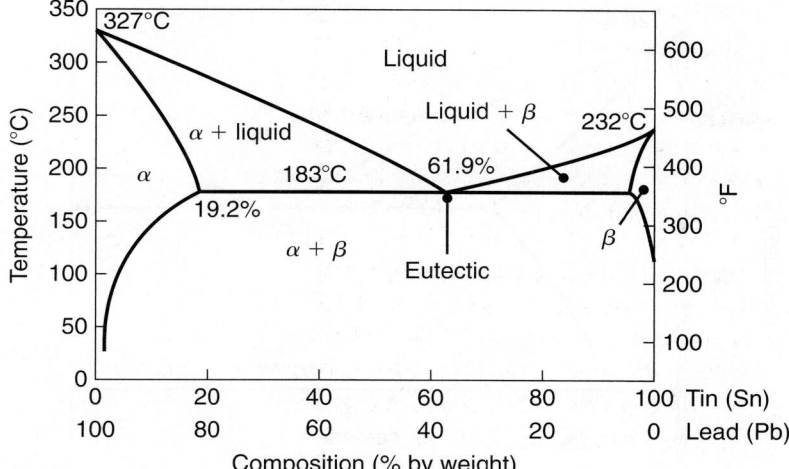

FIGURE 4.7 The lead-tin phase diagram. Note that the composition of the eutectic point for this alloy is 61.9% Sn-38.1% Pb. A composition either lower or higher than this ratio will have a higher liquidus temperature.

general applications; they have composition ranging from 5% Pb-95% Sn to 70% Pb-30% Sn. Each composition has its own melting point.

4.4 | The Iron-Carbon System

Steels and cast irons are represented by the iron-carbon binary system. Commercially pure iron contains up to 0.008% C, steels up to 2.11% C, and cast irons up to 6.67% C, although most cast irons contain less than 4.5% C. In this section, the iron-carbon system is described, including the techniques employed to evaluate and modify the properties of these important materials for specific applications.

The **iron-iron carbide phase diagram** is shown in Fig. 4.8. Although this diagram can be extended to the right—to 100% C (pure graphite) (see Fig. 4.12)—the range that is significant to engineering applications is up to 6.67% C, because Fe_3C is a stable phase. Pure iron melts at a temperature of 1538°C (2798°F), as shown at the left boundary in Fig. 4.8. As iron cools, it first forms delta ferrite, then austenite, and finally alpha ferrite.

Ferrite. Alpha ferrite, or simply **ferrite**, is a solid solution of body-centered cubic iron; it has a maximum solid solubility of 0.022% C at a temperature of 727°C (1341°F). Delta ferrite is stable only at very high temperatures and is of no practical significance in engineering. Just as there is a solubility limit for salt in water (with any extra amount precipitating as solid salt at the bottom of a container), so there is a solid-solubility limit for carbon in iron.

Ferrite is relatively soft and ductile; it is magnetic from room temperature to 768°C (1414°F)—the *Curie temperature* (after M. Curie, 1867–1934). Although very little carbon can dissolve interstitially in bcc iron, the amount of carbon significantly can affect the mechanical properties of ferrite. Furthermore, significant amounts of chromium, manganese, nickel, molybdenum, tungsten, and silicon can be contained in iron in solid solution, thereby imparting desirable properties.

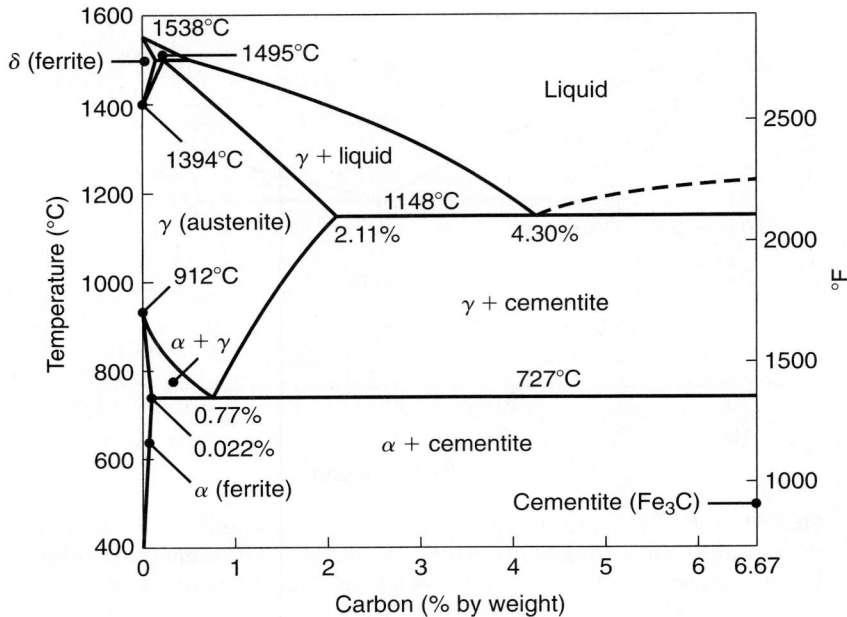

FIGURE 4.8   The iron-iron carbide phase diagram. Because of the importance of steel as an engineering material, this diagram is one of the most important of all phase diagrams.

Austenite.   As shown in Fig. 4.8, within a certain temperature range, iron undergoes a **polymorphic transformation** from the bcc to an fcc structure, becoming *gamma iron* or (more commonly) **austenite** (after W. R. Austen, 1843–1902). This structure has a solid solubility of up to 2.11% C at 1148°C (2098°F). Because the fcc structure has more interstitial positions, the solid solubility of austenite is about two orders of magnitude higher than that of ferrite, with the carbon occupying the interstitial positions (Fig. 4.9a).

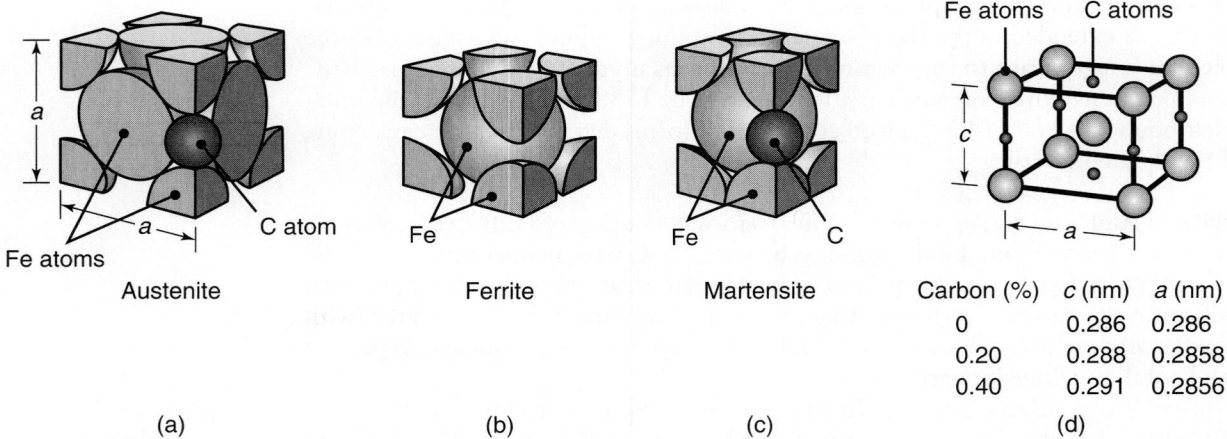

Carbon (%)	c (nm)	a (nm)
0	0.286	0.286
0.20	0.288	0.2858
0.40	0.291	0.2856

(a)   (b)   (c)   (d)

FIGURE 4.9   The unit cells for (a) austenite, (b) ferrite, and (c) martensite. The effect of percentage of carbon (by weight) on the lattice dimensions for martensite is shown in (d). Note the interstitial position of the carbon atoms (see Fig. 1.9). Also note, the increase in dimension *c* with increasing carbon content: this effect causes the unit cell of martensite to be in the shape of a rectangular prism.

Austenite is an important phase in the heat treatment of steels (Section 4.7). It is denser than ferrite, and its single-phase fcc structure is ductile at elevated temperatures. Consequently, it possesses good formability. Large amounts of nickel and manganese also can be dissolved in fcc iron to impart various properties. Steel is nonmagnetic in the austenitic form either at high temperatures or (for austenitic stainless steels) at room temperature.

Cementite.   The right boundary of Fig. 4.8 represents **cementite**, which is 100% iron carbide (Fe_3C) having a carbon content of 6.67%. Cementite, from the Latin *caementum* (meaning stone chips), is also called **carbide**. This iron carbide should not be confused with other carbides used as dies, cutting tools, and abrasives (such as tungsten carbide, titanium carbide, and silicon carbide, described in Chapters 8 and 22). Cementite is a very hard and brittle intermetallic compound and has a significant influence on the properties of steels. It can include other alloying elements, such as chromium, molybdenum, and manganese.

4.5   The Iron-Iron Carbide Phase Diagram and the Development of Microstructures in Steels

The region of the iron-iron carbide phase diagram that is up to 2.11% C and is significant for steels is shown in Fig. 4.10, which is an enlargement of the lower-left portion of Fig. 4.8. Various microstructures can be developed, depending on the carbon content, the amount of plastic deformation (working), and the method of heat treatment. For example, let's consider the eutectic point of iron with a 0.77% C

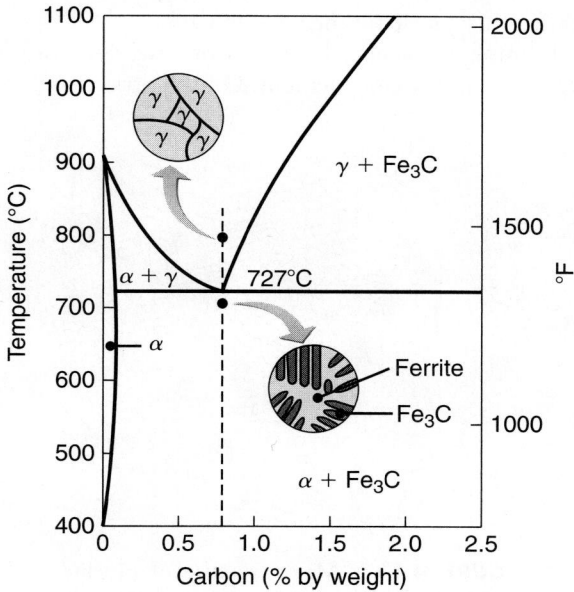

FIGURE   4.10   Schematic   illustration   of   the microstructures for an iron-carbon alloy of eutectoid composition (0.77% carbon) above and below the eutectoid temperature of 727°C (1341°F).

content while it is being cooled very slowly from a temperature of, say, 1100°C (2000°F) in the austenite phase.

The reason for *slow* cooling is to maintain equilibrium. Higher rates of cooling are employed in heat treating, as explained in Section 4.7. At 727°C (1341°F) a reaction takes place in which austenite is transformed into alpha ferrite (bcc) and cementite. Because the solid solubility of carbon in ferrite is only 0.022%, the extra carbon forms cementite. This reaction is called a **eutectoid** (meaning *eutecticlike*) **reaction**. This term means that, at a certain temperature, a single, solid phase (austenite) is transformed into two other solid phases (ferrite and cementite). The structure of eutectoid steel is called **pearlite**, because (at low magnifications) it resembles mother-of-pearl (Fig. 4.11). The microstructure of pearlite consists of alternating layers (**lamellae**) of ferrite and cementite. Consequently, the mechanical properties of pearlite are intermediate between those of ferrite (soft and ductile) and cementite (hard and brittle).

In iron with less than 0.77% C, the microstructure formed consists of a pearlite phase (ferrite and cementite) and a ferrite phase. The ferrite in the pearlite is called **eutectoid ferrite**, and the ferrite phase is called **proeutectoid ferrite** (*pro* meaning before). It forms at a temperature higher than the eutectoid temperature of 727°C (1341°F) in the *alpha + gamma* region. If the carbon content is higher than 0.77%, the austenite transforms into pearlite and cementite. The cementite in the pearlite is called **eutectoid cementite**, and the cementite phase is called **proeutectoid cementite**, because it forms in the alpha + Fe_3C region at a temperature higher than the eutectoid temperature.

4.5.1 Effects of Alloying Elements in Iron

Although carbon is the basic element that transforms iron into steel, other elements are also added to impart a variety of desirable properties. The main effect of these alloying elements on the iron-iron carbide phase diagram is to shift the eutectoid temperature and eutectoid composition (percentage of carbon in steel at the eutectoid point); they shift other phase boundaries as well.

The eutectoid temperature may be raised or lowered from 727°C (1341°F), depending on the particular alloying element. On the other hand, alloying elements

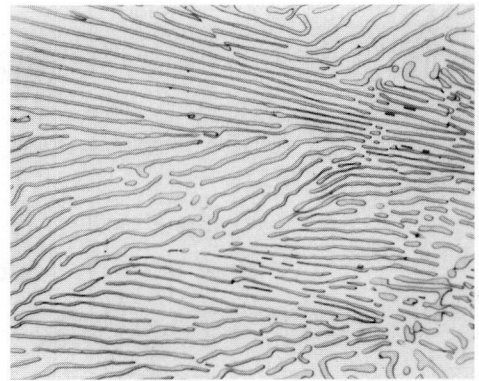

FIGURE 4.11 Microstructure of pearlite in 1080 steel formed from austenite of a eutectoid composition. In this lamellar structure, the lighter regions are ferrite, and the darker regions are carbide. Magnification: 2500×.

always lower the eutectoid composition (that is, its carbon content is lower than 0.77%). Lowering the eutectoid temperature means increasing the austenite range. As result, an alloying element (such as nickel) is known as an **austenite former**. Because nickel has an fcc structure, it favors the fcc structure of austenite. Conversely, chromium and molybdenum have a bcc structure, thus favoring the bcc structure of ferrite. These elements are known as **ferrite stabilizers**.

4.6 | Cast Irons

The term **cast iron** refers to a family of ferrous alloys composed of iron, carbon (ranging from 2.11% to about 4.5%), and silicon (up to about 3.5%). Cast irons usually are classified according to their solidification morphology from the eutectic temperature, as follows (see also Section 12.3):

a. Gray cast iron, or gray iron
b. Ductile cast iron, nodular cast iron, or spheroidal graphite cast iron
c. White cast iron
d. Malleable iron
e. Compacted graphite iron

Cast irons also are classified by their structure: ferritic, pearlitic, quenched and tempered, or austempered.

The equilibrium phase diagram relevant to cast irons is shown in Fig. 4.12, where the right boundary is 100% C (that is, pure graphite). The eutectic temperature is 1154°C (2109°F), and thus, cast irons are completely liquid at temperatures lower than those required for liquid steels. Consequently, iron with a high-carbon content can be cast (see Part II) at lower temperatures than can steels.

Cementite is not completely stable; it is **metastable**, with an extremely low rate of decomposition. It can, however, be made to decompose into alpha ferrite and graphite.

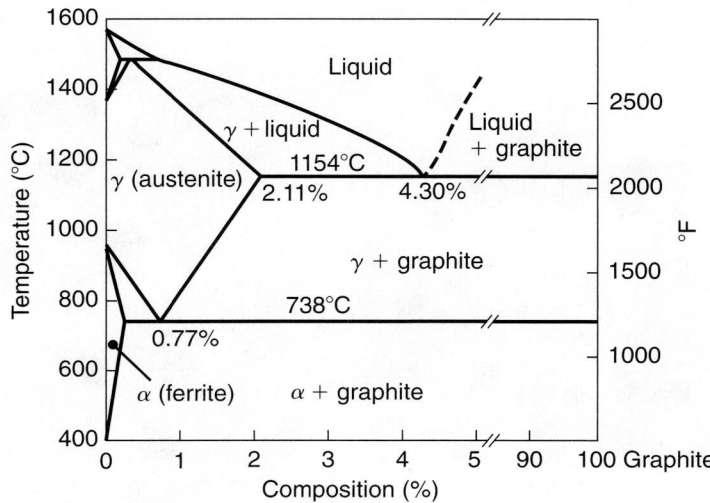

FIGURE 4.12 Phase diagram for the iron-carbon system with graphite (instead of cementite) as the stable phase. Note that this figure is an extended version of Fig. 4.8.

The formation of graphite (**graphitization**) can be controlled, promoted, and accelerated by modifying the composition and the rate of cooling and by the addition of silicon.

Gray cast iron. In this structure, graphite exists largely in the form of *flakes* (Fig. 4.13a). It is called **gray cast iron** or **gray iron**, because when it is broken, the fracture path is along the graphite flakes and has a gray, sooty appearance. These flakes act as stress raisers. As a result, gray iron has negligible ductility and is weak in tension (although strong in compression), as are other brittle materials. On the other hand, the presence of graphite flakes gives this material the capacity to dampen vibrations (by internal friction) and, consequently, the ability to dissipate energy. This capacity makes gray cast iron a suitable and commonly used material for constructing machine-tool bases and machinery structures.

Three types of gray cast iron are **ferritic, pearlitic,** and **martensitic**. Because of the different structures, each has different properties and applications. In ferritic gray iron (also known as *fully gray iron*), the structure consists of graphite flakes in an alpha-ferrite matrix. Pearlitic gray iron has a structure of graphite in a matrix of pearlite; although still brittle, it is stronger than fully gray iron. Martensitic gray iron is obtained by austenitizing a pearlitic gray iron and then rapidly quenching it to produce a structure of graphite in a martensite matrix. As a result, martensitic cast iron is very hard.

Ductile (Nodular) iron. In the ductile-iron structure, graphite is in a **nodular** or **spheroid** form (Fig. 4.13b). This shape permits the material to be somewhat ductile and shock-resistant. The shape of graphite flakes is changed into nodules (spheres) by small additions of magnesium and/or cerium to the molten metal prior to pouring. Ductile iron can be made ferritic or pearlitic by heat treatment. It also can be heat treated to alternatively obtain a structure of tempered martensite (Section 4.7).

White cast iron. The white-cast-iron structure is very hard, wear-resistant, and brittle because of the presence of large amounts of iron carbide (instead of graphite). **White cast iron** is obtained either by cooling gray iron rapidly or by adjusting the composition by keeping the carbon and silicon content low. This type of cast iron is also called *white iron* because of the white crystalline appearance of the fracture surface.

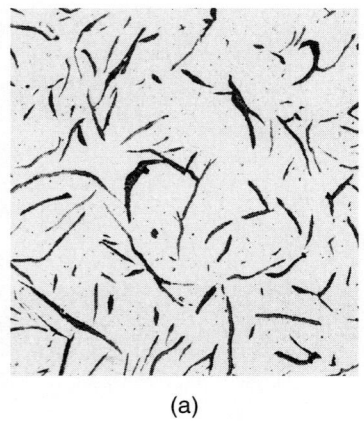 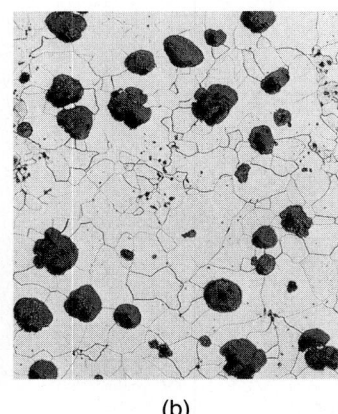 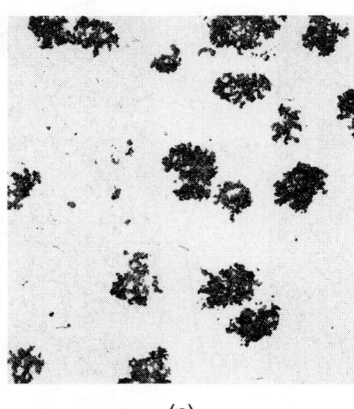

(a)   (b)   (c)

FIGURE 4.13 Microstructure for cast irons. Magnification: 100×. (a) Ferritic gray iron with graphite flakes. (b) Ferritic ductile iron (nodular iron) with graphite in nodular form. (c) Ferritic malleable iron. This cast iron solidified as white cast iron with the carbon present as cementite and was heat treated to graphitize the carbon.

Malleable iron.   Malleable iron is obtained by annealing white cast iron in an atmosphere of carbon monoxide and carbon dioxide, at between 800° and 900°C (1470° and 1650°F), for up to several hours, depending on the size of the part. During this process, the cementite decomposes (*dissociates*) into iron and graphite. The graphite exists as *clusters* or *rosettes* (Fig. 4.13c) in a ferrite or pearlite matrix. Consequently, malleable iron has a structure similar to that of nodular iron. This structure promotes ductility, strength, and shock resistance—hence, the term *malleable* (from the Latin *malleus*, meaning it can be "hammered").

Compacted-graphite iron.   The graphite in this structure is in the form of short, thick, and interconnected flakes having undulating surfaces and rounded extremities. The mechanical and physical properties of this cast iron are intermediate between those of flake-graphite and nodular-graphite cast irons.

4.7 | Heat Treatment of Ferrous Alloys

The various microstructures described thus far can be modified by **heat-treatment** techniques—that is, by controlled heating and cooling of the alloys at various rates. These treatments induce **phase transformations** that greatly influence such mechanical properties as the strength, hardness, ductility, toughness, and wear resistance of the alloys. The effects of thermal treatment depend on the particular alloy, its composition and microstructure, the degree of prior cold work, and the rates of heating and cooling during heat treatment. The processes of recovery, recrystallization, and grain growth (Section 1.6) are examples of thermal treatment, involving changes in the grain structure of the alloy.

   This section will focus on the microstructural changes in the iron-carbon system. Because of their technological significance, the structures considered here are pearlite, spheroidite, bainite, martensite, and tempered martensite. The heat-treatment processes described are annealing, quenching, and tempering.

Pearlite.   If the ferrite and cementite lamellae in the pearlite structure of the eutectoid steel shown in Fig. 4.11 are thin and closely packed, the microstructure is called **fine pearlite**; if they are thick and widely spaced, it is called **coarse pearlite**. The difference between the two depends on the rate of cooling through the eutectoid temperature (the site of a reaction in which austenite is transformed into pearlite). If the rate of cooling is relatively high (as it is in air), fine pearlite is produced; if cooling is slow (as it is in a furnace), coarse pearlite is produced.

Spheroidite.   When pearlite is heated to just below the eutectoid temperature and then held at that temperature for a period of time (**subcritical annealing**, Section 4.11)—such as for a day at 700°C (1300°F)—the cementite lamellae transform to roughly *spherical* shapes (Fig. 4.14). Unlike the lamellar shapes of cementite (which act as stress raisers) **spheroidites** (spherical particles) are less conducive to stress concentration because of their rounded shapes. Consequently, this structure has higher toughness and lower hardness than the pearlite structure. In this form, it can be cold worked, because the ductile ferrite has high toughness, and the spheroidal carbide particles prevent the initiation of cracks within the material.

Bainite.   Visible only by using electron microscopy, **bainite** is a very fine microstructure consisting of ferrite and cementite—somewhat like a pearlitic but having a different morphology. It can be produced in steels with alloying elements and at

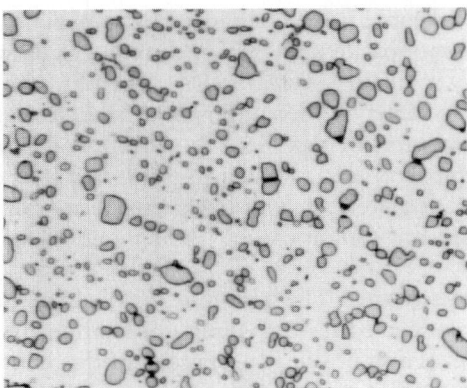

FIGURE 4.14   Microstructure of eutectoid steel. Spheroidite is formed by tempering the steel at 700°C (1292°F). Magnification: 1000×.

cooling rates that are higher than those required for transformation to pearlite. This structure, called **bainitic steel** (after E.C. Bain, 1891–1971), generally is stronger and more ductile than pearlitic steels at the same hardness level.

Martensite.   When austenite is cooled at a high rate (such as by quenching it in water), its fcc structure is transformed into a **body-centered tetragonal** (bct) structure. This structure can be described as a body-centered rectangular prism which is elongated slightly along one of its principal axes (see Fig. 4.9d). This microstructure is called **martensite** (after A. Martens, 1850–1914). Because martensite does not have as many slip systems as a bcc structure (and the carbon is in interstitial positions), it is extremely hard and brittle (Fig. 4.15); it lacks toughness; and therefore, it has limited use. Martensite transformation takes place almost instantaneously because it involves not the diffusion process but a slip mechanism (thus plastic deformation), which is a time-dependent phenomenon that is the mechanism in other transformations as well.

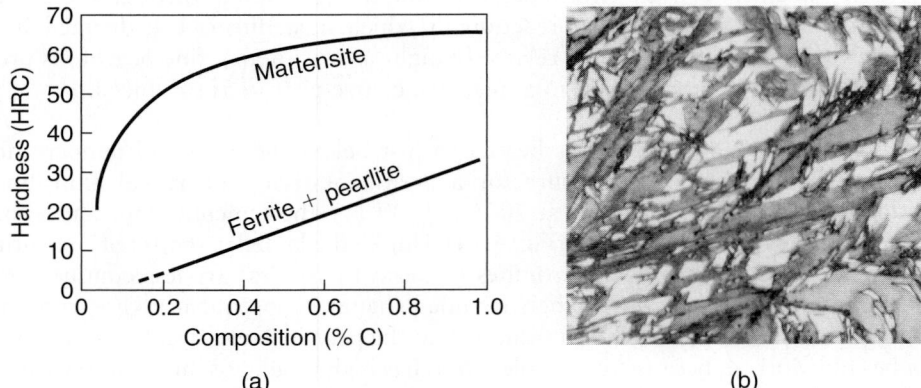

(a)   (b)

FIGURE 4.15   (a) Hardness of martensite as a function of carbon content. (b) Micrograph of martensite containing 0.8% carbon. The gray plate-like regions are martensite; they have the same composition as the original austenite (white regions). Magnification: 1000×.

The material undergoes volume changes because of the differences in densities that result from phase transformations. For example, when austenite transforms to martensite, its volume increases (and hence its density decreases) by as much as 4%. A similar but smaller volume expansion also occurs when austenite transforms to pearlite. These expansions and the thermal gradients present in a quenched part cause internal stresses within the body. They may cause parts to undergo *distortion* or to even crack during heat treatment; **quench cracking** of steels is caused by rapid cooling during quenching.

Distortion is an irreversible dimensional change of the part during heat treatment. It is a general term and may consist of size distortion or shape distortion. *Size distortion* involves changes in the dimensions of the part without a change in shape, whereas *shape distortion* involves bending, twisting, and similar nonsymmetrical dimensional changes. Distortion can be reduced by the proper control of heating and cooling cycles, by improved part design, and by more localized heat treatment of the part.

Retained austenite.   If the temperature to which the alloy is quenched is not sufficiently low, only a portion of the structure is transformed to martensite. The rest is **retained austenite**, which is visible as white areas in the structure along with the dark, needlelike martensite. Retained austenite can cause dimensional instability and cracking, and it lowers the hardness and strength of the alloy.

Tempered martensite.   Martensite is tempered in order to improve its mechanical properties. **Tempering** is a heating process by which hardness is reduced and toughness is improved. The body-centered tetragonal martensite is heated to an intermediate temperature, typically 150° to 650°C (300° to 1200°F), where it decomposes to a two-phase microstructure consisting of body-centered cubic alpha ferrite and small particles of cementite. With increasing tempering time and temperature, the hardness of tempered martensite decreases (Fig. 4.16). The reason is that the cementite particles coalesce and grow, and the distance between the particles in the soft ferrite matrix increases as the less stable and smaller carbide particles dissolve.

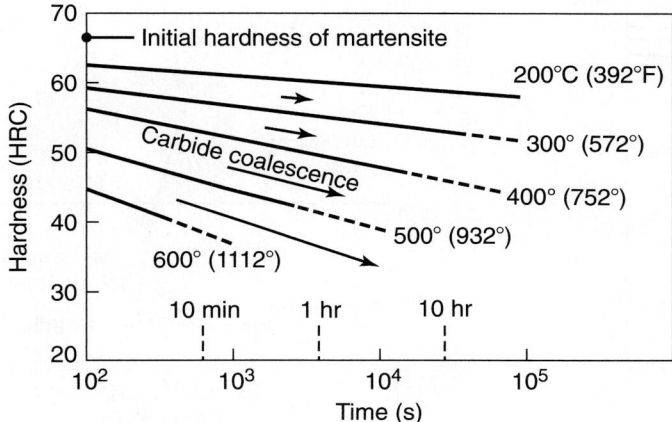

FIGURE 4.16   Hardness of tempered martensite as a function of tempering time for 1080 steel quenched to 65 HRC. Hardness decreases because the carbide particles coalesce and grow in size, thereby increasing the interparticle distance of the softer ferrite.

4.7.1 Time-temperature-transformation diagrams

The transformation from austenite to pearlite (among other structures) is best illustrated by Figs. 4.17b and c. These diagrams are called **isothermal transformation** (IT) **diagrams**, or *time-temperature-transformation (TTT) diagrams*. They are constructed from the data given in Fig. 4.17a, which shows the percentage of austenite

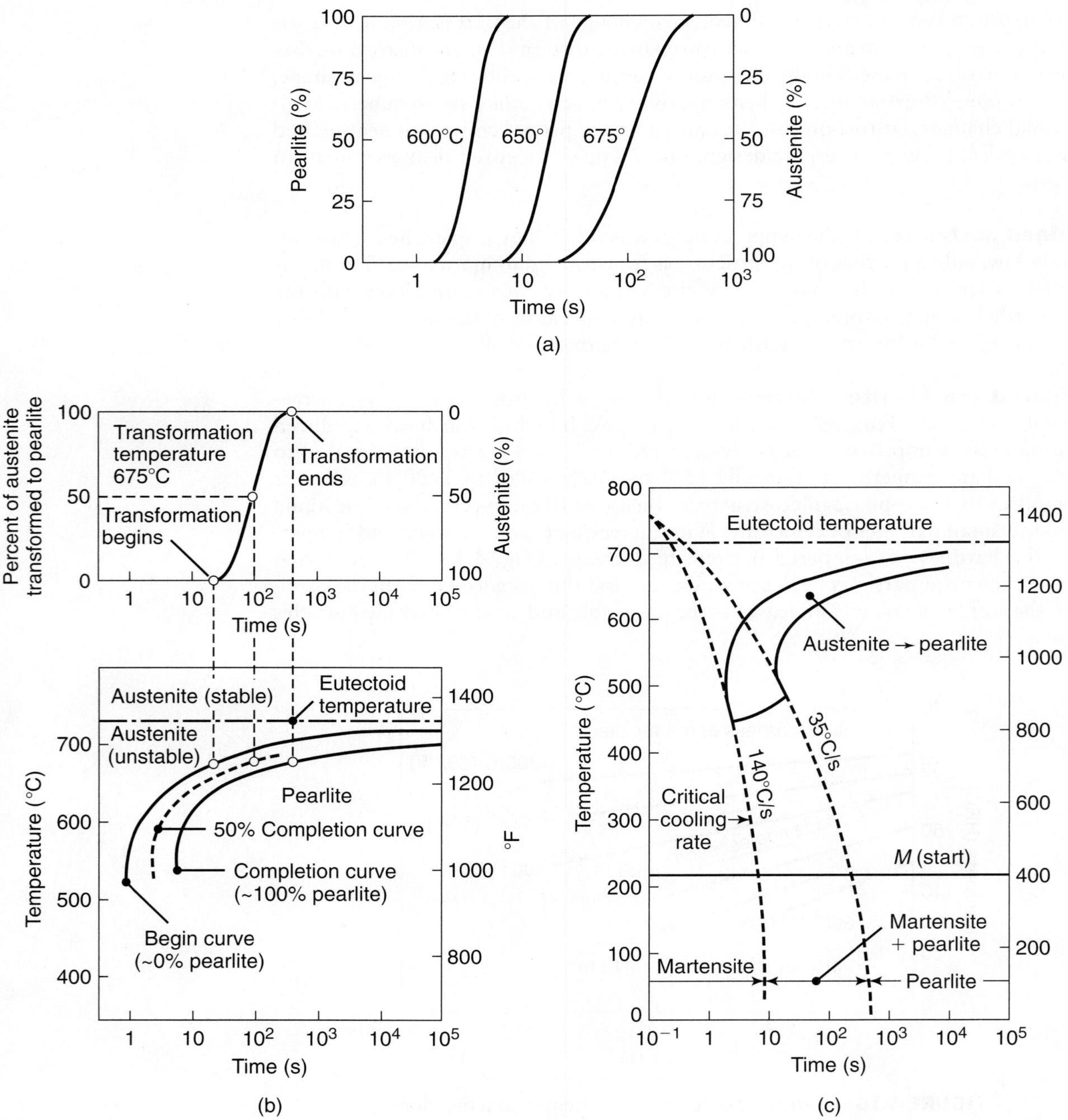

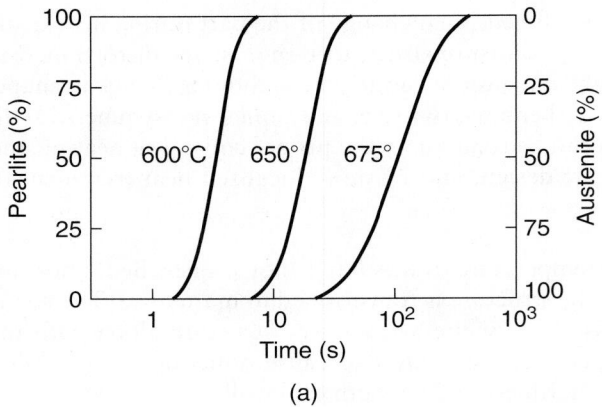

FIGURE 4.17   (a) Austenite-to-pearlite transformation of iron-carbon alloy as a function of time and temperature. (b) Isothermal transformation diagram obtained from (a) for a transformation temperature of 675°C (1247°F). (c) Microstructures obtained for a eutectoid iron-carbon alloy as a function of cooling rate.

transformed into pearlite as a function of temperature and time. The higher the temperature or the longer the time, the higher is the percentage of austenite transformed to pearlite. Note that, for each temperature, there is a minimum time for the transformation to begin. This time period defines the critical cooling rate; with longer times, austenite begins to transform into pearlite. This transformation can be traced in Figs. 4.17b and c.

The TTT diagrams shown allow metallurgists to design heat-treatment schedules to obtain desirable microstructures. For example, consider the TTT curves shown in Fig. 4.17c. The steel can be raised to a very high temperature (above the eutectic temperature) to start with a state of austenite. If the material is cooled very quickly, it can follow the 140°C/s cooling-rate trajectory shown that results in complete martensite. On the other hand, it can be cooled more slowly in a molten salt bath to develop pearlite or bainite containing steels. If tempered martensite is desired, the heat treat and quench stages will be followed by a tempering process.

The differences in the hardness and the toughness of the various structures obtained are shown in Fig. 4.18. Fine pearlite is harder and less ductile than coarse pearlite. The effects of various percentages of carbon, cementite, and pearlite on other mechanical properties of steels are shown in Fig. 4.19.

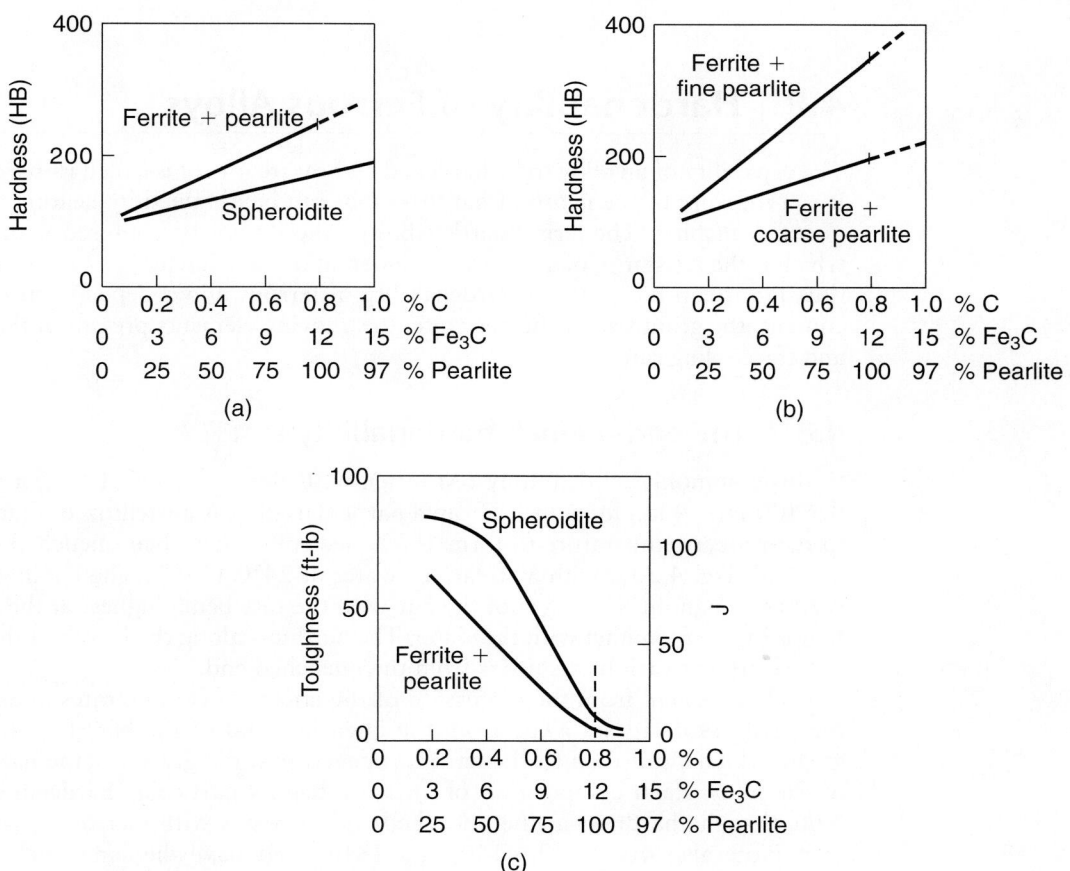

FIGURE 4.18   (a) and (b) Hardness and (c) toughness for annealed plain-carbon steels as a function of carbide shape. Carbides in the pearlite are lamellar. Fine pearlite is obtained by increasing the cooling rate. The spheroidite structure has sphere-like carbide particles.

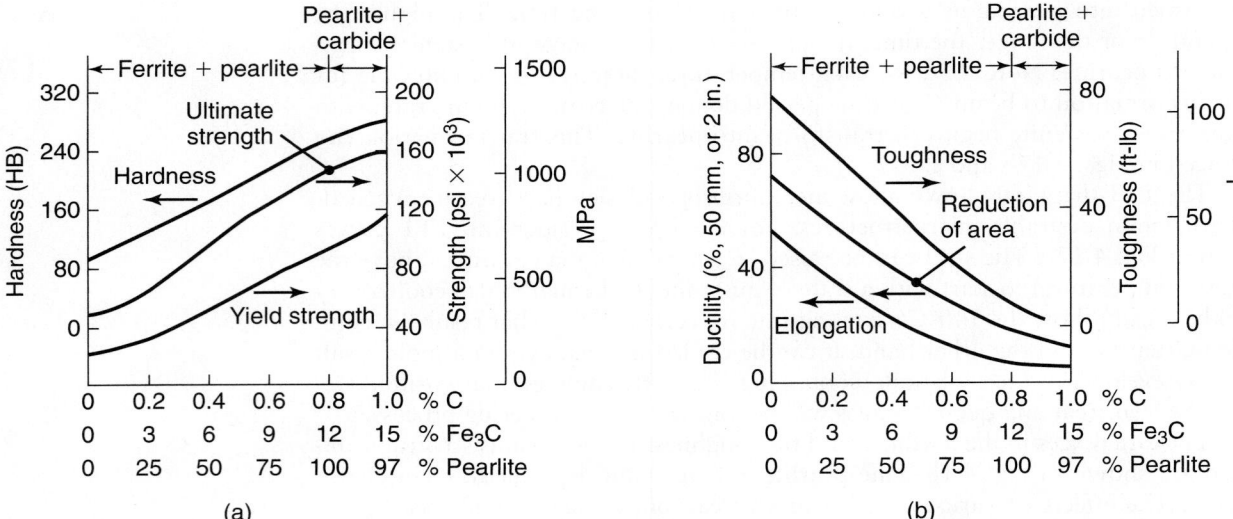

FIGURE 4.19   Mechanical properties of annealed steels as a function of composition and microstructure. Note in (a) the increase in hardness and strength and in (b) the decrease in ductility and toughness with increasing amounts of pearlite and iron carbide.

4.8 | Hardenability of Ferrous Alloys

The capability of an alloy to be hardened by heat treatment is called its **hardenability**. It is a measure of the *depth* of hardness that can be obtained by heating and subsequent quenching. The term "hardenability" should not be confused with hardness, which is the resistance of a material to indentation or scratching. From the discussion thus far, it can be seen that hardenability of ferrous alloys depends on the carbon content, the grain size of the austenite, the alloying elements present in the material, and the cooling rate.

4.8.1 The end-quench hardenability test

In this commonly used **Jominy test** (after W. E. Jominy, 1893–1976), a round test bar 100-mm (4 in.) long, made from a particular alloy, is **austenitized** (that is, heated to the proper temperature to form 100% austenite). It is then quenched directly at one end (Fig. 4.20a) with a stream of water at 24°C (75°F). The cooling rate thus varies throughout the length of the bar with the rate being highest at the lower end that is in direct contact with the water. The hardness along the length of the bar then is measured at various distances from the quenched end.

As expected from the discussion of the effects of cooling rates in Section 4.7, the hardness decreases away from the quenched end of the bar (Fig. 4.20b). The greater the depth to which the hardness penetrates, the greater is the hardenability of the alloy. Each composition of an alloy has its particular **hardenability band**. Note that the hardness at the quenched end increases with increasing carbon content. Note also that 1040, 4140, and 4340 steels have the same carbon content (0.40%), and thus, they have the same hardness (57 HRC) at the quenched end.

Because small variations in composition and in grain size can affect the shape of hardenability curves, each lot of an alloy should be tested individually. The data may be plotted as a band, rather than as a single curve. Hardenability curves are

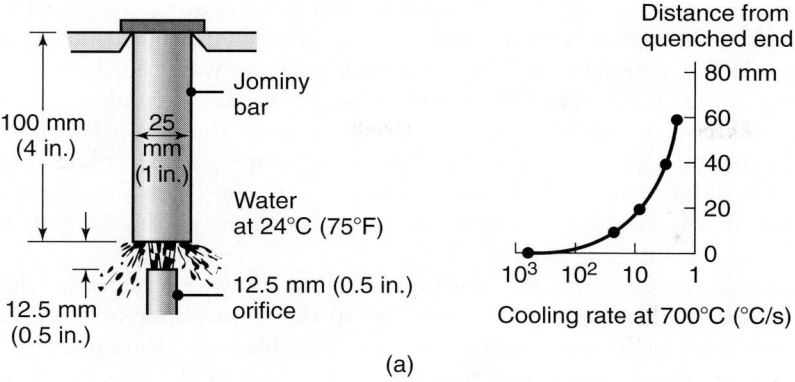

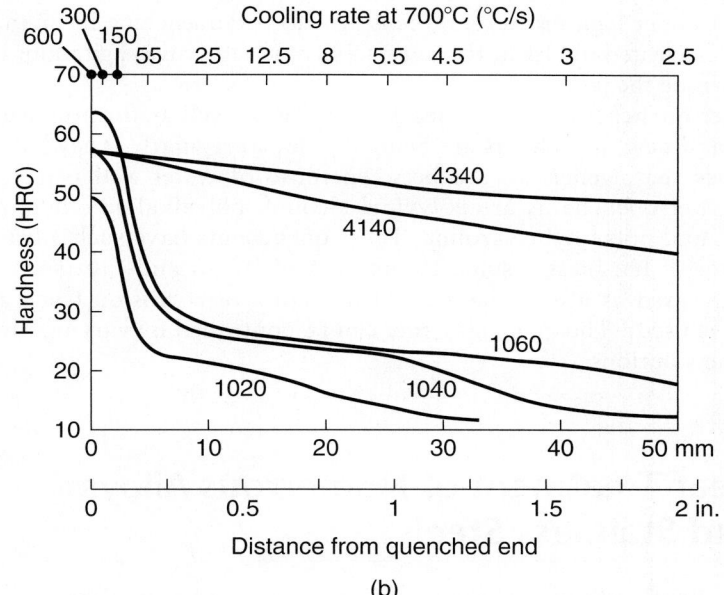

FIGURE 4.20 (a) End-quench test and cooling rate. (b) Hardenability curves for five different steels, as obtained from the end-quench test. Small variations in composition can change the shape of these curves. Each curve is actually a band, and its exact determination is important in the heat treatment of metals for better control of properties.

essential in predicting the hardness of heat-treated parts (such as gears, cams, and various other components) as a function of their composition.

4.8.2 Quenching media

The fluid used for quenching the heated alloy also has an effect on hardenability. Quenching commonly may be carried out in water, brine (saltwater), oils, molten salts, or air. Caustic solutions, polymer solutions, and gases are also used. Because of the differences in the thermal conductivities, specific heats, and heats of vaporization of these media, the rate of cooling of the alloy (**severity of quench**) is also different. In relative terms and in decreasing order, the cooling capacities of several quenching media are as follows: agitated brine, 5; still water, 1; still oil, 0.3; cold gas, 0.1; still air, 0.02.

Agitation is also a significant factor in the rate of cooling. The more vigorous the agitation, the higher is the rate of cooling. In tool steels, the quenching medium is specified by a letter (see Table 5.5), such as W for water hardening, O for oil hardening, and A for air hardening. The cooling rate also depends on the surface-area-to-thickness or surface-area-to-volume ratio of the part. The higher this ratio, the higher is the cooling rate. Thus, for example, a thick plate cools more slowly than a thin plate with the same surface area. These considerations also are significant in the cooling of metals and of plastics in casting and in molding processes.

Water is a common medium for rapid cooling. However, the heated metal may form a **vapor blanket** along its surfaces due to the water-vapor bubbles that form when water boils at the metal–water interface. This blanket creates a barrier to heat conduction because of the lower thermal conductivity of the vapor. Agitating the fluid or the part helps to reduce or eliminate the blanket. Also, water may be sprayed onto the part under high pressure. Brine is an effective quenching medium, because salt helps to nucleate bubbles at the interfaces, which improves agitation. However, brine can corrode the part.

Polymer quenchants can be used for ferrous as well as for nonferrous alloy quenching, and new quenchants are being developed regularly. They have cooling characteristics that, generally, are between those of water and petroleum oils. Typical polymer quenchants are polyvinyl alcohol, polyalkaline oxide, polyvinyl pyrrolidone, and polyethyl oxazoline. These quenchants have such advantages as better control of hardness results, elimination of fumes and fire (such as occur when oils are used as a quenchant), and reduction of corrosion (such as occurs when water is used). The quenching rate can be controlled by varying the concentration of the solutions.

4.9   Heat Treatment of Nonferrous Alloys and Stainless Steels

Nonferrous alloys and some stainless steels generally cannot be heat treated by the techniques used for ferrous alloys. The reason is that nonferrous alloys do not undergo phase transformations like those in steels; the hardening and strengthening mechanisms for these alloys are fundamentally different. Heat-treatable aluminum alloys, copper alloys, martensitic stainless steels, and some other stainless steels are hardened and strengthened by a process called **precipitation hardening**. This heat treatment is a technique in which small particles (of a different phase and called **precipitates**) are dispersed uniformly in the matrix of the original phase (Fig. 4.3a). In this process, precipitate forms because the solid solubility of one element (one component of the alloy) in the other is exceeded.

Three stages are involved in precipitation hardening; they best can be described by reference to the phase diagram for the aluminum-copper system (Fig. 4.21a). For an alloy with the composition 95.5% Al-4.5% Cu, a single-phase (kappa phase) substitutional solid solution of copper (solute) in aluminum (solvent) exists between 500° and 570°C (930° and 1060°F). This *kappa phase* is aluminum-rich, has an fcc structure, and is ductile. Below the lower temperature (that is, below the lower solubility curve) there are two phases: *kappa* and *theta* (a hard intermetallic compound of $CuAl_2$). This alloy can be heat treated, and its properties are modified by two different methods: *solution treatment* and *precipitation*.

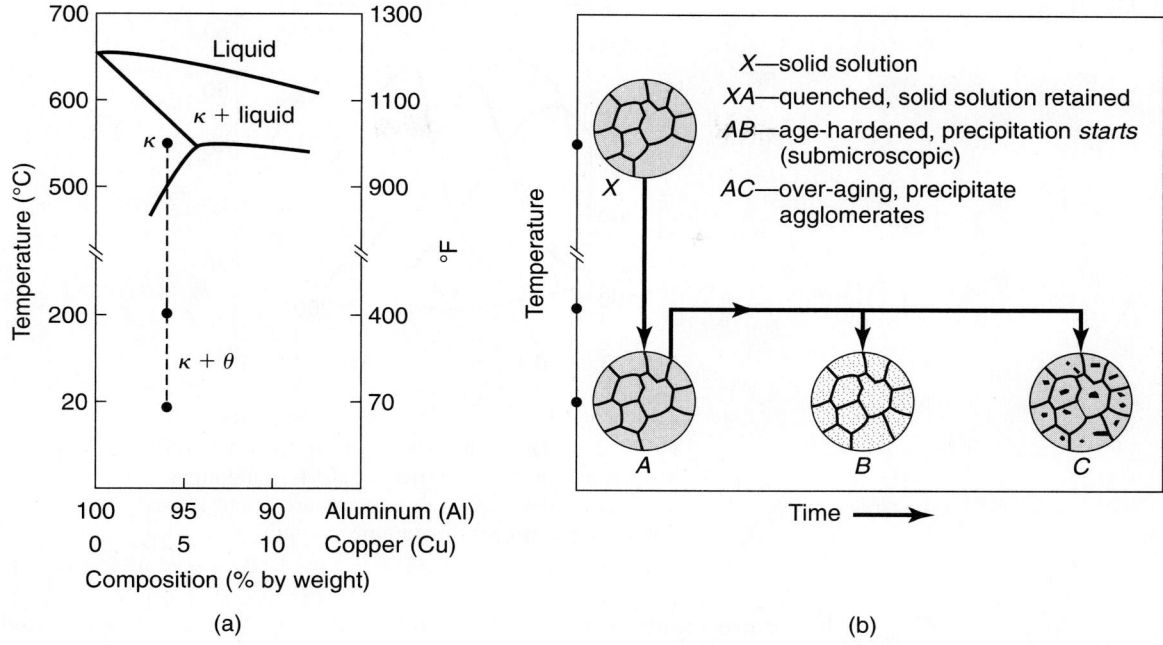

FIGURE 4.21 (a) Phase diagram for the aluminum-copper alloy system. (b) Various microstructures obtained during the age-hardening process.

4.9.1 Solution treatment

In **solution treatment**, the alloy is heated to within the solid-solution kappa phase, say 540°C (1000°F), and then cooled rapidly—for instance, by quenching it in water. The structure obtained soon after quenching (*A* in Fig. 4.21b) consists only of the single-phase kappa. This alloy has moderate strength and considerable ductility.

4.9.2 Precipitation hardening

The structure obtained in *A* in Fig. 4.21b can be made stronger by **precipitation hardening**. The alloy is reheated to an intermediate temperature and held there for a period of time, during which precipitation takes place. The copper atoms diffuse to nucleation sites and combine with aluminum atoms. This process produces the theta phase, which forms as submicroscopic precipitates (shown in *B* by the small dots within the grains of the kappa phase). This structure is stronger than that in *A*, although it is less ductile. The increase in strength is due to increased resistance to dislocation movement in the region of the precipitates.

Aging. Because the precipitation process is one of time and temperature, it is also called *aging*, and the property improvement is known as **age hardening**. If carried out above room temperature, the process is called **artificial aging**. However, several aluminum alloys harden and become stronger over a period of time at room temperature; this process is called **natural aging**. Such alloys are first quenched and then, if it is so desired, they are shaped by plastic deformation at room temperature. Finally, they are allowed to gain strength and hardness by aging naturally. Natural aging can be slowed down by refrigerating the quenched alloy (**cryogenic treatment**).

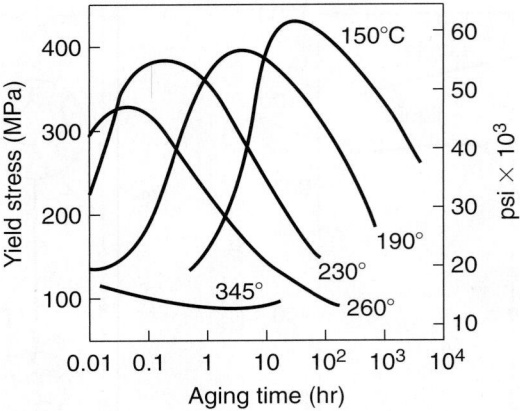

FIGURE 4.22   The effect of aging time and temperature on the yield stress of 2014-T4 aluminum alloy. Note that, for each temperature, there is an optimal aging time for maximum strength.

In the precipitation process, if the reheated alloy is held at the elevated temperature for an extended period of time, the precipitates begin to coalesce and grow. They become larger but fewer, as is shown by the larger dots in C in Fig. 4.21b. This process is called **overaging**, and the resulting alloy is softer and weaker.

There is an optimal time–temperature relationship in the aging process that must be observed in order to obtain desired properties (Fig. 4.22). Obviously, an aged alloy can be used only up to a certain maximum temperature in service; otherwise, it will overage and so lose its strength and hardness. Although weaker, an overaged part has better dimensional stability.

Maraging.   This is a precipitation-hardening treatment for a special group of high-strength iron-based alloys. The word *maraging* is derived from *martensite age hardening*. In this process, one or more intermetallic compounds are precipitated in a matrix of low-carbon martensite. A typical maraging steel may contain 18% Ni in addition to other elements, and aging is completed at 480°C (900°F). Hardening by maraging does not depend on the cooling rate; consequently, uniform and full hardness can be obtained throughout large parts with minimal distortion. Typical uses of maraging steels are in dies and tooling for casting, molding, forging, and extrusion (Parts II and III).

4.10 | Case Hardening

The heat-treatment processes described thus far involve microstructural alterations and property changes in the *bulk* of the material or component by means of *through hardening*. It is not desirable to through harden parts because a hard part lacks the necessary toughness for these applications; a small surface crack could propagate rapidly through such a part and cause total failure. In many cases, however, alteration of only the *surface properties* of a part (hence, the term *case hardening*) is desirable. This method is useful particularly for improving resistance to surface indentation, fatigue, and wear. Typical applications for case hardening are gear teeth, cams, shafts, bearings, fasteners, pins, automotive clutch plates, tools, and dies.

Several surface-hardening processes are available (Table 4.1):

a. **Carburizing** (gas, liquid, and pack carburizing)
b. **Carbonitriding**
c. **Cyaniding**
d. **Nitriding**
e. **Boronizing**
f. **Flame hardening**
g. **Induction hardening**
h. **Laser hardening**

Basically, these are operations in which the component is heated in an atmosphere containing elements (such as carbon, nitrogen, or boron) that alter the composition, microstructure, and properties of surfaces. For steels with sufficiently high-carbon content, surface hardening takes place without using any of these additional elements. Only the heat-treatment processes described in Section 4.7 are needed to alter the microstructures, usually by either flame hardening or induction hardening, as outlined in Table 4.1.

Laser beams and **electron beams** (Sections 27.6 and 27.7) also are used effectively to harden both small and large surfaces, such as gears, valves, punches, and locomotive cylinders. These methods also are used for thorough hardening of relatively small parts. The main advantages of laser surface hardening are close control of power input, low distortion, and the ability to reach areas that would be inaccessible by other means. Capital costs can be high, however, and the depth of the case-hardened layer usually is less than 2.5 mm (0.1 in.).

Because case hardening is a localized heat treatment, case-hardened parts have a hardness gradient. Typically, the hardness is a maximum at the surface and decreases below the surface, with a rate of decrease that depends on the composition of the metal and the process variables. Surface-hardening techniques also can be used for *tempering* (Section 4.11) to modify the properties of surfaces that have been subjected to heat treatment. Various other processes and techniques for surface hardening, such as shot peening and surface rolling, improve wear resistance and various other characteristics.

Decarburization is the phenomenon in which alloys containing carbon lose carbon from their surfaces as a result of heat treatment or of hot working in a medium (usually oxygen) that reacts with the carbon. Decarburization is undesirable because it affects the hardenability of the surfaces of the part by lowering its carbon content. It also adversely affects the hardness, strength, and fatigue life of steels by significantly lowering their endurance limit. Decarburization is best avoided by processing the parts in an inert atmosphere or a vacuum or by using neutral salt baths during heat treatment.

4.11 | Annealing

Annealing is a general term used to describe the restoration of a cold-worked or heat-treated alloy to its original properties. For instance, annealing can be used to increase ductility (hence formability) and reduce hardness and strength or to modify its microstructure. The annealing process also is used to relieve residual stresses in a manufactured part, as well as to improve machinability and dimensional stability. The term annealing also applies to the thermal treatment of glasses and similar products, castings, and weldments.

TABLE 4.1

Outline of Heat Treatment Processes for Surface Hardening

Process	Metals hardened	Element added to surface	Procedure	General characteristics	Typical applications
Carburizing	Low-carbon steel (0.2% C), alloy steels (0.08–0.2% C)	C	Heat steel at 870–950°C (1600–1750°F) in an atmosphere of carbonaceous gases (gas carburizing) or carbon-containing solids (pack carburizing). Then quench.	A hard, high-carbon surface is produced. Hardness 55 to 65 HRC. Case depth < 0.5–1.5 mm (< 0.020 to 0.060 in.). Some distortion of part during heat treatment.	Gears, cams, shafts, bearings, piston pins, sprockets, clutch plates
Carbonitriding	Low-carbon steel	C and N	Heat steel at 700–800°C (1300–1600°F) in an atmosphere of carbonaceous gas and ammonia. Then quench in oil.	Surface hardness 55 to 62 HRC. Case depth 0.07 to 0.5 mm (0.003 to 0.020 in.). Less distortion than in carburizing.	Bolts, nuts, gears
Cyaniding	Low-carbon steel (0.2% C), alloy steels (0.08–0.2% C)	C and N	Heat steel at 760–845°C (1400–1550°F) in a molten bath of solutions of cyanide (e.g., 30% sodium cyanide) and other salts.	Surface hardness up to 65 HRC. Case depth 0.025 to 0.25 mm (0.001 to 0.010 in.). Some distortion.	Bolts, nuts, screws, small gears
Nitriding	Steels (1% Al, 1.5% Cr, 0.3% Mo), alloy steels (Cr, Mo), stainless steels, high-speed tool steels	N	Heat steel at 500–600°C (925–1100°F) in an atmosphere of ammonia gas or mixtures of molten cyanide salts. No further treatment.	Surface hardness up to 1100 HV. Case depth 0.1 to 0.6 mm (0.005 to 0.030 in.) and 0.02 to 0.07 mm (0.001 to 0.003 in.) for high speed steel.	Gears, shafts, sprockets, valves, cutters, boring bars, fuel-injection pump parts
Boronizing	Steels	B	Part is heated using boron-containing gas or solid in contact with part.	Extremely hard and wear resistant surface. Case depth 0.025–0.075 mm (0.001–0.003 in.).	Tool and die steels
Flame hardening	Medium-carbon steels, cast irons	None	Surface is heated with an oxyacetylene torch, then quenched with water spray or other quenching methods.	Surface hardness 50 to 60 HRC. Case depth 0.7 to 6 mm (0.030 to 0.25 in.). Little distortion.	Gear and sprocket teeth, axles, crankshafts, piston rods, lathe beds and centers
Induction hardening	Same as above	None	Metal part is placed in copper induction coils and is heated by high frequency current, then quenched.	Same as above	Same as above

The annealing process consists of the following steps:

1. Heating the workpiece to a specific range of temperature in a furnace.
2. Holding it at that temperature for a period of time (*soaking*).
3. Cooling in air or in a furnace.

The annealing process may be carried out in an inert or a controlled atmosphere, or it may be performed at lower temperatures to prevent or minimize surface oxidation.

An *annealing temperature* may be higher than the material's recrystallization temperature, depending on the degree of cold work. For example, the recrystallization temperature for copper ranges between 200° and 300°C (400° and 600°F), whereas the annealing temperature needed to fully recover the original properties ranges from 260° to 650°C (500° to 1200°F), depending on the degree of prior cold work (see also Section 1.6). **Full annealing** is a term applied to the annealing of ferrous alloys. The steel is heated to above A_1 or A_3 (Fig. 4.23), and the cooling takes place slowly (typically at 10°C (20°F) per hour) in a furnace after it is turned off. The structure obtained through full annealing is coarse pearlite, which is soft and ductile and has small, uniform grains.

To avoid excessive softness from the annealing of steels, the cooling cycle may be done completely in still air. This process is called **normalizing** to indicate that the part is heated to a temperature above A_3 or A_{cm} in order to transform the structure to austenite. Normalizing results in somewhat higher strength and hardness and in lower ductility than does full annealing (Fig. 4.24). The structure obtained is fine pearlite, with small, uniform grains. Normalizing generally is carried out to refine the grain structure, obtain uniform structure (homogenization), decrease residual stresses, and improve machinability. The structure of spheroidizing and the procedure for obtaining it were described in Section 4.7 and shown in Figs. 4.14 and 4.24. *Spheroidizing annealing* improves the cold workability and the machinability of steels.

Process annealing.   During *process annealing* (also called *intermediate annealing*, *subcritical annealing*, or *in-process annealing*), the workpiece is annealed to restore its ductility, part or all of which may have been exhausted by work hardening

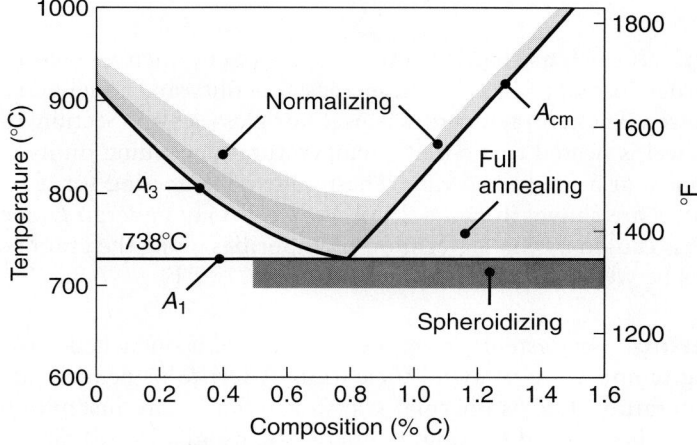

FIGURE 4.23   Heat-treating temperature ranges for plain-carbon steels, as indicated on the iron-iron carbide phase diagram.

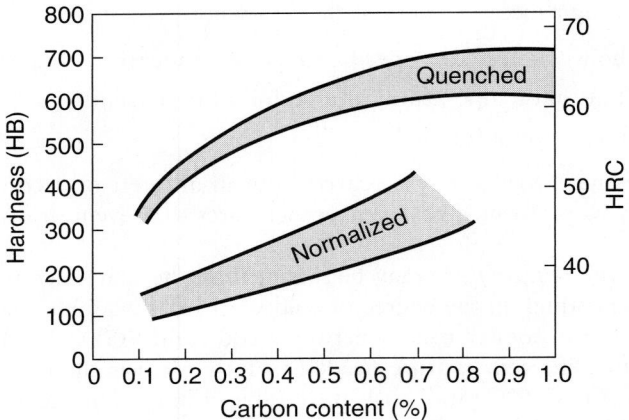

FIGURE 4.24   Hardness of steels in the quenched and normalized conditions as a function of carbon content.

during cold working. Afterwards, the part can be worked further into the final desired shape. If the temperature is high and/or the time of annealing is long, grain growth may result, with adverse effects on the formability of the annealed parts.

Stress-relief annealing.   To reduce or eliminate residual stresses, a workpiece generally is subjected to *stress-relief annealing* (or simply **stress relieving**). The temperature and time required for this process depend on the material and on the magnitude of the residual stresses present. The residual stresses may have been induced during forming, machining, or other shaping processes, or they may have been caused by volume changes during phase transformations.

For steels, the part is not heated to as high as A_1 in Fig. 4.23 in order to avoid phase transformations. Slow cooling (such as occurs in still air) generally is employed. Stress relieving promotes dimensional stability in situations where subsequent relaxing of residual stresses may cause distortion of the part when it is in service over a period of time. It also reduces the tendency toward stress–corrosion cracking (Sections 2.10 and 3.8).

Tempering.   If steels are hardened by heat treatment, then *tempering* or **drawing** is used in order to reduce brittleness, increase ductility and toughness, and reduce residual stresses. The term tempering is used for glasses also (Section 18.4). In tempering, the steel is heated to a specific temperature (depending on its composition) and then cooled at a prescribed rate. The results of tempering for an oil-quenched AISI 4340 steel are shown in Fig. 4.25. Alloy steels may undergo **temper embrittlement**, which is caused by the segregation of impurities along the grain boundaries at temperatures between 480° and 590°C (900° and 1100°F).

Austempering.   In *austempering*, the heated steel is quenched rapidly from the austenitizing temperature to avoid formation of ferrite or pearlite. It is held at a certain temperature until isothermal transformation from austenite to bainite is complete. It is then cooled to room temperature, usually in still air and at a moderate rate in order to avoid thermal gradients within the part. The quenching

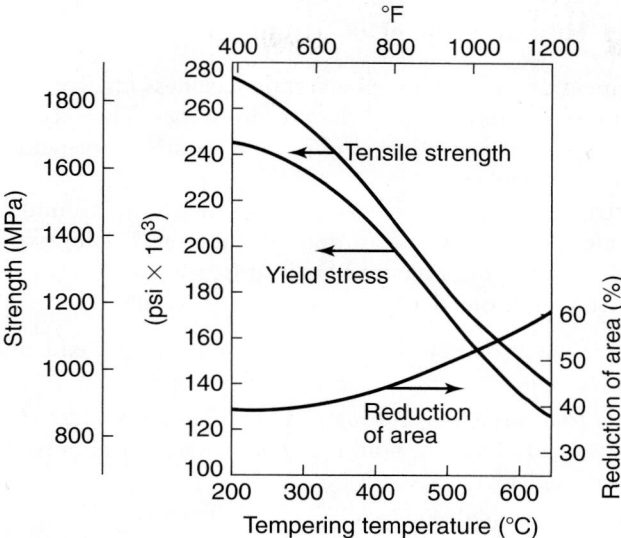

FIGURE 4.25 Mechanical properties of oil-quenched 4340 steel as a function of tempering temperature.

medium most commonly used is molten salt, at temperatures ranging from 160° to 750°C (320° to 1380°F).

Austempering is often substituted for conventional quenching and tempering, either to reduce the tendency toward cracking and distortion during quenching or to improve ductility and toughness while maintaining hardness. Because of the shorter cycle time involved, this process also is economical for many applications. In **modified austempering**, a mixed structure of pearlite and bainite is obtained. The best example of this practice is **patenting**, which provides high ductility and moderately high strength, such as in patented wire (Section 15.8).

Martempering (Marquenching). In *martempering*, the steel or cast iron is first quenched from the austenitizing temperature in a hot-fluid medium, such as hot oil or molten salt. Next, it is held at that temperature until the temperature is uniform throughout the part; then it is cooled at a moderate rate, such as in air, in order to avoid excessive temperature gradients within the part. Usually, the part is then tempered, because the structure obtained is otherwise primarily untempered martensite and, thus, is not suitable for most applications. Martempered steels have less tendency to crack, distort, or develop residual stresses during heat treatment. In **modified martempering**, the quenching temperature is lower, and thus, the cooling rate is higher. The process is suitable for steels with lower hardenability.

Ausforming. In *ausforming* (also called **thermomechanical processing**) the steel is formed into desired shapes within controlled ranges of temperature and time to avoid formation of nonmartensitic transformation products. The part is then cooled at various rates to obtain the desired microstructures. Ausformed parts have superior mechanical properties.

EXAMPLE 4.1 **Heat treatment of an extrusion die**

The heat treatment of parts to obtain a certain hardness involves several considerations regarding the material and its desired properties. The specific heat-treating process has to be planned carefully, and it often requires considerable experience, as is noted in this example.

A hot-extrusion die (see Section 15.3) 200-mm (8-in.) around, 75-mm (3-in.) long, and having a round hole of 75 mm, is made of H21 hot-worked steel (see Tables 5.6 and 5.7). A typical method for heat treating such a die suitable for hot extrusion of metals is outlined as follows:

1. Preheat the die at 815° to 845°C, either in a slightly oxidizing atmosphere or in neutral salt.
2. Transfer it to a furnace operating at 1175°C in a 6 to 12% reducing atmosphere or neutral salt bath; hold it in the furnace for about 20 minutes after the die has reached 1175°C.
3. Cool it in still air to about 65°C.
4. Temper it at 565°C for four hours.
5. Cool it to near room temperature.
6. Retemper it at 650°C for four hours.
7. Cool the die in air.

Source: Courtesy of ASM International.

4.12 | Heat-Treating Furnaces and Equipment

Two basic types of furnaces are used for heat treating: **batch furnaces** and **continuous furnaces**. Because they consume great amounts of energy, their insulation and efficiency are important design considerations, as are their initial cost, the personnel needed for their operation and maintenance, and their safe use.

Uniform temperature and accurate control of temperature–time cycles are important. Modern furnaces are equipped with various electronic controls, including computer-controlled systems programmed to run through a complete heat-treating cycle repeatedly and with reproducible accuracy. Heating-system fuels are usually gas, oil, or electricity (for resistance or induction heating). The type of fuel used affects the furnace's atmosphere. Unlike electric heating, gas or oil introduces combustion products into the furnace (a disadvantage). Electrical heating, however, has a slower start-up time, and it is more difficult to adjust and control.

Batch furnaces.   In a batch furnace the parts to be heat treated are loaded into and unloaded from the furnace in individual batches. The furnace basically consists of an insulated chamber, a heating system, and an access door or doors. Batch furnaces are of the following basic types:

- **Box furnace** is a horizontal rectangular chamber with one or two access doors through which parts are loaded. This type of furnace is used commonly and is versatile, simple to construct and to use, and available in several sizes. A variation of this type is the **car-bottom furnace**. The parts to be heat treated, usually long or large, are loaded onto a flatcar which then moves on rails into the furnace.

- **Pit furnace** is a vertical pit below ground level into which the parts are lowered. This type of furnace is particularly suitable for long parts (such as rods, shafts, and tubing), because they can be suspended by one end and (consequently) are less likely to warp during processing than if positioned horizontally within a box furnace.
- **Bell furnace** is a round or rectangular box furnace without a bottom and is lowered over stacked parts that are to be heat treated. This type of furnace is particularly suitable for coils of wire, rods, and sheet metal.
- **Elevator furnace** is used by loading the parts to be heat treated onto a car platform, rolled into position, and then raised into the furnace. This type of furnace saves space in the plant and can be especially suitable for metal alloys that have to be quenched rapidly, because a quenching tank can be placed directly under the furnace.

Continuous furnaces. In this type of furnace, the parts to be heat treated move continuously through the furnace on conveyors of various designs that use trays, belts, chains, and other mechanisms. Continuous furnaces are suitable for high-production runs, and can be designed and programmed so that complete heat-treating cycles can be performed under tight control.

Salt-bath furnaces. Because of their high-heating rates and better control of uniformity of temperature, *salt baths* commonly are used in various heat-treating operations, particularly for nonferrous strip and wire. Heating rates are high because of the higher thermal conductivity of liquid salts compared to that of air or gases. Depending on the electrical conductivity of the salt, heating may be done externally (for nonconducting salts) or by immersed or submerged electrodes using a low-voltage alternating current. Direct current cannot be used because it subjects the salt to electrolysis. Salt-baths are available for a wide range of temperatures. Lead also can be used as the heating medium.

Fluidized beds. Dry, fine, and loose solid particles (usually aluminum oxide) are heated and suspended in a chamber by an upward flow of hot gas at various speeds. The parts to be heat treated are then placed within the floating particles—hence, the term *fluidized bed*. Because of the constant agitation, the system is efficient, the temperature distribution is uniform, and the heat-transfer rate is high. These furnaces are used for various batch-type applications.

Induction heating. In this method, the part is heated rapidly by the electromagnetic field generated by an *induction coil* carrying an alternating current, which induces eddy currents in the part. The coil, which can be shaped to fit the contour of the part to be heat treated (Fig. 4.26), is made of copper or of a copper-base alloy. The coil, which is usually water cooled, may be designed to quench the part as well after heating it. Induction heating is desirable for localized heat treating, such as that required for gear teeth, cams, and similar parts.

Furnace atmospheres. The atmospheres in furnaces can be controlled so as to avoid (or cause) oxidation, tarnishing, and decarburization of ferrous alloys heated to elevated temperatures. Oxygen causes oxidation (corrosion, rusting, and scaling).

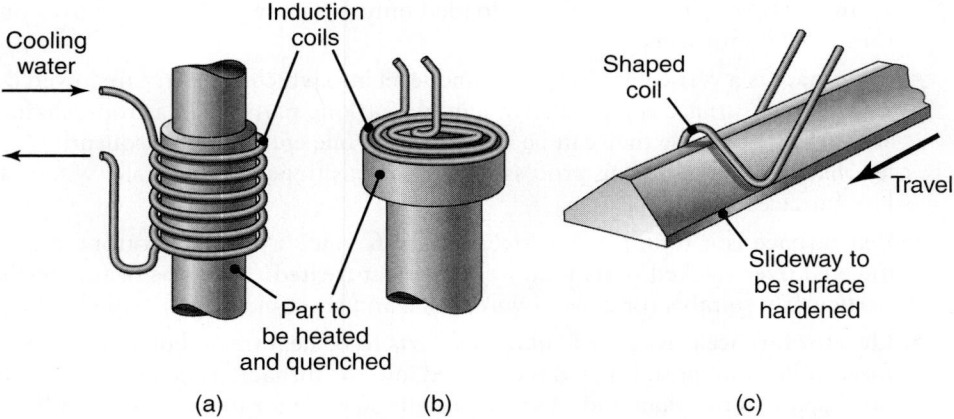

FIGURE 4.26   Types of coils used in the induction heating of various surfaces of parts.

Carbon dioxide has various effects; it may be neutral or decarburizing, depending on its concentration in the furnace atmosphere. Water vapor in the furnace causes the oxidation of steels, resulting in a blue color. Nitrogen is a common neutral atmosphere, and a vacuum provides a completely neutral atmosphere. The term **bluing** is used to describe formation of a thin blue film of oxide on finished parts to improve their appearance and their resistance to oxidation.

4.13 | Design Considerations for Heat Treating

In addition to the metallurgical factors, successful heat treating involves design considerations for avoiding problems such as cracking, distortion, and nonuniformity of the properties throughout the heat-treated part. The rate of cooling during quenching may not be uniform, particularly in complex shapes having varying cross-sections and thicknesses; this nonuniformity may produce severe temperature gradients in the part. Nonuniformity can lead to variations in contraction, resulting in thermal stresses that may cause cracking of the part; furthermore, nonuniform cooling causes residual stresses in the part, which can lead to stress–corrosion cracking. The quenching method selected, the care taken in quenching, and the selection of a proper quenching medium and temperature are, consequently, important considerations.

As a general guideline for part design for heat treating, sharp internal or external corners should be avoided; otherwise, stress concentrations at these corners may raise the level of stresses high enough to cause cracking. The part should have its thicknesses as nearly uniform as possible; also, the transition between regions of different thicknesses should be made smooth. Parts with holes, grooves, keyways, splines, and asymmetrical shapes likewise may be difficult to heat treat, because they may crack during quenching. Large surfaces with thin cross-sections are likely to warp. Hot forgings and hot steel-mill products may have a **decarburized skin** (loss of carbon, Section 4.10); as a result, they may not respond successfully to heat treatment.

SUMMARY

- Commercially pure metals generally do not have sufficient strength for most engineering applications; they must be alloyed with various elements which alter their structures and properties. Important concepts in alloying are the solubility of alloying elements in a host metal and the phases present at various ranges of temperature and composition.

- Alloys basically have two forms: solid solutions and intermetallic compounds. Solid solutions may be substitutional or interstitial. There are certain conditions pertaining to the crystal structure and atomic radii that have to be met in order to develop these structures.

- Phase diagrams show the relationships among the temperature, composition, and the phases present in a particular alloy system. As temperature is decreased at various rates, various transformations take place in microstructures that have widely varying characteristics and properties. Among the binary systems, the most important is the iron-carbon system, which includes a wide range of steels and cast irons. Important components in this system are ferrite, austenite, and cementite. The basic types of cast irons are: gray iron, ductile (nodular) iron, white iron, malleable iron, and compacted-graphite iron.

- The mechanisms for hardening and strengthening metal alloys involve heating the alloy and, subsequently, quenching it at varying cooling rates. As a result, important phase transformations take place, producing structures such as pearlite (fine or coarse), spheroidite, bainite, and martensite. Heat treating of nonferrous alloys and stainless steels involves solution treatment and precipitation hardening.

- The control of the furnace atmosphere, the quenchants used, the characteristics of the equipment, and the shape of the parts to be heat treated are important considerations. Hardenability is the capability of an alloy to be hardened by heat treatment. The end-quench hardenability test (Jominy) is a method commonly used to determine hardenability bands for alloys.

- Case hardening is an important process for improving the wear and fatigue resistance of parts. Several methods are available; among them are carburizing, nitriding, induction hardening, and laser hardening.

- Annealing includes several alternative processes (normalizing, process annealing, stress relieving, tempering, austempering, and martempering), each having the purpose of enhancing the ductility and toughness of heat-treated parts.

KEY TERMS

Age hardening	Cast iron	Eutectoid reaction
Aging	Cementite	Ferrite
Alloy	Curie temperature	Hardenability
Annealing	Decarburization	Heat treatment
Austenite	Distortion	Intermetallic compounds
Bainite	Equilibrium diagram	Jominy test
Case hardening	Eutectic point	Martensite

Normalizing

Overaging

Pearlite

Phase diagram

Phase transformations

Precipitation hardening

Pure metals

Solute

Solution treatment

Solvent

Spheroidites

Stress relieving

Tempering

BIBLIOGRAPHY

ASM Handbook, Vol. 3: *Alloy Phase Diagrams*, ASM International, 1992.

ASM Handbook, Vol. 4: *Heat Treating*, ASM International, 1991.

ASM Handbook, Vol. 9: *Metallography and Microstructures*, ASM International, 2004.

Brooks, C.R., *Principles of the Heat Treatment of Plain Carbon and Low Alloy Steel*, ASM International, 1996.

Bryson, B., *Heat Treatment, Selection, and Application of Tool Steels*, Hanser Gardner, 1997.

Davies, J.R. (ed.), *Surface Hardening of Steels: Understanding the Basics*, ASM International, 2002.

Haimbaugh, R.E., *Practical Induction Heat Treating*. ASM International, 2001.

Heat Treater's Guide: Practices and Procedures for Irons and Steels, ASM International, 1995.

Heat Treater's Guide: Practices and Procedures for Nonferrous Alloys, ASM International, 1996.

Krauss, G., *Steels: Heat Treatment and Processing Principles*, ASM International, 1990.

Totten, G.E., and Howes, M.A.H. (eds.), *Steel Heat Treatment Handbook*, Marcel Dekker, 1997.

Totten, G.E., Bates, C.E., and Clinton, N.A., *Handbook of Quenchants and Quenching Technology*, ASM International, 1992.

REVIEW QUESTIONS

4.1 Describe the difference between a solute and a solvent.

4.2 What is a solid solution?

4.3 What are the conditions for obtaining (a) substitutional and (b) interstitial solid solutions?

4.4 What is the difference between a single-phase and a two-phase system?

4.5 Explain what is meant by "second-phase particle."

4.6 Describe the features of a phase diagram.

4.7 What do the terms "equilibrium" and "constitutional," as applied to phase diagrams, indicate?

4.8 What is the difference between "eutectic" and "eutectoid?"

4.9 What is tempering? Why is it done?

4.10 Explain what is meant by "severity of quenching."

4.11 What are precipitates? Why are they significant in precipitation hardening?

4.12 What is the difference between natural and artificial aging?

4.13 Describe the characteristics of ferrite, austenite, and cementite.

4.14 What is the purpose of annealing?

QUALITATIVE PROBLEMS

4.15 You may have seen some technical literature on products stating that certain parts in those products are "heat treated." Describe briefly your understanding of this term and why the manufacturer mentions it.

4.16 Describe the engineering significance of the existence of a eutectic point in phase diagrams.

4.17 Explain the difference between hardness and hardenability.

4.18 Refer to Table 4.1 and explain why the items listed under "typical applications" are suitable for surface hardening.

4.19 Why is it generally not desirable to use steels in their as-quenched condition?

4.20 Describe the differences between case hardening and through hardening, insofar as engineering applications are concerned.

4.21 Describe the characteristics of (a) an alloy, (b) pearlite, (c) austenite, (d) martensite, and (e) cementite.

4.22 Explain why carbon, of all the elements, is so effective in imparting strength to iron in the form of steel.

4.23 How does the shape of graphite in cast iron affect its properties?

4.24 In Section 4.8, we listed several fluids in terms of their cooling capacity in quenching. Which physical properties of these fluids influence their cooling capacity?

4.25 Why is it important to know the characteristics of heat-treating furnaces?

4.26 Explain why, in the abscissa of Fig. 4.15, the percentage of pearlite begins to go down after 100% carbon content is reached.

4.27 What is the significance of decarburization? Give some examples.

4.28 Explain your understanding of size distortion and of shape distortion in heat-treated parts, and describe their causes.

4.29 Comment on your observations regarding Fig. 4.20.

4.30 Design a heat-treat cycle for carbon steel, including temperature and exposure times, to produce (a) pearlite-martensite steels and (b) bainite-martensite steels.

QUANTITATIVE PROBLEMS

4.31 Using Fig. 4.5, estimate the following quantities for a 20% Cu–80% Ni alloy:

(a) The liquidus temperature

(b) The solidus temperature

(c) The percentage of nickel in the liquid at 1400°C (2550°F)

(d) The major phase at 1400°C

(e) The ratio of solid-to-liquid at 1400°C

4.32 Extrapolating the curves in Fig. 4.16, estimate the time that it would take for 1080 steel to soften to 53 HRC at (a) 200°C and (b) 300°C.

4.33 A typical steel for tubing is AISI 1040, and one for music wire is AISI 1085. Considering their applications, explain the reason for the difference in their carbon contents.

SYNTHESIS, DESIGN, AND PROJECTS

4.34 We stated in this chapter that, in parts design, sharp corners should be avoided in order to reduce the tendency toward cracking during heat treatment. If it is necessary for the part to have sharp corners in order to function properly and it still requires heat treatment, what method would you recommend for manufacturing this part?

4.35 The heat-treatment processes for surface hardening are given in Table 4.1. Each of these processes involves different equipment, procedures, and cycle times; as a result, each incurs different costs. Examine available literature, and contact various facilities. Make a similar table outlining the costs involved in each process.

4.36 We have seen that, as a result of heat treatment, parts can undergo size distortion and shape distortion to various degrees. By referring to the bibliography at the end of this chapter, make a survey of the technical literature, and report quantitative data regarding the distortions of parts having different shapes.

4.37 Figure 4.20 shows hardness distributions in end-quench tests, as measured *along the length* of the round bar. Make a simple qualitative sketch showing the hardness distribution *across the diameter* of the bar. Would the shape of the curve depend on the bar's carbon content? Explain.

4.38   Throughout this chapter, you have seen the importance and the benefits of heat-treating parts (or certain regions of parts), and you have seen some specific examples. Make a survey of the heat-treatment literature available by referring to the bibliography at the end of this chapter. Compile several examples and illustrations of parts that have been heat treated.

4.39   Referring to Fig. 4.26, suggest of a variety of other part shapes to be heat treated, and design appropriate coils for them. Describe how your designs would change if the parts have varying shapes along their length.

4.40   Inspect various parts in your car or home, and identify those that would have been case hardened. Explain your reasons.

Ferrous Metals and Alloys: Production, General Properties, and Applications

Ferrous metals and alloys are the most widely used structural metals. In this chapter, we describe the following:

- Primary production methods for iron and steel.
- Types of ferrous alloys, their properties, and applications.
- Specific applications for various types of steels and cast irons.
- Characteristics of tool and die steels and their selection for specific applications.

5.1 | Introduction

By virtue of their wide range of mechanical, physical, and chemical properties, **ferrous metals and alloys** are among the most useful of all metals. Ferrous metals and alloys contain iron as their base metal; the general categories are *carbon and alloy steels*, *stainless steels*, *tool and die steels*, *cast irons*, and *cast steels*. Ferrous alloys are produced as

- Sheet steel for automobiles, appliances, and containers
- Plates for boilers, ships, and bridges
- Structural members such as I-beams, bar products, axles, crankshafts, and railroad rails
- Gears, tools, dies, and molds
- Music wire
- Fasteners such as bolts, rivets, and nuts

A typical U.S. passenger car contains about 800 kg (1750 lb) of steel, accounting for about 55 to 60% of its weight. As an example of their widespread use, ferrous materials make up 70 to 85% by weight of structural members and mechanical components. Carbon steels are the least expensive of all structural metals. The use of iron and steel as structural materials has been one of the most important modern technological developments. Primitive ferrous tools first appeared in about

4000 to 3000 B.C. They were made from meteoritic iron, obtained from meteorites that had struck the earth. True ironworking began in Asia Minor in about 1100 B.C. and signaled the advent of the *Iron Age*. Invention of the blast furnace, in about 1340 A.D., made possible the production of large quantities of iron and steel.

5.2 | Production of Iron and Steel

5.2.1 Raw materials

The three basic materials used in iron- and steelmaking are **iron ore, limestone**, and **coke**. Although it does not occur in a free state in nature, iron is one of the most abundant elements in the world, making up about 5% of the earth's crust (in the form of various ores). The principal iron ores are *taconite* (a black flint-like rock), *hematite* (an iron-oxide mineral), and *limonite* (an iron oxide containing water). After it is mined, the ore is crushed into fine particles, the impurities are removed by various means (such as magnetic separation), and the ore is formed into pellets, balls, or briquettes using water and various binders. Typically, pellets are about 65% pure iron and about 25 mm (1 in.) in diameter. The concentrated iron ore is referred to as *beneficiated* (as are other concentrated ores). Some iron-rich ores are used directly, without pelletizing.

Coke is obtained from special grades of bituminous coal (a soft coal rich in volatile hydrocarbons and tarry matter) that are heated in vertical ovens to temperatures of up to 1150°C (2100°F) and then cooled with water in quenching towers. Coke has several functions in steelmaking, including (a) generating the high level of heat required for the chemical reactions in ironmaking to take place and (b) producing carbon monoxide (a reducing gas, meaning that it removes oxygen), which is then used to reduce iron oxide to iron. The chemical by-products of coke are used in the making of plastics and of chemical compounds. The gases evolved during the conversion of coal to coke are used as fuel for plant operations.

The function of limestone (calcium carbonate) is to remove impurities from the molten iron. The limestone reacts chemically with impurities, acting like a **flux** (meaning to flow as a fluid) that causes the impurities to melt at a low temperature. The limestone combines with the impurities and forms a **slag** (which is light), floats over the molten metal, and, subsequently, is removed. *Dolomite* (an ore of calcium magnesium carbonate) also is used as a flux. Slag later is used in making cement, fertilizers, glass, building materials, rock-wool insulation, and road ballast.

5.2.2 Ironmaking

The three raw materials described previously are carried to the top of a **blast furnace** (Fig. 5.1) and dumped into it (called *charging the furnace*). This furnace is basically a large steel cylinder lined with refractory (heat-resistant) brick; it has the height of about a ten-story building. The principle of this furnace was developed in Central Europe. The first blast furnace built in the United States began operating in 1621. The charge mixture is melted in a reaction at 1650°C (3000°F), with the air preheated to about 1100°C (2000°F) and *blasted* into the furnace (hence the term "blast furnace") through nozzles (called *tuyeres*). Although a number of reactions may take place, the basic reaction is that of oxygen with carbon to produce carbon monoxide, which, in turn, reacts with the iron oxide and reduces it to **iron**. Preheating the incoming air is necessary because the burning coke alone does not produce sufficiently high temperatures for these reactions to occur.

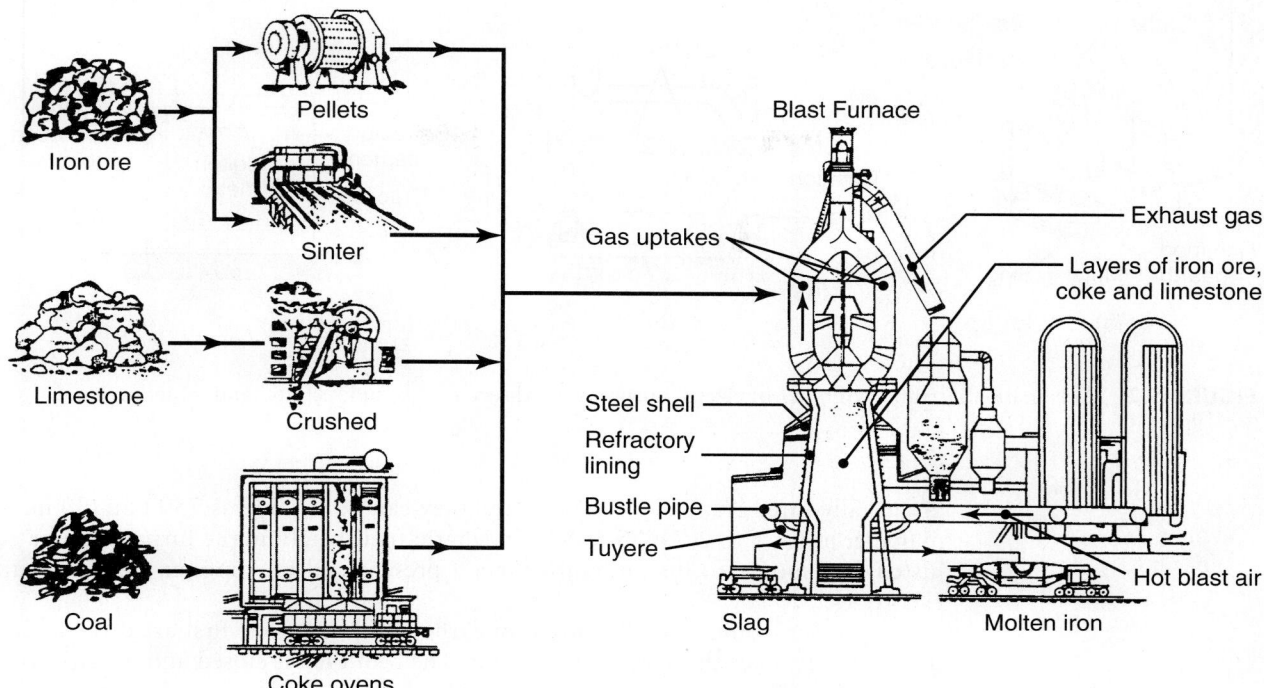

FIGURE 5.1   Schematic illustration of a blast furnace.

The molten metal accumulates at the bottom of the blast furnace, while the impurities float to the top of the metal. At intervals of four to five hours, the molten metal is drawn off (*tapped*) into ladle cars, each holding as much as 160 tons of molten iron. The molten metal at this stage is called **pig iron** or simply **hot metal**; it has a typical composition of 4% C, 1.5% Si, 1% Mn, 0.04% S, 0.4% P, and the rest being pure iron. The word **pig** comes from the early practice of pouring the molten iron into small sand molds arranged like a litter of small pigs around a main channel. The solidified metal (pig) is then used in making iron and steels.

5.2.3 Steelmaking

Steel was first produced in China and Japan in about 600 to 800 A.D. The steelmaking process is essentially one of refining the pig iron by the reduction of the percentage of manganese, silicon, carbon, and other elements and by controlling of the composition of the output with the addition of various elements. The molten metal from the blast furnace is transported into one of three types of furnaces: **open-hearth,** **electric,** or **basic-oxygen**. The name "open-hearth" derives from the shallow hearth shape that is open directly to the flames that melt the metal. Developed in the 1860s, the open-hearth furnace still is important industrially, but it has been replaced by electric furnaces and by the basic-oxygen process, because the latter two are more efficient and produce steels of better quality.

Electric furnace.   The source of heat in this furnace is a continuous electric arc that is formed between the electrodes and the charged metal (Fig. 5.2a and b). Temperatures as high as 1925°C (3500°F) are generated in this type of furnace. There

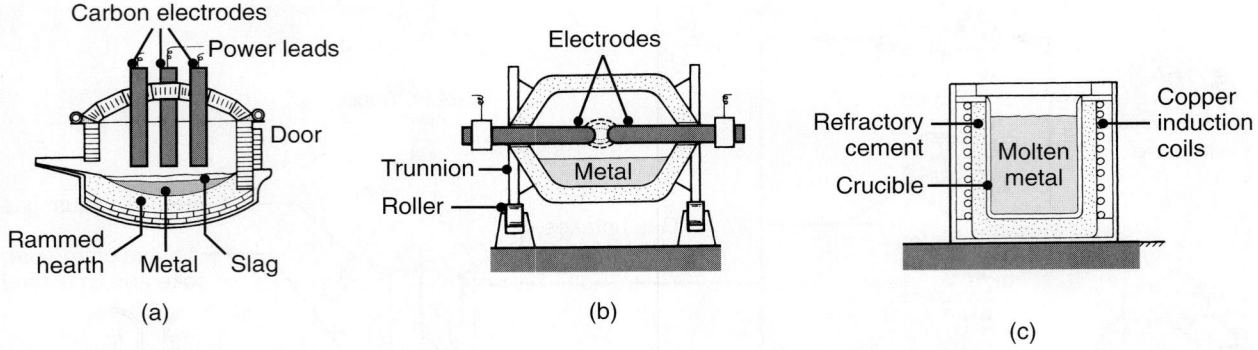

FIGURE 5.2 Schematic illustration of types of electric furnaces: (a) direct arc, (b) indirect arc, and (c) induction.

are usually three graphite electrodes, and they can be as large as 750 mm (30 in.) in diameter and 1.5 to 2.5 m (5 to 8 ft) in length; their height in the furnace can be adjusted in response to the amount of metal present and the amount of wear of the electrodes.

Steel scrap and a small amount of carbon and limestone first are dropped into the electric furnace through the open roof. The roof then is closed and the electrodes are lowered. The power then is turned on, and within a period of about two hours, the metal melts. The current then is shut off, the electrodes are raised, the furnace is tilted, and the molten metal is poured into a *ladle*, which is a receptacle used for transferring and pouring molten metal. Electric-furnace capacities range from 60 to 90 tons of steel per day. The quality of steel produced is better than that from either the open-hearth or the basic-oxygen process.

For smaller quantities, electric furnaces can be of the **induction** type. The metal is placed in a **crucible**—a large pot made of refractory material and surrounded with a copper coil through which alternating current is passed (Fig. 5.2c). The induced current in the charge generates heat and melts the metal. These furnaces also are used for remelting metal for casting.

Basic-oxygen furnace. The basic-oxygen furnace (BOF) is the fastest steel-making process. Typically, 200 tons of molten pig iron and 90 tons of scrap are charged into a vessel (Fig. 5.3). Pure oxygen is then blown into the furnace for about 20 minutes through a water-cooled *lance* (a long tube), under a pressure of about 1250 kPa (180 psi), as shown in Fig. 5.3. Fluxing agents (such as lime) are added through a chute. The vigorous agitation of the oxygen refines the molten metal by an oxidation process in which iron oxide is produced. The oxide reacts with the carbon in the molten metal, producing carbon monoxide and carbon dioxide. The lance is then retracted, and the furnace is tapped by tilting it (note the opening in Fig. 5.3 for the molten metal). The slag is removed by tilting the furnace in the opposite direction. The BOF process is capable of refining 250 tons of steel in 35 to 50 minutes. Most BOF steels, which have low impurity levels and are of better quality than open-hearth furnace steels, are processed into plates, sheets, and various structural shapes, such as I-beams and channels (see Fig. 13.1).

Vacuum furnace. Steel also may be melted in induction furnaces from which the air has been removed, similar to the one shown in Fig. 5.2c. Because the process removes gaseous impurities from the molten metal, vacuum melting produces high-quality steels.

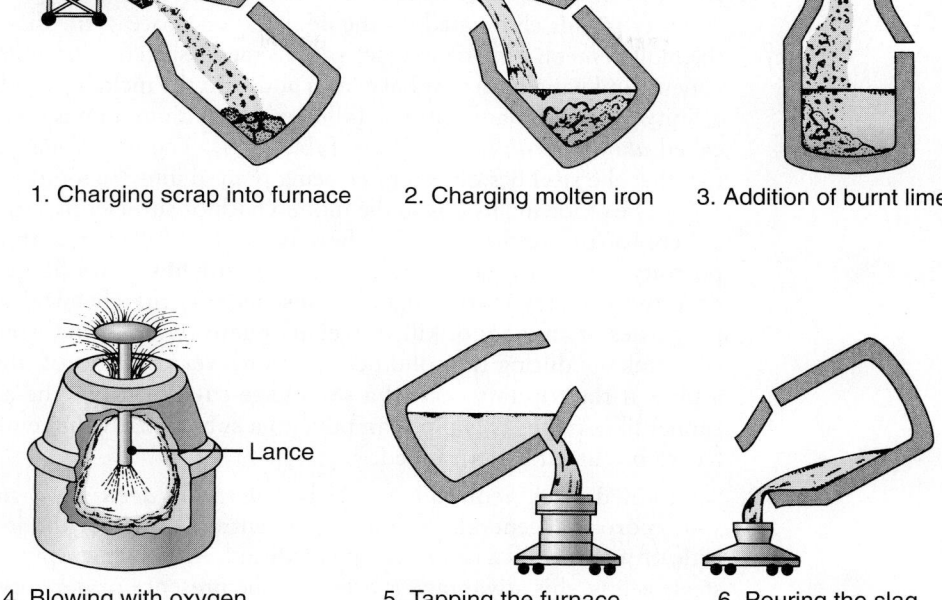

1. Charging scrap into furnace 2. Charging molten iron 3. Addition of burnt lime

4. Blowing with oxygen   —Lance 5. Tapping the furnace 6. Pouring the slag

FIGURE 5.3 Schematic illustrations showing charging, melting, and pouring of molten iron in a basic-oxygen process.

5.3 | Casting of Ingots

Traditionally, the next step in the steelmaking process is the shaping of the molten steel into a solid form (**ingot**) for such further processing as rolling it into shapes, casting it into semifinished forms, or forging it. This process now is replaced largely by **continuous casting**, which improves efficiency by eliminating the need for ingots (see Section 5.4). The molten metal is poured (teemed) from the ladle into ingot molds in which the metal solidifies. Molds usually are made of cupola iron or blast-furnace iron with 3.5% C. They are tapered in order to facilitate the removal of the solidified metal. The bottoms of the molds may be closed or open; if they are open, the molds are placed on a flat surface. The cooled ingots are removed (stripped) from the molds and lowered into **soaking pits**, where they are reheated to a uniform temperature of about 1200°C (2200°F) for subsequent processing by rolling. Ingots may be square, rectangular, or round in cross-section, and their weights range from a few hundred pounds to 40 tons.

Certain reactions take place during the solidification of an ingot. These reactions have an important influence on the quality of the steel produced. For example, significant amounts of oxygen and other gases can dissolve in the molten metal during steelmaking. Most of these gases are rejected during the solidification of the metal, because the solubility limit of the gases in the metal decreases sharply as its temperature decreases (see Fig. 10.15). Rejected oxygen combines with carbon to form carbon monoxide, which causes porosity in the solidified ingot.

Depending on the amount of gas evolved during solidification, three types of steel ingots can be produced: killed, semi-killed, and rimmed.

1. **Killed steel**. Killed steel is a fully deoxidized steel; that is, oxygen is removed and porosity is thus eliminated. In the deoxidation process, the dissolved oxygen in the molten metal is made to react with elements such as aluminum, silicon, manganese, and vanadium that have been added to the melt. These elements have an affinity for oxygen and form metallic oxides. If aluminum is used, the product is called *aluminum-killed steel* (see Table 16.4). The term *killed* comes from the fact that the steel lies quietly after being poured into the mold.

   The oxide inclusions in the molten bath (if sufficiently large) float out and adhere to (or are dissolved in) the slag. A fully killed steel thus is free of any porosity caused by gases; it also is free of any **blowholes** (large spherical holes near the surfaces of the ingot). Consequently, the chemical and mechanical properties of an ingot of killed steel are relatively uniform throughout. Because of shrinkage during the solidification, however, an ingot of this type develops a **pipe** at the top (also called a **shrinkage cavity**). It has the appearance of a funnel-like shape. This pipe can take up a substantial volume of the ingot, as it has to be cut off and scrapped.

2. **Semi-killed steel**. Semi-killed steel is a *partially deoxidized steel*. It contains some porosity (generally in the upper-central section of the ingot), but it has little or no pipe. As a result, scrap is reduced. Although the piping in semi-killed steels is less, this advantage is offset by the presence of porosity in that region. Semi-killed steels are economical to produce.

3. **Rimmed steel**. In a rimmed steel, which generally has a low carbon content (less than 0.15%), the evolved gases are killed (or controlled) partially by the addition of other elements, such as aluminum. The gases produce blowholes along the outer rim of the ingot—hence the term *rimmed*. Rimmed steels have little or no piping, and they have a ductile skin with good surface finish. However, if they are not controlled properly, blowholes may break through the skin. Furthermore, impurities and inclusions tend to segregate toward the center of the ingot. Thus, products made from this steel may be defective and should be inspected.

Refining. The properties and manufacturing characteristics of ferrous alloys are affected adversely by the amount of impurities, inclusions, and other elements present (see Section 2.10). The removal of impurities is known as *refining*. Most refining is done in melting furnaces or in ladles, by means of the addition of various elements. There is an increasing demand for cleaner steels: ones having improved and more uniform properties and a greater consistency of composition. Refining is important particularly in producing high-grade steels and alloys for high-performance and critical applications, such as for aircraft components. Moreover, warranty periods on shafts, camshafts, crankshafts for diesel trucks, and similar parts can be increased significantly by using higher-quality steels. Such steels are subjected to **secondary refining** in ladles (**ladle metallurgy**) and ladle refining (**injection refining**), which generally consists of melting and processing the steel in a vacuum. Several processes using controlled atmospheres (such as electron-beam melting, vacuum-arc remelting, argon-oxygen decarburization, and vacuum-arc double-electrode remelting) have been developed.

5.4 | Continuous Casting

The inefficiencies and the problems involved in making steels traditionally in ingots are alleviated by the *continuous-casting* processes, which produce higher-quality steels at reduced costs (see also Section 13.5.1 on *minimills*). Conceived in the 1860s, continuous or **strand casting** was first developed for casting nonferrous metal strips. The process now is used widely for steel production, with major productivity improvements and cost reduction. One system for continuous casting is shown schematically in Fig. 5.4. The molten metal in the ladle is cleaned, then it is equalized in temperature by blowing nitrogen gas through it for five to ten minutes. The metal then is poured into a refractory-lined intermediate pouring vessel (**tundish**), where impurities are skimmed off. The tundish holds as much as three tons of metal. The molten metal travels downward through water-cooled copper molds and begins to solidify into a path supported by rollers (called *pinch rolls*).

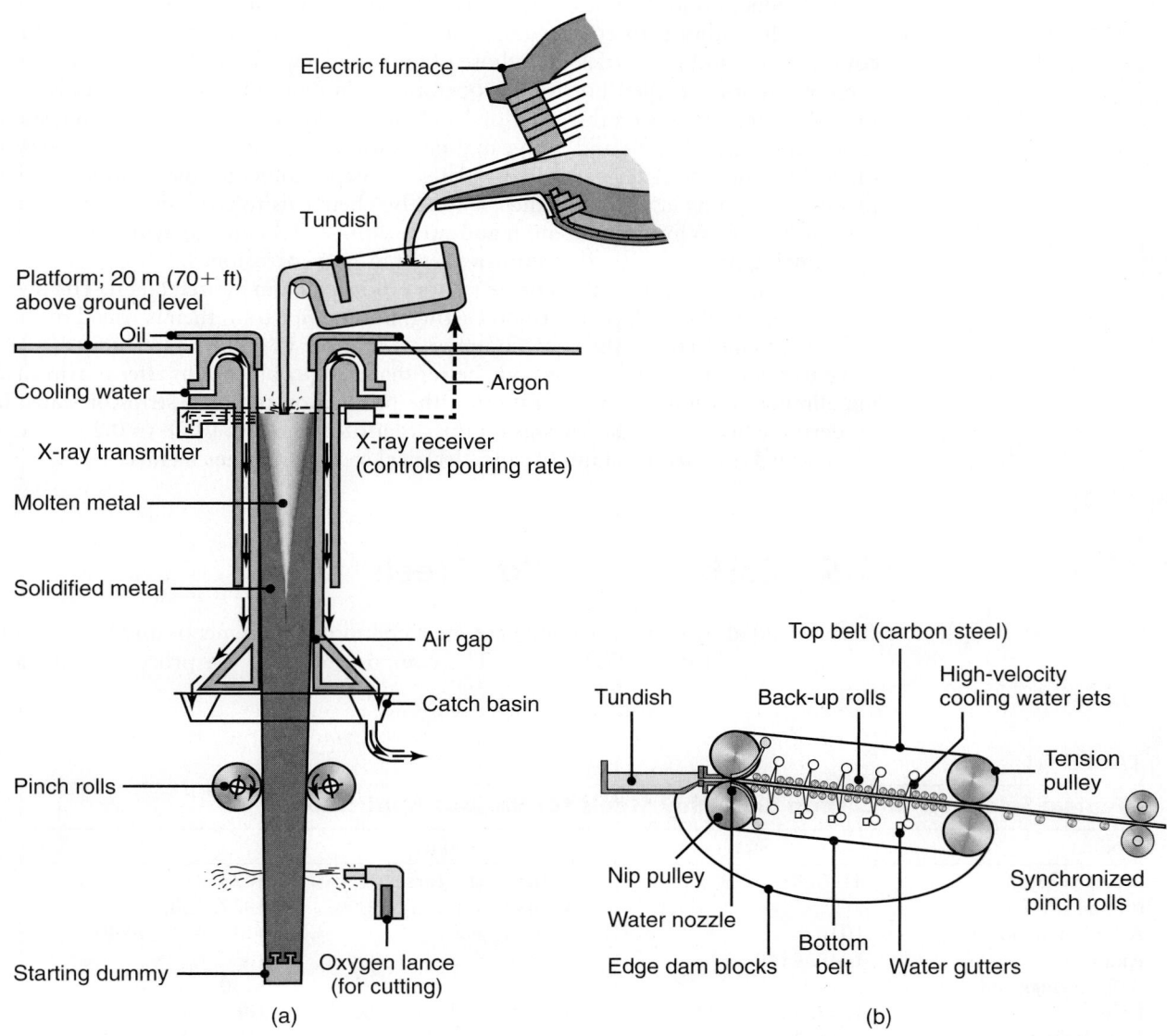

FIGURE 5.4   (a) The continuous-casting process for steel. Typically, the solidified metal descends at a speed of 25 mm/s (1 in./s). Note that the platform is about 20 m (65 ft) above ground level. *Source: Metalcaster's Reference and Guide*, American Foundrymen's Society. (b) Continuous strip casting of nonferrous metal strip. *Source:* Courtesy of Hazelett.

Before starting the casting process, a solid *starter bar* (*dummy bar*) is inserted into the bottom of the mold. When the molten metal is first poured, it freezes onto the dummy bar. The bar is withdrawn at the same rate at which the metal is poured. The cooling rate is such that the metal develops a solidified skin (shell), so as to support itself during its travel downward, typically at speeds of about 25 mm/s (1 in./s). The shell thickness at the exit end of the mold is about 12 to 18 mm (0.5 to 0.75 in.). Additional cooling is provided by water sprays along the travel path of the solidifying metal. The molds generally are coated with graphite or similar solid lubricants in order to reduce both friction and adhesion at the mold–metal interfaces. Also, the molds are vibrated in order to reduce friction and sticking.

The continuously cast metal may be cut into desired lengths by shearing or computer-controlled torch cutting, or it may be fed directly into a rolling mill for further reduction in thickness and for the shaping of products such as channels and I-beams. In addition to costing less, continuously cast metals have more uniform compositions and properties than those obtained by ingot casting. Modern facilities use computer-controlled hot-rolling operations on continuously cast strands with final sheet thicknesses on the order of 2 to 6 mm (0.08 to 0.25 in.) for carbon, stainless, and electrical steels and with capabilities for a rapid switchover from one type of steel to another. Afterwards, steel plates or shapes undergo one or more further processes, such as (a) cleaning and pickling by chemicals to remove surface oxides, (b) cold rolling to improve strength and surface finish, (c) annealing, and (d) coating (galvanizing or aluminizing) to improve resistance to corrosion.

In **strip casting**, thin slabs or strips are produced from molten metal. The metal solidifies in similar fashion to strand casting, but the hot solid then is rolled to form the final shape (Fig. 5.4b). The compressive stresses in rolling (see Section 13.2) serve to reduce porosity and provide better material properties. In effect, strip casting eliminates a hot-rolling operation in the production of metal strips or slabs. In modern facilities, final thicknesses on the order of 2 to 6 mm (0.08 to 0.25 in.) can be obtained for carbon, stainless, and electrical steels and other metals.

5.5 | Carbon and Alloy Steels

Carbon and alloy steels are among the most commonly used metals and have a wide variety of applications (Table 5.1). The compositions and the processing of these

TABLE 5.1

Typical Selection of Carbon and Alloy Steels for Various Applications			
Product	Steel	Product	Steel
Aircraft forgings, tubing, fittings	4140, 8740	Differential gears	4023
		Gears (car and truck)	4027, 4032
Automobile bodies	1010	Landing gear	4140, 4340, 8740
Axles	1040, 4140	Lock washers	1060
Ball bearings and races	52100	Nuts	3130
Bolts	1035, 4042, 4815	Railroad rails and wheels	1080
Camshafts	1020, 1040	Springs (coil)	1095, 4063, 6150
Chains (transmission)	3135, 3140	Springs (leaf)	1085, 4063, 9260, 6150
Coil springs	4063	Tubing	1040
Connecting rods	1040, 3141, 4340	Wire	1045, 1055
Crankshafts (forged)	1045, 1145, 3135, 3140	Wire (music)	1085

steels are controlled in a manner that makes them suitable for numerous applications. They are available in various basic product shapes: plate, sheet, strip, bar, wire, tube, castings, and forgings.

5.5.1 Effects of various elements in steels

Various elements are added to steels in order to impart properties such as hardenability, strength, hardness, toughness, wear resistance, workability, weldability, and machinability. These elements are listed below (in alphabetical order) with summaries of their beneficial and detrimental effects. Generally, the higher the percentages of these elements in steels, the greater are the particular properties that they impart. For example, the higher the carbon content, the greater the hardenability of the steel and the greater its strength, hardness, and wear resistance. On the other hand, ductility, weldability, and toughness are reduced with increasing carbon content.

> **Boron** improves hardenability without the loss of (or even with some improvement in) machinability and formability.
>
> **Calcium** deoxidizes steels, improves toughness, and may improve formability and machinability.
>
> **Carbon** improves hardenability, strength, hardness, and wear resistance; it reduces ductility, weldability, and toughness.
>
> **Cerium** controls the shape of inclusions and improves toughness in high-strength, low-alloy steels; it deoxidizes steels.
>
> **Chromium** improves toughness, hardenability, wear and corrosion resistance, and high-temperature strength; it increases the depth of hardness penetration resulting from heat treatment by promoting carburization.
>
> **Cobalt** improves strength and hardness at elevated temperatures.
>
> **Copper** improves resistance to atmospheric corrosion and, to a lesser extent, increases strength with little loss in ductility; it adversely affects the hot-working characteristics and surface quality.
>
> **Lead** improves machinability; it causes liquid–metal embrittlement.
>
> **Magnesium** has the same effects as cerium.
>
> **Manganese** improves hardenability, strength, abrasion resistance, and machinability; it deoxidizes the molten steel, reduces hot shortness, and decreases weldability.
>
> **Molybdenum** improves hardenability, wear resistance, toughness, elevated-temperature strength, creep resistance, and hardness; it minimizes temper embrittlement.
>
> **Nickel** improves strength, toughness, and corrosion resistance; it improves hardenability.
>
> **Niobium** (*columbium*) imparts fineness of grain size and improves strength and impact toughness; it lowers transition temperature and may decrease hardenability.
>
> **Phosphorus** improves strength, hardenability, corrosion resistance, and machinability; it severely reduces ductility and toughness.
>
> **Selenium** improves machinability.
>
> **Silicon** improves strength, hardness, corrosion resistance, and electrical conductivity; it decreases magnetic-hysteresis loss, machinability, and cold formability.
>
> **Sulfur** improves machinability when combined with manganese; it lowers impact strength and ductility and impairs surface quality and weldability.
>
> **Tantalum** has effects similar to those of niobium.
>
> **Tellurium** improves machinability, formability, and toughness.

Titanium improves hardenability; it deoxidizes steels.

Tungsten has the same effects as cobalt.

Vanadium improves strength, toughness, abrasion resistance, and hardness at elevated temperatures; it inhibits grain growth during heat treatment.

Zirconium has the same effects as cerium.

5.5.2 Residual elements in steels

During steel production, refining, and processing, some *residual elements* (**trace elements**) may still remain. Although the elements in the preceding list also can be considered as residuals, the following generally are considered unwanted residual elements.

Antimony and **arsenic** cause temper embrittlement.

Hydrogen severely embrittles steels; however, heating during processing drives out most of the hydrogen.

Nitrogen improves strength, hardness, and machinability; in aluminum-deoxidized steels, it controls the size of inclusions, improves strength and toughness, and decreases ductility and toughness.

Oxygen slightly increases the strength of rimmed steels; it severely reduces toughness.

Tin causes hot shortness and temper embrittlement.

5.5.3 Designations for steels

Traditionally, the American Iron and Steel Institute (AISI) and the Society of Automotive Engineers (SAE) have designated carbon and alloy steels by using four digits. The first two digits indicate the alloying elements and their percentages, and the last two digits indicate the carbon content by weight. Another numbering system has been the American Society for Testing and Materials (ASTM) designations, which incorporate the AISI and SAE designations and include standard specifications for steel products. For ferrous metals, the designation consists of the letter "A" followed by arbitrary numbers (generally three).

The present numbering system is known as the *Unified Numbering System* (UNS) and has been adopted widely by ferrous and nonferrous industries. It consists of a letter indicating the general class of the alloy, followed by five digits designating its chemical composition. Typical letter designations are:

G—for AISI and SAE carbon and alloy steels

J—for cast steels

K—for miscellaneous steels and ferrous alloys

S—for stainless steels and superalloys

T—for tool steels

Examples are: G41300 for AISI 4130 alloy steel, and T30108 for AISI A-8 tool steel.

5.5.4 Carbon Steels

Carbon steels generally are classified by their proportion (by weight) of carbon content. The general mechanical properties of carbon and alloy steels are shown in Table 5.2. The machinability, formability, and weldability of such steels are described in various chapters throughout this text.

TABLE 5.2

	Typical Mechanical Properties of Selected Carbon and Alloy Steels in the Hot-Rolled, Normalized, and Annealed Condition					
AISI	Condition	Ultimate tensile strength (MPa)	Yield strength (MPa)	Elongation in 50 mm (%)	Reduction of area (%)	Hardness (HB)
1020	As-rolled	448	346	36	59	143
	Normalized	441	330	35	67	131
	Annealed	393	294	36	66	111
1080	As-rolled	1010	586	12	17	293
	Normalized	965	524	11	20	293
	Annealed	615	375	24	45	174
3140	Normalized	891	599	19	57	262
	Annealed	689	422	24	50	197
4340	Normalized	1279	861	12	36	363
	Annealed	744	472	22	49	217
8620	Normalized	632	385	26	59	183
	Annealed	536	357	31	62	149

- **Low-carbon steel**, also called **mild steel**, has less than 0.30% C. It often is used for common industrial products (such as bolts, nuts, sheet, plate, and tubes) and for machine components that do not require high strength.

- **Medium-carbon steel** has 0.30 to 0.60% C. It generally is used in applications requiring higher strength than is available in low-carbon steels, such as in machinery, automotive and agricultural equipment parts (gears, axles, connecting rods, crankshafts), railroad equipment, and parts for metalworking machinery.

- **High-carbon steel** has more than 0.60% C. Generally, high-carbon steel is used for parts requiring strength, hardness, and wear resistance, such as cutting tools, cable, music wire, springs, and cutlery. After being manufactured into shapes, the parts usually are heat treated and tempered. The higher the carbon content of the steel, the higher is its hardness, strength, and wear resistance after heat treatment.

- Carbon steels containing sulfur and phosphorus are known as **resulfurized** carbon steels (11xx series) and **rephosphorized and resulfurized** carbon steels (12xx series). For example, 1112 steel is a resulfurized steel with a carbon content of 0.12%. These steels have improved machinability, as described in Section 21.7.

5.5.5 Alloy steels

Steels containing significant amounts of alloying elements are called **alloy steels**; they usually are made with more care than are carbon steels. **Structural-grade alloy steels** are used mainly in the construction and transportation industries because of their high strength. Other alloy steels are used in applications where strength, hardness, creep and fatigue resistance, and toughness are required. These steels can be heat treated to obtain the desired properties.

5.5.6 High-strength low-alloy steels

In order to improve the strength-to-weight ratio of steels, a number of **high-strength, low-alloy steels (HSLA)** have been developed. These steels have a low carbon content (usually less than 0.30%) and are characterized by a microstructure consisting of fine-grain ferrite as one phase and a hard second phase of martensite and austenite. First developed in the 1930s, HSLA steels usually are produced in sheet form by microalloying and controlled hot rolling. Plates, bars, and structural shapes are made from these steels. The ductility, formability, and weldability of HSLA steels, however, generally are inferior to those of conventional low-alloy steels. To improve these properties, **dual-phase steels** have been developed (see Section 5.5.7).

Sheet products of HSLA steels typically are used for parts of automobile bodies and other transportation equipment (in order to reduce weight and, hence, fuel consumption) and in mining, agricultural, and various other industrial applications. HSLA plates are used in ships, bridges, building construction, and for shapes such as I-beams, channels, and angles used in buildings and in various structures.

Designations. Three categories compose the system of AISI designations for high-strength sheet steel (Table 5.3). *Structural quality* (S) includes the elements C, Mn, P, and N. *Low alloys* (X) contain Nb, Cr, Cu, Mo, Ni, Si, Ti, V, and Zr, either singly or in combination. *Weathering steels* (W) have environmental-corrosion resistance that is approximately four times greater than that of conventional low-carbon steels and contain Si, P, Cu, Ni, and Cr in various combinations. In addition, the formability of these sheet steels is graded by the letters F (excellent), K (good), and O (fair).

Microalloyed steels. These recently developed HSLA steels provide superior properties and can eliminate the need for heat treatment. They have a ferrite-pearlite microstructure with fine dispersed particles of carbonitride. A number of microalloyed steels have been produced, with a typical microalloyed steel containing 0.5% C, 0.8% Mn, and 0.1% V. When subjected to carefully controlled cooling (usually in air) these materials develop improved and uniform strength. Compared to medium-carbon steels, microalloyed steels also can provide cost savings of as much as 10%, since the manufacturing steps of quenching, tempering, and stress relieving are not required.

TABLE 5.3

AISI Designation for High-Strength Sheet Steel			
Yield strength		Chemical composition	Deoxidation practice
psi $\times 10^3$	MPa		
35	240	S = structural quality	F = killed plus sulfide inclusion control
40	275		
45	310		
50	350	X = low alloy	
60	415		K = killed
70	485	W = weathering	
80	550		O = nonkilled
100	690	D = dual phase	
120	830		
140	970		

Example

50	X	F
50 $\times 10^3$ psi min yield strength	Low alloy	Killed plus sulfide inclusion control

Nanoalloyed steels. Now under development, these steels have extremely small grain sizes (10–100 nm), and are produced using metallic glasses (Section 6.14) as a precursor. The metallic glass is subjected to a carefully controlled vitrification (crystallization) process at a high nucleation rate, resulting in fine nanoscale phases. (See also Section 6.16.)

5.5.7 Dual-phase steels

Dual-phase steels, designated with the letter "D" in Table 5.3, are processed specially to have a mixed ferrite and martensite structure. Developed in the late 1960s, these steels have a high work-hardening characteristic (high n value in Eq. 2.8), which improves their ductility and formability. The SAE designations for these steels are similar to those given in Table 5.3, with the exception that another letter is added to indicate the carbon content. Thus, for example, 050XF becomes 050XLF, where L indicates the proportion of carbon (in this case L meaning low carbon).

5.6 | Stainless Steels

Stainless steels are characterized primarily by their corrosion resistance, high strength and ductility, and high chromium content. They are called *stainless* because, in the presence of oxygen (air), they develop a thin, hard, adherent film of chromium oxide that protects the metal from corrosion (*passivation*, see Section 3.8). This protective film builds up again in the event that the surface is scratched; for passivation to occur, the minimum chromium content should be 10 to 12% by weight.

In addition to chromium, other alloying elements in stainless steels typically are nickel, molybdenum, copper, titanium, silicon, manganese, columbium, aluminum, nitrogen, and sulfur. The letter L is used to identify low-carbon stainless steels. The higher the carbon content, the lower is the corrosion resistance of stainless steels. The reason is that the carbon combines with the chromium in the steel and forms chromium carbide; the reduced availability of chromium lowers the passivity of the steel. Also, the chromium carbide introduces a second phase and, thereby, promotes galvanic corrosion. (The statues of soldiers in the Korean War Veterans Memorial in Washington, D.C. are cast in 316L stainless steel.)

Developed in the early 1900s, stainless steels are made by using electric furnaces or the basic-oxygen process and processed by techniques similar to those used in other types of steelmaking. The level of purity is controlled by various refining techniques. Stainless steels are available in a wide variety of shapes, and typical applications include cutlery, kitchen equipment, health care and surgical equipment and are used in the chemical, food-processing, and petroleum industries. A recent trend in the use of stainless steel is as reinforcing bars (rebar) in reinforced concrete for some sections of bridges, highways, buildings, and other forms of construction, especially in a marine environment. The advantages are better mechanical properties and corrosion resistance against chlorides from road salt as well as from the concrete in which the bar is buried. Although the initial cost is high, it is estimated that, because of lower maintenance costs, stainless steel is more economical than carbon-steel bars over the life of the structure.

Stainless steels generally are divided into five types (see Table 5.4).

Austenitic (200 and 300 series). These steels generally are composed of chromium, nickel, and manganese in iron. They are nonmagnetic and have excellent corrosion resistance, but they are susceptible to stress-corrosion cracking. Austenitic

TABLE 5.4

Room-Temperature Mechanical Properties and Typical Applications of Selected Annealed Stainless Steels				
AISI (UNS)	Ultimate tensile strength (MPa)	Yield strength (MPa)	Elongation in 50 mm (%)	Characteristics and typical applications
303 (S30300)	550–620	240–260	53–50	Screw machine products (shafts, valves, bolts, bushings, and nuts) and aircraft fittings (bolts, nuts, rivets, screws, studs)
304 (S30400)	5–620	240–290	60–55	Chemical and food-processing equipment, brewing equipment, cryogenic vessels, gutters, downspouts, and flashings.
316 (S31600)	50–590	210–290	60–55	High corrosion resistance and high creep strength, chemical and pulp handling equipment, photographic equipment, brandy vats, fertilizer parts, ketchup-cooking kettles, and yeast tubs.
410 (S41000)	480–520	240–310	35–25	Machine parts, pump shafts, bolts, bushings, coal chutes, cutlery, tackle, hardware, jet engine parts, mining machinery, rifle barrels, screws, and valves
416 (S41600)	480–520	275	30–20	Aircraft fittings, bolts, nuts, fire extinguisher inserts, rivets, and screws

stainless steels are hardened by cold working. They are the most ductile of all stainless steels and can be formed easily, although with increasing cold work, their formability is reduced. These steels are used in a wide variety of applications, such as kitchenware, fittings, welded construction, lightweight transportation equipment, furnace and heat-exchanger parts, and components for severe chemical environments.

Ferritic (400 series). These steels have a high chromium content—up to 27%. They are magnetic and have good corrosion resistance, but they have lower ductility than austenitic stainless steels. Ferritic stainless steels are hardened by cold working and are not heat treatable. They generally are used for nonstructural applications, such as kitchen equipment and automotive trim.

Martensitic (400 and 500 series). Most martensitic stainless steels do not contain nickel and are hardenable by heat treatment. Their chromium content may be as much as 18%. These steels are magnetic, and they have high strength, hardness, and fatigue resistance, good ductility, and moderate corrosion resistance. Martensitic stainless steels typically are used for cutlery, surgical tools, instruments, valves, and springs.

Precipitation-hardening (PH). These steels contain chromium and nickel along with copper, aluminum, titanium, or molybdenum. They have good corrosion resistance and ductility, and they have high strength at elevated temperatures. Their main application is in aircraft and aerospace structural components.

Duplex structure. These steels have a mixture of austenite and ferrite. They have good strength and have higher resistance to both corrosion (in most environments) and stress–corrosion cracking than do the 300 series of austenitic steels. Typical applications are in water-treatment plants and in heat-exchanger components.

EXAMPLE 5.1 Use of stainless steels in automobiles

The types of stainless steel usually selected by materials engineers for use in automobile parts are 301, 409, 430, and 434. Because of its good corrosion resistance and mechanical properties, type 301 is used for wheel covers. Cold working during the forming process increases its yield strength (by means of strain hardening) and gives the wheel cover a spring-like action.

Type 409 is used extensively for catalytic converters. Type 430 had been used for automotive trim, but it is not as resistant as type 434 is to the de-icing salts used in colder climates in winter. As a result, its use is now limited. In addition to being more corrosion resistant, type 434 closely resembles the color of chromium plating, so it has become an attractive alternative to 430. Stainless steels are well-suited for use in other automobile components as well: exhaust manifolds (replacing cast-iron manifolds to reduce weight, increase durability, provide higher thermal conductivity and reduce emissions), mufflers and tailpipes (to offer better corrosion protection in harsh environments), and brake tubing.

5.7 | Tool and Die Steels

Tool and die steels are specially alloyed steels (Tables 5.5 and 5.6) designed for high strength, impact toughness, and wear resistance at room and elevated temperatures. They commonly are used in the forming and machining of metals (Parts III and IV).

5.7.1 High-speed steels

High-speed steels (HSS) are the most highly alloyed tool and die steels. First developed in the early 1900s, they maintain their hardness and strength at elevated operating temperatures. There are two basic types of high-speed steels: the **molybdenum type** (M-series) and the **tungsten type** (T-series).

TABLE 5.5

Basic Types of Tool and Die Steels	
Type	AISI
High speed	M (molybdenum base)
	T (tungsten base)
Hot work	H1 to H19 (chromium base)
	H20 to H39 (tungsten base)
	H40 to H59 (molybdenum base)
Cold work	D (high carbon, high chromium)
	A (medium alloy, air hardening)
	O (oil hardening)
Shock resisting	S
Mold steels	P1 to P19 (low carbon)
	P20 to P39 (others)
Special purpose	L (low alloy)
	F (carbon-tungsten)
Water hardening	W

TABLE 5.6

Processing and Service Characteristics of Common Tool and Die Steels							
AISI designation	Resistance to decarburization	Resistance to cracking	Approximate hardness (HRC)	Machinability	Toughness	Resistance to softening	Resistance to wear
M2	Medium	Medium	60–65	Medium	Low	Very high	Very high
T1	High	High	60–65	Medium	Low	Very high	Very high
T5	Low	Medium	60–65	Medium	Low	Highest	Very high
H11, 12, 13	Medium	Highest	38–55	Medium to high	Very high	High	Medium
A2	Medium	Highest	57–62	Medium	Medium	High	High
A9	Medium	Highest	35–56	Medium	High	High	Medium to high
D2	Medium	Highest	54–61	Low	Low	High	High to very high
D3	Medium	High	54–61	Low	Low	High	Very high
H21	Medium	High	36–54	Medium	High	High	Medium to high
H26	Medium	High	43–58	Medium	Medium	Very high	High
P20	High	High	28–37	Medium to high	High	Low	Low to medium
P21	High	Highest	30–40	Medium	Medium	Medium	Medium
W1, W2	Highest	Medium	50–64	Highest	High	Low	Low to medium

The **M-series** steels contain up to about 10% molybdenum with chromium, vanadium, tungsten, and cobalt as other alloying elements. The **T-series** steels contain 12 to 18% tungsten with chromium, vanadium, and cobalt as other alloying elements. The M-series steels generally have higher abrasion resistance than the T-series steels, undergo less distortion in heat treatment, and are less expensive. The M-series constitutes about 95% of all the high-speed steels produced in the United States. High-speed steel tools can be coated with titanium nitride and titanium carbide for improved wear resistance.

5.7.2 Die steels

Hot-work steels (H-series) are designed for use at elevated temperatures. They have high toughness as well as high resistance to wear and cracking. The alloying elements are generally tungsten, molybdenum, chromium, and vanadium. **Cold-work steels** (A-, D-, and O-series) are used for cold-working operations. They generally have high resistance to wear and cracking. These steels are available as oil-hardening or air-hardening types. **Shock-resisting steels** (S-series) are designed for impact toughness and are used in applications such as header dies, punches, and chisels. Other properties of these steels depend on the particular composition. Various tool and die materials for a variety of manufacturing applications are presented in Table 5.7.

TABLE 5.7

Typical Tool and Die Materials for Metalworking	
Process	Material
Die casting	H13, P20
Powder metallurgy	
Punches	A2, S7, D2, D3, M2
Dies	WC, D2, M2
Molds for plastics and rubber	S1, O1, A2, D2, 6F5, 6F6, P6, P20, P21, H13
Hot forging	6F2, 6G, H11, H12
Hot extrusion	H11, H12, H13, H21
Cold heading	W1, W2, M1, M2, D2, WC
Cold extrusion	
Punches	A2, D2, M2, M4
Dies	O1, W1, A2, D2
Coining	52100, W1, O1, A2, D2, D3, D4, H11, H12, H13
Drawing	
Wire	WC, diamond
Shapes	WC, D2, M2
Bar and tubing	WC, W1, D2
Rolls	
Rolling	Cast iron, cast steel, forged steel, WC
Thread rolling	A2, D2, M2
Shear spinning	A2, D2, D3
Sheet metals	
Shearing	
Cold	D2, A2, A9, S2, S5, S7
Hot	H11, H12, H13
Pressworking	Zinc alloys, 4140 steel, cast iron, epoxy composites, A2, D2, O1
Deep drawing	W1, O1, cast iron, A2, D2
Machining	Carbides, high-speed steels, ceramics, diamond, cubic boron nitride

Notes: Tool and die materials usually are hardened 55 to 65 HRC for cold working, and 30 to 55 HRC for hot working. Tool and die steels contain one or more of the following major alloying elements: chromium, molybdenum, tungsten, and vanadium. For further details, see the bibliography at the end of this chapter.

SUMMARY

- The major categories of ferrous metals and alloys are carbon steels, alloy steels, stainless steels, and tool and die steels. Their wide range of properties and general low cost have made them among the most useful of all metallic materials.

- Steelmaking processes have been improved upon continuously, notably by the continuous-casting and secondary-refining techniques. These advances have resulted in higher-quality steels and in higher efficiency and productivity.

- Alloying elements greatly influence mechanical, physical, chemical, and manufacturing properties (hardenability, castability, formability, machinability, and weldability) and performance in service.

- Carbon steels generally are classified as low-carbon (mild steel), medium-carbon, and high-carbon steels. Alloy steels contain a variety of alloying elements, particularly chromium, nickel, and molybdenum. Stainless steels generally are classified as austenitic, ferritic, martensitic, and precipitation-hardening.

- Tool and die steels are among the most important materials and are used widely in casting, forming, and machining operations for both metallic and nonmetallic materials. They generally consist of high-speed steels (molybdenum and tungsten types), hot- and cold-work steels, and shock-resisting steels.

KEY TERMS

Alloy steels	Ingot	Refining
Basic-oxygen furnace	Integrated mills	Rimmed steel
Blast furnace	Killed steel	Semi-killed steel
Carbon steels	Microalloyed steels	Stainless steels
Continuous casting	Nanoalloyed steels	Strand casting
Dual-phase steels	Open-hearth furnace	Tool and die steels
Electric furnace	Pig	Trace elements
High-strength low-alloy steels	Pig iron	

BIBLIOGRAPHY

ASM Handbook, Vol. 1: *Properties and Selection: Iron, Steels, and High-Performance Alloys*, ASM International, 1990.

ASM Specialty Handbook: Carbon and Alloy Steels, ASM International, 1995.

ASM Specialty Handbook: Stainless Steels, ASM International, 1994.

ASM Specialty Handbook: Tool Materials, ASM International, 1995.

Beddoes, J., and Parr, J.G., *Introduction to Stainless Steels*, 3rd ed., ASM. International, 1999.

Bryson, B., *Heat Treatment, Selection and Application of Tool Steels*, Hanser Gardner, 1997.

Donachie, N.J., and Donachie, S.J., *Superalloys: A Technical Guide*, 2nd ed., ASM International, 2001.

Farag, M.M., *Materials Selection for Engineering Design*, Prentice Hall, 1997.

Harper, C. (ed.), *Handbook for Materials for Product Design*, 3rd ed., McGraw-Hill, 2001.

Llewellyn, D.T., and Hudd, R.C., *Steels: Metallurgy and Applications*, 3rd ed. Butterworth-Heinemann, 1999.

Roberts, G.A., Krauss, G., and Kennedy, R., *Tool Steels*, 5th ed., ASM International, 1998.

REVIEW QUESTIONS

5.1   What are the major categories of ferrous alloys?

5.2   List the basic raw materials used in making iron and steel and explain their functions.

5.3   List the types of furnaces commonly used in steelmaking and describe their characteristics.

5.4   List and explain the characteristics of the types of steel ingots.

5.5   What does refining mean? How is it done?

5.6   What advantages does continuous casting have over casting into ingots?

5.7   Name the four alloying elements that have the greatest effect on the properties of steels.

5.8   What are trace elements?

5.9   What are the percentage carbon contents of low-carbon, medium-carbon, and high-carbon steels?

5.10   How do stainless steels become stainless?

5.11   What are the major alloying elements in tool and die steels and in high-speed steels?

5.12   How does chromium affect the surface characteristics of stainless steels?

5.13   What kind of furnaces are used to refine steels?

5.14   What is high-speed steel?

5.15   Where does the term "pig iron" come from?

QUALITATIVE PROBLEMS

5.16   Identify several different products that are made of stainless steel and explain why they are made of this material.

5.17   As you may know, professional cooks prefer carbon-steel to stainless-steel knives, even though the latter are more popular with consumers. Explain the reasons for those preferences.

5.18   Why is the control of an ingot structure important?

5.19   Explain why continuous casting has been such an important technological advancement.

5.20   Describe applications in which you would not want to use carbon steels.

5.21   Explain what would happen if the speed of the continuous-casting process (Fig. 5.4) is (a) higher or (b) lower than that indicated (typically 25 mm/s).

5.22   The cost of mill products of metals increases with decreasing thickness and section size. Explain why.

5.23   Describe your observations regarding the information given in Table 5.7.

5.24   How do trace elements affect the ductility of steels?

5.25   Comment on your observations regarding Table 5.1.

5.26   In Table 5.7, D2 steel is listed as a tool and die material for most applications. Why is this so?

5.27   List the common impurities in steel. Which of these are the ones most likely to be minimized if the steel is melted in a vacuum furnace?

5.28   Explain the purpose of the oil in Fig. 5.4 given that the molten-steel temperatures are far above the ignition temperatures of the oil.

5.29   Recent research has identified mold-surface textures that will either (a) inhibit a solidified steel from separating from the mold or (b) force it to stay in contact in continuous casting. What is the advantage of a mold which maintains intimate contact with the steel?

5.30   Identify products which cannot be made of steel and explain why this is so. For example, electrical contacts commonly are made of gold or copper, because their softness results in low contact resistance, while for steel the contact resistance would be very high.

QUANTITATIVE PROBLEMS

5.31   By referring to the available literature, estimate the cost of the raw materials for (a) an aluminum beverage can, (b) a stainless-steel two-quart cooking pot, and (c) the steel hood of a car.

5.32   In Table 5.1, more than one type of steel is listed for some applications. By referring to data in the technical literature listed in the bibliography, determine the range of properties for these steels in various conditions (such as cold worked, hot worked, and annealed).

5.33   Some soft drinks now are available in steel cans (with aluminum tops) that look similar to aluminum cans. Obtain one of each, weigh them when empty, and determine their respective wall thicknesses.

5.34   Using strength and density data, determine the minimum weight of a two-foot long tension member

which must support 1000 lb. manufactured from (a) annealed 303 stainless steel, (b) normalized 8620 steel, (c) as-rolled 1080 steel, (d) any two aluminum alloys, (e) any brass alloy, or (f) pure copper.

5.35   The endurance limit (fatigue life) of steel is approximately one-half the ultimate strength (see Fig. 2.16) but never higher than 100 ksi (700 MPa). For irons, the endurance limit is 40% of the ultimate strength but never higher than 24 ksi (170 MPa). Plot the endurance limit versus the ultimate strength for the steels described in this chapter and for the cast irons in Table 12.3. On the same plot, show the effect of surface finish by plotting the endurance limit, assuming the material is in the as-cast state (see Fig. 2.28).

SYNTHESIS, DESIGN, AND PROJECTS

5.36 Based on the information given in Section 5.5.1, make a table with columns for each improved property (i.e., hardenability, strength, toughness, and machinability). In each column, list the elements that improve that property and identify the element that has the most influence.

5.37 Assume that you are in charge of public relations for a steel-producing company. Outline all of the attractive characteristics of steels that you would like your customers to know about.

5.38 Assume that you are in competition with the steel industry and are asked to list all of the characteristics of steels that are not attractive. Make a list of these characteristics and explain their engineering relevance.

5.39 In Section 5.5.1, we noted the effects of various individual elements (such as lead alone or sulfur alone) on the properties and characteristics of steels. We did not, however, discuss the role of combinations of elements (such as lead and sulfur together). Review the technical literature and prepare a table indicating the combined effects of several elements on steels.

5.40 In the past, waterfowl hunters used lead shot in their shotguns, but this practice resulted in lead poisoning of unshot birds that ingested lead pellets (along with gravel) to help them digest food. Recently, steel and tungsten have been used as replacement materials. If all pellets have the same velocity upon exiting the shotgun barrel, what concerns would you have regarding this material substitution? Consider environmental and performance effects.

5.41 Aluminum has been cited as a possible substitute material for steel in automobiles. What concerns would you have before purchasing an aluminum automobile?

5.42 In the 1940s, the *Yamato* was the largest battleship that had ever been built. Find out the weight of this ship and determine how many automobiles could be built from the steel in this one ship. How long would it take to cast this much steel by continuous casting?

5.43 Search the technical literature and add more parts and materials to those shown in Table 5.1.

Nonferrous Metals and Alloys: Production, General Properties, and Applications

An extensive variety of nonferrous metals and alloys, ranging from aluminum to zinc, are indispensable for a wide variety of applications in engineering. This chapter outlines:

- Production methods for nonferrous metals.
- Physical and mechanical properties of nonferrous metals and alloys and their relevance.
- Applications of these alloys.
- Shape-memory alloys, amorphous alloys, and nanomaterials, and their unique uses.

6.1 | Introduction

Nonferrous metals and alloys cover a wide range, from the more common metals (such as aluminum, copper, and magnesium) to high-strength high-temperature alloys (such as those of tungsten, tantalum, and molybdenum). Although generally more expensive than ferrous metals (Table 6.1), nonferrous metals and alloys have major applications because of properties such as corrosion resistance, high thermal and electrical conductivity, low density, and ease of fabrication (Table 6.2). Typical examples of the applications of nonferrous metals and alloys are aluminum for cooking utensils and aircraft bodies, copper wire for electricity, copper tubing for residential water supply, zinc for galvanized sheet metal for car bodies, titanium for jet-engine turbine blades and for orthopedic implants, and tantalum for rocket engines.

A turbofan jet engine for the Boeing 757 aircraft typically contains the following nonferrous metals and alloys: 38% Ti, 37% Ni, 12% Cr, 6% Co, 5% Al, 1% Nb, and 0.02% Ta. Without these materials, a jet engine (Fig. 6.1) could not be designed, manufactured, and operated at the power and efficiency levels required.

This chapter introduces the general properties, the production methods, and the important engineering applications for nonferrous metals and alloys. The manufacturing properties of these materials (such as formability, machinability, and weldability) are described in various chapters throughout this text.

TABLE 6.1

Approximate Cost-per-unit-volume for Wrought Metals and Plastics Relative to the Cost of Carbon Steel			
Gold	60,000	Magnesium alloys	2–4
Silver	600	Aluminum alloys	2–3
Molybdenum alloys	200–250	High-strength low-alloy steels	1.4
Nickel	35	Gray cast iron	1.2
Titanium alloys	20–40	Carbon steel	1
Copper alloys	5–6	Nylons, acetals, and silicon rubber*	1.1–2
Zinc alloys	1.5–3.5	Other plastics and elastomers*	0.2–1
Stainless steels	2–9		

*As molding compounds.
Note: Costs vary significantly with quantity of purchase, supply and demand, size and shape, and various other factors.

TABLE 6.2

General Characteristics of Nonferrous Metals and Alloys	
Material	Characteristics
Nonferrous alloys	More expensive than steels and plastics; wide range of mechanical, physical, and electrical properties; good corrosion resistance; high-temperature applications
Aluminum	High strength-to-weight ratio; high thermal and electrical conductivity; good corrosion resistance; good manufacturing properties
Magnesium	Lightest metal; good strength-to-weight ratio
Copper	High electrical and thermal conductivity; good corrosion resistance; good manufacturing properties
Superalloys	Good strength and resistance to corrosion at elevated temperatures; can be iron-, cobalt-, and nickel-based alloys
Titanium	Highest strength-to-weight ratio of all metals; good strength and corrosion resistance at high temperatures
Refractory metals	Molybdenum, niobium (columbium), tungsten, and tantalum; high strength at elevated temperatures
Precious metals	Gold, silver, and platinum; generally good corrosion resistance

6.2 | Aluminum and Aluminum Alloys

The important factors in selecting **aluminum** (Al) and its alloys are their high strength-to-weight ratio, resistance to corrosion by many chemicals, high thermal and electrical conductivity, nontoxicity, reflectivity, appearance, and ease of formability and of machinability; they are also nonmagnetic. The principal uses of aluminum and its alloys, in decreasing order of consumption, are in containers and packaging (aluminum cans and foil), buildings and other types of construction, transportation (aircraft and aerospace applications, buses, automobiles, railroad cars, and marine craft), electrical applications (as an economical and nonmagnetic electrical conductor), consumer durables (appliances, cooking utensils, and furniture), and portable tools (Tables 6.3 and 6.4). Nearly all high-voltage transmission wiring is made of aluminum. In its structural (load-bearing) components, 82% of a Boeing 747 aircraft and 70% of a Boeing 777 aircraft is aluminum.

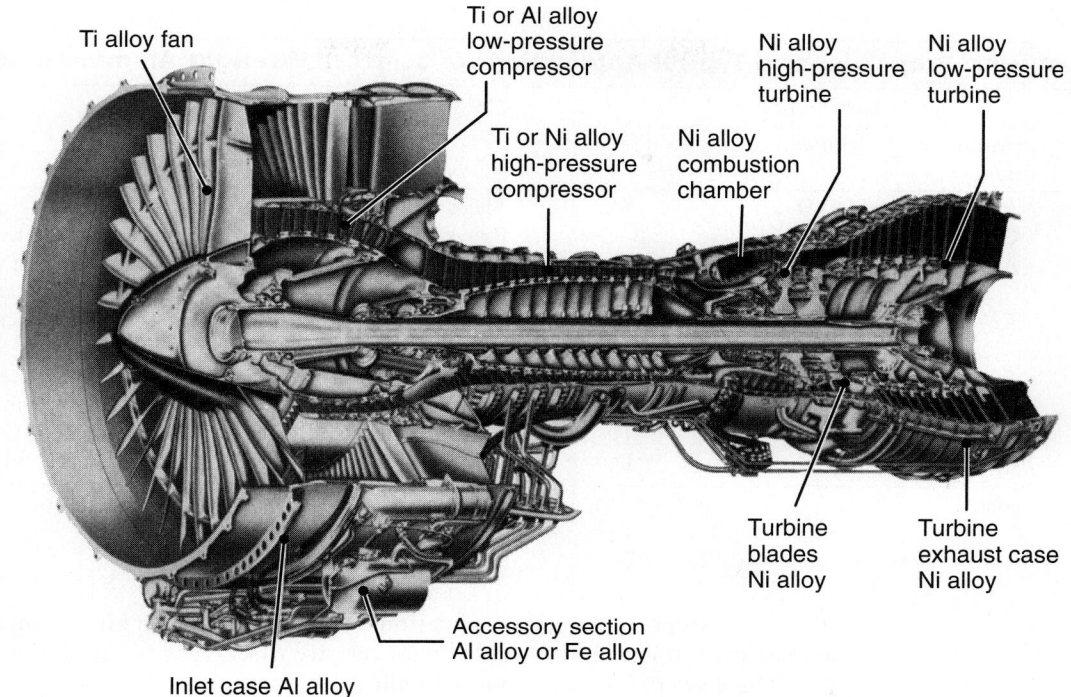

FIGURE 6.1 Cross-section of a jet engine (PW2037) showing various components and the alloys used in manufacturing them. *Source*: Courtesy of United Aircraft Pratt & Whitney.

TABLE 6.3

Properties of Selected Aluminum Alloys at Room Temperature				
Alloy (UNS)	Temper	Ultimate tensile strength (MPa)	Yield strength (MPa)	Elongation in 50 mm (%)
1100 (A91100)	O	90	35	35–45
1100	H14	125	120	9–20
2024 (A92024)	O	190	75	20–22
2024	T4	470	325	19–20
3003 (A93003)	O	110	40	30–40
3003	H14	150	145	8–16
5052 (A95052)	O	190	90	25–30
5052	H34	260	215	10–14
6061 (A96061)	O	125	55	25–30
6061	T6	310	275	12–17
7075 (A97075)	O	230	105	16–17
7075	T6	570	500	11

Aluminum alloys are available as mill products, that is, as wrought products made into various shapes by rolling, extrusion, drawing, and forging (Chapters 13 through 15). Aluminum ingots are available for casting, as is aluminum in powder

TABLE 6.4

Manufacturing Properties and Typical Applications of Selected Wrought Aluminum Alloys				
Alloy	Characteristics*			Typical applications
	Corrosion resistance	Machinability	Weldability	
1100	A	C–D	A	Sheet metal work, spun hollowware, tin stock
2024	C	B–C	B–C	Truck wheels, screw machine products, aircraft structures
3003	A	C–D	A	Cooking utensils, chemical equipment, pressure vessels, sheet metal work, builders' hardware, storage tanks
5052	A	C–D	A	Sheet metal work, hydraulic tubes, and appliances; bus, truck and marine uses
6061	B	C–D	A	Heavy-duty structures where corrosion resistance is needed; truck and marine structures, railroad cars, furniture, pipelines, bridge railings, hydraulic tubing
7075	C	B–D	D	Aircraft and other structures, keys, hydraulic fittings

*A, excellent; D, poor.

form for powder-metallurgy applications (Chapter 17). Most aluminum alloys can be machined, formed, and welded with relative ease.

There are two types of wrought alloys of aluminum:

1. Alloys that can be *hardened by cold working* and are not heat treatable.
2. Alloys that can be *hardened by heat treatment*.

Designation of wrought aluminum alloys. Wrought aluminum alloys are identified by four digits and by a **temper designation** that shows the condition of the material. (See also Unified Numbering System later in this section.) The major alloying element is identified by the first digit.

> 1xxx—Commercially Pure Aluminum: excellent corrosion resistance, high electrical and thermal conductivity, good workability, low strength, not heat-treatable
>
> 2xxx—Copper: high strength-to-weight ratio, low resistance to corrosion, heat-treatable
>
> 3xxx—Manganese: good workability, moderate strength, generally not heat treatable
>
> 4xxx—Silicon: lower melting point, forms an oxide film of a dark-gray to charcoal color, generally not heat-treatable
>
> 5xxx—Magnesium: good corrosion resistance and weldability, moderate to high strength, not heat treatable
>
> 6xxx—Magnesium and Silicon: medium strength; good formability, machinability, weldability, and corrosion resistance; heat treatable
>
> 7xxx—Zinc: moderate to very high strength, heat treatable
>
> 8xxx—Other element

The second digit in these designations indicates modifications of the alloy. For the 1xxx series, the third and fourth digits stand for the minimum amount of aluminum in the alloy. For example, 1050 indicates a minimum of 99.50% Al and 1090 indicates a minimum of 99.90% Al. In other series, the third and fourth digits identify the different alloys in the group and have no numerical significance.

A typical aluminum beverage can may consist of the following aluminum alloys, all in the H19 condition (which is the highest cold-worked state): 3004 or 3104 for the can body, 5182 for the lid, and 5042 for the tab. These alloys are selected for their manufacturing characteristics as well as for economics.

Designation of cast aluminum alloys.   Designations for cast aluminum alloys also consist of four digits. The first digit indicates the major alloy group, as follows:

> 1xx.x—Aluminum (99.00% minimum)
> 2xx.x—Aluminum-Copper
> 3xx.x—Aluminum-Silicon (with copper and/or magnesium)
> 4xx.x—Aluminum-Silicon
> 5xx.x—Aluminum-Magnesium
> 6xx.x—Unused series
> 7xx.x—Aluminum-Zinc
> 8xx.x—Aluminum-Tin

In the 1xx.x series, the second and third digits indicate the minimum aluminum content, as do the third and fourth in wrought aluminum. For the other series, the second and third digits have no numerical significance. The fourth digit (to the right of the decimal point) indicates product form.

Temper designations.   The temper designations for both wrought and cast aluminum are as follows:

- F—As fabricated (by cold or hot working or by casting)
- O—Annealed (from the cold worked or the cast state)
- H—Strain hardened by cold working (for wrought products only)
- T—Heat treated
- W—Solution treated only (unstable temper)

Unified numbering system.   As is the case with steels, aluminum and other nonferrous metals and alloys now are identified internationally by the Unified Numbering System, consisting of a letter indicating the general class of the alloy followed by five digits indicating its chemical composition. For example, A is for aluminum, C for copper, N for nickel alloys, P for precious metals, and Z for zinc. In the UNS designation, 2024 wrought aluminum alloy is A92024.

Production.   Aluminum was first produced in 1825. It is the most abundant metallic element, making up about 8% of the earth's crust, and is produced in a quantity second only to that of iron. The principal ore for aluminum is *bauxite*, which is a hydrous (water-containing) aluminum oxide and includes various other oxides. After the clay and dirt are washed off, the ore is crushed into powder and treated with hot caustic soda (sodium hydroxide) to remove impurities. Alumina (aluminum oxide) is extracted from this solution and then dissolved in a molten sodium-fluoride and aluminum-fluoride bath at 940 to 980°C (1725 to 1800°F). This mixture is then subjected to direct-current electrolysis. Aluminum metal forms at the cathode (negative pole), while oxygen is released at the anode (positive pole). *Commercially pure aluminum* is up to 99.99% Al, also referred to in industry as "four nines" aluminum. The production process consumes a great deal of electricity, which contributes significantly to the cost of aluminum.

Porous aluminum. Blocks of aluminum have been produced that are 37% lighter than solid aluminum and have uniform permeability (*microporosity*). This characteristic allows their use in applications where a vacuum or differential pressure has to be maintained. Examples are the vacuum holding of fixtures for assembly and automation, and the vacuum forming or thermoforming of plastics (Section 19.6). These blocks are 70 to 90% aluminum powder; the rest is epoxy resin. They can be machined with relative ease and can be joined together using adhesives.

EXAMPLE 6.1 An all-aluminum automobile

Aluminum usage in automobiles and in light trucks has been climbing steadily. As recently as 1990, there were no aluminum-structured passenger cars in production anywhere in the world, but in 1997, there were seven of them, including the Plymouth Prowler and the Audi A8 (Fig. 6.2). With weight savings of up to 47% over steel vehicles, such cars use less fuel, create less pollution, and are recyclable.

New alloys and new design and manufacturing methodologies had to be developed. For example, welding and adhesive bonding procedures had to be refined, the structural frame design had to be optimized, and new tooling designs (to allow forming of aluminum) had to be developed. Because of these new technologies, the desired environmental savings were able to be realized without an accompanying drop in performance or safety. In fact, the Audi A8 is the first luxury-class car to earn a dual five-star (highest safety) rating for both driver and front-seat passenger in the National Highway Transportation Safety Administration (NHSTA) New Car Assessment Program.

6.3 | Magnesium and Magnesium Alloys

Magnesium (Mg) is the lightest engineering metal available, and it has good vibration-damping characteristics. Its alloys are used in structural and nonstructural applications wherever weight is of primary importance. Magnesium is also an alloying element in various nonferrous metals.

Typical uses of magnesium alloys are in aircraft and missile components, material-handling equipment, portable power tools, ladders, luggage, bicycles, sporting goods, and general lightweight components. These alloys are available either as castings (such as die-cast camera frames) or as wrought products (such as extruded bars and shapes, forgings, and rolled plate and sheet). Magnesium alloys are also used in printing and textile machinery to minimize inertial forces in high-speed components (Section 3.2).

Because it is not sufficiently strong in its pure form, magnesium is alloyed with various elements (Table 6.5) in order to gain certain specific properties, particularly a high strength-to-weight ratio. A variety of magnesium alloys have good casting, forming, and machining characteristics. Because they oxidize rapidly (*pyrophoric*), a fire hazard exists, and precautions must be taken when machining, grinding, or sand-casting magnesium alloys. Products made of magnesium and its alloys are, nonetheless, not a fire hazard during normal use.

Designation of magnesium alloys. Magnesium alloys are designated with the following:

    a. One or two prefix letters, indicating the principal alloying elements.

    b. Two or three numerals, indicating the percentage of the principal alloying elements and rounded off to the nearest decimal.

(a)

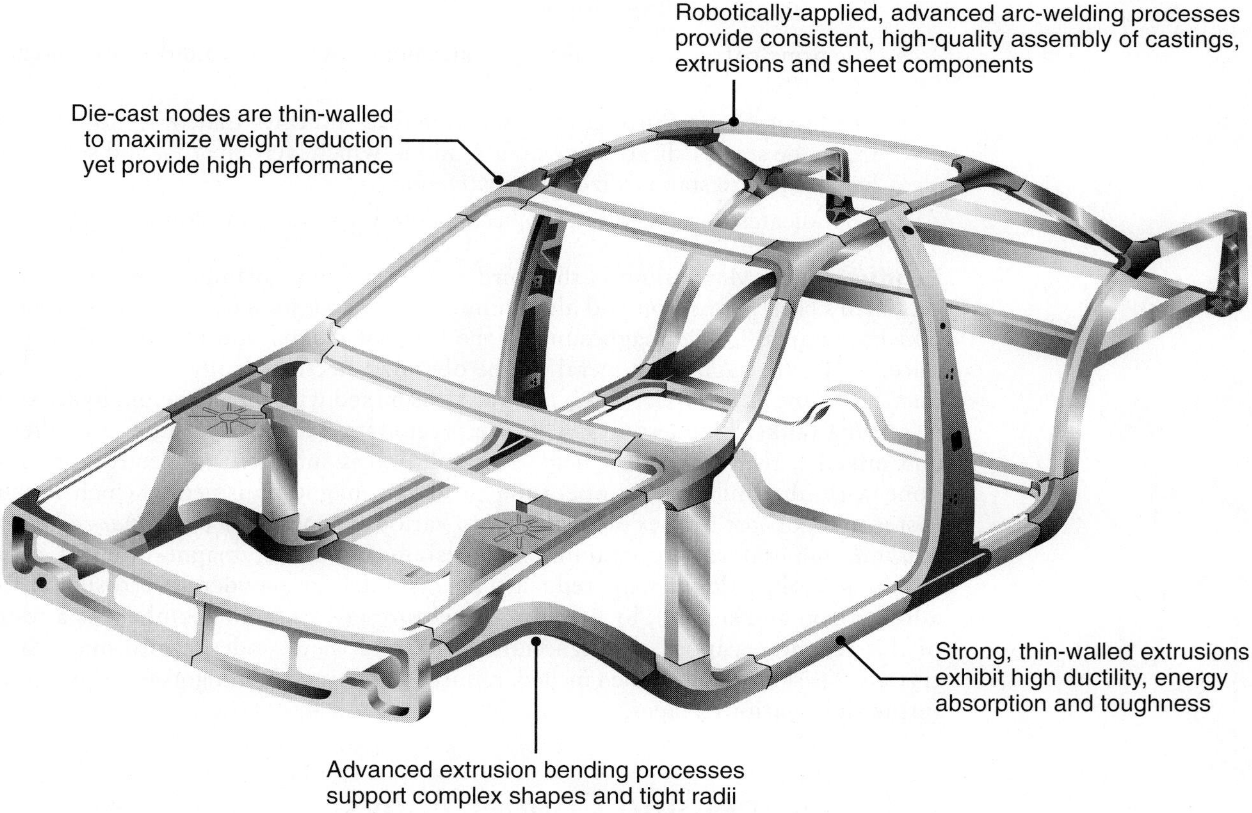

Die-cast nodes are thin-walled to maximize weight reduction yet provide high performance

Robotically-applied, advanced arc-welding processes provide consistent, high-quality assembly of castings, extrusions and sheet components

Strong, thin-walled extrusions exhibit high ductility, energy absorption and toughness

Advanced extrusion bending processes support complex shapes and tight radii

(b)

FIGURE 6.2   (a) The Audi A8 automobile which has an all-aluminum body structure. (b) The aluminum body structure, showing various components made by extrusion, sheet forming, and casting processes. *Source*: Courtesy of ALCOA, Inc.

c. A letter of the alphabet (except the letters I and O) indicating the standardized alloy with minor variations in composition.

d. A symbol for the temper of the material, following the system used for aluminum alloys.

TABLE 6.5

Properties and Typical Forms of Selected Wrought Magnesium Alloys										
Alloy	Composition (%)					Condition	Ultimate tensile strength (MPa)	Yield strength (MPa)	Elongation in 50 mm (%)	Typical forms
	Al	Zn	Mn	Zr	Th					
AZ31 B	3.0	1.0	0.2	—	—	F	260	200	15	Extrusions
						H24	290	220	15	Sheet and plates
AZ80A	8.5	0.5	0.2	—	—	T5	380	275	7	Extrusions and forgings
HK31A	—	—	—	0.7	3	H24	255	200	8	Sheet and plates
ZK60A	—	5.7	—	0.55	—	T5	365	300	11	Extrusions and forgings

For example, take the alloy AZ91C-T6:

- The principal alloying elements are aluminum (A at 9%, rounded off) and zinc (Z at 1%).
- The letter C, the third letter of the alphabet, indicates that this alloy was the third one standardized (later than A and B, which were the first and second alloys that were standardized, respectively).
- T6 indicates that this alloy has been solution treated and artificially aged.

Production. Magnesium is the third most abundant metallic element (2%) in the earth's crust, after iron and aluminum. Most magnesium comes from sea water, which contains 0.13% magnesium in the form of magnesium chloride. First produced in 1808, magnesium metal can be obtained electrolytically or by thermal reduction. In the *electrolytic method*, seawater is mixed with lime (calcium hydroxide) in settling tanks. Magnesium hydroxide precipitates to the bottom, and is filtered and mixed with hydrochloric acid. This solution is subjected to electrolysis (as is done with aluminum); that operation produces magnesium metal, which is then cast into ingots for further processing into various shapes. In the *thermal-reduction method*, mineral rock containing magnesium (dolomite, magnesite, and other rocks) is broken down with reducing agents (such as powdered ferrosilicon, an alloy of iron and silicon), by heating the mixture in a vacuum chamber. As a result of this reaction, vapors of magnesium form, and they condense into magnesium crystals. These crystals are then melted, refined, and poured into ingots to be processed further into various shapes.

6.4 | Copper and Copper Alloys

First produced in about 4000 B.C., **copper** (Cu, from *cuprum*) and its alloys have properties somewhat similar to those of aluminum and its alloys. In addition, they are among the best conductors of electricity and heat (Tables 3.1 and 3.2), and they have good corrosion resistance. They can be processed easily by various forming, machining, casting, and joining techniques.

Copper alloys often are attractive for applications in which a combination of electrical, mechanical, nonmagnetic, corrosion-resistant, thermally conductive, and wear-resistant qualities are required. Applications include electrical and electronic components, springs, cartridges for small arms, plumbing, heat exchangers, marine hardware, and consumer goods (such as cooking utensils, jewelry, and other decorative objects). Although aluminum is the most common material for dies in polymer injection molding (Section 19.3), copper often is used because of its better thermal properties. Pure copper also can be used as a solid lubricant in hot-metal forming operations.

Copper alloys can acquire a wide variety of properties by the addition of alloying elements and by heat treatment, to improve their manufacturing characteristics. The most common copper alloys are brasses and bronzes. **Brass** (which is an alloy of copper and zinc) is one of the earliest alloys developed and has numerous applications, including decorative objects (Table 6.6). **Bronze** is an alloy of copper and tin (Table 6.7). There are also other bronzes, such as *aluminum bronze* (which is an alloy of copper and aluminum) and tin bronzes. **Beryllium copper** (or **beryllium bronze**) and **phosphor bronze** have good strength and hardness for applications such as springs and bearings. Other major copper alloys are *copper nickels* and *nickel silvers*.

Designation of copper alloys.   In the Unified Numbering System, copper is identified with the letter C. For example, cartridge brass is C26200, replacing the outmoded three-digit numbering of CDA 262 (for Copper Development Association). In addition to being identified by their composition, copper and copper alloys are known by various names (Tables 6.5 and 6.6). The temper designations (such as *1/2 hard*, *extra hard*, *extra spring*, and so on) are based on percentage reduction by cold working (such as by rolling or drawing).

Production.   Copper is found in several types of ores, the most common being sulfide ores. The ores are generally of low grade (although some contain up to 15% Cu) and usually are obtained from open-pit mines. The ore first is crushed and then formed into a slurry (a watery mixture with insoluble solid particles). The slurry is ground into fine particles in ball mills (rotating cylinders with metal balls inside to

TABLE 6.6

Properties and Typical Applications of Selected Wrought Copper and Brasses					
Type and UNS number	Nominal composition (%)	Ultimate tensile strength (MPa)	Yield strength (MPa)	Elongation in 50mm (%)	Typical applications
Electrolytic tough pitch copper (C11000)	99.90 Cu, 0.04 O	220–450	70–365	55–4	Downspouts, gutters, roofing, gaskets, auto radiators, bus bars, nails, printing rolls, rivets
Red brass, 85% (C23000)	85.0 Cu, 15.0 Zn	270–725	70–435	55–3	Weather stripping, conduits, sockets, fasteners, fire extinguishers, condenser and heat-exchanger tubing
Cartridge brass, 70% (C26000)	70.0 Cu, 30.0 Zn	300–900	75–450	66–3	Radiator cores and tanks, flashlight shells, lamp fixtures, fasteners, locks, hinges, ammunition components, plumbing accessories
Free-cutting brass (C36000)	61.5 Cu, 3.0 Pb, 35.5 Zn	340–470	125–310	53–18	Gears, pinions, automatic high-speed screw machine parts
Naval brass (C46400 to C46700)	60.0 Cu, 39.25 Zn, 0.75 Sn	380–610	170–455	50–17	Aircraft: turnbuckle barrels, balls, bolts; marine hardware: propeller shafts, rivets, valve stems, condenser plates

TABLE 6.7

Properties and Typical Applications of Selected Wrought Bronzes

Type and UNS number	Nominal composition (%)	Ultimate tensile strength (MPa)	Yield strength (MPa)	Elongation in 50 mm (%)	Typical applications
Architectural bronze (C38500)	57.0 Cu, 3.0 Pb, 40.0 Zn	415 (As extruded)	140	30	Architectural extrusions, store fronts, thresholds, trim, butts, hinges
Phosphor bronze, 5% A (C51000)	95.0 Cu, 5.0 Sn, trace P	325–960	130–550	64–2	Bellows, clutch disks, cotter pins, diaphragms, fasteners, wire brushes, chemical hardware, textile machinery
Free-cutting phosphor bronze (C54400)	88.0 Cu, 4.0 Pb, 4.0 Zn, 4.0 Sn	300–520	130–435	50–15	Bearings, bushings, gears, pinions, shafts, thrust washers, valve parts
Low-silicon bronze, B (C65100)	98.5 Cu, 1.5 Si	275–655	100–475	55–11	Hydraulic pressure lines, bolts, marine hardware, electrical conduits, heat-exchanger tubing
Nickel silver, 65–10 (C74500)	65.0 Cu, 25.0 Zn, 10.0 Ni	340–900	125–525	50–1	Rivets, screws, slide fasteners, hollowware, nameplates

crush the ore (see Fig. 17.6). Chemicals and oil are then added, and the mixture is agitated. The mineral particles form a froth, which is scraped and dried. The dry copper concentrate (as much as one-third of which is copper) is traditionally **smelted** (melted and fused) and refined; this process is known as **pyrometallurgy**, because heat is used to refine the metal. For applications such as for electrical conductors, the copper is refined further electrolytically to a purity of at least 99.95% (*oxygen-free electrolytic copper*). A more recent technique for processing copper is **hydrometallurgy**, a process involving chemical and electrolytic reactions.

6.5 | Nickel and Nickel Alloys

Nickel (Ni) is a silver-white metal discovered in 1751 and a major alloying element that imparts strength, toughness, and corrosion resistance. It is used extensively in stainless steels and in nickel-based alloys (also called **superalloys**). Nickel alloys are used in high-temperature applications (such as jet engine components, rockets, and nuclear power plants), in food-handling and chemical-processing equipment, in coins, and in marine applications. Because nickel is magnetic, nickel alloys also are used in electromagnetic applications, such as solenoids. The principal use of nickel as a metal is in the electroplating of parts for their appearance and for the improvement of their corrosion and wear resistance. Nickel alloys have high strength and corrosion resistance at elevated temperatures. Alloying elements in nickel are chromium, cobalt, and molybdenum. The behavior of nickel alloys in machining, forming, casting, and welding can be modified by various other alloying elements.

A variety of nickel alloys, having a wide range of strengths at different temperatures, have been developed (Table 6.8). Although trade names are still in wide usage, nickel alloys are now identified in the UNS system with the letter N. Thus, Hastelloy G is now N06007. **Monel** is a nickel-copper alloy. **Inconel** is a nickel-chromium alloy with a tensile strength of up to 1400 MPa (200 ksi).

TABLE 6.8

Properties and Typical Applications of Selected Nickel Alloys (All are Trade Names)					
Type and UNS number	Nominal composition (%)	Ultimate tensile strength (MPa)	Yield strength (MPa)	Elongation in 50 mm (%)	Typical applications
Nickel 200 (annealed)	None	380–550	100–275	60–40	Chemical and food processing industry, aerospace equipment, electronic parts
Duranickel 301 (age hardened)	4.4 Al, 0.6 Ti	1300	900	28	Springs, plastics extrusion equipment, molds for glass, diaphragms
Monel R-405 (hot rolled)	30 Cu	525	230	35	Screw-machine products, water meter parts
Monel K-500 (age hardened)	29 Cu, 3 Al	1050	750	30	Pump shafts, valve stems, springs
Inconel 600 (annealed)	15 Cr, 8 Fe	640	210	48	Gas turbine parts, heat-treating equipment, electronic parts, nuclear reactors
Hastelloy C-4 (solution-treated and quenched)	16 Cr, 15 Mo	785	400	54	Parts requiring high-temperature stability and resistance to stress-corrosion cracking

Hastelloy (a nickel-chromium alloy) has good corrosion resistance and high strength at elevated temperatures. **Nichrome** (an alloy of nickel, chromium, and iron) has high electrical resistance and a high resistance to oxidation and is used for electrical heating elements. **Invar** and **Kovar** (alloys of iron and nickel) have relatively low sensitivity to temperature (Section 3.6).

Production.   The main sources of nickel are sulfide and oxide ores, all of which have low concentrations of nickel. Nickel metal is produced by preliminary sedimentary and thermal processes followed by electrolysis; this sequence yields 99.95% pure nickel. Although nickel also is present in the ocean bed in significant amounts, undersea mining of it is not yet economical.

6.6 | Superalloys

Superalloys are important in high-temperature applications; hence, they are also known as **heat-resistant** or **high-temperature alloys.** Superalloys generally have good resistance to corrosion, mechanical and thermal fatigue, mechanical and thermal shock, creep, and erosion at elevated temperatures. Major applications of superalloys are in jet engines and gas turbines. Other applications are in reciprocating engines, rocket engines, tools and dies for hot working of metals, and the nuclear, chemical, and petrochemical industries. Generally, superalloys are identified by trade names or by special numbering systems, and they are available in a variety of shapes. Most superalloys have a maximum service temperature of about 1000°C (1800°F) in structural applications. The temperatures can be as high as 1200°C (2200°F) for nonload-bearing components.

TABLE 6.9

Properties and Typical Applications of Selected Nickel-Based Superalloys at 870°C (1600°F) (All are Trade Names)

Alloy	Condition	Ultimate tensile strength (MPa)	Yield strength (MPa)	Elongation in 50 mm (%)	Typical applications
Astroloy	Wrought	770	690	25	Forgings for high temperature use
Hastelloy X	Wrought	255	180	50	Jet engine sheet parts
IN-100	Cast	885	695	6	Jet engine blades and wheels
IN-102	Wrought	215	200	110	Superheater and jet engine parts
Inconel 625	Wrought	285	275	125	Aircraft engines and structures, chemical processing equipment
Inconel 718	Wrought	340	330	88	Jet engine and rocket parts
MAR-M 200	Cast	840	760	4	Jet engine blades
MAR-M 432	Cast	730	605	8	Integrally cast turbine wheels
René 41	Wrought	620	550	19	Jet engine parts
Udimet 700	Wrought	690	635	27	Jet engine parts
Waspaloy	Wrought	525	515	35	Jet engine parts

Superalloys are referred to as *iron-based*, *cobalt-based*, or *nickel-based*.

- **Iron-based superalloys** generally contain from 32 to 67% Fe, from 15 to 22% Cr, and from 9 to 38% Ni. Common alloys in this group are the *Incoloy* series.
- **Cobalt-based superalloys** generally contain from 35 to 65% Co, from 19 to 30% Cr, and up to 35% Ni. These superalloys are not as strong as nickel-based superalloys, but they retain their strength at higher temperatures.
- **Nickel-based superalloys** are the most common of the superalloys, and they are available in a wide variety of compositions (Table 6.9). The proportion of nickel is from 38 to 76%. They also contain up to 27% Cr and 20% Co. Common alloys in this group are the *Hastelloy, Inconel, Nimonic, René, Udimet, Astroloy,* and *Waspaloy* series.

6.7 | Titanium and Titanium Alloys

Titanium (Ti, named after the giant Greek god Titan) is a silvery white metal, which was discovered in 1791 but not produced commercially until the 1950s. Although expensive, its high strength-to-weight ratio and corrosion resistance at room and elevated temperatures make it attractive for many applications, including aircraft; jet engines (see Fig. 6.1); racing cars; golf clubs; chemical, petrochemical, and marine components; submarine hulls; armor plate; and biomaterials, such as orthopedic implants (Table 6.10). Titanium alloys have been developed for service at 550°C (1000°F) for long periods of time and at up to 750°C (1400°F) for shorter periods.

Unalloyed titanium, known as *commercially pure titanium*, has excellent corrosion resistance for applications where strength considerations are secondary. Aluminum, vanadium, molybdenum, manganese, and other alloying elements impart properties such as improved workability, strength, and hardenability.

TABLE 6.10

Properties and Typical Applications of Selected Wrought Titanium Alloys at Various Temperatures									
Nominal composition (%)	UNS	Condition	Ultimate tensile strength (MPa)	Yield strength (MPa)	Elongation (%)	Reduction of area (%)	Temp. (°C)	Ultimate tensile strength (MPa)	Yield strength (MPa)
99.5 Ti	R50250	Annealed	330	240	30	55	300	150	95
5 Al, 2.5 Sn	R54520	Annealed	860	810	16	40	300	565	450
6 Al, 4 V	R56400	Annealed	1000	925	14	30	300	725	650
		Solution + age	1175	1100	10	20	300	980	900
13 V, 11 Cr, 3 Al	R58010	Solution + age	1275	1210	8	—	425	1100	830

The properties and manufacturing characteristics of titanium alloys are extremely sensitive to small variations in both alloying and residual elements. Therefore, control of composition and processing are important, especially the prevention of surface contamination by hydrogen, oxygen, or nitrogen during processing; these elements cause embrittlement of titanium and, consequently, reduce toughness and ductility.

The body-centered cubic structure of titanium (beta-titanium) is above 880°C (1600°F) and is ductile; whereas its hexagonal close-packed structure (alpha-titanium) is somewhat brittle and is very sensitive to stress corrosion. A variety of other structures (alpha, near-alpha, alpha-beta, and beta) can be obtained by alloying and heat treating, so that the properties can be optimized for specific applications. **Titanium aluminide intermetallics** (TiAl and Ti$_3$Al) have higher stiffness and lower density than conventional titanium alloys, and they can withstand higher temperatures.

Production. Ores containing titanium first are reduced to titanium tetrachloride in an arc furnace, then converted to titanium chloride in a chlorine atmosphere. This compound is reduced further to titanium metal by distillation and leaching (dissolving). This sequence forms *sponge titanium*, which is then pressed into billets, melted, and poured into ingots to be processed later into various shapes. The complexity of these multi-step thermochemical operations (the *Kroll process* developed in the 1940–1950s) adds considerably to the cost of titanium. New developments in electrochemical extraction processes are taking place to reduce the number of steps involved and the energy consumption, thereby reducing the cost of producing titanium.

6.8 | Refractory Metals and Alloys

There are four **refractory metals**: molybdenum, niobium, tungsten, and tantalum. These are called *refractory* because of their high melting point. Although these metals were discovered about 200 years ago and have been used as important alloying elements in steels and superalloys, their use as engineering metals and alloys did not begin until about the 1940s. More than most other metals and alloys, these metals maintain their strength at elevated temperatures. Therefore, they are of great importance and use—in rocket engines, gas turbines, and various other aerospace applications; in the electronic, nuclear-power, and chemical industries; and as tool and die materials. The temperature range for some of these applications is on the order of 1100 to 2200°C (2000 to 4000°F), where strength and oxidation are of major concern.

6.8.1 Molybdenum

Molybdenum (Mo) is a silvery-white metal discovered in the 18th century and has a high melting point, high modulus of elasticity, good resistance to thermal shock, and good electrical and thermal conductivity. Molybdenum is used in greater amounts than is any other refractory metal, with typical applications such as solid-propellant rockets, jet engines, honeycomb structures, electronic components, heating elements, and dies for die casting. The principal alloying elements for molybdenum are titanium and zirconium. It is itself also an important alloying element in cast and wrought alloy steels and in heat-resistant alloys; it imparts strength, toughness, and corrosion resistance. A major limitation of molybdenum alloys is their low resistance to oxidation at temperatures above 500°C (950°F), which necessitates the use of protective coatings.

Production.   The main source for molybdenum is the mineral *molybdenite* (molybdenum disulfide). The ore first is processed and concentrated; later it is reduced by reaction—first with oxygen and then with hydrogen. Powder-metallurgy techniques also are used to produce ingots for further processing into various shapes.

6.8.2 Niobium (Columbium)

Niobium (Nb, for niobium, after Niobe, the daughter of mythical Greek king Tantalus) was first identified in 1801; it is also referred to as **columbium** (after its source mineral, *columbite*). It possesses good ductility and formability and has greater oxidation resistance than other refractory metals. With various alloying elements, niobium alloys can be produced with moderate strength and good fabrication characteristics. These alloys are used in rockets and missiles and in nuclear, chemical, and superconductor applications. Niobium is also an alloying element in various alloys and superalloys. The metal is processed from ores by reduction and refinement and from powder by melting and shaping into ingots.

6.8.3 Tungsten

Tungsten (W, for wolfram, its European name, and from its source mineral, *wolframite*; in Swedish, *tung* means heavy and *sten* means stone) was first identified in 1781; it is the most abundant of all the refractory metals. Tungsten has the highest melting point of any metal (3410°C, 6170°F). As a result, it is characterized by high strength at elevated temperatures. However, it has high density (hence used for balancing weights and counterbalances in mechanical systems, including self-winding watches), is brittle at low temperatures, and offers poor resistance to oxidation. Tungsten imparts strength and hardness to steels at elevated temperatures.

Tungsten alloys are used for applications involving temperatures above 1650°C (3000°F), such as nozzle throat liners in missiles and in the hottest parts of jet and rocket engines, circuit breakers, welding electrodes, tooling for electrical-discharge machining, and spark-plug electrodes. The filament wire in incandescent light bulbs (Section I.1) is made of pure tungsten and is produced by the use of powder-metallurgy and wire-drawing techniques. Tungsten carbide, with cobalt as a binder for the carbide particles, is one of the most important tool and die materials. Tungsten is processed from ore concentrates by chemical decomposition and is then reduced. It is further processed by powder-metallurgy techniques in a hydrogen atmosphere.

6.8.4 Tantalum

Identified in 1802, **tantalum** (Ta, after the mythical Greek king, Tantalus) is characterized by its high melting point (3000°C, 5425°F), high density, good ductility, and resistance to corrosion. However, it has poor resistance to chemicals at temperatures above 150°C (300°F). Tantalum is used extensively in electrolytic capacitors and in various components in the electrical, electronic, and chemical industries; it also is used for thermal applications, such as in furnaces and in acid-resistant heat exchangers. A variety of tantalum-based alloys are available in many forms for use in missiles and aircraft. Tantalum also is used as an alloying element. It is processed by techniques similar to those used for processing niobium.

6.9 | Beryllium

Steel gray in color, **beryllium** (Be, from the ore *beryl*) has a high strength-to-weight ratio. Unalloyed beryllium is used in rocket nozzles, space and missile structures, aircraft disc brakes, and precision instruments and mirrors. It is used in nuclear and x-ray applications because of its low neutron absorption. Beryllium is also an alloying element, and its alloys of copper and nickel are used in various applications including springs (*beryllium copper*), electrical contacts, and nonsparking tools for use in such explosive environments as mines and metal-powder production (Section 17.2). Beryllium and its oxide are toxic; their associated dust and fumes should not be inhaled.

6.10 | Zirconium

Zirconium (Zr) is silvery in appearance; it has good strength and ductility at elevated temperatures and has good corrosion resistance because of an adherent oxide film. This element is used in electronic components and in nuclear-power reactor applications because of its low neutron absorption.

6.11 | Low-Melting Alloys

Low-melting alloys are so named because of their relatively low melting points. The major metals in this category are lead, zinc, tin, and their alloys.

6.11.1 Lead

Lead (Pb, after *plumbum*, the root for the word plumber) has properties of high density, resistance to corrosion (by virtue of the stable lead-oxide layer that forms to protect the surface), softness, low strength, ductility, and good workability. Alloying it with various elements (such as antimony and tin) enhances its desirable properties, making it suitable for piping, collapsible tubing, bearing alloys, cable sheathing, roofing, and lead-acid storage batteries. Lead also is used for damping sound and vibrations, radiation shielding against x-rays, ammunition, as weights, and in the chemical industry.

The oldest known lead artifacts were made in about 3000 B.C. Lead pipes made by the Romans and installed in the Roman baths in Bath, England, two millennia ago are still in use. Lead is also an alloying element in *solders*, steels, and copper alloys; it promotes corrosion resistance and machinability. An additional use of lead is as a solid lubricant for hot-metal forming operations. Because of its toxicity, however, environmental contamination by lead (causing lead poisoning) is a major concern; major efforts are being made to replace lead with other elements (such as *lead-free solders*). The important source mineral for lead is *galena* (PbS). It is mined, smelted, and refined by chemical treatments.

6.11.2 Zinc

Zinc (Zn), is bluish-white in color and is the metal fourth most utilized industrially, after iron, aluminum, and copper. Although its existence was known for many centuries, zinc was not developed until the 18th century. It has two major uses: (1) for galvanizing iron, steel sheet, and wire, (2) as an alloy base for casting.

In **galvanizing**, zinc serves as an anode and protects the steel (cathode) from corrosive attack should the coating be scratched or punctured. Zinc also is used as an alloying element; brass, for example, is an alloy of copper and zinc. Major alloying elements in zinc-based alloys are aluminum, copper, and magnesium; they impart strength and provide dimensional control during casting of the metal. Zinc-based alloys are used extensively in die casting for making such products as fuel pumps and grills for automobiles, components for household appliances such as vacuum cleaners and washing machines, kitchen equipment, various machinery parts, and photoengraving equipment. Another use for zinc is in superplastic alloys which have good formability characteristics by virtue of their capacity to undergo large deformation without failure. A very fine grained 78% Zn-22% Al sheet is a common example of a superplastic zinc alloy which can be formed by methods used for forming plastics or metals.

Production.   A number of minerals containing zinc are found in nature. The principal source mineral is zinc sulfide, also called *zincblende*. The ore first is roasted in air and converted to zinc oxide. It then is reduced to zinc either electrolytically (using sulfuric acid) or by heating it in a furnace with coal (which causes the molten zinc to separate).

6.11.3 Tin

Although used in small amounts, **tin** (Sn, from the Latin *stannum*) is an important metal. The most extensive use of tin (a silvery-white, lustrous metal) is as a protective coating on steel sheet (*tin plate*) used in making containers (*tin cans*) for food and for various other products. The low shear strength of the tin coatings on steel sheet improves its performance in deep drawing and in general presswork. Unlike the case of galvanized steels, if this coating is punctured or destroyed, the steel corrodes because tin is cathodic to it.

Unalloyed tin is used in such applications as a lining material for water distillation plants and as a molten layer of metal over which plate glass is made (Section 18.3). Tin-based alloys (also called **white metals**) generally contain copper, antimony, and lead. The alloying elements impart hardness, strength, and corrosion resistance. Tin is an alloying element for dental alloys and for bronze (copper-tin alloy), titanium, and zirconium alloys. Tin-lead alloys are common soldering materials, with a wide range of compositions and melting points.

Because of their low friction coefficients. (which result from low shear strength and low adhesion) some tin alloys are used as journal-bearing materials. These alloys are known as **babbitts** (after I. Babbitt, 1799–1862) and contain tin, copper, and antimony. **Pewter** is an alloy of tin, copper, and antimony. It was developed in the 15th century and has been used for tableware, hollowware, and decorative artifacts. Organ pipes are made of tin alloys. The most important tin mineral is *cassiterite* (tin oxide), which is of low grade. The ore is mined, concentrated by various techniques, smelted, refined, and cast into ingots for further processing.

6.12 | Precious Metals

The most important *precious* (costly) metals, also called **noble metals**, are described here.

- **Gold** (Au, after *aurum*) is soft and ductile and has good corrosion resistance at any temperature. Typical applications include jewelry, coinage, reflectors, gold leaf for decorative purposes, dental work, electroplating, and electrical contacts and terminals.

- **Silver** (Ag, after *argentum*) is a ductile metal and has the highest electrical and thermal conductivity of any metal (Table 3.1). However, it develops an oxide film that affects its surface characteristics and appearance. Typical applications for silver include tableware, jewelry, coinage, electroplating, photographic film, electrical contacts, solders, bearing linings, and food and chemical equipment. *Sterling silver* is an alloy of silver and 7.5% copper.

- **Platinum** (Pt) is a soft, ductile, grayish-white metal that has good corrosion resistance even at elevated temperatures. Platinum alloys are used as electrical contacts, for spark-plug electrodes, as catalysts for automobile pollution-control devices, in filaments, in nozzles, in dies for extruding glass fibers (Section 18.3), in thermocouples, in the electrochemical industry, as jewelry, and in dental work.

6.13 | Shape-Memory Alloys

Shape-memory alloys are unique in that, after being plastically deformed at room temperature into various shapes, they return to their original shape upon heating. For example, a piece of straight wire made of this material can be wound into the shape of a helical spring; when heated with a match, the spring uncoils and returns to the original straight shape. Shape-memory alloys can be used to generate motion and/or force in temperature-sensitive actuators. Their behavior can also be *reversible*, that is, the shape can switch back and forth repeatedly upon application and removal of heat. A typical shape-memory alloy is 55% Ni-45% Ti (*Nitinol*). Other such alloys are copper-aluminum-nickel, copper-zinc-aluminum, iron-manganese-silicon, and titanium-nickel-hafnium. Shape-memory alloys generally have such properties as good ductility, corrosion resistance, and high electrical conductivity.

Applications of shape-memory alloys include various sensors, eyeglass frames, stents, relays, pumps, switches, connectors, clamps, fasteners, and seals. As an example, a nickel-titanium anti-scald valve has been made to protect people from being scalded in sinks, tubs, and showers. It is installed directly into the piping system and brings the water flow down to a trickle within 3 seconds after the water temperature reaches 47°C (116°F). New developments include thin-film shape-memory alloys deposited on polished silicon substrates for use in microelectromechanical (MEMS) devices (see Chapter 29).

6.14 | Amorphous Alloys (Metallic Glasses)

A class of metal alloys which, unlike metals, do not have a long-range crystalline structure are called *amorphous alloys*; they have no grain boundaries, and the atoms are packed randomly and tightly. The amorphous structure first was obtained in the late 1960s by the extremely **rapid solidification** of the molten alloy (Section 11.5). Because their structure resembles that of glasses, these alloys are also called *metallic glasses*. Amorphous alloys typically contain iron, nickel, and chromium, which are alloyed with carbon, phosphorus, boron, aluminum, and silicon. They are available as wire, ribbon, strip, and powder. One application is for face-plate inserts on golf-club heads; this alloy has a composition of zirconium, beryllium, copper, titanium, and nickel and is made by die casting. Another application is in hollow aluminum baseball bats coated with a composite of amorphous metal by thermal spraying and is said to improve the performance of the bat.

Amorphous alloys exhibit excellent corrosion resistance, good ductility, high strength, and very low loss from magnetic hysteresis. The latter property is utilized in the making of magnetic steel cores for transformers, generators, motors, lamp ballasts, magnetic amplifiers, and linear accelerators. The low magnetic hysteresis loss provides greatly improved efficiency; however, fabrication costs are significant. Amorphous steels are being developed with strengths twice those of high-strength steels, with potential applications in large structures; however, they are presently cost-prohibitive. A major application for the superalloys of rapidly solidified powders is the consolidation into near-net shapes for parts used in aerospace engines.

6.15 | Metal Foams

Metal foams are material structures where the metal consists of only 5 to 20% of the structure's volume. Usually made of aluminum alloys (but also of titanium or tantalum), metal foams can be produced by blowing air into molten metal and tapping the froth that forms at the surface; this froth then solidifies into a foam. Other approaches to producing metal foam include (a) chemical vapor deposition (Section 34.6) into a polymer or carbon foam lattice, and (b) doping molten or powder metals (Chapter 17) with titanium hydride (TiH_2), which then releases hydrogen gas at the elevated casting or sintering temperatures.

Metal foams have unique combinations of strength-to-density and stiffness-to-density ratios, although these ratios are not as high as the base metals themselves. However, they are very lightweight and, thus, are attractive materials for aerospace applications. Because of their porosity, other applications of metal foams include filters and orthopedic implants.

6.16 | Nanomaterials

Important developments involve the production of materials with grains, fibers, films, and composites having particles that are on the order of 1–100 nm in size. First investigated in the early 1980s and generally called *nanomaterials*, they have certain properties that are frequently superior to traditional and commercially available materials. These characteristics include the strength, hardness, ductility, wear resistance and corrosion resistance suitable for structural (load bearing) and nonstructural applications in combination with unique electrical, magnetic, and optical properties.

The composition of a nanomaterial can be any combination of chemical elements. Among the more important compositions are carbides, oxides, nitrides, metals and alloys, organic polymers, and various composites. Synthesis methods include inert-gas condensation, plasma synthesis, electrodeposition, sol-gel synthesis, and mechanical alloying or ball milling.

Synthesized powders are consolidated into bulk materials by various techniques, including compaction and sintering. They are available in a variety of shapes and are called by various names, such as nanocrystalline materials, nanostructured materials, nanophase materials, nanopowders, nanofibers, nanowires, nanotubes, and nanofilms. Because the synthesis of these products is done at atomic levels, their purity (on the order of 99.9999%), their homogeneity, and the uniformity of their microstructure are highly controlled; as a result, their mechanical, physical, and chemical properties also can be controlled precisely.

Among the current and potential applications for nanomaterials are the following:

a. Cutting tools and inserts made of nanocrystalline carbides and other ceramics.

b. Nanophase ceramics that are ductile and machinable.

c. Powders for powder-metallurgy processing.

d. Next-generation computer chips using nanocrystalline starting materials with very high purity, better thermal conductivity, and more durable interconnections.

e. Flat-panel displays for laptop computers and televisions made by synthesizing nanocrystalline phosphorus to improve screen resolution.

f. Spark-plug electrodes, igniters and fuels for rockets, medical implants, high-sensitivity sensors, catalysts for elimination of pollutants, high-power magnets, and high-energy-density batteries.

SUMMARY

- Nonferrous metals and alloys cover a very broad range of materials. They may consist of aluminum, magnesium, and copper and of their alloys, which have a wide range of applications. For higher-temperature service, they consist of nickel, titanium, refractory alloys (molybdenum, niobium, tungsten, tantalum), and superalloys. Other nonferrous material categories include low-melting alloys (lead, zinc, tin) and precious metals (gold, silver, platinum).

- Nonferrous alloys have a wide variety of desirable properties, such as strength, toughness, hardness, and ductility; resistance to high temperature, creep, and oxidation; a wide range of physical, thermal, and chemical properties; and high strength-to-weight and stiffness-to-weight ratios (particularly for aluminum and titanium). They can be heat treated to impart certain desired properties. As in all materials, the selection of a nonferrous material for a particular application requires consideration of many factors, including design and service requirements, long-term effects, chemical affinity to other materials, environmental attack, and cost.

- Shape-memory alloys, amorphous alloys (metallic glasses), and nanomaterials have some properties that are superior to those of conventional materials. Each has several unique applications in product design and manufacturing.

KEY TERMS

Amorphous alloys	Metal foam	Refractory metals
Babbitts	Metallic glasses	Shape-memory alloys
Brass	Nanomaterials	Smelting
Bronze	Pewter	Superalloys
Galvanizing	Precious metals	Temper designation
Low-melting alloys	Pyrometallurgy	

BIBLIOGRAPHY

ASM Handbook, Vol. 2: Properties and Selection: Non-ferrous Alloys and Special-Purpose Materials, ASM International, 1990.
ASM Specialty Handbook: Aluminum and Aluminum Alloys, ASM International, 1993.
ASM Specialty Handbook: Copper and Copper Alloys, ASM International, 2001.
ASM Specialty Handbook: Heat-Resistant Materials, ASM International, 1997.
ASM Specialty Handbook: Magnesium and Magnesium Alloys, ASM International, 1999.
ASM Specialty Handbook: Nickel, Cobalt, and Their Alloys, ASM International 2000.
Bhushan, B. (ed.), *Handbook of Nanotechnology,* Springer, 2004.
Donachie, M.J. (ed.), *Titanium: A Technical Guide,* 2nd ed., ASM International, 2000.

Donachie, M.J., and Donachie, S.J., *Superalloys: A Technical Guide,* 2nd ed., ASM International, 2002.
Edelstein, A.S., and Cammarata, R.C., (eds.), *Nanomaterials: Synsthesis, Properties, and Applications,* Institute of Physics, 1998.
Farag, M.M., *Materials Selection for Engineering Design,* Prentice Hall, 1997.
Fremond, M., and Miyazaki, S., *Shape-Memory Alloys,* Springer Verlag, 1996.
Harper, C. (ed.), *Handbook for Materials for Product Design,* 3rd ed., McGraw-Hill, 2001.
Kaufman, J.G., *Introduction to Aluminum Alloys and Tempers,* ASM International, 2000.
Mitura, S., *Nanomaterials,* Elsevier, 2000.
Superalloys: A Technical Guide, 2nd ed., ASM International, 2002.

REVIEW QUESTIONS

6.1 Given the abundance of aluminum in the earth's crust, explain why it is more expensive than steel.

6.2 Why is magnesium often used as a structural material in power hand tools? Why are its alloys used instead of pure magnesium?

6.3 What are the major uses of copper? What are the alloying elements in brass and bronze, respectively?

6.4 What are superalloys? Why are they so named?

6.5 What properties of titanium make it attractive for use in race-car and jet-engine components? Why is titanium not used widely for engine components in passenger cars?

6.6 What are the individual properties of each of the major refractory metals that define their most useful applications?

6.7 What are metallic glasses? Why is the word "glass" used for these materials?

6.8 What is the composition (a) of babbitts, (b) of pewter, and (c) of sterling silver?

6.9 Which of the materials described in this chapter has the highest (a) density, (b) electrical conductivity, (c) thermal conductivity, (d) strength, and (e) cost?

6.10 What are the major uses of gold, other than in jewelry?

6.11 What are the advantages to using zinc as a coating for steel?

6.12 What are nanomaterials? Why are they being developed?

6.13 Why are aircraft skins made of aluminum alloys, even though magnesium is the lightest metal?

6.14 What are the major uses of lead?

QUALITATIVE PROBLEMS

6.15 Explain why cooking utensils generally are made of stainless steels, aluminum, or copper.

6.16 Would it be advantageous to plot the data in Table 6.1 in terms of cost-per-unit weight rather than of cost-per-unit volume? Explain and give some examples.

6.17 Inspect Table 6.2 and comment on which of the two hardening processes (heat treating and work hardening) is more effective in improving the strength of aluminum alloys.

6.18 Other than mechanical strength, what other factors should be considered in selecting metals and alloys for high-temperature applications?

6.19 Explain why you would want to know the ductility of metals and alloys before selecting them.

6.20 Explain the techniques you would use to strengthen aluminum alloys.

6.21 Assume that, for geopolitical reasons, the price of copper increases rapidly. Name two metals with similar mechanical and physical properties that can be substituted for copper. Comment on your selection and observations.

6.22 If planes (such as a Boeing 757) are made of 79% aluminum, why are automobiles predominantly made of steel?

6.23 Portable (notebook) computers have their housing made of magnesium. Why?

6.24 Table 6.3 lists the manufacturing properties of wrought aluminum alloys. Compare their relative characteristics with those of other metals.

6.25 Most household wiring is made of copper wire. By contrast, grounding wire leading to satellite dishes and the like is made of aluminum. Explain the reason for this.

QUANTITATIVE PROBLEMS

6.26 A simply supported rectangular beam is 25-mm wide and 1-m long, and it is subjected to a vertical load of 30 kg at its center. Assume that this beam could be made of any of the materials listed in Table 6.1. Select three different materials and calculate for each the beam's height that would cause each beam to have the same maximum deflection. Calculate the ratio of the cost for each of the three beams.

6.27 Obtain a few aluminum beverage cans, cut them, and measure their wall thicknesses. Using data in this chapter and simple formulas for thin-walled, closed-end pressure vessels, calculate the maximum internal pressure these cans can withstand before yielding.

6.28 Beverage cans usually are stacked on top of each other in stores. By using information from Problem 6.27 and referring to textbooks on the mechanics of solids, make an estimate of the crushing load each of these cans can withstand.

6.29 Using strength and density data, determine the minimum weight of a two-foot long tension member which must support 750 pounds if it is manufactured from (a) 3003-O aluminum, (b) 5052-H34 aluminum, (c) AZ31B-F magnesium, (d) any brass alloy, and (e) any bronze alloy.

6.30 An automobile engine operates at up to 7000 rpm. If the stroke length for a piston is 6 in. and the piston is made of a 10-lb. steel casting, estimate the inertial stress on the 1-in. diameter connecting rod. If the piston is replaced by the same volume of aluminum alloy, what would be the speed for the same inertia-induced stress?

6.31 Plot the following for the materials described in this chapter: (a) yield strength versus density, (b) modulus of elasticity versus strength, and (c) modulus of elasticity versus relative cost.

SYNTHESIS, DESIGN, AND PROJECTS

6.32 Because of the number of processes involved in making them, the cost of raw materials for metals depends on their condition (hot or cold rolled), shape (plate, sheet, bar, tubing), and size. Make a survey of the technical literature and price lists (or get in touch with suppliers) and prepare a list indicating the cost per 100 kg of the nonferrous materials described in this chapter, as available in different conditions, shapes, and sizes.

6.33 The materials described in this chapter have numerous applications. Make a survey of the available literature in the bibliography and prepare a list of several specific products and applications, indicating the types of materials used.

6.34 Name products that would not have been developed to their advanced stages (as we find them today) if alloys having high strength, high corrosion resistance, and high creep resistance (all at elevated temperatures) had not been developed.

6.35 Assume that you are the technical sales manager of a company that produces nonferrous metals. Choose any one of the metals and alloys described in this chapter and prepare a brochure, including some illustrations, for use as sales literature by your staff in their contact with potential customers.

6.36 Inspect several metal products and components and make an educated guess as to what materials each is made from. Give reasons for your guesses. If you list two or more possibilities, explain your reasoning.

6.37 Give applications for (a) amorphous metals, (b) precious metals, (c) low-melting alloys, and (d) nanomaterials.

6.38 Describe the advantages of making products with multilayer materials. (For example, aluminum bonded to the bottom of stainless-steel pots.)

6.39 Describe applications and designs utilizing shape-memory alloys.

6.40 The Bronze Age is so known because the hardest metals known at the time were bronzes. Therefore, tools, weapons, and armor were made from bronze. Investigate the geographical sources of the metals needed for bronze and identify the known sources in the Bronze Age. (*Note*: Does this explain the Greek interest in the British Isles?)

6.41 Aluminum beverage-can tops are made from 5182 alloy, while the bottoms are made from 3004 alloy. Study the properties of these alloys and explain why they are used for these applications.

6.42 Obtain specimens of pure copper, pure aluminum, and alloys of copper and aluminum. Conduct tension tests on each, plot the stress–strain diagrams, and evaluate the results.

6.43 Comment on your observations regarding the types of materials used in particular sections of the jet engine shown in Fig. 6.1.

6.44 Inspect various small or large appliances in your home and identify the metals and alloys that you think have been used in their construction.

6.45 Referring to recent technical literature, comment on the trends in the use of metallic materials in (a) military vehicles, (b) sports equipment, (c) medical equipment, (d) automotive applications, and (e) aircraft.

Polymers: Structure, General Properties, and Applications

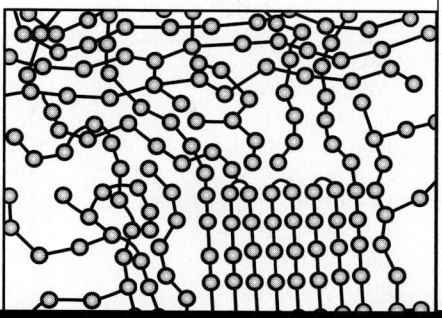

With their widely different properties and applications, plastics continue to be among the most commonly used materials in products. In order to better understand their characteristics, this chapter describes the following:

- Structure of polymers, polymerization processes, crystallinity, and glass-transition temperature.
- How temperature and deformation rate affect the properties of thermoplastics.
- Differences between thermoplastics and thermosets.
- Properties and applications of polymers, their advantages and limitations.

7.1 | Introduction

The word **plastics** first was used as a noun around 1909 and commonly is employed as a synonym for **polymers**. Plastics are one of numerous polymeric materials and have extremely large molecules (*macromolecules* or *giant molecules*). Consumer and industrial products made of polymers include food and beverage containers, packaging, signs, housewares, housings for computers and monitors, textiles, medical devices, foams, paints, safety shields, toys, appliances, lenses, gears, electronic and electrical products, and automobile bodies and components.

Because of their many unique and diverse properties, polymers increasingly have replaced metallic components in applications such as automobiles, civilian and military aircraft, sporting goods, toys, appliances, and office equipment. These substitutions reflect the advantages of polymers in terms of the following characteristics:

- Corrosion resistance and resistance to chemicals
- Low electrical and thermal conductivity
- Low density
- High strength-to-weight ratio (particularly when reinforced)
- Noise reduction
- Wide choice of colors and transparencies

TABLE 7.1

Range of Mechanical Properties for Various Engineering Plastics at Room Temperature				
Material	UTS (MPa)	E (GPa)	Elongation (%)	Poisson's ratio (ν)
ABS	28–55	1.4–2.8	75–5	—
ABS, reinforced	100	7.5	—	0.35
Acetal	55–70	1.4–3.5	75–25	—
Acetal, reinforced	135	10	—	0.35–0.40
Acrylic	40–75	1.4–3.5	50–5	—
Cellulosic	10–48	0.4–1.4	100–5	—
Epoxy	35–140	3.5–17	10–1	—
Epoxy, reinforced	70–1400	21–52	4–2	—
Fluorocarbon	7–48	0.7–2	300–100	0.46–0.48
Nylon	55–83	1.4–2.8	200–60	0.32–0.40
Nylon, reinforced	70–210	2–10	10–1	—
Phenolic	28–70	2.8–21	2–0	—
Polycarbonate	55–70	2.5–3	125–10	0.38
Polycarbonate, reinforced	110	6	6–4	—
Polyester	55	2	300–5	0.38
Polyester, reinforced	110–160	8.3–12	3–1	—
Polyethylene	7–40	0.1–1.4	1000–15	0.46
Polypropylene	20–35	0.7–1.2	500–10	—
Polypropylene, reinforced	40–100	3.5–6	4–2	—
Polystyrene	14–83	1.4–4	60–1	0.35
Polyvinyl chloride	7–55	0.014–4	450–40	—

- Ease of manufacturing and complexity of design possibilities
- Relatively low cost (see Table 6.1)
- Other characteristics that may or may not be desirable (depending on the application), such as low strength and stiffness (Table 7.1), high coefficient of thermal expansion, low useful-temperature range—up to about 350°C (660°F), and lower dimensional stability in service over a period of time.

The word *plastic* is from the Greek word *plastikos*, meaning capable of being molded and shaped. Plastics can be formed, machined, cast, and joined into various shapes with relative ease. Minimal additional surface-finishing operations, if any at all, are required; this characteristic provides an important advantage over metals. Plastics are available commercially as film, sheet, plate, rods, and tubing of various cross-sections.

The word *polymer* first was used in 1866. The earliest polymers were made of **natural organic materials** from animal and vegetable products; *cellulose* is the most common example. By means of various chemical reactions, cellulose is modified into *cellulose acetate*, used in making photographic films (*celluloid*), sheets for packaging, and textile fibers; *cellulose nitrate* for plastics and explosives; *rayon* (a cellulose-based textile fiber); and varnishes. The earliest **synthetic** (manmade) polymer was a phenolformaldehyde, a thermoset developed in 1906 and called *Bakelite* (a trade name, after L. H. Baekeland, 1863–1944).

The development of modern plastics technology began in the 1920s when the raw materials necessary for making polymers were extracted from coal and petroleum

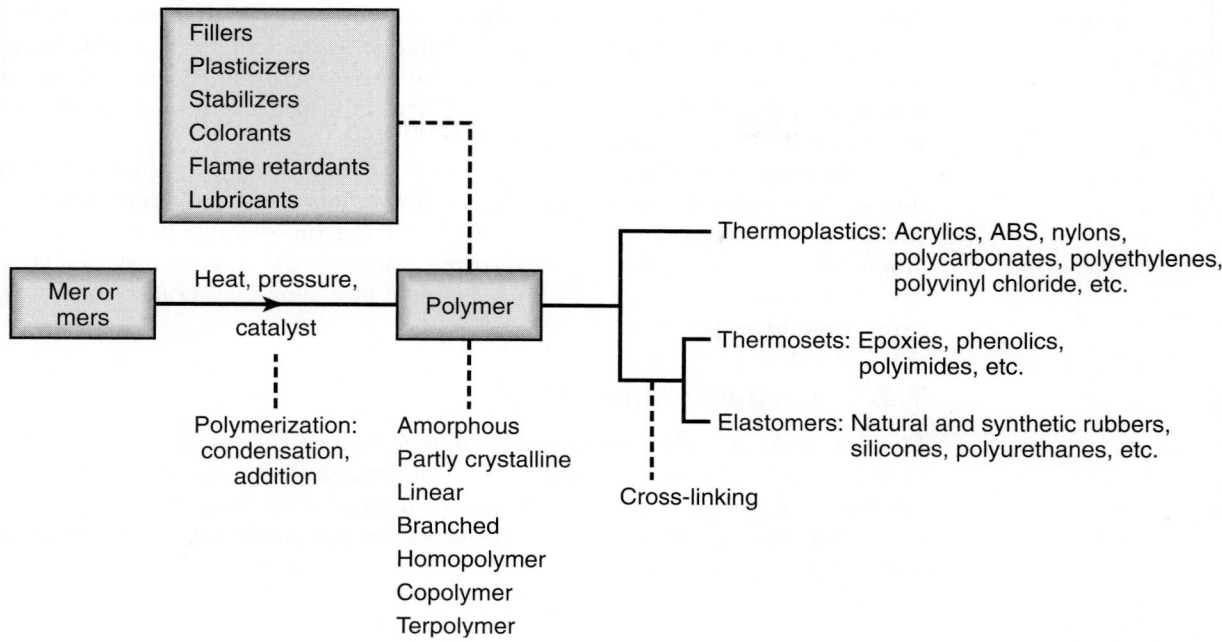

FIGURE 7.1   Outline of the topics described in Chapter 7.

products. Ethylene was the first example of such a raw material; it became the building block for polyethylene. Ethylene is the product of the reaction between acetylene and hydrogen, and acetylene is the product of the reaction between coke and methane. Commercial polymers, such as polypropylene, polyvinyl chloride, polymethylmethacrylate, polycarbonate, and others, are all made in a similar manner; these materials are known as **synthetic organic polymers.**

An outline of the basic process of making various synthetic polymers is given in Fig. 7.1. In polyethylene, only carbon and hydrogen atoms are involved, but other polymer compounds can be obtained by including chlorine, fluorine, sulfur, silicon, nitrogen, and oxygen. As a result, an extremely wide range of polymers—having among them an equally wide range of properties—has been developed.

This chapter describes the relationship of the structure of a polymer to its properties and behavior, during both manufacturing and its service life under various physical and environmental conditions. An outline of the topics to be presented is given in Fig. 7.1. This chapter also describes the properties and engineering applications of plastics, rubbers, and elastomers. Reinforced plastics and composite materials are described in Chapter 9 and processing methods for plastics and reinforced plastics in Chapter 19.

7.2 | The Structure of Polymers

The properties of polymers depend largely on the structures of individual polymer molecules, molecule shape and size, and how molecules are arranged to form a polymer structure. Polymer molecules are characterized by their very large size—a feature that distinguishes them from most other organic chemical compositions.

Polymers are **long-chain molecules** that are formed by *polymerization* (that is, by the linking and cross-linking of different monomers). A **monomer** is the basic building block of a polymer. The word **mer** (from the Greek *meros*, meaning part) indicates the smallest repetitive unit; its use is similar to that of the term *unit cell* in crystal structures of metals (Section 1.2).

The term **polymer** means many mers (or units), generally repeated hundreds or thousands of times in a chainlike structure. Most monomers are *organic materials* in which carbon atoms are joined in *covalent* (electron sharing) bonds with other atoms (such as hydrogen, oxygen, nitrogen, fluorine, chlorine, silicon, and sulfur). An ethylene molecule (Fig. 7.2) is an example of a simple monomer consisting of carbon and hydrogen atoms.

7.2.1 Polymerization

Monomers can be linked into polymers in repeating units to make longer and larger molecules by a chemical process called a **polymerization reaction**. Polymerization processes are complex; they will be described only briefly here. Although there are several variations, two polymerization processes are important: condensation and addition polymerization.

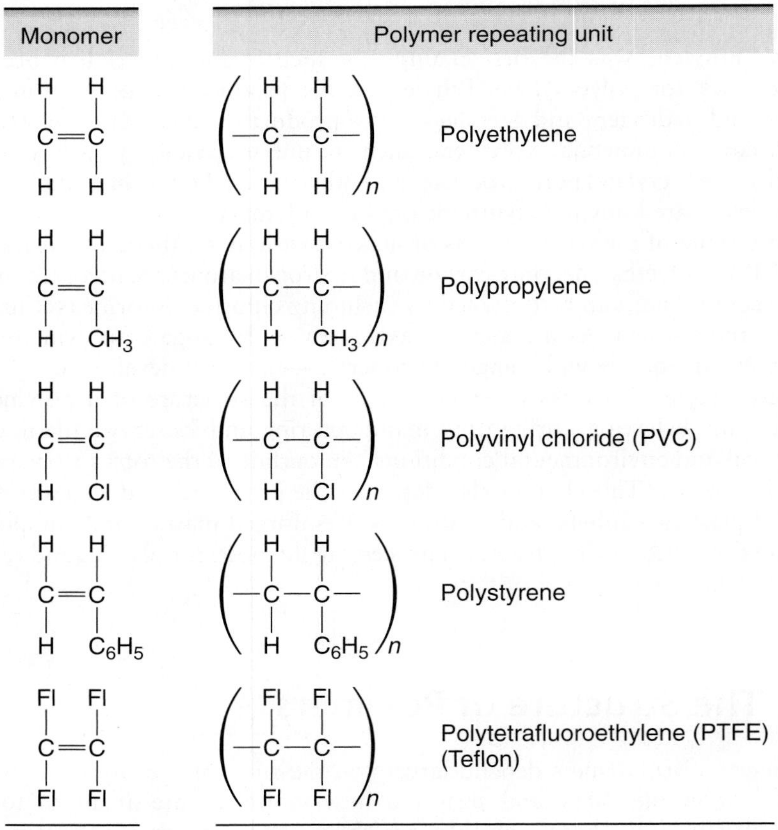

FIGURE 7.2   Molecular structure of various polymers. These are examples of the basic building blocks for plastics.

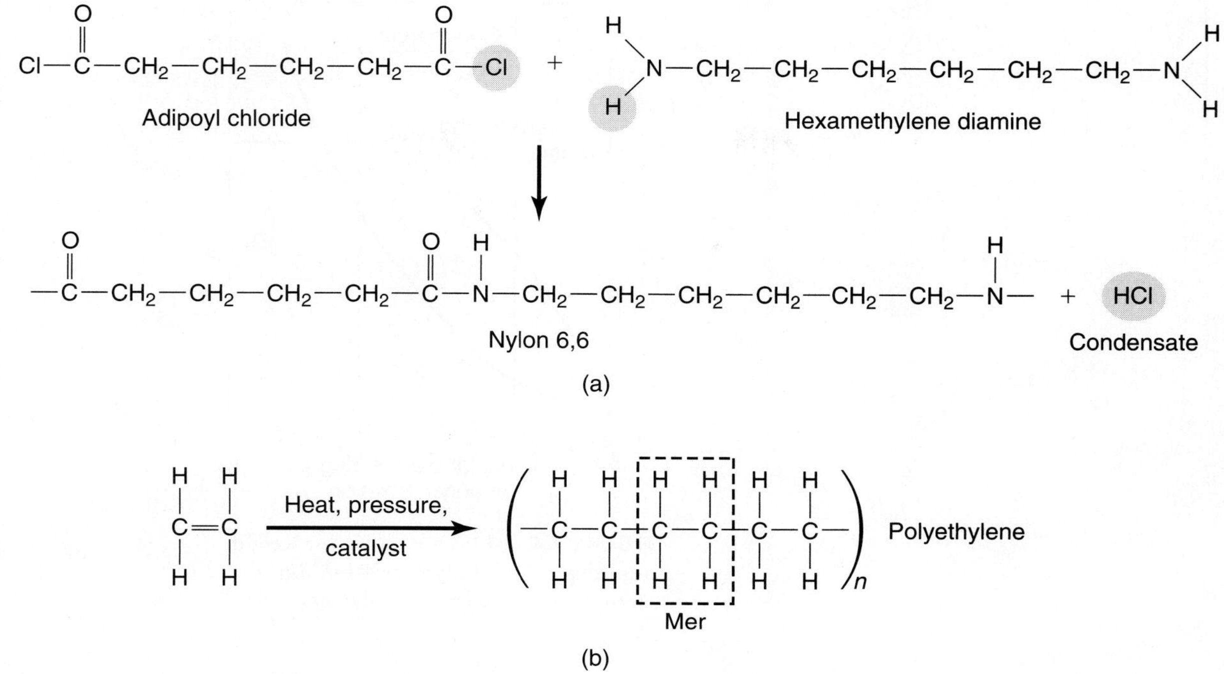

FIGURE 7.3   Examples of polymerization. (a) Condensation polymerization of nylon 6,6 and (b) addition polymerization of polyethylene molecules from ethylene mers.

In **condensation polymerization** (Fig. 7.3), polymers are produced by the formation of bonds between two types of reacting mers. A characteristic of this reaction is that reaction by-products (such as water) are condensed out (hence the name). This process is also known as **step-growth** or **step-reaction polymerization**, because the polymer molecule grows step-by-step until all of one reactant is consumed.

In **addition polymerization** (also called **chain-growth** or **chain-reaction polymerization**), bonding takes place without reaction by-products, as shown in Fig. 7.3b. It is called "chain reaction" because of the high rate at which long molecules form simultaneously, usually within a few seconds. This rate is much higher than that in condensation polymerization. In this reaction, an *initiator* is added to open the double bond between two carbon atoms, which begins the linking process by adding many more monomers to a growing chain. For example, ethylene monomers (Fig. 7.3b) link to produce the polymer *polyethylene*; other examples of addition-formed polymers are shown in Fig. 7.2.

Molecular weight.   The sum of the molecular weights of the mers in a representative chain is known as the *molecular weight* of the polymer. The higher the molecular weight of a given polymer, the greater the average chain length. Most commercial polymers have a molecular weight between 10,000 and 10,000,000. Because polymerization is a random event, the polymer chains produced are not all of equal length, but the chain lengths produced fall into a traditional distribution curve. The average molecular weight of a polymer is determined on a statistical basis by averaging. The spread of the molecular weights in a chain is referred to as the **molecular weight distribution** (MWD). A polymer's molecular weight and its MWD have a strong influence on its properties. For example, the tensile and the impact strength, the resistance to cracking, and the viscosity (in the molten state) of the polymer all increase with increasing molecular weight (Fig. 7.4).

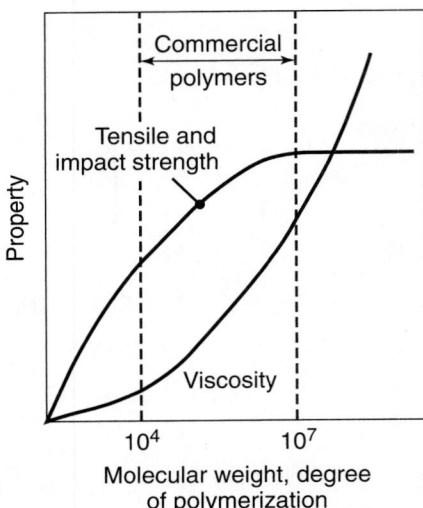

FIGURE 7.4   Effect of molecular weight and degree of polymerization on the strength and viscosity of polymers.

Degree of polymerization.   It is convenient to express the size of a polymer chain in terms of the *degree of polymerization* (DP), which is defined as the ratio of the molecular weight of the polymer to the molecular weight of the repeating unit. For example, polyvinyl chloride (PVC) has a mer weight of 62.5; thus, the DP of PVC with a molecular weight of 50,000 is 50,000/62.5 = 800. In terms of polymer processing (Chapter 19), the higher the DP, the higher is the polymer's viscosity or its resistance to flow (Fig. 7.4). High viscosity adversely affects the ease of shaping and, thus, raises the overall cost of processing.

Bonding.   During polymerization, the monomers are linked together by **covalent bonds**, forming a polymer chain. Because of their strength, covalent bonds also are called **primary bonds**. The polymer chains are, in turn, held together by **secondary bonds**, such as van der Waals bonds, hydrogen bonds, and ionic bonds. Secondary bonds are weaker than primary bonds by one to two orders of magnitude. In a given polymer, the increase in strength and viscosity with molecular weight is due (in part) to the fact that the longer the polymer chain, the greater is the energy needed to overcome the combined strength of the secondary bonds. For example, ethylene polymers having DPs of 1, 6, 35, 140, and 1350 at room temperature are, respectively, in the form of gas, liquid, grease, wax, and hard plastic.

Linear polymers.   The chain-like polymers shown in Fig. 7.2 are called *linear polymers* because of their sequential structure (Fig. 7.5a). However, a linear molecule is not necessarily straight in shape. In addition to those shown in the figure, other linear polymers are polyamides (nylon 6,6) and polyvinyl fluoride. Generally, a polymer consists of more than one type of structure; thus, a linear polymer may contain some branched and cross-linked chains. As a result of branching and cross-linking, the polymer's properties are changed significantly.

Branched polymers.   The properties of a polymer depend not only on the type of monomers but also on their arrangement in the molecular structure. In *branched polymers* (Fig. 7.5b), side-branch chains are attached to the main chain during the synthesis

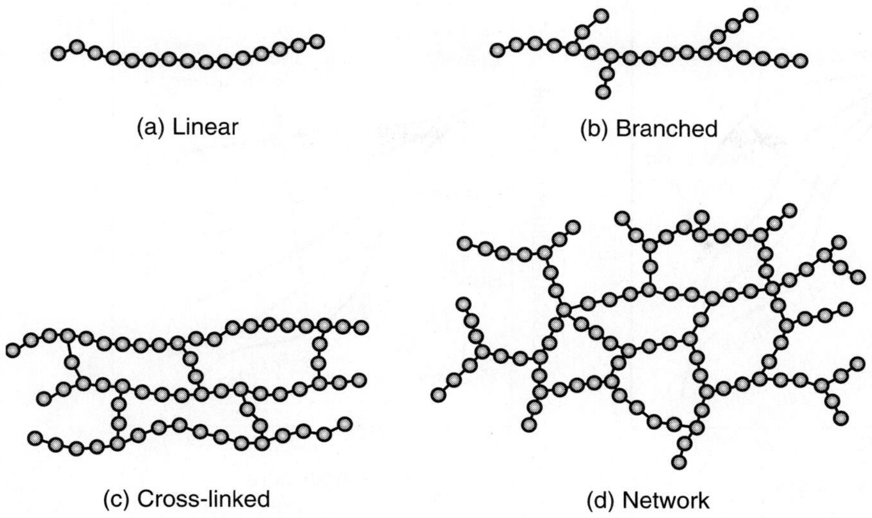

FIGURE 7.5 Schematic illustration of polymer chains. (a) Linear structure—thermoplastics such as acrylics, nylons, polyethylene, and polyvinyl chloride have linear structures. (b) Branched structure, such as in polyethylene. (c) Cross-linked structure—many rubbers or elastomers have this structure, and the vulcanization of rubber produces this structure. (d) Network structure, which is basically highly cross-linked—examples are thermosetting plastics, such as epoxies and phenolics.

of the polymer. Branching interferes with the relative movement of the molecular chains. As a result, their resistance to deformation and stress cracking is increased. The density of branched polymers is lower than that of linear-chain polymers, because the branches interfere with the packing efficiency of polymer chains.

The behavior of branched polymers can be compared to that of linear-chain polymers by making an analogy with a pile of tree branches (branched polymers) and a bundle of straight logs (linear polymers). Note that it is more difficult to move a branch within the pile of branches than to move a log within its bundle. The three-dimensional entanglements of branches make movements more difficult, a phenomenon akin to increased strength.

Cross-linked polymers. Generally three-dimensional in structure, *cross-linked polymers* have adjacent chains linked by covalent bonds (Fig. 7.5c). Polymers with a cross-linked chain structure are called **thermosets**, or **thermosetting plastics**; examples are epoxies, phenolics, and silicones. Cross-linking has a major influence on the properties of polymers (generally imparting hardness, strength, stiffness, brittleness, and better dimensional stability; see Fig. 7.6), as well as in the **vulcanization** of rubber (Section 7.9).

Network polymers. These polymers consist of spatial (three-dimensional) networks of three or more active covalent bonds (Fig. 7.5d). A highly cross-linked polymer also is considered a network polymer. Thermoplastic polymers that already have been formed or shaped can be cross-linked to obtain higher strength by subjecting them to high-energy radiation, such as ultraviolet light, x-rays, or electron beams. However, excessive radiation can cause degradation of the polymer.

Copolymers and terpolymers. If the repeating units in a polymer chain are all of the same type, the molecule is called a *homopolymer*. However, as with solid–solution metal alloys (Section 4.2), two or three different types of monomers

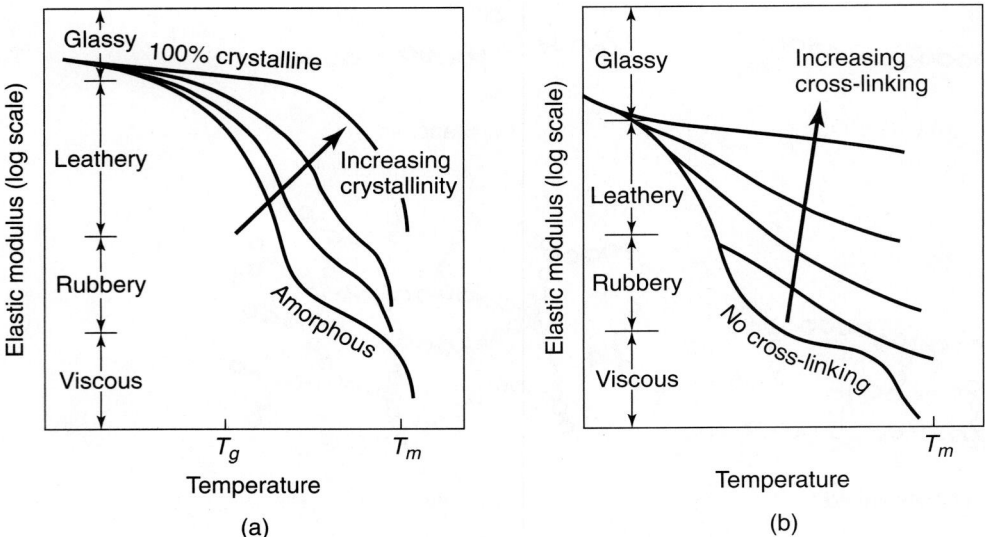

FIGURE 7.6   Behavior of polymers as a function of temperature and (a) degree of crystallinity and (b) cross-linking. The combined elastic and viscous behavior of polymers is known as viscoelasticity.

can be combined to develop certain special properties and characteristics, such as improved strength, toughness, and formability of the polymer. *Copolymers* contain two types of polymers (for example, styrene-butadiene, which is used widely for automobile tires). *Terpolymer*s contain three types (for example, ABS (acrylonitrile-butadiene-styrene), which is used for helmets, telephones, and refrigerator liners).

EXAMPLE 7.1   Dental and medical bone cement

Polymethylmethacrylate (PMMA) is an acrylic polymer commonly used in dental and medical applications as an adhesive and commonly is referred to as bone cement. There are a number of forms of PMMA, but this example describes one common form involving an addition-polymerization reaction. PMMA is delivered in two parts: a powder and a liquid, which are mixed by hand. The liquid wets and partially dissolves the powder, resulting in a liquid with viscosity on the order of 0.1 Ns/m^2 similar to that of vegetable oil. The viscosity increases markedly until a "dough" state is reached in about five minutes and fully hardens from the dough state in an additional five minutes.

The powder consists of high molecular weight poly[(methylmethacrylate)-costyrene] particles of about 50 μm in diameter, containing a small volume fraction of benzoyl peroxide. The liquid consists of a methyl methacrylate (MMA) monomer, with a small amount of dissolved n,n dimethyl-p-toluidine (DMPT). When the liquid and powder are mixed, the MMA wets the particles (dissolving a surface layer of the PMMA particles) and the DMPT cleaves the benzoyl peroxide molecule into two parts to form a catalyst with a free electron (sometimes referred to as a free radical). The resulting catalyst causes rapid growth of PMMA from the MMA mers, so that the final material is a composite of high molecular weight PMMA particles interconnected by PMMA chains. A schematic diagram of fully set bone cement is shown in Fig. 7.7.

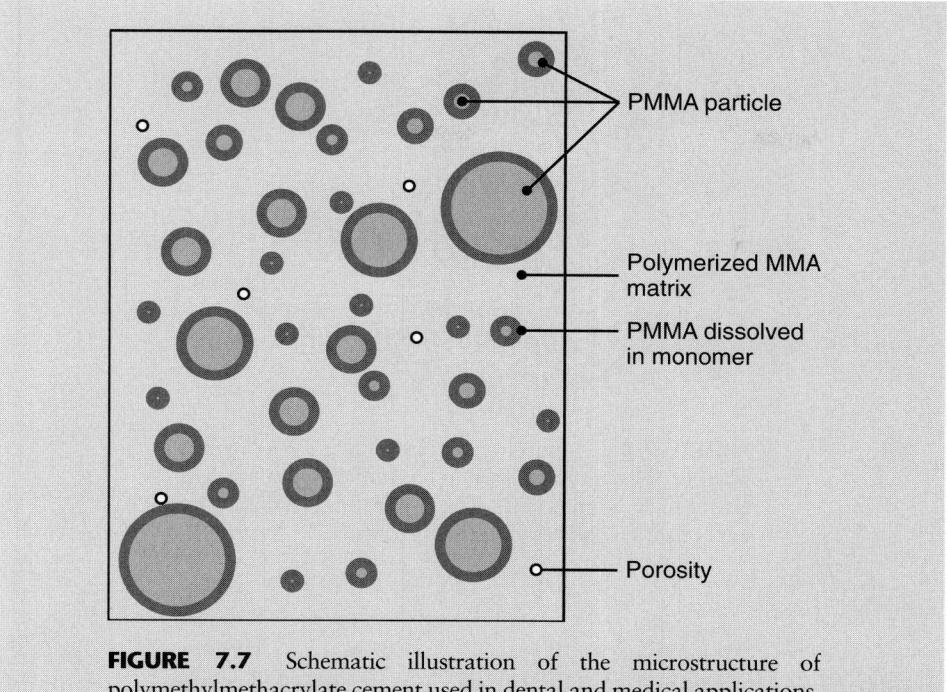

FIGURE 7.7 Schematic illustration of the microstructure of polymethylmethacrylate cement used in dental and medical applications.

7.2.2 Crystallinity

Polymers such as polymethylmethacrylate, polycarbonate, and polystyrene are generally **amorphous**; that is, the polymer chains exist without long-range order (see also *amorphous alloys*, Section 6.14). The amorphous arrangement of polymer chains often is described as being like a bowl of spaghetti or like worms in a bucket —all intertwined with each other. In some polymers, however, it is possible to impart some crystallinity and thereby modify their characteristics. This arrangement may be fostered either during the synthesis of the polymer or by deformation during its subsequent processing.

The crystalline regions in polymers are called **crystallites** (Fig. 7.8). These crystals are formed when the long molecules arrange themselves in an orderly manner, similar to the folding of a fire hose in a cabinet or of facial tissues in a box. A partially crystalline (**semi-crystalline**) polymer can be regarded as a two-phase material, one phase being crystalline and the other amorphous.

By controlling the rate of solidification during cooling and the chain structure, it is possible to impart different **degrees of crystallinity** to polymers, although never 100%. Crystallinity ranges from an almost complete crystal (up to about 95% by volume in the case of polyethylene) to slightly crystallized (mostly amorphous) polymers. The degree of crystallinity also is affected by branching. A linear polymer can become highly crystalline, but a highly branched polymer cannot, although it may develop some low level of crystallinity. It will never achieve a high crystallite content because the branches interfere with the alignment of the chains into a regular crystal array.

Effects of crystallinity.   The mechanical and physical properties of polymers are greatly influenced by the degree of crystallinity: as crystallinity increases, polymers become stiffer, harder, less ductile, more dense, less rubbery, and more resistant

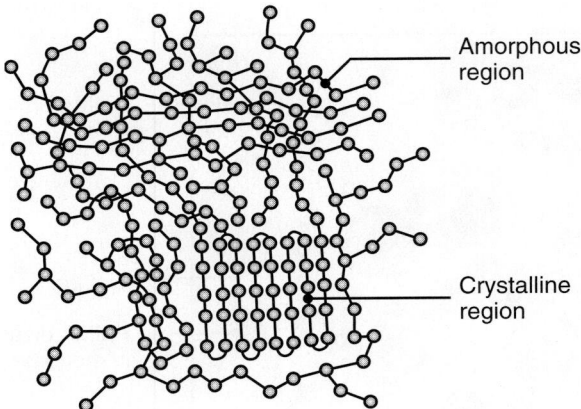

FIGURE 7.8   Amorphous and crystalline regions in a polymer. The crystalline region (crystallite) has an orderly arrangement of molecules. The higher the crystallinity, the harder, stiffer, and less ductile the polymer.

to solvents and heat (Fig. 7.6). The increase in density with increasing crystallinity is called *crystallization shrinkage* and is caused by a more efficient packing of the molecules in the crystal lattice. For example, the highly crystalline form of polyethylene, known as high-density polyethylene (HDPE), has a specific gravity in the range of from 0.941 to 0.970 (80 to 95% crystalline). It is stronger, stiffer, tougher, and less ductile than low-density polyethylene (LDPE), which is about 60 to 70% crystalline and has a specific gravity of about 0.910 to 0.925.

Optical properties of polymers also are affected by the degree of crystallinity. The reflection of light from the boundaries between the crystalline and the amorphous regions in the polymer causes opaqueness. Furthermore, because the index of refraction is proportional to density, the greater the density difference between the amorphous and crystalline phases, the greater is the opaqueness of the polymer. Polymers that are completely amorphous can be transparent, such as polycarbonate and acrylics.

7.2.3 Glass-transition temperature

Although amorphous polymers do not have a specific melting point, they undergo a distinct change in their mechanical behavior across a narrow range of temperature. At low temperatures, they are hard, rigid, brittle, and glassy; at high temperatures, they are rubbery or leathery. The temperature at which a transition occurs is called the **glass-transition temperature** (T_g), also called the *glass point* or *glass temperature*. The term glass is used in this description because glasses, which are amorphous solids, behave in the same manner (see *metallic glasses*, Section 6.14). Although most amorphous polymers exhibit this behavior, an exception is polycarbonate, which is neither rigid nor brittle below its glass-transition temperature. Polycarbonate is tough at ambient temperatures and is used for safety helmets and shields.

To determine T_g, the specific volume of the polymer is determined and plotted against temperature, and marked by a sharp change in the slope of the curve (Fig. 7.9). In the case of highly cross-linked polymers, the slope of the curve changes gradually near T_g, and hence, it can be difficult to determine T_g for these polymers. The

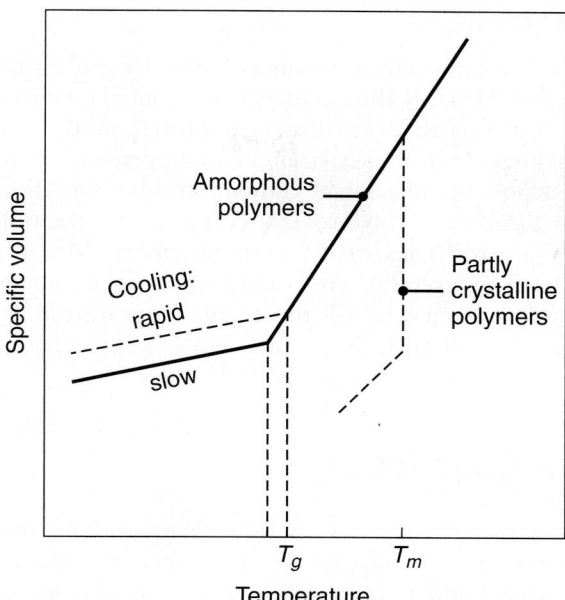

FIGURE 7.9   Specific volume of polymers as a function of temperature. Amorphous polymers, such as acrylic and polycarbonate, have a glass-transition temperature, T_g, but do not have a specific melting point, T_m. Partly crystalline polymers, such as polyethylene and nylons, contract sharply while passing through their melting temperatures during cooling.

glass-transition temperature varies with different polymers (Table 7.2). For example, room temperature is above T_g for some polymers and below it for others. Unlike amorphous polymers, partly crystalline polymers have a distinct melting point, T_m (Fig. 7.9; see also Table 7.2). Because of the structural changes (first-order changes) occurring, the specific volume of the polymer drops suddenly as its temperature is reduced.

TABLE 7.2

Glass-Transition and Melting Temperatures of Some Polymers		
Material	T_g (°C)	T_m (°C)
Nylon 6,6	57	265
Polycarbonate	150	265
Polyester	73	265
Polyethylene		
High density	−90	137
Low density	−110	115
Polymethylmethacrylate	105	—
Polypropylene	−14	176
Polystyrene	100	239
Polytetrafluoroethylene	−90	327
Polyvinyl chloride	87	212
Rubber	−73	—

7.2.4 Polymer blends

The brittle behavior of amorphous polymers below their glass-transition temperature can be reduced by *blending* them—usually with small quantities of an **elastomer** (Section 7.9). These tiny particles are dispersed throughout the amorphous polymer, enhancing its toughness and impact strength by improving its resistance to crack propagation. These polymer blends are known as **rubber modified polymers.**

Advances in blending involve several components, creating **polyblends** that utilize the favorable properties of different polymers. **Miscible blends** (mixing without separation of two phases) are created by a process similar to the alloying of metals that enables polymer blends to become more ductile. Polymer blends account for about 20% of all polymer production.

7.3 | Thermoplastics

It was noted earlier that within each molecule the bonds between adjacent long-chain molecules (secondary bonds) are much weaker than the covalent bonds between mers (primary bonds). It is the strength of the secondary bonds that determines the overall strength of the polymer; linear and branched polymers have weak secondary bonds.

As the temperature is raised above the glass-transition temperature, T_g, or melting point, T_m, certain polymers become easier to form or mold into desired shapes. The increased temperature weakens the secondary bonds (through thermal vibration of the long molecules), and the adjacent chains thus can move more easily when subjected to external shaping forces. When the polymer is cooled, it returns to its original hardness and strength; in other words, the process is reversible. Polymers that exhibit this behavior are known as **thermoplastics** (common examples of which are acrylics, cellulosics, nylons, polyethylenes, and polyvinyl chloride).

The behavior of thermoplastics depends on other variables as well as their structure and their composition. Among the most important are temperature and rate of deformation. Below the glass-transition temperature, most polymers are *glassy* (brittle) and behave like an elastic solid (that is, the relationship between stress and strain is linear (see Fig. 2.2). For example, polymethylmethacrylate (PMMA) is glassy below its T_g, whereas polycarbonate is not glassy below its T_g. The glassy behavior can be represented by a spring whose stiffness is equivalent to the modulus of elasticity of the polymer.

When the applied stress is increased further, the polymer eventually fractures, just as a piece of glass does at ambient temperature. Plastics undergo fatigue and creep phenomena, just as metals do. Typical stress–strain curves for some thermoplastics and thermosets at room temperature are shown in Fig. 7.10. Note that these plastics exhibit various behaviors, which may be described as rigid, soft, brittle, flexible, and so on. The mechanical properties of several polymers listed in Table 7.1 indicate that thermoplastics are about two orders of magnitude less stiff than metals. Their ultimate tensile strength is about one order of magnitude lower than that of metals (see Table 2.1).

Effects of temperature.   If the temperature of a thermoplastic polymer is raised above its T_g, it first becomes *leathery* and then, with increasing temperature, *rubbery* (Fig. 7.6). Finally, at higher temperatures (e.g., above T_m for crystalline thermoplastics), it becomes a *viscous fluid*: its viscosity decreases with increasing temperature. At still higher temperatures, the response of a thermoplastic can be likened to ice

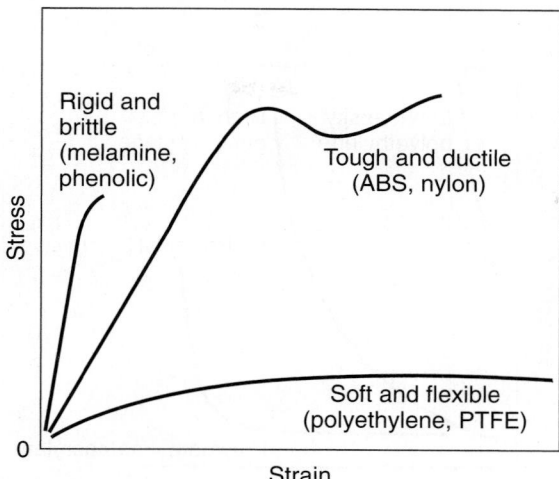

FIGURE 7.10   General terminology describing the behavior of three types of plastics. PTFE (polytetra-fluoroethylene) has *Teflon* as its trade name. *Source*: After R. L. E. Brown.

cream: It can be softened, molded into shapes, refrozen, resoftened, and remolded a number of times. In practice, however, repeated heating and cooling cause **degradation**, or **thermal aging**, of thermoplastics.

The typical effect of temperature on the strength and elastic modulus of thermoplastics is similar to that of metals; with increasing temperature, the strength and the modulus of elasticity decrease and the toughness increases (Fig. 7.11). The effect of temperature on impact strength is shown in Fig. 7.12; note the large difference in the impact behaviors of various polymers.

Effect of rate of deformation.   The behavior of thermoplastics is similar to the strain-rate sensitivity of metals, indicated by the strain-rate sensitivity

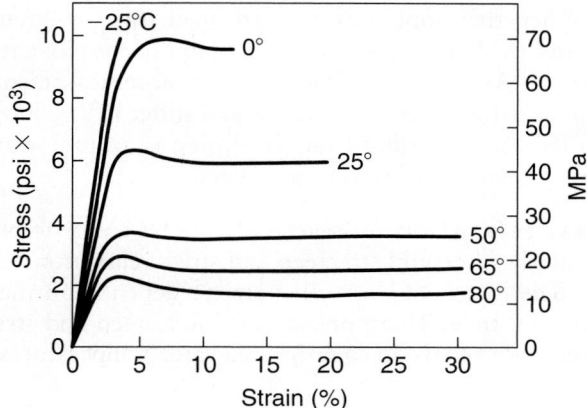

FIGURE 7.11   Effect of temperature on the stress–strain curve for cellulose acetate, a thermoplastic. Note the large drop in strength and the large increase in ductility with a relatively small increase in temperature. *Source*: After T. S. Carswell and H. K. Nason.

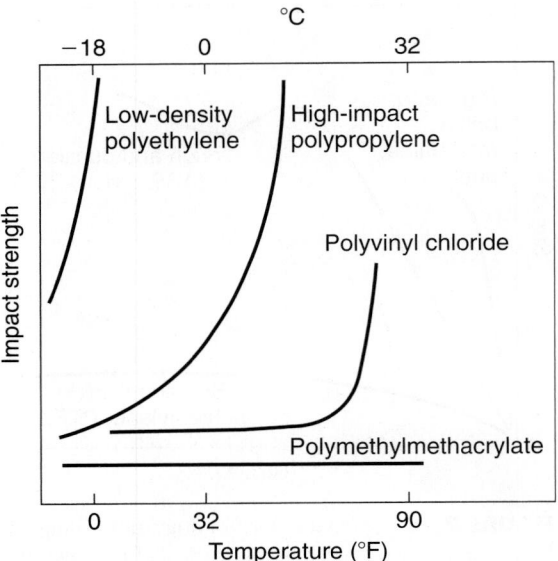

FIGURE 7.12 Effect of temperature on the impact strength of various plastics. Small changes in temperature can have a significant effect on impact strength. *Source*: After P. C. Powell.

exponent m in Eq. (2.9). Thermoplastics in general have high m values, indicating that they can undergo large *uniform deformation* in tension before fracture (Fig. 7.13). Note how (unlike in ordinary metals) the necked region elongates considerably. This phenomenon easily can be demonstrated by stretching a piece of the plastic holder for a 6-pack of beverage cans. Observe the sequence of necking and stretching behavior shown in Fig. 7.13a. This characteristic (which is the same in the superplastic metals) enables the thermoforming of thermoplastics (Section 19.6) into such complex shapes as meat trays, lighted signs, and bottles for soft drinks.

Orientation. When thermoplastics are deformed (say, by stretching) the long-chain molecules tend to align in the general direction of the elongation; this process is called *orientation*. As in metals, the polymer becomes anisotropic (see also Section 1.5), so the specimen becomes stronger and stiffer in the elongated (stretched) direction than in its transverse direction. Stretching is an important technique for enhancing the strength and toughness of polymers.

Creep and stress relaxation. Because of their viscoelastic behavior, thermoplastics are particularly susceptible to creep and stress relaxation and to a larger extent than metals. The extent of these phenomena depends on the polymer, stress level, temperature, and time. Thermoplastics exhibit creep and stress relaxation at room temperature; most metals do so only at elevated temperatures.

Crazing. Some thermoplastics (such as polystyrene and polymethylmethacrylate), when subjected to tensile stresses or to bending, develop localized, wedge-shaped, narrow regions of highly deformed material, called *crazing*. Although they may appear to be like cracks, crazes are spongy material, typically containing about 50% voids. With increasing tensile load on the specimen, these voids coalesce to form a crack, which eventually can lead to a fracture of the polymer. Crazing has been observed both in

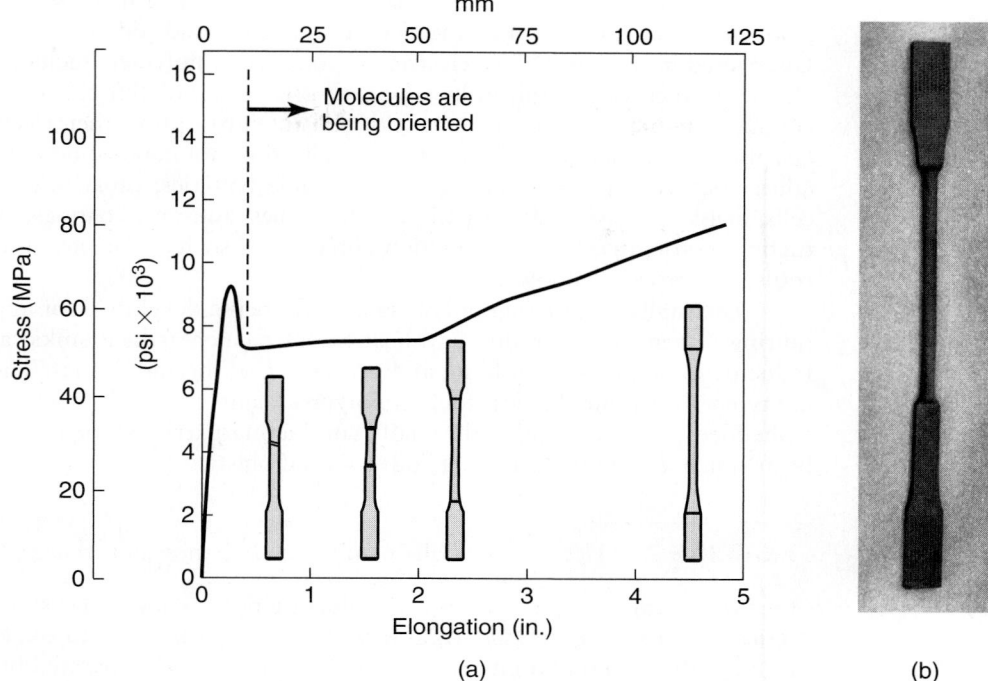

FIGURE 7.13   (a) Load-elongation curve for polycarbonate, a thermoplastic. *Source*: Courtesy of R. P. Kambour and R. E. Robertson. (b) High-density polyethylene tensile-test specimen, showing uniform elongation (the long, narrow region in the specimen).

transparent, glassy polymers and in other types. The environment (particularly the presence of solvents, lubricants, or water vapor) can enhance the formation of crazes (**environmental-stress cracking** and **solvent crazing**). Residual stresses in the material also contribute to crazing and cracking of the polymer; radiation (especially ultraviolet radiation) can increase the crazing behavior in certain polymers.

A phenomenon related to crazing is **stress whitening**. When subjected to tensile stresses (such as those caused by folding or bending), the plastic becomes lighter in color—a phenomenon usually attributed to the formation of microvoids in the material. As a result, the material becomes less translucent (transmits less light) or more opaque. This behavior easily can be demonstrated by bending plastic components commonly found in colored binder strips for report covers, household products, and toys.

Water absorption.   An important characteristic of some polymers, such as nylons, is their ability to absorb water. Water acts as a plasticizing agent: It makes the polymer more plastic (see Section 7.5). In a sense, it lubricates the chains in the amorphous region. With increasing moisture absorption, the glass-transition temperature, the yield stress, and the elastic modulus of the polymer typically are lowered severely. Dimensional changes also occur, especially in a humid environment.

Thermal and electrical properties.   Compared to metals, plastics generally are characterized by low thermal and electrical conductivity, low specific gravity (ranging from 0.90 to 2.2), and a high coefficient of thermal expansion (about an order of magnitude higher; see Tables 3.1 and 3.2.) Because most polymers have low electrical conductivity, they can be used for insulators and as packaging material for electronic components.

The electrical conductivity of some polymers can be increased by **doping** (introducing impurities, such as metal powders, salts, and iodides, into the polymer. Discovered in the late 1970s, **electrically conducting polymers** include polyethylene oxide, polyacetylene, polyaniline, polypyrrole, and polythiophene. The electrical conductivity of polymers increases with moisture absorption; their electronic properties also can be changed by irradiation. Applications for conducting polymers include adhesives, microelectronic devices, rechargeable batteries, capacitors, catalysts, fuel cells, fuel-level sensors, deicer panels, radar dishes, antistatic coatings, and thermoactuating motors (used in linear-motion applications such as for power antennae, sun roofs, and power windows).

Thermally conducting polymers also are being developed for applications requiring dimensional stability and heat transfer (such as heat sinks) as well as for reducing cycle times in molding and processing of thermoplastics. These polymers are typically thermoplastics (such as polypropylene, polycarbonate, nylon) and are embedded with nonmetallic thermally conducting particles; their conductivity can be as much as 100 times that of conventional plastics.

EXAMPLE 7.2 Use of electrically conducting polymers in rechargeable batteries

One of the earliest applications of conducting polymers was in rechargeable batteries. Modern lithium rechargeable batteries use lithium or an oxide of lithium as the cathode and lithium carbide (Li_yC_6) as the anode, separated by a conducting polymer layer. Lithium is used because it is the lightest of all metals and has a high electrochemical potential, so that its energy per volume is highest.

The polymer, usually polyethylene oxide (PEO), with a dissolved lithium salt is placed between the cathode and anode. During discharging, Li_yC_6 is oxidized and discharges free electrons and lithium ions. The electrons drive external electronics, and the Li^+ ions are stored in the polymer. When the cathode is depleted, the battery must be recharged to restore the cathode. During charging, Li+ is transferred through the polymer electrolytes to the cathode. Lithium-ion batteries have good capacity, can generate up to 4.5 V, and can be placed in series to obtain higher voltages. Developments are taking place to make battery cells in which both electrodes are made of conducting polymers; one has been constructed with a capacity of 3.5 V.

7.4 | Thermosetting Plastics

When the long-chain molecules in a polymer are cross-linked in a three-dimensional arrangement, the structure in effect becomes one *giant molecule* with strong covalent bonds. These polymers are called **thermosetting polymers** or **thermosets**, because (during polymerization) the network is completed and the shape of the part is permanently set. This **curing (cross-linking)** reaction, unlike that of thermoplastics, is *irreversible*. The response of a thermosetting plastic to a sufficiently elevated temperature can be likened to what happens in the baking of a cake or in the boiling of an egg; once the cake is baked and cooled, or the egg boiled and cooled, reheating it will not change its shape. Some thermosets (such as epoxy, polyester, and urethane) cure at room temperature, because the heat produced by the exothermic reaction is sufficient to cure the plastic.

The polymerization process for thermosets generally takes place in two stages. The first occurs at the chemical plant, where the molecules are partially polymerized into linear chains. The second stage occurs at the parts-producing plant, where cross-linking is completed under heat and pressure during the molding and shaping of the part (Chapter 19).

Thermosetting polymers do not have a sharply defined glass-transition temperature. Because of the nature of the bonds, the strength and hardness of a thermoset (unlike those of thermoplastics) are not affected by temperature or by rate of deformation. If the temperature is increased sufficiently, the thermosetting polymer instead begins to burn up, degrade, and char. Thermosets generally possess better mechanical, thermal, and chemical properties; electrical resistance; and dimensional stability than do thermoplastics. A typical and common thermoset is **phenolic**, which is a product of the reaction between phenol and formaldehyde. Common products made from this polymer are the handles and knobs on cooking pots and pans and components of light switches and outlets.

7.5 | Additives in Plastics

In order to impart certain specific properties, polymers usually are compounded with *additives*. These additives modify and improve certain characteristics of the polymer, such as stiffness, strength, color, weatherability, flammability, arc resistance (for electrical applications), and ease of subsequent processing.

- **Plasticizers** are added to polymers to impart *flexibility* and *softness* by lowering their glass-transition temperature. Plasticizers are low-molecular-weight solvents with high boiling points (nonvolatile); they reduce the strength of the secondary bonds between the long-chain molecules and, thus, make the polymer flexible and soft. The most common use of plasticizers is in polyvinyl chloride (PVC), which remains flexible during its many uses; other applications are in thin sheets, films, tubing, shower curtains, and clothing materials.

- Most polymers are affected adversely by *ultraviolet radiation* (such as from sunlight) and by *oxygen*; they weaken and break the primary bonds and cause the *scission* (splitting) of the long-chain molecules; the polymer then degrades and becomes stiff and brittle. On the other hand, degradation may be beneficial, as in the disposal of plastic objects by subjecting them to environmental attack (see also Section 7.8). A typical example of protection against ultraviolet radiation is the compounding of certain plastics and rubber with **carbon black** (soot). The carbon black absorbs a high percentage of the ultraviolet radiation. Protection against degradation caused by oxidation, particularly at elevated temperatures, is achieved by adding *antioxidants* to the polymer. Various *coatings* are another means of protecting polymers.

- **Fillers** used in plastics are generally wood flour (fine sawdust), silica flour (fine silica powder), clay, powdered mica, talc, calcium carbonate, and short fibers of cellulose, glass, or asbestos. Because of their low cost, fillers are important in reducing the overall cost of polymers. Depending on their type, fillers also may improve the strength, hardness, toughness, abrasion resistance, dimensional stability, or stiffness of plastics. These properties are greatest at specific percentages of different types of polymer–filler combinations. As with reinforced plastics (Section 9.2), a filler's effectiveness depends on the nature of the bond between the filler material and the polymer chains.

- The wide variety of colors available in plastics is obtained by adding **colorants**— either organic (*dyes*) or inorganic (*pigments*). The selection of a colorant depends on the service temperature and the expected amount of exposure to light. Pigments are dispersed particles; they generally have greater resistance than do dyes to temperature and light.

- If the temperature is sufficiently high, most polymers will ignite and burn; flame colors are usually yellow or blue. The **flammability** (ability to support combustion) of polymers varies considerably, depending on their composition (especially on their chlorine and fluorine content). The flammability of polymers can be reduced either by making them from less-flammable raw materials or by the addition of **flame retardants,** such as compounds of chlorine, bromine, and phosphorus. Cross-linking also reduces polymer flammability.

   The following list gives several common polymers with different burning characteristics:

   1. Plastics that do not burn: *fluorocarbons (Teflon)*
   2. Plastics that do burn but are self-extinguishing: *carbonate, nylon, vinyl chloride*
   3. Plastics that burn and are not self-extinguishing: *acetal, acrylic, acrylonitrile-butadiene-styrene, cellulose, polyester, propylene, styrene*

- **Lubricants** may be added to polymers to reduce friction during their subsequent processing into useful products and to prevent parts from sticking to the molds. Typical lubricants are linseed oil, mineral oil, and waxes (natural and synthetic); metallic soaps such as calcium stearate and zinc stearate also are used. Lubrication also is important in preventing thin polymer films from sticking to each other.

7.6   General Properties and Applications of Thermoplastics

The general characteristics and typical applications of major thermoplastics, particularly as they relate to the manufacturing and service life of plastic products and components, are outlined in this section. General recommendations for various plastics applications are given in Table 7.3, and Table 7.4 lists some of the more common trade names for thermoplastics.

   Acetals (from *acetic* and *alcohol*) have good strength, good stiffness, and good resistance to creep, abrasion, moisture, heat, and chemicals. Typical applications include: mechanical parts and components requiring high performance over a long period (i.e., bearings, cams, gears, bushings, and rollers), impellers, wear surfaces, pipes, valves, shower heads, and housings.

   Acrylics (polymethylmethacrylate, PMMA) possess moderate strength, good optical properties, and weather resistance. They are transparent (but can be made opaque), are generally resistant to chemicals, and have good electrical resistance. Typical applications include: lenses, lighted signs, displays, window glazing, skylights, bubble tops, automotive lenses, windshields, lighting fixtures, and furniture.

   Acrylonitrile-butadiene-styrene (ABS) is rigid and dimensionally stable. It has good impact, abrasion, and chemical resistance; good strength and toughness; good low-temperature properties; and high electrical resistance. Typical applications include: pipes, fittings, chrome-plated plumbing supplies, helmets, tool handles, automotive components, boat hulls, telephones, luggage, housing, appliances, refrigerator liners, and decorative panels.

   Cellulosics have a wide range of mechanical properties, depending on their composition. They can be made rigid, strong, and tough; however, they weather poorly and are affected by heat and chemicals. Typical applications include: tool handles, pens, knobs, frames for eyeglasses, safety goggles, machine guards, helmets,

TABLE 7.3

General Recommendations for Plastic Products		
Design requirement	Typical applications	Plastics
Mechanical strength	Gears, cams, rollers, valves, fan blades, impellers, pistons	Acetals, nylon, phenolics, polycarbonates, polyesters, polypropylenes, epoxies, polyimides
Wear resistance	Gears, wear strips and liners, bearings, bushings, roller blades	Acetals, nylon, phenolics, polyimides, polyurethane, ultrahigh-molecular-weight polyethylene
Frictional properties		
High	Tires, nonskid surfaces, footware, flooring	Elastomers, rubbers
Low	Sliding surfaces, artificial joints	Fluorocarbons, polyesters, polyimides
Electrical resistance	All types of electrical components and equipment, appliances, electrical fixtures	Polymethylmethacrylate, ABS, fluorocarbons, nylon, polycarbonate, polyester, polypropylenes, ureas, phenolics, silicones, rubbers
Chemical resistance	Containers for chemicals, laboratory equipment, components for chemical industry, food and beverage containers	Acetals, ABS, epoxies, polymethylmethacrylate, fluorocarbons, nylon, polycarbonate, polyester, polypropylene, ureas, silicones
Heat resistance	Appliances, cookware electrical components	Fluorocarbons, polyimides, silicones, acetals, polysulfones, phenolics, epoxies
Functional and decorative	Handles, knobs, camera and battery cases, trim moldings, pipe fittings	ABS, acrylics, cellulosics, phenolics, polyethylenes, polypropylenes, polystyrenes, polyvinyl chloride
Functional and transparent	Lenses, goggles, safety glazing, signs, food-processing equipment, laboratory hardware	Acrylics, polycarbonates, polystyrenes, polysulfones
Housings and hollow shapes	Power tools, housings, sport helmets, telephone cases	ABS, cellulosics, phenolics, polycarbonates, polyethylenes, polypropylene, polystyrenes

TABLE 7.4

Trade Names for Thermoplastic Polymers			
Trade name	Type	Trade name	Type
Alathon	Ethylene	Noryl	Polyphenylene oxide
Cycolac	Acrylonitrile-butadiene-styrene	Nylon	Polyamide
Dacron	Polyester	Orlon	Acrylic
Delrin	Acetal	Plexiglas	Acrylic
Dylene	Styrene	Royalite	Acrylonitrile-butadiene-styrene
Envex	Polyimide	Saran	Polyvinyl chloride
Hyzod	Polycarbonate	Sintra	Polyvinyl chloride
Implex	Acrylic (rubber-modified)	Styrofoam	Polystyrene
Kapton	Polyimide	Teflon	Fluorocarbon
Kevlar	Aramid	Torlon	Polyimide
Kodel	Polyester	Tygon	Polyvinyl chloride
Kydex	Acrylic-polyvinyl chloride	Ultem	Polyetherimide
Kynar	Polyvinylidene fluoride	Vespel	Polyimide
Lexan	Polycarbonate	Zerlon	Styrene-methylmethacrylate
Lucite	Acrylic	Zytel	Polyamide
Mylar	Polyester		

tubing and pipes, lighting fixtures, rigid containers, steering wheels, packaging film, signs, billiard balls, toys, and decorative parts.

Fluorocarbons possess good resistance to high temperature (for example, a melting point of 327°C (621°F) for Teflon), chemicals, weather, and electricity. They also have unique nonadhesive properties and low friction. Typical applications include: linings for chemical-processing equipment, nonstick coatings for cookware, electrical insulation for high-temperature wire and cable, gaskets, low-friction surfaces, bearings, and seals.

Polyamides (from the words *poly, amine,* and *carboxyl acid*) are available in two main types: *nylons* and *aramids*.

- **Nylons** (a coined word) have good mechanical properties and abrasion resistance. They are self-lubricating and resistant to most chemicals. All nylons are *hygroscopic* (absorb water); the moisture absorption reduces desirable mechanical properties and increases part dimensions. Typical applications include: gears, bearings, bushings, rollers, fasteners, zippers, electrical parts, combs, tubing, wear-resistant surfaces, guides, and surgical equipment.

- **Aramids** (aromatic polyamides) have very high tensile strength and stiffness. Typical applications include: fibers for reinforced plastics, bulletproof vests, cables, and radial tires.

Polycarbonates are versatile. They have good mechanical and electrical properties, high impact resistance, and they can be made resistant to chemicals. Typical applications include: safety helmets, optical lenses, bullet-resistant window glazing, signs, bottles, food-processing equipment, windshields, load-bearing electrical components, electrical insulators, medical apparatus, business machine components, guards for machinery, and parts requiring dimensional stability.

Polyesters (thermoplastic polyesters; see also Section 7.7) have good mechanical, electrical, and chemical properties; good abrasion resistance; and low friction. Typical applications include: gears, cams, rollers, load-bearing members, pumps, and electromechanical components.

Polyethylenes possess good electrical and chemical properties; their mechanical properties depend on composition and structure. Three major polyethylene classes are: (1) *low density* (LDPE), (2) *high density* (HDPE), and (3) *ultrahigh molecular weight* (UHMWPE). Typical applications for LDPE and HDPE are housewares, bottles, garbage cans, ducts, bumpers, luggage, toys, tubing, bottles, and packaging materials. UHMWPE is used in parts requiring high-impact toughness and resistance to abrasive wear; examples include artificial knee and hip joints.

Polyimides have the structure of a thermoplastic but the nonmelting characteristic of a thermoset. (See also Section 7.7.)

Polypropylenes have good mechanical, electrical, and chemical properties and good resistance to tearing. Typical applications include automotive trim and components, medical devices, appliance parts, wire insulation, TV cabinets, pipes, fittings, drinking cups, dairy-product and juice containers, luggage, ropes, and weather stripping.

Polystyrenes generally have average properties, are somewhat brittle but inexpensive. Typical applications include disposable containers, packaging, trays for meats, cookies and candy, foam insulation, appliances, automotive and radio/TV components, housewares, and toys and furniture parts (as a substitute for wood).

Polysulfones have excellent resistance to heat, water, and steam; they have dielectric properties that virtually are unaffected by humidity, are highly resistant to some chemicals, but are attacked by organic solvents. Typical applications include: steam irons, coffeemakers, hot-water containers, medical equipment that requires sterilization, power-tool and appliance housings, aircraft cabin interiors, and electrical insulators.

Polyvinyl chloride (PVC) has a wide range of properties, is inexpensive and water resistant, and can be made rigid or flexible. It is not suitable for applications requiring strength and heat resistance. *Rigid* PVC is tough and hard; it is used for signs and in the construction industry (for example, in pipes and conduits). *Flexible* PVC is used in wire and cable coatings, in low-pressure flexible tubing and hose, and in footwear, imitation leather, upholstery, records, gaskets, seals, trim, film, sheet, and coatings.

7.7   General Properties and Applications of Thermosetting Plastics

This section outlines the general characteristics and typical applications of the major thermosetting plastics.

Alkyds (from *alkyl*, meaning alcohol, and *acid*) possess good electrical insulating properties, impact resistance, dimensional stability, and have low water absorption. Typical applications are in electrical and electronic components.

Aminos have properties that depend on composition; generally, they are hard, rigid, and resistant to abrasion, creep, and electrical arcing. Typical applications include: small appliance housings, countertops, toilet seats, handles, and distributor caps. **Urea** typically is used for electrical and electronic components, and **melamine** for dinnerware.

Epoxies have excellent mechanical and electrical properties, good dimensional stability, strong adhesive properties, and good resistance to heat and chemicals. Typical applications include: electrical components requiring mechanical strength and high insulation, tools and dies, and adhesives. **Fiber-reinforced epoxies** have excellent mechanical properties and are used in pressure vessels, rocket-motor casings, tanks, and similar structural components.

Phenolics are rigid (although brittle) and dimensionally stable, and they have high resistance to heat, water, electricity, and chemicals. Typical applications include: knobs, handles, laminated panels, telephones; bond material to hold abrasive grains together in grinding wheels; and electrical components (such as wiring devices, connectors, and insulators).

Polyesters (thermosetting polyesters; see also Section 7.6) have good mechanical, chemical, and electrical properties. They generally are reinforced with glass (or other) fibers and also are available as casting resins. Typical applications include: boats, luggage, chairs, automotive bodies, swimming pools, and materials for impregnating cloth and paper.

Polyimides possess good mechanical, physical, and electrical properties at elevated temperatures; they also have good creep resistance, low friction, and low wear characteristics. Polyimides have the nonmelting characteristic of a thermoset but the structure of a thermoplastic. Typical applications include: pump components (bearings, seals, valve seats, retainer rings, and piston rings), electrical connectors for high-temperature use, aerospace parts, high-strength impact-resistant structures, sports equipment, and safety vests.

Silicones have properties that depend on composition. Generally, they weather well, possess excellent electrical properties over a wide range of humidity and temperature, and resist chemicals and heat (see also Section 7.9). Typical applications include: electrical components requiring strength at elevated temperatures, oven gaskets, heat seals, and waterproof materials.

EXAMPLE 7.3 Materials for a refrigerator door liner

In the selection of candidate materials for a refrigerator door liner (where eggs, butter, salad dressings, and small bottles are stored) the following factors should be considered:

1. *Mechanical requirements:* strength, toughness (to withstand impact, door slamming, racking), stiffness, resilience, and resistance to scratching and wear at operating temperatures.

2. *Physical requirements:* dimensional stability and electrical insulation.

3. *Chemical requirements:* resistance to staining, odor, chemical reactions with food and beverages, and cleaning fluids.

4. *Appearance:* color, stability of color, surface finish, texture, and feel.

5. *Manufacturing properties:* methods of manufacturing and assembly, effects of processing on material properties and behavior over a period of time, compatibility with other components in the door, and cost of materials and manufacturing.

An extensive study, considering all of the factors involved, identified two candidate materials for door liners: ABS (acrylonitrile-butadiene-styrene) and HIPS (high-impact polystyrene). One aspect of the study involved the effect of vegetable oils, such as from salad dressing stored in the door shelf, on the strength of these plastics. Experiments showed that the presence of vegetable oils significantly reduced the load-bearing capacity of HIPS. It was found that HIPS becomes brittle in the presence of oils (solvent–stress cracking), whereas ABS is not affected to any significant extent.

7.8 | Biodegradable Plastics

Plastic wastes contribute about 10% of municipal solid waste; by weight; on a volume basis they contribute between two and three times their weight. Only about one-third of plastic production goes into disposable products, such as bottles, packaging, and garbage bags. With the growing use of plastics and great concern over environmental issues regarding the disposal of plastic products and the shortage of landfills, major efforts are underway to develop completely biodegradable plastics. The first attempts were made in the 1980s as a possible solution to roadside litter.

Traditionally, most plastic products have been made from synthetic polymers that are derived from nonrenewable natural resources, are not biodegradable, and are difficult to recycle. **Biodegradability** means that microbial species in the environment (e.g., microorganisms in soil and water) will degrade a portion of (or even the entire) polymeric material, under the proper environmental conditions, and without producing toxic by-products. The end products of the degradation of the biodegradable portion of the material are carbon dioxide and water. Because of the variety of constituents in biodegradable plastics, these plastics can be regarded as composite materials. Consequently, only a portion of these plastics may be truly biodegradable.

Three different *biodegradable plastics* have thus far been developed. They have different degradability characteristics, and they degrade over different periods of time (anywhere from a few months to a few years).

1. The **starch-based system** is the farthest along in terms of production capacity. Starch may be extracted from potatoes, wheat, rice, and corn. The starch

granules are processed into a powder, which is heated and becomes a sticky liquid. The liquid is then cooled, shaped into pellets, and processed in conventional plastic-processing equipment. Various additives and binders are blended with the starch to impart special characteristics to the bioplastic materials. For example, a composite of polyethylene and starch is produced commercially as degradable garbage bags.

2. In the **lactic-based system**, fermenting feed stocks produce lactic acid, which is then polymerized to form a polyester resin. Typical uses include medical and pharmaceutical applications.

3. In **fermentation of sugar** (the third system), organic acids are added to a sugar feed stock. With the use of a specially developed process, the resulting reaction produces a highly crystalline and very stiff polymer, which (after further processing) behaves in a manner similar to polymers developed from petroleum.

Numerous attempts continue to be made to produce fully biodegradable plastics by means of the use of various agricultural waste (*agrowastes*), plant carbohydrates, plant proteins, and vegetable oils. Typical applications include the following:

- Disposable tableware made from a cereal substitute, such as rice grains or wheat flour.

- Plastics made almost entirely from starch extracted from potatoes, wheat, rice, and corn.

- Plastic articles made from coffee beans and rice hulls that are dehydrated and molded under high pressure and temperature.

- Water-soluble and compostable polymers for medical and surgical applications.

- Food and beverage containers (made from potato starch, limestone, cellulose, and water) that can dissolve in storm sewers and oceans without affecting marine life or wildlife.

The long-range performance of biodegradable plastics (both during their useful life-cycle as products and in landfills) has not been assessed fully. There also is concern that the emphasis on biodegradability will divert attention from the issue of the *recyclability* of plastics and the efforts for *conservation* of materials and energy. A major consideration is the fact that the cost of today's biodegradable polymers is substantially higher than that of synthetic polymers. Consequently, a mixture of agricultural waste—such as hulls from corn, wheat, rice, and soy (as the major component)—and biodegradable polymers (as the minor component) is an attractive alternative.

Recycling of plastics. Much effort continues to be expended globally on the collecting and recycling of used plastic products. Thermoplastics are recycled by remelting them and then reforming them into other products. They carry *recycling symbols*, in the shape of a triangle outlined by three clockwise arrows and having a number in the middle. These numbers correspond to the following plastics:

1—PETE (polyethylene)
2—HDPE (high-density polyethylene)
3—V (vinyl)
4—LDPE (low-density polyethylene)
5—PP (polypropylene)
6—PS (polystyrene)
7—Other

Recycled plastics increasingly are being used for a variety of products. For example, a recycled polyester (filled with glass fibers and minerals) is used for the engine cover for an F-series Ford pickup truck, as it has the appropriate stiffness, chemical resistance, and shape retention up to 180°C (350°F).

7.9 | Elastomers (Rubbers)

Elastomers consist of a large family of amorphous polymers having a low glass-transition temperature. They have a characteristic ability to undergo large elastic deformations without rupture; also, they are soft and have a low elastic modulus. The term **elastomer** is derived from the words *elastic* and *mer*.

The structure of elastomers is highly kinked (tightly twisted or curled). They stretch, but then return to their original shape after the load is removed (Fig. 7.14). They can also be cross-linked, the best example of this being the elevated-temperature **vulcanization** of rubber with sulfur, discovered by Charles Goodyear in 1839 and named for Vulcan, the Roman god of fire. Once the elastomer is cross-linked, it cannot be reshaped (for example, an automobile tire, which is one giant molecule, cannot be softened and reshaped).

The terms *elastomer* and *rubber* often are used interchangeably. Generally, an **elastomer** is defined as being capable of recovering substantially in shape and size after the load has been removed. A **rubber** is defined as being capable of recovering from large deformations quickly.

The hardness of elastomers, which is measured with a durometer (Section 2.6), increases with the cross-linking of the molecular chains. As with plastics, a variety of additives can be blended into elastomers to impart specific properties. Elastomers have a wide range of applications in high-friction and nonskid surfaces, protection against corrosion and abrasion, electrical insulation, and shock and vibration insulation. Examples include tires, hoses, weatherstripping, footwear, linings, gaskets, seals, printing rolls, and flooring.

One property of elastomers is their *hysteresis loss* in stretching or compression (Fig. 7.14). The clockwise loop indicates energy loss, whereby mechanical energy is converted into heat. This property is desirable for absorbing vibrational energy (damping) and sound deadening.

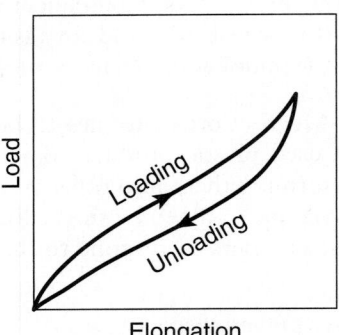

FIGURE 7.14 Typical load-elongation curve for rubbers. The clockwise loop, indicating the loading and the unloading paths, displays the hysteresis loss. Hysteresis gives rubbers the capacity to dissipate energy, damp vibration, and absorb shock loading, as is necessary in automobile tires and in vibration dampers placed under machinery.

Natural rubber. The base for natural rubber is **latex**, a milk-like sap obtained from the inner bark of a tropical tree. Natural rubber has good resistance to abrasion and fatigue, high frictional properties, but low resistance to oil, heat, ozone, and sunlight. Typical applications are tires, seals, shoe heels, couplings, and engine mounts.

Synthetic rubbers. Examples of synthetic rubbers are butyl, styrene butadiene, polybutadiene, and ethylene propylene. Compared to natural rubber, they have better resistance to heat, gasoline, and chemicals, and they have a higher range of useful temperatures. Synthetic rubbers that are resistant to oil are neoprene, nitrile, urethane, and silicone. Typical applications of synthetic rubbers are tires, shock absorbers, seals, and belts.

Silicones. Silicones (see also Section 7.7) have the highest useful temperature range of elastomers (up to 315°C; 600°F), but other properties (such as strength and resistance to wear and oils) generally are inferior to those in other elastomers. Typical applications of silicones are seals, gaskets, thermal insulation, high-temperature electrical switches, and electronic apparatus.

Polyurethane. This elastomer has very good overall properties of high strength, stiffness, and hardness, and it has exceptional resistance to abrasion, cutting, and tearing. Typical applications are seals, gaskets, cushioning, diaphragms for the rubber forming of sheet metals (Section 16.8), and autobody parts.

SUMMARY

- Polymers are a major class of materials and possess a very wide range of mechanical, physical, chemical, and optical properties. Compared to metals, polymers generally are characterized by a lower density, strength, elastic modulus, thermal and electrical conductivity, and cost; by a higher strength-to-weight ratio, higher resistance to corrosion, higher thermal expansion, wider choice of colors and transparencies; and by a greater ease of manufacture into complex shapes.

- Plastics are composed of polymer molecules and various additives. The smallest repetitive unit in a polymer chain is called a mer. Monomers are linked by polymerization processes (condensation and addition) to form larger molecules. The glass-transition temperature separates the region of brittle behavior in polymers from that of ductile behavior.

- The properties of polymers depend on the molecular weight, the structure (linear, branched, cross-linked, or network), the degrees of polymerization and crystallinity, and on additives. Additives have such functions as improving strength, flame retardation, lubrication, imparting flexibility and color, and providing stability against ultraviolet radiation and oxygen. Polymer structures can be modified by several means to impart a wide range of desirable properties to plastics.

- Two major classes of polymers are thermoplastics and thermosets. Thermoplastics become soft and easy to form at elevated temperatures; they return to their original properties when cooled. Their mechanical behavior can be characterized by various spring and damping models. Their behavior includes such phenomena as creep and stress relaxation, crazing, and water absorption. Thermosets, which are obtained by cross-linking polymer chains, do not become soft to any significant extent with increasing temperature. They are much more rigid and harder than thermoplastics, and they offer many fewer choices of color.

- Elastomers have a characteristic ability to undergo large elastic deformations and then return to their original shapes when unloaded. Consequently, they have important applications in tires, seals, footwear, hoses, belts, and shock absorbers.

- Among important considerations in polymers are recyclability and biodegradable plastics, several formulations of which are being developed continually.

KEY TERMS

Additives	Elastomer	Plastics
Biodegradability	Fillers	Polyblends
Bonding	Flame retardants	Polymerization
Branched polymers	Glass-transition temperature	Polymers
Colorants	Latex	Primary bonds
Crazing	Linear polymers	Rubber
Cross-linked polymers	Lubricants	Secondary bonds
Crystallites	Mer	Silicones
Curing	Molecular weight	Stress whitening
Degradation	Monomer	Thermal aging
Degree of crystallinity	Network polymers	Thermoplastics
Degree of polymerization	Orientation	Thermosets
Doping	Plasticizers	Vulcanization

BIBLIOGRAPHY

Berins, M.L., *Plastics Engineering Handbook*, 5th ed., Chapman & Hall, 1995.

Bhowmick, A.K., and Stephens, H.L., *Rubber Products Manufacturing Technology*, Marcel Dekker, 1994.

Buckley, C.P., Bucknall, C.B., and McCrum, N.G., *Principles of Polymer Engineering*, 2nd ed., Oxford University Press, 1997.

Campbell, P., *Plastics Components Design,* Industrial Press, 1996.

Chanda, M., and Roy, S.K., *Plastics Technology Handbook*, 3rd ed., Marcel Dekker, 1998.

Characterization and Failure Analysis of Plastics, ASM International, 2003.

Charrier, J.-M., *Polymeric Materials and Processing: Plastics, Elastomers, and Composites*, Hanser, 1991.

Engineered Materials Handbook, Vol. 2: *Engineering Plastics*, ASM International, 1988.

Engineering Plastics and Composites, 2nd ed., ASM International, 1993.

Fatigue and Tribological Properties of Plastics and Elastomers, William Andrew Inc., 1995.

Feldman, D., and Barbalata, A., *Synthetic Polymers: Technology, Properties, Applications*, Chapman & Hall, 1996.

Harper, C., *Handbook of Plastics, Elastomers, and Composites* 3rd ed., McGraw-Hill, 1996.

Griskey, R.G., *Polymer Process Engineering*, Chapman & Hall, 1995.

Harper, C.A., *Modern Plastics Handbook*, McGraw-Hill, 2000.

MacDermott, C.P., and Shenoy, A.V., *Selecting Thermoplastics for Engineering Applications*, 2nd ed., Marcel Dekker, 1997.

Modern Plastics Encyclopedia, McGraw-Hill, published annually.

Mustafa, N., *Plastics Waste Management: Disposal, Recycling, and Reuse*, Marcel Dekker, 1993.

Nielsen, L.E., and Landel, R.F., *Mechanical Properties of Polymers and Composites*, 2nd ed., Marcel Dekker, 1994.

Rudin, A., *Elements of Polymer Science and Engineering*, 2nd ed., Academic Press, 1999.

Salamone, J.C. (ed.), *Concise Polymeric Materials Encyclopedia*, CRC Press, 1999.

Sperling, L.H., *Polymeric Multicomponent Materials: An Introduction*, Wiley, 1997.

Strong, A.B., *Plastics: Materials and Processing*, 2nd ed., Prentice Hall, 1999.

Ulrich, H., *Introduction to Industrial Polymers*, Hanser, 1994.

Young, R.J., and Lovell, P., *Introduction to Polymers*, Chapman & Hall, 1991.

REVIEW QUESTIONS

7.1 Summarize the important mechanical and physical properties of plastics.

7.2 What are the major differences between (a) the mechanical and (b) the physical properties of plastics and metals?

7.3 What are (a) polymerization and (b) degree of polymerization? What properties are influenced by the degree of polymerization?

7.4 What is the difference between condensation polymerization and addition polymerization?

7.5 What are the differences between linear, branched, and cross-linked polymers?

7.6 Why would we want to synthesize a polymer with a high degree of crystallinity?

7.7 What is a glass-transition temperature?

7.8 What additives are used in plastics? Why?

7.9 What is crazing?

7.10 What are polyblends?

7.11 What are the differences between thermoplastics and thermosets?

7.12 What is an elastomer?

7.13 What is a terpolymer?

7.14 What effects does a plasticizing agent have on a polymer?

7.15 Define the following acronyms: PMMA, PVC, ABS, HDPE, LDPE.

7.16 Describe how a rechargeable lithium battery works.

QUALITATIVE PROBLEMS

7.17 Inspect various plastic components in your automobile and state whether you think that they are made of thermoplastic materials or of thermosetting plastics.

7.18 Give applications for which flammability of plastics would be of major importance.

7.19 What characteristics make polymers advantageous for applications such as gears? What characteristics are drawbacks for such applications?

7.20 What properties do elastomers have that thermoplastics in general do not have?

7.21 Do you think that the substitution of plastics for metals (in products traditionally made of metal) is viewed negatively by the public at large? If so, why?

7.22 Name three plastics that are suitable for use at elevated temperatures.

7.23 Is it possible for a material to have a hysteresis behavior that is the opposite of that shown in Fig. 7.14, so that the arrows run counterclockwise? Explain.

7.24 Observe the behavior of the specimen shown in Fig. 7.13 and state whether the material has a high or a low strain–rate sensitivity exponent m. (See Section 2.2.7.) Explain why it does.

7.25 Add more to the applications column in Table 7.3.

7.26 Discuss the significance of the glass-transition temperature, T_g, in engineering applications.

7.27 Why does cross-linking improve the strength of polymers?

7.28 Describe the methods by which the optical properties of polymers can be altered.

7.29 Can polymers be made to conduct electricity? How?

7.30 Explain the reasons for which elastomers were developed. Are there any substitutes for elastomers? Explain.

7.31 Give several examples of plastic products or components in which creep and stress relaxation are important considerations.

7.32 Describe your opinions regarding the recycling of plastics versus the development of plastics that are biodegradable.

7.33 Explain how you would go about determining the hardness of plastics.

7.34 Compare the values of the elastic modulus from Table 7.1 to the values for metals given in Chapters 2, 5, and 6.

7.35 Why is there so much variation in the stiffness of polymers?

7.36 Explain why thermoplastics are easier to recycle than thermosets.

7.37 Give an example of a process where crazing is desirable.

7.38 Describe how shrink-wrap works.

7.39 List and explain some environmental pros and cons of using plastic shopping bags instead of paper bags.

7.40 List the characteristics required of a polymer for: (a) a total hip replacement insert, (b) a golf ball, (c) an automobile dashboard, (d) clothing, (e) laminated flooring, and (f) fishing nets.

7.41 How can you tell whether a part is made of a thermoplastic?

7.42 As you know, there are plastic paper clips available in various colors. Why are there no plastic staples?

7.43 By incorporating small amounts of a blowing agent, it is possible to manufacture hollow polymer fibers with gas cores. List applications for such fibers.

7.44 In injection molding operations, it is common practice to remove the part from its runner, to place the runner into a shredder, and to recycle the resultant pellets. List the concerns you would have in using such recycled pellets as opposed to so-called "virgin" pellets.

7.45 Based on the subject matter in this chapter, describe how human DNA is similar to a terpolymer.

QUANTITATIVE PROBLEMS

7.46 Calculate the areas under the stress–strain curve (toughness) for the materials in Fig. 7.11, plot them as a function of temperature, and describe your observations.

7.47 Note in Fig. 7.11 that, as expected, the elastic modulus of the polymer decreases as temperature increases. Using the stress–strain curves in the figure, make a plot of the modulus of elasticity versus the temperature. Comment on the shape of the curve.

7.48 A rectangular cantilever beam 120-mm high, 20-mm wide, and 1.5-m long is subjected to a concentrated load of 100 kg at its end. From Table 7.1, select three unreinforced and three reinforced materials and calculate the maximum deflection of the beam in each case. Then select aluminum and steel for the same beam dimensions, calculate the maximum deflection, and compare the results.

7.49 Determine the dimensions of a tubular steel drive shaft for a typical automobile. If you now replace this shaft with an unreinforced and then with a reinforced plastic, what should be its new dimensions in each case in order to transmit the same torque? Choose the materials from Table 7.1 and assume a Poisson's ratio of 0.4.

7.50 Estimate the number of molecules in a typical automobile tire. Estimate the number of atoms.

7.51 Using strength and density data, determine the minimum weight of a 2-ft long tension member which must support a load of 1000 lbs. if it is manufactured from (a) high-molecular-weight polyethylene, (b) polyester, (c) rigid polyvinyl chloride, (d) ABS, (e) polystyrene, and (e) reinforced nylon. Where appropriate, calculate a range of weights for the same polymer.

7.52 Plot the following for any five polymers described in this chapter: (a) UTS versus density and (b) elastic modulus versus UTS. Where appropriate, plot a range of values.

SYNTHESIS, DESIGN, AND PROJECTS

7.53 Describe the design considerations involved in replacing a metal beverage container with one made of plastic.

7.54 Assume that you are manufacturing a product in which all of the gears are made of metal. A salesman visits you and asks you to consider replacing some of these metal gears with plastic ones. Make a list of the questions that you would raise before making a decision.

7.55 Sections 7.6 and 7.7 list several plastics and their applications. Rearrange this information by making a table of products (gears, helmets, luggage, electrical parts, etc.) which shows the types of plastic that can be used to make these products.

7.56 Make a list of products or parts that currently are not made of plastics and offer some reasons why they are not. Support the reasons.

7.57 Review the three curves in Fig. 7.10 and give applications for each type of behavior. Explain your choices.

7.58 Repeat Problem 7.51 for the curves in Fig. 7.12.

7.59 In order to use a steel or aluminum container for an acidic substance, such as tomato sauce, a polymeric barrier must be placed between the container and its contents. Describe methods of producing such a barrier.

7.60 Perform a study of plastics used for some products. Measure the hardness and stiffness of these plastics. (For example, dog chew toys use plastics with a range of properties.)

7.61 Add a column to Table 7.1 which describes the appearance of these plastics, include available colors and opaqueness.

7.62 Perform a literature search and describe the properties and applications of polyaryletherketone (PAEK).

7.63 With Table 7.3 as a guide, inspect various products in both your kitchen and your car and describe the types of plastics that could have been used in making their individual components.

Ceramics, Graphite, and Diamond: Structure, General Properties, and Applications

There are many applications in engineering where the use of metals or polymers is not appropriate; ceramics, diamond, and graphite are among those materials with unique properties. In this chapter, we describe:

- Ceramics: their structure and properties.
- Types of ceramics and their typical applications.
- Glasses: their structure, properties, and applications.
- Diamond: its properties and applications in abrasive machining and micromachining.
- Graphite: its uses as a mold material and as fibers in reinforced plastics.

8.1 | Introduction

The various types of materials described in the preceding chapters is not suitable for certain engineering applications. For example:

a. An electrical insulator to be used at high temperatures
b. Floor tiles to resist spills, scuffing, and abrasion
c. A transparent baking dish
d. Small ball bearings that are light, rigid, hard, and resist high temperatures
e. The space shuttle orbiter, made of aluminum, when its skin temperature reaches 1450°C (2650°F) as it takes off and reenters the atmosphere

From these few examples it is apparent that the properties required are high-temperature strength; hardness; inertness to chemicals, foods, and environment; resistance to wear and corrosion; and low electrical and thermal conductivity.

This chapter describes the general characteristics and applications of those ceramics, glasses, and glass ceramics that are of importance in engineering applications and in manufacturing. Because of their unique characteristics, the properties and uses of two forms of carbon, namely graphite and diamond, are also discussed here. The manufacturing of ceramic and of glass components and various shaping and finishing operations are detailed in Chapter 18. Composites, which are an important group of materials, are described in Chapter 9.

8.2 | The Structure of Ceramics

Ceramics are compounds of metallic and nonmetallic elements. The term *ceramics* (from the Greek words *keramos*, meaning potter's clay, and *keramikos*, meaning clay products) refers both to the material and to the ceramic product itself. Because of the large number of possible combinations of elements, a wide variety of ceramics now is available for a broad range of consumer and industrial applications. The earliest use of ceramics was in pottery and bricks, dating back to before 4000 B.C. Ceramics have been used for many years in automotive spark plugs both as an electrical insulator and for its high-temperature strength. They have become increasingly important in tool and die materials, heat engines, and automotive components (such as exhaust-port liners, coated pistons, and cylinder liners).

Ceramics can be divided into two general categories:

1. **Traditional ceramics**—such as whiteware, tiles, brick, sewer pipe, pottery, and abrasive wheels
2. **Industrial ceramics** (also called **engineering, high-tech,** or **fine ceramics**)—such as turbine, automotive, and aerospace components (Fig. 8.1); heat exchangers; semiconductors; seals; prosthetics; and cutting tools

The structure of ceramic crystals (containing various atoms of different sizes) is among the most complex of all material structures. The bonding between these atoms is generally **covalent** (electron sharing, hence strong bonds) or **ionic** (primary bonding between oppositely charged ions, thus strong bonds). These bonds are much stronger than metallic bonds. Consequently, properties such as hardness and thermal and electrical resistance are significantly higher in ceramics than in metals

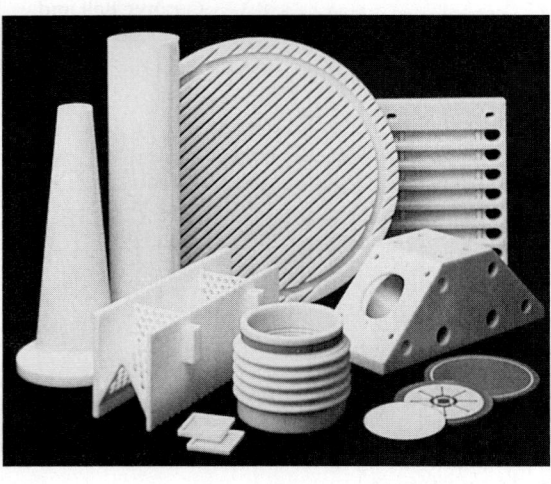

(a)

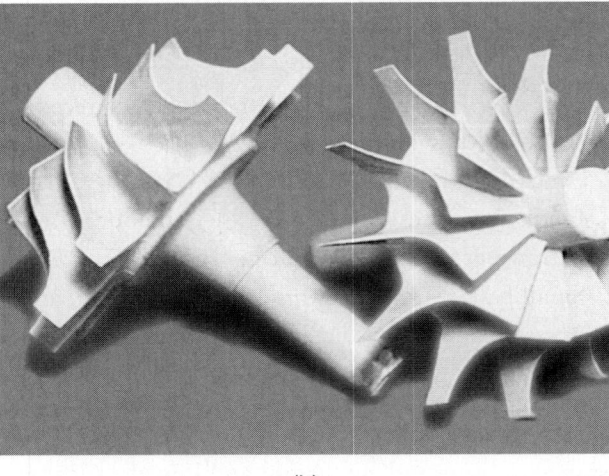

(b)

FIGURE 8.1   A variety of ceramic components. (a) High-strength alumina for high-temperature applications. (b) Gas-turbine rotors made of silicon nitride. *Source:* Courtesy of Wesgo Div., GTE.

(Tables 3.1 and 3.2). Ceramics are available in *single crystal* or in *polycrystalline* form. Grain size has a major influence on the strength and properties of ceramics; the finer the grain size (hence the term **fine ceramics**), the higher the strength and toughness.

8.2.1 Raw materials

Among the oldest of the raw materials used for making ceramics is **clay,** which has a fine-grained sheet-like structure. The most common example is *kaolinite* (from Kaoling, a hill in China). It is a white clay consisting of silicate of aluminum, with alternating weakly-bonded layers of silicon and aluminum ions. When added to kaolinite, water attaches itself to the layers (*adsorption*). This makes the layers slippery and gives wet clay both its well-known softness and the plastic properties (*hydroplasticity*) that make it formable.

Other major raw materials for ceramics that are found in nature are *flint* (a rock composed of very fine-grained silica, SiO_2) and *feldspar* (a group of crystalline minerals consisting of aluminum silicates plus potassium, calcium, or sodium). **Porcelain** is a white ceramic composed of kaolin, quartz, and feldspar; its largest use is in appliances and sanitaryware. In their natural state, these raw materials generally contain impurities of various kinds, which have to be removed prior to further processing of the materials into useful products with reliable performance.

8.2.2 Oxide ceramics

There are two major types of oxide ceramics: alumina and zirconia (Table 8.1).

Alumina. Also called **corundum** or **emery,** *alumina* (aluminum oxide, Al_2O_3) is the most widely used *oxide ceramic*—either in pure form or as a raw material to be blended with other oxides. It has high hardness and moderate strength. Although alumina exists in nature, it contains unknown amounts of impurities and possesses nonuniform properties; as a result, its behavior is unreliable. Aluminum oxide, silicon carbide, and many other ceramics are now manufactured almost totally synthetically, so that their quality can be controlled. First made in 1893, synthetic aluminum oxide is obtained by the fusion of molten bauxite (an aluminum-oxide ore that is the principal source of aluminum), iron filings, and coke in electric furnaces. The cooled product is crushed and graded by size by passing the particles through standard screens. Parts made of aluminum oxide are cold pressed and sintered (*white ceramics*). Their properties are improved by minor additions of other ceramics, such as titanium oxide and titanium carbide.

Structures containing alumina and various other oxides are known as *mullite* and *spinel*; they are used as refractory materials for high-temperature applications. The mechanical and physical properties of alumina are suitable particularly in applications such as electrical and thermal insulation and in cutting tools and abrasives.

Zirconia. *Zirconia* (zirconium oxide, ZrO_2, white in color) has good toughness; good resistance to thermal shock, wear, and corrosion; low thermal conductivity; and a low friction coefficient. **Partially stabilized zirconia** (PSZ) has high strength and toughness and better reliability in performance than does zirconia. It is obtained by doping zirconia with oxides of calcium, yttrium, or magnesium. This process forms a material with fine particles of tetragonal zirconia in a cubic lattice. Typical applications include dies for the hot extrusion of metals and zirconia beads used as grinding and dispersion media for aerospace coatings, for automotive primers and topcoats, and for fine glossy print on flexible food packaging.

TABLE 8.1

Types and General Characteristics of Ceramics

Type	General characteristics
Oxide ceramics	
Alumina	High hardness and moderate strength; most widely used ceramic; cutting tools; abrasives; electrical and thermal insulation.
Zirconia	High strength and toughness; thermal expansion close to cast iron; suitable for high-temperature applications.
Carbides	
Tungsten carbide	Hardness, strength, and wear resistance depend on cobalt binder content; commonly used for dies and cutting tools.
Titanium carbide	Not as tough as tungsten carbide; has nickel and molybdenum as the binder; used as cutting tools.
Silicon carbide	High-temperature strength and wear resistance; used for heat engines and as abrasives.
Nitrides	
Cubic boron nitride	Second-hardest substance known, after diamond; used as abrasives and cutting tools.
Titanium nitride	Gold in color; used as coatings because of low frictional characteristics.
Silicon nitride	High resistance to creep and thermal shock; used in high-temperature applications.
Sialon	Consists of silicon nitrides and other oxides and carbides; used as cutting tools.
Cermets	Consist of oxides, carbides, and nitrides; used in high-temperature applications.
Silica	High-temperature resistance; quartz exhibits piezoelectric effect; silicates containing various oxides are used in high-temperature nonstructural applications.
Glasses	Contain at least 50 percent silica; amorphous structures; several types available with a wide range of mechanical and physical properties.
Glass ceramics	Have a high crystalline component to their structure; good thermal-shock resistance and strong.
Graphite	Crystalline form of carbon; high electrical and thermal conductivity; good thermal-shock resistance.
Diamond	Hardest substance known; available as single crystal or polycrystalline form; used as cutting tools and abrasives and as dies for fine wire drawing.

Two important characteristics of PSZ are its coefficient of thermal expansion (which is only about 20% lower than that of cast iron) and its thermal conductivity (which is about one-third that of other ceramics). Consequently, PSZ is very suitable for heat-engine components, such as cylinder liners and valve bushings, to help keep the cast-iron engine assembly intact. **Transformation-toughened zirconia** (TTZ) has higher toughness because of dispersed tough phases in the ceramic matrix.

EXAMPLE 8.1 Ceramic knives

The use of ceramics now is being extended to ceramics knives generally made of zirconium oxide. Ceramic knives are produced by a process (described in Section 18.2) that starts with a ceramic powder mixed with various binders and compacted (molded) into blanks under high pressure. It then is fired (sintered) at temperatures above 1000°C (1830°F) for several days. The blanks then are ground and polished on a diamond wheel to form a sharp edge, and the handle is attached. The Mohs hardness (Section 2.6) of the zirconium oxide ceramic is 8.2, as compared with 6 for hardened steel and a maximum of 10 for diamond.

Among the advantages of ceramics knives over steel knives: (a) Because of their very high hardness and wear resistance, ceramic knives can last months and even years before sharpening. (b) They are chemically inert. Consequently, they do not stain, food does not stick (hence are easy to clean), and they leave no metallic taste or smell. (c) Because they are lightweight, they are easier to use.

The knives should be stored in wooden knife blocks, and handled carefully. Sharp impact against other objects (such as dishes or dropping it on its edge on a hard surface) should be avoided, as the sharp edges of the knife can chip. Also, they should be used only for cutting (not for prying) and in cutting meat, contact with bones is not advisable. Knives have to be sharpened at the factory to precise edge shape using diamond grinding wheels. Ceramic knives are more expensive than steel knives, typically ranging from $60 for a 3-in. paring knife to $250 for a 6-in. serrated knife. *Source:* Courtesy of Kyocera Corporation

8.2.3 Other ceramics

Major ceramics of other types may be classified as follows:

Carbides.   Typical *carbides* are those made of tungsten and titanium (used as cutting tools and die materials) and of silicon (used as an abrasive, especially in grinding wheels). Some examples of carbides are:

- **Tungsten carbide** (WC) consists of tungsten-carbide particles with cobalt as a binder. The amount of binder has a major influence on the material's properties; toughness increases with cobalt content, whereas hardness, strength, and wear resistance decrease.

- **Titanium carbide** (TiC) has nickel and molybdenum as the binder and is not as tough as tungsten carbide.

- **Silicon carbide** (SiC) has good resistance to wear, thermal shock, and corrosion. It has a low friction coefficient, and it retains strength at elevated temperatures. Thus, it is suitable for high-temperature components in heat engines and also is used as an abrasive. First produced in 1891, synthetic silicon carbide is made from silica sand, coke, and small amounts of sodium chloride and sawdust. The process is similar to that for making synthetic aluminum oxide (Section 8.2).

Nitrides.   Another class of ceramics is the nitrides, examples of which are:

- **Cubic boron nitride** (cBN) is the second hardest known substance (after diamond) and has special applications, such as in cutting tools and as abrasives in grinding wheels. It does not exist in nature and was first made synthetically in the 1970s by means of techniques similar to those used in making synthetic diamond (see Section 8.7).

- **Titanium nitride** (TiN) is used widely as a coating on cutting tools; it improves tool life by virtue of its low frictional characteristics.

- **Silicon nitride** (Si_3N_4) has high resistance to creep at elevated temperatures, low thermal expansion, and high thermal conductivity. Consequently, it resists thermal shock. It is suitable for high-temperature structural applications, such as in automotive-engine and gas-turbine components, cam-follower rollers, bearings, sand-blast nozzles, and components for the paper industry.

Sialon.   Sialon consists of silicon nitride with various additions of aluminum oxide, yttrium oxide, and titanium carbide. It has higher strength and thermal-shock resistance than silicon nitride. Presently, it is used primarily as a cutting-tool material. The word *sialon* is derived from *si*licon, *al*uminum, *o*xygen, and *n*itrogen.

Cermets. *Cermets* are combinations of a *ceramic* phase bonded with a *metallic* phase. Introduced in the 1960s and also called **black ceramics** or **hot-pressed ceramics,** they combine the high-temperature oxidation resistance of ceramics with the toughness, thermal-shock resistance, and ductility of metals. A common application of cermets is in cutting tools with a typical composition being 70% Al_2O_3 and 30% TiC. Other cermets contain various oxides, carbides, and nitrides. They have been developed for high-temperature applications, such as nozzles for jet engines and brakes for aircraft. Cermets can be regarded as composite materials; they can be used in various combinations of ceramics and metals bonded by powder-metallurgy techniques (Chapter 17).

8.2.4 Silica

Abundant in nature, **silica** is a polymorphic material—that is, it can have different crystal structures (for example, the cubic structure is found in refractory bricks used for high-temperature furnace applications). Most glasses contain more than 50% silica. The most common form of silica is **quartz,** which is a hard, abrasive hexagonal crystal used extensively in communications applications as an oscillating crystal of fixed frequency because it exhibits the piezoelectric effect (Section 3.7).

   Silicates are products of the reaction of silica with oxides of aluminum, magnesium, calcium, potassium, sodium, and iron; examples are clay, asbestos, mica, and silicate glasses. **Lithium aluminum silicate** has a very low thermal expansion and thermal conductivity and good thermal-shock resistance. However, it also has a very low strength and fatigue life, thus it is suitable only for nonstructural applications (such as catalytic converters, regenerators, and heat-exchanger components).

8.2.5 Nanoceramics and composites

In order to improve the ductility and manufacturing properties of ceramics, the particle size in ceramics has been reduced using various techniques, such as gas condensation. Called *nanoceramics* or **nanophase ceramics,** these materials consist of atomic clusters containing a few thousand atoms. Control of particle size, distribution, and contamination are important in these ceramics. Nanoceramics exhibit ductility at significantly lower temperatures than do conventional ceramics and are stronger and easier to fabricate and to machine with fewer flaws. Present applications are in the automotive industry (such as valves, rocker arms, turbocharger rotors, and cylinder liners) and in jet-engine components.

   Nanocrystalline *second-phase particles* (on the order of 100 nm or less) and fibers also are used as reinforcements in composites. These composites have enhanced properties, such as tensile strength and creep resistance. (See also *nanomaterials* in Section 6.16.)

8.3   General Properties and Applications of Ceramics

Compared to metals, ceramics typically have the following relative characteristics: brittleness; high strength and hardness at elevated temperatures; high elastic modulus; low toughness, density, thermal expansion; and low thermal and electrical conductivity. However, because of the wide variety of material compositions and grain sizes, the mechanical and physical properties of ceramics vary significantly. For example, the electrical conductivity of ceramics can be modified from poor to good; this change is the principle behind semiconductors.

Because of their sensitivity to flaws, defects, and surface or internal cracks; the presence of different types and levels of impurities; and different methods of manufacturing, ceramics can have a wide range of properties. Although the individual characteristics of ceramics were outlined in Section 8.2, their general mechanical and physical properties are described next.

8.3.1 Mechanical properties

The mechanical properties of several engineering ceramics are presented in Table 8.2. Note that their strength in tension (transverse rupture strength, Section 2.5) is approximately one order of magnitude lower than their compressive strength. The reason for this is their sensitivity to cracks, impurities, and porosity. Such defects lead to the initiation and propagation of cracks under tensile stresses and reduce the tensile strength severely. Thus, reproducibility and reliability (acceptable performance over a specified period of time) is an important aspect in the service life of ceramic components.

The tensile strength of a polycrystalline ceramic increases with decreasing grain size and porosity. This relationship is represented approximately by the expression

$$UTS = UTS_o e^{-nP} \tag{8.1}$$

where P is the volume fraction of pores in the solid (thus, if porosity is 15%, then $P = 0.15$), UTS_o is the tensile strength at zero porosity, and the exponent n ranges between 4 and 7. The modulus of elasticity of ceramics is related approximately to its porosity by the expression

$$E = E_o(1 - 1.9P + 0.9P^2) \tag{8.2}$$

where E_o is the modulus at zero porosity.

Unlike most metals and thermoplastics, ceramics generally lack impact toughness and thermal-shock resistance because of their inherent lack of ductility; once

TABLE 8.2

Properties of Various Ceramics at Room Temperature							
Material	Symbol	Transverse rupture strength (MPa)	Compressive strength (MPa)	Elastic modulus (GPa)	Hardness (HK)	Poisson's ratio (ν)	Density (kg/m^3)
Aluminum oxide	Al_2O_3	140–240	1000–2900	310–410	2000–3000	0.26	4000–4500
Cubic boron nitride	CBN	725	7000	850	4000–5000	—	3480
Diamond	—	1400	7000	830–1000	7000–8000	—	3500
Silica, fused	SiO_2	—	1300	70	550	0.25	—
Silicon carbide	SiC	100–750	700–3500	240–480	2100–3000	0.14	3100
Silicon nitride	Si_3N_4	480–600	—	300–310	2000–2500	0.24	3300
Titanium carbide	TiC	1400–1900	3100–3850	310–410	1800–3200	—	5500–5800
Tungsten carbide	WC	1030–2600	4100–5900	520–700	1800–2400	—	10,000–15,000
Partially stabilized zirconia	PSZ	620	—	200	1100	0.30	5800

Note: These properties vary widely depending on the condition of the material.

initiated, a crack propagates rapidly. In addition to undergoing fatigue failure under cyclic loading, ceramics (particularly glasses) exhibit a phenomenon called **static fatigue.** When subjected to a static tensile load over a period of time, these materials suddenly may fail. This phenomenon occurs in environments where water vapor is present. Static fatigue, which does not occur in a vacuum or in dry air, has been attributed to a mechanism similar to the stress–corrosion cracking of metals.

Ceramic components that are to be subjected to tensile stresses may be *prestressed* in much the same way that concrete is prestressed. Prestressing the shaped ceramic components subjects them to compressive stresses. The methods used include:

- Heat treatment and chemical tempering
- Laser treatment of surfaces
- Coating with ceramics having different thermal-expansion coefficients
- Surface-finishing operations (such as grinding) in which compressive residual stresses are induced on the surfaces

Major advances have been made in improving the toughness and other properties of ceramics, including the development of **machinable** and **grindable ceramics.** Among these advances are the proper selection and processing of raw materials, the control of purity and structure, and the use of reinforcements—with particular emphasis on advanced methods of stress analysis during the design of ceramic components.

8.3.2 Physical properties

Most ceramics have a relatively low specific gravity, ranging from about 3 to 5.8 for oxide ceramics as compared to 7.86 for iron (Table 3.1). They have very high melting or decomposition temperatures.

Thermal conductivity in ceramics varies by as much as three orders of magnitude (depending on their composition), whereas in metals, it varies by only one order. Like that of other materials, the thermal conductivity of ceramics decreases with increasing temperature and porosity, because air is a poor thermal conductor. The thermal conductivity k is related to porosity by

$$k = k_o(1 - P) \tag{8.3}$$

where k_o is the thermal conductivity at zero porosity and P is the porosity as a fraction of the total volume.

Thermal expansion and thermal conductivity induce internal stresses that can lead to thermal shock or to thermal fatigue in ceramics. The tendency toward **thermal cracking** (called **spalling** when a small piece or a layer from the surface breaks off) is lower with the combination of low thermal expansion and high thermal conductivity. For example, fused silica has high thermal-shock resistance because of its virtually zero thermal expansion.

A familiar example that illustrates the importance of low thermal expansion is heat-resistant ceramics for cookware and electric stove tops. (See also *glass ceramics*, Section 8.5.) They can sustain high thermal gradients, from hot to cold and vice versa. Moreover, the similar thermal expansion of both ceramics and metals is an important reason for the use of ceramic components in heat engines. The fact that the thermal conductivity of partially stabilized zirconia components is close to that of the cast iron in engine blocks is an additional advantage to the use of PSZ in heat engines.

Another characteristic is the **anisotropy of thermal expansion** of oxide ceramics (like that exhibited by hexagonal close-packed metals), when the thermal expansion varies with differing direction through the ceramic (by as much as 50% for quartz). This behavior causes thermal stresses that can lead to cracking of the ceramic component.

The *optical* properties of ceramics can be controlled by using various formulations and by controlling the structure. These methods make possible the imparting of different degrees of transparency and translucency and of different colors (for example, single-crystal sapphire is completely transparent; zirconia is white; and fine-grained polycrystalline aluminum oxide is a translucent gray). Porosity influences the optical properties of ceramics much in the same way as air trapped in ice cubes; it makes the material less transparent and gives it a white appearance. Although ceramics are basically resistors, they can be made *electrically conducting* by alloying them with certain elements in order to make the ceramic act like a semiconductor or even like a superconductor.

8.3.3 Applications

Ceramics have numerous consumer and industrial applications. Various types of ceramics are used in the electrical and electronics industries, because they have high electrical resistivity, high dielectric strength (voltage required for electrical breakdown per unit thickness), and magnetic properties suitable for such applications as magnets for speakers.

The capability of ceramics to maintain their strength and stiffness at elevated temperatures makes them very attractive for high-temperature applications. The higher operating temperatures made possible by the use of ceramic components mean more efficient burning of fuel and reduction of emissions in automobiles. Currently, internal combustion engines are only about 30% efficient, but with the use of ceramic components, the operating performance can be improved by at least 30%.

Much research has been conducted on developing materials and techniques for an all-ceramic heat engine capable of operating at temperatures up to 1000°C (1830°F). Progress, however, has been slower than expected because of such problems as unreliability, lack of sufficient toughness, difficulty with lubricating bearings and hot components, an unmet need for reliable nondestructive evaluation techniques, and a lack of the capability for structural ceramics (such as silicon nitride and silicon carbide) to be produced economically in near-net shape, as weighed against the need for the machining and finishing processes required for dimensional accuracy of the engine. Ceramics that are being used successfully, especially in automotive gas-turbine engine components (such as rotors), are silicon nitride, silicon carbide, and partially stabilized zirconia.

Coating metal with ceramics is another application; it may be done to reduce wear, prevent corrosion, or provide a thermal barrier. The tiles on the *space shuttles*, for example, are made of silica fibers having an open cellular structure that consists of 5% silica. The rest of the tile structure is air, so the tile is not only very light in weight but is also an excellent heat barrier. The tiles (34,000 on each shuttle) are bonded to the aluminum skin of the space shuttle with several layers of silicon-based adhesives. The skin temperature on the shuttle reaches 1400°C (2550°F) because of frictional heat from contact with the atmosphere.

Other attractive properties of ceramics are their low density and high elastic modulus. They enable engine weight to be reduced and, in other applications, allow the inertial forces generated by moving parts to be lower. Ceramic turbochargers, for example, are about 40% lighter than conventional ones. High-speed components for machine tools also are candidates for ceramics (Section 25.3). Furthermore, the high elastic modulus of ceramics makes them attractive for improving the stiffness of machines, while reducing the weight. Their high resistance to wear makes them suitable for applications such as cylinder liners, bushings, seals, bearings, and liners for cylinders and gun barrels.

EXAMPLE 8.2 Ceramic gun barrels

The wear resistance and low density of ceramics have lead to investigation regarding their use as liners for gun barrels. Their limited success has lead to more recent developments in making composite ceramic gun barrels, which have improved performance over traditional steel barrels. The 50 caliber zirconia ceramic barrel is formed in several separate segments (each 150–200 mm (6–8 in.) long and with a wall thickness of 3.75 mm. (0.150 in.) by the shaping and sintering processes described in Section 17.9.

The segments subsequently are machined to the required dimensions and surface finish. Zirconia has been chosen for its high toughness, flexural strength, specific heat, operating temperature, and very low thermal conductivity. The thermal properties are important to the performance of the barrel and the bullet.

The separate ceramic segments are then joined, and the barrel is wrapped with a carbon-fiber/polymer-matrix composite which subjects the ceramic barrel to a compressive stress of 100,000 psi, thus greatly improving its capacity to withstand tensile stresses developed during firing. The inside of the barrel is then rifled (cutting of internal spiral grooves to give rotation to the exiting bullet for gyroscopic stability) and fitted to a breech.

Source: Courtesy of K.H. Kohnken, Surface Conversion Technologies, Inc., Cumming, Georgia.

EXAMPLE 8.3 Ceramic ball and roller bearings

Silicon-nitride ceramic ball and roller bearings are used when high temperature, high speed, or marginally lubricated conditions occur. The bearings can be made entirely from ceramics or just the ball and rollers are ceramic and the races are metal, in which case they are referred to as hybrid bearings (Fig. 8.2). Examples of machines utilizing ceramic and hybrid bearings include high-performance machine

(a)

FIGURE 8.2   A selection of ceramic bearings and races.
Source: Courtesy of The Timken Company.

(b)

FIGURE 8.2   (*continued*)

tool spindles, metal-can seaming heads, high-speed flow meters and the Space Shuttle's main booster rocket's liquid oxygen and hydrogen pumps.

The ceramic spheres have a diameter tolerance of 0.13 μm (5 μin.) and a surface roughness of 0.02 μm (0.8 μin.). They have high wear resistance, high fracture toughness, perform well with little or no lubrication, and have low density. The balls have a coefficient of thermal expansion one-fourth that of steel, and they can withstand temperatures of up to 1400°C (2550°F). Produced from titanium and carbon nitride by using powder-metallurgy techniques, the full-density titanium carbonitride (TiCN) or silicon nitride (Si_3N_4) bearing-grade material can be twice as hard as chromium steel and 40% lighter. Components up to 300 mm (12 in.) in diameter can be produced.

Bioceramics.   Because of their strength and inertness, ceramics are used as bio-materials (*bioceramics*) to replace joints in the human body, as prosthetic devices, and in dental work. Commonly used bioceramics are aluminum oxide, hydroxy ap-atite, tricalcium phosphate, silicon nitride, and various compounds of silica. Ceramic implants can be made porous, whereby bone can grow into the porous structure (likewise with porous titanium implants) and develop a strong bond with structural integrity.

8.4 | Glasses

Glass is an amorphous solid with the structure of a liquid. It has been *supercooled* (cooled at a rate too high to allow crystals to form). Technically, glass is defined as an inorganic product of fusion that has cooled to a rigid condition without crystallizing. Glass has no distinct melting or freezing point, thus its behavior is similar to that of amorphous alloys (see *metallic glasses*, Section 6.14) and amorphous polymers (Section 7.2).

Glass beads first were produced in about 2000 B.C., and the art of glass-blowing followed in about 200 B.C. Silica was used for all glass products until the late 1600s. Rapid developments in glasses began in the early 1900s. Currently, there are some 750 different types of commercially available glasses with applications ranging from window glass to glass for containers, cookware, lighting, TV tubes and CRTs, and to glasses with special mechanical, electrical, high-temperature, anti-chemical, corrosion, and optical characteristics. Special glasses are used in fiber optics (for communication by light with little loss in signal power) and in glass fibers with very high strength (for use in reinforced plastics).

All glasses contain at least 50% silica, which is known as a **glass former**. The composition and properties of glasses can be modified greatly by the addition of oxides of aluminum, sodium, calcium, barium, boron, magnesium, titanium, lithium, lead, and potassium. Depending on their function, these oxides are known as **intermediates** (or **modifiers**).

8.4.1 Types of glasses

Almost all *commercial glasses* are categorized by type (Table 8.3).

- **Soda-lime glass** (the most common type)
- **Lead-alkali glass**
- **Borosilicate glass**
- **Aluminosilicate glass**
- **96%-silica glass**
- **Fused silica glass**

Glasses also are classified as colored, opaque (white and translucent), multiform (variety of shapes), optical, photochromatic (darkens when exposed to light, as in sunglasses), photosensitive (changing from clear to opaque), fibrous (drawn into long fibers, as in fiberglass), and foam or cellular (containing bubbles, thus a good thermal insulator). Glasses also can be referred to as **hard** or **soft,** usually in the sense of a thermal rather than mechanical property (see also hardness of glasses, Section 8.4.2). Thus, a soft glass softens at a lower temperature than does a hard glass. Soda-lime and lead-alkali glasses are considered soft, the rest hard.

TABLE 8.3

Properties of Various Glasses					
	Soda-lime glass	Lead glass	Borosilicate glass	96% silica	Fused silica
Density	High	Highest	Medium	Low	Lowest
Strength	Low	Low	Moderate	High	Highest
Resistance to thermal shock	Low	Low	Good	Better	Best
Electrical resistivity	Moderate	Best	Good	Good	Good
Hot workability	Good	Best	Fair	Poor	Poorest
Heat treatability	Good	Good	Poor	None	None
Chemical resistance	Poor	Fair	Good	Better	Best
Impact-abrasion resistance	Fair	Poor	Good	Good	Best
Ultraviolet-light transmission	Poor	Poor	Fair	Good	Good
Relative cost	Lowest	Low	Medium	High	Highest

8.4.2 Mechanical properties

The behavior of glass, like that of most ceramics, generally is regarded as perfectly elastic and brittle. The modulus of elasticity for commercial glasses ranges mostly from 55 to 90 GPa (8 to 13 million psi), and their Poisson's ratios from 0.16 to 0.28. Hardness of glasses, as a measure of resistance to scratching, ranges from 5 to 7 on the Mohs scale; that is equivalent to a range approximately from 350 to 500 HK. (See Fig. 2.16.)

Glass in *bulk* form generally has a strength of less than 140 MPa (20 ksi). The relatively low strength of bulk glass is attributed to the presence of small flaws and microcracks on its surface, some or all of which may be introduced during normal handling of the glass by inadvertent abrading. These defects reduce the strength of glass by two to three orders of magnitude, compared to its ideal (defect free) strength. Glasses can be strengthened by thermal or chemical treatments to obtain high strength and toughness (Section 18.4). The strength of glass theoretically can reach 35 GPa (5 million psi). When molten glass is freshly drawn into fibers (**fiberglass**), its tensile strength ranges from 0.2 to 7 GPa (30 to 1000 ksi), with an average value of about 2 GPa (300 ksi). These glass fibers are stronger than steel; they are used to reinforce plastics in such applications as boats, automobile bodies, furniture, and sports equipment (Tables 2.1 and 9.1).

The strength of glass usually is measured by bending it. The surface of the glass is thoroughly abraded (roughened) to ensure that the test gives a strength level that is reliable for actual service under adverse conditions. The phenomenon of **static fatigue**, observed in ceramics, also is exhibited by glasses. As a guide, if a glass item must withstand a load for 1000 hours or longer, the maximum stress that can be applied to it is approximately one-third the maximum stress that the same item can withstand during the first second of loading.

8.4.3 Physical properties

Glasses have low thermal conductivity, high electrical resistivity, and dielectric strength. Their thermal expansion coefficients are lower than those for metals and plastics, and may even approach zero. For example, titanium silicate glass (a clear, synthetic high-silica glass) has a near-zero coefficient of thermal expansion. Fused silica (a clear, synthetic amorphous silicon dioxide of very high purity) also has a near-zero coefficient of expansion. The optical properties of glasses (such as reflection, absorption, transmission, and refraction) can be modified by varying their composition and treatment. Glasses generally are resistant to chemical attack and are ranked by their resistance to corrosion by acids, alkalis, or water.

8.5 | Glass Ceramics

Although glasses are amorphous, **glass ceramics** (such as *Pyroceram*, a trade name) have a high crystalline component to their microstructure. Glass ceramics contain large proportions of several oxides, and thus, their properties are a combination of those for glass and those for ceramics. Most glass ceramics are stronger than glass. These products first are shaped and then heat treated, with **devitrification** (recrystallization) of the glass occurring. Unlike most glasses, which are clear, glass ceramics are generally white or gray in color.

The hardness of glass ceramics ranges approximately from 520 to 650 HK. They have a near-zero coefficient of thermal expansion; as a result, they have high thermal-shock resistance. They are strong, because of the absence of the porosity usually found in conventional ceramics. The properties of glass ceramics can be improved by modifying their composition and by heat-treatment techniques. First developed in 1957, glass ceramics are suitable for cookware, heat exchangers in gas turbine engines, radomes (housings for radar antennas), and electrical and electronics applications.

8.6 | Graphite

Graphite is a crystalline form of carbon having a *layered structure* with basal planes or sheets of close-packed carbon atoms (see Fig. 1.4). Consequently, graphite is weak when sheared along the layers. This characteristic, in turn, gives graphite its low frictional properties as a solid lubricant. However, its frictional properties are low only in an environment of air or moisture; in a vacuum, graphite is abrasive and a poor lubricant. Unlike in other materials, strength and stiffness of graphite increase with temperature. Amorphous graphite is known as **lampblack** (black soot) and is used as a pigment.

Although brittle, graphite has high electrical and thermal conductivity and good resistance to thermal shock and to high temperature (although it begins to oxidize at 500°C (930°F). It is, therefore, an important material for applications such as electrodes, heating elements, brushes for motors, high-temperature fixtures and furnace parts, mold materials (such as crucibles for the melting and casting of metals), and seals (Fig. 8.3). A characteristic of graphite is its resistance to chemicals; thus, for example, it is used in filters for corrosive fluids. Also, its low absorption cross-section and high scattering cross-section for thermal neutrons make graphite suitable for nuclear applications. Ordinary pencil "lead" is a mixture of graphite and clay.

Graphite is available commercially in square, rectangular, and round shapes of various sizes and generally is graded in decreasing order of grain size: *industrial, fine-grain,* and *micrograin.* As in ceramics, the mechanical properties of graphite improve with decreasing grain size. Micrograin graphite can be impregnated with copper. In

(a)

(b)

FIGURE 8.3   (a) Various engineering components made of graphite. *Source:* Courtesy of Poco Graphite, Inc., a Unocal Co. (b) Examples of graphite electrodes for electrical discharge machining. *Source:* Courtesy of Unicor, Inc.

this form, it is used for electrodes in electrical-discharge machining (Section 27.5) and for furnace fixtures. Graphite usually is processed first by molding or forming, then by oven baking, and finally by machining to the final shape.

Graphite fibers.   An important use of graphite is as fibers in reinforced plastics and composite materials, as described Section 9.2.

Carbon and graphite foams.   These foams have important properties of high service temperatures, chemical inertness, low thermal expansion, and thermal and electrical properties that can be tailored for specific applications. Carbon foams are available either in graphitic or nongraphitic structures. Graphitic foams (typically produced from petroleum, coal tar, and synthetic pitches) have low density, high thermal and electrical conductivity (but lower mechanical strength), and are much more expensive than nongraphitic foams (produced from coal or organic resins), which are highly amorphous.

These foams have a cellular microstructure with interconnected pores, thus their mechanical properies depend on density (see also Section 8.3). They easily can be machined into various complex shapes with appropriate tooling. Applications of carbon foams include their use as core materials for aircraft and ship interior panels, structural insulation, sound absorption panels, substrates for space-borne mirrors, lithium-ion batteries, and fire and thermal protection.

Buckyballs.   A more recent development is the production of carbon molecules (usually C60) in the shape of a soccer ball, called **buckyballs,** after Buckminster Fuller (1895–1983), the inventor of the geodesic dome. Also called **fullerenes,** these chemically-inert spherical molecules are produced from soot and act much like solid lubricant particles. Fullerenes can become low-temperature superconductors when mixed with metals. Despite their promise and significant research investment, no commercial applications of buckyballs currently exist.

Nanotubes.   Carbon nanotubes have a geometric structure similar to that of graphite sheets; one can picture a nanotube as a rolled-up sheet of graphite. Carbon nanotubes are a few nanometers in diameter and typically a few micrometers in length. While nanotubes have been suggested for a number of applications, none have been commercially successful to date. Nanotubes most often are cited as a natural building material for new microelectromechanical systems (see Chapter 29).

8.7 | Diamond

Diamond is a principal form of carbon with a covalently bonded structure. It is the hardest substance known (7000 to 8000 HK). However, it is brittle and begins to decompose in air at about 700°C (1300°F), but it resists higher temperatures in nonoxidizing environments.

Synthetic (also called **industrial**) **diamond** was first made in 1955. A common method of manufacturing it is to subject graphite to a hydrostatic pressure of 14 GPa (2 million psi) and a temperature of 3000°C (5400°F). Synthetic diamond is identical to natural diamond. It has superior properties because of its lack of impurities and is used extensively in industrial applications. It is available in various sizes and shapes; for abrasive machining, the most common grit size is 0.01 mm (0.004 in.) in diameter. **Diamond-like carbon** (DLC) also has been developed and is used as a diamond film coating, as described in Section 34.13. Diamond particles also can be *coated* with nickel, copper, or titanium for improved performance in grinding operations.

In addition to its use in jewelry, gem-quality synthetic diamond has possible applications as heat sinks for computers, in telecommunications and integrated-circuit industries, and as windows for high-power lasers. Its electrical conductivity is 50 times higher than that of natural diamond, and it is 10 times more resistant to laser damage.

Because of its favorable characteristics, diamond has many important applications, such as:

- Cutting-tool materials, as a single crystal or in polycrystalline form
- Abrasives in grinding wheels, for grinding hard materials
- Dressing of grinding wheels (i.e., sharpening of the abrasive grains)
- Dies for drawing wire less than 0.06 mm (0.0025 in.) in diameter
- Coatings for cutting tools and dies

SUMMARY

- Several nonmetallic materials are of major importance both in engineering applications and in manufacturing processes. Ceramics, which are compounds of metallic and nonmetallic elements, generally are characterized by high hardness, high compressive strength, high elastic modulus, low thermal expansion, high temperature resistance, good chemical inertness, low density, and low thermal and electrical conductivity. On the other hand, they are brittle and have low toughness. Nanophase ceramics have better properties than common ceramics. Porosity in ceramics has important effects on their properties.

- Ceramics generally are categorized as either traditional or industrial (or high-tech) ceramics; the latter are particularly attractive for applications such as heat-engine components, cutting tools, and components requiring resistance against wear and corrosion. Ceramics of importance in design and manufacturing are the oxide ceramics (alumina and zirconia), tungsten and silicon carbides, nitrides, and cermets.

- Glasses are supercooled liquids and are available in a wide variety of compositions and of mechanical, physical, and optical properties. Glass ceramics predominantly are crystalline in structure and have properties that are more desirable than those of glasses.

- Glass in bulk form has relatively low strength, but glasses can be strengthened by thermal and chemical treatments in order to obtain high strength and toughness. Glass fibers are used widely as a reinforcing material in composite materials, such as in fiber-reinforced plastics.

- Graphite, fullerenes, and diamond are forms of carbon which display unusual combinations of properties. These materials have several unique and emerging applications in engineering and manufacturing. Graphite has high-temperature and electrical applications; graphite fibers are used to reinforce plastics and other composite materials. Diamond (both natural and synthetic) is used as cutting tools for fine machining operations, as dies for thin-wire drawing, and as abrasives for grinding wheels. Diamond-like carbon has applications as a coating material giving improved wear resistance.

KEY TERMS

Alumina	Feldspar	Nitrides
Bioceramics	Flint	Oxide ceramics
Buckyballs	Fullerenes	Partially stabilized zirconia
Carbides	Glass	Porcelain
Carbon	Glass ceramics	Porosity
Carbon foam	Glass fibers	Sialon
Ceramics	Glass former	Silica
Cermets	Graphite	Static fatigue
Clay	Industrial ceramics	Transformation-toughened zirconia
Devitrification	Industrial diamond	White ceramics
Diamond	Ion implantation	Zirconia
Diamond-like carbon	Nanophase ceramics	

BIBLIOGRAPHY

Barsoum, M.W., *Fundamentals of Ceramics,* Iop Institute of Physics, 2003.

Bioceramics: Materials and Applications, American Ceramic Society, Vol. I, 1995; Vol. II, 1996, Vol. III, 1999.

Brook, R.J. (ed.), *Concise Encyclopedia of Advanced Ceramic Materials,* Pergamon, 1991.

Cheremisinoff, N.P., *Handbook of Ceramics and Composites,* 3 vols., Marcel Dekker, 1991.

Chiang, Y.-M., *Physical Ceramics: Principles for Ceramics Science and Engineering,* Wiley, 1995.

Doresmus, R.H., *Glass Science,* 2nd ed., Wiley, 1994.

Ellis, W.S., *Glass,* Avon Books, 1998.

Edinsinghe, M.J., *An Introduction to Structural Engineering Ceramics,* Ashgate Pub. Co., 1997.

Engineered Materials Handbook, Vol. 4: *Ceramics and Glasses,* ASM International, 1991.

Green, D.J., *An Introduction to the Mechanical Properties of Ceramics,* Cambridge Univ. Press, 1998.

Harper, C.A. (ed.), *Handbook of Ceramics, Glasses, and Diamonds,* McGraw-Hill, 2001.

Hench, L.L., and Wilson, J. (eds.), *An Introduction to Bioceramics,* World Scientific Pub., 1993.

Holand, W., and Beall, G.H., *Design and Properties of Glass Ceramics,* American Chemical Society, 2001.

Lu, H.Y., *Introduction to Ceramic Science,* Marcel Dekker, 1996.

Musikant, S., *What Every Engineer Should Know About Ceramics,* Marcel Dekker, 1991.

Nanoceramics, Institute of Materials, 1993.

Pfaender, H.G. (ed.), *Schott Guide to Glass,* Chapman & Hall, 1996.

Phillips, G.C., *A Concise Introduction to Ceramics,* Van Nostrand Rheinhold, 1991.

Pierson, H.O., *Handbook of Carbon, Graphite, Diamond and Fullerenes: Properties, Processing and Applications,* Noyes Pub., 1993.

Prelas, M.A., Popovichi, G., and Bigelow, L.K. (eds.), *Handbook of Industrial Diamonds and Diamond Films,* Marcel Dekker, 1998.

Richerson, D.W., *Modern Ceramic Engineering: Properties, Processing, and Use in Design,* 2nd ed., Marcel Dekker, 1992.

Schwartz, M.M. (ed.), *Handbook of Structural Ceramics,* McGraw-Hill, 1992.

Vincenzini, P., *Fundamentals of Ceramic Engineering,* Elsevier, 1991.

Wilks, J., and Wilks, E., *Properties and Applications of Diamond,* Butterworth-Heinemann, 1991.

REVIEW QUESTIONS

8.1 Compare the major differences between the properties of ceramics and those of metals and plastics.

8.2 List the major types of ceramics that are useful in engineering applications.

8.3 What do the following materials consist of? (a) carbides, (b) cermets, and (c) sialon.

8.4 List the major limitations of ceramics.

8.5   What is porcelain?

8.6   What is glass? Why is it called a supercooled material?

8.7   What is devitrification?

8.8   List the major types of glasses and their applications.

8.9   What is static fatigue?

8.10   Describe the major uses of graphite.

8.11   What is the significance of Al_2O_3 in this chapter?

8.12   How are alumina ceramics produced?

8.13   What is the difference between a carbide and a nitride?

8.14   What features of partially stabilized zirconia differentiate it from other ceramics?

8.15   Is diamond a ceramic? Why or why not?

8.16   What is a buckyball?

8.17   What are the major uses of diamond?

QUALITATIVE PROBLEMS

8.18   Explain why ceramics are weaker in tension than in compression.

8.19   What are the advantages of cermets? Suggest applications in addition to those given in the text.

8.20   Explain why the electrical and thermal conductivity of ceramics decreases with increasing porosity.

8.21   Explain why the mechanical property data in Table 8.2 have such a broad range. What is the significance in engineering practice?

8.22   What reasons can you think of that encouraged the development of synthetic diamond?

8.23   Explain why the mechanical properties of ceramics generally are better than those of metals.

8.24   How are ceramics made tougher?

8.25   Mention and describe applications in which static fatigue can be important.

8.26   How does porosity affect the mechanical properties of ceramics, and why does it do so?

8.27   What properties are important in making heat-resistant ceramics for use on oven tops? Why?

8.28   Describe the differences between the properties of glasses and those of ceramics.

8.29   A large variety of glasses is now available. Why is this so?

8.30   What is the difference between the structure of graphite and that of diamond? Is it important? Explain.

8.31   What materials are suitable for use as a coffee cup? Explain.

8.32   Aluminum oxide and partially stabilized zirconia are described as white in appearance. Can they be colored? If they can, how would you accomplish this?

8.33   Both ceramics and metal castings (Part II) are known to be stronger in compression than in tension. What similar reasons exist for these behaviors?

8.34   Why does the strength of a ceramic part depend on its size?

8.35   In old castles and churches in Europe, the glass windows display pronounced ripples and are thicker at the bottom than at the top. Explain why this has occurred.

8.36   Ceramics are hard and strong in both compression and shear. Why, then, are they not used as nails or other fasteners?

8.37   It was stated in the text that ceramics have a wider range of strengths in tension than metals. List reasons why this is so, with respect to both the ceramic properties that cause variations and the difficulties in obtaining repeatable results.

QUANTITATIVE PROBLEMS

8.38   If a fully dense ceramic has the properties that $UTS_o = 180$ MPa and $E_o = 300$ GPa, what are these properties at 20% porosity for values of $n = 4, 5, 6,$ and 7, respectively?

8.39   Plot the UTS, E, and k values for ceramics as a function of porosity P, describe, and explain the trends that you observe in their behavior.

8.40 What would be the tensile strength and the modulus of elasticity of the ceramic in Problem 8.38 for porosities of 10% and 30% for the four n values given?

8.41 Calculate the thermal conductivities for ceramics at porosities of 10%, 20%, and 30% for $k_o = 0.7$ W/m·K.

8.42 A ceramic has $k_o = 0.65$ W/m·K. If this ceramic is shaped into a cylinder with a porosity distribution given by $P = 0.1(x/L)(1 - x/L)$, where x is the distance from one end of the cylinder and L is the total cylinder length, plot the porosity as a function of distance, evaluate the average porosity, and calculate the average thermal conductivity.

8.43 It can be shown that the minimum weight of a column which will support a given load depends on the ratio of the material's stiffness to the square root of its density. Plot this property for a ceramic as a function of porosity.

SYNTHESIS, DESIGN, AND PROJECTS

8.44 Make a list of the ceramic parts that you can find around your house and in your car. Explain why those parts are made of ceramics.

8.45 Assume that you are working in technical sales and are fully familiar with all the advantages and limitations of ceramics. Which of the markets traditionally using nonceramic materials do you think ceramics can penetrate? What would you say to your potential customers during your sales visits? What kind of questions do you think they will ask?

8.46 Describe applications in which a ceramic material with a near-zero coefficient of thermal expansion would be desirable.

8.47 The modulus of elasticity of ceramics is maintained largely at elevated temperatures. What engineering applications could benefit from this characteristic?

8.48 List and discuss the factors that you would take into account when replacing a metal component with a ceramic component.

8.49 Obtain some data from the available technical literature in the bibliography and quantitatively show the effect of temperature on the strength and the modulus of elasticity of several ceramics. Comment on how the shape of these curves differs from those of metals.

8.50 Assume that the cantilever beam in Quantitative Problem 3.16 in Chapter 3 is made of ceramic. How different would your answer be when compared to that for a beam made of metal? Explain clearly, giving numerical examples.

8.51 It was noted in Section 8.4.1 that there are several basic types of glasses available. Make a survey of the available technical literature in the bibliography and prepare a table for these glasses indicating various mechanical, physical, and optical properties.

8.52 Ceramic pistons are being considered for a high-speed combustion engine. List the benefits and concerns that you would have regarding this application.

8.53 Pyrex cookware displays a unique phenomenon: It functions well for a large number of cycles and then shatters into many pieces. Investigate this phenomenon, list the probable causes, and discuss the manufacturing considerations that may alleviate or contribute to such failures.

8.54 It has been noted that the strength of brittle materials (such as ceramics and glasses) is very sensitive to surface defects such as scratches (notch sensitivity). Obtain some pieces of these materials, make scratches on them, and test them by carefully clamping them in a vise and bending them. Comment on your observations.

CHAPTER

9

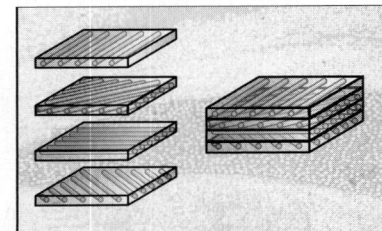

Composite Materials: Structure, General Properties, and Applications

With their attractive properties, especially high strength-to-weight and stiffness-to-weight ratios, reinforced plastics and composites are among the most important developments in materials. This chapter describes:

- Structure of reinforced plastics.
- Types of reinforcing fibers and matrix materials and their role in overall properties.
- Structure and properties of metal-matrix and ceramic-matrix composites.
- Selection and applications of reinforced plastics and composites.

9.1 | Introduction

A composite material is a combination of two or more chemically distinct and insoluble phases with a recognizable interface, in such a manner that its properties and structural performance are superior to those of the constituents acting independently. These combinations are known as **metal-matrix** and **ceramic-matrix composites.** As shown in Table 7.1, fiber reinforcements significantly improve the strength, stiffness, and creep resistance of plastics—particularly their strength-to-weight and stiffness-to-weight ratios. Composite materials have found increasingly wider applications in aircraft (Fig. 9.1), space vehicles, offshore structures, piping, electronics, automobiles, boats, and sporting goods.

The oldest example of composites, dating back to 4000 B.C., is the addition of straw to clay in the making of mud huts and of bricks for structural use. In this combination, the straws are the reinforcing fibers and the clay is the matrix. Another example of a composite material is reinforced concrete, which was developed in the 1800s. In fact, concrete itself is a composite material, consisting of cement, sand, and gravel. However, by itself, concrete is brittle and has little or no useful tensile strength; the steel rods impart the necessary tensile strength to the composite.

Composites, in the most general sense, can be thought of as a wide variety of materials, such as cermets (Section 8.2), two-phase alloys (Section 4.2), natural materials such as wood and bone, and general reinforced or combined materials such as steel-wire reinforced automobile tires. These materials also should be recognized as composite materials, even if they are not emphasized in this chapter.

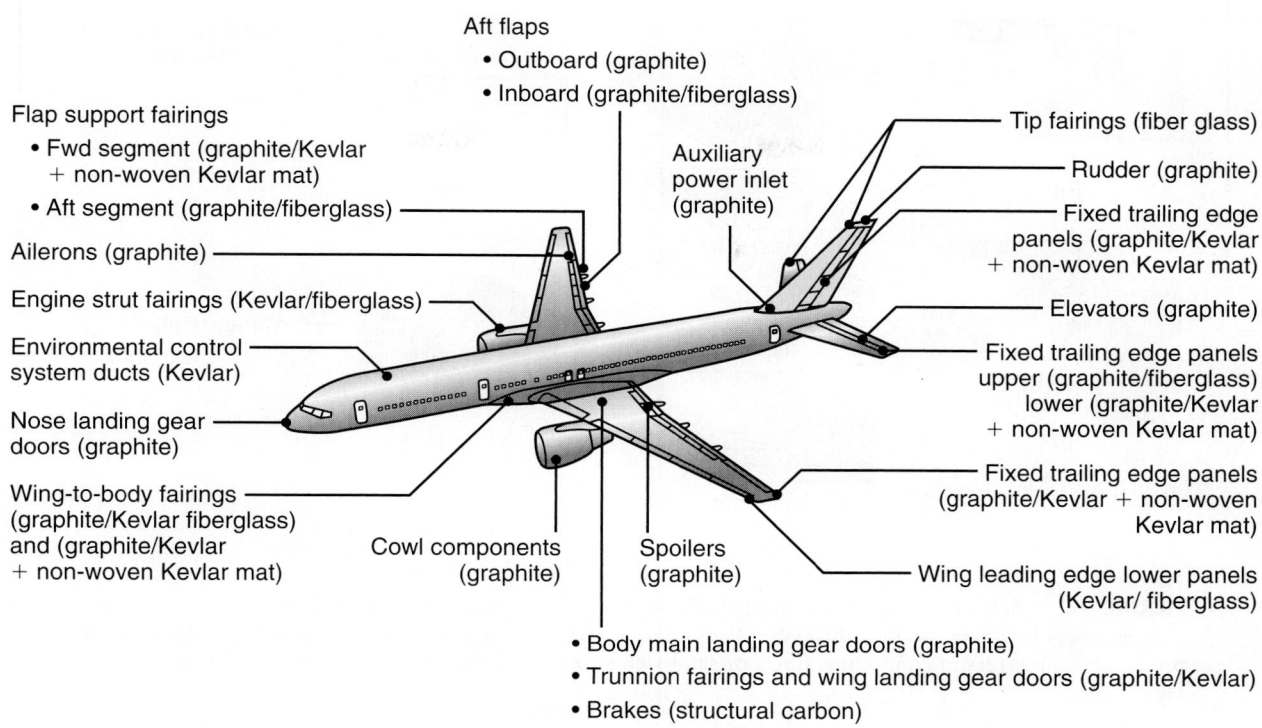

Aft flaps
• Outboard (graphite)
• Inboard (graphite/fiberglass)

Flap support fairings
• Fwd segment (graphite/Kevlar + non-woven Kevlar mat)
• Aft segment (graphite/fiberglass)

Ailerons (graphite)

Engine strut fairings (Kevlar/fiberglass)

Environmental control system ducts (Kevlar)

Nose landing gear doors (graphite)

Wing-to-body fairings (graphite/Kevlar fiberglass) and (graphite/Kevlar + non-woven Kevlar mat)

Auxiliary power inlet (graphite)

Cowl components (graphite)

Spoilers (graphite)

Tip fairings (fiber glass)

Rudder (graphite)

Fixed trailing edge panels (graphite/Kevlar + non-woven Kevlar mat)

Elevators (graphite)

Fixed trailing edge panels upper (graphite/fiberglass) lower (graphite/Kevlar + non-woven Kevlar mat)

Fixed trailing edge panels (graphite/Kevlar + non-woven Kevlar mat)

Wing leading edge lower panels (Kevlar/ fiberglass)

• Body main landing gear doors (graphite)
• Trunnion fairings and wing landing gear doors (graphite/Kevlar)
• Brakes (structural carbon)

FIGURE 9.1   Application of advanced composite materials in Boeing 757-200 commercial aircraft. *Source*: Courtesy of Boeing Commercial Airplane Company.

This chapter describes the structure of composite materials, the types and characteristics of reinforcing fibers used, and typical major applications of these materials. The processing and the shaping of composite materials are described in Chapter 19.

9.2 | The Structure of Reinforced Plastics

Reinforced plastics, also known as **polymer-matrix composites** (PMC) and **fiber-reinforced plastics** (FRP), consist of **fibers** (the discontinuous or dispersed phase) in a polymer **matrix** (the continuous phase), as shown in Fig. 9.2. These fibers are strong and stiff (Table 9.1), and they have high specific strength (strength-to-weight ratio) and specific stiffness (stiffness-to-weight ratio), as shown in Fig. 9.3. In addition to high specific strength and specific stiffness, reinforced-plastic structures have improved fatigue resistance, greater toughness, and higher creep resistance than those made of unreinforced plastics. Such structures are relatively easy to design, fabricate, and repair.

The fibers in reinforced plastics by themselves have little structural value; they have stiffness in their longitudinal direction but no transverse stiffness or strength. The plastic matrix is less strong and less stiff than the fiber, but it is tougher and often more chemically inert than the fibers. Reinforced plastics possess the advantages of each of the two constituents. The percentage of fibers (by volume) in reinforced plastics usually ranges between 10% and 60%. Practically, the percentage of fiber in a matrix is limited by the average distance between adjacent fibers or particles.

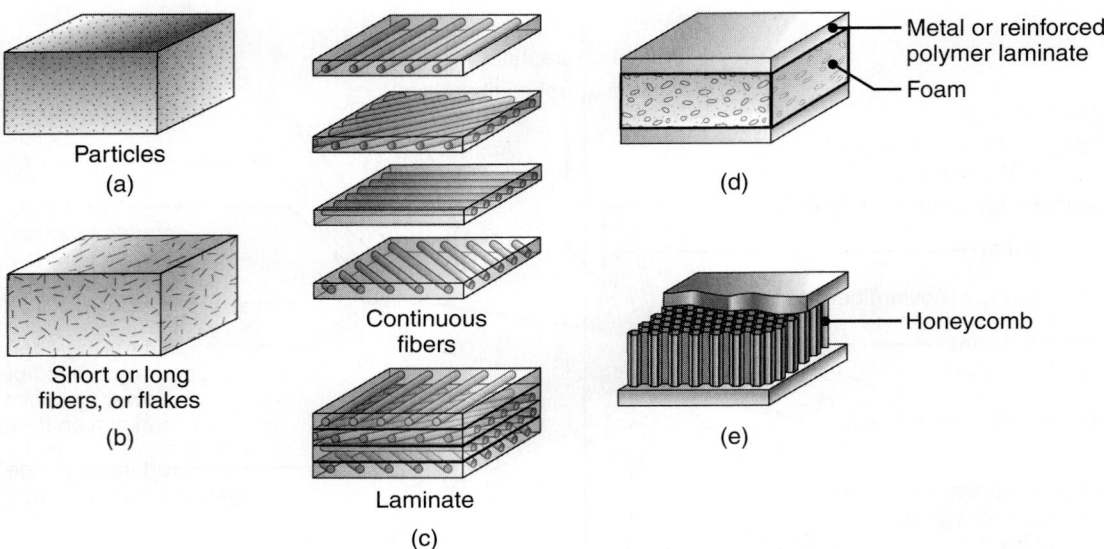

FIGURE 9.2   Schematic illustration of methods of reinforcing plastics (matrix) with (a) particles, (b) short or long fibers or flakes, and (c) through (e) continuous fibers. The laminate structures shown in (d) can be produced from layers of continuous fibers or sandwich structures using a foam or honeycomb core (see also Fig. 16.50).

TABLE 9.1

Types and General Characteristics of Composite Materials	
Material	Characteristics
Fibers	
Glass	High strength, low stiffness, high density; lowest cost; E (calcium aluminoborosilicate) and S (magnesia-aluminosilicate) types commonly used
Graphite	Available as high modulus or high strength; low cost; less dense than glass.
Boron	High strength and stiffness; highest density; highest cost; has tungsten filament at its center
Aramids (Kevlar)	Highest strength-to-weight ratio of all fibers; high cost
Other fibers	Nylon, silicon carbide, silicon nitride, aluminum oxide, boron carbide, boron nitride, tantalum carbide, steel, tungsten, molybdenum
Matrix materials	
Thermosets	Epoxy and polyester, with the former most commonly used; others are phenolics, fluorocarbons, polyethersulfone, silicon, and polyimides
Thermoplastics	Polyetheretherketone; tougher than thermosets but lower resistance to temperature
Metals	Aluminum, aluminum-lithium, magnesium, and titanium; fibers are graphite, aluminum oxide, silicon carbide, and boron
Ceramics	Silicon carbide, silicon nitride, aluminum oxide, and mullite; fibers are various ceramics

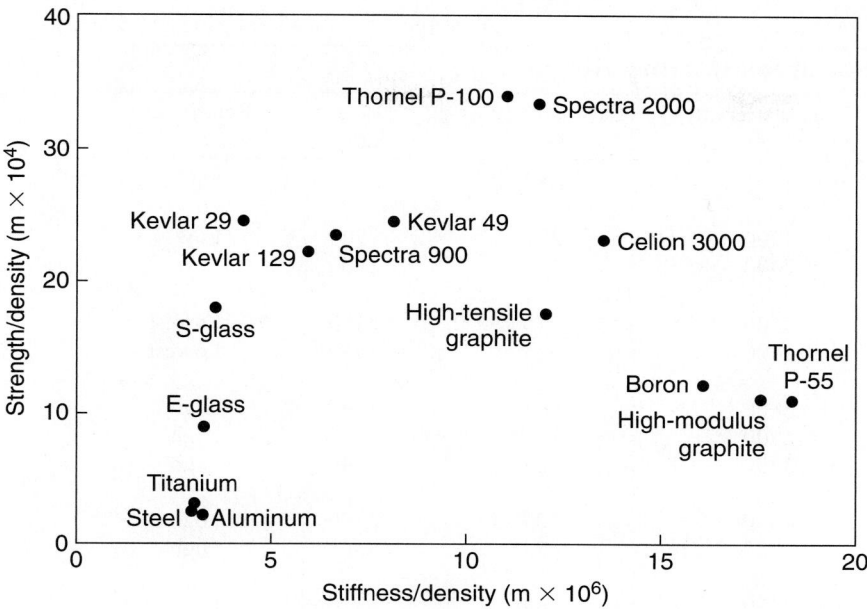

FIGURE 9.3   Specific tensile strength (tensile strength-to-density ratio) and specific tensile modulus (modulus of elasticity-to-density ratio) for various fibers used in reinforced plastics. Note the wide range of specific strengths and stiffnesses available.

The highest practical fiber content is 65%; higher percentages generally result in lower structural properties.

When more than one type of fiber is used in a reinforced plastic, the composite is called a **hybrid**. Hybrids generally have even better properties, but they are more costly. Glass or carbon fiber-reinforced hybrid plastics are developed for high-temperature applications with continuous use ranging up to about 300°C (550°F). Generally, however, these fibers are brittle, abrasive, lack toughness, and can degrade chemically when exposed to the environment. Also, as noted in the bottom of Table 9.2, the properties of fibers can vary significantly, depending on the quality of the material and on the method of processing.

9.2.1 Reinforcing fibers

Reinforcing fibers for polymer-matrix composites are glass, graphite, aramids, and boron (Table 9.2). Their properites are described next.

Glass fibers.   Glass fibers are used the most widely and are the least expensive of all fibers. The composite material is called **glass-fiber reinforced plastic** (GFRP) and may contain between 30 and 60% glass fibers by volume. Fibers are made by drawing molten glass through small openings in a platinum die (see Section 17.10.3). The molten glass then is elongated mechanically, cooled, and wound on a roll. A protective coating or sizing may be applied to facilitate their passage through the machinery. The glass fibers are treated with **silane** (a silicon hydride) for improved wetting and bonding between the fiber and the matrix.

The principal types of glass fibers are:

- **E-type:** a calcium aluminoborosilicate glass, the type most commonly used.

- **S-type:** a magnesia-aluminosilicate glass, offering higher strength and stiffness but at higher cost.

- **E-CR-type:** a high-performance glass fiber, offering higher resistance to elevated temperatures and acid corrosion than does the E glass.

TABLE 9.2

Typical Properties of Reinforcing Fibers			
Type	Tensile strength (MPa)	Elastic modulus (GPa)	Density (kg/m³)	Relative cost
Boron	3500	380	2600	Highest
Carbon				
High strength	3000	275	1900	Low
High modulus	2000	415	1900	Low
Glass				
E-type	3500	73	2480	Lowest
S-type	4600	85	2540	Lowest
Kevlar				
29	2800	62	1440	High
49	2800	117	1440	High
129	3200	85	1440	High
Nextel				
312	1630	135	2700	High
610	2770	328	3960	High
Spectra				
900	2270	64	970	High
1000	2670	90	970	High
2000	3240	115	970	High
Alumina (Al_2O_3)	1900	380	3900	High
Silicon carbide	3500	400	3200	High

Note: These properties vary significantly depending on the material and method of preparation.

Graphite fibers. Graphite fibers (Fig. 9.4a), although more expensive than glass fibers, have a combination of low density, high strength, and high stiffness; the product is called **carbon-fiber reinforced plastic** (CFRP). All graphite fibers are made by **pyrolysis** of organic **precursors**, commonly of *polyacrylonitrile* (PAN) because of its low cost. *Rayon* and *pitch* (the residue from catalytic crackers in petroleum refining) also can be used as precursors. Pyrolysis is the process of inducing chemical changes by heat—for instance, by burning a length of yarn and causing the material to carbonize and become black in color. With PAN, the fibers are partially cross-linked at a moderate temperature (in order to prevent melting during subsequent processing steps) and are elongated simultaneously. At this point, the fibers

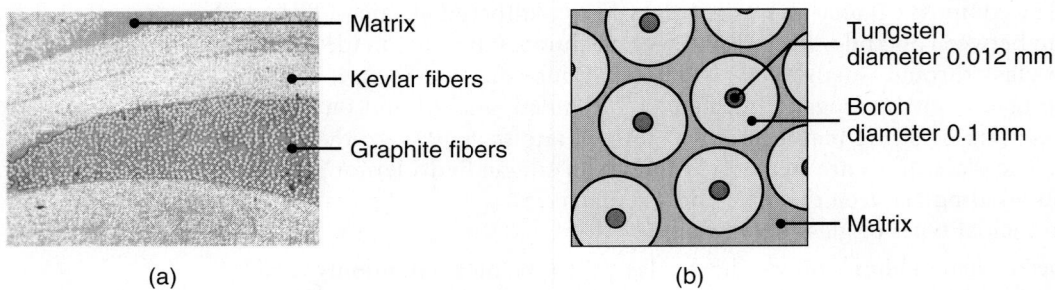

(a)   (b)

FIGURE 9.4 (a) Cross-section of a tennis racket, showing graphite and aramid (Kevlar) reinforcing fibers. *Source:* Courtesy of J. Dvorak, Mercury Marine Corporation and F. Garrett, Wilson Sporting Goods Co. (b) Cross-section of boron fiber-reinforced composite material.

are *carburized*; they are exposed to elevated temperature to expel the hydrogen (de-hydrogenation) and the nitrogen (denitrogenation) from the PAN. The temperatures for carbonizing range up to about 1500°C (2730°F) and for graphitizing up to 3000°C (5400°F).

The difference between *carbon* and *graphite*, although the words often are used interchangeably, depends on the temperature of pyrolysis and the purity of the material. Carbon fibers are generally 80 to 95% carbon; graphite fibers are usually more than 99% carbon. A typical carbon fiber contains amorphous (noncrystalline) carbon and graphite (crystalline carbon). These fibers are classified by their elastic modulus, which ranges from 35 to 800 GPa: *low*, *intermediate*, *high*, and *very high modulus*. Tensile strengths range from 250 to 2600 MPa (Table 9.2).

Conductive graphite fibers.   These fibers are produced to make it possible to en-hance the electrical and thermal conductivity of reinforced plastic components. These fibers are coated with a metal (usually nickel) using a continuous electroplating process; the coating is typically 0.5-μm thick on a 7-μm-diameter graphite fiber core. Available in a chopped or a continuous form, the conductive fibers are incorporated directly into injection-molded plastic parts. Applications include electromagnetic and radio-frequency shielding and lightning-strike protection.

Polymer fibers.   Polymer fibers may be made of nylon, rayon, acrylics, and aramids; the most common are **aramid fibers**. Aramids (Section 7.6) are among the toughest fibers (**Kevlar**) and have very high specific strengths (Fig. 9.3). Aramids can undergo some plastic deformation prior to fracture and, hence, have higher toughness than brittle fibers. However, aramids absorb moisture (hygroscopic), degrading their properties and complicating their application.

Another high-performance polyethylene fiber is *Spectra* (a trade name); it has an ultra-high molecular weight and high molecular-chain orientation. Spectra has better abrasion resistance and flexural-fatigue resistance than aramid fibers and at a similar cost. In addition, because of its lower density (970 kg/m^3), it has a higher specific strength and specific stiffness than aramid fibers. However, a low melting point and poor adhesion characteristics as compared to other poly-mers are its major limitations to applications. Among newer fibers is *Nextel*, a trade name.

Polymer fibers are made by two processes: melt spinning and dry spinning. **Melt spinning** involves extruding a liquid polymer through small holes in a die (*spinnerettes*). The fibers are then cooled before being gathered and wound onto bobbins. These fibers may be stretched to further orient and strenghten the polymer in the fibers. In **dry spinning**, the polymer is dissolved in a liquid solution to form a partially oriented liquid–crystal form. As the polymer passes through the spin-nerette, it is further oriented, and at this point, the fibers are washed, dried, and wound. Aramids are oriented in solution and are oriented fully when they pass through the spinnerette and, therefore, do not need to be drawn further.

Boron fibers.   These fibers consist of boron deposited (by chemical vapor-deposition techniques,) onto tungsten fibers (Fig. 9.4b). Boron also can be de-posited onto carbon fibers. Boron fibers have desirable properties, such as high strength and stiffness, both in tension and in compression, and resistance to high temperatures. However, because of the high density of tungsten, they are heavy and also are expensive.

Other fibers.   A variety of other fibers used in composites are silicon carbide, sili-con nitride, aluminum oxide, sapphire, steel, tungsten, molybdenum, boron carbide, boron nitride, and tantalum carbide. **Whiskers** also are used as reinforcing fibers (see also Section 22.10). Whiskers are tiny needle-like single crystals that grow from 1 to 10 μm (40 to 400 μin.) in diameter with high aspect ratios (the ratio of fiber length to its diameter) ranging from 100 to 15,000. Because of their small size, whiskers are either free of imperfections or the imperfections they contain do not significantly affect their strength, which approaches the theoretical strength of the material (size effect). The elastic moduli of whiskers range between 400 and 700 GPa, and their tensile strength is on the order of 15 to 20 GPa.

9.2.2 Fiber size and length

Fibers are very strong and stiff in tension. The reason is that the molecules in the fibers are oriented in the longitudinal direction, and their cross-sections are so small (usually less than 0.01 mm (0.0004 in.) in diameter), that the probability is low for any defects to exist in the fiber. Glass fibers, for example, can have tensile strengths as high as 4600 MPa (650 ksi), whereas the strength of glass in bulk form (Section 8.4) is much lower.

Fibers generally are classified as **short** (**discontinuous**) or **long** (**continuous**) fibers. The designations "short" and "long" fiber are, in general, based on the following distinction: In a given type of fiber, if the mechanical properties improve as a result of increasing the average fiber length, then it is called a *short fiber*. If no such improvement in properties occurs, it is called a *long fiber*. Short fibers typically have aspect ratios between 20 and 60, while long fibers have between 200 and 500.

Reinforcing elements also may be in the form of *chopped fibers*, *particles*, or *flakes* or in the form of continuous *roving* (slightly twisted strand of fibers), *woven fabric* (similar to cloth), *yarn* (twisted strand), and *mats* of various combinations. Various hybrid yarns also are available.

9.2.3 Matrix materials

The matrix in reinforced plastics has three principal functions:

1. Support the fibers in place and transfer the stresses to them while they carry most of the load.
2. Protect the fibers against physical damage and the environment.
3. Reduce the propagation of cracks in the composite by virtue of the greater ductility and toughness of the plastic matrix.

Matrix materials are usually *thermoplastics* or *thermosets* and commonly consist of epoxy, polyester, phenolic, fluorocarbon, polyethersulfone, or silicon. The most commonly used are the epoxies (80% of all reinforced plastics) and the polyesters (less expensive than the epoxies). Polyimides, which resist exposure to temperatures in excess of 300°C (575°F), continue to be developed for use as a matrix with graphite fibers. Some thermoplastics, such as polyetheretherketone (PEEK) are also used as matrix materials; they generally have higher toughness than thermosets, but their resistance to temperature is lower, being limited to 100° to 200°C (200° to 400°F).

9.3 | Properties of Reinforced Plastics

The mechanical and physical properties of reinforced plastics depend on the type, shape, and orientation of the reinforcing material, the length of the fibers, and the volume fraction (percentage) of the reinforcing material. Short fibers are less

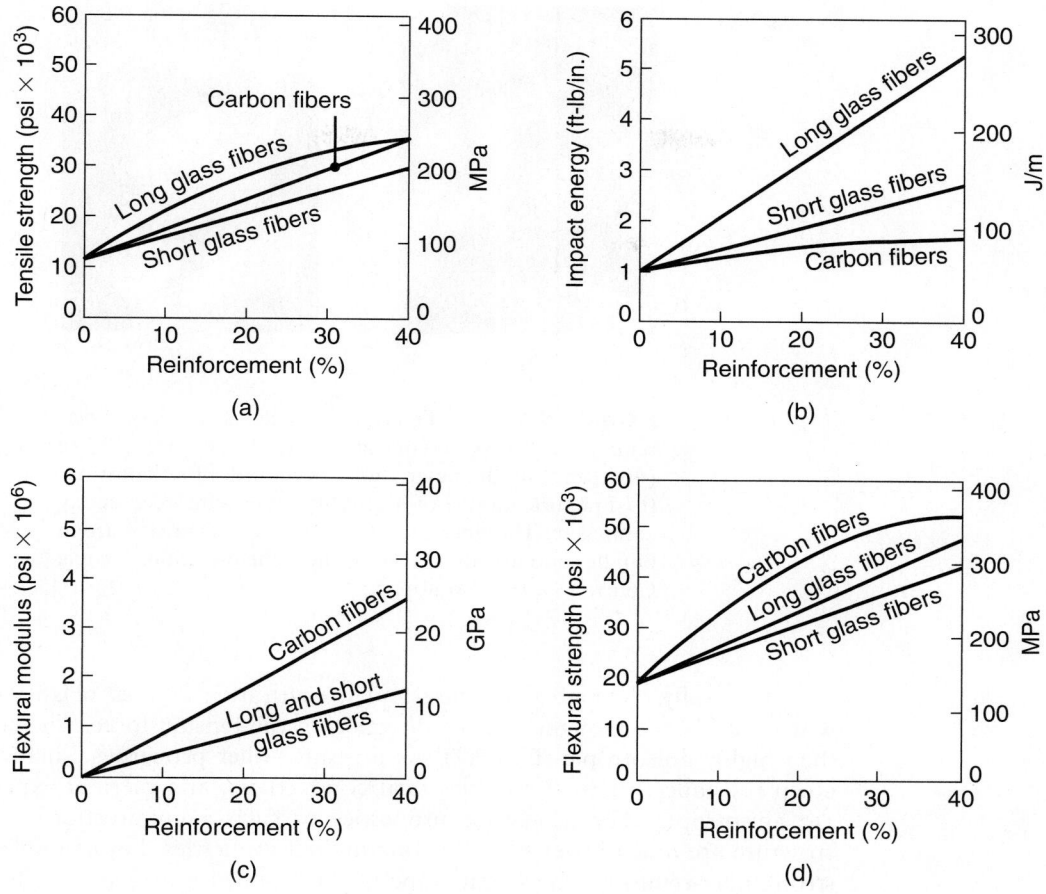

FIGURE 9.5 The effect of type of fiber on various properties of fiber-reinforced nylon (6,6). *Source:* Courtesy of NASA.

effective than long fibers (Fig. 9.5), and their properties are strongly influenced by temperature and time under load. Long fibers transmit the load through the matrix better, thus they are commonly used in critical applications, particularly at elevated temperatures. Physical properties of reinforced plastics and their resistance to fatigue, creep, and wear greatly depend on the type and amount of reinforcement. Composites can be tailored in order to impart specific properties (such as permeability and dimensional stability), to make processing easier, and to reduce production costs.

A critical factor in reinforced plastics is the strength of the bond between the fiber and the polymer matrix, because the load is transmitted through the fiber-matrix interface. Weak interfacial bonding causes **fiber pullout** and **delamination** of the structure, particularly under adverse environmental conditions. Adhesion at the interface can be improved by special surface treatments, such as coatings and coupling agents. Glass fibers, for example, are treated with **silane** (a silicon hydride) for improved wetting and bonding between the fiber and the matrix. The importance of proper bonding can be appreciated by inspecting the fracture surfaces of reinforced plastics, shown in Figs. 9.6a and b. Note, for example, the separation between the fibers and the matrix; obviously, better adhesion between them improves the overall strength of the composite.

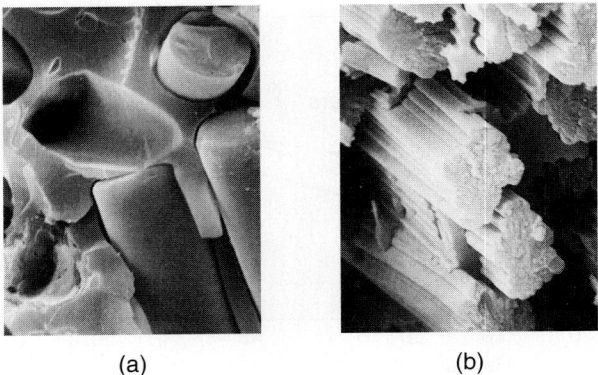

(a)   (b)

FIGURE 9.6 (a) Fracture surface of a glass fiber-reinforced epoxy composite. The fibers are 10 μm (400 μin.) in diameter and have random orientation. (b) Fracture surface of a graphite fiber-reinforced epoxy composite. The fibers, 9 μm to 11 μm in diameter, are in bundles and are all aligned in the same direction. *Source:* Courtesy of L. J. Broutman.

Generally, the highest stiffness and strength in reinforced plastics is obtained when the fibers are aligned in the direction of the tension force. The composite is then highly anisotropic (Fig. 9.7). As a result, other properties, such as stiffness, creep resistance, thermal and electrical conductivity, and thermal expansion, also are anisotropic. The transverse properties of such a unidirectionally reinforced structure are much lower than the longitudinal properties. For example, note how strong fiber-reinforced packaging tape is when pulled in tension, yet how easily it can split when pulling in the width direction.

Because it is an engineered material, a reinforced plastic part can be given an optimal configuration for a specific service condition. For example, if the part is to be subjected to forces in different directions (such as in thin-walled, pressurized vessels),

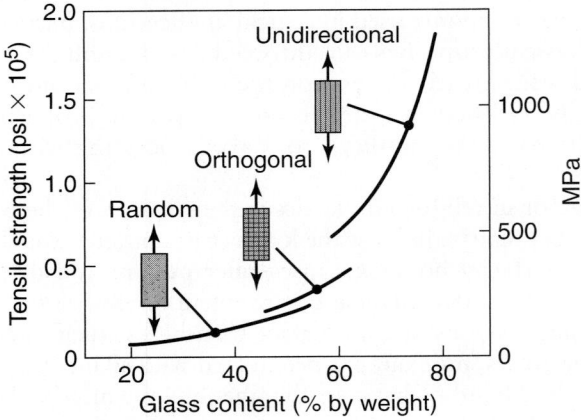

FIGURE 9.7 The tensile strength of glass-reinforced polyester as a function of fiber content and fiber direction in the matrix.

(a) the fibers can be criss-crossed in the matrix, or (b) layers of fibers oriented in different directions can be built up into a laminate having improved properties in more than one direction. (See *filament winding*, Section 18.12.2.) Also, a composite flywheel rotor has been produced using a special weaving technique in which the reinforcing fibers (E-glass) are aligned in the radial direction as well as in the hoop direction. Designed for mechanical-energy storage systems in low-emissions electric and hybrid vehicles, the flywheel can operate at speeds up to 50,000 rpm.

9.3.1 Strength and elastic modulus of reinforced plastics

The strength and elastic modulus of a reinforced plastic with unidirectional fibers can be determined in terms of the strength and moduli, both of the fibers and of the matrix, and in terms of the volume fraction of fibers in the composite. In the following equations, c refers to the composite, f to the fiber, and m to the matrix. The total load, P_c, on the composite is shared by the fiber (P_f) and the matrix (P_m). Thus,

$$P_c = P_f + P_m \tag{9.1}$$

which can be written as

$$\sigma_c A_c = \sigma_f A_f + \sigma_m A_m \tag{9.2}$$

where A_c, A_f, and A_m are the cross-sectional areas of the composite, the fiber, and the matrix, respectively; thus, $A_c = A_f + A_m$. Let's now denote x as the area fraction of the fibers in the composite. (Note that x also represents the volume fraction, because the fibers are uniformly longitudinal in the matrix.) Equation (9.2) can be written as follows:

$$\sigma_c = x\sigma_f + (1 - x)\sigma_m \tag{9.3}$$

The fraction of the total load carried by the fibers now can be calculated. First, note that in the composite under a tension load, the strains sustained by the fibers and the matrix are the same (that is, $e_c = e_f = e_m$). Next, recall from Section 2.2, that

$$e = \frac{\sigma}{E} = \frac{P}{AE}$$

Consequently,

$$\frac{P_f}{P_m} = \frac{A_f E_f}{A_m E_m} \tag{9.4}$$

Since the relevant quantities for a specific situation are known, by using Eq. (9.1), the fraction P_f/P_c can be found. Then, using the foregoing relationships, the elastic modulus, E_c, of the composite can be calculated by replacing σ in Eq. (9.3) with E. Thus,

$$E_c = xE_f + (1 - x)E_m \tag{9.5}$$

As an example, let's assume that a graphite-epoxy reinforced plastic with longitudinal fibers contains 20% graphite fibers. The elastic modulus of the fiber is 300 GPa, and that of the epoxy matrix is 100 GPa. Let's calculate the elastic modulus of the composite and the fraction of the load supported by the fibers. The data given are $x = 0.2$, $E_f = 300$ GPa, and $E_m = 100$ GPa. Using Eq. (9.5),

$$E_c = 0.2(300) + (1 - 0.2)100 = 60 + 80 = 140 \text{ GPa.}$$

The load fraction P_f/P_m can be obtained from Eq. (9.4) as:

$$\frac{P_f}{P_m} = 0.2\frac{(300)}{0.8(100)} = 0.75$$

Because

$$P_c = P_f + P_m \quad \text{and} \quad P_m = \frac{P_f}{0.75}$$

Therefore,

$$P_c = P_f + \frac{P_f}{0.75} = 2.33P_f, \quad \text{or} \quad P_f = 0.43P_c$$

Thus, the fibers support 43% of the load, even though they occupy only 20% of the cross-sectional area (and hence volume) of the composite.

9.4 | Applications of Reinforced Plastics

The first engineering application of reinforced plastics was in 1907 for an acid-resistant tank made of a phenolic resin with asbestos fibers. In the 1920s, *formica*, a trade name, was developed and used commonly for counter tops. Epoxies first were used as a matrix material in the 1930s. Beginning in the 1940s, boats were made with fiberglass, and reinforced plastics were used for aircraft, electrical equipment, and sporting goods. Major developments in composites began in the 1970s, resulting in materials that are now called **advanced composites**. Glass or carbon fiber-reinforced hybrid plastics are available for high-temperature applications, with continuous use ranging up to about 300°C (550°F).

Reinforced plastics typically are used in commercial and military aircraft, rocket components, helicopter blades, automobile bodies, leaf springs, drive shafts, pipes, ladders, pressure vessels, sporting goods, helmets, boat hulls, and various other structures and components. Applications include components in the DC-10, the L-1011, and the Boeing 727, 757, 767, and 777 commercial aircraft. The Boeing 777 is made of about 9% composites by total weight; that proportion is triple the composite content of previous Boeing transport aircraft. The floor beams and panels and most of the vertical and horizontal tail are made of composite materials.

By virtue of the resulting weight savings, reinforced plastics have reduced fuel consumption by about 2%. The newly designed Airbus jumbo jet A380, with a capacity of 550 to 700 passengers and to be in service in 2006, will have horizontal stabilizers, ailerons, wing boxes and leading edges, secondary mounting brackets of the fuselage, and the deck structure made of composites with carbon fibers, thermosetting resins, and thermoplastics. The upper fuselage will be made of alternating layers of aluminum and glass fiber-reinforced epoxy prepregs.

The structure of the Lear Fan 2100 passenger aircraft is almost totally made of graphite-epoxy reinforced plastic. Nearly 90% of the structure of the lightweight *Voyager* aircraft, which circled the earth without refueling, was made of carbon-reinforced plastic. The contoured frame of the Stealth bomber is made of composites consisting of carbon and glass fibers, epoxy-resin matrices, high-temperature polyimides, and other advanced materials. Boron fiber-reinforced composites are used in military aircraft, golf-club shafts, tennis rackets, fishing rods, and sailboards (Fig. 9.8). A more recent example is the development of a small, all-composite

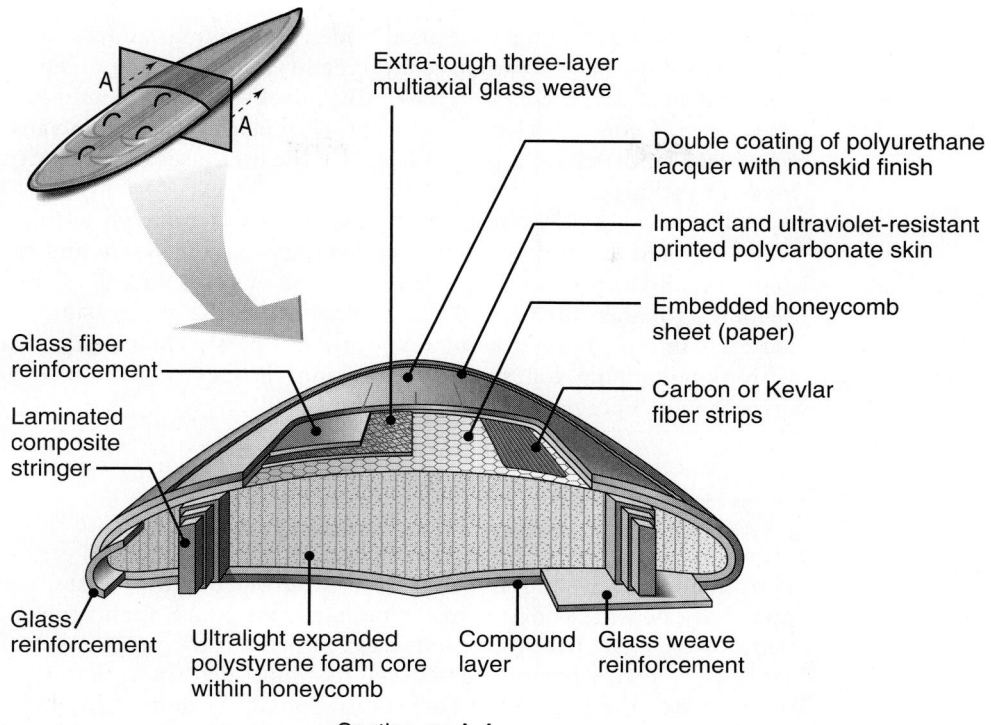

Extra-tough three-layer multiaxial glass weave

Double coating of polyurethane lacquer with nonskid finish

Impact and ultraviolet-resistant printed polycarbonate skin

Embedded honeycomb sheet (paper)

Carbon or Kevlar fiber strips

Glass fiber reinforcement

Laminated composite stringer

Glass reinforcement

Ultralight expanded polystyrene foam core within honeycomb

Compound layer

Glass weave reinforcement

Section on A-A

FIGURE 9.8   Cross-section of a composite sailboard, an example of advanced materials construction. *Source:* K. Easterling, *Tomorrow's Materials* (2nd ed.), p. 133. Institute of Metals, 1990.

ship (twin-hull catamaran design) for the U.S. Navy capable of speeds of 50 knots (58 mph).

The processing of reinforced plastics can present significant challenges. As a result, several innovative techniques have been developed for manufacturing both large and small parts, particularly by molding, forming, cutting, and assembly. Careful inspection and testing of reinforced plastics is essential in critical applications, in order to ensure that good bonding between the reinforcing fiber and the matrix has been obtained throughout the whole structure. In some instances, however, the cost of inspection can be as high as one-quarter of the total cost of the composite product.

EXAMPLE 9.1 Fiberglass ladders

The traditional material for ladders has been wood or aluminum. Partly because of its weight, a wood ladder gives the user a sense of security. However, wood can have internal and external defects that, if undetected, can reduce significantly the strength of the ladder and, therefore, compromise safety. When dry, wood does not conduct electricity; when wet, however, it does so. Thus, a wet wood ladder should not come in contact with electrical wiring. Aluminum ladders are lightweight, and they can be designed for high strength and stiffness (in spite of the relatively low elastic modulus). Although they conduct electricity, they are longer-lasting than wood ladders, and they require little maintenance.

Composite materials are used widely in making ladders for various uses. Fiberglass is the most common reinforcing fiber, with epoxies and polyesters as the matrix material. These ladders have advantages similar to aluminum, and because they do not conduct electricity, they are preferred by electricians. Fiberglass-reinforced ladders have a rugged feel, give the user a sense of security, and can be made in various colors.

Fiberglass ladders can absorb moisture and undergo surface weathering when exposed to outdoor environments, especially in warm and moist climates and in sunlight (Section 7.3). There is a loss of color and gloss, and the surface becomes rougher because of fiber prominence due to erosion of the matrix material and to fiber *blooming* (exposure of fibers). These ladders can be coated with polyurethane for surface protection. Proper maintenance and periodic inspection of fiberglass ladders are important.

EXAMPLE 9.2 Composite military helmets and body armor

Personnal protective equipment in the form of body armor and composite helmets have become widespread for military and police applications. Body armor depends on high-strength woven fibers to prevent the penetration of projectiles. To stop a bullet, a composite material first must deform or flatten it; this process occurs when the bullet's tip comes into contact with as many individual fibers of the composite as possible without the fibers being pushed aside. The momentum associated with projectiles is felt, of course, by the user of the armor, but successful designs will contain bullets and shrapnel and prevent serious and fatal injuries.

There are two main types of body armor: *soft armor*, which relies upon many layers of woven, high-strength fibers and is designed mainly to defeat handguns; and *hard armor*, which uses a metal, ceramic, or polymer plate in addition to the woven fiber and is intended to provide protection against rifle rounds and shrapnel. A schematic of body armor is shown in Figure 9.9.

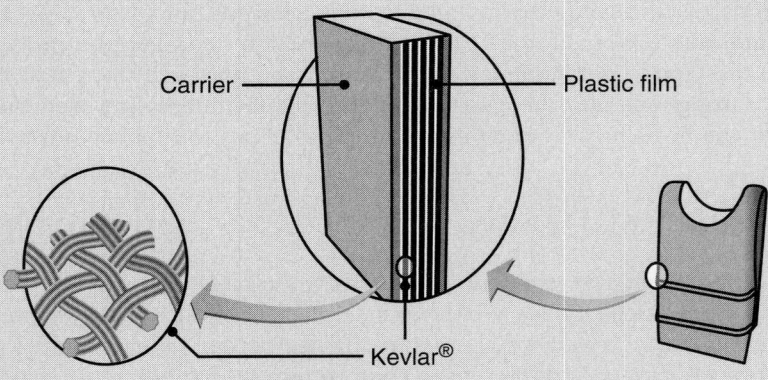

FIGURE 9.9 (a) Schematic illustration of body armor showing the layers of woven fibers; (b) Blunting and containment of a handgun bullet by body armor.

A number of fiber meshes have been used in body armor applications. Different suppliers use different combinations of fiber meshes and may include additional layers to provide protection agains blunt trauma. Historically, the first fiber used for flexible body armor was Kevlar 29 (an aramid), which has been improved through a number of versions. Other forms include Kevlar 49, Kevlar 129, and Kevlar Protera—a form whose tensile strength and energy-absorbing capabilities have been improved through the development of advanced spinning processes to produce the fibers. Aramid fibers are used very commonly in flexible body armor. Honeywell also produces an aramid fiber-based body armor, but other designs, such as Akzo Noble's TWARON aramid fiber, use over a thousand finely-spun filaments that interact with each other to dissipate the impact energy.

Spectra fiber is used to make the Spectra Shield composite for use in body armor. A layer of Spectra Shield composite consists of two unidirectional layers of Spectra fiber arranged to cross each other at 0- and 90-degree angles and held in place by a flexible resin. Both the fiber and resin layers are sealed between two thin sheets of polyethylene film, which is similar in appearance to plastic food wrap.

Hard armor uses a number of designs, but typically consists of a steel, ceramic (usually aluminum oxide and silica), or polyethylene plates strategically located to prevent penetration of ballistic particles to critical areas. Designs currently being evaluated utilize fluids with suspended nanoparticles of silica. At low strain rates, these fluids are inviscid and flow readily. At high strain rates typical of ballistic particles, these fluids are very resistant to deformation and can provide additional protection. The fluid is contained by the woven fiber mesh (it acts like a sponge holding the fluid in place) and is contained by the outer fabric.

In addition, a composite military helmet has been developed which, although weighing about the same as a conventional manganese-steel helmet, covers more of the head and offers twice the ballistic and fragmentation protection. A composite helmet has a nonwoven fiber construction made with Spectra fibers in a thermo-setting polymer matrix, which effectively stops the bullet by flattening it as it strikes the first layer of material.

Source: Courtesy of Pinnacle Armor, AlliedSignal Corp., and CGS Gallet SA.

9.5 | Metal-Matrix Composites

The advantages of a *metal matrix* over a *polymer matrix* are higher elastic modulus, toughness, ductility, and higher resistance to elevated temperatures. The limitations are higher density and a greater difficulty in processing parts. Matrix materials in *metal-matrix composites* (MMC) are usually aluminum, aluminum-lithium alloy (lighter than aluminum), magnesium, copper, titanium, and superalloys (Fig. 9.10). Fiber materials are graphite, aluminum oxide, silicon carbide, boron, molybdenum, and tungsten. The elastic modulus of nonmetallic fibers ranges between 200 and 400 GPa with tensile strengths being in the range from 2000 to 3000 MPa.

Typical compositions and applications for metal-matrix composites are given in Table 9.3. Because of their high specific stiffness, light weight, and high thermal conductivity, boron fibers in an aluminum matrix have been used for structural tubular supports in the space shuttle orbiter. Other applications are in bicycle frames and sporting goods.

FIGURE 9.10 Examples of metal-matrix composite parts. *Source:* Courtesy of Metal Matrix Cast Composites, LLC.

TABLE 9.3

Metal-Matrix Composite Materials and Applications

Fiber	Matrix	Applications
Graphite	Aluminum	Satellite, missile, and helicopter structures
	Magnesium	Space and satellite structures
	Lead	Storage-battery plates
	Copper	Electrical contacts and bearings
Boron	Aluminum	Compressor blades and structural supports
	Magnesium	Antenna structures
	Titanium	Jet-engine fan blades
Alumina	Aluminum	Superconductor restraints in fission power reactors
	Lead	Storage-battery plates
	Magnesium	Helicopter transmission structures
Silicon carbide	Aluminum, titanium	High-temperature structures
	Superalloy (cobalt-base)	High-temperature engine components
Molybdenum, tungsten	Superalloy	High-temperature engine components

EXAMPLE 9.3 Aluminum-matrix composite brake calipers

One of the trends in automobile design and manufacture is the increased push towards lighter weight designs in order to realize improved performance and/or fuel economy. This can be seen in the development of metal-matrix composite brake calipers. Traditional brake calipers are made from cast iron and can weigh around 3 kg (6.6 lb) each in a small car, and up to 14 kg (30 lb) in a truck. The cast-iron caliper could be redesigned completely using aluminum to achieve weight savings, but this would require a larger volume, and the space available between the wheel and rotor is highly constrained.

A new brake caliper was designed using an aluminum alloy locally reinforced with precast composite inserts using continuous ceramic fiber. This is a nanocrystalline alumina fiber, with a diameter of 10 to 12 μm, and fiber volume fraction of 65%. The fiber and material properties are summarized in Table 9.4. Finite element analysis confirmed the placement and amount of reinforcement, leading to a design that exceeded minimum design requirements and that matched deflections of cast-iron calipers in a packaging-constrained environment. The new brake caliper is shown in Fig. 9.11. It produces a weight savings of 50% and brings added benefits of easy recyclability and corrosion resistance.

TABLE 9.4

Summary of Fiber and Material Properties for an Automotive Brake Caliper

Property	Alumina fiber	Aluminum-reinforced composite material
Tensile strength	3100 MPa (450 ksi)	1.5 GPa (220 ksi)
Elastic modulus	380 GPa (55 Mpsi)	270 GPa (39 Mpsi)
Density	3.9 g/cm^3	3.48 g/cm^3

FIGURE 9.11 Aluminum-matrix composite brake caliper using nanocrystallyne alumina fiber reinforcement. *Source:* Courtesy of 3M Speciality Materials Division.

9.6 | Ceramic-Matrix Composites

Ceramic-matrix composites (CMC) are important because of their resistance to high temperatures and corrosive environments. As described in Section 8.3, ceramics are strong and stiff, they resist high temperatures, but they generally lack toughness.

Matrix materials that retain their strength up to 1700°C (3000°F) are silicon carbide, silicon nitride, aluminum oxide, and mullite (a compound of aluminum, silicon, and oxygen). Carbon/carbon-matrix composites retain much of their strength (up to 2500°C; 4500°F), although they lack oxidation resistance at high temperatures. Fiber materials are usually carbon and aluminum oxide. Applications of CMC include jet and automotive engine components, deep-sea mining equipment, pressure vessels, structural components, cutting tools, and dies for the extrusion and the drawing of metals.

9.7 | Other Composites

Composites also may consist of *coatings* of various types on base metals or substrates (Chapter 34). Examples are

- Plating of aluminum or other metals over plastics, generally for decorative purposes
- Enamels
- Vitreous (glasslike) coatings on metal surfaces for various functional or ornamental purposes

Composites also are made into cutting tools and dies, such as cemented carbides and cermets. Other composites are grinding wheels made of aluminum oxide, silicon carbide, diamond, or cubic-boron-nitride abrasive particles—held together with various organic, inorganic, or metallic binders (Chapter 26). A composite of granite particles in an epoxy matrix has high strength, good vibration-damping capacity (better than gray cast iron), and good frictional characteristics. It is used in machine-tool beds for some precision grinders.

SUMMARY

- Composites are an important class of engineered materials with numerous attractive properties. Three major categories are fiber-reinforced plastics, metal-matrix composites, and ceramic-matrix composites. They have a wide range of applications in the aircraft, aerospace, and transportation industries; sporting goods; and structural components.

- In fiber-reinforced plastics, the fibers usually are glass, graphite, aramids, or boron. Polyester and epoxies commonly are used as the matrix material. These composites have particularly high toughness and high strength-to-weight and stiffness-to-weight ratios.

- In metal-matrix composites, the fibers usually are graphite, boron, aluminum oxide, silicon carbide, molybdenum, or tungsten. Matrix materials generally consist of aluminum, aluminum-lithium alloy, magnesium, copper, titanium, and superalloys.

- For ceramic-matrix composites, the fibers are usually carbon and aluminum oxide, and the matrix materials are silicon carbide, silicon nitride, aluminum oxide, carbon, or mullite (a compound of aluminum, silicon, and oxygen).

- In addition to the type and quality of the materials used, important factors in the structure of composite materials are the size and length of the fibers, their volume percentage compared with that of the matrix, the strength of the bond at the fiber-matrix interface, and the orientation of the fibers in the matrix.

KEY TERMS

Advanced composites	Fiber pullout	Precursor
Ceramic matrix	Fibers	Pyrolysis
Composite materials	Hybrid	Reinforced plastics
Delamination	Matrix	Silane
Engineered materials	Metal matrix	Whiskers

BIBLIOGRAPHY

Agarwal, B.D., and Broutman, L.J., *Analysis and Performance of Fiber Composites*, 2nd ed., Wiley, 1990.

ASM Handbook, Vol. 21: Composites, ASM International, 2001

Belitskus, D.L., *Fiber and Whisker Reinforced Ceramics for Structural Applications*, Marcel Dekker, 1993.

Bertholet, J.-M., *Composite Materials: Mechanical Behavior and Structural Analysis*, Springer, 1999.

Bittence, J.C. (ed.), *Engineering Plastics and Composites*, ASM International, 1990.

Chawla, K.K., *Composite Materials: Science and Engineering*, 2nd ed., Springer, 1998.

Cheremisinoff, N.P., and Cheremisinoff, P.N., *Fiberglass Reinforced Plastics*, Noyes Publications, 1995.

Daniel, I.M., and Ishai, O., *Engineering Mechanics of Composite Materials*, Oxford, 1994.

Delmonte, J. *Metal-Polymer Composites*, Van Nostrand Reinhold, 1990.

Engineered Materials Handbook, Vol. 1: Composites, ASM International, 1987.

Engineering Plastics and Composites, 2nd ed., ASM International, 1993.

Fitzer, E., and Manocha, L.M., *Carbon Reinforcements and Carbon/Carbon Composites*, Springer, 1998.

Gay, D., Hoa, S.V., and Tsai, S.W., *Composite Materials: Design and Applications*, CRC Press, 2002.

Gutowski, T.G., *Advanced Composites Manufacturing*, Wiley, 1997.

Handbook of Ceramics and Composites, 3 vols. Marcel Dekker, 1991.

Harper, C.A, *Handbook of Plastics, Elastomers, and Composites*, 3rd ed., McGraw-Hill, 1996.

Hull, D., and Clyne, T.W., *An Introduction to Composite Materials*, Cambridge Univ. Press, 1996.

Jang, B.Z., *Advanced Polymer Composites: Principles and Applications*, ASM International, 1994.

Kelley, A., Cahn, R.W., and Bever, M.B. (eds.), *Concise Encyclopedia of Composite Materials,* Rev. ed., Pergamon, 1994.

Mallick, P.K., *Composites Engineering Handbook*, Marcel Dekker, 1997.

Mileiko, S.T., *Metal and Ceramic Based Composites*, Elsevier, 1997.

Miller, E., *Introduction to Plastics and Composites: Mechanical Properties and Engineering Applications*, Marcel Dekker, 1995.

Nielsen, L.E., and Landel, R.F., *Mechanical Properties of Polymers and Composites*, Marcel Dekker, 1994.

Peters, S. (ed.), *Handbook of Composites*, Chapman & Hall, 1997.

Pilato, L.A., and Michno, M.J., *Advanced Composite Materials*, Springer, 1994.

Potter, K., *Introduction to Composite Products: Design, Development and Manufacture*, Kluwer, 1997.

Rosato, D.V., DiMattia, D.P., and Rosato, D.V., *Designing with Plastics and Composites: A Handbook*, Van Nostrand Reinhold, 1991.

Schwartz, M., *Composite Materials*, Vol. 1: *Properties, Nondestructive Testing, and Repair*; Vol. 2: *Processing, Fabrication, and Applications*. Prentice Hall, 1997.

REVIEW QUESTIONS

9.1 Distinguish between composites and metal alloys.

9.2 Describe the functions of the matrix and of the reinforcing fibers. What fundamental differences are there in the characteristics of the two materials?

9.3 What reinforcing fibers generally are used to make composites? Which type of fiber is the strongest? Which type is the weakest?

9.4   What is the range in length and diameter of reinforcing fibers?

9.5   List the important factors that determine the properties of reinforced plastics.

9.6   Compare the advantages and limitations of metal-matrix composites, reinforced plastics, and ceramic-matrix composites.

9.7   What are the most commonly used matrix materials?

9.8   What is a hybrid composite?

9.9   What material properties are improved by the addition of reinforcing fibers?

9.10   What is the purpose of the matrix material?

9.11   What are the most common types of glass fibers?

9.12   Is there a difference between a carbon fiber and a graphite fiber? Explain.

9.13   How can a graphite fiber be made electrically and thermally conductive?

9.14   What is a whisker? What is the difference between a whisker and a fiber?

9.15   What do boron fibers consist of? Why are they heavy?

9.16   Why are metal-matrix composites of interest?

9.17   What characteristics do metal-matrix materials have in common?

QUALITATIVE PROBLEMS

9.18   How do you think the use of straw in clay originally came about in making brick for dwellings?

9.19   What products have you personally seen that are made of reinforced plastics? How can you tell?

9.20   What applications are not well suited for composite materials?

9.21   What is the difference between a composite material and a coated material?

9.22   Identify metals and alloys that have strengths comparable to those of reinforced plastics.

9.23   The many advantages of composite materials were described in this chapter. What limitations or disadvantages do these materials have? What suggestions would you make to overcome these limitations?

9.24   What factors contribute to the cost of reinforcing fibers? (See Table 9.2.)

9.25   Give examples of composite materials other than those stated in this chapter.

9.26   A hybrid composite is defined as one containing two or more different types of reinforcing fibers. What advantages would such a composite have over others?

9.27   Explain why the behavior of the materials given in Fig. 9.5 is as shown.

9.28   Why are fibers capable of supporting a major portion of the load in composite materials?

9.29   Do metal-matrix composites have advantages over reinforced plastics? Explain.

9.30   Give reasons for the development of ceramic-matrix composites. Name some possible applications.

9.31   Explain how you would go about determining the hardness of reinforced plastics and of composite materials. Are hardness measurements on these types of materials meaningful? Does the size of the indentation make a difference? Explain.

9.32   How would you go about trying to determine the strength of a fiber?

9.33   What are the advantages of whiskers as a reinforcing material?

9.34   It has been stated that glass fibers are much stronger than bulk glass. Why is this so?

9.35   Under what circumstances could a glass be used as a matrix?

9.36   When the American Plains states were settled, no trees existed for the construction of housing. Pioneers cut bricks from sod—basically, prairie soil as a matrix and grass and its root system as reinforcement. Explain why this worked. Also, if you were a pioneer, would you stack the bricks with the grass horizontally or vertically? Explain.

9.37   Compare the advantages and disadvantages of metal-matrix composites, reinforced plastics, and ceramic-matrix composites.

9.38   You studied the many advantages of composite materials in this chapter. What limitations or disadvantages do these materials have? What suggestions would you make to overcome these limitations?

9.39   By incorporating small amounts of blowing agent, it is possible to manufacture hollow polymer fibers with gas cores. List applications for such fibers.

QUANTITATIVE PROBLEMS

9.40 Calculate the average increase in the properties of the plastics given in Table 7.1 as a result of their reinforcement and describe your observations.

9.41 In the example in Section 9.3, what would be the percentage of the load supported by the fibers if their strength is 1100 MPa and the matrix strength is 200 MPa? What if the fiber stiffness is doubled and the matrix stiffness is halved?

9.42 Make a survey of the recent technical literature and present data indicating the effects of fiber length on such mechanical properties as the strength, the elastic modulus, and impact energy of reinforced plastics.

9.43 Calculate the percent increase in the mechanical properties of reinforced nylon from the data shown in Fig. 9.4.

9.44 Plot E/ρ and $E/\rho^{0.5}$ for the composite materials listed in Table 9.1 and make a comparison to the properties of the materials described in Chapters 4 through 8. (See also Table 9.2.)

9.45 Calculate the stress in the fibers and in the matrix for the example in Section 9.3.1. Assume that the cross-sectional area is 0.1 in^2 and P_c = 500 lb.

9.46 Repeat the calculations in the example in Section 9.3.1 (a) if a high-modulus carbon fiber is used and (b) if Kevlar 29 is used.

9.47 Refer to the properties listed in Table 7.1. If acetal is reinforced with E-type glass fibers, what is the range of fiber content in glass-reinforced acetal?

9.48 Plot the elastic modulus and strength of an aluminum metal-matrix composite with high-modulus carbon fibers as a function of fiber content.

9.49 For the data in the numerical example at the end of Section 9.3.1, what should be the fiber content so that the fibers and the matrix fail simultaneously? Use an allowable fiber stress of 200 MPa and a matrix strength of 50 MPa.

SYNTHESIS, DESIGN, AND PROJECTS

9.50 What applications for composite materials can you think of other than those listed in Section 9.4? Why do you think your applications are suitable for these materials?

9.51 Using the information given in this chapter, develop special designs and shapes for new applications of composite materials.

9.52 Would a composite material with a strong and stiff matrix and a soft and flexible reinforcement have any practical uses? Explain.

9.53 Make a list of products for which the use of composite materials could be advantageous because of their anisotropic properties.

9.54 Inspect Fig. 9.1 and explain what other components of an aircraft, including parts in the cabin, could be made of composites.

9.55 Name applications in which both specific strength and specific stiffness (Fig. 9.2) are important.

9.56 What applications for composite materials can you think of in which high thermal conductivity would be desirable?

9.57 As with other materials, the mechanical properties of composites are obtained by preparing appropriate specimens and then testing them. Explain what problems you might encounter in preparing specimens for testing in tension. Suggest methods for making appropriate specimens, including their shape and how they are clamped into the jaws of testing machines.

9.58 Design and describe a test method to determine the mechanical properties of reinforced plastics in their thickness direction.

9.59 Developments are taking place in techniques for three-dimensional reinforcement of plastics. Describe (a) applications in which strength in the thickness direction of the composite is important and (b) your ideas on how to achieve this strength. Include simple sketches of the structure utilizing such reinforced plastics.

9.60 As described in this chapter, reinforced plastics can be affected adversely by environment—in particular by moisture, chemicals, and temperature variations. Design and describe test methods to determine the mechanical properties of composite materials subjected to these conditions.

9.61 Comment about your observations on the design of the sailboard shown in Fig. 9.8.

9.62 Describe the similarities and differences between ordinary corrugated cardboard and a honeycomb structure.

9.63 Suggest product designs in which corrugated cardboard can be used. Comment on the advantages and limitations.

9.64 Suggest consumer-product designs that could utilize honeycomb structures. For example, an elevator can use a honeycomb laminate as a stiff and lightweight floor material.

9.65 Make a survey of various sports equipment and identify the components made of composite materials. Explain the reasons for and the advantages of using composites in these specific applications.

9.66 Several material combinations and structures were described in this chapter. In relative terms, identify those that would be suitable for applications involving one of the following: (a) very low temperatures, (b) very high temperatures, (c) vibrations, and (d) high humidity.

9.67 Obtain a textbook on composite materials and investigate the effective stiffness of a continuous fiber-reinforced polymer. Plot the stiffness of such a composite as a function of orientation with respect to the fiber direction.

9.68 Derive a general expression for the coefficient of thermal expansion for a continuous fiber-reinforced composite in the fiber direction.

9.69 Instead of a constant cross-section, it is possible to make fibers or whiskers with a varying cross-section or a "wavy" fiber. What advantages would such fibers have?

9.70 Describe how you can produce some simple composite materials using raw materials available around your home. Explain.

Metal-Casting Processes and Equipment

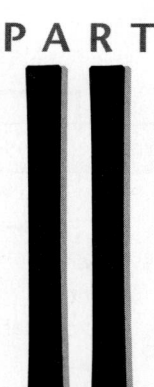

P A R T

II

As described throughout the rest of this book, several different methods are available to shape metals into useful products. One of the oldest processes is **casting**, which basically involves pouring molten metal into a mold cavity where (upon solidification) it takes the shape of the cavity. Casting first was used around 4000 B.C. to make ornaments, copper arrowheads, and various other objects. A wide variety of products can be cast, and the process is capable of producing intricate shapes in one piece, including those with internal cavities, such as engine blocks. Figure II.1 shows cast components in a typical automobile, a product that was used in the introduction to Part I to illustrate the selection and use of a variety of materials. The casting processes developed over the years are shown in Fig. II.2.

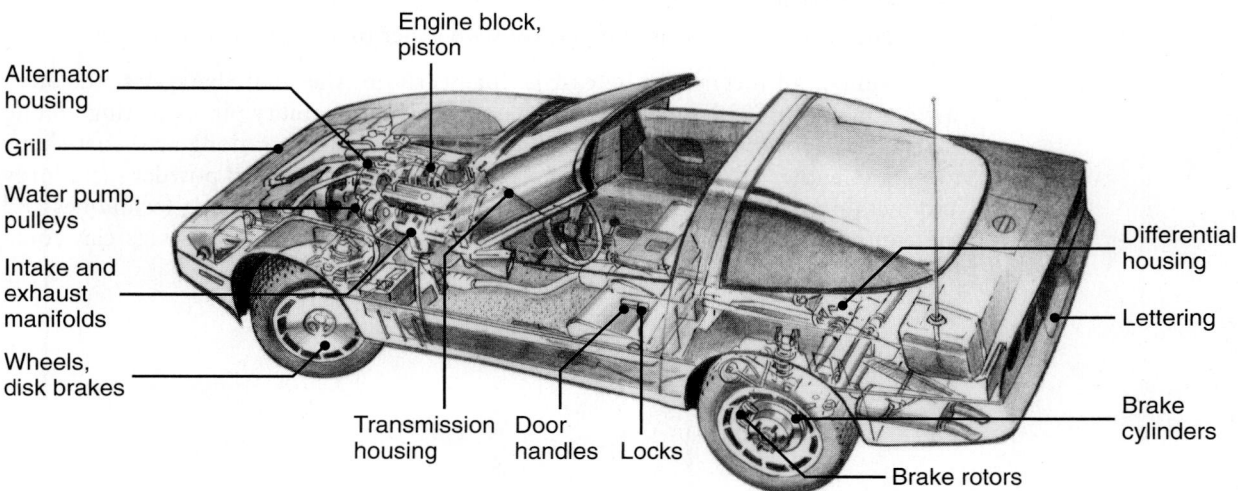

FIGURE II.1   Cast parts in a typical automobile.

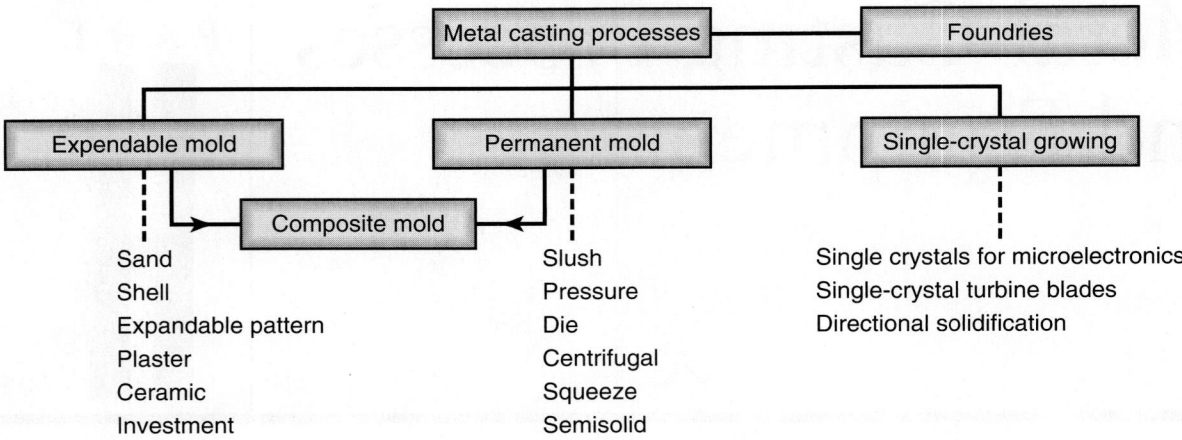

FIGURE II.2   Outline of metal-casting processes described in Part II.

As in all manufacturing, each casting process has its own characteristics, applications, advantages, limitations, and costs. Casting processes are most often selected over other manufacturing methods, for the following reasons:

- Casting can produce complex shapes and with internal cavities or hollow sections.
- Very large parts can be produced in one piece.
- Casting can utilize materials that are difficult or uneconomical to process by other means.
- The casting process is competitive with other manufacturing processes.

Almost all metals can be cast in (or nearly in) the final shape desired, often with only minor finishing operations required. This capability places casting among the most important *net-shape manufacturing* technologies, along with net-shape forging (Chapter 14), stamping of sheet metal (Chapter 16), and powder metallurgy and metal-injection molding (Chapter 17). With modern processing techniques and the control of chemical composition, mechanical properties of castings can equal those made by other manufacturing processes.

Fundamentals of Metal Casting

First used about six thousand years ago, casting continues to be an important manufacturing process for producing very small (as well as very large) parts. In order to understand the fundamental aspects of this process, this chapter describes:

- Mechanisms of solidification in metals and their alloys.
- Significance of solidification patterns in casting.
- Characteristics of fluid flow and heat transfer in molds and their effects.
- Role of gases and shrinkage in defect formation in casting.

10.1 | Introduction

The **casting** process basically involves (a) pouring molten metal into a mold patterned after the part to be manufactured, (b) allowing it to solidify, and (c) removing the part from the mold. As with all other manufacturing processes, an understanding of the fundamentals is essential for the production of good quality and economical castings and for establishing proper techniques for mold design and casting practice.

Important considerations in casting operations are as follows:

- Flow of the molten metal into the mold cavity
- Solidification and cooling of the metal in the mold
- Influence of the type of mold material

This chapter describes the relationship among the many factors involved in casting. The flow of molten metal into the mold cavity first is discussed in terms of mold design and fluid-flow characteristics. Solidification and cooling of metals in the mold are affected by several factors, including the metallurgical and thermal properties of the metal. The type of mold also has an important influence because it affects the rate of cooling of the metal in the mold. Finally, factors influencing defect formation also are described.

Industrial metal-casting processes, design considerations, and casting materials are described in Chapters 11 and 12. The casting of ceramics and of plastics, which involves somewhat similar methods and procedures for metal, are described in Chapters 18 and 19, respectively.

10.2 | Solidification of Metals

After molten metal is poured into a **mold,** a series of events takes place during the solidification of the metal and its cooling to ambient temperature. These events greatly influence the size, shape, uniformity, and chemical composition of the grains formed throughout the casting, which in turn influence its overall properties. The significant factors affecting these events are the type of metal, the thermal properties of both the metal and the mold, the geometric relationship between volume and surface area of the casting, and the shape of the mold.

10.2.1 Pure metals

Because a pure metal has a clearly defined melting (or freezing) point, it solidifies at a constant temperature, as shown in Fig. 10.1. Pure aluminum, for example, solidifies at 660°C (1220°F), iron at 1537°C (2798°F), and tungsten at 3410°C (6170°F). (See also Table 3.1 and Fig. 4.4.) After the temperature of the molten metal drops to its freezing point, its temperature remains constant while the *latent heat of fusion* is given off. The *solidification front* (solid–liquid interface) moves through the molten metal from the mold walls in toward the center. The solidified metal, called the casting, is taken out of the mold and allowed to cool to ambient temperature.

The grain structure of a pure metal cast in a square mold is shown in Fig. 10.2a. At the mold walls, which are at ambient temperature or at least are much cooler than the molten metal, the metal cools rapidly and produces a solidified **skin** or *shell* of fine equiaxed grains. The grains grow in a direction opposite to that of the heat transfer out through the mold. Those grains that have favorable orientation grow preferentially and are called **columnar grains** (Fig. 10.3). As the driving force of the heat transfer is reduced away from the mold walls, the grains become equiaxed and coarse. Those grains that have substantially different orientations are blocked from

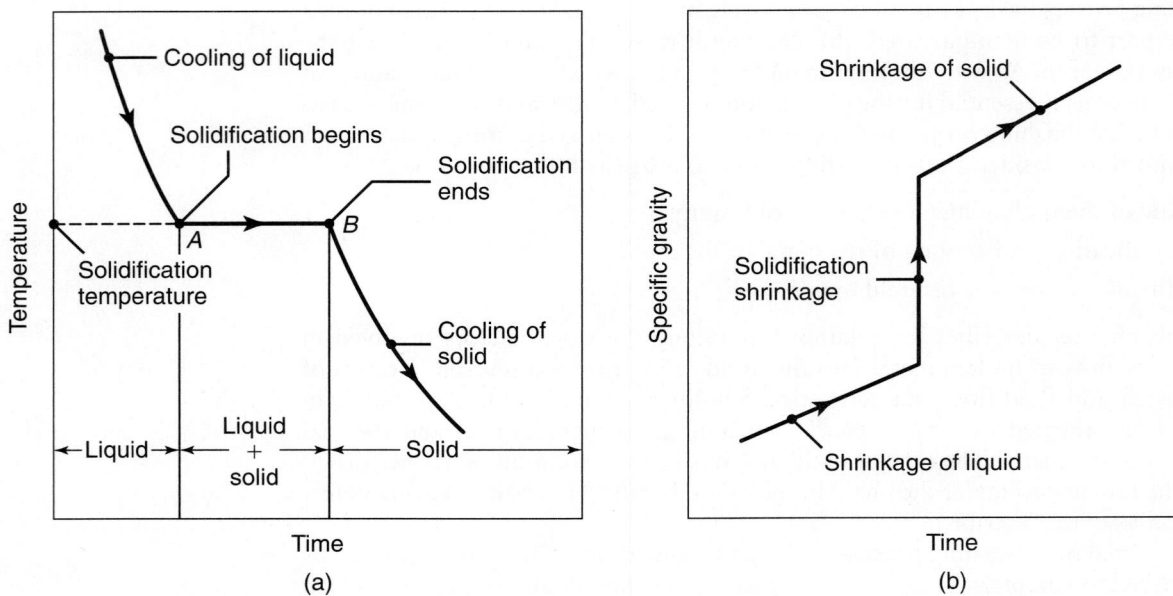

FIGURE 10.1   (a) Temperature as a function of time for the solidification of pure metals. Note that freezing takes place at a constant temperature. (b) Density as a function of time.

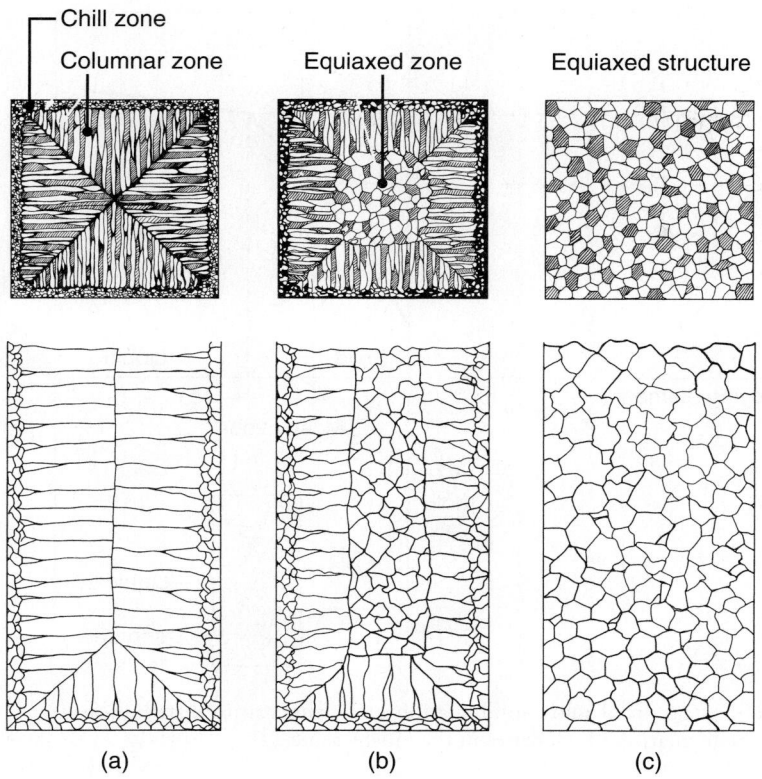

FIGURE 10.2   Schematic illustration of three cast structures of metals solidified in a square mold: (a) pure metals; (b) solid-solution alloys; and (c) structure obtained by using nucleating agents. *Source:* After G. W. Form, J. F. Wallace, J. L. Walker, and A. Cibula.

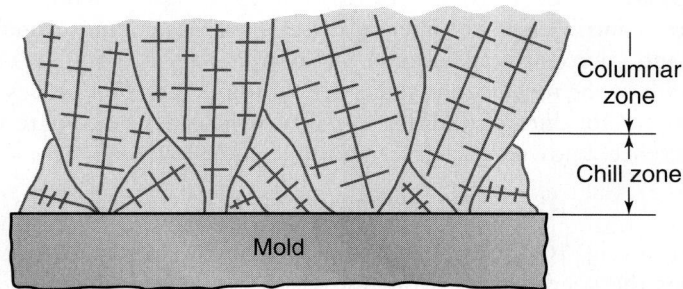

FIGURE 10.3   Development of a preferred texture at a cool mold wall. Note that only favorably oriented grains grow away from the surface of the mold.

further growth. Such grain development is known as **homogenous nucleation,** meaning that the grains (crystals) grow upon themselves, starting at the mold wall.

10.2.2 Alloys

Solidification in alloys begins when the temperature drops below the liquidus, T_L, and is complete when it reaches the solidus, T_S (Fig. 10.4). Within this temperature

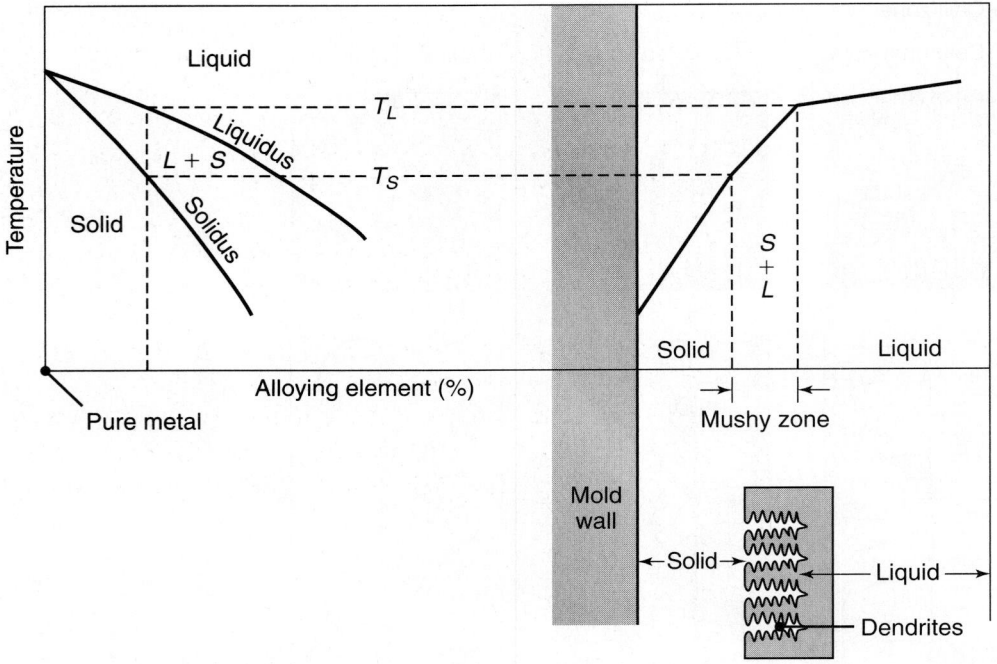

FIGURE 10.4 Schematic illustration of alloy solidification and temperature distribution in the solidifying metal. Note the formation of dendrites in the mushy zone.

range, the alloy is in a *mushy* or *pasty* state consisting of **columnar dendrites** (from the Greek *dendron*, meaning "akin to," and *drys*, meaning "tree"). Note the presence of liquid metal between the dendrite arms. Dendrites have three-dimensional arms and branches (secondary arms) which eventually interlock, as can be seen in Fig. 10.5. The study of dendritic structures (although complex) is important, because such structures contribute to detrimental factors, such as compositional variations, segregation, and microporosity within a cast part.

The width of the **mushy zone** (where both liquid and solid phases are present) is an important factor during solidification. This zone is described in terms of a temperature difference, known as the **freezing range**, as follows:

$$\text{Freezing range} = T_L - T_S \tag{10.1}$$

It can be seen in Fig. 10.4 that pure metals have a freezing range that approaches zero, and that the solidification front moves as a plane front without forming a mushy zone. *Eutectics* (Section 4.3) solidify in a similar manner, with an approximate plane front. The type of structure developed after solidification depends on the composition of the eutectic. In alloys with a nearly symmetrical phase diagram, the structure is generally lamellar, with two or more solid phases present, depending on the alloy system. When the volume fraction of the minor phase of the alloy is less than about 25%, the structure generally becomes fibrous. These conditions are particularly important for cast irons.

For alloys, a *short freezing range* generally involves a temperature difference of less than 50°C (90°F), and a *long freezing range* greater than 110°C (200°F). Ferrous castings generally have narrow mushy zones, whereas aluminum and magnesium alloys have wide mushy zones. Consequently, these alloys are in a mushy state throughout most of the solidification process.

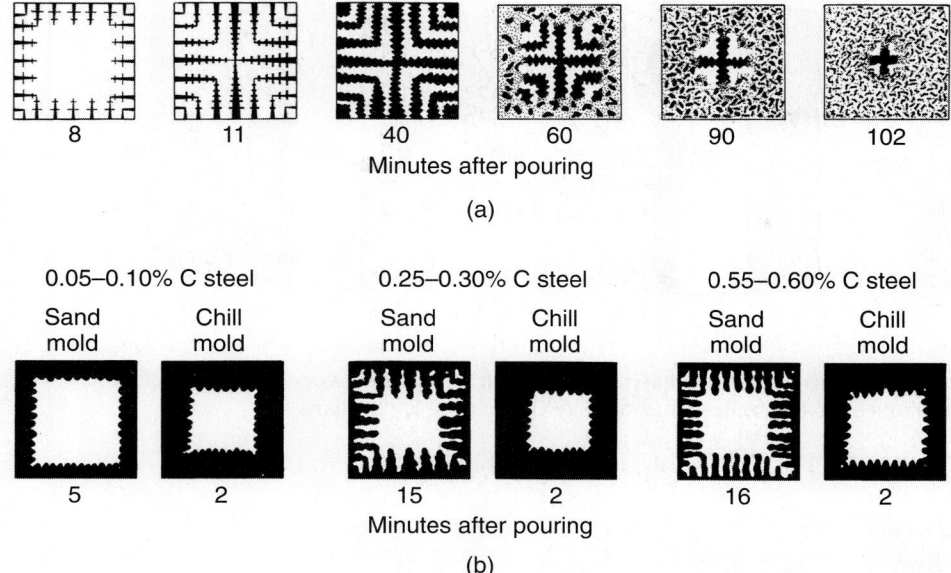

8   11   40   60   90   102
Minutes after pouring

(a)

0.05–0.10% C steel   0.25–0.30% C steel   0.55–0.60% C steel

Sand   Chill   Sand   Chill   Sand   Chill
mold   mold   mold   mold   mold   mold

5   2   15   2   16   2
Minutes after pouring

(b)

FIGURE 10.5   (a) Solidification patterns for gray cast iron in a 180-mm (7-in.) square casting. Note that after 11 minutes of cooling, dendrites reach each other, but the casting is still mushy throughout. It takes about two hours for this casting to solidify completely. (b) Solidification of carbon steels in sand and chill (metal) molds. Note the difference in solidification patterns as the carbon content increases. *Source:* After H. F. Bishop and W. S. Pellini.

Effects of cooling rates.   Slow cooling rates (on the order of 10^2 K/s) or long, local solidification times result in *coarse* dendritic structures with large spacing between dendrite arms. For higher cooling rates (on the order of 10^4 K/s) or short, local solidification times, the structure becomes *finer* with smaller dendrite arm spacing. For still higher cooling rates (on the order of from 10^6 to 10^8 K/s) the structures developed are *amorphous*, as described in Section 6.14.

The structures developed and the resulting grain size influence the properties of the casting. As grain size decreases, the strength and the ductility of the cast alloy increase, microporosity (interdendritic shrinkage voids) in the casting decreases, and the tendency for the casting to crack (*hot tearing*, see Fig. 10.12) during solidification decreases. Lack of uniformity in grain size and grain distribution results in castings with *anisotropic properties*.

A criterion describing the kinetics of the liquid–solid interface is the ratio G/R, where G is the *thermal gradient* and R is the *rate* at which the liquid–solid interface moves. Typical values for G range from 10^2 to 10^3 K/m, and for R range from 10^{-3} to 10^{-4} m/s. Dendritic-type structures (Fig. 10.6a and b) typically have a G/R ratio in the range of 10^5 to 10^7, whereas ratios of 10^{10} to 10^{12} produce a plane-front, nondendritic liquid–solid interface (Fig. 10.7).

10.2.3 Structure–property relationships

Because all castings are expected to possess certain properties to meet design and service requirements, the relationships between properties and the structures developed during solidification are important aspects of casting. This section describes

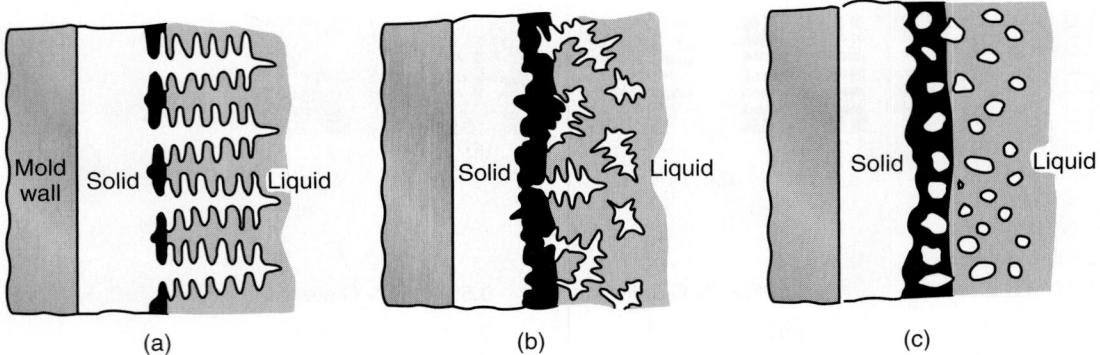

FIGURE 10.6   Schematic illustration of three basic types of cast structures: (a) columnar dendritic; (b) equiaxed dendritic; and (c) equiaxed nondendritic. *Source:* Courtesy of D. Apelian.

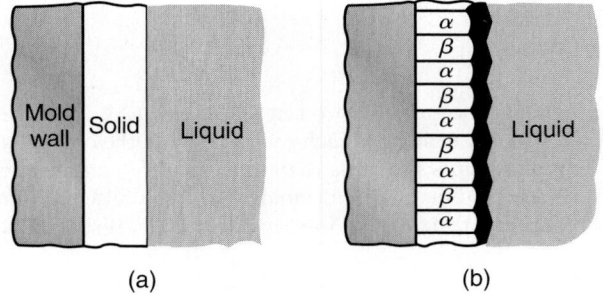

FIGURE 10.7   Schematic illustration of cast structures in (a) plane front, single phase, and (b) plane front, two phase. *Source:* Courtesy of D. Apelian.

these relationships in terms of dendrite morphology and the concentration of alloying elements in various regions within the metal.

The compositions of dendrites and the liquid metal are given by the *phase diagram* of the particular alloy. When the alloy is cooled very slowly, each dendrite develops a uniform composition. However, under the normal (faster) cooling rates encountered in practice, **cored dendrites** are formed. Cored dendrites have a surface composition different from that at their centers, a difference referred to as a *concentration gradient*. The surface of the dendrite has a higher concentration of alloying elements than does its core due to solute rejection from the core towards the surface during solidification of the dendrite (**microsegregation**). The darker shading in the interdendritic liquid near the dendrite roots shown in Fig. 10.6 indicates that these regions have a higher solute concentration; microsegregation in these regions is much more pronounced than in others.

There are several types of **segregation**. In contrast with microsegregation, **macrosegregation** involves differences in composition throughout the casting itself. In situations where the solidifying front moves away from the surface of a casting as a plane front (Fig. 10.7), lower melting-point constituents in the solidifying alloy are driven towards the center (**normal segregation**). Consequently, such a casting has a higher concentration of alloying elements at its center than at its surfaces. In dendritic structures such as those found in solid–solution alloys (Fig. 10.2b), the opposite occurs; that is, the center of the casting has a lower concentration of alloying elements (**inverse segregation**) than does its surface. The reason is that liquid metal (having a higher concentration of alloying elements) enters the cavities developed from solidification shrinkage in the dendrite arms, which have solidified sooner.

Another form of segregation is due to gravity. **Gravity segregation** describes the process whereby higher-density inclusions or compounds sink, and lighter elements (such as antimony in an antimony–lead alloy) float to the surface.

A typical cast structure of a solid–solution alloy with an inner zone of equiaxed grains is shown in Fig. 10.2b. This inner zone can be extended throughout the casting, as shown in Fig. 10.2c, by adding an **inoculant** (*nucleating agent*) to the alloy. The inoculant induces nucleation of the grains throughout the liquid metal (**heterogeneous nucleation**).

Because of the presence of *thermal gradients* in a solidifying mass of liquid metal and gravity and the resultant density differences, *convection* has a strong influence on the structures developed. Convection promotes the formation of an outer chill zone, refines grain size, and accelerates the transition from columnar to equiaxed grains. The structure shown in Fig. 10.6b also can be obtained by increasing convection within the liquid metal, whereby dendrite arms separate (**dendrite multiplication**). Conversely, reducing or eliminating convection results in coarser and longer columnar dendritic grains.

The dendrite arms are not particularly strong and can be broken up by agitation or mechanical vibration in the early stages of solidification (**semisolid metal forming** and **rheocasting**, see Section 11.3.7). This process results in finer grain size with equiaxed nondendritic grains distributed more uniformly throughout the casting (Fig. 10.6c). A side benefit is the **thixotropic** behavior of alloys (that is, the viscosity decreases when the liquid metal is agitated), leading to improved castability. Another form of semisolid metal forming is *thixotropic casting*, where a solid billet is heated to the semisolid state and then injected into a die-casting mold (Section 11.3.5). The heating is usually by convection in a furnace but can be enhanced by the use of mechanical or electromagnetic methods.

Experiments are being conducted during space flights concerning the effects of gravity on the microstructure of castings. Lack of gravity (**microgravity**) means that, unlike on earth, there are no significant density differences or thermal gradients (and therefore no convection) during solidification. This lack of convection affects the solidification structure and the distribution of impurities. Recent experiments involve the growth of crystals for production of cadmium-zinc telluride, mercury-zinc telluride, and selenium-doped gallium arsenide semiconductor samples (see Section 28.3).

10.3 | Fluid Flow

To emphasize the importance of fluid flow in casting, let's briefly describe a basic gravity casting system, as shown in Fig. 10.8. The molten metal is poured through a **pouring basin** or **cup**; it then flows through the **gating system** (sprue, runners and gates) into the mold cavity. As illustrated in Fig. 11.3, the **sprue** is a tapered vertical channel through which the molten metal flows downward in the mold. **Runners** are the channels that carry the molten metal from the sprue into the mold cavity or connect the sprue to the **gate** (that portion of the runner through which the molten metal enters the mold cavity). **Risers** (also called **feeders**) serve as reservoirs of molten metal to supply any molten metal necessary to prevent porosity due to shrinkage during solidification. (See also Fig. 11.3.)

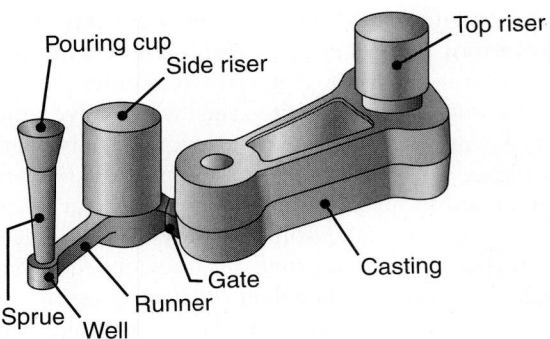

FIGURE 10.8   Schematic illustration of a typical riser-gated casting. Risers serve as reservoirs, supplying molten metal to the casting as it shrinks during solidification.

Although such a gating system appears to be relatively simple, successful casting requires proper design and control of the solidification process to ensure adequate fluid flow in the system. For example, an important function of the gating system in sand casting is to *trap contaminants* (such as oxides and other inclusions) in the molten metal by having the contaminants adhere to the walls of the gating system, thereby preventing them from reaching the mold cavity. Furthermore, a properly designed gating system helps avoid or minimize problems (such as premature cooling, turbulence, and gas entrapment). Even before it reaches the mold cavity, the molten metal must be handled carefully to avoid the formation of oxides on molten-metal surfaces from exposure to the environment or the introduction of impurities into the molten metal.

Two basic principles of fluid flow are relevant to gating design: Bernoulli's theorem and the law of mass continuity.

Bernoulli's theorem.   This theorem is based on the principle of the conservation of energy and relates pressure, velocity, the elevation of the fluid at any location in the system, and the frictional losses in a system that is full of liquid. This is according to the equation

$$h + \frac{p}{\rho g} + \frac{v^2}{2g} = \text{constant} \tag{10.2}$$

where h is the elevation above a certain reference plane, p is the pressure at that elevation, v is the velocity of the liquid at that elevation, ρ is the density of the fluid (assuming that it is incompressible), and g is the gravitational constant. Conservation of energy requires that, at a particular location in the system, the following relationship be satisfied as

$$h_1 + \frac{p_1}{\rho g} + \frac{v_1^2}{2g} = h_2 + \frac{p_2}{\rho g} + \frac{v_2^2}{2g} + f \tag{10.3}$$

where the subscripts 1 and 2 represent two different elevations, respectively. Here f represents the frictional loss in the liquid as it travels downward through the system. The frictional loss includes such factors as energy loss at the liquid-mold wall interfaces and turbulence in the liquid.

Mass continuity.   The law of mass continuity states that for incompressible liquids and in a system with impermeable walls the rate of flow is constant. Thus,

$$Q = A_1 v_1 = A_2 v_2 \qquad (10.4)$$

where Q is the volume rate of flow (such as m^3/s), A is the cross-sectional area of the liquid stream, and v is the average velocity of the liquid in that cross-sectional location. The subscripts 1 and 2 refer to two different locations in the system. According to this law, the flow rate must be maintained anywhere in the system. The permeability of the walls of the system is important, because otherwise, some liquid will permeate through the walls (such as occurs in sand molds). Thus, the flow rate will decrease as the liquid moves through the system. Coatings often are used to inhibit such behavior in sand molds.

Sprue design.   An application of the two principles just stated is the traditional tapered design of sprues (shown in Fig. 10.8). The shape of the sprue can be determined by using Eqs. (10.3) and (10.4). Assuming that the pressure at the top of the sprue is equal to the pressure at the bottom, and that there are no frictional losses, the relationship between height and cross-sectional area at any point in the sprue is given by the parabolic relationship

$$\frac{A_1}{A_2} = \sqrt{\frac{h_2}{h_1}} \qquad (10.5)$$

where, for example, the subscript 1 denotes the top of the sprue and 2 denotes the bottom. Moving downward from the top, the cross-sectional area of the sprue must therefore decrease. Depending on the assumptions made, expressions other than Eq. (10.5) can also be obtained. For example, we may assume a certain molten-metal velocity, V_1, at the top of the sprue. Then, using Eqs. (10.3) and (10.4), an expression can be obtained for the ratio A_1/A_2 as a function of h_1, h_2, and V_1.

Note that in a free-falling liquid (such as water from a faucet), the cross-sectional area of the stream decreases as it gains velocity downward. If we design a sprue with a constant cross-sectional area and pour the molten metal into it, regions may develop where the liquid loses contact with the sprue walls. As a result, **aspiration** (a process whereby air is sucked in or entrapped in the liquid) may take place. A common alternative to tapered sprues is the use of straight-sided sprues with a **choking** mechanism at the bottom, consisting of either a choke core or a runner choke (as shown in Fig. 11.4). The choke slows flow sufficiently to prevent aspiration in the sprue.

Modeling.   Another application of the foregoing equations is in the *modeling of mold filling*. For example, consider the situation shown in Fig. 10.7 where molten metal is poured into a pouring basin; it flows through a sprue to a runner and a gate and fills the mold cavity. If the pouring basin has a much larger cross-sectional area than the sprue bottom, then the velocity of the molten metal at the top of the pouring basin is very low and can be taken as zero. If frictional losses are due to a viscous dissipation of energy, then f in Eq. (10.3) can be taken as a function of the vertical distance and is often approximated as a linear function. Therefore, the velocity of the molten metal leaving the gate is obtained from Eq. (10.3) as

$$v = c\sqrt{2gh}$$

where h is the distance from the sprue base to the liquid metal height and c is a friction factor. For frictionless flow, c equals unity and is always between 0 and 1.

The magnitude of c varies with mold material, runner layout, and channel size and can include energy losses due to turbulence, as well as viscous effects.

If the liquid level has reached a height of x, then the gate velocity is

$$v = c\sqrt{2g}\sqrt{h-x}$$

The flow rate through the gate will be the product of this velocity and the gate area according to Eq. (10.4). The shape of the casting will determine the height as a function of time. Integrating Eq. (10.4) gives the mean fill time and flow rate and dividing the casting volume by this mean flow rate gives the mold fill time.

Simulation of mold filling assists designers in the specification of the runner diameter, as well as the size and number of sprues and pouring basins. To ensure that the runners stay open, the fill time must be a small fraction of the solidification time, but the velocity should not be so high as to erode the mold material (referred to as *mold wash*) or to result in too high of a Reynolds number (see the following). Otherwise, turbulence and associated air entrainment results. Fortunately, many computational tools now are available to evaluate gating designs and assist in the sizing of components.

Flow characteristics. An important consideration of the fluid flow in gating systems is the presence of **turbulence**, as opposed to the *laminar flow* of fluids. The *Reynolds number*, Re, is used to quantify this aspect of fluid flow. It represents the ratio of the *inertia* to the *viscous* forces in fluid flow and is defined as

$$\text{Re} = \frac{vD\rho}{\eta} \tag{10.6}$$

where v is the velocity of the liquid, D is the diameter of the channel, and ρ and η are the density and viscosity of the liquid, respectively. The higher the Reynolds number, the greater the tendency for turbulent flow to occur.

In gating systems, Re typically ranges from 2000 to 20,000, where a value of up to 2000 represents laminar flow. Between 2000 and 20,000, it represents a mixture of laminar and turbulent flow. Such a mixture generally is regarded as harmless in gating systems. However, Re values in excess of 20,000 represent severe turbulence, resulting in air entrainment and the formation of *dross* (the scum that forms on the surface of molten metal) from the reaction of the liquid metal with air and other gases. Techniques for minimizing turbulence generally involve avoidance of sudden changes in flow direction and in the geometry of channel cross-sections in gating system design.

Dross or slag can be eliminated almost completely only by *vacuum casting* (Section 11.3.2). Conventional atmospheric casting mitigates dross or slag by (a) skimming, (b) using properly designed pouring basins and runner systems, or (c) using filters, which also can eliminate turbulent flow in the runner system. Filters usually are made of ceramics, mica, or fiberglass; their proper location and placement are important for effective filtering of dross and slag.

10.4 | Fluidity of Molten Metal

The capability of the molten metal to fill mold cavities is called *fluidity*, which consists of two basic factors: (1) characteristics of the molten metal and (2) casting parameters. The following characteristics of molten metal influence fluidity.

Viscosity. As viscosity and its sensitivity to temperature (*viscosity index*) increase, fluidity decreases.

Surface tension.   A high surface tension of the liquid metal reduces fluidity. Because of this, oxide films on the surface of the molten metal have a significant adverse effect on fluidity. For example, an oxide film on the surface of pure molten aluminum triples the surface tension.

Inclusions.   Because they are insoluble, inclusions can have a significant adverse effect on fluidity. This effect can be verified by observing the viscosity of a liquid (such as oil) with and without sand particles in it; the liquid with sand in it has a higher viscosity and, hence, lower fluidity.

Solidification pattern of the alloy.   The manner in which solidification takes place (Section 10.2) can influence fluidity. Moreover, fluidity is inversely proportional to the freezing range. The shorter the range (as in pure metals and eutectics), the higher the fluidity. Conversely, alloys with long freezing ranges (such as solid–solution alloys) have lower fluidity.

The following casting parameters influence fluidity and also influence the fluid flow and thermal characteristics of the system.

Mold design.   The design and dimensions of the sprue, runners, and risers all influence fluidity.

Mold material and its surface characteristics.   The higher the thermal conductivity of the mold and the rougher its surfaces, the lower the fluidity of the molten metal. Although heating the mold improves fluidity, it slows down solidification of the metal. Thus the casting develops coarse grains and hence has lower strength.

Degree of superheat.   Superheat (defined as the increment of temperature of an alloy above its melting point) improves fluidity by delaying solidification. The **pouring temperature** often is specified instead of the degree of superheat, because it is specified more easily.

Rate of pouring.   The slower the rate of pouring molten metal into the mold, the lower the fluidity because of the higher rate of cooling when poured slowly.

Heat transfer.   This factor directly affects the viscosity of the liquid metal (see below).

Although complex, the term **castability** generally is used to describe the ease with which a metal can be cast to produce a part with good quality. This term includes not only fluidity but the nature of casting practices as well.

10.4.1 Tests for fluidity

Several tests have been developed to quantify fluidity, although none is accepted universally. In one such common test, the molten metal is made to flow along a channel that is at room temperature (Fig. 10.9); the distance the metal flows before it solidifies and stops flowing is a measure of its fluidity. Obviously, this length is a function of the thermal properties of the metal and the mold, as well as of the design of the channel. Still, such fluidity tests are useful and simulate casting situations to a reasonable degree.

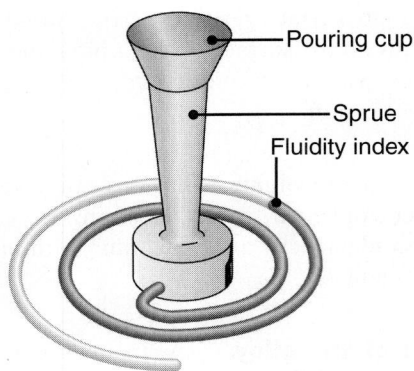

FIGURE 10.9 A test method for fluidity using a spiral mold. The *fluidity index* is the length of the solidified metal in the spiral passage. The greater the length of the solidified metal, the greater is its fluidity.

10.5 | Heat Transfer

The heat transfer during the complete cycle (from pouring, to solidification, and to cooling to room temperature) is another important consideration in metal casting. Heat flow at different locations in the system is a complex phenomenon and depends on several factors relating to the material cast and the mold and process parameters. For instance, in casting thin sections, the metal flow rates must be high enough to avoid premature chilling and solidification. On the other hand, the flow rate must not be so high as to cause excessive turbulence—with its detrimental effects on the casting process.

A typical temperature distribution at the mold liquid–metal interface is shown in Fig. 10.10. Heat from the liquid metal is given off through the mold wall and to the surrounding air. The temperature drop at the air–mold and mold–metal interfaces is caused by the presence of boundary layers and imperfect contact at these interfaces. The shape of the curve depends on the thermal properties of the molten metal and the mold.

10.5.1 Solidification time

During the early stages of solidification, a thin, solidified skin begins to form at the cool mold walls, and as time passes, the thickness of the skin increases (Fig. 10.11). With flat mold walls, this thickness is proportional to the square root of time. Thus, doubling the time will make the skin $\sqrt{2} = 1.41$ times or 41% thicker.

The **solidification time** is a function of the volume of a casting and its surface area (*Chvorinov's rule*):

$$\text{Solidification time} = C \left(\frac{\text{Volume}}{\text{Surface area}} \right)^n \qquad (10.7)$$

where C is a constant that reflects (a) the mold material, (b) the metal properties (including latent heat), and (c) the temperature. The parameter n has a value between 1.5 and 2 but usually is taken as 2. Thus, a large, solid sphere will solidify and cool to ambient temperature at a much slower rate than will a smaller solid sphere.

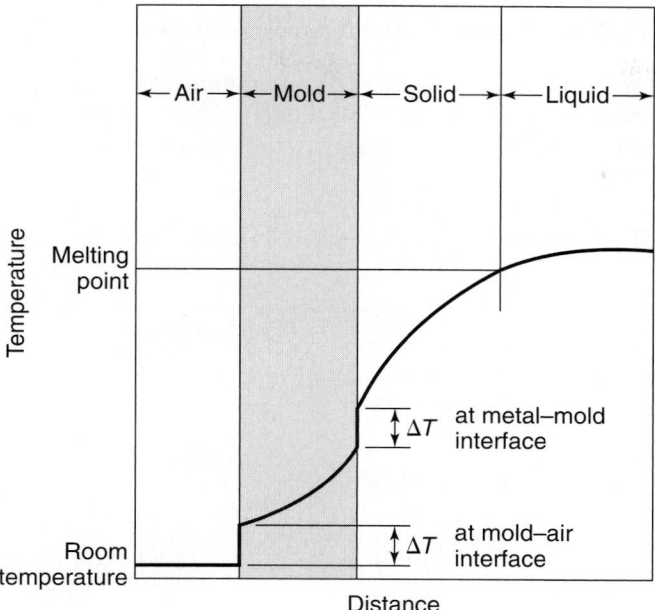

FIGURE 10.10 Temperature distribution at the interface of the mold wall and the liquid metal during the solidification of metals in casting.

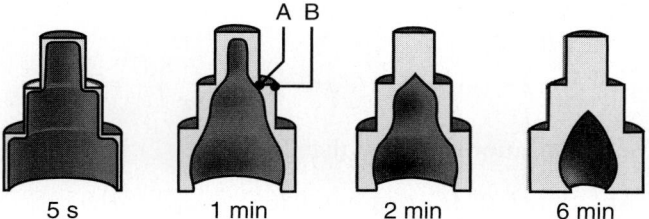

FIGURE 10.11 Solidified skin on a steel casting. The remaining molten metal is poured out at the times indicated in the figure. Hollow ornamental and decorative objects are made by a process called slush casting, which is based on this principle. *Source:* After H. F. Taylor, J. Wulff, and M. C. Flemings.

The reason for this is that the volume of a sphere is proportional to the cube of its diameter, and the surface area is proportional to the square of its diameter. Similarly, it can be shown that molten metal in a cube-shaped mold will solidify faster than in a spherical mold of the same volume (see Example 10.1).

The effects of mold geometry and elapsed time on skin thickness and shape are shown in Fig. 10.11. As illustrated, the unsolidified molten metal has been poured from the mold at different time intervals ranging from five seconds to six minutes. Note that (as expected) the skin thickness increases with elapsed time, and the skin is thinner at internal angles (location *A* in the figure) than at external angles (location *B*). This latter condition is caused by slower cooling at internal angles than at external angles.

EXAMPLE 10.1 Solidification times for various shapes

Three metal pieces being cast have the same volume but different shapes: One is a sphere, one a cube, and the other a cylinder with its height equal to its diameter. Which piece will solidify the fastest, and which one the slowest? Assume that $n = 2$.

Solution   The volume of the piece is taken as unity. Thus from Eq. (10.7),

$$\text{Solidification time} \propto \frac{1}{(\text{Surface area})^2}$$

The respective surface areas are as follows:

Sphere:

$$V = \left(\frac{4}{3}\right)\pi r^3, \; r = \left(\frac{3}{4\pi}\right)^{1/3}$$

$$A = 4\pi r^2 = 4\pi\left(\frac{3}{4\pi}\right)^{2/3} = 4.84$$

Cube:

$$V = a^3, \; a = 1, \text{ and } A = 6a^2 = 6$$

Cylinder:

$$V = \pi r^2 h = 2\pi r^3, \; r = \left(\frac{1}{2\pi}\right)^{1/3}$$

$$A = 2\pi r^2 + 2\pi rh = 6\pi r^2 = 6\pi\left(\frac{1}{2\pi}\right)^{2/3} = 5.54$$

The respective solidification times are therefore

$$t_{\text{sphere}} = 0.043C, \; t_{\text{cube}} = 0.028C, \; t_{\text{cylinder}} = 0.033C$$

Hence, the cube-shaped piece will solidify the fastest, and the spherical piece will solidify the slowest.

10.5.2 Shrinkage

Because of their thermal expansion characteristics, metals usually shrink (contract) during solidification and while cooling to room temperature. *Shrinkage*, which causes dimensional changes and (sometimes) cracking, is the result of the following three sequential events:

1. Contraction of the molten metal as it cools prior to its solidification.
2. Contraction of the metal during phase change from liquid to solid (latent heat of fusion).
3. Contraction of the solidified metal (the casting) as its temperature drops to ambient temperature.

The largest potential amount of shrinkage occurs during the cooling of the casting to ambient temperature. The amount of contraction during the solidification

TABLE 10.1

Volumetric Solidification Contraction or Expansion for Various Cast Metals			
Contraction (%)		Expansion (%)	
Aluminum	7.1	Bismuth	3.3
Zinc	6.5	Silicon	2.9
Al-4.5% Cu	6.3	Gray iron	2.5
Gold	5.5		
White iron	4–5.5		
Copper	4.9		
Brass (70–30)	4.5		
Magnesium	4.2		
90% Cu-10% Al	4		
Carbon steels	2.5–4		
Al-12% Si	3.8		
Lead	3.2		

of various metals is shown in Table 10.1. Note that some metals (such as gray cast iron) expand (the reason is that graphite has a relatively high specific volume, and when it precipitates as graphite flakes during solidification of the gray cast iron, it causes a net expansion of the metal). Shrinkage is further discussed in Section 12.2.1 in connection with design considerations in casting.

10.6 | Defects

As will be seen in this section (as well as other sections throughout Parts II through VI), various defects can develop in manufacturing processes depending on factors such as materials, part design, and processing techniques. While some defects affect only the appearance of the parts made, others can have major adverse effects on the structural integrity of the parts.

Several defects can develop in castings (Figs. 10.12 and 10.13). Because different names have been used in the past to describe the same defect, the International Committee of Foundry Technical Associations has developed a standardized nomenclature, consisting of seven basic categories of casting defects identified with boldface capital letters:

A—Metallic projections, consisting of fins, flash, or projections such as swells and rough surfaces.

B—Cavities, consisting of rounded or rough internal or exposed cavities including blowholes, pinholes, and shrinkage cavities (see *porosity*, Section 10.6.1).

C—Discontinuities, such as cracks, cold or hot tearing, and cold shuts. If the solidifying metal is constrained from shrinking freely, cracking and tearing may occur. Although several factors are involved in tearing, coarse grain size and the presence of low melting-point segregates along the grain boundaries (*intergranular*) increase the tendency for hot tearing. *Cold shut* is an interface in a casting that lacks complete fusion because of the meeting of two streams of liquid metal from different gates.

D—Defective surface, such as surface folds, laps, scars, adhering sand layers, and oxide scale.

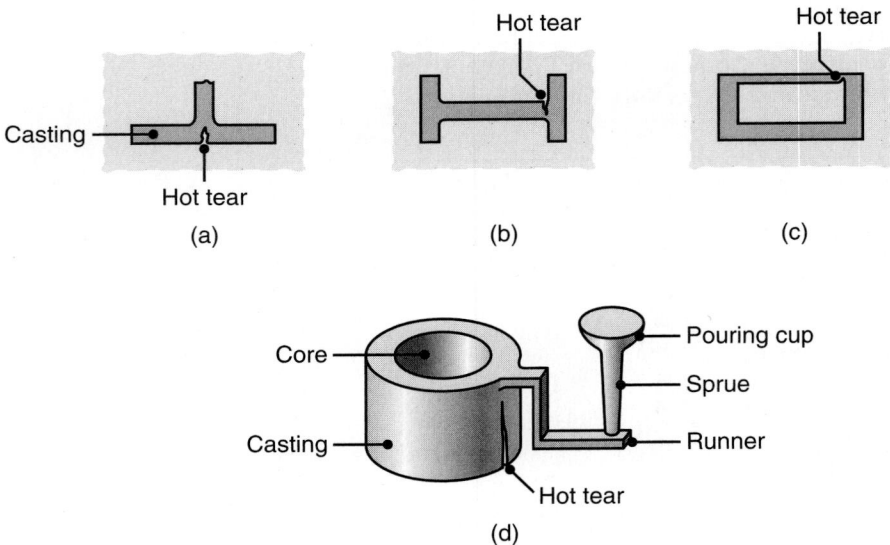

FIGURE 10.12   Examples of hot tears in castings. These defects occur because the casting cannot shrink freely during cooling, owing to constraints in various portions of the molds and cores. Exothermic (heat-producing) compounds may be used (as exothermic padding) to control cooling at critical sections to avoid hot tearing.

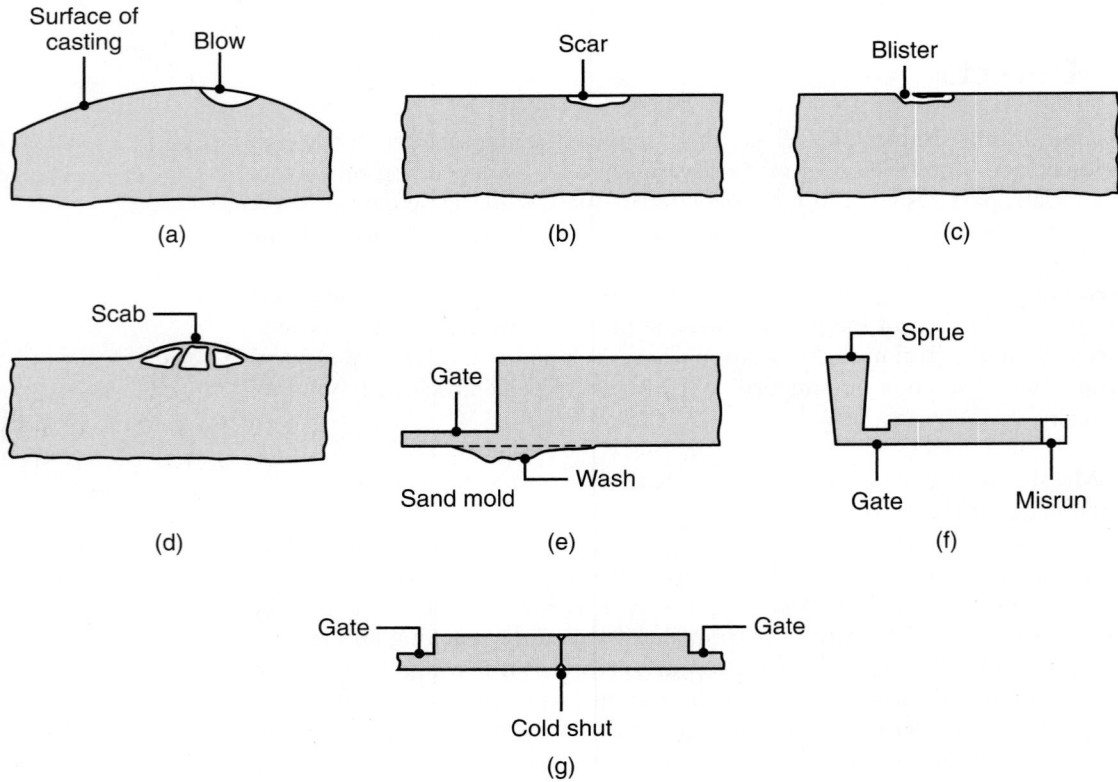

FIGURE 10.13   Examples of common defects in castings. These defects can be minimized or eliminated by proper design and preparation of molds and control of pouring procedures. *Source:* After J. Datsko.

E—**Incomplete casting**, such as misruns (due to premature solidification), insufficient volume of the metal poured, and runout (due to loss of metal from mold after pouring). Incomplete castings also can result from the molten metal being at too low a temperature or from pouring the metal too slowly.

F—**Incorrect dimensions or shape**, due to factors such as improper shrinkage allowance, pattern-mounting error, irregular contraction, deformed pattern, or warped casting.

G—**Inclusions**, which form during melting, solidification, and molding; generally nonmetallic. They are regarded as harmful because they act as stress raisers and, thus, reduce the strength of the casting. Particles as small as 30 μm can be filtered out during processing of the molten metal. Inclusions may form during melting when the molten metal reacts with the environment (usually oxygen) or with the crucible or mold material. Chemical reactions among components in the molten metal itself may produce inclusions; slags and other foreign material entrapped in the molten metal also become inclusions. Spalling of the mold and core surfaces also can produce inclusions, thus indicating the importance of the quality of molds and of their maintenance.

10.6.1 Porosity

Porosity in a casting may be caused by *shrinkage*, or *gases*, or both. Porous regions can develop in castings because of **shrinkage** of the solidified metal. Thin sections in a casting solidify sooner than thicker regions; as a result, molten metal flow into the thicker regions that have not yet solidified. Porous regions may develop at their centers because of contraction as the surfaces of the thicker region begin to solidify first. *Microporosity* also can develop when the liquid metal solidifies and shrinks between dendrites and between dendrite branches.

Porosity is detrimental to the ductility of a casting and its surface finish, making it permeable and, thus, affecting the pressure tightness of a cast pressure vessel. Porosity caused by shrinkage can be reduced or eliminated by various means such as:

- Adequate liquid metal should be provided to avoid cavities caused by shrinkage.
- Internal or external **chills**, as those used in sand casting (Fig. 10.14), also are an effective means of reducing shrinkage porosity. The function of chills is to increase the rate of solidification in critical regions. Internal chills usually are made of the same material as the casting and are left in the casting. However, there may be problems involved in proper fusion of the internal chills with the casting, thus foundries generally avoid use of internal chills for this reason. External chills may be made of the same material or may be iron, copper, or graphite.
- With alloys, porosity can be reduced or eliminated by making the temperature gradient steep. For example, mold materials that have higher thermal conductivity may be used.
- Subjecting the casting to *hot isostatic pressing* is another method of reducing porosity (see Section 17.3.2).

Because liquid metals have much greater solubility for **gases** than do solid metals (Fig. 10.15), when a metal begins to solidify, the dissolved gases are expelled from the solution. Gases also may result from reactions of the molten metal with the

00

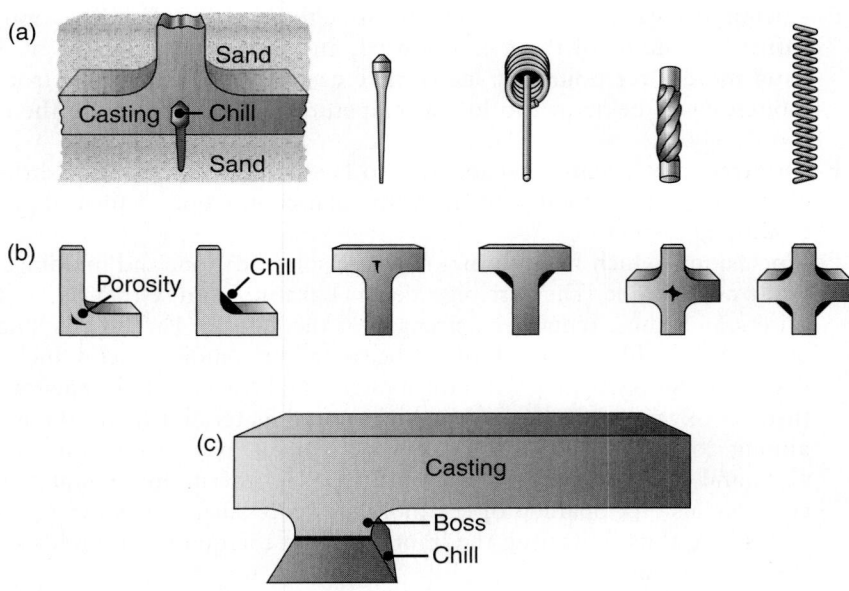

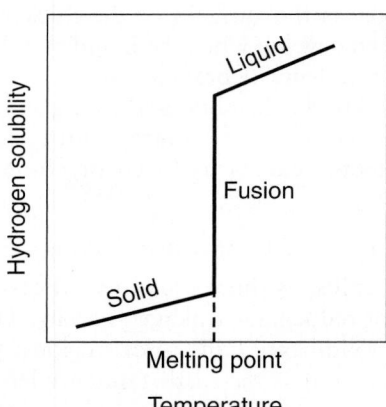

FIGURE 10.14 Various types of (a) internal and (b) external chills (dark areas at corners) used in castings to eliminate porosity caused by shrinkage. Chills are placed in regions where there is a larger volume of metal, as shown in (c).

FIGURE 10.15 Solubility of hydrogen in aluminum. Note the sharp decrease in solubility as the molten metal begins to solidify.

mold materials. Gases either accumulate in regions of existing porosity (such as in interdendritic regions) or cause microporosity in the casting, particularly in cast iron, aluminum, and copper. Dissolved gases may be removed from the molten metal by *flushing* or *purging* with an inert gas or by melting and pouring the metal in a vacuum. If the dissolved gas is oxygen, the molten metal can be *deoxidized*. Steel usually is deoxidized with aluminum, silicon, copper-based alloys with phosphorus copper, titanium, and zirconium-bearing materials.

Whether microporosity is a result of shrinkage or is caused by gases may be difficult to determine. If the porosity is spherical and has smooth walls (similar to the shiny holes in Swiss cheese), it is generally from gases. If the walls are rough and angular, porosity is likely from shrinkage between dendrites. Gross porosity is from shrinkage and usually is called a **shrinkage cavity**.

EXAMPLE 10.2 Casting of aluminum automotive pistons

Figure 10.16 shows an aluminum piston used in automotive internal-combustion engines. These products must be manufactured at very high rates with very tight dimensional tolerances and strict material requirements in order to achieve proper operation. Economic concerns are obviously paramount, and it is essential that pistons be produced with a minimum of expensive finishing operations and with few rejected parts.

Aluminum pistons are manufactured through casting because of its capability to produce near-net shaped parts at the required production rates. However, with poorly designed molds, underfills or excess porosity can cause parts to be rejected, adding to the cost. These defects were traditionally controlled through the use of large machining allowances coupled with the intuitive design of molds based on experience.

The pistons are produced from high-silicon alloys, such as 413.0 aluminum alloy. This alloy has high fluidity and can create high-definition surfaces through permanent mold casting; it also has high resistance to corrosion, good weldability, and low specific gravity. The universal acceptance of aluminum pistons for internal combustion engine applications is due mainly to their light weight and high thermal conductivity. Their low inertia allows for higher engine speeds and reduced counterweighting in the crankshaft, and the higher thermal conductivity allows for more efficient heat transfer from the engine.

The H13 tool-steel mold is preheated 200° to 450°C, depending on the cast alloy and part size. Initially, the preheat is achieved with a hand-held torch,

FIGURE 10.16   Aluminum piston for an internal combustion engine: (a) as-cast and (b) after machining.

but after a few castings, the mold reaches a steady–state temperature profile. The molten aluminum is heated to between 100° to 200°C above its liquidus temperature, and then a shot is placed into the infeed section of the mold. Once the molten metal shot is in place, a piston drives the metal into the mold. Because of the high thermal conductivity of the mold material, heat extraction from the molten metal is rapid, and the metal can solidify in small channels before filling the mold completely. Solidification usually starts at one end of the casting before the mold is filled fully.

As with most alloys, it is desired to begin solidification at one extreme end of the casting and have the solidification front proceed across the volume. This results in a directionally solidified microstructure and the elimination of gross porosity that arises when two solidification fronts meet inside a casting. Regardless, casting defects such as undercuts, hot spots, porosity, cracking, and entrapped air zone defects (such as blow holes and scabs) can occur. With poor mold design, 5% of the castings can be defective.

In order to improve the reliability and reduce the costs associated with permanent mold casting, computer-based modeling of mold filling takes place to suggest potential causes of defects. The computer models use the Bernoulli and continuity equations—coupled with heat transfer and solidification—to model the casting process and identify potential shortcomings. For example, Fig. 10.17 shows a result from a simulation of mold filling where a volume of entrapped air remains in the mold. This is corrected by placing a vent in the area of concern to allow air to escape during casting. Computer simulation allows designers to evaluate mold features and geometries before purchasing expensive tooling and has become an indispensible process to reduce costs and eliminate defects in casting.

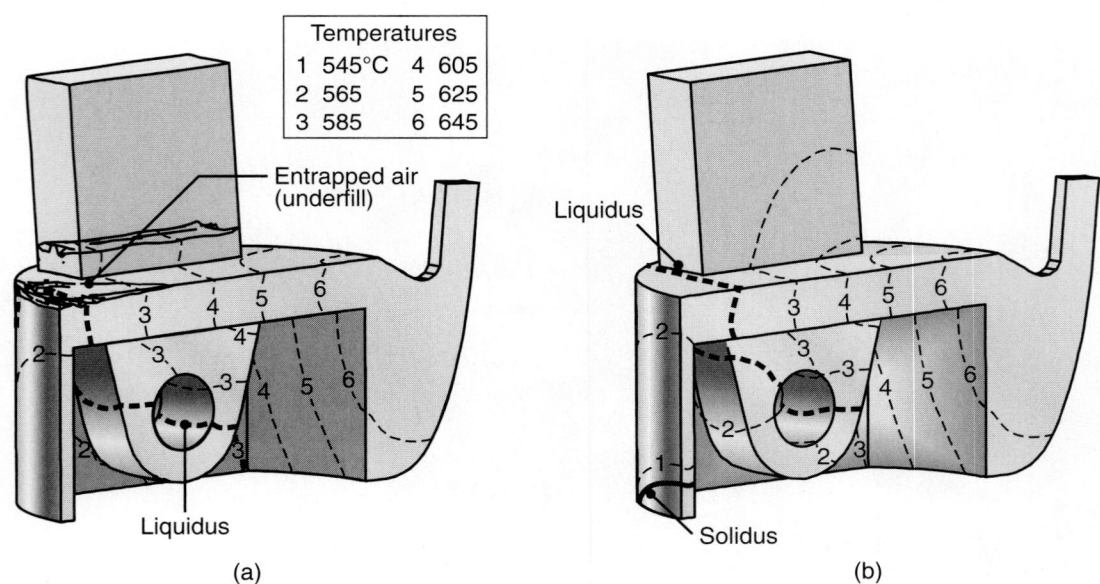

(a)   (b)

FIGURE 10.17   Simulation of mold filling and solidification. (a) 3.7 seconds after start of pour. Note that the mushy zone has been established before the mold is filled completely. (b) Using a vent in the mold for removal of entrapped air, 5 seconds after pour.

SUMMARY

- Casting is a solidification process in which molten metal is poured into a mold and allowed to cool. The metal may flow through a variety of passages (pouring basins, sprues, runners, risers, and gating systems) before reaching the final mold cavity. Bernoulli's theorem, the continuity law, and the Reynolds number are the analytical tools used in designing castings, with the goals of an appropriate flow rate and the elimination of defects associated with fluid flow.

- Solidification of pure metals takes place at a constant temperature, whereas solidification of alloys occurs over a range of temperatures. Phase diagrams are important tools for identifying the solidification point or points for technologically important metals.

- Composition and cooling rates of the molten metal affect the size and shape of the grains and the dendrites in the solidifying alloy. In turn, the size and structure of grains and dendrites influence properties of the solidified casting. Solidification time is a function of the volume of a casting and its surface area (Chvorinov's rule).

- The grain structure of castings can be controlled by various means to obtain the desired properties. Because metals contract during solidification and cooling, cavities can form in the casting. Porosity caused by gases evolved during solidification can be a significant problem, particularly because of its adverse effect on the mechanical properties of castings. Various defects also can develop in castings from lack of control of material and process variables.

- Although most metals shrink during solidification, gray cast iron and some aluminum alloys actually expand. Dimensional changes and cracking (hot tearing) are difficulties which can arise during solidification and cooling. Seven basic categories of casting defects have been classified.

- Melting practices have a direct effect on the quality of castings, as do foundry operations, such as pattern and mold making, pouring of the melt, removal of cast parts from molds, cleaning, heat treatment, and inspection.

KEY TERMS

Aspiration	Gate	Pouring basin
Bernoulli's theorem	Gating system	Reynolds number
Casting	Heterogeneous nucleation	Risers
Chills	Homogeneous nucleation	Runners
Columnar dendrites	Inoculant	Segregation
Columnar grains	Macrosegregation	Shrinkage
Cored dendrites	Microsegregation	Skin
Dendrites	Mold	Solidification
Fluidity	Mushy zone	Sprue
Freezing range	Porosity	Turbulence

BIBLIOGRAPHY

Alexiades, V., *Mathematical Modeling of Melting and Freezing Processes*, Hemisphere, 1993.

Analysis of Casting Defects, American Foundrymen's Society, 1974.

ASM Handbook, Vol. 15: *Casting*, ASM International, 1988.

Bradley, E.F., *High-Performance Castings: A Technical Guide*, Edison Welding Institute, 1989.

Campbell, J., *Castings*, Butterworth-Heinemann, 1991.

Casting in *Tool and Manufacturing Engineers Handbook*, Volume II: *Forming*, Society of Manufacturing Engineers, 1984.

Casting Defects Handbook, American Foundrymen's Society, 1972.

Flemings, M.C., *Solidification Processing*, McGraw-Hill, 1974.

Heine, R., *Principles of Metal Casting*, McGraw-Hill, 1999.

Liebermann, H.H. (ed.), *Rapidly Solidified Alloys*, Marcel Dekker, 1993.

Mikelonis, P.J. (ed.), *Foundry Technology: A Source Book*, ASM International, 1982.

Minkoff, I., *Solidification and Cast Structure*, Wiley, 1986.

Steel Castings Handbook, 6th ed., Steel Founders' Society of America, 1995.

Szekely, J., *Fluid Flow Phenomena in Metals Processing*, Academic Press, 1979.

REVIEW QUESTIONS

10.1 Why is casting an important manufacturing process?

10.2 What is the difference between the solidification of pure metals and metal alloys?

10.3 What are dendrites?

10.4 State the difference between short and long freezing ranges. How is range determined?

10.5 What is superheat?

10.6 Define shrinkage and porosity. How can you tell whether cavities in a casting are due to porosity or to shrinkage?

10.7 What is the function of chills?

10.8 What is the Reynolds number? Why is it important in casting?

10.9 How is fluidity defined? Why is it important?

10.10 Explain the reasons for hot tearing in castings.

10.11 Why is it important to remove dross or slag during the pouring of molten metal into the mold? What methods are used to remove them?

10.12 What are the effects of mold materials on fluid flow and heat transfer?

10.13 Why is Bernoulli's equation important in casting?

10.14 Describe thixocasting and rheocasting.

QUALITATIVE PROBLEMS

10.15 Describe the stages involved in the contraction of metals during casting.

10.16 Explain the reasons why heat transfer and fluid flow are important in metal casting.

10.17 We know that pouring metal at a high rate into a mold has certain disadvantages. Are there any disadvantages to pouring it very slowly?

10.18 Describe the events depicted in Fig. 10.5.

10.19 Would you be concerned about the fact that parts of internal chills are left within the casting? What materials do you think chills should be made of and why?

10.20 What practical demonstrations can you offer to indicate the relationship of the solidification time to the volume and surface area?

10.21 Explain why you may want to subject a casting to various heat treatments.

10.22 Why does porosity have detrimental effects on the mechanical properties of castings? Would physical properties (such as thermal and electrical conductivity) also be affected by porosity? Explain.

10.23 A spoked handwheel is to be cast in gray iron. In order to prevent hot tearing of the spokes, would you insulate the spokes or chill them? Explain.

10.24 Which of the following considerations is/are important for a riser to function properly? Must it: (a) have a surface area larger than the part being cast, (b) be kept open to atmospheric pressure, and/or (c) solidify first? Why?

10.25   Explain why the constant C in Eq. (10.7) depends on mold material, metal properties, and temperature.

10.26   Are external chills as effective as internal chills? Explain.

10.27   Explain why gray cast iron undergoes expansion rather than contraction during solidification, as shown in Table 10.1.

10.28   Referring to Fig. 10.11, explain why internal corners (such as A) develop a thinner skin than external corners (such as B) during solidification.

10.29   Note the shape of the two risers in Fig. 10.8, and discuss your observations with respect to Eq. (10.7).

10.30   Is there any difference in the tendency for shrinkage void formation in metals with short and long freezing ranges, respectively? Explain.

10.31   What is the influence of the cross-sectional area of the spiral channel in Fig. 10.9 on fluidity test results? What is the effect of sprue height? If this test is run with the test setup heated to elevated temperatures, would the test results be more useful? Explain.

10.32   It has long been observed by foundrymen and ingot casters that low pouring temperatures (i.e., low superheat) promote the formation of equiaxed grains over columnar grains. Also, equiaxed grains become finer as the pouring temperature decreases. Explain these phenomena.

10.33   What would you expect to occur (in casting metal alloys) if the mold was agitated aggressively (vibrated) after the molten metal had been in the mold for a sufficient amount of time to form a skin?

10.34   If you examine a typical ice cube, you will see pockets and cracks in the cube. However, some ice cubes are tubular in shape and do not have noticeable air pockets or cracks in their structure. Explain this phenomenon.

10.35   How can you tell whether cavities in a casting are due to shrinkage or entrained air bubbles?

10.36   Describe the drawbacks to having a riser that is (a) too large and (b) too small.

10.37   What are the benefits and drawbacks to having a pouring temperature that is much higher than the metal's melting temperature? What are the advantages and disadvantages in having the pouring temperature remain close to the melting temperature?

QUANTITATIVE PROBLEMS

10.38   Sketch a graph of specific volume versus temperature for a metal that shrinks as it cools from the liquid state to room temperature. On the graph, mark the area where shrinkage is compensated for by risers.

10.39   A round casting is 0.2 m (7.9 in.) in diameter and 0.5 m (19.7 in.) in length. Another casting of the same metal is elliptical in cross-section with a major-to-minor axis ratio of 2 and has the same length and cross-sectional area as the round casting. Both pieces are cast under the same conditions. What is the difference in the solidification times of the two castings?

10.40   A 100-mm (4-in.) thick square plate and a right circular cylinder with a radius of 100 mm (4 in.) and a height of 50 mm each have the same volume. If each is to be cast using a cylindrical riser, will each part require the same size riser to ensure proper feeding? Explain.

10.41   Assume that the top of a round sprue has a diameter of 3 in. (75 mm) and is at a height of 8 in. (200 mm) from the runner. Based on Eq. (10.5), plot the profile of the sprue diameter as a function of its height. Assume that the sprue wall has a diameter of 0.25 in. (6 mm) at the bottom.

10.42   Pure aluminum is poured into a sand mold. The metal level in the pouring basin is 8 in. above the metal level in the mold, and the runner is circular with a 0.5-in. diameter. What is the velocity and rate of the flow of the metal into the mold? Is the flow turbulent or laminar?

10.43   A cylinder with a diameter of 1 in. and height of 3 in. solidifies in three minutes in a sand casting operation. What is the solidification time if the cylinder height is doubled? What is the time if the diameter is doubled?

10.44   The volume flow rate of metal into a mold is 0.01 m^3/s. The top of the sprue has a diameter of 20 mm, and its length is 200 mm. What diameter should be specified at the bottom of the sprue to prevent aspiration? What is the resultant velocity and Reynolds number at the bottom of the sprue if the metal being cast is aluminum with a viscosity of 0.004 Ns/m^2?

10.45   A rectangular mold with dimensions 100 mm × 200 mm × 400 mm is filled with aluminum with no superheat. Determine the final dimensions of the part as it cools to room temperature. Repeat the analysis for gray cast iron.

10.46 The constant C in Chvorinov's rule is given as 3 s/mm^2 and is used to produce a cylindrical casting with a diameter of 75 mm and height of 125 mm. Estimate the time for the casting to fully solidify. The mold can be broken safely when the solidified shell is at least 20 mm. Assuming the cylinder cools evenly, how much time must transpire after pouring the molten metal before the mold can be broken?

10.47 Assume that you are an instructor covering the topics described in this chapter, and you are giving a quiz on the numerical aspects to test the understanding of the students. Prepare two quantitative problems and supply the answers.

SYNTHESIS, DESIGN, AND PROJECTS

10.48 Can you devise fluidity tests other than that shown in Fig. 10.9? Explain the features of your test methods.

10.49 Figure P10.49 indicates various defects and discontinuities in cast products. Review each one and offer solutions to avoid them.

10.50 The fluidity test shown in Fig. 10.9 only illustrates the principle of this test. Design a setup for such a test, showing the type of materials and the equipment to be used. Explain the method by which you would determine the length of the solidified metal in the spiral passage.

10.51 Utilizing the equipment and materials available in a typical kitchen, design an experiment to reproduce results similar to those shown in Fig. 10.11. Comment on your observations.

10.52 One method of relieving stress concentrations in a part is to apply a small, uniform plastic deformation to it. Make a list of your concerns and recommendations if such an approach is suggested for a casting.

10.53 If a casting of a given shape is to be doubled in volume, describe the effects on mold design, including the required change in the size of the risers, runners, chokes, and sprues.

10.54 Small amounts of slag often persist after skimming and are introduced into the molten-metal flow in casting. Recognizing that the slag is much less dense than the metal, design mold features that will remove small amounts of slag before the metal reaches the mold cavity.

10.55 Figure II.1 shows a variety of components in a typical automobile that are produced by casting. Think of other products, such as power tools and small appliances, and prepare a similar illustration as done in that figure.

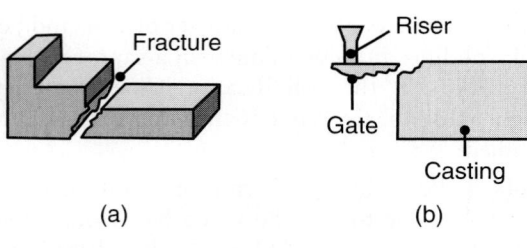  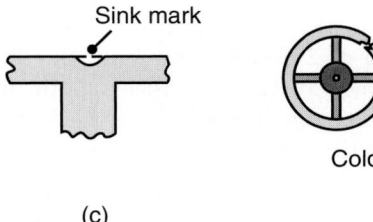 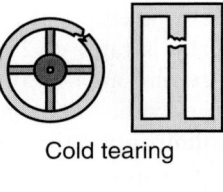

(a) (b) (c) (d)

FIGURE P10.49

Metal-Casting Processes

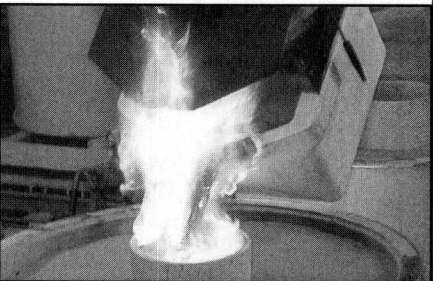

Following a description of the fundamentals of solidification of metals in the preceding chapter and the roles of fluid flow and heat transfer in molds, we now describe in detail:

- Characteristics of expendable-mold and permanent-mold processes.
- Applications, advantages, and limitations of common casting processes.
- Casting of single crystals.
- Inspection techniques for castings.
- Brief review of foundries and their automation.

Typical products made by casting: engine blocks, crankshafts, hubcaps, power tools, turbine blades, plumbing, zipper teeth, dies and molds, gears, railroad wheels, propellers, office equipment, statues, and housings.

Alternative processes: forging, powder metallurgy, machining, and fabrication.

11.1 | Introduction

The first metal castings were made during the period from 4000 to 3000 B.C., using stone and metal molds for casting copper. Various casting processes have been developed over time, each with its own characteristics and applications (see also Fig. I.7a), to meet specific engineering and service requirements (Table 11.1). A large variety of parts and components are made by casting, such as engine blocks, crankshafts, automotive components and powertrains (Fig. 11.1), agricultural and railroad equipment, pipes and plumbing fixtures, power tools, gun barrels, frying pans, office equipment, and very large components for hydraulic turbines.

Two trends have had a major impact on the casting industry. The first is the mechanization and automation of the casting process, which has led to significant changes in the use of equipment and labor. Advanced machinery and automated process-control systems have replaced traditional methods of casting. The second major trend has been the increasing demand for high-quality castings with close dimensional tolerances.

This chapter is organized around the major classifications of casting practices (see Fig. II.2 in the Introduction to Part II). These classifications are related to mold materials, molding processes, and methods of feeding the mold with molten metal.

TABLE 11.1

Summary of Casting Processes		
Process	Advantages	Limitations
Sand	Almost any metal cast; no limit to part size, shape, or weight; low tooling cost	Some finishing required; relatively coarse surface finish; wide tolerances
Shell mold	Good dimensional accuracy and surface finish; high production rate	Part size limited; expensive patterns and equipment
Evaporative pattern	Most metals cast with no limit to size; complex part shapes	Patterns have low strength and can be costly for low quantities
Plaster mold	Intricate part shapes; good dimensional accuracy and surface finish; low porosity	Limited to nonferrous metals; limited part size and volume of production; mold-making time relatively long
Ceramic mold	Intricate part shapes; close-tolerance parts; good surface finish.	Limited part size
Investment	Intricate part shapes; excellent surface finish and accuracy; almost any metal cast	Part size limited; expensive patterns, molds, and labor
Permanent mold	Good surface finish and dimensional accuracy; low porosity; high production rate	High mold cost; limited part shape and complexity; not suitable for high-melting-point metals
Die	Excellent dimensional accuracy and surface finish; high production rate	High die cost; limited part size; generally limited to nonferrous metals; long lead time
Centrifugal	Large cylindrical or tubular parts with good quality; high production rate	Expensive equipment; limited part shape

The major categories are as follows:

1. **Expendable molds,** which typically are made of sand, plaster, ceramics, and similar materials and generally are mixed with various binders (*bonding agents*) for improved properties. A typical sand mold consists of 90% sand, 7% clay, and 3% water. As described in Chapter 8, these materials are *refractories* (that is, they are capable of withstanding the high temperatures of molten metals). After the casting has solidified, the mold is broken up to remove the casting.

2. **Permanent molds,** which are made of metals that maintain their strength at high temperatures. As the name implies, they are used repeatedly and are designed in such a way that the casting can be removed easily and the mold used for the next casting. Metal molds are better heat conductors than expendable nonmetallic molds (see Table 3.1); hence, the solidifying casting is subjected to a higher rate of cooling, which in turn affects the microstructure and grain size within the casting.

3. **Composite molds,** which are made of two or more different materials (such as sand, graphite, and metal) combining the advantages of each material. These molds have a permanent and an expendable portion and are used in various casting processes to improve mold strength, control the cooling rates, and optimize the overall economics of the casting process.

The general characteristics of sand casting and other casting processes are given in Table 11.2. Almost all commercially used metals can be cast. The surface finish obtained is largely a function of the mold material; although, as expected, sand castings generally have rough, grainy surfaces. Dimensional tolerances generally are not as good as those in machining and other net-shape processes. However, intricate shapes can be made by casting, such as cast-iron engine blocks and very large propellers for ocean liners.

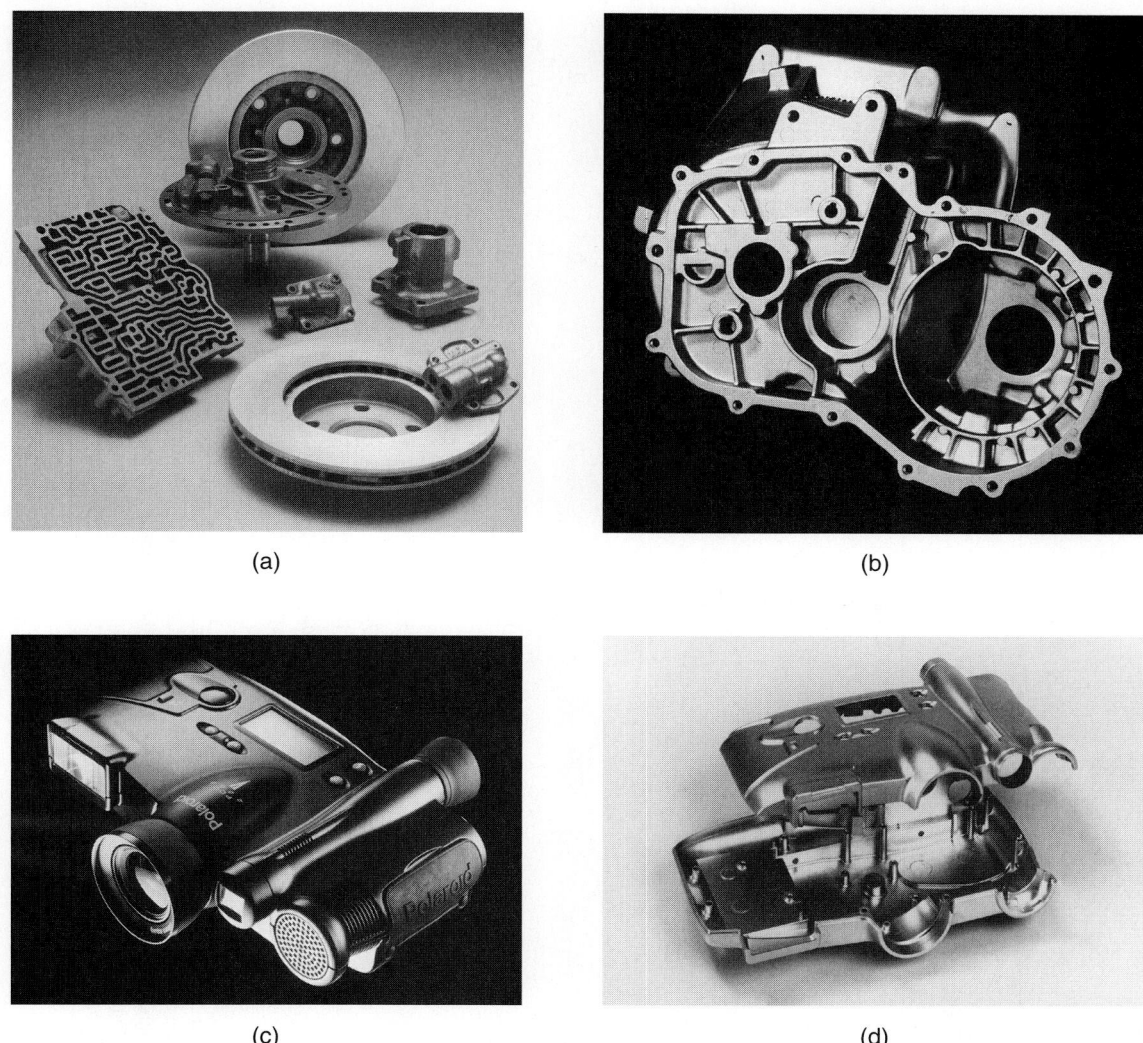

(a)

(b)

(c)

(d)

FIGURE 11.1   (a) Typical gray-iron castings used in automobiles, including the transmission valve body (left) and the hub rotor with disk-brake cylinder (front). *Source:* Courtesy of Central Foundry Division of General Motors Corporation. (b) A cast transmission housing. (c) The Polaroid PDC-2000 digital camera with a AZ191D die-cast, high-purity magnesium case. (d) A two-piece Polaroid camera case made by the hot-chamber die-casting process. *Source:* Courtesy of Polaroid Corporation and Chicago White Metal Casting, Inc.

Because of their unique characteristics and applications, particularly in manufacturing microelectronic devices (Part V), basic crystal-growing techniques also are described in this chapter, which concludes with a brief overview of modern foundries.

11.2 | Expendable-Mold Casting Processes

The major categories of expendable-mold casting are sand, shell mold, plaster mold, ceramic mold, evaporative pattern, and investment casting.

TABLE 11.2

General Characteristics of Casting Processes

	Sand	Shell	Evaporative pattern	Plaster	Investment	Permanent mold	Die	Centrifugal
Typical materials cast	All	All	All	Nonferrous (Al, Mg, Zn, Cu)	All	All	Nonferrous (Al, Mg, Zn, Cu)	All
Weight (kg):								
minimum	0.01	0.01	0.01	0.01	0.001	0.1	<0.01	0.01
maximum	No limit	100+	100+	50+	100+	300	50	5000+
Typ. surface finish (R_a in μm)[1]	5–25	1–3	5–25	1–2	0.3–2	2–6	1–2	2–10
Porosity[1]	3–5	4–5	3–5	4–5	5	2–3	1–3	1–2
Shape complexity[1]	1–2	2–3	1–2	1–2	1	2–3	3–4	3–4
Dimensional accuracy[1]	3	2	3	2	1	1	1	3
Section thickness (mm):								
Minimum	3	2	2	1	1	2	0.5	2
Maximum	No limit	—	—	—	75	50	12	100
Typ. dimensional tolerance (mm/mm)	1.6–4 mm (0.25 mm for small parts)	±0.003		±0.005–0.010	±0.005	±0.015	±0.001–0.005	0.015
Cost[1,2]								
Equipment	3–5	3	2–3	3–5	3–5	2	1	1
Pattern/die	3–5	2–3	2–3	3–5	2–3	2	1	1
Labor	1–3	3	3	1–2	1–2	3	5	5
Typical lead time[2]	Days	Weeks	Weeks	Days	Weeks	Weeks	Weeks–months	Months
Typical production rate[2] (parts/mold-hour)	1–20	5–50	1–20	1–10	1–1000	5–50	2–200	1–1000
Minimum quantity[2]	1	100	500	10	10	1000	10,000	10–10,000

Notes: 1. Relative rating, from 1 (best) to 5 (worst). For example, die casting has relatively low porosity, mid to low shape complexity, high dimensional accuracy, high equipment and die costs, and low labor costs. These ratings are only general; significant variations can occur, depending on the manufacturing methods used.

2. Approximate values without the use of rapid prototyping technologies.

Source: Data taken from J. A. Schey, *Introduction to Manufacturing Processes*, 3d. ed., McGraw-Hill 2000.

11.2.1 Sand casting

The traditional method of casting metals is in sand molds and has been used for millennia. Sand casting is still the most prevalent form of casting; in the United States alone, about 15 million tons of metal are cast by this method each year. Typical applications of sand casting include machine bases, large turbine impellers, propellers, plumbing fixtures, and numerous components for agricultural and railroad equipment. The capabilities of sand casting are given in Table 11.2.

Basically, *sand casting* consists of (a) placing a pattern (having the shape of the desired casting) in sand to make an imprint, (b) incorporating a gating system, (c) removing the pattern and filling the mold cavity with molten metal, (d) allowing the metal to cool until it solidifies, (e) breaking away the sand mold, and (f) removing the casting (Fig. 11.2).

Sands. Most sand-casting operations use silica sand (SiO_2) as mold material. Sand is inexpensive and is suitable as mold material because of its high-temperature characteristics and high melting point. There are two general types of sand: **naturally bonded** (*bank sand*) and **synthetic** (*lake sand*). Because its composition can be controlled more accurately, synthetic sand is preferred by most foundries. For proper functioning, mold sand must be clean and preferably new.

Several factors are important in the selection of sand for molds, and it involves certain tradeoffs with respect to properties. Sand having fine, round grains can be packed closely and, thus, forms a smooth mold surface. Although fine-grained sand enhances mold strength, the fine grains also lower mold *permeability* (penetrating through pores). Good permeability of molds and cores allows gases and steam evolved during the casting to escape easily. The mold also should have good *collapsibility* to allow for the casting to shrink while cooling and, thus, to avoid defects in the casting, such as hot tearing and cracking (see Fig. 10.12).

Types of sand molds. Sand molds (Fig. 11.3) are characterized by the types of sand that comprise them and by the methods used to produce them. There are three

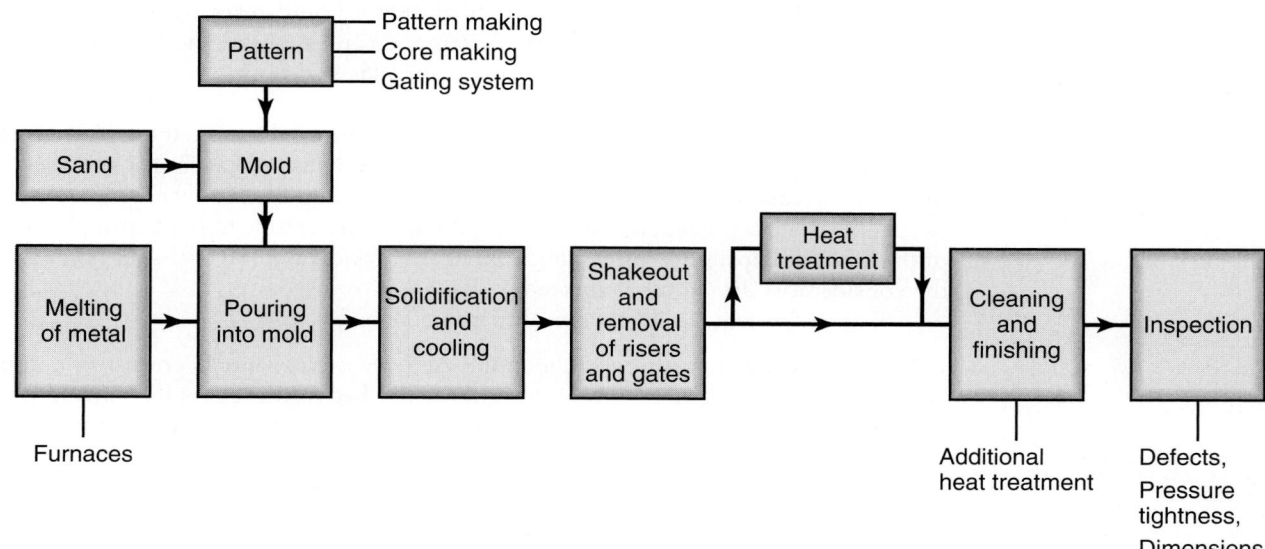

FIGURE 11.2 Outline of production steps in a typical sand-casting operation.

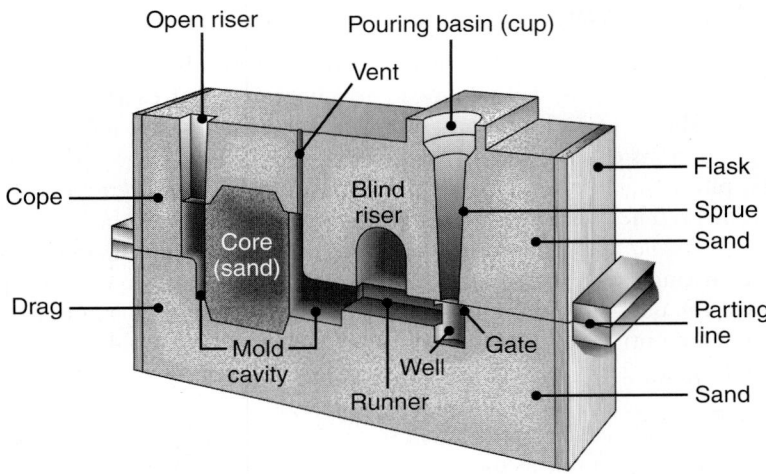

FIGURE 11.3 Schematic illustration of a sand mold, showing various features.

basic types of sand molds: green-sand, cold-box, and no-bake molds. The most common mold material is **green molding sand**, which is a mixture of sand, clay, and water. The term "green" refers to the fact that the sand in the mold is moist or damp while the metal is being poured into it. Green-sand molding is the least expensive method of making molds, and the sand is recycled easily for subsequent use. In the **skin-dried** method, the mold surfaces are dried, either by storing the mold in air or by drying it with torches. Because of their higher strength, these molds generally are used for large castings

In the **cold-box mold** process, various organic and inorganic *binders* are blended into the sand to bond the grains chemically for greater strength. These molds are more accurate dimensionally than green-sand molds but are more expensive. In the **no-bake mold** process, a synthetic liquid resin is mixed with the sand, and the mixture hardens at room temperature. Because the bonding of the mold in this and in the cold-box process takes place without heat, they are called **cold-setting processes.**

Sand molds are oven dried (*baked*) prior to pouring the molten metal; they are stronger than green-sand molds and impart better dimensional accuracy and surface finish to the casting. However, this method has the following drawbacks: (a) distortion of the mold is greater, (b) the castings are more susceptible to hot tearing because of the lower collapsibility of the mold, and (c) the production rate is lower because of the considerable drying time required.

The major features of sand molds are described next.

1. The **flask**, which supports the mold itself. Two-piece molds consist of a **cope** on top and a **drag** on the bottom; the seam between them is the *parting line*. When more than two pieces are used in a sand mold, the additional parts are called *cheeks*.
2. A **pouring basin** or **pouring cup**, into which the molten metal is poured.
3. A **sprue**, through which the molten metal flows downward.
4. The **runner system**, which has channels that carry the molten metal from the sprue to the mold cavity. **Gates** are the inlets into the mold cavity.

5. **Risers,** which supply additional molten metal to the casting as it shrinks during solidification. Two types of risers, a *blind riser* and an *open riser*, are shown in Fig. 11.3.

6. **Cores,** which are inserts made from sand. They are placed in the mold to form hollow regions or otherwise define the interior surface of the casting. Cores also are used on the outside of the casting to form features such as lettering on the surface of a casting or deep external pockets.

7. **Vents,** which are placed in molds to carry off gases produced when the molten metal comes into contact with the sand in the mold and the core. Vents also exhaust air from the mold cavity as the molten metal flows into the mold.

Patterns.   *Patterns* are used to mold the sand mixture into the shape of the casting and may be made of wood, plastic, or metal. The selection of a pattern material depends on the size and shape of the casting, the dimensional accuracy and the quantity of castings required, and the molding process. Because patterns are used repeatedly to make molds, the strength and durability of the material selected for a pattern must reflect the number of castings that the mold will produce. Patterns may be made of a combination of materials to reduce wear in critical regions, and they usually are coated with a **parting agent** to facilitate the removal of the casting from the molds.

Patterns can be designed with a variety of features to fit specific applications and economic requirements. **One-piece patterns**, also called *loose* or *solid patterns*, generally are used for simpler shapes and low-quantity production; they generally are made of wood and are inexpensive. **Split patterns** are two-piece patterns, made such that each part forms a portion of the cavity for the casting; in this way, castings with complicated shapes can be produced. **Match-plate patterns** are a common type of mounted pattern in which two-piece patterns are constructed by securing each half of one or more split patterns to the opposite sides of a single plate (Fig. 11.4). In such constructions, the gating system can be mounted on the drag side of the pattern. This type of pattern is used most often in conjunction with molding machines and large production runs to produce smaller castings.

An important development in molding and pattern making is the application of **rapid prototyping** (Chapter 20). In sand casting, for example, a pattern can be fabricated in a rapid prototyping machine and fastened to a backing plate at a fraction of the time and cost of machining a pattern. There are several rapid prototyping techniques with which these tools can be produced quickly.

Pattern design is a critical aspect of the total casting operation. The design should provide for **metal shrinkage**, ease of removal from the sand mold by means of a taper or draft (Fig. 11.5), and proper metal flow in the mold cavity. These topics are described in greater detail in Chapter 12.

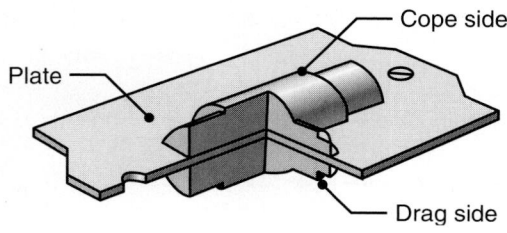

FIGURE 11.4   A typical metal match-plate pattern used in sand casting.

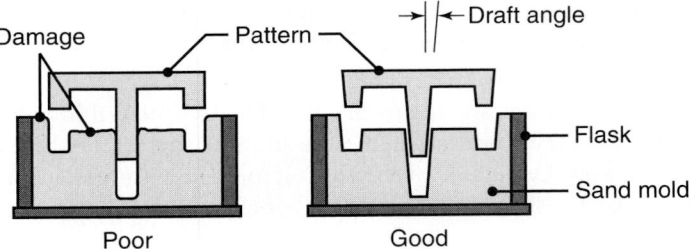

FIGURE 11.5 Taper on patterns for ease of removal from the sand mold.

Cores. For castings with internal cavities or passages, such as those found in an automotive engine block or a valve body, *cores* are utilized. Cores are placed in the mold cavity to form the interior surfaces of the casting and are removed from the finished part during shakeout and further processing. Like molds, cores must possess strength, permeability, ability to withstand heat, and collapsibility; hence, cores are made of sand aggregates. The core is anchored by **core prints,** which are recesses added to the pattern to support the core and to provide vents for the escape of gases (Fig. 11.6). A common problem with cores is that (for some casting requirements, as in the case where a recess is required) they may lack sufficient structural support in the cavity. To keep the core from shifting, metal supports (**chaplets**) may be used to anchor the core in place (Fig. 11.6b).

Cores generally are made in a manner similar to that used in moldmaking; the majority are made with shell (see Section 11.2.2), no-bake, or cold-box processes. Cores are shaped in *core boxes*, which are used in much the same way that patterns are used to form sand molds.

Sand-molding machines. The oldest known method of molding, which is still used for simple castings, is to compact the sand by hand hammering (*tamping*) or ramming it around the pattern. For most operations, however, the sand mixture is compacted around the pattern by *molding machines*. These machines eliminate arduous labor, offer high-quality casting by improving the application and distribution of forces, manipulate the mold in a carefully controlled manner, and increase production rate.

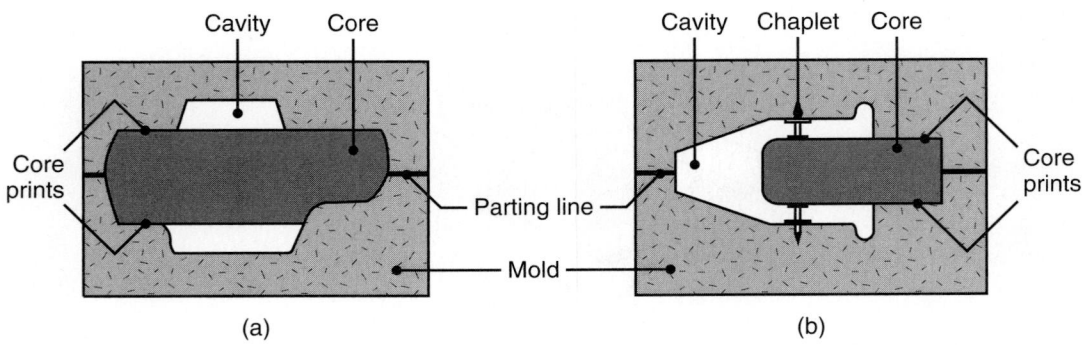

FIGURE 11.6 Examples of sand cores showing core prints and chaplets to support cores.

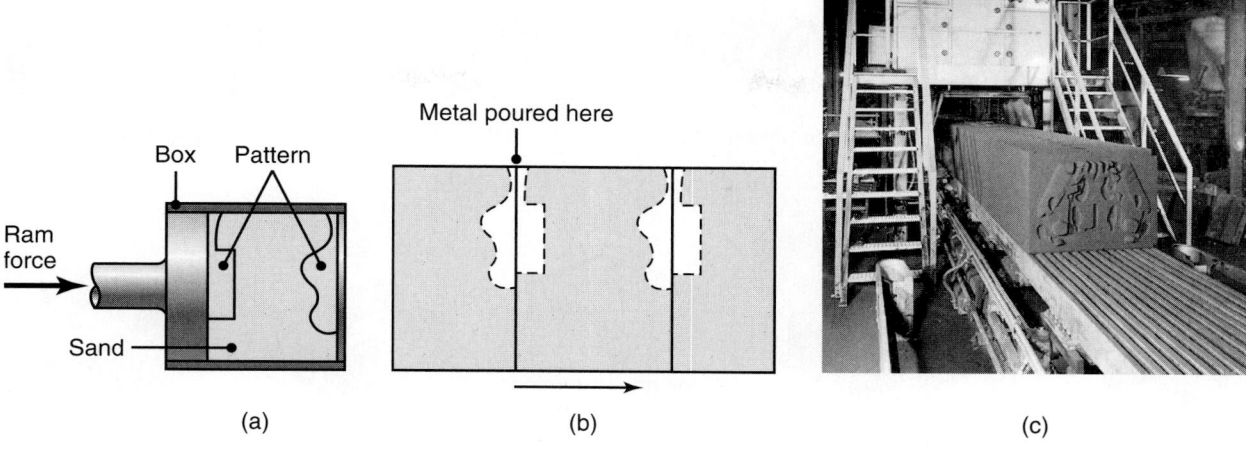

Box Pattern

Metal poured here

Ram
force

Sand

(a)   (b)   (c)

FIGURE 11.7   Vertical flaskless molding. (a) Sand is squeezed between two halves of the pattern. (b) Assembled molds pass along an assembly line for pouring. (c) A photograph of a vertical flaskless molding line. *Source:* Courtesy of American Foundry Society.

In **vertical flaskless molding**, the halves of the pattern form a vertical chamber wall against which sand is blown and compacted (Fig. 11.7). Then the mold halves are packed horizontally with the parting line oriented vertically and moved along a pouring conveyor. This operation is simple and eliminates the need to handle flasks, allowing for very high-production rates, particularly when other aspects of the operation (such as coring and pouring) are automated.

Sandslingers fill the flask uniformly with sand under a high-pressure stream; they are used to fill large flasks and are operated typically by machine. An impeller in the machine throws sand from its blades (or cups) at such high speeds that the machine not only places the sand but also rams it appropriately.

In **impact molding**, the sand is compacted by a controlled explosion or instantaneous release of compressed gases. This method produces molds with uniform strength and good permeability.

In **vacuum molding** (also known as the *V process*), the pattern is covered tightly with a thin sheet of plastic. A flask is placed over the coated pattern and is filled with dry, binderless sand. A second sheet of plastic then is placed on top of the sand, and a vacuum action compacts the sand so that the pattern can be withdrawn. Both halves of the mold are made this way and assembled. During pouring, the mold remains under a vacuum but the casting cavity does not. When the metal has solidified, the vacuum is turned off, and the sand falls away, releasing the casting. Vacuum molding produces castings with high-quality surface detail and dimensional accuracy; it is suited especially well for large, relatively flat (plane) castings.

The sand-casting operation.   After the mold has been shaped and the cores have been placed in position, the two halves (cope and drag) are closed, clamped, and weighted down to prevent the separation of the mold sections under the pressure exerted when the molten metal is poured into the mold cavity. A complete sequence of operations in sand casting is shown in Fig. 11.8.

After solidification, the casting is shaken out of its mold, and the sand and oxide layers adhering to the casting are removed by vibration (using a shaker) or by sand blasting. Castings also are cleaned by blasting with steel shot or grit

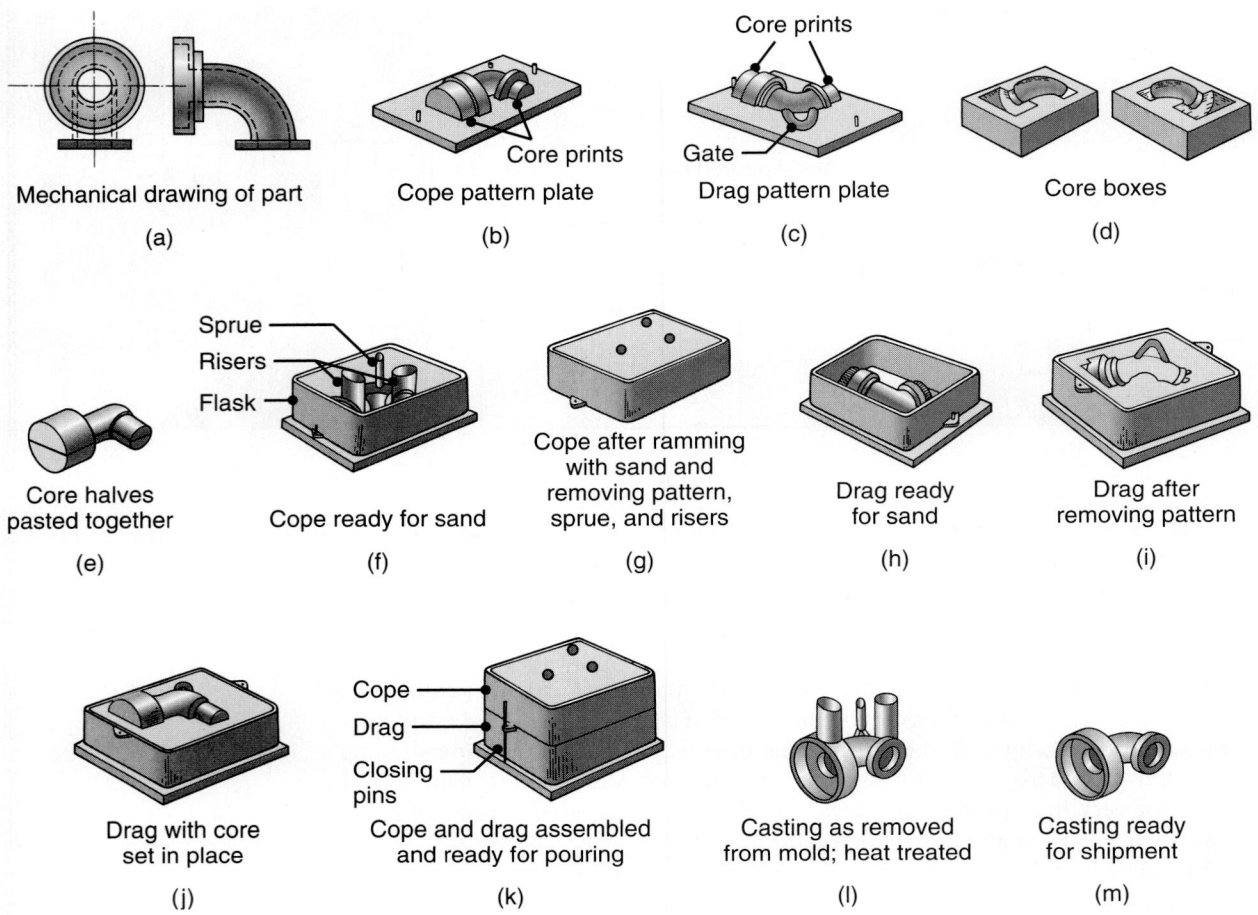

Mechanical drawing of part
(a)

Cope pattern plate
Core prints
(b)

Core prints
Gate
Drag pattern plate
(c)

Core boxes
(d)

Core halves pasted together
(e)

Sprue
Risers
Flask
Cope ready for sand
(f)

Cope after ramming with sand and removing pattern, sprue, and risers
(g)

Drag ready for sand
(h)

Drag after removing pattern
(i)

Drag with core set in place
(j)

Cope
Drag
Closing pins
Cope and drag assembled and ready for pouring
(k)

Casting as removed from mold; heat treated
(l)

Casting ready for shipment
(m)

FIGURE 11.8 Schematic illustration of the sequence of operations for sand casting. (a) A mechanical drawing of the part is used to generate a design for the pattern. Considerations such as part shrinkage and draft must be built into the drawing. (b–c) Patterns have been mounted on plates equipped with pins for alignment. Note the presence of core prints designed to hold the core in place. (d–e) Core boxes produce core halves, which are pasted together. The cores will be used to produce the hollow area of the part shown in (a). (f) The cope half of the mold is assembled by securing the cope pattern plate to the flask with aligning pins and attaching inserts to form the sprue and risers. (g) The flask is rammed with sand and the plate and inserts are removed. (h) The drag half is produced in a similar manner with the pattern inserted. A bottom board is placed below the drag and aligned with pins. (i) The pattern, flask, and bottom board are inverted; and the pattern is withdrawn, leaving the appropriate imprint. (j) The core is set in place within the drag cavity. (k) The mold is closed by placing the cope on top of the drag and securing the assembly with pins. The flasks then are subjected to pressure to counteract buoyant forces in the liquid, which might lift the cope. (l) After the metal solidifies, the casting is removed from the mold. (m) The sprue and risers are cut off and recycled, and the casting is cleaned, inspected, and heat treated (when necessary). *Source*: Courtesy of Steel Founders' Society of America.

(*shot blasting*; Section 26.8). The risers and gates are cut off by oxyfuel-gas cutting, sawing, shearing, and abrasive wheels; or they are trimmed in dies. Gates and risers on steel castings also may be removed with air carbon-arc (Section 30.8) or torches. Castings may be cleaned further by electrochemical means or by pickling with chemicals to remove surface oxides.

The casting subsequently may be *heat treated* to improve certain properties required for its intended service use; these processes are important, particularly for steel castings. *Finishing operations* may involve machining, straightening, or forging with dies (sizing) to obtain final dimensions. *Inspection* is an important final step and is carried out to ensure that the casting meets all design and quality-control requirements.

Rammed-graphite molding.   In this process, rammed graphite (Section 8.6) is used to make molds for casting reactive metals, such as titanium and zirconium. Sand cannot be used because these metals react vigorously with silica. The molds are packed like sand molds, air dried, baked at 175°C (350°F), fired at 870°C (1600°F), and then stored under controlled humidity and temperature. The casting procedures are similar to those for sand molds.

11.2.2 Shell molding

Shell molding first was developed in the 1940s and has grown significantly because it can produce many types of castings with close dimensional tolerances and a good surface finish at low cost. Shell-molding applications include small mechanical parts requiring high precision, such as gear housings, cylinder heads, and connecting rods. The process also is used widely in producing high-precision molding cores. The capabilities of shell-mold casting are given in Table 11.2.

In this process, a mounted pattern made of a ferrous metal or aluminum is (a) heated to a range of 175° to 370°C (350° to 700°F), (b) coated with a parting agent (such as silicone), and (c) clamped to a box or chamber. The box contains fine sand, mixed with 2.5 to 4% of a thermosetting resin binder (such as phenol-formaldehyde) that coats the sand particles. Either the box is rotated upside down (Fig. 11.9) or the sand mixture is blown over the pattern, allowing it to coat the pattern.

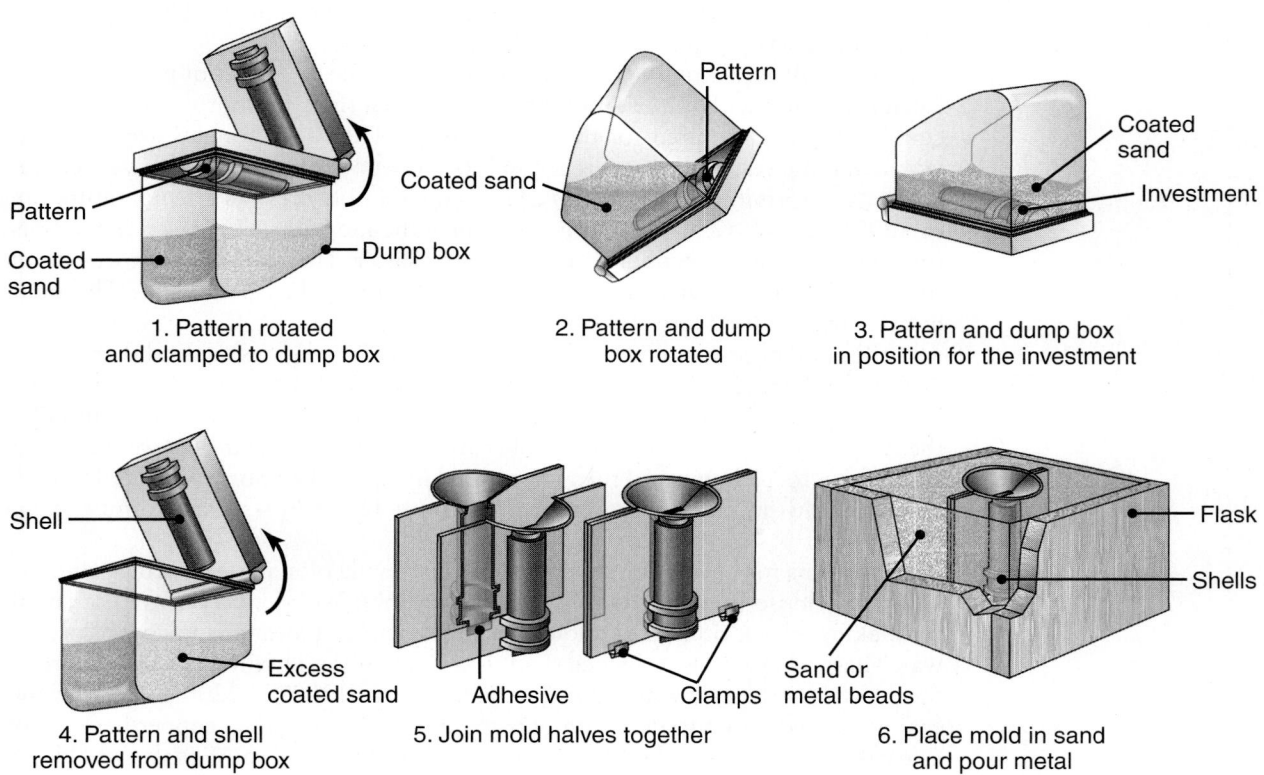

1. Pattern rotated and clamped to dump box

2. Pattern and dump box rotated

3. Pattern and dump box in position for the investment

4. Pattern and shell removed from dump box

5. Join mold halves together

6. Place mold in sand and pour metal

FIGURE 11.9   The shell-molding process, also called the *dump-box* technique.

The assembly then is placed in an oven for a short period of time to complete the curing of the resin. In most shell-molding machines, the oven consists of a metal box with gas-fired burners that swing over the shell mold to cure it. The shell hardens around the pattern and is removed from the pattern using built-in ejector pins. Two half-shells are made in this manner and are bonded or clamped together to form a mold.

The thickness of the shell can be determined accurately by controlling the time that the pattern is in contact with the mold. In this way, the shell can be formed with the required strength and rigidity to hold the weight of the molten liquid. The shells are light and thin (usually 5 to 10 mm, 0.2 to 0.4 in.), and consequently, their thermal characteristics are different from those for thicker molds.

Shell sand has a much lower permeability than the sand used for green-sand molding, because a sand of much smaller grain size is used for shell molding. The decomposition of the shell-sand binder also produces a high volume of gas. Consequently, unless the molds are vented properly, trapped air and gas can cause serious problems in the shell molding of ferrous castings. The high quality of the finished casting can reduce cleaning, machining, and other finishing costs significantly. Complex shapes can be produced with less labor, and the process can be automated fairly easily.

11.2.3 Plaster-mold casting

This process, and the ceramic-mold and investment casting processes described in Sections 11.2.4 and 11.2.6, are known as **precision casting** because of the high dimensional accuracy and good surface finish obtained. Typical parts made are lock components, gears, valves, fittings, tooling, and ornaments. The castings usually weigh less than 10 kg (22 lb) and are typically in the range of 125 to 250 g (1/4 to 1/2 lb), although parts as light as 1 g (0.035 oz) have been made. The capabilities of plaster-mold casting are given in Table 11.2.

In the *plaster-molding* process, the mold is made of plaster of paris (gypsum or calcium sulfate) with the addition of talc and silica flour to improve strength and to control the time required for the plaster to set. These components are mixed with water, and the resulting slurry is poured over the pattern. After the plaster sets (usually within 15 minutes) it is removed, and the mold is dried at a temperature range of 120° to 260°C (250° to 500°F) to remove the moisture. Higher drying temperatures may be used, depending on the type of plaster. The mold halves are assembled to form the mold cavity and are preheated to about 120°C (250°F). The molten metal is then poured into the mold.

Because plaster molds have very low permeability, gases evolved during solidification of the metal cannot escape. Consequently, the molten metal is poured either in a vacuum or under pressure. Mold permeability can be increased substantially by the *Antioch process*, in which the molds are dehydrated in an *autoclave* (pressurized oven) for 6 to 12 hours and then rehydrated in air for 14 hours. Another method of increasing the permeability of the mold is to use foamed plaster containing trapped air bubbles.

Patterns for plaster molding generally are made of materials such as aluminum alloys, thermosetting plastics, brass, or zinc alloys. Wood patterns are not suitable for making a large number of molds, because they are repeatedly in contact with the water-based plaster slurry. Since there is a limit to the maximum temperature that the plaster mold can withstand (generally about 1200°C; 2200°F), plaster-mold casting is used only for aluminum, magnesium, zinc, and some copper-based alloys. The castings have a good surface finish with fine details. Because plaster molds have lower thermal conductivity than others, the castings cool slowly, and thus, a more uniform grain structure is obtained with less warpage. Wall thickness of the cast parts can be 1 to 2.5 mm (0.04 to 0.1 in.).

11.2.4 Ceramic-mold casting

The *ceramic-mold casting* process (also called *cope-and-drag investment casting*) is similar to the plaster-mold process, with the exception that it uses refractory mold materials suitable for high-temperature applications. Typical parts made are impellers, cutters for machining operations, dies for metalworking, and molds for making plastic and rubber components. Parts weighing as much as 700 kg (1500 lb) have been cast by this process.

The slurry is a mixture of fine-grained zircon ($ZrSiO_4$), aluminum oxide, and fused silica, which are mixed with bonding agents and poured over the pattern (Fig. 11.10), which has been placed in a flask.

The pattern may be made of wood or metal. After setting, the molds (ceramic facings) are removed, dried, burned off to remove volatile matter, and baked. The molds are clamped firmly and used as all-ceramic molds. In the *Shaw process*, the ceramic facings are backed by fireclay (which resists high temperatures) to give strength to the mold. The facings then are assembled into a complete mold, ready to be poured.

The high-temperature resistance of the refractory molding materials allows these molds to be used for casting ferrous and other high-temperature alloys, stainless steels, and tool steels. Although the process is somewhat expensive, the castings have good dimensional accuracy and surface finish over a wide range of sizes and intricate shapes.

11.2.5 Evaporative-pattern casting (lost-foam process)

Evaporative-pattern casting (and investment casting discussed next), is sometimes referred to as the *expendable-pattern* casting processes or the *expendable mold–expendable pattern* processes. It is unique in that a mold and a pattern must be produced for every casting, whereas the patterns in the processes described thus far are reusable. Typical applications are cylinder heads, engine blocks, crankshafts, brake components, manifolds, and machine bases. The capabilities of evaporative-pattern casting are given in Table 11.2.

The *evaporative-pattern casting* process uses a polystyrene pattern, which evaporates upon contact with molten metal to form a cavity for the casting; this process is also known as *lost-foam casting* and falls under the trade name *Full-Mold* process. It

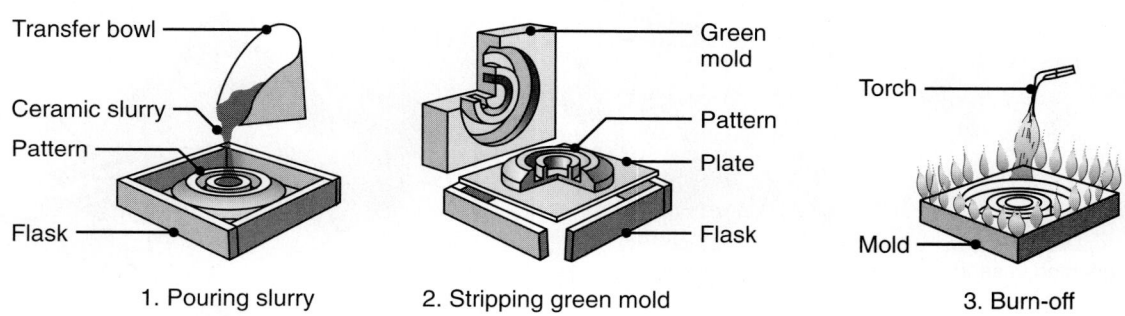

1. Pouring slurry 2. Stripping green mold 3. Burn-off

FIGURE 11.10 Sequence of operations in making a ceramic mold. *Source: Metals Handbook*, Vol. 5, 8th ed.

has become one of the more important casting processes for ferrous and nonferrous metals, particularly for the automotive industry.

In this process, raw, expendable polystyrene (EPS) beads containing 5 to 8% pentane (a volatile hydrocarbon) are placed in a preheated die which usually is made of aluminum. The polystyrene expands and takes the shape of the die cavity. Additional heat is applied to fuse and bond the beads together. The die then is cooled and opened, and the polystyrene pattern is removed. Complex patterns also may be made by bonding various individual pattern sections using hot-melt adhesive (Section 32.4.1).

The pattern is coated with a water-based refractory slurry, dried, and placed in a flask. The flask then is filled with loose, fine sand, which surrounds and supports the pattern (Fig 11.11) and may be dried or mixed with bonding agents to give it additional strength. The sand periodically is compacted, without removing the polystyrene pattern; then the molten metal is poured into the mold. This action immediately vaporizes the pattern and fills the mold cavity, completely replacing the space previously occupied by the polystyrene pattern. The heat degrades (depolymerizes) the polystyrene, and the degradation products are vented into the surrounding sand.

The flow velocity of the molten metal in the mold depends on the rate of degradation of the polymer. Studies have shown that the flow of the metal basically is laminar, with Reynolds numbers in the range of 400 to 3000. The velocity of the molten metal at the metal–polymer pattern front (interface) is in the range of 0.1 to 1.0 m/s and can be controlled by producing patterns with cavities or hollow sections. Thus, the velocity will increase as the molten metal crosses these hollow regions, similar to pouring the metal into an empty cavity.

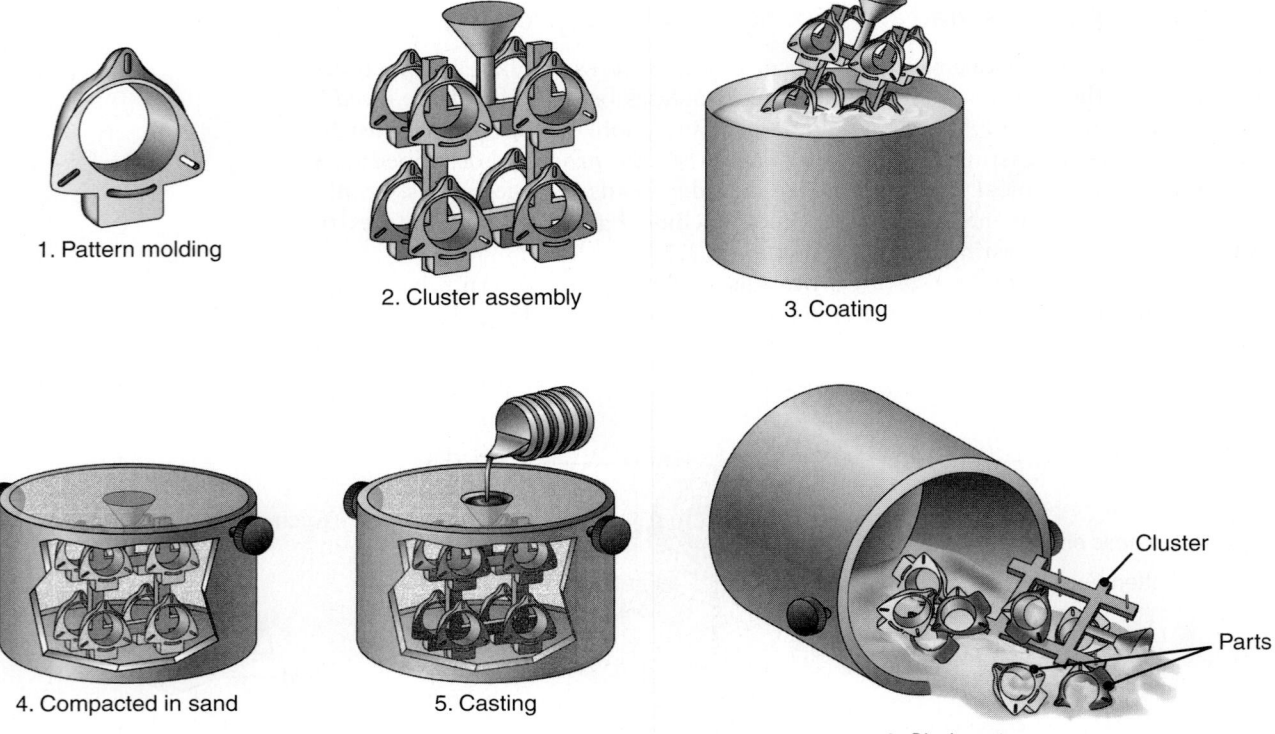

1. Pattern molding

2. Cluster assembly

3. Coating

4. Compacted in sand

5. Casting

6. Shakeout

Cluster

Parts

FIGURE 11.11   Schematic illustration of the expendable-pattern casting process, also known as lost-foam or evaporative–pattern casting.

Because the polymer requires considerable energy to degrade, large thermal gradients are present at the metal–polymer interface. In other words, the molten metal cools faster than it would if it were poured directly into an empty cavity. Consequently, fluidity is less than in sand casting. This has important effects on the microstructure throughout the casting and also leads to directional solidification of the metal. Polymethylmethacrylate (PMMA) and polyalkylene carbonate also may be used as pattern materials for ferrous castings.

The evaporative-pattern process has a number of advantages over other casting methods:

- The process is relatively simple because there are no parting lines, cores, or riser systems. Hence, it has design flexibility.

- Inexpensive flasks are satisfactory for the process.

- Polystyrene is inexpensive and can be processed easily into patterns having complex shapes, various sizes, and fine surface detail.

- The casting requires minimal finishing and cleaning operations.

- The process can be automated and is economical for long production runs. However, a major factor is the cost to produce the die used for expanding the polystyrene beads to make the pattern and the need for two sets of tooling. Also, the process is used for new casting designs and not for existing designs.

In a modification of the evaporative-pattern process, called the *Replicast C-S process*, a polystyrene pattern is surrounded by a ceramic shell; then the pattern is burned out prior to pouring the molten metal into the mold. Its principal advantage over investment casting (using wax patterns, Section 11.2.6) is that carbon pickup into the metal is avoided entirely. Further developments in evaporative-pattern casting include the production of metal-matrix composites (Sections 9.5 and 19.14). During the molding of the polymer pattern, fibers or particles are embedded throughout, which then become an integral part of the casting. Additional techniques include the modification and grain refinement of the casting by using grain refiners and modifier-master alloys (Section 11.7) within the pattern while it is being molded.

CASE STUDY 11.1 | Lost-Foam Casting of Engine Blocks

One of the most important parts in an internal combustion engine is the engine block. The engine block forms the enclosure for the engine, provides the basic structure that encloses the pistons and cylinders, and encounters significant pressure during operation.

Recognizing the industry pressures on high-quality, low-cost and lightweight designs, Mercury Castings built a lost-foam casting line to produce aluminum engine blocks and cylinder heads. One example of a part produced through lost-foam casting is a 60-hp 3-cylinder engine block used for marine applications and illustrated in Fig. 11.12. Previously manufactured as eight separate die castings, the block was converted to a single, 22-lb lost-foam casting with a weight and cost savings of 2 lb and $25 on each block, respectively. Lost-foam casting also allowed consolidation of the engine's cylinder head and exhaust and cooling systems into the block and eliminated the associated machining and fasteners required in sand-cast or die-cast designs. In addition, since the pattern contained holes and these could be cast without the use of cores, numerous drilling operations were eliminated.

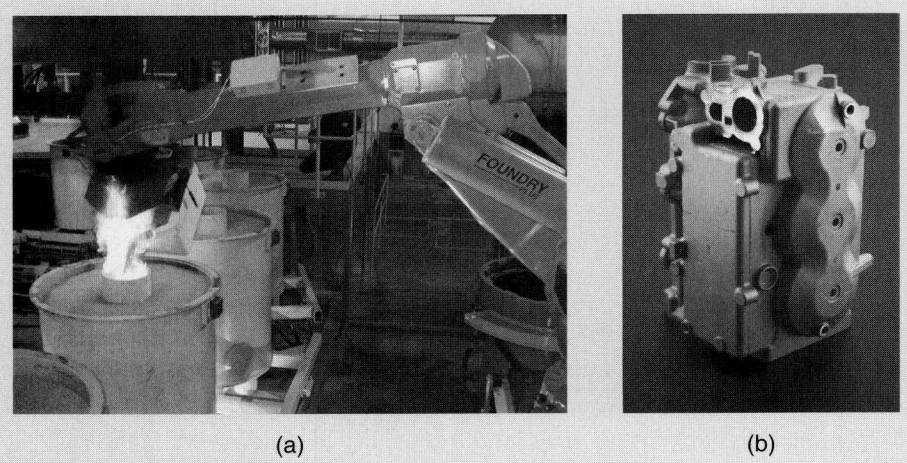

FIGURE 11.12   (a) Metal is poured into mold for lost-foam casting of a 60-hp, 3-cylinder marine engine; (b) finished engine back. *Source:* Courtesy of Mercury Marine.

Mercury Marine also was in the midst of developing a new V6 engine utilizing a new corrosion-resistant aluminum alloy with increased wear resistance. This engine design also required a cylinder block and head integration. It features hollow sections for water-jacket cooling that couldn't be cored out in die-casting or semi-permanent molding (the processes used for its other V6 blocks). Based on the success the foundry had with the 3-cylinder lost-foam block, engineers applied lost-foam casting for the V6 die block. The new engine block now involves only one casting, which is lighter and cheaper than the previous designs. Produced with an integrated cylinder head and exhaust and cooling system, this component is cast hollow to create a more efficient water-jacket cooling of the engine during operation.

The company also has developed a pressurized lost-foam process. A foam pattern is made, placed in a flask, and surrounded by sand. The flask then is placed in a pressure vessel where a robot pours molten aluminum onto the polystyrene pattern. A lid on the pressure vessel is closed and a pressure of 150 psi is applied to the casting until it solidifies. The result is a casting with better dimensional accuracy, lower porosity, and improved strength as compared to conventional lost-foam casting.

Source: Courtesy of Mercury Marine.

11.2.6 Investment casting

The *investment-casting* process, also called the **lost-wax process**, was first used during the period from 4000 to 3000 B.C. Typical parts made are components for office equipment as well as mechanical components, such as gears, cams, valves, and ratchets.

Parts up to 1.5 m (60 in.) in diameter and weighing as much as 1140 kg (2500 lb) have been cast successfully by this process. The capabilities of invstment casting are given in Table 11.2.

The sequences involved in investment casting are shown in Fig. 11.13. The pattern is made of wax or of a plastic such as polystyrene by molding or rapid pro-totyping techniques. The pattern then is dipped into a slurry of refractory material such as very fine silica and binders, including water, ethyl silicate, and acids. After this initial coating has dried, the pattern is coated repeatedly to increase its thickness for better strength.

The term *investment* derives from the fact that the pattern is invested (sur-rounded) with the refractory material. Wax patterns require careful handling because they are not strong enough to withstand the forces encountered during mold making; however, unlike plastic patterns, wax can be recovered and reused.

The one-piece mold is dried in air and heated to a temperature of 90° to 175°C (200° to 375°F). It is held in an inverted position for about 12 hours to melt out the wax. The mold is then fired to 650° to 1050°C (1200° to 1900°F) for about four hours (depending on the metal to be cast) to drive off the water of crystallization (chemically combined water) and to burn off any residual wax. After the metal has been poured and has solidified, the mold is broken up and the casting is removed. A number of patterns can be joined to make one mold, called a **tree** (Fig. 11.13),

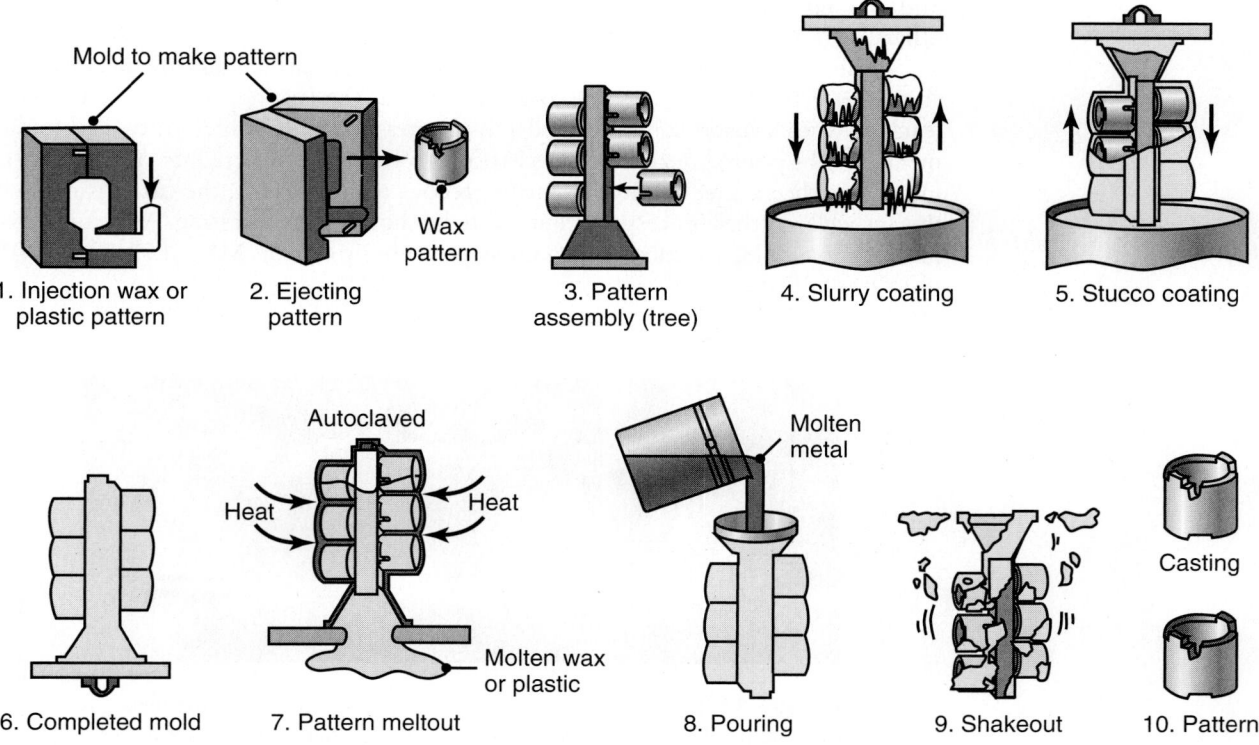

FIGURE 11.13   Schematic illustration of the investment casting (lost-wax) process. Castings produced by this method can be made with very fine detail and from a variety of metals. *Source*: Courtesy of Steel Founders' Society of America.

significantly increasing the production rate. For small parts, the tree can be inserted into a permeable flask and filled with a liquid slurry investment. The investment then is placed into a chamber and evacuated (to remove the air bubbles in it) until the mold solidifies. The flask usually is placed in a vacuum-casting machine, so that molten metal is drawn into the permeable mold and onto the part, producing fine detail.

Although the mold materials and labor involved make the lost-wax process costly, it is suitable for casting high-melting-point alloys with good surface finishes and close dimensional tolerances; few or no finishing operations, which otherwise would add significantly to the total cost of the casting, are required. This process is capable of producing intricate shapes, with parts weighing from 1 g to 35 kg (0.035 oz to 75 lb) from a wide variety of ferrous and nonferrous metals and alloys. Recent advances include the casting of titanium aircraft-engine and structural airframe components with wall thicknesses on the order of 1.5 mm (0.060 in.), thus competing with previously used sheet-metal structures.

Ceramic-shell investment casting.   A variation of the investment-casting process is *ceramic-shell casting*. It uses the same type of wax or plastic pattern, which is dipped first in ethyl silicate gel and subsequently into a fluidized bed (see Section 4.12) of fine-grained fused silica or zircon flour. The pattern then is dipped into coarser-grained silica to build up additional coatings and develop a proper thickness so that the pattern can withstand the thermal shock due to pouring. The rest of the procedure is similar to investment casting. This process is economical and is used extensively for the precision casting of steels and high-temperature alloys.

The sequence of operations involved in making a turbine disk by this method is shown in Fig. 11.14. If ceramic cores are used in the casting, they are removed by leaching with caustic solutions under high pressure and temperature. The molten metal may be poured in a vacuum to extract evolved gases and reduce oxidation, thus improving the casting quality. To further reduce microporosity, the castings made by this (as well as other processes) are subjected to hot isostatic pressing. Aluminum castings, for example, are subjected to a gas pressure up to 100 MPa (15 ksi) at 500°C (900°F).

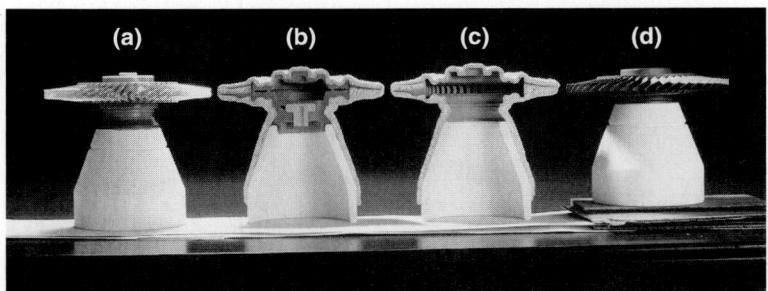

FIGURE 11.14   Investment casting of an integrally cast rotor for a gas turbine. (a) Wax pattern assembly. (b) Ceramic shell around wax pattern. (c) Wax is melted out and the mold is filled, under a vacuum, with molten superalloy. (d) The cast rotor, produced to net or near-net shape. *Source*: Courtesy of Howmet Corporation.

EXAMPLE 11.1 Investment-cast superalloy components for gas turbines

Since the 1960s, investment-cast superalloys have been replacing wrought counterparts in high-performance gas turbines. Much development has been taking place in producing cleaner superalloys (nickel-based and cobalt-based). Improvements have been made in melting and casting techniques, such as vacuum-induction melting and using microprocessor controls. Impurity and inclusion levels have continually been reduced, improving the strength and ductility of these components. Such control is essential, because these parts operate at a temperature only about 50°C (90°F) below the solidus temperature of the alloy (Section 4.3).

The microstructure of an integrally investment-cast, gas-turbine rotor is shown in the upper half of Fig. 11.15. Note the fine, uniform, equiaxed grains throughout the rotor cross-section. The procedures include the use of a nucleant addition to the molten metal, as well as close control of its superheat, pouring techniques, and control of cooling rate of the casting. In contrast, note the coarse-grain structure in the lower half of Fig. 11.15 showing the same type of rotor cast conventionally. This rotor has inferior properties compared with the fine-grained rotor. Due to developments in these processes, the proportion of cast parts to other parts in aircraft engines has increased from 20% to about 45% by weight.

FIGURE 11.15 Cross-section and microstructure of two rotors: (top) investment-cast; (bottom) conventionally cast. *Source: Advanced Materials and Processes*, October 1990, p. 25. ASM International.

11.3 | Permanent-Mold Casting Processes

Permanent-mold casting processes have certain advantages over other casting processes, as described here.

11.3.1 Permanent-mold casting

In *permanent-mold casting*, (also called *hard-mold casting*), two halves of a mold are made from materials with high resistance to erosion and thermal fatigue, such as cast iron, steel, bronze, graphite, or refractory metal alloys. Typical parts made are automobile pistons, cylinder heads, connecting rods, gear blanks for appliances, and kitchenware. Parts that can be made economically generally weigh less than 25 kg (55 lb), although special castings weighing a few hundred kilograms have been made using this process. The capabilities of permanent-mold casting are given in Table 11.2.

The mold cavity and gating system are machined into the mold and, thus, become an integral part of it. To produce castings with internal cavities, cores made of metal or sand aggregate are placed in the mold prior to casting. Typical core materials are oil-bonded or resin-bonded sand, plaster, graphite, gray iron, low-carbon steel, and hot-work die steel. Gray iron is used most commonly, particularly for large molds for aluminum and magnesium casting. Inserts also are used for various parts of the mold.

In order to increase the life of permanent molds, the surfaces of the mold cavity usually are coated with a refractory slurry (such as sodium silicate and clay) or sprayed with graphite every few castings. These coatings also serve as parting agents and as thermal barriers, thus controlling the rate of cooling of the casting. Mechanical ejectors (such as pins located in various parts of the mold) may be required for removal of complex castings; ejectors usually leave small round impressions.

The molds are clamped together by mechanical means and heated to about 150° to 200°C (300° to 400°F) to facilitate metal flow and reduce thermal damage to the dies due to high-temperature gradients. Molten metal then is poured through the gating system. After solidification, the molds are opened and the casting is removed. Special means that are employed to cool the mold include water or the use of fins similar to those found on motorcycle or lawnmower engines that cool the engine block.

Although the permanent-mold casting operation can be performed manually, it can be automated for large production runs. This process is used mostly for aluminum, magnesium, copper alloys, and gray iron because of their generally lower melting points, although steels also can be cast using graphite or heat-resistant metal molds. It produces castings with a good surface finish, close dimensional tolerances, uniform and good mechanical properties, and at high production rates. Although equipment costs can be high because of high die costs, labor costs are kept low by automating the process. The process is not economical for small production runs and intricate shapes, because of the difficulty in removing the casting from the mold. However, easily collapsable sand cores can be used, which are then removed from castings, leaving intricate internal cavities. This process then is called **semipermanent-mold casting**.

11.3.2 Vacuum casting

A schematic illustration of the *vacuum-casting* process or *counter-gravity low-pressure* (CL) *process* (not to be confused with the vacuum-molding process described in Section 11.2.1) is shown in Fig. 11.16. Vacuum casting is an alternative to investment, shell-mold, and green-sand casting and is suitable particularly for thin-walled (0.75 mm; 0.03 in.) complex shapes with uniform properties. Typical parts made are gas-turbine components from superalloys. These parts, which are often in the shape of superalloys for gas turbines, have walls as thin as 0.5 mm (0.02 in.).

In this process, a mixture of fine sand and urethane is molded over metal dies and cured with amine vapor. The mold then is held with a robot arm and immersed partially into molten metal held in an induction furnace. The metal may be melted in air (*CLA process*) or in a vacuum (*CLV process*). The vacuum reduces the air pressure inside the mold to about two-thirds of atmospheric pressure, thus drawing the molten metal into the mold cavities through a gate in the bottom of the mold. The

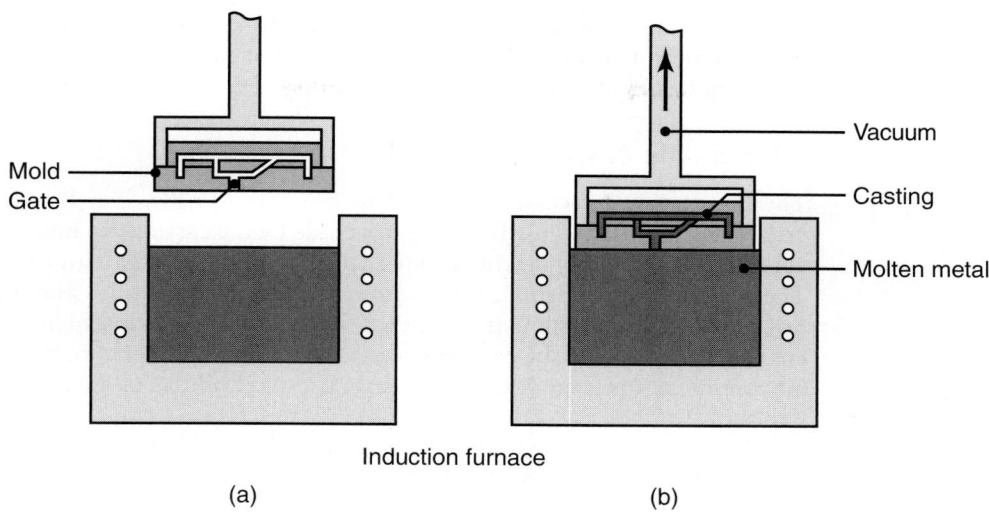

FIGURE 11.16   Schematic illustration of the vacuum-casting process. Note that the mold has a bottom gate. (a) Before and (b) after immersion of the mold into the molten metal. *Source: After R. Blackburn.*

metal in the furnace is at a temperature of usually 55°C (100°F) above the liquidus temperature of the alloy. Consequently, it begins to solidify within a very short time. After the mold is filled, it is withdrawn from the molten metal.

The process can be automated, and production costs are similar to those for green-sand casting. Carbon, low- and high-alloy steel, and stainless steel parts weighing as much as 70 kg (155 lb) have been vacuum cast by this method. CLA parts are made easily at high volume and relatively low cost. CLV parts usually involve reactive metals, such as aluminum, titanium, zirconium, and hafnium.

11.3.3 Slush casting

It was noted in Fig. 10.11 that a solidified skin develops first in a casting, and this skin then becomes thicker with time. Hollow castings with thin walls can be made by permanent-mold casting using this principle: a process called *slush casting*. This process is suitable for small production runs and generally is used for making ornamental and decorative objects (such as lamp bases and stems) and toys from low-melting-point metals such as zinc, tin, and lead alloys. The molten metal is poured into the metal mold. After the desired thickness of solidified skin is obtained, the mold is inverted (or slung) and the remaining liquid metal is poured out. The mold halves then are opened, and the casting is removed. Note that this operation is very similar to making hollow chocolate shapes, eggs, and other confectionaries.

11.3.4 Pressure casting

In the two permanent-mold processes described previously, the molten metal flows into the mold cavity by gravity. In *pressure casting* (also called *pressure pouring* or *low-pressure casting*), the molten metal is forced upward by gas pressure into a graphite or metal mold. The pressure is maintained until the metal has solidified completely in the mold. The molten metal also may be forced upward by a vacuum, which also removes dissolved gases and produces a casting with lower porosity.

Pressure casting generally is used for high-quality castings (such as steel railroad-car wheels), although these wheels also may be cast in sand molds or semipermanent molds made of graphite and sand.

11.3.5 Die casting

The *die casting* process, developed in the early 1900s, is a further example of permanent-mold casting. Typical parts made by die casting are motor housings, engine blocks, business-machine and appliance components, hand tools, and toys. The weight of most castings ranges from less than 90 g (3 oz) to about 25 kg (55 lb). Equipment costs, particularly the cost of dies, are somewhat high, but labor costs are generally low, because the process is now semi- or fully automated. Die casting is economical for large production runs. The capabilities of die casting are given in Table 11.2.

In this process, the molten metal is forced into the die cavity at pressures ranging from 0.7 to 700 MPa (0.1–100 ksi). The European term *pressure-die casting* (or simply die casting), which is described in this section, is not to be confused with the term *pressure casting* described in Section 11.3.4. There are two basic types of die-casting machines: hot-chamber and cold-chamber.

The **hot-chamber process** (Fig. 11.17) involves the use of a piston, which traps a certain volume of molten metal and forces it into the die cavity through a gooseneck and nozzle. Pressures range up to 35 MPa (5000 psi) with an average of about 15 MPa (2000 psi). The metal is held under pressure until it solidifies in the die. To improve die life and to aid in rapid metal cooling (thereby reducing cycle time) dies usually are cooled by circulating water or oil through various passageways in the die block. Low-melting-point alloys (such as zinc, magnesium, tin, and lead) commonly are cast using this process. Cycle times usually range from 200 to 300 shots (individual injections) per hour for zinc, although very small components such as zipper teeth can be cast at rates of 18,000 shots per hour.

In the **cold-chamber process** (Fig. 11.18), molten metal is poured into the injection cylinder (*shot chamber*). The chamber is not heated, hence the term *cold chamber*. The metal is forced into the die cavity at pressures usually ranging from 20 to 70 MPa (3 to 10 ksi), although they may be as high as 150 MPa (20 ksi).

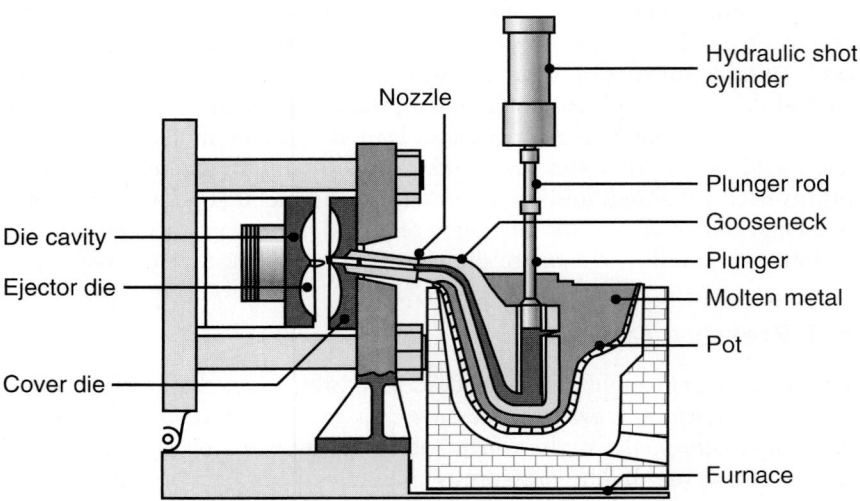

FIGURE 11.17   Schematic illustration of the hot-chamber die-casting process.

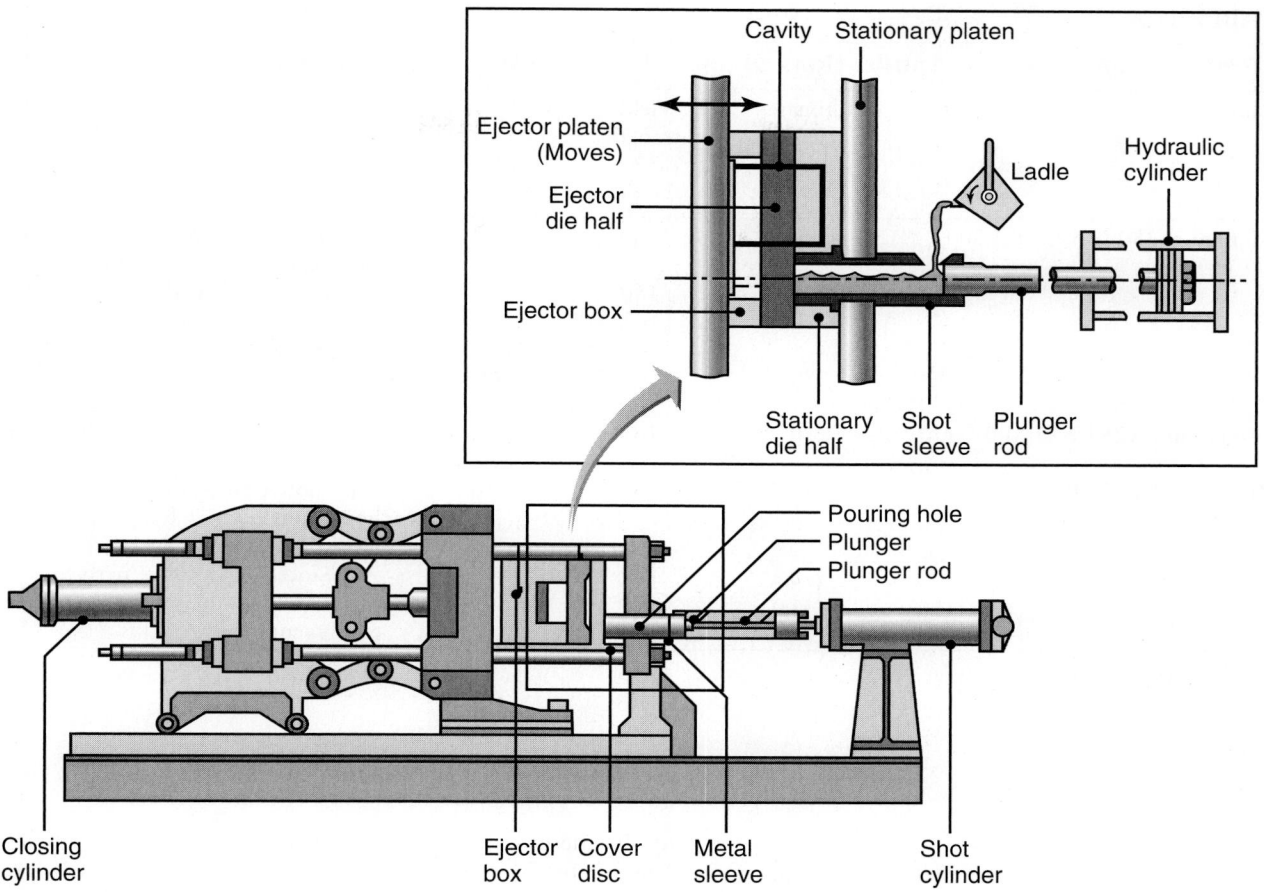

FIGURE 11.18 Schematic illustration of the cold-chamber die-casting process. These machines are large compared to the size of the casting, because high forces are required to keep the two halves of the dies closed under pressure.

The machines may be horizontal (as in the figure) or vertical, in which case the shot chamber is vertical. High-melting-point alloys of aluminum, magnesium, and copper normally are cast using this method, although other metals (including ferrous metals) also can be cast. Molten-metal temperatures start at about 600°C (1150°F) for aluminum and some magnesium alloys, and increase considerably for copper-based and iron-based alloys.

Process capabilities and machine selection. Die casting has the capability for rapid production of strong, high-quality parts with complex shapes, especially with aluminum, brass, magnesium, and zinc (Table 11.3). It also produces good dimensional accuracy and surface details, so that parts require little or no subsequent machining or finishing operations (net-shape forming). Because of the high pressures involved, walls as thin as 0.38 mm (0.015 in.) are produced, which are thinner than those obtained by other casting methods. However, ejector marks remain, as may small amounts of flash (thin material squeezed out between the dies) at the die parting line.

TABLE 11.3

Properties and Typical Applications of Some Common Die-Casting Alloys				
Alloy	Ultimate tensile strength (MPa)	Yield strength (MPa)	Elongation in 50 mm (%)	Applications
Aluminum 380 (3.5 Cu-8.5 Si)	320	160	2.5	Appliances, automotive components, electrical motor frames and housings
13 (12 Si)	300	150	2.5	Complex shapes with thin walls, parts requiring strength at elevated temperatures
Brass 858 (60 Cu)	380	200	15	Plumbing fixtures, lock hardware, bushings ornamental castings
Magnesium AZ91 B (9 Al-0.7 Zn)	230	160	3	Power tools, automotive parts, sporting goods
Zinc No. 3 (4 Al)	280	—	10	Automotive parts, office equipment, household utensils, building hardware, toys
No. 5 (4 Al-1 Cu)	320	—	7	Appliances, automotive parts, building hardware, business equipment

Source: American Die Casting Institute.

A typical part made by die casting is shown in Fig. 11.1d; note the intricate shape and fine surface detail. In the fabrication of certain parts, die casting can compete favorably with other manufacturing methods (such as sheet-metal stamping and forging) or other casting processes. In addition, because the molten metal chills rapidly at the die walls, the casting has a fine-grained, hard skin with high strength. Consequently, the strength-to-weight ratio of die-cast parts increases with decreasing wall thickness. With a good surface finish and dimensional accuracy, die casting can produce smooth surfaces for bearings that otherwise normally would be machined.

Components such as pins, shafts, and threaded fasteners can be die cast integrally. Called **insert molding**, this process is similar to placing wooden sticks in popsicles prior to freezing (see also Section 19.3). For good interfacial strength, insert surfaces may be knurled (see Fig. 23.1f), grooved, or splined. Steel, brass, and bronze inserts are used commonly in die-casting alloys. In selecting insert materials, the possibility of galvanic corrosion should be taken into account. To avoid this potential problem, the insert can be insulated, plated, or surface-treated.

Because of the high pressures involved, dies for die casting have a tendency to part unless clamped together tightly. Die-casting machines are thus rated according to the clamping force that can be exerted to keep the dies closed. The capacities of commercially available machines range from about 25 to 3000 tons. Other factors involved in the selection of die-casting machines are die size, piston stroke, shot pressure, and cost.

Die-casting dies (Fig. 11.19) may be *single cavity, multiple cavity* (with several identical cavities), *combination cavity* (with several different cavities), or *unit dies* (simple small dies that can be combined in two or more units in a master holding die). Typically, the ratio of die-weight to part-weight is 1000 to 1, thus the die for a

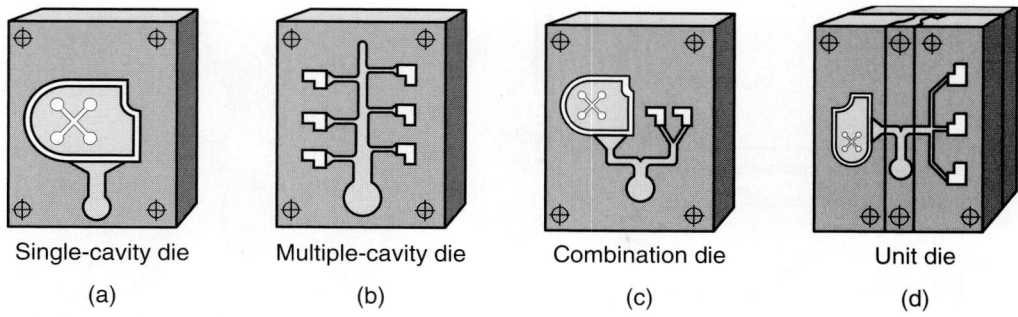

Single-cavity die	Multiple-cavity die	Combination die	Unit die
(a)	(b)	(c)	(d)

FIGURE 11.19   Various types of cavities in a die-casting die. *Source*: Courtesy of American Die Casting Institute.

casting weighing 2 kg would weigh about 2000 kg. The dies usually are made of hot-work die steels or mold steels (see Section 5.7). Die wear increases with the temperature of the molten metal. **Heat checking** of dies (surface cracking from repeated heating and cooling of the die, discussed in Section 3.6) can be a problem. When die materials are selected and maintained properly, dies may last more than a half a million shots before any significant die wear takes place.

Die design includes the taper (draft) to allow the removal of the casting. The sprues and runners may be removed either manually or by using trim dies in a press. The entire die casting and finishing processes are now highly automated. Lubricants (parting agents) often are applied as thin coatings on die surfaces; they usually are water-based lubricants with graphite or other compounds in suspension. Because of the high cooling capacity of water, these fluids also are effective in keeping die temperatures low for improved die life.

11.3.6 Centrifugal casting

As its name implies, the *centrifugal-casting* process utilizes inertial forces (caused by rotation) to distribute the molten metal into the mold cavities—a method that was first suggested in the early 1800s. There are three types of centrifugal casting: true centrifugal casting, semicentrifugal casting, and centrifuging.

True centrifugal casting.   In *true centrifugal casting*, hollow cylindrical parts (such as pipes, gun barrels, bushings, engine-cylinder liners, bearing rings with or without flanges, and streetlamp posts) are produced by the technique shown in Fig. 11.20. In this process, molten metal is poured into a rotating mold. The axis of rotation is usually horizontal but can be vertical for short workpieces. Molds are made of steel, iron, or graphite and may be coated with a refractory lining to increase mold life. The mold surfaces can be shaped so that pipes with various external designs can be cast. The inner surface of the casting remains cylindrical, because the molten metal is distributed uniformly by the centrifugal forces. However, because of density differences, lighter elements (such as dross, impurities, and pieces of the refractory lining) tend to collect on the inner surface of the casting. Consequently, the properties of the casting can vary throughout its thickness.

Cylindrical parts ranging from 13 mm (0.5 in.) to 3 m (10 ft) in diameter and 16 m (50 ft) long can be cast centrifugally with wall thicknesses ranging from 6 to 125 mm (0.25 to 5 in.). The pressure generated by the centrifugal force is high (as much as 150 g); such high pressure is necessary for casting thick-walled parts. Castings

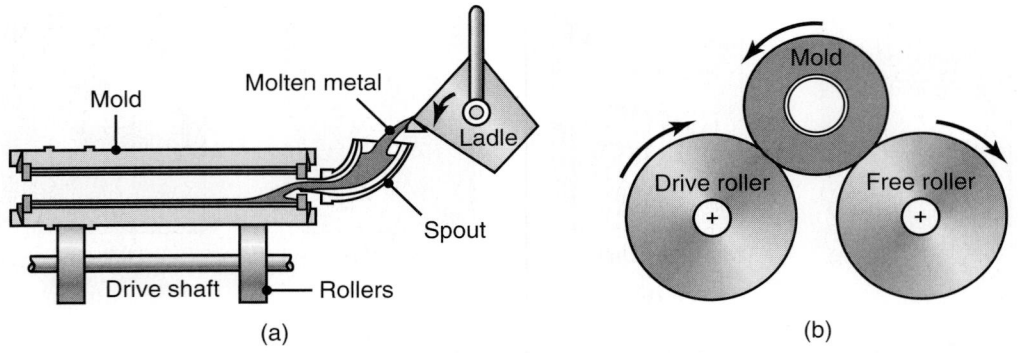

FIGURE 11.20   (a) Schematic illustration of the centrifugal-casting process. Pipes, cylinder liners, and similarly shaped parts can be cast with this process. (b) Side view of the machine.

with good quality, dimensional accuracy, and external surface detail are produced by this process. The capabilities of centrifugal casting are given in Table 11.2.

Semicentrifugal casting.   An example of *semicentrifugal casting* is shown in Fig. 11.21a. This method is used to cast parts with rotational symmetry, such as a wheel with spokes.

Centrifuging.   In *centrifuging* (also called *centrifuge casting*), mold cavities of any shape are placed at a certain distance from the axis of rotation. The molten metal is poured from the center and is forced into the mold by centrifugal forces (Fig. 11.21b). The properties of the castings can vary by distance from the axis of rotation, as in true centrifugal casting.

11.3.7 Squeeze casting and semisolid metal forming

Two casting processes that basically are combinations of casting and forging (Chapter 14) are *squeeze casting* and *semisolid-metal forming*.

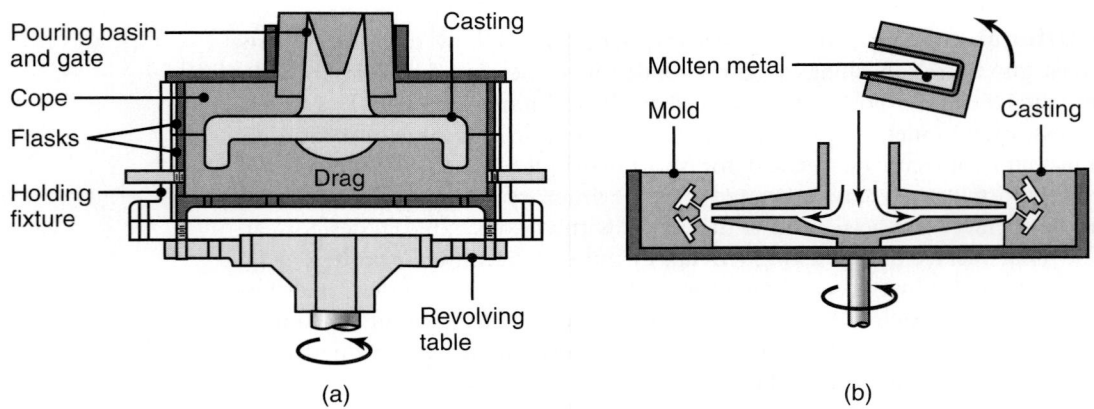

FIGURE 11.21   (a) Schematic illustration of the semicentrifugal casting process. Wheels with spokes can be cast by this process. (b) Schematic illustration of casting by centrifuging. The molds are placed at the periphery of the machine, and the molten metal is forced into the molds by centrifugal force.

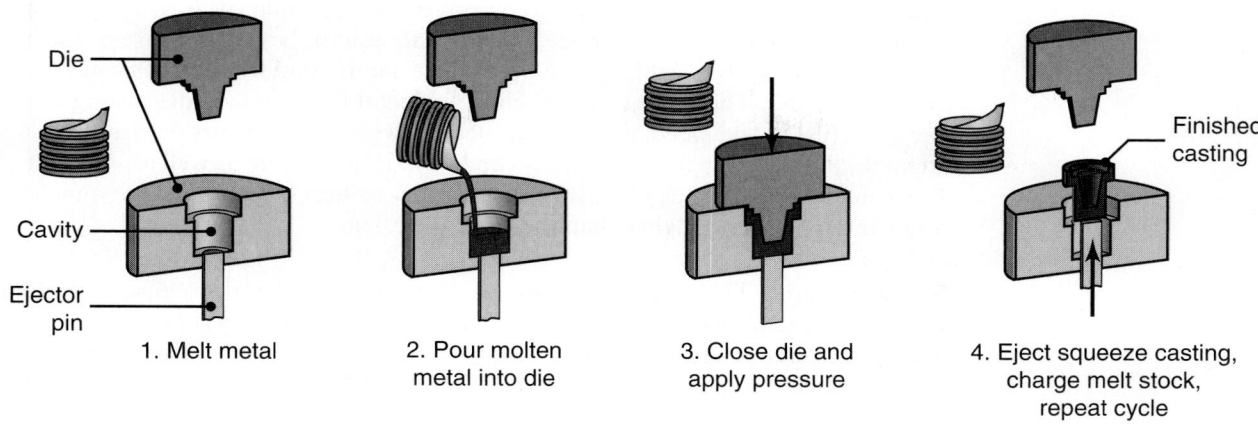

Die

Cavity

Ejector
pin

1. Melt metal

2. Pour molten
metal into die

3. Close die and
apply pressure

Finished
casting

4. Eject squeeze casting,
charge melt stock,
repeat cycle

FIGURE 11.22   Sequence of operations in the squeeze-casting process. This process combines the advantages of casting and forging.

Squeeze casting.   The *squeeze casting* (or *liquid-metal forging*) process was developed in the 1960s and involves the solidification of molten metal under high pressure (Fig. 11.22). Typical products made are automotive components and mortar bodies (a short-barrelled cannon). The machinery includes a die, punch, and ejector pin. The pressure applied by the punch keeps the entrapped gases in solution, and the contact under high pressure at the die–metal interface promotes rapid heat transfer, thus resulting in a fine microstructure with good mechanical properties.

The application of pressure also overcomes feeding difficulties that may arise when casting metals with a long freezing range (Section 10.2.2). The pressures required in squeeze casting are lower than those for hot or cold forging. Complex parts can be made to near-net shape with fine surface detail from both nonferrous and ferrous alloys.

Semisolid-metal forming.   Semisolid-metal forming (also called *mushy-state processing*, see Fig. 10.4) was developed in the 1970s and put into commercial production by 1981. When it enters the die, the metal (consiting of liquid and solid components) is stirred so that all of the dendrites are crushed into fine solids, and when cooled in the die, it developes into a fine-grained structure. The alloy exhibits *thixotropic* behavior (hence the process also is called **thixoforming**), meaning its viscosity decreases when agitated. Thus, for example, at rest and above its solidus temperature, the alloy has the consistency of butter, but when agitated vigorously, its consistency becomes more like motor oil. Processing metals in their mushy state also has lead to developments in *mushy-state extrusion*, *forging*, and *rolling* (hence the term *semisolid metalworking*). These processes also are used in making parts with specially designed casting or wrought alloys and metal-matrix composites.

Thixotropic behavior has been utilized in developing technologies that combine casting and forging of parts using cast billets that are forged when 30 to 40% liquid. Magnesium-alloy automotive parts, for example, have been made in machines with a feature that combines the processes of die casting and the injection molding of plastics using reciprocating screws (see Section 19.3). Parts made include control arms, brackets, and steering components. Processing steels by thixoforming has not yet reached the same stage as with magnesium, largely because of the high temperatures involved which adversely affect die life and the difficulty in making complex shapes. The advantages of semisolid metal forming over die casting are (a) the structures developed are homogeneous, with uniform properties and high strength, (b) both thin and thick parts can be made, (c) casting as well as wrought alloys can be used, and (d) parts subsequently can be heat treated. However, material and overall costs are higher than those for die casting.

Rheocasting. This is another technique, first investigated in the 1960s, for forming metals in the semisolid state. The metal is heated to just above its solidus temperature and poured into a vessel to cool it down to the semisolid state. The slurry then is mixed and delivered to the mold or die. This process is being used successfully with aluminum and magnesium alloys.

11.3.8 Composite mold casting operations

Composite molds are made of two or more different materials and are used in shell molding and other casting processes. They generally are employed in casting complex shapes, such as impellers for turbines. Composite molds increase the strength of the mold, improve the dimensional accuracy and surface finish of castings, and can help reduce overall costs and processing time. Molding materials commonly used are shells (made as described previously), plaster, sand with binder, metal, and graphite. These molds also may include cores and chills to control the rate of solidification in critical areas of castings.

11.4   Casting Techniques for Single-Crystal Components

The characteristics of single-crystal and polycrystalline structures in metals were described in Section 1.2. This section describes the techniques used to cast single-crystal components (such as gas turbine blades) which generally are made of nickel-based superalloys and used in the hot stages of the engine. The procedures involved also can be used for other alloys and components.

Conventional casting of turbine blades. The *conventional-casting process* uses a ceramic mold. The molten metal is poured into the mold and begins to solidify at the ceramic walls. The grain structure developed is polycrystalline, similar to that shown in Fig. 10.2c. However, the presence of grain boundaries makes this structure susceptible to creep and cracking along the boundaries under the centrifugal forces and elevated temperatures commonly encountered in an operating gas turbine.

Directionally solidified blades. The *directional-solidification process* (Fig. 11.23a) was first developed in 1960. The ceramic mold is preheated by radiant heating, and the mold is supported by a water-cooled chill plate. After the metal is poured into the mold, the chill-plate assembly is lowered slowly. Crystals begin to grow at the chill-plate surface and on upward, like the columnar grains shown in Fig. 10.3. The blade thus is solidified directionally with longitudinal but no transverse grain boundaries. Consequently, the blade is stronger in the direction of centrifugal forces developed in the gas turbine.

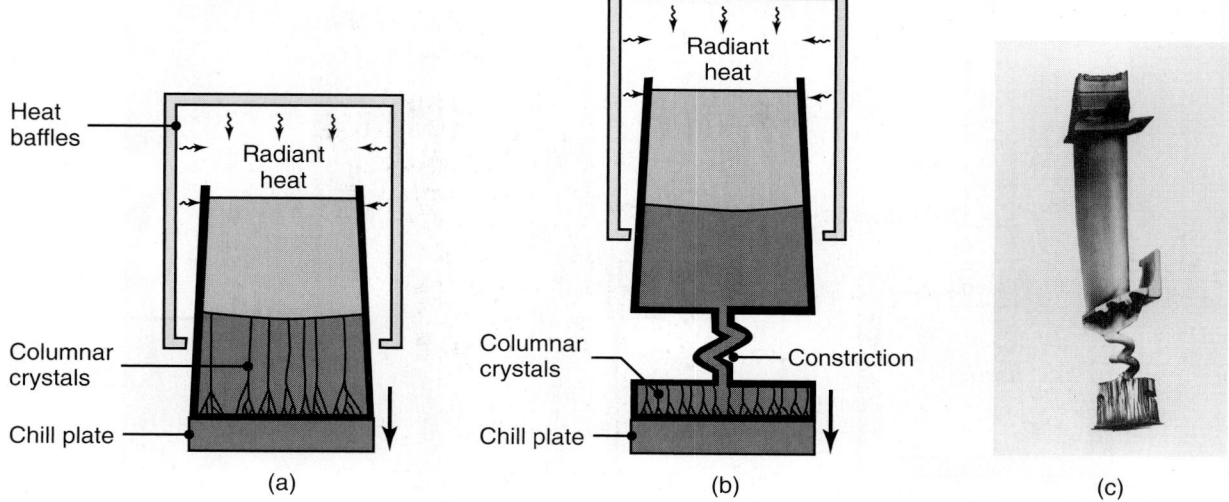

FIGURE 11.23 Methods of casting turbine blades: (a) directional solidification; (b) method to produce a single-crystal blade; and (c) a single-crystal blade with the constriction portion still attached. *Source:* (a) and (b) After B. H. Kear, (c) Courtesy of ASM International.

Single-crystal blades. In *crystal growing*, developed in 1967, the mold has a constriction in the shape of a corkscrew or helix (Figs. 11.23b and c). The cross-section is so small that it allows only one crystal to fit through. The mechanism of crystal growth is such that only the most favorably oriented crystals are able to grow (a situation similar to that shown in Fig. 10.3) through the helix because all others are intercepted by the walls of the helical passage.

As the assembly is lowered slowly, a single crystal grows upward through the constriction and begins to grow in the mold. Strict control of the rate of movement is important. The solidified mass in the mold is a single-crystal blade. Although these blades are more expensive than other types, the lack of grain boundaries makes them resistant to creep and thermal shock, so they have a longer and more reliable service life.

Single-crystal growing. Single-crystal growing is a major activity in the semiconductor industry in the manufacture of the silicon wafers in microelectronic devices (Chapter 28). There are two basic methods of crystal growing.

- In the **crystal-pulling method**, also known as the **Czochralski (CZ) process** (Fig. 11.24a), a seed crystal is dipped into the molten metal and then pulled out slowly (at a rate of about 10 μm/s) while being rotated. The liquid metal begins to solidify on the seed, and the crystal structure of the seed is continued throughout. *Dopants* (alloying elements) may be added to the liquid metal to impart special electrical properties. Single crystals of silicon, germanium, and various other elements are grown with this process. Single-crystal ingots with typical dimensions of 50 to 150 mm (2 to 6 in.) in diameter and over 1 m (40 in.) in length can be produced with this technique.

- The second technique for crystal growing is the **floating-zone method** (Fig. 11.24b). Starting with a rod of polycrystalline silicon resting on a single crystal, an induction coil heats these two pieces while the coil moves slowly upward. The single crystal grows upward while maintaining its orientation. Thin wafers then are cut from the rod, cleaned, and polished for use in microelectronic device fabrication.

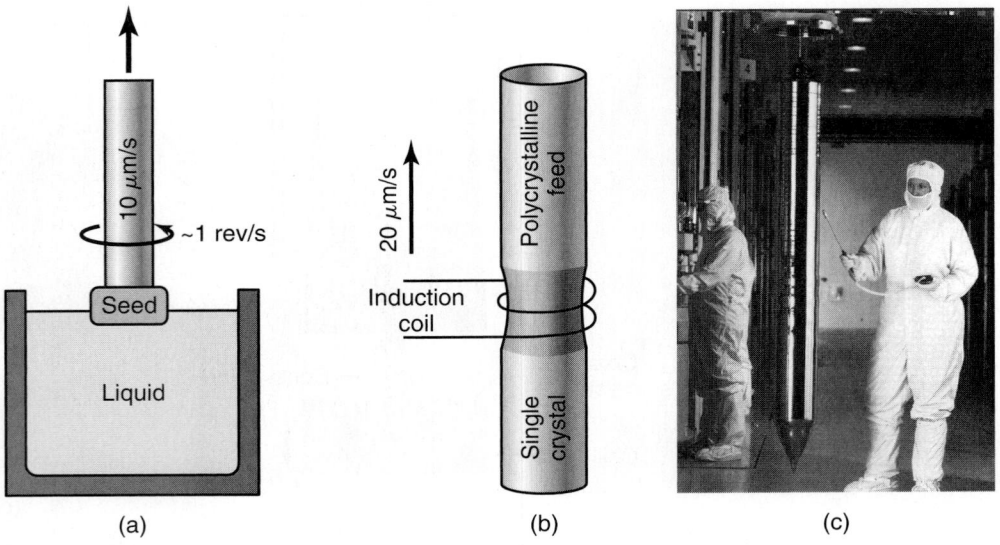

FIGURE 11.24   Two methods of crystal growing: (a) crystal pulling (Czochralski process) and (b) the floating-zone method. Crystal growing is important especially in the semiconductor industry. (c) A single-crystal ingot produced by the Czochralski process. *Source:* Courtesy of Intel Corp.

11.5 | Rapid Solidification

The properties of *amorphous alloys* (also known as *metallic glasses*) were described in Section 6.14. The technique for making these alloys (called *rapid solidification*) involves cooling the molten metal at rates as high as 10^6 K/s, so that it does not have sufficient time to crystallize (see also Fig. 1.11). Rapid solidification results in a significant extension of solid solubility, grain refinement, and reduced microsegregation (see Section 10.2.3), among other effects.

In a common method called **melt spinning** (Fig. 11.25), the alloy is melted by induction in a ceramic crucible. It is then propelled under high gas pressure at very high speed against a rotating copper disk (chill block), which chills the alloy rapidly (**splat cooling**).

11.6 | Inspection of Castings

The control of all stages during casting—from mold preparation to the removal of castings from molds or dies—is important in maintaining good quality. Several methods can be used to inspect castings to determine their quality and the presence and types of any possible defects. Castings can be inspected *visually* or *optically* for surface defects. Subsurface and internal defects are investigated using various *nondestructive* techniques (Section 36.10). In *destructive* testing (Section 36.11), test specimens are removed from various sections of a casting to test for strength, ductility, and other mechanical properties and to determine the presence, location, and distribution of porosity and any other defects.

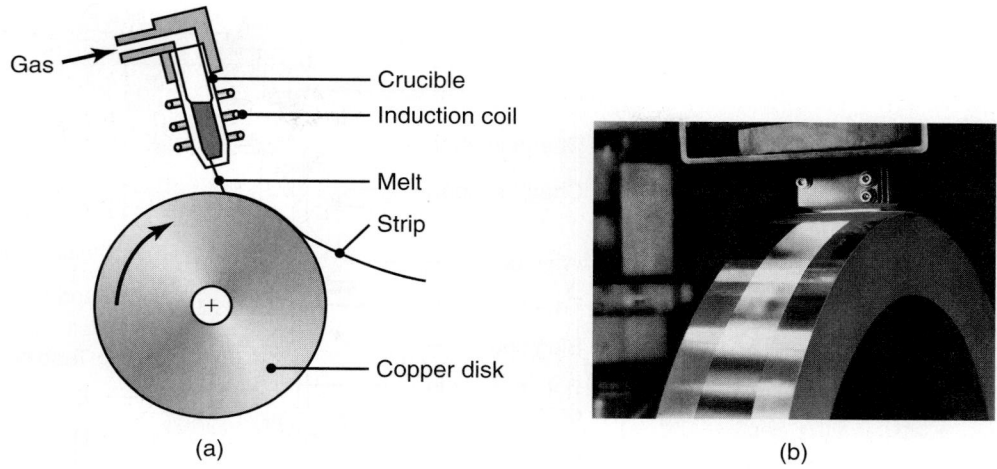

FIGURE 11.25 (a) Schematic illustration of melt-spinning to produce thin strips of amorphous metal. (b) Photograph of nickel-alloy production through melt-spinning *Source:* Courtesy of Siemens AG.

Pressure tightness of cast components (valves, pumps, and pipes) usually is determined by sealing the openings in the casting and pressurizing it with water, oil, or air. (Because air is compressible, its use is very dangerous in such tests because of the possibility of a sudden explosion due to a major flaw in the casting.) For extreme leak-tightness requirements in critical applications, pressurized helium or specially scented gases with detectors (sniffers) are used. The casting then is inspected for leaks while the pressure is maintained. Unacceptable or defective castings are remelted for reprocessing.

11.7 | Melting Practice and Furnaces

The melting practice is an important aspect of casting operations, because it has a direct bearing on the quality of castings. Furnaces are charged with melting stock, consisting of metal, alloying elements, and various other materials (such as **flux** and slag-forming constituents). Fluxes are inorganic compounds that refine the molten metal by removing dissolved gases and various impurities. They may be added manually or can be injected automatically into the molten metal.

Melting furnaces. The melting furnaces commonly used in foundries are electric-arc furnaces, induction furnaces, crucible furnaces, and cupolas.

- **Electric-arc** furnaces, described in Section 5.2.3 and illustrated in Fig. 5.2, are used extensively in foundries and have such advantages as a high rate of melting (and thus high-production rate), much less pollution than other types of furnaces, and the ability to hold the molten metal (keep it at a constant temperature for a period of time) for alloying purposes.

- **Induction** furnaces (Fig. 5.2c) are useful especially in smaller foundries and produce smaller composition-controlled melts. There are two basic types. The *coreless induction furnace* consists of a crucible completely surrounded with a

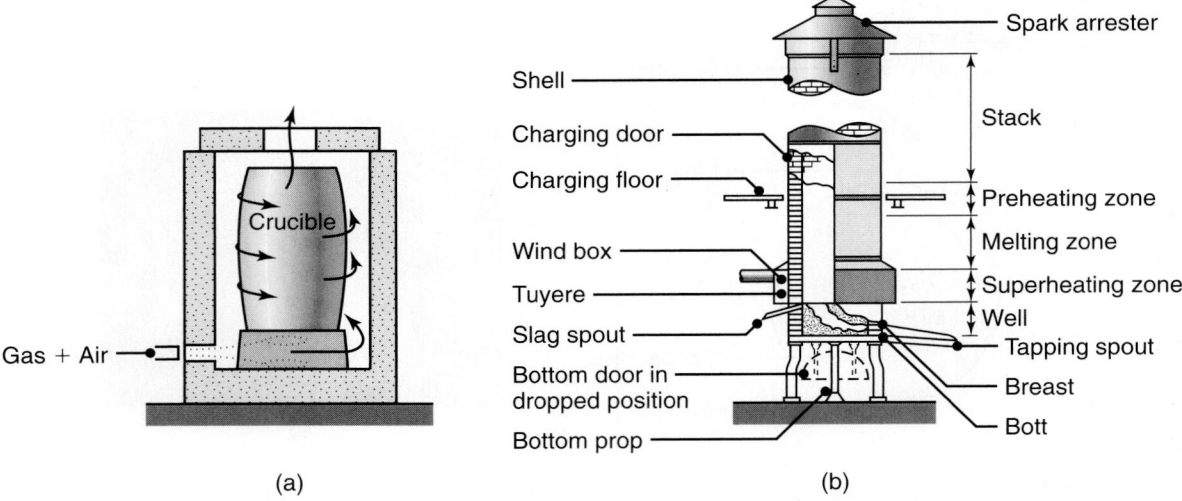

FIGURE 11.26   Two types of melting furnaces used in foundries: (a) crucible, and (b) cupola.

water-cooled copper coil through which a high-frequency current passes. Because there is a strong electromagnetic stirring action during induction heating, this type of furnace has excellent mixing characteristics for alloying and adding a new charge of metal.

The other type, called a *core* or *channel furnace*, uses a low-frequency current (as low as 60 Hz) and has a coil that surrounds only a small portion of the unit. These furnaces commonly are used in nonferrous foundries and are particularly suitable for (a) superheating (that is, heating above normal casting temperature to improve fluidity), (b) holding (which makes it suitable for die-casting applications), and (c) duplexing (using two furnaces to, for instance, melt the metal in one furnace and transfer it to another).

- **Crucible** furnaces (Fig. 11.26a), which have been used extensively throughout history, are heated with various fuels, such as commercial gases, fuel oil, and fossil fuel, as well as electricity. They may be stationary, tilting, or movable.

- **Cupolas** are basically vertical, refractory-lined steel vessels charged with alternating layers of metal, coke, and flux (Fig. 11.26b). Although they require major investments and increasingly are being replaced by induction furnaces, cupolas operate continuously, have high melting rates, and produce large amounts of molten metal.

- **Levitation melting** involves *magnetic suspension* of the molten metal. An induction coil simultaneously heats a solid billet and stirs and confines the melt, thus eliminating the need for a crucible (which could be a source of contamination with oxide inclusions). The molten metal flows downward into an investment-casting mold placed directly below the coil. Investment castings made with this method are free of refractory inclusions and of gas porosity and have uniform fine-grained structure.

11.8 | Foundries and Foundry Automation

Casting operations usually are carried out in **foundries** (from the Latin *fundere*, meaning melting and pouring). Although these operations traditionally have involved much manual labor, modern foundries have automated and computer-integrated facilities for all aspects of their operations. They produce a wide variety and sizes of castings at high-production rates, with good quality control, and at low cost.

As outlined in Fig. 11.2, foundry operations initially involve two separate groups of activities. The first group is pattern and mold making. Computer-aided design and manufacturing and rapid prototyping techniques now are used to minimize trial and error and, thus, improve efficiency. A variety of automated machinery is used to minimize labor costs, which can be significant in the production of castings. The second group of activities is melting the metals, controlling their composition and impurities, and pouring them into molds.

The rest of the operations, such as pouring into molds carried along conveyors, shakeout, cleaning, heat treatment, and inspection, also are automated. Automation minimizes labor, reduces the possibility of human error, increases the production rate, and attains higher quality levels. Industrial robots (Section 37.7) now are used extensively in foundry operations, such as cleaning, riser cutting, mold venting, mold spraying, pouring, sorting, and inspection. Automatic storage and retrieval systems for cores and patterns using automated guided vehicles (Section 37.6) are other operations.

The level of automation in foundries is an important economic consideration, particularly in view of the fact that many foundries are small businesses. The extent of automation depends on the type of products made. For example, a die-casting facility or a foundry making parts for the automotive industry may involve production runs in the hundreds of thousands; thus, a high level of automation is desirable. On the other hand, a jobbing foundry producing short production runs may not be automated to the same extent. Furthermore, foundries in the past have been viewed as somewhat dirty, hot, and labor-intensive operations. Today, it can be difficult to find qualified personnel to work in such an environment. Consequently, automation increasingly has become necessary to compensate for the decline in worker competency.

SUMMARY

- Casting processes generally are classified as expendable-mold or permanent-mold casting. The most common expendable-mold processes are sand, shell-mold, plaster-mold, ceramic-mold, and investment casting. Common permanent-mold processes include slush casting, pressure casting, die casting, and centrifugal casting. Expendable-mold casting usually involves lower mold and equipment costs but produces less dimensional accuracy.

- The molds used in permanent-mold casting are made of metal or graphite and are used repeatedly to produce a large number of parts. Because metals are good heat conductors but do not allow gases to escape, permanent molds fundamentally have different effects on casting than sand or other aggregate mold materials.

- In permanent-mold casting, die and equipment costs are relatively high, but the processes are economical for large production runs. Scrap loss is low and dimensional accuracy is relatively high—with good surface details.

- Other casting processes include squeeze casting (combination of casting and forging), semisolid-metal forming, rapid solidification for the production of amorphous alloys, and the casting of single-crystal components (such as turbine blades) for the hot stages of jet engines.

- Melting processes and their control also are important factors in casting operations. They include proper melting of the metals; preparation for alloying; removal of slag and dross; and pouring the molten metal into the molds. Inspection of castings for possible internal or external defects also is important.
- Castings subsequently may be subjected to further processing (such as heat treatment and machining operations) to produce the final desired shapes and surface characteristics.

KEY TERMS

Binders	Green molding sand	Pressure casting
Centrifugal casting	Insert molding	Rammed-graphite molding
Ceramic-mold casting	Investment casting	Rapid prototyping
Chaplets	Levitation melting	Rapid solidification
Composite molds	Lost-foam process	Rheocasting
Core prints	Lost-wax process	Sand casting
Cores	Master alloys	Semisolid-metal forming
Crystal growing	Parting agent	Shell-mold casting
Die casting	Patterns	Slush casting
Evaporative-pattern casting	Permanent-mold casting	Sodium-silicate process
Expendable-pattern casting	Permanent molds	Squeeze casting
Expendable molds	Plaster-mold casting	Thixotropy
Fluxes	Precision casting	Vacuum casting
Foundry		

BIBLIOGRAPHY

Allsop, D.F., and Kennedy, D., *Pressure Die Casting—Part II: The Technology of the Casting and the Die*, Pergamon, 1983.
An Introduction to Die Casting, American Die Casting Institute, 1981.
ASM Handbook, Vol. 15: *Casting*, ASM International, 1988.
Bradley, E.F., *High-Performance Castings: A Technical Guide*, Edison Welding Institute, 1989.
Campbell, J., *Castings*, 2nd ed., Butterworth-Heinemann, 2003.
Clegg, A.J., *Precision Casting Processes*, Pergamon, 1991.
Kaye, A., and Street, A.C., *Die Casting Metallurgy*, Butterworth, 1982.

Liebermann, H.H. (ed.), *Rapidly Solidified Alloys*, Marcel Dekker, 1993.
Mikelonis, P.J. (ed.), *Foundry Technology: A Source Book*, ASM International, 1982
Product Design for Die Casting, Diecasting Development Council, 1988.
Steel Castings Handbook, 6th ed., ASM International, 1995.
Street, A.C., *The Diecasting Book*, 2nd ed., Portcullis Press, 1986.
The Metallurgy of Die Castings, Society of Die Casting Engineers, 1986.
Young, K.P., *Semi-solid Processing*, Chapman & Hall, 1997.

REVIEW QUESTIONS

11.1 Describe the differences between expendable and permanent molds.

11.2 Name the important factors in selecting sand for molds.

11.3 What are the major types of sand molds? What are their characteristics?

11.4 List important considerations when selecting pattern materials.

11.5 What is the function of a core?

11.6 What is the difference between sand- and shell-mold casting?

11.7 What are composite molds? Why are they used?

11.8 Describe the features of plaster-mold casting.

11.9 Why is the investment-casting process capable of producing fine surface detail on castings?

11.10 Name the type of materials used for permanent-mold casting processes.

11.11 What are the advantages of pressure casting?

11.12 List the advantages and limitations of die casting.

11.13 What is the purpose of a riser? A vent?

11.14 Give reasons for using die inserts.

11.15 What is squeeze casting? What are its advantages?

11.16 What are the advantages of the lost-foam casting process?

QUALITATIVE PROBLEMS

11.17 If you need only a few units of a particular casting, which process(es) would you use? Why?

11.18 What are the reasons for the large variety of casting processes that have been developed over the years? Explain with specific examples.

11.19 Why does die casting produce the smallest cast parts?

11.20 What differences, if any, would you expect in the properties of castings made by permanent mold versus sand casting?

11.21 Would you recommend preheating the molds used in permanent-mold casting? Would you remove the casting soon after it has solidified? Explain your reasons.

11.22 Referring to Fig. 11.3, do you think it is necessary to weigh down or clamp the two halves of the mold? Explain your reasons. Do you think that the kind of metal cast, such as gray cast iron versus aluminum, should make a difference in the clamping force? Explain.

11.23 Explain why squeeze casting produces parts with better mechanical properties, dimensional accuracy, and surface finish than do expendable-mold processes.

11.24 How would you attach the individual wax patterns on a "tree" in investment casting?

11.25 Describe the measures that you would take to reduce core shifting in sand casting.

11.26 You have seen that, even though die casting produces thin parts, there is a limit to how thin they can be. Why can't even thinner parts be made by this process?

11.27 How are hollow parts with various cavities made by die casting? Are cores used? If so, how? Explain.

11.28 It was stated that the strength-to-weight ratio of die-cast parts increases with decreasing wall thickness. Explain why.

11.29 How are risers and sprues placed in sand molds? Explain with appropriate sketches.

11.30 In shell-mold casting, the curing process is critical to the quality of the finished mold. In this part of the process, the shell-mold assembly and cores are placed in an oven for a short period of time to complete the curing of the resin binder. List probable causes of unevenly cured cores or of uneven core thicknesses.

11.31 Why does the die-casting machine shown in Fig. 11.17 have such a large mechanism to close the dies? Explain.

11.32 Chocolate is available in hollow shapes. What process is used to make these candies?

11.33 What are the benefits and drawbacks to heating the mold in investment casting before pouring in the molten metal?

11.34 The slushy state of alloys refers to that state between the solidus and liquidus temperatures, as described in Section 10.2. Pure metals do not have such a slushy state. Does this mean that pure metals cannot be slush cast? Explain.

11.35 Can a chaplet also be a chill? Explain.

11.36 Rank the casting processes described in this chapter in terms of their solidification rate. (That is, which processes extract heat the fastest from a given volume of metal?)

QUANTITATIVE PROBLEMS

11.37 Estimate the clamping force for a die-casting machine in which the casting is rectangular with projected dimensions of 125 mm × 175 mm (5 in. × 7 in.). Would your answer depend on whether it is a hot-chamber or cold-chamber process? Explain.

11.38 The blank for the spool shown in Figure P11.38 is to be sand cast out of A-319—an aluminum casting alloy. Make a sketch of the wooden pattern for this part and include all necessary allowances for shrinkage and machining.

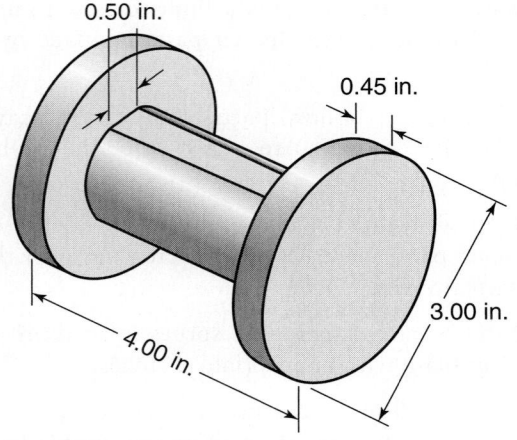

0.50 in.

0.45 in.

3.00 in.

4.00 in.

FIGURE P11.38

11.39 Repeat Problem 11.38 but assume that the aluminum spool is to be cast using expendable-pattern casting. Explain the important differences between the two patterns.

11.40 In sand casting, it is important that the cope-mold half be weighted down with sufficient force to keep it from floating when the molten metal is poured in. For the casting shown in Figure P11.40, calculate the minimum amount of weight necessary to keep the cope from floating up as the molten metal is poured in. (*Hint*: The buoyancy force exerted by the molten metal on the cope is dependent on the effective height of the metal head above the cope.)

11.41 If an acceleration of 100 g is necessary to produce a part in true centrifugal casting and the part has an inner diameter of 10 in., a mean outer diameter of 14 in., and a length of 25 ft, what rotational speed is needed?

11.42 A jeweler wishes to produce twenty gold rings in one investment-casting operation. The wax parts are attached to a wax central sprue of a 0.5 in. diameter. The rings are located in four rows, each 0.5 in. from the other on the sprue. The rings require a 0.125-in. diameter, 0.5-in. long runner to the sprue. Estimate the weight of gold needed to completely fill the rings, runners, and sprues. The specific gravity of gold is 19.3.

11.43 Assume that you are an instructor covering the topics described in this chapter, and you are giving a quiz on the numerical aspects to test the understanding of the students. Prepare two quantitative problems and supply the answers.

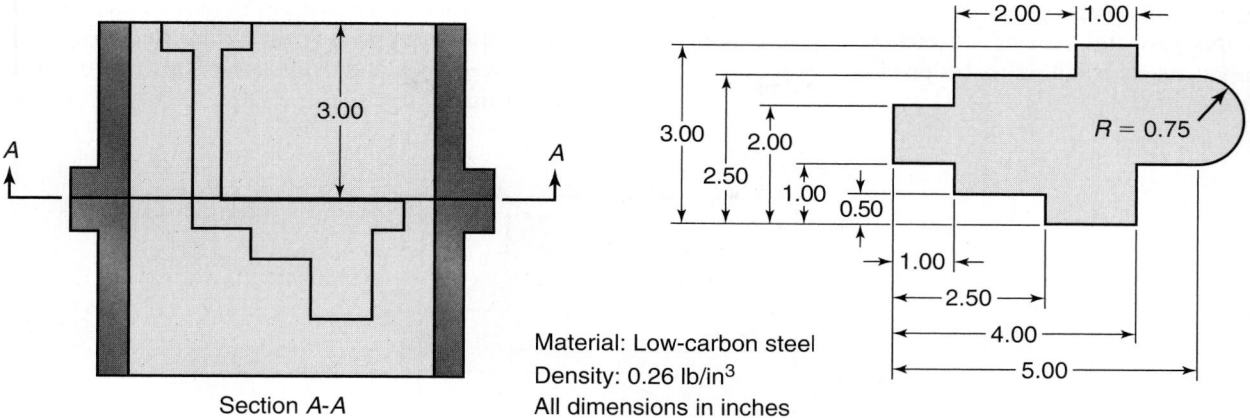

Section *A-A*

Material: Low-carbon steel
Density: 0.26 lb/in³
All dimensions in inches

FIGURE P11.40

SYNTHESIS, DESIGN, AND PROJECTS

11.44 Make a list of the mold and die materials used in the casting processes described in this chapter. Under each type of material, list the casting processes that are employed and explain why these processes are suitable for that particular mold or die material.

11.45 The optimum shape of a riser is spherical to ensure that it cools more slowly than the casting it feeds. However, spherically shaped risers are difficult to cast. (a) Sketch the shape of a blind riser that is easy to mold but also has the smallest possible surface area-to-volume ratio. (b) Compare the solidification time of the riser in part (a) to that of a riser shaped like a right circular cylinder. Assume that the volume of each riser is the same and the height for each is equal to the diameter. (See Example 10.1)

11.46 Sketch an automated casting line consisting of machinery, conveyors, robots, sensors, etc., that automatically could perform the expendable-pattern casting process.

11.47 Which of the casting processes would be most suitable for making small toys? Why?

11.48 Describe the procedures that would be involved in making a large bronze statue. Which casting process(es) would be suitable? Why?

11.49 Write a brief report on the permeability of molds and the techniques that are used to determine permeability.

11.50 Light metals commonly are cast in vulcanized rubber molds. Conduct a literature search and describe the mechanics of this process.

11.51 It sometimes is desirable to cool metals more slowly than they would be if the molds were maintained at room temperature. List and explain the methods you would use to slow down the cooling process.

11.52 The part shown in Figure P11.52 is a hemispherical shell used as an acetabular (mushroom shaped)

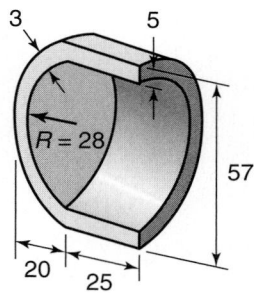

Dimensions in mm

FIGURE P11.52

cup in a total hip replacement. Select a casting process for this part and provide a sketch of all the patterns or tooling needed if it is to be produced from a cobalt-chrome alloy.

11.53 Porosity developed in the boss of a casting is illustrated in Figure P11.53. Show that this problem can be eliminated by simply repositioning the parting line of this casting.

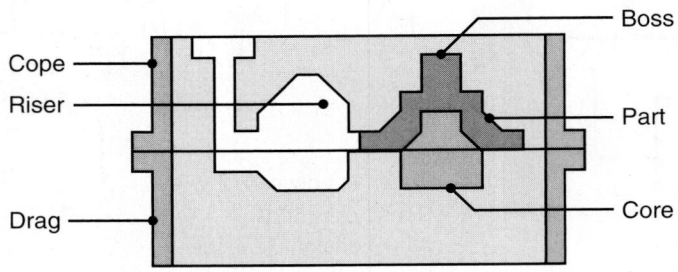

FIGURE P11.53

Metal Casting: Design, Materials, and Economics

This final chapter on metal casting serves as a guide to the important considerations regarding the interrelationships among product design, material and process selection, and economical methods of casting for specific applications. We describe in detail:

- Design considerations for expendable and permanent mold casting.
- General guidelines for successful casting.
- Characteristics and applications of nonferrous and ferrous casting alloys.
- Economic considerations in metal casting.

12.1 | Introduction

It was noted in the preceding two chapters that successful casting practice requires the proper control of a large number of variables. These variables pertain to the particular characteristics of the metals and alloys cast, method of casting, mold and die materials, mold design, and various process parameters. The flow of the molten metal in the mold cavities, the gating systems, the rate of cooling, and the gases evolved all influence the quality of a casting.

This chapter describes general design considerations and guidelines for metal casting and presents suggestions for avoiding defects. It also describes the characteristics of the alloys that are commonly cast, together with their typical applications. Because the economics of casting operations are just as important as the technical aspects, this chapter also briefly outlines the basic economic factors relevant to casting operations.

12.2 | Design Considerations in Casting

As in all manufacturing operations, certain guidelines and design principles pertaining to casting have been developed over many years. Although these principles have been established primarily through experience, analytical methods, process simulation and modeling, and computer-aided design and manufacturing techniques have all come into wide use as well, thus improving productivity and the quality of castings and resulting in significant cost savings.

All casting operations share some characteristics, such as phase change and thermal shrinkage during the casting cycle. Consequently, a number of design considerations apply equally to, for example, sand casting and die casting. However, each process will have its own particular design considerations; sand casting will require the consideration of mold erosion and associated sand inclusions in the casting, whereas die casting will not have this concern (although its has its own problems, such as heat checking of dies which reduce die life).

Troubleshooting the causes of defects very often is complicated, and the considerations presented in this chapter are by no means an exhaustive list. Also, defects often are random and difficult to reproduce, further complicating the implementation of corrective measures. In most cases, a given mold design will produce mostly good parts and some defective ones. It is very difficult for a mold to produce no defective parts; for these reasons, quality control procedures must be implemented for critical applications of castings.

12.2.1 General design considerations for castings

There are two types of design issues in casting: (a) geometric features, tolerances, etc., that should be incorporated into the part and (b) mold features that are needed to produce the desired casting. Robust design of castings usually involves the following steps:

1. Design the part so that the shape is cast easily. A number of important design considerations are given in this chapter to assist in such design efforts.

2. Select a casting process and a material suitable for the part, size, mechanical properties, and so on. Often, the design of the part will not be independent of the first step given, and the part, material and process have to be specified simultaneously.

3. Locate the parting line of the mold in the part.

4. Locate and design the gates to allow uniform feeding of the mold cavity with molten metal.

5. Select an appropriate runner geometry for the system.

6. Locate mold features such as sprue, screens, and risers, as appropriate.

7. Make sure proper controls and good practices are in place

We will now examine these rules for general casting conditions and then discuss specific rules applicable to particular casting operations.

Design of cast parts. The following considerations are important in designing castings, as outlined in Fig. 12.1.

- **Corners, angles, and section thickness.** Sharp corners, angles, and fillets should be avoided as much as possible, because they act as stress raisers and may cause cracking and tearing of the metal (as well as of the dies) during solidification. Fillet radii should be selected to reduce stress concentrations and to ensure proper liquid-metal flow during pouring. Fillet radii usually range from 3 to 25 mm (1/8 to 1 in.), although smaller radii may be permissible in small castings and in specific applications. However, if the fillet radii are too large, the volume of the material in those regions also is large, and consequently, the rate of cooling is lower.

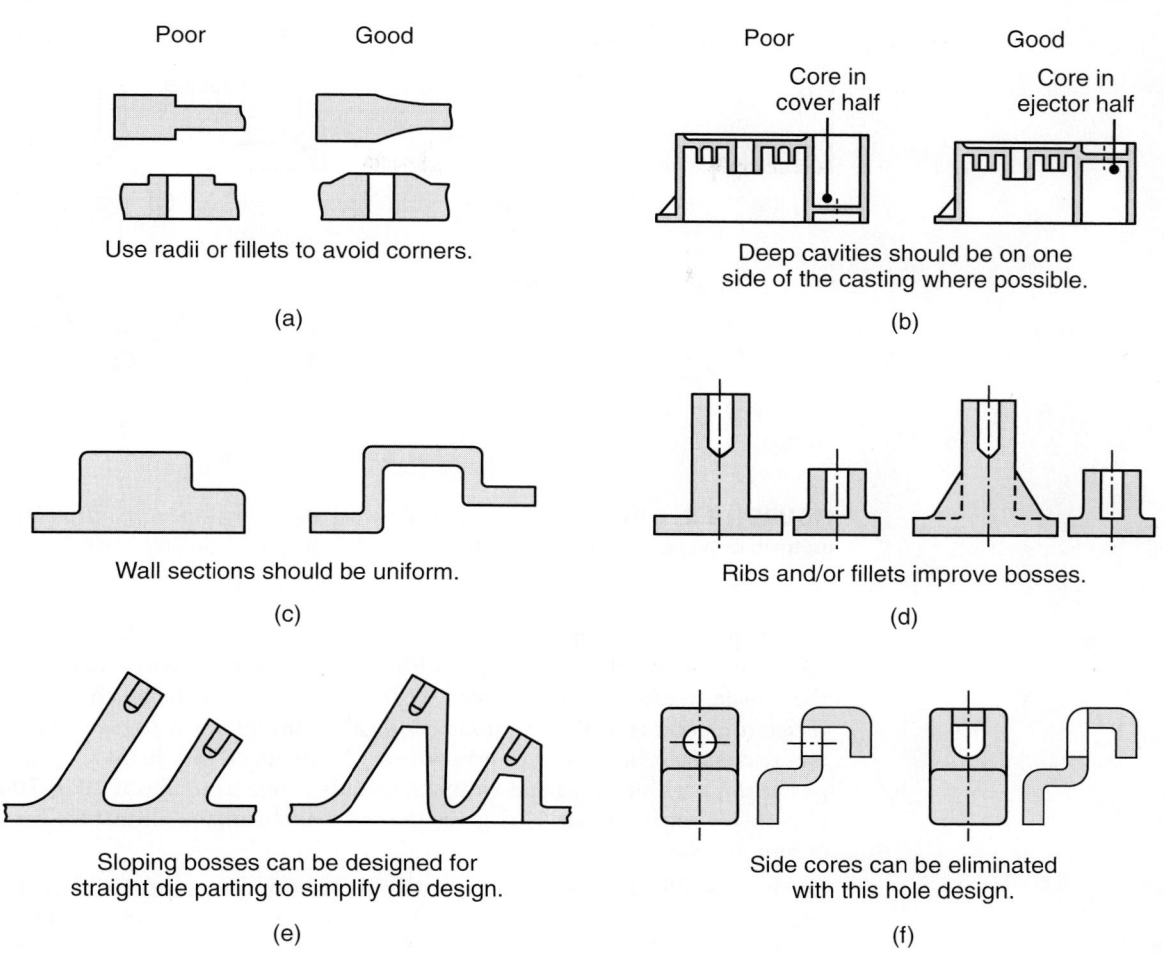

Poor   Good   Poor   Good

Use radii or fillets to avoid corners.

(a)

Deep cavities should be on one
side of the casting where possible.

(b)

Wall sections should be uniform.

(c)

Ribs and/or fillets improve bosses.

(d)

Sloping bosses can be designed for
straight die parting to simplify die design.

(e)

Side cores can be eliminated
with this hole design.

(f)

FIGURE 12.1 Suggested design modifications to avoid defects in castings. *Source*: Courtesy of the American Die Casting Institute.

Section changes in castings should be blended smoothly into each other. The location of the largest circle that can be inscribed in a particular region is critical so far as shrinkage cavities are concerned (Figs. 12.2a and b). Because the cooling rate in regions with larger circles is lower, they are called **hot spots.** These regions can develop **shrinkage cavities** and **porosity** (Figs. 12.2c and d). Cavities at hot spots can be eliminated by using small cores. Although they produce cored holes in the casting (Fig. 12.2e), these holes do not affect its strength significantly. It is important to maintain (as much as possible) uniform cross-sections and wall thicknesses throughout the casting to avoid or minimize shrinkage cavities. Although they increase the cost of production, *metal paddings* or *chills* in the mold can eliminate or minimize hot spots (see Fig. 10.14).

- **Flat areas.** Large flat areas (plain surfaces) should be avoided, since they may warp during cooling because of temperature gradients, or they develop poor surface finish because of an uneven flow of metal during pouring. One of the common techniques to resolve this is to break up flat surfaces with staggered ribs and serrations.

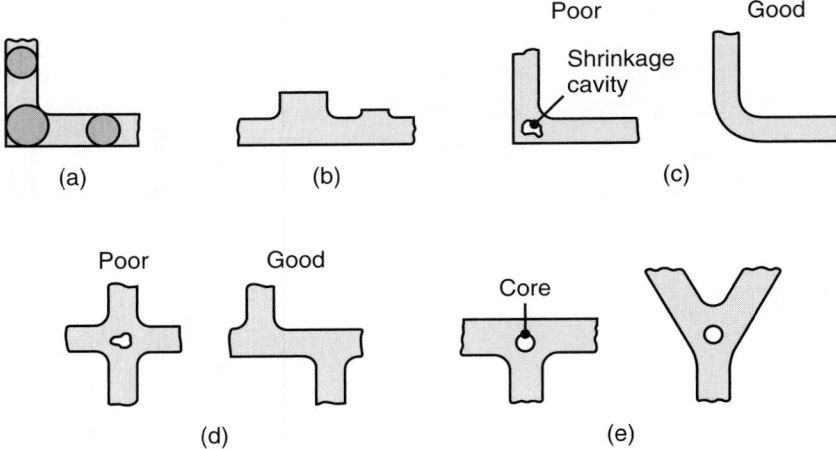

FIGURE 12.2 Examples of designs showing the importance of maintaining uniform cross-sections in castings to avoid hot spots and shrinkage cavities.

- **Shrinkage.** To avoid cracking of the casting during cooling, there should be allowances for shrinkage during solidification. In castings with intersecting ribs, the tensile stresses can be reduced by staggering the ribs or by changing the intersection geometry. Pattern dimensions also should allow for shrinkage of the metal during solidification and cooling. Allowances for shrinkage, known as **patternmaker's shrinkage allowances**, usually range from about 10 to 20 mm/m (1/8 to 1/4 in./ft). Table 12.1 gives the normal shrinkage allowance for metals commonly sand cast.

- **Draft.** A small draft (taper) typically is provided in sand-mold patterns to enable removal of the pattern without damaging the mold (see Fig. 11.5). Drafts generally range from 5 to 15 mm/m (1/16 to 3/16 in./ft). Depending on the quality of the pattern, draft angles usually range from 0.5° to 2°. The angles on inside surfaces typically are twice this range; they have to be higher than those for outer surfaces because the casting shrinks inward towards the core.

- **Dimensional tolerances.** Dimensional tolerances depend on the particular casting process, size of the casting, and type of pattern used. Tolerances should be

TABLE 12.1

Normal Shrinkage Allowance for Some Metals Cast in Sand Molds	
Metal	%
Gray cast iron	0.83–1.3
White cast iron	2.1
Malleable cast iron	0.78–1.0
Aluminum alloys	1.3
Magnesium alloys	1.3
Yellow brass	1.3–1.6
Phosphor bronze	1.0–1.6
Aluminum bronze	2.1
High-manganese steel	2.6

as wide as possible, within the limits of good part performance; otherwise, the cost of the casting increases. In commercial practice, tolerances are usually in the range of ±0.8 mm (1/32 in.) for small castings and increase with the size of the castings. Tolerances for large castings, for instance, may be as much as ±6 mm (0.25 in.).

- **Lettering and markings.** It is common practice to include some form of part identification (such as lettering or corporate logos) in castings. These features can be sunk into the casting or can protrude from the surface; which one is most desirable depends on the method of producing the molds. For example, in sand casting, a pattern plate is produced by machining on a CNC mill, and it is simpler to machine letters into the pattern plate. On the other hand, in die casting, it is simpler to machine letters into the mold.

- **Finishing operations.** In designing a casting, it is important to consider the subsequent machining and finishing operations that often have to take place. For example, if a hole is to be drilled in a casting, it is better to locate the hole on a flat surface than on a curved surface in order to prevent the drill from wandering. An even better design would incorporate a small dimple as a starting point for the drilling operations. Castings should include features that allow them to be clamped easily into machine tools.

Selecting the casting process. Casting process selection cannot be separated from a discussions of economics (see Section 12.4). However, Table 11.1 lists some of the advantages and limitations of casting processes that have an impact on casting design. Specific design rules for expendable and permanent mold operations are discussed next.

Locating the parting line. A part should be oriented in a mold so that the large portion of the casting is relatively low and the height of the casting is minimized. Part orientation also determines the distribution of porosity. For example, in casting aluminum, hydrogen is soluble in liquid metal but is not soluble as the aluminum solidifies (see Fig. 10.15). Thus, hydrogen bubbles can form during the casting of aluminum, and these will float upwards due to buoyancy and there will be a higher porosity in the top parts of castings. Therefore, critical surfaces should be oriented so that they face downwards.

A properly oriented part then can have the parting line specified. The parting line is the line or plane separating the upper (cope) and lower (drag) halves of molds (see Fig. 11.4). In general, the parting line should be along a flat plane rather than be contoured. Whenever possible, the parting line should be at the corners or edges of castings rather than on flat surfaces in the middle of the casting, so that the *flash* at the parting line (material squeezing out between the two halves of the mold) will not be as visible. The location of the parting line is important because it influences mold design, ease of molding, number and shape of cores required, method of support, and the gating system.

The parting line should be placed as low as possible (relative to the casting) for less dense metals (such as aluminum alloys) and located at around mid-height for denser metals (such as steels). However, the metal should not be allowed to flow vertically, especially when unconstrained by a sprue. The placement of the parting line has a large effect on the remainder of the mold design. For example, in sand casting, it is typical that the runners, gates, and sprue well are placed in the drag on the parting line. Also, the placement of the parting line and orientation of the part determines the number of cores needed, and it is preferable to avoid the use of cores whenever practical.

Locating and designing gates.   The gates are the connections between the runners and the part to be cast. Some of the considerations in designing gating systems are:

- Multiple gates often are preferable and are necessary for large parts. Multiple gates have the benefits of allowing lower pouring temperature and reducing the temperature gradients in the casting.
- Gates should feed into thick sections of castings.
- A fillet should be used where a gate meets a casting; this feature produces less turbulence than abrupt junctions.
- The gate closest to the sprue should be placed sufficiently far away so that the gate can be easily removed. This distance may be as small as a few millimeters for small castings and up to 500 mm for large parts.
- The minimum gate length should be three to five times the gate diameter, depending on the metal being cast. The cross-section should be large enough to allow the filling of the mold cavity and should be smaller than the runner cross-section.
- Curved gates should be avoided, but when necessary, a straight section in the gate should be located immediately adjacent to the casting.

Runner design.   The runner is a horizontal distribution channel that accepts molten metal from the sprue and delivers it to the gates. One runner is used for simple parts, but two-runner systems can be specified for more complicated castings. Runners are used to trap dross (dross is a mixture of oxide and metal and forms on the surface of metals) and keep it from entering the gates and mold cavity. Commonly, dross traps are placed at the ends of runners, and the runner projects above the gates to ensure that the metal in the gates is tapped from below the surface.

Designing other mold features.   The main goal in designing a *sprue* (described in Section 10.3) is to achieve the required metal flow rates while preventing aspiration or excessive dross formation. Flow rates are determined such that turbulence is avoided, but the mold is filled quickly compared to the solidification time required. A *pouring basin* can be used to ensure that the metal flow into the sprue is uninterrupted; also, if molten metal is maintained in the pouring basin during pouring, then the dross will float and will not enter the mold cavity. *Filters* are used to trap large contaminants, and these also serve to slow the metal velocity and make the flow more laminar. *Chills* can be used to speed solidification of the metal in a particular region of a casting.

Establishing good practices.   It has been observed widely that a given mold design can produce acceptable parts as well as defective ones and rarely will produce good or only defective parts. To check for defective castings, quality control procedures are necessary. Some of the common concerns are the following:

- Starting with a high-quality molten metal is essential for producing superior castings. Pouring temperature, metal chemistry, gas entrainment, and handling procedures all can affect the quality of metal being poured into a mold.
- The pouring of metal should not be interrupted, since this can lead to dross entrainment and turbulence. The meniscus of the molten metal in the mold cavity should experience a continuous, uninterrupted, and upward advance.

- The different cooling rates within the body of a casting cause residual stresses. Stress relieving (Section 4.11) thus may be necessary to avoid distortions of castings in critical applications.

12.2.2 Design for expendable-mold casting

Expendable-mold processes have certain specific design considerations, which mainly are attributed to the mold material, size of parts, and manufacturing method. Clearly, a casting in an expendable-mold process (such as investment casting) will cool much slower than in, say, die casting, which has important implications in the layout of molds.

Important design considerations for expendable-mold casting are the following:

Mold layout.   The features in the mold must be placed logically and compactly, with gates as necessary. One of the most important goals in mold layout is to have solidification initiate at one end of the mold and progress in a uniform front across the casting, with the risers solidifying last. Traditionally, mold layout has been based on experience and on considerations of fluid flow and heat transfer. More recently, commercial computer programs based on finite-difference algorithms have become available. These allow the simulation of mold filling and the rapid evaluation of mold layouts.

Riser design.   A major concern in the design of castings is the size and placement of risers. Risers are extremely useful in affecting the solidification-front progression across a casting and are an essential feature in the mold layout described previously. Blind risers are good design features and maintain heat longer than open risers. Risers are designed according to six basic rules:

1. The riser must not solidify before the casting. This usually is enforced by avoiding the use of small risers and by using cylindrical risers with small aspect ratios (small ratios of height to cross-section). Spherical risers are the most efficient shape but are difficult to work with.

2. The riser volume must be large enough to provide a sufficient amount of liquid metal to compensate for shrinkage in the casting.

3. Junctions between the casting and feeder should not develop a hot spot where shrinkage porosity can occur.

4. Risers must be placed so that the liquid metal can be delivered to locations where it is most needed.

5. There must be sufficient pressure to drive the liquid metal into locations in the mold where it is needed. Risers therefore are not as useful for metals with low density (such as aluminum alloys) as they are for those with a higher density (such as steel and cast iron).

6. The pressure head from the riser should suppress cavity formation and encourage complete cavity filling.

Machining allowance.   Because most expendable-mold castings require some additional finishing operations, such as machining and grinding, allowances should be made in casting design for these operations. Machining allowances, which are included in pattern dimensions, depend on the type of casting and increase with the size and section thickness of castings. Allowances usually range from about 2 to 5 mm (0.1 to 0.2 in.) for small castings to more than 25 mm (1 in.) for large castings.

12.2.3 Design for permanent-mold casting

Typical design guidelines and examples for permanent-mold casting are discussed in Example 12.1. Special considerations are involved in designing tooling for die casting. Although designs may be modified to eliminate the draft for better dimensional accuracy, a draft angle of 1/2° or even 1/4° usually is required; otherwise, galling (localized seizure or sticking of material) may take place between the part and the dies and cause distortion.

Die-cast parts are nearly-net shaped, requiring only the removal of gates and minor trimming to remove flashing and other minor defects. The surface finish and dimensional accuracy of die-cast parts are very good (see Table 11.2), and in general, they do not require a machining allowance.

EXAMPLE 12.1 Illustrations of poor and good casting designs

Several examples of poor and good designs in permanent-mold and die casting are shown in Fig. 12.3. The significant differences in design are outlined here for each example.

a. The lower portion of the design on the left has a thin wall with no apparent functional role. This location of the part thus may fracture if subjected to high forces or impact. The good design eliminates this problem and also may simplify die and mold manufacturing.

b. Large flat surfaces always present difficulties in casting metals (as well as nonmetallic materials), as they tend to warp and develop uneven surfaces. A common practice to avoid this situation is to break up the surface with ribs and serrations on the reverse side of the casting. This approach greatly reduces distortion and, furthermore, does not adversely affect the appearance and function of the flat surface.

c. This example of poor and good design is relevant not only to castings but also to parts that are machined or ground. It is difficult to produce sharp internal radii or corners which may be required for functional purposes, such as inserts that are designed to reach the bottom of the part cavity. Also, in the case of lubricated cavities, the lubricant can accumulate at the bottom and, being incompressible, will prevent full insertion of a part into the cavity. Placement of a small radius at the corners or periphery at the bottom of the part eliminates this problem.

d. The function of such a part could be, for instance, a knob to be gripped and rotated—hence the outer features along its periphery. Note in the design on the left that the inner periphery of the knob also has features which are not functional but help save material. The casting die for the good design is easier to manufacture.

e. Note that the poor design has sharp fillets at the base of the longitudinal grooves, which means that the die has sharp (knife edge) protrusions. Because of their sharpness, it is possible that over extended use of the die these edges can chip off.

f. The poor design on the left has threads reaching the right face of the casting. It then is possible that during casting some molten metal can penetrate this region, thus forming a flash and interfering with the function of the threaded insert, such as when a nut is used. The good design uses an offset on the threaded rod, eliminating this problem. This design consideration also is applicable for the injection-molding of plastics, an example of which is shown in Fig. 19.9.

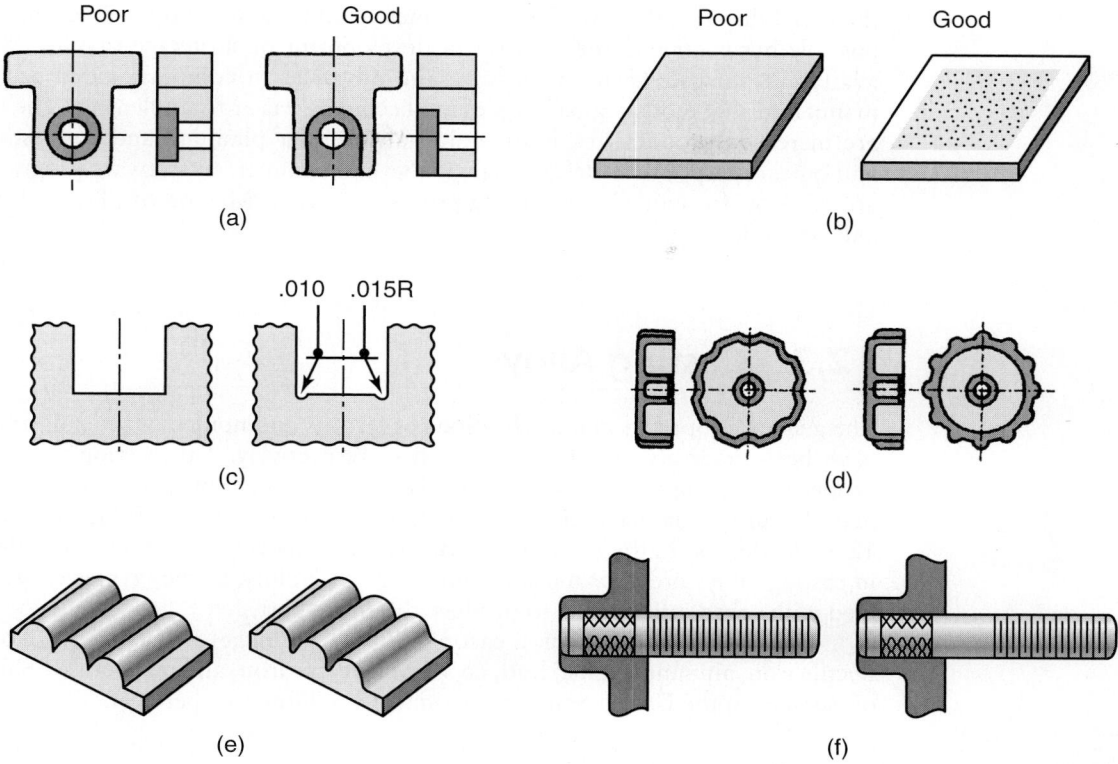

FIGURE 12.3 Examples of undesirable (poor) and desirable (good) casting designs. *Source:* Courtesy of American Die Casting Institute.

12.2.4 Computer modeling of casting processes

Because casting involves complex interactions among material and process variables, a quantitative study of these interactions is essential to the proper design of castings and the production of high-quality castings. Although in the past such studies have presented major difficulties, rapid advances in computers and modeling techniques have led to important innovations in modeling various aspects of casting—including fluid flow, heat transfer, and the microstructures developed during solidification—under various casting-process conditions.

Modeling of *fluid flow in molds* is based on Bernoulli's and the continuity equations (Section 10.3). It predicts the behavior of the metal during pouring into the gating system and its travel into the mold cavity as well as the velocity and pressure distributions in the system. Progress also is being made in the modeling of *heat transfer* in casting. Modern software can couple fluid flow and heat transfer and the effects of surface conditions, thermal properties of the materials involved, and natural and forced convection on cooling. Note that the surface conditions vary during solidification, as a layer of air develops between the casting and the mold wall due to shrinkage. Similar studies are being conducted on modeling the development of *microstructures* in casting. These studies encompass heat flow, temperature gradients, nucleation and growth of crystals, formation of dendritic and equiaxed structures, impingement of grains on each other, and movement of the liquid–solid interface during solidification.

Such models now are capable of predicting, for example, the width of the mushy zone (see Fig. 10.4) during solidification and the grain size in castings. Similarly, the capability to calculate isotherms (lines of equal temperature) gives insight into possible hot spots and the subsequent development of shrinkage cavities. With the availability of user-friendly software and advances in computer-aided design and manufacturing, modeling techniques are becoming easier to implement. The benefits are increased productivity, improved quality, easier planning and cost estimating, and quicker response to design changes. Several commercial software programs now are available for modeling of casting processes, such as Magmasoft, ProCast, Solidia, and AFSsolid.

12.3 | Casting Alloys

The general properties and applications of ferrous and nonferrous metals and alloys have been described in Chapters 5 and 6, respectively. This section describes the properties and applications of cast metals and alloys; their properties and casting and manufacturing characteristics are summarized in Fig. 12.4 and Tables 12.2 through 12.5. In addition to their casting characteristics, important additional considerations in casting alloys are their machinability and weldability, as they typically are assembled with other components to produce the entire part.

   The most commonly used casting alloy (in tonnage) is gray iron, followed by ductile iron, aluminum, zinc, lead, copper, malleable iron, and magnesium. Shipments of castings in the United States are around 13 million tons per year.

12.3.1 Nonferrous casting alloys

Common nonferrous casting alloys are the following:

Aluminum-based alloys.   Alloys with an aluminum base have a wide range of mechanical properties, mainly because of various hardening mechanisms and heat treatments that can be used with them (Section 4.9). These alloys have high electrical conductivity and generally good atmospheric corrosion resistance. However, their resistance to some acids and all alkalis is poor and care must be taken to prevent galvanic corrosion. They are nontoxic, lightweight, and have good machinability. Except for alloys with silicon, they generally have low resistance to wear and abrasion. Aluminum-based alloys have numerous applications, including architectural and decorative use. An increasing trend is their use in automobiles, for components such as engine blocks, cylinder heads, intake manifolds, transmission cases, suspension components, wheels and brakes. Parts made of aluminum-based and magnesium-based alloys are known as **light-metal** castings.

Magnesium-based alloys.   The lowest density of all commercial casting alloys are those in the magnesium-based group. They have good corrosion resistance and moderate strength, depending on the particular heat treatment used. Typical applications include automotive wheels, housings, and air-cooled engine blocks.

Copper-based alloys.   Although somewhat expensive, copper-based alloys have the advantages of good electrical and thermal conductivity, corrosion resistance, and nontoxicity, as well as wear properties suitable for bearing materials. A wide variety of copper-based alloys are available, including brasses, aluminum bronzes, phosphore bronzes, and tin bronzes.

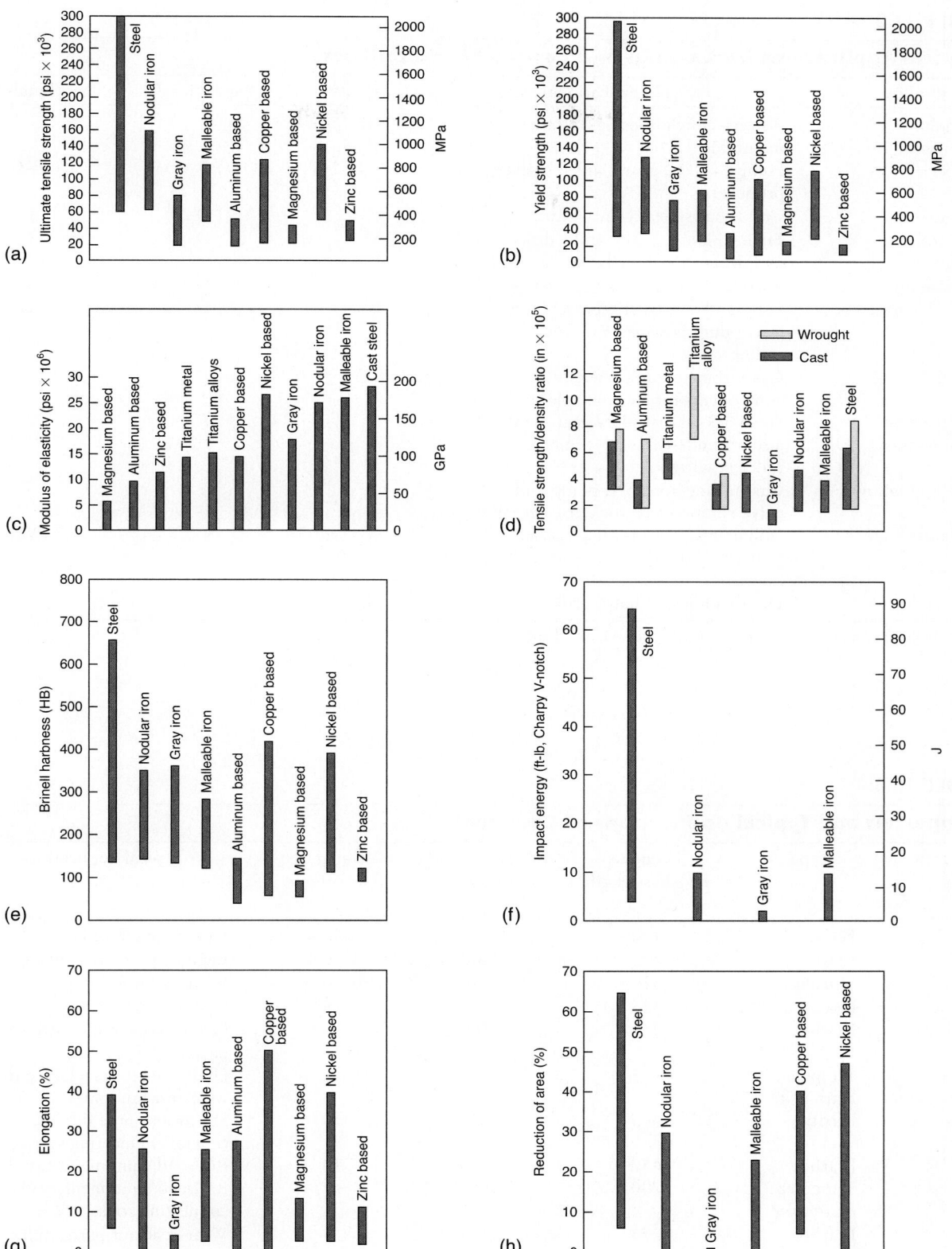

FIGURE 12.4 Mechanical properties for various groups of cast alloys. Note that even within the same group, the properties vary over a wide range, particularly for cast steels. *Source:* Courtesy of Steel Founders' Society of America.

TABLE 12.2

Typical Applications for Castings and Casting Characteristics

Type of alloy	Typical applications	Castability*	Weldability*	Machinability*
Aluminum	Pistons, clutch housings, intake manifolds	E	F	G–E
Copper	Pumps, valves, gear blanks, marine propellers	F–G	F	F–G
Ductile iron	Crankshafts, heavy-duty gears	G	D	G
Gray iron	Engine blocks, gears, brake disks and drums, machine bases	E	D	G
Magnesium	Crankcase, transmission housings	G–E	G	E
Malleable iron	Farm and construction machinery, heavy-duty bearings, railroad rolling stock	G	D	G
Nickel	Gas turbine blades, pump and valve components for chemical plants	F	F	F
Steel (carbon and low-alloy)	Die blocks, heavy-duty gear blanks, aircraft undercarriage members, railroad wheels	F	E	F
Steel (high-alloy)	Gas-turbine housings, pump and valve components, rock-crusher jaws	F	E	F
White iron	Mill liners, shot-blasting nozzles, railroad brake shoes, crushers, and pulverizers	G	VP	VP
Zinc	Door handles, radiator grills	E	D	E

*E = excellent; G = good; F = fair; VP = very poor; D = difficult.

TABLE 12.3

Properties and Typical Applications of Cast Irons

Cast iron	Type	Ultimate tensile strength (MPa)	Yield strength (MPa)	Elongation in 50 mm (%)	Typical applications
Gray	Ferritic	170	140	0.4	Pipe, sanitary ware
	Pearlitic	275	240	0.4	Engine blocks, machine tools
	Martensitic	550	550	0	Wear surfaces
Ductile (Nodular)	Ferritic	415	275	18	Pipe, general service
	Pearlitic	550	380	6	Crankshafts, highly stressed parts
	Tempered martensite	825	620	2	High-strength machine parts, wear-resistant parts
Malleable	Ferritic	365	240	18	Hardware, pipe fittings, general engineering service
	Pearlitic	450	310	10	Railroad equipment, couplings
	Tempered martensite	700	550	2	Railroad equipment, gears, connecting rods
White	Pearlitic	275	275	0	Wear-resistant parts, mill rolls

TABLE 12.4

Mechanical Properties of Gray Cast Irons

ASTM class	Ultimate tensile strength (MPa)	Compressive strength (MPa)	Elastic modulus (GPa)	Hardness (HB)
20	152	572	66–97	156
25	179	669	79–102	174
30	214	752	90–113	210
35	252	855	100–119	212
40	293	965	110–138	235
50	362	1130	130–157	262
60	431	1293	141–162	302

TABLE 12.5

Properties and Typical Applications of Nonferrous Cast Alloys

Alloys (UNS)	Condition	Ultimate tensile strength (MPa)	Yield strength (MPa)	Elongation in 50 mm (%)	Typical applications
Aluminum alloys					
195 (AO1950)	Heat treated	220–280	110–220	8.5–2	Sand castings
319 (AO3190)	Heat treated	185–250	125–180	2–1.5	Sand castings
356 (AO3560)	Heat treated	260	185	5	Permanent mold castings
Copper alloys					
Red brass (C83600)	Annealed	235	115	25	Pipe fittings, gears
Yellow brass (C86400)	Annealed	275	95	25	Hardware, ornamental
Manganese bronze (C86100)	Annealed	480	195	30	Propeller hubs, blades
Leaded tin bronze (C92500)	Annealed	260	105	35	Gears, bearings, valves
Gun metal (C90500)	Annealed	275	105	30	Pump parts, fittings
Nickel silver (C97600)	Annealed	275	175	15	Marine parts, valves
Magnesium alloys					
AZ91A	F	230	150	3	Die castings
AZ63A	T4	275	95	12	Sand and permanent mold castings
AZ91C	T6	275	130	5	High-strength parts
EZ33A	T5	160	110	3	Elevated-temperature parts
HK31A	T6	210	105	8	Elevated-temperature parts
QE22A	T6	275	205	4	Highest-strength parts

Zinc-based alloys. A low-melting-point alloy group, zinc-based alloys have good corrosion resistance, good fluidity, and sufficient strength for structural applications. These alloys commonly are used in die casting, particularly for parts with thin walls and intricate shapes.

Tin-based alloys. Although low in strength, these alloys have good corrosion resistance and typically are used for bearing surfaces.

Lead-based alloys. These alloys have applications similar to tin-based alloys, but the toxicity of lead is a major drawback to their wider application.

High-temperature alloys. High-temperature alloys have a wide range of properties and typically require temperatures of up to 1650°C (3000°F) for casting titanium and superalloys and higher for refractory alloys (Mo, Nb, W, and Ta). Special techniques are used to cast these alloys for nozzles and various jet- and rocket-engine components. Some of these alloys are more suitable and economical for casting than for shaping by other manufacturing methods, such as forging.

12.3.2 Ferrous casting alloys

Commonly cast ferrous alloys are as follows:

Cast irons. Cast irons represent the largest quantity of all metals cast, and they can be cast easily into intricate shapes. They generally possess several desirable properties, such as wear resistance, hardness, and good machinability. The term *cast iron* refers to a family of alloys, and as described in Section 4.6, they are classified as gray cast iron (gray iron), ductile (nodular or spheroidal) iron, white cast iron, malleable iron, and compacted-graphite iron. Their general properties and typical applications are given in Tables 12.3 and 12.4.

a. **Gray cast iron.** Castings of gray cast iron have relatively few shrinkage cavities and low porosity. Various forms of gray cast iron are *ferritic, pearlitic*, and *martensitic*. Because of differences in their structures, each type has different properties. The mechanical properties for several classes of gray cast iron are given in Table 12.4. Typical uses of gray cast iron are in engine blocks, electric-motor housings, pipes, and wear surfaces for machines. Also, its high damping capacity has made gray iron a common material for machine-tool bases. Gray cast irons are specified by a two-digit ASTM designation. For example, class 20 specifies that the material must have a minimum tensile strength of 20 ksi (140 MPa).

b. **Ductile (nodular) iron.** Typically used for machine parts, housings, gears, pipe, rolls for rolling mills, and automotive crankshafts, ductile irons are specified by a set of two-digit numbers. For example, class or grade 80-55-06 indicates that the material has a minimum tensile strength of 80 ksi (550 MPa), a minimum yield strength of 55 ksi (380 MPa), and 6% elongation in 2 in. (50 mm).

c. **White cast iron.** Because of its extreme hardness and wear resistance, white cast iron is used mainly for rolls for rolling mills, railroad-car brake shoes, and liners in machinery for processing abrasive materials.

d. **Malleable iron.** The principal use of malleable iron is for railroad equipment and various types of hardware, fittings, and components for electrical applications. Malleable irons are specified by a five-digit designation. For example, 35018 indicates that the yield strength of the material is 35 ksi (240 MPa) and its elongation is 18% in 2 in.

e. **Compacted-graphite iron.** First produced commercially in 1976, compacted-graphite iron (CGI) has properties that are between those of gray and ductile irons. Gray iron has good damping and thermal conductivity but low ductility, whereas ductile iron has poor damping and thermal conductivity but high tensile strength and fatigue resistance. Compacted-graphite iron has damping and thermal properties similar to gray iron and strength and stiffness comparable to those of ductile iron. Because of its strength, parts made of CGI can be smaller and, thus, lighter. It is easy to cast and has consistent properties throughout the casting, and its machinability is better than that of ductile iron (which is an important consideration, since it is used for automotive engine blocks and cylinder heads). New casting techniques are being developed to further improve the machinability of CGI.

Cast steels. Because of the high temperatures required to melt cast steels (up to about 1650°C, 3000°F), casting them requires considerable experience. The high temperatures involved present difficulties in the selection of mold materials, particularly in view of the high reactivity of steels with oxygen during the melting and pouring of the metal. Steel castings possess properties that are more uniform (isotropic) than those made by mechanical working processes (Part III). Cast steels can be welded; however, welding alters the cast microstructure in the heat-affected zone (see Fig. 30.17), thus influencing the strength, ductility, and toughness of the base metal. Subsequent heat treatment must be performed to restore the mechanical properties of the casting. Cast weldments have gained importance for assembling large machines and structures where complex configurations or the size of the casting may prevent casting of the part economically in one location. Cast steels have important applications in equipment for railroads, mining, chemical plants, oil fields, and heavy construction.

Cast stainless steels. Casting of stainless steels involves considerations similar to those for steels. Stainless steels generally have long freezing ranges and high melting temperatures. They can develop several structures, depending on their composition and processing parameters. Cast stainless steels are available in various compositions, and they can be heat treated and welded. These products have high heat and corrosion resistance, especially in the chemical and food industries. Nickel-based casting alloys are used for severely corrosive environments and for very high-temperature service.

12.4 | Economics of Casting

As is the case with all manufacturing processes, the cost of each cast part (**unit cost**) depends on several factors, including materials, equipment, and labor. Upon reviewing the various casting processes in Chapter 11, it is noted that some require more labor than others, some require expensive dies and machinery, and some require a great deal of time to produce the castings (Table 12.6). Each of these individual factors

TABLE 12.6

General Cost Characteristics of Casting Processes				
Casting process	Cost*			Production rate (pieces/hr)
	Die	Equipment	Labor	
Sand	L	L	L–M	<20
Shell mold	L–M	M–H	L–M	<50
Plaster	L–M	M	M–H	<10
Investment	M–H	L–M	H	<1000
Permanent mold	M	M	L–M	<60
Die	H	H	L–M	<200
Centrifugal	M	H	L–M	<50

*L = low; M = medium; H = high.

thus affects (to varying degrees) the overall cost of a casting operation. As described in greater detail in Section 40.7, the cost of a product includes the costs of materials, labor, tooling, and equipment. Preparations for casting a product include the production of molds and dies that require raw materials, time, and effort—all of which also influence product cost.

As can be seen in Table 12.6, relatively little cost is involved in molds for sand casting. On the other hand, molds for various processes and die-casting dies require expensive materials and a great deal of preparation. There are also major costs involved in making patterns for casting, although (as stated in Section 11.2.1) much progress is being made in utilizing rapid prototyping techniques to reduce costs and time.

Costs also are involved in melting and pouring the molten metal into molds and in heat treating, cleaning, and inspecting the castings. Heat treating is an important part of the production of many alloy groups (especially ferrous castings) and may be necessary to produce improved mechanical properties. However, heat treating also introduces another set of production problems (such as scale formation on casting surfaces and warpage of the part) which can be a significant aspect of production costs. The labor and skills required for these operations can vary considerably, depending on the particular process and level of automation in the foundry. Investment casting, for example, requires much labor because of the many steps involved in the operation, although some automation is possible, such as in the use of robots (Fig. 12.5). On the other hand, operations such as a highly automated die-casting process can maintain high-production rates with little labor required.

FIGURE 12.5   A robot generates a ceramic shell on wax patterns (trees) for investment casting. The robot is programmed to dip the trees and then place them in an automated drying system. With many layers, a thick ceramic shell suitable for investment casting is formed. *Source:* Courtesy of Wisconsin Precision Casting Corporation.

It also should be noted that the equipment cost per casting will decrease as the number of parts cast increases. Sustained high-production rates, therefore, can justify the high cost of dies and machinery. However, if the demand is relatively small, the cost-per-casting increases rapidly. It then becomes more economical to manufacture the parts by sand casting or other casting processes described in this chapter or by other manufacturing processes described in detail in Parts III and IV.

SUMMARY

- General guidelines have been established to aid designers in producing castings free from defects and meeting dimensional tolerances, service requirements, and various specifications and standards. These principles concern the shape of the casting and various techniques to minimize hot spots that could lead to shrinkage cavities. Because of the large number of variables involved, close control of all parameters is essential, particularly those related to the nature of liquid-metal flow into the molds and dies and the rate of cooling in different regions of the mold.

- Numerous nonferrous and ferrous casting alloys are available with a wide range of properties, casting characteristics, and applications. Because many castings are designed and produced to be assembled with other mechanical components and structures (subassemblies), various other considerations (such as weldability, machinability, and surface characteristics) also are important.

- Within the limits of good performance, the economic aspects of casting are just as important as technical considerations. Factors affecting the overall cost are the cost of materials, molds, dies, equipment, and labor—each of which varies with the particular casting operation. An important parameter is the cost per casting, which can justify large expenditures for large production runs using automated machinery and computer controls.

KEY TERMS

Cast iron	Flash	Patternmaker's shrinkage allowance
Compacted-graphite iron	Hot spots	Porosity
Design principles	Machining allowance	Shrinkage cavities
Draft	Parting line	Unit cost

BIBLIOGRAPHY

Abrasion-Resistant Cast Iron Handbook, American Foundry Society, 2000.

Alexiades, V., *Mathematical Modeling of Melting and Freezing Processes*, Hemisphere, 1993.

Allsop, D.F., and Kennedy, D., *Pressure Die Casting-Part II: The Technology of the Casting and the Die*, Pergamon, 1983.

An Introduction to Die Casting, American Die Casting Institute, 1981.

Analysis of Casting Defects, American Foundrymen's Society, 1974.

ASM Handbook, Vol. 15: *Casting*, ASM International, 1988.

ASM Specialty Handbook: Cast Irons, ASM International, 1996.

Bradley, E.F., *High-Performance Castings: A Technical Guide*, Edison Welding Institute, 1989.

Campbell, J., *Castings*, 2nd ed., Butterworth-Heinemann, 2003.

Casting Defects Handbook, American Foundrymen's Society, 1972.

Clegg, A.J., *Precision Casting Processes*, Pergamon, 1991.

Davis, J.R. (ed.), *Cast Irons*, ASM International, 1996.

Elliott, R., *Cast Iron Technology*, Butterworths, 1988.

Heine, R.W., Loper, C.R., Jr., and Rosenthal, C., *Principles of Metal Casting*, 2nd ed., McGraw-Hill, 1967.

Investment Casting Handbook, Investment Casting Institute, 1997.

Johns, R., *Casting Design*, American Foundrymen's Society, 1987.

Karlsson, L. (ed.), *Modeling in Welding, Hot Powder Forming and Casting*, ASM International, 1997.

Kaye, A., and Street, A.C., *Die Casting Metallurgy*, Butterworth, 1982.

Krauss, G., *Steels: Heat Treatment and Processing Principles*, ASM International, 1990.

Kurz, W., and Fisher, D.J., *Fundamentals of Solidification*, Trans Tech Pub., 1994.

Liebermann, H.H. (ed.), *Rapidly Solidified Alloys*, Marcel Dekker, 1993.

The Metallurgy of Die Castings, Society of Die Casting Engineers, 1986.

Powell, G.W., Cheng, S.-H., and Mobley, C.E., Jr., *A Fractography Atlas of Casting Alloys*, Battelle Press, 1992.

Rowley, M.T. (ed.), *International Atlas of Casting Defects*, American Foundrymen's Society, 1974.

Steel Castings Handbook, 6th ed., ASM International, 1995.

Walton, C.F., and Opar, T.J. (eds.), *Iron Castings Handbook*, 3rd ed., Iron Castings Society, 1981.

REVIEW QUESTIONS

12.1 Describe the general design considerations in metal casting.

12.2 What are hot spots? What is their significance in metal casting?

12.3 What is shrinkage allowance? Machining allowance?

12.4 Are drafts necessary in all molds?

12.5 What are light-metal castings? Where are they used most commonly?

12.6 Name the types of cast irons available and list their major characteristics and applications.

12.7 Why are steels more difficult to cast than cast irons? What is the consequence of this?

12.8 Describe the important factors involved in the economics of casting operations.

12.9 What is the difference between a runner and a gate?

12.10 What is the difference between machining allowance and dimensional tolerance?

12.11 List the rules for locating parting lines in casting.

QUALITATIVE PROBLEMS

12.12 Describe the procedure you would follow to determine whether a defect in a casting is a shrinkage cavity or a porosity caused by gases.

12.13 Explain how you would go about avoiding hot tearing.

12.14 Describe your observation concerning the design changes shown in Fig. 12.1.

12.15 If you need only a few castings of the same design, which three processes would be the most expensive per piece cast?

12.16 Do you generally agree with the cost ratings in Table 12.6? If so, why?

12.17 Add more examples to those shown in Fig. 12.2.

12.18 Explain how ribs and serrations are helpful in casting flat surfaces that otherwise may warp. Give a specific illustration.

12.19 Describe the nature of the design changes made in Fig. 12.3. What general principles do you observe in this figure?

12.20 Note in Fig. 12.4 that the ductility of some cast alloys is very low. Do you think this should be a significant concern in engineering applications of castings? Explain.

12.21 Do you think there will be fewer defects in a casting made by gravity pouring versus one made by pouring under pressure? Explain.

12.22 Explain the difference in the importance of drafts in green-sand casting versus permanent-mold casting.

12.23 What type of cast iron would be suitable for heavy-machine bases, such as presses and machine tools? Why?

12.24 Explain the advantages and limitations of sharp and rounded fillets, respectively, in casting design.

12.25 Explain why the elastic modulus, E, of gray cast iron varies so widely, as shown in Table 12.4.

12.26 Why are risers not as useful in die casting as compared with sand casting?

12.27 Describe the drawbacks to having a riser that is (a) too large and (b) too small.

12.28 Why can blind risers be smaller than open-top risers?

12.29 If you were to incorporate lettering or numbers on a sand-cast part, would you make them protrude from the surface or recess them into the surface? What if the part were to be made by investment casting? Explain your answer.

12.30 The general design recommendations for a well in sand casting (see Fig. 11.3) are that (a) its diameter should be at least twice the exit diameter of the sprue and (b) its depth should be approximately twice the depth of the runner. Explain the consequences of deviating from these guidelines.

12.31 The heavy regions of parts typically are placed in the drag in sand casting and not in the cope. Explain why.

QUANTITATIVE PROBLEMS

12.32 When designing patterns for casting, patternmakers use special rulers that automatically incorporate solid shrinkage allowances into their designs. For example, a 12-in. patternmaker's ruler is longer than one foot. How long should a patternmaker's ruler be for making patterns for (a) aluminum castings and (b) high-manganese steel?

12.33 Using the data given in Table 12.2, develop approximate plots of (a) castability versus weldability and (b) castability versus machinability for at least five of the materials listed in the table.

SYNTHESIS, DESIGN, AND PROJECTS

12.34 List casting processes that are suitable for making hollow parts with (a) complex external features, (b) complex internal features, and (c) both external and internal features. Explain your choices.

12.35 Small amounts of slag and dross often persist after skimming and are introduced into the molten metal flow in casting. Recognizing that slag and dross are less dense than the molten metal, design mold features that will remove small amounts of slag before the metal reaches the mold cavity.

12.36 For the cast metal wheel illustrated in Fig. P12.36, show how (a) riser placement, (b) core placement, (c) padding, and (d) chills may be used to help feed molten metal and eliminate porosity in the isolated hub boss.

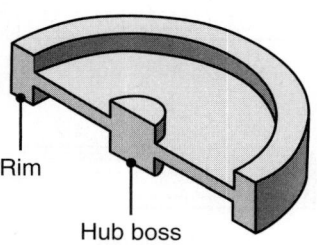

Rim

Hub boss

FIGURE P12.36

12.37   Assume that the introduction to this chapter is missing. Write a brief introduction to highlight the importance of the topics covered in it.

12.38   In Fig. P12.38, the original casting design shown in (a) was modified to the design shown in (b).

The casting is round and has a vertical axis of symmetry. As a functional part, what advantages do you think the new design has over the old one?

12.39   An incorrect and a correct design for casting are shown in Fig. P12.39. Review the changes made and comment on their advantages.

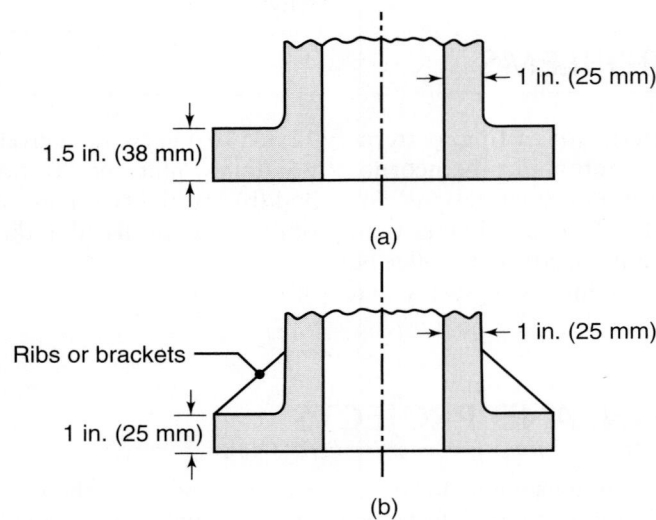

FIGURE P12.38

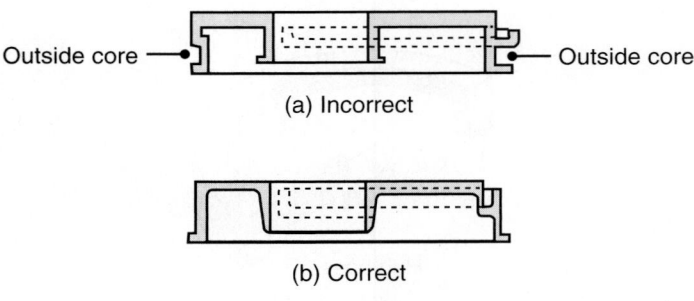

FIGURE P12.39

12.40 Three sets of designs for die casting are shown in Fig. P12.40. Note the changes made to die design 1 and comment on the reasons.

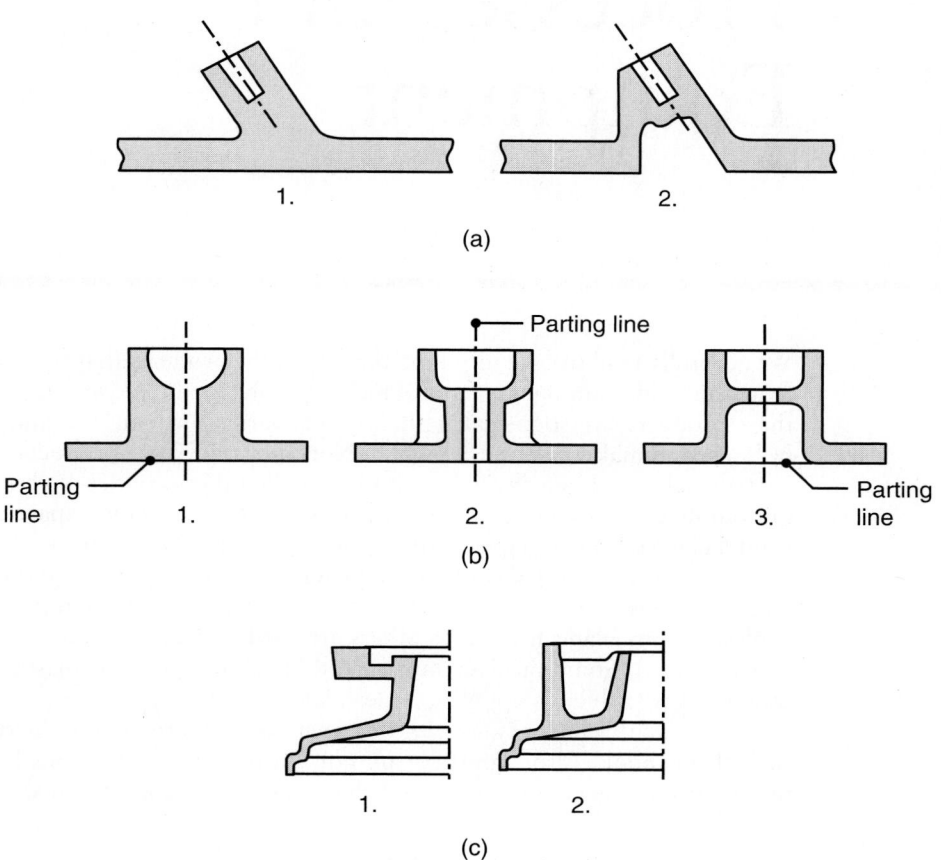

FIGURE P12.40

Forming and Shaping Processes and Equipment

We generally tend to take for granted many of the products that we use today and the materials and components from which they are made. However, when we inspect these products, we soon realize that a wide variety of materials and processes have been used in making them (Fig. III.1). Note also that some products consist of a few parts (mechanical pencils, light fixtures), while others consist of thousands of parts (automobiles, computers) or even millions of parts (airplanes, space shuttles). Some products have simple shapes with smooth curvatures (ball bearings, bicycle handles), but others have complex configurations (engine blocks, pumps) and detailed surface features (coins, silverware). Some products are used in critical applications (elevator cables, turbine blades), whereas others are used in routine applications (paper clips, forks, knives). Some products are very thin (aluminum foil, plastic film), whereas others are very thick (ship hulls, boiler plates).

Note that the words *forming* and *shaping* both are used in the title for this section of this book. Although there are not always clear distinctions between the two terms, forming generally indicates "changing" the shape of an existing solid body.

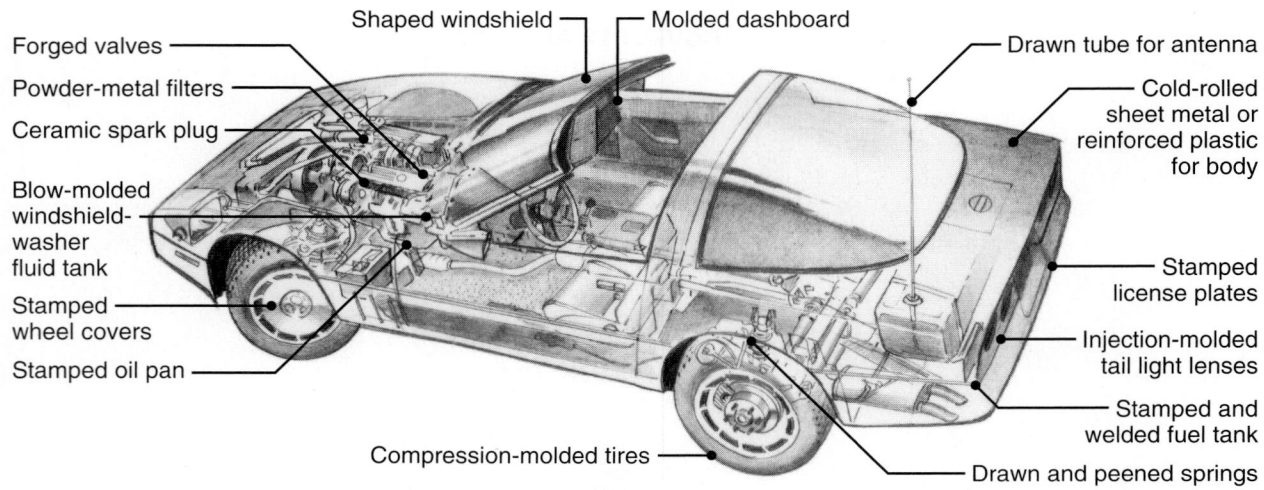

FIGURE III.1   Formed and shaped parts in a typical automobile.

Thus, in **forming processes,** the starting material (usually called the workpiece, stock, or blank) may be in the shape of a plate, sheet, bar, rod, wire, or tubing of various cross-sections. For example, an ordinary wire coat hanger is made by forming a straight piece of wire by bending and twisting it into the shape of a hanger. Also for example, the metal body for an automobile typically is made of cold-rolled, flat steel sheet which then is formed into various shapes (hood, roof, trunk, door panels) using a pair of large dies.

Shaping processes typically involve the molding and casting of soft or molten materials, and the finished product is usually at or near the final desired shape. It may require little or no further finishing. A plastic coat hanger, for example, is made by forcing molten plastic into a two-piece mold with a cavity in the shape of the hanger. Telephone receivers, refrigerator-door liners, computer housings, and countless other plastic products likewise are shaped by forcing the molten polymer into a mold and letting it solidify. As another example, the white ceramic insulator for an automotive spark plug is made by shaping clay in a mold, then drying it and firing it in a furnace.

Some of the forming and shaping operations produce long *continuous products,* such as plates, sheets, tubing, wire, and bars with various cross-sections. Rolling, extrusion, and drawing processes (Chapters 13 and 15) are capable of making such products, which then are cut into desired lengths. On the other hand, processes such as forging (Chapter 14), sheet metals (Chapter 16), powder metallurgy (Chapter 17), ceramics and glasses (Chapter 18), and plastics and reinforced plastics (Chapter 19) typically produce *discrete products.*

The initial material used in forming and shaping metals is usually molten metal, which is *cast* into individual *ingots,* or *continuously cast* into slabs, rods, or pipes. Cast structures are converted to *wrought structures* by the plastic-deformation processes described in various other chapters. The raw material used also may consist of *metal powders,* which then are pressed and sintered (heated without melting) into individual products. For plastics, the starting material is usually pellets, flakes, or powder, and for ceramics, it is clays and oxides obtained from ores or produced synthetically.

The important factors involved in each forming and shaping process are described in Part III, along with how material properties and processes affect product quality (Table III.1). We also explain why some materials can be processed only by certain manufacturing methods and why parts with particular shapes can be processed only by certain techniques and not by others. The characteristics of the machinery and the equipment used in these processes also significantly affect product quality, production rate, and the economics of a particular manufacturing operation.

TABLE III.1

General Characteristics of Forming and Shaping Processes	
Process	Characteristics
Rolling	
Flat	Production of flat plate, sheet, and foil at high speeds; good surface finish, especially in cold rolling; very high capital investment; low-to-moderate labor cost
Shape	Production of various structural shapes (such as I-beams and rails) at high speeds; includes thread rolling; requires shaped rolls and expensive equipment; low-to-moderate labor cost; moderate operator skill
Forging	Production of discrete parts with a set of dies; some finishing operations usually required; usually performed at elevated temperatures, but also cold for smaller parts; die and equipment costs are high; moderate-to-high labor cost; moderate-to-high operator skill
Extrusion	Production of long lengths of solid or hollow shapes with constant cross-section; product is then cut into desired lengths; usually performed at elevated temperatures; cold extrusion has similarities to forging and is used to make discrete products; moderate-to-high die and equipment cost; low-to-moderate labor cost; low-to-moderate operator skill
Drawing	Production of long rod and wire with various cross-sections; good surface finish; low-to-moderate die, equipment, and labor costs; low-to-moderate operator skill
Sheet-metal forming	Production of a wide variety of shapes with thin walls and with simple or complex geometries; generally low-to-moderate die, equipment, and labor costs; low-to-moderate operator skill
Powder metallurgy	Production of simple or complex shapes by compacting and sintering metal powders; moderate die and equipment cost; low labor cost and skill
Processing of plastics and composite materials	Production of a wide variety of continuous or discrete products by extrusion, molding, casting, and fabricating processes; moderate die and equipment costs; high operator skill in processing of composite materials
Forming and shaping of ceramics	Production of discrete products by various shaping, drying, and firing processes; low-to-moderate die and equipment cost; moderate-to-high operator skill

Rolling of Metals

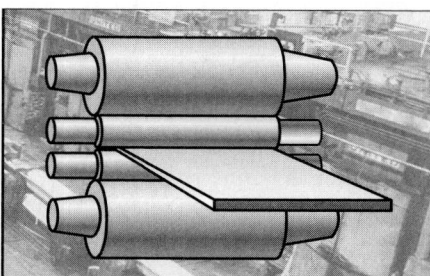

In this first chapter of Part III on the forming and shaping of metallic and non-metallic materials, we describe the rolling of metals, an important primary process. Specifically, we will discuss:

- Fundamentals of flat- and shape-rolling processes and their capabilities.
- The parameters involved in determining quantities such as forces, torque, and power.
- Types of defects that can develop, and how to avoid or minimize them.
- Characteristics of rolling mills and the manner in which rolls can be arranged in various configurations for specific purposes.

Typical products made by various rolling processes: Plates for ships, bridges, structures, machines; sheet metal for car bodies, aircraft fuselages, appliances, containers; foil for packaging; I-beams, railroad rails, channels, rings, seamless pipe and tubing; bolts, screws, and threaded components.

Alternative processes: Continuous casting, extrusion, drawing, machining of threaded components

13.1 | Introduction

Rolling is the process of reducing the thickness or changing the cross-section of a long workpiece by compressive forces applied through a set of **rolls** (Fig. 13.1). This process is similar to rolling dough with a rolling pin to reduce its thickness. Rolling, which accounts for about 90% of all metals produced by metalworking processes, was first developed in the late 1500s. Modern steelmaking practices and the production of various ferrous and nonferrous metals and alloys now generally involve combining continuous casting with rolling processes. This greatly improves productivity and lowers production costs, as described in Section 5.4. Nonmetallic materials also are rolled to reduce their thickness and enhance their properties. Typical applications are in the rolling of plastics, powder metals, ceramic slurry, and hot glass, as described in detail in their appropriate sections of this book.

Rolling first is carried out at elevated temperatures (hot rolling). During this phase, the coarse-grained, brittle, and porous structure of the ingot (or the continuously cast metal) is broken down into a *wrought structure* having a finer grain size

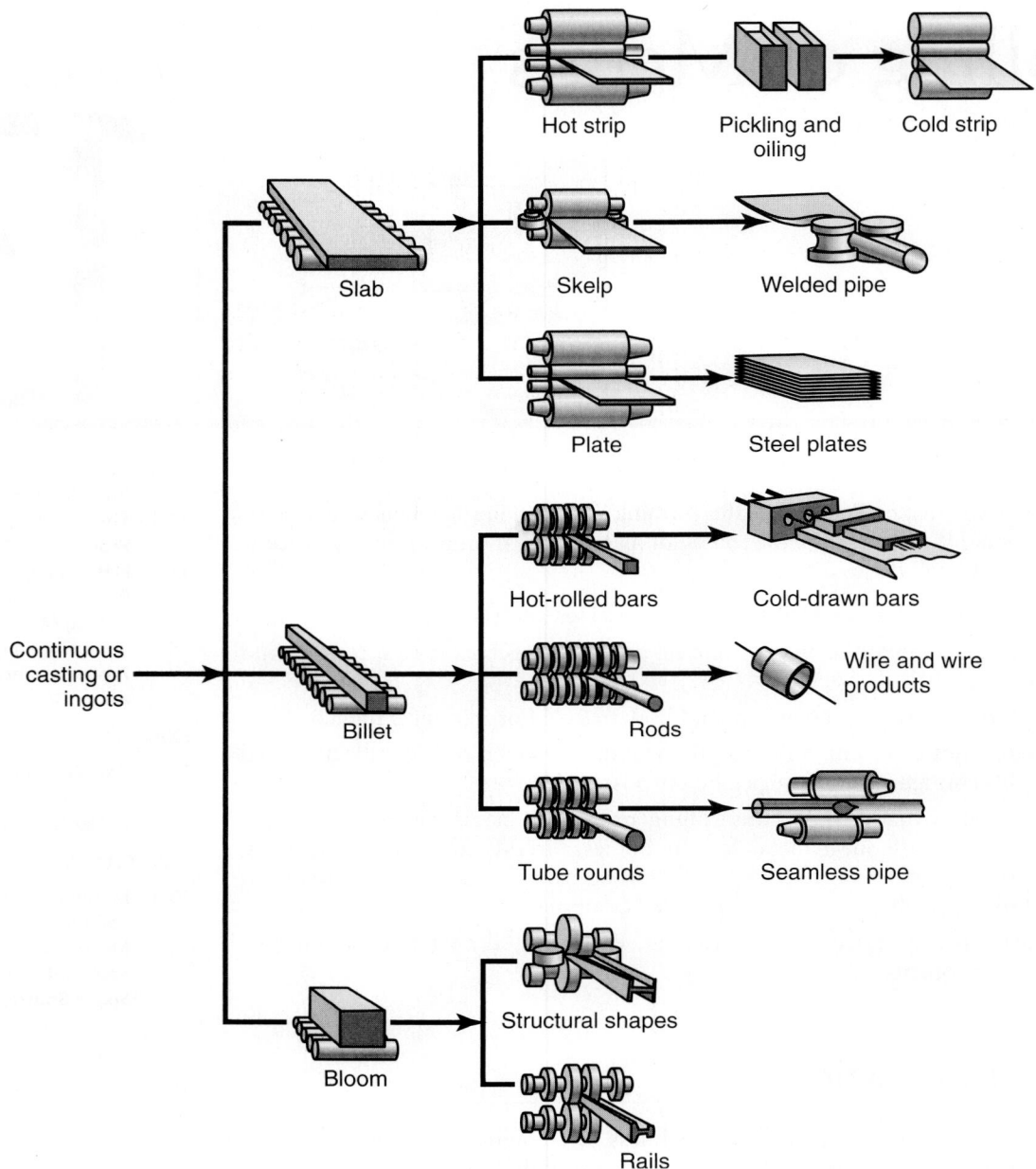

FIGURE 13.1 Schematic outline of various flat-rolling and shape-rolling processes. *Source:* After American Iron and Steel Institute.

and enhanced properties, such as strength and hardness. Subsequently, rolling typically is carried out at room temperature (cold rolling), whereby the rolled product has higher strength and hardness and better surface finish. However, it requires more energy (because of the higher strength of the material at room temperature) and will result in a product with anisotropic properties (due to preferred orientation or mechanical fibering, see Section 1.5).

Plates generally have a thickness of more than 6 mm (1/4 in.) and are used for structural applications, such as ship hulls, boilers, bridges, machinery, and nuclear vessels. Plates can be as much as 300 mm (12 in.) thick for large structural supports, 150 mm (6 in.) thick for reactor vessels, and 100 to 125 mm (4 to 5 in.) thick for warships and tanks.

Sheets generally are less than 6 mm thick and typically are provided to manufacturing facilities as coils—weighing as much as 30,000 kg (33 tons)—or as flat sheets for further processing into various products. Sheets typically are used for automobile and aircraft bodies, appliances, food and beverage containers, and kitchen and office equipment. Commercial aircraft fuselages and trailer bodies usually are made of a minimum of 1 mm (0.04 in.) thick aluminum-alloy sheets. For example, the skin thickness of a Boeing 747 fuselage is 1.8 mm (0.07 in.) and of a Lockheed L1011 is 1.9 mm (0.075 in.). Steel sheets used for automobile and appliance bodies are typically about 0.7 mm (0.03 in.) thick. Aluminum beverage cans are made from sheets 0.28 mm (0.01 in.) thick. After processing into a can, this sheet metal becomes a cylindrical body with a wall thickness of 0.1 mm (0.004 in.). Aluminum **foil** (typically used for wrapping and for candy and chewing gum) has a thickness of 0.008 mm (0.0003 in.), although thinner foils down to 0.003 mm (0.0001 in.) also can be produced with a variety of metals.

This chapter describes the fundamentals of **flat-rolling** and various **shape-rolling** operations, the production of seamless tubing and pipe, and discusses the important factors involved in rolling practices.

13.2 | The Flat-Rolling Process

A schematic illustration of the *flat-rolling* process is shown in Fig. 13.2a. A metal strip of thickness h_0 enters the **roll gap** and is reduced to thickness h_f by a pair of rotating rolls—each roll being powered individually by electric motors. The surface speed of the rolls is V_r. The velocity of the strip increases from its entry value of V_o as it moves through the roll gap in the same manner in which an incompressible fluid must flow faster as it moves through a converging channel. The velocity of the strip is highest at the exit from the roll gap and is denoted as V_f.

Because the surface speed of the rigid roll is constant, there is *relative sliding* between the roll and the strip along the arc of contact in the roll gap, L. At one point

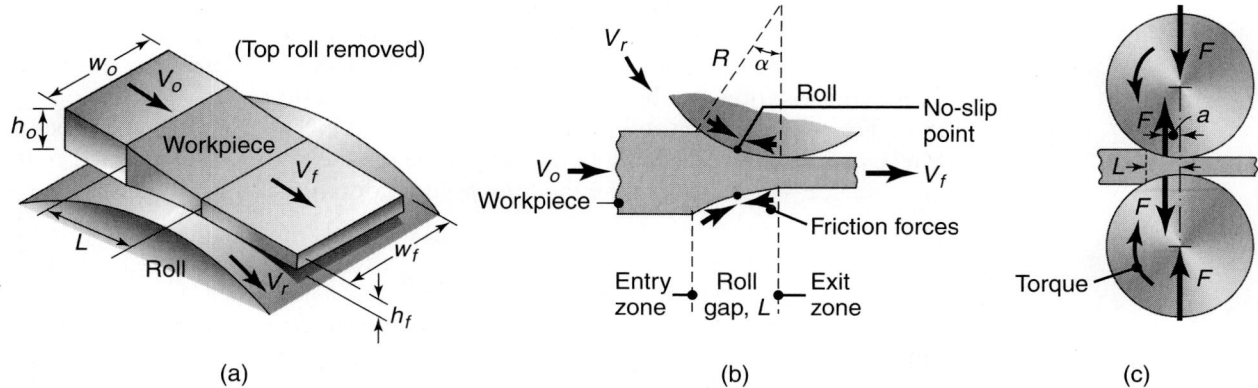

FIGURE 13.2   (a) Schematic illustration of the flat-rolling process. (b) Friction forces acting on strip surfaces. (c) Roll force, F, and the torque, T, acting on the rolls. The width of the strip, w, usually increases during rolling, as shown later in Fig. 13.5.

along the contact length (called the **neutral point** or **no-slip point**) the velocity of the strip is the same as that of the roll. To the left of this point, the roll moves faster than the strip; to the right of this point, the strip moves faster than the roll. Consequently, the frictional forces—which oppose motion between two sliding bodies—act on the strip, as shown in Fig. 13.2b.

The rolls pull the material into the roll gap through a *net frictional force* on the material. Thus, the net frictional force must be to the right in Fig. 13.2b. This also means that the frictional force to the left of the neutral point must be higher than the friction force to the right. Although friction is necessary for rolling materials (just as it is in driving a car on a road), energy is dissipated in overcoming friction. Thus, increasing friction also increases rolling forces and power requirements. Furthermore, high friction could damage the surface of the rolled product (or cause sticking, as it can occur in rolling dough). Thus, a compromise is made in practice—one which induces low coefficients of friction by using effective lubricants.

The maximum possible **draft** is defined as the difference between the initial and final strip thicknesses or $(h_o - h_f)$. It can be shown that this is a function of the coefficient of friction, μ, between the strip and the roll and the roll radius, R, by the following relationship:

$$h_o - h_f = \mu^2 R \qquad (13.1)$$

Thus, as expected, the higher the friction and the larger the roll radius, the greater the maximum possible draft becomes. Note that this situation is similar to the use of large tires (high R) and rough treads (high μ) used on farm tractors and off-road, earth-moving equipment, thus permitting the vehicles to travel over rough terrain without skidding.

13.2.1 Roll force, torque, and power requirement

The rolls apply pressure on the flat strip in order to reduce its thickness, resulting in a *roll force*, F, as shown in Fig. 13.2c. Note that this force is shown in the figure as perpendicular to the plane of the strip rather than at an angle. This is because, in practice, the arc of contact is very small compared with the roll radius, so we can assume that the roll force is perpendicular to the workplace without causing significant error in calculations. The roll force in flat rolling can be estimated from the formula

$$F = L w Y_{\text{avg}} \qquad (13.2)$$

where L is the roll-strip contact length, w is the width of the strip, and Y_{avg} is the average true stress (see Section 2.2) of the strip in the roll gap. Equation (13.2) is for a *frictionless* situation; however, an estimate of the *actual roll force* may be made by increasing this force by about 20% to account for the effect of friction.

The *torque* on the roll is the product of F and a. The power required per roll can be estimated by assuming that F acts in the middle of the arc of contact; thus in Fig. 13.2c, $a = L/2$. Therefore, the *total power* (for two rolls) in S.I. units is

$$\text{Power} = \frac{2\pi F L N}{60,000} \text{ kW} \qquad (13.3)$$

where F is in newtons, L is in meters, and N is the revolutions per minute of the roll. In traditional English units, the total power can be expressed as

$$\text{Power} = \frac{2\pi F L N}{33,000} \text{ hp} \qquad (13.4)$$

where F is in pounds and L is in feet.

EXAMPLE 13.1 Calculation of roll force and torque in flat rolling

An annealed copper strip, 9 in. (228 mm) wide and 1.00 in. (25 mm) thick, is rolled to a thickness of 0.80 in. (20 mm) in one pass. The roll radius is 12 in. (300 mm), and the rolls rotate at 100 rpm. Calculate the roll force and the power required in this operation.

Solution The roll force is determined from Eq. (13.2) in which L is the roll-strip contact length. It can be shown from simple geometry that this length is given approximately by

$$L = \sqrt{R(h_o - h_f)} = \sqrt{12(1.00 - 0.80)} = 1.55 \text{ in.}$$

The average true stress, Y_{avg}, for annealed copper is determined as follows. First note that the absolute value of the true strain that the strip undergoes in this operation is

$$\varepsilon = \ln\left(\frac{1.00}{0.80}\right) = 0.223$$

Referring to Fig. 2.6, note that annealed copper has a true stress of about 12,000 psi in the unstrained condition, and at a true strain of 0.223, the true stress is 40,000 psi. Thus, the average true stress is $(12,000 + 40,000)/2 = 26,000$ psi. We can now define the roll force as

$$F = LwY_{avg} = (1.55)(9)(26,000) = 363,000 \text{ lb} = 1.6 \text{ MN}$$

The total power is calculated from Eq. (13.4), noting that $N = 100$ rpm. Thus,

$$\text{Power} = \frac{2\pi FLN}{33,000} = \frac{2\pi(363,000)(1.55/12)(100)}{33,000}$$

$$= 898 \text{ hp} = 670 \text{ kW}$$

Exact calculation of the force and the power requirements in rolling is difficult because of the uncertainties involved in (a) determining the exact contact geometry between the roll and the strip and (b) accurately estimating both the coefficient of friction and the strength of the material in the roll gap, particularly for hot rolling because of the sensitivity of the strength of the material to strain rate (see Section 2.2.7).

Reducing roll force.   Roll forces can cause significant deflection and flattening of the rolls (as it does in a rubber tire). Such changes in turn will affect the rolling operation. Also, the columns of the **roll stand** (including the housing, chocks, and bearings, as shown in Fig. 13.3) may deflect under high roll forces to such an extent that the roll gap can open up significantly. Consequently, the rolls have to be set closer than originally calculated in order to compensate for this deflection and to obtain the desired final thickness.

Roll forces can be reduced by the following means:

- Reducing friction at the roll-workpiece interface
- Using smaller-diameter rolls to reduce the contact area
- Taking smaller reductions-per-pass to reduce the contact area
- Rolling at elevated temperatures to lower the strength of the material
- Applying tensions to the strip

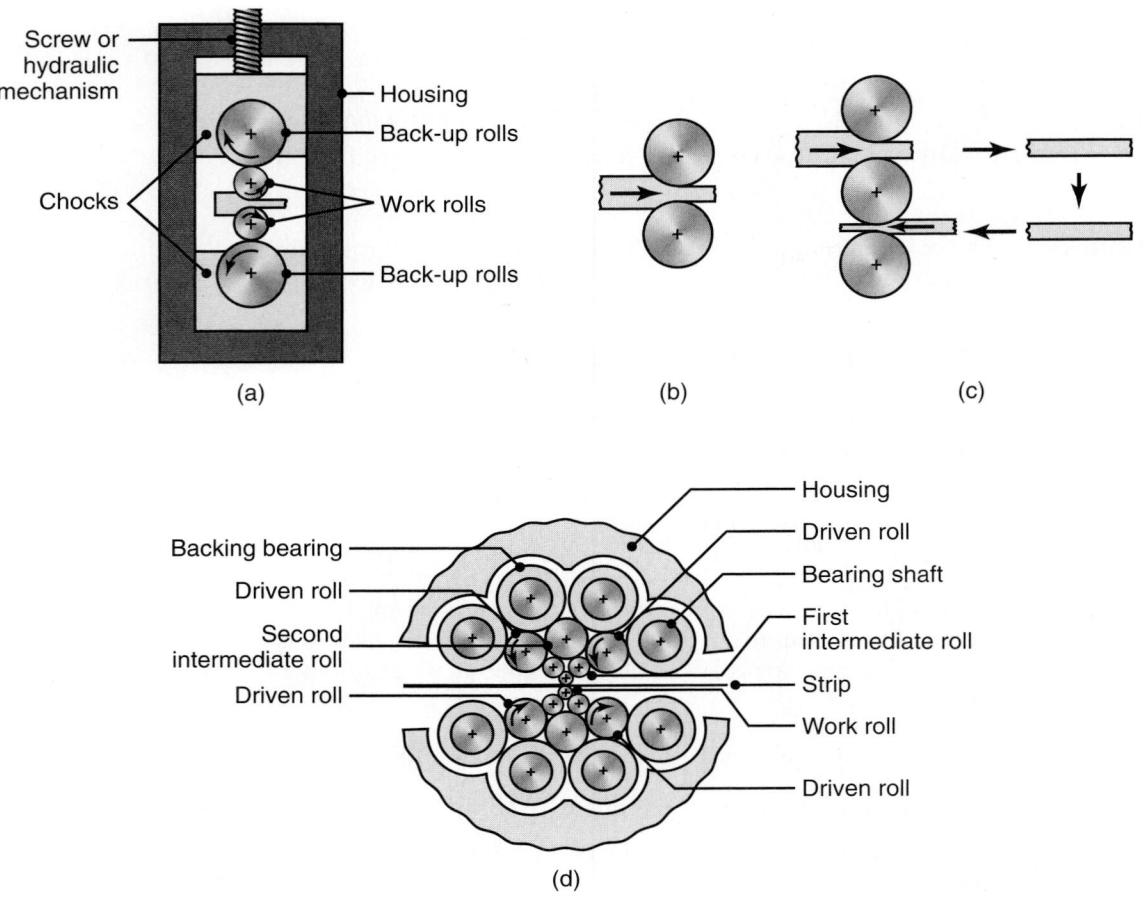

FIGURE 13.3 Schematic illustration of various roll arrangements: (a) four-high rolling mill showing various features. The stiffness of the housing, the rolls, and the roll bearings are all important in controlling and maintaining the thickness of the rolled strip; (b) two-high mill; (c) three-high mill; and (d) cluster (or Sendzimir) mill.

Another effective method of reducing roll forces is to apply longitudinal **tension** to the strip during rolling (as a result of which the compressive stresses required to plastically deform the material become smaller). Because they require high roll forces, tensions are important particularly in rolling high-strength metals. Tensions can be applied to the strip either at the entry zone (**back tension**), at the exit zone (**front tension**), or both. Back tension is applied to the sheet by applying a braking action to the reel that supplies the sheet into the roll gap (*pay-off reel*) by some suitable means. Front tension is applied by increasing the rotational speed of the *take-up reel*. Although it has limited and specialized applications, rolling also can be carried out by front tension only, with no power supplied to the rolls—a process known as **Steckel rolling**.

13.2.2 Geometric considerations

Because of the forces acting on them, rolls undergo shape changes during rolling. Just as a straight beam deflects under a transverse load, roll forces tend to bend the rolls *elastically* during rolling (Fig. 13.4a). As expected, the higher the elastic modulus of the roll material, the smaller the roll deflection.

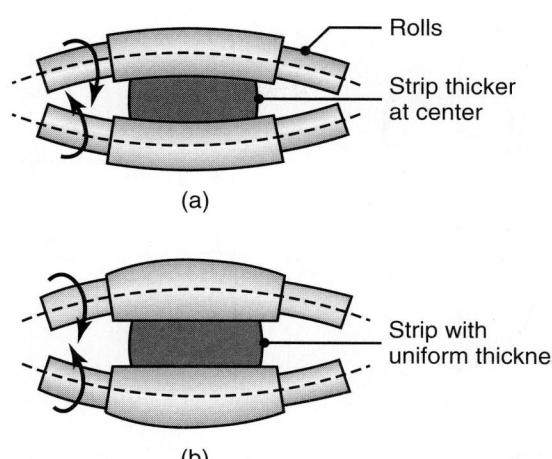

(a)

(b)

FIGURE 13.4   (a) Bending of straight cylindrical rolls caused by roll forces. (b) Bending of rolls ground with camber, producing a strip with uniform thickness through the strip width. Deflections have been exaggerated for clarity.

As a result of roll bending, the rolled strip tends to be thicker at its center than at its edges (**crown**). The usual method of avoiding this problem is to grind the rolls in such a way that their diameter at the center is slightly larger than at their edges (**camber**). Thus, when the roll bends, the strip being rolled now has a constant thickness along its width (Fig. 13.4b). For rolling sheet metals, the radius of the maximum camber point is generally 0.25 mm (0.01 in.) greater than that at the edges of the roll. However, as expected, a particular camber is correct only for a certain load and strip width. To reduce the effects of deflection, the rolls also can be subjected to external bending by applying moments at their bearings (a technique demonstrated simply by bending a wooden stick at its ends, a manipulation that simulates camber).

Because of the heat generated by plastic deformation during rolling, rolls can become slightly barrel shaped (**thermal camber**). Unless compensated for by some means, this condition can produce strips that are thinner at the center than at the edges. Consequently, the total (or final) camber can be controlled by adjusting the location and the flow rate of the coolant along the length of the rolls during hot rolling.

Roll forces also tend to *flatten* the rolls elastically, producing an effect much like the flattening of automobile tires under load. Flattening of the rolls is undesirable as it results, in effect, in a larger roll radius. This, in turn, means a larger contact area for the same draft, and the roll force increases because of the now larger contact area.

Spreading.   In rolling plates and sheets with high width-to-thickness ratios, the width of the strip remains effectively constant during rolling. However, with smaller ratios (such as a strip with a square cross-section), its width increases significantly as it passes through the rolls (an effect commonly observed in the rolling of dough with a rolling pin). This increase in width is called *spreading* (Fig. 13.5). In the calculation of the roll force, the width w in Eq. (13.2a) then is taken as an average width.

It can be shown that spreading increases with (a) decreasing width-to-thickness ratio of the entering strip (because of reduction in the width constraint), (b) increasing

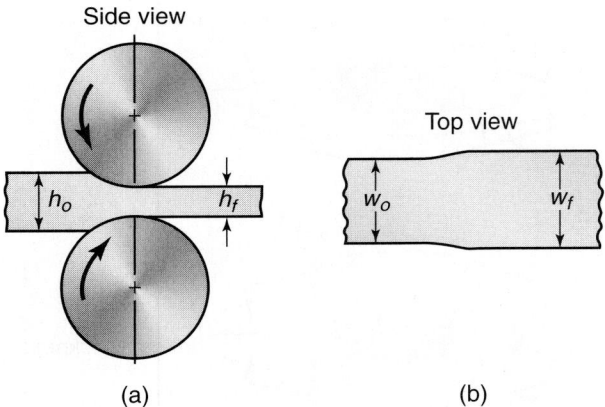

FIGURE 13.5   Increase in strip width (spreading) in flat rolling. Note that similar spreading can be observed when dough is rolled with a rolling pin.

friction, and (c) decreasing ratio of the roll radius to the strip thickness. The last two effects are due to the increased longitudinal constraint of the material flow in the roll gap. Spreading can be prevented also by using additional rolls (with vertical axes) in contact with the edges of the rolled product in the roll gap (*edger mills*), thus providing a physical constraint to spreading.

13.2.3 Vibration and chatter

Vibration and *chatter* can have significant effects on product quality and the productivity of metalworking operations. Chatter, generally defined as *self-excited vibration*, can occur in rolling as well as in extrusion, drawing, machining, and grinding operations. In rolling, it leads to periodic variations in the thickness of the rolled sheet and in its surface finish and, consequently, can lead to excessive scrap (see Table 40.6). Chatter in rolling has been found to occur predominantly in *tandem mills*. Chatter is very detrimental to productivity; it has been estimated, for example, that modern rolling mills could operate at up to 50% higher speeds were it not for chatter.

Chatter is a very complex phenomenon (see also Section 25.4) and results from interactions between the structural dynamics of the mill stand and the dynamics of the rolling operation. Rolling speed and lubrication are found to be the two most important parameters. Although not always practical to implement, it also has been suggested that chatter can be reduced by increasing the distance between the stands of the rolling mill (see Fig. 13.11), increasing the strip width, decreasing the reduction-per-pass (draft), increasing the roll radius, increasing the strip-roll friction, and incorporating dampers in the roll supports.

13.3 | Flat-Rolling Practice

The initial rolling steps (*breaking down*) of the material typically is done by **hot rolling** (above the recrystallization temperature of the metal, Section 1.6). As described in Section 10.2 and illustrated in Fig. 10.2, a **cast structure** typically is dendritic, and it includes coarse and nonuniform grains; this structure usually is brittle and may be porous. Hot rolling converts the cast structure to a **wrought structure** (Fig. 13.6) with finer grains and enhanced ductility, both of which result from the breaking up of brittle grain boundaries and the closing up of internal defects (especially porosity). Typical temperature ranges for hot rolling are about 450°C (850°F) for aluminum alloys, up to 1250°C (2300°F) for alloy steels, and up to 1650°C (3000°F) for refractory alloys (see also Table 14.3).

The product of the first hot-rolling operation is called a **bloom** or **slab** (see Fig. 13.1). A bloom usually has a square cross-section, at least 150 mm (6 in.) on the side; a slab usually is rectangular in cross-section. Blooms are processed further by *shape rolling* into structural shapes, such as I-beams and railroad rails (Section 13.5). Slabs are rolled into plates and sheets. **Billets** usually are square (with a cross-sectional area smaller than blooms) and later are rolled into various shapes, such as round rods and bars, using shaped rolls. Hot-rolled round rods (**wire rod**) are used as the starting material for rod- and wire-drawing operations (Chapter 15).

In the hot rolling of blooms, billets, and slabs, the surface of the material usually is **conditioned** (prepared for a subsequent operation) prior to rolling them. Conditioning is done by means such as using a torch (*scarfing*) to remove heavy scale or by rough grinding to smoothen surfaces. Prior to cold rolling, the scale developed during hot rolling may be removed by *pickling* with acids (acid etching), by such mechanical means as blasting with water, or by grinding to remove other defects as well.

Cold rolling is carried out at room temperature and, compared with hot rolling, produces sheets and strips with a much better surface finish (because of lack of scale), dimensional tolerances, and mechanical properties (because of strain hardening).

Pack rolling is a flat-rolling operation in which two or more layers of metal are rolled together, thus improving productivity. *Aluminum foil*, for example, is pack rolled in two layers, so only the top and bottom outer layers have been in contact with the rolls. Note that, one side of aluminum foil is matte, while the other side is shiny. The foil-to-foil side has a matte and satiny finish, but the foil-to-roll side is shiny and bright because it has been in contact under high contact stresses with the polished rolls during rolling.

Rolled mild steel, when subsequently stretched during sheet-forming operations, undergoes *yield-point elongation* (Section 16.3)—a phenomenon that causes surface irregularities called *stretcher strains* or *Lüder's bands*. To correct this situation, the sheet metal is subjected to a final, light pass of 0.5 to 1.5% reduction known as **temper rolling** or **skin pass**.

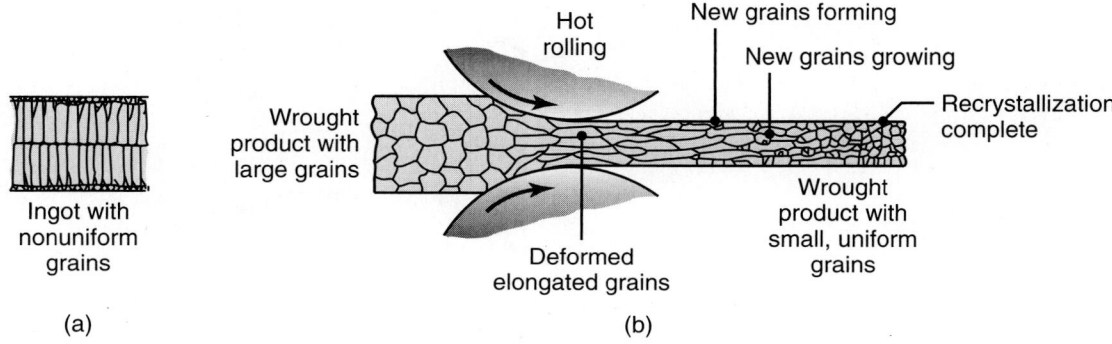

FIGURE 13.6 Changes in the grain structure of cast or of large-grain wrought metals during hot rolling. Hot rolling is an effective way to reduce grain size in metals for improved strength and ductility. Cast structures of ingots or continuous castings are converted to a wrought structure by hot working.

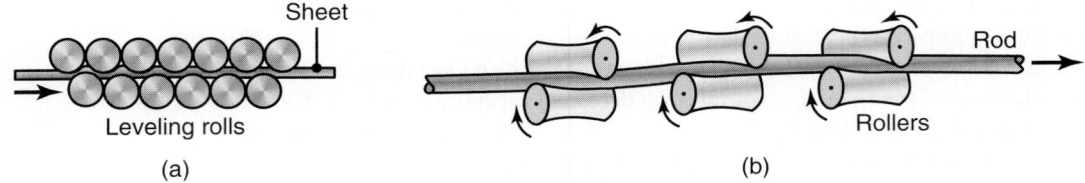

FIGURE 13.7 (a) A method of roller leveling to flatten rolled sheets. (b) Roller leveling to straighten drawn bars.

A rolled sheet may not be sufficiently flat as it leaves the roll gap, due to factors such as variations in the incoming material or in the processing parameters during rolling. To improve flatness, the rolled strip typically goes through a series of **leveling rolls**. Several roller arrangements are used, as shown in Fig. 13.7. The work-piece is flexed in opposite directions as it passes through the sets of rollers, where each roll usually is driven separately by an individual electric motor.

13.3.1 Defects in rolled plates and sheets

Defects may be present on the surfaces of rolled plates and sheets, or there may be internal structural defects. Defects are undesirable not only because they degrade surface appearance but also because they may adversely affect the strength, forma-bility, and other manufacturing characteristics. Several surface defects (such as scale, rust, scratches, gouges, pits, and cracks) have been identified for sheet metals. These defects may be caused by inclusions and impurities in the original cast material or by various other conditions related to material preparation and to the rolling operation.

Wavy edges on sheets (Fig. 13.8a) are the result of roll bending. The strip is thinner along its edges than at its center (see Fig. 13.4a), thus the edges elongate more than the center. Consequently, the edges buckle because they are constrained by the central region from expanding freely in the longitudinal (rolling) direction. The **cracks** shown in Fig. 13.8b and c are usually the result of poor material ductili-ty at the rolling temperature. Because the quality of the edges of the sheet may affect sheet-metal forming operations, edge defects in rolled sheets often are removed by

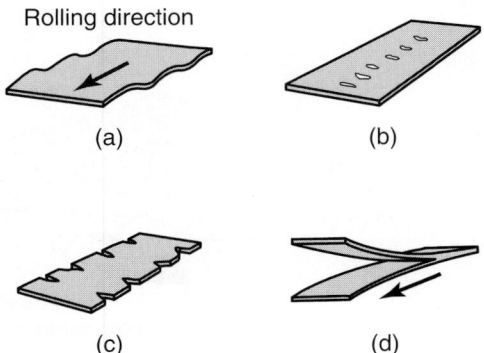

FIGURE 13.8 Schematic illustration of typical defects in flat rolling: (a) wavy edges; (b) zipper cracks in the center of the strip; (c) edge cracks; and (d) alligatoring.

shearing and slitting operations (Section 16.2). **Alligatoring** (Fig. 13.8d) is a complex phenomenon and typically is caused by nonuniform bulk deformation of the billet during rolling or by the presence of defects in the original cast material.

13.3.2 Other characteristics of rolled metals

Residual stresses.   Because of nonuniform deformation of the material in the roll gap, residual stresses can develop in rolled plates and sheets, especially during cold rolling. Small-diameter rolls or small thickness reductions-per-pass tend to deform the metal plastically more at its surfaces than in the bulk (Fig. 13.9a). This situation results in compressive residual stresses on the surfaces and tensile stresses in the bulk. Conversely, large-diameter rolls or high reductions-per-pass tend to deform the bulk more than the surfaces (Fig. 13.9b). This is due to the higher frictional constraint at the surfaces along the arc of contact—a situation that produces residual stress distributions that are the opposite of those in the case of small-diameter rolls.

Dimensional tolerances.   Thickness tolerances for cold-rolled sheets usually range from ±0.1 to 0.35 mm (±0.004 to 0.014 in.), depending on the thickness. Tolerances are much greater for hot-rolled plates because of thermal effects. *Flatness tolerances* are usually within ±15 mm/m (±3/16 in./ft) for cold rolling and ±55 mm/m (5/8 in./ft) for hot rolling.

Surface roughness.   The ranges of surface roughness in cold and hot rolling are given in Fig. 23.13 which, for comparison, includes ranges for other manufacturing processes. Note that cold rolling can produce a very fine surface finish, hence products made of cold-rolled sheets may not require additional finishing operations, depending on the application. Note also that hot rolling and sand casting produce the same range of surface roughness.

Gage numbers.   The thickness of a sheet usually is identified by a *gage number:* the smaller the number, the thicker the sheet. Several numbering systems are used in

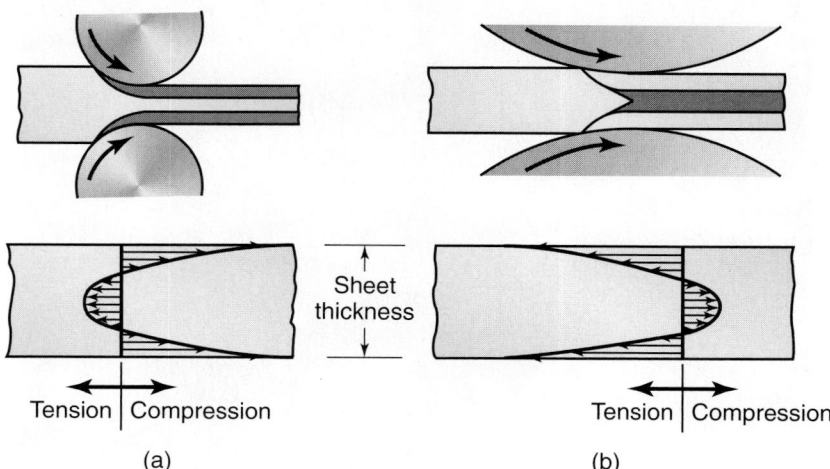

FIGURE 13.9   (a) Residual stresses developed in rolling with small-diameter rolls or at small reductions in thickness per pass. (b) Residual stresses developed in rolling with large-diameter rolls or at high reductions-per-pass. Note the reversal of the residual stress patterns.

industry, depending on the type of sheet metal being classified. Rolled sheets of copper and of brass also are identified by thickness changes during rolling, such as 1/4 hard, 1/2 hard, and so on.

13.4 | Rolling Mills

Several types of *rolling mills* and equipment are available with diverse roll arrangements. Although the equipment for hot and cold rolling is essentially the same, there are important differences in the roll materials, process parameters, lubricants, and cooling systems. The design, construction, and operation of rolling mills (Fig. 13.10) require major investments. Highly automated mills produce close-tolerance, high-quality plates and sheets at high production rates and low cost-per-unit weight, particularly when integrated with continuous casting. Rolling speeds may range up to 40 m/s (130 ft/s). The width of rolled products may range up to 5 m (200 in.).

Two-high rolling mills (Figs. 13.3b) are used for hot rolling in initial breakdown passes (*primary roughing* or **cogging mills**) on cast ingots or in continuous casting, with roll diameters ranging from 0.6 to 1.4 m (24 to 55 in.). In the **three-high mill** (*reversing mill*, Fig. 13.3c) the direction of material movement is reversed after each pass, using elevator mechanisms and various manipulators. The plate being rolled, which may weigh as much as 160 tons, is raised repeatedly to the upper roll gap, rolled, and then lowered to the lower roll gap and rolled, and so on.

Four-high mills (Fig. 13.3a) and **cluster mills** (Sendzimir or **Z mill**, Fig. 13.3d) are based on the principle that small-diameter rolls lower roll forces (because of small roll-strip contact area) and power requirements and reduce spreading. Moreover, when worn or broken, small rolls can be replaced at lower cost than can large ones.

FIGURE 13.10   A general view of a rolling mill. *Source:* Courtesy of Ispat Inland.

On the other hand, small rolls deflect more under roll forces and have to be supported by other large-diameter rolls, as is done in four-high and cluster mills. Although the cost of a Sendzimir mill facility can be very high, it is particularly suitable for cold rolling thin sheets of high-strength metals. Common rolled widths in this mill are 0.66 m (26 in.) with a maximum of 1.5 m (60 in.).

In **tandem rolling** the strip is rolled continuously through a number of **stands** to thinner gages with each pass (Fig 13.11). Each stand consists of a set of rolls with its own housing and controls; a group of stands is called a *train*. The control of the strip thickness and the speed at which it travels through each roll gap is critical. Extensive electronic and computer controls are used in these operations, particularly in precision rolling.

Roll materials.   The basic requirements for roll materials are strength and resistance to wear. Common roll materials are cast iron, cast steel, and forged steel; tungsten carbides are also used for small-diameter rolls, such as the working roll in the cluster mill (Fig. 13.3d). Forged-steel rolls, although more costly than cast rolls, have higher strength, stiffness, and toughness than cast-iron rolls. Rolls for cold rolling are ground to a fine finish. For special applications, they also are polished. Rolls made for cold rolling should not be used for hot rolling because they may crack from thermal cycling (*heat checking*) and *spalling* (cracking or flaking of surface layers). Recall also from earlier discussions that the elastic modulus of the roll influences roll deflection and flattening.

Note that the bottom surface of an aluminum beverage can, for example, has what appear to be longitudinal scratches on it. This is explained by the fact that the

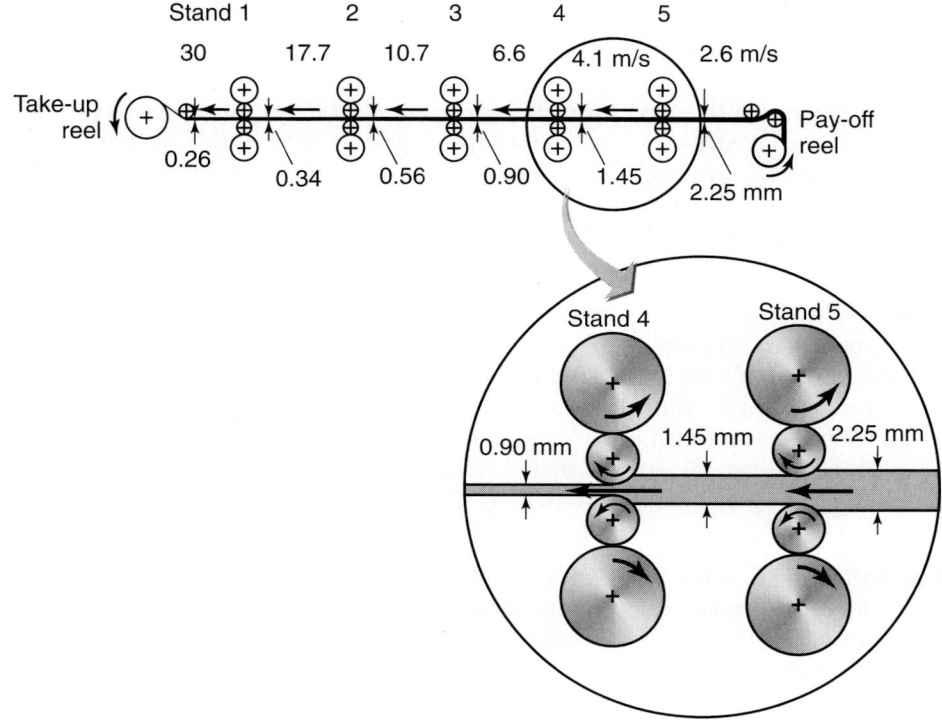

FIGURE 13.11   An example of a tandem-rolling operation.

surface is a replica of the surface finish of the roll, which is produced by grinding (see Fig. 26.2a). In this way, the rolling direction of the original aluminum sheet also can be determined easily.

Lubricants.   Hot rolling of ferrous alloys usually is carried out without lubricants, although graphite may be used. Water-based solutions are used to cool the rolls and to break up the scale on the rolled material. Nonferrous alloys are hot rolled with a variety of compounded oils, emulsions, and fatty acids. Cold rolling is carried out with water-soluble oils or low-viscosity lubricants, such as mineral oils, emulsions, paraffin, and fatty oils. The heating medium, such as that used in heat treating billets and slabs, may also act as a lubricant. Residual salts from molten-salt baths (Section 4.12) also offer effective lubrication during rolling.

13.5 | Various Rolling Processes and Mills

Several rolling processes and mills have been developed to produce a specific family of product shapes.

Shape rolling.   Straight and long structural shapes (such as channels, I-beams, railroad rails, and solid bars) are formed at elevated temperatures by *shape rolling* (*profile rolling*), in which the stock goes through a set of specially designed rolls (Fig. 13.12; see also Fig. 13.1). *Cold shape rolling* also can be done with the starting materials in the shape of wire with various cross-sections. Because the material's cross-section is reduced nonuniformly, the design of a series of rolls (**roll-pass design**) requires considerable experience in order to avoid external and internal defects, hold dimensional tolerances, and reduce roll wear.

Roll forging.   In this operation (also called *cross rolling*), the cross-section of a round bar is shaped by passing it through a pair of rolls with profiled grooves (Fig. 13.13). Roll forging typically is used to produce tapered shafts and leaf springs, table knives, and hand tools; it also may be used as a preliminary forming operation, to be followed by other forging processes.

Skew rolling.   A process similar to roll forging is *skew rolling*, typically used for making ball bearings (Fig. 13.14a). Round wire or rod is fed into the roll gap, and roughly spherical blanks are formed continuously by the action of the rotating rolls. Another method of forming near-spherical blanks for ball bearings is to shear pieces from a round bar and then upset them in headers (see also Fig. 14.10) between two dies with hemispherical cavities (Fig. 13.14b). The balls subsequently are ground and polished in special machinery (see Fig. 26.15).

Ring rolling.   In *ring rolling*, a thick ring is expanded into a large-diameter thinner one. The ring is placed between sets of two rolls, one of which is driven while the other is idle (Fig. 13.15a). Its thickness is reduced by bringing the rolls closer together as they rotate. Since the volume of the ring material remains constant during plastic deformation (volume constancy), the reduction in ring thickness is compensated by an increase in its diameter. Depending on its size, the ring-shaped blank may be produced by such means as cutting from a plate, piercing, or cutting a thick-walled

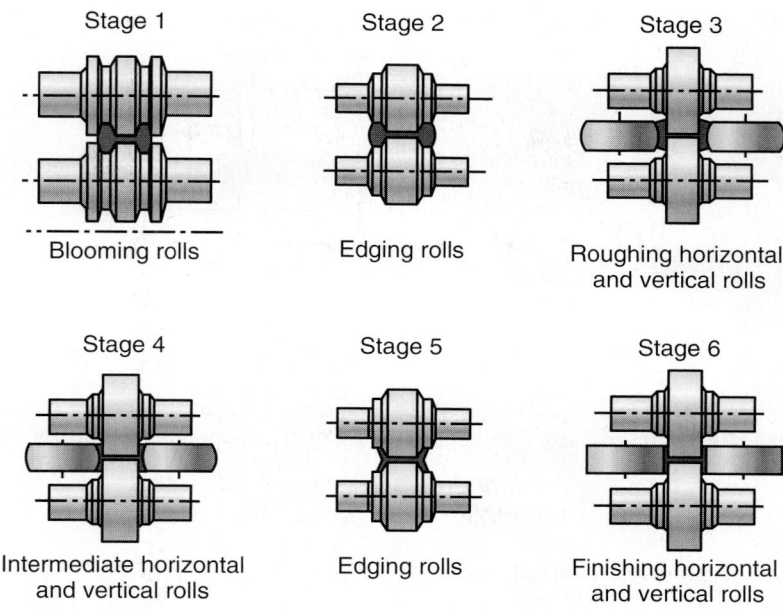

FIGURE 13.12   Steps in the shape rolling of an I-beam part. Various other structural sections, such as channels and rails, also are rolled by this kind of process.

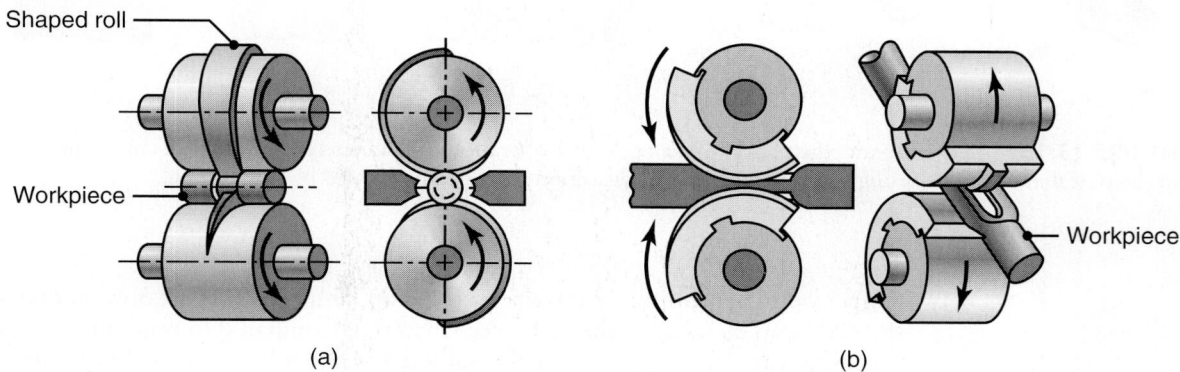

FIGURE 13.13   Two examples of the roll-forging operation, also known as *cross-rolling*. Tapered leaf springs and knives can be made by this process. *Source:* After J. Holub.

pipe. Various shapes can be ring rolled using shaped rolls (Fig. 13.15). It should be noted that the thickness of rings also can be reduced by an open-die forging process, as illustrated in Fig. 14.4c; however, dimensional control and surface finish will not be as good as in ring rolling.

Typical applications of ring rolling are large rings for rockets and turbines, jet-engine cases, gearwheel rims, ball-bearing and roller-bearing races, flanges, and reinforcing rings for pipes. The process can be carried out at room or at an elevated

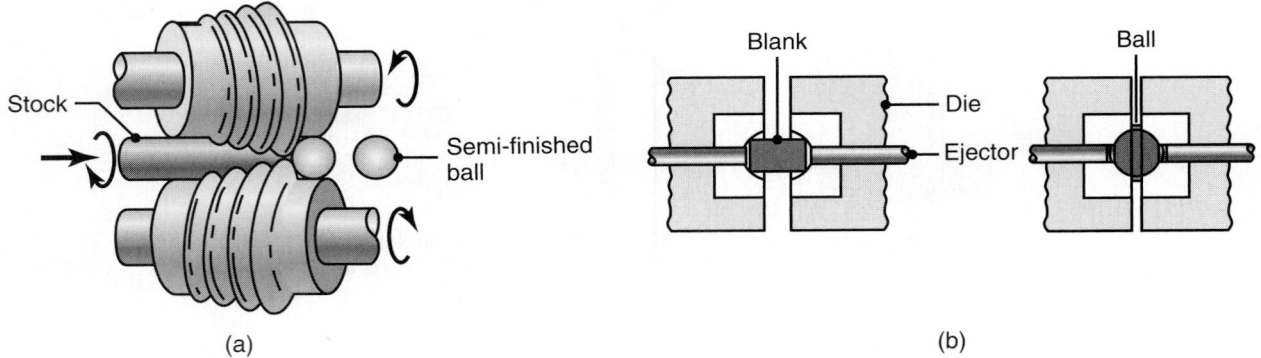

FIGURE 13.14   (a) Production of steel balls by the skew-rolling process. (b) Production of steel balls by upsetting a cylindrical blank. Note the formation of flash. The balls made by these processes subsequently are ground and polished for use in ball bearings.

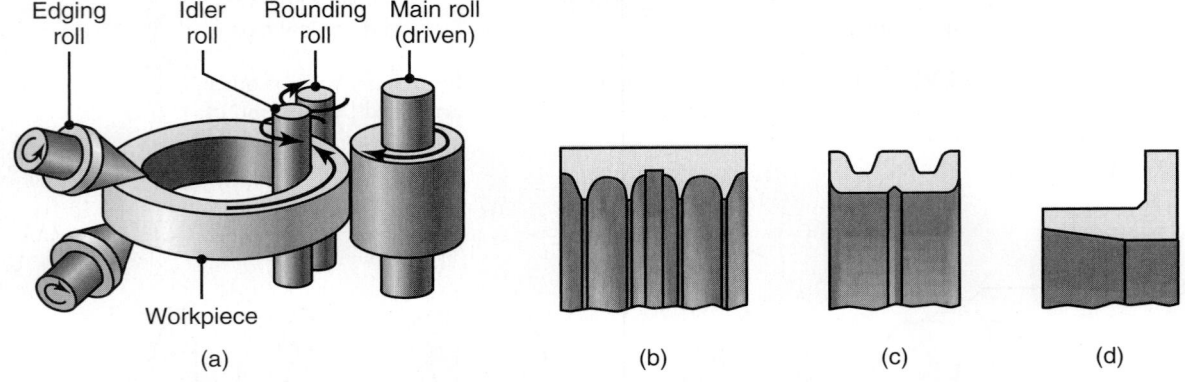

FIGURE 13.15   (a) Schematic illustration of a ring-rolling operation. Thickness reduction results in an increase in the part diameter. (b) through (d) Examples of cross-sections that can be formed by ring rolling.

temperature, depending on the size (which can be up to 10 ft, or 3 m, in diameter), strength, and ductility of the workpiece material. Compared to other manufacturing processes capable of producing the same part, the advantages of ring rolling are short production times, material savings, close dimensional tolerances, and favorable grain flow in the product, thus enhancing its strength in the desired direction. The design of the profile rolls requires considerable experience. Analytical techniques are being developed to rely less on established practice and help minimize defects in rolled products.

Thread rolling.   *Thread rolling* is a cold-forming process by which straight or tapered threads are formed on round rods or wire by passing them between dies. Threads are formed on the rod or wire with each stroke of a pair of flat reciprocating dies (Fig. 13.16a). In another method, threads are formed with *rotary dies* (Fig. 13.16c), at production rates as high as 80 pieces per second. Typical products are screws, bolts, and similar threaded parts. Depending on die design, the major diameter of a rolled thread may or may not be larger than a machined thread (Fig. 13.17a)—that is, the same as the blank diameter. The thread-rolling process

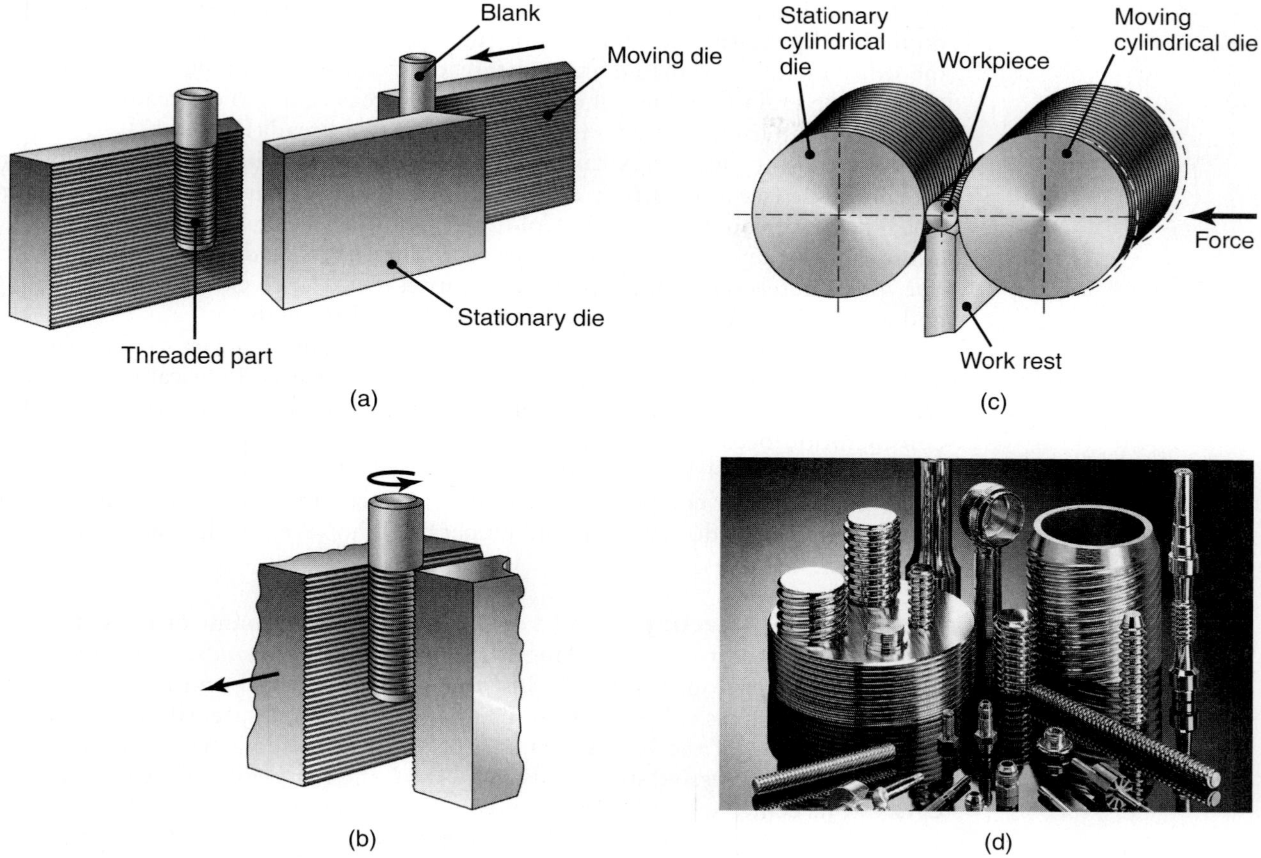

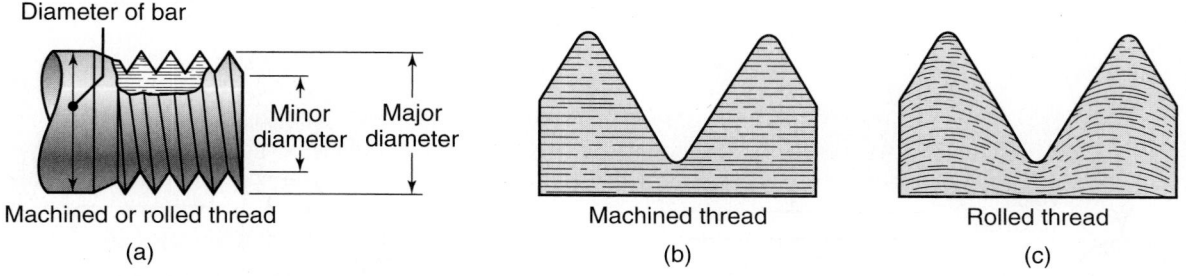

Blank
Moving die
Threaded part
Stationary die
(a)

Stationary cylindrical die
Workpiece
Moving cylindrical die
Force
Work rest
(c)

(b)

(d)

FIGURE 13.16 Thread rolling processes: (a) and (b) reciprocating flat dies; (c) two-roller dies; (d) A collection of thread-rolled parts made economically at high production rates. *Source:* Courtesy of Tesker Manufacturing Corp.

Diameter of bar
Minor diameter
Major diameter
Machined or rolled thread
(a)

Machined thread
(b)

Rolled thread
(c)

FIGURE 13.17 (a) Features of a machined or rolled thread. Grain flow in (b) machined and (c) rolled threads. Unlike machining, which cuts through the grains of the metal, the rolling of threads imparts improved strength because of cold working and favorable grain flow.

is capable of generating other shapes as well, such as grooves and various gear forms, and it is used in the production of almost all threaded fasteners at high production rates.

The thread-rolling process has the advantages of generating threads with good strength (due to cold working) and without any loss of material (scrap). The surface finish produced is very smooth, and the process induces compressive residual stresses on the workpiece surfaces, thus improving fatigue life. Thread rolling is superior to the other methods of manufacturing threads, because machining the threads cuts through the grain-flow lines of the material, whereas rolling the threads results in a grain-flow pattern that improves the strength of the thread (Fig. 13.17).

Spur and helical gears can be produced by cold-rolling processes similar to thread rolling (see also Section 24.7). The process may be carried out on solid cylindrical blanks or on precut gears. Cold rolling of gears has extensive applications in automatic transmissions and in power tools. **Internal thread rolling** can be carried out with a fluteless **forming tap**. This operation is similar to external thread rolling, and it produces accurate internal threads with good strength.

Lubrication is important in thread-rolling operations in order to obtain a good surface finish and surface integrity and to minimize defects. Lubrication affects the manner in which the material deforms during deformation, which is an important consideration because of the possibility of internal defects being developed (for example, see Fig. 14.16). Typically made of hardened steel, rolling dies are expensive because of their complex shape. They usually cannot be reground after they are worn. With proper die materials and preparation, however, die life may range up to millions of pieces.

Rotary tube piercing.   Also known as the **Mannesmann process**, this is a hot-working operation for making long, thick-walled *seamless pipe and tubing* (Fig. 13.18). Developed in the 1880s, this process is based on the principle that when a round bar is subjected to radial compressive forces, tensile stresses develop at the center of the bar (see Fig. 2.9). When it is subjected continuously to these cyclic compressive stresses (Fig. 13.18b), the bar begins to develop a small cavity at its center which begins to grow. (This phenomenon can be demonstrated with a short piece of round eraser by rolling it back and forth on a hard flat surface, as shown in Fig. 13.18b.)

Rotary tube piercing is carried out using an arrangement of rotating rolls (Fig. 13.18c). The axes of the rolls are *skewed* in order to pull the round bar through the rolls by the axial component of the rotary motion. An internal mandrel assists the operation by expanding the hole and sizing the inside diameter of the tube. The mandrel may be held in place by a long rod, or it may be a floating mandrel

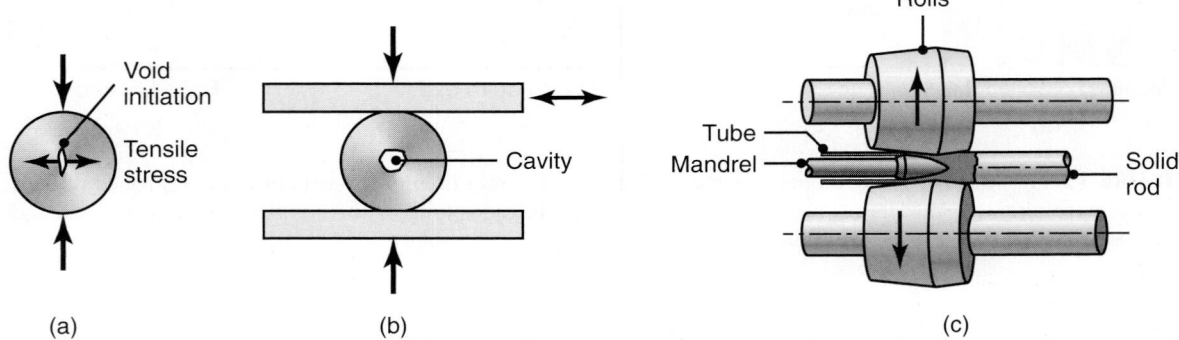

FIGURE 13.18   Cavity formation in a solid, round bar and its utilization in the rotary tube-piercing process for making seamless pipe and tubing. (See also Fig. 2.9.)

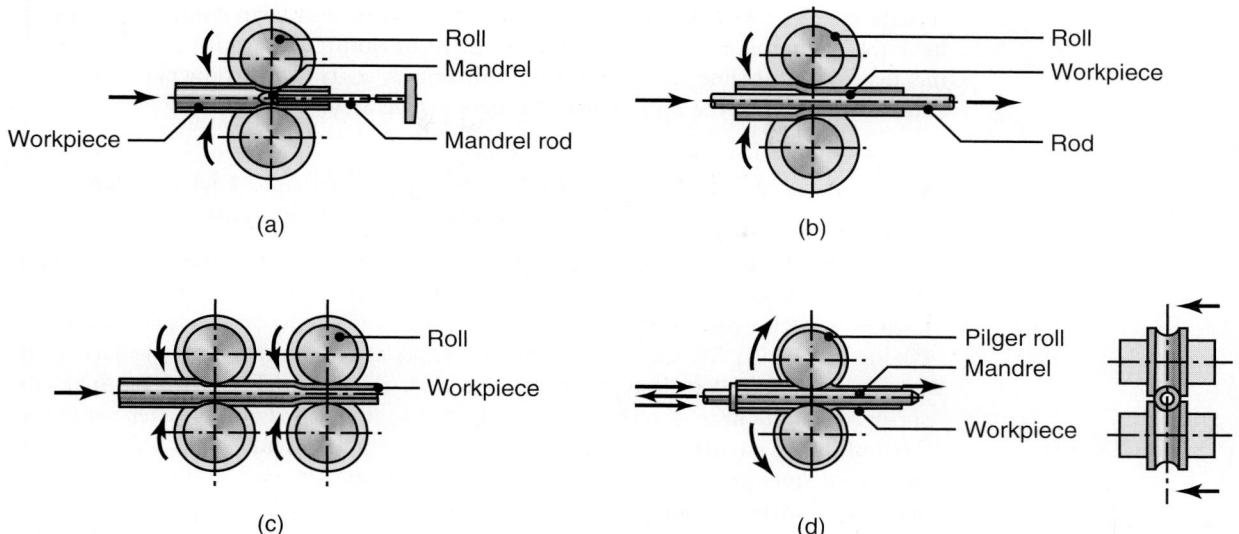

FIGURE 13.19 Schematic illustration of various tube-rolling processes: (a) with a fixed mandrel; (b) with a floating mandrel; (c) without a mandrel; and (d) pilger rolling over a mandrel and a pair of shaped rolls. Tube diameters and thicknesses also can be changed by other processes, such as drawing, extrusion, and spinning.

without a support. Because of the severe deformation that the bar undergoes, the material must be high in quality and free from defects (since internal defects may propagate rapidly and cause premature failure of the part during forming).

Tube rolling.   The diameter and thickness of pipes and tubing can be reduced by *tube rolling*, which utilizes shaped rolls (Fig. 13.19). Some of these operations can be carried out either with or without an internal mandrel. In the *pilger mill*, the tube and an internal mandrel undergo a reciprocating motion; the rolls are specially shaped and are rotated continuously. During the gap cycle on the roll, the tube is advanced and rotated, starting another cycle of tube reduction. As a result, the tube undergoes a reduction both in diameter and wall thickness. Steel tubing of 265 mm (10.5 in.) in diameter have been produced by this process. Other operations for tube manufacturing are described in Chapter 15.

13.5.1 Various Mills

Integrated mills.   These mills are large facilities that involve complete integration of the activities—from the production of hot metal in a blast furnace to the casting and rolling of finished products ready to be shipped to the customer.

Minimills.   Competition in the steel industry has led to the development of *minimills*, in which scrap metal is (a) melted in electric-arc furnaces, (b) cast continuously, and (c) rolled directly into specific lines of products. Each minimill produces essentially one type of rolled product (rod, bar, or structural sections such as angle iron) from basically one type of metal or alloy. The scrap metal, which is obtained locally (to reduce transportation costs), is typically old machinery, cars, and farm equipment. Minimills have the economic advantages of low-investment optimal operations for each type of metal and product line and of low labor and energy costs. The products usually are oriented to markets in the mill's particular geographic area.

CASE STUDY 13.1 | Manufacturing Solid-Rocket Motor-Case Segments for a Space Shuttle

The United States space shuttles are very complex spacecraft that require extensive engineering to design and manufacture. The components on these space-launch vehicles are highly specialized and are subjected to extreme conditions during flight. To lift the orbiter into space, a space shuttle employs two large, strapped-on cylindrical solid-rocket boosters, as seen in Fig. 13.20, which also shows the launching of the space shuttle *U.S.S. Atlantis*. Each booster provides a 3.3-million lb thrust force for the initial lift-off and assent. The rocket boosters consist of steel case segments which contain the solid-rocket fuel propellant, and act as very large pressure vessels during launch.

The manufacture of the steel case segments is very challenging and requires a series of special processes and equipment (Fig. 13.21). The case segments are each approximately 3 m (120 in.) in diameter, 3 m (120 in.) long, and less than 13 mm (0.5 in) thick. The dimensional tolerance on thickness is 0.25 mm (0.010 in.). A very thin and tightly controlled wall section is required for this pressure-vessel application. However, the case segments are so large that the final component weight is very close to the weight limit in facilities for the melting and casting of specialty-grade steel into ingots to be used as the starting stock (blank). Thus, it is imperative that very little material be wasted or discarded during the manufacturing process.

FIGURE 13.20   The Space Shuttle *U.S.S. Atlantis* is launched by two strapped-on solid-rocket boosters. *Source:* Courtesy of NASA.

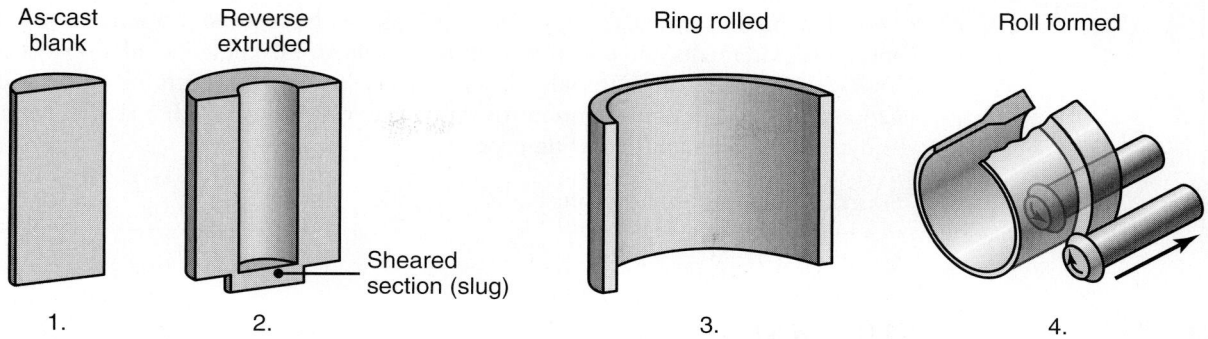

FIGURE 13.21   The forming processes involved in the manufacture of solid rocket casings for the space shuttles.

The mechanical property requirements for these components dictates that ultra-clean steel be utilized to eliminate melt-related material defects. Ladish D6AC steel, with an ultimate tensile strength of 1500–1700 MPa (220–250 ksi), is first cast into large, premium-quality, and vacuum-melted ingots. Because of the need to save material, the size of the steel ingots is very close to the weight of the finished components. Thus, material-efficient manufacturing steps have to be incorporated. To transform the cast ingot into its final cylinder components, several different manufacturing steps are utilized: (a) upsetting, (b) extrusion, (c) ring rolling, (d) roll forming, (e) machining, and (f) heat treating.

Indirect (or backward) extrusion (see Fig. 15.3) is used to first produce a rough cylindrical shape for the case segments from the cast ingot. In this process, the ingot is upset (see Fig. 14.3), and then a large central punch is pressed into the upset steel blank. As the punch presses into the center of the heated metal, the metal is extruded backward between the moving punch and the container. The rough cylinder is removed from the container, and the thin slug of material that remains underneath the punch at the end of the operation is removed by hot shearing (see Fig. 16.2).

The hot cylinder is then placed on a large ring-rolling machine where it is ring rolled (see Fig. 13.15) to achieve the proper shape. This process reduces the wall thickness of the cylinder while increasing its diameter. The length of the cylinder is maintained by containment rolls or pinch rollers on the end faces to keep the material from flowing out in that direction. The rolled cylinder has a contour that includes the end flanges and the central pressure vessel. After this operation is completed, the cylinder is machined to remove all scale (from hot working) and to produce a smooth, uniform, and bright surface finish. This step is designed to remove a minimum amount of material to produce a shape that can be ready for the final forming operation.

The machined cylinder is processed further by roll forming (also known as flow forming or shear forming, see Section 16.9.) to produce the final pressure-vessel walls. This process is performed at room temperature by means of two opposing rollers, whereby the wall section is reduced, the cylinder diameter remains constant, and its length is increased. The cylinder wall is thus cold reduced to a net final thickness. It is then heat treated by normalizing, austenitizing, quenching, and tempering. Only the end flanges of the case segments require some final machining for proper assembly into a rocket.

The manufacturing steps described above have been optimized for the special requirements and constraints of this application. These include: (a) a clean, high tensile-strength material, (b) a defect-free and inspectable component geometry, (c) a minimum amount of input material loss, and (d) a stable, robust, and reproducible manufacturing process.

Source: Courtesy of D. Furrer, Ladish Co., Inc.

SUMMARY

- Rolling is the process of reducing the thickness or changing the cross-section of a long strip by compressive forces applied through a set of rolls. In addition to flat rolling, shape rolling is used to make products with various cross-sections. Products made by rolling include: (a) plate, sheet, foil, rod, seamless pipe, and tubing; (b) shape-rolled products, such as I-beams and structural shapes; and (c) bars of various cross-section. Other rolling operations include ring rolling and thread rolling.

- Rolling may be carried out at room temperature (cold rolling) or at elevated temperatures (hot rolling). The process involves several material and process variables, including roll diameter (relative to material thickness), reduction-per-pass, speed, lubrication, and temperature. Spreading, bending, and flattening are important considerations for controlling the dimensional accuracy of the rolled stock.

- Rolling mills have a variety of roll configurations, such as two-high, three-high, four-high, cluster (Sendzimir), and tandem. Front and/or back tension may be applied to the material to reduce roll forces.

- Continuous casting and rolling of ferrous and of nonferrous metals into semifinished products have become a common practice because of their economic benefits.

- Integrated mills are large facilities involving the complete sequence of activities, from the production of hot metal in a blast furnace to the casting and the rolling of finished products ready to be shipped to the customer. On a much smaller scale, minimills utilize scrap metal which is melted in electric-arc furnaces, cast and continuously rolled into specific lines of products.

KEY TERMS

Alligatoring	Flat rolling	Pilger mill
Back tension	Foil	Plate
Billet	Front tension	Ring rolling
Bloom	Gage number	Roll
Camber	Hot rolling	Roll forging
Cast structure	Mannesmann process	Roll stand
Cogging mill	Minimill	Rolling
Cold rolling	Neutral point	Rolling mill
Crown	Osprey process	Rotary tube piercing
Draft	Pack rolling	Sendzimir mill

Shape rolling

Sheet

Skew rolling

Slab

Spray casting

Spreading

Stand

Steckel rolling

Tandem rolling

Temper rolling

Thread rolling

Tube rolling

Wrought structure

BIBLIOGRAPHY

Blazynski, T.Z., *Plasticity and Modern Metal-forming Technology*, Elsevier, 1989.

Ginzburg, V.B., *High-Quality Steel Rolling: Theory and Practice*, Marcel Dekker, 1993.

_____, *Steel-Rolling Technology: Theory and Practice*, Marcel Dekker, 1989.

Hosford, W.F., and Caddell, R.M., *Metal Forming: Mechanics and Metallurgy*, 2d ed., Prentice Hall, 1993.

Lange, K. (ed.), *Handbook of Metal Forming*, McGraw-Hill, 1985.

Roberts, W.L., *Cold Rolling of Steel*, Marcel Dekker, 1978.

_____, *Hot Rolling of Steel*, Marcel Dekker, 1983.

REVIEW QUESTIONS

13.1 What is the difference between a plate and a sheet?

13.2 Define (a) roll gap, (b) neutral point, and (c) draft.

13.3 What factors contribute to spreading in flat rolling?

13.4 Explain the types of deflection that rolls undergo.

13.5 Describe the difference between a bloom, a slab, and a billet.

13.6 Why may roller leveling be necessary?

13.7 List the defects commonly observed in flat rolling.

13.8 What are the advantages of tandem rolling? Pack rolling?

13.9 Make a list of some parts that can be made by (a) shape rolling and (b) thread rolling.

13.10 How are seamless tubes produced?

13.11 Why is the surface finish of a rolled product better in cold rolling than in hot rolling?

13.12 What is a Sendzimir mill? What are its important features?

13.13 What typically is done to make sure the product in flat rolling is not crowned?

QUALITATIVE PROBLEMS

3.14 Explain why the rolling process was invented and developed.

13.15 Flat rolling reduces the thickness of plates and sheets. It is possible, instead, to reduce their thickness by simply stretching the material. Would this be a feasible process? Explain.

13.16 Explain how the residual stress patterns shown in Fig. 13.9 become reversed when the roll radius or reduction-per-pass is changed.

13.17 Explain whether it would be practical to apply the roller-leveling technique shown in Fig. 13.7 to thick plates.

13.18 Describe the factors that influence the magnitude of the roll force, F, in Fig. 13.2c.

13.19 Explain how you would go about applying front and back tensions to sheet metals during rolling. How would you go about controlling these tensions?

13.20 We noted that rolls tend to flatten under roll forces. Describe the methods by which flattening can be reduced. Which property or properties of the roll material can be increased to reduce flattening?

13.21 It was stated that spreading in flat rolling increases with the (a) decreasing width-to-thickness ratio of the entering material, (b) decreasing friction, and (c) decreasing ratio of the roll radius to the strip thickness. Explain why.

13.22 As stated in this chapter, flat rolling can be carried out by front tension only, using idling rolls (Steckel rolling). Since the torque on the rolls is now zero, where, then, is the energy coming from to supply the work of deformation in rolling?

13.23   What is the consequence of applying too high a back tension in rolling?

13.24   Note in Fig. 13.3d that the driven rolls (powered rolls) are the third set from the work roll. Why isn't power supplied through the work roll itself? Is it even possible? Explain.

13.25   Describe the importance of controlling roll speeds, roll gaps, temperature, and other process variables in a tandem-rolling operation, as shown in Fig. 13.11.

Explain how you would go about determining the distance between the stands.

13.26   In Fig. 13.9a, if you remove the top compressive layer by, say, grinding, will the strip remain flat? If not, which way will it curve and why?

13.27   Name several products that can be made by each of the operations shown in Fig. 13.1.

13.28   List the possible consequences of rolling at (a) too high of a speed and (b) too low of a speed.

13.29   Describe your observations concerning Fig. 13.12.

QUANTITATIVE PROBLEMS

13.30   Using simple geometric relationships and the inclined-plane principle for friction, prove Eq. (13.1).

13.31   Estimate the roll force, F, and the torque for an AISI 1020 carbon-steel strip that is 200 mm wide and 10 mm thick and rolled to a thickness of 7 mm. The roll radius is 200 mm, and it rotates at 200 rpm.

13.32   In Example 13.1, calculate the roll force and the power for the case in which the workpiece material is 1100-O aluminum and the roll radius R is 10 in.

13.33   Calculate the individual drafts in each of the stands in the tandem-rolling operation shown in Fig. 13.11.

13.34   Assume that you are an instructor covering the topics described in this chapter, and you are giving a quiz on the numerical aspects to test the understanding of the students. Prepare two quantitative problems and supply the answers.

SYNTHESIS, DESIGN, AND PROJECTS

13.35   A simple sketch for a four-high mill stand is shown in Fig. 13.3a. Make a survey of the technical literature and present a more detailed sketch for such a stand, showing the major components.

13.36   Obtain a piece of soft, round rubber eraser, such as that at the end of pencils, and duplicate the process shown in Fig. 13.18b. Note how the central portion of the eraser will begin to erode away, producing a hole.

13.37   If you repeat the experiment in Problem 13.36 with a harder eraser, such as that used for erasing ink, you will note that the whole eraser will begin to crack and crumble. Explain why.

13.38   Design a set of rolls to produce cross-sections other than those shown in Fig. 13.12.

Forging of Metals

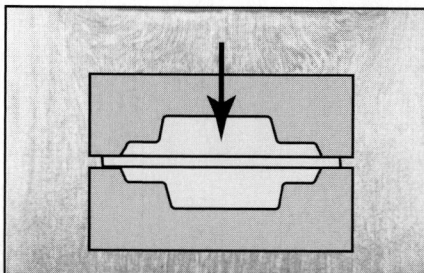

CHAPTER

14

In this chapter, we examine the fundamentals of forging and related processes, including design and economic considerations. Topics covered include:

- Open-die forging operations for producing simple shapes.
- Impression-die and closed-die forging operations for producing more intricate shapes.
- Various forging operations, such as heading, piercing, coining, swaging, and cold extrusion.
- Forging defects and die failures and their causes.
- Economics of forging as it relates to process selection.
- Design of forged parts, die design and manufacturing, and selection of materials and lubricants.

Typical parts made by forging and related processes: Shafts, gears, bolts, turbine blades, hand tools, dies, and components for machinery, transportation, and farm equipment.

Alternative processes: Casting, powder metallurgy, machining, and fabrication.

14.1 | Introduction

Forging is a basic process in which the workpiece is shaped by compressive forces applied through various dies and tooling. One of the oldest and important metalworking operations, dating back at least to 4000 B.C., forging first was used to make jewelry, coins, and various implements by hammering metal with tools made of stone. Forged parts now include large rotors for turbines; gears; bolts and rivets; cutlery (Fig. 14.1a); hand tools; numerous structural components for machinery, aircraft (Fig. 14.1b and c), and railroads; and a variety of other transportation equipment.

Unlike rolling operations described in Chapter 13 which generally produce continuous plates, sheets, strips, or various structural cross-sections, forging operations produce *discrete parts*. Because the metal flow in a die and the material's grain structure can be controlled, forged parts have good strength and toughness, and are very reliable for highly stressed and critical applications (Fig. 14.2). Simple forging

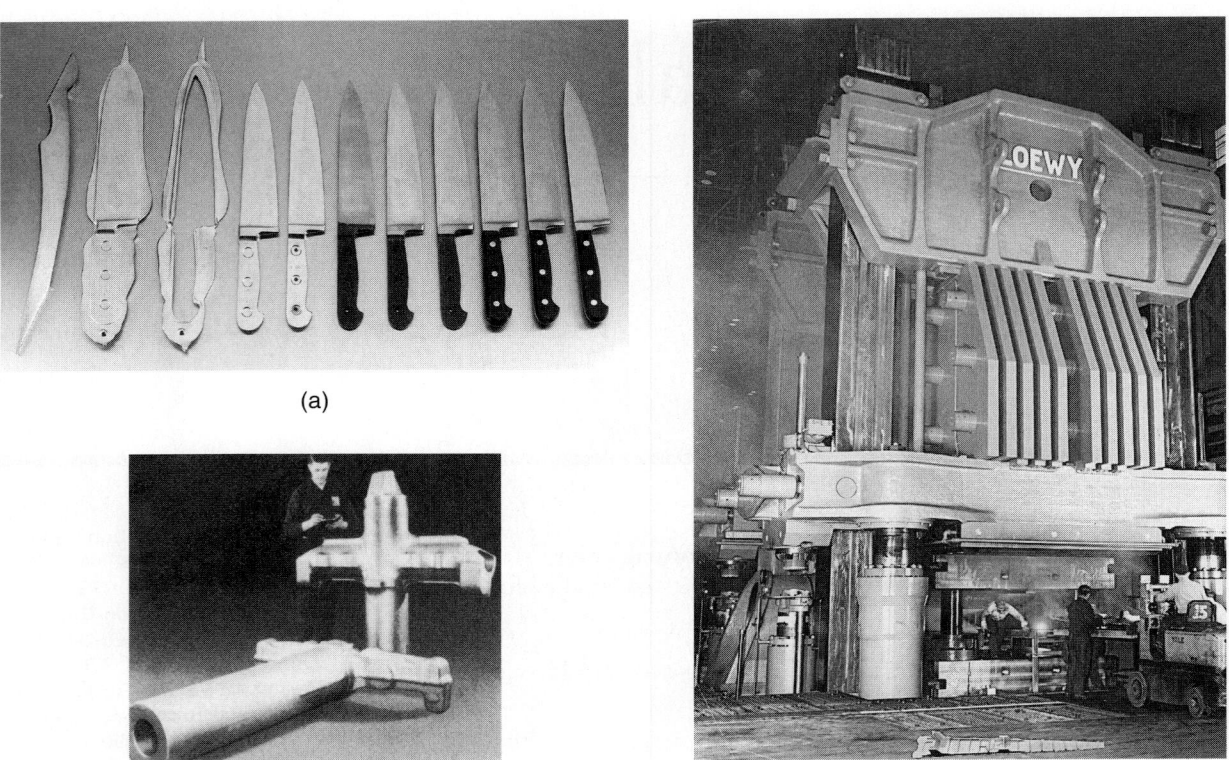

(a)

(b)

(c)

FIGURE 14.1   (a) Schematic illustration of the steps involved in forging a knife. (b) Landing-gear components for the C5A and C5B transport aircraft, made by forging. (c) General view of a 445 MN (50,000 ton) hydraulic press. *Source*: (a) Courtesy of Mundial LLC. (b) and (c) Courtesy of Wyman-Gordon Company.

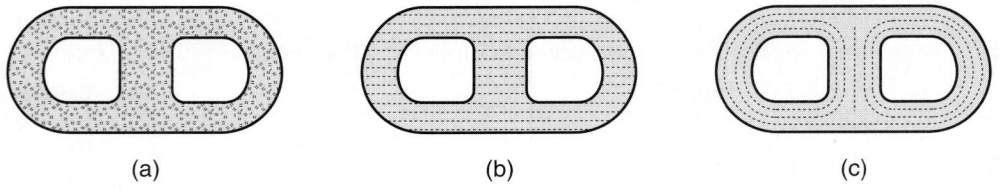

(a)   (b)   (c)

FIGURE 14.2   Schematic illustration of a part made by three different processes showing grain flow. (a) Casting by the processes described in Chapter 11. (b) Machining from a blank, described in Part IV this book, and (c) forging. Each process has its own advantages and limitations regarding external and internal characteristics, material properties, dimensional accuracy, surface finish, and the economics of production. *Source*: Courtesy of Forging Industry Association.

operations can be performed with a heavy hammer and an anvil, as has been done traditionally by blacksmiths. However, most forgings require a set of dies and such equipment as a press or a powered forging hammer.

Forging may be carried out at room temperature (*cold forging*) or at elevated temperatures (*warm* or *hot forging*), depending on the temperature; (see Section 1.7). Cold forging requires higher forces (because of the higher strength of the workpiece material), and the workpiece material must possess sufficient ductility at room temperature to undergo the necessary deformation without cracking. Cold-forged parts have a good surface finish and dimensional accuracy. Hot forging requires lower forces, but the dimensional accuracy and surface finish of the parts are not as high as in cold forging.

Forgings generally are subjected to additional finishing operations, such as heat treating to modify properties and machining to obtain accurate final dimensions and surface finish. These finishing operations can be minimized by *precision forging*, which is an important example of *net-shape* or *near-net shape* forming processes. As we shall see throughout this book, components that can be forged successfully also may be manufactured economically by other methods, such as by casting (Chapter 11), powder metallurgy (Chapter 17), or machining (Part IV). Each of these will produce a part having different characteristics, particularly with regard to strength, toughness, dimensional accuracy, surface finish, and the possibility of internal or external defects.

14.2 | Open-Die Forging

Open-die forging is the simplest forging operation (Table 14.1). Although most open-die forgings generally weigh 15 to 500 kg (30 to 1000 lb), forgings as heavy as 300 tons have been made. Part sizes may range from very small parts such as nails, pins, and bolts, to very large (up to 23-m (75-ft) long shafts for ship propellers). Open-die forging can be depicted by a solid workpiece placed between two flat dies and reduced in height by compressing it (Fig. 14.3a), a process which is also called **upsetting** or **flat-die forging**. The die surfaces also may have shallow cavities or incorporate features to produce relatively simple forgings.

The deformation of the workpiece under *frictionless conditions* is shown in Fig. 14.3b. Because constancy of volume is maintained, any reduction in height increases the diameter of the forged part. Note in Fig. 14.3b that the workpiece is

TABLE 14.1

General Characteristics of Forging Processes

Process	Advantages	Limitations
Open die	Simple and inexpensive dies; wide range of part sizes; good strength characteristics; generally for small quantities	Limited to simple shapes; difficult to hold close tolerances; machining to final shape necessary; low production rate; relatively poor utilization of material; high degree of skill required
Closed die	Relatively good utilization of material; generally better properties than open-die forgings; good dimensional accuracy; high production rates; good reproducibility	High die cost, not economical for small quantities; machining often necessary
Blocker	Low die costs; high production rates	Machining to final shape necessary; parts with thick webs and large fillets
Conventional	Requires much less machining than blocker type; high production rates; good utilization of material	Higher die cost than blocker type
Precision	Close dimensional tolerances; very thin webs and flanges possible; machining generally not necessary; very good material utilization	High forging forces, intricate dies, and provision for removing forging from dies

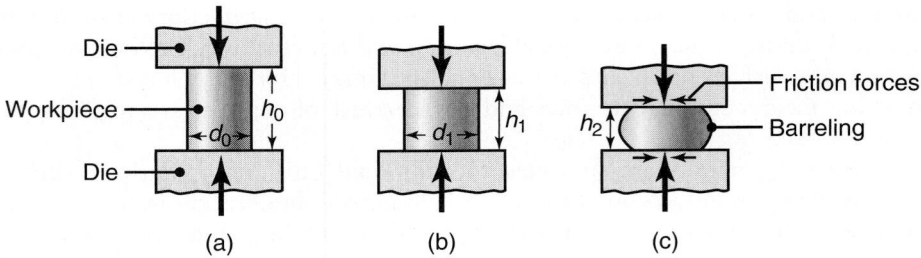

FIGURE 14.3 (a) Solid cylindrical billet upset between two flat dies. (b) Uniform deformation of the billet without friction. (c) Deformation with friction. Note barreling of the billet caused by friction forces at the billet–die interfaces.

deformed *uniformly*. In actual operations, however, there is friction, and the part develops a *barrel* shape (Fig. 14.3c)—a deformation also known as *pancaking*.

Barreling is caused primarily by frictional forces at the die–workpiece interfaces that oppose the outward flow of the materials at these interfaces and, thus, can be minimized by using an effective lubricant. Barreling also can develop in upsetting hot workpieces between cold dies. The material at and near the die surfaces cools rapidly, while the rest of the workpiece remains relatively hot. Consequently, the material at the top and bottom of the workpiece has higher resistance to deformation than the material at its center. As a result, the central portion of the workpiece expands laterally to a greater extent than do its ends. Barreling from thermal effects can be reduced or eliminated by using heated dies. Thermal barriers, such as glass cloth, at the die–workpiece interfaces also can be used for this purpose.

Cogging (also called *drawing out*) is basically an open-die forging operation in which the thickness of a bar is reduced by successive forging steps (*bites*) at specific intervals (Fig. 14.4a). The thickness of bars and rings can be reduced by similar open-die forging techniques, as shown in Fig. 14.4b and c. Because the contact area between the die and the workpiece is small, a long section of a bar can be reduced in thickness without requiring large forces or heavy machinery. Blacksmiths perform

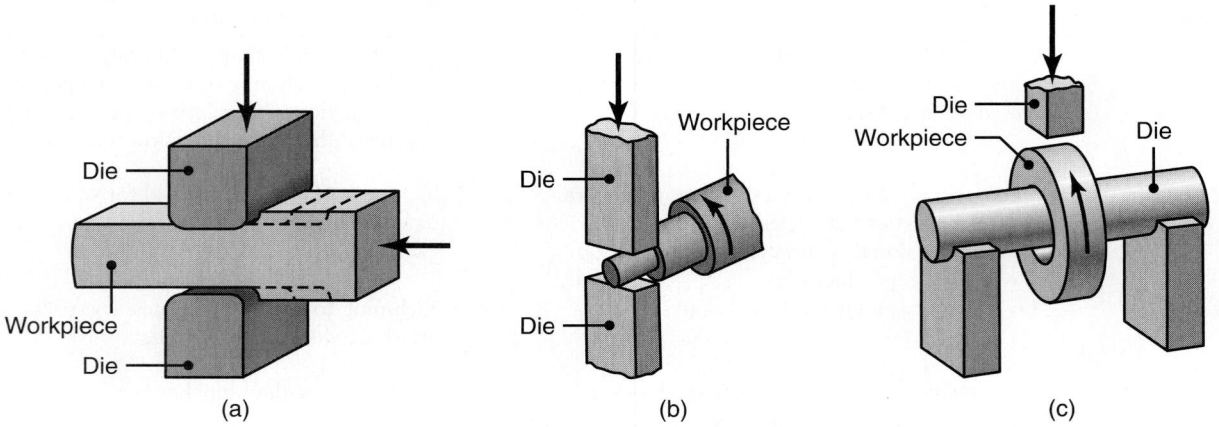

FIGURE 14.4 (a) Schematic illustration of a cogging operation on a rectangular bar. Blacksmiths use this process to reduce the thickness of bars by hammering the part on an anvil. Reduction in thickness is accompanied by barrelling, as in Fig. 14.3c. (b) Reducing the diameter of a bar by open-die forging; note the movements of the dies and the workpiece. (c) The thickness of a ring being reduced by open-die forging.

such operations with a hammer and an anvil using hot pieces of metal—typical product examples being iron fences of various designs. Thus, it is noted that cogging can be a rough substitute for rolling operations. Cogging of larger workpieces usually is done using mechanized equipment and computer controls in which lateral and vertical movements are coordinated to produce the desired part.

Forging force.   The *forging force*, F, in an *open-die forging* operation on a solid cylindrical workpiece can be estimated from the formula

$$F = Y_f \pi r^2 \left(1 + \frac{2\mu r}{3h} \right) \tag{14.1}$$

where Y_f is the *flow stress* of the material (see example below), μ is the coefficient of friction between the workpiece and the die, and r and h are the radius and height of the workpiece, respectively. (Derivation of this formula and of others for various forging processes are given in references lised in the Bibliography at the end of the chapter.)

EXAMPLE 14.1 Calculation of forging force in upsetting

A solid cylindrical slug made of 304 stainless steel is 150 mm (6 in.) in diameter and 100 mm (4 in.) high. It is reduced in height by 50% at room temperature by open-die forging with flat dies. Assuming that the coefficient of friction is 0.2, calculate the forging force at the *end* of the stroke.

Solution   The forging force at the end of the stroke is calculated using Eq. (14.1), in which the dimensions pertain to the final dimensions of the forging. Thus, the final height is $h = 100/2 = 50$ mm, and the final radius, r, is determined from volume constancy by equating the volumes before and after deformation. Hence,

$$(\pi)(75)^2(100) = (\pi)(r)^2(50)$$

And therefore, $r = 106$ mm (4.17 in.).

The quantity Y_f in Eq. (14.1) is the flow stress of the material, which is the stress required to continue plastic deformation of the workpiece at a particular true strain. The absolute value of the true strain that the workpiece has undergone at the end of the stroke in this operation is

$$\varepsilon = \ln\left(\frac{100}{50}\right) = 0.69$$

We can determine the flow stress by referring to Eq. (2.8) and noting from Table 2.3 that for 304 stainless steel $K = 1275$ MPa, and $n = 0.45$. Thus, for a true strain of 0.69, the flow stress is calculated to be 1100 MPa. Another method is to refer to Fig. 2.6 and note that the flow stress for 304 stainless steel at a true strain of 0.69 is about 1000 MPa (140 ksi). The small difference between the two values is due to the fact that the data in Table 2.3 and Figure 2.6 are from different sources. Taking the latter value, the forging force now can be calculated, noting that in this problem the units in Eq. (14.1) must be in N and m. Thus,

$$F = (1000)(10^6)(\pi)(0.106)^2 (1) + \frac{(2)(0.2)(0.106)}{(3)(0.050)}$$

$$= 4.5 \times 10^7 \text{ N} = 45 \text{ MN} = 10^7 \text{ lb} = 5000 \text{ tons}$$

14.3 | Impression-Die and Closed-Die Forging

In *impression-die forging*, the workpiece takes the shape of the die cavity while being forged between two shaped dies (Fig. 14.5a through c). This process usually is carried out at elevated temperatures for enhanced ductility of the metals and to lower the forces. Note in Fig. 14.5c that, during deformation, some of the material flows outward and forms a **flash**. The flash has an important role in impression-die forging. The high pressure and the resulting high frictional resistance in the flash presents a severe constraint to the material in the die for outward flow. Thus, based on the principle that in plastic deformation the material flows in the direction of least resistance (because it requires less energy), the material begins to flow into the die cavity, ultimately filling it completely.

The standard terminology for a typical forging die is shown in Fig. 14.5d. Instead of being made as one piece, dies may be made of several pieces (segmented), including **die inserts** (Fig. 14.6) and particularly for complex shapes. The inserts can be replaced easily in the case of wear or failure in a particular section of the die and usually are made of stronger and harder materials. Of course, dies should allow the removal of the forged workpiece without difficulty (see also Section 14.6).

The blank to be forged is prepared by such means as (a) *cropping* from an extruded or drawn bar stock; (b) a preform from operations such as *powder metallurgy*; (c) *casting*; or (d) using a preform blank from a prior forging operation. The blank is placed on the lower die, and as the upper die begins to descend, the blank's shape gradually changes—as is shown for the forging of a connecting rod in Fig. 14.7a.

Preforming operations (Figs. 14.7b and c) typically are used to distribute the material properly into various regions of the blank using simple shaped dies of various contours. In **fullering**, material is *distributed away* from an area. In **edging**, it is

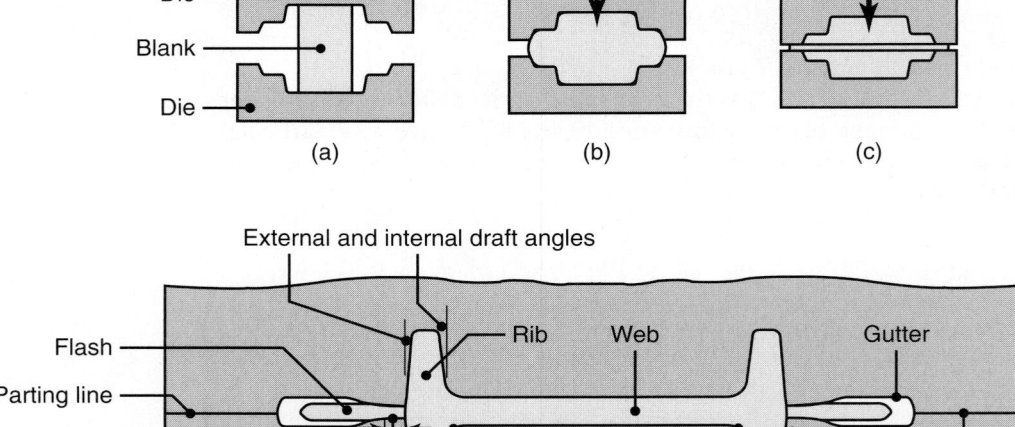

FIGURE 14.5   (a) through (c) Stages in impression-die forging of a solid round billet. Note the formation of flash, which is excess metal that is subsequently trimmed off (see Fig. 14.7). (d) Standard terminology for various features of a forging die.

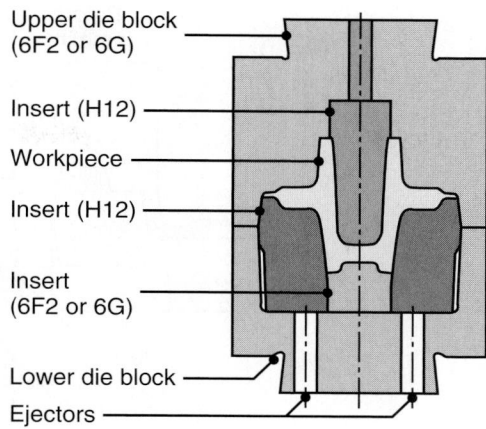

FIGURE 14.6   Die inserts used in forging an automotive axle housing.

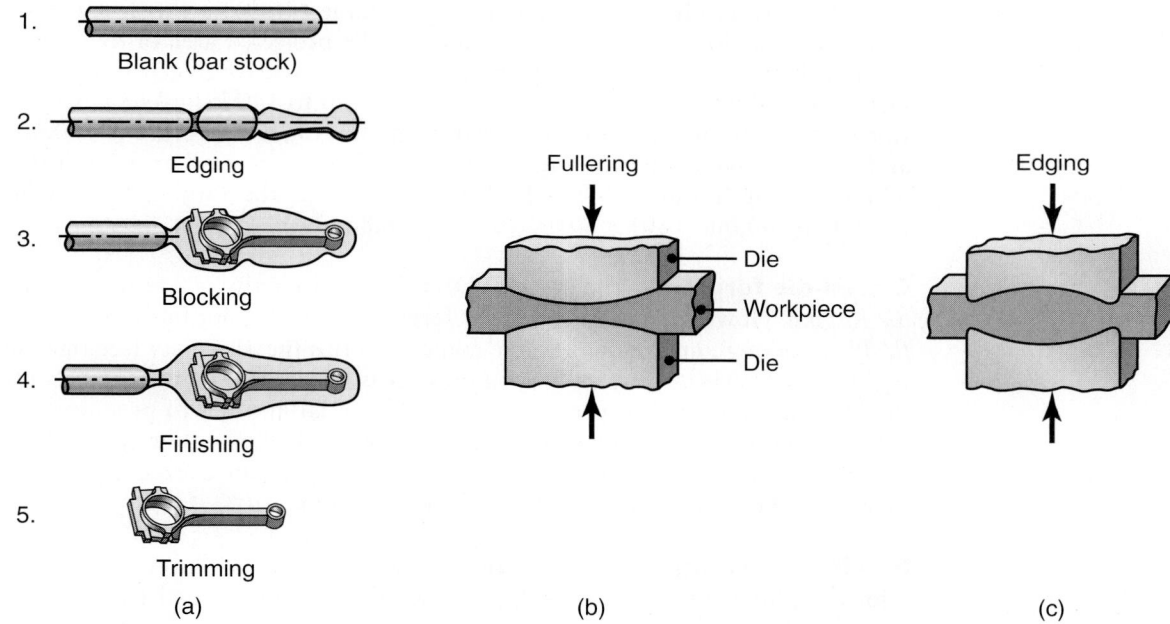

FIGURE 14.7   (a) Stages in forging a connecting rod for an internal combustion engine. Note the amount of flash required to ensure proper filling of the die cavities. (b) Fullering and (c) edging operations to properly distribute the material when preshaping the blank for forging.

gathered into a localized area. The part then is formed into the rough shape (say, a connecting rod) by a process called **blocking**, using *blocker dies*. The final operation is the finishing of the forging in *impression dies* that give the forging its final shape. The flash is removed later by a trimming operation (Fig. 14.8).

Forging force.   The *forging force*, *F*, required to carry out an *impression-die forging* operation can be estimated from the formula

$$F = kY_f A \tag{14.2}$$

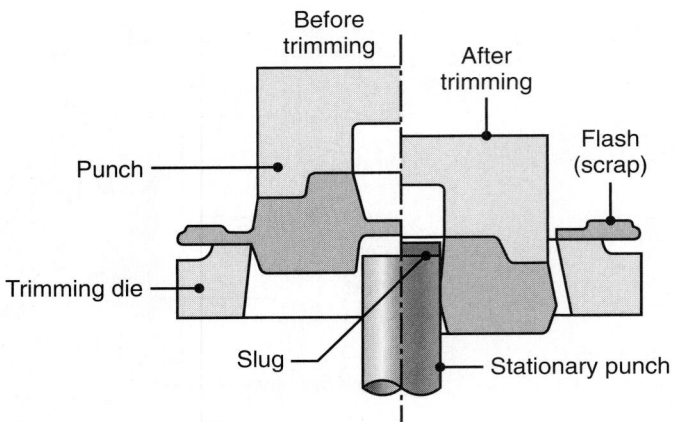

FIGURE 14.8   Trimming flash from a forged part. Note that the thin material at the center is removed by punching.

where k is a multiplying factor obtained from Table 14.2, Y_f is the flow stress of the material at the forging temperature, and A is the projected area of the forging, including the flash. In hot-forging operations, the actual forging pressure for most metals typically ranges from 550 to 1000 MPa (80 to 140 ksi). As an example, assume that the flow stress of a material at the forging temperature is 100,000 psi, and a part (such as that shown in Fig. 14.7a) has a projected area (with flash) of 60 in². Taking a value of $k = 10$ from Table 14.2, the forging force would be $F = (10)(100,000)(60) = 60 \times 10^6$ lb = 26.8 tons.

Closed-die forging.   The process shown in Fig. 14.5 also is referred to as *closed-die forging*. However, in true closed-die forging, flash does not form (hence the term *flashless forging*), and the workpiece completely fills the die cavity (see right side of Fig. 14.9b). Consequently, the forging pressure is very high. Thus, accurate control of the blank volume and proper die design are essential in order to produce a forging with the desired dimensional tolerances. Undersized blanks prevent the complete filling of the die cavity; conversely, oversized blanks generate excessive pressures and may cause dies to fail prematurely or the machine to jam.

Precision forging.   In order to reduce the number of additional finishing operations required (net-shape forming)—hence the cost—the trend has been toward greater precision in forged products. Typical precision-forged products are gears, connecting rods, and turbine blades. Precision forging requires (a) special and more complex dies, (b) precise control of the blank's volume and shape, and (c) accurate positioning of the blank in the die cavity. Also, because of the higher forces required to obtain fine details on the part, this process requires higher-capacity equipment. Aluminum and magnesium alloys are suitable, particularly for precision

TABLE 14.2

Range of k Values for Eq. (14.2)	
Simple shapes, without flash	3–5
Simple shapes, with flash	5–8
Complex shapes, with flash	8–12

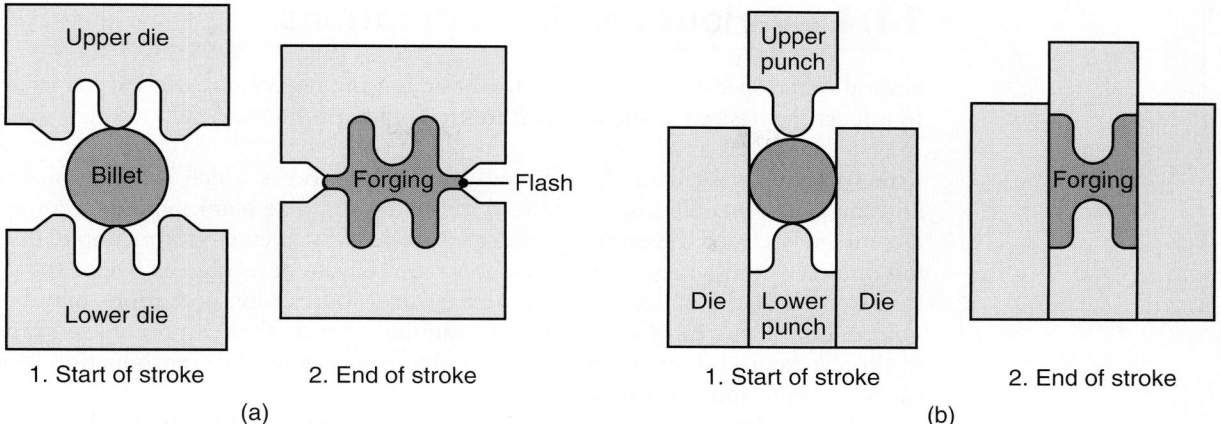

FIGURE 14.9 Comparison of (a) closed-die forging with flash and (b) precision or flashless forging of a round billet. *Source*: After H. Takemasu, V. Vazquez, B. Painter, and T. Altan.

forging because of the relatively low forging loads and temperatures that they require; steels and titanium also can be precision forged.

A typical forging operation and product quality. A forging operation typically involves the following sequence of steps:

1. Prepare a slug, billet, or preform by processes such as shearing (cropping), sawing, or cutting off. If necessary, clean surfaces by such means as shot blasting.

2. For hot forging, heat the workpiece in a suitable furnace and then, if necessary, descale it after heating with a wire brush, water jet, steam, or by scraping. Some descaling also may occur during the initial stages of forging, when the scale (which is brittle) falls off during deformation in forging.

3. For hot forging, preheat and lubricate the dies; for cold forging, lubricate the blank.

4. Forge the billet in appropriate dies and in the proper sequence. If necessary, remove any excess material (such as flash) by trimming, machining, or grinding.

5. Clean the forging, check its dimensions, and (if necessary) machine it to final dimensions and specified tolerances.

6. Perform additional operations, such as straightening and heat treating (for improved mechanical properties). Also perform any finishing operations that may be required, such as machining and grinding.

7. Inspect the forging for any external and internal defects.

The quality, dimensional tolerances, and surface finish of a forging depend on how well these operations are performed and controlled. Generally, dimensional tolerances range between ±0.5 and ±1% of the dimensions of the forging. In good practice, tolerances for hot forging of steel are usually less than ±6 mm (1/4 in.); in precision forging, they can be as low as ±0.25 mm (0.01 in.). Other factors that contribute to dimensional inaccuracies are draft angles, radii, fillets, die wear, die closure (whether the dies have closed properly), and mismatching of the dies. The surface finish of the forging depends on blank preparation, die surface finish, die wear, and the effectiveness of the lubricant.

14.4 | Various Forging Operations

Several other operations related to the basic forging process are carried out in order to impart the desired shape and features to forged products.

Coining.   This is essentially a closed-die forging process which is used typically in minting coins, medallions, and jewelry (Figs. 14.10). The blank or slug is coined in a completely closed die cavity. In order to produce fine details (for example, the detail on newly minted coins), the pressures required can be as high as five or six times the strength of the material. On some parts, several coining operations may be required. Lubricants cannot be applied in coining, because they can become entrapped in the die cavities and (being incompressible) prevent the full reproduction of die-surface details and surface finish.

   Marking parts with letters and numbers also can be done rapidly through this process. The coining process also is used with forgings and with other products to improve surface finish and to impart the desired dimensional accuracy with little or no change in part size. Called **sizing**, this process requires high pressures.

Heading.   Also called **upset forging**, *heading* is essentially an upsetting operation, usually performed on the end of a round rod or wire in order to increase the cross-section. Typical examples are nails, bolt heads, screws, rivets, and various other fasteners (Fig. 14.11a). Heading can be carried out cold, warm, or hot. An important consideration in heading is the tendency for the bar to buckle if its unsupported length-to-diameter ratio is too high. This ratio usually is limited to less than 3:1, but with appropriate dies, it can be higher. For example, higher ratios can be accommodated if the diameter of the die cavity is not more than 1.5 times the bar diameter.

   Heading operations are performed on machines called **headers**, which usually are highly automated with production rates of hundreds of pieces per minute for small parts. Hot heading operations on larger parts typically are performed on *horizontal upsetters*. These machines tend to be noisy; soundproof enclosure or the use of ear protectors is required. Heading operations can be combined with cold-extrusion processes to make various parts, as described in Section 15.4.

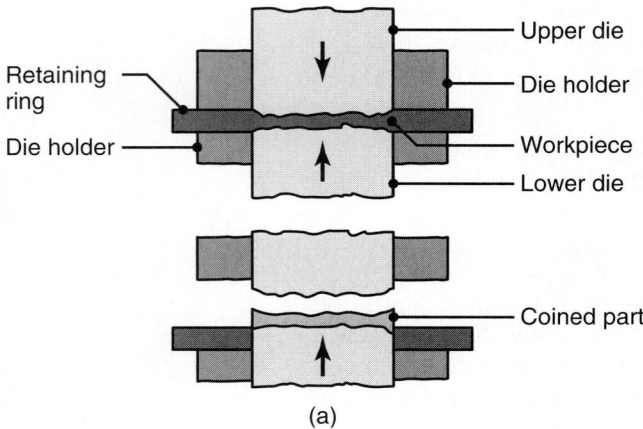

(a)   (b)

FIGURE 14.10   (a) Schematic illustration of the coining process. The earliest coins were made by open-die forging and lacked precision and sharp details. (b) An example of a modern coining operation, showing the workpiece and tooling. Note the detail and superior surface finish that can be achieved in this process. *Source*: Courtesy of C & W Steel Stamp Co., Inc.

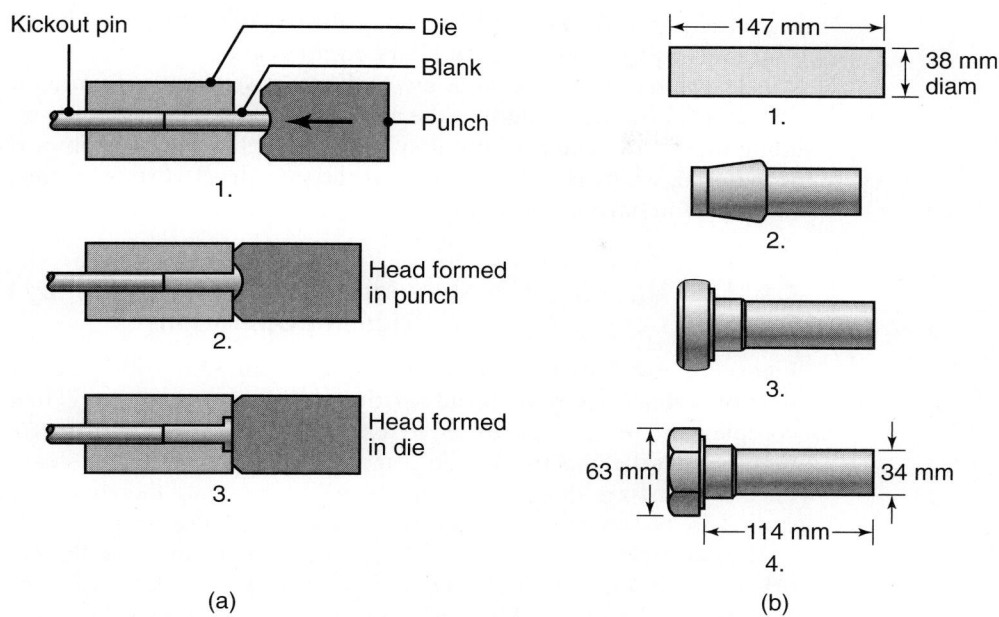

FIGURE 14.11   (a) Heading operation to form heads on fasteners, such as nails and rivets. (b) Sequence of operations to produce a typical bolt head by heading.

Piercing.   This is a process of indenting the surface of a workpiece (but not breaking through) with a punch in order to produce a cavity or an impression (Fig. 14.12). The workpiece may be confined in a container (such as a die cavity) or may be unconstrained. The deformation of the workpiece will depend on how much it is constrained from flowing freely as the punch descends. A common example of piercing is the indentation of heads of bolts, such as the hexagonal cavity in bolt heads. Piercing may be followed by punching to produce a hole in the part (for a similar depiction of this situation, see the *slug* above the stationary punch in

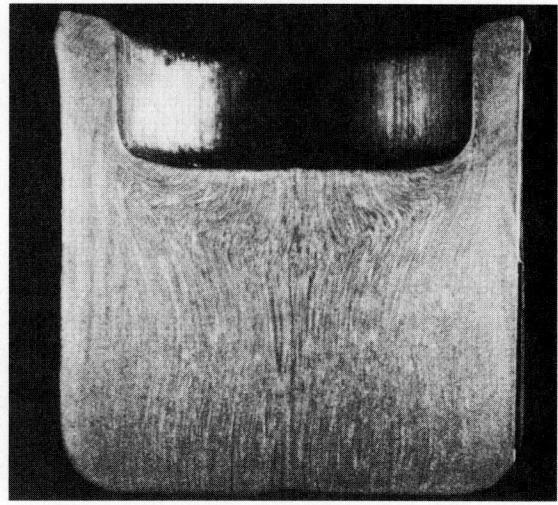

FIGURE 14.12   A pierced round billet showing grain-flow pattern (see also Fig. 14.2c). *Source*: Courtesy of Ladish Co., Inc.

the central portion of Fig. 14.8). Piercing also is performed to produce hollow regions in forgings using side-acting auxiliary equipment.

The *piercing force* depends on (a) the cross-sectional area and the tip geometry of the punch, (b) the strength of the material, and (c) the magnitude of friction at the sliding interfaces. The pressure may range from three to five times the strength of the material, which is approximately at the same level of stress required to make an indentation in hardness testing.

CASE STUDY 14.1   Manufacture of a Stepped Pin by Heading and Piercing Operations

Figure 14.13a shows a stepped pin made from SAE 1008 steel and used as a part of a roller assembly to adjust the position of a car seat. The part is fairly complex and must be produced in a progressive manner in order to produce the required details and fill the die completely. The cold-forging steps used to produce this part are shown in Figure 14.13b. First, a solid, cylindrical blank is extruded in two operations, followed by an upsetting operation. The upsetting operation uses a conical cross-section in the die to produce the preform and is oriented such that material is concentrated at the top of the part in order to ensure proper die filling. After the impression-die forming, a piercing operation is performed to form the bore. The part is made to net shape on a cold forming machine at a rate of 240 parts per minute.

Hubbing.   This process consists of pressing a hardened punch with a particular tip geometry into the surface of a block of metal. The cavity produced subsequently is used as a die for forming operations, such as those occuring in the making of tableware. The die cavity usually is shallow, but for deeper cavities, some material may be removed from the surface by machining prior to hubbing (see Fig. 23.3c). The *hubbing force* can be estimated from the equation

$$\text{Hubbing force} = 3(\text{UTS})(A) \tag{14.3}$$

where UTS is obtained from Table 2.2, and A is the projected area of the impression. Thus, for example, for high-strength steel with UTS = 1500 MPa and a part with a

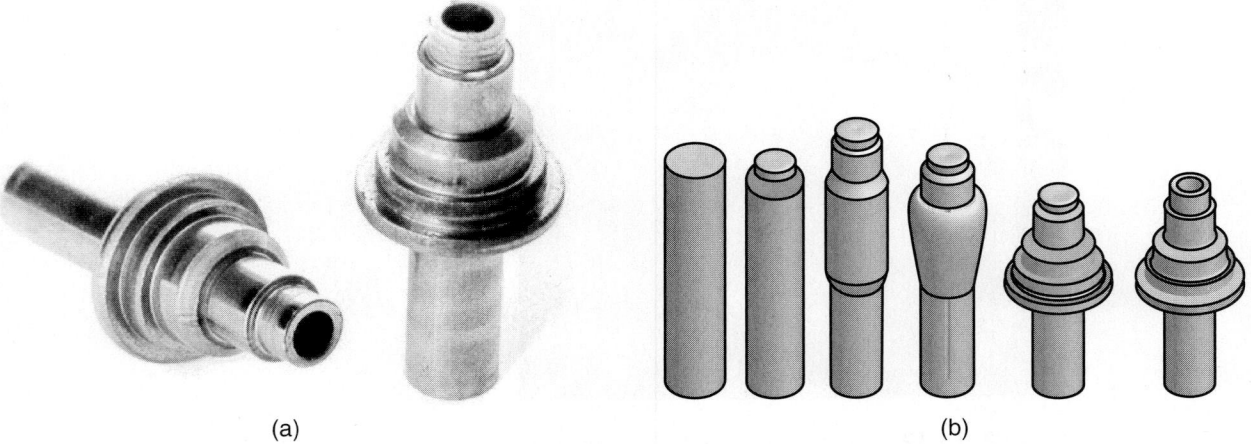

(a)   (b)

FIGURE 14.13   (a) The stepped pin used in Case Study 14.1. (b) Illustration of the manufacturing steps used to produce the stepped pin. *Source*: Courtesy of National Machinery, LLC.

projected area of 400 mm^2, the hubbing force would be $(3)(1500 \text{ N/mm}^2)$ $(400 \text{ mm}^2) = 1.8 \text{ MN} = 179$ tons.

Orbital forging. This is a process in which the upper die moves along an orbital path and forms the part *incrementally*. The operation is similar to the action of a mortar and pestle used for crushing herbs and seeds. Although not in common use, typical components that may be forged by this process are disk-shaped and conical parts, such as bevel gears and gear blanks. The forging force is relatively small, because at any particular instant, the die contact is concentrated onto a small area of the workpiece (see also *incremental forging* below). The operation is relatively quiet, and parts can be formed within 10 to 20 cycles of the orbiting die.

Incremental forging. In this process, a blank is forged into a shape with a tool that forms the blank in several small steps. Note that this operation is somewhat similar to cogging (see Fig. 14.4), in which the die penetrates the blank to different depths along the surface. Because of the smaller die-contact area, this process requires much lower forces when compared with conventional impression-die forging, and the tools are simpler and less costly.

Isothermal forging. Also known as **hot-die forging**, the dies in this process are heated to the same temperature as that of the hot workpiece. Because it remains hot, the low strength and high ductility of the workpiece are maintained during forging. Also, the forging load is low, and material flow within the die cavity is improved. Complex parts with good dimensional accuracy can be isothermally forged to near-net shape by one stroke in a hydraulic press. The dies for hot forging of high-temperature alloys usually are made of nickel or molybdenum alloys (because of their resistance to high temperature), but steel dies can be used for aluminum alloys. Isothermal forging is expensive, and production rate is low. However, it can be economical for specialized, intricate forgings made of materials such as aluminum, titanium, and superalloys, provided that the quantity required is sufficiently high to justify the die costs.

Rotary swaging. In this process (also known as *radial forging, rotary forging* or simply *swaging*), a solid rod or tube is subjected to radial impact forces by a set of reciprocating dies of the machine (Fig. 14.14a and b). The die movements are obtained by means of a set of rollers in a cage in an action similar to that of a roller bearing. The workpiece is stationary and the dies rotate (while moving radially in their slots), striking the workpiece at rates as high as 20 strokes per second. In *die-closing swaging machines*, die movements are obtained through the reciprocating motion of wedges (Fig. 14.14c). The dies can be opened wider than those in rotary swagers, thereby accommodating large-diameter or variable-diameter parts. In another type of machine, the dies do not rotate but move radially in and out. Typical products made are screwdriver blades and soldering-iron tips.

The swaging process also can be used to *assemble* fittings over cables and wire; in such cases, the tubular fitting is swaged directly onto the cable. This process also is used for operations such as *pointing* (tapering the tip of a cylindrical part) and *sizing* (finalizing the dimensions of a part).

Swaging generally is limited to a maximum workpiece diameter of about 150 mm (6 in.); parts as small as 0.5 mm (0.02 in.) have been swaged. Dimensional tolerances range from ±0.05 to ±0.5 mm (0.002 to 0.02 in.). The process is suitable for medium-to-high rates of production where rates as high as 50 parts-per-minute are possible, depending on part complexity. It is a versatile process, and it is limited in length only by the length of the bar supporting the mandrel (if one is needed).

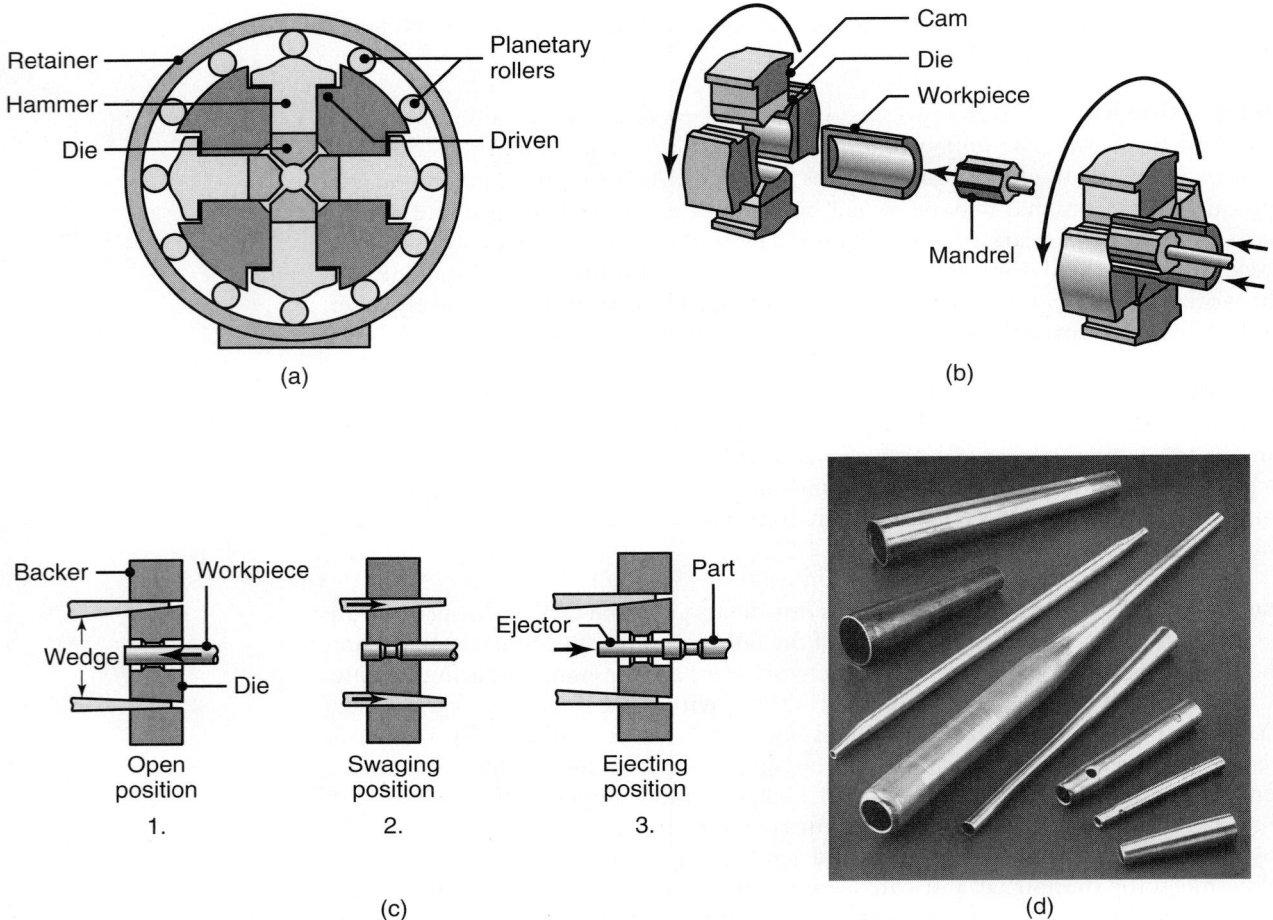

FIGURE 14.14 (a) Schematic illustration of the rotary-swaging process. (b) Forming internal profiles on a tubular workpiece by swaging. (c) A die-closing swaging machine showing forming of a stepped shaft. (d) Typical parts made by swaging. *Source*: (d) Courtesy of J. Richard Industries.

Tube swaging. In this process, the internal diameter and/or the thickness of the tube is reduced with or without the use of *internal mandrels* (Figs. 14.15a and b). For small-diameter tubing, high-strength wire can be used as a mandrel. Mandrels also can be made with longitudinal grooves, to allow swaging of internally shaped tubes (Fig. 14.15c). For example, the rifling in gun barrels (internal spiral grooves to give gyroscopic effect to bullets) can be produced by swaging a tube over a mandrel with spiral grooves. Special machinery has been built to swage gun barrels and other parts with starting diameters as large as 350 mm (14 in.).

14.5 | Forgeability of Metals—Forging Defects

Forgeability generally is defined as the capability of a material to undergo deformation without cracking. Various tests have been developed and recommended to quantify forgeability; however, because of their complex nature, only two simple tests have had general acceptance: upsetting and hot-twist.

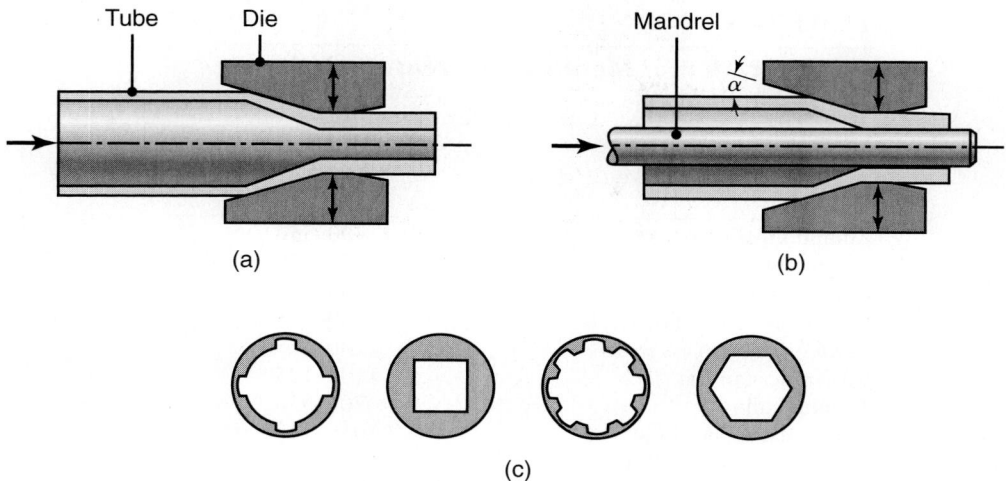

FIGURE 14.15   (a) Swaging of tubes without a mandrel; note the increase in wall thickness in the die gap. (b) Swaging with a mandrel; note that the final wall thickness of the tube depends on the mandrel diameter. (c) Examples of cross-sections of tubes produced by swaging on shaped mandrels. Rifling (internal spiral grooves) in small gun barrels can be made by this process.

In the **upsetting test**, a solid, cylindrical specimen is upset between flat dies, and the reduction in height at which cracking on the barreled surfaces begins is noted (see also Fig. 2.19d). The greater the deformation prior to cracking, the greater the forgeability of the metal. The second method is the **hot-twist test**, in which a round specimen is twisted continuously in the same direction until it fails. This test is performed on a number of specimens and at different temperatures, and the number of complete turns that each specimen undergoes before failure at each temperature is plotted. The temperature at which the maximum number of turns occurs then becomes the forging temperature for maximum forgeability. The hot-twist test has been found to be useful, particularly for steels.

The forgeability of various metals and alloys is given in Table 14.3 in decreasing order. They are based on considerations such as ductility and strength of the material, forging temperature required, frictional behavior, and the quality of the forgings produced. These ratings should be regarded only as general guidelines. Typical *hot-forging temperature* ranges for various metals and alloys are included in Table 14.3. Note that higher forging temperature does not necessarily indicate greater difficulty in forging that material. For *warm* forging, temperatures range from 200° to 300°C (400° to 600°F) for aluminum alloys and from 550° to 750°C (1000° to 1400°F) for steels.

Forging defects.   In addition to surface cracking during forging, other defects also can develop as a result of the material flow pattern in the die, as described next in Section 14.6 regarding die design. For example, if there is an insufficient volume of material to fill the die cavity completely, the web may buckle during forging and develop laps (Fig. 14.16a). On the other hand, if the web is too thick, the excess material flows past the already formed portions of the forging and develops internal cracks (Fig. 14.16b).

TABLE 14.3

Classification of Metals in Decreasing Order of Forgeability	
Metal or alloy	Approximate range of hot-forging temperatures (°C)
Aluminum alloys	400–550
Magnesium alloys	250–350
Copper alloys	600–900
Carbon-and low-alloy steels	850–1150
Martensitic stainless steels	1100–1250
Austenitic stainless steels	1100–1250
Titanium alloys	700–950
Iron-based superalloys	1050–1180
Cobalt-based superalloys	1180–1250
Tantalum alloys	1050–1350
Molybdenum alloys	1150–1350
Nickel-based superalloys	1050–1200
Tungsten alloys	1200–1300

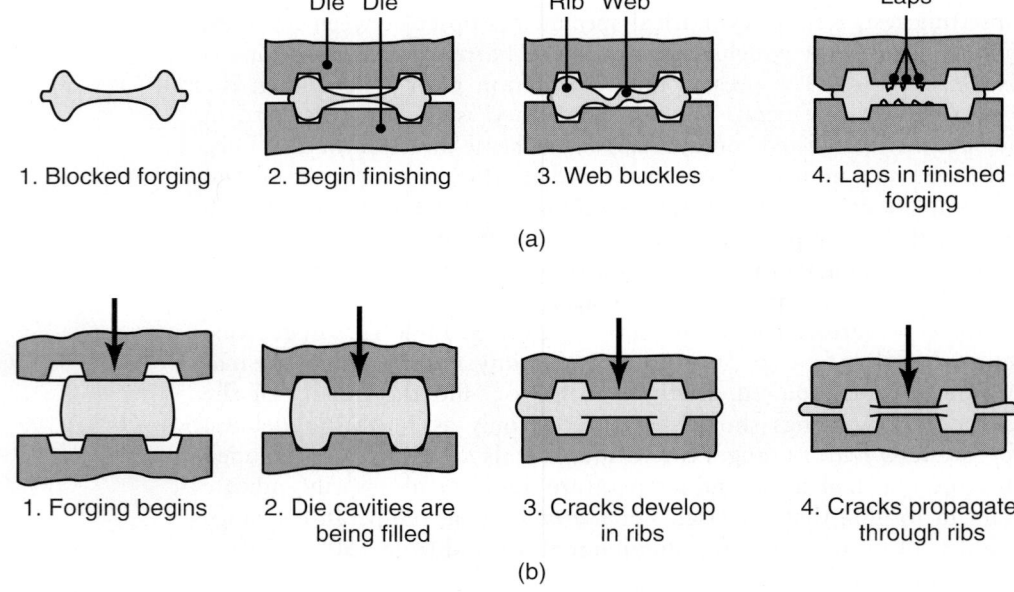

FIGURE 14.16 Examples of defects in forged parts. (a) Laps formed by web buckling during forging; web thickness should be increased to avoid this problem. (b) Internal defects caused by an oversized billet. Die cavities are filled prematurely, and the material at the center flows past the filled regions as the dies close.

The various radii in the forging-die cavity can influence significantly the formation of such defects. Internal defects also may develop from (a) nonuniform deformation of the material in the die cavity, (b) temperature gradients throughout the workpiece during forging, and (c) microstructural changes caused by phase transformations. The grain-flow pattern of the material in forging also is important. The flow lines may reach a surface perpendicularly, as shown in Fig. 14.12. This condition, known as **end grains**, exposes the grain boundaries directly to the environment and can be attacked by it, developing a rough surface and also acting as stress raisers.

Forging defects can cause fatigue failures, and they also may lead to such other problems as corrosion and wear during the service life of the forged component. The importance of inspecting forgings prior to their placement in service, particularly in critical applications such as in aircraft, is obvious. Inspection techniques for manufactured parts are described in Chapter 36.

14.6 | Die Design, Die Materials, and Lubrication

The design of forging dies requires considerable knowledge and experience regarding the shape and complexity of the workpiece, its ductility, its strength and sensitivity to deformation rate and temperature, and its frictional characteristics. Die distortion under high forging loads also is an important design consideration, particularly if close dimensional tolerances are required. The most important rule in die design is the fact that the part will flow in the direction of least resistance. Thus, the workpiece *intermediate shape*s should be planned so that they properly fill the die cavities. An example of the intermediate shapes for a connecting rod are shown in Fig. 14.6a.

With continuing advances in developing reliable simulation of all types of metalworking operations, sofware is available to help predict material flow in forging die cavities under various conditions such as temperature and heat transfer, frictional conditions at die-workpiece contact surfaces, and forging speed. Such software can be very helpful in die design and in eliminating future problems with defective forgings (see also Section 38.7).

Preshaping. In a properly preshaped workpiece, the material should not flow easily into the flash (as otherwise die filling will be incomplete), the grain flow pattern should be favorable for the products' strength and reliability, and excessive sliding at the workpiece–die interfaces should be minimized in order to reduce die wear. The selection of preshapes requires considerable experience and involves calculations of cross-sectional areas at each location in the forging. Computer modeling and simulation techniques are very useful in such calculations.

Die design features. The terminology for forging dies is shown in Fig. 14.5d, and the significance of various features is described next. Some of these considerations are similar to those for casting (Section 12.2). For most forgings, the **parting line** is at the largest cross-section of the part. For simple symmetrical shapes, the parting line is normally a straight line at the center of the forging, but for more complex shapes, the line may not lie in a single plane. The dies then are designed in such a way that they lock during engagement, in order to avoid side thrust, balance forces, and maintain die alignment during forging.

After sufficiently constraining lateral flow to ensure proper die filling, the flash material is allowed to flow into a *gutter*, so that the extra flash does not increase the forging load unnecessarily. A general guideline for flash clearance between dies is 3% of the maximum thickness of the forging. The length of the *land* is usually two to five times the flash thickness.

Draft angles are necessary in almost all forging dies in order to facilitate the removal of the part from the die. Upon cooling, the forging shrinks both radially and longitudinally, so internal draft angles are made larger than external ones. Internal angles are about 7 to 10°, external angles about 3 to 5°.

Selection of the proper radii for corners and fillets is important in order to ensure smooth flow of the metal into the die cavity and to improve die life. Small radii generally are undesirable because of their adverse effect on metal flow and their tendency to wear rapidly (as a result of stress concentration and thermal cycling). Small fillet radii also can cause fatigue cracking of the dies. As a general rule, these radii should be as large as can be permitted by the design of the forging. As with the patterns used in casting, *allowances* are provided in forging-die design because machining the forging may be necessary to obtain final desired dimensions and surface finish. Machining allowance should be provided at flanges, at holes, and at mating surfaces.

Die materials.   Most forging operations (particularly for large parts) are carried out at elevated temperatures. General requirements for die materials therefore are

- Strength and toughness at elevated temperatures.
- Hardenability and ability to harden uniformly.
- Resistance to mechanical and thermal shock.
- Wear resistance (particularly resistance to abrasive wear) because of the presence of scale in hot forging.

Common die materials are tool and die steels containing chromium, nickel, molybdenum, and vanadium (see Tables 5.5 and 5.6). Dies are made from die blocks, which themselves are forged from castings and then machined and finished to the desired shape and surface finish. Die-manufacturing methods are described in Section 14.7.

Lubrication.   A wide variety of metalworking fluids can be used in forging, as described in Section 33.7. Lubricants greatly influence friction and wear. Consequently, they affect the forces required (see Eq. 14.1) and the manner in which the material flows into the die cavities. They also can act as a thermal barrier between the hot workpiece and the relatively cool dies—thus slowing the rate of cooling of the workpiece and improving metal flow. Another important role of the lubricant is to act as a *parting agent*—preventing the forging from sticking to the dies and helping in its release from the die.

14.7 | Die Manufacturing Methods—Die Failures

From the topics described thus far, it should be evident that dies, their quality, and die life are highly significant aspects of the total manufacturing operation, including the quality of the parts produced. This is particularly noteworthy in view of the fact that the vast majority of discrete parts that are produced in large quantities (such as gears, shafts, bolts, and numerous components of machinery, transportation and farm equipment), as well as castings of all types of products, are made in individual dies and molds. Dies also have an impact on the overall economics of manufacturing, because of their cost and the lead time needed to produce them, as some dies require months to manufacture. Equally important considerations are the maintenance of dies and their modifications as parts are first produced.

Several manufacturing methods, either singly or in combination, can be used to make dies for forging as well as for other metalworking processes. These processes include (a) casting, (b) forging, (c) machining, (d) grinding, (e) electrical and electrochemical methods—particularly electrical-discharge machining (EDM) and wire EDM—and (f) lasers for small dies. An important and continuing development is the production of tools and dies by **rapid tooling** using rapid protoyping techniques, as described in Section 20.5.

The process of producing a die cavity in a **die block** is called **die sinking**. The process of *hubbing* (Section 14.4), either cold or hot, also may be used to make small dies with shallow cavities. Dies are usually heat treated for higher hardness and wear resistance (Chapter 31). If necessary, their surface profile and finish are improved further by finish grinding and polishing either by hand or by programmable industrial robots.

The choice of a die manufacturing method depends on its size and shape and the particular operation in which the die is to be used, such as casting, forging, extrusion, powder metallurgy, and plastics molding. As in all manufacturing operations, cost often dictates the process selected because tool and die costs can be significant in manufacturing operations. Dies of various sizes and shapes can be **cast** from steels, cast irons, and nonferrous alloys. The processes used for preparing them may range from sand casting (for large dies weighing several tons), to shell molding (for casting small dies). Cast steels generally are preferred for large dies because of their strength and toughness, as well as the ease with which the steel composition, grain size, and properties can be controlled and modified.

Most commonly, dies are *machined* from *forged die blocks* by processes such as high-speed milling, turning, grinding, and electrical discharge and electrochemical machining. Such an operation is shown in Fig. I.14b for making molds for eyeglass frames. For high-strength and wear-resistant die materials that are hard or are heat treated (thus difficult to machine), processes such as hard machining (Section 25.6) and electrical and elctrochemical machining are a common practice. Typically, a die is machined by milling on computer-controlled machine tools with various software (see Fig. 24.2) that have the capability (economically) of optimizing the cutting-tool path. Thus, the best surface finish can be obtained in the least possible machining time. Equally important is the setup for machining, because dies should be machined as much as possible in one setup without having to remove them from their fixtures and reorient them for subsequent machining operations.

After using heat treatment to achieve the desired mechanical properties, dies usually are subjected to *finishing operations* (Section 26.7), such as grinding, polishing, and chemical and electrical processes, to obtain the desired surface finish and dimensional accuracy. This also may include *laser surface treatments* and *coatings* (Chapter 34) to improve die life. Lasers also are used for die repair and reconfiguration of the worn regions of dies (see also Fig. 33.11).

Die costs. From the preceding discussion, it is evident that the cost of a die will depend greatly on its size, shape complexity, application, and the surface finish required, as well as the die material and manufacturing, heat treating, and finishing methods employed. Consequently, specific die costs cannot be categorized easily. Some qualitative ranges of tool and die costs are given throughout this book, such as in Table 12.6. Even small and relatively simple dies can cost hundreds of dollars to make, and the cost of a set of dies for automotive body panels can be as much as $2 million. On the other hand, because a large number of parts usually are made from one set of dies, *die cost per piece made* is generally a small portion of a part's manufacturing cost (see also Section 40.11). The lead time required to produce dies also can have a significant impact on the overfall manufacturing cost, particularly in a global and competitive marketplace.

Die failures.   Failure of dies in manufacturing operations generally results from one or more of the following causes:

- Improper die design
- Defective or improper selection of die material
- Improper manufacturing and improper heat-treatment and finishing operations
- Overheating and heat checking (i..e., cracking caused by temperature cycling)
- Excessive wear
- Overloading (i.e., excessive force on the die)
- Improper alignment of the die components with respect to their movements
- Misuse
- Improper handling of the die

Although these factors typically apply to dies made of tool and die steels, many also apply to other die materials, such as carbides, ceramics, and diamond.

The proper design of dies is as important as the proper selection of die materials. In order to withstand the forces involved, a die must have sufficiently large cross-sections and clearances (to prevent jamming). Abrupt changes in cross-section, sharp corners, radii, fillets, and a coarse surface finish (including grinding marks and their orientation on die surfaces) act as stress raisers and, thus, can have detrimental effects on die life. For improved strength and to reduce the tendency for cracking, dies may be made in segments and assembled into a complete die with rings that prestress the dies. Proper handling, installation, assembly, and alignment of dies are essential. Overloading of tools and dies can cause premature failure. A common cause of failure in cold-extrusion dies is that of the operator (or of a programmable robot) to remove a formed part from the die before loading it with another blank.

14.8 | Forging Machines

A variety of forging machines are available with a range of capacities (tonnage), speeds, and speed-stroke characteristics (Table 14.4).

TABLE 14.4

Typical Speed Ranges of Forging Equipment	
Equipment	m/s
Hydraulic press	0.06–0.30
Mechanical press	0.06–1.5
Screw press	0.6–1.2
Gravity drop hammer	3.6–4.8
Power drop hammer	3.0–9.0
Counterblow hammer	4.5–9.0

Hydraulic presses.   These presses operate at constant speeds and are *load limited* or load restricted. In other words, a press stops if the load required exceeds its capacity. Large amounts of energy can be transmitted to a workpiece by a constant load throughout a stroke—the speed of which can be controlled. Because forging in a hydraulic press takes longer than in the other types of forging machines described next, the workpiece may cool rapidly unless the dies are heated (see *isothermal forging*, Section 14.4). Compared to mechanical presses, hydraulic presses are slower and involve higher initial costs, but they require less maintenance.

A hydraulic press typically consists of a frame with two or four columns, pistons, cylinders (Fig. 14.17a), rams, and hydraulic pumps driven by electric motors. The ram speed can be varied during the stroke. Press capacities range up to 125 MN (14,000 tons) for open-die forging and up to 450 MN (50,000 tons) in North America, 640 MN (72,000 tons) in France, and 730 MN (82,000 tons) in Russia for closed-die forging. The main landing-gear support beam for the Boeing 747 aircraft is forged in a 450-MN (50,000-ton) hydraulic press, shown in Fig. 14.1c (with the part in the forefront). This part is made of a titanium alloy and weighs approximately 1350 kg (3000 lb.).

Mechanical presses.   These presses are basically of either the crank or the eccentric type (Fig. 14.17b). The speed varies from a maximum at the center of the stroke to zero at the bottom of the stroke, thus they are *stroke limited*. The energy in a mechanical press is generated by a large flywheel powered by an electric motor. A clutch engages the flywheel to an eccentric shaft. A connecting rod translates the rotary motion into a reciprocating linear motion. A *knuckle-joint* mechanical press is shown in Fig. 14.17c. Because of the linkage design, very high forces can be applied in this type of press (see also Fig. 11.18).

The force available in a mechanical press depends on the stroke position and becomes extremely high at the bottom "dead" center. Thus, proper setup is essential to avoid breaking the dies or equipment components. Mechanical presses have high production rates, are easier to automate, and require less operator skill than do other types of machines. Press capacities generally range from 2.7 to 107 MN (300 to 12,000 tons). Mechanical presses are preferred for forging parts with high prescision.

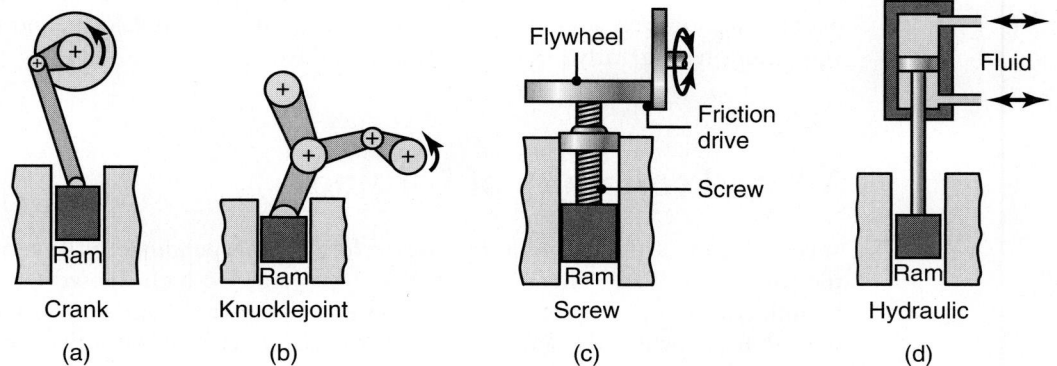

FIGURE 14.17   Schematic illustration of the principles of various forging machines. (a) Mechanical press with an eccentric drive; the eccentric shaft can be replaced by a crankshaft to give the up-and-down motion to the ram. (b) Knuckle-joint press. (c) Screw press. (d) Hydraulic press.

Screw presses.   These presses (Fig. 14.17d) derive their energy from a flywheel; hence, they are *energy limited*. The forging load is transmitted through a large vertical screw, and the ram comes to a stop when the flywheel energy is dissipated. If the dies do not close at the end of the cycle, the operation is repeated until the forging is completed. Screw presses are used for various open-die and closed-die forging operations. They are suitable particularly for small production quantities, especially thin parts with high precision, such as turbine blades. Press capacities range from 1.4 to 280 MN (160 to 31,500 tons).

Hammers.   Hammers derive their energy from the potential energy of the ram, which is converted into kinetic energy; hence, they are *energy limited*. Unlike hydraulic presses, hammers (as the name implies) operate at high speeds, and the resulting low forming time minimizes the cooling of a hot forging. Low cooling rates then allow the forging of complex shapes, particularly those with thin and deep recesses. To complete the forging, several successive blows usually are made in the same die. Hammers are available in a variety of designs and are the most versatile and the least expensive type of forging equipment.

Drop hammers.   In *power drop hammers*, the ram's downstroke is accelerated by steam, air, or hydraulic pressure at about 750 kPa (100 psi). Ram weights range from 225 to 22,500 kg (500 to 50,000 lb) with energy capacities ranging up to 1150 kJ (850,000 ft-lb). In the operation of *gravity drop hammers* (a process called **drop forging**), the energy is derived from the free-falling ram (potential energy). The available energy of a drop hammer is the product of the ram's weight and the height of its drop. Ram weights range from 180 to 4500 kg (400 to 10,000 lb) with energy capacities ranging up to 120 kJ (90,000 ft-lb).

Counterblow hammers.   This hammer has two rams that simultaneously approach each other horizontally or vertically to forge the part. As in open-die forging operations, the part may be rotated between blows for proper shaping of the workpiece during forging. Counterblow hammers operate at high speeds and transmit less vibration to their bases. Capacities range up to 1200 kJ (900,000 ft-lb).

High-energy-rate forging (HERF) machines.   In this type of machine, the ram is accelerated rapidly by inert gas at high pressure, and the part is forged in one blow at a very high speed. Although there are several types of these machines, various problems associated with their operation, maintenance, die breakage, and safety considerations have greatly limited their use in industry.

14.9 | Economics of Forging

Several factors are involved in the cost of forgings. Depending on the complexity of the forging, tool and die costs range from moderate to high. However, as in other manufacturing operations, this cost is spread out over the number of parts forged with that particular die set. Thus, even though the cost of workpiece material per piece made is constant, setup and tooling costs-per-piece decrease as the number of pieces forged increases (Fig. 14.18).

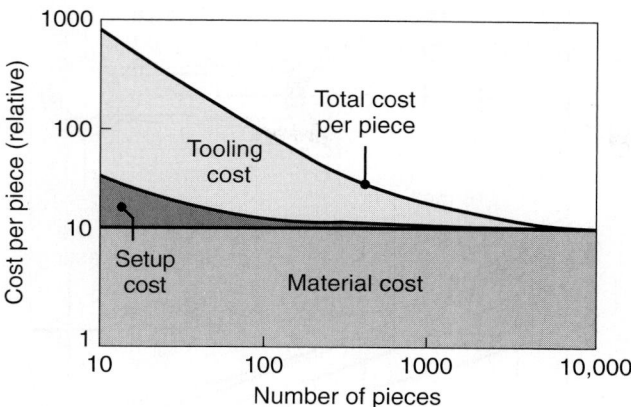

FIGURE 14.18 Typical (cost-per-piece) in forging; note how the setup and the tooling costs-per-piece decrease as the number of pieces forged increases if all pieces use the same die.

The ratio of the cost of the die material to the total cost of forging the part increases with the weight of forgings; the more expensive the material, the higher the cost of the material relative to the total cost. Because dies must be made and forging operations must be performed regardless of the size of forging, the cost of dies and of the forging operation relative to material cost is high for small parts, and conversely, die material costs are relatively low.

The size of forgings also has some effect on cost. Sizes range from small forgings (such as utensils and small automotive components) to large ones (such as gears and crankshafts and connecting rods for large engines). As forging size increases, the share of material cost in the total cost also increases but at a lower rate. This occurs because (a) the incremental increase in die cost for larger dies is relatively small, (b) the machinery and operations involved are essentially the same regardless of forging size, and (c) the labor involved per piece made is not that much higher.

The total cost involved in a forging operation is not influenced to any major extent by the type of materials forged. Because they have been reduced significantly by automated and computer-controlled operations, labor costs in forging generally are moderate. Also, die design and manufacturing are now performed by computer-aided design and manufacturing techniques (Chapter 38), which result in major savings in time and effort.

The cost of forging a part compared to that of making it by various casting techniques, powder metallurgy, machining, or other methods is an important consideration in a competitive global marketplace.

For example, all other factors being the same and depending on the number of pieces required, manufacturing a certain part by, say, expendable-mold casting may well be more economical than producing it by forging for shorter production runs (Fig. 14.19). This casting method does not require expensive molds and tooling, whereas forging requires expensive dies. Competitive aspects of manufacturing and process selection are discussed in greater detail in Chapter 40.

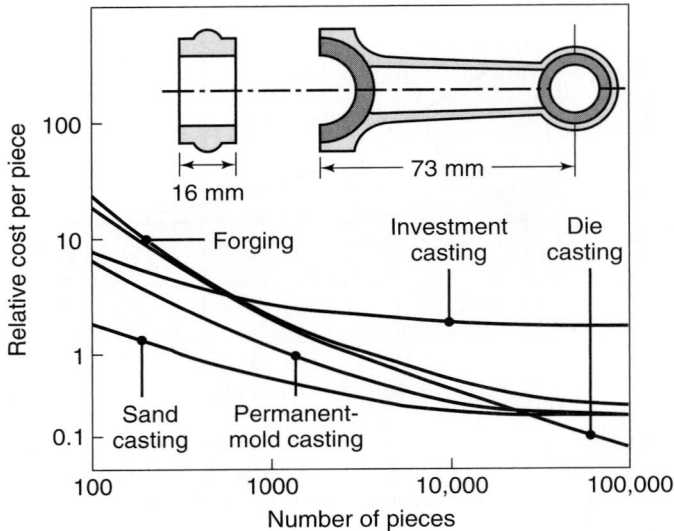

FIGURE 14.19   Relative unit costs of a small connecting rod made by various forging and casting processes. Note that, for large quantities, forging is more economical. Sand casting is the most economical process for fewer than about 20,000 pieces.

CASE STUDY 14.2 | Suspension Components for the Lotus Elise Automobile

The automotive industry increasingly has been subjected to a demanding set of performance, cost, fuel efficiency, and environmental regulations. One of the main strategies in improving vehicle design with respect to all of these (seemingly) contradictory constraints is to reduce vehicle weight while using advanced materials and manufacturing processes to preserve performance and safety. Previous design optimization has shown that weight savings of up to 34% could be realized on suspension system components, which is significant since suspensions make up approximately 12% of a car's mass. These weight savings could be achieved largely by developing optimum designs, using advanced analytical tools, and using net-shape or near-net shape steel forgings instead of cast-iron components. In addition, significant cost savings have been demonstrated in many parts when using optimized steel forgings compared to aluminum castings and extrusions.

The Lotus Elise is a high-performance sports car designed for superior ride and handling. The Lotus group investigated the use of steel forgings instead of extruded-aluminum suspension uprights in order to reduce cost and improve reliability and performance. Their development efforts consisted of two phases, as shown in Fig. 14.20. The first phase involved the development of a forged-steel component that can be used on the existing Elise sports car; the second phase involved the production of a suspension upright for a new model.

A new design was developed using an iterative process with advanced software tools to reduce the number of components and to determine the optimum geometry. The material selected for the upright was an air-cooled forged steel, which gives uniform grain size and microstructure and uniform high strength without the need for heat treatment. These materials also have approximately 20% higher fatigue strengths than traditional carbon steels, such as AISI 1548-HT, which is used for similar applications.

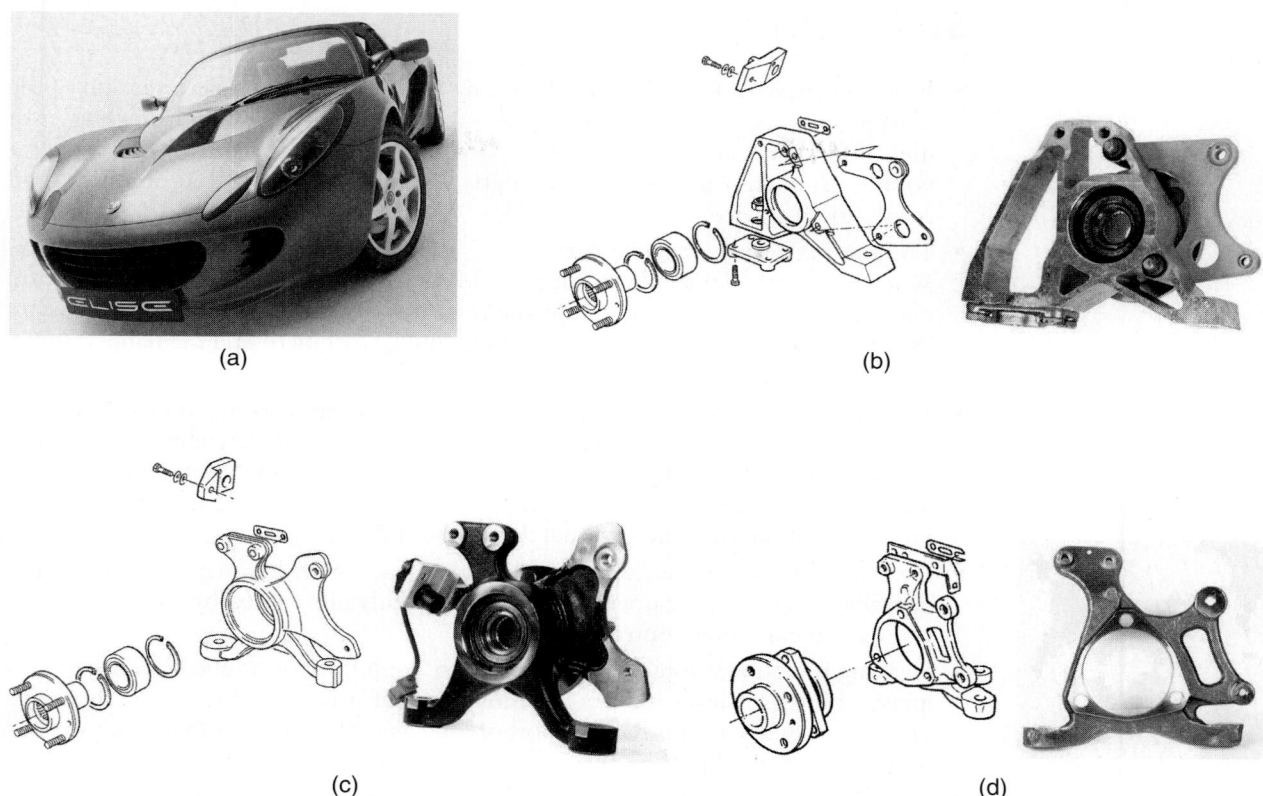

FIGURE 14.20   (a) The Lotus Elise Series 2 automobile; (b) illustration of the original design for the vertical suspension uprights, using an aluminum extrusion; (c) retrofit design, using a steel forging; (d) optimized steel forging design for new car models. *Source*: (a) Courtesy of Fox Valley Motorcars, (b) through (d) Courtesy of Lotus Engineering and the American Iron and Steel Institute.

TABLE 14.5

Comparison of Suspension Upright Designs for the Lotus Elise Automobile				
Figure 14.20 sketch	Material	Application	Mass (kg)	Cost ($)
(b)	Aluminum extrusion, steel bracket, steel bushing, housing	Original design	2.105	85
(c)	Forged steel	Phase I	2.685 (+28%)	27.7 (−67%)
(d)	Forged steel	Phase II	2.493 (+18%)	30.8 (−64%)

The revised designs are summarized in Table 14.5. As can be seen, the optimized new forging design (Fig 14.20d) resulted in significant cost savings. Although it also resulted in a small weight increase when compared to the aluminum-extrusion design, the weight penalty is recognized as quite small, and the use of forged steel for such components is especially advantageous in fatigue-loading conditions constantly encountered by suspension components. The new design also had certain performance advantages in that the component stiffness is now higher, which registered as improved customer satisfaction and better "feel" during driving. Furthermore, the new design reduced the number of parts required, thus satisfying another fundamental principle in design.

Source: Courtesy of Lotus Engineering and the American Iron and Steel Institute.

SUMMARY

- Forging denotes a family of metalworking processes in which deformation of the workpiece is carried out by compressive forces typically applied through a set of dies. Forging is capable of producing a wide variety of structural parts with favorable characteristics, such as strength, toughness, dimensional accuracy, and reliability in service.

- The forging process can be carried out at room, warm, or high temperatures. Workpiece material behavior during deformation, friction, heat transfer, and material-flow characteristics in the die cavity are important considerations, as are the proper selection of die materials, lubricants, workpiece and die temperatures, forging speeds, and equipment.

- Various defects can develop if the forging process is not controlled properly, especially in workpiece quality, billet or preform shape, and die geometry. Computer-aided design and manufacturing techniques now are used extensively in die design and manufacturing, preform design, predicting material flow, and in avoiding the possibility of internal and external defects during forging.

- A variety of forging machines are available, each with its own capabilities and characteristics. Forging operations are now highly automated by using industrial robots and computer controls.

- Swaging is a type of rotary forging in which a solid rod or a tube is reduced in diameter by the reciprocating radial movement of a set of two or four dies. The process is suitable for producing short or long lengths of bar or tubing with various internal or external profiles.

- Because die failure has a major economic impact, die design, material selection, and production method are of major importance. A variety of die materials and manufacturing methods are available, including advanced material-removal and finishing processes.

KEY TERMS

Barreling	Fullering	Open-die forging
Closed-die forging	Hammers	Orbital forging
Cogging	Heading	Piercing
Coining	Hot-twist test	Precision forging
Edging	Hubbing	Presses
End grain	Impression-die forging	Sizing
Flash	Incremental forging	Swaging
Forgeability	Isothermal forging	Upsetting
Forging	Net-shape forging	

BIBLIOGRAPHY

Altan, T., Ngaile, G., and Shen, G. (eds), *Cold and Hot Forging: Fundamentals and Applications*, ASM International, 2004.

ASM Handbook, Vol. 14: *Forming and Forging*, ASM International, 1988.

Blazynski, T.Z., *Plasticity and Modern Metal-forming Technology*, Elsevier, 1989.

Byrer, T.G. (ed.), *Forging Handbook*, Forging Industry Association, 1985.

Dieter, G.E., Kuhn, H.A., and Semiatin, S.L. (eds.), *Handbook of Workability and Process Design*, ASM International, 2003.

Lange, K. (ed.), *Handbook of Metal Forming*, McGraw-Hill, 1985.

Open Die Forging Manual, 3rd ed., Forging Industry Association, 1982.

Prasad, Y.V.R.K., and Sasidhara, S. (eds.), *Hot Working Guide: A Compendium of Processing Maps*, ASM International, 1997.

Product Design Guide for Forging, Forging Industry Association, 1997.

Reid, D.T. (ed.), *Fundamentals of Tool Design*, 3rd ed., Society of Manufacturing Engineers, 1991.

Thomas, A., *DFRA Forging Handbook: Die Design*, Drop Forging Research Association, 1980.

Tool and Manufacturing Engineers Handbook, 4th ed., Vol. 2, *Forming*, Society of Manufacturing Engineers, 1984.

REVIEW QUESTIONS

14.1 What is the difference between cold, warm, and hot forging?

14.2 Explain the difference between open-die and impression-die forging.

14.3 Explain the difference between fullering, edging, and blocking.

14.4 What factors are involved in precision forging?

14.5 What type of parts can rotary swaging produce?

14.6 Explain the features of a typical forging die.

14.7 Explain what is meant by "load limited," "energy limited," and "stroke limited," as these terms pertain to forging machines.

14.8 What is flash?

14.9 Why is hubbing an attractive alternative to producing simple dies?

14.10 What is the difference between piercing and punching?

QUALITATIVE PROBLEMS

14.11 How can you tell whether a certain part is forged or cast? Explain the features that you would investigate.

14.12 Why is the intermediate shape of a part important in forging operations?

14.13 Explain the functions of flash in impression-die forging.

14.14 Why is control of the volume of the blank important in closed-die forging?

14.15 Why are there so many types of forging machines? Describe the capabilities and limitations of each.

14.16 What are the advantages and limitations of (a) a cogging operation and (b) isothermal forging?

14.17 Describe your observations concerning Fig. 14.16.

14.18 What are the advantages and limitations of using die inserts?

14.19 Review Fig. 14.5d and explain why inner draft angles are larger than outer angles. Is this also true for permanent-mold casting?

14.20 Comment on your observations regarding the grain-flow pattern in Fig. 14.12.

14.21 Describe your observations concerning the control of the final tube thickness in Fig. 14.15.

14.22 By inspecting some forged products (such as a pipe wrench) you can see that the lettering on them is raised rather than sunk. Offer an explanation as to why they are made that way.

14.23 Describe the difficulties involved in precisely defining the term "forgeability."

14.24 Identify casting design rules (described in Section 12.2) that also can be applied to forging.

QUANTITATIVE PROBLEMS

14.25   Calculate the forging force for a solid, cylindrical workpiece made of 1020 steel that is 3.5 in. high and 5 in. in diameter and is to be reduced in height by 30%. Let the coefficient of friction be 0.2.

14.26   Using Eq. (14.2), estimate the forging force for the workpiece in Problem 14.25, assuming that it is a complex forging and that the projected area of the flash is 40% greater than the projected area of the forged workpiece.

14.27   Take two solid, cylindrical specimens of equal diameter but different heights and compress them (frictionless) to the same percent reduction in height. Show that the final diameters will be the same.

14.28   In Example 14.1, calculate the forging force, assuming that the material is 1100-O aluminum and that the coefficient of friction is 0.2.

14.29   Using Eq. (14.1), make a plot of the forging force, F, as a function of the radius, r, of the workpiece. Assume that the flow stress, Y_f, of the material is constant. Remember that the volume of the material remains constant during forging, thus as h decreases, r increases.

14.30   How would you go about calculating the punch force required in a hubbing operation, assuming that the material is mild steel and the projected area of the impression is 0.5 in^2? Explain clearly. (*Hint*: See Section 2.6 on hardness.)

14.31   A mechanical press is powered by a 30-hp motor and operates at forty strokes per minute. It uses a flywheel, so that the crankshaft speed does not vary appreciably during the stroke. If the stroke is 6 in., what is the maximum constant force that can be exerted over the entire stroke length?

14.32   To what thickness can a cylinder of 5052-O aliminum that is 3 in. in diameter and 2 in. high be forged in a press that can generate 100,000 lb?

14.33   Assume that you are an instructor covering the topics described in this chapter, and you are giving a quiz on the numerical aspects to test the understanding of the students. Prepare two quantitative problems and supply the answers.

SYNTHESIS, DESIGN, AND PROJECTS

14.34   Devise an experimental method whereby you can measure only the force required for forging the flash in impression-die forging.

14.35   Assume that you represent the forging industry and that you are facing a representative of the casting industry. What would you tell that person about the merits of forging processes?

14.36   Figure P14.36 shows a round impression-die forging made from a cylindrical blank, as shown on the left. As described in this chapter, such parts are made in a sequence of forging operations. Suggest a sequence of intermediate forging steps to make the part on the right and sketch the shape of the dies needed.

14.37   Gears can be made by forging, especially bevel gears. Make a survey of the technical literature and describe the sequence of manufacturing steps involved. Comment on the quality of such a gear as compared with one made by the casting processes described in Chapter 11.

14.38   Forging is one method of producing turbine blades for jet engines (in addition to casting or machining). Study the design of such blades and the relevant technical literature in the Bibliography, then prepare a step-by-step procedure to produce such blades by forging. Comment on the possible difficulties that may be encountered and offer solutions.

14.39   In comparing forged parts with those that are cast, we have noted that the same part may be made by either process. Comment on the pros and cons of each process, considering factors such as part size and shape complexity, design flexibility, mechanical properties developed, and performance in service.

14.40   From the data given in Table 14.3, obtain the approximate value of the yield strength of these materials at hot-forging temperatures. Plot a bar chart showing the

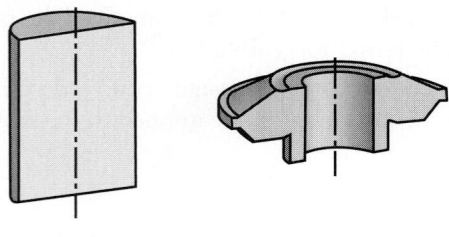

FIGURE P14.36

maximum diameter of a hot-forged part produced on a press with a 60 ton capacity as a function of the material.

14.41 Obtain a number of bolts, nails, and screws of different sizes. Measure the volume of the heads and calculate the original unsupported length-to-diameter ratio for these parts. Discuss these numbers with respect to the discussion in the text.

14.42 Review the sequence of operations in the production of the stepped pin shown in Fig. 14.13. If the conical-upsetting step is not performed, what would be the consequences on the final part?

14.43 Perform simple cogging operations on pieces of clay using a flat piece of wood and make observations regarding the spread of the pieces as a function of the original cross-sections of the clay specimens (for example, square, or rectangular with different thickness-to-width ratios).

14.44 Discuss the possible environmental concerns regarding the operations described in this chapter.

CHAPTER

15

Extrusion and Drawing of Metals

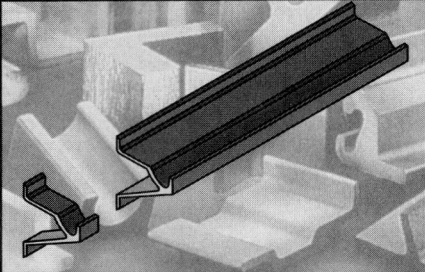

This chapter examines the fundamentals of extrusion and drawing and their applications in manufacturing. We specifically describe:

• Characteristics of direct, indirect, and hydrostatic extrusion.

• Forces involved in extrusion as a function of material and processing parameters.

• Extrusion practice and die design to avoid common defects.

• Drawing of rod, wire, and tubing.

• Die design for product integrity.

Typical parts made by extrusion and drawing: Long pieces with a wide variety of constant cross-sections, rods, shafts, bars for machinery and automotive power-train applications, aluminum ladders, collapsible tubes, wire for numerous electrical and mechanical applications.

Alternative processes: Machining, powder metallurgy, and casting.

15.1 | Introduction

Extrusion and drawing have numerous applications in the manufacture of continuous as well as discrete products from a wide variety of metals and alloys. In **extrusion**, a cylindrical billet is forced through a die (Fig. 15.1) in a manner similar to squeezing toothpaste from a tube or extruding Play-Doh® in various cross-sections in a toy press. A wide variety of solid or hollow cross-sections may be produced by extrusion, which essentially are semifinished parts. A characteristic of extrusion (from the Latin *extrudere*, meaning "to force out") is that large deformations can take place without fracture (see Section 2.2.8), because the material is under high triaxial compression during extrusion. Since the die geometry remains unchanged throughout the operation, extruded products typically have a constant cross-section. Lead pipes were made by extrusion in the 1700s. Plastics are extruded extensively, as described in Section 19.2.

Typical products made by extrusion are railings for sliding doors, window frames, tubing having various cross-sections, aluminum ladders, and numerous structural and architectural shapes. Extrusions can be cut into desired lengths, which then become discrete parts, such as brackets, gears, and coat hangers (Fig. 15.2). Commonly

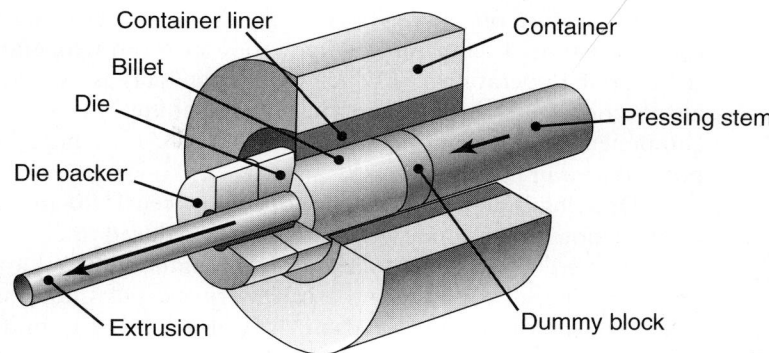

FIGURE 15.1 Schematic illustration of the direct-extrusion process.

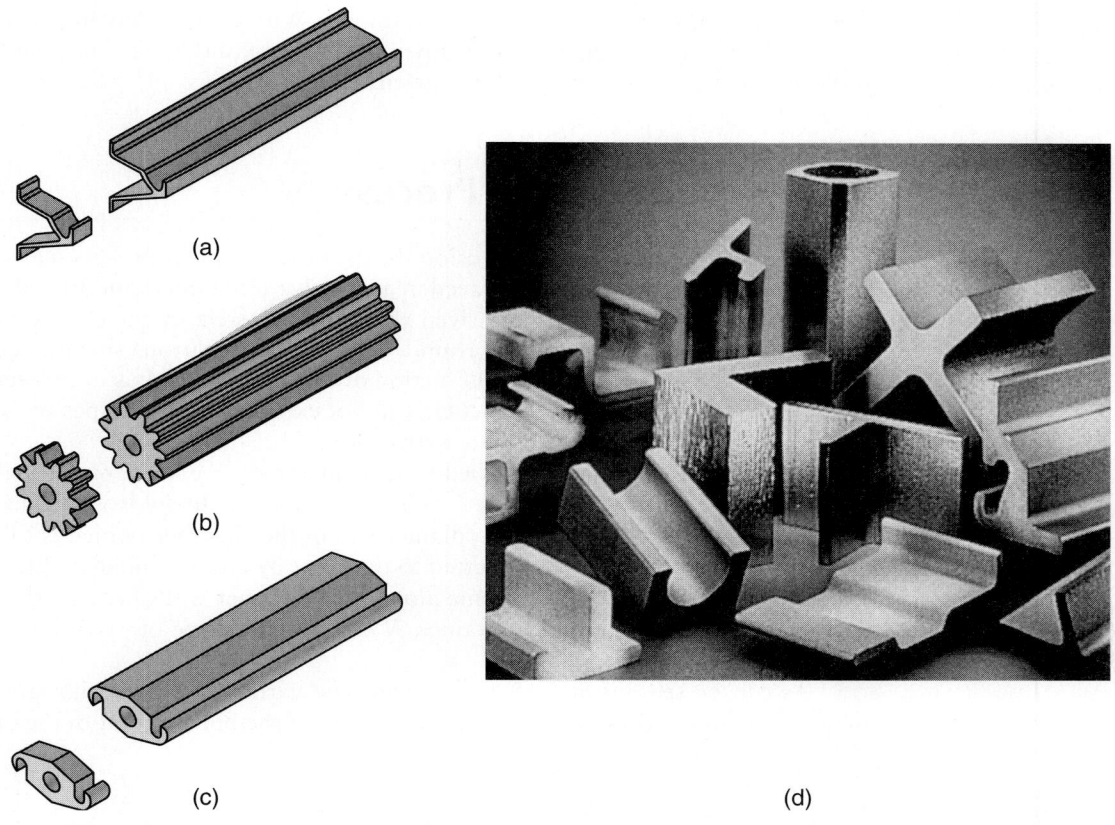

FIGURE 15.2 Extrusions and examples of products made by sectioning off extrusions. *Source:* Courtesy of Plymouth Extruded Shapes.

extruded materials are aluminum, copper, steel, magnesium, and lead; other metals and alloys also can be extruded, with various levels of difficulty.

Because a chamber is involved, each billet is extruded individually, thus extrusion is a batch or semicontinuous process. Extrusion can be economical for large production runs as well as for short ones. Tool costs generally are low, particularly for producing simple, solid cross-sections.

Depending on the ductility of the material, extrusion may be carried out at room or elevated temperatures. Extrusion at room temperature often is combined with forging operations, in which case it generally is known as **cold extrusion** (see also Section 14.4). It has numerous important applications, including fasteners and components for automobiles, bicycles, motorcycles, heavy machinery, and transportation equipment.

Drawing is an operation developed between 1000 and 1500 A.D., in which the cross-section of solid rod, wire, or tubing is reduced or changed in shape by pulling it through a die. Drawn rods are used for shafts, spindles, and small pistons and as the raw material for fasteners (such as rivets, bolts, and screws). In addition to round rods, various profiles also can be drawn. The term *drawing* also is used to refer to making cup-shaped parts by sheet-metal forming operations, as described in Section 16.9.

The distinction between the terms **rod** and **wire** is somewhat arbitrary, rod being larger in cross-section than wire. In industry, wire generally is defined as a rod that has been drawn through a die at least once. Wire drawing involves smaller diameters than rod drawing with sizes down to 0.01 mm (0.0005 in.) for magnet wire and even smaller for use in very low-current fuses.

15.2   The Extrusion Process

There are three basic types of extrusion. In the most common process (called **direct** or **forward extrusion**), a billet is placed in a *chamber* (container) and forced through a die opening by a hydraulically driven ram (pressing stem or punch), as shown in Fig. 15.1. The die opening may be round, or it may have various shapes, depending on the extruded shape desired. The function of the dummy block is to protect the tip of the pressing stem (punch), particularly in hot extrusion. Other types of extrusion are indirect, hydrostatic, and impact extrusion.

In **indirect** extrusion (also called reverse, inverted, or backward extrusion), the die moves toward the unextruded billet (Fig. 15.3a). In **hydrostatic** extrusion (Fig. 15.3b), the billet is smaller in diameter than the chamber (which is filled with a fluid), and the pressure is transmitted to the billet by a ram. Unlike in direct extrusion, there is no friction to overcome along the container walls because the billet is stationary with respect to the container. A less common type of extrusion is *lateral* (or *side*) *extrusion* (Fig. 15.3c).

As can be seen in Fig. 15.4, the geometric variables in extrusion are the die angle, α, and the ratio of the cross-sectional area of the billet to that of the extruded

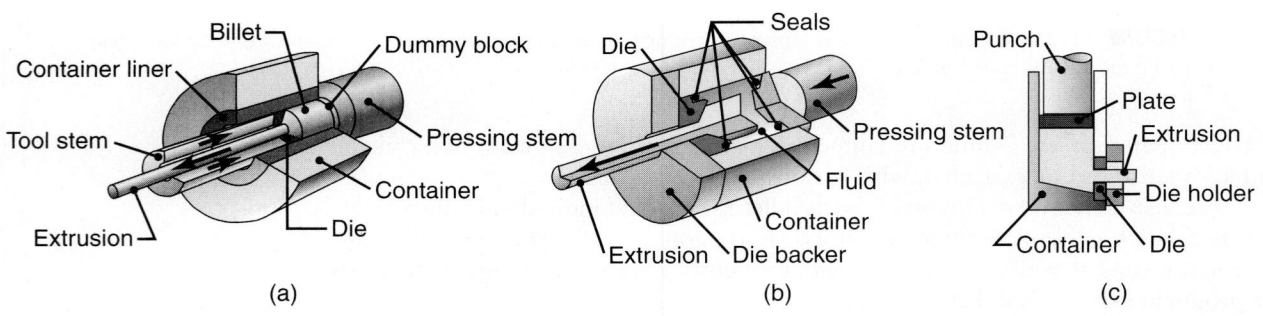

FIGURE 15.3   Types of extrusion: (a) indirect; (b) hydrostatic; (c) lateral.

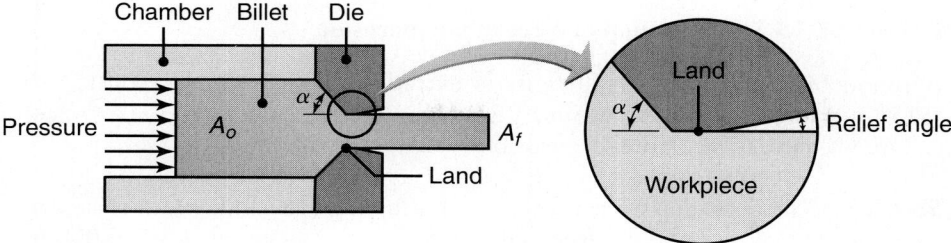

FIGURE 15.4 Process variables in direct extrusion. The die angle, reduction in cross-section, extrusion speed, billet temperature, and lubrication all affect the extrusion pressure.

product, A_o/A_f, called the **extrusion ratio**, R. Other variables are the temperature of the billet, the speed at which the ram travels, and the type of lubricant used.

Extrusion force. The force required for extrusion depends on (a) the strength of the billet material, (b) the extrusion ratio, (c) friction between the billet and the chamber and die surfaces, and (d) process variables, such as the temperature of the billet and the speed of extrusion. The *extrusion force, F,* can be estimated from the formula

$$F = A_o k \ln\left(\frac{A_o}{A_f}\right) \tag{15.1}$$

where k is the **extrusion constant** (which is determined experimentally) and A_o and A_f are the billet and extruded product areas, respectively. The k value in Eq. (15.1) is thus a measure of the strength of the material being extruded and the frictional conditions. Figure 15.5 gives the k values for several metals for a range of extrusion temperatures.

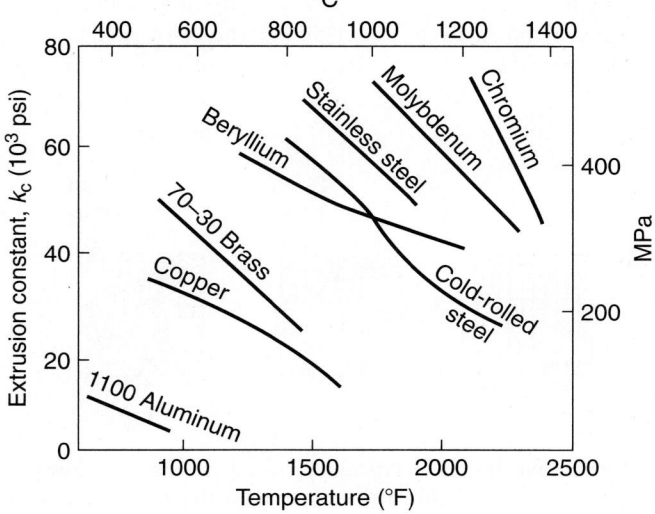

FIGURE 15.5 Extrusion constant k for various metals at different temperatures. *Source:* After P. Loewenstein.

EXAMPLE 15.1 Calculation of force in hot extrusion

A round billet made of 70–30 brass is extruded at a temperature of 1250°F (675°C). The billet diameter is 5 in. (125 mm), and the diameter of the extrusion is 2 in. (50 mm). Calculate the extrusion force required.

Solution The extrusion force is calculated using Eq. (15.1), in which the extrusion constant, k, is obtained from Fig. 15.5. For this material, k = 35,000 psi (250 MPa) at the given extrusion temperature. Thus,

$$F = \pi(2.5)^2(35,000) \ln\left[\frac{\pi(2.5)^2}{\pi(1.0)^2}\right] = 1.26 \times 10^6 \text{ lb}$$

$$= 630 \text{ tons} = 5.5 \text{ MN}$$

(See Section 15.6 for capacities of extrusion presses.)

Metal flow in extrusion. The metal flow pattern in extrusion, as in other forming processes, is important because of its influence on the quality and the mechanical properties of the extruded product. The flow pattern is determined by the basic principle that energy is minimized in any such process. The material flows longitudinally, much like incompressible fluid flow in a channel; thus, extruded products have an elongated grain structure (preferred orientation). Section 15.5 describes how improper metal flow during extrusion can produce various defects in the extruded product.

A common technique for investigating the flow pattern is to section the round billet in half lengthwise and then mark one face with a square grid pattern. The two halves are placed in the chamber together and are extruded. The two pieces then are taken apart and studied. Figure 15.6 shows typical flow patterns obtained by this technique for the case of direct extrusion with *square dies* (90° die angle). The flow pattern is a function of several variables, including friction.

The conditions under which these different flow patterns occur are described in the caption of Fig. 15.6. Note the **dead-metal zones** in Fig. 15.6b and c, where the metal at the corners essentially is stationary. This situation is similar to the stagnation of fluid flow in channels that have sharp angles or turns. Aluminum, for example, typically is hot extruded with square dies and without any lubrication. The

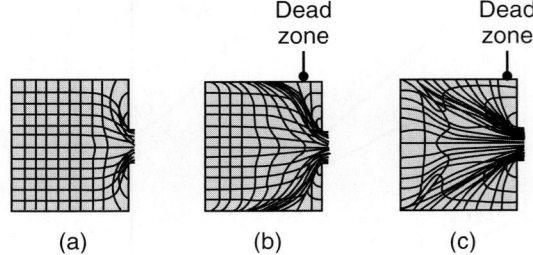

(a)   (b)   (c)

FIGURE 15.6 Types of metal flow in extruding with square dies. (a) Flow pattern obtained at low friction or in indirect extrusion. (b) Pattern obtained with high friction at the billet–chamber interfaces. (c) Pattern obtained at high friction or with cooling of the outer regions of the billet in the chamber. This type of pattern, observed in metals whose strength increases rapidly with decreasing temperature, leads to a defect known as pipe (or extrusion) defect.

shiny, burnished surface produced is a result of the material flowing past the interface with the dead-metal zone.

Process parameters.   Because they have high ductility, wrought aluminum, copper, magnesium and their alloys, steels, and stainless steels are extruded with relative ease into numerous shapes. Other metals (such as titanium and refractory metals) also can be extruded but only with some difficulty and considerable die wear.

In practice, extrusion ratios, R, usually range from about 10 to 100. They may be higher for special applications (400 for softer nonferrous metals) or lower for less ductile materials, although the ratio usually has to be at least 4 to deform the material plastically through the bulk of the workpiece. Extruded products usually are less than 7.5 m (25 ft) long because of the difficulty in handling greater lengths, but they can be as long as 30 m (100 ft). Ram speeds range up to 0.5 m/s (100 ft/min). Generally, lower speeds are preferred for aluminum, magnesium, and copper with higher speeds for steels, titanium, and refractory alloys. Dimensional tolerances in extrusion are usually in the range of ±0.25–2.5 mm (±0.01–0.1 in.), and they increase with increasing cross-section.

Most extruded products, particularly those with small cross-sections, generally require straightening and twisting. This is accomplished typically by stretching and twisting the extruded product, usually in a hydraulic stretcher equipped with jaws. The presence of a die angle causes a small portion of the end of the billet to remain in the chamber after the operation has been completed. This portion (called scrap or the *butt end*) subsequently is removed by cutting off the extrusion at the die exit and removing the scrap from the chamber. Alternatively, another billet or a graphite block may be placed in the chamber to extrude the piece remaining from the previous extrusion.

In **coaxial extrusion** or **cladding**, coaxial billets are extruded together—provided that the strength and ductility of the two metals are compatible. An example is copper clad with silver. *Stepped extrusions* are produced by extruding the billet partially in one die and then in one or more larger dies (see also *cold extrusion*, Section 15.4). Lateral extrusion (Fig. 15.3c) is used for the sheathing of wire and the coating of electric wire with plastic.

15.3 | Hot Extrusion

For metals and alloys that do not have sufficient ductility at room temperature (or in order to reduce the forces required), extrusion is carried out at elevated temperatures (Table 15.1). As in all other hot-working operations, hot extrusion has special requirements because of the high operating temperatures. For example, die wear can be excessive, and cooling of the surfaces of the hot billet (in the cooler chamber) and the die can result in highly nonuniform deformation (Fig. 15.6c). To reduce cooling of the billet and to prolong die life, extrusion dies may be preheated, as is done in hot-forging operations.

Because the billet is hot, it develops an oxide film unless it is heated in an inert-atmosphere furnace. This film can be abrasive (see Section 33.2), and it can affect the flow pattern of the material. It also results in an extruded product that may be unacceptable when a good surface finish is important. In order to avoid the formation of oxide films on the hot extruded product, the dummy block placed ahead of the ram (Fig. 15.1) is made a little smaller in diameter than the container. As a result, a thin shell (*skull*) consisting mainly of the outer oxidized layer of the billet is left in the container. The skull is removed later from the chamber.

TABLE 15.1

Typical Extrusion Temperature Ranges for Various Metals and Alloys	
	°C
Lead	200–250
Aluminum and its alloys	375–475
Copper and its alloys	650–975
Steels	875–1300
Refractory alloys	975–2200

Die design.   Die design requires considerable experience, as can be appreciated by reviewing Fig. 15.7. *Square dies* (*shear dies*) are used in extruding nonferrous metals, especially aluminum. Square dies develop dead-metal zones, which in turn form a "die angle" (see Fig. 15.6b and c) along which the material flows in the deformation zone. The dead-metal zones produce extrusions with bright finishes because of the burnishing that takes places as the material flows past the "die angle" surface.

Tubing is extruded from a solid or hollow billet. Wall thickness usually is limited to 1 mm (0.040 in.) for aluminum, 3 mm (0.125 in.) for carbon steels, and 5 mm (0.20 in.) for stainless steels. When using solid billets, the ram is fitted with a mandrel that pierces a hole into the billet. Billets with a previously pierced hole also may be extruded in this manner. Because of friction and the severity of deformation, thin-walled extrusions are more difficult to produce than those with thick walls.

Hollow cross-sections (Fig. 15.8a) can be extruded by welding-chamber methods and using various dies known as a **spider die, porthole die,** and **bridge die** (Fig. 15.9b). During extrusion, the metal divides and flows around the supports for the internal mandrel into strands. (This condition is much like that of air flowing around a moving car and rejoining downstream or water flowing around large rocks in a river and rejoining.) The strands being extruded then become rewelded under the high pressures existing in the welding chamber before they exit through the die. The rewelded surfaces have good strength because they have not been exposed to the environment, which otherwise would develop oxides on the surfaces inhibiting good welding. However the welding-chamber process is suitable only for aluminum and some of its

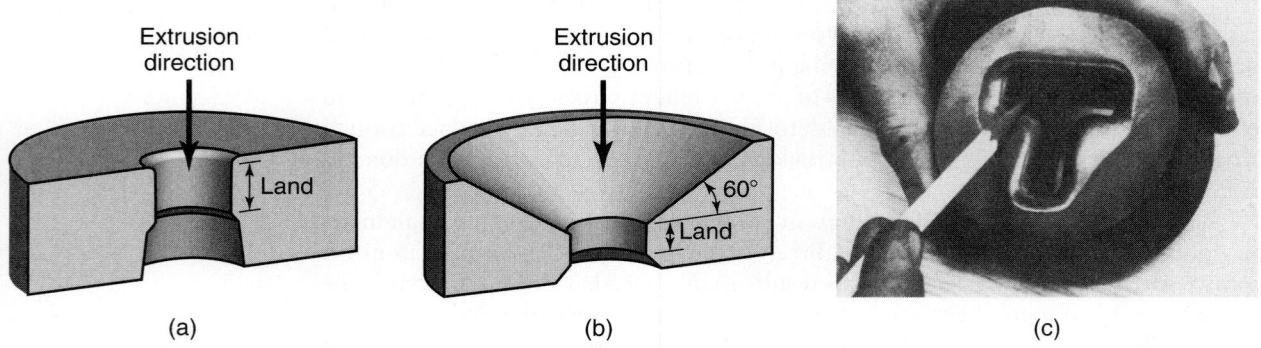

(a)   (b)   (c)

FIGURE 15.7   Typical extrusion–die configurations: (a) die for nonferrous metals; (b) die for ferrous metals; (c) die for a T-shaped extrusion made of hot-work die steel and used with molten glass as a lubricant. *Source*: (c) Courtesy of LTV Steel Company.

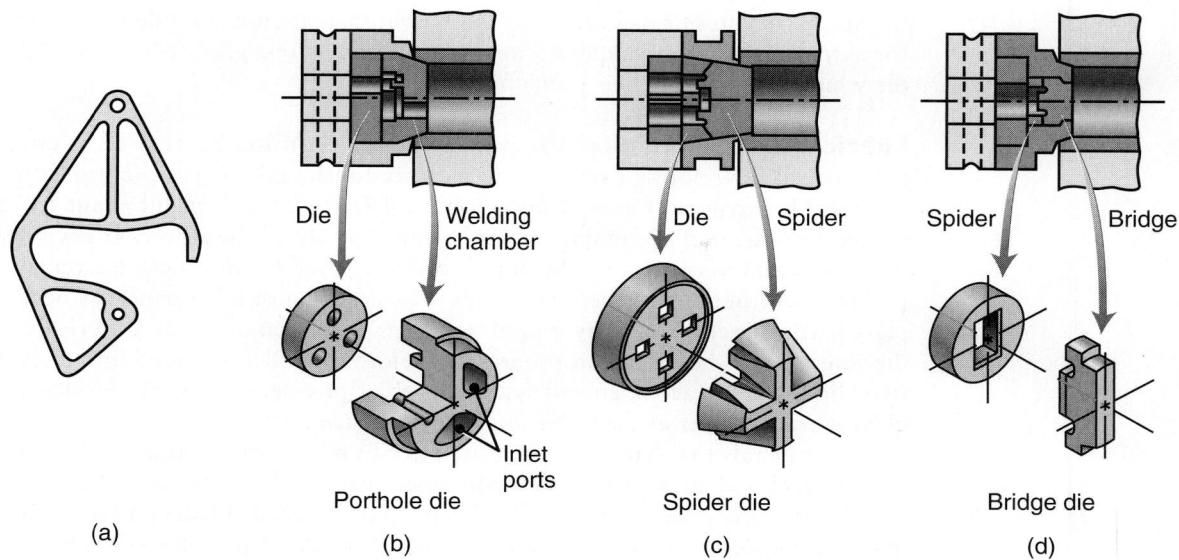

FIGURE 15.8 (a) An extruded 6063-T6 aluminum-ladder lock for aluminum extension ladders. This part is 8 mm (5/16 in.) thick and is sawed from the extrusion (see Fig. 15.2). (b) through (d) Components of various dies for extruding intricate hollow shapes. *Source*: (b) through (d) after K. Laue and H. Stenger.

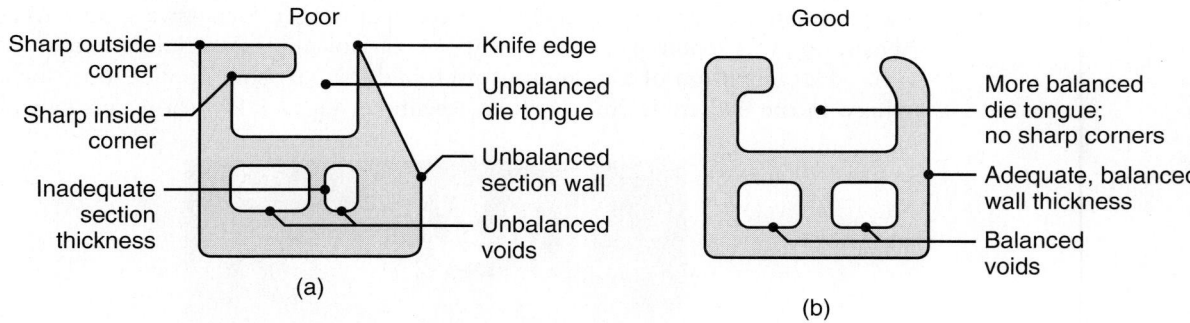

FIGURE 15.9 Poor and good examples of cross-sections to be extruded. Note the importance of eliminating sharp corners and of keeping section thicknesses uniform. *Source*: J. G. Bralla (ed.), *Handbook of Product Design for Manufacturing*. New York: McGraw-Hill Publishing Company, 1986. Used with permission.

alloys, because of their capacity for developing a strong weld under pressure (as described in Section 30.2). Lubricants, of course, cannot be used, because they prevent rewelding of the metal in the die.

Guidelines for proper die design in extrusion are illustrated in Fig. 15.9. Note the (a) importance of symmetry of cross-section, (b) avoidance of sharp corners, and (c) avoidance of extreme changes in die dimensions within the cross-section.

Die materials. Die materials for hot extrusion usually are hot-worked die steels (Section 5.7). Coatings (such as partially stabilized zirconia) may be applied to the dies to extend their life. Partially stabilized zirconia dies (Section 8.2.2) also are used for hot extrusion of tubes and rods. However, they are not suitable for making dies for extruding complex shapes because of the severe stress gradients developed in the die which may lead to their premature failure.

Lubrication.   Lubrication is important in hot extrusion because of its effects on (a) material flow during extrusion, (b) surface finish and integrity, (c) product quality, and (d) extrusion forces. *Glass* (Section 8.4) is an excellent lubricant for steels, stainless steels, and high-temperature metals and alloys. In a process developed in the 1940s and known as the **Séjournet process** (after J. Séjournet), a circular, glass pad is placed in the chamber at the die entrance. The hot billet conducts heat to the glass pad, whereby a thin layer of glass begins to melt and acts as a lubricant at the die interface as the extrusion progresses. Before the billet is placed in the chamber, its cylindrical surface is coated with a layer of powdered glass to develop a thin, glass lubricant layer at the billet–chamber interface.

For metals that have a tendency to stick to the container and the die, the billet can be enclosed in a thin-walled container made of a softer and lower-strength metal, such as copper or mild steel. This procedure is called **jacketing** or **canning**. In addition to acting as a low-friction interface, this jacket prevents contamination of the billet by the environment. Also, if the billet material is toxic or radioactive, the jacket prevents it from contaminating the environment. This technique also can be used for extruding reactive metal powders (Section 17.3).

EXAMPLE 15.2 Manufacture of aluminum heat sinks

Aluminum is used widely to transfer heat for both cooling and heating applications because of its very high thermal conductivity. In fact, on a weight-to-cost basis, no other material conducts heat as economically as does aluminum.

Hot extrusion of aluminum is preferred for heat-sink applications, such as those in the electronics industry. For example, Fig. 15.10a shows an extruded

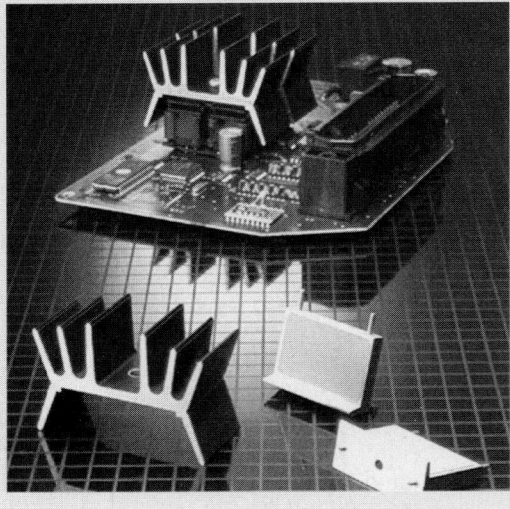

(a)

FIGURE 15.10   (a) Aluminum extrusion used as a heat sink for a printed circuit board. *Source*: Courtesy of Aluminum Extruders Council.

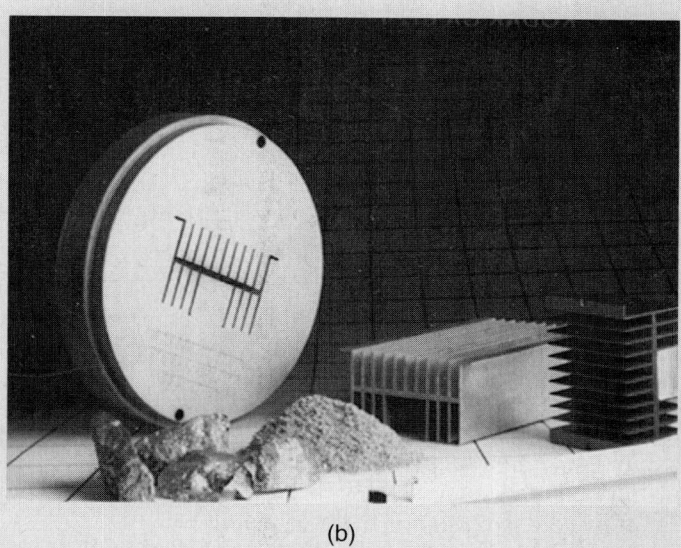

(b)

FIGURE 15.10 (*Continued*) (b) Die and resulting heat-sink profiles. *Source:* Countesy of Aluminum Extruders Council.

heat sink used to remove heat from a transformer on a printed circuit board. Heat sinks usually are designed with a large number of fins that are intended to maximize the surface area and are evaluated from a thermodynamics standpoint using computer simulations. These fins are very difficult and expensive to machine, forge, or roll form. However, the tooling for hot extrusion can be produced through electrical-discharge machining, so that this process is favorable economically.

Figure 15.10b shows a die and hot-extruded cross-section suitable to serve as a heat sink. The shapes shown could be produced through a casting operation, but extrusion is preferred, since there is no internal porosity and the thermal conductivity is slightly higher.

15.4 | Cold Extrusion

Developed in the 1940s, *cold extrusion* is a general term that often denotes a *combination* of operations, such as direct and indirect *extrusion and forging* (Fig. 15.11).

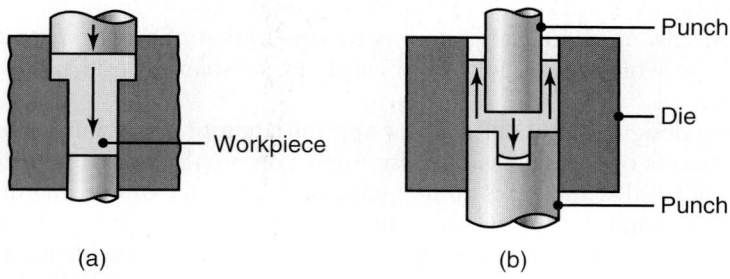

(a)   (b)

FIGURE 15.11   Two examples of cold extrusion. Thin arrows indicate the direction of metal flow during extrusion.

Cold extrusion is used widely for components in automobiles, motorcycles, bicycles, and appliances and in transportation and farm equipment.

This process uses slugs cut from cold-finished or hot-rolled bars, wire, or plates. Slugs that are less than about 40 mm (1.5 in.) in diameter are sheared (*cropped*), and if necessary, their ends are squared off by processes such as upsetting, machining, or grinding. Larger-diameter slugs are machined from bars into specific lengths. Cold-extruded parts weighing as much as 45 kg (100 lb) and having lengths of up to 2 m (80 in.) can be made, although most parts weigh much less. Powder-metal slugs (preforms) also may be cold extruded.

The *force, F*, in cold extrusion may be estimated from the formula

$$F = 1.7 A_o Y_{avg} \varepsilon \tag{15.2}$$

where A_o is the cross-sectional area of the blank, Y_{avg} is the average flow stress of the metal, and ε is the true strain that the piece undergoes based on its original and final cross-sectional area; i.e., $\ln(A_o/A_f)$. For example, assume that a round slug of 10 mm in diameter and made of a metal with $Y_{avg} = 50{,}000$ psi is reduced to a final diameter of 7 mm by cold extrusion. The force would be

$$F = 1.7(\pi)(10^2/4)(50{,}000)[\ln(10/7)^2] = 4.8 \times 10^6 \text{ lb} = 2140 \text{ tons}$$

Cold extrusion has the following advantages over hot extrusion:

- Improved mechanical properties resulting from work-hardening, provided that the heat generated by plastic deformation and friction does not recrystallize the extruded metal.

- Good control of dimensional tolerances, reducing the need for subsequent machining or finishing operations.

- Improved surface finish, due partly to lack of an oxide film and provided that lubrication is effective.

- Production rates and costs that are competitive with those of other methods of producing the same part, such as machining. Some machines are capable of producing more than 2000 parts per hour.

The magnitude of the stresses on the tooling in cold extrusion, however, is very high (especially with steel and specialty-alloy workpieces), being on the order of the hardness of the workpiece material. The punch hardness usually ranges between 60 and 65 HRC and the die hardness between 58 and 62 HRC. Punches are a critical component, as they must possess not only sufficient strength but also sufficient toughness and resistance to wear and fatigue failure.

Lubrication is critical, especially with steels, because of the possibility of sticking (*seizure*) between the workpiece and the tooling (in the case of lubricant breakdown). The most effective means of lubrication is the application of a *phosphate-conversion coating* on the workpiece followed by a coating of soap or wax, as described in Section 34.9.

Tooling design and the selection of appropriate tool and die materials are crucial to the success of cold extrusion. Also important are the selection and control of workpiece material with regard to its quality, the accuracy of the slug dimensions, and its surface condition. Several specialty alloys have been developed (particularly for critical applications requiring high performance) that are suitable for a variety of cold-extrusion and cold-forming operations with good properties, dimensional tolerances, and at low cost.

EXAMPLE 15.3 Cold-extruded part

A typical cold-extruded product, similar to the metal component of an automotive spark plug, is shown in Fig. 15.12. First, a slug is sheared off the end of a round rod (Fig. 15.12, left). It then is cold extruded (Fig. 15.12, middle) in an operation similar to those shown in Fig. 15.11 but with a blind hole. Then the material at the bottom of the blind hole is punched out, producing the small slug shown. Note the diameters of the slug and of the hole at the bottom of the sectioned part, respectively.

Investigating material flow during the deformation of the slug helps avoid defects and leads to improvements in punch and die design. Furthermore, the part usually is sectioned in the midplane then polished and etched to show the grain flow, as shown in Fig. 15.13 (see also Fig. 14.11).

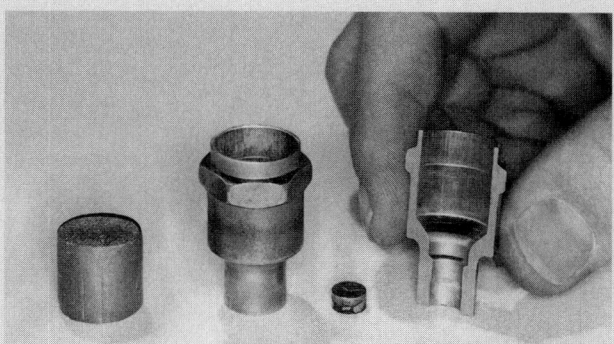

FIGURE 15.12   Production steps for a cold-extruded spark plug. *Source:* Courtesy of National Machinery Company.

FIGURE 15.13   A cross-section of the metal part in Fig. 15.12 showing the grain-flow pattern. *Source:* Courtesy of National Machinery Company.

15.4.1 Impact Extrusion

Impact extrusion is similar to indirect extrusion, and the process often is included in the cold-extrusion category. The punch descends rapidly on the blank (slug), which is extruded backward (Fig. 15.14). Because of volume constancy, the thickness of the tubular extruded section is a function of the clearance between the punch and the die cavity.

Typical products made by this process are shown in Fig. 15.15a and b. Other examples of impact extrusion are the production of collapsible tubes (similar to those used for toothpaste), light fixtures, automotive parts, and small pressure vessels. Most nonferrous metals can be impact extruded in vertical presses and at production rates as high as two parts per second.

The maximum diameter of the parts made is about 150 mm (6 in.). The impact-extrusion process can produce thin-walled tubular sections—ones having thickness-to-diameter ratios as small as 0.005. Consequently, the symmetry of the part and the concentricity of the punch and the blank are important.

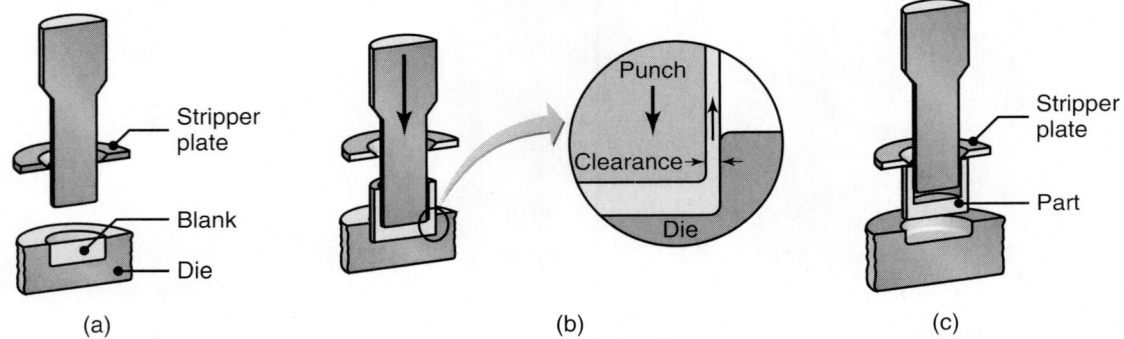

FIGURE 15.14 Schematic illustration of the impact-extrusion process. The extruded parts are stripped by the use of a stripper plate, because they tend to stick to the punch.

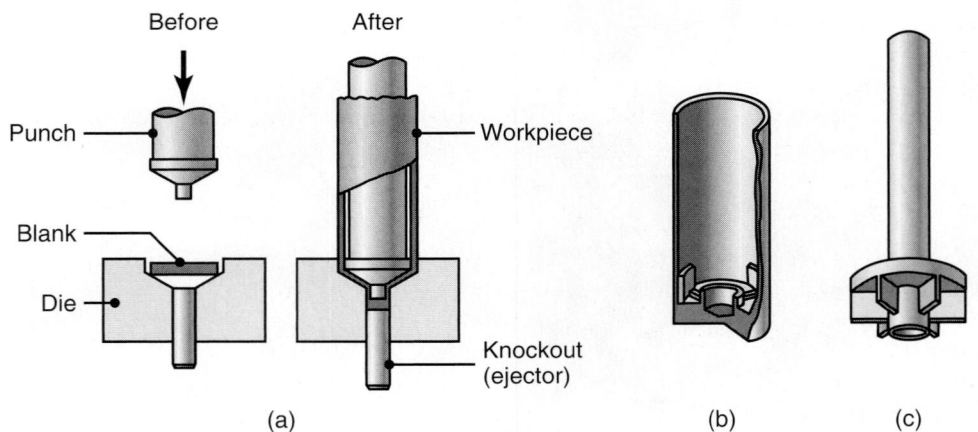

FIGURE 15.15 (a) Impact extrusion of a collapsible tube by the *Hooker process*. (b) and (c) Two examples of products made by impact extrusion. These parts also may be made by casting, forging, or machining. The choice of process depends on the materials involved, part dimensions and wall thickness, and the product properties desired. Economic considerations also are important in final process selection.

15.4.2 Hydrostatic extrusion

In *hydrostatic extrusion*, the pressure required in the chamber is supplied via a piston through an incompressible fluid medium surrounding the billet (Fig. 15.3b). Pressures are typically on the order of 1400 MPa (200 ksi). The high pressure in the chamber transmits some of the fluid to the die surfaces, where it significantly reduces friction. Hydrostatic extrusion usually is carried out at room temperature, typically using vegetable oils as the fluid (particularly castor oil, because it is a good lubricant and its viscosity is not influenced significantly by pressure).

Brittle materials can be extruded successfully by this method, because the hydrostatic pressure (along with low friction and the use of small die angles and high extrusion ratios) increases the ductility of the material. Long wires also have been extruded from an aluminum billet at room temperature and at an extrusion ratio of 14,000, which means that a 1-m billet becomes a 14-km long wire. In spite of the success obtained, hydrostatic extrusion has had limited industrial applications, mainly because of the somewhat complex nature of the tooling, the experience needed with high pressures and the design of specialized equipment, and the long cycle times required—all of which make the process uneconomical.

15.5 | Extrusion Defects

Depending on workpiece material condition and process variables, extruded products can develop several types of defects that can affect significantly their strength and product quality. Some defects are visible to the naked eye, while others can be detected only by the techniques described in Section 36.10. There are three principal *extrusion defects*: surface cracking, pipe, and internal cracking.

Surface cracking.   If extrusion temperature, friction, or speed is too high, surface temperatures can rise significantly, which may cause surface cracking and tearing (*fir-tree cracking* or *speed cracking*). These cracks are intergranular (i.e., along the grain boundaries; see Fig. 2.26) and usually are caused by **hot shortness** (Section 1.4.2). These defects occur especially in aluminum, magnesium, and zinc alloys, although they may also occur in high-temperature alloys. This situation can be avoided by lowering the billet temperature and the extrusion speed.

Surface cracking also may occur at lower temperatures, where it has been attributed to periodic sticking of the extruded product along the die land. Because of the similarity in appearance to the surface of a bamboo stem, it is known as a **bamboo defect.** When the product being extruded temporarily sticks to the *die land* (see Fig. 15.7), the extrusion pressure increases rapidly. Shortly thereafter, the product moves forward again, and the pressure is released. The cycle then is repeated continually, producing periodic circumferential cracks on the surface.

Pipe.   The type of metal-flow pattern in extrusion shown in Fig. 15.6c tends to draw surface oxides and impurities toward the center of the billet—much like a funnel. This defect is known as *pipe defect*, *tailpipe*, or *fishtailing*. As much as one-third of the length of the extruded product may contain this type of defect and, thus, has to be cut off as scrap. Piping can be minimized by modifying the flow pattern to be more uniform, such as by controlling friction and minimizing temperature gradients. Another method is to machine the billet's surface prior to extrusion so that scale and surface impurities are removed. These impurities also can be removed by the chemical etching of the surface oxides prior to extrusion.

Internal cracking.   The center of the extruded product can develop cracks, called *center cracking*, *center-burst*, *arrowhead fracture*, or *chevron cracking*

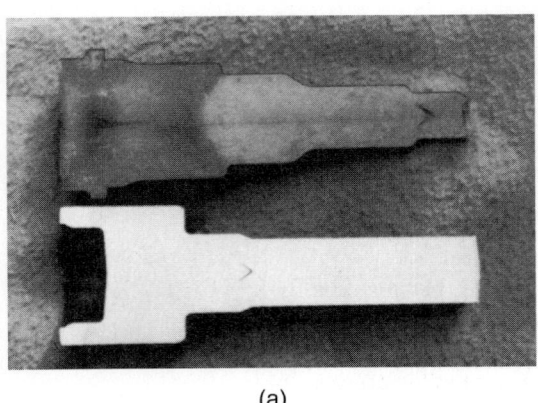

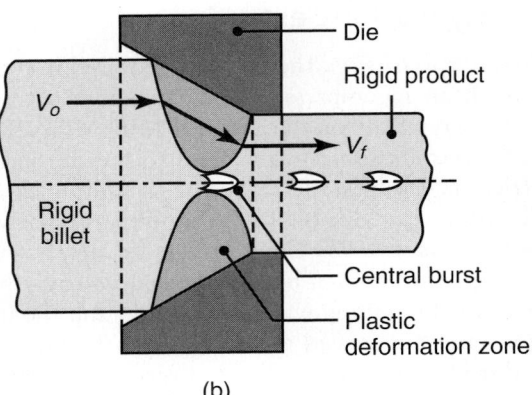

(a)   (b)

FIGURE 15.16   (a) Chevron cracking (central burst) in extruded round steel bars. Unless the products are inspected, such internal defects may remain undetected and later cause failure of the part in service. This defect can also develop in the drawing of rod, of wire, and of tubes. (b) Schematic illustration of rigid and plastic zones in extrusion. The tendency toward chevron cracking increases if the two plastic zones do not meet. Note that the plastic zone can be made larger either by decreasing the die angle or by increasing the reduction in cross-section (or both). *Source:* After B. Avitzur.

(Fig. 15.16a). These cracks are attributed to a state of hydrostatic tensile stress at the centerline in the deformation zone in the die (Fig. 15.16b)—a situation similar to the necked region in a tensile-test specimen (see Fig. 2.22). These cracks also have been observed in tube extrusion and in tube spinning (see Fig. 16.43b and c); they appear on the inside surfaces of tubes. The tendency for center cracking (a) increases with increasing die angle, (b) increases with increasing amount of impurities, and (c) decreases with increasing extrusion ratio and friction.

15.6 | Extrusion Equipment

The basic equipment for extrusion is a *horizontal hydraulic press* (Fig. 15.17). These presses (see also Fig. 14.17) are suitable for extrusion because the stroke and speed of the operation can be controlled, depending on the particular application. They are capable of applying a constant force over a long stroke. Consequently, long billets can be used, and the production rate thus increased. Hydraulic presses with a

FIGURE 15.17   General view of a 9-MN (1000-ton) hydraulic-extrusion press. *Source:* Courtesy of Jones & Laughlin Steel Corporation.

ram-force capacity as high as 120 MN (14,000 tons) have been built, particularly for hot extrusion of large-diameter billets.

Vertical hydraulic presses typically are used for cold extrusion. They generally have less capacity than those used for hot extrusion, but they take up less floor space. In addition to such presses, *crankjoint* and *knucklejoint* mechanical presses also are used for cold extrusion and for impact extrusion to mass produce small components. Multistage operations, where the cross-sectional area is reduced in a number of individual operations, are carried out on specially designed presses.

15.7 | The Drawing Process

In *drawing*, the cross-section of a long rod or wire typically is reduced or changed by pulling (hence the term drawing) it through a die called a *draw die* (Fig. 15.18). Thus, the difference between drawing and extrusion is that in extrusion the material is pushed through a die, whereas in drawing it is pulled through it. Rod and wire products cover a very wide range of applications, including rods for shafts for power and motion transmissions, machine and structural components, and as blanks for bolts and rivets, electrical wiring, cables, tension-loaded structural members, welding electrodes, springs, paper clips, spokes for bicycle wheels, and stringed musical instruments.

The major processing variables in drawing are similar to those in extrusion—that is, reduction in cross-sectional area, die angle, friction along the die-workpiece interfaces, and drawing speed. The die angle influences the drawing force and the quality of the drawn product.

Drawing force.   The expression for the *drawing force, F*, under *ideal and frictionless* conditions is similar to that for extrusion. It is given by the expression

$$F = Y_{\text{avg}} A_f \ln\left(\frac{A_o}{A_f}\right) \tag{15.3}$$

where Y_{avg} is the average true stress of the material in the die gap. Because more work has to be done to overcome friction, the force increases with increasing friction. Furthermore, because of the nonuniform deformation that occurs within the die zone, additional energy also is required (known as the *redundant work of deformation*). Although various equations have been developed to estimate the force (as described in greater detail in advanced texts), a useful formula that includes friction and the redundant work is

$$F = Y_{\text{avg}} A_f \left[\left(1 + \frac{\mu}{\alpha}\right) \ln\left(\frac{A_o}{A_f}\right) + \frac{2}{3}\alpha\right] \tag{15.4}$$

where α is the die angle in radians.

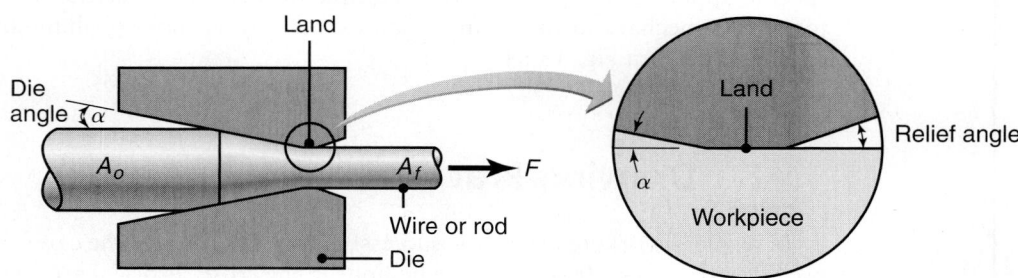

FIGURE 15.18   Process variables in wire drawing. The die angle, the reduction in cross-sectional area per pass, the speed of drawing, the temperature, and the lubrication all affect the drawing force, *F*.

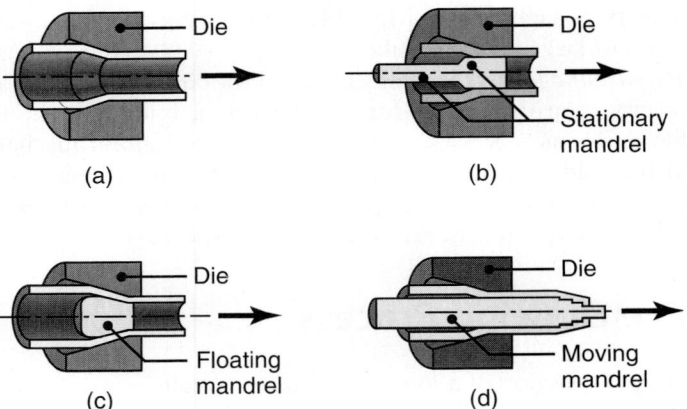

FIGURE 15.19   Examples of tube-drawing operations, with and without an internal mandrel. Note that a variety of diameters and wall thicknesses can be produced from the same initial tube stock (which has been made by other processes).

As can be seen from these equations, the drawing force increases as reduction increases. However, there has to be a limit to the magnitude of the force, because when the tensile stress reaches the yield stress of the metal being drawn, the workpiece will simply yield and, eventually, break. It can be shown that, ideally and without friction, the maximum reduction in cross-sectional area per pass is 63%. Thus, for example, a 10-mm diameter rod can be reduced (at most) to a diameter of 6.1 mm in one pass without failure.

It can be shown that, for a certain reduction in diameter and a certain frictional condition, there is an *optimum die angle* at which the drawing force is a minimum. This does not mean that the process should be carried out at this optimum angle, because there are other product quality considerations.

Drawing of other shapes.   Various solid cross-sections can be produced by drawing through dies with different profiles. Proper die design and the proper selection of reduction sequence per pass require considerable experience to ensure proper material flow in the die, reduce internal or external defects, and improve surface quality.

The wall thickness, diameter, or shape of tubes that have been produced by extrusion or by other processes described in this book can be reduced further by *tube-drawing* processes (Fig. 15.19). Tubes as large as 0.3 m (12 in.) in diameter can be drawn by these techniques. Mandrels of various profiles are available for these operations.

Wedge-shaped dies are used for the drawing of flat strips and are used only in specific applications. However, the principle of this process is the fundamental deformation mechanism in **ironing**, used extensively in making aluminum beverage cans, as shown in Fig. 16.31.

15.8 | Drawing Practice

As in all metalworking processes, successful drawing requires the careful selection of process parameters. In drawing, reductions in the cross-sectional area per pass range up to about 45%. Usually, the smaller the initial cross-section, the smaller the reduction per pass. Fine wires usually are drawn at 15 to 25% reduction per pass and larger sizes at 20 to 45%. Reductions of higher than 45% may result in lubricant breakdown, leading to surface-finish deterioration. Although most drawing is done at room temperature, drawing large solid or hollow sections can be done at elevated temperatures in order to reduce forces.

A light reduction (**sizing pass**) also may be taken on rods to improve surface finish and dimensional accuracy. However, because they basically deform only the surface layers, light reductions usually produce highly nonuniform deformation of the material and its microstructure. Consequently, the properties of the material will vary with location within the cross-section.

Note in Fig. 15.18 that a rod or wire has to have its tip reduced in cross-section in order to be fed through the die opening and be pulled. This typically is done by *swaging* the tip of the rod or wire in a manner similar to that shown in Fig. 14.15a and b; this operation is called *pointing*. Drawing speeds depend on the material and on the reduction in cross-sectional area. They may range from 1 to 2.5 m/s (200 to 500 ft/min) for heavy sections to as much as 50 m/s (165 ft/s) for very fine wire, such as that used for electromagnets. Because the product does not have sufficient time to dissipate the heat generated, temperatures can rise substantially at high drawing speeds and can have detrimental effects on product quality.

Drawn copper and brass wires are designated by their *temper* (such as 1/4 hard, 1/2 hard, etc.) because of work hardening. Intermediate annealing between passes may be necessary to maintain sufficient ductility of the material during cold drawing. High-carbon steel wires for springs and for musical instruments are made by heat treating (**patenting**) the drawn wire; the microstructure obtained is fine pearlite (see Fig. 4.11). These wires have ultimate tensile strengths as high as 5 GPa (700 ksi) and a tensile reduction of area of about 20%.

Bundle drawing.   Although very fine wire can be produced by drawing, the cost can be high. One method employed to increase productivity is to draw many wires (a hundred or more) simultaneously as a *bundle*. The wires are separated from one another by a suitable metallic material with similar properties but lower chemical resistance (so that it subsequently can be leached out from the drawn-wire surfaces).

Bundle drawing produces wires that are somewhat polygonal in cross-section rather than round. In addition to continuous lengths, techniques have been developed to produce fine wire that is broken or chopped into various sizes and shapes. These wires then are used in applications such as electrically conductive plastics, heat-resistant and electrically conductive textiles, filter media, radar camouflage, and medical implants. The wires produced can be as small as 4 μm (0.00016 in.) in diameter and can be made from such materials as stainless steels, titanium, and high-temperature alloys.

Die design.   The characteristic features of a typical die for drawing are shown in Fig. 15.20. Die angles usually range from 6° to 15°. Note, however, that there are two angles (entering and approach) in a typical die. The basic design for this type of die was developed through many years of trial and error. The purpose of the bearing surface (*land*) is to set the final diameter of the product (*sizing*). Furthermore, when a worn die is reground, the land maintains the exit dimension of the die opening.

A set of dies is required for **profile drawing**, which involves various stages of deformation to produce the final profile. The dies may be made in one piece or (depending on the complexity of the cross-sectional profile) with several segments held together in a retaining ring. Computer-aided design techniques are being implemented to design dies for smooth material flow through the dies as well as to minimize defects. A set of idling cylindrical or shaped rolls also may be used in drawing rods or bars of various shapes. Such an arrangement (*Turk's head*) is more versatile than

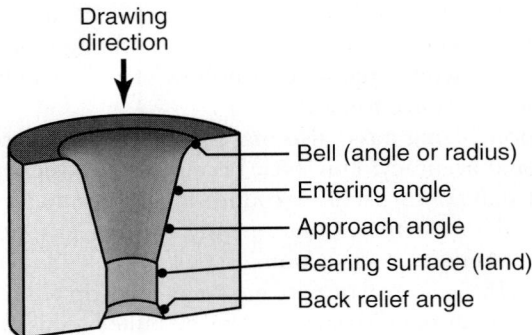

FIGURE 15.20   Terminology of a typical die used for drawing a round rod or wire.

that in ordinary draw dies, because the rolls can be adjusted to different positions and angles for specific profiles.

Die materials.   Die materials for drawing (Table 5.7) typically are tool steels and carbides. For hot drawing, cast-steel dies are used because of their high resistance to wear at elevated temperatures. Diamond dies are used for drawing fine wire with diameters ranging from 2 μm to 1.5 mm (0.0001 to 0.06 in.). They may be made from a single-crystal diamond or in polycrystalline form with diamond particles in a metal matrix (*compacts*). Because of their lack of tensile strength and toughness, carbide and diamond dies typically are used as **inserts** or **nibs**, which are supported in a steel casing (Fig. 15.21).

Lubrication.   Proper lubrication is essential in drawing in order to improve die life and product surface finish and to reduce drawing forces and temperature. Lubrication is critical, particularly in tube drawing, because of the difficulty of maintaining a sufficiently thick lubricant film at the mandrel–tube interface. In the drawing of rods, a common method of lubrication uses phosphate coatings (see Section 15.4).

The following are the basic methods of lubrication used in wire drawing (see also Chapter 33):

- **Wet drawing** in which the dies and the rod are immersed completely in the lubricant
- **Dry drawing** in which the surface of the rod to be drawn is coated with a lubricant by passing it through a box filled with the lubricant (*stuffing box*)

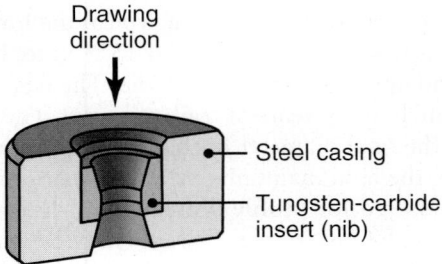

FIGURE 15.21   Tungsten-carbide die insert in a steel casing. Diamond dies used in drawing thin wire are encased in a similar manner.

- **Metal coating** in which the rod or wire is coated with a soft metal, such as copper or tin, that acts as a solid lubricant
- **Ultrasonic vibration** of the dies and mandrels, where vibrations reduce forces, improve surface finish and die life, and allow larger reductions per pass without failure.

15.9 | Drawing Defects and Residual Stresses

Typical defects in a drawn rod or wire are similar to those observed in extrusion, especially *center cracking* (see Fig. 15.15). Another major type of defect in drawing is **seams**, which are longitudinal scratches or folds in the material. Seams may open up during subsequent forming operations (such as upsetting, heading, thread rolling, or bending of the rod or wire), and they can cause serious quality-control problems. Various other surface defects (such as scratches and die marks) also can result from improper selection of the process parameters, poor lubrication, or poor die condition.

Because they undergo nonuniform deformation during drawing, cold-drawn products usually have *residual stresses*. For light reductions, such as only a few percent, the longitudinal-surface residual stresses are compressive (while the bulk is in tension), and thus fatigue life is improved. Conversely, heavier reductions induce tensile surface stresses (while the bulk is in compression). Residual stresses can be significant in causing stress-corrosion cracking of the part over a period of time. Moreover, they cause the component to *warp* if a layer of material subsequently is removed (see Fig. 2.30), such as by slitting, machining, or grinding.

Rods and tubes that are not sufficiently straight (or are supplied as coil) can be straightened by passing them through an arrangement of rolls placed at different axes—a process similar to roller leveling (see Fig. 13.7).

15.10 | Drawing Equipment

Although they are available in several designs, the equipment for drawing is basically of two types: draw bench and bull block.

A **draw bench** contains a single die, and its design is similar to that of a long, horizontal tension-testing machine (Fig. 15.22). The pulling force is supplied by a

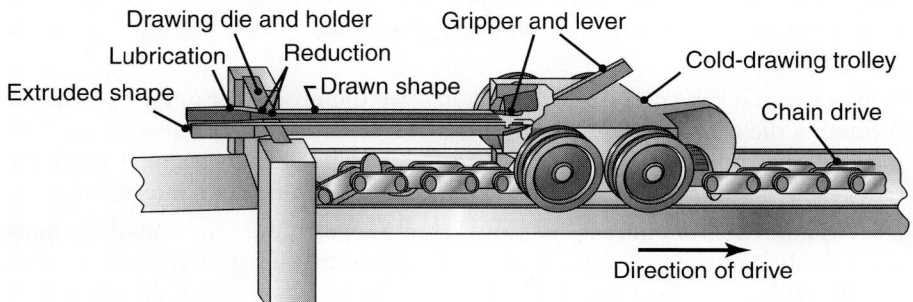

FIGURE 15.22   Cold drawing of an extruded channel on a draw bench to reduce its cross-section. Individual lengths of straight rods or of cross-sections are drawn by this method.

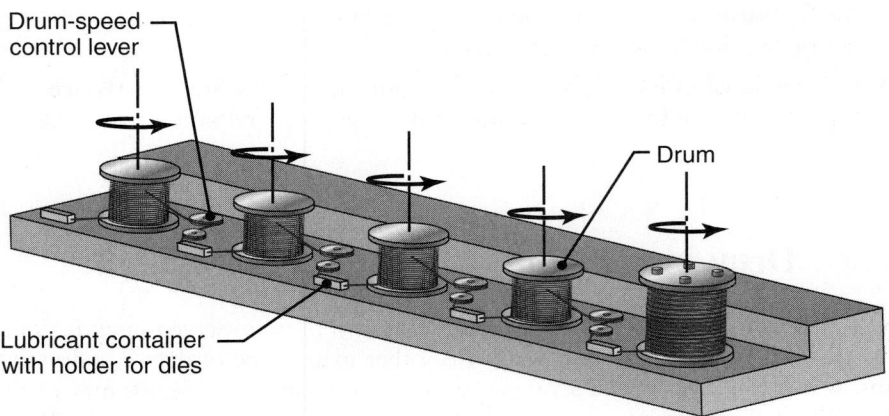

FIGURE 15.23 An illustration of multistage wire drawing typically used to produce copper wire for electrical wiring. *Source*: After H. Auerswald.

chain drive or is activated hydraulically. Draw benches are used for a single-length drawing of straight rods and tubes with diameters larger than 20 mm (0.75 in.) and lengths up to 30 m (100 ft.). Machine capacities reach 1.3 MN (300 klb) of pulling force with a speed range of 6 to 60 m/min (20 to 200 ft/min).

Very long rods and wire (many kilometers) and wire of smaller cross-sections, usually less than 13 mm (0.5 in.), are drawn by a rotating **drum** (**bull block** or **capstan**, Fig. 15.23). The tension in this setup provides the force required for drawing the wire, usually through multiple dies (*tandem drawing*).

SUMMARY

- Extrusion is the process of forcing a billet through a die to reduce its cross-section or to produce various solid or hollow cross-sections. This process generally is carried out at elevated temperatures in order to reduce forces and improve the ductility of the material.

- Important factors in extrusion are die design, extrusion ratio, billet temperature, lubrication, and extrusion speed. Although the term "cold extrusion" applies to extrusion at room temperature, it is also the name for a combination of extrusion and forging operations. Cold extrusion is capable of economically producing discrete parts in various shapes with good mechanical properties and dimensional tolerances.

- Rod, wire, and tube drawing basically involve the process of pulling the material through a die or a set of dies in tandem. The cross-sections of most drawn products are round, but other shapes also can be drawn. Drawing tubular products to reduce either their diameter or their thickness usually requires internal mandrels.

- Die design, reduction in cross-sectional area per pass, and selection of die materials and lubricants are all important parameters in making drawn products of high quality with a good surface finish. External as well as internal defects (chevron cracking) can develop both in extrusion and in drawing. Their minimization or avoidance depends principally on the die angle, the reduction per pass, and the quality of the workpiece material.

KEY TERMS

Bamboo defect
Bridge die
Bull block
Bundle drawing
Canning
Capstan
Center cracking
Chevron cracking
Cold extrusion
Conversion coating
Dead-metal zone
Draw bench

Drawing
Extrusion
Extrusion constant
Extrusion defects
Extrusion ratio
Fir-tree cracking
Hydrostatic extrusion
Impact extrusion
Ironing
Jacketing
Patenting
Pipe defect

Porthole die
Rod
Seam
Séjournet process
Shear die
Sizing pass
Speed cracking
Spider die
Turk's head
Wire

BIBLIOGRAPHY

Altan, T., Ngaile, G., and Shen, G. (eds.), *Cold and Hot Forging: Fundamentals and Applications*, ASM International, 2004
ASM Handbook, Vol. 14: *Forming and Forging*, ASM International, 1988.
Blazynski, T.Z., *Plasticity and Modern Metal-forming Technology*, Elsevier, 1989.
Hosford, W.F., and Caddell, R.M., *Metal Forming: Mechanics and Metallurgy*, 2nd ed., Prentice Hall, 1993.
Inoue, N., and Nishihara, M. (eds.), *Hydrostatic Extrusion: Theory and Applications*, Elsevier, 1985.

Lange, K. (ed.), *Handbook of Metal Forming*, McGraw-Hill, 1985.
Laue, K., and Stenger, H., *Extrusion—Processes, Machinery, Tooling*, ASM International, 1981.
Michaeli, W., *Extrusion Dies*, 2nd ed., Hanser, 1992.
Nonferrous Wire Handbook, 2 vols. The Wire Association International, Inc., 1977 and 1981.
Sheppard, T., *Extrusion of Aluminum Alloys*, Chapman & Hall, 1997.

REVIEW QUESTIONS

15.1 How does extrusion differ from rolling and forging processes?

15.2 What is the difference between extrusion and drawing?

15.3 What is a spider die? What is it used for?

15.4 Why are wires sometimes drawn in bundles?

15.5 What is a dead-metal zone?

15.6 Define (a) cladding, (b) dummy block, (c) shear dies, (d) skull, and (e) canning.

15.7 Why is glass a good lubricant in hot extrusion?

15.8 Explain why cold extrusion is an important manufacturing process.

15.9 What types of defects may occur in (a) extrusion and (b) drawing?

15.10 What is the difference between direct and reverse extrusion?

15.11 What is a land? What is its function in a die?

15.12 How are tubes extruded? Can they also be drawn? Explain.

15.13 Is it possible to extrude straight gears as well as helical gears? What further processing may be necessary?

15.14 What materials are used in making dies for extrusion?

15.15 What is the difference between piping and bambooing?

QUALITATIVE PROBLEMS

15.16 Explain why extrusion is a batch or a semicontinuous process. Do you think it can be made into a continuous process? Explain.

15.17 Explain the different ways by which changing the die angle affects the extrusion process.

15.18 Glass is a good lubricant in hot extrusion. Would you use glass for impression-die forging also? Explain.

15.19 How would you go about avoiding center-cracking defects in extrusion? Explain why your methods would be effective.

15.20 What is the purpose of a stripper plate in impact extrusion?

15.21 Table 15.1 gives temperature ranges for extruding various metals. Describe the possible consequences of extruding at a temperature (a) below and (b) above these ranges.

15.22 Will the force in direct extrusion vary as the billet becomes shorter? If so, why?

15.23 Comment on the significance of grain-flow patterns, such as those shown in Fig. 15.6.

15.24 In which applications could you use the type of impact-extruded parts shown in Fig. 15.15?

15.25 Can spur gears be made by (a) drawing and (b) extrusion? Can helical gears? Explain.

15.26 We have seen in Chapter 13 that applying back tension in rolling reduces the roll force. What advantages, if any, would applying back tension in rod or wire drawing have? Explain.

15.27 How would you prepare the end of a wire in order to be able to feed it through a die so that a drawing operation can commence?

15.28 What is the purpose of a dummy block in extrusion?

15.29 Describe your observations concerning Fig. 15.8.

15.30 Occasionally, steel wire drawing will take place within a sheath of a soft metal, such as copper or lead. Why would this sheath be useful?

15.31 What are the advantages to bundle drawing?

15.32 Under what circumstances is backwards extrusion preferable to direct extrusion?

15.33 Why is lubrication detrimental when extruding with a porthole die?

15.34 What is the purpose of a container liner in direct extrusion? (See Fig. 15.1.)

QUANTITATIVE PROBLEMS

15.35 Estimate the force required in extruding 70-30 brass at 700°C, if the billet diameter is 125 mm and the extrusion ratio is 20.

15.36 Assuming an ideal drawing process, what is the smallest final diameter to which a 100-mm diameter rod can be drawn?

15.37 If you include friction in Problem 15.36, would the final diameter be different? Explain.

15.38 Calculate the extrusion force for a round billet 200 mm in diameter, made of stainless steel, and extruded at 1000°C to a diameter of 50 mm.

15.39 Show that, for a perfectly plastic material with a yield stress, Y, and under frictionless conditions, the pressure, p, in direct extrusion is

$$p = Y \ln\left(\frac{A_o}{A_f}\right)$$

15.40 Show that, for the same conditions stated in Problem 15.39, the drawing stress, σ_d, in wire drawing is

$$\sigma_d = Y \ln\left(\frac{A_o}{A_f}\right)$$

15.41 Plot the equations given in Problems 15.39 and 15.40 as a function of the percent reduction in area of the workpiece. Describe your observations.

15.42 A planned extrusion operation involves steel at 1000°C with an initial diameter of 120 mm and a final diameter of 20 mm. Two presses, one with capacity of 20 MN and the other with a capacity of 10 MN, are available for the operation. Is the smaller press sufficient for this operation? If not, what recommendations would you make to allow the use of the smaller press?

15.43 A round wire made of a perfectly plastic material with a yield stress of 30,000 psi is being drawn from a diameter of 0.1 to 0.07 in. in a draw die of 15°. Let the coefficient of friction be 0.1. Using both Eqs. (15.3) and (15.4), estimate the drawing force required. Comment on the differences in your answer.

15.44 Use Problem 15.43, but for 304 stainless steel (see Table 2.3).

15.45 Assume that you are an instructor covering the topics described in this chapter, and you are giving a quiz on the numerical aspects to test the understanding of the students. Prepare two quantitative problems and supply the answers.

SYNTHESIS, DESIGN, AND PROJECTS

15.46 Assume that you are the technical director of trade associations of (a) extruders and (b) rod-and wire-drawing operations. Prepare a technical leaflet for potential customers stating all of the advantages of these processes, respectively.

15.47 Assume that the summary to this chapter is missing. Write a one-page summary of the highlights to the wire-drawing process.

15.48 Review the technical literature and make a detailed list of the manufacturing steps involved in the manufacture of common metallic hypodermic needles.

15.49 Figure 15.2 shows examples of discrete parts that can be made by cutting extrusions into individual pieces. Name several other products that can be made in a similar fashion.

15.50 Survey the technical literature and explain how external vibrations can be applied to wire drawing to reduce friction. Comment also on the possible directions of vibration, such as longitudinal or torsional.

15.51 A popular child's toy is a miniature extrusion press using a soft dough or a putty to make various shapes. Obtain such a toy and demonstrate the surface defects that may develop.

15.52 In Case Study 14.2 on suspension uprights in Chapter 14, it was stated that there was a significant cost improvement using forgings when compared to extrusions. List and explain the reasons why you think these cost savings were possible.

15.53 It was stated that in bundle drawing the wires produced were more polygonal in cross-section than round. Produce some thin pieces of round clay or obtain cooked spaghetti, bundle them together, and compress them radially in a suitable fixture—a manner similar to tightening one's belt. Comment on the shapes produced.

16 Sheet-Metal Forming Processes

This chapter describes the important characteristics of sheet metals and the wide variety of manufacturing processes employed to produce numerous sheet-metal products. Topics covered include:

- Specific material properties of sheet metals that affect their formability.
- Shearing operations where blanks are cut from large sheets and strips.
- Various bending operations for sheets, plates, and tubes.
- Fundamentals of forming operations, including stretch forming, rubber forming, bulging, spinning, peen forming, and superplastic forming.
- Deep drawing and ironing for the production of containers with thin walls.
- Design and economic considerations for sheet-metal forming.

Typical parts made by sheet-metal forming: Car bodies, office furniture, appliances, aircraft fuselages, trailers, fuel tanks, and cookware.

Alternative process: Die casting (relatively small, thin-walled parts), thermoforming.

16.1 | Introduction

Products made of **sheet metals** are all around us. They include a very wide range of consumer and industrial products, such as beverage cans, cookware, file cabinets, metal desks, appliances, car bodies, trailers, and aircraft fuselages (Fig. 16.1). Sheet forming dates back to about 5000 B.C., when household utensils and jewelry were made by hammering and stamping gold, silver, and copper. Compared to those made by casting and by forging, sheet-metal parts offer the advantages of light weight and versatile shape.

As described throughout this chapter, there are numerous processes employed for making sheet-metal parts. However, the term **pressworking** or **press forming** is used commonly in industry to describe general sheet-forming operations, because they typically are performed on *presses* (described in Sections 14.8 and 16.14) using a set of dies. A sheet-metal part produced in presses is called a

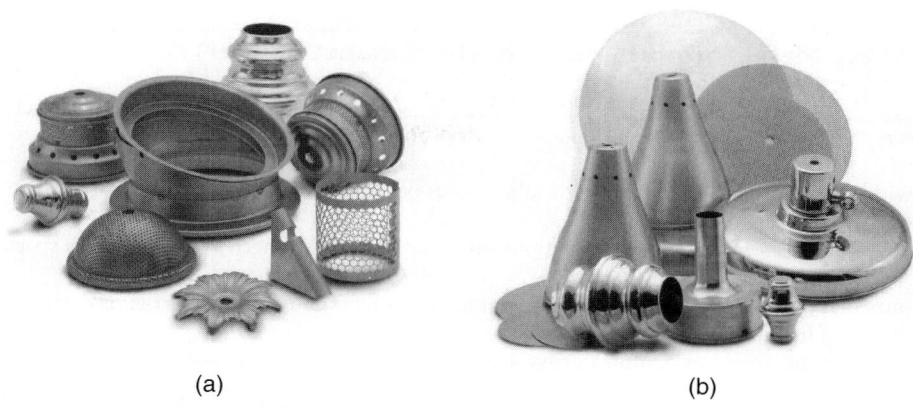

(a)   (b)

FIGURE 16.1   Examples of sheet-metal parts. (a) Stamped parts. (b) Parts produced by spinning. *Source:* (a) Courtesy of Williamsburg Metal Spinning & Stamping Corp.

stamping (after the word *stamp*, first used around 1200 A.D. and meaning "to force downward" or "to pound"). Note that this is a term similar to a *forging* or a *casting* commonly used to describe parts made by those individual processes using dies or molds, respectively.

Low-carbon steel is the most commonly used sheet metal because of its low cost and generally good strength and formability characteristics. Aluminum is the most common material for such sheet-metal applications as beverage cans, packaging, kitchen utensils, and applications for corrosion resistance. The common metallic materials for aircraft and aerospace applications are aluminum and titanium, although they are being replaced increasingly with composite materials, as described in Chapters 9 and 19.

This chapter first describes the methods by which blanks are cut from large rolled sheets then processed further into desired shapes by a wide variety of methods. The chapter also includes discussions on the characteristic features of sheet metals, the techniques employed to determine their formability, and the construction of forming-limit diagrams. All of the major processes of sheet forming and the equipment used to make sheet-metal products (as outlined in Table 16.1) are also described.

16.2 | Shearing

Before a sheet-metal part is made, a *blank* of suitable dimensions first is removed from a large sheet (usually from a coil) by *shearing*. This sheet is cut by subjecting it to shear stresses, generally using a punch and a die (Fig. 16.2a). The typical features of the sheared edges of the sheet and of the slug are shown in Fig. 16.2b and c, respectively. Note that the edges are not smooth nor are they perpendicular to the plane of the sheet.

Shearing generally starts with the formation of cracks on both the top and bottom edges of the workpiece (at points *A* and *B*, and *C* and *D* in Fig. 16.2a). These cracks eventually meet each other and complete separation occurs. The rough *fracture surfaces* are due to these cracks; the smooth and shiny *burnished surfaces*

TABLE 16.1

General Characteristics of Sheet-Metal Forming Processes (In alphabetic order)	
Forming process	Characteristics
Drawing	Shallow or deep parts with relatively simple shapes, high production rates, high tooling and equipment costs
Explosive	Large sheets with relatively simple shapes, low tooling costs but high labor cost, low-quantity production, long cycle times
Magnetic-pulse	Shallow forming, bulging, and embossing operations on relatively low-strength sheets, requires special tooling
Peen	Shallow contours on large sheets, flexibility of operation, generally high equipment costs, process also used for straightening formed parts
Roll	Long parts with constant simple or complex cross-sections, good surface finish, high production rates, high tooling costs
Rubber	Drawing and embossing of simple or relatively complex shapes, sheet surface protected by rubber membranes, flexibility of operation, low tooling costs
Spinning	Small or large axisymmetric parts, good surface finish, low tooling costs but labor costs can be high unless operations are automated
Stamping	Includes a wide variety of operations, such as punching, blanking, embossing, bending, flanging, and coining; simple or complex shapes formed at high production rates; tooling and equipment costs can be high, but labor cost is low
Stretch	Large parts with shallow contours, low-quantity production, high labor costs, tooling and equipment costs increase with part size
Superplastic	Complex shapes, fine detail and close dimensional tolerances, long forming times (hence production rates are low), parts not suitable for high-temperature use

on the hole and the slug are from the contact and rubbing of the sheared edge against the walls of the punch and die, respectively.

The major processing parameters in shearing are

- The shape of the punch and die
- The speed of punching
- Lubrication
- The **clearance**, c, between the punch and the die

The clearance is a major factor in determining the shape and the quality of the sheared edge. As the clearance increases, the zone of deformation (Fig. 16.3a) becomes larger, and the sheared edge becomes rougher. The sheet tends to be pulled into the clearance region, and the perimeter or edges of the sheared zone become rougher. Unless such edges are acceptable as produced, secondary operations may be required to make them smoother (which will increase the production cost).

Edge quality can be improved with increasing punch speed; speeds may be as high as 10 to 12 m/s (30 to 40 ft/s). As shown in Fig. 16.3b, sheared edges can undergo severe cold working due to the high shear strains involved. Work hardening of the edges then will reduce the ductility of the edges and thus adversely affect the formability of the sheet during subsequent operations, such as bending and stretching.

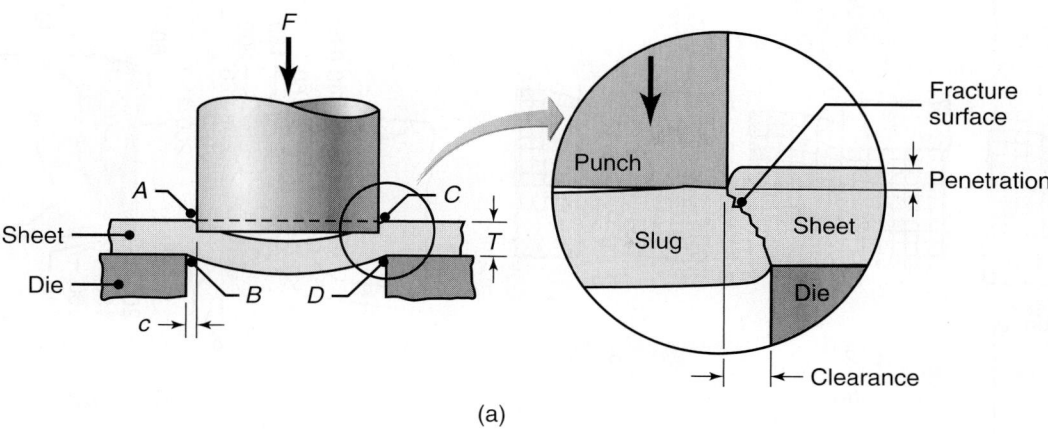

(a)

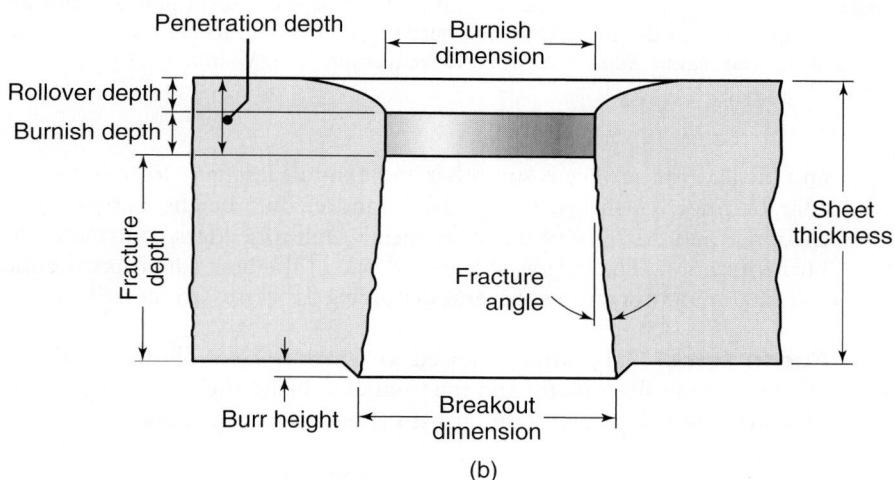

(b)

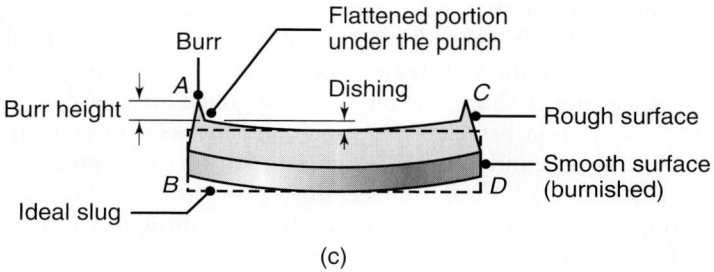

(c)

FIGURE 16.2   (a) Schematic illustration of shearing with a punch and die, indicating some of the process variables. Characteristic features of (b) a punched hole; (c) the slug. (*Note:* The scales of (b) and (c) are different.)

The ratio of the burnished area to the rough areas along the sheared edge (a) increases with increasing ductility of the sheet metal and (b) decreases with increasing sheet thickness and clearance. The extent of the deformation zone in Fig. 16.3 depends on the punch speed. With increasing speed, the heat generated by plastic deformation is confined to a smaller and smaller zone. Consequently, the sheared zone is narrower,

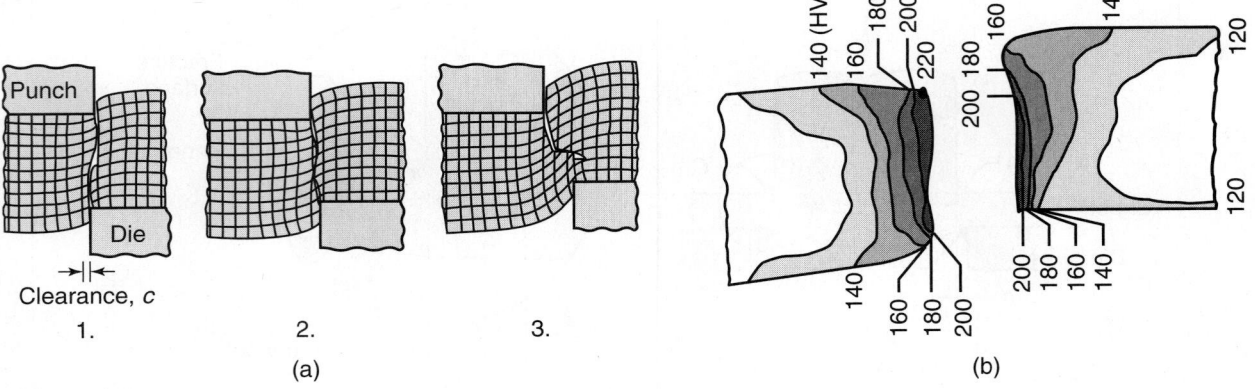

FIGURE 16.3   (a) Effect of the clearance, *c*, between punch and die on the deformation zone in shearing. As the clearance increases, the material tends to be pulled into the die rather than be sheared. In practice, clearances usually range between 2 and 10% of the thickness of the sheet. (b) Microhardness (HV) contours for a 6.4-mm (0.25-in.) thick AISI 1020 hot-rolled steel in the sheared region. *Source:* After H. P. Weaver and K. J. Weinmann.

and the sheared surface is smoother and exhibits less burr formation. A **burr** is a thin edge or ridge, as shown in Fig. 16.2b and c. Burr height increases with increasing clearance and ductility of the sheet metal. Dull tool edges contribute greatly to large burr formation. The height, shape, and size of the burr can affect significantly subsequent forming operations. Several **deburring** processes are described in Section 26.8.

Punch force.   The force required to punch is basically the product of the shear strength of the sheet metal and the total area being sheared along the periphery. The *maximum punch force, F*, can be estimated from the equation

$$F = 0.7TL(\text{UTS}) \qquad (16.1)$$

where *T* is the sheet thickness, *L* is the total length sheared (such as the perimeter of a hole), and UTS is the ultimate tensile strength of the material. As the clearance increases, the punch force decreases, and the wear on dies and punches also is reduced. The effects of punch shape and die shape on punch forces are described in Section 16.2.3.

Friction between the punch and the workpiece can, however, increase punch force significantly. Furthermore, in addition to the punch force, a force is required to strip the punch from the sheet during its return stroke. This second force, which is in opposite direction of the punch force, is difficult to estimate because of the many factors involved in the operation.

EXAMPLE 16.1 Calculation of punch force

Estimate the force required for punching a 1-in. (25-mm) diameter hole through a $\frac{1}{8}$-in. (3.2-mm) thick annealed titanium-alloy Ti-6Al-4V sheet at room temperature.

Solution   The force is estimated from Eq. (16.1) where the UTS for this alloy is found from Table 6.10 to be 1000 MPa or 140,000 psi. Thus,

$$F = 0.7\left(\frac{1}{8}\right)(\pi)(1)(140{,}000) = 38{,}500 \text{ lb} = 19.25 \text{ tons} = 0.17 \text{ MN}$$

16.2.1 Shearing operations

The most common shearing operations are **punching**—where the sheared slug is scrap (Fig. 16.4a) or may be used for some other purpose—and **blanking**—where the slug is the part to be used and the rest is scrap. The operations described next, as well as those described throughout the rest of this chapter, generally are carried out on computer-numerical-controlled machines with quick-change toolholders. Such machines are useful, particularly in making prototypes of sheet-metal parts requiring several operations to produce.

Die cutting.   This is a shearing operation that consists of the following basic processes (Fig. 16.4b):

- **Perforating:** punching a number of holes in a sheet
- **Parting:** shearing the sheet into two or more pieces
- **Notching:** removing pieces (or various shapes) from the edges
- **Lancing:** leaving a tab without removing any material

Parts produced by these processes have various uses, particularly in assembly with other components. Perforated sheet metals with hole diameters ranging from around 1 mm (0.040 in.) to 75 mm (3 in.) have uses as filters, as screens, in ventilation, as guards for machinery, in noise abatement, and in weight reduction of fabricated parts and structures. They are punched in crank presses (see Fig. 14.17a) at rates as high as 300,000 holes per minute, using special dies and equipment.

Fine blanking.   Very smooth and square edges can be produced by *fine blanking* (Fig. 16.5a). One basic die design is shown in Fig. 16.5b. A V-shaped stinger or impingement mechanically locks the sheet tightly in place and prevents the type of distortion of the material shown in Figs. 16.2b and 16.3. The fine-blanking process, which was developed in the 1960s, involves clearances on the order of 1% of the sheet thickness and may range from 0.5 to 13 mm (0.02 to 0.5 in.) in most cases. Dimensional tolerances are on the order of $+/-0.05$ mm (0.002 in.) and less than $+/-0.025$ mm (0.001 in.) in the case of edge perpendicularity.

Slitting.   Shearing operations can be carried out by means of a pair of circular blades similar to those in a can opener (Fig. 16.6). In *slitting*, the blades follow either a straight line, a circular path, or a curved path. A slit edge normally has a burr, which may be folded over the sheet surface by rolling it (flattening) between two rolls. If not performed properly, slitting operations can cause various distortions of the sheared edges.

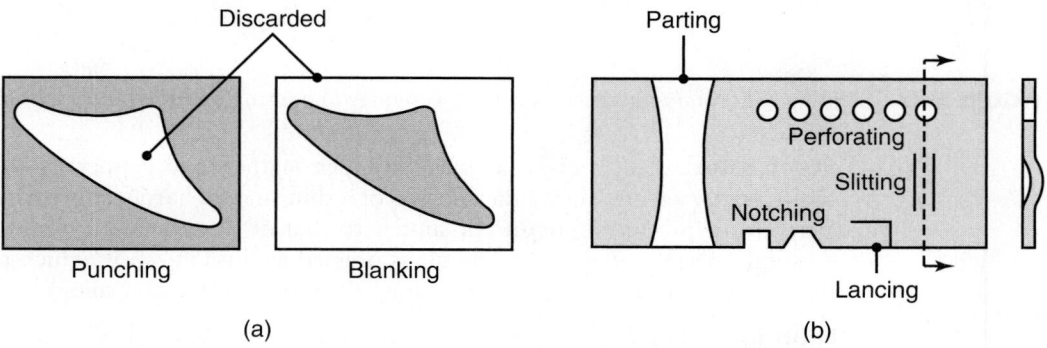

FIGURE 16.4   (a) Punching (piercing) and blanking. (b) Examples of various die-cutting operations on sheet metal.

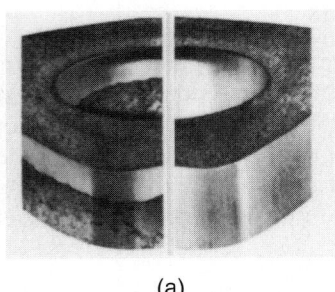

(a)

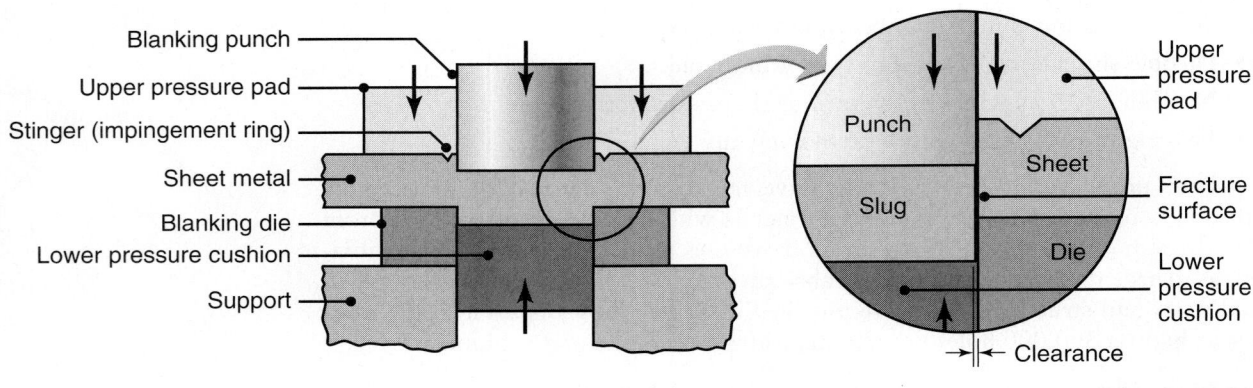

(b)

FIGURE 16.5   (a) Comparison of sheared edges produced by conventional (left) and by fine-blanking (right) techniques. (b) Schematic illustration of one setup for fine blanking. *Source*: Courtesy of Feintool U.S. Operations.

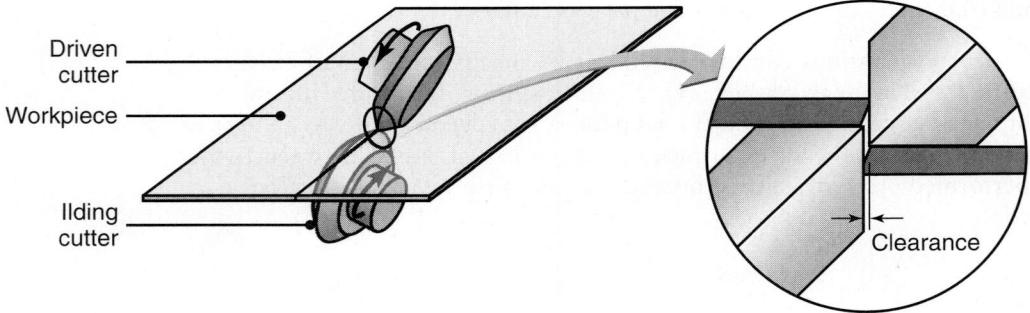

FIGURE 16.6   Slitting with rotary knives. This process is similar to opening cans.

Steel rules.   Soft metals (as well as paper, leather, and rubber) can be blanked with a *steel-rule die*. Such a die consists of a thin strip of hardened steel bent into the shape to be produced (a concept similar to that of a cookie cutter) and held on its edge on a flat wooden base. The die is pressed against the sheet which rests on the flat surface, and it shears the sheet along the shape of the steel rule.

Nibbling.   In *nibbling*, a machine called a *nibbler* moves a small straight punch up and down rapidly into a die. A sheet is fed through the gap and many overlapping holes are made. Using manual or automatic control, sheets can be cut along any desired path. In addition to its flexibility, an advantage of nibbling is that intricate slots and notches, such as those shown in Fig. 16.4b, can be produced using standard punches. The process is economical for small production runs because no special dies are required.

Scrap in shearing. The amount of scrap (*trim loss*) produced in shearing operations can be significant and can be as high as 30% on large stampings (see Table 40.3). Scrap can be a significant factor in manufacturing cost, and it can be reduced substantially by efficient arrangement of the shapes on the sheet to be cut (**nesting**, see Fig. 16.51). Computer-aided design techniques have been developed to minimize the scrap from shearing operations.

16.2.2 Tailor-welded blanks

In the sheet-metal forming processes to be described throughout this chapter, the blank is usually a one-piece sheet of one thickness cut from a large sheet. An important variation from these conditions involves *laser-beam butt welding* (see Section 30.7) of two or more pieces of sheet metal with different shapes and thicknesses. Because of the small thicknesses involved, the proper alignment of the sheets prior to welding is important. The welded assembly subsequently is formed into a final shape (see Example 16.2).

This technique is becoming increasingly important, particularly to the automotive industry. Because each sub-piece now can have a different thickness, grade, coating, or other property, tailor-welded blanks possess the needed properties in the desired locations in the blank. The result is

- Reduction in scrap
- Elimination of the need for subsequent spot welding (i.e., in the making of the car body)
- Better control of dimensions
- Improved productivity

EXAMPLE 16.2 Tailor-welded sheet metal for automotive applications

An example of the use of tailor-welded sheet metals in automobile bodies is shown in Fig. 16.7. Note that five different pieces are blanked first, which includes cutting by laser beams. Four of these pieces are 1 mm thick, and one is 0.8 mm thick. These pieces are laser butt-welded and then stamped into the final shape. In this manner, the blanks can be tailored for a particular application, not only as to shape and thickness but also by using different quality sheets—with or without coatings.

Laser-welding techniques are highly developed; as a consequence, weld joints are very strong and reliable. The growing trend toward welding and forming sheet-metal pieces makes possible significant flexibility in the product design, structural stiffness, formability, and crash behavior of an automobile. It also makes possible the use of different materials in one component, weight savings, and cost reduction in materials, scrap, equipment, assembly, and labor.

There are increasing applications for this type of production in automotive companies. The various components shown in Fig. 16.8 utilize the advantages outlined above. For example, note in Fig. 16.8(b) that the strength and stiffness required for the support of the shock absorber are achieved by welding a round piece onto the surface of the large sheet. The sheet thickness in such components is varied (depending on its location and on its contribution to such characteristics as stiffness and strength) and, thereby, makes possible significant weight savings without loss of structural strength and stiffness.

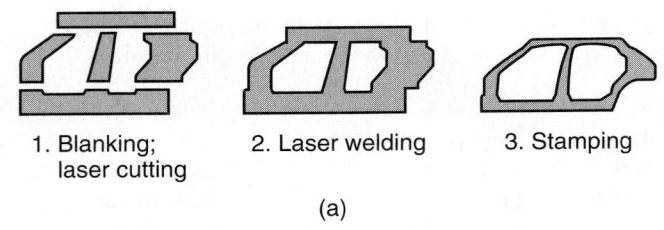

1. Blanking; laser cutting 2. Laser welding 3. Stamping

(a)

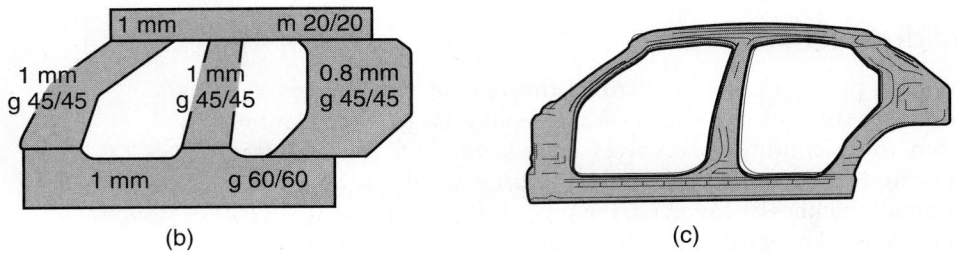

(b)   (c)

Legend

g 60/60 (45/45)   Hot-galvanized alloy steel sheet. Zinc amount: 60/60 (45/45) g/m².
m 20/20   Double-layered iron-zinc alloy electroplated steel sheet. Zinc amount: 20/20 g/m².

FIGURE 16.7 Production of an outer side panel of a car body by laser butt-welding and stamping. *Source:* After M. Geiger and T. Nakagawa.

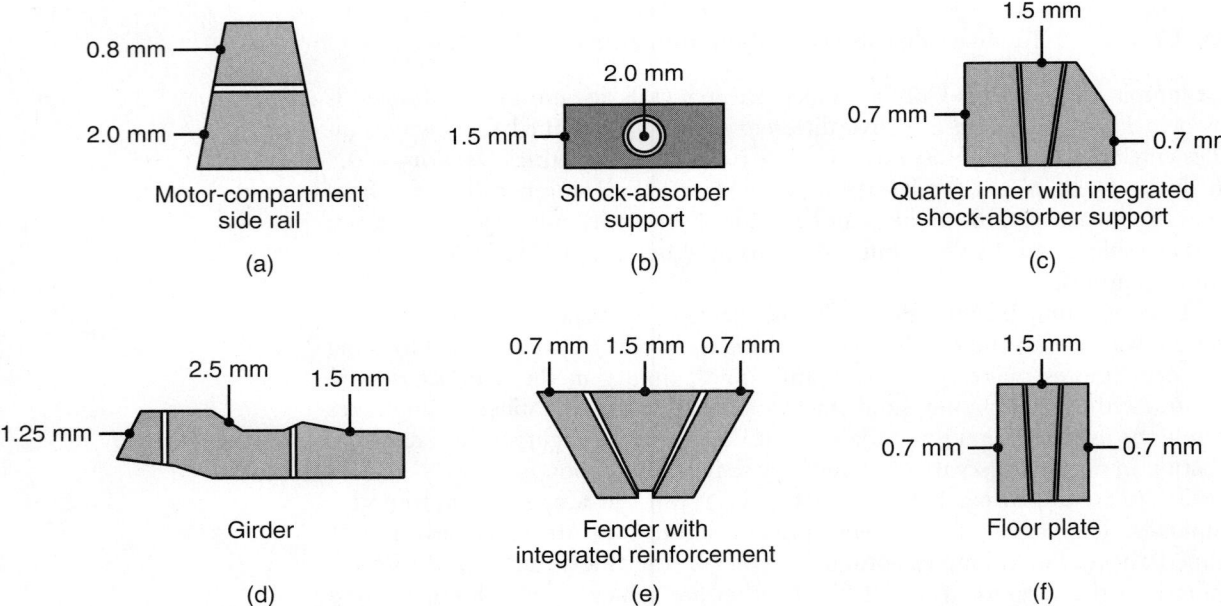

FIGURE 16.8 Examples of laser butt-welded and stamped automotive-body components. *Source:* After M. Geiger and T. Nakagawa.

16.2.3 Characteristics and type of shearing dies

The features and types of various shearing dies are described in this section.

Clearance.   Because the formability of the sheared part can be influenced by the quality of its sheared edges, clearance control is important. The appropriate clearance depends on

- The type of material and its temper
- The thickness and size of the blank
- Its proximity to the edges of other sheared edges or the edges of the original blank

Clearances generally range between 2 and 8% of the sheet thickness, but they may be as small as 1% (as in fine blanking) or as large as 30%. The smaller the clearance, the better is the quality of the edge. If the sheared edge is rough and not acceptable, it can be subjected to a process called **shaving** (Fig. 16.9a), whereby the extra material from the edge is trimmed by cutting, as also depicted in Fig. 21.3.

As a general guideline, (a) clearances for soft materials are less than those for harder grades; (b) the thicker the sheet, the larger the clearance must be; and (c) as the ratio of hole diameter to sheet thickness decreases, clearances should be larger. In using larger clearances, attention must be paid to the rigidity and the alignment of the presses, the dies, and their setups.

Punch and die shape.   Note in Fig. 16.2a that both the surfaces of the punch and of the die are flat. Because the entire thickness is sheared at the same time, the punch force increases rapidly during shearing. The location of the regions being sheared at any particular instant can be controlled by *beveling* the punch and die surfaces (Fig. 16.10). This shape is similar to that of some paper punches, which you can observe by looking closely at the tip of the punch. Beveling is suitable particularly for shearing thick sheets because it reduces the force at the beginning of the stroke. It also reduces the operation's noise level, because the operation is smoother.

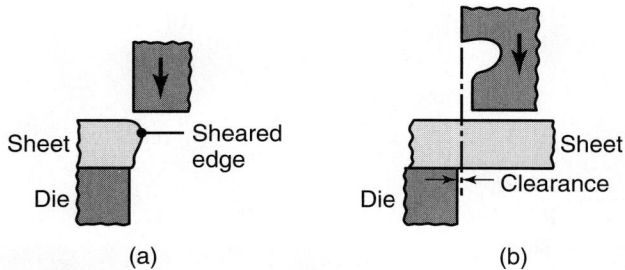

(a)   (b)

FIGURE 16.9   Schematic illustrations of the shaving process. (a) Shaving a sheared edge. (b) Shearing and shaving combined in one stroke.

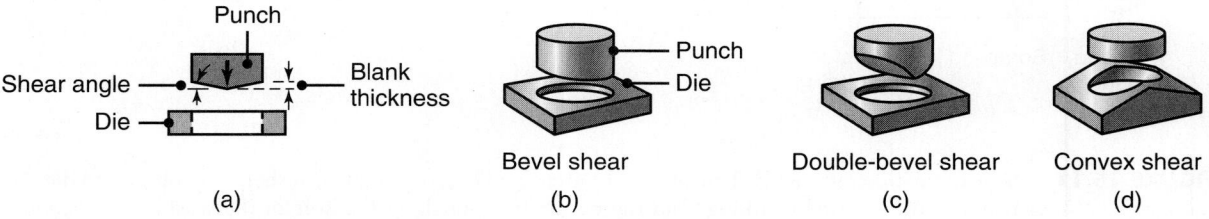

(a)   (b)   (c)   (d)

FIGURE 16.10   Examples of the use of shear angles on punches and dies.

Note in Fig. 16.10c that the punch tip is symmetrical, and in Fig. 16.10d, the die is symmetrical. Hence, there are no lateral forces acting on the punch to cause distortion. By contrast, the punch in Fig. 16.10b has a single taper, and thus, the punch is subjected to a lateral force. Consequently, the punch and press setups in this latter case must both have sufficient lateral stiffness so that they neither produce a hole that is located improperly nor allow the punch to hit the edge of the lower die (as it might at points *B* or *D* in Fig. 16.2a), causing damage.

Compound dies.   Several operations on the same sheet may be performed in one stroke at one station with a *compound die* (Fig. 16.11). Such combined operations usually are limited to relatively simple shapes, because (a) the process is somewhat slow and (b) the dies rapidly become much more expensive to produce than those for individual shearing operations, especially for complex dies.

Progressive dies.   Parts requiring multiple operations to produce can be made at high production rates in *progressive dies*. The sheet metal is fed through as a coil

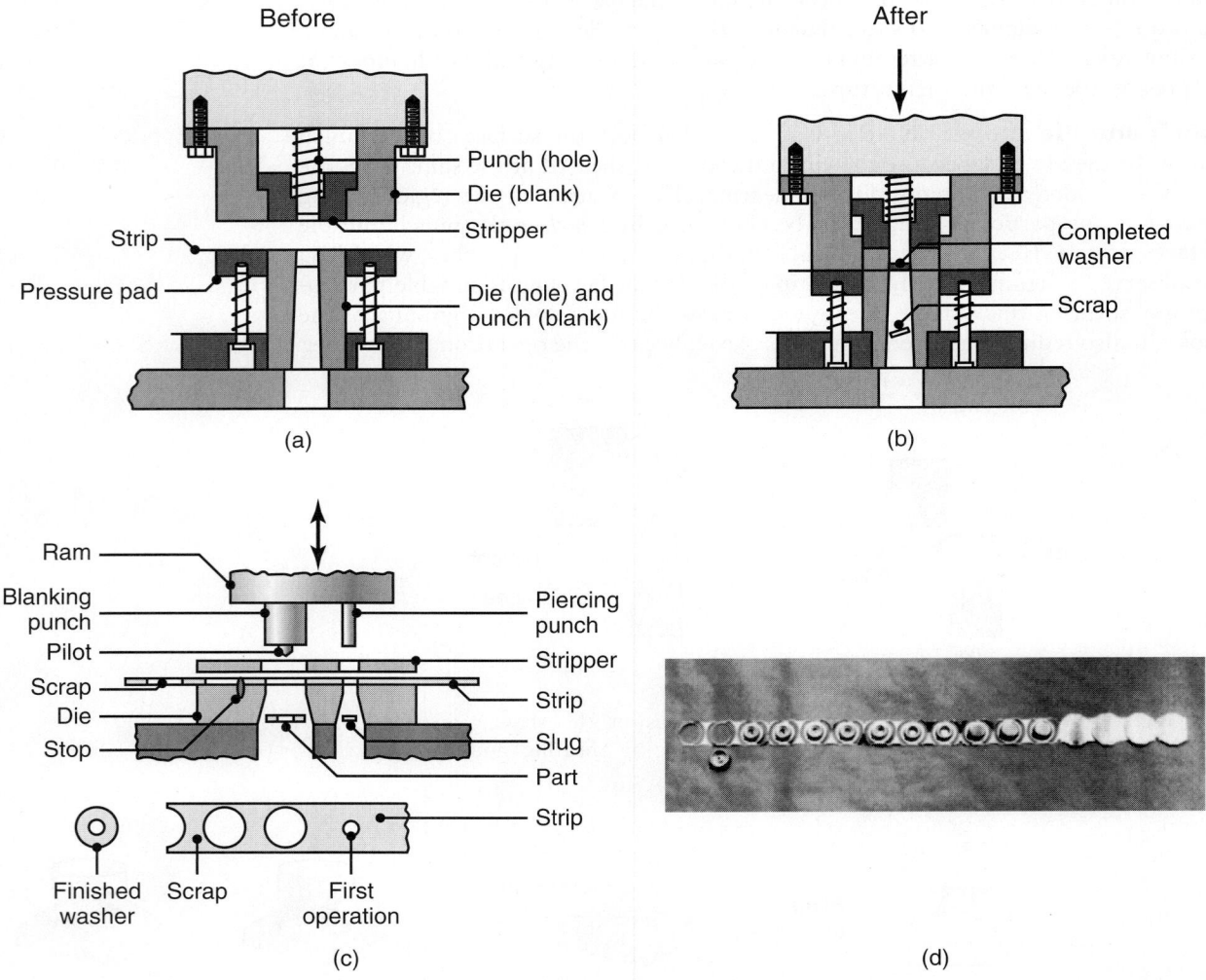

FIGURE 16.11   Schematic illustrations: (a) before and (b) after blanking a common washer in a compound die. Note the separate movements of the die (for blanking) and the punch (for punching the hole in the washer). (c) Schematic illustration of making a washer in a progressive die. (d) Forming of the top piece of an aerosol spray can in a progressive die. Note that the part is attached to the strip until the last operation is completed.

strip, and a different operation (such as punching, blanking, and notching) is performed at the same station of the machine with each stroke of a series of punches (Fig. 16.11c). An example of a part made in progressive dies is shown in Fig. 16.11d; the part is the small round piece that supports the plastic tip in spray cans.

Transfer dies.   In a *transfer-die* setup, the sheet metal undergoes different operations at different stations of the machine which are arranged along a straight line or a circular path. After each step in a station, the part is transferred to the next station for further operations.

Tool and die materials.   Tool and die materials for shearing generally are tool steels and (for high production rates) carbides (see Table 5.7). Lubrication is important for reducing tool and die wear, thus improving edge quality.

16.2.4 Miscellaneous methods of cutting sheet metal

There are several other methods of cutting sheets and, particularly, plates:

- **Laser-beam cutting** is an important process (Section 26.7) typically used with computer-controlled equipment to cut a variety of shapes consistently, in various thicknesses, and without the use of any dies. Laser-beam cutting also can be combined with punching and shearing. These processes cover different and complementary ranges. Parts with certain features can be produced best by one process; some with other features can be produced best by the other process. Combination machines incorporating both capabilities have been designed and built (see also Example 27.1).
- **Water-jet cutting** is a cutting process that is effective on many metallic as well as nonmetallic materials (Section 27.8).
- Cutting with a **band saw**; this method is a chip-removal process.
- **Friction sawing** involves a disk or blade which rubs against the sheet or plate at high surface speeds (Section 24.5).
- **Flame cutting** is another common method, particularly for thick plates; it is used widely in ship building and on heavy structural components.

16.3 | Sheet-Metal Characteristics and Formability

After a blank is cut from a larger sheet or coil, it is formed into various shapes by several processes described in the rest of this chapter. We will now briefly review those characteristics of sheet metals that have important effects on these forming operations, as outlined in Table 16.2.

Elongation.   Sheet-metal-forming processes rarely involve simple uniaxial stretching like that in a tension test. However, observations from tensile testing are useful and necessary for understanding the behavior of metals in these operations. Recall from Section 2.2 that a specimen subjected to tension first undergoes **uniform elongation**, and that when the load exceeds the ultimate tensile strength of the material, the specimen begins to neck and thus elongation is no longer uniform.

Because the material usually is being stretched in sheet forming, high uniform elongation is desirable for good formability. The true strain at which necking begins is equal numerically to the *strain-hardening exponent* (n) shown in Eq. (2.8). Thus, a high n value indicates large uniform elongation (see also Table 2.3). Necking may be *localized* or it may be *diffuse*, depending on the *strain-rate sensitivity* (m) of the material; this relationship is given in Eq. (2.9). The higher the value of m, the more diffuse the neck becomes. A diffuse neck is desirable in sheet-forming operations. In addition to

TABLE 16.2

Characteristics of Metals Important in Sheet-Forming Operations

Characteristic	Importance
Elongation	Determines the capability of the sheet metal to stretch without necking and failure; high strain-hardening exponent (n) and strain-rate sensitivity exponent (m) are desirable
Yield-point elongation	Typically observed with mild-steel sheets (also called Lüder's bands or stretcher strains), flamelike depressions on the sheet surface, can be eliminated by temper rolling but sheet must be formed within a certain time after rolling
Anisotropy (planar)	Exhibits different behavior in different planar directions, present in cold-rolled sheets because of preferred orientation or mechanical fibering, causes earing in deep drawing, can be reduced or eliminated by annealing but at lowered strength
Anisotropy (normal)	Determines thinning behavior of sheet metals during stretching, important in deep drawing
Grain size	Determines surface roughness on stretched sheet metal, the coarser the grain—the rougher the appearance (orange peel), also affects material strength.
Residual stresses	Typically caused by nonuniform deformation during forming, results in part distortion when sectioned, can lead to stress-corrosion cracking, reduced or eliminated by stress relieving.
Springback	Due to elastic recovery of the plastically deformed sheet after unloading, causes distortion of part and loss of dimensional accuracy, can be controlled by techniques such as overbending and bottoming of the punch
Wrinkling	Caused by compressive stresses in the plane of the sheet, can be objectionable, depending on its extent, can be useful in imparting stiffness to parts by increasing their section modulus, can be controlled by proper tool and die design
Quality of sheared edges	Depends on process used; edges can be rough, not square, and contain cracks, residual stresses, and a work-hardened layer, which are all detrimental to the formability of the sheet; edge quality can be improved by fine blanking, reducing the clearance, shaving, and improvements in tool and die design and lubrication
Surface condition of sheet	Depends on sheet rolling practice; important in sheet forming as it can cause tearing and poor surface quality

uniform elongation and necking, the **total elongation** of the specimen (in terms of that for a 50-mm gage length) is also a significant factor in the formability of sheet metals.

Yield-point elongation. Low-carbon steels and some aluminum-magnesium alloys exhibit a behavior called *yield-point elongation*—having both upper and lower yield points (Fig. 16.12a). This behavior results in **Lüder's bands** (also called *stretcher-strain marks* or *worms*) on the sheet (Fig. 16.12b). These are elongated depressions on the surface of the sheet, such as can be found on the bottom of cans containing common household products (Fig. 16.12c). These marks may be objectionable in the final product, because coarseness in the surface degrades appearance and may cause difficulties in subsequent coating and painting operations.

The usual method of avoiding these marks is to eliminate or reduce yield-point elongation by reducing the thickness of the sheet 0.5 to 1.5% by cold rolling (**temper** or **skin rolling**). Because of strain aging, however, the yield-point elongation reappears after a few days at room temperature or after a few hours at higher temperatures. To prevent this undesirable occurrence, the material should be formed within a certain time limit (which depends on the type of the steel).

Anisotropy. An important factor that influences sheet-metal forming is *anisotropy* (*directionality*) of the sheet. Recall that anisotropy is acquired during the thermomechanical processing of the sheet, and that there are two types of anisotropy *crystallographic anisotropy* (preferred orientation of the grains) and *mechanical fibering* (alignment of impurities, inclusions, and voids throughout the thickness of the sheet). The relevance of this subject is discussed further in Section 16.4.

Grain size. As described in Section 1.4, grain size affects mechanical properties and influences the surface appearance of the formed part (*orange peel*). The smaller

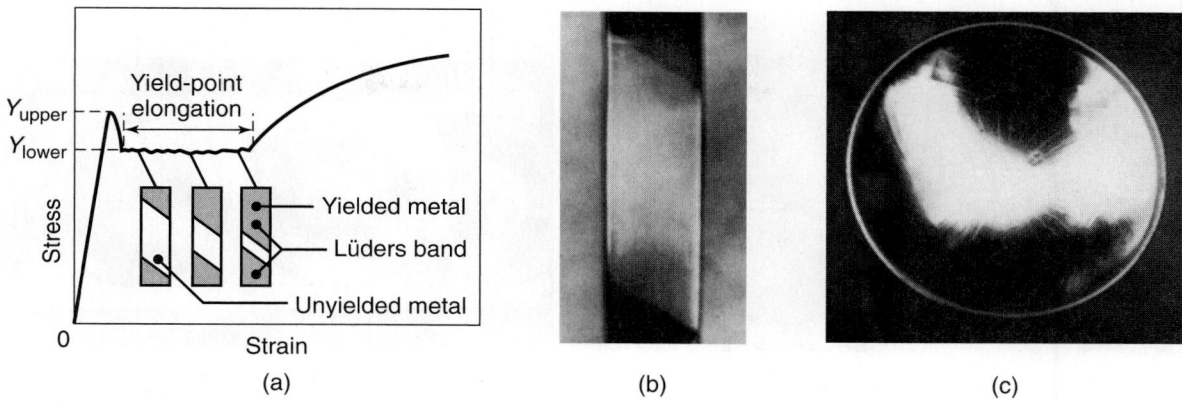

FIGURE 16.12   (a) Yield-point elongation in a sheet-metal specimen. (b) Lüder's bands in a low-carbon steel sheet. (c) Stretcher strains at the bottom of a steel can for household products. *Source:* (b) Courtesy of Caterpillar Inc.

the grain size, the stronger is the metal; and the coarser the grain, the rougher is the surface appearance. An ASTM grain size of 7 or finer (Table 1.1) is preferred for general sheet forming operations.

Dent resistance of sheet metals.   Dents commonly are found on cars, appliances, and office furniture. Dents usually are caused by dynamic forces from moving objects that hit the sheet metal. In typical automotive panels, for example, velocities at impact range up to 45 m/s (150 ft/s). Thus, it is the *dynamic yield stress* (yield stress under high rates of deformation) rather than the static yield stress that is the significant strength parameter.

Dynamic forces tend to cause *localized* dents, whereas static forces tend to *diffuse* the dented area. This phenomenon may be demonstrated by trying to dent a piece of flat sheet metal, first by pushing a ball-peen hammer against it and then by striking it with the hammer. Note how localized the dent will be in the latter case. *Dent resistance* of sheet-metal parts has been found to (a) increase as the sheet thickness and its yield-stress increases, and (b) decrease as its elastic modulus and its overall panel stiffness increase. Consequently, panels rigidly held at their edges have lower dent resistance because of their higher stiffness.

16.4 | Formability Tests for Sheet Metals

Sheet-metal formability is of great technological and economic interest, and it generally is defined as the ability of the sheet metal to undergo the desired shape change without failure, such as by necking, cracking, or tearing. As we shall see throughout the rest of this chapter, sheet metals (depending in part on geometry) may undergo two basic modes of deformation: (1) *stretching* and (2) *drawing*. There are important distinctions between these two modes, and different parameters are involved in determining formability under these different conditions. This section describes the methods that generally are used to predict formability.

Cupping tests.   The earliest tests developed to predict sheet-metal formability were cupping tests (Fig. 16.13a). In the *Erichsen test*, the sheet specimen is clamped between two circular, flat dies, and a steel ball or round punch is forced into the sheet until a crack begins to appear on the stretched specimen. The *punch depth, d*, at which failure occurs is a measure of the formability of the sheet. Although this and similar tests are easy to perform, they do not simulate the exact conditions of actual forming operations, and hence are not particularly reliable, especially for complex parts.

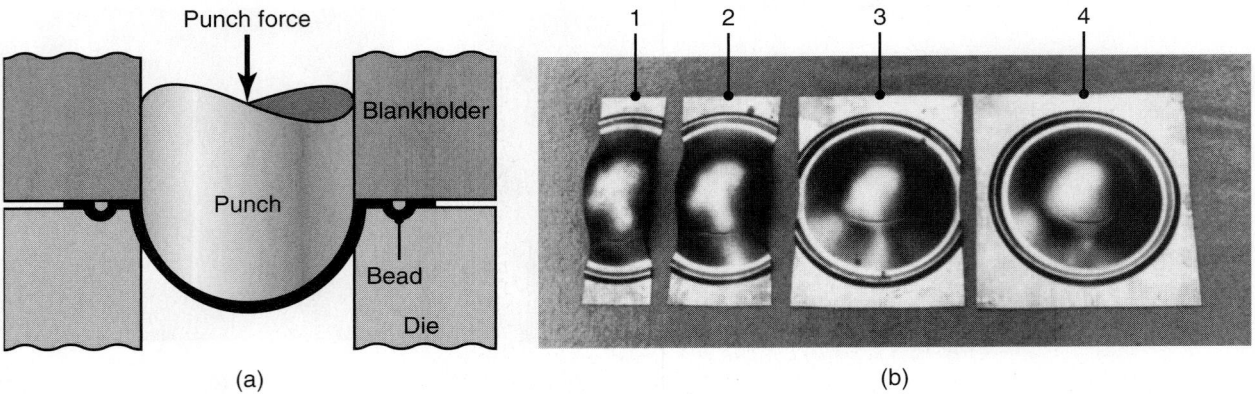

FIGURE 16.13 (a) A cupping test (the Erichsen test) to determine the formability of sheet metals. (b) Bulge-test results on steel sheets of various widths. The specimen farthest left is subjected to, basically, simple tension. The specimen farthest right is subjected to equal biaxial stretching. *Source:* Courtesy of Inland Steel Company.

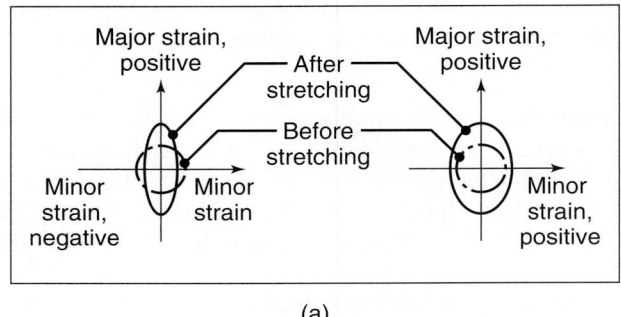

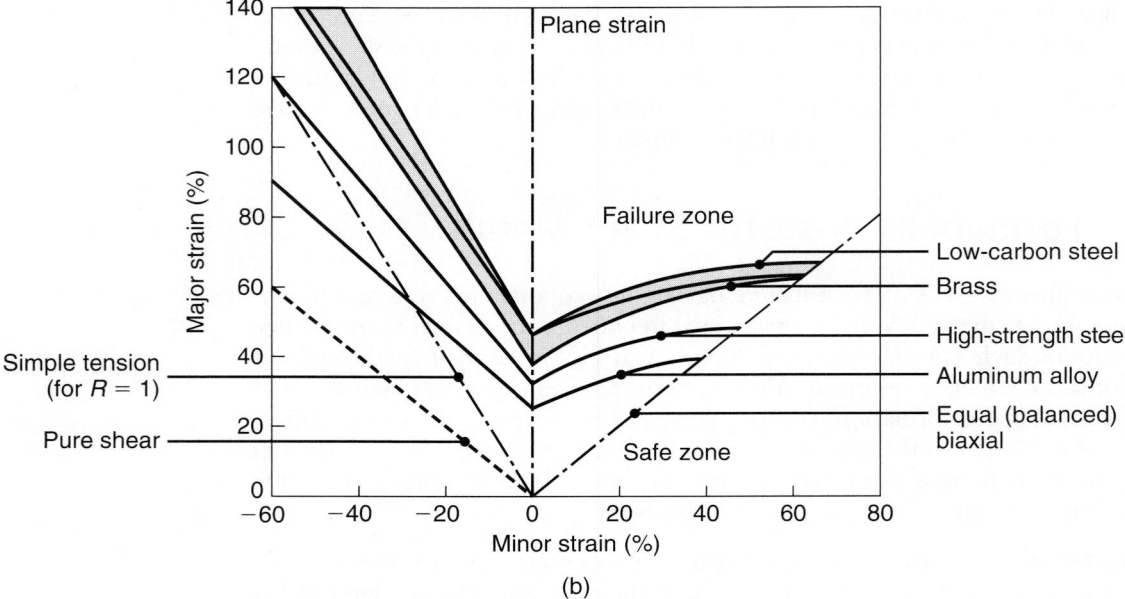

FIGURE 16.14 (a) Strains in deformed circular grid patterns. (b) Forming-limit diagrams (FLD) for various sheet metals. Although the major strain is always positive (stretching), the minor strain may be either positive or negative. R is the normal anisotropy of the sheet, as described in Section 16.4. *Source:* After S. S. Hecker and A. K. Ghosh.

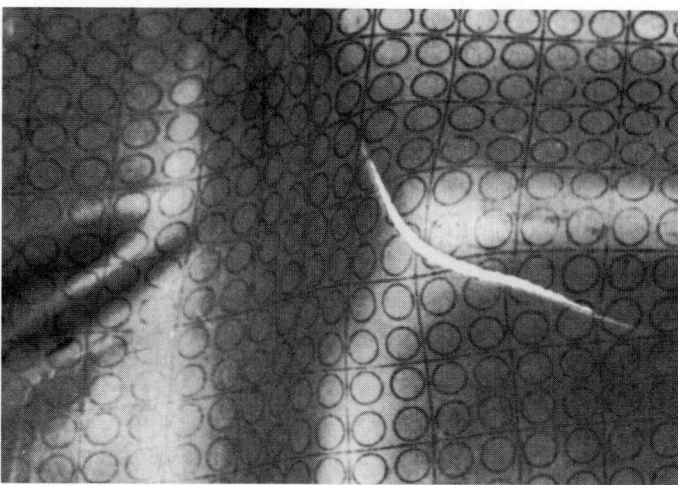

FIGURE 16.15   The deformation of the grid pattern and the tearing of sheet metal during forming. The major and minor axes of the circles are used to determine the coordinates on the forming-limit diagram in Fig. 16.14b. *Source:* After S. P. Keeler.

Forming-limit diagrams.   An important advance in testing the formability of sheet metals is the development of *forming-limit diagrams*, as shown in Fig. 16.14. A forming-limit diagram (FLD) for a particular metal is constructed by first marking the flat sheet with a grid pattern of circles (see Fig. 16.15), using electrochemical or photoprinting techniques. The blank then is stretched over a punch (Fig. 16.13a), and the deformation of the circles is observed and measured in regions where failure (*necking* and *tearing*) has occurred. Although they typically are 2.5 to 5 mm (0.1 to 0.2 in.) in diameter, for improved accuracy of measurement, the circles should be made as small as is practical.

In order to develop unequal stretching to simulate actual sheet-forming operations, the flat specimens are cut to varying widths (Fig. 16.13b) and then tested. Note that a square specimen (farthest right in the figure) produces *equal biaxial stretching* (such as that achieved in blowing up a spherical balloon), whereas a narrow specimen (farthest left in the figure) approaches the state of *uniaxial stretching* (that is, simple tension). After a series of such tests is performed on a particular sheet metal and at different widths, a forming-limit diagram is constructed showing the boundaries between *failure* and *safe* regions (Fig. 16.14b).

In order to develop a forming-limit diagram, the major and minor engineering strains as measured from the deformation of the original circles are obtained. Note in Fig. 16.14a that the original circle has deformed into an ellipse. The *major axis* of the ellipse represents the major direction and magnitude of stretching. The major strain is the *engineering strain* in this direction and is always *positive*, because the sheet is being stretched. The *minor axis* of the ellipse represents the magnitude of the stretching or shrinking in the *transverse* direction.

Note, however, that the minor strain can be either *positive* or *negative*. For example, if a circle is placed in the center of a tensile-test specimen and then stretched uniaxially (simple tension), the specimen becomes narrower as it is stretched (due to the Poisson effect), and thus, the minor strain is negative. (This behavior can be demonstrated easily by stretching a rubber band and observing the dimensional changes it undergoes.) On the other hand, if we place a circle on a spherical rubber balloon and inflate it, the minor and major strains are both positive and equal in magnitude.

By comparing the surface areas of the original circle and the deformed circle on the formed sheet, we also can determine whether the thickness of the sheet has changed during deformation. Because the volume remains constant in plastic deformation, we know that if the area of the deformed circle is larger than the original circle, the sheet has become thinner. This phenomenon can be demonstrated easily by blowing up a balloon and noting that it becomes more translucent as it is stretched (because it is getting thinner).

The data thus obtained from different locations in each of the samples shown in Fig. 16.13b are then plotted as shown in Fig. 16.14b. The curves represent the boundaries between *failure zones* and *safe zones* for each type of metal, and as can be noted, the higher the curve, the better the formability of that particular metal. As expected, different materials and conditions (such as cold worked or heat treated) have different forming-limit diagrams.

Taking the aluminum alloy in Fig. 16.14b as an example, if a circle in a particular location on the sheet has undergone major and minor strains of plus 20% and minus 10%, respectively, there would be no tear in that location of the specimen. On the other hand, if the major and minor strains were plus 80% and minus 40%, respectively, at another location, there would be a tear in that particular location of the specimen. An example of a formed sheet-metal part with a grid pattern is shown in Fig. 16.15. Note the deformation of the circular patterns in the vicinity of the tear on the formed sheet.

It is important to note in forming-limit diagrams that a compressive minor strain of, say, 20% is associated with a higher major strain than is a tensile (positive) minor strain of the same magnitude. In other words, it is desirable for the minor strain to be negative (that is, shrinking in the minor direction). In the forming of complex parts, special tooling can be designed to take advantage of the beneficial effect of negative minor strains on formability.

The effect of sheet thickness on forming-limit diagrams is to raise the curves in Fig. 16.14b. The thicker the sheet, the higher its formability curve, and thus the more formable it is. On the other hand, in actual forming operations, a thick blank may not bend as easily around small radii without cracking (as described in Section 16.5 on bending). Friction and lubrication at the interface between the punch and the sheet metal are also important factors in the test results. With well-lubricated interfaces, the strains in the sheet are distributed more uniformly over the punch. Also, as expected and depending on the material and its notch sensitivity, surface scratches, deep gouges, and blemishes can reduce formability significantly and thereby lead to premature tearing and failure of the part.

16.5 | Bending Sheets, Plates, and Tubes

Bending is one of the most common industrial forming operations. We merely have to look at an automobile body, appliance, paper clip, or file cabinet to appreciate how many parts are shaped by bending. Furthermore, bending also imparts stiffness to the part by increasing its moment of inertia. Note, for example, how corrugations, flanges, beads, and seams improve the stiffness of structures without adding any weight. As a specific example, observe the diametral stiffness of a metal can with and without circumferential beads (see also *beading*).

The terminology used in the bending of a sheet or plate is shown in Fig. 16.16. Note that the outer fibers of the material are in tension, while the inner fibers are in compression. Because of the Poisson effect, the width of the part (*bend length*, L) has become smaller in the outer region and larger in the inner region than the original width (as can be seen in Fig. 16.17c). This phenomenon may be observed easily by bending a rectangular rubber eraser and observing the changes in its shape.

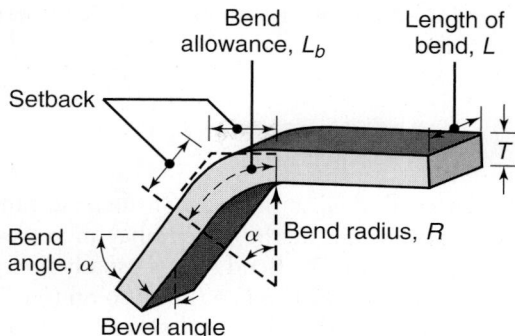

FIGURE 16.16 Bending terminology. Note that the bend radius is measured to the inner surface of the bent part.

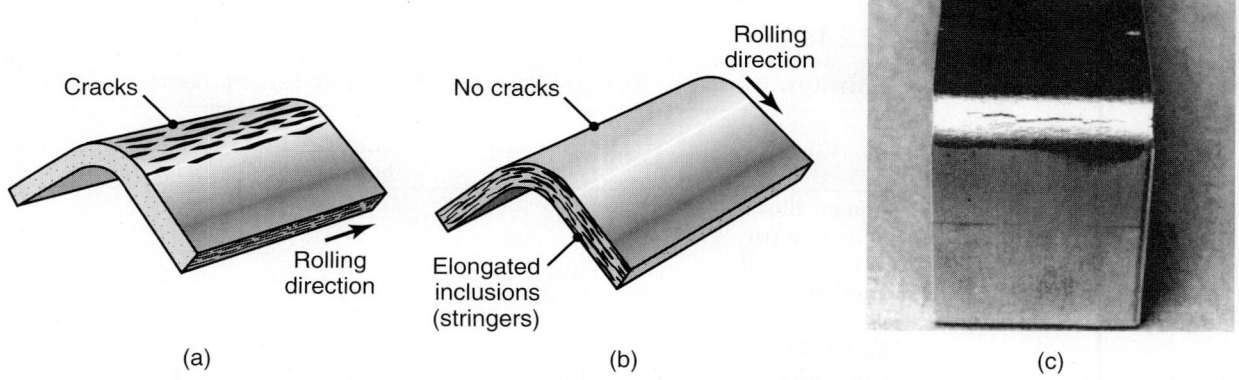

FIGURE 16.17 (a) and (b) The effect of elongated inclusions (stringers) on cracking as a function of the direction of bending with respect to the original rolling direction of the sheet. (c) Cracks on the outer surface of an aluminum strip bent to an angle of 90°. Note also the narrowing of the top surface in the bend area (due to the Poisson effect).

As shown in Fig. 16.16, the **bend allowance**, L_b, is the length of the *neutral axis* in the bend and is used to determine the length of the blank for a part to be bent. The position of the neutral axis, however, depends on the radius and the bend angle (as described in texts on mechanics of materials). An approximate formula for the bend allowance is given by

$$L_b = \alpha(R + kT) \qquad (16.2)$$

where α is the bend angle (in radians), T is the sheet thickness, R is the bend radius, and k is a constant. In practice, k values typically range from 0.33 (for $R < 2T$) to 0.5 (for $R > 2T$). Note that for the ideal case, the neutral axis is at the center of the sheet thickness, $k = 0.5$, and, hence,

$$L_b = \alpha\left[R + \left(\frac{T}{2}\right)\right] \qquad (16.3)$$

Minimum bend radius. The radius at which a crack first appears at the outer fibers of a sheet being bent is referred to as the *minimum bend radius*. It can be shown that the engineering strain on the outer and inner fibers of a sheet during bending is given by the expression

$$e = \frac{1}{(2R/T) + 1} \tag{16.4}$$

Thus, as R/T decreases (that is, as the ratio of the bend radius to the thickness becomes smaller), the tensile strain at the outer fiber increases, and the material eventually develops cracks (Fig. 16.17). Bend radius usually is expressed (reciprocally) in terms of the thickness, such as $2T$, $3T$, $4T$, and so on (see Table 16.3). Thus, a $3T$ minimum bend radius indicates that the *smallest radius* to which the sheet can be bent without cracking is three times its thickness.

There is an inverse relationship between *bendability* and the tensile reduction of the area of the material (Fig. 16.18). The *minimum bend radius*, R, is, approximately,

$$R = T\left(\frac{50}{r} - 1\right) \tag{16.5}$$

TABLE 16.3

Minimum Bend Radius for Various Metals at Room Temperature		
	Condition	
Material	Soft	Hard
Aluminum alloys	0	6T
Beryllium copper	0	4T
Brass (low-leaded)	0	2T
Magnesium	5T	13T
Steels		
Austenitic stainless	0.5T	6T
Low-carbon, low-alloy, and HSLA	0.5T	4T
Titanium	0.7T	3T
Titanium alloys	2.6T	4T

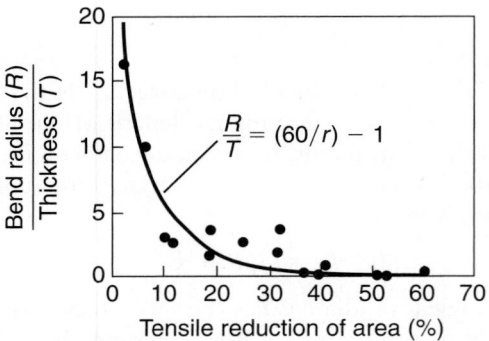

FIGURE 16.18 Relationship between R/T ratio and tensile reduction of area for sheet metals. Note that sheet metal with a 50% tensile reduction of area can be bent over itself in a process like the folding of a piece of paper without cracking. *Source*: After J. Datsko and C.T. Yang.

where r is the tensile reduction of area of the sheet metal. Thus, for $r = 50$, the minimum bend radius is zero; that is, the sheet can be folded over itself (see *hemming*) in much the same way as a piece of paper is folded. To increase the bendability of metals, we may increase their tensile reduction of area of the metal either by heating or by bending in a high-pressure environment. Bendability also depends on the *edge condition* of the sheet. Since rough edges are points of stress concentration, bendability decreases as edge roughness increases.

Another significant factor in edge cracking is the amount, shape, and hardness of *inclusions* present in the sheet metal and the amount of cold working that the edges undergo during shearing. Because of their pointed shape, inclusions in the form of stringers are more detrimental than globular-shaped inclusions (see also Fig. 2.23). The resistance to edge cracking during bending can be improved greatly by removing the cold-worked regions by shaving or machining the edges of the part (see Fig. 16.9) or by annealing it to improve its ductility.

Anisotropy of the sheet is another important factor in bendability. Cold rolling results in anisotropy by *preferred orientation* or by *mechanical fibering* due to the alignment of any impurities, inclusions, and voids that may be present, as shown in Fig. 1.13. Prior to laying out (on such a sheet) the blanks to be bent (*nesting*, see Fig. 16.51), caution should be exercised in cutting it in the proper direction from a rolled sheet; however, this choice may not always be possible in practice.

Springback.   Because all materials have a finite modulus of elasticity, plastic deformation always is followed by some elastic recovery when the load is removed (see Fig. 2.3). In bending, this recovery is called *springback*, which can be observed easily by bending and then releasing a piece of sheet metal or wire. Springback occurs not only in flat sheets and plates, but also in solid or hollow bars and tubes of any cross-section. As noted in Fig. 16.19, the final bend angle after springback is smaller than the angle to which the part was bent, and the final bend radius is larger than before springback occurs.

Springback can be calculated approximately in terms of the radii R_i and R_f (Fig. 16.19), as

$$\frac{R_i}{R_f} = 4\left(\frac{R_i Y}{ET}\right)^3 - 3\left(\frac{R_i Y}{ET}\right) + 1 \qquad (16.6)$$

Note from this formula that springback increases (a) as the R/T ratio and the yield stress, Y, of the material increase, and (b) as the elastic modulus, E, decreases.

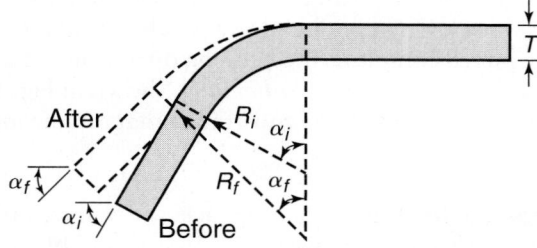

FIGURE 16.19   Springback in bending. The part tends to recover elastically after bending, and its bend radius becomes larger. Under certain conditions, it is possible for the final bend angle to be smaller than the original angle (negative springback).

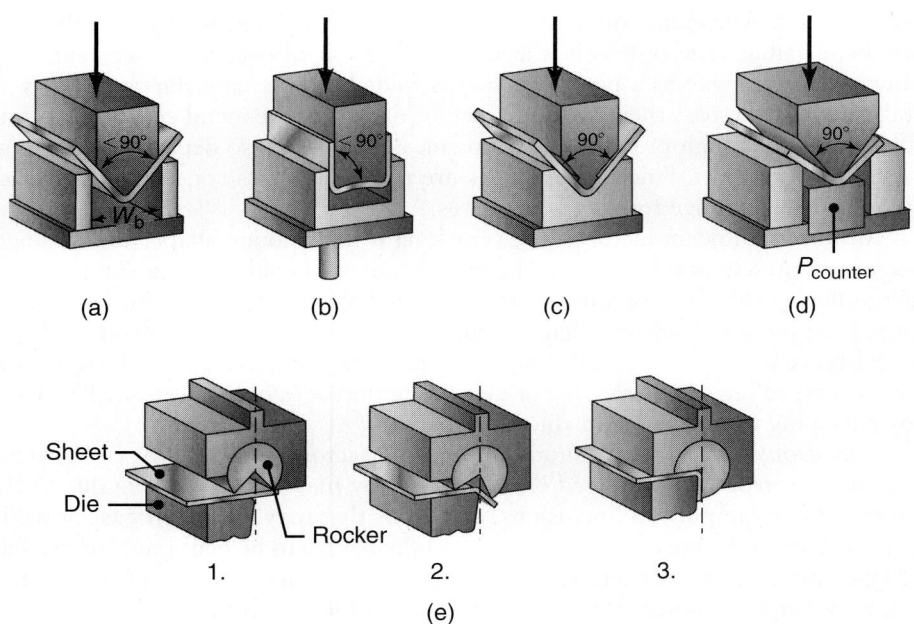

FIGURE 16.20   Methods of reducing or eliminating springback in bending operations.

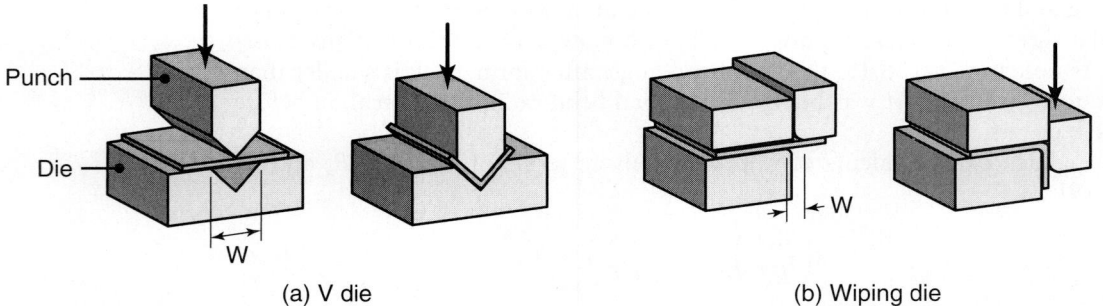

FIGURE 16.21   Common die-bending operations showing the die-opening dimension, W, used in calculating bending forces.

In V-die bending (Figs. 16.20 and 16.21), it is possible for the material to also exhibit *negative springback*. This condition is caused by the nature of the deformation occurring just as the punch completes the bending operation at the end of the stroke. Negative springback does not occur in *air bending*, shown in Fig. 16.22a (also called *free bending*), because of the lack of the constraints that a V-die imposes on the bend area.

Compensation for springback.   Springback in forming operations usually is compensated for by *overbending* the part (Fig. 16.20a and b). Several trials may be necessary to obtain the desired results. Another method is to *coin* the bend area by subjecting it to highly localized compressive stresses between the tip of the punch and the die surface (Fig. 16.20c and d)—a technique known as *bottoming* the punch. Another method is *stretch bending*, in which the part is subjected to tension while being bent (see also *stretch forming*, Section 16.6).

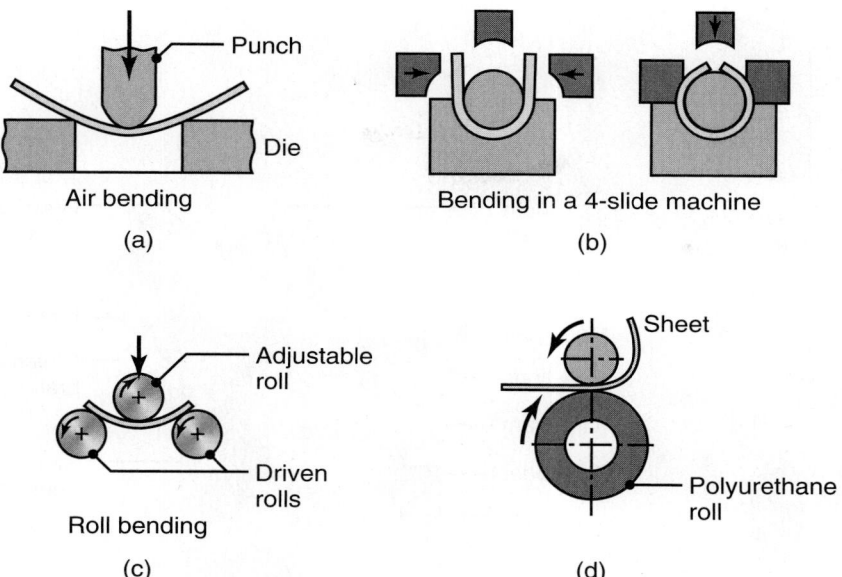

Air bending
(a)

Bending in a 4-slide machine
(b)

Roll bending
(c)

(d)

FIGURE 16.22 Examples of various bending operations.

Bending force. The bending force for sheets and plates can be estimated by assuming that the process is one of the simple bending of a rectangular beam, as described in texts on mechanics of solids. Thus, the bending force is a function of the strength of the material, the length, L, of the bend, the thickness, T, of the sheet, and the die opening, W, as shown in Fig. 16.21. Excluding friction, the *maximum bending force, P*, is

$$P = \frac{kYLT^2}{W} \tag{16.7}$$

where the factor k ranges from about 0.3 for a wiping die, to about 0.7 for a U-die, to about 1.3 for a V-die, and Y is the yield stress of the material.

For a V-die, Eq. (16.7) is often modified as

$$P = \frac{(\text{UTS})LT^2}{W} \tag{16.8}$$

where UTS is the ultimate tensile strength of the material. This equation applies well to situations in which the punch-tip radius and the sheet thickness are relatively small compared to the die opening, W.

The force in die bending varies throughout the bending cycle. It increases from zero to a maximum, and it may even decrease as the bend is completed. The force then increases sharply as the punch reaches the bottom of its stroke and the part touches the bottom of the die. In air bending (Fig. 16.22a), however, the force does not increase again after it begins to decrease, as it has no resistance to its free movement downward.

16.6 | Miscellaneous Bending and Related Operations

Press-brake forming. Sheet metal or plate can be bent easily with simple fixtures using a press. Sheets or narrow strips that are 7 m (20 ft) or even longer usually are bent in a *press brake* (Fig. 16.23). The machine utilizes long dies in a mechanical or

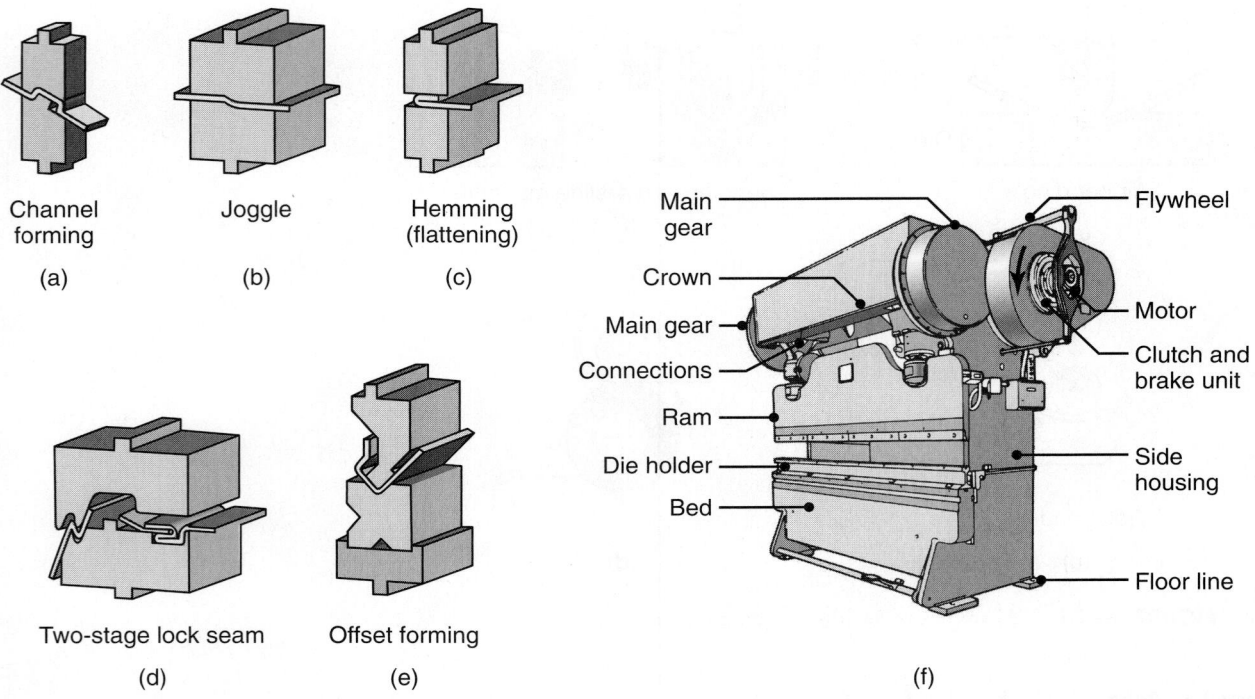

FIGURE 16.23 (a) through (e) Schematic illustrations of various bending operations in a press brake. (f) Schematic illustration of a press brake. *Source*: Courtesy of Verson Allsteel Company.

hydraulic press and is particularly suitable for small production runs. As can be seen in Fig. 16.23, the tooling is simple, their motions are only up and down, and they easily are adaptable to a wide variety of shapes. Also, the process can be automated easily for low-cost, high-production runs. *Die materials* for press brakes may range from hardwood (for low-strength materials and small-production runs) to carbides for strong and abrasive sheet materials and also are chosen to improve die life. For most applications, however, carbon-steel or gray-iron dies generally are used.

Bending in a four-slide machine. Bending relatively short pieces can be done on a machine such as that shown in Fig. 16.22b. In these machines, the lateral movements of the dies are controlled and synchronized with the vertical die movement to form the part into desired shapes. This process is useful in making seamed tubing and conduits, bushings, fasteners, and various machinery components.

Roll bending. In this process (Fig. 16.22c), plates are bent using a set of rolls. By adjusting the distance between the three rolls, various curvatures can be obtained. This process is flexible and is used widely for bending plates for applications such as boilers, cylindrical pressure vessels, and various curved structural members. Figure 16.22d shows the bending of a strip with a compliant roll made of polyurethane, which conforms to shape change as the hard upper roll presses upon the strip.

Beading. In *beading*, the periphery of the sheet metal is bent into the cavity of a die (Fig. 16.24). The bead imparts stiffness to the part by increasing the moment of inertia of that section. Also, beads improve the appearance of the part and eliminate exposed sharp edges that can be hazardous.

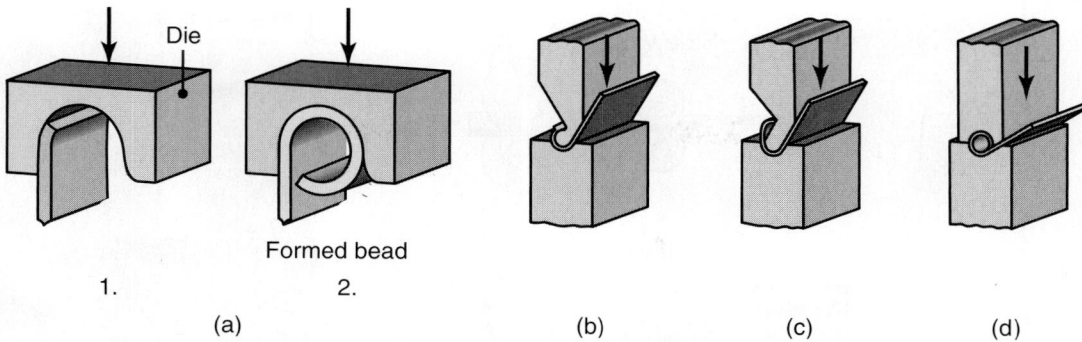

FIGURE 16.24   (a) Bead forming with a single die. (b) though (d) Bead forming with two dies in a press brake.

Flanging.   This is a process of bending the edges of sheet metals, usually to 90°. In **shrink flanging** (Fig. 16.25a), the flange is subjected to compressive hoop stresses which, if excessive, can cause the flange periphery to wrinkle. The wrinkling tendency increases with decreasing radius of curvature of the flange. In **stretch flanging**, the flange periphery is subjected to tensile stresses that, if excessive, can lead to cracking along the periphery.

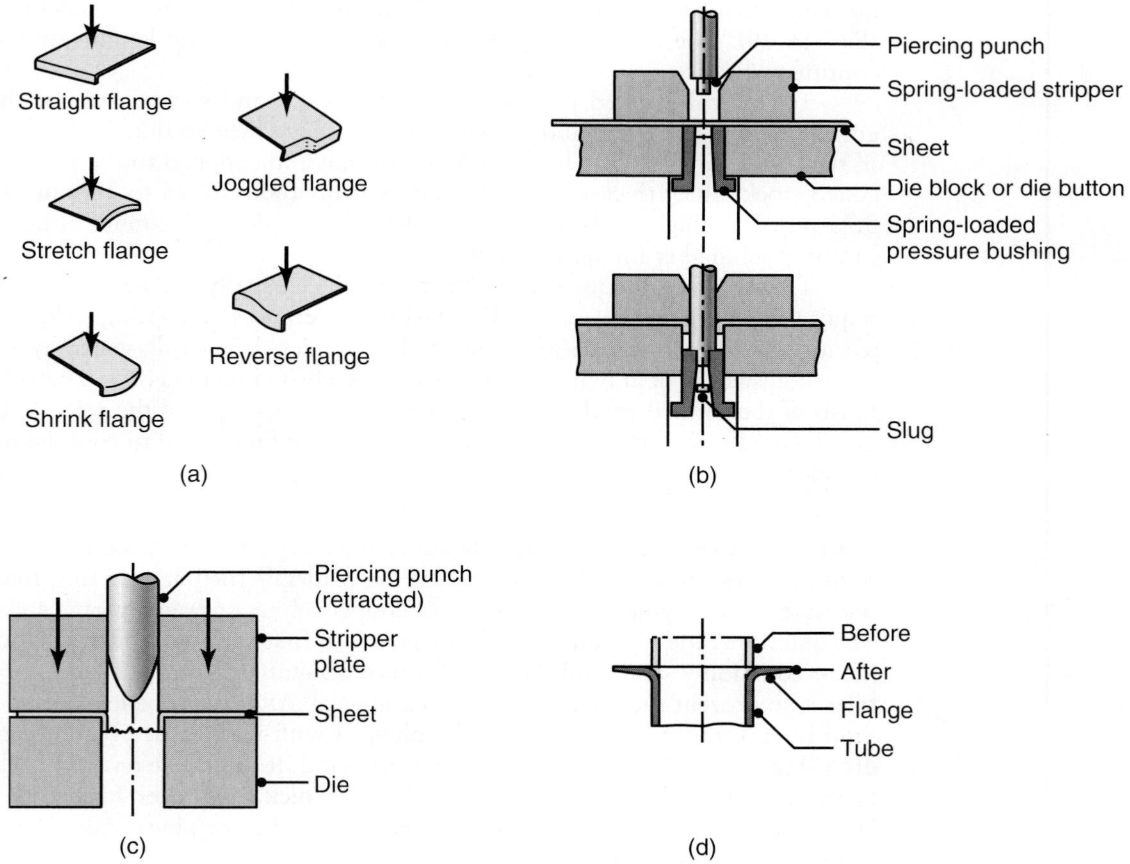

FIGURE 16.25   Various flanging operations. (a) Flanges on flat sheet. (b) Dimpling. (c) The piercing of sheet metal to form a flange. In this operation, a hole does not have to be pre-punched before the punch descends. Note, however, the rough edges along the circumference of the flange. (d) The flanging of a tube. Note the thinning of the edges of the flange.

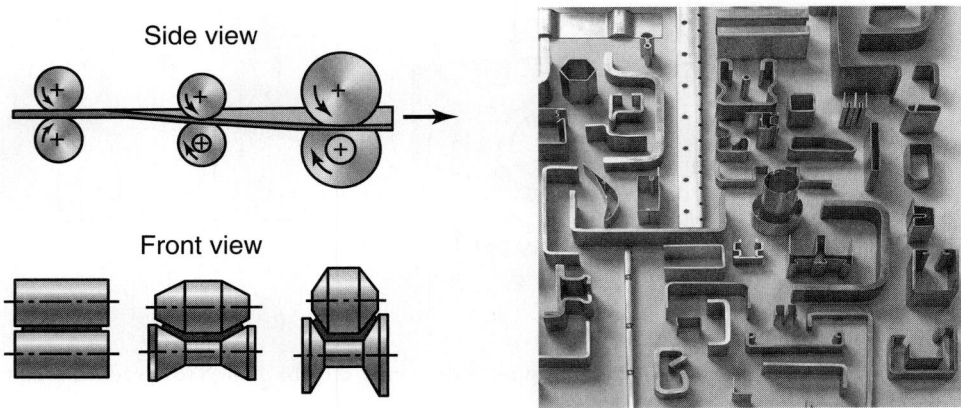

Side view

Front view

FIGURE 16.26 (a) Schematic illustration of the roll-forming process. (b) Examples of roll-formed cross-sections. *Source:* (b) Courtesy of Sharon Custom Metal Forming, Inc.

Roll forming. This process, which is also called *contour-roll forming* or *cold-roll forming*, is used for forming continuous lengths of sheet metal and for large production runs. As it passes through a set of rolls, the metal strip is bent in consecutive stages (Fig. 16.26). The formed strip is then sheared into specific lengths and stacked continuously.

Typical roll-formed products are panels, door and picture frames, channels, gutters, siding, and pipes and tubing with lock seams (see Section 32.5). The length of the part is limited only by the amount of material supplied to the rolls from the coiled stock. Sheet thickness usually ranges from about 0.125 to 20 mm (0.005 to 0.75 in.). Forming speeds are generally below 1.5 m/s (300 ft/min), although they can be much higher for special applications.

The design and sequencing of the rolls (which usually are mechanically driven) require considerable experience. Dimensional tolerances and springback, as well as tearing and buckling of the strip, have to be considered. The rolls generally are made of carbon steel or of gray iron, and they may be chromium plated for a better surface finish of the formed product and for better wear resistance of the rolls. Lubricants may be used to reduce roll wear, to improve surface finish, and to cool the rolls and the sheet being formed.

Tube bending and forming. Bending and forming tubes and other hollow sections requires special tooling because of the tendency for buckling and folding, as one notes when trying to bend a piece of copper tubing or even a plastic soda straw. The oldest method of bending a tube or pipe is to first pack its inside with loose particles (commonly sand) and then bend it into a suitable fixture. The function of the filler is to prevent the tube from buckling inward. After the tube has been bent, the sand is shaken out. Tubes also can be plugged with various flexible internal mandrels (Fig. 16.27) for the same purpose as the sand. It should be noted that (because of its lower tendency for buckling) a relatively thick tube to be formed to a large bend radius can be bent safely without the use of fillers or plugs. (See also *tube hydroforming*, Section 16.8.)

Forming tubes and tubular shapes such as exhaust pipes, fuel filler tubes, and exhaust manifolds also can be done using internal *fluid pressure* (replacing the polyurethane plug) with the ends of the tubes sealed by mechanical means. In this process (*tube hydroforming*), the part is expanded in a split-female die at pressures

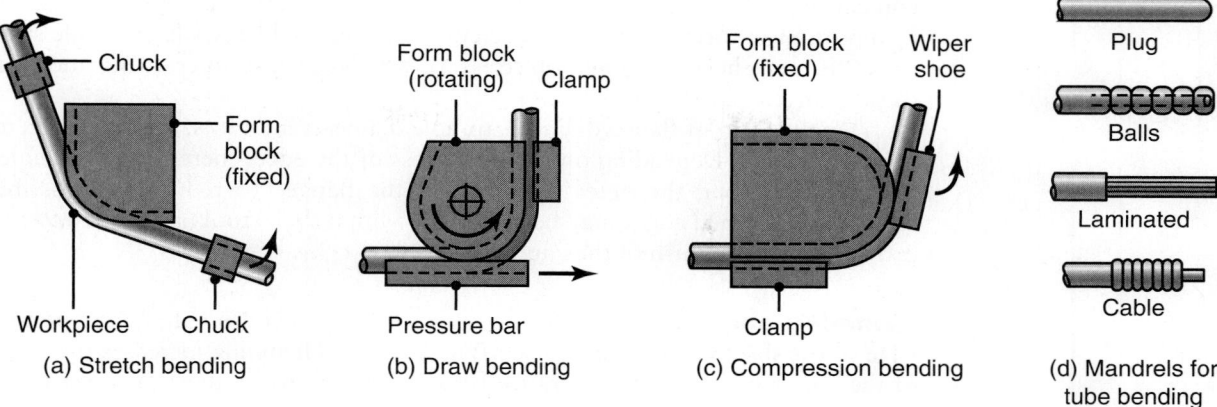

(a) Stretch bending   (b) Draw bending   (c) Compression bending   (d) Mandrels for tube bending

FIGURE 16.27   Methods of bending tubes. Internal mandrels or filling of tubes with particulate materials such as sand are often necessary to prevent collapse of the tubes during bending. Tubes also can be bent by a technique consisting of a stiff, helical tension spring slipped over the tube. The clearance between the OD of the tube and the ID of the spring is small, thus the tube cannot kink and the bend is uniform.

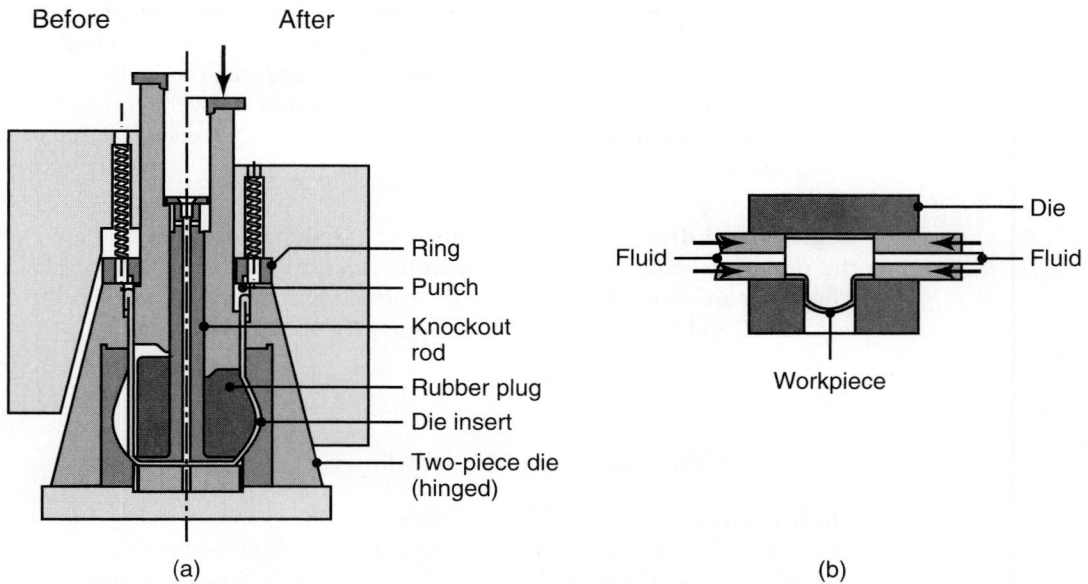

FIGURE 16.28   (a) The bulging of a tubular part with a flexible plug. Water pitchers can be made by this method. (b) Production of fittings for plumbing by expanding tubular blanks under internal pressure. The bottom of the piece is then punched out to produce a "T." *Source*: After J. A. Schey.

on the order of 600 MPa (90,000 psi). The die then is opened to remove the formed part (Fig. 16.28b).

Dimpling, piercing, and flaring.   In *dimpling* (Fig. 16.25b), a hole first is punched and then expanded into a flange. Flanges also may be produced by *piercing* with a shaped punch (Fig. 16.25c). Tube ends also can be flanged by a similar process (Fig. 16.25d). When the bend angle is less than 90° (as in fittings with conical ends) the process is called *flaring*. The condition of the edges (see Fig. 16.3) is important in these operations. Stretching the material causes high tensile stresses along the periphery (tensile hoop stresses), which can lead to cracking and tearing of the flange.

As the ratio of flange diameter to hole diameter increases, the strains increase proportionately. Depending on the roughness of the edge, there will be a tendency for cracking along the outer periphery of the flange. To reduce this possibility, sheared or punched edges may be shaved off with a sharp tool (see Fig. 16.9) to improve the surface finish of the edge.

Hemming and seaming.   In the *hemming* process (also called *flattening*), the edge of the sheet is folded over itself (Fig. 16.23c). Hemming increases the stiffness of the part, improves its appearance, and eliminates sharp edges. *Seaming* involves joining two edges of sheet metal by hemming (Fig. 16.23d). Double seams are made by a similar process using specially shaped rollers for watertight and airtight joints, such as are needed in food and beverage containers.

Bulging.   This process involves placing a tubular, conical, or curvilinear part into a split-female die and then expanding it, usually with a polyurethane plug (Fig. 16.28a). The punch then is retracted, the plug returns to its original shape (by total elastic recovery), and the formed part is removed by opening the split dies. Typical products made are coffee or water pitchers, beer barrels, and beads on oil drums. For parts with complex shapes, the plug (instead of being cylindrical) may be shaped in order to apply higher pressures at critical regions of the part. The major advantages of using polyurethane plugs is that they are very resistant to abrasion, wear, and lubricants; furthermore, they do not damage the surface finish of the part being formed.

Segmented dies.   These dies consist of individual segments that are placed inside the part to be formed and expanded mechanically in a generally radial direction. They then are retracted to remove the formed part. Segmented dies are relatively inexpensive, and they can be used for large production runs.

EXAMPLE 16.3 Manufacturing of bellows

Bellows are manufactured by a bulging process, as shown in Fig. 16.29. After the tube is bulged at several equidistant locations, it is compressed axially to uniformly collapse the bulged regions, thus forming bellows. The tube material must be able to undergo the large strains involved during the collapsing process without developing cracks.

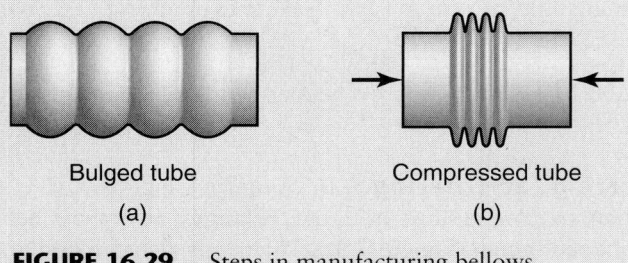

Bulged tube   Compressed tube

(a)   (b)

FIGURE 16.29   Steps in manufacturing bellows.

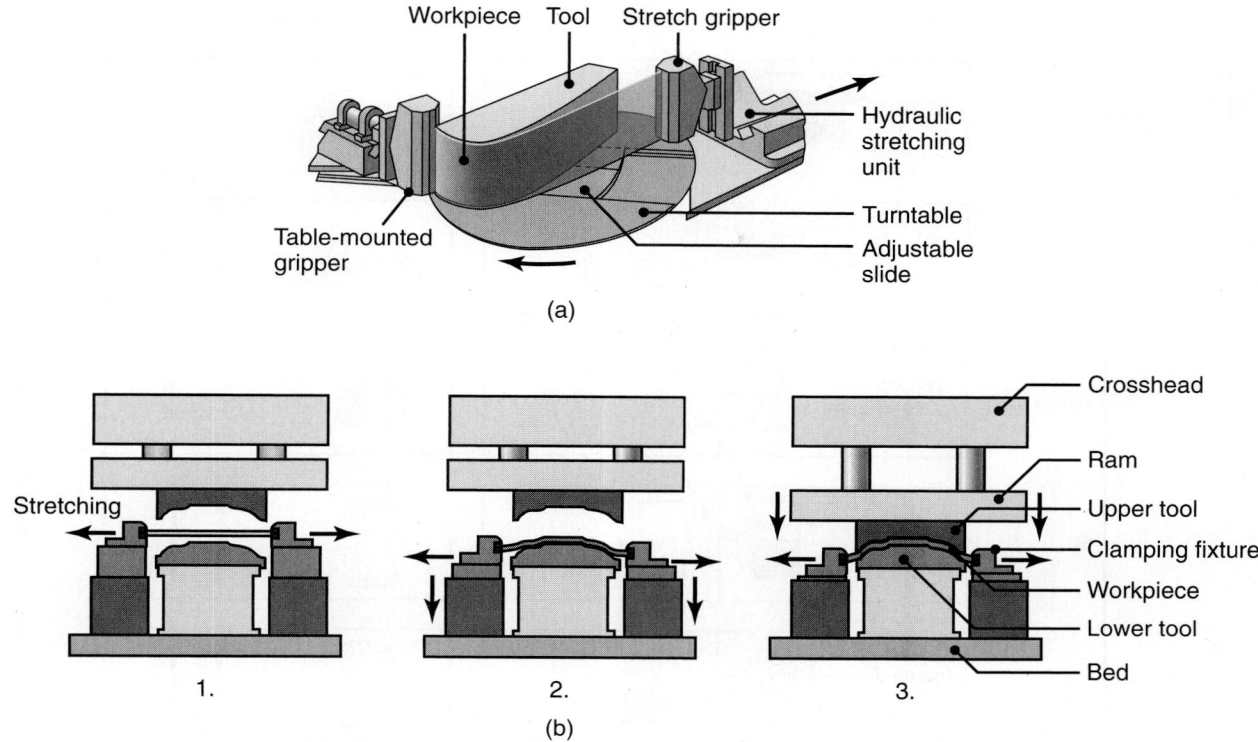

FIGURE 16.30 Schematic illustration of a stretch-forming process. Aluminum skins for aircraft can be made by this method. *Source*: Courtesy of Cyril Bath Co.

Stretch forming. In *stretch forming*, the sheet metal is clamped along its edges and then stretched over a male die (*form block* or *form punch*). The die moves upward, downward, or sideways, depending on the particular design of the machine (Fig. 16.30). Stretch forming is used primarily to make aircraft wing-skin panels, fuselages, and boat hulls. Aluminum skins for the Boeing 767 and 757 aircraft, for example, are made by stretch forming—with a tensile force of 9 MN (2 million lb). The rectangular sheets are 12 m × 2.5 m × 6.4 mm (40 ft × 8.3 ft × 0.25 in.). Although this process generally is used for low-volume production, it is versatile and economical, particularly for the aerospace industry.

In most operations, the blank is a rectangular sheet clamped along its narrower edges and stretched lengthwise, thus allowing the material to shrink in width. Controlling the amount of stretching is important in order to prevent tearing. Stretch forming cannot produce parts with sharp contours or with re-entrant corners (depressions on the surface of the die). Various accessory equipment can be used in conjunction with stretch forming, including further forming with both male and female dies while the part is under tension. Dies for stretch forming generally are made of zinc alloys, steel, plastics, or wood. Most applications require little or no lubrication.

16.7 | Deep Drawing

Numerous parts made of sheet metal are cylindrical or box shaped, such as pots and pans, all types of containers for food and beverages (Fig. 16.31), stainless-steel kitchen sinks, canisters, and automotive fuel tanks. Such parts usually are made by

Process	Process illustration	Result
1. Blanking		
2. Deep drawing		
3. Redrawing		
4. Ironing		
5. Doming		
6. Necking		
7. Seaming		

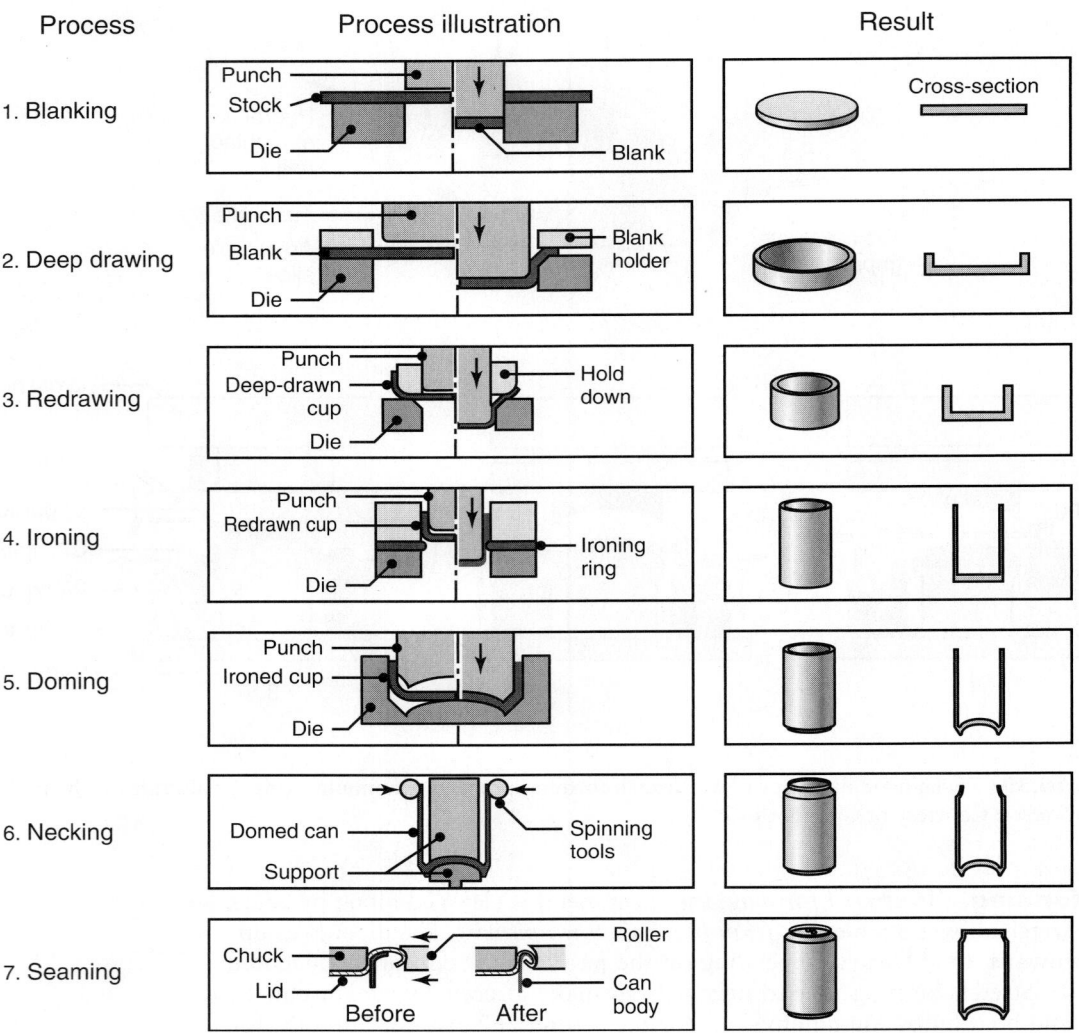

FIGURE 16.31 The metal-forming processes involved in manufacturing a two-piece aluminum beverage can.

a process in which a punch forces a flat sheet-metal blank into a die cavity (Fig. 16.32a). Although the process generally is called *deep drawing* (because of its capability for producing deep parts), it also is used to make parts that are shallow or have moderate depth. It is one of the most important metalworking processes because of the wide use of products made.

In the basic deep-drawing process, a round sheet-metal blank is placed over a circular die opening and is held in place with a **blankholder** or *hold-down ring* (Fig. 16.32b). The punch travels downward and forces the blank into the die cavity, forming a cup. The important variables in deep drawing are the properties of the sheet metal, the ratio of blank diameter, D_o; the punch diameter, D_p; the clearance, c, between punch and die; the punch radius, R_p; the die-corner radius, R_d; the blankholder force; and friction and lubrication between all contacting surfaces.

During the drawing operation, the movement of the blank into the die cavity induces compressive circumferential (hoop) stresses in the flange, which tend to cause the flange to *wrinkle* during drawing. This phenomenon can be demonstrated

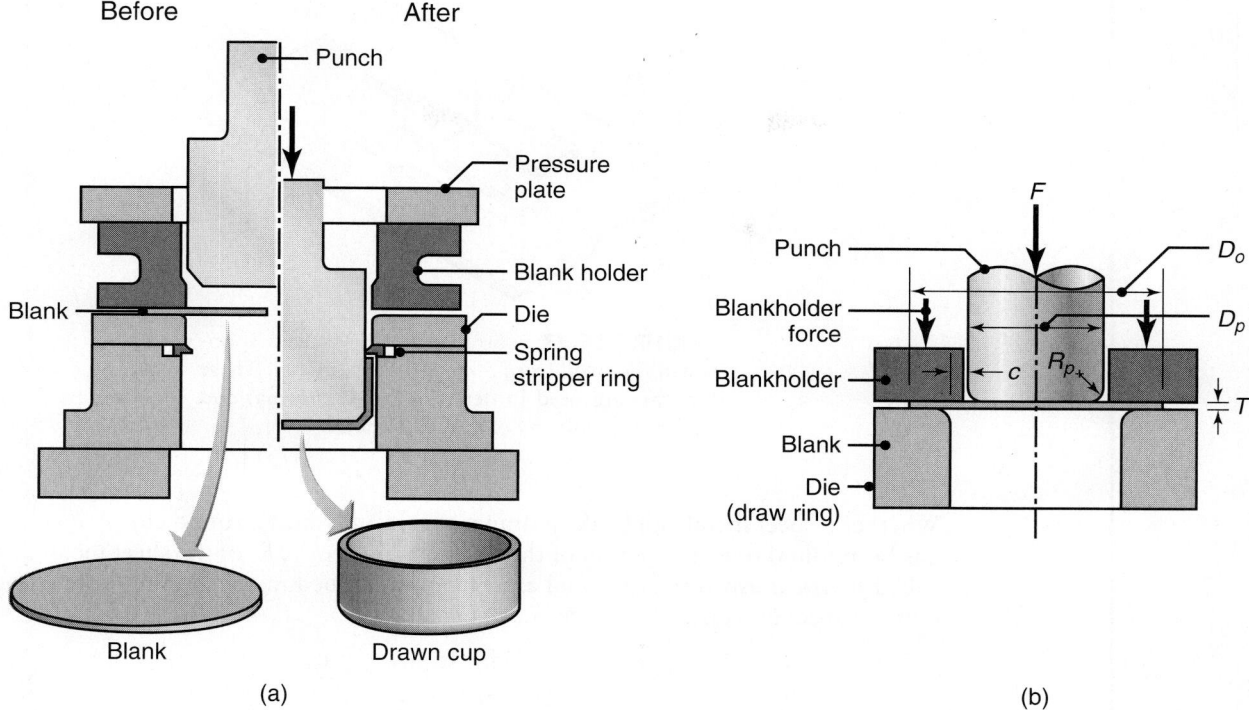

FIGURE 16.32   (a) Schematic illustration of the deep-drawing process on a circular sheet-metal blank. The stripper ring facilitates the removal of the formed cup from the punch. (b) Process variables in deep drawing. Except for the punch force, F, all the parameters indicated in the figure are independent variables.

simply by trying to force a circular piece of paper into a round cavity, such as a drinking glass. Wrinkling can be reduced or eliminated if a blankholder is kept under the effect of a certain force. In order to improve performance, the magnitude of this force can be controlled as a function of punch travel.

Because of the many variables involved, the *punch force*, F, is difficult to calculate directly. It has been shown, however, that the *maximum punch force*, F_{max}, can be estimated from the formula

$$F_{max} = \pi D_p T (\text{UTS})[(D_o/D_p) - 0.7] \qquad (16.9)$$

where the nomenclature is the same as that in Fig. 16.32b. It can be seen that the force increases with increasing blank diameter, thickness, strength, and the ratio (D_o/D_p). The wall of the cup is subjected principally to a longitudinal (vertical) tensile stress due to the punch force. Elongation under this stress causes the cup wall to become thinner and, if excessive, can cause *tearing* of the cup.

16.7.1 Deep drawability

In a deep-drawing operation, failure generally results from the *thinning* of the cup wall under high longitudinal tensile stresses. If we follow the movement of the material as it flows into the die cavity, it can be seen that the sheet metal (a) must be capable of undergoing a reduction in width due to reduction in diameter, and (b) it must also resist thinning under the longitudinal tensile stresses in the cup wall. *Deep drawability* generally is expressed by the **limiting drawing ratio** (LDR), as

$$\text{LDR} = \frac{\text{Maximum blank diameter}}{\text{Punch diameter}} = \frac{D_o}{D_p} \qquad (16.10)$$

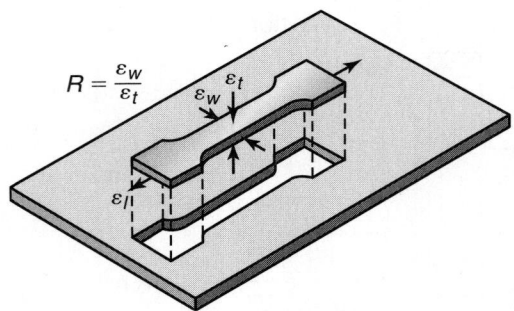

FIGURE 16.33   Strains on a tensile-test specimen removed from a piece of sheet metal. These strains are used in determining the normal and planar anisotropy of the sheet metal.

Whether a sheet metal can be deep drawn successfully into a round cup-shaped part has been found to be a function of the **normal anisotropy**, R, of the sheet metal (also called *plastic anisotropy*). Normal anisotropy is defined in terms of the true strains that the specimen undergoes in tension (Fig. 16.33):

$$R = \frac{\text{Width strain}}{\text{Thickness strain}} = \frac{\varepsilon_w}{\varepsilon_t} \qquad (16.11)$$

In order to determine the magnitude of R, a tensile-test specimen first is prepared and subjected to an elongation of 15 to 20%. The true strains that the sheet undergoes are calculated in the manner discussed in Section 2.2. Because cold-rolled sheets generally have anisotropy in their *planar* direction, the R value of a specimen cut from a rolled sheet will depend on its orientation with respect to the rolling direction of the sheet. For this condition, an average value, R_{avg}, is calculated from the equation

$$R_{\text{avg}} = \frac{R_0 + 2R_{45} + R_{90}}{4} \qquad (16.12)$$

where the subscripts are the angles with respect to the rolling direction of the sheet. Some typical R_{avg} values are given in Table 16.4.

The experimentally determined relationship between R_{avg} and the limiting drawing ratio is shown in Fig. 16.34. It has been established that no other mechanical property of sheet metal shows as consistent a relationship to LDR as does R_{avg}.

TABLE 16.4

Typical Ranges of Average Normal Anisotropy, R_{avg}, for Various Sheet Metals	
Zinc alloys	0.4–0.6
Hot-rolled steel	0.8–1.0
Cold-rolled, rimmed steel	1.0–1.4
Cold-rolled, aluminum-killed steel	1.4–1.8
Aluminum alloys	0.6–0.8
Copper and brass	0.6–0.9
Titanium alloys (α)	3.0–5.0
Stainless steels	0.9–1.2
High-strength, low-alloy steels	0.9–1.2

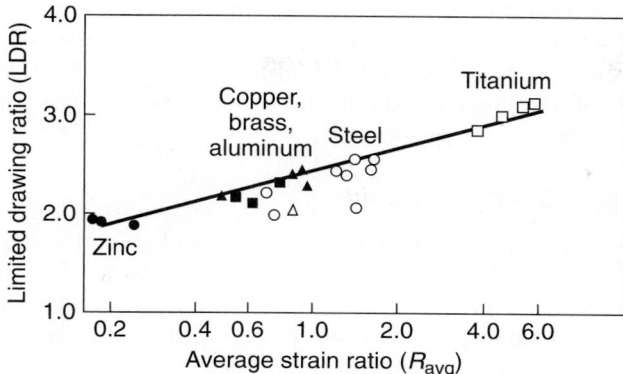

FIGURE 16.34 The relationship between average normal anisotropy and the limiting drawing ratio for various sheet metals. *Source*: After M. Atkinson.

Thus, by using a simple tensile-test result and obtaining the normal anisotropy of the sheet metal, the limiting drawing ratio of a material can be determined.

Earing. In drawing, the edges of cups may become wavy—a phenomenon called *earing* (Fig. 16.35). Ears are objectionable on drawn cups because they have to be trimmed off as they serve no useful purpose and interfere with further processing of the cup, which results in scrap. Earing is caused by the **planar anisotropy** of the sheet, and the number of ears produced may be two, four, or eight, depending on the processing history and microstructure of the sheet. If the sheet is stronger in the rolling direction than transverse to the rolling direction and the strength varies uniformly with respect to orientation, then two ears will form. If the sheet has high strength at different orientations, then more ears will form.

The planar anisotropy of the sheet is indicated by ΔR. It is defined in terms of directional R values from the equation

$$\Delta R = \frac{R_0 - 2R_{45} + R_{90}}{2} \tag{16.13}$$

When $\Delta R = 0$, no ears form. The height of the ears increases as ΔR increases.

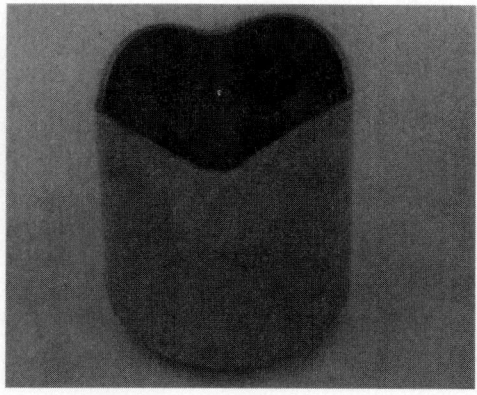

FIGURE 16.35 Earing in a drawn steel cup caused by the planar anisotropy of the sheet metal.

It can be seen that deep drawability is enhanced by a high R_{avg} value and a low ΔR. Generally, however, sheet metals with high R_{avg} also have high ΔR values. Sheet-metal textures are being developed continually to improve drawability by controlling the type of alloying elements in the material as well as various processing parameters during rolling of the sheet.

16.7.2 Deep-drawing practice

Certain guidelines have been established for successful deep-drawing practice. The blankholder pressure is chosen generally as 0.7 to 1.0% of the sum of the yield strength and the ultimate tensile strength of the sheet metal. Too high a blankholder force increases the punch force and causes the cup wall to tear. On the other hand, if the blankholder force is too low, wrinkling will occur.

Clearances are usually 7 to 14% greater than sheet thickness. If the clearance is too small, the blank may be pierced or sheared by the punch. The corner radii of the punch and of the die are also important parameters. If they are too small, they can cause fracture at the corners; if they are too large, the cup wall may wrinkle—a phenomenon called *puckering*.

Draw beads (Fig. 16.36) often are necessary to control the flow of the blank into the die cavity. Beads restrict the free flow of the sheet metal by bending and unbending it during the drawing cycle; they thereby increase the force required to pull the sheet into the die cavity. This phenomenon can be demonstrated simply by placing a strip of paper or aluminum foil through one's fingers in an arrangement similar to that shown in Fig. 13.36a. You will note that a certain force now is required to pull the strip through your fingers. Draw beads also help to reduce the required blankholder forces, because the beaded sheet has a higher stiffness (because of it higher moment of inertia) and, hence, a lower tendency to wrinkle. Draw bead diameters may range from 13 to 20 mm (0.50 to 0.75 in.)—the latter being applicable for large stampings, such as automotive panels.

Draw beads also are useful in drawing *box-shaped* and *nonsymmetric* parts, because they can present significant difficulties in practice (Fig. 16.36b and c). Note in Fig. 16.36c, for example, that various regions of the part undergo different types

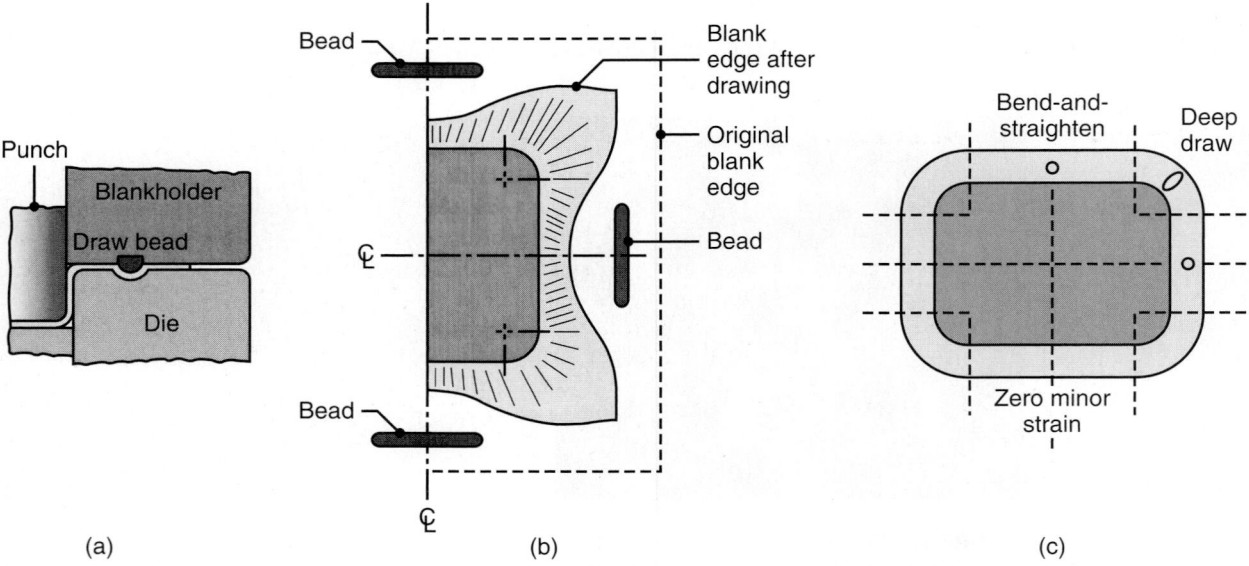

| (a) | (b) | (c) |

FIGURE 16.36   (a) Schematic illustration of a draw bead. (b) Metal flow during the drawing of a box-shaped part while using beads to control the movement of the material. (c) Deformation of circular grids in the flange in deep drawing.

of deformation during drawing. (Recall also the fundamental principle that the material flows in the direction of least resistance.)

In order to avoid tearing of the sheet metal during forming, it often is necessary to incorporate the following factors:

- Proper design and location of draw beads
- Large die radii
- Effective lubrication
- Proper blank size and shape
- The cutting off of corners of square or rectangular blanks at 45° to reduce tensile stresses developed during drawing
- The use of blanks free of internal and external defects

Ironing.   Note in Fig. 16.32 that if the clearance between the punch and the die is sufficiently large, the drawn cup will have thicker walls at its rim than at its base. The reason for this is that the cup rim consists of material from the outer diameter of the blank, hence it has been reduced in diameter more (and thus becomes thicker) than the material constituting the rest of the cup wall. As a result, the cup will develop a nonuniform wall thickness. The thickness of the cup wall can be controlled by a process called *ironing*, in which a drawn cup is pushed through one or more ironing rings (see Fig. 16.31). The clearance between the ironing rings and the punch is less than the cup wall thickness, so the cup after ironing has essentially constant wall thickness (and is equal to the clearance, except for some small elastic recovery). Aluminum beverage cans, for example, typically undergo two or three ironing operations in one stroke in which the drawn cup is pushed through a set of ironing rings.

Redrawing.   Containers that are too difficult to draw in one operation generally undergo *redrawing* (see Fig. 16.31). Because of the volume constancy of the metal, the cup becomes longer as it is redrawn to smaller diameters. In *reverse redrawing*, the cup is placed upside down in the die and thus is subjected to bending in the direction opposite to its original configuration.

Drawing without blankholder.   Deep drawing also may be carried out successfully without a blankholder, provided that the sheet metal is sufficiently thick to prevent wrinkling. A typical range of the diameter is

$$D_o - D_p < 5T \tag{16.14}$$

where T is the sheet thickness. The dies are contoured specially for this operation.

Embossing.   This is an operation consisting of shallow or moderate draws made with male and female matching shallow dies (Fig. 16.37). Embossing is used principally

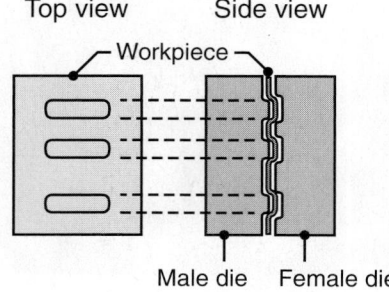

FIGURE 16.37 An embossing operation with two dies. Letters, numbers, and designs on sheet-metal parts can be produced by this process.

for the stiffening of flat sheet-metal panels and for purposes of decorating, numbering, and lettering, such as those on the lid of aluminum beverage cans.

Tooling and equipment for drawing.   The most common tool and die materials for deep drawing are tool steels and cast irons and include dies produced from ductile-iron castings made by the lost-foam process. Other materials such as carbides and plastics also may be used (see Table 5.7). Die manufacturing methods are described in detail in Section 14.7. Because of the generally axisymmetric shape of the punch and die components (such as for making cylindrical cans and containers), they can be manufactured on equipment such as high-speed machining on computer-controlled lathes.

The equipment for deep drawing is usually a *double-action hydraulic press* or a *mechanical press*, the latter generally being favored because of its higher operating speed. In the double-action hydraulic press, the punch and the blankholder are controlled independently. Punch speeds generally range between 0.1 and 0.3 m/s (20 and 60 ft/min).

EXAMPLE 16.4 Manufacturing of food and beverage cans

Can manufacturing is a major and competitive industry worldwide with approximately 100-billion beverage cans and 30-billion food cans produced each year in the United States alone. These containers are strong and lightweight (typically weighing less than 0.5 oz), and they are under an internal pressure of 90 psi—reliably and without leakage of its contents. There are stringent requirements for the surface finish of the can, since brightly decorated and shiny cans are preferred over dull-looking containers. Considering all of these features, metal cans are very inexpensive. Can makers charge approximately $40 per 1000 cans or about 4 cents per can. Thus, the cost of empty cans alone in a six-pack is 24 cents, which also indicates the importance of recycling cans.

Food and beverage cans may be produced in a number of ways, the most common ones being two-piece and three-piece cans. Two-piece cans consist of the can body and the lid (Fig. 16.38a). The body is made of one piece which has been drawn and ironed, hence the industry practice of referring to this style as D&I (drawn and ironed) cans. Three-piece cans are produced by attaching a lid and a bottom to a sheet-metal cylindrical body.

(a)

FIGURE 16.38   (a) Aluminum beverage cans. Note the excellent surface finish.

Scored region   Integral rivet   Pop-top cantilever

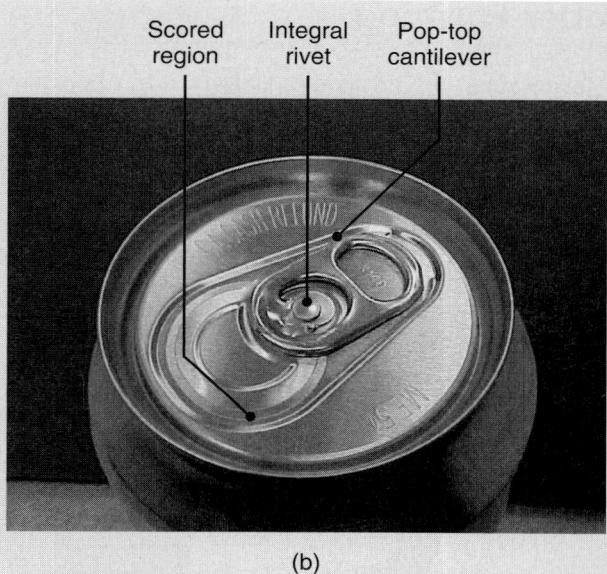

(b)

FIGURE 16.38 (*Continued*) (b) Detail of the can lid showing integral rivet and scored edges for the pop-top.

Drawn and ironed can bodies are produced from a number of alloys, but the most common are 3004-H19 aluminum (see Section 6.2) and electrolytic tin-plated ASTM A623 steel. Aluminum lids are used for both steel and aluminum cans and are produced from 5182-H19 or 5182-H48. The lid presents a demanding set of design requirements, as can be appreciated by reviewing Fig. 16.38b. Not only must the can lid be scored easily (curved grooves around the tab), but an integral rivet is formed and headed in the lid to hold the tab in place. Aluminum alloy 5182 has the unique characteristics of having sufficient formability to enable forming of the integral rivet without cracking and has the ability to be scored. The lids basically are stamped from 5182 aluminum sheet, the pop-top is scored, and a plastic seal is placed around the periphery of the lid. This polymer layer seals the can's contents after the lid is seamed to the can body, as described next.

The traditional method of manufacturing the can bodies is shown in Fig. 16.31. The process starts with 5.5-in. diameter blanks produced from rolled sheet stock. These blanks are (a) deep drawn to a diameter of around 3.5 in., (b) redrawn to the final diameter of around 2.6 in., (c) ironed through two or three ironing rings in one pass, and (d) domed for the can bottom. The deep-drawing and ironing operations are performed in a special type of press that typically produces cans at speeds over 400 strokes per minute. Following this series of operations, a number of additional processes take place.

Necking of the can body is performed through either spinning (Section 16.9) or by die necking (a forming operation similar to that shown in Fig. 15.19a, where a thin-walled tubular part is pushed into the die), and it is then spin-flanged. The reason for necking the can top is that the 5182 aluminum used for the lid is relatively expensive. Thus, by tapering the top of the can, a smaller volume of material is needed, thereby reducing the cost. Also, it should be noted that the cost of a can often is calculated to millionths of a dollar, hence any design feature that reduces its cost will be exploited by this competitive industry.

Source: Courtesy of J. E. Wang, Texas A&M University.

16.8 | Rubber Forming

In the processes described in the preceding sections, it has been noted that the dies generally are made of solid materials, such as steels and carbides. However, in *rubber forming* (also known as the *Guerin process*), one of the dies in a set is made of a flexible material, typically a polyurethane membrane. Polyurethanes are used widely because of their

- Resistance to abrasion
- Resistance to cutting or tearing by burrs or other sharp edges on the sheet metal
- Long fatigue life

In the bending and embossing of sheet metal by this process, the female die is replaced with a rubber pad (Fig. 16.39). Note that the outer surface of the sheet is

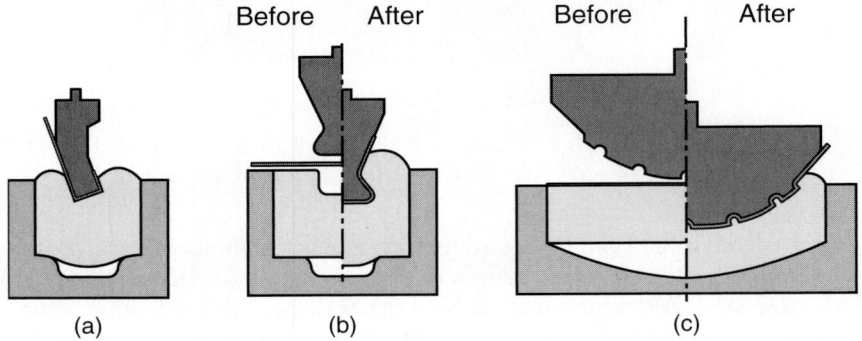

FIGURE 16.39   Examples of the bending and the embossing of sheet metal with a metal punch and with a flexible pad serving as the female die. *Source*: Courtesy of Polyurethane Products Corporation.

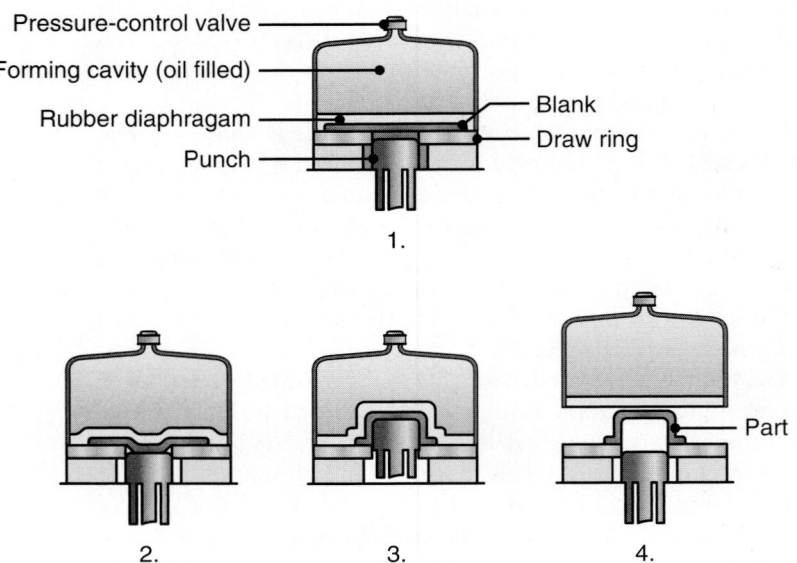

FIGURE 16.40   The hydroform (or fluid-forming) process. Note that in contrast to the ordinary deep-drawing process, the pressure in the dome forces the cup walls against the punch. The cup travels with the punch; in this way, deep drawability is improved.

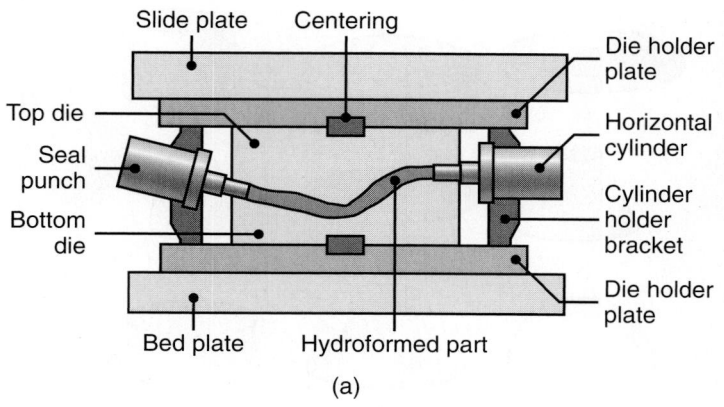

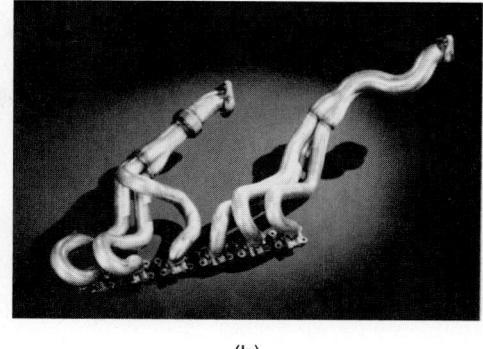

FIGURE 16.41 (a) Schematic illustration of the tube-hydroforming process. (b) Example of tube-hydroformed parts. Automotive-exhaust and structural components, bicycle frames, and hydraulic and pneumatic fittings are produced through tube hydroforming. *Source:* Courtesy of Schuler GmBH.

protected from damage or scratches, because it is not in contact with a hard metal surface during forming. Pressures in rubber forming are typically on the order of 10 MPa (1500 psi).

In the **hydroform** or *fluid-forming process* (Fig. 16.40), the pressure over the rubber membrane is controlled throughout the forming cycle with a maximum pressure of up to 100 MPa (15,000 psi). This procedure allows close control of the part during forming and prevents wrinkling or tearing. Deeper draws are obtained than in conventional deep drawing, because the pressure around the rubber membrane forces the cup against the punch. As a result, the friction at the punch–cup interface increases, which then reduces the longitudinal tensile stresses in the cup and, thus, delays fracture.

The control of frictional conditions in rubber forming, as well as in other sheet-forming operations, can be a critical factor in making parts successfully. The use of proper lubricants and their application methods is also important. In *tube hydroforming* (Fig. 16.41), metal tubing is formed in a die and pressurized internally by a fluid. This process, which now is being applied more widely, can form simple tubes as well as various intricate hollow shapes (Fig. 16.41b). Parts made by this process include automotive-exhaust and tubular structural components.

When selected properly, rubber-forming and hydroforming processes have the advantages of (a) the capability to form complex shapes, (b) forming parts with laminated sheets of various materials and coatings, (c) flexibility and ease of operation, (d) the avoidance of damage to the surfaces of the sheet, (e) low die wear, and (f) low tooling cost.

16.9 | Spinning

Spinning is a process which involves the forming of axisymmetric parts over a mandrel by the use of various tools and rollers—a process that is similar to that of forming clay on a potter's wheel.

Conventional spinning. In *conventional spinning*, a circular blank of flat or preformed sheet metal is placed and held against a mandrel and rotated while a rigid tool deforms and shapes the material over the mandrel (Fig. 16.42a). The tool may be activated either manually or (for higher production rates) by computer-controlled

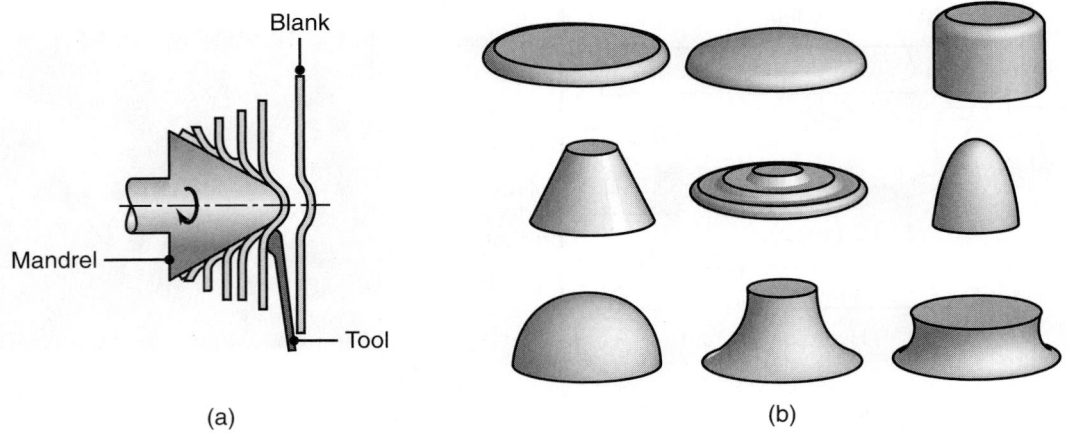

FIGURE 16.42 (a) Schematic illustration of the conventional spinning process. (b) Types of parts conventionally spun. All parts are axisymmetric.

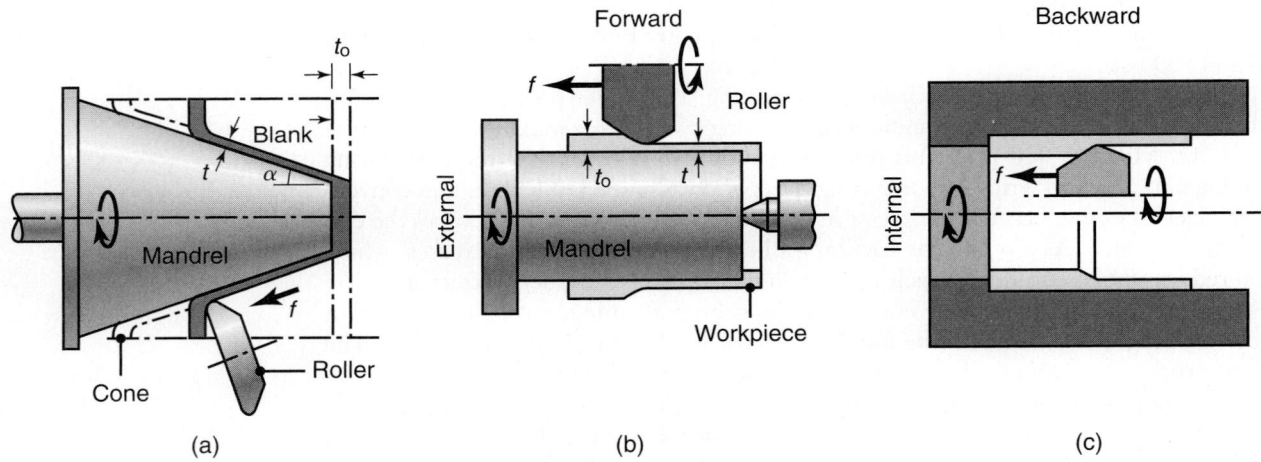

FIGURE 16.43 (a) Schematic illustration of the shear-spinning process for making conical parts. The mandrel can be shaped so that curvilinear parts can be spun. (b) and (c) Schematic illustrations of the tube-spinning process.

mechanisms. The process typically involves a sequence of passes, and it requires considerable skill. Conventional spinning is suitable particularly for conical and curvilinear shapes (Fig. 16.42b), which otherwise would be difficult or uneconomical to produce. Part diameters may range up to 6 m (20 ft). Although most spinning is performed at room temperature, thick parts and metals with high strength or low ductility require spinning at elevated temperatures.

Shear spinning. Also known as *power spinning*, *flow turning*, *hydrospinning*, and *spin forging*, this operation produces an axisymmetric conical or curvilinear shape, reducing the sheet's thickness while maintaining its maximum (blank) diameter (Fig. 16.43a). A single forming roller can be used, but two rollers are preferable in order to balance the forces acting on the mandrel. Typical parts made are rocket motor casings and missile nose cones. Parts up to 3 m (10 ft) in diameter can be formed by shear spinning. This operation wastes little material, and it can be completed in a relatively short time, some within as little as a few seconds. Various shapes can be spun with fairly simple tooling, which generally is made of tool steel.

The *spinnability* of a metal in this process generally is defined as the maximum reduction in thickness to which a part can be subjected by spinning without fracture. Spinnability is found to be related to the tensile reduction of area of the material, just as is bendability (see Fig. 16.18.) Thus, if a metal has a tensile reduction of area of 50% or higher, its thickness can be reduced by as much as 80% in just one spinning pass. For metals with low ductility, the operation is carried out at elevated temperatures by heating the blank in a furnace and transferring it rapidly to the mandrel.

Tube spinning.   In *tube spinning*, the thickness of hollow, cylindrical blanks is reduced or shaped by spinning them on a solid, round mandrel using rollers (Fig. 16.43). The reduction in wall thickness results in a longer tube. This operation may be carried out externally or internally, thus various external and internal profiles can be produced from cylindrical blanks with constant wall thickness. The parts may be spun *forward* or *backward*; this nomenclature is similar to that of direct and indirect extrusion, as described in Section 15.2. The maximum thickness reduction per pass in tube spinning is related to the tensile reduction of area of the material, as it is in shear spinning. Tube spinning can be used to make rocket, missile, and jet engine parts, pressure vessels, and automotive components, such as car and truck wheels.

16.10 | Superplastic Forming

The superplastic behavior of certain metals and alloys was described in Section 2.2.7, where tensile elongations on the order of 2000% were obtained within certain temperature ranges. Common examples of such materials are zinc-aluminum and titanium alloys, which have very fine grains—typically less than 10 to 15 μm (see Table 1.1). Superplastic alloys can be formed into complex shapes by *superplastic forming*—a process that employs common metalworking techniques—as well as by polymer-processing techniques (such as thermoforming, vacuum forming, and blow molding, to be described in Chapter 19). The behavior of the material in superplastic forming is similar to that of bubble gum or hot glass, which when blown expands many times its original diameter before it bursts.

Superplastic alloys, particularly Zn-22Al and Ti-6Al-4V, also can be formed by bulk deformation processes, including closed-die forging, coining, hubbing, and extrusion. Commonly used die materials in superplastic forming are low-alloy steels, cast tool steels, ceramics, graphite, and plaster of paris. Selection depends on the forming temperature and the strength of the superplastic alloy.

The very high ductility and relatively low strength of superplastic alloys offer the following advantages:

- Complex shapes can be formed out of one piece, with fine detail, close tolerances, and elimination of secondary operations.

- Weight and material savings can be realized because of the good formability of the materials.

- Little or no residual stresses develop in the formed parts.

- Because of the low strength of the material at forming temperatures, the tooling can be made of materials that have lower strength than those in other metalworking processes, hence tooling costs are lower.

Superplastic forming has, on the other hand, the following limitations:

- The material must not be superplastic at service temperatures, as otherwise the part will undergo shape changes.

- Because of the high strain-rate sensitivity of the superplastic material, it must be formed at sufficiently low strain rates, typically 10^{-4} to 10^{-2}/s. Forming times range anywhere from a few seconds to several hours, thus cycle times are much longer than those of conventional forming processes. Consequently, superplastic forming is a batch-forming process.

Diffusion bonding/superplastic forming.   Fabricating complex sheet-metal structures by combining *diffusion bonding* with *superplastic forming* (SPF/DB) is an important process, particularly in the aerospace industry. Typical structures made are shown in Fig. 16.44, in which flat sheets are diffusion bonded (see Section 31.7) and formed. In this process, selected locations of the sheets are first diffusion bonded while the rest remains unbonded, using a layer of material (*stop-off*) to prevent bonding. The structure then is expanded in a mold, typically by using pressurized neutral (argon) gas, thus taking the shape of the mold. These structures have high stiffness-to-weight ratios because they are thin, and by design, they have high section modulus. This important feature makes this process particularly attractive in aircraft and aerospace applications.

The SPF/DB process improves productivity by eliminating mechanical fasteners, and it produces parts with good dimensional accuracy and low residual stresses. The technology is well advanced for titanium structures for aerospace applications. In addition to various aluminum alloys being developed using this technique, other metals for superplastic forming include various nickel alloys.

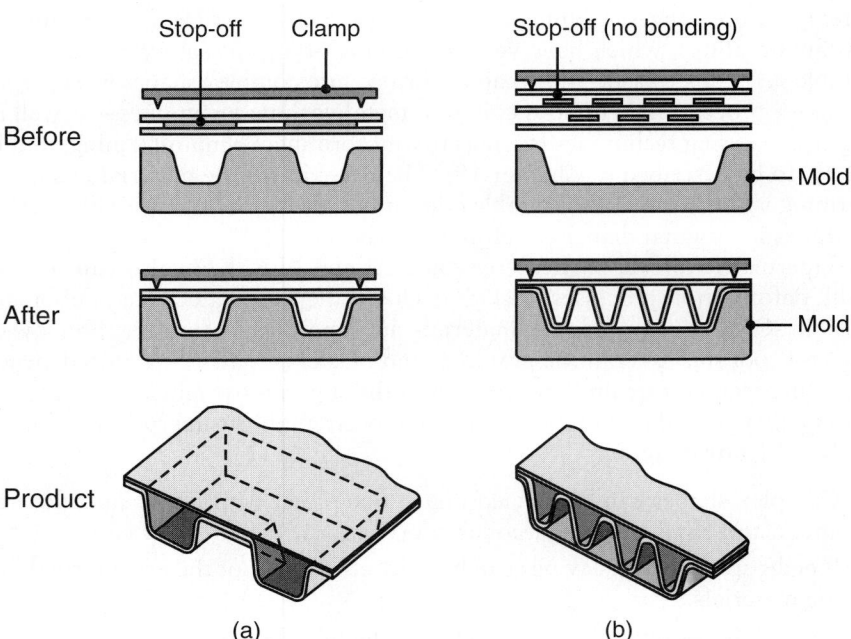

FIGURE 16.44   Types of structures made by diffusion bonding and superplastic forming of sheet metals. Such structures have a high stiffness-to-weight ratio. *Source:* Courtesy of Rockwell International Corp.

EXAMPLE 16.5 Applications of superplastic forming/diffusion bonding

The majority of applications for SPF/DB involve titanium parts for military aircraft, such as the Toronado and the Mirage 2000. The components made include fuselage bulkheads, leading-edge slats, heat-exchanger ducts, and cooler-outlet ducts. The nozzle fairing of the F-15 fighter aircraft also is made by this process. In civilian applications, the Airbus A340 has its water-closet, drain, and fresh-water maintenance panels (made of Ti-6Al-4V) manufactured by this process.

The superplastic forming process usually is carried out at about 900°C (1650°F) for titanium alloys and at about 500°C (930°F) for aluminum alloys; temperatures for diffusion bonding are similar. However, the presence of an oxide layer on aluminum sheets is a significant problem that degrades the bond strength in diffusion bonding. To illustrate cycle times, 718 nickel-alloy sheets 2 mm (0.080 in.) in thickness were, in one application, superplastically formed in ceramic dies at 950°C (1740°F) using argon gas at a pressure of 2 MPa (300 psi). The cycle time was four hours.

16.11 | Specialized Forming Processes

Although not as commonly used as the other processes covered thus far, several other sheet-forming processes that are used for specialized applications are described in this section.

Explosive forming. Explosives generally are used for demolition in construction, road building, and for many destructive purposes. However, by controlling their quantity and shape, explosives can be used as a source of energy for sheet-metal forming. First utilized to form metals in the early 1900s, in *explosive forming*, the sheet-metal blank is clamped over a die, and the entire assembly is lowered into a tank filled with water (Fig. 16.45a). The air in the die cavity then is evacuated, an explosive charge is placed at a certain height, and the charge is detonated.

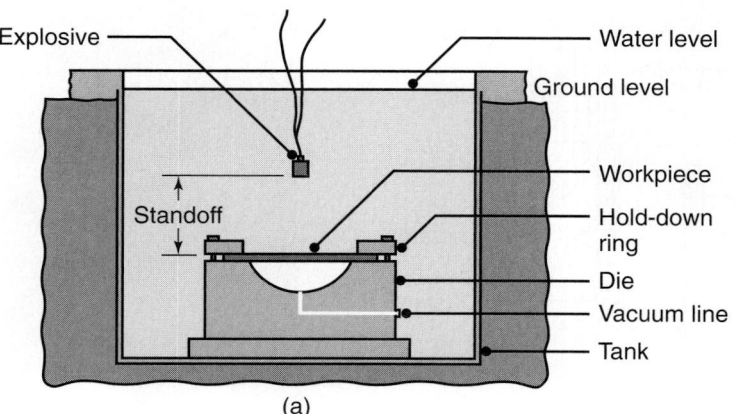
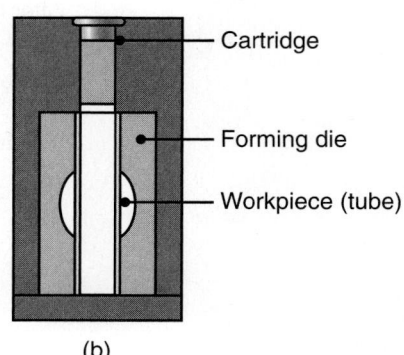

(a)   (b)

FIGURE 16.45 (a) Schematic illustration of the explosive forming process. (b) Illustration of the confined method of the explosive bulging of tubes.

The explosive generates a shock wave with a pressure which is sufficient to form sheet metals. The *peak pressure*, *p*, generated in water is given by the expression

$$p = K\left(\frac{\sqrt[3]{W}}{R}\right)^a \tag{16.15}$$

where *p* is in psi, *K* is a constant which depends on the type of explosive, such as 21,600 for TNT (trinitrotoluene), *W* is the weight of the explosive in pounds, *R* is the distance of the explosive from the sheet-metal surface (called the *standoff*) in feet, and *a* is a constant, generally taken as 1.15.

A variety of shapes can be formed through explosive forming, provided that the material is ductile at the high rates of deformation characteristic of this process (see Table 2.4). This process is versatile, as there is virtually no limit to the size of the sheet or plate. It is suitable particularly for low-quantity production runs of large parts, such as those used in aerospace applications. Steel plates 25 mm (1 in.) thick, and 3.6 m (12 ft) in diameter have been formed by this method. Tubes with wall thicknesses as much as 25 mm (1 in.) have been bulged by this process. The explosive-forming method also can be used at a much smaller scale, as shown in Fig. 16.45b. In this case, a *cartridge* (canned explosive) is used as the source of energy. The process can be useful in the bulging and expanding of thin-walled tubes for specialized applications.

The mechanical properties of parts made by explosive forming are basically similar to those of others made by conventional forming methods. Depending on the number of parts to be produced, dies may be made of aluminum alloys, steel, ductile iron, zinc alloys, reinforced concrete, wood, plastics, or composite materials.

Magnetic-pulse forming.   In *magnetic-pulse forming* or *electromagnetic forming*, the energy stored in a capacitor bank is discharged rapidly through a magnetic coil. In a typical example, a ring-shaped coil is placed over a tubular workpiece. The tube then is collapsed by magnetic forces over a solid piece, thus making the assembly an integral part (Fig. 16.46).

The mechanics of this process is based on the fact that a magnetic field produced by the coil (Fig. 16.46a) crosses the metal tube (which is an electrical conductor) and generates *eddy currents* in the tube. In turn, these currents produce their

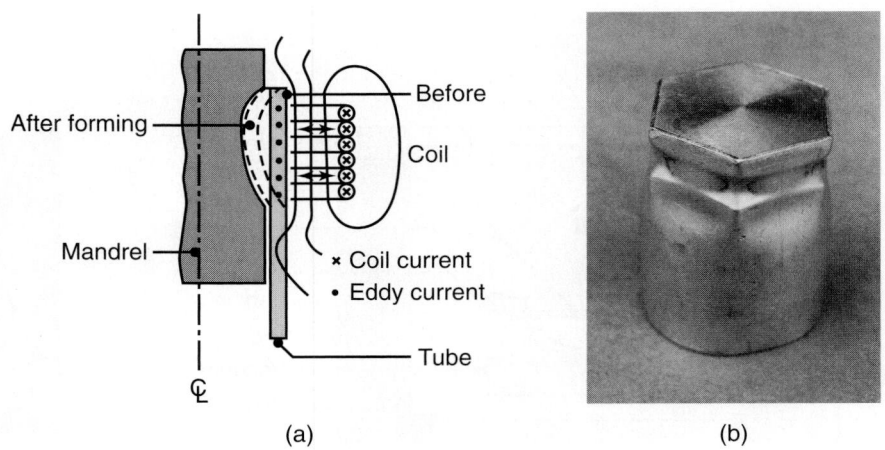

(a)   (b)

FIGURE 16.46   (a) Schematic illustration of the magnetic-pulse forming process used to form a tube over a plug. (b) Aluminum tube collapsed over a hexagonal plug by the magnetic-pulse forming process.

own magnetic field. The forces produced by the two magnetic fields oppose each other. The repelling force generated between the coil and the tube then collapses the tube over the inner piece. The higher the electrical conductivity of the workpiece, the higher the magnetic forces. It is not necessary for the workpiece material to have magnetic properties.

Magnetic-pulse forming is used for collapsing thin-walled tubes over rods, cables, and plugs; for compression-crimp sealing of automotive oil filter canisters; for bulging and flaring operations; and for swaging end fittings onto torque tubes for the Boeing 777 aircraft. Flat, magnetic coils also can be made for operations such as embossing and the shallow drawing of sheet metals.

Peen forming. Peen forming is used to produce curvatures on thin sheet metals by *shot peening* (see Section 34.2) one surface of the sheet. As a result, the surface of the sheet is subjected to compressive stresses, which tend to expand the surface layer. Because the material below the peened surface remains rigid, the surface expansion causes the sheet to develop a curvature. The process also induces compressive surface residual stresses, which improve the fatigue strength of the sheet.

Peening is done with cast-iron or steel shot discharged either from a rotating wheel or by an air blast from a nozzle. Peen forming is used by the aircraft industry to generate smooth and complex curvatures on aircraft wing skins. Cast-steel shot about 2.5 mm (0.1 in.) in diameter, traveling at speeds of 60 m/s (200 ft/s), have been used to form wing panels 25 m (80 ft) long. For heavy sections, shot diameters as large as 6 mm (1/4 in.) may be used. The peen-forming process also is used for straightening twisted or bent parts, including out-of-round rings to make them round.

Laser forming. This process involves the application of laser beams as a heat source in specific regions of the sheet metal. The steep thermal gradients developed through the thickness of the sheet causes thermal stresses, which are sufficiently high to cause localized plastic deformation of the sheet. Using this method, for example, a sheet can be bent permanently without using dies. In **laser-assisted forming**, the laser acts as a localized heat source, thus reducing the strength of the sheet metal at specific locations, improving formability, and increasing process flexibility. Applications include straightening, bending, embossing, and forming of complex tubular or flat components.

Microforming. This is a more recent development and describes a family of processes that are used to produce very small metallic parts and components. Examples of *minitaturized products* include a wristwatch with an integrated digital camera and one gigabyte of a computer storage component. Typical components made by microforming include small shafts for micromotors, springs, screws, and a variety of cold-headed, extruded, bent, embossed, coined, punched or deep-drawn parts. Dimensions are typically in the submillimeter range, and weights are on the order of milligrams.

Electrohydraulic forming. Also called *underwater spark* or *electric-discharge forming*, the source of energy in this forming process is a spark between electrodes that are connected with a short, thin wire. The rapid discharge of the energy from a capacitor bank through the wire generates a shock wave, similar to those created by explosives. The energy levels are lower than those in explosive forming, being typically a few kJ. The pressure developed in the water medium is sufficiently high to form the part. Electrohydraulic forming is a batch process and can be used in making various small parts.

Gas mixtures. As an energy source in this process, a gas mixture in a closed container is ignited. The pressure generated is sufficiently high to form sheet-metal parts. Although not used in practice, the principle of this process is similar to that used for the generation of pressure in an internal combustion engine.

Liquefied gases. Liquefied gases (such as liquid nitrogen) also have been used to develop pressures sufficiently high to form sheet metals. When allowed to reach room temperature in a closed container, liquefied nitrogen becomes gaseous and expands, developing the necessary pressure. Although not used in practice, the process is capable of forming relatively shallow parts.

CASE STUDY 16.1 | Cymbal Manufacture

Cymbals (Fig. 16.47a) are an essential percussion instrument for all forms of music. Modern drum-set cymbals cover a wide variety of sounds—from deep, dark, and warm, to bright, high-pitched, and cutting. Some cymbals sound musical, while some are "trashy." A wide variety of sizes, shapes, weights, hammerings, and surface finishes (Fig. 16.47b) are available to achieve the desired performance.

Cymbals are produced from metals—such as B20 bronze (80% Cu-20% Sn with a trace of silver), B8 bronze (92% Cu-8% Sn), nickel-silver alloy, and brass—using various methods of processing. The manufacturing sequence for producing B20 bronze cymbals is shown in Fig. 16.48. The metal first is cast into mushroom-shaped ingots and then cooled by ambient temperature. It then is rolled successively (up to 14 times) with water cooling the metal with each pass through the rolling mill. Special care is taken to roll the bronze at a different angle with each pass to minimize anisotropy, impart preferred grain orientation, and develop an even, round shape. The as-rolled blanks then are reheated and stretch formed (pressed) into the cup or bell shape that determines the cymbal's overtones. The cymbals then are center-drilled or punched to create hang-holes and trimmed on a rotary shear to approximate final diameters. This operation is followed by another stretch-forming step to achieve the characteristic "Turkish dish" form that controls pitch.

(a)

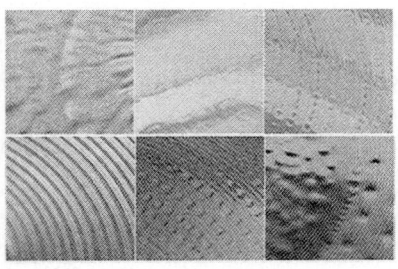

(b)

FIGURE 16.47   (a) A selection of common cymbals. (b) Detailed view of different surface textures and finishes of cymbals. *Source:* Courtesy of W. Blanchard, Sabian Ltd.

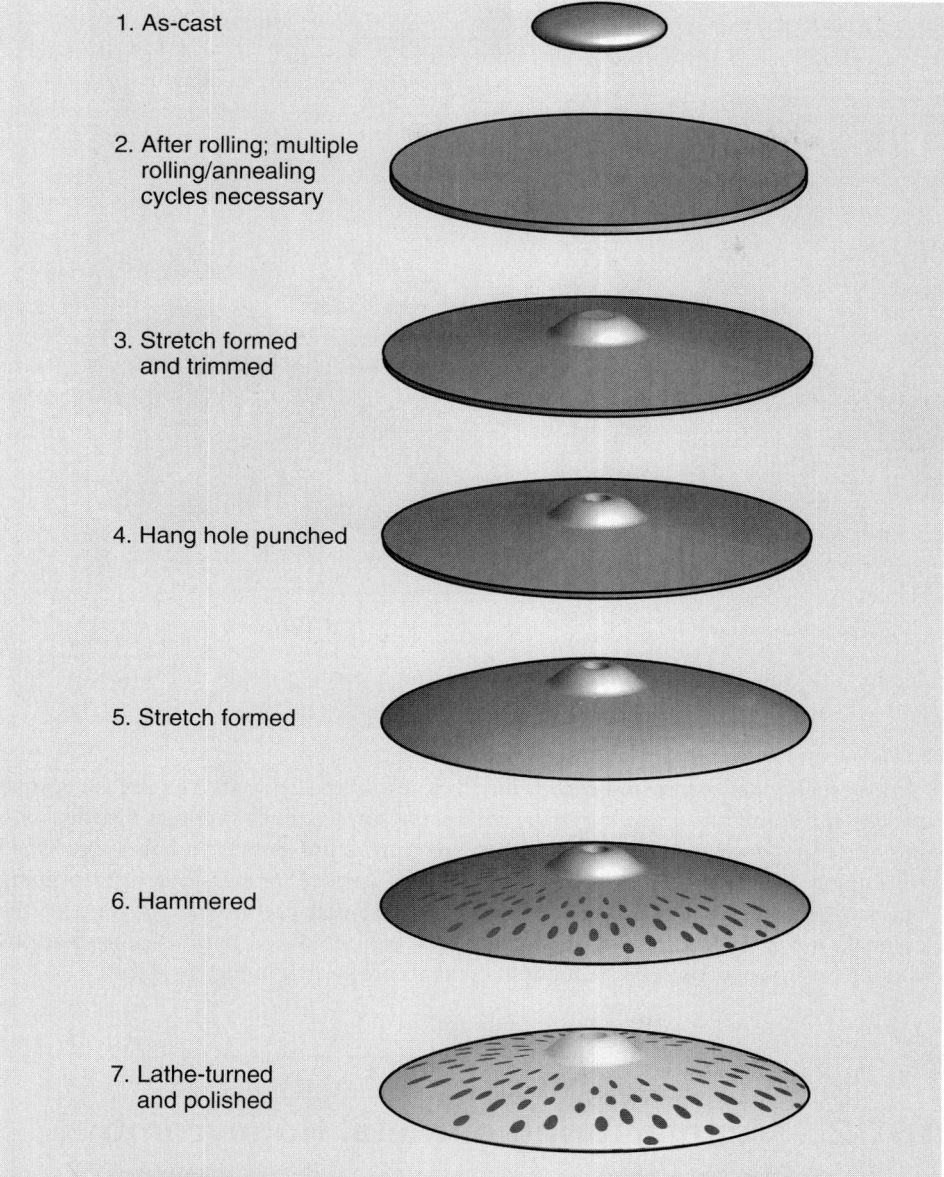

1. As-cast

2. After rolling; multiple
   rolling/annealing
   cycles necessary

3. Stretch formed
   and trimmed

4. Hang hole punched

5. Stretch formed

6. Hammered

7. Lathe-turned
   and polished

FIGURE 16.48   Manufacturing sequence for the production of cymbals. *Source:* Courtesy of W. Blanchard, Sabian Ltd.

The cymbals then are hammered to impart a distinctive character to each instrument. Hammering can be done by hand (see Fig. 16.49b) or in automatic peen-forming machines (Fig. 16.49a). Hand hammering involves placing the bronze blank on a steel anvil, where the cymbals then are struck manually by hand hammers. Automatic peen-forming is done on machinery without templates, since the cymbals have already been pressed into shape, but the pattern is controllable and uniform. The size and pattern of the peening operations depends on the desired response, such as tone, sound, response, and pitch of the cymbal.

A number of finishing operations are performed on the cymbals. This merely can involve cleaning and printing of identifying information, as some musicians prefer the natural surface appearance and sound of formed, hot-rolled bronze. More commonly, the cymbals are lathe-turned (without any machining fluid) in order to

(a)

(b)

FIGURE 16.49   Hammering of cymbals. (a) Automated hammering on a peening machine; (b) hand hammering of cymbals. *Source:* Courtesy of W. Blanchard, Sabian Ltd.

remove the oxide surface and reduce the thickness of the cymbal to create the desired weight and sound. As a result of this process, the surface finish becomes lustrous and, in some cases, develops a favorable surface microstructure. Some cymbals are polished to a glossy "brilliant finish." In many cases, the surface indentations from peening persist after finishing; this is recognized as an essential performance feature of the cymbal, and it is also an aesthetic feature that is appreciated by musicians. Various surface finishes associated with modern cymbals are shown in Fig. 16.47b.

Source: Courtesy of W. Blanchard, Sabian Ltd.

16.12   Manufacturing of Metal Honeycomb Structures

A *honeycomb structure* consists basically of a core of honeycomb or other corrugated shapes bonded to two thin outer skins (Fig. 16.50). The most common example of such a structure is corrugated cardboard, which has a high stiffness-to-weight ratio and is used extensively in packaging for shipping consumer and industrial goods. Because of their light weight and high resistance to bending forces, metal honeycomb structures are used for aircraft and aerospace components, in buildings, and in transportation equipment. Metal honeycomb manufacturing methods are described in this section because they involve operations that are best classified under sheet-metal forming processes. It should be noted, however, that honeycomb structures also may be made of nonmetallic materials, such as cardboard, polymers, and a variety of composite materials.

Honeycomb structures are made most commonly of 3000-series aluminum. However, they also are made of titanium, stainless steels, and nickel alloys for

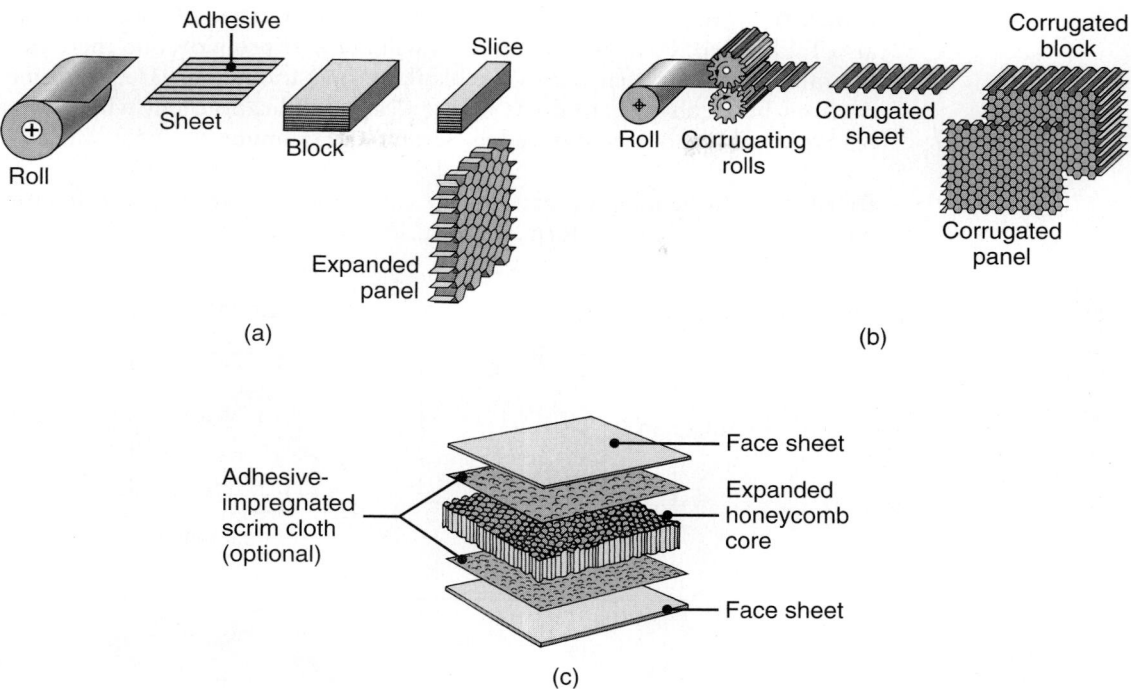

FIGURE 16.50   Methods of manufacturing honeycomb structures: (a) expansion process; (b) corrugation process; (c) assembling a honeycomb structure into a laminate.

specialized applications and corrosion resistance. More recent developments include making honeycomb structures using reinforced plastics, such as aramid-epoxy.

There are two basic methods of manufacturing honeycomb materials. In the **expansion process**, which is the more common method (Fig. 16.50a), sheets first are cut from a coil, and an *adhesive* (see Section 32.4) is applied at intervals (node lines) on their surfaces. The sheets then are stacked and cured in an oven, developing strong bonds at the adhesive joints. The block finally is cut into slices of the desired dimension and stretched to produce a honeycomb structure.

The **corrugation process** (Fig. 16.50b) is similar to the process used in making corrugated cardboard. The sheet metal first passes through a pair of specially designed rolls, becoming a corrugated sheet; it is then cut into desired lengths. Adhesive is applied to the node lines, the corrugated sheets are assembled into a block, and the block then is cured. Because the sheets are preformed already, no expansion process is involved. The honeycomb material finally is made into a sandwich structure (Fig. 16.50c) using face sheets which are joined by adhesives (or *brazed*, see Section 32.2) to the top and bottom surfaces.

16.13   Design Considerations in Sheet-Metal Forming

As with most other processes described throughout this book, certain design guidelines and practices have evolved with time. Careful design using the best established design practices, computational tools, and manufacturing techniques is the best approach to achieving high-quality designs and realizing cost savings. The following guidelines apply to sheet-metal forming operations, with the most significant design issues identified.

Blank design. Material scrap is the primary concern in blanking operations (see also Table 40.6). Poorly designed parts will not *nest* properly, and there can be considerable scrap between successive blanking operations (Fig. 16.51). Some restrictions on blank shapes are made by the design application, but whenever possible, blanks should be designed to reduce scrap to a minimum.

Bending. In bending operations, the main concerns are material fracture, wrinkling, and the inability to form the bend. As shown in Fig. 16.52, a sheet-metal part

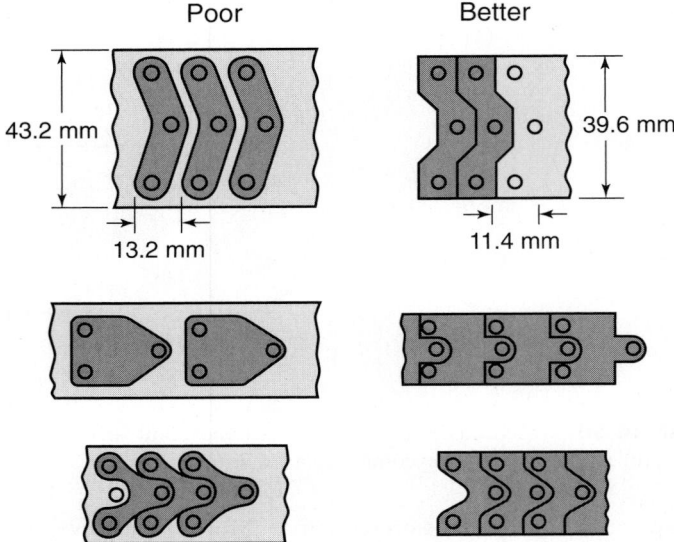

FIGURE 16.51 Efficient nesting of parts for optimum material utilization in blanking. *Source:* Courtesy of Society of Manufacturing Engineers.

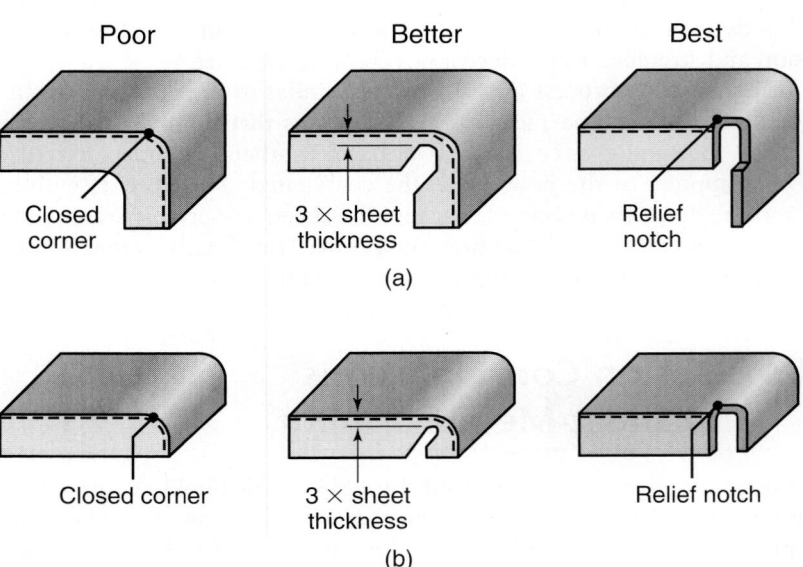

FIGURE 16.52 Control of tearing and buckling of a flange in a right angle bend. *Source:* Courtesy of Society of Manufacturing Engineers.

with a flange that is to be bent will force the flange to undergo compression, which can cause buckling (see also *flanging*, Section 16.6). This problem can be controlled with a relief notch cut to limit the stresses from bending, or else a design modification as shown in the figure can be made to eliminate the problem. Right-angle bends have similar difficulties, and relief notches also can be used to avoid tearing (Fig. 16.53).

Because the bend radius is a highly stressed area, all stress concentrations should be removed from the bend-radius location. An example is parts with holes near bends. It is advantageous to move the hole away from the bend area, but when this is not possible, a crescent slot or ear can be used (Fig. 16.54a). Similarly, when bending flanges, tabs and notches should be avoided, since these stress concentrations will greatly reduce the formability. When tabs are necessary, large radii should be used to reduce stress concentration (Fig. 16.54b).

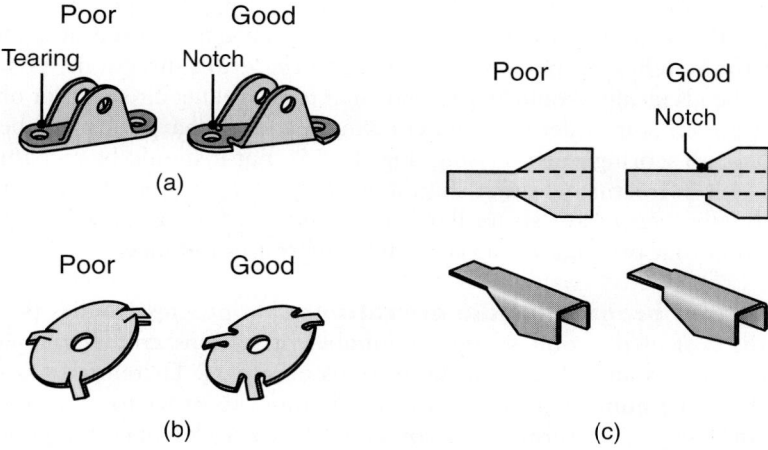

FIGURE 16.53   Application of notches to avoid tearing and wrinkling in right-angle bending operations. *Source:* Courtesy of Society of Manufacturing Engineers.

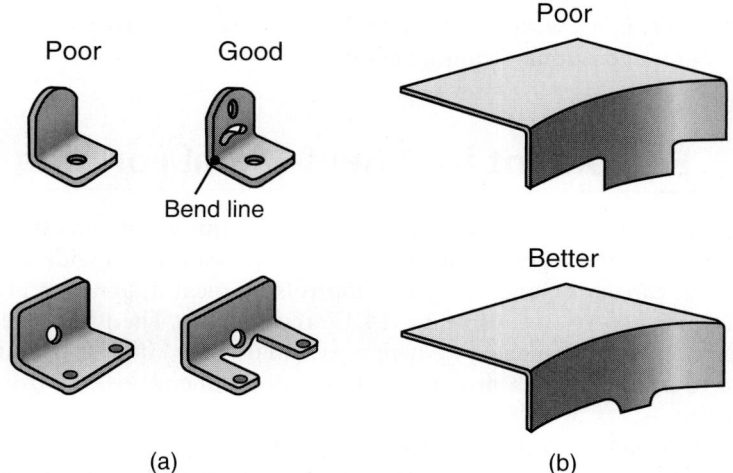

FIGURE 16.54   Stress concentrations near bends. (a) Use of a crescent or ear for a hole near a bend. (b) Reduction of severity of tab in flange. *Source:* Courtesy of Society of Manufacturing Engineers.

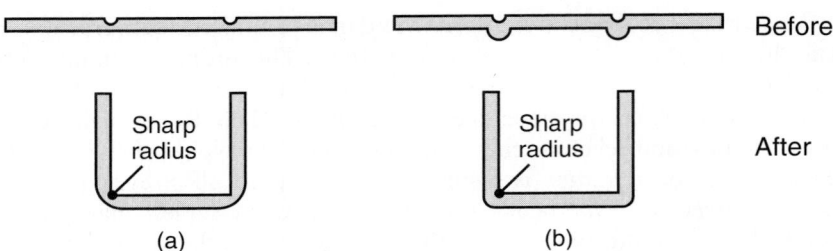

FIGURE 16.55 Application of (a) scoring or (b) embossing to obtain a sharp inner radius in bending. Unless properly designed, these features can lead to fracture. *Source:* Courtesy of Society of Manufacturing Engineers.

When the bending and production of notches are to be used, it is important to orient the notches properly with respect to the grain direction. As shown in Fig. 16.17, bends ideally should be perpendicular to the rolling direction (or oblique, if this is not possible) in order to avoid cracking. Bending sharp radii can be accomplished through scoring or embossing (Fig. 16.55), but it should be recognized that this can result in fracturing. Burrs should not be present in a bend allowance, since they are brittle (because of strain hardening) and can fracture, leading to a stress concentration that propagates the crack into the rest of the sheet.

Stamping and progressive-die operations. In progressive dies (see Section 16.2.3), the cost of the tooling and the number of stations are determined by the number of features and spacing of the features on a part. Therefore, it is advantageous to hold the number of features to a minimum in order to minimize tooling costs. Closely spaced features may provide insufficient clearance for punches and may require two punches. Narrow cuts and protrusions also are problematic for forming with a single punch and die.

Deep drawing. After a deep-drawing operation, a cup invariably will spring back towards its original shape. For this reason, designs that use a vertical wall in a deep-drawn cup may be difficult to form. Relief angles of at least 3° on each wall are easier to produce. Cups with sharp internal radii are difficult to produce, and deep cups often require additional ironing operations.

16.14 | Equipment for Sheet-Metal Forming

For most general pressworking operations, the basic equipment consists of mechanical, hydraulic, pneumatic, or pneumatic-hydraulic presses with a wide variety of designs, features, capacities, and computer controls. Typical designs for press frames are shown in Fig. 16.56 (see also Figs. 14.17 and 16.23f). The proper design, stiffness, and construction of such equipment is essential to the efficient operation of the system and to achieve a high production rate, good dimensional control, and high product quality.

The traditional **C-frame** structure (Fig. 16.56a) has been used widely for ease of tool and workpiece accessibility, but it is not as stiff as the **box-type pillar** (Fig. 16.56e) or the **double-column frame** structure (Fig. 16.56f). Furthermore, accessibility has become less important due to advances in automation and in the use of industrial robots and computer controls.

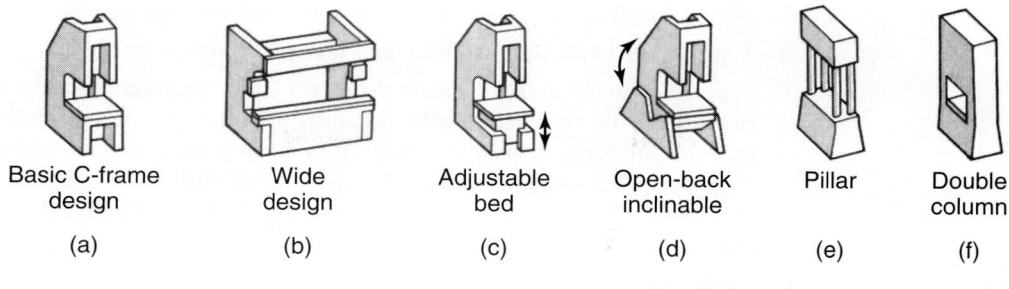

Basic C-frame design (a) Wide design (b) Adjustable bed (c) Open-back inclinable (d) Pillar (e) Double column (f)

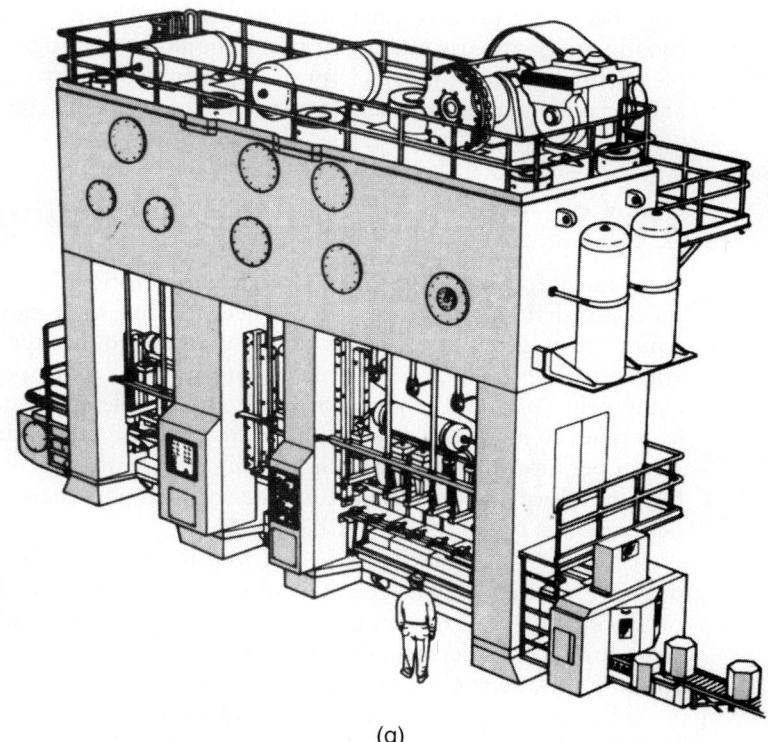

(g)

FIGURE 16.56 (a) through (f) Schematic illustrations of types of press frames for sheet-forming operations. Each type has its own characteristics of stiffness, capacity, and accessibility. (g) A large stamping press. *Source:* (a) through (f) *Engineer's Handbook*, VEB Fachbuchverlag, 1965. (g) Verson Allsteel Company.

Press selection for sheet-metal forming operations depends on several factors:

1. Type of forming operation, the size and shape of the dies, and the tooling required.
2. Size and shape of workpieces.
3. Length of stroke of the slide (slides), the number of strokes per minute, the operating speed, and the *shut height* (the distance from the top of the bed to the bottom of the slide with the stroke down).
4. Number of slides. Single-action presses have one reciprocating slide. Double-action presses have two slides, reciprocating in the same direction. They typically are used for deep drawing—one slide for the punch and the other for the blankholder. Triple-action presses have three slides; they are typically used for reverse redrawing and for other complicated forming operations.

5. Maximum force required (press capacity and tonnage rating).

6. Type of mechanical, hydraulic, and computer controls.

7. Features for changing dies. Because the time required for changing dies in presses can be significant (which can be as much as a few hours) and thus affect productivity, rapid die-changing systems have been developed. Following a system called *single-minute exchange of die* (SMED), die setups can be changed in less than ten minutes by using computer-controlled hydraulic or pneumatic systems.

8. Safety features.

Because a press is a major capital investment, its present and future use for a broad variety of parts and applications should be investigated. Versatility and multiple use are important factors in press selection, particularly for product modifications and for making new products to respond to continually changing global markets.

16.15 | Economics of Sheet-Forming Operations

Sheet-metal forming involves economic considerations similar to those for the other processes that have been described. Sheet-forming operations are very versatile, and a number of different processes can be used to produce the same part. The costs involved (see also Chapter 40) depend on the particular operations (such as die and equipment costs and labor). For small and simple sheet-metal parts, die costs and lead times to make the dies are rather low. On the other hand, these costs for large-scale operations (such as stretch forming of aircraft panels and boat hulls) are very high. Furthermore, because the number of such parts needed is rather low, the cost-per-piece can be very high (see Fig. 14.18).

Similar considerations are involved in other sheet-forming operations. Deep drawing requires expensive dies and tooling, but a very high number of parts are produced with the same setup, such as containers, cans, and similar products. These costs for other processes such as punching, blanking, bending, and spinning vary considerably, depending on part size and thickness.

Equipment costs vary widely and depend largely on the complexity of the forming operation, part loading and unloading features, part size and shape, and level of automation and computer control required. Automation, in turn, directly affects the amount of labor required and the skill level. The higher the extent of automation, the lower the skill level required. Furthermore, many sheet-metal parts generally require some finishing operations—one of the most common being deburring of the edges of the part, which generally is labor intensive, although some advances have been made in automated deburring (which itself requires computer-controlled equipment; hence, it can be costly).

As an example of the versatility of sheet-forming operations and the costs involved, note that a cup-shaped part can be formed by deep drawing, spinning, rubber forming, or explosive forming. Moreover, it also can be formed by impact extrusion, casting, or fabricating it from different pieces. The part shown in Fig. 16.57 can be made either by deep drawing or by conventional spinning, but the die costs for the two processes are significantly different.

Deep-drawing dies have many components, and they cost much more than the relatively simple mandrels and tools employed in spinning. Consequently, the die cost-per-part in drawing will be high if only a few parts are needed. On the other hand, this part can be formed by deep drawing in a much shorter time than by spinning, even if the latter operation is automated and computer controlled. Furthermore, spinning

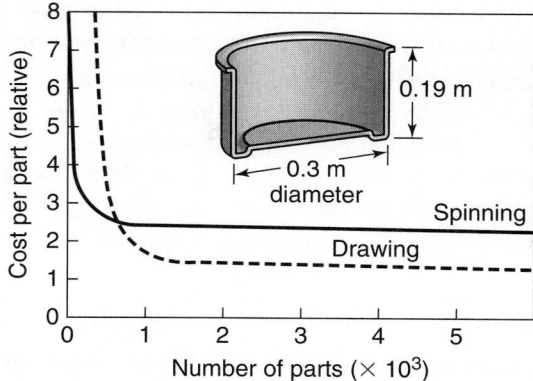

FIGURE 16.57 Cost comparison for manufacturing a round sheet-metal container either by conventional spinning or by deep drawing. Note that for small quantities, spinning is more economical.

generally requires more skilled labor. Considering these factors, the breakeven point can be seen as around 700 parts, and deep drawing is more economical for quantities greater than that. Chapter 40 describes further details of the economics of manufacturing.

SUMMARY

- Sheet-metal forming processes are among the most versatile of all operations. They generally are used on workpieces having high ratios of surface area to thickness. Unlike bulk deformation processes such as forging and extrusion, the thickness of the material in sheet forming often is prevented from being reduced (to avoid necking and tearing).

- Important material parameters are the quality of the sheared edge of the sheet metal prior to forming, the capability of the sheet to stretch uniformly, the material's resistance to thinning, its normal and planar anisotropy, its grain size, and its yield-point elongation (for low-carbon steels).

- The forces and energy required in sheet-metal forming processes are transmitted to the workpiece through solid tools and dies, flexible rubber or polyurethane members, or by electrical, chemical, magnetic, and gaseous means.

- Because of the relatively thin materials used, springback, buckling, and wrinkling are significant problems in sheet forming. Springback is a function of the yield stress, the elastic modulus, and the ratio of bend radius to thickness. These problems can be reduced or eliminated by proper tool and die design, by minimizing the unsupported length of the sheet during processing, and by controlling the thickness of the incoming sheet and its mechanical properties.

- Among important developments is the superplastic forming of diffusion-bonded sheets. The process is capable of producing complex sheet-metal structures, particularly for aerospace applications (which require particularly high stiffness-to-weight ratios).

- Several test methods have been developed for predicting the formability of sheet metals. In bending operations, the tensile reduction of the area of the sheet gives an indication of its bendability (minimum bend radius); this also applies to the parameter spinnability of metals (maximum reduction in thickness per pass).

- For general stamping operations, forming-limit diagrams are very useful, because they establish quantitative relationships among the major and minor principal strains that limit safe forming. For deep-drawing operations, the important parameter is the normal or plastic anisotropy of the sheet (the ratio of width strain to thickness strain in tensile testing).

KEY TERMS

Beading	Explosive forming	Plastic anisotropy
Bendability	Fine blanking	Press brake
Bend allowance	Flanging	Progressive dies
Bending	Formability	Punching
Blankholder	Forming-limit diagram	Redrawing
Blanking	Hemming	Roll forming
Bulging	Honeycomb structures	Rubber forming
Burnished surface	Hydroform process	Shaving
Burr	Ironing	Shearing
Clearance	Laser forming	Slitting
Compound dies	Limiting drawing ratio	Spinning
Deburring	Lüder's bands	Springback
Deep drawing	Magnetic-pulse forming	Steel rule
Dent resistance	Microforming	Stretch forming
Dimpling	Minimum bend radius	Superplastic forming
Drawbead	Nesting	Tailor-welded blanks
Drawing	Nibbling	Transfer dies
Earing	Normal anisotropy	Wrinkling
Electrohydraulic forming	Peen forming	
Embossing	Planar anisotropy	

BIBLIOGRAPHY

ASM Handbook, Vol. 14: *Forming and Forging*, ASM International, 1988.

Benson, S., *Press Brake Technology*, Society of Manufacturing Engineers, 1997.

Bitzer, T., *Honeycomb Technology: Materials, Design, Manufacturing, Applications and Testing*, Chapman & Hall, 1997.

Blazynski, T.Z., *Plasticity and Modern Metal-forming Technology*, Elsevier, 1989.

Boljanovic, V., *Sheet Metal Forming Process and Die Design*, Industrial Press, 2004.

Fundamentals of Tool Design, 4th ed., Society of Manufacturing Engineers, 1998.

Gillanders, J., *Pipe and Tube Bending Manual*, FMA International, 1994.

Hosford, W.F., and Caddell, R.M., *Metal Forming: Mechanics and Metallurgy*, 2nd ed., Prentice Hall, 1993.

Iliescu, C., *Cold-Pressing Technology*, Elsevier, 1990.

Morgan, E., *Tinplate and Modern Canmaking Technology*, Pergamon, 1985.

Pearce, R., *Sheet Metal Forming*, Chapman & Hall, 1997.

Progressive Dies, Society of Manufacturing Engineers, 1994.

Smith, D., *Die Design Handbook*, 3rd ed., Society of Manufacturing Engineers, 1990.

_____, *Fundamentals of Pressworking*, Society of Manufacturing Engineers, 1994.

Suchy, I., *Handbook of Die Design*, McGraw-Hill, 1997.

Tool and Manufacturing Engineers Handbook, 4th ed., Vol. 2: *Forming*. Society of Manufacturing Engineers, 1984.

REVIEW QUESTIONS

16.1 How does sheet-metal forming differ from rolling, forging, and extrusion?

16.2 What causes burrs? How can they be reduced or eliminated?

16.3 Explain the difference between punching and blanking.

16.4 List the various operations performed by die cutting. What types of applications do these processes have in manufacturing?

16.5 Describe the difference between compound, progressive, and transfer dies.

16.6 Describe the characteristics of sheet metals that are important in sheet-forming operations. Explain why they are important.

16.7 Describe the features of forming-limit diagrams (FLDs).

16.8 List the properties of materials that influence springback. Explain why and how they do so.

16.9 Give one specific application for each of the common bending operations described in this chapter.

16.10 Why do tubes buckle when bent? What is the effect of the tube thickness-to-diameter ratio?

16.11 Define normal anisotropy and explain why it is important in determining the deep drawability of a material.

16.12 Describe earing and why it occurs.

16.13 What are the advantages of rubber forming? Which processes does it compete with?

16.14 What is the difference between deep drawing and redrawing?

16.15 What are the differences and similarities between conventional spinning and shear spinning?

16.16 How is roll forming fundamentally different from rolling?

16.17 What is nesting?

16.18 What is microforming? What is its significance in modern manufacturing?

16.19 What are the advantages of superplastic forming?

QUALITATIVE PROBLEMS

16.20 Outline the differences that you have observed between products made of sheet metals and those made by casting and forging.

16.21 Describe the cutting process that takes place when a pair of scissors cuts through aluminum foil.

16.22 Identify the material and process variables that influence the punch force in shearing and explain how each of these affects this force.

16.23 Explain why springback in bending depends on yield stress, elastic modulus, sheet thickness, and bend radius.

16.24 What is the significance of the size of the circles in the grid patterns shown in Fig. 16.15? What is the significance of the thickness of the lines?

16.25 Explain why cupping tests may not predict well the formability of sheet metals in actual forming processes.

16.26 It was stated that the thicker the sheet metal, the higher the curves in Fig. 16.14b become. Why do you think this effect occurs?

16.27 Identify the factors that influence the deep-drawing force, F, in Fig. 16.32b and explain why they do so.

16.28 Why are the beads in Fig. 16.36b placed in those particular locations?

16.29 A general rule for dimensional relationships for successful drawing without a blankholder is given by Eq. (16.14). Explain what would happen if this limit were exceeded.

16.30 Section 16.2.1 stated that the punch stripping force is difficult to estimate because of the many factors involved. Make a list of these factors with brief explanations about why they would affect the stripping force.

16.31 Is it possible for the forming-limit diagram shown in Fig. 16.14b to have a negative major strain? Explain.

16.32 Inspect Fig. 16.14b and explain clearly whether in a sheet-forming operation you would like to develop a state of strain that is in the left half or in the right half of the forming-limit diagram.

16.33 Is it possible to have ironing take place in an ordinary deep-drawing operation? What is the most important factor?

16.34 Note the roughness of the periphery of the flanged hole in Fig. 16.25c and comment on its possible effects when the part is used in a product.

16.35   What recommendations would you make in order to eliminate the cracking of the bent piece shown in Fig. 16.17c? Explain your reasons.

16.36   As you can see, the forming-limit diagram axes pertain to engineering strains given as a percentage. Describe your thoughts about whether the use of true strains, as in Eq. (2.7), would have any significant advantage.

16.37   It has been stated that the drawability of a material is higher in the hydroform process than in the deep-drawing process. Explain why.

16.38   Give several specific examples from this chapter in which friction is desirable and several in which it is not desirable.

16.39   As you can see, some of the operations described in this chapter produce considerable scrap. Describe your thoughts regarding the reuse, recycling,

or disposal of this scrap. Consider its size, its shape, and its contamination by metalworking fluids during processing.

16.40   In the manufacture of automotive body panels from carbon-steel sheet, stretcher strains (or Lüder's bands) are observed, which can detrimentally can affect the surface finish. How can these be eliminated?

16.41   A coil of sheet metal is taken to a furnace and annealed in order to improve its ductility. It is found, however, that the sheet has a lower, limiting drawing ratio than it had before annealing. Explain why this effect has occurred.

16.42   Through changes in clamping or die design, it is possible for a sheet metal to undergo a negative minor strain. Explain how this effect can be advantageous.

16.43   How would you produce the part shown in Fig. 16.41b other than by tube hydroforming?

QUANTITATIVE PROBLEMS

16.44   Calculate R_{avg} for a metal where the R values for the 0°, 45°, and 90° directions are 0.9, 1.6, and 1.75, respectively. What is the limiting drawing ratio (LDR) for this material?

16.45   Calculate the value of ΔR in Problem 16.44. Will any ears form when this material is deep drawn? Explain.

16.46   Estimate the limiting drawing ratio for the materials listed in Table 16.4.

16.47   Prove Eq. (16.4).

16.48   Regarding Eq. (16.4), it has been stated that (in bending) actual values of the strain in the outer fibers (i.e., in tension) are higher than those on the inner fibers (in compression), the reason being that the neutral axis shifts during bending. With an appropriate sketch, explain this phenomenon.

16.49   Using Eq. (16.15) and the K value for TNT, plot the pressure as a function of weight (W) and R, respectively. Describe your observations.

16.50   Section 16.5 states that the k values in bend allowance depend on the relative magnitudes of R and T. Explain why this relationship exists.

16.51   In explosive forming, calculate the peak pressure in water for 0.3 lb of TNT at a standoff distance of 3 ft. Comment on whether or not the magnitude of this pressure is sufficiently high to form sheet metals.

16.52   Why is the bending force, P, proportional to the square of the sheet thickness, as seen in Eqs. (16.7) and (16.8) ?

16.53   In Fig. 16.14a, measure the respective areas of the solid outlines and compare them with the areas of the original circles. Calculate the final thicknesses of the sheets, assuming that the original sheet is 1 mm thick.

16.54   With the aid of a free-body diagram, prove the existence of compressive hoop stresses in the flange in a deep-drawing operation.

16.55   Plot Eq. (16.6) in terms of the elastic modulus, E, and the yield stress, Y, of the material and describe your observations.

16.56   What is the minimum bend radius for a 2-mm thick sheet metal with a tensile reduction of area of 30%? Does the bend angle affect your answer? Explain.

16.57   When a round sheet-metal blank is deep drawn, it is found that it does not exhibit any earing. Its R values in the 0° and 90° directions to rolling are 1.4 and 1.8, respectively. What is the R value in the 45° direction?

16.58   Survey the technical literature and explain the mechanism by which negative springback can occur in V-die bending. Show that negative springback does not occur in air bending.

16.59   Using the data in Table 16.3 and referring to Eq. (16.5), calculate the tensile reduction of area for the materials and the conditions listed in the table.

16.60   What is the force required to punch a square hole, 100 mm on each side, in a 1-mm thick 5052-O aluminum sheet by using flat dies? What would be your answer if beveled dies are used?

16.61 In Example 16.4 it was stated that the reason for reducing the top of cans (necking) is to save material for making the lid. How much material will be saved if the lid diameter is reduced by 10%? By 15%?

16.62 Estimate the percentage of scrap in producing round blanks if the clearance between blanks is one-tenth of the blank radius. Consider single and two-row blanking, as sketched in Fig. P16.62.

16.63 Assume that you are an instructor covering the topics described in this chapter, and you are giving a quiz on the numerical aspects to test the understanding of the students. Prepare two quantitative problems and supply the answers.

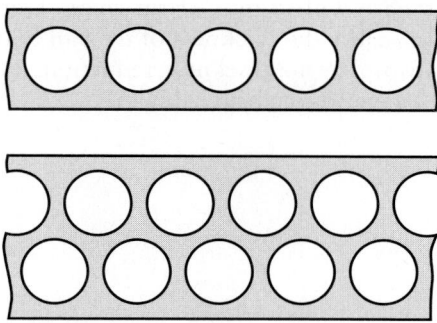

FIGURE P16.62

SYNTHESIS, DESIGN, AND PROJECTS

16.64 Examine some of the products in your home that are made of sheet metal and discuss the process or combination of processes by which you think they were made.

16.65 Consider several shapes to be blanked from a large sheet (such as oval, triangular, L-shaped, and so forth) by laser-beam cutting and sketch a nesting layout to minimize scrap generation.

16.66 Give several product applications for (a) hemming and (b) seaming.

16.67 Many axisymmetric missile bodies are made by spinning. What other methods could you use if spinning processes were not available?

16.68 Give several structural designs and applications in which diffusion bonding and superplastic forming can be used jointly.

16.69 Inspect sheet-metal parts in an automobile and describe your thoughts as to which of the processes or combinations of processes were used in making them. Comment on the reasons why more than one process may have to be used.

16.70 Name several parts that can be made in compound dies and several others that can be made in transfer dies.

16.71 On the basis of experiments, it has been suggested that concrete—either plain or reinforced—can be a suitable material for dies in metal-forming operations, especially for large parts. Describe your thoughts regarding this suggestion, considering die geometry and any other factors that may be relevant.

16.72 Metal cans are either two-piece (in which the bottom and sides are integral) or three-piece (in which the sides, the bottom, and the top are each separate pieces). For a three-piece can, should the vertical seam in the can body be (a) in the rolling direction, (b) normal to the rolling direction, or (c) oblique to the rolling direction? Prove your answer.

16.73 Investigate methods for determining optimum blank shapes for deep-drawing operations. Sketch the optimally shaped blanks for rectangular cups, and optimize their layout on a large metal sheet.

16.74 The design shown in Fig. P16.74 is proposed for a metal tray—the main body of which is made from cold-rolled sheet steel. Noting its features and that the sheet is bent in two different directions, comment on various manufacturing considerations. Include factors such as anisotropy of the rolled sheet, its surface texture, the bend directions, the nature of the sheared edges, and the way the handle is snapped in for assembly.

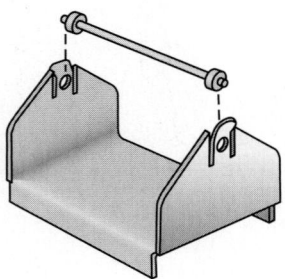

FIGURE P16.74

16.75 Using a ball-peen hammer, strike the surface of aluminum sheets of various thicknesses until they develop a curvature. Describe your observations about the shapes produced.

16.76 Inspect a common paper punch and observe the shape of the punch tip. Compare it with those shown in Fig. 16.10 and comment on your observations.

16.77 Obtain an aluminum beverage can and slit it in half lengthwise with a pair of tin snips. Using a micrometer, measure the thickness of the can bottom and the wall. Estimate the thickness reductions in ironing and the diameter of the original blank.

Processing of Metal Powders

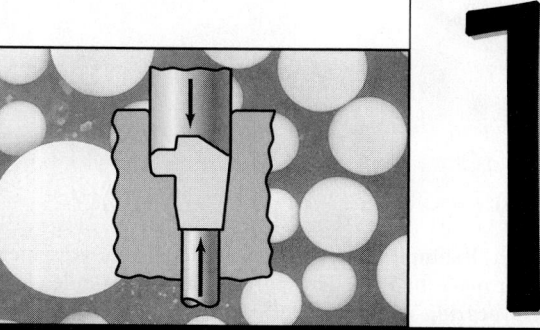
This chapter explores the manufacturing processes and technologies for producing net-shape parts from metal powders. Specifically, we describe:

- Methods of producing metal powders.
- Processes used to consolidate and sinter powders into desired shapes.
- Additional processes and finishing operations to enhance properties.
- Competitive aspects of this unique process.

Typical products made: Connecting rods, piston rings, gears, cams, bushings, bearings, cutting tools, surgical implants, magnets, and metal filters.

Alternative processes: Casting, forging, and machining.

17.1 | Introduction

In the manufacturing processes we have described thus far, the raw materials used have been metals and alloys either in a molten state (casting) or in solid form (metalworking). This chapter describes the **powder metallurgy** (P/M) process, in which metal powders are compacted into desired and often complex shapes and sintered (heated without melting) to form a solid piece. This process first was used by the Egyptians in about 3000 B.C. to make iron tools. One of its first modern uses was in the early 1900s to make the tungsten filaments for incandescent light bulbs. The availability of a wide range of metal–powder compositions, the ability to produce parts to net dimensions (**net-shape forming**), and the overall economics of the operation give this unique process its numerous attractive and expanding applications.

A wide range of parts and components are made by powder-metallurgy techniques (Fig. 17.1): (a) balls for ball-point pens; (b) automotive components (which now constitute about 70% of the P/M market) such as piston rings, connecting rods, brake pads, gears, cams, and bushings; (c) tool steels, tungsten carbides, and cermets as tool and die materials; (d) graphite brushes impregnated with copper for electric motors; (e) magnetic materials; (f) metal filters and oil-impregnated bearings with controlled porosity; (g) metal foams; (h) surgical implants, and (i) several others for aerospace, nuclear, and industrial applications. Advances in this technology now permit *structural* parts of aircraft, such as landing gear components, engine-mount supports, engine disks, impellers, and engine nacelle frames to be made by P/M.

(a)

(b)

(c)

FIGURE 17.1   (a) Examples of typical parts made by powder-metallurgy processes. (b) Upper trip lever for a commercial irrigation sprinkler made by P/M. This part is made of an unleaded brass alloy; it replaces a die-cast part with a 60% cost savings. (c) Main-bearing metal-powder caps for 3.8 and 3.1 liter General Motors automotive engines. *Source*: (a) and (b) Reproduced with permission from *Success Stories on P/M Parts*, 1998. Metal Powder Industries Federation, Princeton, New Jersey, 1998. (c) Courtesy of Zenith Sintered Products, Inc., Milwaukee, Wisconsin.

Powder metallurgy has become competitive with processes (such as casting, forging, and machining), particularly for relatively complex parts made of high-strength and hard alloys. Although most parts weigh less than 2.5 kg (5 lb), they can weigh as much as 50 kg (100 lb). It has been shown that P/M parts can be mass-produced economically in quantities as small as 5000 per year and as much as 100 million per year for vibrator weights for cell phones.

The most commonly used metals in P/M are iron, copper, aluminum, tin, nickel, titanium, and the refractory metals. For parts made of brass, bronze, steels, and stainless steels, *prealloyed powders* are used, where each powder particle itself is an alloy. Metal sources are generally bulk metals and alloys, ores, salts, and other compounds.

17.2 | Production of Metal Powders

The powder-metallurgy process consists of the following operations, in sequence (Fig. 17.2):

1. Powder production
2. Blending
3. Compaction
4. Sintering
5. Finishing operations

17.2.1 Methods of powder production

There are several methods of producing metal powders, and most of them can be produced by more than one method. The choice depends on the requirements of the end product. The microstructure, bulk and surface properties, chemical purity, porosity, shape, and size distribution of the particles depend on the particular process used (Figs. 17.3 and 17.4). These characteristics are important because they significantly affect the flow and permeability during compaction and in subsequent sintering operations. Particle sizes produced range from 0.1 to 1000 μm (4 to 0.04 in.).

Atomization.   Atomization produces a liquid-metal stream by injecting molten metal through a small orifice. The stream is broken up by jets of inert gas or air (Fig. 17.5a) or water (Fig. 17.5b) known as *gas* and *water atomization*, respectively.

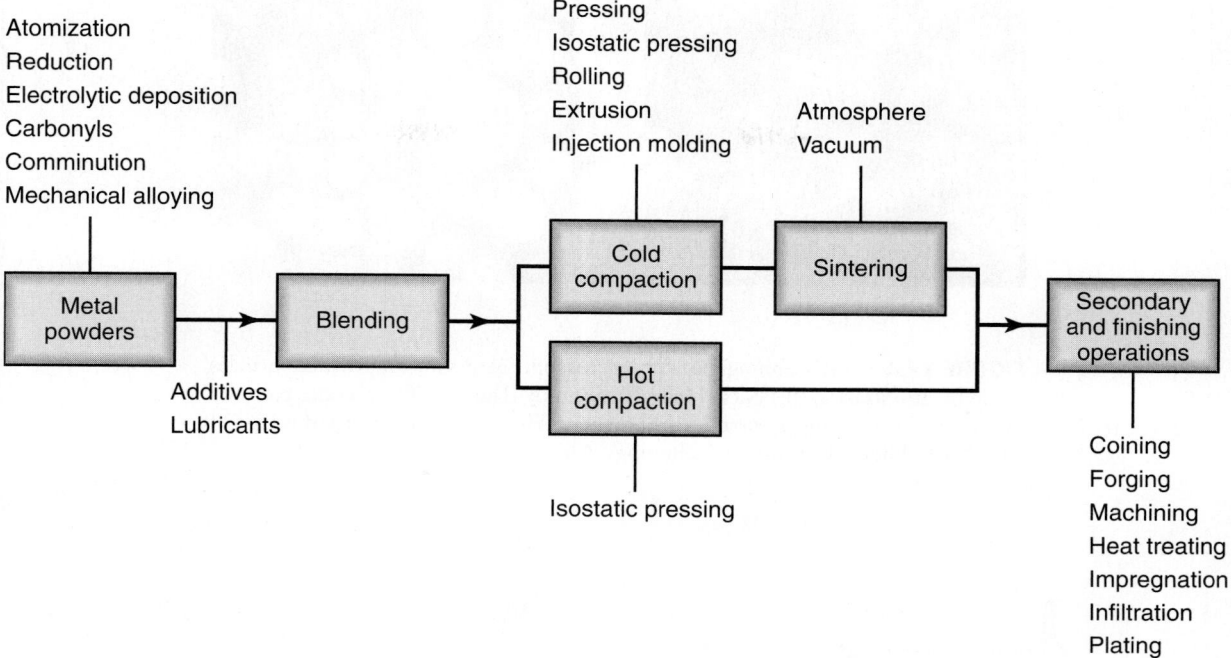

FIGURE 17.2   Outline of processes and operations involved in making powder-metallurgy parts.

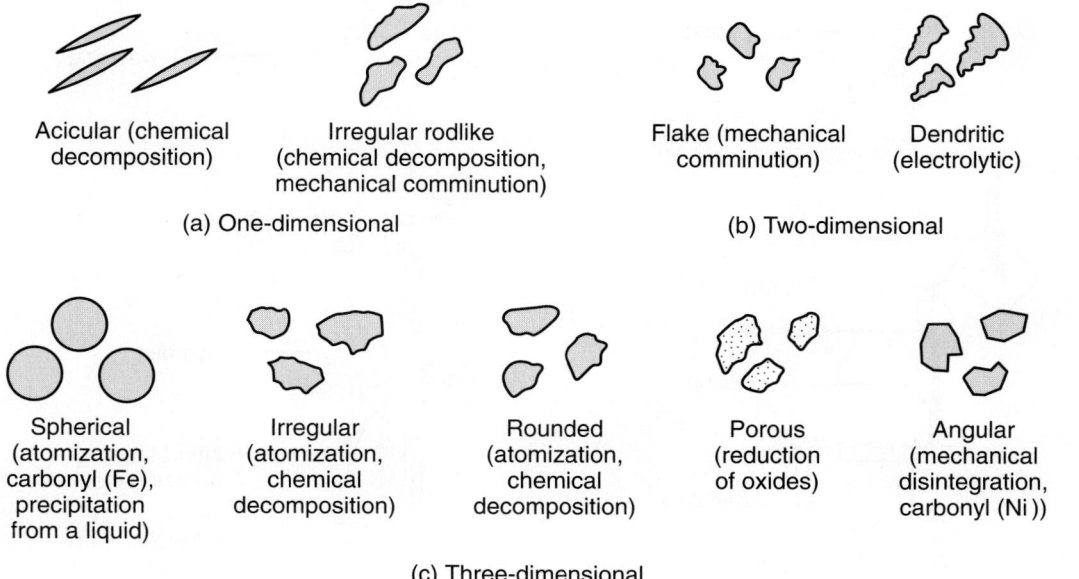

(a) One-dimensional

(b) Two-dimensional

(c) Three-dimensional

FIGURE 17.3   Particle shapes in metal powders, and the processes by which they are produced. Iron powders are produced by many of these processes.

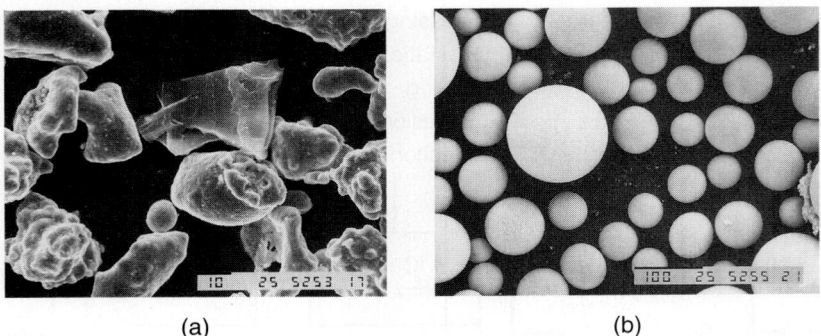

(a)   (b)

FIGURE 17.4 (a) Scanning-electron microscope image of iron-powder particles made by atomization. (b) Nickel-based superalloy (Udimet 700) powder particles made by the rotating electrode process; see Fig. 17.5c. *Source*: Courtesy of P.G. Nash, Illinois Institute of Technology, Chicago.

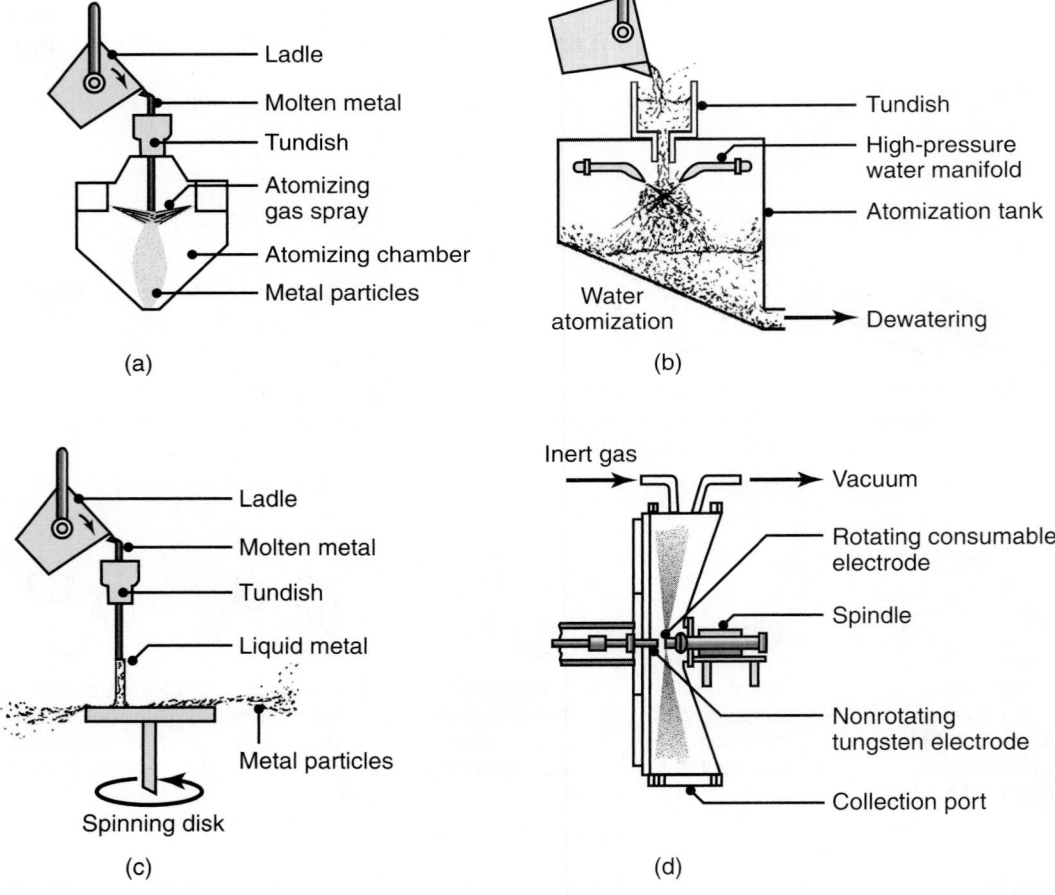

FIGURE 17.5 Methods of metal-powder production by atomization: (a) gas atomization; (b) water atomization; (c) atomization with a rotating consumable electrode; and (d) centrifugal atomization with a spinning disk or cup.

The size and shape of the particles formed depends on the temperature of the molten metal, rate of flow, nozzle size, and jet characteristics. The use of water results in a slurry of metal powder and liquid at the bottom of the atomization chamber. Although the powders must be dried before they can be used, the water allows for more rapid cooling of the particles and higher production rates. On the other hand, gas atomization usually results in more spherical particles (see Fig. 17.3c).

In *centrifugal atomization*, the molten-metal stream drops onto a rapidly rotating disk or cup, so that centrifugal forces break up the molten-metal stream and generate particles (Fig. 17.5c). In another variation of this method, a consumable electrode is rotated rapidly (about 15,000 rev/min) in a helium-filled chamber (Fig. 17.5d). The centrifugal force breaks up the molten tip of the electrode into metal particles.

Reduction.   The *reduction* of metal oxides (i.e., removal of oxygen) uses gases, such as hydrogen and carbon monoxide, as reducing agents. By this means, very fine metallic oxides are reduced to the metallic state. The powders produced are spongy and porous and have uniformly sized spherical or angular shapes.

Electrolytic deposition.   *Electrolytic deposition* utilizes either aqueous solutions or fused salts. The powders produced are among the purest available.

Carbonyls.   *Metal carbonyls*, such as iron carbonyl ($Fe(CO)_5$) and nickel carbonyl ($Ni(CO)_4$), are formed by letting iron or nickel react with carbon monoxide. The reaction products are then decomposed to iron and nickel, and they turn into small, dense, uniformly spherical particles of high purity.

Comminution.   *Mechanical comminution* (*pulverization*) involves crushing (Fig. 17.6), milling in a ball mill, or grinding of brittle or less ductile metals into small particles. A *ball mill* (Fig. 17.6b) is a machine with a rotating hollow cylinder partly filled with steel or white cast-iron balls. With brittle materials, the powder particles produced have angular shapes; with ductile metals, they are flaky and are not particularly suitable for powder-metallurgy applications.

Mechanical alloying.   In *mechanical alloying*, powders of two or more pure metals are mixed in a ball mill, as illustrated in Fig. 17.7. Under the impact of the hard balls, the powders fracture and bond together by diffusion, forming alloy powders. The dispersed phase can result in strengthening of the particles or can impart special electrical or magnetic properties of the powder.

Miscellaneous methods.   Other less commonly used methods for making powders are:

- **Precipitation** from a chemical solution
- Production of fine metal chips by **machining**
- **Vapor condensation**

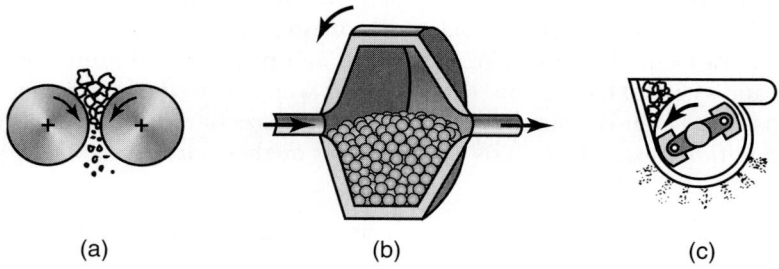

(a)   (b)   (c)

FIGURE 17.6   Methods of mechanical comminution to obtain fine particles: (a) roll crushing, (b) ball mill, and (c) hammer milling.

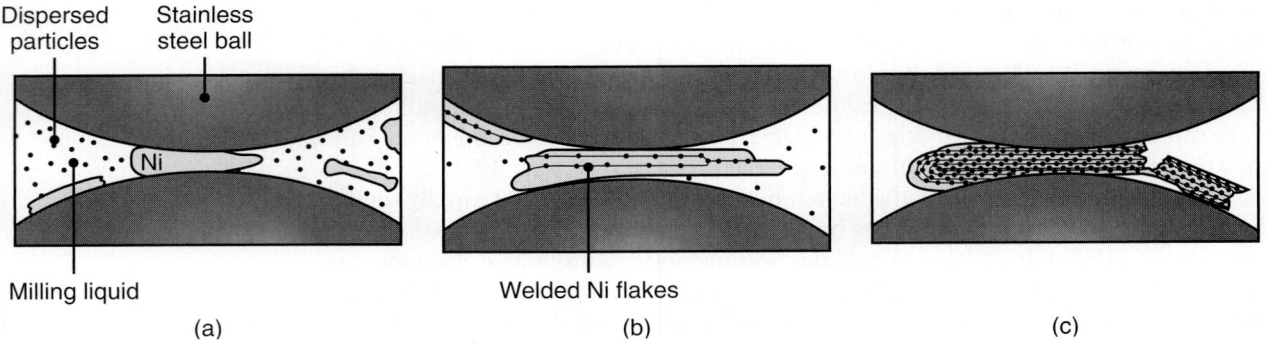

FIGURE 17.7   Mechanical alloying of nickel particles with dispersed smaller particles. As nickel particles are flattened between two balls, the second smaller phase is impressed into the nickel surface and eventually is dispersed throughout the particle due to successive flattening, fracture, and welding events.

More recent developments include techniques based on *high-temperature extractive metallurgical processess*—based on the reaction of volatile halides (a compound of halogen and an electropositive element) with liquid metals and the controlled reduction and reduction/carburization of solid oxides.

Nanopowders.   Most recent developments include the production of *nanopowders* of copper, aluminum, iron, titanium, and various other metals (see also *nanomaterials* in Section 6.16). Because these powders are pyrophoric (ignite spontaneously) or are contaminated readily when exposed to air, they are shipped as thick slurries under hexane gas (which itself is highly volatile and combustible). When the material is subjected to large plastic deformation by compression and shear at stress levels of 5500 MPa (800 ksi) during processing of the powders, the particle size is reduced, and the material becomes pore-free and possesses enhanced properties.

Microencapsulated powders.   These metal powders are coated completely with a binder. For electrical applications (such as magnetic components of ignition coils and other pulsed AC and DC applications), the binder acts like an insulator, preventing electricity from flowing between particles and thus reducing eddy-current losses. The powders are compacted by warm pressing, and they are used with the binder still in place. (See also *powder-injection molding*, Section 17.3.3.)

17.2.2 Particle size, shape, and distribution

Particle size usually is measured by *screening*—that is, by passing the metal powder through screens (*sieves*) of various mesh sizes. Screen analysis is achieved by using a vertical stack of screens with the mesh size becoming finer as the powder flows downward through the screens. The larger the mesh size, the smaller is the opening in the screen. For example, a mesh size of 30 has an opening of 600 μm, size 100 has 150 μm, and size 400 has 38 μm. (This method is similar to the numbering of abrasive grains. The larger the number, the smaller is the size of the abrasive particle.)

In addition to screen analysis, several other methods are available for particle-size analysis:

1. **Sedimentation,** which involves measuring the rate at which particles settle in a fluid.

2. **Microscopic analysis,** which may include the use of transmission and scanning electron microscopy.

3. **Light scattering** from a laser that illuminates a sample consisting of particles suspended in a liquid medium. The particles cause the light to be scattered, and a detector then digitizes the signals and computes the particle-size distribution.
4. **Optical** (such as particles blocking a beam of light), which is then sensed by a photocell.
5. **Suspending particles** in a liquid and then detecting particle size and distribution by electrical sensors.

Particle shape.   A major influence on processing characteristics, particle shape usually is described in terms of aspect ratio or shape factor. *Aspect ratio* is the ratio of the largest dimension to the smallest dimension of the particle. This ratio ranges from unity (for a spherical particle) to about 10 for flake-like or needle-like particles.

Shape factor (SF).   Also called the *shape index*, this is a measure of the ratio of the surface area of the particle to its volume—normalized by reference to a spherical particle of equivalent volume. Thus, the shape factor for a flake is higher than that for a sphere.

Size distribution.   The size distribution of particles is an important consideration, because it affects the processing characteristics of the powder. The distribution of particle size is given in terms of a *frequency-distribution* plot (see Section 36.7 for details.) The maximum is called the *mode size*.

Other properites of metal powders that have an effect on their behavior in processing them are: (a) *flow properties* when filled into dies, (b) *compressibility* when being compacted, (c) *density*, as defined in various terms such as theoretical density, apparent density, and the density when the powder is shaken or tapped in the die cavity.

17.2.3 Blending metal powders

Blending (mixing) powders is the next step in powder-metallurgy processing. It is carried out for the following purposes:

- Powders of different metals and other materials can be mixed in order to impart special physical and mechanical properties and characteristics to the P/M product. Note that mixtures of metals can be produced by alloying the metal before producing a powder, or else blends can be produced. Proper mixing is essential to ensure the uniformity of mechanical properties throughout the part.
- Even when a single metal is used, the powders may vary significantly in size and shape, hence they must be blended to obtain uniformity from part to part. The ideal mix is one in which all of the particles of each material (and of each size and morphology) are distributed uniformly.
- *Lubricants* can be mixed with the powders to improve their flow characteristics. They reduce friction between the metal particles, improve flow of the powder metals into the dies, and improve die life. Lubricants typically are stearic acid or zinc stearate in a proportion of from 0.25 to 5% by weight.
- Other additives—*binders* (as in sand molds) are used to develop sufficient *green strength* (see Section 17.3), and additives also can be used to facilitate sintering.

Powder mixing must be carried out under controlled conditions in order to avoid contamination or deterioration. Deterioration is caused by excessive mixing, which may alter the shape of the particles and work-harden them and, thus, make the subsequent compacting operation more difficult. Powders can be mixed in air, in

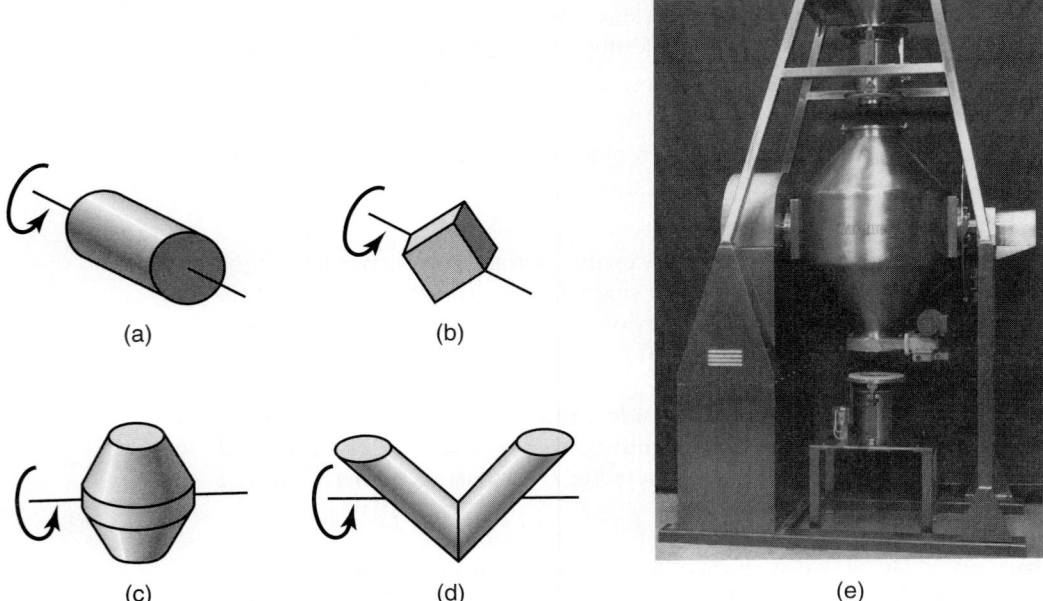

FIGURE 17.8   (a) through (d) Some common bowl geometries for mixing or blending powders. (e) A mixer suitable for blending metal powders. Since metal powders are abrasive, mixers rely on the rotation or tumbling of enclosed geometries as opposed to using aggressive agitators. *Source: Courtesy of Kemutec Group, Inc.*

inert atmospheres (to avoid oxidation), or in liquids (which act as lubricants and make the mix more uniform). Several types of blending equipment are available (Fig. 17.8).

Hazards.   Because of their high surface area-to-volume ratio, metal powders can be explosive, particularly aluminum, magnesium, titanium, zirconium, and thorium. Great care must be exercised both during blending and in storage and handling. Precautions include (a) grounding equipment, (b) preventing sparks (by using non-sparking tools) and avoiding friction as a source of heat, and (c) avoiding dust clouds, open flames, and chemical reactions.

17.3 | Compaction of Metal Powders

Compaction is the step in which the blended powders are pressed into various shapes in dies, as shown in sequence in Fig. 17.9. The purposes of compaction are to obtain the required shape, density, and particle-to-particle contact and to make the part sufficiently strong for further processing. The powder (*feedstock*) is fed into the die by a feed show, and the upper punch descends into the die. The presses used are actuated either hydraulically or mechanically, and the process generally is carried out at room temperature, although it can be done at elevated temperatures.

The pressed powder is known as **green compact**, since it has a low strength just as is seen in green parts in slip casting (Section 18.2.1). The green parts are very fragile (similar to chalk) and can crumble or become damaged very easily; this situation is exacerbated by poor pressing practices. To obtain higher green strengths, the powder must be fed properly into the die cavity, and proper pressures must be developed throughout the part.

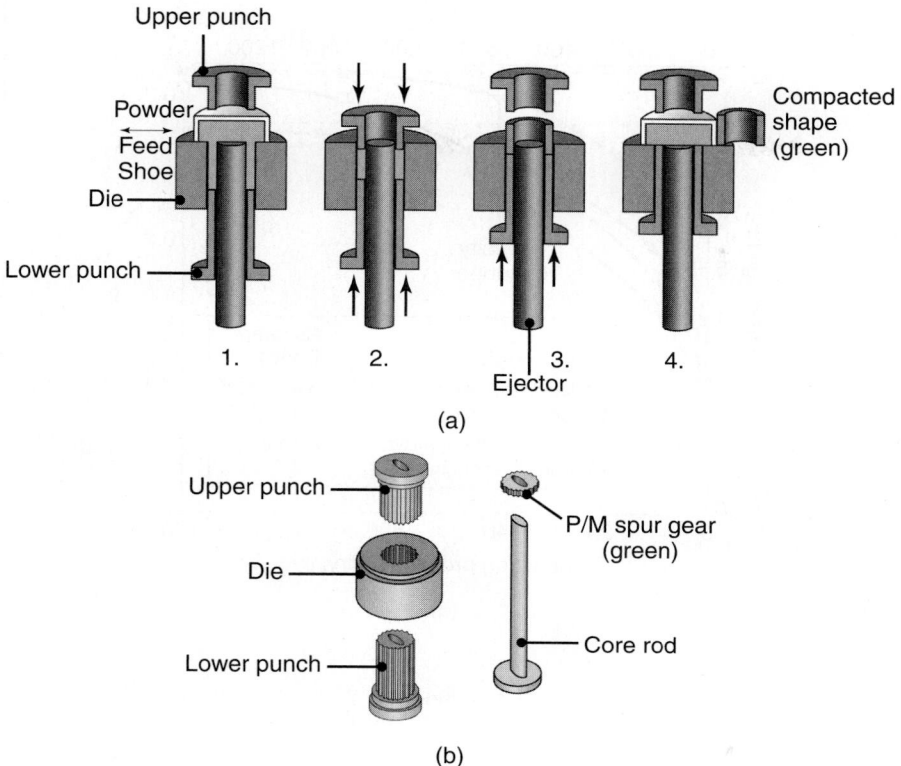

FIGURE 17.9   (a) Compaction of metal powder to form a bushing. The pressed-powder part is called green compact. (b) Typical tool and die set for compacting a spur gear. *Source*: Courtesy of Metal Powder Industries Federation.

The *density* of the green compact depends on the pressure applied (Fig. 17.10a). As the compacting pressure is increased, the compact density approaches that of the metal in its bulk form. An important factor in density is the size distribution of the particles. If all of the particles are of the same size, there always will be some porosity when they are packed together, theoretically a porosity of at least 24% by volume. (Observe, for example, a box filled with tennis balls and note that there are always open spaces between the balls.) Introducing smaller particles into the powder mix will fill the spaces between the larger powder particles and, thus, result in a higher density of the compact (see also *porous aluminum*, Section 6.2).

The higher the density of the compacted part, the higher is its strength and elastic modulus (Fig. 17.10b). The reason is that the higher the density, the higher the amount of solid metal in the same volume, and hence the greater its strength (resistance to external forces). Because of friction between (a) the metal particles in the powder and (b) the punch surfaces and die walls, the density within the part can vary considerably. This variation can be minimized by proper punch and die design and by control of friction. Thus, for example, it may necessary to use multiple punches with separate movements, in order to ensure that the density is more uniform throughout the part (Fig. 17.11). Recall a similar discussion regarding the compaction of sand in mold making (see Fig. 11.7). However, density variation in components such as gears, cams, bushings, and structural parts may be desirable. For example, densities can be increased in critical locations where high strength and wear resistance are important and reduced where they are not.

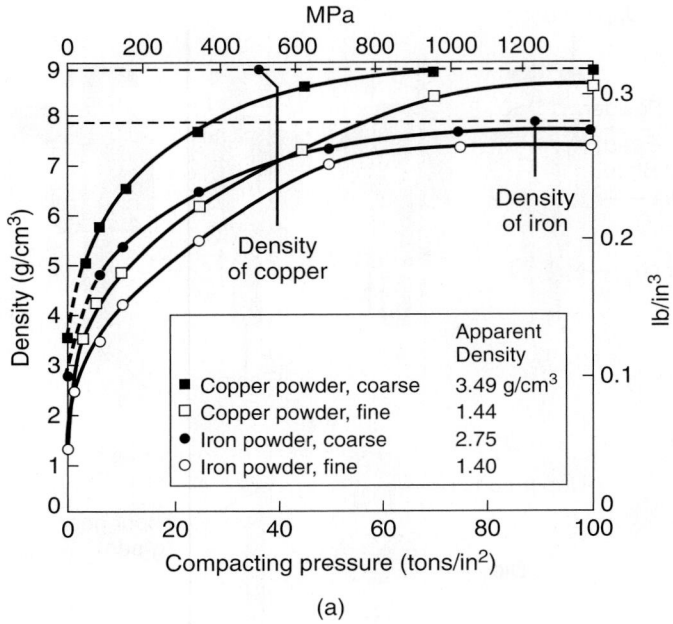

(a)

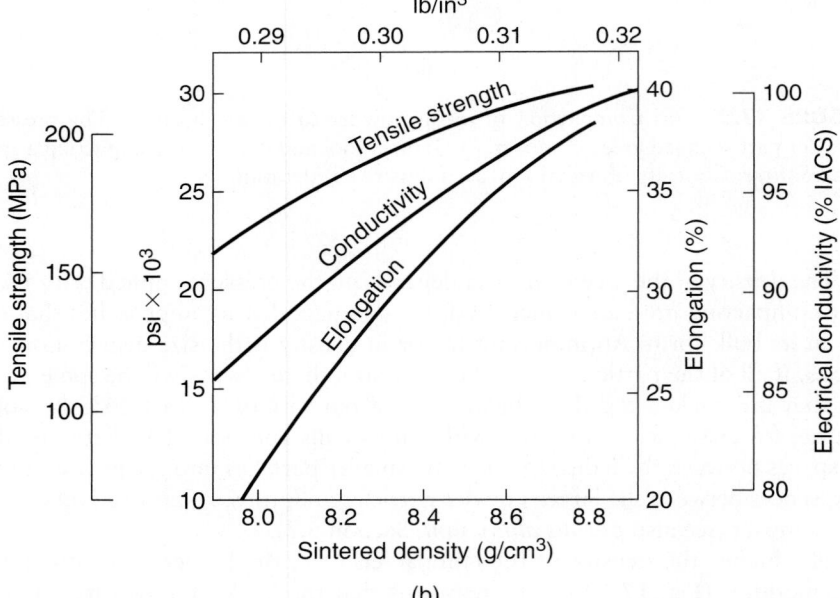

(b)

FIGURE 17.10   (a) Density of copper- and iron-powder compacts as a function of compacting pressure. Density greatly influences the mechanical and physical properties of P/M parts. (b) Effect of density on tensile strength, elongation, and electrical conductivity of copper powder. *Source*: (a) After F. V. Lenel, (b) IACS: International Annealed Copper Standard (for electrical conductivity).

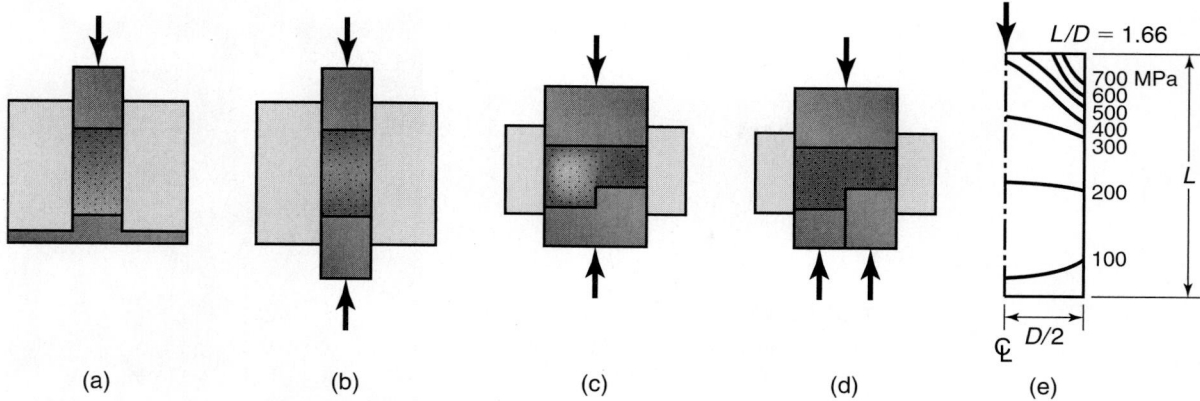

FIGURE 17.11 Density variation in compacting metal powders in various dies: (a) and (c) single-action press; (b) and (d) double-action press. Note in (d) the greater uniformity of density from pressing with two punches with separate movements when compared with (c). (e) Pressure contours in compacted copper powder in a single-action press. *Source*: After P. Duwez and L. Zwell.

17.3.1 Equipment

The pressure required for pressing metal powders ranges from 70 MPa (10 ksi) (for aluminum) to 800 MPa (120 ksi) (for high-density iron parts) (see Table 17.1). The compacting pressure required depends on the characteristics and shape of the particles, on the method of blending, and on the lubricant.

Press capacities are on the order of 1.8 to 2.7 MN (200 to 300 tons), although presses with much higher capacities are used for special applications. Most applications require less than 100 tons. For small tonnage, crank- or eccentric-type mechanical presses are used; for higher capacities, toggle or knucklejoint presses are employed (see Fig. 14.17c). Hydraulic presses (Fig. 17.12) with capacities as high as 45 MN (5000 tons) can be used for large parts. Press selection depends on part size and its configuration, density requirements, and production rate. However, the higher the pressing speed, the greater the tendency for the press to trap air in the die cavity and, thus, prevent proper compaction.

TABLE 17.1

Compacting Pressures for Various Powders	
Metal	Pressure (MPa)
Aluminum	70–275
Brass	400–700
Bronze	200–275
Iron	350–800
Tantalum	70–140
Tungsten	70–140
Other materials	
Aluminum oxide	110–140
Carbon	140–165
Cemented carbides	140–400
Ferrites	110–165

FIGURE 17.12 A 7.3-MN (825-ton) mechanical press for compacting metal powder. *Source*: Courtesy of Cincinnati Incorporated.

17.3.2 Isostatic pressing

Green compacts may be subjected to *hydrostatic pressure* in order to achieve more uniform compaction and, hence, density.

In **cold isostatic pressing** (CIP), the metal powder is placed in a flexible rubber mold typically made of neoprene rubber, urethane, polyvinyl chloride, or another elastomer (Fig. 17.13). The assembly then is pressurized hydrostatically in a chamber, usually using water. The most common pressure is 400 MPa (60 ksi), although pressures of up to 1000 MPa (150 ksi) may be used. The ranges for CIP and other compacting methods in terms of the size and complexity of a part are shown in Fig. 17.14. A typical application is automotive cylinder liners.

In **hot isostatic pressing** (HIP), the container generally is made of a high-melting-point sheet metal, and the pressurizing medium is high-temperature inert gas or a vitreous (glasslike) fluid (Fig. 17.15). Common conditions for HIP are pressures as high as 100 MPa (15 ksi)—although it can be three times as high—and at a temperature

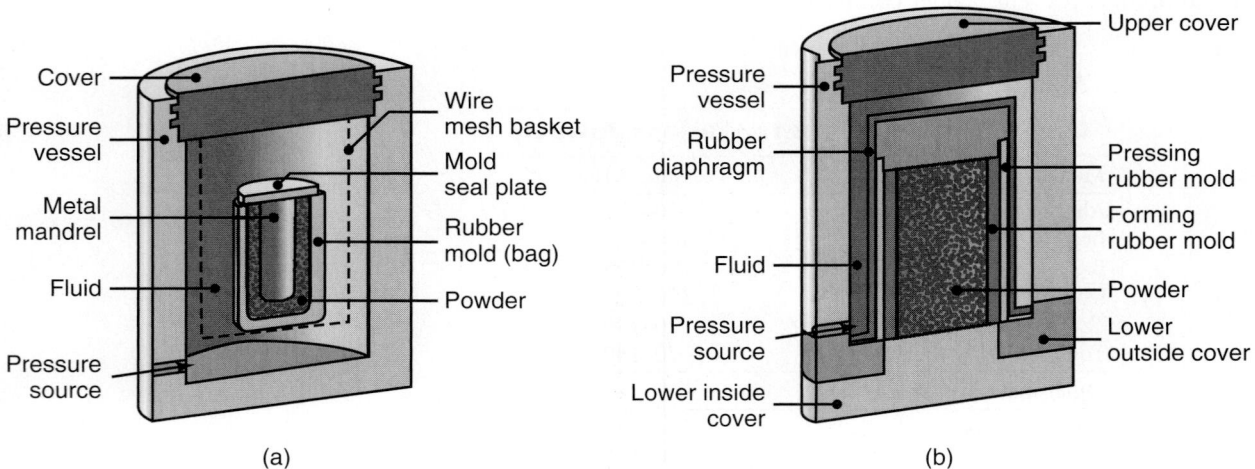

(a)   (b)

FIGURE 17.13 Schematic diagram of cold isostatic pressing, as applied to forming a tube. The powder is enclosed in a flexible container around a solid-core rod. Pressure is applied isostatically to the assembly inside a high-pressure chamber. *Source*: Reprinted with permission from R.M. German, *Powder Metallurgy Science*. Metal Powder Industries Federation, Princeton, NJ; 1984.

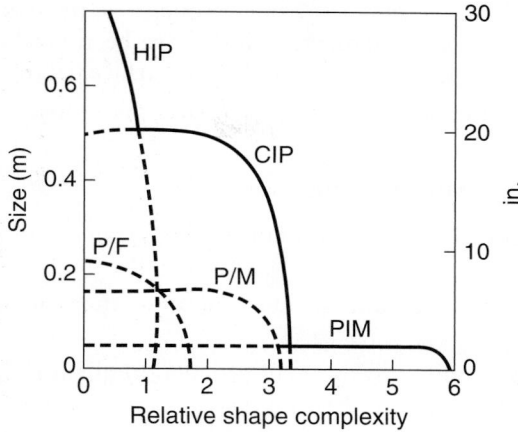

FIGURE 17.14 Capabilities, with respect to part size and shape complexity, available from various P/M operations. P/F means powder forging. *Source:* Courtesy of Metal Powder Industries Federation.

of 1200°C (2200°F). The main advantage of HIP is its ability to produce compacts having almost 100% density, good metallurgical bonding of the particles, and good mechanical properties. Consequently, it has gained wide acceptance in making high-quality parts.

The HIP process is used mainly in making superalloy components for the aircraft and aerospace industries and in military, medical, and chemical applications. It also is used (a) to close internal porosity, (b) to improve properties in superalloy and titanium-alloy castings for the aerospace industry, and (c) as a final densification step for tungsten-carbide cutting tools and P/M tool steels.

The main advantages of isostatic pressing are the following:

• Because of the uniformity of pressure from all directions and the absence of die-wall friction, it produces fully dense compacts of practically uniform grain

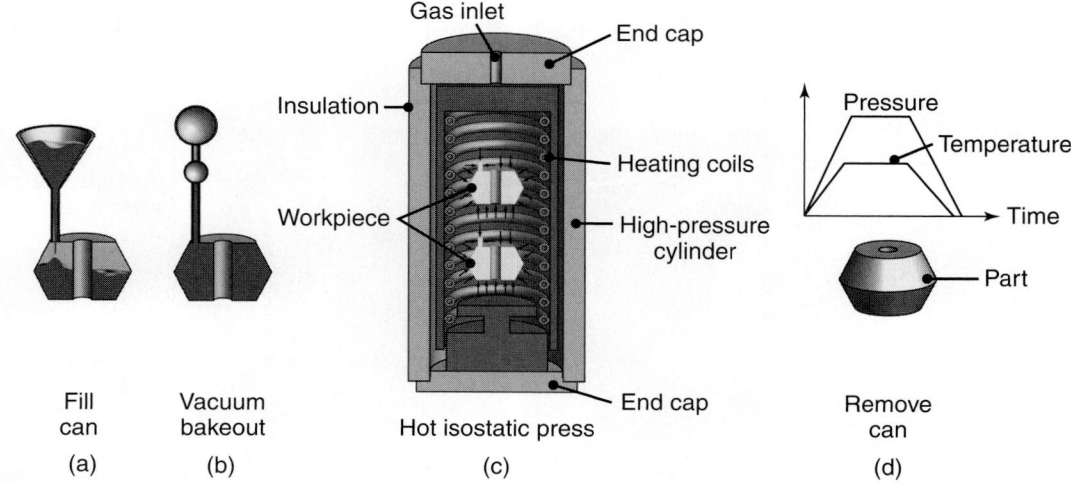

FIGURE 17.15 Schematic illustration of hot isostatic pressing. The pressure and temperature variation versus time are shown in the diagram.

structure and density (hence, isotropic properties), irrespective of part shape. Parts with high length-to-diameter ratios have been produced with very uniform density, strength, toughness, and good surface detail.

- HIP is capable of handling much larger parts than those in other compacting processes.

The limitations of HIP are as follows:

- Wider dimensional tolerances than those obtained in other compacting processes.

- Higher equipment cost and production time than are required by other processes.

- Applicability only to relatively small production quantities, typically less than 10,000 parts per year.

EXAMPLE 17.1 Hot isostatic pressing of a valve lifter

An HIP-clad valve lifter used in a full range of medium- to heavy-duty truck diesel engines is shown in Fig. 17.16. The 0.2-kg (0.45-lb) valve lifter rides on the camshaft and opens and closes the engine valves. As such, it is desired to have a tungsten-carbide face for wear resistance and a steel shaft for fatigue resistance. Before the HIP valve lifter was developed, parts were produced through furnace brazing but resulted in occasional field failures and relatively high scrap rates. The required annual production of these parts is over 400,000, so high scrap rates are particularly objectionable.

The HIP-clad product consists of a 9% Co bonded tungsten-carbide (WC) face made from powder (pressed and sintered), a steel sheet-metal cap fitted over the WC disk, a copper-alloy foil interlayer, and a steel shaft. The steel cap is electron-beam welded to the steel shaft and then the assembly is hot isostatically pressed to provide a very strong bond. The HIPing takes place at 1010°C (1850°F) at a pressure of 100 MPa (15,000 psi). The tungsten-carbide surface has a density of 14.52 to 14.72 g/cm^3, a hardness of 90.8 ± 5 HRA, and a minimum transverse rupture strength of 2450 MPa (355,000 psi).

Secondary operations are limited to grinding the face to remove any protruding sheet-metal cap and to expose the wear-resistant tungsten-carbide

FIGURE 17.16   A valve lifter for heavy-duty diesel engines produced from a hot-isostatic-pressed carbide cap on a steel shaft. *Source*: Courtesy of Metal Powder Industries Federation and Bodycote, Inc.

face. The high reliability of the HIP bond drastically reduced scrap rates to under 0.2%. No field failures have been experienced in over four years of full production. Production costs also were substantially reduced because of the hot isostatic pressing step

Source: Courtesy of Metal Powder Industry Federation.

17.3.3 Miscellaneous compacting and shaping processes

Powder-injection molding (PIM).   In this process (also called **metal-injection molding** (MIM)), very fine metal powders ($<10~\mu$m) are blended with a 25 to 45% polymer or a wax-based binder. The mixture then undergoes a process similar to die casting (Section 11.3.5; see also *injection molding of plastics* in Section 19.3); it is injected into the mold at a temperature of 135° to 200°C (275° to 400°F). The molded green parts are placed in a low-temperature oven to burn off the plastic (*debinding*), or the binder is removed by solvent extraction. The parts then are sintered in a furnace at temperatures as high as 1375°C (2500°F). Subsequent operations (such as hole tapping and heat treating) also may be performed.

Generally, metals that are suitable for powder-injection molding are those that melt at temperatures above 1000°C (1830°F); examples are carbon and stainless steels, tool steels, copper, bronze, and titanium. Typical parts made are components for watches, small-caliber gun barrels, scope rings for rifles, door hinges, impellers for sprinkler systems, and surgical knives.

The major advantages of powder-injection molding over conventional compaction are:

- Complex shapes having wall thicknesses as small as 5 mm (0.2 in.) can be molded and then removed easily from the dies.
- Mechanical properties are nearly equal to those of wrought products.
- Dimensional tolerances are good.
- High production rates can be achieved by using multicavity dies.
- Parts produced by the PIM process compete well against small investment-cast parts, small forgings, and complex machined parts. However, it does not compete well with zinc and aluminum die casting or with screw machining.

The major limitations of PIM are the high cost and limited availablitity of fine metal powders.

Rolling.   In *powder rolling* (also called **roll compaction**), the metal powder is fed into the roll gap in a two-high rolling mill (Fig. 17.17) and is compacted into a continuous strip at speeds of up to 0.5 m/s (100 ft/min). The rolling process can be carried out at room or at elevated temperature. Sheet metal for electrical and electronic components and for coins can be made by this process.

Extrusion.   Powders can be compacted by *extrusion*, whereby the powder is encased in a metal container and hot extruded. After sintering, preformed P/M parts may be reheated and forged in a closed die to their final shape. Superalloy powders, for example, are hot extruded for enhanced properties.

Pressureless compaction.   In *pressureless compaction*, the die is filled with metal powder by gravity, and the powder is sintered directly in the die. Because of

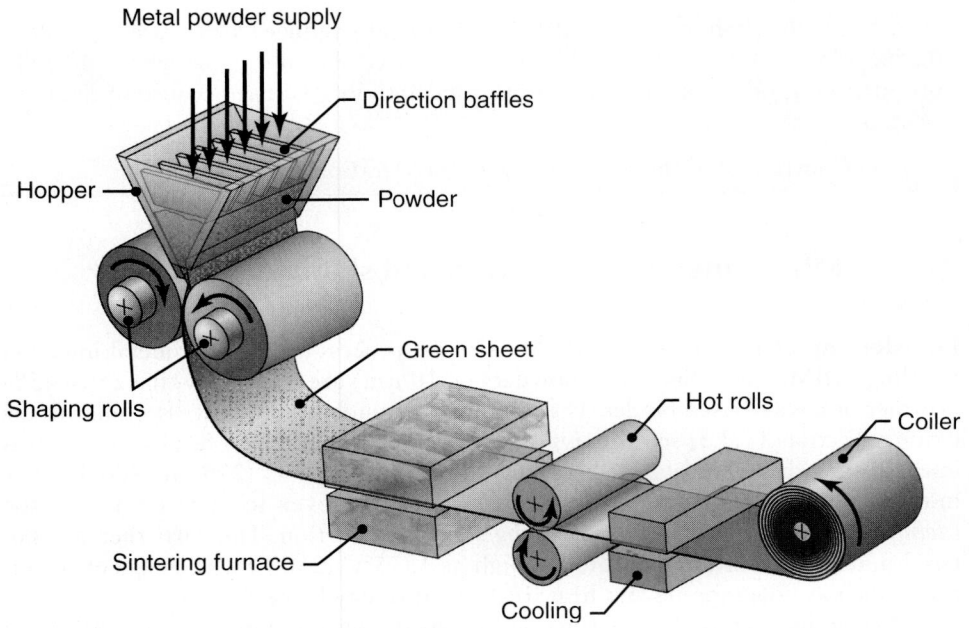

FIGURE 17.17 An illustration of powder rolling.

the resulting low density, pressureless compaction is used principally for porous metal parts, such as filters.

Spray deposition. Spray deposition is a *shape-generation process* (Fig. 17.18). The basic components of the spray-deposition process for metal powders are (a) an atomizer, (b) a spray chamber with inert atmosphere, and (c) a mold for producing preforms. The mold may be made in various shapes, such as billets, tubes, disks, and cylinders.

Although there are several variations, the best known is the *Osprey process* shown in Fig. 17.18. After the metal is atomized, it is deposited onto a cooled preform mold, usually made of copper or ceramic, where it solidifies. The metal particles bond together, developing a density that normally is above 99% of the solid-metal density. Spray-deposited forms may be subjected to additional shaping and consolidation processes, such as forging, rolling, and extrusion. The grain size is

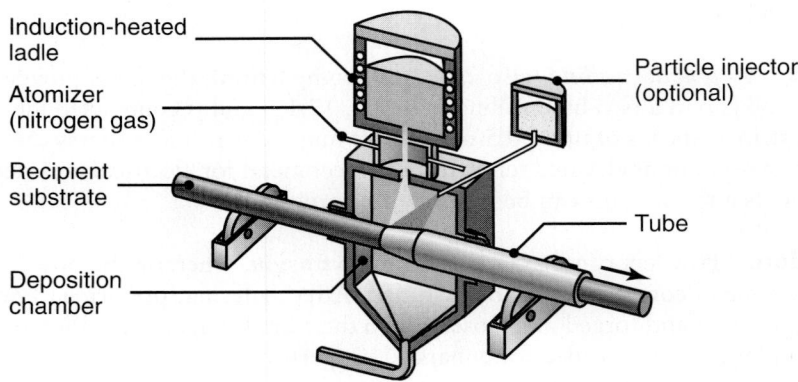

FIGURE 17.18 Spray deposition (*Osprey process*) in which molten metal is sprayed over a rotating mandrel to produce seamless tubing and pipe.

fine, and the mechanical properties are comparable to those for wrought products made of the same alloy.

Ceramic molds.   Ceramic molds for shaping metal powders are made by the technique used in investment casting. After the mold is made, it is filled with metal powder and placed in a steel container. The space between the mold and the container is filled with particulate material. The container then is evacuated, sealed, and subjected to hot isostatic pressing. Titanium-alloy compressor rotors for missile engines have been made by this process.

17.3.4 Punch and die materials

The selection of punch and die materials for powder metallurgy depends on the abrasiveness of the powder metal and the number of parts to be produced. Most common die materials are air- or oil-hardening tool steels (such as D2 or D3), with a hardness range of 60 to 64 HRC (Table 5.7). Because of their higher hardness and wear resistance, tungsten-carbide dies are used for more severe applications. Punches generally are made of similar materials.

Close control of die and punch dimensions is essential for proper compaction and die life. Too large a clearance between the punch and the die will allow the metal powder to enter the gap, where it will severely interfere with the operation and cause eccentric parts. Diametral clearances are generally less than 25 μm (0.001 in.). Die and punch surfaces must be lapped or polished (in the direction of tool movements in the die) for improved die life and overall performance.

17.4 | Sintering

Sintering is the process whereby green compacts are heated in a controlled-atmosphere furnace to a temperature below the melting point but sufficiently high to allow bonding (fusion) of the individual particles. As we have seen, the green compact is brittle, and its strength (*green strength*) is low. The nature and strength of the bond between the particles and, hence, that of the sintered compact, depend on the complex mechanisms of diffusion, plastic flow, evaporation of volatile materials in the compact, recrystallization, grain growth, and pore shrinkage.

The principal variables in sintering are temperature, time, and the furnace atmosphere. Sintering temperatures (Table 17.2) are generally within 70 to 90% of

TABLE 17.2

Sintering Temperature and Time for Various Metals		
Material	Temperature (°C)	Time (min)
Copper, brass, and bronze	760–900	10–45
Iron and iron-graphite	1000–1150	8–45
Nickel	1000–1150	30–45
Stainless steels	1100–1290	30–60
Alnico alloys (for permanent magnets)	1200–1300	120–150
Ferrites	1200–1500	10–600
Tungsten carbide	1430–1500	20–30
Molybdenum	2050	120
Tungsten	2350	480
Tantalum	2400	480

the melting point of the metal or alloy (see Table 3.1). Sintering times (Table 17.2) range from a minimum of about 10 minutes for iron and copper alloys to as much as 8 hours for tungsten and tantalum.

Continuous-sintering furnaces, which are used for most production, have three chambers:

1. *Burn-off chamber* for volatilizing the lubricants in the green compact in order to improve bond strength and prevent cracking.

2. *High-temperature chamber* for sintering.

3. *Cooling chamber.*

To obtain optimum properties, proper control of the furnace atmosphere is important for successful sintering. An oxygen-free atmosphere is essential to control the carburization and decarburization of iron and iron-based compacts and to prevent oxidation of the powders. A vacuum generally is used for sintering refractory-metal alloys and stainless steels. The gases most commonly used for sintering are hydrogen, dissociated or burned ammonia, partially combusted hydrocarbon gases, and nitrogen.

Sintering mechanisms are complex and depend on the composition of the metal particles as well as on the processing parameters. The sintering mechanisms are *diffusion, vapor-phase transport*, and *liquid-phase sintering*. As temperature increases, two adjacent powder particles begin to form a bond by a **diffusion mechanism** (*solid-state bonding*, Fig. 17.19a); as a result of this, the strength, density, ductility, and thermal and electrical conductivities of the compact increase. At the same time, however, the compact shrinks. Hence, allowances should be made for shrinkage as are done in casting.

A second sintering mechanism is **vapor-phase transport** (Fig. 17.19b). Because the material is heated to very close to its melting temperature, metal atoms will release to the vapor phase from the particles. At convergent geometries (the interface of two particles), the melting temperature is locally higher, and the vapor phase resolidifies. Thus, the interface grows and strengthens while each particle shrinks as a whole.

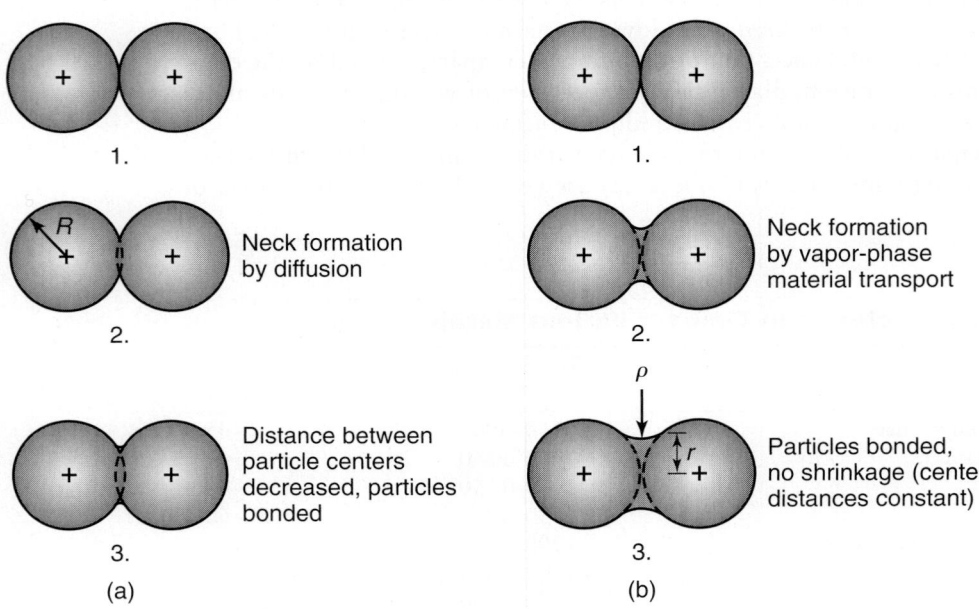

FIGURE 17.19   Schematic illustration of two mechanisms for sintering metal powders: (a) solid-state material transport; and (b) vapor-phase material transport. R = particle radius, r = neck radius, and ρ = neck-profile radius.

If two adjacent particles are of different metals, *alloying* can take place at the interface of the two particles. If one of the particles has a lower melting point than the other, the particle will melt and (because of surface tension) surround the particle that has not melted. Known as **liquid-phase sintering**, an example of this mechanism is cobalt in tungsten-carbide tools and dies (see Example 17.4 and also Section 22.4). Stronger and denser parts can be obtained in this way. In **spark sintering** (an experimental process), loose metal powders are placed in a graphite mold, heated by electric current, subjected to a high-energy discharge, and compacted—all in one step.

Mechanical properties.   Depending on temperature, time, and the processing history, different structures and porosities can be obtained in a sintered compact and, thus, affect its properties. Porosity cannot be eliminated completely because (a) voids remain after compaction and (b) gases evolve during sintering. Porosity may consist either of a *network* of interconnected pores or of *closed holes*. Generally, if the density of the material is less than 80% of its theoretical density, the pores are interconnected. Although porosity reduces the strength of the P/M product, it is an important characteristic for making metal filters and bearings and to allow for infiltration with liquid lubricants by surface tension.

Typical mechanical properties for several sintered P/M alloys are given in Table 17.3. The differences in mechanical properties of wrought versus P/M metals are given in Table 17.4. To further evaluate the differences between the properties of P/M, wrought, and cast metals and alloys, compare this table to the ones given in Parts I and II.

TABLE 17.3

Mechanical Properties of Selected P/M Materials							
Designation	MPIF type	Condition	Ultimate tensile strength (MPa)	Yield strength (MPa)	Elastic modulus (GPa)	Hardness	Elongation in 25 mm (%)
Ferrous					70		
FC-0208	N	AS	225	205	70	45 HRB	<0.5
		HT	295	—	110	95 HRB	<0.5
	R	AS	415	330	110	70 HRB	1
		HT	550	—	130	35 HRC	<0.5
	S	AS	550	395	130	80 HRB	1.5
		HT	690	655	145	40 HRC	<0.5
FN-0405	S	AS	425	240	145	72 HRB	4.5
		HT	1060	880	160	39 HRC	1
	T	AS	510	295	160	80 HRB	6
		HT	1240	1060		44 HRC	1.5
Aluminum							
601 AB, pressed bar		AS	110	48	—	60 HRH	6
		HT	252	241	—	75 HRH	2
Brass							
CZP-0220	T	—	165	76	—	55 HRH	13
	U	—	193	89	—	68 HRH	19
	W	—	221	103	—	75 HRH	23
Titanium							
Ti-6Al-4V		HIP	917	827	—	—	13
Superalloys							
Stellite 19		—	1035	—	—	49 HRC	<1

MPIF: Metal Powder Industries Federation. AS = as sintered, HT = heat treated, HIP = hot isostatically pressed.

TABLE 17.4

Comparison of Mechanical Properties of Some Wrought and Equivalent P/M Metals (as Sintered)

Metal	Condition	Density (%)	UTS (MPa)	Elongation in 50 mm (%)
Aluminum				
2014-T6	Wrought (W)	—	480	20
	P/M	94	330	2
6061-T6	W	—	310	15
	P/M	94	250	2
Copper, OFHC	W, annealed	—	235	50
	P/M	89	160	8
Brass, 260	W, annealed	—	300	65
	P/M	89	255	26
Steel, 1025	W, hot rolled	—	590	25
	P/M	84	235	2
Stainless steel, 303	W, annealed	—	620	50
	P/M	82	360	2

Note: The density and strength of P/M materials greatly increase with further processing, such as forging, isostatic pressing, and heat treatments.

The effects of various manufacturing processes on the mechanical properties of a titanium alloy are shown in Table 17.5. Note that hot-isostatic-pressed (HIP) titanium has properties that are similar to those for cast and forged titanium. It should be remembered, however, that unless they are precision forged, forgings generally require some additional machining or finishing processes that a P/M component may not require.

TABLE 17.5

Mechanical Property Comparisons for Ti-6AL-4V Titanium Alloy

Process (*)	Density (%)	Yield strength (MPa)	Ultimate tensile strength (MPa)	Elongation (%)	Reduction of area (%)
Cast	100	840	930	7	15
Cast and forged	100	875	965	14	40
Blended elemental (P + S)	98	786	875	8	14
Blended elemental (HIP)	>99	805	875	9	17
Prealloyed (HIP)	100	880	975	14	26

(*) P + S = pressed and sintered, HIP = hot isostatically pressed.
Source: Courtesy of R.M. German.

EXAMPLE 17.2 Shrinkage in sintering

In solid-state bonding during the sintering of a metal-powder green compact, the linear shrinkage is 4%. If the desired sintered density is 95% of the theoretical density of the metal, what should be the density of the green compact?

Solution Linear shrinkage is defined as $\Delta L/L_o$, where L_o is the original length. The volume shrinkage during sintering is

$$V_{\text{sint}} = V_{\text{green}} \left(1 - \frac{\Delta L}{L_o} \right)^3 \qquad (17.1)$$

The volume of the green compact must be higher than that of the sintered part, but the mass does not change during sintering. Thus, we can rewrite this expression in terms of the density ρ:

$$\rho_{green} = \rho_{sint}\left(1 - \frac{\Delta L}{L_o}\right)^3 \qquad (17.2)$$

Therefore,

$$\rho_{green} = 0.95(1 - 0.04)^3 = 0.84 \text{ or } 84\%$$

17.5 | Secondary and Finishing Operations

In order to further improve the properties of sintered P/M products, or to impart special characteristics, several additional operations may be carried out after sintering.

1. **Coining** and **sizing** are compacting operations, performed under high pressure in presses. The purposes of these operations are to impart dimensional accuracy to the sintered part and to improve its strength and surface finish by further densification.

2. *Preformed and sintered* alloy-powder compacts subsequently may be cold or hot **forged** to the desired final shapes and sometimes by *impact forging*. These products have a good surface finish, good dimensional tolerances, and a uniform and fine grain size. The superior properties obtained make this technology particularly suitable for such applications as highly stressed automotive and jet-engine components.

3. Powder-metal parts may be subjected to other finishing operations such as:
   - **Machining:** for producing various geometric features by milling, drilling, and tapping (to produce threaded holes).
   - **Grinding:** for improved dimensional accuracy and surface finish.
   - **Plating:** for improved appearance and resistance to wear and corrosion.
   - **Heat treating:** for improved hardness and strength.

4. The inherent porosity of P/M components can be utilized by **impregnating** them with a fluid. Bearings and bushings that are lubricated internally with up to 30% oil by volume are made by immersing the sintered bearing in heated oil. These bearings have a continuous supply of lubricant (due to capillary action) during their service lives (also referred to as *permanently lubricated*). Universal joints also are made by means of grease-impregnated P/M techniques, thus no longer requiring traditional grease fittings.

5. **Infiltration** is a process whereby a slug of a lower-melting-point metal is placed in contact the sintered part. The assembly then is heated to a temperature sufficiently high to melt the slug. The molten metal infiltrates the pores by capillary action and produces a relatively pore-free part having good density and strength. The most common application is the infiltration of iron-based compacts by copper. The advantages of infiltration are that the hardness and tensile strength of the part are improved and the pores are filled, thus preventing moisture penetration (which could cause corrosion). Furthermore, since some porosity is desirable when an infiltrant is used, the part may be sintered only partially, resulting in lower thermal warpage.

EXAMPLE 17.3 Powder-metallurgy gears for a garden tractor

Gears have been manufactured competitively using high-quality powders having high compressibility and, thus, requiring low compacting pressures. These parts have from medium-to-high density and are suitable for severe applications with high loads and for high-wear surfaces. In one application, a reduction gear for a garden tractor was made from iron powder and infiltrated with copper. Although its strength was acceptable, the wear rate under high loads was too high. Wear resulted in loss of tooth profile, side loading on the bearings, and high noise level. To improve strength and wear resistance, a new powder was selected containing 2.0% Ni, 0.5% C, and 0.5% Mo; the balance was atomized iron powder. Because of the size of the part, the pressing loads were very high.

The part was redesigned, and the tooling was made with three punches to compact the part to three different densities. The entire part required a compacting load of 4 MN (450 tons). By this method, high density was obtained in those sections of the part that required high strength and wear resistance. The high density also permitted carburization for hardness improvement. The presintered part was pressed again (to a higher density of 7.3 to 7.5 g/cm^3 in the tooth area) for improved strength, while the hub of the gear remained at its pressed density of 6.4 to 6.6 g/cm^3. The weight of the gear was 1 kg (2.25 lb).

EXAMPLE 17.4 Production of tungsten carbide for tools and dies

Tungsten carbide is an important tool and die material, mostly because of its hardness, strength, and wear resistance over a wide range of temperatures. Powder-metallurgy techniques are used in making such carbides. First, powders of tungsten and carbon are blended together in a ball mill or a rotating mixer. The mixture, which is basically 94% W and 6% C (by weight), is heated to approximately 1500°C (2800°F) in a vacuum-induction furnace. As a result, the tungsten is carburized and forms tungsten carbide in a fine powder form. A binding agent (usually cobalt) then is added to the tungsten carbide (together with an organic fluid, such as hexane), and the mixture is ball milled to produce a uniform and homogenous mix. This process can take several hours or even days.

The mixture is dried and consolidated, usually by cold compaction and under pressures of about 200 MPa (30,000 psi). The part then is sintered in a hydrogen atmosphere or vacuum furnace at a temperature of from 1350° to 1600°C (2500° to 2900°F), depending on its composition. At this temperature, the cobalt is in a liquid phase and acts as a binder for the carbide particles. Powders also may be hot pressed at the sintering temperature using graphite dies.

The tungsten carbide undergoes a linear shrinkage of about 16% during sintering—corresponding to a volume shrinkage of about 40%. Consequently, control of size and shape is important for producing tools with the desired dimensions and tolerances. A combination of other carbides (such as titanium carbide and tantalum carbide) also can be produced by this method using mixtures made by the method described. The trend is to use finer and finer particles, in order to further enhance the mechanical properties of such composite structures.

17.6 | Design Considerations

Because of the unique properties of metal powders, their flow characteristics in the die, and the brittleness of green compacts, there are certain design principles that should be followed (Figs. 17.20 through 17.22).

1. The shape of the compact must be kept as simple and uniform as possible. Sharp changes in contour, thin sections, variations in thickness, and high length-to-diameter ratios should be avoided.

2. Provision must be made for ejection of the green compact from the die without damaging the compact. Thus, holes or recesses should be parallel to the axis of punch travel. Chamfers also should be provided to avoid damage to the edges during ejection.

3. P/M parts should be made with the widest acceptable dimensional tolerances (consistent with their intended applications) in order to increase tool and die life and reduce production costs.

4. Part walls generally should not be less than 1.5 mm (0.060 in.) thick; however, with special care, walls as thin as 0.34 mm (0.0135 in.) can be pressed successfully on components 1 mm (0.04 in.) in length. Walls with length-to-thickness ratios greater than 8:1 are difficult to press, and density variations are virtually unavoidable.

5. Steps in parts can be produced if they are simple and their size doesn't exceed 15% of the overall part length. Larger steps can be pressed, but they require more complex, multiple-motion tooling.

6. Letters can be pressed if they are oriented perpendicular to the direction of pressing and can be raised or recessed. Raised letters are more susceptible to damage in the green stage and also prevent stacking during sintering.

7. Flanges or overhangs can be produced by a step in the die. However, long flanges can be broken upon ejection and may require more elaborate tooling. A long flange should incorporate a draft around the flange, a radius at the bottom edge, and a radius at the juncture of the flange and/or component body to reduce stress concentrations and the likelihood of fracture.

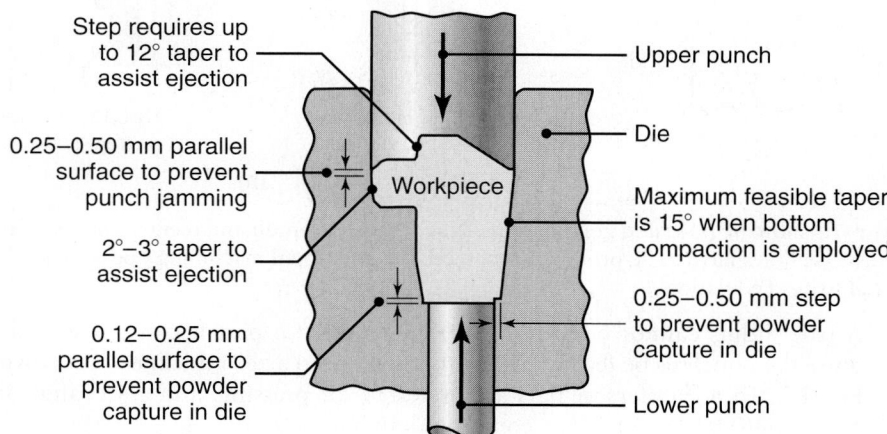

FIGURE 17.20 Die geometry and design features for powder-metal compaction. *Source:* Courtesy of Metal Powder Industries Federation.

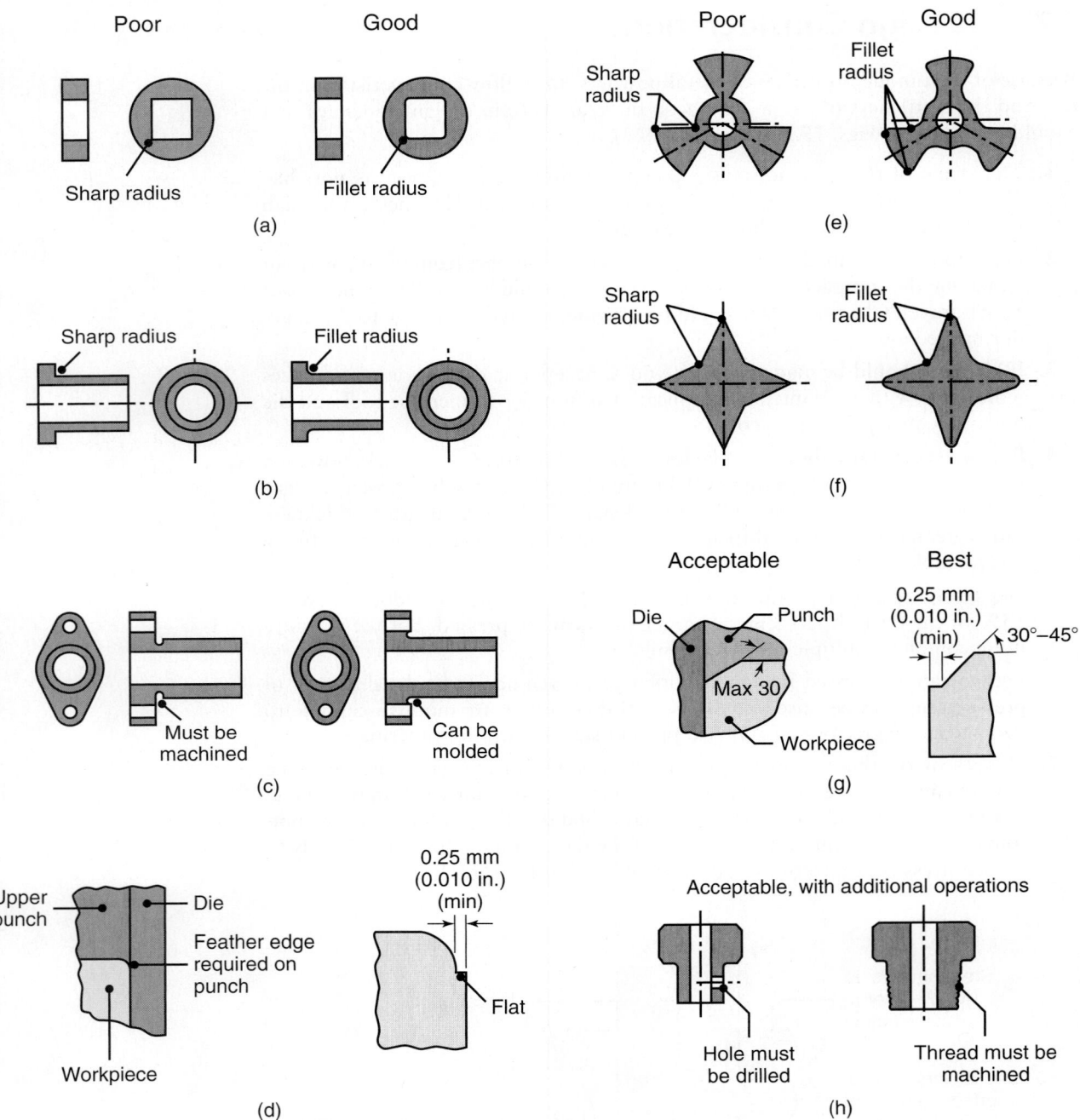

FIGURE 17.21   Examples of P/M parts showing poor and good designs. Note that sharp radii and reentry corners should be avoided and that threads and transverse holes have to be produced separately by additional machining operations. *Source:* Courtesy of Metal Powder Industries Federation.

8. A true radius cannot be pressed into the edge of a part because it would require the punch to be feathered (gently tapered) to a zero thickness as shown in Fig. 17.21d. Chamfers or flats are preferred for pressing, and a 45° angle in a 0.25 mm (0.010 in.) flat is a common design approach.

9. Keys, keyways, and holes used for transmitting torques on gears and pulleys can be formed during powder compaction. Bosses can be produced provided that proper drafts are used and their length is small compared to the overall component.

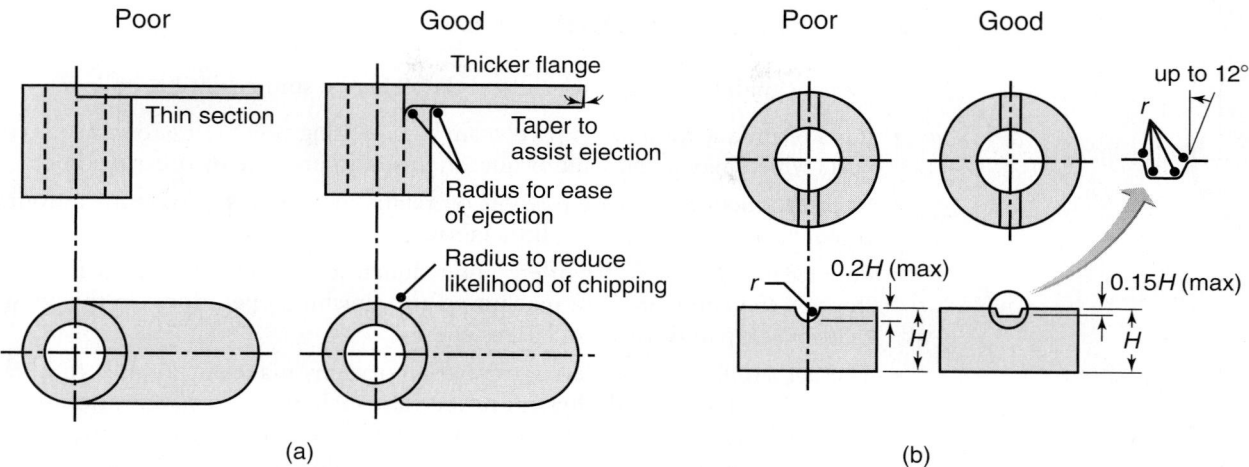

Poor   Good   Poor   Good

Thin section
Thicker flange
Taper to assist ejection
Radius for ease of ejection
Radius to reduce likelihood of chipping

up to 12°
r
0.2H (max)
r
H
0.15H (max)
H

(a)   (b)

FIGURE 17.22   (a) Design features to use with unsupported flanges. (b) Design features for use with grooves. *Source*: Courtesy of Metal Powder Industries Federation.

10. Notches and grooves can be made if they are oriented perpendicular to the pressing direction. It is recommended that circular grooves not exceed a depth of 20% of the overall component, and rectangular grooves should not exceed 15%.

11. Parts produced through powder-injection molding have similar design constraints as with injection molding of polymers. With PIM, wall thicknesses should be uniform to minimize distortion during sintering. Also, molds should be designed with smooth transitions to prevent powder accumulation and to allow uniform distribution of metal powder (Fig. 17.23).

12. Dimensional tolerances of sintered P/M parts are usually on the order of ±0.05 to 0.1 mm (±0.002 to 0.004 in.). Tolerances improve significantly with additional operations, such as sizing, machining, and grinding.

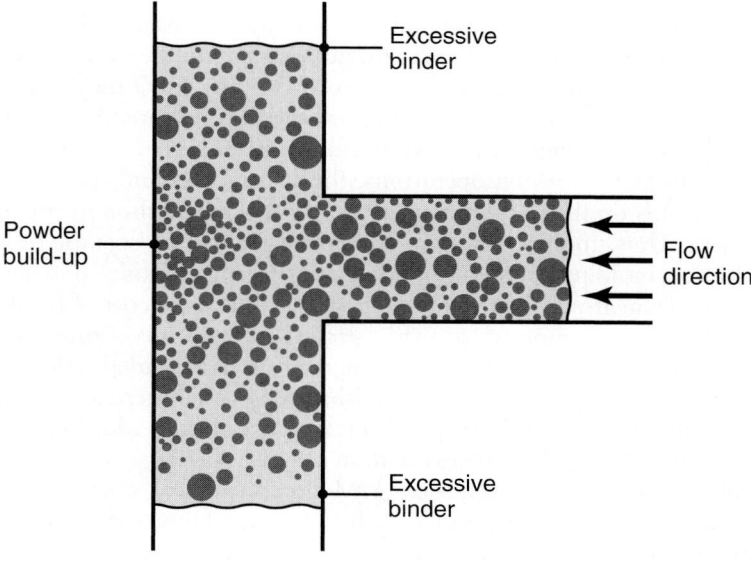

Excessive binder

Powder build-up

Flow direction

Excessive binder

FIGURE 17.23   Illustration of non-uniform powder distribution due to abrupt channel transitions in metal injection molding.

17.7 | Process Capabilities

The process capabilities of powder metallurgy may be summarized as follows:

- It is a technique for making parts from high-melting-point refractory metals, and parts which may be difficult or uneconomical to produce by other methods.
- High production rates are possible on relatively complex parts using automated equipment and requiring little labor.
- Powder-metal processing offers good dimensional control and (in many instances) the elimination of machining and finishing operations; in this way, it reduces scrap and waste and saves energy.
- The availability of a wide range of compositions makes it possible to obtain special mechanical and physical properties, such as stiffness, vibration damping, hardness, density, toughness, and specific electrical and magnetic properties. Some of the newer highly alloyed superalloys can be manufactured into parts only by P/M processing.
- It offers the capability of impregnation and infiltration for specific applications.

The limitations to P/M are:

- The high cost of metal powder, particularly those for powder-injection molding, as compared to that of raw materials to be cast or wrought.
- The high cost of tooling and equipment for small production runs.
- Limitations on part size and shape complexity.
- Mechanical properties, such as strength and ductility, that generally are lower than those obtained by forging. However, the properties of full-density P/M parts made by HIP or by additional forging operations can be as high as those made by other processes.

17.8 | Economics of Powder Metallurgy

Because powder metallurgy can produce parts at net or near-net shape and thus eliminate many secondary manufacturing and assembly operations, it increasingly has become competitive with casting, forging, and machining. On the other hand, the high initial cost of punches, dies, and equipment for P/M processing means that production volume must be sufficiently high to warrant this expenditure. Although there are exceptions, the process generally is economical for quantities over 10,000 pieces.

As in other metalworking operations, the cost of dies and tooling in powder metallurgy depends on the part shape complexity and the method of processing the metal powders. Thus, tooling costs for processes such as hot isostatic pressing and powder-injection molding are higher than the more conventional powder processing. Because it is a near-net shape manufacturing method, the cost of finishing operations in P/M is low compared to other processes, such as casting and forging. However, if there is certain features to the part (such as threaded holes, undercuts, and transverse cavities and holes), then finishing costs will increase. Consequently, following design guidelines in P/M to minimize or avoid such additional operations can be more important in this process than in others.

Equipment costs for conventional P/M processing are somewhat similar to bulk deformation processing of metals, such as forging. However, the cost increases significantly when using methods such as HIP and PIM. Labor costs are not as high

TABLE 17.6

Forged and P/M Titanium Parts and Cost Savings				
Part	Weight (kg)			
	Forged billet	P/M	Final part	Cost saving (%)
F-14 Fuselage brace	2.8	1.1	0.8	50
F-18 Engine mount support	7.7	2.5	0.5	20
F-18 Arrestor hook support fitting	79.4	25	12.9	25
F-14 Nacelle frame	143	82	24.2	50

in other processes, primarily because the individual operations (such as powder blending, compaction, and sintering) are performed on highly automated equipment. Thus, the skills required are not as high.

The near-net-shape capability of P/M significantly reduces or eliminates scrap. For example, weight comparisons of aircraft components produced by forging and by P/M processes are shown in Table 17.6. Note that the P/M parts are subjected to further machining processes, thus the final parts weigh less than those made by either of the two processes alone.

SUMMARY

- Powder metallurgy is a net-shape forming process consisting of producing metal powders, blending them, compacting them in dies, and sintering them to impart strength, hardness, and toughness. Compaction also may be carried out by cold or hot isostatic pressing for improved properties. Although the size and the weight of its products are limited, the P/M process is capable of producing relatively complex parts economically, in net-shape form, to close dimensional tolerances, and from a wide variety of metal and alloy powders.

- Secondary and finishing operations may be performed on P/M parts to improve their dimensional accuracy, surface finish, mechanical and physical properties, and appearance. These operations include forging, heat treating, machining, grinding, plating, infiltration (such as with oil), and impregnation (such as with low-melting-point metals).

- Control of powder shape and quality, process variables, and sintering atmospheres are important considerations in product quality. Density and mechanical and physical properties can be controlled by tooling design and by compacting pressure.

- An important P/M process is powder-injection molding, which involves mixing the very fine metal powders with polymers to make them flow more easily into molds of complex shape during compaction. The polymers subsequently evaporate during sintering.

- Design considerations for powder metallurgy include the shape of the part, the ability to eject the green compact from the die, and the dimensional tolerances acceptable for the particular application. The P/M process is suitable for medium- to high-volume production runs and for relatively small parts. It has some competitive advantages over other methods of production, such as casting, forging, and machining.

KEY TERMS

Atomization	Hot isostatic pressing	Powder metallurgy
Blending	Impregnation	Pressing
Carbonyls	Infiltration	Pressureless compaction
Cold isostatic pressing	Injection molding	Reduction
Comminution	Mechanical alloying	Screening
Compaction	Metal injection molding	Shape factor
Diffusion	Oxide powder in tube process	Sintering
Electrolytic deposition	Powder injection molding	Spark sintering
Green compact		
Green strength		

BIBLIOGRAPHY

ASM Handbook, Vol. 7: *Powder Metal Technologies and Applications*, ASM International, 1998.

Atkinson, H.V., and Rickinson, B.A., *Hot Isostatic Pressing*, Adam Hilger, 1991.

Dowson, G., *Powder Metallurgy*: *The Process and its Products*, Adam Hilger, 1990.

Fayed, M., and Otten L., (eds.), *Handbook of Powder Science and Technology*, 2nd ed., Chapman & Hall, 1997.

German, R.M., *Powder Injection Molding*, Metal Powder Industries Federation, 1990.

_____, *Sintering Technology and Practice*, Wiley, 1996.

_____, and Bose, A., *Injection Molding of Metals and Ceramics*, Metal Powder Industries Federation, 1997.

Hausner, H.H., and Mall, M.K., *Handbook of Powder Metallurgy*, Chemical Publishing Company, 1982.

Kuhn, H.A., and Ferguson, B.L., *Powder Forging*, Metal Powder Industries Federation, 1990.

Lawley, A., *Atomization: The Production of Metal Powders*, Metal Powder Industries Federation, 1992.

Lenel, F.V., *Powder Metallurgy: Principles and Applications*, American Powder Metallurgy Institute, 1980.

Powder Metallurgy Design Manual, 2nd ed., Metal Powder Industries Federation, 1995.

REVIEW QUESTIONS

17.1 Describe briefly the production steps involved in making powder-metallurgy parts.

17.2 Name the various methods of powder production and explain the types of powders produced.

17.3 Explain why metal powders are blended.

17.4 Is green strength important? Explain.

17.5 Name the methods used in metal powder compaction.

17.6 What hazards are involved in P/M processing? Explain their causes.

17.7 What requirements should punches and dies meet in P/M processing?

17.8 Describe what occurs to metal powders during sintering.

17.9 Why might secondary and finishing operations be performed on P/M parts?

17.10 Explain the difference between impregnation and infiltration. Give some applications for each.

17.11 What is mechanical alloying? What are its advantages over the conventional alloying of metals?

17.12 Why are protective atmospheres necessary in sintering? What would be the effects on the properties of P/M parts if such atmospheres were not used?

17.13 How are spherical powders produced? How are sharp-edged powders produced?

17.14 What is "screening" of metal powders? Why is it done?

QUALITATIVE PROBLEMS

17.15 Why is there density variation in the compacting of powders? How is it reduced?

17.16 What is the magnitude of the stresses and forces involved in powder compaction?

17.17 Give the reasons why powder-injection molding is an important process.

17.18 How does the equipment used for powder compaction vary from that used in other metalworking operations?

17.19 Why do mechanical and physical properties depend on the density of P/M parts?

17.20 What are the effects of the different shapes and sizes of metal particles in P/M processing?

17.21 Describe the relative advantages and limitations of cold and hot isostatic pressing.

17.22 Are the requirements for punch and die materials in powder metallurgy different from those for forging and extrusion? Explain.

17.23 We have stated that P/M can be competitive with processes such as casting and forging. Explain why this is so, giving ranges of applications.

17.24 Explain the reasons for the shapes of the curves shown in Fig. 17.10 and for their relative positions on the charts.

17.25 Should green compacts be brought up to the sintering temperature slowly or rapidly? Explain your reasoning.

17.26 Because they undergo special processing, metal powders are more expensive than the same metals in bulk form, especially powders used in powder-injection molding. How is the additional cost justified in processing powder-metallurgy parts?

17.27 In Fig. 17.11e, we noted that the pressure is not uniform across the diameter of the compact at a particular distance from the punch. What is the reason for this variation?

17.28 Why do the compacting pressure and the sintering temperature depend on the type of powder metal?

17.29 Comment on the shapes and the ranges of the curves of process capabilities in Fig. 17.14.

QUANTITATIVE PROBLEMS

17.30 Estimate the maximum tonnage required to compact a brass slug 2.5 in. in diameter. Would the height of the slug make any difference in your answer? Explain your reasoning.

17.31 Refer to Fig. 17.10a. What should be the volume of loose, fine iron powder in order to make a solid cylindrical compact 25 mm in diameter and 15 mm high?

17.32 Determine the shape factors for (a) a cylinder with dimensional ratios of 1:1:1 and (b) a flake with ratios of 1:10:10.

17.33 Estimate the number of particles in a 400 g sample of iron powder, if the particle size is 75 μm.

17.34 Assume that the surface of a copper particle is covered by an oxide layer 0.1 mm in thickness. What is the volume (and the percentage of volume) occupied by this layer if the copper particle itself is 60 μm in diameter?

17.35 Survey the technical literature to obtain data on shrinkage during the sintering of P/M parts. Comment on your observations.

17.36 Plot the total surface area of a 100 g sample of aluminum as a function of the natural log of particle size.

17.37 A coarse copper powder is compacted in a mechanical press at a pressure of 20 tons/in^2. During sintering, the green part shrinks an additional 8%. What is the final density?

17.38 A gear is to be manufactured from iron powders. It is desired that it have a final density 90% that of cast iron, and it is known that the shrinkage in sintering will be approximately 5%. For a gear that is 2.5 in. in diameter and has a 0.75 in. hub, what is the required press force?

17.39 Assume that you are an instructor covering the topics described in this chapter, and you are giving a quiz on the numerical aspects to test the understanding of the students. Prepare two quantitative problems and supply the answers.

SYNTHESIS, DESIGN, AND PROJECTS

17.40   Make sketches of P/M products in which density variations (see Fig. 17.11) would be desirable. Explain why in terms of the functions of these parts.

17.41   Compare the design considerations for P/M products to those for (a) casting and (b) forging. Describe your observations.

17.42   Are there applications in which you, as a manufacturing engineer, would not recommend a P/M product? Explain.

17.43   Describe in detail other methods of manufacturing the parts shown in Fig. 17.1.

17.44   How large is the grain size of metal powders that can be produced in atomization chambers? Conduct a literature search to determine the answer.

17.45   Plot the opening size versus the mesh size for screens used in powder-size sorting.

17.46   Use the Internet to locate suppliers of metal powders and compare the cost of the powder to the cost of ingots for five different materials.

17.47   It is known that in the design of P/M gears the hub outside diameter should be as far as possible from the root of the gear. Explain why this is the case.

17.48   Explain why powder-metal parts commonly are used for machine elements requiring good frictional and wear characteristics and for mass-produced parts.

17.49   It was stated that powder-injection molding competes well with investment casting and small forgings for various materials but not with zinc and aluminum die castings. Explain why.

17.50   Describe how the information given in Fig. 17.14 would be helpful to you in designing P/M parts.

17.51   We have stated that in the process shown in Fig. 17.18 shapes produced are limited to axisymmetric parts. Do you think it would be possible to produce other shapes as well? Describe how you would modify the design of the setup to make those shapes and explain the difficulties that may be encountered.

Processing of Ceramics, Glass, and Super-conductors

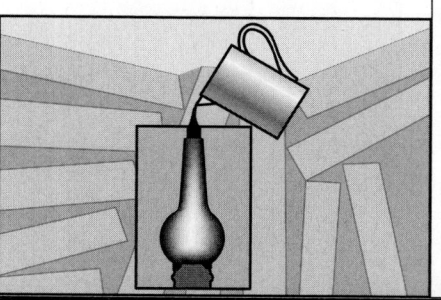

In this chapter, we describe the manufacturing processes used for producing discrete parts from ceramics and glass and making superconducting wire and tape. Specifically:

- Preparation of raw materials as fine ceramic particles.
- Casting, pressing, extrusion, and molding to produce discrete product shapes.
- Drying and firing to induce strength and hardness.
- Finishing operations to improve size and surface finish.
- Production of superconductors into wire and tapes.

Typical products made: Ceramics: insulators, rotors for gas turbines, lightweight components for high-speed machines, ball and roller bearings, seals, ovenware, tiles, and glasses: glazing, laminated glass, bulletproof glass, bulbs, lenses, bottles, glass fibers, rods, and tubing.

18.1 | Introduction

The properties and various applications of ceramics and glasses were described in Chapter 8. These materials have important characteristics, such as high-temperature strength and hardness, low electrical and thermal conductivity, inertness to chemicals, and resistance to wear and corrosion. The wide range of applications for these materials include parts as simple as floor tiles and dishes, electrical insulators and spark plugs, and ball bearings and thermal insulation for the space shuttle orbiter.

In this chapter, we describe the techniques that are available for processing ceramics into numerous useful products (Fig. 18.1). The methods employed for ceramics consist of crushing the raw materials, shaping them by various means, drying, firing, and finishing operations. For glasses, the processes involves mixing and melting the raw materials in a furnace and shaping them in molds and by various techniques—depending on the shape and size of the part—such as for discrete products (bottles) and continuous products (flat glass, rods, tubing, and fibers). Glasses also are strengthened by thermal and chemical means as well as by lamination with polymer sheets (as done with windshields and bulletproof glass).

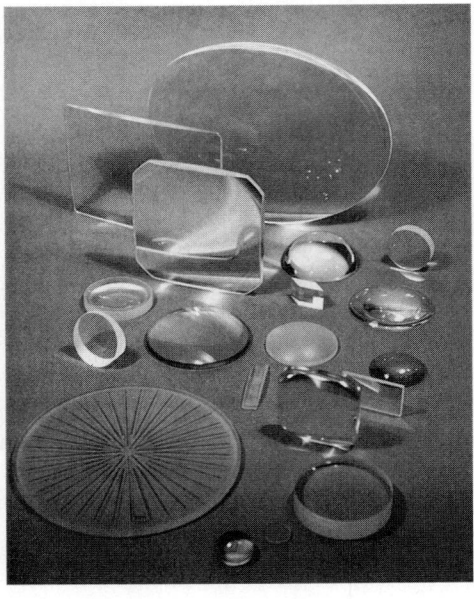

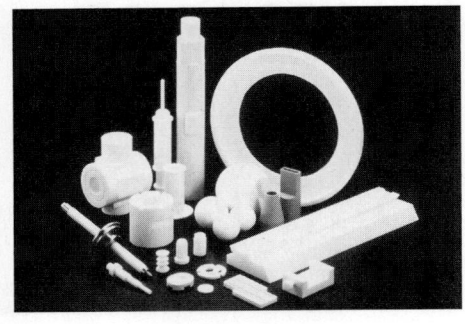

(a)   (b)

FIGURE 18.1   Example of typical glass parts. *Source*: (a) Courtesy of Commercial Optical Manufacturing, Inc. (b) Courtesy of Kyocera.

18.2 | Shaping Ceramics

Several techniques are available for processing ceramics into useful products (Table 18.1), depending on the type of ceramics involved and their shapes. Production of some ceramic parts (such as pottery, ovenware, or floor tiles) generally does not involve the same level of control of materials and processes as do high-tech components (made of such structural ceramics as silicon nitride and silicon carbide) and cutting tools (using, for example, aluminum oxide). Generally, however, the procedure involves the following steps (Fig. 18.2):

1. Crushing or grinding the raw materials into very fine particles
2. Mixing them with additives to impart certain desirable characteristics
3. Shaping, drying, and firing the material

TABLE 18.1

General Characteristics of Ceramics Processing		
Process	Advantages	Limitations
Slip casting	Large parts, complex shapes, low equipment cost	Low production rate, limited dimensional accuracy
Extrusion	Hollow shapes and small diameters, high production rate	Parts have constant cross-section, limited thickness
Dry pressing	Close tolerances, high production rates (with automation)	Density variation in parts with high length-to-diameter ratios, dies require abrasive-wear resistance, equipment can be costly
Wet pressing	Complex shapes, high production rate	Limited part size and dimensional accuracy, tooling costs can be high
Hot pressing	Strong, high-density parts	Protective atmospheres required, die life can be short
Isostatic pressing	Uniform density distribution	Equipment can be costly
Jiggering	High production rate with automation, low tooling cost	Limited to axisymmetric parts, limited dimensional accuracy
Injection molding	Complex shapes, high production rate	Tooling can be costly

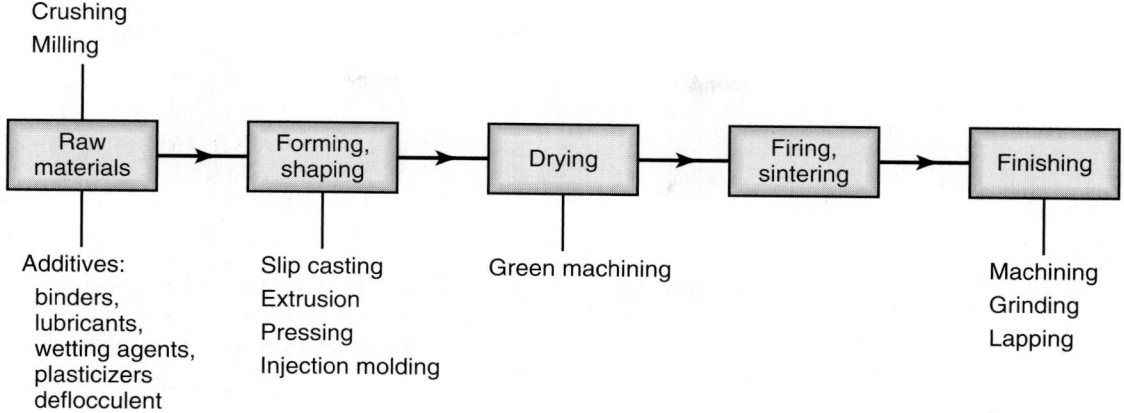

FIGURE 18.2   Processing steps involved in making ceramic parts.

They may be subjected to additional processing, such as machining and grinding, for better control of dimensions and surface finish.

The first step in processing ceramics is the *crushing* (also called *comminution* or *milling*) of the raw materials. Crushing generally is done in a *ball mill* (see Fig. 17.6), either dry or wet. Wet crushing is more effective, because it keeps the particles together and also prevents the suspension of fine particles in the air. The particles then may be *sized* (passed through a sieve), filtered, and washed.

The ground particles are then mixed with *additives*—the functions of which are one or more of the following:

- **Binder:** for holding ceramic particles together.
- **Lubricant:** to reduce internal friction between particles during molding and to help remove the part from the mold.
- **Wetting agent:** to improve mixing.
- **Plasticizer:** to make the mix more plastic and formable.
- **Agents:** to control foaming and sintering.
- **Deflocculent:** to make the ceramic-water suspension more uniform by changing the electrical charges on the particles of clay (so that the particles repel rather than attract each other). Water is added to make the mixture more pourable and less viscous. Typical deflocculents are Na_2CO_3 and Na_2SiO_3 in amounts of less than 1%.

The three basic shaping processes for ceramics are casting, plastic forming, and pressing, as described next.

18.2.1 Casting

The most common casting process is **slip casting** (also called **drain casting**), as in Fig. 18.3. A **slip** is a suspension of *colloidal* (small particles that do not settle) ceramic particles in an *immiscible liquid* (insoluble in each other), which is generally water. The slip is poured into a porous mold, typically made of plaster of paris. Molds also may consist of several components.

The slip must have sufficient fluidity and low enough viscosity to flow easily into the mold, much like the fluidity of molten metals described in Section 10.3. Pouring the slip must be done properly, as air entrapment can be a significant problem during casting. Iron and other magnetic materials are removed using in-line separators.

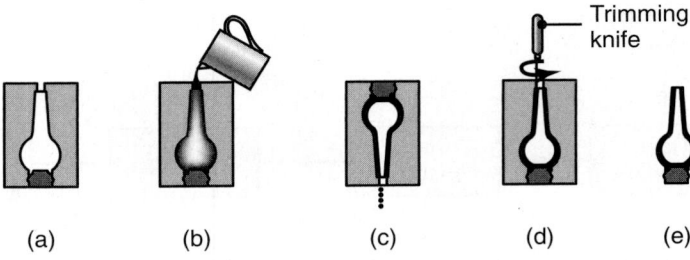

(a) (b) (c) (d) (e)

FIGURE 18.3 Sequence of operations in slip-casting a ceramic part. After the slip has been poured, the part is dried and fired in an oven to give it strength and hardness. *Source*: After F. H. Norton.

After the mold has absorbed some of the water from the outer layers of the suspension, it is inverted, and the remaining suspension is poured out (for the making of hollow objects, as in the slush casting of metals described in Section 11.3.3). The top of the part is then trimmed (note the trimming tool in Fig. 18.3d), the mold is opened, and the part is removed.

Large and complex parts (such as plumbing ware, art objects, and dinnerware) can be made by slip casting. Although mold and equipment costs are low, dimensional control is poor, and the production rate is low. In some applications, components of the product (such as handles for cups and pitchers) are made separately and then joined, using the slip as an adhesive. Molds also may consist of multicomponents. Iron and other magnetic materials are removed using in-line magnetic separators.

For solid-ceramic parts, the slip is supplied continuously into the mold to replenish the absorbed water; otherwise, the part will shrink. At this stage, the part is described as either a soft solid or semirigid. The higher the concentration of solids in the slip, the less water has to be removed. The part (called *green*, as in powder metallurgy) then is fired.

While the ceramic parts are still green, they may be machined to produce certain features or to give dimensional accuracy to the parts. Because of the delicate nature of the green compacts, however, machining usually is done manually or with simple tools. For example, the flashing in a slip casting may be removed gently with a fine wire brush, or holes can be drilled in the mold. Detailed work (such as the tapping of threads) generally is not done on green compacts because warpage (due to firing) makes such machining not viable.

Doctor-blade process. Thin sheets of ceramics (less than 1.5 mm (0.06 in.) thick) can be made by a casting technique called the *doctor-blade process* (Fig. 18.4). The slip is cast over a moving plastic belt while its thickness is controlled by a blade. Ceramic sheets also may be produced by other methods, including: (a) **rolling** the slip between pairs of rolls and (b) **casting** the slip over a paper tape which subsequently burns off during firing.

18.2.2 Plastic forming

Plastic forming (also called *soft, wet,* or *hydroplastic forming*) can be carried out by various methods, such as extrusion, injection molding, or molding and jiggering (Fig. 18.5). Plastic forming tends to orient the layered structure of clay along the direction of material flow and, hence, tends to cause anisotropic behavior of the material both in subsequent processing and in the final properties of the ceramic product.

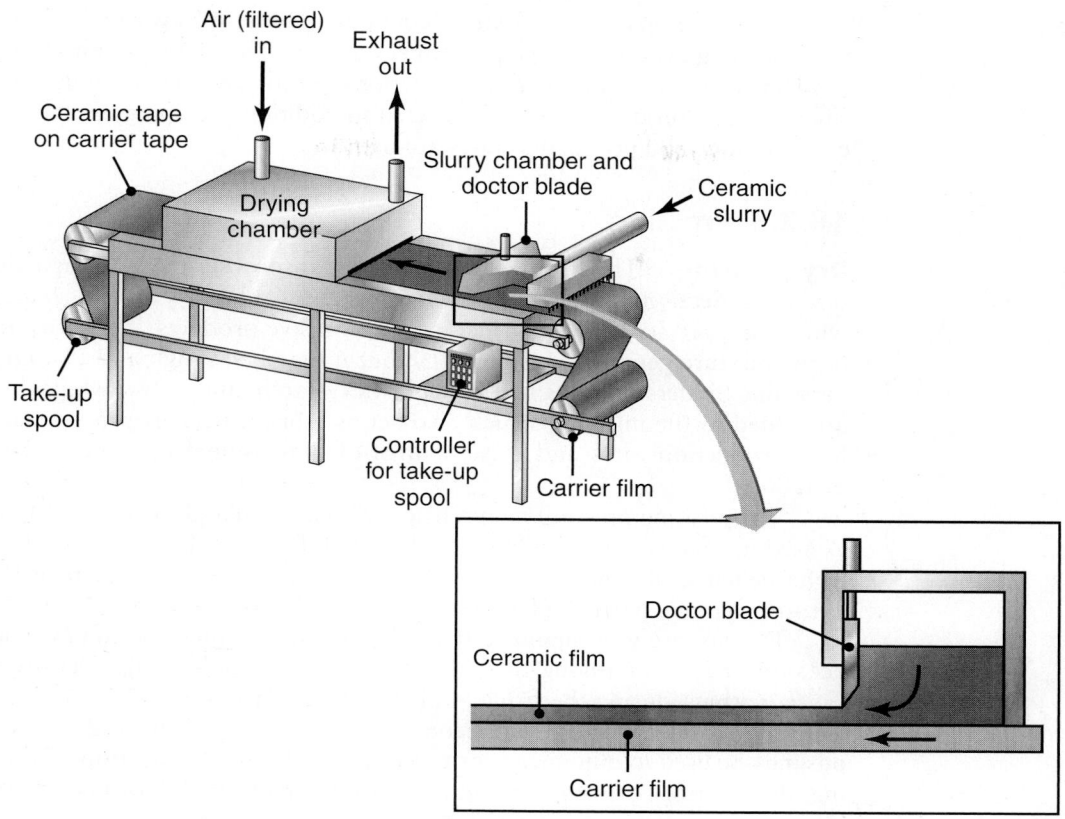

FIGURE 18.4   Production of ceramic sheets through the doctor-blade process.

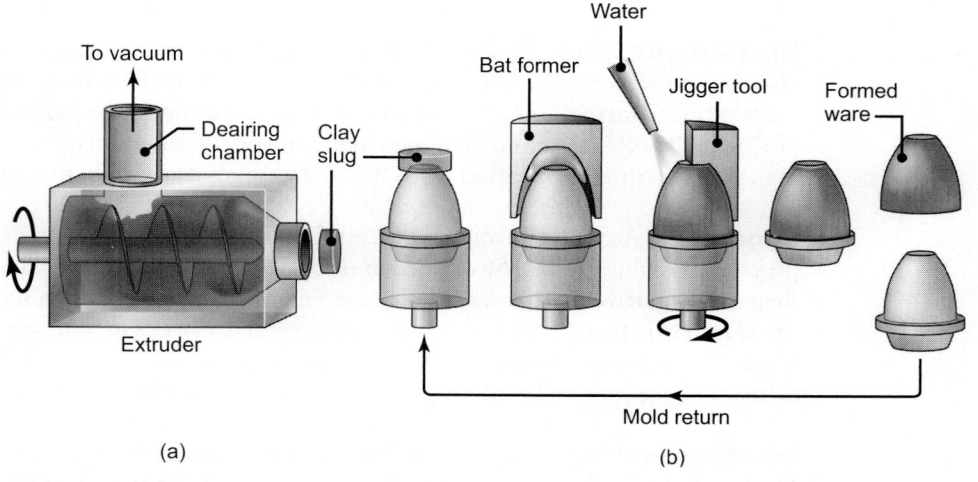

FIGURE 18.5   (a) Extruding and (b) jiggering operations. *Source*: After R. F. Stoops.

In *extrusion*, the clay mixture (containing 20 to 30% water) is forced through a die opening by a screw-type piece of equipment. The cross-section of the extruded product is constant, and there are limitations to wall thickness for hollow extrusions. The extruded products may be subjected to additional shaping operations. Tooling costs are low, and production rates are high.

18.2.3 Pressing

Dry pressing.   This is a technique similar to powder-metal compaction, as described in Section 17.3. *Dry pressing* is used for relatively simple shapes, such as whiteware, refractories for furnaces, and abrasive products. The moisture content of the mixture generally is below 4%, but it may be as high as 12%. Organic and inorganic binders (such as stearic acid, wax, starch, and polyvinyl alcohol) usually are added to the mixture, which also act as lubricants. This process has the same high production rates and close control of dimensional tolerances as in powder metallurgy.

The pressing pressure ranges from 35 to 200 MPa (5 to 30 ksi). Modern presses used for dry pressing are highly automated. Dies usually are made of carbides or of hardened steel. They must have high wear resistance in order to withstand the abrasive ceramic particles; hence, they can be expensive.

Density can vary significantly in dry-pressed ceramics (as in P/M compaction) because of friction among the particles and at the mold walls. Density variations cause warping during the firing, which is particularly severe for parts having high length-to-diameter ratios. The recommended maximum ratio is 2:1. Several methods may be used to minimize density variations, including: (a) proper design of tooling, (b) vibratory pressing and impact forming (particularly for nuclear-reactor fuel elements), and (c) isostatic pressing.

Wet pressing.   In *wet pressing*, the part is formed in a mold while under high pressure in a hydraulic or mechanical press. This process generally is used to make intricate shapes. Moisture content usually ranges from 10 to 15%. Production rates are high; however, (a) part size is limited, (b) dimensional control is difficult to achieve because of shrinkage during drying, and (c) tooling costs can be high.

Isostatic pressing.   Used extensively in powder metallurgy, *isostatic pressing* also is used for ceramics in order to obtain uniform density distribution throughout the part during compaction. For example, automotive spark-plug insulators are made by this method at room temperature, while silicon-nitride vanes for high-temperature applications (see Fig. 8.1) are made by *hot isostatic pressing*.

Jiggering.   A combination of processes is used to make ceramic plates. In this process, clay slugs first are extruded and formed into a *bat* over a plaster mold. They then are jiggered on a rotating mold (see Fig. 18.5b). *Jiggering* is a motion in which the clay bat is formed by means of templates or rollers. The part then is dried and fired. The jiggering process is confined to axisymmetric parts and has limited dimensional accuracy. The operation is automated for improved productivity.

Injection molding.   *Injection molding* is used extensively for the precision forming of ceramics in high-technology applications, such as for rocket-engine components. The raw material is mixed with a binder, such as a thermoplastic polymer (polypropylene, low-density polyethylene, or ethylene vinyl acetate) or wax. The binder usually is removed by pyrolysis (inducing chemical changes by heat); the part then is sintered by firing.

The injection-molding process can produce thin sections (typically less than 10 to 15 mm (0.4 to 0.6 in.) thick) from most engineering ceramics, such as alumina, zirconia, silicon nitride, silicon carbide, and sialon. Thicker sections require careful control of the materials used and of the processing parameters in order to avoid defects, such as internal voids and cracks—especially those due to shrinkage.

Hot pressing.   In this process (also called *pressure sintering*), the pressure and the heat are applied simultaneously. This method reduces porosity and, thus, makes the part denser and stronger. Graphite commonly is used as a punch and die material, and protective atmospheres usually are employed during pressing.

Hot isostatic pressing (Section 17.3.2) also may be used, particularly to improve shape accuracy and the quality of high-technology ceramics, such as silicon carbide and silicon nitride. Glass-encapsulated HIP processing has been shown to be effective for this purpose.

18.2.4 Drying and firing

The next step in ceramic processing is to dry and fire the part to give it the proper strength and hardness. *Drying* is a critical stage because of the tendency for the part to warp or crack from variations in the moisture content and in the thickness of the part. Control of atmospheric humidity and of ambient temperature is important in order to reduce warping and cracking.

Loss of moisture during drying causes shrinkage of the part by as much as 15 to 20% from the original, moist size (Fig. 18.6). In a humid environment, the evaporation rate is low, and consequently, the moisture gradient across the thickness of the part is lower than that in a dry environment. This gradient, in turn, prevents a large, uneven gradient in shrinkage from the surface to the interior during drying.

A ceramic part that has been shaped by any of the methods described thus far is in the *green* state. This part can be machined in order to bring it closer to a near-net shape. Although the green part should be handled carefully, machining it is not particularly difficult because of its relative softness.

Firing (also called **sintering**) involves heating the part to an elevated temperature in a controlled environment. Some shrinkage occurs during firing. Firing gives the ceramic part its strength and hardness. This improvement in properties results from (a) the development of a strong bond between the complex oxide particles in the ceramic and (b) reduced porosity. A more recent technology (although not yet commercialized) involves the **microwave sintering** of ceramics in furnaces operating at more than 2 GHz. Its cost effectiveness depends on the availability of inexpensive furnace insulation.

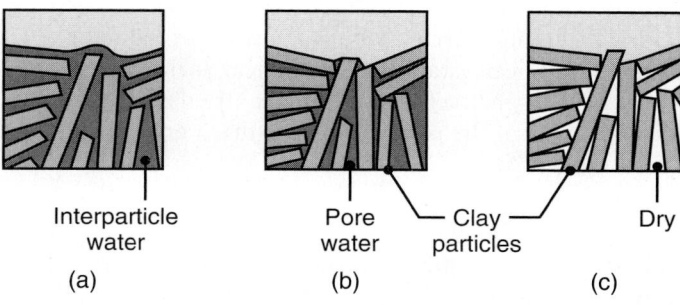

Interparticle
water

Pore
water

Clay
particles

Dry

(a)   (b)   (c)

FIGURE 18.6   Shrinkage of wet clay caused by the removal of water during drying. Shrinkage may be as much as 20% by volume. *Source*: After F. H. Norton.

Nanophase ceramics (described in Section 8.2.5) can be sintered at lower temperatures than those used for conventional ceramics. They are easier to fabricate, because they can be compacted at room temperature to high densities, hot pressed to theoretical density, and formed into net-shaped parts without using binders or sintering aids.

18.2.5 Finishing operations

Because firing causes dimensional changes, additional operations may be performed to (a) give the ceramic part its final shape, (b) improve its surface finish and dimensional tolerances, and (c) remove any surface flaws. Although they are hard and brittle, major advances have been made in producing **machinable ceramics** and **grindable ceramics**, thus enabling the production of ceramic components with high dimensional accuracy and a good surface finish. An example is silicon carbide, which can be machined into final shapes from sintered blanks.

The finishing processes employed can be one or more of the following operations, which are described in detail in various sections in Part IV of this book:

1. *Grinding* (using a diamond wheel)
2. *Lapping* and *honing*
3. *Ultrasonic machining*
4. *Drilling* (by using a diamond-coated drill)
5. *Electrical-discharge machining*
6. *Laser-beam machining*
7. *Abrasive water-jet cutting*
8. *Tumbling* (to remove sharp edges and grinding marks)

Process selection is an important consideration because of the brittle nature of most ceramics and the additional costs involved in some of these processes. The effect of the finishing operation on the properties of the product also must be considered. For instance, because of notch sensitivity, the finer the finish, the higher the part's strength and load-carrying capacity—particularly its fatigue strength (see Fig. 2.28). Ceramic parts also may undergo static fatigue, as described for glass in Section 18.5.

To improve their appearance and strength and to make them impermeable, ceramic products often are coated with a **glaze** or **enamel**, which forms a glassy coating after firing.

EXAMPLE 18.1 Dimensional changes during the shaping of ceramic components

A solid, cylindrical ceramic part is to be made with a final length, L, of 20 mm. For this material, it has been established that linear shrinkages during drying and firing are 7 and 6%, respectively, based on the dried dimension, L_d. Calculate (a) the initial length, L_o, of the part and (b) the dried porosity, P_d, if the porosity of the fired part, P_f, is 3%.

Solution

a. On the basis of the information given and noting that firing is preceded by drying, we can write

$$\frac{(L_d - L)}{L_d} = 0.06$$

or

$$L = (1 - 0.06)L_d$$

Hence,

$$L_d = \frac{20}{0.94} = 21.28 \text{ mm}$$

$$L_o(1 + 0.07)L_d = (1.07)(21.28) = 22.77 \text{ mm}$$

b. Since the final porosity is 3%, the actual volume, V_a, of the solid material in the part is

$$V_a = (1 - 0.03)V_f = 0.97V_f$$

where V_f is the fired volume of the part. Because the linear shrinkage during firing is 6%, we can determine the dried volume, V_d, of the part as:

$$V_d = \frac{V_f}{(1 - 0.06)^3} = 1.2V_f$$

Hence,

$$\frac{V_a}{V_d} = \frac{0.97}{1.2} = 0.81, \text{ or } 81\%$$

Therefore, the porosity, P_d, of the dried part is 19%.

18.3 | Forming and Shaping of Glass

Glass is processed by melting and then shaping it, either in molds and various devices or by blowing. Glass shapes produced include flat sheets and plates, rods, tubing, glass fibers, and discrete products such as bottles and headlights. Glass products may be as thick as those for large telescope mirrors and as thin as those for holiday tree ornaments. The strength of glass can be improved by thermal and chemical treatments (which induce compressive surface residual stresses) or by laminating it with a thin sheet of tough plastic.

Glass products generally can be categorized as follows:

1. **Flat sheets** or **plates** ranging in thickness from about 0.8 to 10 mm (0.03 to 0.4 in.), such as window glass, glass doors, and table tops.
2. **Rods** and **tubing** used for chemicals, neon lights, and decorative artifacts.
3. **Discrete products** such as bottles, vases, headlights, and television tubes.
4. **Glass fibers** to reinforce composite materials (Section 9.2.1) and for use in fiber optics.

All glass forming and shaping processes begin with molten glass, typically in the range of 1000° to 1200°C (1830° to 2200°F). The glass has the appearance of a red-hot, viscous syrup and is supplied from a melting furnace or tank.

18.3.1 Flat-sheet and plate glass

Flat-sheet glass can be made by the float-glass method or by drawing or rolling it from the molten state. All three methods are continuous processes.

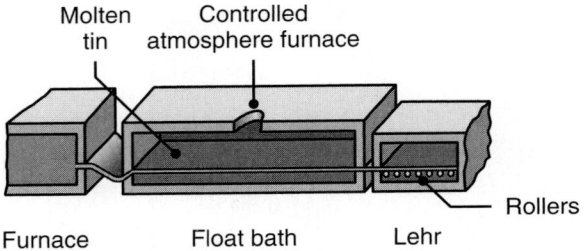

FIGURE 18.7   The float method of forming sheet glass.
Source: Courtesy of Corning Glass Works.

1. In the **float method** (Fig. 18.7), molten glass from the furnace is fed into a long bath in which the glass—under a controlled atmosphere and at a temperature of 1150°C (2100°F)—floats over a bath of molten tin. The glass then moves at a temperature of about 650°C (1200°F) over rollers into another chamber (*lehr*), where it solidifies. *Float glass* has smooth surfaces (*fire-polished*), and further grinding or polishing is not necessary. The widths can be as much as 4 m (13 ft). Both thin and plate glass are made by this process.

2. The **drawing** process for making flat sheets or plates involves a machine in which the molten glass passes through a pair of rolls (Fig. 18.8a) in an arrangement similar to an old-fashioned clothes wringer. The solidifying glass is squeezed between these two rolls (forming it into a sheet) and then moved forward over a set of smaller rolls.

3. In the **rolling** process (Fig. 18.8b), the molten glass is squeezed between rollers, forming a sheet. The surfaces of the glass may be embossed with a pattern by using textured roller surfaces. In this way, the glass surface becomes a replica of the roll surface. Thus, glass sheet produced by drawing or rolling has a rough surface appearance.

18.3.2 Tubing and rods

Glass tubing is manufactured by the process shown in Fig. 18.9. Molten glass is wrapped around a rotating hollow mandrel (cylindrical or cone-shaped) and is drawn out by a set of rolls. Air is blown through the mandrel to prevent the glass

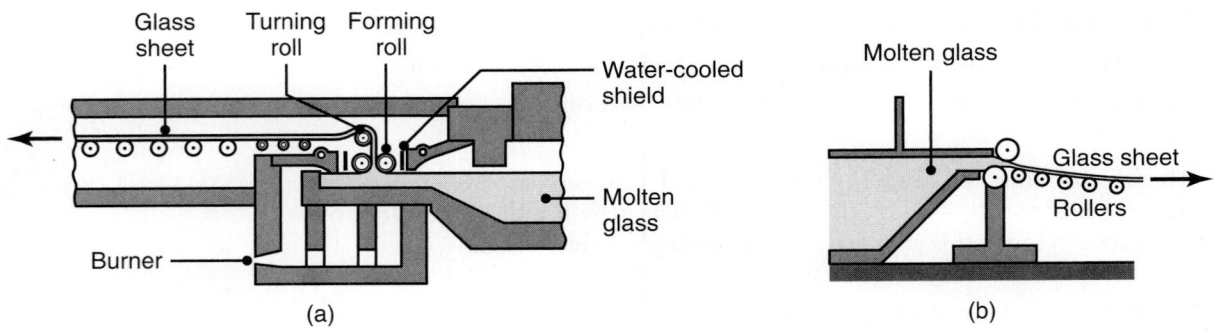

FIGURE 18.8   (a) Drawing process for drawing sheet glass from a molten bath. (b) Rolling process. *Source*: After W. D. Kingery.

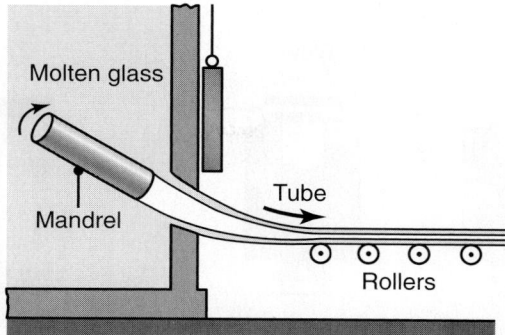

FIGURE 18.9 Manufacturing process for glass tubing. Air is blown through the mandrel to keep the tube from collapsing. Glass tubes for fluorescent bulbs are made by this method.

tube from collapsing. These machines may be horizontal, vertical, or slanted downward. This is the method used in making the glass tubes for fluorescent bulbs, using machines (such as the Corning Ribbon Machine) capable of producing bulbs at rates of 2000 per minute. Glass rods are made in a similar manner, but air is not blown through the mandrel. The drawn product becomes a solid glass rod.

18.3.3 Discrete glass products

Several processes are used to make discrete glass objects, as described here.

Blowing.   Hollow and thin-walled glass items (such as bottles, vases, and flasks) are made by *blowing*—a process that is similar to the blow molding of thermoplastics (Section 19.4). The steps involved in the production of an ordinary glass bottle by the blowing process are shown in Fig. 18.10. Blown air expands a hollow gob of heated glass against the inner walls of the mold. The mold usually is coated with a parting agent (such as oil or emulsion) to prevent the glass from sticking to the mold. Blowing may be followed by a second blowing operation for finalizing product shape, called the *blow and blow* process.

The surface finish of products made by the blowing process is acceptable for most applications, such as bottles and jars. It is difficult to control the wall thickness of the product, but this process is economical for high-rate production. Incandescent light bulbs (see Fig. I.3) are made in highly-automated blowing machines at a rate of 2000 bulbs per minute.

Pressing.   In the *pressing* process, a gob of molten glass is placed into a mold and pressed into a confined shape with a plunger. Thus, the process is similar to closed-die forging. The mold may be made in one piece (such as that shown in Fig. 18.11) or it may be a split mold (Fig. 18.12). After being pressed, the solidifying glass acquires the shape of the mold-plunger cavity. Because of the confined environment, the product has a higher dimensional accuracy than can be obtained with blowing. Pressing in one-piece molds cannot be used for (a) shapes of products from which the plunger cannot be retracted, or (b) thin-walled items. For example, split molds are used for bottles, while for thin-walled items, pressing can be combined with blowing. Known as *press and blow*, the pressed part is subjected to air pressure (hence the term blow), which further expands the glass into the mold.

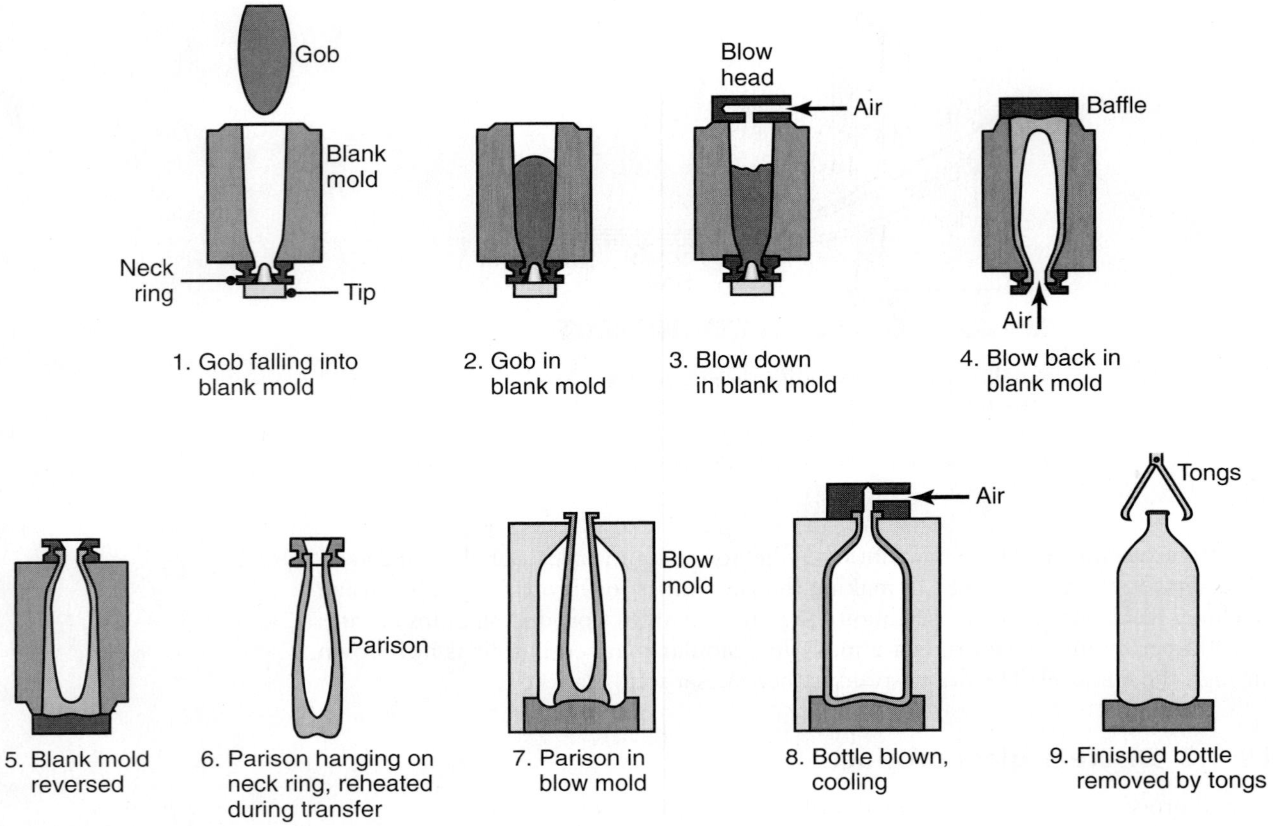

Gob
Blank mold
Neck ring
Tip
1. Gob falling into blank mold
2. Gob in blank mold
Blow head
Air
3. Blow down in blank mold
Baffle
Air
4. Blow back in blank mold
5. Blank mold reversed
Parison
6. Parison hanging on neck ring, reheated during transfer
Blow mold
7. Parison in blow mold
Air
8. Bottle blown, cooling
Tongs
9. Finished bottle removed by tongs

FIGURE 18.10 Steps in manufacturing an ordinary glass bottle. *Source*: After F. H. Norton.

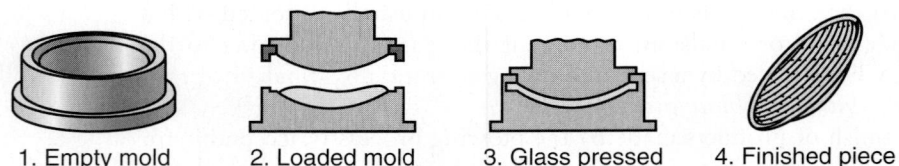

1. Empty mold
2. Loaded mold
3. Glass pressed
4. Finished piece

FIGURE 18.11 Manufacturing a glass item by pressing glass into a mold. *Source*: Courtesy of Corning Glass Works.

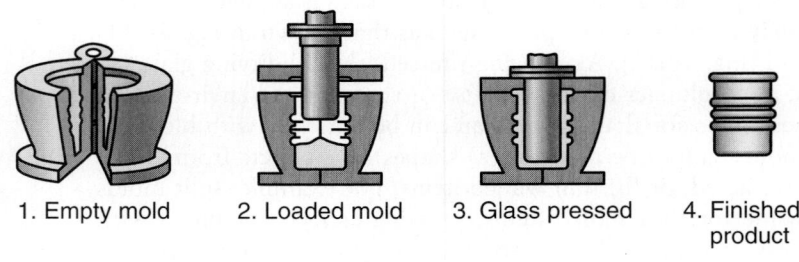

1. Empty mold
2. Loaded mold
3. Glass pressed
4. Finished product

FIGURE 18.12 Pressing glass into a split mold. *Source*: After E. B. Shand.

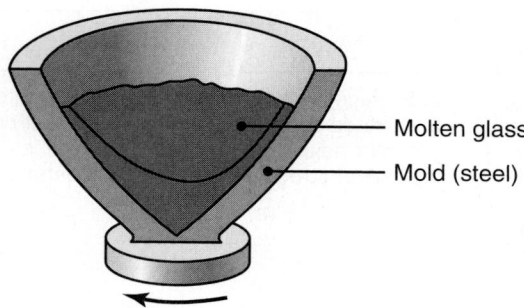

FIGURE 18.13   Centrifugal casting of glass. Large telescope lenses and television-tube funnels are made by this process. *Source*: Courtesy of Corning Glass Works.

Centrifugal casting.   Also known in the glass industry as **spinning** (Fig. 18.13), this process is similar to that used for metals. The centrifugal force pushes the molten glass against the mold wall, where it solidifies. Typical products made are TV picture tubes and missile nose cones.

Sagging.   Shallow dish-shaped or lightly embossed glass parts can be made by the *sagging* process. A sheet of glass is placed over the mold and heated. The glass sags by its own weight and takes the shape of the mold. The process is similar to the thermoforming of thermoplastics (Section 19.6), but no pressure or vacuum is involved. Typical applications are dishes, sunglass lenses, mirrors for telescopes, and lighting panels.

Glass ceramics manufacture.   Glass ceramics (trade names: *Pyroceram, Corningware*) contain large proportions of several oxides, as noted in Section 8.5. Thus, their manufacturing involves a combination of the methods used for ceramics and glasses. Glass ceramics are shaped into discrete products (such as dishes and baking pans) and then heat treated, whereby glass becomes *devitrified* (recrystallized).

18.3.4 Glass fibers

Continuous glass fibers are drawn through multiple orifices (200 to 400 holes) in heated platinum plates at speeds as high as 500 m/s (1700 ft/s). Fibers as small as 2 μm (80 μin.) in diameter can be produced by this method. In order to protect their surfaces, fibers subsequently are coated with chemicals. Short fibers (*chopped*) are produced by subjecting long fibers to compressed air or steam as they leave the orifice.

    Glass wool (short glass fibers)—used as a thermal-insulating material or for acoustic insulation—is made by a *centrifugal spraying* process in which molten glass is ejected (spun) from a rotating head. The diameter of the fibers is typically in the range of 20 to 30 μm (800 to 1200 μin.).

18.4   Techniques for Strengthening and Annealing Glass

Glass can be strengthened by the processes described next, and discrete glass products may be subjected to annealing and to other finishing operations to impart their desired properties and surface characteristics.

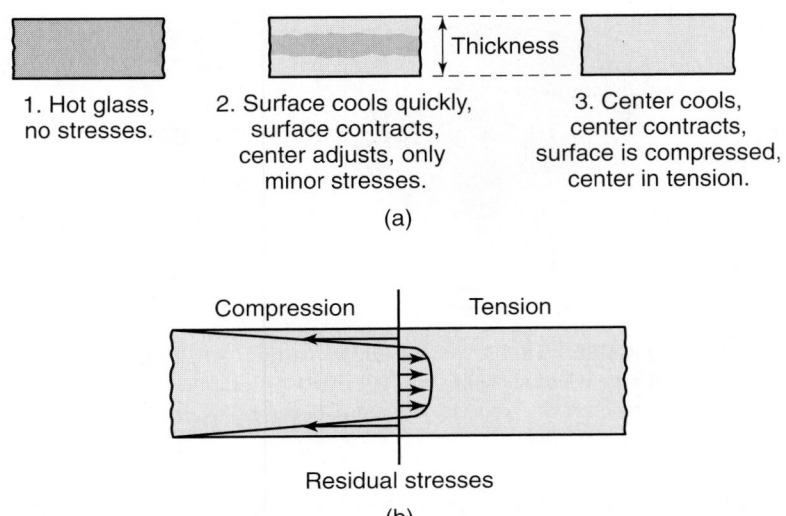

FIGURE 18.14   (a) The stages involved in inducing compressive surface residual stresses for improved strength. (b) Residual stresses in a tempered glass plate.

Thermal tempering.   In this process (also called *physical tempering* or *chill tempering*), the surfaces of the hot glass are cooled rapidly by a blast of air (Fig. 18.14). As a result, the surfaces shrink, and (at first) tensile stresses develop on the surfaces. As the bulk of the glass begins to cool, it contracts. The already solidified surfaces of the glass are then forced to contract, and consequently, they develop residual compressive surface stresses, while the interior develops tensile stresses (see also Section 2.11). Compressive surface stresses improve the strength of the glass in the same way that they do in metals and other materials.

The higher the coefficient of thermal expansion of the glass and the lower its thermal conductivity, the higher will be the level of residual stresses developed, and hence, the stronger the glass becomes. Thermal tempering takes a relatively short time (minutes), and can be applied to most glasses. Because of the high amount of energy stored in residual stresses, **tempered glass** shatters into a large number of pieces when broken. The broken pieces are not as sharp and hazardous as those from ordinary window glass.

Chemical tempering.   In this process, the glass is heated in a bath of molten KNO_3, K_2SO_4, or $NaNO_3$, depending on the type of glass. Ion exchanges then take place with larger atoms replacing the smaller atoms in the surface of the glass. As a result, residual compressive stresses develop on the surface. This condition is similar to that created by forcing a wedge between two bricks in a brick wall.

The time required for chemical tempering is about one hour longer than that for thermal tempering. It may be performed at various temperatures. At low temperatures, part distortion is minimal, therefore complex shapes can be treated. At elevated temperatures, there may be some distortion of the part, but the product then can be used at higher temperatures without loss of strength.

Laminated glass.   Laminated glass is a product of another strengthening method called *laminate strengthening*. It consists of two pieces of flat glass with a thin sheet of tough plastic in between. When laminated glass is cracked, its pieces are held together by the plastic sheet—a phenomenon commonly observed in a shattered automobile windshield.

Traditionally, flat glass for glazing windows and doors has been strengthened with wire netting (such as *chicken wire*—with a hexagonal mesh) embedded in the glass during its production. When a hard object strikes the surface, the glass shatters, but the pieces are held together because of the toughness of the wire, which has both strength and ductility.

Bulletproof glass.   Laminated glass has considerable resistance and can prevent the full penetration of solid objects because of the presence of a tough polymer film in between the two layers of glass. *Bulletproof glass* (used in some automobiles, armored bank vehicles, and buildings) is a more challenging design. This is due to the very high speed and energy level of the bullet and the small size and the shape of the bullet tip—a small contact area and high localized stresses. Depending on the caliber of the weapon, bullet speeds range from about 350 to 950 m/s (1150 to 3100 ft/s).

Bulletproof glass (also called *bullet-resistant glass*) ranges in thickness from 7 to 75 mm (0.3 to 3 in.). The thinner ones are for handguns, and thicker ones are for rifles. Although there are several variations, bulletproof glass basically consists of glass laminated with a polymer sheet. The capacity of a bulletproof glass to stop a bullet depends on (a) the type and thickness of the glass; (b) the size, shape, weight, and speed of the bullet; and (c) the properties and thickness of the polymer sheet.

Polycarbonate sheets commonly are used for bulletproof glass. As a material widely used for safety helmets, windshields, and guards for machinery, polycarbonate is a tough and flexible polymer. Combined with a thick glass, it can stop a bullet, although the glass develops a circular shattered region. Proper bonding of these layers over the glass surface is also an important consideration, as there usually is more than one round fired during such encounters. Also, in order to maintain the transparency of the glass and minimize distortion, the index of refraction of the glass and the polymer must be nearly identical.

If a polymer sheet is only on one side of the glass, it is known as a *one-way* bulletproof glass. In a vehicle, the polymer layer is on the inside surface of the glass. An external bullet will not penetrate the window, because the bullet will strike the glass first, shattering it, absorbing some of the energy of the bullet, thus slowing it down. The remaining energy is dissipated in the polymer sheet which then stops the bullet. This arrangement allows someone inside the vehicle to fire back. A bullet from inside penetrates the polymer sheet and forces the glass to break outwards, allowing the bullet to go through. Thus, a one-way glass stops a bullet fired from outside but allows a bullet to be fired from inside.

A more recent design for bulletproof glass consists of two adjacent layers of thermoplastic polymer sheet over the same surface of the glass and is based on a somewhat different principle.

- The outermost layer (the side where the bullet enters) is an acrylic sheet (polymethylmethacrylate, PMMA). This sheet *dulls* the tip of the bullet—thus slowing down the bullet's speed and its ability to penetrate easily because of the now blunt tip. Additionally, the acrylic film has high weather resistance, making it suitable as the outer layer, which is exposed to the elements.
- The next layer is a polycarbonate sheet. Because it has high toughness, the polycarbonate layer stops the bullet, which has already been dulled when penetrating the acrylic sheet first. The glass shatters in the same manner as in other designs.

18.4.1 Finishing operations

As in metal products, residual stresses can develop also in glass products if they are not cooled at a sufficiently low rate. In order to ensure that the product is free from these stresses, it is **annealed** by a process similar to the stress-relief annealing of metals. The glass is heated to a certain temperature and then cooled gradually. Depending on the size, the thickness, and the type of the glass, annealing times may range from a few minutes to as long as 10 months, as in the case of a 600-mm (24-in.) mirror for a telescope in an observatory.

In addition to annealing, glass products may be subjected to further operations, such as cutting, drilling, grinding, and polishing. Sharp edges and corners can be smoothed by (a) grinding (as seen in glass tops for desks and shelves) or (b) holding a torch against the edges (**fire polishing**), which rounds them by localized softening of the glass and surface tension.

In all finishing operations on glass and other brittle materials, care should be exercised to ensure that there is no surface damage, especially stress raisers such as rough surface finish and scratches. Because of their notch sensitivity, even a single scratch can cause premature failure of the part, especially if the scratch is in a direction where the tensile stresses are a maximum.

18.5   Design Considerations for Ceramics and Glasses

Ceramic and glass products require careful selection of composition, processing methods, finishing operations, and methods of assembly with other components. The potential consequences of part failure are always a significant factor in designing ceramic and glass products. The knowledge of limitations (such as general lack of tensile strength, sensitivity to internal and external defects, and low impact toughness) is important. On the other hand, these limitations have to be balanced against such desirable characteristics as hardness, scratch resistance, compressive strength at room and elevated temperatures, and a wide range of diverse and desirable physical properties.

The control of processing parameters and of the quality and level of impurities in the raw materials is important. As in all design decisions, there are priorities and limitations, and several factors should be considered simultaneously, including the number of parts needed and the costs of tooling, equipment, and labor.

Dimensional changes and warping and the possibility of cracking during processing and service life are significant factors in selecting methods for shaping these materials. When a ceramic or glass component is part of a larger assembly, its compatibility with other components is another important consideration. Particularly important are the type of loading and its thermal expansion (such as in seals and windows with metal frames) Recall from Table 3.1, the wide range of coefficients of thermal expansion for various metallic and nonmetallic materials. Thus, when a plate glass fits tightly in a metal window frame, temperature variations (such as the sun shining on only a portion of the window) can cause thermal stresses which may lead to cracking—a phenomenon that commonly has been observed in some tall buildings. A common solution is placing rubber seals between the glass and the frame to avoid stresses due to differences in thermal expansion.

As noted in Sections 8.3.1 and 8.4.2, ceramics and glasses undergo a phenomenon called *static fatigue*, whereby they can suddenly break under a static load after a period of time. Although it does not occur in a vacuum or in dry air, provisions must be made to avoid such failure. A general guide is that, in order for a glass item to withstand a load of 1000 hours or longer, the maximum stress that can be applied is about one-third of the maximum stress that it can withstand during the first second of loading.

18.6 | Processing of Superconductors

Although *superconductors* (see Section 3.7) have major energy-saving potential in the generation, storage, and distribution of electrical power, their processing into useful shapes and sizes for practical applications has presented significant difficulties. Two basic types of superconductors are:

- Metals, called **low-temperature superconductors** (LTSC), include combinations of niobium, tin, and titanium.
- Ceramics, called **high-temperature superconductors** (HTSC), include various copper oxides. Here, "high" temperature means closer to ambient temperature, and hence HTSCs are of greater pratical use.

Ceramic superconducting materials are available in powder form. The fundamental difficulty in manufacturing them is their (a) inherent brittleness and (b) anisotropy, making it difficult to align the grains in the proper direction to achieve high efficiency. The smaller the grain size, the more difficult it is to align the grains.

The basic manufacturing process for superconductors consists of the following steps:

1. Preparing the powder, mixing it, and grinding it in a ball mill down to a grain size of 0.5 to 10 μm
2. Forming the powder into shape
3. Heat treating it

The most common forming process is the **oxide powder in tube** (OPIT) method. The powder first is packed into silver tubes (because silver has the highest electrical conductivity of any metal, Table 3.2) and sealed at both ends. The tubes then are worked mechanically by such processes as swaging, drawing, extrusion, isostatic pressing, and rolling. The final shapes may be wire, tape, coil, or bulk.

Other principal methods of processing are: (a) coating silver wire with superconducting material, (b) the deposition of superconductor films by laser ablation, (c) the doctor-blade process, (d) explosive cladding, and (e) chemical spraying. The formed part subsequently may be heat treated to improve the grain alignment of the superconducting powder.

CASE STUDY 18.1 | **Production of High-Temperature Superconducting Tapes**

Significant progress has been made in recent years in understanding high-temperature superconducting materials and their potential use as electrical conductors. Two bismuth-based oxides are superconducting ceramic materials of choice for various military and commercial applications, such as electrical propulsion for ships and submarines, shallow water and ground minesweeping systems, transmission cable generators, and superconducting magnetic energy storage (SMES). A variety of processing methods have been explored to produce wires and multifilament tapes. The powder-in-tube process (Fig. 18.15) has been used successfully to fabricate long lengths of bismuth-based wires and tapes with desirable properties. The following example demonstrates this method for the production of high-temperature superconducting multifilament tapes.

Production of the multifilament tapes involves the following steps:

1. A composite billet first is produced using a silver casing and ceramic powder. The casing is an annealed high-purity silver that is filled with the bismuth-ceramic powder in an inert atmosphere. It is compacted in several increments

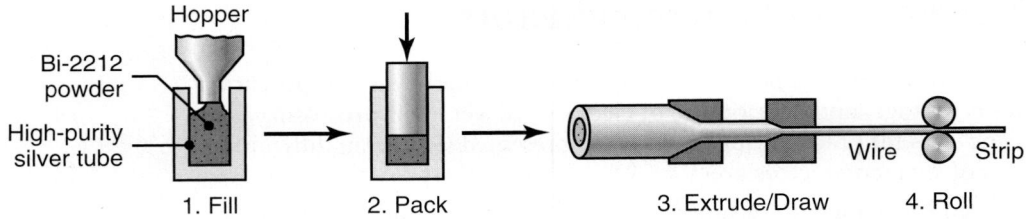

FIGURE 18.15 Schematic illustration of the powder-in-tube process.

to a 30% relative density using a steel ram. In order to minimize density gradients (such as those shown in Fig. 17.11), about one gram of powder is added to the billet for each stroke of the ram. Each billet is weighed and measured to verify the initial packing density. The billet ends then are sealed with a silver alloy to avoid contamination during subsequent deformation processing.

2. The billet is extruded and drawn to reduce its diameter and increase the powder density. Billets are drawn to a final diameter of 1.63 mm on a draw bench using 12 passes with a 20.7% reduction per pass. The dies have a semi-cone angle of 8°, and the drawing speed is approximately 1.4 m/min. A semi-soluble oil and zinc-stearate spray are used as lubricants.

3. Following the drawing process, the wire is transformed progressively into tape in a single-stand rolling mill in two-high and four-high configurations. For the four-high case, the diameter of the backup rolls (which are the work rolls for the two-high configuration) is 213 mm, and the diameter of the work rolls is 63.5 mm. The final tape dimensions are 100 to 200 μm in thickness, 2 to 3 mm in width with a ceramic core ranging from 40 to 80 μm in thickness and 1.0 to 1.5 mm in width.

Source: Courtesy of S. Vaze and M. Pradheeradhi, Concurrent Technologies Corporation.

SUMMARY

- Ceramic products are shaped by various casting, plastic forming, or pressing techniques. The parts then are dried and fired to impart strength and hardness. Finishing operations (such as machining and grinding) may be performed to give the part its final shape and dimensional accuracy or to subject it to surface treatments. Because of their inherent brittleness, ceramics are processed with due consideration of distortion and cracking. The control of raw-material quality and processing parameters also is important.

- Glass products are made by several shaping processes that are similar to those used for ceramics and plastics. They are available in a wide variety of forms, compositions, and mechanical, physical, and optical properties. Their strength can be improved by thermal and chemical treatments.

- Continuous methods of glass processing are drawing, rolling, and floating. Discrete glass products can be manufactured by blowing, pressing, centrifugal casting, or sagging. The parts subsequently may be annealed to relieve residual stresses.

- Design considerations for ceramics and glasses are guided by such factors as their general lack of tensile strength and toughness and their sensitivity to external and internal defects. Warping and cracking during production are important considerations.

- Manufacturing superconductors into useful products is challenging because of the anisotropy and inherent brittleness of the materials involved. Although new processes also are being developed, the basic process consists of packing the powder into a silver tube and deforming it plastically into desired shapes.

KEY TERMS

Binder

Blow and blow

Blowing

Bulletproof glass

Centrifugal casting

Chemical tempering

Deflocculent

Doctor-blade process

Drawing

Fire polishing

Firing

Float glass

Gob

High-temperature superconductors

Hot pressing

Injection molding

Jiggering

Laminated glass

Low-temperature superconductors

Microwave sintering

Oxide-powder-in-tube process

Plastic forming

Plasticizer

Press and blow

Pressing

Sagging

Slip

Slip casting

Tempered glass

Thermal tempering

Wetting agent

BIBLIOGRAPHY

Bourdillon, A., *High Temperature Superconductors: Processing and Science*, Academic Press, 1994.

Concise Encyclopedia of Advanced Ceramics, The MIT Press, 1991.

Engineered Materials Handbook, Vol. 4: *Ceramics and Glasses*, ASM International, 1991.

German, R.M., and Bose, A., *Injection Molding of Metals and Ceramics*, Metal Powder Industries Federation, 1997.

Ghosh, A., Hiremath B., and Halloran, J. (eds.), *Design for Manufacturability of Ceramic Components*, American Ceramic Society, 1995.

Green, D.J., *An Introduction to the Mechanical Properties of Ceramics*, Cambridge Univ. Press, 1998.

Handbook of Ceramics and Composites, 3 vols., Marcel Dekker, 1991.

Hiremath, B., Gupta, T., and Nair, K.M. (eds.), *Ceramic Manufacturing Practices and Technologies*. American Ceramic Society, 1997.

Hiremath, B., Bruce, A., and Ghosh, A. (eds.), *Manufacture of Ceramic Components*, American Ceramic Society, 1995.

Hlavac, J., *The Technology of Glass and Ceramics*, Elsevier, 1983.

McHale, A.E. (ed.), *Phase Diagrams and Ceramic Processes*, Chapman & Hall, 1997.

McLellen, G.W., and Shand, E.B., *Glass Engineering Handbook*, McGraw-Hill, 1984.

Mistler, R.E., and Twiname, E.R., *Tape Casting: Theory and Practice*, American Ceramic Society, 1999.

Musikant, S., *What Every Engineer Should Know About Ceramics*, Marcel Dekker, 1991.

Mutsuddy, B.C., *Ceramic Injection Molding*, Chapman & Hall, 1995.

Rahaman, M.N., *Ceramics Processing and Sintering*, Marcel Dekker, 1998.

Reed, J.S., *Principles of Ceramics Processing*, 2nd ed., Wiley, 1995.

Reimanis, I., Henager, C., and Tomsia, A. (eds.), *Ceramic Joining*, American Ceramic Society, 1997.

Richerson, D.W., *Modern Ceramic Engineering*, 2nd ed., Marcel Dekker, 1992.

Sabia, R., Greenhurt, V., and Pantano, C. (eds.), *Finishing of Advanced Ceramics and Glasses*, American Ceramic Society, 1999.

Schwartz, M.M. (ed.), *Handbook of Structural Ceramics*, McGraw-Hill, 1992.

Wachtman, J.B. (ed.), *Ceramic Innovations in the 20th Century*, American Ceramic Society, 1999.

Weimer, A. (ed.), *Carbide, Nitride and Boride Materials: Synthesis and Processing*, Chapman & Hall, 1997.

REVIEW QUESTIONS

18.1   Outline the steps involved in processing (a) ceramics and (b) glasses.

18.2   List and describe the functions of additives in ceramics.

18.3   Describe the sequence involved in the slip-casting process.

18.4   What is the doctor-blade process?

18.5   Explain the advantages of isostatic pressing.

18.6   What is jiggering? What shapes does it produce?

18.7   Name the parameters that are important in drying ceramic products.

18.8   What types of finishing operations are used on ceramics? On glass? Why?

18.9   Describe the methods by which sheet glass is made.

18.10   How is glass tubing produced? Are there limitation to their maximum length?

18.11   Describe the glass-blowing process.

18.12   What is the difference between physical and chemical tempering of glass?

18.13   What is the structure of laminated glass? Bulletproof glass?

18.14   How are glass fibers made? What are their sizes?

QUALITATIVE PROBLEMS

18.15   Inspect various products at home, work, or in stores. Noting their shape, color, and transparency, identify those that are made of (a) ceramic, (b) glass, and (c) glass ceramics.

18.16   Describe the differences and similarities in processing metal powders versus ceramics.

18.17   What should be the property requirements for the metal balls in a ball mill?

18.18   Which property of glasses allows them to be expanded to large dimensions by blowing? Can metals undergo such behavior? Explain.

18.19   Explain why ceramic parts may distort or warp during drying. What precautions should be taken to avoid this situation?

18.20   What properties should plastic sheets have to be used in laminated glass? Why?

18.21   It is stated that the higher the coefficient of thermal expansion of a glass and the lower its thermal conductivity, the higher the level of the residual stresses developed. Explain why.

18.22   Are any of the processes used for making discrete glass products similar to ones described in preceding chapters? Describe them.

18.23   Injection molding is a process that is used for powder metals, polymers, and ceramics. Explain why is this so.

18.24   Are there any similarities between the strengthening mechanisms used on glass and those used on metallic materials? Explain.

18.25   Explain the phenomenon of static fatigue and how it affects the service life of a ceramic or glass component.

18.26   Describe and explain the differences in the manner in which each of the following flat surfaces would fracture when struck with a heavy piece of rock: (a) ordinary window glass, (b) tempered glass, and (c) laminated glass.

18.27   Is there any flash that develops in slip casting? How would you propose to remove such flash?

18.28   Are there similarities between slip casting and shell-mold casting? Explain.

QUANTITATIVE PROBLEMS

18.29 For Example 18.1, calculate (a) the porosity of the dried part if the porosity of the fired part is to be 9% and (b) the initial length, L_o, of the part if the linear shrinkages during drying and firing are 8 and 7%, respectively.

18.30 What would be the answers to Problem 18.29 if the quantities given were halved?

18.31 Assume that you are an instructor covering the topics described in this chapter, and you are giving a quiz on the numerical aspects to test the understanding of the students. Prepare two quantitative problems and supply the answers.

SYNTHESIS, DESIGN, AND PROJECTS

18.32 Describe similarities and differences between the processes described in this chapter and those (a) in Part II on metal casting and (b) in Part III on forming.

18.33 Consider some ceramic products with which you are familiar and outline a sequence of processes that you think were used to manufacture them.

18.34 Make a survey of the technical literature and describe the differences (if any) between the quality of glass fibers made for use in reinforced plastics and those made for use in fiber-optic communications. Comment on your observations.

18.35 How different are the design considerations for ceramics from those for other materials? Explain.

18.36 Locate a ceramics/pottery shop and investigate the different techniques used for coloring and decorating a ceramic part. What are the different methods of applying a metallic finish to the part?

18.37 Give examples of designs and applications where static fatigue should be taken into account.

18.38 Perform a literature search and make a list of automotive parts made of ceramics. Explain why they are made of ceramics.

18.39 Describe your thoughts on the processes that can be used to make (a) small ceramic statues, (b) whiteware for bathrooms, (c) common bricks, and (d) floor tiles.

18.40 As we have seen, one method of producing superconducting wire and strip is by compacting powders of these materials, placing them into a tube, and drawing them through dies or rolling them. Describe your thoughts concerning the steps and possible difficulties involved in each step of this production.

18.41 We have explained briefly the characteristics of bulletproof glass. Describe your own thoughts on possible new designs for this type of glass. Explain your reasoning.

CHAPTER 19

Forming and Shaping Plastics and Composite Materials

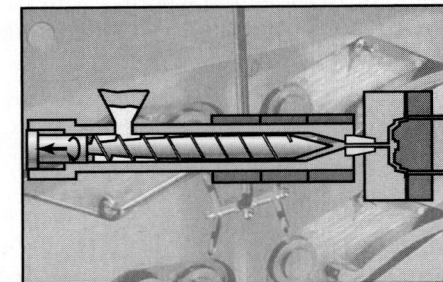

In this chapter, we describe the manufacturing processes involved in producing polymers and composite materials for consumer and industrial products. We specifically describe:

- Extrusion as a basic process for producing rods, tubing, and pellets for subsequent use.
- How discrete plastic parts (such as bottles, automotive parts, and electrical parts with metal components) are made.
- Production of plastic sheets and films.
- How reinforced plastics and various composite materials are produced.
- Characteristics of machinery used, mold design, and economic considerations.

Typical parts made: Extremely wide variety of consumer and industrial products with a wide variety of colors and characteristics, but with characteristically limited strength and temperature resistance.

Alternative processes: Casting, forming, powder metallurgy, and machining, depending on material-substitution possibilities.

19.1 | Introduction

The processing of plastics and elastomers involves operations similar to those used in the forming and shaping of metals, described in preceding chapters. The processing of rubbers and elastomers began in the 1800s with the discovery of vulcanization by C. Goodyear in 1839. Plastics began to be developed in the 1920s, and rapid progress in the 1940s and onward led to important advances in the design and manufacture of various processing equipment to make numerous consumer and industrial products in large quantities. In the 1970s, reinforced plastics began to be introduced, leading the way for rapid progress in the use of composite materials—with unique properties and attendant challenges in producing them.

As we have noted in Chapter 7, thermoplastics melt and thermosets cure at relatively low temperatures. Hence, unlike metals, they are relatively easy to handle and require much less force and energy to process. Plastics in general can be molded, cast,

TABLE 19.1

General Characteristics of Forming and Shaping Processes for Plastics and Composite Materials	
Process	Characteristics
Extrusion	Continuous, uniformly solid or hollow, and complex cross-sections; high production rates; relatively low tooling costs; wide tolerances
Injection molding	Complex shapes of various sizes; thin walls; very high production rates; costly tooling; good dimensional accuracy
Structural foam molding	Large parts with high stiffness-to-weight ratio; less expensive tooling than in injection molding; low production rates
Blow molding	Hollow, thin-walled parts and bottles of various sizes; high production rates; relatively low tooling costs
Rotational molding	Large, hollow items of relatively simple shape; relatively low tooling costs; relatively low production rates
Thermoforming	Shallow or relatively deep cavities; low tooling costs; medium production rates
Compression molding	Parts similar to impression-die forging; expensive tooling; medium production rates
Transfer molding	More complex parts than compression molding; higher production rates; high tooling costs; some scrap loss
Casting	Simple or intricate shapes made with rigid or flexible low-cost molds; low production rates
Processing of composite materials	Long cycle times; expensive operation; tooling costs depend on process

formed, and machined into complex shapes in few operations, with relative ease, and at high production rates (Table 19.1). They also can be joined by various means (Section 32.6) and coated (generally for improved appearance) by various techniques (described in Chapter 32). Plastics are shaped into discrete products or as sheets, plates, rods, and tubing which then may be formed by secondary processes into a variety of discrete products. The types and properties of polymers and the shape and complexity of components that can be produced are influenced greatly by their method of manufacture and processing parameters.

Plastics usually are shipped to manufacturing plants as pellets, granules, or powders and are melted (for thermoplastics) just before the shaping process. Liquid plastics that cure into solid form are used especially in the making of thermosets and reinforced-plastic parts. With increasing awareness of our environment, raw materials also may consist of reground or chopped plastics obtained from recycling centers. As expected, however, product quality is not as high for such materials.

In this chapter, we follow the outline shown in Fig. 19.1, which describes the basic processes and economics of forming and shaping plastics and reinforced plastics. We also describe processing techniques for metal-matrix and ceramic-matrix composites, which have become increasingly important in various applications with critical requirements. We begin with *melt processing* techniques (starting with extrusion) and continue on to molding processes—both categories involving the application of external pressure during processing.

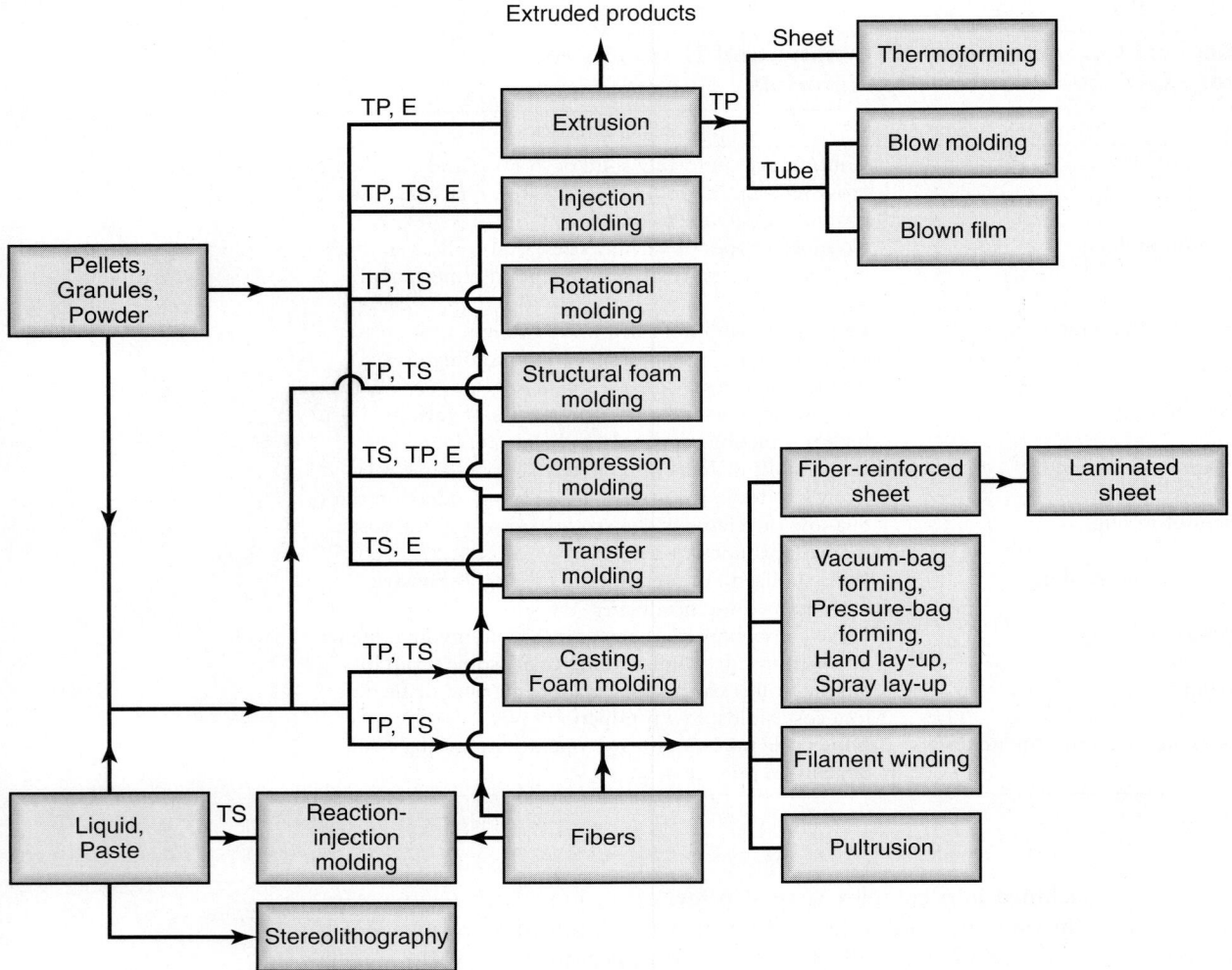

FIGURE 19.1 Outline of forming and shaping processes for plastics, elastomers, and composite materials. (TP = Thermoplastic; TS = Thermoset; E = Elastomer.)

19.2 | Extrusion

In *extrusion*, which constitutes the largest volume of plastics produced, raw materials in the form of thermoplastic pellets, granules, or powder are placed into a *hopper* and fed into the barrel of a *screw extruder* (Fig. 19.2). The barrel is equipped with a helical screw that blends the pellets and conveys them down the barrel towards the die. The internal friction from the mechanical action of the screw, heats the pellets and liquifies them. The screw action also builds up pressure in the barrel.

Screws have three distinct sections:

1. *Feed section:* Conveys the material from the hopper into the central region of the barrel.
2. *Melt section* (also called *compression* or *transition section*): Where the heat generated by the viscous shearing of the plastic pellets and by the external heaters causes melting to begin.
3. *Metering* or *pumping section*: Where additional shearing (at a high rate) and melting occur with pressure building up at the die.

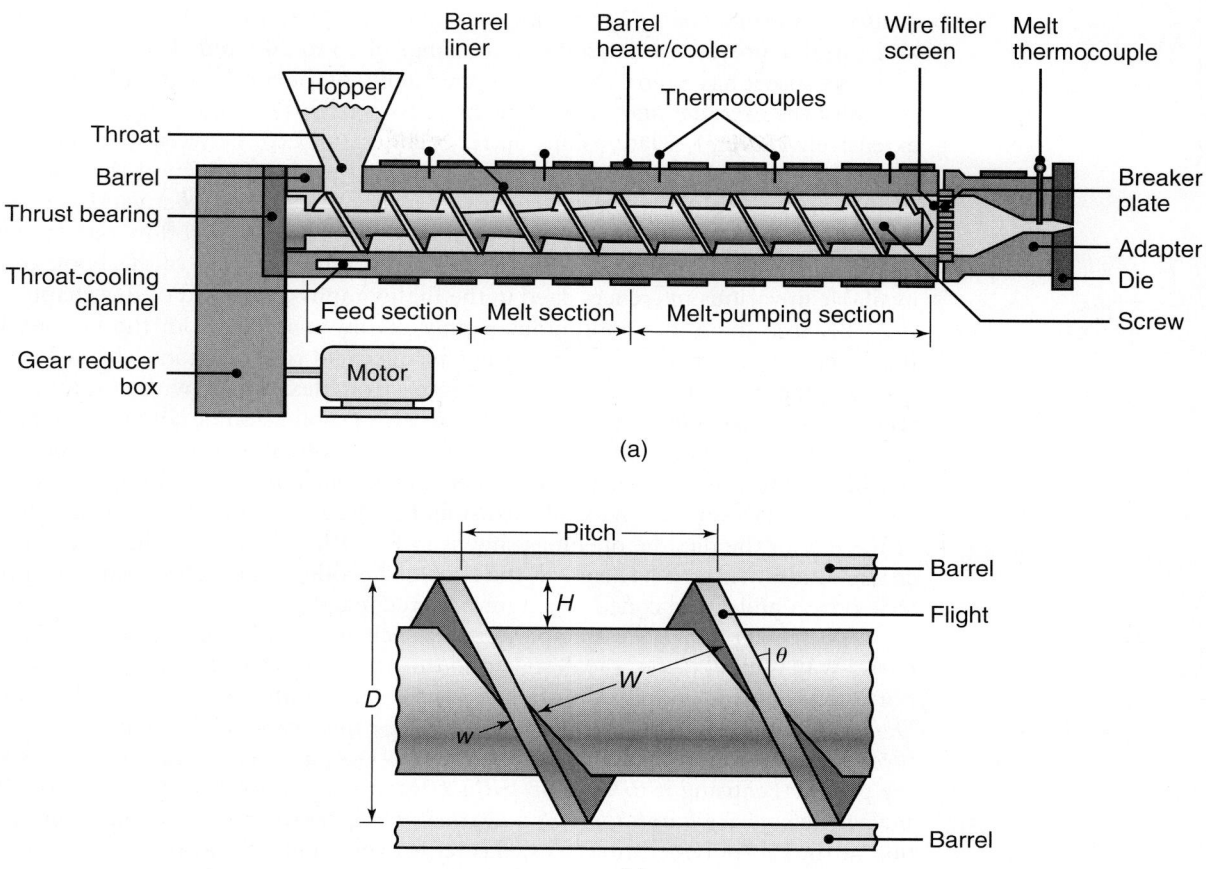

FIGURE 19.2 (a) Schematic illustration of a typical screw extruder. (b) Geometry of an extruder screw. Complex shapes can be extruded with relatively simple and inexpensive dies.

The lengths of these individual sections can be changed to accommodate the melting characteristics of different types of plastics. The molten plastic is forced through a die in a process similar to that of extruding metals. A metal-wire screen (Fig. 19.2a) usually is placed just before the die to filter out unmelted or congealed resin. This screen also helps build up back pressure in the barrel; it is replaced periodically. Between the screen and the die is a *breaker plate*, which has several small holes in it and helps improve mixing of the polymer prior to entering the die. The extruded product is then cooled, generally by exposing it to blowing air or by passing it through a water-filled channel (trough).

Controlling the rate and uniformity of cooling is important to minimize product shrinkage and distortion. In addition to single-screw extruders, other designs include *twin* (two parallel screws side by side) and *multiple screws* for polymers that are difficult to extrude (see also *reciprocating screw*, Section 19.3).

A typical helical screw is shown in Fig. 19.2b and indicates the important parameters that affect the mechanics of polymer extrusion. At any point in time, the molten plastic is in the shape of a helical ribbon with thickness, H, and width, W, and is conveyed towards the extruder outlet by the rotating screw flights. The shape, pitch, and flight angle of the helical screw are important parameters, as they affect the flow of the polymer through the extruder. The ratio of the barrel length, L, to its diameter, D, is also important. The *L/D* ratio in typical commercial extruders ranges from 5 to 30, and barrel diameters are generally in the range of 25 to 200 mm (1 to 8 in.).

Because it has a direct bearing on the quality of the product extruded and on the design of the extruder and the die, the mechanics of this operation have been studied extensively. Several relationships have been established between the dimensions shown in Fig. 19.2b, the screw rotational speed, and the viscosity of the polymer to describe what are known as *extruder characteristics* and *die characteristics*. These characteristics then determine such quantities as the pressure and flow rate at any location during the operation. Further details are beyond the scope of this book and are available in various references cited in the Bibliography at the end of this chapter.

Because there is a continuous supply of raw material from the hopper, long products can be extruded continuously (such as solid rods, sections, channels, sheet, tubing, pipe, and architectural components). Complex shapes with constant cross-section can be extruded with relatively inexpensive tooling. Some common die profiles are shown in Fig. 19.3. Note that some of the die profiles are not intuitive, but this is attributable to the polymer usually experiencing much greater and uneven shape recovery than encountered in metal extrusion. Furthermore, since the polymer will *swell* at the exit of the die, the openings shown in Fig. 19.3 are smaller than the extruded cross-sections. After it has cooled, the extruded product may subsequently be *drawn* (*sized*) by a puller and coiled or cut into desired lengths.

The control of processing parameters such as extruder-screw rotational speed, barrel-wall temperatures, die design, and rate of cooling and drawing speeds are important in order to ensure product integrity and uniform dimensional accuracy. *Defects* observed in extruding plastics are similar to those observed in metal extrusion (described in Section 15.5). Die shape is important, as it can induce high stresses in the product, causing it to develop surface fractures (as also occur in metals). Other surface defects are *bambooing* and *sharkskin effects*—due to a combination of friction at the die–polymer interfaces, elastic recovery, and nonuniform deformation of the outer layers of the product with respect to its bulk during extrusion.

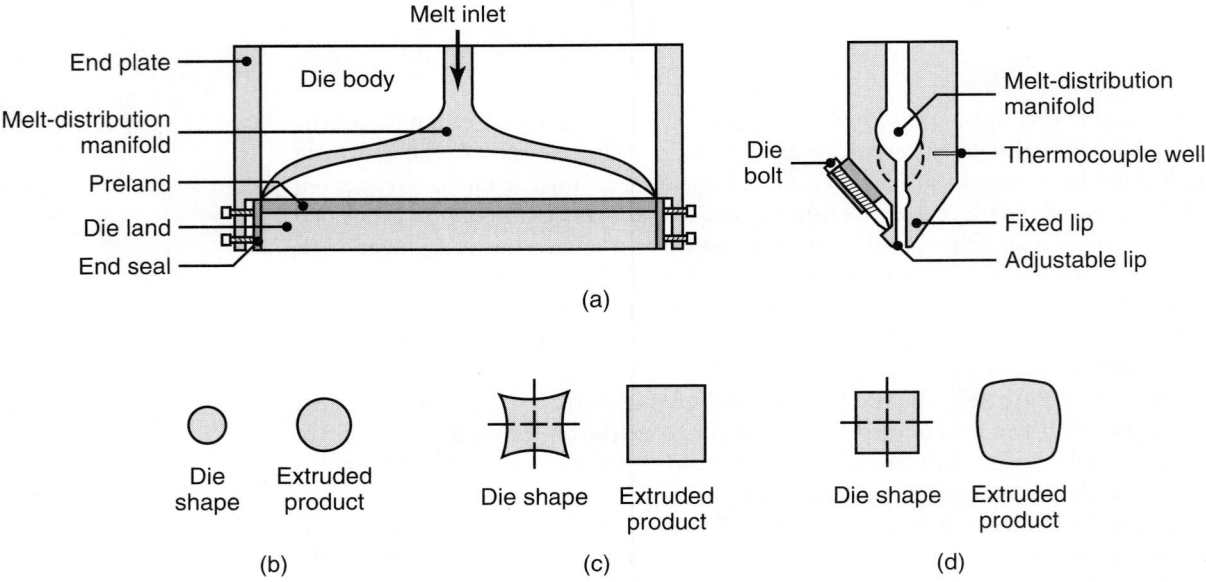

FIGURE 19.3   Common extrusion die geometries: (a) coat-hanger die for extruding sheet; (b) round die for producing rods; and (c) and (d) nonuniform recovery of the part after it exits the die. *Source:* (a) *Encyclopedia of Polymer Science and Engineering* (2nd ed.). Copyright © 1985. Reprinted by permission of John Wiley & Sons, Inc.

Extruders generally are rated by the diameter, D, of the barrel and the length-to-diameter (L/D) ratio of the barrel. Machinery costs can be on the order of $300,000, including the cost for the equipment for downstream cooling and winding of the extruded product.

19.2.1 Miscellaneous extrusion processes

There are several variations of the basic extrusion process for producing the following types of products.

Plastic tubes and pipes.   These are produced in an extruder with a *spider die*, as shown in Fig. 19.4a (see also Fig. 15.8 for details). Woven fiber or wire reinforcements also may be fed through specially designed dies in this operation for the production of

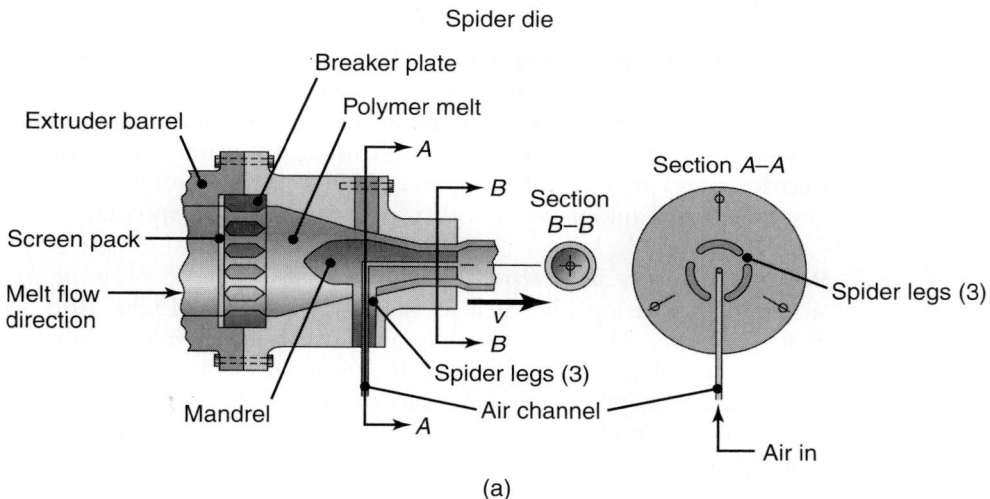

(a)

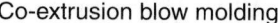

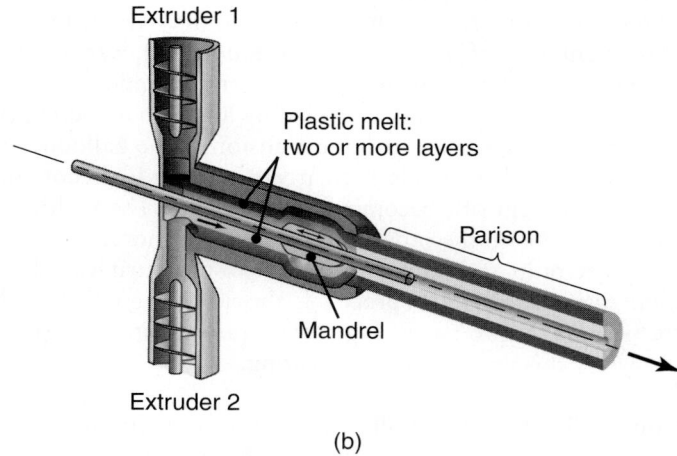

(b)

FIGURE 19.4   Extrusion of tubes. (a) Extrusion using a spider die (see also Fig. 15.8) and pressurized air. (b) Coextrusion for producing a bottle.

reinforced hoses that need to withstand higher pressures. The extrusion of tubes is also a necessary first step for related processes, such as extrusion blow molding and blown film.

Rigid plastic tubing. Extruded by a process in which the die is *rotated*, rigid plastic tubing causes the polymer to be sheared and biaxially oriented during extrusion. As a result, the tube has a higher crushing strength and a higher strength-to-weight ratio than conventionally extruded tubes.

Coextrusion. Shown in Fig. 19.4b, coextrusion involves simultaneous extrusion of two or more polymers through a single die. The product cross-section thus contains different polymers—each with its own characteristics and function. Coextrusion commonly is performed in shapes such as flat sheets, films, and tubes, and is used especially in food packaging where different layers of polymers have different functions, such as (a) inertness for food, (b) serving as barriers to fluids such as water or oil, and (c) labeling of the product.

Plastic-coated electrical wire. Electrical wire cable, and strips also are extruded and coated with plastic by this process. The wire is fed into the die opening at a controlled rate with the extruded plastic in order to produce a uniform coating. *Plastic-coated paper clips* also are made by coextrusion. To ensure proper insulation, extruded electrical wires are checked continuously for their resistance as they exit the die; they also are marked automatically with a roller to identify the specific type of wire.

Polymer sheets and films. These can be produced by using a specially designed flat-extrusion die, such as that shown in Fig. 19.3a. Also known as the *coathanger die*, it is designed to distribute the polymer melt evenly throughout the width. The polymer is extruded by forcing it through the die, after which the extruded sheet is taken up—first on water-cooled rolls and then by a pair of rubber-covered pull-off rolls. Generally, polymer *sheet* is considered to be thicker than 0.5 mm, and *film* is thinner than 0.5 mm.

Thin polymer films. Common *plastic bags* and other thin polymer film products are made from **blown film**, which is made from a thin-walled tube produced by an extruder (Fig. 19.5). In this process, a tube is extruded continuously vertically upward and then expanded into a balloon shape by blowing air through the center of the extrusion die until the desired film thickness is reached. Because of the molecular orientation of thermoplastics (Section 7.3), a *frost line* develops on the balloon in this process from which its transparency is reduced.

The balloon usually is cooled by air from a cooling ring around it, which can also act as a barrier to further expansion of the balloon, thus controlling its dimensions. The cooled bubble is then *slit* lengthwise, becoming *wrapping film*, or it is pinched and cut off, becoming a *plastic bag*. The width of the film produced after slitting can be on the order of 6 m (20 ft) or more.

The ratio of the blown diameter to the extruded tube diameter is known as the *blow ratio*, which in this process is about 3:1. Note that, as described in Section 2.2.7, the polymer has to have a high strain-rate sensitivity exponent, m, to successfully be blown by this process without tearing.

Plastic films. Plastic films, especially polytetrafluoroethylene (PTFE; trade name: *Teflon*), can be produced by *shaving* the circumference of a solid, round plastic billet using specially designed knives in a manner similar to producing veneer from a large piece of round wood. The process is called **skiving** (see also Section 24.4).

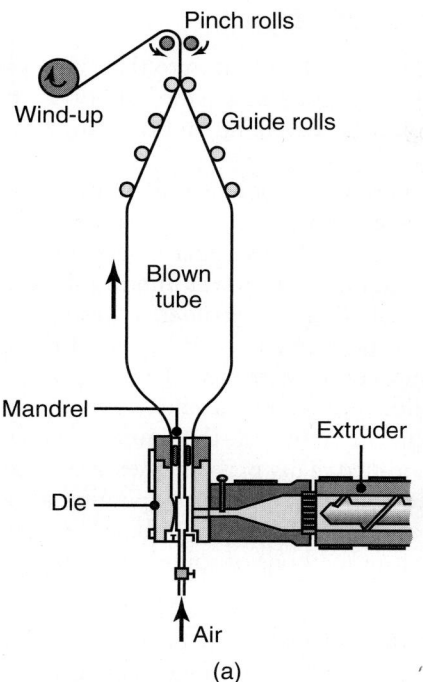

(a)   (b)

FIGURE 19.5   (a) Schematic illustration of the production of thin film and plastic bags from tube—first produced by an extruder and then blown by air. (b) A blown-film operation. This process is well developed, producing inexpensive and very large quantities of plastic film and shopping bags. *Source:* Courtesy of Windmoeller & Hoelscher Corp.

Pellets.   Used as raw material for other plastics-processing methods described in this chapter, pellets also are made by extrusion. A small-diameter, solid rod is extruded continuously and then *chopped* into short lengths (*pellets*). With some modifications, extruders also can be used as simple melters for other shaping processes, such as for injection molding and blow molding.

EXAMPLE 19.1   Blown film

Assume that a typical plastic shopping bag made by blown film has a lateral dimension (width) of 400 mm. (a) What should be the extrusion-die diameter? (b) These bags are relatively strong in use. How is this strength achieved?

Solution

a. The perimeter of the flat bag is $(2)(400) = 800$ mm. Since the original cross-section of the film is round, the blown diameter should be $\pi D = 800$, thus $D = 255$ mm. Recall that in this process, a tube is expanded from 1.5 to 2.5 times the extrusion-die diameter. Taking the maximum value of 2.5, we calculate the die diameter as $255/2.5 = 100$ mm.

b. Note in Fig. 19.5a that after being extruded, the balloon is being pulled upward by the pinch rolls. Thus, in addition to diametral stretching and the attendant molecular orientation, the film is *stretched* and *oriented* in the longitudinal direction. The resulting biaxial orientation of the polymer molecules significantly improves the strength and toughness of the plastic bag.

19.2.2 Production of polymer reinforcing fibers

Polymer fibers have numerous important applications. In addition to their use as reinforcement in composite materials, these fibers are used in a wide variety of consumer and industrial products, including clothing, carpeting, fabrics, rope, and packaging.

Most synthetic fibers used in reinforced plastics are polymers that are extruded through the tiny holes of a device called a **spinneret** (resembling a shower head) to form continuous filaments of semi-solid polymer. The extruder forces the polymer through the spinneret, which may have from one to several hundred holes. If the polymers are thermoplastics, they first are melted in the extruder, as described in Section 19.2. Thermosetting polymers also can be formed into fibers by first dissolving or chemically treating them so that they can be extruded. These operations are performed at high production rates and with very high reliability.

As the filaments emerge from the holes in the spinneret, the liquid polymer is converted first to a rubbery state and then solidified. This process of extrusion and solidification of continuous filaments is called **spinning**. The term spinning also is used for the production of natural textiles (such as cotton or wool), where short pieces of fiber are twisted into yarn. There are four methods of spinning fibers: melt, wet, dry, and gel spinning.

1. **In melt spinning** (shown in Fig. 19.6) the polymer is melted for extrusion through the spinneret and then solidified directly by cooling. A typical spinneret for this operation has about 50 holes of around 0.25 mm (0.01 in.) in diameter and is about 5 mm (0.2 in.) thick. The fibers that emerge from the spinneret are cooled by forced-air convection and are simultaneously pulled, so that their final diameter becomes much smaller than the spinneret opening. Polymers (such as nylon, olefin, polyester, and PVC) are produced in this matter. Because of the important applications of nylon and polyester fibers, melt spinning is the most important fiber-manufacturing process.

   Melt-spun fibers also can be extruded from the spinneret in various other cross-sections, such as trilobal (triangle with curved sides), pentagonal, octagonal, and hollow shapes. Hollow fibers trap air and, thus, provide additional thermal insulation, while other cross-sections have specific applications.

2. **Wet spinning** is the oldest process for fiber production and is used for polymers that have been dissolved in a solvent. The spinnerets are submerged in a chemical bath. As the filaments emerge, they precipitate in the chemical bath, producing a fiber that is then wound onto a bobbin. The term "wet spinning" refers to the use of a precipitating liquid bath, resulting in wet fibers that require drying before they can be used. Acrylic, rayon, and aramid fibers can be produced by this process.

3. **Dry spinning** is used for thermosets carried by a solvent. However, instead of precipitating the polymer by dilution as in wet spinning, solidification is achieved by evaporating the solvent in a stream of air or inert gas. The filaments do not come in contact with a precipitating liquid, thus eliminating the need for drying. Dry spinning may be used for the production of acetate, triacetate, polyether-based elastane, and acrylic fibers.

4. **Gel spinning** is a special process used to obtain high-strength or special fiber properties. The polymer is not melted completely or dissolved in liquid, but the molecules are bound together at various points in liquid-crystal form. This operation produces strong inter-chained forces in the resulting filaments that

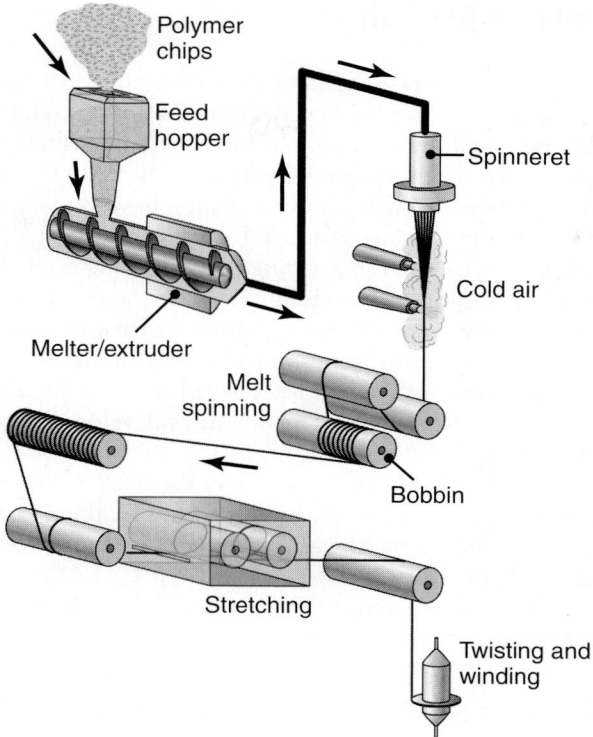

FIGURE 19.6 The melt-spinning process for producing polymer fibers. The fibers are then used in a variety of applications, including fabrics and as reinforcements for composite materials.

can increase significantly the tensile strength of the fibers. In addition, the liquid crystals are aligned along the fiber axis by the strain encountered during extrusion. The filaments emerge from the spinneret with an unusually high degree of orientation relative to each other—further enhancing their strength. This process also is called *dry-wet spinning*, because the filaments first pass through air and then are cooled further in a liquid bath. Some high-strength polyethylene and aramid fibers are produced by gel spinning.

A necessary step in the production of most fibers is the application of significant *stretching* to induce orientation of the polymer molecules in the fiber direction. This orientation is the main reason for the high strength of the fibers compared to the polymer in bulk form. The stretching can be done while the polymer is still pliable— just after extrusion from the spinneret—or it can be performed as a cold-drawing operation. The strain induced can be as high as 800%.

Graphite fibers are produced from different polymer fibers by *pyrolysis*. In this operation, controlled heat in the range of 1500° to 3000°C (2730° to 5400°F) is applied to the polymer fiber (typically polyacrylonitrile, PAN) to drive off all elements except the carbon. The fiber is under tension in order to develop a high degree of orientation in the resulting fiber structure. (See also Section 9.2.1 on the properties of graphite fibers and other details.)

19.3 | Injection Molding

Injection molding is similar to hot-chamber die casting (Fig. 19.7; see also Section 11.3.5). The pellets or granules are fed into the heated cylinder, and the melt is forced into the mold either by a hydraulic *plunger* or by the *rotating screw* system of an extruder. As in plastics extrusion, the barrel (cylinder) is heated externally to promote melting of the polymer. In injection-molding machines, however, a far greater portion of the heat transferred to the polymer is due to frictional heating.

Modern machines are of the *reciprocating* or *plasticating screw type* (Fig. 19.7b) with the sequence of operation shown in Fig. 19.8. As the pressure builds up at the mold entrance, the rotating screw begins to move backwards under pressure to a predetermined distance. This movement controls the volume of material to be injected. The screw then stops rotating and is pushed forward hydraulically, forcing the molten plastic into the mold cavity. The pressures developed usually range from 70 to 200 MPa (10,000 to 30,000 psi).

Some injection-molded products are shown in Fig. 19.9. Other products include cups, containers, housings, tool handles, knobs, toys, plumbing fixtures, telephone receivers, and electrical and communication-equipment components. For thermoplastics, the molds are kept relatively cool at about 90°C (190°F). Thermoset parts are molded in heated molds at about 200°C (400°F), where *polymerization* and *cross-linking* take place.

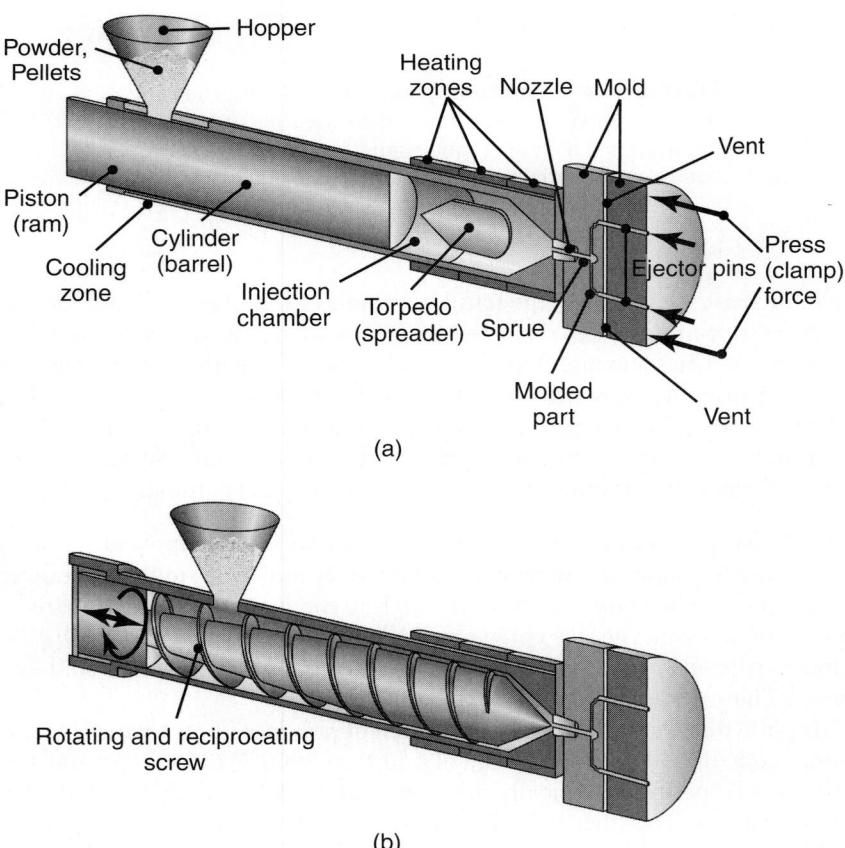

(a)

(b)

FIGURE 19.7   Schematic illustration of injection molding with (a) plunger and (b) reciprocating rotating screw.

1. Build up polymer in front of sprue bushing; pressure pushes the screw backwards. When sufficient polymer has built-up, rotation stops.

2. When the mold is ready, the screw is pushed forward by a hydraulic cylinder, filling the sprue bushing, sprue and mold cavity with polymer. The screw begins rotating again to build up more polymer.

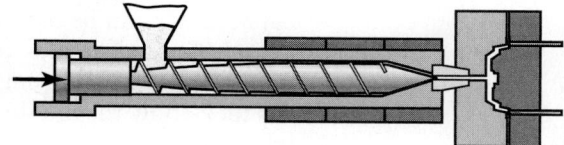

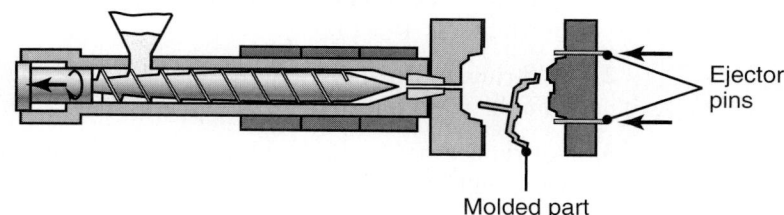

Rotating and reciprocating screw

3. After polymer is solidified/cured, the mold opens, and ejector pins remove the molded part.

Ejector pins

Molded part

FIGURE 19.8   Sequence of operations in the injection molding of a part with a reciprocating screw. This process is used widely for numerous consumer and commercial products, such as toys, containers, knobs, and electrical equipment (see Fig. 19.9).

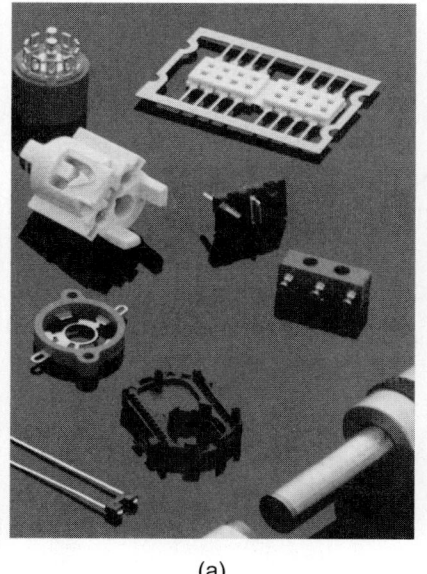

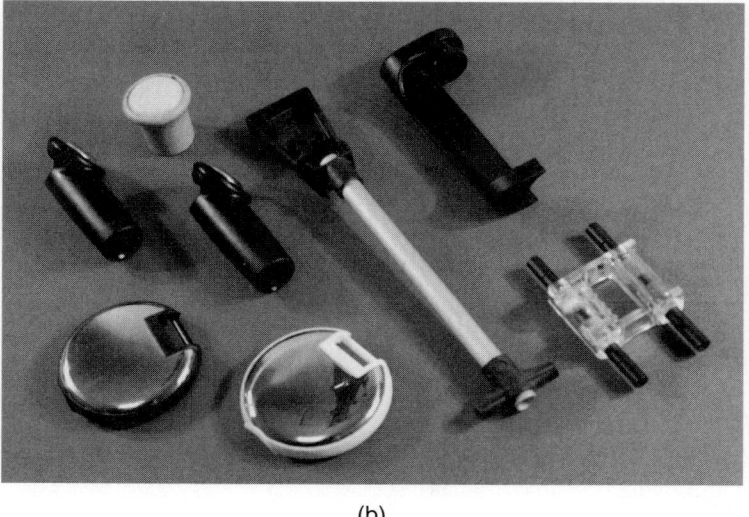

(a)   (b)

FIGURE 19.9   Typical products made by injection molding, including examples of insert molding. *Source:* (a) Courtesy of Plainfield Molding, Inc. (b) Courtesy of Rayco Mold and Mfg. LLC.

After the part has cooled sufficiently (for thermoplastics) or cured (for thermosets), the molds are opened and the part is removed from the mold using *ejectors*. The molds then are closed, and the process is repeated automatically. Elastomers also are injection molded into discrete products by these processes. Because the material is molten when injected into the mold, complex shapes with good dimensional accuracy can be obtained. However, because of uneven cooling of the part in the mold, residual stresses develop in the part.

Molds with moving and unscrewing mandrels also are used in injection molding, as they allow the molding of parts having multiple cavities or internal and external threaded features. To accommodate part design, molds may have several components (Fig. 19.10), including runners (such as those used in metal-casting dies), cores, cavities, cooling channels, inserts, knockout pins, and ejectors. There are three basic types of molds:

1. **Cold-runner, two-plate mold:** this design is the simplest and most common, as shown in Fig. 19.11a.

2. **Cold-runner, three-plate mold** (Fig. 19.11b): the runner system is separated from the part when the mold is opened.

3. **Hot-runner mold** (Fig. 19.11c), also called **runnerless mold:** the molten plastic is kept hot in a heated runner plate.

In cold-runner molds, the solidified plastic remaining in the channels connecting the mold cavity to the end of the barrel must be removed, which usually is done by trimming. Later, this scrap can be chopped and recycled. In hot-runner molds (which are more expensive), there are no gates, runners, or sprues attached to the

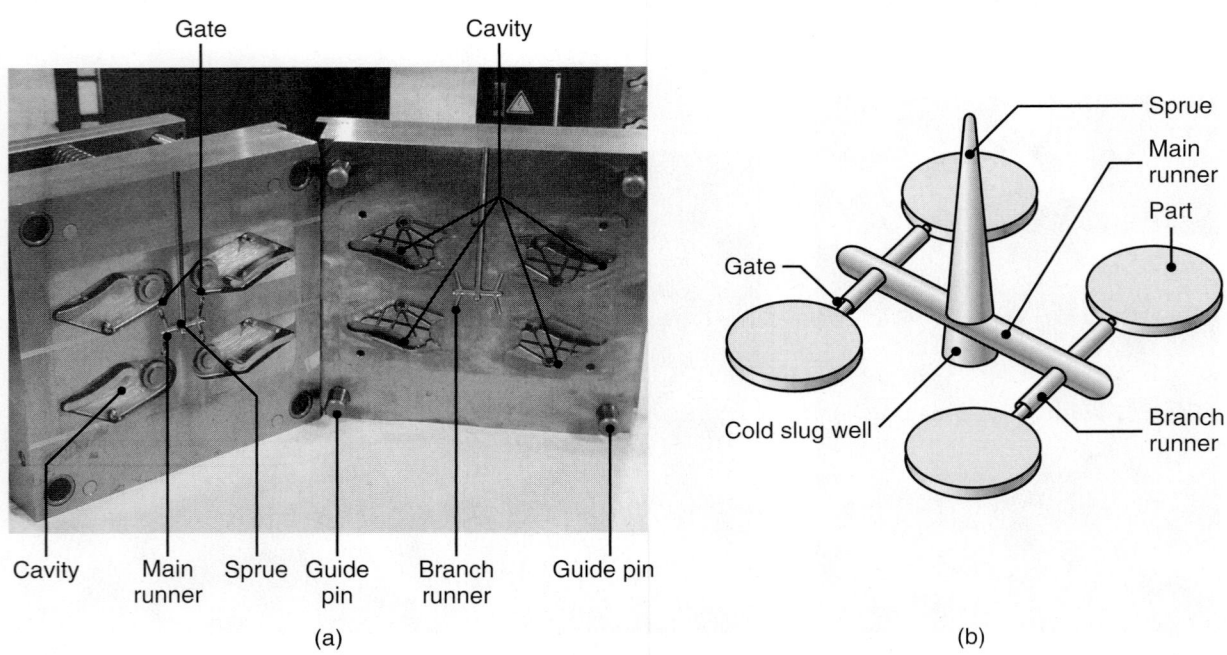

(a)   (b)

FIGURE 19.10   Illustration of mold features for injection molding. (a) Two-plate mold with important features identified. (b) Schematic illustration of the features in a mold. *Source:* Courtesy of Tooling Molds West, Inc.

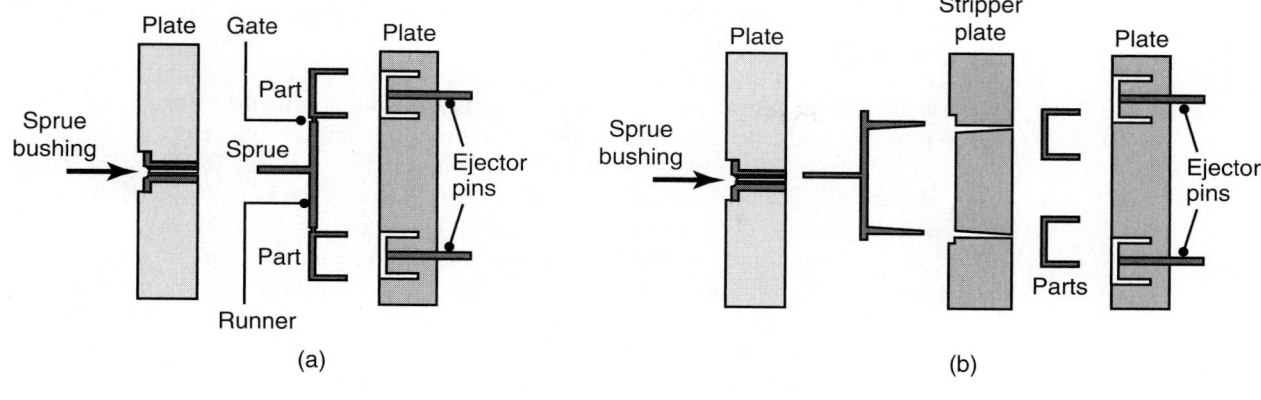

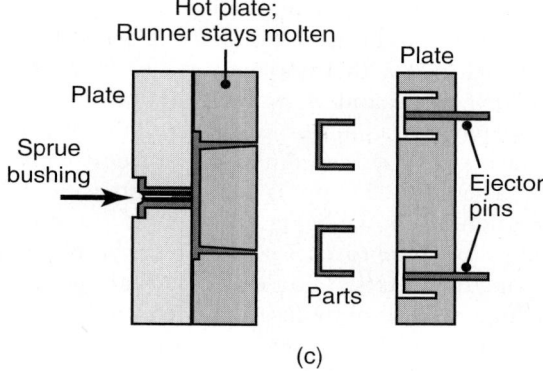

FIGURE 19.11 Types of molds used in injection molding: (a) two-plate mold; (b) three-plate mold; and (c) hot-runner mold.

molded part. Cycle times are shorter, because only the molded part must be cooled and ejected.

   Multicomponent injection molding (also called *coinjection* or *sandwich molding*) allows the forming of parts with a combination of various colors and shapes. An example is the molding of rear-light covers for automobiles made of different materials and colors, such as red, amber, and white. Also, for some parts, printed film can be placed in the mold cavity, so they need not be decorated or labeled after molding.

   Insert molding involves metallic components (such as screws, pins, and strips) that are placed in the mold cavity prior to injection and then become an integral part of the molded product (Fig. 19.9). The most common examples of such combinations are electrical and automotive components and faucet parts.

CASE STUDY 19.1 | EPOCH Hip Replacement

Each year, over one million artificial joints are replaced world-wide, thus eliminating pain and greatly improving the quality of life of their recipients. A total hip replacement usually is comprised of two major components: (a) a femoral component placed in the leg and (b) an acetabular component for the hip. The manufacture of hip stems was described in Example I.5 of the General

Introduction; the EPOCH hip stem is a recent development based on polymer processing technology.

One of the problems that has been encountered with total hip replacements is that of bone resorption. To keep it from breaking down and being reassimilated into the body, healthy bone must be stressed to maintain its strength and vigor. However, metal implants that are very stiff tend to transfer loads to the femur differently than the patient's natural hip, and much of the bone towards the top of the femur is subjected to very little stress. The body reacts to this low-stress state by resorbing the bone. This loss of bone mass can result in failure of the bone and lead to fatigue failure of the implant because of the new and harmful loading environment. This is serious, because repairing or revising an implant is significantly more painful for the patient and involves significantly more rehabilitation, and revisions usually are not as successful as the first implant.

A new and innovative implant design is the EPOCH system, shown in Fig. 19.12. In this system, the hip consists of three layers: (1) an outermost layer comprised of a diffusion-bonded (see Section 31.7) titanium-alloy mesh for bone in-growth, (2) a thermoplastic polymer layer intended to provide a stiffness similar to bone, and (3) a cobalt-alloy core to provide static and fatigue strength. The polymer used in this hip is from the polyaryletherketone (PAEK) family, and the layers are sized to mimic bone stiffness while maximizing fatigue strength.

The EPOCH hip is manufactured through an insert injection-molding operation, as shown in Fig. 19.13. A 85-ton vertical injection-molding machine is used (as opposed to a horizontal injection-molding machine) in order to facilitate locating the inserts. Two flexible pads are placed into seats on the mold halves. A machined cobalt-alloy insert is preheated and placed in the bottom half of the mold. The mold is closed, and the polymer is injected into the cavity, thus adhering to the insert, partially permeating into the porous pad, and integrally locking all components together. After the part is ejected from the mold, the flash

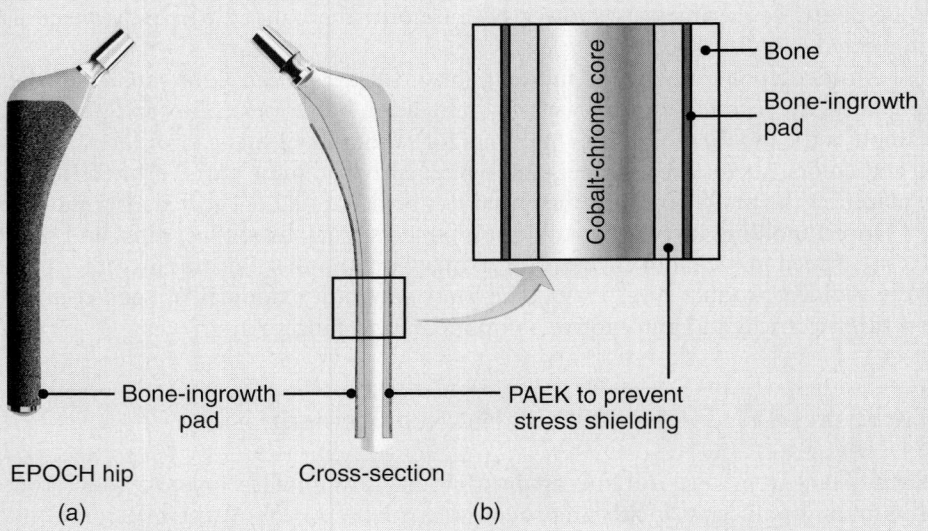

EPOCH hip

(a)

Cross-section

(b)

FIGURE 19.12   The EPOCH hip stem. This design uses a PAEK (polyaryletherketone) layer and bone-ingrowth pad around a cobalt-chrome core in order to maximize bone ingrowth. *Source:* Courtesy of Zimmer, Inc.

FIGURE 19.13   An EPOCH hip is removed from the mold after an insert injection-molding operation. *Source:* Courtesy of Zimmer, Inc.

is trimmed, and the hip stem is finished to obtain the desired surface finish. A cross-section of the resulting structure is shown in Fig. 19.12(b).

The EPOCH hip has improved greatly the ability for bone to remain healthy around the implant, leading to longer implant life, fewer revisions, and less pain and suffering for artificial joint recipients.

Source: Courtesy of M. Hawkins, Zimmer, Inc.

Overmolding.   This is a process for making products (such as hinge joints and ball-and-socket joints) in one operation and without the need for post-molding assembly. Two different plastics usually are used to ensure that no bonds will form between the molded halves of the joint, as otherwise motion would be impeded.

In **ice-cold molding**, the same type of plastic is used to form both components of the joint. The operation is carried out in a standard injection-molding machine and in one cycle. A two-cavity mold is used with cooling inserts positioned in the area of contact between the first and the second molded component of the joint. In this way, no bonds develop between the two pieces, and thus, the two components have free movements, such as in a hinge or a sliding mechanism.

Process capabilities.   Injection molding is a high-rate production process and permits good dimensional control. Although most parts generally weigh from 100 to 600 g (3 to 20 oz), they can be much heavier, such as in automotive-body panels and exterior components. Typical cycle times range from 5 to 60 seconds, although they can be several minutes for thermosetting materials.

Injection molding is a versatile process capable of producing complex shapes with good dimensional accuracy and at high production rates. As in other forming processes, mold design and the control of material flow in the die cavities are important factors in the quality of the product and thus to avoid defects. Because of the basic similarities to metal casting regarding material flow and heat transfer, *defects* observed in injection molding are somewhat of the same nature, as outlined next. Methods of avoiding defects consist of the proper control of temperatures, pressures, and mold design modifications using simulation software.

- For example, in Fig. 10.13g, the molten metal flows in from two opposite runners and then meets in the middle of the mold cavity. Thus, a cold shut in casting is equivalent to *weld lines* in injection molding.

- If the runner cross-sections are too small, the polymer may solidify prematurely, thus preventing full filling of the mold cavity. Solidification of the outer layers in thick sections can cause *porosity* or *voids* due to shrinkage, as in the metal parts shown in Fig. 12.2.

- If for some reason the dies do not close completely or due to die wear, a *flash* will form in a manner similar to flash formation in impression-die forging (see Figs. 14.5 and 19.19c).

- A defect known as *sink marks* (pull-in) also is observed in injection-molded parts, similar to that shown in Fig. 19.31c.

Much progress has been made in the analysis and design of molds and material flow in injection molding. *Modeling techniques* and *simulation software* have been developed for studying optimum gating systems, mold filling, mold cooling, and part distortion. *Software programs* now are available to expedite the design process for molding parts with good dimensions and characteristics. The programs take into account such factors as injection pressure, temperature, heat transfer, and the condition of the resin.

Machines. Injection-molding machines are usually horizontal (Fig. 19.14). Vertical machines are used for making small, close-tolerance parts and for insert molding.

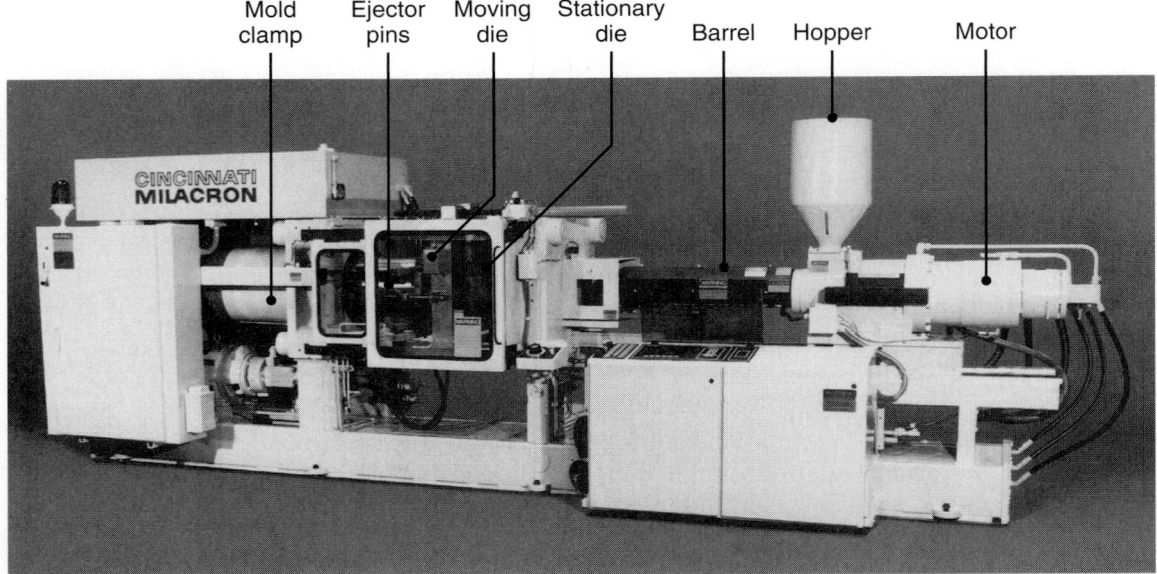

FIGURE 19.14 A 2.2-MN (250-ton) injection-molding machine. The tonnage is the force applied to keep the dies closed during the injection of molten plastic into the mold cavities and hold it there until the parts are cool and stiff enough to be removed from the die. *Source:* Courtesy of Cincinnati Milacron, Plastics Machinery Division.

The clamping force on the dies generally is supplied by hydraulic means, although electrical means also are used (which weigh less and are quieter than hydraulic machines). Modern machines are equipped with microprocessors and microcomputers in a control panel and monitor all aspects of the operation.

The machines are rated according to the capacity of the mold and the clamping force. In most machines, this force ranges from 0.9 to 2.2 MN (100 to 250 tons). The largest machine in operation has a capacity of 45 MN (5000 tons), and it can produce parts weighing 25 kg (55 lb). The cost of a 100-ton machine ranges from about $60,000 to about $90,000 and of a 300-ton machine from about $85,000 to about $140,000. Die costs typically range from $20,000 to $200,000. Consequently, high-volume production is essential to justify such high expenditure.

The molds generally are made of tool steels, beryllium-copper, or aluminum. They may have multiple cavities, so that more than one part can be made in one cycle (see also Fig. 11.19). Mold costs can be on the order of $100,000 for large ones. Mold life may be on the order of 2 million cycles for steel molds, but it can be about only 10,000 cycles for aluminum molds.

EXAMPLE 19.2 Force required in injection molding

A 250-ton injection-molding machine is to be used to make spur gears 4.5 in. in diameter and 0.5 in. thick. The gears have a fine-tooth profile. How many gears can be injection-molded in one set of molds? Does the thickness of the gears influence your answer?

Solution Because of the fine detail involved (fine gear teeth), let's assume that the pressures required in the mold cavity will be on the order of 100 MPa (15 ksi). The cross-sectional (projected) area of the gear is $\pi(4.5)^2/4 = 15.9$ in^2. If we assume that the parting plane of the two halves of the mold is in the middle of the gear, the force required is $(15.9)(15,000) = 238,500$ lb.

Since the capacity of the machine is 250 tons, we have $(250)(2000) = 500,000$ lb of clamping force available. Hence, the mold can accommodate two cavities and produce two gears per cycle. Because it does not influence the cross-sectional area of the gear, the thickness of the gear does not influence directly the pressures involved, and thus, it does not change the answer.

19.3.1 Reaction-injection molding

In the *reaction-injection molding* (RIM) process, a monomer and two or more reactive fluids are forced at high speed into a mixing chamber at a pressure of 10 to 20 MPa (1400 to 2800 psi) and then into the mold cavity (Fig. 19.15). Chemical reactions take place rapidly in the mold, and the polymer solidifies. Typical polymers are polyurethane, nylon, and epoxy. Cycle times may range up to about 10 minutes, depending on the materials, part size, and shape.

Major applications of this process include automotive parts (such as bumpers and fenders, steering wheels and instrument panels), thermal insulation for refrigerators and freezers, water skis, and stiffeners for structural components. Parts made may range up to about 50 kg (110 lb). Reinforcing fibers (such as glass or graphite) also may be used to improve the product's strength and stiffness. Depending on the number of parts to be made and the part quality required, molds can be made of common materials, such as steel or aluminum.

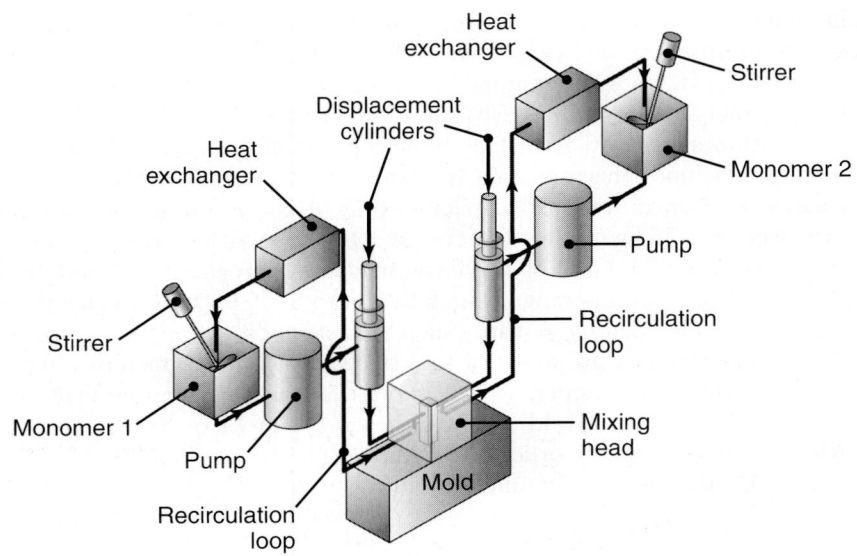

FIGURE 19.15 Schematic illustration of the reaction-injection molding process. Typical parts made are automotive-body panels, water skis, and thermal insulation for refrigerators and freezers.

19.4 | Blow Molding

Blow molding is a modified extrusion- and injection-molding process. In **extrusion blow molding**, a tube or preform (usually oriented so that it is vertical) is first extruded. It is clamped into a mold with a cavity much larger than the tube diameter and blown outward to fill the mold cavity (Fig. 19.16a). Depending on the material, the blow ratio may be as high as 7:1. Blowing usually is done with a hot-air blast at a pressure ranging from 350 to 700 kPa (50 to 100 psi). Drums with a volume as large as 2000 liters (530 gallons) can be made by this process. Typical die materials are steel, aluminum, and beryllium copper.

In some operations, the extrusion is continuous, and the molds move with the tubing. The molds close around the tubing, sealing off one end, breaking the long tube into individual sections, and moving away as air is injected into the tubular piece. The part then is cooled and ejected from the mold. Corrugated-plastic pipe and tubing are made by continuous blow molding in which the pipe or tubing is extruded horizontally and blown into moving molds.

In **injection blow molding**, a short tubular piece (**parison**) first is injection-molded (Fig. 19.16b) into cool dies (parisons may be made and stored for later use). The dies open, and the parison is transferred to a blow-molding die by an indexing mechanism (Fig. 19.16c). Hot air is injected into the parison, expanding it to the walls of the mold cavity. Typical products made are plastic beverage bottles (typically made of polyethylene or polyetheretherketone, PEEK) and small, hollow containers. A related process is **stretch blow molding**, where the parison is expanded and elongated simultaneously, subjecting the polymer to biaxial stretching and thus enhancing its properties.

Multilayer blow molding involves the use of coextruded tubes or parisons and thus permits the production of a multilayer structure (see Fig. 19.4b). A typical

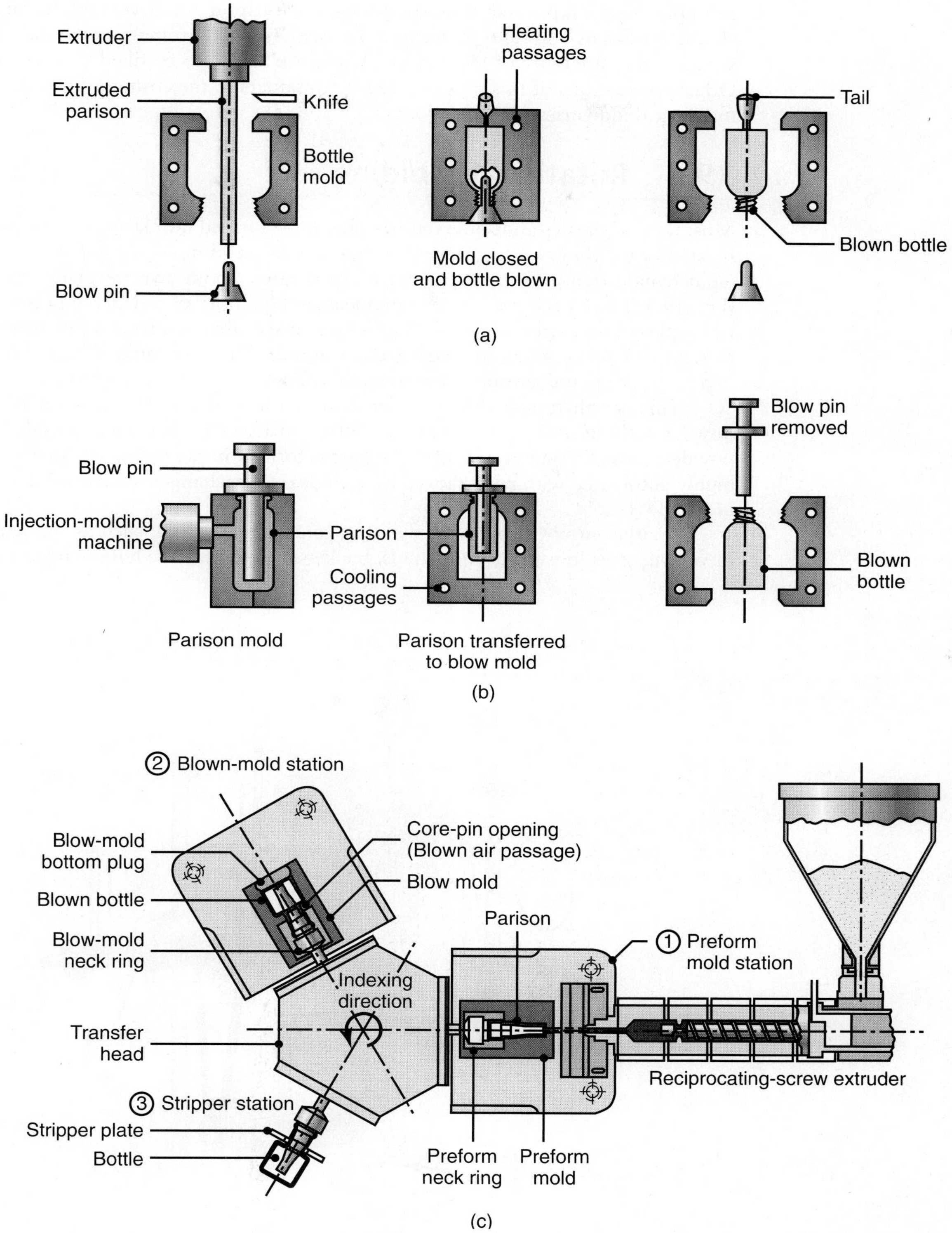

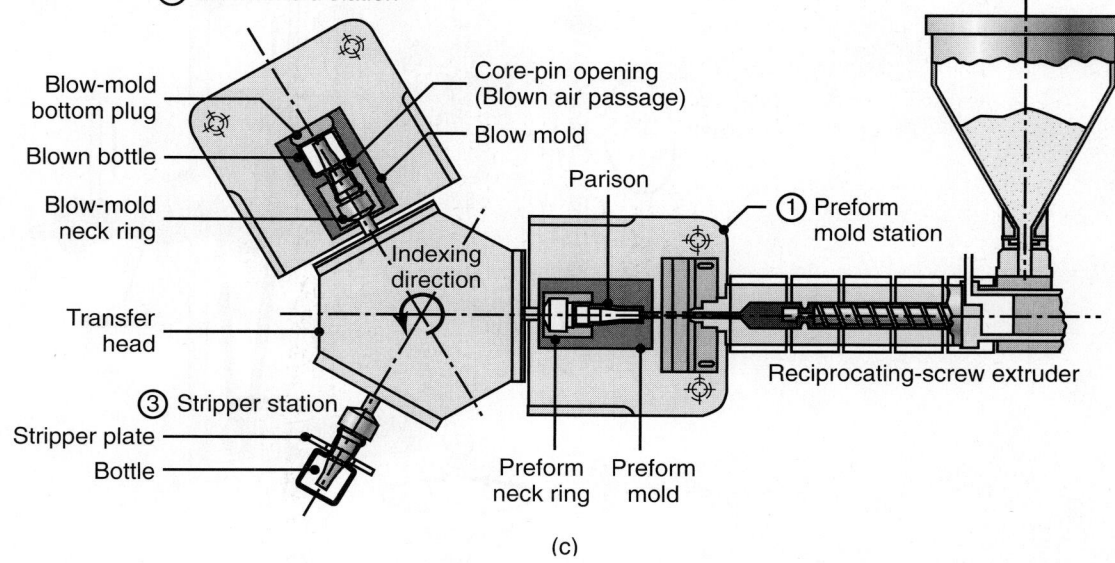

FIGURE 19.16   Schematic illustrations of (a) the extrusion blow-molding process for making plastic beverage bottles; (b) the injection blow-molding process; and (c) a three-station injection blow-molding machine for making plastic bottles.

example of such a product is plastic packaging for food and beverages, having such characteristics as odor and permeation barrier, taste and aroma protection, scuff resistance, the capability of being printed, and the ability to be filled with hot fluids. Other applications of this process are for containers in the cosmetics and the pharmaceutical industries.

19.5 | Rotational Molding

Most thermoplastics and some thermosets can be formed into large, hollow parts by *rotational molding*. In this process, a thin-walled metal mold is made in two pieces (split-female mold) and is designed to be rotated about two perpendicular axes (Fig. 19.17). For each part cycle, a premeasured quantity of powdered plastic material is placed inside the warm mold. (The powder is obtained from a polymerization process that precipitates a powder from a liquid.) Then the mold is heated (usually in a large oven) and is rotated continuously about the two principal axes.

This action tumbles the powder against the mold, where the heat fuses the powder without melting it. For thermosetting parts, a chemical agent is added to the powder; cross-linking occurs after the part is formed in the mold. The machines are highly automated with parts moved by an indexing mechanism similar to that shown in Fig. 19.16c.

A large variety of parts are made by rotational molding, such as storage tanks of various sizes, trash cans, boat hulls, buckets, housings, large hollow toys, carrying

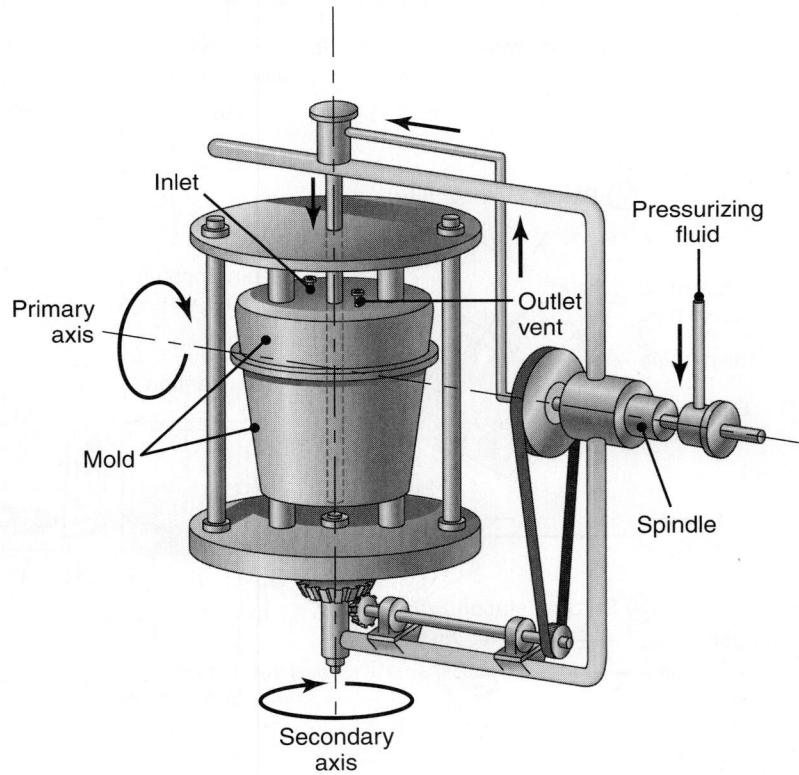

FIGURE 19.17 The rotational molding (rotomolding or rotocasting) process. Trash cans, buckets, and plastic footballs can be made by this process.

cases, and footballs. Various metallic or plastic inserts or components also may be molded integrally into the parts being made by this process.

In addition to powders, liquid polymers (**plastisols**) also can be used in rotational molding—PVC plastisols being the most common material. In this operation (called **slush molding** or *slush casting*), the mold is heated and rotated simultaneously. Due to the tumbling action, the polymer is forced against the inside walls of the mold, where it melts and coats the mold walls. The part is cooled while it is still rotating and removed by opening the mold. Parts made are typically thin-walled products, such as boots and toys.

Process capabilities.   Rotational molding can produce parts with complex, hollow shapes with wall thicknesses as small as 0.4 mm (0.016 in.). Parts as large as 1.8 m × 1.8 m × 3.6 m (6 ft × 6 ft × 12 ft) with a volume as large as 80,000 liters (21,000 gallons) have been produced. The outer surface finish of the part is a replica of the surface finish of the inside mold walls. Cycle times are longer than in other molding processes. Quality-control considerations usually involve accurate weight of the powder, proper rotational speed of the mold, and temperature–time relationships during the oven cycle.

19.6 | Thermoforming

Thermoforming is a process for forming thermoplastic sheets or films over a mold by means of the application of heat and pressure (Fig. 19.18). In this process, a sheet is (a) clamped and heated to the *sag point* (above the *glass-transition temperature*, T_g, of the polymer; Table 7.2), usually by radiant heating, and (b) forced against the mold surfaces through the application of a vacuum or air pressure. The sheets used in thermoforming usually are made by sheet extrusion and are available as a coiled strip or as lengths and widths of various sizes. They also are available filled with various materials for making parts with specific applications.

The mold is generally at room temperature, thus the shape produced becomes set upon contact with the mold. Because of the low strength of the materials formed, the pressure differential caused by a vacuum usually is sufficient for forming. However, thicker and more complex parts require air pressure, which may range from about 100 to 2000 kPa (15 to 300 psi), depending on the type of material and thickness

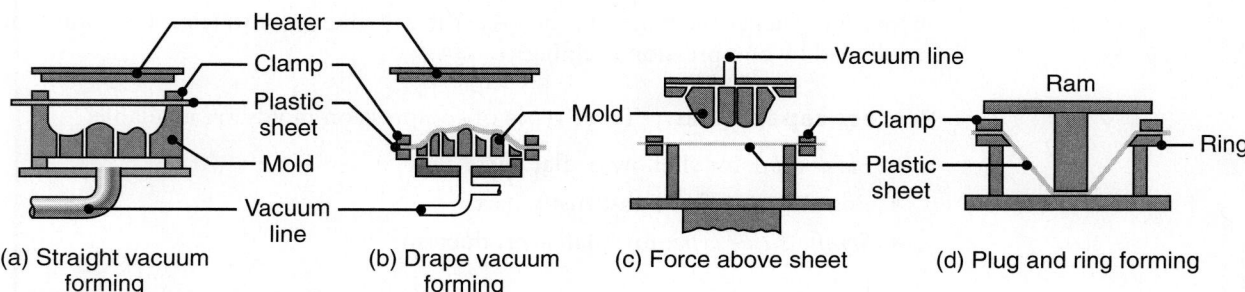

FIGURE 19.18   Various thermoforming processes for a thermoplastic sheet. These processes commonly are used in making advertising signs, cookie and candy trays, panels for shower stalls, and packaging.

of the sheet. Mechanical means, such as the use of *plugs*, also may be employed to help form the parts. Variations of the basic thermofoming process are shown in Fig. 19.18.

Process capabilities.   Typical parts made by thermoforming are packaging, trays for cookies and candy, advertising signs, refrigerator liners, appliance housings, and panels for shower stalls. Parts with openings or holes cannot be formed by this process because the pressure differential cannot be maintained during forming. Because thermoforming is a combination of *drawing* and *stretching* operations (much like in some sheet-metal forming), the material must exhibit high, uniform elongation; otherwise, it will neck and tear. Thermoplastics have high capacities for uniform elongation by virtue of their high strain-rate sensitivity exponent, m, as described in Section 2.2.7.

Molds for thermoforming usually are made of aluminum because high strength is not required, hence tooling is relatively inexpensive. Thermoforming molds have small holes in order to pull a vacuum. These holes typically are less than 0.5 mm (0.02 in.) in diameter; otherwise, they would leave marks on the parts formed. Quality considerations include tearing of the sheet during forming, nonuniform wall thickness, improperly filled molds, poor part definition, and lack of surface details.

19.7 | Compression Molding

In *compression molding*, a preshaped charge of material, premeasured volume of powder, or viscous mixture of liquid-resin and filler material is placed directly into a heated mold cavity that typically is around 200°C (400°F) but can be much higher. Forming is done under pressure from a plug or from the upper half of the die (Fig. 19.19), thus the process is somewhat similar to closed-die forging of metals.

Pressures range from about 10 to 150 MPa (1400 to 22,000 psi). As seen in Fig. 19.19, there is a flash formed, which subsequently is removed by trimming or by some other means. Typical parts made are dishes, handles, container caps, fittings, electrical and electronic components, washing-machine agitators, and housings. Fiber-reinforced parts with chopped fibers also are formed exclusively by this process.

Compression molding is used mainly with thermosetting plastics with the original material being in a partially polymerized state. However, thermoplastics also may be formed. Cross-linking is completed in the heated die. Curing times range from about 0.5 to 5 minutes, depending on the material and on part thickness and shape. The thicker the material, the longer it will take for it to cure. Elastomers also are shaped by compression molding.

Process capabilities.   Three types of compression molds are available:

* *Flash type*: for shallow or flat parts
* *Positive type*: for high-density parts
* *Semipositive type*: for quality production

Undercuts in parts are not recommended; however, dies can be designed to open sideways (Fig. 19.19d) to allow removal of the molded part. In general, the complexity of parts produced is less than that from injection molding, but the dimensional control is better. Surface areas of compression-molded parts may range up to about 2.5 m² (8 ft²). Because of their relative simplicity, dies for compression

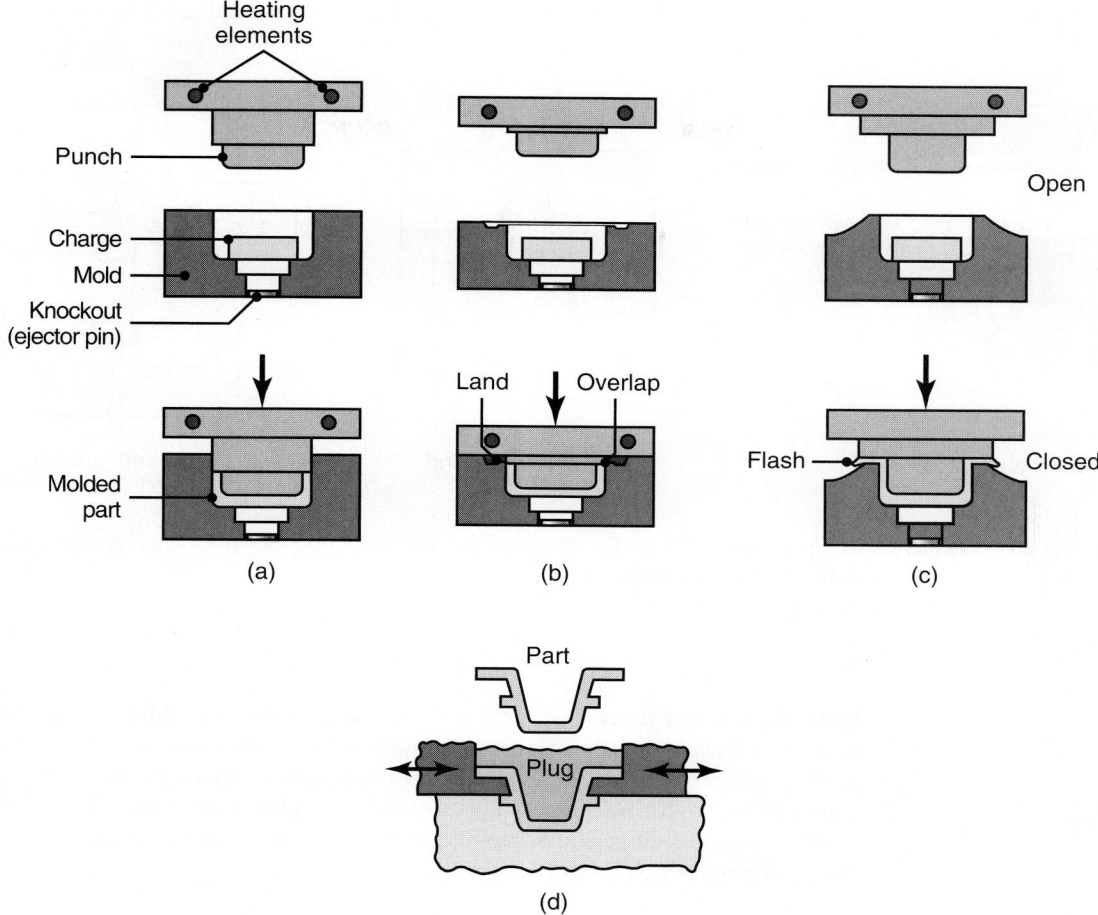

FIGURE 19.19 Types of compression molding—a process similar to forging: (a) positive, (b) semipositive, and (c) flash, which is later trimmed off. (d) Die design for making a compression-molded part with external undercuts.

molding generally are less costly than those used in injection molding. They typically are made of tool steels and may be chrome plated or polished for an improved surface finish of the molded product.

19.8 | Transfer Molding

Transfer molding represents a further development of the process of compression molding. The uncured thermosetting resin is placed in a heated transfer pot or chamber (Fig. 19.20), and after the material is heated, it is injected into heated closed molds. Depending on the type of machine used, a ram, plunger, or rotating-screw feeder forces the material to flow through the narrow channels into the mold cavity at pressures up to 300 MPa (43,000 psi). This viscous flow generates considerable heat, which raises the temperature of the material and homogenizes it. Curing takes place by cross-linking. Because the resin is in a molten state as it enters the molds, the complexity of the parts and the dimensional control approach those of injection molding.

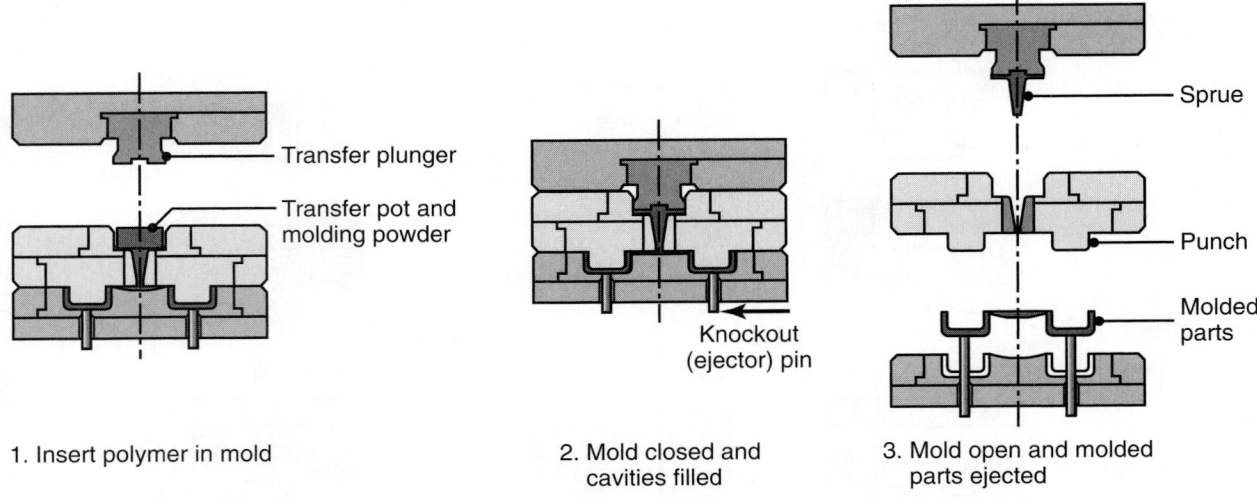

FIGURE 19.20 Sequence of operations in transfer molding for thermosetting plastics. This process is suitable particularly for intricate parts with varying wall thickness.

Process capabilities. Typical parts made by transfer molding are electrical connectors and electronic components, rubber and silicone parts, and the encapsulation of microelectronic devices. The process is suitable particularly for intricate shapes with varying wall thicknesses. The molds tend to be more expensive than those for compression molding, and some excess material is left in the channels of the mold during filling, which is later removed.

19.9 | Casting

Some thermoplastics (such as nylons and acrylics) and thermosetting plastics (epoxies, phenolics, polyurethanes, and polyester) can be *cast* into a variety of shapes using either rigid or flexible molds (Fig. 19.21). Compared to other methods of processing plastics, casting is a slow but simple and inexpensive process. However, the polymer must have sufficiently low viscosity in order to flow easily into the mold.

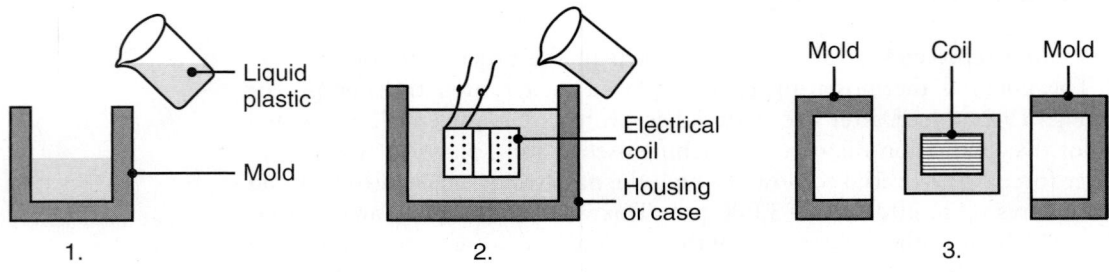

FIGURE 19.21 Schematic illustration of (1) casting, (2) potting, and (3) encapsulation processes for plastics and electrical assemblies, where the surrounding plastic serves as a dielectric.

Typical parts cast are gears (especially nylon), bearings, wheels, thick sheets, lenses, and components requiring resistance to abrasive wear.

In the basic conventional casting of thermoplastics, a mixture of monomer, catalyst, and various additives (activators) is heated to above its melting point, T_m, and poured into the mold. The part is formed after polymerization takes place at ambient pressure. Degassing may be necessary for product integrity. Intricate shapes can be produced using *flexible molds*, which are then peeled off (in a manner similar to using rubber gloves) and reused. As with metals, thermoplastics also may be *cast continuously*, where the polymer is carried over continuous stainless-steel belts and polymerized by external heat.

Centrifugal casting. This is a process similar to centrifugal metal casting (Section 11.3.6) and is used with thermoplastics, thermosets, and reinforced plastics with short fibers.

Potting and encapsulation. As a variation of casting that is important, particularly to the electrical and electronics industry, potting and encapsulation involve casting the plastic material (typically a liquid resin, such as expoxy) around an electrical component (such as a transformer) to embed it in the plastic. *Potting* (Fig. 19.21b) is carried out in a housing or case, which becomes an integral part of the component and fixes it in position. In *encapsulation* (Fig. 19.21c), the component is coated with a layer of the plastic, surrounding it completely, and then solidifying.

In both of these processes the plastic material can serve as a *dielectric* (nonconductor); consequently, it must be free of moisture and porosity, which would require processing in a vacuum. Mold materials may be metal, glass, or various polymers. Small structural members (such as hooks, studs, and similar parts) may be encapsulated partially by dipping them in a hot thermoplastic using polymers of various colors.

19.10 | Foam Molding

Products such as styrofoam cups, food containers, insulating blocks, and shaped packaging materials (such as for shipping appliances, computers, and electronics) are made by *foam molding*, using expandable **polystyrene beads** as the raw material. As it readily can be seen upon close inspection, these products have a **cellular structure**. The structure may have *open and interconnected porosity* (for polymers with low viscosity) or have *closed cells* (for polymers with high viscosity).

There are several techniques that can be used in foam molding. In the basic operation, polystyrene beads obtained by polymerization of styrene monomer are placed in a mold with a blowing agent—typically pentane (a volatile hydrocarbon) or inert gas (nitrogen)—and exposed to heat—usually by steam. As a result, the beads expand to as much as 50 times their original size and take the shape of the mold cavity. The amount of expansion can be controlled by varying the temperature and time. Various other particles also may be added, including hollow glass beads or plastic spheres, to impart specific structural characteristics to the foam produced.

Polystyrene beads are available in three sizes: (a) small: for cups with a finished part density of about 50 kg/m^3 (3 lb/ft^3), (b) medium: for molded shapes, and (c) large: for molding insulating blocks with a finished part density of about 15–30 kg/m^3 (which can then be cut to size). The bead size selected also depends on the minimum wall thickness of the product: the smaller the size, the thinner the part. The beads also can be colored prior to expansion, thus the part becomes integrally colored. Both thermoplastics and thermosets can be used for foam molding, but thermosets are in a liquid-processing form and thus are similar to the condition of polymers in reaction-injection molding.

A common method of foam molding is to use *pre-expanded polystyrene beads* in which the beads are expanded partially by steam (hot air, hot water, or an oven also can be used) in an open-top chamber. The beads then are placed in a storage bin and allowed to stabilize for a period of 3 to 12 hours. They then can be molded into desired shapes in the manner described previously.

Structural foam molding.   This is a molding process used to make plastic products with a *solid outer skin* and a *cellular core structure*. Typical products made are furniture components, computer and business-machine housings, and moldings (replacing more expensive wood moldings). In this process, thermoplastics are mixed with a blowing agent (usually an inert gas such as nitrogen) and injection molded into cold molds of desired shapes. The rapid cooling against the cold-mold surfaces produces a skin that is rigid (which can be as much as 2 mm (0.08 in.) thick) and a core of the part that is cellular in structure. The overall part density can be as low as 40% of the density of the solid plastic. Thus, with a rigid skin and a less-dense bulk, molded parts have a high stiffness-to-weight ratio (see also Fig. 3.1).

Polyurethane foam processing.   Products such as furniture cushions and insulating blocks are made by this process. Basically, the operation starts with the mixing of two or more components; chemical reactions then take place after the mixture is (a) poured into molds of various shapes or (b) sprayed over surfaces with a spray gun to provide sound and thermal insulation. Various low-pressure and high-pressure machines are available, having computer controls for proper mixing. The mixture solidifies with a cellular structure, the characteristics of which depend on the type and proportion of the components used.

19.11 | Cold Forming and Solid-Phase Forming

Processes that have been used in the cold working of metals (such as rolling, closed-die forging, coining, deep drawing, and rubber forming—described in Part III) also can be used to form thermoplastics at room temperature (*cold forming*). Typical materials formed are polypropylene, polycarbonate, ABS, and rigid PVC. Important considerations regarding this process are: (a) the polymer must be sufficiently ductile at room temperature (thus polystyrenes, acrylics, and thermosets cannot be formed) and (b) its deformation must be nonrecoverable (in order to minimize springback and creep of the formed part).

The advantages of the cold forming of plastics over other methods of shaping are:

* Strength, toughness, and uniform elongation are increased.
* Plastics with high molecular weights can be used to make parts with superior properties.
* Forming speeds are not affected by part thickness because (unlike other plastic-processing methods) there is no heating or cooling involved. Cycle times generally are shorter than in molding processes.

Solid-phase forming.   Also called *solid-state forming*, this process is carried out at a temperature of 10° to 20°C (20° to 40°F) *below* the melting temperature of the plastic (for a crystalline polymer). Thus, the forming operation takes place while the polymer is still in a solid state. The main advantages of solid-phase forming over cold forming are that forming forces and springback are lower. These processes are not used as widely as hot-processing methods and generally are restricted to special applications.

19.12 | Processing Elastomers

We have described the properties, characteristics, and applications of elastomers and rubbers in Section 7.9. Recall that (in terms of its processing characteristics) a thermoplastic *elastomer* is a polymer. In terms of its function and performance, it is a *rubber*. The raw material to be processed into various shapes is basically a compound of rubber and various additives and fillers. The additives include *carbon black*—an important element that enhances properties such as tensile and fatigue strength, abrasion and tear resistance, ultraviolet protection, and resistance to chemicals.

These materials then are mixed (Banbury mixers) to break them down and lower the viscosity; the mixture subsequently is *vulcanized* using sulfur as the vulcanizing agent. This compound then is ready for further processing (such as calendering, extrusion, and various molding processes), which may also include reinforcements in such forms as fibers and fabric. During processing, the part becomes cross-linked, imparting the desirable properties that we all associate with rubber products, ranging from rubber boots to pneumatic tires.

Elastomers can be shaped by a variety of processes that also are used for shaping thermoplastics. Thermoplastic elastomers commonly are shaped by extrusion or injection molding—extrusion being the more economical and the faster process. They also can be formed by blow molding or thermoforming. Thermoplastic polyurethane, for example, can be shaped by all conventional methods. It also can be blended with thermoplastic rubbers, polyvinyl chloride compounds, ABS, and nylon to obtain specific properties.

The temperatures for elastomer extrusion are typically in the range of 170° to 230°C (340° to 440°F) and for molding are up to 60°C (140°F). Dryness of the materials is important for product integrity. Reinforcements also are used in conjunction with extrusion to impart greater resistance to failure of the part. Examples of extruded elastomer products are tubing, hoses, moldings, and inner tubes. Injection-molded elastomer products cover a broad range of applications, such as numerous components for automobiles and appliances.

Rubber and some thermoplastic sheets are formed by the **calendering** process (Fig. 19.22), wherein a warm mass of the compound is fed into a series of rolls and is **masticated**. The thickness produced is typically 0.3 to 1 mm (0.01 to 0.40 in.) but can be less by stretching the material. It then is stripped off in the form of a sheet, which may be as wide as 3 m (10 ft) and at speeds on the order of 2 m/s (6.5 ft/s).

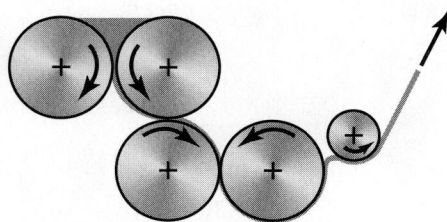

FIGURE 19.22   Schematic illustration of calendering. Sheets produced by this process subsequently are used in thermoforming. The process also is used in the production of various elastomer and rubber products.

The calendered rubber then may be processed into various products, such as tires and belts for machinery. The rubber or thermoplastics also may be formed over both surfaces of a tape, paper, fabric, or various soft or rigid plastics, thus making them permanently *laminated*. Roll surfaces also may be textured to produce a rubber sheet with various patterns and designs.

Discrete rubber products, such as gloves, are made by *dipping* a metal form (such as in the shape of a hand for making gloves) repeatedly into a liquid compound that adheres to the form. A typical compound is *latex*, which is a milklike sap obtained from the inner bark of a tropical tree. The compound is then vulcanized (cross-linked), usually in steam, and stripped from the form and becomes a discrete product.

19.13 | Processing Polymer-Matrix Composites

As we have described in Chapter 9, polymer-matrix composites (PMCs) (also called **reinforced plastics**) are *engineered materials* with unique mechanical properties, especially high strength-to-weight ratio, stiffness-to-weight ratio, creep resistance, and directional properties. Because of their complex structure, reinforced plastics require special methods to shape them into consumer and industrial products (Fig. 19.23).

Polymer-matrix composites can be fabricated by various methods, which we describe in this chapter. Fabrication to ensure reliable properties in composite parts and structures, particularly over the long range of their service life, can be challenging because of the presence of two or more types of materials. The matrix and the reinforcing fibers in the composite have—*by design*—very different properties and characteristics and consequently have different responses to the methods of processing (Section 9.2).

The several steps involved in manufacturing reinforced plastics and the time and care required make the processing costs very substantial, and generally, they are not competitive with traditional materials and shapes. This situation has necessitated the careful assessment and integration of the design and manufacturing processes (concurrent engineering) in order to take advantages of the unique properties of these composites. This is done while minimizing manufacturing costs and maintaining long-range product integrity, reliability, and production rate. Additonally, an important safety and environmental concern in reinforced plastics is the dust generated during processing. For example, airborne carbon fibers are known to remain in the work area long after the fabrication of parts has been completed.

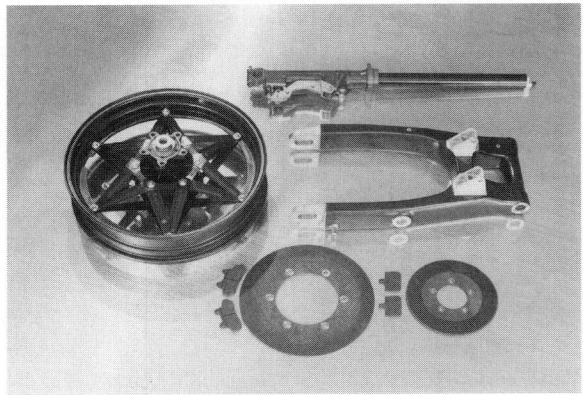

FIGURE 19.23   Reinforced plastic components for a Honda motorcycle. The parts shown are front and rear forks, rear swing-arm, wheel, and brake disks.

19.13.1 Fiber impregnation

In order to obtain good bonding between the reinforcing fibers and the polymer matrix and to protect them during handling, fibers are surface treated by impregnation (*sizing*). When impregnation is carried out as a separate step, the resulting partially cured sheets are called by various terms, as described in this section.

Prepregs. In a typical procedure for making fiber-reinforced plastic *prepregs* (meaning preimpregnated with resin), the continuous fibers first are aligned and subjected to a surface treatment to enhance the adhesion to the polymer matrix (Fig. 19.24a). They then are coated by dipping them in a resin bath and are made into a *tape* (Fig. 19.24b), typically in widths of 75 to 150 mm (3 to 6 in.). Individual segments of prepreg tape then are cut and assembled into *laminated structures* (Fig. 19.25a), such as the horizontal stabilizer for the F-14 fighter aircraft.

Typical products made from prepregs are flat or corrugated architectural paneling, panels for construction and electrical insulation, and structural components of aircraft, requiring good property retention over a period of time and under adverse conditions typically encountered by military aircraft (including fatigue strength under hot or wet conditions).

Because the process of laying prepreg tapes is a time-consuming and labor-intensive operation, special and high-automated *computer-controlled tape-laying machines* have been built for this purpose (Fig. 19.25b). The prepreg tapes automatically are cut from a reel and placed on a mold in the desired patterns with much better dimensional control than can be achieved by hand. The layout patterns can be modified easily and quickly for a variety of parts by computer control and with high repeatability.

Sheet-molding compound (SMC). In making this compound, continuous strands of reinforcing fiber first are chopped into short fibers (Fig. 19.26) and deposited in random orientations over a layer of resin paste. Generally, the paste is a polyester mixture (which may contain fillers, such as various mineral powders) and

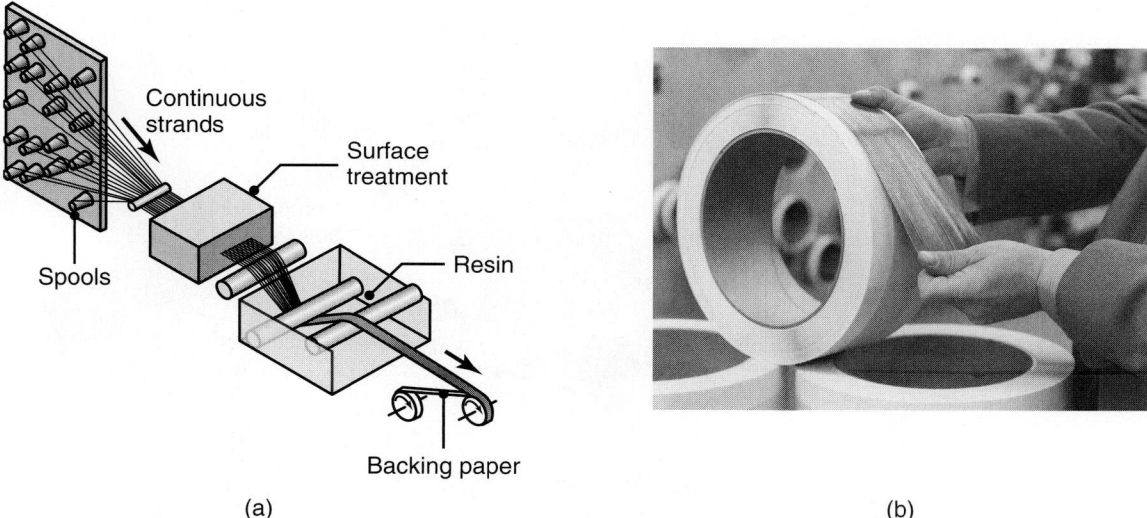

(a)   (b)

FIGURE 19.24 (a) Manufacturing process for polymer-matrix composite tape. (b) Boron-epoxy prepreg tape. These tapes are then used in making reinforced plastic parts and components with high strength-to-weight and stiffnes-to-weight ratios, particularly important for aircraft and aerospace applications and sports equipment. *Source:* (a) Courtesy of T.-W. Chou, R.L. McCullough, and R.B. Pipes. (b) Courtesy of Avco Specialty Materials/Textron.

(a)　　　　　　　　　　　　　　　　(b)

FIGURE 19.25 (a) Single-ply layup of boron-epoxy tape for the horizontal stabilizer for an F-14 fighter aircraft. (b) A 10-axis computer-numerical-controlled tape-laying system. This machine is capable of laying up 75- and 150-mm (3- and 6-in.) wide tapes on contours of up to ±30° and at speeds of up to 0.5 m/s (1.7 ft/s). *Source*: (a) Courtesy of Grumman Aircraft Corporation. (b) Courtesy of The Ingersoll Milling Machine Company.

is carried on a polymer film (such as polyethylene). A second layer of resin paste then is deposited on top, and the sheet is pressed between rollers.

The product then is gathered into rolls (or placed into containers in several layers) and stored until it has undergone a maturation period and has reached the desired viscosity. The maturing process takes place under controlled conditions of temperature and humidity and usually takes one day.

The molding compounds should be stored at a temperature sufficiently low to delay curing. They have a limited shelf life (usually around 30 days) and must be processed within this period. Alternatively, the resin and the fibers can be mixed together only at the time they are placed into the mold.

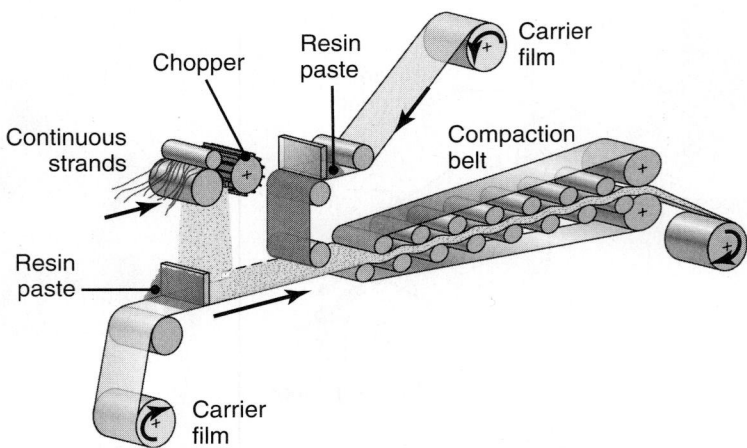

FIGURE 19.26 Schematic illustration of the manufacturing process for producing fiber-reinforced plastic sheets. The sheet still is viscous at this stage and later can be shaped into various products. *Source*: After T.-W. Chou, R. L. McCullough, and R. B. Pipes.

Bulk-molding compound (BMC). These compounds are in the shape of bil-lets, (hence the term bulk) and generally are up to 50 mm (2 in.) in diameter. They are made in the same manner as SMCs and extruded to produce a bulk form. When processed into products, BMCs have flow characteristics that are similar to those of dough, thus they also are called *dough-molding compounds* (DMC).

Thick-molding compound (TMC). Thick-molding compounds combine a characteristic of BMC (lower cost) with one of SMC (higher strength). They are usu-ally injection molded, using chopped fibers of various lengths. One application is in electrical components because of the high dielectric strength of TMC.

19.13.2 Molding of reinforced plastics

There are several molding processes used for reinforced plastics, as described in this section.

Compression molding. The material is placed between two molds, and pres-sure is applied. The molds either may be at room temperature or heated to acceler-ate hardening of the part. The material may be a bulk-molding compound, which is a viscous, sticky mixture of polymers, fibers, and additives. Generally, it is molded into the shape of a log, which subsequently is cut or sliced into the desired shape. Fiber lengths generally range from 3 to 50 mm (0.125 to 2 in.), although longer fibers of 75 mm (3 in.) also may be used.

Sheet-molding compounds also can be processed by compression molding. These compounds are similar to bulk-molding compounds, except that the resin–fiber mixture is laid between plastic sheets to make a sandwich that can be handled easily. The sheets have to be removed prior to placing the SMC in the mold.

Vacuum-bag molding. In this process (Fig. 19.27a), prepregs are laid in a mold to form the desired shape. The pressure required to shape the product and to devel-op good bonding is obtained by covering the lay-up with a plastic bag and creating a vacuum. Curing takes place at room temperature or in an oven.

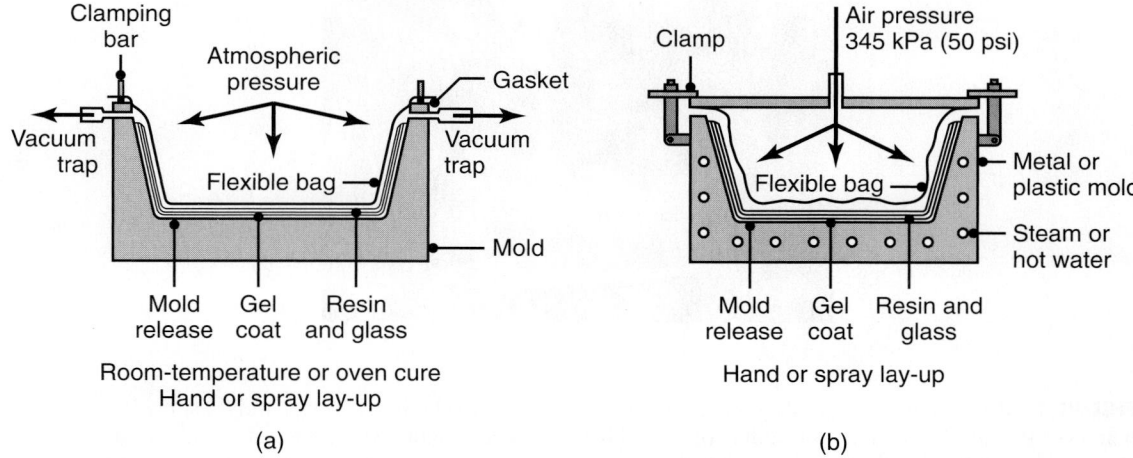

FIGURE 19.27 Schematic illustration of (a) vacuum-bag forming, and (b) pressure-bag forming. These processes are used in making discrete reinforced plastic parts. *Source*: After T. H. Meister.

A variation of this process is **pressure-bag molding** (Fig. 19.27b). A flexible bag is placed over the resin and reinforcing fiber mixture, and pressure is applied over the mold at a range typically of 200 to 400 kPa (30 to 50 psi). If higher heat and pressure are needed to produce parts with higher density and fewer voids, the entire assembly is put into an *autoclave* (a chamber under heat and pressure).

Care should be exercised to maintain fiber orientation if specific directional properties are desired. In chopped-fiber materials, no specific orientation is intended. In order to prevent the resin from sticking to the vacuum bag and also to facilitate removal of excess resin, several sheets of various materials (*release cloth* or *bleeder cloth*) are placed on top of the prepreg sheets.

The molds can be made of metal (usually aluminum), but they more often are made from the same resin (with reinforcement) as the material to be cured. This practice eliminates any difficulties caused by the difference in thermal expansion between the mold and the part.

Contact molding.   This is a series of processes that use a single male or female mold (Fig. 19.28), hence it also is called *open-mold processing,* which can be made of such materials as reinforced plastics, wood, metal, or plaster. This is a *wet* method

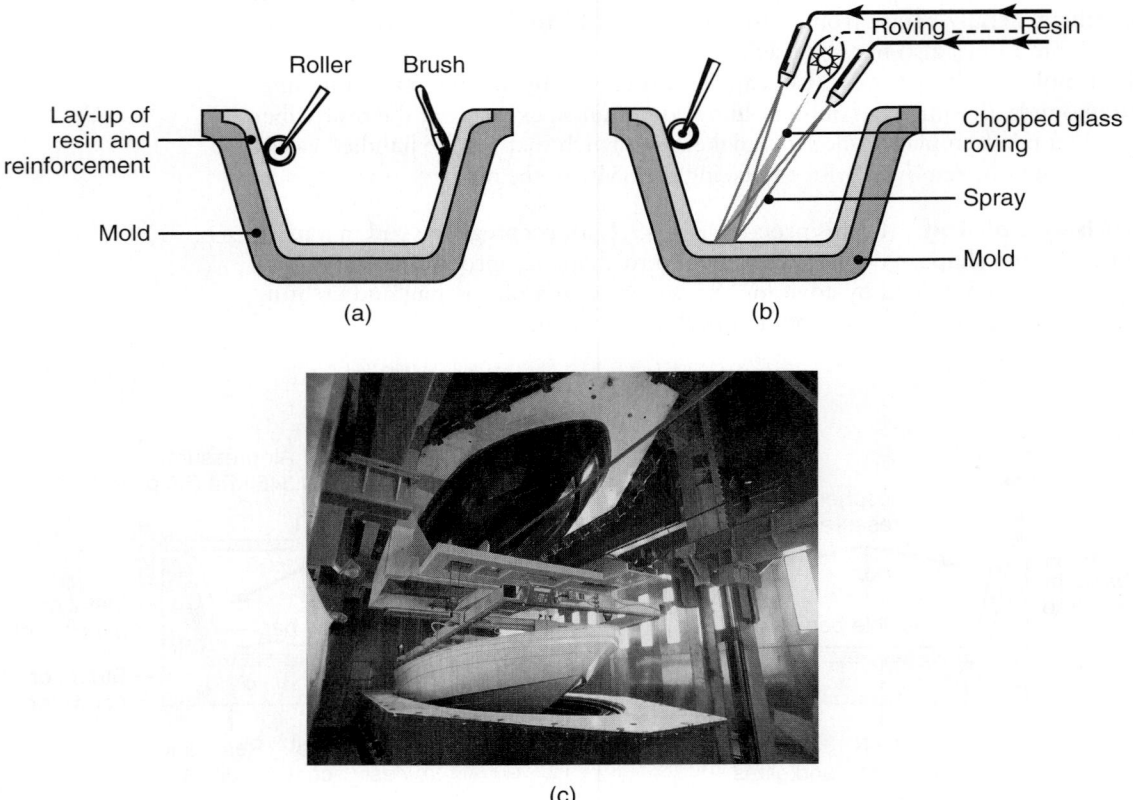

FIGURE 19.28   Manual methods of processing reinforced plastics: (a) hand lay-up, and (b) spray lay-up. Note that, even though the process is slow, only one mold is required. The figures show a female mold, but male molds also are used. These methods also are called *open-mold processing.* (c) A boat hull made by these processes. *Source*: Courtesy of Genmar Holdings, Inc.

in which the materials are applied in layers and the reinforcement is impregnated with the resin at the time of molding. Contact molding is used in making *laminated products* with high surface area-to-thickness ratios, hence the process also is called *contact lamination*. Typical examples of products are swimming pools, boat hulls, automotive-body panels, tub and shower units, and housings.

The simplest method of contact molding is **hand lay-up**. The materials are placed in proper order (resins and reinforcements), brushed, and shaped in the mold by hand with a roller (Fig. 19.28a). The squeezing action of the roller expels any trapped air bubbles and compacts the part. The reinforcements placed in the mold may consist of various shapes, including prepregs. Consequently, their orientation in the final product can be controlled.

In **spray lay-up**, molding is done by spraying the materials into the mold. As seen in Fig. 19.28b, both the resin and the chopped fibers are sprayed over the mold surfaces. Rolling the deposited materials (as in hand lay-up) to remove any porosity may be necessary in this process. Since the chopped fibers have random orientations, directional properties cannot be imparted in products made by this process.

These two processes are relatively slow, have high labor costs, and require significant time and labor in finishing operations. Also, the choice of materials that can be used is limited. However, they are simple to perform, and the tooling is inexpensive. Note also that only the mold-side surface of the formed part is smooth from being in contact with the mold surfaces.

Resin-transfer molding. This process is based on transfer molding (Section 19.8). A resin is mixed with a catalyst and is forced by a piston-type, positive-displacement pump into the mold cavity that is filled with a fiber reinforcement. This process is a viable alternative to hand lay-up, spray-up, and compression molding for low- or intermediate-volume production.

Transfer/injection molding. This is an automated operation that combines compression-molding, injection-molding, and transfer-molding processes. This combination has the good surface finish, dimensional stability, and mechanical properties obtained in compression molding and the high-automation capability and low cost of injection molding and transfer molding.

EXAMPLE 19.3 Tennis rackets made of composite materials

In order to impart certain desirable characteristics (such as light weight and stiffness), composite-material tennis rackets are manufactured with graphite, fiberglass, boron, ceramic (silicon carbide), and Kevlar as the reinforcing fibers. These rackets have a foam core, and some have a unidirectional (while others have a braided) reinforcement.

Rackets with boron fibers have the highest stiffness, followed by those with graphite (carbon), glass, and Kevlar fibers. The racket with the lowest stiffness has 80% fiberglass, while the stiffest has 95% graphite and 5% boron. Thus, it has the highest percentage of inexpensive reinforcing fiber and the smallest percentage of the most expensive fiber.

A new development is the use of piezoelectric fibers integrated into the composite material of the racket frame. The purpose of the piezoelectric material is to modify the stiffness of the racket—depending on whether a power or control shot is desired. Also, the piezoelectric material can be used to dampen vibrations in order to improve the feel of the tennis racket.

Two major manufacturing strategies are used for composite tennis rackets. They can be compression molded (see Fig. 19.19), where fibers or cloths are impregnated with epoxy and placed in the mold. An inner tube is inflated or injected with an expanding foam to produce a hollow cross-section—the shape of which is defined by the mold. Some tennis rackets are made by injection molding by using a nylon matrix and short reinforcing fibers (usually graphite).

19.13.3 Filament winding, pultrusion, and pulforming

Filament winding.   This is a process in which the resin and fibers are combined at the time of curing (Fig. 19.29a) in order to develop a composite structure. Axisymmetric parts, (such as pipes and storage tanks) and even nonsymmetric parts are produced on a rotating mandrel. The reinforcing filament, tape, or roving is wrapped continuously around the form. The reinforcements are impregnated by passing them through a polymer bath. This process can be modified by wrapping the mandrel with a prepreg material.

The products made by filament winding are very strong because of their highly reinforced structure. Parts as large as 4.5 m (15 ft) in diameter and 20 m (65 ft) long have been made by this process. Filament winding also has been used for strengthening cylindrical or spherical pressure vessels (Fig. 19.29b) made of materials such as aluminum and titanium, where the presence of a metal inner lining makes the part impermeable. Filament winding also can be used directly over solid-rocket propellant forms. Seven-axis computer-controlled machines have

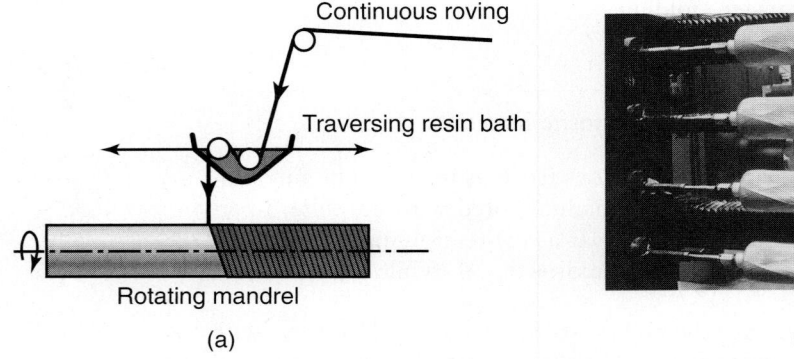

(a)   (b)

FIGURE 19.29   (a) Schematic illustration of the filament-winding process; (b) fiberglass being wound over aluminum liners for slide-raft inflation vessels for the Boeing 767 aircraft. The products made by this process have high strength-to-weight ratio and also serve as lightweight pressure vessels. *Source:* Courtesy of Brunswick Corporation.

been developed that automatically dispense several unidirectional prepregs to also make nonsymmetric parts, such as aircraft engine ducts, fuselages, propellers, blades, and struts.

Pultrusion.   Long parts with various uniform cross-sections (such as rods, profiles, flat strips, and tubing) are made continuously by the *pultrusion* process. This sequence of operations is shown in Fig. 19.30. The continuous reinforcement, glass roving, or fabric, (typically made of calcium aluminosilicate glass fiber—E type, see Section 9.2.1) is supplied through several bobbins. The bundle is pulled first through a thermosetting polymer bath (usually polyesters), then through a performing die, and through a heated steel die.

The product is cured during its travel through the heated die, which has a length of up to about 1.5 m (5 ft.) and a speed slow enough to allow sufficient time for the polymer to set. (Note that this is an operation similar to continuously baking bread and cookies or making resin-bonded grinding wheels.) The exiting material then is cut into desired lengths. Typical products made by pultrusion (which may contain up to about 75% reinforcing fiber) are golf clubs, ski poles, fishing poles, drive shafts, and such structural members as ladders, walkways, and handrails. Cross-sections as large as 1.5 m × 0.3 m (60 in. × 12 in.) have been made by this process.

Pulforming.   Continuously reinforced products other than constant cross-sectional profiles are made by *pulforming*. After being pulled through the polymer bath, the composite is clamped between the two halves of a die and cured into a finished product. The dies recirculate and shape the products successively. Commonly made products are hammer handles reinforced by glass fibers and curved automotive leaf springs.

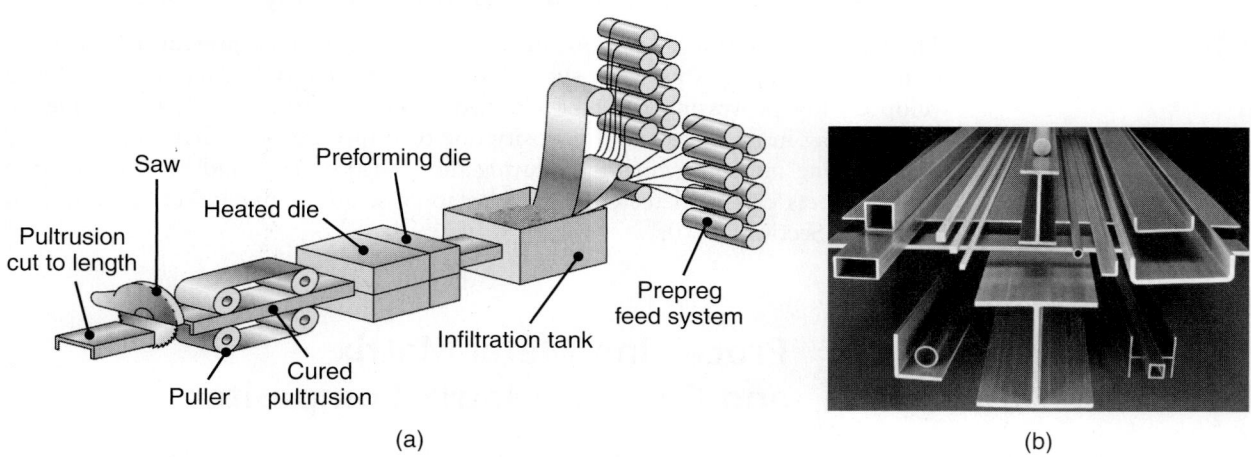

(a)   (b)

FIGURE 19.30   (a) Schematic illustration of the pultrusion process. (b) Examples of parts made by pultrusion. The major components of fiberglass ladders (used especially by electricians) are made by this process. They are available in different colors but are heavier because of the presence of glass fibers. *Source*: Courtesy of Strongwell Corporation.

EXAMPLE 19.4 **Polymer automotive-body panels shaped by various processes**

The trend towards the use of polymeric materials for automobile bodies has been increasing at an accelerating rate; this example outlines typical applications of polymers. Three commonly used and competing processing methods are: (a) injection-molded thermoplastics and elastomers; (b) reaction-injection molded polyurea/polyurethanes; and (c) compression-molded sheet-molding compound (SMC) with resin-transfer-molded polyester and vinylester.

Typical examples of parts made for automobiles are: (a) body panels and other large exterior components made by injection molding; (b) front fenders and rear quarter panels made of polyphenylene-ether/nylon or thermoplastric polyester; (c) outer door panels made of polycarbonate/ABS, and (d) fascias made of thermoplastic polyolefin. These materials are selected for design flexibility, impact strength and toughness, corrosion resistance, high durability, and low mass. Vertical panels and fascias are made in multicavity molds on large injection-molding machines, then assembled mechanically to a steel frame.

Large exterior-body parts also are made of reaction-injection molded (RIM) polyurethane, although polyureas have become important for body panels and bumpers. Thermoset fascias are made of reinforced RIM polyurethane and (more recently) new polyureas because of their higher thermal stability, low-temperature toughness, and lower cycle times required. Large horizontal exterior-body panels (such as hoods, roofs, and rear decks) are made of reinforced polyester or vinylester in the form of compression-molded sheet-molding compounds. Lower-volume parts are made by transfer molding (RTM).

Environmental and recycling considerations in material and process selection for automobiles have become increasingly important. For example, polyphenylene oxide is being replaced with polycarbonate, which is made out of 100% recycled or reclaimed materials.

19.13.4 Quality considerations in processing reinforced plastics

The major quality considerations in the processes described previously concern internal voids and gaps between successive layers of material. Volatile gases that develop during processing must be allowed to escape from the lay-up through the vacuum bag in order to avoid porosity due to trapped gases. Microcracks may develop during improper curing or during the transportation and handling of parts. These defects can be detected using ultrasonic scanning and other techniques described in Section 36.10.

19.14 Processing Metal-Matrix and Ceramic-Matrix Composites

Metal-matrix composites (MMC) can be made into near-net shaped parts by the following processes.

- **Liquid-phase processing** basically consists of casting together the liquid-matrix material (such as aluminum or titanium) and the solid reinforcement (such as graphite, aluminum oxide, or silicon carbide) by *conventional casting* processes or by *pressure-infiltration casting*. In the latter process, pressurized gas is used to force the liquid-metal matrix into a preform (usually shaped out of wire or sheet and made of reinforcing fibers).

- **Solid-phase processing** consists basically of *powder-metallurgy* techniques, including cold and hot isostatic pressing. Proper mixing is important in order to obtain homogeneous distribution of the fibers. An example of this technique is the production of tungsten-carbide tools and dies with cobalt as the matrix material, as described in Example 17.4.

- **Two-phase (liquid–solid) processing** involves technologies which consist of *rheocasting* (Section 11.3.7) and the techniques of *spray atomization* and *deposition*. In the latter two processes, the reinforcing fibers are mixed with a matrix that contains both liquid and solid phases of the metal.

In making complex metal-matrix composite parts with whisker or fiber reinforcement, die geometry and control of process variables are very important for ensuring the proper distribution and orientation of the fibers within the part. MMC parts made by powder-metallurgy techniques generally are heat treated for optimum properties.

EXAMPLE 19.5 Metal-matrix composite brake rotors and cylinder liners

Some brake rotors currently are being made of composites consisting of an aluminum-based matrix reinforced with 20% silicon-carbide particles. The particles are stirred into molten aluminum alloys, and the mixture is cast into ingots. The ingots then are remelted and cast into shapes (such as brake rotors and drums) using casting processes (such as green-sand, bonded-sand, investment, permanent-mold, and squeeze casting. These rotors (a) are about one-half the weight of those made of gray cast iron, (b) conduct heat three times faster, (c) add the stiffness and wear-resistance characteristics of ceramics, and (d) reduce noise and vibration because of internal damping.

To improve the wear- and heat-resistance of cast-iron cylinder liners in aluminum engine blocks, aluminum-matrix liners also are being developed. The metal-matrix layer consists of 12% aluminum-oxide fiber and 9% graphite fiber and has a thickness that ranges from 1.5 to 2.5 mm (0.06 to 0.1 in.).

19.14.1 Processing ceramic-matrix composites

Several processes are used in making ceramic-matrix composites (CMC), including more recent techniques such as melt infiltration, controlled oxidation, and hot-press sintering. Although the latter are largely in the experimental stage, they continue to be developed to improve the properties and performance of these composites.

- **Slurry infiltration** is the most common process to make ceramic-matrix composites. It involves the preparation of a fiber preform that first is hot pressed and then impregnated with a combination of slurry (which contains the matrix powder), a carrier liquid, and an organic binder. High strength, toughness, and uniform structure are obtained by this process, but the product has limited high-temperature properties. A further improvement of this process is **reaction bonding** or **reaction sintering** of the slurry.

- **Chemical-synthesis** processes involve the sol-gel and the polymer-precursor techniques. In the **sol-gel process**, a *sol* (a colloidal fluid having the liquid as its continuous phase) that contains fibers is converted to a *gel*. The gel then is subjected to heat treatment to produce a ceramic-matrix composite. The **polymer-precursor method** is analogous to the process used in making ceramic fibers, using aluminum oxide, silicon nitride, and silicon carbide.

- In **chemical-vapor infiltration**, a porous fiber preform is infiltrated with the matrix phase using the chemical vapor deposition technique (Section 34.6). The product has very good high-temperature properties, but the process is time consuming and costly.

19.15 | Design Considerations

Design considerations in the forming and shaping of plastics are similar to those for casting metals. The selection of appropriate materials from an extensive list requires considerations of (a) service requirements, (b) possible long-range effects on their properties and behavior (such as dimensional stability and wear), and (c) ultimate disposal of the product after its life cycle. Some of these issues were described in Section I.6 in the General Introduction and Section 7.8.

Outlined here are the general design guidelines for the production of plastic and composite-material parts.

1. As we have seen, processes for plastics have inherent flexibility to produce a wide variety of part shapes and sizes. Complex parts with internal and external features can be produced with relative ease and at high production rates. Consequently, a process such as injection molding competes well with powder-injection molding and die casting, all of which are capable of producing complex shapes and having thin walls. In considering possible process substitutions, it is necessary to bear in mind that the materials involved and their characteristics are quite different—each having its own properties that are important to a particular application.

2. Compared to metals, plastics have much lower stiffness and strength. Therefore, section sizes and shapes should be selected accordingly. Depending on the application, a high section modulus can be achieved based on design principles common to I-beams and tubes. Large, flat surfaces can be stiffened by such simple means as prescribing curvatures on parts. For example, observe the stiffness of very thin but gently curved slats in venetian blinds. Such simple design aspects can be ascertained by inspecting the differences between the hoods of garden tractors—for example, one made of sheet metal and the other of plastic. Reinforcement with fibers or particles also are effective in achieving this objective.

3. The overall part shape and thickness often determine the particular shaping or molding process to be selected. Even after a particular process is chosen, the designs of the part and the die should be such that they will not present difficulties concerning proper shape generation (Fig. 19.31), dimensional control, and surface finish. Because of low stiffness and thermal effects, dimensional tolerances (especially for thermoplastics) are not as small as in metalworking processes. Dimensional tolerances are much smaller in injection molding, for example, than they are in thermoforming. As in the casting of metals and alloys, the control of material flow in the mold cavities is important. The effects of molecular orientation during the processing of the polymer also should be considered, especially in extrusion, thermoforming, and blow molding.

4. Large variations in cross-sectional areas, section thicknesses, and abrupt changes in geometry should be avoided to achieve proper shape generation. Note for example, the *sink marks* (pull-in) shown in the top piece in Fig. 19.31c are due to the fact that thick sections solidify last. Furthermore, *contraction in larger*

Original   Distortion   Modified
design   design

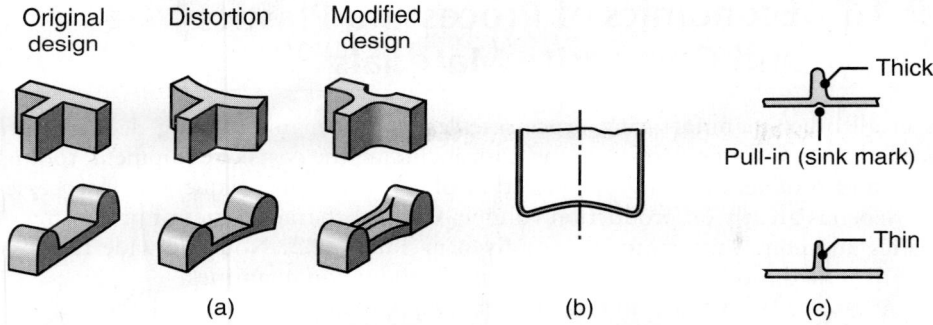

Thick

Pull-in (sink mark)

Thin

(a)   (b)   (c)

FIGURE 19.31   Examples of design modifications to eliminate or minimize distortion in plastic parts: (a) suggested design changes to minimize distortion; (b) stiffening the bottoms of thin plastic containers by doming—a technique similar to the process used to shape the bottoms of aluminum beverage cans; and (c) design change in a rib to minimize pull-in (sink mark) caused by shrinkage during the cooling of thick sections in molded parts.

cross-sections tends to cause *porosity* in plastic parts, as it does in metal casting (see Fig. 12.2), thus affecting product quality. On the other hand, a lack of stiffness may make it more difficult to remove thin parts from molds after shaping them.

5. The low elastic modulus of plastics further requires that shapes be selected properly for improved stiffness of the component (Fig. 19.31b), particularly when saving material is an important factor. Note that these considerations are similar to those in the design for metal castings and forgings as is the need for drafts (typically less than one degree for polymers) to enable removal of the part from molds and dies. Generally, the recommended part thickness ranges from about 1 mm (0.04 in.) for small parts to about 3 mm (0.12 in.) for large parts.

6. Physical properties (especially a high coefficient of thermal expansion) are important considerations. Improper part design or assembly can lead to distortion (warping) and uneven shrinking (Fig. 19.31a). Plastics can be molded easily around metallic parts and inserts; however, their interfacial strength and compatibility with metals when so assembled is an important consideration.

7. The properties of the final product depend on the original material and its processing history. For example, the cold working of polymers improves their strength and toughness. On the other hand, because of the nonuniformity of deformation (even in simple rolling), residual stresses develop in polymers just as they do in metals. These stresses also can be due to the thermal cycling of the part during processing. The magnitude and direction of residual stresses (however produced) are important factors, such as in stress cracking. Furthermore, these stresses can relax over a period of time and cause distortion of the part during its service life.

8. A major design advantage of reinforced plastics is the directional nature of the strength of the composite (for example, see Fig. 9.7). External forces applied to the material are transferred by the resin matrix to the fibers, which are much stronger and stiffer than the matrix. When all of the fibers are oriented in one direction, the resulting composite material is exceptionally strong in the fiber direction and weak in others.

   To achieve strength in two principal directions, individual unidirectional layers are laid at the corresponding angles to each other as is done in filament winding. If strength in the third (thickness) direction is required, a different type of reinforcement (such as woven fiber) is used to form a sandwich structure.

19.16 Economics of Processing Plastics and Composite Materials

As in all other manufacturing processes, design and manufacturing decisions are based ultimately on performance and cost, including the costs of equipment, tooling labor, and production. The final selection of a process or a sequence of processes also depends greatly on production volume. General characteristics of processing of plastics and composite materials are given in Table 19.2. Note the wide range of equipment and tooling costs and economical production quantities.

It also can be seen from this table that equipment costs and tool and die costs are related somewhat, and for processes such as thermoforming, casting, and rotational molding, these costs are relatively low. The most expensive are injection-molding machines, followed by compressing-molding and transfer-molding machines. Tool and die costs are also high. Thus, in an operation like injection molding, the size of the die and the optimum number of cavities in the die for producing more and more parts in one cycle are important considerations, as they are in die casting.

Larger dies may be considered in order to accommodate a larger number of cavities (with runners to each cavity) but at the expense of increasing die cost even further. On the other hand, more parts will be produced per machine cycle, thus production rate will increase. Therefore, a detailed analysis must be made to determine the overall die size, the number of cavities in the die, and the machine capacity required in order to optimize the total operation and to produce parts at minimum cost. Similar considerations also apply to the other plastics processing methods described throughout this chapter.

For the production of composite materials, equipment and tooling costs for most molding operations are generally high, production rates are low, and economical production quantities vary widely. High equipment and tooling costs can be acceptable if the production run is large.

TABLE 19.2

Comparative Production Characteristics of Various Molding Methods			
Molding method	Equipment and tooling cost	Production rate	Economical production quantity
Extrusion	M–L	VH–H	VH
Injection molding	VH	VH	VH
Rotational molding	M	M–L	M
Blow molding	M	H–M	H
Compression molding	H–M	M	H–M
Transfer molding	H	M	VH
Thermoforming	M–L	M–L	H–M
Casting	M–L	M–L	L
Centrifugal casting	H–M	M–L	M–L
Pultrusion	H–M	H	H
Filament winding	H–M	L	L
Spray lay-up and hand lay-up	L–VL	L–VL	L

VH = very high; H = high; M = medium; L = low; VL = very low.

SUMMARY

- Thermoplastics can be shaped by a variety of processes, including extrusion, molding, casting, and thermoforming, as well as by some of the processes used in metalworking. The raw material usually is in the form of pellets, granules, and powders. The high strain-rate sensitivity of thermoplastics allows extensive stretching in forming operations, thus complex and deep parts can be produced easily. Thermosetting plastics generally are molded or cast, and they have better dimensional accuracy than in forming thermoplastics.

- Fiber-reinforced plastics are processed into structural components using liquid plastics, prepregs, and bulk- and sheet-molding compounds. Fabricating techniques include various molding methods, filament winding, pultrusion, and pulforming. The type and orientation of the fibers and the strength of the bond between fibers and matrix and between layers of materials are important considerations.

- The design of plastic parts must take into account their low strength and stiffness and their physical properties, such as high thermal expansion and generally low resistance to temperature. Inspection techniques are available to determine the integrity of these products.

- Because of their expanding use in critical applications, the processing of metal-matrix and ceramic-matrix composites has undergone important developments to ensure product integrity and reduce costs. Metal-matrix composites are processed by liquid-phase, solid-phase, and two-phase processes. Ceramic-matrix composites can be processed by slurry infiltration, chemical synthesis, or chemical-vapor infiltration.

- Because of the wide variety of low-cost materials and manufacturing techniques available, the economics of processing plastics and composite materials is an important consideration, particularly when compared to those of metal components. These factors include machinery and die costs, cycle times, and production rate and volume.

KEY TERMS

Blow molding	Extrusion	Parison
Blow ratio	Extrusion blow molding	Pellets
Bulk-molding compound	Filament winding	Plastisols
Calendering	Foam molding	Potting
Casting	Hand lay-up	Prepregs
Chemical synthesis	Ice-cold forming	Pulforming
Chemical-vapor infiltration	Injection molding	Pultrusion
Coathanger die	Insert molding	Reaction-injection molding
Coextrusion	Liquid-phase processing	Resin transfer molding
Cold forming	Masticated	Rotational molding
Compression molding	Melt spinning	Sheet-molding compound
Contact molding	Open-die processing	Sink marks
Encapsulation	Overmolding	Sizing

Slurry infiltration

Slush molding

Solid-phase forming

Solid-phase processing

Spinneret

Spinning

Spray lay-up

Structural foam molding

Swell

Thermoforming

Thick-molding compound

Transfer molding

Two-phase processing

Vacuum-bag molding

BIBLIOGRAPHY

Baird, D.G. and Collias, D.I., *Polymer Processing: Principles and Design*, Wiley, 1998.

Berins, M.L. (ed.), *Plastics Engineering Handbook,* 5th ed., Chapman & Hall, 1994.

Bhowmick, A.K., Hall M.M., and Benarey, H.A., *Rubber Products Manufacturing Technology*, Marcel Dekker, 1994.

Bryce, D., *Plastic Injection Molding*, Vol. I, 1996; Vol. II, 1997; Vol. III, 1998; Vol. IV, 1999, Society of Manufacturing Engineers.

Buckley, C.P., Bucknall, C.B., and McCrum, N.G., *Principles of Polymer Engineering*, 2nd ed., Oxford, 1997.

Campbell, P.D.O., *Plastic Component Design*, Industrial Press, 1996.

Chabot, J.F., Jr., *The Development of Plastics Processing Machinery and Methods*, Wiley, 1992.

Chanda, J.M. and Roy, S.K., *Plastics Technology Handbook*, 3rd ed., Marcel Dekker, 1998.

Cracknell, P.S. and Dyson, R.W., *Handbook of Thermoplastics Injection Mould Design*, Kluwer, 1993.

Engineering Materials Handbook, Vol. 1: *Composites*, ASM International, 1987.

Feldman, D. and Barbalata, A., *Synthetic Polymers: Technology, Properties, Applications*, Kluwer, 1996.

Gastrow, H., *Injection Molds: 130 Proven Designs*, Hanser Gardner, 2002.

Griskey, R.G., *Polymer Process Engineering*, Kluwer, 1995.

Gutowski, T.G., *Advanced Composites Manufacturing*, Wiley, 1997.

Harper, C., *Handbook of Plastics, Elastomers, and Composites*, 3rd ed., McGraw-Hill, 1996.

Isayev, A.I., *Injection and Compression Molding Fundamentals*, Marcel Dekker, 1987.

Johnson, P.S., *Rubber Processing: An Introduction*, Hanser Gardner, 2001.

Karszes, W.M., *Extrusion Coating: Equipment and Materials*, Tappi Press, 1997.

Lee, N.C., *Blow Molding Design Guide*, Hanser Gardner, 1998.

Macosko, C.W., *Fundamentals of Reaction-Injection Molding*, Oxford University Press 1989.

Mazumdar, S.K., *Composites Manufacturing: Materials, Products and Process Engineering*, CRC, 2001.

Mallick, P.K. (ed.), *Composites Engineering Handbook*, Marcel Dekker, 1997.

Malloy, R.A., *Plastic Part Design for Injection Molding—An Introduction*, Hanser Gardner, 1994.

Mills, N., *Mechanical Design of Polymer Foam Products*, Chapman & Hall, 1997.

Modern Plastics Encyclopedia, McGraw-Hill, annual.

Morton-Jones, D.H., *Polymer Processing*, Chapman & Hall, 1989.

Muccio, E.A., *Plastic Part Technology*, ASM International, 1991.

_____, *Plastics Processing Technology*, ASM International, 1994.

Potter, K., *Resin Transfer Moulding*, Chapman & Hall, 1997.

Pye, R.G.W., *Injection Mold Design*, 4th ed., Wiley, 1989.

Quinn, J.A., *Composites Design Manual*, Technomic, 1999.

Rauwendaal, C., *Polymer Extrusion*, 4th ed., Hanser Gardner, 2001.

Rosato, D.V., *Plastics Processing Data Handbook*, 2nd ed., Chapman & Hall, 1997.

Rosen, S. R., *Thermoforming: Improving Process Performance*, Society of Manufacturing Engineers, 2002.

Schwartz, M.M., *Composite Materials*, Vol. 2: *Processing, Fabrication, and Applications*, Prentice Hall, 1999.

Starr, T.F., *Pultrusion for Engineers*, CRC Press, 2000.

Stevens, M.J., and Covas, J.A., *Extruder Principles and Operation*, 2nd ed., Kluwer, 1995.

Strong, A.B., *Plastics: Materials and Processing*, Prentice Hall, 1996.

_____, *Fundamentals of Composites Manufacturing*, Society of Manufacturing Engineers, 1989.

Sweeney, F.M., *Reaction Injection Molding Machinery and Processes*, Marcel Dekker, 1987.

Tool and Manufacturing Engineers Handbook, 4th ed., Vol. 8: *Plastic Part Manufacturing*, Society of Manufacturing Engineers, 1996.

Wilkinson, A.N. and Ryan, A., *Polymer Processing and Structure Development*, Kluwer, 1999.

Winchaeli, W., *Extrusion Dies for Plastics and Rubber*, 2nd ed., Hanser Gardner 1992.

REVIEW QUESTIONS

19.1 What are the forms of raw materials for processing plastics into products?

19.2 How are injection-molding machines rated?

19.3 Describe the blow-molding process.

19.4 What is (a) a parison, (b) a plastisol, and (c) a prepreg?

19.5 How is thin plastic film produced?

19.6 List several common products that can be made by thermoforming.

19.7 What similarities and differences are there between compression molding and closed-die forging?

19.8 Explain the difference between potting and encapsulation.

19.9 Describe the advantages of cold-forming plastics over other processing methods.

19.10 Name the major methods used in processing reinforced plastics.

19.11 What are the characteristics of filament-wound products? Explain why they are desirable.

19.12 Describe the methods that can be used to make tubular plastic products.

19.13 List the major design considerations in forming and shaping reinforced plastics.

19.14 What is pultrusion? Pulforming?

19.15 Describe the principal manufacturing processes used to make (a) metal-matrix and (b) ceramic-matrix composites.

19.16 How are plastic sheet and plastic film produced?

19.17 What process is used to make foam drinking cups?

19.18 If a polymer is in the form of a thin sheet, is it a thermoplastic or thermoset? Why?

19.19 How are polymer fibers made? Why are they much stronger than bulk forms of the polymer?

19.20 What is a spinneret?

19.21 What are the advantages of coextrusion?

QUALITATIVE PROBLEMS

19.22 Describe the features of a screw extruder and its functions.

19.23 Explain why injection molding is capable of producing parts with complex shapes and fine detail.

19.24 Describe the advantages of applying the traditional metal-forming techniques described in Chapters 13 through 16 to form (a) thermoplastics and (b) thermosets.

19.25 Explain the reasons why some plastic-forming processes are more suitable for certain polymers than for others.

19.26 Describe the problems involved in recycling products made from reinforced plastics.

19.27 Can thermosetting plastics be used in injection molding? Explain.

19.28 By inspecting plastic containers, such as ones for talcum powder, you can see that the integral lettering on them is raised rather than depressed. Can you offer an explanation as to why they are molded in that way?

19.29 Outline the precautions that you would take in shaping (a) reinforced plastics and (b) other composites.

19.30 Describe the factors that contribute to the cost of each forming and shaping process described in this chapter.

19.31 An injection-molded nylon gear is found to contain small pores. It is recommended that the material be dried before molding it. Explain why drying will solve this problem.

19.32 Explain why operations such as blow molding and film-bag making are performed vertically. Why is the movement of the material upward?

19.33 Comment on the principle of operation of the tape-laying machine shown in Fig. 19.25b. How is the cost for such a machine justified?

19.34 Typical production volumes are given in Table 19.2. Comment on your observations and explain why there is such a wide range.

19.35   What determines the cycle time for (a) injection molding, (b) thermoforming, and (c) compression molding?

19.36   Does the pull-in defect (sink marks) shown in Fig. 19.31c also occur in metal-forming and casting processes? Explain.

19.37   Comment on the differences between the barrel section of an extruder and that of an injection-molding machine.

19.38   What determines the time intervals at which the indexing head in Fig. 19.16c rotates from station to station?

19.39   Identify processes that are suitable for making small production runs on plastic parts, of say, 100.

19.40   Identify processes that are capable of producing parts with the following fiber orientations in each: (a) uniaxial, (b) cross-ply, (c) in-plane random, and (d) three-dimensional random.

QUANTITATIVE PROBLEMS

19.41   Estimate the die-clamping force required for injection molding five identical 6-in. diameter disks in one die. Include the runners of appropriate length and diameter.

19.42   A two-liter plastic beverage bottle is made from a parison with a diameter that is the same as that of the threaded neck of the bottle and is 5 in. long. Assuming uniform deformation during blow molding, estimate the wall thickness of the tubular portion of the parison.

19.43   Make a survey of a variety of sports equipment and identify the components made of composite materials. Explain the reasons for and advantages of using composites for these specific applications.

19.44   Consider a styrofoam drinking cup. Measure the volume of the cup and its weight. From this information, estimate the percent increase in volume that the polystyrene beads have undergone.

19.45   During the sterilization process for producing intravenous (IV) bags for medical applications, the polymer bags are subjected to an internal pressure of 30 psi. If the bag diameter is 4 in. and its shape can be approximated as a thin-walled, closed-end, cylindrical pressure vessel, what should be the wall thickness to ensure that the bag does not burst during sterilization? Assume that the allowable tensile stress of the polymer is 10 ksi.

SYNTHESIS, DESIGN, AND PROJECTS

19.46   Give examples of several parts suitable for insert molding. How would you manufacture these parts if insert molding were not available?

19.47   Give other examples of design modifications in addition to those shown in Fig. 19.31.

19.48   With specific examples, discuss the design issues involved in making products out of plastics, when compared to reinforced plastics.

19.49   Inspect various plastic components in your car and identify the processes that could have been used in making them.

19.50   Explain the design considerations involved in replacing a metal beverage can with one made completely of plastic.

19.51   Inspect several similar products that are made either from metals or from plastics, such as a metal bucket and a plastic bucket of similar shape and size.

Comment on their respective shapes and thicknesses and explain the reasons for their differences, if any.

19.52   Write a brief paper on how plastic coatings are applied to (a) electrical wiring, (b) sheet-metal panels, (c) wire baskets, racks, and similar structures, and (d) handles for electrician's tools, such as wire cutters and pliers requiring electrical insulation.

19.53   Inspect several electrical components (such as light switches, outlets, and circuit breakers) and describe the processes used in making them.

19.54   Based on experiments, it has been suggested that polymers (either plain or reinforced) can be a suitable material for dies in sheet-metal forming operations described in Chapter 16. Describe your thoughts regarding this suggestion, considering die geometry and any other factors that may be relevant.

19.55   Think of plastic parts that are made using two or more of the processes described in this chapter.

19.56 As we know, plastic forks, spoons, and knives are not particularly strong or rigid. What suggestions would you have to make them better? Describe processes that could be used for producing them and comment on the production costs involved.

19.57 For ease of sorting for recycling, a rapidly increasing number of plastic products are now identified with a triangular symbol with a single-digit number at its center and two or more letters under it. Explain what these numbers indicate and why they are used.

19.58 Explain the similarities and differences between the product design principles for the processes described in this chapter and those for the chapters in Part III. Describe your observations.

19.59 Make a survey of the technical literature and describe how different types of (a) pneumatic tires, (b) automotive hoses, and (c) garden hoses are manufactured.

19.60 Obtain a boxed kit for assembling a model car or airplane. Examine the injection-molded parts provided and describe your thoughts on the layout of the molds to produce these parts.

19.61 Using the Internet, obtain the following information:

a. Cost of raw polystyrene beads and raw polyethylene pellets.

b. Melt properties (such as glass–transition temperatures and melting temperatures) of plastics.

c. Availability of polymer additives, such as coloring agents and flow property additives.

d. Sizes, features, and costs of injection-molding machines.

19.62 In injection-molding operations, it is common practice to remove the part from its runner, place the runner in a shredder, and recycle it by producing pellets. List the concerns you may have in using such recycled pellets for products, as against "virgin" pellets.

19.63 An increasing environmental concern is the very long period required for the degradation of polymers in landfills. Noting the information given in Section 7.8 on biodegradable plastics, perform a literature search on the trends and developments in the production of these plastics.

19.64 Inspect some plastic slats for venetian blinds and describe your thoughts on how to produce them. Compare them with sheet-metal slats and comment on their relative cross-sections, curvature, and performance characteristics.

19.65 Die swell in extrusion is radially uniform for circular cross-sections but is not uniform for other cross-sections, as shown in Figs. 19.3b through d. Recognizing this, make a qualitative sketch of a die profile that will produce (a) triangular and (b) gear-shaped cross-sections of extruded plastics.

19.66 Examine some common plastic poker chips and give an opinion on how they were manufactured.

19.67 Inspect various plastic products around your home and observe sink marks (see Fig. 19.31c) that typically exist on the surface opposite the thicker sections. Make suggestions as to how they can be prevented during (a) product design and (b) manufacturing.

19.68 Obtain different kinds of toothpaste tubes, carefully cut them across, and comment on your observations regarding the type of materials used and how the tube could be produced.

20 Rapid-Prototyping Operations

This chapter describes the technologies associated with rapid prototyping, which share the characteristics of computer integration, production without the use of traditional tools and dies, and the ability to rapidly produce a single part on demand. Topics covered include:

- Principles of processes using polymers, ceramics, and metals as raw materials.
- Unique applications of these operations.
- Rapid tooling for the production of molds and various tooling.
- Production of engineering parts by rapid manufacturing.

Typical parts made: Wide variety of components for design and product evaluation, tools, and finished products in limited quantity for industrial and medical applications.

Alternative processes: Machining and fabricating.

20.1 | Introduction

In the development of a new product, there is invariably a need to produce a single example or **prototype** of a designed part or system before the allocation of large amounts of capital to new production facilities or assembly lines. The main reasons for this need are that the capital cost is very high and production tooling takes considerable time to prepare. Consequently, a working prototype is needed for design evaluation and troubleshooting before a complex product or system is ready to be produced and marketed.

A typical product development process was outlined in Fig. I.4 in the General Introduction. An iterative process naturally occurs when (a) errors are discovered or (b) more efficient or better design solutions are gleaned from the study of an earlier-generation prototype. The main problem with this approach, however, is that the production of a prototype can be extremely time consuming. Tooling can take several months to prepare, and the production of a single complicated part by conventional manufacturing operations can be very difficult. Furthermore, while waiting for a prototype to be prepared, facilities and staff still generate costs.

An even more important concern is the speed with which a product flows from concept to a marketable item. In a competitive marketplace, it is well known that products that are introduced before those of their competitors generally are more profitable and enjoy a larger share of the market. At the same time, there are important concerns regarding the production of high-quality products. For these reasons, there is a concerted effort to bring high-quality products to market quickly.

A technology which considerably speeds the iterative product-development process is the concept and practice of **rapid prototyping** (RP)—also called **solid freeform fabrication.** Examples of rapid-prototyped parts are shown in Fig. 20.1.

Developments in rapid prototyping began in the mid-1980s. The advantages of this technology include the following:

- Physical models of parts produced from CAD data files can be manufactured in a matter of hours and allow the rapid evaluation of manufacturability and design effectiveness. In this way, rapid prototyping serves as an important tool for visualization and for concept verification.

- With suitable materials, the prototype can be used in subsequent manufacturing operations to produce the final parts. In this way, rapid prototyping itself serves as an important manufacturing technology.

- Rapid-prototyping operations can be used in some applications to produce actual tooling for manufacturing operations (**rapid tooling**). In this way, one can obtain tooling in a matter of a few days.

Rapid-prototyping processes can be classified into three major groups: **subtractive, additive,** and **virtual.** As the names imply, subtractive processes involve material removal from a workpiece that is larger than the final part. Additive processes build up a part by adding material incrementally to produce the part. Virtual processes use advanced computer-based visualization technologies.

Almost all materials can be used through one or more rapid-prototyping operations, as outlined in Table 20.1. However, because their properties are more suitable for these operations, polymers are the workpiece material most commonly used today, followed by ceramics and metals. However, new processes are being introduced continually, and, thus, existing processes and materials are improved. The more common materials used in rapid-prototyping operations are summarized in Table 20.2. This chapter is intended to serve as a general introduction to the most common rapid-prototyping operations, describe their advantages and limitations, and explore the present and future applications of these processes.

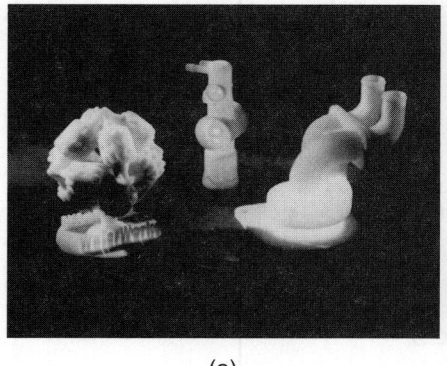

(a)   (b)   (c)

FIGURE 20.1   Examples of parts made by rapid-prototyping processes: (a) selection of parts from fused-deposition modeling; (b) stereolithography model of cellular phone; and (c) selection of parts from three-dimensional printing. *Source:* (a) Courtesy of Stratasys, Inc., (b) and (c) Courtesy of 3D Systems, Inc.

TABLE 20.1

Characteristics of Additive Rapid-Prototyping Technologies				
Supply phase	Process	Layer creation technique	Phase change type	Materials
Liquid	Stereolithography	Liquid layer curing	Photopolymerization	Photopolymers (such as acrylates, epoxies, colorable resins, and filled resins)
	Fused-deposition modeling	Extrusion of melted polymer	Solidification by cooling	Polymers (such as ABS and polyacrylate), wax, metal and ceramic powders with binder
	Ballistic-particle manufacturing	Droplet deposition	Solidification by cooling	Polymers and wax
Powder	Three-dimensional printing	Layer of powder and binder droplet deposition	No phase change	Ceramic, polymer, metal powder, and sand
	Selective laser sintering[a]	Layer of powder	Laser-driven sintering, melting, and solidification	Polymers, metals with a binder, metals, ceramics and sand with a binder
Solid	Laminated-object manufacturing	Deposition of sheet material	No phase change	Paper and polymers

[a] Some selective laser sintering systems are sold under the trade name Three-Dimensional Printers.

TABLE 20.2

Mechanical Properties of Selected Materials for Rapid Prototyping				
Process	Material	Tensile strength (MPa)	Elastic modulus (GPa)	Elongation in 50 mm (%)
Stereolithography	SL 5180[a]	55–65	2.4–2.6	6–11
	SL 5195[a]	46.5	2.1	11
	SL 5510[b]	73	2.8	7.9
	SL 7940[b]	37–39	1.3	18–21
Fused-deposition modeling	Polycarbonate	62	—	100
	ABS	35	2.5	50
Selective laser sintering	Nylon	36	1.4	32
	polycarbonate	23.4	1.2	5
	polyamide	44	1.6	9
	SOMOS 201	17.3	14	130
	ST-100[c]	305	137	10
Three-dimensional printing	S3 stainless steel	406	148	8.00
	S4 stainless steel	682	147	8.00

[a] After a 90-min UV cure; [b] after a 90-min UV cure at 80°C; [c] sintered and bronze-infiltrated steel powder.

20.2 | Subtractive Processes

Making a prototype traditionally has involved a series of processes using a variety of tooling and machines, and it usually takes anywhere from weeks to months, depending on part complexity and size. This approach requires skilled operators using *material removal by machining and finishing operations* (as described in detail in Part IV)—one by one—until the prototype is completed. To speed this process, subtractive processes increasingly use computer-based technologies, such as:

- **Computer-based drafting packages,** which can produce three-dimensional representations of parts.
- **Interpretation software,** which can translate the CAD file into a format usable by manufacturing software.
- **Manufacturing software,** which is capable of planning the operations required to produce the desired shape.
- **Computer-numerical-control machinery** with the capabilities necessary to produce the parts.

When a prototype is required only for the purpose of shape verification, a soft material (usually a polymer or a wax) is used as the workpiece in order to reduce or avoid any machining difficulties. The material intended for use in the actual application also can be machined, but this operation may be more time consuming, depending on the machinability of the material. Depending on part complexity and machining capabilities, prototypes can be produced in a matter of from a few days to a few weeks. Subtractive systems can take many forms; they are similar in approach to the manufacturing cells described in Section 39.2. Operators may or may not be involved, although the handling of parts is usually a human task.

20.3 | Additive Processes

Additive rapid-prototyping operations all build parts in *layers*, and as summarized in Table 20.1, they consist of *stereolithography, fused-deposition modeling, ballistic-particle manufacturing, three-dimensional printing, selective laser sintering and laminated-object manufacturing.* In order to visualize the methodology used, it is beneficial to think of constructing a loaf of bread by stacking and bonding individual slices on top of each other. All of the processes described in this section *build parts slice-by-slice.* The main difference between the various additive processes lies in the method of producing the individual slices, which are typically 0.1 to 0.5 mm (0.004 to 0.020 in.) thick and can be thicker for some systems.

All additive operations require elaborate software. As an example, note the solid part shown in Fig. 20.2a. The first step is to obtain a CAD file description of the part. The computer then constructs slices of the three-dimensional part (Fig. 20.2b). Each slice is analyzed separately, and a set of instructions is compiled in order to provide the rapid-prototyping machine with detailed information regarding the manufacture of the part. Figure 20.2d shows the paths of the extruder in one slice, using the fused-deposition-modeling operation, described in Section 20.3.1.

This approach requires operator input in the setup of the proper computer files and in the initiation of the production process. Following this stage, the machines generally operate unattended and provide a rough part after a few hours. The part then is subjected to a series of manual finishing operations (such as sanding and painting) in order to complete the rapid-prototyping process.

It should be recognized that the setup and finishing operations are very labor intensive and that the production time is only a portion of the time required to obtain a prototype. In general, however, additive processes are much faster than subtractive processes, taking as little as a few minutes to a few hours to produce a part.

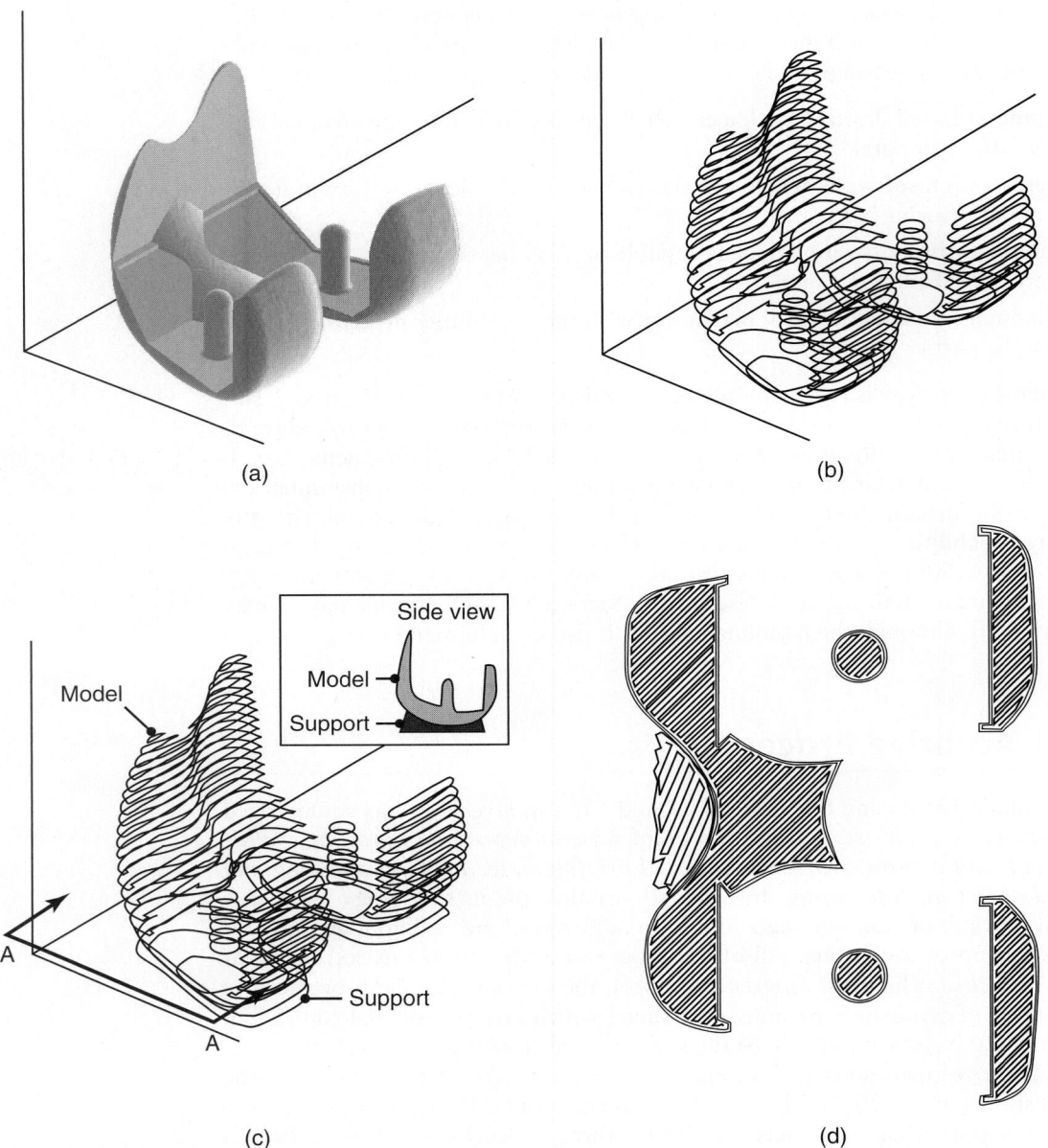

FIGURE 20.2   The computational steps in producing a stereolithography file. (a) Three-dimensional description of part. (b) The part is divided into slices (only one in 10 is shown). (c) Support material is planned. (d) A set of tool directions is determined to manufacture each slice. Also shown is the extruder path at section A–A from (c) for a fused-deposition-modeling operation.

20.3.1 Fused-deposition modeling

In the *fused-deposition-modeling* (FDM) process (Fig. 20.3), a gantry-robot controlled extruder head moves in two principal directions over a table; the table can be raised and lowered as needed. A thermoplastic or wax filament is extruded through the small orifice of a heated die. The initial layer is placed on a foam foundation by extruding the filament at a constant rate while the extruder head follows

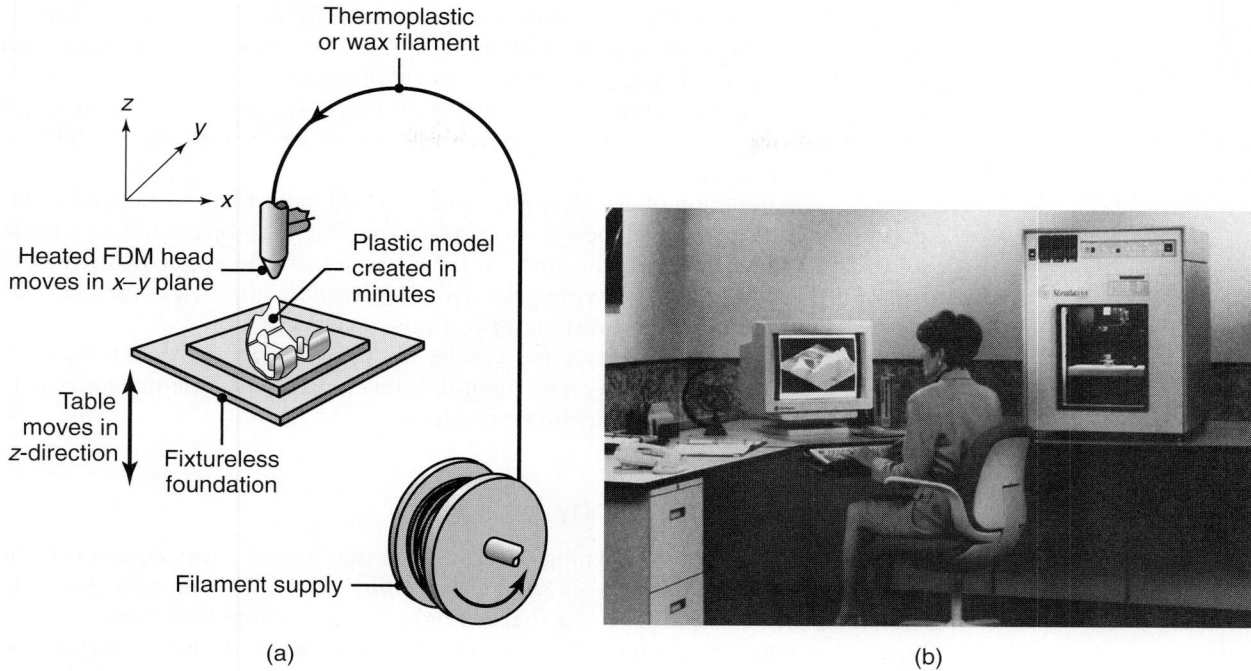

FIGURE 20.3 (a) Schematic illustration of the fused-deposition-modeling process. (b) The FDM 5000, a fused-deposition-modeling machine. *Source*: Courtesy of Stratysis, Inc.

a predetermined path (see Fig. 20.2d). When the first layer is completed, the table is lowered so that subsequent layers can be superimposed.

Occasionally, complicated parts are required, such as the one shown in Fig. 20.4a. This part is difficult to manufacture directly, because once the part has been constructed up to height a, the next slice would require the filament to be placed in a location where no material exists beneath to support it. The solution is to extrude a support material separately from the modeling material, as shown in Fig. 20.4b. Note that the use of such support structures allows all of the layers to be supported by the material directly beneath them. The support material is produced with a less-dense filament spacing on a layer, so it is weaker than the model material and can be broken off easily after the part is completed.

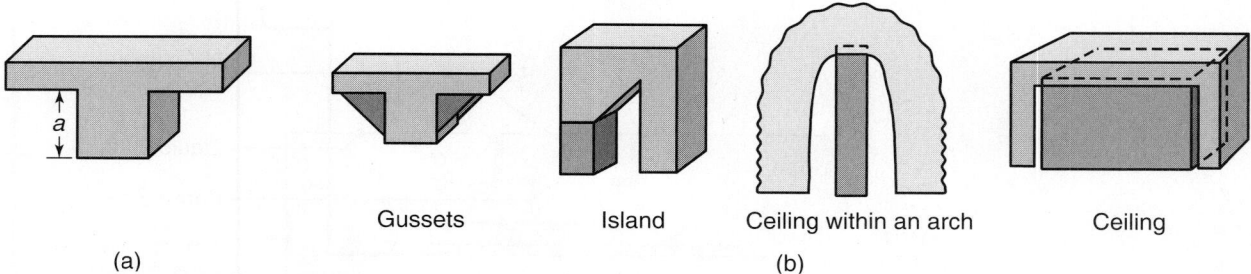

FIGURE 20.4 (a) A part with a protruding section which requires support material. (b) Common support structures used in rapid-prototyping machines. *Source*: After P.F. Jacobs, *Rapid Prototyping & Manufacturing: Fundamentals of Stereolithography*. Society of Manufacturing Engineers, 1992.

The layers in an FDM machine are determined by the extruder-die diameter, which are typically from 0.50 to 0.25 mm (0.02 to 0.01 in.). This thickness represents the best achievable tolerance in the vertical direction. In the x–y plane, dimensional accuracy can be as fine as 0.025 mm (0.001 in.)—as long as a filament can be extruded into the feature. A wide variety of polymers and waxes are available for different applications.

Close examination of an FDM-produced part will indicate that a stepped surface exists on oblique exterior planes. If this surface roughness is objectionable, a heated tool can be used to smooth the surface, or a coating can be applied (often in the form of a polishing wax). However, the overall tolerances are then compromised—unless care is taken in these finishing operations.

Although some FDM machines can be obtained for around $30,000, others can cost as much as $150,000. The main difference between these machines is the maximum size of the parts that can be produced.

20.3.2 Stereolithography

A very common rapid-prototyping process—one that actually was developed prior to fused-deposition modeling—is *stereolithography* (SLA). This process (Fig. 20.5) is based on the principle of *curing* (hardening) a liquid photopolymer into a specific shape. A vat containing a mechanism whereby a platform can be lowered and raised is filled with a photocurable liquid-acrylate polymer. The liquid is a mixture of acrylic monomers, oligomers (polymer intermediates), and a photoinitiator (a compound which undergoes a reaction upon the absorption of light).

At its highest position (depth a in Fig. 20.5), a shallow layer of liquid exists above the platform. A *laser* generating an ultraviolet beam is focused upon a selected surface area of the photopolymer and then moved around in the x–y plane. The beam cures that portion of the photopolymer (say, a ring-shaped portion) and, thereby, produces a solid body. The platform then is lowered sufficiently to cover the cured polymer with another layer of liquid polymer, and the sequence is repeated. The process is repeated until level b in Fig. 20.5 is reached. Thus far, we have generated a cylindrical part with a constant wall thickness. Note that the platform is now lowered by a vertical distance ab.

At level b, the x–y movements of the beam define a wider geometry, so we now have a flange-shaped portion that is being produced over the previously

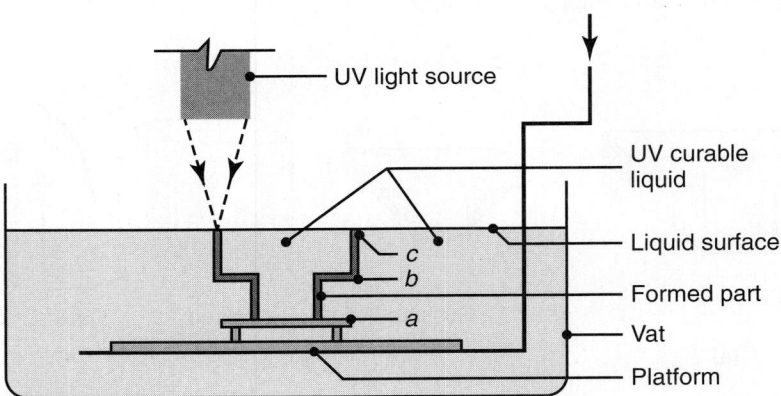

FIGURE 20.5   Schematic illustration of the stereolithography process.

formed part. After the proper thickness of the liquid has been cured, the process is repeated, producing another cylindrical section between levels *b* and *c*. Note that the surrounding liquid polymer is still fluid (because it has not been exposed to the ultraviolet beam) and that the part has been produced from the bottom up in individual "slices." The unused portion of the liquid polymer can be used again to make another part or another prototype.

Note that the term "stereolithography," as used to describe this process, comes from the facts that the movements are three-dimensional and the process is similar to lithography (see Section 28.7), in which the image to be printed on a flat surface is ink-receptive and the blank areas are ink-repellent. Also note that stereolithography can utilize (like FDM) a weaker support material. In stereolithography, this support takes the form of perforated structures. After its completion, the part is removed from the platform, blotted, and cleaned ultrasonically and with an alcohol bath. Then the support structure is removed, and the part is subjected to a final curing cycle in an oven. The smallest tolerance that can be achieved in stereolithography depends on the sharpness of the focus of the laser; typically, it is around 0.0125 mm (0.0005 in). Oblique surfaces also can be of very high quality.

Solid parts can be produced by applying special laser-scanning patterns to speed up production. For example, by spacing scan lines in stereolithography, volumes or pockets of uncured polymer can be formed within cured shells. When the part later is placed in a post-processing oven, the pockets cure and a solid part forms. Similarly, parts to be investment-cast will have a drainable honeycomb structure which permits a significant fraction of the part to remain uncured.

Total cycle times in stereolithography range from a few hours to a day—without post-processing such as sanding and painting. Depending on their capacity, the cost of the machines is in the range of $100,000 to $400,000. The cost of the liquid polymer is on the order of $300 per gallon. The maximum part size that can be produced is 0.5 m × 0.5 m × 0.6 m (19 in. × 19 in. × 24 in.).

Stereolithography has been used with highly focused lasers to produce parts with micrometer-sized features. The use of optics required to produce such features necessitates thinner layers and lower volumetric cure rates. However, this process has been used for the fabrication of micromechanical systems (see Chapter 29), and in such cases is called **microstereolithography**.

EXAMPLE 20.1 Computer mouse design

When contacted by a major computer manufacturer (and potential first-time customer), Logitech Company quoted a bid for building a two-button computer mouse of unique design within two weeks (Fig. 20.6). Stereolithography was employed efficiently within concurrent engineering (see Section I.3 in the General Introduction). The customer gave Logitech control drawings on the new mouse, and concurrent-engineering development took place with two design teams: an electrical-engineering team emphasizing control circuitry and a mechanical-engineering team emphasizing casing layout and button geometry.

On the first day, both engineering teams began their design work. Overnight on the first day, the stereolithography machine produced the first prototype of the mouse button. Design work on the mouse top and the complicated button mechanism was completed during the second day, and the parts were sent to the

FIGURE 20.6   A two-button computer mouse. *Source: Courtesy of 3D Systems, Inc.*

stereolithography machine and produced before the end of the third day. By this time, completed prototype circuitry boards had been constructed and were ready to be assembled into the prototype. The total elapsed time between the launch and the completed functional prototype was seven days. The company was able to demonstrate the design with working prototypes within the two-week deadline. The company now ships over one million units each year to the computer manufacturer.

Source: Courtesy of Logitech and 3D Systems, Inc.

20.3.3 Selective laser sintering

Selective laser sintering (SLS) is a process based on the sintering of nonmetallic or (less commonly) metallic powders selectively into an individual object. Selective-laser-sintering machines marketed by 3D Systems are called **three-dimensional printers**, although this term historically has been associated with ballistic-particle manufacturing (Section 20.3.4). The basic elements in this process are shown in Fig. 20.7. The bottom of the processing chamber is equipped with two cylinders:

1. A powder feed cylinder, which is raised incrementally to supply powder to the part-build cylinder through a roller mechanism.

2. A part-build cylinder, which is lowered incrementally as the part is being formed.

A thin layer of powder first is deposited in the part-build cylinder. A laser beam guided by a process-control computer using instructions generated by the three-dimensional CAD program of the desired part is then focused on that layer, tracing and sintering a particular cross-section into a solid mass. The powder in other areas remains loose, yet it supports the sintered portion. Another layer of powder then is deposited; this cycle is repeated again and again until the entire three-dimensional part has been produced. The loose particles are shaken off, and the part is recovered.

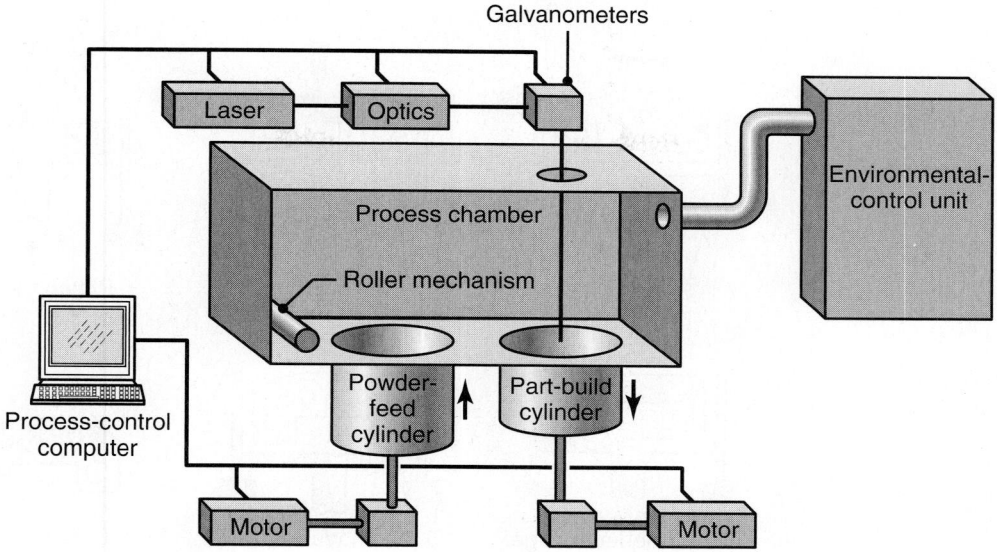

FIGURE 20.7 Schematic illustration of the selective-laser-sintering process. *Source:* After C. Deckard and P.F. McClure.

The part does not require further curing—unless it is a ceramic, which has to be fired to develop strength.

A variety of materials can be used in this process, including polymers (such as ABS, PVC, nylon, polyester, polystyrene, and epoxy), wax, metals, and ceramics with appropriate binders. It is most common to use polymers because of the smaller, less expensive, and less complicated lasers required for sintering. With ceramics and metals, it is common to sinter only a polymer binder which has been blended with the ceramic or metal powders.

20.3.4 Ballistic-particle manufacturing

In the *ballistic-particle manufacturing* process, a stream of a material (such as plastic, ceramic, metal, or wax) is ejected through a small orifice at a surface (target) using an ink-jet type mechanism. The mechanism uses a piezoelectric pump, which operates when an electric charge is applied, generating a shock wave that propels 50-μm droplets at a rate of 10,000 per second. The operation is repeated in a manner similar to other processes to form the part with layers deposited on top of each other. The ink-jet head is guided by a three-axis robot.

Three-dimensional printing (3DP) is related to ballistic-particle manufacturing with the exception that (instead of depositing the prototype material) the print head deposits an inorganic-binder material. This binder (glue or adhesive) represents the ballistic particles and is directed onto a layer of ceramic or metal powders, as shown in Fig. 20.8. A piston (supporting the powder bed) is lowered incrementally. With each step, a layer is deposited and then unified by the binder. Commonly used powder materials are aluminum oxide, silicon carbide, silica, and zirconia.

A common part produced by 3DP is a ceramic casting shell (see Section 11.2), in which an aluminum-oxide or aluminum-silica powder is fused with a silica binder. The molds have to be post-processed in two steps: (1) curing at around 150°C (300°F) and (2) firing at 1000° to 1500°C (1840° to 2740°F).

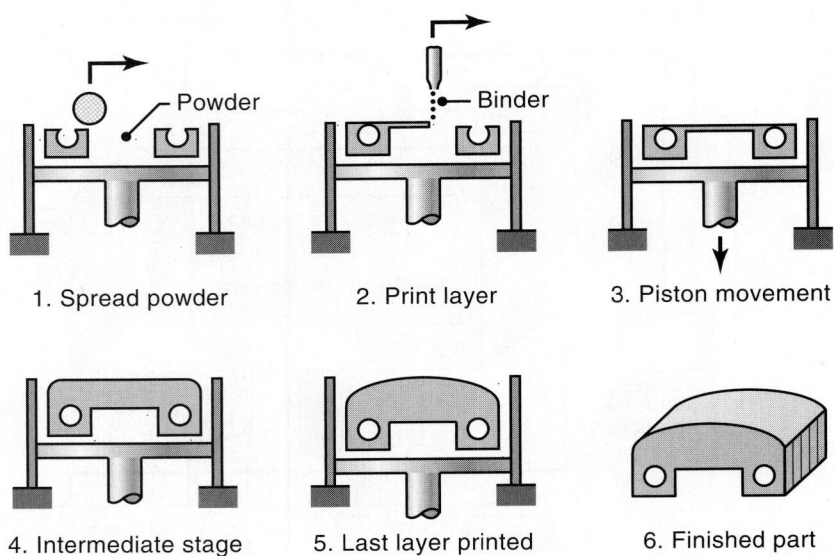

1. Spread powder   2. Print layer   3. Piston movement

4. Intermediate stage   5. Last layer printed   6. Finished part

FIGURE 20.8   Schematic illustration of the three-dimensional-printing process. *Source:* After E. Sachs and M. Cima.

The parts produced through 3DP are somewhat porous and, therefore, may lack strength. Three-dimensional printing can be combined with sintering and metal infiltration to produce fully dense parts using the sequence shown in Fig. 20.9. Here, the part is produced (as before) by directing a binder onto powders to produce the part, but the build sequence then is followed by sintering to burn off the binder and partially fuse the metal powders—just as in metal-injection molding, described in Section 17.3. Common materials used in 3DP are stainless steels, aluminum, titanium and silicon dioxide.

For metallic parts, sintering can be followed by infiltration with a lower-melting-point metal to produce fully dense parts; this approach represents an efficient strategy for *rapid tooling* (see Section 20.5). Common infiltration materials are copper and bronze, which provide good heat-transfer capabilities as well as wear resistance.

EXAMPLE 20.2 Fuselage fitting for helicopters

Sikorsky Aircraft Company needed to produce a limited number of the fuselage fittings shown in Fig. 20.10a. Sikorsky wanted to produce the dies for forging using three-dimensional-printing technologies. A die was designed using the CAD part description. Forging allowances were incorporated and flashing accommodated by the die design.

The dies were printed using a three-dimensional printer produced by ProMetal and are shown in Fig. 20.10b. The dies were made by producing 0.178 mm (0.007 in.) layers with stainless-steel powder as the workpiece media. The total time spent in the 3DP machine was just under 45 hours. This was followed by curing of

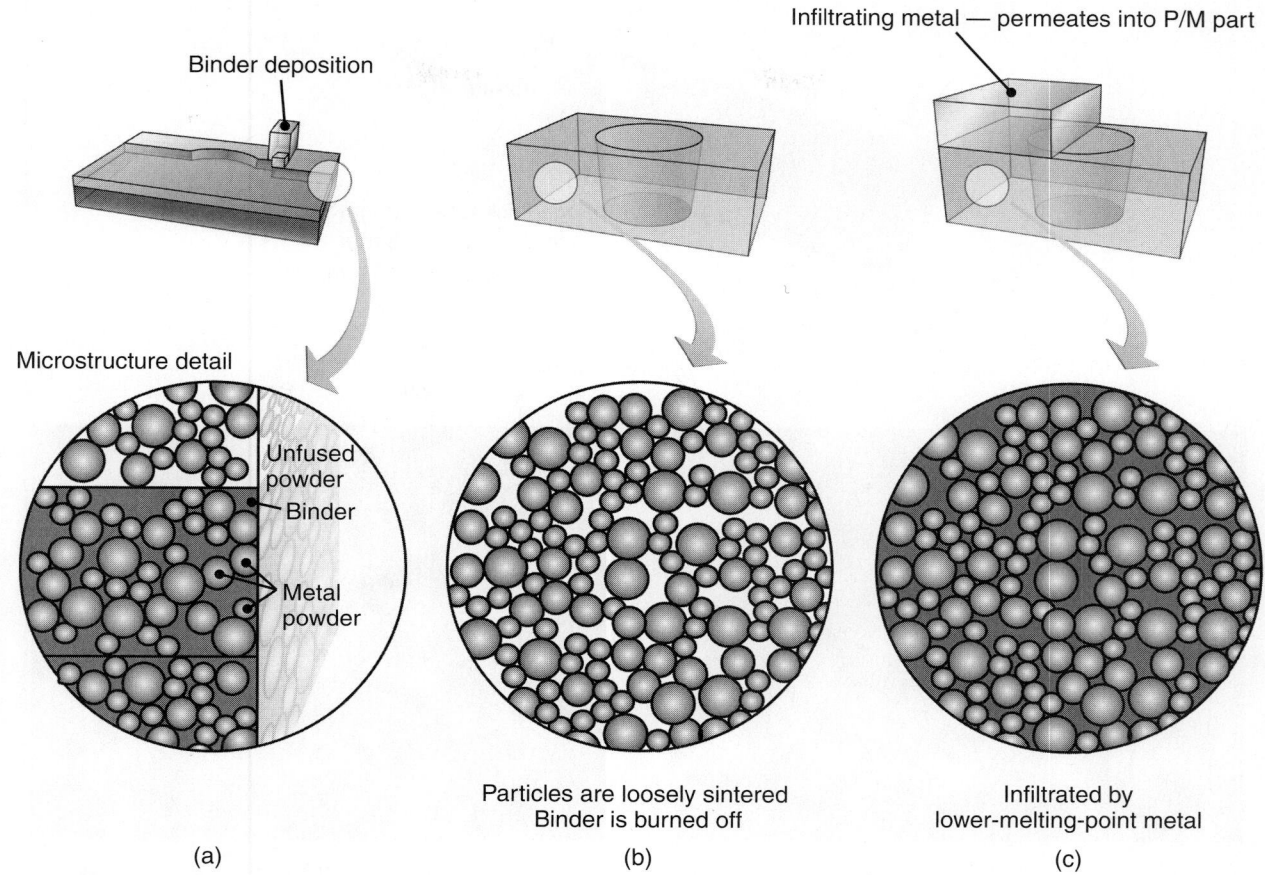

Binder deposition

Infiltrating metal — permeates into P/M part

Microstructure detail

Unfused powder

Binder

Metal powder

Particles are loosely sintered
Binder is burned off

Infiltrated by
lower-melting-point metal

(a)　　　　　　　　　　(b)　　　　　　　　(c)

FIGURE 20.9 Three-dimensional printing using (a) part-build, (b) sinter, and (c) infiltration steps to produce metal parts. (d) An example of a bronze-infiltrated stainless-steel part produced through three-dimensional printing. *Source:* Courtesy of the ProMetal Division of Ex One Corporation.

the binder (10 hours, plus five hours for cool-down), sintering (40 hours, plus 17 hours for cool-down), and infiltration (27 hours, plus 15 hours for cool-down). The dies then were finished and positioned in a die holder, and the part was forged in a 800-ton hydraulic press with a die temperature of around 300°C. An as-forged part is shown in Fig. 20.10c and requires trimming of the flash before it can be used. The dies were produced in just over six days—compared to the many months required for conventional die production.

Source: Courtesy of the ProMetal Division of Ex One Corporation.

20.3.5 Laminated-object manufacturing and solid-ground curing

Additional technologies for producing parts through additive rapid-prototyping operations are available in industry but are less common than the processes described thus far. Three of the more important processes are described next.

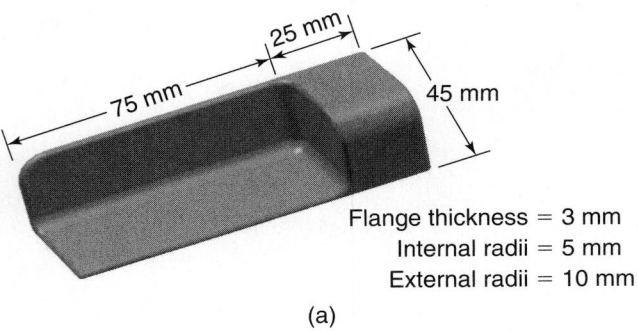

Flange thickness = 3 mm
Internal radii = 5 mm
External radii = 10 mm

(a)

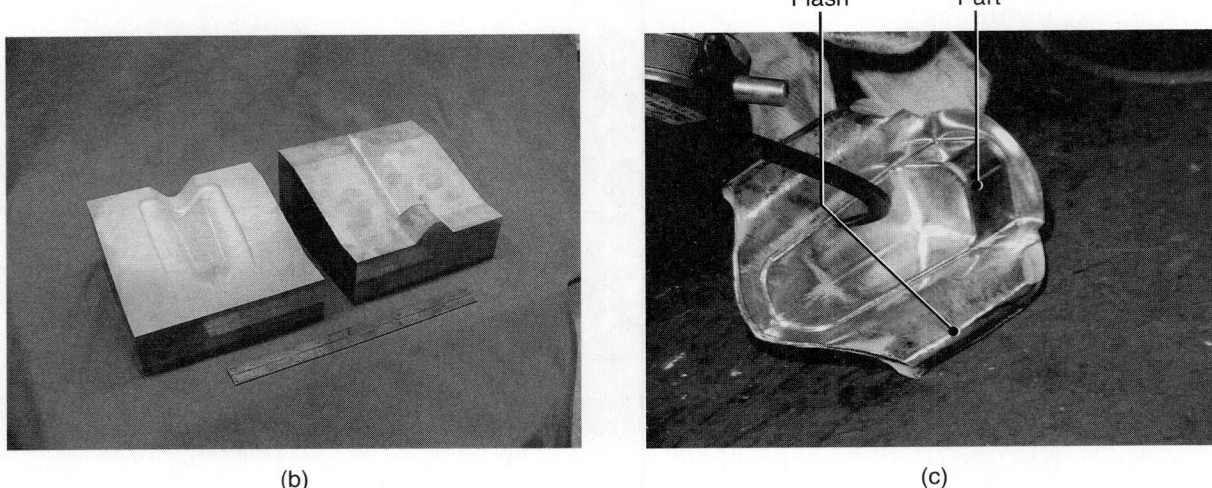

(b)   (c)

FIGURE 20.10   A fitting required for a helicopter fuselage. (a) CAD representation with added dimensions. (b) Dies produced by three-dimensional printing. (c) Final forged workpiece. *Source:* Courtesy of the ProMetal Division of Ex One Corporation.

Laminated-object manufacturing.   Lamination implies a laying down of layers which are bonded adhesively to one another. *Laminated-object manufacturing* (LOM) uses layers of paper or plastic sheets with a heat-activated glue on one side to produce parts. The desired shapes are burned into the sheet with a laser, and the parts are built layer by layer (Fig. 20.11). Once the part is completed, the excess material must be removed manually. This process is simplified by programming the laser to burn perforations in crisscrossed patterns. The resulting grid lines make the part appear as if it had been constructed from gridded paper (with squares printed on them similar to graph paper).

Solid-ground curing.   This process is unique in that entire slices of a part are manufactured at one time. As a result, a large throughput is achieved compared to that from other rapid-prototyping processes. However, *solid-ground curing* (SGC) is among the most expensive processes, hence its adoption has been much less common than that of other types of rapid prototyping, and new machines are not available. The basic method consists of the following steps:

1. Once a slice is created by the computer software, a mask of the slice is printed on a glass sheet by using an electrostatic printing process similar to that used

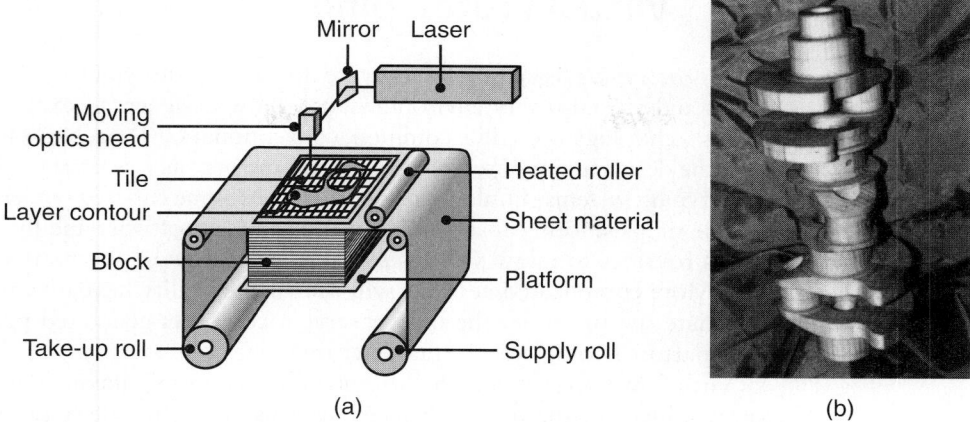

FIGURE 20.11 (a) Schematic illustration of the laminated-object-manufacturing process. (b) Crankshaft-part example made by LOM. *Source:* (a) Courtesy of Helysis, Inc. (b) After L. Wood.

in laser printers. A mask is required, because the area of the slice where the solid material is desired remains transparent.

2. While the mask is being prepared, a thin layer of photoreactive polymer is deposited on the work surface and is spread evenly.

3. The photomask is placed over the work surface, and an ultraviolet floodlight is projected through the mask. Wherever the mask is clear, the light shines through to cure the polymer and causes the desired slice to be hardened.

4. The unaffected resin (still liquid) is vacuumed off the surface.

5. Water-soluble liquid wax is spread across the work area, filling the cavities previously occupied by the unexposed liquid polymer. Since the workpiece is on a chilling plate and the workspace remains cool, the wax hardens quickly.

6. The layer then is milled to achieve the correct thickness and flatness.

7. This process is repeated—layer by layer—until the part is completed.

Solid-ground curing has the advantage of a high production rate, because entire slices are produced at once and two glass screens are used concurrently. That is, while one mask is being used to expose the polymer, the next mask already is being prepared, and it is ready as soon as the milling operation is completed.

Laser-engineered net shaping. More recent developments in additive manufacturing processes involve the principle of using a laser beam to melt and deposit metal powder or wire—layer by layer—over a previously molten layer. The patterns of deposited layers are controlled by a CAD file. This near-net-shaping process is called *laser-engineered net shaping* (LENS, a trade name) and is based on the technologies of laser welding and cladding. The heat input and cooling is controlled precisely to obtain a favorable microstructure.

The deposition process is carried out inside a closed area in an argon environment to avoid the adverse effects of oxidation (particularly for aluminum). It is suitable for a wide variety of metals and specialty alloys for the direct manufacturing of parts, including fully dense tools and molds. Also, it can be used for repairing thin and delicate components. There are other, similar processing methods using lasers, including *controlled-metal buildup* (CMB) and *precision-metal deposition* (PMD, a trade name).

20.4 | Virtual Prototyping

Virtual prototyping is a purely software form of prototyping and uses advanced graphics and virtual-reality environments to allow designers to examine a part. In a way, this technology is used by common, conventional CAD packages to render a part so that the designer can observe and evaluate the part as it is drawn. However, virtual-prototyping systems should be recognized as extreme cases of rendering detail.

The simplest forms of such systems use complex software and three-dimensional graphics routines to allow viewers to change the view of the parts on a computer screen. More complicated versions will use virtual-reality headgear and gloves with appropriate sensors to let the user observe a computer-generated prototype of the desired part in a completely virtual environment.

Virtual prototyping has the advantage of the instantaneous rendering of parts for evaluation, but the more advanced systems are costly. Because familiarity with software interfaces is a necessary prerequisite to their application, these systems have very steep learning curves. Furthermore, many manufacturing and design practitioners prefer a physical prototype to evaluate, rather than a video-screen rendering. They often perceive virtual-reality prototypes to be inferior to mechanical prototypes, even though designers debug as many or more errors in the virtual environment.

There have been some important examples of complicated products which have been produced without any physical prototype whatsoever (**paperless design**). Perhaps the best-known example is the Boeing 777 aircraft, where mechanical fits and interferences were evaluated on a CAD system and difficulties were corrected before manufacture of the first production model (see Section 38.5).

20.5 | Direct Manufacturing and Rapid Tooling

While extremely beneficial as a demonstration and visualization tool, rapid-prototyping processes also have been used as a manufacturing step in production. There are two basic methodologies used:

1. Direct production of engineering metal, ceramic, and polymer components or parts by rapid prototyping.
2. Production of tooling by rapid prototyping for use in further manufacturing operations.

The polymer parts that can be obtained from various rapid-prototyping operations are useful not only for design evaluation and troubleshooting, but occasionally these processes can be used to manufacture parts directly—referred to as *direct manufacturing*. Thus, the component is generated directly to a near-net shape from a computer file on part geometry. The main limitations to widespread use of rapid prototyping for direct manufacturing or *rapid manufacturing* (RM) are:

• Economic, as raw-material costs are high and the time required to produce each part is too high to be viable for large production runs. However, there are many applications where production runs are small enough to justify direct manufacturing through rapid-prototyping technologies.

• The long-term and consistent performance of the RM parts (as compared with the more traditional methods of manufacturing them) and their well-established performance characteristics such as wear and life cycle.

Much progress is being made to address these concerns to make rapid manufacturing a more competitive and viable option in manufacturing. The future of these processes remains challenging and promising, especially in view of the fact that RM is now being regarded as a method of product on demand. Customers will be able to order a particular part, which will be produced within a relatively short waiting time.

CASE STUDY 20.1 | Invisalign Orthodontic Aligners

Orthodontic braces have been available to straighten teeth for more than fifty years. These involve metal, ceramic, or plastic brackets that are bonded adhesively to teeth with fixtures for attachment to a wire which then forces compliance on the teeth and straightens them to the desired shape within a few years. Conventional orthodontic braces are a well-known and successful technique for long-term dental health. However, there are several drawbacks to conventional braces, including: (a) they are aesthetically unappealing, (b) the sharp wires and brackets can be painful, (c) they trap food leading to premature tooth decay, (d) brushing and flossing teeth are far more difficult and less effective with braces in place, and (e) certain foods must be avoided because they will damage the braces.

One solution is the Invisalign system, made by Align Technology, Inc. It consists of a series of aligners, each of which the person wears for approximately two weeks. Each aligner (see Fig. 20.12) consists of a precise geometry which incrementally moves the teeth to the desired positions. Because the aligners can be removed for eating, brushing, and flossing, most of the drawbacks of conventional braces are eliminated. Furthermore, since they are produced from a transparent plastic, they do not seriously affect the person's appearance.

The Invisalign product uses a combination of advanced technologies in the production process shown in Fig. 20.13. The treatment begins with an orthodontist or a general dentist creating a polymer impression of the patient's teeth (Fig 20.13a). These impressions then are used to create a three-dimensional CAD representation of the patient's teeth, as shown in Fig. 20.13b. Proprietary computer-aided design software then assists in the development of a treatment strategy for moving the teeth in an optimal manner.

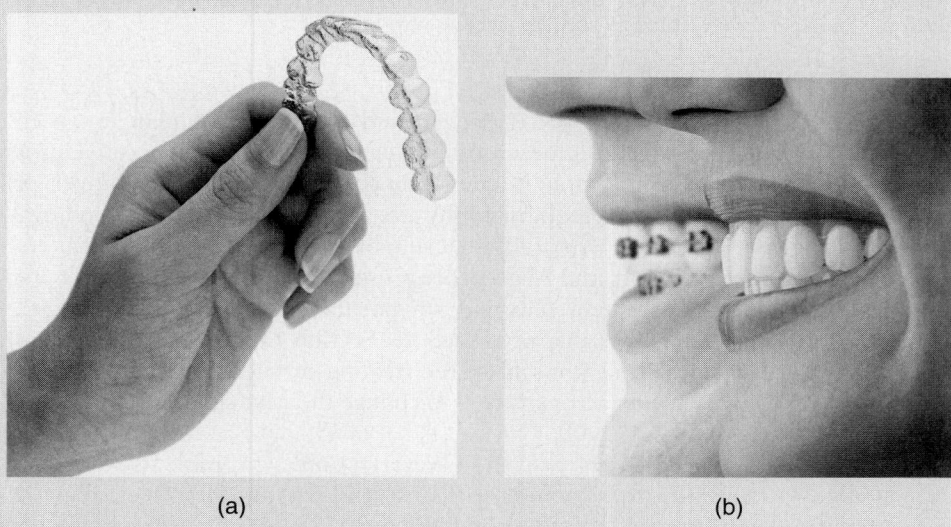

(a)   (b)

FIGURE 20.12 (a) An aligner for orthodontic use manufactured using a combination of rapid tooling and thermoforming. (b) Comparison of conventional orthodontic braces to the use of transparent aligners. *Source:* Courtesy of Align Technology, Inc.

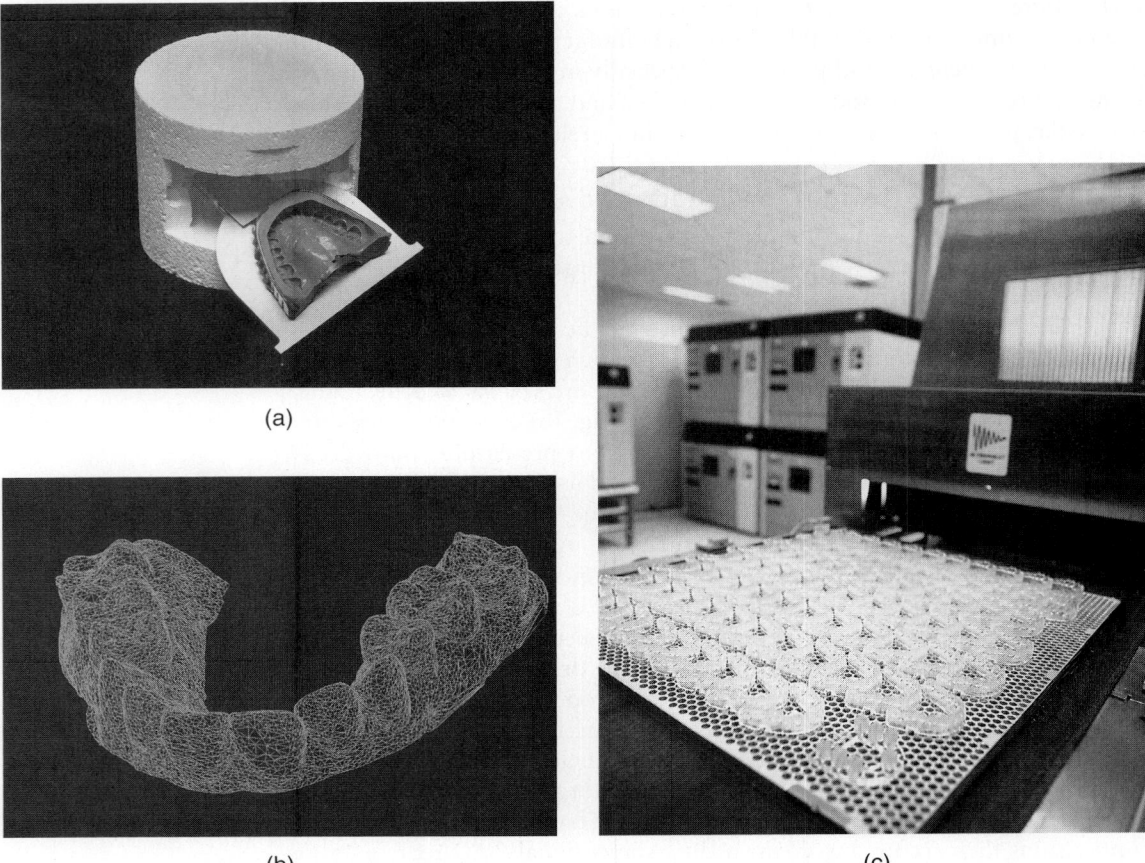

(a)

(b)

(c)

FIGURE 20.13   The manufacturing sequence for Invisalign orthodontic aligners. (a) Creation of a polymer impression of the patient's teeth. (b) Computer modeling to produce CAD representations of desired tooth profiles. (c) Production of incremental models of desired tooth movement. An aligner is produced by thermoforming a transparent plastic sheet against this model. *Source:* Courtesy of Align Technology, Inc.

Once the treating doctor has approved and the treatment plan has been developed, the computer-based information is used to produce the aligners. This is done through a novel application of stereolithography. Although a number of materials are available for stereolithography, they have a characteristic yellow-brown shade to them and therefore are unsuitable for direct application as an orthodontic product. Instead, the Align process uses a stereolithography machine that produces patterns of the desired incremental positions of the teeth (Fig. 20.13c). A sheet of clear polymer is then thermoformed (see Section 19.6) over these patterns to produce the aligners. These are sent to the treating orthodontist, and with the doctor's supervision, patients are instructed to change the next set of aligners every two weeks.

The Invisalign product has proven to be very popular for patients who wish to promote dental health and to preserve their teeth long into their lives. The use of stereolithography to produce accurate tools quickly and inexpensively allows this orthodontic treatment to be viable economically.

Source: Courtesy of Align Technology, Inc.

20.5.1 Rapid tooling

While the number of materials available in rapid-prototyping operations continues to increase, there is an inherent limit to material selection when these processes are considered. For example, it often is desirable to use metallic parts, while some of the highly developed and most widely available rapid-prototyping operations involve polymeric materials. The solution is to use the components manufactured through rapid prototyping as aids in further processing.

Known as *rapid tooling* (RT), the simplest example of this approach is the direct manufacture of sand molds through three-dimensional printing using foundry sand as the workpiece material. As another example, consider the investment-casting operation shown in Fig. 20.14. Here the individual patterns are made in a rapid-prototyping operation (in this case, stereolithography) and then used as the blanks in assembling a "tree" for investment casting. It should be noted that this method requires a polymer that will melt and burn from the ceramic mold completely; such polymers are available

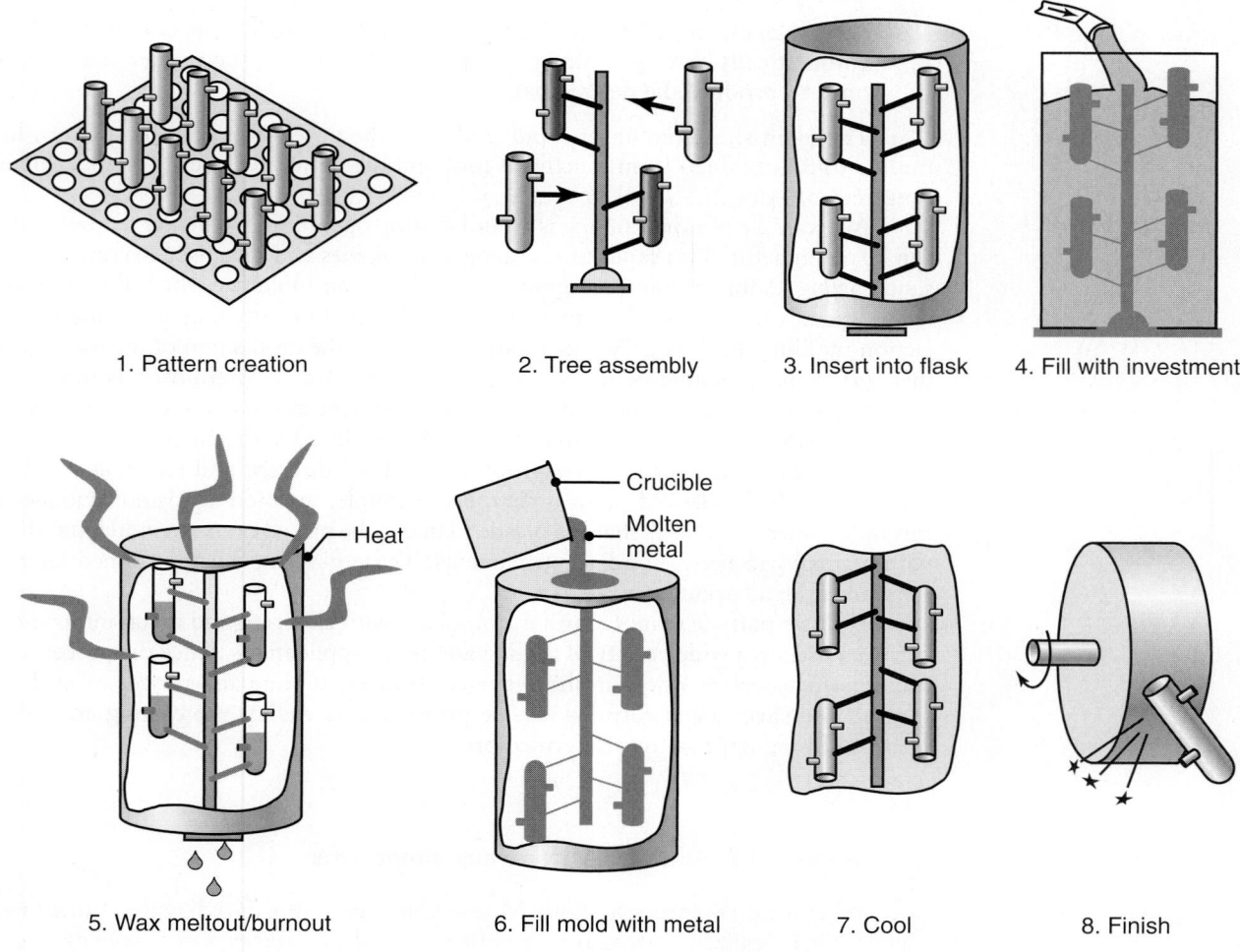

1. Pattern creation   2. Tree assembly   3. Insert into flask   4. Fill with investment

5. Wax meltout/burnout   6. Fill mold with metal   7. Cool   8. Finish

FIGURE 20.14   Manufacturing steps for investment casting that uses rapid-prototyped wax parts as blanks. This method uses a flask for the investment, but a shell method also can be used. *Source*: Courtesy of 3D Systems, Inc.

for all forms of polymer rapid-prototyping operations. Furthermore, the parts drawn in CAD programs usually are software-modified to account for shrinkage; it is the modified part that is produced in the rapid-prototyping machinery. Note also that the mold can be produced directly from three-dimensional printing when using foundry sand as the workpiece material.

Several methods have been devised for the rapid production of tooling by means of rapid-prototyping processes. The advantages to RT include the following:

1. The high cost of labor and short supply of skilled patternmakers can be overcome.
2. There is a major reduction in lead time.
3. Hollow designs can be adopted easily so that light-weight castings can be produced more easily.
4. The integral use of CAD technologies allows the use of modular dies with base-mold tooling (match plates) and specially fabricated inserts. This modular technique can further reduce tooling costs.
5. Chill- and cooling-channel placement in molds can be optimized more easily, leading to reduced cycle times.
6. Shrinkage due to solidification or thermal contraction can be compensated for automatically through software to produce tooling of the proper size and, in turn, to produce the desired parts.

The main shortcoming of rapid tooling is the potentially reduced pattern life (unlike those obtained from machined tool and die materials, such as tool steels or tungsten carbides).

An example of rapid tooling is a sand-casting operation, where the pattern plates can be manufactured via rapid-prototyping technologies and then used in conventional sand casting. Another common application is injection molding, where the mold (or more typically a *mold insert*) is manufactured via rapid prototyping and is used in injection-molding machines. Two methods are used for the production of the tooling: either (a) a high-melting-point thermoplastic or a stable thermoset is used for low-temperature injection molding or (b) an investment-cast metal insert is produced.

Molds for slip-casting ceramics also can be produced in this manner. To make individual molds, rapid-prototyping processes are used directly, and the molds will be produced with the desired permeability. For example, in fused-deposition modeling, this requirement mandates that the plastic filaments be placed onto the individual slices with a small gap between adjacent filaments. These filaments are positioned later at right angles in adjacent layers.

Ballistic-particle manufacturing, combined with sintering and metal infiltration, has been used in a wide variety of tooling and mold applications. Since the process can use stainless-steel powders infiltrated with bronze, tooling for processes such as forging and sheet-metal forming can be produced, as well as the casting and polymer-processing applications described previously.

EXAMPLE 20.3 Automotive rear-wiper motor cover

Manufacturing engineers at Ford Motor Company were faced with a dilemma when it was realized that, to meet production goals, a prototype rear-wiper motor cover was needed for the 1994 Explorer within a six-week time frame (Fig. 20.15). Ford's tooling supplier could not meet this deadline, because to machine the part, then to grind it, and finally to verify its dimensions usually took many times longer.

FIGURE 20.15   Rapid tooling for a rear-wiper motor cover.
Source: Courtesy of 3D Systems, Inc.

From a CAD file, a prototype was created, then (as an example of classical rapid-prototyping utility) was produced by stereolithography, and was fitted over the wiper motor. As often occurs with new designs and assemblies, an interference problem was discovered, one requiring a redesign. The wax prototype was machined by hand, the new fit was verified, and the modified part served as the template for adjusting the CAD drawing. From the revised CAD drawing, a new prototype was produced and the fit was assured.

Using mold-design software, "negative" patterns of the parts were produced that incorporated shrink factors to compensate for the prototype resin—the A2 die steel (Section 5.7) to be used as the tooling, and the polypropylene end material. These mold patterns were produced using an investment-casting-compatible wax on a stereolithography system and then cast using A2 steel. The first molds then were evaluated, and improvements were made in the cooling channels and the injector-pin components of the tool. A second set of tools were investment cast to produce 30,000 pieces per month. Total turnaround time was four weeks.

Source: Courtesy of Ford Motor Company and 3D Systems.

SUMMARY

- Rapid prototyping has grown into a unique manufacturing discipline within the past two decades. As a physical-model producing technology, it is a useful technique for identifying and correcting design errors. Several techniques have been developed for producing parts through rapid prototyping.

- Fused-deposition modeling consists of a computer-controlled extruder through which a polymer filament is deposited to produce a part slice-by-slice.

- Stereolithography involves a computer-controlled laser-focusing system that cures a liquid thermosetting polymer containing a photosensitive curing agent.

- Solid-ground curing involves curing an entire layer at one time with an ultraviolet lamp and then wiping it with a water-soluble wax to fill in the uncured locations. This results in higher throughput than other rapid-prototyping processes.

- Laminated-object manufacturing uses a laser beam to first cut the slices on paper or plastic sheets (laminations), then it applies an adhesive layer, and finally stacks the sheets to produce the part.
- Ballistic-particle manufacturing and the related process of three-dimensional printing use an inkjet mechanism to deposit liquid droplets of the prototyping material or to deposit a liquid binder onto powders, respectively.
- Selective laser sintering uses a high-powered laser beam to sinter powders in a desired pattern.
- Rapid-prototyping techniques have made possible much faster product development times, and they are having a major effect on other manufacturing processes. When appropriate materials are used, rapid-prototyping machinery can produce blanks for investment casting or similar processes, so that metallic parts now can be obtained quickly and inexpensively, even for lot sizes as small as one part. Such technologies also can be applied to producing molds for operations (such as injection molding), thereby, significantly reducing the lead time between design and manufacture.

KEY TERMS

Additive processes	Laminated-object manufacturing	Stereolithography
Ballistic-particle manufacturing	Photopolymer	Subtractive processes
Desktop machines	Prototype	Three-dimensional printing
Direct Manufacturing	Rapid tooling	Virtual prototyping
Free-form fabrication	Selective laser sintering	
Fused-deposition modeling	Solid-ground curing	

BIBLIOGRAPHY

Beaman, J.J., Barlow, J.W., Bourell, D.L., and Crawford, R., *Solid Freeform Fabrication*, Kluwer 1997.

Bennett, G. (ed.), *Developments in Rapid Prototyping and Tooling*, Institution of Mechanical Engineers, 1997.

Burns, M., *Automated Fabrication*, Prentice Hall, 1993.

Chua, C.K., and Fua, L.K., *Rapid Prototyping: Principles and Applications in Manufacturing*, Wiley, 1997.

Hilton, P.D., and Jacobs, P.F., *Rapid Tooling: Technologies and Industrial Applications*, Marcel Dekker, 2000.

Jacobs, P.F., *Rapid Prototyping and Manufacturing: Fundamentals of StereoLithography*, McGraw-Hill, 1993.

_____ , *StereoLithography and Other RP&M Technologies: From Rapid Prototyping to Rapid Tooling*, Society of Manufacturing Engineers, 1995.

Kocran, D., *Solid Preform Manufacturing*, Elsevier, 1993.

Pham, D.T., and Dimov, S.S., *Rapid Tooling: The Technologies and Applications of Rapid Prototyping and Rapid Tooling*, Springer-Verlag, 2001.

Wood, L., *Rapid Automated Prototyping: An Introduction*, Industrial Press, 1993.

REVIEW QUESTIONS

20.1   What is stereolithography?

20.2   Can rapid-prototyped parts be made of paper? Explain.

20.3   What is virtual prototyping, and how does it differ from additive methods?

20.4   What is fused-deposition modeling?

20.5 What is meant by rapid tooling? What is its significance in manufacturing?

20.6 Why are photopolymers essential for stereolithography?

20.7 What are (a) 3DP, (b) LOM, (c) SLS, (d) SLA, (e) FDM, and (f) LENS?

20.8 What starting materials can be used in fused-deposition modeling? In three-dimensional printing?

20.9 Why are the metal parts in three-dimensional printing often infiltrated by another metal?

QUALITATIVE PROBLEMS

20.10 Examine a coffee cup and determine in which orientation you would choose to produce the part if using (a) fused-deposition manufacturing or (b) laminated-object manufacturing processes.

20.11 How would you quickly manufacture tooling for injection molding? Explain any difficulties that may be encountered.

20.12 Summarize the rapid-prototyping processes and the materials which can be used for them.

20.13 Which processes described in this chapter are best suited for the production of ceramic parts? Why?

20.14 Why are so few parts in commercial products directly manufactured through rapid-prototyping operations?

20.15 Outline methods of producing metal parts that use the processes described in this chapter.

20.16 What are the cleaning and finishing operations in rapid-prototyping processes? Why are they necessary?

20.17 Careful analysis of a rapid-prototyped part indicates that it is made up of layers with a clear filament outline visible on each layer. Is the material a thermoset or a thermoplastic? Explain.

20.18 What is the main difference between additive manufacturing and rapid prototyping?

20.19 Make a list of the advantages and limitations of each of the rapid-prototyping operations described in this chapter.

20.20 If you are making a prototype of a toy automobile, list the post-rapid-prototyping finishing operations you would perform and why.

QUANTITATIVE PROBLEMS

20.21 Using an approximate cost of $500 per gallon for the liquid polymer, estimate the material cost of a rapid-prototyped rendering of a typical computer mouse.

20.22 The extruder head in a fused-deposition modeling setup has a diameter of 1.25 mm (0.05 in.) and produces layers that are 0.25 mm (0.01 in.) thick. If the extruder head and polymer extrudate velocities are both 50 mm/s, estimate the production time for the generation of a 50-mm (2-in.) solid cube. Assume that there is a 15 second delay between layers as the extruder head is moved over a wire brush for cleaning.

20.23 Using the data for Problem 20.22 and assuming that the porosity for the support material is 50%, calculate the production rate for making a 100-mm (4-in.) high cup with an outside diameter of 88 mm (3.5 in.) and wall thickness of 6 mm (0.25 in.). Consider both the case with the closed-end (a) up and (b) down.

SYNTHESIS, DESIGN, AND PROJECTS

20.24 A current topic of research is to produce parts by rapid-prototyping operations and then to use them in experimental stress analysis in order to infer the strength of the final parts produced by conventional manufacturing operations. List your concerns with this approach and outline means of addressing these concerns.

20.25   Because of relief of residual stresses during curing, long unsupported overhangs in parts from stereolithography will tend to curl. Suggest methods of controlling or eliminating this problem.

20.26   Because rapid-prototyping machines represent a large capital investment, few companies can justify the purchase of their own system. Consequently, service companies which produce parts based on their customers' drawings have become common. Conduct an informal survey of such service companies, determine the classes of rapid-prototyping machines that are used, and determine their percentages.

20.27   One of the major advantages of stereolithography is that it can use semi-transparent polymers, so that internal details of parts can be discerned readily. List and describe a few parts in which this feature is valuable.

20.28   A manufacturing technique is being proposed which uses a variation of fused-deposition modeling, where there are two polymer filaments that are melted and mixed before being extruded in order to produce the workpiece. What advantages does this method have?

20.29   Identify the RP processes described in this chapter that you can perform (on a modest scale) with the materials available in your home or you can purchase easily at low cost. Explain how you would go about it. Consider materials such as thin plywood, thick paper, glue, and butter and the use of various tools and energy sources.

20.30   Design a machine that uses rapid-prototyping technologies to produce ice sculptures. Describe its basic features, commenting on the effect of size and shape complexity on your design.

Machining Processes and Machine Tools

Parts manufactured by the casting, forming, and shaping processes described in Parts II and III often require further operations before the product is ready for use, including many parts made by near-net or net-shape methods. Consider, for example, the following features on parts, and whether they could be produced by the processes discussed thus far:

- Smooth and shiny surfaces, such as the bearing surfaces of the crankshaft shown in Fig. IV.1.
- Small-diameter deep holes in a part, such as an injector nozzle shown in Fig. IV.2.

Before   After

FIGURE IV.1 A forged crankshaft before and after machining the bearing surfaces. The shiny bearing surfaces of the part on the right cannot be made to its final dimensions and surface finish by any of the processes described in previous chapters of this book. *Source:* Courtesy of Wyman-Gordon Company.

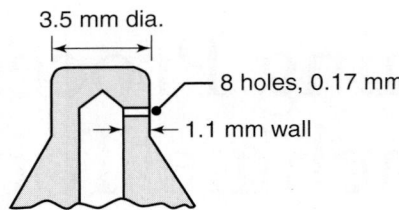

FIGURE IV.2 Cross-section of a fuel-injection nozzle showing a small hole made by the electrical-discharge machining process, as described in Section 27.5. The material is heat-treated steel.

- Parts with sharp features, a threaded section, and with the close dimensional tolerances specified, such as that shown in Fig. IV.3.
- A threaded hole or holes on different surfaces of a part for mechanical assembly with other components.
- Special surface finish and textures for functional purposes or for appearance.

A brief review will indicate that none of the forming and shaping processes described thus far is capable of producing parts with such specific characteristics, and that the parts will require further manufacturing operations. **Machining** is a general term describing a group of processes that consist of the *removal* of material and *modification* of the surfaces of a workpiece after it has been produced by various methods. Thus, machining involves *secondary* and *finishing* operations.

The very wide variety of shapes produced by machining can be seen clearly in an automobile, as shown in Fig. IV.4. It also should be recognized that some parts may be produced to final shape (net shape) and at high quantities by forming and shaping processes, such as die casting and powder metallurgy. However, machining

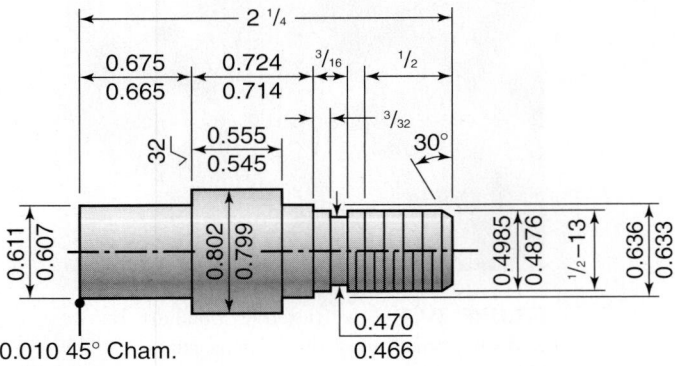

FIGURE IV.3 A machined and threaded part showing various dimensions and tolerances; all dimensions are in inches. Note that some tolerances are only a few thousandths of an inch.

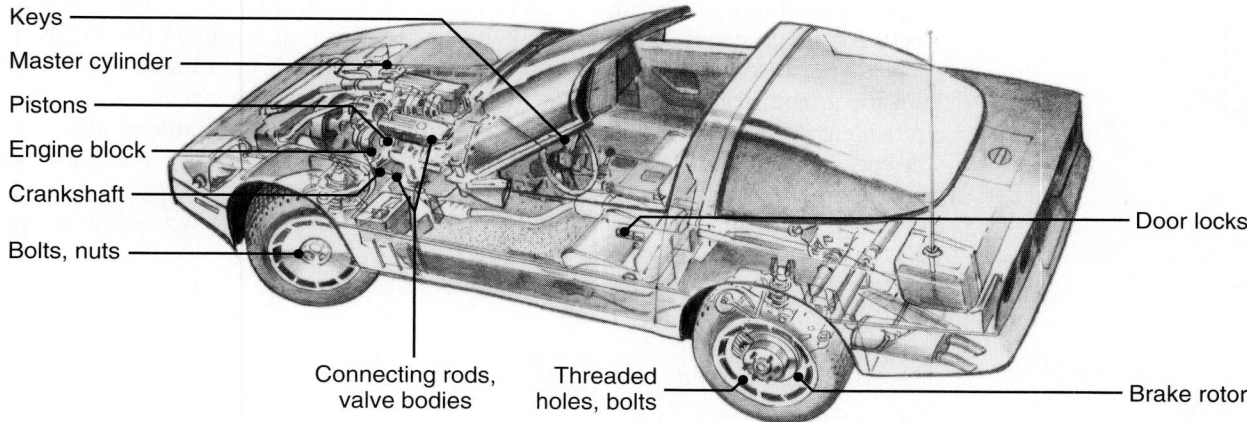

Keys
Master cylinder
Pistons
Engine block
Crankshaft
Bolts, nuts

Connecting rods, valve bodies
Threaded holes, bolts

Door locks
Brake rotor

FIGURE IV.4   Typical parts on an automobile that require machining operations to impart desirable surface characteristics, dimensions, and tolerances.

may be more economical provided that the number of parts required is relatively small or the material and part shape allows them to be machined at high rates and quantities and with high dimensional accuracy. A good example is the production of brass screw-machine parts on multiple-spindle automatic screw machines.

In general, however, resorting to machining suggests that a part could not have been produced to the final desired specifications by the primary processes used in making them and that additional operations are necessary. We again emphasize the importance of *net-shape manufacturing*, as described in Section I.5, to avoid these additional steps and reduce production costs.

Furthermore, in spite of their advantages, material-removal processes have the following disadvantages

- They *waste material* (although the amount may be relatively small).
- The processes generally takes *longer* than it does to shape by other processes.
- They generally require *more energy* than do forming and shaping operations.
- They can have *adverse effects* on the surface quality and properties of the product.

As outlined in Fig. I.7e in the General Introduction, machining consists of several major types of material-removal processes:

- **Cutting,** typically involving single-point or multipoint cutting tools, each with a clearly defined shape (Chapters 23 through 25).
- **Abrasive processes,** such as grinding and related processes (Chapter 26).
- **Advanced machining processes** utilizing electrical, chemical, laser, thermal, and hydrodynamic methods to accomplish this task (Chapter 27).

The machines on which these operations are performed are called **machine tools**. As described throughout Part IV, their construction and characteristics greatly influence these operations, as well as product quality, surface finish, and dimensional accuracy.

As can be seen in Table I.2 in the General Introduction, the first primitive tools (dating back many millennia) were made for the main purpose of chipping away and cutting all types of natural materials (such as wood, stone, vegetation, and live-stock) for the purpose of food and shelter. Note also that it was in the 1500s that progress began on manufacturing products by machining operations, particularly with the introduction of lathes. We now have available a wide variety of computer-controlled machine tools and modern techniques (using various materials and energy sources) and are capable of making functional parts as small as tiny insects and with cross-sections smaller than a human hair.

As in other manufacturing operations, it is important to view machining operations as a *system*, consisting of the

- *Workpiece*
- *Cutting tool*
- *Machine tool* and
- *Production personnel*

Machining cannot be carried out efficiently or economically and also meet stringent part specifications without a thorough knowledge of the interactions among these four elements.

In the following seven chapters, we will describe first the basic mechanics of chip formation in machining: tool forces, power, temperature, tool wear, surface finish and integrity, cutting tools, and cutting fluids. We then discuss specific machining processes—their capabilities, limitations, and typical applications—and identify important machine-tool characteristics for operations such as turning, milling, boring, drilling, and tapping. The features of *machining centers*—versatile machine tools controlled by computers and capable of efficiently performing a variety of operations—also are described.

We then describe processes in which the removal of material (to a very high dimensional accuracy and surface finish) is carried out by **abrasive processes** and related operations. Common examples are grinding wheels, coated abrasives, honing, lapping, buffing, polishing, shot-blasting, and ultrasonic machining.

For technical and economic reasons, some parts cannot be machined satisfactorily by cutting or abrasive processes. Since the 1940s, important developments have taken place in **advanced machining processes**, such as chemical, electrochemical, electrical-discharge, laser-beam, electron-beam, abrasive-jet, and hydrodynamic machining.

The knowledge gained in Part IV will enable us to assess the capabilities and limitations of material-removal processes; machine tools and related equipment; their proper selection for maximum efficiency, productivity, and low cost; and how these processes fit into the broader scheme of manufacturing operations.

Fundamentals of Machining

As an introduction to the machining processes to be covered in subsequent chapters, we describe in this chapter the mechanism by which materials are cut. Specifically:

- How chips are produced during machining.
- Force and power requirements in machining.
- Factors involved in temperature rise and its effects.
- How cutting tools wear and fail.
- Surface finish and integrity of parts produced by machining.
- Machinability of materials.

21.1 | Introduction

Cutting processes remove material from the surface of a workpiece by producing **chips.** Some of the more common cutting processes are illustrated in Fig. 21.1 (see also Fig. I.7e).

- **Turning,** in which the workpiece is rotated and a cutting tool removes a layer of material as it moves to the left, as in Fig. 21.1a.
- **Cutting-off** operation, where the cutting tool moves radially inward and separates the right piece from the bulk of the blank.
- **Slab-milling** operation, in which a rotating cutting tool removes a layer of material from the surface of the workpiece.
- **End-milling** operation, in which a rotating cutter travels along a certain depth in the workpiece and produces a cavity.

In the turning process, illustrated in greater detail in Fig. 21.2, the cutting tool is set at a certain *depth of cut* (mm or in.) and travels to the left with a certain velocity as the workpiece rotates. The *feed* or *feed rate* is the distance the tool travels horizontally per unit revolution of the workpiece (mm/rev or in./rev). This movement of the tool produces a chip, which moves up the face of the tool.

In order to analyze this process in detail, a two-dimensional model of it is presented in Fig. 21.3a. In this idealized model, a cutting tool moves to the left along

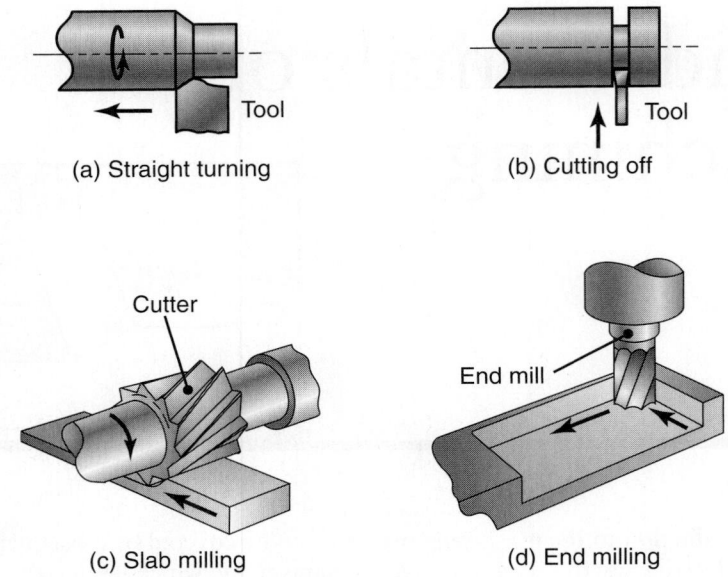

(a) Straight turning

(b) Cutting off

(c) Slab milling

(d) End milling

FIGURE 21.1   Some examples of common machining operations.

the workpiece at a constant velocity, V, and a depth of cut, t_o. A chip is produced ahead of the tool by plastically deforming and shearing the material continuously along the shear plane. This phenomenon can be demonstrated by slowly scraping the surface of a stick of butter lengthwise with a sharp knife and observing the formation of a chip. It is noted that chocolate shavings to make decorations on cakes and pastries also are produced in a similar manner.

In comparing Figs. 21.2 and 21.3, note that the feed in turning is equivalent to t_o, and the depth of cut in turning is equivalent to the width of cut (dimension perpendicular to the page) in the idealized model. These relationships can be visualized by rotating Fig. 21.3 clockwise by 90°. With this brief introduction as a background, we now will describe the cutting process in greater detail.

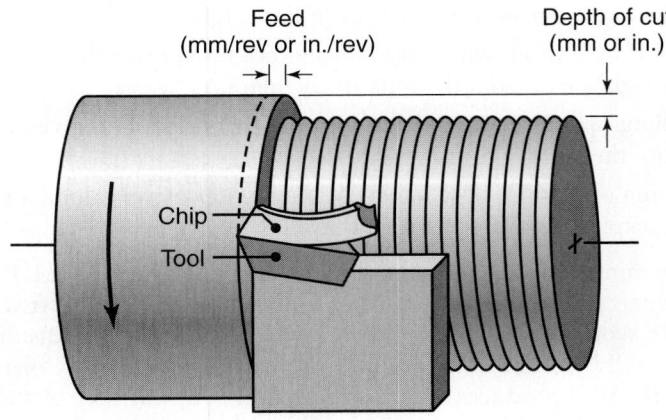

FIGURE 21.2   Schematic illustration of the turning operation showing various features.

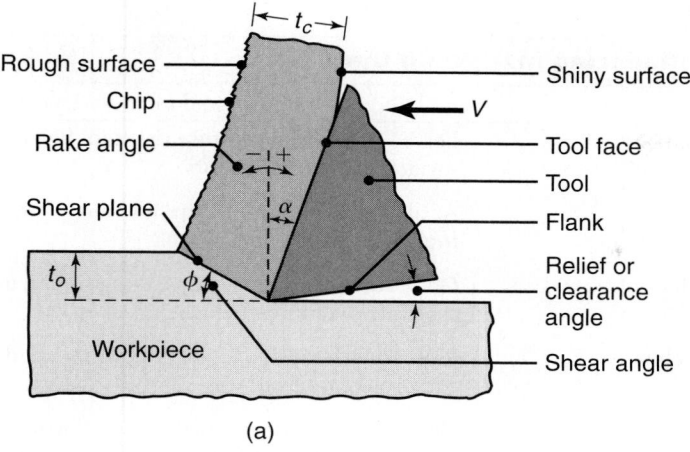

(a)

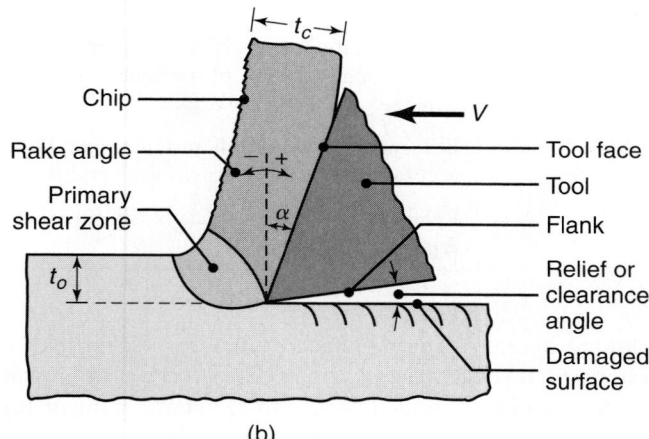

(b)

FIGURE 21.3 Schematic illustration of a two-dimensional cutting process, also called orthogonal cutting: (a) Orthogonal cutting with a well-defined shear plane, also known as the Merchant model. Note that the tool shape, depth of cut, t_o, and the cutting speed, V, are all independent variables: (b) Orthogonal cutting without a well-defined shear plane.

21.2 | Mechanics of Cutting

The factors that influence the cutting process are outlined in Table 21.1. In order to appreciate the contents of this table, let's now identify the major *independent variables* in the cutting process as follows: (a) tool material and coatings; (b) tool shape, surface finish, and sharpness; (c) workpiece material and condition; (d) cutting speed, feed, and depth of cut; (e) cutting fluids; (f) characteristics of the machine tool; and (g) workholding and fixturing.

Dependent variables in cutting are those that are influenced by changes in the independent variables listed above, and include: (a) type of chip produced, (b) force and energy dissipated during cutting, (c) temperature rise in the workpiece, the tool, and the chip, (d) tool wear and failure, and (e) surface finish and surface integrity of the workpiece.

TABLE 21.1

Factors Influencing Machining Operations	
Parameter	Influence and interrelationship
Cutting speed, depth of cut, feed, cutting fluids	Forces, power, temperature rise, tool life, type of chip, surface finish and integrity
Tool angles	As above; influence on chip flow direction; resistance to tool wear and chipping
Continuous chip	Good surface finish; steady cutting forces; undesirable, especially in automated machinery
Built-up edge chip	Poor surface finish and integrity; if thin and stable, edge can protect tool surfaces
Discontinuous chip	Desirable for ease of chip disposal; fluctuating cutting forces; can affect surface finish and cause vibration and chatter
Temperature rise	Influences tool life, particularly crater wear and dimensional accuracy of workpiece; may cause thermal damage to workpiece surface
Tool wear	Influences surface finish and integrity, dimensional accuracy, temperature rise, forces and power
Machinability	Related to tool life, surface finish, forces and power, and type of chip

When machining operations yield unacceptable results, normal troubleshooting requires a systematic investigation. A typical question posed is which of the independent variables should be changed first, and to what extent, if (a) the surface finish of the workpiece being cut is poor and unacceptable, (b) the cutting tool wears rapidly and becomes dull, (c) the workpiece becomes very hot, and (d) the tool begins to vibrate and chatter.

In order to understand these phenomena and respond to these questions, let's first study the mechanics of chip formation—a subject that has been studied extensively since the early 1940s. Several models (with varying degrees of complexity) have been proposed. As is being done in other manufacturing processes (such as casting, molding, shaping, and forming), advanced machining models also are being continuously developed. These methods include *computer simulation* of the machining process with the purpose of studying the complex interactions of the many variables involved in the process, while developing capabilities to *optimize* machining operations.

The simple model shown in Fig. 21.3a (referred as the M.E. Merchant model, developed in the early 1940s) is sufficient for our purposes. This model is known as **orthogonal cutting**, because it is two dimensional and the forces involved (as we later show in Fig. 21.11) are perpendicular to each other. The cutting tool has a **rake angle** of α (positive, as shown in the figure) and a **relief** or **clearance angle**.

Microscopic examination of chips obtained in actual machining operations have revealed that they are produced by *shearing* (as modeled in Fig. 21.4a)—similar to the movement in a deck of cards sliding against each other. Shearing takes place along a **shear zone** (usually along a well-defined plane referred to as the **shear plane**) at an angle, ϕ (called the **shear angle**). Below the shear plane, the workpiece remains undeformed; above it, the chip that is already formed moves up the rake face of the tool. The dimension d in the figure is highly exaggerated to show the mechanism involved. In reality, this dimension is only on the order of 10^{-2} to 10^{-3} mm (10^{-3} to 10^{-4} in.).

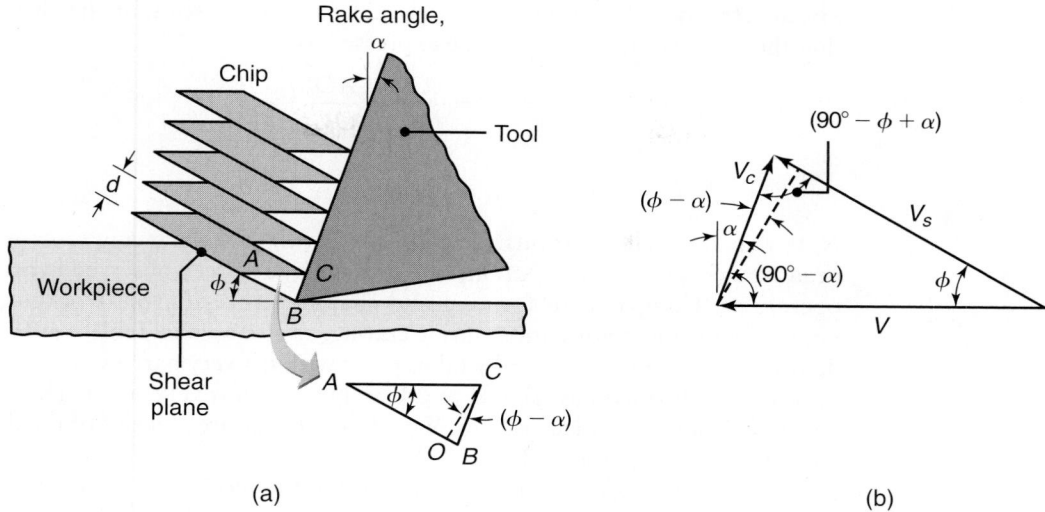

FIGURE 21.4 (a) Schematic illustration of the basic mechanism of chip formation by shearing. (b) Velocity diagram showing angular relationships among the three speeds in the cutting zone.

Some materials (notably cast irons at low speeds) do not shear along a well-defined plane but instead shear in a zone, as shown in Fig. 21.3b. Shearing in such a volume is not in itself objectionable, but it can lead to surface defects in the workpiece (as will be discussed later).

Cutting ratio. It can be seen that the chip thickness, t_c, can be determined by knowing the depth of cut, t_o, and α and ϕ. The ratio of t_o/t_c is known as the **cutting ratio** (or chip-thickness ratio), r, and is related to the two angles by the following relationships:

$$\tan \phi = \frac{r \cos \alpha}{1 - r \sin \alpha} \tag{21.1a}$$

and

$$r = \frac{t_o}{t_c} = \frac{\sin \phi}{\cos(\phi - \alpha)} \tag{21.1b}$$

Because the chip thickness always is greater than the depth of cut, the value of r always is less than unity. The reciprocal of r is known as the *chip-compression ratio* or *factor* and is thus a measure of how thick the chip has become compared to the depth of cut; hence, the chip-compression ratio always is greater than unity. The depth of cut also is referred to as the *undeformed chip thickness*, as may be visualized by reviewing Fig. 21.3.

The cutting ratio is an important and useful parameter for evaluating cutting conditions. Since the undeformed chip thickness, t_o, is a machine setting and is therefore known, the cutting ratio can be calculated easily by measuring the chip thickness with a micrometer. With the rake angle also known for a particular cutting operation (it is a function of the tool and workpiece geometry in use), Eq. (21.1) allows calculation of the shear angle.

Although we have referred to t_o as the *depth of cut*, note that in a machining process (such as turning, as shown in Fig. 21.2), this quantity is the *feed*. To visualize this situation, assume, for instance, that the workpiece in Fig. 21.2 is a thin-walled tube, and the width of the cut is the same as the thickness of the tube. Then by rotating Fig. 21.3 clockwise by 90°, we note that it is now similar to the view in Fig. 21.2.

Shear strain.   Referring now to Fig. 21.4a, we can see that the **shear strain**, γ, that the material undergoes can be expressed as

$$\gamma = \frac{AB}{OC} = \frac{AO}{OC} + \frac{OB}{OC}$$

or

$$\gamma = \cot \phi + \tan(\phi - \alpha) \qquad (21.2)$$

Note that large shear strains are associated with low shear angles or with low or negative rake angles. Shear strains of 5 or higher have been observed in actual cutting operations. Compared to forming and shaping processes, the workpiece material undergoes greater deformation during cutting, as also seen in Table 2.4. Furthermore, deformation in cutting generally takes place within a very narrow deformation zone. In other words, the dimension $d = OC$ in Fig. 21.4a is very small. Thus, the rate at which shearing takes place is high. We discuss the nature and size of the deformation zone further in Section 21.3.

The shear angle has great significance in the mechanics of cutting operations. It influences force and power requirements, chip thickness, and temperature. Consequently, much attention has been focused on determining the relationships among shear angle, cutting process variables, and workpiece material properties. One of the earliest analyses was based on the assumption that the shear angle adjusts itself to minimize the cutting force or that the shear plane is a plane of maximum shear stress. This analysis yielded the expression

$$\phi = 45° + \frac{\alpha}{2} - \frac{\beta}{2} \qquad (21.3)$$

where β is the **friction angle** and is related to the *coefficient of friction*, μ, at the tool–chip interface by the expression $\mu = \tan \beta$. Among the many shear–angle relationships developed, another useful formula that generally is applicable is

$$\phi = 45° + \alpha - \beta \qquad (21.4)$$

The coefficient of friction in metal cutting generally ranges from about 0.5 to 2, indicating that the chip encounters considerable frictional resistance while moving up the tool's rake face. Experiments have shown that μ varies considerably along the tool–chip interface because of steep variations in contact pressure and temperature. Consequently, μ also is called the *apparent mean coefficient of friction.*

Equation (21.3) indicates that: (a) as the rake angle decreases and/or the friction at the tool–chip interface (rake face) increases, the shear angle decreases and the chip becomes thicker; (b) thicker chips mean more energy dissipation because the shear strain is higher [see Eq. (21.2)]; and (c) because work done during cutting is converted into heat, the temperature rise is also higher. The effects of these phenomena are described throughout the rest of this chapter.

Velocities in the cutting zone.   Note in Fig. 21.3 that (since chip thickness is greater than the depth of cut) the velocity of the chip, V_c, has to be lower than the cutting speed, V.

Since mass continuity has to be maintained,

$$V t_o = V_c t_c \quad \text{or} \quad V_c = Vr$$

Hence,

$$V_c = \frac{V \sin \phi}{\cos(\phi - \alpha)} \qquad (21.5)$$

A *velocity diagram* also can be constructed, as shown in Fig. 21.4b, where from trigonometric relationships, we obtain the equation

$$\frac{V}{\cos(\phi - \alpha)} = \frac{V_s}{\cos \alpha} = \frac{V_c}{\sin \phi} \qquad (21.6a)$$

where V_s is the velocity at which shearing takes place in the shear plane. Note also that

$$r = \frac{t_o}{t_c} = \frac{V_c}{V} \qquad (21.6b)$$

These velocity relationships will be utilized further in Section 21.3 when describing power requirements in cutting operations.

21.2.1 Types of chips produced in metal cutting

The types of metal chips commonly observed in practice and their photomicrographs are shown in Fig. 21.5. The four main types are:

- **Continuous**
- **Built-up edge**
- **Serrated** or **segmented**
- **Discontinuous**

Let's first note that a chip has two surfaces. One surface has been in contact with the rake face of the tool and has a shiny and burnished appearance caused by rubbing as the chip moves up the tool face. The other surface is from the original surface of the workpiece; it has a jagged, rough appearance (as can be seen on the chips in Figs. 21.3 and 21.5) caused by the shearing mechanism shown in Fig. 21.4a. This surface is exposed to the environment and has not come into any contact with any solid body.

Continuous chips.   *Continuous chips* usually are formed with ductile materials, machined at high cutting speeds and/or high rake angles (Fig. 21.5a). The deformation of the material takes place along a narrow shear zone called the *primary shear zone*. Continuous chips may develop a *secondary shear zone* (Fig. 21.5b) because of high friction at the tool–chip interface; this zone becomes thicker as friction increases.

Deformation in continuous chips also may take place along a wide primary shear zone with *curved boundaries* (see Fig. 21.3b), unlike that shown in Fig. 21.5a. Note that the lower boundary of the deformation zone is *below* the machined surface, subjecting it to distortion, as depicted by the distorted vertical lines in the machined subsurface. This situation occurs generally in machining soft metals at low speeds and low rake angles. It usually results in a poor surface finish and induces surface residual stresses, which may be detrimental to the properties of the machined part in their surface life.

Although they generally produce a good surface finish, continuous chips are not necessarily desirable, particularly with computer-controlled machine tools in wide use, as they tend to become tangled around the tool holder, the fixturing, the workpiece, as well as the chip-disposal systems. The operation may have to be stopped to clear away the chips. This problem can be alleviated with **chip breakers** (to follow) or by changing parameters, such as cutting speed, feed, depth of cut, and by using cutting fluids.

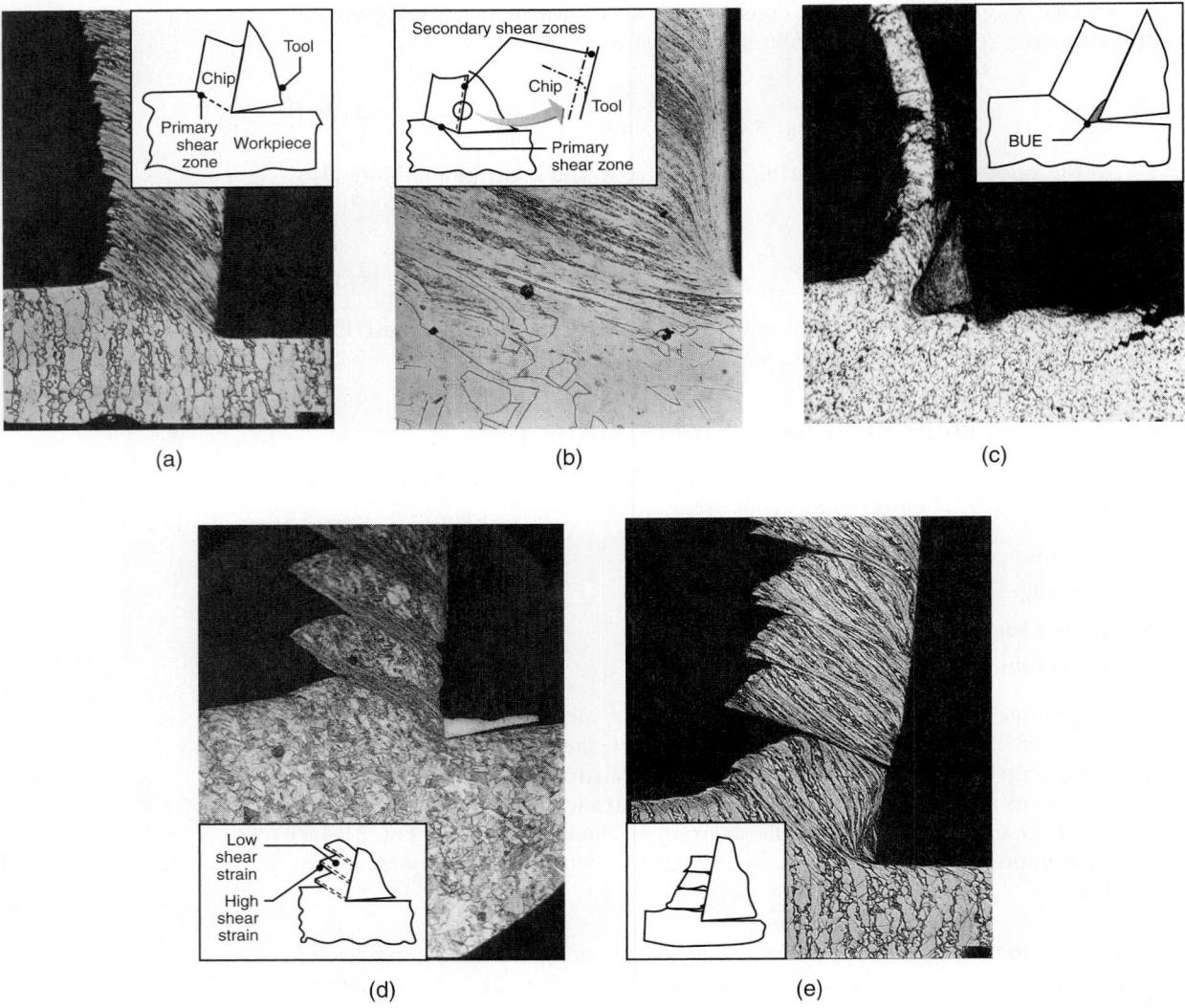

(a) (b) (c)

(d) (e)

FIGURE 21.5 Basic types of chips produced in orthogonal metal cutting, their schematic representation, and photomicrographs of the cutting zone: (a) continuous chip with narrow, straight, and primary shear zone; (b) continuous chip with secondary shear zone at the chip–tool interface; (c) built-up edge; (d) segmented or nonhomogeneous chip; and (e) discontinuous chip. *Source:* After M. C. Shaw, P.K. Wright, and S. Kalpakjian.

Built-up edge chips. A *built-up edge* (BUE) consists of layers of material from the workpiece that gradually are deposited on the tool tip—hence the term *built-up* (Fig. 21.5c). As it grows larger, the BUE becomes unstable and eventually breaks apart. Part of the BUE material is carried away by the tool-side of the chip; the rest is deposited randomly on the workpiece surface. The cycle of BUE formation and destruction is repeated continuously during the cutting operation until corrective measures are taken. In effect, built-up edge changes the geometry of the cutting edge and dulls it (Fig. 21.6a).

Built-up edge commonly is observed in practice. It is a major factor that adversely affects surface finish, as can be seen in Figs. 21.5c and 21.6b and c. However, a thin, stable BUE usually is regarded as desirable because it reduces tool wear by

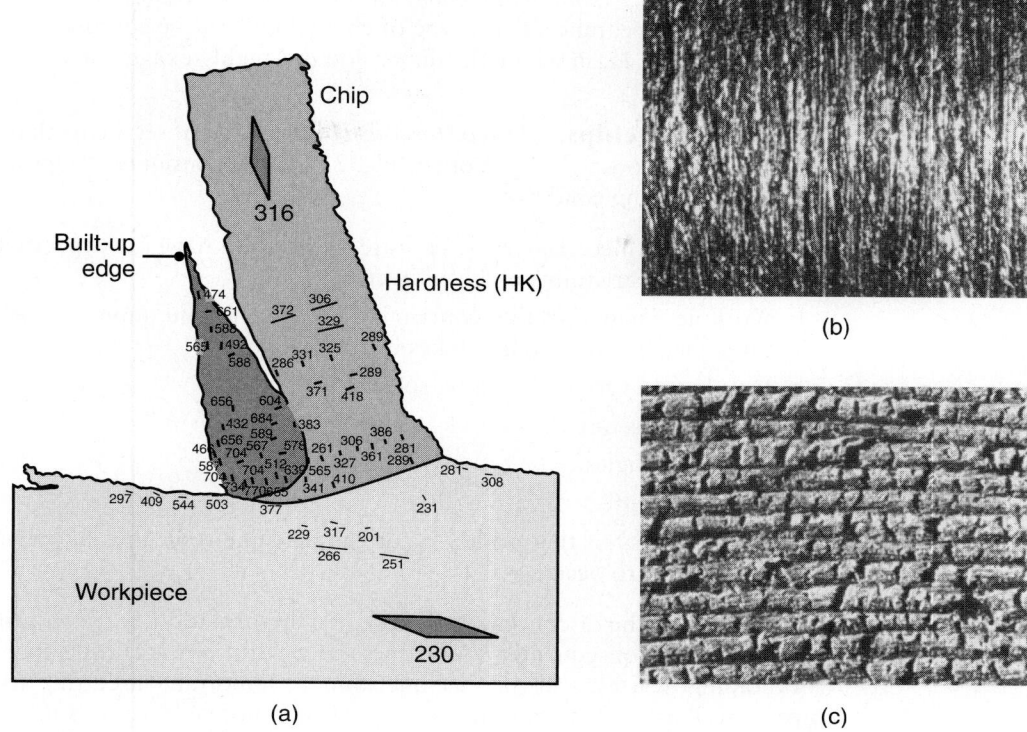

FIGURE 21.6 (a) Hardness distribution with a built-up edge in the cutting zone (material, 3115 steel). Note that some regions in the built-up edge are as much as three times harder than the bulk metal of the workpiece. (b) Surface finish produced in turning 5130 steel with a built-up edge. (c) Surface finish on 1018 steel in face milling. Magnifications: 15×. *Source*: Courtesy of Metcut Research Associates, Inc.

protecting its rake face. Cold-worked metals generally have less of a tendency to form BUE than in their annealed condition. Because of work hardening and deposition of successive layers of material, the BUE hardness increases significantly (Fig. 21.6a). As the cutting speed increases, the size of the BUE decreases; in fact it may not form at all.

The tendency for BUE formation can be reduced by one or more of the following means:

- Increase the cutting speeds
- Decrease the depth of cut
- Increase the rake angle
- Use a sharp tool
- Use an effective cutting fluid
- Use a cutting tool that has lower chemical affinity for the workpiece material

Serrated chips. *Serrated chips* (also called *segmented* or *nonhomogeneous* chips, see Fig. 21.5d) are semicontinuous chips with large zones of low shear strain and small zones of high shear strain, hence the latter zone is called *shear localization*. Metals with low thermal conductivity and strength that decreases sharply with temperature (thermal softening) exhibit this behavior, most notably titanium. The chips have a sawtooth-like appearance. (This type of chip should not be confused with the illustration in Fig. 21.4a, in which the dimension d is highly exaggerated.)

Discontinuous chips.   *Discontinuous chips* consist of segments that may be attached firmly or loosely to each other (Fig. 21.5e). Discontinuous chips usually form under the following conditions:

- Brittle workpiece materials, because they do not have the capacity to undergo the high shear strains involved in cutting.
- Workpiece materials that contain hard inclusions and impurities or have structures such as the graphite flakes in gray cast iron.
- Very low or very high cutting speeds.
- Large depths of cut.
- Low rake angles.
- Lack of an effective cutting fluid.
- Low stiffness of the toolholder or the machine tool, thus allowing vibration and chatter to occur.

Because of the discontinuous nature of chip formation, forces continually vary during cutting. Consequently, the stiffness or rigidity of the cutting-tool holder, the workholding devices, and the machine tool are important in cutting with serrated chips as well as with discontinuous chips. If it is not sufficiently stiff, the machine tool may begin to vibrate and chatter, as discussed in detail in Section 25.4. This, in turn, adversely affects the surface finish and dimensional accuracy of the machined part and may cause premature wear or damage to the cutting tool—even to the components of the machine tool, if excessive.

Chip curl.   In all cutting operations performed on metals, as well as nonmetallic materials such as plastics and wood, chips develop a curvature (*chip curl*) as they leave the workpiece surface (Fig. 21.5). Among factors affecting the chip curl are:

- The distribution of stresses in the primary and secondary shear zones.
- Thermal effects.
- Work-hardening characteristics of the workpiece material.
- The geometry of the cutting tool.
- Cutting fluids.

Process variables also affect chip curl. Generally, as the depth of cut decreases, the radius of curvature decreases; that is, the chip becomes curlier. Also, cutting fluids can make chips become more curly (the radius of curvature decreases), thus reducing the tool–chip contact area and concentrating the heat closer to the tip of the tool. As a result, tool wear increases.

Chip breakers.   As stated previously, continuous and long chips are undesirable as they tend to become entangled and severely interfere with machining operations and also become a potential safety hazard. If all of the process variables are under control, the usual procedure employed to avoid such a situation is to break the chip intermittently with cutting tools that have *chip breaker* features, as shown in Fig. 21.7.

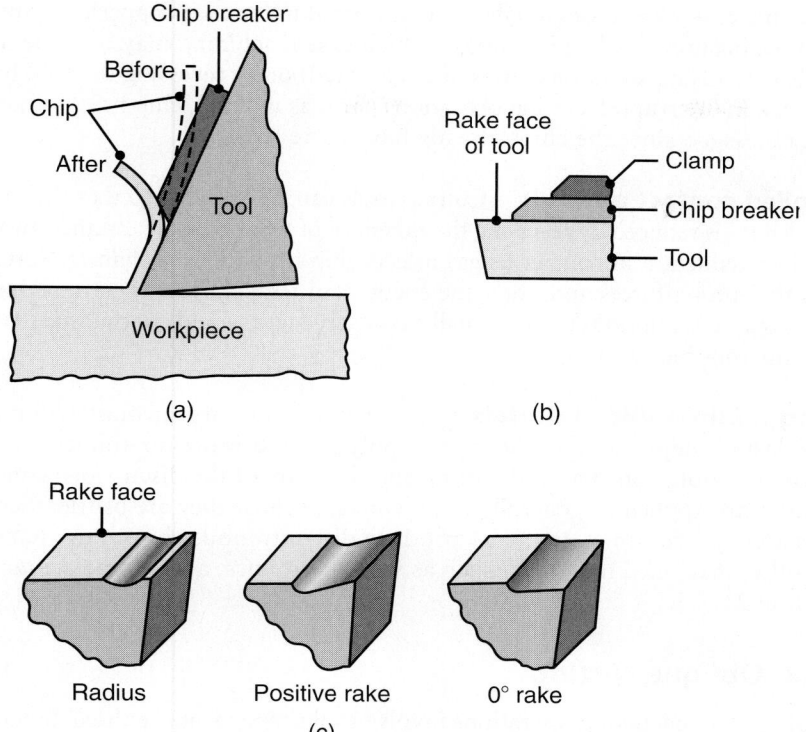

FIGURE 21.7 (a) Schematic illustration of the action of a chip breaker. Note that the chip breaker decreases the radius of curvature of the chip and eventually breaks it. (b) Chip breaker clamped on the rake face of a cutting tool. (c) Grooves in cutting tools acting as chip breakers. Most cutting tools used now are *inserts* with built-in chip-breaker features.

Chip breakers have traditionally been a piece of metal clamped to the tool's rake face, which bend and break the chip. However, most modern cutting tools and inserts now have built-in chip-breaker features (see Fig. 22.2) of various designs (Fig. 21.7). Chips also can be broken by changing the tool geometry to control chip flow, as in the turning operations shown in Fig. 21.8. Experience has indicated that the ideal chip size to be broken is in the shape of either the letter C or the number 9 and fits within a 25-mm (1-in.) square space.

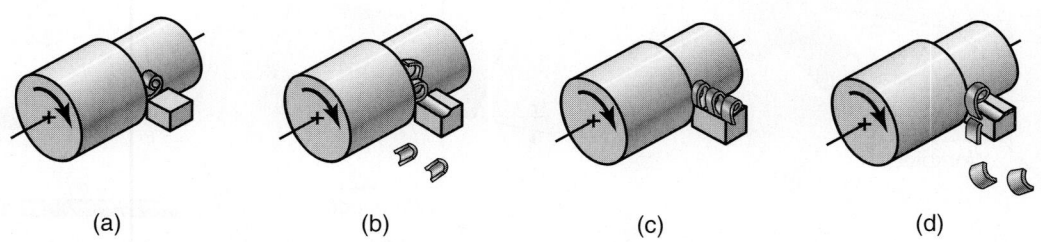

FIGURE 21.8 Chips produced in turning: (a) tightly curled chip; (b) chip hits workpiece and breaks; (c) continuous chip moving radially away from workpiece; and (d) chip hits tool shank and breaks off. *Source*: After G. Boothroyd.

With soft workpiece materials (such as pure aluminum or copper), chip breaking by such means may not be effective, in which case machining may be done in small increments (pausing so that a chip is not generated) or by reversing the feed by small increments. In interrupted-cutting operations (such as milling), chip breakers generally are not necessary, since the chips already have finite lengths.

Controlled contact on tools.   Cutting tools can be designed so that the tool–chip contact length is reduced by recessing the rake face of the tool some distance away from its tip. This reduction in contact length affects chip-formation mechanics. Primarily, it reduces the cutting forces and, thus, the energy and temperature. Determination of an optimum length is important as too small a contact length would concentrate the heat at the tool tip, thus increasing wear.

Cutting nonmetallic materials.   A variety of chips are encountered in cutting thermoplastics, depending on the type of polymer and process parameters, such as depth of cut, tool geometry, and cutting speed. Many of the discussions concerning metals also are applicable generally to polymers. Because they are brittle, thermosetting plastics and ceramics generally produce discontinuous chips. For characteristics of other machined materials (such as wood, ceramics, and composite materials) see Section 21.7.3.

21.2.2 Oblique cutting

The majority of machining operations involve tool shapes that are three-dimensional, thus the cutting is *oblique*. The basic difference between oblique and orthogonal cutting can be seen in Fig. 21.9a. Whereas in orthogonal cutting, the chip slides directly up the face of the tool, in oblique cutting, the chip is *helical* and at an angle i, called the **inclination angle** (Fig. 21.9b). Note the lateral direction of chip movement in oblique cutting—a situation that is similar to a snow-plow blade which throws the snow sideways. It can be seen that such a helical chip moves sideways and away from the cutting zone and doesn't obstruct it as it would in orthogonal cutting.

Note that the chip in Fig. 21.9a flows up the rake face of the tool at angle α_c (**chip flow angle**), which is measured in the plane of the tool face. Angle α_i is the

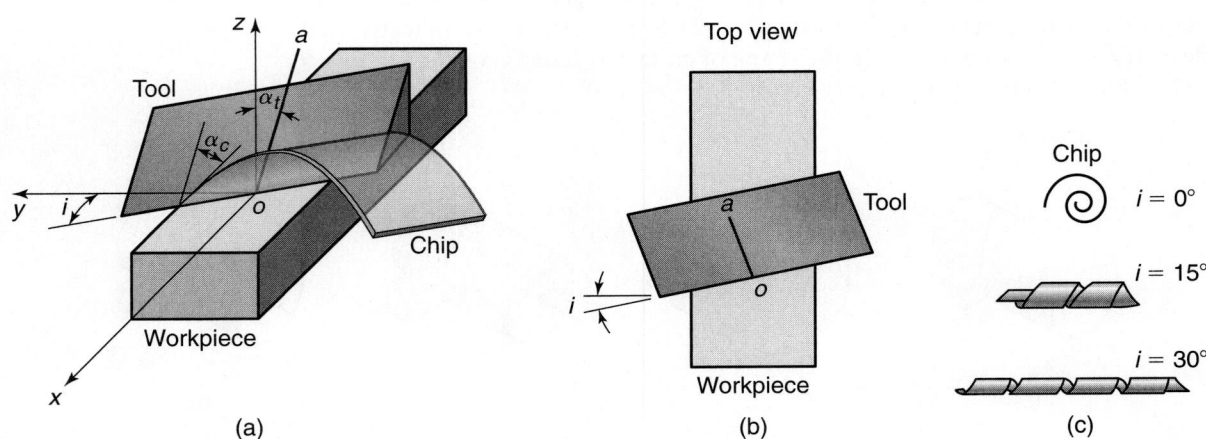

FIGURE 21.9   (a) Schematic illustration of cutting with an oblique tool. Note the direction of chip movement. (b) Top view, showing the inclination angle, *i*. (c) Types of chips produced with tools at increasing inclination angles.

normal rake angle, and it is a basic geometric property of the tool. This is the angle between line oz normal to the workpiece surface and line oa on the tool face.

The workpiece material approaches the cutting tool at a velocity, V, and leaves the surface (as a chip) with a velocity, V_c. The effective rake angle, α_e, is calculated in the plane of these two velocities. Assuming that the chip flow angle, α_c, is equal to the inclination angle, i (and this assumption has been found, through experiments, to be appropriate), the effective rake angle, α_e, is

$$\alpha_e = \sin^{-1}(\sin^2 i + \cos^2 i \sin \alpha_n). \tag{21.7}$$

Since both i and α_n can be measured directly, the effective rake angle can be calculated. Note that, as i increases, the effective rake angle increases and the chip becomes thinner and longer, and as a consequence, the cutting force decreases. The influence of the inclination angle on chip shape is shown in Fig. 21.9c.

A typical single-point turning tool used on a lathe is shown in Fig. 21.10a. Note the various angles involved, each of which has to be selected properly for efficient cutting. Although these angles usually can be produced by grinding, the majority of cutting tools are now available as **inserts**, as shown in Fig. 21.10b and to be described in detail in Chapter 22. Various three-dimensional cutting tools are described in greater detail in Chapters 23 and 24, including those for drilling, tapping, milling, planing, shaping, broaching, sawing, and filing.

Shaving and skiving.   Thin layers of material can be removed from straight or curved surfaces by a process similar to the use of a plane to shave wood. *Shaving* is useful particularly in improving the surface finish and dimensional accuracy of sheared parts and punched slugs (Fig. 16.9). Another application of shaving is in finishing gears with a cutter that has the shape of the gear tooth (see Section 24.7). Parts that are long or have a combination of shapes are shaved by *skiving* with a specially-shaped cutting tool, where it moves tangentially across the length of the workpiece.

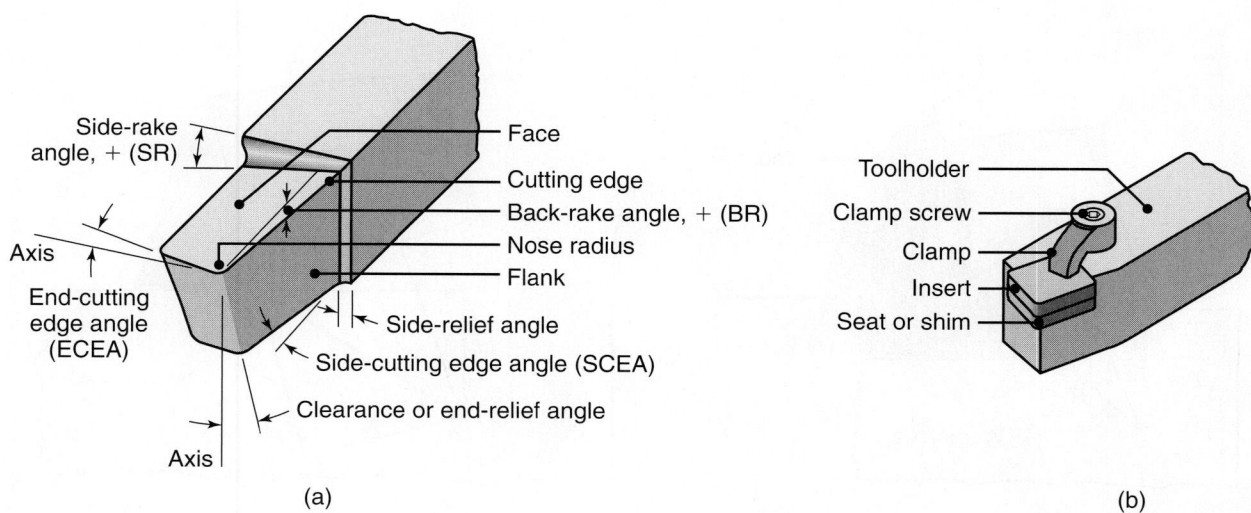

(a)

(b)

FIGURE 21.10   (a) Schematic illustration of a right-hand cutting tool. The various angles on these tools and their effects on machining are described in Section 23.2 Although these tools traditionally have been produced from solid tool-steel bars, they have been replaced largely with (b) inserts made of carbides and other materials of various shapes and sizes.

21.3 | Cutting Forces and Power

Knowledge of the *cutting forces* and *power* involved in machining operations is important for the following reasons:

- Data on cutting forces is essential so that:
  a. Machine tools can be properly designed to minimize distortion of the machine components, maintain the desired dimensional accuracy of the machined part, and help select appropriate toolholders and workholding devices.
  b. The workpiece is capable of withstanding these forces without excessive distortion.
- Power requirements must be known in order to enable the selection of a machine tool with adequate electric power.

The forces acting in orthogonal cutting are shown in Fig. 21.11a. The **cutting force**, F_c, acts in the direction of the cutting speed, V, and supplies the energy required for cutting. The ratio of the cutting force to the cross-sectional area being cut (i.e., the product of width of cut and depth of cut) is referred to as the *specific cutting force*.

The **thrust force**, F_t, acts in a direction normal to the cutting speed. These two forces produce the **resultant force**, R, as can be seen from the force circle shown in Fig. 21.11b. Note that the resultant force can be resolved into two components on the tool face: a **friction force**, F, along the tool-chip interface and a **normal force**, N, perpendicular to it. It can also be shown that

$$F = R \sin \beta \qquad (21.8a)$$

and

$$N = R \cos \beta \qquad (21.8b)$$

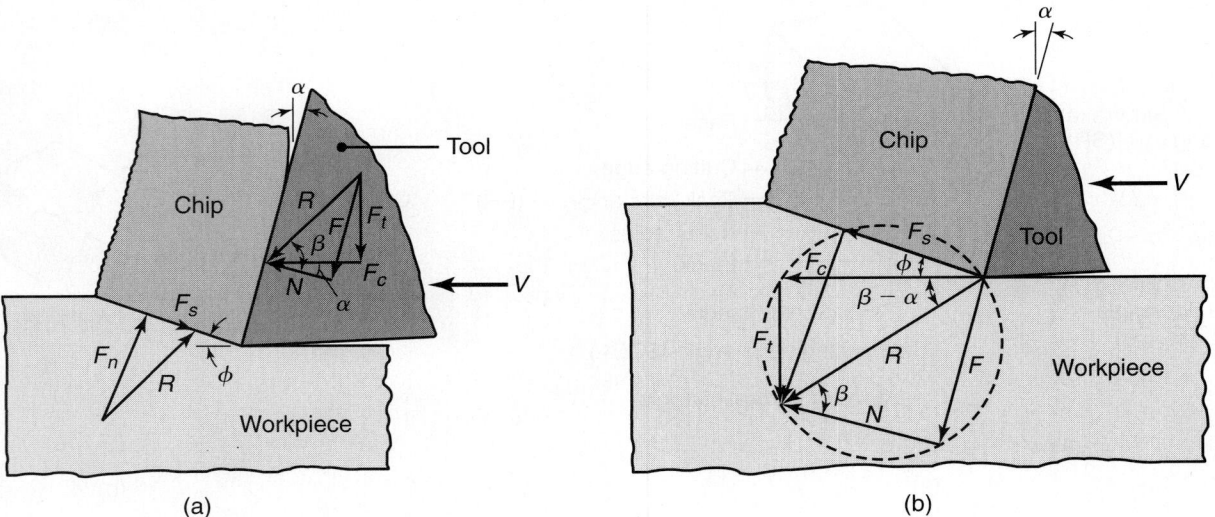

FIGURE 21.11   (a) Forces acting in the cutting zone during two-dimensional cutting. Note that the resultant force, R, must be colinear to balance the forces. (b) Force circle to determine various forces acting in the cutting zone.

Note also that the resultant force is balanced by an equal and opposite force along the shear plane and is resolved into a **shear force**, F_s and a **normal force**, F_n. It can be shown that these forces can be expressed as follows:

$$F_s = F_c \cos \phi - F_t \sin \phi \tag{21.9}$$

and

$$F_n = F_c \sin \phi + F_t \cos \phi \tag{21.10}$$

Because the area of the shear plane can be calculated by knowing the shear angle and the depth of cut, the shear and normal stresses in the shear plane can be determined.

The ratio of F to N is the **coefficient of friction**, μ, at the tool–chip interface, and the angle β is the **friction angle** (as in Fig. 21.11). The magnitude of μ can be determined as

$$\mu = \frac{F}{N} = \frac{F_t + F_c \tan \alpha}{F_c - F_t \tan \alpha} \tag{21.11}$$

Although the magnitude of forces in actual cutting operations is generally on the order of a few hundred newtons, the *local stresses* in the cutting zone and the *pressures* on the tool are very high because the contact areas are very small. For example, the tool-chip contact length (see Fig. 21.3) is typically on the order of 1 mm (0.04 in.). Consequently, the tool tip is subjected to very high stresses, which lead to wear and sometimes chipping and fracture of the tool.

Thrust force. A knowledge of the *thrust force* in cutting is important because the tool holder, the workholding devices, and the machine tool must be sufficiently stiff to support this force with minimal deflections. For example, if the thrust force is too high or if the machine tool is not sufficiently stiff, the tool will be pushed away from the workpiece surface being machined. This movement will, in turn, reduce the depth of cut, resulting in a lack of dimensional accuracy in the machined part.

We also can show the effect of rake angle and friction angle on the direction of thrust force by noting from Fig. 21.11b that

$$F_t = R \sin(\beta - \alpha) \tag{21.12a}$$

or

$$F_t = F_c \tan(\beta - \alpha) \tag{21.12b}$$

Note that the magnitude of the cutting force, F_c, is always positive, as shown in Fig. 21.11, because it is this force that supplies the work required in cutting. However, the sign of the thrust force, F_t, can be either positive or negative, depending on the relative magnitudes of β and α. When $\beta > \alpha$, the sign of F_t is positive (that is *downward*), and when $\beta < \alpha$, the sign is negative (that is *upward*). Therefore, it is possible to have an upward thrust force under the conditions of (a) high rake angles, (b) low friction at the tool–chip interface, or (c) both. A negative thrust force can have important implications in the design of machine tools and workholders and in the stability of the cutting process.

Power. *Power* is the product of force and velocity. Thus, referring to Fig. 21.11, the power input in cutting is

$$\text{Power} = F_c V \tag{21.13}$$

This power is dissipated mainly in the shear zone (due to the energy required to shear the material) and on the rake face of the tool (due to tool–chip interface friction).

From Figs. 21.4b and 21.11, it can be seen that the power dissipated in the shear plane is

$$\text{Power for shearing} = F_s V_s \qquad (21.14)$$

Letting w be the width of cut, the **specific energy for shearing**, u_s, is given by

$$u_s = \frac{F_s V_s}{w t_o V} \qquad (21.15)$$

Similarly, the power dissipated in friction is

$$\text{Power for friction} = F V_c \qquad (21.16)$$

and the **specific energy for friction**, u_f, is

$$u_f = \frac{F V_c}{w t_o V} = \frac{F r}{w t_o} \qquad (21.17)$$

The **total specific energy**, u_t, thus is

$$u_t = u_s + u_f \qquad (21.18)$$

Because of the many factors involved, reliable prediction of cutting forces and power still is based largely on experimental data, such as those given in Table 21.2. The wide range of values shown can be attributed to differences in strength within each material group and to various other factors, such as friction, use of cutting fluids, and processing variables.

The sharpness of the tool tip also influences forces and power. Because it rubs against the machined surface and makes the deformation zone ahead of the tool larger, duller tools require higher forces and power.

Measuring cutting forces and power. Cutting forces can be measured using a **force transducer** (typically with quartz piezoelectric sensors), a **dynamometer** or a **load cell** (with resistance-wire strain gages placed on octagonal rings) mounted on the cutting-tool holder. Transducers have much higher natural frequency and stiffness

TABLE 21.2

Approximate Range of Energy Requirements in Cutting Operations at the Drive Motor of the Machine Tool (For Dull Tools, Multiply by 1.25)		
Material	Specific energy	
	$W \cdot s/mm^3$	$hp \cdot min/in^3$
Aluminum alloys	0.4–1	0.15–0.4
Cast irons	1.1–5.4	0.4–2
Copper alloys	1.4–3.2	0.5–1.2
High-temperature alloys	3.2–8	1.2–3
Magnesium alloys	0.3–0.6	0.1–0.2
Nickel alloys	4.8–6.7	1.8–2.5
Refractory alloys	3–9	1.1–3.5
Stainless steels	2–5	0.8–1.9
Steels	2–9	0.7–3.4
Titanium alloys	2–5	0.7–2

than dynamometers, which are prone to excessive deflection and vibration. Also, it is possible to calculate cutting force from the **power consumption** during cutting, provided that the mechanical efficiency of the machine tool is known or can be determined. The power can be measured easily by a power monitor, such as a wattmeter. The *specific energy* in cutting (such as that shown in Table 21.2) also can be used to calculate cutting forces.

EXAMPLE 21.1 Relative energies in cutting

In an orthogonal cutting operation, $t_o = 0.005$ in., $V = 400$ ft/min, $\alpha = 10°$, and the width of cut $= 0.25$ in. It is observed that $t_c = 0.009$ in., $F_c = 125$ lb, and $F_t = 50$ lb. Calculate the percentage of the total energy that goes into overcoming friction at the tool-chip interface.

Solution The percentage of the energy can be expressed as

$$\frac{\text{Friction energy}}{\text{Total energy}} = \frac{FV_c}{F_cV} = \frac{Fr}{F_c}$$

where

$$r = \frac{t_o}{t_c} = \frac{5}{9} = 0.555$$

$$F = R \sin \beta$$

$$F_c = R \cos(\beta - \alpha)$$

and

$$R = \sqrt{F_t^2 + F_c^2} = \sqrt{50^2 + 125^2} = 135 \text{ lb}$$

Thus,

$$125 = 135 \cos(\beta - 10)$$

so

$$\beta = 32°$$

and

$$F = 135 \sin 32° = 71.5 \text{ lb}$$

Hence,

$$\text{Percentage} = \frac{(71.5)(0.555)}{125} = 0.32 \quad \text{or} \quad 32\%$$

21.4 | Temperatures in Cutting

As in all metalworking processes where plastic deformation is involved, the energy dissipated in cutting is converted into heat which, in turn, raises the temperature in the cutting zone. *Temperature rise* is a very important factor in machining because of its major adverse effects such as:

- Excessive temperature lowers the strength, hardness, stiffness, and wear resistance of the cutting tool; tools also may soften and undergo plastic deformation; thus tool shape is altered.

- Increased heat causes uneven dimensional changes in the part being machined, making it difficult to control its dimensional accuracy and tolerances.
- Excessive temperature rise can induce thermal damage and metallurgical changes in the machined surface, adversely affecting its properties.

From the preceding sections, it can be seen that the main sources of heat in machining are: (a) the work done in shearing in the primary shear zone, (b) energy dissipated as friction at the tool–chip interface, and (c) heat generated as the tool rubs against the machined surface, especially for dull or worn tools. Much effort has been expended in establishing relationships among temperature and various material and process variables in cutting. A comprehensive expression for the *mean temperature*, T_{mean}, in *orthogonal cutting* is

$$T = \frac{1.2Y_f}{pc}\sqrt[3]{\frac{Vt_o}{K}}, \tag{21.19a}$$

where the mean temperature is in °F, Y_f is the flow stress in psi, pc is the volumetric specific heat in in.-lb/in³-°F, and K is the thermal diffusivity (ratio of thermal conductivity to volumetric specific heat) in in²/s. Because the material parameters in this equation depend on temperature, it is important to use appropriate values that are compatible with the predicted temperature range. It can be seen from Eq. (21.19a) that the mean cutting temperature increases with workpiece strength, cutting speed, and depth of cut; it decreases with increasing specific heat and thermal conductivity of the workpiece material.

An expression for the *mean temperature in turning* on a lathe is given by

$$T_{\text{mean}} \ \propto \ V^a f^b \tag{21.19b}$$

where V is the cutting speed, and f is the feed of the tool, as shown in Fig. 21.2. Approximate values of the exponents a and b are: for *carbide* tools, $a = 0.2$ and $b = 0.125$; for *high-speed steel* tools, $a = 0.5$ and $b = 0.375$.

Temperature distribution.   Because the sources of heat generation in machining are concentrated in the primary shear zone and at the tool–chip interface, it is to be expected that there will be severe temperature gradients in the cutting zone. A typical temperature distribution is shown in Fig. 21.12. Note the presence of severe gradients, and that the *maximum* temperature is about halfway up the tool–chip interface. From the preceding discussions thus far, it will be apparent that the particular temperature pattern depends on several factors pertaining to material properties and cutting conditions, including the type of cutting fluid used (if any) during machining.

The temperatures developed in a *turning* operation on 52100 steel are shown in Fig. 21.13. The temperature distribution along the *flank surface* of the tool is shown in Fig. 21.13a, for V of 60, 90, and 170 m/min, respectively, as a function of the distance from the tip of the tool. The temperature distributions at the *tool–chip interface* for the same three cutting speeds are shown in Fig. 21.13b, as a function of the fraction of the contact length. Thus, zero on the abscissa represents the tool tip, and 1.0 represents the end of the tool–chip contact length.

Note that the temperature increases with cutting speed and that the highest temperature is almost 1100°C (2000°F). The presence of such high temperatures in machining can be verified by simply observing the dark-bluish color of the chips (caused by oxidation) produced at high cutting speeds. Chips can become red hot and create a safety hazard for the operator.

From Eq. (21.19b) and the values for the exponent a, it can be seen that the cutting speed, V, greatly influences temperature. The explanation is that (as speed

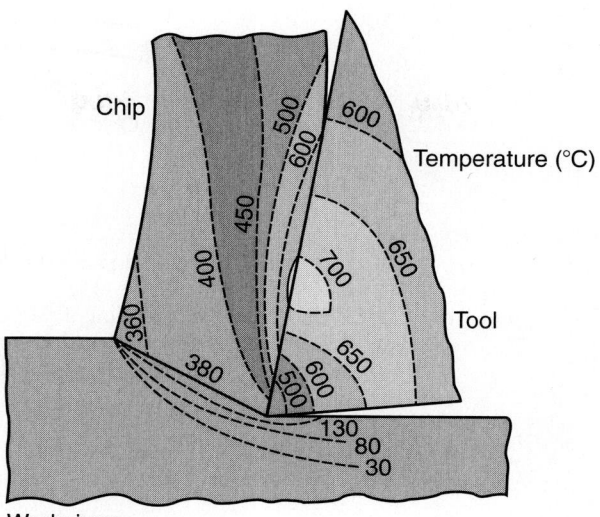

FIGURE 21.12 Typical temperature distribution in the cutting zone. Note the severe temperature gradients within the tool and the chip, and that the workpiece is relatively cool. *Source:* After G. Vieregge.

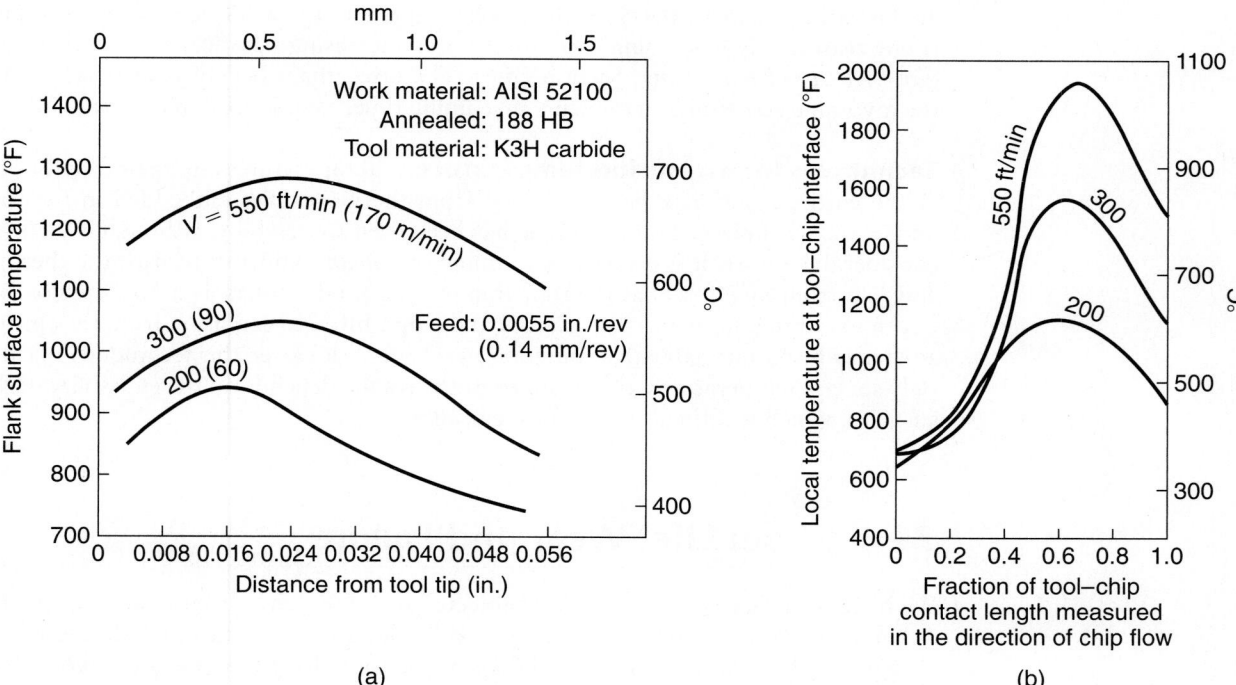

(a)

(b)

FIGURE 21.13 Temperatures developed in turning 52100 steel: (a) flank temperature distribution and (b) tool–chip interface temperature distribution. *Source:* After B. T. Chao and K. J. Trigger.

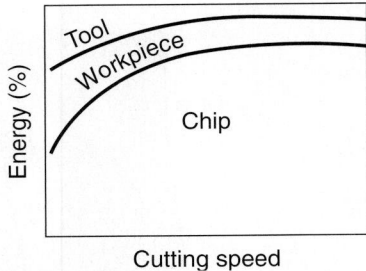

FIGURE 21.14   Proportion of the heat generated in cutting transferred into the tool, workpiece, and chip as a function of the cutting speed. Note that the chip removes most of the heat.

increases the time for heat dissipation decreases and hence temperature rises (eventually becoming an almost adiabatic process). This effect of speed can be demonstrated easily by rubbing your hands together faster and faster.

As can be seen from Fig. 21.14, the chip carries away most of the heat generated. It has been estimated that 90% of the energy is dissipated in the chip during a typical machining operation, with the rest in the tool and the workpiece. Note in this figure that (as the cutting speed increases) a larger proportion of the total heat generated is carried away by the chip, and less heat goes into the workpiece or the tool. This is one reason why machining speeds have been increasing significantly over the years (see *high-speed machining*, Section 25.5). The other main benefit is associated with the favorable economics in reducing machining time, as described in Section 25.8.

Techniques for measuring temperature.   Temperatures and their distribution in the cutting zone may be determined from **thermocouples** embedded in the tool and/or the workpiece. This technique has been used successfully, although it involves considerable effort. It is easier to determine the *mean* temperature with the **thermal emf** (electromotive force) at the tool-chip interface, which acts as a *hot junction* between two different materials (i.e., tool and chip). **Infrared radiation** from the cutting zone may be monitored with a *radiation pyrometer*. However, this technique indicates only surface temperatures; the accuracy of the results depends on the emissivity of the surfaces, which is difficult to determine accurately.

21.5 | Tool Life: Wear and Failure

We have seen that cutting tools are subjected to (a) high localized stresses at the tip of the tool, (b) high temperatures, especially along the rake face, (c) sliding of the chip along the rake face, and (d) sliding of the tool along the newly cut workpiece surface. These conditions induce **tool wear**, which is a major consideration in all machining operations, as are mold and die wear in casting and metalworking. Tool wear adversely affects tool life, the quality of the machined surface and its dimensional accuracy, and consequently, the economics of cutting operations.

Wear is a gradual process, much like the wear of the tip of an ordinary pencil. The rate of tool wear depends on tool and workpiece materials, tool geometry, process parameters, cutting fluids, and the characteristics of the machine tool. Tool wear and the changes in tool geometry during cutting manifest themselves in different ways, generally classified as *flank wear, crater wear, nose wear, notching, plastic deformation of the tool tip, chipping,* and *gross fracture* (Fig. 21.15).

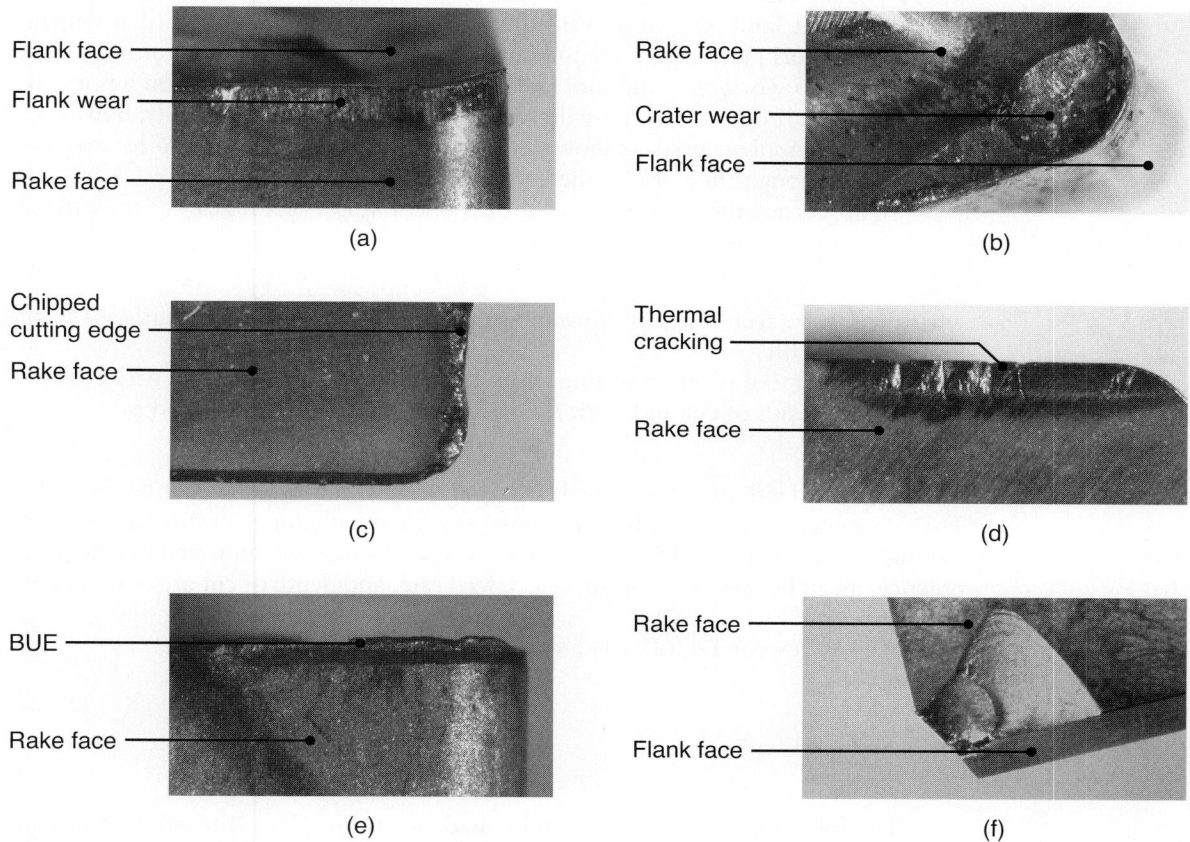

FIGURE 21.15 (a) Flank wear and crater wear in a cutting tool; the tool moves to the left, as in Fig. 21.3. (b) View of the rake face of a turning tool, showing various wear patterns. (c) View of the flank face of a turning tool, showing various wear patterns. (d) Types of wear on a turning tool: 1. flank wear; 2. crater wear; 3. chipped cutting edge; 4. thermal cracking on rake face; 5. built-up edge; 6. catastrophic failure. (See also Fig. 21.18.) *Source:* Courtesy of Kennametal, Inc.

21.5.1 Flank wear

Flank wear occurs on the relief (flank) face of the tool (Fig. 21.15a, c, and d). It generally is attributed to (a) rubbing of the tool along the machined surface, causing adhesive and/or abrasive wear (see Section 33.5) and (b) high temperatures, which adversely affect tool-material properties.

In a classic study by F. W. Taylor on the machining of steels conducted in the early 1890s, the following approximate relationship was established:

$$VT^n = C \qquad (21.20a)$$

where V is the cutting speed, T is the time (in minutes) that it takes to develop a certain flank **wear land** (shown as VB in Fig. 21.15c), n is an exponent (that depends on tool and workpiece materials and cutting conditions), and C is a constant. Each combination of workpiece and tool materials and each cutting condition has its own n and C values, both of which are determined experimentally. Generally, however, n depends on the tool material, as shown in Table 21.3, and C on the workpiece material. Note that the magnitude of C is the cutting speed at $T = 1$ min.

To appreciate the importance of the exponent n, Eq. (21.20) can be rewritten as

$$T = \left(\frac{C}{V}\right)^{\frac{1}{n}} \qquad (21.20b)$$

where it can be seen that for constant values of C, the smaller the value of n, the lower the tool life.

Cutting speed is the most important process variable associated with tool life, followed by depth of cut and feed, f. Equation (21.20) can be modified as

$$VT^n d^x f^y = C \qquad (21.21)$$

where d is the depth of cut and f is the feed in mm/rev or in./rev, as shown in Fig. 21.2. The exponents x and y must be determined experimentally for each cutting condition. Taking $n = 0.15$, $x = 0.15$, and $y = 0.6$ as typical values encountered in machining practice, it can be seen that cutting speed, feed rate, and depth of cut are of decreasing importance.

We can rewrite Eq. (21.21) as

$$T = C^{1/n} V^{-1/n} d^{-x/n} f^{-y/n} \qquad (21.22)$$

or, using typical values, as

$$T \simeq C^7 V^{-7} d^{-1} f^{-4} \qquad (21.23)$$

The following observations can be made from Eq. (21.23) to obtain the same tool life: (a) if the feed or the depth of cut is increased, the cutting speed must be decreased (and vice versa) and (b) depending on the exponents, a reduction in speed can result in an increase in the volume of the material removed because of the increased feed and/or depth of cut.

Tool-life curves. *Tool-life curves* are plots of experimental data obtained by performing cutting tests on various materials and under different cutting conditions,

TABLE 21.3

Ranges of n Values for the Taylor Eq. (21.20a) for Various Tool Materials	
High-speed steels	0.08–0.2
Cast alloys	0.1–0.15
Carbides	0.2–0.5
Coated carbides	0.4–0.6
Ceramics	0.5–0.7

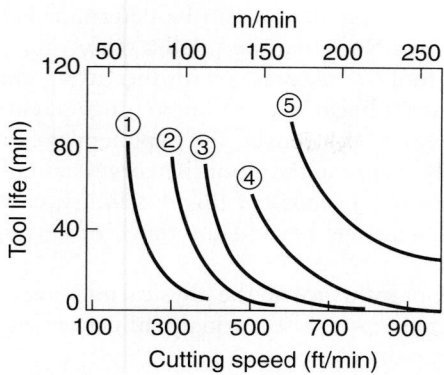

FIGURE 21.16   Effect of workpiece hardness and microstructure on tool life in turning ductile cast iron. Note the rapid decrease in tool life (approaching zero) as the cutting speed increases. Tool materials have been developed that resist high temperatures, such as carbides, ceramics, and cubic boron nitride, as will be described in Chapter 22.

such as cutting speed, feed, depth of cut, tool material and geometry, and cutting fluids. Note in Fig. 21.16, for example, that (a) tool life decreases rapidly as the cutting speed increases; (b) the condition of the workpiece material has a strong influence on tool life; and (c) there is a large difference in tool life for different workpiece-material microstructures.

Heat treatment of the workpiece is important, largely due to increasing workpiece hardness. For example, ferrite has a hardness of about 100 HB, pearlite 200 HB, and martensite 300 to 500 HB. Impurities and hard constituents in the material or on the surface of the workpiece (such as rust, scale, slag, etc.) also are important factors because their abrasive action reduces tool life.

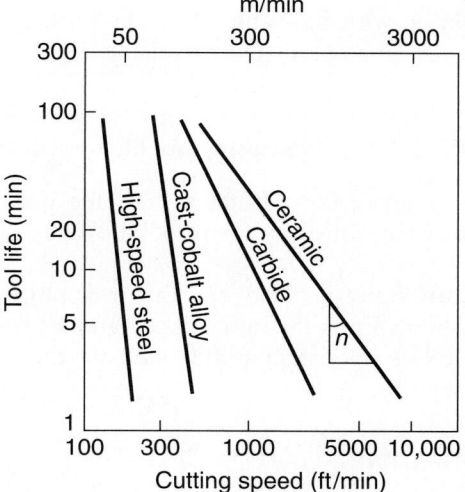

FIGURE 21.17   Tool-life curves for a variety of cutting-tool materials. The negative inverse of the slope of these curves is the exponent n in the Taylor tool-life equation, and C is the cutting speed at $T = 1$ min, ranging from about 200 to 10,000 ft/min in this figure.

Tool-life curves, from which the exponent n can be determined (Fig. 21.17), generally are plotted on log–log paper. Note that the smaller the n value, the steeper the curve, and thus the faster the tool life decreases with increasing cutting speed. Although tool-life curves are somewhat linear over a limited range of cutting speeds, they rarely are linear over a wide range. Moreover, the exponent n can indeed become *negative* at low cutting speeds, meaning that tool-life curves actually can reach a maximum and then curve downward. Because of this possibility, caution should be exercised when using tool-life equations beyond the range of cutting speeds to which they are applicable.

Because temperature has a major influence on the physical and mechanical properties of materials, it can be expected that it also strongly influences wear. Thus, as temperature increases, flank wear rapidly increases.

EXAMPLE 21.2 Determining the coefficients C and n in the Taylor equation

Determine the C and n values in the Taylor equation for the four different tool materials shown in Fig. 21.17.

Solution We have noted that the value of C corresponds to the cutting speed for a tool life of one minute. By extrapolation, the four lines in Fig. 21.17 have a tool life of one minute (which in this case is the abscissa). We find that ranging from HSS to ceramics, C values are about 200, 400, 3,000, and 11,000, respectively.

The n values are obtained from the negative inverse slopes of the lines on a log–log paper. We then estimate that:

HSS is 6° or $n = 0.11$
Cast-cobalt alloy is 8° or $n = 0.14$
Carbide is 25° or $n = 0.47$
Ceramic is 36° or $n = 0.73$

Comparing with the values given in Table 21.3, we note that our estimates are within the range of n values given.

EXAMPLE 21.3 Increasing tool life by reducing the cutting speed

Using the Taylor Eq. (21.20a) for tool life and letting $n = 0.5$ and $C = 400$, calculate the percentage increase in tool life when the cutting speed is reduced by 50%.

Solution Since $n = 0.5$, the Taylor equation can be rewritten as $VT^{0.5} = 400$. Let's denote V_1 as the initial speed and V_2 the reduced speed; thus $V_2 = 0.5V_1$. Because C is a constant at 400, we have the relationship

$$0.5V_1\sqrt{T_2} = V_1\sqrt{T_1}$$

Simplifying this expression, $T_2/T_1 = 1/0.25 = 4$. This indicates that the tools life change is

$$\frac{T_2 - T_1}{T_1} = \left(\frac{T_2}{T_1}\right) - 1 = 4 - 1 = 3$$

or that tool life is increased by 300%. Thus, reduction in cutting speed has resulted in a major increase in tool life. Note also that, for this problem, the magnitude of C is not relevant.

TABLE 21.4

Allowable Average Wear Land (see *VB* in Fig. 21.15c) for Cutting Tools in Various Machining Operations		
Operation	Allowable wear land (mm)	
	High-speed-steel tools	Carbide tools
Turning	1.5	0.4
Face milling	1.5	0.4
End milling	0.3	0.3
Drilling	0.4	0.4
Reaming	0.15	0.15

Note: Allowable wear for ceramic tools is about 50% higher. Allowable notch wear, VB_{max}, is about twice that for *VB*.

Allowable wear land. We realize that we have to sharpen a knife or a pair of scissors when the quality of the cut deteriorates or the forces required are too high. Similarly, cutting tools need to be replaced (or resharpened) when (a) the surface finish of the machined workpiece begins to deteriorate, (b) cutting forces increase significantly, or (c) temperature rises significantly. The *allowable wear land* (VB in Fig. 21.15c) for various machining conditions is given in Table 21.4. For improved dimensional accuracy, tolerances, and surface finish, the allowable wear land may be smaller than the values given in the table.

The *recommended cutting speed* for a high-speed steel tool is generally the one that yields a tool life of 60 to 120 min, and for a carbide tool, it is 30 to 60 min. However, depending on the particular workpiece, the operation, and the high-productivity considerations due to the use of modern, computer-controlled machine tools, the cutting speeds selected can vary significantly from these values.

Optimum cutting speed. We have noted that as cutting speed increases, tool life is reduced rapidly. On the other hand, if the cutting speed is low, tool life is long, but the *rate* at which material is removed is also low. Thus, there is an *optimum cutting speed*. Because it involves several other parameters, we will describe this topic further in Section 25.8.

EXAMPLE 21.4 Effect of cutting speed on material removal

The effect of cutting speed on the volume of metal removed between tool changes (or resharpenings) can be appreciated by analyzing Fig. 21.16. Assume that a material is being machined in the "one" condition (that is, as cast with a hardness of 265 HB). We note that when the cutting speed is 60 m/min, tool life is about 40 min. Thus, the tool travels a distance of 60 m/min × 40 min = 2400 m before it has to be replaced. However, when the cutting speed is increased to 120 m/min, the tool life is reduced to about 5 min, and the tool travels 120 m/min × 5 min = 600 m before it has to be replaced.

Since the volume of material removed is proportional directly to the distance the tool has traveled, it can be seen that by *decreasing* the cutting speed *more* material is removed between tool changes. It is important to note, however, that the lower the cutting speed, the longer the time required to machine a part, which has a significant economic impact on the operation (see Section 25.8).

21.5.2 Crater wear

Crater wear occurs on the rake face of the tool, as shown in Fig. 21.15a, b, and d, and Fig. 21.18, which shows various types of tool wear and failures. It readily can be seen that crater wear changes the tool–chip interface contact geometry. The most significant factors influencing crater wear are (a) temperature at the tool–chip interface and (b) the chemical affinity between the tool and workpiece materials. Additionally, the factors influencing flank wear also may influence crater wear.

Crater wear generally is attributed to a **diffusion mechanism**, that is, the movement of atoms across the tool–chip interface. Since diffusion rate increases with increasing temperature, crater wear increases as temperature increases. Note in Fig. 21.19, for example, how rapidly crater wear increases within a narrow range of temperature. *Coating* tools is an effective means of slowing the diffusion process and thus reducing crater wear. Typical coatings are titanium nitride, titanium carbide, titanium carbonitride, and aluminum oxide and are described in greater detail in Section 22.6.

When comparing Figs. 21.12 and 21.15a, it can be seen that the location of the maximum depth of crater wear, *KT*, coincides with the location of the *maximum temperature* at the tool–chip interface. An actual cross-section of this interface, when cutting steel at high speeds, is shown in Fig. 21.20. Note that the crater-wear pattern on the tool coincides with its discoloration pattern, which is an indication of the presence of high temperatures.

21.5.3 Other types of wear, chipping, and fracture

We now describe the factors involved in other types of cutting tool wear and fracture.

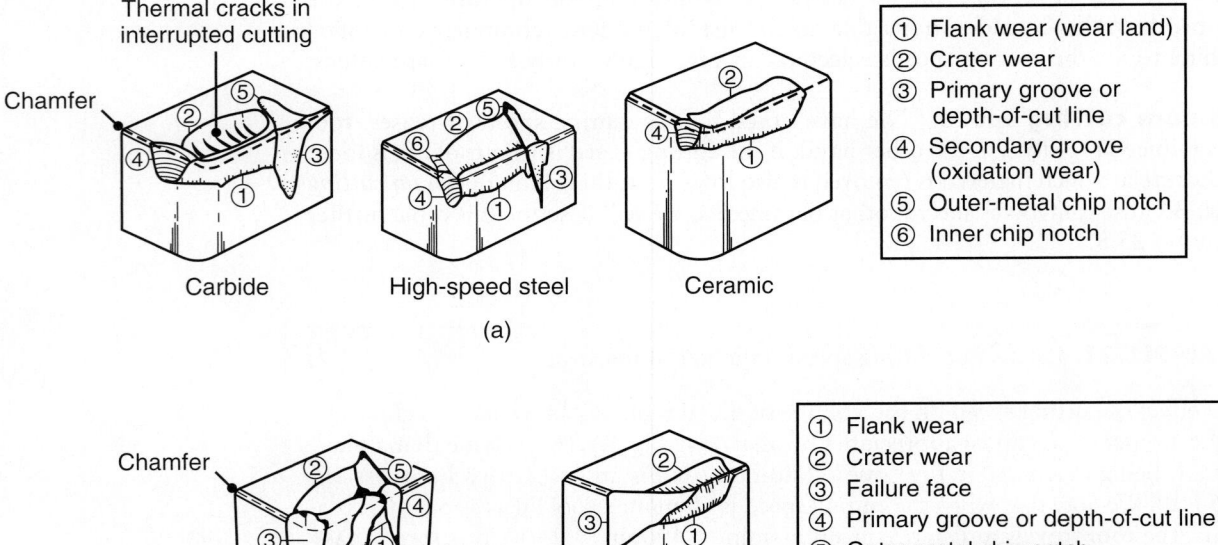

FIGURE 21.18 (a) Schematic illustrations of types of wear observed on various cutting tools. (b) Schematic illustrations of catastrophic tool failures. A wide range of parameters influence these wear and failure patterns. *Source:* Courtesy of V. C. Venkatesh.

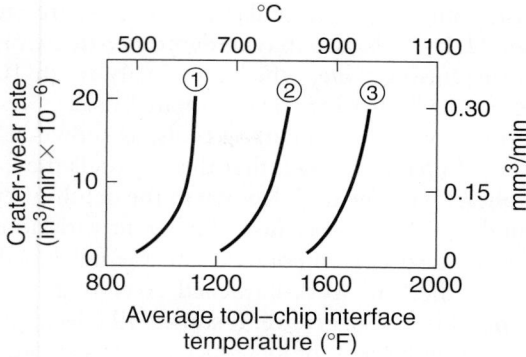

FIGURE 21.19   Relationship between crater-wear rate and average tool–chip interface temperature: (1) High-speed steel, (2) C1 carbide, and (3) C5 carbide (see Table 22.4). Note how rapidly crater-wear rate increases with an incremental increase in temperature. *Source*: After B. T. Chao and K. J. Trigger.

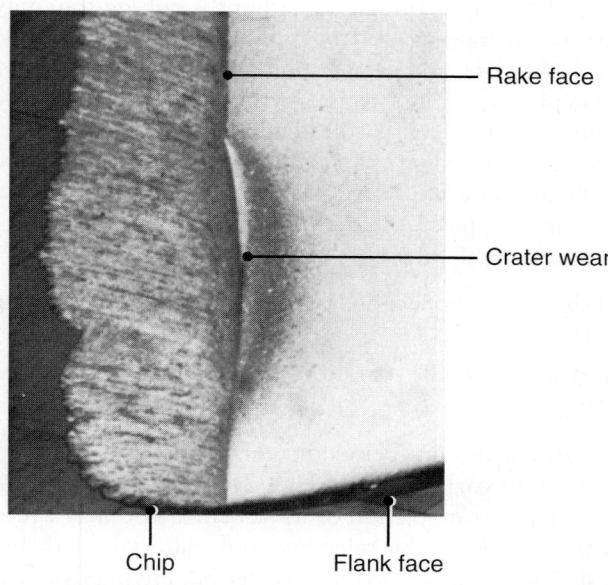

FIGURE 21.20   Interface of a cutting tool (right) and chip (left) in machining plain-carbon steel. The discoloration of the tool indicates the presence of high temperatures. Compare this figure with the temperature profiles shown in Fig. 21.12. *Source*: Courtesy of P. K. Wright.

Nose wear (Fig. 21.15b) is the rounding of a sharp tool, due to mechanical and thermal effects. It dulls the tool, affects chip formation, and causes rubbing of the tool over the workpiece, raising its temperature and possibly inducing residual stresses on the machined surface. A related phenomenon is *edge rounding*, as shown in Fig. 21.15b and c.

Temperature rise is important particularly for high-speed-steel tools, as can be appreciated from Fig. 22.1. Tools also may undergo *plastic deformation* because of temperature rises in the cutting zone, which can easily reach 1000°C (1800°F) in machining steels and can be higher for stronger materials.

The *notch* or *groove* observed on cutting tools, as shown in Figs. 21.15b and c and 21.18, has been attributed to the fact that this region is the boundary where the chip is no longer in contact with the tool. Known as the **depth-of-cut line** (DOC) with a depth VN, this boundary oscillates because of inherent variations in the cutting operation. Furthermore, this region is in contact with the machined surface generated during the previous cut; the thin work-hardened layer that can develop will contribute to the formation of the wear groove. If sufficiently deep, the notch can lead to gross chipping of the tool tip because of its reduced cross-section, as well as its notch sensitivity of the tool material.

Scale and oxide layers on a workpiece surface also contribute to notch wear, because these layers are hard and abrasive. Thus, light cuts should not be taken on rusted workpieces, and the depth-of-cut should be greater than the thickness of the oxide film or the work-hardened layer. In Fig. 21.3, for example, the depth of cut, t_o, should be greater than the thickness of the scale on the workpiece.

In addition to wear, tools also may undergo *chipping*, in which a small fragment from the cutting edge of the tool breaks away. This phenomenon, which typically occurs in brittle tool materials such as ceramics, is similar to chipping the tip of a pencil if it is too sharp. The chipped fragments from the cutting tool may be very small (**microchipping** or **macrochipping**), or they may be relatively large, variously called **gross chipping**, **gross fracture**, and **catastrophic failure** (Fig. 21.15d and 21.18).

Chipping also may occur in a region of the tool where a small crack or defect already exists. Unlike wear, which is a gradual process, chipping is a sudden loss of tool material, and a corresponding change in its shape. As can be expected, chipping has a major detrimental effect on surface finish, surface integrity, and the dimensional accuracy of the workpiece.

Two main causes of chipping are:

- **Mechanical shock** (i.e., impact due to interrupted cutting, as in turning a splined shaft on a lathe).
- **Thermal fatigue** (i.e., cyclic variations in the temperature of the tool in interrupted cutting).

Thermal cracks usually are perpendicular to the cutting edge of the tool, as shown on rake face of the carbide tool in Figs. 21.15d and 21.18a. Major variations in the workpiece material composition or its structure also may cause chipping.

Chipping can be reduced by selecting tool materials with high impact and thermal-shock resistance, as described in Chapter 22. High, positive rake angles can contribute to chipping because of the small included angle of the tool tip, as can be visualized from Fig. 21.3. Also, it is possible for the crater-wear region to progress toward the tool tip, thus weakening the tip because of reduced material volume and causing chipping.

21.5.4   Tool-condition monitoring

With computer-controlled machine tools and automated manufacturing, the reliable and repeatable performance of cutting tools is a critical consideration. As described in Chapters 23 through 25, modern machine tools operate with little direct supervision by a machine operator and generally are enclosed, making it impossible or difficult to monitor the machining operation and the condition of the tool. It is therefore essential to continuously and indirectly monitor the condition of the cutting tool so as to note, for example: wear, chipping, or gross failure.

In modern machine tools, tool-condition monitoring systems are integrated into computer numerical control and programmable logic controllers. Techniques for tool-condition monitoring typically fall into two general categories: direct and indirect.

The **direct method** for observing the condition of a cutting tool involves optical measurements of wear, such as the periodic observation of changes in the tool profile. This is a common and reliable technique and is done using a microscope (**toolmakers' microscope**). However, this requires that the cutting operation be stopped for tool observation. Another direct method involves programming the tool to contact a sensor after every machining cycle; this allows the detection of broken tools. Usually, the sensor has the appearance of a pin that must be depressed by the tool tip.

Indirect methods of observing tool conditions involve the correlation of the tool condition with parameters such as cutting forces, power, temperature rise, workpiece surface finish, vibration, and chatter. A powerful technique is **acoustic emission technique** (AE), which utilizes a piezoelectric transducer mounted on a tool holder. The transducer picks up acoustic emissions (typically above 100 kHz), which result from the stress waves generated during cutting. By analyzing the signals, tool wear and chipping can be monitored. This technique is effective particularly in precision-machining operations where cutting forces are low because of the small amounts of material removed. Another effective use of AE is in detecting the fracture of small carbide tools at high cutting speeds.

A similar indirect tool-condition monitoring system consists of **transducers** that are installed in original machine tools or are retrofitted on existing machines. They continually monitor torque and forces during cutting. The signals are preamplified, and a microprocessor analyzes and interprets their content. The system is capable of differentiating the signals that come from different sources, such as tool breakage, tool wear, a missing tool, overloading of the machine tool, or colliding with machine components. The system also can compensate automatically for tool wear and, thus, improve dimensional accuracy of the part being machined.

The design of transducers must be such that they are: (a) nonintrusive to the machining operation, (b) accurate and repeatable in signal detection, (c) resistant to abuse and shop-floor environment, and (d) cost effective. Continued progress is being made in the development of **sensors**, including the use of *infrared* and *fiber-optic techniques* for temperature measurement during machining.

In lower-cost computer numerical-control machine tools, monitoring is done by **tool-cycle time**. In a production environment, once the life expectancy of a cutting tool or insert has been determined, it can be entered in the machine control unit, so that the operator is prompted to make a tool or cutter change when this time is reached. This process is inexpensive and fairly reliable, although not totally so because of the inherent statistical variation in tool life.

21.6 | Surface Finish and Integrity

Surface finish influences not only the dimensional accuracy of machined parts but also their properties and their performance in service. The term *surface finish* describes the geometric features of a surface (see Chapter 33), and *surface integrity* pertains to material properties, such as fatigue life and corrosion resistance, which are strongly influenced by the nature of the surface produced.

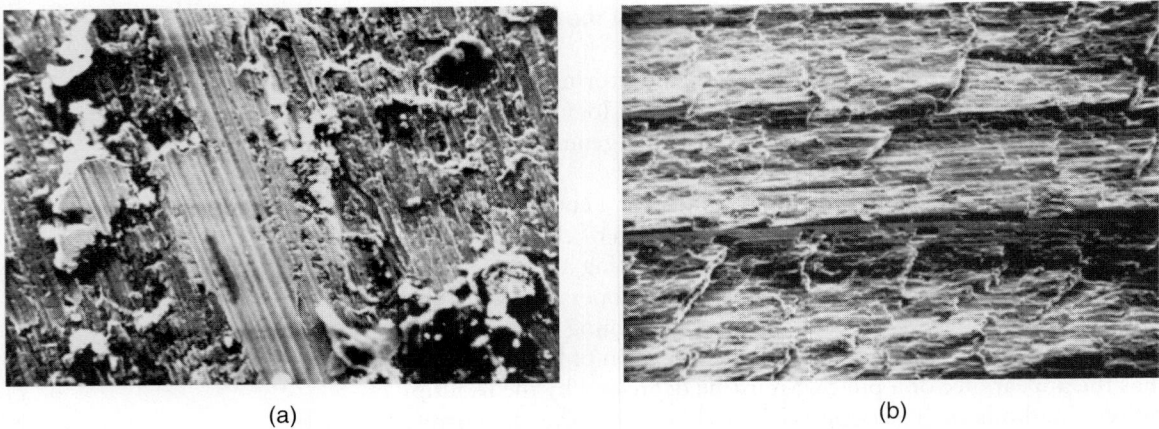

(a)   (b)

FIGURE 21.21   Machined surfaces produced on steel (highly magnified), as observed with a scanning electron microscope: (a) turned surface and (b) surface produced by shaping. *Source*: Courtesy of J. T. Black and S. Ramalingam.

With its significant effect on the tool-tip profile, the *built-up edge* has the greatest influence on surface finish. Figure 21.21 shows the surfaces obtained in two different cutting operations. Note the considerable damage to the surfaces from BUE; its damage is manifested in the scuffing marks, which deviate from the straight grooves that would result from normal machining, as seen in Fig. 21.2. Ceramic and diamond tools generally produce better surface finish than other tools, largely because of their much lower tendency to form a BUE.

A *dull* tool has a large radius along its edges, just as the tip of a dull pencil or the cutting edge of a knife. Figure 21.22 illustrates the relationship between the radius of the cutting edge and the depth of the cut in orthogonal cutting. Note that at small depths of cut, the rake angle effectively can become negative, and the tool simply

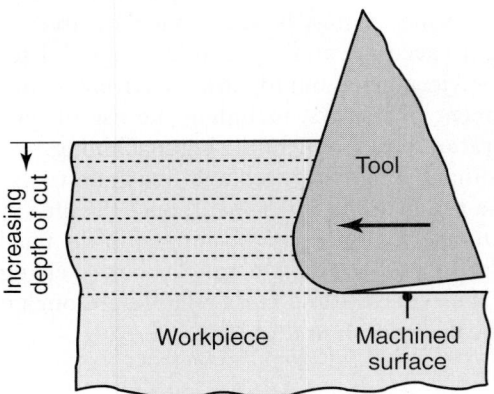

FIGURE 21.22   Schematic illustration of a dull tool with respect to the depth of cut in orthogonal machining (exaggerated). Note that the tool has a positive rake angle, but as the depth of cut decreases, the rake angle effectively can become negative. The tool then simply rides over the workpiece (without cutting) and burnishes its surface; this action raises the workpiece temperature and causes surface residual stresses.

may ride over the workpiece surface instead of cutting it and producing chips. This is a phenomenon similar to trying to scrape a thin layer from the surface of a stick of butter with a dull knife.

If the *tip radius* of the tool (not to be confused with the radius R in Fig. 21.15b) is large in relation to the depth of cut, the tool simply will rub over the machined surface. Rubbing will generate heat and induce residual surface stresses, which in turn may cause surface damage, such as tearing and cracking. Consequently, the depth of cut should be greater than the radius on the cutting edge.

In a turning operation, as in other cutting processes, the tool leaves a spiral profile (**feed marks**) on the machined surface as it moves across the workpiece, as shown in Figs. 21.2 and 21.23. We can see that the higher the feed, f, and the smaller the tool-nose radius, R, the more prominent these marks will be. It can be shown that the surface roughness for such a case is given by

$$R_a = \frac{f^2}{8R} \qquad (21.24)$$

where R_a is the arithmetic mean value, as described in Section 33.3. Although not significant in rough machining operations, feed marks are important in finish machining. Further details on surface roughness are given for individual machining processes as they are discussed.

Vibration and chatter are described in detail in Section 25.4. For now, it should be recognized that, if the tool vibrates or chatters during cutting, it will affect the workpiece surface finish adversely. The reason is that a vibrating tool periodically changes the dimensions of the cut. Excessive chatter also can cause chipping and premature failure of the more brittle cutting tools, such as ceramics and diamond.

Factors influencing *surface integrity* are:

- Temperatures generated during processing and possible metallurgical transformations.
- Surface residual stresses.
- Severe plastic deformation and strain hardening of the machined surfaces, tearing, and cracking.

Each of these factors can have major adverse effects on the machined part but can be taken care of by careful selection and maintenance of cutting tools and control of process variables.

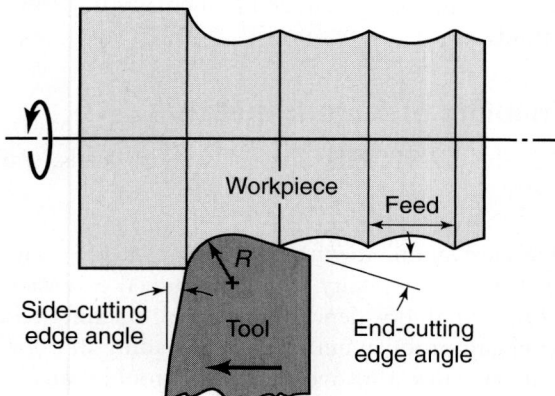

FIGURE 21.23 Schematic illustration of feed marks on a surface being turned (exaggerated).

The difference between *finish machining* and *rough machining* should be emphasized. In finish machining, it is important to consider the surface finish to be produced; whereas, in rough machining, the main purpose is to remove a large amount of material at a high rate. Surface finish is not a primary consideration, since it will be improved during finish machining. Of course, it is important that there be no subsurface-damage results from rough machining that cannot be removed during finish machining (see Fig. 21.21).

21.7 | Machinability

The *machinability* of a material is usually defined in terms of four factors:

1. Surface finish and surface integrity of the machined part.
2. Tool life.
3. Force and power required.
4. The level of difficulty in chip control.

Thus, good machinability indicates good surface finish and surface integrity, long tool life, and low force and power requirements. As for chip control and as stated earlier regarding continuous chips, long, thin, stringy, and curled chips can interfere severely with the cutting operation by becoming entangled in the cutting zone.

Because of the complex nature of cutting operations, it is difficult to establish relationships that quantitatively define the machinability of a particular material. In machining practice, tool life and surface roughness generally are considered to be the most important factors in machinability. Although not used much any more due to their qualitative and misleading nature approximate **machinability ratings (index)** have been available for many years for each type of material and its condition.

In these ratings, the standard material is AISI 1112 steel (resulfurized), with a rating of 100. This means that for a tool life of 60 min, this steel should be machined at a cutting speed of 100 ft/min (30 m/min). Examples of typical ratings are: 3140 steel at 55; free-cutting brass at 300; 2011 wrought aluminum at 200; pearlitic gray iron at 70; and precipitation-hardening 17-7 steel at 20.

These qualitative aspects of machinability are not sufficient (of course) to guide a machine operator in determining the machining parameters in order to produce an acceptable part economically. Hence, in subsequent chapters, we present several tables where, for various groups of materials, *specific recommendations* are given regarding such parameters as cutting speed, feed, depth of cut, cutting tools and their shape, and type of cutting fluids.

21.7.1 Machinability of ferrous metals

This section describes the machinability of steels, alloy steels, stainless steels, and cast irons.

Steels. Because steels are among the most important engineering materials (as also noted in Chapter 5) their machinability has been studied extensively. Carbon steels have a wide range of machinability, depending on their ductility and hardness. If too ductile, chip formation can produce built-up edge, leading to poor surface finish; if the steel is too hard, it can cause abrasive wear of the tool because of the presence of carbides in the steel. Cold-worked carbon steels are desirable from a machinability standpoint.

An important group of steels is **free-machining steels**, containing sulfur and phosphorus. *Sulfur* forms manganese-sulfide inclusions (second-phase particles), which act as stress raisers in the primary shear zone. As a result, the chips produced break up easily and are small, thus improving machinability. The size, shape, distribution, and concentration of these inclusions significantly influence machinability. Elements such as *tellurium* and *selenium*, both of which are similar to sulfur chemically, act as *inclusion modifiers* in resulfurized steels.

Phosphorus in steels has two major effects: (a) it strengthens the ferrite, causing increased hardness and resulting in better chip formation and surface finish, and (b) it increases hardness and thus causes the formation of short chips instead of continuous stringy ones, thereby improving machinability. Note that soft steels can be difficult to machine because of their tendency for built-up edge formation and the resulting poor surface finish.

In **leaded steels,** a high percentage of lead solidifies at the tips of manganese-sulfide inclusions. In non-resulfurized grades of steel, lead takes the form of dispersed fine particles. Lead is insoluble in iron, copper, and aluminum and their alloys. Because of its low shear strength, lead acts as a *solid* lubricant, and is smeared over the tool–chip interface during cutting.

When the temperature developed is sufficiently high, such as at high cutting speeds and feeds, the lead melts directly in front of the tool, acting as a *liquid* lubricant. In addition to this effect, lead lowers the shear stress in the primary shear zone, thus reducing cutting forces and power consumption. Lead can be used with every grade of steel and is identified by the letter "L" between the second and third numerals in steel identification (for example, 10L45). (Note that in stainless steels, a similar use of the letter *L* means "low carbon," a condition that improves their corrosion resistance.)

However, because lead is a well-known *toxin* and a pollutant, there are serious environmental concerns about its use in steels (estimated at 4500 tons of lead consumption every year in the production of steels). Consequently, there is a continuing trend toward eliminating the use of lead in steels (**lead-free steels**). *Bismuth* and *tin* are substitutes for lead in steels, although their performance is not as high.

Calcium-deoxidized steels contain oxide flakes of calcium silicates (CaSO) formed as flakes which, in turn, reduce the strength of the secondary shear zone and decrease tool–chip interface friction and wear. Temperature rise also is reduced correspondingly. Consequently, these steels produce less crater wear, especially at high cutting speeds.

Alloy steels can have a wide variety of compositions and hardnesses. Consequently, their machinability cannot be generalized, although they have higher levels of hardness and other mechanical properties. An important trend in machining these steels is *hard turning*, as described in detail in Section 25.6. Using polycrystalline cubic-boron-nitride cutting tools, alloy steels at hardness levels of 45 to 65 HRC can be machined with good surface finish, integrity, and dimensional accuracy.

Effects of various elements in steels. The presence of *aluminum* and *silicon* in steels is always harmful, because these elements combine with oxygen to form aluminum oxide and silicates, which are hard and abrasive. As a result, tool wear increases and machinability is reduced.

Carbon and *manganese* have various effects on the machinability of steels, depending on their composition. Plain low-carbon steels (less than 0.15% C) can produce poor surface finish by forming a built-up edge. Cast steels are more abrasive, although their machinability is similar to that of wrought steels. Tool and die steels are very difficult to machine and usually require annealing prior to machining.

Machinability of most steels is improved by cold working, which hardens the material and reduces the tendency for built-up edge formation.

Other alloying elements (such as *nickel, chromium, molybdenum,* and *vanadium*), which improve the properties of steels, generally reduce machinability. The effect of *boron* is negligible. Gaseous elements such as *hydrogen* and *nitrogen* can have particularly detrimental effects on the properties of steel. *Oxygen* has been shown to have a strong effect on the aspect ratio of the manganese-sulfide inclusions: the higher the oxygen content, the lower the aspect ratio, and the higher the machinability.

In improving the machinability of steels, however, it is important to consider the possible detrimental effects of the alloying elements on the properties and strength of the machined part in service. At elevated temperatures, for example, lead causes *embrittlement* of steels (liquid-metal embrittlement and hot shortness; see Section 1.4.2), although at room temperature it has no effect on mechanical properties.

Sulfur can reduce the hot workability of steels severely because of the formation of iron sulfide (unless sufficient manganese is present to prevent such formation). At room temperature, the mechanical properties of resulfurized steels depend on the orientation of the deformed manganese-sulfide inclusions (anisotropy). Rephosphorized steels are significantly less ductile and are produced solely to improve machinability.

Stainless steels.   Austenitic (300 series) steels generally are difficult to machine. Chatter can be a problem, necessitating the need for machine tools with high stiffness. Ferritic stainless steels (also 300 series) have good machinability. Martensitic (400 series) steels are abrasive, tend to form a built-up edge, and require tool materials with high hot hardness and crater-wear resistance. *Precipitation-hardening stainless steels* are strong and abrasive, thus requiring hard and abrasion-resistant tool materials.

Cast irons.   *Gray irons* generally are machinable, but they can be abrasive depending on composition, especially pearlite. Free carbides in castings reduce their machinability and cause tool chipping or fracture. *Nodular* and *malleable irons* are machinable using hard tool materials.

21.7.2 Machinability of nonferrous metals

A summary of the machinability of nonferrous metals and alloys is given next, in alphabetic order.

Aluminum is generally very easy to machine, although the softer grades tend to form a built-up edge, resulting in poor surface finish. Thus, high cutting speeds, high rake angles, and high relief angles are recommended. Wrought aluminum alloys with high *silicon* content and cast aluminum alloys are generally abrasive, hence they require harder tool materials. Dimensional tolerance control may be a problem in machining aluminum, because it has a high thermal expansion coefficient and a relatively low elastic modulus.

Beryllium generally is machinable, but because the fine particles produced during machining are toxic, it requires machining in a controlled environment.

Cobalt-based alloys are abrasive and highly work-hardening. They require sharp, abrasion-resistant tool materials and low feeds and speeds.

Copper in the wrought condition can be difficult to machine because of built-up edge formation, although cast copper alloys are easy to machine. Brasses are easy to machine, especially with the addition of lead (*leaded free-machining brass*). Note, however, the toxicity of lead. Bronzes are more difficult to machine than brass.

Magnesium is very easy to machine, with good surface finish and prolonged tool life. However, care should be exercised because of its high rate of oxidation (*pyrophoric*) and the danger of fire.

Molybdenum is ductile and work-hardening. It can produce poor surface finish, thus sharp tools are essential.

Nickel-based alloys and superalloys are work-hardening, abrasive, and strong at high temperatures. Their machinability depends on their condition and improves with annealing.

Tantalum is very work-hardening, ductile, and soft. It produces a poor surface finish, and tool wear is high.

Titanium and its alloys have very poor thermal conductivity (the lowest of all metals), causing significant temperature rise and built-up edge. They are highly reactive and can be difficult to machine.

Tungsten is brittle, strong, and very abrasive, hence its machinability is low, although it greatly improves at elevated temperatures.

Zirconium has good machinability, but it requires a coolant-type cutting fluid because of the danger of explosion and fire.

21.7.3 Machinability of miscellaneous materials

Thermoplastics generally have a low thermal conductivity, a low elastic modulus, and are thermally softening. Consequently, machining them requires sharp tools with positive rake angles (to reduce cutting forces), large relief angles, small depths of cut and feed, relatively high speeds, and proper support of the workpiece. External cooling of the cutting zone may be necessary to keep the chips from becoming gummy and sticking to the tools. Cooling usually can be achieved with a jet of air, vapor mist, or water-soluble oils.

Thermosetting plastics are brittle and sensitive to thermal gradients during cutting; machining conditions generally are similar to those of thermoplastics.

Polymer-matrix composites are very abrasive because of the fibers present, hence they are difficult to machine. Fiber tearing, pulling, and edge delamination are significant problems and can lead to severe reduction in the load-carrying capacity of the machined component. Machining of these materials requires careful handling and removal of debris to avoid contact with and inhaling of the fibers.

Metal-matrix and ceramic-matrix composites can be difficult to machine, depending on the properties of the matrix material and the reinforcing fibers.

Graphite is abrasive; it requires sharp, hard, and abrasion-resistant tools.

Ceramics have a steadily improved machinability, particularly with the development of *machinable ceramics* and *nanoceramics* (Section 8.2.5) and with the selection of appropriate processing parameters, such as *ductile-regime cutting* (described in Section 25.7).

Wood is an orthotropic material with properties varying with respect to its grain direction. Consequently, the type of chips and the surfaces produced vary significantly, depending also on the type of wood and its condition. Woodworking, which dates back to 3000 B.C. remains largely an art. Two basic requirements are generally sharp tools and high cutting speeds.

21.7.4 Thermally assisted machining

Metals and alloys that are difficult to machine at room temperature can be machined more easily at elevated temperatures. In *thermally assisted machining* (also called **hot machining**), a source of heat (such as a torch, induction coil, electric current, laser-beam, electron-beam, and plasma arc) is focused onto an area just ahead of the cutting tool. First investigated in the early 1940s, this operation typically is carried out above the homologous temperature of $T/T_m = 0.5$ (see Section 1.7, and Tables 1.2 and 3.1). Thus, for example, steels are hot machined above a temperature range of 650° to 750° (1200° to 1400°F). Although difficult and complicated to perform in production plants, the general advantages of hot machining are: (a) reduced cutting forces, (b) increased tool life, (c) higher material-removal rates, and (d) reduced tendency for vibration and chatter.

SUMMARY

- Machining processes are often necessary in order to impart the desired dimensional accuracy, geometric features, and surface-finish characteristics to components, particularly those with complex shapes that cannot be produced economically or with other shaping techniques. On the other hand, these processes generally take more time, waste some material in the form of chips, and may have adverse effects on surfaces produced.

- Commonly observed chip types in machining are continuous, built-up edge, discontinuous, and serrated. Important process variables in machining are tool shape and tool material; cutting conditions such as speed, feed, and depth of cut; use of cutting fluids; and the characteristics of the workpiece material and the machine tool. Parameters influenced by these variables are forces and power consumption; tool wear; surface finish and surface integrity; temperature rise; and dimensional accuracy of the workpiece.

- Temperature rise is an important consideration, since it can have adverse effects on tool life as well as on the dimensional accuracy and surface integrity of the machined part.

- Two principal types of tool wear are flank wear and crater wear. Tool wear depends on workpiece and tool material characteristics; cutting speed, feed, depth-of-cut, and cutting fluids; and the characteristics of the machine tool. Tool failure also may occur by notching, chipping, and gross fracture.

- Surface finish of machined components can affect product integrity adversely. Important variables are the geometry and condition of the cutting tool, type of chip produced, and process variables.

- Machinability generally is defined in terms of surface finish, tool life, force and power requirements, and chip control. Machinability of materials depends on their composition, properties, and microstructure. Thus, proper selection and control of process variables are important

KEY TERMS

Acoustic emission	Chipping of tool	Depth-of-cut line
Allowable wear land	Clearance angle	Diffusion
Built-up edge	Continuous chip	Discontinuous chip
Chip	Crater wear	Feed marks
Chip breaker	Cutting force	Flank wear
Chip curl	Cutting ratio	Friction angle

Hot machining
Inclination angle
Machinability
Machinability ratings
Machine tool
Machining
Notch wear
Oblique cutting
Orthogonal cutting
Primary shear zone

Rake angle
Relief angle
Rephosphorized steel
Resulfurized steel
Secondary shear zone
Serrated chip
Shaving
Shear angle
Shear plane
Skiving

Specific energy
Surface finish
Surface integrity
Taylor equation
Thrust force
Tool-condition monitoring
Tool life
Turning
Wear land

BIBLIOGRAPHY

Astakhov, V.P., *Metal Cutting Mechanics*, CRC Press, 1998.
ASM Handbook, Vol. 16: *Machining*, ASM International, 1989.
Boothroyd, G., and Knight, W.A., *Fundamentals of Machining and Machine Tools*, 2nd ed., Marcel Dekker, 1989.
Brown, J., *Advanced Machining Technology Handbook*, McGraw-Hill, 1998.
Davis, J.R. (ed.), *Tool Materials*, ASM International, 1995.
DeVries, W.R., *Analysis of Material Removal Processes*, Springer, 1992.
Juneja, B.L., *Fundamentals of Metal Cutting and Machine Tools*, Wiley, 1987.
Kalpakjian, S. (ed.), *Tool and Die Failures: Source Book*, ASM International, 1982.
Metal Cutting Tool Handbook, 7th ed., Industrial Press, 1989.

Meyers, A.L., and Slattery, T., *Basic Machining Reference Handbook*, Industrial Press, 1988.
Oxley, P.L.B., *Mechanics of Machining: An Analytical Approach to Assessing Machinability*, Wiley, 1989.
Shaw, M.C., *Metal Cutting Principles*, 2nd ed., Oxford, 2005.
Stephenson, D.A., and Agapiou, J.S., *Metal Cutting Theory and Practice*, Marcel Dekker, 1997.
Stout, K.J., Davis, E.J., and Sullivan, P.J., *Atlas of Machined Surfaces*, Chapman and Hall, 1990.
Tool and Manufacturing Engineers Handbook, 4th ed., Vol. 1: *Machining*. Society of Manufacturing Engineers, 1983.
Trent, E.M., and Wright, P.K., *Metal Cutting*, 4th ed., Butterworth Heinemann, 2000.
Walsh, R.A., *McGraw-Hill Machining and Metalworking Handbook*, McGraw-Hill, 1994.

REVIEW QUESTIONS

21.1 Explain why continuous chips may not necessarily be desirable.

21.2 Name the factors that contribute to the formation of discontinuous chips.

21.3 Explain the difference between positive and negative rake angles. What is the importance of the rake angle?

21.4 Comment on the role and importance of the relief angle.

21.5 Explain the difference between discontinuous chips and segmented chips.

21.6 Why should we be interested in the magnitude of the thrust force in cutting?

21.7 What are the differences between orthogonal and oblique cutting?

21.8 Is there any advantage to having a built-up edge on a tool? Explain.

21.9 What is the function of chip breakers? How do they function?

21.10 Identify the forces involved in a cutting operation. Which of these forces contributes to the power required?

21.11 Explain the characteristics of different types of tool wear.

21.12 List the factors that contribute to poor surface finish in cutting

21.13 Explain what is meant by the term *machinability* and what it involves. Why does titanium have poor machinability?

QUALITATIVE PROBLEMS

21.14 Are the locations of maximum temperature and crater wear related? If so, explain why.

21.15 Is material ductility important for machinability? Explain.

21.16 Explain why studying the types of chips produced is important in understanding cutting operations.

21.17 Why do you think the maximum temperature in orthogonal cutting is located at about the middle of the tool–chip interface? (*Hint*: Note that the two sources of heat are (a) shearing in the primary shear plane and (b) friction at the tool–chip interface.)

21.18 Tool life can be almost infinite at low cutting speeds. Would you then recommend that all machining be done at low speeds? Explain.

21.19 Explain the consequences of allowing temperatures to rise to high levels in cutting.

21.20 The cutting force increases with the depth-of-cut and decreasing rake angle. Explain why.

21.21 Why is it not always advisable to increase cutting speed in order to increase production rate?

21.22 What are the consequences if a cutting tool chips?

21.23 What are the effects of performing a cutting operation with a dull tool? A very sharp tool?

21.24 To what factors do you attribute the difference in the specific energies when machining the materials shown in Table 21.2? Why is there a range of energies for each group of materials?

21.25 Explain why it is possible to remove more material between tool resharpenings by lowering the cutting speed.

21.26 Noting that the dimension d in Fig. 21.4a is very small, explain why the shear strain rate in metal cutting is so high.

21.27 Explain the significance of Eq. (21.7).

21.28 Comment on your observations regarding Figs. 21.12 and 21.13.

21.29 Describe the consequences of exceeding the allowable wear land (Table 21.4) for various cutting-tool materials.

21.30 Comment on your observations regarding the hardness variations shown in Fig. 21.6a.

21.31 Why does the temperature in cutting depend on the cutting speed, feed, and depth-of-cut? Explain in terms of the relevant process variables.

21.32 You will note that the values of a and b in Eq. (21.19b) are higher for high-speed steels than for carbides. Why is this so?

21.33 As shown in Fig. 21.14, the percentage of the total cutting energy carried away by the chip increases with cutting speed. Why?

21.34 Describe the effects that a dull tool can have on cutting operations.

21.35 Explain whether it is desirable to have a high or low (a) n value and (b) C value in the Taylor tool-life equation.

21.36 The tool-life curve for ceramic tools in Fig. 21.17 is to the right of those for other tool materials. Why?

21.37 Why are tool temperatures low at low cutting speeds and high at high cutting speeds?

21.38 Can high-speed machining be performed without the use of a cutting fluid?

21.39 Given your understanding of the basic metal-cutting process, what are the important physical and chemical properties of a cutting tool?

QUANTITATIVE PROBLEMS

21.40 Let $n = 0.5$ and $C = 300$ in the Taylor equation for tool wear. What is the percent increase in tool life if the cutting speed is reduced by (a) 30% and (b) 50%?

21.41 Assume that, in orthogonal cutting, the rake angle is 15° and the coefficient of friction is 0.2. Using Eq. (21.3), determine the percentage increase in chip thickness when the friction is doubled.

21.42 Derive Eq. (21.11).

21.43 Taking carbide as an example and using Eq. (21.19b), determine how much the feed should be reduced in order to keep the mean temperature constant when the cutting speed is tripled.

21.44 Using trigonometric relationships, derive an expression for the ratio of shear energy to frictional energy in orthogonal cutting, in terms of angles α, β, and ϕ only.

21.45 An orthogonal cutting operation is being carried out under the following conditions: $t_o = 0.1$ mm, $t_c = 0.2$ mm, width of cut = 5 mm, $V = 2$ m/s, rake angle = 10°, $F_c = 500$ N, and $F_t = 200$ N. Calculate the percentage of the total energy that is dissipated in the shear plane.

21.46 Explain how you would go about estimating the C and n values for the four tool materials shown in Fig. 21.17.

21.47 Derive Eqs. (21.1).

21.48 Assume that, in orthogonal cutting, the rake angle, α, is 25° and the friction angle, β, is 30° at the chip–tool interface. Determine the percentage change in chip thickness when the friction angle is 50°. (*Note:* do not use Eqs. (21.3) or (21.4)).

21.49 Show that, for the same shear angle, there are two rake angles that give the same cutting ratio.

21.50 With appropriate diagrams, show how the use of a cutting fluid can change the magnitude of the thrust force, F_t, in Fig. 21.11.

21.51 For a turning operation using a ceramic cutting tool, if the speed is increased by 50%, by what factor must the feed rate be modified to obtain a constant tool life? Use $n = 0.5$ and $y = 0.6$.

21.52 In Example 21.3, if the cutting speed, V, is doubled, will the answer be different? Explain.

21.53 Using Eq. (21.24) select an appropriate feed for $R = 1$ mm and a desired roughness of 1 μm. How would you adjust this feed to allow for nose wear of the tool during extended cuts? Explain your reasoning.

21.54 Using a carbide cutting tool, the temperature in a cutting operation with a speed of 300 ft/min and feed of 0.002 in./rev is measured to be 1200°F. What is the approximate temperature if the speed is doubled? What speed is required to lower the maximum cutting temperature to 900°F?

21.55 Assume that you are an instructor covering the topics described in this chapter, and you are giving a quiz on the numerical aspects to test the understanding of the students. Prepare two quantitative problems and supply the answers.

SYNTHESIS, DESIGN, AND PROJECTS

21.56 As we have seen, chips carry away the majority of the heat generated during machining. If chips did not have this capacity, what suggestions would you make in order to be able to carry out machining processes without excessive heat? Explain.

21.57 Tool life is increased greatly when an effective means of cooling and lubrication is implemented. Design methods of delivering this fluid to the cutting zone and discuss the advantages and limitations of your design.

21.58 Design an experimental setup whereby orthogonal cutting can be simulated in a turning operation on a lathe.

21.59 Describe your thoughts on whether chips produced during machining can be used to make useful products. Give some examples of possible products and comment on their characteristics and differences if the

same products were made by other manufacturing processes. Which types of chips would be desirable for this purpose?

21.60 We have stated that cutting tools can be designed so that the tool–chip contact length is reduced by recessing the rake face of the tool some distance away from its tip. Explain the possible advantages of such a tool.

21.61 We have stated that the chip formation mechanism also can be observed by scraping the surface of a stick of butter with a sharp knife. Using butter at different temperatures, including frozen, conduct such an experiment. Keep the depth of cut constant and hold the knife at different angles (to simulate the tool rake angle), including oblique scraping. Describe your observations regarding the type of chips produced. Also comment on the force that your hand feels while scraping and whether you observe any chatter when the butter is very cold.

21.62 Experiments have shown that it is possible to produce thin, wide chips, such as 0.08 mm (0.003 in.) thick and 10 mm (4 in.) wide, which would be similar to a rolled sheet. Materials have been aluminum, magnesium, and stainless steel. A typical setup would be similar to orthogonal cutting, by machining the periphery of a solid round bar with a straight tool moving radially inward. Describe your thoughts on producing thin metal sheet by this method, its surface characteristics, and its properties.

21.63 Describe your thoughts on the recycling of chips produced during machining in a plant. Include considerations regarding chips produced by dry cutting versus those produced by machining with a cutting fluid.

Cutting-Tool Materials and Cutting Fluids

Continuing our coverage of the fundamentals of cutting in the preceding chapter, we now describe two essential elements in machining: cutting-tool materials and cutting fluids. Specifically:

- Types and characteristics of cutting-tool materials.
- Properties and applications of high-speed steels, carbides, ceramics, cubic boron nitride, and diamond.
- Coatings on tools, their composition, and how they function.
- Types of cutting fluids and their applications.
- Trends in near-dry and dry machining.

22.1 | Introduction

The selection of cutting-tool materials for a particular application is among the most important factors in machining operations, as is the selection of mold and die materials for forming and shaping processes. We will discuss throughout this chapter the relevant properties and performance characteristics of all major types of cutting-tool materials, which will help us in tool selection. However, as it will become apparent, the complex nature of this subject does not always render itself to the determination of appropriate tool materials, hence we also must rely on guidelines and *recommendations* that have been accumulated in industry over many years. Beginning with Chapter 23, information on tool material recommendations for specific workpiece materials and machining operations will be presented.

As noted in the preceding chapter, the cutting tool is subjected to (a) high temperatures, (b) high contact stresses, and (c) rubbing along the tool–chip interface and along the machined surface. Consequently, the cutting-tool material must possess the following characteristics:

- **Hot hardness:** so that the hardness, strength, and wear resistance of the tool are maintained at the temperatures encountered in machining operations. This ensures that the tool does not undergo any plastic deformation and, thus, retains its shape and sharpness. Tool-material hardness as a function of temperature is shown in Fig. 22.1. Note the wide response of these materials and (not surprisingly) how well ceramics maintain their hardness at high temperatures.

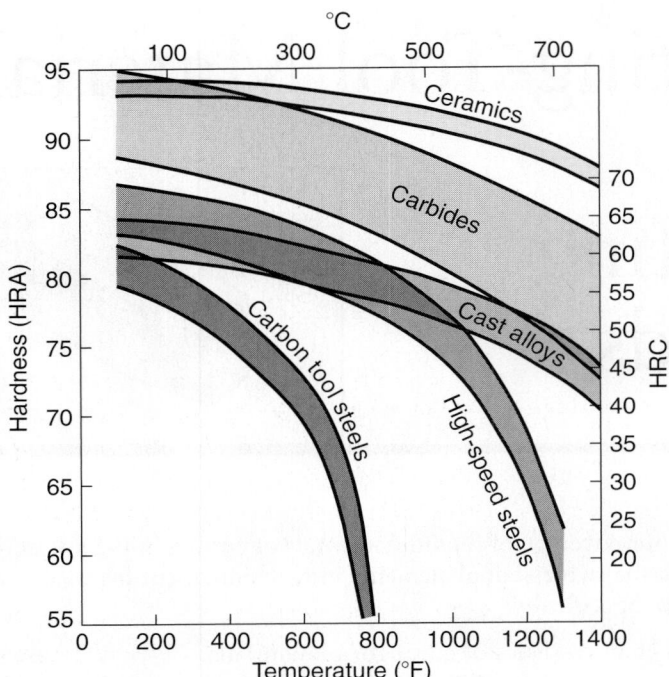

FIGURE 22.1 The hardness of various cutting-tool materials as a function of temperature (hot hardness). The wide range in each group of materials is due to the variety of tool compositions and treatments available for that group.

Carbon tool steels commonly were used as tool materials until the development of high-speed steels in the early 1900s. Note that carbon tool steels rapidly begin to lose their hardness even at moderate temperatures, meaning that they cannot be used for machining at high speeds (thus high temperatures). Consequently, production rate will be low and costs will be high.

- **Toughness** and **impact strength** (mechanical shock): so that impact forces on the tool encountered repeatedly in interrupted cutting operations (such as milling, turning a splined shaft on a lathe, or due to vibration and chatter during machining) do not chip or fracture the tool.

- **Thermal shock resistance:** to withstand the rapid temperature cycling encountered in interrupted cutting.

- **Wear resistance:** so that an acceptable tool life is obtained before the tool has to be replaced.

- **Chemical stability** and **inertness:** with respect to the material being machined, to avoid or minimize any adverse reactions, adhesion, and tool–chip diffusion that would contribute to tool wear.

To respond to these demanding requirements, various cutting-tool materials with a wide range of mechanical, physical, and chemical properties have been developed over the years, as shown in Table 22.1. The properties listed in the first column

TABLE 22.1

General Properties of Tool Materials

Property	High-speed steels	Cast-cobalt alloys	Carbides		Ceramics	Cubic boron nitride	Single-crystal diamond *
			WC	TiC			
Hardness	83–86 HRA	82–84 HRA 46–62 HRC	90–95 HRA 1800–2400 HK	91–93 HRA 1800–3200 HK	91–95 HRA 2000–3000 HK	4000–5000 HK	7000–8000 HK
Compressive strength, MPa psi $\times 10^3$	4100–4500 600–650	1500–2300 220–335	4100–5850 600–850	3100–3850 450–560	2750–4500 400–650	6900 1000	6900 1000
Transverse rupture strength, MPa psi $\times 10^3$	2400–4800 350–700	1380–2050 200–300	1050–2600 150–375	1380–1900 200–275	345–950 50–135	700 105	1350 200
Impact strength, J in.-lb	1.35–8 12–70	0.34–1.25 3–11	0.34–1.35 3–12	0.79–1.24 7–11	<0.1 <1	<0.5 <5	<0.2 <2
Modulus of elasticity, GPa psi $\times 10^6$	200 30	— —	520–690 75–100	310–450 45–65	310–410 45–60	850 125	820–1050 120–150
Density, kg/m³ lb/in³	8600 0.31	8000–8700 0.29–0.31	10,000–15,000 0.36–0.54	5500–5800 0.2–0.22	4000–4500 0.14–0.16	3500 0.13	3500 0.13
Volume of hard phase, %	7–15	10–20	70–90	—	100	95	95
Melting or decomposition temperature, °C °F	1300 2370	— —	1400 2550	1400 2550	2000 3600	1300 2400	700 1300
Thermal conductivity, W/m K	30–50	—	42–125	17	29	13	500–2000
Coefficient of thermal expansion, $\times 10^{-6}$/°C	12	—	4–6.5	7.5–9	6–8.5	4.8	1.5–4.8

*The values for polycrystalline diamond are generally lower, except impact strength, which is higher.

are useful in determining desirable tool-material characteristics for a particular application. Thus, for example:

- Hardness and strength are important with regard to the hardness and strength of the workpiece material to be machined.
- Impact strength is important in making interrupted cuts in machining, such as milling.
- Melting temperature of the tool material is important versus the temperatures developed in the cutting zone.
- The physical properties of thermal conductivy and coefficient of thermal expansion are important in determining the resistance of the tool materials to thermal fatigue and shock.

It must be recognized, however, that the desirable tool properties in a particular machining operation can appear contradictory with regard to the properties of the cutting tool itself. This situation readily can be seen from Table 22.2 by noting the opposite directions of the long horizontal arrows. As examples: (a) high-speed steels are tough, but they have limited hot hardness; and (b) ceramics have high resistance to temperature and wear, but they are brittle and can chip. Note also that the cost of tools increases as we move to the right in Table 22.2—from HSS to diamond (which is the most expensive).

Table 22.3 shows the operating characteristics of tool materials in machining. Tool materials generally are divided into the following categories, listed in the order in which they were developed and implemented in industry. Note that many of these

TABLE 22.2

General Characteristics of Cutting-Tool Materials (These Tool Materials Have a Wide Range of Compositions and Properties. Overlapping Characteristics Exist in Many Categories of Tool Materials.)

	High-speed steels	Cast-cobalt alloys	Uncoated carbides	Coated carbides	Ceramics	Polycrystalline cubic boron nitride	Diamond
Hot hardness	→						
Toughness	←						
Impact strength	←						
Wear resistance	→						
Chipping resistance	←						
Cutting speed	→						
Thermal-shock resistance	←						
Tool material cost	→						
Depth of cut	Light to heavy	Light to heavy	Light to heavy	Light to heavy	Light to heavy	Light to heavy	Very light for single-crystal diamond
Processing method	Wrought, cast, HIP* sintering	Cast and HIP sintering	Cold pressing and sintering	CVD or PVD†	Cold pressing and sintering or HIP sintering	High-pressure, high-temperature sintering	High-pressure, high-temperature sintering

Source: After R. Komanduri.
*Hot-isostatic pressing.
†Chemical-vapor deposition, physical-vapor deposition.

TABLE 22.3

General Operating Characteristics of Cutting-Tool Materials			
Tool materials	General characteristics	Modes of tool wear or failure	Limitations
High-speed steels	High toughness, resistance to fracture, wide range of roughing and finishing cuts, good for interrupted cuts	Flank wear, crater wear	Low hot hardness, limited hardenability, and limited wear resistance
Uncoated carbides	High hardness over a wide range of temperatures, toughness, wear resistance, versatile and wide range of applications	Flank wear, crater wear	Cannot use at low speeds because of cold welding of chips and microchipping
Coated carbides	Improved wear resistance over uncoated carbides, better frictional and thermal properties	Flank wear, crater wear	Cannot use at low speeds because of cold welding of chips and microchipping
Ceramics	High hardness at elevated temperatures, high abrasive wear resistance	Depth-of-cut line notching, microchipping, gross fracture	Low strength, and low thermo-mechanical fatigue strength
Polycrystalline cubic boron nitride (cBN)	High hot hardness, toughness, cutting-edge strength	Depth-of-cut line notching, chipping, oxidation, graphitization	Low strength, and low chemical stability at higher temperature
Diamond	High hardness and toughness, abrasive wear resistance	Chipping, oxidation, graphitization	Low strength, and low chemical stability at higher temperatures

Source: After R. Komanduri and other sources.

materials also are used for dies and molds in casting, forming, and shaping metallic and nonmetallic materials.

1. High-speed steels
2. Cast-cobalt alloys
3. Carbides
4. Coated tools
5. Alumina-based ceramics
6. Cubic boron nitride
7. Silicon-nitride-based ceramics
8. Diamond
9. Whisker-reinforced materials and nanomaterials

Carbon steels are the oldest of tool materials and have been used widely for drills, taps, broaches, and reamers since the 1880s. Low-alloy and medium-alloy steels were developed later for similar applications but with longer tool life. Although inexpensive and easily shaped and sharpened, these steels do not have sufficient hot hardness and wear resistance for cutting at high speeds when the temperature rises significantly. Their use is limited to very low-speed cutting operations, hence they are not of any particular significance in modern manufacturing.

In this chapter the following are described:

- The characteristics, applications, and limitations of these tool materials in machining operations, including the required characteristics we outlined and including costs.

- The applicable range of process variables for optimal performance.
- The types and characteristics of cutting fluids and their specific applications in a wide variety of machining operations.

22.2 | High-Speed Steels

High-speed steel (HSS) tools are so named because they were developed to machine at higher speeds than was previously possible. First produced in the early 1900s, *high-speed steels* are the most highly alloyed of the tool steels. They can be hardened to various depths, have good wear resistance, and are relatively inexpensive. Because of their toughness (hence high resistance to fracture), high-speed steels are suitable especially for (a) high, positive rake-angle tools (i.e., those with small included angles), (b) interrupted cuts, (c) machine tools with low stiffness that are subject to vibration and chatter, and (d) complex and single-piece tools, such as drills, reamers, taps, and gear cutters. Their most important limitation (due to their lower hot hardness), is that cutting speeds are low, compared to those of carbide tools as can be seen in Fig. 22.1.

There are two basic types of high-speed steels: **molybdenum** (*M-series*) and **tungsten** (*T-series*). The M-series contains up to about 10% Mo with Cr, V, W, and Co as alloying elements. The T-series contains 12 to 18% W with Cr, V, and Co as alloying elements. Carbides formed in the steel constiute about 10 to 20% by volume. The M-series generally has higher abrasion resistance than the T-series, undergoes less distortion during heat treating, and is less expensive. Consequently, 95% of all high-speed steel tools are made of the M-series. Table 5.6 includes three of these steels and their characteristics.

High-speed-steel tools are available in wrought (rolled or forged), cast, and powder-metallurgy (sintered) forms. They can be **coated** for improved performance, as described in Section 22.5. High-speed-steel tools also may be subjected to *surface treatments* (such as case hardening for improved hardness and wear resistance, see Section 4.10) or steam treatment at elevated temperatures to develop a hard, black oxide layer (*bluing*) for improved performance, including a reduced tendency for built-up edge formation.

EXAMPLE 22.1 Role of alloying elements in high-speed-steel cutting tools

List the major alloying elements in high-speed steels and describe their effects in cutting tools.

Solution   The major alloying elements in HSS are chromium, vanadium, tungsten, cobalt, and molybdenum.
To appreciate their role in cutting tools, refer to Section 5.5.1 on the effects of various elements in steels and note the following:

> *Chromium*: improves toughness, wear resistance, and high-temperature strength.
> *Vanadium*: improves toughness, abrasion resistance, and hot hardness.
> *Tungsten* and *cobalt* have similar effects: improved strength and hot hardness.
> *Molybdenum*: improves wear resistance, toughness, and high-temperature strength and hardness.

22.3 | Cast-Cobalt Alloys

Introduced in 1915, *cast-cobalt alloys* have the following composition ranges: 38 to 53% Co, 30 to 33% Cr, and 10 to 20% W. Because of their high hardness (typically 58 to 64 HRC), they have good wear resistance and can maintain their hardness at elevated temperatures. They are not as tough as high-speed steels and are sensitive to impact forces. Consequently, they are less suitable than high-speed steels for interrupted cutting operations. Commonly known as *Stellite* tools, these alloys are cast and ground into relatively simple shapes. They now are used only for special applications that involve deep, continuous **roughing cuts** at relatively high feeds and speeds—as much as twice the rates possible with high-speed steels. As described in Section 23.2, roughing cuts usually involve high feed rates and large depths of cut with the primary purpose of removing large amounts of material with little regard for surface finish. Conversely, **finishing cuts** are performed at a lower feeds and depths of cut, and the surface finish produced is a priority.

22.4 | Carbides

The two groups of tool materials just described possess the required toughness, impact strength, and thermal shock resistance, but they also have important limitations, particularly with respect to strength and hot hardness. Consequently, they cannot be used as effectively where high cutting speeds (hence high temperatures) are involved. However, this condition often is necessary to improve plant productivity.

To meet the challenge for increasingly higher cutting speeds, *carbides* (also known as *cemented* or *sintered carbides*) were introduced in the 1930s. Because of their high hardness over a wide range of temperatures (Fig. 22.1), high elastic modulus, high thermal conductivity, and low thermal expansion, carbides are among the most important, versatile, and cost-effective tool and die materials for a wide range of applications. The two major groups of carbides used for machining are *tungsten carbide* and *titanium carbide*. In order to differentiate them from the coated tools described in Section 22.6, plain-carbide tools usually are referred to as **uncoated carbides**.

22.4.1 Tungsten carbide

Tungsten carbide (WC) typically consists of tungsten-carbide particles bonded together in a cobalt matrix. These tools are manufactured using powder-metallurgy techniques (hence the term *sintered carbides* or *cemented carbides*), as described in Example 17.4. Tungsten-carbide particles are first combined with *cobalt* in a mixer, resulting in a composite material with a cobalt matrix surrounding the carbide particles. These particles, which are 1 to 5 μm (40 to 200 μin.) in size, are then pressed and sintered into the desired *insert* shapes. Tungsten carbides frequently are compounded with *titanium carbide* and *niobium carbide* to impart special properties to the material.

The amount of cobalt present, ranging typically from 6 to 16%, significantly affects the properties of tungsten-carbide tools. As the cobalt content increases, the strength, hardness, and wear resistance of WC decrease, while its toughness increases because of the higher toughness of cobalt. Tungsten-carbide tools generally are used for cutting steels, cast irons, and abrasive nonferrous materials and largely have replaced HSS tools because of their better performance.

Micrograin carbides.   Cutting tools also are made of submicron and ultra fine-grained (*micrograin*) carbides, including tungsten carbide, titanium carbide, and tantalum carbide. The grain size is in the range of 0.2 to 0.8 μm (8 to 30 μin.). Compared to the traditional carbides described previously, these tool materials are stronger, harder, and more wear resistant, thus improving productivity. In one application, micro-drills with diameters on the order of 100 μm (0.004 in.) are being made from these materials and used in the fabrication of microelectronic circuit boards (Chapter 28).

Functional gradient carbides.   In these tools, the composition of the carbide in the insert has a gradient through its near-surface depth—instead of being uniform, as is in common carbide inserts. The gradient has a smooth distribution of compositions and phases with functions similar to those described as the desirable properties of coatings on cutting tools. Graded mechanical properties eliminate stress concentrations and promote tool life and performance. However, they are more expensive and cannot be justified for all applications.

22.4.2 Titanium carbide

Titanium carbide (TiC) consists of a nickel-molybdenum matrix. It has higher wear resistance than tungsten carbide but is not as tough. Titanium carbide is suitable for machining hard materials (mainly steels and cast irons) and for cutting at speeds higher than those appropriate for tungsten carbide.

22.4.3 Inserts

High-speed steel tools are shaped in one piece and ground to impact various geometric features (Fig 21.10a); such tools include drill bits and milling and gear cutters. After the cutting edge wears, the tool has to be removed from its holder and reground.

   Although a supply of sharp or resharpened tools usually is available from tool rooms, tool-changing operations are time-consuming and inefficient. The need for a more effective method has led to the development of *inserts*, which are individual cutting tools with several cutting points (Fig. 22.2). A square insert has eight cutting points, and a triangular insert has six. Inserts usually are clamped on the *toolholder*

FIGURE 22.2   Typical carbide inserts with various shapes and chip-breaker features: Round inserts also are available, as can be seen in Figs. 22.3c and 22.4. The holes in the inserts are standardized for interchangeability in toolholders. *Source:* Courtesy of Kyocera Engineered Ceramics, Inc.

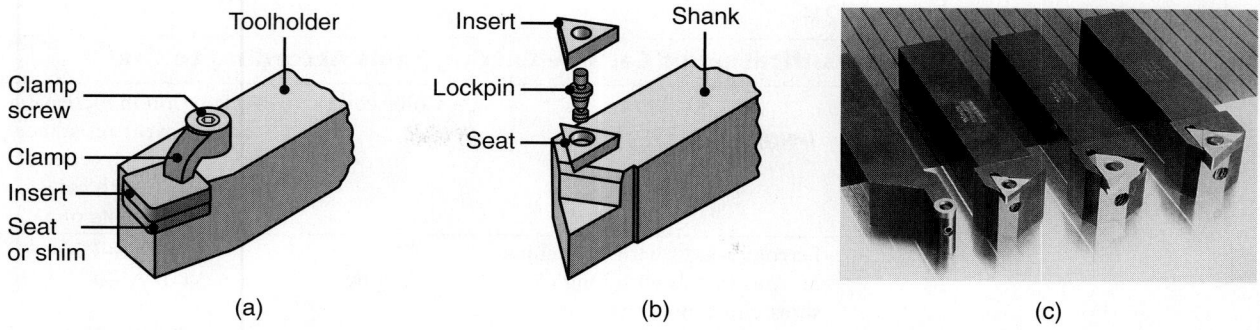

FIGURE 22.3   Methods of mounting inserts on toolholders: (a) clamping and (b) wing lockpins. (c) Examples of inserts mounted with threadless lockpins, which are secured with side screws. *Source*: Courtesy of Valenite.

with various locking mechanisms (Fig. 22.3). Although not used as commonly, inserts also may be *brazed* to the tool shank, but this practice largely has been abandoned.

Clamping is the preferred method of securing an insert because each insert has a number of cutting points, and after one edge is worn, it is **indexed** (rotated in its holder) to make available another cutting point. In addition to the examples in this figure, a wide variety of other toolholders is available for specific applications, including those with quick insertion and removal features.

Carbide inserts are available in a variety of shapes, such as square, triangle, diamond, and round. The strength of the cutting edge of an insert depends on its shape. The smaller the included angle (Fig. 22.4), the lower the strength of the edge. In order to further improve edge strength and prevent chipping, all insert edges usually are honed, chamfered, or produced with a negative land (Fig. 22.5). Most inserts are honed to a radius of about 0.025 mm (0.001 in.).

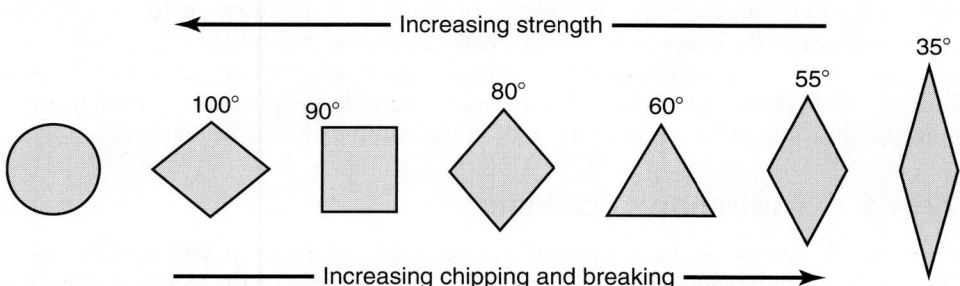

FIGURE 22.4   Relative edge strength and tendency for chipping of inserts with various shapes. Strength refers to the cutting edge indicated by the included angles. *Source*: Courtesy of Kennametal, Inc.

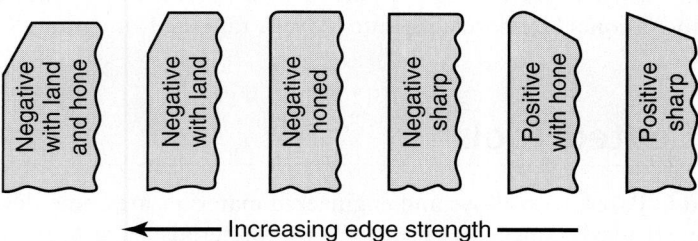

FIGURE 22.5   Edge preparation for inserts to improve edge strength. *Source*: Courtesy of Kennametal, Inc.

TABLE 22.4

ISO Classification of Carbide Cutting Tools According to Use			
Symbol	Workpiece material	Color code	Designation in increasing order of wear resistance and decreasing order of toughness in each category (in increments of 5)
P	Ferrous metals with long chips	Blue	P01, P05–P50
M	Ferrous metals with long or short chips, nonferrous metals	Yellow	M10–M40
K	Ferrous metals with short chips, nonferrous metals, nonmetallic materials	Red	K01, K10–K40

    Chip breaker features on inserts (described in Section 21.2.1) are for the purposes of (a) controlling chip flow during machining, (b) eliminating long chips, and (c) reducing vibration and heat generated. Carbide inserts are available commercially with a wide variety of complex chip-breaker features, typical examples of which are shown in Fig. 22.2. The selection of a particular chip-breaker feature depends on the feed and depth of cut of the operation, the workpiece material, the type of chip produced during cutting, and whether it is a roughing or finishing cut. Optimum chip-breaker geometries are being developed using computer-aided design and finite-element analysis techniques.

    Stiffness of the machine tool (Section 25.3) is of major importance when using carbide tools. Light feeds, low speeds, and chatter are detrimental because they tend to damage the tool's cutting edge. Light feeds, for example, concentrate the forces and temperature closer to the edges of the tool, increasing the tendency for the edges to chip off.

    Low cutting speeds tend to encourage cold welding of the chip to the tool. Cutting fluids should be applied continuously and in large quantities if used to minimize the heating and cooling of the tool in interrupted cutting operations.

22.4.4 Classification of carbides

With rapidly increasing global manufacturing and wider use of ISO (International Organization for Standardization) standards, carbide grades are classified using the letters P, M, and K (as shown in Tables 22.4 and 22.5) for a range of applications, including the traditional C grades used in the U.S. Because of the wide variety of carbide compositions available and the broad range of machining applications and workpiece materials involved, efforts for ISO classification continue to be a difficult task. This is true especially when comparing ISO grades with the traditional grades classified by the American National Standards Institute (ANSI), ranging from grades C1 to C8.

22.5 | Coated Tools

As described in Part I, new alloys and engineered materials are being developed continuously, particularly since the 1960s. These materials have high strength and toughness but generally are abrasive and chemically reactive with tool materials. The difficulty of machining these materials efficiently and the need for improving

TABLE 22.5

Classification of Tungsten Carbides According to Machining Applications

ISO standard	ANSI Classification number (Grade)	Materials to be machined	Machining operation	Type of carbide	Characteristics of	
					Cut	Carbide
K30–K40	C1	Cast iron, nonferrous metals, and nonmetallic materials requiring abrasion resistance	Roughing	Wear-resistant grades; generally straight WC-Co with varying grain sizes	Increasing cutting speed ↓ ↑ Increasing feed rate	Increasing hardness and wear resistance ↓ ↑ Increasing strength and binder content
K20	C2		General purpose			
K10	C3		Light finishing			
K01	C4		Precision finishing			
P30–P50	C5	Steels and steel alloys requiring crater and deformation resistance	Roughing	Crater-resistant grades; various WC-Co compositions with TiC and/or TaC alloys	Increasing cutting speed ↓ ↑ Increasing feed rate	Increasing hardness and wear resistance ↓ ↑ Increasing strength and binder content
P20	C6		General purpose			
P10	C7		Light finishing			
P01	C8		Precision finishing			

Note: The ISO and ANSI comparisons are approximate.

the performance in machining the more common engineering materials have led to important developments in *coated tools*. Coatings have unique properties, such as

- Lower friction
- Higher adhesion
- Higher resistance to wear and cracking
- Acting as a diffusion barrier
- Higher hot hardness and impact resistance

Coated tools can have tool lives 10 times longer than those of uncoated tools, allowing for high cutting speeds and thus reducing both the time required for machining operations and production costs. As can be seen in Fig. 22.6, machining time has been reduced steadily by a factor of more than 100 since 1900. This improvement has had a major impact on the economics of machining operations in conjuction with continued improvements in the design and construction of modern machine tools and their computer controls. As a result, coated tools now are used in 40 to 80% of all machining operations, particularly in turning, milling, and drilling. Surveys have indicated that the use of coated tools is more prevalent in larger companies than in smaller ones.

22.5.1 Coating materials and coating methods

Commonly used coating materials are titanium nitride (TiN), titanium carbide (TiC), titanium carbonitride (TiCN), and aluminum oxide (Al_2O_3). These coatings, generally in the thickness range of 2 to 15 μm (80 to 600 μin.), are applied on cutting tools and inserts by two techniques, as described in greater detail in Section 34.6.

1. **Chemical-vapor deposition** (CVD), including **plasma-assisted chemical-vapor deposition.**
2. **Physical-vapor deposition** (PVD).

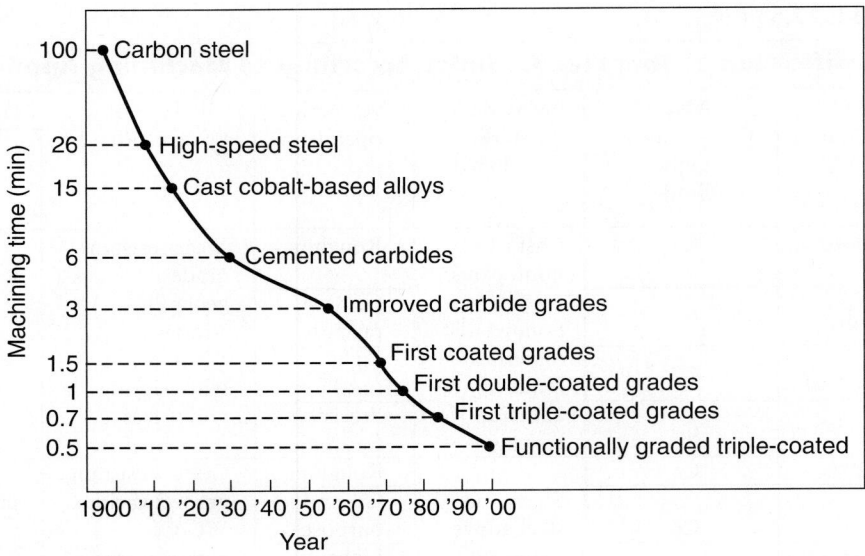

FIGURE 22.6   Relative time required to machine with various cutting-tool materials, indicating the year the tool materials were first introduced. Note that machining time has been reduced by two orders of magnitude within a hundred years. *Source:* Courtesy of Sandvik.

The CVD process is the most commonly used method for carbide tools with multiphase and ceramic coatings, both of which are described later in this section. The PVD-coated carbides with TiN coatings, on the other hand, have higher cutting-edge strength, lower friction, lower tendency to form a built-up edge, and are smoother and more uniform in thickness (which is generally in the range of 2 to 4 μm (80 to 160 μin.). Another technology (used particularly for multiphase coatings) is **medium-temperature chemical-vapor deposition** (MTCVD), developed to machine ductile (nodular) iron and stainless steels and to provide higher resistance to crack propagation than CVD coatings provide.

Coatings for cutting tools and dies should have the following general characteristics:

- **High hardness** at elevated temperatures, to resist wear.
- **Chemical stability** and **inertness** to the workpiece material, to reduce wear.
- **Low thermal conductivity** to prevent temperature rise in the substrate.
- **Compatibility** and **good bonding** to the substrate to prevent flaking or spalling.
- **Little or no porosity** in the coating to maintain its integrity and strength.

The effectiveness of coatings is enhanced by the hardness, toughness, and high thermal conductivity of the substrate (which may be carbide or high-speed steel). Honing of the cutting edges is an important procedure for the maintainance of coating strength; otherwise, the coating may peel or chip off at sharp edges and corners.

Titanium-nitride coatings.   Titanium-nitride coatings have low friction coefficients, high hardness, resistance to high temperature, and good adhesion to the substrate. Consequently, they greatly improve the life of high-speed steel tools, as well as the lives of carbide tools, drill bits, and cutters. Titanium-nitride coated tools (gold in color) perform well at higher cutting speeds and feeds. Flank wear is significantly

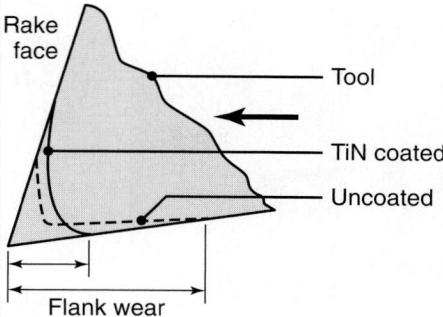

FIGURE 22.7 Schematic illustration of typical wear patterns on uncoated high-speed-steel and titanium-nitride coated tools. Note that flank wear is significantly lower for the coated tool.

lower than that of uncoated tools (Fig. 22.7), and flank surfaces can be reground after use, since regrinding the tool does not remove the coating on the rake face of the tool. However, coated tools do not perform as well at low cutting speeds because the coating can be worn off by chip adhesion. Therefore, the use of appropriate cutting fluids to discourage adhesion is important.

Titanium-carbide coatings. Titanium-carbide coatings on tungsten-carbide inserts have high flank-wear resistance in machining abrasive materials.

Ceramic coatings. Because of their chemical inertness, low thermal conductivity, resistance to high temperature, and resistance to flank and crater wear, ceramics are suitable coating materials for tools. The most commonly used ceramic coating is aluminum oxide (Al_2O_3). However, because they are very stable (not chemically reactive), oxide coatings generally bond weakly to the substrate.

Multiphase coatings. The desirable properties of the coatings just described can be combined and optimized with the use of *multiphase coatings*. Carbide tools now are available with two or three layers of such coatings and are effective particularly in machining cast irons and steels.

For example, one could first desposit TiC over the substrate, followed by Al_2O_3, and then TiN. The first layer should bond well with the substrate; the outer layer should resist wear and have low thermal conductivity. The intermediate layer should bond well and be compatible with both layers.

Typical applications of multiple-coated tools are as follows:

1. High-speed, continuous cutting: TiC/Al_2O_3
2. Heavy-duty, continuous cutting: TiC/Al_2O_3/TiN
3. Light, interrupted cutting: TiC/TiC + TiN/TiN

Coatings also are available in **alternating multiphase layers**. The thickness of these layers is on the order of 2 to 10 μm—thinner than regular multiphase coatings (Fig. 22.8). The reason for using thinner coatings is that coating hardness increases with decreasing grain size—a phenomenon that is similar to the increase in strength of metals with decreasing grain size. Therefore, thinner layers are harder than thicker layers.

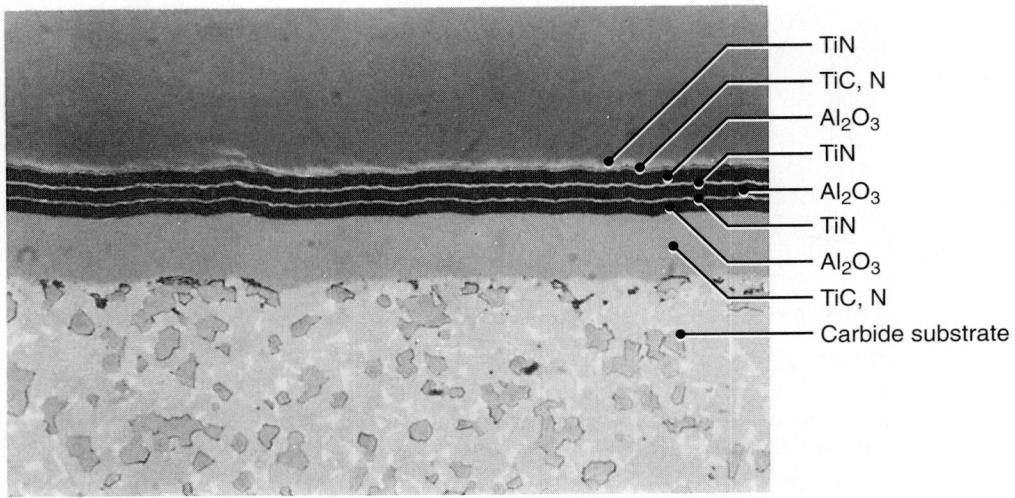

- TiN
- TiC, N
- Al$_2$O$_3$
- TiN
- Al$_2$O$_3$
- TiN
- Al$_2$O$_3$
- TiC, N
- Carbide substrate

FIGURE 22.8   Multiphase coatings on a tungsten-carbide substrate. Three alternating layers of aluminum oxide are separated by very thin layers of titanium nitride. Inserts with as many as thirteen layers of coatings have been made. Coating thicknesses are typically in the range of 2 to 10 μm. *Source*: Courtesy of Kennametal, Inc.

A typical multiphase-coated carbide tool may consist of the following layers, starting from the top; along with their primary functions:

1. TiN: low friction
2. Al$_2$O$_3$: high thermal stability
3. TiCN: fiber reinforced with a good balance of resistance to flank wear and crater wear, particularly for interrupted cutting
4. A thin-carbide substrate: high fracture toughness
5. Thick-carbide substrate: hard and resistant to plastic deformation at high temperatures

Diamond coatings.   The properties and applications of diamond, diamond coatings, and *diamond-like carbon* are described in Sections 8.7 and 34.13, respectively, and their use as cutting tools is given in Section 22.9. Polycrystalline diamond is being used widely as a coating for cutting tools, particularly on tungsten-carbide and silicon-nitride inserts. Diamond-coated tools are effective particularly in machining nonferrous metals, and abrasive materials such as aluminum alloys containing silicon, fiber-reinforced and metal-matrix composite materials, and graphite. As many as tenfold improvements in tool life have been obtained over other coated tools.

Diamond-coated inserts are available commercially. Thin films are deposited on substrates with PVD and CVD techniques. Thick films are obtained by growing a large sheet of pure diamond, which is then laser cut to shape and brazed to a carbide insert. *Multilayer nanocrystal diamond coatings* also are being developed, with interlocking layers of diamond that give strength to the coating. As in all coatings, it is important to develop good adherence of the diamond film to the substrate and minimize the difference in thermal expansion between the diamond and substrate materials.

22.5.2 Miscellaneous coating materials

Major advances have been taking place in further improving the performance of coated tools. **Titanium carbonitride** (TiCN) and **titanium-aluminum nitride** (TiAlN) are effective in cutting stainless steels. TiCN (which is deposited with physical-vapor deposition techniques) is harder and tougher than TiN and can be used on carbides and high-speed steel tools. TiAlN is effective in machining aerospace alloys. Chromium-based coatings, such as **chromium carbide** (CrC), have been found to be effective in machining softer metals that tend to adhere to the cutting tool, such as aluminum, copper, and titanium. Other new materials include **zirconium nitride** (ZrN) and **hafnium nitride** (HfN). Considerable experimental data is required before these coatings and their behavior can be assessed fully for proper applications in machining.

More recent developments include (a) **nanolayer coatings** including carbide, boride, nitride, oxide, or some combination (see also Section 6.16), and (b) **composite coatings**, using a variety of materials. The hardness of some of these coatings approaches that of cubic boron nitride. Although they are still in experimental stages, it is expected that these coatings will have the combined benefits of various types of coatings as well as wider applications in machining operations.

22.5.3 Ion implantation

In this process, ions are introduced into the surface of the cutting tool, improving its surface properties (Section 34.6). The process does not change the dimensions of tools. **Nitrogen-ion** implanted carbide tools have been used successfully on alloy steels and stainless steels. **Xenon-ion** implantation of tools is also under development.

22.6 | Alumina-Based Ceramics

Ceramic tool materials, introduced in the early 1950s, primarily consist of fine-grained, high-purity **aluminum oxide** (see also Section 8.2). They are cold pressed into insert shapes under high pressure and sintered at high temperature; the end product is referred to as **white** (**cold-pressed**) **ceramics**. Additions of titanium carbide and zirconium oxide help improve properties such as toughness and thermal-shock resistance.

Alumina-based ceramic tools have very high abrasion resistance and hot hardness (Fig. 22.9). Chemically, they are more stable than high-speed steels and carbides, so they have less tendency to adhere to metals during cutting and a correspondingly lower tendency to form a built-up edge. Consequently, in cutting cast irons and steels, good surface finish is obtained with ceramic tools. However, ceramics lack

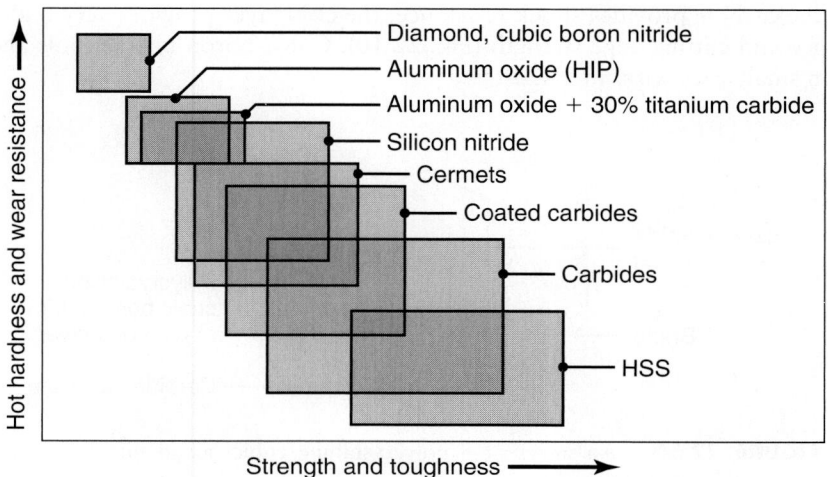

FIGURE 22.9 Ranges of mechanical properties for various groups of tool materials. See also Tables 22.1 through 22.5.

toughness, and their use may result in premature tool failure by chipping or catastrophic failure.

Ceramic inserts are available in shapes similar to the shapes of carbide inserts. They are effective in high-speed, uninterrupted cutting operations, such as finishing or semifinishing by turning. To reduce thermal shock, cutting should be performed either dry or with a copious amount of cutting fluid applied in a steady stream. Improper or intermittent applications of the fluid can cause thermal shock and fracture of the ceramic tool.

Ceramic tool shape and setup are important. Negative rake angles (large included angles) generally are preferred in order to avoid chipping due to the poor tensile strength of ceramics. Tool failure can be reduced by increasing the stiffness and damping capacity of machine tools, mountings, and workholding devices, thus reducing vibration and chatter.

Cermets. *Cermets* (from the words *cer*amic and *met*al) were first used in the early 1950s and consist of ceramic particles in a metallic matrix. The were introduced in the 1960s and are *black* or *hot-pressed ceramics* (carboxides). A typical cermet consists of 70% aluminum oxide and 30% titanium carbide; other cermets contain molybdenum carbide, niobium carbide, and tantalum carbide. Although they have chemical stability and resistance to built-up edge formation, the brittleness and high cost of cermets have been a limitation to their wider use. Further refinements of these tools have resulted in improved strength, toughness, and reliability. Their performance is somewhere between that of ceramics and carbides and has been particularly suitable for light roughing cuts and high-speed finishing cuts. Chip-breaker features are important for cermet inserts. Although they can be coated, the benefits of coated cermets are somewhat controversial, as the improvement in wear resistance appears to be marginal.

22.7 | Cubic Boron Nitride

Next to diamond, *cubic boron nitride* (cBN) is the hardest material presently available. Introduced in 1962 under the trade name Borazon, cubic boron nitride is made by bonding a 0.5 to 1 mm (0.02 to 0.04 in.) layer of **polycrystalline cubic boron nitride** to a carbide substrate by sintering under high pressure and high temperature. While the carbide provides shock resistance, the cBN layer provides very high wear resistance and cutting-edge strength (Fig. 22.10). Cubic-boron-nitride tools also are made in small sizes without a substrate.

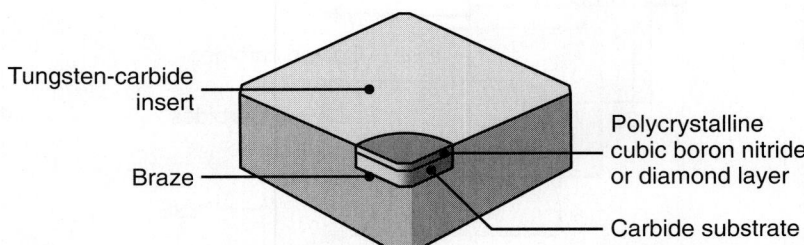

Tungsten-carbide insert

Braze

Polycrystalline cubic boron nitride or diamond layer

Carbide substrate

FIGURE 22.10 An insert of a polycrystalline cubic boron nitride or a diamond layer on tungsten carbide.

At elevated temperatures, cBN is chemically inert to iron and nickel (hence no wear due to diffusion). Its resistance to oxidation is high and, thus, is particularly suitable for cutting hardened ferrous and high-temperature alloys (see *hard machining*, Section 25.6). It also is used as an abrasive. Because cBN tools are brittle, stiffness of the machine tool and the fixturing is important to avoid vibration and chatter. Furthermore, in order to avoid chipping and cracking due to thermal shock, machining generally should be performed dry (that is, cutting fluids should be avoided) particularly in interrupted cutting operations (such as milling), which repeatedly subject the tool to thermal cycling.

22.8 | Silicon-Nitride-Based Ceramics

Developed in the 1970s, *silicon-nitride* (SiN) *based ceramic* tool materials consist of silicon nitride with various additions of aluminum oxide, yttrium oxide, and titanium carbide. These tools have toughness, hot hardness, and good thermal-shock resistance. An example of a SiN-based material is **sialon**, named after the elements of which it is composed: *si*licon, *al*uminum, *o*xygen, and *n*itrogen. It has higher thermal-shock resistance than silicon nitride and is recommended for machining cast irons and nickel-based superalloys at intermediate cutting speeds. Because of their chemical affinity to iron at elevated temperatures, however, SiN-based tools are not suitable for machining steels.

22.9 | Diamond

Of all known materials, the hardest substance is diamond, as also described in Section 8.7. As a cutting tool, it has highly desirable properties, such as low friction, high wear resistance, and the ability to maintain a sharp cutting edge. Diamond is used when good surface finish and dimensional accuracy are required, particularly with soft nonferrous alloys, and abrasive nonmetallic and metallic materials (especially some aluminum-silicon alloys). *Synthetic* or industrial *diamonds* now are used widely because natural diamond has flaws and its performance can be unpredictable, as is the case with abrasives used in grinding wheels.

Single-crystal diamond of various carats can be used for special applications. However, they have been replaced largely by **polycrystalline diamond** (PCD) tools, called *compacts*, which also are used as dies for fine wire drawing. These diamond tools consist of very small synthetic crystals fused by a high-pressure, high-temperature process to a thickness of about 0.5 to 1 mm (0.02 to 0.04 in.) and bonded to a carbide substrate; this product is similar to cBN tools (Fig. 22.11). The random orientation of

FIGURE 22.11 Inserts with polycrystalline cubic boron nitride tips (top row), and solid-polycrystalline cBN inserts (bottom row). *Source:* Courtesy of Valenite.

the diamond crystals prevents the propagation of cracks through the structure, significantly improving its toughness.

Because diamond is brittle, tool shape and sharpness are important. Low rake angles generally are used to provide a strong cutting edge (because of the larger included angles). Special attention should be given to proper mounting and crystal orientation in order to obtain optimum tool life. Wear may occur through microchipping (caused by thermal stresses and oxidation) and through transformation to carbon (caused by the heat generated during cutting). Diamond tools can be used satisfactorily at almost any speed, but are most suitable for light, uninterrupted finishing cuts. In order to minimize tool fracture, the single-crystal diamond must be resharpened as soon as it becomes dull. Because of its strong chemical affinity at elevated temperatures (resulting in diffusion), diamond is not recommended for machining plain-carbon steels or titanium, nickel, and cobalt-based alloys. Diamond also is used as an abrasive in grinding and polishing operations and as *coatings*.

22.10 | Whisker-Reinforced Tool Materials

In order to further improve the performance and wear resistance of cutting tools (particularly in machining new materials and composites), continued progress is being made in developing new tool materials with enhanced properties, such as:

- High fracture toughness
- Resistance to thermal shock
- Cutting-edge strength
- Creep resistance
- Hot hardness

Advances include the use of *whiskers* as reinforcing fibers in composite cutting-tool materials. Examples of *whisker-reinforced cutting tools* include (a) silicon-nitride-based tools reinforced with silicon-carbide whiskers and (b) aluminum-oxide-based tools reinforced with 25 to 40% silicon-carbide whiskers, sometimes with the addition of *zirconium oxide* (ZrO_2). Silicon-carbide whiskers are typically 5 to 100 μm long and 0.1 to 1 μm in diameter. The high reactivity of silicon carbide with ferrous metals makes SiC-reinforced tools unsuitable for machining irons and steels.

22.11 | Tool Costs and Reconditioning of Tools

Tool costs vary widely, depending on the tool material, size, shape, chip-breaker features, and quality. The cost for a typical 0.5 in. (12.5 mm) *insert* is approximately: (a) $2 to $10 for uncoated carbides, (b) $6 to $10 for coated carbides, (c) $8 to $12 for ceramics, (d) $50 to $60 for diamond-coated carbides, (e) $60 to $90 for cubic boron nitride, and (f) $90 to $100 for a diamond-tipped insert.

After reviewing the costs involved in machining and considering all of the aspects involved in the total operation, it can be seen that the cost of an individual insert is relatively insignificant. Tooling costs in machining have been estimated to be on the order of 2 to 4% of the manufacturing costs. This small amount is due to the fact that a single cutting tool, for example, can perform a large amount of material removal before it is indexed and eventually recycled. Recall from Section 21.5 that the expected tool life can be in the range of 30 to 60 minutes. Thus, considering that a square insert has eight cutting edges, it indicates that the tool lasts a long time before it is removed from the machine and replaced.

Cutting tools can be **reconditioned** by resharpening them, using tool and cutter grinders in toolrooms having special fixtures. This operation may be carried out by hand or on computer-controlled tool and cutter grinders. Advanced methods of machining also are used, as described in Chapter 27. Reconditioning of coated tools also is done by recoating them, usually in special facilities for these purposes. It is important that reconditioned tools have the same geometric features as the original tools. Often a decision has to be made whether reconditioning or further reconditioning of tools is economical, especially when the costs of typical small inserts are not a major item in the total operation. Recycling tools is always a significant consideration, especially if they contain expensive and strategically important materials, such as tungsten and cobalt.

22.12 | Cutting Fluids

Cutting fluids have been used extensively in machining operations to achieve the following results:

- Reduce friction and wear, thus improving tool life and the surface finish of the workpiece.
- Cool the cutting zone, thus improving tool life and reducing the temperature and thermal distortion of the workpiece.
- Reduce forces and energy consumption.
- Flush away the chips from the cutting zone, and thus prevent the chips from interfering with the cutting process, particularly in operations such as drilling and tapping.
- Protect the machined surface from environmental corrosion.

Depending on the type of machining operation, the cutting fluid needed may be a **coolant**, a **lubricant, or both**. The effectiveness of cutting fluids depends on a number of factors, such as the type of machining operation, tool and workpiece materials, cutting speed, and the method of application. Water is an excellent coolant and can reduce effectively the high temperatures developed in the cutting zone. However, water is not an effective lubricant, hence it does not reduce friction. Furthermore, it causes the rusting of workpieces and machine-tool components. On the other hand, as we have seen, effective lubrication is an important factor in machining operations.

The need for a cutting fluid depends on the *severity* of the particular machining operation, which may be defined as the level of temperatures and forces encountered, the tendency for built-up edge formation, the ease with which chips produced can be removed from the cutting zone, and how effectively the fluids can be applied to the proper region at the tool–chip interface. The relative severity of specific machining processes in increasing order of severity are: sawing, turning, milling, drilling, gear cutting, thread cutting, tapping, and internal broaching.

There are operations, however, in which the cooling action of cutting fluids can be detrimental. It has been shown that cutting fluids may cause the chip to become *more curly*, and thus concentrating the heat closer to the tool tip, which reduces tool life. More importantly, in interrupted cutting operations, such as milling with multiple-tooth cutters, cooling of the cutting zone leads to thermal cycling of the cutter teeth, which can cause *thermal cracks* by thermal fatigue or thermal shock.

However, beginning with the mid-1990s, there has been a major trend for **near-dry machining,** meaning a minimal use of cutting fluids, as well as *dry machining*; we describe these trends in Section 22.12.1.

Cutting-fluid action.   The basic mechanisms of lubrication in metalworking operations are described in greater detail in Section 33.6. Here, we briefly describe the mechanisms by which cutting fluids influence machining operations. It is not immediately clear how a cutting fluid can penetrate the important rake face of the tool and influence the cutting process. Studies have shown that the cutting fluid gains access to the tool–chip interface by seeping from the sides of the chip by the *capillary action* of the interlocking network of surface asperities in the interface. Because of the small size of this capillary network, the cutting fluid should have a small molecular size and possess proper wetting (surface tension) characteristics. Thus, for example, grease cannot be an effective lubricant in maching, but low molecular weight oils suspended in water (emulsions) are very effective.

EXAMPLE 22.2 Effects of cutting fluids on machining

A machining operation is being carried out with a cutting fluid that is an effective lubricant. What will be the changes in the mechanics of the cutting operation if the fluid is shut off?

Solution   Since the cutting fluid is a good lubricant, the following chain of events will take place after the fluid is shut off:

   a. Friction at the tool–chip interface will increase.

   b. The shear angle will decreases, according to Eq. (21.3).

   c. The shear strain will increase, as seen from Eq. (21.2).

   d. The chip will become thicker.

   e. A built-up edge is likely to form.

As a result of these changes, the following events will occur:

   a. The shear energy in the primary zone will increase.

   b. The frictional energy in the secondary zone will increase.

   c. The total energy will increase.

   d. The temperature in the cutting zone will rise, causing greater tool wear.

   e. Surface finish will begin to deteriorate and dimensional tolerances may be difficult to maintain because of the increased temperature and thermal expansion of the workpiece during machining.

Types of cutting fluids.   The characteristics and applications of metalworking fluids are described in Section 33.7. Briefly, four general types of cutting fluids commonly are used in machining operations:

1. **Oils** (also called *straight oils*) including mineral, animal, vegetable, compounded, and synthetic oils typically are used for low-speed operations where temperature rise is not significant.

2. **Emulsions** (also called *soluble oils*) are a mixture of oil and water and additives, generally are used for high-speed operations because temperature rise is significant. The presence of water makes emulsions very effective coolants.

3. **Semisynthetics** are chemical emulsions containing little mineral oil, diluted in water, and with additives that reduce the size of oil particles, making them more effective.

4. **Synthetics** are chemicals with additives, diluted in water, and contain no oil.

Because of the complex interaction of the cutting fluid, the workpiece materials, temperature, time, and cutting-process variables, the application of fluids cannot be generalized. In Chapters 23 and 24, we have given cutting fluid recommendations for specific machining operations; these recommendations also are available from numerous suppliers of metalworking fluids.

Methods of cutting-fluid application.   There are four basic methods of cutting-fluid applications in machining:

1. **Flooding.** This is the most common method (shown in Fig. 22.12) indicating good and poor flooding practices. Flow rates typically range from 10 L/min (3 gal/min) for single-point tools to 225 L/min (60 gal/min) per cutter for multiple-tooth cutters, such as in milling. In some operations, such as drilling and milling, fluid pressures in the range of 700 to 14,000 kPa (100 to 2000 psi) are used to flush away the chips produced to prevent interfering with the operation.

2. **Mist.** This type of cooling supplies fluid to inaccessible areas, similar to using an aerosol can, and provides better visibility of the workpiece being machined (as compared to flood cooling). It is effective particularly with water-based fluids at air pressures of 70 to 600 kPa (10 to 80 psi). However, it has limited

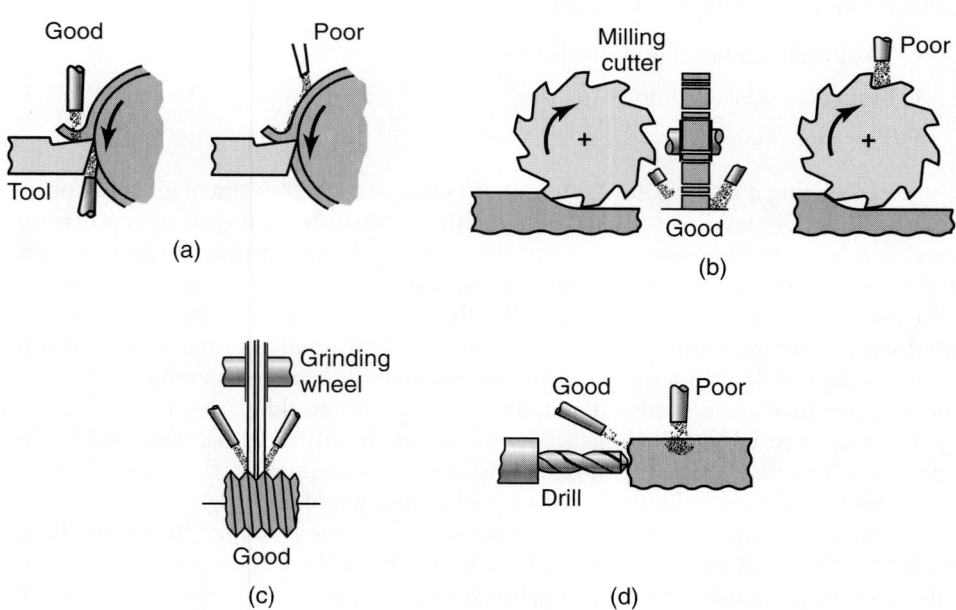

FIGURE 22.12   Schematic illustration of the proper methods of applying cutting fluids (flooding) in various machining operations: (a) turning, (b) milling, (c) thread grinding, and (d) drilling.

cooling capacity. Mist application requires venting to prevent the inhalation of airborn fluid particles by the machine operator and others nearby.

3. **High-pressure systems.** With the increasing speed and power of modern, computer-controlled machine tools, heat generation in machining has become a significant factor. Particularly effective is the use of high-pressure refrigerated coolant systems to increase the rate of heat removal from the cutting zone. High pressures also are used in delivering the cutting fluid using specially designed nozzles that aim a powerful jet of fluid to the zone, particularly into the *clearance* or *relief face* of the tool (see Fig. 21.3). The pressures employed, which are usually in the range of 5.5 to 35 MPa (800 to 5000 psi), act as a chip breaker in situations where the chips produced would otherwise be long and continuous, interfering with the cutting operation. In order to avoid damage to the workpiece surface by impact from any particles present in the high-pressure jet, contaminant size in the coolant should not exceed 20 μm (800 μin.). Proper and continuous filtering of the fluid also is essential to maintain quality.

4. **Through the cutting tool system.** We have pointed out the severity of various machining operations in terms of the difficulty of supplying fluids into the cutting zone and flushing away the chips. For a more effective application, narrow passages can be produced in cutting tools, as well as in toolholders, through which cutting fluids can be applied under high pressure. We have shown two applications of this method: (a) gun drilling, shown in Fig. 23.24, with a long, small hole through the body of the drill itself, and (b) boring bars, shown in Fig. 23.17a, where there is a long hole through the shank (toolholder) to which an insert is clamped. Similar designs have been developed for cutting tools and inserts and for delivering cutting fluids through the spindle of the machine tool.

Effects of cutting fluids. The selection of a cutting fluid also should include considerations such as its effects on

- Workpiece material and machine tools
- Biological considerations
- The environment

In selecting a cutting fluid, one should consider whether the machined component will be subjected to stress and adverse effects, possibly leading to stress-corrosion cracking. This consideration is important particularly for cutting fluids with sulfur and chlorine additives. For example, (a) cutting fluids containing sulfur should not be used with nickel-base alloys, and (b) fluids containing chlorine should not be used with titanium. Cutting fluids also may affect adversely the machine tool components, thus their compatibility with various metallic and nonmetallic materials in the machine tool must be considered. Machined parts should be cleaned and washed in order to remove any cutting-fluid residue, as described in Section 34.16. This operation can be significant in time and cost. Consequently, the trend is to use water-based, low-viscosity fluids for ease of cleaning and filtering.

Because the machine-tool operator is usually in close proximity to cutting fluids, the health effects of operator contact with fluids should be of primary concern. Mist, fumes, smoke, and odors from cutting fluids can cause severe skin reactions and respiratory problems, especially in using fluids with chemical constituents such as sulfur, chlorine, phosphorus, hydrocarbons, biocides, and various additives. Much progress has been made in ensuring the safe use of cutting fluids in manufacturing plants, keeping in view the reducing or eliminating their use by adopting the more recent trends in dry or near-dry machining techniques (see Section 21.12.1).

Cutting fluids (as well as other metalworking fluids used in manufacturing operations) may undergo chemical changes as they are used repeatedly over time. These changes may be due to environmental effects or to contamination from various sources, including metal chips and fine particles produced during machining and tramp oil (from leaks in hydraulic systems, oils on sliding members of machines, and lubricating systems for the machine tools). These changes involve the growth of microbes (bacteria, molds, and yeast), particularly in the presence of water, becoming an environmental hazard and also adversely affecting the characteristics and effectiveness of the cutting fluids.

Several techniques (such as settling, skimming, centrifuging, and filtering) are available for clarifying used cutting fluids. Recycling involves treatment of the fluids with various additives, agents, biocides, and deodorizers, as well as water treatment (for water-based fluids). Disposal practices for these fluids must comply with federal, state, and local laws and regulations.

22.12.1 Near-dry and dry machining

For economic and environmental reasons, there has been a continuing worldwide trend since the mid-1990s to minimize or eliminate the use of metalworking fluids. This trend has lead to the practice of *near-dry machining* (NDM) with major benefits such as:

- Alleviating the environmental impact of using cutting fluids, improving air quality in manufacturing plants, and reducing health hazards.
- Reducing the cost of machining operations, including the cost of maintenance, recycling, and disposal of cutting fluids.
- Further improving surface quality.

The significance of this approach is apparent when noting that, in the United States alone, millions of gallons of metalworking fluids are consumed each year. Furthermore, it has been estimated that metalworking fluids constitute about 7 to 17% of the total machining costs.

The principle behind near-dry cutting is the application of a fine mist of air-fluid mixture containing a very small amount of cutting fluid, including vegetable oil. The mixture is delivered to the cutting zone through the spindle of the machine tool, typically through a 1-mm diameter nozzle and under a pressure of 600 kPa (85 psi). It is used at rates on the order of 1 to 100 cc/hr, which is estimated to be (at most) one ten-thousandth of that used in flood cooling. Consequently, the process also is known as *minimum quantity lubrication* (MQL).

Dry machining also is a viable alternative. With major advances in cutting tools, dry machining has been shown to be effective in various machining operations (especially turning, milling, and gear cutting) on steels, steel alloys, and cast irons, but generally not for aluminum alloys.

One of the functions of a metal-cutting lubricant is to flush chips from the cutting zone. This seems to be problematic with dry machining. However, tool designs have been created that allow application of pressurized air, often through the tool shank. The compressed air doesn't serve a lubrication purpose, provides only limited cooling, but is very effective at clearing chips from the cutting interface.

Cryogenic machining.   More recent developments in machining include the use of cryogenic gases such as *nitrogen* or *carbon dioxide* as a coolant in machining. With small-diameter nozzles and at a temperature of $-200°C$ ($-320°F$), liquid nitrogen is injected into the cutting zone. Because of reduction in temperature, tool hardness is maintained and tool life is enhanced, thus allowing higher cutting speeds. Also, the chips are more brittle, hence machinability is increased. Furthermore, the nitrogen simply evaporates and thus has no adverse environmental impact.

SUMMARY

- A wide variety of cutting-tool materials have been developed over the past century, the most commonly used ones being high-speed steels, carbides, ceramics, cubic boron nitride, and diamond. Tool materials have a broad range of mechanical and physical properties, such as hot hardness, toughness, chemical stability and inertness, and resistance to chipping and wear.
- Various tool coatings have been developed with major improvements in tool life, surface finish, and the economics of machining operations. Common coating materials are titanium nitride, titanium carbide, titanium carbonitride, and aluminum oxide; diamond coatings also are gaining acceptance. The trend is toward multiphase coatings for even better performance.
- The selection of appropriate tool materials depends not only on the material to be machined but also on process parameters and the characteristics of the machine tool.
- Cutting fluids are important in machining operations and reduce friction, wear, cutting forces, and power requirements. Generally, slower cutting operations and those with high cutting-tool pressures require a fluid with good lubricating characteristics. In high-speed operations, where temperature rise can be significant, fluids with good cooling capacity and some lubricity are required. The selection of cutting fluids must include considerations of their possible adverse effects on the machined parts, machine tools and their components, personnel, and the environment.

KEY TERMS

Alumina-based ceramics	Dry machining	Roughing cuts
Carbides	Finishing cuts	Sialon
Cast-cobalt alloys	Flooding	Silicon-nitride-based ceramics
Ceramics	High-speed steels	Stellite
Cermets	Inserts	Titanium carbide
Chemical stability	Lubricants	Titanium nitride
Chip breaker	Micrograin carbides	Tool reconditioning
Coated tools	Mist	Tool costs
Coolants	Multiphase coatings	Toughness
Cryogenic machining	Nanocrystalline	Tungsten carbide
Cubic boron nitride	Near-dry machining	Uncoated carbides
Cutting fluids	Polycrystalline cubic boron nitride	Wear resistance
Diamond tools	Polycrystalline diamond	Whisker-reinforced tools
Diamond coatings	Reconditioning of tools	

BIBLIOGRAPHY

ASM Handbook, Vol. 16: *Machining*. ASM International, 1989.

ASM Handbook, Vol. 3: *Properties and Selection: Stainless Steels, Tool Materials and Special Purpose Metals*, ASM International, 1980.

ASM Specialty Handbook: Tool Materials, ASM International, 1995.

Byers, J.P. (ed.), *Metalworking Fluids*, Marcel Dekker, 1994.

Kalpakjian, S. (ed.), *Tool and Die Failures: Source Book*, ASM International, 1982.

Komanduri, R., *Tool Materials*, in *Kirk-Othmer Encyclopedia of Chemical Technology*, 4th ed., Vol. 24, 1997.

Machinery's Handbook, Industrial Press, revised periodically.

Machining Data Handbook, 3rd ed., 2 vols. Machinability Data Center, 1980.

Nachtman, E.S., and Kalpakjian, S., *Lubricants and Lubrication in Metalworking Operations*. Marcel Dekker, 1985.

Roberts, G.A., Krauss, G., and Kennedy, R., *Tool Steels*, 5th ed., ASM International, 1997.

Shaw, M.C., *Metal Cutting Principles*, 2nd ed., Oxford, 2005.

Schey, J.A., *Tribology in Metalworking–Friction: Wear and Lubrication*, ASM International, 1983.

Sluhan, C. (ed.), *Cutting and Grinding Fluids: Selection and Application*, Society of Manufacturing Engineers, 1992.

Stephenson, D.A., and Agapiou, J.S., *Metal Cutting Theory and Practice*, Marcel Dekker, 1996.

Trent, E.M., and Wright, P.K., *Metal Cutting*, 4th ed., Butterworth-Heinemann, 1999.

Tool and Manufacturing Engineers Handbook, 4th ed., Vol. 1: *Machining*, Society of Manufacturing Engineers, 1983.

REVIEW QUESTIONS

22.1 What are the major properties required of cutting-tool materials?

22.2 What are the differences in composition and properties between carbon steel and high-speed steel tools?

22.3 What is the composition of a typical carbide tool?

22.4 Why were cutting-tool inserts developed?

22.5 Why are tools coated? What are the common coating materials? What is a multiphase coating? What are its advantages?

22.6 Explain the applications and limitations of ceramic tools.

22.7 What is the composition of sialon?

22.8 How are cutting tools reconditioned?

22.9 List the major functions of cutting fluids.

22.10 Explain how cutting fluids penetrate the cutting zone.

22.11 List the methods by which cutting fluids are applied in machining operations.

22.12 Describe the advantages and limitations of (a) single-crystal and (b) polycrystalline diamond tools.

22.13 What is the hardest known material next to diamond?

22.14 What is a cermet? What are its advantages?

22.15 Explain the difference between M-series and T-series high-speed steels.

QUALITATIVE PROBLEMS

22.16 Explain why so many different types of cutting-tool materials have been developed over the years. Why are they still being developed further?

22.17 Which tool-material properties are suitable for interrupted cutting operations? Why?

22.18 Describe the reasons for coating cutting tools with multiple layers of different materials.

22.19 Make a list of the alloying elements used in high-speed steels. Explain why they are so effective in cutting tools.

22.20 As we have stated in Section 22.1, tool materials can have conflicting properties in machining operations. Describe your observations regarding this matter.

22.21 Comment on the purposes of chamfers on inserts and their design features

22.22 Explain the economic impact of the trend shown in Fig. 22.6.

22.23 Why does temperature have such an important effect on tool life?

22.24 Ceramic and cermet cutting tools have certain advantages over carbide tools. Why, then, are they not completely replacing carbide tools?

22.25 Can cutting fluids have any adverse effects? If so, what are they?

22.26 Describe the trends you observe in Table 22.2.

22.27 Why are chemical stability and inertness important in cutting tools?

22.28 How would you go about measuring the effectiveness of cutting fluids? Explain any difficulties that you might encounter.

22.29 Titanium-nitride coatings on tools reduce the coefficient of friction at the tool–chip interface. What is the significance of this?

22.30 Describe the necessary conditions for optimal utilization of the capabilities of diamond and cubic-boron-nitride cutting tools.

22.31 List and comment on the advantages of coating high-speed steel tools.

22.32 Explain the limits of application when comparing tungsten-carbide and titanium-carbide cutting tools.

22.33 Negative rake angles generally are preferred for ceramic, diamond, and cubic-boron-nitride tools. Why?

22.34 Do you think that there is a relationship between the cost of a cutting tool and its hot hardness? Explain.

22.35 Survey the technical literature and give some typical values of cutting speeds for high-speed-steel tools and for a variety of workpiece materials.

22.36 In Table 22.1, the last two properties listed can be important to the life of the cutting tool. Why?

22.37 It has been stated that titanium-nitride coatings allow cutting speeds and feeds to be higher than those for uncoated tools. Survey the technical literature and prepare a table showing the percentage increase of speeds and feeds that would be made possible by coating the tools.

22.38 You will note in Fig. 22.1 that all tool materials have a wide range of hardnesses for a particular temperature, especially carbides. Describe all of the factors that are responsible for this wide range.

22.39 List and explain the considerations involved in the decision to recondition, recycle, or discard a cutting tool.

22.40 Referring to Table 22.1, state which tool materials would be suitable for interrupted cutting operations. Explain.

22.41 Which of the properties listed in Table 22.1 is, in your opinion, the least important in cutting tools? Explain.

22.42 If a drill bit is intended for woodworking applications, what material is it most likely to be made from? (*Hint*: Temperatures rarely rise to 400°C in woodworking.) Are there any reasons why such a drill bit cannot be used to drill a few holes in a metal? Explain.

22.43 What are the consequences of a coating having a different coefficient of thermal expansion than the substrate?

22.44 Discuss the relative advantages and limitations of near-dry machining. Consider all relevant technical and economic aspects.

QUANTITATIVE PROBLEMS

22.45 Review the contents of Table 22.1. Plot several curves to show range relationships, if any, among parameters such as hardness, transverse rupture strength, and impact strength. Comment on your observations.

22.46 Obtain data on the thermal properties of various cutting fluids. Identify those that are basically effective coolants (such as water-based fluids) and those that are basically good lubricants (such as oils).

22.47 The first column in Table 22.2 shows seven properties that are important to cutting tools. For each of the tool materials listed in the table, add numerical data for each of these properties. Describe your observations, including any data that overlaps.

SYNTHESIS, DESIGN, AND PROJECTS

22.48 Describe in detail your thoughts regarding the technical and economic factors involved in tool material selection.

22.49 One of the principal concerns with coolants is degradation due to biological attack by bacteria. To prolong life, chemical biocides often are added, but these biocides greatly complicate the disposal of coolants. Conduct a literature search regarding the latest developments in the use of environmentally benign biocides in cutting fluids.

22.50 Contact several different suppliers of cutting tools or search their web sites. Make a list of the costs of typical cutting tools of various sizes, shapes, and features.

22.51 As you can see, there are several types of cutting-tool materials available today for machining operations.

Yet, there is much research and development that is being carried out on these materials. Make a list of the reasons why you think such studies are being conducted. Comment on each with a specific application or example.

22.52 Assume that you are in charge of a laboratory for developing new or improved cutting fluids. On the basis of the topics presented in this and in Chapter 21, suggest a list of topics for your staff to investigate. Explain why you have chosen those topics.

22.53 Survey the technical literature and describe the trends in new cutting-tool materials and coatings. Which of these are becoming available to industry?

CHAPTER

23

Machining Processes Used to Produce Round Shapes: Turning and Hole Making

With the preceding two chapters as background, this chapter describes machining processes that are capable of generating round shapes. These could be externally round (or axisymmetric) shapes or holes. This chapter emphasizes:

- Characteristics of lathes and turning operations.
- External and internal material-removal processes.
- The type of cutting tools used in these operations.

Typical parts made: Machine components, engine blocks and heads, parts with complex shapes and close tolerances, and externally and internally threaded parts.

Alternative processes: Precision casting, powder metallurgy, powder injection molding, abrasive machining, and thread rolling.

23.1 | Introduction

This chapter describes machining processes with the capability of producing parts that basically are round in shape. Typical products made are as small as miniature screws for the hinges of eyeglass frames and as large as turbine shafts for hydroelectric power plants, rolls for rolling mills, cylinders, and gun barrels.

One of the most basic machining processes is *turning*, meaning that the part is rotated while it is being machined. The starting material is generally a workpiece that has been made by other processes, such as casting, forging, extrusion, drawing, or powder metallurgy, as described in Parts II and III. Turning processes, which typically are carried out on a *lathe* or similar *machine tools*, are outlined in Fig. 23.1 and Table 23.1. These machines are very versatile and capable of producing a wide variety of shapes as outlined below:

- **Turning:** to produce straight, conical, curved, or grooved workpieces (Fig. 23.1a through d), such as shafts, spindles, and pins.

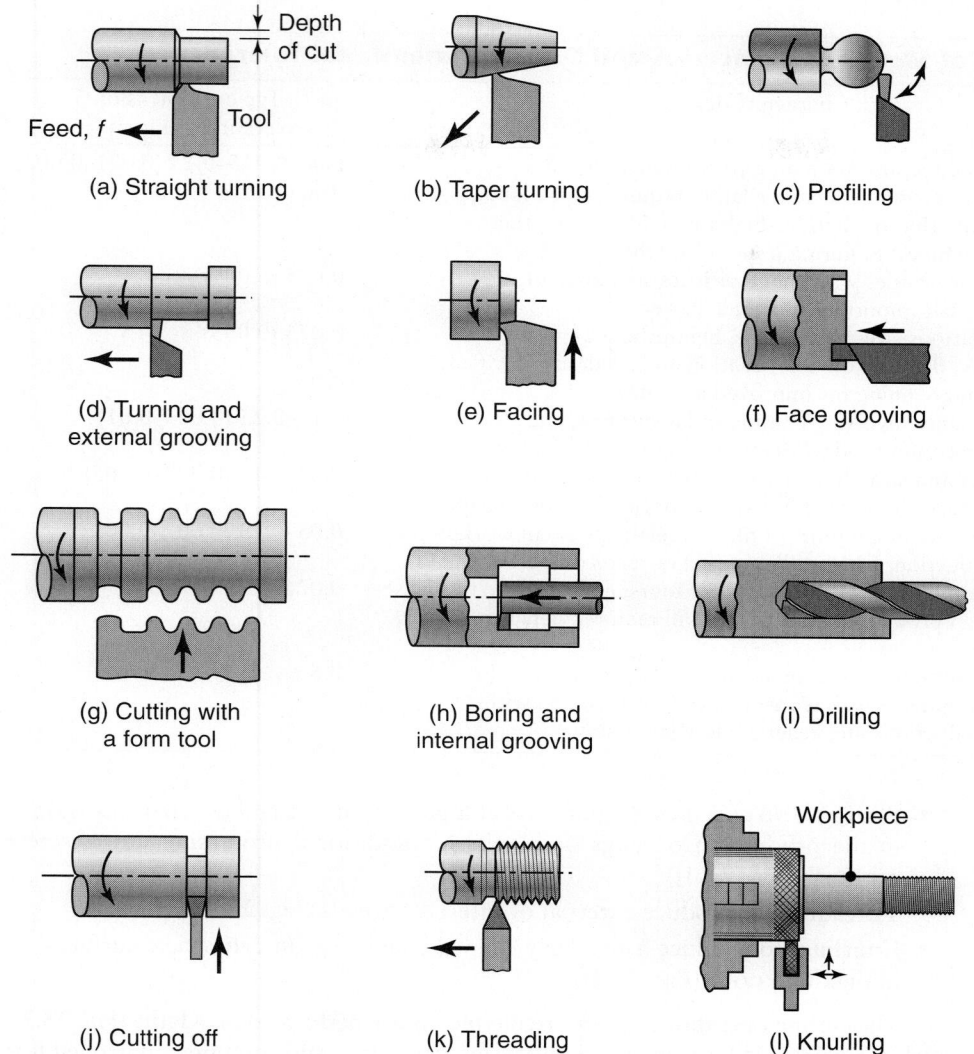

FIGURE 23.1 Miscellaneous cutting operations that can be performed on a lathe. Note that all parts are circular—a property known as axisymmetry. The tools used, their shape, and the processing parameters are described throughout this chapter.

- **Facing:** to produce a flat surface at the end of the part and perpendicular to its axis (Fig. 23.1e), useful for parts that are assembled with other components. *Face grooving* produces grooves for applications such as O-ring seats (Fig. 23.1f).

- **Cutting with form tools:** (Fig. 23.1g) to produce various axisymmetric shapes for functional or aesthetic purposes.

- **Boring:** to enlarge a hole or cylindrical cavity made by a previous process or to produce circular internal grooves (Fig. 23.1h).

- **Drilling:** to produce a hole (Fig. 23.1i), which may be followed by boring to improve its dimensional accuracy and surface finish.

TABLE 23.1

General Characteristics of Machining Processes and Typical Dimensional Tolerances

Process	Characteristics	Typical dimensional tolerances, ± mm (in.)
Turning	Turning and facing operations on all types of materials, uses single-point or form tools, engine lathes require skilled labor, low production rate (but medium-to-high rate with turret lathes and automatic machines) requiring less-skilled labor	Fine: 0.025–0.13 (0.001–0.005) Rough: 0.13 (0.005)
Boring	Internal surfaces or profiles with characteristics similar to turning, stiffness of boring bar important to avoid chatter	0.025 (0.001)
Drilling	Round holes of various sizes and depths, high production rate, labor skill required depends on hole location and accuracy specified, requires boring and reaming for improved accuracy	0.075 (0.003)
Milling	Wide variety of shapes involving contours, flat surfaces, and slots; versatile; low-to-medium production rate; requires skilled labor	0.13–0.25 (0.005–0.01)
Planing	Large flat surfaces and straight contour profiles on long workpieces, low-quantity production, labor skill required depends on part shape	0.08–0.13 (0.003–0.005)
Shaping	Flat surfaces and straight contour profiles on relatively small workpieces, low-quantity production, labor skill required depends on part shape	0.05–0.13 (0.002–0.003)
Broaching	External and internal surfaces, slots, and contours; good surface finish; costly tooling; high production rate; labor skill required depends on part shape	0.025–0.15
Sawing	Straight and contour cuts on flat or structural shapes, not suitable for hard materials unless saw has carbide teeth or is coated with diamond, low production rate, generally low labor skill	0.8

- **Parting:** also called **cutting off,** to cut a piece from the end of a part, as is done in the production of slugs or blanks for additional processing into discrete products (Fig. 23.1j).
- **Threading:** to produce external or internal threads (Fig. 23.1k).
- **Knurling:** to produce a regularly shaped roughness on cylindrical surfaces, as in making knobs (Fig. 23.1l).

The cutting operations summarized typically are performed on a **lathe** (Fig. 23.2), which is available in a variety of designs, sizes, capacities, and computer-controlled features (as discussed in Section 23.3 and Chapter 25). As shown in Figs. 21.2 and 23.3, turning is performed at various (1) rotational speeds, N, of the workpiece clamped in a spindle, (2) depths-of-cut, d, and (3) feeds, f, depending on the workpiece materials, cutting-tool materials, surface finish and dimensional accuracy required, and the characteristics of the machine tool.

This chapter describes the process parameters, cutting tools, process capabilities, and the characteristics of the machine tools that are used to produce a variety of parts with round shapes. Finally, design considerations to improve productivity for each group of processes are given.

23.2 | The Turning Process

The majority of turning operations involve the use of simple single-point cutting tools, with the geometry of a typical right-hand cutting tool shown in Figs. 21.10 and 23.4. As can be seen, such tools are described by a standardized nomenclature. Each group of workpiece materials has an optimum set of tool angles, which have been developed largely through experience (Table 23.2).

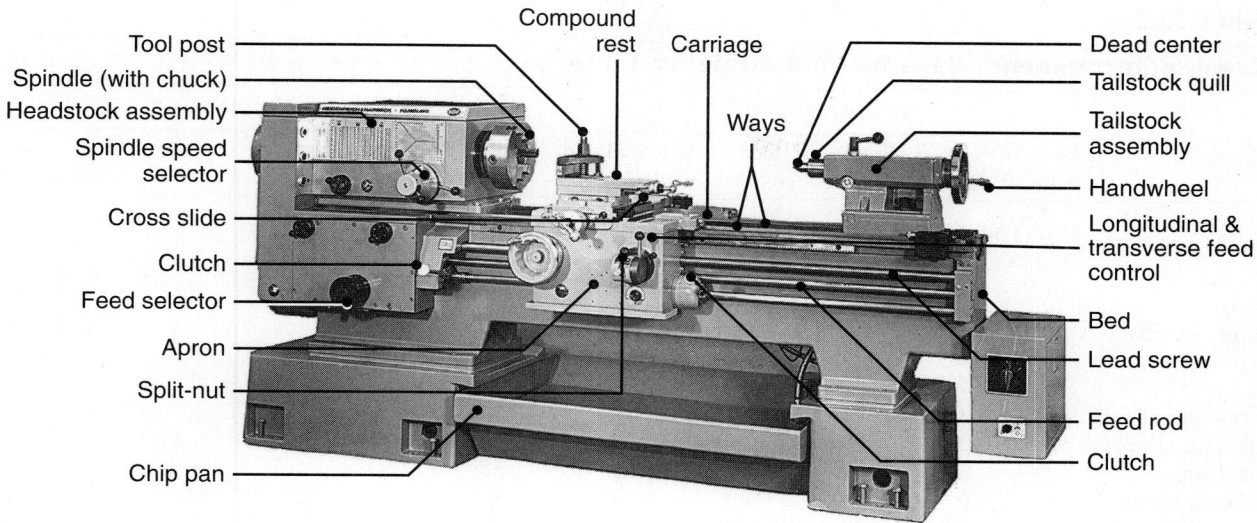

FIGURE 23.2 General view of a typical lathe, showing various components. *Source*: Courtesy of Heidenreich & Harbeck.

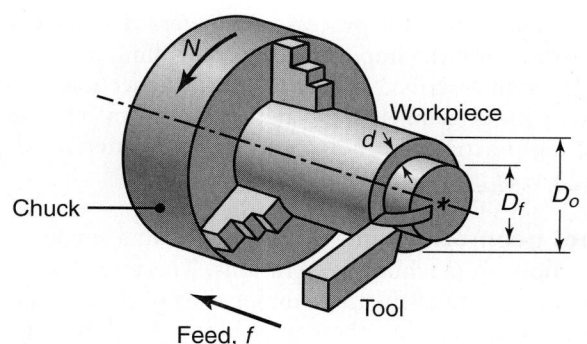

FIGURE 23.3 Schematic illustration of the basic turning operation, showing depth-of-cut, d; feed, f; and spindle rotational speed, N in rev/min. Cutting speed is the surface speed of the workpiece at the tool tip.

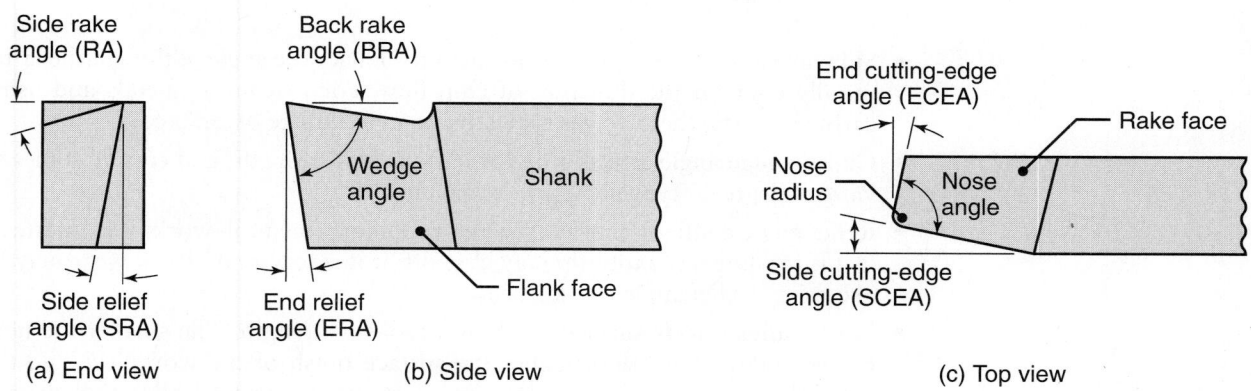

(a) End view　　　(b) Side view　　　(c) Top view

FIGURE 23.4 Designations for a right-hand cutting tool. Right-hand means that the tool travels from right to left, as shown in Fig. 23.3.

TABLE 23.2

General Recommendations for Tool Angles in Turning

Material	High-speed steel					Carbide inserts				
	Back rake	Side rake	End relief	Side relief	Side and end cutting edge	Back rake	Side rake	End relief	Side relief	Side and end cutting edge
Aluminum and magnesium alloys	20	15	12	10	5	0	5	5	5	15
Copper alloys	5	10	8	8	5	0	5	5	5	15
Steels	10	12	5	5	15	−5	−5	5	5	15
Stainless steels	5	8–10	5	5	15	−5–0	−5–5	5	5	15
High-temperature alloys	0	10	5	5	15	5	0	5	5	45
Refractory alloys	0	20	5	5	5	0	0	5	5	15
Titanium alloys	0	5	5	5	15	−5	−5	5	5	5
Cast irons	5	10	5	5	15	−5	−5	5	5	15
Thermoplastics	0	0	20–30	15–20	10	0	0	20–30	15–20	10
Thermosets	0	0	20–30	15–20	10	0	15	5	5	15

The important process parameters that have a direct influence on machining processes and the importance of controlling these parameters for optimized productivity were described in Chapter 21. This section outlines turning-process parameters such as tool geometry and material-removal rate and gives data for recommended cutting practices, including cutting-tool materials, depth-of-cut, feed, cutting speed, and use of cutting fluids.

Tool geometry. The various angles in a single-point cutting tool have important functions in machining operations. These angles are measured in a coordinate system consisting of the three major axes of the tool shank, as can be seen in Fig. 23.4. Note, however, that these angles may be different, with respect to the workpiece, after the tool is installed in the toolholder.

- **Rake angle** is important in controlling both the direction of chip flow and the strength of the tool tip. Positive rake angles improve the cutting operation by reducing forces and temperatures. However, positive angles also result in a small included angle of the tool tip (as in Figs. 21.3 and 23.4) which may lead to premature tool chipping and failure, depending on the toughness of the tool material.

- **Side rake angle** is more important than the **back rake angle**, although the latter usually controls the direction of chip flow. For machining metals and using carbide inserts, these angles typically are in the range of −5° to 5°.

- **Cutting-edge angle** affects chip formation, tool strength, and cutting forces to various degrees. Typically they are around 15°.

- **Relief angle** controls interference and rubbing at the tool–workpiece interface. If it is too large, the tool tip may chip off; if it is too small, flank wear may be excessive. Relief angles typically are 5°.

- **Nose radius** affects surface finish and tool-tip strength. The smaller the nose radius (sharp tool), the rougher the surface finish of the workpiece and the lower the strength of the tool. However, large nose radii can lead to tool *chatter*, as described in Section 25.4.

Material-removal rate.   The *material-removal rate* (MRR) in turning is the volume of material removed per unit time with the units of mm^3/min or in^3/min. Referring to Figs. 21.2 and 23.3, note that for each revolution of the workpiece, a ring-shaped layer of material is removed, which has a cross-sectional area that equals the product of the distance the tool travels in one revolution (feed, f) and the depth of cut, d. The volume of this ring is the product of the cross-sectional area $(f)(d)$ and the average circumference of the ring, πD_{avg}, where

$$D_{avg} = \frac{D_o + D_f}{2}$$

For light cuts on large-diameter workpieces, the average diameter may be replaced by D_o.

The rotational speed of the workpiece is N, and the material removal rate per revolution is $(\pi)(D_{avg})(d)(f)$. Since there are N revolutions per minute, the removal rate is

$$\text{MRR} = \pi\, D_{avg}\, d\, f\, N \qquad (23.1a)$$

The dimensional accuracy of this equation can be checked by substituting dimensions into the right-hand side. For instance, (mm)(mm)(mm/rev)(rev/min) = mm^3/min, which indicates volume rate of removal. Note that Eq. (23.1a) also can be written as

$$\text{MRR} = d\, f\, V \qquad (23.1b)$$

where V is the cutting speed and MRR has the same unit of mm^3/min.

The cutting time, t, for a workpiece of length l can be calculated by noting that the tool travels at a feed rate of fN = (mm/rev)(rev/min) = mm/min. Since the distance traveled is l mm, the cutting time is

$$t = \frac{l}{fN} \qquad (23.2)$$

The cutting time does not include the time required for *tool approach* and *retraction*. Because the time spent in noncutting cycles of a machining operation is unproductive and adversely affects the overall economics, the time involved in approaching and retracting tools to and from the workpiece is an important consideration. Machine tools are designed and built to minimize this time. One method of accomplishing this is to rapidly traverse the tools during noncutting cycles followed by a slower movement as the tool engages the workpiece.

The foregoing equations and the terminology used are summarized in Table 23.3.

Forces in turning.   The three principal forces acting on a cutting tool are shown in Fig. 23.5. These forces are important in the design of machine tools, as well as in the deflection of tools and workpieces for precision-machining operations. The machine tool and its components must be able to withstand these forces without causing significant deflections, vibrations, and chatter in the overall operation.

The **cutting force**, F_c, acts downward on the tool tip and, thus, tends to deflect the tool downward and the workpiece upward. The cutting force supplies the energy required for the cutting operation, and it can be calculated using the data given in Table 21.2, from the energy per unit volume, described in Section 21.3. The product of the cutting force and its radius from the workpiece center determines the *torque* on the spindle. The product of the torque and the spindle speed determines the *power* required in the turning operation.

TABLE 23.3

Summary of Turning Parameters and Formulas

N = Rotational speed of the workpiece, rpm

f = Feed, mm/rev or in./rev

v = Feed rate, or linear speed of the tool along workpiece length, mm/min or in./min

   = fN

V = Surface speed of workpiece, m/min or ft/min

   = $\pi D_o N$ (for maximum speed)

   = $\pi D_{avg} N$ (for average speed)

l = Length of cut, mm or in.

D_o = Original diameter of workpiece, mm or in.

D_f = Final diameter of workpiece, mm or in.

D_{avg} = Average diameter of workpiece, mm or in.

   = $(D_o + D_f)/2$

d = Depth of cut, mm or in.

   = $(D_o + D_f)/2$

t = Cutting time, s or min

   = l/fN

MRR = mm^3/min or in^3/min

   = $\pi D_{avg} df N$

Torque = $N \cdot m$ or $lb \cdot ft$

   = $F_c D_{avg}/2$

Power = kW or hp

   = $(Torque)(\omega)$, where $\omega = 2\pi N$ rad/min

Note: The units given are those that are used commonly; however, appropriate units must be used and checked in the formulas.

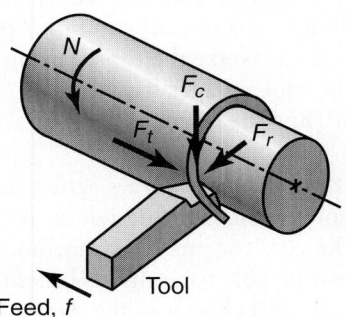

FIGURE 23.5 Forces acting on a cutting tool in turning. F_c is the cutting force, F_t is the thrust or feed force (in the direction of feed), and F_r is the radial force that tends to push the tool away from the workpiece being machined.

The **thrust force**, F_t, acts in the longitudinal direction. It also is called the **feed force** because it is in the feed direction of the tool. This force tends to push the tool towards the right, and away from the chuck. The **radial force**, F_r, acts in the radial direction and tends to push the tool away from the workpiece. Because of the many factors involved in the cutting process, forces F_t and F_r are difficult to calculate directly; they usually are determined experimentally if desired.

Roughing and finishing cuts.   In machining, the usual procedure is to first perform one or more *roughing cuts* at high feed rates and large depths-of-cut (and therefore high material-removal rates) but with little consideration of dimensional tolerance and surface roughness. These cuts then are followed by a *finishing cut*, at a lower feed and depth-of-cut in order to produce a good surface finish.

Tool materials, feeds, and cutting speeds.   The general characteristics of cutting-tool materials have been described in Chapter 22. A broad range of applicable cutting speeds and feeds for these tool materials is given in Fig. 23.6 as a general guideline in turning operations. Specific recommendations regarding turning-process parameters for various workpiece materials and cutting tools are given in Table 23.4.

Cutting fluids.   Many metallic and nonmetallic materials can be machined without a cutting fluid, but in most cases, the application of a cutting fluid can improve the operation significantly. General recommendations for cutting fluids appropriate to various workpiece materials are given in Table 23.5. However, there is a major current trend towards **near-dry** and **dry machining**, with important benefits, as described in Section 22.12.1.

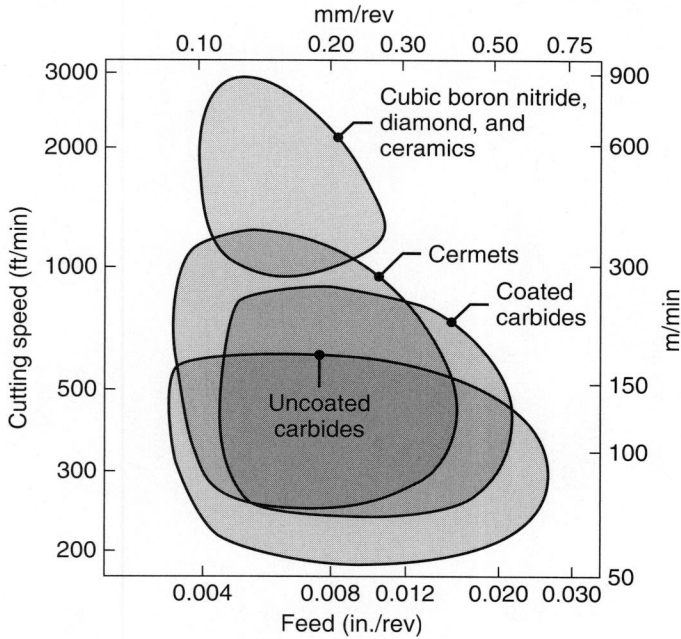

FIGURE 23.6   The range of applicable cutting speeds and feeds for a variety of tool materials.

TABLE 23.4

General Recommendations for Turning Operations

Workpiece material	Cutting tool	General-purpose starting conditions			Range for roughing and finishing		
		Depth of cut, mm (in.)	Feed, mm/rev (in./rev)	Cutting speed, m/min (ft/min)	Depth of cut, mm (in.)	Feed, mm/rev (in./rev)	Cutting speed, m/min (ft/min)
Low-C and free machining steels	Uncoated carbide	1.5–6.3 (0.06–0.25)	0.35 (0.014)	90 (300)	0.5–7.6 (0.02–0.30)	0.15–1.1 (0.006–0.045)	60–135 (200–450)
	Ceramic-coated carbide	"	"	245–275 (800–900)	"	"	120–425 (400–1400)
	Triple-coated carbide	"	"	185–200 (600–650)	"	"	90–245 (300–800)
	TiN-coated carbide	"	"	105–150 (350–500)	"	"	60–230 (200–750)
	Al_2O_3 ceramic	"	0.25 (0.010)	395–440 (1300–1450)	"	"	365–550 (1200–1800)
	Cermet	"	0.30 (0.012)	215–290 (700–950)	"	"	105–455 (350–1500)
Medium and high-C steels	Uncoated carbide	1.2–4.0 (0.05–0.20)	0.30 (0.012)	75 (250)	2.5–7.6 (0.10–0.30)	0.15–0.75 (0.006–0.03)	45–120 (150–400)
	Ceramic-coated carbide	"	"	185–230 (600–750)	"	"	120–410 (400–1350)
	Triple-coated carbide	"	"	120–150 (400–500)	"	"	75–215 (250–700)
	TiN-coated carbide	"	"	90–200 (300–650)	"	"	45–215 (150–700)
	Al_2O_3 ceramic	"	0.25 (0.010)	335 (1100)	"	"	245–455 (800–1500)
	Cermet	"	0.25 (0.010)	170–245 (550–800)	"	"	105–305 (350–1000)
Cast iron, gray	Uncoated carbide	1.25–6.3 (0.05–0.25)	0.32 (0.013)	90 (300)	0.4–12.7 (0.015–0.5)	0.1–0.75 (0.004–0.03)	75–185 (250–600)
	Ceramic-coated carbide	"	"	200 (650)	"	"	120–365 (400–1200)
	TiN-coated carbide	"	"	90–135 (300–450)	"	"	60–215 (200–700)
	Al_2O_3 ceramic	"	0.25 (0.010)	455–490 (1500–1600)	"	"	365–855 (1200–2800)
	SiN ceramic	"	0.32 (0.013)	730 (2400)	"	"	200–990 (650–3250)

(Continued)

Workpiece material	Cutting tool	Depth of cut mm (in.)	Feed mm/rev (in.)	Cutting speed m/min (ft/min)	Depth of cut mm (in.)	Feed mm/rev (in.)	Cutting speed m/min (ft/min)
Stainless steel, austenitic	Triple-coated carbide	1.5–4.4 (0.06–0.175)	0.35 (0.014)	150 (500)	0.5–12.7 (0.02–0.5)	0.08–0.75 (0.003–0.03)	75–230 (250–750)
	TiN-coated carbide	"	"	85–160 (275–525)	"	"	55–200 (175–650)
	Cermet	"	0.30 (0.012)	185–215 (600–700)	"	"	105–290 (350–950)
High-temperature alloys, nickel based	Uncoated carbide	2.5 (0.10)	0.15 (0.006)	25–45 (75–150)	0.25–6.3 (0.01–0.25)	0.1–0.3 (0.004–0.012)	15–30 (50–100)
	Ceramic-coated carbide	"	"	45 (150)	"	"	20–60 (65–200)
	TiN-coated carbide	"	"	30–55 (95–175)	"	"	20–85 (60–275)
	Al$_2$O$_3$ ceramic	"	"	260 (850)	"	"	185–395 (600–1300)
	SiN ceramic	"	"	215 (700)	"	"	90–215 (300–700)
	Polycrystalline cBN	"	"	150 (500)	"	"	120–185 (400–600)
Titanium alloys	Uncoated carbide	1.0–3.8 (0.04–0.15)	0.15 (0.006)	35–60 (120–200)	0.25–6.3 (0.01–0.25)	0.1–0.4 (0.004–0.015)	10–75 (30–250)
	TiN-coated carbide	"	"	30–60 (100–200)	"	"	10–100 (30–325)
Aluminum alloys Free machining	Uncoated carbide	1.5–5.0 (0.06–0.20)	0.45 (0.018)	490 (1600)	0.25–8.8 (0.01–0.35)	0.08–0.62 (0.003–0.025)	200–670 (650–2000)
	TiN-coated carbide	"	"	550 (1800)	"	"	60–915 (200–3000)
	Cermet	"	"	490 (1600)	"	"	215–795 (700–2600)
	Polycrystalline diamond	"	"	760 (2500)	"	"	305–3050 (1000–10,000)
High silicon	Polycrystalline diamond	"	"	530 (1700)	"	"	365–915 (1200–3000)

(Continued)

TABLE 23.4 (*Continued*)

General Recommendations for Turning Operations

Workpiece material	Cutting tool	General-purpose starting conditions			Range for roughing and finishing		
		Depth of cut, mm (in.)	Feed, mm/rev (in./rev)	Cutting speed, m/min (ft/min)	Depth of cut, mm (in.)	Feed, mm/rev (in./rev)	Cutting speed, m/min (ft/min)
Copper alloys	Uncoated carbide	1.5–5.0 (0.06–0.20)	0.25 (0.010)	260 (850)	0.4–7.51 (0.015–0.3)	0.15–0.75 (0.006–0.03)	105–535 (350–1750)
	Ceramic-coated carbide	"	"	365 (1200)	"	"	215–670 (700–2200)
	Triple-coated carbide	"	"	215 (700)	"	"	90–305 (300–1000)
	TiN-coated carbide	"	"	90–275 (300–900)	"	"	45–455 (150–1500)
	Cermet	"	"	245–425 (800–1400)	"	"	200–610 (650–2000)
	Polycrystalline diamond	"	"	520 (1700)	"	"	275–915 (900–3000)
Tungsten alloys	Uncoated carbide	2.5 (0.10)	0.2 (0.008)	75 (250)	0.25–5.0 (0.01–0.2)	0.12–0.45 (0.005–0.018)	55–120 (175–400)
	TiN-coated carbide	"	"	85 (275)	"	"	60–150 (200–500)
Thermoplastics and thermosets	TiN-coated carbide	1.2 (0.05)	0.12 (0.005)	170 (550)	0.12–5.0 (0.005–0.20)	0.08–0.35 (0.003–0.015)	90–230 (300–750)
	Polycrystalline diamond	"	"	395 (1300)	"	"	150–730 (500–2400)
Composites, graphite reinforced	TiN-coated carbide	1.9 (0.075)	0.2 (0.008)	200 (650)	0.12–6.3 (0.005–0.25)	0.12–1.5 (0.005–0.06)	105–290 (350–950)
	Polycrystalline diamond	"	"	760 (2500)	"	"	550–1310 (1800–4300)

Source: Based on data from Kennametal, Inc.

Note: Cutting speeds for high-speed-steel tools are about one-half those for uncoated carbides.

TABLE 23.5

General Recommendations for Cutting Fluids for Machining (see also Chapter 33)	
Material	Type of fluid
Aluminum	D, MO, E, MO + FO, CSN
Beryllium	MC, E, CSN
Copper	D, E, CSN, MO + FO
Magnesium	D, MO, MO + FO
Nickel	MC, E, CSN
Refractory metals	MC, E, EP
Steels	
carbon and low-alloy	D, MO, E, CSN, EP
stainless	D, MO, E, CSN
Titanium	CSN, EP, MO
Zinc	C, MC, E, CSN
Zirconium	D, E, CSN

Note: CSN = chemicals and synthetics; D = dry; E = emulsion; EP = extreme pressure; FO = fatty oil; and MO = mineral oil.

EXAMPLE 23.1 Material-removal rate and cutting force in turning

A 6-in.-long, 0.5-in.-diameter 304 stainless-steel rod is being reduced in diameter to 0.480 in. by turning on a lathe. The spindle rotates at $N = 400$ rpm, and the tool is traveling at an axial speed of 8 in./min. Calculate the cutting speed, material-removal rate, cutting time, power dissipated, and cutting force.

Solution The cutting speed is the tangential speed of the workpiece. The maximum cutting speed is at the outer diameter, D_o, and is obtained from the expression

$$V = \pi D_o N$$

Thus,

$$V = (\pi)(0.500)(400) = 628 \text{ in./min} = 52 \text{ ft/min}$$

The cutting speed at the machined diameter is

$$V = (\pi)(0.480)(400) = 603 \text{ in./min} = 50 \text{ ft/min}$$

From the information given, note that the depth-of-cut is

$$d = \frac{0.500 - 0.480}{2} = 0.010 \text{ in.}$$

and the feed is

$$f = \frac{8}{400} = 0.02 \text{ in./rev}$$

According to Eq. (23.1a), the material-removal rate is then

$$\text{MRR} = (\pi)(0.490)(0.010)(0.02)(400) = 0.12 \text{ in}^3/\text{min}$$

Equation (23.1b) also can be used, where we find MRR = $(0.01)(0.02)(52)$ $(12) = 0.12$ in^3/min. The actual time to cut, according to Eq. (23.2), is

$$t = \frac{6}{(0.02)(400)} = 0.75 \text{ min}$$

The power required can be calculated by referring to Table 21.2 and taking an average value for stainless steel as 4 W·s/mm^3 = 4/2.73 = 1.47 hp·min/in^3. Therefore, the power dissipated is

$$\text{Power} = (1.47)(0.123) = 0.181 \text{ hp}$$

Since 1 hp = 396,000 in.-lb/min, the power dissipated is 71,700 in.-lb/min.

The cutting force, F_c, is the tangential force exerted by the tool. Power is the product of torque, T, and the rotational speed in radians per unit time; hence,

$$T = \frac{(71,700)}{(2\pi)(400)} = 29 \text{ lb-in.}$$

Since $T = F_c D_{avg}/2$,

$$F_c = \frac{29}{0.490/2} = 118 \text{ lb}$$

23.3 | Lathes and Lathe Operations

Lathes generally are considered to be the oldest machine tools. Although woodworking lathes originally were developed during the period from 1000 to 1001 B.C., metalworking lathes with lead screws were not built until the late 1700s. The most common lathe originally was called an *engine lathe*, because it was powered with overhead pulleys and belts from nearby engines on the factory floor. Today, these lathes are all equipped with individual electric motors.

The maximum spindle speed of lathes typically is around 4000 rpm but only may be about 200 rpm for large lathes. For special applications, speeds may range to 10,000 rpm, 40,000 rpm, or higher for very high-speed machining. The cost of lathes ranges from about $2000 for bench types to over $100,000 for larger units.

Although simple and versatile, an engine lathe requires a skilled machinist, because all controls are manipulated by hand. Consequently, it is inefficient for repetitive operations and for large production runs. The rest of this section will describe the various types of automation that usually are added to improve efficiency.

23.3.1 Lathe components

Lathes are equipped with a variety of components and accessories, as shown in Fig. 23.2. Their features and functions are described next.

Bed.   The bed supports all major components of the lathe. Beds have a large mass and are built rigidly, usually from gray or nodular cast iron. (See also Section 25.3 on new materials for machine-tool structures.) The top portion of the bed has two **ways** with various cross-sections that are hardened and machined for wear resistance and dimensional accuracy during turning. In *gap-bed lathes*, a section of the bed in front of the headstock can be removed to accommodate larger-diameter workpieces.

Carriage. The carriage or carriage assembly slides along the ways and consists of an assembly of the *cross-slide, tool post*, and *apron*. The cutting tool is mounted on the *tool post*, usually with a *compound rest* that swivels for tool positioning and adjustment. The *cross-slide* moves radially in and out, controlling the radial position of the cutting tool in operations such as facing (Fig. 23.1e). The *apron* is equipped with mechanisms for both manual and mechanized movement of the carriage and the cross-slide by means of the *lead screw.*

Headstock. The headstock is fixed to the bed and is equipped with motors, pulleys, and V-belts that supply power to a *spindle* at various rotational speeds. The speeds can be set through manually-controlled selectors or by electrical controls. Most headstocks are equipped with a set of gears, and some have various drives to provide a continuously variable speed range to the spindle. Headstocks have a *hollow spindle* to which workholding devices (such as *chucks* and *collets*, see Section 23.3.2) are mounted and long bars or tubing can be fed through them for various turning operations. The accuracy of the spindle is important for precision in turning, particularly in high-speed machining; preloaded tapered or ball bearings typically are used to rigidly support the spindle.

Tailstock. The tailstock, which can slide along the ways and be clamped at any position, supports the other end of the workpiece. It is equipped with a center that may be fixed (*dead center*), or it may be free to rotate with the workpiece (*live center*). Drills and reamers can be mounted on the tailstock *quill* (a hollow cylindrical part with a tapered hole) to drill axial holes in the workpiece.

Feed rod and lead screw. The feed rod is powered by a set of gears through the headstock. The rod rotates during the lathe operation and provides movement to the carriage and the cross-slide by means of gears, a friction clutch, and a keyway along the length of the rod. Closing a *split nut* around the lead screw engages it with the carriage; it is also used for cutting threads accurately.

Lathe specifications. A lathe generally is specified by:

- Its *swing*, the maximum diameter of the workpiece that can be machined (Table 23.6).
- The maximum distance between the headstock and tailstock centers.
- The length of the bed.

For example, a lathe may have the following size: 360 mm (14 in.) swing by 760 mm (30 in.) between centers by 1830 mm (6 ft) length of bed. As we describe next, lathes are available in a variety of styles and types of construction and power. Maximum workpiece diameters may be as much as 2 m (78 in.).

23.3.2 Workholding devices and accessories

Workholding devices are important, particularly in machine tools and machining operations, as they must hold the workpiece securely. As shown in Fig. 23.3, one end of the workpiece is clamped to the spindle of the lathe by a chuck, collet, face plate (see Fig. 23.7d), or mandrel.

A **chuck** usually is equipped with three or four *jaws*. *Three-jaw* chucks generally have a geared-scroll design that makes the jaws self-centering. They are used for round workpieces (such as bar stock, pipes, and tubing), which can be centered to within 0.025 mm (0.001 in.). *Four-jaw* (independent) chucks have jaws that can be

TABLE 23.6

Typical Capacities and Maximum Workpiece Dimensions for Machine Tools			
Machine tool	Maximum dimension (m)	Power (kW)	Maximum rpm
Lathes (swing/length)			
Bench	0.3/1	<1	3000
Engine	3/5	70	4000
Turret	0.5/1.5	60	3000
Automatic screw	0.1/0.3	20	10,000
Boring machines (work diameter/length)			
Vertical spindle	4/3	200	300
Horizontal spindle	1.5/2	70	1000
Drilling machines			
Bench and column (drill diameter)	0.1	10	12,000
Radial (column to spindle distance)	3	—	—
Numerical control (table travel)	4	—	—

Note: Larger capacities are available for special applications.

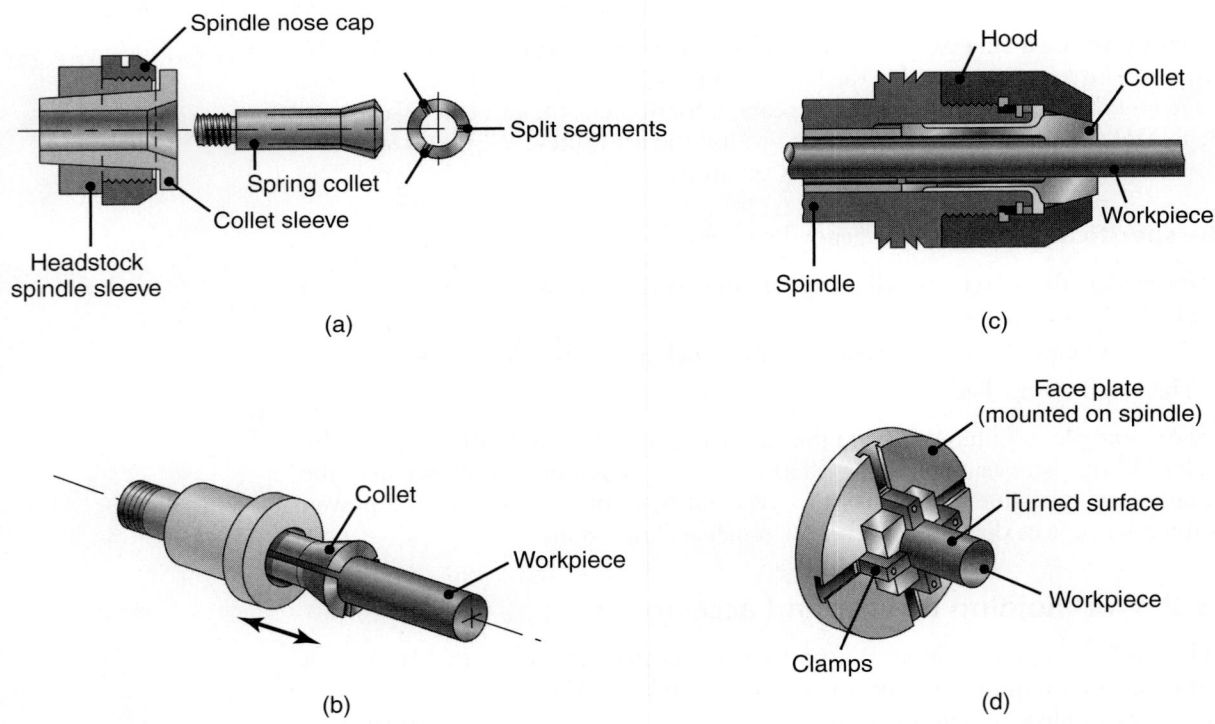

(a)

(b)

(c)

(d)

FIGURE 23.7   (a) and (b) Schematic illustrations of a draw-in type collet. The workpiece is placed in the collet hole, and the conical surfaces of the collet are forced inward by pulling it with a draw bar into the sleeve. (c) A push-out type collet. (d) Workholding of a workpiece on a face plate.

moved and adjusted independently of each other. Thus, they can be used for square, rectangular, or odd-shaped workpieces. Because they are constructed more ruggedly than three-jaw chucks, four-jaw chucks are used for heavy workpieces or for work requiring multiple chuckings where concentricity is important.

The jaws in some types of chucks can be reversed to permit clamping of the workpieces either on the outside surfaces or on the inside surfaces of hollow workpieces, such as pipes and tubing. Also available are jaws that are made of low-carbon steel (*soft jaws*) which can be machined into desired shapes. Because of their low strength and hardness, soft jaws conform to small irregularities on workpieces and, therefore, result in better clamping. Chucks can be *power* or *manually actuated* using a chuck wrench. Because they take longer to operate, manually actuated chucks generally are used only for toolroom and limited production runs.

Chucks are available in various designs and sizes. Their selection depends on the type and speed of operation, workpiece size, production and dimensional accuracy requirements, and the jaw forces required. By controlling the magnitude of jaw forces, an operator can ensure that the part does not slip in the chuck during machining. High spindle speeds can reduce jaw (clamping) forces significantly due to the effect of *centrifugal forces*; this effect is important particularly in precision tube turning. Modern jaw-actuating mechanisms permit a higher clamping force for roughing and lower force for finishing operations.

To meet the increasing demands for stiffness, precision, versatility, power, and high cutting speeds in modern machine tools, major advances have been made in the design of workholding devices. **Power chucks,** actuated pneumatically or hydraulically, are used in automated equipment for high production rates, including the loading of parts using industrial robots. Also available are several types of power chucks with lever- or wedge-type mechanisms to actuate the jaws; these chucks have jaw movements (stroke) that usually are limited to about 13 mm (0.5 in.).

A **collet** is basically a longitudinally-split, tapered bushing. The workpiece (generally with a maximum diameter of 1 in.) is placed inside the collet, and the collet is pulled (*draw-in collet*; Fig. 23.7a and b) or pushed (*push-out collet*; Fig. 23.7c) mechanically into the spindle. The tapered surfaces shrink the segments of the collet radially, tightening onto the workpiece. Collets are used for round workpieces as well as for other shapes (e.g., square or hexagonal workpieces) and are available in a wide range of incremental sizes.

One advantage to using a collet (rather than a three- or four-jaw chuck) is that the collet grips nearly all of the part circumference, making it well suited particularly for parts with small cross-sections. Because the radial movement of the collet segments is small, workpieces generally should be within 0.125 mm (0.005 in.) of the nominal size of the collet.

Face plates are used for clamping irregularly shaped workpieces. The plates are round and have several slots and holes through which the workpiece is bolted or clamped (Fig. 23.7d). **Mandrels** (Fig. 23.8) are placed inside hollow or tubular workpieces and are used to hold workpieces that require machining on both ends or on their cylindrical surfaces. Some mandrels are mounted between centers on the lathe.

Accessories. Several devices are available as accessories and attachments for lathes. Among these devices are the following:

- Carriage and cross-slide stops, with various designs to stop the carriage at a predetermined distance along the bed.
- Devices for turning parts with various tapers.
- Milling, sawing, gear-cutting, and grinding attachments.
- Various attachments for boring, drilling, and thread cutting.

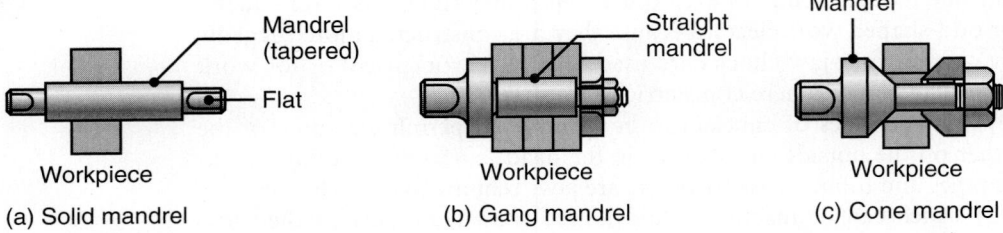

FIGURE 23.8 Various types of mandrels to hold workpieces for turning. These mandrels usually are mounted between centers on a lathe. Note that in (a), both the cylindrical and the end faces of the workpiece can be machined, whereas in (b) and (c), only the cylindrical surfaces can be machined.

23.3.3 Lathe operations

In a typical turning operation, the workpiece is clamped by any one of the workholding devices described previously. Long and slender parts must be supported by a *steady rest* and *follow rest* placed on the bed, as otherwise the part will deflect under the cutting forces. These rests usually are equipped with three adjustable fingers or rollers that support the workpiece while allowing it to rotate freely. Steady rests are clamped directly on the *ways* of the lathe (as in Fig. 23.2), whereas *follow rests* are clamped on the carriage and travel with it.

The cutting tool is attached to the tool post, which is driven by the lead screw, and removes material by traveling along the bed. A *right-hand* tool travels toward the headstock, and a *left-hand* tool travels toward the tailstock. Facing operations are done by moving the tool radially with the cross-slide and also clamping the carriage for better dimensional accuracy.

Form tools are used to produce various shapes on solid, round workpieces (Fig. 23.1g) by moving the tool radially inward while the part is rotating. Form cutting is not suitable for deep and narrow grooves or sharp corners because vibration and chatter may result and cause poor surface finish. As a rule, (a) the formed length of the part should not be greater than about 2.5 times the minimum diameter of the part, (b) the cutting speed should be set properly, and (c) cutting fluids should be used. The stiffnesses of the machine tools and workholding devices also are important considerations.

Boring on a lathe is similar to turning. It is performed inside hollow workpieces or in a hole made previously by drilling or other means. Out-of-shape holes can be straightened by boring. The workpiece is held in a chuck or in some other suitable workholding device. Boring large workpieces is described in Section 23.4.

Drilling (Section 23.5) can be performed on a lathe by mounting the drill bit in a chuck in the tailstock quill. The workpiece is clamped in a workholder on the headstock, and the drill bit is advanced by rotating the hand wheel of the tailstock. Holes drilled in this manner may not be sufficiently concentric because of the tendency for the drill to drift radially. The concentricity of the hole can be improved by subsequently boring the drilled hole. Drilled holes may be **reamed** (Section 23.6) on lathes in a manner similar to drilling, thus improving hole dimensional tolerances and surface finish.

The tools for *parting, grooving, thread cutting,* and various other operations are specially shaped for their particular purpose or are available as inserts. *Knurling* is performed on a lathe with hardened rolls (see Fig. 23.1l), in which the surface of the rolls is a replica of the profile to be generated. The rolls are pressed radially against the rotating workpiece, while the tool moves axially along the part.

23.3.4 Types of lathes

There are a number of other lathe types, brief descriptions of which are given here.

Bench lathes.   As the name suggests, these lathes are placed on a workbench or a table. They have low power, are usually operated by hand feed, and are used to machine small workpieces. *Toolroom lathes* have high precision, enabling the machining of parts to close dimensional tolerances.

Special-purpose lathes.   These lathes are used for applications (such as railroad wheels, gun barrels, and rolling-mill rolls) with workpiece sizes as large as 1.7 m in diameter by 8 m in length (66 in. to 25 ft) and capacities of 450 kW (600 hp).

Tracer lathes.   These lathes have special attachments that are capable of turning parts with various contours. Also called *duplicating lathes* or *contouring lathes*, the cutting tool follows a path that duplicates the contour of the template, similar to a pencil following the shape of a plastic stencil. However, operations typically performed on a tracer lathe have been replaced largely by *numerical-control lathes* and *turning centers*, as described in Section 25.2.

Automatic lathes.   Lathes have been automated increasingly over the years; manual machine controls have been replaced by various mechanisms that enable machining operations to follow a certain prescribed sequence. In a *fully automatic lathe*, parts are fed and removed automatically, whereas in *semiautomatic* machines, these functions are performed by the operator. Automatic lathes may have a horizontal or vertical spindle and are suitable for medium- to high-volume production.

Lathes that do not have tailstocks are called *chucking machines* or *chuckers*. They are used for machining individual pieces of regular or irregular shapes and are either single- or multiple-spindle types. In another type of automatic lathe, the bar stock is fed periodically into the lathe, and a part is machined and cut off from the end of the bar stock.

Automatic bar machines.   Also called **automatic screw machines**, these machine tools are designed for high-production-rate machining of screws and similar threaded parts. All operations on these machines are performed automatically with tools attached to a special turret. After each part or screw is machined to finished dimensions, the bar stock is fed forward automatically through the hole in the spindle and then cut off. These machines may be equipped with single or multiple spindles. Capacities range from 3- to 150-mm (1/8- to 6-in.) diameter bar stock. The long stock is supported by special fixtures as they enter the spindle hole.

Single-spindle automatic bar machines are similar to turret lathes and are equipped with various cam-operated mechanisms. There are two types of single-spindle machines. In *Swiss-type automatics,* the cylindrical surface of the solid-bar stock is machined with a series of tools that move radially and in the same plane—toward the workpiece. The bar stock is clamped close to the headstock spindle, which minimizes deflections due to cutting forces. These machine tools are capable of high-precision machining of small-diameter parts.

The other single-spindle machine (called the *American type*) is similar to a small, automatic turret lathe. The turret is on a vertical plane, and all motions of the machine components are controlled by cams. Automatic bar machines now are equipped with computer numerical controls, eliminating the use of cams, and the operation is programmed for a specific product. (See Section 37.3).

Multiple-spindle automatic bar machines typically have from four to eight spindles arranged in a circle on a large drum with each carrying an individual workpiece. The cutting tools are arranged in various positions in the machine and move in both axial and radial directions. Each part is machined in stages, as it moves from one station to the next. Because all operations are carried out simultaneously, the cycle time per part is reduced.

Turret lathes. These machine tools are capable of performing multiple cutting operations, such as turning, boring, drilling, thread cutting, and facing (Fig. 23.9). Several cutting tools (usually as many as six) are mounted on the hexagonal *main turret*, which is rotated after each specific cutting operation is completed. The lathe usually has a *square turret* on the cross-slide, mounting as many as four cutting tools. The workpiece (generally a long, round bar stock) is advanced a preset distance through the chuck. After the part is machined, it is cut off by a tool mounted on the square turret, which moves radially into the workpiece. The rod then is advanced the same preset distance, and the next part is machined.

Turret lathes (either bar type or chucking type) are versatile, and the operations may be carried out either by hand, using the turnstile (capstan wheel), or automatically. Once set up properly, these machines do not require highly skilled operators. *Vertical turret lathes* also are available; they are more suitable for short, heavy workpieces with diameters as large as 1.2 m (48 in.).

The turret lathe shown in Fig. 23.9 is known as a **ram-type** turret lathe—one in which the ram slides in a separate base on the saddle. The short stroke of the turret slide limits this machine to relatively short workpieces and light cuts in both small- and medium-quantity production. In another style (called the **saddle type**), the main turret is installed directly on the saddle, which slides along the bed. The

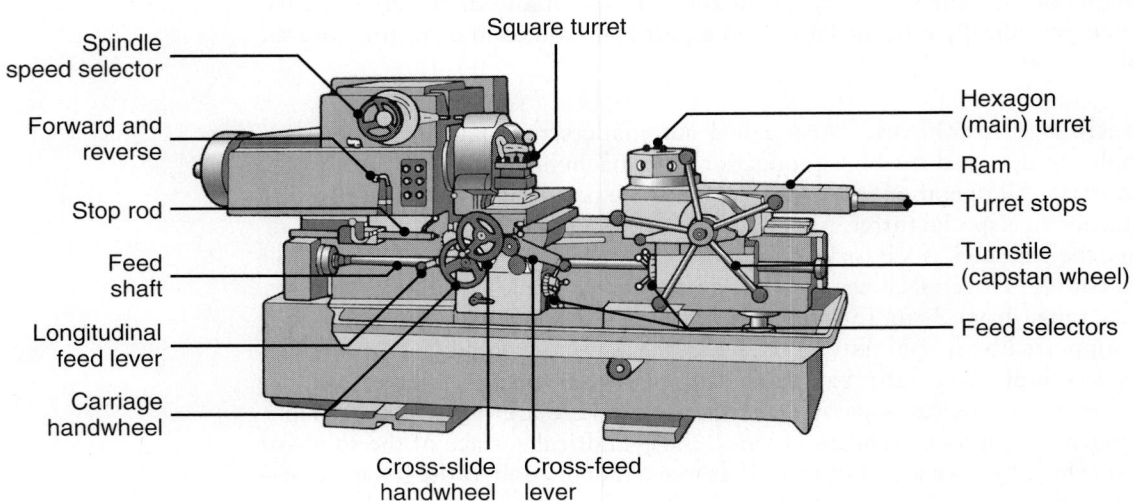

FIGURE 23.9 Schematic illustration of the components of a turret lathe. Note the two turrets: square and hexagonal (main).

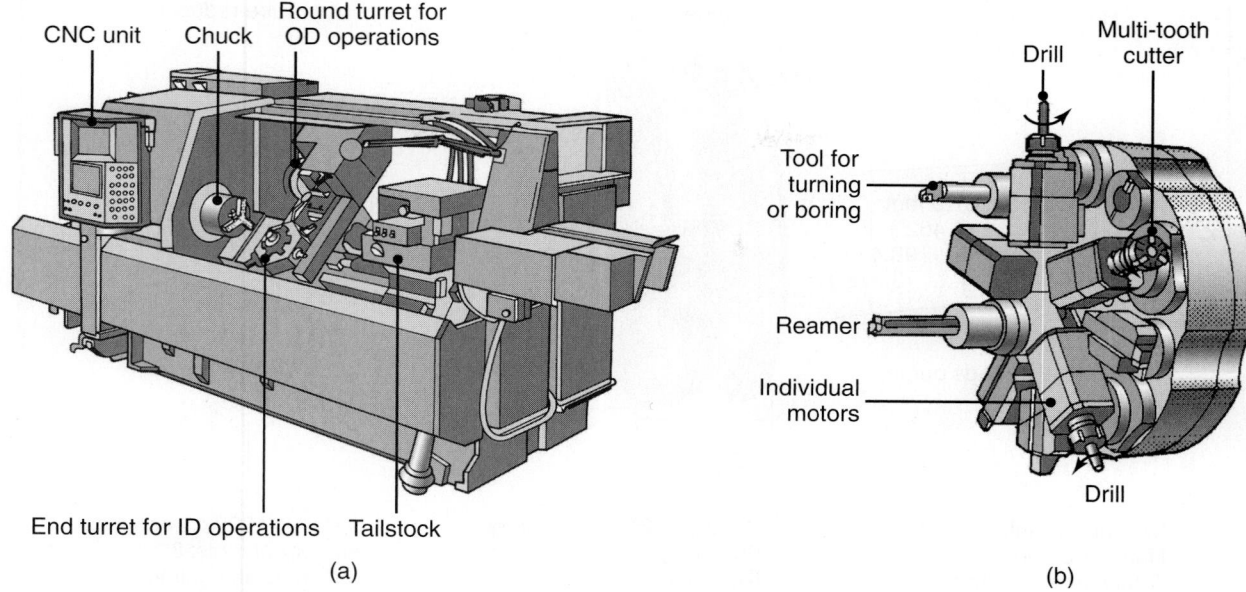

FIGURE 23.10   (a) A computer numerical-control lathe. Note the two turrets on this machine. These machines have higher power and spindle speed than other lathes in order to take advantage of new cutting tools with enhanced properties. (b) A typical turret equipped with ten tools, some of which are powered.

length of the stroke is limited only by the length of the bed. This type of lathe is constructed more heavily and is used to machine large workpieces. Because of the heavy weight of the components, saddle-type lathe operations are slower than ram type.

Computer-controlled lathes.   In the most advanced lathes, movement and control of the machine tool and its components are achieved by *computer numerical controls* (CNC). The features of such a lathe are shown in Fig. 23.10a. These lathes generally are equipped with one or more turrets, and each turret is equipped with a variety of tools and performs several operations on different surfaces of the workpiece (Fig. 23.10b). Workpiece diameters may be as much as 1 m (36 in.).

To take advantage of new cutting-tool materials, computer-controlled lathes are designed to operate faster and have higher power available as compared to other lathes. They are equipped with *automatic tool changers* (ATC). Their operations are reliably repetitive, maintain the desired dimensional accuracy, and require less-skilled labor (once the machine is set up). They are suitable for low- to medium-volume production.

EXAMPLE 23.2 Typical parts made on CNC turning-machine tools

The capabilities of CNC turning-machine tools are illustrated in Fig. 23.11. Material and number of cutting tools used and machining times are indicated for each part. These parts also can be made on manual or turret lathes, although not as effectively or consistently.

Source: Courtesy of Monarch Machine Tool Company

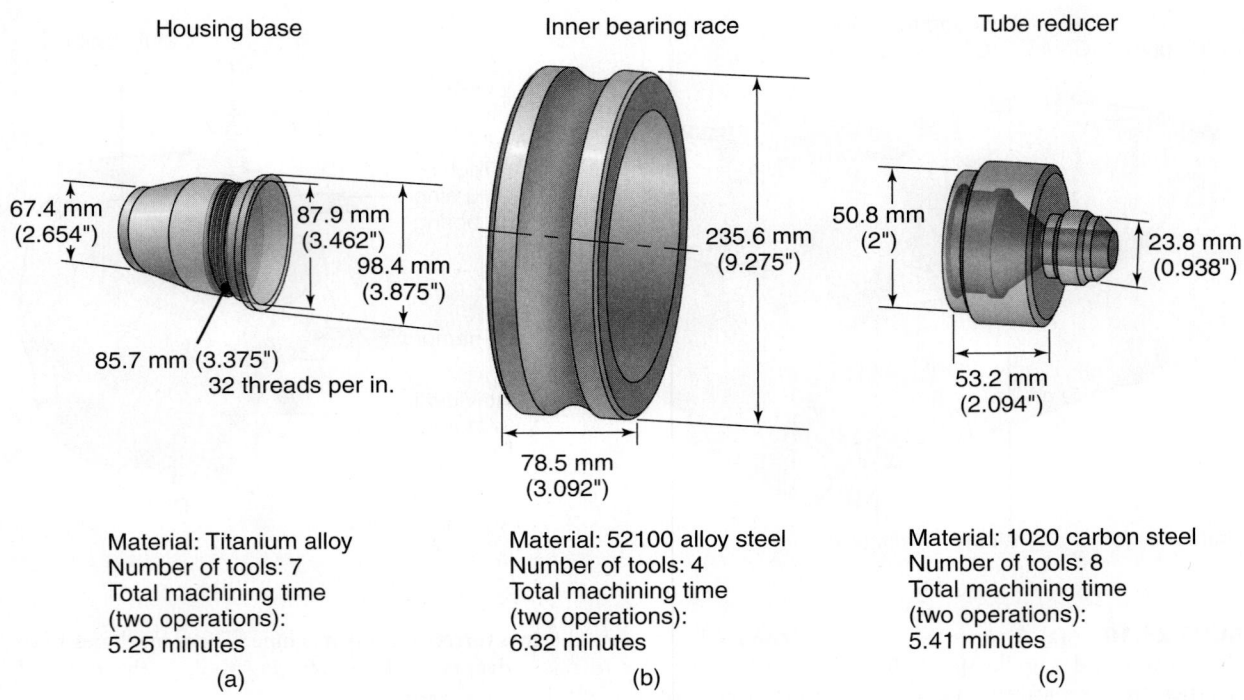

FIGURE 23.11   Typical parts made on CNC lathes.

EXAMPLE 23.3   Machining of complex shapes

Note that, in Example 23.2, the parts are axisymmetric. The capabilities of CNC turning can be further illustrated by referring to Figs. 23.12, which shows three additional, more complex parts: a pump shaft, a crankshaft, and a tubular part with internal rope thread. As in most operations, the machining of these parts consists of both roughing and finishing cuts.

1. *Pump shaft* (Fig. 23.12a). This part, as well as a wide variety of similar parts with external and internal features, including camshafts, was produced on a CNC lathe with two turrets, similar in construction to the machine shown in Fig. 23.10a. Each turret can hold as many as eight tools. To produce this particular shape, the upper turret is programmed in such a manner that its radial movement is synchronized with shaft rotation (Fig. 23.12b).

    The spindle turning angle is monitored directly, a processor performs a high-speed calculation, and the CNC issues a command to the cam turret in terms of that angle. It has absolute-position feedback using a high-accuracy scale system. The CNC compares the actual value with the commanded one and performs an automatic compensation using a built-in learning function. The turret has a lightweight design for smooth operation by reducing inertial forces.

    This shaft may be made of aluminum or stainless steel. The machining parameters for aluminum are given in Table 23.7 (see Part (a) in the first column of the table). These parameters may be compared with the data given in Table 23.4, which has only a broad and approximate range as a

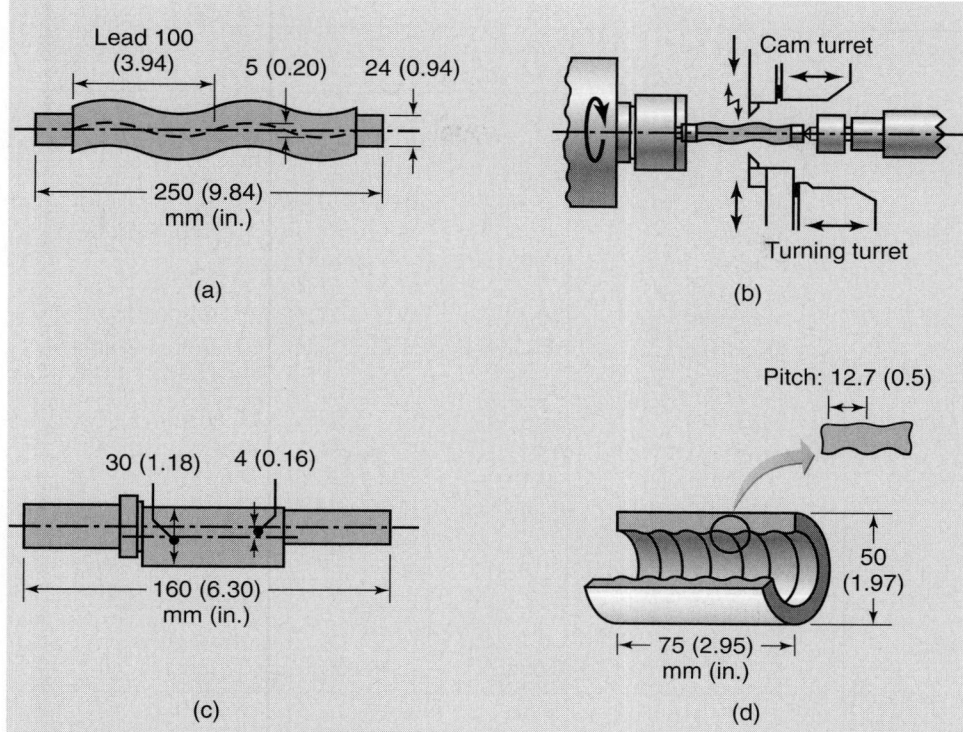

FIGURE 23.12   Examples of more complex shapes that can be produced on a CNC lathe.

guideline. The inserts were a K10 (C3) uncoated carbide with a compacted polycrystalline diamond (see Fig. 22.10). The OD machining in the table shown refers to the two, straight cylindrical ends of the part. The total machining time for an aluminum shaft was 24 min. For stainless steel, it was 55 min, because the cutting speed for stainless steel is considerably lower than that for aluminum.

2. *Crankshaft* (Fig. 23.12c). This part is made of ductile (nodular) cast iron, and the machining parameters are shown in Part (b) of Table 23.7. The insert was K10 carbide. The machining time was 25 min; note that this time is on the same order of magnitude as that for the part described above.

3. *Tubular part with internal rope threads* (Fig. 23.12d). This part, made of 304 stainless steel, was machined under the conditions given for Part (c) in Table 23.7. The starting blank was a straight tubular piece similar to a bushing. The cutting tools were coated carbide and cermet. The boring bar was made of tungsten carbide for increased stiffness and, hence, improved dimensional accuracy and surface finish. For the threaded portion, the dimensional accuracy was ±0.05 mm (0.002 in.), with a surface finish of R_a = 2.5 μm (100 μin.).

   The machining time for this part was 1.5 min, which is much shorter than those for the previous two parts. The reason is that (a) this part is shorter, (b) less material is removed, (c) it does not have the eccentricity features of the first two parts (so the radial movement of the cutting tool is not a function of the angular position of the part), and (d) the cutting speed is higher.

Source: Based on technical literature supplied by Okuma Corp.

TABLE 23.7

Operation	Speed rpm	Cutting speed	Depth of cut	Feed	Tool
Part a and b: **OD**					
Roughing	1150	160 m/min (525 fpm)	3 mm (0.12 in.)	0.3 mm/rev (0.012 in./rev)	K10 (C3)
Finishing	1750	250 (820)	0.2 (0.008)	0.15 (0.006)	K10 (C3)
Lead					
Roughing	300	45 (148)	3 (0.12)	0.15 (0.006)	K10 (C3)
Finishing	300	45 (148)	0.1 (0.004)	0.15 (0.006)	Diamond compact
Part c: Eccentric shaft					
Roughing	200	5–11 (16–136)	1.5 (0.059)	0.2 (0.008)	K10 (C3)
Finishing	200	5–11 (16–136)	0.1 (0.004)	0.05 (0.0020)	K10 (C3)
Part d: Internal thread					
Roughing	800	70 (230)	1.6 (0.063)	0.15 (0.006)	Coated carbide
Finishing	800	70 (230)	0.1 (0.004)	0.15 (0.006)	Cermet

23.3.5 Turning-process capabilities

Relative *production rates* in turning (as well as in other machining operations described in the rest of this chapter and in Chapter 24) are shown in Table 23.8. These rates have an important bearing on productivity in machining operations. Note that

TABLE 23.8

Typical Production Rates for Various Machining Operations	
Operation	Rate
Turning	
Engine lathe	Very low to low
Tracer lathe	Low to medium
Turret lathe	Low to medium
Computer-controlled lathe	Low to medium
Single-spindle chuckers	Medium to high
Multiple-spindle chuckers	High to very high
Boring	Very low
Drilling	Low to medium
Milling	Low to medium
Planing	Very low
Gear cutting	Low to medium
Broaching	Medium to high
Sawing	Very low to low

Note: Production rates indicated are relative: *Very low* is about one or more parts per hour; *medium* is approximately 100 parts per hour; and *very high* is 1000 or more parts per hour.

there are major differences in the production rate among these processes. These differences are due not only to the inherent characteristics of the processes and machine tools but also to various other factors, such as the setup times and the types and sizes of the workpieces to be machined.

The ratings given in Table 23.8 are relative, and there can be significant variations in special applications. For example, heat-treated, high-carbon, cast-steel rolls (for rolling mills) can be machined on special lathes using cermet tools at material-removal rates as high as 6000 cm^3/min (370 in^3/min). Also called *high removal-rate machining*, important requirements are (a) very high rigidity of the machine tool to avoid tool breakage due to chatter and (b) high power of up to 450 kW (600 hp).

The surface finish (Figs. 23.13) and dimensional accuracy (Fig. 23.14) obtained in turning and related operations depend on several factors, such as the characteristics and condition of the machine tool, stiffness, vibration and chatter, process parameters, tool geometry and wear, the use of cutting fluids, the machinability of the workpiece material, and operator skill. As a result (and not surprisingly), a wide range of surface finishes can be obtained, as shown in Fig. 23.13. (See also Fig. 27.4.)

23.3.6 Design considerations and guidelines for turning operations

Several considerations are important in designing parts to be machined economically by turning operations. Because machining in general (a) takes considerable time, thus increasing the production cost, (b) wastes material, and (c) is not as economical as forming or shaping parts, it must be avoided as much as possible. When turning operations are necessary, the following general design guidelines should be followed:

1. Parts should be designed so that they can be fixtured and clamped easily in workholding devices. Thin, slender workpieces are difficult to support properly so that they can withstand clamping and cutting forces. (See also *flexible fixturing*, Section 37.8.)

2. The dimensional accuracy and surface finish specified should be as wide as permissible for the part to still function properly.

3. Sharp corners, tapers, steps, and major dimensional variations in the part should be avoided.

4. Blanks to be machined should be as close to final dimensions as possible (such as by near-net-shape forming), so as to reduce production cycle time.

5. Parts should be designed so that cutting tools can travel across the workpiece without obstruction.

6. Design features should be such that commercially available standard cutting tools, inserts, and toolholders can be used.

7. Workpiece materials should be selected for their machinability (Section 21.7) as much as possible.

Guidelines for turning operations.   A general guide to the probable causes of problems in turning operations is given in Table 23.9. Recall that Chapters 21 and 22 described the factors influencing the parameters listed.

In addition to the various recommendations concerning tools and process parameters described thus far, an important factor is the presence of *vibration* and *chatter* (Section 25.4). Vibration during cutting can cause poor surface finish, poor dimensional accuracy, excessive tool wear, and premature tool failure. The following

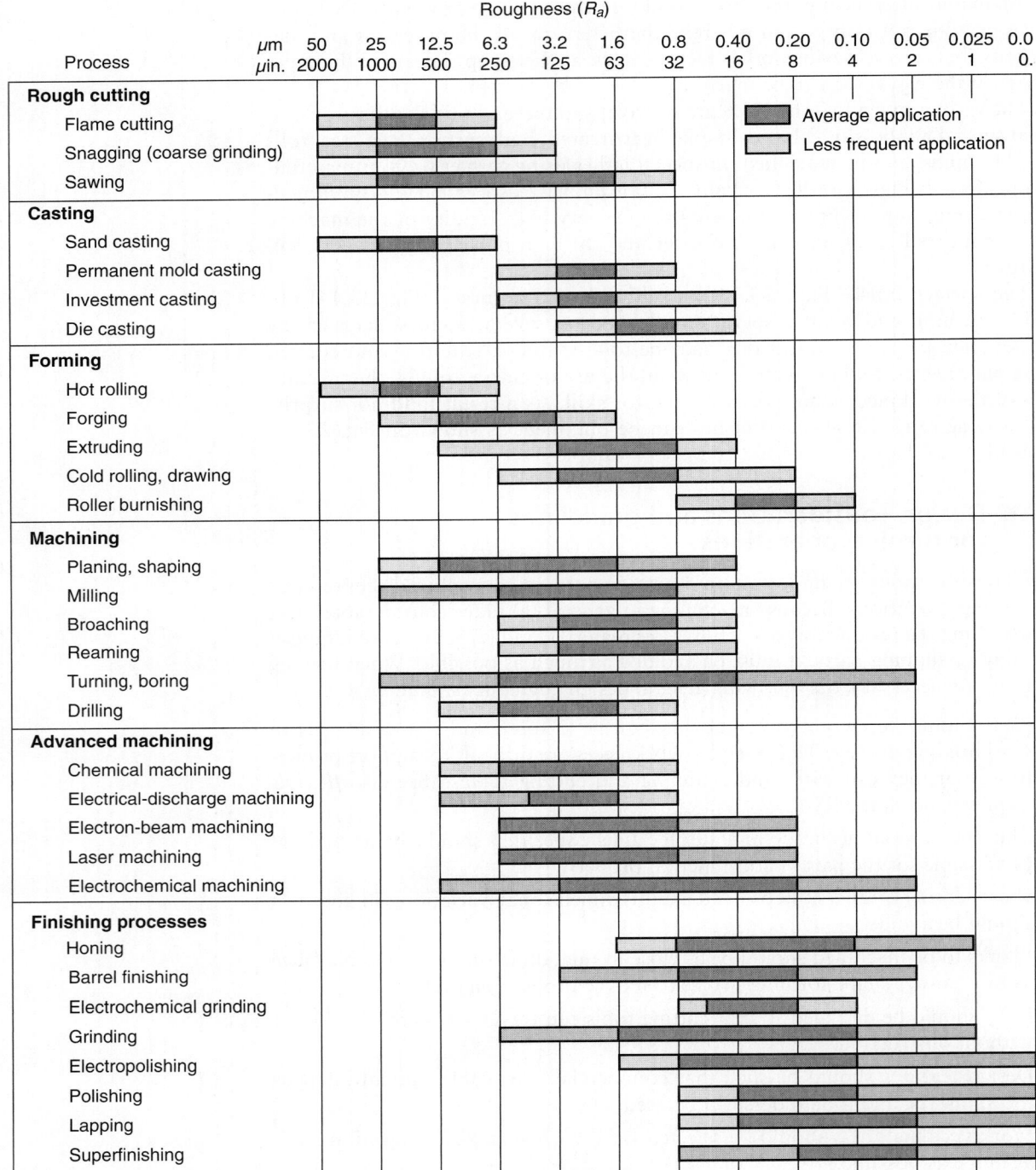

FIGURE 23.13   The range of surface roughnesses obtained in various processes. Note the wide range within each group, especially in turning and boring.

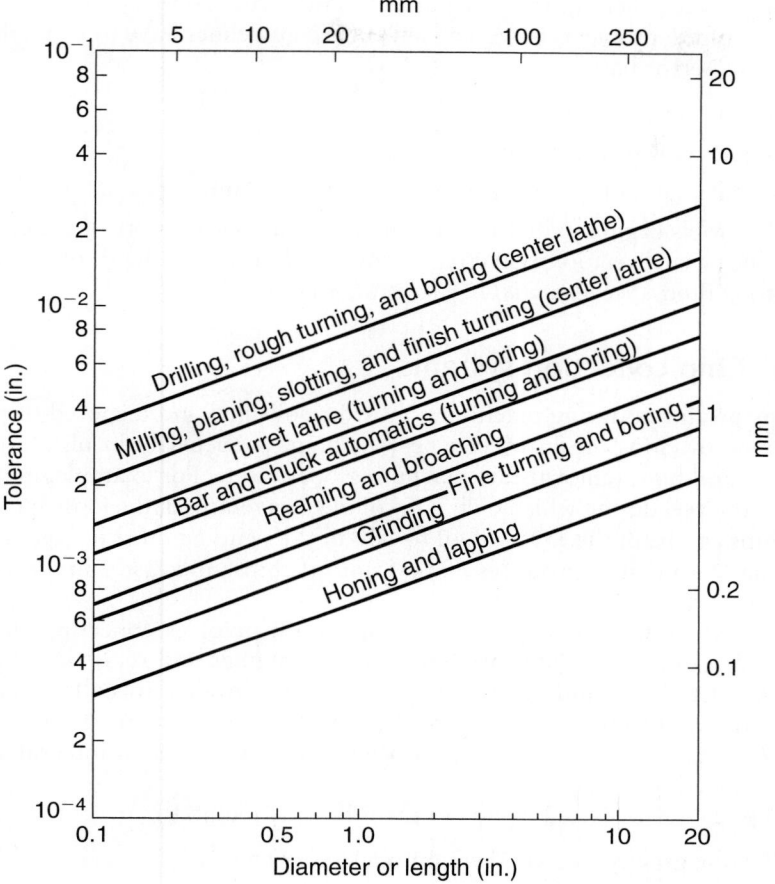

FIGURE 23.14   Range of dimensional tolerances obtained in various machining processes as a function of workpiece size. Note that there is an order of magnitude difference between small and large workpieces.

TABLE 23.9

General Troubleshooting Guide for Turning Operations	
Problem	Probable causes
Tool breakage	Tool material lacks toughness, improper tool angles, machine tool lacks stiffness, worn bearings and machine components, machining parameters too high
Excessive tool wear	Machining parameters too high, improper tool material, ineffective cutting fluid, improper tool angles
Rough surface finish	Built-up edge on tool; feed too high; tool too sharp, chipped, or worn; vibration and chatter
Dimensional variability	Lack of stiffness of machine tool and workholding devices, excessive temperature rise, tool wear
Tool chatter	Lack of stiffness of machine tool and workholding devices, excessive tool overhang, machining parameters not set properly

list outlines some generally accepted guidelines for turning operations. Because of the complexity of the problem, however, some of the guidelines have to be implemented on a trial-and-error basis.

1. Minimize tool overhang.
2. Support workpiece rigidly.
3. Use machine tools with high stiffness and high damping capacity.
4. When tools begin to vibrate and chatter, modify one or more of the process parameters, such as tool geometry, cutting speed, feed rate, depth-of-cut, or use of cutting fluid. (See also *adaptive control*, Section 37.4.)

23.3.7 Chip collection systems

The chips produced during machining must be collected and disposed of properly. The volume of chips produced can be very high, particularly in ultra-high-speed machining and high-removal-rate machining operations. For example, in a drilling operation on steel during which only 1 in^3 of metal is removed, the loose bulk volume of the chips can be in the range of 40 to 800 in^3, depending on chip type. Likewise, the milling of 1 in^3 steel produces 30 to 45 in^3 of chips, and cast iron produces 7 to 15 in^3 of chips.

Also called **chip management**, the operation involves collecting chips from their source in the machine tool in an efficient manner and removing them from the work area. Long and stringy chips are more difficult to collect than short chips, which are produced by using tools with chipbreaker features (see Figs. 21.7 and 22.2). Thus, the type of chip produced (*chip control*) is an integral aspect of the chip-collecting system.

Chips can be collected by any of the following methods:

- Allowing gravity to drop them onto a steel conveyor belt.
- Dragging the chips from a settling tank.
- Using augers with feed screws (similar to those in meat grinders).
- Using magnetic conveyors (for ferrous chips).
- Employing vacuum methods of chip removal.

Modern machine tools are designed with automated chip-handling features. It should be noted that there may be a considerable amount of cutting fluid mixed with and adhering to the chips produced, hence proper filtration or draining is important. The cutting fluid and sludge can be separated using *chip wringers* (centrifuges). Chip-processing systems usually require considerable floor space and can cost from $60,000 for small shops to over $1 million for large plants.

The collected chips may be recycled if it is economical to do so. Prior to their removal from a manufacturing plant, the large volume of chips can be reduced to as little as one-fifth of the loose volume by *compaction* (crushing) into briquettes or by *shredding*. Dry chips are more valuable for recycling because of reduced environmental contamination. The method chosen for chip disposal depends on economics as well as on compliance with local, state, and federal regulations. The trend is to recycle all chips as well as the used cutting fluids and the sludge.

23.3.8 Cutting screw threads

Screw threads are extremely common features. A *screw thread* may be defined as a ridge of uniform cross-section that follows a helical or spiral path on the outside or inside of a cylindrical (*straight thread*) or tapered surface (*tapered thread*). Machine screws, bolts, and nuts have straight threads, as do threaded rods for applications such as the lead screw in lathes and various machinery components (Fig. 23.2). Threads may be *right-handed* or *left-handed*. Tapered threads commonly are used for water or gas pipes and plumbing supplies, which require a watertight or airtight connection.

Threads traditionally have been machined, but increasingly, they are formed by **thread rolling** (described in Section 13.5). Rolled threads now constitute the largest quantity of threaded parts produced. It also may be possible to cast threaded parts, but there are limitations to dimensional accuracy, surface finish, and minimum dimensions. Production rates are not as high as those obtained in other processes.

Threads can be machined externally or internally with a cutting tool with a process called *thread cutting* or *threading*. Internal threads also can be produced with a special threaded tool, called a *tap*, and the process is called *tapping* (Section 23.7). External threads also may be cut with a *die* or by milling. Although it adds considerably to the cost, threads subsequently may be ground for high dimensional accuracy and surface finish for applications such as screw drives in machines.

Screw-thread cutting on a lathe.   A typical thread-cutting operation on a lathe is shown in Fig. 23.15a. The cutting tool, the shape of which depends on the type of

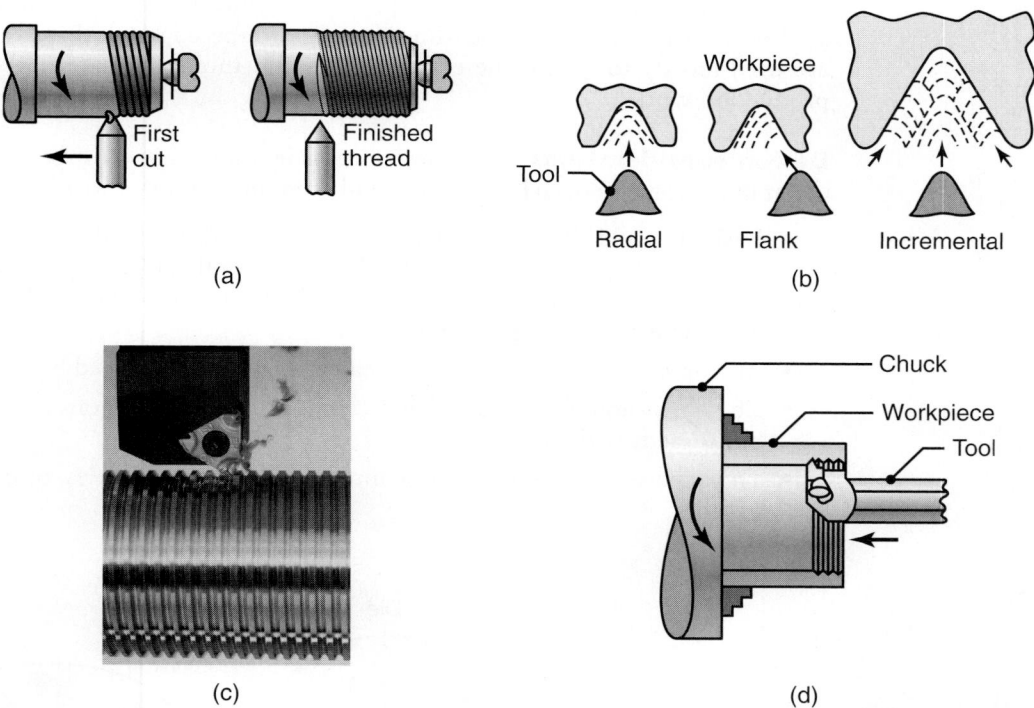

FIGURE 23.15   (a) Cutting screw threads on a lathe with a single-point cutting tool. (b) Cutting screw threads with a single-point tool in several passes, normally utilized for large threads. The small arrows in the figures show the direction of feed, and the broken lines show the position of the cutting tool as time progresses. Note that in *radial cutting*, the tool is fed directly into the workpiece. In *flank cutting*, the tool is fed into the piece along the right face of the thread. In *incremental cutting*, the tool is first fed directly into the piece at the center of the thread, then at its sides, and finally into the root. (c) A typical coated-carbide insert in the process of cutting screw threads on a round shaft. (d) Cutting internal screw threads with a carbide insert. *Source*: (c): Courtesy of Iscar Metals Inc.

thread to be cut, is mounted on a holder and moved along the length of the workpiece by the *lead screw* on the lathe. This movement is achieved by the engagement of a *split nut* (also called a *half nut*) inside the apron of the lathe (see Fig. 23.2).

The axial movement of the tool in relation to the rotation of the workpiece determines the *lead* of the screw thread (that is, the axial distance moved in one complete revolution of the screw). For a fixed spindle speed, the slower the tool movement, the finer the thread will be. In thread cutting, the cutting tool may be fed radially into the workpiece, thus cutting both sides of the thread at the same time, as in form cutting described earlier. However, this method usually produces a poor surface finish.

A number of passes in the sequence shown in Fig. 23.15b generally are required to produce threads with good dimensional accuracy and surface finish. Figure 23.15c shows a carbide insert for screw-thread cutting (*threading insert*) machining threads on a round shaft, and Fig. 23.15d shows an internal screw-thread cutting process. Cutting threads on a lathe is an old and versatile method, but it requires considerable operator skill and is a slow process. Consequently, except for small production runs, the process largely has been replaced by other methods, such as thread rolling, automatic screw machining, and the use of CNC lathes.

The production rate in cutting screw threads can be increased with tools called *die-head chasers* (Fig. 23.16a and b). These tools typically have four cutters with multiple teeth and can be adjusted radially. After the threads are cut, the cutters open automatically (thus the alternative name *self-opening die heads*) by rotating around their axes to allow the part to be removed. *Solid-threading dies* (Fig. 23.16c) also are available for cutting straight or tapered screw threads. These dies are used mostly to thread the ends of pipes and tubing and are not suitable for production work.

Design considerations.   The design considerations that must be taken into account in order to produce high-quality and economical screw threads are as follows:

- Designs should allow for the termination of threads before they reach a shoulder. Internal threads in blind holes should have an unthreaded length at the bottom. The term *blind hole* refers to a hole that does not go through the thickness of the workpiece. (For example, see Fig. 23.1i.)
- Attempts should be made to eliminate shallow, blind tapped holes.
- Chamfers should be specified at the ends of threaded sections to minimize fin-like threads with burrs.
- Threaded sections should not be interrupted with slots, holes, or other discontinuities.

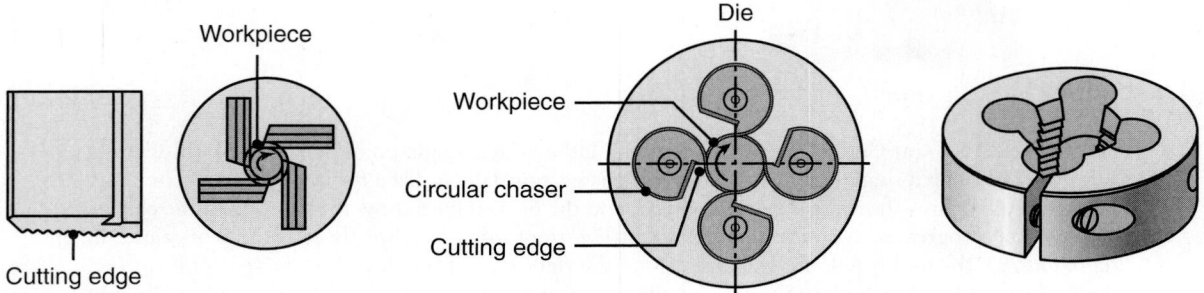

FIGURE 23.16   (a) Straight chasers for cutting threads on a lathe. (b) Circular chasers. (c) A solid threading die.

- Standard threading tooling and inserts should be used as much as possible.

- Thin-walled parts should have sufficient thickness and strength to resist the clamping and cutting forces. A common rule of thumb is that the minimum engagement length of a fastener should be 1.5 times the diameter.

- Parts should be designed so that all cutting operations can be completed in one setup.

23.4 | Boring and Boring Machines

Boring is performed to enlarge a hole made previously by some other process or to produce circular internal profiles in hollow workpieces (Fig. 23.1h). The cutting tools are similar to those used in turning and are mounted on a *boring bar* (Fig. 23.17a) to reach the full length of the bore. The boring bar must be sufficiently stiff to minimize tool deflection, thus maintain dimensional accuracy and avoid vibration and chatter. A material with a high elastic modulus (such as tungsten carbide) would be desirable. Boring bars have been designed and built with capabilities for damping vibration (Fig. 23.17b).

Boring operations on relatively small workpieces can be carried out on a lathe; large workpieces are machined on **boring mills**. These machine tools are either horizontal or vertical and are capable of performing various operations, such as turning, facing, grooving, and chamfering. In **horizontal boring machines**, the workpiece is mounted on a table that can move horizontally in both the axial and radial directions. The cutting tool is mounted on a spindle that rotates in the headstock, which is capable of both vertical and longitudinal movements. Drills, reamers, taps, and milling cutters also can be mounted on the machine spindle. A **vertical boring mill** (Fig. 23.18) is similar to a lathe, has a vertical axis of workpiece rotation, and can accommodate workpieces with diameters as much as 2.5 m (98 in.).

The cutting tool is usually a single point, made of M2 and M3 high-speed steel or P10 (C7) and P01 (C8) carbide. It is mounted on the tool head, which is capable of vertical movement (for boring and turning) and radial movement (for facing) guided by the cross-rail. The head can be swiveled to produce conical (tapered) holes. Cutting speeds and feeds for boring are similar to those for turning. (For capabilities of boring operations, see Table 23.8.)

Boring machines are available with a variety of features. Machine capacities range up to 150 kW (200 hp) and are available with computer numerical controls,

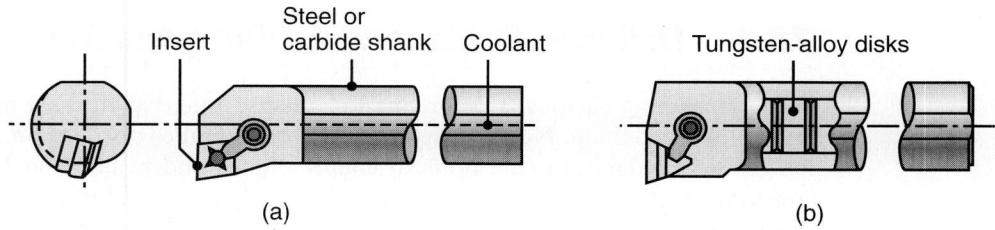

FIGURE 23.17 (a) Schematic illustration of a steel boring bar with a carbide insert. Note the passageway in the bar for cutting fluid application. (b) Schematic illustration of a boring bar with tungsten-alloy "inertia disks" sealed in the bar to counteract vibration and chatter during boring. This system is effective for boring bar length-to-diameter ratios of up to 6.

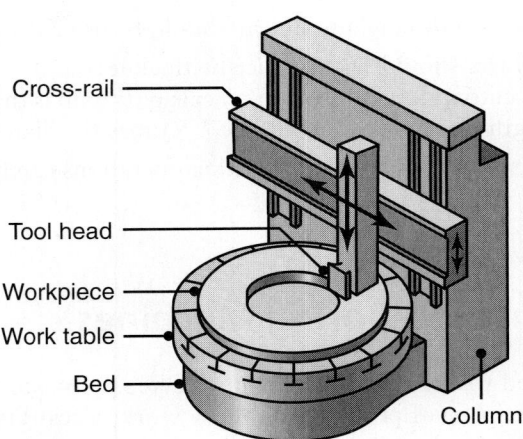

FIGURE 23.18 Schematic illustration of a vertical boring mill. Such a machine can accommodate workpiece sizes as large as 2.5 m (98 in.) in diameter.

allowing all movements of the machine to be programmed. Little operator involvement is required and consistency and productivity are improved.

Jig borers are vertical boring machines with high-precision bearings. Although they are available in various sizes and are used in tool rooms for making jigs and fixtures, they now are being replaced by more versatile numerical control machines.

Design considerations for boring. Guidelines for efficient and economical boring operations are similar to those for turning. Additionally, the following factors should be considered:

- Whenever possible, through holes rather than blind holes should be specified. Recall that the term *blind hole* refers to a hole that does not go through the thickness of the workpiece.
- The greater the length-to-bore-diameter ratio, the more difficult it is to hold dimensions because of the deflections of the boring bar due to cutting forces, as well as the higher tendency for vibration and chatter.
- Interrupted internal surfaces—such as internal splines or radial holes that go through the tickness of the part—should be avoided.

23.5 | Drilling, Drills, and Drilling Machines

When inspecting various large or small products, note that the vast majority have several holes in them. Note, for example, (a) the number of rivets on an airplane's wings and fuselage, (b) the bolts in engine blocks and heads, and (c) numerous consumer and industrial products. Holes typically are used for assembly with fasteners (such as bolts, screws, and rivets, each of which requires a hole) or for design purposes (such as weight reduction, ventilation, access to the inside of parts, or for appearance).

Hole making is among the most important operations in manufacturing, and **drilling** is a major and common hole-making process. Other processes for producing holes are punching (as described in Section 16.2) and various advanced machining processes (Chapter 27). The cost of hole making is among the highest machining costs in automotive engine production.

23.5.1 Drills

Drills typically have high length-to-diameter ratios (Fig. 23.19), hence they are capable of producing relatively deep holes. However, drills are somewhat flexible and should be used with care in order to drill holes accurately and to prevent breakage. Furthermore, the chips that are produced within the hole move in a direction opposite to the forward movement of the drill. Thus, chip disposal and the effectiveness of cutting fluids can present significant difficulties in drilling.

Drills generally leave a *burr* on the bottom surface upon breakthrough, necessitating deburring operations (Section 26.8). Also, because of its rotary motion, drilling produces holes with walls that have circumferential marks. In contrast, punched holes have longitudinal marks (see Fig. 16.5a). This difference is significant in terms of the hole's fatigue properties, as we describe in Section 33.2.

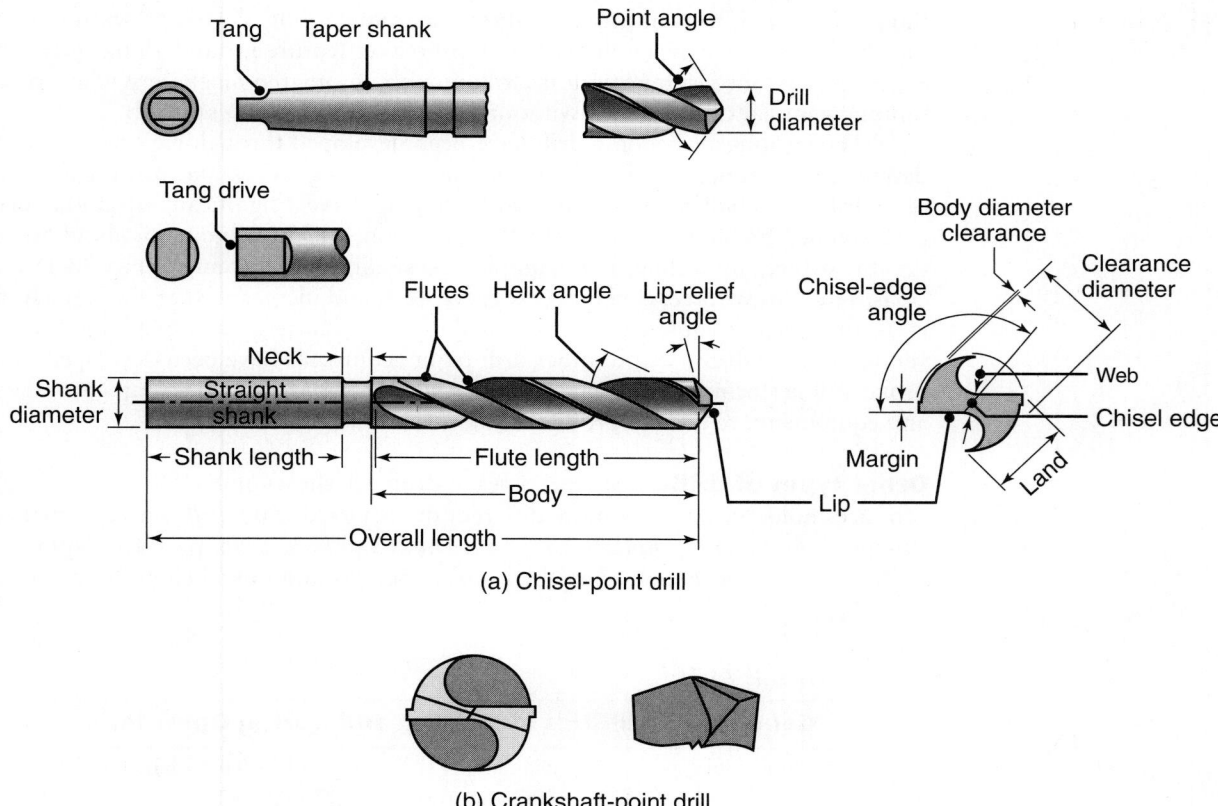

(a) Chisel-point drill

(b) Crankshaft-point drill

FIGURE 23.19   Two common types of drills: (a) Chisel-point drill. The function of the pair of margins is to provide a bearing surface for the drill against walls of the hole as it penetrates into the workpiece. Drills with four margins (*double-margin*) are available for improved drill guidance and accuracy. Drills with chip-breaker features also are available. (b) Crankshaft drill. These drills have good centering ability, and because chips tend to break up easily, these drills are suitable for producing deep holes.

The diameter of a hole produced by drilling is slightly larger than the drill diameter (*oversize*), as one can note by observing that a drill can be removed easily from the hole it has just produced. The amount of oversize depends on the quality of the drill and of the equipment used, as well as on the machining practices employed. Furthermore, depending on their thermal properties, some metals and nonmetallic materials expand significantly due to the heat produced by drilling, thus the final hole diameter could be smaller than the drill diameter. For better surface finish and dimensional accuracy, drilled holes may be subjected to subsequent operations, such as *reaming* and *honing*. The capabilities of drilling and boring operations are shown in Table 23.10.

Twist drill.   The most common drill is the conventional *standard-point twist drill* (Fig. 23.19a). The geometry of the drill point is such that the normal rake angle and velocity of the cutting edge vary with the distance from the center of the drill. The main features of this drill are (with typical ranges of angles given in parentheses): (a) *point angle* (118° to 135°), (b) *lip-relief angle* (7° to 15°), (c) *chisel-edge angle* (125° to 135°), and (d) *helix angle* (15° to 30°).

Two spiral grooves (*flutes*) run the length of the drill, and the chips produced are guided upward through these grooves. The grooves also serve as passageways to enable the cutting fluid to reach the cutting edges. Some drills have internal longitudinal holes (see, for example, the drill shown in Fig. 23.22a) through which cutting fluids are forced, thus improving lubrication and cooling as well as washing away the chips. Drills are available with a **chip-breaker** feature ground along the cutting edges. This feature is important in drilling with automated machinery where a continuous removal of long chips without operator assistance is essential.

The various angles on a drill have been developed through experience and are designed to produce accurate holes, minimize drilling forces and torque, and optimize drill life. Small changes in drill geometry can have a significant effect on a drill's performance, particularly in the chisel-edge region, which accounts for about 50% of the thrust force in drilling. For example, too small a lip relief angle (Fig. 23.19a) increases the thrust force, generates excessive heat, and increases wear. Conversely, too large an angle can cause chipping or breaking of the cutting edge. In addition to conventional point drills, several other drill-point geometries have been developed to improve drill performance and increase the penetration rate. Special grinding techniques and equipment are used to produce these geometries.

Other types of drills.   Several types of drills are shown in Fig. 23.20. A *step drill* produces holes with two or more different diameters. A *core drill* is used to make an existing hole larger. *Counterboring* and *countersinking drills* produce depressions on the surface to accommodate the heads of screws and bolts below the workpiece

TABLE 23.10

General Capabilities of Drilling and Boring Operations			
Cutting tool type	Diameter range (mm)	Hole depth/diameter	
		Typical	Maximum
Twist	0.5–150	8	50
Spade	25–150	30	100
Gun	2–50	100	300
Trepanning	40–250	10	100
Boring	3–1200	5	8

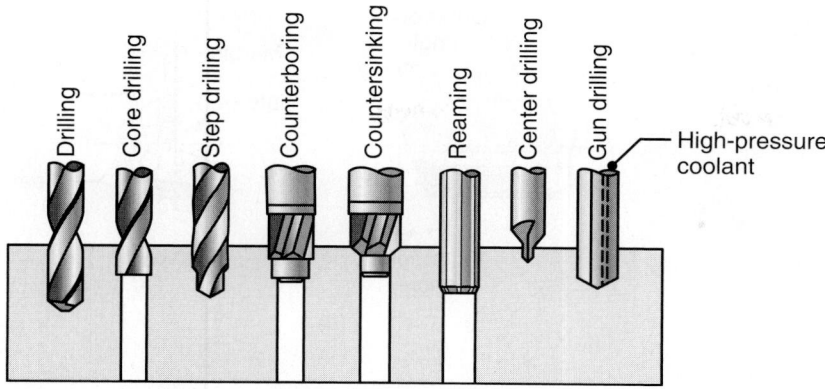

FIGURE 23.20   Various types of drills and drilling and reaming operations.

surface. A *center drill* is short and is used to produce a hole at the end of a piece of stock, so that it may be mounted between centers of the headstock and the tailstock of a lathe (Fig. 23.2). A *spot drill* is used to spot (to start) a hole at the desired location on a surface.

Spade drills (Fig. 23.21a) have removable tips or bits and are used to produce large-diameter and deep holes. These drills have the advantages of higher stiffness (because of the absence of flutes in the body of the drill), ease of grinding the cutting edges, and lower cost. A similar drill is the *straight-flute drill* (Fig. 23.21b).

Solid carbide and carbide-tipped drills (Fig. 23.21c and d) are available for drilling hard materials (such as cast irons), high-temperature metals, abrasive materials (such as concrete and brick—*masonry drills*), and composite materials with abrasive fiber reinforcements (such as glass and graphite).

Gun drilling.   Developed originally for drilling gun barrels, *gun drilling* is used for drilling deep holes and requires a special drill (Fig. 23.22). The depth-to-diameter

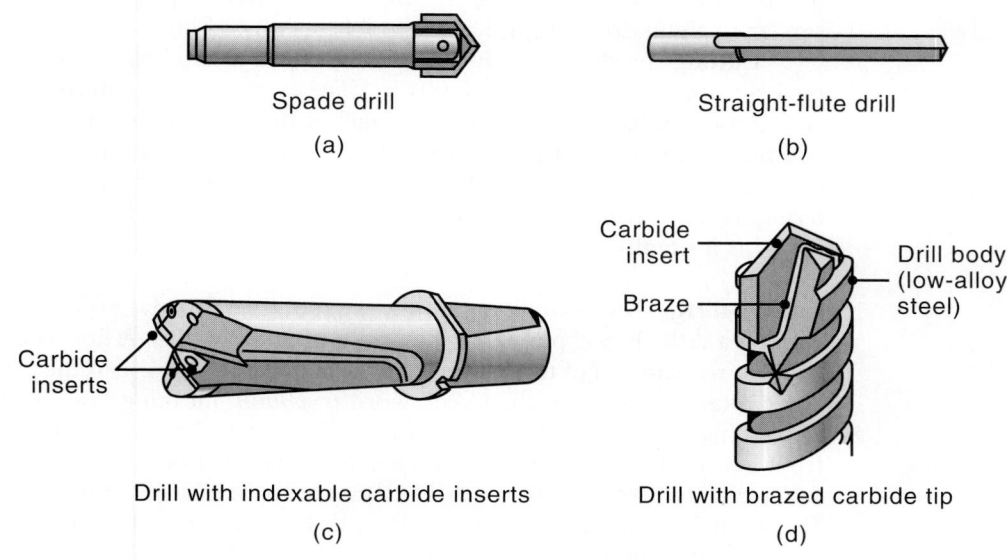

FIGURE 23.21   Various types of drills.

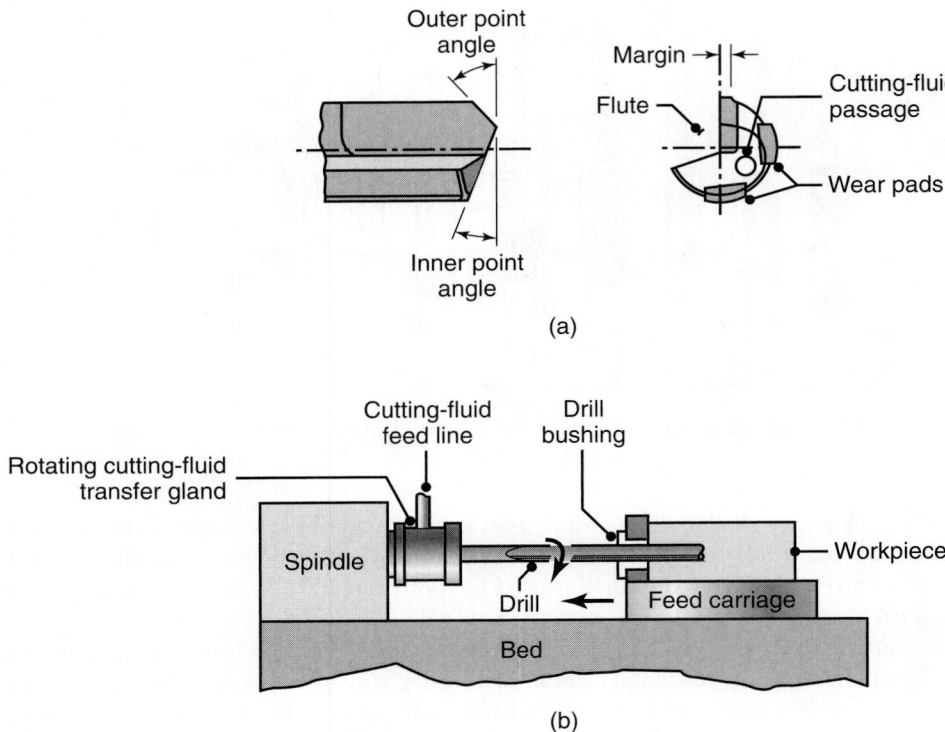

FIGURE 23.22   (a) A gun drill showing various features. (b) Schematic illustration of the gun-drilling operation.

ratios of holes produced can be 300:1 or even higher. The thrust force (the radial force that tends to push the drill sideways) is balanced by bearing pads on the drill that slide along the inside surface of the hole. Consequently, a gun drill is self-centering—an important feature when drilling straight, deep holes. A variation of this process is **gun trepanning** (see the next subsection), which uses a cutting tool similar to a gun drill except that the tool has a central hole.

Cutting speeds in gun drilling are usually high, and feeds are low. Tolerances typically are about 0.025 mm (0.001 in.) The cutting fluid is forced under high pressure through a longitudinal hole (passage) in the body of the drill (Fig. 23.22a). In addition to its function to lubricate and cool the workpiece, the fluid also flushes out chips that otherwise would be trapped in the deep hole being drilled, thus interfering with the drilling operation. The tool does not have to be retracted to clear the chips, as it usually is done with twist drills.

Trepanning.   In *trepanning*, the cutting tool (Fig. 23.23a) produces a hole by removing a disk-shaped piece (*core*), usually from flat plates. A hole is thus produced without reducing all of the material being removed to chips, as is the case in drilling. The process is based on the Greek word *trypanon*, meaning "boring a hole." The trepanning process can be used to make disks up to 250 mm (10 in.) in diameter from flat sheets, plates, or structural members such as I-beams. It also can be used to make circular grooves in which O-rings are to be placed (similar to Fig. 23.1f). Trepanning can be carried out on lathes, drill presses, or other machine tools using single-point or multipoint tools, as shown in Fig. 23.23b.

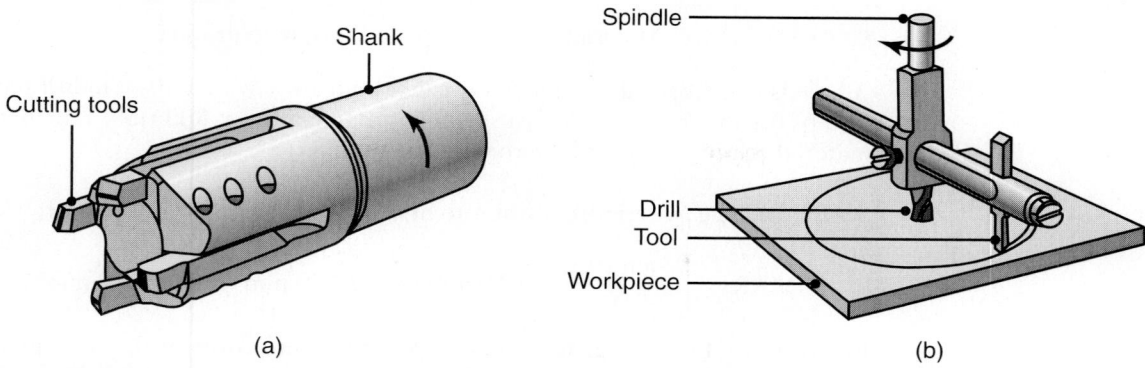

Shank

Cutting tools

(a)

Spindle

Drill

Tool

Workpiece

(b)

FIGURE 23.23   (a) Trepanning tool. (b) Trepanning with a drill-mounted single cutter.

23.5.2 Material-removal rate in drilling

The *material-removal rate* (MRR) in drilling is the volume of material removed by per unit time. For a drill with a diameter D, the cross-sectional area of the drilled hole is $\pi D^2/4$. The velocity of the drill perpendicular to the workpiece is the product of the feed, f (the distance the drill penetrates per unit revolution), and the rotational speed, N, where $N = V/\pi D$. Thus,

$$\text{MRR} = \left(\frac{\pi D^2}{4}\right) fN \tag{23.3}$$

The dimensional accuracy of this equation can be checked, as was done for Eq. (23.1), by noting that MRR = $(mm^2)(mm/rev)(rev/min)$ = mm^3/min, which is the correct unit for volume removed per unit time.

23.5.3 Thrust force and torque

The *thrust force* in drilling acts perpendicular to the hole axis; if this force is excessive, it can cause the drill to bend or break. An excessive thrust force also can distort the workpiece, particularly if it does not have sufficient stiffness (for example, thin sheet-metal structures), or it can cause the workpiece to slip into the workholding fixture.

The thrust force depends on factors such as (a) the strength of the workpiece material, (b) feed, (c) rotational speed, (d) drill diameter, (e) drill geometry, and (f) cutting fluids. Accurate calculation of the thrust force on the drill is difficult. Thrust forces typically range from a few newtons for small drills to as high as 100 kN (23.5 klb) for drilling high-strength materials with large drills. Experimental data are available as an aid in designing and using drills and drilling equipment.

Torque.   A knowledge of the magnitude of the *torque* in drilling is essential for estimating the power requirement; however, because of the many factors involved, it is difficult to calculate. Torque can be calculated from the data given in Table 21.2 by noting that the power dissipated during drilling is the product of torque and rotational speed and that we first have to calculate the material-removal rate. Torque in drilling can be as high as 4000 N·m (3000 lb-ft).

EXAMPLE 23.4 Material-removal rate and torque in drilling

A hole is being drilled in a block of magnesium alloy with a 10-mm drill bit, at a feed of 0.2 mm/rev, and with the spindle running at $N = 800$ rpm. Calculate the material-removal rate and the torque on the drill.

Solution The material-removal rate first is calculated from Eq. (23.3):

$$\text{MRR} = \left[\frac{(\pi)(10)^2}{4}\right](0.2)(800) = 12,570 \text{ mm}^3/\text{min} = 210 \text{ mm}^3/\text{s}$$

Referring to Table 21.2, let's take an average unit power of $0.5 \text{ W}\cdot\text{s/mm}^3$ for magnesium alloys. The power required is then

$$\text{Power} = (210)(0.5) = 105 \text{ W}$$

Power is the product of the torque on the drill and the rotational speed, which in this case is $(800)(2\pi)/60 = 83.8$ radians per second. Noting that $W = J/s$ and $J = N\cdot m$, we find that

$$T = \frac{105}{83.8} = 1.25 \text{ N}\cdot\text{m}$$

23.5.4 Drill materials and sizes

Drills usually are made of high-speed steels (M1, M7, and M10) and solid carbides or with carbide tips (typically made of K20 (C2) cabide), like those shown in Fig. 23.21c and d. Drills now are coated commonly with titanium nitride or titanium carbonitride for increased wear resistance (Section 22.5). Polycrystalline-diamond-coated drills are used for producing fastener holes in fiber-reinforced plastics. Because of their high wear resistance, several thousand holes can be drilled with little damage to the material.

Although there are continued developments, standard twist-drill sizes basically consist of the following series:

Numerical: No. 97 (0.0059 in.) to No. 1 (0.228 in.)
Letter: A (0.234 in.) to Z (0.413 in.)
Fractional: Straight shank from $\frac{1}{64}$ to $1\frac{1}{4}$ in. (in $\frac{1}{64}$ in. increments) to $1\frac{1}{2}$ in. (in $\frac{1}{32}$ in. increments), and larger drills in larger increments. Taper shank from $\frac{1}{8}$ to $1\frac{3}{4}$ in. (in $\frac{1}{64}$ in. increments) to 3.5 in. (in $\frac{1}{16}$ in. increments).
Millimeter: From 0.05 mm (0.002 in.) in increments of 0.01 mm.

23.5.5 Drilling practice

Drills and similar holemaking tools usually are held in *drill chucks*, which may be tightened with or without keys. Special chucks and collets with various quick-change features that do not require stopping the spindle are available for use on production machinery.

Because it does not have a centering action, a drill tends to "walk" on the workpiece surface at the beginning of the operation. This problem is particularly severe with small-diameter long drills and can lead to failure. To start a hole properly, the drill should be guided, using fixtures (such as a bushing) to keep it from deflecting laterally. A small starting hole can be made with a *center drill* (usually with a point angle of 60°), or the drill point may be ground to an *S* shape (*helical* or *spiral point*). This shape has a self-centering characteristic—thus eliminating the need for center drilling—and produces accurate holes and with improved drill life. These factors are important particularly in automated production with CNC machines in which the usual practice is to use a spot drill. To keep the drill more centered, the point angles of the spot drill and of the drill are matched. Other alternatives for minimizing walking of the drill bit are to use a centering punch to produce an initial impression in which drilling starts or else to incorporate dimples or other features into the cast or forged blank.

Drilling recommendations.   Recommended ranges for drilling speeds and feeds are given in Table 23.11. The speed is the *surface speed* of the drill at its periphery. Thus, a 0.5 in. (12.7 mm) drill rotating at 300 rpm has a surface speed of $\left(\frac{0.5}{2}\text{ in.}\right)(300\text{ rev/min})(2\pi\text{ rad/rev})(\frac{1}{12}\text{ ft/in.}) = 39\text{ ft/min} = 12\text{ m/min}$. When drilling holes smaller than 1 mm (0.040 in.) in diameter, rotational speeds can range up to 30,000 rpm, depending on the workpiece material.

The *feed* in drilling is the distance the drill travels into the workpiece per revolution. For example, Table 23.11 recommends that for most workpiece materials, a drill 1.5 mm (0.060 in.) in diameter should have a feed of 0.025 mm/rev. If the speed column in the table indicates that the drill should rotate at, say, 2000 rpm, then the drill should travel into the workpiece at a linear speed of (0.025 mm/rev) (2000 rev/min) = 50 mm/min = 2 in./min.

Chip removal during drilling can be difficult, especially for deep holes in soft and ductile workpiece materials. The drill should be retracted periodically (*pecking*) to remove chips that may have accumulated along the flutes. Otherwise, it may break because of excessive torque, or it may "walk" off location and produce a misshaped hole. A general guide to the probable causes of problems in drilling operations is given in Table 23.12.

Drill reconditioning.   Drills are reconditioned by grinding them either manually or using special fixtures. Proper reconditioning of drills is important, particularly

TABLE 23.11

General Recommendations for Speeds and Feeds in Drilling

Workpiece material	Surface speed		Feed, mm/rev (in./rev) — Drill diameter		rpm — Drill diameter	
	m/min	ft/min	1.5 mm (0.060 in.)	12.5 mm (0.5 in.)	1.5 mm	12.5 mm
Aluminum alloys	30–120	100–400	0.025 (0.001)	0.30 (0.012)	6400–25,000	800–3000
Magnesium alloys	45–120	150–400	0.025 (0.001)	0.30 (0.012)	9600–25,000	1100–3000
Copper alloys	15–60	50–200	0.025 (0.001)	0.25 (0.010)	3200–12,000	400–1500
Steels	20–30	60–100	0.025 (0.001)	0.30 (0.012)	4300–6400	500–800
Stainless steels	10–20	40–60	0.025 (0.001)	0.18 (0.007)	2100–4300	250–500
Titanium alloys	6–20	20–60	0.010 (0.0004)	0.15 (0.006)	1300–4300	150–500
Cast irons	20–60	60–200	0.025 (0.001)	0.30 (0.012)	4300–12,000	500–1500
Thermoplastics	30–60	100–200	0.025 (0.001)	0.13 (0.005)	6400–12,000	800–1500
Thermosets	20–60	60–200	0.025 (0.001)	0.10 (0.004)	4300–12,000	500–1500

Note: As hole depth increases, speeds and feeds should be reduced. Selection of speeds and feeds also depends on the specific surface finish required.

TABLE 23.12

General Troubleshooting Guide for Drilling Operations	
Problem	Probable causes
Drill breakage	Dull drill, drill seizing in hole because of chips clogging flutes, feed too high, lip relief angle too small
Excessive drill wear	Cutting speed too high, ineffective cutting fluid, rake angle too high, drill burned and strength lost when sharpened
Tapered hole	Drill misaligned or bent, lips not equal, web not central
Oversize hole	Same as above, machine spindle loose, chisel edge not central, side force on workpiece
Poor hole surface finish	Dull drill, ineffective cutting fluid, welding of workpiece material on drill margin, improperly ground drill, improper alignment

with automated manufacturing on computer numerical control machines. Hand grinding is difficult and requires considerable skill in order to produce symmetric cutting edges. Grinding on fixtures is accurate and is done on special computer-controlled grinders. Coated drills can be recoated.

Measuring drill life. Drill life, as well as tap life, usually is measured by the number of holes drilled before they become dull. The test procedure consists of clamping a block of material on a suitable dynamometer or force transducer and drilling a number of holes while recording the torque or thrust force during each successive operation. After a number of holes have been drilled, the torque and force begin to increase because the tool is becoming dull. *Drill life* is defined as the number of holes drilled until this transition begins. Other techniques, such as monitoring vibration and acoustic emissions (Section 21.5.4), also may be used to determine drill life.

23.5.6 Drilling machines

Drilling machines are used for drilling holes, tapping, reaming, and small-diameter boring operations. The most common machine is the **drill press**, the major components of which are shown in Fig. 23.24a. The workpiece is placed on an adjustable table, either by clamping it directly into the slots and holes on the table or by using a vise, which in turn is clamped to the table. The drill is lowered manually by a hand wheel or by power feed at preset rates. Manual feeding requires some skill in judging the appropriate feed rate.

Drill presses usually are designated by the largest workpiece diameter that can be accommodated on the table and typically range from 150 to 1250 mm (6 to 50 in.). In order to maintain proper cutting speeds at the cutting edges of drills, the spindle speed on drilling machines has to be adjustable to accommodate different drill sizes. Adjustments are made by means of pulleys, gear boxes, or variable-speed motors.

The types of drilling machines range from simple *bench-type drills* used to drill small-diameter holes to large *radial drills* (Fig. 23.24b), which can accommodate large workpieces. The distance between the column and the spindle center can be as much as 3 m (10 ft). The drill head of *universal drilling machines* can be swiveled to drill holes at an angle. Developments in drilling machines include numerically controlled three-axis machines, in which the operations are performed automatically and in the desired sequence with the use of a turret (Fig. 23.25). Note that the turret holds several different drilling tools.

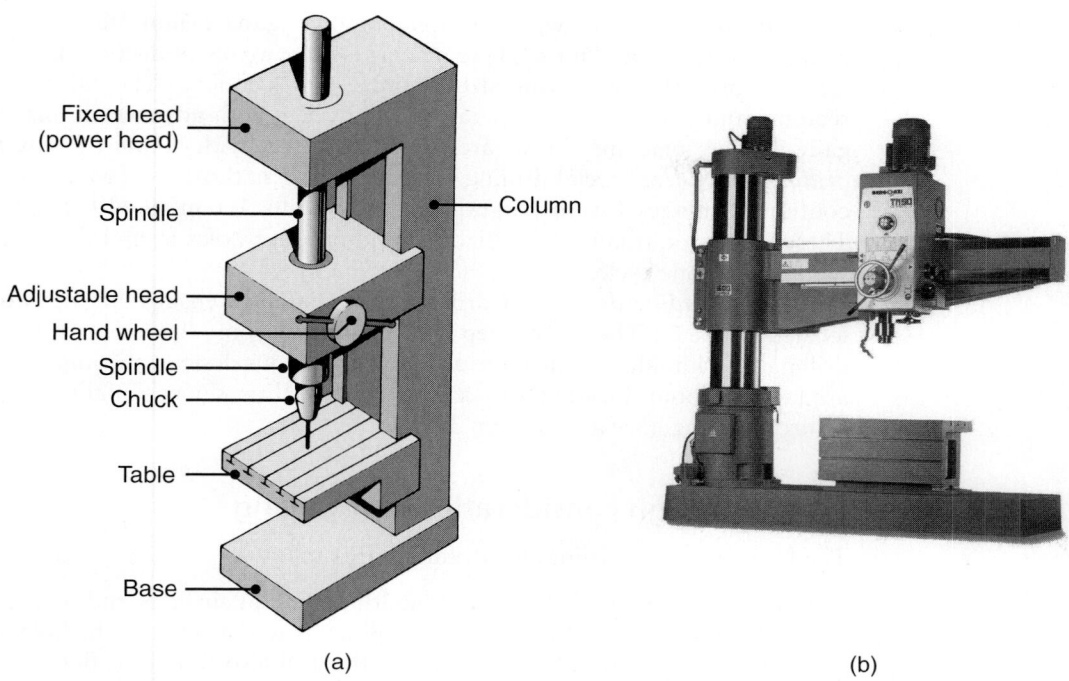

Fixed head
(power head)

Column

Spindle

Adjustable head

Hand wheel

Spindle

Chuck

Table

Base

(a)

(b)

FIGURE 23.24   (a) Schematic illustration of the components of a vertical drill press. (b) A radial drilling machine. *Source:* (b) Courtesy of Willis Machinery and Tools.

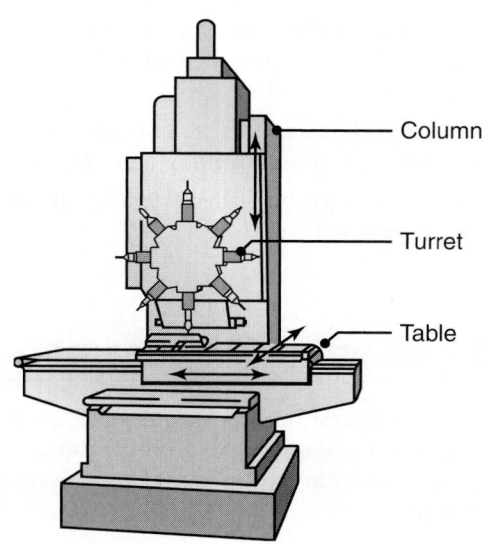

Column

Turret

Table

FIGURE 23.25   A three-axis computer numerical-control drilling machine. The turret holds as many as eight different tools, such as drills, taps, and reamers.

Drilling machines with multiple spindles (**gang drilling**) are used for high-production-rate operations. These machines are capable of drilling, in one cycle, as many as 50 holes of varying sizes, depths, and locations. They also are used for reaming and counterboring operations. However, with advances in machine tools, gang-drilling machines now are being replaced with *numerical-control turret drilling machines*. Special drilling machines, such as those used to produce holes in continuous hinges (*piano hinges*), use twist drills 1 mm (0.040 in.) in diameter. These machines usually are horizontal and produce holes in up to 3-m (10-ft) long segments on one cycle.

Workholding devices for drilling are essential to ensure that the workpiece is located properly. They also keep the workpiece from slipping or rotating during drilling. Workholding devices are available in various designs; the important features are (a) three-point locating for accuracy and (b) three-dimensional workholding for secure fixturing. (See also Section 37.8.)

23.5.7 Design considerations for drilling

The basic design guidelines for drilling are as follows:

- Designs should allow holes to be drilled on flat surfaces and perpendicular to the drill motion. Otherwise, the drill tends to deflect, and the hole will not be located accurately. Exit surfaces for the drill also should be flat.
- Interrupted hole surfaces should be avoided or minimized for improved dimensional accuracy, drill life, and to avoid vibrations.
- Hole bottoms should match, if possible, standard drill-point angles; flat bottoms or odd shapes should be avoided.
- Through holes are preferred over blind holes. If holes with large diameters are required, the workpiece should have a preexisting hole, preferably made during fabrication of the part (such as by casting, powder metallurgy, or forming). Dimples should be provided when holes are not practical, in order to reduce the tendency of the drill to walk.
- Parts should be designed so that all drilling can be performed with a minimum of fixturing and without having to reposition the workpiece.
- Blind holes must be drilled deeper than subsequent reaming or tapping operations that may be performed.

23.6 | Reaming and Reamers

Reaming is an operation used to (a) make an existing hole dimensionally more accurate than can be obtained by drilling alone, and (b) improve its surface finish. The most accurate holes in workpieces generally are produced by the following sequence of operations:

1	Centering	3	Boring
2	Drilling	4	Reaming

For even better accuracy and surface finish, holes may be *burnished* or internally *ground* and *honed* (Sections 26.6 and 26.10).

A *reamer* (Fig. 23.26a) is a multiple-cutting-edge tool with straight or helically fluted edges that remove very little material. For soft metals, a reamer typically removes a minimum of 0.2 mm (0.008 in.) on the diameter of a drilled hole and for harder metals about 0.13 mm (0.005 in.). Attempts to remove smaller layers can be detrimental, as the reamer may be damaged or the hole surface may become burnished (see also Fig. 21.22 as an analogy). In this case, honing would be preferred. In general, reamer speeds are one-half those of the same-sized drill and three times the feed rate.

Hand reamers are straight or have a tapered end in the first third of their length. Various *machine reamers* (also called *chucking reamers*, because they are mounted in a chuck and operated by a machine) are available in two types: (1) *Rose reamers* have cutting edges with wide margins and no relief (Fig. 23.26a). They remove considerable material and true up a hole for flute reaming. (2) *Fluted reamers* have small margins and relief with a rake angle of about 5°. They usually are used for light cuts of about 0.1 mm (0.004 in.) on the hole diameter.

Shell reamers (which are hollow and are mounted on an arbor) generally are used for holes larger than 20 mm (0.75 in.). *Expansion reamers* are adjustable for small variations in hole size and also to compensate for wear of the reamer's cutting edges. *Adjustable reamers* (Fig. 23.26b) can be set for specific hole diameters and therefore are versatile.

Reamers may be held rigidly (as in a chuck), or they may *float* in their holding fixtures to ensure alignment or be *piloted* in guide bushings placed above and below the workpiece. A further development in reaming consists of the *dreamer*—a tool which combines drilling and reaming. The tip of the tool produces a hole by drilling and the rest of the same tool performs a reaming operation. A similar development involves drilling and tapping in one stroke using a single tool.

Reamers typically are made of high-speed steels (M1, M2, and M7) or solid carbides (K20; C2), or have carbide-cutting edges. Reamer maintenance and reconditioning are important for hole accuracy and surface finish.

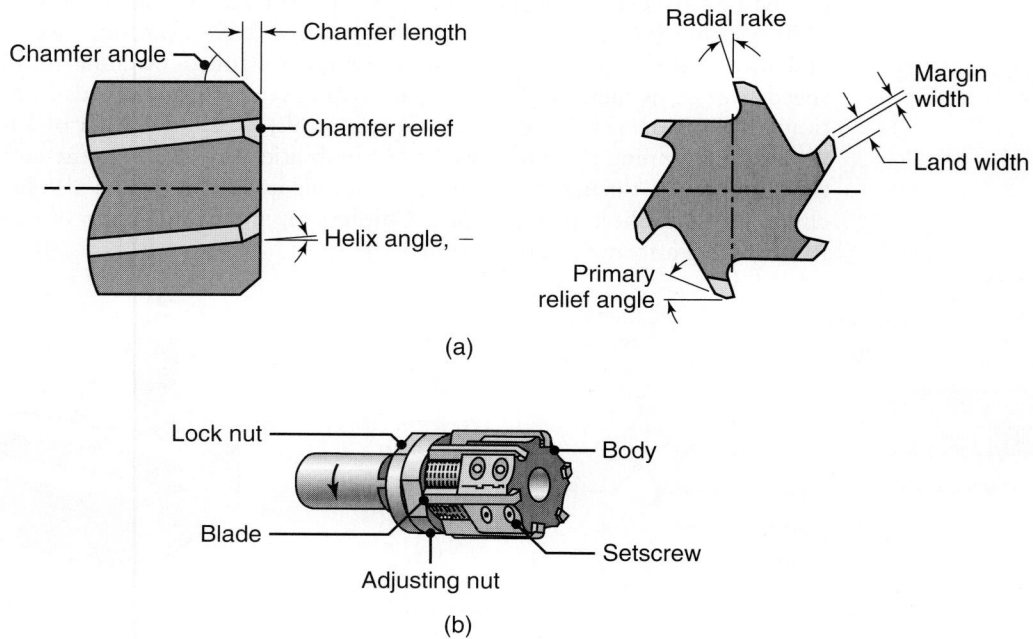

(a)

(b)

FIGURE 23.26   (a) Terminology for a helical reamer. (b) Inserted-blade adjustable reamer.

23.7 | Tapping and Taps

Internal threads in workpieces can be produced by *tapping*. A *tap* is a chip-producing threading tool with multiple cutting teeth (Fig. 23.27a). Taps generally are available with two, three, or four flutes. The most common production tap is the two-flute spiral-point tap. The two-flute tap forces the chips into the hole so that the tap needs to be retracted only at the end of the cut. Three-fluted taps are stronger because more material is available in the flute. Tap sizes range up to 100 mm (4 in.).

Tapered taps are designed to reduce the torque required for the tapping of through holes. *Bottoming taps* are for tapping blind holes to their full depth. *Collapsible taps* are used in large-diameter holes; after tapping has been completed, the tap is collapsed mechanically and is removed from the hole without rotation.

Chip removal can be a significant problem during tapping because of the small clearances involved. If chips aren't removed properly, the excessive torque that results can break the tap. The use of a cutting fluid and periodic reversal and removal of the tap from the hole are effective means of chip removal and of improving the quality of the tapped hole. For higher tapping productivity, drilling and tapping can be combined in a single tool (*drapping*). The tool has a drilling section at its tip, followed by a tapping section.

Tapping may be done by *hand* or with machines such as (a) drilling machines, (b) lathes, (c) automatic screw machines, and (d) vertical CNC milling machines combining the correct relative rotation and the longitudinal feed. Special tapping machines are available with features for multiple tapping operations. Multiple-spindle tapping heads are used extensively, particularly in the automotive industry where 30 to 40% of machining operations involve the tapping of holes. One system for the automatic tapping of nuts is shown in Fig. 23.27b.

With proper lubrication, tap life may be as high as 10,000 holes. Tap life can be determined with the same technique used to measure drill life. Taps usually are made of high-speed steels (M1, M2, M7, and M10). Productivity in tapping operations can be improved by *high-speed tapping* with surface speeds as high as 100 m/min (350 ft/min). *Self-reversing* tapping systems also have been improved significantly and now are in use with modern computer-controlled machine tools. Operating speeds can be as high as 5000 rpm, although actual cutting speeds in most applications are considerably lower. Cycle times typically are on the order of 1 to 2 seconds.

Some tapping systems now have capabilities for directing the cutting fluid to the cutting zone through the spindle and a hole in the tap, which also helps flush the chips out of the hole being tapped. **Chipless tapping** is a process of internal thread rolling using a forming tap (Section 13.5).

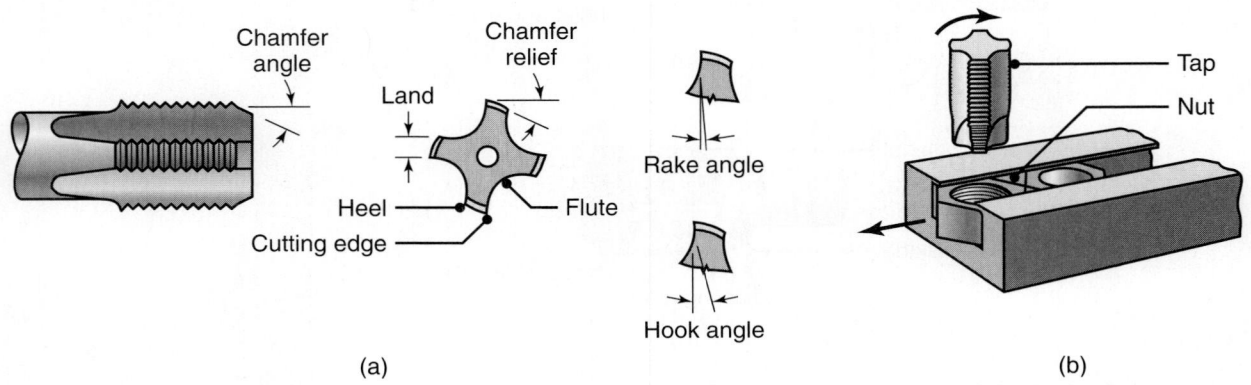

(a)   **(b)**

FIGURE 23.27   (a) Terminology for a tap. (b) Tapping of steel nuts in production.

CASE STUDY 23.1 | Bone Screw Retainer

A cervical spine implant is shown in Fig. 23.28a. In the event that a patient requires cervical bone fusion at one or more vertebral levels, this implant can act as an internal stabilizer by decreasing the amount of motion in this region to help promote a successful fusion. The plate affixes to the anterior aspect of the spine with bone screws that go through the plate and into the bone. The undersurface of the plate has a very rough surface that helps hold the plate in place while the bone screws are being inserted.

One concern with this type of implant is the possibility of the bone screws loosening with time due to normal, repetitive loading from the patient. In extreme cases, this can result in a screw backing out, with the head of the screw no longer flush with the plate—a condition that obviously is undesirable. This implant uses a retainer to prevent the bone screw from backing out away from the plate. The part drawing for the retainer is shown in the left half of Fig. 23.28b.

Screw and retainer
inserted in plate

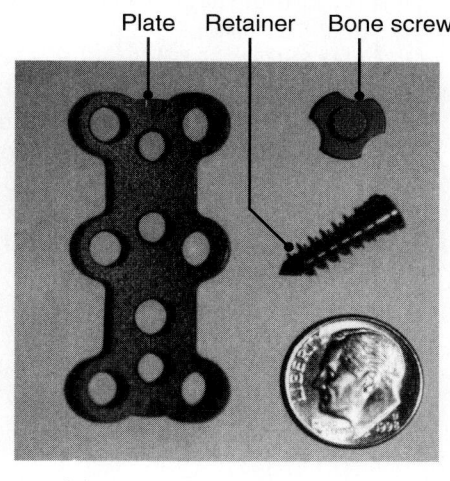

Plate   Retainer   Bone screw

(a)

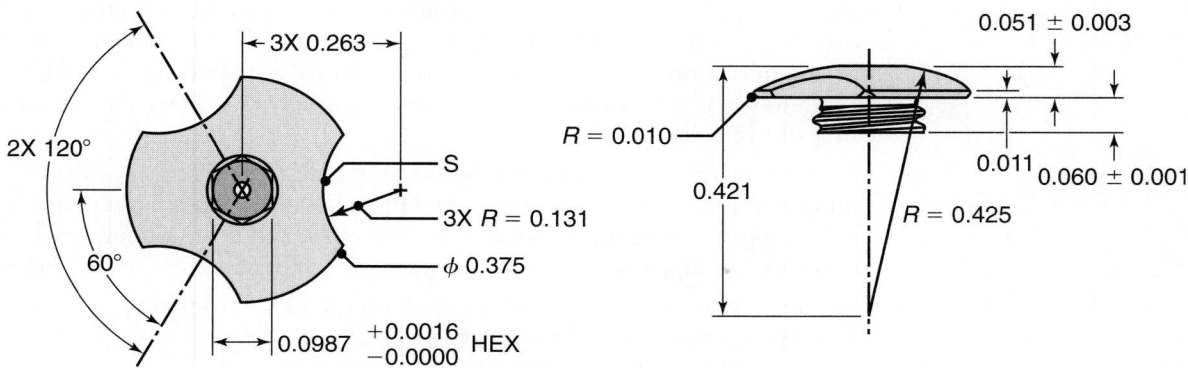

Note: Thread must start at point S to assure retainer interferes with bone screw.

(b)

FIGURE 23.28   A cervical spine implant. All dimensions in inches.

The retainer has a number of design features that are essential for it to function correctly and without complicating the surgical procedure. To ease its use in surgery, the plate is provided with the retainers already in place, with the circular notches aligned with the bone screw holes. This allows the surgeon to insert the bone screws without interference from the retainer. Once the screws are inserted, the surgeon turns the retainer a few degrees so that each screw head is then captured. In order to ensure the retainer's proper orientation in the plate, the thread of its shank must start in the same axial location as point S in Fig. 23.28b.

The manufacturing steps to produce this part are shown in Fig. 23.28b. First, a 0.5-in. diameter Ti-6Al-4V rod is placed in a CNC lathe and faced, then the threaded area is turned to the diameter necessary to machine the threads. The thread is turned on the shank—but over a longer length than ultimately required because of difficulties in obtaining high-quality threads at the start of machining. The cap then is turned to the required diameter and the 0.10 in. radius is machined on the underside of the head. The part is removed, inspected, and placed in another CNC lathe, where it is faced to length. The spherical radius in the cap is machined, the center hole is drilled, and the hex head is broached. The cap is removed and inspected. If the desired length has not been achieved, it is lapped (Section 26.7) to the final dimension.

At this point, the retainer is placed in a CNC milling machine using a specially designed fixture which consists basically of a threaded, tapered hole. By carefully applying a pre-determined torque on the retainer when placing it into the fixture, the starting location of the threads can be controlled accurately. Once the cap is located in the fixture, the three circular notches are machined as per the drawing. The retainer then is deburred and tumbled to remove all sharp corners, and the bottom is heavy-grit blasted to match that of the underside of the plate. Finally, the parts are anodized (Section 34.10) and passivated to obtain the desired biocompatability.

Source: Courtesy of J. Mankowski and B. Pyszka, Master Metal Engineering Inc., and C. Lyle and M. Handwerker, Wright Medical Technology, Inc.

SUMMARY

- Machining processes that typically produce external and internal circular profiles are turning, boring, drilling, and tapping. Because of the three-dimensional nature of these operations, chip movement and its control are important considerations. Chip removal can be a significant problem (especially in drilling and tapping) and can lead to tool breakage.

- Optimization of each machining process requires an understanding of the interrelationships among design parameters (part shape, dimensional accuracy, and surface finish) and process parameters (cutting speed, feed, and depth-of-cut), tool material and shape, use of cutting fluids, and the sequence of operations to be performed.

- High-speed machining, hard turning, and ultraprecision machining, are important developments that have helped reduce machining costs and produce parts with exceptional surface finish and dimensional accuracy.

- The parts to be machined may have been produced by casting, forging, extrusion, or powder metallurgy. The closer the blank to be machined to the final shape desired (near-net shape), the fewer the number and extent of the subsequent machining processes required.

KEY TERMS

Automatic bar machine	Facing	Parting
Back rake angle	Feed force	Power chuck
Bed	Feed rod	Rake angle
Boring	Finishing cuts	Reamer
Boring mill	Form tools	Reaming
Carriage	Gun drilling	Reconditioning
Chip management	Hard machining	Relief angle
Chuck	Headstock	Roughing cuts
Collet	High-speed machining	Screw threads
Cutting-edge angle	Hole making	Side rake angle
Diamond turning	Jig borer	Tailstock
Drill life	Knurling	Tapping
Drilling	Lathes	Threading
Drill press	Lead screw	Trepanning
Dry machining	Mandrel	Turning
Ductile-regime cutting	Material-removal rate	Turret lathe
Engine lathe	Near-dry machining	Twist drill
Face plate	Nose radius	Ultraprecision machining

BIBLIOGRAPHY

Arnone, M., *High Performance Machining*, Hanser, 1998.

ASM Handbook, Vol. 16: *Machining*, ASM International, 1989.

Boothroyd, G., and Knight, W.A., *Fundamentals of Machining and Machine Tools*, 2nd ed., Marcel Dekker, 1989.

Brown, J., *Advanced Machining Technology Handbook*, McGraw-Hill, 1998.

Byers, J.P. (ed.), *Metalworking Fluids*, Marcel Dekker, 1994.

Handbook of High-Speed Machining Technology, Chapman and Hall, 1985.

Hoffman, E.G., *Jigs and Fixture Design*, 4th ed., Industrial Press, 1996.

Juneja, B.L., *Fundamentals of Metal Cutting and Machine Tools*, Wiley, 1987.

Krar, S.F., and Check, A.F., *Technology of Machine Tools,* 5th ed., Glencoe Macmillan/McGraw-Hill, 1996.

Machinery's Handbook, Industrial Press, revised periodically.

Machining Data Handbook, 3rd ed., 2 vols. Machinability Data Center, 1980.

Metal Cutting Tool Handbook, 7th ed., Industrial Press, 1989.

Meyers, A.L., and Slattery, T., *Basic Machining Reference Handbook*, Industrial Press, 1988.

Nachtman, E.S., and Kalpakjian, S., *Lubricants and Lubrication in Metalworking Operations*, Marcel Dekker, 1985.

Rechetov, D.N., and Portman, V.T., *Accuracy of Machine Tools*, ASME International, 1989.

Sluhan, C. (ed.), *Cutting and Grinding Fluids: Selection and Application*, Society of Manufacturing Engineers, 1992.

Stephenson, D.A., and Agapiou, J.S., *Metal Cutting Theory and Practice*, Marcel Dekker, 1996.

Stout, K.J., Davis, E.J., and Sullivan, P.J., *Atlas of Machined Surfaces*, Chapman and Hall, 1990.

Tool and Manufacturing Engineering Handbook, 4th ed., Vol. 1: *Machining*. Society of Manufacturing Engineers, 1983.

Walsh, R.A., *McGraw-Hill Machining and Metalworking Handbook*, McGraw-Hill, 1994.

Weck, M., *Handbook of Machine Tools*, 4 vols. Wiley, 1984.

REVIEW QUESTIONS

23.1 Describe the types of machining operations that can be performed on a lathe.

23.2 Explain the functions of different angles on a single-point cutting tool.

23.3 Why were power chucks developed?

23.4 Why can (a) boring on a lathe and (b) tapping be difficult operations?

23.5 Why is there more than one turret in turret lathes?

23.6 Describe the differences between boring a workpiece on a lathe and boring it on a horizontal boring mill.

23.7 How is drill life determined?

23.8 Why are reaming operations performed?

23.9 Describe the difference between a steady rest and a follow rest. Give an application of each.

23.10 Explain the functions of the saddle on a lathe.

23.11 Describe the relative advantages of (a) self-opening and (b) solid-die heads for threading.

23.12 How is a boring mill different from a lathe?

23.13 Explain how external threads are cut on a lathe.

23.14 What is the difference between a blind hole and a through hole? What is the significance of it?

QUALITATIVE PROBLEMS

23.15 Explain the reasoning behind the various design guidelines for turning.

23.16 You will note that we have used both the terms "tool strength" and "tool-material strength." Do you think there is a difference between them? Explain.

23.17 Explain why the sequence of drilling, boring, and reaming produces a hole that is more accurate than just drilling and reaming it.

23.18 Explain why machining operations may be necessary even on net-shape or near-net-shape parts made by precision casting, forming, or powder-metallurgy products.

23.19 What are the consequences of drilling with a drill bit that has not been sharpened properly?

23.20 A badly oxidized and uneven round bar is being turned on a lathe. Would you recommend a small or a large depth-of-cut? Explain your reasons.

23.21 Describe the problems, if any, that may be encountered in clamping a workpiece made of a soft metal in a three-jaw chuck.

23.22 Does the force or torque in drilling change as the hole depth increases? Explain.

23.23 Explain the similarities and differences in the design guidelines for turning and for boring.

23.24 What are the advantages and applications of having a hollow spindle in the headstock of a lathe?

23.25 Explain how you would go about producing a taper on a round workpiece on a lathe.

23.26 Assume that you are asked to perform a boring operation on a large-diameter hollow workpiece. Would you use a horizontal or a vertical boring mill? Explain.

23.27 Explain the reasons for the major trend in producing threads by thread rolling versus thread cutting. What would be the differences (if any) in the types of threads produced and their performance characteristics?

23.28 In some materials the hole drilled can be smaller than the diameter of the drill. Explain this phenomenon and identify the relevant material properties that could influence it.

23.29 Describe your observations concerning the contents of Tables 23.2 and 23.4 and explain why those particular recommendations are made.

23.30 We have seen that cutting speed, feed, and depth-of-cut are the main parameters in a turning operation. In relative terms, at what values should these parameters be set for a (a) roughing operation and (b) finishing operation?

23.31 Explain the economic justification for purchasing a turret lathe instead of a conventional lathe.

23.32 The footnote to Table 23.11 states that (in drilling) as the hole diameter increases, speeds and feeds should be reduced. Explain why.

23.33 In modern manufacturing with computer-controlled machine tools, which types of metal chips would be undesirable and why?

23.34 List and explain the factors that contribute to poor surface finish in the processes described in this chapter.

23.35 The operational severity for reaming is much less than that for tapping, even though they are both internal material-removal processes which can be difficult. Why?

23.36 Review Fig. 23.6 and comment on the factors involved in determining the height of the zones (cutting speed) for various tool materials.

23.37 We have stated that some chucks are power actuated. Make a survey of the technical literature and describe the basic design of such chucks.

23.38 What operations typically can be performed on a drill press but not on a lathe?

23.39 Explain how gun drills remain centered during drilling. Why is there a hollow, longitudinal channel in a gun drill?

23.40 Comment on the magnitude of the wedge angle on the tool shown in Fig. 23.4.

23.41 If inserts are used in a drill bit, is the shank material important? If so, what properties are important? Explain.

23.42 Refer to Fig. 23.10b and (in addition to the tools shown) describe other types of cutting tools that can be placed in toolholders to perform other machining operations.

QUANTITATIVE PROBLEMS

23.43 Calculate the same quantities as in Example 23.1 for high-strength titanium alloy and at $N = 800$ rpm.

23.44 Estimate the machining time required to rough turn a 0.4-m long, annealed copper-alloy round bar from 60-mm diameter to 55-mm diameter using a high-speed-steel tool. (See Table 23.4.) Estimate the time required for an uncoated carbide tool.

23.45 A high-strength cast-iron bar 5 in. in diameter is being turned on a lathe at a depth-of-cut $d = 0.050$ in. The lathe is equipped with a 15-hp electric motor and has a mechanical efficiency of 80%. The spindle speed is 500 rpm. Estimate the maximum feed that can be used before the lathe begins to stall?

23.46 A 0.4-in. diameter drill is used on a drill press operating at 300 rpm. If the feed is 0.005 in./rev, what is the MRR? What is the MRR if the drill diameter is doubled?

23.47 In Example 23.4, assume that the workpiece material is high-strength aluminum alloy and the spindle is running at $N = 600$ rpm. Estimate the torque required for this operation.

23.48 In a drilling operation, a 0.5-in. drill bit is being used in a low-carbon steel workpiece. The hole is a blind hole, which will be tapped to a depth of 1 in. The drilling operation takes place with a feed of 0.010 in./rev and a spindle speed of 700 rpm. Estimate the time required to drill the hole prior to tapping.

23.49 A 3-in. diameter low-strength, stainless-steel cylindrical part is to be turned on a lathe at 600 rpm with a depth-of-cut of 0.2 in. and a feed of 0.025 in./rev. What should be the minimum horsepower of the lathe?

23.50 A 6-in. diameter aluminum cylinder 10 in. in length is to have its diameter reduced to 4.5 in. Using the typical machining conditions given in Table 23.4, estimate the machining time if a TiN-coated carbide tool is used.

23.51 For the data in Problem 23.50, calculate the power required.

23.52 Assume that you are an instructor covering the topics described in this chapter, and you are giving a quiz on the numerical aspects to test the understanding of the students. Prepare two quantitative problems and supply the answers.

SYNTHESIS, DESIGN, AND PROJECTS

23.53 Would you consider the machining processes described in this chapter as net-shape processing (requiring no further processing)? Near-net-shape processing? Explain with appropriate examples.

23.54 If a bolt breaks in a hole so that the head is no longer present, it is removed by first drilling a hole in the bolt shank and then using a special tool to remove the bolt. Inspect such a tool and explain how it works.

Can you think of any other means of removing the broken bolt from the hole? Explain.

23.55 Describe the machine tool and the fixtures needed to machine baseball bats from wooden cylindrical stock.

23.56 An important trend in machining operations is the increased use of flexible fixturing. Conduct a search on the Internet regarding flexible fixturing and comment on their design and operation.

23.57 Review Fig. 23.7d and explain if it would be possible to machine eccentric shafts (like that shown in Fig. 23.12c) on such a setup. What if the part is long compared to its cross-section?

23.58 We have seen that boring bars can be designed with internal damping capabilities in order to reduce or eliminate vibration and chatter during boring machining. Referring to the technical and manufacturers' literature, describe details of designs for such boring bars.

23.59 Would it be difficult to use the machining processes described in this chapter on various soft nonmetallic or rubber-like materials? Explain your thoughts, commenting on the role of the physical and mechanical properties of such materials on the machining operation and any difficulties in producing the desired shapes and dimensional accuracies.

23.60 With appropriate sketches, describe the principles of various fixturing methods and workholding devices that can be used for the processes described in this chapter. Include three-point locating and three-dimensional workholding for drilling and similar operations.

23.61 Make a comprehensive table of the process capabilities of the machining processes described in this chapter. Using several columns, describe the machine tools involved, type of cutting tools and tool materials used, shapes of parts produced, typical maximum and minimum sizes, surface finish, dimensional tolerances, and production rates.

23.62 Inspect various utensils in a typical kitchen and identify the ones that have some similarity to the processes described in this chapter. Comment on their principle of operation.

23.63 In Fig. 23.14, we have shown tolerances that can be obtained by the process listed in the figure. Give specific examples of parts and applications where such ultra-high dimensional accuracies are essential.

Machining Processes Used to Produce Various Shapes: Milling, Broaching, Sawing, and Filing; Gear Manufacturing

In this chapter, the processes for producing various complex shapes are described. Specifically, this chapter presents the following topics:

- Milling operations.
- Machining of long, flat surfaces and profiles.
- Broaching for producing external and internal profiles.
- Sawing and filing operations.
- Production of gears by machining.

Typical parts made: Parts with complex external and internal features on various surfaces, splines, and gears.

Alternative processes: Die casting, precision casting, precision forging, powder metallurgy, and powder-injection molding.

24.1 | Introduction

In addition to producing parts with various external or internal round profiles, machining operations can produce many other parts with more complex shapes (Fig. 24.1). In this chapter, several cutting processes and machine tools that are capable of producing these shapes using single-point, multi-tooth, and cutting tools are described (see also Table 23.1). This chapter begins with one of the most versatile processes—*milling*—in which a rotating cutter removes material while traveling along various axes with respect to the workpiece.

Other machining processes are then described (such as *planing, shaping,* and *broaching*), in which the tool or workpiece travels along a straight path producing flat and various shaped surfaces. Next, *sawing* processes are covered, which generally are used to prepare blanks from rods and flat plate which subsequently may be subjected

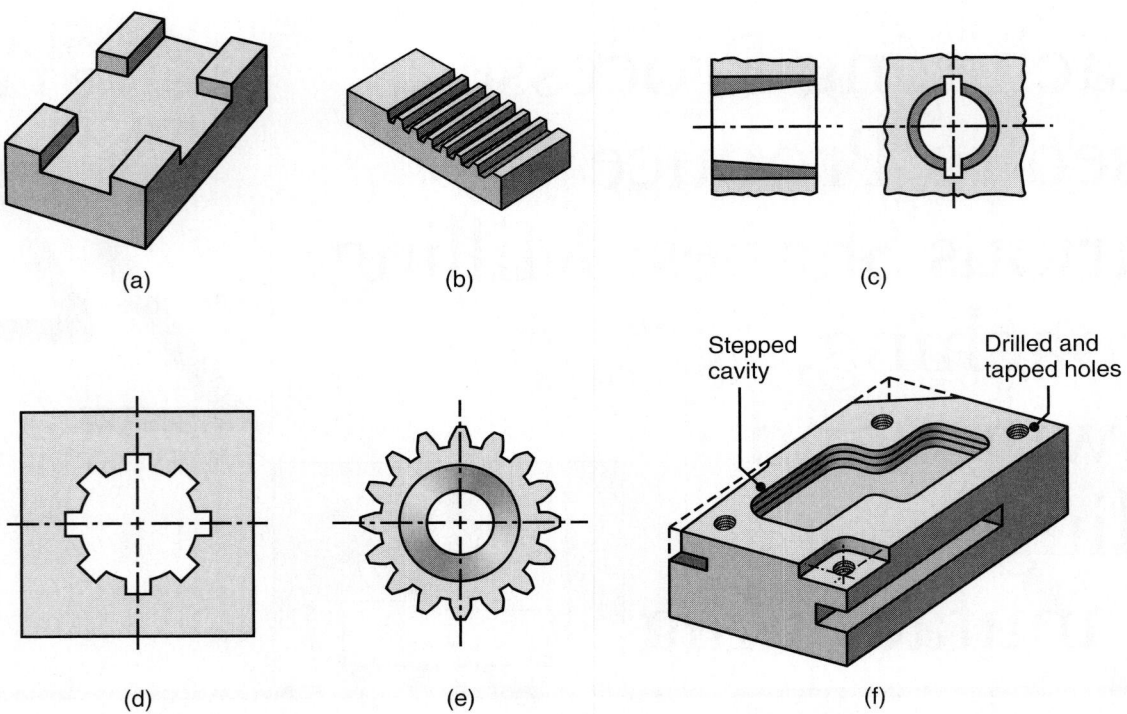

(a)   (b)   (c)

(d)   (e)   Stepped cavity   Drilled and tapped holes (f)

FIGURE 24.1   Typical parts and shapes that can be produced with the machining processes described in this chapter.

to forming, machining, and welding operations. Also briefly discussed is *filing*, which is used to remove small amounts of material, usually from edges and corners.

Finally, *gear manufacturing* is discussed, through processes in which gear teeth are produced by casting, powder-metallurgy, and by plastic deformation, such as by forging. This chapter describes gear production by several machining processes that use special cutters and the quality and properties of the gears made by these processes.

24.2 | Milling and Milling Machines

Milling includes a number of highly versatile machining operations, taking place in a variety of configurations (Fig. 24.2) with the use of a **milling cutter**—a multi-tooth tool that produces a number of chips in one revolution.

24.2.1 Peripheral milling

In this process (also called *plain milling*), the axis of cutter rotation is parallel to the workpiece surface, as shown in Fig. 24.2a. The cutter body, which generally is made of high-speed steel, has a number of teeth along its circumference; each tooth acts like a single-point cutting tool. When the cutter is longer than the width of the cut, the process is called *slab milling*. Cutters for peripheral milling may have *straight* or *helical teeth* (as shown in Fig. 24.2a), resulting in orthogonal or oblique cutting action, respectively. Helical teeth generally are preferred over straight teeth because the tooth is partially engaged with the workpiece as it rotates. Consequently, the cutting force and the torque on the cutter are lower, resulting in a smoother operation and reduced chatter. (See also Fig. 21.9.)

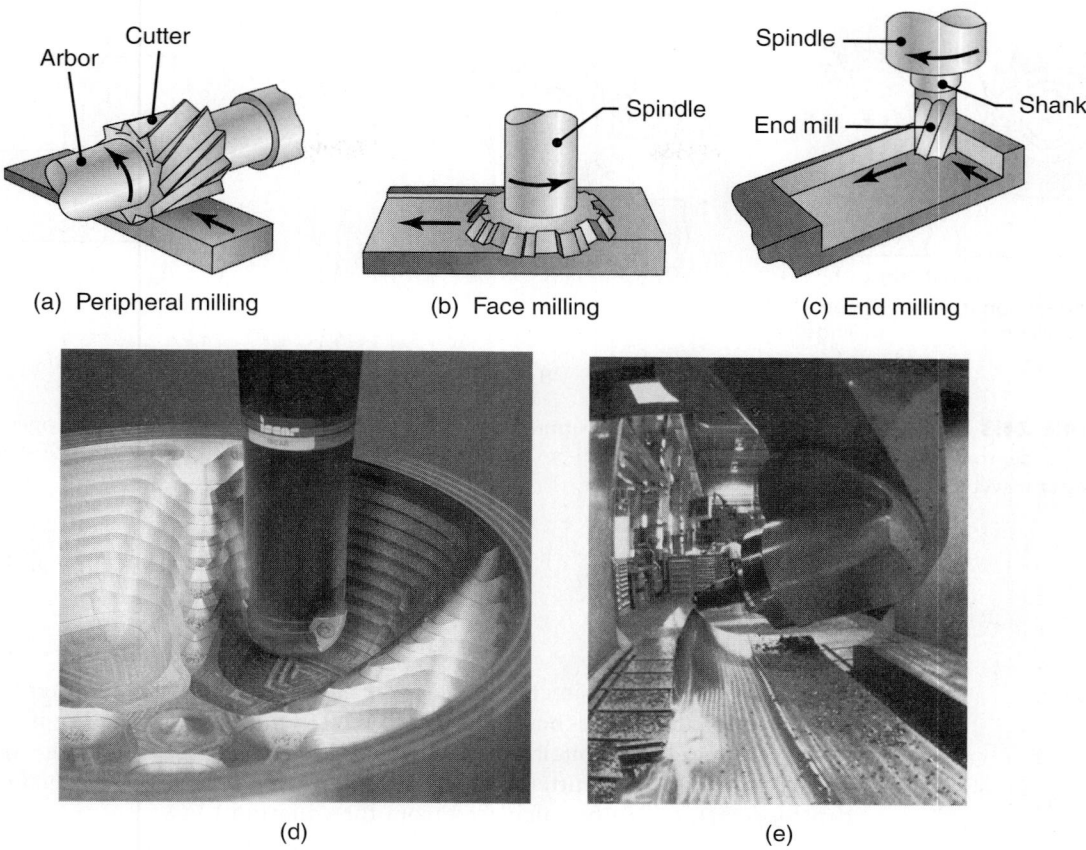

FIGURE 24.2 Some basic types of milling cutters and milling operations. (a) Peripheral milling. (b) Face milling. (c) End milling. (d) Ball-end mill with indexable coated-carbide inserts machining a cavity in a die block. (e) Milling a sculptured surface with an end mill, using a five-axis numerical control machine. *Source:* (d) Courtesy of Iscar Metals, Inc. (e) Courtesy of The Ingersoll Milling Machine Co.

Conventional milling and climb milling. Note in Fig. 24.3a that the cutter rotation can be either clockwise or counter-clockwise; this is significant in the operation. In *conventional milling* (also called *up milling*), the maximum chip thickness is at the end of the cut as the tooth leaves the workpiece surface. The advantages to conventional milling are that (a) tooth engagement is not a function of workpiece surface characteristics and (b) contamination or scale (oxide layer) on the surface does not adversely affect tool life. This is the more common method of milling. The cutting process is smooth, provided that the cutter teeth are sharp. Otherwise, the tooth will rub against the surface before it begins to cut. Also, there may be a tendency for the tool to chatter, and the workpiece has a tendency to be pulled upward (because of the cutter rotation direction), necessitating proper clamping.

In *climb milling* (also called *down milling*), cutting starts at the surface of the workpiece where the chip is thickest. The advantage is that the downward component of the cutting force holds the workpiece in place, particularly for slender parts. However, because of the resulting impact forces when the teeth engage the workpiece, this operation must have a rigid workholding setup, and gear backlash must be eliminated in the table feed mechanism. Climb milling is not suitable for the machining of workpieces having surface scale, such as hot-worked metals, forgings, and castings. The scale is hard and abrasive and causes excessive wear and damage to the cutter teeth, thus shortening tool life.

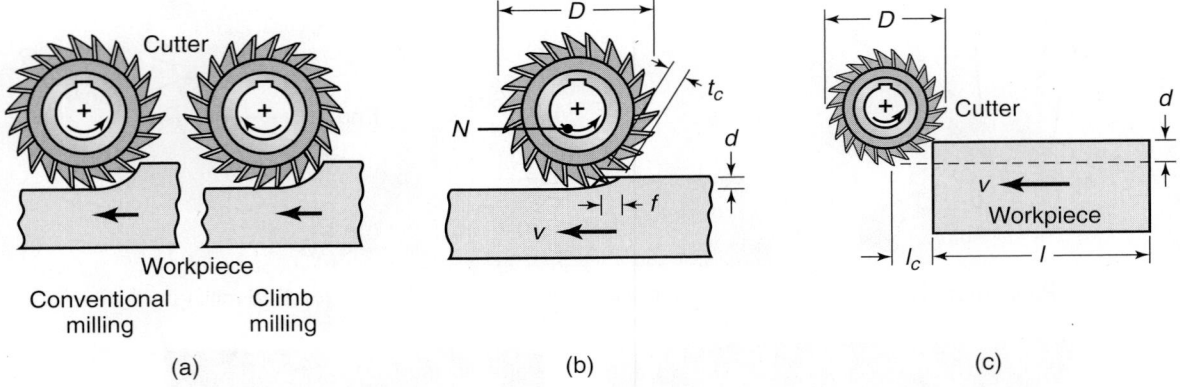

FIGURE 24.3   (a) Schematic illustration of conventional milling and climb milling. (b) Slab-milling operation showing depth-of-cut, d; feed per tooth, f; chip depth of cut, t_c; and workpiece speed, v. (c) Schematic illustration of cutter travel distance, l_c, to reach full depth-of-cut.

Milling parameters.   The cutting speed, V, in peripheral milling is the surface speed of the cutter, or

$$V = \pi D N \tag{24.1}$$

where D is the cutter diameter and N is the rotational speed of the cutter (Fig. 24.4). Note that the thickness of the chip in slab milling varies along its length because of the relative longitudinal motion between the cutter and the workpiece. For a straight-tooth cutter, the approximate *undeformed chip thickness* (*chip depth-of-cut*), t_c, can be calculated from the equation

$$t_c = 2f\sqrt{\frac{d}{D}} \tag{24.2}$$

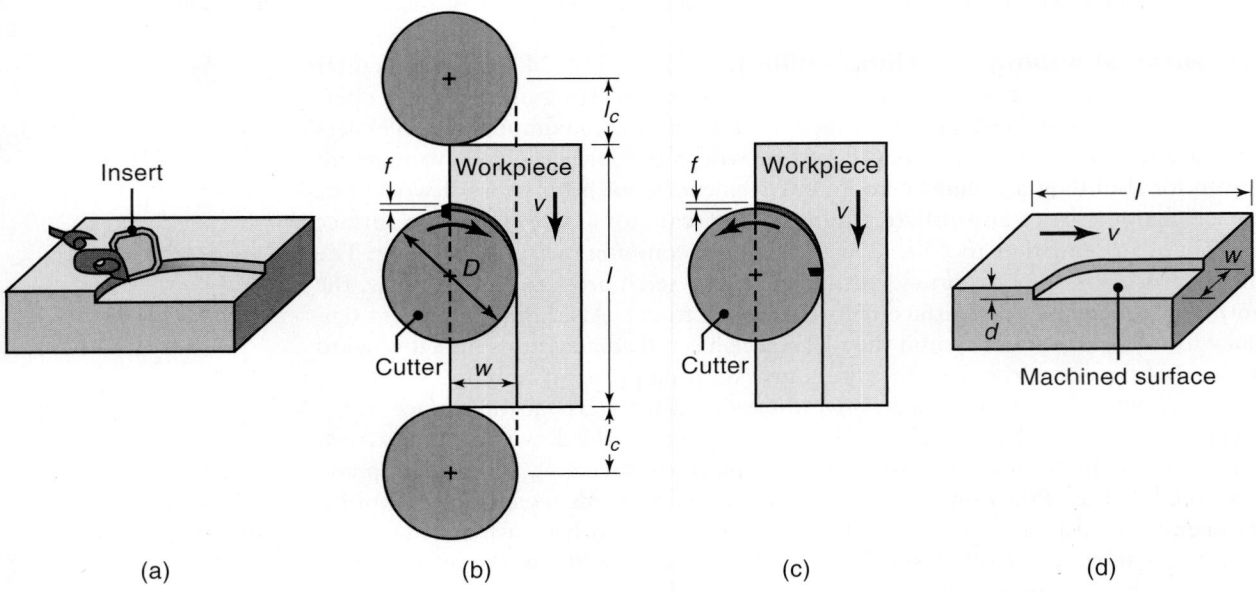

FIGURE 24.4   Face-milling operation showing (a) action of an insert in face milling; (b) climb milling; (c) conventional milling; (d) dimensions in face milling. The width of cut, w, is not necessarily the same as the cutter radius.

where f is the feed-per-tooth of the cutter, that is, the distance the workpiece travels per tooth of the cutter, in mm/tooth or in./tooth, and d is the depth-of-cut. As t_c becomes larger, the force on the cutter tooth increases.

Feed-per-tooth is determined from the equation

$$f = \frac{v}{Nn} \tag{24.3}$$

where v is the linear speed (*feed rate*) of the workpiece and n is the number of teeth on the cutter periphery. The dimensional accuracy of this equation can be checked by using appropriate units for the individual terms; thus, for instance, (mm/tooth) = (m/min)(10^3 mm/m)/(rev/min)(number of teeth/rev).

The cutting time, t, is given by the expression

$$t = \frac{l + l_c}{v} \tag{24.4}$$

where l is the length of the workpiece (Fig. 24.3c) and l_c is the horizontal extent of the cutter's first contact with the workpiece. Based on the assumption that $l_c \ll l$ (although this generally is not the case), the *material-removal rate* (MRR) is

$$\text{MRR} = \frac{lwd}{t} = wdv \tag{24.5}$$

where w is the width of the cut which (in slab milling) is the same as the width of the workpiece. As stated in Section 23.2, the distance that the cutter travels in the non-cutting cycle of the milling operation is an important economic consideration, and should be minimized, by means such as faster travel of the machine tool components. The foregoing equations and the terminology used are summarized in Table 24.1.

TABLE 24.1

Summary of Peripheral Milling Parameters and Formulas

N = Rotational speed of the milling cutter, rpm

F = Feed, mm/tooth or in./tooth

D = Cutter diameter, mm or in.

n = Number of teeth on cutter

v = Linear speed of the workpiece or feed rate, mm/min or in./min

V = Surface speed of cutter, m/min or ft/min
      = DN

f = Feed per tooth, mm/tooth or in./tooth
      = v/Nn

l = Length of cut, mm or in.

t = Cutting time, s or min
      = $(l + l_c)/v$, where l_c = extent of the cutter's first contact with workpiece

MRR = mm^3/min or in^3/min
      = wdv, where w is the width of cut

Torque = $\text{N} \cdot \text{m}$ or $\text{lb} \cdot \text{ft}$
      = $F_c D/2$

Power = kW or hp
      = (Torque)(ω), where $\omega = 2\pi N$ radians/min

The *power requirement* in peripheral milling can be measured and calculated, but the *forces* acting on the cutter (tangential, radial, and axial; see also Fig. 23.5) are difficult to calculate because of the many variables involved, particularly cutting-tool geometry. These forces can be measured experimentally for a variety of conditions. However, the *torque* on the cutter spindle also can be calculated from the power. (See the following example.) Although the torque is the product of the tangential force on the cutter and the cutter's radius, the tangential force per tooth will depend on how many teeth are engaged during the cut.

EXAMPLE 24.1 Material-removal rate, power, torque, and cutting time in slab milling

A slab-milling operation is being carried out on a 12-in. long, 4-in. wide annealed mild-steel block at a feed $f = 0.01$ in./tooth and a depth of cut $d = \frac{1}{8}$ in. The cutter is $D = 2$ in. in diameter, has 20 straight teeth, rotates at $N = 100$ rpm, and by definition is wider than the block to be machined. Calculate the material-removal rate, estimate the power and torque required for this operation, and calculate the cutting time.

Solution From the information given, the linear speed of the workpiece, v, can be calculated from Eq. (24.3):

$$v = fNn = (0.01)(100)(20) = 20 \text{ in./min}$$

From Eq. (24.5), the material-removal rate is calculated to be

$$\text{MRR} = (4)\left(\frac{1}{8}\right)(20) = 10 \text{ in}^3/\text{min}$$

Since the workpiece is annealed mild steel, let's estimate unit power from Table 21.2 as 1.1 hp.min/in^3. Hence, the power required can be estimated as

$$\text{Power} = (1.1)(10) = 11 \text{ hp}$$

The torque acting on the cutter spindle also can be calculated by noting that power is the product of torque and the spindle rotational speed (in radians per unit time). Therefore,

$$\text{Torque} = \frac{\text{Power}}{\text{Rotational speed}}$$

$$= \frac{(11 \text{ hp})(33,000 \text{ lb-ft/min.hp})}{(100 \text{ rpm})(2\pi)} = 578 \text{ lb-ft.}$$

The cutting time is given by Eq. (24.4), in which the quantity l_c can be shown, from simple geometric relationships and for $D \gg d$, to approximate

$$l_c = \sqrt{Dd} = \sqrt{(2)\left(\frac{1}{8}\right)} = 0.5 \text{ in.}$$

Thus, the cutting time is

$$t = \frac{12 + 0.5}{20} = 0.625 \text{ min} = 37.5 \text{ s}$$

24.2.2 Face milling

In *face milling* (Fig 24.4), the cutter is mounted on a spindle having an axis of rotation perpendicular to the workpiece surface (Fig. 24.2b) and removes material in the manner shown in Fig. 24.4a. The cutter rotates at a rotational speed N, and the workpiece moves along a straight path at a linear speed v. When the cutter rotation is as shown in Fig. 24.4b, the operation is *climb milling*; when it is in the opposite direction (Fig. 24.4c), the operation is *conventional milling*. The cutting teeth, such as carbide inserts, are mounted on the cutter body as shown in Fig. 24.5 (see also Fig. 22.3c).

Because of the relative motion between the cutter teeth and the workpiece, face milling leaves *feed marks* on the machined surface (Fig. 24.6), similar to those left by turning operations (Fig. 21.2). Note that surface roughness of the workpiece depends on the corner geometry of the insert and feed per tooth.

The terminology for a face-milling cutter, as well as the various angles, are shown in Fig. 24.7. As can be seen from the side view of the insert in Fig. 24.8, the *lead angle* of the insert in face milling has a direct influence on the *undeformed chip thickness*, as it does in turning operations. As the lead angle (positive, as shown in Fig. 24.8b) increases, the undeformed chip thickness decreases (as does chip thickness), and the length of contact (hence chip width) increases. The lead angle also influences the forces in milling. It can be seen that as the lead angle decreases, there is a smaller and smaller vertical-force component (axial force on the cutter spindle). The range of *lead angles* for most face-milling cutters is typically from 0° to 45°. Note in Fig. 24.8 that the cross-sectional area of the undeformed chip remains constant.

A wide variety of milling cutters is available. The cutter diameter should be chosen such that it will not interfere with fixtures and other components in the setup. In a typical face-milling operation, the ratio of the cutter diameter, D, to the width of cut, w, should be no less than 3:2.

FIGURE 24.5 A face-milling cutter with indexable inserts. *Source:* Courtesy of Ingersoll Cutting Tool Company.

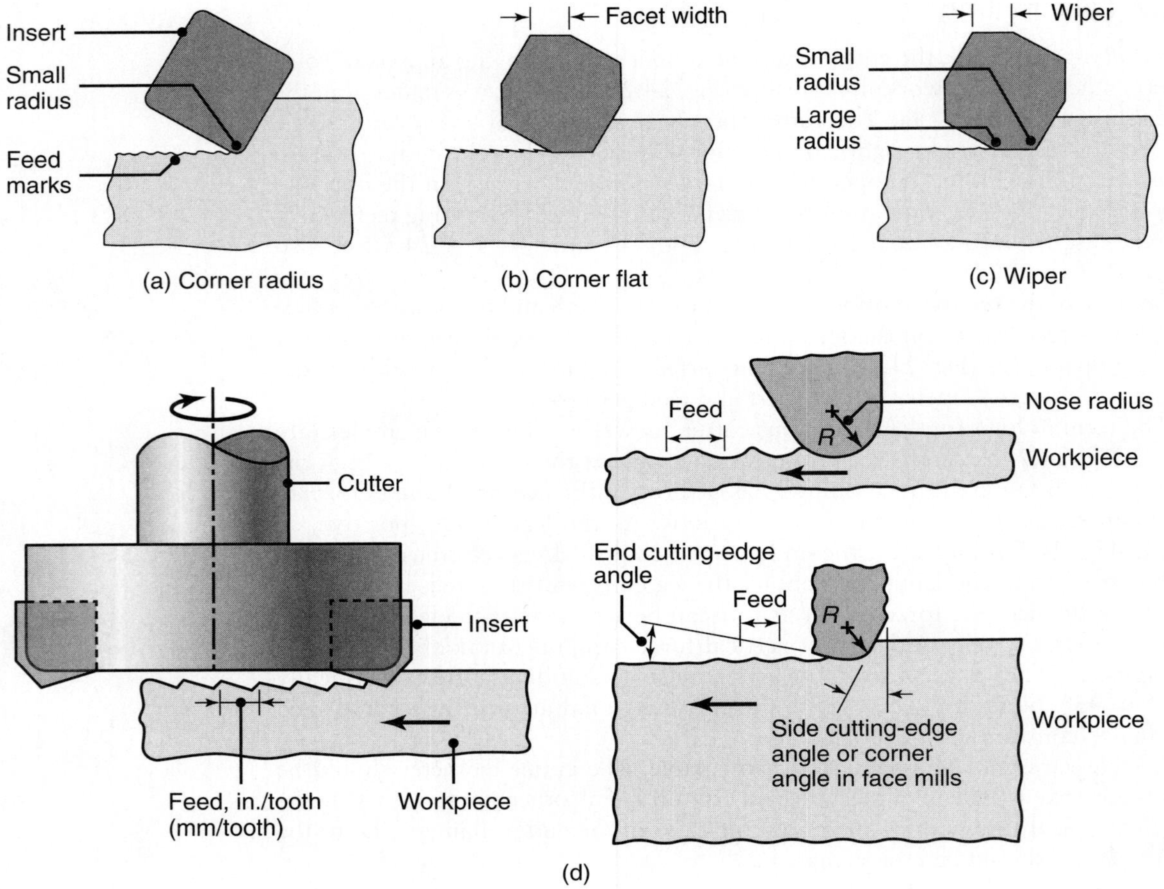

Insert
Small radius
Feed marks

(a) Corner radius

Facet width

(b) Corner flat

Wiper
Small radius
Large radius

(c) Wiper

Cutter
Insert
Feed, in./tooth (mm/tooth)
Workpiece

Feed
Nose radius
Workpiece
R

End cutting-edge angle
Feed
R
Side cutting-edge angle or corner angle in face mills
Workpiece

(d)

FIGURE 24.6   Schematic illustration of the effect of insert shape on feed marks on a face-milled surface: (a) small corner radius, (b) corner flat on insert, and (c) wiper, consisting of a small radius followed by a large radius which leaves smoother feed marks. (d) Feed marks due to various insert shapes.

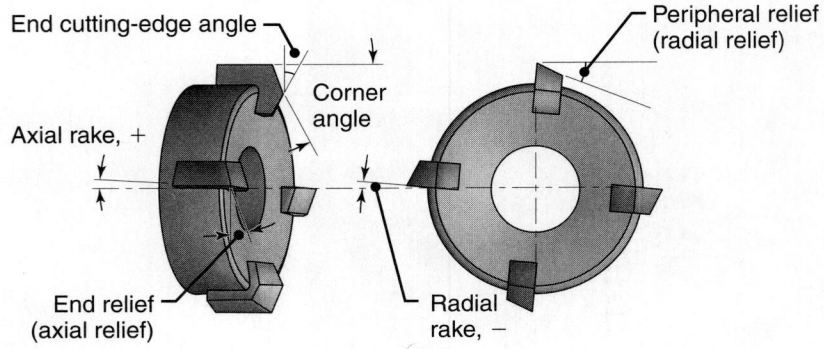

End cutting-edge angle
Corner angle
Axial rake, +
End relief (axial relief)
Peripheral relief (radial relief)
Radial rake, −

FIGURE 24.7   Terminology for a face-milling cutter.

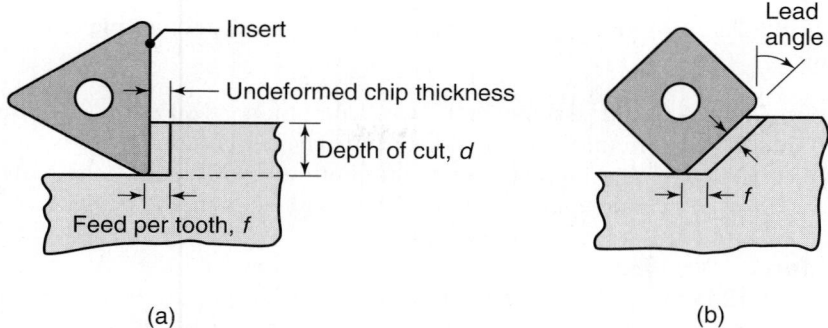

(a)

(b)

FIGURE 24.8 The effect of the lead angle on the undeformed chip thickness in face milling. Note that as the lead angle increases, the chip thickness decreases, but the length of contact (i.e., chip width) increases. The edges of the insert must be sufficiently large to accommodate the contact length increase.

The relationship of cutter diameter to insert angles and their position relative to the surface to be milled is important in that it will determine the angle at which an insert *enters* and *exits* the workpiece. Note in Fig. 24.4b for climb milling that (if the insert has zero axial and radial rake angles, see Fig. 24.7) the rake face of the insert engages the workpiece directly. As seen in Fig. 24.9a and b, however, the same insert may engage the workpiece at different angles, depending on the relative positions of the cutter and the workpiece width. Note in Fig. 24.9a, that the tip of the insert makes the first contact, so there is a possibility for the cutting edge to chip off. In Fig. 24.9b, on the other hand, the first contacts (at entry, reentry, and the two exits) are at an angle and away from the tip of the insert. Therefore, there is a lower tendency for the insert to fail, because the forces on the insert vary more slowly. Note from Fig. 24.7 that the radial and axial rake angles also will have an effect on this operation.

Figure 24.9c shows the exit angles for various cutter positions. Note that in the first two examples, the insert exits the workpiece at an angle, thus causing the force on the insert to reduce to zero at a slower rate (desirable) than in the third example, where the insert exits the workpiece suddenly (which is undesirable for tool life).

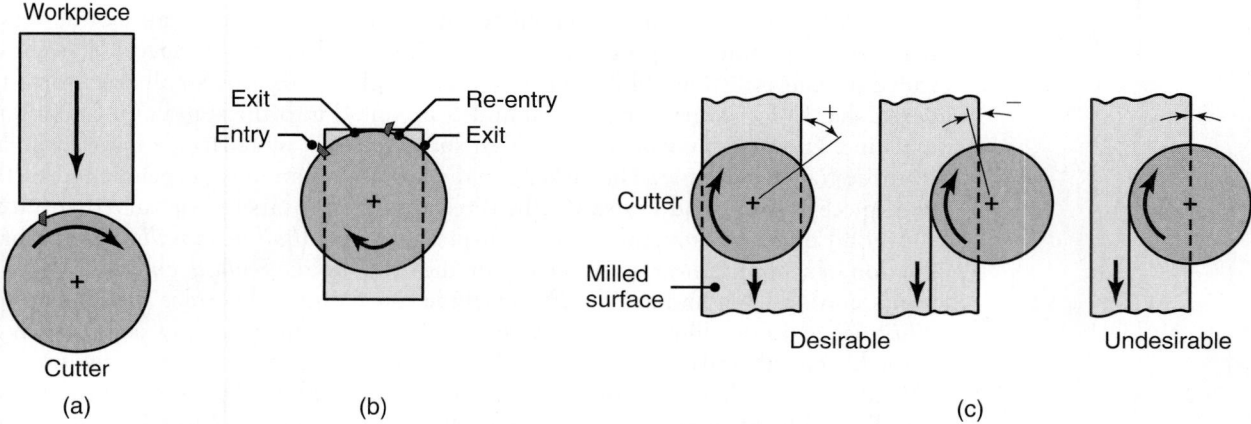

(a)

(b)

(c)

FIGURE 24.9 (a) Relative position of the cutter and insert as they first engage the workpiece in face milling. (b) Insert positions towards the end of cut. (c) Examples of exit angles of insert, showing desirable (positive or negative angle) and undesirable (zero angle) positions. In all figures, the cutter spindle is perpendicular to the page and rotates clockwise.

EXAMPLE 24.2 Material-removal rate, power required, and cutting time in face milling

Refer to Fig. 24.4 and assume that $D = 150$ mm, $w = 60$ mm, $l = 500$ mm, $d = 3$ mm, $v = 0.6$ m/min, and $N = 100$ rpm. The cutter has 10 inserts, and the workpiece material is a high-strength aluminum alloy. Calculate the material-removal rate, cutting time, and feed-per-tooth, and estimate the power required.

Solution   We first note that the cross-section of the cut is $wd = (60)(3) = 180$ mm^2. Noting that the workpiece speed, v, is 0.6 m/min = 600 mm/min, the material-removal rate (MRR) can be calculated as

$$\text{MRR} = (180)(600) = 108,000 \text{ mm}^3/\text{min}$$

The cutting time is given by

$$t = \frac{l + 2l_c}{v}$$

Note from Fig. 24.4 that, for this problem, $l_c = \dfrac{D}{2} = 75$ mm. The cutting time is then

$$t = \frac{500 + 150}{10} = 65 \text{ s} = 1.08 \text{ min}$$

The feed-per-tooth can be obtained from Eq. (24.3). Noting that $N = 100$ rpm =1.67 rev/s, we find that

$$f = \frac{10}{(1.67)(10)} = 0.6 \text{ mm/tooth}$$

For this material, let's estimate the unit power from Table 21.2 to be 1.1 W·s/mm^3. Thus, the power is

$$\text{Power} = (1.1)(1800) = 1980 \text{ W} = 1.98 \text{ kW}$$

24.2.3 End milling

End milling is an important and common machining operation because of its versatility and capability to produce various profiles and curved surfaces. The cutter, called an **end mill** (Fig. 24.2c) has either a straight shank (for small sizes) or a tapered shank (for larger cutter sizes) and is mounted into the spindle of the milling machine. End mills may be made of high-speed steels or with carbide inserts, similar to those for face milling. The cutter usually rotates on an axis perpendicular to the workpiece surface, and it also can be tilted to machine tapered or curved surfaces.

End mills are available with hemispherical ends (*ball nose mills*) for the production of sculptured surfaces, such as on dies and molds. *Hollow end mills* have internal cutting teeth and are used to machine the cylindrical surface of solid, round workpieces. End milling can produce a variety of surfaces at any depth, such as curved, stepped, and pocketed (Fig. 24.2d). The cutter can remove material on both its end and its cylindrical cutting edges, as can be seen in Fig. 24.2c. Both vertical spindle and horizontal spindle machines, as well as machining centers, can be used for end milling workpieces of various sizes and shapes. The machines can be programmed such that the cutter can follow a complex set of paths that optimize the whole machining operation for productivity and minimum cost.

High-speed end milling. High-speed machining and its applications are described in Section 25.5. *High-speed end milling* has become an important process with numerous applications, such as the milling of large aluminum-alloy aerospace components and honeycomb structures with spindle speeds in the range of 20,000 to 60,000 rpm. The machines must have high stiffness and accuracy, usually requiring hydrostatic (air) bearings and high-quality workholding devices. The spindles have a rotational accuracy of 10 μm, thus the workpiece surface is also very accurate. At such high rates of material removal, chip collection and disposal can be a significant problem, as discussed in Section 23.3.7.

The production of cavities in metalworking dies (**die sinking**)—such as in forging or in sheet-metal forming—also is done by high-speed end milling often using TiAlN-coated ball nose end mills (Fig. 24.10). The machines have *four-axis* or *five-axis* movements (see, for example, Fig. 24.18) and are able to accommodate dies as large as 3 m × 6 m (9 ft × 18 ft) and weighing 60 tons. It is not surprising, then, that such dies can cost around $2 million. The advantages of five-axis machines are that they (a) are capable of machining very complex shapes in a single setup, (b) can use shorter cutting tools (thus reducing the tendency for vibration), and (c) enable drilling of holes at various compound angles.

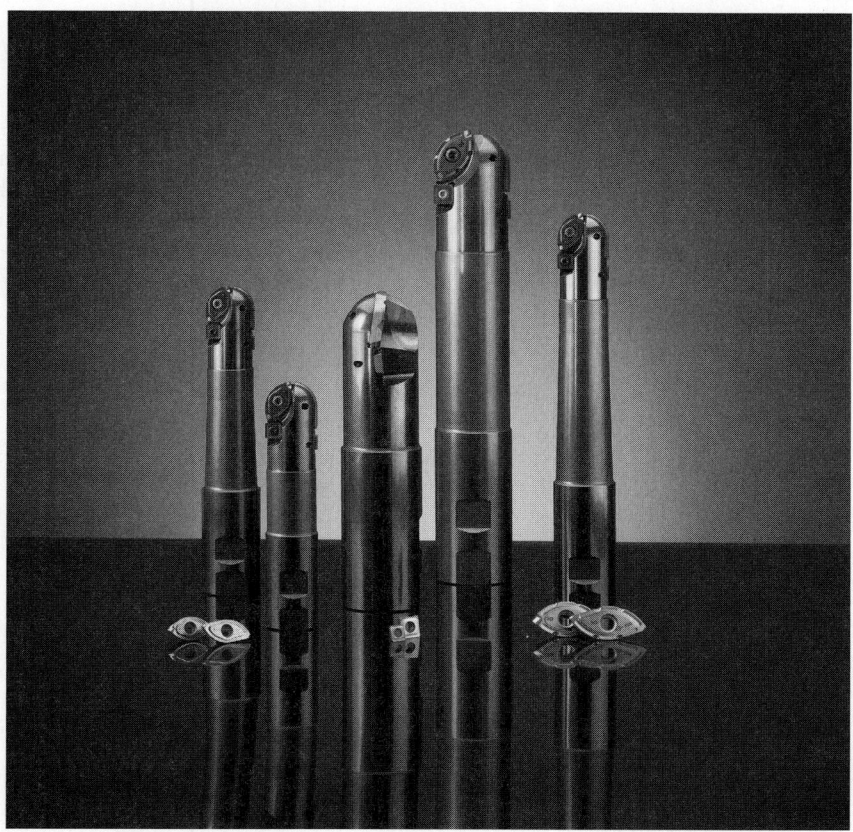

FIGURE 24.10 Ball nose end mills. These cutters are able to produce elaborate contours and are used often in the machining of dies and molds. (See also Fig. 24.2d.) *Source:* Courtesy of Dijet, Inc.

24.2.4 Other milling operations and milling cutters

Several other milling operations and cutters are used to machine workpieces. In **straddle milling**, two or more cutters are mounted on an arbor and are used to machine two parallel surfaces on the workpiece (Fig. 24.11a). **Form milling** produces curved profiles using cutters that have specially shaped teeth (Fig. 24 to 11b). Such cutters also are used for cutting gear teeth, as described in Section 24.7. **Slotting** and **slitting** operations are done using *circular cutters*, as shown in Fig. 24.11c and d, respectively. The teeth may be staggered slightly, like those in a saw blade (Section 24.5), to provide clearance for the cutter when making deep slots. *Slitting saws* are relatively thin, usually less than 5 mm $\left(\frac{3}{16}\,\text{in.}\right)$. *T-slot cutters* are used to mill T-slots, such as those found in machine-tool work tables for clamping workpieces. As shown in Fig. 24.12a, a slot first is milled with an end mill; then the cutter machines the complete profile of the T-slot in one pass.

Key seat cutters are used to make the semicylindrical key seats (Woodruff) for shafts. *Angle milling cutters* (single-angle or double-angle) are used to produce tapered surfaces with various angles. *Shell mills* (Fig. 24.12b) are hollow inside and are mounted on a shank; this allows the same shank to be used for different-sized cutters. The use of shell mills is similar to that of end mills. Milling with a single cutting tooth mounted on a high-speed spindle is known as *fly cutting*; generally it is used in simple face-milling and boring operations. This tool can be shaped as a single-point cutting tool and can be placed in various radial positions on the spindle in an arrangement similar to that shown in Fig. 23.23b.

24.2.5 Toolholders

Milling cutters are classified as either arbor cutters or shank cutters. **Arbor cutters** are mounted on an *arbor* (see Figs. 24.11 and 24.15a), for operations such as peripheral, face, straddle, and form milling. In **shank-type cutters**, the cutter and the shank are made in one piece—the most common examples being end mills. Small end mills have straight shanks, but larger end mills have tapered shanks for better mounting in

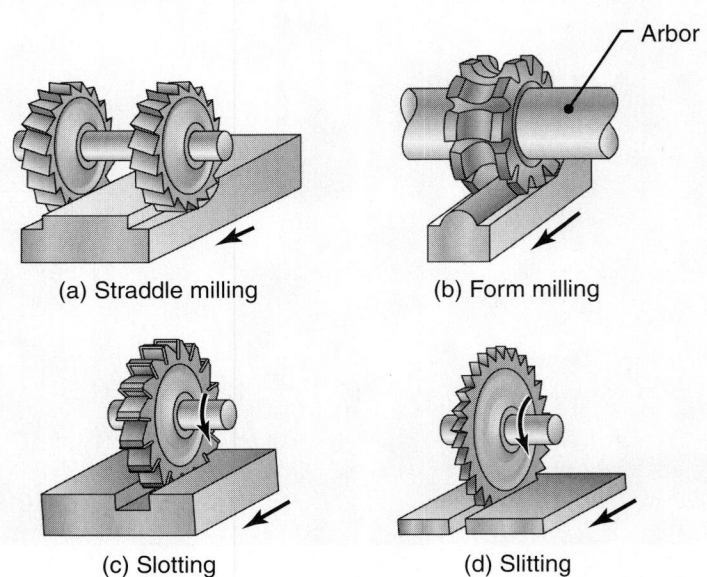

(a) Straddle milling   (b) Form milling

(c) Slotting   (d) Slitting

FIGURE 24.11   Cutters for (a) straddle milling, (b) form milling, (c) slotting, and (d) slitting with a milling cutter.

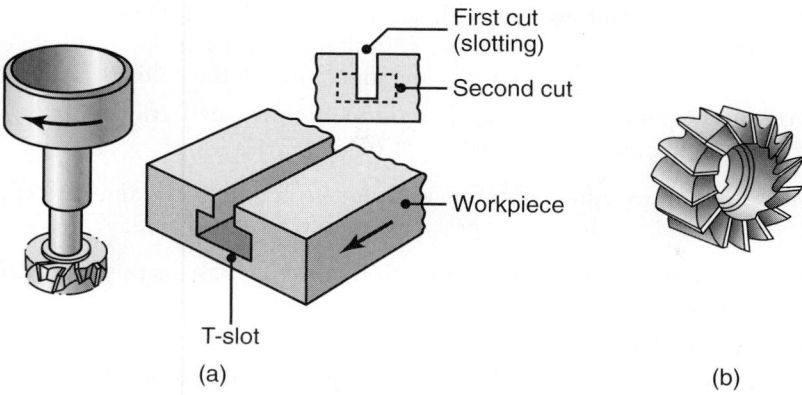

First cut
(slotting)

Second cut

Workpiece

T-slot

(a)

(b)

FIGURE 24.12   (a) T-slot cutting with a milling cutter. (b) A shell mill.

the machine spindle in order to resist the higher forces and torque involved during cutting. Cutters with straight shanks are mounted in collet chucks or special end-mill holders; those with tapered shanks are mounted in tapered toolholders.

In addition to mechanical types, hydraulic toolholders and arbors also are available. The stiffness of cutters and toolholders is important for surface quality and to reduce vibration and chatter during milling operations.

24.2.6 Milling process capabilities

In addition to the various characteristics of the milling processes described thus far, milling process capabilities include parameters such as surface finish, dimensional tolerances, production rate, and cost considerations. Data on process capabilities are presented in Tables 23.1 and 23.8, Figs. 23.13 and 23.14, and in Chapter 40.

The conventional ranges of cutting speeds and feeds for milling are given in Table 24.2. Depending on the workpiece material and its condition, cutting-tool material, and process parameters, cutting speeds vary widely in the range of 30 to 3000 m/min (90 to 10,000 ft/min). Feed-per-tooth typically ranges from about 0.1 mm (0.004 in.) to 0.5 mm (0.02 in.), and depths-of-cut are usually 1 to 8 mm (0.04 to 0.30 in.). For cutting-fluid recommendations, see Table 23.5.

A general **troubleshooting guide** for milling operations is given in Table 24.3; the last four items in this table are illustrated in Figs. 24.13 and 24.14. *Back striking* involves double feed marks made by the trailing edge of the cutter. Note from Table 24.3 that some recommendations (such as changing milling parameters or cutting tools) are easier to accomplish than others (such as changing tool angles, cutter geometry, and the stiffness of spindles and workholding devices).

24.2.7 Design and operating guidelines for milling

The guidelines for turning and boring, given in Sections 23.3.6 and 23.4, also generally are applicable to milling operations. Additional factors relevant to milling operations include the following:

- Standard milling cutters should be used as much as possible, depending on part design features. Costly special cutters should be avoided.
- Chamfers should be specified instead of radii; it is difficult to smoothly match various intersecting surfaces if radii are specified.

- Internal cavities and pockets with sharp corners should be avoided because of the difficulty of milling them, since cutting teeth or inserts have a finite edge radius. When possible, the corner radius should match the milling cutter geometry.
- Workpieces should be sufficiently rigid to minimize deflections that may result from clamping and cutting forces.

Guidelines for avoiding vibration and chatter in milling are similar to those for turning. In addition, the following practices should be considered:

- Cutters should be mounted as close to the spindle base as possible in order to reduce tool deflections.
- Tool holders and fixturing devices should be as rigid as possible.
- In cases of vibration and chatter, tool shape and process conditions should be modified, and cutters with fewer cutting teeth or with random tooth spacing should be used.

TABLE 24.2

General Recommendations for Milling Operations

Material	Cutting tool	General-purpose starting conditions		Range of conditions	
		Feed mm/tooth (in./tooth)	Speed m/min (ft/min)	Feed mm/tooth (in./tooth)	Speed m/min (ft/min)
Low-carbon and- **free machining steels**	Uncoated carbide, coated carbide, cermets	0.13–0.20 (0.005–0.008)	120–180 (400–600)	0.085–0.38 (0.003–0.015)	90–425 (300–1400)
Alloy steels					
Soft	Uncoated, coated, cermets	0.10–0.18 (0.004–0.007)	90–170 (300–550)	0.08–0.30 (0.003–0.012)	60–370 (200–1200)
Hard	Cermets, PcBN	0.10–0.15 (0.004–0.006)	180–210 (600–700)	0.08–0.25 (0.003–0.010)	75–460 (250–1500)
Cast iron, gray					
Soft	Uncoated, coated, cermets, SiN	0.10–10.20 (0.004–0.008)	120–760 (400–2500)	0.08–0.38 (0.003–0.015)	90–1370 (300–4500)
Hard	Cermets, SiN, PcBN	0.10–0.20 (0.004–0.008)	120–210 (400–700)	0.08–0.38 (0.003–0.015)	90–460 (300–1500)
Stainless steel, Austenitic	Uncoated, coated, cermets	0.13–0.18 (0.005–0.007)	120–370 (400–1200)	0.08–0.38 (0.003–0.015)	90–500 (300–1800)
High-temperature alloys Nickel based	Uncoated, coated, cermets, SiN, PcBN	0.10–0.18 (0.004–0.007)	30–370 (100–1200)	0.08–0.38 (0.003–0.015)	30–550 (90–1800)
Titanium alloys	Uncoated, coated, cermets	0.13–0.15 (0.005–0.006)	50–60 (175–200)	0.08–0.38 (0.003–0.015)	40–140 (125–450)
Aluminum alloys					
Free machining	Uncoated, coated, PCD	0.13–0.23 (0.005–0.009)	610–900 (2000–3000)	0.08–0.46 (0.003–0.018)	300–3000 (1000–10,000)
High silicon	PCD	0.13 (0.005)	610 (2000)	0.08–0.38 (0.003–0.015)	370–910 (1200–3000)
Copper alloys	Uncoated, coated, PCD	0.13–0.23 (0.005–0.009)	300–760 (1000–2500)	0.08–0.46 (0.003–0.018)	90–1070 (300–3500)
Plastics	Uncoated, coated, PCD	0.13–0.23 (0.005–0.009)	270–460 (900–1500)	0.08–0.46 (0.003–0.018)	90–1370 (300–4500)

Source: Based on data from Kennametal, Inc.
Note: Depths-of-cut, *d*, usually are in the range of 1 to 8 mm (0.04 to 0.3 in.). PcBN: polycrystalline cubic-boron nitride. PCD: polycrystalline diamond. See also Table 23.4 for range of cutting speeds within tool material groups.

TABLE 24.3

General Troubleshooting Guide for Milling Operations

Problem	Probable causes
Tool breakage	Tool material lacks toughness, improper tool angles, machining parameters too high
Excessive tool wear	Machining parameters too high, improper tool material, improper tool angles, improper cutting fluid
Rough surface finish	Feed per tooth too high, too few teeth on cutter, tool chipped or worn, built-up edge, vibration and chatter
Tolerances too broad	Lack of spindle and workholding stiffness, excessive temperature rise, dull tool, chips clogging cutter
Workpiece surface burnished	Dull tool, depth-of-cut too low, radial relief angle too small
Back striking	Dull cutting tools, tilt in cutter spindle, negative tool angles
Chatter marks	Insufficient stiffness of system; external vibrations; feed, depth of cut, and width of cut too large
Burr formation	Dull cutting edges or too much honing, incorrect angle of entry or exit, feed and depth of cut too high, incorrect insert shape
Breakout	Lead angle too low, incorrect cutting edge geometry, incorrect angle of entry or exit, feed and depth of cut too high

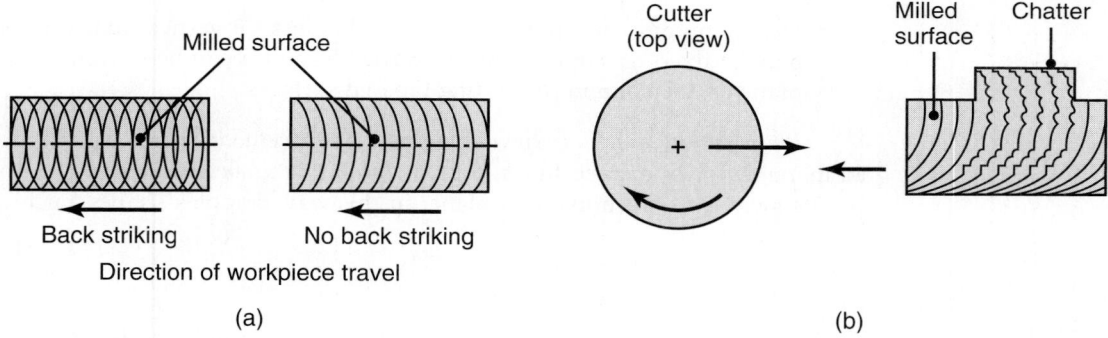

FIGURE 24.13   Machined surface features in face milling. See also Fig. 24.6.

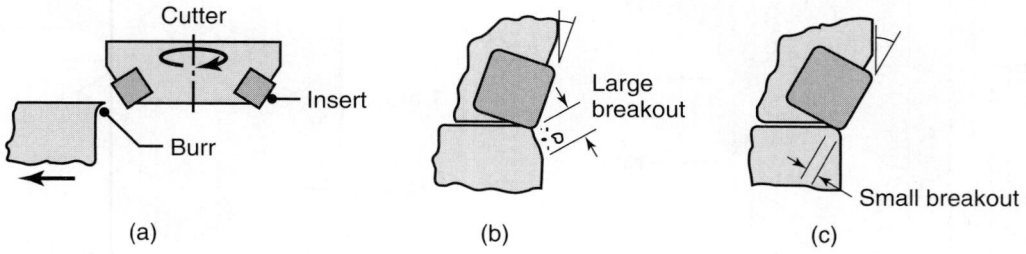

FIGURE 24.14   Edge defects in face milling: (a) burr formation along workpiece edge, (b) breakout along workpiece edge, and (c) how it can be avoided by increasing the lead angle (see also last row in Table 24.3).

24.2.8 Milling machines

Because they are capable of performing a variety of cutting operations, milling machines are among the most versatile and useful machine tools. The first milling machine was built in 1820 by Eli Whitney (1765–1825). A wide selection of milling machines with numerous features is now available. The features of typical standard milling machines are described next. It should be noted, however, that many of these machines and operations are now being replaced with *computer controls* and *machining centers*.

Column-and-knee type machines.   Used for general-purpose milling operations, *column-and-knee type machines* are the most common milling machines. The spindle on which the milling cutter is mounted may be *horizontal* (Fig. 24.15a) for peripheral milling or *vertical* for face and end milling, boring, and drilling operations (Fig. 24.15b). The basic components of these machines are:

- *Work table*: on which the workpiece is clamped using T-slots. The table moves longitudinally relative to the saddle.
- *Saddle*: supports the table and can move in the transverse direction.
- *Knee*: supports the saddle and gives the table vertical movement so that the depth of cut can be adjusted and workpieces with various heights can be accommodated.
- *Overarm*: used on horizontal machines; it is adjustable to accommodate different arbor lengths.
- *Head*: contains the spindle and cutter holders. In vertical machines, the head may be fixed or can be adjusted vertically, and it can be swiveled in a vertical plane on the column for cutting tapered surfaces.

Plain milling machines have three axes of movement, which usually are imparted manually or by power. In *universal column-and-knee milling machines*, the table can be swiveled on a horizontal plane. In this way, complex shapes (such as helical

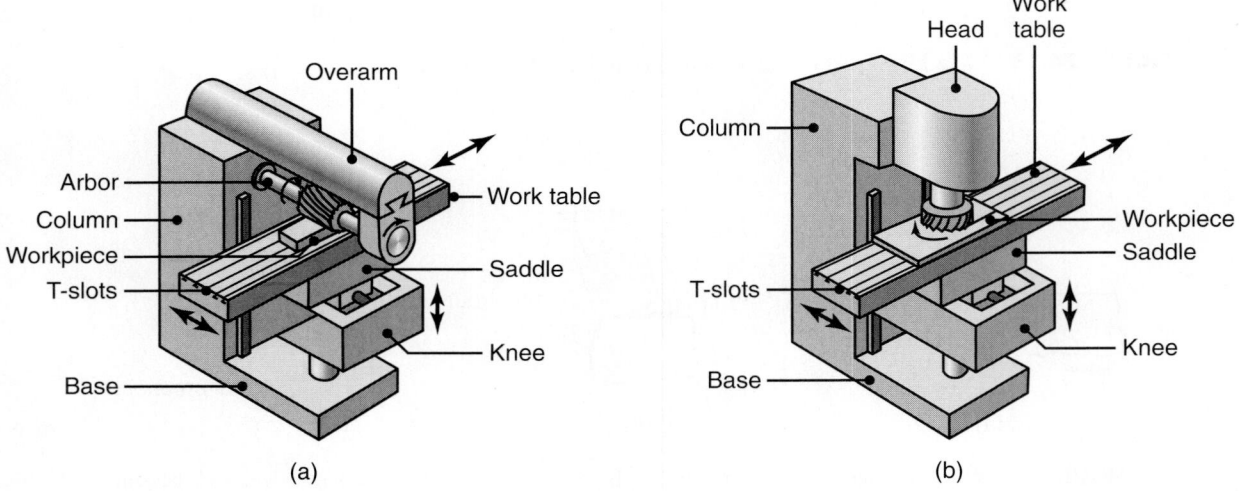

(a)   (b)

FIGURE 24.15   Schematic illustration of (a) a horizontal-spindle column-and-knee type milling machine and (b) vertical-spindle column-and-knee type milling machine. *Source*: After G. Boothroyd.

grooves at various angles) can be machined to produce parts such as gears, drills, taps, and cutters.

Bed-type milling machines.   In *bed-type machines*, the work table is mounted directly on the bed, which replaces the knee and can move only longitudinally (Fig. 24.16). These machines are not as versatile as other types, but they have high stiffness and typically are used for high-production work. The spindles may be horizontal or vertical and of duplex or triplex types (with two or three spindles) for the simultaneous machining of two or three workpiece surfaces.

Other types of milling machines.   Several other types of milling machines are available (see also *machining centers*, Section 25.2). *Planer-type milling machines*, which are similar to bed-type machines, are equipped with several heads and cutters to mill different surfaces. They are used for heavy workpieces and are more efficient than planers (Section 24.3) when used for similar purposes. *Rotary-table machines* are similar to vertical milling machines and are equipped with one or more heads for face-milling operations.

Milling machines are being replaced rapidly by *computer numerical-control machines* (CNC). These machine tools are versatile and capable of milling, drilling, boring, and tapping with repetitive accuracy (Fig. 24.17). Also available are *profile milling machines*, which have five axes of movement (Fig. 24.18); note the three linear and two angular movements of the machine components.

Workholding devices and accessories.   The workpiece to be milled must be clamped securely to the work table in order to resist cutting forces and prevent slipping during milling. Various fixtures and vises generally are used for this purpose. (See also Section 37.8 on *flexible fixturing*.) They are mounted and clamped to the work table using the T-slots seen in Figs. 24.15a and b. Vises are used for small production work on small parts. Fixtures are used for higher-production work and can be automated by various mechanical and hydraulic means.

Accessories for milling machines include various fixtures and attachments for the machine head (as well as the work table) designed to adapt them to different milling operations. The accessory that has been used most commonly in the past, typically in job shops, is the *universal dividing (index) head*. Manually operated, this fixture rotates (indexes) the workpiece to specified angles between individual machining steps. Typically, it has been used to mill parts with polygonal surfaces and

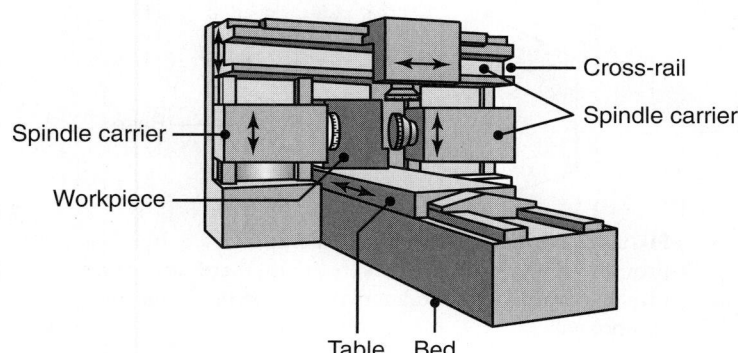

FIGURE 24.16   Schematic illustration of a bed-type milling machine.

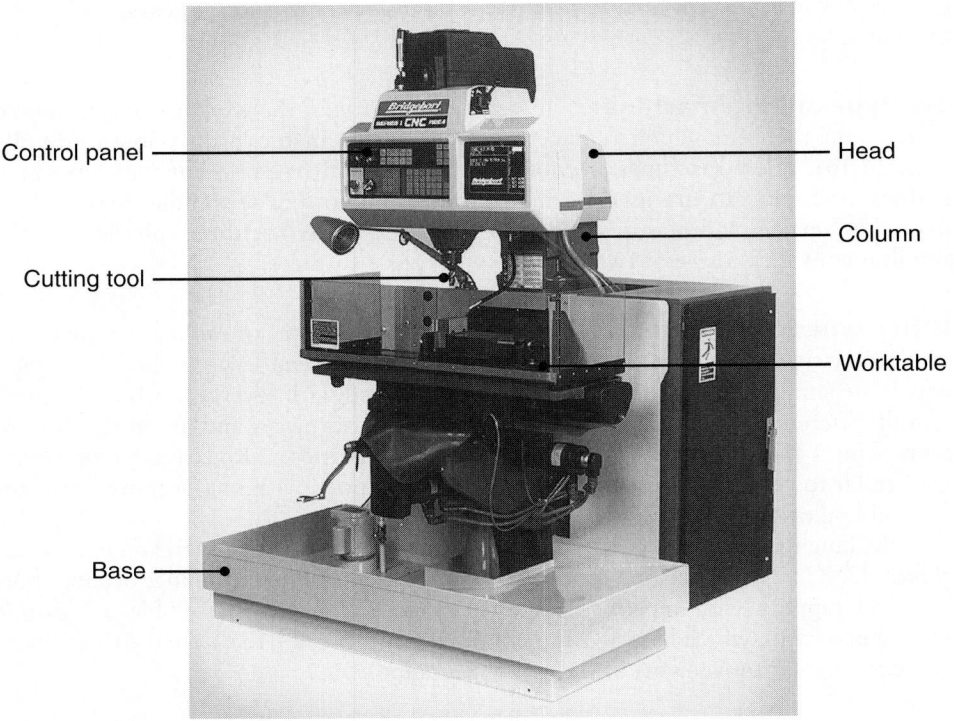

FIGURE 24.17 A computer numerical-control (CNC) vertical-spindle milling machine. This machine is one of the most versatile machine tools. The original vertical-spindle milling machine used in job shops is still referred to as a "Bridgeport," after its manufacturer in Bridgeport, Connecticut. *Source*: Courtesy of Bridgeport Machines Division, Textron Inc.

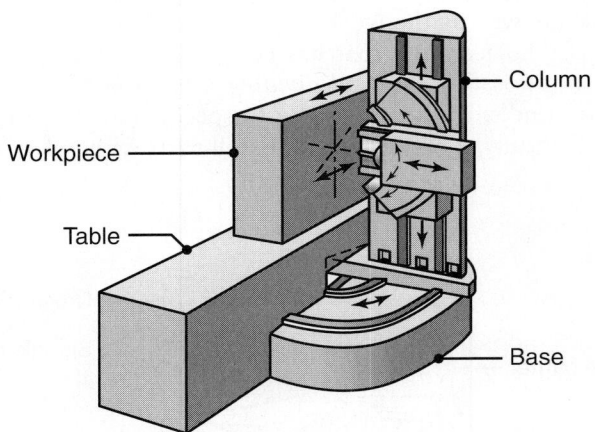

FIGURE 24.18 Schematic illustration of a five-axis profile milling machine. Note that there are three principal linear and two angular movements of machine components.

to machine gear teeth. Dividing heads now are used only for low-volume, job-shop quantities; they have been replaced by CNC controls and machining centers.

24.3 | Planing and Shaping

Planing.   This is a relatively simple machining operation by which flat surfaces, as well as cross-sections with grooves and notches, are produced along the length of the workpiece (Fig. 24.19). Planing usually is done on large workpieces, as large as 25 m × 15 m (75 ft × 40 ft), although a length of 10 m is more typical. In a **planer**, the workpiece is mounted on a table that travels back and forth along a straight path. A horizontal cross-rail, which can be moved vertically along the ways of the column, is equipped with one or more tool heads. The cutting tools are mounted on the heads, and the machining is done along a straight path. In order to prevent tool-cutting edges from chipping when they rub along a workpiece during the return stroke, tools are either tilted or lifted mechanically or hydraulically.

Because of the reciprocating motion of the workpiece, elapsed non-cutting time during the return stroke is significant. Consequently, these operations are neither efficient nor economical (except for low-quantity production, which is generally the case for large and long workpieces). The efficiency of the operation can be improved by equipping planers with tool holders and tools that cut in both directions of table travel. Also, because of the length of the workpiece, it is essential to equip cutting tools with chip breakers. Otherwise, the chips produced can be very long, interfering with the operation and becoming a safety hazard.

Cutting speeds in planers can range up to 120 m/min (400 ft/min) with power capacities of up to 110 kW (150 hp). Recommended speeds for cast irons and stainless steels are in the range of 3 to 6 m/min (10 to 20 ft/min) and up to 90 m/min (300 ft/min) for aluminum and magnesium alloys. Feeds usually are in the range of 0.5 to 3 mm/stroke (0.02 to 0.125 in./stroke). The most common tool materials are M2 and M3 high-speed steels and K20 (C2) and P20 (C6) carbides.

Shaping.   Machining by *shaping* is basically the same as by planing, except that (a) it is the tool and not the workpiece that travels, and (b) workpieces are smaller, typically less than 1 m × 2 m (3 ft. × 6 ft.) of surface area. In a **horizontal shaper**, the cutting tool travels back and forth along a straight path. The tool is attached to the tool head, which is mounted on the ram. The ram has a reciprocating motion. In

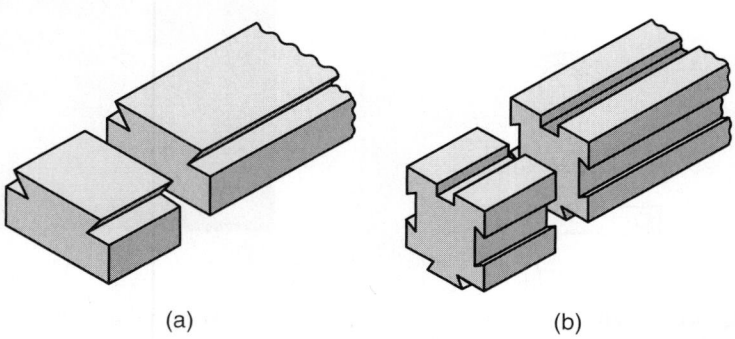

(a)   (b)

FIGURE 24.19   Typical parts that can be made on a planer.

most machines, cutting is done during the forward movement of the ram (*push cut*); in others, it is done during the return stroke of the ram (*draw cut*). Vertical shapers (**slotters**) are used to machine notches, keyways, and dies. Because of low production rates, only special-purpose shapers (such as gear shapers, Section 24.7.2) are in common use today.

24.4 | Broaching and Broaching Machines

The *broaching* operation is similar to shaping with a long multiple-tooth cutter and is used to machine internal and external surfaces, such as holes of circular, square, or irregular section; keyways; the teeth of internal gears; multiple spline holes; and flat surfaces (Fig. 24.20). In a **broach** (Fig. 24.21a), the total depth of material removed in one stroke is the sum of the depths-of-cut of each tooth of the broach. A large broach can remove material as deep as 38 mm (1.5 in.) in one stroke. Broaching is an important production process and can produce parts with good surface finish and dimensional accuracy. It competes favorably with other processes (such as boring, milling, shaping, and reaming) to produce similar shapes. Although broaches can be expensive, the cost is justified with high-quantity production runs.

Broaches.   The terminology for a typical broach is given in Fig. 24.21b. The *rake* (hook) *angle* depends on the material cut (as it does in turning and other cutting operations) and usually ranges between 0° and 20°. The *clearance angle* is typically 1° to 4°; finishing teeth have smaller angles. Too small a clearance angle causes rubbing of the teeth against the broached surface. The *pitch* of the teeth depends on factors such as the length of the workpiece (length of cut), tooth strength, and size and shape of chips.

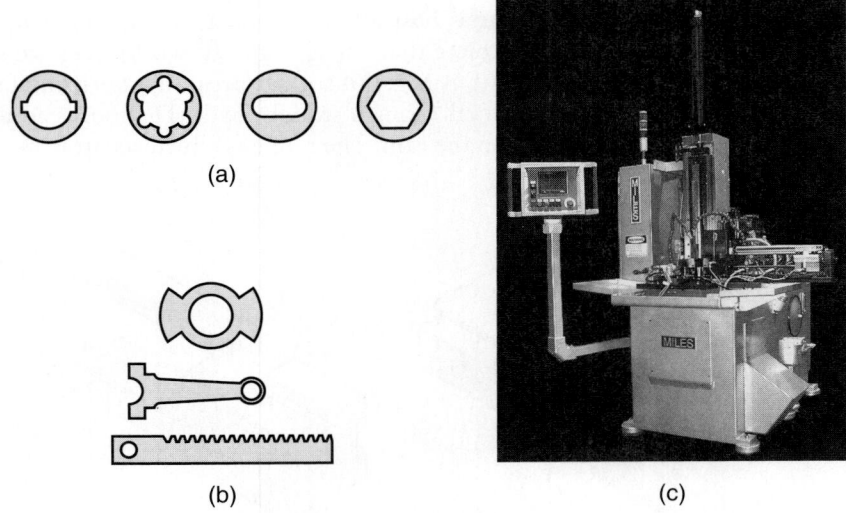

(a)

(b)   (c)

FIGURE 24.20   (a) Typical parts made by internal broaching. (b) Parts made by surface broaching. (c) Vertical broaching machine. *Source*: (a) and (b) Courtesy of General Broach and Engineering Company, (c) Courtesy of Ty Miles, Inc.

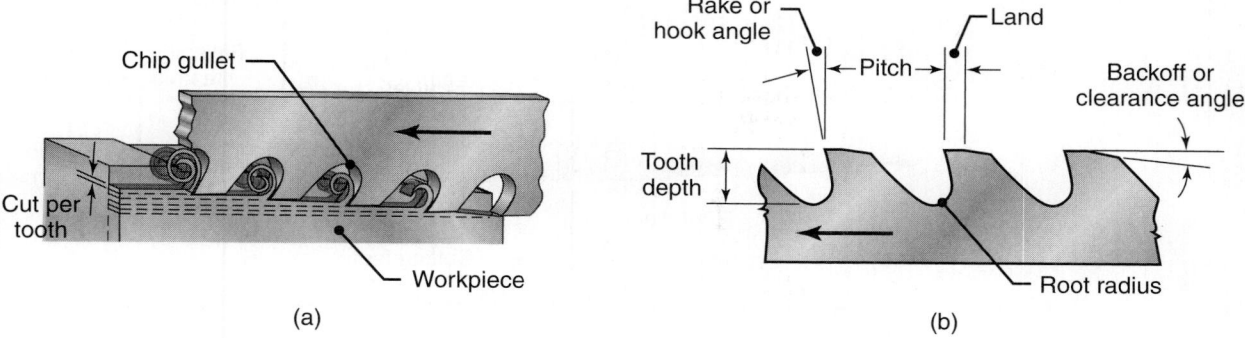

Chip gullet

Cut per tooth

Workpiece

(a)

Rake or hook angle

Land

Pitch

Backoff or clearance angle

Tooth depth

Root radius

(b)

FIGURE 24.21   (a) Cutting action of a broach showing various features. (b) Terminology for a broach.

The tooth depth and pitch must be sufficiently large to accommodate the chips produced during broaching, particularly for long workpieces. At least two teeth should be in contact with the workpiece at all times. The following formula may be used to obtain the pitch for a broach to cut a surface of length l:

$$\text{Pitch} = k\sqrt{l} \tag{24.6}$$

where k is a constant, equal to 1.76 when l is in mm and 0.35 when l is in inches. An average pitch for small broaches is in the range of 3.2 to 6.4 mm (0.125 to 0.25 in.), and for large ones, it is in the range of 12.7 to 25 mm (0.5 to 1 in.). The depth-of-cut per tooth depends on the workpiece material and the surface finish required. It is usually in the range of 0.025 to 0.075 mm (0.001 to 0.003 in.) for medium-sized broaches but can be larger than 0.25 mm (0.01 in.) for larger broaches.

Broaches are available with various tooth profiles, including some with *chip breakers* (Fig. 24.22). The variety of *surface* broaches include *slab* (for cutting flat surfaces), *slot, contour, dovetail, pot* (for precision external shapes), and *straddle*. *Internal* broach types include *hole* (for close-tolerance holes, round shapes, and other shapes; Fig. 24.23), *keyway, internal gear*, and *rifling* (for gun barrels). Irregular internal shapes usually are broached by starting with a round hole drilled or bored in the workpiece.

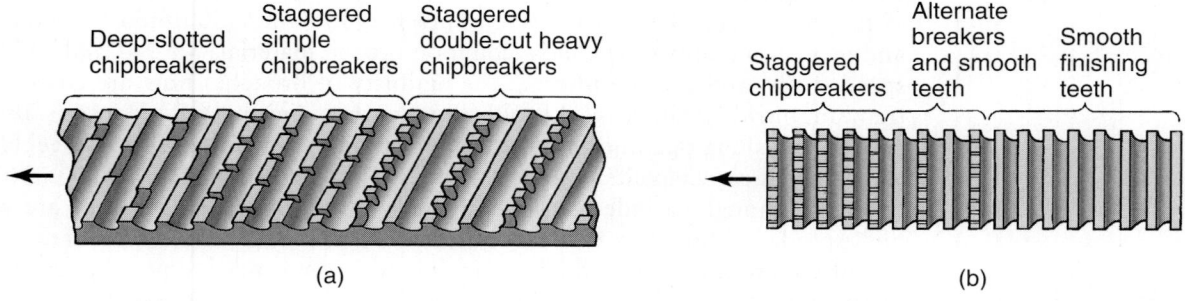

FIGURE 24.22   Chipbreaker features on (a) a flat broach and (b) a round broach.

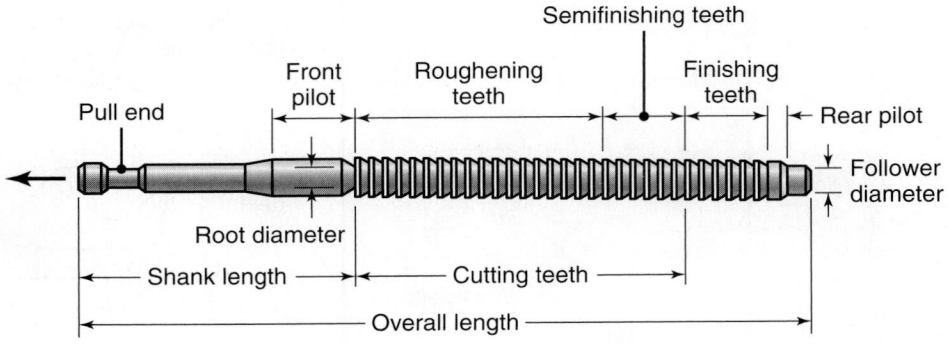

FIGURE 24.23   Terminology for a pull-type internal broach used for enlarging long holes.

Turn broaching.   This process is used typically for broaching the bearing surfaces of crankshafts and similar parts. The crankshaft is rotated between centers, and the broach, which is equipped with multiple carbide inserts, passes tangentially across the bearing surfaces and removes material. Turn broaching is a combination of *shaving* and *skiving* (removing a thin layer of material with a specially-shaped cutting tool). Straight as well as circular broaches are used successfully in turn broaching. Machines that broach a number of crankshafts simultaneously have been built.

Broaching machines.   The machines for broaching are relatively simple in construction, have only linear motions, and usually are actuated hydraulically, although some are moved by crank, screw, or rack. Many styles of broaching machines are available, and sizes range from machines for making needle-like parts to those used for broaching gun barrels, including rifled (internal spiral grooves) gun barrels. Broaching machines either pull or push the broaches and are either horizontal or vertical. *Push broaches* usually are shorter, generally in the range of 150 to 350 mm (6 to 14 in.). *Pull broaches* tend to straighten the hole, whereas pushing permits the broach to follow any irregularity of the leader hole. Horizontal machines are capable of longer strokes. The *force* required to pull or push the broach depends on the (a) strength of the workpiece material, (b) total depth and width of cut, (c) cutting speed, (d) tooth profile, and (e) use of cutting fluids. The pulling force capacities of broaching machines are as high as 0.9 MN (100 tons).

Process parameters.   Cutting speeds for broaching may range from 1.5 m/min (5 ft/min) for high-strength alloys to as much as 15 m/min (50 ft/min) for aluminum and magnesium alloys. The most common broach materials are M2 and M7 high-speed steels, and carbide inserts. The majority of broaches now are coated with titanium nitride for improved tool life and surface finish. Ceramic inserts also are used for finishing operations in some applications. Smaller, high-speed steel blanks for broaches can be made with powder-metallurgy techniques for better control of quality. Although carbide or ceramic inserts can be indexed after they are worn, high-speed steel broach teeth have to be resharpened by grinding, which reduces the size of the broach. Cutting fluids generally are recommended, especially for internal broaching.

Design considerations.   Broaching, as with other machining processes, requires that certain guidelines be followed in order to obtain economical and high-quality production. The major requirements are as follows:

- Parts should be designed so that they can be clamped securely in broaching machines. Parts should have sufficient structural strength and stiffness to withstand the cutting forces during broaching.
- Blind holes, sharp corners, dovetail splines, and large flat surfaces should be avoided.
- Chamfers are preferable to round corners.

EXAMPLE 24.3 **Broaching internal splines**

The part shown in Fig. 24.24 is made of nodular iron (65-45-15; Section 12.3.2) with internal splines—each 50 mm (2 in.) long. The splines have 19 involute teeth

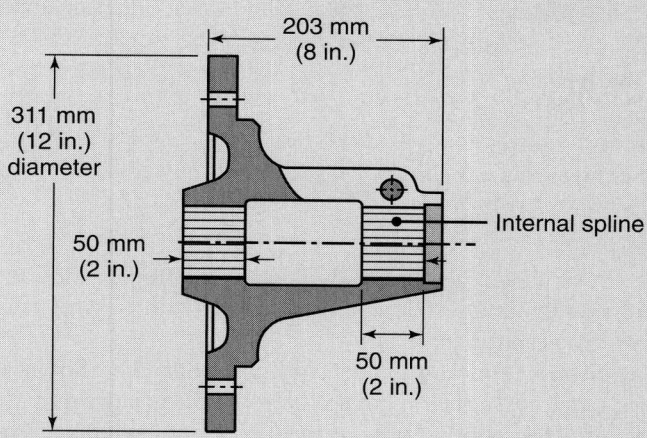

FIGURE 24.24   Example of a part with internal splines produced by broaching.

with a pitch diameter of 63.52 mm (2.5009 in.). An M2 high-speed steel broach with 63 teeth, a length of 1.448 m (57 in.), and a diameter the same as the pitch diameter was used to produce the splines. The cut per tooth was 0.116 mm (0.00458 in.). The production rate was 63 pieces per hour. The number of parts per grind was 400, with a total broach life of about 6000 parts.

Source: ASM International.

24.5 | Sawing

Sawing is a common and old cutting operation dating back to around 1000 B.C., in which the cutting tool is a blade having a series of small teeth (**saw**). Each tooth removes a small amount of material with each stroke of the saw. This process can be used for all metallic and nonmetallic materials and is capable of producing various

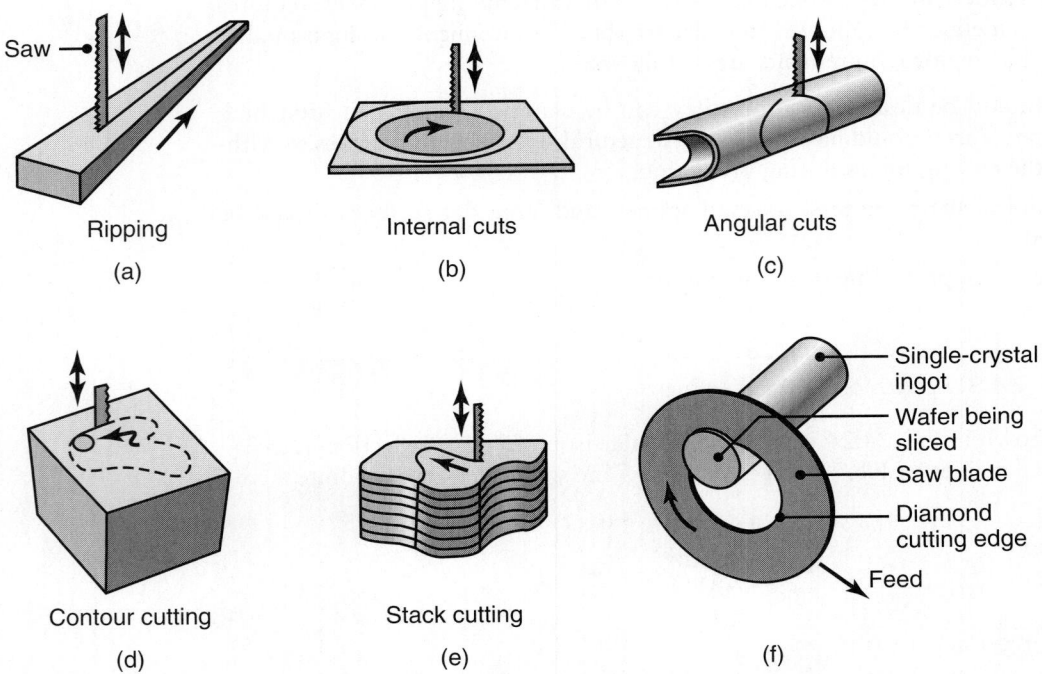

FIGURE 24.25   Examples of various sawing operations.

shapes (Fig. 24.25). Sawing is an efficient bulk-removal process and can produce near-net shapes from raw materials. The width-of-cut (**kerf**) in sawing usually is narrow. Thus, the process wastes little material.

Typical saw-tooth and saw-blade configurations are shown in Fig. 24.26. Tooth spacing is generally in the range of 0.08 to 1.25 teeth per mm (2 to 32 per in.). A wide variety of sizes, tooth forms, tooth spacing, and blade thicknesses and widths are available. Saw blades generally are made from high-carbon and high-speed steels (M2 and M7). Carbide or high-speed steel-tipped steel blades are used to saw harder materials (Fig. 24.27).

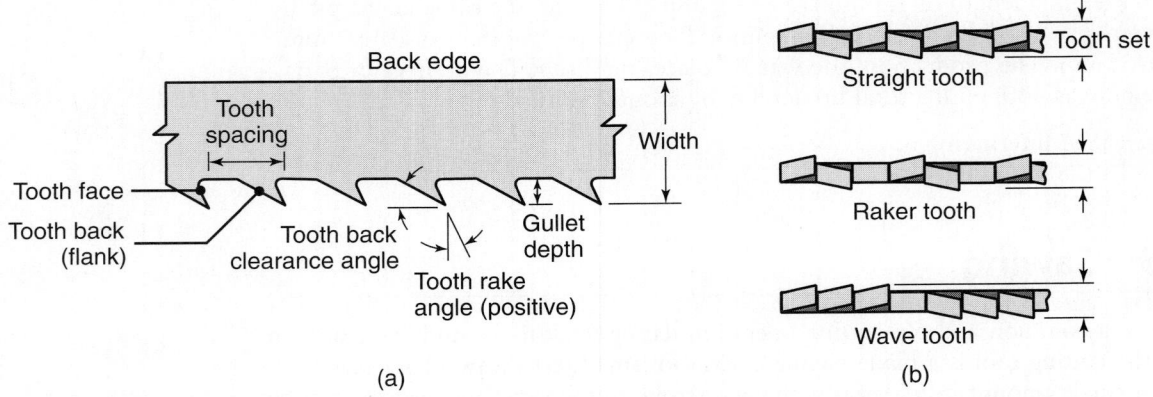

FIGURE 24.26   (a) Terminology for saw teeth. (b) Types of tooth sets on saw teeth staggered to provide clearance for the saw blade to prevent binding during sawing.

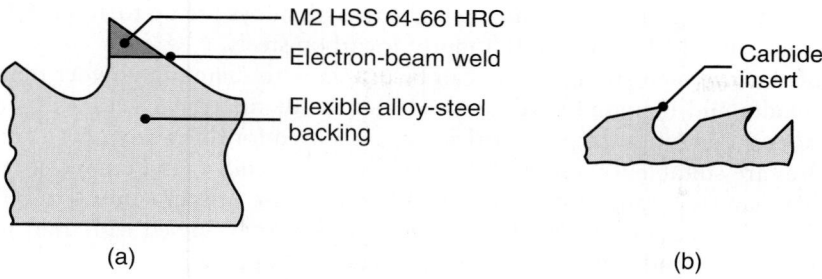

FIGURE 24.27 (a) High-speed-steel teeth welded on a steel blade. (b) Carbide inserts brazed to blade teeth.

The **tooth set** in a saw (Fig. 24.26b) is important in that it provides a sufficiently wide kerf for the blade to move freely in the workpiece without binding and excessive frictional resistance, thus reducing the heat generated. Heat can have adverse effects on the cut, especially when cutting thermoplastics which soften rapidly when heated. The tooth set also allows the blade to track a path accurately, following the pattern to be cut without wandering. At least two or three teeth always should be engaged with the workpiece in order to prevent *snagging* (catching of the saw tooth on the workpiece). This is why thin materials, especially sheet metals, can be difficult to saw. The thinner the stock, the finer the saw teeth should be, and the greater the number of teeth per unit length of the saw. Cutting fluids generally are used to improve the quality of cut and the life of the saw.

Types of saws. *Hacksaws* have straight blades and reciprocating motions. Developed in the 1650s, they generally are used to cut off bars, rods, and structural shapes. They may be manual or power operated. Because cutting takes place during only one of the two reciprocating strokes, hacksaws are not as efficient as band saws (described later). *Power hacksaw* blades are usually 1.2 to 2.5 mm (0.05 to 0.10 in.) thick and up to 610 mm (24 in.) long. The rate of strokes ranges from 30 per minute for high-strength alloys to 180 per minute for carbon steels. The hacksaw frame in power hacksaws is weighted by various mechanisms, applying as much as 1.3 kN (300 lb) of force to the workpiece to improve the cutting rate. *Hand hacksaw* blades are thinner and shorter than power hacksaw blades, which have as many as 1.2 teeth per mm (32 per in.) for sawing sheet metal and thin tubing.

   Circular saws (also called *cold saws* when cutting metal) generally are used for high-production-rate sawing, a process called *cutting off*. Cutting-off operations also can be carried out with thin, abrasive disks, as we describe in Section 26.4. Cold sawing is used in industry very commonly, particularly for cutting large cross-sections. Cold saws are available with a variety of tooth profiles and sizes and can be fed at any angle into the workpiece. In modern machines, cutting off with circular saws produces relatively smooth surfaces with good thickness control and dimensional accuracy due to the stiffness of the machines and of the saws. The inner-diameter cutting saw shown in Fig. 24.25f is used widely to cut single-crystal silicon wafers in microelectronic devices (see also Section 28.4).

   Band saws have continuous, long, flexible blades and, thus, have a continuous cutting action. Vertical band saws are used for straight as well as *contour cutting* of flat sheets and other parts supported on a horizontal table (Fig. 24.25d). Also available are computer-controlled band saws with the capability of guiding the contour path automatically. Power band saws also are available; they have higher productivity than power hacksaws because of their continuous cutting action. With high-speed steel blades, cutting speeds for sawing high-strength alloys are up to about 60 m/min (200 ft/min) and 120 m/min (400 ft/min) for carbon steels.

Blades and high-strength wire can be *coated* with diamond powder (**diamond-edged blades** and **diamond-wire saws**) so that the diamond particles act as cutting teeth (abrasive cutting); carbide particles also are used for this purpose. These blades and wires are suitable for sawing hard metallic, nonmetallic, and composite materials. Wire diameters range from 13 mm (0.5 in.) for use in rock cutting to 0.08 mm (0.003 in.) for precision cutting. Hard materials also can be sawed with thin, abrasive disks and with the advanced machining processes (Chapter 27).

Friction sawing.   *Friction sawing* is a process in which a mild-steel blade or disk rubs against the workpiece at speeds of up to 7600 m/min (25,000 ft/min). The frictional energy is converted into heat, which rapidly softens a narrow zone in the workpiece. The action of the blade, which can have teeth or notches for higher cutting efficiency, pulls and ejects the softened metal from the cutting zone. The heat generated in the workpiece produces a *heat-affected zone* (Section 30.9) on the cut surfaces. Thus, the workpiece properties along the cut edges can be affected adversely by this process. Because only a small portion of the blade is engaged with the workpiece at any time, the blade itself cools rapidly as it passes through the air. This friction-sawing process is suitable for hard, ferrous metals and reinforced plastics but not for nonferrous metals because of their tendency to stick to the blade. Friction-sawing disks as large as 1.8 m (6 ft) in diameter are used to cut off large steel sections. Friction sawing also is used commonly to remove flash from castings.

24.6 | Filing

Filing involves the small-scale removal of material from a surface, corner, edge, or hole—including the removal of burrs. First developed around 1000 B.C., files usually are made of hardened steel and are available in a variety of cross-sections, such as flat, round, half-round, square, and triangular. Files can have many tooth forms and grades of coarseness. Although filing usually is done by hand, filing machines with automatic features are available for high production rates, with files reciprocating at up to 500 strokes/min.

Band files consist of file segments, each about 75 mm (3 in.) long, which are riveted to a flexible steel band and are used in a manner similar to band saws. Disk-type files also are available.

Rotary files and **burs** (Fig. 24.28) are used for such applications as deburring, scale removal from surfaces, producing chamfers on parts, and removing small amounts

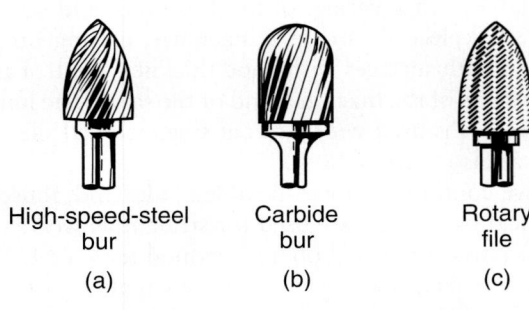

High-speed-steel   Carbide   Rotary
bur   bur   file

(a)   (b)   (c)

FIGURE 24.28   Types of burs used in burring operations.

of material in die making. These cutters generally are conical, cylindrical, or spherical in shape and have various tooth profiles. Their cutting action (similar to that of reamers) removes small amounts of material at high rates. The rotational speed of burs ranges from 1500 rpm for cutting steels using large burs to as high as 45,000 rpm for magnesium using small burs.

24.7 | Gear Manufacturing by Machining

Several processes for making gears or producing gear teeth on various components were described in Parts II and III, such as by casting, forging, extrusion, drawing, thread rolling, and powder metallurgy. Blanking of sheet metal also can be used for making thin gears, such as those used in mechanical watches, clocks, and similar mechanisms. Plastic gears can be made by such processes as injection molding and casting.

Gears may be as small as those used in watches or as large as 9 m (30 ft) in diameter for rotating mobile crane superstructures. The dimensional accuracy and surface finish required for gear teeth depend on the intended use. Poor gear-tooth quality contributes to inefficient energy transmission, increased vibration and noise, and adversely affects the gear's frictional and wear characteristics. Submarine gears, for example, have to be of extremely high quality so as to reduce noise levels, thus helping to avoid detection.

The standard nomenclature for an involute spur gear is shown in Fig. 24.29. Starting with a wrought or cast gear blank, there are two basic methods of making such gear teeth: *form cutting* and *generating*.

24.7.1 Form cutting

In *form cutting*, the cutting tool is similar to a form-milling cutter made in the shape of the space between the gear teeth (Fig. 24.30a). The gear-tooth shape is reproduced

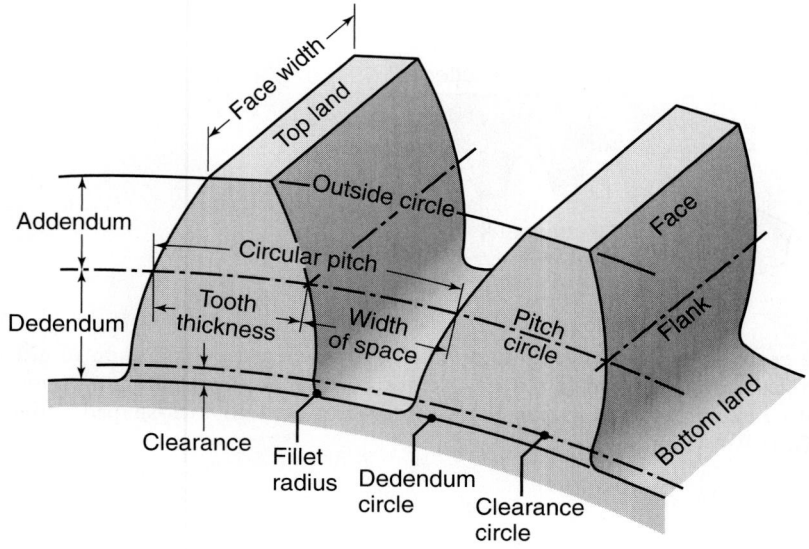

FIGURE 24.29   Nomenclature for an involute spur gear.

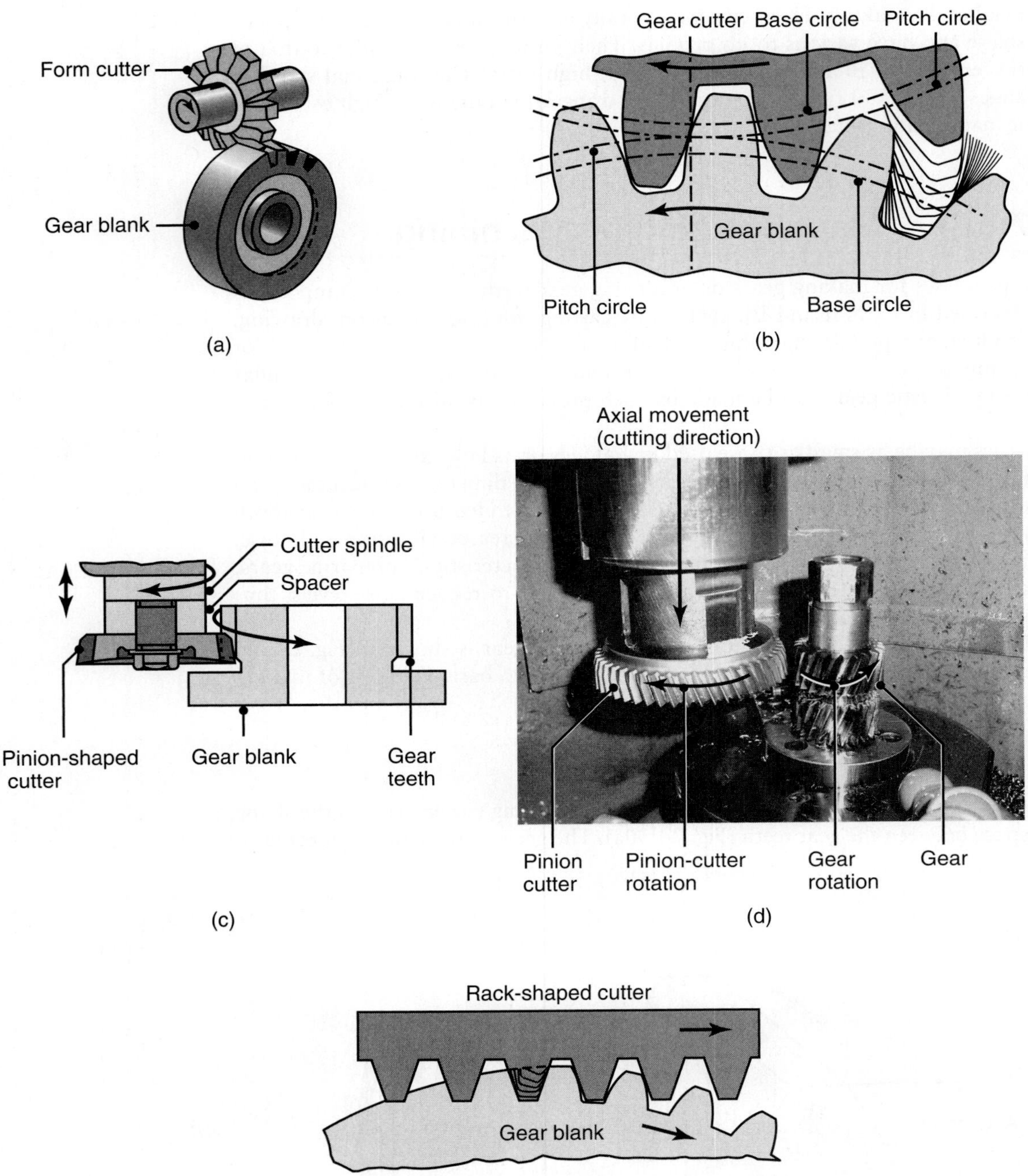

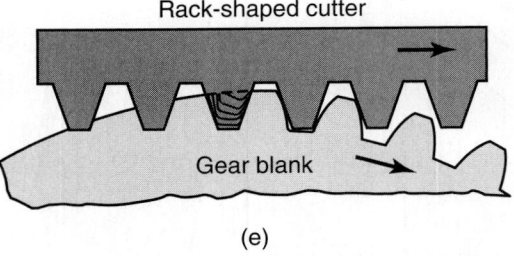

FIGURE 24.30   (a) Producing gear teeth on a blank by form cutting. (b) Schematic illustration of gear generating with a pinion-shaped gear cutter. (c) and (d) Gear generating in a gear shaper using a pinion-shaped cutter. Note that the cutter reciprocates vertically. (e) Gear generating with rack-shaped cutter. *Source:* (d) Courtesy of Schafer Gear Works, Inc.

by machining the gear blank around its periphery. The cutter travels axially along the length of the gear tooth and at the appropriate depth to produce the gear-tooth profile. After each tooth is cut, the cutter is withdrawn, the gear blank is rotated (*indexed*), and the cutter proceeds to cut another tooth. This process continues until all teeth are machined.

Each cutter is designed to cut a range of number of teeth. The precision of the form-cut tooth profile depends on the accuracy of the cutter and on the machine and its stiffness. Although inefficient, form cutting can be done on milling machines with the cutter mounted on an arbor and the gear blank mounted in a dividing head. Because the cutter has a fixed geometry, form cutting can be used only to produce gear teeth that have a constant width—that is on spur or helical gears but not on bevel gears. Internal gears and gear teeth on straight surfaces (such as in a rack and pinion) are form cut with a shaped cutter using a machine similar to a shaper. Form cutting is a relatively simple process and can be used for cutting gear teeth with various profiles. On the other hand, it is a slow operation; furthermore, some types of machines require skilled labor. Machines with semiautomatic features can be used economically for form cutting on a limited-production basis. Generally, however, form cutting is suitable only for low-quantity production.

Broaching also can be used to machine gear teeth and is particularly suitable for producing internal teeth. The broaching process is rapid and produces fine surface finish with high dimensional accuracy. However, because a different broach is required for each gear size (and broaches are expensive), this method is suitable almost exclusively for high-quantity production.

Gear teeth also may be cut on special machines with a single-point cutting tool that is guided by a *template* in the shape of the gear-tooth profile. Because the template can be made much larger than the gear tooth, dimensional accuracy is improved.

24.7.2 Gear generating

The cutting tool used in *gear generating* may be one of the following: (a) pinion-shaped cutter, (b) rack-shaped straight cutter, and (c) hob.

- A **pinion-shaped cutter** can be considered as one of the two gears in a conjugate pair and the other as the gear blank (Fig. 24.30b). This type of cutter is used on *gear shapers* (Fig. 24.30c and d). The cutter has an axis parallel to that of the gear blank and rotates slowly with the blank at the same pitch–circle velocity and in an axial-reciprocating motion. A train of gears provides the required relative motion between the cutter shaft and the gear-blank shaft. Cutting may take place at either the downstroke or the upstroke of the machine. Because the clearance required for the cutter travel is small, gear shaping is suitable for gears that are located close to obstructing surfaces, such as a flange in the gear blank in Fig. 24.30c and d. The process can be used for low-quantity as well as high-quantity production.

- On a **rack shaper**, the generating tool is a *segment* of a rack (Fig. 24.30e), which reciprocates parallel to the axis of the gear blank. Because it is not practical to have more than 6 to 12 teeth on a rack cutter, the cutter must be disengaged at suitable intervals and returned to the starting point. The gear blank remains fixed during the operation.

- A **hob** (Fig. 24.31) is basically a gear-cutting worm, or screw, made into a gear-generating tool by a series of longitudinal slots or gashes machined into it to form the cutting teeth. When hobbing a spur gear, the angle between the hob

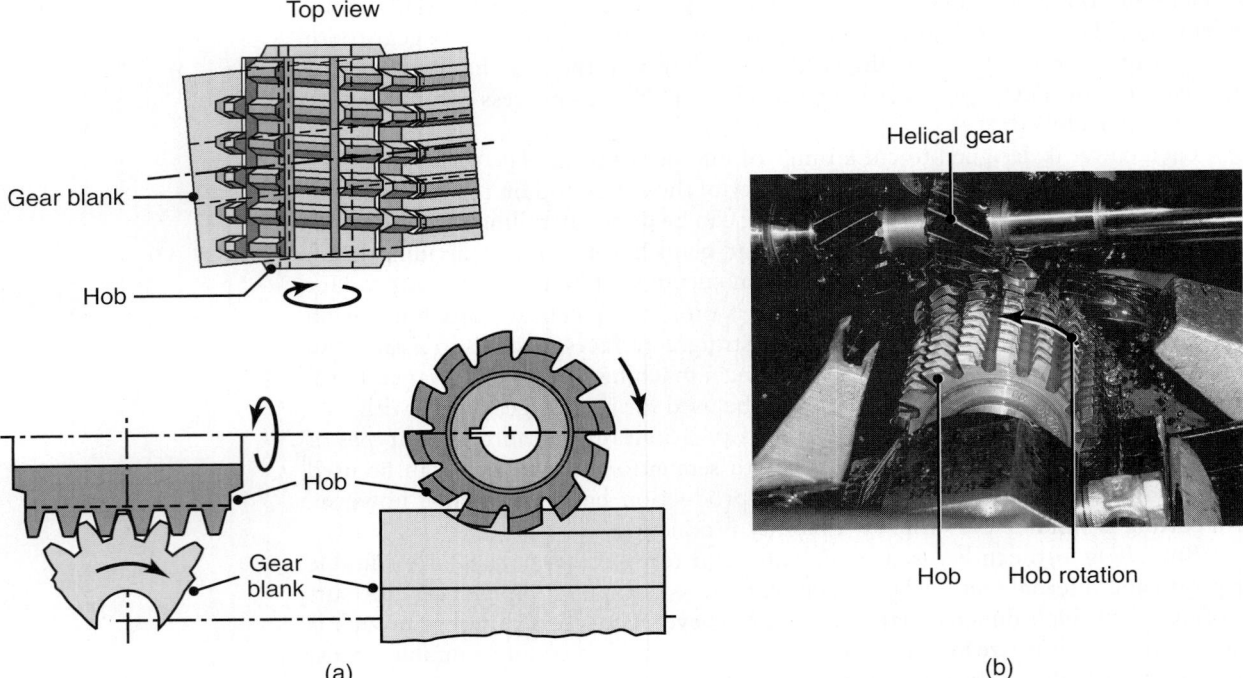

Top view

Gear blank

Hob

Hob

Gear blank

Helical gear

Hob   Hob rotation

(a)   (b)

FIGURE 24.31   (a) Schematic illustration of gear cutting with a hob. (b) Production of a worm gear by hobbing. *Source:* Courtesy of Schafer Gear Works, Inc.

and gear-blank axes is 90° minus the lead angle at the hob threads. All motions in hobbing are rotary, and the hob and gear blank rotate continuously—much as two gears in mesh—until all of the teeth are cut.

Hobs are available with one, two, or three threads. For example, if the hob has a single thread and the gear is to have 40 teeth, the hob and the gear spindle must be geared together such that the hob makes 40 revolutions while the gear blank makes one revolution. Similarly, if a double-threaded hob is used, the hob would make 20 revolutions to the gear blank's one revolution. In addition, the hob must be fed parallel to the gear axis for a distance greater than the face width of the gear tooth (Fig. 24.29) in order to produce straight teeth on spur gears. The same hobs and machines can be used to cut helical gears by tilting the axis of the hob spindle.

Because it produces a variety of gears at high rates and with good dimensional accuracy, gear hobbing is used extensively in industry. Although the process also is suitable for low-quantity production, it is most economical for medium- to high-quantity production. Gear-generating machines also can produce spiral-bevel and hypoid gears. Like most other machine tools, modern gear-generating machines are computer controlled. *Multiaxis computer-controlled machines* are capable of generating many types and sizes of gears using indexable milling cutters.

24.7.3 Cutting bevel gears

Straight bevel gears generally are roughed out in one cut with a form cutter on machines that index automatically. The gear then is finished to the proper shape on a gear generator. The generating method is analogous to the rack-generating method

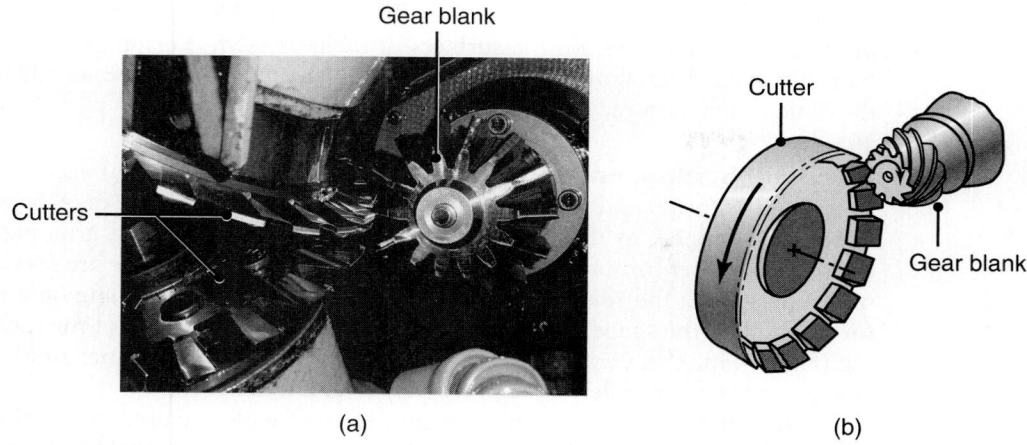

Gear blank

Cutters

Cutter

Gear blank

(a)

(b)

FIGURE 24.32 (a) Cutting a straight bevel-gear blank with two cutters. (b) Cutting a helical bevel gear. *Source*: Courtesy of Schafer Gear Works, Inc.

described previously. The cutters reciprocate across the face of the bevel gear as does the tool on a shaper (Fig. 24.32a). The machines for spiral bevel gears essentially operate on the same principle, and the spiral cutter is basically a face-milling cutter with a number of straight-sided cutting blades protruding from its periphery (Fig. 24.32b).

24.7.4 Gear-finishing processes

As produced by any of the processes described previously, the surface finish and dimensional accuracy of gear teeth may not be sufficient for certain applications. Moreover, the gears may be noisy or their mechanical properties (especially fatigue life and wear resistance) may not be sufficiently high. Several *finishing processes* are available to improve the surface quality of the gears. The choice of process is dictated by the method of gear manufacture and whether the gears have been hardened by heat treatment. As described in Chapter 4, heat treating can cause distortion of parts. Consequently, for a precise gear-tooth profile, heat-treated gears should be subjected to appropriate finishing operations.

Shaving. The gear-shaving process involves a cutter made in the exact shape of the finished tooth profile which removes small amounts of metal from the surface of the gear teeth. The cutter teeth are slotted or gashed at several points along its width, making the process similar to fine broaching. The cutter has a reciprocating motion. Shaving and burnishing (described next) can be performed only on gears with a hardness of 40 HRC or lower.

Although the tools are expensive and special machines are required, shaving is rapid and is the most commonly used process for gear finishing. It produces gear teeth with improved surface finish and a good dimensional accuracy of the tooth profile. Shaved gears subsequently may be heat treated and then ground for improved hardness, wear resistance, and a more accurate tooth profile.

Burnishing. The surface finish of gear teeth also can be improved by burnishing. Introduced in the 1960s, burnishing is basically a surface plastic-deformation process (see Section 34.2) using a special hardened, gear-shaped burnishing die that subjects the tooth surfaces to a surface-rolling action (**gear rolling**). The resulting cold working of the tooth surfaces not only improves the surface finish, but it also induces compressive residual stresses on the surfaces of the gear teeth, thus improving their fatigue life. It has been shown, however, that burnishing does not significantly improve the dimensional accuracy of the gear tooth.

Grinding, honing, and lapping.   For the highest dimensional accuracy, tooth spacing and form, and surface finish, gear teeth subsequently may be ground, honed, and lapped, as described in Chapter 26. Specially-dressed grinding wheels are used for either forming or generating gear-tooth surfaces. There are several types of grinders, with the single-index form grinder being the most commonly used. In **form grinding**, the shape of the grinding wheel is identical to that of the tooth spacing (Fig. 24.33a). In **generating**, the grinding wheel acts in a manner similar to the gear-generating cutter described previously (Fig. 24.32b).

The **honing** tool is a plastic gear impregnated with fine abrasive particles. The honing process is faster than grinding and is used to improve surface finish. To further improve the surface finish, ground gear teeth are **lapped**, using abrasive compounds with either (a) a gear-shaped lapping tool made of cast iron or bronze or (b) a pair of

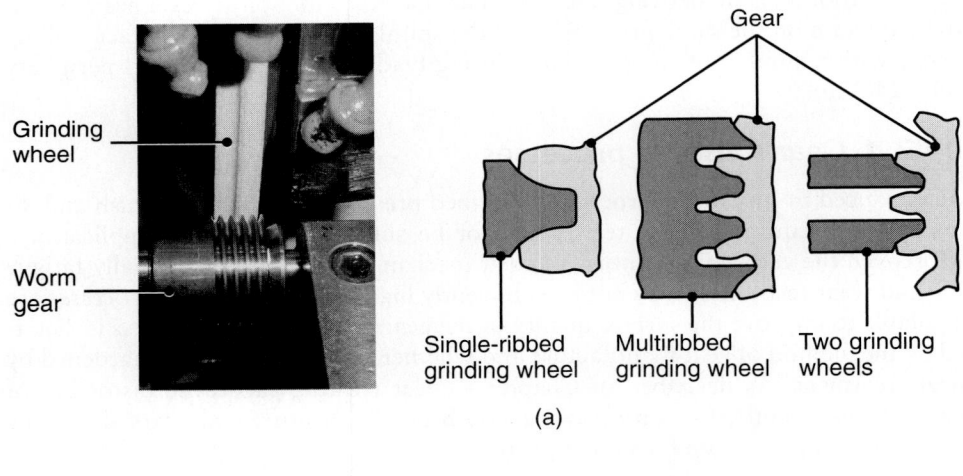

(a)

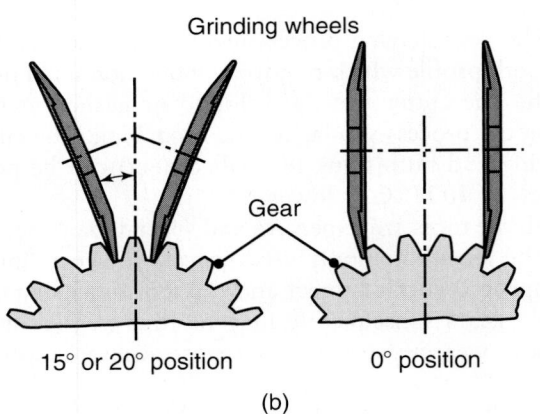

(b)

FIGURE 24.33   Finishing gears by grinding: (a) form grinding with shaped grinding wheels; (b) grinding by generating using two wheels.

mating gears that are run together. Although production rates are lower and costs are higher, these finishing operations are suitable, particularly for producing hardened gears of very high quality, long life, and quiet operation.

24.7.5 Design considerations and economics of gear machining

Design considerations for gear-cutting operations may be summarized as follows:

- The design of gear blanks is important, especially for complex gear teeth; provision should be made for clamping blanks securely on the machine.
- The specified dimensional accuracy and surface finish on gear teeth should be as broad as possible to reduce production costs.
- Wide gears are more difficult to machine than narrow ones.
- Gears should be machined prior to their assembly on shafts.
- Sufficient clearance should be provided between gear teeth and flanges, shoulders, and other features, so that the cutting tool can machine without interference.
- Provision should be made for using standard cutters wherever possible.

Economics. As in all machining operations, the cost of gears increases rapidly with improved surface finish and quality. Figure 24.34 shows the relative manufacturing cost of gears as a function of quality, as specified by AGMA (American Gear Manufacturers Association) and DIN (Deutsches Institut für Normung) numbers. The higher the number, the higher the dimensional accuracy of the gear teeth. As noted in this figure, the manufacturing cost can vary by an order of magnitude depending on dimensional tolerances.

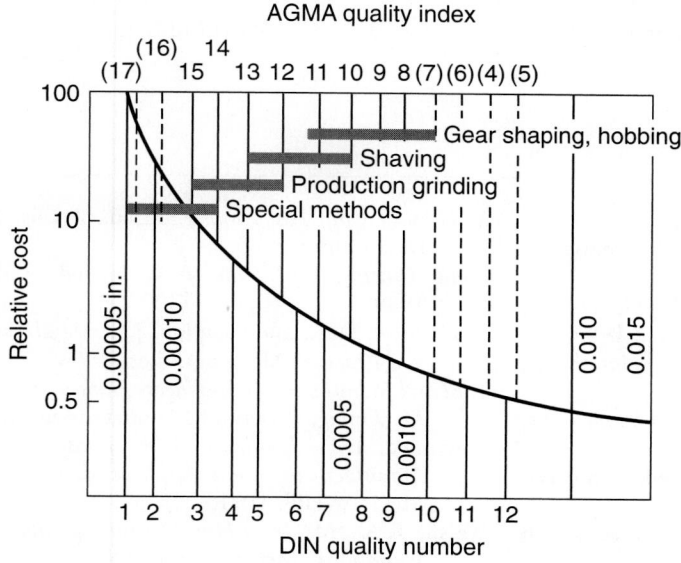

FIGURE 24.34 Gear manufacturing cost as a function of gear quality. The numbers along the vertical lines indicate tolerances.

SUMMARY

- Various complex shapes can be machined by the processes described in this chapter. Milling is one of the most common machining processes, because it is capable of economically producing a variety of shapes on workpieces.

- Although these processes are similar to those such as turning, drilling, and boring, most of the latter processes utilize multitooth tools and cutters at various axes with respect to the workpiece.

- The machine tools for these operations now are mostly computer controlled, have various features, and have much more flexibility in operation.

- In addition to the various forming and shaping processes described in other chapters, gears also are produced by machining, either by form cutting or generating; the latter produces gears with better surface finish and higher dimensional accuracy. The quality of the gear-tooth profile is improved further by finishing processes, such as shaving, burnishing, grinding, honing, and lapping.

KEY TERMS

Arbor	Form cutting	Pull broach
Broaching	Friction sawing	Push broach
Bur	Gear generating	Rack shaper
Burnishing	High-speed milling	Sawing
Climb milling	Hob	Shaping
Die sinking	Honing	Shaving
End milling	Kerf	Slab milling
Face milling	Lapping	Tooth set
Filing	Milling	Turn broaching
Fly cutting	Planing	Workholding

BIBLIOGRAPHY

Arnone, M., *High Performance Machining*, Hanser, 1998.

ASM Handbook, Vol. 16: *Machining*, ASM International, 1989.

Boothroyd, G., and Knight, W.A., *Fundamentals of Machining and Machine Tools*, 2nd ed., Marcel Dekker, 1989.

Brown, J., *Advanced Machining Technology Handbook*, McGraw-Hill, 1998.

Ewert, R.H., *Gears and Gear Manufacture: The Fundamentals*, Chapman and Hall, 1997.

Hoffman, E.G., *Jigs and Fixture Design*, 4th ed., Industrial Press, 1996.

Krar, S.F., and Check, A.F., *Technology of Machine Tools*, 5th ed., Glencoe Macmillan/McGraw-Hill, 1996.

Machinery's Handbook, Industrial Press, revised periodically.

Machining Data Handbook, 3rd ed., 2 vols. Machinability Data Center, 1980.

Metal Cutting Tool Handbook, 7th ed., Industrial Press, 1989.

Stephenson, D.A., and Agapiou, J.S., *Metal Cutting: Theory and Practice*, Marcel Dekker, 1996.

Tool and Manufacturing Engineers Handbook, 4th ed., Vol. 1: *Machining*, Society of Manufacturing Engineers, 1983.

Townsend, D.P., *Dudley's Gear Handbook: The Design, Manufacturing, and Application of Gears*, 2nd ed., McGraw-Hill, 1991.

Walsh, R.A., *McGraw-Hill Machining and Metalworking Handbook*. McGraw-Hill, 1994.

Weck, M., *Handbook of Machine Tools*, 4 vols., Wiley, 1984.

REVIEW QUESTIONS

24.1 Why is milling such a versatile machining process?

24.2 Describe the different types of cutters used in milling operations and give an application of each type.

24.3 What are the advantages of helical teeth over straight teeth on cutters for slab milling?

24.4 Explain the relative characteristics of climb milling and up milling.

24.5 Describe the geometric features of a broach and explain their functions.

24.6 Why is sawing a commonly used process? Does it have any limitations? Explain.

24.7 Why do some saw blades have staggered teeth?

24.8 Explain the difference between shaving and burnishing.

24.9 What advantages do bed-type milling machines have over column-and-knee type machines for production operations?

24.10 Why is the axis of a hob tilted with respect to the axis of the gear blank?

24.11 Describe the difference between finishing by form grinding and by generating.

24.12 What is a shell mill? Why are they used?

24.13 Why is it difficult to saw very thin sections or sheet metals?

QUALITATIVE PROBLEMS

24.14 Explain why broaching crankshaft bearings is an attractive alternative to other machining processes.

24.15 Several guidelines are presented in this chapter for various cutting operations. Discuss the reasoning behind these guidelines.

24.16 Explain why hacksaws are not as productive as band saws.

24.17 In milling operations with horizontal- and vertical-spindle machines, which one is likely to hold dimensional accuracy better? Why?

24.18 What similarities and differences are there in slitting with a milling cutter and with a saw?

24.19 Why do machined gears have to be subjected to finishing operations? Which of the finishing processes are not suitable for hardened gear teeth? Why?

24.20 How would you reduce the surface roughness shown in Fig. 24.6?

24.21 Why are machines such as the one shown in Fig. 24.17 so useful?

24.22 Comment on your observations concerning the designs shown in Fig. 24.20b and on the usefulness of broaching operations.

24.23 Explain how contour cutting could be started in a band saw, as shown in Fig. 24.25d.

24.24 In Fig. 24.27a, high-speed steel cutting teeth are welded to a steel blade. Would you then recommend that the whole blade be made of high-speed steel? Explain your reasons.

24.25 Describe the parts and conditions under which broaching would be the preferred method of machining.

24.26 With appropriate sketches, explain the differences between and similarities among shaving, broaching, and turn-broaching operations.

24.27 Would you consider the machining processes described in this chapter to be near-net or net-shape processing? Explain with appropriate examples.

24.28 Why is end milling such a versatile process? Explain with examples.

24.29 List and explain factors that contribute to poor surface finish in the processes described in this chapter.

24.30 Explain the possible reasons why a knife cuts well when it is moved back and forth. Consider factors such as the material cut, friction, and the dimensions of the cut.

24.31 Are there size limitations to parts to be sawed? Explain.

24.32 Why is it difficult to use friction sawing on nonferrous metals?

24.33 Would you recommend broaching a keyway on a gear blank before or after machining the teeth? Why?

QUANTITATIVE PROBLEMS

24.34 In milling operations, the total cutting time can be significantly influenced by (a) the magnitude of the noncutting distance, l_c, shown in Figs. 24.3 and 24.4 and (b) the ratio of width of cut, w, to the cutter diameter, D. Sketch several combinations of these parameters, give dimensions, select feeds and cutting speeds, etc., and determine the total cutting time. Comment on your observations.

24.35 You are performing a slab-milling operation at a specified cutting speed (surface speed of the cutter) and feed-per-tooth. Explain the procedure for determining the table speed required.

24.36 Show that the distance l_c in slab milling is approximately equal to $\sqrt{Dd}$ for situations where $D \gg d$. See Fig. 24.3c.

24.37 In Example 24.1, which of the quantities will be affected when the feed is increased to $f = 0.02$ in./tooth?

24.38 Calculate the chip depth of cut, t_c, and the torque for Example 24.1.

24.39 Estimate the time required to face mill a 10-in.-long, 2-in.-wide brass block with a 6-in. diameter cutter with 10 high-speed steel inserts.

24.40 A 10-in.-long, 1-in.-thick plate is being cut on a band saw at 150 ft/min. The saw has 12 teeth per in. If the feed-per-tooth is 0.003 in., how long will it take to saw the plate along its length?

24.41 A single-thread hob is used to cut 40 teeth on a spur gear. The cutting speed is 120 ft/min and the hob is 3 in. in diameter. Calculate the rotational speed of the spur gear.

24.42 Assume that in the face-milling operation shown in Fig. 24.4 the workpiece dimensions are 5 in. by 10 in. The cutter is 6 in. in diameter, has 8 teeth, and rotates at 300 rpm. The depth of cut is 0.125 in. and the feed is 0.005 in./tooth. Assume that the specific energy requirement for this material is 2 hp·min/in³ and that only 75% of the cutter diameter is engaged during cutting. Calculate (a) the power required and (b) the material-removal rate.

24.43 A slab-milling operation will take place on a part 250 mm long and 50 mm wide. A helical cutter 75 mm in diameter with 10 teeth will be used. If the feed-per-tooth is 0.2 mm/tooth and the cutting speed is 0.75 m/s, find the machining time and metal-removal rate for removing 6 mm from the surface of the part.

24.44 Calculate the ranges of typical machining times for face milling a 10-in.-long, 2-in.-wide cutter with a depth-of-cut of 0.1 in. for the following workpiece materials: (a) low-carbon steel, (b) titanium alloys, (c) aluminum alloys, and (d) thermoplastics.

24.45 Explain whether the feed marks left on the workpiece by a face-milling cutter (as shown in Fig. 24.13a) are segments of true circles. Describe the parameters you consider in answering this question.

24.46 In describing the broaching operations and the design of broaches, we have not given equations regarding feeds, speeds, and material-removal rates, as we have done in turning and milling operations. Review Fig. 24.21 and develop such equations.

SYNTHESIS, DESIGN, AND PROJECTS

24.47 The part shown in Fig. 24.1f is to be machined from a rectangular blank. Suggest the machine tool(s) required, the fixturing needed, and the types and sequence of operations to be performed. Discuss your answer in terms of the workpiece material, such as aluminum versus stainless steel.

24.48 Referring to Fig. 24.1f, would you prefer to machine this part from a preformed blank (near-net shape) rather than a rectangular blank? If so, how would you prepare such a blank? How would the number of parts required influence your answer?

24.49 Assume that you are an instructor covering the topics described in this chapter, and you are giving a quiz on the numerical aspects to test the understanding of the students. Prepare two quantitative problems and supply the answers.

24.50 Suggest methods whereby milling cutters of various designs (including end mills) can incorporate carbide inserts.

24.51 Some handbooks include tables of "do's and don'ts" regarding machining operations and the equipment used. Survey the available literature and prepare such a table for milling operations.

24.52 Make a comprehensive table of the process capabilities of the machining processes described in this chapter. Using several columns, list the machines involved, types of tools and tool materials used, shapes of blanks and parts produced, typical maximum and minimum sizes, surface finish, dimensional tolerances, and production rates.

24.53 Based on the data developed in Problem 24.52, describe your thoughts regarding the procedure to be followed in determining what type of machine tool to select when machining a particular part.

24.54 Using the Internet, obtain specifications on the smallest and largest milling machines available and compare your results with the results obtained by your classmates.

24.55 Make a list of all the processes that can be used in manufacturing gears, including those described in Parts II and III of this text. For each process, describe the advantages, limitations, and the quality of gears produced.

24.56 If expanded honeycomb panels (see Section 16.12) were to be machined in a form-milling operation, what precautions would you take to keep the sheet metal from buckling due to tool forces? Think up as many solutions as you can.

Machining Centers, Advanced Machining Concepts and Structures, and Machining Economics

In this chapter, advances in machine-tool design and construction for improved characteristics, accuracy, productivity, and reduced manufacturing costs are presented. Specifically, the following are described:

- The concepts and characteristics of machining and turning centers.
- The structure of machine tools. Their design, construction, and materials used.
- The design and operation of hexapods and reconfigurable machining centers.
- Why vibration and chatter occur in machining operations, and how they can be avoided.
- The continued importance of high-speed, hard, and ultraprecision machining.
- Factors involved in machining costs and how to minimize them.

25.1 | Introduction

In the preceding chapters, the characteristics and features of machine tools that typically are used for specific sets of machining operations were described. Also, the desirability of having flexibility in these machines was emphasized, so that more operations could be performed without having to refixture the workpieces on different machines. This chapter describes the design and capabilities of computer-controlled **machining centers,** which have the required flexibility and the versatility that other individual machine tools do not have. Consequently, they often have become the first choice in machine-tool selection.

Also described is a more recent development in the flexibility of manufacturing processes, namely the concept of **reconfigurable machines.** Here, the design of the machines consists of modular components, which can be arranged and rearranged quickly in a number of configurations—hence the term *reconfigurable*. The system uses advanced computer hardware and software, communications systems, and reconfigurable controllers.

Trends in the design and materials for machine tools are presented in this chapter. These developments require an understanding of the performance of machine tools, modules, and their components, particularly with regard to their stiffness, vibration, chatter, and damping characteristics. These considerations are important, not only for dimensional accuracy and the quality of the surfaces produced, but also because of their influence on tool life, productivity, and overall machining economics. The final sections in this chapter describe the economics of machining operations, identifying the factors that contribute to machining costs. High-speed, hard, and ultraprecision machining are described with the tacit understanding that these topics are tied strongly to machining economics considerations. A simple method of cost analysis is presented, which can help determine the conditions under which machining parameters can be selected so that minimum machining cost or minimum machining time-per-piece can be achieved.

25.2 | Machining Centers

In describing the individual machining processes and machine tools in the preceding chapters, it was noted that each machine, regardless of how highly it is automated, is designed to perform basically the same type of operations, such as turning, boring, drilling, milling, broaching, planing, or shaping. Also, it was shown that most parts manufactured by the methods described throughout this book require further operations on their various surfaces before they are completed. Note, for example, that the parts shown in Fig. 25.1 have a variety of complex geometric features and that all of the surfaces on these parts require different types of machining operations (such as milling, facing, boring, drilling, reaming, or threading) to meet certain specific

(a)

(b)

FIGURE 25.1   Examples of parts that can be machined on machining centers using various processes such as turning, facing, milling, drilling, boring, reaming, and threading. Such parts ordinarily would require the use of a variety of machine tools to complete. (a) Forged motorcycle wheel, finish machined to tolerance, and subsequently polished and coated; (b) Detailed view of an engine block, showing complex cavities, threaded holes and planar surfaces. *Source*: (a) Courtesy of R.C. Components, (b) Courtesy of Donovan Engineering, programming by Norman Woodruff, and Photography by Ed Dellis, Powersports photography.

requirements concerning shapes, features, dimensional tolerances, and surface finish. A brief review will lead to the following observations:

- Some possibilities exist in *net-shape* or *near-net shape* production of these parts, depending on specific requirements regarding shapes, dimensional tolerances, detailed surface features, surface finish, and various mechanical and other properties to meet service requirements. Shaping processes that are candidates for these parts are precision casting, powder metallurgy, powder-injection molding, and precision forging. Even then, however, it is very likely that the parts still will need some additional finishing operations. For example, deep small-diameter holes, threaded holes, flat surfaces for sealing with gaskets, very close dimensional tolerances, sharp corners and edges, and flat or curved surfaces with different surface-finish requirements will require further machining.

- If some form of machining is required or if it is shown to be more economical to finish machine these parts to their final shapes, then it is obvious that none of the machine tools described in Chapters 23 and 24 *individually* could produce these parts completely. We also note that, traditionally, machining operations are performed by moving the workpiece from one machine tool to another until all of the required machining operations are completed.

The concept of machining centers.   The traditional method of machining parts using different types of machine tools has been and continues to be a viable and efficient manufacturing method. It can be highly automated in order to increase productivity, and it is indeed the principle behind **transfer lines** (also called *dedicated manufacturing lines,* DML), as described in Section 37.2. Commonly used in *high-volume* or *mass production,* transfer lines consist of several specific machine tools arranged in a logical sequence.

The workpiece (such as an automotive engine block) is moved from station to station with a specific machining operation performed at each station, after which it is transferred to the next machine for another specific machining operation, and so on. There are situations, however, where transfer lines are not feasible or economical, particularly when the types of products to be processed are changed rapidly due to factors such as product demand or changes in product shape or style. It is a very expensive and time-consuming process to rearrange these machine tools to respond to the needs for the next production cycle. An important concept developed in the late 1950s is that of **machining centers.**

A machining center (Fig. 25.2) is an advanced, computer-controlled machine tool that is capable of performing a variety of machining operations on different surfaces and different orientations of a workpiece without having to remove it from its workholding device or fixture. The workpiece generally is stationary, and the cutting tools rotate as they do in milling, drilling, honing, tapping, and similar operations. Whereas in transfer lines or in typical shops and factories the workpiece is brought *to the machine,* note that in machining centers, it is the machining operation that is brought *to the workpiece.*

In referring to the word *workpiece,* we also should point out that the workpiece in a machining center also can consist of all types of **tooling.** Tooling includes forming and cutting tools, cutters and tool holders, tool shanks for holding tool inserts, molds for casting, male and female dies for forming, punches for metalworking and powder metallurgy, rams for extrusion, workholding devices, and fixturing—all of which also have to be manufactured. Since the geometries often are quite complicated and a variety of machining operations are performed, these tools commonly are produced in machinine centers.

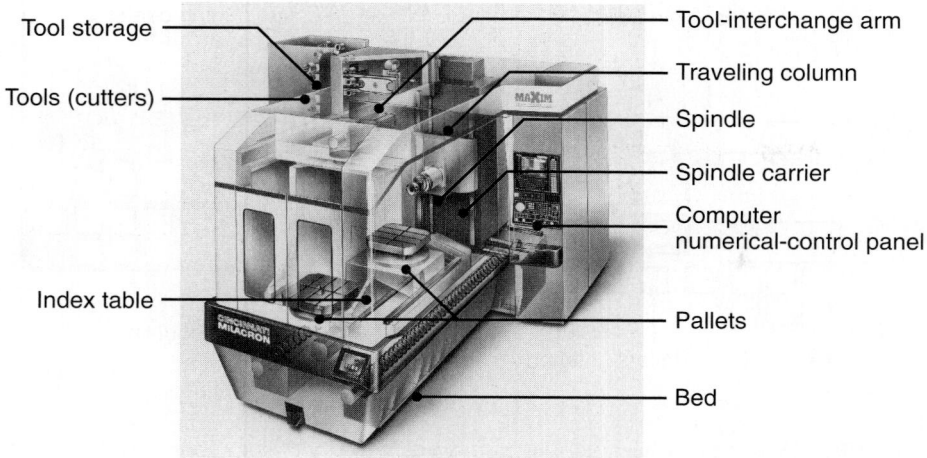

Tool storage

Tools (cutters)

Tool-interchange arm

Traveling column

Spindle

Spindle carrier

Computer numerical-control panel

Index table

Pallets

Bed

FIGURE 25.2   A horizontal-spindle machining center equipped with an automatic tool changer. Tool magazines can store up to 200 cutting tools of various functions and sizes. *Source:* Courtesy of Cincinnati Milacron, Inc.

The development of machining centers is related closely to advances in automation and computer control of machine tools, the details of which are described in Chapter 37. Recall that as an example of the advances in modern lathes, Fig. 23.10 illustrates a numerically controlled lathe (**turning center**) with two turrets carrying several cutting tools.

Components of a machining center.   The workpiece in a machining center is placed on a **pallet** or *module* that can be moved and swiveled (oriented) in various directions (Fig. 25.3). After a particular machining operation has been completed,

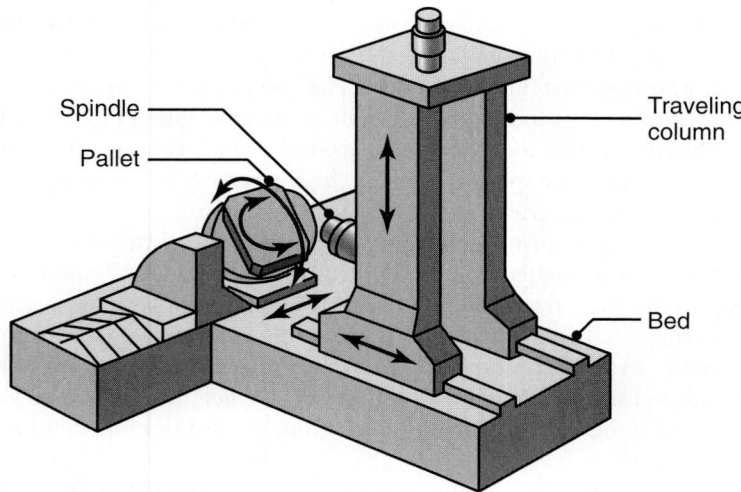

Spindle

Pallet

Traveling column

Bed

FIGURE 25.3   Schematic illustration of the principle of a five-axes machining center. Note that in addition to the three linear movements (three axes), the pallet, which supports the workpiece, can be swiveled around two axes (hence a total of five axes), allowing the machining of complex shapes, such as those shown in Fig. 25.1. *Source:* Courtesy of Toyoda Machinery.

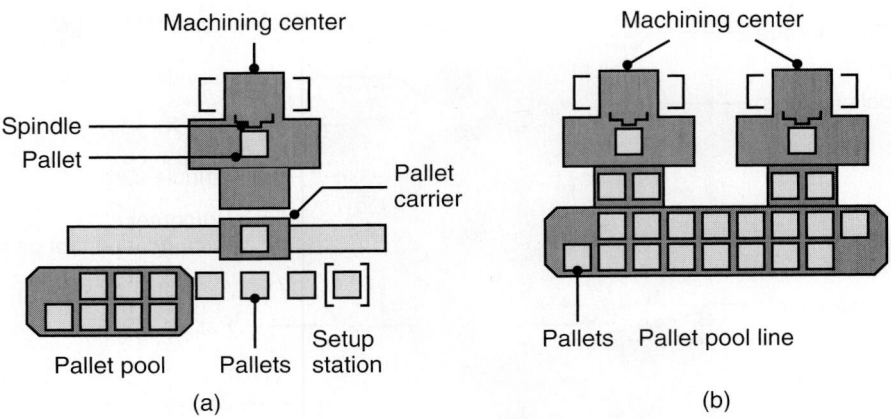

FIGURE 25.4   (a) Schematic illustration of the top view of a horizontal-spindle machining center showing the pallet pool, set-up station for a pallet, pallet carrier, and an active pallet in operation (shown directly below the spindle of the machine). (b) Schematic illustration of two machining centers with a common pallet pool. Various other pallet arrangements are possible in such systems. *Source:* Courtesy of Hitachi Seiki Co., Ltd.

another operation begins, which may require reindexing of the workpiece on its pallet. After all of the machining operations have been completed, the pallet automatically moves away with the finished part, and another pallet (carrying another workpiece to be machined) is brought into position by an **automatic pallet changer** (Fig. 25.4). All movements are computer controlled, and pallet-changing cycle times are on the order of only 10 to 30 seconds. Pallet stations are available with several pallets serving the machining center. The machines also can be equipped with various automatic features, such as part loading and unloading devices.

A machining center is equipped with a programmable **automatic tool changer** (ATC). Depending on the particular design, up to 200 cutting tools can be stored in a magazine, drum, or chain (*tool storage*). Auxiliary tool storage also is available on some special and large machining centers. The cutting tools are selected automatically with random access for the shortest route to the machine spindle. The maximum dimensions that the cutting tools can reach around a workpiece in a machining center is known as the **work envelope**—a term that first was used in connection with industrial robots, as will be described in Section 37.6.

The **tool-exchange arm** shown in Fig. 25.5 is a common design; it swings around to pick up a particular tool and places it in the spindle. Note that each tool has its own toolholder, thus making the transfer of cutting tools to the machine spindle very efficient. Tools are identified by bar codes, coded tags, or memory chips attached directly to their toolholders. *Tool-changing times* are typically between 5 and 10 seconds but may be up to 30 seconds for tools weighing up to 110 kg (250 lb). Because tool changing is a non-cutting operation, the continuing trend is to reduce the times even more.

Machining centers may be equipped with a **tool-checking** and/or **part-checking station** that feeds information to the machine control system so that it can compensate for any variations in tool settings or tool wear. **Touch probes** (Fig. 25.6) can be installed into a tool holder to determine workpiece-reference surfaces for

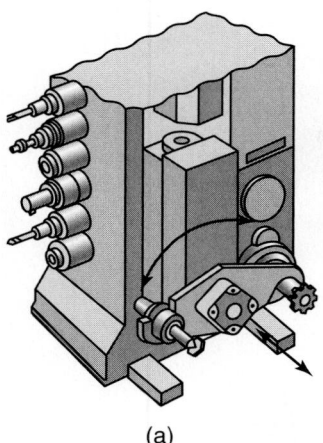 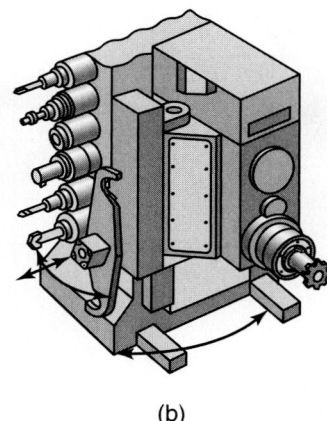

(a)   (b)

FIGURE 25.5   Swing-around tool changer on a horizontal-spindle machining center. (a) The tool-exchange arm is placing a toolholder with a cutting tool into the machine spindle. Note the axial and rotational movement of the arm. (b) The arm is returning to its home position. Note its rotation along a vertical axis after placing the tool and the two degrees of freedom in its home position.

the selection of tool settings and for the on-line inspection of parts being machined. Note in Figure 25.6 that several surfaces can be contacted (see also *sensor technology,* Section 37.7) and that their relative positions are determined and stored in the database of the computer software. The data then are used to program tool paths and to compensate for tool length, tool diameter, as well as for tool wear in more advanced machine tools.

25.2.1 Types of machining centers

There are various designs for machining centers. The two basic types are vertical spindle and horizontal spindle, but many machines are capable of operating along both axes.

Vertical-spindle machining centers   Also called *vertical machining centers* (VMC), these are capable of performing various machining operations on parts with deep cavities, such as in mold and die making. A vertical-spindle machining center (which is similar to a vertical-spindle milling machine) is shown in Fig. 25.7. The tool magazine is on the left of the figure, and all operations and movements are directed and modified through the computer-control panel shown on the right. Because the thrust forces in vertical machining are directed downward, such machines have high stiffness and produce parts with good dimensional accuracy. These machines generally are less expensive than horizontal-spindle machines.

Horizontal-spindle machining centers.   Also called *horizontal machining centers* (HMC), these are suitable for large as well as tall workpieces that require machining on a number of their surfaces. The pallet can be swiveled on different

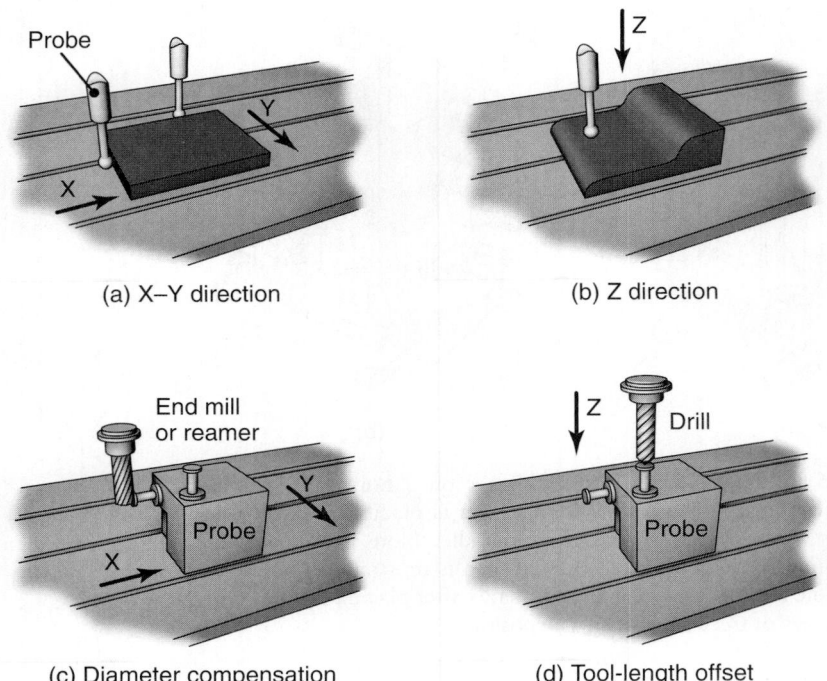

(a) X–Y direction

(b) Z direction

(c) Diameter compensation

(d) Tool-length offset

FIGURE 25.6   Touch probes used in machining centers for determining workpiece and tool positions and surfaces relative to the machine table or column. Touch probe (a) determining the X–Y (horizontal) position of a workpiece, (b) determining the height of a horizontal surface, (c) determining the planar position of the surface of a cutter (for instance, for cutter–diameter compensation), and (d) determining the length of a tool for tool-length offset.

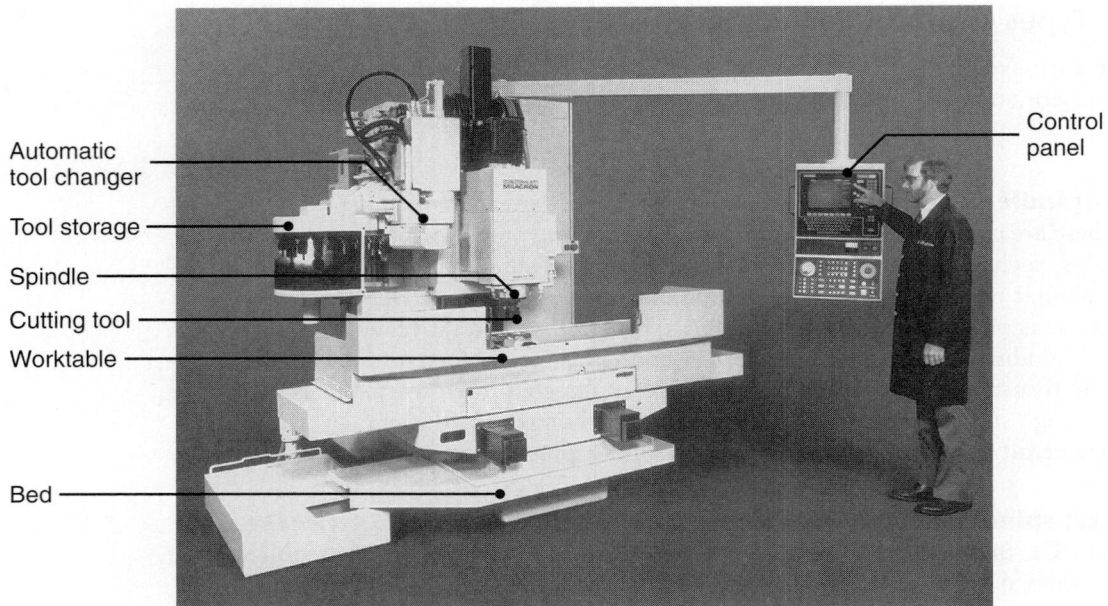

FIGURE 25.7   A vertical-spindle machining center. The tool magazine is on the left of the machine. The control panel on the right can be swiveled by the operator. *Source:* Courtesy of Cincinnati Milacron, Inc.

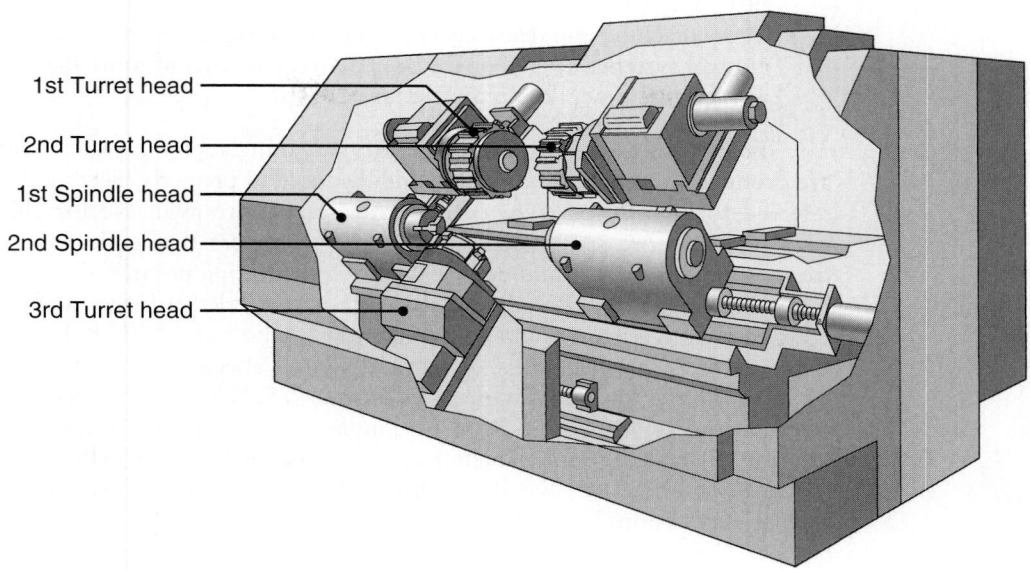

1st Turret head

2nd Turret head

1st Spindle head

2nd Spindle head

3rd Turret head

FIGURE 25.8   Schematic illustration of a computer numerical-controlled turning center. Note that the machine has two spindle heads and three turret heads, making the machine very flexible in its machining capabilities. *Source:* Courtesy of Hitachi Seiki Co., Ltd.

axes (for example, see Fig. 25.3) to various angular positions. Another category of horizontal-spindle machines is **turning centers,** which are computer-controlled *lathes* with several features. A three-turret turning center is shown in Fig. 25.8. It is constructed with two horizontal spindles and three turrets equipped with a variety of cutting tools used to perform several operations on a rotating workpiece.

Universal machining centers are equipped with both vertical and horizontal spindles. They have a variety of features and are capable of machining all of the surfaces of a workpiece (that is, vertical and horizontal and at a wide range of angles).

25.2.2 Characteristics and capabilities of machining centers

The major characteristics of machining centers are summarized here:

- Machining centers are capable of handling a wide variety of part sizes and shapes efficiently, economically, repetitively, and with high dimensional accuracy—with tolerances on the order of ± 0.0025 mm (0.0001 in.).

- These machines are versatile and capable of quick changeover from one type of product to another.

- The time required for loading and unloading workpieces, changing tools, gaging of the part, and troubleshooting is reduced. Therefore productivity is improved, thus reducing labor requirements (particularly skilled labor) and minimizing production costs.

- These machines are equipped with tool-condition monitoring devices for the detection of tool breakage and wear as well as probes for tool-wear compensation and for tool positioning.

- In-process and post-process gaging and inspection of machined workpieces are now features of machining centers.
- These machines are relatively compact and highly automated and have advanced control systems, so one operator can attend to two or more machining centers at the same time, thus reducing labor costs.

Because of the high productivity of machining centers, large amounts of chips are produced and must be collected and disposed of properly (Section 23.3.7); this is referred to as *chip management*. Several designs are available for **chip collection** with one or more spiral (screw) or chain conveyors that collect the chips along troughs in the machine and deliver them to a collecting point.

Machining centers are available in a wide variety of sizes and features, and their costs range from about \$50,000 to \$1 million and higher. Typical capacities range up to 75 kW (100 hp). Maximum spindle speeds are usually in the range of 4000 to 8000 rpm, and some are as high as 75,000 rpm for special applications using small-diameter cutters. Modern spindles can accelerate to a speed of 20,000 rpm in only 1.5 seconds. Some pallets are capable of supporting workpieces weighing as much as 7000 kg (15,000 lb), although even higher capacities are available for special applications.

25.2.3 Selection of machining centers

Machining centers generally require significant capital expenditure, so to be cost effective, they may have to be used for more than one shift per day. Consequently, there must be a sufficient and continued demand for parts to justify this purchase. Because of their inherent versatility, however, machining centers can be used to produce a wide range of products, particularly for *just-in-time manufacturing*, as will be described in Section 39.5.

The selection of the type and size of machining centers depends on several factors, especially:

- Type of products, their size, and their shape complexity
- Type of machining operations to be performed, and the type and number of cutting tools required
- Dimensional accuracy required
- Production rate required

EXAMPLE 25.1 Machining outer bearing races on a turning center

Outer bearing races (Fig. 25.9) are machined on a turning center. The starting material is a hot-rolled 52100 steel tube with 91 mm (3.592 in.) OD and 75.5 mm (2.976 in.) ID. The cutting speed is 95 m/min (313 ft/min) for all operations. All tools are carbide, including the cutoff tool (last operation), which is 3.18 mm $\left(\frac{1}{8} \text{in.}\right)$ instead of 4.76 mm $\left(\frac{3}{16} \text{in.}\right)$ for the high-speed steel cutoff tool that formerly was used.

The material saved by this change is significant because the race width is small. The turning center was able to machine these races at high speeds and with repeatable tolerances of ±0.025 mm (0.001 in.). (See also Example 23.2.)

Source: Courtesy of McGill Manufacturing Company.

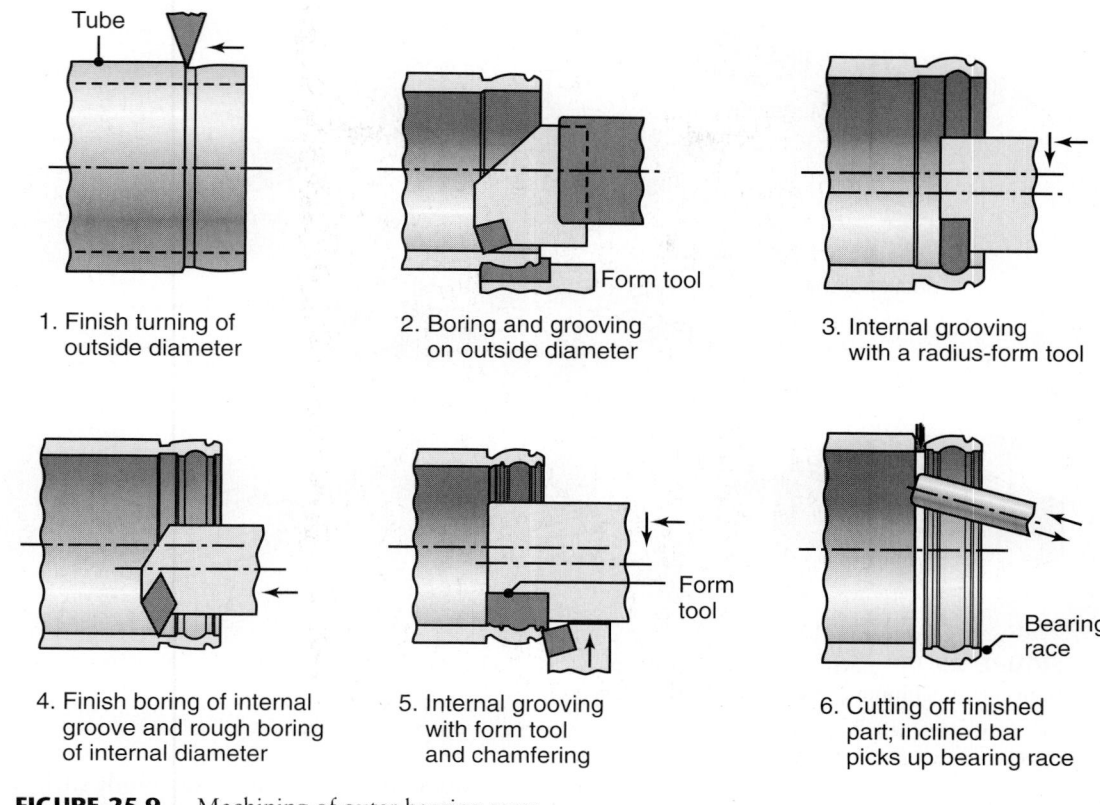

1. Finish turning of outside diameter

2. Boring and grooving on outside diameter

Form tool

3. Internal grooving with a radius-form tool

4. Finish boring of internal groove and rough boring of internal diameter

5. Internal grooving with form tool and chamfering

Form tool

6. Cutting off finished part; inclined bar picks up bearing race

Bearing race

FIGURE 25.9 Machining of outer bearing races.

25.2.4 Reconfigurable machines and systems

The need for the flexibility of manufacturing processes has led to the more recent concept of *reconfigurable machines*, consisting of various modules. The term *reconfigurable* stems from the fact that (using advanced computer hardware and reconfigurable controllers and utilizing advances in information management technologies) the machine components can be arranged and rearranged quickly in a number of configurations to meet specific production demands. Based on the basic machine-tool structure of a three-axis machining center, Fig. 25.10 shows an example of how it

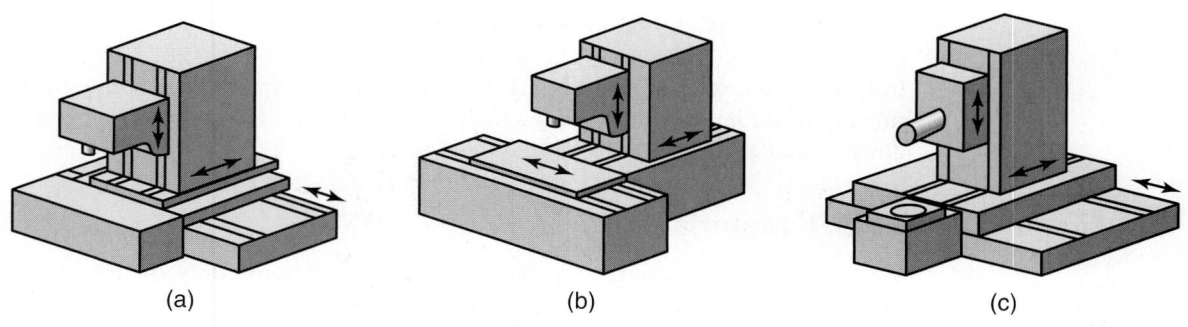

(a)　　　　　(b)　　　　　(c)

FIGURE 25.10 Schematic illustration of a reconfigurable modular machining center capable of accommodating workpieces of different shapes and sizes and requiring different machining operations on their various surfaces. *Source:* After Y. Koren.

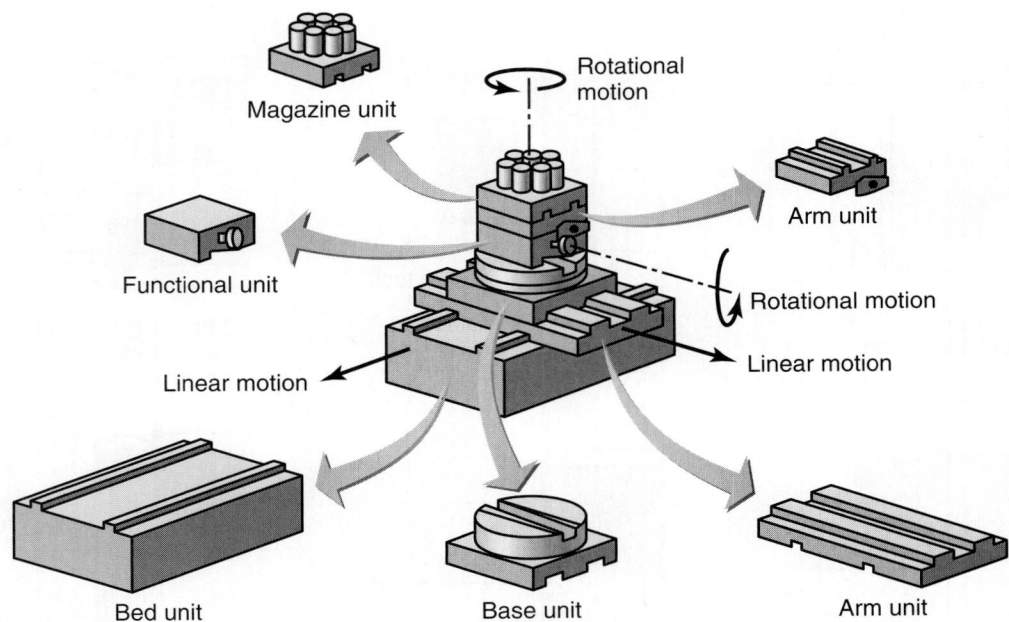

FIGURE 25.11   Schematic illustration of the assembly of different components of a reconfigurable machining center. *Source:* After Y. Koren.

can be reconfigured to become a modular machining center. With such flexibility, the machine can perform different machining operations while accommodating various workpiece sizes and part geometries. Another example is given in Fig. 25.11, where a five-axis (three linear and two rotational movements) machine can be reconfigured by assembling different modules.

Reconfigurable machines have the promise of (a) improving the productivity and efficiency of manufacturing operations, (b) reducing lead time for production, and (c) providing cost-effective and rapid response to market demands (see also Chapter 39). These capabilities are significant, especially in view of the frequent introduction of new products into a highly competitive global marketplace, fluctuations in product demand and product mix, and unpredictable modifications in product design.

25.3 | Machine-Tool Structures

In this section, we describe the materials and design aspects of machine-tool structures that are important in producing parts with acceptable geometric features and dimensional and surface finish characteristics.

25.3.1 Materials

We outline here the materials that have been used commonly or are suitable for machine-tool structures and describe their characteristics.

- **Gray cast iron,** the first material used in machine tool structures, has the advantages of a good damping capacity and low cost, but the disadvantage of being heavy. Most machine-tool structures are made of class 40 cast iron;

some are made of class 50 (see Table 12.4). Each casting requires a pattern, the cost of which increases significantly with the size of the part.

- **Welded steel** structures are lighter than cast-iron structures. Wrought steels (a) are available in a wide range of section sizes and shapes (such as channels, angles, and tubes), (b) have desirable mechanical properties, (c) have good manufacturing characteristics (such as formability, machinability, and weldability), and (d) have low cost. Structures made of steels can have high stiffness-to-weight ratios, using cross-sections such as tubes and channels. On the other hand, their damping capacity is very low.

- **Ceramic** components are used in advanced machine tools for their strength, stiffness, corrosion resistance, surface finish, and good thermal stability. Ceramic machine-tool components were first introduced in the 1980s. Spindles and bearings now can be made of silicon nitride, which has better friction and wear characteristics than traditional metallic materials. Furthermore, the low density of ceramics makes them suitable as the components of high-speed machinery that undergo rapid reciprocating or rotating movements in which low inertial forces are desirable to maintain the system's stability, reduce inertial forces, and reduce the noncutting time in high-speed machining operations.

- **Composites** may consist of a polymer-matrix, metal-matrix, or ceramic-matrix with various reinforcing materials. The compositions can be tailored to provide appropriate mechanical properties in selected axes of the machine tool. Although they are expensive and (presently) limited in use, composites are likely to become significant materials for high-accuracy, high-speed machining applications.

- **Granite-epoxy composites** (with a typical composition of 93% crushed granite and 7% epoxy binder) were first used in precision centerless and internal grinders in the early 1980s. These composite materials have several favorable properties: (a) good castability (which allows design versatility in machine tools), (b) high stiffness-to-weight ratio, (c) thermal stability, (d) resistance to environmental degradation, and (e) good damping capacity.

- **Polymer concrete** is a mixture of crushed concrete and plastic (typically poly-methylmethacrylate) and can be cast easily into desired shapes for machine bases and various components. It was first introduced in the 1980s, and several new compositions are being developed. Although it has good damping capacity, polymer concrete has low stiffness (about one-third that of class 40 cast iron) and poor thermal conductivity. It also can be used for sandwich construction with cast irons, thus combining the advantages of each type of material. Plain concrete also can be poured into cast-iron machine-tool structures to increase their mass and improve their damping capacity. Filling the cavities of machine-tool bases with loose sand also has been shown to be an effective means of improving damping capacity.

25.3.2 Machine-tool design considerations

Important considerations in machine tools generally involve the following factors:

- Design, materials, and construction.
- Spindle materials and construction.
- Thermal distortion of machine components.
- Error compensation and the motion control of moving components along slideways.

Stiffness.   Stiffness is a major factor in the dimensional accuracy of a machine tool. It is a function of (a) the elastic modulus of the materials used and (b) the geometry of the structural components—including the spindle, bearings, drive train, and slideways. Machine stiffness can be enhanced by design improvements, such as using diagonally arranged interior ribs.

Damping.   Damping is a very critical factor in reducing or eliminating vibration and chatter in machining operations. Principally, it involves the (a) types of materials used and (b) type and number of joints (for example, welded versus bolted) in the machine-tool structure. As described previously, cast irons and polymer-matrix composites have much better damping capacity than metals or ceramics. Also, the greater the number of joints in a structure, the more the damping.

Thermal distortion.   An important factor in machine tools is the thermal distortion of its components, which contributes significantly to its lack of precision. It has been estimated that about 50% of machine-tool errors are due to temperature. Machine components expand or contract to various extents in different directions.

There are two sources of heat in machine tools:

1. *Internal sources*—such as from bearings, ballscrews, machine ways, spindle motors, pumps, servomotors, as well as from the cutting zone during machining.

2. *External sources*—such as from cutting fluids, nearby furnaces, heaters, other nearby machines, sunlight, and fluctuations in ambient temperature from sources—such as air-conditioning units, vents, or even someone opening or closing a door or a window.

These considerations are important, particularly in **precision** and **ultraprecision machining** where dimensional tolerances and surface finish are now approaching the nanometer range. The machines used for these applications are equipped with the following:

- Various thermal and geometric real-time error-compensating features, including the modeling of heating and cooling and electronic compensation for accurate ballscrew positions.
- Gas or fluid hydrostatic spindle bearings.
- New designs for traction or friction drives for linear motion.
- Extremely fine feed and position controls using microactuators.
- Fluid-circulation channels in the machine-tool base for the maintainance of thermal stability.

The structural components of the machine tool can be made of materials with high dimensional stability and a low coefficient of thermal expansion, such as Super-Invar, granite, ceramics, and composites. *Retrofitting* also is a viable option for enhancing the performance of older machines.

Assembly techniques for machine-tool components.   Traditionally, machine-tool components have been assembled using threaded fasteners and welding. Advanced assembly techniques now include integral casting and resin bonding. Using a hybrid casting technology, steel guideways (with their higher stiffness) can be cast integrally over a cast-iron bed. *Resin bonding* is being used to assemble machine tools, replacing mechanical fastening. Adhesives (described in Section 32.4) have favorable characteristics for machine-tool construction, as they do not require special preparation and are suitable for assembling both the nonmetallic and the metallic components.

Guideways. The preparation of guideways in machine tools traditionally has required significant effort. The plain cast-iron ways in machines (which is the most common) require much hand scraping to achieve the required precision and service life. Other materials also are being investigated, including molded epoxies and various polymer-based coatings. The movements of various components in a machine tool along various axes usually have utilized conventional *ballscrews, rotating-screw drives,* and *rotary motors.* This system of mechanical and electrical components has several disadvantages, such as speed limitations, length restrictions, inertia effects, gear backlash and other errors, wear of the components, and low efficiency.

Linear motor drives. A *linear motor* is like a typical rotary electric motor that has been rolled out (opened) flat. This is the same principle used in some high-speed ground transportation systems in which the cars are levitated by magnetic forces. The sliding surfaces in these drives are separated by an air gap and as a result have very low friction.

Linear motor drives in machine tools have the following important advantages:

- Simplicity and minimal maintenance—since there is one moving part and no mechanical linkages.

- Smooth operation, better positioning accuracy, and repeatability—at as low as submicron range.

- A wide range of linear speeds—from 1 μm/s to 5 m/s.

- Acceleration rates of about 1 to 2 g (10 to 20 m/s^2) and as high as 4 to 10 g for smaller units.

- The moving components undergo no wear, because there is no physical contact between the sliding surfaces of the machine.

Machine foundations. Foundation materials, their mass, and the manner in which they are installed in a plant are important considerations, as they help reduce vibration and do not affect adversely the performance of other nearby machinery. For example, in the installation of a special grinder for high-precision grinding of 2.75-m (9-ft) diameter marine-propulsion gears, the concrete foundation was 6.7 m (22 ft) deep. The large mass of concrete combined with the machine base reduced the amplitude of vibrations. Even better results can be obtained when the machine is installed on an *independent* concrete slab, which is isolated from the rest of the plant floor with shock-isolation devices.

25.3.3 Hexapod machines

Developments in the design and materials used for machine-tool structures and components are taking place continually. The important goals are (a) imparting machining flexibility to machine tools, (b) increasing their *machining envelope* (that is, the space within which machining can be done), and (c) making them lighter. A truly innovative machine-tool structure is a self-contained octahedral (eight-sided) machine frame. Referred to as **hexapods** (Fig. 25.12) or *parallel kinematic linked machines,* this design is based on a mechanism called the *Stewart platform* (after D. Stewart)—an invention developed in 1966 and first used to position aircraft cockpit simulators. The main advantage is that the links in the hexapod are loaded axially; the bending stresses and deflections are minimal, resulting in an extremely stiff structure.

(a)   (b)

FIGURE 25.12   (a) A hexapod machine tool, showing its major components. (b) A detailed view of the cutting tool in a hexapod machining center. *Source:* Courtesy of National Institute of Standards and Technology.

The workpiece is mounted on a stationary table. Three pairs of *telescoping tubes* (struts or legs), each with its own motor and equipped with ballscrews, are used to maneuver a rotating cutting-tool holder. During machining of a part with various features and curvatures, the machine controller automatically shortens some tubes and extends others, so that the cutter can follow a specified path around the workpiece. Six sets of coordinates are involved in these machines (hence the term *hexapod*, meaning "six legged"): three linear sets and three rotational sets. Every motion of the cutter (even a simple linear motion) is translated into six coordinated leg lengths moving in real time. The motions of the legs are rapid. Consequently, high accelerations and decelerations with high inertia forces are involved.

These machines (a) have high stiffness, (b) are not as massive as machining centers, (c) have about one-third fewer parts than machining centers, (d) have a large machining envelope (hence greater access to the work zone), (e) are capable of maintaining the cutting tool perpendicular to the surface being machined (thus improving the machining operation), and (f) have high flexibility (with six degrees of freedom) in the production of parts with various geometries and sizes without the need to refixture the work in progress. Unlike most machine tools, these machines basically are portable. In fact, *hexapod attachments* now are available so that a conventional machining center easily can be converted into a hexapod machine.

A limited number of these machines have been built. In view of their potential as efficient machine tools, their performance is being evaluated continually regarding stiffness, thermal distortion, friction within the struts, dimensional accuracy, speed of operation, repeatability, and reliability. The cost of a hexapod machine is currently about $500,000 but is likely to come down significantly as they become more accepted and used.

25.4   Vibration and Chatter in Machining Operations

In describing machining processes and machine tools, it was noted on several occasions that *machine stiffness* is as important as any other parameter in machining. Low stiffness can cause *vibration and chatter* of the cutting tools and the machine components and, thus, have adverse effects on product quality. Uncontrolled vibration and chatter can result in the following:

- Poor surface finish, as shown in the right central region of Fig. 25.13.
- Loss of dimensional accuracy of the workpiece.
- Premature wear, chipping, and failure of the cutting tool—a critical consideration with brittle tool materials, such as ceramics, some carbides, and diamond.
- Possible damage to the machine-tool components from excessive vibration.
- Objectionable noise, particularly if it is of high frequency, such as the squeal heard when turning brass on a lathe.

Vibration and chatter in machining are complex phenomena. There are two basic types of vibration in machining: forced vibration and self-excited vibration.

Forced vibration.   Forced vibration generally is caused by some *periodic* applied force present in the machine tool, such as that from gear drives, imbalance of the machine-tool components, misalignment, and motors and pumps. In processes such as the milling or turning of a splined shaft or a shaft with a keyway or hole, forced vibrations are caused by the periodic engagement of the cutting tool with entry to and exit from the workpiece surface. (For example, see Figs. 24.9 and 24.14.)

The basic solution to forced vibration is to isolate or remove the forcing element. For example, if the forcing frequency is at or near the natural frequency of a machine-tool system component, one of the frequencies may be raised or lowered. The amplitude of vibration can be reduced by increasing the stiffness or by damping the system. Although changing the cutting parameters generally does not appear to greatly influence the magnitude of forced vibrations, changing the cutting speed and the tool geometry can be helpful. It also is recognized that the source of vibrations

FIGURE 25.13   Chatter marks (right of center of photograph) on the surface of a turned part. *Source:* Courtesy of General Electric Company.

can be minimized by changing the configuration of the machine-tool components—such as if the driving forces are close to or act through the *center of gravity* of that component. This approach will reduce the bending moment on the component, reduce deflections, and thus improve accuracy.

Self-excited vibration.   Generally called **chatter**, self-excited vibration is caused by the interaction of the chip-removal process and the structure of the machine tool. Self-excited vibrations usually have a very high amplitude. Chatter typically begins with a disturbance in the cutting zone and includes (a) the type of chips produced, (b) a lack of homogeneity in the workpiece material or its surface condition, and (c) variations in the frictional conditions at the tool–chip interface, as influenced by cutting fluids and their effectiveness.

The most important type of self-excited vibration is **regenerative chatter**, which is caused when a tool is cutting a surface that has a roughness or geometric disturbances developed from the previous cut (for example, see Figs. 21.2 and 21.21). Consequently, the depth of cut varies, and the resulting variations in the cutting force subject the tool to vibrations. This process continues repeatedly (hence the term regenerative). This type of vibration can be observed easily while driving a car over a rough road (the so-called *washboard effect*).

Self-excited vibrations generally can be controlled by

- Increasing the **stiffness** and, especially, the **dynamic stiffness** of the system.
- **Damping.**

Dynamic stiffness is defined as the ratio of the applied-force amplitude to the vibration amplitude. For example, recall in Fig. 23.23b on trepanning that there are four components involved in the deflections and hence causing the vibration: the spindle, the supporting arm for the cutting tool, the drill, and the cutting tool. Experience and analysis would indicate that unless all of the components are sufficiently stiff, such an operation likely will lead to chatter, beginning with torsional vibration around the spindle axis and twisting of the arm. Examples are long and slender drills that may undergo torsional vibrations and cutting tools that are long or are not well-supported (such as that shown schematically in Fig. 23.3) that would also lead to vibration and chatter.

Factors influencing chatter.   It has been observed that the tendency for chatter during machining is proportional to the cutting forces and the depth and width of the cut. Because cutting forces increase with strength (hence the hardness of the workpiece material), the tendency to chatter generally increases as hardness increases. Thus, aluminum and magnesium alloys have lower tendency to chatter than do martensitic and precipitation-hardening stainless steels, nickel alloys, and high-temperature and refractory alloys.

Another important factor in chatter is the type of chip produced during cutting operations. Continuous chips involve steady cutting forces. Consequently, such chips generally do not cause chatter. Discontinuous chips and serrated chips (Fig. 21.5), on the other hand, may do so. These types of chips are periodic, and the resulting force variations during cutting can cause chatter. Other factors that also may contribute to chatter are the use of dull cutters, a lack of cutting fluids, and worn machine-tool ways and components.

Damping.   Damping is defined as the rate at which vibrations decay. This effect can be demonstrated on an automobile's shock absorbers by pushing down on the car's front or rear end and observing how soon the vertical motion stops. Damping is a major factor in controlling machine-tool vibration and chatter and consists of internal and external damping.

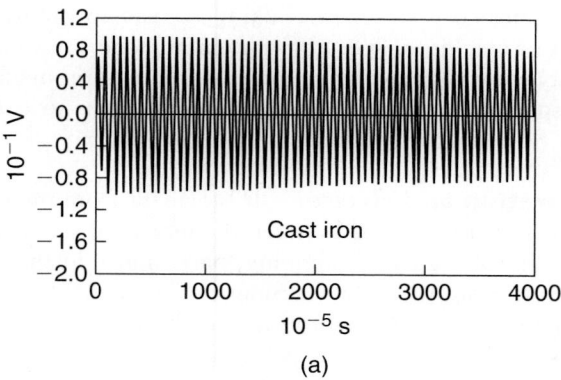 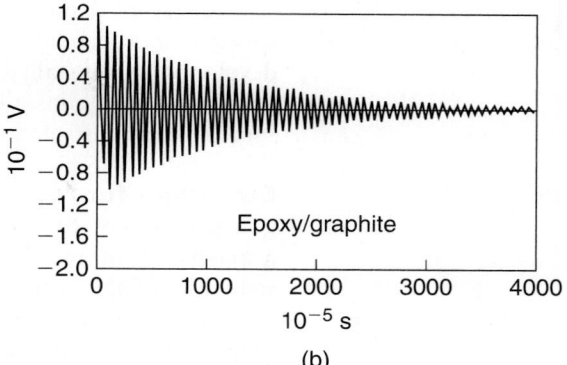

FIGURE 25.14 The relative damping capacity of (a) gray cast iron and (b) an epoxy-granite composite material. The vertical scale is the amplitude of vibration and the horizontal scale is time.

Internal damping results from the energy loss in materials during vibration. For example, steels have a lower damping capacity than gray cast iron, and composite materials have a higher damping capacity than gray cast iron (Fig. 25.14). The difference in the damping capacity of materials can be observed by striking them with a gavel and listening to the sound. For example, try striking a brass cymbal, a piece of concrete, and a piece of wood and listen to the variations in sound.

Bolted joints in the structure of a machine tool also are a source of damping; their effectiveness depends on size, position, and the number of joints. Because friction dissipates energy, small relative movements along dry (unlubricated) joints dissipate energy and, thus, help improve damping. Because all machine tools consist of a number of large and small components assembled by various means, this type of damping is cumulative. Note in Fig. 25.15 how overall damping increases as the number of components on a lathe and their contact areas increase. However, as expected, the overall system stiffness of the machine generally is reduced as the number of joints increases.

Another method of damping in order to reduce or eliminate vibration and chatter is implemented in some boring bars and tool holders. As described and illustrated in Fig. 23.17b, damping can be accomplished by mechanical means—dissipating energy by the frictional resistance of the components within the structure of the boring bar.

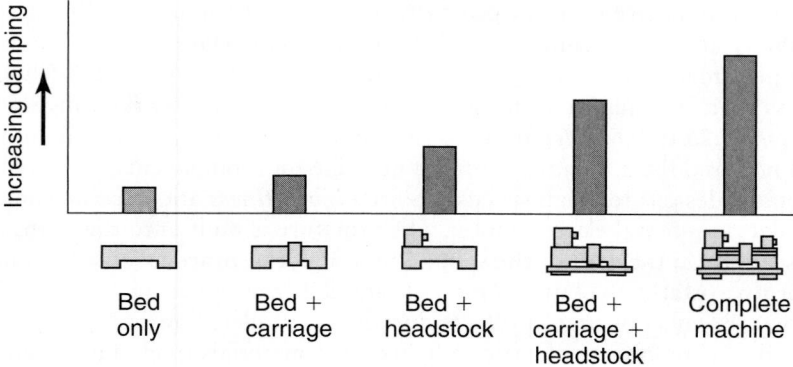

FIGURE 25.15 The damping of vibrations as a function of the number of components on a lathe. Joints dissipate energy; the greater the number of joints, the higher the damping capacity of the machine. *Source:* After J. Peters.

External damping is accomplished with external dampers that are similar to shock absorbers on automobiles or machines. Special vibration absorbers have been developed and installed on machine tools for this purpose. Furthermore, machine tools also can be installed on specially prepared floors and foundations to isolate forced vibrations, such as those from nearby machines on the same floor.

Guidelines for reducing vibration and chatter. It is evident from the foregoing discussion that a balance must be achieved between the increased stiffness of a machine tool and the desirability of increased damping, particularly in the construction of high-precision machine tools. In various sections of Chapters 23 and 24, we have outlined general guidelines for reducing vibration and chatter in machining operations. The basic guidelines may be summarized as follows:

- Minimize tool overhang.
- Improve the stiffness of workholding devices and support workpieces rigidly.
- Modify tool and cutter geometry to minimize forces or make them uniform.
- Change process parameters, such as cutting speed, feed, depth-of-cut, and cutting fluids.
- Increase the stiffness of the machine tool and its components by improving design, using larger cross-sections and materials with a higher elastic modulus.
- Improve the damping capacity of the machine tool.

25.5 | High-Speed Machining

With continuing demands for higher productivity and lower production costs, investigations have been carried out since the late 1950s to increase the cutting speed and the material-removal rate in machining, particularly for applications in the aerospace and automotive industries.

The term "high" in *high-speed machining* (HSM) is somewhat relative; as a general guide, an approximate range of cutting speeds may be defined as follows:

High speed: 600 to 1800 m/min (2000 to 6000 ft/min)
Very high speed: 1800 to 18,000 m/min (6000 to 60,000 ft/min)
Ultrahigh speed: >18,000 m/min

Spindle rotational speeds in machine tools now range up to 50,000 rpm, although the automotive industry generally has limited them to 15,000 rpm for better reliability and less downtime should a failure occur. The *spindle power* required in high-speed machining is generally on the order of 0.004 W/rpm (0.005 hp/rpm)—much less than in traditional machining, which is typically in the range of 0.2 to 0.4 W/rpm (0.25 to 0.5 hp/rpm). Feed rates in high-speed machining are now up to 1 m/s (3 ft/s), and the acceleration rates of machine-tool components are very high.

Spindle designs for high speeds require *high stiffness* and *accuracy* and generally involve an integral electric motor. The armature is built onto the shaft, and the stator is placed in the wall of the spindle housing. The bearings may be rolling elements or hydrostatic; the latter is more desirable because it requires less space than the former. Because of *inertia effects* during the acceleration and deceleration of machine-tool components, the use of lightweight materials (including ceramics and composite materials) is an important consideration.

The selection of appropriate cutting-tool materials is, of course, a major consideration. Based on our discussions on tools and their selection in Chapter 22 (especially by reviewing Table 22.2), it is apparent that (depending on the workpiece material) multiphase coated carbides, ceramics, cubic-boron nitride, and diamond are all candidate tool materials for this operation.

It also is important to note that high-speed machining should be considered primarily for operations in which **cutting time** is a significant portion of the time in the overall machining operation. As we will describe in Section 38.6 and Chapter 40, **non-cutting time** and various other factors are important considerations in the overall assessment of the benefits of high-speed machining for a particular application.

Research and development work continues on high-speed machining, especially in the turning, milling, boring, and drilling of aluminum alloys, titanium alloys, steels, and superalloys. Much data have been collected regarding the effect of high cutting speed on (a) the type of chips produced, (b) cutting forces and power, (c) temperatures generated, (d) tool wear and failure, (e) surface finish and integrity, and (f) the economics of the process.

These studies have indicated that high-speed machining is economical for certain specific applications. For example, it has been implemented in machining (a) aluminum structural components for aircraft; (b) submarine propellers of 6 m (20 ft) in diameter, made of a nickel-aluminum-bronze alloy, and weighing 55,000 kg (50 tons); and (c) automotive engines, with five to ten times the productivity of traditional machining. High-speed machining of complex three- and five-axis contours has been made possible by advances in CNC control technology, as described regarding *machining centers* in this chapter and in Chapter 37.

Another major factor in the adoption of high-speed machining has been the requirement to further improve dimensional tolerances in cutting operations. As we have seen in Fig. 21.14, as the cutting speed increases, more and more of the heat generated is removed by the chip, thus the tool and (more importantly) the workpiece remain close to ambient temperature. This is beneficial, because there is no thermal expansion or warping of the workpiece during machining.

The important machine-tool characteristics in high-speed machining may be summarized as follows:

1. Spindle design for stiffness, accuracy, and balance at very high rotational speeds.
2. Bearing characteristics.
3. Inertia of the machine-tool components.
4. Fast feed drives.
5. Selection of appropriate cutting tools.
6. Processing parameters and their computer control.
7. Workholding devices that can withstand high centrifugal forces.
8. Chip removal systems effective at very high rates of material removal.

CASE STUDY 25.1 | High-Speed Dry Machining of Cast-Iron Engine Blocks

High-speed machining has clear economic advantages as long as spindle damage and chatter are avoided. Similarly, dry machining is recognized as a promising approach to solving environmental problems associated with cutting fluids. Unfortunately, it often has been perceived that the two approaches cannot be applied simultaneously because the higher speeds would result in low tool life. However, high-speed dry machining is possible with aluminum and cast iron and is examined here for roughing, finishing, and holemaking operations on a cast-iron cylinder head for a 4.3 liter engine.

High-speed dry machining is a demanding operation. Heat dissipation without a liquid coolant requires high-performance coatings and heat-resistant tool materials. The cutting-tool materials are either ceramic or cBN because of the intense heat generated in the process. In addition, machine tools with high power are required, and spindles with high stiffness and power ratings are needed. Chatter can be a problem with high-speed machining, hence CNC programs must be developed that maintain process parameters within a window of operation that is known to be chatter-free.

The key to dry machining the cast-iron head is the use of two advanced techniques. The first is *red-crescent machining,* where high temperatures are generated in the workpiece in front of the tool, creating a visible red arc (hence the name of the process). The second technique is flush-fine machining, using in-spindle high-pressure air to remove chips which normally would be carried away by a cutting fluid. Red-crescent machining refers to milling operations that take place at high feed rates and high spindle speeds. This combination reduces the thrust force against the workpiece, as compared to a standard milling operation. Red-crescent machining concentrates heat in the material in front of the tool and can reach 600° to 700°C (1100° to 1300°F) in some cases. This is because (at the high speeds) the heat generated is localized in the shear zone and does not have time to conduct into the workpiece or the tool. This high-temperature zone generates a red glow. This is beneficial to cast irons, because the heat increases the ductility and reduces the strength of the cast iron, allowing for machining at much higher efficiency. Part distortion does not occur, because most of the heat is retained in the chip and removed before the heat can conduct into the workpiece. A comparison of red-crescent machining with conventional machining for this particular application is included in Table 25.1.

Finish milling is accomplished by using low-thrust machining strategies, usually by using cutters with fewer inserts per cutter, at lower feed-per-revolution, and higher surface speeds than conventional machining. Low thrust-force machining has an additional advantage in that workholder design is simplified, often allowing better access to complicated parts. cBN inserts are used for finish machining of the cast-iron engine blocks. Although cBN is more expensive than the alternatives of carbide or ceramic inserts, tool life is longer and fewer setups are needed, thus the insert cost per machined part can be comparable or lower than the conventional tool materials.

Flush-fine machining is a process that uses high-pressure through-spindle air for chip disposal and also removes heat from the cutting tool. The air pressure required is in the range of 1100 to 1380 kPa (160 to 200 psi) in order to reduce the temperature of the workpiece by approximately 50%. Flush-fine machining is suitable for most shallow holes (depth-to-diameter ratios of five or less), which

TABLE 25.1

Comparison of Conventional versus Red-Crescent Machining Conditions

Parameter	Conventional	Red crescent
Surface speed (ft/min)	250	5000
Spindle speed (rpm)	160	3200
Feed (in./rev)	0.020	0.005
Number of inserts	10	5
Radial thrust force (lb)	4300	262
Metal-removal rate (in^3/min)	32	80

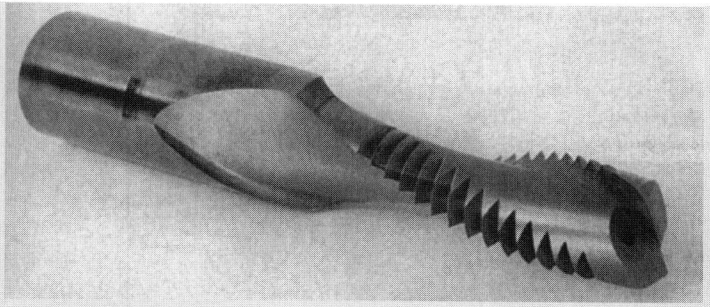

FIGURE 25.16 A high-speed tool for single-point milling, chamfering, counterboring, and threading of holes. *Source:* Courtesy of Makino, Inc.

applies to most automotive applications. Deeper holes are produced using more conventional drilling operations. Shallow holes, single-tool milling, chamfering, counterboring, and threading of holes can be accomplished using tools such as the one shown in Fig. 25.16. This process uses high-speed, high-accuracy CNC control to move a small-diameter tool in a helical pattern to shape and thread the holes. In addition, this allows a single tool to machine holes of different diameters.

Source: Adapted from "High Speed, Dry Machining Can Cut Cycle Times and Cost," by G. Hyatt, *Manufacturing Engineering,* Vol. 119, No. 3, 1997.

25.6 | Hard Machining

We have noted that as the hardness of the workpiece increases, its machinability decreases, and tool wear and fracture, surface finish, and surface integrity can become significant problems. As will be described in Chapters 26 and 27, there are several other mechanical (including abrasive machining), nonmechanical, and advanced techniques of removing material economically from hard or hardened metals and alloys. However, it is still possible to apply traditional machining processes to hard metals and alloys by selecting an appropriate hard-tool material and using machine tools with high stiffness, power, and precision.

A common example is the finish machining of heat-treated steel (45 to 65 HRC) shafts, gears, pinions, and various automotive components using polycrystalline cubic-boron nitride (PcBN), cermet, or ceramic cutting tools. Called *hard machining* or *hard turning,* this process produces machined parts with good dimensional accuracy, surface finish (as low as 0.25 μm or 10 μin.), and surface integrity. The (a) available power, (b) static and dynamic stiffness of the machine tool and its spindle, and (c) workholding devices and fixturing are important factors.

As we described in Section 25.3, trends in the design and construction of modern machine tools include the use of hydrostatic bearings for the spindles and slideways. The headstock and the slanted bed in the machines (see Fig. 23.10a) can be made of *granite composite materials* with unique properties. Cutting-tool selection and edge preparation also are important to avoid premature failure in hard machining.

From both technical, economic, and ecological considerations, hard turning can compete successfully with the *grinding* process and with the quality and characteristics approaching those produced in grinding machines. For instance, in some specific cases, hard turning has been shown to be three times faster than grinding, requires fewer operations to finish the part, and utilizes five times less energy. A detailed comparative example of hard turning versus grinding will be presented in Example 26.4.

25.7 | Ultraprecision Machining

Beginning in the 1960s, increasing demands were being made for the precision manufacturing of components for computer, electronic, nuclear, and defense applications. Some examples include optical mirrors, computer memory disks, and drums for photocopying machines. Surface-finish requirements are in the tens of *nanometer* (10^{-9} m, 0.001 μm, or 0.04 μin.) range, and dimensional tolerances and form accuracies are in the μm and sub-μm range.

This trend for ultraprecision manufacturing continues to grow. Modern **ultraprecision machine tools** with advanced computer controls can position a cutting tool within an accuracy approaching 1 nm, as can be seen from Fig. 25.17. Also note in this figure that higher precision now is being achieved by processes such as abrasive machining, ion-beam machining, and molecular manipulation. We will describe *abrasive machining* and *advanced machining processes* in Chapters 26 and 27, respectively. We will cover the topics of *micromachining, fabrication of microelectronic* and *microelectromechanical systems* (MEMS), and *nanomanufacturing* in Chapters 28 and 29.

The cutting tool for *ultraprecision machining* applications is almost exclusively a single-crystal diamond, where the process is called **diamond turning.** The diamond tool has a polished cutting edge with a radius as small as a few nm. Wear of the diamond can be a significant problem, and recent advances include *cryogenic diamond turning,* in which the tooling system is cooled by liquid nitrogen to a temperature of about $-120°C$ ($-184°F$).

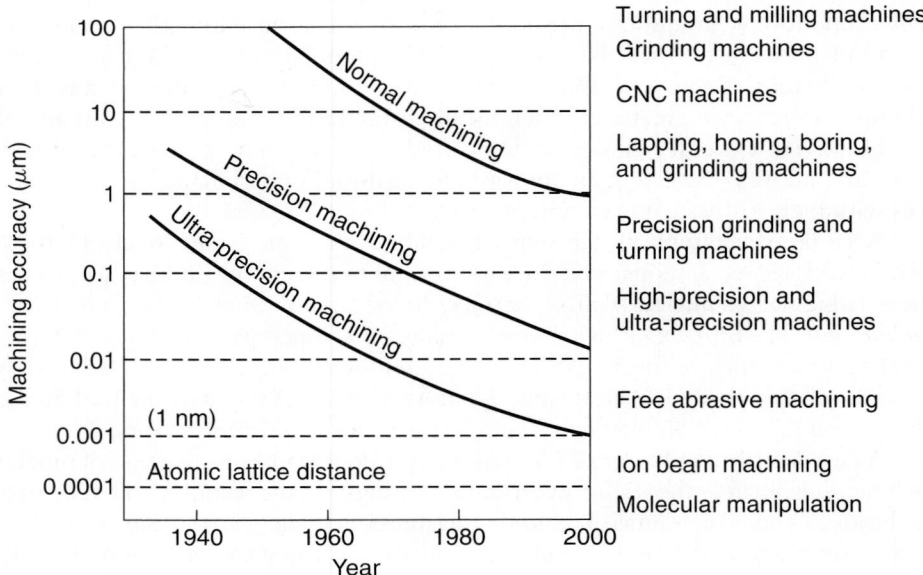

FIGURE 25.17   Improvements in machining accuracy over the years using ultraprecision machining technologies. *Source:* After C.J. McKeown, N. Taniguchi, G. Byrne, D. Dornfeld, and B. Denkena.

The workpiece materials for ultraprecision machining include copper alloys, aluminum alloys, silver, gold, electroless nickel, infrared materials, and plastics (acrylics). With depths of cut in the nm range, hard and brittle materials produce continuous chips in a process known as **ductile-regime cutting.** (See also *ductile-regime grinding* in Section 26.3.4.) Deeper cuts in brittle materials produce discontinuous chips.

The machine tools for these applications are built with very high precision and high machine, spindle, and workholding-device stiffnesses. These ultraprecision machines (parts of which are made of structural materials with low thermal expansion and good dimensional stability, see Section 25.3) are located in a dust-free environment (*clean rooms*) where the temperature is controlled to within a fraction of one degree.

Vibrations from internal machine sources—as well as from external sources, such as nearby machines on the same floor—must be avoided as much as possible. Laser metrology is used for feed and position control, and the machines are equipped with highly advanced computer-control systems and with thermal- and geometric error-compensating features.

General considerations for precision machining.   There are several important factors in precision and ultraprecision machining and machine tools, somewhat similar to those in high-speed machining:

1. Machine-tool design, construction, and assembly—providing stiffness, damping, and geometric accuracy.
2. Motion control of various components—both linear and rotational.
3. Spindle technology.
4. Thermal growth of the machine tool, compensation for thermal growth, and control of the machine-tool environment.
5. Cutting-tool selection and application.
6. Machining parameters.
7. Real-time performance and control of the machine tool and implementation of a tool-condition monitoring system.

25.8 | Machining Economics

In the Introduction to Part IV, it was stated that the important limitations of machining operations include (a) the relatively longer time required to machine a part, as compared to forming and shaping it, (b) the need to reduce noncutting time, and (c) the fact that material is inevitably wasted. Despite these drawbacks, machining is indispensable, particularly for producing complex workpiece shapes with external and internal features and obtaining high dimensional accuracy and surface finish.

We have outlined the material and process parameters that are relevant to efficient machining operations. When analyzing the *economics* of machining, however, several other factors have to be considered. As will be described in greater detail in Chapter 40, these factors include the costs involved in (a) machine tools, workholding devices and fixtures, and cutting tools; (b) labor and overhead; (c) the time consumed in setting up the machine for a particular operation; (d) material handling and movement, such as loading the blank and unloading the machined part; (e) gaging for dimensional accuracy and surface finish; and (f) cutting times and noncutting times.

Actual machining time is an important consideration. Recall also the discussion in Section 25.5 on the importance of noncutting time in assessing the economic relevance of high-speed machining. Unless noncutting time is a significant portion of the floor-to-floor time, high-speed machining should not be considered unless its other benefits are relevant.

Minimizing machining cost-per-piece.   As in all manufacturing processes and operations, the relevant parameters in machining can be selected and specified in such a manner that the *machining cost-per-piece* (as well as *machining time-per-piece*) is minimized. Various methods and approaches have been developed to accomplish this goal. With the increasing use of software and user-friendly computers, this task now has become easier. However, in order for the results to be reliable, it is essential that input data be accurate and up-to-date. Described next is one of the simpler and more common methods of analyzing machining costs in a turning operation.

In machining a part by turning, the total machining cost-per-piece, C_p, consists of

$$C_p = C_m + C_s + C_l + C_t \tag{25.1}$$

where

C_m = Machining cost
C_s = Cost of setting up for machining; mounting the cutter, setting up fixtures, and preparing the machine tool for the operation
C_l = Cost of loading, unloading, and machine handling
C_t = Tooling cost, often only about 5% of the total cutting operation. Consequently, using the least expensive tool is not always an effective way of reducing machining costs

The **machining cost** is given by

$$C_m = T_m(L_m + B_m) \tag{25.2}$$

where T_m is the machining time-per-piece, L_m is the labor cost of production operator per hour, and B_m is the burden rate or overhead charge of the machine—including depreciation, maintenance, indirect labor, and the like. The **setup cost** is a fixed figure in dollars-per-piece. The **loading, unloading,** and **machine handling cost** is

$$C_l = T_l(L_m + B_m) \tag{25.3}$$

where T_l is the time involved in loading and unloading the part, changing speeds, feed rates, and so on. The **tooling cost** is

$$C_t = \frac{1}{N_p}\left[T_c(L_m + B_m) + T_g(L_g + B_g) + D_c \right] \tag{25.4}$$

where N_p is the number of parts machined per tool grind, T_c is the time required to change the tool, T_g is the time required to grind the tool, L_g is the labor cost of tool grinder operator per hour, B_g is the burden rate of tool grinder per hour, and D_c is the depreciation of the tool in dollars per grind.

The **time** needed to produce one part is

$$T_p = T_l + T_m + \frac{T_c}{N_p} \tag{25.5}$$

where T_m has to be calculated for a particular operation. For example, let's consider a turning operation. The machining time is

$$T_m = \frac{l}{fN} = \frac{\pi l d}{fV} \tag{25.6}$$

where l is the length of cut, f is the feed, N is the rpm of the workpiece, d is the workpiece diameter, and V is the cutting speed. Note that appropriate units must be used in all of these equations. From the Taylor tool-life equation (Eq. 21.20a) we have

$$VT^n = C$$

Hence,

$$T = \left(\frac{C}{V}\right)^{1/n} \qquad (25.7)$$

where T is the time (in minutes) required to reach a flank wear of certain dimension, after which the tool has to be reground or changed. Thus the number of pieces per tool grind is simply

$$N_p = \frac{T}{T_m} \qquad (25.8)$$

The combination of Eqs. (25.6), (25.7), and (25.8) gives

$$N_p = \frac{fC^{1/n}}{\pi l d V^{(1/n)-1}} \qquad (25.9)$$

The **cost-per-piece**, C_p, in Eq. (25.1) now can be defined in terms of several variables. To find the optimum cutting speed and also the optimum tool life for **minimum cost**, we have to differentiate C_p with respect to V and set it to zero. Thus,

$$\frac{\partial C_p}{\partial V} = 0 \qquad (25.10)$$

We then find that the optimum cutting speed, V_o, is

$$V_o = \frac{C(L_m + B_m)^n}{\{[(\frac{1}{n}) - 1][T_c(L_m + B_m) + T_g(L_g + B_g) + D_c]\}^n} \qquad (25.11)$$

and the optimum tool life, T_o, is

$$T_o = [(\tfrac{1}{n}) - 1]\frac{T_c(L_m + B_m) + T_g(L_g + B_g) + D_c}{L_m + B_m} \qquad (25.12)$$

To find the optimum cutting speed and also the optimum tool life for **maximum production,** we have to differentiate T_p with respect to V and set it to zero. Thus,

$$\frac{\partial T_p}{\partial V} = 0 \qquad (25.13)$$

The **optimum cutting speed** now becomes

$$V_o = \frac{C}{\{[(\frac{1}{n}) - 1]T_c\}^n} \qquad (25.14)$$

and the **optimum tool life** is

$$T_o = [(\tfrac{1}{n}) - 1]T_c \qquad (25.15)$$

A qualitative plot of minimum cost-per-piece and the minimum time-per-piece (that is, the maximum production rate) is given in Fig. 25.18. The cost of a machined surface also depends on the finish required; the machining cost increases rapidly with finer surface finish.

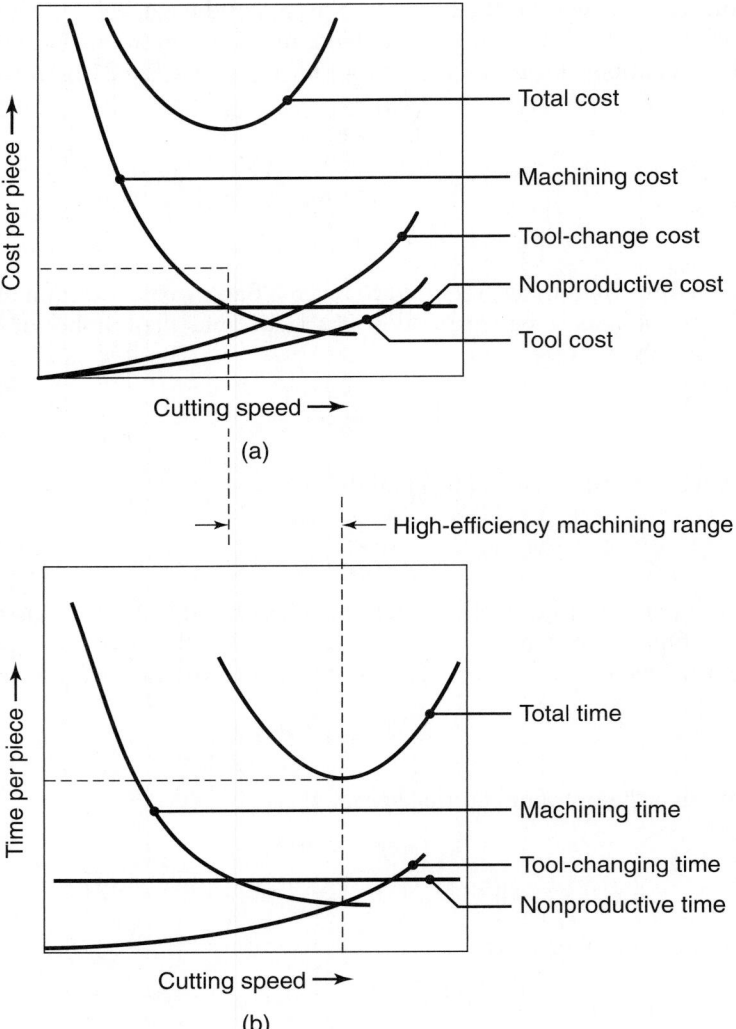

FIGURE 25.18   Graphs showing (a) cost-per-piece and (b) time-per-piece in machining. Note the optimum speeds for both cost and time. The range between the two is known as the high-efficiency machining range.

This analysis indicates the importance of identifying all relevant parameters in a machining operation, determining various cost factors, obtaining relevant tool-life curves for the particular operation, and properly measuring the various time intervals involved in the overall operation. The importance of obtaining accurate data is shown clearly in Fig. 25.18, as small changes in cutting speed can have a significant effect on the minimum cost- or time-per-piece.

SUMMARY

- Because they are versatile and capable of performing a variety of machining operations on small and large workpieces of various shapes, machining centers are now among the most important machine tools. Their selection depends on factors such as part complexity, the number and type of machining operations to be performed, the number of cutting tools required, and the dimensional accuracy and production rate required.

- Vibration and chatter in machining are among important considerations for workpiece dimensional accuracy, surface finish, and tool life. Stiffness and damping capacity of machine tools are important factors in controlling vibration and chatter. New materials are being developed continually and used in the construction of machine-tool structures.

- The economics of machining processes depend on factors such as nonproductive costs, machining costs, tool-change costs, and tool costs. Optimum cutting speeds can be determined for both minimum machining time-per-piece and minimum cost-per-piece.

KEY TERMS

Automatic pallet changer

Automatic tool changer

Chatter

Chip collection

Damping

Dynamic stiffness

Forced vibration

Hexapods

High-efficiency machining range

Linear-motor drive

Machining center

Modular construction

Pallet

Reconfigurable machines

Regenerative chatter

Self-excited vibration

Stiffness

Tool-exchange arm

Tool- and part-checking station

Touch probes

Turning center

Universal machining center

Work envelope

BIBLIOGRAPHY

ASM Handbook, Vol. 16, *Machining,* ASM International, 1989.

Boothroyd, G., and Knight, W.A., *Fundamentals of Machining and Machine Tools,* 2nd ed., Marcel Dekker, 1989.

Brown, J., *Advanced Machining Technology Handbook,* McGraw-Hill, 1998.

Erdel, B., *High-Speed Machining,* Society of Manufacturing Engineers, 2003.

Krar, S.F., and Check, A.F., *Technology of Machine Tools,* 5th ed., Glencoe/Macmillan McGraw-Hill, 1996.

Nakazawa, H., *Principles of Precision Engineering,* Oxford, 1994.

Reshetov, D.N., and Portman, V.T., *Accuracy of Machine Tools,* American Society of Mechanical Engineers, 1989.

Rivin, E.I., *Stiffness and Damping in Mechanical Design,* Marcel Dekker, 1999.

Tool and Manufacturing Engineers Handbook, 4th ed., Vol. 1, *Machining,* Society of Manufacturing Engineers, 1983.

Weck, M., *Handbook of Machine Tools,* 4 vols. Wiley, 1984.

REVIEW QUESTIONS

25.1 Describe the distinctive features of machining centers and explain why these machines so versatile.

25.2 Why are pallet changers and tool changers integral parts of machining centers?

25.3 Explain how the tooling system in a machining center operates.

25.4 Explain the trends in materials used for machine-tool structures.

25.5 Is there any difference between chatter and vibration? Explain.

25.6 Explain the importance of foundations for installing machine tools.

25.7 What are the typical tool-changing times in a machining center?

25.8 What types of materials are machine-tool bases made from? Why?

25.9 What factors contribute to costs in machining operations?

25.10 What is meant by "modular" construction of machine tools?

25.11 What is the high-efficiency machining range? Why is it so called?

QUALITATIVE PROBLEMS

25.12 Explain the technical and economic factors that led to the development of machining centers.

25.13 We have noted that spindle speeds in machining centers vary over a wide range. Explain the reasons, giving specific applications.

25.14 Explain the importance of stiffness and damping of machine tools. How are they accomplished?

25.15 Are there machining operations described in Chapters 23 and 24 that cannot be performed in machining and turning centers? Explain, with specific examples.

25.16 Describe how the touch probes shown in Fig. 25.6 operate.

25.17 How important is the control of cutting-fluid temperature in operations performed in machining centers? Explain.

25.18 Review Fig. 25.10 on modular machining centers and explain workpieces and operations that would be suitable on such machines.

25.19 Describe the adverse effects of vibration and chatter in machining regarding both the workpiece and the cutting tool.

25.20 We know that we may be able to decrease machining costs by changing the cutting speed. Explain which costs are likely to change and how, as the cutting speed is changed.

25.21 Explain the differences in the functions of a turret and a spindle in turning centers.

25.22 Explain how the pallet arrangements shown in Fig. 25.4a and b would be operated in the use of these machines.

25.23 Review the tool changer shown in Fig. 25.5 and explain any constraints to making their operations faster, thus reducing the tool changing time.

25.24 In addition to the number of joints in a machine tool (see Fig. 25.15), what other factors influence the rate at which damping increases?

25.25 Describe workpieces that would not be suitable for machining on a machining center. Give specific examples.

25.26 Other than the fact that they each have a minimum, are the overall shapes and slopes of the total-cost and total-time curves in Fig. 25.18 important? Explain.

25.27 In Section 21.3, we stated that the thrust force in cutting can be negative at high rake angles and/or low friction at the chip–tool interface. This can have an adverse effect on the stability of the cutting process. Explain why.

25.28 Explain the advantages and disadvantages machine-tool frames made of gray-iron castings?

25.29 What are the advantages and disadvantages of welded-steel frames? Bolted-steel frames? Using adhesive bonding to assemble components of machine tools?

25.30 Why have concrete or polymer concretes been used in some machine tools?

25.31 Describe the economic considerations involved in selecting machining and turning centers.

25.32 Give examples of forced vibration or self-excited vibration in general engineering practice.

25.33 Explain specific situations where thermal distortion of machine-tool components would be important.

25.34 What are the workpiece requirements that would make a machining center preferable to conventional milling machines, lathes, and drill presses?

25.35 Explain how you would go about reducing each of the cost factors in machining operations. What difficulties would you encounter in doing so?

QUANTITATIVE PROBLEMS

25.36 A machining-center spindle and tool extend 12 in. from its machine-tool frame. What temperature change can be tolerated to maintain a tolerance of 0.0001 in. in machining? A tolerance of 0.001 in.? Assume that the spindle is made of steel.

25.37 Calculate the time to manufacture the parts in Example 25.1 using conventional machining and high-speed machining, using the data in Case Study 25.1.

25.38 In the production of a machined valve, the labor rate is $19.00 per hour, the tool-grinder labor rate is $25.00 per hour, and the general overhead rate is $15.00 per hour. The tool is made of high-speed steel and costs $25.00, takes five minutes to change, ten minutes to grind, and can be reground four times before it needs to be replaced. The workpiece is a free-machining aluminum. Estimate the optimum cutting speed from a cost perspective. Use $C = 100$ for V_o in m/min.

SYNTHESIS, DESIGN, AND PROJECTS

25.39 If you were the chief engineer in charge of the design of machining and turning centers, what changes and improvements would you recommend on existing models?

25.40 Study the technical literature and outline the trends in the design of modern machine tools. Explain why there are those trends.

25.41 Make a list of components of machine tools that could be made of ceramics and explain why ceramics would be suitable.

25.42 Survey the literature from various machine-tool manufacturers and prepare a comprehensive table, indicating the capabilities, sizes, power, and costs of machining and turning centers. Comment on your observations.

25.43 As you can appreciate, the cost of machining and turning centers is considerably higher than for traditional machine tools. Since many operations performed by machining centers also can be done on conventional machines, how would you go about justifying the high cost of these centers? Explain with appropriate examples.

25.44 Conduct a literature search or contact manufacturers to describe the details of how hexapod machines operate.

25.45 In Part III of this text, we described forming and shaping processes for metallic and nonmetallic materials. Based on this knowledge, explain whether it would be feasible to design a machine that could be named a "forming center," capable of performing two or more different shaping processes. Give specific examples of parts that can be made and describe your ideas on the design of such a machine or machines.

25.46 Explain if it could be possible to design and build machining centers without computer controls.

25.47 In your experience using tools or other devices, have you come across situations where you experienced vibration and chatter? If so, give details and explain how you would go about minimizing them.

25.48 Describe your thoughts on whether or not it is feasible to include grinding operations (see Chapter 26) in machining centers. Explain the nature of any difficulties that may be encountered.

25.49 The following experiment is designed to better demonstrate the effect of tool overhang in vibration and chatter. With a sharp tool, scrape the surface of a piece of soft metal by holding the tool with your arm fully stretched. Repeat the experiment by holding the tool with your hand as close to the metal as possible. Describe your observations regarding the tendency for the tool to vibrate. Repeat the experiment with different types of metallic and nonmetallic materials.

25.50 A large bolt is to be produced from extruded hexagonal stock by placing the hex stock into a chuck and machining the cylindrical shank of the bolt by turning. List the difficulties that may be encountered in this operation.

26

Abrasive Machining and Finishing Operations

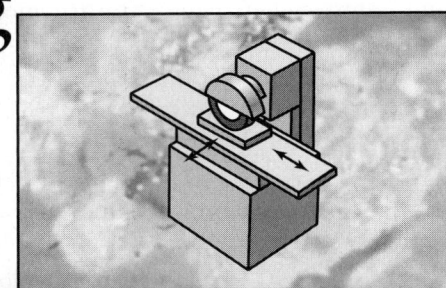

Abrasive machining is among the final steps in the production of parts and is an important series of processes because of its capability to impart high dimensional accuracy and surface finish. In this chapter, we specifically describe:

- Fundamentals and capabilities of the grinding process.
- Characteristics of typical abrasives.
- Structure of grinding wheels and various abrasive tools.
- Grinding operations and their capabilities.
- Finishing processes that are based on abrasive action.
- How abrasive machining can compete with other processes.

Typical parts made: Any part requiring high dimensional accuracy and surface finish, such as ball and roller bearings, piston rings, valves, cams, gears, and dies.

Alternative processes: Precision machining, electrical-discharge machining, electrochemical machining and grinding, and abrasive-jet machining.

26.1 | Introduction

There are many situations in manufacturing where the processes described thus far cannot produce the required dimensional accuracy and surface finish for a part, or the workpiece material is too hard or too brittle to process. Consider, for example, the accuracy and fine surface finish on ball and roller bearings, pistons, valves, cylinders, cams, gears, dies, and numerous precision components used in instrumentation. One of the best methods for producing such demanding characteristics on parts is abrasive machining.

An **abrasive** is a small, hard particle having sharp edges and an irregular shape, unlike the cutting tools described earlier. Abrasives are capable of removing small amounts of material from a surface through a cutting process that produces tiny chips. Most of us are familiar with using *grinding wheels* (bonded abrasives), as shown in Fig. 26.1, to sharpen knives and tools, as well as using *sandpaper* or *emery cloth* to smoothen surfaces and remove sharp corners. As described in this chapter, abrasives also are used to hone, lap, buff, and polish workpieces. With the use of computer-controlled machines, abrasive processes now are capable of producing (a) a wide variety of workpiece geometries, as can be seen in Fig. 26.2 and (b) very

FIGURE 26.1   A variety of bonded abrasives used in abrasive-machining processes. *Source*: Courtesy of Norton Company.

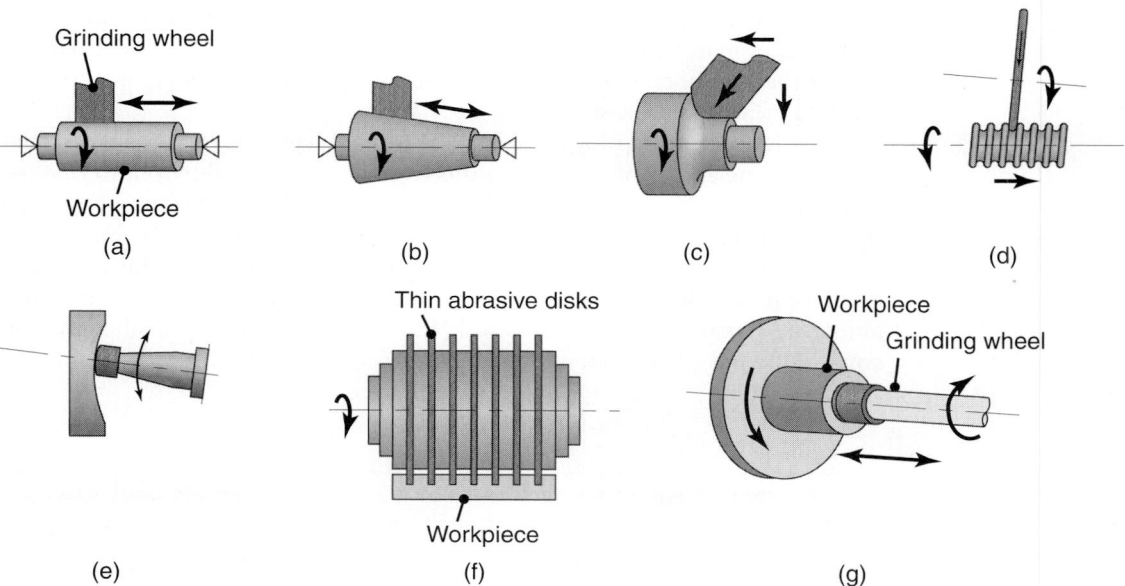

FIGURE 26.2   The types of workpieces and operations typical of grinding: (a) cylindrical surfaces, (b) conical surfaces, (c) fillets on a shaft, (d) helical profiles, (e) concave shape, (f) cutting off or slotting with thin wheels, and (g) internal grinding.

fine dimensional accuracy and surface finishes, as shown in Figs. 23.13, 23.14, and 27.4. For example, dimensional tolerances can be less than 1 μm (40 μin.), and surface roughnesses can be as fine as 0.025 μm (1 μin.).

Because they are hard, abrasives also are used in *finishing processes* for heat-treated metals and alloys and for very hard parts in applications such as (a) finishing of ceramics and glasses, (b) cutting off lengths of bars, structural shapes, masonry, and concrete, (c) removing unwanted weld beads and spatter, and (d) cleaning surfaces with jets of air or water containing abrasive particles.

In this chapter, the characteristics of abrasives are presented along with their use in various material-removal processes. As with cutting operations, the mechanics of abrasive-machining operations first are described. This knowledge will then assist in establishing the interrelationships of workpiece material and process variables and the dimensional accuracy, surface finish, and surface integrity of the parts produced.

26.2 | Abrasives and Bonded Abrasives

Abrasives that are used most commonly in abrasive-machining operations are:

Conventional abrasives

- *Aluminum oxide* (Al_2O_3)
- *Silicon carbide* (SiC)

Superabrasives

- *Cubic boron nitride* (cBN)
- *Diamond*

As described in Chapter 8, these abrasives are much harder than conventional cutting-tool materials, as seen by comparing Tables 22.1 and 26.1. Because the last two of the four abrasives listed above are the two hardest materials known, they are referred to as *superabrasives*.

In addition to hardness, an important characteristic of abrasives is **friability**—defined as the ability of abrasive grains to fracture (break down) into smaller pieces. This is an important property and gives abrasives their *self-sharpening* characteristics, which are essential in maintaining their sharpness during use. High friability indicates low strength or low fracture resistance of the abrasive. Thus, a highly friable abrasive grain fragments more rapidly under grinding forces than one with low friability. For example, aluminum oxide has lower friability than silicon carbide and, correspondingly, less tendency to fragment.

TABLE 26.1

Ranges of Knoop Hardness for Various Materials and Abrasives			
Common glass	350–500	Titanium nitride	2000
Flint, quartz	800–1100	Titanium carbide	1800–3200
Zirconium oxide	1000	Silicon carbide	2100–3000
Hardened steels	700–1300	Boron carbide	2800
Tungsten carbide	1800–2400	Cubic boron nitride	4000–5000
Aluminum oxide	2000–3000	Diamond	7000–8000

The *shape* and *size* of the abrasive grain also affect its friability. For example, blocky grains (which are analogous to a negative rake angle in single-point cutting tools, as shown in Fig. 21.3) are less friable than less-blocky or plate-like grains. Furthermore, because the probability of defects is lower as the grain size becomes smaller (due to the *size effect*), smaller grains are stronger and less friable than larger ones.

Abrasive types.   The abrasives commonly found in nature are *emery*, *corundum* (alumina), *quartz*, *garnet*, and *diamond*. Because these natural abrasives generally contain unknown amounts of impurities and possess nonuniform properties, their performance is inconsistent and unreliable. Consequently, abrasives have been made synthetically for many years, as described next.

- **Aluminum oxide** was first made in 1893 and is produced by fusing bauxite, iron filings, and coke. Fused aluminum oxides are categorized as *dark* (less friable), *white* (very friable), and *single crystal*.

- **Seeded gel** was first introduced in 1987 and is the purest form of *unfused aluminum oxide*. It also is known as *ceramic aluminum oxide*. It has a grain size on the order of 0.2 μm, which is much smaller than other types of commonly used abrasive grains. These grains are sintered to form larger sizes. Because they are harder than fused alumina and have relatively high friability, seeded gels maintain their sharpness and are used especially for difficult-to-grind materials.

- **Silicon carbide** was first discovered in 1891 and is made with silica sand and petroleum coke. Silicon carbides are divided into *black* (less friable) and *green* (more friable) and generally have higher friability than aluminum oxides. Hence, they have a greater tendency to fracture and remain sharp.

- **Cubic-boron nitride** was first developed in the 1970s. Its properties and characteristics are described in Sections 8.2.3 and 22.7.

- **Diamond** (also known as synthetic or industrial diamond) was first used as an abrasive in 1955. Its properties and characteristics are described in Sections 8.7 and 22.9.

Abrasive grain size.   As used in manufacturing operations, abrasives generally are very small when compared to the size of cutting tools and inserts that we described in Chapters 21 and 22. Also, abrasives have sharp edges, allowing the removal of very small quantities of material from the workpiece surface. Consequently, a very fine surface finish and dimensional accuracy can be obtained using abrasives as tools.

The size of an abrasive *grain* is identified by a **grit number**, which is a function of sieve size: the smaller the grain size, the larger the grit number. For example, grit number 10 is regarded as very coarse, 100 as fine, and 500 as very fine. Sandpaper and emery cloth also are identified in this manner, as you can observe readily by noting the grit number printed on the back of the abrasive paper or cloth.

Abrasive workpiece-material compatibility.   As in selecting cutting-tool materials for machining particular workpiece materials, the affinity of an abrasive grain to the workpiece material is an important consideration. The less the reactivity of the two materials, the less wear and dulling of the grains occur during grinding, both of which would make the operation less efficient and cause damage to the workpiece surface (see Section 26.3.1 for details). Recall that because of its chemical affinity, diamond cannot be used for grinding steels because diamond dissolves in iron at the high temperatures encountered in grinding.

Generally, the following recommendations are made for abrasive selection:

- *Aluminum oxide:* Carbon steels, ferrous alloys, and alloy steels.
- *Silicon carbide:* Nonferrous metals, cast irons, carbides, ceramics, glass, and marble.
- *Cubic boron nitride:* Steels and cast irons above 50 HRC hardness and high-temperature alloys.
- *Diamond:* Ceramics, cemented carbides, and some hardened steels.

26.2.1 Grinding wheels

Because each abrasive grain typically removes only a very small amount of material at a time, high rates of material removal can be achieved only if a large number of these grains act together. This is done by using **bonded abrasives**, typically in the form of a *grinding wheel*, in which the abrasive grains are distributed and oriented randomly.

As shown schematically in Fig. 26.3, the abrasive grains in a grinding wheel are held together by a **bonding material** (Section 26.2.2) which acts as supporting posts or braces between the grains. In bonded abrasives, *porosity* is essential in order to provide clearance for the chips being produced and to provide cooling; otherwise, the chips would severely interfere with the grinding operation. Porosity can be observed by looking closely at the surface of any grinding wheel. As you can appreciate, it is impossible to utilize a grinding wheel with no porosity, one that is fully dense and solid; there simply is no room for any chips to form.

A very wide variety of types and sizes of abrasive wheels are made today. Some of the more commonly used types of grinding wheels for conventional abrasives are shown in Fig. 26.4. Superabrasive wheels are shown in Fig. 26.5. Note that due to their high cost, only a small volume of these wheels consists of superabrasives.

Bonded abrasives are marked with a standardized system of letters and numbers indicating: the type of abrasive, grain size, grade, structure, and bond type. Figure 26.6 shows the marking system for aluminum-oxide and silicon-carbide bonded abrasives, and Fig. 26.7 shows the marking system for diamond and cubic-boron-nitride bonded abrasives.

The *cost* of grinding wheels depends on the type and size of the wheel. Small wheels (up to about 25 mm (1 in.) in diameter) cost approximately $2 to $10 for conventional abrasives, $30 to $100 for diamond, and $50 to $200 for cubic-boron-nitride wheels. For a large wheel of about 500 mm in diameter and 250 mm in width

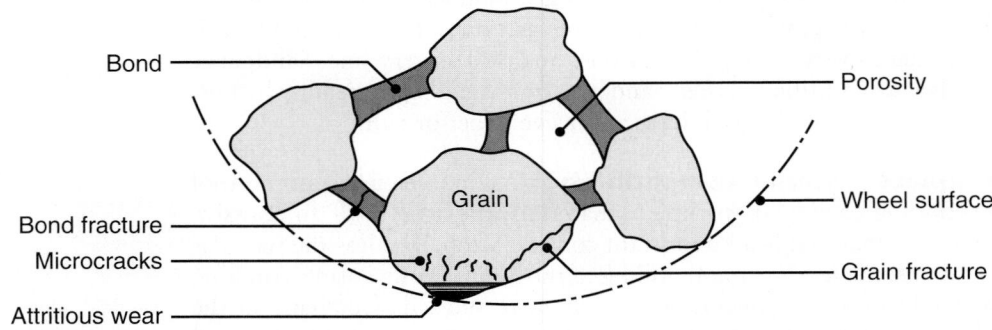

FIGURE 26.3   Schematic illustration of a physical model of a grinding wheel showing its structure and wear and fracture patterns.

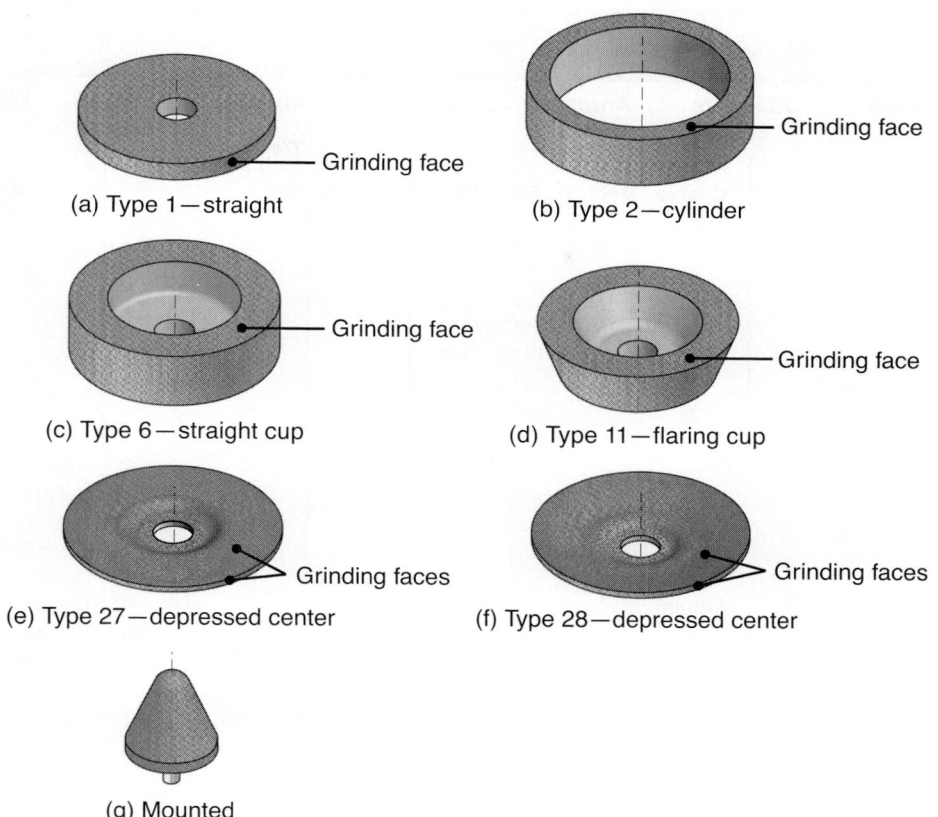

(a) Type 1—straight

Grinding face

(b) Type 2—cylinder

Grinding face

(c) Type 6—straight cup

Grinding face

(d) Type 11—flaring cup

Grinding face

(e) Type 27—depressed center

Grinding faces

(f) Type 28—depressed center

Grinding faces

(g) Mounted

FIGURE 26.4   Common types of grinding wheels made with conventional abrasives. Note that each wheel has a specific grinding face; grinding on other surfaces is improper and unsafe.

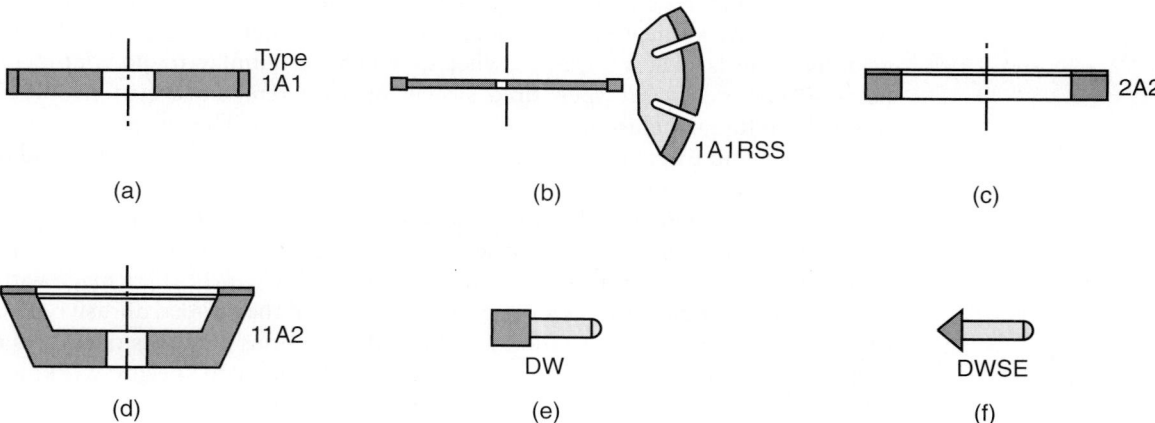

Type 1A1

(a)

1A1RSS

(b)

2A2

(c)

11A2

(d)

DW

(e)

DWSE

(f)

FIGURE 26.5   Examples of superabrasive wheel configurations. The annular regions (rim) are superabrasive grinding surfaces, and the wheel itself (core) generally is made of metal or composites. The bonding materials for the superabrasives are: (a), (d), and (e) resinoid, metal, or vitrified; (b) metal; (c) vitrified; and (f) resinoid.

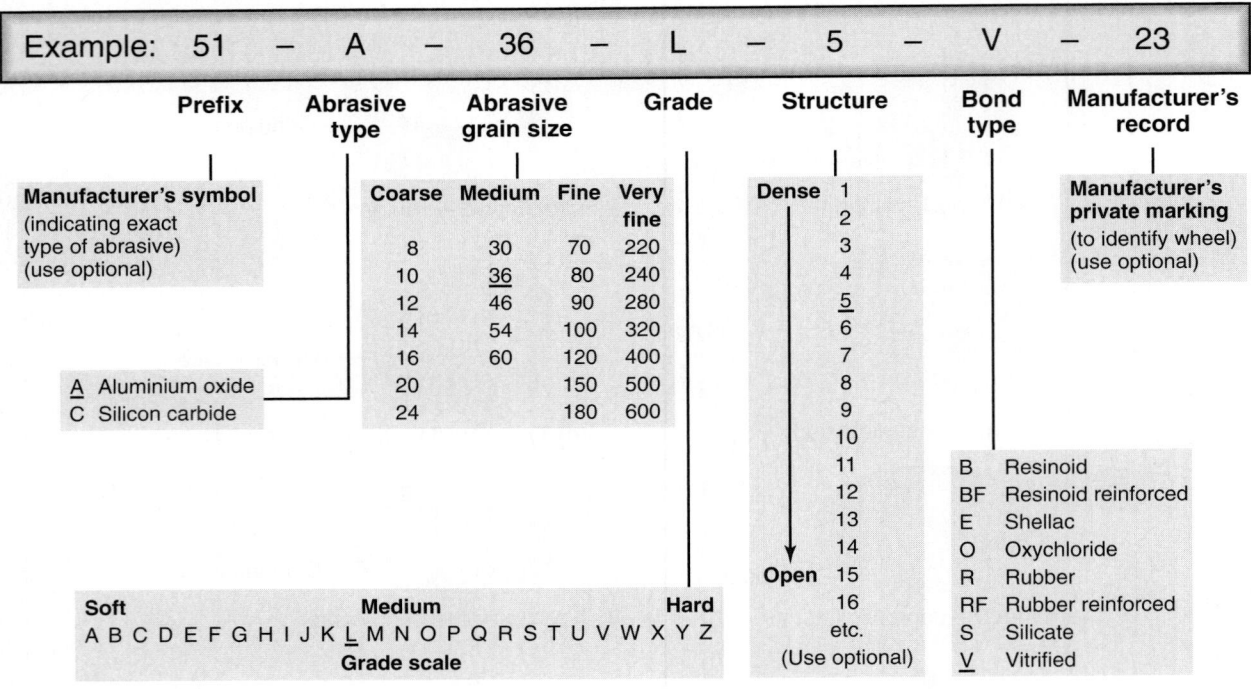

FIGURE 26.6   Standard marking system for aluminum-oxide and silicon-carbide bonded abrasives.

(20 in. × 10 in.), the wheel costs are $500 for conventional abrasives, $5000 to $8000 for diamond, and almost $20,000 for cubic boron nitride.

26.2.2 Bond types

The common types of bonds used in bonded abrasives are described next.

Vitrified.   Essentially a glass, a *vitrified bond* (also called a *ceramic bond*, particularly outside the United States) is the most common and widely used bond material. The raw materials consist of feldspar (a crystalline mineral) and clays. They are mixed with the abrasives, moistened, and molded under pressure into the shape of grinding wheels. These "green" wheels, which are similar to powder-metallurgy parts (Chapter 17), are then fired slowly up to a temperature of about 1250°C (2300°F) to fuse the glass and develop structural strength. The wheels are then cooled slowly (to avoid thermal cracking), finished to size, inspected for quality and dimensional accuracy, and tested for defects.

Wheels with vitrified bonds are strong, stiff, porous, and resistant to oils, acids, and water. However, they are brittle and lack resistance to mechanical and thermal shock. To improve their strength during use, vitrified wheels also are made with steel-backing plates or cups for better structural support of the bonded abrasive. The color of the grinding wheel can be modified by adding various elements during its manufacture; in this way, wheels can be color coded for use with specific workpiece materials, such as ferrous, nonferrous, ceramic, and so on.

Resinoid.   Resinoid bonding materials are *thermosetting resins* and are available in a wide range of compositions and properties (Sections 7.4 and 7.7). Because the bond is an organic compound, wheels with *resinoid bonds* also are called **organic wheels**.

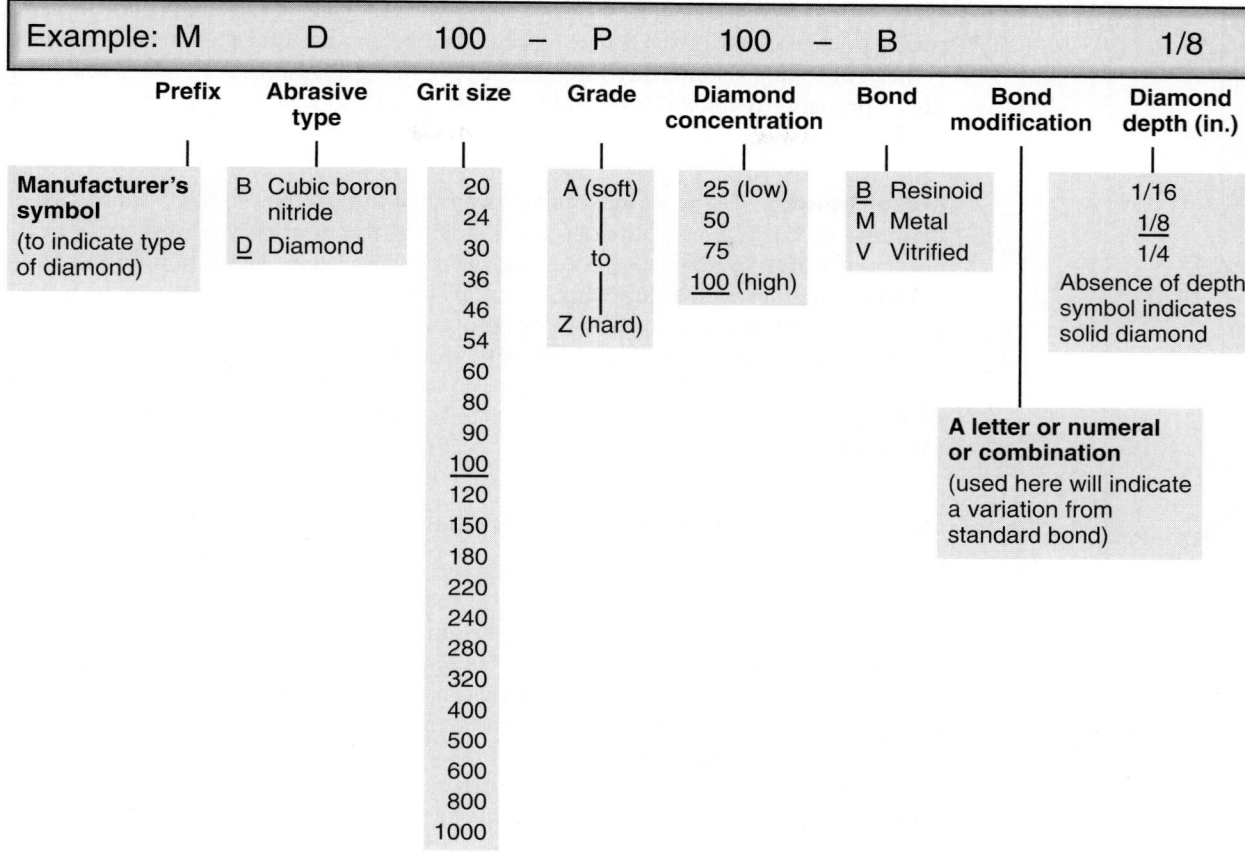

FIGURE 26.7 Standard marking system for cubic-boron-nitride and diamond bonded abrasives.

The manufacturing technique for producing them basically consists of (a) mixing the abrasive with liquid or powdered phenolic resins and additives, (b) pressing the mixture into the shape of a grinding wheel, and (c) curing it at temperatures of about 175°C (350°F).

Because the elastic modulus of thermosetting resins is lower than that of glasses, resinoid wheels are more flexible than vitrified wheels. A more recent development is the use of *polyimide* (Section 7.7) as a substitute for the phenolic in resinoid wheels; it is tougher and more resistant to higher temperatures. In addition to pressing them, these grinding wheels also are manufactured by *injection molding* (see Sections 17.3 and 19.3)

Reinforced wheels. These wheels typically consist of one or more layers of *fiberglass mats* of various mesh sizes. The fiberglass in this laminate structure provides reinforcement in resinoid wheels by way of retarding the disintegration of the wheel should it break for some reason during use, rather than improving its strength. Large-diameter resinoid wheels can be supported additionally with one or more internal rings made of round steel bars which are inserted during the molding of the wheel.

Thermoplastic. In addition to thermsetting resins, thermoplastic bonds also are being used in grinding wheels. Wheels are available with sol-gel abrasives bonded with thermoplastics.

Rubber.   The most flexible matrix used in abrasive wheels is rubber. The manufacturing process consists of (a) mixing crude rubber, sulfur, and the abrasive grains together, (b) rolling the mixture into sheets, (c) cutting out circles, and (d) heating them under pressure to vulcanize the rubber. Thin wheels can be made in this manner and are used like saws for cutting-off operations (*cut-off blades*).

Metal bonds.   Using powder-metallurgy techniques, the abrasive grains (usually diamond or cubic boron nitride) are bonded to the periphery of a metal wheel to depths of 6 mm (0.25 in.) or less, as shown in Fig. 26.5. Metal bonding is carried out under high pressure and temperature. The wheel itself (the core) may be made of aluminum, bronze, steel, ceramics, or composite materials—depending on the wheel requirements, such as strength, stiffness, and dimensional stability. Superabrasive wheels may be *layered* so that a single abrasive layer is plated or brazed to a metal wheel with a particular desired shape. These wheels are lower in cost and are used for small-production quantities.

26.2.3 Wheel grade and structure

The *grade* of a bonded abrasive is a measure of its bond strength, thus it includes both the *type* and the *amount* of bond in the wheel. Because strength and hardness are directly related (see Section 2.6.2), the grade also is referred to as the *hardness* of a bonded abrasive. Thus, a hard wheel has a stronger bond and/or a larger amount of bonding material between the grains than a soft wheel. The *structure* of a bonded abrasive is a measure of its *porosity* (that is, the spacing between the grains, as shown in Fig. 26.3). The structure of bonded abrasives ranges from *dense* to *open*, as shown in Fig. 26.6. As we have stated, some porosity is essential to provide clearance for the grinding chips; otherwise, they would interfere with the grinding operation.

26.3 | The Grinding Process

Grinding is a chip-removal process that uses an individual abrasive grain as the cutting tool (Fig. 26.8a). The major differences between the action of an abrasive grain and that of a single-point cutting tool can be summarized as follows:

- The individual abrasive grains have *irregular shapes* and are spaced *randomly* along the periphery of the wheel (Fig. 26.9).
- The average rake angle of the grains is highly negative, typically $-60°$ or even lower. Consequently, grinding chips undergo much larger plastic deformation than they do in other machining processes. (See Section 21.2.)
- The radial positions of the grains over the peripheral surface of a wheel vary, thus not all grains are active during grinding.
- Surface speeds (that is, cutting speeds) in grinding are very high, typically 20 to 30 m/s (4000 to 6000 ft/min), and may be as high as 150 m/s (30,000 ft/min) in high-speed grinding using specially designed and manufactured wheels.

The grinding process and its parameters can be observed best in the *surface-grinding* operation shown schematically in Fig. 26.10. A straight grinding wheel (Fig. 26.4a) with a diameter D removes a layer of metal at a depth d (**wheel depth of cut**). An individual grain on the periphery of the wheel moves at a tangential velocity, V, while the workpiece moves at a velocity, v. Each abrasive grain removes a small chip which has an *undeformed thickness* (**grain depth-of-cut**), t, and an *undeformed length, l*.

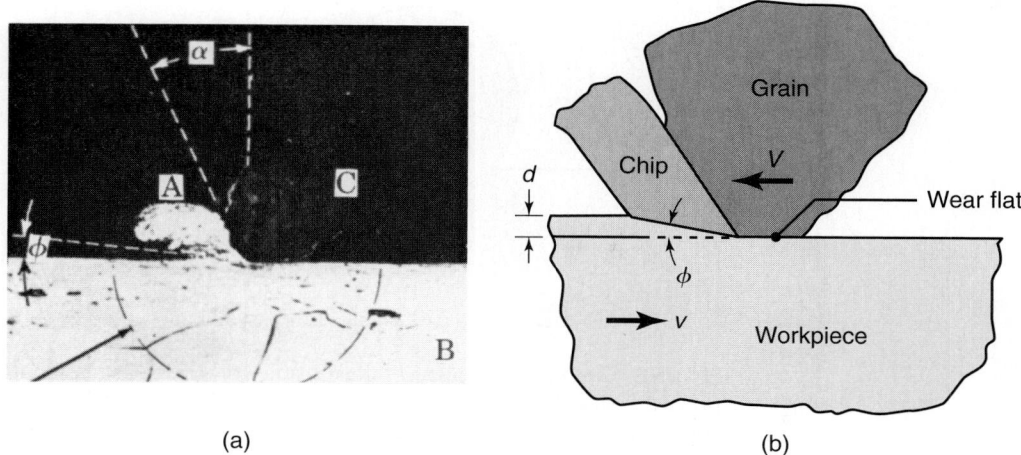

(a)   (b)

FIGURE 26.8 (a) Grinding chip being produced by a single abrasive grain: (A) chip, (B) workpiece, (C) abrasive grain. Note the large negative rake angle of the grain. The inscribed circle is 0.065 mm (0.0025 in.) in diameter. (b) Schematic illustration of chip formation by an abrasive grain with a wear flat. Note the negative rake angle of the grain and the small shear angle. *Source:* (a) After M. E. Merchant.

FIGURE 26.9 The surface of a grinding wheel (A46-J8V) showing abrasive grains, wheel porosity, wear flats on grains, and metal chips from the workpiece adhering to the grains. Note the random distribution and shape of the abrasive grains. Magnification: 50×. *Source:* S. Kalpakjian.

From geometric relationships, it can be shown that the undeformed chip length, l, in surface grinding (Fig. 26.10) is given approximately by the expression

$$l = \sqrt{Dd} \tag{26.1}$$

and the undeformed chip thickness, t, by the expression

$$t = \sqrt{\left(\frac{4v}{VCr}\right)\sqrt{\left(\frac{d}{D}\right)}} \tag{26.2}$$

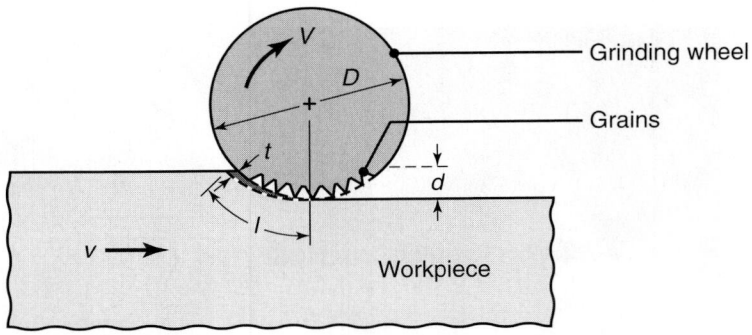

FIGURE 26.10 Schematic illustration of the surface-grinding process, showing various process variables. The figure depicts conventional (up) grinding.

where C is the number of cutting points per unit area of the periphery of the wheel; generally, it is estimated to be in the range of 0.1 to 10 per mm² (10^2 to 10^3 per in²). The quantity r is the ratio of chip width to average undeformed chip thickness and has an estimated value typically of between 10 and 20.

As an example, l and t can be calculated for the following process parameters. Let $D = 200$ mm, $d = 0.05$ mm, $v = 30$ m/min, and $V = 1800$ m/min. Using the preceding formulas,

$$l = \sqrt{(200)(0.05)} = 3.2 \text{ mm} = 0.13 \text{ in.}$$

Assuming that $C = 2$ per mm² and that $r = 15$,

$$t = \sqrt{\frac{(4)(30)}{(1800)(2)(15)}\sqrt{\frac{0.05}{200}}} = 0.006 \text{ mm} = 0.00023 \text{ in.}$$

Because of plastic deformation, the actual chip will be shorter and thicker than the values calculated. (See Fig. 26.8.) It can be seen that grinding chips are much smaller than those typically obtained in metal-cutting operations, as described in Chapter 21.

Grinding forces.   A knowledge of grinding forces is essential for:

- Estimating power requirements.
- Designing grinding machines and workholding fixtures and devices.
- Determining the deflections that the workpiece as well as the grinding machine itself may undergo. Note that deflections, in turn, adversely affect dimensional accuracy and are especially critical in precision and ultraprecision grinding.

If it is assumed that the cutting force on the grain is proportional to the cross-sectional area of the undeformed chip, it can be shown that the **grain force** (which is tangential to the wheel) is proportional to the process variables.

$$\text{Grain force} \propto \left(\frac{v}{V}\sqrt{\frac{d}{D}}\right)(\text{strength of the material}) \tag{26.3}$$

Forces in grinding are usually much smaller than those in the machining operations described in Chapters 23 and 24 because of the small dimensions involved. Grinding forces should be kept low in order to avoid distortion and to maintain the high dimensional accuracy of the workpiece.

Specific energy. The *energy* dissipated in producing a grinding chip consists of the energy required for the following actions:

- Chip formation.
- Plowing, as shown by the ridges formed in Fig. 26.11.
- Friction, caused by rubbing of the grain along the workpiece surface.

Note that after some use, the grains along the periphery of the wheel develop a **wear flat** (Fig. 26.8b), a phenomenon that is similar to *flank wear* in cutting tools (Fig. 21.15). The wear flat continuously rubs along the ground surface, dissipates energy (because of friction), and makes the grinding operation less efficient.

Specific-energy requirements in grinding are defined as the energy per unit volume of material ground from the workpiece surface and are shown in Table 26.2. Note that these specific-energy levels are much higher than those in the machining operations given in Table 21.2. This difference has been attributed to factors such as the presence of a wear flat, high negative rake angles of the grains (which require more energy), and possible contribution of the size effect (the smaller the chip, the higher the energy required to produce it). Also, it has been observed that with effective lubrication, specific energies in grinding can be reduced by a factor of four or more.

As shown in Example 26.1, the *grinding force* and the *thrust force* in grinding can be calculated from the specific-energy data.

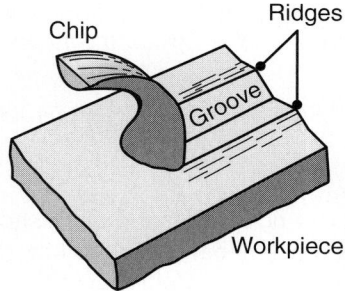

FIGURE 26.11 Chip formation and plowing of the workpiece surface by an abrasive grain.

TABLE 26.2

Approximate Specific-Energy Requirements for Surface Grinding

Workpiece material	Hardness	Specific energy	
		$W \cdot s/mm^3$	$hp \cdot min/in^3$
Aluminum	150 HB	7–27	2.5–10
Cast iron (class 40)	215 HB	12–60	4.5–22
Low-carbon steel (1020)	110 HB	14–68	5–25
Titanium alloy	300 HB	16–55	6–20
Tool steel (T15)	67 HRC	18–82	6.5–30

EXAMPLE 26.1 Forces in surface grinding

A surface-grinding operation is being performed on low-carbon steel with a wheel of diameter $D = 10$ in. that is rotating at $N = 4000$ rpm and a width-of-cut of $w = 1$ in. The depth-of-cut is $d = 0.002$ in. and the feed rate of the workpiece, v, is 60 in./min. Calculate the cutting force (force tangential to the wheel), F_c, and the thrust force (normal to the workpiece surface), F_n.

Solution The material-removal rate (MRR) first is determined as

$$\text{MRR} = dwv = (0.002)(1)(60) = 0.12 \text{ in}^3/\text{min}$$

The power consumed is given by

$$\text{Power} = (u)(\text{MRR})$$

where u is the specific energy, as obtained from Table 26.2. (See also Section 21.3.) For low-carbon steel, it is estimated to be 15 hp · min/in³. Thus,

$$\text{Power} = (15)(0.12) = 1.8 \text{ hp}$$

By noting that 1 hp = 33,000 ft · lb/min = 396,000 in. · lb/min,

$$\text{Power} = (1.8)(396,000) = 712,800 \text{ in.} \cdot \text{lb/min}$$

Since power is defined as

$$\text{Power} = T\omega$$

where T is the torque and is equal to $T = F_c D/2$ and ω is the rotational speed of the wheel in radians per minute ($\omega = 2\pi N$). It follows that

$$712,800 = (F_c)\left(\frac{10}{2}\right)(2\pi)\,(4000)$$

and therefore, $F_c = 5.7$ lb.

The thrust force, F_n, can be calculated directly; however, it also can be estimated by noting from experimental data in the technical literature that it is about 30% higher than the cutting force, F_c. Consequently,

$$F_n = (1.3)(5.7) = 7.4 \text{ lb}$$

Temperature. The temperature rise in grinding is an important consideration because:

- It can adversely affect the surface properties, including metallurgical changes.
- The temperature rise can cause residual stresses on the workpiece.
- Temperature gradients in the workpiece cause distortions due to thermal expansion and contraction of the workpiece surface, thus making it difficult to control dimensional accuracy.

The *surface temperature rise* in grinding has been found to be related to process variables by the following expression:

$$\text{Temperature rise} \propto D^{1/4} d^{3/4} \left(\frac{V}{v} \right)^{1/2} \tag{26.4}$$

Thus, temperature increases with increasing depth of cut, d, wheel diameter, D, and wheel speed, V, and decreases with increasing workpiece speed, v. Note from this equation that the depth of cut has the largest exponent, hence it has the greatest influence on temperature.

Peak temperatures during grinding can reach 1600°C (3000°F). However, the time involved in producing a chip is extremely short (microseconds), so the chip produced may or may not melt. Because the chips carry away much of the heat generated (as do chips formed in high-speed machining processes, Section 25.5), only a fraction of the heat produced in grinding is conducted to the workpiece. If this was not the case, it would be very difficult to grind workpieces with sufficient dimensional accuracy.

Sparks. The sparks produced when grinding metals are actually chips that glow, due to the *exothermic* (heat producing) reaction of the hot chips with oxygen in the atmosphere. Sparks do not occur when grinding in an oxygen-free environment. The color, intensity, and shape of the sparks depend on the composition of the metal being ground. Charts are available that help identify the type of metal being ground from the appearance of its sparks. If the heat generated due to exothermic reaction is sufficiently high, chips can melt, acquire a spherical shape (because of surface tension), and solidify as metal particles.

Tempering. Excessive temperature rise in grinding can cause *tempering* and *softening* of the workpiece surface. Process variables must be selected carefully in order to avoid excessive temperature rise. The use of grinding fluids is an effective means of controlling temperature.

Burning. Excessive temperature during grinding may burn the surface being ground. A *burn* is characterized by a bluish color on ground steel surfaces—an indication that high temperature caused oxidation. It can be detected by etching and metallurgical techniques. A burn may not be objectionable in itself unless surface layers have undergone *phase transformations* (Chapter 4). For example, martensite forming in higher-carbon steels from rapid cooling is called a *metallurgical burn*. This condition will adversely influence the surface properties of ground parts, reducing surface ductility and toughness.

Heat checking. High temperatures in grinding may cause the workpiece surface to develop cracks; this condition is known as *heat checking*. The cracks usually are perpendicular to the grinding direction. Under severe conditions, however, parallel cracks also may appear. As expected, such a surface lacks toughness and has low fatigue and corrosion resistance. (Note that heat checking also occurs in dies for die casting, as described in Section 11.3.5.)

Residual stresses. Temperature gradients within the workpiece during grinding are primarily responsible for *residual stresses*. Grinding fluids, their method of application, as well as process parameters such as depth of cut and speeds significantly influence the magnitude and type of residual stresses developed (tension or compression). Because of the adverse effect of tensile residual stresses on fatigue strength, process variables should be selected carefully. Residual stresses usually can be reduced by lowering wheel speed and increasing workpiece speed (**low-stress grinding** or *gentle grinding*). Softer-grade wheels (known as **free-cutting wheels**) also may be used.

26.3.1 Grinding-wheel wear

Grinding-wheel wear is an important consideration because it adversely affects the shape and dimensional accuracy of ground surfaces—similar to the wear on cutting tools. Grinding-wheel wear is caused by three different mechanisms: attritious grain wear, grain fracture, and bond fracture.

Attritious grain wear.   In *attritious wear*, the cutting edges of an originally sharp grain become dull and develop a *wear flat* (Fig. 26.8b) that is similar to flank wear in cutting tools. Wear is caused by the interaction of the grain with the workpiece material, involving both physical and chemical reactions. These reactions are complex and involve diffusion, chemical degradation or decomposition of the grain, fracture at a microscopic scale, plastic deformation, and melting.

Attritious wear is low when the two materials (grain and workpiece) are *chemically inert* with respect to each other, much like what has been observed with cutting tools. The more inert the materials, the lower the tendency for reaction and adhesion to occur between the grain and the workpiece. For example, because aluminum oxide is relatively inert with respect to iron, its rate of attritious wear when used to grind steels is much lower than that of silicon carbide and diamond. On the other hand, silicon carbide can dissolve in iron, so it is not suitable for grinding steels. Cubic-boron nitride has a higher inertness with respect to steels, hence it is suitable for use as an abrasive.

Therefore, the selection of the type of abrasive for low attritious wear is based on the *reactivity* of the grain with the workpiece and on their relative mechanical properties, such as hardness and toughness. The environment and the type of grinding fluid used also have an influence on grain–workpiece interactions.

Grain fracture.   Because abrasive grains are brittle, their fracture characteristics in grinding are important. If the wear flat caused by attritious wear is excessive, the grain becomes dull and grinding becomes inefficient and produces undesirably high temperatures.

Ideally, the grain should fracture or fragment at a moderate rate so that new sharp cutting edges are produced continuously during grinding. This is equivalent to breaking a dull piece of chalk or a stone into two or more pieces in order to expose new sharp edges. Section 26.2 described the *friability* of abrasives (the extent to which they are self-sharpening) as an important factor in effective grinding.

The selection of grain type and size for a particular application also depends on the attritious-wear rate. A grain–workpiece material combination with high attritious wear and low grain friability dulls grains and develops a large wear flat. Grinding then becomes inefficient, and surface damage (such as burning) is likely to occur.

Bond fracture.   The strength of the bond (grade) is a significant parameter in grinding. If the bond is too strong, dull grains cannot be dislodged. This prevents other sharp grains along the circumference of the grinding wheel from contacting the workpiece to remove chips, and the grinding process becomes inefficient. On the other hand, if the bond is too weak, the grains are dislodged easily, and the wear rate of the wheel increases. In this case, maintaining dimensional accuracy becomes difficult.

In general, softer bonds are recommended for harder materials and for reducing residual stresses and thermal damage to the workpiece. Hard-grade wheels are used for softer materials and for removing large amounts of material at high rates.

26.3.2 Grinding ratio

Grinding-wheel wear generally is correlated with the amount of workpiece material ground by a parameter called the *grinding ratio, G,* and defined as

$$G = \frac{\text{Volume of material removed}}{\text{Volume of wheel wear}} \tag{26.5}$$

Grinding ratios in practice vary widely, ranging from 2 to 200 and even higher, depending on the type of wheel, workpiece material, grinding fluid, and process parameters (such as depth of cut and speeds of wheel and workpiece). It has been shown that effective grinding fluids can increase the grinding ratio by a factor of ten or more, thus greatly improving wheel life.

During grinding, a particular wheel may **act soft** (wheel wear is high) or **act hard** (wear is low), regardless of the wheel grade. Note, for example, that an ordinary pencil *acts* soft when writing on rough paper and acts hard when writing on soft paper—even though it is the same pencil. Acting hard or soft is a function of the force on the individual grain on the periphery of the wheel. The higher the force, the greater the tendency for the grains to fracture or be dislodged from the wheel surface, the higher the wheel wear, and the lower the grinding ratio. From Eq. (26.3), note that the grain force increases with the strength of the workpiece material, work speed, and depth of cut and decreases with increasing wheel speed and wheel diameter. Thus, a grinding wheel acts soft when v and d increase or when V and D decrease.

It should be noted that attempting to obtain a high grinding ratio in practice (so as to extend wheel life) isn't always desirable, because high ratios may indicate grain dulling and, thus, possible surface damage of the workpiece. A lower ratio may be acceptable when an overall technical and economic analysis justifies it.

EXAMPLE 26.2 Action of a grinding wheel

A surface-grinding operation is being carried out with the wheel running at a constant spindle speed. Will the wheel act soft or hard as the wheel wears down over a period of time? Assume that the depth of cut, d, remains constant and the wheel is dressed periodically (see Section 26.3.3).

Solution   Referring to Eq. (26.3), note that the parameters that change over time in this operation are the wheel diameter, D, and its surface speed, V. As D becomes smaller, the relative grain force increases, thus the wheel acts softer. To accommodate the changes due to the reduction of wheel diameter over time or to make provisions for using wheels of different diameters, some grinding machines are equipped with variable-speed spindle motors.

26.3.3 Dressing, truing, and shaping of grinding wheels

Dressing is the process of

- *Conditioning* worn grains on the surface of a grinding wheel by producing *sharp new edges* on grains so that they cut more effectively.

- *Truing*, which is producing a true circle on a wheel that has become out of round.

Dressing is necessary when excessive attritious wear dulls the wheel (called **glazing** because of the shiny appearance of the wheel surface) or when the wheel becomes loaded. For softer wheels, truing and dressing are done separately, but for harder wheels (such as cBN), both are done in one operation.

Loading of a grinding wheel occurs when the porosities on the wheel surfaces (Fig. 26.9) become filled or clogged with chips from the workpiece. Loading can occur in grinding soft materials or by improper selection of wheels or process parameters. A loaded wheel cuts inefficiently and generates much frictional heat, which results in surface damage and loss of dimensional accuracy of the workpiece.

The techniques used to dress grinding wheels are:

- A specially shaped *diamond-point tool* or *diamond cluster* is moved across the width of the grinding face of a rotating wheel and removes a small layer from the wheel surface with each pass. This method can be performed either dry or wet, depending on whether the wheel is to be used dry or wet, respectively. In practice, however, the wear of the diamond can be significant with harder wheels, in which case a diamond disk or cup wheel can be used.

- A set of *star-shaped steel disks* is pressed manually against the wheel. Material is removed from the wheel surface by crushing the grains. As a result, this method produces a coarse surface on the wheel and is used only for rough grinding operations on bench or pedestal grinders.

- *Abrasive sticks* may be used to dress grinding wheels, particularly softer wheels. However, this technique is not appropriate for precision grinding operations.

- Dressing techniques for metal-bonded diamond wheels involve the use of *electrical-discharge* and *electrochemical machining* techniques, as will be described in Chapter 27. These processes erode away very thin layers of the metal bond and, thus, expose new diamond cutting edges.

- Dressing for form grinding involves *crush dressing* or *crush forming*. The process consists of pressing a metal roll on the surface of the grinding wheel, which typically is a vitrified wheel. The roll (which usually is made of high-speed steel, tungsten carbide, or boron carbide) has a machined or ground profile (form) on its periphery. Thus, it reproduces a replica of this profile on the surface of the grinding wheel being dressed. (See Section 26.4.)

Dressing techniques and their frequency are important for quality, because they affect grinding forces and workpiece surface finish. Modern computer-controlled grinders are equipped with automatic dressing features, which dress the wheel continually as grinding progresses. The first contact of the dressing tool on the grinding wheel is very important, as it determines the nature of the new surface produced. This action usually is monitored precisely by using piezoelectric or acoustic-emission sensors (Section 38.8). Furthermore, features such as vibration sensors, power monitors, and strain gauges also are used in the dressing setup of high-precision grinding machines.

For a typical aluminum-oxide wheel, the depth removed during dressing is on the order of 5 to 15 μm (200 to 600 μin.), but for a cBN wheel, it would be 2 to 10 μm (80 to 400 μin.). Consequently, modern dressing systems have a resolution as low as 0.25 to 1 μm (10 to 40 μin.).

Grinding wheels can be **shaped** to the form to be ground on the workpiece (Section 26.4). The grinding face on the Type 1 straight wheel shown in Fig. 26.4a is cylindrical; therefore, it produces a flat surface. However, this surface can be shaped into various forms by dressing it (Fig. 26.12a). Although templates have been used for this purpose, modern grinders commonly are equipped with computer-controlled shaping features. Unless it already has the desired form, the diamond dressing tool traverses the wheel face automatically along a certain prescribed path (Fig. 26.12b) and produces very accurate surfaces. Note in Fig. 26.12b that the axis of the diamond dressing tool remains normal to the grinding-wheel face at the point of contact.

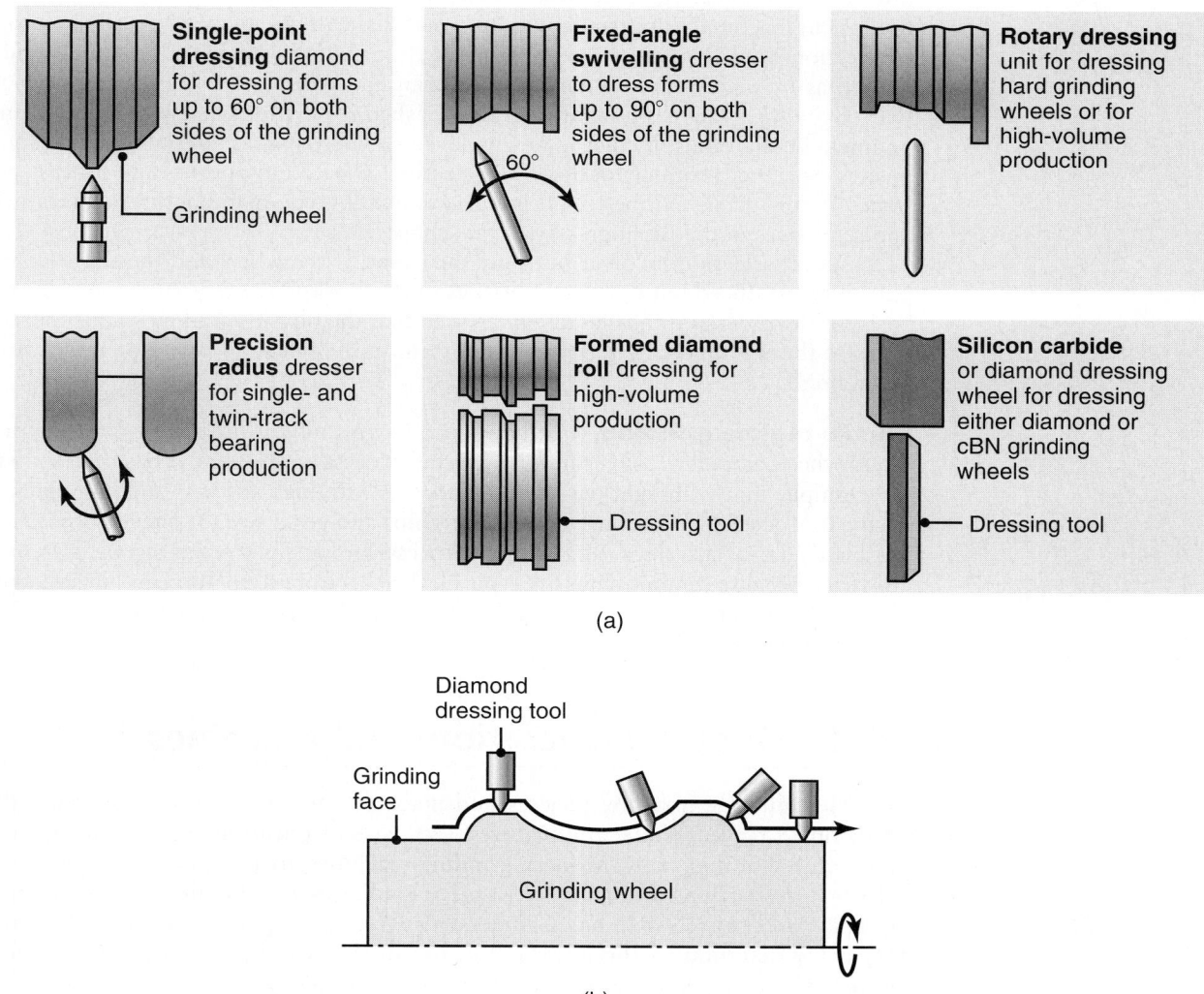

Single-point dressing diamond for dressing forms up to 60° on both sides of the grinding wheel

Grinding wheel

Fixed-angle swivelling dresser to dress forms up to 90° on both sides of the grinding wheel

60°

Rotary dressing unit for dressing hard grinding wheels or for high-volume production

Precision radius dresser for single- and twin-track bearing production

Formed diamond roll dressing for high-volume production

Dressing tool

Silicon carbide or diamond dressing wheel for dressing either diamond or cBN grinding wheels

Dressing tool

(a)

Diamond dressing tool

Grinding face

Grinding wheel

(b)

FIGURE 26.12   (a) Forms of grinding-wheel dressing. (b) Shaping the grinding face of a wheel by dressing it by computer control. Note that the diamond dressing tool is normal to the surface at point of contact with the wheel. *Source*: Courtesy of Okuma Machinery Works Ltd.

TABLE 26.3

Typical Ranges of Speeds and Feeds for Abrasive Processes				
Process variable	Grinding, conventional	Grinding, creep-feed	Polishing	Buffing
Wheel speed (m/min)	1500–3000	1500–3000	1500–2400	1800–3500
Work speed (m/min)	10–60	0.1–1	—	—
Feed (mm/pass)	0.01–0.05	1–6	—	—

26.3.4 Grindability of materials and wheel selection

The *grindability* of materials is (like *machinability* described in Section 21.7) difficult to define precisely. Grindability is a general indicator of how easy it is to grind a material and includes considerations such as the quality of the surface produced, surface finish, surface integrity, wheel wear, cycle time, and overall economics of the operation. As in machinability, grindability of a material can be enhanced greatly by proper selection of process parameters (Table 26.3), grinding wheels and grinding fluids, as well as by machine characteristics, fixturing methods, and workholding devices.

Grinding practices now are well established for a wide variety of metallic and nonmetallic materials, including newly developed aerospace materials such as composites. Specific recommendations for selecting wheels and appropriate process parameters for metals can be found in various handbooks, manufacturers' literature, and references in the Bibliography of this chapter.

Wheel selection involves not only the shape of the wheel and the shape of the part to be produced but the characteristics of the workpiece material as well. On the basis of the discussion thus far, it can be seen that the physical and mechanical properties of the workpiece material are important in the selection of a type of abrasive and a bond.

Ductile-regime grinding. Ceramics can be ground with relative ease using diamond wheels as well as using carefully selected process parameters. It has been shown, for example, that with light passes and machines with high stiffness and damping capacity, it is possible to produce continuous chips and good surface integrity by grinding brittle materials (Fig. 26.11) in a process known as *ductile-regime grinding*. However, because ceramic chips are typically 1 to 10 μm (40 to 400 μin.) in size, they are more difficult to remove from grinding fluids than metal chips and require fine filters and special techniques.

26.4 | Grinding Operations and Machines

The selection of a grinding process and machine for a particular application depends on workpiece shape and features, size, ease of fixturing, and the production rate required (Table 26.4). Modern grinding machines are computer controlled and have features such as automatic workpiece loading and unloading, part clamping, dressing, and wheel shaping. Grinders also can be equipped with probes and gauges for determining the relative position of the wheel and workpiece surfaces (see also Fig. 25.6) as well as with tactile-sensing features, whereby diamond dressing-tool breakage (for example) can be detected readily during the dressing cycle. The basic types of grinding operations are described in this section.

The relative movement of the wheel may be along the surface of the workpiece (*traverse* grinding, *through-feed* grinding, *cross-feeding*), or it may move radially into

TABLE 26.4

General Characteristics of Abrasive Machining Processes and Machines

Process	Characteristics	Typical maximum dimensions, length and diameter (m)*
Surface	Flat surfaces on most materials; production rate depends on table size and level of automation; labor skill depends on part complexity; production rate is high on vertical-spindle rotary-table machines	Reciprocating table L: 6 Rotary table D: 3
Cylindrical	Round workpieces with stepped diameters; low production rate unless automated; low to medium labor skill	Workpiece D: 0.8, roll grinders D: 1.8, universal grinders D: 2.5
Centerless	Round and slender workpieces; high production rate; low to medium labor skill	Workpiece D: 0.8
Internal	Holes in workpiece; low production rate; low to medium labor skill	Hole D: 2
Honing	Holes in workpiece; low production rate; low labor skill	Spindle D: 1.2
Lapping	Flat, cylindrical, or curved; high production rate; low labor skill	Table D: 3.7
Ultrasonic machining	Holes and cavities with various shapes; suitable for hard and brittle materials; medium labor skill	—

*Larger capacities are available for special applications.

the workpiece (*plunge* grinding). Surface grinders comprise the largest percentage of grinders used in industry followed by bench grinders (usually with two wheels at each end of the spindle), cylindrical grinders, and tool and cutter grinders—the least common being internal grinders.

Surface grinding.   *Surface grinding* is one of the most common operations (Fig. 26.13), generally involving the grinding of flat surfaces. Typically, the workpiece is secured on a *magnetic chuck* attached to the work table of the *grinder* (Fig. 26.14); nonmagnetic materials are held by vises, vacuum chucks, or some other fixture. A straight wheel is mounted on the horizontal spindle of the surface grinder. Traverse grinding occurs as the table reciprocates longitudinally and is fed laterally (in the direction of the spindle axis) after each stroke. In *plunge grinding*, the wheel is moved radially *into* the workpiece, as it is when grinding a groove (Fig. 26.13b).

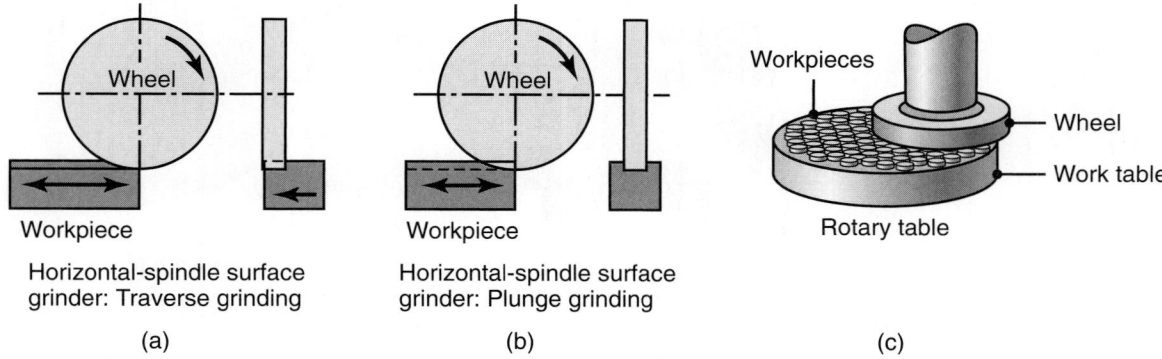

FIGURE 26.13   Schematic illustrations of various surface-grinding operations. (a) Traverse grinding with a horizontal-spindle surface grinder. (b) Plunge grinding with a horizontal-spindle surface grinder, producing a groove in the workpiece. (c) A vertical-spindle rotary-table grinder (also known as the *Blanchard* type).

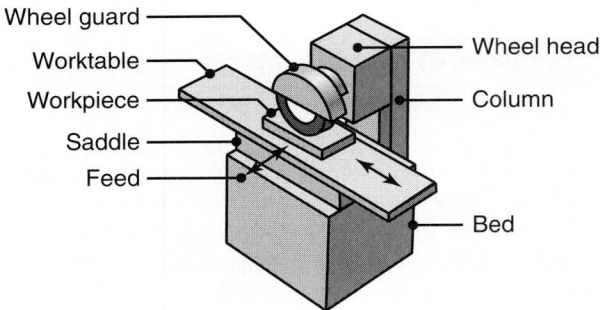

FIGURE 26.14   Schematic illustration of a horizontal-spindle surface grinder.

In addition to the surface grinder shown in Fig. 26.14, other types include *vertical spindles* and *rotary tables* (referred to as the *Blanchard* type, Fig. 26.13c). These configurations allow a number of pieces to be ground in one setup. Steel balls for ball bearings, for example, are ground in special setups and at high production rates (Fig. 26.15).

Cylindrical grinding.   In *cylindrical grinding* (also called *center-type grinding*, Fig. 26.16; see also Fig. 26.2), the external cylindrical surfaces and shoulders of the workpiece are ground, such as for crankshaft bearings, spindles, pins, and bearing rings. The rotating cylindrical workpiece reciprocates laterally along its axis to cover the width to be ground. In *roll grinders* used for large and long workpieces such as rolls for rolling mills (see Fig. 13.1), the grinding wheel reciprocates. These machines are capable of grinding rolls as large as 1.8 m (72 in.) in diameter.

The workpiece in cylindrical grinding is held between centers or in a chuck, or it is mounted on a faceplate in the headstock of the grinder. For straight cylindrical

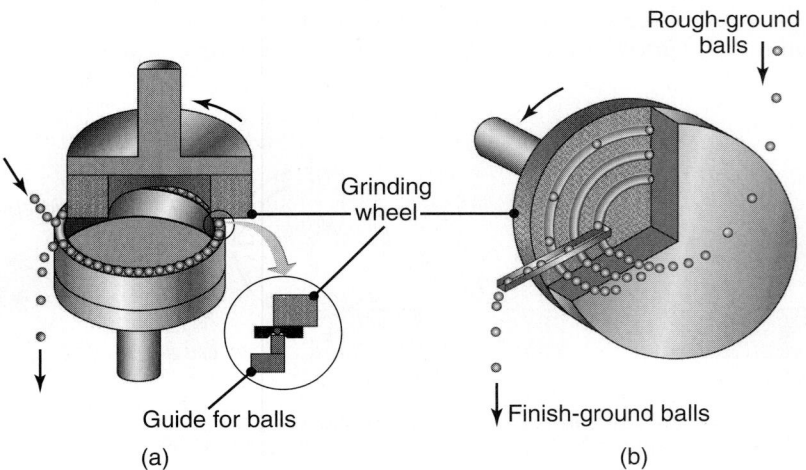

(a)   (b)

FIGURE 26.15   (a) Rough grinding of steel balls on a vertical-spindle grinder. The balls are guided by a special rotary fixture. (b) Finish grinding of balls in a multiple-groove fixture. The balls are ground to within 0.013 mm (0.0005 in.) of their final size.

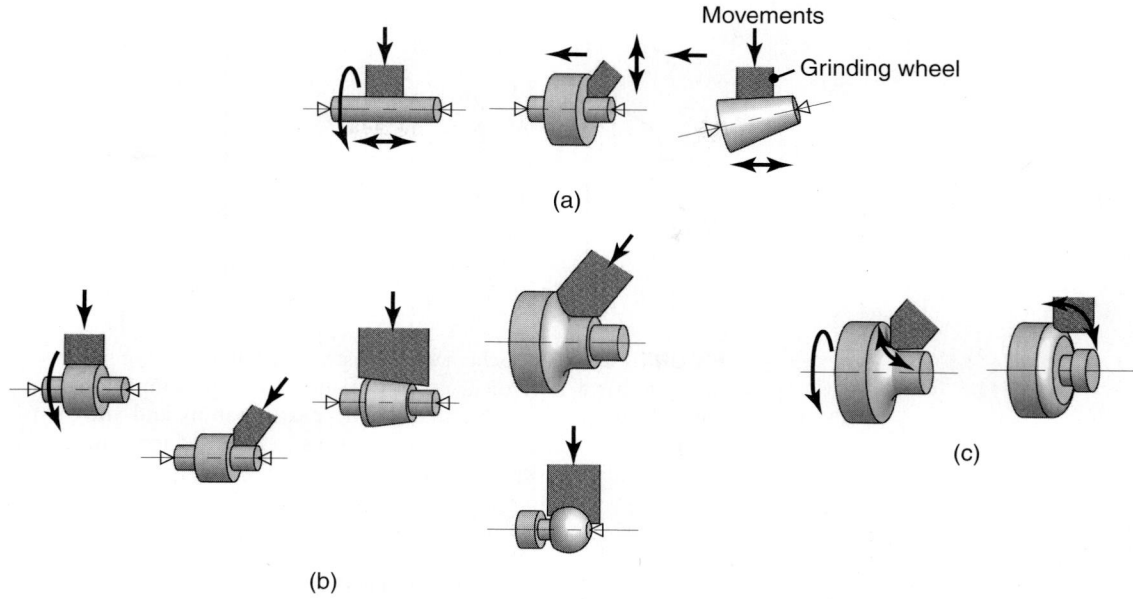

FIGURE 26.16   Examples of various cylindrical-grinding operations: (a) traverse grinding, (b) plunge grinding, and (c) profile grinding. *Source*: Courtesy of Okuma Machinery Works Ltd.

surfaces, the axes of rotation of the wheel and workpiece are parallel. The wheel and workpiece are each driven by separate motors and at different speeds. Long workpieces with two or more diameters also can be ground on cylindrical grinders. Cylindrical grinding also can produce shapes (*form grinding* and *plunge grinding*) in which the wheel is dressed to the workpiece form to be ground (Fig. 26.17).

Cylindrical grinders are identified by the maximum diameter and length of the workpiece that can be ground—similar to engine lathes. In *universal grinders*, both the workpiece and the wheel axes can be moved and swiveled around a horizontal plane, thus permitting the grinding of tapers and other shapes.

With computer-control features, *noncylindrical* parts such as cams can be ground on rotating workpieces. As Fig. 26.18 illustrates, the workpiece spindle speed is synchronized such that the radial distance between the workpiece and the wheel axes is varied continuously to produce a particular shape, such as the one shown.

Thread grinding is done on cylindrical grinders using specially dressed wheels matching the shape of the threads, as shown in Fig. 26.19. (See also *centerless grinding*.)

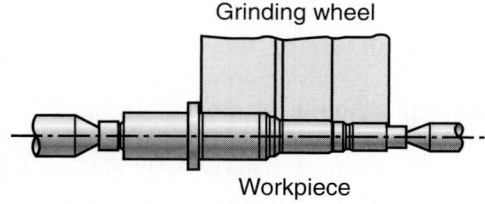

FIGURE 26.17   Plunge grinding of a workpiece on a cylindrical grinder with the wheel dressed to a stepped shape.

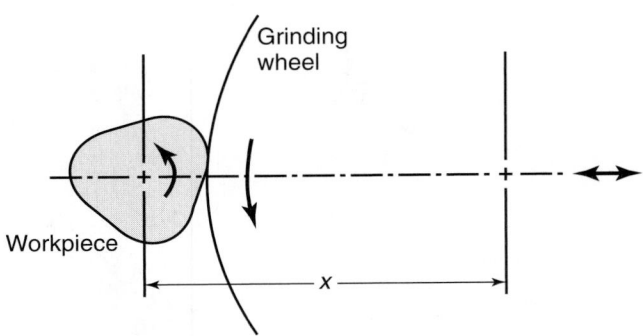

FIGURE 26.18 Schematic illustration of grinding a noncylindrical part on a cylindrical grinder with computer controls to produce the shape. The part rotation and the distance x between centers is varied and synchronized to grind the particular workpiece shape.

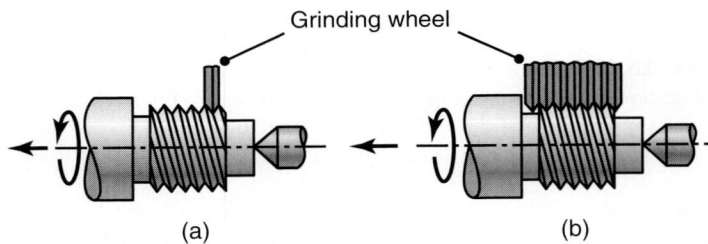

(a)   (b)

FIGURE 26.19 Thread grinding by (a) traverse and (b) plunge grinding.

Although expensive, threads produced by grinding are the most accurate of any manufacturing process and have a very fine surface finish with applications such as ball-screw mechanisms used for precise movement of machine components. The workpiece and wheel movements are synchronized to produce the pitch of the thread, usually in about six passes.

EXAMPLE 26.3 Cycle patterns in cylindrical grinding

As in most grinding operations, the grinding wheel in cylindrical grinding typically makes several passes along a path in order to produce the final geometry on the workpiece. Figure 26.20 illustrates the cycle patterns for producing various shapes on a multifunctional, computer-controlled precision grinder. The downward arrowheads with numbers indicate the beginning of the grinding cycle.

The determination of the optimum and most economical pattern for minimum cycle time depends on the amount of material to be removed, part shape, and the process parameters chosen. These patterns are generated automatically by the software in the computer controls of the grinder.

Source: Courtesy of Toyoda Machinery.

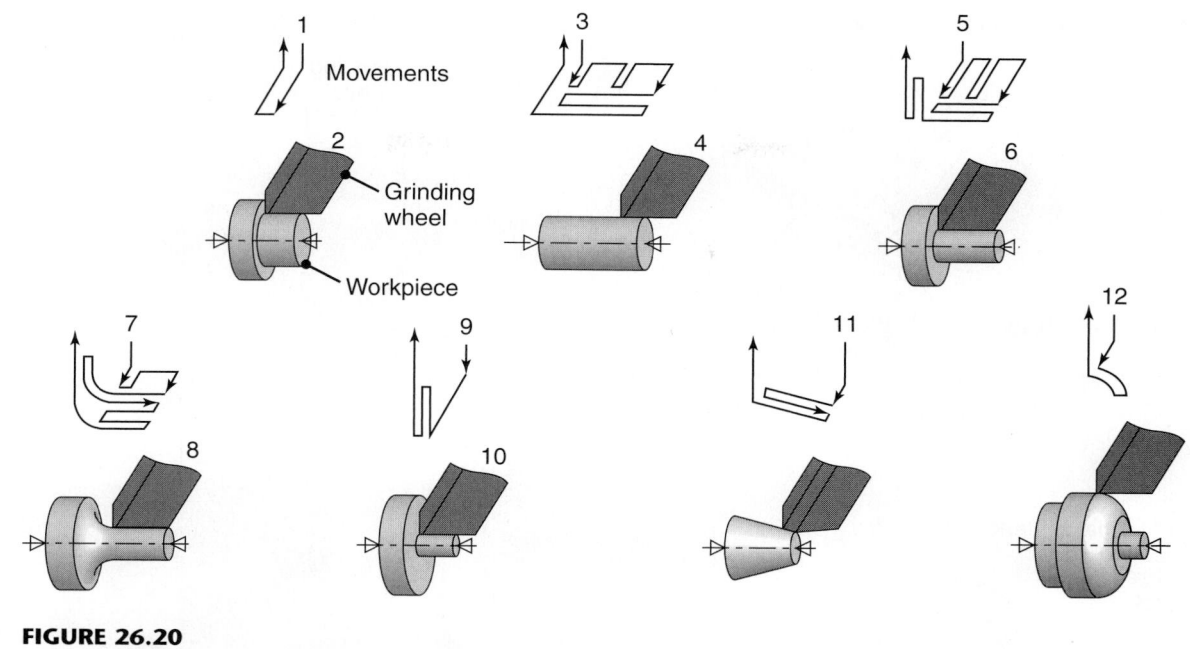

FIGURE 26.20

Internal grinding. In *internal grinding* (Fig. 26.21), a small wheel is used to grind the inside diameter of the part, such as in bushings and bearing races. The workpiece is held in a rotating chuck and the wheel rotates at 30,000 rpm or higher. Internal profiles also can be ground with profile-dressed wheels that move radially into the workpiece. The headstock of internal grinders can be swiveled on a horizontal plane to grind tapered holes.

Centerless grinding. *Centerless grinding* is a high-production process for continuously grinding cylindrical surfaces in which the workpiece is supported not by centers (hence the term "centerless") or chucks but by a *blade* (Figs. 26.22a and b). Typical parts made by centerless grinding are roller bearings, piston pins, engine valves, camshafts, and similar components. Parts with diameters as small as 0.1 mm (0.004 in.) can be ground. Centerless grinders (Fig. 26.22d) are capable of wheel surface speeds on the order of 10,000 m/min (35,000 ft/min), typically using cubic-boron-nitride abrasive wheels.

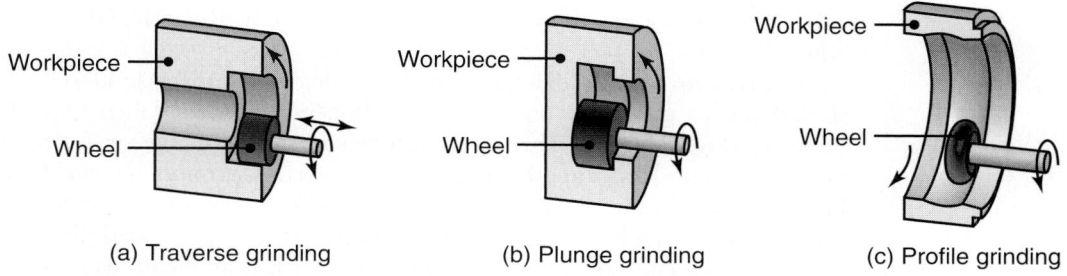

(a) Traverse grinding   (b) Plunge grinding   (c) Profile grinding

FIGURE 26.21 Schematic illustrations of internal grinding operations: (a) traverse grinding, (b) plunge grinding, and (c) profile grinding.

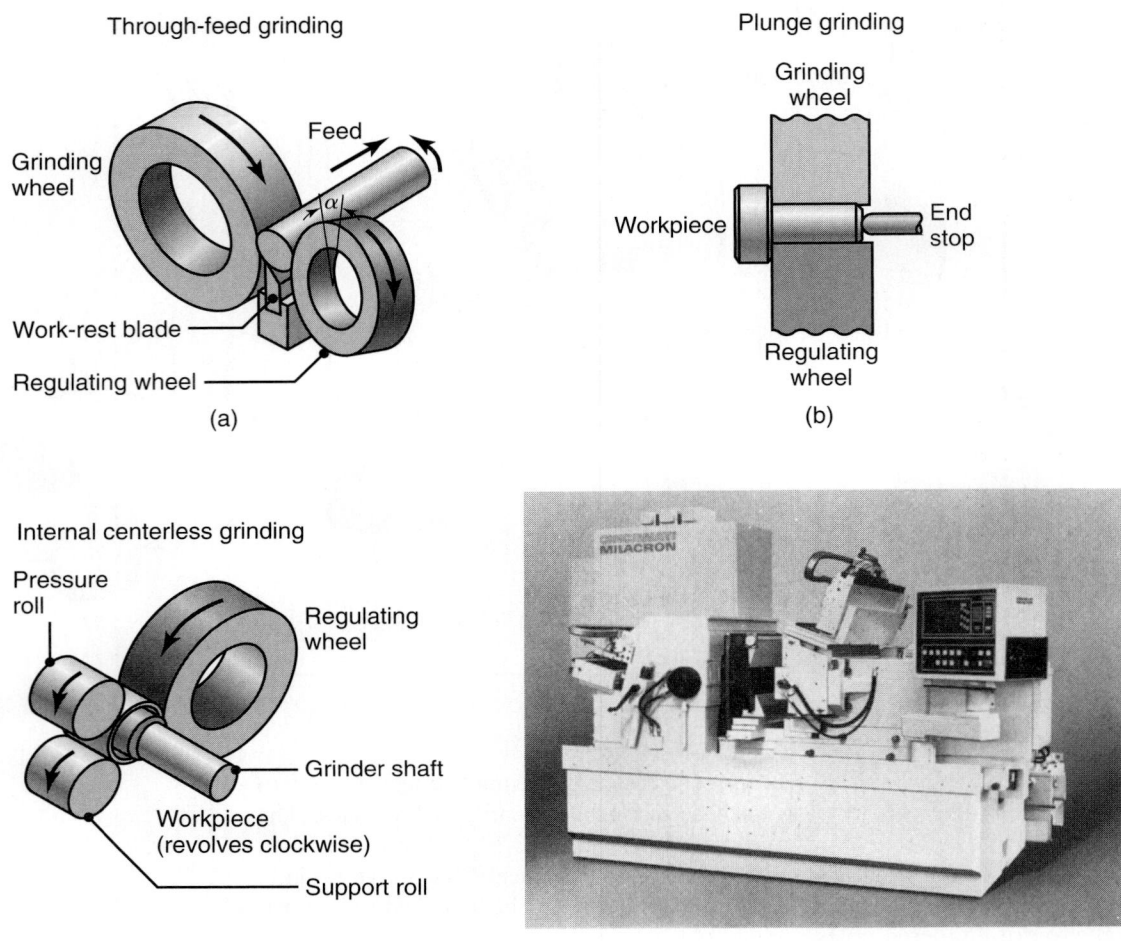

Through-feed grinding

Plunge grinding

Internal centerless grinding

FIGURE 26.22 Schematic illustrations of centerless grinding operations: (a) through-feed grinding, (b) plunge grinding, (c) internal grinding, and (d) a computer numerical-control cylindrical-grinding machine. *Source*: Courtesy of Cincinnati Milacron, Inc.

In *through-feed grinding*, the workpiece is supported on a work-rest blade and is ground between two wheels. Grinding is done by the larger wheel, while the smaller wheel regulates the axial movement of the workpiece. The *regulating wheel* (which is rubber bonded) is tilted and runs at a speed of only about 1/20th of the grinding-wheel speed.

Parts with variable diameters (such as bolts, valve tappets, and multiple-diameter shafts) can be ground by centerless grinding in a process called *in-feed* or *plunge grinding* (Fig. 26.22b). This operation is similar to plunge or form grinding on cylindrical grinders. Tapered pieces are centerless ground by *end-feed grinding*. High-production-rate thread grinding also can be done with centerless grinders using specially dressed wheels. In *internal centerless grinding*, the workpiece is supported between three rolls and is ground internally; typical applications are sleeve-shaped parts and rings (Fig. 26.22c).

Creep-feed grinding.   Grinding traditionally has been associated with small rates of material removal (Table 26.4) and fine finishing operations. However, grinding can be used also for large-scale metal-removal operations similar to and competing with milling, broaching, and planing. In *creep-feed grinding*, the wheel depth of cut, d, is as much as 6 mm (0.25 in.), and the workpiece speed is low (Fig. 26.23). To keep workpiece temperatures low and improve surface finish, the wheels are softer-grade resin bonded and have an open structure.

The machines used for creep-feed grinding have special features, such as power-up to 225 kW (300 hp), high stiffness (because of the high forces due to the large depth of material removed), high damping capacity, variable spindle and work-table speeds, and ample capacity for grinding fluids. Grinders are equipped with features for continuously dressing the wheel using a diamond roll as the dressing tool.

Creep-feed grinding can be competitive with other machining processes and economical for specific applications, such as grinding shaped punches, key seats, twist-drill flutes, the roots of turbine blades (Fig. 26.23c), and various complex, superalloy parts. Because the wheel is dressed to the shape of the workpiece to be produced, the workpiece does not have to be previously shaped by milling, shaping, or broaching. Near-net-shape castings and forgings thus are suitable for creep-feed grinding. Although a single grinding pass generally is sufficient, a second pass may be necessary for improved surface finish.

Heavy stock removal by grinding.   Grinding can be used also for heavy stock removal by increasing process parameters. This operation can be economical in specific applications and can compete favorably with machining processes, particularly milling, turning, and broaching. In this operation, surface finish is of secondary importance and the grinding wheel (or belt) is utilized to its fullest while minimizing cost-per-piece. The dimensional tolerances in this process are on the same order as those obtained by most machining processes. Heavy stock removal by grinding is performed also on welds, castings, and forgings to smoothen weld beads and remove flash.

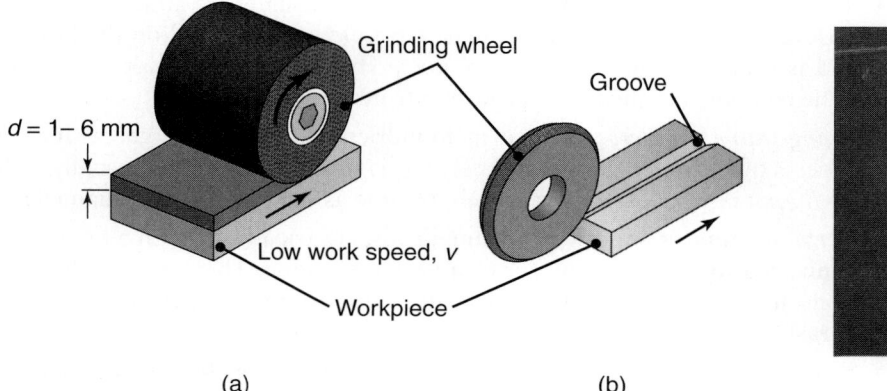
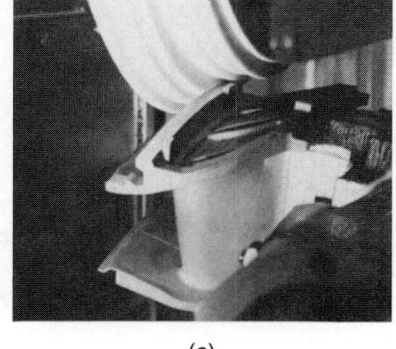

FIGURE 26.23   (a) Schematic illustration of the creep-feed grinding process. Note the large wheel depth-of-cut, d. (b) A shaped groove produced on a flat surface by creep-feed grinding in one pass. Groove depth is typically on the order of a few mm. (c) An example of creep-feed grinding with a shaped wheel. This operation also can be performed by some of the processes described in Chapter 27. *Source*: Courtesy of Blohm, Inc.

EXAMPLE 26.4 Grinding versus hard turning

In Section 25.6, *hard turning* was described. An example of this is the machining of heat-treated steels (usually above 45 HRC) using a single-point polycrystalline cubic-boron-nitride cutting tool. In view of the discussions presented thus far, it is evident that grinding and hard turning can be competitive in specific applications. Consequently, there has been considerable debate regarding the respective merits of the two processes.

Hard turning continues to be increasingly competitive with grinding, and dimensional tolerances and surface finish are beginning to approach those obtained with grinding. As seen in Tables 21.2 and 26.2, turning requires much less energy than grinding. Also, thermal and other types of damage to the workpiece surface is less likely to occur; cutting fluids may not be necessary; and the machine tools are less expensive. In addition, finishing the part while still chucked in the lathe eliminates the need for material handling and setting the part in the grinder.

However, workholding devices for large and slender workpieces during hard turning can present significant problems because the cutting forces used are higher than grinding forces. Furthermore, tool wear and its control can be a significant problem when compared to the automatic dressing of grinding wheels. It is evident that the competitive position of hard turning versus grinding has to be evaluated individually for each application in terms of product surface finish, integrity, quality, and overall economics.

Other grinding operations.   Various operations that are carried out on special-purpose grinders are summarized as follows.

- *Universal tool and cutter grinders* are used for grinding single-point or multi-point tools and cutters, including drills. They are equipped with special work-holding devices for accurate positioning of the tools to be ground. A variety of CNC tool grinders is now available, making the operation simpler and faster and with consistent results. However, the cost of these grinders ranges from about \$150,000 to \$400,000.

- *Tool-post grinders* are self-contained units and usually are attached to the tool post of a lathe (as in Fig. 23.2). The workpiece is mounted on the headstock and is ground by moving the tool post. These grinders are versatile, but the lathe components should be protected from abrasive debris.

- *Swing-frame grinders* are used in foundries for grinding large castings. The process of the rough grinding of castings is called **snagging** and usually is done on *floorstand grinders* using wheels as large as 0.9 m (36 in.) in diameter.

- *Portable grinders* are driven pneumatically, electrically, or with a flexible shaft connected to an electric motor or a gasoline engine. They are used for operations such as grinding off weld beads and *cutting-off* operations using thin abrasive disks.

- *Bench and pedestal grinders* are used for the routine off-hand grinding of tools and small parts. They usually are equipped with two wheels mounted on the two ends of the shaft of an electric motor. Generally, one wheel is coarse for rough grinding, and the other is fine for finish grinding.

Grinding fluids.   The functions of grinding fluids are similar to those of cutting fluids, as described in Section 22.12. Although grinding and other abrasive-removal processes can be performed dry, the use of a fluid is important because it:

- Prevents temperature rise in the workpiece.
- Improves part surface finish and dimensional accuracy.
- Improves the efficiency of the operation by reducing wheel wear and loading, and by lowering power consumption.

Grinding fluids are typically water-based emulsions for general grinding and oils for thread grinding (Table 26.5). They may be applied as a stream (flood) or as mist (which is a mixture of fluid and air). Because of the high surface speeds involved, an airstream (*air blanket*) around the periphery of the wheel usually prevents the fluid from reaching the wheel–workpiece interface. Special nozzles that conform to the shape of the cutting surface of the grinding wheel have been designed in which the grinding fluid is applied under high pressure.

There can be a significant rise in the temperature of water-based grinding fluids, as they remove heat from the grinding zone. This can cause the workpiece to expand, making it difficult to control dimensional tolerances. To maintain uniform workpiece temperature, a common method is to use refrigerating systems (chillers) through which the grinding fluid is circulated continuously and maintained at the same temperature. As described in Section 22.12, the biological and ecological aspects of metalworking fluids regarding their disposal, treatment, and recycling are important considerations in their selection and use. The practices employed must comply with federal, state, and local laws and regulations.

Grinding chatter.   *Chatter* is particularly important in grinding operations because it adversely affects surface finish and wheel performance. Studying **chatter marks** on ground surfaces often can identify their source, which may include (a) the bearings and spindles of the grinding machine, (b) nonuniformities in the grinding wheel (as manufactured), (c) uneven wheel wear, (d) poor dressing techniques, (e) using grinding wheels that are not balanced properly, and (f) external sources (such as nearby machinery). The grinding operation itself can cause *regenerative chatter*, as it does in machining.

The important factors in controlling chatter are the stiffness of the grinding machine, stiffness of workholding devices, and damping. General guidelines have been established to reduce the tendency for chatter in grinding, especially (a) using soft-grade wheels, (b) dressing the wheel frequently, (c) changing dressing techniques, (d) reducing the material-removal rate, and (e) supporting the workpiece rigidly.

Safety in grinding operations.   Because grinding wheels are brittle and rotate at high speeds, certain procedures must be followed carefully in their handling, storage,

TABLE 26.5

General Recommendations for Grinding Fluids	
Material	Grinding fluid
Aluminum	E, EP
Copper	CSN, E, MO + FO
Magnesium	D, MO
Nickel	CSN, EP
Refractory metals	EP
Steels	CSN, E
Titanium	CSN, E

D = dry; E = emulsion; EP = extreme pressure; CSN = chemicals and synthetics; MO = mineral oil; FO = fatty oil.

and use. Failure to follow these procedures and the instructions and warnings printed on individual wheel labels may result in serious injury or fatality. Grinding wheels should be stored properly and protected from environmental extremes, such as high temperature or humidity. They should be inspected visually for cracks and damage prior to installing them on grinders. Vitrified wheels should be tested prior to use by *ringing* them (that is, supporting them at the hole, tapping them gently, and listening to the sound). A damaged wheel will have a flat ring to it—similar to a cracked dinner plate.

Damage to a grinding wheel can reduce its **bursting speed** severely. Defined as the surface speed at which a freely rotating wheel bursts (explodes), the bursting speed is calculated in terms of bursting rpm for a particular wheel and depends on the type of wheel—its bond, grade, and structure. In diamond and cBN wheels (Fig. 26.5), which are operated at high surface speeds, the type of core material used in the wheel affects the bursting speed. Metal cores, for example, have the highest bursting speed on the order of about 250 m/s (800 ft/s).

26.5 | Design Considerations for Grinding

Design considerations for grinding operations are similar to those for machining that were described in various sections in Chapters 23 and 24. In addition, specific attention should be given to the following:

- Parts to be ground should be designed so that they can be mounted securely, either in chucks, magnetic tables, or suitable fixtures and workholding devices. Thin, straight, or tubular workpieces may distort during grinding, thus requiring special attention.
- If high dimensional accuracy is required, interrupted surfaces (such as holes and keyways) should be avoided, as they can cause vibrations and chatter.
- Parts for cylindrical grinding should be balanced, and long and slender designs should be avoided to minimize deflections. Fillets and corner radii should be as large as possible, or relief should be provided for these by prior machining.
- In centerless grinding, short pieces may be difficult to grind accurately because of their lack of support on the blade. In through-feed grinding, only the largest diameter on the parts can be ground.
- Designs requiring accurate form grinding should be kept simple to avoid frequent form dressing of the wheel.
- Deep and small holes and blind holes requiring internal grinding should be avoided, or they should include a relief.

In general, designs should require that a minimum amount of material be removed by grinding—except in creep-feed grinding. Moreover, in order to maintain good dimensional accuracy, designs preferably should allow for all grinding to be done without having to reposition the workpiece. (This guideline also is applicable to all manufacturing processes and operations.)

26.6 | Ultrasonic Machining

In *ultrasonic machining* (UM), material is removed from a surface by microchipping and erosion with loose, fine abrasive grains in a water slurry (Fig. 26.24a).

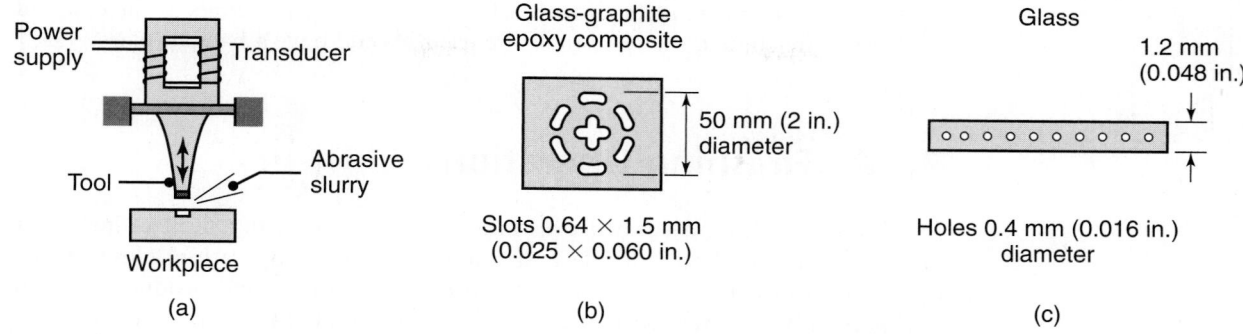

FIGURE 26.24   (a) Schematic illustration of the ultrasonic machining process. (b) and (c) Types of parts made by this process. Note the small size of holes produced.

The tip of the tool (*sonotrode*) vibrates at a frequency of 20 kHz and low amplitude of 0.0125 to 0.075 mm (0.0005 to 0.003 in.). This vibration, in turn, imparts a high velocity to abrasive grains between the tool and the workpiece. The stress produced by the impact of abrasive particles on the workpiece surface is high because (a) the time of contact between the particle and the surface is very short (10 to 100 μs) and (b) the area of contact is very small. In brittle materials, these impact stresses are sufficiently high to cause microchipping and erosion of the workpiece surface.

Ultrasonic machining is best suited for materials that are hard and brittle, such as ceramics, carbides, precious stones, and hardened steels. Two applications of ultrasonic machining are shown in Fig. 26.24b and c. A special tool is required for each shape to be produced, hence it is also called a *form tool*. The tip of the tool (which is attached to a transducer through the toolholder) usually is made of mild steel.

The abrasive grains are typically boron carbide, although aluminum oxide or silicon carbide are also used, with grain sizes ranging from grit number 100 for roughing to grit number 1000 for finishing operations. The grains are carried in a water slurry with concentrations of 20 to 60% by volume; the slurry also carries the debris away from the cutting zone.

Rotary ultrasonic machining (RUM).   In this process, the abrasive slurry is replaced by a tool with metal-bonded diamond abrasives either impregnated or electroplated on the tool surface. The tool is vibrated ultrasonically and rotated at the same time while being pressed against the workpiece surface at a constant pressure. The process is similar to a face-milling operation (Fig. 24.5) with the inserts being replaced by abrasives. The chips produced are washed away by a coolant pumped through the core of the rotating tool. The rotary ultrasonic-machining process is effective particularly in producing deep holes and high material-removal rates in ceramic parts.

Design considerations for ultrasonic machining.   The basic design guidelines for UM include the following:

- Avoid sharp profiles, corners, and radii, because these are eroded by the abrasive slurry.
- Holes produced will have some taper.

- Because of the tendency for the chipping of brittle materials at the exit end of holes machined, the bottom of the parts should have a backup plate.

26.7 | Finishing Operations

In addition to those described thus far, several processes that utilize fine abrasive grains also are used on workpieces as the final finishing operation. However, these operations can contribute significantly to production time and product cost. Thus, they should be specified with due consideration to their costs and benefits.

Coated abrasives.   Common examples of *coated abrasives* are sandpaper and emery cloth. The majority of coated abrasives are made using aluminum oxide. Silicon carbide and zirconia alumina make up the rest. They usually have a much more open structure than the abrasives on grinding wheels, and have grains that are more pointed. The grains are deposited electrostatically on flexible backing materials, such as paper, cotton, rayon polyester, polynylon, and various blends of these materials. As shown in Fig. 26.25, the bonding material (matrix) typically is resin, which first is applied to the backing (*make coat*); then the grains are bonded with a second layer (*size coat*). The grains have their long axes perpendicular to the plane of the backing, thus improving their cuttting action.

Coated abrasives are available as sheets, belts, and disks. They are used extensively to finish flat or curved surfaces of metallic and nonmetallic parts, metallographic specimens, and in woodworking. The precision of the surface finish obtained depends primarily on grain size.

Belt grinding.   Coated abrasives also are used as *belts* for high-rate material removal with good surface finish. *Belt grinding* has become an important production process, in some cases replacing conventional grinding operations. Belts with grit numbers ranging from 16 to 1500 are available. Belt speeds are in the range of 700 to 1800 m/min (2500 to 6000 ft/min). Machines for abrasive-belt operations require proper belt support and have rigid construction to minimize vibrations.

Conventional coated abrasives have randomly placed abrasives on their surface and may consist of single or multiple layers of abrasives. Another technique is *microreplication*, in which abrasives in the shape of tiny aluminum-oxide pyramids are placed in a predetermined orderly arrangement on the belt surface. Used on stainless steels and superalloys, their performance is more consistent and the temperatures involved are lower. Typical applications include belt grinding of surgical implants, golf clubs, firearms, turbine blades, and medical and dental instruments.

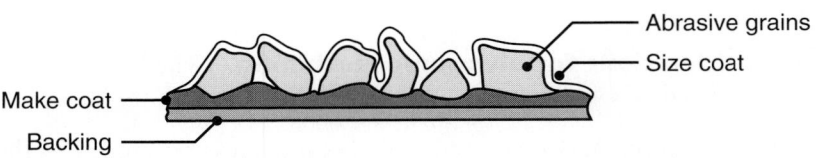

FIGURE 26.25   Schematic illustration of the structure of a coated abrasive. Sandpaper (developed in the 16th century) and emery cloth are common examples of coated abrasives.

EXAMPLE 26.5 Belt grinding of turbine nozzle vanes

The turbine nozzle vane shown in Fig. 26.26 was investment cast (Section 11.8) from a cobalt-based superalloy. To remove a thin diffusion layer from the root

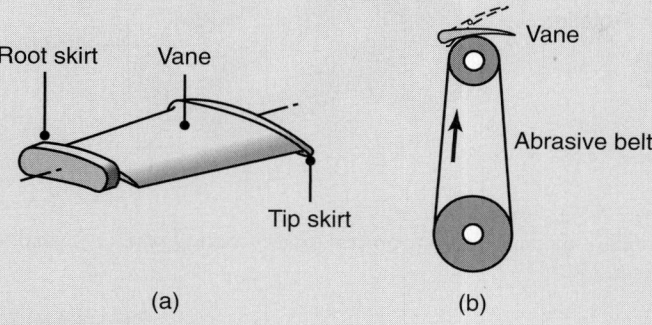

(a)   (b)

FIGURE 26.26

skirt and tip skirt sections of the vane, it was ground on a cloth-backed abrasive belt (60 grit aluminum oxide). The vanes were mounted on a fixture and ground dry at a belt surface speed of 1800 m/min (6000 ft/min). Production rate was 93 seconds per piece. Each vane weighed 21.65 g before and 20.25 g after belt grinding, a reduction in weight of about 6.5%.

Source: Courtesy of ASM International.

Wire brushing. In this process, also called *power brushing*, the workpiece is held against a circular wire brush that rotates at speeds ranging from 1750 rpm for large wheels to 3500 rpm for small wheels. As they rub, the tips of the wire produce longitudinal scratches on the workpiece surface. Wire brushing is used to produce a fine or controlled surface texture. Performed under the proper conditions, wire brushing also may be considered as a light material-removal process. In addition to metal wires, polymeric wires (nylon) embedded with abrasives also can be used effectively. (See also *diamond wire saws*, Section 24.5.)

Honing. *Honing* is an operation that is used primarily to improve the surface finish of holes produced by processes such as boring, drilling, and internal grinding. The honing tool consists of a set of aluminum-oxide or silicon-carbide bonded abrasive sticks, usually called *stones* (Fig. 26.27). They are mounted on a mandrel that rotates in the hole at surface speeds of 45 to 90 m/min (150 to 300 ft/min), applying a radial force. The tool has a reciprocating axial motion, which produces a cross-hatched pattern on the surface of the hole. The stones can be adjusted radially for different

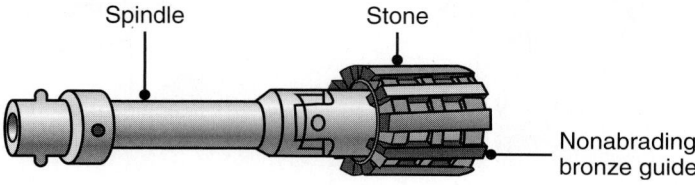

FIGURE 26.27   Schematic illustration of a honing tool used to improve the surface finish of bored or ground holes.

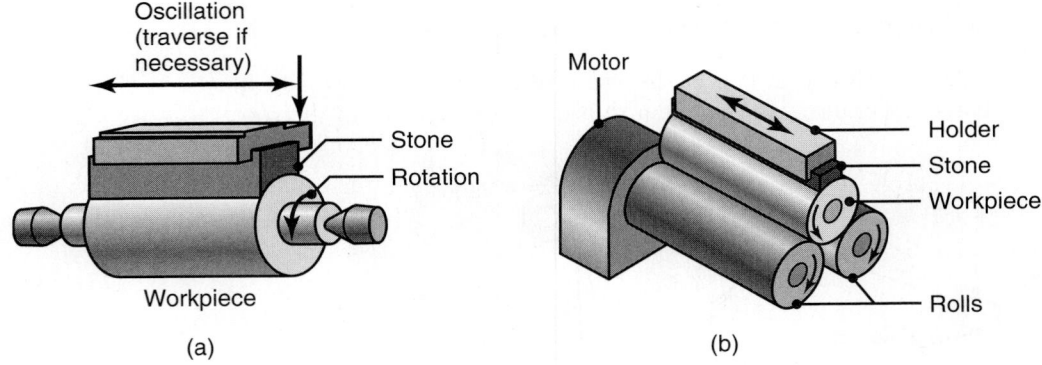

FIGURE 26.28 Schematic illustrations of the superfinishing process for a cylindrical part. (a) Cylindrical microhoning. (b) Centerless microhoning.

hole sizes. Honing fluids generally are used, which are oil- or water-based and help flush away the debris. Honing also is done on external cylindrical or flat surfaces and to manually remove sharp edges on cutting tools and inserts. (See Fig. 22.5.)

The quality of the surface finish produced by honing can be controlled by the type and size of abrasive used, the pressure applied, and speed. A fluid is used to remove chips and to keep temperatures low. If not performed properly, honing can produce holes that are neither straight nor cylindrical but rather shapes that are bell-mouthed, wavy, barrel-shaped, or tapered.

Superfinishing. In this process, the pressure applied is very light and the motion of the honing stone has a short stroke. Its motion is controlled so that the grains do not travel along the same path on the surface of the workpiece. Examples of external superfinishing of a round part are shown in Fig. 26.28.

Lapping. This is an operation used for finishing flat, cylindrical, or curved surfaces. The *lap* (Fig. 26.29a) is relatively soft and porous and usually is made of cast iron,

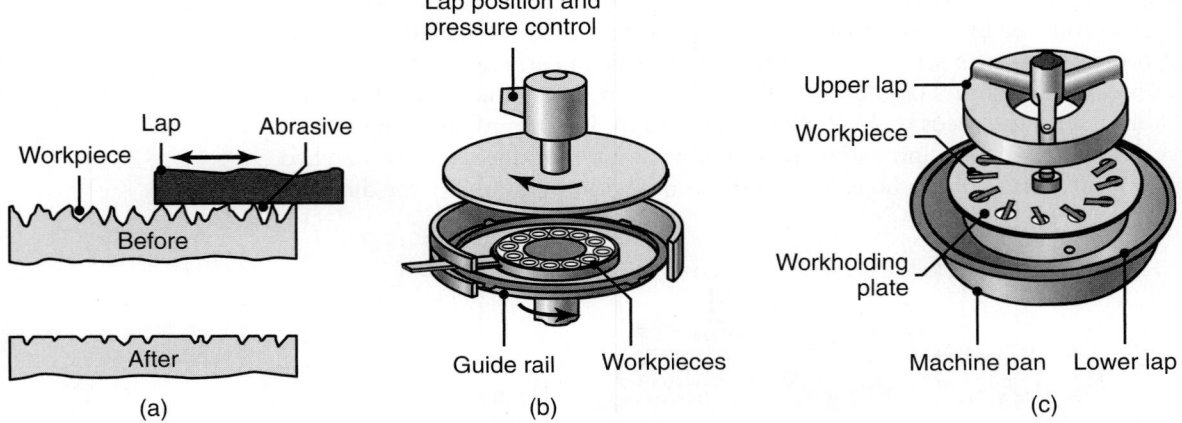

FIGURE 26.29 (a) Schematic illustration of the lapping process. (b) Production lapping on flat surfaces. (c) Production lapping on cylindrical surfaces.

copper, leather, or cloth. The abrasive particles are embedded either in the lap or may be carried through a slurry. Lapping of spherical objects and glass lenses is done using specially shaped laps. *Running-in* of mating gears can be done by lapping, such as on the hypoid gears for rear axles. Depending on the type and hardness of the workpiece material, lapping pressures range from 7 to 140 kPa (1 to 20 psi).

Dimensional tolerances on the order of ±0.0004 mm (0.000015 in.) can be obtained in lapping using fine abrasives up to grit size 900, and the surface finish can be as smooth as 0.025 to 0.1 μm (1 to 4 μin.). Production lapping on flat or cylindrical pieces is done on machines similar to those shown in Fig. 26.29b and c.

Polishing.   *Polishing* is a process that produces a smooth, lustrous surface finish. The basic mechanism involved in the polishing process is the softening and smearing of surface layers by frictional heating developed during polishing and some very-fine scale abrasive removal from the workpiece surface. The shiny appearance of polished surfaces results from a smearing action.

Polishing is done with disks or belts made of fabric, leather, or felt that are coated typically with fine powders of aluminum oxide or diamond. In *double-sided polishing*, pairs of pads are fixed on the faces of platens that rotate horizontally and in opposite directions. Parts with irregular shapes, sharp corners, deep recesses, and sharp projections can be difficult to polish.

Chemical–mechanical polishing.   *Chemical–mechanical polishing* (CMP) is extremely important in the semiconductor industry. The process is shown in Fig. 26.30 and uses a suspension of abrasive particles in a water-based solution with a chemistry selected to give controlled corrosion. This process therefore removes material from the workpiece through combined abrasion and corrosion effects. The results are a surface of exceptionally fine surface finish and a workpiece that is especially flat. For this reason, the process is often referred to as **chemical–mechanical planarization**.

A major application of this process is the polishing of silicon wafers (Section 28.4). In this case, the primary function of CMP is to polish a wafer at the micro-level. Therefore, to evenly remove material across the whole wafer, the wafer is held on a rotating carrier face down and is pressed against a polishing pad attached to a rotating disk, as shown in Fig. 26.30. The relative velocity of the pad across the disk can be shown to be constant, whereas if the wafer were not rotating, the velocity

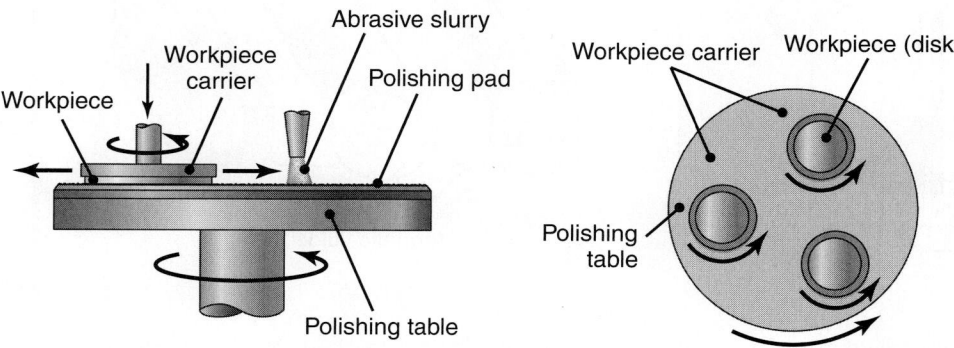

FIGURE 26.30   Schematic illustration of the chemical–mechanical polishing (CMP) process. This process is used widely in the manufacture of silicon wafers and integrated circuits and also is known as *chemical–mechanical planarization*. For other materials and applications, more carriers and more disks per carrier are possible.

would change with the radius. The pad contains grooves intended to uniformly supply slurry to all wafers. Also, pad rotation ensures that a linear layer does not develop (see Section 33.3).

Specific abrasive and solution-chemistry combinations have been developed for the polishing of copper, silicon, silicon dioxide, aluminum, tungsten and other metals. For example, for silicon dioxide or silicon polishing, an alkaline slurry of colloidal silica (SiO_2 particles in a KOH solution or NH4OH) is fed continuously to the pad/wafer interface.

Electropolishing.   Mirror-like finishes can be obtained on metal surfaces by *electropolishing*, a process that is the reverse of electroplating (to be described in Section 34.9). Because there is no mechanical contact with the workpiece, this process is suitable particularly for polishing irregular shapes. The electrolyte attacks projections and peaks on the workpiece surface at a higher rate than the rest of the surface, producing a smooth surface. Electropolishing also is used for deburring operations.

Polishing in magnetic fields.   In this technique, abrasive slurries are supported with magnetic fields. There are two basic methods.

In the *magnetic-float polishing* of ceramic balls illustrated schematically in Fig. 26.31a, a magnetic fluid (containing abrasive grains and extremely fine ferromagnetic particles in a carrier fluid, such as water or kerosene) is filled in the chamber within a guide ring. The ceramic balls are located between a drive shaft and a float. The abrasive grains, ceramic balls, and the float (which is made of a nonmagnetic material) all are suspended by magnetic forces. The balls are pressed against the rotating drive shaft and are polished by the abrasive action. The forces applied by the abrasive particles on the balls are extremely small and controllable, hence the polishing action is very fine. Because polishing times are much lower than those involved in other polishing methods, this process is very economical and the surfaces produced have few, if any, significant defects.

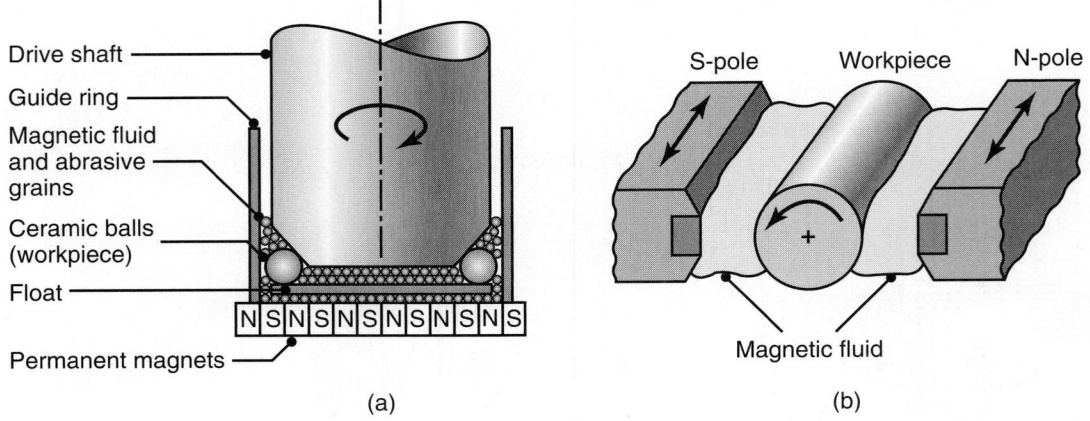

(a)   (b)

FIGURE 26.31   Schematic illustration of polishing of balls and rollers using magnetic fields. (a) Magnetic-float polishing of ceramic balls. (b) Magnetic-field-assisted polishing of rollers. *Source*: After R. Komanduri, M. Doc, and M. Fox.

Magnetic-field-assisted polishing of ceramic rollers is illustrated in Fig. 26.31b. A ceramic or steel roller (as the workpiece) is clamped and rotated on a spindle. The magnetic poles are oscillated, introducing a vibratory motion to the magnetic-abrasive conglomerate. This action polishes the cylindrical roller surface. Bearing steels of 63 HRC have been mirror-finished in 30 seconds using this process.

Buffing. *Buffing* is similar to polishing (with the exception that an even finer surface finish is obtained) using very fine abrasives on soft disks that typically are made of cloth or hide. The abrasive is supplied externally from a stick of abrasive compound.

26.8 | Deburring Operations

Burrs are thin ridges, usually triangular in shape, that develop along the edges of a workpiece from operations such as machining, shearing sheet metals (as in Figs. 16.2 and 16.3), and trimming forgings and castings. Burrs can be detected by simple means such as with a finger, toothpick, or cotton swab. Visual inspection of burrs includes the use of magnifiers and microscopes. Although efforts are being made, there are no widely accepted standards for the definition of burrs, partly because there can be a variety of burrs developed on parts.

Burrs have several disadvantages: (a) They may interfere with the mechanical assembly of parts and can cause the jamming of parts, misalignment, and short circuits in electrical components. (b) Because they are usually sharp, they can be a safety hazard to personnel. (c) Burrs may reduce the fatigue life of components. On the other hand, burrs on thin drilled or tapped components (such as the tiny parts in mechanical watches and mechanisms) can provide additional thickness and thus improve the holding torque of screws.

Several **deburring processes** are available. Their cost-effectiveness depends on factors such as the extent of deburring required, part complexity and burr location, the number of parts to be deburred, floor space available, labor costs, and safety and environmental considerations. Deburring operations include:

1. Manual deburring with files and scrapers. It is estimated that manual deburring can add up to 10% to the cost of manufacturing the part.

2. Mechanical deburring by machining pieces such as cylindrical parts on a rotating spindle.

3. Wire brushing or using rotary nylon brushes consisting of filaments embedded with abrasive grits.

4. Abrasive belts.

5. Ultrasonic machining.

6. Electropolishing.

7. Electrochemical machining.

8. Magnetic-abrasive finishing.

9. Vibratory finishing.

10. Shot blasting or abrasive blasting.

11. Abrasive-flow machining (such as extruding a semi-solid abrasive slurry over the edges of the part).

12. Thermal energy machining using lasers or plasma.

The last four processes are described next; other processes are covered elsewhere throughout this book.

Vibratory and barrel finishing.   These processes are used to remove burrs from large numbers of relatively small workpieces. In this batch-type operation, specially shaped *abrasive pellets* of nonmetallic or metallic media (stones or balls) are placed in a container along with the parts to be deburred. The container is either *vibrated* or *tumbled* using various mechanical means. The impact of individual abrasives and metal particles removes the burrs and sharp edges from the parts. Depending on the application, this can be a *dry* or a *wet* process. Liquid compounds may be added for purposes such as decreasing and adding corrosion resistance to the part being deburred. When using chemically active fluids and abrasives, this is a form of chemical–mechanical polishing.

Shot blasting.   Also called **grit blasting**, this involves abrasive particles (usually sand) that are propelled by a high-velocity jet of air, or by a rotating wheel, onto the surface of the workpiece. Shot blasting is particularly useful in deburring metallic and nonmetallic materials, and in stripping, cleaning, and removing surface oxides. The surfaces produced have a matte finish, but surface damage can result if the process parameters are not properly controlled. **Microabrasive blasting** consists of small-scale polishing and etching, using very fine abrasives, on bench-type units.

Abrasive-flow machining.   This involves the use of abrasive grains, such as silicon carbide or diamond, that are mixed in a putty-like matrix, and then forced back and forth through the openings and passageways in the workpiece. The movement of the abrasive matrix under pressure erodes away both burrs and sharp corners and polishes the part. This process is particularly suitable for workpieces with internal cavities, such as those produced by casting, that are inaccessible by other means. Pressures applied range from 0.7 to 22 MPa (100 to 3200 psi). External surfaces also can be deburred with this method by containing the workpiece within a fixture that directs the abrasive media to the edges and the areas to be deburred. The deburring of a turbine impeller by abrasive-flow machining is illustrated in Fig. 26.32.

Thermal energy deburring.   This process consists of placing the part in a chamber, which is then injected with a mixture of natural gas and oxygen. When the mixture is ignited, a burst of heat is produced, at a temperature of about 3300°C (6000°F). The burrs are instantly heated and melt, while the temperature of the part reaches only about 150°C (300°F). However, there are drawbacks to this process: (a) larger burrs tend to form beads after melting. (b) thin and slender parts may distort, and (c) it does not polish or buff the workpiece surfaces as do several other deburring processes.

Robotic deburring.   Deburring and flash removal from finished products are performed increasingly by *programmable robots* (Section 37.6) using a force-feedback system for controlling the path and rate of burr removal. This method eliminates tedious and expensive manual labor and results in more consistent and repeatable

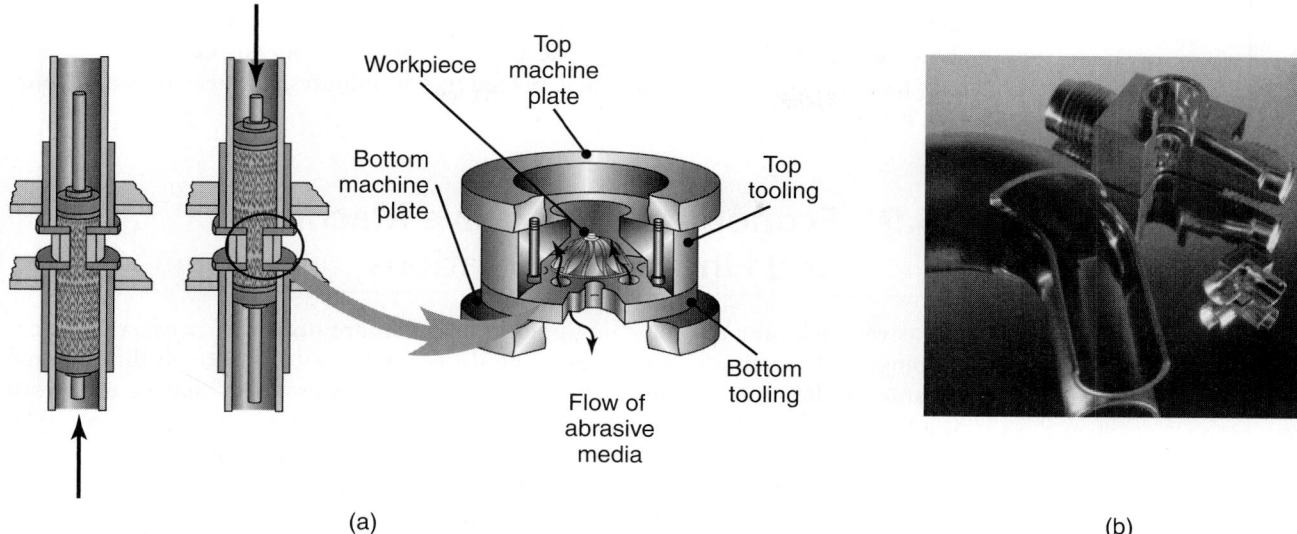

(a)   (b)

FIGURE 26.32 (a) Schematic illustration of abrasive-flow machining to deburr a turbine impeller. The arrows indicate movement of the abrasive media. Note the special fixture, which is usually different for each part design. (b) Valve fittings treated by abrasive-flow machining to eliminate burrs and improve surface quality. *Source*: Courtesy of Extrude Hone Corp.

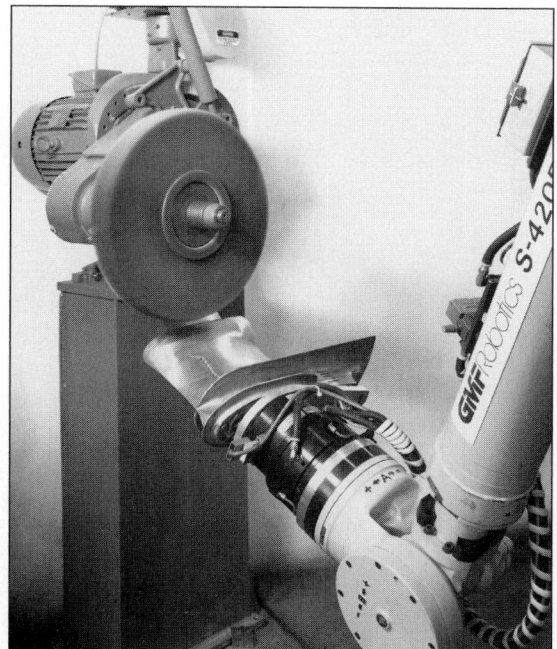

FIGURE 26.33 A deburring operation on a robot-held die-cast part for an outboard motor housing using a grinding wheel. Abrasive belts (Fig. 26.26) or flexible abrasive radial-wheel brushes also can be used for such operations. *Source*: Courtesy of Acme Manufacturing Company.

deburring. An example is the robotic deburring of a die-cast outboard motor housing shown in Fig. 26.33. In another application, the manual deburring of a double-helical gear for a helicopter gearbox was deburred in 150 minutes, whereas robotic deburring required 15 minutes.

26.9 Economics of Abrasive Machining and Finishing Operations

Abrasive machining and finishing operations often are necessary because forming, shaping, and machining processes alone do not achieve high enough dimensional accuracy and/or good quality surface finishes. Abrasive processes may be used both as a finishing process and as a large-scale material-removal operation. For example, creep-feed grinding is an economical alternative to machining operations (such as milling), even though wheel wear is high. Also, grinding and hard turning have become competitive for certain specific applications.

Much progress has been made in automating the equipment involved in these operations, including the use of computer controls, process optimization, and robotic handling of parts. Consequently, labor costs and production times have been reduced, even though such machinery generally requires significant capital investment.

Because they are additional operations, the processes described in this chapter can contribute significantly to product cost. Furthermore, as the surface-finish requirement increases, more operations are necessary, hence the cost increases

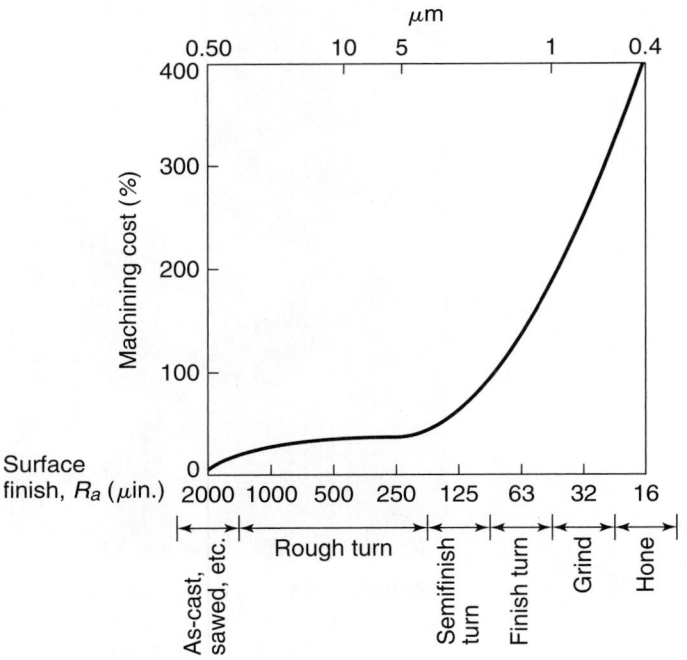

FIGURE 26.34   Increase in the cost of machining and finishing a part as a function of the surface finish required. This is the main reason that the surface finish specified on parts should not be any finer than necessary for the part to function properly.

further, as can be seen in Fig. 26.34. Note how rapidly the cost increases as surface finish is improved by additional processes, such as grinding and honing.

The cost components (such as equipment, tooling, labor, and operator skill) in abrasive processes vary widely. Whereas, for grinding, machinery costs can be high, the costs for other finishing processes are rather low. Grinding wheel costs are generally low compared to other aspects of the overall grinding operation. As discussed previously, however, the cost of wheels can run into hundreds or even thousands of dollars, depending on their composition and size. The costs of finishing tools, such as for honing and lapping, vary widely. Labor costs and operator skill depend greatly on how well the equipment is automated.

If finishing is likely to be an important factor in manufacturing a particular product, the conceptual and design stages should involve an analysis of the level of surface finish and dimensional accuracy required and of whether they can be relaxed. Furthermore, all processes that precede finishing operations should be analyzed for their capability to produce a more acceptable surface characteristics. As we have described throughout this book, this can be accomplished through the proper selection of tools, process parameters, metalworking fluids, and the characteristics of the machine tools and the workholding devices involved.

SUMMARY

- Abrasive machining often is necessary and economical when workpiece hardness is high, the materials are brittle, and surface finish and dimensional tolerance requirements are high.

- Conventional abrasives consist of aluminum oxide and silicon carbide. Superabrasives consist of cubic boron nitride and diamond. The friability of abrasive grains is an important factor in their performance—as are the shape and size of the grains.

- Grinding wheels consist of a combination of abrasive grains and bonding agents. Important characteristics of wheels are abrasive grain and bond types, grade, and hardness. Wheels also may be reinforced to maintain their integrity if and when a crack develops during their use.

- Grinding-wheel wear is an important consideration in the surface quality and integrity of the ground product. Dressing and truing of wheels are important operations and are done by various techniques.

- A variety of abrasive-machining processes and machinery are available for surface, external, and internal grinding. Abrasive machining also is used for large-scale material-removal processes (such as creep-feed grinding), making it competitive with machining processes (such as milling and turning).

- The selection of abrasives and process variables, including grinding fluids, is important in order to obtain the desired surface finish and dimensional accuracy. Otherwise, damage to surfaces (such as burning, heat checking, detrimental residual stresses, and chatter) may develop.

- Several finishing operations are available for improving surface finish. Because they can contribute significantly to product cost, however, the appropriate selection and implementation of these operations are important.

- Deburring may be necessary for some finished components. Commonly used methods are vibratory finishing, barrel finishing, and shot blasting, although thermal energy and other methods are also available.

KEY TERMS

Abrasive-flow machining

Abrasives

Aluminum oxide

Attritious wear

Barrel finishing

Belt grinding

Bonded abrasives

Buffing

Burning

Burr

Chatter marks

Chemical–mechanical polishing

Coated abrasives

Creep-feed grinding

Cubic boron nitride

Deburring

Diamond

Dressing

Electropolishing

Finishing

Free-cutting wheels

Friability

Glazing

Grade

Grain depth-of-cut

Grain size

Grindability

Grinding

Grinding ratio

Grit number

Hardness of wheel

Heat checking

Honing

Lapping

Loading

Low-stress grinding

Magnetic-float polishing

Magnetic-field-assisted polishing

Metallurgical burn

Microreplication

Polishing

Reinforced wheels

Resinoid bond

Robotic deburring

Rotary ultrasonic machining

Seeded gel

Shot blasting

Silicon nitride

Snagging

Sonotrode

Sparks

Specific energy

Structure of wheel

Superabrasives

Superfinishing

Tempering

Truing

Ultrasonic machining

Vibratory finishing

Vitrified bond

Wear flat

Wheel depth of cut

Wire brushing

BIBLIOGRAPHY

Andrew, C., Howes, T.D., and Pearce, T.R.A., *Creep Feed Grinding*, Industrial Press, 1985.

ASM Handbook, Vol. 16: *Machining*, ASM International, 1989.

Brown, J., *Advanced Machining Technology Handbook*, McGraw-Hill, 1998.

Farago, F.T., *Abrasive Methods Engineering*, Vol. 1 and Vol. 2, Industrial Press, 1976 and 1980.

Gillespie, L.K., *Deburring and Edge Finishing Handbook*, Society of Manufacturing Engineers/American Society of Mechanical Engineers, 2000.

———, *Hand Deburring—Increasing Shop Productivity*, Society of Manufacturing Engineers, 2003.

Hwa, L.S., *Chemical Mechanical Polishing in Silicon Processing*, Academic Press, 1999.

King, R.I., and Hahn, R.S., *Handbook of Modern Grinding Technology*, Chapman & Hall/Methuen, 1987.

Krar, S., *Grinding Technology*, 2nd ed., Delmar Publishers, 1995

Krar, S., and Ratterman, E., *Superabrasives: Grinding and Machining with CBN and Diamond*, McGraw-Hill, 1990.

Machinery's Handbook, Industrial Press, revised periodically.

Machining Data Handbook, 3rd ed., 2 vols. Machinability Data Center, 1980.

Malkin, S., *Grinding Technology: Theory and Applications of Machining with Abrasives*, Wiley, 1989.

Marinescu, I.D. (ed.), *Handbook of Ceramics Grinding & Polishing*, Noyes, 1998.

Maroney, M. L., *A Guide to Metal and Plastic Finishing*, Industrial Press, 1991.

Metzger, J.L., *Superabrasive Grinding*, Butterworth, 1986.

Nachtman, E.S., and Kalpakjian, S., *Lubricants and Lubrication in Metalworking Operations*, Marcel Dekker, 1985.

Salmon, S.C., *Modern Grinding Process Technology*, McGraw-Hill, 1992.

Schneider, A.F., *Mechanical Deburring and Surface Finishing Technology*, Marcel Dekker, 1990.

Shaw, M.C., *Principles of Abrasive Processing*, Oxford, 1996.

Sluhan, C. (ed.), *Cutting and Grinding Fluids: Selection and Application*, Society of Manufacturing Engineers, 1992.

Szymanski, A., and Borkowski, J., *Technology of Abrasives and Abrasive Tools*, Ellis Horwood, 1992.

Tool and Manufacturing Engineers Handbook, 4th ed., Vol. 1: *Machining*, Society of Manufacturing Engineers, 1983.

Webster, J.A., Marinescu I.D., and Trevor, T.D., *Abrasive Processes*, Marcel Dekker, 1999.

REVIEW QUESTIONS

26.1 What is an abrasive? What are superabrasives?

26.2 How is the size of an abrasive grain related to its number?

26.3 Why are most abrasives made synthetically?

26.4 Describe the structure of a grinding wheel and its features.

26.5 Explain the characteristics of each type of bond used in bonded abrasives.

26.6 Describe the (a) grade and (b) structure of bonded abrasives.

26.7 What causes grinding sparks? Is it useful to observe them? Explain.

26.8 Define metallurgical burn.

26.9 Explain the mechanisms by which grinding wheels wear.

26.10 Define (a) friability, (b) wear flat, (c) grinding ratio, (d) truing, and (e) dressing.

26.11 Explain what is meant by a grinding wheel acting "soft" or acting "hard."

26.12 What is creep-feed grinding and what are its advantages?

26.13 What is the principle of ultrasonic machining?

26.14 List the finishing operations commonly used in manufacturing operations. Why are they necessary? Explain why they should be minimized.

26.15 How is centerless grinding different from cylindrical grinding?

26.16 What are the differences between coated and bonded abrasives?

26.17 What is heat checking in grinding? What is its significance?

26.18 What is the purpose of the slurry in chemical mechanical polishing? What about the liquid?

QUALITATIVE PROBLEMS

26.19 Why are grinding operations necessary for components that have previously been machined by the processes described in Chapters 23 and 24?

26.20 Explain why there are so many different types, shapes, and sizes of grinding wheels.

26.21 Explain the reasons for the large difference between the specific energies involved in machining (Table 21.2) and in grinding (Table 26.2).

26.22 We have stated that ultrasonic machining is best suited for hard and brittle materials. Explain why.

26.23 Explain why parts with irregular shapes, sharp corners, deep recesses, and sharp projections can be difficult to polish.

26.24 Explain the reasons why so many deburring operations have been developed over the years.

26.25 What precautions should you take when grinding with high precision? Comment on the machine, process parameters, the grinding wheel, and grinding fluids.

26.26 Why does grinding temperature decrease with increasing work speed (Eq. 26.4)? Does this mean that for a work speed of zero, the temperature is infinite? Explain.

26.27 Describe the similarities and differences in the action of metalworking fluids in machining versus grinding.

26.28 What factors could contribute to chatter in grinding? Explain why.

26.29 The grinding ratio, G, depends on the following: the type of grinding wheel, workpiece hardness, wheel depth of cut, wheel and workpiece speeds, and the type of grinding fluid. Explain why.

26.30 Generally, it is recommended that, when grinding hardened steels, the grinding wheel should be of a relatively soft grade. Explain the reason.

26.31 In Fig. 26.4, the proper grinding faces are shown with an arrow for each type of wheel. Explain why the other surfaces of the wheels should not be used for grinding, and what the consequences may be in doing so.

26.32 Explain the factors involved in selecting the appropriate type of abrasive for a particular grinding operation.

26.33 Describe the effects of a wear flat on the overall grinding operation.

26.34 What difficulties, if any, could you encounter in grinding thermoplastics? Thermosets? Ceramics?

26.35 Are there any similarities among grinding, honing, polishing, and buffing? Explain.

26.36 Is the grinding ratio important in evaluating the economics of a grinding operation? Explain.

26.37 We know that grinding can produce a very fine surface finish on a workpiece. Is this necessarily an indication of the quality of a part? Explain.

26.38 What are the consequences of allowing the temperature to rise during grinding?

26.39 If not done properly, honing can produce holes that are bell-mouthed, wavy, barrel-shaped, or tapered. Explain how this is possible.

26.40 Jewelry applications require the grinding of diamonds into desired shapes. How is this done, since diamond is the hardest material known?

26.41 Why are speeds so much higher in grinding than in cutting?

QUANTITATIVE PROBLEMS

26.42 Calculate the chip dimensions in surface grinding for the following process variables: $D = 8$ in., $d = 0.001$ in., $v = 100$ ft/min, $V = 6000$ ft/min, $C = 500$ per in^2, and $r = 20$.

26.43 If the strength of the workpiece material is doubled, what should be the percentage decrease in the wheel depth-of-cut, d, in order to maintain the same grain force, with all other variables being the same?

26.44 Assume that a surface-grinding operation is being carried out under the following conditions: $D = 200$ mm, $d = 0.1$ mm, $v = 40$ m/min, and $V = 3000$ m/min. These conditions then are changed to the following: $D = 150$ mm, $d = 0.1$ mm, $v = 30$ m/min, and $V = 2500$ m/min. How different is the temperature rise from the rise that occurs with the initial conditions?

26.45 Estimate the percent increase in the cost of the grinding operation if the specification for the surface finish of a part is changed from 63 to 16 μin.

26.46 Assume that the energy cost for grinding an aluminum part with a specific energy requirement of 8 W·s/mm^3 is $0.90 per piece. What would be the energy cost of carrying out the same operation if the workpiece material were T15 tool steel?

26.47 On the basis of the information given in Chapter 23 and this chapter, comment on the feasibility of producing a 10 mm hole that is 100 mm deep in a copper alloy (a) by conventional drilling and (b) by internal grinding.

26.48 In describing grinding processes, we have not given the type of equations regarding feeds, speeds, material-removal rates, total grinding time, etc., as we have done in the turning and milling operations in Chapters 23 and 24. Study the quantitative relationships involved and develop such equations for grinding operations.

26.49 What would be the answers to Example 26.1 if the workpiece is high-strength titanium and the width-of-cut is $w = 0.75$ in.? Give your answers in newtons.

26.50 It is known that, in grinding, heat checking occurs when grinding with a spindle speed of 4000 rpm, a wheel diameter of 10 in., and a depth-of-cut of 0.0015 in. for a feed rate of 50 ft/min. For this reason, the standard operating procedure is to keep the spindle speed at 3500 rpm. If a new, 8-in. diameter wheel is used, what spindle speed can be used before heat checking occurs? What spindle speed should be used to keep the same grinding temperatures as those encountered with the existing operating conditions?

26.51 It is desired to grind a hard aerospace aluminum alloy. A depth of 0.003 in. is to be removed from a cylindrical section 10 in. long and 4 in. in diameter. If each part is to be ground in not more than one minute, what is the approximate power requirement for the grinder? What if the material is changed to a hard titanium alloy?

26.52   A grinding operation takes place with a 10 in. grinding wheel with a spindle speed of 3000 rpm. The workpiece feed rate is 60 ft/min and the depth-of-cut is 0.002 in. Contact thermometers record an approximate maximum temperature of 1800°F. If the workpiece is steel, what is the temperature if the speed is increased to 4000 rpm? What if the speed is 10,000 rpm?

26.53   Derive an expression for the angular velocity of the wafer, as shown in Fig. 26.30b as a function of the radius and angular velocity of the pad in chemical–mechanical polishing.

SYNTHESIS, DESIGN, AND PROJECTS

26.54   List and explain factors that contribute to poor surface finish in the processes described in this chapter.

26.55   With appropriate sketches, describe the principles of various fixturing methods and devices that can be used for the processes described in this chapter.

26.56   Explain the major design guidelines for grinding.

26.57   Which of the processes described in this chapter are suitable particularly for workpieces made of (a) ceramics, (b) thermoplastics, and (c) thermosets? Why?

26.58   Make a comprehensive table of the process capabilities of abrasive-machining processes. Using several columns, describe the machines involved, the type of abrasive tools used, the shapes of blanks and parts produced, typical maximum and minimum sizes, surface finish, tolerances, and production rates.

26.59   Based on the data developed in Problem 26.58, describe your thoughts regarding the procedure to be followed in determining what type of machine tool to select for a particular part to be machined by abrasive means.

26.60   Vitrified grinding wheels (also called ceramic wheels) use a glass-like bond to hold the abrasive grains together. Given your understanding of ceramic-part manufacture (as described in Chapter 18), list methods of producing vitrified wheels.

26.61   A somewhat controversial subject in grinding is size effect; that is, there is an apparent increase in the strength of a workpiece when the depth of penetration by grinding abrasives is reduced. Design an experimental setup whereby this size effect can be examined.

26.62   Describe as many parameters as you can that could affect the final surface finish in grinding. Include process parameters as well as setup and equipment effects.

26.63   Assume that you are an instructor covering the topics described in this chapter, and you are giving a quiz on the numerical aspects to test the understanding of the students. Prepare three quantitative problems and supply the answers.

26.64   Conduct a literature search and explain how observing the color, brightness, and shape of sparks produced in grinding can be a useful guide to identifying the type of material being ground and its condition.

26.65   Review Fig. 26.4 and give possible grinding applications for each type of wheels shown in the figure.

26.66   Visit a large hardware store and inspect the grinding wheels on display. Make a note of the markings on the wheels and, based on the marking system shown in Fig. 26.6, comment on your observations, including the most common types of wheels available in the store.

26.67   Obtain a small grinding wheel or a piece of a large wheel. (a) Observe its surfaces using a magnifier or a microscope and compare these with Fig. 26.9. (b) Rub the abrasive wheel by pressing it hard against a variety of flat metallic and nonmetallic materials. Describe your observations regarding the surfaces produced.

26.68   Explain how you would produce the steel balls that subsequently are ground on the type of setups shown in Fig. 26.15.

26.69   In reviewing the abrasive machining processes in this chapter, you will note that some use bonded abrasives while others involve loose abrasives. Make two separate lists for these processes and comment on your observations.

26.70   Do you think the plunge-grinding operation shown in Fig. 26.17 could compete with a turning operation to produce the same type of part? Explain.

26.71   Observe the cycle patterns shown in Fig. 26.20 and comment on why they follow those particular patterns.

26.72   Obtain pieces of sand paper and emery cloth of different coarseness. Using a magnifier or a microscope, observe its surface features and compare them with Fig. 26.25.

26.73 In this chapter, we described the principles of machining and turning centers, and we will describe similar centers in Chapter 27 on advanced machining operations. Based on the contents of this chapter, describe your thoughts on whether or not it would be possible to design and build a "grinding center." Explain if such a center could perform all types of grinding operations or would be confined to specific types. Comment also on any difficulties that have to be dealt with in such machines and operations.

Advanced Machining Processes

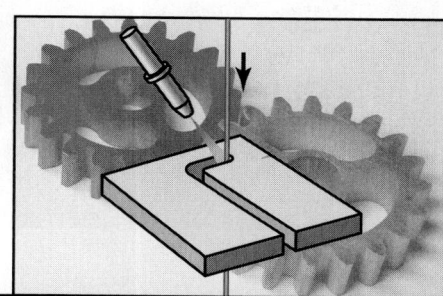

In this chapter, advanced machining processes are described that are based on non-mechanical means of material removal. Specifically, the following are described:

- Fundamentals of processes that utilize chemical, electrical, and high-energy beams.
- Characteristics of each process and its applications.
- Competitive aspects compared to other processes.

Typical parts made: Skin panels for missiles and aircraft, turbine blades, nozzles, parts with complex cavities and small-diameter deep holes, dies, laser cutting of sheet metals, cutting of thick metallic and nonmetallic parts.

Alternative methods: Abrasive machining, ultrasonic machining, and precision machining.

27.1 | Introduction

The machining processes described in the preceding chapters involved material removal by mechanical means by either chip formation, abrasion, or microchipping. However, there are situations where mechanical methods are not satisfactory, economical, or even possible for the following reasons:

- The **strength** and **hardness** of the workpiece material is very high, typically above 400 HB.
- The workpiece material is too **brittle** to be machined without damage to the workpiece. This is typically the case with highly heat-treated alloys, glass, ceramics, and powder-metallurgy parts.
- The workpiece is too **flexible** or too slender to withstand forces in machining or grinding, or the parts are difficult to clamp in fixtures and workholding devices.
- The **shape** of the part is *complex* (Fig. 27.1), including such features as internal and external profiles or holes with high length-to-diameter ratios in very hard materials.
- Special **surface finish** and **dimensional tolerance** requirements exist that cannot be obtained by other manufacturing processes or are uneconomical through alternative processes.

(a)   (b)

FIGURE 27.1   Examples of parts made by advanced machining processes. (a) Samples of parts produced by water-jet cutting. (b) Turbine blade, produced by plunge electrical–discharge machining, in a fixture to produce the holes by EDM. *Source:* (a) Courtesy of OMAX Corporation (b) Courtesy of HI-TEK Mfg., Inc.

- **Temperature rise** during processing and **residual stresses** developed in the workpiece are not desirable or acceptable.

These requirements led to the development of chemical, electrical, laser, and high-energy beams as energy sources for removing material from metallic or nonmetallic workpieces, as outlined in Table 27.1. These advanced methods, which in the past have been called *nontraditional* or *unconventional machining*, began to be introduced in the 1940s. As shall be seen, these processes remove material not by producing chips as in machining and grinding but by means such as chemical dissolution, etching, melting, evaporation, and hydrodynamic action—sometimes with the assistance of fine abrasive particles. A major advantage of these processes is that the hardness of the workpiece is not relevant.

When selected and applied properly, advanced machining processes offer major technical and economic advantages over more traditional machining methods. This chapter describes these processes, including their characteristics, typical applications, limitations, product quality, dimensional accuracy, surface finish, and economics.

27.2 | Chemical Machining

Chemical machining (CM) was developed from the observation that chemicals attack and etch most metals, stones, and some ceramics, thereby removing small amounts of material from the surface. The CM process is carried out by chemical dissolution using **reagents** or **etchants**, such as acids and alkaline solutions. Chemical machining is the oldest of the advanced machining processes and has been used in engraving metals and hard stones, in deburring, and in the production of printed-circuit boards and microelectronic devices (see Chapters 28 and 29).

Chemical milling.   In *chemical milling*, shallow cavities are produced on plates, sheets, forgings, and extrusions, generally for the overall reduction of weight (Fig. 27.2). This process has been used on a wide variety of metals with depths of metal removal as large as 12 mm (0.5 in.). Selective attack by the chemical reagent on different areas of the workpiece surfaces is controlled by removable layers of material (called **masking**, Fig. 27.3a) or by partial immersion in the reagent.

TABLE 27.1

General Characteristics of Advanced Machining Processes

Process	Characteristics	Process parameters and typical material-removal rate or cutting speed
Chemical machining (CM)	Shallow removal on large flat or curved surfaces; blanking of thin sheets; low tooling and equipment cost; suitable for low production runs	0.0025–0.1 mm/min.
Electrochemical machining (ECM)	Complex shapes with deep cavities; highest rate of material removal among other nontraditional processes; expensive tooling and equipment; high power consumption; medium-to-high production quantity	V: 5–25 DC; A: 1.5–8 A/mm^2; 2.5–12 mm/min, depending on current density
Electrochemical grinding (ECG)	Cutting off and sharpening hard materials, such as tungsten-carbide tools; also used as a honing process; higher removal rate than grinding	A: 1–3 A/mm^2; Typically 25 mm^3/s per 1000 A
Electrical-discharge machining (EDM)	Shaping and cutting complex parts made of hard materials; some surface damage may result; also used as a grinding and cutting process; expensive tooling and equipment	V: 50–380; A: 0.1–500; Typically 300 mm^3/min
Wire EDM	Contour cutting of flat or curved surfaces; expensive equipment	Varies with material and thickness
Laser-beam machining (LBM)	Cutting and hole making on thin materials; heat-affected zone; does not require a vacuum; expensive equipment; consumes much energy	0.50–7.5 m/min
Electron-beam machining (EBM)	Cutting and hole making on thin materials; very small holes and slots; heat-affected zone; requires a vacuum; expensive equipment	1–2 mm^3/min
Water-jet machining (WJM)	Cutting all types of nonmetallic materials; suitable for contour cutting of flexible materials; no thermal damage; noisy	Varies considerably with material
Abrasive water-jet machining (AWJM)	Single or multilayer cutting of metallic and nonmetallic materials	Up to 7.5 m/min
Abrasive-jet machining (AJM)	Cutting, slotting, deburring, etching, and cleaning of metallic and nonmetallic materials; tends to round off sharp edges; can be hazardous	Varies considerably with material

The procedure for chemical milling consists of the following:

1. If the part to be machined has residual stresses from prior processing, the stresses first should be relieved in order to prevent warping after chemical milling.

2. The surfaces are degreased and cleaned thoroughly to ensure both good adhesion of the masking material and uniform material removal. Scale from heat treatment also should be removed.

3. The masking material is applied. Masking with tapes or paints (**maskants**) is a common practice, although elastomers (rubber and neoprene) and plastics (polyvinyl chloride, polyethylene, and polystyrene) also are used. The maskant material should not react with the chemical reagent.

4. If required, the maskant that covers various regions that require etching is peeled off by the scribe-and-peel technique.

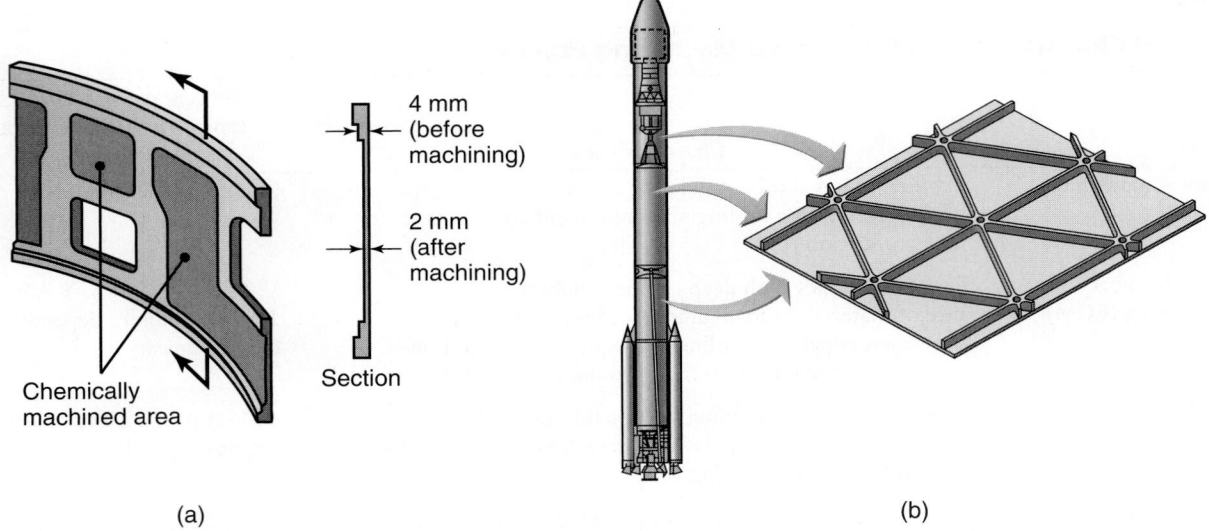

(a)

(b)

FIGURE 27.2 (a) Missile skin-panel section contoured by chemical milling to improve the stiffness-to-weight ratio of the part. (b) Weight reduction of space-launch vehicles by the chemical milling of aluminum-alloy plates. These panels are chemically milled after the plates first have been formed into shape by a process such as roll forming or stretch forming. The design of the chemically machined rib patterns can be modified readily at minimal cost.

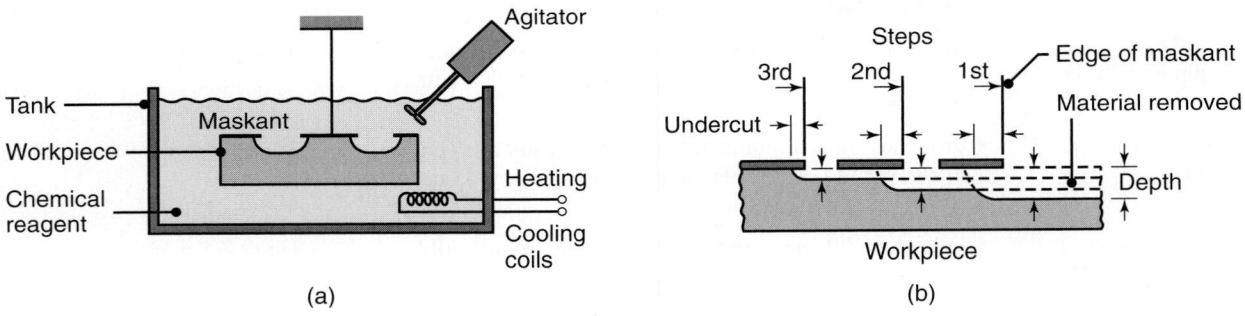

(a)

(b)

FIGURE 27.3 (a) Schematic illustration of the chemical-machining process. Note that no forces or machine tools are involved in this process. (b) Stages in producing a profiled cavity by chemical machining; note the undercut.

5. The exposed surfaces are machined chemically with etchants, such as sodium hydroxide (for aluminum), solutions of hydrochloric and nitric acids (for steels), or iron chloride (for stainless steels). Temperature control and agitation (stirring) during chemical milling is important in order to obtain a uniform depth from the material removed.

6. After machining, the parts should be washed thoroughly to prevent further reactions with or exposure to any etchant residues.

7. The rest of the masking material is removed and the part is cleaned and inspected. Note that the masking material is unaffected by the reagent but usually is dissolved easily by a different type of solvent.

8. Additional finishing operations may be performed on chemically milled parts.

9. This sequence of operations can be repeated to produce stepped cavities and various contours (Fig. 27.3b).

Chemical milling is used in the aerospace industry to remove shallow layers of material from large aircraft components, missile skin panels, and extruded parts for airframes. Tank capacities for reagents are as large as 3.7 m × 15 m (12 ft × 50 ft). This process also is used to fabricate microelectronic devices and often is referred to as **wet etching** for these products. The ranges of surface finish and tolerance obtained by chemical machining and other machining processes are shown in Fig. 27.4.

Some surface damage may result from chemical milling because of *preferential etching* and *intergranular attack*, which adversely affect surface properties. The chemical milling of welded and brazed structures may result in uneven material removal. The chemical milling of castings may result in uneven surfaces caused by porosity in and nonuniformity of the material.

Chemical blanking. *Chemical blanking* is similar to the blanking of sheet metals in that it is used to produce features which penetrate through the thickness of the material (as shown in Fig. 16.4), with the exception that the material is removed by chemical dissolution rather than by shearing. Typical applications for

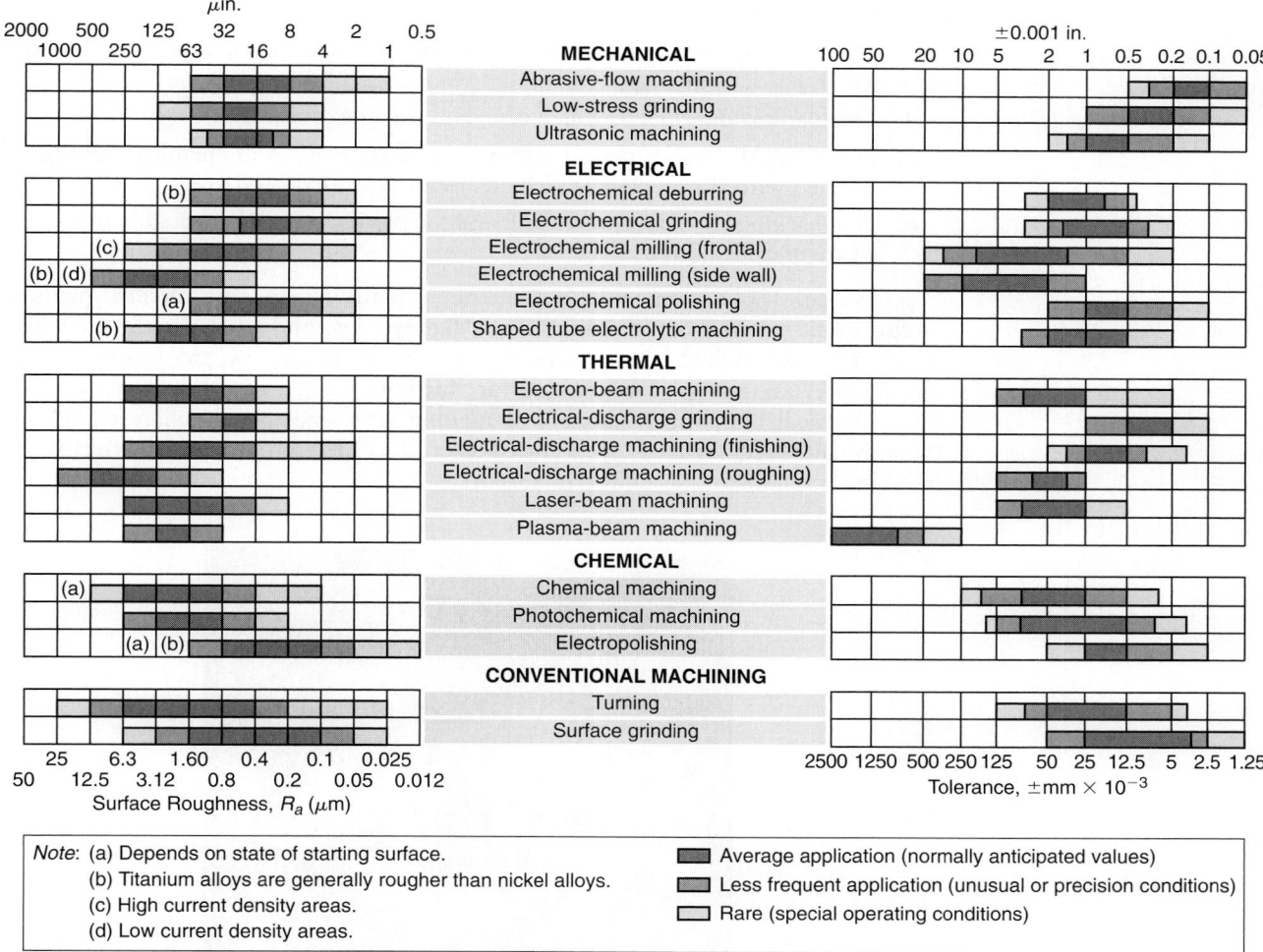

FIGURE 27.4 Surface roughness and tolerances obtained in various machining processes. Note the wide range within each process (see also Fig. 23.13). *Source: Machining Data Handbook*, 3rd ed. Copyright © 1980. Used by permission of Metcut Research Associates, Inc.

chemical blanking are the burr-free etching of printed-circuit boards, decorative panels, and thin sheet-metal stampings, as well as the production of complex or small shapes.

Photochemical blanking. *Photochemical blanking* (also called photoetching) is a modification of chemical milling. Material is removed (usually from a flat thin sheet) by photographic techniques. Complex, burr-free shapes can be blanked (Fig. 27.5) on metals as thin as 0.0025 mm (0.0001 in.). Sometimes called **photochemical machining**, this process also is used for etching.

The procedure in photochemical blanking consists of the following steps:

1. The design of the part to be blanked is prepared at a magnification of up to 100×. A photographic negative then is made and reduced to the size of the finished part. The reduced negative of the design is called **artwork**. The original (enlarged) drawing allows inherent design errors to be reduced by the amount of reduction (such as 100×) for the final artwork image.

2. The sheet blank is coated with a photosensitive material (**photoresist**) by dipping, spraying, spincasting, or roller coating and dried in an oven. This coating is often called the *emulsion*.

3. The negative is placed over the coated blank and exposed to ultraviolet light, which hardens the exposed areas.

4. The blank is developed, which dissolves the unexposed areas.

5. The blank then is immersed into a bath of reagent (as in chemical milling) or sprayed with the reagent, which etches away the exposed areas.

6. The masking material is removed, and the part is washed thoroughly to remove all chemical residues.

Typical applications for photochemical blanking are fine screens, printed-circuit cards, electric-motor laminations, flat springs, and masks for color televisions. Although skilled labor is required, tooling costs are low; the process can be automated; and it is economical for medium- to high-production volume. Photochemical blanking is capable of making very small parts where traditional blanking dies (Section 16.2) are difficult to produce. The process is also effective for blanking fragile workpieces and materials.

FIGURE 27.5 Various parts made by chemical blanking. Note the fine detail. *Source:* Courtesy of Buckbee-Mears, St. Paul.

The handling of chemical reagents requires precautions and special safety considerations to protect the workers against exposure to both liquid chemicals and volatile chemicals. Furthermore, the disposal of chemical by-products from this process is a major drawback, although some by-products can be recycled.

Design considerations for chemical machining. Design guidelines for chemical machining are:

- Because the etchant attacks all exposed surfaces continuously, designs involving sharp corners, deep and narrow cavities, severe tapers, folded seams, or porous workpiece materials should be avoided.

- Because the etchant attacks the material in both vertical and horizontal directions, *undercuts* may develop (as shown by the areas under the edges of the maskant in Fig. 27.3). Typically, tolerances of $\pm 10\%$ of the material thickness can be maintained in chemical blanking.

- In order to improve the production rate, the bulk of the workpiece should be shaped by other processes (such as by machining) prior to chemical machining.

- Dimensional variations can occur because of size changes in artwork due to humidity and temperature. This variation can be minimized by properly selecting artwork media and by controlling the environment in the artwork generation and the production area in the plant.

- Many product designs now are made with computer-aided design systems (Chapter 38). However, product drawings must be translated into a protocol that is compatible with the equipment for photochemical artwork generation.

27.3 | Electrochemical Machining

Electrochemical machining (ECM) is basically the reverse of electroplating (Section 34.9). An **electrolyte** acts as current carrier (Fig. 27.6), and the high rate of electrolyte movement in the tool-workpiece gap (typically 0.1 to 0.6 mm) washes metal ions away from the workpiece (*anode*) before they have a chance to plate onto the tool (*cathode*). Note that the cavity produced is the female mating image of the tool shape.

The shaped tool, either a solid or tubular form, generally is made of brass, copper, bronze, or stainless steel. The electrolyte is a highly conductive inorganic

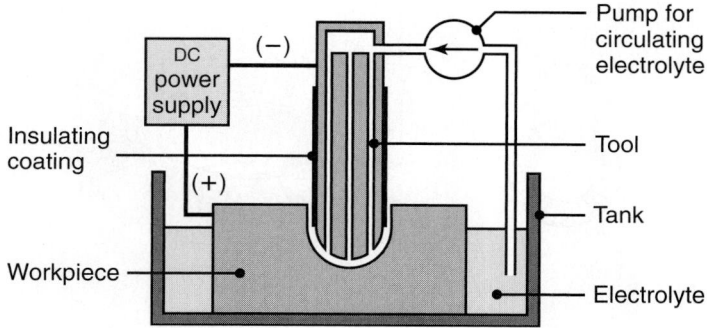

FIGURE 27.6 Schematic illustration of the electrochemical machining process.

fluid, such as an acqueous solution of sodium nitrate. It is pumped through the passages in the tool, at rates of 10 to 16 m/s (3 to 50 ft/s). A DC power supply in the range of 10 to 25 V maintains current densities, which for most applications are 20 to 200 A/cm^2 (130 to 1300 A/in^2) of active machined surface.

The *material-removal rate* (MRR) in electrochemical machining for a current efficiency of 100% may be estimated from the equation

$$\text{MRR} = CI \qquad (27.1)$$

where MRR is in mm^3/min, I is the current in amperes, and C is a material constant, in mm^3/A-min. For pure metals, C depends on the valence; the higher the valence, the lower is its value.

Machines having current capacities as high as 40,000 A and as low as 5 A are available. The penetration rate of the tool is proportional to the current density. Material removal rate typically ranges between 1.5 and 4 mm^3 per A-min. Because the metal removal rate is only a function of ion exchange rate, it is not affected by the strength, hardness, or toughness of the workpiece.

Process capabilities. The basic concept of electrochemical machining was patented in 1929, and developed rapidly during the 1950–1960s whereby it has become an important manufacturing process, although not as widely used as other processes described in this chapter. It is generally used to machine complex cavities and shapes in high-strength materials, particularly in the aerospace industry for the mass production of turbine blades, jet-engine parts, and nozzles (Fig. 27.7), as well as in the automotive (engines castings and gears) and medical industries.

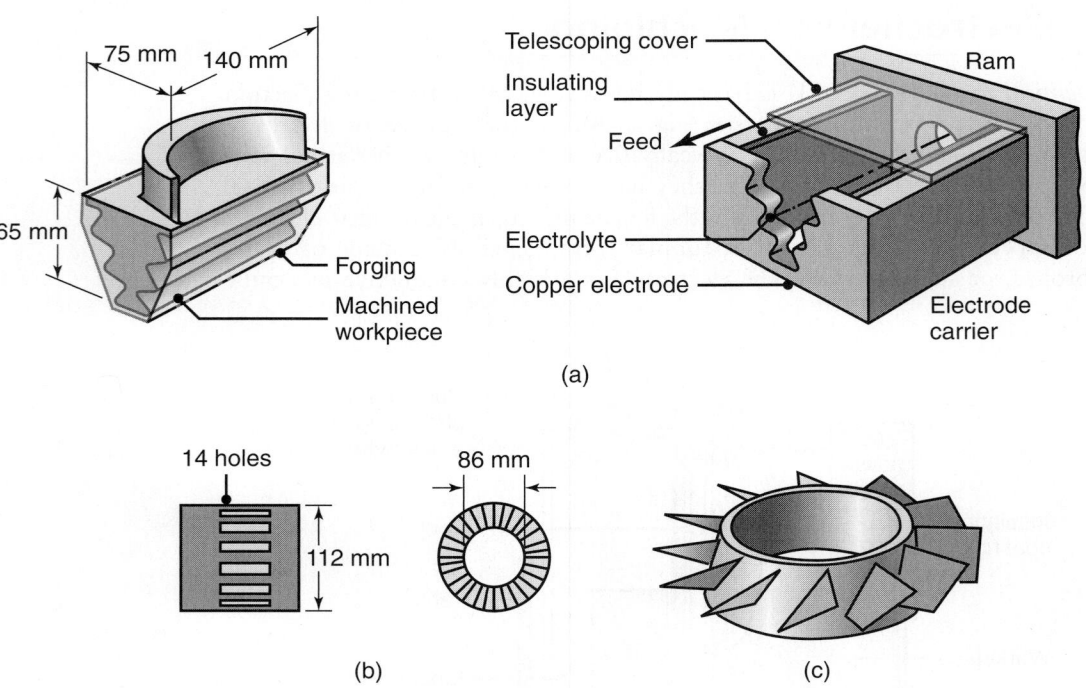

FIGURE 27.7 Typical parts made by electrochemical machining. (a) Turbine blade made of a nickel alloy of 360 HB. Note the shape of the electrode on the right. (b) Thin slots on a 4340-steel roller-bearing cage. (c) Integral airfoils on a compressor disk.

Electrochemical machining also is used for machining and finishing forging-die cavities (*die sinking*) and to produce small holes. Versions of this process are used for turning, facing, milling, slotting, drilling, trepanning, and profiling, as well as in the production of continuous metal strips and webs. More recent applications of ECM include *micromachining* (Chapter 29) for the electronics industry.

A modification of ECM is the *shaped-tube electrolytic machining* (STEM), used for drilling small-diameter deep holes, as in turbine blades. The tool is a titanium tube, coated with an electrically insulating resin. Holes as small as 0.5 mm can be drilled, at depth-to-diameter ratios of as high as 300:1.

The ECM process leaves a burr-free, bright surface; in fact, it can also be used as a deburring process. It does not cause any thermal damage to the part, and the lack of tool forces prevents part distortion. Furthermore, there is no tool wear (since only hydrogen is generated at the cathode), and the process is capable of producing complex shapes. However, the mechanical properties of components made by the ECM process should be compared carefully to those of other material-removal methods.

Electrochemical-machining systems now are available as *numerically-controlled machining centers* with the capability of high production rates, high flexibility, and the maintenance of close dimensional tolerances. The ECM process also can be combined with electrical-discharge machining (EDM) on the same machine. Further developments include hybrid-machining systems, in which ECM and other advanced machining processes (as described throughout this chapter) are combined to take advantage of the capabilities of each process.

Design considerations for electrochemical machining. The following are general design guidelines for electrochemical machining:

- Because of the tendency for the electrolyte to erode away sharp profiles, electrochemical machining is not suited for producing sharp square corners or flat bottoms.
- Controlling the electrolyte flow may be difficult, so irregular cavities may not be produced to the desired shape with acceptable dimensional accuracy.
- Designs should make provision for a small taper for holes and cavities to be machined.

CASE STUDY 27.1 Electrochemical Machining of a Biomedical Implant

A total knee-replacement system consists of a femoral and tibial implant combined with an ultra-high molecular-weight polyethylene (UHMWPE) insert, as shown in Fig. 27.8a. The polyethylene has superior wear resistance and low friction against the cobalt-chrome alloy femoral implant. The UHMWPE insert is compression molded (Section 19.7), and the metal implant is cast and ground on its external mating surfaces.

Designers of implants, manufacturing engineers, and clinicians have been concerned particularly with the contact surface in the cavity of the metal implant that mates with a protrusion on the polyethylene insert. As the knee articulates during normal motion, the polyethylene slides against the metal part, becoming a potentially serious wear site. This geometry is necessary to ensure lateral stability of the knee (that is, to prevent the knee from buckling sideways).

In order to produce a smooth surface, the grinding of the bearing surfaces of the metal implant using both hand-held and cam-mounted grinders was a procedure that had been followed for many years. However, grinding produced

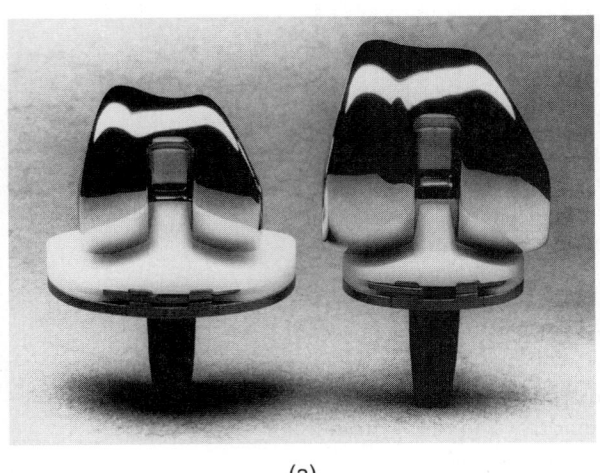

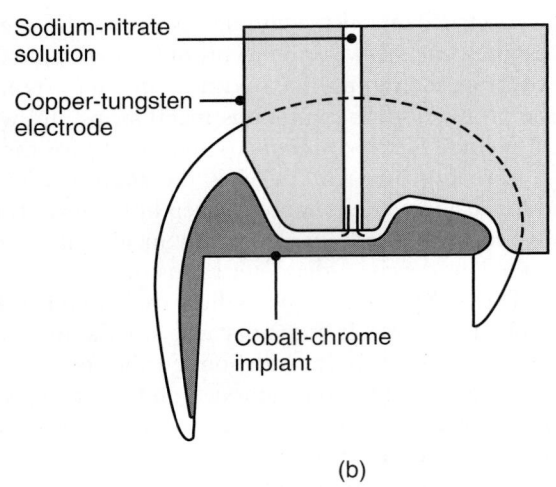

Sodium-nitrate solution

Copper-tungsten electrode

Cobalt-chrome implant

(a)   (b)

FIGURE 27.8   (a) Two total knee-replacement systems showing metal implants (top pieces) with an ultra-high molecular-weight polyethylene insert (bottom pieces). (b) Cross-section of the ECM process as applied to the metal implant. *Source:* Courtesy of Biomet, Inc.

marginal repeatability and quality. The interior surfaces of this part are extremely difficult to access for grinding, and the cobalt-chrome alloy is difficult to grind. Consequently, advanced machining processes (particularly electrochemical machining) were ideal candidates for this operation.

As shown in Fig. 27.8b, the current procedure consists of placing the metal implant in a fixture and bringing a tungsten electrode of the desired final contour in close proximity to the implant. The electrolyte is a sodium nitrate and water mixture and is pumped through the tool, filling the gap between the tool and implant. A power source (typically 10 V and 225 A) is applied, causing local electrochemical machining of the high spots on the implant surface and producing a polished surface.

The electrolyte flow volume can be controlled to maximize surface quality. When the flow rate is too low, defects appear on the machined surface as localized dimples; if the rate is too high, machining times become longer. Typical machining times for this part are four to six minutes.

Source: Courtesy of T. Hershberger and R. Redman, Biomet, Inc., Warsaw, Indiana.

27.3.1   Pulsed electrochemical machining (PECM)

This process is a refinement of ECM; it uses very high current densities (on the order of 1 A/mm^2), but the current is *pulsed* rather than direct current. The purpose of pulsing is to eliminate the need for high electrolyte flow rates, which limit the usefulness of ECM in die and mold making (die sinking). Investigations have shown that PECM improves fatigue life, and the process has been proposed as a possible method for eliminating the recast layer left on die and mold surfaces by electrical-discharge machining. The tolerances obtained typically are in the range of 20 to 100 μm.

Machines can perform a combination of both EDM and PECM, thus the need to move the tool and workpiece between the two processes is eliminated. However, it is difficult to maintain precise alignment when moving the workpiece from the EDM to the PECM. If misaligned significantly, all polishing will occur where the gap is smallest, and passivation will occur where the gap is largest. Also, the process leaves metal residues suspended in the acqueous solution, which is harmful to the environment if discarded without treatment.

The ECM process can be effective for micromachining as well. The complete lack of tool wear implies that the process can be used for precision-electronic components, although the erosion problem due to stray current has to be overcome. ECM machines now have increasing flexibility by implementation of numerical controls.

27.4 | Electrochemical Grinding

Electrochemical grinding (ECG) combines electrochemical machining with conventional grinding. The equipment used is similar to a conventional grinder, except that the wheel is a rotating cathode embedded with abrasive particles (Fig. 27.9a). The wheel is metal bonded with diamond or aluminum-oxide abrasives and rotates at a surface speed of from 1200 to 2000 m/min (4000 to 7000 ft/min).

The abrasives have two functions: (1) to serve as insulators between the wheel and the workpiece and (2) to mechanically remove electrolytic products from the working area. A flow of electrolyte solution (usually sodium nitrate) is provided for the electrochemical machining phase of the operation. Current densities range from 1 to 3 A/mm^2 (500 to 2000 A/in^2). The majority of metal removal in ECG is by electrolytic action and typically, less than 5% of the metal is removed by the abrasive action of the wheel. Therefore, wheel wear is very low, and the workpiece remains cool. Finishing cuts usually are made by the grinding action but only to produce a surface with good finish and dimensional accuracy.

The ECG process is suitable for applications similar to those for milling, grinding, and sawing (Fig. 27.9b), but it is not adaptable to cavity-sinking operations. This process can be applied successfully to carbides and high-strength alloys. It offers a distinct advantage over traditional diamond-wheel grinding when processing very hard materials where wheel wear can be high. ECG machines are available with numerical controls, which improve dimensional accuracy, repeatability, and increase productivity.

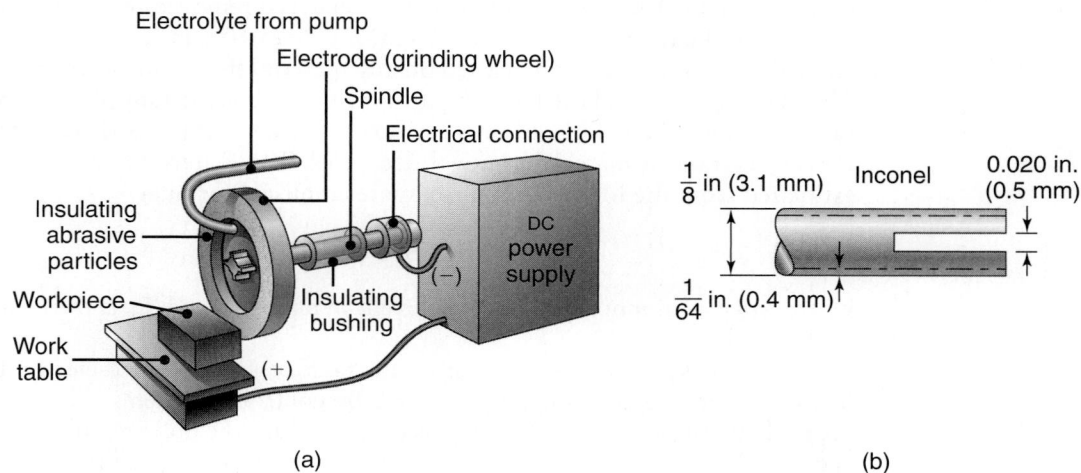

(a)   (b)

FIGURE 27.9   (a) Schematic illustration of the electrochemical-grinding process. (b) Thin slot produced on a round nickel-alloy tube by this process.

Electrochemical honing combines the fine abrasive action of honing with electrochemical action. Although the equipment is costly, this process is as much as five times faster than conventional honing, and the tool lasts as much as ten times longer. It is used primarily for finishing internal cylindrical surfaces.

Design considerations for electrochemical grinding.   In addition to those already listed for electrochemical machining, ECG requires two additional design considerations:

- Designs should avoid sharp inside radii.
- If flat surfaces are to be produced, the electrochemically ground surface should be narrower than the width of the grinding wheel.

27.5 | Electrical-Discharge Machining

The principle of *electrical-discharge machining* (EDM) (also called *electrodischarge* or *spark-erosion machining*) is based on the erosion of metals by spark discharges. We know that when two current-conducting wires are allowed to touch each other an arc is produced. If we look closely at the point of contact between the two wires, we note that a small portion of the metal has been eroded away, leaving a small crater.

Although this phenomenon has been known since the discovery of electricity, it was not until the 1940s that a machining process based on this principle was developed. The EDM process has become one of the most important and widely used production technologies in manufacturing.

Principle of operation.   The basic EDM system consists of a shaped tool (*electrode*) and the workpiece connected to a DC power supply and placed in a **dielectric** (electrically nonconducting) fluid (Fig. 27.10a). When the potential difference between the tool and the workpiece is sufficiently high, the dielectric breaks down and a transient spark discharges through the fluid, removing a very small amount of metal from the workpiece surface. The capacitor discharge is repeated at rates between 200 and 500 kHz with voltages usually ranging between 50 and 380 V and currents from 0.1 to 500 A. The volume of material removed per spark discharge is typically in the range of 10^{-6} to 10^{-4} mm^3 (10^{-10} to 10^{-8} in^3).

The EDM process can be used on any material that is an electrical conductor. The melting point and the latent heat of melting are important physical properties that determine the volume of metal removed per discharge. As these quantities increase, the rate of material removal decreases. The material-removal rate can be estimated from the following approximate empirical formula

$$\text{MRR} = 4 \times 10^4 \, I T_w^{-1.23} \tag{27.2}$$

where MRR is in mm^3/min, I is the current in amperes, and T_w is the melting point of the workpiece in °C.

The workpiece is fixtured within the tank containing the dielectric fluid, and its movements are controlled by numerically controlled systems. The gap between the tool and the workpiece (*overcut*) is critical. Thus, the downward feed of the tool is controlled by a servomechanism, which automatically maintains a constant gap.

Because the process doesn't involve mechanical energy, the hardness, strength, and toughness of the workpiece material do not necessarily influence the removal

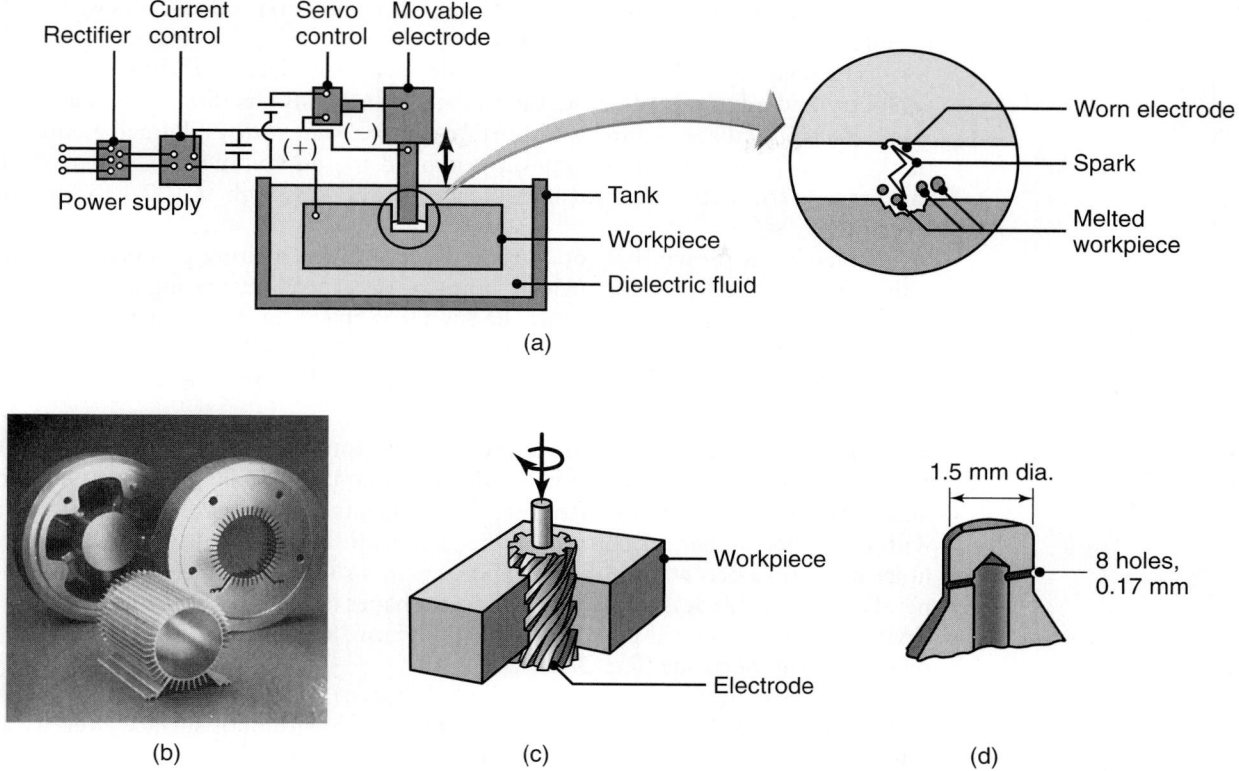

FIGURE 27.10 (a) Schematic illustration of the electrical-discharge machining process. This is one of the most widely used machining processes, particularly for die-sinking applications. (b) Examples of cavities produced by the electrical-discharge machining process, using shaped electrodes. Two round parts (rear) are the set of dies for extruding the aluminum piece shown in front (see also Fig. 15.9b). (c) A spiral cavity produced by EDM using a slowly rotating electrode similar to a screw thread. (d) Holes in a fuel-injection nozzle made by EDM; the material is heat-treated steel. *Source*: (b) Courtesy of AGIE USA Ltd.

rate. The frequency of discharge or the energy per discharge, the voltage, and the current usually are varied to control the removal rate. The removal rate and surface roughness increase with (a) increasing current density and (b) decreasing frequency of sparks.

Dielectric fluids. The functions of the dielectric fluid are to:

1. Act as an insulator until the potential is sufficiently high.
2. Provide a cooling medium.
3. Act as a flushing medium and carry away the debris in the gap.

The most common dielectric fluids are mineral oils, although kerosene and distilled and deionized water also are used in specialized applications. Clear, low-viscosity fluids also are available; although more expensive, these fluids make cleaning easier. The machines are equipped with a pump and filtering system for the dielectric fluid.

Electrodes. Electrodes for EDM usually are made of graphite, although brass, copper, or copper-tungsten alloys also are used. The tools can be shaped by forming, casting, powder metallurgy, or CNC machining techniques. Tungsten-wire electrodes as small as 0.1 mm (0.005 in.) in diameter have been used to produce holes with depth-to-hole diameter ratios of up to 400:1.

The sparks in this process also erode away the electrode, thus changing its geometry and adversely affecting the shape produced and its dimensional accuracy. *Tool (electrode) wear* is thus an important factor. *Wear ratio* is defined as the ratio of the volume of workpiece material removed to the volume of tool wear. This ratio ranges from about 3:1 for metallic electrodes to as high as 100:1 for graphite electrodes.

It has been shown that tool wear is related to the melting points of the materials involved: the lower the melting point of the electrode, the higher is the wear rate. Also, the higher the current, the higher is the wear. Consequently, graphite electrodes have the highest wear resistance. Tool wear can be minimized by reversing the polarity and using copper tools—a process called **no-wear EDM.**

Process capabilities. Electrical-discharge machining has numerous applications, such as the production of dies for forging, extrusion, die casting, injection molding, and large sheet-metal automotive-body components (**die-sinking machining centers** with computer numerical control). Other applications include deep, small-diameter holes using tungsten wire as the electrode, narrow slots in parts, cooling holes in superalloy turbine blades, and various intricate shapes (see Figs. 27.10b and c). Stepped cavities can be produced by controlling the relative movements of the workpiece in relation to the electrode (Fig. 27.11).

Because of the molten and resolidified (recast) surface structure, high rates of material removal produce a very rough surface finish with poor surface integrity and low fatigue properties. Therefore, finishing cuts are made at low removal rates, or the recast layer is removed subsequently by finishing operations. It has been shown that surface finish can be improved by oscillating the electrode in a planetary motion at amplitudes of 10 to 100 μm.

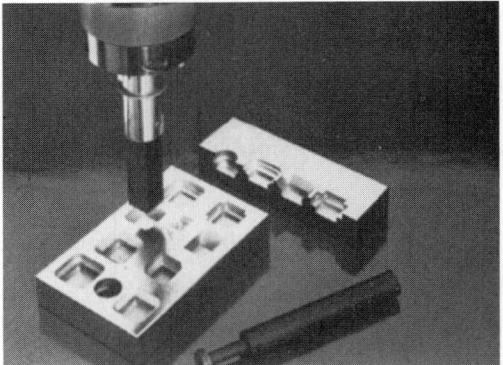

FIGURE 27.11 Stepped cavities produced with a square electrode by the EDM process. The workpiece moves in the two principal horizontal directions and its motion is synchronized with the downward movement of the electrode to produce these cavities. Also shown is a round electrode capable of producing round or elliptical cavities. *Source:* Courtesy of AGIE USA Ltd.

Design considerations for EDM.   The general design guidelines for electrical-discharge machining are as follows:

- Parts should be designed so that the required electrodes can be shaped properly and economically.
- Deep slots and narrow openings should be avoided.
- For economic production, the surface finish specified should not be too fine.
- In order to achieve a high production rate, the bulk of material removal should be done by conventional processes (roughing out).

27.5.1   Wire EDM

A variation of EDM is *wire EDM* (Fig. 27.12) or *electrical-discharge wire cutting*. In this process, which is similar to contour cutting with a band saw (see Fig. 24.25), a slowly moving wire travels along a prescribed path, cutting the workpiece. This process is used to cut plates as thick as 300 mm (12 in.) and to make punches, tools, and dies from hard metals. It also can cut intricate components for the electronics industry. Figure 27.13a shows a thick plate being cut by this process, and a wire EDM machine is shown in Fig. 27.13b.

The wire usually is made of brass, copper, tungsten, or molybdenum; zinc- or brass-coated and multi-coated wires also are used. The wire diameter is typically about 0.30 mm (0.012 in.) for roughing cuts and 0.20 mm (0.008 in.) for finishing cuts. The wire should have high electrical conductivity and tensile strength, as the tension on it is typically 60% of its tensile strength.

The wire usually is used only once, as it is relatively inexpensive compared to the type of operation it performs. It travels at a constant velocity in the range of 0.15 to 9 m/min (6 to 360 in./min), and a constant gap (kerf) is maintained during the cut. The trend in the use of dielectric fluids is toward clear, low-viscosity fluids.

The cutting speed generally is given in terms of the cross-sectional area cut-per-unit time. Typical examples are: 18,000 mm²/hr (28 in²/hr) for 50-mm (2-in.) thick

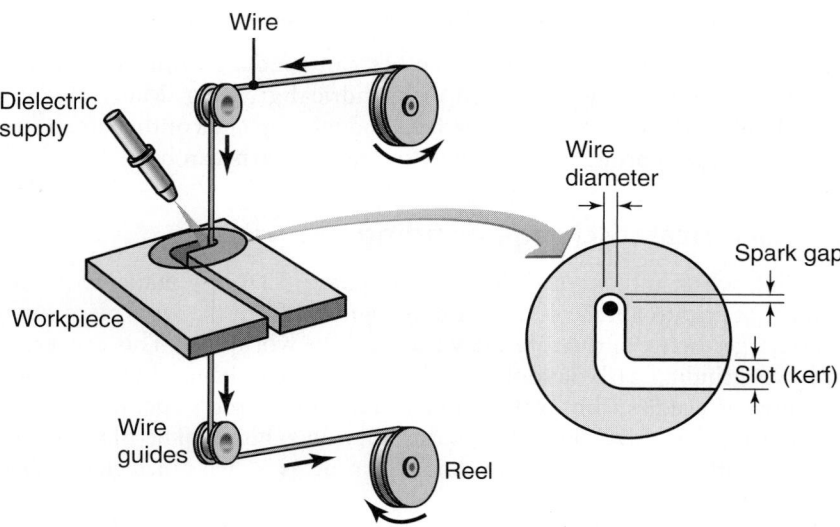

FIGURE 27.12   Schematic illustration of the wire EDM process. As many as 50 hours of machining can be performed with one reel of wire, which is then discarded.

(a)   (b)

FIGURE 27.13   (a) Cutting a thick plate with wire EDM. (b) A computer-controlled wire EDM machine. *Source:* Courtesy of AGIE USA Ltd.

D2 tool steel and 45,000 mm^2/hr (70 in^2/hr) for 150-mm (6-in.) thick aluminum. These removal rates indicate a linear cutting speed of 18,000/50 = 360 mm/hr = 6 mm/min and 45,000/150 = 300 mm/hr = 5 mm/min, respectively.

Modern wire EDM machines (**multiaxis EDM wire-cutting machining centers**) are capable of producing three-dimensional shapes and are equipped with such features as:

- Computer controls for controlling the cutting path of the wire (Fig. 27.13b) and its angle with respect to the workpiece plane.
- Multiheads for cutting two parts at the same time.
- Features such as controls for preventing wire breakage.
- Automatic self-threading features in case of wire breakage.
- Programmed machining strategies to optimize the operation.

Two-axis computer-controlled machines can produce cylindrical shapes in a manner similar to a turning operation or cylindrical grinding. Many modern wire EDM machines allow the control of the feed and take-up ends of the wire to traverse independently in two principal directions, so tapered parts can be made.

27.5.2 Electrical-discharge grinding

The grinding wheel in *electrical-discharge grinding* (EDG) is made of graphite or brass and contains no abrasives. Material is removed from the workpiece surface by spark discharges between the rotating wheel and the workpiece. This process is used primarily for grinding carbide tools and dies but also can be used with fragile parts, such as surgical needles, thin-walled tubes, and honeycomb structures. The ECDG process is faster than EDG, but power consumption is higher. The EDG process can be combined with electrochemical grinding. The process is then called **electrochemical-discharge grinding** (ECDG).

Material is removed by chemical action (with the electrical discharges from the graphite wheel breaking up the oxide film) and is washed away by the electrolyte flow. The material-removal rate in EDG can be estimated from the equation:

$$\text{MRR} = KI \qquad\qquad (27.3)$$

where MRR is in mm^3/min, I is the current in amperes, and K is a workpiece material factor in mm^3/A-min. For example, $K = 4$ for tungsten carbide, and $K = 16$ for steel.

In **sawing** with EDM, a setup similar to a band or circular saw (but without any teeth) is used with the same electrical circuit for EDM. Narrow cuts can be made at high rates of metal removal. Because cutting forces are negligible, this process can be used on thin and slender components as well.

27.6 | Laser-Beam Machining

In *laser-beam machining* (LBM), the source of energy is a **laser** (an acronym for *light amplification by stimulated emission of radiation*), which focuses optical energy on the surface of the workpiece (Fig. 27.14a). The highly focused, high-density energy source melts and evaporates portions of the workpiece in a controlled manner. This process (which does not require a vacuum) is used to machine a variety of metallic and nonmetallic materials.

There are several types of lasers used in manufacturing operations (Table 27.2):

CO$_2$ (*pulsed* or *continuous wave*)
Nd:YAG (neodymium: yttrium-aluminum-garnet)
Nd:glass, ruby
Excimer lasers (from the words *exci*ted and di*mer*, meaning two mers or two molecules of the same chemical composition)

Important physical parameters in LBM are the *reflectivity* and *thermal conductivity* of the workpiece surface and its specific heat and latent heats of melting and evaporation. The lower these quantities, the more efficient the process. The cutting depth may be expressed as

$$t = \frac{CP}{vd} \tag{27.4}$$

TABLE 27.2

General Applications of Lasers in Manufacturing	
Application	Laser type
Cutting	
Metals	PCO$_2$, CWCO$_2$, Nd:YAG, ruby
Plastics	CWCO$_2$
Ceramics	PCO$_2$
Drilling	
Metals	PCO$_2$, Nd:YAG, Nd:glass, ruby
Plastics	Excimer
Marking	
Metals	PCO$_2$, Nd:YAG
Plastics	Excimer
Ceramics	Excimer
Surface treatment, metals	CWCO$_2$
Welding, metals	PCO$_2$, CWCO$_2$, Nd:YAG, Nd:glass, ruby

Note: P = pulsed, CW = continuous wave, Nd:YAG = neodymium: yttrium-aluminum-garnet.

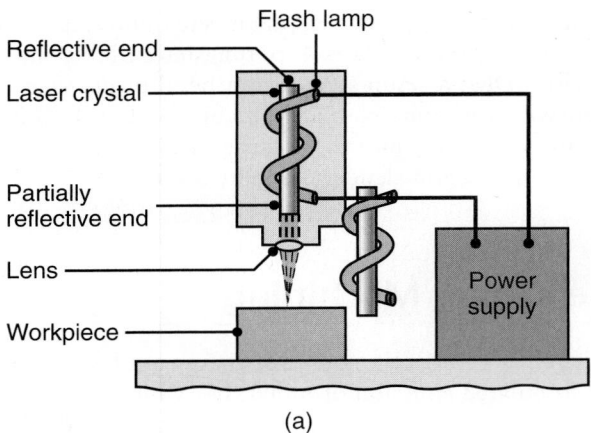

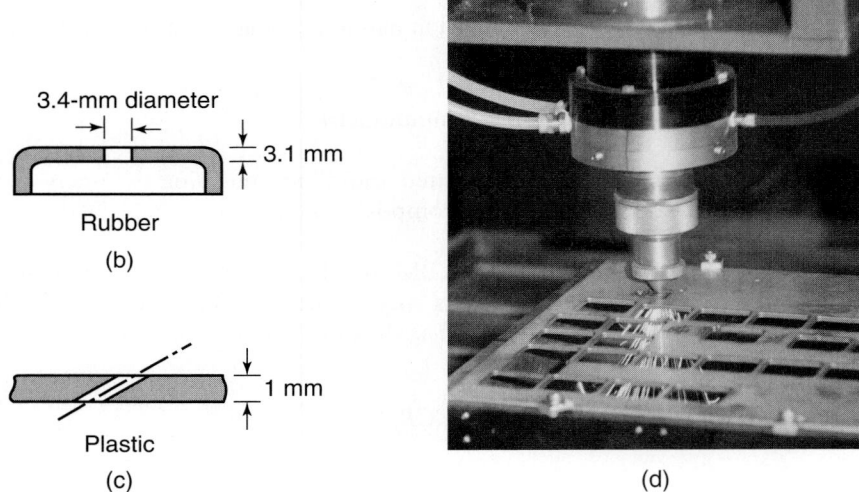

FIGURE 27.14 (a) Schematic illustration of the laser-beam machining process. (b) and (c) Examples of holes produced in nonmetallic parts by LBM. (d) Cutting sheet metal with a laser beam. *Source*: (d) Courtesy of Rofin-Sinar, Inc.,

where t is the depth, C is a constant for the process, P is the power input, v is the cutting speed, and d is the laser-spot diameter. Peak energy densities of laser beams are in the range of 5 to 200 kW/mm^2.

The surface produced by LBM is usually rough and has a *heat-affected zone* (as described in Section 30.9), which in critical applications may have to be removed or heat treated. Kerf width is an important consideration, as it is in other cutting processes, such as sawing, wire EDM, and electron-beam machining.

Laser beams may be used in combination with a gas stream (such as oxygen) to increase energy absorption (**laser-beam torch**) for cutting sheet metals. High-pressure, inert-gas (nitrogen or argon) assisted laser cutting is used for stainless steel and aluminum; it leaves an oxide-free edge that can improve weldability. Gas streams also have the important function of blowing away molten and vaporized material from the workpiece surface.

Process capabilities.   Laser-beam machining is used widely for drilling, trepanning, and cutting metals, nonmetallic materials, ceramics, and composite materials (Fig. 27.14b and c). The abrasive nature of composite materials and the cleanliness of the operation have made laser-beam machining an attractive alternative to traditional machining methods. Holes as small as 0.005 mm (0.0002 in.) with hole depth-to-diameter ratios of 50:1 have been produced in various materials, although a more practical minimum is 0.025 mm (0.001 in.). Steel plates as thick as 32 mm (1.25 in.) can be cut with laser beams.

Laser-beam machining is being used increasingly in the electronics and automotive industries. Bleeder holes for fuel-pump covers and lubrication holes in transmission hubs are being drilled with lasers, for example. The cooling holes in the first stage of producing vanes for the Boeing 747 jet engines also are produced by lasers. Significant cost savings have been achieved by laser-beam machining—a process that is competing with electrical-discharge machining.

Laser beams also are used for the following applications:

- **Welding.**
- Small-scale and localized **heat treating** of metals and ceramics to modify their surface mechanical and tribological properties.
- **Marking** of parts, such as letters, numbers, codes, etc. Marking also can be done by processes such as (a) with ink, (b) with mechanical devices such as punches, pins, styluses, scroll rolls, or by stamping, and (c) by etching. Although the equipment is more expensive than that used in other methods, marking and engraving with lasers has become increasingly common due to its accuracy, reproducibility, flexibility, ease of automation, and on-line application in manufacturing.

The inherent *flexibility* of the laser-cutting process—including its fiber-optic beam delivery, simple fixturing, low setup times, and the availability of multi-kW machines and two- and three-dimensional computer-controlled robotic laser-cutting systems—is an attractive feature. Therefore, laser cutting for sheet can compete successfully with the traditional punching processes described in Chapter 16. There are efforts now to combine the two processes for improved overall efficiency. (See Example 27.1.)

Extreme caution should be exercised with lasers. Even low-power lasers can cause damage to the retina of the eye if proper safety precautions are not observed.

Design considerations for LBM.   General design guidelines for laser-beam machining are:

- Designs with sharp corners should be avoided, since they can be difficult to produce.
- Deep cuts will produce tapered walls.
- Reflectivity of the workpiece surface is an important consideration in laser-beam machining; because they reflect less, dull and unpolished surfaces are preferable.
- Any adverse effects on the properties of the machined materials caused by the high local temperatures and heat-affected zone should be investigated.

EXAMPLE 27.1 Combining laser cutting and punching of sheet metal

As has been seen, laser cutting and punching processes have their respective advantages and limitations regarding both technical and economic aspects. The advantages of laser cutting are generally (a) smaller batches, (b) the flexibility of the operation, (c) a wide range of thicknesses, (d) prototyping capability, (e) materials and composites that otherwise may be cut with difficulty, and (f) complex geometries that can be programmed.

Punching advantages and drawbacks include (a) required large lot sizes for economic justification of purchasing tooling, (b) relatively simple parts, (c) a small range of part thicknesses, (d) fixed and limited punch geometries (even when using turrets), (e) rapid production, and (f) integration with subsequent processing after punching.

It is evident that the two processes cover different but complementary ranges. It is not difficult to visualize parts with some features that can be produced best by one process, and other features that are best produced by the other process.

Machines have been designed and built in such a manner that the processes and fixturing can be utilized jointly to their full extent but without interfering with each other's operational boundaries. The purpose of combining them is to increase the overall efficiency and productivity of the manufacturing process for parts that are within the capabilities of each of the two processes—similar to the concept of the machining centers described in Section 25.2. For example, turret-punch presses have been equipped with an integrated laser head; the machine can punch or laser cut, but it cannot do both simultaneously.

A number of factors have to be taken into account in such a combination with respect to the characteristics of each operation: (1) the ranges of sizes, thicknesses, and shapes to be produced and how they are to be nested; (2) processing and setup times, including the loading, fixturing, and unloading of parts; (3) programming for cutting; and (4) the process capabilities of each method, including dynamic characteristics, vibrations, and shock from punching (and isolation) that may disturb adjustments and alignments of the laser components.

27.7 | Electron-Beam Machining

The energy source in *electron-beam machining* (EBM) is high-velocity electrons, which strike the workpiece surface and generate heat (Fig. 27.15). The machines utilize voltages in the range of 50 to 200 kV to accelerate the electrons to speeds of 50 to 80% of the speed of light. Applications of this process are similar to those of laser-beam machining, except that EBM requires a vacuum. Consequently, it is used much less than laser-beam machining.

Electron-beam machining can be used for very accurate cutting of a wide variety of metals. Surface finish is better and kerf width is narrower than those for other thermal cutting processes. (See also Section 30.7 on *electron-beam welding*.) The interaction of the electron beam with the workpiece surface produces hazardous x-rays. Therefore, the equipment should be used only by highly trained personnel.

Plasma-arc cutting. In plasma-arc cutting (PAC) *plasma beams* (ionized gas) are used to rapidly cut ferrous and nonferrous sheets and plates. The temperatures generated are very high (9400°C; 17,000°F in the torch for oxygen as a plasma gas). Consequently, the process is fast, the kerf width is small, and the surface finish is

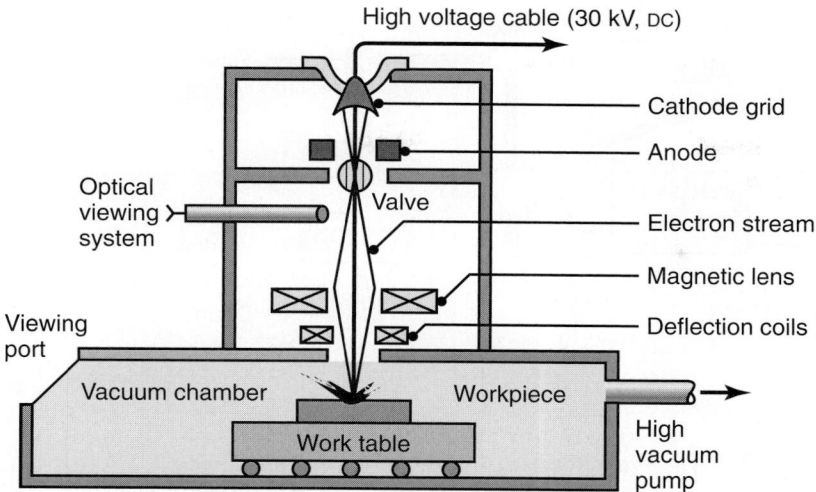

High voltage cable (30 kV, DC)

Cathode grid

Anode

Optical viewing system

Valve

Electron stream

Magnetic lens

Deflection coils

Viewing port

Vacuum chamber

Workpiece

Work table

High vacuum pump

FIGURE 27.15   Schematic illustration of the electron-beam machining process. Unlike LBM, this process requires a vacuum, so workpiece size is limited to the size of the vacuum chamber.

good. Parts as thick as 150 mm (6 in.) can be cut. Material-removal rates are much higher than those associated with the EDM and LBM processes, and parts can be machined with good reproducibility. Plasma-arc cutting is highly automated today, using programmable controllers.

Design considerations for EBM.   The guidelines for EBM generally are similar to those for LBM. Additional considerations are:

- Because vacuum chambers have limited capacity, parts or batches should match closely the size of the vacuum chamber for a high-production–rate per cycle.

- If a part requires electron-beam machining on only a small portion of the workpiece, consideration should be given to manufacturing it as a number of smaller components and assembling them after electron-beam machining.

27.8 | Water-Jet Machining

When we put our hand across a jet of water or air, we feel a considerable concentrated force acting on it. This force results from the momentum change of the stream and, in fact, is the principle on which the operation of water or gas turbines is based. In *water-jet machining* (WJM) (also called **hydrodynamic machining**), this force is utilized in cutting and deburring operations (Fig. 27.16). (See also *water-jet peening*, Section 34.2.)

The water jet acts like a saw and cuts a narrow groove in the material. A pressure level of about 400 MPa (60 ksi) generally is used for efficient operation, although pressures as high as 1400 MPa (200 ksi) can be generated. Jet-nozzle diameters range between 0.05 and 1 mm (0.002 and 0.040 in.). A water-jet cutting machine and its operation are shown in Fig. 27.16b. A variety of materials can be cut, including plastics, fabrics, rubber, wood products, paper, leather, insulating materials, brick, and composite materials (Fig. 27.16c).

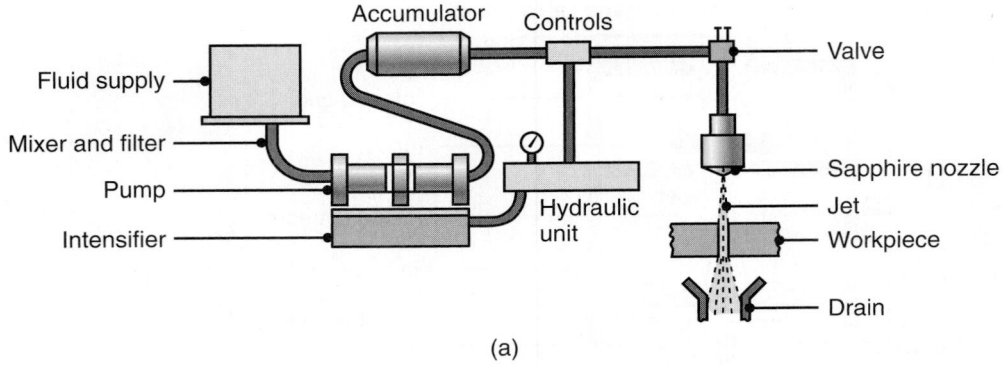

(a)

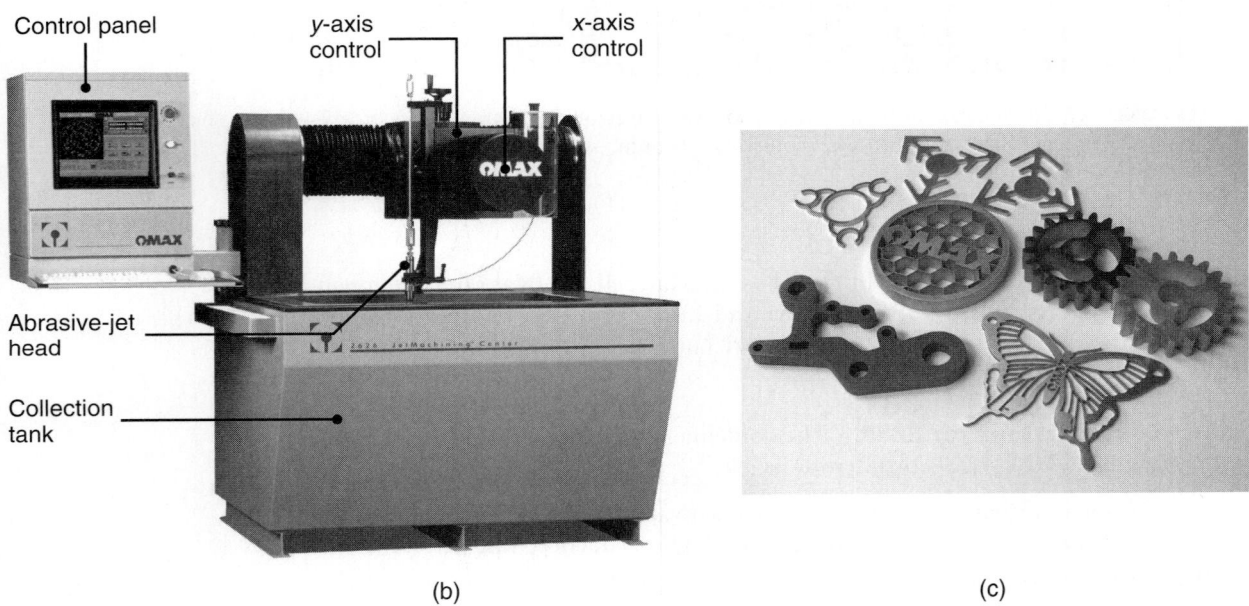

(b)   (c)

FIGURE 27.16 (a) Schematic illustration of the water-jet machining process. (b) A computer-controlled water-jet cutting machine. (c) Examples of various nonmetallic parts produced by the water-jet cutting process. *Source:* Courtesy of OMAX Corporation.

Depending on the materials, thicknesses can range up to 25 mm (1 in.) and higher. Vinyl and foam coverings for automobile dashboards (as well as some body panels) are being cut using multiple-axis, robot-guided water-jet machining equipment. Because it is an efficient and clean operation compared to other cutting processes, it also is used in the food-processing industry for cutting and slicing food products.

The advantages of this process are:

- Cuts can be started at any location without the need for predrilled holes.
- No heat is produced.
- No deflection of the rest of the workpiece takes place, thus the process is suitable for flexible materials.
- Little wetting of the workpiece takes place.
- The burr produced is minimal.
- It is an environmentally safe manufacturing process.

Abrasive water-jet machining. In *abrasive water-jet machining* (AWJM), the water jet contains abrasive particles (such as silicon carbide or aluminum oxide), which increase the material-removal rate above that of water-jet machining. Metallic, nonmetallic, and advanced composite materials of various thicknesses can be cut in single or multilayers.

This process is suitable particularly for heat-sensitive materials that cannot be machined by processes in which heat is produced. Cutting speeds can be as high as 7.5 m/min (25 ft/min) for reinforced plastics but much lower for metals. Consequently, this process may not be acceptable for situations requiring high production rates.

The minimum hole size that can be produced satisfactorily to date is about 3 mm (0.12 in.); maximum hole depth is on the order of 25 mm (1 in.). With multiple-axis and robotic-controlled machines, complex three-dimensional parts can be machined to finish dimensions. The optimum level of abrasives in the jet stream is controlled automatically in modern AWJM systems. Nozzle life has been improved by making nozzles from rubies, sapphires, and carbide-based composite materials (Fig. 27.16a).

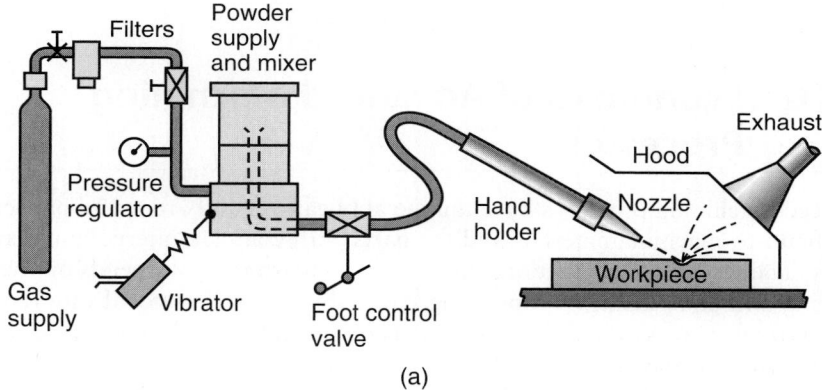

(a)

(b)

FIGURE 27.17 (a) Schematic illustration of the abrasive-jet machining process. (b) Examples of parts produced through abrasive-jet machining, produced in 50-mm (2-in.) thick 304 stainless steel. *Source*: Courtesy of OMAX Corporation.

27.9 | Abrasive-Jet Machining

In *abrasive-jet machining* (AJM), a high-velocity jet of dry air, nitrogen, or carbon dioxide containing abrasive particles is aimed at the workpiece surface under controlled conditions (Fig. 27.17). The impact of the particles develops a sufficiently concentrated force (see also Section 26.6) to perform operations such as (a) cutting small holes, slots, or intricate patterns in very hard or brittle metallic and nonmetallic materials, (b) deburring or removing small flash from parts, (c) trimming and beveling, (d) removing oxides and other surface films, and (e) generally cleaning components with irregular surfaces.

The gas-supply pressure is on the order of 850 kPa (125 psi), and the abrasive-jet velocity can be as high as 300 m/s (100 ft/s) and is controlled by a valve. The nozzles are usually made of tungsten carbide or sapphire, both of which have abrasive wear resistance. The abrasive size is in the range of from 10 to 50 μm (400 to 2000 μin.). Because the flow of the free abrasives tends to round off corners, designs for abrasive-jet machining should avoid sharp corners. Also, holes made in metal parts tend to be tapered. There is some hazard involved in using this process because of airborne particulates. This problem can be avoided by using the abrasive water-jet machining process.

27.10 | Economics of Advanced Machining Processes

Advanced machining processes have unique applications and are useful particularly for difficult-to-machine materials and for parts with complex internal and external profiles. The economic production run for a particular process depends on the costs of tooling and equipment, the operating costs, the material-removal rate required, the level of operator skill required, as well as on secondary and finishing operations that subsequently may be necessary.

In chemical machining (which is inherently a slow process), an important factor is the cost of reagents, maskants, and disposal—together with the cost of cleaning the parts. In electrical-discharge machining, the cost of electrodes and the need to periodically replace them can be significant.

The rate of material removal and the production rate can vary significantly in these processes, as can be seen in Table 27.1. The cost of tooling and equipment also varies considerably, as does the operator skill required. The high capital investment for machines (such as for electrical and high-energy beam machining, especially when equipped with robotic controls) has to be justified in terms of the production runs and the feasibility of manufacturing the same part by other means, if at all possible.

CASE STUDY 27.2 | Manufacture of Stents

Heart attacks, strokes, and other cardiovascular diseases claim one life every 33 seconds in the United States alone.[1] These illnesses are attributed mostly to coronary artery disease, which is the gradual buildup of fat (cholesterol) within the artery wall causing the coronary arteries to become narrowed or blocked. This condition reduces the blood flow to the heart muscle and eventually leads to a

---

[1] 2002 Heart and Stroke Statistical Update, American Heart Association, p. 4

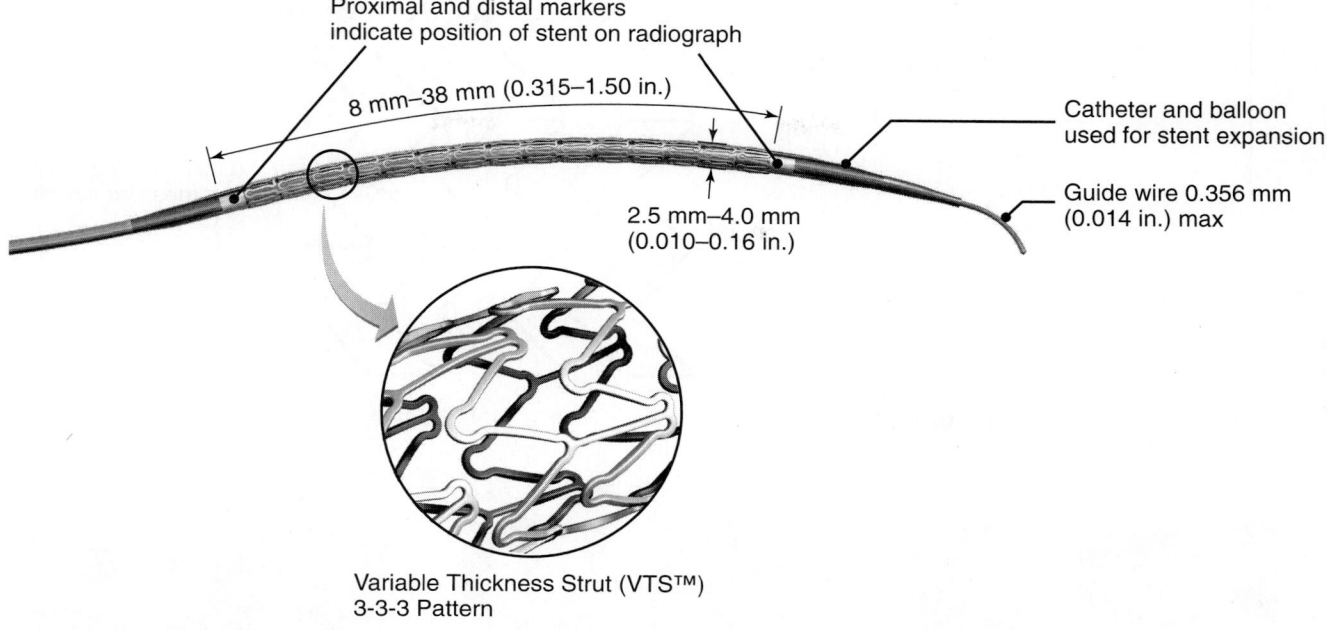

Proximal and distal markers
indicate position of stent on radiograph

8 mm–38 mm (0.315–1.50 in.)

Catheter and balloon
used for stent expansion

Guide wire 0.356 mm
(0.014 in.) max

2.5 mm–4.0 mm
(0.010–0.16 in.)

Variable Thickness Strut (VTS™)
3-3-3 Pattern

FIGURE 27.18 The Guidant MULTI-LINK TETRA™ coronary stent system.

heart attack, stroke, or other cardiovascular diseases. One of the most popular methods today for keeping blocked arteries open is to implant a stent into the artery. The MULTI-LINK TETRA™ (shown in Fig. 27.18) consists of a tiny mesh tube that is expanded using a balloon dilatation catheter and implanted into a blocked or partially blocked coronary artery. A stent serves as a scaffolding or mechanical brace to keep the artery open. A stent offers the patient a minimally invasive method for treating coronary heart disease for which the alternative is usually open-heart bypass surgery—a procedure with greater risk, pain, rehabilitation time, and cost to the patient.

Stent manufacturing is extremely demanding, and the accuracy and precision of its design are paramount for proper performance. The manufacture of the stent must allow satisfaction of all of the design constraints and provide an extremely reliable device; thus, strict quality control procedures must be in place throughout the manufacturing process. When designing a stent, there are many material-selection factors to consider, such as radial strength, corrosion resistance, endurance limit, flexibility, and biocompatibility. Radial strength is important, because the stent must withstand the pressure the artery exerts on the stent upon expansion.

Because the stent is implantable in the body, it must be corrosion resistant; also, it must be able to withstand the stress undulations caused by heart-beats. Consequently, the material also must have high resistance to fatigue. In addition, the stent must have the appropriate wall thickness in order to be flexible enough to negotiate through the tortuous anatomy of the heart. Above all, the material must be biocompatible to the body, because it remains there for the rest of the patient's life. A standard material for stents—one that satisfies all of these requirements—is 316L stainless steel.

0.12 mm (0.0049 in.) section thickness to provide radiopacity

0.091 mm (0.0036 in.) thickness for flexibility

FIGURE 27.19   Detail of the 3-3-3 MULTI-LINK TETRA™ pattern.

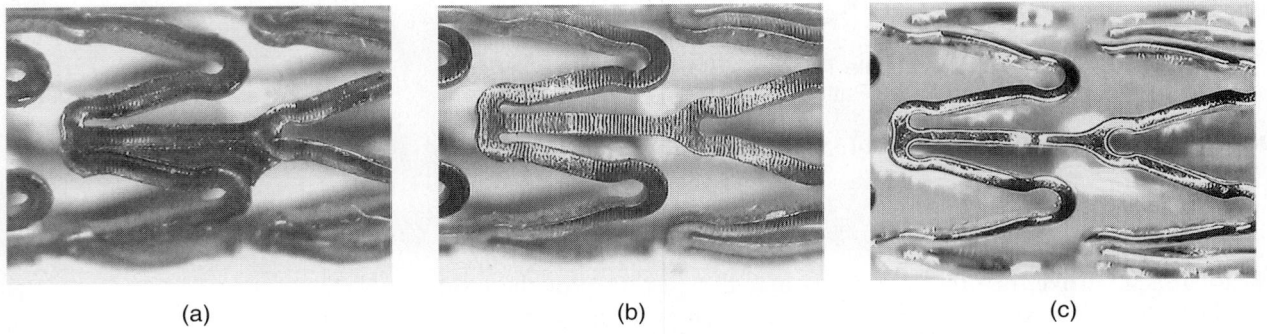

(a)   (b)   (c)

FIGURE 27.20   Evolution of the stent surface. (a) MULTI-LINK TETRA™ after lasing. Note that a metal slug still is attached. (b) After removal of slag. (c) After electropolishing.

The pattern of a stent also affects its performance. A finite-element analysis of the stent's design first is performed to determine how the stent will perform under the stresses induced by the beating heart. The strut pattern, tube thickness, and strut-leg width all affect the performance of the stent. Many different patterns are possible, such as the MULTI-LINK TETRA™ stent, which is shown in Fig. 27.19 and includes some of the critical dimensions.

A stent starts out as a drawn stainless-steel tube with an outer diameter that matches the final stent dimension and a wall thickness selected to provide proper strength when expanded. The drawn tube then is laser machined to achieve the desired pattern (Fig. 27.20a). This method has proven to be very effective because of the small, intricate pattern of the stent and the tight dimensional tolerances that it must have. As the laser cuts the pattern of the stent, it leaves small metal slugs that need to be removed; therefore, it is very important that the laser cut all the way through the tube wall. A thick oxide layer or slag inevitably develops on the stainless steel due to the thermal and chemical attack from the air. In addition, weld splatter, burrs, and other surface defects result from laser machining. Therefore, finishing operations are required to remove the splatter and oxide layer.

The first finishing operation after laser machining involves chemical etching to remove as much of the slag on the stent as possible; this operation typically is done in an acid solution, and the resulting surface is shown in Fig. 27.20b. Once the slag has been sufficiently removed, any residual burrs from the laser-machining process need to be removed, and the stent surface must be finished. It is very important that the surface finish of the stent be smooth to prevent possible formation of clots (thrombi) on the stent. To remove the sharp points, the stent is electropolished by passing an electrical current through an electrochemical solution to ensure a proper surface finish (Fig. 27.20c) that is both shiny and smooth. The stent then is placed onto a balloon catheter assembly, sterilized, and packaged for delivery to the surgeon.

Source: Courtesy of K. L. Graham, Guidant Corporation.

SUMMARY

- Advanced machining processes have unique capabilities and utilize chemical, electrochemical, electrical, and high-energy-beam sources of energy. The mechanical properties of the workpiece material are not significant, because these processes rely on mechanisms that do not involve the strength, hardness, ductility, or toughness of the material. Rather, they involve physical, chemical, and electrical properties.

- Chemical and electrical methods of machining are suitable particularly for hard materials and complex shapes. They do not produce forces (and therefore can be used for thin, slender, and flexible workpieces), significant temperatures, or residual stresses. However, the effects of these processes on surface integrity must be investigated, as they can damage surfaces considerably, thus reducing the fatigue life of the product.

- High-energy-beam machining processes basically utilize laser beams, electron beams, and plasma beams. They have important industrial applications, possess a high flexibility of operation with robotic controls, and are competitive economically with various other processes.

- Water-jet machining, abrasive water-jet machining, and abrasive-jet machining processes can be used for cutting as well as deburring operations. Because they do not utilize hard tooling, they have an inherent flexibility of operation.

KEY TERMS

Abrasive-jet machining	Electrochemical honing	Photochemical blanking
Abrasive water-jet machining	Electrochemical machining	Photochemical machining
Chemical blanking	Electrode	Photoetching
Chemical machining	Electrolyte	Plasma beams
Chemical milling	Electron-beam machining	Plasma-arc cutting
Die sinking	Etchant	Pulsed electrochemical machining
Dielectric	Hydrodynamic machining	Reagent
Electrical-discharge grinding	Laser	Shaped-tube electrolytic machining
Electrical-discharge machining	Laser-beam machining	Undercut
Electrochemical grinding	No-wear EDM	Wire EDM
Electrochemical-discharge grinding	Photoresist	Water-jet machining

BIBLIOGRAPHY

ASM Handbook, Vol. 16: *Machining*, ASM International, 1989.

Brown, J., *Advanced Machining Technology Handbook*, McGraw-Hill, 1998.

Chryssolouris, G., and Sheng, P., *Laser Machining, Theory and Practice*, Springer, 1991.

Crafer, R.C., and Oakley, P.J., *Laser Processing in Manufacturing*, Chapman & Hall, 1993.

El-Hofy, H.A.-G., *Advanced Machining Processes*, McGraw-Hill, 2005.

Gillespie, L., *Deburring and Edge Finishing Handbook*, Society of Manufacturing Engineers, 1999.

Guitran, E.B., *The EDM Handbook*, Hanser-Gardner, 1997.

Jain, V.K., and Pandey, P.C., *Theory and Practice of Electrochemical Machining*, Wiley, 1993.

Jameson, E.C., *Electrical Discharge Machining*, Society of Manufacturing Engineers, 2001.

Lange, K. (ed.), *Handbook of Metal Forming* (Chapter 32, *Die Manufacture*), McGraw-Hill, 1985.

Machinery's Handbook, Industrial Press, revised periodically.

Machining Data Handbook, 3rd ed., 2 vols., Machinability Data Center, 1980.

McGeough, J.A., *Advanced Methods of Machining*, Chapman & Hall, 1988.

Migliore, L., *Laser Materials Processing*, Marcel Dekker, 1996.

Momber, A.W., and Kovacevic, R., *Principles of Abrasive Water Jet Machining*, Springer, 1998.

Powell, J., *CO_2 Laser Cutting*, 2nd ed., Springer, 1998.

Sommer, C., and Sommer, S., *Wire EDM Handbook*, Technical Advanced Publishing Co., 1997.

Steen, W.M., *Laser Material Processing*, Springer, 1991.

Taniguchi, N., *Energy-Beam Processing of Materials*, Oxford, 1989.

Tool and Manufacturing Engineers Handbook, 4th ed., Vol. 1: *Machining*, Society of Manufacturing Engineers, 1983.

REVIEW QUESTIONS

27.1 Name the processes involved in chemical machining. Describe briefly their principles.

27.2 What should be the properties of maskants? Explain.

27.3 Describe the similarities and differences between chemical blanking and conventional blanking using dies.

27.4 Explain the difference between chemical machining and electrochemical machining.

27.5 What is the underlying principle of electrochemical grinding?

27.6 Explain how the EDM process is capable of producing complex shapes.

27.7 What are the capabilities of wire EDM? Could this process be used to make tapered parts? Explain.

27.8 Describe the advantages of water-jet machining.

27.9 What is the difference between photochemical blanking and chemical blanking?

27.10 Can contoured cavities be machined chemically?

27.11 What type of workpiece is not suitable for laser-beam machining?

27.12 What is an undercut? Why must it be considered in chemical machining?

QUALITATIVE PROBLEMS

27.13 Give possible technical and economic reasons why the processes described in this chapter might be preferred over those described in the preceding chapters.

27.14 Why has electrical-discharge machining become so widely used in industry?

27.15 Explain why the mechanical properties of workpiece materials are not significant in most of the processes described in this chapter.

27.16 In which types of products or parts is the wire EDM process most applicable?

27.17 Why may different advanced machining processes affect the fatigue strength of materials to different degrees?

27.18 Explain why it is difficult to produce sharp profiles and corners with some of the processes described in this chapter.

27.19 Which of the advanced machining processes causes thermal damage? What is the consequence of such damage to workpieces?

27.20 In abrasive water-jet machining, at what stage

is the abrasive introduced in the water jet? Survey the available literature, then draw a schematic illustration of the equipment involved.

27.21 Describe your thoughts regarding the laser-beam machining of nonmetallic materials. Give several possible applications, including their advantages as compared to other processes.

27.22 Why is the preshaping or premachining of parts sometimes desirable in the processes described in this chapter?

27.23 Are deburring operations necessary for parts made by advanced machining processes? Explain and give several specific examples.

27.24 Do you think it should be possible to produce spur gears by advanced machining processes, starting with a round blank? Explain.

27.25 List and explain factors that contribute to a poor surface finish in the processes described in this chapter.

27.26 Conduct a survey of the available technical literature and describe the types of surfaces produced by electron-beam, plasma-arc, and laser cutting.

27.27 It was stated that graphite is the preferred material for EDM tooling. Would graphite be useful in wire EDM? Explain.

27.28 What is the purpose of the abrasives in electrochemical grinding?

27.29 Why are lasers increasingly used to mark parts?

27.30 Which of the processes described in this chapter are suitable for producing very small and deep holes? Why?

27.31 Is kerf width important in wire EDM? Explain.

27.32 Are there similarities between photochemical machining and solid ground curing? (See Section 20.3.6.)

QUANTITATIVE PROBLEMS

27.33 A 100-mm deep hole that is 20 mm in diameter is being produced by electrochemical machining. A high production rate is more important than machined surface quality. Estimate the maximum current and the time required to perform this operation.

27.34 If the operation in Problem 27.33 were performed on an electrical-discharge machine, what would be the estimated machining time?

27.35 A cutting-off operation is being performed with a laser beam. The workpiece being cut is $\frac{3}{4}$ in.

thick and 8 in. long. If the kerf is $\frac{3}{32}$ in. wide, estimate the time required to perform this operation.

27.36 A 1.0-in.-thick copper plate is being machined through wire EDM. The wire moves at a speed of 5 ft/min and the kerf width is $\frac{1}{16}$ in. What is the required power? Note that it takes 1550 J (2100 ft-lb) to melt one gram of copper.

SYNTHESIS, DESIGN, AND PROJECTS

27.37 Would you consider designing a machine tool that combines (in one machine) two or more of the processes described in this chapter? Explain. For what types of parts would such a machine be useful? Make a preliminary sketch for such a machine.

27.38 Repeat Problem 27.37, combining processes described in (a) Chapters 13 through 16, (b) Chapters 23 and 24, and (c) Chapters 26 and 27. Give a preliminary sketch of a machine for each of the three groups. How would you convince a prospective customer of the merits of such machines?

27.39 Make a list of machining processes that may be suitable for each of the following materials: (a) ceramics, (b) cast iron, (c) thermoplastics, (d) thermosets, (e) diamond, and (f) annealed copper.

27.40 Which of the processes described in this chapter require a vacuum? Explain why?

27.41 How would you manufacture a large-diameter, conical, round disk with a thickness that decreases from the center outward?

27.42   Describe the similarities and differences among the various design guidelines presented in this chapter.

27.43   Describe any size limitations in advanced machining processes. Give examples.

27.44   We have seen that there are several hole-making methods. Based on the topics covered in Parts III and IV, make a comprehensive table of hole-making processes. Describe the advantages and limitations of each method, comment on the quality and surface integrity of the holes produced, and give examples of specific applications.

27.45   An example of combining laser cutting and the punching of sheet metal is given in Example 27.1. Considering the relevant parameters involved, design a system whereby both processes can be used in combination to produce parts from sheet metal.

27.46   Marking surfaces with numbers and letters for part-identification purposes can be done not only with labels but by various mechanical and nonmechanical methods. Based on the processes described throughout this book thus far, make a list of these methods explaining their advantages, limitations, and typical applications.

27.47   Precision engineering is a term that is used to describe manufacturing high-quality parts with close dimensional tolerances and good surface finish. Based on their process capabilities, make a list of advanced machining processes with decreasing order of the quality of parts produced. Include a brief commentary on each method.

27.48   With appropriate sketches, describe the principles of various fixturing methods and workholding devices that can be used for the processes described in this chapter.

27.49   Make a table of the process capabilities of the advanced machining processes described in this chapter. Use several columns describing the machines involved, the type of tools and tool materials used, the shapes of blanks and parts produced, the typical maximum and minimum sizes, surface finish, tolerances, and production rates.

27.50   One of the general concerns regarding advanced machining processes is that, in spite of their many advantages, they generally are slower than conventional machining processes. Conduct a survey of the speeds, machining times, and production rates involved and prepare a table comparing their respective process capabilities.

27.51   We have seen that several of the processes described in Part IV of this book can be employed (either singly or in combination) to make or finish dies for metalworking operations. Write a brief technical paper on these methods, describing their advantages, limitations, and typical applications.

27.52   Would the processes described in this chapter be difficult to perform on various nonmetallic or rubber-like materials? Explain your thoughts, commenting on the influence of various physical and mechanical properties of workpiece materials, part geometries, etc.

27.53   Make a list of the processes described in this chapter in which the following properties are relevant: (a) mechanical, (b) chemical, (c) thermal, and (d) electrical. Are there processes in which two or more of these properties are important? Explain.

Fabrication of Microelectronic Devices and Micromanufacturing

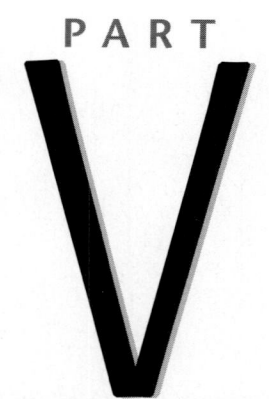

Consider the manufacture of a simple, metal spur gear; this can be produced from a forged blank and machining operations if the gear is ten centimeters in diameter. A gear that is two millimeters in diameter is very difficult to produce. Such a small gear would require extremely fine blanking or chemical etching from sheet metal or a form of electroforming. While possible to produce, this type of gear is too small for conventional forging and cutting strategies. On the other hand, a gear that is a few tens of micrometers in diameter can be produced using techniques involving optical lithography, wet and dry chemical etching, as well as related processes described in the following two chapters. Note that a gear that is only a few nanometers in diameter would be extremely difficult to produce; indeed, such a gear would have only (at most) a few tens of atoms across its face. The example of such a gear is extremely informative and can be put in perspective with the illustration of length scales in Fig. V.1.

For most of the profession's history, engineers have emphasized the design and manufacture of relatively large components. Conventional manufacturing processes (as described in Chapters 11 through 27) typically produce parts that are larger than a millimeter or so or can be described as visible to the naked eye. These parts generally are referred to as *macroscale*, the word macro being derived from *makros* in Greek (meaning long). The processing of such parts is known as *macromanufacturing*. Numerous examples can be given, ranging from products found in a hardware store, to castings and forgings used in machinery, and to products as large as automobiles and aircraft. This is the most developed and best understood size range from a design and manufacturing standpoint with a wide variety of processes suitable for producing components of this size.

The gear that was a few tens of micrometers across fits into the realm of micromanufacture, as shown in Fig. V.1. *Micromanufacturing* by definition refers to manufacturing on a microscopic scale—not visible to the naked eye. The field of micromanufacturing has been developed mostly for electronic devices of all kinds, including computer processors and memory chips, sensors, magnetic storage devices, etc. For the most part, this form of manufacturing relies heavily on lithography approaches, wet and dry etching, and coating operations. In addition, the micromanufacturing of semiconductors exploits the unique ability of silicon to form oxides. Examples of products that rely upon micromanufacturing approaches are a wide variety of sensors and probes (see Fig. V.2), inkjet printing heads, microactuators and associated devices, magnetic hard-drive heads, and microelectronic devices such as computer processors and memory chips. Modern manufacturing methods allow the production of a wide variety of features at these length scales, but most experience is with electronic devices. Microscale mechanical devices are still a relatively new technology, but a technology that has developed with surprising speed.

The gear at nanometric length scales and the gear at length scales between the macro- and microscales both are difficult to produce. No established manufacturing processes have been developed for either situation, although promising alternatives are under development.

Examples of mesomanufacturing are extremely small motors and bearings and components for miniature devices such as hearing aids, medical devices such as stents and valves, and mechanical watches (which use exactly the same gear in our example). Note that mesomanufacturing overlaps both macro- and micromanufacturing examples in Fig. V.1.

In *nanomanufacturing*, parts are produced at nanometer length scales (that is, one billionth of a meter). The term usually refers to manufacturing approaches below the micrometer scale—between 10^{-6} and 10^{-9} meters in length. Many of the features in integrated circuits are at this length scale, but very little else; molecularly engineered medicines and other forms of biomanufacturing are the only commercial examples. However, it has been recognized that many physical and biological processes act at this length scale and that this approach holds much promise for future innovations.

This part of the book emphasizes micro- and nanomanufacturing. These subdisciplines within the broad range of manufacturing engineering are over five decades old, but they have been developed rapidly in the past two decades or so. Products using these processes have become very common and pervasive in modern society. Computers, communications, video, and control hardware of all types depend upon these manufacturing and material approaches.

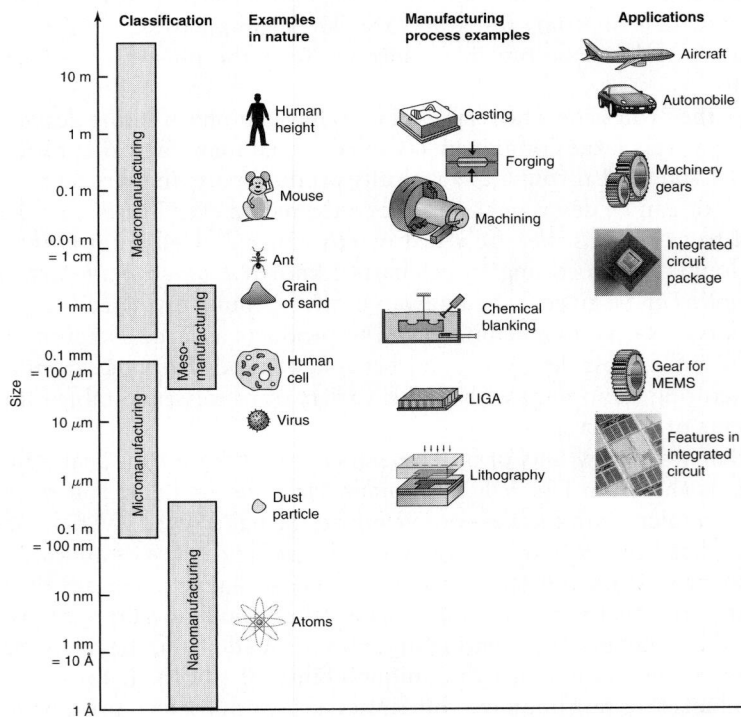

FIGURE V.1 Illustration of the regimes of macro-, meso-, micro-, and nanomanufacturing and the range of common sizes of parts and the capabilities of manufacturing processes in producing these parts.

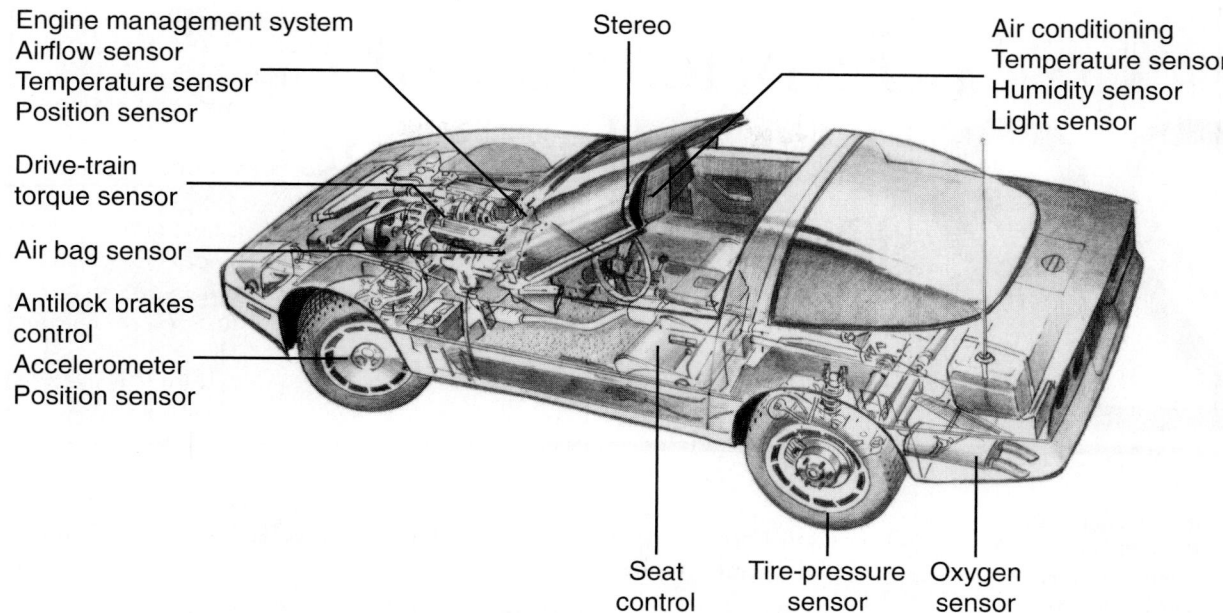

Engine management system
Airflow sensor
Temperature sensor
Position sensor

Drive-train
torque sensor

Air bag sensor

Antilock brakes
control
Accelerometer
Position sensor

Stereo

Air conditioning
Temperature sensor
Humidity sensor
Light sensor

Seat
control

Tire-pressure
sensor

Oxygen
sensor

FIGURE V.2   Microelectronic and microelectromechanical devices and parts used in a typical automobile.

In Chapter 28, we describe the manufacture of silicon wafers and microelectronic devices, which include a wide variety of computer processors, memory devices, and integrated circuits. Computers and microelectronic devices are very widespread in modern society and have had a profound impact on culture. Communication, entertainment, control, transportation, engineering design and manufacturing, and medicine all have been changed greatly by the ready availability of metal-oxide-semiconductor (MOS) devices, usually based on single-crystal silicon. Microelectronics is the best known and commercially important example of micromanufacturing, with some aspects of the applications exemplifying nanomanufacturing. The techniques used in packaging and assembling the integrated circuits onto printed circuit boards also are presented.

The production of microscale devices that are mechanical and electrical in nature then are described in Chapter 29. Depending on their level of integration, these are called micromechanical devices or microelectromechanical systems (MEMS). While the historical origins of MEMS manufacture stem from the same processes used for microelectronic systems and identical processes and production sequences are still used, a number of unique processes have been developed for the manufacture of microscale electromechanical devices and systems.

CHAPTER 28

Fabrication of Microelectronic Devices

This chapter presents the science and technology involved in the production of integrated circuits. We describe:

• Special properties of silicon for producing oxide and dopants.
• Treatment of a cast ingot and machining operations to produce a disk.
• Lithography, etching, and doping processes.
• Wet and dry etching for circuit manufacturing.
• Connections from a transistor to a computer level.
• Packages for the integrated circuits and manufacturing methods for printed circuit boards.

Typical parts produced: Computer processors, memory chips, printed circuit boards, and integrated circuits of all types.

28.1 | Introduction

Although semiconducting materials have been used in electronics for a long time, it was the invention of the *transistor* in 1947 that set the stage for what would become one of the greatest technological advancements in all of history. *Microelectronics* have played an ever-increasing role in our lives since **integrated circuit** (IC) technology became the foundation for calculators, wrist watches, home-appliances and automotive controls, information systems, telecommunications, robotics, space travel, weaponry, and personal computers.

The major advantages of today's ICs are their very small size and low cost. As fabrication technology becomes more advanced, the size of devices continues decreasing. Consequently, more components can be put onto a **chip**—a small piece of semiconducting material on which the circuit is fabricated. In addition, mass production and automation have helped reduce the cost of each completed circuit. The components fabricated include transistors, diodes, resistors, and capacitors.

Typical chips produced today have sizes that range from 0.5 mm × 0.5 mm to more than 50 mm × 50 mm. In the past, no more than 100 devices could be fabricated on a single chip; new technology now allows densities in the range of 10 million

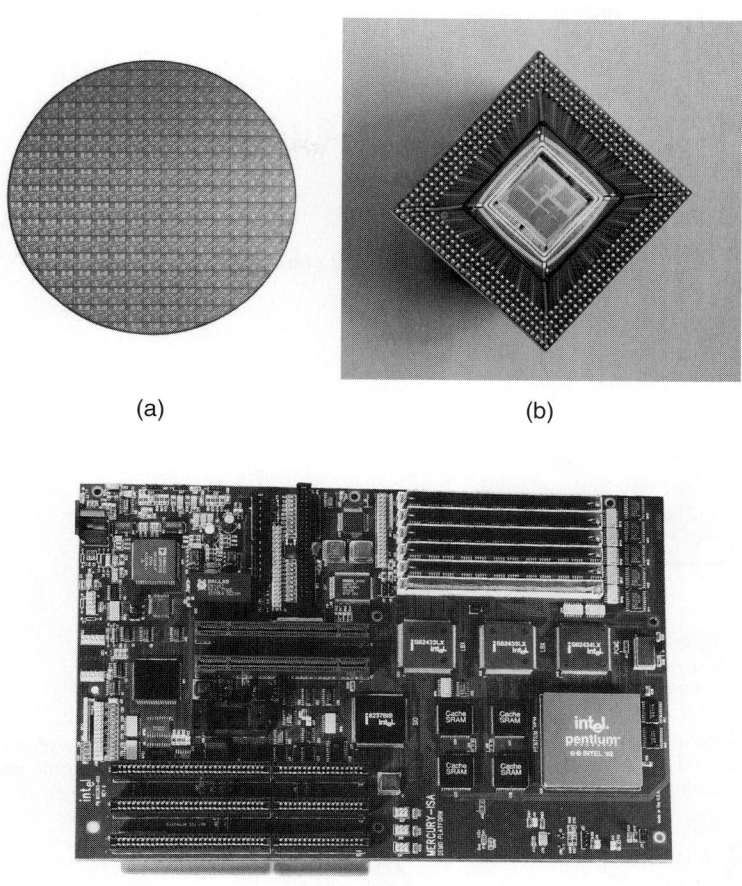

(a)   (b)

(c)

FIGURE 28.1   (a) A completed eight-inch wafer with completed dice. (b) A single chip in a ball-grid array (BGA) with cover removed. (c) A printed circuit board. *Source*: Courtesy of Intel Corporation.

devices per chip (Fig. 28.1). This magnitude of integration has been termed **very large-scale integration** (VLSI). Some of the most advanced ICs may contain more than 100 million devices, termed **ultra large-scale integration** (ULSI).

In this chapter, we describe the processes that currently are used in the fabrication of microelectronic devices and integrated circuits and follow the outline shown in Fig. 28.2. The major steps in fabricating a **metal-oxide-semiconductor field-effect transistor** (MOSFET)—which is one of the dominant devices used in modern IC technology—are shown in Fig. 28.3.

This chapter first introduces the basic properties of semiconductors and properties of silicon and then describes each of the major fabrication steps. Packaging of the integrated circuits and assembling them onto circuit boards also are discussed. Finally, we will describe current trends and forecasts in the microelectronics industry.

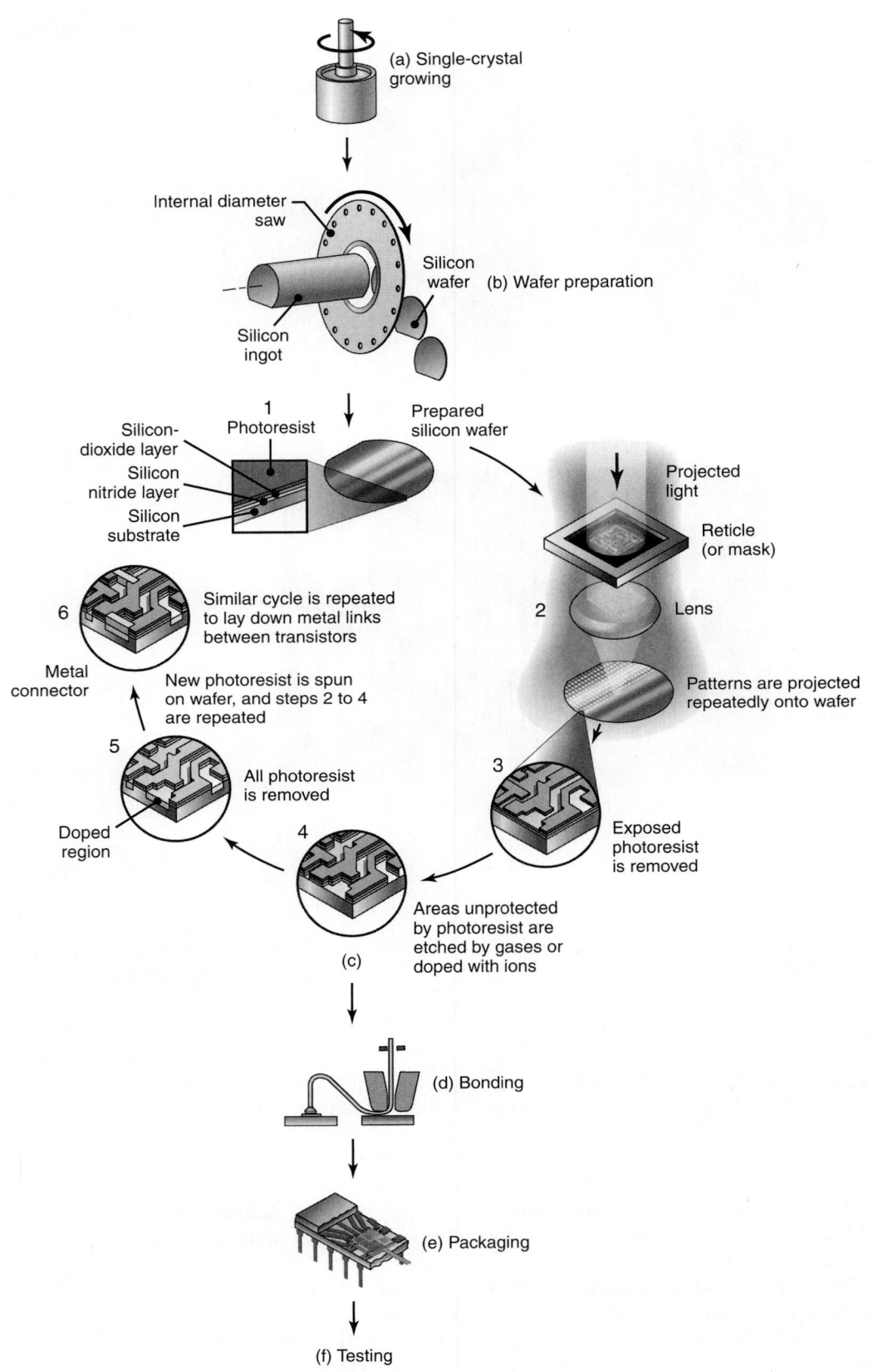

(a) Single-crystal growing

Internal diameter saw

Silicon wafer   **(b) Wafer preparation**

Silicon ingot

1 Photoresist

Silicon-dioxide layer

Silicon nitride layer

Silicon substrate

Prepared silicon wafer

Projected light

Reticle (or mask)

2   Lens

Patterns are projected repeatedly onto wafer

6   Similar cycle is repeated to lay down metal links between transistors

Metal connector

New photoresist is spun on wafer, and steps 2 to 4 are repeated

5   All photoresist is removed

Doped region

4   3   Exposed photoresist is removed

Areas unprotected by photoresist are etched by gases or doped with ions

(c)

(d) Bonding

(e) Packaging

(f) Testing

FIGURE 28.2   Outline of the general fabrication sequence for integrated circuits.

870

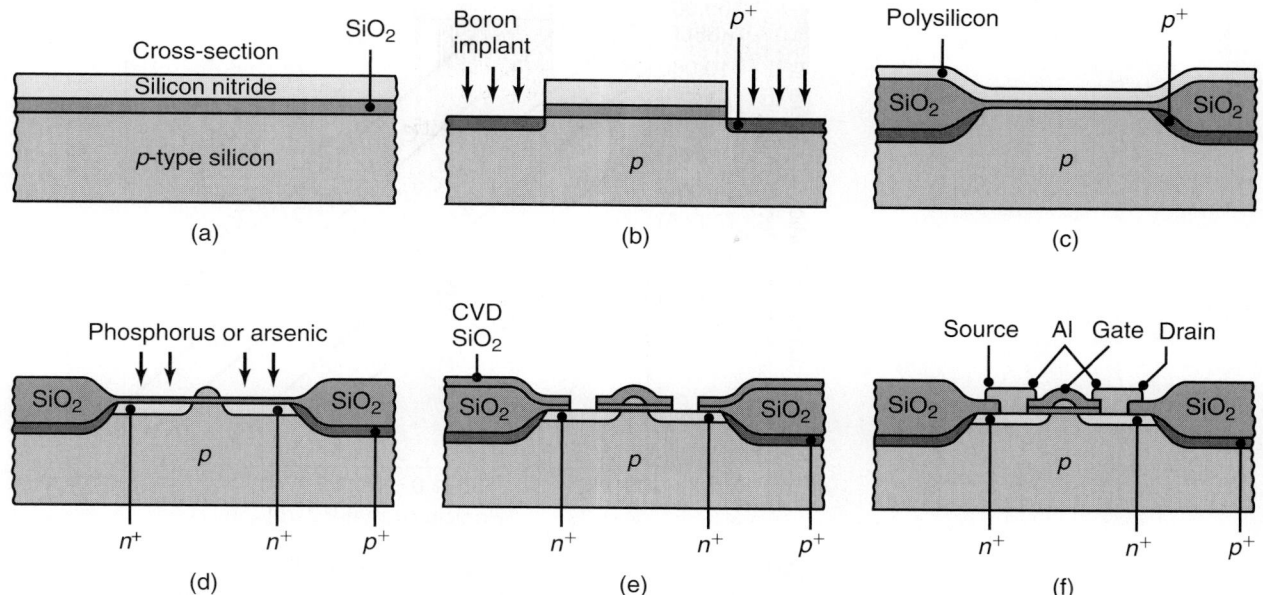

FIGURE 28.3   Cross-sectional views of the fabrication of a MOS transistor. *Source*: After R.C. Jaeger.

28.2 | Clean Rooms

Clean rooms are essential for the production of most integrated circuits. This can be appreciated by noting the scale of manufacturing that is to be performed. Integrated circuits are typically a few millimeters in length, and the smallest features in a transistor on the circuit may be as small as a few tens of nanometers (nano = 10^{-9}). This size range is smaller than particles that normally we don't consider harmful, such as dust, smoke, perfume, and bacteria. However, if these contaminants are present on a silicon wafer during processing, they can compromise the performance of the entire device. Thus, it is essential that all of the potentially harmful particles be eliminated from the integrated-circuit-manufacturing environment.

There are various levels of clean rooms, which are defined by the **class** of the clean rooms. The classification system refers to the number of 0.5 μm or larger particles within a cubic foot of air. Thus, a Class-10 clean room has 10 or fewer such particles per cubic foot. Clearly, the size and the number of particles are important to define the class of a clean room, as shown in Fig. 28.4. Most clean rooms for microelectronics manufacuring range from Class 1 to Class 10. In comparison, the contamination level in modern hospitals is on the order of 10,000 particles per cubic foot.

To obtain controlled atmospheres that are free from particulate contamination, all ventilating air is passed through a *high-performance particulate air* (HEPA) filter. In addition, the air usually is conditioned so that it is at 21°C (70°F) and 45% relative humidity.

The largest source of contaminants in a clean room is people themselves. Skin particles, hair, perfume and makeup, clothing, bacteria, and viruses are given off naturally by people and in large enough numbers to quickly compromise a Class-100 clean room. For these reasons, most clean rooms require special coverings, such as white laboratory coats, gloves and hair nets, as well as the avoidance of perfumes

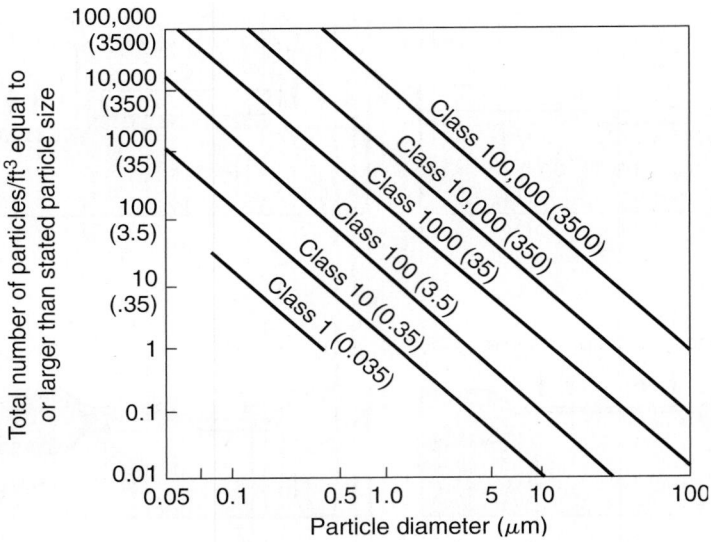

FIGURE 28.4   Allowable particle size counts for different clean room classes. The numbers in parenthesis are particle counts per cubic meter.

and makeup. The most stringent clean rooms require full body coverings, called bunny suits. There are other stringent precautions, as well. For example, the use of a pencil instead of a ball point pen can produce objectionable graphite particles, and special clean-room paper is required to prevent the accumulation of paper particles in the air.

Clean rooms are designed such that the cleanliness at critical processing locations is greater than in the clean room in general. This is accomplished by directing the filtered ventilating air so that it displaces ambient air and directs dust particles away from the process. This can be facilitated by laminar-flow hooded work areas.

28.3 | Semiconductors and Silicon

As the name suggests, **semiconductor materials** have electrical properties that lie between those of conductors and insulators and exhibit resistivities between 10^{-3} and 10^8 Ω-cm.

Semiconductors have become the foundation for electronic devices because their electrical properties can be altered when controlled amounts of selected impurity atoms are added to their crystal structures. These impurity atoms (also known as **dopants**) have either one more valence electron (*n*-type or negative dopant) or one less valence electron (*p*-type or positive dopant) than the atoms in the semiconductor lattice.

For silicon, which is a Group IV element in the Periodic Table, typical *n*-type and *p*-type dopants include phosphorus (Group V) and boron (Group III), respectively. The electrical operation of semiconductor devices can be controlled through the creation of regions with different doping types and concentrations.

Although the earliest electronic devices were fabricated on *germanium*, **silicon** has become the industry standard. The abundance of alternative forms of silicon in the Earth is second only to that of oxygen, making it attractive economically. Silicon's main advantage over germanium is its large energy gap (1.1 eV) compared to that of germanium (0.66 eV). This energy gap allows silicon-based devices to operate at temperatures of about 150°C (270°F) higher than devices fabricated on germanium, which operate at about 100°C (180°F).

Another important processing advantage of silicon is that its oxide (*silicon dioxide*) is an excellent electrical insulator and can be used for both isolation and passivation purposes. Conversely, germanium oxide is water soluble and unsuitable for electronic devices. Furthermore, the oxidized form of silicon, SiO_2, allows the production of *metal-oxide-semiconductor* (MOS) devices, which are the basis for MOS-transistors. These materials make up memory devices, processors, and the like and are by far the largest volume of semiconductor material produced worldwide.

Structure of silicon. The crystallographic structure of silicon is a diamond-type fcc structure, as shown in Fig. 28.5, along with the Miller indices of an fcc material. Miller indices are a useful notation for identifying planes and directions within a unit cell. A crystallographic plane is defined by the reciprocal of its intercepts on the three axes. Since anisotropic etchants preferentially remove material in certain crystallographic planes, the orientation of the silicon crystal in a wafer is an important consideration.

In spite of its advantages, silicon has a larger energy gap (1.1 eV) than germanium oxide and therefore, has a higher maximum operating temperature (about 200°C; 400°F). This limitation has encouraged the development of compound semiconductors, specifically **gallium arsenide**. Its major advantage over silicon is its ability

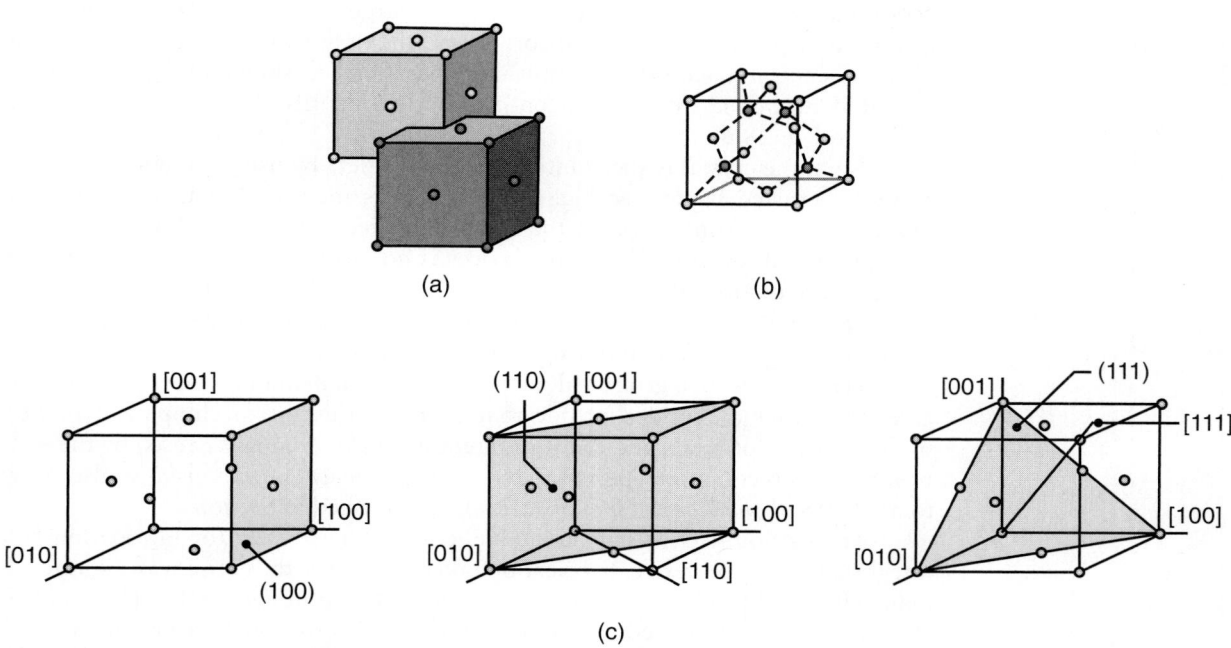

FIGURE 28.5 Crystallographic structure and Miller indices for silicon. (a) Construction of a diamond-type lattice from interpenetrating face-centered cubic cells; one of eight penetrating cells is shown. (b) Diamond-type lattice of silicon; the interior atoms have been shaded darker than the surface atoms. (c) Miller indices for a cubic lattice.

to emit light, thus allowing the fabrication of devices such as lasers and light-emitting diodes (LEDs).

Devices fabricated on gallium arsenide also have much higher operating speeds than those fabricated on silicon. Some of gallium arsenide's disadvantages include its considerably higher cost, greater processing complications, and the difficulty of growing high-quality oxide layers, the need for which is emphasized throughout the rest of this chapter.

28.4 | Crystal Growing and Wafer Preparation

Silicon occurs naturally in the forms of silicon dioxide and various silicates. It must, however, undergo a series of purification steps in order to become the high-quality, defect-free, single-crystal material that is required for semiconductor device fabrication. The process begins by heating silica and carbon together in an electric furnace, which results in a 95 to 98% pure polycrystalline silicon. This material is converted to an alternative form (commonly trichlorosilane), which in turn is purified and decomposed in a high-temperature hydrogen atmosphere. The resulting product is extremely high-quality **electronic-grade silicon** (EGS).

Single-crystal silicon usually is obtained through the **Czochralski** or **CZ process,** as we have described in Section 11.4. This process utilizes a seed crystal that is dipped into a silicon melt and then is pulled out slowly while being rotated. At this point, controlled amounts of impurities can be added to obtain a uniformly doped crystal.

The result of the CZ process is a cylindrical single-crystal ingot, typically 150 to 300 mm (6 to 12 in.) in diameter and over 1 m (40 in.) in length. Unfortunately, this technique does not allow for exact control of the ingot diameter. Therefore, ingots commonly are grown a few millimeters larger than the required size and are ground to a precise diameter. Silicon wafers then are produced from silicon ingots by a sequence of machining and finishing operations, as shown in Fig. 28.6. A notch or flat is machined into the silicon cylinder to identify its crystal orientation, as shown in Fig. 28.6b.

Next, the crystal is sliced into individual **wafers** by using an inner-diameter diamond-encrusted blade (see Fig. 24.25f). In this method, a rotating, ring-shaped blade with its cutting edge on the inner diameter is utilized. While the substrate depth required for most electronic devices is no more than several microns, wafers typically are cut to a thickness of about 0.5 mm (0.02 in.). This thickness provides the necessary physical support for the absorption of temperature variations and the mechanical support needed during subsequent fabrication.

The wafer is then ground along its edges with a diamond wheel. This operation gives the wafer a rounded profile that is more resistant to chipping. Finally, the wafers must be polished and cleaned to remove surface damage caused by the sawing process. This is commonly performed by *chemical–mechanical polishing* (also referred to as *chemical–mechanical planarization*), as described in Section 26.7.

At this point, the single-crystal silicon wafer is ready for fabrication of the integrated circuit or device. Fabrication takes place over the entire wafer surface, and many chips are produced at the same time, as shown in Fig. 28.1a. The number of chips that can be produced is dependent on the cross-sectional area of the wafer. For this reason, a number of advanced circuit manufacturers have moved towards using larger single-crystal cylinders, and 300 mm (12 in.) diameter wafers now are increasingly common.

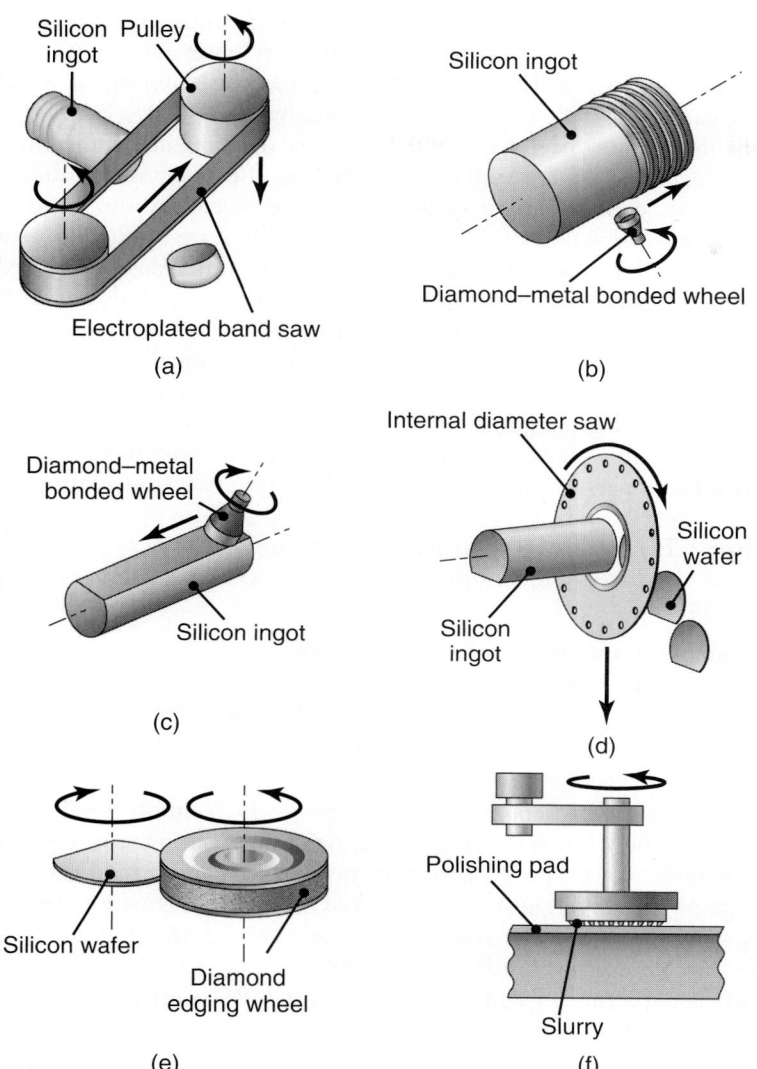

FIGURE 28.6 Finishing operations on a silicon ingot to produce wafers: (a) sawing the ends off the ingot; (b) grinding of the end and cylindrical surfaces of a silicon ingot; (c) machining of a notch or flat; (d) slicing of wafers; (e) end grinding of wafers; (f) chemical–mechanical polishing of wafers.

28.5 | Film Deposition

Films of many different types are used extensively in microelectronic-device processing, particularly insulating and conducting films. Commonly deposited films include polysilicon, silicon nitride, silicon dioxide, tungsten, titanium, and aluminum. In some instances, the wafers merely serve as a mechanical support on which custom *epitaxial layers* are grown.

Epitaxy. Defined as the growth of a vapor deposit, *epitaxy* or electrodeposit occurs when the crystal orientation of the deposit is related directly to the crystal orientation in the underlying crystalline substrate. The advantages of processing on these deposited films instead of on the actual wafer surface include: fewer impurities (notably carbon and oxygen), improved device performance, and the tailoring of material properties (which cannot be done on the wafers themselves).

Some of the major functions of deposited films are **masking** and protecting the semiconductor surface. In masking applications, the film effectively must inhibit the passage of dopants and concurrently display an ability to be etched into patterns of high resolution. Upon completion of device fabrication, films are applied to protect the underlying circuitry. Films used for masking and protecting include silicon dioxide, phosphosilicate glass (PSG), and silicon nitride. Each of these materials has distinct advantages, and they often are used in combination.

Conductive films are used primarily for device interconnection. These films must have a low resistivity, be capable of carrying large currents, and be suitable for connection to terminal packaging leads with wire bonds. Generally, aluminum and copper are used for this purpose. Increasing circuit complexity has required up to six levels of conductive layers, all of which must be separated by insulating films.

Film deposition.   Films may be deposited using a number of techniques, which involve a variety of pressures, temperatures, and vacuum systems, as described here (see also Chapter 34):

- One of the simplest and oldest methods is **evaporation**, which is used primarily for depositing metal films. In this process, the metal is heated in a vacuum to its point of vaporization. Upon evaporation, the metal forms a thin layer on the substrate surface. The heat for evaporation usually is generated by a heating filament or electron beam.

- Another method of metal deposition is **sputtering**, which entails bombarding a target with high-energy ions (usually argon, Ar^+) in a vacuum. Sputtering systems usually include a DC power source to produce the energized ions. As the ions impinge on the target, atoms are knocked off and subsequently deposited on wafers mounted within the system. Although some argon may be trapped within the film, sputtering results in very uniform coverage. Advances in this field include using a radio-frequency power source (**RF sputtering**) and introducing magnetic fields (**magnetron sputtering**).

- In one of the most common techniques, **chemical vapor deposition** (CVD), film depositing is achieved by way of the reaction and/or decomposition of gaseous compounds. Using this technique, silicon dioxide is deposited routinely by the oxidation of silane or a chlorosilane. Figure 28.7a shows a continuous CVD reactor that operates at atmospheric pressure.

  A similar method that operates at lower pressures is referred to as **low-pressure chemical vapor deposition** (LPCVD) and is shown in Fig. 28.7b. Capable of coating hundreds of wafers at a time, this method results in a much higher production rate than that of an atmospheric-pressure CVD and provides superior film uniformity with less consumption of carrier gases. This technique commonly is used for depositing polysilicon, silicon nitride, and silicon dioxide.

- **Plasma-enhanced chemical vapor deposition** (PECVD) involves the processing of wafers in an RF plasma containing the source gases. This method has the advantage of maintaining low wafer temperature during deposition.

Silicon **epitaxy** layers (in which the crystalline layer is formed using the substrate as a seed crystal) can be grown using a variety of methods. If the silicon is deposited from the gaseous phase, the process is known as **vapor-phase epitaxy** (VPE).

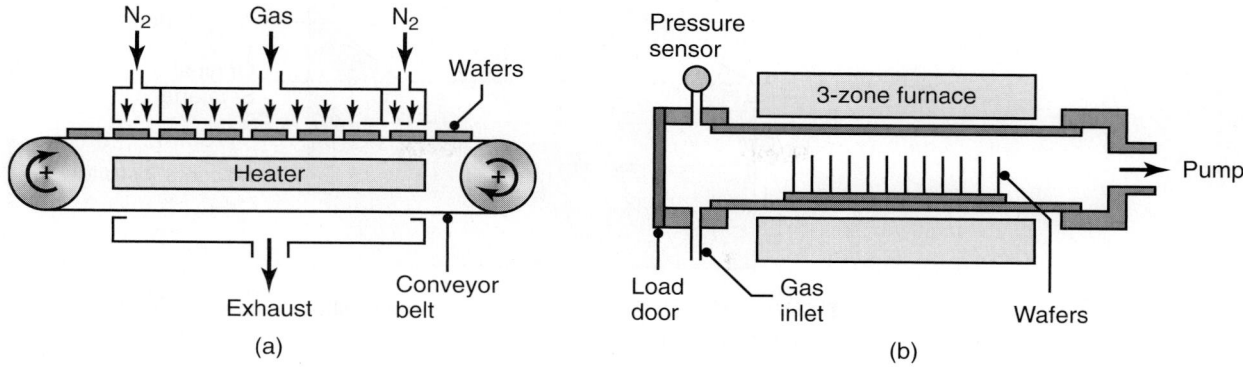

FIGURE 28.7   Schematic diagrams of (a) a continuous, atmospheric-pressure CVD reactor and (b) a low-pressure CVD. *Source:* After S.M. Sze.

In another variation, the heated substrate is brought into contact with a liquid solution containing the material to be deposited, called **liquid-phase epitaxy** (LPE).

Another high-vacuum process utilizes evaporation to produce a thermal beam of molecules that are deposited on the heated substrate. Called **molecular-beam epitaxy** (MBE), this process results in a very high degree of purity. In addition, since the films are grown one atomic layer at a time, it is possible to have excellent control over doping profiles. This level of control is important especially in gallium-arsenide technology. Unfortunately, the MBE process has relatively low growth rates as compared to other conventional film-deposition techniques.

28.6 | Oxidation

The term *oxidation* refers to the growth of an oxide layer as a result of the reaction of oxygen with the substrate material. Oxide films also can be formed using the previously described deposition techniques. The thermally grown oxides, described in this section, display a higher level of purity than deposited oxides, because they are grown directly from the high-quality substrate. However, deposition methods must be used if the composition of the desired film is different from that of the substrate material.

Silicon dioxide is the most widely used oxide in IC technology today, and its excellent characteristics are one of the major reasons for the widespread use of silicon. Aside from its effectiveness in dopant masking and device isolation, silicon dioxide's most critical role is that of the "gate oxide" material. Silicon surfaces have an extremely high affinity for oxygen, and a freshly sawed slice of silicon will grow a native oxide of 30 to 40 Å thickness ($\text{Å} = 10^{-7}$ mm) quickly.

- **Dry oxidation** is a relatively simple process and is accomplished by elevating the substrate temperature, typically to about 750° to 1100°C (1380° to 2020°F), in an oxygen-rich environment. As a layer of oxide forms, the oxidizing agents must be able to pass through the oxide and reach the silicon surface where the actual reaction takes place. Thus, an oxide layer does not continue to grow on top of itself, but rather, it grows from the silicon surface outward. Some of the silicon substrate is consumed in the oxidation process (Fig. 28.8).

  The ratio of oxide thickness to the amount of silicon consumed is found to be 1:0.44. Thus, to obtain an oxide layer 1000 Å thick, approximately 440

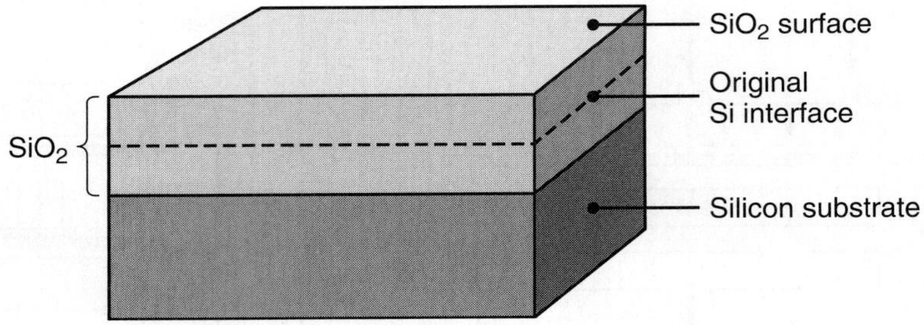

FIGURE 28.8   Growth of silicon dioxide showing consumption of silicon. *Source*: After S.M. Sze.

Å of silicon will be consumed. This requirement does not present a problem, as substrates always are grown sufficiently thick.

One important effect of this consumption of silicon is the rearrangement of dopants in the substrate near the interface. As different impurities have different segregation coefficients or mobilities in silicon dioxide, some dopants deplete away from the oxide interface while others pile up. Hence, processing parameters must be adjusted to compensate for this effect.

- Another oxidation technique utilizes a water-vapor atmosphere as the agent, hence it is called **wet oxidation**. This method effects a considerably higher growth rate than that of dry oxidation, but it suffers from a lower oxide density and, therefore, a lower dielectric strength. The common practice in industry is to combine both dry and wet oxidation methods by growing an oxide in a three-part layer: dry–wet–dry. This approach combines the advantages of wet oxidation's much higher growth rate and dry oxidation's high quality.

- The foregoing two oxidation methods are useful primarily for coating the entire silicon surface with oxide, but it also may be necessary to oxidize only certain portions of the surface. The procedure of oxidizing only certain areas is termed **selective oxidation** and uses silicon nitride, which inhibits the passage of oxygen and water vapor. Thus, by covering certain areas with silicon nitride, the silicon under these areas remains unaffected, but the uncovered areas are oxidized.

28.7 | Lithography

Lithography is the process by which the geometric patterns that define devices are transferred to the substrate surface. A summary of lithographic techniques is given in Table 28.1. There are many forms of lithography, but the most common form used today is *photolithography*. Electron-beam and x-ray lithography are of great interest because of their ability to transfer patterns of higher resolution, which is a necessary feature for the increased miniaturization of integrated circuits. However, most integrated-circuit applications can be manufactured successfully with photolithography. A comparison of the basic lithography methods is shown in Fig. 28.9.

Photolithography.   Photolithography is the most widely used form of lithography. It uses a **reticle** (also called a **photomask** or **mask**), which is a glass or quartz plate with a pattern of the chip deposited onto it with a chromium film. The reticle image can be the same size as the desired structure on the chip, but it is often an enlarged image (usually 5 to 20× larger). Enlarged images then are focused onto a wafer through a lens system; this is referred to as *reduction lithography*.

TABLE 28.1

General Characteristics of Lithography Techniques		
Method	Wavelength (nm)	Finest feature size (nm)
Ultraviolet (Photolithography)	365	350
Deep UV	248	250
Extreme UV	10–20	30–100
X-ray	0.01–1	20–100
Electron beam	—	80

Source: P.K. Wright, *21ˢᵗ Century Manufacturing*, Prentice Hall, 2001

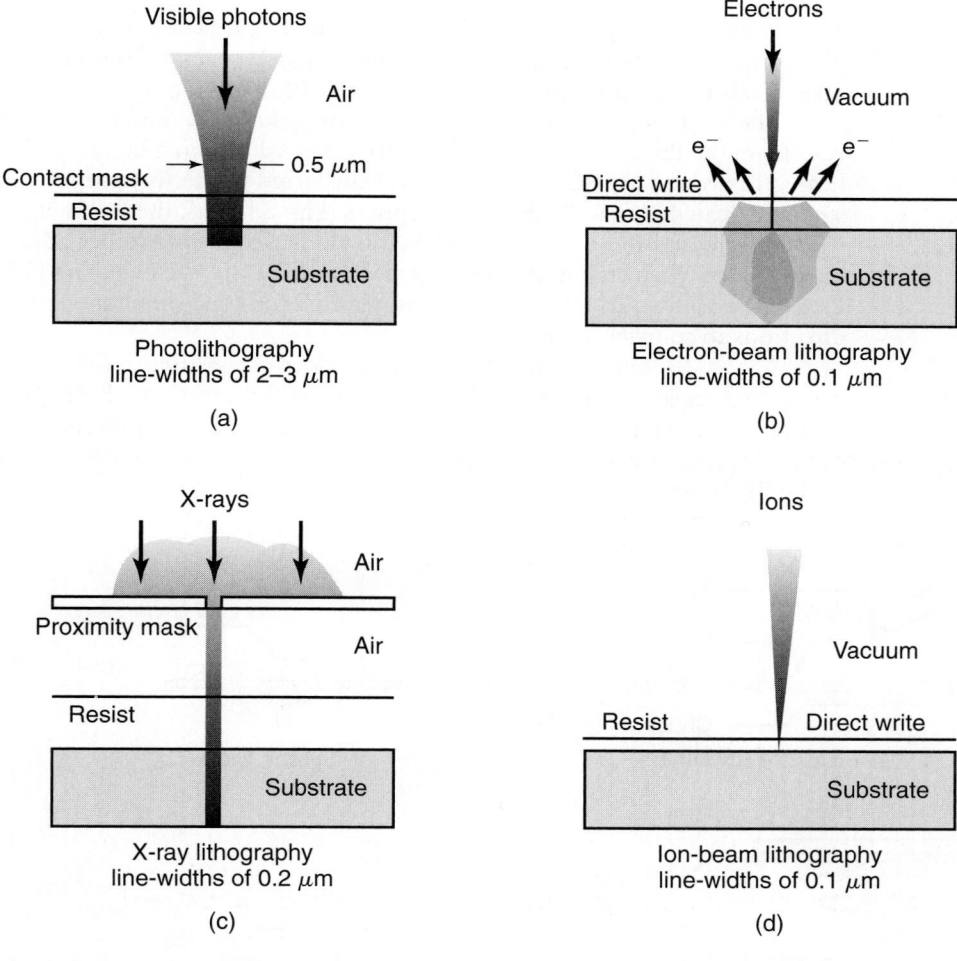

FIGURE 28.9   Comparison of lithography techniques.

In current practice, the *lithographic process* is applied to each microelectronic circuit as many as 25 times—each time using a different reticle to define the different areas of the working devices. Typically designed at several thousand times their final size, reticle patterns undergo a series of reductions before being applied permanently to a defect-free quartz plate. Computer-aided design (CAD; see Section 38.4) has had a major impact on reticle design and generation.

Cleanliness is especially important in lithography, and many manufacturers now are using robotics and specialized wafer-handling apparatus in order to minimize dust and dirt contamination. Once the film-deposition process is completed and the desired reticle patterns have been generated, the wafer is cleaned and coated with an organic polymer known as a **photoresist** (PR).

A photoresist consists of three principal components:

1. A polymer that changes structure when exposed to radiation.
2. A sensitizer that controls the reactions in the polymer.
3. A solvent, necessary to deliver the polymer in liquid form.

Photoresist layers of 0.5 to 2.5 μm (20 to 100 μin.) thick are obtained by applying the PR to the substrate and then spinning it at several thousand rpm for 30 or 60 seconds to give uniform coverage (Fig. 28.10).

The next step in photolithography is **prebaking** the wafer to remove the solvent from the PR and harden it. This step is carried out on a hot plate that has been heated to around 100°C. The pattern is transferred to the wafer through stepper or step-and-scan systems. With wafer steppers (Fig. 28.11a), the full image is exposed in one flash, and the reticle pattern then is refocused onto another adjacent section of the wafer. With step-and-scan systems (Fig. 28.11b), the exposing light source is focused into a line, and the reticle and wafer are translated simultaneously in opposite directions to transfer the pattern.

The wafer must be aligned carefully under the desired reticle. In this crucial step called **registration**, the reticle must be aligned correctly with the previous layer on the wafer. Once the reticle is aligned, it is subjected to UV radiation. Upon development and removal of the exposed PR, a duplicate of the reticle pattern will appear in the PR layer.

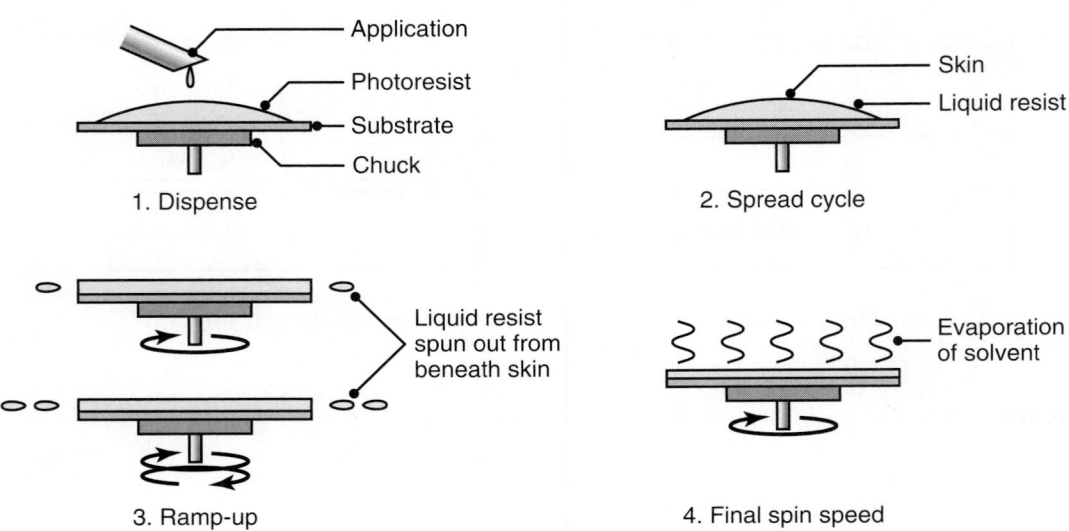

FIGURE 28.10   Spinning of an organic coating on a wafer.

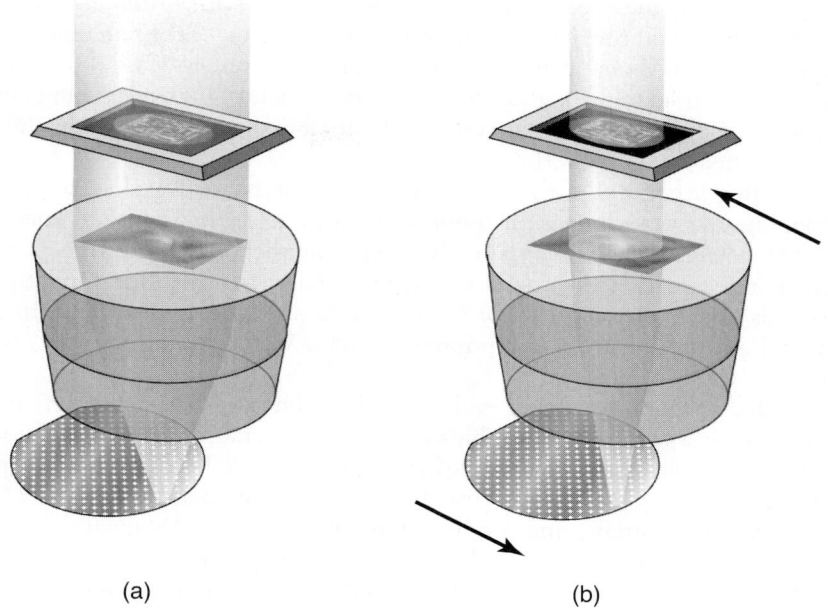

(a)   (b)

FIGURE 28.11   Schematic illustration of (a) wafer stepper technique to pattern transfer and (b) step-and-scan technique.

As seen in Fig. 28.12, the reticle can be a negative image or a positive image of the desired pattern. A positive reticle uses the UV radiation to break down the chains in the organic film, so that these films are removed preferentially by the developer. Positive masking is more common than negative masking, because with negative masking the photoresist can swell and distort, making it unsuitable for small geometries. Newer negative photoresist materials do not have this problem.

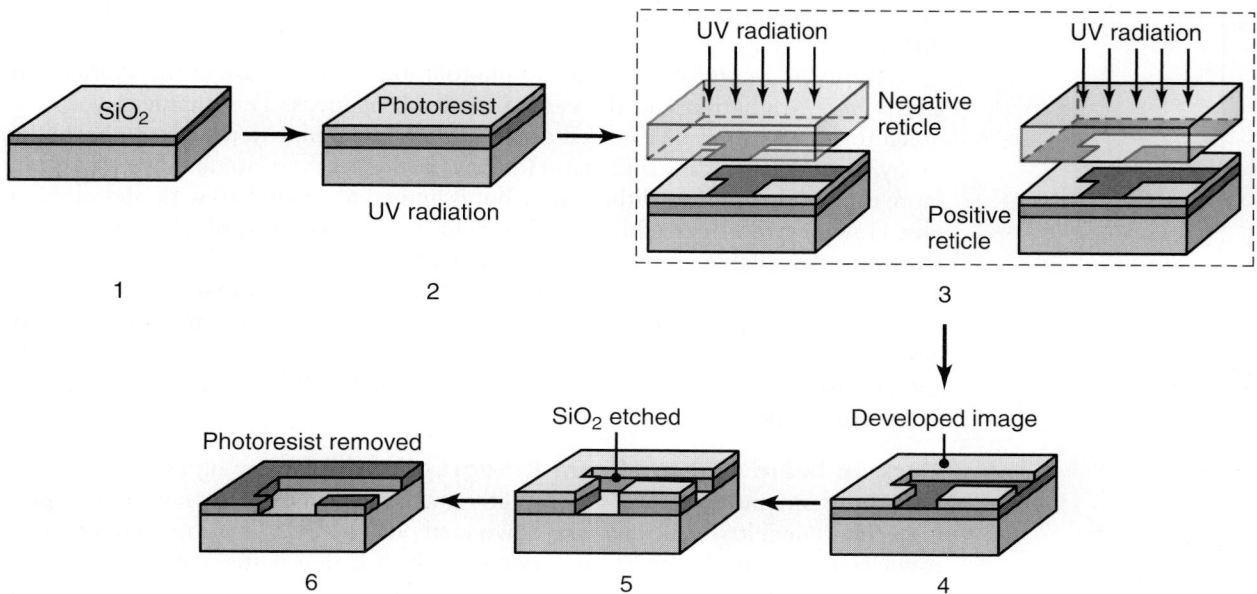

FIGURE 28.12   Pattern transfer by photolithography. Note that the mask in Step 3 can be a positive or negative image of the pattern.

Following the exposure and development sequence, **postbaking** the wafer drives off the solvent and toughens and improves the adhesion of the remaining resist. In addition, a deep UV treatment (baking the wafer to about 150° to 200°C in ultraviolet light) can be used to further strengthen the resist against high-energy implants and dry etches. The underlying film not covered by the PR is then etched away (Section 28.8) or implanted (Section 28.9).

After lithography, the developed PR must be removed in a process called **stripping**. In *wet stripping*, the photoresist is dissolved by solutions such as acetone or strong acids. *Dry stripping* involves exposing the PR to an oxygen plasma, which also is referred to as **ashing**. Dry stripping has become more popular, because it (a) does not involve disposal of consumed hazardous chemicals and (b) is easier to control and can result in exceptional surfaces. Wet stripping solutions tend to lose potency in use.

One of the major issues in lithography is **line width**, which refers to the width of the smallest feature imprintable on the silicon surface. As circuit densities have escalated over the years, device sizes and features have become smaller and smaller. Today, minimum commercially feasible line widths are between 90 and 150 nm with considerable research being conducted in regard to 70 nm line widths.

As pattern resolution and device miniaturization have been limited by the wavelength of the radiation source used, the need has arisen to move to wavelengths shorter than those in the ultraviolet range, such as "deep" UV wavelengths, "extreme" UV wavelengths, electron beams, and x-rays. In these technologies, the photoresist is replaced by a similar resist that is sensitive to a specific range of shorter wavelengths.

Extreme ultraviolet lithography.   The pattern resolution in photolithography is limited by light diffraction. One of the means of reducing the effects of diffraction is to use ever shorter wavelengths. *Extreme ultraviolet lithography* (EUV) uses light at a wavelength of 13 nm in order to obtain features in the range of 30 to 100 nm in size. The waves are focused through highly reflective molybdenum-silicon mirrors (instead of glass lenses that absorb EUV light) through the mask to the wafer surface.

X-ray lithography.   Although photolithography is the most widely used lithography technique, it has fundamental resolution limitations associated with light diffraction.

X-ray lithography is superior to photolithography because of the shorter wavelength of the radiation and the very large depth of focus. This characteristic allows much finer patterns to be resolved and is far less susceptible to dust than photolithography. Furthermore, the aspect ratio (defined as the depth to lateral dimension) can be more than 100 with x-ray lithography but is limited to around 10 with photolithography. However, to achieve this benefit, synchrotron radiation is required, which is expensive and available at only a few research laboratories.

Given the large capital investment required for a manufacturing facility, industry has preferred to refine and improve optical lithography instead of investing new capital into x-ray based production. Currently, x-ray lithography is not widespread; however, the LIGA process (see Section 29.3) fully exploits the benefits of x-ray lithography.

Electron-beam and ion-beam lithography.   Like x-ray lithography, *electron-beam* (e-beam) and *ion-beam* (i-beam) lithography are superior to photolithography in terms of attainable resolutions. These two methods involve high current density in narrow electron or ion beams (*pencil sources*), which scan a pattern one pixel at a time onto a wafer. The masking is done by controlling the point-by-point transfer of the stored pattern; therefore, the masking is done using software. These techniques have the advantages of accurate control of exposure over small areas of the wafer, large depth of focus, and low defect densities. Resolutions are limited to about 10 nm because of electron scatter, although 2 nm resolutions have been reported for some materials.

It should be noted that the scan time increases significantly as the resolution increases, because more highly focused beams are required. The main drawback of these techniques is that electron and ion beams have to be maintained in a vacuum, which significantly increases equipment complexity and production cost. Furthermore, the scan time for a wafer is much slower than that for other lithographic methods.

SCALPEL.   Figure 28.13 shows the SCALPEL (*s*cattering with *a*ngular *l*imitation *p*rojection *e*lectron-beam *l*ithography) process, where a mask is produced from about an 0.1-μm thick membrane of silicon nitride and patterned with an approximately 50-nm thick coating of tungsten. High-energy electrons pass through both the silicon nitride and the tungsten, but the tungsten scatters the electrons widely, whereas the silicon nitride results in very little scattering. An aperture blocks the scattered electrons, resulting in a high-quality image at the wafer.

The limitation to this process is the small-sized masks that are currently in use. However, this process has high potential. Perhaps its most significant advantage is that energy does not need to be absorbed by the reticle. Instead, it is blocked by the aperture, which is not as fragile or expensive as the reticle.

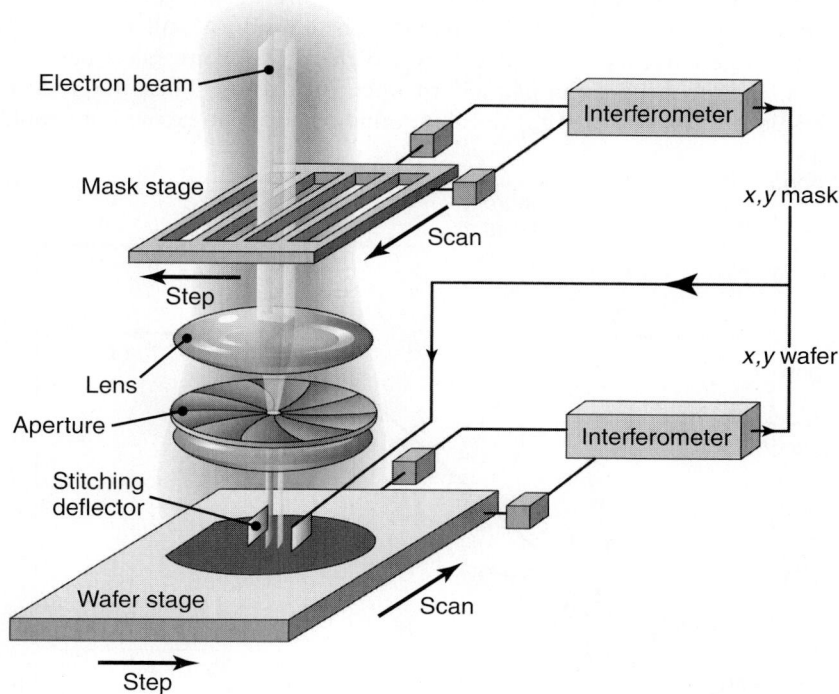

FIGURE 28.13   Schematic illustration of the SCALPEL process.

EXAMPLE 28.1 Moore's law

Gordon Moore, an inventor of the IC and past chairman of Intel Corporation, observed in 1965 that the surface area of a single transistor is reduced by 50% every 12 months. In 1975, he revised this estimate to every two years, and the resulting estimate is now widely known as Moore's law. This law has been remarkably accurate. Figure 28.14 shows the historical progression of feature size in *dynamic read-only memory* (DRAM) bits as well as projected future developments. Looking ahead, there are some major impediments to the continued reliability of Moore's law, some of the more important of which are the following:

- To produce ever smaller features in the transistor requires that even more stringent manufacturing tolerances be achieved. For example, 180-nm line widths require ±14-nm dimensional tolerances, whereas 50-nm line widths require ±4 nm. This is especially problematic for the metal connection lines within the transistor.

- Smaller transistors only can operate if the dopant concentration is increased. However, above a certain limit, the doping atoms cluster together. The result is that *p*-type and *n*-type silicon cannot be produced reliably at small length scales.

- The gate-switching energy of transistors has not been reduced at the same rate as their size; the result is increased power consumption in integrated circuits. This has a serious consequence in that it is very difficult to dissipate the heat produced.

- At smaller length scales, microprocessors require smaller voltages for proper operation. However, since the power consumption is still relatively high, very large currents are needed between the power-conversion devices and central-processing units of modern microprocessors. These large currents result in resistive heating, thus compounding the heat-extraction problems.

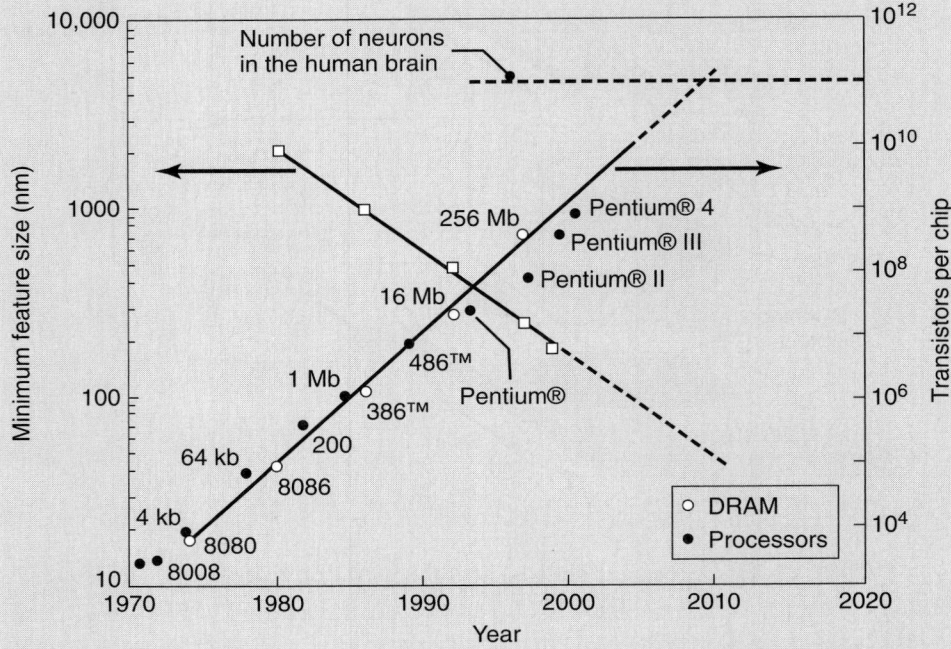

FIGURE 28.14   Illustration of Moore's law. *Source:* After M. Madou.

- Quantum effects play a large role at the small length scales, and system stability becomes an issue.

Much research is being directed towards overcoming these limitations. Moore's law was intended as a prediction of the short-term future of the semiconductor industry and was put forward at a time when photolithography was the only option. During the four decades since it was first stated, researchers often have identified insurmountable problems which, in turn, have been overcome.

28.8 | Etching

Etching is the process by which entire films or particular sections of films are removed, and it plays an important role in the fabrication sequence. One of the most important criteria in this process is **selectivity**, which refers to the ability to etch one material without etching another. A summary of etching processes and etchants are given in Tables 28.2 and 28.3.

In silicon technology, an etching process must etch the silicon-dioxide layer effectively and with minimal removal of either the underlying silicon or the resist material. In addition, polysilicon and metals must be etched into high-resolution lines with vertical wall profiles and also with minimal removal of either the underlying insulating film or photoresist. Typical etch rates range from hundreds to several thousands of angstroms per minute, and selectivities (defined as the ratio of the etch rates of the two films) can range from 1:1 to 100:1.

28.8.1 Wet etching

Wet etching involves immersing the wafers in a liquid solution, usually acidic. One primary feature of most wet-etching operations is that they are *isotropic* (that is, they etch in all directions of the workpiece at the same rate). This results in undercuts beneath the mask material (for example, see Fig. 28.15a) and limits the resolution of geometric features in the substrate.

Effective etching requires the following conditions:

1. Etchant transport to the surface.
2. A chemical reaction.

TABLE 28.2

General Characteristics of Silicon Etching Operations						
	Temperature (°C)	Etch rate (μm/min)	{111}/{100} selectivity	Nitride etch rate (nm/min)	SiO_2 etch rate (nm/min)	p^{++} etch stop
Wet etching						
$HF : HNO_3 : CH_3COOH$	25	1–20	—	Low	10–30	No
KOH	70–90	0.5–2	100:1	<1	10	Yes
Ethylene-diamine pyrochatechol (EDP)	115	0.75	35:1	0.1	0.2	Yes
$N(CH_3)_4OH$ (TMAH)	90	0.5–1.5	50:1	<0.1	<0.1	Yes
Dry (plasma) etching						
SF_6	0–100	0.1–0.5	—	200	10	No
SF_6/C_4F_8 (DRIE)	20–80	1–3	—	200	10	No

Source: Adapted from N. Maluf, *An Introduction to Microelectromechanical Systems Engineering,* Artech House, 2000.

TABLE 28.3

Comparison of Etch Rates

Etchant	Target material	Etch rate (nm/min)[a]							
		Polysilicon n^+	Polysilicon, undoped	Silicon dioxide	Silicon nitride	Phosphosilicate glass, annealed	Aluminum	Titanium	Photoresist (OCG-820PR)
Wet etchants									
Concentrated HF (49%)	Silicon oxides	0	—	2300	14	3600	4.2	>1000	0
25:1 HF:H_2O	Silicon oxides	0	0	9.7	0.6	150	—	—	0
5:1 BHF[b]	Silicon oxides	9	2	100	0.9	440	140	>1000	0
Silicon etchant (126 HNO_3:60 H_2O:5 NH_4F)	Silicon	310	100	9	0.2	170	400	300	0
Aluminum etchant (16 H_3PO_4:1 HNO_3:1 HAc:2 H_2O)	Aluminum	<1	<1	0	0	<1	660	0	0
Titanium etchant (20 H_2O:1 H_2O_2:1 HF)	Titanium	1.2	—	12	0.8	210	>10	880	0
Piranha (50 H_2SO_4:1 H_2O_2)	Cleaning off metals and organics	0	0	0	0	0	180	240	>10
Acetone (CH_3COOH)	Photoresist	0	0	0	0	0	0	0	>4000
Dry Etchants									
CF_4 + CHF_3 + He, 450W	Silicon oxides	190	210	470	180	620	—	>1000	220
SF_6 + He, 100W	Silicon nitrides	73	67	31	82	61	—	>1000	69
SF_6, 12.5W	Thin silicon nitrides	170	280	110	280	140	—	>1000	310
O_2, 400W	Ashing photoresist	0	0	0	0	0	0	0	340

Notes:

[a] Results are for fresh solutions at room temperature unless noted. Actual etch rates will vary with temperature and prior use of solution, area of exposure of film, other materials present, film impurities, and microstructure.

[b] Buffered hydrofluoric acid, 33% NH_4F and 8.3% HF by weight.

Source: After K. Williams and R. Muller, *J. Microelectromechanical Systems*, Vol. 5, 1996, pp. 256–269.

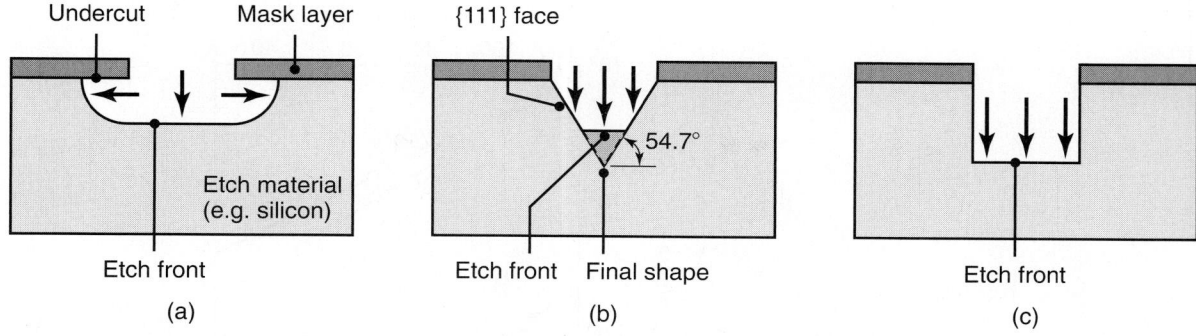

FIGURE 28.15 Etching directionality. (a) Isotropic etching: etch proceeds vertically and horizontally at approximately the same rate, with significant mask undercut. (b) Orientation-dependant etching (ODE): etch proceeds vertically, terminating on {111} crystal planes with little mask undercut. (c) Vertical etching: etch proceeds vertically with little mask undercut. *Source:* Courtesy of K.R. Williams, Agilent Laboratories.

3. Transport of reaction products away from the surface.
4. Ability to stop the etching process rapidly in order to obtain superior pattern transfer (**etch stop**), usually using an underlying layer with high selectivity.

If the first or third condition limits the speed of the process, agitation or stirring of the solution can increase etching rates. If the second condition limits the speed of the process, the etching rate will depend strongly on temperature, etching material, and solution composition. Therefore, reliable etching requires both good temperature control and repeatable stirring capability.

Isotropic etchants. These are used widely for:

* Removal of damaged surfaces.
* Rounding of sharp etched corners to avoid stress concentrations.
* Reducing roughness after anisotropic etching.
* Creating structures in single-crystal slices.
* Evaluating defects.

Fabrication of microelectronic devices and microelectromechanical systems, described in Chapter 29, require the precise machining of structures done through masking. However, masking is a challenge with isotropic etchants. The strong acids (a) etch aggressively at a rate of up to 50 μm/min with an etchant of 66% HNO_3 and 34%HF, although etch rates of 0.1 to 1 μm/s are more typical and (b) produce rounded cavities. Furthermore, the etch rate is very sensitive to agitation, and therefore, lateral and vertical features are difficult to control.

The size of the features in an integrated circuit determines its performance, and for this reason, there is a strong desire to produce well-defined, extremely small structures. Such small features cannot be attained through isotropic etching because of the poor definition which results from undercutting of masks.

Anisotropic etching. This takes place when etching is strongly dependent on compositional or structural variations in the material. There are two basic kinds of anisotropic etching: *orientation-dependent etching* (ODE) and *vertical etching*, although most vertical etching is done with dry plasmas and is discussed later. Orientation-dependent etching commonly occurs in a single crystal when etching takes place at different rates in different directions, as shown in Fig. 28.15b.

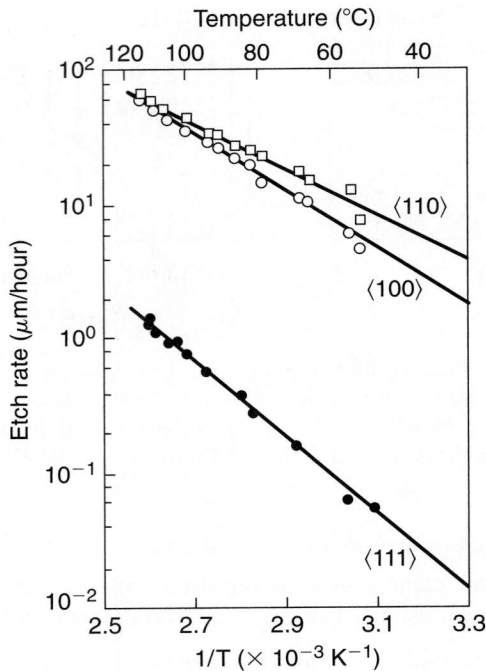

FIGURE 28.16 Etch rates of silicon in different crystallographic orientations using ethylene-diamine/pyrocatechol-in-water as the solution. *Source:* After Seidel, H., *et al.*, Journal Electrochemical Society, 1990, pp. 3612–3626.

When performed properly, these etchants produce geometric shapes with walls defined by the crystallographic planes that resist the etchants. For example, Fig. 28.16 shows the vertical etch rate for silicon as a function of temperature. As can be seen, etching is more than one order of magnitude slower in the [111] crystal direction than in other directions; therefore, well-defined walls can be obtained along the [111] crystal direction.

The **anisotropy ratio** for etching is defined by

$$AR = \frac{E_1}{E_2} \qquad (28.1)$$

where E is the etch rate in a crystallographic direction of interest, and the subscripts refer to two directions or materials of interest. **Selectivity** is defined in a similar manner but refers to the etch rates between materials of interest. The anisotropy ratio is unity for isotropic etchants and can be as high as 400/200/1 for (110)/(100)/(111) silicon. The <111> planes always etch the slowest, but the <100> and <110> planes can be controlled through etchant chemistry.

Masking is also a concern for anisotropic etching but for different reasons than with isotropic etching. Anisotropic etching is slow (typically 3 μm/min), thus anisotropic etching through a wafer may take several hours. Silicon oxide may etch too rapidly to use as a mask, hence a high-density silicon-nitride mask may be needed.

Often, it is important to rapidly halt the etching process, referred to as *etch stop*. This is typically the case when thin membranes are to be manufactured or features with very precise thickness control are needed. Conceptually, this can be accomplished by removing the wafer from the etching solution. However, etching depends to a great extent on the ability to circulate fresh etchants to the desired locations. Since the circulation varies across a wafer's surface, this strategy for halting the etching process would lead to large variations in the etched depth.

The most common approach to obtain uniform feature sizes across a wafer is to use a *boron etch stop*, where a boron layer is diffused or implanted into silicon. Examples of common etch stops are the placement of a boron-doped layer beneath silicon or the placement of silicon dioxide (SiO_2) beneath silicon nitride (Si_3N_4). Since anisotropic etchants do not attack boron-doped silicon as aggressively as they do undoped silicon, surface features or membranes can be created by **back etching**. An example of the boron etch-stop approach is shown in Fig. 28.17.

Several etchant formulations have been developed, including hydrofluoric acid, phosphoric acid, mixtures of nitric acid and hydrofluoric acid, potassium hydrochloride, and mixtures of phosphoric acid, nitric acid, acetic acid, and water. Wafer cleaning is done using a solution consisting of sulfuric acid and peroxide (Piranha solution, a trade name). Photoresist can be removed with these solutions, although acetone is used more commonly for this purpose.

28.8.2 Dry etching

Modern integrated circuits are etched exclusively with *dry etching*, which involves the use of chemical reactants in a low-pressure system. In contrast to the wet process, dry etching can have a high degree of directionality, resulting in highly anisotropic-etching profiles (Fig. 28.15c). Also, the dry-etching process requires only small amounts of the reactant gases, whereas the solutions used in the wet-etching process have to be refreshed periodically. Dry etching usually involves a plasma or discharge in areas of high-electric and magnetic fields; any gases that are present are dissociated to form ions, photons, electrons, or highly reactive molecules.

There are several specialized dry etching techniques, as described next.

Sputter etching.   *Sputter etching* removes material by bombarding it with noble gas ions, usually Ar^+. The gas is ionized in the presence of a cathode and anode (Fig. 28.18). If a silicon wafer is the target, the momentum transfer associated with the bombardment of atoms causes bond breakage and material to be ejected or sputtered. If the silicon chip is the substrate, then the material in the target is deposited onto the silicon after it has been sputtered by the ionized gas. Some of the concerns with sputter etching are the following:

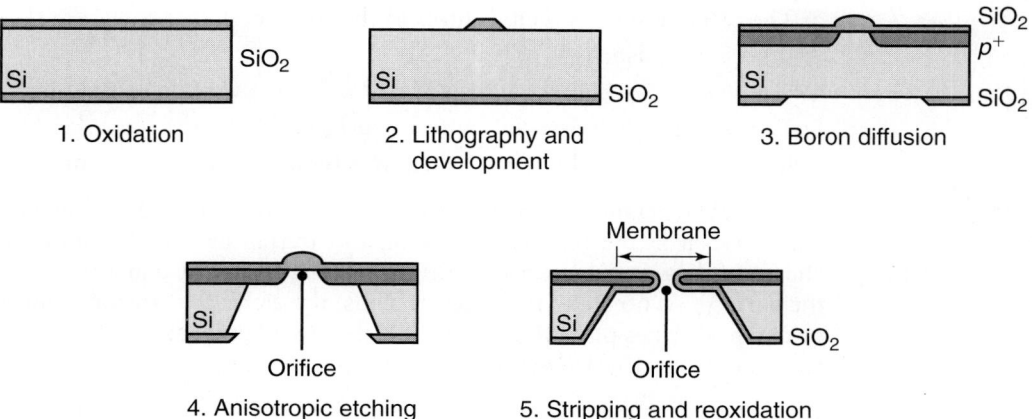

FIGURE 28.17   Application of a boron etch stop and back etching to form a membrane and orifice. *Source:* After Brodie, I., and Murray, J.J., *The Physics of Microfabrication*, Plenum Press, 1982.

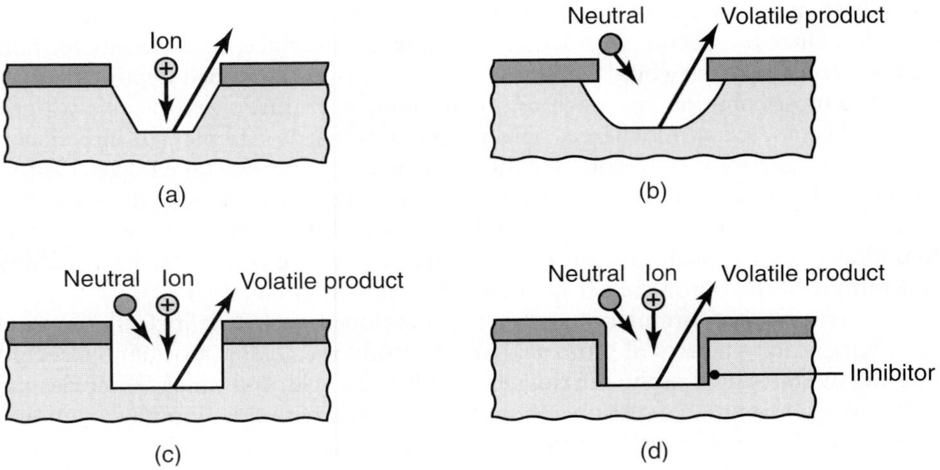

FIGURE 28.18 Machining profiles associated with different dry-etching techniques: (a) sputtering; (b) chemical; (c) ion-enhanced energetic; (d) ion-enhanced inhibitor. *Source:* After M. Madou.

- The ejected material can be redeposited onto the target, especially with large aspect ratios.
- Sputter etching is not material selective, because most materials sputter at about the same rate; therefore, masking is difficult.
- Sputter etching is slow with etch rates limited to tens of nm/min.
- Sputtering can cause damage or excessive erosion of the material.
- The photoresist is difficult to remove.

Reactive plasma etching. Also referred to as *dry chemical etching*, *reactive plasma etching* involves chlorine or fluorine ions (generated by RF excitation) and other molecular species that diffuse to and chemically react with the substrate, forming a volatile compound which is removed by the vacuum system. The mechanism of reactive plasma etching is shown in Fig. 28.19. Here:

1: A reactive species is produced (such as CF_4), dissociating upon impact with energetic electrons to produce fluorine atoms.
2: The reactive species then diffuses to the surface.
3: It becomes adsorbed.
4: The reactive species chemically reacts to form a volatile compound.
5: The reactant then desorbs from the surface.
6: It diffuses into the bulk gas, where it is removed by the vacuum system.

Some reactants polymerize on the surface and require additional removal either with oxygen in the plasma reactor or an external ashing operation. The electrical charge of the reactive species is not great enough to cause damage through impact on the surface, so no sputtering occurs. Thus, the etching is isotropic and undercutting of the mask takes place (Fig. 28.15a). Table 28.2 lists some of the more common dry etchants, their target materials, and typical etch rates.

Physical–chemical etching. Processes such as *reactive ion-beam etching* (RIBE) and *chemically assisted ion-beam etching* (CAIBE) combine the advantages of physical and chemical etching. These processes use a chemically reactive species

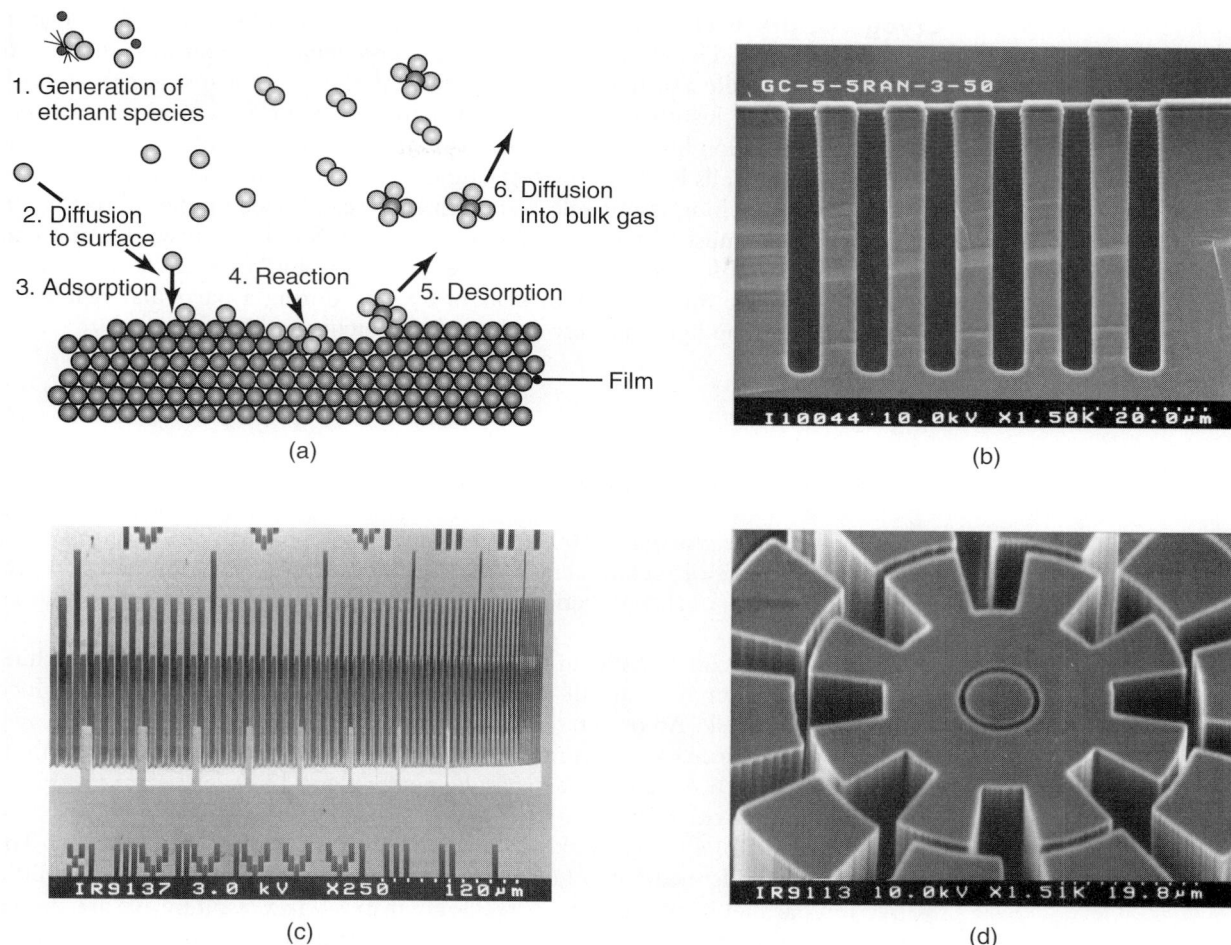

FIGURE 28.19 (a) Schematic illustration of reactive plasma etching. (b) Example of a deep reactive-ion etched trench. Note the periodic undercuts or scallops. (c) Near-vertical sidewalls produced through DRIE with an anisotropic-etching process. (d) An example of cryogenic dry etching showing a 145-μm deep structure etched into silicon using a 2.0-μm thick oxide masking layer. The substrate temperature was $-140°C$ during etching. *Source:* (a) After M. Madou. (b) through (d) After R. Kassing and I.W. Rangelow, University of Kassel, Germany.

to drive material removal, but this is assisted physically by the impact of ions onto the surface. In RIBE, also known as *deep reactive-ion etching* (DRIE), vertical trenches hundreds of nanometers deep can be produced by periodically interrupting the etching process and depositing a polymer layer, as shown in Fig. 28.19d.

In CAIBE, ion bombardment can assist dry chemical etching by:

- Making the surface more reactive.
- Clearing the surface of reaction products and allowing the chemically reactive species access to the cleared areas.
- Providing the energy to drive surface chemical reactions; however, the neutral species do most of the etching.

Physical–chemical etching is extremely useful because the ion bombardment is directional—so that etching is anisotropic. Also, the ion-bombardment energy is low and does not contribute much to mask removal. This allows the generation of near-vertical walls with very large aspect ratios. Since the ion bombardment does not remove material directly, masks can be used.

Cryogenic dry etching. This is an approach used to obtain very deep features with vertical walls. The workpiece is lowered to cryogenic temperatures, and then the process of chemically assisted ion-beam etching takes place. The low temperatures involved ensure that insufficient energy is available for a surface chemical reaction to take place, unless ion bombardment is normal to the surface. Oblique impacts (such as occur on side walls in deep crevices) cannot drive the chemical reactions.

Since dry etching is not selective, etch stops cannot be applied directly. Dry-etching reactions must be terminated when the target film is removed. Optical emission spectroscopy often is used to determine the "end point" of a reaction. Filters can be used to capture the wavelength of light emitted during a particular reaction. A noticeable change in light intensity at the point of etching will be detected.

EXAMPLE 28.2 Comparison of wet and dry etching

Consider the case where a <100> wafer has an oxide mask placed on it in order to produce square or rectangular holes. The sides of the square are oriented precisely within the <110> direction (see Fig. 28.5) of the wafer surface, as shown in Fig. 28.20.

Isotropic etching results in the cavity, as shown in Fig. 28.20a. Since etching occurs at constant rates in all directions, a rounded cavity is produced which undercuts the mask. An orientation-dependent etchant produces the cavity shown in Fig. 28.20b. Since etching is much faster in the <100> and <110> directions than in the <111> direction, sidewalls defined by the <111> plane are generated. For silicon, these sidewalls are at an angle of 54.74° to the surface.

The effect of a larger mask or shorter etch time is shown in Fig. 28.20c. The resultant pit is defined by <111> sidewalls and by a bottom in the <100> direction parallel to the surface. A rectangular mask and resultant pit are shown in Fig. 28.20d. Deep reactive-ion etching is depicted in Fig. 28.20e. Note that a polymer layer is deposited periodically onto the hole sidewalls to allow for deep pockets, but scalloping (greatly exaggerated in the figure) is unavoidable. A hole resulting from chemically-reactive ion etching is shown in Fig. 28.20f.

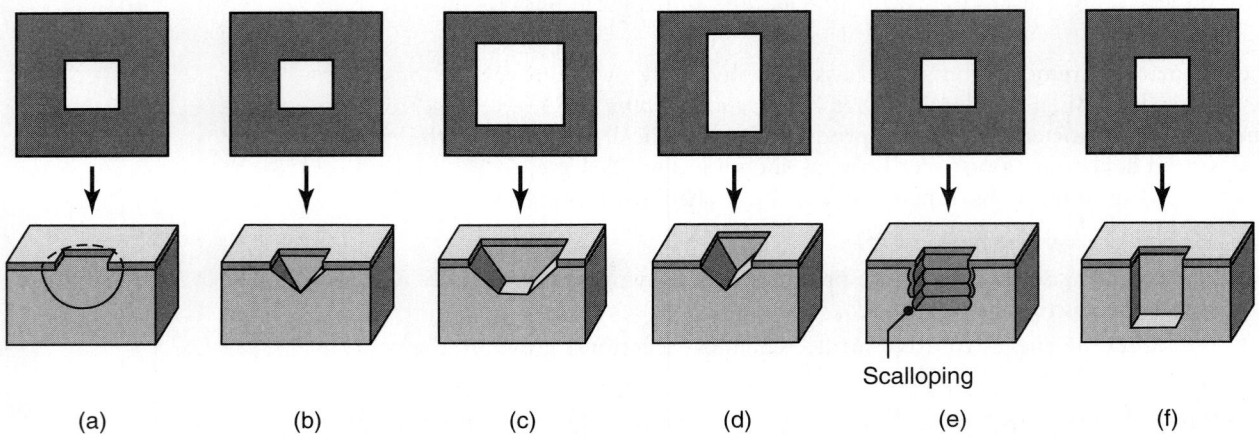

Scalloping

| (a) | (b) | (c) | (d) | (e) | (f) |

FIGURE 28.20 Various holes generated from a square mask in: (a) isotropic (wet) etching; (b) orientation-dependant etching (ODE); (c) ODE with a larger hole; (d) ODE of a rectangular hole, (e) deep reactive-ion etching; and (f) vertical etching. *Source:* After M. Madou.

28.9 | Diffusion and Ion Implantation

We should recall again that the electrical operation of microelectronic devices depends on regions that have different doping types and concentrations. The electrical character of these regions is altered through the introduction of dopants into the substrate, which is accomplished by the *diffusion* and *ion-implantation processes*. This step in the fabrication sequence is repeated several times, since many different regions of microelectronic devices must be defined.

In the *diffusion* process, the movement of atoms is a result of thermal excitation. Dopants can be introduced to the substrate surface in the form of a deposited film, or the substrate can be placed in a vapor containing the dopant source. The process takes place at elevated temperatures, usually 800° to 1200°C (1500° to 2200°F). Dopant movement within the substrate is strictly a function of temperature, time, the diffusion coefficient (or diffusivity) of the dopant species, as well as the type and quality of the substrate material.

Because of the nature of diffusion, the dopant concentration is very high at the substrate surface and drops off sharply when away from the surface. To obtain a more uniform concentration within the substrate, the wafer is heated further to drive in the dopants in a process called **drive-in diffusion**. Diffusion (desired or undesired) always will occur at high temperatures; this fact always is taken into account during subsequent processing steps. Although the diffusion process is relatively inexpensive, it is highly isotropic.

Ion implantation is a much more extensive process and requires specialized equipment (Fig. 28.21; see also Section 34.7). Implantation is accomplished by accelerating the ions through a high-voltage field of as much as one million electron volts and then by choosing the desired dopant by means of a mass separator. In a manner similar to that of cathode-ray tubes, the beam is swept across the wafer by sets of deflection plates, thus ensuring uniform coverage of the substrate. The complete implantation system must be operated in a vacuum.

The high-velocity impact of ions on the silicon surface damages the lattice structure and results in lower electron mobilities. This condition is undesirable, but the damage can be repaired by an annealing step, which involves heating the substrate to relatively low temperatures, usually 400° to 800°C (750° to 1500°F), for 15 to 30 min. This provides the energy that the silicon lattice needs to rearrange and mend itself.

Another important function of annealing is driving in the implanted dopants. Implantation alone imbeds the dopants less than half a micron below the silicon surface; annealing enables the dopants to diffuse to a more desirable depth of a few microns.

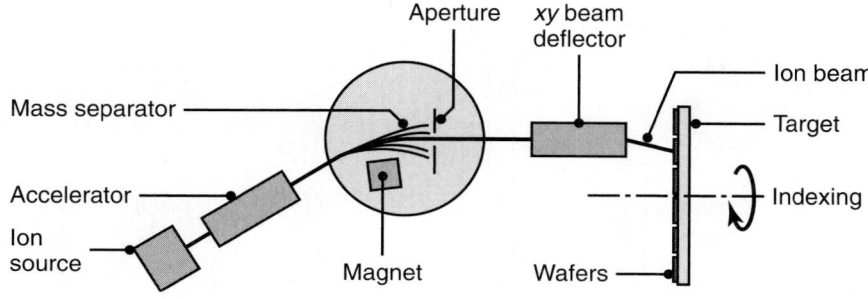

FIGURE 28.21   Schematic illustration of an apparatus for ion implantation.

EXAMPLE 28.3 Processing of a *p*-type region in *n*-type silicon

Assume that we wish to create a *p*-type region within a sample of *n*-type silicon. Draw cross-sections of the sample at each processing step in order to accomplish this.

Solution   See Fig. 28.22. This simple device is known as a *pn-junction diode*, and the physics of its operation are the foundation for most semiconductor devices.

Processing step

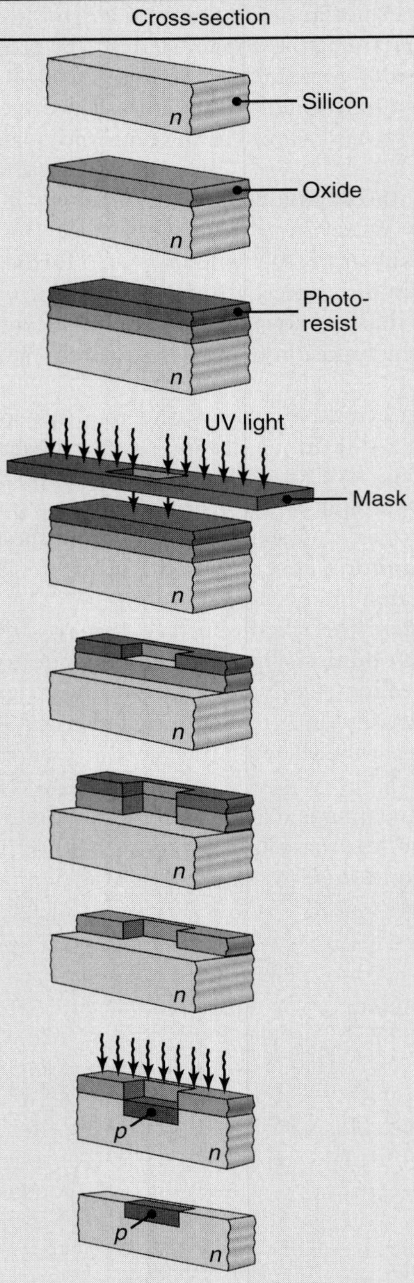

Cross-section	Description
1.	Sample of *n*-type silicon
2.	Grow silicon dioxide by oxidation
3.	Apply photoresist
4.	Expose photoresist using appropriate lithographic mask
5.	Develop photoresist
6.	Etch silicon dioxide
7.	Remove photoresist
8.	Implant boron
9.	Remove silicon dioxide

FIGURE 28.22   Fabrication sequence for a *pn*-junction diode.

28.10 | Metallization and Testing

In the preceding sections, we focused only on device fabrication. However, generating a complete and functional integrated circuit requires these devices to be interconnected, and this must take place on a number of levels (Fig. 28.23). **Interconnections** are made with metals that exhibit low electrical resistance and good adhesion to dielectric insulator surfaces. Aluminum and aluminum-copper alloys remain the most commonly used materials for this purpose in VLSI technology today.

However, as device dimensions continue to shrink, electromigration has become more of a concern with aluminum interconnects. **Electromigration** is the process by which aluminum atoms are moved physically by the impact of drifting electrons under high-current conditions. In extreme cases, this can lead to severed and/or shorted metal lines. Solutions to the electromigration problem include (a) the addition of sandwiched metal layers such as tungsten and titanium and more recently (b) the use of pure copper, which displays lower resistivity and significantly better electromigration performance than aluminum.

Metals are deposited using standard deposition techniques, and interconnection patterns are generated through lithographic and etching processes, as previously

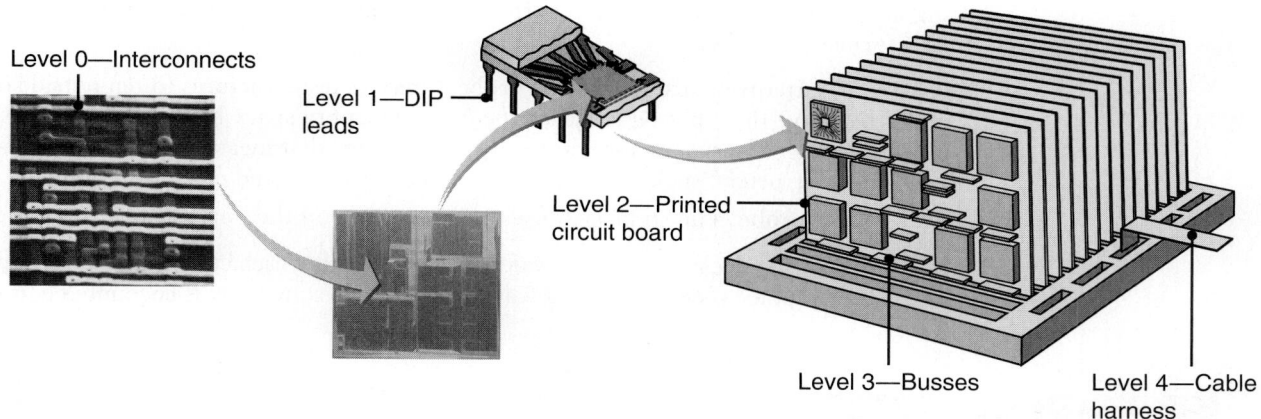

Level	Element example	Interconnection method
Level 0	Transistor within an IC	IC metallization
Level 1	ICs, other discrete components	Package leads or module interconnections
Level 2	IC packages	Printed circuit board
Level 3	Printed circuit boards	Connectors (busses)
Level 4	Chassis or box	Connectors/cable harnesses
Level 5	System, e.g., computer	

FIGURE 28.23　Connections between elements in the hierarchy for integrated circuits.

described. Modern ICs typically have one to six layers of metallization, in which case each layer of metal is insulated by a dielectric.

Planarization (producing a planar surface) of these inter-layer dielectrics is critical to the reduction of metal shorts and the linewidth variation of the interconnect. A common method used to achieve a planar surface is a uniform oxide-etch process that smoothens out the "peaks" and "valleys" of the dielectric layer.

However, today's standard for planarizing high-density interconnects quickly has become *chemical–mechanical polishing* (CMP). This process entails physically polishing the wafer surface in a manner similar to that by which a disc or belt sander flattens the ridges in a piece of wood. A typical CMP process combines an abrasive medium with a polishing compound (or slurry) and can polish a wafer to within 300 Å (1.2×10^{-6} in.) of being perfectly flat.

Layers of metal are connected together by **vias**; access to the devices on the substrate is achieved through **contacts** (Fig. 28.24). In recent years, as devices have become smaller and faster, the size and speed of some chips have become limited by the metallization process itself.

Wafer processing is completed upon application of a *passivation layer*, usually silicon nitride (Si_3N_4). The silicon nitride acts as a barrier for sodium ions and also provides excellent scratch resistance.

The next step is to test each of the individual circuits on the wafer. Each chip (also known as a **die**) is tested by a computer-controlled probe platform containing needle-like probes which access the bonding pads on the die. The probes are of two forms:

1. **Test patterns or structures**. The probe measures test structures (often outside of the active dice) placed in the *scribe line* (the empty space between dies). These consist of transistors and interconnect structures that measure various processing parameters, such as resistivity, contact resistance, and electromigration.

2. **Direct probe**. This approach uses 100% testing on the bond pads of each die.

The platform scans across the wafer and tests whether each circuit functions properly with computer-generated timing waveforms. If a defective chip is encountered, it is

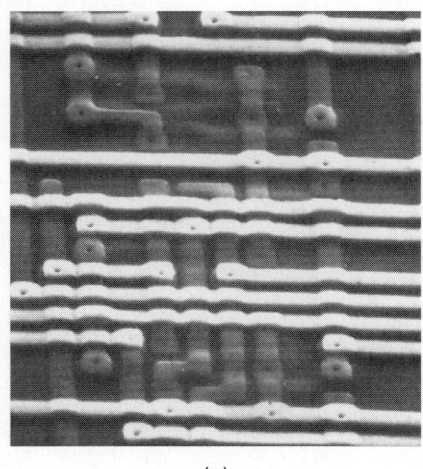

(a)

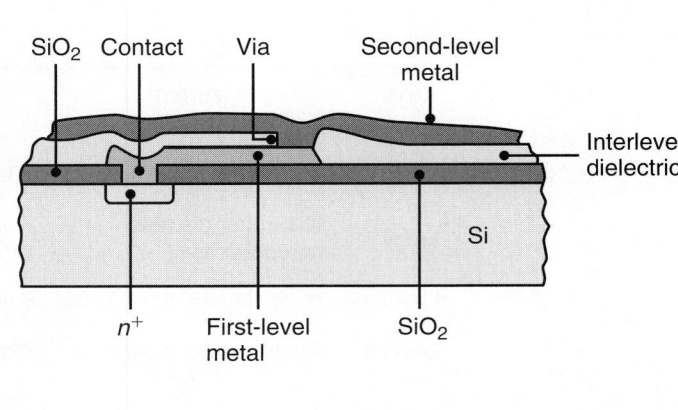

(b)

FIGURE 28.24   (a) Scanning electron microscope (SEM) photograph of a two-level metal interconnect. Note the varying surface topography. (b) Schematic illustration of a two-level metal interconnect structure. *Source:* (a) Courtesy of National Semiconductor Corporation. (b) After R.C. Jaeger.

marked with a drop of ink. Up to one third of the cost of a microelectronic circuit can be incurred during this testing.

After the wafer-level testing is completed, back grinding may be done to remove a large amount of the original substrate. Final die thickness is dependent on the packaging requirement, but anywhere from 25 to 75% of the wafer thickness can be removed. After back grinding, each die is separated from the wafer. *Diamond sawing* is a commonly used separation technique and results in very straight edges with minimal chipping and cracking damage. The chips then are sorted; the functional dice are sent on for packaging, and the inked dice are discarded.

28.11 | Wire Bonding and Packaging

The working dice must be attached to a more rugged foundation to ensure reliability. One simple method is to fasten a die to its packaging material with an epoxy cement. Another method makes use of a eutectic bond, which is made by heating metal-alloy systems (see Section 4.3). One widely used mixture is 96.4% Au and 3.6% Si, and it has a eutectic point at 370°C (700°F).

Once the chip has been attached to its substrate, it must be connected electrically to the package leads. This is accomplished by **wire bonding** very thin (25 μm diameter; 0.001 μin.) gold wires from the package leads to bonding pads located around the perimeter or down the center of the die (Fig. 28.25). The bonding pads on the die are drawn typically at 75 to 100 μm (0.003 to 0.004 in.) per side, and the bond wires are attached using thermocompression, ultrasonic, or thermosonic techniques (Fig. 28.26).

The connected circuit now is ready for final packaging. The *packaging* process largely determines the overall cost of each completed IC, since the circuits are mass produced on the wafer but then are packaged individually. Packages are available in a variety of styles; the appropriate one must reflect the operating requirements.

Consideration of a circuit's package includes: chip size, number of external leads, operating environment, heat dissipation, and power requirements. For example, ICs that are used for military and industrial applications require packages of particularly high strength, toughness, and temperature resistance.

Packages are produced from polymers, metals, or ceramics. *Metal* containers are produced from alloys such as Kovar (an iron-cobalt-nickel alloy with a low coefficient of expansion), which provide a hermetic seal and good thermal conductivity

 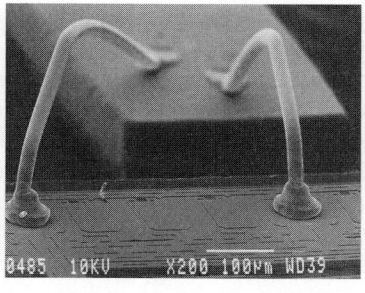 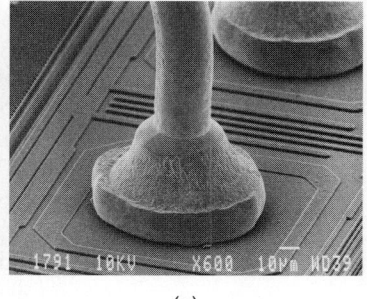

| (a) | (b) | (c) |

FIGURE 28.25   (a) SEM photograph of wire bonds connecting package leads (left-hand side) to die bonding pads. (b) and (c) Detailed views of (a). *Source*: Courtesy of Micron Technology, Inc.

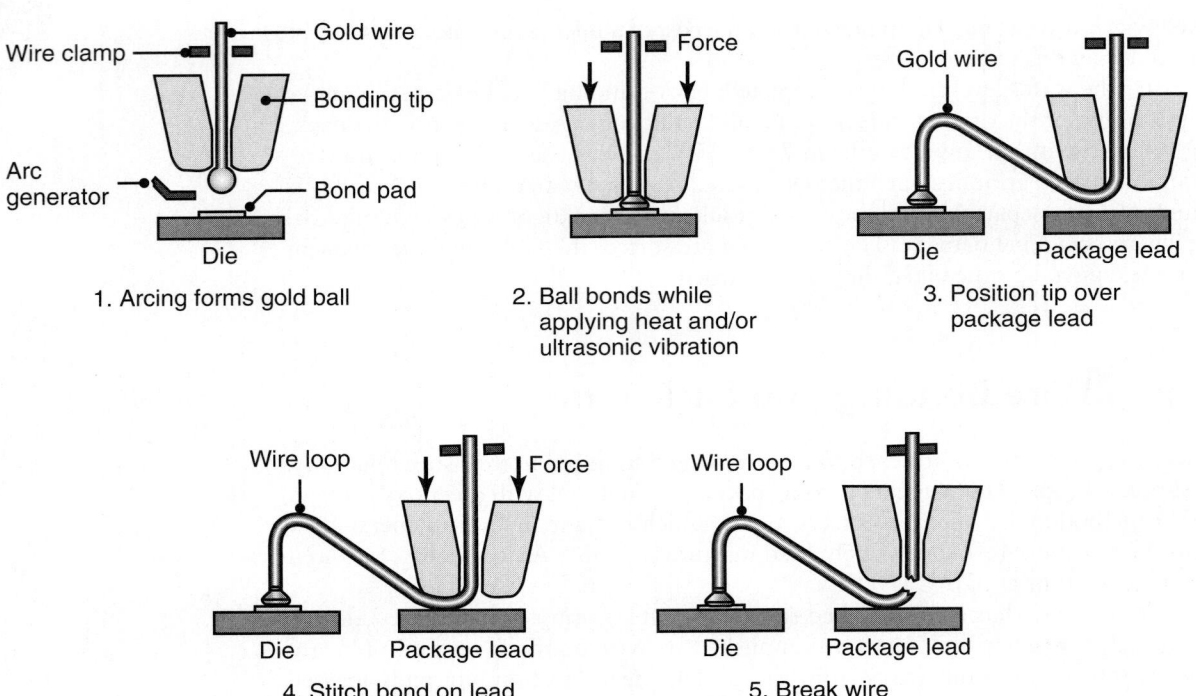

FIGURE 28.26   Schematic illustration of thermosonic welding of gold wires from package leads to bonding pads.

but are limited in the number of leads that can be used. *Ceramic* packages usually are produced from Al_2O_3 and also are hermetic and have good thermal conductivity but have higher lead counts than metal packages. However, they also are more expensive. *Plastic* packages are inexpensive and have high lead counts, but they have high thermal resistance and are not hermetic.

An older style of packaging is the **dual-in-line package** (DIP), shown schematically in Fig. 28.27a. Characterized by low cost and ease of handling, DIP packages are made of thermoplastics, epoxies, or ceramics, and can have from 2 to 500 external leads. Ceramic packages are designed for use over a broader temperature range and in high-reliability and military applications, thus costing considerably more than plastic packages.

Figure 28.27b shows a flat, ceramic package in which the package and all of the leads are in the same plane. This package style does not offer the ease of handling or the modular design of the DIP package. For this reason, it usually is affixed permanently to a multiple-level circuit board in which the low profile of the flat package is necessary.

Surface-mount packages have become the standard for today's integrated circuits. Some common examples are shown in Fig. 28.27c, where it can be seen that the main difference among them is in the shape of the connectors.

The DIP connection to the surface board is via prongs (which are inserted into corresponding holes), while a surface mount is soldered onto a specially fabricated pad or *land*. A *land* is a raised-solder platform for component interconnections in a printed circuit board. Package size and layouts are selected from standard patterns and usually require adhesive bonding of the package to the board followed by **wave soldering** of the connections. (See Section 32.3.3.)

Faster and more versatile chips require increasingly tightly-spaced connections. **Pin-grid arrays** (PGAs) use tightly packed pins which connect by way of through-holes

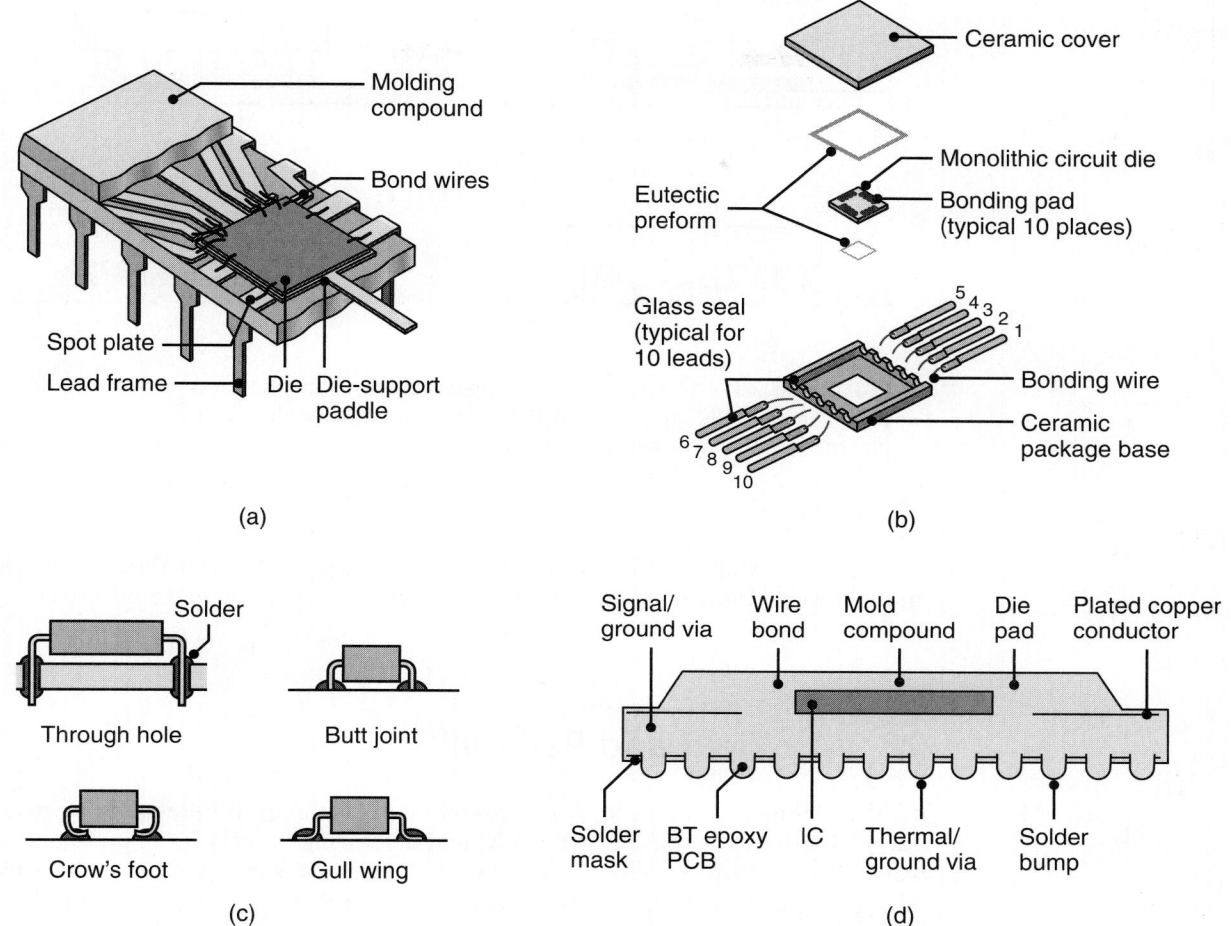

FIGURE 28.27   Schematic illustrations of various IC packages: (a) dual-in-line package (DIP); (b) flat, ceramic package; (c) common surface-mount configurations; (d) ball-grid arrays.

onto printed circuit boards. However, PGAs and other in-line and surface-mount packages are extremely susceptible to plastic deformation of the wires and legs, especially with small-diameter, closely-spaced wires. One way of achieving tight packing of connections and avoiding the difficulties of slender connections is through **ball-grid arrays** (BGA), as shown in Fig. 28.27d. This type of array has a solder-plated coating on a number of closely-spaced metal balls on the underside of the package. The spacing between the balls can be as small as 50 μm (2000 μin.), but more commonly it is standardized as 1.0 mm (0.040 in.), 1.27 mm (0.050 in.), or 1.5 mm (0.060 in.).

Although BGAs can be designed with over 1000 connections, this is extremely rare and usually 200 to 300 connections are sufficient for demanding applications. By using **reflow soldering** (Section 323.3), the solder serves to center the BGA by surface tension, resulting in well-defined electrical connections for each ball. **Flip-chip technology** (FCT) refers to the attachment procedure of ball-grid arrays and is depicted in Fig. 28.28. The final encapsulation with an epoxy is necessary not only to more securely attach the IC package to the printed circuit board but also to transfer heat evenly during its operation.

After the chip has been sealed in the package, it undergoes final testing. Because one of the main purposes of packaging is isolation from the environment,

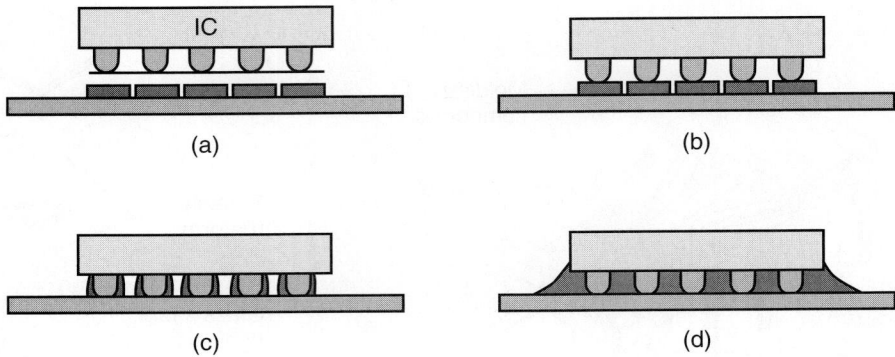

FIGURE 28.28 Illustration of flip-chip technology. Flip-chip package with (a) solder-plated metal balls and pads on the printed circuit board; (b) flux application and placement; (c) reflow soldering; (d) encapsulation.

testing at this stage usually involves heat, humidity, mechanical shock, corrosion, and vibration. Destructive tests also are performed to investigate the effectiveness of sealing.

28.12 | Yield and Reliability

Yield is defined as the ratio of functional chips to the total number of chips produced. The overall yield of the total IC manufacturing process is the product of the wafer yield, bonding yield, packaging yield, and test yield. This can range from only a few percent for new processes to more that 90% for mature manufacturing lines.

Most yield loss occurs during wafer processing due to the more complex nature of this type of processing. Wafers commonly are separated into regions of good and bad chips. Failures at this stage can arise from point defects (such as oxide pinholes), film contamination, metal particles, and area defects (such as uneven film deposition or etch nonuniformity).

A major concern about completed ICs is their **reliability** and **failure rate**. Since no device has an infinite lifetime, statistical methods are used to characterize the expected lifetimes and failure rates of microelectronic devices. The unit for failure rate is the FIT, defined as "one failure per 1 billion device-hours." However, complete systems may have millions of devices, so the overall failure rate in entire systems is correspondingly higher.

Equally important in failure analysis is determining the *failure mechanism* (that is, the actual process that causes the device to fail). Common failures due to processing involve the following:

- Diffusion regions (nonuniform current flow and junction breakdown).
- Oxide layers (dielectric breakdown and accumulation of surface charge).
- Lithography (uneven definition of features and mask misalignment).
- Metal layers (poor contact and electromigration resulting from high current densities).

- Other failures can originate in improper chip mounting, poorly formed wire bonds, and loss of package hermeticity.

Because device lifetimes are very long, it is impractical to study device failure under normal operating conditions. One method of studying failures efficiently is called **accelerated life testing** and involves accelerating the conditions whose effects cause device breakdown. Cyclic variations in temperature, humidity, voltage, and current are used to stress the components. Chip mounting and packaging are strained by cyclical temperature variations. Statistical data taken from these tests are then used to predict device-failure modes and device life under normal operating conditions.

28.13 | Printed Circuit Boards

Packaged integrated circuits seldom are used alone; rather, they usually are combined with other ICs to serve as building blocks of a yet larger system. A *printed circuit board* (PCB) is the substrate for the final interconnections among all of the completed chips and serves as the communication link between the outside world and the microelectronic circuitry within each packaged IC. In addition to the ICs, circuit boards also usually contain discrete circuit components (such as resistors and capacitors) which take up too much "real estate" on the limited silicon surface, have special power-dissipation requirements, or cannot be implemented on a chip. Other common discrete components are inductors (which cannot be integrated onto the silicon surface), high-performance transistors, large capacitors, precision resistors, and crystals (for frequency control).

A printed circuit board is basically a plastic (resin) material containing several layers of copper foil (Fig. 28.29). *Single-sided* PCBs have copper tracks on only one side of an insulating substrate; *double-sided* boards have copper tracks on both sides. Multilayered boards also can be constructed from alternating layers of copper and insulator. Single-sided boards are the simplest form of circuit board.

Double-sided boards usually must have locations where electrical connectivity is established between the features on both sides of the board. This is accomplished with vias, as shown in Fig. 28.29. Multilayered boards can have partial, buried, or through-hole vias to allow for extremely flexible PCBs. Double-sided and multilayered boards are beneficial, because IC packages can be bonded to both sides of the board, allowing for more compact designs.

The insulating material is usually an epoxy resin 0.25 to 3 mm (0.01 to 0.12 in.) thick reinforced with an epoxy/glass fiber and is referred to as E-glass (see Section 9.2.1). The assembly is produced by impregnating sheets of glass fiber with epoxy and pressing the layers together between hot plates or rolls. The heat and pressure cure the board, resulting in a stiff and strong basis for printed circuit boards.

Boards are sheared to a desired size, and about 3-mm diameter locating holes then are drilled or punched into the board corners to permit alignment and proper location of the board within the chip-insertion machines. Holes for vias and connections are punched or produced through CNC drilling; stacks of boards can be drilled simultaneously to increase production rates.

The conductive patterns on circuit boards are defined by lithography, although originally, they were produced through screen-printing technologies—hence the term

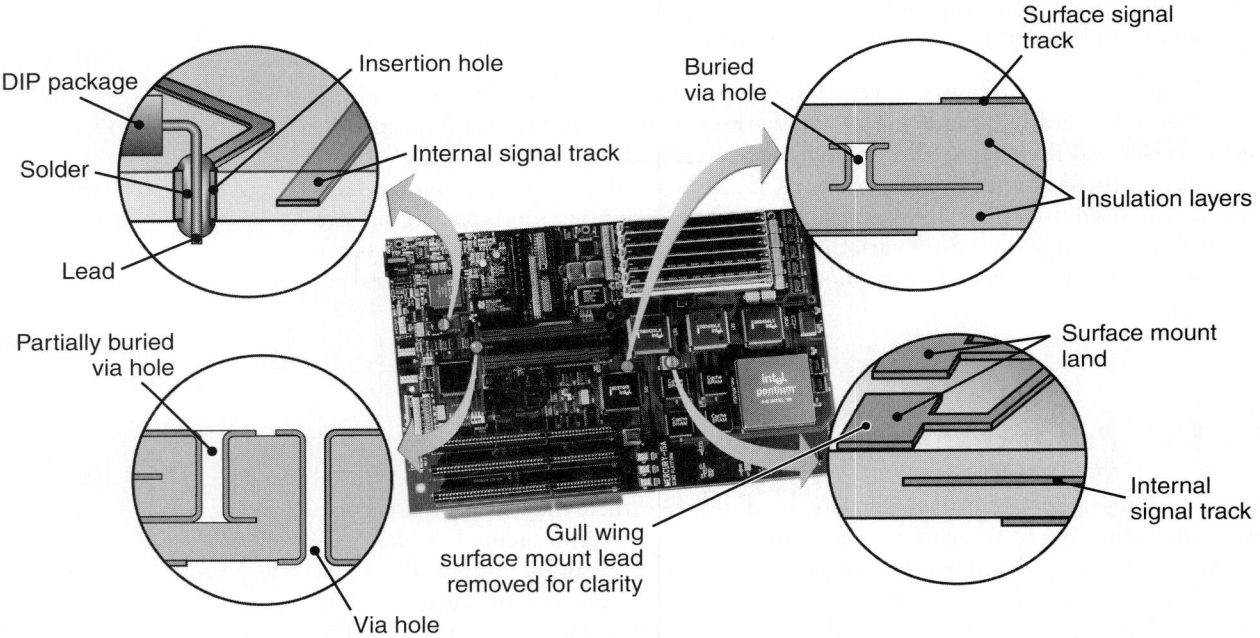

FIGURE 28.29   Printed circuit board structures and design features.

printed circuit board or *printed wiring board* (PWB). In the *subtractive method*, a copper foil is bonded to the circuit board. The desired pattern on the board is defined by a positive mask developed through photolithography, and the remaining copper is removed through wet etching. In the *additive method*, a negative mask is placed directly onto an insulator substrate to define the desired shape. Electroless plating and electroplating of copper serve to define the connections, tracks, and lands on the circuit board.

The ICs and other discrete components then are fastened to the board by soldering. This is the final step in making both the integrated circuits and the microelectronic devices they contain into larger systems through connections on printed circuit boards. *Wave soldering* and *reflow paste soldering* (see Section 32.3 and Example 32.1) are the preferred methods of soldering ICs onto circuit boards.

Some of the design considerations in laying out PCBs are the following:

1. Wave soldering should be used only on one side of the board; thus, all through-hole mounted components should be inserted from the same side of the board. Surface-mount devices placed on the insertion side of the board must be reflow soldered in place; surface-mount devices on the lead side can be wave soldered.

2. To allow good solder flow in wave soldering, IC packages should be laid out carefully on the printed circuit board. Inserting the packages in the same direction is advantageous for automated placing, because random orientations can cause problems in the flow of solder across all of the connections.

3. The spacing of ICs is determined mainly by the need to remove heat during the operation. Sufficient clearance between packages and adjacent boards is required to allow forced air flow and heat convection.

4. There also should be sufficient space around each IC package to allow for reworking and repairing without disturbing adjacent devices.

SUMMARY

- The microelectronics industry is developing rapidly. The possibilities for new device concepts and circuit designs appear to be endless. The fabrication of microelectronic devices and integrated circuits involves several different types of processes, many of which have been adapted from those of other fields of manufacturing.

- A rough shape of single-crystal silicon is obtained from the Czochralski process. This shape is ground to a cylinder of well-controlled dimensions, and a notch or flat is machined into the cylinder. The cylinder then is sliced into wafers which are ground on their edges and subjected to chemical–mechanical polishing to complete the wafer.

- After bare wafers have been prepared, they undergo repeated oxidation or film deposition and lithographic or etching steps to open windows in the oxide layer in order to access the silicon substrate.

- Wet etching is isotropic and relatively fast. However, dry etching (using gas plasmas) is anisotropic and allows for more accurate lithography and large-scale integration of integrated circuits.

- After each of the processing cycles is completed, dopants are introduced into various regions of the silicon structure through diffusion and ion implantation. The devices then are interconnected by multiple metal layers, and the completed circuit is packaged and made accessible through electrical connections.

- Finally, the packaged circuit and other discrete devices are soldered to a printed circuit board for final installation.

KEY TERMS

Accelerated-life testing	Diffusion	Evaporation
Bonding	Dopants	Failure rate
Chemical vapor deposition	Dry etching	Film deposition
Chemical–mechanical polishing	Dry oxidation	Gallium arsenide
Chip	Dual-in-line package	Integrated circuit
Contacts	Electromigration	Ion implantation
Czochralski process	Epitaxy	Line width
Die	Etching	Lithography

Masking

Metal-oxide-semiconductor
  field-effect transistor

Metallization

Oxidation

Packaging

Photoresist

Planarization

Postbaking

Prebaking

Printed circuit board

Registration

Reliability

Reticle

Selective oxidation

Selectivity

Semiconductor

Silicon

Sputtering

Surface-mount package

Very large-scale integration

Vias

Yield

Wafer

Wet etching

Wet oxidation

Wire bonding

Yield

BIBLIOGRAPHY

Bakoglu, H.B., *Circuits, Interconnections, and Packaging for VLSI*, Addison-Wesley, 1990.

Berger, L.I., *Semiconductor Materials*, CRC Press, 1997.

Brar, A.S., and Narayan, P.B., *Materials and Processing Failures in the Electronics and Computer Industries: Analysis and Prevention*, ASM International, 1993.

Campbell, S.A., *The Science and Engineering of Microelectronic Fabrication*, Oxford University Press, 1996.

Capillo, C., *Surface Mount Technology*, McGraw-Hill, 1990.

Chandrakasan, A., and Brodersen, R. (eds.), *Low Power CMOS Design*, IEEE, 1998.

Chandrakasan, A., and Brodersen, R., *Low Power Digital CMOS Design*, Kluwer Academic Pub., 1995.

Coombs, C.F., Jr. (ed.), *Printed Circuits Handbook*, 4th ed., McGraw-Hill, 1995.

Edwards, P.R., *Manufacturing Technology in the Electronics Industry*, Chapman & Hall, 1991.

Electronic Materials Handbook, Vol. 1: *Packaging*. ASM International, 1989.

Ghandhi, S.K., *VLSI Fabrication Principles: Silicon and Gallium Arsenide*, 2nd ed., Wiley, 1993.

Harper, C.A. (ed.), *Electronic Packaging and Interconnection Handbook*, 2nd ed., McGraw-Hill, 1996.

Humpston, G., and Jacobson, D.M., *Principles of Soldering*, ASM International, 2004.

Hwang, J.S., *Modern Solder Technology for Competitive Electronics Manufacturing*, McGraw-Hill, 1996.

Jackson, K.A., and Schroter, W. (eds.), *Handbook of Semiconductor Technology*, Vol. 2: *Processing of Semiconductors*, Wiley, 2000.

Judd, M., and Brindley, K., *Soldering in Electronics Assembly*, 2nd ed., Newnes, 1999.

Lambert, L.P., *Soldering for Electronic Assemblies*, Marcel Dekker, 1988.

Lau, J.H. (ed.), *Electronic Packaging: Design, Materials, Process, and Reliability*, McGraw-Hill, 1998.

Lee, H.H., *Fundamentals of Microelectronics Processing*, McGraw-Hill, 1990.

Madou, M.J., *Fundamentals of Microfabrication*, CRC Press, 2nd ed., 2002.

Mahajan, S., and Harsha, K.S.S., *Principles of Growth and Processing of Semiconductors*, McGraw-Hill, 1998.

Manko, H.H., *Soldering Handbook for Printed Circuits and Surface Mounting*, Van Nostrand Reinhold, 1995.

Marks, L., *Printed Circuit Assembly Design*, McGraw-Hill, 2000.

Matisoff, B.S., *Handbook of Electronics Manufacturing*, 3rd ed., Chapman & Hall, 1996.

Mroczkowski, R.S., *Electronic Connector Handbook: Theory and Applications*, McGraw-Hill, 1997.

Pecht, M. (ed.), *Electronic Packaging Materials and Their Properties*. CRC Press, 1998.

Prasad, R.P., *Surface Mount Technology*, 2nd ed., Chapman & Hall, 1997.

Runyan, W.R., and Shaffner, T.J., *Semiconductor Measurements and Instrumentation*, 2nd ed., McGraw-Hill, 1998.

Schroder, D.K., *Semiconductor Material and Device Characterization*, 2nd ed., Wiley-Interscience, 1998.

Taur, Y., and Ning, T.H., *Fundamentals of Modern VLSI Devices*, Cambridge, 1998.

Tummala, R., Rymaszewski, E.J., and Klopfenstein, A.G., *Microelectronics Packaging Fundamentals*, Chapman & Hall, 1999.

Van Zant, P., *Microchip Fabrication*, 4th ed., McGraw-Hill, 2000.

Wolf, W.H., *Modern VLSI Design: Systems on Silicon*, 2nd ed., Prentice Hall, 1998

Sze, S.M. (ed.), *High Speed Semiconductor Devices*, Wiley, 1993.

————, *Modern Semiconductor Device Physics*, Wiley, 1997.

Wolf, S., and Tauber, R.N., *Silicon Processing for the VLSI Era*, Vol. 1: *Process Technology*, 1986; Vol. 2: *Process Integration*, Lattice Press, 1990.

Yeap, G., *Practical Low Power Digital VLSI Design*, Kluwer Academic Pub., 1997.

REVIEW QUESTIONS

28.1 Define the terms: wafer, chip, die, device, integrated circuit, line width, registration, surface mount, accelerated-life testing, and yield.

28.2 Why is silicon the semiconductor most used in IC technology?

28.3 What do VLSI, IC, CVD, CMP, and DIP stand for?

28.4 How do *n*-type and *p*-type dopants differ?

28.5 How is epitaxy different from other techniques used for deposition?

28.6 Explain the differences between wet and dry oxidation.

28.7 How is silicon nitride used in oxidation?

28.8 What are the purposes of prebaking and postbaking in lithography?

28.9 Define selectivity and isotropy and their importance in relation to etching.

28.10 Compare diffusion and ion implantation.

28.11 What is the difference between evaporation and sputtering?

28.12 What do BJT and MOSFET stand for?

28.13 What are the levels of interconnection?

28.14 Which is cleaner, a Class-10 or a Class-1 clean room?

QUALITATIVE PROBLEMS

28.15 In a horizontal epitaxial reactor (Fig. P28.15), the wafers are placed on a stage (susceptor) that is tilted by a small amount, usually 1 to 3°. Why is this done?

28.16 The table below describes three wafer-manufacturing changes: increasing the wafer diameter, reducing the chip size, and increasing the process complexity. Complete the table by filling in "increase," "decrease," or "no change," and indicate the effect that each change would have on the wafer yield and on the overall number of functional chips.

Effects of Manufacturing Changes

Change	Wafer yield	Number of functional chips
Increase wafer diameter		
Reduce chip size		
Increase process complexity		

28.17 The speed of a transistor is proportional directly to the width of its polysilicon gate: a narrower gate results in a faster transistor and a wider gate results in a slower transistor. Knowing that the manufacturing process has a certain variation for the gate width (say ± 0.1 μm), how might a designer alter the gate sizing of a critical circuit in order to minimize its speed variation? Are there any negative effects of this change?

28.18 A common problem in ion implantation is channeling, in which the high-velocity ions travel deep into the material via channels along the crystallographic planes before finally being stopped. What is one simple way to stop this effect?

28.19 Examine the hole profiles in Fig. P28.19 and explain how they might be produced.

28.20 Referring to Fig. 28.20, sketch the holes generated from a circular mask.

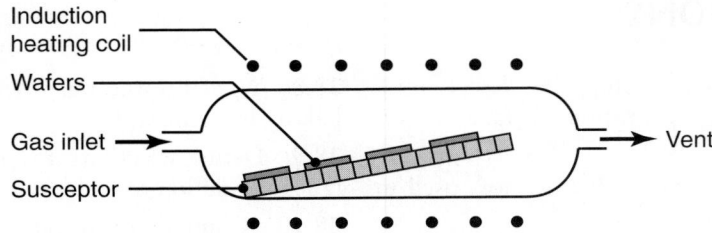

FIGURE P28.15

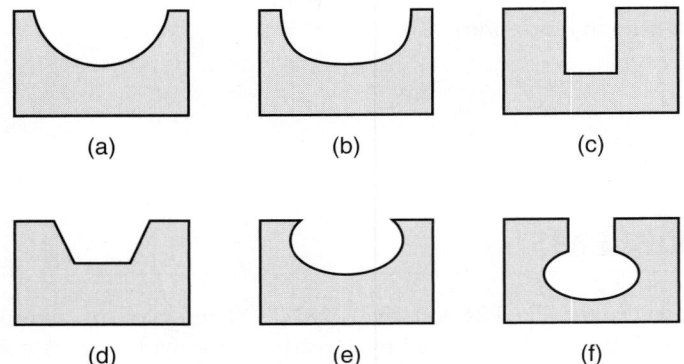

(a)   (b)   (c)

(d)   (e)   (f)

FIGURE P28.19

QUANTITATIVE PROBLEMS

28.21 A certain wafer manufacturer produces two equal-sized wafers, one containing 500 chips and the other containing 300 chips. After testing, it is observed that 50 chips on each wafer are defective. What are the yields of these two wafers? Can any relationship be drawn between chip size and yield?

28.22 A chlorine-based polysilicon etching process displays a polysilicon-to-resist selectivity of 4:1 and a polysilicon-to-oxide selectivity of 50:1. How much resist and exposed oxide will be consumed in etching 3500 Å of polysilicon? What would the polysilicon-to-oxide selectivity have to be in order to reduce the loss to only 40 Å of exposed oxide?

28.23 During a processing sequence, three silicon-dioxide layers are grown by oxidation to: 2500 Å, 4000 Å, and 1500 Å. How much of the silicon substrate is consumed?

28.24   A certain design rule calls for metal lines to be no less than 2 μm wide. If a 1-μm-thick metal layer is to be wet-etched, what is the minimum photoresist width allowed (assuming that the wet etching is perfectly isotropic)? What would be the minimum photoresist width if a perfectly anisotropic dry-etching process is used?

SYNTHESIS, DESIGN, AND PROJECTS

28.25   Inspect various electronic and computer equipment, take them apart as much as you can, and identify components that may have been manufactured by the techniques described in this chapter.

28.26   Do any aspects of this chapter's contents and the processes described bear any similarity to the processes described throughout previous parts of this book? Explain and describe what they are.

28.27   Describe your understanding of the important features of clean rooms, and how they are maintained.

28.28   Describe products that would not exist without the knowledge and techniques described in this chapter. Explain.

28.29   Review the technical literature and give more details regarding the type and shape of the abrasive wheel used in the wafer-cutting process shown in Step 2 in Fig. 28.2.

28.30   As you know, microelectronic devices may be subjected to hostile environments (such as high temperature, humidity, and vibration) as well as physical abuse (such as being dropped on a hard surface) Describe your thoughts on how you would go about testing these devices for their endurance under these conditions.

28.31   Review the specific devices indicated in Fig. V.2. Choose any one of these devices and investigate what they are, their characteristics, how they are manufactured, and at what costs.

29

Fabrication of Microelectromechanical Devices and Systems (MEMS)

Many of the processes and materials used for microelectronic devices also are used for manufacturing micromechanical devices and microelectromechanical systems. In this chapter, we describe:

• Specialized processes for manufacturing MEMS devices.
• Micromachining of three-dimensional features.
• Surface micromachining.
• Dry etching and wet etching.
• Electroforming.

Typical parts made: Sensors, actuators, accelerometers, optical switches, inkjet printing mechanisms, micromirrors, micromachines, and microdevices.

29.1 | Introduction

The preceding chapter dealt with the manufacture of integrated circuits and products that operate purely on electrical or electronic principles, called *microelectronic devices*. These semiconductor-based devices often have the common characteristic of extreme miniaturization. A large number of devices exist that are mechanical in nature and are of a similar size as microelectronic devices. A *micromechanical device* is a product that is purely mechanical in nature and has dimensions between a few millimeters and atomic length scales, such as some very small gears and hinges.

Microelectromechanical devices (MEMS devices) are products that combine mechanical and electrical or electronic elements at these very small length scales. *Microelectromechanical systems* (MEMS) are microelectromechanical devices that also incorporate an integrated electrical system into one product. Common examples of micromechanical devices are sensors of all types (Fig. 29.1). Microelectromechanical systems are rarer, but typical examples are airbag sensors and digital micromirror devices.

Many of the materials and manufacturing methods and systems described in Chapter 28 also apply to the manufacture of microelectromechanical devices and systems. However, it should be noted that microelectronic devices are semiconductor based, whereas microelectromechanical devices and portions of MEMS do not

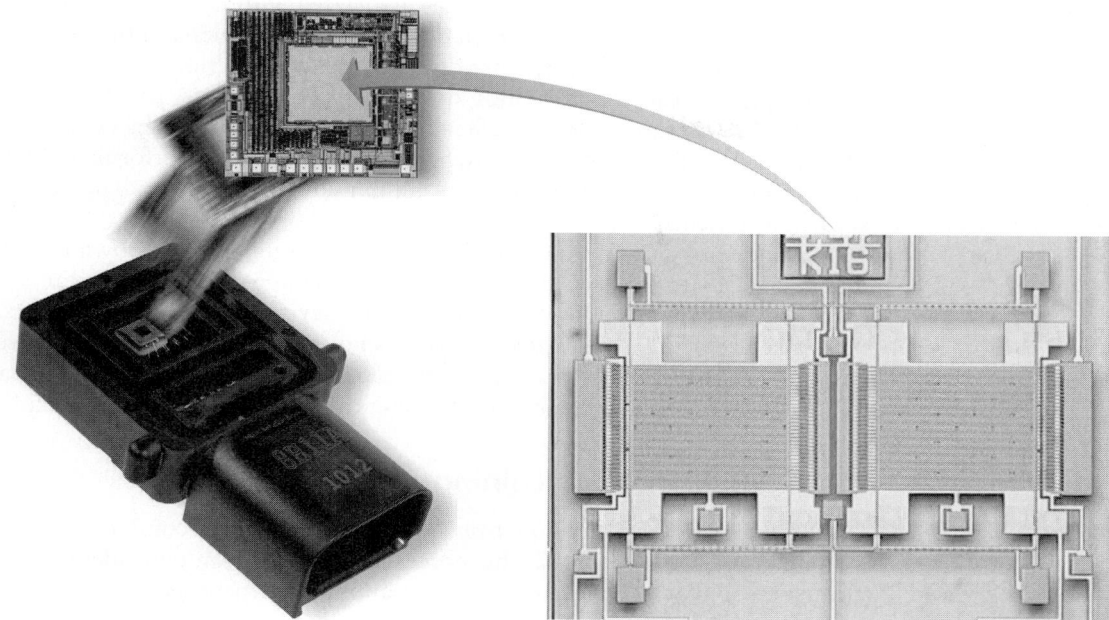

FIGURE 29.1 A gyroscope sensor used for automotive applications that combines mechanical and electronic systems. Perhaps the most widespread use of MEMS devices is in sensors of all kinds. *Source:* Courtesy of Motorola Corporation.

have this material restriction. This allows the use of many more materials and the development of processes suitable for these materials. Regardless, silicon often is used because several highly advanced and reliable manufacturing processes have been developed for microelectronic applications. This chapter emphasizes the manufacturing processes that are applicable specifically to microelectromechanical devices and systems, but it should be realized that processes and concepts such as lithography, metallization, etching, coating, and packaging still apply.

The field of MEMS and MEMS devices is advancing rapidly, and new processes or variations of existing processes are being developed continually. It has been suggested that MEMS technology can have widespread industrial applications; however, only a few industries have exploited MEMS thus far. Many of the processes described in this chapter have not yet become widely applied but are of interest to researchers and practitioners in MEMS.

29.2 | Micromachining of MEMS Devices

In the manufacture of MEMS devices, lithography and etching can be used to obtain two-dimensional or two-and-one-half-dimensional features on wafer surfaces; however, three-dimensional features often are required. The production of features ranging from micrometers to millimeters is called *micromachining*. As described in Section 28.8, the use of anisotropic-etching techniques allows the fabrication of devices with well-defined walls and high aspect ratios. For this reason, some single-crystal silicon MEMS devices have been fabricated.

One of the recognized difficulties associated with using silicon for MEMS devices is the high adhesion encountered at small length scales and the associated rapid wear. Most commercial devices are designed to avoid friction, for example, by using flexing springs instead of bushings; however, this complicates the designs and makes some MEMS devices not feasible. Significant research now is being conducted to identify materials and lubricants that provide reasonable life and performance. Silicon carbide, diamond, and metals (such as aluminum, tungsten, and nickel) are being studied as potential MEMS materials.

Lubricants also have been investigated. For example, it is known that surrounding the MEMS device in a silicone oil practically eliminates adhesive wear (see Section 33.5), but it also limits the performance of the device. Self-assembling layers of polymers also are being investigated, as well as novel and new materials with self-lubricating characteristics. However, the tribology (friction, wear, and lubrication) of MEMS devices remains the main technological barrier to their widespread use.

29.2.1 Bulk micromachining

Until the early 1980s, *bulk micromachining* was the most common method of machining at micrometer scales. This process uses orientation-dependent etches on single-crystal silicon (see Fig. 28.15b). This approach depends on etching down into a surface and stopping on certain crystal faces, doped regions, and etchable films to form a desired structure. As an example of this process, consider the fabrication of the silicon cantilever shown in Fig. 29.2. Using the masking techniques described in Section 28.7, a rectangular patch of the *n*-type silicon substrate is changed to *p*-type silicon through boron doping. Etchants such as potassium hydroxide will not be able to remove heavily boron-doped silicon, hence this patch will not be etched.

A mask then is produced, such as with silicon nitride on silicon. When etched with potassium hydroxide, the undoped silicon will be removed rapidly, while the mask and the doped patch essentially will be unaffected. Etching progresses until the (111) planes are exposed in the *n*-type silicon substrate; they undercut the patch, leaving a suspended cantilever as shown.

29.2.2 Surface micromachining

Bulk micromachining is useful for producing very simple shapes. It is restricted to single-crystal materials, since polycrystalline materials will not machine at different rates in different directions when using wet etchants. Many MEMS applications

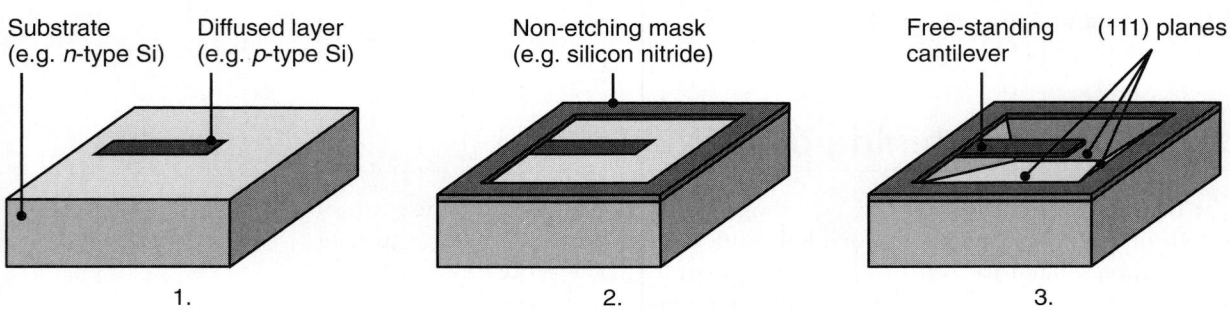

Substrate (e.g. *n*-type Si) Diffused layer (e.g. *p*-type Si) Non-etching mask (e.g. silicon nitride) Free-standing cantilever (111) planes

1.   2.   3.

FIGURE 29.2 Schematic illustration of bulk micromachining. (1) Diffuse dopant in desired pattern. (2) Deposit and pattern masking film. (3) Orientation-dependent etching (ODE) leaves behind a freestanding structure. *Source*: Courtesy of K. R. Williams, Agilent Laboratories.

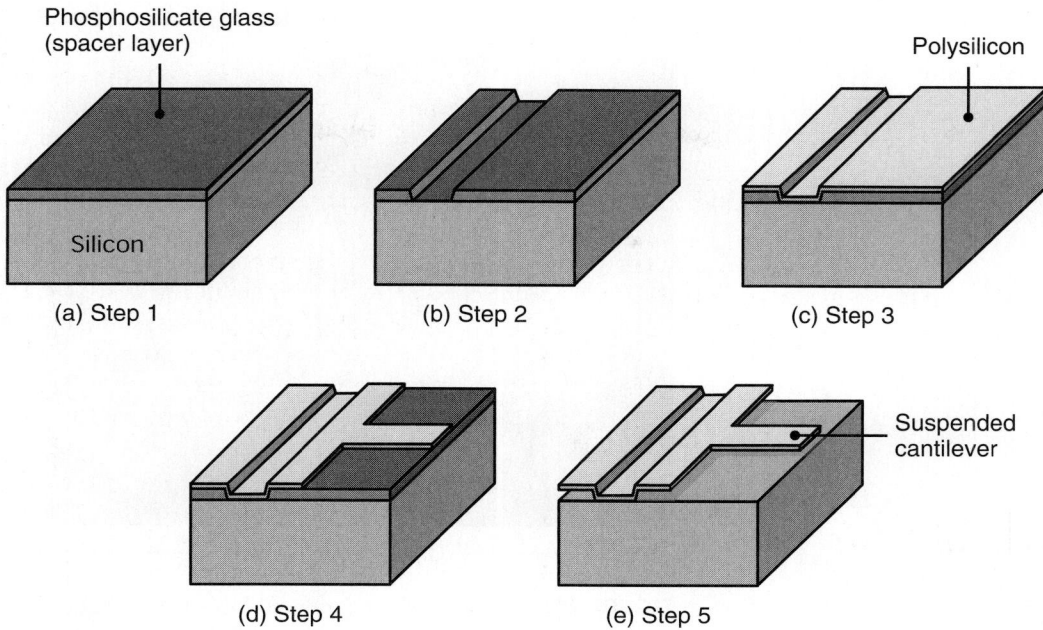

Phosphosilicate glass
(spacer layer)

Polysilicon

Silicon

(a) Step 1

(b) Step 2

(c) Step 3

Suspended
cantilever

(d) Step 4

(e) Step 5

FIGURE 29.3   Schematic illustration of the steps in surface micromachining: (a) deposition of a phosphosilicate glass (PSG) spacer layer; (b) etching of spacer layer; (c) deposition of polysilicon; (d) etching of polysilicon; (e) selective wet etching of PSG, leaving the silicon substrate and deposited polysilicon unaffected.

require the use of other materials, so alternatives to bulk micromachining are needed. One such alternative is *surface micromachining*. The basic steps in surface micromachining are illustrated in Fig. 29.3 for silicon devices. A spacer or sacrificial layer is deposited onto a silicon substrate coated with a thin dielectric layer (called an *isolation* or *buffer layer*).

Phosphosilicate glass deposited by chemical vapor deposition is the most common material for a spacer layer, because it etches very rapidly in hydrofluoric acid. Figure 29.3b shows the spacer layer after the application of masking and etching. At this stage, a structural thin film is deposited onto the spacer layer; the film can be polysilicon, metal, metal alloy, or a dielectric (Fig 29.3c). The structural film is then patterned, usually through dry etching in order to maintain vertical walls and tight dimensional tolerances. Finally, wet etching of the sacrificial layer leaves a free-standing, three-dimensional structure, as shown in Fig. 29.3e. It should be noted that the wafer must be annealed to remove the residual stresses in the deposited metal before it is patterned. If this is not done, the structural film will severely warp once the spacer layer is removed.

Figure 29.4 shows a microlamp that emits a white light when current is passed through it. It has been produced through a combination of surface and bulk micromachining. The top patterned layer is a 2.2-μm thick layer of plasma-etched tungsten, forming a meandering filament and bond pad. The rectangular overhang is dry-etched silicon nitride. The steeply sloped layer is wet-HF-etched phosphosilicate glass. The substrate is silicon, which is ODE etched.

The etchant used to remove the spacer layer must be chosen carefully. It must preferentially dissolve the spacer layer while leaving the dielectric, silicon, and structural film as intact as possible. With large features and narrow spacer layers,

Film 2 μm thick

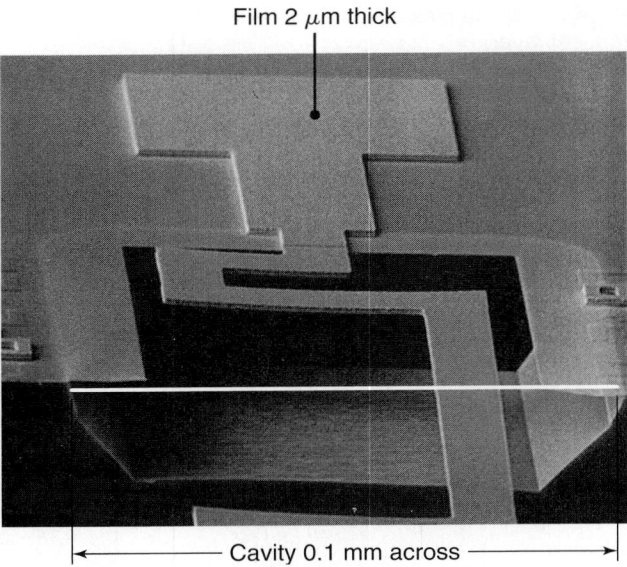

Cavity 0.1 mm across

FIGURE 29.4   A microlamp produced from a combination of bulk and surface micromachining processes. *Source:* Courtesy of K.R. Williams, Agilent Technologies.

this becomes very difficult to do, and etching can take many hours. To reduce the etching time, additional etched holes can be designed into the microstructures to increase access of the etchant to the spacer layer.

Another difficulty that must be overcome is **stiction** after wet etching. Consider the situation illustrated in Fig. 29.5. After the spacer layer has been removed, the liquid etchant is dried from the wafer surface. A meniscus formed between the layers then results in capillary forces that can deform the film and cause it to contact the substrate as the liquid evaporates. Since adhesive forces are more significant at small length scales, it is possible that the film may stick permanently to the surface, and thus the desired three-dimensional features will not be produced.

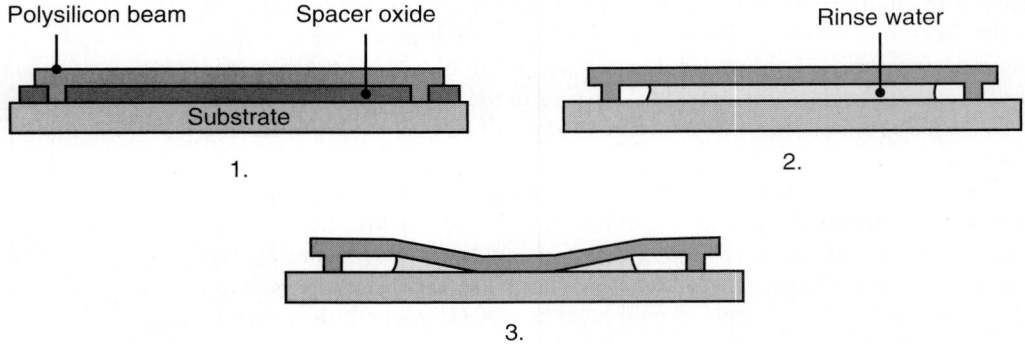

FIGURE 29.5   Stiction after wet etching: (1) unreleased beam; (2) released beam before drying; (3) released beam pulled to the surface by capillary forces during drying. Once contact is made, adhesive forces prevent the beam from returning to its original shape. *Source:* After B. Bhushan.

EXAMPLE 29.1   Surface micromachining of a hinge

Surface micromachining is a very widespread technology for the production of MEMS. Applications include accelerometers, pressure sensors, micropumps, micromotors, actuators, and microscopic locking mechanisms. Often, these devices require very large vertical walls, which cannot be manufactured directly because the high vertical structure is difficult to deposit. This is overcome by machining large, flat structures horizontally and then rotating or folding them into an upright position, as shown in Fig. 29.6.

Figure 29.6a shows a micromirror that has been inclined with respect to the surface on which it was manufactured. Such systems can be used for reflecting light (that is oblique to a surface) onto detectors or towards other sensors. It is apparent that a device which has such depth and has the aspect ratio of the deployed mirror is very difficult to machine directly. Instead, it is easier to surface micromachine the mirror along with a linear actuator and then to fold the mirror into a deployed position. In order to do so, special hinges (as shown in Fig. 29.6b) are integrated into the design.

Figure 29.7 shows the cross-section of a hinge during its manufacture. The following steps are involved in the production of the hinges:

1. A 2-μm thick layer of phosphosilicate glass (PSG) is deposited onto the substrate material.

2. A 2-μm thick layer of polysilicon (Poly1 in Fig. 29.7a) is deposited onto the PSG, patterned by photolithography, and dry etched to form the desired structural elements, including the hinge pins.

3. A second layer of sacrificial PSG with a thickness of 0.5 μm is deposited (Fig. 29.7b).

4. The connection locations are etched through both layers of PSG (Fig. 29.7c).

5. A second layer of polysilicon (Poly2 in Fig. 29.7d) is deposited, patterned, and etched.

6. The sacrificial layers of PSG are removed by wet etching.

(a)

(b)

FIGURE 29.6   (a) SEM image of a deployed micromirror. (b) Detail of the micromirror hinge. *Source*: Courtesy of Sandia National Laboratories.

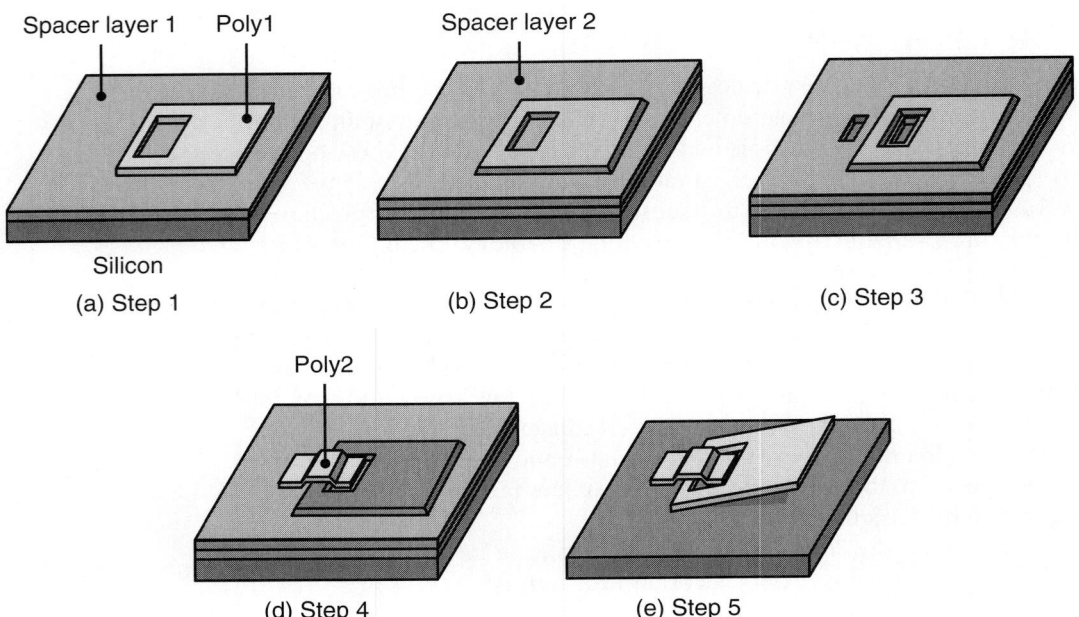

FIGURE 29.7   Schematic illustration of the steps required to manufacture a hinge. (a) Deposition of a phosphosilicate glass (PSG) spacer layer and polysilicon layer (see Fig. 29.3). (b) Deposition of a second spacer layer. (c) Selective etching of the PSG. (d) Deposition of polysilicon to form a staple for the hinge. (e) After selective wet etching of the PSG, the hinge can rotate.

Hinges such as these have very high friction. If mirrors (as shown in Fig. 29.6) are manipulated manually and carefully with probe needles, they will remain in position. Often, such mirrors will be combined with linear actuators to precisely control their deployment.

EXAMPLE 29.2 Digital micromirror device

An example of a commercial MEMS-based product is the digital pixel technology (DPT™) device, illustrated in Fig. 29.8. This device uses an array of *digital micromirror devices* (DMD™) to project a digital image, such as in computer-driven projection systems. The aluminum mirrors can be tilted so that light is directed into or away from the optics that focus light onto a screen. This way each mirror can represent a pixel of an image's resolution. The mirror allows light or dark pixels to be projected, but levels of gray also can be accommodated. Since the switching time is about 15 μs (which is much faster than the human eye can respond to), the mirror will switch between the on and off states in order to reflect the proper light dose to the optics.

The fabrication steps for producing the DMD device are shown in Fig. 29.9. This sequence is similar to other surface micromachining examples but has the following important differences:

- All micromachining steps take place at temperatures below 400°C—low enough to ensure that no damage occurs to the electronic circuit.

- A thick silicon-dioxide layer is deposited and is chemical–mechanical polished to provide an adequate foundation for the MEMS device.

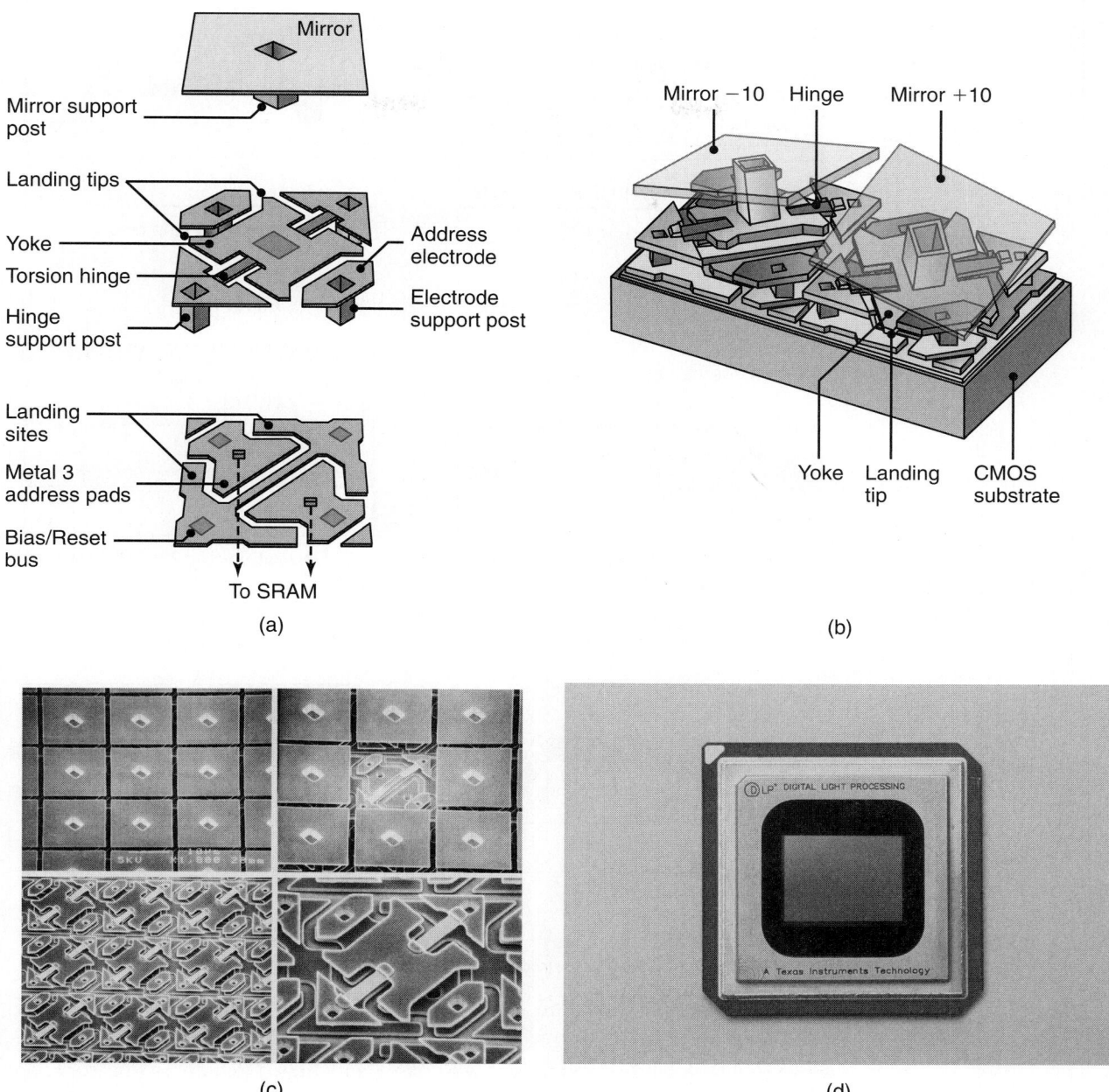

FIGURE 29.8   The Texas Instruments digital pixel technology (DPT™) device. (a) Exploded view of a single digital micromirror device (DMD™). (b) View of two adjacent DMD pixels. (c) Images of DMD arrays with some mirrors removed for clarity; each mirror measures approximately 17 μm (670 μin.) on a side. (d) A typical DPT device used for digital projection systems, high definition televisions, and other image display systems. The device shown contains 1,310,720 micromirrors and measures less than 50 mm (2 in.) per side. *Source*: Courtesy of Texas Instruments Corp.

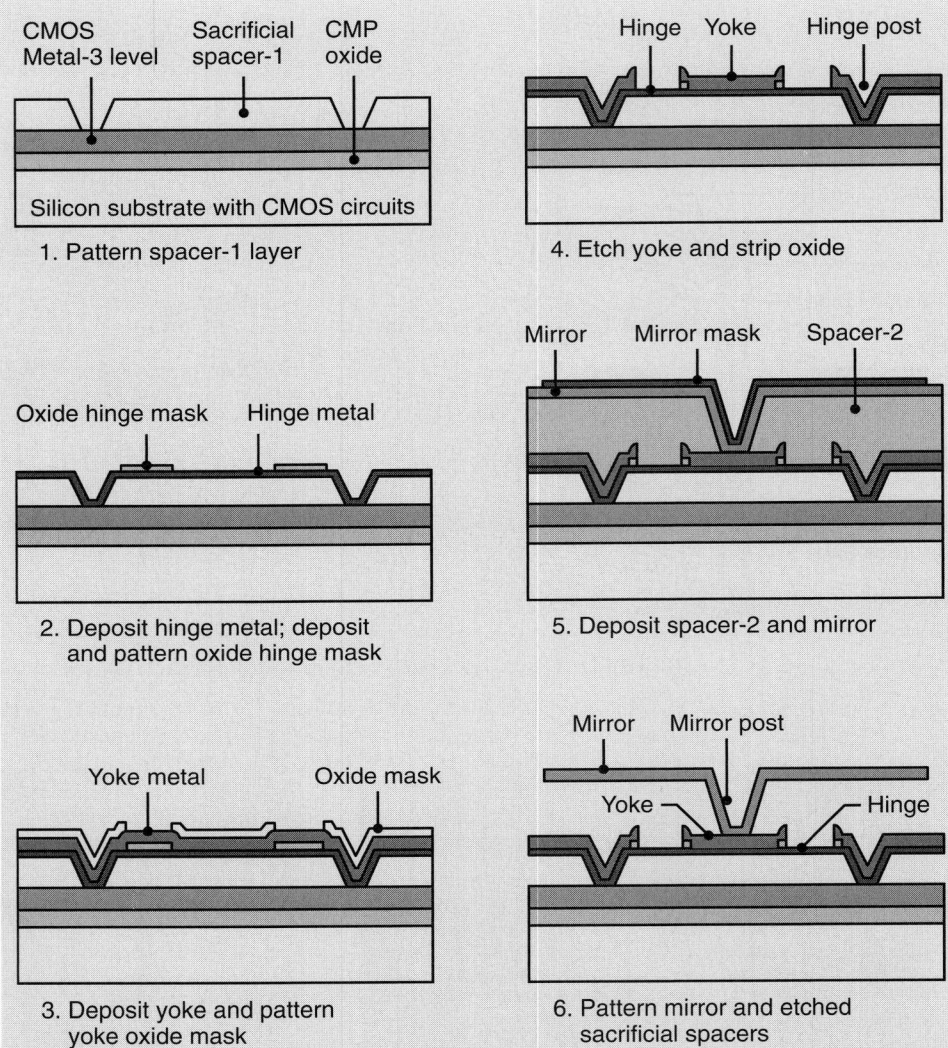

FIGURE 29.9   Manufacturing sequence for the Texas Instruments DMD device.

- The landing pads and electrodes are produced from aluminum, which is deposited by sputtering.
- High reliability requires low stresses and high strength in the torsional hinge, which is produced from a proprietary aluminum alloy.
- The MEMS portion of the DMD is very delicate, and special care must be taken in separating the dies. When completed, a wafer saw (see Fig. 28.6c) cuts a trench along the edges of the DMD, which allows the individual dice to be broken apart at a later stage.
- A special step deposits a layer that prevents adhesion between the yoke and landing pads.
- The DMD is placed in a hermetically sealed ceramic package (Fig. 29.10) with an optical window.

An array of such mirrors represents a grayscale screen. Using three mirrors (one each for red, green, and blue light) for each pixel results in a color image with

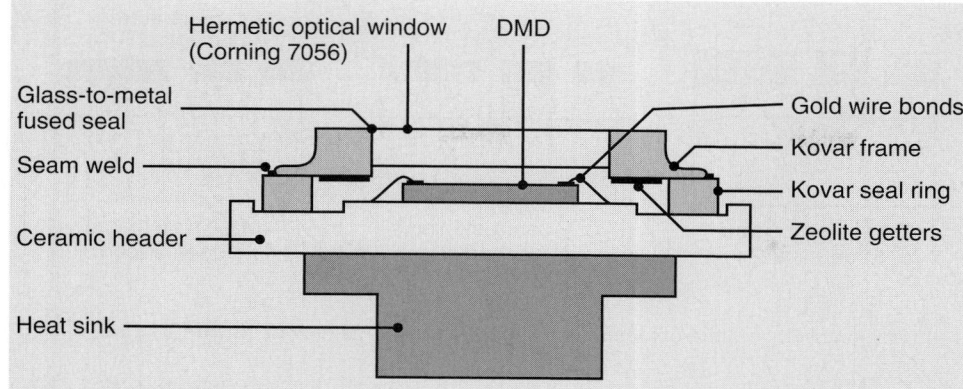

FIGURE 29.10   Ceramic flat-package construction used for the DMD device.

millions of discrete colors. Digital pixel technology is applied widely in digital projection systems, high-definition television, and other optical applications. However, to produce the device shown in Fig. 29.8 requires much more than two-and-one-half-dimensional features, so full three-dimensional, multi-part assemblies have to be manufactured.

SCREAM.   Another method for making very deep MEMS structures is the SCREAM process (*single-crystal silicon reactive etching and metallization*), depicted in Fig. 29.11. In this technique, standard lithography and etching processes produce trenches 10 to 50 μm (400 to 2000 μin.) deep, which are then protected by a layer of chemically vapor-deposited silicon oxide. An anisotropic-etching step removes the oxide only at the bottom of the trench, and the trench then is extended through dry etching. An isotropic etching (using sulfur hexafluoride, SF_6) laterally etches the exposed sidewalls at the bottom of the trench. This undercut (when it overlaps adjacent undercuts) releases the machined structures.

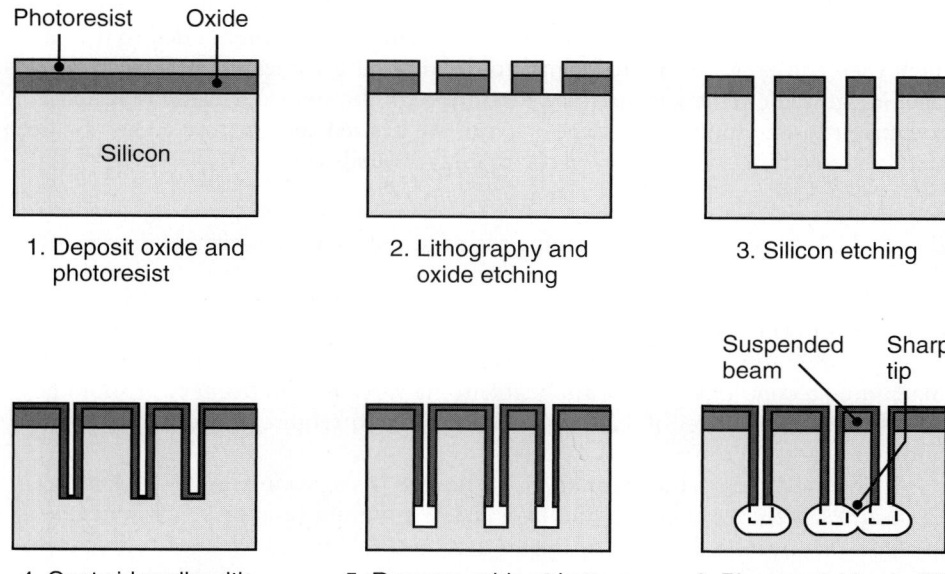

FIGURE 29.11   The SCREAM process. *Source*: After N. Maluf.

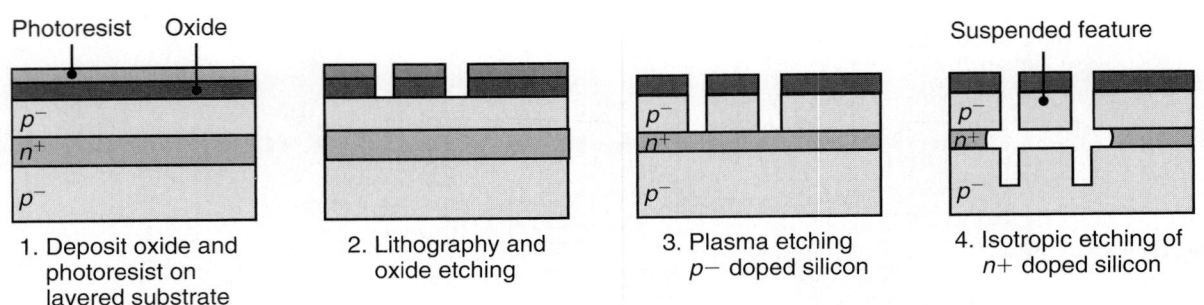

Photoresist   Oxide

Suspended feature

1. Deposit oxide and
photoresist on
layered substrate

2. Lithography and
oxide etching

3. Plasma etching
p− doped silicon

4. Isotropic etching of
n+ doped silicon

FIGURE 29.12   Schematic illustration of silicon micromachining by single-step plasma etching (SIMPLE) process.

SIMPLE.   An alternative to SCREAM is the SIMPLE technique (*silicon micromachining by single-step plasma etching*), as depicted in Fig. 29.12. This technique uses a chlorine-gas based plasma-etching process that machines p-doped or lightly-doped silicon anisotropically, but heavily n-doped silicon isotropically. A suspended MEMS device thus can be produced in one plasma-etching device, as shown in the figure.

Some of the concerns with the SIMPLE process are:

- The oxide mask is machined (although at a slower rate) by the chlorine-gas plasma. Therefore, relatively thick oxide masks are required.
- The isotropic etch rate is low, typically 50 nm/min. Consequently, this is a very slow process.
- The layer beneath the structures will have developed deep trenches, which may affect the motion of free-hanging structures.

Etching combined with diffusion bonding.   Very tall structures can be produced in crystalline silicon through a combination of *silicon-diffusion bonding* and *deep reactive-ion etching* (SFB-DRIE), as illustrated in Fig. 29.13. First, a silicon wafer is prepared with an insulating oxide layer with the deep trench areas defined by a standard lithography procedure. This step is followed by conventional wet or dry etching to form a large cavity. A second layer of silicon is fusion bonded to this layer, which then can be ground and lapped to the desired thickness, if necessary. At this stage, integrated circuitry is manufactured through the steps outlined in Fig. 28.2. A protective resist is applied and exposed, and the desired trenches are etched by deep reactive-ion etching to the cavity in the first layer of silicon.

EXAMPLE 29.3 Operation and fabrication sequence for a thermal inkjet printer

Thermal inkjet printers are perhaps the most successful application of MEMS to date. These printers operate by ejecting nano- or pico-liters (10^{-12} liters) of ink from a nozzle towards paper. Inkjet printers use a variety of designs, but silicon-machining technology is most applicable to high-resolution printers. It should be noted that a resolution of 1200 dpi (dots per inch) requires a nozzle spacing of approximately 20 μm.

The mode of operation of an inkjet printer is shown in Fig. 29.14. When an ink droplet is to be generated and expelled, a tantalum resistor below a nozzle is heated. This heats a thin film of ink so that a bubble forms within five microseconds (which expands rapidly) with internal pressures reaching 1.4 MPa (200 psi). As a result, fluid is forced rapidly out of the nozzle. Within 24 μs, the tail of the inkjet droplet separates because of surface tension, the heat source is removed

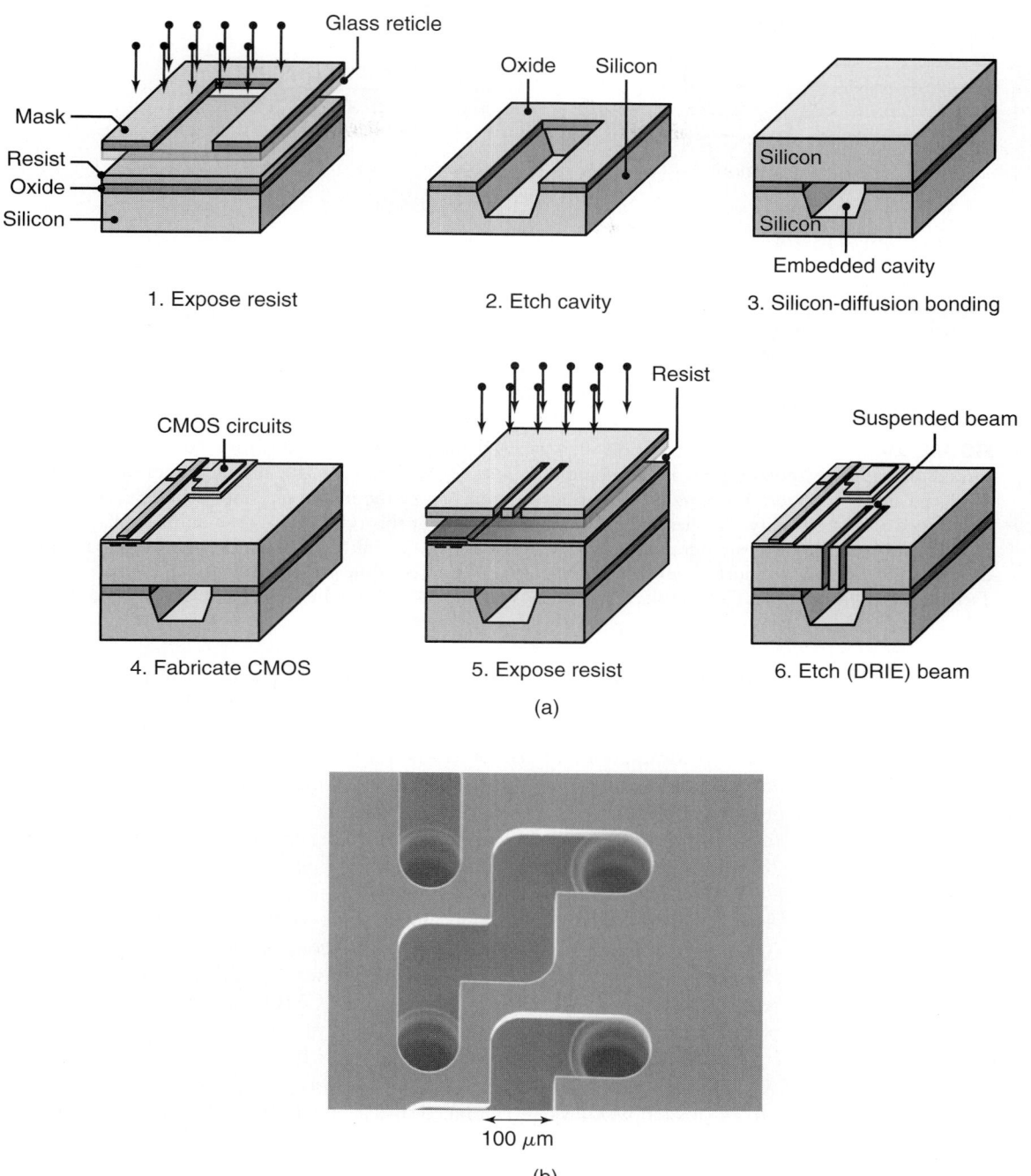

1. Expose resist

2. Etch cavity

3. Silicon-diffusion bonding

4. Fabricate CMOS

5. Expose resist

6. Etch (DRIE) beam

(a)

100 μm

(b)

FIGURE 29.13 (a) Schematic illustration of silicon-diffusion bonding combined with deep reactive-ion etching to produce large, suspended cantilevers. (b) A microfluid-flow device manufactured by DRIE etching two separate wafers then aligning and silicon-fusion bonding them together. Afterward, a Pyrex layer (not shown) is anodically bonded over the top to provide a window to observe fluid flow. *Source*: (a) After N. Maluf. (b) Courtesy of K.R. Williams, Agilent Technologies.

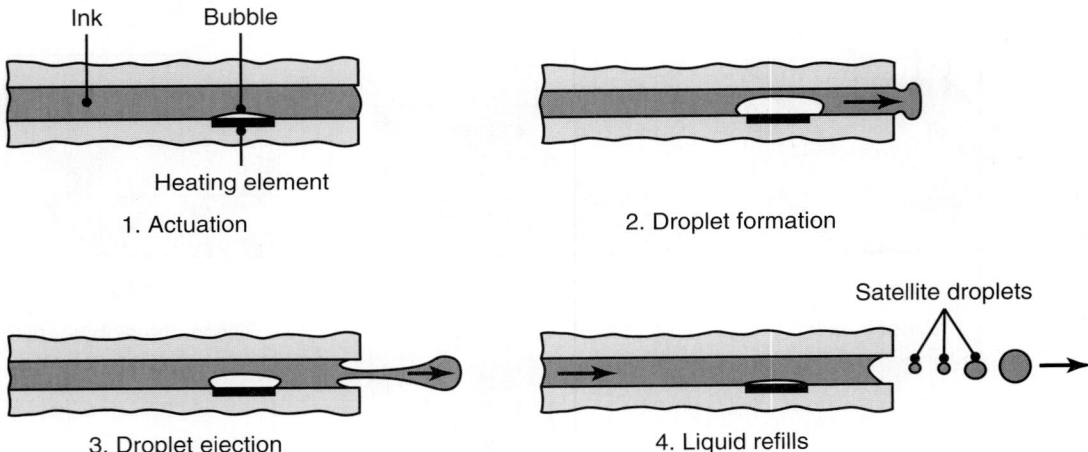

1. Actuation

2. Droplet formation

3. Droplet ejection

4. Liquid refills

FIGURE 29.14   Sequence of operation of a thermal inkjet printer. (1) Resistive heating element is turned on, rapidly vaporizing ink and forming a bubble. (2) Within 5 μm, the bubble has expanded and displaced liquid ink from the nozzle. (3) Surface tension breaks the ink stream into a bubble, which is discharged at high velocity. The heating element is turned off at this time, so that the bubble collapses as heat is transferred to the surrounding ink. (4) Within 24 μs, an ink droplet (and undesirable satellite droplets) are ejected, and surface tension of the ink draws more liquid from the reservoir. *Source*: From Tseng, F-G, "Microdroplet Generators," *The MEMS Handbook*, M. Gad-el-Hak (ed.), CRC Press, 2002.

(turned off), and the bubble collapses inside the nozzle. Within 50 s, sufficient ink has been drawn into the nozzle from a reservoir to form the desired meniscus for the next droplet.

Traditional inkjet printer heads were made with electroformed nickel nozzles, produced separately from the integrated circuitry, and required a bonding operation to attach these two components. With increasing printer resolution, it is more difficult to bond the components with a tolerance of less than a few micrometers. For this reason, single component or monolithic fabrication is of interest.

The fabrication sequence for a monolithic inkjet printer head is shown in Fig. 29.15. A silicon wafer is prepared and coated with a phosphosilicate-glass (PSG) pattern and low-stress silicon-nitride coating. The ink reservoir is obtained by isotropically etching the back side of the wafer, followed by PSG removal and enlargement of the reservoir. The required CMOS (complimentary metal-oxide semiconductor) controlling circuitry then is produced and a tantalum heater pad is deposited. The aluminum interconnection between the tantalum pad and the CMOS circuit is formed, and the nozzle is produced through laser ablation. An array of such nozzles can be placed inside an inkjet printing head, and resolutions of 2400 dpi or higher can be achieved.

29.3 | The LIGA Microfabrication Process

LIGA is a German acronym for the combined process of x-ray lithography, electrodeposition, and molding (**x-ray *li*thographie, *g*alvanoformung und *a*bformung**). A schematic illustration of this process is given in Fig. 29.16.

The LIGA process involves the following steps:

1. A very thick (up to hundreds of microns) resist layer of polymethylmethacrylate (PMMA) is deposited onto a primary substrate.

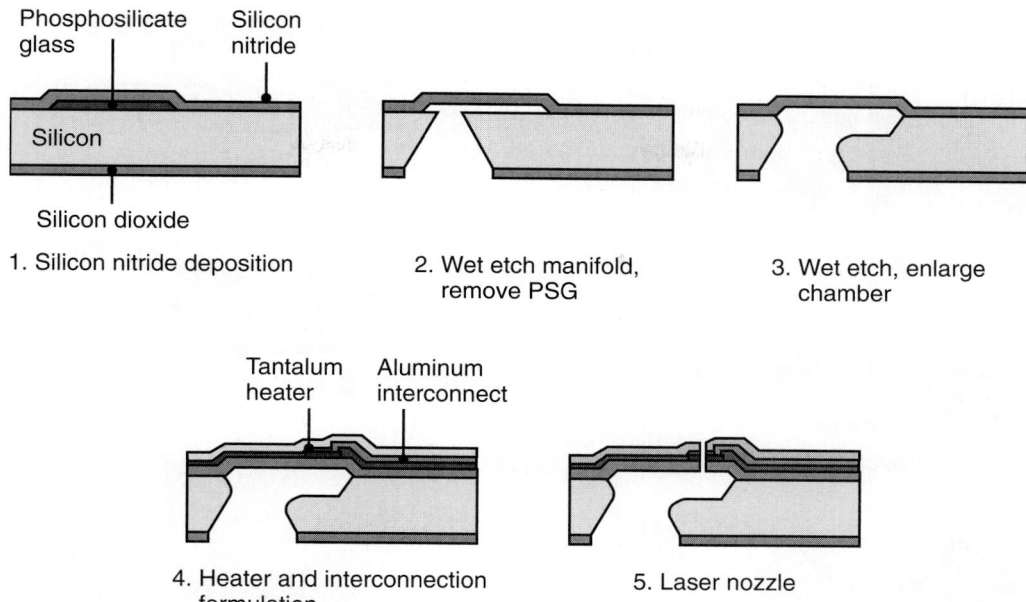

FIGURE 29.15   The manufacturing sequence for producing thermal inkjet printer heads. *Source*: From Tseng, F-G, "Microdroplet Generators," *The MEMS Handbook*, M. Gad-el-Hak (ed.), CRC Press, 2002.

2. The PMMA is exposed to columnated x-rays and developed.

3. Metal is electrodeposited onto the primary substrate.

4. The PMMA is removed or stripped, resulting in a free-standing metal structure.

5. Plastic injection molding takes place.

Depending on the application, the final product from a LIGA process may consist of:

• A free-standing metal structure resulting from the electrodeposition process.

• A plastic injection-molded structure.

• An investment-cast metal part using the injection-molded structure as a blank.

• A slip-cast ceramic part produced with the injection-molded parts as the molds.

The substrate used in LIGA is a conductor or a conductor-coated insulator. Examples of primary substrate materials include austenitic steel plate, silicon wafers with a titanium layer, and copper plated with gold, titanium or nickel. Metal-plated ceramic and glass also have been used. The surface may be roughened by grit blasting to encourage good adhesion of the resist material.

Resist materials must have high x-ray sensitivity, dry- and wet-etching resistance when unexposed, and thermal stability. The most common resist material is polymethylmethacrylate, which has a very high molecular weight (more than 10^6 grams per mole). The x-rays break the chemical bonds, leading to the production of free radicals and to a significantly reduced molecular weight in the exposed region. Organic solvents then preferentially dissolve the exposed PMMA in a wet-etching process. After development, the remaining three-dimensional structure is rinsed and dried, or it is spun and blasted with dry nitrogen.

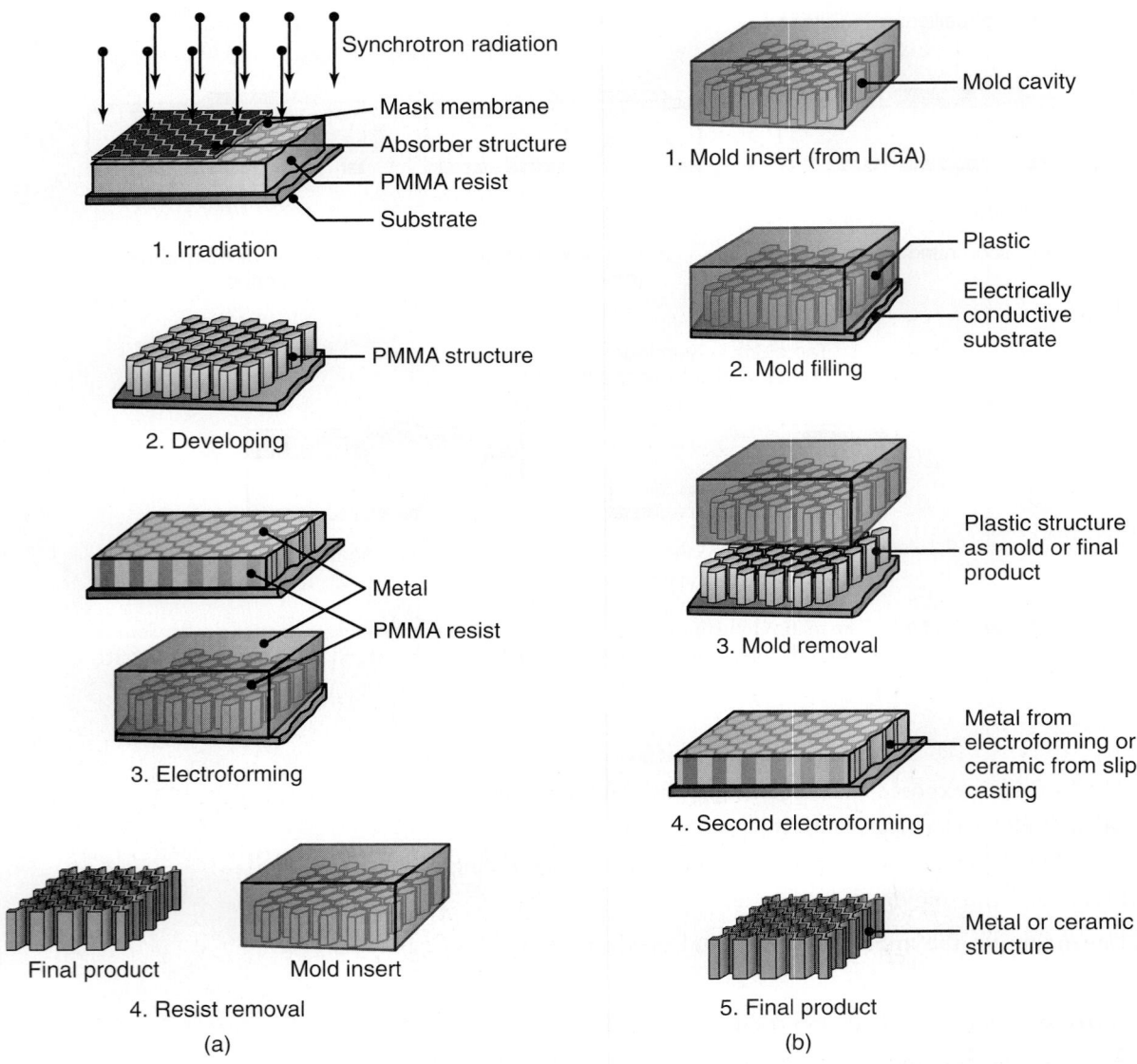

FIGURE 29.16 The LIGA (lithography, electrodeposition and molding) technique. (a) Primary production of a metal final product or mold insert. (b) Use of the primary part for secondary operations or replication. *Source*: Courtesy of IMM Institut für Mikrotechnik.

The electrodeposition of metal usually involves the electroplating of nickel. The nickel is deposited onto exposed areas of the substrate; it fills the PMMA structure and even can coat the resist (Fig. 29.16a). Nickel is the material of choice because of the relative ease in electroplating with well-controlled deposition rates. Electroless plating of nickel also is possible, and it can be deposited directly onto electrically insulating substrates. However, because nickel displays high wear rates in MEMS, significant research has been directed towards the use of other materials or coatings.

After the metal structure has been deposited, precision grinding removes either the substrate material or a layer of the deposited nickel referred to as *planarization* (producing a planar surface; see Section 28.10). The need for planarization is obvious when it is recognized that three-dimensional MEMS devices require micrometer tolerances on layers many hundreds of micrometers thick. Planarization is difficult to achieve; conventional lapping leads to preferential removal of the soft PMMA and smearing of the metal. Planarization usually is accomplished with a diamond-lapping procedure referred to as *nanogrinding*. Here, a diamond-slurry-loaded, soft-metal plate is used to remove material in order to maintain flatness within 1 μm (40 μin.) over a 75-mm (3-in.) diameter substrate.

If cross-linked, the PMMA resist then is exposed to synchrotron x-ray radiation and removed by exposure to an oxygen plasma or through solvent extraction. The result is a metal structure, which may be used for further processing. Examples of free-standing metal structures produced through the electrodeposition of nickel are shown in Fig. 29.17.

The processing steps used to make free-standing metal structures are extremely time consuming and expensive. The main advantage of LIGA is that these structures serve as molds for the rapid replication of submicron features through molding operations. The processes that can be used for producing micromolds are shown and compared in Table 29.1; it can be seen that LIGA provides some clear advantages. Reaction injection molding, injection molding, and compression molding (see Chapter 19) also have been used to make these micromolds.

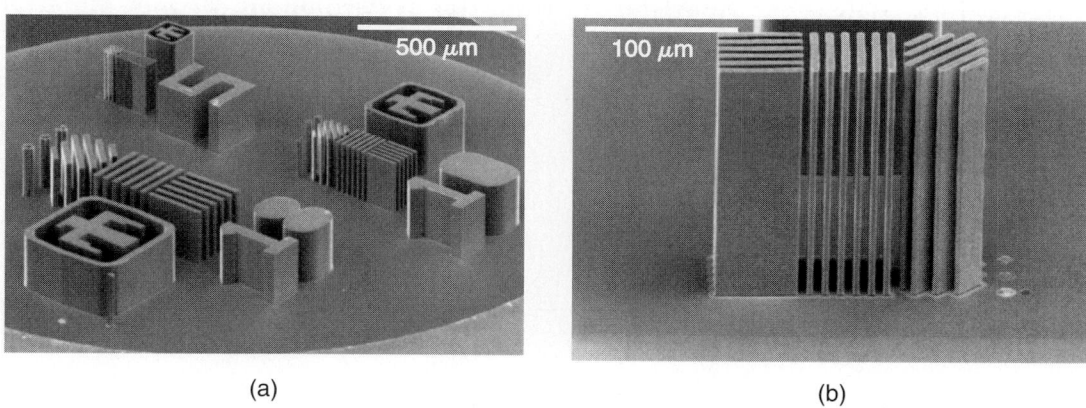

(a)   (b)

FIGURE 29.17   (a) Electroformed, 200-μm-tall nickel structures; (b) Detail of 5-μm-wide nickel lines and spaces. *Source*: After T. Christenson, *MEMS Handbook*, CRC Press, 2002.

TABLE 29.1

Comparison of Micromold Manufacturing Techniques		
	Production technique	
	LIGA	Laser machining	EDM
Aspect ratio	10-50	10	up to 100
Surface roughness	<50 nm	100 nm	0.3–1 μm
Accuracy	<1 μm	1–3 μm	1–5 μm
Mask required?	Yes	No	No
Maximum height	1–500 μm	200–500 μm	microns to millimeters

Source: Weber, L., Ehrfeld, W., Freimuth, H., Lacher, M., Lehr, M., and Pech, P., *SPIE Micromachining and Microfabrication Process Technology II*, Austin, Texas, 1996, and K.R. Williams, Agilent Laboratories.

EXAMPLE 29.4 Production of rare-earth magnets

A number of scaling issues in electromagnetic devices indicate that there is an advantage in using rare-earth magnets from the samarium cobalt (SmCo) and neodymium iron boron (NdFeB) families, which are available in powder form. They are of interest because they can produce magnets that have an order of magnitude more powerful than conventional magnets (Table 29.2). These materials can be used when effective miniature electromagnetic transducers are to be produced.

The processing steps to manufacture these magnets are shown in Fig. 29.18. The polymethylmethacrylate mold is produced by exposure to x-ray radiation and solvent extraction. The rare-earth powders are mixed with a binder of epoxy and applied to the PMMA mold through a combination of calendering (see Fig. 19.22) and pressing. After curing in a press at a pressure around 70 MPa (10 ksi), the substrate is planarized. The substrate then is subjected to a magnetizing field of at least 35 kilo-oersteds (kOe) in the desired orientation. Once the material has been magnetized, the PMMA substrate is dissolved, leaving behind the rare-earth magnets, as shown in Fig. 29.19.

Source: Courtesy of T. Christenson, Sandia National Laboratories.

TABLE 29.2

Comparison of Properties of Permanent Magnet Materials	
Material	Energy product (Gauss-Oersted $\times 10^{-6}$)
Carbon steel	0.20
36% Cobalt steel	0.65
Alnico I	1.4
Vicalloy I	1.0
Platinum-cobalt	6.5
$Nd_2Fe_{14}B$, fully dense	40
$Nd_2Fe_{14}B$, bonded	9

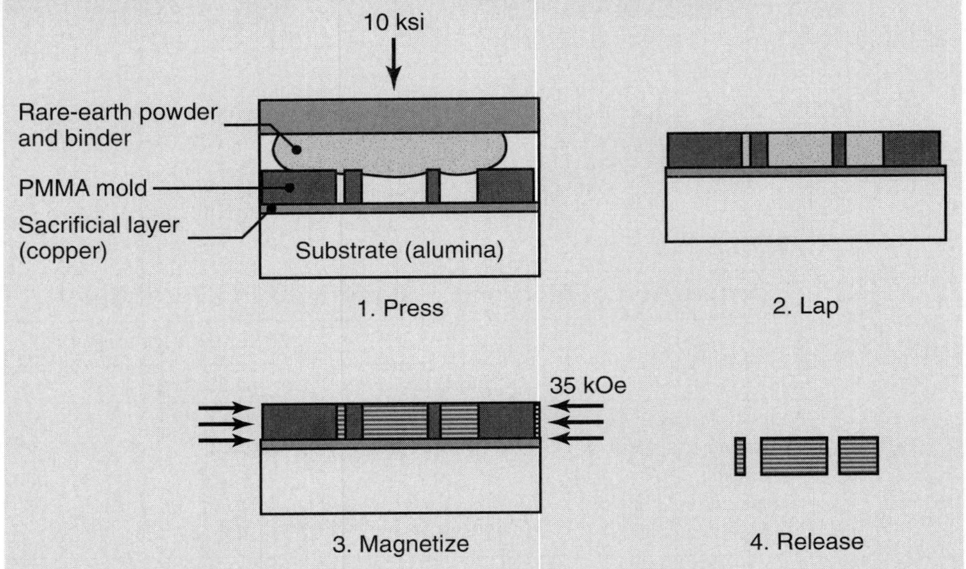

FIGURE 29.18 Fabrication process used to produce rare-earth magnets for microsensors.
Source: Courtesy of T. Christenson, Sandia National Laboratories.

(a)   (b)

FIGURE 29.19   SEM images of $Nd_2Fe_{14}B$ permanent magnets. Powder particle size ranges from 1 to 5 μm, and the binder is a methylene-chloride resistant epoxy. Mild distortion is present in the image due to magnetic perturbation of the imaging electrons. Maximum energy products of 9 MGOe have been obtained with this process. *Source*: Courtesy of T. Christenson, Sandia National Laboratories.

Multilayer X-ray lithography.   The LIGA technique is very powerful for producing MEMS devices with large aspect ratios and reproducible shapes. Often, it is useful to obtain a multilayer stepped structure that cannot be made directly through LIGA. For non-overhanging geometries, direct plating can be applied. In this technique, a layer of electrodeposited metal with surrounding PMMA is produced as described previously. A second layer of PMMA resist then is bonded to this structure and x-ray exposed with an aligned x-ray mask.

Often, it is useful to have overhanging geometries within complex MEMS devices. A batch diffusion-bonding and release procedure has been developed for this purpose, as schematically illustrated in Fig. 29.20a. This process involves the

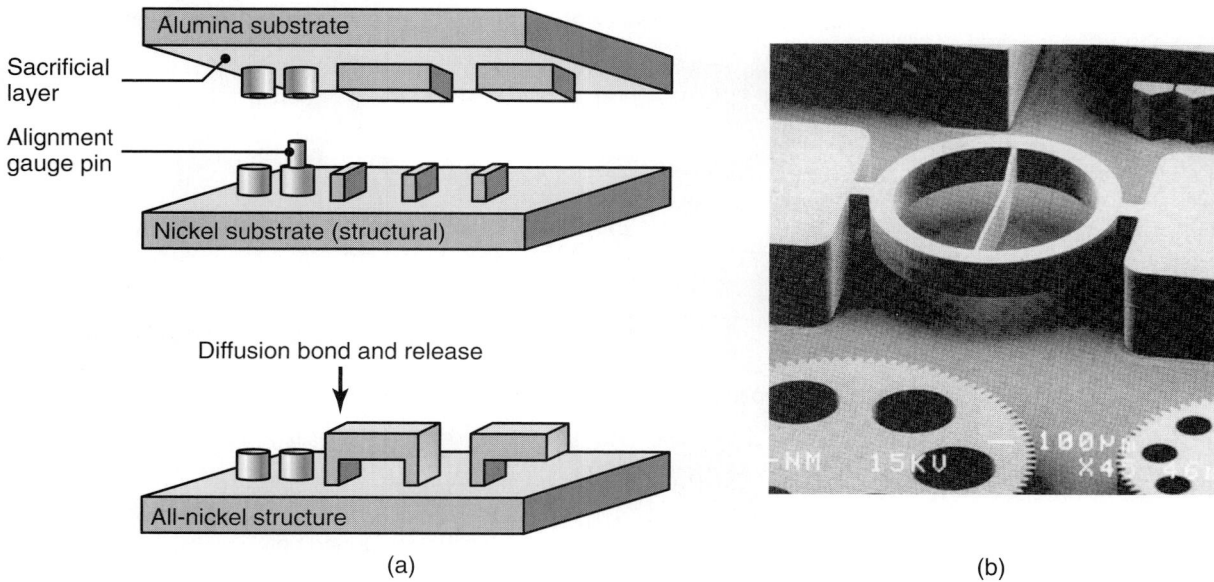

(a)   (b)

FIGURE 29.20   (a) Schematic of multilevel MEMS fabrication through wafer-scale diffusion bonding. (b) A suspended ring structure for measurement of tensile strain, formed by two-layer wafer-scale diffusion bonding. *Source*: (b) Courtesy of T. Christenson, Sandia National Laboratories.

preparation of two PMMA patterned and electroformed layers with the PMMA subsequently removed. The wafers then are aligned face to face with guide pins that press-fit into complementary structures on the opposite surface. The substrates then are joined in a hot press, and a sacrificial layer on one substrate is etched away, leaving one layer bonded to the other. Figure 29.20b shows an example of such a structure.

HEXSIL.   The *HEXSIL* process, illustrated in Fig. 29.21, combines *hex*agonal honeycomb structures, *sil*icon micromachining, and thin-film deposition to produce high aspect-ratio, free-standing structures. HEXSIL can produce tall structures with a shape definition that rivals LIGA.

In HEXSIL, a deep trench first is produced in single-crystal silicon using dry etching and followed by a shallow wet etching to make the trench walls smoother. The depth of the trench matches the desired structure height and is limited to practically around 100 μm. An oxide layer is then grown or deposited onto the silicon followed by an undoped-polycrystalline silicon layer, which leads to good mold filling and good shape definition. A doped-silicon layer follows, providing a resistive portion of the microdevice. Electroplated or electroless nickel plate then is deposited.

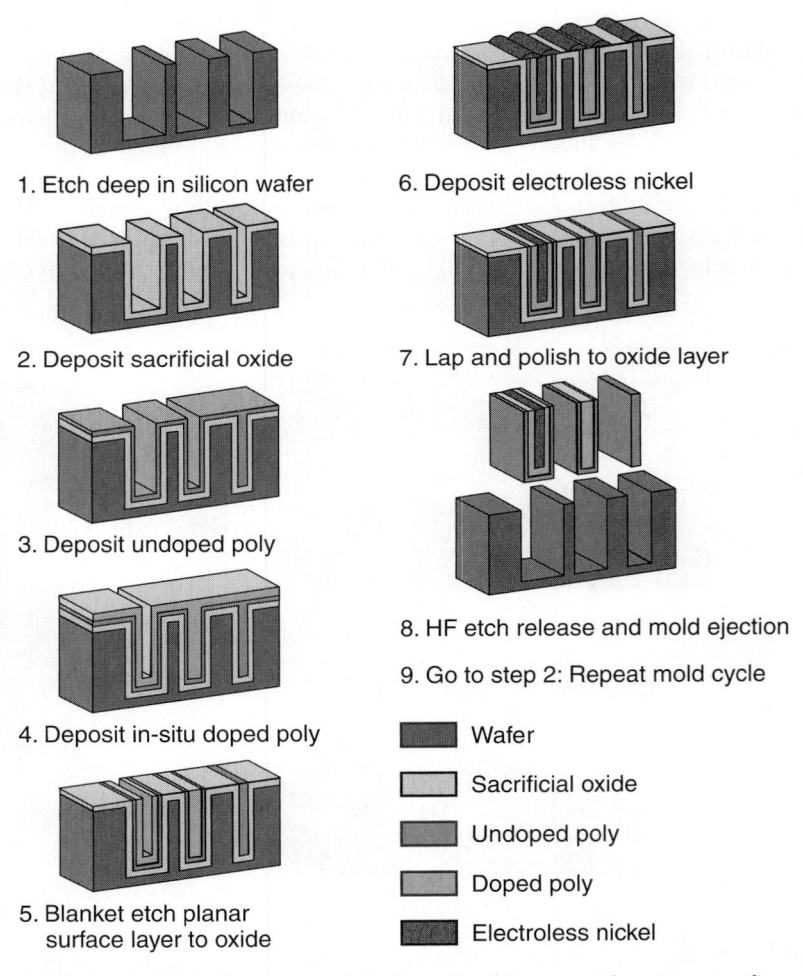

1. Etch deep in silicon wafer

2. Deposit sacrificial oxide

3. Deposit undoped poly

4. Deposit in-situ doped poly

5. Blanket etch planar surface layer to oxide

6. Deposit electroless nickel

7. Lap and polish to oxide layer

8. HF etch release and mold ejection

9. Go to step 2: Repeat mold cycle

- Wafer
- Sacrificial oxide
- Undoped poly
- Doped poly
- Electroless nickel

FIGURE 29.21   Illustration of the *hex*agonal honeycomb structure, *sil*icon micromachining, and thin-film deposition—the HEXSIL process.

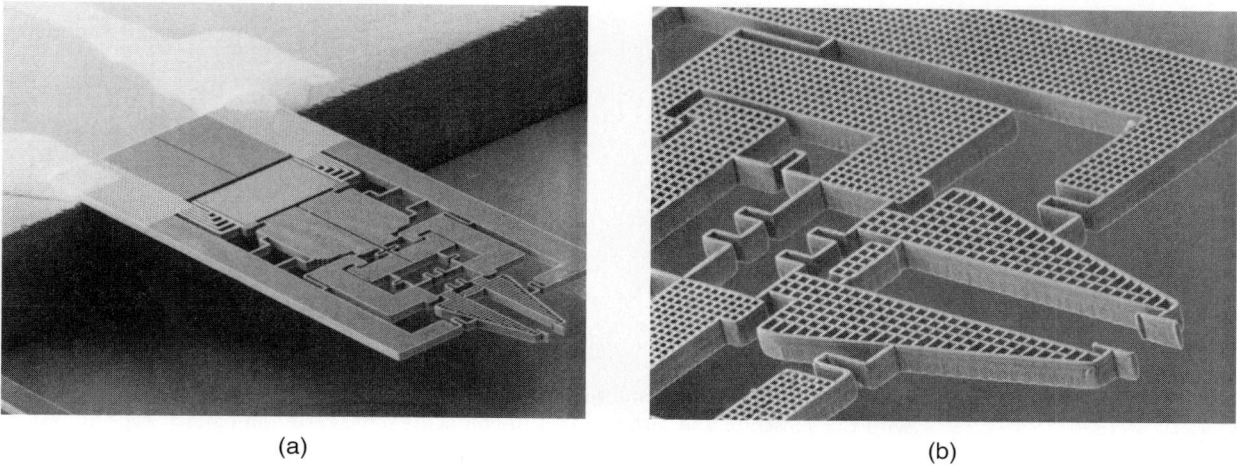

(a)   (b)

FIGURE 29.22   (a) SEM image of microscale tweezers used in microassembly and microsurgery applications. (b) Detailed view of gripper. *Source:* Courtesy of MEMS Precision Instruments (memspi.com).

Figure 29.21 shows various trench widths to demonstrate the different structures that can be produced in HEXSIL.

Figure 29.22 shows microscale tweezers produced through the HEXSIL process. These tweezers use a thermally activated bar that activates the tweezers. They have been used for microassembly and microsurgery applications.

29.4 | Solid Free-Form Fabrication of Devices

Solid free-form fabrication is another term for rapid prototyping, as described in Chapter 20. This method is unique in that complex three-dimensional structures are produced through additive manufacturing, as opposed to material removal. Many of the advances in rapid prototyping also are applicable to MEMS manufacture for processes with sufficiently high resolution. Recall that *stereolithography* involves curing a liquid thermosetting polymer using a photoinitiator and a highly focused light source. Conventional stereolithography uses layers between 75 and 500 μm in thickness with a laser dot focused to a 0.05–0.25 mm diameter.

Microstereolithography.   In *microstereolithography*, the process uses the same basic approach as in stereolithography (see Section 20.3.2). However, there are some important differences between microstereolithography and stereolithography, including:

- The laser is more highly focused (to a diameter as small as 1 μm), as compared to 10 to over 100 μm in stereolithography.
- Layer thicknesses are around 10 μm, which is an order of magnitude smaller than in stereolithography.
- The photopolymers used must have much lower viscosities to ensure the formation of uniform layers.
- Support structures are not needed in microstereolithography, since the smaller structures can be supported adequately by the fluid.

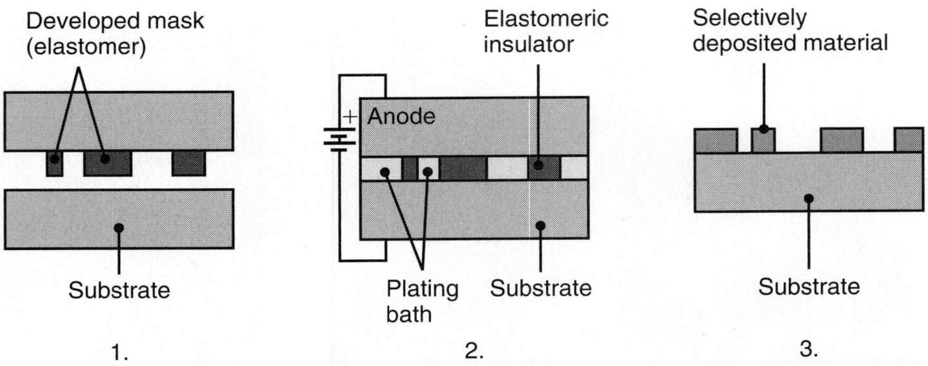

FIGURE 29.23   The instant-masking process: (1) bare substrate; (2) during deposition with the substrate and instant mask in contact; (3) the resulting pattern deposit. *Source:* Courtesy of A. Cohen, MEMGen Corporation.

- Parts with significant metal and ceramic content can be produced by suspending nanoparticles in the liquid photopolymer.

The microstereolithography technique has a number of cost advantages, but the MEMS devices made by this method are difficult to integrate with the controlling circuitry.

Electrochemical fabrication.   Instant masking is a technique for producing MEMS devices (Fig. 29.23). The solid free-form fabrication of MEMS devices using instant masking is known as *electrochemical fabrication* (EFAB). A mask of elastomeric material first is produced through conventional photolithography techniques, described in Section 28.7. The mask is pressed against the substrate in an electrodeposition bath, so that the elastomer conforms to the substrate and excludes the plating solution in contact areas. Electrodeposition takes place in areas that are not masked, eventually producing a mirror image of the mask. By using a sacrificial filler made of a second material, complex three-dimensional shapes can be produced complete with overhangs, arches, and other features, using instant-masking technology.

CASE STUDY 29.1 | Accelerometer for Automotive Air Bags

Accelerometers based on lateral resonators represent the largest commercial application of surface micromachining today and are used widely as sensors for automotive air-bag deployment systems. The sensor portion of such an accelerometer is shown in Fig. 29.24. A central mass is suspended over the substrate but anchored through four slender beams, which act as springs to center the mass under static-equilibrium conditions. An acceleration causes the mass to deflect, reducing or increasing the clearance between the fins on the mass and the stationary fingers on the substrate. By measuring the electrical capacitance between the mass and fins, the deflection of the mass (and therefore the acceleration or deceleration of the system) can be directly measured. Figure 29.24 shows an arrangement for the measurement of acceleration in one direction, but commercial sensors employ several masses so that accelerations can be measured in multiple directions simultaneously.

Figure 29.25 shows the 50-g surface micromachined accelerometer (ADXL-50) with on-board signal conditioning and self-diagnostic electronics. The polysilicon sensing element (visible in the center of the die) occupies only 5% of

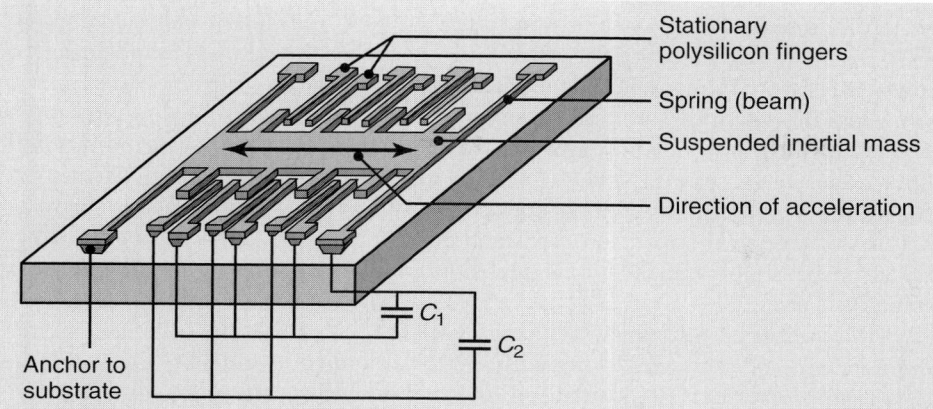

Stationary polysilicon fingers

Spring (beam)

Suspended inertial mass

Direction of acceleration

C_1

C_2

Anchor to substrate

FIGURE 29.24   Schematic illustration of a micro-acceleration sensor. *Source*: After N. Maluf.

the total die area, and the whole chip measures 500×625 μm (20×25 μin). The mass is approximately 0.3 μg, and the sensor has a measurement accuracy of 5% over the ± 50 g range.

Fabrication of the accelerometer proved to be a challenge, since it required a *complementary metal-oxide-semiconductor* (CMOS) fabrication sequence to be integrated closely with a surface micromachining approach. Analog Devices, Inc. was able to modify a CMOS production technique to directly incorporate surface micromachining. In the sensor design, $n+$ underpasses connect the sensor area to the electronic circuitry, replacing the usual heat-sensitive aluminum connect lines. Most of the sensor processing is inserted into the fabrication process right after a borosilicate-glass planarization.

After the planarization, a designated sensor region or *moat* is cleared in the center of the die (Fig. 29.26a). A thin oxide then is deposited to passivate the $n+$ underpass connects and followed by a thin, low-pressure chemical-vapor deposited

FIGURE 29.25   Photograph of Analog Devices' ADXL-50 accelerometer with a surface micromachined capacitive sensor (center), on-chip excitation, self-test and signal-conditioning circuitry. The entire chip measures 0.500×0.625 mm. *Source*: From R.A. Core, *et al.*, *Solid State Technology*, pp. 39–47, October 1993.

(LPCVD) nitride to act as an etch stop for the final polysilicon released etching (Fig. 29.26b). The spacer or sacrificial oxide used is a 1.6 μm (64 μin.) densified low-temperature oxide (LTO) deposited over the whole die (Fig. 29.26c).

In a first etching, small depressions that will form bumps or dimples on the underside of the polysilicon sensor are created in the LTO layer. These bumps will limit adhesive forces and sticking in case the sensor comes in contact with the substrate. A subsequent etching cuts anchors into the spacer layer to provide regions of electrical and mechanical contact (Fig. 29.26c). The 2-μm (80-μin.) thick sensor of polysilicon layer is deposited, implanted, annealed, and patterned (Fig. 29.27a).

Metallization follows, which starts with the removal of the sacrificial spacer oxide from the circuit area along with the LPCVD nitride and LTO layer. A low-temperature oxide is deposited on the polysilicon-sensor part, and contact openings appear in the integrated-circuit part of the die where platinum is deposited to form platinum silicide (Fig. 29.27b). The trimmable thin-film

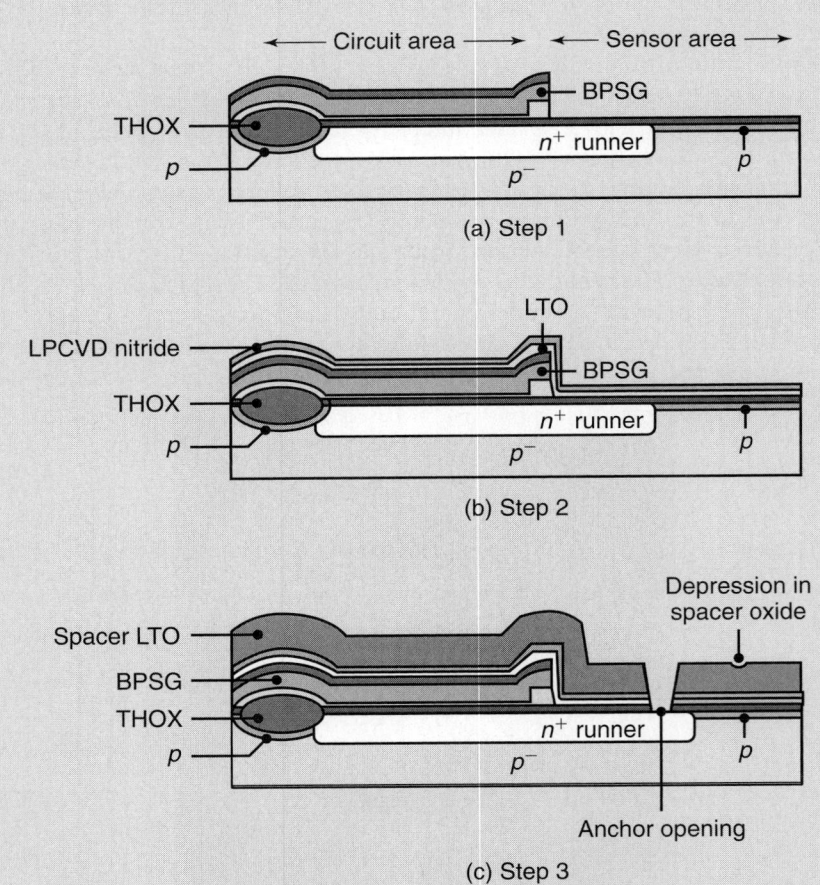

FIGURE 29.26 Preparation of IC chip for polysilicon. (a) Sensor area post-BPSG planarization and moat mask. (b) Blanket deposition of thin oxide and thin nitride layer. (c) Bumps and anchors made in an LTO spacer layer. *Source*: From R.A. Core, *et al.*, *Solid State Technology*, pp. 39–47, October 1993.

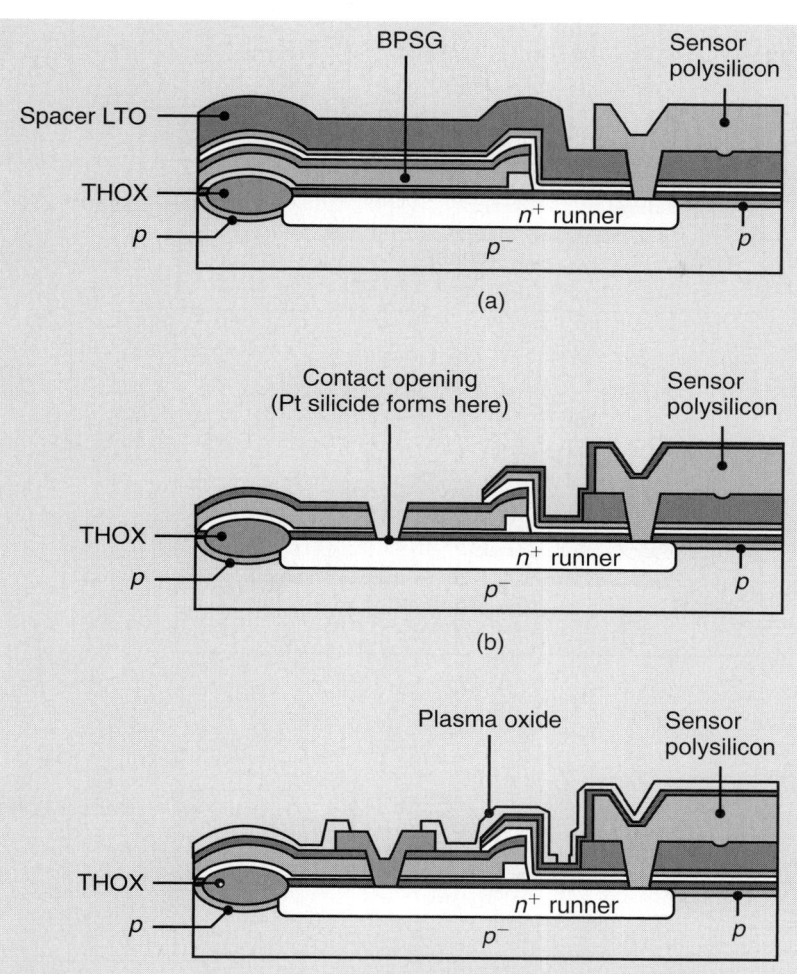

FIGURE 29.27 Polysilicon deposition and IC metallization. (a) Cross sectional view after polysilicon deposition, implanting, annealing, and patterning. (b) Sensor area after removal of dielectrics from circuit area, contact mask, and platinum silicide. (c) Metallization scheme and plasma-oxide passivation and patterning. *Source:* From R.A. Core, *et al.*, *Solid State Technology*, pp. 39–47, October 1993.

material, TiW barrier metal, and Al/Cu interconnect metal are sputtered on and patterned in the integrated circuit area.

The circuit area is then passivated in two separate deposition steps. First, plasma oxide is deposited and patterned (Fig. 29.27c). Then it is followed by a plasma nitride (Fig. 29.28a) to form a seal with the previously deposited LCVD nitride. The nitride acts as a hydrofluoric-acid barrier in the subsequent etch release in surface micromachining. The plasma oxide left on the sensor acts as an etch stop for the removal of the plasma nitride (Fig. 29.28a). The sensor area then is prepared for the final release etch. The dielectrics are removed from the sensor, and the final protective resist mask is applied. The photoresist protects the circuit

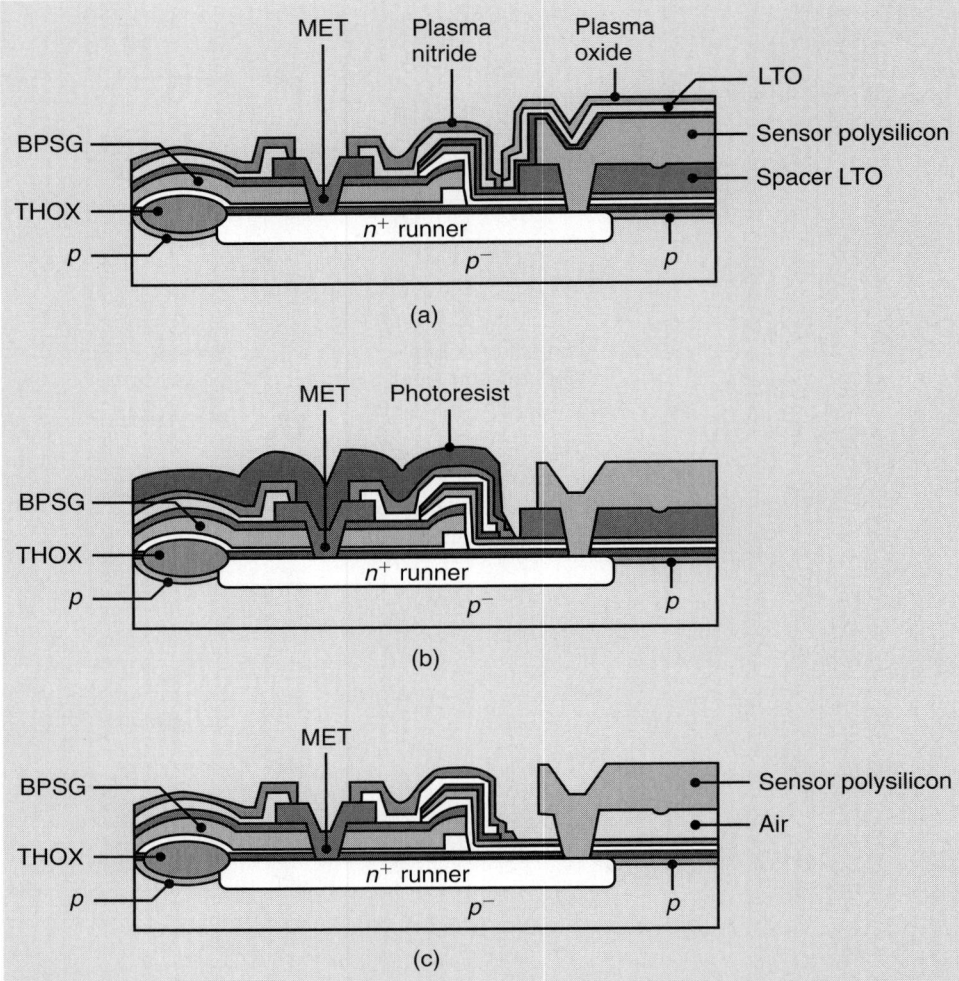

FIGURE 29.28   Pre-release preparation and release. (a) Post-plasma nitride passivation and patterning. (b) Photoresist protection of the IC. (c) Free-standing, released, polysilicon beam. *Source*: From R.A. Core, *et al.*, *Solid State Technology*, pp. 39–47, October 1993.

area from the long-term buffered oxide etch (Fig. 29.28b). The final device cross-section is shown in Fig. 29.28c.

Source: Adapted from M. Madou, *Fundamentals of Microfabrication*, 2nd ed., CRC Press, 2002.

SUMMARY

- The MEMS field is relatively new and developing rapidly. Most successful commercial MEMS applications are in the optics, printing, and sensor industries. The possibilities for new device concepts and circuit designs appear to be endless.

- MEMS devices are manufactured through techniques and with materials that (for the most part) have been pioneered in the microelectronics industry. Bulk and surface micromachining are processes that are well developed for single-crystal silicon.

- Specialized processes for MEMS include variations of machining, such as DRIE, SIMPLE, and SCREAM. These processes produce free-standing mechanical structures in silicon.
- Polymer MEMS can be manufactured through LIGA or microstereolithography. LIGA combines x-ray lithography and electroforming to produce three-dimensional structures. Related processes include multilayer x-ray lithography and HEXSIL.

KEY TERMS

Bulk micromachining
Diffusion bonding
EFAB
Electrochemical fabrication
HEXSIL
LIGA

MEMS
Micromachining
Microstereolithography
Multilayer x-ray lithography
Planarization
Sacrificial layer

SCREAM
SIMPLE
Stiction
Surface micromachining

BIBLIOGRAPHY

Bakoglu, H. B., *Circuits, Interconnections, and Packaging for VLSI*, Addison-Wesley, 1990.

Berger, L.I., *Semiconductor Materials*, CRC Press, 1997.

Brar, A.S., and Narayan, P.B., *Materials and Processing Failures in the Electronics and Computer Industries: Analysis and Prevention*, ASM International, 1993.

Campbell, S.A., *The Science and Engineering of Microelectronic Fabrication*, Oxford, 1996.

Chandrakasan, A., and Brodersen, R. (eds.), *Low Power CMOS Design*, IEEE, 1998.

Chandrakasan, A., and Brodersen, R. *Low Power Digital CMOS Design*, Kluwer, 1995.

Chang, C.-Y., and Sze, S.M. (eds.), *ULSI Devices*, Wiley-Interscience, 2000.

Colclaser, R.A., *Microelectronics: Processing and Device Design*, Wiley, 1980.

Electronic Materials Handbook, Vol. 1: *Packaging*, ASM International, 1989.

Elwenspoek, M., and Jansen, H., *Silicon Micromachining*, Cambridge University Press, 1998.

Elwenspoek, M., and Wiegerink, R., *Mechanical Microsensors*, Springer, 2001.

Gad-el-Hak, M. (ed.), *The MEMS Handbook*, CRC Press, 2002.

Ghandhi, S.K., *VLSI Fabrication Principles,* 2nd ed., Wiley-Interscience, 1993.

Griffin, P.B., Plummer, J. D., and Deal, M. D., *Silicon VLSI Technology: Fundamentals, Practice and Modeling*, Prentice Hall, 2000.

Harper, C.A. (ed.), *Electronic Packaging and Interconnection Handbook,* 3rd ed., McGraw-Hill, 2000.

_____, *High-Performance Printed Circuit Boards*, McGraw-Hill, 2000.

Hwang, J.S., *Modern Solder Technology for Competitive Electronics Manufacturing*, McGraw-Hill, 1996.

Javits, M.W. (ed.), *Printed Circuit Board Materials Handbook*, McGraw-Hill, 1997.

Judd, M., and Brindley, K., *Soldering in Electronics Assembly*, 2nd ed., Newnes, 1999.

Kovacs, G.T.A., *Micromachined Transducers Sourcebook*, McGraw-Hill, 1998.

Madou, M.J., *Fundamentals of Microfabrication*, CRC Press, 2nd ed., 2002.

Mahajan, S., and Harsha, K.S.S., *Principles of Growth and Processing of Semiconductors*, McGraw-Hill, 1998.

Maluf, N., *An Introduction to Microelectromechanical Systems Engineering*, Artech House, 2000.

Manko, H.H., *Soldering Handbook for Printed Circuits and Surface Mounting*, Van Nostrand Reinhold, 1995.

Matisoff, B.S., *Handbook of Electronics Manufacturing*, 3rd ed., Chapman & Hall, 1996.

Nishi, Y., and Doering, R. (eds.), *Handbook of Semiconductor Manufacturing Technologies*, Marcel Dekker, 2000.

Pierret, R.F., and Neudeck, G.W., *Semiconductor Device Fundamentals*, Addison-Wesley, 1996.

Quirk, M., and Serda, J., *Semiconductor Manufacturing Technology*, Prentice Hall, 2000.

Sze, S.M. (ed.), *Modern Semiconductor Devices Physics*, Wiley, 1997.

_____, *Semiconductor Devices: Physics and Technology*, Wiley, 2001.

Taur, Y., and Ning, T.H., *Fundamentals of Modern VLSI Devices*, Cambridge, 1998.

van Zant, P., *Microchip Fabrication: A Practical Guide to Semiconductor Processing*, 4th ed., McGraw-Hill, 2000

Varadan, V.K., Jiang, X., and Varadan, V., *Microstereolithography and other Fabrication Techniques for 3D MEMS*, Wiley, 2001.

Wise, K.D. (ed.), *Special Issue on Integrated Sensors, Microactuators and Microsystems (MEMS)*, Proceedings of the IEEE, 1998.

Wolf, S., and Tauber, R.N., *Silicon Processing for the VSLI Era: Process Technology*, 2nd ed., Lattice Press, 1999.

Yeap, G., *Practical Low Power Digital VLSI Design*. Kluwer, 1997.

REVIEW QUESTIONS

29.1 Define MEMS.

29.2 Why is silicon often used with MEMS and MEMS devices?

29.3 Describe bulk and surface micromachining.

29.4 What is the purpose of a spacer layer in surface micromachining?

29.5 What do SIMPLE and SCREAM stand for?

29.6 Compare wet and dry etching.

29.7 What is the main limitation to successful application of MEMS?

29.8 What is Moore's Law?

29.9 What are common applications for MEMS and MEMS devices?

29.10 What is LIGA? What are its advantages?

29.11 Which process(es) in this chapter allow the fabrication of products from polymers?

29.12 What is a sacrificial layer?

29.13 What is HEXSIL?

29.14 What are the differences between stereolithography and microstereolithography?

QUALITATIVE PROBLEMS

29.15 What is the difference between isotropic and anisotropic etching?

29.16 Lithography produces projected shapes, so that true three-dimensional shapes are more difficult to produce. What lithography processes are best able to produce three-dimensional shapes, such as lenses?

29.17 What is the difference between chemically reactive ion etching and dry-plasma etching?

29.18 The MEMS devices discussed in this chapter apply to macroscale machine elements, such as spur gears, hinges, and beams. Which of the following machine elements can and cannot be applied to MEMS and why? (a) ball bearings, (b) bevel gears, (c) worm gears, (d) cams, (e) helical springs, (f) rivets, and (g) bolts.

29.19 Referring to Fig. 28.20, sketch the holes generated from a circular mask.

29.20 Explain how you would produce a spur gear if its thickness was one-tenth of its diameter and its diameter was (a) 10 μm, (b) 100 μm, (c) 1 mm, (d) 10 mm, and (e) 100 mm.

29.21 List the advantages and disadvantages of surface micromachining compared to bulk micromachining.

29.22 What are the main limitations to the LIGA process?

29.23 What other process can be used to make the microtweezers in Fig. 29.22 other than HEXSIL?

QUANTITATIVE PROBLEMS

29.24 A polymide photoresist needs 100 mJ/cm^2 per micrometer of thickness to develop properly. How long does a 150 μm film need to develop when exposed by a 1000 W/m^2 light source?

29.25 Calculate the undercut in etching a 10-μm deep trench if the anisotropy ratio is (a) 200, (b) 2, and (c) 0.5. What is the sidewall slope for these cases?

29.26 How many levels are needed to produce the micromotor shown in Fig. 28.19d?

29.27 Conduct an Internet search and determine the smallest diameter hole that can be produced by (a) drilling, (b) punching, (c) water–jet cutting, (d) laser machining, (e) chemical etching, and (f) EDM.

SYNTHESIS, DESIGN, AND PROJECTS

29.28 List similarities and differences between IC technologies covered in Chapter 28 and miniaturization technologies presented in this chapter.

29.29 Figure I.11 in the General Introduction shows a mirror which is suspended on a torsional beam and can be inclined through electrostatic attraction by applying a voltage on either side of the mirror at the bottom of the trench. Make a flow chart of the manufacturing operations required to produce this device.

29.30 Referring to Fig. 29.5, design an experimental study to find the critical dimensions of an overhanging cantilever which will not stick to the substrate.

29.31 Design an accelerometer using the (a) SCREAM process and (b) HEXSIL process.

29.32 Design a micromachine or device that allows the direct measurement of the mechanical properties of a thin film.

29.33 Perform a literature search and write a one-page summary of applications in biomems.

PART VI

Joining Processes and Equipment

When inspecting various products around us, we note that some products are made of only one component, such as paper clips, nails, steel balls for bearings, staples, screws, and bolts. Almost all products, however, are assembled from components that have been manufactured as individual parts. Even relatively simple products consist of at least two different parts joined by various means. For example: (a) some kitchen knives have wooden or plastic handles that are attached to the metal blade with fasteners; (b) cooking pots and pans have metal, plastic, or wooden handles and knobs that are attached to the pot by various methods; and (c) the eraser of an ordinary pencil is attached with a brass sleeve.

On a much larger scale, observe computers, motorcycles, washing machines, power tools, and airplanes and how their numerous components are assembled and joined so that they can function reliably and also are economical to produce. As also shown in Table I.1 in the General Introduction, a rotary lawn mower has about 300 parts, and a grand piano has 12,000 parts. A C-5A transport plane has more than 4 million parts, and a Boeing 747-400 aircraft has more than 6 million parts. A typical automobile consists of 15,000 components, some of which are shown in Fig. VI.1.

Joining is an all-inclusive term, covering processes such as welding, brazing, soldering, adhesive bonding, and mechanical fastening. These processes are

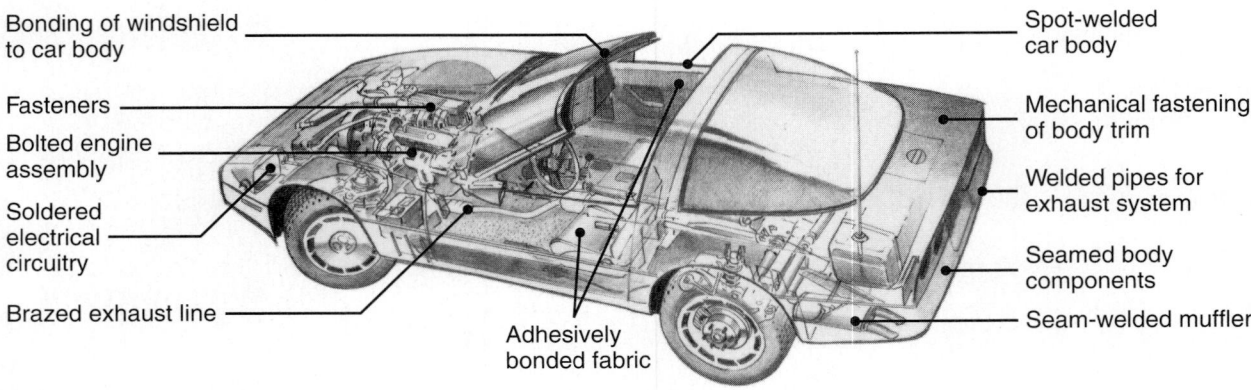

FIGURE VI.1   Various parts in a typical automobile assembled using the processes described in Part VI.

an essential and important aspect of manufacturing and *assembly* operations, for one or more of the following reasons:

- Even a relatively simple product may be impossible to manufacture as a *single piece*. Consider, for example, the tubular construction shown in Fig. VI.2a. Assume that each of the arms of this product is 5 m (15 ft) long, the tubes are 100 mm (4 in.) in diameter, and their wall thickness is 1 mm (0.04 in.). After reviewing all of the manufacturing processes described in the preceding chapters, we would conclude that manufacturing this product in one piece would be impossible or uneconomical.

- The product (such as a cooking pot with a handle) is *more economical* to manufacture as individual components, which are then assembled.

- Products such as automobile engines, hair dryers, and printers need to be designed to be able to be *taken apart* for maintenance or replacement of their parts.

- *Different properties* may be desirable for functional purposes of the product. For example, surfaces subjected to friction, wear, corrosion, or environmental attack generally require characteristics that differ significantly from those of the component's bulk. Examples are: (a) masonry drills with carbide cutting tips brazed to the shank of a drill (Fig. VI.2b), (b) automotive brake shoes, and (c) grinding wheels bonded to a metal backing (Section 26.2).

- *Transporting* the product in individual components and assembling them at home or at the customer's plant may be easier and less costly than transporting the completed item. Most bicycles, large toys, metal or wood shelving, and machine tools and presses are assembled after the components or subassemblies have been transported to the appropriate site.

Although there are different ways of categorizing the wide variety of available joining processes, we will follow the classification by the American Welding Society

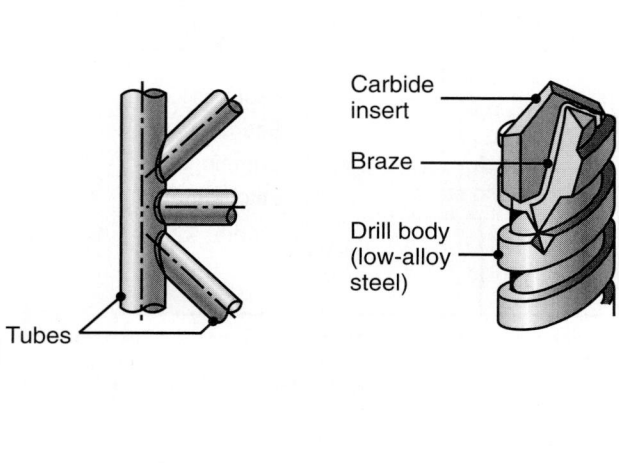

(a)   (b)   (c)

FIGURE VI.2   Examples of parts utilizing joining processes. (a) A tubular part fabricated by joining individual components. This product cannot be manufactured in one piece by any of the methods described in previous chapters if it consists of thin-walled, large-diameter, tubular-shaped long arms. (b) A drill bit with a carbide cutting insert brazed to a steel shank—an example of a part where two materials need to be joined for performance reasons. (c) Spot welding of automobile bodies. Source: (c) Courtesy of Ford Motor Co.

(AWS). Accordingly, joining processes fall into three major categories (see Figs. VI.3 and I.7f):

- *Welding*
- *Adhesive bonding*
- *Mechanical fastening*

Table VI.1 lists the general relative characteristics of various joining processes. Welding processes, in turn, generally are classified into three basic categories:

- *Fusion welding*
- *Solid-state welding*
- *Brazing and soldering*

As will be seen, some types of welding processes can be classified into both the fusion and the solid-state categories.

Fusion welding is defined as the *melting together and coalescing* of materials by means of heat usually supplied by chemical or electrical means. Filler metals may or may not be used. Fusion welding is comprised of consumable- and nonconsumable-electrode arc welding and high-energy-beam welding processes. The welded joint

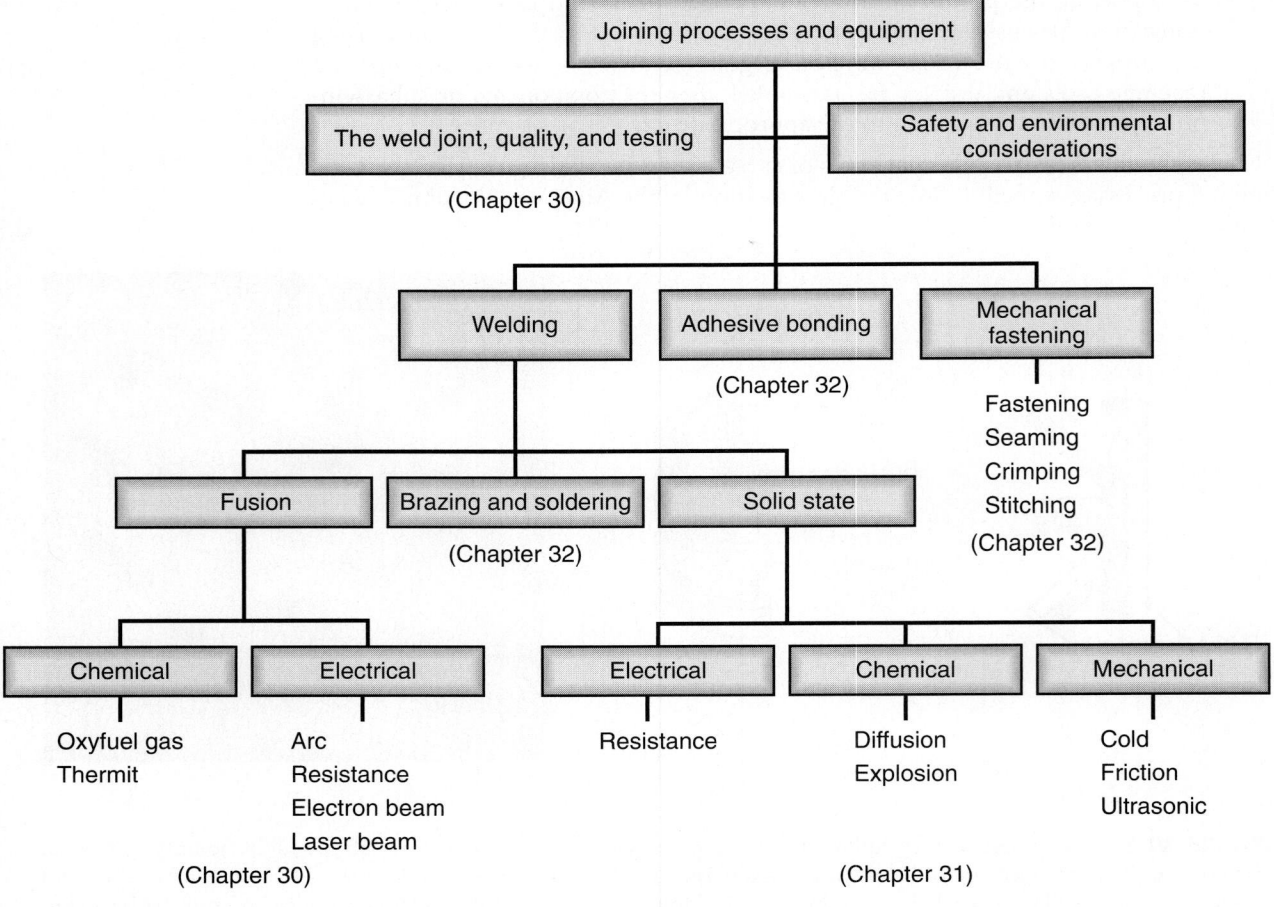

FIGURE VI.3   Outline of topics described in Part VI.

TABLE VI.1

Comparison of Various Joining Methods

Method	Strength	Design variability	Small parts	Large parts	Tolerances	Reliability	Ease of maintenance	Visual inspection	Cost
Arc welding	1	2	3	1	3	1	2	2	2
Resistance welding	1	2	1	1	3	3	3	3	1
Brazing	1	1	1	1	3	1	3	2	3
Bolts and nuts	1	2	3	1	2	1	1	1	3
Riveting	1	2	3	1	1	1	3	1	2
Fasteners	2	3	3	1	2	2	2	1	3
Seaming and crimping	2	2	1	3	3	1	3	1	1
Adhesive bonding	3	1	1	2	3	2	3	3	2

Note: 1 = very good; 2 = good; 3 = poor. For cost, 1 is the lowest.

undergoes important metallurgical and physical changes, which, in turn, have a major effect on the properties and performance of the welded component or structure. Some simple welded joints are shown in Fig. VI.4.

In **solid-state welding**, joining takes place *without fusion*. Consequently, there is no liquid (molten) phase in the joint. The basic processes in this category are diffusion bonding and cold, ultrasonic, friction, resistance, and explosion welding. **Brazing** uses filler metals and involves lower temperatures than welding. **Soldering** also uses similar filler metals (solders) and involves even lower temperatures.

Adhesive bonding has unique applications requiring strength, sealing, thermal and electrical insulating, vibration damping, and resistance to corrosion between dissimilar metals. **Mechanical fastening** involves traditional methods of using various fasteners, especially bolts, nuts, and rivets. The **joining of plastics** can be accomplished by adhesive bonding, fusion by various external or internal heat sources, and mechanical fastening.

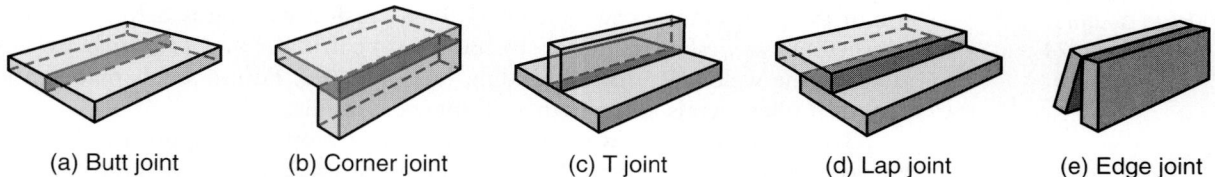

(a) Butt joint   (b) Corner joint   (c) T joint   (d) Lap joint   (e) Edge joint

FIGURE VI.4   Examples of joints that can be made through the various joining processes described in Chapters 30 through 32.

CHAPTER

30

Fusion-Welding Processes

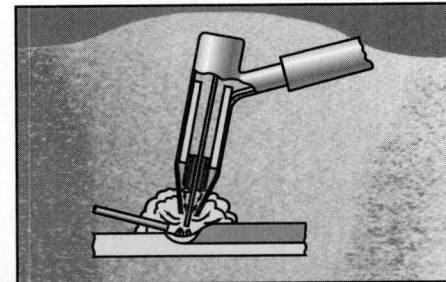

This chapter describes fusion-welding processes in which two pieces are joined together by the application of heat, which then melts and fuses the interface. We describe in detail their principles, characteristics, and applications. The topics covered in this chapter include:

- Oxyfuel-gas welding, where acetylene and oxygen provide the energy needed for welding.
- Arc-welding processes that use electrical energy and nonconsumable and consumable electrodes to form the weld.
- High-energy-beam welding processes such as laser- and electron-beam welding.
- A description of how these processes are used in cutting metals.
- The nature and characteristics of the weld joint.
- Factors involved in the weldability of metals.
- Good joint design practices and process selection.

30.1 | Introduction

The welding processes described in this chapter involve the partial melting and fusion of the joint between two members. Here, **fusion welding** is defined as *melting* together and *coalescing* materials by means of heat. *Filler metals* (which are metals added to the weld area during welding) may be used. Fusion welds made without the use of filler metals are known as *autogenous* welds.

This chapter describes the major classes of fusion-welding processes. It covers the basic principles of each process; the equipment used; its relative advantages, limitations, and capabilities; and the economic considerations affecting the process selection (Table 30.1). These processes include the oxyfuel-gas, arc, and high-energy-beam (laser-beam and electron-beam) welding processes, which have important and unique applications in modern manufacturing.

The chapter continues with a description of weld-zone features and the wide variety of discontinuities and defects that can exist in welded joints. The weldability of various ferrous and nonferrous metals and alloys then are reviewed. The chapter ends with a discussion of design guidelines for welding, giving several examples of

TABLE 30.1

General Characteristics of Fusion Welding Processes							
Joining process	Operation	Advantage	Skill level required	Welding position	Current type	Distortion*	Typical cost of equipment ($)
Shielded metal arc	Manual	Portable and flexible	High	All	AC, DC	1 to 2	Low (1500+)
Submerged arc	Automatic	High deposition	Low to medium	Flat and horizontal	AC, DC	1 to 2	Medium (5000+)
Gas metal arc	Semiautomatic or automatic	Works with most metals	Low to high	All	DC	2 to 3	Medium (3000+)
Gas tungsten arc	Manual or automatic	Works with most metals	Low to high	All	AC, DC	2 to 3	Medium (5000+)
Flux-cored arc	Semiautomatic or automatic	High deposition	Low to high	All	DC	1 to 3	Medium (2000+)
Oxyfuel	Manual	Portable and flexible	High	All	—	2 to 4	Low (500+)
Electron beam, laser beam	Semiautomatic or automatic	Works with most metals	Medium to high	All	—	3 to 5	High (100,000–1 million)

*1=highest; 5=lowest.

good weld-design practices. As in all manufacturing processes, the economics of welding is an equally significant aspect of the overall operation. Welding processes, equipment, and labor costs will be discussed in Section 31.8.

30.2 | Oxyfuel-Gas Welding

Oxyfuel-gas welding (OFW) is a general term used to describe any welding process that uses a **fuel gas** combined with *oxygen* to produce a flame. This flame is the source of the heat that is used to melt the metals at the joint. The most common gas-welding process uses *acetylene*; this process is known as *oxyacetylene-gas welding* (OAW) and is used typically for structural sheet-metal fabrication, automotive bodies, and various repair work.

Developed in the early 1900s, the OAW process utilizes the heat generated by the combustion of acetylene gas (C_2H_2) in a mixture with oxygen. The heat is generated in accordance with a pair of chemical reactions. The primary combustion process, which occurs in the inner core of the flame (Fig. 30.1), involves the following reaction:

$$C_2H_2 + O_2 \longrightarrow 2CO + H_2 + heat \qquad (30.1)$$

This reaction dissociates the acetylene into carbon monoxide and hydrogen and produces about one-third of the total heat generated in the flame. The secondary combustion process is

$$2CO + H_2 + 1.5O_2 \longrightarrow 2CO_2 + H_2O + heat \qquad (30.2)$$

This reaction consists of the further burning of both the hydrogen and the carbon monoxide and produces about two-thirds of the total heat. Note that the reaction also produces water vapor. The temperatures developed in the flame can reach 3300°C (6000°F).

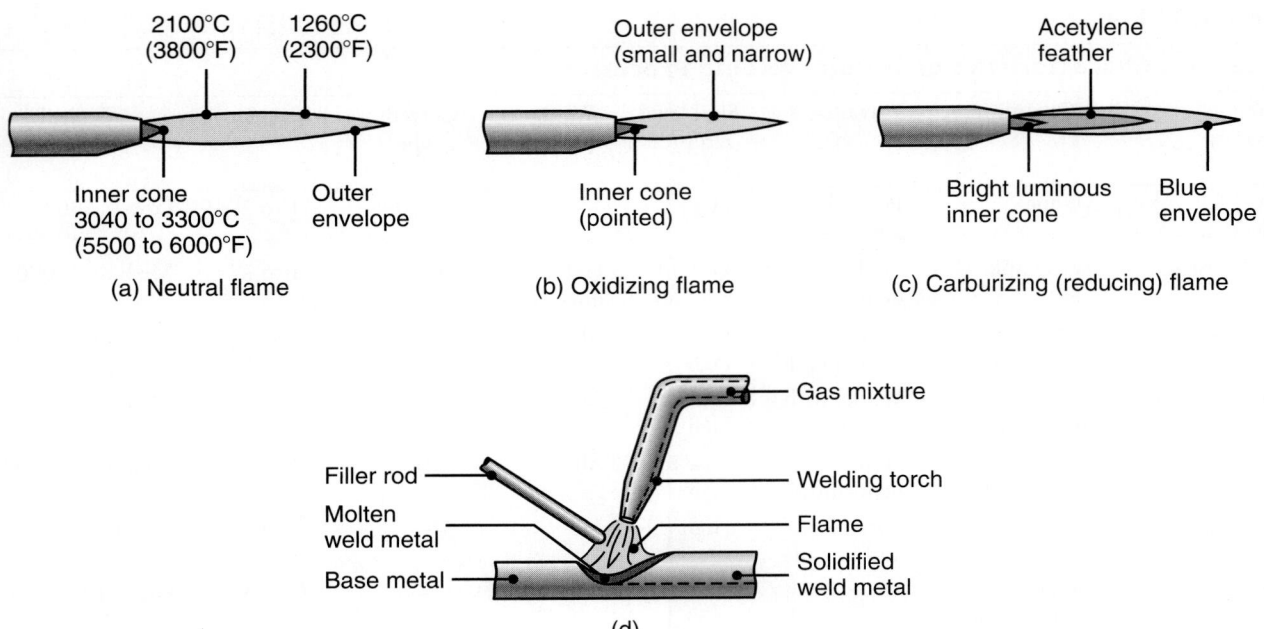

FIGURE 30.1   Three basic types of oxyacetylene flames used in oxyfuel-gas welding and cutting operations: (a) neutral flame; (b) oxidizing flame; (c) carburizing or reducing flame. The gas mixture in (a) is basically equal volumes of oxygen and acetylene. (d) The principle of the oxyfuel-gas welding operation.

Flame types.   The proportion of acetylene and oxygen in the gas mixture is an important factor in oxyfuel-gas welding. At a ratio of 1:1 (that is, when there is no excess oxygen), the flame is considered to be *neutral* (Fig. 30.1a). With a greater oxygen supply, the flame can be harmful (especially for steels), because it oxidizes the metal. For this reason, a flame with excess oxygen is known as an **oxidizing flame** (Fig. 30.1b). Only in the welding of copper and copper-based alloys is an oxidizing flame desirable, because in those cases, a thin protective layer of *slag* (compounds of oxides) forms over the molten metal. If the oxygen is insufficient for full combustion, the flame is known as a **reducing** (one having excess acetylene) or **carburizing flame** (Fig. 30.1c). The temperature of a reducing flame is lower, hence it is suitable for applications requiring low heat, such as brazing, soldering, and flame-hardening operations.

Other fuel gases (such as hydrogen and methylacetylene propadiene) also can be used in oxyfuel-gas welding. However, the temperatures developed by these gases are low. Hence, they are used for welding (a) metals with low melting points (such as lead) and (b) parts that are thin and small. The flame with pure hydrogen gas is colorless, hence it is difficult to adjust the flame by eyesight.

Filler metals.   Filler metals are used to supply additional metal to the weld zone during welding. They are available as **filler rods** or *wire* (Fig. 30.1d) and may be bare or coated with **flux.** The purpose of the flux is to retard oxidation of the surfaces of the parts being welded by generating a *gaseous shield* around the weld zone. The flux also helps to dissolve and remove oxides and other substances from the weld zone, thus contributing to the formation of a stronger joint. The slag developed (compounds of oxides, fluxes, and electrode-coating materials) protects the molten puddle of metal against oxidation as it cools.

Welding practice and equipment.   Oxyfuel-gas welding can be used with most ferrous and nonferrous metals for almost any workpiece thickness, but the relatively low heat input limits the process to thicknesses of less than 6 mm (0.25 in.). The basic steps can be summarized as follows:

1. Prepare the edges to be joined and establish and maintain their proper position by using clamps and fixtures.
2. Open the acetylene valve and ignite the gas at the tip of the torch. Open the oxygen valve and adjust the flame for that particular operation (Fig. 30.2).
3. Hold the torch at about 45° from the plane of the workpiece with the inner flame near the workpiece and the filler rod at about 30° to 40°.
4. Touch the filler rod to the joint and control its movement along the joint length by observing the rate of melting and filling of the joint.

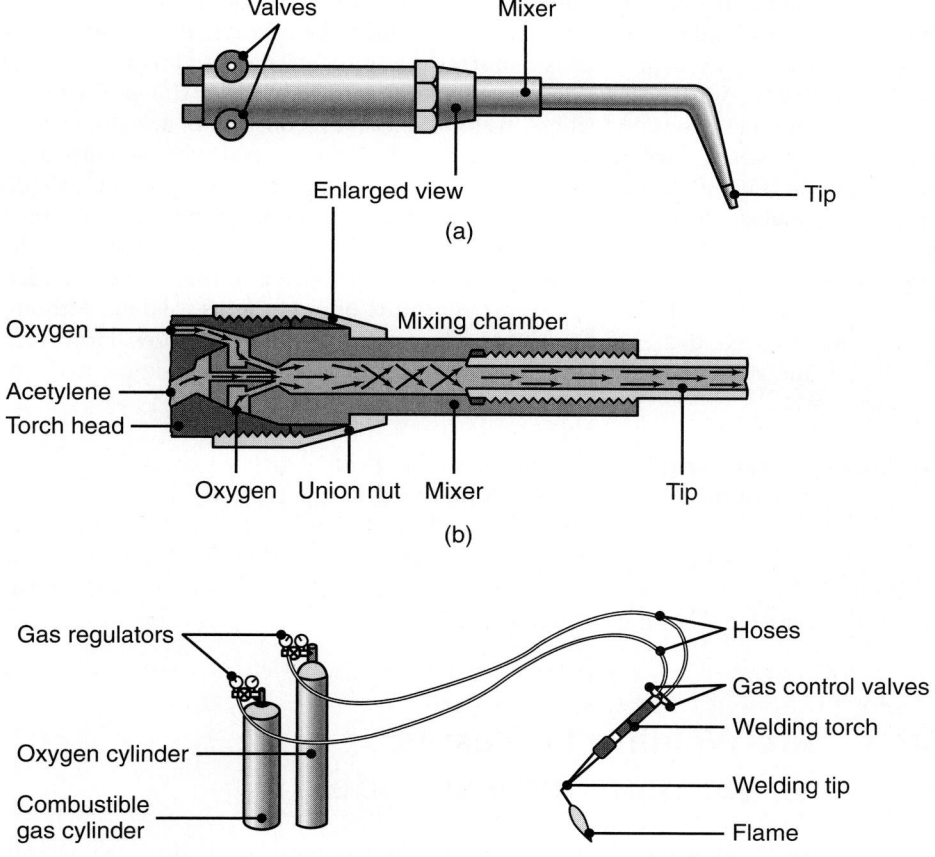

FIGURE 30.2   (a) General view of and (b) cross-section of a torch used in oxyacetylene welding. The acetylene valve is opened first; the gas is lit with a spark lighter or a pilot light; then the oxygen valve is opened; and the flame adjusted. (c) Basic equipment used in oxyfuel-gas welding. To ensure correct connections, all threads on acetylene fittings are left-handed, whereas those for oxygen are right-handed. Oxygen regulators usually are painted green and acetylene regulators red.

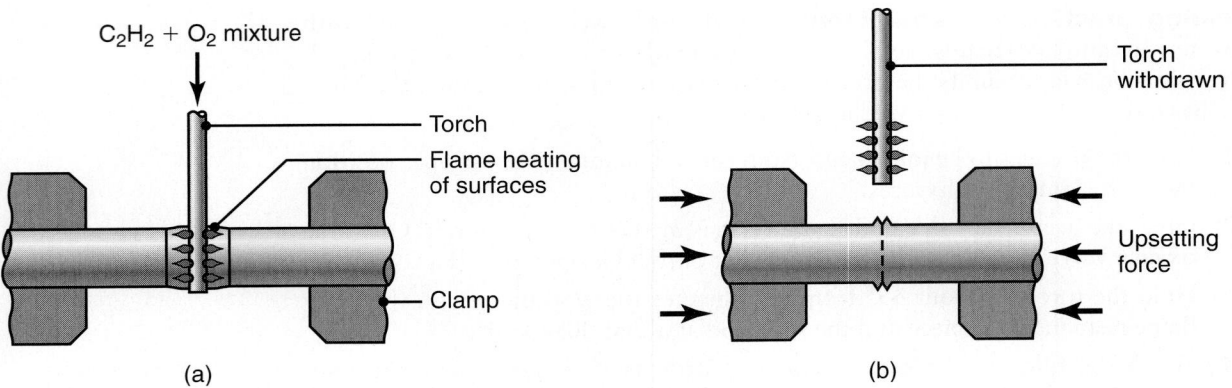

FIGURE 30.3   Schematic illustration of the pressure-gas welding process: (a) before and (b) after. Note the formation of a flash at the joint, which later can be trimmed off.

Small joints made by this process may consist of a single-weld bead. Deep-V groove joints are made in multiple passes. Cleaning the surface of each weld bead prior to depositing a second layer is important for joint strength and to avoid defects (see Section 30.9). Wire brushes (hand or power) may be used for this purpose.

The equipment for oxyfuel-gas welding basically consists of a **welding torch** connected by hoses to high-pressure gas cylinders and equipped with pressure gages and regulators (Fig. 30.2c). The use of safety equipment (such as goggles with shaded lenses, face shields, gloves, and protective clothing) is essential. Proper connection of the hoses to the cylinders is an important factor in safety. Oxygen and acetylene cylinders have different threads, so the hoses cannot be connected to the wrong cylinders.

The low equipment cost is an attractive feature of oxyfuel-gas welding. Although it can be mechanized, this operation essentially is manual and, hence, slow. However, it has the advantages of being portable, versatile, and economical for simple and low-quantity work.

Pressure-gas welding.   In this method, the welding of two components starts with the heating of the interface by means of a torch using (typically) an oxyacetylene-gas mixture (Fig. 30.3a). After the interface begins to melt, the torch is withdrawn. A force is applied to press the two components together (Fig. 30.3b) and is maintained until the interface solidifies. Note the formation of a flash due to the upsetting of the joined ends of the two components.

30.3 Arc-Welding Processes: Nonconsumable Electrode

In *arc welding*, developed in the mid-1800s, the heat required is obtained from electrical energy. The process involves either a *consumable* or a *nonconsumable electrode*. An arc is produced between the tip of the electrode and the workpiece to be welded, by using an AC or a DC power supply. This arc produces temperatures of about 30,000°C (54,000°F), which are much higher than those developed in oxyfuel-gas welding.

In *nonconsumable-electrode* welding processes, the electrode is typically a **tungsten electrode** (Fig. 30.4). An externally supplied shielding gas is necessary because of the high temperatures involved in order to prevent oxidation of the weld

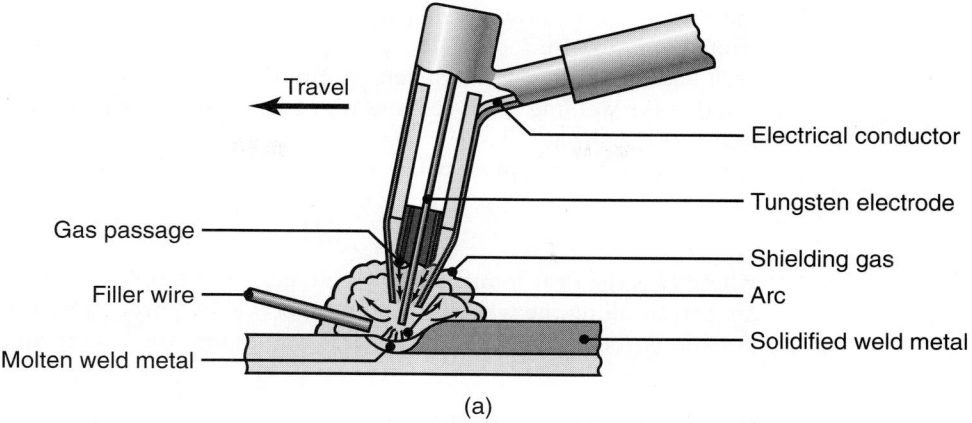

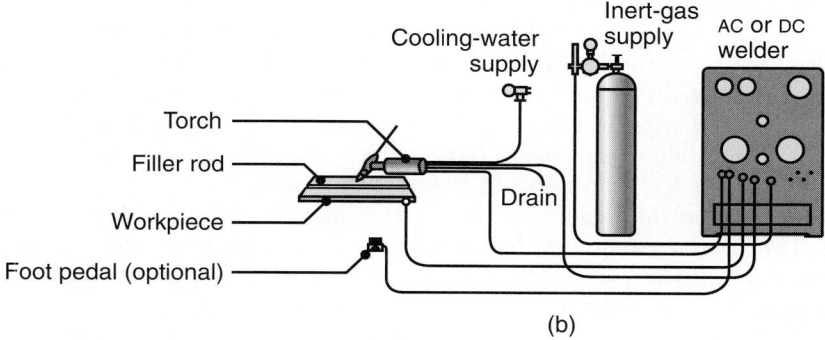

FIGURE 30.4   (a) The gas tungsten-arc welding process formerly known as TIG (for tungsten inert gas) welding. (b) Equipment for gas tungsten-arc welding operations.

zone. Typically, direct current is used, and its **polarity** (that is the direction of current flow) is important. Its selection depends on such factors as the type of electrode, metals to be welded, and depth and width of the weld zone.

In **straight polarity**—also known as *direct-current electrode negative* (DCEN)—the workpiece is positive (anode), and the electrode is negative (cathode). It generally produces welds that are narrow and deep (Fig. 30.5a). In **reverse polarity**—also known as *direct-current electrode positive* (DCEP)—the workpiece is negative, and

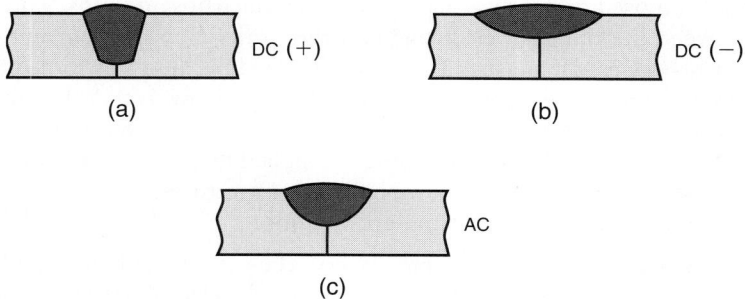

FIGURE 30.5   The effect of polarity and current type on weld beads: (a) DC current with straight polarity; (b) DC current with reverse polarity; (c) AC current.

the electrode is positive. Weld penetration is less, and the weld zone is shallower and wider (Fig. 30.5b). Hence, DCEP is preferred for sheet metals and for joints with very wide gaps. In the AC **current** method, the arc pulsates rapidly. This method is suitable for welding thick sections and for using large-diameter electrodes at maximum currents (Fig. 30.5c).

The heat input in electric-arc welding is given by the expression

$$H = \frac{EI}{v} \qquad (30.3)$$

where H is the heat input, E is the voltage, I is the current, and v is the velocity the arc travels along the weld line. As in other welding processes, however, only a small portion of the theoretical heat generated goes into the immediate weld area.

Gas tungsten-arc welding.   In *gas tungsten-arc welding* (GTAW), formerly known as *TIG welding* (for "tungsten inert gas"), the filler metal is supplied from a **filler wire** (Fig. 30.4a). Because the tungsten electrode is not consumed in this operation, a constant and stable *arc gap* is maintained at a constant current level. The filler metals are similar to the metals to be welded, and flux is not used. The shielding gas is usually argon or helium (or a mixture of the two). Welding with GTAW may be done without filler metals—for example, in the welding of close-fit joints.

Depending on the metals to be welded, the power supply is either DC at 200 A, or AC at 500 A (Fig. 30.4b). In general, AC is preferred for aluminum and magnesium, because the cleaning action of AC removes oxides and improves weld quality. Thorium or zirconium may be used in the tungsten electrodes to improve their electron emission characteristics. Power supply ranges from 8 to 20 kW. Contamination of the tungsten electrode by the molten metal can be a significant problem, particularly in critical applications, because it can cause discontinuities in the weld. Therefore, contact of the electrode with the molten-metal pool should be avoided.

The GTAW process is used for a wide variety of metals and applications, particularly aluminum, magnesium, titanium, and the refractory metals. It is suitable especially for thin metals. The cost of the inert gas makes this process more expensive than SMAW but provides welds with very high quality and surface finish. It is used in a variety of critical applications with a wide range of workpiece thicknesses and shapes. The equipment is portable.

Plasma-arc welding.   In *plasma-arc welding* (PAW), developed in the 1960s, a concentrated plasma arc is produced and directed towards the weld area. The arc is stable and reaches temperatures as high as 33,000°C (60,000°F). A **plasma** is ionized hot gas composed of nearly equal numbers of electrons and ions. The plasma is initiated between the tungsten electrode and the orifice by a low-current pilot arc. Unlike other processes, the plasma arc is concentrated because it is forced through a relatively small orifice. Operating currents usually are below 100 A, but they can be higher for special applications. When a filler metal is used, it is fed into the arc, as is done in GTAW. Arc and weld-zone shielding is supplied by means of an outer-shielding ring and the use of gases, such as argon, helium, or mixtures.

There are two methods of plasma-arc welding:

- In the **transferred-arc** method (Fig. 30.6a), the workpiece being welded is part of the electrical circuit. The arc transfers from the electrode to the workpiece— hence the term *transferred*.

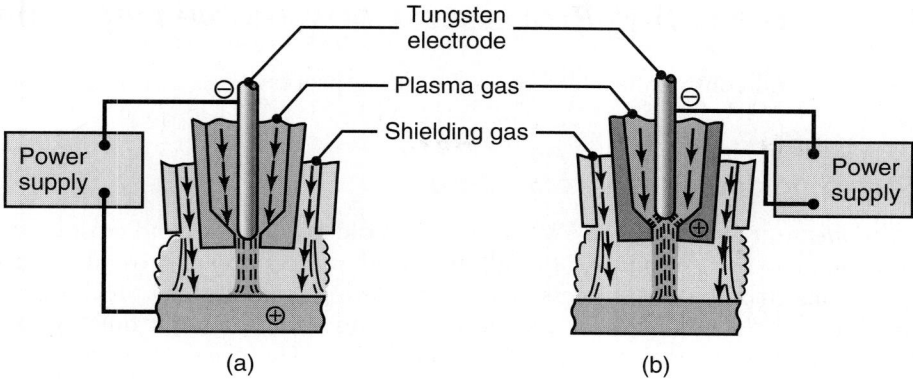

FIGURE 30.6 Two types of plasma-arc welding processes: (a) transferred and (b) nontransferred. Deep and narrow welds can be made by this process at high welding speeds.

- In the **nontransferred** method (Fig. 30.6b), the arc occurs between the electrode and the nozzle, and the heat is carried to the workpiece by the plasma gas. This thermal-transfer mechanism is similar to that for an oxyfuel flame (see Section 30.2).

Compared to other arc-welding processes, plasma-arc welding has better arc stability, less thermal distortion, and higher energy concentration, thus permitting deeper and narrower welds. In addition, higher welding speeds from 120 to 1000 mm/min (5 to 40 in./min) can be achieved. A variety of metals can be welded with part thicknesses generally less than 6 mm (0.25 in.).

The high heat concentration can penetrate completely through the joint (**keyhole technique**) with thicknesses as much as 20 mm (0.75 in.) for some titanium and aluminum alloys. In the keyhole technique, the force of the plasma arc displaces the molten metal and produces a hole at the leading edge of the weld pool. Plasma-arc welding (rather than the GTAW process) often is used for butt and lap joints because of its higher energy concentration, better arc stability, and higher welding speeds. Proper training and skill are essential for operators who use this equipment. Safety considerations include protection against glare, spatter, and noise from the plasma arc.

Atomic-hydrogen welding. In *atomic-hydrogen welding* (AHW), an arc is generated between two tungsten electrodes in a shielding atmosphere of hydrogen gas. The arc is maintained independently of the workpiece or parts being welded. The hydrogen gas normally is diatomic (H_2), but where the temperatures are over 6000°C (11,000°F) near the arc, the hydrogen breaks down into its atomic form, simultaneously absorbing a large amount of heat from the arc. When the hydrogen strikes a relatively cold surface (i.e., the weld zone), it recombines into its diatomic form and rapidly releases the stored heat. The energy in AHW can be varied easily by changing the distance between the arc stream and the workpiece surface. This process is being replaced by shielded metal-arc welding, mainly because of the availability of inexpensive inert gases.

30.4 | Arc-Welding Processes: Consumable Electrode

There are several consumable-electrode arc-welding processes, as described below.

30.4.1 Shielded metal-arc welding

Shielded metal-arc welding (SMAW) is one of the oldest, simplest, and most versatile joining processes. About 50% of all industrial and maintenance welding currently is performed by this process. The electric arc is generated by touching the tip of a *coated electrode* against the workpiece and withdrawing it quickly to a distance sufficient to maintain the arc (Fig. 30.7a). The electrodes are in the shapes of thin, long rods (hence this process also is known as **stick welding**) that are held manually.

The heat generated melts a portion of the electrode tip, its coating, and the base metal in the immediate arc area. The molten metal consists of a mixture of the base metal (workpiece), the electrode metal, and substances from the coating on the electrode; this mixture forms the weld when it solidifies. The electrode coating deoxidizes the weld area and provides a shielding gas to protect it from oxygen in the environment.

A bare section at the end of the electrode is clamped to one terminal of the power source, while the other terminal is connected to the workpiece being welded (Fig. 30.7b). The current, which may be DC or AC, usually ranges between 50 and 300 A. For sheet-metal welding, DC is preferred because of the steady arc it produces. Power requirements generally are less than 10 kW.

The SMAW process has the advantages of being relatively simple, versatile and requiring a smaller variety of electrodes. The equipment consists of a power supply, cables, and an electrode holder. The SMAW process commonly is used in general construction, shipbuilding, pipelines, and maintenance work. It is useful especially for work in remote areas where a portable fuel-powered generator can be used as the power supply. This process is suited best for workpiece thicknesses of 3 to 19 mm (0.12 to 0.75 in.), although this range can be extended easily by skilled operators using multiple-pass techniques (Fig. 30.8).

The multiple-pass approach requires that the slag be cleaned after each weld bead. Unless removed completely, the solidified slag can cause severe corrosion of the weld area and lead to failure of the weld, but it also prevents fusion of weld layers and, therefore, compromises the weld strength. Before another weld is applied, the slag should be removed completely, such as by wire brushing or weld chipping. Consequently, both labor costs and material costs are high.

30.4.2 Submerged-arc welding

In *submerged-arc welding* (SAW), the weld arc is shielded by a *granular flux* consisting of lime, silica, manganese oxide, calcium fluoride, and other compounds. The flux is fed into the weld zone from a hopper by gravity flow through a nozzle (Fig. 30.9). The thick layer of flux completely covers the molten metal. It prevents spatter and

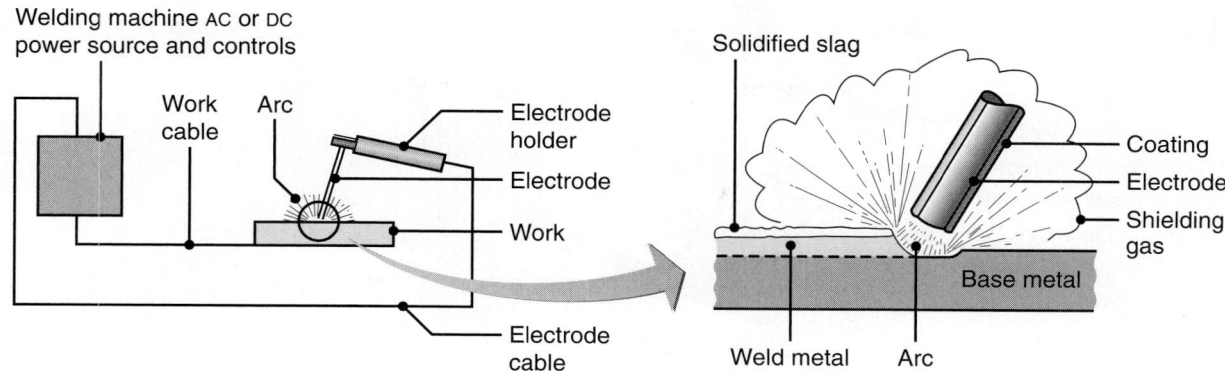

FIGURE 30.7 Schematic illustration of the shielded metal-arc welding process. About 50% of all large-scale industrial-welding operations use this process.

sparks and suppresses the intense ultraviolet radiation and fumes characteristic of the SMAW process. The flux also acts as a thermal insulator by promoting deep penetration of heat into the workpiece. The unused flux can be recovered (using a recovery tube), treated, and reused.

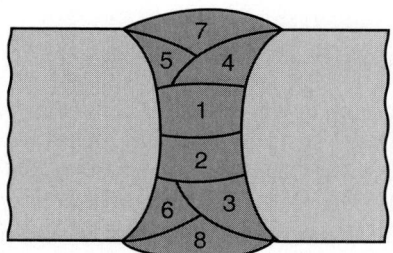

FIGURE 30.8 A deep weld showing the buildup sequence of eight individual weld beads.

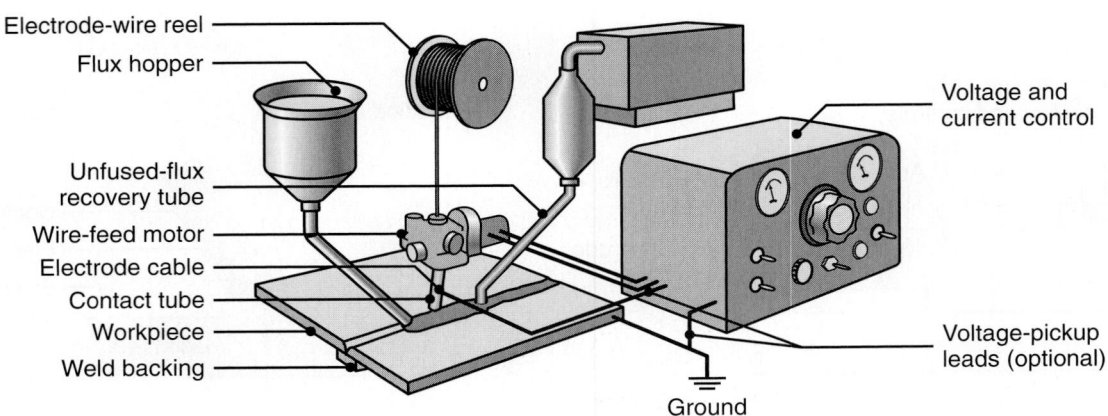

Electrode-wire reel
Flux hopper
Voltage and current control
Unfused-flux recovery tube
Wire-feed motor
Electrode cable
Contact tube
Workpiece
Weld backing
Voltage-pickup leads (optional)
Ground

FIGURE 30.9 Schematic illustration of the submerged-arc welding process and equipment. The unfused flux is recovered and reused.

The consumable electrode is a coil of bare round wire 1.5 to 10 mm $\left(\frac{1}{16} \text{ to } \frac{3}{8} \text{ in.}\right)$ in diameter; it is fed automatically through a tube (**welding gun**). Electric currents typically range between 300 and 2000 A. The power supplies usually are connected to standard single- or three-phase power lines with a primary rating up to 440 V.

Because the flux is gravity fed, the SAW process is limited largely to welds in a flat or horizontal position having a backup piece. Circular welds can be made on pipes and cylinders—provided that they are rotated during welding. As Fig. 30.9 shows, the unfused flux can be recovered, treated, and reused. This process is automated and is used to weld a variety of carbon and alloy steel and stainless-steel sheets or plates at speeds as high as 5 m/min (16 ft/min). The quality of the weld is very high—with good toughness, ductility, and uniformity of properties. The SAW process provides very high welding productivity, depositing 4 to 10 times the amount of weld metal per hour as the SMAW process. Typical applications include thick-plate welding for shipbuilding and for pressure vessels.

30.4.3 Gas metal-arc welding

In *gas metal-arc welding* (GMAW), developed in the 1950s and formerly called *metal inert-gas* (MIG) *welding*, the weld area is shielded by an effectively inert atmosphere of argon, helium, carbon dioxide, or various other gas mixtures (Fig. 30.10a). The consumable bare wire is fed automatically through a nozzle into the weld arc by a wire-feed drive motor (Fig. 30.10b). In addition to using inert shielding gases, deoxidizers usually are present in the electrode metal itself in order to prevent oxidation of the molten-weld puddle. Multiple-weld layers can be deposited at the joint.

Metal can be transferred by three methods in the GMAW process:

1. In **spray transfer**, small, molten metal droplets from the electrode are transferred to the weld area at a rate of several hundred droplets per second. The transfer is spatter-free and very stable. High DC current and voltages and large-diameter electrodes are used with argon or an argon-rich gas mixture used as the shielding gas. The average current required in this process can be reduced

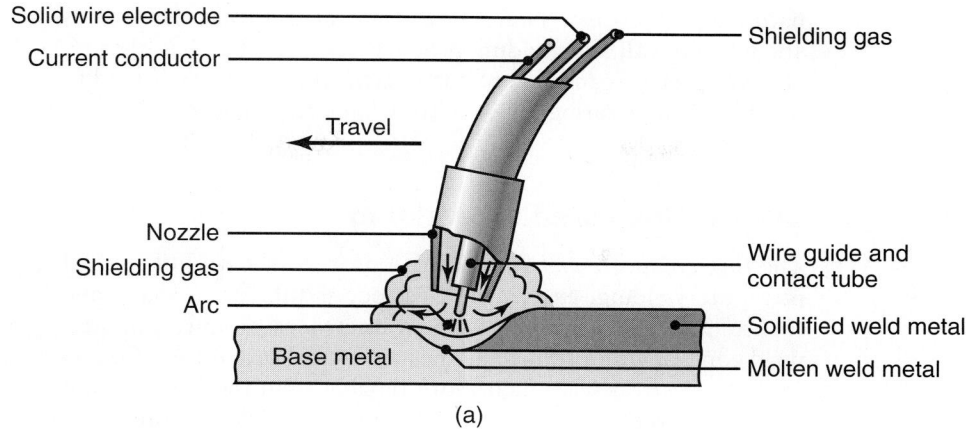

Solid wire electrode
Current conductor
Shielding gas

Travel

Nozzle
Shielding gas
Arc
Wire guide and contact tube
Solidified weld metal
Base metal
Molten weld metal

(a)

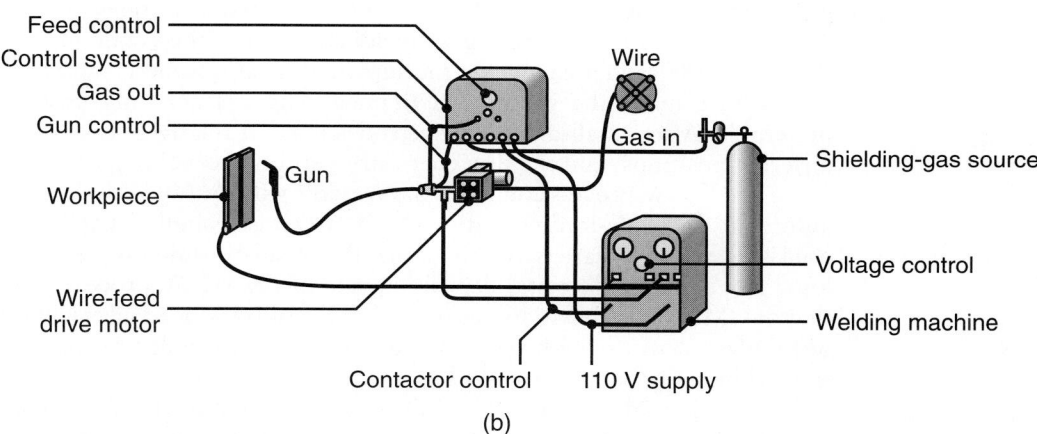

Feed control
Control system
Gas out
Gun control
Wire
Gas in

Workpiece
Gun

Shielding-gas source

Wire-feed drive motor

Voltage control
Welding machine

Contactor control   110 V supply

(b)

FIGURE 30.10 (a) Schematic illustration of the gas metal-arc welding process, formerly known as MIG (for metal inert gas) welding. (b) Basic equipment used in gas metal-arc welding operations.

by using a **pulsed arc,** which superimposes high-amplitude pulses onto a low, steady current. The process can be used in all welding positions.

2. In **globular transfer,** carbon-dioxide-rich gases are utilized, and globules are propelled by the forces of the electric-arc transfer of the metal, resulting in considerable spatter. High welding currents are used, making it possible for greater weld penetration and higher welding speed than are achieved in spray transfer. Heavier sections commonly are joined by this method.

3. In **short circuiting,** the metal is transferred in individual droplets (more than 50 per second), as the electrode tip touches the molten weld metal and short circuits. Low currents and voltages are utilized with carbon-dioxide-rich gases and electrodes made of small-diameter wire. The power required is about 2 kW.

The temperatures generated in GMAW are relatively low. Consequently, this method is suitable only for thin sheets and sections of less than 6 mm (0.25 in.), otherwise incomplete fusion may occur. The operation is easy to handle and is used very commonly for welding ferrous metals in thin sections. Pulsed-arc systems are used for thin ferrous and nonferrous metals.

This process is suitable for welding most ferrous and nonferrous metals and is used extensively in the metal-fabrication industry. Because of the relatively simple nature of the process, the training of operators is easy. The process is versatile, rapid, and economical, and welding productivity is double that of the SMAW process. The GMAW process can be automated easily and lends itself readily to robotics and to flexible manufacturing systems (see Chapters 37 and 39).

30.4.4 Flux-cored arc welding

The *flux-cored arc welding* (FCAW) process (shown in Fig. 30.11) is similar to gas metal-arc welding, except the electrode is tubular in shape and is filled with flux (hence the term *flux-cored*). Cored electrodes produce a more stable arc, improve weld contour, and produce better mechanical properties of the weld metal. The flux in these electrodes is much more flexible than the brittle coating used on SMAW electrodes, so the tubular electrode can be provided in long coiled lengths.

The electrodes are usually 0.5 to 4 mm (0.020 to 0.15 in.) in diameter, and the power required is about 20 kW. Self-shielded cored electrodes also are available. They do not require external gas shielding, because they contain emissive fluxes that shield the weld area against the surrounding atmosphere. Small-diameter electrodes have made the welding of thinner materials not only possible but often preferable. Also, small-diameter electrodes make it relatively easy to weld parts in different positions, and the flux chemistry permits the welding of many metals.

The FCAW process combines the versatility of SMAW with the continuous and automatic electrode-feeding feature of GMAW. It is economical and versatile, so it is used for welding a variety of joints, mainly on steels, stainless steels, and nickel alloys. The higher weld-metal deposition rate of the FCAW process (compared with that of GMAW) has led to its use in the joining of sections of all thicknesses. The use of *tubular electrodes* with very small diameters has extended the use of this process to workpieces of smaller section size.

A major advantage of FCAW is the ease with which specific weld-metal chemistries can be developed. By adding alloying elements to the flux core, virtually

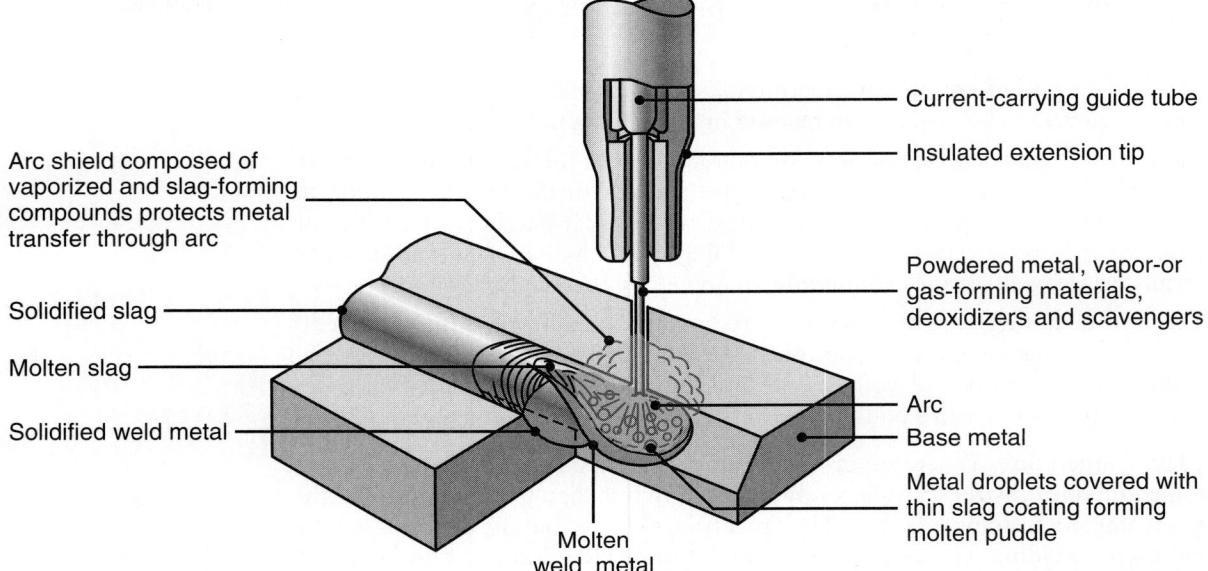

FIGURE 30.11   Schematic illustration of the flux-cored arc-welding process. This operation is similar to gas metal-arc welding, shown in Fig. 30.10.

any alloy composition can be produced. This process is easy to automate and is readily adaptable to flexible manufacturing systems and robotics.

30.4.5 Electrogas welding

Electrogas welding (EGW) is used primarily for welding the edges of sections vertically and in one pass with the pieces placed edge to edge (butt joint). It is classified as a *machine-welding process*, because it requires special equipment (Fig. 30.12). The weld metal is deposited into a weld cavity between the two pieces to be joined. The space is enclosed by two water-cooled copper *dams* (*shoes*) to prevent the molten slag from running off. Mechanical drives move the shoes upward. Circumferential welds (such as on pipes) also are possible, with the workpiece rotating.

Single or multiple electrodes are fed through a conduit, and a continuous arc is maintained using flux-cored electrodes at up to 750 A or solid electrodes at 400 A. Power requirements are about 20 kW. Shielding is done by means of an inert gas, such as carbon dioxide, argon, or helium—depending on the type of material being welded. The gas may be provided from an external source, produced from a flux-cored electrode, or from both.

The equipment for electrogas welding is reliable and training for operators is relatively simple. Weld thickness ranges from 12 to 75 mm (0.5 to 3 in.) on steels, titanium, and aluminum alloys. Typical applications are in the construction of bridges, pressure vessels, thick-walled and large-diameter pipes, storage tanks, and ships.

30.4.6 Electroslag welding

Electroslag welding (ESW) and its applications are similar to electrogas welding (Fig. 30.13). The main difference is that the arc is started between the electrode tip and the bottom of the part to be welded. Flux is added, which then is melted by the heat of the arc. After the molten slag reaches the tip of the electrode, the arc is extinguished. Heat is produced continuously by the electrical resistance of the molten slag. Because the arc is extinguished, ESW is not strictly an arc-welding process.

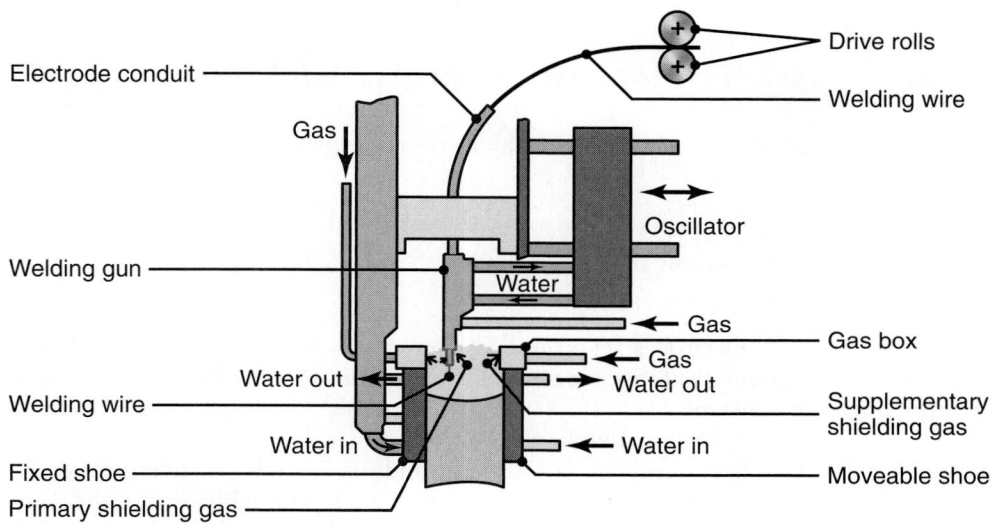

FIGURE 30.12   Schematic illustration of the electrogas-welding process.

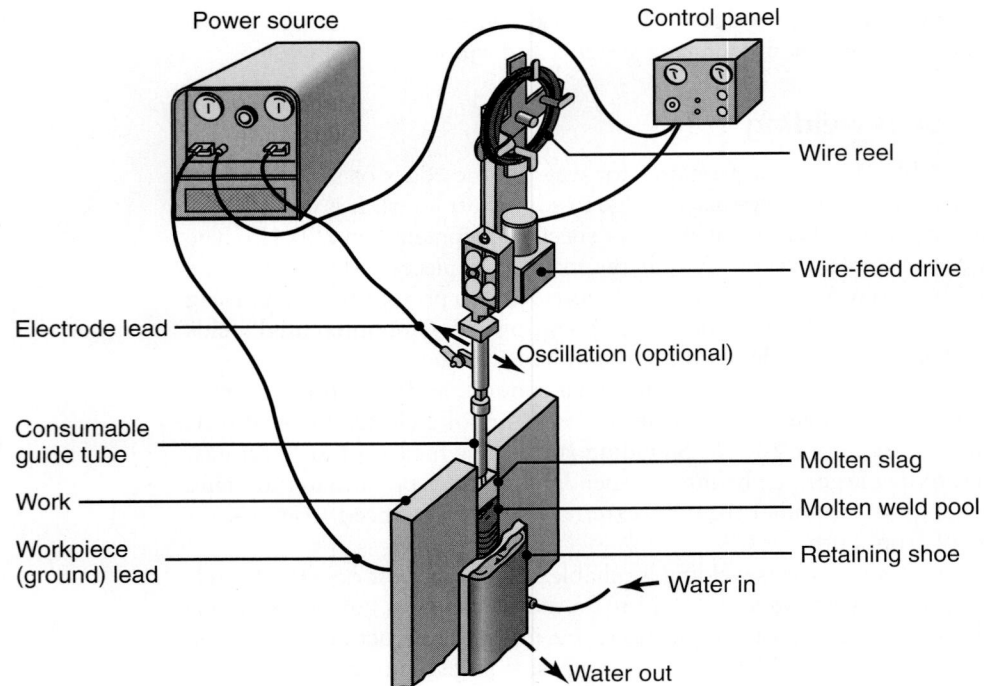

FIGURE 30.13 Equipment used for electroslag-welding operations.

Single or multiple solid as well as flux-cored electrodes may be used. The guide may be nonconsumable (conventional method) or consumable.

Electroslag welding is capable of welding plates with thicknesses ranging from 50 mm to more than 900 mm (2 to 36 in.), and welding is done in one pass. The current required is about 600 A at 40 to 50 V, although higher currents are used for thick plates. Travel speed of the weld is in the range of 12 to 36 mm/min (0.5 to 1.5 in./min). Weld quality is good. This process is used for large structural-steel sections, such as heavy machinery, bridges, oil rigs, ships, and nuclear-reactor vessels.

30.5 | Electrodes for Arc Welding

Electrodes for the consumable arc-welding processes described are classified according to the

- Strength of the deposited weld metal
- Current (AC or DC)
- Type of coating

Electrodes are identified by numbers and letters (Table 30.2) or by color code, particularly if they are too small to imprint with identification. Typical coated-electrode dimensions are in the range of 150 to 460 mm (6 to 18 in.) in length, and 1.5 to 8 mm $\left(\frac{1}{16} \text{ to } \frac{5}{16} \text{ in.}\right)$ in diameter.

Specifications for electrodes and filler metals (including dimensional tolerances, quality control procedures, and processes) are published by the American Welding Society (AWS) and the American National Standards Institute (ANSI). Some specifications appear in the Aerospace Materials Specifications (AMS) by the Society of

TABLE 30.2

| Designations for Mild–Steel Coated Electrodes |

The prefix "E" designates arc welding electrode.

The first two digits of four-digit numbers and the first three digits of five-digit numbers indicate minimum tensile strength:

E60XX	60,000 psi minimum tensile strength
E70XX	70,000 psi minimum tensile strength
E110XX	110,000 psi minimum tensile strength

The next-to-last digit indicates position:

| EXX1X | All positions |
| EXX2X | Flat position and horizontal fillets |

The last two digits together indicate the type of covering and the current to be used.
The suffix (Example: EXXXX-A1) indicates the approximate alloy in the weld deposit:

—A1	0.5% Mo
—B1	0.5% Cr, 0.5% Mo
—B2	1.25% Cr, 0.5% Mo
—B3	2.25% Cr, 1% Mo
—B4	2% Cr, 0.5% Mo
—B5	0.5% Cr, 1% Mo
—C1	2.5% Ni
—C2	3.25% Ni
—C3	1% Ni, 0.35% Mo, 0.15% Cr
—D1 and D2	0.25–0.45% Mo, 1.75% Mn
—G	0.5% min. Ni, 0.3% min. Cr, 0.2% min. Mo, 0.1% min. V, 1% min. Mn (only one element required)

Automotive Engineers (SAE). Electrodes are sold by weight and are available in a wide variety of sizes and specifications. Selection and recommendations for electrodes for a particular metal and its application can be found in supplier literature and in the various handbooks and references listed at the end of this chapter.

Electrode coatings. Electrodes are *coated* with clay-like materials (that include silicate binders) and powdered materials (such as oxides, carbonates, fluorides, metal alloys, and cellulose (cotton cellulose and wood flour)). The coating (which is brittle and takes part in complex interactions during welding) has the following basic functions:

- Stabilize the arc.
- Generate gases to act as a shield against the surrounding atmosphere; the gases produced are carbon dioxide and water vapor (and carbon monoxide and hydrogen in small amounts).
- Control the rate at which the electrode melts.
- Act as a flux to protect the weld against the formation of oxides, nitrides, and other inclusions and (with the resulting slag) to protect the molten-weld pool.
- Add alloying elements to the weld zone to enhance the properties of the joint—among these are deoxidizers to prevent the weld from becoming brittle.

The deposited electrode coating or slag must be removed after each pass in order to ensure a good weld; a wire brush (manual or power) can be used for this purpose. Bare electrodes and wire made of stainless steels and aluminum alloys also are available. They are used as filler metals in various welding operations.

30.6 | Electron-Beam Welding

In *electron-beam welding* (EBW), developed in the 1960s, heat is generated by high-velocity narrow-beam electrons. The kinetic energy of the electrons is converted into heat as they strike the workpiece. This process requires special equipment to focus the beam on the workpiece, typically in a vacuum. The higher the vacuum, the more the beam penetrates, and the greater is the depth-to-width ratio; thus, the methods are called EBW-HV (for high vacuum) and EBW-MV (for medium vacuum). Welding some materials also may be done by EBW-NV (for no vacuum).

Almost any metal can be welded by EBW, and workpiece thicknesses can range from foil to plate. The intense energy also is capable of producing holes in the workpiece (**keyhole technique**; Section 30.3). Generally, no shielding gas, flux, or filler metal is required. Capacities of electron guns range up to 100 kW.

This process has the capability of making high-quality welds that are almost parallel-sided, are deep and narrow, and have small heat-affected zones (see Section 30.9). Depth-to-width ratios range between 10 and 30. The sizes of welds made by EBW are much smaller than those of welds made by conventional processes. Using automation and servo controls, parameters can be controlled accurately at welding speeds as high as 12 m/min (40 ft/min).

Almost any metal can be butt- or lap-welded with this process at thicknesses up to 150 mm (6 in.). Distortion and shrinkage in the weld area is minimal. Weld quality is good and of very high purity. Typical applications include the welding of aircraft, missile, nuclear and electronic components, and gears and shafts for the automotive industry. Electron-beam welding equipment generates x-rays, hence proper monitoring and periodic maintenance are essential.

30.7 | Laser-Beam Welding

Laser-beam welding (LBW) utilizes a high-power laser beam as the source of heat, to produce a fusion weld. Because the beam can be focused onto a very small area, it has high energy density and deep-penetrating capability. The beam can be directed, shaped, and focused precisely on the workpiece. Consequently, this process is suitable particularly for welding deep and narrow joints (Fig. 30.14) with depth-to-width ratios typically ranging from 4 to 10.

In the automotive industry, welding transmission components is the most widespread application. Among numerous other applications is the welding of thin parts for electronic components. The laser beam may be **pulsed** (in milliseconds) for applications (such as the spot welding of thin materials) with power levels up to 100 kW. **Continuous** multi-kW laser systems are used for deep welds on thick sections.

Laser-beam welding produces welds of good quality with minimum shrinkage and distortion. Laser welds have good strength and generally are ductile and free of porosity. The process can be automated to be used on a variety of materials with thicknesses of up to 25 mm (1 in.); it is effective particularly on thin workpieces. As stated in Section 16.2.2, *tailor-welded* sheet-metal blanks are joined principally by laser-beam welding using robotics for precise control of the beam path during welding.

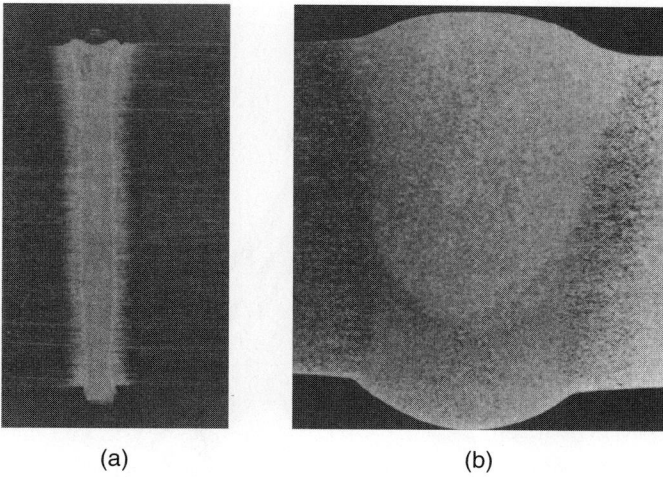

(a)   (b)

FIGURE 30.14   Comparison of the size of weld beads: (a) laser-beam or electron-beam welding and (b) tungsten-arc welding. *Source*: Courtesy of American Welding Society, *Welding Handbook*, 8th ed., 1991.

Typical metals and alloys welded include aluminum, titanium, ferrous metals, copper, superalloys, and the refractory metals. Welding speeds range from 2.5 m/min (8 ft/min) to as high as 80 m/min (250 ft/min) for thin metals. Because of the nature of the process, welding can be done in otherwise inaccessible locations. As in other and similar automated welding systems, the operator skill required is minimal. Safety particularly is important in laser-beam welding due to the extreme hazards to the eye as well as the skin; solid-state (YAG) lasers also are dangerous. (See Table 27.2 on types of lasers.)

The major advantages of LBW over EBW are the following:

- A vacuum is not required, and the beam can be transmitted through air.
- Laser beams can be shaped, manipulated, and focused optically (using fiber optics), so the process can be automated easily.
- The beams do not generate x-rays.
- The quality of the weld is better than in EBW with less tendency for incomplete fusion, spatter, porosity, and less distortion.

EXAMPLE 30.1   Laser welding of razor blades

A closeup of the Gillette Sensor™ razor cartridge is shown in Fig. 30.15. Each of the two narrow, high-strength blades has 13 pinpoint welds—11 of which can be seen (as darker spots, about 0.5 mm in diameter) on each blade in the photograph. You can inspect the welds on actual blades with a magnifying glass or a microscope.

The welds are made with a Nd:YAG laser equipped with fiber-optic delivery. This equipment provides very flexible beam manipulation and can target exact locations along the length of the blade. With a set of these machines, production is at a rate of 3 million welds per hour with accurate and consistent weld quality.

Source: Courtesy of Lumonics Corporation, Industrial Products Division.

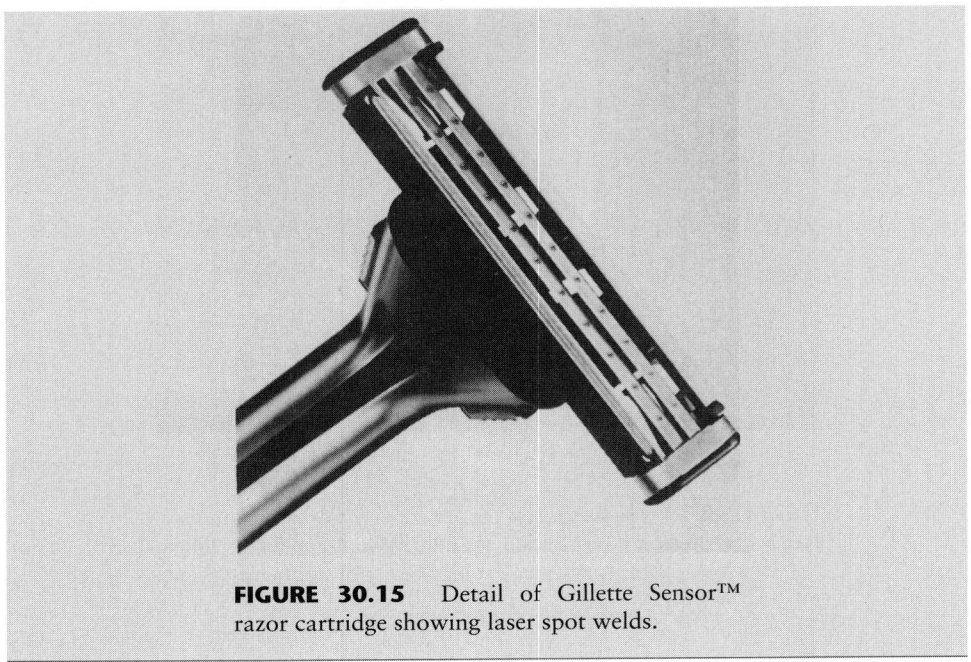

FIGURE 30.15 Detail of Gillette Sensor™ razor cartridge showing laser spot welds.

30.8 | Cutting

In addition to mechanical means, a piece of material can be separated into two or more parts or into various contours by the use of a heat source that melts and removes a narrow zone in the workpiece. The sources of heat can be torches, electric arcs, or lasers.

Oxyfuel-gas cutting.   *Oxyfuel-gas cutting* (OFC) is similar to oxyfuel welding, but the heat source now is used to remove a narrow zone from a metal plate or sheet (Fig. 30.16a). This process is suitable particularly for steels. The basic reactions with steel are

$$Fe + O \longrightarrow FeO + heat \tag{30.4}$$

$$3Fe + 2O_2 \longrightarrow Fe_3O_4 + heat \tag{30.5}$$

and

$$4Fe + 3O_2 \longrightarrow 2\ Fe_2O_3 + heat \tag{30.6}$$

The greatest heat is generated by the second reaction, and it can produce a temperature rise to about 870°C (1600°F). However, this temperature is not sufficiently high to cut steels, therefore the workpiece is *preheated* with fuel gas, and oxygen is introduced later (see nozzle cross-section in Fig. 30.16a). The higher the carbon content of the steel, the higher is the preheating temperature required. Cutting takes place mainly by the oxidation (burning) of the steel; some melting also takes place. Cast irons and steel castings also can be cut by this method. This process generates a **kerf** similar to that produced by sawing with a saw blade or by wire EDM (see Fig. 27.12).

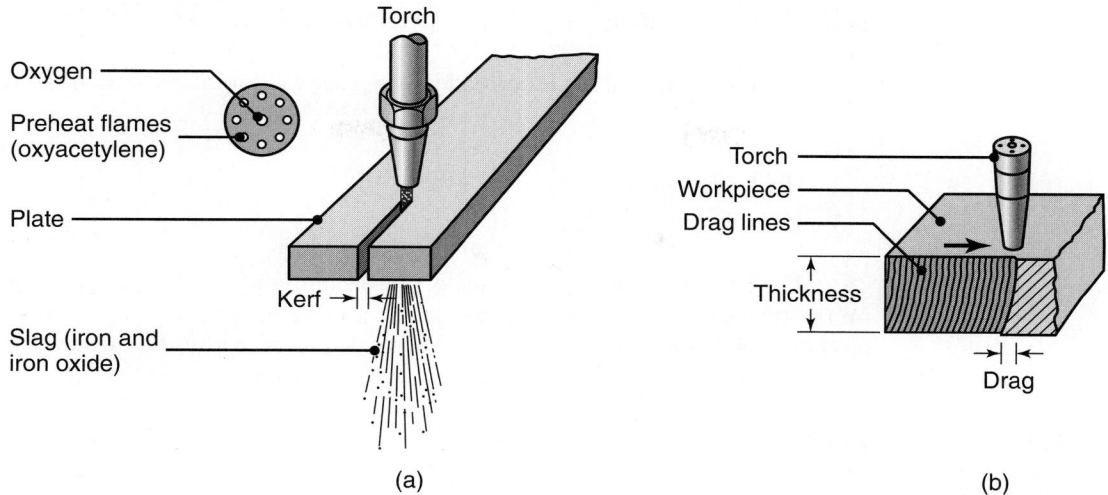

FIGURE 30.16   (a) Flame cutting of a steel plate with an oxyacetylene torch, and a cross-section of the torch nozzle. (b) Cross-section of a flame-cut plate showing drag lines.

The maximum thickness that can be cut by OFC depends mainly on the gases used. With oxyacetylene gas, the maximum thickness is about 300 mm (12 in.); with oxyhydrogen, about 600 mm (24 in.). Kerf widths range from about 1.5 to 10 mm (0.06 to 0.4 in.) with reasonably good control of tolerances. The flame leaves **drag lines** on the cut surface (Fig. 30.16b), which results in a rougher surface than that produced by processes such as sawing, blanking, or other operations using mechanical cutting tools. Distortion caused by uneven temperature distribution can be a problem in OFC.

Although long used for salvage and repair work, oxyfuel-gas cutting can be used in manufacturing as well. Torches may be guided along various paths manually, mechanically, or automatically by machines using programmable controllers and robots. *Underwater cutting* is done with specially designed torches that produce a blanket of compressed air between the flame and the surrounding water.

Arc cutting.   *Arc cutting* processes are based on the same principles as arc welding processes. A variety of materials can be cut at high speeds by arc cutting. As in welding, these processes also leave a heat-affected zone which needs to be taken into account, particularly in critical applications.

In **air carbon-arc cutting** (CAC-A), a carbon electrode is used, and the molten metal is blown away by a high-velocity air jet. Thus, the metal being cut doesn't have to oxidize. This process is used especially for gouging and scarfing (removal of metal from a surface). However, this process is noisy, and the molten metal can be blown substantial distances and cause safety hazards.

Plasma-arc cutting (PAC) produces the highest temperatures. It is used for the rapid cutting of nonferrous and stainless-steel plates. The cutting productivity of this process is higher than that of oxyfuel-gas methods. It produces a good surface finish, narrow kerfs, and is the most popular cutting process utilizing programmable controllers employed in manufacturing today.

Electron beams and **lasers** also are used for very accurately cutting a wide variety of metals, as was described in Sections 27.6 and 27.7. The surface finish is better than that of other thermal cutting processes, and the kerf is narrower.

30.9 | The Weld Joint, Quality, and Testing

Three distinct zones can be identified in a typical weld joint, as shown in Fig. 30.17:

1. *Base metal*
2. *Heat-affected zone*
3. *Weld metal*

The metallurgy and properties of the second and third zones strongly depend on the type of metals joined, the particular joining process, the filler metals used (if any), and welding process variables. A joint produced without a filler metal is called *autogenous*, and its weld zone is composed of the *resolidified base metal*. A joint made with a filler metal has a central zone called the *weld metal* and is composed of a mixture of the base and the filler metals.

Solidification of the weld metal.   After the application of heat and the introduction of the filler metal (if any) into the weld zone, the weld joint is allowed to cool to ambient temperature. The *solidification* process is similar to that in casting and begins with the formation of columnar (dendritic) grains. (See Fig. 10.3.) These grains are relatively long and form parallel to the heat flow. Because metals are much better heat conductors than the surrounding air, the grains lie parallel to the plane of the two components being welded (Fig. 30.18a). In contrast, the grains in a shallow weld are shown in Fig. 30.18b and c.

Grain structure and grain size depend on the specific metal alloy, the particular welding process employed, and the type of filler metal. Because it began in a molten state, the weld metal basically has a *cast* structure, and since it has cooled slowly, it has coarse grains. Consequently, this structure generally has low strength, toughness, and ductility. However, the proper selection of filler–metal composition or of heat treatments following welding can improve the mechanical properties of the joint.

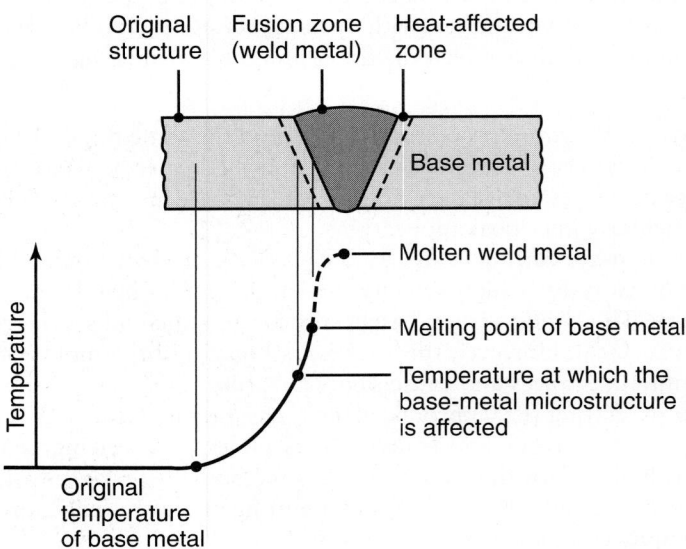

FIGURE 30.17   Characteristics of a typical fusion-weld zone in oxyfuel-gas and arc welding.

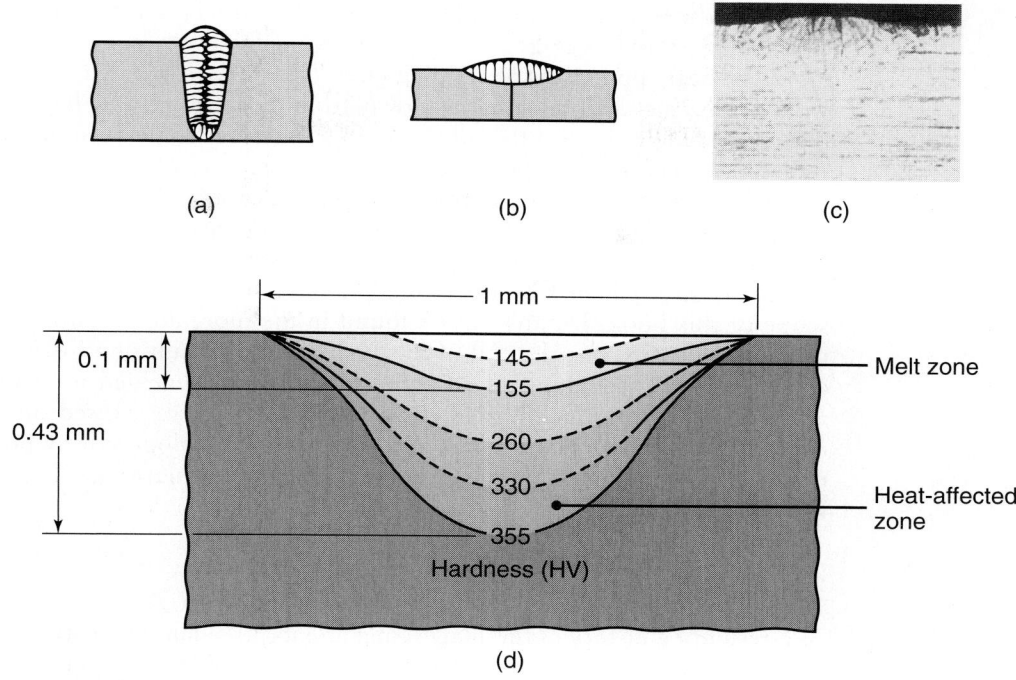

FIGURE 30.18 Grain structure in a (a) deep weld and (b) shallow weld. Note that the grains in the solidified weld metal are perpendicular to their interface with the base metal. (c) Weld bead on a cold-rolled nickel strip produced by a laser beam. (d) Microhardness (HV) profile across a weld bead.

The resulting structure depends on the particular alloy, its composition, and the thermal cycling to which the joint is subjected. For example, cooling rates may be controlled and reduced by *preheating* the general weld area prior to welding. Preheating is important particularly for metals having high thermal conductivity, such as aluminum and copper. Without preheating, the heat produced during welding dissipates rapidly through the rest of the parts being joined.

Heat-affected zone. The *heat-affected zone* (HAZ) is within the base metal itself. It has a microstructure different from that of the base metal prior to welding, because it has been subjected temporarily to elevated temperatures during welding. The portions of the base metal that are far enough away from the heat source do not undergo any structural changes during welding because of the far lower temperature to which they are subjected.

The properties and microstructure of the HAZ depend on (a) the rate of heat input and cooling and (b) the temperature to which this zone was raised. In addition to metallurgical factors (such as original grain size, grain orientation, and degree of prior cold work), physical properties (such as the specific heat and thermal conductivity of the metals) also influence the size and characteristics of this zone.

The strength and hardness of the heat-affected zone (Fig. 30.18d) depend partly on how the original strength and hardness of the base metal was developed prior to the welding. As was described in Chapters 2 and 4, they may have been developed by (a) cold working, (b) solid-solution strengthening, (c) precipitation hardening, or (d) various heat treatments. The effects of these strengthening methods are complex, and the simplest to analyze are those in the base metal that has been cold-worked, such as by cold rolling or cold forging.

The heat applied during welding *recrystallizes* the elongated grains of the cold-worked base metal. Grains that are away from the weld metal will recrystallize into fine, equiaxed grains. On the other hand, grains close to the weld metal have been subjected to elevated temperatures for a longer period of time. Consequently, the grains will grow in size (grain growth), and this region will be softer and have lower strength. Such a joint will be weakest at its heat-affected zone.

The effects of heat on the HAZ for joints made from dissimilar metals and for alloys strengthened by other methods are so complex as to be beyond the scope of this book. Details can be found in the more advanced references listed in the Bibliography at the end of this chapter. As a result of a history of thermal cycling and its attendant microstructural changes, a welded joint may develop various **discontinuities**. Welding discontinuities can be caused also by inadequate or careless application of proper welding technologies or by poor operator training. The major discontinuities that affect weld quality are described in the next section.

30.9.1 Weld quality

As a result of a history of thermal cycling and its attendant microstructural changes, a welded joint may develop various **discontinuities**. Welding discontinuities also can be caused by an inadequate or careless application of proper welding technologies or by poor operator training. The major discontinuities that affect weld quality are described here.

Porosity.   *Porosity* in welds is caused by

- Gases released during melting of the weld area but trapped during solidification.
- Chemical reactions during welding.
- Contaminants.

Most welded joints contain some porosity, which is generally in the shape of spheres or of elongated pockets. (See also Section 10.6.1.) The distribution of porosity in the weld zone may be random, or the porosity may be concentrated in a certain region in the zone.

Porosity in welds can be reduced by the following practices:

- Proper selection of electrodes and filler metals.
- Improved welding techniques, such as preheating the weld area or increasing the rate of heat input.
- Proper cleaning and the prevention of contaminants from entering the weld zone.
- Reduced welding speeds to allow time for gas to escape.

Slag inclusions.   *Slag inclusions* are compounds such as oxides, fluxes, and electrode-coating materials that are trapped in the weld zone. If shielding gases are not effective during welding, contamination from the environment also may contribute to such inclusions. Welding conditions also are important; with control of welding process parameters, the molten slag will float to the surface of the molten weld metal and thus will not become entrapped.

Slag inclusions can be prevented by the following practices:

- Cleaning the weld-bead surface before the next layer is deposited by means of a wire brush (hand or power) or chipper.
- Providing sufficient shielding gas.
- Redesigning the joint to permit sufficient space for proper manipulation of the puddle of molten weld metal.

Incomplete fusion and penetration. *Incomplete fusion* (lack of fusion) produces poor weld beads, such as those shown in Fig. 30.19. A better weld can be obtained by the use of the following practices:

- Raising the temperature of the base metal.
- Cleaning the weld area before welding.
- Modifying the joint design and changing the type of electrode used.
- Providing sufficient shielding gas.

Incomplete penetration occurs when the depth of the welded joint is insufficient. Penetration can be improved by the following practices:

- Increasing the heat input.
- Reducing the travel speed during the welding.
- Modifying the joint design.
- Ensuring that the surfaces to be joined fit each other properly.

Weld profile. *Weld profile* is important not only because of its effects on the strength and appearance of the weld, but also because it can indicate incomplete fusion or the presence of slag inclusions in multiple-layer welds.

- **Underfilling** results when the joint is not filled with the proper amount of weld metal (Fig. 30.20a).
- **Undercutting** results from the melting away of the base metal and the consequent generation of a groove in the shape of a sharp recess or notch (Fig. 30.20b). If it

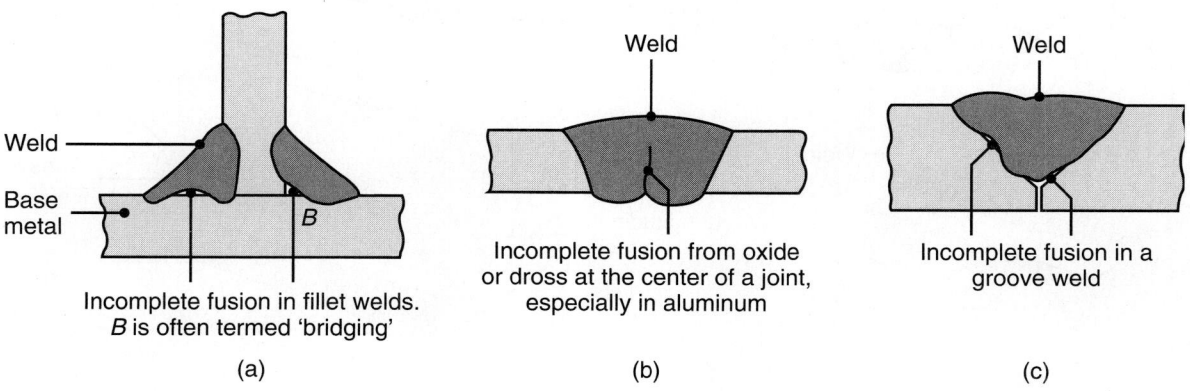

Weld

Base metal

B

Incomplete fusion in fillet welds. *B* is often termed 'bridging'

(a)

Weld

Incomplete fusion from oxide or dross at the center of a joint, especially in aluminum

(b)

Weld

Incomplete fusion in a groove weld

(c)

FIGURE 30.19   Examples of various discontinuities in fusion welds.

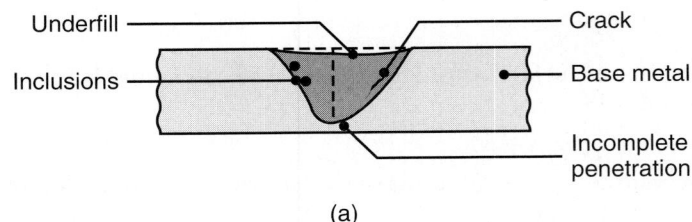

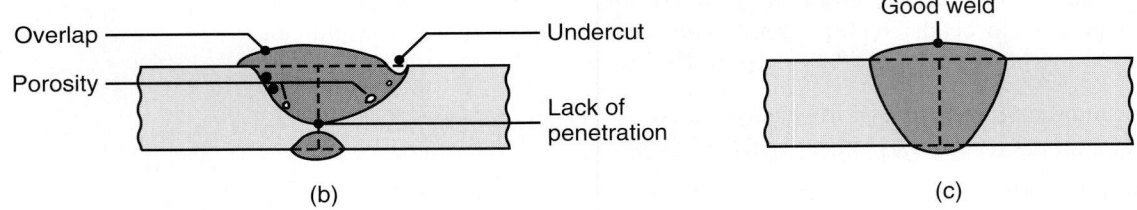

FIGURE 30.20   Examples of various defects in fusion welds.

is deep or sharp, an undercut can act as a stress raiser and can reduce the fatigue strength of the joint; in such cases, it may lead to premature failure.

- **Overlap** is a surface discontinuity (Fig. 30.20b) usually caused by poor welding practice or by the selection of improper materials. A good weld is shown in Fig. 30.20c.

Cracks.   Cracks may occur in various locations and directions in the weld area. Typical types of cracks are longitudinal, transverse, crater, underbead, and toe cracks (Fig. 30.21).

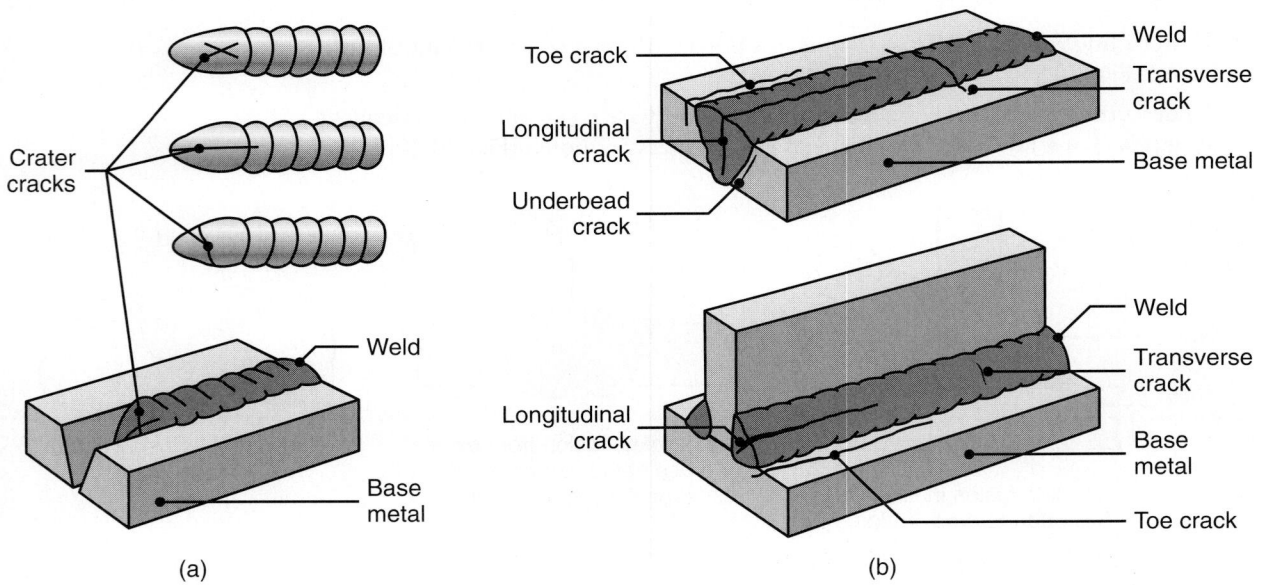

FIGURE 30.21   Types of cracks developed in welded joints. The cracks are caused by thermal stresses, similar to the development of hot tears in castings, as shown in Fig. 10.12.

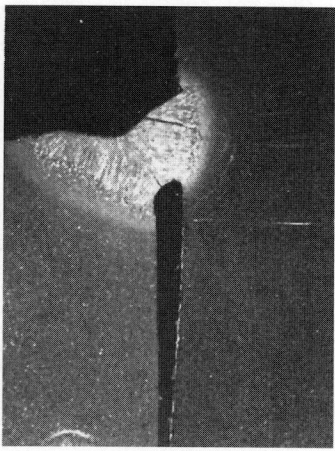

FIGURE 30.22   Crack in a weld bead. The two welded components were not allowed to contract freely after the weld was completed. *Source:* Courtesy of Packer Engineering.

These cracks generally result from a combination of the following factors:

- Temperature gradients that cause thermal stresses in the weld zone.
- Variations in the composition of the weld zone that cause different rates of contraction during cooling.
- Embrittlement of grain boundaries (Section 1.4), caused by the segregation of such elements as sulfur to the grain boundaries, and when the solid–liquid boundary moves when the weld metal begins to solidify.
- Hydrogen embrittlement (Section 2.10.2).
- Inability of the weld metal to contract during cooling (Fig. 30.22). This is a situation similar to *hot tears* that develop in castings (Fig 10.12) and is related to excessive restraint of the workpiece during the welding operation.

Cracks also are classified as **hot cracks** that occur while the joint is still at elevated temperatures and **cold cracks** that develop after the weld metal has solidified. The basic crack-prevention measures in welding are the following:

- Modify the joint design to minimize stresses developed from shrinkage during cooling.
- Change the parameters, procedures, and sequence of the welding operation.
- Preheat the components to be welded.
- Avoid rapid cooling of the welded components.

Lamellar tears.   In describing the anisotropy of plastically deformed metals in Section 1.5, it was stated that the workpiece is weaker when tested in its thickness direction because of the alignment of nonmetallic impurities and inclusions (stringers). This condition is evident particularly in rolled plates and in structural shapes. In welding such components, *lamellar tears* may develop because of shrinkage of the restrained components of the structure during cooling. Such tears can be avoided by providing for shrinkage of the members or by modifying the joint design to make the weld bead penetrate the weaker component more deeply.

Surface damage. Some of the metal may spatter during welding and be deposited as small droplets on adjacent surfaces. In arc-welding processes, the electrode inadvertently may touch the parts being welded at places other than the weld zone (**arc strikes**). Such surface discontinuities may be objectionable for reasons of appearance or of subsequent use of the welded part. If severe, these discontinuities adversely may affect the properties of the welded structure, particularly for notch-sensitive metals. Using proper welding techniques and procedures is important in avoiding surface damage.

Residual stresses. Because of localized heating and cooling during welding, the expansion and contraction of the weld area causes *residual stresses* in the workpiece. (See also Section 2.11.) Residual stresses can lead to the following defects:

- Distortion, warping, and buckling of the welded parts (Fig. 30.23).
- Stress-corrosion cracking (Section 2.10.2).
- Further distortion, if a portion of the welded structure is subsequently removed, such as by machining or sawing.
- Reduced fatigue life of the welded structure.

The type and distribution of residual stresses in welds is described best by reference to Fig. 30.24a. When two plates are being welded, a long, narrow zone is subjected to elevated temperatures, while the plates, as a whole, are essentially at ambient temperature. After the weld is completed and as time elapses, the heat from the weld zone dissipates laterally into the plates, while the weld area cools. Thus, the plates begin to expand longitudinally, while the welded length begins to contract (Fig. 30.22a).

If the plate is not constrained, it will warp, as shown in Fig. 30.22a. However, if the plate is not free to warp, it will develop residual stresses that typically are distributed like those shown in Fig. 30.24. Note that the magnitude of the compressive residual stresses in the plates diminishes to zero at a point far

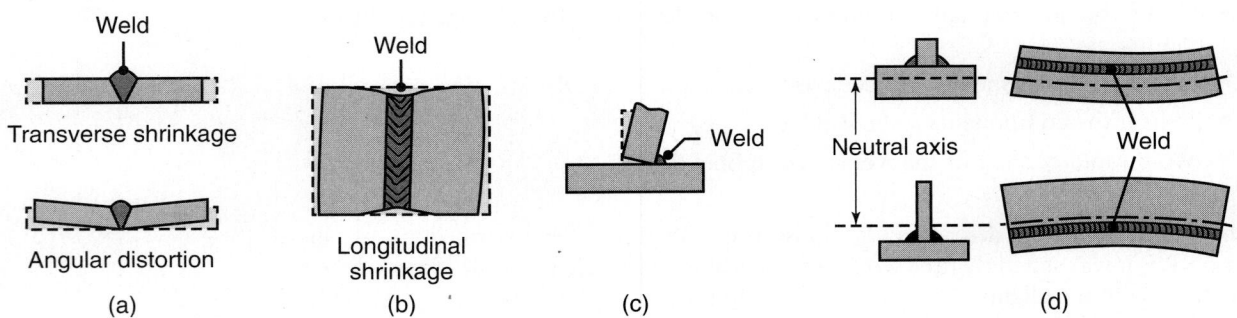

FIGURE 30.23 Distortion of parts after welding. Distortion is caused by differential thermal expansion and contraction of different regions of the welded assembly.

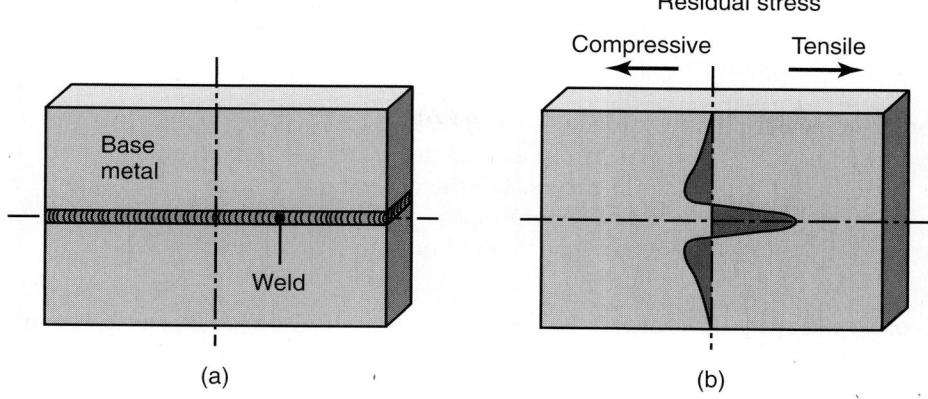

FIGURE 30.24 Residual stresses developed in (a) a straight-butt joint. Note that the residual stresses shown in (b) must be balanced internally. (See also Fig. 2.29.)

away from the weld area. Because no external forces are acting on the welded plates, the tensile and compressive forces represented by these residual stresses must balance each other.

Events leading to the distortion of a welded structure is shown in Fig. 30.25. Before welding, the structure is stress-free, as shown in Fig. 30.25a. The shape may be fairly rigid, and fixturing also may be present to support the structure. When the weld bead is placed, the molten metal fills the gap between the surfaces to be joined, and flows outward to form the weld bead. At this point, the weld is not under any stress. Afterward, the weld bead solidifies, and both the weld bead and the surrounding material cool to room temperature. As these materials cool, they try to contract but are constrained by the bulk of the weldment. The result is that the weldment distorts (Fig. 30.25c) and residual stresses develop.

The residual-stress distribution shown places the weld and the HAZ in a state of residual tension, which is harmful from a fatigue standpoint. Many welded structures will use cold-worked materials (such as extruded or roll-formed shapes), and these are relatively strong and fatigue-resistant. The weld itself may have porosity (see Fig. 30.20b), which can act as a stress riser and aid fatigue-crack growth, or there could be other cracks that can grow in fatigue. In general, the HAZ is less fatigue-resistant than

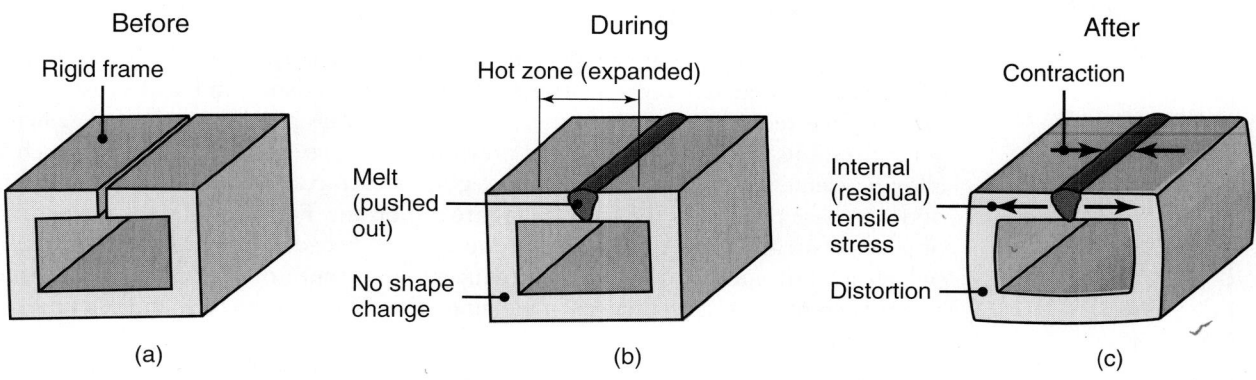

FIGURE 30.25 Distortion of a welded structure. *Source*: After J.A. Schey.

the base metal. Thus, the residual stresses developed can be very harmful, and it is not unusual to further treat welds in highly stressed or fatigue-susceptible applications, as discussed next.

In complex welded structures, residual-stress distributions are three-dimensional and, consequently, difficult to analyze. The previous discussion involved two plates that were not restrained from movement. In other words, the plates were not an integral part of a larger structure. On the other hand, if they are restrained, reaction stresses will be generated, because the plates are not free to expand or contract. This situation arises particularly in structures with high stiffness.

Stress relieving of welds.   The problems caused by residual stresses (such as distortion, buckling, and cracking) can be reduced by **preheating** the base metal or the parts to be welded. Preheating reduces distortion by reducing the cooling rate and the level of thermal stresses developed (by lowering the elastic modulus). This technique also reduces shrinkage and possible cracking of the joint.

For optimum results, preheating temperatures and cooling rates must be controlled carefully in order to maintain acceptable strength and toughness in the welded structure. The workpieces may be heated in several ways, including: (a) in a furnace, (b) electrically (resistively or inductively), or (c) by radiant lamps or a hot-air blast for thin sections. The temperature and time required for *stress relieving* depend on the type of material and on the magnitude of the residual stresses developed.

Other methods of stress relieving include peening, hammering, or surface rolling of the weld-bead area. These techniques induce compressive residual stresses, which, in turn, lower or eliminate tensile residual stresses in the weld. For multilayer welds, the first and last layers should not be peened in order to protect them against possible peening damage.

Residual stresses also can be relieved or reduced by plastically deforming the structure by a small amount. For instance, this technique can be used in welded pressure vessels by pressurizing the vessels internally (*proof-stressing*). In order to reduce the possibility of sudden fracture under high internal pressure, the weld must be made properly and must be free of notches and discontinuities, which could act as points of stress concentration.

In addition to being preheated for stress relieving, welds may be *heat treated* by various other techniques in order to modify other properties. These techniques include the annealing, normalizing, quenching, and tempering of steels and the solution treatment and aging of various alloys as described in Chapter 4.

30.9.2 Weldability

The *weldability* of a metal usually is defined as its capacity to be welded into a specific structure that has certain properties and characteristics and will satisfactorily meet service requirements. Weldability involves a large number of variables, hence generalizations are difficult. As noted previously, the material characteristics (such as alloying elements, impurities, inclusions, grain structure, and processing history) of both the base metal and the filler metal are important. For example, the weldability of steels decreases with increasing carbon content because of martensite formation (which is hard and brittle) and thus reduces the strength of the weld. Coated steel sheets present various challenges in welding, depending on the type and thickness of the coating.

Because of the effects of melting and solidification and of the consequent microstructural changes, a thorough knowledge of the phase diagram and the response of the metal or alloy to sustained elevated temperatures is essential. Also influencing weldability are mechanical and physical properties: strength, toughness, ductility, notch sensitivity, elastic modulus, specific heat, melting point, thermal expansion, surface-tension characteristics of the molten metal, and corrosion resistance.

Preparation of surfaces for welding is important, as are the nature and properties of surface-oxide films and of adsorbed gases. The particular welding process employed significantly affects the temperatures developed and their distribution in the weld zone. Other factors that affect weldability are shielding gases, fluxes, moisture content of the coatings on electrodes, welding speed, welding position, cooling rate, and level of preheating, as well as such post-welding techniques as stress relieving and heat treating.

Weldability of ferrous materials:

- *Plain-carbon steels:* Weldability is excellent for low-carbon steels, fair to good for medium-carbon steels, poor for high-carbon steels.
- *Low-alloy steels:* Weldability is similar to that of medium-carbon steels.
- *High-alloy steels:* Weldability generally is good under well-controlled conditions.
- *Stainless steels:* These generally are weldable by various processes.
- *Cast irons:* These generally are weldable, although their weldability varies greatly.

Weldability of nonferrous materials:

- *Aluminum alloys:* These are weldable at a high rate of heat input. An inert shielding gas and lack of moisture are important. Aluminum alloys containing zinc or copper generally are considered unweldable.
- *Copper alloys:* Depending on composition, these generally are weldable at a high rate of heat input. An inert shielding gas and lack of moisture are important.
- *Magnesium alloys:* These are weldable with the use of a protective shielding gas and fluxes.
- *Nickel alloys:* Weldability is similar to that of stainless steels. Lack of sulfur is important.
- *Titanium alloys:* These are weldable with the proper use of shielding gases.
- *Tantalum:* Weldability is similar to that of titanium.
- *Tungsten:* Weldable under well-controlled conditions.
- *Molybdenum:* Weldability is similar to that of tungsten.
- *Niobium (Columbium):* Weldability is good.

30.9.3 Testing of welds

As in all manufacturing processes, the quality of a welded joint is established by testing. Several standardized tests and test procedures have been established. They are available from many organizations, such as the American Society for Testing and Materials (ASTM), the American Welding Society (AWS), the American Society of Mechanical Engineers (ASME), the American Society of Civil Engineers (ASCE), and various federal agencies.

Welded joints may be tested either *destructively* or *nondestructively*. (See also Sections 36.10 and 36.11.) Each technique has certain capabilities and limitations, as well as sensitivity, reliability, and requirements for special equipment and operator skill.

Destructive testing techniques:

- **Tension test:** Longitudinal and transverse tension tests are performed on specimens removed from actual welded joints and from the weld-metal area. Stress–strain curves then are obtained by the procedures described in Section 2.2. These curves indicate the yield strength, Y, ultimate tensile strength, UTS, and ductility of the welded joint (elongation and reduction of area) in different locations and directions.

- **Tension-shear test:** The specimens in the tension-shear test (Fig. 30.26a and b) are prepared to simulate conditions to which actual welded joints are subjected. These specimens are subjected to tension, so the shear strength of the weld metal and the location of fracture can be determined.

- **Bend test:** Several bend tests have been developed to determine the ductility and strength of welded joints. In one common test, the welded specimen is bent around a fixture (*wrap-around* bend test; Fig. 30.26c). In another test, the specimens are tested in *three-point transverse bending* (Fig. 30.26d; see also Fig. 2.11a). These tests help to determine the relative ductility and strength of welded joints.

- **Fracture toughness test:** Fracture toughness tests commonly utilize the impact testing techniques described in Section 2.9. *Charpy V-notch* specimens first are prepared and then tested for toughness. Another toughness test is the *drop-weight test*, in which the energy is supplied by a falling weight.

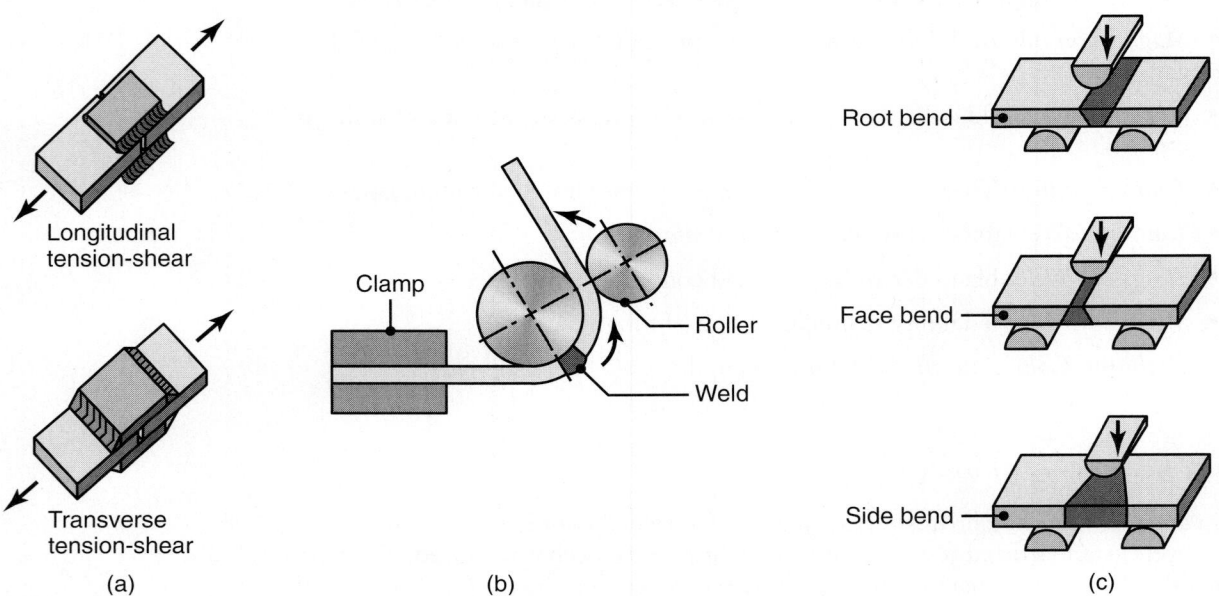

FIGURE 30.26   (a) Specimens for longitudinal tension-shear testing and for transfer tension-shear testing. (b) Wrap–around bend-test method. (c) Three-point transverse bending of welded specimens.

- **Corrosion and creep tests:** In addition to mechanical tests, welded joints also may be tested for their resistance to corrosion and creep. Because of the difference in the composition and microstructure of the materials in the weld zone, *preferential corrosion* may take place in the zone. Creep tests are important in determining the behavior of welded joints and structures when subjected to elevated temperatures.

Nondestructive testing techniques.   Welded structures often have to be tested nondestructively, particularly for critical applications where weld failure can be catastrophic, such as in pressure vessels, load-bearing structural members, and power plants. Nondestructive testing techniques for welded joints generally consist of the following methods. (These tests are described later in Section 36.10.)

- **Visual**
- **Radiographic (x-rays)**
- **Magnetic-particle**
- **Liquid-penetrant**
- **Ultrasonic**

Testing for hardness distribution in the weld zone also may be a useful indicator of weld strength and microstructural changes.

30.10 | Joint Design and Process Selection

In describing individual welding processes, we have given several examples of the types of welds and joints produced and their applications in numerous consumer and industrial products of various designs. Typical types of joints produced by welding and their terminology are shown in Fig. 30.27. Standardized symbols commonly used in engineering drawings to describe the types of welds are shown in Fig. 30.28. These symbols identify the type of weld, the groove design, the weld size and length, the welding process, the sequence of operations, and other necessary information.

The general design guidelines for welding are summarized next, with some examples given in Fig. 30.29. Various other types of joint design will be given in Chapters 31 and 32.

- Product design should minimize the number of welds because welding can be costly (unless automated).
- Weld location should be selected to avoid excessive stresses or stress concentrations in the welded structure and for appearance.
- Weld location should be selected so as not to interfere with any subsequent processing of the joined parts or with its intended use.
- Components should fit properly prior to welding. The method used to prepare edges (such as sawing, machining, or shearing) can affect weld quality.
- The need for edge preparation should be avoided or minimized.
- Weld-bead size should be kept to a minimum to conserve weld metal and for better appearance.

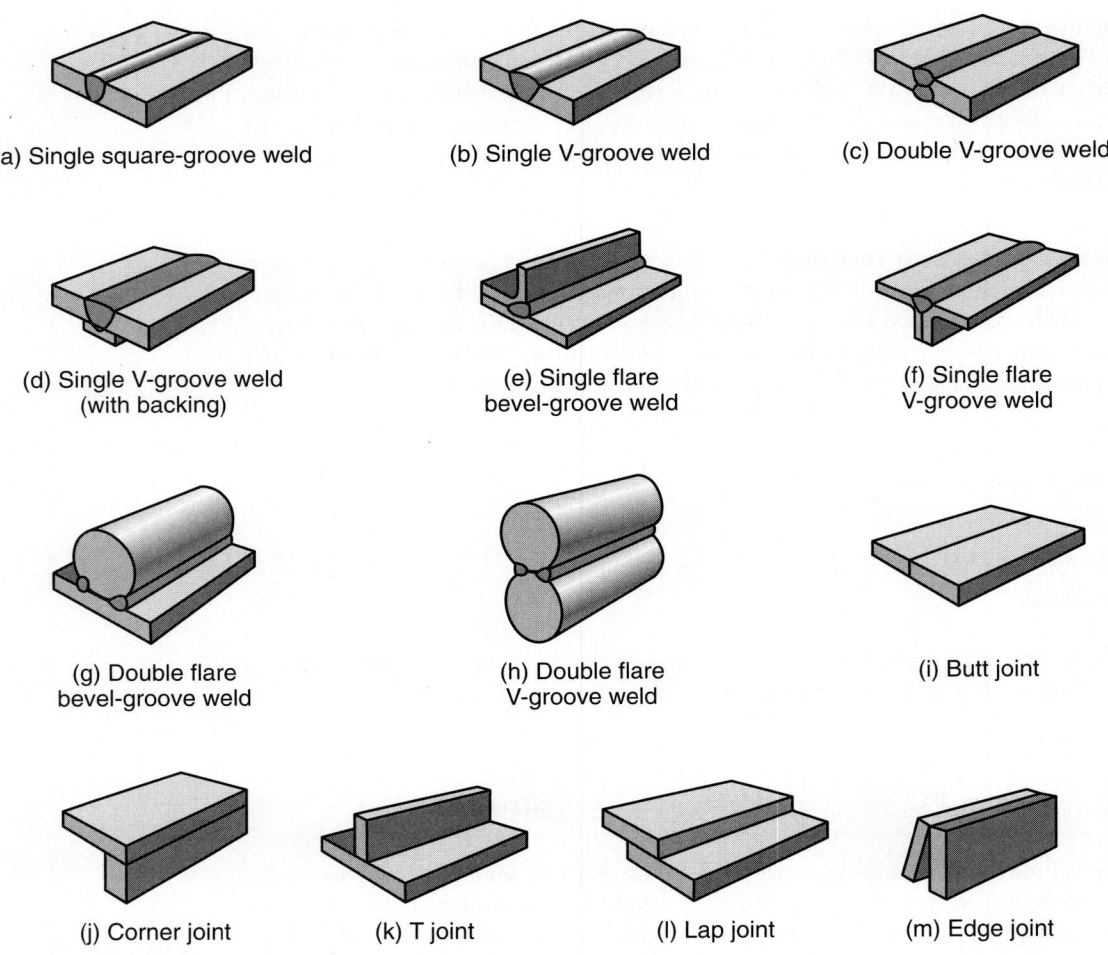

(a) Single square-groove weld   (b) Single V-groove weld   (c) Double V-groove weld

(d) Single V-groove weld   (e) Single flare   (f) Single flare
(with backing)   bevel-groove weld   V-groove weld

(g) Double flare   (h) Double flare   (i) Butt joint
bevel-groove weld   V-groove weld

(j) Corner joint   (k) T joint   (l) Lap joint   (m) Edge joint

FIGURE 30.27   Examples of welded joints and their terminology.

Process selection.   In addition to the process characteristics, capabilities, and material considerations described thus far in this chapter, the selection of a weld joint and an appropriate welding process involve the following considerations (see also Chapters 31 and 32).

- Configuration of the parts or structure to be joined, joint design, thickness and size of the components, and number of joints required.
- The methods used in manufacturing the components to be joined.
- Type of materials involved, which may be metallic or nonmetallic.
- Location, accessibility, and ease of joining.
- Application and service requirements, such a type of loading, stresses generated, and the environment.
- Effects of distortion, warping, discoloration of appearance and service.
- Costs involved in edge preparation, joining, and postprocessing (including machining, grinding, and finishing operations).
- Costs of equipment, materials, labor and skills required, and the joining operation.

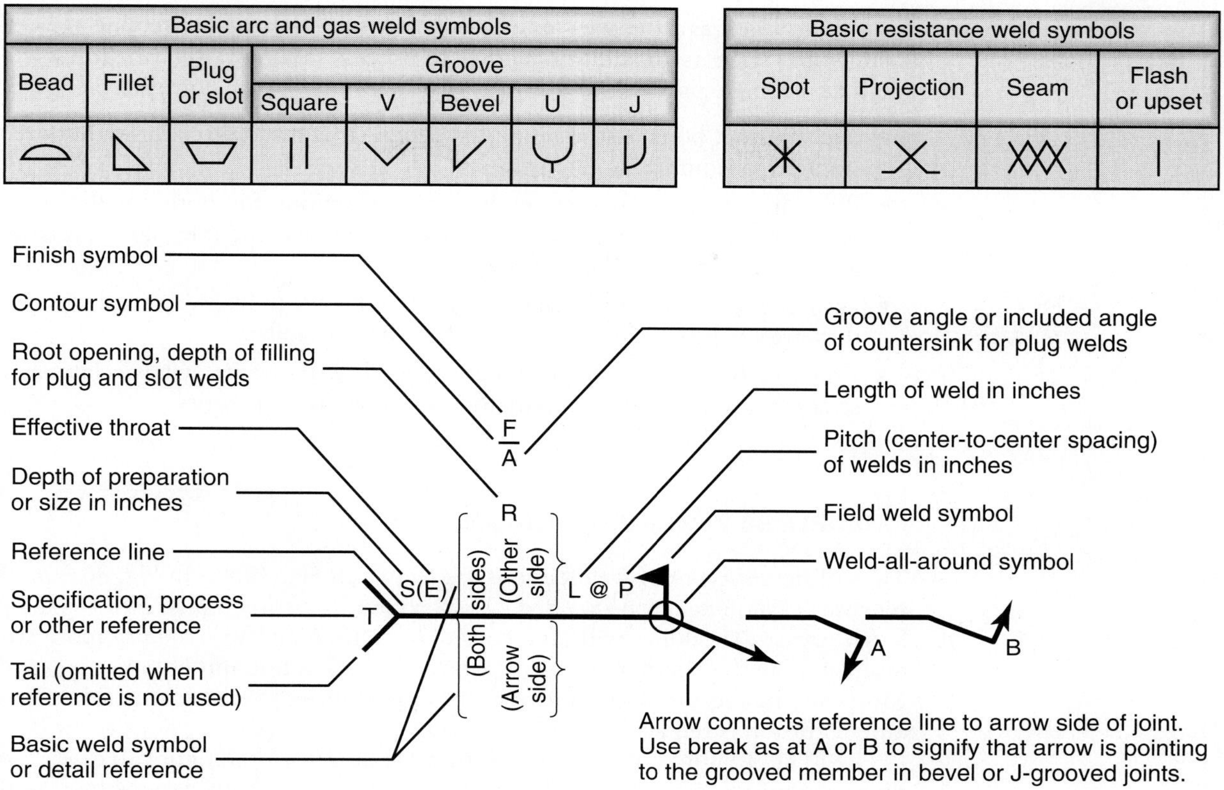

Basic arc and gas weld symbols							
Bead	Fillet	Plug or slot	Groove				
			Square	V	Bevel	U	J
⌒	◺	�⌣	||	∨	⩔	⩒	⫫

Basic resistance weld symbols			
Spot	Projection	Seam	Flash or upset
✳	✕	⋙	|

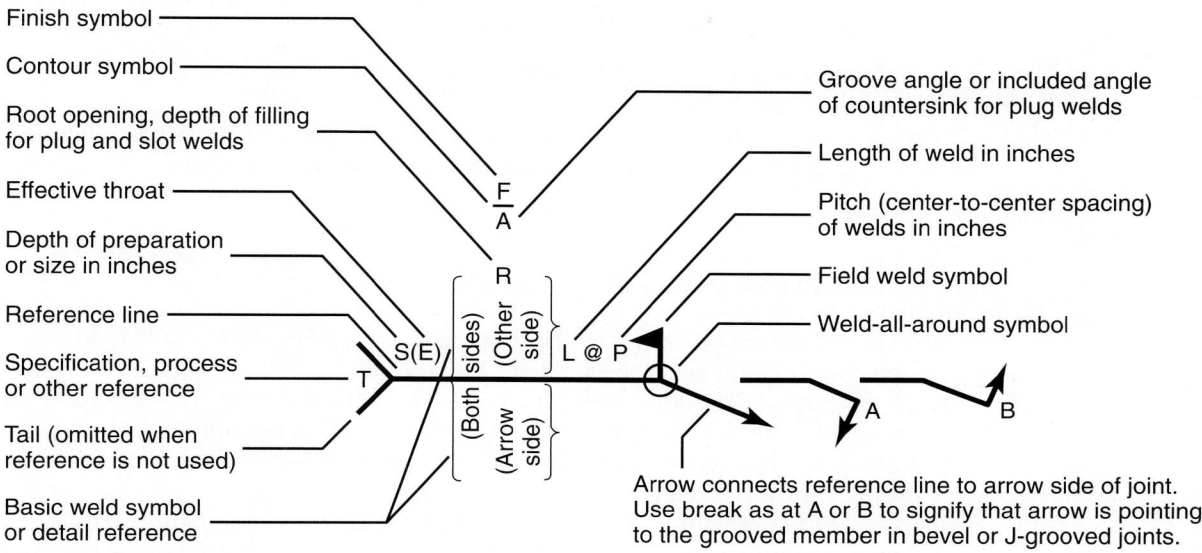

Finish symbol
Contour symbol
Root opening, depth of filling for plug and slot welds
Effective throat
Depth of preparation or size in inches
Reference line
Specification, process or other reference
Tail (omitted when reference is not used)
Basic weld symbol or detail reference

Groove angle or included angle of countersink for plug welds
Length of weld in inches
Pitch (center-to-center spacing) of welds in inches
Field weld symbol
Weld-all-around symbol

F
A
R
S(E)
T
(Other side)
(Arrow side)
(Both sides)
L @ P
A
B

Arrow connects reference line to arrow side of joint. Use break as at A or B to signify that arrow is pointing to the grooved member in bevel or J-grooved joints.

FIGURE 30.28   Standard identification and symbols for welds.

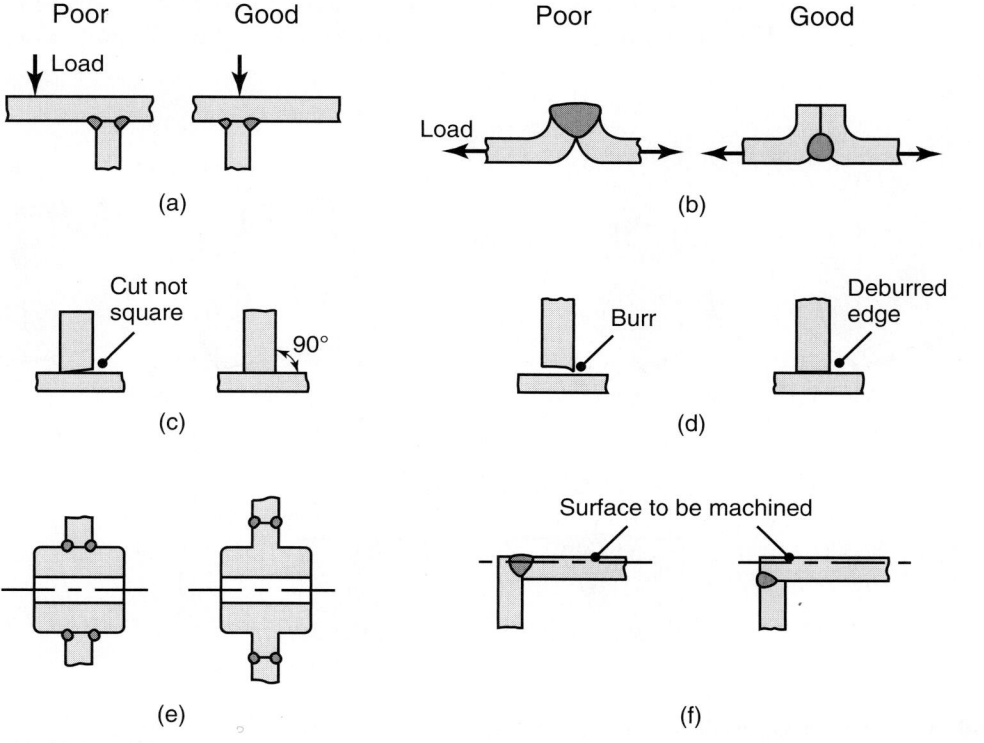

Poor   Good   Poor   Good

Load

(a)

Load

(b)

Cut not square

90°

(c)

Burr

Deburred edge

(d)

(e)

Surface to be machined

(f)

FIGURE 30.29   Some design guidelines for welds. *Source:* After J.G. Bralla.

Table VI.1 gave the various characteristics of individual welding processes, which would serve as an additional guide to process selection. Refering to this table, note that no single process has a high rating in all categories. For example:

- Arc welding, bolts, and riveting have high strength and reliability but are not suitable for joining small parts.
- Resistance welding has strength and applications for both small and large parts. It is not easy to inspect visually for reliability, and it has lower tolerances and reliability than other processes.
- Fasteners are useful for large parts and can be easy to inspect visually. However, they are costly and do not have much design variability.
- Adhesive bonding has high design variability. However, it has relatively low strength and is difficult to visually inspect for joint integrity.

EXAMPLE 30.2 Weld design selection

Three different types of weld designs are shown in Fig. 30.30. In Fig. 30.30a, the two vertical joints can be welded either externally or internally. Note that full-length external welding will take considerable time and will require more weld material than the alternative design, which consists of intermittent internal welds. Moreover, by the alternative method, the appearance of the structure is improved and distortion is reduced.

In Fig. 30.30b, it can be shown that the design on the right can carry three times the moment M of the one on the left. Note that both designs require the same amount of weld metal and welding time. In Fig. 30.30c, the weld on the left requires about twice the amount of weld material than does the design on the right. Also note that because more material must be machined, the design on the left will require more time for edge preparation, and more base metal will be wasted as a result.

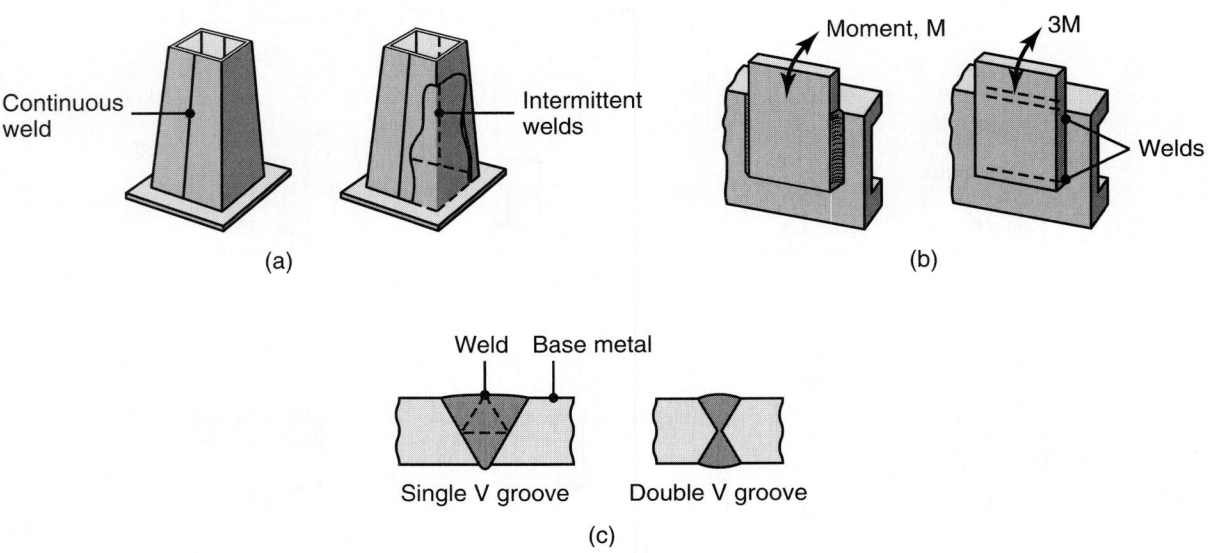

FIGURE 30.30   Examples of weld designs used in Example 30.2.

SUMMARY

- Oxyfuel-gas, arc, and high-energy-beam welding are among the most commonly used joining operations. Gas welding uses chemical energy; to supply the necessary heat, arc and high-energy-beam welding use electrical energy instead.

- In all of these processes, heat is used to bring the joint being welded to a liquid state. Shielding gases are used to protect the molten-weld pool and the weld area against oxidation. Filler rods may or may not be used in oxyfuel-gas and arc welding to fill the weld area.

- The selection of a welding process for a particular operation depends on the workpiece material, on its thickness and size, on its shape complexity, on the type of joint, on the strength required, and on the change in product appearance caused by welding.

- A variety of welding equipment is available—much of which is now robotics and computer controlled with programmable features.

- The cutting of metals also can be done by processes, of which the principles are based on oxyfuel-gas and arc welding. The cutting of steels occurs mainly through oxidation (burning). The highest temperatures for cutting are obtained by plasma-arc cutting.

- The metallurgy of the welded joint is an important aspect of all welding processes, because it determines the strength and toughness of the joint. The welded joint consists of solidified metal and a heat-affected zone; each has a wide variation in microstructure and properties, depending on the metals joined and on the filler metals.

- The metallurgy of the welded joint is an important aspect of all welding processes, because it determines the strength and toughness of the joint. The welded joint consists of solidified metal and a heat-affected zone; each has a wide variation in microstructure and properties, depending on the metals joined and on the filler metals.

- Discontinuities can develop in the weld zone (such as porosity, inclusions, incomplete welds, tears, surface damage, and cracks). Residual stresses and relieving them also are important considerations in welding.

- The weldability of metals and alloys depends greatly on their composition, the type of welding operation and process parameters employed, and on the control of welding parameters.

- General guidelines are available to help in the initial selection of suitable and economical welding methods for a particular application.

KEY TERMS

Arc cutting	Electrode	Gas tungsten-arc welding
Arc welding	Electrogas welding	Heat-affected zone
Atomic-hydrogen welding	Electron-beam welding	Inclusions
Base metal	Electroslag welding	Joining
Carburizing flame	Filler metal	Kerf
Coated electrode	Flux	Keyhole technique
Consumable electrode	Flux-cored arc welding	Laser-beam welding
Discontinuities	Fusion welding	Neutral flame
Drag lines	Gas metal-arc welding	Nonconsumable electrode

Oxidizing flame

Reducing flame

Tears

Oxyfuel-gas cutting

Residual stresses

Weld profile

Oxyfuel-gas welding

Shielded metal-arc welding

Weld metal

Plasma-arc welding

Slag

Weldability

Polarity

Stick welding

Welding gun

Porosity

Submerged-arc welding

BIBLIOGRAPHY

ASM Handbook, Vol. 6: *Welding, Brazing, and Soldering*, ASM International, 1993.

Bowditch, W.A., and Bowditch, K.E., *Welding Technology Fundamentals*, Goodheart-Willcox, 1997.

Cary, H.B., *Modern Welding Technology*, 4th ed., Prentice Hall, 1997.

Croft, D., *Heat Treatment of Welded Structures*, Woodhead Publishing, 1996

Davies, A.C., *The Science and Practice of Welding*, 10th ed., 2 vols., Cambridge University Press, 1993.

Evans, G.M., and Bailey, N., *Metallurgy of Basic Weld Metal*, Woodhead Publishing, 1997.

Galyen, J., Sear, G., and Tuttle, C., *Welding: Fundamentals and Procedures*, Wiley, 1984.

Granjon, H., *Fundamentals of Welding Metallurgy*, Woodhead Publishing, 1991.

Hicks, J.G., *Welded Joint Design*, 2nd ed., Abington, 1997.

Houldcroft, P.T., *Welding and Cutting: A Guide to Fusion Welding and Associated Cutting Processes*, Industrial Press, 1988.

Introduction to the Nondestructive Testing of Welded Joints, 2nd ed., American Society of Mechanical Engineers, 1996.

Jeffus, L.F., *Welding: Principles and Applications*, 4th ed., Delmar Publishers, 1997.

Jellison, R., *Welding Fundamentals*, Prentice Hall, 1995.

Krou, S., *Welding Metallurgy*, Wiley, 1987.

Lancaster, J.F., *The Metallurgy of Welding*, Chapman & Hall, 1993.

Linnert, J.E., *Welding Metallurgy*, 4th ed., Vol. 1, American Welding Society, 1994.

Messler, R.W., Jr., *Joining of Advanced Materials*, Butterworth-Heinemann, 1993.

————, *Flux Cored Arc Welding Handbook*, 1998.

Minnick, W.H., *Gas Metal Arc Welding Handbook*, Goodheart-Willcox, 2000.

Mouser, J.D., *Welding Codes, Standards, and Specifications*, McGraw-Hill, 1997.

North, T.H., *Advanced Joining Technologies*, Chapman & Hall, 1990.

Powell, J., CO_2 *Laser Cutting*, Springer, 1992.

Schultz, H., *Electron Beam Welding*, Woodhead Publishing, 1994.

Steen, W.M., *Laser Material Processing*, 2nd ed., Springer, 1998.

Stout, R.D., *Weldability of Steels*, Welding Research Council, 1987.

Tool and Manufacturing Engineers Handbook, Vol. 4: *Quality Control and Assembly*, Society of Manufacturing Engineers, 1986.

Welding Handbook, 8th ed., 3 vols., American Welding Society, 1987.

Welding Inspection, American Welding Society, 1980.

Weld Integrity and Performance, ASM International, 1997.

REVIEW QUESTIONS

30.1 Describe fusion as it relates to welding operations.

30.2 Explain the features of neutralizing, reducing, and oxidizing flames. Why is a reducing flame so called?

30.3 Explain the basic principles of arc-welding processes.

30.4 Why is shielded metal-arc welding a commonly used process? Why is it also called stick welding?

30.5 Describe the functions and characteristics of electrodes. What functions do coatings have? How are electrodes classified?

30.6 What are the similarities and differences between consumable and nonconsumable electrodes?

30.7 Explain how cutting takes place when an oxyfuel-gas torch is used. How is underwater cutting done?

30.8 What is the purpose of flux? Why is it not needed in gas tungsten-arc welding?

30.9 What is meant by weld quality? Discuss the factors that influence it.

30.10 Explain why some joints may have to be preheated prior to welding.

30.11   How is weldability defined?

30.12   Describe the common types of discontinuities in welded joints.

30.13   What is meant by point design?

30.14   What types of destructive tests are performed on welded joints?

QUALITATIVE PROBLEMS

30.15   Explain the reasons why so many different welding processes have been developed.

30.16   What is the effect of the thermal conductivity of the workpiece on kerf width in oxyfuel-gas cutting?

30.17   Describe the differences between oxyfuel-gas cutting of ferrous and of nonferrous alloys. Which properties are significant?

30.18   Could you use oxyfuel-gas cutting for a stack of sheet metals? (*Note:* For stack cutting, see Fig. 24.25e.) Explain.

30.19   What are the advantages of electron-beam and laser-beam welding when compared to arc welding?

30.20   Discuss the need for and role of fixtures for holding workpieces in the welding operations described in this chapter.

30.21   Describe the common types of discontinuities in welds and explain the methods by which they can be avoided.

30.22   Explain the significance of the stiffness of the components being welded on both weld quality and part shape.

30.23   How would you go about detecting underbead cracks in a weld?

30.24   Could plasma-arc cutting be used for non-metallic materials? If so, would you select a transferred or a nontransferred type of arc? Explain.

30.25   What factors influence the size of the two weld beads shown in Fig. 30.14?

30.26   Which of the processes described in this chapter are not portable? Can they be made so? Explain.

30.27   Describe your observations concerning the contents of Table 30.1.

30.28   What determines whether a certain welding process can be used for workpieces in horizontal, vertical, or upside-down positions (or any position)? Explain your answer and give examples of appropriate applications.

30.29   Explain the factors involved in electrode selection in arc-welding processes.

30.30   In Table 30.1, there is a column on the distortion of welded components that is ordered from lowest distortion to highest. Explain why the degree of distortion varies among different welding processes.

30.31   Explain the significance of residual stresses in welded structures.

30.32   Comment on your observations regarding the shape of the weld beads shown in Fig. 30.5. Which ones would you recommend for thin sheet metals?

30.33   Why is oxyacetylene welding limited to rather thin sections?

30.34   Rank the processes described in this chapter in terms of (a) cost and (b) weld quality.

30.35   What are the sources of weld spatter? How can spatter be controlled?

30.36   Must the filler metal be made of the same composition as the base metal to be welded? Explain your response.

30.37   Describe your observations concerning Fig. 30.18

30.38   In Fig. 30.24b, assume that most of the top portion of the top piece is cut horizontally with a sharp saw. The residual stresses now will be disturbed and the part will undergo shape change, as was described in Section 2.11. For this case, how do you think the part will distort: curved downward or upward? Explain your response. (See also Fig. 2.29d.)

30.39   Describe the reasons that fatigue failures generally occur in the heat-affected zones of welds instead of through the weld bead itself.

30.40   If the materials to be welded are preheated, is the likelihood for porosity increased or decreased? Explain.

30.41   Make a list of welding processes that are suitable for producing (a) butt joints (where the weld is in the form of a line or line segment), (b) spot welds, and (c) both butt joints and spot welds.

QUANTITATIVE PROBLEMS

30.42   A welding operation takes place on an aluminum-alloy plate. A pipe 50-mm in diameter with a 4-mm wall thickness and a 60-mm length is butt-welded onto a section of 15 × 15 × 5 mm angle iron. The angle iron is of an L-shape and has a length of 0.3 m. If the weld zone in a gas tungsten-arc welding process is approximately 8 mm wide, what would be the temperature increase of the entire structure due to the heat input from welding only? What if the process were an electron-beam welding operation with a bead width of 6 mm? Assume that the electrode requires 1500 J and the aluminum alloy requires 1200 J to melt one gram.

30.43   A welding operation will take place on carbon steel. The desired welding speed is around 0.7 in./s. If an arc-welding power supply is used with a voltage of 10 V, what current is needed if the weld width is to be 0.2 in.?

30.44   In oxyacetylene, arc, and laser-beam cutting, the processes basically involve melting the workpiece. If an 80-mm-diameter hole is to be cut from a 250-mm-diameter and 12-mm-thick plate, plot the mean temperature rise in the blank as a function of kerf. Assume that one-half of the energy goes into the blank.

30.45   Plot the hardness in Fig. 30.18d as a function of the distance from the top surface and discuss your observations.

SYNTHESIS, DESIGN, AND PROJECTS

30.46   Comment on workpiece size and shape limitations (if any) for each of the processes described in this chapter.

30.47   Review the types of welded joints shown in Fig. 30.27 and give an application for each.

30.48   Comment on the design guidelines given in this chapter.

30.49   Make a summary table outlining the principles of the processes described in this chapter, together with examples of their applications.

30.50   Prepare a table of the processes described in this chapter and give the range of welding speeds as a function of workpiece material and thicknesses.

30.51   Assume that you are asked to inspect a welded structure for a critical application. Describe the procedure that you would follow.

30.52   Explain the factors that contribute to any differences in properties across a welded joint.

30.53   Explain why preheating the components to be welded is effective in reducing the likelihood of developing cracks.

30.54   Review the poor and good joint designs shown in Fig. 30.29 and explain why they are labeled so.

30.55   In building large ships, there is a need to weld large sections of steel together to form a hull. For this application, consider each of the welding operations discussed in this chapter and list the benefits and drawbacks of that particular operation for this application. Which welding process would you select? Why?

30.56   Perform a literature search and describe the relative advantages and limitations of CO_2 and Nd: YAG lasers.

30.57   Inspect various parts and components in an automobile and explain if any of the processes described in this chapter has been used in joining them.

30.58   Similar to Problem 30.57 but for kitchen utensils and appliances. Are there any major differences between these two types of product lines? Explain.

30.59   Make an outline of the general guidelines for safety in welding operations. For each of the operations described in this chapter, prepare a poster which effectively and concisely gives specific instructions for safe practices in welding (or cutting). Review the various publications of the National Safety Council and other similar organizations.

30.60   Are there common factors affecting the weldability, castability, formability, and machinability of metals? Explain with appropriate examples.

30.61   If you find a flaw in a welded joint during inspection, how would you go about determining whether the flaw is important?

30.62   Lattice booms for cranes are constructed from extruded cross-sections that are welded together. Any warpage which causes such a boom to deviate from straightness severely reduces its lifting capacity. Perform a literature search on the approaches used to minimize distortion due to welding and to correct it, specifically in the construction of lattice booms.

30.63 A common practice in repairing expensive broken or worn parts (such as may occur when a fragment is broken from a forging) is to fill the area with layers of weld beads and then to machine the part back to its original dimensions. Make a list of the precautions that you would suggest to someone who uses this approach.

30.64 A welded frame needs first to be disassembled and then to be repaired (by rewelding the members). What procedures would you recommend for disassembly of the frame and in preparation for rewelding?

30.65 Assume that you are asked to give a quiz to students on the contents of this chapter. Prepare three qualitative questions and supply the answers.

Solid-State Welding Processes

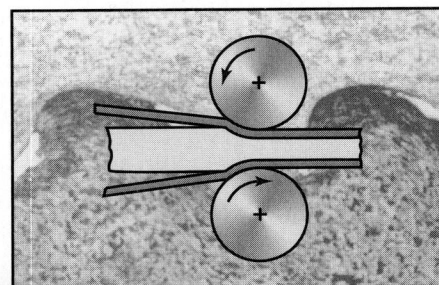

Solid-state welding processes comprise a family of important joining processes. This chapter describes:

• The principles of joining processes where no external heat is applied.

• Welding processes where heat is involved but is generated internally or else is insufficient to cause a phase change in the workpiece, and where no filler is used.

• Advantages, limitations, and applications of these processes.

• Unique applications of diffusion bonding combined with superplastic forming.

• Economic considerations in the selection of welding processes.

31.1 | Introduction

This chapter describes **solid-state welding** processes in which joining takes place without fusion at the interface of the two parts to be welded. Unlike in the fusion-welding processes described in Chapter 30, no liquid or molten phase is present in the joint. The principle of solid-state welding is demonstrated best with the following example. If two clean surfaces are brought into atomic contact with each other under sufficient pressure, they form bonds and produce a joint. To form a strong bond, it is essential that the interface be free of oxide films, residues, metalworking fluids, other contaminants, and even adsorbed layers of gas.

Solid-state bonding involves one or more of the following phenomena:

• **Diffusion:** The transfer of atoms across an interface; thus, applying external heat improves the strength of the bond between the two surfaces being joined, as occurs in *diffusion bonding*. Heat may be generated internally by friction (as utilized in *friction welding*), through electrical-resistance heating (as in *resistance welding* processes, such as *spot welding*), and externally by induction heating (as in *butt welding* tubes).

• **Pressure:** The higher the pressure, the stronger is the interface (as in *roll bonding* and *explosion welding*), where plastic deformation also occurs at the interface. Pressure and resistance heating may be combined, as in *flash welding*, *stud welding*, and *resistance projection welding*.

- **Relative interfacial movements:** When movements of the contacting surfaces (*faying surfaces*) occur (as in *ultrasonic welding*), even very small amplitudes will disturb the mating surfaces, break up any oxide films, and generate new, clean surfaces—thus improving the strength of the bond.

Most of the joining processes outlined here now are automated by using *robotics, vision systems, sensors*, and *adaptive* and *computer controls* for cost reduction, consistency, reliability of weld quality, and higher productivity. The costs involved in the joining process are outlined in Section 31.8.

31.2 | Cold Welding and Roll Bonding

In *cold welding* (CW), pressure is applied to the workpieces through dies or rolls. Because of the *plastic deformation* involved, it is necessary that at least one (but preferably both) of the mating parts be ductile. Prior to welding, the interface is degreased, wire-brushed, and wiped to remove oxide smudges. Cold welding can be used to join small workpieces made of soft, ductile metals. Applications include wire stock and electrical connections.

During the joining of two *dissimilar* metals that are mutually soluble, brittle *intermetallic compounds* may form (Section 4.2.2); these will produce a weak and brittle joint. An example occurs in the bonding of aluminum and steel, where a brittle intermetallic compound is formed at the interface. The best bond strength is obtained with two similar materials.

Roll bonding.   The pressure required for welding can be applied through a pair of rolls (Fig. 31.1); this process is called *roll bonding* or *roll welding* (ROW). Developed in the 1960s, roll bonding is used for manufacturing some U.S. coins (see Example 31.1). This process can be carried out at elevated temperatures (*hot roll bonding*). Surface preparation is important for interfacial strength.

Typical examples are the *cladding* of (a) pure aluminum over precipitation-hardened aluminum-alloy sheet (Alclad) and (b) stainless steel over mild steel (for corrosion resistance). A common application of roll bonding is the production of bimetallic strips for thermostats and similar controls using two layers of materials with different thermal-expansion coefficients. Bonding in only selected regions in the interface can be achieved by depositing a parting agent, such as graphite or ceramic, called *stop-off* (see Section 31.7).

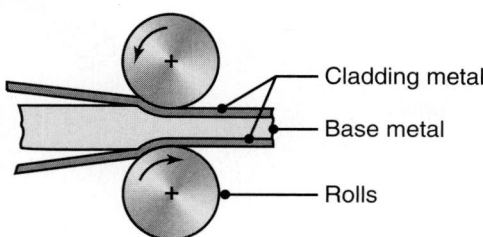

FIGURE 31.1   Schematic illustration of the roll bonding or cladding process.

EXAMPLE 31.1 Roll bonding of the U.S. quarter

The technique used for manufacturing composite U.S. quarters is the roll bonding of two outer layers of 75% Cu-25% Ni (cupronickel), where each is 1.2 mm (0.048 in.) thick with an inner layer of pure copper 5.1 mm (0.20 in.) thick. To obtain good bond strength, the faying surfaces are cleaned chemically and wire-brushed. First, the strips are rolled to a thickness of 2.29 mm (0.090 in.); a second rolling operation reduces the thickness to 1.36 mm (0.0535 in.). The strips thus undergo a total reduction in thickness of 82%.

Because volume constancy is maintained in plastic deformation, there is a major increase in the surface area between the layers, and it causes the generation of clean interfacial surfaces. This extension in surface area under the high pressure of the rolls combined with the solid solubility of nickel in copper (see Section 4.2.1) produces a strong bond.

31.3 | Ultrasonic Welding

In *ultrasonic welding* (USW), the faying surfaces of the two components are subjected to a static normal force and oscillating shearing (tangential) stresses. The shearing stresses are applied by the tip of a **transducer** (Fig. 31.2a), which is similar to that used for ultrasonic machining. (See Fig. 26.24a.) The frequency of oscillation is generally in the range of 10 to 75 kHz, although an even lower or higher frequency can be employed. Proper coupling between the transducer and the tip (called—by analogy with *electrode*—a *sonotrode*, from the word *"sonic"*) is important for efficient operation.

The shearing stresses cause plastic deformation at the interface of the two components, breaking up oxide films and contaminants, thus allowing good contact and producing a strong solid-state bond. The temperature generated in the weld zone is usually in the range of one-third to one-half of the melting point (absolute scale) of the metals joined. Consequently, neither melting nor fusion takes place.

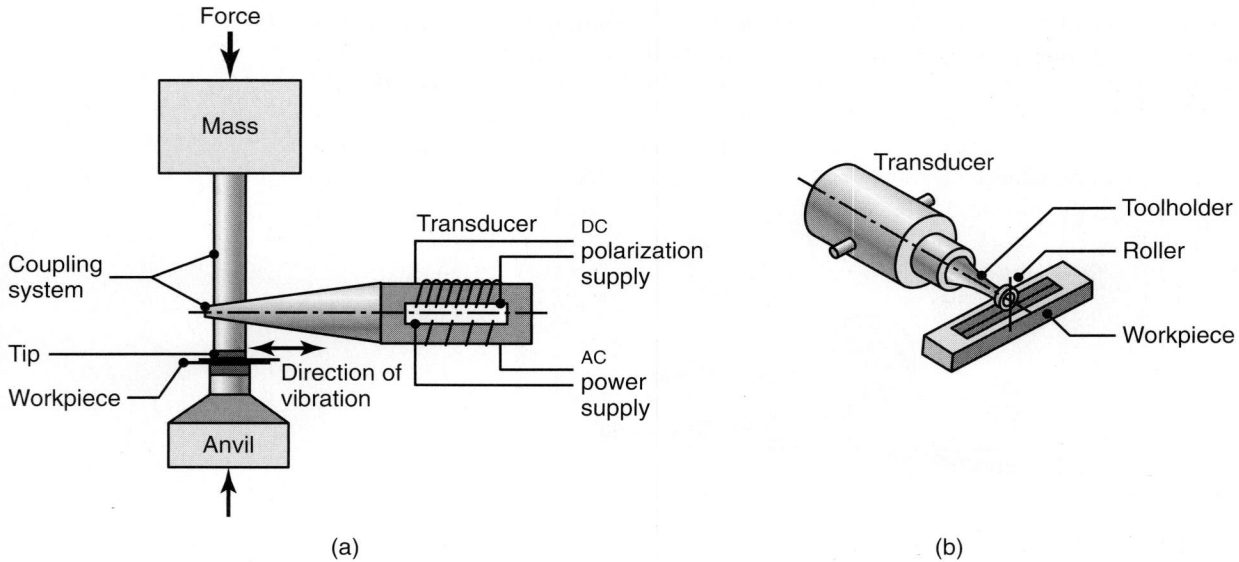

FIGURE 31.2   (a) Components of an ultrasonic-welding machine for making lap welds. The lateral vibrations of the tool tip cause plastic deformation and bonding at the interface of the workpieces. (b) Ultrasonic seam welding using a roller as the *sonotrode*.

In certain situations, however, the temperature generated can be sufficiently high to cause metallurgical changes in the weld zone. The mechanism responsible for the joining of *thermoplastics* by ultrasonic welding is different from that for metals, and melting does take place at the interface, because plastics have much lower melting temperatures. (See Table 7.2.)

The ultrasonic-welding process is versatile and reliable. It can be used with a wide variety of metallic and nonmetallic materials, including dissimilar metals (*bimetallic strips*). It is used extensively for the joining of plastics, for packaging with foils, and (in the automotive and consumer electronics industries) for the lap welding of sheet, foil, and thin wire. The welding tip can be replaced with *rotating disks* (Fig. 31.2b) for the seam welding of structures in which one component is sheet, foil, or polymer-woven material (a process similar *to resistance seam welding*, Section 31.5.2). Moderate skill is required to operate the equipment.

31.4 | Friction Welding

In the joining processes described thus far, the energy required for welding (typically chemical, electrical, or ultrasonic) is supplied from external sources. In *friction welding* (FRW), the heat required for welding is generated through (as the name implies) friction at the interface of the two components being joined. You can demonstrate the significant rise in temperature caused by friction by rubbing your hands together or by sliding down a rope rapidly.

In friction welding, developed in the 1940s, one of the workpiece components remains stationary while the other is placed in a chuck or collet and rotated at a high constant speed. The two members to be joined then are brought into contact under an axial force (Fig. 31.3). After sufficient contact is established, the rotating member is brought to a quick stop (so that the weld is not destroyed by shearing) while the axial force is increased. Oxides and other contaminants at the interface are removed by the radially outward movement of the hot metal at the interface.

The rotating member must be clamped securely to the chuck or collet to resist both torque and axial forces without slipping. The pressure at the interface and the resulting friction produce sufficient heat for a strong joint to form.

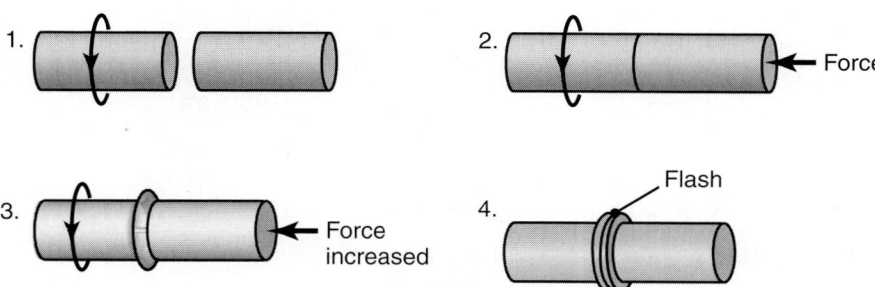

FIGURE 31.3   Sequence of operations in the friction-welding process: 1. Left-hand component is rotated at high speed. 2. Right-hand component is brought into contact under an axial force. 3. Axial force is increased; flash begins to form. 4. Left-hand component stops rotating; weld is completed. The flash subsequently can be removed by machining or grinding.

The weld zone usually is confined to a narrow region; its size depends on the following parameters:

- Amount of heat generated.
- Thermal conductivity of the materials.
- Mechanical properties of the materials at elevated temperature.

The shape of the welded joint depends on the rotational speed and on the axial pressure applied (Fig. 31.4). These factors must be controlled to obtain a uniform, strong joint. The radially outward movement of the hot metal at the interface pushes oxides and other contaminants out of the interface.

Friction welding can be used to join a wide variety of materials, provided that one of the components has some rotational symmetry. Solid or tubular parts can be joined by this method with good joint strength. Solid steel bars up to 100 mm (4 in.) in diameter and pipes up to 250 mm (10 in.) in outside diameter have been welded successfully by this process.

The surface speed of the rotating member may be as high as 15 m/s (3000 ft/min). Because of the combined heat and pressure, the interface in FRW develops a flash by plastic deformation (upsetting) of the heated zone. This flash (if objectionable) easily can be removed by machining or grinding. Friction-welding machines are automated fully, and the operator skill required is minimal—once individual cycle times for the complete operation are set properly.

Inertia friction welding.   This process is a modification of FRW, although the term has been used interchangeably with friction welding. The energy required for frictional heating in inertia friction welding is supplied by the kinetic energy of a flywheel. The flywheel is accelerated to the proper speed, the two members are brought into contact, and an axial force is applied. As friction at the interface slows the flywheel, the axial force is increased. The weld is completed when the flywheel has come to a stop. The timing of this sequence is important for good weld quality.

The rotating mass in inertia-friction-welding machines can be adjusted for applications requiring different levels of energy (these levels depend on workpiece size and properties). In one application of inertia friction welding, 10-mm- (0.4-in.)-diameter shafts are welded to automotive turbocharger impellers at a rate of one joint every 15 seconds.

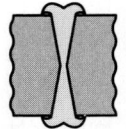   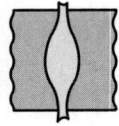   

(a) High pressure   (b) Low pressure   (c) Optimum
or low speed   or high speed

FIGURE 31.4   Shape of the fusion zones in friction welding as a function of the axial force applied and the rotational speed.

Linear friction welding.   In a further development of friction welding, the interface of the two components to be joined is subjected to a *linear* reciprocating motion, as opposed to a rotary motion. In *linear friction welding*, the components do not have to be circular or tubular in their cross-section. The process is capable of welding square or rectangular components (as well as round parts) made of metals or plastics. In this process, one part is moved across the face of the other part using a balanced reciprocating mechanism.

In one application, a rectangular titanium-alloy part was friction welded at a linear frequency of 25 Hz with an amplitude of ± 2 mm (0.08 in.) under a pressure of 100 MPa (15,000 psi) acting on a 240 mm^2 (0.38 in^2) interface. Various other metal parts have been welded successfully, with rectangular cross-sections as large as 50 mm $\times$ 20 mm (2 in. $\times$ 0.8 in.).

Friction stir welding.   In conventional friction welding, heating interfaces is achieved through friction by rubbing two contacting surfaces. In the friction stir welding (FSW) process, developed in 1991, a third body is rubbed against the two surfaces to be joined. A rotating nonconsumable probe (typically 5 to 6 mm in diameter and 5 mm high) is plunged into the joint (Fig. 31.5). The contact pressures cause frictional heating, raising the temperature to the range of 230° to 260°C (450° to 500°F). The probe at the tip of the rotating tool forces heating and mixing (or stirring) of the material in the joint.

Materials such as aluminum, copper, steel, and titanium have been welded successfully, and developments are taking place to extend FSW applications to polymers and composite materials. This process now is being applied to aerospace, automotive, shipbuilding, and military vehicles, using sheet or plates. With developments in rotating-tool design, other possible applications include inducing microstructural changes, grain refinement in materials, and localized toughness improvement in castings.

The welding equipment can be a conventional, vertical-spindle milling machine (Fig. 24.15b), and the process is relatively easy to implement. The thickness of the welded material can be as little as 1 mm and as much as 50 mm (2 in.) welded in a single pass. FSW welds have high quality, minimal pores, and uniform material structure. The welds are produced with low heat input and, therefore, low

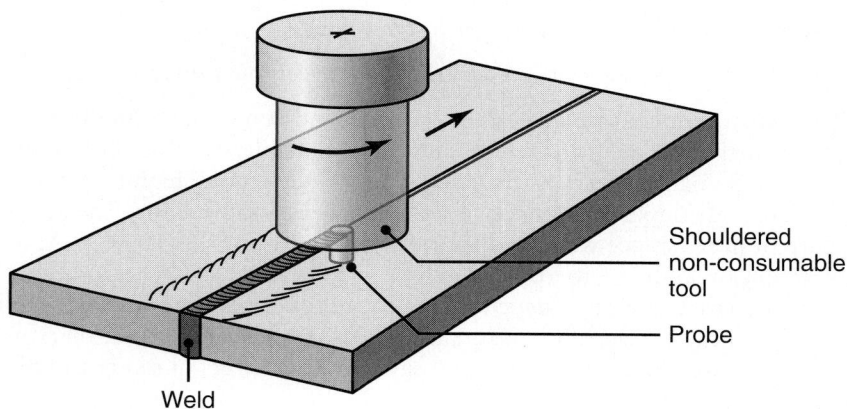

Shouldered
non-consumable
tool

Probe

Weld

FIGURE 31.5   The principle of the friction stir-welding process. Aluminum-alloy plates up to 75 mm (3 in.) thick have been welded by this process.

distortion and little microstructural changes. No shielding gases or surface cleaning are required.

31.5 | Resistance Welding

The category of *resistance welding* (RW) covers a number of processes in which the heat required for welding is produced by means of **electrical resistance** across the two components to be joined. These processes have major advantages, such as not requiring consumable electrodes, shielding gases, or flux.

The heat generated in resistance welding is given by the general expression

$$H = I^2 R t \tag{31.1}$$

where

H = Heat generated in joules (watt-seconds)
I = Current (in amperes)
R = Resistance (in ohms)
t = Time of current flow (in seconds)

Equation (31.1) often is modified so that it represents the actual heat energy available in the weld by including a factor K, which represents the energy losses through conduction and radiation. The equation then becomes

$$H = I^2 R t K \tag{31.2}$$

where the value of K is less than unity.

Referring to the resistance spot-welding operation shown in Fig. 31.6, the *total resistance* is the sum of the following properties:

- Resistances of the electrodes
- Electrode–workpiece contact resistance
- Resistances of the individual parts to be welded
- Workpiece–workpiece contact resistance (between the *faying surfaces*)

The actual temperature rise at the joint depends on the specific heat and the thermal conductivity of the metals to be joined. For example, metals such as aluminum and copper have high thermal conductivity, and so they require high heat concentrations. Similar or dissimilar metals can be joined by resistance welding. The magnitude of the current in resistance welding operations may be as high as 100,000 A, but the voltage is typically only 0.5 to 10 V.

The strength of the bond depends on surface roughness and on the cleanliness of the mating surfaces. Oil films, paint, and thick oxide layers should, therefore, be removed before welding. The presence of uniform, thin layers of oxide and of other contaminants is not as critical.

Developed in the early 1900s, resistance welding processes require specialized machinery. Much of it is now operated by programmable computer control. Generally, the machinery is not portable, and the process is suitable primarily for

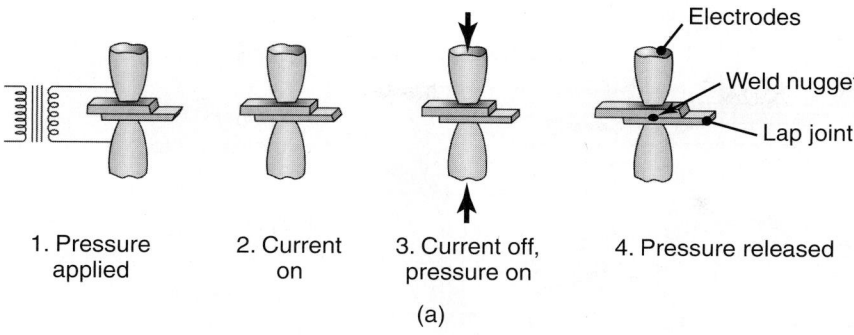

1. Pressure   2. Current   3. Current off,   4. Pressure released
   applied   on   pressure on

(a)

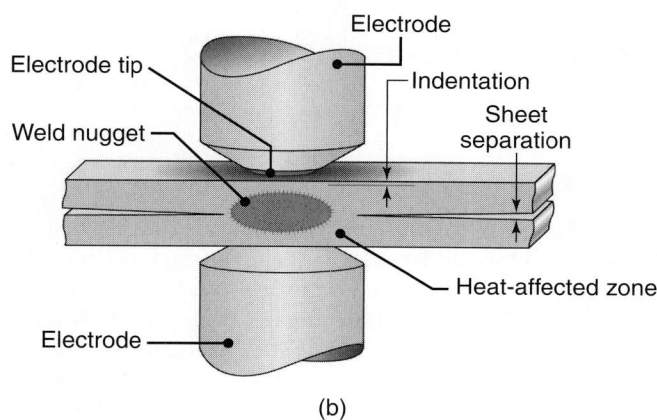

(b)

FIGURE 31.6   (a) Sequence of events in resistance spot welding. (b) Cross-section of a spot weld showing the weld nugget and the indentation of the electrode on the sheet surfaces. This is one of the most commonly used processes in sheet-metal fabrication and in automotive-body assembly.

use in manufacturing plants and machine shops. The operator skill required is minimal, particularly with modern machinery.

31.5.1 Resistance spot welding

In *resistance spot welding* (RSW), the tips of two opposing solid, cylindrical electrodes touch a lap joint of two sheet metals, and resistance heating produces a spot weld (Fig. 31.6a). In order to obtain a strong bond in the **weld nugget**, pressure is applied until the current is turned off and the weld has solidified. Accurate control and timing of the electric current and of the pressure are essential in resistance welding. In the automotive industry, for example, the number of cycles range up to about 30 at a frequency of 60 Hz. (See also *high-frequency resistance welding* in Section 31.5.3.)

The weld nugget (Fig. 31.6b) is generally 6 to 10 mm (0.25 to 0375 in.) in diameter. The surface of the spot weld has a slightly discolored indentation. Currents range from 3000 to 40,000 A. The current level depends on the materials being welded and on their thicknesses. For example, the current is typically 10,000 A for steels and 13,000 A for aluminum. Electrodes generally are made of copper alloys and must have sufficient electrical conductivity and hot strength to maintain their shape.

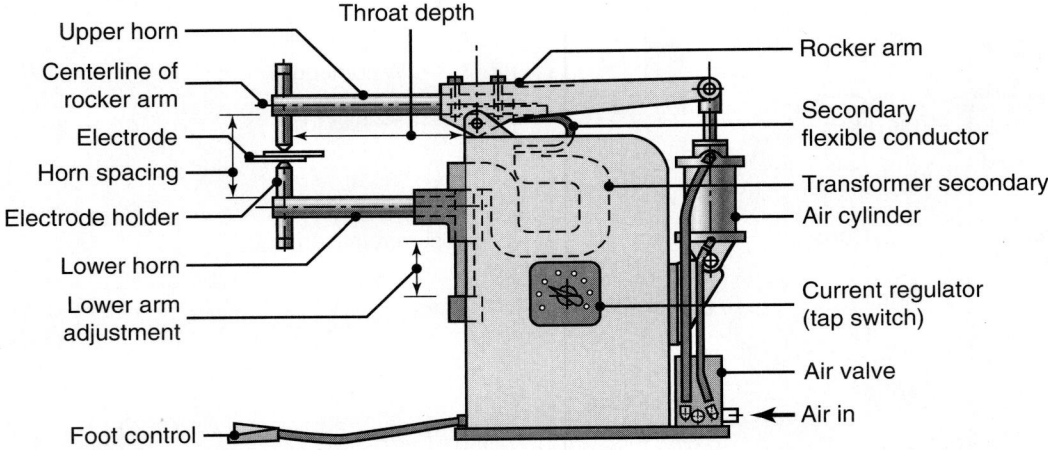

FIGURE 31.7 Schematic illustration of an air-operated rocker-arm spot-welding machine.

Spot welding is the simplest and most commonly used resistance-welding process. Welding may be performed by means of single (most common) or multiple pairs of electrodes (as many as a hundred or more), and the required pressure is supplied through mechanical or pneumatic means (Fig. 31.7). **Rocker-arm type** spot-welding machines normally are used for smaller parts; **press type** machines are used for larger workpieces. The shape and surface condition of the electrode tip and the accessibility of the site are important factors in spot welding. A variety of electrode shapes are used to spot-weld areas that are difficult to reach (Fig. 31.8).

Spot welding is used widely for fabricating sheet-metal parts. Examples range from attaching handles to stainless-steel cookware (Fig. 31.9a) to the spot welding of mufflers (Fig. 31.9b) and to large sheet-metal structures. Modern spot-welding equipment is computer controlled for optimum timing of current and pressure; its spot-welding guns are manipulated by programmable robots (Fig. 31.9c). Automobile bodies can have as many as 10,000 spot welds; they are welded at high rates using multiple electrodes (see Fig. I.13).

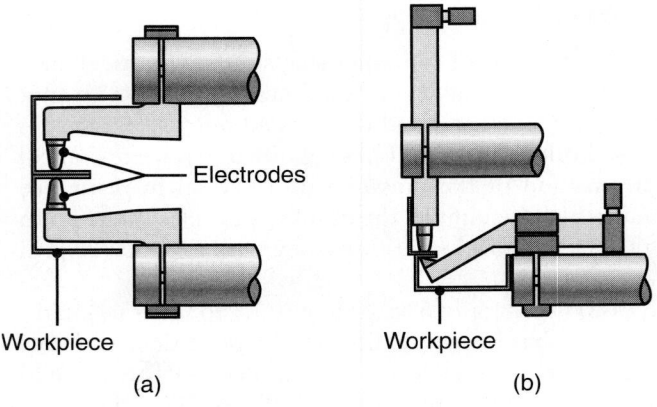

FIGURE 31.8 Two electrode designs for easy access to the components to be welded.

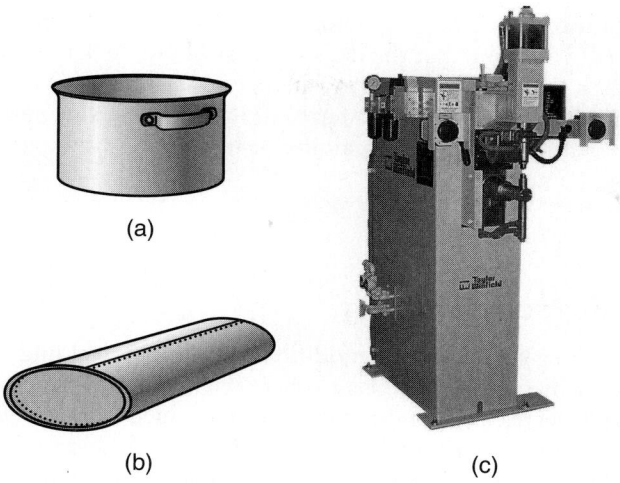

(a)

(b)   (c)

FIGURE 31.9 (a) and (b) Spot-welded cookware and muffler. (c) An automated spot-welding machine. The welding tip can move in three principal directions. Sheets as large as 2.2 × 0.55 m (88 × 22 in.) can be accommodated in this machine with proper workpiece supports. *Source*: Courtesy of Taylor–Winfield Corporation.

Testing spot welds. Spot-welded joints may be tested for weld-nugget strength using the following techniques (Fig. 31.10):

- Tension-shear
- Cross-tension
- Twist
- Peel

Because they are easy to perform and are inexpensive, tension-shear tests commonly are used in fabricating facilities. The cross-tension and twist tests are capable of revealing flaws, cracks, and porosity in the weld area. The peel test commonly is used for thin sheets. After the joint has been bent and peeled, the shape and size of the torn-out weld nugget are evaluated.

EXAMPLE 31.2 Heat generated in spot welding

Assume that two steel sheets 1-mm (0.04-in.) thick are being spot-welded at a current of 5000 A and over a current flow time of 0.1 second by means of electrodes 5 mm (0.2 in.) in diameter. Estimate the heat generated and its distribution in the weld zone if the effective resistance in the operation is 200 $\mu\Omega$.

Solution According to Eq. (31.1),

$$\text{Heat} = (5000)^2(0.0002)(0.1) = 500 \text{ J}$$

From the information given, the weld-nugget volume can be estimated to be 30 mm³ (0.0018 in³). Assume that the density for steel (Table 3.1) is 8000 kg/m³ (0.008 g/mm³); then the weld nugget has a mass of 0.24 g. The heat required to melt 1 g of steel is about 1400 J, so the heat required to melt the weld nugget is (1400)(1400)(0.24) = 336 J. The remaining heat (164 J) is dissipated into the metal surrounding the nugget.

31.5.2 Resistance seam welding

Resistance seam welding (RSEW) is a modification of spot welding wherein the electrodes are replaced by rotating wheels or rollers (Fig. 31.11a). Using a continuous AC power supply, the electrically conducting rollers produce a spot weld whenever the current reaches a sufficiently high level in the AC cycle. With a high enough frequency or slow enough traverse speed, these spot welds actually overlap into a continuous seam and produce a joint that is liquid tight and gas tight (Fig. 31.11b).

In **roll spot welding**, current to the rollers is applied only intermittently, resulting in a series of spot welds at specified intervals along the length of the seam (Fig. 31.11c). In **mash seam welding** (Fig. 31.11d), the overlapping welds are about one to two times the sheet thickness, and the welded seam thickness is only about 90% of the original

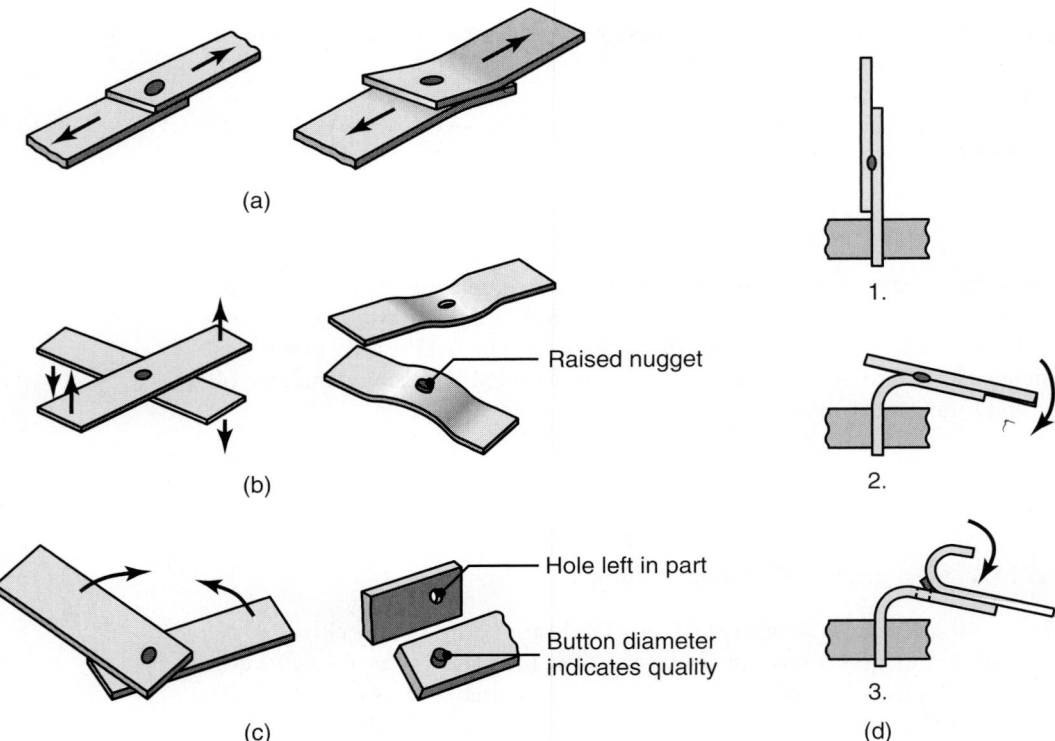

FIGURE 31.10   Test methods for spot welds: (a) tension-shear test, (b) cross-tension test, (c) twist test, (d) peel test (see also Fig. 32.9).

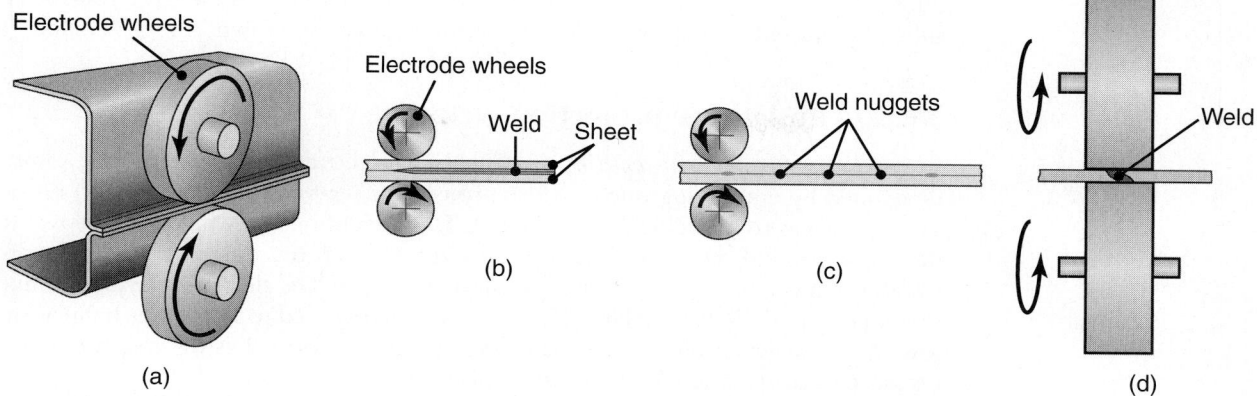

(a)　　　(b)　　　(c)　　　(d)

FIGURE 31.11　　(a) Seam-welding process in which rotating rolls act as electrodes. (b) Overlapping spots in a seam weld. (c) Roll spot welds and (d) Mash seam welding.

sheet thickness. This process also is used in producing *tailor-welded* sheet-metal blanks, which also can be made by laser welding (see Section 16.2.2).

The RSEW process is used to make the longitudinal (side) seam of cans for household products, mufflers, gasoline tanks, and other containers. The typical welding speed is 1.5 m/min (60 in./min) for thin sheets.

31.5.3 High-frequency resistance welding

High-frequency resistance welding (HFRW) is similar to seam welding, except that high-frequency current (up to 450 kHz) is employed. A typical application is the production of *butt-welded* tubing or pipe where the current is conducted through two sliding contacts (Fig. 31.12a) to the edges of roll-formed tubes. The heated edges then are pressed together by passing the tube through a pair of squeeze rolls. Any flash formed then is trimmed off.

Structural sections (such as I-beams) can be fabricated by HFRW by welding the webs and flanges made from long, flat pieces. Spiral pipe and tubing, finned tubes for heat exchangers, and wheel rims also may be made by these techniques. In

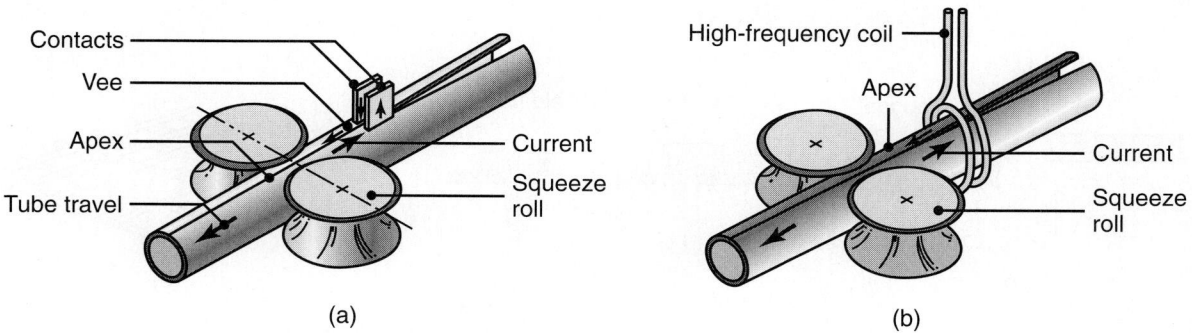

(a)　　　(b)

FIGURE 31.12　　Two methods of high-frequency continuous butt welding of tubes.

another method, called **high-frequency induction welding** (HFIW), the roll-formed tube is subjected to high-frequency induction heating, as shown in Fig. 31.12b.

31.5.4 Resistance projection welding

In *resistance projection welding* (RPW), high electrical resistance at the joint is developed by embossing one or more projections (*dimples*, see Fig. 16.37) on one of the surfaces to be welded (Figs. 31.13). The projections may be round or oval for design or strength purposes. High, localized temperatures are generated at the projections, which are in contact with the flat mating part. The electrodes (typically made of copper-based alloys) are large and flat and water cooled to keep their temperature low. Weld nuggets similar to those in spot welding are formed as the electrodes exert pressure to soften and compress the projections.

Spot-welding equipment can be used for RPW by modifying the electrodes. Although the embossing of the workpieces adds expense, this process produces a number of welds in one pass, extends electrode life, and is capable of welding metals of different thicknesses, such as a sheet welded over a plate. Nuts and bolts can be welded to sheets and plates by this process (Fig. 31.13c and d) with projections that are produced by machining or forging. Joining a network of rods and wires (such as the ones making up metal baskets, grills (Fig. 31.13e) oven racks, and shopping carts) also is considered resistance projection welding because of the many small contact areas between crossing wires (grids).

31.5.5 Flash welding

In *flash welding* (FW), also called **flash butt welding**, heat is generated very rapidly from the arc as the ends of the two members begin to make contact and develop an

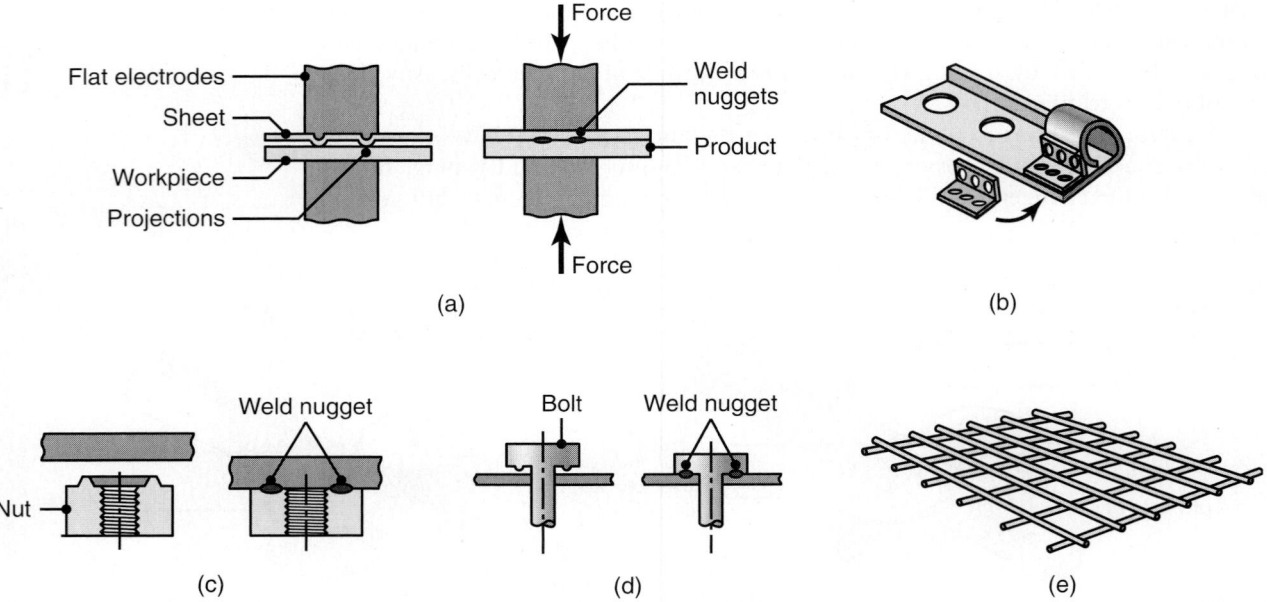

FIGURE 31.13   (a) Schematic illustration of resistance projection welding. (b) A welded bracket. (c) and (d) Projection welding of nuts or threaded bosses and studs. (e) Resistance-projection-welded grills.

electrical resistance at the joint (Fig. 31.14a). After the proper temperature is reached and the interface begins to soften, an axial force is applied at a controlled rate, and a weld is formed by plastic deformation of the joint. The mechanism is called *hot up-setting*, and the term *upset welding* (UW) also is used for this process. Some molten metal is expelled from the joint as a shower of sparks during the process—hence the name flash welding. Because of the presence of an arc, this process also can be classified as arc welding.

Impurities and contaminants are squeezed out during this operation, therefore the quality of the weld is good. However, a significant amount of material may be burned off during the welding process. The joint may be machined later to improve its appearance. The machines for FW usually are automated and large and have a variety of power supplies ranging from 10 to 1500 kVA.

The flash-welding process is suitable for end-to-end or edge-to-edge joining sheets of similar or dissimilar metals 0.2 to 25 mm (0.01 to 1 in.) thick and for end joining bars 1 to 75 mm (0.05 to 3 in.) in diameter. Thinner sections have a tendency to buckle under the axial force applied during welding. Rings made by forming processes (such as those shown in Fig. 16.22) also can be flash butt welded. This process also is used to repair broken band-saw blades with the use of fixtures that are mounted on the band-saw frame.

The flash-welding process can be automated for reproducible welding operations. Typical applications are the joining of pipe and of tubular shapes for metal

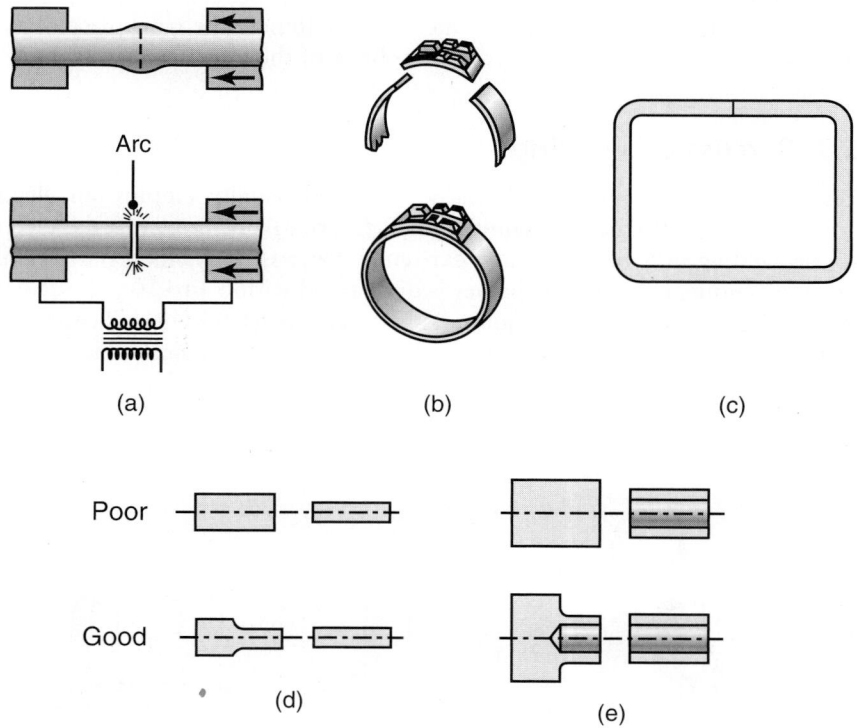

FIGURE 31.14   (a) Flash-welding process for end-to-end welding of solid rods or tubular parts. (b) and (c) Typical parts made by flash welding. (d) and (e) Some design guidelines for flash welding.

furniture and windows. It also is used for welding the ends of sheets or coils of wire in continuously operating rolling mills (Chapter 13) and in the feeding of wire-drawing equipment (Chapter 15). Once the appropriate process parameters are established, the required operator skill is minimal. Some design guidelines for mating surfaces in flash welding are shown in Fig. 31.14d and e. Note the importance of uniform cross-sections at the joint.

31.5.6 Stud welding

Stud welding (SW) also is called *stud arc welding* and is similar to flash welding. The stud (which may be a small part or, more commonly, a threaded rod, hanger, or handle) serves as one of the electrodes while being joined to another component, which is usually a flat plate (Fig. 31.15). Polarity for aluminum is usually direct-current electrode positive (DCEP), and for steel it is direct-current electrode negative (DCEN).

In order to concentrate the heat generated, prevent oxidation, and retain the molten metal in the weld zone, a disposable ceramic ring (*ferrule*) is placed around the joint. The equipment for stud welding can be automated with various controls for arcing and for applying pressure. Portable stud-welding equipment also is available. Typical applications of stud welding include automobile bodies, electrical panels, and shipbuilding; it is used also in building construction.

In **capacitor-discharge stud welding**, a DC arc is produced from a capacitor bank. No ferrule or flux is required, because the welding time is very short—on the order of 1 to 6 milliseconds. The choice between this process and stud arc welding depends on such factors as the types of metals to be joined, the workpiece thickness and cross-section, the stud diameter, and the shape of the joint.

31.5.7 Percussion welding

The resistance-welding processes already described usually employ an electrical transformer to meet the power requirements. Alternatively, however, the electrical energy for welding may be stored in a capacitor. *Percussion welding* (PEW) utilizes this latter technique, in which the power is discharged within 1 to 10 milliseconds to develop localized high heat at the joint. This process is useful where heating of the components adjacent to the joint is to be avoided, such as in electronic assemblies and electrical wires.

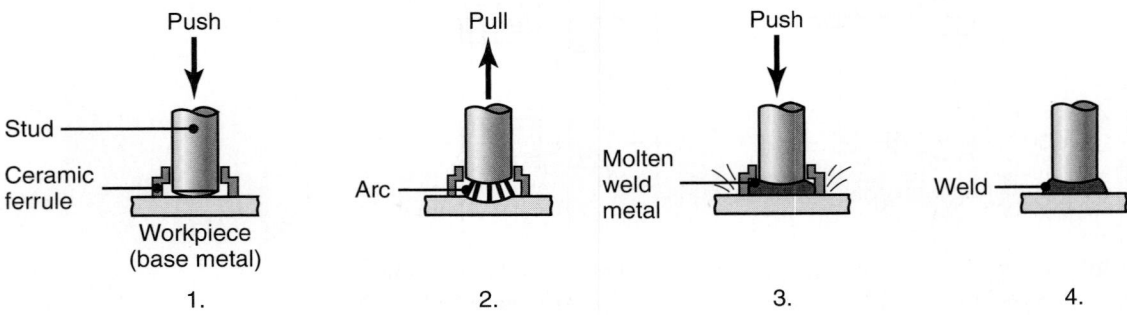

FIGURE 31.15   The sequence of operations in stud welding commonly used for welding bars, threaded rods, and various fasteners onto metal plates.

EXAMPLE 31.3 Resistance welding versus laser-beam welding in the can-making industry

The cylindrical bodies of cans for food and for household products have been resistance seam welded (a lap joint up the side of the can) for many years. Beginning in about 1987, laser-beam welding technology was introduced into the can-making industry. The joints are welded by lasers with the same productivity as resistance welding, but with the following advantages.

- As opposed to the lap joints suitable for resistance welding, laser welding utilizes butt joints. Thus, some material is saved. Multiplied by the billions of cans made each year, this amount becomes a very significant savings.

- Because laser welds have a very narrow zone (Fig. 31.16; see also Fig. 30.14), the unprinted area on the can surface (printing margin) is reduced greatly. As a result, the can's appearance and its customer acceptance are improved.

- The resistance lap-welded joint can be subject to corrosion by the contents of the can (tomato juice, for example). This effect may change the taste and can cause potential liability risk. A butt joint made by laser-beam welding eliminates this problem.

Source: Courtesy of G.F. Benedict.

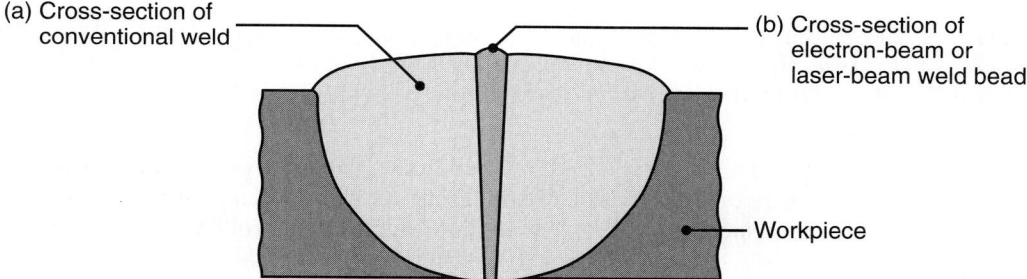

(a) Cross-section of conventional weld

(b) Cross-section of electron-beam or laser-beam weld bead

Workpiece

FIGURE 31.16   The relative sizes of the weld beads obtained by tungsten-arc and by electron-beam or laser-beam welding.

31.6 | Explosion Welding

In *explosion welding* (EXW), pressure is applied by detonating a layer of explosive that has been placed over one of the components being joined, called the *flyer plate* (Fig. 31.17a and b). The contact pressures developed are extremely high, and the kinetic energy of the plate striking the mating component causes a wavy interface.

This impact mechanically interlocks the two surfaces (Fig. 31.17c and d), so that pressure welding by plastic deformation also takes place. The flyer plate is placed at an angle, and any oxide films present at the interface are broken up and propelled out of the interface. As a result, the bond strength from explosion welding is very high.

The explosive may be a flexible plastic sheet or cord or in granulated or liquid form, which is cast or pressed onto the flyer plate. Detonation speed is usually in the range of from 2400 to 3600 m/s (8000 to 12,000 ft/s); it depends on the type of explosive, the thickness of the explosive layer, and its packing density. There is a minimum denotation speed necessary for welding to occur in this process. Detonation is carried out using a standard commercial blasting cap.

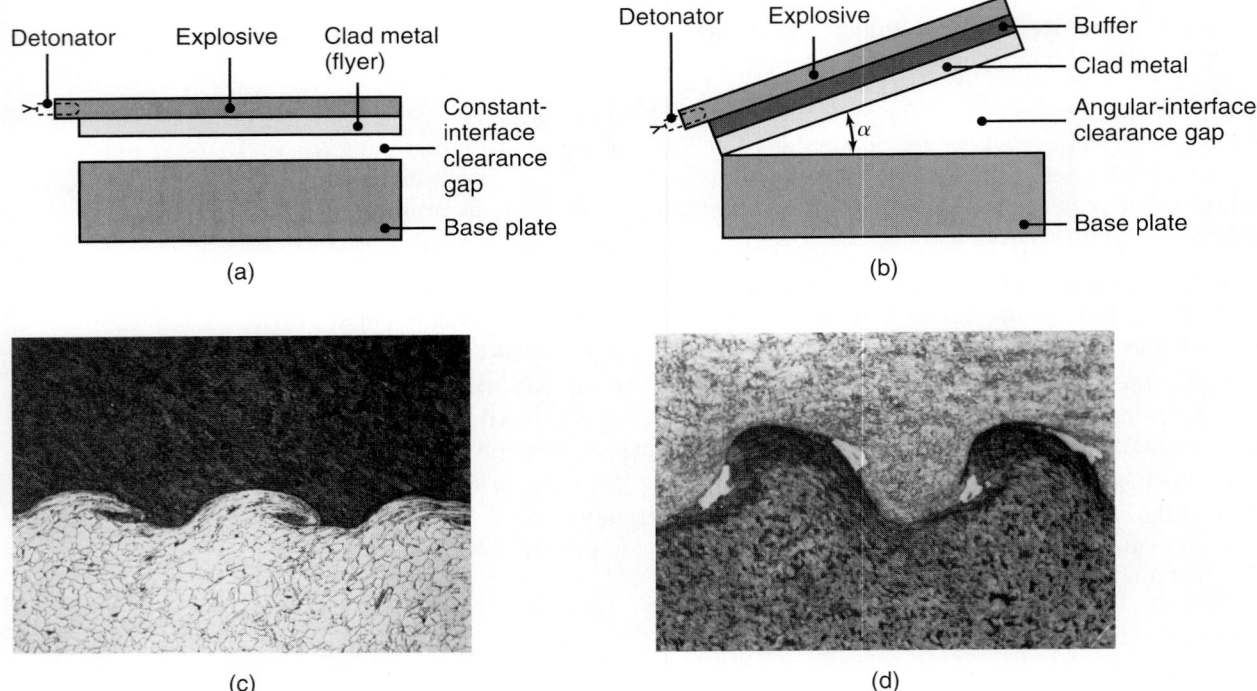

FIGURE 31.17   Schematic illustration of the explosion-welding process: (a) constant-interface clearance gap and (b) angular-interface clearance gap. (c) Cross-section of explosion-welded joint: titanium (top) and low-carbon steel (bottom). (d) Iron-nickel alloy (top) and low-carbon steel (bottom).

This process was developed in the 1940s. It is suitable particularly for cladding a plate or a slab with a dissimilar metal. Plates as large as 6×2 m (20×7 ft) have been clad explosively. These plates then may be rolled into thinner sections. Tubes and pipes can be joined to the holes in the header plates of boilers and heat exchangers by placing the explosive inside the tube; the explosion expands the tube. Explosion welding is inherently dangerous, so it requires safe handling by well-trained and experienced personnel.

31.7 | Diffusion Bonding

Diffusion bonding or **diffusion welding** (DFW) is a process in which the strength of the joint results primarily from diffusion (movement of atoms across the interface) and secondarily from plastic deformation of the faying surfaces. This process requires temperatures of about $0.5\ T_m$ (where T_m is the melting point of the metal on the absolute scale) in order to have a sufficiently high diffusion rate between the parts being joined.

The bonded interface in DFW essentially has the same physical and mechanical properties as the base metal. Its strength depends on (a) pressure, (b) temperature, (c) time of contact, and (d) how clean the faying surfaces are. These requirements can be relaxed by using a filler metal at the interface. Depending on the materials joined, brittle intermetallic compounds may form at the interface. These may be avoided by electroplating the surfaces with suitable metal alloys.

In diffusion bonding, pressure may be applied by dead weights, a press, differential gas pressure, or by the thermal expansion of the parts to be joined. The parts usually are heated in a furnace or by electrical resistance. High-pressure autoclaves also are used for bonding complex parts.

Although this process was developed in the 1970s as a modern welding technology, the principle of diffusion bonding dates back centuries to when goldsmiths bonded gold over copper to create a product called **filled gold**. First, a thin layer of gold foil is produced by hammering. Then the gold is placed over copper, and a weight is placed on top of it. Finally, assembly is placed in a furnace and left until a strong bond is obtained, hence this process is also called *hot-pressure welding* (HPW).

Diffusion bonding generally is most suitable for joining dissimilar metals. It also is used for reactive metals (such as titanium, beryllium, zirconium, and refractory metal alloys) and for composite materials such as metal-matrix composites (Section 9.5). Diffusion bonding is also an important mechanism of sintering in powder metallurgy (Section 17.4). Because diffusion involves migration of the atoms across the joint, this process is slower than other welding processes.

Although DFW is used for fabricating complex parts in low quantities for the aerospace, nuclear, and electronics industries, it has been automated to make it suitable and economical for moderate-volume production. Unless the process is highly automated, considerable operator training and skill is required. Equipment cost is related approximately to the diffusion-bonded area and is in the range of $3 to $6/mm^2 ($2000 to $4000/in^2).

EXAMPLE 31.4 Diffusion-bonding applications

Diffusion bonding is suitable especially for such metals as titanium and the superalloys used in military aircraft. Design possibilities allow the conservation of expensive strategic materials and the reduction of manufacturing costs. The military aircraft illustrated in Fig. 31.18 has more than 100 diffusion-bonded parts, some of which are shown.

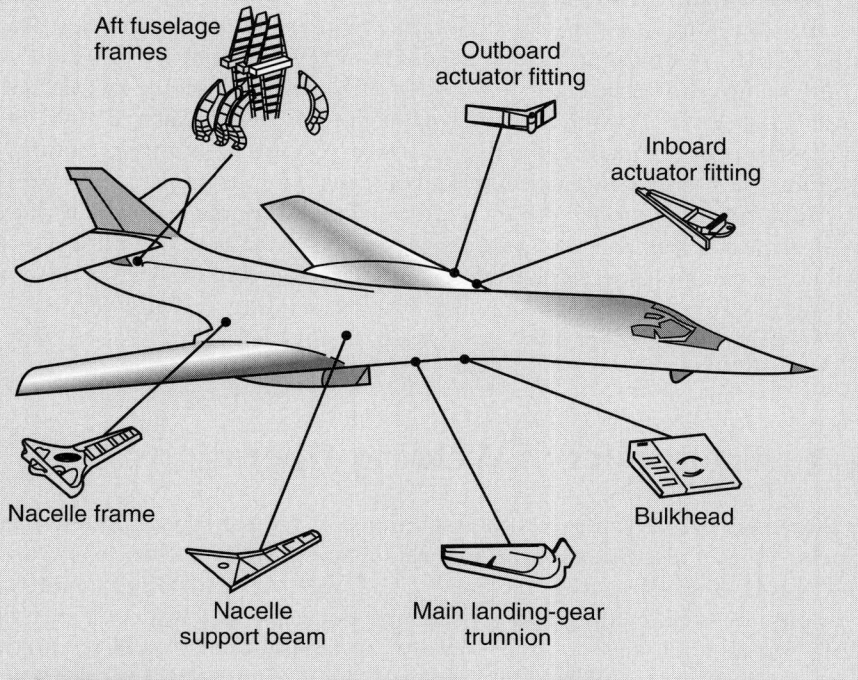

FIGURE 31.18   Aerospace diffusion-bonding application.

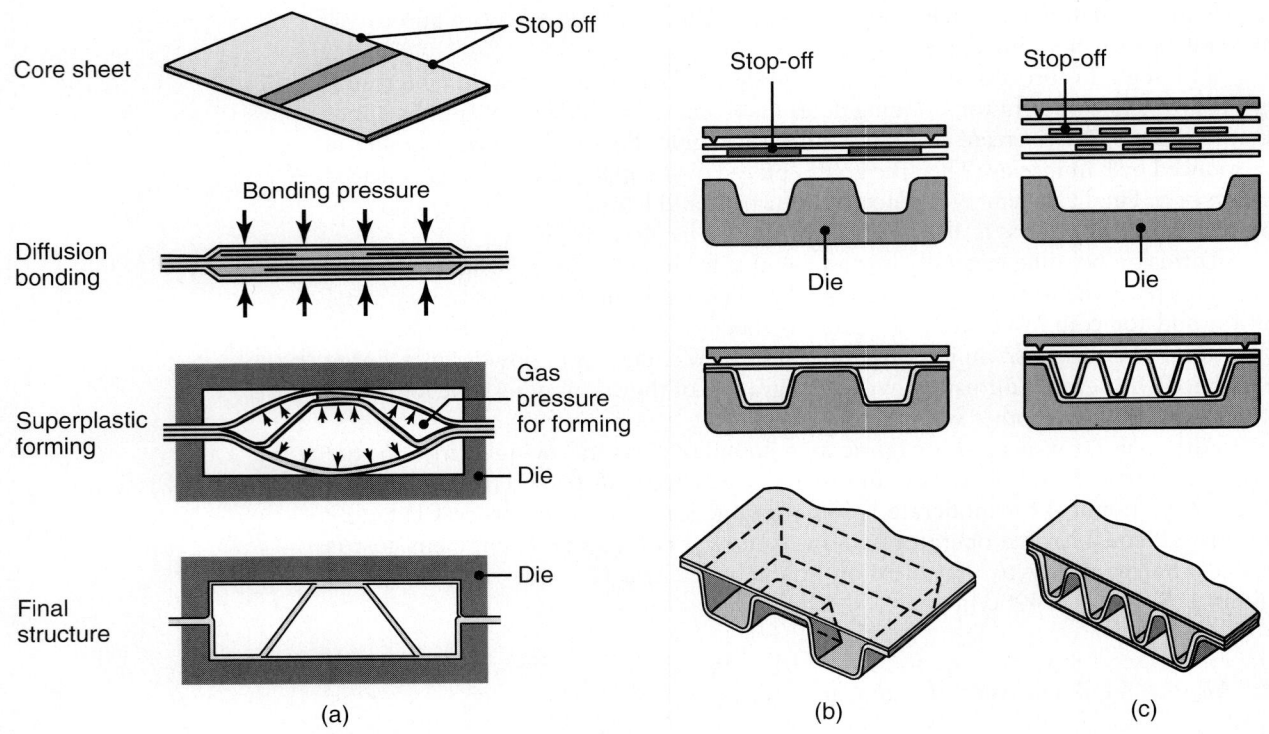

FIGURE 31.19 The sequence of operations in the fabrication of a structure by the diffusion bonding and superplastic forming of three originally flat sheets. *Sources*: (a) After D. Stephen and S.J. Swadling. (b) and (c) Courtesy of Rockwell International Corp.

Diffusion bonding/superplastic forming. Sheet-metal structures can be fabricated by combining *diffusion bonding* with *superplastic forming* (see also Section 16.10). Typical structures in which flat sheets (usually) are diffusion bonded and formed are shown in Fig. 31.19. After the diffusion bonding of selected locations of the sheets, the unbonded (*stop-off*) regions are expanded in a mold by air or fluid pressure. These structures are thin and have high stiffness-to-weight ratios, hence they are useful particularly in aircraft and aerospace applications.

This process improves productivity by eliminating the number of parts in a structure, mechanical fasteners, labor, and cost. It produces parts with good dimensional accuracy and low residual stresses. First developed in the 1970s, this technology now is well advanced for titanium structures (using typically Ti-6Al-4V and 7475-T6) and various other alloys for aerospace applications.

31.8 | Economics of Welding Operations

The characteristics, advantages, and limitations of welding processes described thus far included brief discussions regarding welding costs. The relative costs of some selected processes are shown in Tables 30.1 and VI.1. As in all other manufacturing operations, costs in welding and joining processes can vary widely, depending on such factors as equipment capacity, level of automation, labor skill required, weld quality, production rate, preparation required, and various other considerations specific to a particular joining process.

The general *welding and joining costs* for some selected and common operations (all described throughout Chapters 30 through 32) can be summarized as follows:

- *High*: brazing and fasteners (such as bolts and nuts) as they require holemaking operations and fastener costs.
- *Intermediate*: arc welding, riveting, adhesive bonding.
- *Low*: resistance welding, seaming, and crimping, as they are relatively simple to perform and automate.

Equipment costs for welding can be summarized as follows, noting that these costs can vary widely:

- *High* ($100,000 to $200,000): electron-beam and laser-beam welding.
- *Intermediate* ($5000 to $50,000+): spot, submerged arc, gas metal-arc, gas tungsten-arc, flux-cored arc, electrogas, electroslag, plasma arc, and ultrasonic welding.
- *Low* ($1000+): shielded metal-arc and oxyfuel-gas welding.

Labor costs in welding generally are higher than other metalworking operations because of the operator skill, welding time, and preparation required. However, much depends on the automation of the equipment employed, including the use of robotics and computer controls programmed to follow a prescribed path (*seam tracking*) during welding. It has been observed that in systems with robotic controls, the number reaches 80%, whereas in manual welding operations (see Table 30.1), the actual welding time spent by the operator is only about 30% of the overall time.

Labor costs may be summarized as follows:

- *High to intermediate:* oxyfuel-gas welding and shielded metal-arc welding.
- *High to low:* electron-beam and laser-beam welding and flux-cored arc welding.
- *Intermediate to low:* submerged arc welding.

SUMMARY

- In addition to the traditional joining processes of oxyfuel-gas and arc welding, a number of other joining processes that are based on producing a strong joint under pressure and/or heat are available.
- Surface preparation and cleanliness are important in some of these processes. Pressure is applied mechanically or by explosives. Heat may be supplied externally (electrical resistance or furnaces), or it may be generated internally (as in friction welding).
- Among important developments is the combining of diffusion-bonding and super-plastic-forming processes. Productivity is improved, as is the capability to make complex parts economically.
- As in all manufacturing operations, certain hazards are inherent in welding operations. Some relate to the machinery and equipment used, others to the nature of the process. Proper safety precautions always must be taken in work areas where welding is done.

KEY TERMS

Cold welding	Friction welding	Roll bonding (welding)
Diffusion bonding (welding)	High-frequency resistance welding	Roll spot welding
Explosion welding	Inertia friction welding	Solid-state welding
Faying surfaces	Linear friction welding	Stud welding
Ferrule	Percussion welding	Superplastic forming
Filled gold	Resistance projection welding	Ultrasonic welding
Flash welding	Resistance seam welding	Weld nugget
Flyer plate	Resistance spot welding	
Friction stir welding	Resistance welding	

BIBLIOGRAPHY

ASM Handbook, Vol. 6: *Welding, Brazing, and Soldering*, ASM International, 1993.

Bowditch, W.A., and Bowditch, K.E., *Welding Technology Fundamentals*, Goodheart–Willcox, 1997.

Cary, H.B., *Modern Welding Technology*, 4th ed., Prentice Hall, 1997.

Davies, A.C., *The Science and Practice of Welding*, 10th ed., 2 vols., Cambridge University Press, 1993.

Easterling, K.E., *Introduction to the Physical Metallurgy of Welding*, Butterworth, 1983.

Evans, G.M., and Bailey, N., *Metallurgy of Basic Weld Metal*, Woodhead Publishing, 1997.

Galyen, J., Sear, G., and Tuttle, C., *Welding: Fundamentals and Procedures*, Wiley, 1984.

Granjon, H., *Fundamentals of Welding Metallurgy*, Woodhead Publishing, 1991.

Hicks, J.G., *Welded Joint Design*, 2nd ed., Abington, 1997.

Jeffus, L.F., *Welding: Principles and Applications*, 4th ed., Delmar Publishers, 1997.

Jellison, R., *Welding Fundamentals*, Prentice Hall, 1995.

Lancaster, J.F., *The Metallurgy of Welding*, Chapman & Hall, 1993.

Messler, R.W., Jr., *Joining of Advanced Materials*, Butterworth–Heinemann, 1993.

Mouser, J.D., *Welding Codes, Standards, and Specifications*, McGraw-Hill, 1997.

Nicholas, M.G., *Joining Processes: Introduction to Brazing and Diffusion Bonding*, Chapman & Hall, 1998.

North, T.H., *Advanced Joining Technologies*, Chapman & Hall, 1990.

Tool and Manufacturing Engineers Handbook, Vol. 4: *Quality Control and Assembly*, Society of Manufacturing Engineers, 1986.

Welding Handbook, 8th ed., 3 vols., American Welding Society, 1987.

REVIEW QUESTIONS

31.1 Explain what is meant by solid-state welding.

31.2 What is cold welding? Why is it so called?

31.3 What are faying surfaces in solid-state welding processes? What is their significance?

31.4 What is the basic principle of (a) ultrasonic welding and (b) diffusion bonding?

31.5 What advantages does friction welding have over other methods described in this and in the preceding chapter?

31.6 Describe the advantages and limitations of explosion welding.

31.7 Describe the principle of resistance-welding processes.

31.8 What type of products are suitable for stud welding? Why?

31.9 What is the advantage of linear friction welding over inertia friction welding?

31.10 Describe how high-frequency butt welding operates.

QUALITATIVE PROBLEMS

31.11 Explain the similarities and differences between the joining processes described in this chapter and those described in Chapter 30.

31.12 Explain the reasons why the processes described in this chapter were developed.

31.13 Describe your observations concerning Fig. 31.17c and d.

31.14 Would you be concerned regarding the size of the weld beads shown in Fig. 31.16? Explain.

31.15 Discuss the factors that influence the strength of (a) a diffusion-bonded and (b) a cold-welded component.

31.16 Describe the sources of heat for the processes described in this chapter.

31.17 Comment on the feasibility of applying explosion welding in a factory environment.

31.18 Can the roll-bonding process be applied to a variety of part configurations? Explain.

31.19 Why is diffusion bonding (when combined with the superplastic forming of sheet metals) an attractive fabrication process? Does it have any limitations?

31.20 Describe the features of a weld nugget. What does its strength depend on?

31.21 Make a list of some products that can be fabricated by resistance-welding processes.

31.22 Give some of the reasons that spot welding is used commonly in automotive bodies and in home appliances.

31.23 Explain the significance of the magnitude of the pressure applied through the electrodes during a spot-welding operation.

31.24 Give some applications for (a) flash welding, (b) stud welding, and (c) percussion welding.

31.25 Discuss the need and role of fixtures in holding workpieces in the welding operations described in this chapter.

31.26 Inspect Fig. 31.4 and explain why those particular fusion-zone shapes are developed as a function of pressure and speed. Comment on the influence of the material's properties.

31.27 Discuss your observations concerning the welding design guidelines illustrated in Fig. 31.14d and e.

31.28 Can you friction weld if one of the pieces to be welded is noncircular? Explain, giving some examples.

31.29 Could the process shown in Fig. 31.12 also be applicable to part shapes other than round? Explain your answer and give specific examples.

31.30 What applications could be suitable for the roll-spot-welding process shown in Fig. 31.11?

31.31 Survey the available technical literature on friction welding and prepare a table of the similar and dissimilar metals and nonmetallic materials that can be friction-welded.

31.32 Could the projection-welded parts shown in Fig. 31.13 be made by any of the processes described in other chapters in this book? Explain.

31.33 Explain the difference between resistance seam welding and resistance spot welding.

31.34 Referring to Fig. 14.11b, could you use any of the processes described in Chapters 30 and 31 to make a large bolt by welding the head to the shank? Explain the advantages and limitations of this approach.

QUANTITATIVE PROBLEMS

31.35 Two flat copper sheets (each 1.5 mm thick) are being spot welded by the use of a current of 7000 A and a current flow time of 0.3 s. The electrodes are 5 mm in diameter. Estimate the heat generated in the weld zone. Assume that the resistance is 200 $\mu\Omega$.

31.36 Calculate the temperature rise in Problem 31.35 assuming that the heat generated is confined to the volume of material directly between the two round electrodes and the temperature is distributed uniformly.

31.37 Calculate the range of allowable currents for Problem 31.35 if the temperature should be between 0.7 and 0.85 times the melting temperature of copper. Repeat this problem for carbon steel.

31.38 The energy applied in friction welding is given by the formula $E = IS^2/C$, where I is the moment of inertia of the flywheel, S is the spindle speed in rpm, and C is a constant of proportionality (5873, when the moment of inertia is given in lb-ft^2). For a spindle speed of

600 rpm and an operation in which a steel tube (3.5 in. OD, 0.25 in. wall thickness) is welded to a flat frame, what is the required moment of inertia of the flywheel if all of the energy is used to heat the weld zone (approximated as the material $\frac{1}{4}$ in. deep and directly below the tube)? Assume that 1.4 ft-lbm is needed to melt the electrode.

31.39   The energy required in ultrasonic welding is known to be related to the product of the thickness and hardness of the workpiece. Explain why this relationship exists.

SYNTHESIS, DESIGN, AND PROJECTS

31.40   Explain how you would fabricate the structures shown in Fig. 31.19 by methods other than diffusion bonding and superplastic forming.

31.41   Comment on workpiece size and shape limitations (if any) for each of the processes described in this chapter.

31.42   Describe part shapes that cannot be joined by the processes described in this chapter. Gives specific examples.

31.43   Prepare a table giving the welding speeds (as a function of the relevant parameters) for the processes described in Chapters 30 and 31. Comment on your observations.

31.44   Make a comprehensive outline of this chapter; include sketches of possible welded-joint designs (other than those given) and of their engineering applications. Give specific examples for each type of joint.

31.45   Using a magnifier, inspect the cross-section of coins (such as the U.S. dime and nickel) and comment on your observations.

31.46   In spot-weld tests, what would be the reason for weld failure to occur at the locations shown in Fig. 31.10?

31.47   Describe the methods you would use for removing the flash from welds, such as those shown in Fig. 31.3. How would you automate these methods for high production rates?

31.48   For each of the operations described in this chapter, prepare a poster which effectively and concisely gives specific instructions for safe practices in welding. Consult various publications of the National Safety Council and similar organizations.

31.49   In the roll-bonding process shown in Fig. 31.1, how would you go about ensuring that the interfaces are clean and free of contaminants so that a good bond is developed? Explain.

31.50   Inspect several metal containers for household products and for food and beverages. Identify those that have utilized any of the processes described in this chapter. Describe your observations.

31.51   Discuss various other processes that can be used in attaching tubes to head plates in boilers.

31.52   Describe part designs that cannot be joined by any of the friction-welding processes described in this chapter.

31.53   Inspect the sheet-metal body of an automobile, and comment on the size and frequency of the spot welds applied. How would you go about estimating the number of welds?

31.54   Alclad stock is made from 5182 aluminum alloy and has both sides coated with a thin layer of pure aluminum. The 5182 provides high strength, while the outside layers of pure aluminum provide good corrosion resistance because of their stable oxide film. Hence, Alclad is used commonly in aerospace structural applications. Investigate other common roll-bonded materials and their uses, and prepare a summary table.

31.55   Perform an Internet survey of available spot-welding machines, their capabilities, and their costs. Describe your observations.

31.56   What kind of applications could the structures shown in Fig. 31.19b and c have? Explain.

Brazing, Soldering, Adhesive-Bonding, and Mechanical-Fastening Processes

In this last chapter on joining, we will describe various joining, bonding, and fastening processes that involve mechanisms unlike those in the preceding two chapters. Specific topics covered are:

- Brazing, soldering, and their applications.
- The use of adhesives for joining materials.
- Methods of mechanical fastening, especially using bolts and rivets.
- Methods for joining thermoplastics, thermosets, ceramics, and glasses.
- Economics of these joining processes.

32.1 | Introduction

In most of the joining processes described in Chapters 30 and 31, the mating surfaces of the components were heated to elevated temperatures by various external or internal means to cause fusion or bonding at the joint. What if we want to join materials that cannot withstand high temperatures, such as electronic components? What if the parts to be joined are delicate, intricate, or are made of two or more materials with very different characteristics, properties, sizes, thicknesses, and cross-sections?

This chapter first describes two joining processes—*brazing* and *soldering*—that require lower temperatures than those for welding. Filler metals are placed in or supplied to the joint and are melted using an external source of heat. Upon solidification, a strong joint is obtained. Brazing and soldering are distinguished arbitrarily by temperature. Temperatures for soldering are lower than those for brazing, and the strength of a soldered joint is much lower.

We also describe the principles and types of *adhesive bonding*. The ancient method of joining parts with animal-derived glues (typically employed in bookbinding, labeling, and packaging) has been developed into an important joining technology for metallic and nonmetallic materials. This process has wide applications in numerous consumer and industrial products, as well as in the aircraft and aerospace industries. Bonding materials (such as thermoplastics, thermosets, ceramics, and glasses) either to each other or to other materials presents various challenges.

Although all of the joints described thus far are of a permanent nature, in many applications, joined components have to be taken apart for replacement, maintenance, repair, or adjustment. How do we take apart a product without destroying the joint? If we need joints that are truly nonpermanent but are as strong as welded joints, the obvious solution is to use *mechanical fastening*, such as using bolts, screws, nuts, or a variety of other fasteners. We also describe the advantages and limitations of these techniques.

32.2 | Brazing

Brazing is a joining process in which a filler metal is placed between the faying surfaces to be joined (or at their periphery) and the temperature is raised sufficiently to melt the filler metal (but not the components (base metal)—as would be the case in fusion welding). Thus, brazing is a liquid–solid-state bonding process. Upon cooling and solidification of the filler metal, a strong joint is obtained. Filler metals used for brazing typically melt above 450°C (840°F), which is below the melting point (*solidus temperature*) of the metals to be joined (for example, see Fig. 4.5). Brazing comes from the word *brass* (an archaic word meaning "to harden") and was first used as far back as 3000 to 2000 B.C. Typical applications of brazing are shown in Fig. 32.1.

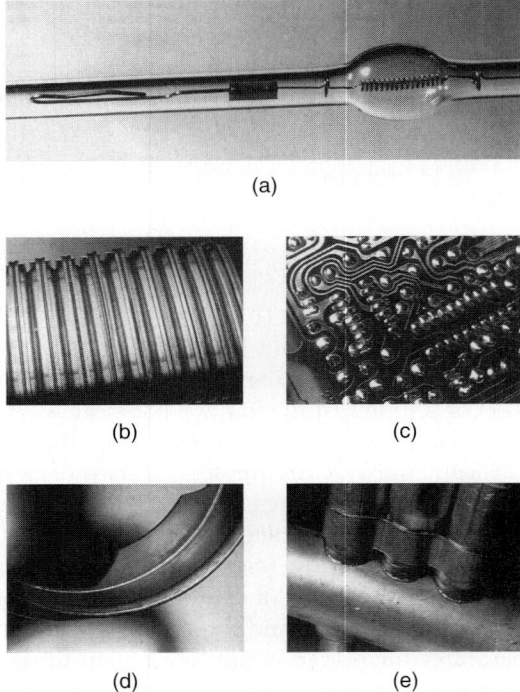

(a)

(b)   (c)

(d)   (e)

FIGURE 32.1   Examples of brazed and soldered parts. (a) Resistance brazed light-bulb filament; (b) brazed radiator heat exchanger; (c) soldered circuit board; (d) brazed ring housing; and (e) brazed heat exchanger *Source:* Courtesy of Edison Welding Institute.

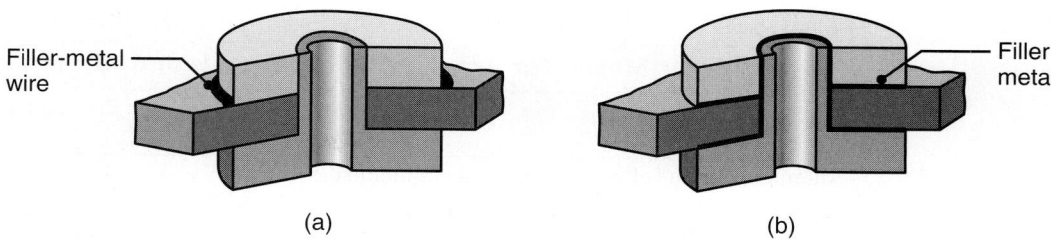

FIGURE 32.2 An example of furnace brazing: (a) before and (b) after brazing. The filler metal is a shaped wire and moves into the interfaces by capillary action with the application of heat.

In a typical brazing operation, a filler (braze) metal wire is placed along the periphery of the components to be joined (Fig. 32.2a). Heat is applied by various external means, which melts the braze metal and fills (by *capillary* action) the closely fitting space (joint clearance) at the interfaces (Fig. 32.2b).

In **braze welding**, filler metal (typically brass) is deposited at the joint with a technique similar to oxyfuel-gas welding (see Fig. 30.1d). The major difference is that the base metal does not melt. Its main application is in repair work on cast steels and irons. Because of the wider gaps between the components being welded (as in oxyfuel-gas welding), more braze metal is used than in conventional brazing.

In general, dissimilar metals can be assembled with good joint strength. Examples of joints made are shown in Fig. 32.3. Intricate, lightweight shapes can be joined rapidly and with little distortion.

Filler metals. Several filler metals (*braze metals*) are available with a range of brazing temperatures (Table 32.1). Note that, unlike those for other welding operations, filler metals for brazing generally have a composition significantly different from the metals to be joined. They come in a variety of shapes, such as wire, rod, rings, shims, and filings. The selection of the type of filler metal and its composition

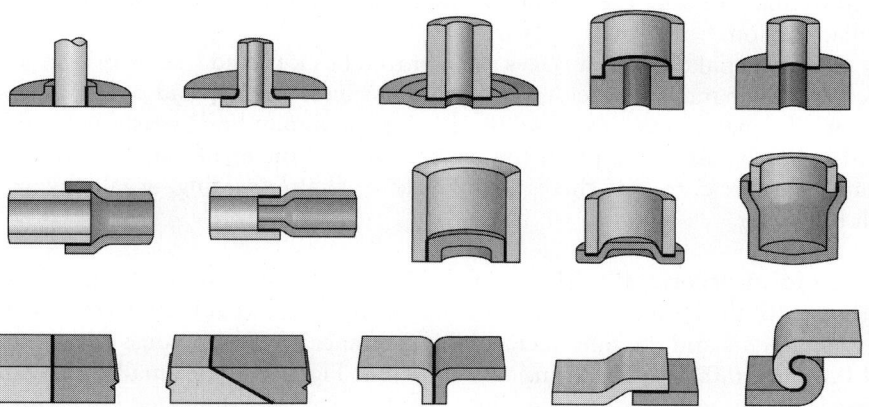

FIGURE 32.3 Joint designs commonly used in brazing operations. The clearance between the two parts being brazed is an important factor in joint strength. If the clearance is too small, the molten braze metal will not penetrate the interface fully. If it is too large, there will be insufficient capillary action for the molten metal to fill the interface.

TABLE 32.1

Typical Filler Metals for Brazing Various Metals and Alloys		
Base metal	Filler metal	Brazing temperature (°C)
Aluminum and its alloys	Aluminum-silicon	570–620
Magnesium alloys	Magnesium-aluminum	580–625
Copper and its alloys	Copper-phosphorus	700–925
Ferrous and nonferrous (except aluminum and magnesium)	Silver and copper alloys, copper-phosphorus	620–1150
Iron-, nickel-, and cobalt-based alloys	Gold	900–1100
Stainless steels, nickel- and cobalt-based alloys	Nickel-silver	925–1200

are important in order to avoid *embrittlement* of the joint by (a) grain-boundary penetration of liquid metal, as described in Section 1.4.2; (b) formation of *brittle intermetallic compounds* at the joint, as described in Section 4.2.2; and (c) *galvanic corrosion* in the joint.

Because of diffusion between the filler metal and the base metal, the mechanical and metallurgical properties of a joint can change in subsequent processing or during the service life of a brazed part. For example, when titanium is brazed with pure tin as the filler metal, it is possible for the tin to diffuse completely into the titanium base metal when subjected to subsequent aging or to heat treatment. Consequently, the joint no longer exists.

Fluxes.   The use of a *flux* is essential in brazing in order to prevent oxidation and to remove oxide films from workpiece surfaces. Brazing fluxes generally are made of borax, boric acid, borates, fluorides, and chlorides. *Wetting agents* also may be added to improve both the wetting characteristics of the molten filler metal and the capillary action.

It is essential that the surfaces to be brazed be clean and free from rust, oil, and other contaminants in order to (a) obtain proper wetting and spreading of the molten filler metal in the joint and (b) develop maximum bond strength. Sand blasting also may be used to improve the surface finish of the faying surfaces for brazing. Because they are corrosive, fluxes must be removed after brazing, usually by washing with hot water.

Brazed joint strength.   The strength of the brazed joint depends on (a) the joint clearance, (b) the joint area, and (c) the nature of the bond at the interfaces between the components and the filler metal. Joint clearances typically range between 0.025 and 0.2 mm (0.001 to 0.008 in.). As shown in Fig. 32.4, the smaller the gap, the higher is the *shear strength* of the joint. The shear strength of brazed joints can reach 800 MPa (120 ksi) by using brazing alloys containing silver (*silver solder*). Also, there is an optimum gap to achieve maximum *tensile strength*.

Because clearances are very small, surface roughness of the mating surfaces becomes important. The surfaces to be brazed must be cleaned chemically or mechanically to ensure full capillary action. Consequently, the use of a flux also is important.

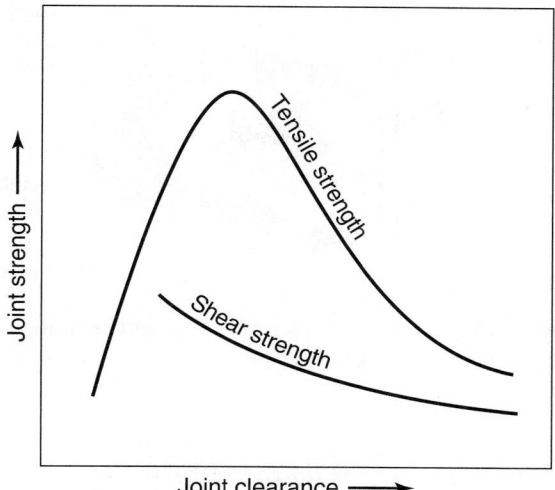

FIGURE 32.4   The effect of joint clearance on tensile and shear strength of brazed joints. Note that unlike tensile strength, shear strength continually decreases as the clearance increases.

32.2.1 Brazing methods

The heating methods used in brazing also identify the various processes, as described next.

Torch brazing.   The heat source in *torch brazing* (TB) is oxyfuel gas with a carburizing flame (see Fig. 30.1c). Brazing is performed by first heating the joint with the torch and then depositing the brazing rod or wire in the joint. Suitable part thicknesses are typically in the range of 0.25 to 6 mm (0.01 to 0.25 in.). Torch brazing is difficult to control and requires skilled labor; however, it can be automated as a production process by using multiple torches. This process also can be used for repair work.

Furnace brazing.   The parts in *furnace brazing* (FB) first are cleaned and preloaded with brazing metal in appropriate configurations; the assembly is then placed in a furnace, where it is heated uniformly. Furnaces may be either a batch-type for complex shapes or a continuous-type for high production runs—especially for small parts with simple joint designs. Vacuum furnaces or neutral atmospheres are used for metals that react with the environment. Skilled labor is not required, and complex shapes can be brazed because the whole assembly is heated uniformly in the furnace.

Induction brazing.   The source of heat in *induction brazing* (IB) is induction heating by high-frequency AC current. Parts are preloaded with filler metal and are placed near the induction coils for rapid heating. Unless a protective (neutral) atmosphere is utilized, fluxes generally are used. Part thicknesses usually are less than 3 mm (0.125 in.). Induction brazing is suitable particularly for brazing parts continuously (Fig. 32.5).

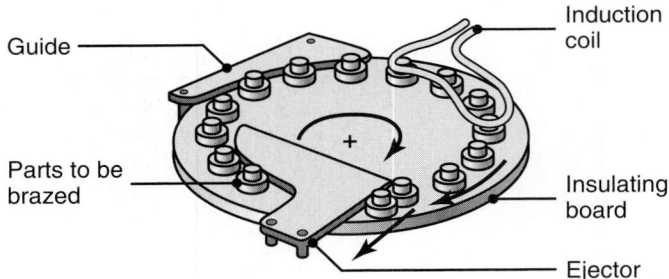

FIGURE 32.5   Schematic illustration of a continuous induction-brazing setup for increased productivity.

Resistance brazing.   In *resistance brazing* (RB), the source of heat is the electrical resistance of the components to be brazed. Electrodes are utilized in this method, as they are in resistance welding. Parts either are preloaded with filler metal or supplied with the metal externally during brazing. Parts commonly brazed have thicknesses of 0.1 to 12 mm (0.004 to 0.5 in.). As in induction brazing, the process is rapid, heating zones can be confined to very small areas, and the process can be automated to produce reliable and uniform quality.

Dip brazing.   *Dip brazing* (DB) is carried out by dipping the assemblies to be brazed into either a molten filler-metal bath or a molten salt bath (Section 4.12) at a temperature just above the melting point of the filler metal. Thus, all component surfaces are coated with the filler metal. Consequently, dip brazing in metal baths typically is used for small parts (such as sheet, wire, and fittings) usually less than 5 mm (0.2 in.) in thickness or diameter. Molten salt baths (which also act as fluxes) are used for complex assemblies of various thicknesses. Depending on the size of the parts and the bath size, as many as 1000 joints can be made at one time by dip brazing.

Infrared brazing.   The heat source in *infrared brazing* (IRB) is a high-intensity quartz lamp. This process is suitable particularly for brazing very thin components usually less than 1 mm (0.04 in.) thick, including honeycomb structures (Section 16.12). The radiant energy is focused on the joint, and brazing can be carried out in a vacuum. *Microwave heating* also may be used.

Diffusion brazing.   *Diffusion brazing* (DFB) is carried out in a furnace where (with proper control of temperature and time) the filler metal diffuses into the faying surfaces of the components to be joined. The brazing time required may range from 30 minutes to 24 hours. This process is used for strong lap or butt joints and for difficult joining operations. Because the rate of diffusion at the interface does not depend on the thickness of the components, part thicknesses may range from foil to as much as 50 mm (2 in.).

High-energy beams.   For specialized and high-precision applications and with high-temperature metals and alloys, *electron-beam* or *laser-beam* heating may be used.

Braze welding.   The joint in *braze welding* is prepared as it is in fusion welding, as described in Chapter 30. While an oxyacetylene torch with an oxidizing flame is used, filler metal is deposited at the joint rather than by capillary action. As a result, considerably more filler metal is used than in brazing. However, temperatures in

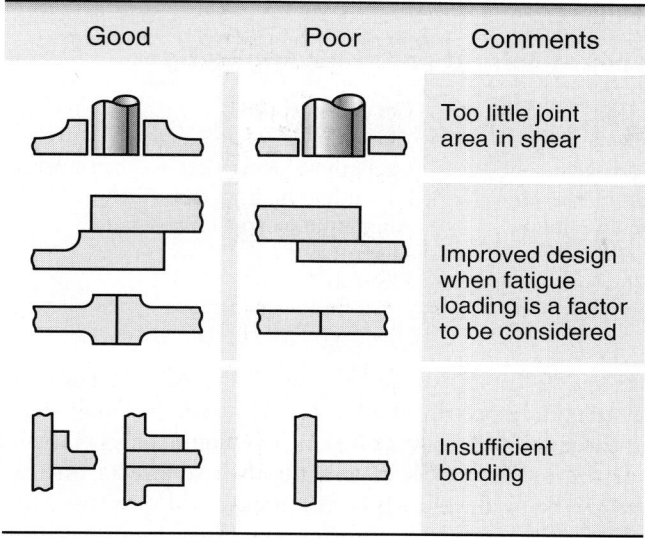

Good	Poor	Comments
		Too little joint area in shear
		Improved design when fatigue loading is a factor to be considered
		Insufficient bonding

FIGURE 32.6   Examples of good and poor design for brazing. *Source:* American Welding Society.

braze welding generally are lower than in fusion welding, and part distortion is minimal. The use of a flux is essential in this process. The principal use of braze welding is for maintenance and repair work, such as on ferrous castings and steel components, although this process also can be automated for mass production.

32.2.2 Design for brazing

As in all joining processes, *joint design* is important in brazing. Some design guidelines are given in Fig. 32.6. Strong joints require a larger contact area for brazing than for welding. A variety of special fixtures may be required to hold the parts together during brazing; some will have provision for thermal expansion and contraction during the operation.

32.3 | Soldering

In *soldering*, the filler metal (called **solder**) melts at a relatively low temperature. As in brazing, the solder fills the joint by capillary action between closely fitting or closely placed components. Important characteristics for solders are low surface tension and high wetting capability. Heat sources for soldering are usually soldering irons, torches, or ovens. Soldering with copper-gold and tin-lead alloys first was practiced as far back as 4000 to 3000 B.C. The word "solder" is derived from the Latin *solidare*, meaning "to make solid."

32.3.1 Types of solders and fluxes

Solders melt at a temperature that is the *eutectic point* of the solder alloy (for example, see Fig. 4.7.) Solders traditionally have been tin-lead alloys in various proportions. For example, a solder of 61.9% Sn-38.1% Pb composition melts at 188°C (370°F),

TABLE 32.2

Types of Solders and their Applications	
Tin-lead	General purpose
Tin-zinc	Aluminum
Lead-silver	Strength at higher than room temperature
Cadmium-silver	Strength at high temperatures
Zinc-aluminum	Aluminum, corrosion resistance
Tin-silver	Electronics
Tin-bismuth	Electronics

whereas tin melts at 232°C (450° F) and lead at 327°C (621°F). For special applications and higher joint strength (especially at elevated temperatures), other solder compositions are tin-zinc, lead-silver, cadmium-silver, and zinc-aluminum alloys (Table 32.2).

Because of the toxicity of lead and its adverse effects on the environment, **lead-free solders** are being developed continuously and now are coming into wider use. Among the various candidate materials are silver, indium, and bismuth eutectic alloys in combination with tin. Three typical compositions are 96.5% Sn-3.5% Ag, 42% Sn-58% Bi, and 48% Sn-52% In. However, none of these combinations are suitable for every application.

Fluxes are used in soldering and for the same purposes as they are in welding and brazing, as described in Section 32.2. Fluxes for soldering are generally of two types:

1. *Inorganic acids* or *salts*, such as zinc-ammonium-chloride solutions which clean the surface rapidly. After soldering, the flux residues should be removed by washing the joint thoroughly with water to avoid corrosion.

2. Noncorrosive *resin-based fluxes* used in electrical applications.

32.3.2 Solderability

Solderability may be defined in a manner similar to weldability (Section 30.9.2). Special fluxes have been developed to improve the solderability of many metals and alloys. As a general guide: (a) Copper, silver, and gold are easy to solder. (b) Iron and nickel are more difficult to solder. (c) Aluminum and stainless steels are difficult to solder because of their thin, strong oxide films. (d) Titanium, magnesium, cast irons, and steels, as well as ceramics and graphite, can be soldered by first plating them with suitable metallic elements that induce interfacial bonding. This method is similar to that used for joining carbides and ceramics (see Section 32.6.3). A common example of this method is *tinplate*, which is steel sheet coated with tin, thus making it very easy to solder. Tinplate is a common material for making cans for food.

32.3.3 Soldering techniques

Some soldering techniques are similar to brazing methods, such as:

Torch soldering (TS)
Furnace soldering (FS)
Iron soldering (INS) (with the use of a soldering iron)
Induction soldering (IS)
Resistance soldering (RS)
Dip soldering (DS)
Infrared soldering (IRS)

Other soldering techniques are:

Ultrasonic soldering (in which a transducer subjects the molten solder to ultrasonic cavitation. This action removes the oxide films from the surfaces to be joined and thus eliminates the need for a flux—hence the term *fluxless soldering*)
Reflow (paste) **soldering** (RS)
Wave soldering (WS)

The last two soldering techniques are significantly different from other soldering methods and are used widely for bonding and packaging in *surface-mount technology*, as described in Section 28.11. They are described in more detail in the following sections.

Reflow soldering.   *Solder pastes* are solder-metal particles held together by flux, binding, and wetting agents. The pastes are semi-solid in consistency, have high viscosity, but are able to maintain a solid shape for relatively long periods of time. The paste is placed directly onto the joint or on flat objects for finer detail, and it can be applied via a *screening* or *stenciling* process, as shown in Fig. 32.7a. Stenciling is

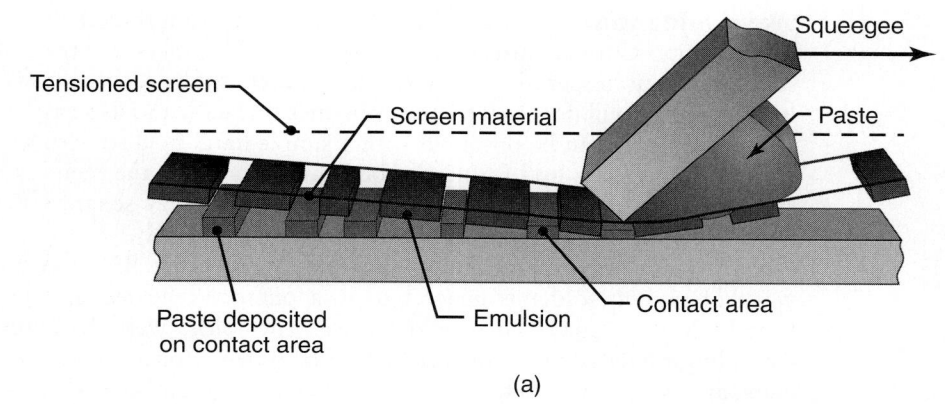

(a)

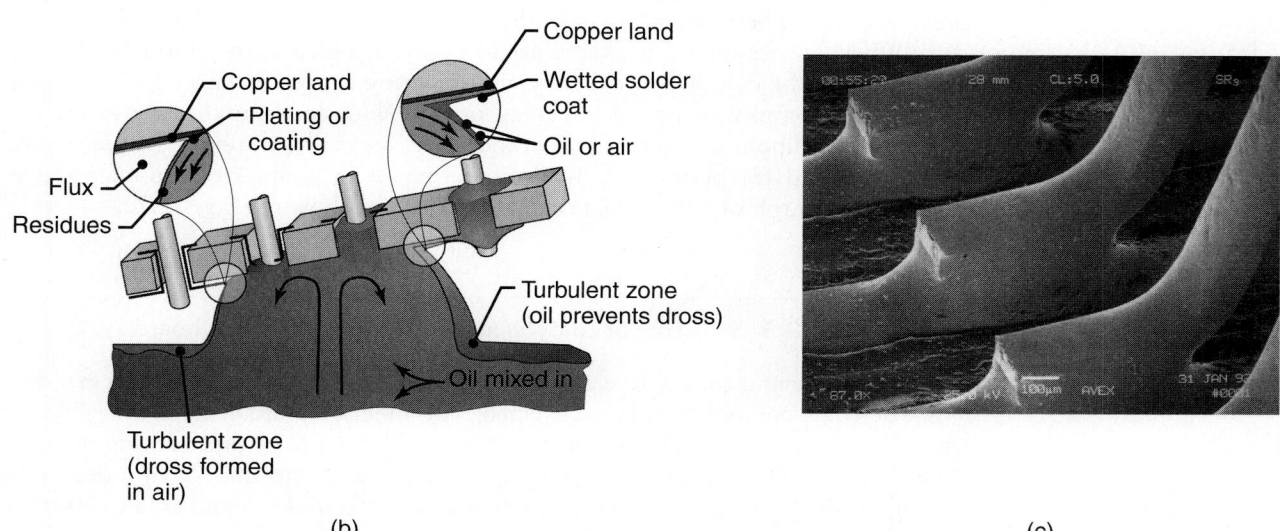

(b)   (c)

FIGURE 32.7   (a) Screening solder paste onto a printed circuit board in reflow soldering. (b) Schematic illustration of the wave-soldering process. (c) SEM image of wave-soldered joint on surface-mount device. *Source:* (a) After V. Solberg.

used very commonly during the attachment of electrical components to printed circuit boards. An additional benefit of reflow soldering is that the surface tension of the paste helps keep surface-mount packages aligned on their pads; this feature improves the reliability of the solder joints.

Once the paste has been placed and the joint assembled, the paste is heated in a furnace, and soldering takes place. In reflow soldering, the product is heated in a controlled manner, so that the following sequence of events occurs:

1. Solvents present in the paste are evaporated.
2. The flux in the paste is activated, and fluxing action occurs.
3. The components are preheated carefully.
4. The solder particles are melted, and they wet the joint.
5. The assembly is cooled at a low rate to prevent thermal shock and fracture of the solder joint.

While this process appears to be straightforward, there are several process variables for each stage, and good control over temperatures and exposures must be maintained at each stage in order to ensure proper joint strength.

Wave soldering. *Wave soldering* is a very common technique for attaching circuit components to their boards (see Section 28.11). To understand the principle of wave soldering, it is imperative to note that molten solder does not wet all surfaces. Indeed, solder will not stick to most polymer surfaces, and it is easy to remove while molten. Also, as can be observed with a simple hand-held soldering iron, the solder wets metal surfaces and forms a good bond only when the metal is preheated to a certain temperature. Therefore, wave soldering requires separate fluxing and preheating operations before it can be completed successfully.

A typical wave-soldering operation is shown in Fig. 32.7b. A standing laminar wave of molten solder is generated by a pump. Preheated and prefluxed circuit boards then are conveyed over the wave. The solder wets the exposed metal surfaces, but it does not remain attached to the polymer package for the integrated circuits, and it does not stick to the polymer-coated circuit boards. An *air knife* (which is basically a high-velocity jet of hot air) blows excess solder away from the joint to prevent bridging between adjacent leads.

When surface-mount packages are to be wave soldered, they must be bonded adhesively to the circuit board before the soldering can commence. This bonding usually is accomplished by (a) screening or stenciling epoxy onto the boards, (b) placing the components in their proper locations, (c) curing the epoxy, (d) inverting the board, and (e) performing wave soldering. A scanning-electron-microscope (SEM) photograph of a typical surface-mount joint is shown in Fig. 32.7c.

EXAMPLE 32.1 Soldering of components onto a printed circuit board

The computer and consumer electronics industries place extremely high demands on electronic components. It is expected that integrated circuits and other electronic devices will function reliably for extended periods of time, during which they can be subjected to significant temperature variations and to vibration. In recognition of this requirement, it is essential that the solder joints used to attach such devices to circuit boards be sufficiently strong and reliable, and also that the solder joints be applied extremely rapidly with automated equipment.

A continuing trend in the computer and the consumer electronics industries is toward the reduction of chip sizes and increasing compactness of circuit boards. Further space savings are achieved by mounting integrated circuits into surface-mount packages, which allow tighter packing on a circuit board and (more importantly) the mounting of components on both sides of a circuit board.

A challenging problem arises when a printed circuit board has both surface-mount and in-line circuits on the same board and it is desired to solder all of the joints via a reliable automated process. A subtle point should be recognized: all of the in-line circuits should be restricted to insertion from one side of the board. Indeed, there is no performance requirement which would dictate otherwise, and this restriction greatly simplifies manufacturing.

The basic steps in soldering the connections on such a board are as follows (Fig. 32.7b and c):

1. Apply solder paste to one side.
2. Place the surface-mount packages onto the board and insert in-line packages through the primary side of the board.
3. Reflow the solder.
4. Apply adhesive to the secondary side of the board.
5. Attach the surface-mount devices on the secondary side using the adhesive.
6. Cure the adhesive.
7. Perform a wave-soldering operation on the secondary side to produce an electrical attachment of the surface mounts and the in-line circuits to the board.

Applying solder paste is done with chemically etched stencils or screens so that the paste is placed only onto the designated areas of a circuit board. (Stencils are used more widely for fine-pitched devices and produce a more uniform paste thickness.) Surface-mount circuit components then are placed on the board, and the board is heated in a furnace to around 200°C (400°F) to reflow the solder and form strong connections between the surface mount and the circuit board.

At this point, the components with leads are inserted into the primary side of the board, their leads are crimped, and the board is flipped over. An adhesive pattern is printed onto the board, using a dot of epoxy at the center of a surface-mount component location. The surface-mount packages are then placed onto the adhesive, usually by high-speed automated, computer-controlled systems. The adhesive is then cured, the board is flipped, and wave soldering is done.

The wave-soldering operation simultaneously joins the surface-mount components to the secondary side and solders the leads of the in-line components from the board's primary side. The board then is cleaned and inspected prior to performing electronic quality checks.

32.3.4 Soldering applications and design guidelines

Soldering is used extensively in the electronics industry. Because soldering temperatures are relatively low, a soldered joint has very limited service use at elevated temperatures. Moreover, because solders generally do not have much strength, the process cannot be used for load-bearing (structural) members. However, joint strength can be improved significantly by *mechanical interlocking* of the joint (Fig. 32.8).

Soldering can be used to join various metals and thicknesses. Copper and such precious metals as silver and gold are easy to solder. Aluminum and stainless steels are difficult to solder because of their strong, thin oxide film. However, these and other metals can be soldered with the aid of special fluxes that modify surfaces.

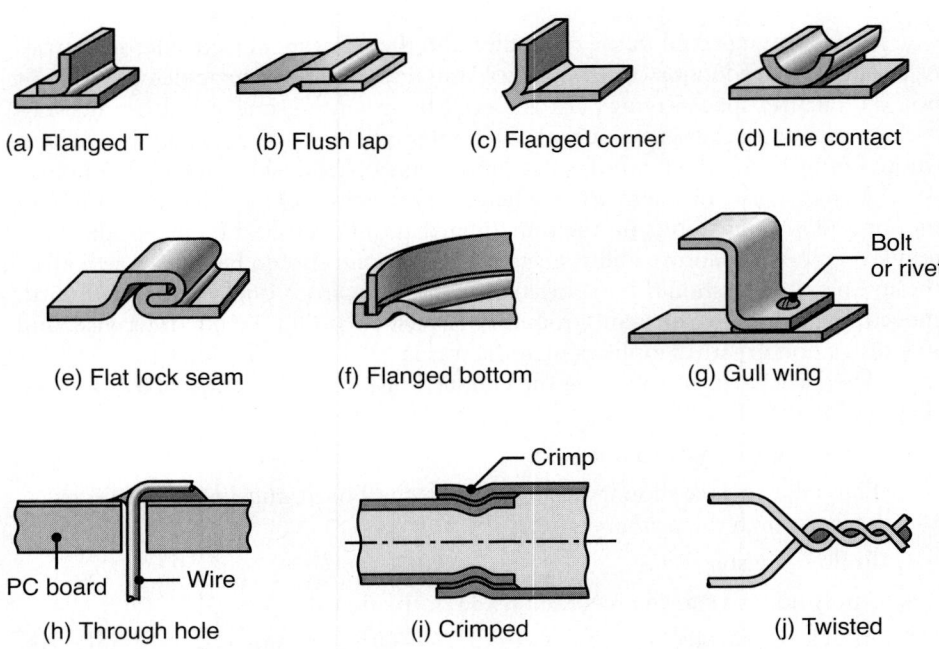

(a) Flanged T   (b) Flush lap   (c) Flanged corner   (d) Line contact

(e) Flat lock seam   (f) Flanged bottom   (g) Gull wing

(h) Through hole   (i) Crimped   (j) Twisted

FIGURE 32.8   Joint designs commonly used for soldering.

Although manual operations require skill and are time-consuming, soldering speeds can be high with automated equipment.

Design guidelines for soldering are similar to those for brazing. Some frequently used joint designs are shown in Fig. 32.8. Note the importance of large contact surfaces (because of the low strength of solders) for developing sufficient joint strength in soldered products. Since the faying surfaces generally would be small, solders rarely are used to make butt joints.

32.4 | Adhesive Bonding

Numerous parts and components can be joined and assembled by **adhesives** rather than by one or more of the joining methods described thus far. *Adhesive bonding* has been a common method of joining and assembly in such applications as labeling, packaging, bookbinding, home furnishings, and footware. Developed in 1905, plywood is a typical example of adhesive bonding with several layers of wood bonded with wood glue.

Adhesive bonding has gained increased acceptance in manufacturing ever since its first use on a large scale: the assembly of load-bearing components in aircraft during World War II (1939–1945). Adhesives are available in various forms: liquid, paste, solution, emulsion, powder, tape, and film. When applied, adhesives typically are about 0.1 mm (0.004 in.) thick.

To meet the requirements of a particular application, an adhesive may require one or more of the following properties (Table 32.3):

- Strength: shear and peel
- Toughness
- Resistance to various fluids and chemicals
- Resistance to environmental degradation, including heat and moisture
- Capability to wet the surfaces to be bonded

TABLE 32.3

Typical Properties and Characteristics of Chemically Reactive Structural Adhesives					
	Epoxy	Polyurethane	Modified acrylic	Cyanoacrylate	Anaerobic
Impact resistance	Poor	Excellent	Good	Poor	Fair
Tension-shear strength, MPa (10^3 psi)	15.4 (2.2)	15.4 (2.2)	25.9 (3.7)	18.9 (2.7)	17.5 (2.5)
Peel strength, N/m (lbf/in.)	<525 (3)	14,000 (80)	5250 (30)	<25 (3)	1750 (10)
Max. service temperature, °C (°F)	120 (250)	80 (175)	120 (250)	80 (175)	150 (300)

32.4.1 Types of adhesives and adhesive systems

Several types of adhesives are available, and more continue to be developed that provide adequate joint strength—including fatigue strength (Table 32.4). The three basic types of adhesives are the following:

1. **Natural adhesives**—such as starch, dextrin (a gummy substance obtained from starch), soya flour, and animal products.

2. **Inorganic adhesives**—such as sodium silicate and magnesium oxychloride.

3. **Synthetic organic adhesives**—which may be thermoplastics (used for non-structural and some structural bonding) or thermosetting polymers (used primarily for structural bonding).

Because of their strength, synthetic organic adhesives are the most important in manufacturing processes, particularly for load-bearing applications. They are classified as follows.

- **Chemically reactive:** polyurethanes, silicones, epoxies, cyanoacrylates, modified acrylics, phenolics, and polyimides. Also included are anaerobics, which cure in the absence of oxygen, such as Loctite® for threaded fasteners (see also Case Study 32.1).

- **Pressure sensitive:** natural rubber, styrene-butadiene rubber, butyl rubber, nitrile rubber, and polyacrylates.

- **Hot melt:** thermoplastics (such as ethylene-vinyl acetate copolymers, polyolefins, polyamides, and polyester) and thermoplastic elastomers.

- **Reactive hot-melt:** a thermoset portion (based on urethane's chemistry) with improved properties.

- **Evaporative** or **diffusion:** vinyls, acrylics, phenolics, polyurethanes, synthetic rubbers, and natural rubbers.

- **Film** and **tape:** nylon epoxies, elastomer epoxies, nitrile phenolics, vinyl phenolics, and polyimides.

- **Delayed tack:** styrene-butadiene copolymers, polyvinyl acetates, polystyrenes, and polyamides.

- **Electrically** and **thermally conductive:** epoxies, polyurethanes, silicones, and polyimides. Electrical conductivity is obtained by the addition of fillers, such as silver (used most commonly), copper, aluminum, and gold. Fillers that improve the electrical conductivity of adhesives generally also improve their thermal conductivity.

Adhesive systems.   These may be classified on the basis of their specific chemistries.

- **Epoxy-based systems:** These systems have high strength and high-temperature properties to as high as 200°C (400°F). Typical applications include automotive brake linings and as a bonding agent for sand molds for casting.

TABLE 32.4

General Characteristics of Adhesives		
Type	Comments	Applications
Acrylic	Thermoplastic; quick setting; tough bond at room temperature; two component; good solvent chemical and impact resistance; short work life; odorous; ventilation required	Fiberglass and steel sandwich bonds, tennis racquets, metal parts, and plastics
Anaearobic	Thermoset; easy to use; slow curing; bonds at room temperature; curing occurs in absence of air; will not cure where air contacts adherents; one component; and not good on permeable surfaces	Close fitting machine parts such as shafts and pulleys, nuts and bolts, and bushings and pins
Epoxy	Thermoset; one or two component; tough bond; strongest of engineering adhesives; high tensile and low peel strengths; resists moisture and high temperature; difficult to use	Metal, ceramic and rigid plastic parts
Cyanoacrylate	Thermoplastic; quick setting; tough bond at room temperature; easy to use; colorless	"Crazy glue™"
Hot melt	Thermoplastic; quick setting; rigid or flexible bonds; easy to apply; brittle at low temperatures; based on ethylene vinyl acetate, polyolefins, polyamides and polyesters	Bonds most materials Packaging, book binding, and metal can joints
Pressure sensitive	Thermoplastic variable strength bonds; primer anchors adhesive to roll tape backing material—a release agent on the back of web permits unwinding; and made of polyacrylate esters and various natural and synthetic rubbers	Tapes, labels, and stickers
Phenolic	Thermoset; oven cured; strong bond; high tensile and low impact strength; brittle; easy to use; cures by solvent evaporation	Acoustical padding, brake lining and clutch pads, abrasive grain bonding, and honeycomb structures
Silicone	Thermoset; slow curing; flexible; bonds at room temperature; high impact and peel strength; rubber-like	Gaskets and sealants
Formaldehyde Urea Melamine Phenol Resorcinol	Thermoset; strong with wood bonds; urea is inexpensive, available as powder or liquid and requires a catalyst; melamine is more expensive, cures with heat and the bond is waterproof; resorcinol forms waterproof bond at room temperature. Types can be combined	Wood joints, plywood, and bonding
Urethane	Thermoset; bonds at room temperature or oven cure; good gap-filling qualities.	Fiberglass body parts, rubber, and fabric
Water-based Animal Vegetable Rubbers	Inexpensive, nontoxic, nonflammable	Wood, paper, fabric, leather, and dry seal envelopes.

- **Acrylics:** These are suitable for applications with substrates that are not clean.
- **Anaerobic systems:** The curing of these adhesives is done under oxygen deprivation, and the bond usually is hard and brittle. Curing times can be reduced by external heat or by ultraviolet (UV) radiation.
- **Cyanoacrylate:** The bond lines are thin and the bond sets within from 5 to 40 s.

- **Urethanes:** These have high toughness and flexibility at room temperature, and they are used widely as sealants.
- **Silicones:** Highly resistant to moisture and solvents, these have high impact and peel strength; however, curing times are typically in the range of 1 to 5 days.

Many of these adhesives can be combined to optimize their properties, such as *epoxy-silicon*, *nitrile-phenolic*, and *epoxy-phenolic*. The least expensive adhesives are epoxies and phenolics, which are followed in affordability by polyurethanes, acrylics, silicones, and cyanoacrylates. Adhesives for high-temperature applications in a range up to about 260°C (500°F) (such as polyimides and polybenzimidazoles) are generally the most expensive. Most adhesives have an optimum temperature (ranging from about room to about 200°C) for maximum shear strength.

32.4.2 Electrically conducting adhesives

Although the majority of the usage of adhesive bonding is for mechanical strength, a relatively recent advance is the development and application of electrically conducting adhesives to replace lead-based solder alloys, particularly in the electronics industry. They require curing or setting temperatures that are lower than those required for soldering.

In these adhesives, the polymer is the matrix and contains conducting metals (fillers) in forms such as flakes and particles (see also Section 7.3 on *electrically conducting polymers*). There is a minimum proportion (by volume) of fillers necessary in order to make the adhesive electrically conducting; typically, it is in the range of 40 to 70%.

The size, shape, and distribution of the metallic particles, the heat and pressure application method, and the individual conducting particle contact geometry can be controlled to impart isotropic or anisotropic electrical conductivity to the adhesive. Metals used are typically silver, nickel, copper, and gold, as well as carbon. Recent developments include polymeric particles (such as polystyrene) coated with thin metallic films of silver or gold. Matrix materials are generally epoxies, although thermoplastics also are used and are available as film or as paste. Applications of electrically conducting adhesives include calculators, remote controls, and control panels. In addition, there are high-density uses in electronic assemblies, liquid-crystal displays, pocket TVs, and electronic games.

32.4.3 Surface preparation, process capabilities, and applications

Surface preparation is very important in adhesive bonding. Joint strength depends greatly on the absence of dirt, dust, oil, and various other contaminants. This dependence can be observed when one is attempting to apply an adhesive tape over a dusty or oily surface. Contaminants also affect the wetting ability of the adhesive and prevent even spreading of the adhesive over the interface. Thick, weak, or loose oxide films on workpiece surfaces are detrimental to adhesive bonding. On the other hand, a porous or a thin and strong oxide film may be desirable, particularly one with some surface roughness to improve adhesion or to introduce mechanical locking. However, the roughness must not be too high, because air may be trapped and the joint strength is reduced. Various compounds and primers are available which modify surfaces to improve adhesive-bond strength. Liquid adhesives may be applied by brushes, sprayers, and rollers.

Process capabilities. Adhesives can be used for bonding a wide variety of similar and dissimilar metallic and nonmetallic materials and components with different shapes, sizes, and thicknesses. Adhesive bonding also can be combined with

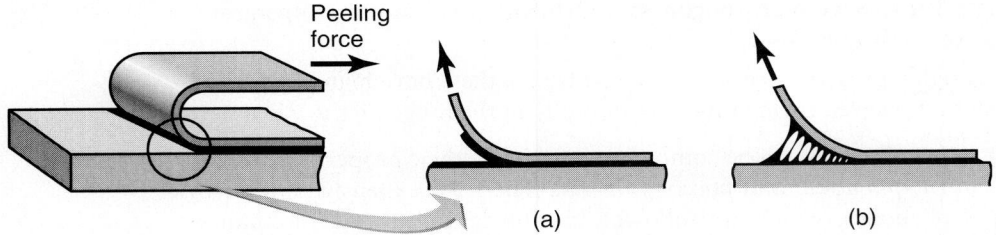

FIGURE 32.9   Characteristic behavior of (a) brittle and (b) tough adhesives in a peeling test. This test is similar to the peeling of adhesive tape from a solid surface.

mechanical joining methods (Section 32.5) to further improve the strength of the bond. Joint design and bonding methods require care and skill. Special equipment is usually required, such as fixtures, presses, tooling, and autoclaves and ovens for curing.

Adhesive joints are designed to withstand shear, compressive, and tensile forces, but they should not be subjected to *peeling* (Fig. 32.9). Note, for example, how easily you can peel adhesive tape from a surface yet be unable to slide it along the surface. During peeling, the behavior of an adhesive either may be brittle or ductile and tough—requiring high forces to peel it.

Applications.   Major industries that use adhesive bonding extensively are aerospace, automotive, appliances, and building products. Applications include: the attachment of rear-view mirrors to windshields, automotive brake-lining assemblies, laminated windshield glass, appliances, helicopter blades, honeycomb structures, and aircraft bodies and control surfaces.

An important consideration in the use of adhesives in production is *curing time*, which can range from a few seconds (at high temperatures) to several hours (at room temperature), particularly for thermosetting adhesives. Thus, production rates can be low compared to those of other joining processes. Furthermore, adhesive bonds for structural applications rarely are suitable for service above 250°C (500°F).

Nondestructive inspection of the quality and strength of adhesively bonded components can be difficult. Some of the techniques described in Section 36.10 (such as acoustic impact (tapping), holography, infrared detection, and ultrasonic testing) are effective nondestructive-testing methods for adhesives.

The major advantages of adhesive bonding are outlined below.

- It provides a bond at the interface either for structural strength or for nonstructural applications, such as sealing, insulation, the prevention of electrochemical corrosion between dissimilar metals, and the reduction of vibration and of noise (by means of internal damping at the joints).

- It distributes the load at an interface and thereby eliminates localized stresses that usually result from joining the components with mechanical fasteners, such as bolts and screws. Moreover, structural integrity of the sections is maintained (because no holes are required).

- The external appearance of the bonded components is unaffected.

- Very thin and fragile components can be bonded without significant increase in their weight.

- Porous materials and materials of very different properties and sizes can be joined.

- Because it usually is carried out at a temperature between room temperature and about 200°C (400°F), there is no significant distortion of the components or change in their original properties. Avoiding distortion is important, particularly for materials that are heat sensitive.

The major limitations of adhesive bonding are the following:

- A limited range of service temperatures.
- Bonding time can be long.
- The need for great care in surface preparation.
- The difficulty of testing bonded joints nondestructively, particularly for large structures.
- The limited reliability of adhesively bonded structures during their service life and under hostile environmental conditions (such as *degradation* by temperature, oxidation, stress corrosion, radiation, and dissolution) may be a significant concern.

The cost of adhesive bonding depends on the particular operation. In many cases, however, the overall economics of the process make adhesive bonding an attractive joining process. Sometimes it is the only one that is feasible or practical. The cost of equipment varies greatly, depending on the size and type of operation.

32.4.4 Design for adhesive bonding

- Designs for adhesive bonding should ensure that joints are subjected only to compressive, tensile, and shear forces and not to peeling or cleavage.
- Several joint designs for adhesive bonding are shown in Figs. 32.10 through 32.12. They vary considerably in strength; hence, selection of the appropriate design is important and should include considerations such as the type of loading and the environment.
- Butt joints require large bonding surfaces. Simple lap joints tend to distort under tension because of the force couple at the joint. (See Fig. 31.10a.)
- The coefficients of thermal expansion of the components to be bonded preferably should be close in order to avoid internal stresses during adhesive bonding. Also, situations in which thermal cycling can cause differential movement across the joint should be avoided.

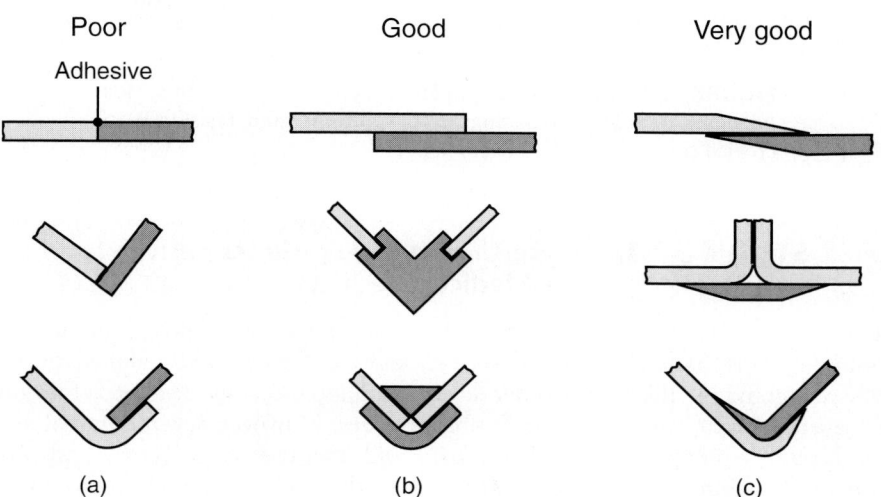

FIGURE 32.10   Various joint designs in adhesive bonding. Note that good designs require large contact areas between the members to be joined.

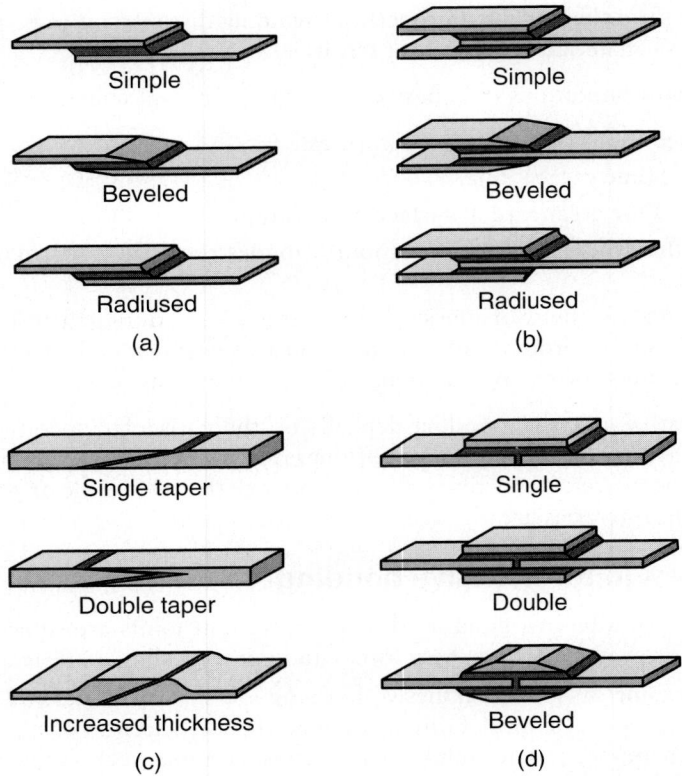

FIGURE 32.11 Desirable configurations for adhesively bonded joints: (a) single lap, (b) double lap, (c) scarf, and (d) strap.

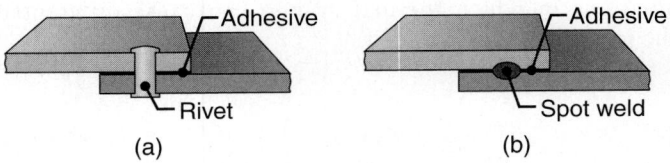

FIGURE 32.12 Two examples of combination joints, for purposes of improved strength, air tightness, and resistance to crevice corrosion.

CASE STUDY 32.1 | **Light Curing of Acryclic Adhesives for Medical Products**

Cobe Cardiovascular, Inc. is a leading manufacturer of blood collection and processing systems, as well as extracorporeal systems for cardiovascular surgery. In 1993, the company (like many other device manufacturers) used solvents for bonding several device components and subassemblies. However, several federal agencies began to encourage industries to avoid the use of solvents, and Cobe particularly wanted to eliminate their use of methylene chloride for environmental and occupational safety reasons. Towards this goal, the company began to redesign most of its assemblies to accept light-curing (ultraviolet or visible) adhesives. Most of the company's devices were made of transparent plastics. Consequently, their engineers required clear adhesive bonds for aesthetic purposes and with no tendency for stress cracking or crazing.

As an example of a typical product, Cobe's blood salvage or collection reservoir is an oval polycarbonate device approximately 300 mm (12 in.) tall, 200 mm (8 in.) in major diameter, and 100 mm (4 in.) deep (Fig. 32.13). The reservoir is a one-time use or disposable device; its purpose is to collect and hold the blood during open-heart and chest surgery or for arthroscopic and emergency room procedures. Up to 3000 cc of blood may be stored in the reservoir while awaiting passage into a 250 cc centrifuge, which cleans the blood and returns it to the patient after the surgical procedure is completed. The collection reservoir consists of a clear, polycarbonate lid joined to a polycarbonate bucket. The joint is a tongue-and-groove configuration; the goal was to create a strong, elastic joint that could withstand repeated stresses with no chance of leakage.

Light-cured acrylic adhesives offer a range of performance properties that make them well suited for this application. First and foremost, they achieve high bond strength to the thermoplastics typically used to form medical-device housings. For example, Loctite® 3211 (see *anaerobic adhesives*, Section 32.4.1), achieves shear strengths of 11 MPa (1600 psi) on polycarbonate. As important as the initial shear strength may be, it is even more important that the adhesive be able to maintain the high bond strength after sterilization. Fortunately, disposable medical devices typically are subjected to very few sterilization cycles during manufacture. Also, these adhesives can endure a limited number of cycles of gamma irradiation, electron beam irradiation, autoclaving, ethylene oxide, or chemical immersion.

Another consideration that makes light-cured adhesives well suited for this application is their availability in formulations that allow them to withstand large strains prior to yielding; Loctite® 3211 yields at elongations in excess of 200%. This flexibility

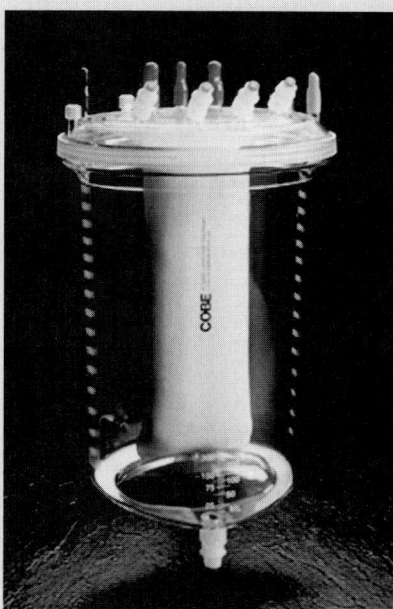

FIGURE 32.13   The Cobe Laboratories blood reservoir. The lid is bonded to the bowl with an airtight adhesive joint and tongue-in-groove joint. *Source*: Courtesy of Cobe Laboratories.

is critical, because the bonded joints typically are subjected to large amounts of bending and flexing when the devices are pressurized during qualification testing and use. If an adhesive is too rigid, it will fail this type of testing, even if it offers higher shear strength than a comparable and more flexible adhesive. Finally, light-cured acrylics are widely available in formulations that meet international quality standard certification (ISO; see Section 36.6), which means that when processed properly they will not cause biocompatibility problems in the final assembly.

While these performance features are attractive, the adhesive also must meet certain processing characteristics during manufacturing. Light-cured acrylic adhesives have found wide use in medical-device assembly/joining operations, because their processing characteristics are compatible with the high-speed automated manufacturing processes employed. These adhesives are available in a wide variety of viscosities and are dispensed easily through either pressure-time or positive-displacement dispensing systems. Once dispensed on the part, they can remain in contact with even highly stressed plastic parts for several minutes or longer without causing stress cracking or degradation of the plastic. For example, Liquid Loctite® 3211 can remain in contact with polycarbonate that has been bent to induce stresses up to 17 MPa (2500 psi) for more than 15 minutes without stress cracking. Finally, the adhesive can be converted completely from a liquid to a solid state in seconds when exposed to light of the proper intensity and wavelength.

Since Loctite® 3211 absorbs light in the visible as well as the ultraviolet range, it can be used successfully on plastics that contain UV blockers, such as many grades of polycarbonate. The ability to have a long open time when parts can be positioned yet cure the adhesive on demand is a unique benefit to light-curing adhesives, dramatically reducing scrap costs. The equipment used to irradiate the part with high-intensity light typically requires a space of 10×20 ft^2 on a production line, which generally is much less than that required for the ovens used by heat-cured adhesives or the racking shelves required for slower-curing adhesives. Since floor space carries a cost premium in clean-room environments, this is a significant benefit.

It also is important that the joint be designed properly to maximize joint performance. If the enclosure is bonded with a joint consisting of two flat faces in intimate contact, peel stresses (see Fig. 32.9) will be acting on the bond when the vessel is pressurized. Peel stresses are the most difficult type for an adhesive joint to withstand, due to the fact that the entire load is concentrated on the leading edge of the joint. The tongue-and-groove design used addressed this concern, whereby the groove acts as a reservoir for holding the adhesive during the dispensing operation. When the parts are mated and the adhesive is cured, this design allows much of the load on the joint (when the device is pressurized) to be translated into shear and tensile forces, which the adhesive is much better suited to withstand. The gap between the tongue and the groove can vary widely because most light-cured adhesives quickly can be cured to depths in excess of 5 mm (0.20 in.). This allows the manufacturer to have a robust joining process (meaning wide dimensional tolerances can be accommodated).

With the new design and by using this adhesive, the environmental concerns and the issues associated with solvent bonding thus were eliminated with the benefit of a safer, faster, and more consistent bond. The light-curing adhesive provided the aesthetic-bond line the company wanted—one that was clear and barely perceptible. It also provided the structural strength needed and, thus, maintained a competitive edge in the marketplace.

Source: Courtesy of P.J. Courtney, Loctite Corporation.

32.5 | Mechanical Fastening

Two or more components may have to be joined or fastened in such a way that they can be taken apart sometime during the product's service life or life cycle. Numerous products (including mechanical pencils, watches, computers, appliances, engines, and bicycles) have components that are fastened mechanically. *Mechanical fastening* may be preferred over other methods for the following reasons:

- Ease of manufacturing.
- Ease of assembly and transportation.
- Ease of disassembly, maintenance, parts replacement, or repair.
- Ease in creating designs that require movable joints, such as hinges, sliding mechanisms, and adjustable components and fixtures.
- Lower overall cost of manufacturing the product.

The most common method of mechanical fastening is by the use of bolts, nuts, screws, pins, and a variety of other **fasteners**; these operations are known also as **mechanical assembly**. Mechanical fastening generally requires that the components have *holes* through which the fasteners are inserted. These joints may be subjected to both shear and tensile stresses and should be designed to resist these forces.

Hole preparation. An important aspect of mechanical fastening is hole preparation. As described in Chapters 16, 23, and 27, a hole in a solid body can be produced by several processes, such as punching, drilling, chemical and electrical means, and high-energy beams. The selection of these depends on the type of material, its properties, and its thickness. Recall from Parts II and III that holes also may be *produced integrally* in the product during casting, forging, extrusion, and powder metallurgy. For improved accuracy and surface finish, many of these hole-making operations may be followed by finishing operations, such as shaving, deburring, reaming, and honing, as was described in various sections in Part IV.

Because of the fundamental differences in their characteristics, each of the hole-making operations produces holes with different surface finishes, surface properties, and dimensional characteristics. The most significant influence of a hole in a solid body is its tendency to reduce the component's fatigue life by stress concentration. For holes, fatigue life can be improved best by inducing compressive residual stresses on the cylindrical surface of the hole. These stresses usually are developed by pushing a round rod (*drift pin*) through the hole and expanding it by a very small amount. This operation plastically deforms the surface layers of the hole in a manner similar to that seen in shot peening or in roller burnishing (Section 34.2).

Threaded fasteners. Bolts, screws, and nuts are among the most commonly used *threaded fasteners*. Numerous standards and specifications (including thread dimensions, dimensional tolerances, pitch, strength, and the quality of the materials used to make these fasteners) are described in the references at the end of this chapter. Bolts and screws may be secured with nuts, or they may be *self-tapping*—whereby the screw either cuts or forms the thread into the part to be fastened. The self-tapping method is particularly effective and economical in plastic products where fastening does not require a tapped hole or a nut.

If the joint is to be subjected to vibration (such as in aircraft, engines, and machinery), several specially designed nuts and lock washers are available. They increase the frictional resistance in the torsional direction and so inhibit any vibrational loosening of the fasteners.

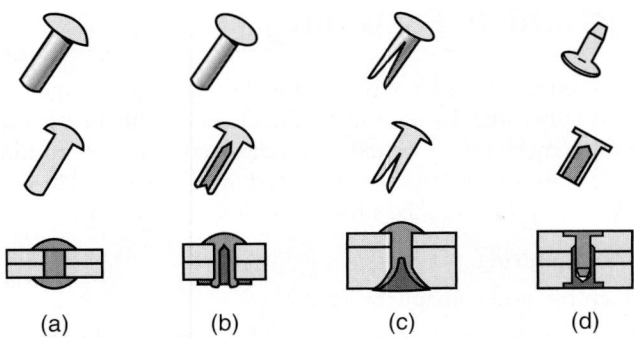

FIGURE 32.14   Examples of rivets: (a) solid, (b) tubular, (c) split or bifurcated, and (d) compression.

Rivets.   The most common method of permanent or semipermanent mechanical joining is by *riveting* (Fig. 32.14). Hundreds of thousands of rivets may be used in the construction and assembly of one large commercial aircraft. Rivets may be solid or tubular. Installing a solid rivet takes two steps: placing the rivet in the hole (usually punched or drilled) and deforming the end of its shank by upsetting it (*heading*). A hollow rivet is installed by flaring its smaller end. Explosives can be placed within the rivet cavity and detonated to expand the end of the rivet. Riveting may be done either at room or at elevated temperature. This operation also may be performed by hand or by mechanized means, including the use of programmable robots. Some design guidelines for riveting are illustrated in Fig. 32.15.

32.5.1 Other fastening methods

Numerous types of fasteners are used in joining and assembly applications. The most common types are described here.

Metal stitching and stapling.   The process of *metal stitching and stapling* (Fig. 32.16) is much like that of the ordinary stapling of papers. This operation is fast, and it is suitable particularly for joining thin metallic and nonmetallic materials, including wood. A common example is the stapling of cardboard containers. In

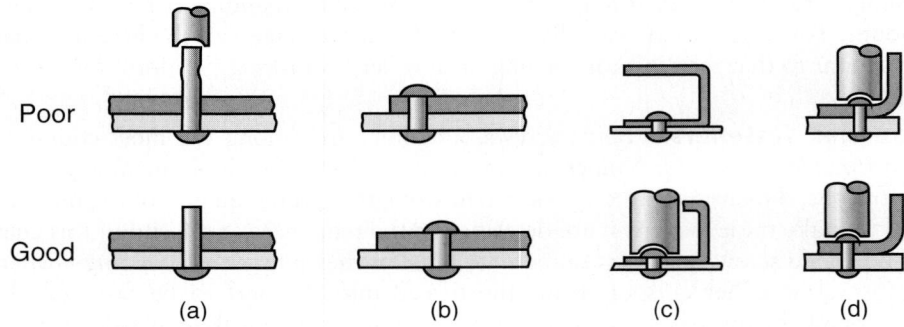

FIGURE 32.15   Design guidelines for riveting. (a) Exposed shank is too long; the result is buckling instead of upsetting. (b) Rivets should be placed sufficiently far from edges to avoid stress concentrations. (c) Joined sections should allow ample clearance for the riveting tools. (d) Section curvature should not interfere with the riveting process. *Source*: After J.G. Bralla.

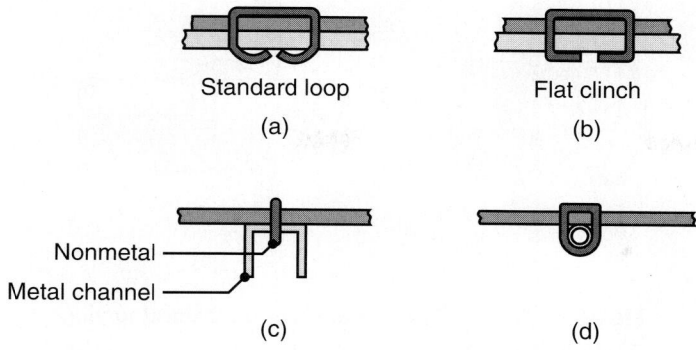

Standard loop

(a)

Flat clinch

(b)

Nonmetal ——

Metal channel ——

(c)

(d)

FIGURE 32.16   Typical examples of metal stitching.

clinching, the fastener material must be sufficiently thin and ductile to withstand the large localized deformation during sharp bending.

Seaming.   *Seaming* is based on the simple principle of folding two thin pieces of material together. It is a process much like the joining of two pieces of paper (when a paper clip is not available) by folding them at the corner (Fig. 32.17). Common examples of seaming are seen at the tops of beverage cans (see last illustration in Fig. 16.31), in containers for food and household products, and in heating and air-conditioning ducts. In seaming, the materials should be capable of undergoing bending and folding at very small radii; otherwise, they will crack. The performance and reliability of seams may be improved, (as well as making them impermeable) by the addition of adhesives, polymeric coatings and seals, or by soldering.

Crimping.   The *crimping* process is a method of joining without using fasteners. It can be done with beads or dimples (Fig. 32.18), which can be produced by shrinking or swaging operations. Crimping can be done on both tubular and flat parts, provided that the materials are sufficiently thin and ductile to withstand the large localized deformations. Caps are fastened to bottles by crimping just as some connectors are to electrical wiring.

Snap-in fasteners.   Several types of spring or snap-in fasteners are shown in Fig. 32.19. Such fasteners are used widely in automotive bodies and household appliances. They are economical, and they permit easy and rapid component assembly.

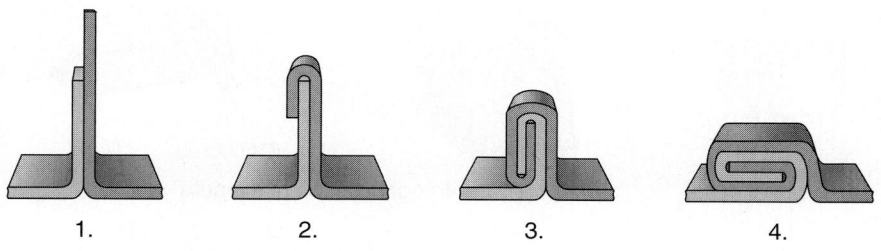

1.   2.   3.   4.

FIGURE 32.17   Stages in forming a double-lock seam.

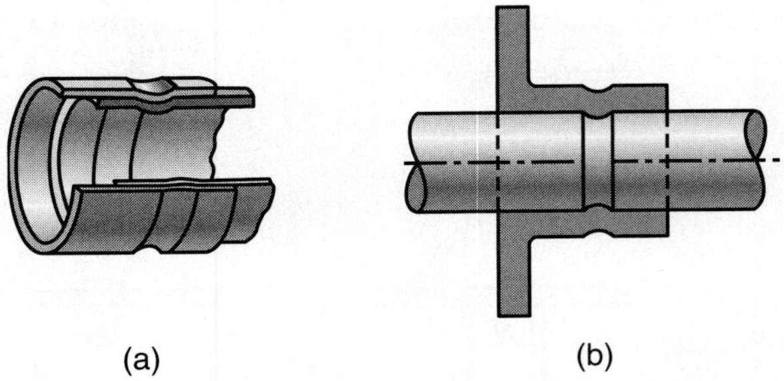

(a)   (b)

FIGURE 32.18   Two examples of mechanical joining by crimping.

Shrink and press fits.   Components also may be assembled by shrink fitting and press fitting. *Shrink fitting* is based on the thermal contractions of two components. Typical applications are the assembly of die components and mounting gears and cams onto shafts. In *press fitting*, one component is forced over another; this process results in high joint strength.

Shape-memory alloys.   The characteristics of these materials were described in Section 6.13, where we noted their use as fasteners because of their capability to recover their shape upon heating. Several advanced applications include their use as coupling in the assembly of titanium-alloy tubing for aircraft.

32.5.2 Design for mechanical fastening

The design of mechanical joints requires consideration of the type of loading to which the structure will be subjected and of the size and spacing of holes. *Compatibility* of the fastener material with that of the components to be joined is important.

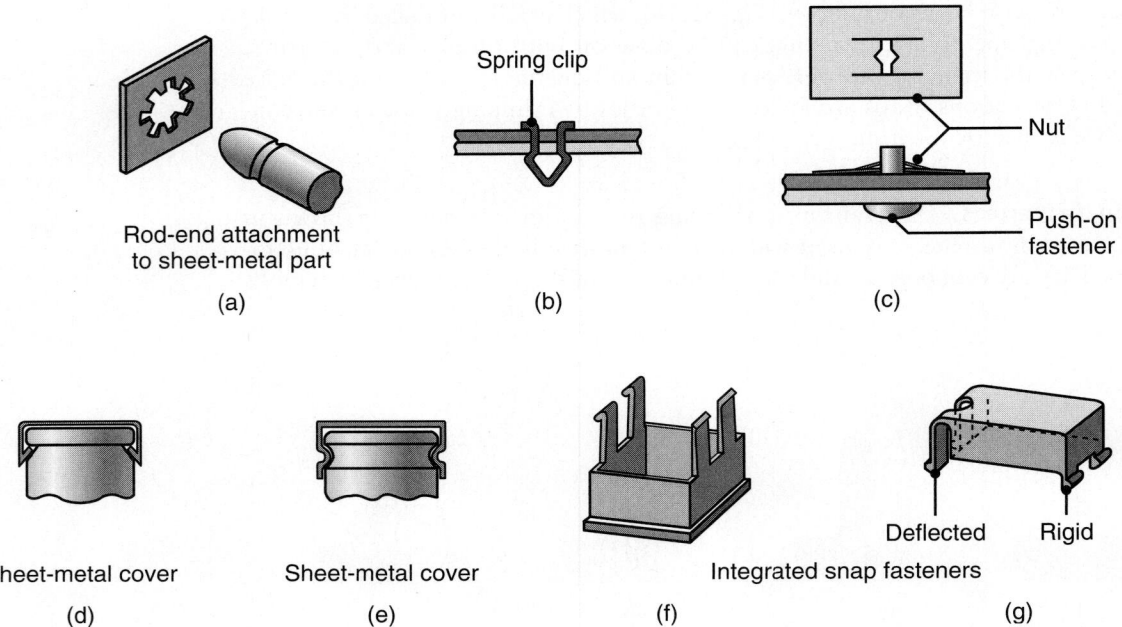

FIGURE 32.19   Examples of spring and snap-in fasteners used to facilitate assembly.

Incompatibility may lead to *galvanic corrosion*, also known as *crevice corrosion* (Section 3.8). For example, in a system where a steel bolt or rivet is used to fasten copper sheets, the bolt is anodic and the copper plate is cathodic; this combination causes rapid corrosion and loss of joint strength. Aluminum or zinc fasteners on copper products will react in a similar manner.

Other general design guidelines for mechanical joining include the following (see also Section 37.10):

- It generally is less costly to use fewer (but larger) fasteners than to use a large number of small ones.
- Part assembly should be accomplished with a minimum number of fasteners.
- The fit between parts to be joined should be as loose as possible to reduce costs and to facilitate the assembly process.
- Fasteners of standard size should be used whenever possible.
- Holes should not be too close to edges or corners to avoid the possibility of tearing the material when it is subjected to external forces.

32.6 | Joining Plastics, Ceramics, and Glasses

Plastics can be joined by many of the methods already described for joining metals and nonmetallic materials, especially by adhesive bonding and mechanical fastening.

32.6.1 Joining thermoplastics

Thermoplastics can be joined by the following methods: thermal means, adhesive bonding, solvent bonding, and mechanical fastening.

Thermal methods. Thermoplastics soften and melt as temperature is increased. Consequently, they can be joined by means whereby heat (either from an external or internal source) is generated at the interface, allowing **fusion** to take place. The heat softens the thermoplastic at the interface to a viscous or molten state and ensures a good bond with the application of pressure.

Because of the low thermal conductivity of thermoplastics and if applied at too high a rate, the heat source may burn or char the surfaces of the components, possibly causing difficulties in obtaining sufficiently deep fusion for joint strength. *Oxidation* also can be a problem in joining some polymers (such as polyethylene) causing degradation. Typically, an inert shielding gas (such as nitrogen) is used to prevent oxidation.

External heat sources may be chosen from among the following; the choice depends on the compatibility of the polymers to be joined:

- *Hot air, inert gases*, or a filler material of the same type also is used.
- In a process known as *hot-tool welding* or *hot-plate welding*, heated tools and dies are pressed against and heat the surfaces to be joined by the interdiffusion of molecular chains. This process commonly is used in butt-welded pipes and tubing (end-to-end).
- *Infrared radiation* (using high-intensity quartz heat lamps) is focused into a narrow beam onto the surface to be joined.
- *Radio-frequency* is useful particularly for thin films; frequencies are in the range of 100 to 500 Hz.

- *Dielectric heating* at frequencies of up to 100 MHz are effecive for the through–heating of polymers, such as nylon, polyvinyl chloride, polyurethane, and rubber.

- *Electrical-resistance* elements (such as wire or braids, or carbon-based tapes, sheets, and ropes) are placed at the interface to create heat by the passing of electrical current—a process known as *resistive-implant welding*. Alternatively, in *induction welding*, these elements at the interface may be subjected to a radio-frequency field. In both cases, because they are left in the weld zone, the elements at the interface must be compatible with the use of the joined product.

- *Lasers* emitting defocused beams at low power prevent degradation of the polymer.

Internal heat sources are developed by the following means:

- *Ultrasonic welding* is the most commonly used process for thermoplastics, particularly amorphous polymers such as ABS and high-impact polystyrene; frequencies are in the range of 20 to 40 kHz.

- *Friction welding* (also called *spin welding* for polymers) and *linear friction welding* (also called *vibration welding*) are useful particularly for joining polymers with a high degree of crystallinity, such as acetal, polyethylene, nylons, and polypropylene.

- *Orbital welding* is similar to friction welding, except that the rotary motion of one component is in an orbital path.

The fusion method is effective particularly with plastics that cannot be bonded easily by means of adhesives. Plastics (such as polyvinyl chloride, polyethylene, polypropylene, acrylics, and acrylonitrile butadiene styrene (ABS)) can be joined in this manner. For example, specially designed portable fusion-sealing systems are used to allow in-field joining of plastic pipe (usually made of polyethylene and used for natural-gas delivery).

Coextruded multiple food wrappings consist of different types of films, which are bonded by heat during extrusion (Section 19.2.1). Each film has a different function—for example, one film may keep out moisture, another may keep out oxygen, and a third film may facilitate heat sealing during the packaging process. Some wrappings have as many as seven layers—all bonded together during production of the film.

Adhesive bonding.   This method is best illustrated in the joining of sections of polyvinyl chloride pipe (used extensively in plumbing systems) and ABS pipe (used in drain, waste, and vent systems). The adhesive is applied to the connecting sleeve and pipe surfaces using a primer to improve adhesion (a step much like that using primers in painting), and then the pieces are pushed together.

Adhesive bonding of polyethylene, polypropylene, and polytetrafluoroethylene (Teflon) can be difficult, because adhesives do not bond readily to them. The surfaces of parts made of these materials usually have to be treated chemically to improve bonding. The use of adhesive primers or double-sided adhesive tapes also is effective.

Mechanical fastening.   This method is effective particularly for most thermoplastics (because of their inherent toughness and resilience) and for joining plastics to metals. Plastic or metal screws may be used. The use of self-tapping metal screws is a common practice. *Integrated snap fasteners* have gained wide acceptance for simplifying assembly operations; fastener geometries are shown in Fig. 32.19f and g. Because the fastener can be molded directly at the same time as the plastic, it adds very little to the cost of the assembly. This technique is very cost-effective, because it reduces assembly time and minimizes the number of parts required.

Solvent bonding.   This method consists of the following sequence of steps:

1. Roughening the surfaces with an abrasive.
2. Wiping and cleaning the surfaces with a solvent appropiate for the particular polymer.
3. Pressing the surfaces together and holding them together until sufficient joint strength is developed.

Electromagnetic bonding.   Thermoplastics also may be joined by magnetic means by embedding tiny particles on the order of 1 μm (40 μin.) in the polymer. A high-frequency field then causes induction heating of the polymer and melts it at the interfaces to be joined.

32.6.2 Joining thermosets

Thermosetting plastics (such as epoxy and phenolics) can be joined by the following techniques:

- *Threaded* or other molded-in *inserts*.
- *Mechanical fasteners*, particularly using self-tapping screws and integrated snap fasteners.
- **Solvent bonding.**
- *Co-curing*, where the two components to be joined are placed together and cured simultaneously.
- *Adhesive bonding*.

32.6.3 Joining ceramics and glasses

A wide variety and types of ceramics and glasses now are available with unique and important properties. Ceramics and glasses are used as products, as components of products, or as tools, molds, and dies. These materials often are assembled into components or subassemblies and are joined either with the same type of material or with different metallic or nonmetallic materials. Generally, ceramics, glasses, and many materials can be joined by adhesive bonding. A typical example is assembling broken ceramic pieces using a two-component epoxy, which is dispensed from two separate tubes, and is mixed just prior to application. Other joining methods include mechanical means, such as fasteners and spring or press fittings.

Ceramics.   As we described in Chapter 8, ceramics have properties that are very different from metallic and nonmetallic materials, including stiffness, hardness, brittleness, resistance to high temperature, and chemical inertness. Thus, joining them to each other or to other metallic or nonmetallic materials requires special considerations; and several highly specialized joining processes have been developed.

A common technique that is effective in joining difficult-to-bond material combinations consists of first applying a coating of a material that bonds itself well to one or both components—thus acting as a bonding agent. For example, the surface of *alumina ceramics* can be *metallized*, as described in Section 34.5. In this technique known as the *Mo-Mn process*, the ceramic part first is coated with a slurry of oxides of molybdenum and manganese. After being fired, this forms a glassy layer on the part's surface. This layer then is plated with nickel, and since the part now has a metallic surface, it can be brazed to a metal surface using an appropriate filler metal.

Tungsten carbide and titanium carbide can be brazed easily to other metals because both have a metallic matrix: WC has a matrix of cobalt, and TiC has nickel-molybdenum alloy as a matrix. Common applications include brazing cubic-boron-nitride or diamond tips to carbide inserts (Figs. 22.10 and 22.11) and carbide tips to masonry drills (Figs. 23.22). Depending on their particular structure, ceramics and metals also can be joined by *diffusion bonding*. It may be necessary to place a metallic layer at the joint to make it stronger.

Ceramic components also can be joined or assembled together during their primary shaping process; a common example is attaching handles to coffee mugs prior to firing them. Thus, the shaping of the whole product is done integrally rather than as an additional operation after the part is already made.

Glasses.   As evidenced by the availability of numerous glass objects, glasses can be bonded easily to each other. This commonly is done by first sofening the surface to be joined, then pressing the two pieces together, and cooling them. Bonding glass to metals also is possible because of the diffusion of metal ions into the amorphous surface structure of glass. However, the differences in their coefficients of thermal expansion must be taken into account.

32.7 | Economics of Joining Operations

As in the economics of welding operations (described in Section 31.8), the joining processes described in this chapter also greatly depend on several considerations. Reviewing Table VI.1, it can be noted that in relative terms the cost distribution for some of these processes is as follows:

- *Highest:* brazing, bolts, nuts, and other fasteners.
- *Intermediate:* riveting and adhesive bonding.
- *Low:* seaming and crimping.

Because of the variety of processes involved, below we outline briefly the general costs involved.

Brazing

- *Manual brazing:* The basic equipment costs about $300, but it can run $50,000+ for automated systems.
- *Furnace brazing:* Costs vary widely, ranging from about $2000 for simple batch furnaces to $300,000+ for continuous-vacuum furnaces.
- *Induction brazing:* For small units the cost is about $10000.
- *Resistance brazing:* Equipment costs range from $1000 for simple units to $10,000+ for larger, more complex units.
- *Dip brazing:* The cost of equipment varies widely, from $2000 to $200,000+; the more expensive equipment includes various computer-control features.
- *Infrared brazing:* Equipment cost ranges from $500 to $30,000.
- *Diffusion brazing:* The cost of equipment ranges from $50,000 to $300,000.

Soldering.   The cost of soldering equipment depends on its complexity and on the level of automation. The cost ranges from less than $100 for manual soldering irons to $50,000+ for automated equipment.

SUMMARY

- Joining processes that do not rely on fusion or pressure at the interfaces include brazing and soldering. These processes instead utilize filler material that requires some temperature rise in the joint. They can be used to join dissimilar metals of intricate shapes and various thicknesses.

- Adhesive bonding has gained increased acceptance in major industries, such as aerospace and automotive. In addition to good bond strength, adhesives have other favorable characteristics, such as the ability to seal, to insulate, to prevent electrochemical corrosion between dissimilar metals, and to reduce vibration and noise by means of internal damping in the bond.

- Surface preparation and joint design are important factors in adhesive bonding.

- Mechanical fastening is one of the oldest and most common joining methods. Bolts, screws, and nuts are common fasteners for machine components and structures which are likely to be taken apart for maintenance, for ease of transportation, or for various other reasons.

- Rivets are semipermanent or permanent fasteners used in buildings, bridges, and transportation equipment. A wide variety of other fasteners and fastening techniques is available for numerous permanent or semipermanent applications.

- Thermoplastics can be joined by fusion-welding techniques, by adhesive bonding, or by mechanical fastening. Thermosets usually are joined by mechanical means (such as molded-in inserts and fasteners) or by solvent bonding. Ceramics can be joined by adhesive bonding and metallizing techniques. Glasses are joined by heating the interface and by using adhesives.

KEY TERMS

Adhesive bonding	Integrated snap fastener	Shrink fitting
Braze welding	Hole preparation	Snap-in fastener
Brazing	Head-free solders	Soldering
Crimping	Mechanical fastening	Solvent bonding
Electrically conducting adhesives	Press fitting	Stapling
Fasteners	Reflow soldering	Stitching
Filler metal	Rivet	Threaded fasteners
Flux	Seaming	Wave soldering

BIBLIOGRAPHY

Adams, R.D., Comyn, J., and Wake, W.C., *Structural Adhesive Joints in Engineering*, Chapman & Hall, 1997.

ASM Handbook, Vol. 6: *Welding, Brazing, and Soldering*, ASM International, 1993.

Baghdachi, J., *Adhesive Bonding Technology*, Marcel Dekker, 1996.

Bickford, J.H., and Nassar, S. (eds.), *Handbook of Bolts and Bolted Joints*, Marcel Dekker, 1998.

Blake, A., *What Every Engineer Should Know about Threaded Fasteners*, Marcel Dekker, 1983.

Brazing Handbook, 4th ed., American Welding Society, 1991.

Engineered Materials Handbook, Vol. 3, *Adhesives and Sealants*, ASM International, 1991.

Hamrock, B.J., Jacobson, B., and Schmid, S.R., *Fundamentals of Machine Elements*, McGraw-Hill, 1998.

Handbook of Plastics Joining: A Practical Guide, William Andrew, Inc., 1996.

Humpston, G., and Jacobson, D.M., *Principles of Soldering*, ASM International, 2004.

Hwang, J.S., *Modern Solder Technology for Competitive Electronics Manufacturing*, McGraw-Hill, 1996.

Jacobson, D.M., and Humpston, G., *Principles of Brazing*, ASM International, 2005.

Lambert, L., *Soldering for Electronic Assemblies*, Marcel Dekker, 1988.

Landrock, A.H., *Adhesives Technology Handbook*, Noyes, 1985.

Lee, L.-H., *Adhesive Bonding*, Plenum, 1991.

Lincoln, B., Gomes, K.J., and Braden, J.F., *Mechanical Fastening of Plastics*, Marcel Dekker, 1984.

Manko, H.H., *Soldering Handbook for Printed Circuits and Surface Mounting*, Van Nostrand Reinhold, 1995.

———, *Solders and Soldering—Materials, Design, Production, and Analysis for Reliable Bonding*, McGraw-Hill, 1992.

Messler, R.W., Jr., *Joining of Advanced Materials*, Butterworth–Heinemann, 1993.

Nicholas, M.G., *Joining Processes: Introduction to Brazing and Diffusion Bonding*, Chapman & Hall, 1998.

Parmley, R.O. (ed.), *Standard Handbook of Fastening and Joining*, 3rd ed., McGraw-Hill, 1997.

Peaslee, R.L., *Brazing Footprints: Case Studies in High-Temperature Brazing*, Wall Colmonoy Corp., 2003.

Pecht, M.G., *Soldering Processes and Equipment*, Wiley, 1993.

Petrie, E.M., *Handbook of Adhesives and Sealants*, McGraw-Hill, 1999.

Rahn, A., *The Basics of Soldering*, Wiley Interscience, 1993.

Sadek, M.M., *Industrial Applications of Adhesive Bonding*, Elsevier, 1987.

Satas, D. (ed.), *Handbook of Pressure Sensitive Adhesive Technology*, 3rd ed., Satas & Associates, 1999.

Schwartz, M.M., *Brazing*, 2nd ed., ASM International, 2003.

———, *Ceramic Joining*, ASM International, 1990.

———, *Joining of Composite-Matrix Materials*, ASM International, 1994.

Shields, J., *Adhesives Handbook*, 3rd ed., Butterworths, 1984.

Skeist, I., *Handbook of Adhesives*, 3rd ed., Van Nostrand Reinhold, 1990.

Speck, J.A., *Mechanical Fastening, Joining, and Assembly*, Marcel Dekker, 1997.

Welding Handbook, 8th ed., 3 vols., American Welding Society, 1987.

Woodgate, R.W., *Handbook of Machine Soldering*, Wiley, 1996.

REVIEW QUESTIONS

32.1   Explain the principle underlying brazing.

32.2   What is the difference between brazing and braze welding?

32.3   Are fluxes necessary in brazing? If so, why?

32.4   Do you think that it is acceptable to differentiate brazing from soldering arbitrarily, by temperature of the application? Comment.

32.5   Why is surface preparation important in adhesive bonding?

32.6   Why have mechanical joining methods been developed? Give several specific examples of their applications.

32.7   Describe the similarities and differences between the functions of a bolt and those of a rivet.

32.8   What precautions should be taken in the mechanical joining of dissimilar metals?

32.9   What difficulties are involved in joining plastics? Why?

32.10   What is wave soldering?

32.11   What is a peel test? Why is it useful?

32.12   How is the filler metal applied in furnace brazing?

32.13   Why is welding not used in attaching electrical components to circuit boards?

32.14   Describe some applications in manufacturing for single-sided and some for double-sided adhesive tapes.

QUALITATIVE PROBLEMS

32.15   Comment on your observations concerning the joints shown in Figs. 32.3 and 32.6.

32.16   How different is adhesive bonding from other joining methods? What limitations does it have?

32.17   Discuss the need for fixtures for holding workpieces in the joining processes described in this chapter.

32.18   Soldering generally is applied to thinner components. Explain why.

32.19 Explain why adhesively bonded joints tend to be weak in peeling.

32.20 It is common practice to tin plate electrical terminals to facilitate soldering. Why is it tin that is used?

32.21 How important is a close fit for two parts that are to be brazed?

32.22 If you are designing a joint that needs to be strong and also needs to be disassembled a few times during the product life, what kind of joint would you use? Explain.

32.23 Loctite® is an adhesive used to keep bolts from vibrating loose; it basically glues the bolt to the nut and threaded hole. Explain how it functions.

QUANTITATIVE PROBLEMS

32.24 Refer to the simple butt and lap joints shown in the top left of Fig. 32.10. (a) Assuming the area of the butt joint is 3 mm × 20 mm and referring to the adhesive properties given in Table 32.3, estimate the minimum and maximum tensile force that this joint can withstand. (b) Estimate these forces for the lap joint assuming that its area is 15 mm × 15 mm.

32.25 In Fig. 32.12a, assume that the cross-section of the lap joint is 20 mm × 20 mm, and that the solid rivet diameter is 5 mm and it is made of copper. Using the strongest adhesive shown in Table 32.3, estimate the maximum tensile force that this joint can withstand.

32.26 As shown in Fig. 32.15a, a rivet can buckle if it is too long. Referring to Chapter 14 on forging, determine the maximum length-to-diameter ratio of a rivet so that it would not buckle during riveting.

32.27 Figure 32.4 shows qualitatively the tensile and shear strength in brazing as a function of joint clearance. Search the technical literature, obtain data, and plot these curves quantitatively. Comment on your observations.

SYNTHESIS, DESIGN, AND PROJECTS

32.28 Examine various household products and describe how their components are joined and assembled. Explain why those particular processes were used.

32.29 Name several products that have been assembled by (a) seaming, (b) stitching, and (c) soldering.

32.30 Suggest methods of attaching a round bar (made of a thermosetting plastic) perpendicularly to a flat metal plate. Discuss their strengths.

32.31 Describe the tooling and equipment that are necessary to perform the double-lock seaming operation shown in Fig. 32.17. Start with a flat sheet.

32.32 Prepare a list of design guidelines for joining by the processes described in this chapter. Would these guidelines be common to most processes? Explain.

32.33 What joining methods would be suitable for assembling a thermoplastic cover over a metal frame? Assume that the cover is removed periodically, like the top of a coffee can.

32.34 Answer Problem 32.33, but for a cover made of (a) a thermoset, (b) a metal, and (c) a ceramic. Describe the factors involved in your selection of methods.

32.35 Comment on workpiece size and shape limitations (if any) for each of the processes described in this chapter.

32.36 Describe part shapes that cannot be joined by the processes explained in this chapter. Give specific examples.

32.37 Describe the similarities and differences between the processes described in this chapter and those described in Chapters 30 and 31.

32.38 Give examples of products in which rivets in a structure or assembly may have to be removed and later replaced by new rivets.

32.39 Using the Internet, investigate the geometry of the heads of screws that are permanent fasteners—that is, ones that can be screwed in but not out.

32.40 Obtain a soldering iron and attempt to solder two wires together. First, try to apply the solder at the same time as you first put the soldering iron tip to the wires. Second, preheat the wires before applying the solder. Repeat the same procedure for a cool surface and a heated surface. Record your results and explain your findings.

32.41 Perform a literature search to determine the properties and types of adhesives used to affix artificial hips onto the human femur.

32.42 Using two strips of steel 1 in. wide and 8 in. long, design and fabricate a joint which gives the highest strength in a tension test in the longitudinal direction.

PART VII

Surface Technology

Our first visual or tactile contact with the objects around us is through their **surfaces**. We can see or feel surface roughness, texture, waviness, color, reflectivity, and other features like scratches and nicks. The preceding chapters described the properties of materials and manufactured components basically in terms of their bulk characteristics, such as strength, ductility, hardness, and toughness. Also included were some descriptions of the influences of surfaces on these properties—influences such as the effect of surface preparation on fatigue life and the sensitivity of brittle materials to surface scratches.

Machinery and accessories have many members that slide against each other: pistons and cylinders, slideways, bearings, and tools and dies for cutting and forming. Close examination reveals that (a) some of these surfaces are smooth while others are rough, (b) some are lubricated while others are dry, (c) some are subjected to heavy loads while others support light loads, (d) some are subjected to elevated temperatures (hot-working dies) while others are at room temperature, and (e) some surfaces slide against each other at high relative speeds (high cutting speeds) while others move slowly (the saddle or carriage on a machine-tool bed).

In addition to its geometric features, a surface constitutes a thin layer on the bulk material. A surface's physical, chemical, metallurgical, and mechanical properties depend not only on the material and its processing history but also on the environment to which the surface is exposed. The term **surface integrity** is used to describe its physical, chemical, and mechanical characteristics.

Because of the various mechanical, physical, thermal, and chemical effects which result from its processing history, the surface of a manufactured part generally possesses properties and behavior that are significantly different from those of its bulk. Although the bulk material generally determines the component's overall mechanical properties, the component's surfaces directly influence the part performance in the following areas (Fig. VII.1):

- Friction and wear of tools, molds, dies, and of the products made.
- Effectiveness of lubricants during the manufacturing process and throughout the part's service life.
- Appearance and geometric features of the part and their role in subsequent operations, such as welding, soldering, adhesive bonding, painting, and coating, as well as resistance to corrosion.
- Crack initiation as a result of surface defects like roughness, scratches, seams, and heat-affected zones, which can lead to weakening and premature failure of the part, such as through fatigue.

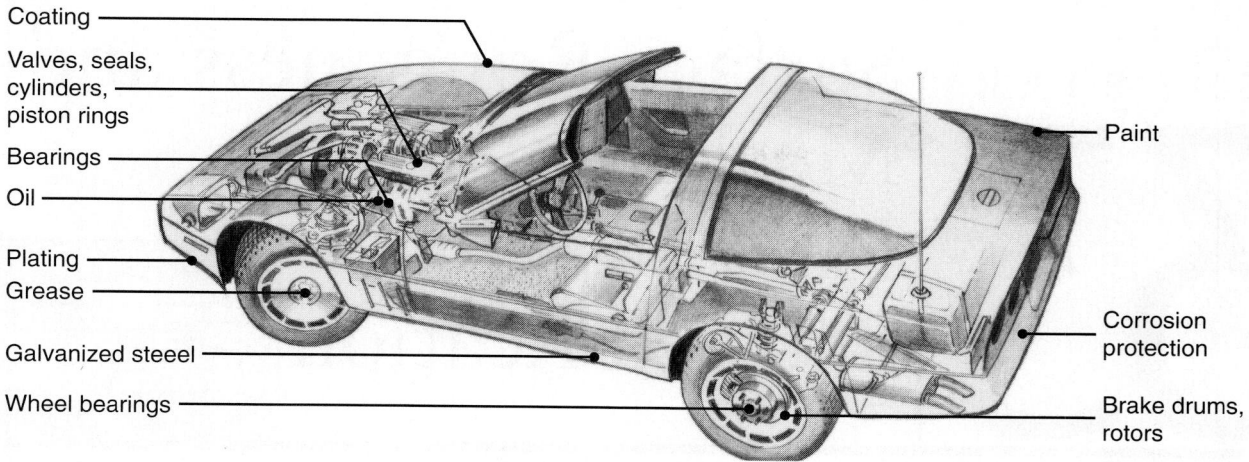

Coating

Valves, seals, cylinders, piston rings

Bearings

Oil

Plating

Grease

Galvanized steeel

Wheel bearings

Paint

Corrosion protection

Brake drums, rotors

FIGURE VII.1   Components in a typical automobile that are related to the topics described in Part VII.

• Thermal and electrical conductivity of contacting bodies. For example, rough surfaces have higher thermal and electrical resistances than smooth surfaces.

Following the outline shown in Fig. VII.2, this part of the book will present surface characteristics in terms of their structure and topography. The material and process variables that influence the friction and wear of materials then will be described. Several mechanical, thermal, electrical, and chemical methods can be used to modify surfaces for improved frictional behavior, effectiveness of lubricants, resistance to wear and corrosion, and surface finish and appearance.

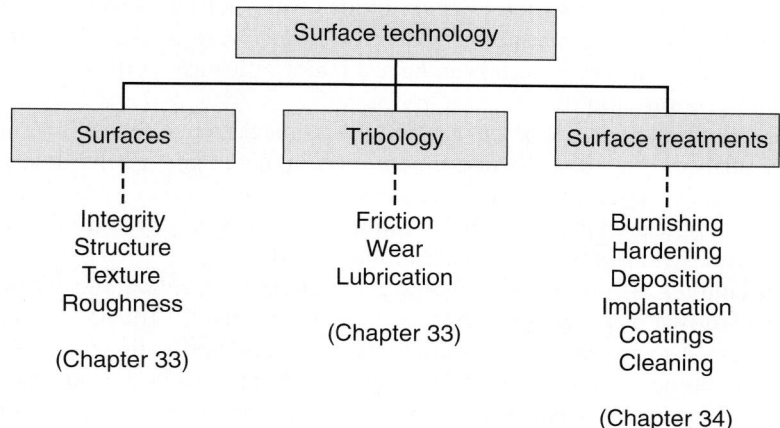

Surface technology

Surfaces

Tribology

Surface treatments

Integrity
Structure
Texture
Roughness

(Chapter 33)

Friction
Wear
Lubrication

(Chapter 33)

Burnishing
Hardening
Deposition
Implantation
Coatings
Cleaning

(Chapter 34)

FIGURE VII.2   An outline of topics covered in Part VII.

33
Surface Roughness and Measurement; Friction, Wear, and Lubrication

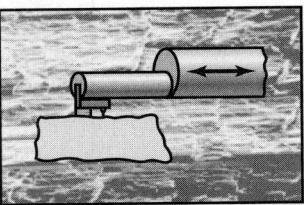

This chapter describes the features of surfaces that have a direct effect on both manufacturing processes and the service life of the parts produced. Specifically:

- Surface features such as roughness, texture, and lay.
- The nature of friction and its quantification.
- Factors involved in wear and wear reduction.
- Types of lubrication and their effects on manufacturing operations.
- Characteristics, selection, and applications of metalworking fluids.

33.1 | Introduction

This chapter begins with a description of the nature of surfaces, which are distinct entities and have properties significantly different from those of the bulk, particularly because of the presence of various surface oxide layers. Several defects can exist on a surface, depending on the manner in which the surface was generated. These defects (as well as various surface textures) can have a major influence on the surface integrity of workpieces, tools, and dies.

Two common methods of surface-roughness measurement in engineering practice are described, including the instrumentation involved and a brief description of surface-roughness requirements in engineering practice. Because of their increasing importance in precision manufacturing and nanofabrication, three-dimensional surface measurements also are discussed and illustrated.

This chapter also describes those aspects of *friction, wear,* and *lubrication*—collectively known as *tribology*—that are relevant to manufacturing processes and operations and to the service life of products. We first describe friction and wear for metallic and nonmetallic materials and how they are influenced by various material and process variables. An understanding of these relationships is necessary for the proper selection of tool and die materials and of appropriate metalworking fluids for a particular operation. Wear has a major economic impact, as it is estimated that (in the United States alone) the total cost of replacing worn parts is more than $100 billion per year.

This chapter then describes the fundamentals of metalworking fluids, including the types, characteristics, and application of commonly used liquid and solid lubricants and the lubrication practices employed. We also highlight the importance of biological and environmental considerations in the use, application, recycling, and ultimate disposal of metalworking fluids.

33.2 | Surface Structure and Integrity

Upon close examination of the surface of a piece of metal, it is found that it generally consists of several layers (Fig. 33.1):

1. The *bulk* metal (also known as the metal *substrate*) has a structure that depends on the composition and processing history of the metal.

2. Above this bulk metal is a layer that usually has been deformed plastically and work hardened to a greater extent than the bulk during the manufacturing process. The depth and properties of the work-hardened layer (*surface structure*) depend on such factors as the processing method used and how much frictional sliding the surface undergoes. For example, if the surface is produced by machining using a dull and worn tool or is ground with a dull grinding wheel, this layer will be relatively thick and usually will have residual stresses.

3. Unless the metal is processed and kept in an inert (oxygen-free) environment or is a noble metal (such as gold or platinum), an *oxide layer* forms over the work-hardened layer. The oxide on a metal surface generally is much harder than the base metal, hence it is more abrasive. As a result, it has important effects on friction, wear, and lubrication. For example,

   • *Iron* has an oxide structure with FeO adjacent to the bulk metal, followed by a layer of Fe_3O_4, and then a layer of Fe_2O_3 (which is exposed to the environment).

   • *Aluminum* has a dense, amorphous (without crystalline structure) layer of Al_2O_3 with a thick, porous, and hydrated aluminum-oxide layer over it.

   • *Copper* has a bright, shiny surface when freshly scratched or machined. Soon after, however, it develops a Cu_2O layer, which then is covered with a layer of CuO. This gives copper its somewhat dull color.

   • *Stainless steels* are "stainless" because they develop a protective layer of chromium oxide (*passivation*), as was described in Section 3.8.

4. Under normal environmental conditions, surface oxide layers generally are covered with *adsorbed* layers of gas and moisture.

5. Finally, the outermost surface of the metal may be covered with *contaminants* such as dirt, dust, grease, lubricant residues, cleaning-compound residues, and pollutants from the environment.

It can be seen that surfaces have properties that generally are very different from those of the substrate material. The factors which pertain to the surface structures of

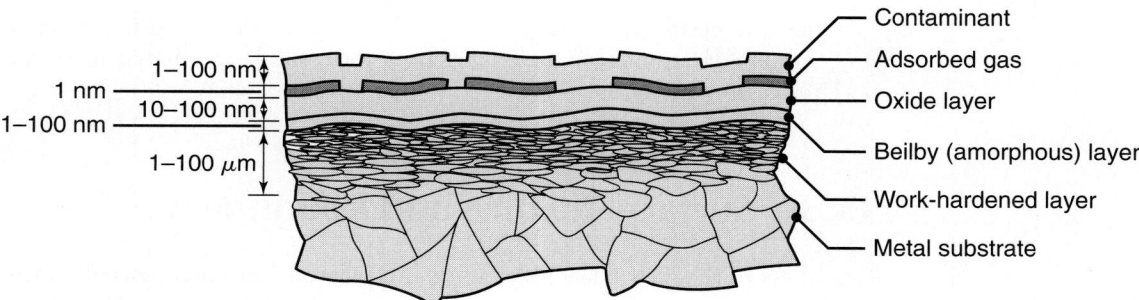

FIGURE 33.1   Schematic illustration of a cross-section of the surface structure of metals. The thickness of the individual layers depends on both processing conditions and processing environment. *Source:* After E. Rabinowicz and B. Bhushan.

the metals just described also are factors in the surface structure of plastics and ceramics. The surface texture of these materials depends (as with metals) on the method of production.

Surface integrity.   Surface integrity describes not only the topological (geometric) features of surfaces and their physical and chemical properties but their mechanical and metallurgical properties and characteristics as well. Surface integrity is an important consideration in manufacturing operations, because it influences such properties as fatigue strength, resistance to corrosion, and service life (for example, see Fig. 2.28).

Several surface defects caused by and produced during component manufacturing can be responsible for inadequate surface integrity. These defects usually are caused by a combination of factors, such as (a) defects in the original material, (b) the method by which the surface is produced, and (c) the lack of proper control of the process parameters (which can result in excessive stresses, temperatures, or surface deformation).

The following are general definitions of the major **surface defects** (listed in alphabetical order) found in practice:

- **Cracks** may be external or internal; cracks that require a magnification of $10\times$ or higher to be seen by the naked eye are called **microcracks**.
- **Craters** are shallow depressions.
- **Heat-affected zone** is the portion of a metal which is subjected to thermal cycling without melting, such as that shown in Fig. 30.17.
- **Inclusions** are small, nonmetallic elements or compounds in the material.
- **Intergranular attack** is the weakening of grain boundaries through liquid-metal embrittlement and corrosion.
- **Laps, folds,** and **seams** are surface defects resulting from the overlapping of material during processing.
- **Metallurgical transformations** involve microstructural changes caused by temperature cycling of the material; these may consist of phase transformations, recrystallization, alloy depletion, decarburization, and molten and then recast, resolidified, or redeposited material.
- **Pits** are shallow surface depressions, usually the result of chemical or physical attack.
- **Residual stresses** (tension or compression) on the surface are caused by nonuniform deformation and nonuniform temperature distribution.
- **Splatter** is small resolidified molten metal particles deposited on a surface, such as during welding.
- **Surface plastic deformation** is a severe surface deformation caused by high stresses due to factors such as friction, tool and die geometry, worn tools, and processing methods.

33.3 | Surface Texture and Roughness

Regardless of the method of production, all surfaces have their own characteristics, which collectively are referred to as **surface texture**. Although the description of surface texture as a geometrical property is complex, certain guidelines have been established for identifying surface texture in terms of well defined and measurable quantities (Fig. 33.2).

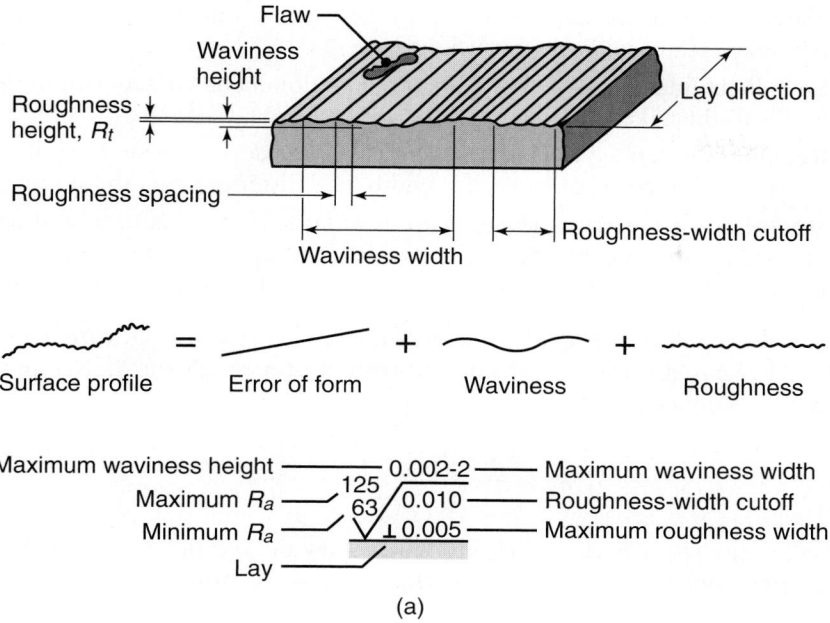

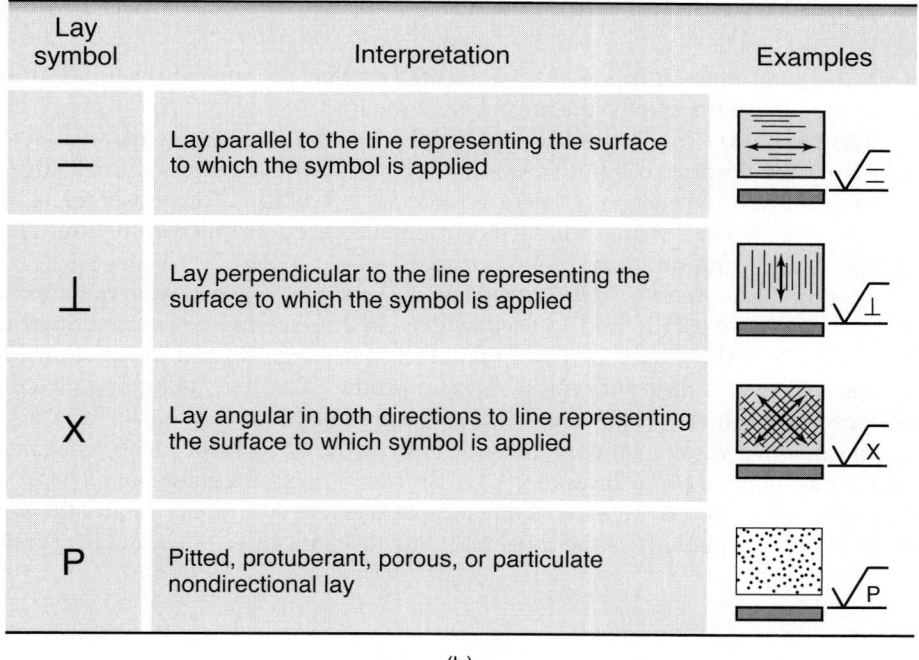

(b)

FIGURE 33.2   (a) Standard terminology and symbols to describe surface finish. The quantities are given in μin. (b) Common surface lay symbols.

- **Flaws** or **defects** are random irregularities, such as scratches, cracks, holes, depressions, seams, tears, or inclusions.
- **Lay** (**directionality**) is the direction of the predominant surface pattern, usually visible to the naked eye.
- **Roughness** is defined as closely spaced, irregular deviations on a small scale; it is expressed in terms of its height, width, and distance along the surface.
- **Waviness** is a recurrent deviation from a flat surface; it is measured and described in terms of the space between adjacent crests of the waves (*waviness width*) and height between the crests and valleys of the waves (*waviness height*).

Surface roughness generally is described by two methods. The **arithmetic mean value** (R_a) is based on the schematic illustration of a rough surface, as shown in Fig. 33.3. It is defined as

$$R_a = \frac{a + b + c + d + \cdots}{n} \tag{33.1}$$

where all ordinates, $a, b, c, \ldots$, are absolute values and n is the number of readings.

The **root-mean-square roughness** (R_q, formerly identified as *RMS*) is defined as

$$R_q = \sqrt{\frac{a^2 + b^2 + c^2 + d^2 + \cdots}{n}} \tag{33.2}$$

The datum line AB in Fig. 33.3 is located so that the sum of the areas above the line is equal to the sum of the areas below the line.

The **maximum roughness height** (R_t) also can be used and is defined as the height from the deepest trough to the highest peak. It indicates how much material has to be removed in order to obtain a smooth surface, such as by polishing.

The units generally used for surface roughness are μm (micron) or μin. Note that 1 μm = 40 μin., or 1 μin. = 0.025 μm.

Because of its simplicity, the arithmetic mean value (R_a) was adopted internationally in the mid-1950s and is used widely in engineering practice. Equations (33.1) and (33.2) show that there is a relationship between R_q and R_a, as shown by the ratio R_q/R_a. This ratio for typical surfaces produced by machining and finishing processes are: cutting, 1.1; grinding, 1.2; and lapping and honing, 1.4.

In general, a surface cannot be described by its R_a or R_q value alone, since these values are averages. Two surfaces may have the same roughness value but have actual topography which is very different. For example, a few deep troughs on an otherwise smooth surface will not affect the roughness values significantly. However, the type of

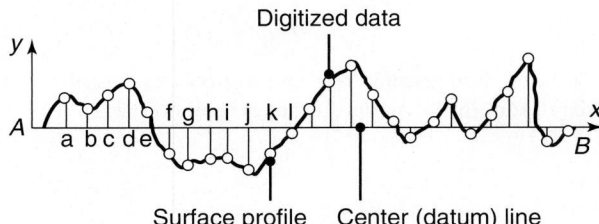

FIGURE 33.3   Coordinates used for surface-roughness measurement using Eqs. (33.1) and (33.2).

surface profile can be significant in terms of friction, wear, and fatigue characteristics of a manufactured product. Consequently, it is important to analyze a surface in great detail, particularly for parts to be used in critical applications.

Symbols for surface roughness.   Acceptable limits for surface roughness are specified on technical drawings by symbols, typically shown around the check mark in the lower portion of Fig. 33.2a, and the values of these limits are placed to the left of the check mark. The symbols and their meanings concerning the lay are given in Fig. 33.2b. Note that the symbol for the lay is placed at the lower right of the check mark. Symbols used to describe a surface specify only its roughness, waviness, and lay; they do not include flaws. Therefore, whenever necessary, a special note is included in technical drawings to describe the method which should be used to inspect for surface flaws.

Measuring surface roughness.   Typically, instruments called **surface profilometers** are used to measure and record surface roughness. A profilometer has a *diamond*

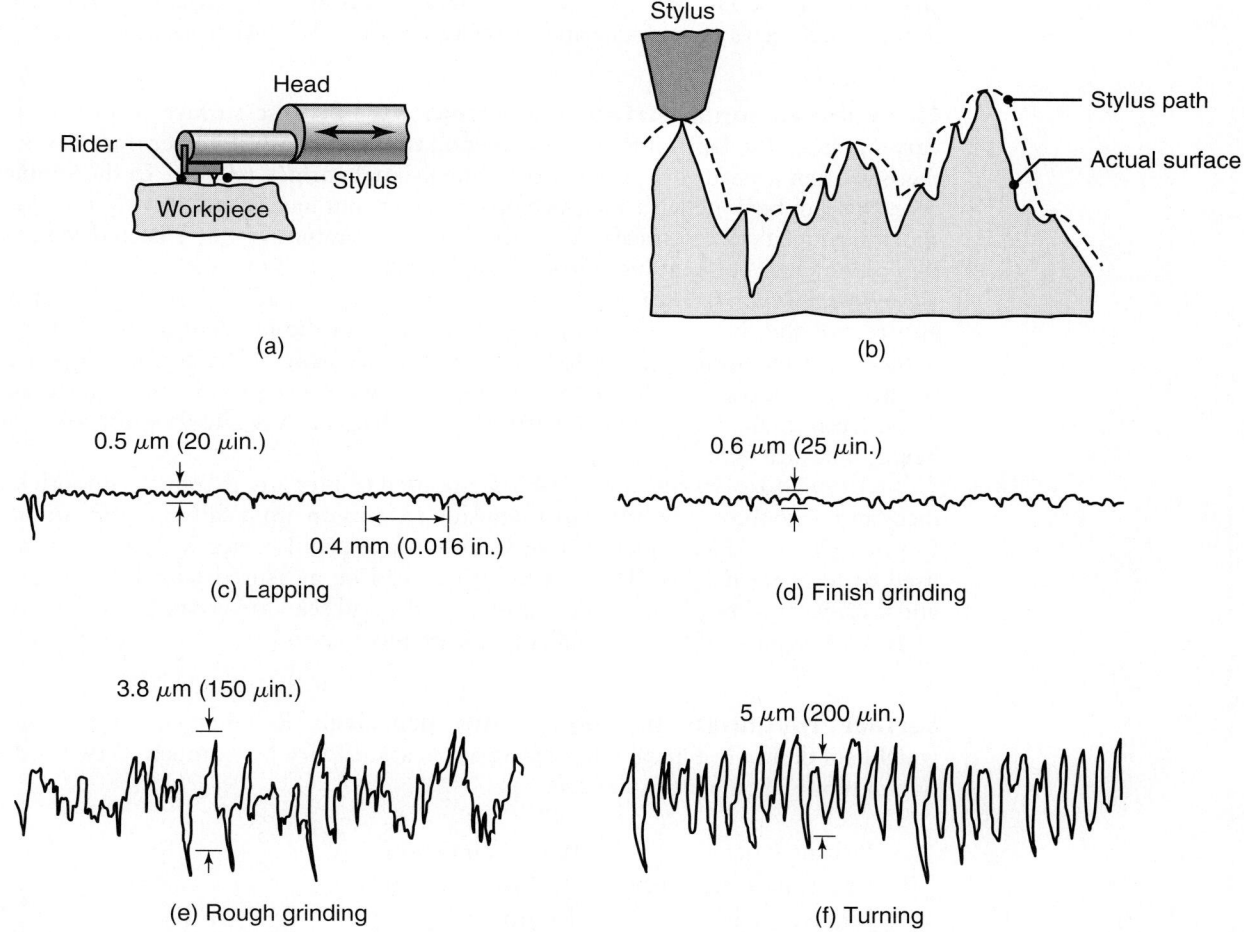

FIGURE 33.4   (a) Measuring surface roughness with a stylus. The rider supports the stylus and guards against damage. (b) Path of the stylus in surface-roughness measurements (broken line) compared to the actual roughness profile. Note that the profile of the stylus path is smoother than that of the actual surface. (c) through (f) Typical surface profiles produced by various machining and surface-finishing processes. Note the difference between the vertical and horizontal scales.

stylus that travels along a straight line over the surface (Fig. 33.4a). The distance that the stylus travels is called the **cutoff**, which generally ranges from 0.08 to 25 mm (0.003 to 1 in.). A cutoff of 0.8 mm (0.03 in.) is typical for most engineering applications. The rule of thumb is that the cutoff must be large enough to include 10 to 15 roughness irregularities, as well as all surface waviness.

In order to highlight the roughness, profilometer traces are recorded on an exaggerated vertical scale (a few orders of magnitude larger than the horizontal scale; see Fig. 33.4c through f); the magnitude of the scale is called **gain** on the recording instrument. Thus, the recorded profile is distorted significantly, and the surface appears to be much rougher than it actually is. The recording instrument compensates for any surface waviness; it indicates only roughness.

Because of the finite radius of the diamond stylus tip, the path of the stylus is different than the actual surface (note the path with the broken line in Fig. 33.4b), and the measured roughness is lower. The most commonly used stylus-tip diameter is 10 μm (400 μin.). The smaller the stylus diameter and the smoother the surface, the closer is the path of the stylus to the actual surface profile.

Surface roughness can be observed directly through an *optical* or *scanning electron microscope*. Stereoscopic photographs are useful particularly for three-dimensional views of surfaces and also can be used to measure surface roughness.

Three-dimensional surface measurement.   Because surface properties can vary significantly depending on the direction in which a profilometer trace is taken, there is often a need to measure three-dimensional surface profiles. In the simplest case, this can be done with a surface profilometer that has a capability of indexing a short distance between traces. A number of other alternatives have been developed, two of which are optical interferometers and atomic-force microscopes.

Optical-interference microscopes shine a light against a reflective surface and record the interference fringes that result from the incident and its reflected waves. This technique allows for a direct measurement of the surface slope over the area of interest. As the vertical distance between the sample and interference objective is changed, the fringe patterns also change, thus allowing for a surface height measurement.

Atomic-force microscopes (AFMs) are used to measure extremely smooth surfaces and have the capability of distinguishing atoms on atomically smooth surfaces. In principle, an AFM is merely a very fine surface profilometer with a laser that is used to measure heights. The surface profile can be measured with high accuracy and with vertical resolution on the atomic scale, and scan areas can be on the order of 100 μm square, although smaller areas are more common.

Surface roughness in engineering practice.   Requirements for surface-roughness design in typical engineering applications vary by as much as two orders of magnitude. Some examples are:

Bearing balls	0.025 μm (1 μin.)
Crankshaft bearings	0.32 μm (13 μin.)
Brake drums	1.6 μm (63 μin.)
Clutch-disk faces	3.2 μm (125 μin.)

Because of the many material and process variables involved, the range of roughness produced even within the same manufacturing process can be significant.

33.4 | Friction

Friction plays an important role in manufacturing processes because of the relative motion and the forces that always are present on tools, dies, and workpieces. Friction (a) dissipates energy (thus generating heat, which can have detrimental effects on an operation) and (b) impedes free movement at interfaces (thus friction can affect significantly the flow and deformation of materials in metalworking processes). On the other hand, friction is not always undesirable. For example, without friction, it would be impossible to roll metals, clamp workpieces on machines, or hold drill bits in chucks.

There are a number of explanations for the phenomenon of friction. A commonly accepted theory of friction is the **adhesion theory**, based on the observation that two clean and dry surfaces (regardless of how smooth they are) contact each other (*junction*) at only a fraction of their apparent contact area (Fig. 33.5). The maximum slope of the surface ranges typically between 5° and 15°. In such a situation, the normal (contact) load, N, is supported by the minute **asperities** (small projections from the surface) that are in contact with each other. Therefore, the normal stresses at these asperities are high; this causes *plastic deformation* at the junctions. Their contact creates an *adhesive bond*—the asperities form *microwelds*. Cold-pressure welding (see Section 31.2) is based on this principle on a large scale.

Another theory of friction is the **abrasion theory**, which is based on the notion that an asperity from a hard surface (such as a tool) penetrates and plows through a softer surface (workpiece). Plowing may (a) cause displacement of the material and/or (b) produce small chips or slivers, as in cutting and abrasive processes. Numerous other explanations for frictional behavior have been suggested, but for most applications in manufacturing, adhesion and abrasion mechanisms are most important.

The sliding motion between two bodies which have such an interface is possible only if a *tangential force* is applied. This tangential force is required to *shear* the junctions or plow through the softer material; it is called the **friction force**, F. The ratio of F/N (Fig. 33.5a) is the **coefficient of friction**, μ. Depending on the materials and processes involved, coefficients of friction in manufacturing vary significantly. For example, in metal-forming processes, it ranges from about 0.03 for cold working to 0.7 for hot working, and from 0.5 to as much as 2 for machining.

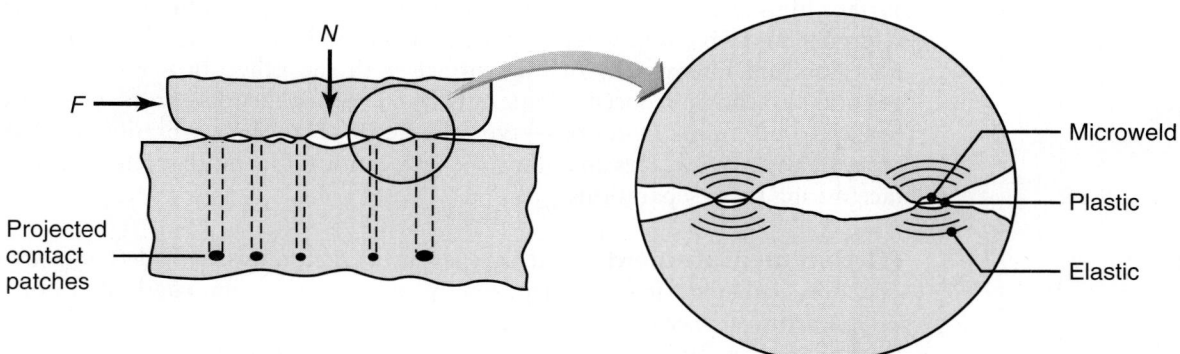

FIGURE 33.5 Schematic illustration of the interface of two bodies in contact showing real areas of contact at the asperities. In engineering surfaces, the ratio of the apparent-to-real areas of contact can be as high as 4 to 5 orders of magnitude.

Almost all of the energy dissipated in overcoming friction is converted into heat (a small fraction becomes stored energy in the plastically deformed regions; see Section 1.6), raising the interface temperature. The temperature increases with friction, sliding speed, decreasing thermal conductivity, and decreasing specific heat of the sliding materials. The interface temperature may be high enough to soften and even melt the surfaces and, sometimes, to cause microstructural changes. Temperature also affects the viscosity and other properties of lubricants, causing their breakdown. Note, for example, how butter and oil burn and are degraded when temperatures are excessive. These results, in turn, adversely affect the operations involved.

Friction in plastics and ceramics.   Although their strength is low compared to that of metals (see Tables 2.2 and 7.1), *plastics* generally possess low frictional characteristics. This property makes plastics better than metals for bearings, gears, seals, prosthetic joints, and general friction-reducing applications. In fact, polymers sometimes are described as *self-lubricating*.

The factors involved in the friction and wear of metals also generally are applicable to polymers. In sliding, the plowing component of friction in thermoplastics and elastomers is a significant factor because of their viscoelastic behavior (they exhibit both viscous and elastic behavior) and subsequent hysteresis loss (see Fig. 7.14). This condition can be simulated by dragging a dull nail across the surface of a rubber tire and observing how the tire surface quickly recovers its shape.

An important factor in plastics applications is the effect of temperature rise at the sliding interfaces caused by friction. As described in Section 7.3, thermoplastics lose their strength and become soft as temperature increases. Their low thermal conductivity and low melting points are significant when considering the heat generated by friction. If the temperature rise is not controlled, sliding surfaces can undergo permanent deformation and thermal degradation.

The frictional behavior of various polymers on metals is similar to that of metals on metals. The well known low friction of PTFE (Teflon) is attributed to its molecular structure, which has no reactivity with metals. Thus, its adhesion is poor, and its friction is low. The frictional behavior of *ceramics* is similar to that of metals; thus, adhesion and plowing at interfaces contribute to the friction force in ceramics as well.

Reducing friction.   Friction can be reduced through the selection of materials that have low adhesion (such as carbides and ceramics) and through the use of surface films and coatings. *Lubricants* (such as oils) or solid films (such as graphite) interpose an adherent film between the tool, die, and workpiece. This film minimizes adhesion and interactions of one surface with the other, thus reducing friction. Friction also can be reduced significantly by subjecting the tool- or die-workpiece interface to *ultrasonic vibrations*—typically at 20 kHz. These vibrations periodically separate the two surfaces and allow the lubricant to flow more freely into the interface during these separations.

Friction measurement.   The coefficient of friction usually is determined experimentally, either during manufacturing processes or in simulated laboratory tests using small-scale specimens of various shapes. A test that has gained wide acceptance—particularly for bulk deformation processes—is the *ring-compression test*. A flat ring is upset plastically between two flat platens (Fig. 33.6a). As its height is reduced, the ring expands radially outward. If friction at the interfaces is zero, both the inner and outer diameters of the ring expand as if it were a solid disk. With increasing friction, however, the internal diameter becomes smaller. For a particular reduction in height, there is a critical friction value at which the internal diameter increases from the original if μ is lower and decreases if μ is higher (Fig. 33.6b).

Good lubrication   Poor lubrication

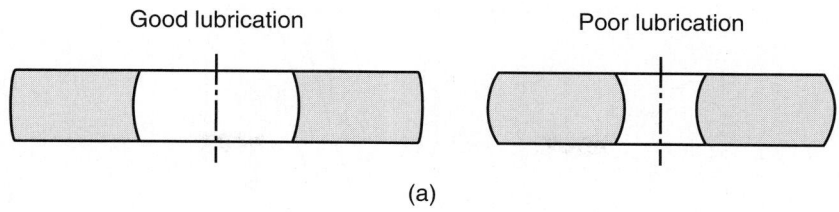

(a)

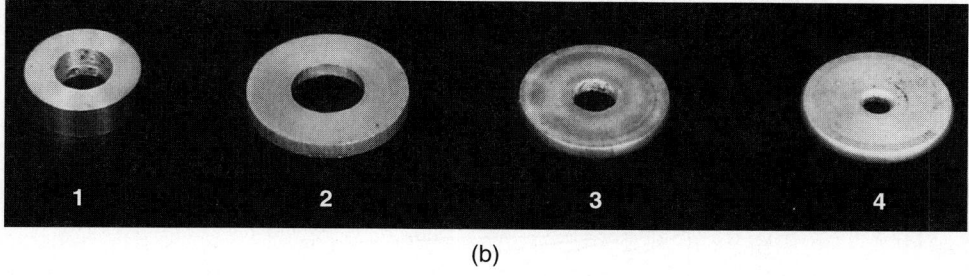

1   2   3   4

(b)

FIGURE 33.6   Ring-compression test between flat dies. (a) Effect of lubrication on type of ring-specimen barreling. (b) Test results: (1) original specimen and (2) to (4) increasing friction. *Source*: After A.T. Male and M.G. Cockcroft.

By measuring the change in the specimen's internal diameter and using the curves shown in Fig. 33.7 (which are obtained through theoretical analyses), the coefficient of friction can be determined. Each ring geometry and material has its own specific set of curves. The most common geometry of a specimen has an outer diameter, inner diameter, and height proportions of 6:3:2, respectively. The actual size of the specimen usually is not relevant in these tests. Thus, once the percentage of reduction in internal diameter and height is known, the magnitude of μ can be determined using the appropriate chart.

EXAMPLE 33.1 Determination of the coefficient of friction

In a ring-compression test, a specimen 10 mm in height with an outside diameter (OD) of 30 mm and inside diameter (ID) of 15 mm is reduced in thickness by 50%. Determine the coefficient of friction, μ, if the OD is 38 mm after deformation.

Solution It first is necessary to determine the new ID (which is obtained using volume constancy), as follows:

$$\text{Volume} = \frac{\pi}{4}(30^2 - 15^2)10 = \frac{\pi}{4}(38^2 - \text{ID}^2)5$$

Using this equation, the new ID is calculated as 9.7 mm.
    Thus, the change in internal diameter is

$$\Delta \text{ID} = \frac{(15 - 9.7)}{15} = 0.35 \text{ or } 35\% \text{ (decrease)}$$

With a 50% reduction in height and a 35% reduction in internal diameter, the fiction coefficient can be obtained from Fig. 33.7 as

$$\mu = 0.21$$

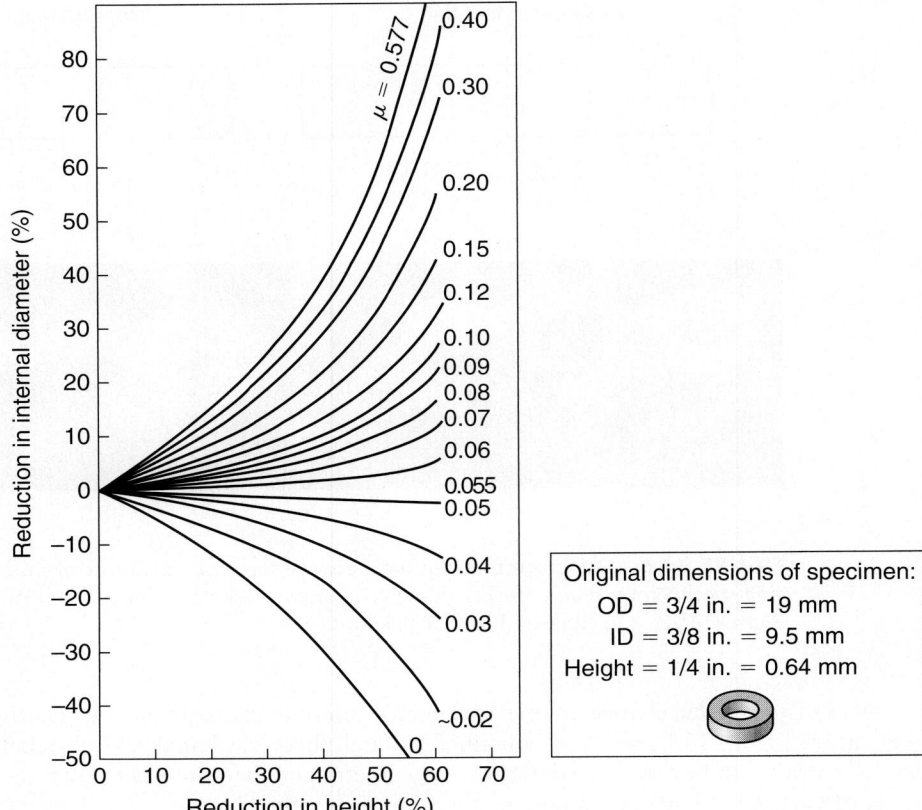

FIGURE 33.7   Chart to determine friction coefficient from a ring-compression test. Reduction in height and change in internal diameter of the ring are measured; then μ is read directly from this chart. For example, if the ring specimen is reduced in height by 40% and its internal diameter decreases by 10%, the coefficient of friction is 0.10.

33.5 | Wear

Wear has great technological and economic significance because it changes the shapes of tools and dies and, consequently, affects the tool life, tool size, and quality of the parts produced. The importance of wear is evident in the number of parts and components that continually have to be replaced or repaired. Examples of wear in manufacturing processes include dull drill bits that have to be reground, worn cutting tools that have to be indexed, tools and dies that have to be repaired or replaced, and countless others. *Wear plates* (which are subjected to high loads) are an important component in some metalworking machinery. These plates (also known as *wear parts*) are expected to wear and can be replaced easily.

Although wear generally alters a part's surface topography and may result in severe surface damage, it also can have a beneficial effect. The *running-in* period for various machines and engines produces this type of wear by removing the peaks from asperities (Fig. 33.8). Thus, under controlled conditions, wear may be regarded as a type of smoothing or polishing process. The major types of wear encountered in manufacturing operations are described next.

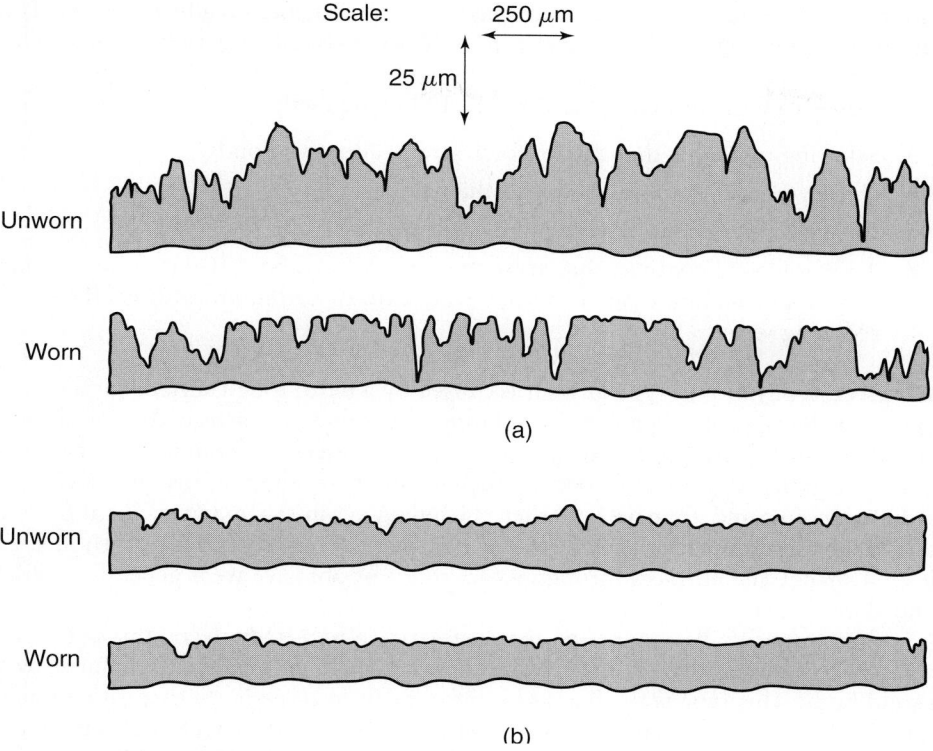

FIGURE 33.8 Changes in original (a) wire-brushed and (b) ground-surface profiles after wear. *Source*: After E. Wild and K.J. Mack.

Adhesive wear. If a tangential force is applied to the model shown in Fig. 33.9, shearing can take place either (a) at the original interface or (b) along a path below or above the interface, causing *adhesive wear*. Because of factors such as strain hardening at the asperity contact, diffusion, and mutual solid solubility, the adhesive bonds often are stronger than the base metals. Thus, during sliding, fracture usually follows a path in the weaker or softer component; this is how a wear fragment is generated. Although this fragment is attached to the harder component (upper surface in Fig. 33.9c), it eventually becomes detached during further rubbing at the interface and develops into a loose **wear particle**. This process is known as adhesive wear or sliding wear.

In more severe cases, such as ones with high loads and strongly bonded asperities, adhesive wear is described as scuffing, smearing, tearing, galling, or seizure

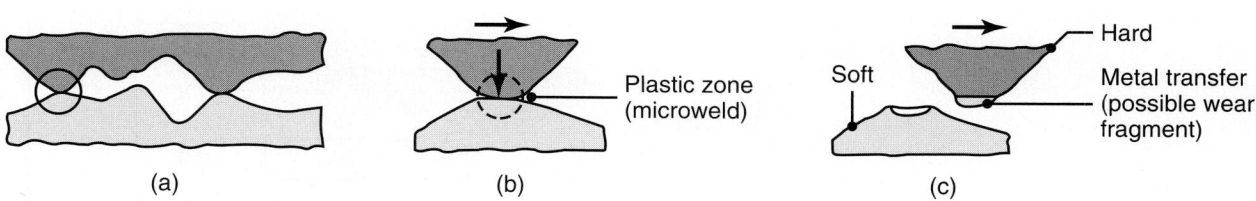

FIGURE 33.9 Schematic illustration of (a) two contacting asperities, (b) adhesion between two asperities, and (c) the formation of a wear particle.

(**severe wear**). Oxide layers on surfaces have a large influence on adhesive wear. They can act as a protective film, resulting in **mild wear**—which consists of small wear particles.

Adhesive wear can be reduced by the following methods:

a. Selecting materials that do not form strong adhesive bonds.

b. Using a harder material as one of the pair.

c. Using materials that oxidize more easily.

d. Applying hard coatings that serve methods **a** to **c** above. Coating one surface with a soft material (such as tin, silver, lead, or cadmium) also is effective in reducing wear.

Abrasive wear.   This type of wear is caused by a hard, rough surface (or a surface containing hard, protruding particles) sliding across another surface. As a result, *microchips* or *slivers* are produced, thereby leaving grooves or scratches on the softer surface (Fig. 33.10). In fact, processes such as filing, grinding, ultrasonic machining, and abrasive-jet and abrasive water-jet machining act in this manner. The difference is that, in these operations, the process parameters are controlled to produce the desired shapes and surfaces through wear; whereas, abrasive wear generally is unintended and unwanted.

There are two basic types of abrasive wear. In *two-body wear*, abrasive action takes place between two sliding surfaces or a hard, abrasive particle in contact with a solid body. This type is the basis of *erosive* wear. In *three-body wear*, an abrasive particle is present between two sliding solid bodies, such as a wear particle carried by a lubricant. Such a situation indicates the importance of properly filtering lubricants in metalworking operations to remove any contaminants.

The *abrasive wear resistance* of pure metals and ceramics has been found to be directly proportional to their hardness. Thus, abrasive wear can be reduced by increasing the hardness of materials (usually by heat treating) or by reducing the normal load. Elastomers and rubbers resist abrasive wear as well, because they deform elastically and then recover when abrasive particles cross over their surfaces. The best examples are automobile tires, which last a long time even though they are operated on road surfaces that generally are rough and abrasive. Even hardened steels would not last long under such conditions.

Corrosive wear.   Also known as *oxidation* or *chemical wear*, this type of wear is caused by chemical or electrochemical reactions between the surfaces and the environment. The fine corrosive products on the surface constitute the wear particles in this type of wear. When the corrosive layer is destroyed or removed through sliding or abrasion, another layer begins to form, and the process of removal and corrosive-layer formation is repeated. Among corrosive media are water, sea water, oxygen,

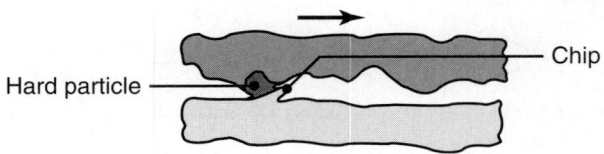

FIGURE 33.10   Schematic illustration of abrasive wear in sliding. Longitudinal scratches on a surface usually indicate abrasive wear.

acids and chemicals, and atmospheric hydrogen sulfide and sulfur dioxide. Corrosive wear can be reduced by:

- Selecting materials that will resist environmental attack.
- Controlling the environment.
- Reducing operating temperatures to lower the rate of chemical reaction.

Fatigue wear.   Fatigue wear (also called *surface fatigue* or *surface-fracture* wear) is caused when the surface of a material is subjected to cyclic loading; one example of this is the rolling contact in bearings. The wear particles usually are formed through *spalling* or *pitting*. Another type of fatigue wear is by *thermal fatigue*. Cracks on the surface are generated by thermal stresses from thermal cycling, such as when a cool die repeatedly contacts hot workpieces (heat checking). These cracks then join, and the surface begins to spall, producing fatigue wear.

This type of wear usually occurs in hot-working and die-casting dies. Fatigue wear can be reduced by:

- Lowering contact stresses.
- Reducing thermal cycling.
- Improving the quality of materials by removing impurities, inclusions, and various other flaws that may act as local points for crack initiation.

Other types of wear.   Several other types of wear can be seen in manufacturing operations.

- **Erosion** is caused by loose abrasive particles abrading a surface.
- **Fretting corrosion** occurs at interfaces that are subjected to very small reciprocal movements.
- **Impact wear** is the removal (by impacting particles) of small amounts of material from a surface.

In many cases, component wear is the result of a combination of different types of wear. Note in Fig. 33.11, for example, that even in the same forging die,

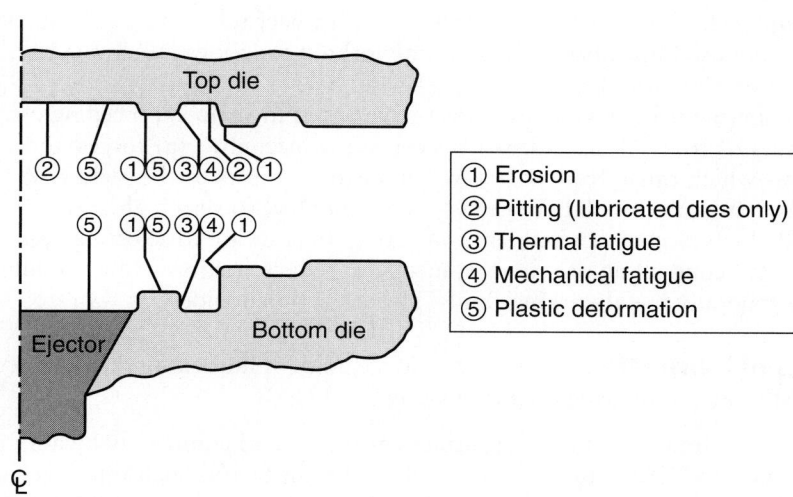

FIGURE 33.11   Types of wear observed in a single die used for hot forging. *Source*: After T.A. Dean.

various types of wear take place in different locations; a similar situation can exist in cutting tools, as shown in Fig. 21.18.

Wear of plastics and ceramics.   The wear behavior of plastics is similar to that of metals. Their abrasive wear behavior depends partly on the ability of the polymer to deform and recover elastically, as in elastomers. Typical polymers with good wear resistance are polyimides, nylons, polycarbonate, polypropylene, acetals, and high-density polyethylene. These polymers are molded or machined to make gears, pulleys, sprockets, and similar mechanical components. Because plastics can be made with a wide variety of compositions, they also can be blended with internal lubricants, such as PTFE (Teflon), silicon, graphite, molybdenum disulfide, and rubber particles that are interspersed within the polymer matrix.

Wear of reinforced plastics.   The wear resistance of reinforced plastics depends on the type, amount, and direction of reinforcement in the polymer matrix. Carbon, glass, and aramid fibers all improve wear resistance. Wear takes place when fibers are pulled out of the matrix (*fiber pullout*). Wear is highest when the sliding direction is parallel to the fibers, because they can be pulled out more easily in this case. Long fibers increase the wear resistance of composites, because they are more difficult to pull out and they prevent cracks in the matrix from propagating to the surface as easily.

Wear of ceramics.   When ceramics slide against metals, wear is caused by (a) small-scale plastic deformation and brittle surface fracture, (b) surface chemical reactions, (c) plowing, and (d) fatigue. Metals can be transferred to the oxide-type ceramic surfaces, forming metal oxides. Thus, sliding actually takes place between the metal and the metal-oxide surface.

33.6 | Lubrication

There is evidence that the lubrication of surfaces to reduce friction and wear dates back about four millenia to lubricate various linear and rotary moving components. For example, chariot wheels were lubricated with beef tallow in 1400 B.C. Various oils also were used for lubrication in metalworking processes (see Table I.2), beginning in about 600 A.D.

In manufacturing processes (as we have noted in various preceding chapters), the surfaces of tools, dies, and workpieces are subjected to (a) forces and contact pressures—which range from very low values to multiples of the yield stress of the workpiece material, (b) relative speeds—from very low to very high, and (c) temperatures—0which range from ambient to melting. In addition to selecting appropriate materials and controlling process parameters to reduce friction and wear, **lubricants** (or more generally **metalworking fluids**) also are applied widely.

Regimes of lubrication.   There are four regimes of lubrication that are generally of interest in manufacturing operations (Fig. 33.12):

1. In **thick-film lubrication**, the surfaces are separated completely by a film of lubricant, and lubricant viscosity is an important factor. Such films can develop in some regions of the workpiece in high-speed operations and also can develop from high-viscosity lubricants that become trapped at die-workpiece interfaces. A thick lubricant film results in a dull, grainy surface appearance on the

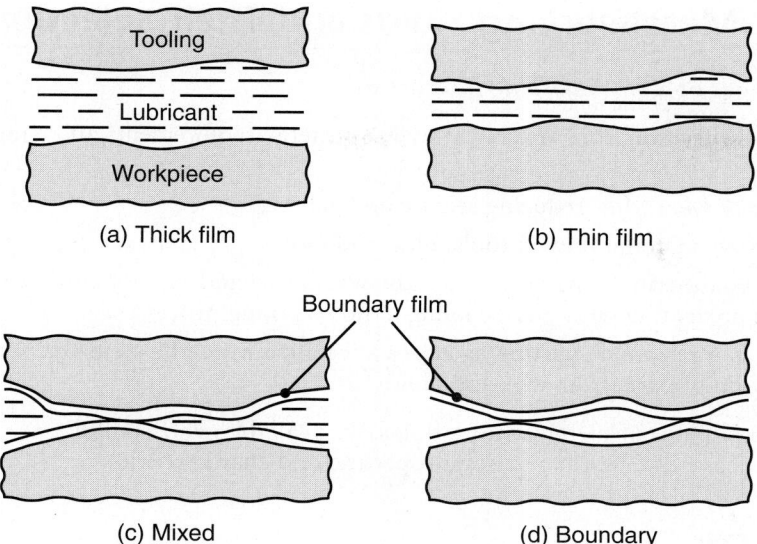

FIGURE 33.12   Types of lubrication generally occurring in metalworking operations. *Source*: After W.R.D. Wilson.

workpiece after forming operations, whereby the degree of roughness depends on grain size. In operations such as coining and precision forging, trapped lubricants are undesirable, because they prevent accurate shape generation.

2. As the load between the die and the workpiece increases or as the speed and viscosity of the metalworking fluid decrease, the lubricant film becomes thinner (**thin-film lubrication**). This condition raises the friction at the sliding interfaces and results in slight wear.

3. In **mixed lubrication**, a significant portion of the load is carried by the physical contact (asperities) of the two surfaces. The rest is carried by the fluid film trapped in pockets, such as the valleys between asperities.

4. In **boundary lubrication**, the load is supported by contacting surfaces covered with a *boundary film* of lubricant (Fig. 33.12d). This is a thin (molecular) lubricant layer that is attracted physically to the metal surfaces, thus preventing direct metal-to-metal contact of the two bodies and reducing wear. Boundary lubricants are typically natural oils, fats, fatty acids, esters, or soaps. However, boundary films can *break down* as a result of (a) desorption caused by high temperatures at the sliding interfaces or (b) being rubbed off during sliding. Deprived of this protective film, the sliding metal surfaces then may wear and score severely.

Other considerations.   The valleys in the surface roughnesses of the contacting bodies can serve as local reservoirs or pockets for lubricants, thereby supporting a substantial portion of the load. The workpiece (not the die) should have the rougher surface; otherwise, the rougher and harder die surface (acting as a file) may damage the workpiece surface. The recommended surface roughness on most dies is about 0.4 μm (15 μin.). The overall *geometry* of the interacting bodies is another important consideration in ensuring proper lubrication. The movement of the workpiece into the deformation zone (as during wire drawing, extrusion, and rolling) should allow a supply of lubricant to be carried into the die–workpiece interface. With the proper selection of process parameters, a relatively thick lubricant film can be entrained and maintained.

33.7 | Metalworking Fluids and Their Selection

The functions of a *metalworking fluid* are to:

- **Reduce friction,** thus reducing force and energy requirements and temperature rise.
- **Reduce wear,** thus reducing seizure and galling.
- **Improve material flow** in tools, dies, and molds.
- Act as a **thermal barrier** between the workpiece and its tool and die surfaces, thus preventing workpiece cooling in hot-working processes.
- Act as a **release** or **parting agent**—a substance which helps in the removal or ejection of parts from dies and molds.

Several types of metalworking fluids are available which fulfill these requirements and have diverse chemistries, properties, and characteristics.

33.7.1 Oils

Oils maintain high film strength on the surface of a metal, as we can observe when trying to clean an oily surface. Although they are very effective in the reduction of friction and wear, oils have low thermal conductivity and low specific heat. Consequently, they do not conduct away the heat generated by friction and plastic deformation effectively. Furthermore, it is difficult and costly to remove oils from component surfaces that are to be painted or welded, and it is difficult to dispose of them.

The sources of oils can be **mineral** (*petroleum* or *hydrocarbon*), **animal,** or **vegetable** in nature. Oils may be *compounded* with any number of additives or with other oils. Compounding is used to change such properties as viscosity–temperature behavior, surface tension, heat resistance, and boundary-layer characteristics. Mineral oils (with or without additives) used undiluted are known as **neat oils.** Oils can be contaminated by the lubricants used for the slideways and various components of the machine tools and metalworking machinery. These oils have different characteristics than those used for the process itself and thus can have adverse effects. When present in the metalworking fluid itself, these oils are known as **tramp oil.**

33.7.2 Emulsions

An *emulsion* is a mixture of two immiscible liquids (usually of oil and water in various proportions) along with additives. *Emulsifiers* are substances that prevent the dispersed droplets in a mixture from joining together (hence the term immiscible). Milky in appearance, emulsions also are known as **water-soluble oils** or **water-based coolants** and are of two types: direct and indirect. In a *direct emulsion,* mineral oil is dispersed in water in the form of very small droplets. In an *indirect emulsion,* water droplets are dispersed in the oil. Direct emulsions are important metalworking fluids, because the presence of water gives them high cooling capacity. They are effective particularly in high-speed machining where temperature rise has detrimental effects on tool life, the surface integrity of workpieces, and the dimensional accuracy of parts.

33.7.3 Synthetic and semisynthetic solutions

Synthetic solutions are chemical fluids that contain inorganic and other chemicals dissolved in water; they do not contain any mineral oils. Various chemical agents are added to a particular solution to impart different properties. *Semisynthetic* solutions are basically synthetic solutions to which small amounts of emulsifiable oils have been added.

33.7.4 Soaps, greases, and waxes

Soaps are typically reaction products of sodium or potassium salts with fatty acids. Alkali soaps are soluble in water, but other metal soaps generally are insoluble. Soaps are effective *boundary lubricants* and can form thick film layers at die–workpiece interfaces, particularly when applied on conversion coatings for cold metalworking applications.

Greases are solid or semisolid lubricants and generally consist of soaps, mineral oil, and various additives. They are highly viscous and adhere well to metal surfaces. Although used extensively in machinery, greases are of limited use in manufacturing processes.

Waxes may be of animal or plant (*paraffin*) origin. Compared to greases, they are less "greasy" and are more brittle. Waxes are of limited use in metalworking operations—except as lubricants for copper and (as chlorinated paraffin) as lubricants for stainless steels and high-temperature alloys.

33.7.5 Additives

Metalworking fluids usually are blended with various additives, such as the following:

- Oxidation inhibitors
- Rust-preventing agents
- Foam inhibitors
- Wetting agents
- Odor-controlling agents
- Antiseptics

Sulfur, chlorine, and *phosphorus* are important oil additives. Known as **extreme-pressure** (EP) **additives** and used singly or in combination, they react chemically with metal surfaces and form adherent surface films of metallic sulfides and chlorides. These films have low shear strength and good anti-weld properties and, thus, effectively can reduce friction and wear. On the other hand, they may preferentially attack the cobalt binder in tungsten-carbide tools and dies (*selective leaching*), causing changes in the surface roughness and integrity of those tools.

33.7.6 Solid lubricants

Because of their unique properties and characteristics, several solid materials are used as lubricants in manufacturing operations. Described here are four of the most commonly used *solid lubricants.*

Graphite. The general properties of graphite were described in Section 8.6. Graphite is weak in shear along its basal planes (see Fig. 1.4), thus it has a low coefficient of friction in that direction. It can be an effective solid lubricant, particularly at elevated temperatures. However, friction is low only in the presence of air or moisture. In a vacuum or an inert-gas atmosphere, friction is very high; in fact, graphite can be abrasive in these situations. Graphite can be applied either by rubbing it on surfaces or by making it part of a *colloidal* (dispersion of small particles) *suspension* in a liquid carrier such as water, oil, or an alcohol.

Molybdenum disulfide.   This is a widely used lamellar solid lubricant; it is somewhat similar in appearance to graphite. However, unlike graphite, it has a high friction coefficient in an ambient environment. Oils commonly are used as carriers for molybdenum disulfide (MoS_2) and as a lubricant at room temperature. Molybdenum disulfide can be rubbed onto the surfaces of a workpiece.

Metallic and polymeric films.   Because of their low strength, thin layers of soft metals and polymer coatings also are used as solid lubricants. Suitable metals include lead, indium, cadmium, tin, silver, polymers such as PTFE (Teflon), polyethylene, and methacrylates. However, these coatings have limited applications because of their lack of strength under high contact stresses and at elevated temperatures.

Soft metals also are used to coat high-strength metals, such as steels, stainless steels, and high-temperature alloys. For example, copper or tin is chemically deposited on the surface of a metal before it is processed. If the oxide of a particular metal has low friction and is sufficiently thin, the oxide layer can serve as a solid lubricant, particularly at elevated temperatures.

Glasses.   Although it is a solid material, glass becomes viscous at elevated temperatures and, hence, can serve as a liquid lubricant. Viscosity is a function of temperature (but not of pressure) and depends on the type of glass. Poor thermal conductivity also makes glass attractive, since it acts as a thermal barrier between hot workpieces and relatively cool dies. Glass lubrication typically is used in such applications as hot extrusion and forging.

Conversion coatings.   Lubricants may not always adhere properly to workpiece surfaces, particularly under high normal and shearing stresses. This property has the greatest effects in forging, extrusion, and the wire drawing of steels, stainless steels, and high-temperature alloys. For these applications, the workpiece surfaces first are transformed through a chemical reaction with acids (hence the term *conversion*). This reaction leaves a somewhat rough and spongy surface, which acts as a carrier for the lubricant. After treatment, any excess acid from the surface is removed using borax or lime. A liquid lubricant (such as a soap) then is applied to the surface. The lubricant film adheres to the surface and cannot be scraped off easily. *Zinc-phosphate* conversion coatings often are used on carbon and low-alloy steels. *Oxalate* coatings are used for stainless steels and high-temperature alloys.

Fullerenes or buckyballs.   Discussed in Section 8.6, these are carbon molecules in the shape of soccer balls. When placed between sliding surfaces, these molecules act like tiny ball bearings. They perform well as solid lubricants and are effective particularly in aerospace applications as bearings.

33.7.7 Selection of metalworking fluids

Selecting a metalworking fluid for a particular application and workpiece material involves consideration of several factors:

1. Particular manufacturing process.
2. Workpiece material.
3. Tool or die material.
4. Processing parameters.
5. Compatibility of the fluid with the tool and die materials and the workpiece.
6. Required surface preparation.
7. Method of fluid application.

8. Removal of the fluid and cleaning of the workpiece after processing.

9. Contamination of the fluid by other lubricants, such as those used to lubricate machinery.

10. Storage and maintenance of fluids.

11. Treatment of waste lubricant.

12. Biological and environmental considerations.

13. Costs involved in all of the aspects listed here.

When selecting an oil as a lubricant, it is necessary to investigate its viscosity, temperature, and pressure characteristics. Low viscosity can have a significantly detrimental effect and cause high friction and wear. The specific function of a metal-working fluid (whether primarily a *lubricant* or a *coolant*) also must be taken into account. Water-based fluids are very effective coolants, but as lubricants, they are not as effective as oils. It is estimated that water-based fluids are used in 80 to 90% of all machining applications.

Metalworking fluids should:

- Not leave any harmful residues that could interfere with machinery operations.

- Not stain or corrode the workpiece or equipment.

- Be checked periodically for deterioration caused by bacterial growth, accumulation of oxides, chips, wear debris, and general degradation and breakdown due to temperature and time. The presence of wear particles is particularly important, because they cause damage to the system; proper inspection and filtering are thus essential.

After the completion of manufacturing operations, workpiece surfaces usually have lubricant residues; these should be removed prior to further processing, such as welding or painting. Oil-based lubricants are more difficult and expensive to remove than water-based fluids. Various cleaning solutions and techniques used for this purpose will be described in Section 34.16.

Biological and environmental considerations.   Biological and environmental considerations are important factors in metalworking-fluid selection. Potential hazards may be involved by contacting or inhaling some of these fluids, such as inflammation of the skin (*dermatitis*) and long-term exposure to *carcinogens*. The improper disposal of metalworking fluids may cause adverse effects on the environment as well. To prevent or restrict the growth of microorganisms such as bacteria, yeasts, molds, algae, and viruses, chemicals (*biocides*) are added to metalworking fluids.

Much progress has been made in developing environmentally safe fluids and the technology and equipment for their proper treatment, recycling, and disposal. Laws and regulations concerning the manufacture, transportation, use, and disposal of metalworking fluids are promulgated by the Occupational Safety and Health Administration (OSHA), the National Institute for Occupational Safety and Health (NIOSH), and the Environmental Protection Agency (EPA) of the United States.

SUMMARY

- Surfaces and their properties are as important as the bulk properties of the materials. A surface has not only a particular shape, roughness, and appearance but also has properties that differ significantly from those of the bulk material as well.

- Surfaces are exposed to the environment and thus are subject to environmental attack. They also may come into contact with tools and dies (during processing) or with other components (during their service life).

- The geometric and material properties of surfaces can affect their friction, wear, fatigue, corrosion, and electrical and thermal conductivity significantly.

- The measurement and description of surface features (including their characteristics) are important aspects of manufacturing. The most common surface-roughness measurement is the arithmetic mean value. The instrument usually used to measure surface roughness is a profilometer.

- Friction and wear are among the most significant factors in processing materials. Much progress has been made in understanding these phenomena and identifying the factors that govern them.

- Other important factors are the affinity and solid solubility of the two materials in contact, the nature of surface films, the presence of contaminants, and process parameters such as load, speed, and temperature.

- A wide variety of metalworking fluids is available for specific applications, including oils, emulsions, synthetic solutions, and solid lubricants.

- The selection and use of lubricants require careful consideration of many factors regarding the workpiece and die materials and the particular manufacturing process.

- Metalworking fluids have various lubricating and cooling characteristics. Biological and environmental considerations also are important factors in their selection.

KEY TERMS

Abrasive wear	Lay	Substrate
Additives	Lubricant	Surface defects
Adhesion	Lubrication	Surface finish
Adhesive wear	Maximum roughness height	Surface integrity
Arithmetic mean value	Metalworking fluids	Surface profilometer
Asperities	Microwelds	Surface roughness
Boundary lubrication	Mixed lubrication	Surface structure
Coefficient of friction	Oils	Surface texture
Compounded oils	Oxide layer	Thick-film lubrication
Conversion coatings	Pit	Thin-film lubrication
Coolant	Plowing	Tribology
Emulsion	Ring-compression test	Ultrasonic vibrations
Extreme-pressure additives	Root-mean-square average	Water-soluble oils
Fatigue wear	Running-in	Waviness
Flaw	Selective leaching	Waxes
Fretting corrosion	Self-lubricating	Wear coefficient
Friction force	Severe wear	Wear parts
Greases	Soaps	
Impact wear	Solid lubricants	

BIBLIOGRAPHY

Arnell, R.D., Davies, P.B., Halling, J., and Whomes, T.L., *Tribology: Principles and Design Applications*, Springer, 1991.

ASM Handbook, Vol. 5: *Surface Engineering*, ASM International, 1994.

ASM Handbook, Vol. 18: *Friction, Lubrication, and Wear Technology*, ASM International, 1992.

Bayer, R.G., *Mechanical Wear Prediction and Prevention*, Marcel Dekker, 1994.

Bhushan, B., *Principles and Applications of Tribology*, Wiley, 1999.

_____ (ed.), *Handbook of Micro/Nanotribology*, 2nd ed., CRC Press, 1998.

_____, *Modern Tribology Handbook*, CRC Press, 2001.

Blau, P.J., *Friction Science and Technology*, Marcel Dekker, 1995.

Booser, E.R. (ed.), *Tribology Data Handbook*, CRC Press, 1998.

Budinski, K.G., *Surface Engineering for Wear Resistance*, Prentice Hall, 1988.

Burakowski, T., and Wiershon, T., *Surface Engineering of Metals: Principles, Equipment, Technologies*, CRC Press, 1998.

Byers, J.P. (ed.), *Metalworking Fluids*, Marcel Dekker, 1994.

Chattopadhyay, R., *Surface Wear: Analysis, Treatment, and Prevention*, ASM International, 2001.

Glasser, W.A., *Materials for Tribology*, Elsevier, 1992.

Hutchins, I.M., *Tribology: Friction and Wear of Engineering Materials*, CRC Press, 1992.

Lansdown, A.R., and Price, A. L., *Materials to Resist Wear: A Guide to Their Selection and Use*, Pergamon Press, 1986.

Ludema, K.C., *Friction, Wear, Lubrication: A Textbook in Tribology*, CRC Press, 1996.

Miller, R.W., *Lubricants and Their Applications*, McGraw-Hill, 1993.

Nachtman, E.S., and Kalpakjian, S., *Lubricants and Lubrication in Metalworking Operations*, Marcel Dekker, 1985.

Neele, M.J. (ed.), *The Tribology Handbook*, 2nd ed., Butterworth-Heinemann, 1995.

Peterson, M.R., and Winer, W.O. (eds.), *Wear Control Handbook*, American Society of Mechanical Engineers, 1980.

Rabinowicz, E., *Friction and Wear of Materials*, 2nd ed., Wiley, 1995.

Schey, J.A., *Tribology in Metalworking—Friction, Lubrication and Wear*, ASM International, 1983.

Seireg, A.A., *Friction and Lubrication in Mechanical Design*, Marcel Dekker, 1998.

Stachowiak, G.W., and Batchelor, A.W., *Engineering Tribology*, Butterworth-Heinemann, 2001.

Williams, J.A., *Engineering Tribology*, Oxford University Press, 1995.

REVIEW QUESTIONS

33.1 What is meant by surface texture? Surface integrity?

33.2 List and explain the types of defects found on surfaces.

33.3 Explain the terms (a) roughness, and (b) waviness.

33.4 What are the causes of lay on surfaces?

33.5 Why are the results from a profilometer not a true depiction of the actual surface?

33.6 What is the significance of a rise in surface temperature resulting from friction?

33.7 Describe the features of the ring-compression test. Does it require the measurement of forces?

33.8 List the types of wear generally observed in engineering practice.

33.9 How can adhesive wear be reduced? Abrasive wear?

33.10 What functions should a lubricant perform in manufacturing processes?

33.11 Describe the factors involved in lubricant selection.

QUALITATIVE PROBLEMS

33.12 Give several examples that show the importance of friction in manufacturing processes.

33.13 What is the significance of the fact that hardness of metal oxides is generally much higher than that of the base metals themselves?

33.14 What factors would you consider in specifying the lay of a surface for a part?

33.15 Explain why identical surface-roughness values do not necessarily represent the same type of surface.

33.16   How does the wear of molds, tools, and dies affect a manufacturing operation?

33.17   Comment on the surface roughness of various parts and components with which you are familiar.

33.18   What is the significance of the fact that the path of the stylus and the actual profile of the surface are not necessarily the same?

33.19   Give two examples for each category in which surface waviness would be (a) desirable and (b) undesirable.

33.20   Same as Problem 33.19 but for surface roughness.

33.21   For each of the surface lays shown in Fig. 33.2b, give an example of a manufacturing process which will produce that lay.

33.22   Give several reasons why an originally round specimen in a ring-compression test may become oval after it is upset.

33.23   Explain how a shoe horn (metal or plastic) facilitates the process of putting on shoes.

33.24   Describe what takes place in the running-in process.

33.25   Why is the abrasive-wear resistance of a material a function of its hardness?

33.26   Make a list of parts and components in consumer and industrial products that have to be replaced because of wear.

33.27   List manufacturing operations where high friction is desirable and those where low friction is desirable.

QUANTITATIVE PROBLEMS

33.28   Figure 33.1 shows various layers in the surface structure of metals. Consult the Bibliography at the end of this chapter and obtain data on the thicknesses for each of these layers. Comment on your observations.

33.29   Refer to the profile shown in Fig. 33.3 and place some reasonable numerical values for the vertical distances from the center line. Calculate the R_a and R_q values. Then, give another set of values for the same general profile and calculate the same two quantities. Comment on your observations.

33.30   Obtain several different parts made of various materials, inspect their surfaces under an optical microscope at different magnifications, and make an educated guess as to what manufacturing process or finishing process was used to produce each of these parts. Explain your reasoning.

33.31   Refer to Fig. 33.6b and make measurements of the external and internal diameters (in the horizontal direction in the photograph) of the four specimens shown. Remembering that in plastic deformation the volume of the rings remains constant, calculate (a) the reduction in height and (b) the coefficient of friction for each of the three compressed specimens.

33.32   Using Fig. 33.7, make a plot of the coefficient of friction versus the change in internal diameter for a constant reduction in height of 40%.

33.33   Assume that in Example 33.1 the coefficient of friction is 0.15. If all other parameters remain the same, what is the new internal diameter of the specimen?

SYNTHESIS, DESIGN, AND PROJECTS

33.34   Would it be desirable to integrate surface-roughness measuring instruments into the machine tools described in Parts III and IV? How would you go about doing so, taking into consideration the factory environment in which they are to be used? Make some preliminary sketches.

33.35   Section 33.2 listed major surface defects. How would you go about determining whether or not each of these defects is a significant factor in a particular application?

33.36   Why are the requirements for surface-roughness design in engineering applications so broad? Explain with specific examples.

33.37   Describe the tribological differences between ordinary machine elements (such as gears and bearings) and metalworking processes. Consider such factors as load, speed, and temperature.

33.38   Explain why the types of wear in Fig. 33.11 occur in those particular locations in the forging die.

33.39   Wear can have detrimental effects in manufacturing operations. Can you visualize situations in which wear could be beneficial? Give examples.

33.40   You have undoubtedly replaced parts in various appliances and automobiles because they were worn. Describe the methodology you would follow in determining the type(s) of wear these components have undergone.

Surface Treatments, Coatings, and Cleaning

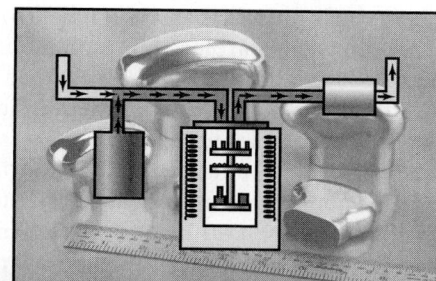

The previous chapters have described methods of producing desired shapes from a wide variety of materials. While material and process selection is extremely important, often the surface properties of a component determine its performance, and various surface finishing operations may be performed for technical and aesthetic reasons subsequent to manufacturing a part. Specifically, we describe the processes for:

- Enhancing appearance and surface texture of parts.
- Improving resistance to corrosion, chemicals, and the environment.
- Increasing hardness, wear resistance, and fatigue life and decreasing friction.
- Methods used to clean surfaces of manufactured parts.

34.1 | Introduction

After a part is manufactured, some of its surfaces may have to be processed further in order to ensure certain properties and characteristics. It may be necessary to perform **surface treatments** in order to:

- *Improve resistance to wear, erosion, and indentation* (machine-tool ways, wear surfaces of machinery, and shafts, rolls, cams, and gears)
- *Control friction* (sliding surfaces of tools, dies, bearings, and machine ways)
- *Reduce adhesion* (electrical contacts)
- *Improve lubrication* (surface modification to retain lubricants)
- *Improve resistance to corrosion and oxidation* (sheet metals for automobile bodies, gas-turbine components, food packaging, and medical devices)
- *Improve fatigue resistance* (bearings and shafts with fillets)
- *Rebuild surfaces* (worn tools, dies, and machine components)
- *Modify surface texture* (appearance, dimensional accuracy, and frictional characteristics)
- *Impart decorative features* (color and texture)

Numerous techniques are used to impart these characteristics to various types of metallic, nonmetallic, and ceramic materials based on mechanisms that involve (a) plastic deformation of the workpiece surface, (b) chemical reactions, (c) thermal means, (d) deposition, (e) implantation, and (f) organic coatings and paints. We begin with surface-hardening techniques and then continue with descriptions of different types of coatings that are applied to surfaces using various means. Some of these techniques also are used in the manufacture of semiconductor devices, as described in Chapter 28.

We then describe techniques that are used to impart texture on workpiece surfaces and types of organic coatings for various purposes. The chapter ends with methods used for cleaning manufactured surfaces before the components are assembled into the completed product and ready for service. Environmental considerations regarding the fluids used and the waste material from various surface-treatment processes also are included, as they are an important factor to be considered.

34.2 | Mechanical Surface Treatments

Several techniques are used to mechanically improve the surface properties of manufactured components. The more common methods are described here.

Shot peening. In shot peening, the workpiece surface is impacted repeatedly with a large number of cast steel, glass, or ceramic shot (small balls), which make overlapping indentations on the surface. This action causes plastic surface deformation at depths up to 1.25 mm (0.05 in.) using shot sizes that range from 0.125 to 5 mm (0.005 to 0.2 in.) in diameter. Because the plastic deformation is not uniform throughout the part's thickness, shot peening causes compressive residual stresses on the surface, thus improving the fatigue life of the component by delaying fatigue crack inititiation. Unless the process parameters are controlled properly, the plastic deformation of the surface can be so severe that it can damage the surface. The extent of deformation can be reduced by *gravity peening*, which involves larger shot sizes but with fewer impacts on the workpiece surface.

Shot peening is used extensively on shafts, gears, springs, oil-well drilling equipment, and jet-engine parts (such as turbine and compressor blades). However, it should be noted that (if these parts are subjected to high temperatures) the residual stress will begin to relax (thermal relaxation) and their beneficial effects will be diminished greatly. An example is gas-turbine blades at their operating temperatures.

Laser shot peening. In this process, also called *laser shock peening* and first developed in the mid-1960s (but not commercialized until recently), the workpiece surface is subjected to planar laser shocks (pulsed) from high-power lasers. This surface-treatment process produces compressive residual-stress layers that are typically 1 mm (0.04 in.) deep with less than 1% of cold working of the surface. Laser peening has been applied successfully and reliably to jet-engine fan blades and to materials such as titanium, nickel alloys, and steels for improved fatigue resistance and some corrosion resistance. Laser intensities necessary for this process are on the order of 100 to 300 J/cm² and have a pulse duration of about 30 nanoseconds. Currently, the basic limitation of this process for industrial, cost-effective applications is the high cost of the high-power lasers (up to 1 kW) that must operate at energy levels of 100 J/pulse.

Water-jet peening.   In this relatively recent process, a water jet at pressures as high as 400 MPa (60,000 psi) impinges on the surface of the workpiece, inducing compressive residual stresses and surface and subsurface hardening at the same level as in shot peening. The water-jet peening process has been used successfully on steels and aluminum alloys. The control of process variables (jet pressure, jet velocity, the design of the nozzle, and its distance from the surface) is important in order to avoid excessive surface roughness and surface damage.

Ultrasonic peening.   This process uses a hand tool based on a piezoelectric transducer. Operating at a frequency of 22 kHz, it can have a variety of heads for different applications.

Roller burnishing.   In this process, also called *surface rolling*, the surface of the component is cold worked by a hard and highly polished roller or set of rollers. This process is used on various flat, cylindrical, or conical surfaces (Fig. 34.1). Roller burnishing improves surface finish by removing scratches, tool marks, and pits and induces beneficial compressive-surface residual stresses. Consequently, corrosion re-sistance is improved, since corrosive products and residues cannot be entrapped. In a variation of this process, called *low-plasticity burnishing*, the roller travels only once over the surface, inducing residual stresses and minimal plastic deformation of the surface.

Internal cylindrical surfaces also are burnished by a similar process, called **ballizing** or **ball burnishing**. In this process, a smooth ball (slightly larger than the bore diameter) is pushed through the length of the hole.

Roller burnishing is used to improve the mechanical properties of surfaces as well as their surface finish. It can be used either by itself or in combination with other finishing processes, such as grinding, honing, and lapping. The equipment can be mounted on various CNC machine tools for improved productivity and consis-tency of performance. All types of metals (soft or hard) can be roller burnished. Roller burnishing typically is used on hydraulic-system components, seals, valves, spindles, and fillets on shafts.

Explosive hardening.   In explosive hardening, the surfaces are subjected to high transient pressures through the placement and detonation of a layer of an explosive sheet directly on the workpiece surface. The contact pressures that develop as a result

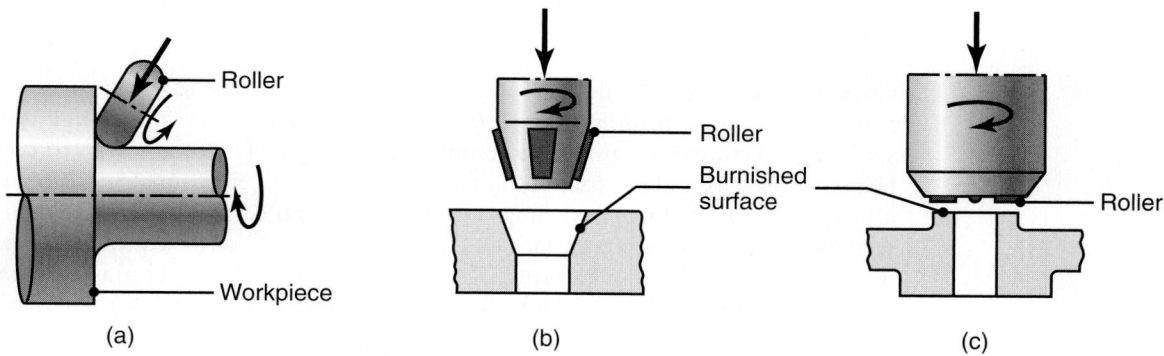

FIGURE 34.1   Burnishing tools and roller burnishing of (a) the fillet of a stepped shaft to induce compressive-surface residual stresses for improved fatigue life; (b) a conical surface; and (c) a flat surface.

can be as high as 35 GPa (5×10^6 psi) and can last about 2 to 3 μs. Major increases in surface hardness can be effected using this method with very little change (less than 5%) in the shape of the component. Railroad rail surfaces, for example, can be hardened by this method.

34.3 | Mechanical Plating and Cladding

Mechanical plating.   In mechanical plating (also called *mechanical coating, impact plating,* or *peen plating*), fine metal particles are compacted over the workpiece surfaces by glass, ceramic, or porcelain beads that are propelled by rotary means (such as by tumbling). This process, which is basically one of cold-welding particles, typically is used for hardened-steel parts for automobiles with plating thickness usually less than 25 μm (0.001 in.).

Cladding.   In this process, also called *clad bonding*, metals are bonded with a thin layer of corrosion-resistant metal through the application of pressure using rolls or other means. A typical application is the cladding of aluminum (*Alclad*), in which a corrosion-resistant layer of aluminum alloy (usually in sheet or tubular form) is clad over an aluminum-alloy body (core). The cladding layer is anodic to the core and usually has a thickness which is less than 10% of the total thickness.

Examples of cladding are 2024 aluminum clad with 1230 aluminum and 3003, 6061, and 7178 aluminum clad with 7072 aluminum. Other applications are steels clad with stainless-steel or nickel alloys. The cladding material also may be applied using dies (as in cladding steel wire with copper) or explosives. Multiple-layer cladding also is utilized in special applications.

Laser cladding consists of the fusion of a different material over the substrate. It has been applied successfully to metals and ceramics, especially for enhanced friction and good wear behavior of the components.

34.4 | Case Hardening and Hard Facing

Surfaces may be hardened by thermal means in order to improve their frictional and wear properties as well as their resistance to indentation, erosion, abrasion, and corrosion. The most common methods are described next.

Case hardening.   Traditional methods of case hardening (*carburizing, carbonitriding, cyaniding, nitriding, flame hardening,* and *induction hardening*) were described in Section 4.10 and were summarized in Table 4.1. In addition to common heat sources (gas and electricity), an electron beam or laser beam also can be used as a heat source in surface hardening of both metals and ceramics. Case hardening, as well as some of the other surface-treatment processes described in this chapter, induces residual stresses on surfaces. The formation of martensite during case hardening causes compressive residual stresses on surfaces. Such stresses are desirable, because they improve the fatigue life of components by delaying the initiation of fatigue cracks.

Hard facing.   In this process, a relatively thick layer, edge, or point of wear-resistant hard metal is deposited on the workpiece surface, using the fusion-welding techniques described in Chapter 30. Numerous layers (known as *weld overlay*) can be deposited to repair worn parts. Hard facing enhances the wear resistance of the materials, hence such materials are used in the manufacture of tools, dies, and various industrial components. Worn parts also can be hard faced for extended use.

Spark hardening.   Hard coatings of tungsten, chromium, or molybdenum carbides can be deposited using an electric arc in a process called *spark hardening, electric spark hardening,* or *electrospark deposition.* The deposited layer is typically 250 μm thick. Hard-facing alloys can be used as electrodes, rods, wires, or powder in spark hardening. Typical applications for these alloys are as valve seats, oil-well drilling tools, and dies for hot metalworking.

34.5 | Thermal Spraying

Thermal spraying is a series of important processes in which coatings of various metals, alloys, carbides, ceramics, and polymers are applied to metal surfaces by a spray gun with a stream of oxyfuel flame, electric arc, or plasma arc. The earliest applications of thermal spraying (in the 1910s) involved metals (hence the term **metallizing** also has been used), and these processes are under continuous development. The surfaces to be sprayed first are cleaned of oil and dirt and roughened to improve bond strength, such as by grit blasting (see Section 26.8). The coating material can be in the form of wire, rod, or powder, and when the droplets or particles impact the workpiece, they solidify and bond to the surface.

Depending on the process, particle velocities typically range from a low of about 150 to 1000 m/s but can be higher for special applications. Temperatures are in the range of 3000° to 8000°C (5500° to 14,000°F). The coating is hard and wear resistant with a layered structure of deposited material. However, the coating can have a porosity of as high as 20% due to entrapped air and oxide particles because of the high temperatures involved. Bond strength depends on the particular process and techniques used, is mostly mechanical in nature (hence the importance of roughening the surface prior to spraying), but can be metallurgical in some cases. Bond strength generally ranges between 7 and 80 MPa (1 and 12 ksi), depending on the particular process used.

Typical applications of thermal spraying include aircraft engine components (such as in rebuilding worn parts), structures, storage tanks, tank cars, rocket motor nozzles, and components which require resistance to wear and corrosion. In an automobile, thermal spraying typically can be applied to the following components: crankshaft, valves, fuel-injection nozzle, piston rings, and engine block. The process also is used in gas and metrochemical industries, for the repair of worn parts, and to restore dimensional accuracy to parts that have not been machined or formed properly.

The source of energy in thermal-spraying processes are of two types: combustion and electrical.

Combustion spraying

- **Thermal wire spraying** (Fig. 34.2a): The oxyfuel flame melts the wire and deposits it on the surface. Its bond is of medium strength, and the process is relatively inexpensive.

- **Thermal metal-powder spraying** (Fig. 34.2b): Similar to flame wire spraying but uses a metal powder instead of the wire.

- **Detonation gun:** Controlled and repeated explosions take place using an oxyfuel-gas mixture. The detonation gun has a performance similar to that of plasma.

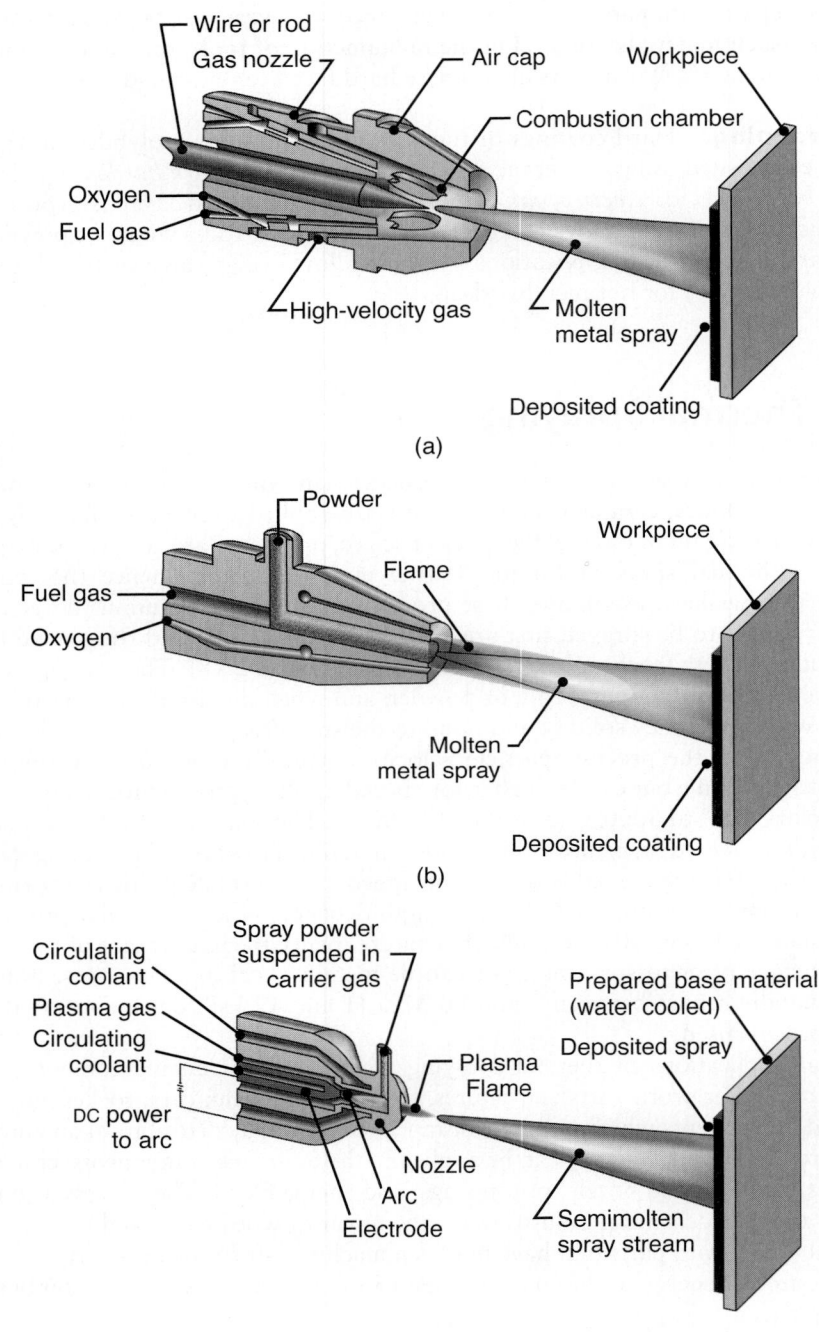

FIGURE 34.2 Schematic illustrations of thermal-spray operations: (a) thermal wire spray, (b) thermal metal-powder spray, and (c) plasma spray.

- **High-velocity oxyfuel-gas spraying** (HVOF): Produces a high performance similar to the detonation gun but is less expensive.

Electrical spraying

- **Twin-wire arc:** An arc is formed between two consumable wire electrodes. The resulting bond has good strength, and the process is the least expensive.
- **Plasma:** Either conventional, high-energy, or vacuum (Fig. 34.2c) plasma produces temperatures on the order of 8300°C (15,000°F) and results in good bond strength with very low oxide content. **Low-pressure plasma spray** (LPPS) and **vacuum plasma spray** both produce coatings with high bond strength and with very low levels of porosity and surface oxides.

Cold spraying.   This is a more recent process in which the particles to be sprayed are at a lower temperature and are not melted. Thus, in cold spraying, the oxidation is minimal. The spray jet is narrow, highly focused, and has very high impact velocities.

EXAMPLE 34.1 Repair of a worn turbine-engine shaft by thermal spraying

The shaft of the helical gear for a GE T-38 gas turbine engine had two worn regions on its nitrided surfaces. The case-hardened depth was 0.3 mm (0.012 in.). Even though the helical gears were in good condition, the part was considered scrap because there was no approved method of repair. The worn regions first were machined undersize, grit blasted, and coated with tungsten carbide (12% cobalt content; see Section 22.5) using the high-velocity oxyfuel-gas thermal-spraying (HVOF) technique. The part then was finish machined to the dimensions of the original shaft. The total cost of repair was a fraction of the projected cost of replacing the part.

Source: Courtesy of Plasma Technology, Inc.

34.6 | Vapor Deposition

Vapor deposition is a process in which the substrate (workpiece surface) is subjected to chemical reactions by gases that contain chemical compounds of the material to be deposited. The coating thickness is usually a few μm, which is much less than the thicknesses that result from the techniques described in Sections 34.2 and 34.3. The deposited materials can consist of metals, alloys, carbides, nitrides, borides, ceramics, or oxides. Control of coating composition, thickness, and porosity are important. The substrate may be metal, plastic, glass, or paper. Typical applications for vapor deposition are the coating of cutting tools, drills, reamers, milling cutters, punches, dies, and wear surfaces.

There are two major vapor-deposition processes: physical vapor deposition and chemical vapor deposition.

34.6.1 Physical vapor deposition

The three basic types of physical vapor deposition (PVD) processes are (1) vacuum deposition or arc evaporation, (2) sputtering, and (3) ion plating. These processes are carried out in a high vacuum and at temperatures in the range of 200° to 500°C (400° to 900°F). In physical vapor deposition, the particles to be deposited are carried physically to the workpiece rather than carried by chemical reactions (as in chemical vapor deposition).

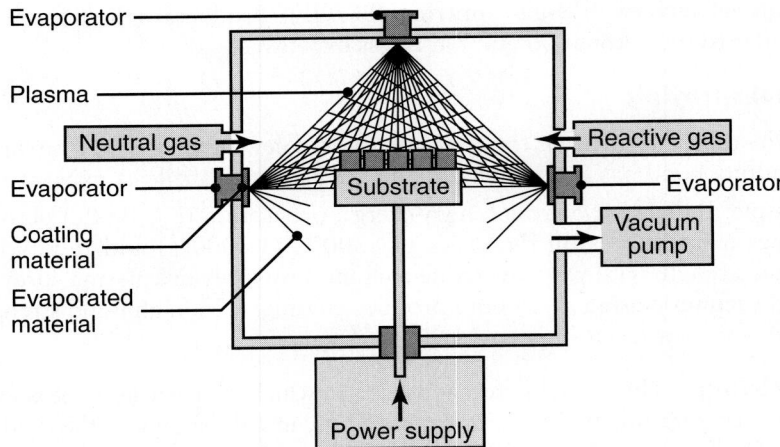

FIGURE 34.3   (a) Schematic illustration of the physical-vapor-deposition process. Note that there are three arc evaporators, and the parts to be coated are placed on a tray inside the chamber.

Vacuum deposition.   In vacuum deposition (or evaporation), the metal is evaporated at a high temperature in a vacuum and is deposited on the substrate (which is usually at room temperature or slightly higher for improved bonding). Coatings of uniform thickness can be deposited, even on complex shapes. In **arc deposition** (PV/ARC), the coating material (cathode) is evaporated by several arc evaporators (Fig. 34.3) using highly localized electric arcs. The arcs produce a highly reactive plasma, which consists of ionized vapor of the coating material. The vapor condenses on the substrate (anode) and coats it. Applications of this process are both functional (oxidation-resistant coatings for high temperature applications, electronics, and optics) and decorative (hardware, appliances, and jewelry). **Pulsed-laser deposition** is a more recent related process in which the source of energy is a pulsed laser.

Sputtering.   In sputtering, an electric field ionizes an inert gas (usually argon). The positive ions bombard the coating material (cathode) and cause sputtering (ejecting) of its atoms. These atoms then condense on the workpiece, which is heated to improve bonding (Fig. 34.4). In **reactive sputtering**, the inert gas is replaced by

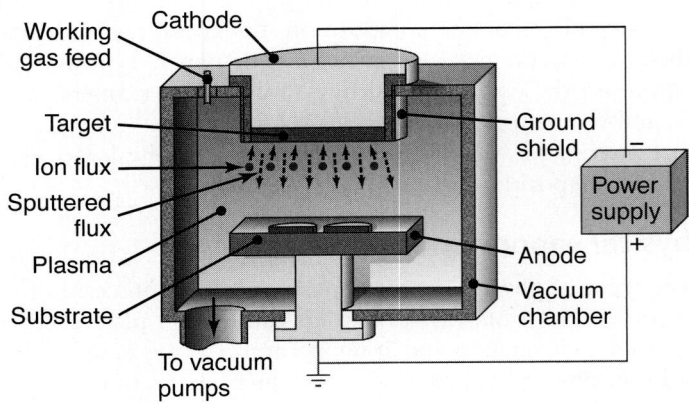

FIGURE 34.4   Schematic illustration of the sputtering process.

a reactive gas (such as oxygen) in which case the atoms are oxidized and the oxides are deposited. Carbides and nitrides also are deposited by reactive sputtering. Very thin polymer coatings also can be deposited on metal and polymeric substrates with a reactive gas, causing polymerization of the plasma. **Radio-frequency** (RF) sputtering is used for nonconductive materials, such as electrical insulators and semiconductor devices.

Ion plating. Ion plating is a generic term that describes various combined processes of sputtering and vacuum evaporation. An electric field causes a glow discharge, generating a plasma (Fig. 34.5). The vaporized atoms in this process are ionized only partially. **Ion-beam-enhanced (assisted) deposition** is capable of producing thin films as coatings for semiconductor, tribological, and optical applications. Bulky parts can be coated in large chambers using high-current power supplies of 15 kW and voltages of 100,000 DC. **Dual ion-beam deposition** is a hybrid coating technique that combines physical vapor deposition with simultaneous ion-beam bombardment. This technique results in good adhesion on metals, ceramics, and polymers. Ceramic bearings and dental instruments are examples of its applications.

34.6.2 Chemical vapor deposition

Chemical vapor deposition (CVD) is a *thermochemical* process (Fig. 34.6). In a typical application, such as coating cutting tools with titanium nitride (TiN), the tools are placed on a graphite tray and heated at 950° to 1050°C (1740° to 1920°F) at atmospheric pressure in an inert atmosphere. Titanium tetrachloride (a vapor), hydrogen, and nitrogen then are introduced into the chamber. The chemical reactions form titanium nitride on the tool surfaces. For a coating of titanium carbide, methane is substituted for the other gases.

Deposited coatings usually are thicker than those obtained using PVD. A typical cycle for CVD is long, consisting of (a) three hours of heating, (b) four hours of coating, and (c) six to eight hours of cooling to room temperature. The thickness of the coating depends on the flow rates of the gases used, the time, and the temperature.

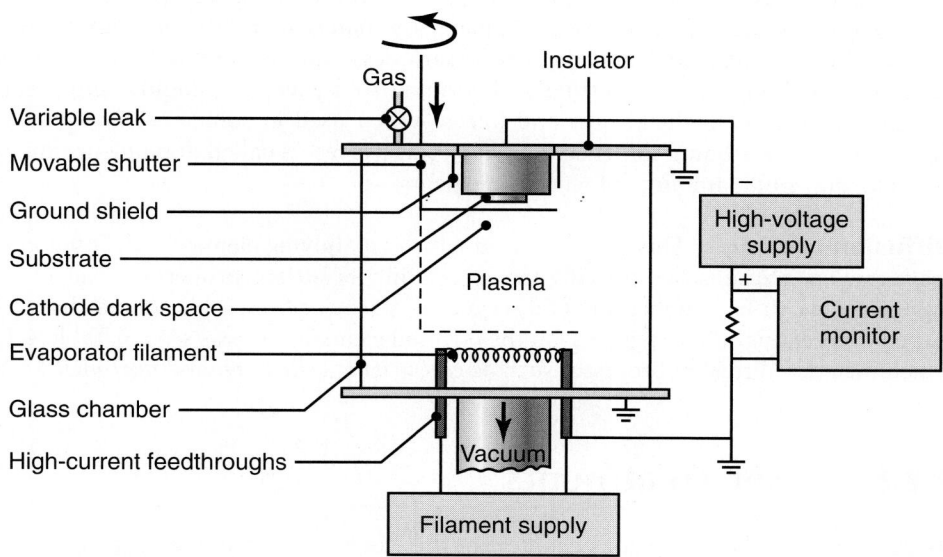

FIGURE 34.5   Schematic illustration of an ion-plating apparatus.

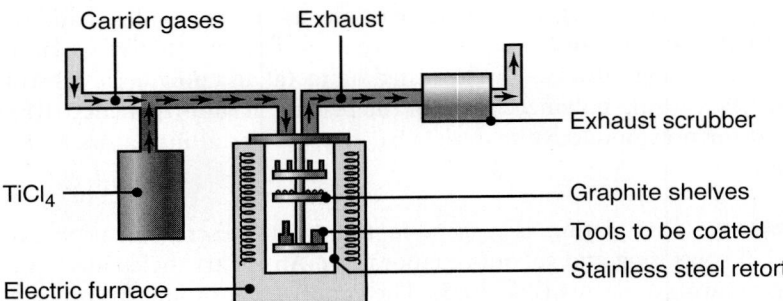

FIGURE 34.6   Schematic illustration of the chemical-vapor-deposition process. Note that parts and tools to be coated are placed on trays inside the chamber.

The types of coatings and workpiece materials allowable are fairly unrestricted in CVD. Almost any material can be coated and any material can serve as a substrate, although bond strength may vary. The CVD process also is used to produce diamond coatings without using binders, unlike polycrystalline diamond films which use 1 to 10% binder materials. A more recent development in chemical vapor deposition is **medium-temperature** CVD (MTCVD). This technique results in a higher resistance to crack propagation than CVD coatings.

34.7 | Ion Implantation and Diffusion Coating

In **ion implantation**, ions (charged atoms) are introduced into the surface of the workpiece material. The ions are accelerated in a vacuum to such an extent that they penetrate the substrate to a depth of a few μm. Ion implantation (not to be confused with ion plating) modifies surface properties by increasing surface hardness and improving resistance to friction, wear, and corrosion. This process can be controlled accurately, and the surface can be masked to prevent ion implantation in unwanted locations.

Ion implantation is effective particularly on materials such as aluminum, titanium, stainless steels, tool and die steels, carbides, and chromium coatings. This process typically is used on cutting and forming tools, dies and molds, and metal prostheses, such as artificial hips and knees. When used in some specific applications, such as semiconductors (Section 28.3), this process is called **doping**—meaning alloying with small amounts of various elements.

Diffusion coating.   This is a process in which an alloying element is diffused into the surface of the substrate (usually steels), altering its surface properties. The alloying elements can be supplied in solid, liquid, or gasous states. This process has acquired different names (depending on the diffused element), as was seen in Table 4.1, which included diffusion processes such as *carburizing, nitriding,* and *boronizing*.

34.8 | Laser Treatments

As we have described in various chapters of this book, lasers are having increasingly broader applications in manufacturing processes (laser machining, forming, joining,

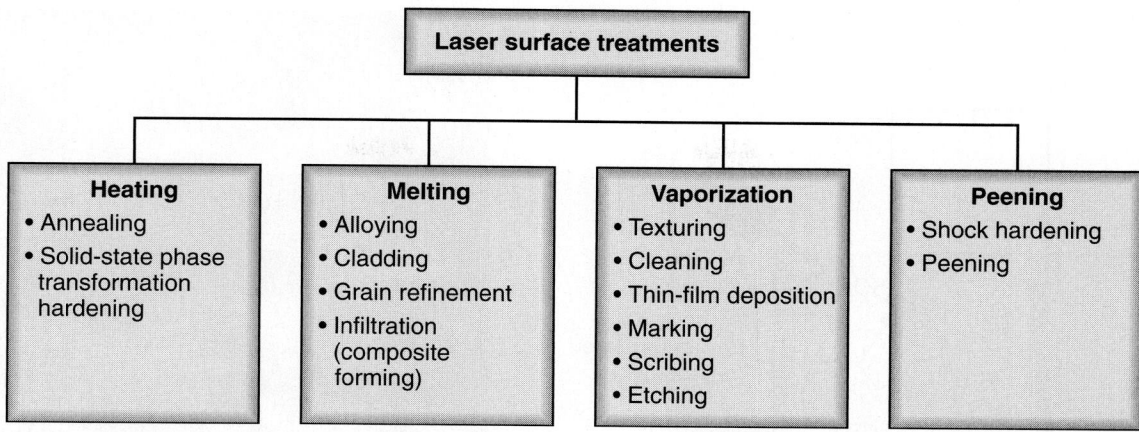

FIGURE 34.7 An outline of laser surface-engineering processes. *Source:* After N.B. Dahotre

rapid prototyping, and metrology) and surface engineering (laser peening, alloying, surface treatments, and texturing). Powerful, efficient, reliable, and less-expensive lasers now are available for a variety of cost-effective surface treatments, as outlined in Fig. 34.7.

EXAMPLE 34.2 Typical current and potential applications of laser surface engineering

In this example, we give several applications of lasers in engineering practice. The most commonly used lasers are Nd:YAG and CO_2; excimer lasers generally are used for surface texturing (see also Table 27.2).

Localized surface hardening—Cast irons: diesel-engine cylinder liners, automobile steering assemblies, and camshafts. Carbon steels: gears and electromechanical parts.
Surface alloying—Alloy steels: bearing components. Stainless steels: diesel-engine valves and seat inserts. Tool and die steels: dies for forming and die casting.
Cladding—Alloy steels: automotive valves and valve seats. Superalloys: turbine blades.
Ceramic coating—Aluminum-silicon alloys: automotive-engine bore.
Surface texturing—Metals, plastics, ceramics, and wood: all types of products.

34.9 Electroplating, Electroless Plating, and Electroforming

Plating, as with other coating processes, imparts the properties of resistance to wear and corrosion, high electrical conductivity, and better appearance and reflectivity, as well as similar desirable properties.

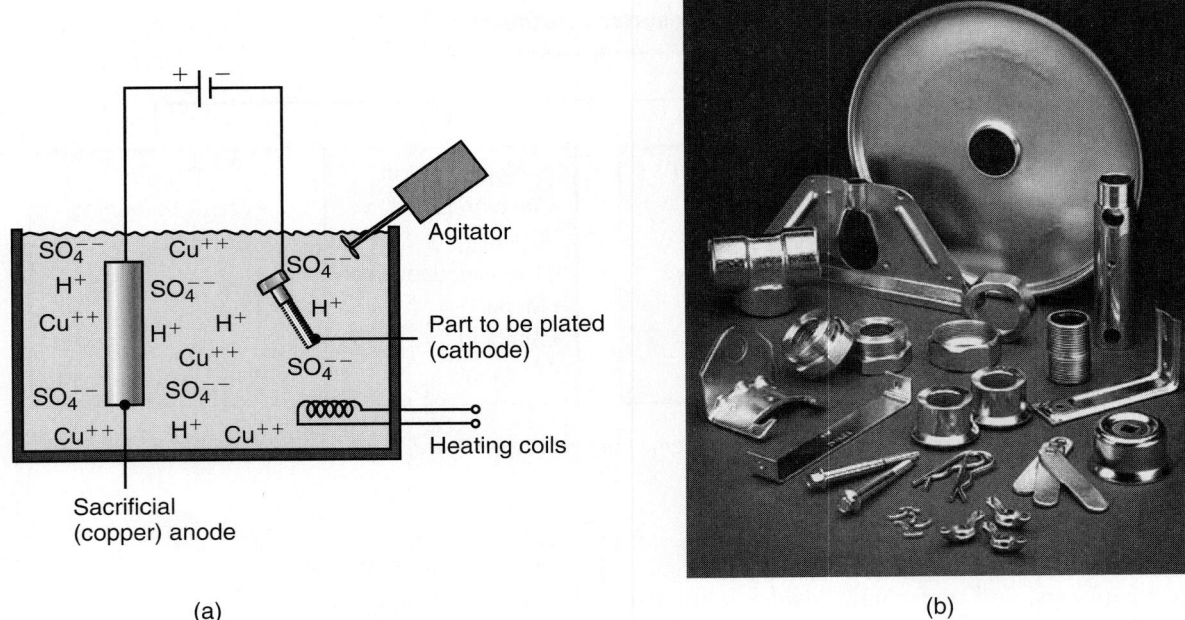

FIGURE 34.8 (a) Schematic illustration of the electroplating process. (b) Examples of electroplated parts. *Source*: Courtesy of BFG Electroplating.

Electroplating. In electroplating, the workpiece (cathode) is plated with a different metal (anode), which is transferred through a water-based electrolytic solution (Fig. 34.8). Although the plating process involves a number of reactions, basically, the process consists of the following sequence:

1. The metal ions from the anode are discharged using the potential energy from the external source of electricity.
2. The metal ions are dissolved into the solution.
3. They are deposited on the cathode.

The volume of the plated metal can be calculated from the expression

$$\text{Volume of metal plated} = cIt \tag{34.1}$$

where I is the current in amperes, t is time, and c is a constant that depends on the plate metal, the electrolyte, and the efficiency of the system and, typically, is in the range of 0.03 to 0.1 mm³/amp–s. Note that, for the same volume of material deposited, the larger the workpiece surface plated, the thinner is the layer. The time required for elecroplating typically is long, because the deposition rate is typically on the order of 75 μm/hour. Thin-plated layers are typically on the order of 1 μm (40 μin.), and for thick layers, it can be as much as 500 μm.

The plating solutions are either strong acids or cyanide solutions. As the metal is plated from the solution, it needs to be periodically replenished, and this is accomplished through two principal methods: additional salts of metals occasionally are added to the solution or a *sacrificial anode* of the metal to be plated is used in the electroplating tank, which dissolves at the same rate that the metal is deposited.

There are three main forms of electroplating:

1. In **rack plating**, parts to be plated are placed in a rack, which then is conveyed through a series of process tanks.

2. In **barrel plating**, small parts are placed inside a permeable barrel, which is placed inside the process tank(s). This form of electroplating commonly is performed with small parts, such as bolts, nuts, gears, fittings, etc. Electrolytic fluid can penetrate through the barrel and provide the metal for plating, and electrical contact is provided through the barrel and contact with other parts.

3. In **brush processing**, the electrolytic fluid is pumped through a hand-held brush with metal bristles. The workpiece can be very large in this circumstance, and this process is suitable for field repair or plating and can be used to apply coatings on large equipment without disassembly.

Simple electroplating can be achieved in a single-process bath or tank, but more commonly, a sequence of operations is used in a plating line. For example, the following process tanks may be part of an electroplating operation:

- Chemical cleaning and degreasing tanks will be used to remove surface contaminents and enhance surface adhesion of the plated coating.

- The workpieces may be exposed to a strong acid bath (pickling solution) to reduce or eliminate the thickness of the oxide coating on the workpiece.

- A base coating may be applied. This may involve the same or a different metal than the ultimate surface. For example, if the desired metal coating will not adhere well to the substrate, an intermediate coating can be applied. Also, if thick films are desired, a plating tank can be used to quickly develop a film, and a subsequent tank with brightener additives in the electrolytic solution is used to develop the ultimate surface finish.

- A tank where final electroplating is performed.

- Rinse tanks will be used throughout the sequence.

Rinse tanks are necessary for a number of reasons. Some plating is performed using cyanide salts to deliver the required metal ions. If any residue acid (such as from a pickling tank) is conveyed to the cyanide-solution tank, poisonous hydrogen-cyanide gas is exhausted. (This is a significant safety concern, and environmental controls are essential in plating facilities.) Also, residue plating solution will contain some metal ions, and it is often desireable to recover those ions by capturing them in a rinse tank.

The rate of film deposition depends on the local current density and is not necessarily uniform on a part. Workpieces with complex shapes may require an altered geometry because of varying plating thicknesses, as shown in Fig. 34.9.

Common plating metals are chromium, nickel (corrosion protection), cadmium, copper (corrosion resistance and electrical conductivity), and tin and zinc (corrosion protection, especially for sheet steel). **Chromium plating** is done by first plating the metal with copper, then with nickel, and finally with chromium. **Hard chromium plating** is done directly on the base metal and results in a surface hardness of up to 70 HRC (see Fig. 2.14) and a thickness of about 0.05 mm (0.002 in.) or more. This method is used to improve the resistance to wear and corrosion of tools, valve stems, hydraulic shafts, and diesel- and aircraft-engine cylinder liners. It also is used to rebuild worn parts.

Examples of electroplating include copper-plating aluminum wire and phenolic boards for printed circuits, chrome-plating hardware, tin-plating copper electrical terminals (for ease of soldering), galvanizing sheet metal (see also Section 34.11), and plating components such as metalworking dies that require resistance to wear

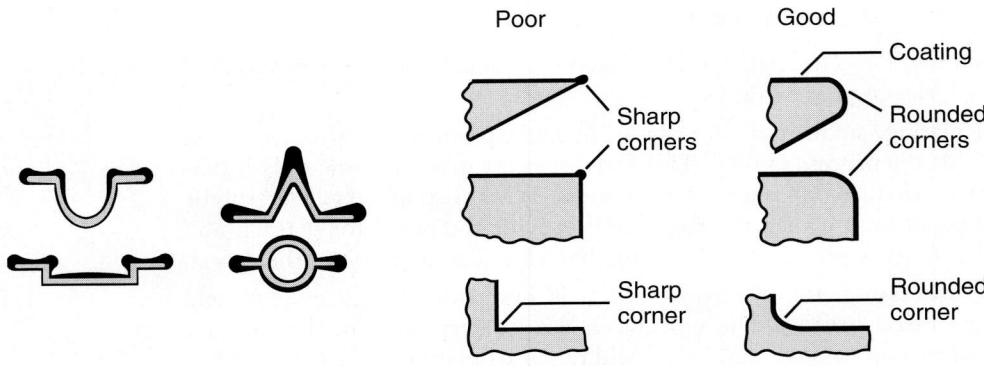

FIGURE 34.9   (a) Schematic illustration of nonuniform coatings (exaggerated) in electroplated parts. (b) Design guidelines for electroplating. Note that sharp external and internal corners should be avoided for uniform plating thickness.

and galling (cold welding of small pieces from the workpiece surface). Metals such as gold, silver, and platinum are important electroplating materials in the electronics and jewelry industries for decorative purposes.

Plastics (such as ABS, polypropylene, polysulfone, polycarbonate, polyester, and nylon) also can be electroplating substrates. Because they are not electrically conductive, plastics must be preplated using a process such as electroless nickel plating. Parts to be coated may be simple or complex, and size is not a limitation.

Electroless plating.   This process is carried out by a chemical reaction and without the use of an external source of electricity. The most common application utilizes nickel as the plating material, although copper also is used. In electroless nickel plating, nickel chloride (a metallic salt) is reduced (using sodium hypophosphite as the reducing agent) to nickel metal, which then is deposited on the workpiece. The hardness of nickel plating ranges between 425 and 575 HV and subsequently can be heat treated to 1000 HV. The coating has excellent wear and corrosion resistance.

Cavities, recesses, and the inner surfaces of tubes can be plated successfully. Electroless plating also can be used with nonconductive materials, like plastics and ceramics. This process is more expensive than electroplating. However, unlike that of electroplating, the coating thickness of electroless plating is always uniform.

Electroforming.   A variation of electroplating, electroforming actually is a metal-fabricating process. Metal is electrodeposited on a *mandrel* (also called a *mold* or a *matrix*), which then is removed; thus, the coating itself becomes the product (Fig. 34.10). Both simple and complex shapes can be produced by electroforming with wall thicknesses as small as 0.025 mm (0.001 in.). Parts may weigh from a few grams to as much as 270 kg (600 lb). Production rates can be increased through the use of multiple mandrels.

Mandrels are made from a variety of materials: metallic (zinc or aluminum) or nonmetallic (which can be made electrically conductive with the proper coatings). Mandrels should be able to be removed physically without damaging the electroformed part. They also may be made of low-melting alloys, wax, or plastics, which can be melted away or dissolved with suitable chemicals.

The electroforming process is suitable particularly for low production quantities or intricate parts (such as molds, dies, waveguides, nozzles, and bellows) made of nickel, copper, gold, and silver. It also is suitable for aerospace, electronics, and electrooptics applications.

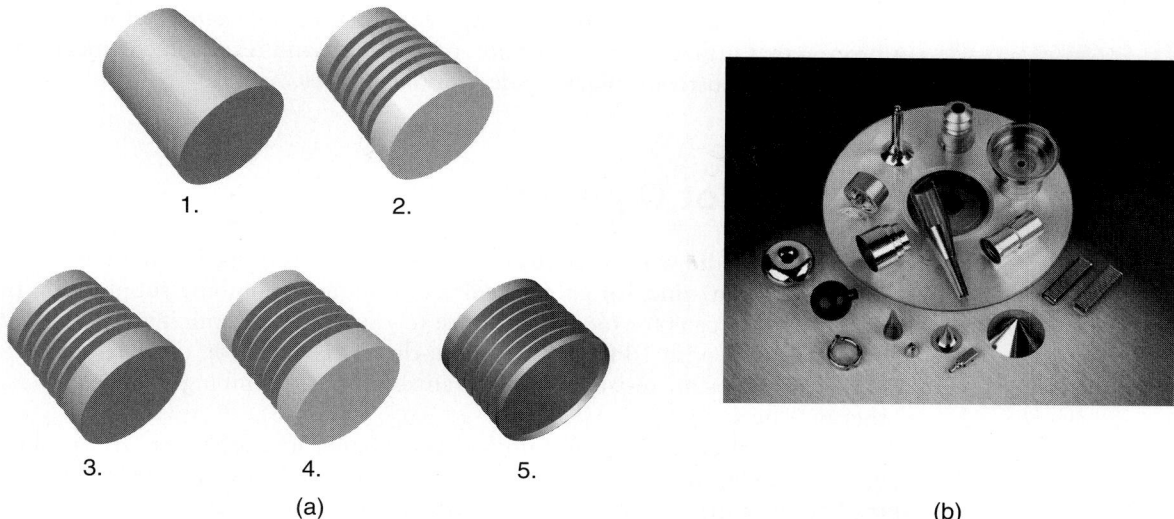

FIGURE 34.10   (a) Typical sequence in electroforming. (1) A mandrel is selected with the correct nominal size. (2) The desired geometry is machined into the mandrel (in this case a bellows). (3) The desired metal is electroplated onto the mandrel. (4) The plated material is trimmed, if necessary. (5) The mandrel is dissolved through chemical machining. (b) A collection of electroformed parts. *Source*: Courtesy of Servometer® Cedar Grove, NJ.

34.10 | Conversion Coatings

Conversion coating (also called *chemical-reaction priming*) is the process of producing a coating that forms on metal surfaces as a result of chemical or electrochemical reactions. Various metals (particularly steel, aluminum, and zinc) can be conversion-coated. Oxides that naturally form on their surfaces are a form of conversion coating.

Phosphates, chromates, and oxalates are used to produce these coatings. They are used for purposes such as providing corrosion protection, prepainting, and decorative finishing. An important application is in the conversion coating of workpieces to serve as lubricant carriers in cold-forming operations, particularly zinc-phosphate and oxalate coatings (see Section 33.7.6). The two common methods of coating are *immersion* and *spraying*. The equipment required depends on the method of application, the type of product, and quality considerations.

Anodizing.   This is an oxidation process (*anodic oxidation*) in which the workpiece surfaces are converted to a hard and porous oxide layer that provides corrosion resistance and a decorative finish. The workpiece is the anode in an electrolytic cell immersed in an acid bath, which results in chemical adsorption of oxygen from the bath. Organic dyes of various colors (typically black, red, bronze, gold, or gray) can be used to produce stable, durable surface films. Typical applications for anodizing are aluminum furniture and utensils, architectural shapes, automobile trim, picture frames, keys, and sporting goods. Anodized surfaces also serve as a good base for painting, especially on aluminum, which otherwise is difficult to paint.

Coloring.   As the name implies, coloring involves processes that alter the color of metals, alloys, and ceramics. This change is caused by the conversion of surfaces (by chemical, electrochemical, or thermal processes) into chemical compounds such as oxides, chromates, and phosphates. An example is the *blackening* of iron and steels, a process that utilizes solutions of hot, caustic soda and results in chemical reactions that produce a lustrous, black oxide film on surfaces.

34.11 | Hot Dipping

In hot dipping, the workpiece (usually steel or iron) is dipped into a bath of molten metal, such as (a) zinc, for galvanized-steel sheet and plumbing supplies; (b) tin, for tinplate and tin cans for food containers; (c) aluminum (aluminizing); and (d) *terne*, an alloy of lead with 10 to 20% tin. Hot-dipped coatings on discrete parts provide long-term corrosion resistance to galvanized pipes, plumbing supplies, and many other products.

A typical continuous *hot-dipped galvanizing* line for sheet steel is shown in Fig. 34.11. The rolled sheet first is cleaned electrolytically and scrubbed by brushing. The sheet then is annealed in a continuous furnace with controlled atmosphere and temperature and dipped in molten zinc at about 450°C (840°F). The thickness of the zinc coating is controlled by a wiping action from a stream of air or steam, called an *air knife* (similar to air-drying in car washes). Proper draining for the removal of excess coating materials is important.

The coating thickness usually is given in terms of coating weight per unit surface area of the sheet, typically 150 to 900 g/m^2 (0.5 to 3 oz/ft^2). Service life depends on the thickness of the zinc coating and the environment to which it is exposed. Various **precoated sheet steels** are used extensively in automobile bodies. Proper draining to remove excess coating materials is an important consideration.

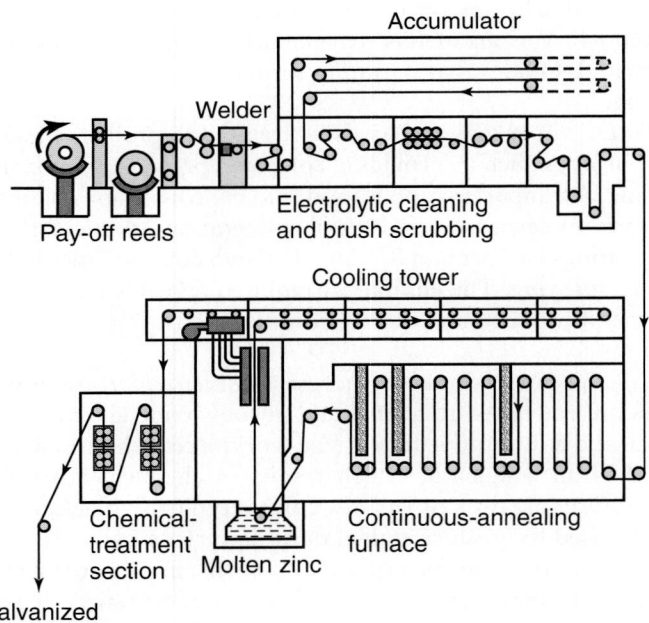

FIGURE 34.11   Flow line for the continuous hot-dipped galvanizing of sheet steel. The welder (upper left) is used to weld the ends of coils to maintain continuous material flow. *Source*: Courtesy of the American Iron and Steel Institute.

34.12 | Porcelain Enameling; Ceramic and Organic Coatings

Metals can be coated with a variety of glassy (vitreous) coatings to provide corrosion and electrical resistance and to provide protection at elevated temperatures. These coatings usually are classified as **porcelain enamels** and generally include enamels and ceramics. The root of the word "porcelain" is *porcellana*, in Italian meaning "marine shell." Note that the word *enamel* also is used as a term for glossy paints, indicating a smooth, hard coating.

Enamels.   Porcelain enamels are glassy inorganic coatings that consist of various metal oxides and are available in various colors and transperancies. *Enameling* (which was a fully developed art by the Middle Ages) involves fusing the coating material to the substrate by heating them both at 425° to 1000°C (800° to 1800°F) to liquefy the oxides. The coating may be applied by dipping, spraying, or electrodeposition, and thicknesses are usually 0.05 to 0.6 mm (0.002 to 0.025 in.). The viscosity of the material can be controlled using binders so that the coating adheres to vertical surfaces during application. Depending on their composition, enamels have varying resistances to alkali, acids, detergents, cleansers, and water.

Typical applications for porcelain enameling are for household appliances, plumbing fixtures, chemical-processing equipment, signs, cookware, and jewelry. Porcelain enamels also are used as protective coatings on jet-engine components. Metals coated are typically steels, cast iron, and aluminum. Glasses are used as a lining (for chemical resistance) where the thickness of the glass is much greater than that of the enamel. **Glazing** is the application of glassy coatings onto ceramic wares to give them decorative finishes and to make them impervious to moisture.

Ceramic coatings.   Materials such as powders of hard metals, aluminum oxide, and zirconium oxide are applied to the substrate at room temperature using binders. These coatings act as thermal barriers and have been applied (usually by thermal spraying techniques) to hot-extrusion dies, turbine blades, diesel-engine components, and nozzles for rocket motors to extend the life of these components. They also are used for electrical-resistance applications to withstand repeated arcing. Plasma arcs are used where temperatures may reach 15,000°C (27,000°F), which is much higher than those obtained by flames.

Organic coatings.   Metal surfaces can be coated or precoated with a variety of organic coatings, films, and laminates to improve appearance and corrosion resistance. Coatings are applied to the coil stock on continuous lines with thicknesses generally of 0.0025 to 0.2 mm (0.0001 to 0.008 in.). Such coatings have a wide range of properties: flexibility, durability, hardness, resistance to abrasion and chemicals, color, texture, and gloss. Coated sheet metal subsequently is formed into various products, such as TV cabinets, appliance housings, paneling, shelving, residential-building siding, gutters, and metal furniture.

More critical applications involve, for example, the protection of naval aircraft, which are subjected to high humidity, rain, sea water, pollutants (such as from ship exhaust stacks), aviation fuel, deicing fluids, and battery acid, as well as being impacted by particles such as dust, gravel, stones, and deicing salts. For aluminum structures, organic coatings consist typically of an epoxy primer and a polyurethane topcoat with a lifetime of four to six years. Primer performance is an important factor in the durability of the coating.

EXAMPLE 34.3 Ceramic coatings for high-temperature applications

Certain product characteristics (such as wear resistance and thermal and electrical insulation—particularly at elevated temperatures) can be imparted through ceramic coatings rather than to the base metals or materials themselves. Selecting materials with such bulk properties can be expensive and may not meet the structural strength requirements in a particular application.

For example, a wear-resistance component does not have to be made completely from a wear-resistant material, since the properties of only a thin layer of its surface are relevant to wear. Consequently, coatings have important applications. Table 34.1 shows various ceramic coatings and their typical applications at elevated temperatures. These coatings may be applied either singly or in layers—as is done in multiple-layer coated cutting tools (see Fig. 22.8)

TABLE 34.1

Property	Type of ceramic	Application
Wear resistance	Chromium oxide, aluminum oxide, aluminum titania	Pumps, turbine shafts, seals, compressor rods for the petroleum industry; plastics extruder barrels; extrusion dies
Thermal insulation	Zirconium oxide (yttria stabilized), zironium oxide (calcia stabilized), magnesium zirconate	Fan blades, compressor blades, and seals for gas turbines; valves, pistons, and combustion heads for automotive engines
Electrical insulation	Magnesium aluminate, aluminum oxide	Induction coils, brazing fixtures, general electrical applications

34.13 | Diamond Coating and Diamond-Like Carbon

The properties of *diamond* that are relevant to manufacturing engineering were described in Section 8.7. Important advances have been made in the diamond coating of metals, glass, ceramics, and plastics using various techniques, such as chemical vapor deposition (CVD), plasma-assisted vapor deposition, and ion-beam-enhanced deposition.

Examples of diamond-coated products are: scratchproof windows (such as those used in aircraft and missile sensors for protection against sandstorms); sunglasses; cutting tools (such as inserts, drills, and end mills); wear faces of micrometers and calipers; surgical knives; razors; electronic and infrared heat seekers and sensors; light-emitting diodes; diamond-coated speakers for stereo systems; turbine blades; and fuel-injection nozzles.

Techniques also have been developed to produce **free-standing diamond films** on the order of 1 mm (0.04 in.) thick and up to 125 mm (5 in.) in diameter. These include smooth, optically clear diamond film, unlike the hazy gray diamond film formerly produced. This film is then laser cut to desired shapes and brazed onto cutting tools (for example).

The development of these techniques combined with the important properties of diamond (hardness, wear resistance, high thermal conductivity, and transparency to ultraviolet light and microwave frequencies) have enabled the production of various aerospace and electronic parts and components.

Studies also are continuing regarding the growth of diamond films on crystalline-copper substrate by the implantation of carbon ions. An important application is in making computer chips (see Chapter 28). Diamond can be doped to form *p*- and *n*-type ends on semiconductors to make transistors, and its high thermal conductivity allows the closer packing of chips than would be possible with silicon or gallium-arsenide chips, significantly increasing the speed of computers.

Diamond-like carbon. Diamond-like carbon (DLC) coatings, a few nanometers in thickness, are produced by a low-temperature, ion-beam-assisted deposition process. Its structure is between those of diamond and graphite. Less expensive than diamond films but with similar properties (such as low friction, high hardness, and chemical inertness, as well as having a smooth surface), DLC has applications in such areas as tools and dies, gears, engine components, bearings, microelectromechanical systems, and microscale probes. As a coating on cutting tools, its hardness is about 5000 HV (as compared to double that for diamond).

34.14 | Surface Texturing

As we have seen throughout the preceding chapters, each manufacturing process (such as casting, forging, powder metallurgy, injection molding, machining, grinding, polishing, electrical-discharge machining, grit blasting, and wire brushing), produces a certain surface texture and appearance. Obviously, some of these processes can be used to modify the surface produced by a previous process; for example, grinding the surface of a cast part. However, manufactured surfaces can be modified further by secondary operations for technical, functional, optical, or aesthetic reasons.

Called *surface texturing*, these additional processes generally consist of the following techniques:

- **Etching:** Using chemicals or sputtering techniques.
- **Electric arcs.**
- **Lasers:** Using excimer lasers with pulsed beams; applications include molds for permanent-mold casting, rolls for temper mills, golf-club heads, and computer hard disks.
- **Atomic oxygen:** Reacting with surfaces to produce a fine, conelike surface texture.

The possible adverse effects of these processes on material properties and the performance of the textured parts are important considerations.

34.15 | Painting

Because of its decorative and functional properties (such as environmental protection, low cost, relative ease of application, and the range of available colors), **paint** has been used widely as a surface coating. The engineering applications of painting range from appliances and machine tools to automobile bodies and aircraft fuselages. Paints generally are classified as

- **Enamels:** Produce a smooth coat and dry with a glossy or semiglossy appearance.
- **Lacquers:** Form a film by evaporation of a solvent.
- **Water-based paints:** These are applied easily but have a porous surface and absorb water, making them more difficult to clean than the first two types.

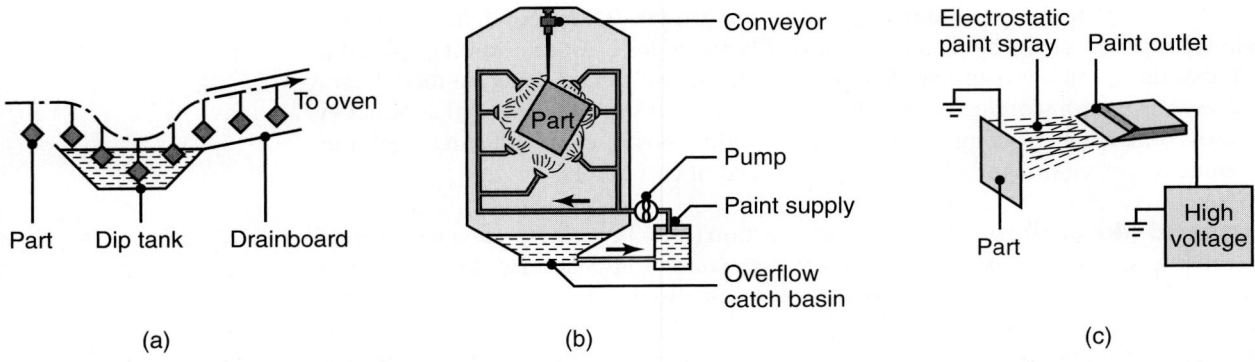

FIGURE 34.12   Methods of paint application: (a) dip coating, (b) flow coating, and (c) electrostatic spraying (used particularly for automotive bodies)

Paints are available with good resistance to abrasion, temperature extremes, and fading; are easy to apply; and dry quickly. Selection of a particular paint depends on specific requirements. Among these are resistance to mechanical actions (abrasion, marring, impact, and flexing) or to chemical actions (acids, solvents, detergents, alkalis, fuels, staining, and general environmental attack).

Common methods of applying paint are dipping, brushing, rolling, and spraying (Fig. 34.12). In **electrocoating** or **electrostatic spraying**, paint particles are charged *electrostatically* and are attracted to surfaces to be painted, producing a uniformly adherent coating. Unlike conventional spraying, in which as much as 70% of the paint may be lost, the loss can be as little as 10% in electrostatic spraying. However, deep recesses and corners can be difficult to coat with this method. The use of *robotic controls* for guiding the spray nozzles are common.

34.16 | Cleaning of Surfaces

We have stressed the importance of surfaces in manufacturing and the effects of deposited or adsorbed layers of various elements and contaminants on surface characteristics. A clean surface can have both beneficial and detrimental effects. Although a surface that is not clean may reduce the tendency for adhesion and galling, cleanliness generally is essential for a more effective application of coatings, painting, adhesive bonding, welding, brazing, and soldering, as well as for the reliable functioning of manufactured parts in machinery, assembly operations, and food and beverage containers.

Cleaning involves the removal of solid, semisolid, or liquid contaminants from a surface, and it is an important part of manufacturing operations and the economics of production. The word *clean* or the degree of cleanliness of a surface is somewhat difficult to define. Two simple and common tests are as follows:

1. Wiping with a clean cloth the surface of, say, a dinner plate and observing any residues on the cloth.

2. Observing whether water continuously coats the surface of the plate (*waterbreak test*). If water collects as individual droplets, the surface is not clean. (You can test this phenomenon by wetting dinner plates that have been cleaned to different degrees.)

The type of cleaning process required depends on the type of *metalworking-fluid residues* and *contaminants* to be removed. For example, water-based fluids are easier and less expensive to remove than oil-based fluids. Contaminants (also called *soils*) may consist of rust, scale, chips (and other metallic and nonmetallic debris), metalworking fluids, solid lubricants, pigments, polishing and lapping compounds, and general environmental elements.

Basically, there are three types of cleaning methods, as described here.

Mechanical cleaning. This operation consists of physically disturbing the contaminants, often with wire or fiber brushing, abrasive blasting (jets), tumbling, or steam jets. Many of these processes are effective particularly in removing rust, scale, and other solid contaminants. **Ultrasonic cleaning** also is placed in this category.

Electrolytic cleaning. In this process, a charge is applied the part to be cleaned in an acqueous (often alkaline) cleaning solution. This charge results in bubbles of hydrogen or oxygen (depending on polarity) being released at the part's surface. The bubbles are abrasive and aid in the removal of contaminants.

Chemical cleaning. Chemical cleaning usually involves the removal of oil and grease from surfaces. The operation consists of one or more of the following processes:

- **Solution:** The soil dissolves in the cleaning solution.
- **Saponification:** A chemical reaction that converts animal or vegetable oils into a soap that is soluble in water.
- **Emulsification:** The cleaning solution reacts with the soil or lubricant deposits and forms an emulsion; the soil and the emulsifier then become suspended in the emulsion.
- **Dispersion:** The concentration of soil on the surface is decreased by surface-active elements in the cleaning solution.
- **Aggregation:** Lubricants are removed from the surface by various agents in the cleanser and are collected as large dirt particles.

Cleaning fluids. Common cleaning fluids used in conjunction with electrochemical processes for more effective cleaning include the following:

- **Alkaline solutions:** A complex combination of water-soluble chemicals. They are the least expensive and most widely used in manufacturing operations. Small parts may be cleaned in rotating drums or barrels. Most parts are cleaned on continuous conveyors by spraying them with the solution and rinsing them with water.
- **Emulsions:** Generally consist of kerosene and oil in water and various types of emulsifiers.
- **Solvents:** Typically petroleum solvents, chlorinated hydrocarbons, and mineral spirits. They generally are used for short runs. Fire and toxicity are major hazards.
- **Hot vapors:** Chlorinated solvents can be used to remove oil, grease, and wax. The solvent is boiled in a container and then condensed. This hot-vapor process is simple, and the cleaned parts are dry.
- **Acids, Salts,** and **Organic compound mixtures:** These are effective in cleaning parts covered with heavy paste or oily deposits and rust.

Design guidelines for cleaning. Cleaning discrete parts with complex shapes can be difficult. Some design guidelines are as follows:

- Avoid deep, blind holes.

- Make several smaller components instead of one large component, which may be difficult to clean.
- Provide appropriate drain holes in the parts to be cleaned.

The *treatment* and *disposal* of cleaning fluids, as well as of various fluids and waste materials from the processes described in this chapter, are among the most important considerations for environmentally safe manufacturing operations. See also Section 1.6.

SUMMARY

- Surface treatments are an important aspect of all manufacturing processes. They are used to impart specific physical and mechanical properties, such as appearance, and corrosion, friction, wear, and fatigue resistance. Several techniques are available for modifying surfaces.
- The processes used include mechanical working and surface treatments, such as heat treatment, deposition, and plating and additional surface coatings, such as enamels, nonmetallic materials, and paints.
- Clean surfaces can be important in the further processing and use of the product (e.g., coating, painting, or welding). Cleaning can have a significant economic impact on manufacturing operations. Various mechanical and chemical cleaning methods may be utilized.

KEY TERMS

Anodizing	Electroless plating	Painting
Ballizing	Electroplating	Physical vapor deposition
Blackening	Enamel	Porcelain enamel
Case hardening	Explosive hardening	Roller burnishing
Chemical cleaning	Free-standing diamond film	Shot peening
Chemical vapor deposition	Glazing	Spraying
Cladding	Hard-chromium plating	Sputtering
Cleaning fluids	Hard facing	Surface texturing
Coloring	Hot dipping	Thermal spraying
Conversion coating	Ion implantation	Vacuum evaporation
Diamond coating	Ion plating	Vapor deposition
Diamond-like carbon	Laser peening	Waterbreak test
Diffusion coating	Mechanical plating	Water-jet peening
Electroforming	Metallizing	

BIBLIOGRAPHY

ASM Handbook, Vol. 5, *Surface Engineering*, ASM International, 1994.

Bhushan, B., and Gupta, B.K., *Handbook of Tribology: Materials, Coatings, and Surface Treatments*, McGraw-Hill, 1991.

Budinski, K.G., *Surface Engineering for Wear Resistance*, Prentice Hall, 1988.

Burakowski, T., and Wierschon, T., *Surface Engineering of Metals: Principles, Equipment, Technologies*, CRC Press, 1998.

Davis, J.R. (ed.), *Surface Engineering for Corrosion and Wear Resistance*, IOM Communications and ASM International, 2001.

—— (ed.), *Handbook of Thermal Spray Technology*, Thermal Spray Society and ASM International, 2004.

—— (ed.), *Surface Hardening of Steels*, ASM International, 2002.

Dennis, J.K., and Such, T.E., *Nickel and Chromium Plating*, 3rd ed., ASM International, 1993.

Edwards, J., *Coating and Surface Treatment Systems for Metals: A Comprehensive Guide to Selection*, ASM International, 1997.

Handbook of Surface Treatments and Coatings, ASME Press, 2003.

Hocking, M.G., Vasanastre, V., and Sidky, P.S., *Metallic and Ceramic Coatings*, Addison-Wesley, 1989.

Holmberg, K., and Matthews, A., *Coating Tribology: Properties, Techniques, and Applications*. Elsevier, 1994.

Inagaki, N., *Plasma Surface Modification and Plasma Polymerization*, Technomic, 1996.

Kawai, S. (ed.), *Anodizing and Coloring of Aluminum Alloys*, Metal Finishing Information Services and ASM International, 2002.

Liberto, N., *User's Guide to Powder Coating*, 4th ed., Society of Manufacturing Engineers, 2003.

Lambourne, R., and Strivens, T.A., *Paint and Surface Coatings: Theory and Practice*, 2nd ed., Woodhead, 1999.

Lindsay, J.H. (ed.), *Coatings and Coating Processes for Metals*, ASM International, 1998.

Park, J., *Chemical Vapor Deposition*, ASM International, 2001.

Peterson, M.B., and Winer, W.O. (eds.), *Wear Control Handbook*, American Society of Mechanical Engineers, 1980.

Prelas, M.A., Popovichi, G., and Bigelow, L.K. (eds.), *Handbook of Industrial Diamonds and Diamond Films*, Marcel Dekker, 1998.

Rausch, W., *The Phosphating of Metals*, 2nd ed., ASM International, 1990.

Rudzki, G.J., *Surface Finishing Systems: Metal and Non-Metal Finishing Handbook Guide*, Finishing Publications Ltd., 1983.

Schey, J.A., *Tribology in Metalworking—Friction, Lubrication and Wear*, ASM International, 1983.

Stern, K.H. (ed.), *Metallurgical and Ceramic Protective Coatings*, Chapman & Hall, 1997.

Sudarshan, T.S. (ed.), *Surface Modification Technologies*, ASM International, 1998.

Tamarin, Y., *Protective Coatings for Turbine Blades*, ASM International, 2002.

Tool and Manufacturing Engineers Handbook, 4th ed., Vol. 3, *Materials, Finishing and Coating*, Society of Manufacturing Engineers, 1985.

Utech, B., *A guide to High Performance Coating*, Society of Manufacturing Engineers, 2002.

van Ooij, W.J., Bierwagen, G.P., Skerry, B.S., and Mills, D., *Corrosion Control of Metals by Organic Coatings*, CRC Press, 1999.

REVIEW QUESTIONS

34.1 Explain why surface treatments may be necessary for manufactured products.

34.2 What are the advantages of roller burnishing?

34.3 Explain the difference between case hardening and hard facing.

34.4 Describe the principles of physical and chemical vapor deposition. What applications do they have?

34.5 What is the principle of electroforming? What are its advantages?

34.6 Explain the difference between electroplating and electroless plating.

34.7 How is hot dipping performed?

34.8 What is an air knife? How does it function?

34.9 What tests are there to determine the cleanliness of surfaces?

34.10 Describe the common painting systems presently in use in industry.

34.11 What is a conversion coating? Why is it so called?

34.12 What are the similarities and differences between electroplating and anodizing?

34.13 Describe the difference between thermal spraying and plasma spraying.

34.14 What is cladding, and why is it performed?

QUALITATIVE PROBLEMS

34.15 Describe how roller-burnishing processes induce compressive residual stresses on the surfaces of parts.

34.16 Explain why you might want to coat parts with ceramics.

34.17 Give examples of part designs that are suitable for hot-dip galvanizing.

34.18 Repeat Problem 34.17, but for cleaning.

34.19 Give some applications of mechanical surface treatment.

34.20 It has been observed in practice that a thin layer of chrome plating (such as that on older-model automobile bumpers) is better than a thick layer. Explain why, considering the effect of thickness on the tendency for cracking.

34.21 It is well known that coatings may be removed or depleted during the service life of components, particularly at elevated temperatures. Describe the factors involved in the strength of coatings and their durability.

34.22 Roller burnishing typically is applied to steel parts. Why is this so?

34.23 Make a list of the coating processes described in this chapter and classify them as "thick" or "thin."

34.24 Which of the processes described in this chapter are used only for small parts? Why is this so?

34.25 Shiny, metallic balloons commonly are made with festive printed patterns which are produced by printing screens and then plated on the balloons. How can metallic coatings be plated onto a rubber sheet?

34.26 Why is galvanizing important for automotive-body sheet-metals?

34.27 Explain the principles involved in various techniques for applying paints.

QUANTITATIVE PROBLEMS

34.28 Taking a simple example, such as the parts shown in Fig. 34.1, estimate the force required for roller burnishing. (See Sections 2.6 and 14.4.)

34.29 Estimate the plating thickness in electroplating a 50-mm solid-metal ball using a current of 10 A and a plating time of 2 hours. Assume that $c = 0.08$ in Eq. (34.1).

SYNTHESIS, DESIGN, AND PROJECTS

34.30 Which surface treatments are functional, and which are decorative? Are there any that serve both functions? Explain.

34.31 An artificial implant has a porous surface area where it is expected that the bone will attach and grow into the implant. Make recommendations for producing a porous surface and then review the literature to determine the actual processes used.

34.32 If one is interested in obtaining a textured surface on a coated piece of metal, should one apply the coating first or apply the texture first?

34.33 It is known that a mirror-like surface finish can be obtained by plating workpieces that are ground (that is, the surface finish improves after coating). Explain how this occurs.

34.34 Outline the reasons why the topics described in this chapter are important in manufacturing processes and operations.

34.35 Because they evaporate, solvents and similar cleaning solutions have adverse environmental effects.

Describe your thoughts on what modifications could be made to render cleaning solutions more environmentally friendly.

34.36 When an electrically insulating specimen is to be placed in a scanning electron microscope for examination, the specimen first is coated with a thin layer of gold. How would you produce this layer of gold?

34.37 A roller-burnishing operation is performed on a shaft shoulder to increase fatigue life. It is noted that the resultant surface finish is poor, and a proposal is made to machine the surface layer to further improve fatigue life. Will this be advisable? Explain.

34.38 You can simulate the shot-peening process by using a ball-peen hammer (in which one of the heads is round). Using such a hammer, make numerous indentations on the surface of a piece of aluminum sheet (a) 2-mm and (b) 10-mm thick, respectively, which is placed on a hard flat surface. You will note that both pieces will develop curvatures but one will become concave and the other become convex. Describe your observations and explain why this happens. (*Hint*: See Fig. 2.13)

34.39 Obtain several pieces of small metal parts (such as bolts, nuts, rods, and sheet metal) and perform the waterbreak test on them. Then clean the surfaces using various cleaning fluids and repeat the test. Describe your observations.

34.40 Inspect various products around your house (such as small and large appliances, silverware, metal vases and boxes, kitchen utensils, and hand tools) and comment on the type of coatings they may have. Also comment on those products that do not appear to have any coatings, stating the possible reasons for this.

34.41 Survey the technical literature and prepare a brief report on the environmental considerations regarding the application of the processes described in this chapter.

PART VIII

Common Aspects of Manufacturing

The preceding chapters have described the techniques used to modify the surfaces of components and products to obtain certain desirable properties, and we have discussed the advantages and limitations of each technique. Although dimensional accuracies obtained in individual manufacturing processes were described, we have not yet explained how parts are *measured* and *inspected* before they are assembled.

Dimensions and other surface features of a part are measured to ensure that it is manufactured consistently and within the specified range of dimensional tolerances. The vast majority of manufactured parts are components or a subassembly of a product, and they must fit and be assembled properly so that the product performs its intended function during its service life. For example: (a) A piston should fit into a cylinder within specified tolerances; (b) A turbine blade should fit properly into its slot on a turbine disk; and (c) The slideways of a machine tool must be produced with a certain accuracy so that the parts produced on that machine are accurate within their desired specifications.

Measurement of the relevant dimensions and features of parts is an integral aspect of **interchangeable parts manufacturing**, the basic concept of standardization and mass production. For example, if a ball bearing in a machine is worn and has to be replaced, all one has to do is purchase a similar one with the same specification or part number. The same is now done with all products, ranging from bolts and nuts to gears and to electric motors.

In the following two chapters, we will describe the principles involved in and the various instruments and modern machines used for measuring dimensional features, such as length, angle, flatness, and roundness. Testing and inspecting parts are important aspects of manufacturing operations, thus the methods used for the nondestructive and destructive testing of parts also are described.

One of the most important aspects of manufacturing is **product quality**. We will explain the technological and economic importance of *building quality into a product* rather than inspecting the product after it is made (as has been done traditionally). The significance of this concept should be clear in view of competitive manufacturing in a global economy.

Engineering Metrology and Instrumentation

In this part of the book we describe the importance of the measurement of manufactured parts. Specifically:

- Traditional methods of measurement.
- Modern machines to measure geometric features of parts.
- Features of measuring instruments and machines and how to select them.
- Importance of automated measurement.
- Principles of dimensioning and tolerancing.

35.1 | Introduction

This chapter presents the principle methods of measurement and the characteristics of the instruments used in manufacturing. **Engineering metrology** is defined as the measurement of dimensions such as length, thickness, diameter, taper, angle, flatness, profile, and others. Consider, for example, the slideways for machine tools (Fig. 35.1); these components must have specific dimensions, angles, and flatness in order for the machine to function properly and with the desired dimensional accuracy.

Traditionally, measurements have been made *after* the part has been produced—known as **post-process inspection**. Here, the term "inspection" means checking the dimensions of what has been produced or is being produced and determining whether it complies with the specified dimensional tolerances and other specifications. Today, however, measurements are being made while the part is being produced on the machine—known as **in-process, on-line,** or **real-time inspection**.

An important aspect of metrology in manufacturing processes is the **dimensional tolerance** (that is, the permissible variation in the dimensions of a part). Tolerances are important because of their impact on the proper functioning of a product, part interchangeability, and manufacturing costs. Generally, the smaller the tolerance, the higher are the production costs. This chapter ends with a discussion on dimensional limits and fits used in engineering practice.

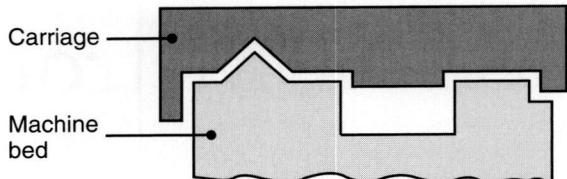

FIGURE 35.1   Cross-section of a machine-tool slide-way. The width, depth, angles, and other dimensions all must be produced and measured accurately for the machine tool to function as expected.

35.2 | Measurement Standards

Our earliest experience with measurement is usually with a simple *ruler* to measure lengths (linear dimensions). Rulers are used as a *standard* against which dimensions are measured. Traditionally, in English-speaking countries, the units of *inch* and *foot* have been used, which originally were based on parts of the human body. Consequently, it was common to find significant variations in the length of one foot.

In most of the world, however, the *meter* has been used as a length standard. Originally, one meter was defined as one ten-millionth of the distance between the north pole and the equator. The original meter length subsequently was standardized as the distance between two scratches on a platinum-iridium bar kept under controlled conditions in a building outside Paris. In 1960, the meter officially was defined as 1,650,763.73 wavelengths (in a vacuum) of the orange light given off by electrically excited krypton 86 (a rare gas). The precision of this measurement was set as 1 part in 10^9. The meter is now a Système International d'Unites, (SI) unit of length, and is the international standard. The smallest SI dimensions are measured in nanometers (1 nm $= 10^{-9}$ m).

Numerous measuring instruments and devices are used in engineering metrology, each of which has its own application, resolution, precision, and other features. Two terms that commonly are used to describe the type and quality of an instrument are:

1. **Resolution** is the smallest difference in dimensions that the measuring instrument can detect or distinguish. A wooden yardstick, for example, has far less resolution than a micrometer.

2. **Precision** (sometimes incorrectly called accuracy) is the degree to which the instrument gives repeated measurements of the same standard. For example, an aluminum ruler will expand or contract (depending on temperature variations in the environment in which it is used) or is held by hand, thus subjecting it to a higher temperature than the surrounding air, which affects the precision.

In engineering metrology, the words **instrument** and **gage** often are used interchangeably. Temperature control is very important, particularly for making measurements with precision instruments. The standard measuring temperature is 20°C (68°F), and all gages are calibrated at this temperature. In the interest of accuracy, measurements should be taken in controlled environments maintaining this temperature usually within ±0.3°C (0.5°F).

EXAMPLE 35.1   Length measurements throughout history

Many standards for length measurement have been developed during the past 6000 years. A common standard in Egypt around 4000 B.C. was the King's Elbow, which was equivalent to 0.4633 m. One *elbow* was equal to 1.5 feet (or 2 hand-spans, 6 hand-widths, or 24 finger-thicknesses). In 1101 A.D., King Henry I declared a new standard called the *yard* (0.9144 m), which was the distance from his nose to the tip of his thumb.

During the Middle Ages, almost every kingdom and city established its own length standard—some with identical names. In 1528, French physician J. Fernel proposed the distance between Paris and Amiens (a city 120 km north of Paris) as a general length reference. During the 17th century, some scientists suggested that the length of a certain pendulum be used as a standard. In 1661, British architect Sir Christopher Wren suggested a pendulum with a period of 1/2 second be used. Dutch mathematician C. Huygens proposed a pendulum which had a length one-third of Wren's and which had a period of 1 second.

To put an end to the confusion of length measurement, a definitive length standard began to be developed in 1790 in France with the concept of a *métre* (from the Greek word *metron*, meaning "measure"). A gage block one meter long was made of pure platinum with a rectangular cross-section and was placed in the National Archives in Paris in 1799. Copies of this gage were made for other countries over the years.

During the three years of 1870 to 1872, international committees met and decided on an international meter standard. The new bar was made of 90% platinum and 10% iridium with an x-shaped cross-section and overall dimensions of 20 mm × 20 mm. Three marks were engraved at each end of the bar. The standard meter is the distance between the central marks at each end, measured at 0°C. Extremely accurate measurement is now based on the speed of light in a vacuum, which is calculated by multiplying the wavelength of the standardized infrared beam of a laser by its frequency.

35.3   Geometric Features of Parts;
Analog and Digital Measurements

In this section, we list the most common quantities and geometric features that typically are measured in engineering practice and in products made by the manufacturing processes described throughout this book.

- *Length*—including all linear dimensions of parts.
- *Diameter*—outside and inside, including parts with different outside and inside diameters (steps) along their length.
- *Roundness*—including out-of-roundness, concentricity, and eccentricity.
- *Depth*—such as drilled or bored holes and cavities in dies and molds.
- *Straightness*—such as shafts, bars, and tubing.
- *Flatness*—such as machined and ground surfaces.
- *Parallelism*—such as two shafts or slideways in machines.
- *Perpendicularity*—such as a threaded bar inserted into a flat plate.

- *Angles*—including internal and external angles.
- *Profile*—such as curvatures in castings, forgings, and on car bodies.

A wide variety of instruments and machines are available to accurately and rapidly measure these quantities on stationary parts or on parts that are in continuous production. Because of major and continuing trends in automation and the computer control of manufacturing operations, modern measuring equipment are now an *integral* part of production machines. The implementation of digital instrumentation and developments in computer-integrated manufacturing (described in Part IX of this book) has led to the total integration of measurement technologies within the manufacturing systems.

It is important to recognize the advantages of *digital* over *analog* instruments. As will be obvious from our description of traditional measuring equipment in Section 35.4, accurate measurement on an analog instrument, such as a vernier caliper or micrometer (Fig. 35.2a), relies on the skill of the operator to properly interpolate and read the graduated scales. In contrast, a digital caliper does not require any particular skills, because measurements are indicated directly (Fig. 35.2b). More

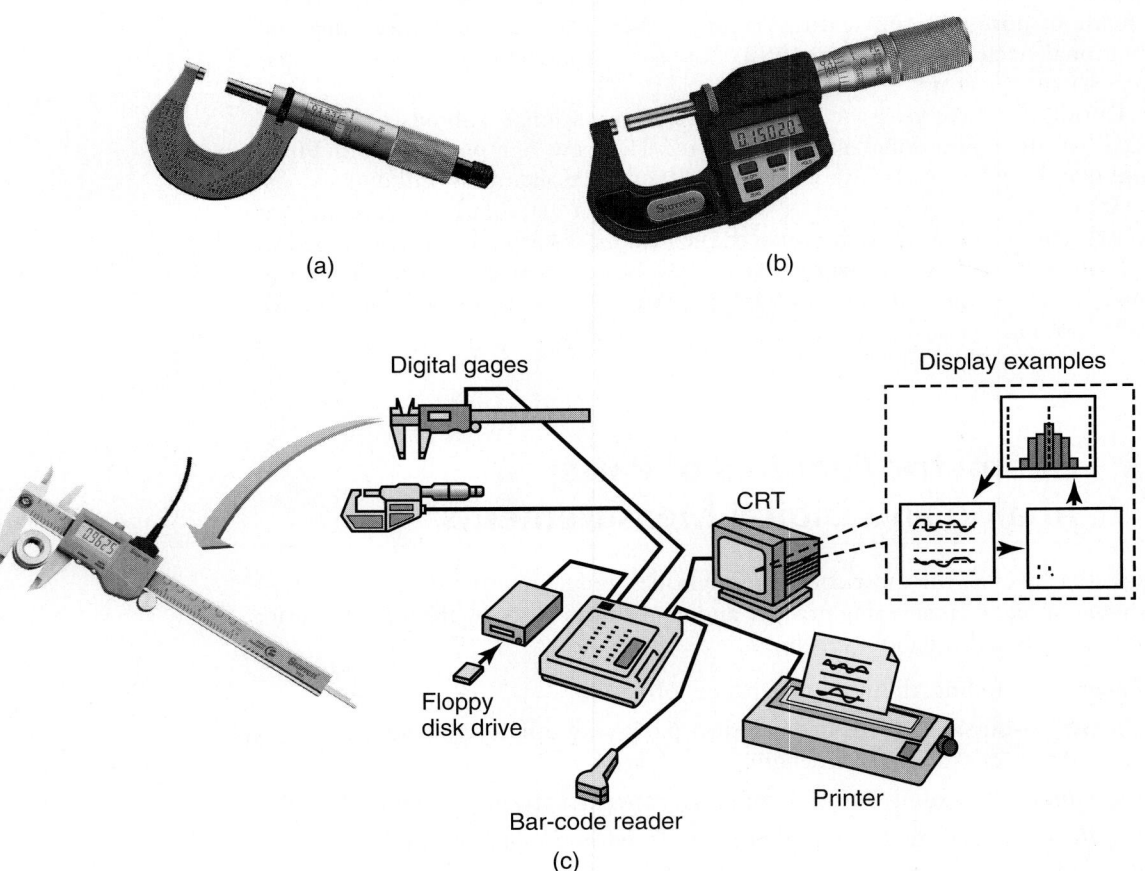

(a)

(b)

Digital gages

Display examples

CRT

Floppy
disk drive

Bar-code reader

Printer

(c)

FIGURE 35.2   (a) A vernier (analog) micrometer. (b) A digital micrometer with a range of 0 to 1 in. (0 to 25 mm) and a resolution of 50 μin. (1.25 μm). Generally, it is easier to read dimensions on this instrument than on analog micrometers. (c) Schematic illustration showing the integration of digital gages with miroprocessors for real-time data acquisition for statistical process control. *Source:* (a) Courtesy of L.C. Starrett Co. (b) Courtesy of Mitutoyo Corp.

importantly, however, digital equipment can be integrated easily into other equipment (Fig. 35.2c), including production machinery and systems for statistical process control (SPC), as described in detail in Chapter 36.

35.4   Traditional Measuring Methods and Instruments

In this section, we describe the characteristics of traditional measuring methods and instruments that have been used over many years and are still used extensively in many parts of the world. However, these instruments rapidly are being replaced with more efficient and advanced instruments and measuring machines, as described in Section 35.5.

35.4.1 Line-graduated instruments

These instruments are used for measuring lengths or angles. *Graduated* means "marked to indicate a certain quantity."

Linear measurement (direct reading)

- **Rules:** The simplest and most commonly used instrument for making linear measurements is a steel rule (*machinist's rule*), bar, or tape with fractional or decimal graduations. Lengths are measured directly to an accuracy that is limited to the nearest division, usually 1 mm (0.040 in.).

- **Calipers:** These instruments can be used to measure inside or outside lengths. Also called *caliper gages* and *vernier calipers* (named for P. Vernier, who lived in the 1600s), they have a graduated beam and a sliding jaw. *Digital calipers* are in increasingly wider use.

- **Micrometers:** These instruments commonly are used for measuring the thickness and inside or outside dimensions of parts. *Digital micrometers* are equipped with digital readouts (Fig. 35.2b) in metric or English units. Micrometers also are available for measuring internal diameters (*inside micrometer*) and depths (*micrometer depth gage*, Fig. 35.3). The anvils on micrometers can be equipped with conical or ball contacts to measure recesses, threaded-rod diameters, and wall thicknesses of tubes and curved sheets.

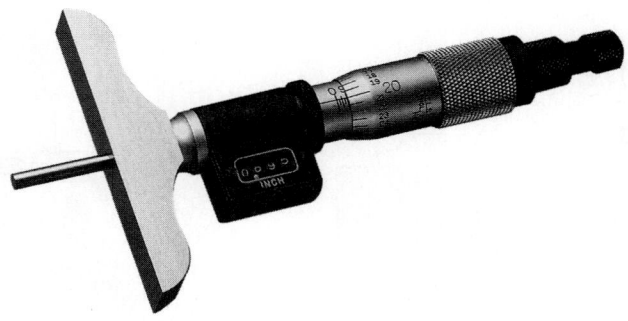

FIGURE 35.3   A digital-micrometer depth gage. *Source:* Courtesy of Starrett Co.

Linear measurement (indirect reading). These instruments typically are *calipers* and *dividers* without any graduated scales. They are used to transfer the measured size to a direct-reading instrument, such as a rule. Because of the experience required to use them and their dependence on graduated scales, the accuracy of indirect-measurement tools is limited. *Telescoping gages* can be used for the indirect measurement of holes or cavities.

Angle measurement

- **Bevel protractor:** This is a direct-reading instrument similar to a common protractor, except that it has a movable element. The two blades of the protractor are placed in contact with the part being measured, and the angle is read directly on the vernier scale. Another common type of bevel protractor is the *combination square*, which is a steel rule equipped with devices for measuring 45° and 90° angles.

- **Sine bar:** Measuring with this method involves placing the part on an inclined bar (sine bar) or plate and adjusting the angle by placing gage blocks on a surface plate. After the part is placed on the sine bar, a dial indicator is used to scan the top surface of the part. Gage blocks (see Section 35.4.4) are added or removed as necessary until the top surface is parallel to the surface plate. The angle on the part then is calculated using trigonometric relationships.

- **Surface plates:** These plates are used to place parts to be measured and the measuring instruments. They typically are made of cast iron or natural stones (such as granite) and are used extensively in engineering metrology. Granite surface plates have the desirable properties of being resistant to corrosion and nonmagnetic, and of having low thermal expansion, thereby minimizing thermal distortion.

Comparative length measurement. Instruments used for measuring comparative lengths (also called *deviation-type* instruments) amplify and measure variations or deviations in the distance between two or more surfaces. These instruments compare dimensions (hence the word *comparative*), of which the most common example is a **dial indicator** (Fig. 35.4). These are simple mechanical devices that convert linear displacements of a pointer to the amount of rotation of an indicator on a circular dial. The indicator is set to zero at a certain reference surface, and the instrument or the surface to be measured (either external or internal) is brought into contact with the pointer. The movement of the indicator is read directly on the circular dial (either plus or minus) to accuracies as high as 1 μm (40 μin.). Dial indicators with electrical and fluidic amplification mechanisms and with digital readout also are available.

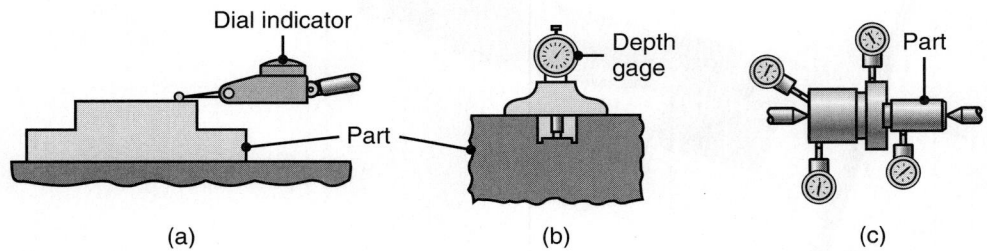

FIGURE 35.4 Three uses of dial indicators: (a) roundness, (b) depth, and (c) multiple-dimension gaging of a part.

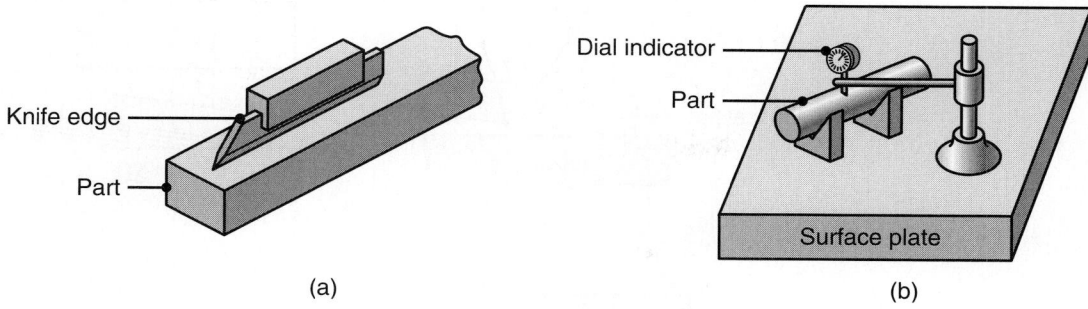

FIGURE 35.5 Measuring straightness manually with (a) a knife-edge rule and (b) a dial indicator. *Source:* After F.T. Farago.

35.4.2 Measuring geometric features

Straightness. Straightness commonly can be checked using a straightedge or a dial indicator (Fig. 35.5). An *autocollimator* (which resembles a telescope with a light beam that bounces back from the object) is used to accurately measure small angular deviations on a flat surface. *Laser beams* now are used commonly to align individual machine elements in the assembly of machine components.

Flatness. Flatness can be measured by mechanical means using a surface plate and a dial indicator. This method can be used to measure perpendicularity, which also can be measured using precision-steel squares.

Another method for measuring flatness is **interferometry**, using an *optical flat*. This device is a glass disk or fused-quartz disk with parallel flat surfaces, which is placed on the surface of the workpiece (Fig. 35.6a). When a *monochromatic* light beam (a light beam with one wavelength) is aimed at the surface at an angle, the optical flat splits the light beam into two beams, appearing as light and dark bands to the naked eye (Fig. 35.6b). The number of fringes that appear is related to the distance between the surface of the part and the bottom surface of the optical flat (Fig. 35.6c). Consequently, a truly flat workpiece surface (that is, one in which the angle between the two surfaces is zero) will not split the light beam and fringes will not appear. When surfaces are not flat, the fringes are curved (Fig. 35.6d). The interferometry method also is used for observing surface textures and scratches (Fig. 35.6e).

Diffraction gratings consist of two optical-flat glasses of different lengths with closely spaced parallel lines scribed on their surfaces. The grating on the shorter glass is inclined slightly. As a result, *interference fringes* develop when it is viewed over the longer glass. The position of these fringes depends on the relative position of the two sets of glasses. With modern equipment and using electronic counters and photoelectric sensors, a resolution of 2.5 μm (0.0001 in.) can be obtained with gratings having 40 lines/mm (1000 lines/in.).

Roundness. This feature usually is described as a deviation from true roundness (which mathematically is manifested in a circle). The term *out-of-roundness* (ovality) is actually more descriptive of the shape of the part (Fig. 35.7a) than the word roundness. True roundness is essential to the proper functioning of rotating shafts, bearing races, pistons, cylinders, and steel balls in bearings.

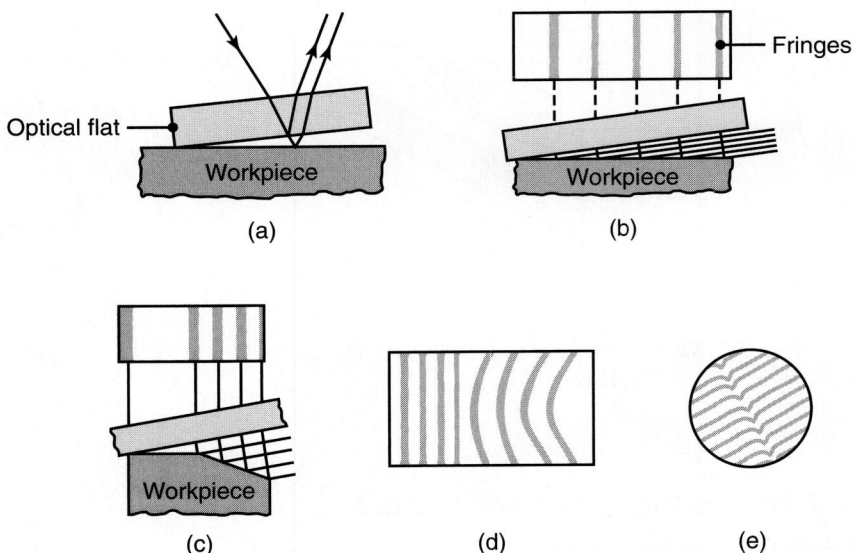

FIGURE 35.6   (a) Interferometry method for measuring flatness using an optical flat. (b) Fringes on a flat, inclined surface. An optical flat resting on a perfectly flat workpiece surface will not split the light beam, and no fringes will be present. (c) Fringes on a surface with two inclinations. *Note:* the greater the incline, the closer together are the fringes. (d) Curved fringe patterns indicate curvatures on the workpiece surface. (e) Fringe pattern indicating a scratch on the surface.

Methods of measuring roundness generally fall into two categories.

1. The round part is placed on a V-block or between centers (Fig. 35.7b and c, respectively) and is rotated while the point of a dial indicator is in contact with the part surface. After a full rotation of the workpiece, the difference between the maximum and minimum readings on the dial is noted. This difference is called the **total indicator reading** (TIR) or *full indicator movement.* This method also can be used to measure the straightness (squareness) of end faces of shafts that are machined, such as the facing operation shown in Fig. 23.1e.

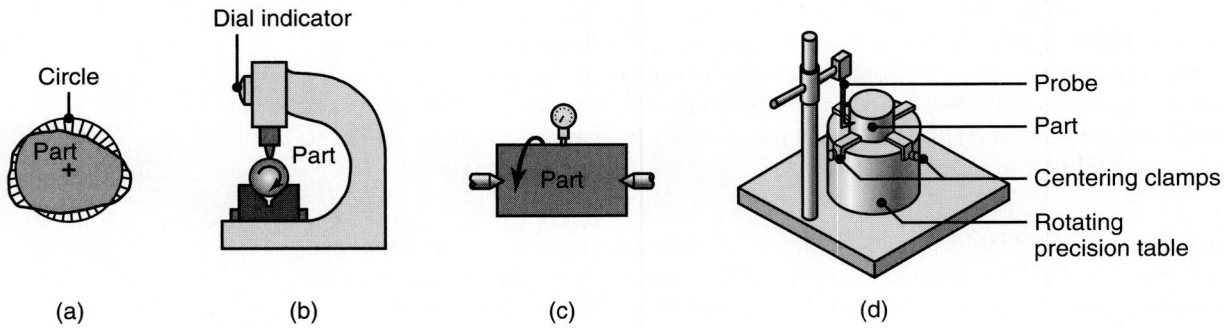

FIGURE 35.7   (a) Schematic illustration of out-of-roundness (exaggerated). Measuring roundness using (b) a V-block and dial indicator, (c) a round part supported on centers and rotated, and (d) circular tracing. *Source:* After F.T. Farago.

2. In *circular tracing,* the part is placed on a platform, and its roundness is measured by rotating the platform (Fig. 35.7d). Conversely, the probe can be rotated around a stationary part to take the measurement.

Profile. Profile may be measured by means such as (a) comparing the surface with a template or profile gage (such as radii and fillets) for conformity and (b) using a number of dial indicators or similar instruments. The best method, however, is using the advanced measuring machines described in Section 35.5.

Screw threads and gear teeth. Threads can be measured by means of *thread gages* of various designs that compare the produced thread against a standard thread. Some of the gages used are *threaded plug gages, screw-pitch gages,* micrometers with cone-shaped points, and *snap gages* (see Section 35.4.4) with anvils in the shape of threads. Gear teeth are measured using (a) instruments that are similar to dial indicators, (b) calipers (Fig. 35.8a), and (c) micrometers using pins or balls of various diameters (Fig. 35.8b). Advanced methods include the use of optical projectors and coordinate measuring machines.

35.4.3 Optical contour projectors

These instruments (also called optical comparators) first were developed in the 1940s to check the geometry of cutting tools for machining screw threads, but they now are used for checking all profiles (Fig. 35.9). The part is mounted on a table or between centers, and the image is projected onto a screen at magnifications of $100\times$ or higher. Linear and angular measurements are made directly on the screen, which is marked with reference lines and circles. The screen can be rotated to allow angular measurements.

35.4.4 Gages

This section describes several common gages that have simple solid shapes and cannot be classified as instruments.

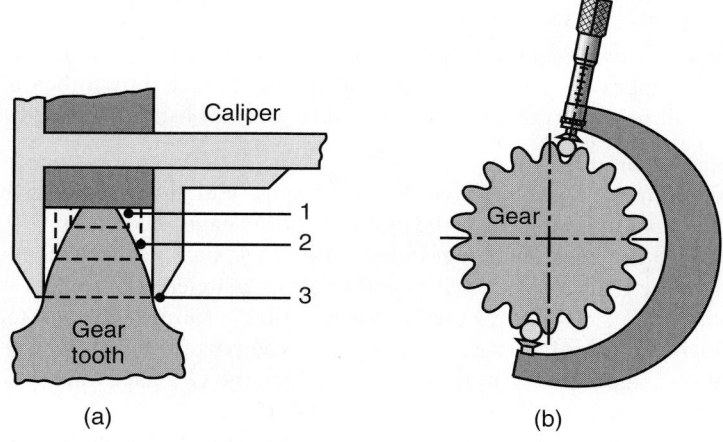

(a)

(b)

FIGURE 35.8 Measuring gear-tooth thickness and profile with (a) a gear-tooth caliper and (b) pins or balls and a micrometer. *Source*: Courtesy of American Gear Manufacturers Association.

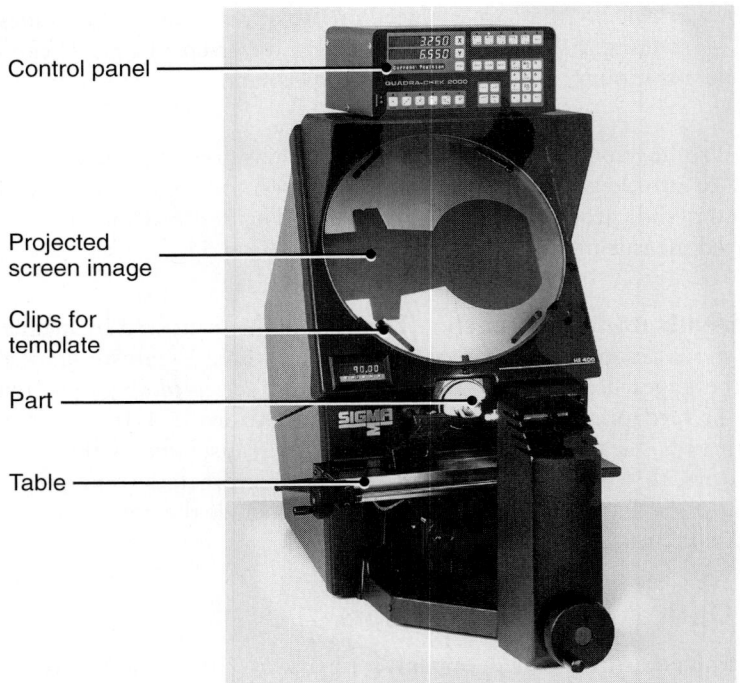

Control panel

Projected
screen image

Clips for
template

Part

Table

FIGURE 35.9   A bench-model horizontal-beam contour projector
with a 16-in.-diameter screen with 150-W tungsten halogen illumination.
Source: Courtesy of L.S. Starrett Company, Precision Optical Division.

Gage blocks.   Gage blocks are individual square, rectangular, or round blocks of
various sizes. For general use, they are made from heat-treated and stress-relieved
alloy steels. Better gage blocks are made of ceramics (often zirconia) and chromium carbide—unlike steels, these materials do not rust, but they are brittle and must
be handled carefully. *Angle blocks* are made the same way and are used for angular
gaging. Gage blocks have a flatness within 1.25 μm (50 μin.). Environmental temperature control is important when gages are used for high-precision measurements.

Fixed gages.   These gages are replicas of the shapes of the parts to be measured.
Although fixed gages are easy to use and inexpensive, they only indicate whether a
part is too small or too large when compared to an established standard.

- **Plug gages** commonly are used for holes (Fig. 35.10a and b). The *GO gage* is
  smaller than the *NOT GO* (or *NO GO*) *gage* and slides into any hole which
  has a dimension smaller than the diameter of the gage. The *NOT GO* gage must
  not go into the hole. Two gages are required for such measurements, although
  both may be on the same device—either at opposite ends or in two steps at one
  end (*step-type gage*). Plug gages also are available for measuring internal tapers
  (in which deviations between the gage and the part are indicated by the looseness of the gage), splines, and threads (in which the *GO* gage must screw into the
  threaded hole).

- **Ring gages** (Fig. 35.10c) are used to measure shafts and similar round parts.
  Ring thread gages are used to measure external threads. The *GO* and *NOT*

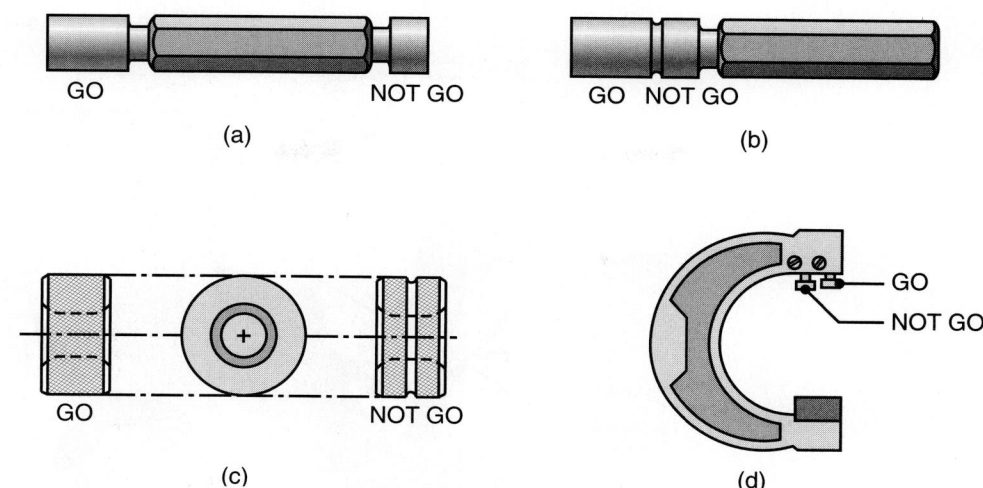

FIGURE 35.10   (a) Plug gage for holes with *GO* and *NOT GO* on opposite ends. (b) Plug gage with *GO* and *NOT GO* on one end. (c) Plain ring gages for gaging round rods. Note the difference in knurled surfaces to identify the two gages. (d) Snap gage with adjustable anvils.

GO features on these gages are identified by the type of knurling on the outside diameters of the rings.

- **Snap gages** (Fig. 35.10d) commonly are used to measure external dimensions. They are made with adjustable gaging surfaces for use with parts which have different dimensions. One of the gaging surfaces can be set at a different gap from the other, thus making it a one-unit *GO* and *NOT GO* gage.

Air gages.   The basic operation of an air gage (also called *pneumatic gage*) is shown in Fig. 35.11a. The gage head (air plug) has two or more holes, typically 1.25 mm (0.05 in.) in diameter, through which pressurized air (supplied by a constant-pressure line) escapes. The smaller the gap between the gage and the hole, the more difficult it is for the air to escape, and hence, the higher the back pressure. The back pressure (which is sensed and indicated by a pressure gage) is calibrated to measure the dimensional variations of holes.

The air gage can be rotated during use to indicate and measure any out-of-roundness of the hole. The outside diameters of parts (such as pins and shafts) also can be measured when the air plug is in the shape of a ring slipped over the part. In cases where a ring is not suitable to use, a fork-shaped gage head (with the air holes at the tips) can be used (Fig. 35.11b). Various shapes of air heads can be prepared for use in specialized applications on parts with different geometric features, such as the conical head shown in Fig. 35.11c.

Air gages are easy to use and the resolution can be as fine as 0.125 μm (5 μin.). If the surface roughness of the part is too high, the readings may be unreliable. The compressed-air supply must be clean and dry for proper operation. The part being measured does not have to be free of dust, metal particles, or similar contaminants, because the air will blow them away. The noncontacting nature and the low pressure of an air gage has the benefit of not distorting or damaging the measured part, as could be the case with mechanical gages—thus giving erroneous readings.

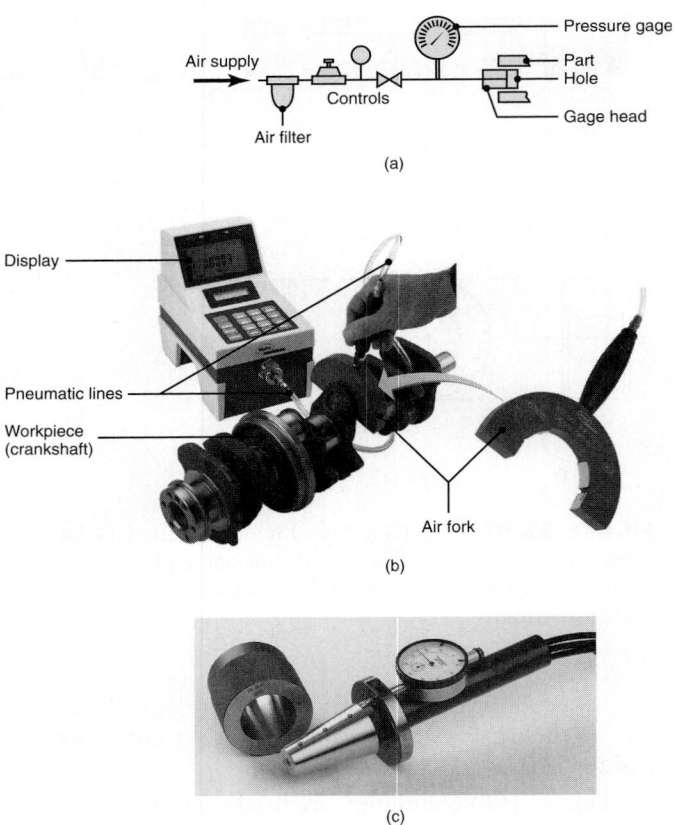

FIGURE 35.11 (a) Schematic illustration of the principle for an air gage. (b) Illustration of an air-gage system used to measure the main bearing dimension on a crank shaft. (c) A conical head for air gaging; note the small air holes on the conical surface. *Source:* (b) Courtesy of Mahr Federal, Inc. (c) Courtesy of Stotz Gaging Co.

35.5 Modern Measuring Instruments and Machines

We begin with descriptions and characteristics of modern measuring instruments, because they are far more accurate and reliable than the traditional instruments used, such as micrometers, calipers, dial indicators, and protractors.

Electronic gages. Unlike mechanical systems, *electronic gages* sense the movement of the contacting pointer through changes in the electrical resistance of a strain gage, inductance, or capacitance. The electrical signals then are converted and displayed digitally as linear dimensions with a digital readout. A hand-held electronic gage for measuring bore diameters is shown in Fig. 35.12. When its handle is squeezed slightly, the tool can be inserted into the bore, and the bore diameter is read directly. A microprocessor-assisted electronic gage for measuring vertical length is shown in Fig. 35.13.

A commonly used electronic gage is the **linear-variable differential transformer** (LVDT) used extensively for measuring small displacements. Electronic caliper gages with diamond-coated edges are available. The chemical vapor deposition (CVD) coating on these gages has a wear resistance superior to that of steel or tungsten-carbide edges; it also resists chemicals.

FIGURE 35.12   An electronic gage for measuring bore diameters. The measuring head is equipped with three carbide-tipped steel pins for wear resistance. The LED display reads 29.158 mm. *Source:* Courtesy of TESA SA.

Although they are more expensive than other types of gages, electronic gages have advantages in ease of operation, rapid response, digital readout, less possibility of human error, versatility, flexibility, and the capability to be integrated into automated systems through microprocessors and computers.

Laser micrometers.   In this instrument, a laser beam scans the workpiece (Fig. 35.14), typically at a rate of 350 times per second. Laser micrometers are capable of resolutions as high as 0.125 μm (5 μin.). They are suitable not only for stationary parts but also for in-line measurement of stationary, rotating, or vibrating

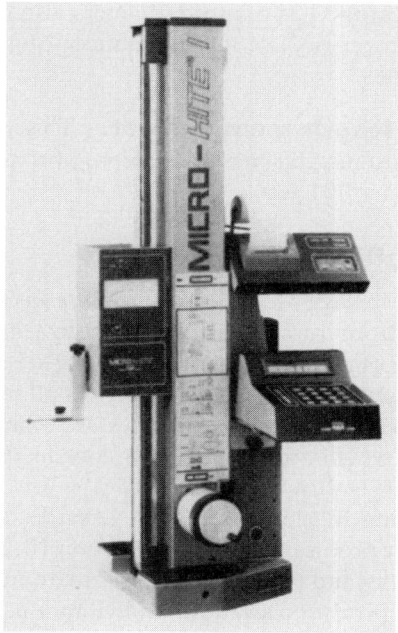

FIGURE 35.13   An electronic vertical-length measuring instrument with a resolution of 1 μm (40 min). *Source:* Courtesy of TESA SA.

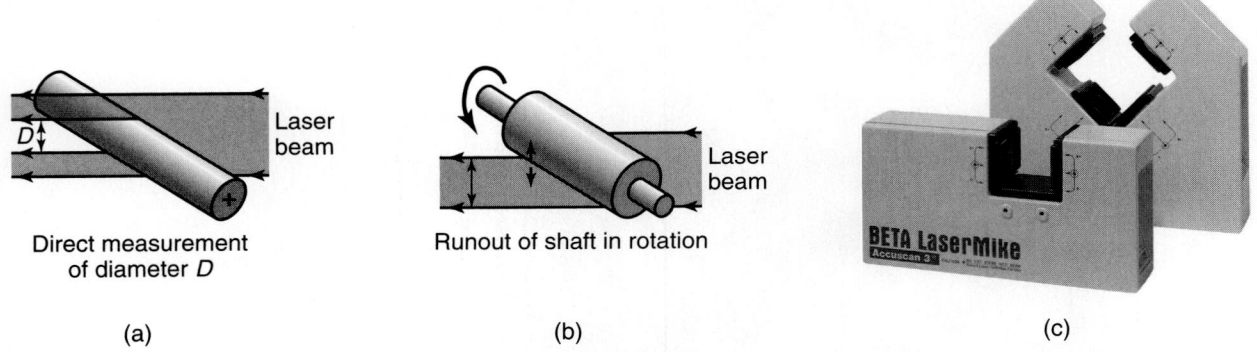

FIGURE 35.14   (a) and (b) Two types of measurements made with a laser scan micrometer. (c) Two types of laser micrometers. Note that the instrument in the front scans the part (placed in the opening) in one dimension; the larger instrument scans the part in two dimensions. *Source:* Courtesy of BETA LaserMike.

parts, as well as parts in continuous, high-speed production. Because there is no physical contact, they also can measure parts that are at elevated temperatures or are too flexible to be measured by other means. The laser beams can be of various types (such as scanning or rastoring for stationary parts), yielding **point cloud** descriptions of part surfaces. They are of the shadow type or are CCD-based for in-line measurement while in production.

Laser micrometers are available with various capacities and features. They can be hand-held for manual operation, or they can be mounted on and integrated with computer-controlled machines and statistical-process control units.

Laser interferometry.   This technique is used to check and calibrate machine tools for various geometric features during assembly. This method has better accuracies than gages or indicators. Laser interferometers also are used to automatically compensate for positioning errors in coordinate-measuring machines and computer-numerical control machines.

Photoelectric digital length measurement.   This is an instrument that can measure the overall dimensions, thickness, and depth of a variety of parts. Resolution settings can range from 5 to 0.01 μm.

35.5.1 Coordinate-measuring machines

As schematically shown in Fig. 35.15a, a *coordinate-measuring machine* (CMM) basically consists of a platform on which the workpiece being measured is placed and then is moved linearly or rotated. A probe (Fig. 35.15b; see also Fig. 25.6) is attached to a head (capable of various movements) and records all measurements. In addition to the tactile probe shown, other types of probes are scanning, laser (Fig. 35.15c), and vision probes, all of which are nontactile. A coordinate-measuring machine for inspection of a typical part is shown in Fig. 35.15d.

Coordinate-measuring machines are very versatile and capable of recording measurements of complex profiles with high resolution (0.25 μm; 10 μin.) and high speed. They are built rigidly and ruggedly to resist environmental effects in manufacturing plants, such as temperature variations and vibration. They can be placed close to machine tools for efficient inspection and rapid feedback; this way, processing parameters are corrected before the next part is made. Although large CMMs can be expensive, most machines with a touch probe and computer controlled three-dimensional movements are suitable for use in small shops and generally cost under $20,000.

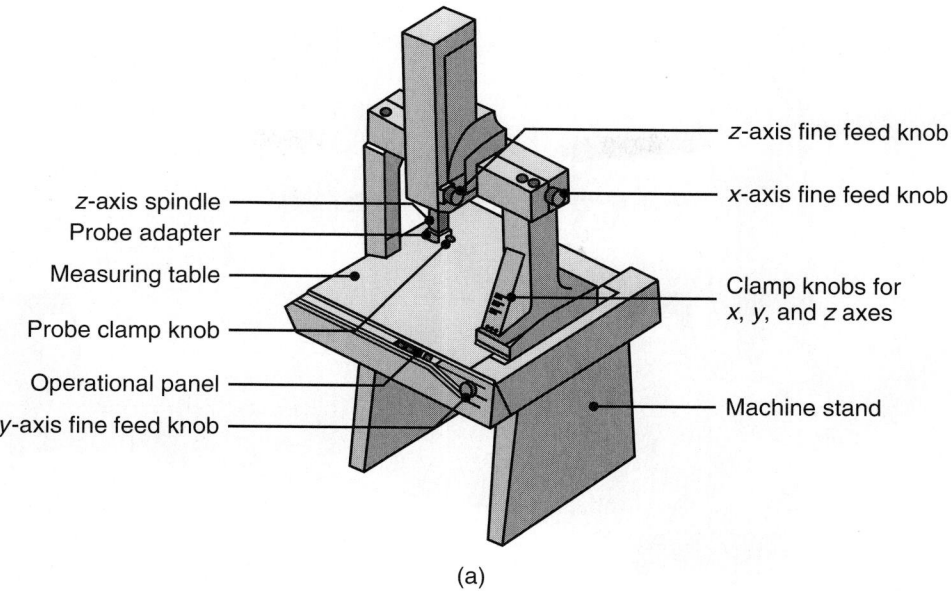

(a)

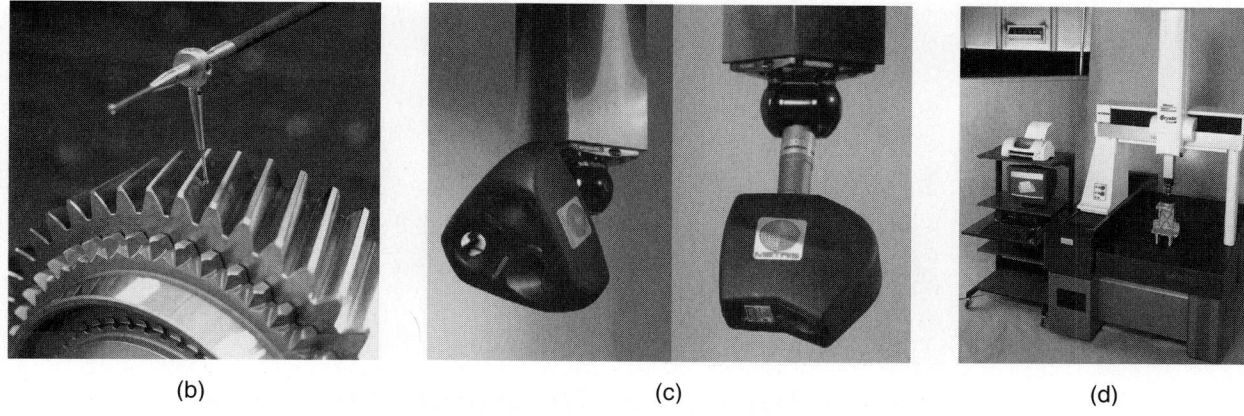

(b)   (c)   (d)

FIGURE 35.15 (a) Schematic illustration of a coordinate-measuring machine. (b) A touch signal probe. (c) Examples of laser probes. (d) A coordinate-measuring machine with a complex part being measured. *Source:* (b) through (d) Courtesy of Mitutoyo America Corp.

EXAMPLE 35.2 Coordinate-measuring machine for car bodies

A large, horizontal CNC coordinate-measuring machine used to measure all dimensions of a car body is shown in Fig. 35.16. This machine has a measuring range of 6 m × 1.6 m × 2.4 m high (236 in. × 63 in. × 94 in.), and a resolution of 0.1 μm (4 μin.). The system has temperature compensation within a range of 16° to 26°C (60° to 78°F) to maintain measurement accuracy. For efficient measurements, this machine has two heads with touch-trigger probes that are controlled simultaneously and have full three-dimensional movements. The measuring speed is 5 mm/s (0.2 in./s). The probes are software controlled, and the machine is equipped with safety devices to prevent the probes from inadvertently hitting any part of the car body during their movements. The equipment

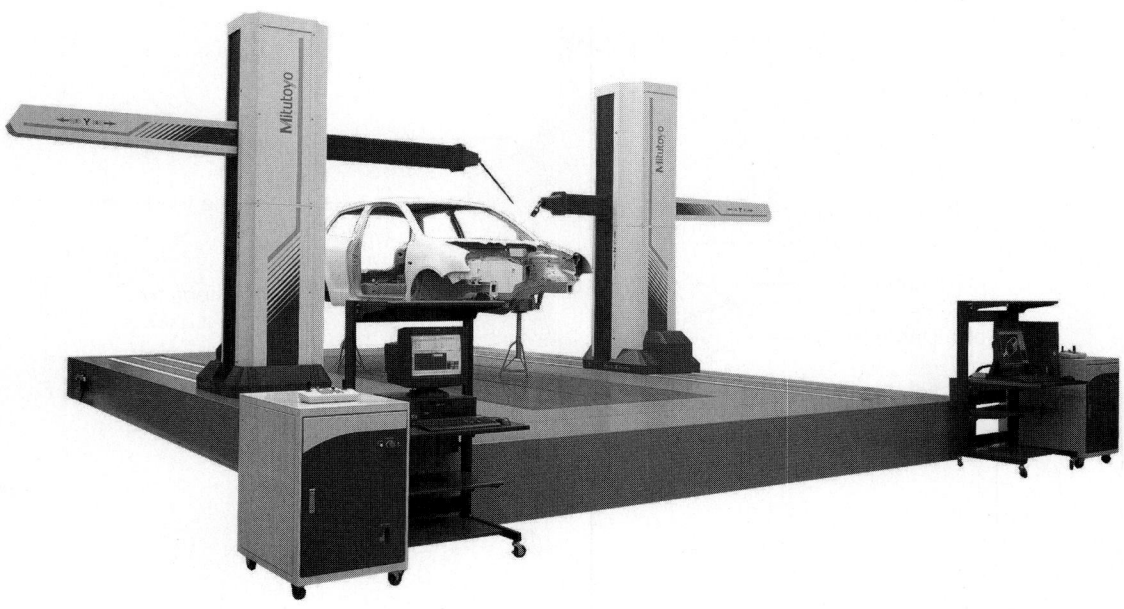

FIGURE 35.16   A large coordinate-measuring machine with two heads measuring various dimensions on a car body. *Source:* Courtesy of Mitutoyo America Corp.

shown around the base of the machine includes supporting hardware and software which controls all movements and records all measurements.

Source: Courtesy of Mitutoyo America Corporation.

35.6 | Automated Measurement

Automated measurement and inspection is based on various **on-line sensor systems** that monitor the dimensions of parts while they are being made and, if necessary, use these measurements as input to make corrections. (See Section 37.7.) Manufacturing cells and flexible manufacturing systems (Chapter 39) have led to the adoption of advanced measuring techniques and systems.

To appreciate the importance of on-line monitoring of dimensions, let's briefly consider the following question: If a machine has been producing a certain part with acceptable dimensions, what factors contribute to the subsequent deviation in the dimensions of the same part produced by the same machine? There are several technical as well as human factors involved:

- Static and dynamic deflections of the machine because of vibrations and fluctuating forces are caused by machine characteristics and variations in the properties and dimensions of the incoming material.

- Distortion of the machine because of thermal effects caused by such changes as in temperature of the environment, of metalworking fluids, and of machine bearings and various components.

- Wear of tools, dies, and molds can affect the dimensional accuracy of the parts produced.
- Human errors and miscalculations.

As a result of these factors, the dimensions of parts will vary, thus making continuous monitoring during production necessary.

35.7   General Characteristics and Selection of Measuring Instruments

The characteristics and quality of measuring instruments generally are described by various specific terms, defined as follows (in alphabetical order):

- **Accuracy:** The degree of agreement of the measured dimension with its true magnitude.
- **Amplification:** See Magnification.
- **Calibration:** Adjusting or setting an instrument to give readings that are accurate within a reference standard.
- **Drift:** See Stability.
- **Linearity:** The accuracy of the readings of an instrument over its full working range.
- **Magnification:** The ratio of instrument output to the input dimension (also called *amplification*).
- **Precision:** Degree to which an instrument gives repeated measurement of the same standard.
- **Repeat accuracy:** The same as accuracy, but repeated many times.
- **Resolution:** Smallest dimension that can be read on an instrument.
- **Rule of 10** *(gage maker's rule)*: An instrument or gage should be 10 times more accurate than the dimensional tolerances of the part being measured. (A factor of 4 is known as the *mil standard rule*).
- **Sensitivity:** Smallest difference in dimension that an instrument can distinguish or detect.
- **Speed of response:** How rapidly an instrument indicates the measurement, particularly when a number of parts are measured in rapid succession.
- **Stability:** An instrument's capability to maintain its calibration over a period of time (also called *drift*).

Selection of an appropriate measuring instrument for a particular application also depends on (a) the size and type of parts to be measured, (b) the environment (temperature, humidity, dust, and so on), (c) operator skills required, and (d) the cost of equipment.

35.8   Geometric Dimensioning and Tolerancing

Individually manufactured parts and components eventually are assembled into products. We take it for granted that when a thousand lawnmowers are manufactured and assembled, each part of the mower will mate properly with another component. For example, the wheels of the lawnmower will slip easily into their axles, or the pistons will fit properly into the cylinders, being neither too tight nor too loose.

Likewise, when we have to replace a broken or worn bolt on an old machine, we purchase an identical bolt. We are confident from similar experiences in the past that the new bolt will fit properly in the machine. The reason we feel confident is that the bolt is manufactured according to certain standards and the dimensions of all similar bolts will vary by only a small, specified amount.

In other words, all bolts are manufactured within a certain range of dimensional tolerance; thus, all similar bolts are *interchangeable*. We also expect that the new bolt will function satisfactorily for a period of time, unless abused or misused. Bolts periodically are subjected to various tests during their production to make sure that their quality is within certain specifications.

Dimensional tolerance.   This is defined as the permissible or acceptable variation in the dimensions (height, width, depth, diameter, and angles) of a part. The root of the word "tolerance" is the Latin *tolerare*, meaning "to endure" or "put up with." Tolerances are unavoidable, because it is virtually impossible (and unnecessary) to manufacture two parts that have precisely the same dimensions.

Furthermore, because close dimensional tolerances can increase the product cost significantly, a narrow tolerance range is undesirable economically. However, for some parts, close tolerances are necessary for proper functioning and are worth the added expense associated with narrow tolerance ranges. Examples are precision measuring instruments and gages, hydraulic pistons, and bearings for aircraft engines.

Measuring dimensional tolerances and features of parts rapidly and reliably can be a challenging task. For example, each of the 6 million parts on a Boeing 747-400 aircraft requires the measurement of about 25 features, representing a total of 150 million measurements.

Surveys have shown that the dimensional tolerances on state-of-the-art manufactured parts are shrinking by a factor of 3 every 10 years, and that this trend will continue. It is estimated that accuracies of (a) conventional turning and milling machines will rise from the present 7.5 to 1 μm, (b) diamond-wheel wafer-slicing machines for semiconductor fabrication to 0.25 μm, (c) precision diamond turning machines to 0.01 μm, and (d) ultraprecision ion-beam machining to less than 0.001 μm. (See also Fig. 25.17.)

Importance of tolerance control.   Dimensional tolerances become important only when a part is to be assembled or mated with another part. Surfaces that are free and not functional do not need close tolerance control. Thus, for example, the accuracy of the holes and the distance between the holes for a connecting rod are far more critical than the rod's width and thickness at various locations along its length (see Fig. 14.7).

To appreciate the importance of dimensional tolerances, let's assemble a simple round shaft (axle) and a wheel with a round hole. Assume that we want the axle's diameter to be 1 in. (Fig. 35.17). We go to a hardware store and purchase a 1-in. round rod and a wheel with a 1-in. hole. Will the rod fit into the hole without forcing it or will it be loose in the hole? The 1-in. dimension is the **nominal size** of the shaft. If we purchase such a rod from a different store, at a different time, and select one randomly from a large lot, the chances are that each rod will have a slightly different diameter. Machines with the same setup may produce rods of slightly different diameters, depending on a number of factors, such as speed of operation, temperature, lubrication, and variations in the properties of the incoming material. If we now specify a *range* of diameters for both the rod and the hole of the wheel, we can predict the type of fit correctly.

Certain terminology has been established to clearly define these geometric quantities, such as the International Organization for Standardization (ISO) system shown in Fig. 35.17. Note that both the shaft and the hole have minimum and maximum diameters, respectively, the difference being the tolerance for each member. A proper engineering drawing would specify these parameters with numerical values, as shown in Fig. 35.18.

The range of dimensional tolerances possible in manufacturing processes is given in various figures and tables throughout this book. There is a general relationship between tolerances and part size (Fig. 35.19) and between tolerances and the surface finish of parts manufactured by various processes (Fig. 35.20). Note the wide range of tolerances and surface finishes obtained. Also, the larger the part, the greater is its obtainable tolerance range.

Definitions.   Several terms are used to describe features of dimensional relationships between mating parts. Details of these definitions are available in the ANSI B4.2, ANSI Y14.5, and ISO/TC10/SC5 standards. Because of the complex geometric relationships involved among all of the parts to be assembled, the definitions of these terms can be somewhat confusing.

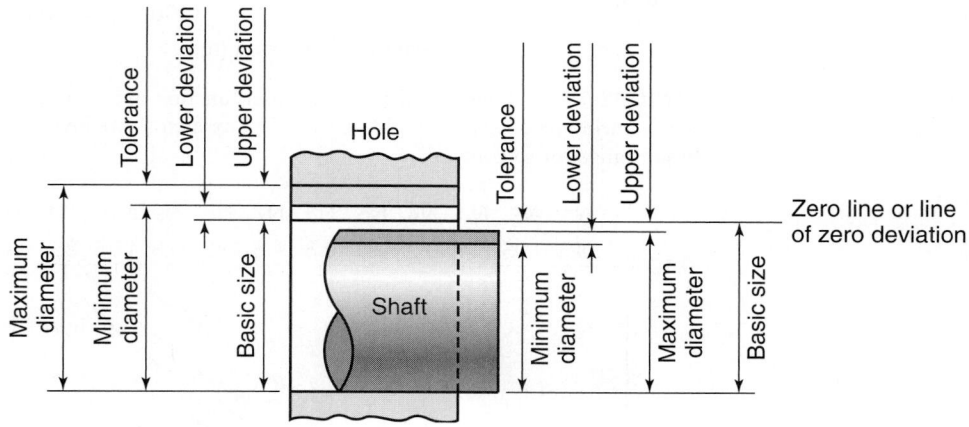

FIGURE 35.17   Basic size, deviation, and tolerance on a shaft, according to the ISO system.

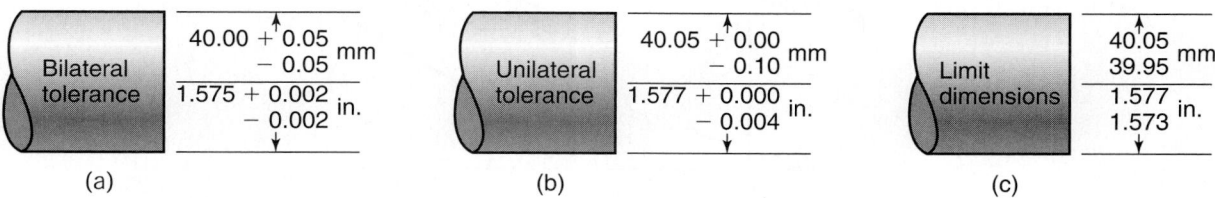

FIGURE 35.18   Various methods of assigning tolerances on a shaft: (a) bilateral tolerance, (b) unilateral tolerance, and (c) limit dimensions.

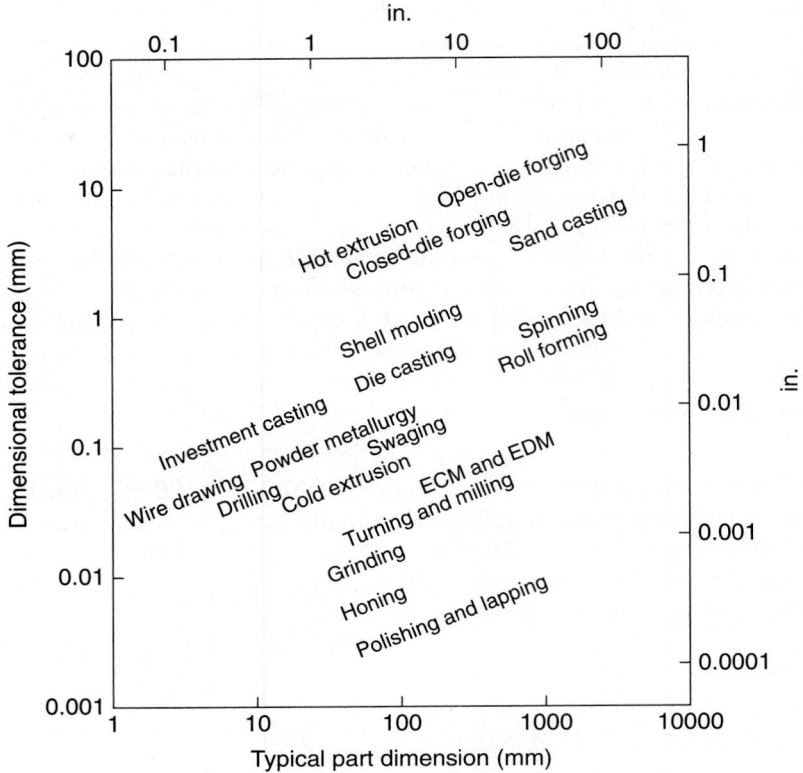

FIGURE 35.19   Dimensional tolerances as a function of part size for various manufacturing processes. Note that because many factors are involved, there is a broad range for tolerances.

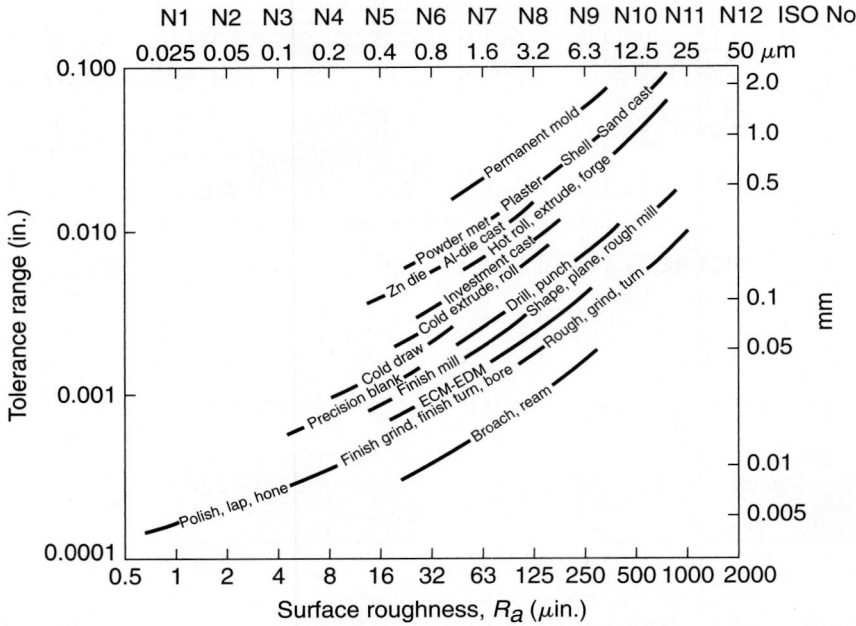

FIGURE 35.20   Dimensional tolerance range and surface roughness obtained in various manufacturing processes. These tolerances apply to a 25-mm (1-in.) workpiece dimension. *Source*: After J.A. Schey.

The commonly used terms for geometric characteristics are defined briefly as follows, in alphabetical order:

- **Allowance:** The specified difference in dimensions between mating parts; also called *functional dimension* or *sum dimension.*

- **Basic size:** Dimension from which limits of size are derived using tolerances and allowances.

- **Bilateral tolerance:** Deviation (plus or minus) from the basic size.

- **Clearance:** The space between mating parts.

- **Clearance fit:** Fit that allows for rotation or sliding between mating parts.

- **Datum:** A theoretically exact axis, point, line, or plane.

- **Feature:** Physically identifiable portion of a part, such as hole, slot, pin, or chamfer.

- **Fit:** The range of looseness or tightness that can result from the application of a specific combination of allowance and tolerance in the design of mating part features.

- **Geometric tolerancing:** Tolerances that involve shape features of the part.

- **Hole-basis system:** Tolerances based on a zero line on the hole; also called *standard hole practice* or *basic hole system.*

- **Interference:** Negative clearance.

- **Interference fit:** A fit having limits of size so prescribed that an interference always results when mating parts are assembled.

- **International tolerance grade** (IT): A group of tolerances that vary depending on the basic size but provide the same relative level of accuracy within a grade.

- **Limit dimensions:** The maximum and minimum dimensions of a part; also called *limits.*

- **Maximum material condition** (MMC): The condition where a feature of a certain size contains the maximum amount of material within the stated limits of that size.

- **Nominal size:** An approximate dimension that is used for the purpose of general identification.

- **Positional tolerancing:** A system of specifying the true position, size, and form of the features of a part, including allowable variations.

- **Shaft-basis system:** Tolerances based on a zero line on the shaft; also called *standard shaft practice* or *basic shaft system.*

- **Standard size:** Nominal size in integers and common subdivisions of length.

- **Transition fit:** Fit with small clearance or interference that allows for accurate location of mating parts.

- **Unilateral tolerancing:** Deviation in one direction only from the nominal dimension.

- **Zero line:** Reference line along the basic size from which a range of tolerances and deviations are specified.

Because the dimensions of holes are more difficult to control than those of shafts, the hole-basis system commonly is used for specifying tolerances in shaft and hole assemblies. The symbols used to indicate geometric characteristics are shown in Fig. 35.21a and b.

Type of feature	Type of tolerance	Characteristic	Symbol
Individual (no datum reference)	Form	Flatness	▱
		Straightness	—
		Circularity (roundness)	○
		Cylindricity	⌀
Individual or related	Profile	Profile of a line	⌒
		Profile of a surface	⌓
Related (datum reference required)	Orientation	Perpendicularity	⊥
		Angularity	∠
		Parallelism	//
	Location	Position	⊕
		Concentricity	◎
	Runout	Circular runout	╱
		Total runout	⫽

(a)

.605 Basic or exact dimension	Ⓟ Projected tolerance zone
-A- Datum feature symbol	⌀ Diametrical (cylindrical) tolerance zone or feature
Ⓜ Maximum material condition	⊕ ⌀.005 Ⓜ A Feature control frame
Ⓢ Regardless of feature size	
Ⓛ Least material condition	Ⓐ1 Datum target symbol

(b)

FIGURE 35.21 Geometric characteristic symbols to be indicated on engineering drawings of parts to be manufactured. *Source:* Courtesy of The American Society of Mechanical Engineers.

Limits and fits. Limits and fits are essential in specifying dimensions for holes and shafts. There are two standards on limits and fits, as described by the American National Standards Institute (see ANSI B4.1, B4.2, and B4.3). One standard is based on the traditional inch unit. The other is based on the metric unit and has been developed in greater detail. In these standards, capital letters always refer to the hole and lowercase letters to the shaft.

SUMMARY

- In modern manufacturing technology, many parts are processed using a high degree of precision and, thus, require measuring instrumentation with several features and characteristics.

- Many devices are available for inspection—from simple gage blocks to electronic gages with high resolution. The selection of a particular measuring instrument depends on factors such as the type of measurement for which it will be used, the environment in which it will be used, and the accuracy of measurement required.

- Major advances have been made in automated measurement, linking measuring devices to microprocessors and computers for accurate in-process control of manufacturing operations. Reliable linking, monitoring, display, distribution, and manipulation of data are important factors, as are the significant costs involved in implementing them.

- Dimensional tolerances and their selection are important factors in manufacturing. Tolerances not only affect the accuracy and operation of all types of machinery and equipment but also can influence product cost significantly.

- The smaller (tighter) the range of tolerances specified, the higher is the cost of production. Tolerances should be as broad as possible but should also maintain the functional requirements of the product.

KEY TERMS

Air gage	Fits	Pneumatic gage
Analog instruments	Fixed gage	Precision
Autocollimator	Gage block	Resolution
Bevel protractor	Interferometry	Ring gage
Comparative length-measuring	Laser micrometer	Sensitivity
instruments	Limits	Snap gage
Coordinate-measuring machine	Line-graduated instruments	Tolerance
Dial indicator	Measurement standards	Toolmaker's microscope
Diffraction gratings	Micrometer	Total indicator reading
Digital instruments	Optical contour projector	Vernier caliper
Dimensional tolerance	Optical flat	
Electronic gages	Plug gage	

BIBLIOGRAPHY

Bentley, J.P., *Principles of Measurement Systems*, 3rd ed., Addison-Wesley, 1995.

Bjorke, O., *Computer-Aided Tolerancing*, 2nd ed., ASME Press, 1992.

Bosch, J.A. (ed.), *Coordinate Measuring Machines and Systems*, Marcel Dekker, 1995.

Creveling, C.M., *Tolerance Control: A Handbook for Developing Optimal Specifications*, Addison-Wesley, 1996.

Drake, P.J., *Dimensioning and Tolerancing Handbook*, McGraw-Hill, 1999.

Farago, F.T., and Curtis, M.A., *Handbook of Dimensional Measurement*, 3rd ed., Industrial Press, 1994.

Gooldy, G., *Geometric Dimensioning and Tolerancing*, Rev. ed., Prentice Hall, 1995.

Handbook of Metrology, Brown & Sharpe, 1992.

Henzhold, G., *Handbook of Geometric Tolerancing: Design, Manufacturing and Inspection*, Wiley, 1995.

Kennedy, C.W., Hoffman, E.G., and Bond, S.D., *Inspection and Gaging*, 6th ed., Industrial Press, 1987.

Krulikowski, A., *Fundamentals of Geometric Dimensioning and Tolerancing*, Delmar, 1997.

Liggett, J.V., *Dimensional Variation Management Handbook: A Guide for Quality, Design, and Manufacturing Engineers*, Prentice Hall, 1993.

Machinery's Handbook, revised periodically, Industrial Press.

Meadows, J.D., *Geometric Dimensioning and Tolerancing*, Marcel Dekker, 1995.

———, *Measurement of Geometric Tolerances in Manufacturing*, Marcel Dekker, 1998.

Morris, A.S., *Measurement and Calibration for Quality Assurance*, Wiley, 1998.

Peterson, M.B., and Winer, W.O. (eds.), *Wear Control Handbook*, American Society of Mechanical Engineers, 1980.

Puncochar, D.E., *Interpretation of Geometric Dimensioning and Tolerancing*, 2nd ed., Industrial Press, 1997.

Tool and Manufacturing Engineers Handbook, 4th ed., Vol. 4, *Quality Control and Assembly*, Society of Manufacturing Engineers, 1987.

Whitehouse, D.J., *Handbook of Surface Metrology*, IOP Publishing, 1994.

Wilson, B.A., *Design Dimensioning and Tolerancing*. Goodheart-Wilcox, 1996.

———, *Dimensioning and Tolerancing Handbook*. Genium, 1998.

Winchell, W., *Inspection and Measurement in Manufacturing*, Society of Manufacturing Engineers, 1996.

REVIEW QUESTIONS

35.1 Explain what is meant by standards for measurement.

35.2 Why is it important to control temperature during the measurement of dimensions?

35.3 Explain the difference between direct-reading and indirect-reading linear measurements. Name the instruments used in each category.

35.4 What is meant by comparative length measurement?

35.5 Describe the characteristics of electronic gages.

35.6 Explain how flatness is measured. What is an optical flat?

35.7 Describe the principle of an optical comparator.

35.8 Why have coordinate measuring machines become important instruments?

35.9 What is the difference between a plug gage and a ring gage?

35.10 What are dimensional tolerances? Why is their control important?

35.11 Explain the difference between tolerance and allowance.

35.12 What is the difference between bilateral and unilateral tolerance?

35.13 How is straightness measured?

QUALITATIVE PROBLEMS

35.14 Why are the words "accuracy" and "precision" often incorrectly interchanged?

35.15 Why do manufacturing processes produce parts with a wide range of tolerances?

35.16 Explain the need for automated inspection.

35.17 Dimensional tolerances for nonmetallic parts usually are wider than for metallic parts. Explain why. Would this also be true for ceramics parts?

35.18   Comment on your observations regarding Fig. 35.20. Why does dimensional tolerance increase with increasing surface roughness?

35.19   What are the advantages and limitations of gage blocks made of zirconia?

35.20   Review Fig. 35.19 and comment on the range of tolerances and part dimensions produced by various manufacturing processes.

35.21   In the game of darts, is it better to be accurate or to be precise? Explain.

35.22   What are the advantages and limitations of *GO* and *NOT GO* gages?

QUANTITATIVE PROBLEMS

35.23   Assume that a steel rule expands by 0.08% due to an increase in environmental temperature. What will be the indicated diameter of a shaft whose diameter at room temperature was 1.500 in.?

35.24   If the same steel rule as in Problem 35.23 is used to measure aluminum extrusions, what will be the indicated diameter of a part that was 1.500 in. at room temperature? What is the actual dimension? What if the part were a thermoplastic?

35.25   A shaft must meet a design requirement of being at least 1.25 in. in diameter, but it can be 0.01 in. oversized. Express the shaft's tolerance as it would appear on an engineering drawing.

SYNTHESIS, DESIGN, AND PROJECTS

35.26   Describe your thoughts on the merits and limitations of digital measuring equipment over analog. Give specific examples.

35.27   Take an ordinary vernier micrometer (Fig. 35.2a) and a simple round rod. Ask five of your classmates to measure the rod diameter with this micrometer. Comment on your observations.

35.28   Obtain a digital micrometer and a steel ball of, say, 1/4 in. diameter. Measure its diameter for the following conditions when the ball has been placed in (a) a freezer, (b) boiling water, and (c) held in your hand for different periods of time. Note the variations, if any, of measured dimensions, and comment on it.

35.29   Repeat Problem 35.28 but with the following parts: (a) plastic lid of a small jar, (b) thermoset part such as the knob or handle from the lid of a sauce pan, (c) small juice glass, (d) ordinary rubber eraser.

35.30   What is the sigificance of the tests described in Problems 35.28 and 35.29?

35.31   Explain the relative advantages and limitations of a tactile probe versus a laser probe.

35.32   Make simple sketches of some forming-and-cutting-machine tools (as described in Parts III and IV of this book) and integrate them with the various types of measuring equipment discussed in this chapter. Comment on the possible difficulties involved in doing so.

35.33   What method would you use to measure the thickness of a foam-rubber part? Explain.

35.34   Obtain one or more of the following parts and describe how you would measure as many of the key dimensions as possible. Include the type of instruments to be used and the measurement method.

   a. An automotive brake pad.

   b. A plastic soft-drink bottle.

   c. A compact disc or floppy disk.

35.35   Inspect various parts and components in consumer products (including small appliances) and comment on how tight dimensional tolerances have to be in order for these products to function properly.

35.36   As you know, very thin sheet-metal parts can distort differently when held from various locations and edges of the part, just as a piece of paper or aluminum foil does. How then could you use a coordinate measuring machine for "accurate" measurements? Explain.

Quality Assurance, Testing, and Inspection

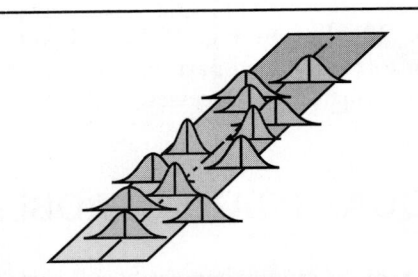

This chapter outlines the procedures used to ensure the manufacturing of high-quality products. The specific topics covered are:

- The nature of quality and the factors that influence it in manufactured products.
- Total quality-management approaches and Taguchi methods.
- Statistical methods of quality control and the use of control charts to make sure that the production of parts meets quality standards.
- Destructive- and nondestructive-testing methods for product inspection.

36.1 | Introduction

Throughout this book, we repeatedly have observed that manufactured products develop certain external and internal characteristics, which result (in part) from the production processes used. External characteristics most commonly involve dimensions, size, and surface finish and integrity issues such as surface damage from cutting tools or friction during the processing of the workpiece. Internal characteristics include defects, such as porosity, impurities, inclusions, phase transformations, embrittlement, cracks, debonding of laminations, and residual stresses.

Some of these defects may exist in the original stock, and some are introduced or induced during the particular manufacturing operation. Before they are marketed, manufactured parts and products are inspected for several characteristics. This inspection routine is important in order to:

- Ensure dimensional accuracy so that parts properly fit into other components during assembly.
- Identify products whose failure or malfunction has potentially serious implications, such as bodily injury or fatality. Typical examples are the failure of cables, switches, brakes, grinding wheels, railroad wheels, turbine blades, pressure vessels, and welded joints.

This chapter identifies and describes the various methods that commonly are used to inspect manufactured products.

Product quality always has been one of the most important aspects of manufacturing operations. In view of a global competitive market, *continuous improvement in* *quality* is now a major priority, particularly for large corporations in industrialized countries. In Japan, the single term **kaizen** is used to signify never-ending improvement.

The prevention of defects in products and on-line inspection of parts are major goals in all manufacturing activities. We again emphasize that *quality must be built into a product* and not merely checked for after the product already has been made. Thus, close cooperation and communication among design and manufacturing engineers and direct involvement and encouragement of the company management are vital.

Major advances in quality engineering and productivity have been made, largely because of the efforts of quality experts such as Deming, Taguchi, and Juran. The importance of the quality, reliability, and safety of products in a global economy now is recognized internationally by the establishment of various ISO and QSO standards and by the Malcolm Baldrige National Quality Award in the United States.

36.2 | Product Quality

We all have used terms like "poor quality" or "high quality" to describe a particular product or the products of a particular company. What is quality? Although we may recognize it when we see or use a product, quality (unlike most technical terms) is difficult to define precisely. Simply and generally, quality may be defined as a *product's fitness for use.*

Several dimensions of quality generally are identified; these include characteristics such as performance, durability, reliability, robustness, and serviceability, as well as aesthetics and perceived quality. Thus, quality is a broadbased characteristic or property, and its factors consist not only of well-defined technical considerations but also of subjective opinions.

For example, consider the following: (a) the handle on a kitchen utensil is installed crooked, or the handle discolors or cracks, (b) a weighing scale functions erratically, (c) a plastic toy breaks easily, (d) a vacuum cleaner requires frequent repairs, (e) the stem of a Phillips screwdriver wears rapidly, and (f) a machine tool cannot maintain the dimensional tolerances of the workpiece because of lack of stiffness or due to poor construction. These examples all lead us to believe that the product is of low quality. The general perception is that a high-quality product is one which performs its functions reliably over a long period of time without breaking down or requiring repairs. Some examples of this type of product are "good-quality" refrigerators, washing machines, kitchen knives, automobiles, and bicycles.

Note that, in describing good- or poor-quality products, this book has not yet stated the lifetimes of products or any of their technical specifications. We have seen that design and manufacturing engineers have the responsibility of selecting and specifying materials for the components of the products to be made. Thus, when selecting the metal for a screwdriver stem, we can specify materials that have high strength and high resistance to wear and corrosion and process them using appropriate manufacturing techniques and possibly heat treatments and coatings.

However, it is important to note that materials possessing better properties generally are more expensive and may be more difficult to process than those with poorer properties. The level of quality that a manufacturer chooses for its products may depend on the market for which the products are intended. For example, low-cost, low-quality tools have their own market niche.

As we will describe in Chapter 40, the total product cost depends on several variables, including the level of automation in the manufacturing plant. Thus, there are many ways for the engineer to review and modify overall product design and

TABLE 36.1

Life Expectancy of Some Products	
Dollar bills	18 months
Car batteries	4 years
Hair dryers	5 years
Water heaters for homes	10 years
Vacuum cleaners	10 years
Air conditioning units	15 years
Nuclear reactors	40 years
Automobile disk brakes	65,000 km (40,000 mi)
Mufflers	50,000 km (30,000 mi)
Tires	65,000–100,000 km (40,000–60,000 miles)

manufacturing processes to minimize a product's cost without affecting its quality. Contrary to general public perception, high quality products do not necessarily cost more, especially when considering the fact that poor-quality products:

- Present difficulties in assembling and maintaining components.
- Result in the need for in-field repairs.
- Have the significant built-in cost of customer dissatisfaction.

Quality standards are essentially a balance between several considerations; this balance is also called **return on quality** (ROQ) and usually includes some limit on the expected life of the product. Typical life expectancies of some products are given in Table 36.1 (See also Table 40.3).

36.3 | Quality Assurance

Quality assurance is the total effort made by a manufacturer to ensure that its products conform to a detailed set of specifications and standards. It can be defined as all actions necessary to ensure that quality requirements will be satisfied; whereas, *quality control* is the set of operational techniques used to fulfill quality requirements.

These standards cover several types of parameters, such as dimensions, surface finish, tolerances, composition, and color, as well as mechanical, physical, and chemical properties. In addition, standards usually are written to ensure proper assembly, using interchangeable defect-free components and resulting in a product that performs as originally intended by its designers.

Quality assurance is the responsibility of everyone involved with design and manufacturing. The often-repeated statement that quality must be built into a product reflects this important concept. Although a finished product can be inspected for quality, quality cannot be inspected *into* a finished product.

An important aspect of quality assurance is the capability to (a) analyze defects as they occur on the production line and (b) promptly eliminate them or reduce them to acceptable levels. In an even broader sense, quality assurance involves evaluating the product and customer satisfaction. The sum total of all these activities is referred to as *total quality control* and, in a larger sense, *total quality management*.

It is clear that, in order to control quality, we have to be able to:

- Measure quantitatively the level of quality.
- Identify all of the material and process variables that can be controlled.

The quality level built in during production then can be checked by continuously inspecting the product to determine whether it meets the specifications for dimensional tolerances, surface finish, defects, and other characteristics.

36.4 | Total Quality Management

Total quality management (TQM) is a system that emphasizes the concept that quality must be designed and built into a product. It is a systems approach in that both management and employees make a concerted effort to consistently manufacture high-quality products. Defect *prevention* (rather than defect *detection*) is the major goal here.

Leadership and *teamwork* in the organization are essential. They ensure that the goal of **continuous improvement** in manufacturing operations is imperative, because they reduce product variability and they improve customer satisfaction. The TQM concept also requires us to control the *processes* and not the *parts produced*, so that process variability is reduced and no defective parts are allowed to continue through the production line.

Quality circle.   This concept consists of regular meetings by groups of employees (workers, supervisors, and managers) who discuss how to improve and maintain product quality at all stages of the manufacturing operation. Worker involvement, responsibility, creativity, and team effort are emphasized. Comprehensive training is provided so that the worker can become conscious of quality and be capable of analyzing statistical data, identifying causes of poor quality, and taking immediate action to correct the situation. Experience has indicated that quality circles are more effective in *lean manufacturing* environments (see Section 39.6).

Quality engineering as a philosophy.   Experts in quality control have put into a larger perspective many of the quality-control concepts and methods. Notable among these experts are Deming, Juran, and Taguchi, whose philosophies of quality and product cost had a major impact on modern manufacturing. Their philosophies of quality engineering are outlined next.

36.4.1 Deming methods

During World War II, W.E. Deming (1900–1993) and several others developed new methods of *statistical process control* for wartime-industry manufacturing plants. The methods of statistical control arose from the recognition that there were variations in the (a) performance of machines and people and (b) quality and dimensions of raw materials. Their efforts involved not only statistical methods of analysis but also a new way of looking at manufacturing operations—that is, from the perspective of improving quality while lowering costs.

Deming recognized that manufacturing organizations are systems of management, workers, machines, and products. His basic ideas are summarized in the now-famous *14 points*, given in Table 36.2. These points are not to be seen as a checklist or menu of tasks; they are the characteristics that Deming recognized in companies that produce high-quality goods. He placed great emphasis on communication, direct worker involvement, and education in statistics and modern manufacturing technology. His ideas have been accepted widely since the end of World War II.

TABLE 36.2

Deming's 14 Points	
1. Create constancy of purpose toward improvement of product and service.	
2. Adopt the new philosophy.	
3. Cease dependence on mass inspection to achieve quality.	
4. End the practice of awarding business on the basis of price tag.	
5. Improve constantly and forever the system of production and service, to improve quality and productivity, and thus constantly decrease cost.	
6. Institute training for the requirements of a particular task, and document it for future training.	
7. Institute leadership, as opposed to supervision.	
8. Drive out fear so that everyone can work effectively.	
9. Break down barriers between departments.	
10. Eliminate slogans, exhortations, and targets for zero defects and new levels of productivity.	
11. Eliminate quotas and management by numbers, numerical goals. Substitute leadership.	
12. Remove barriers that rob the hourly worker of pride of workmanship.	
13. Institute a vigorous program of education and self-improvement.	
14. Put everybody in the company to work to accomplish the transformation.	

36.4.2 Juran methods

A contemporary of Deming, J.M. Juran (1904–), emphasized the following:

- Recognizing quality at all levels of an organization, including upper management.
- Fostering a responsive corporate culture.
- Training all personnel in how to *plan*, *control*, and *improve* quality.

The concern of the top management in an organization is with business and management, whereas those in quality control are concerned with technology. These different worlds have often been at odds, and their conflicts have led to quality problems.

Planners determine who the customers are and their needs. An organization's customers may be external (the end users who purchase the product or service), or they may be internal (the different parts of an organization that rely on other segments of the organization to supply them with products and services). The planners then develop product and process designs to respond to the customer's needs. The plans are turned over to those in charge of operations, who then are responsible for implementing both quality control and continued improvement in quality.

36.5 | Taguchi Methods

In the G. Taguchi (1924–) methods, high quality and low costs are achieved by combining engineering and statistical methods to optimize product design and manufacturing processes. *Taguchi methods* is a term that refers to the approaches developed by Taguchi to manufacture high-quality products. One fundamental viewpoint forwarded is the quality challenge facing manufacturers: provide products that delight your customers. To satisfy customers, manufacturers should provide products with the following characteristics:

- High reliability
- Perform the desired functions well

- Good appearance
- Inexpensive
- Upgradeable
- Available in the quantities desired when needed
- Robust over their intended life (see Section 36.5.1)

These product characteristics clearly are the goal of manufacturers striving to provide high-quality products. Although it is quite challenging to actually meet all of these characteristics, *excellence in manufacturing* is certainly a prerequisite.

Taguchi also contributed to the approaches that are used to document quality. It must be recognized that any deviation from the optimum state of a product represents a financial loss because of such factors as reduced product life, performance, and economy. Loss of quality is defined as the financial loss to society after the product is shipped, with the following results:

- Poor quality leads to customer dissatisfaction.
- Costs are incurred in servicing and repairing defective products, some in the field.
- The manufacturer's credibility in the marketplace is diminished.
- The manufacturer eventually loses its share of the market.

The Taguchi methods of *quality engineering* emphasize the importance of the following:

- **Enhancing cross-functional team interaction:** Design engineers and process or manufacturing engineers communicate with each other in a common language. They quantify the relationships between design requirements and manufacturing process selection. (See also Section I.3 in the General Introduction).
- **Implementing experimental design:** The factors involved in a process or operation and their interactions are studied simultaneously.

In *experimental design*, the effects of controllable and uncontrollable variables on the product are identified. This approach minimizes variations in product dimensions and properties and ultimately brings the mean to the desired level.

The methods used for *experimental design* are complex. They involve the use of *factorial design* and *orthogonal arrays*, which reduce the number of experiments required. These methods also are capable of identifying the effects of variables that cannot be controlled (called *noise*), such as changes in environmental conditions.

The use of these methods results in (a) the rapid identification of the controlling variables (observing *main effects*) and (b) the ability to determine the best method of process control. The control of these variables sometimes requires new equipment or major modifications to existing equipment. For example, variables affecting dimensional tolerances in machining a particular component can be identified readily and (whenever possible) the correct cutting speed, feed, cutting tool, and cutting fluids can be specified.

An important concept introduced by Taguchi is that any deviation from a design objective constitutes a loss in quality. Consider, for example, the tolerancing standards in Fig. 35.18. There is a range of dimensions over which a part is acceptable. On the other hand, the Taguchi philosophy calls for a minimization of deviation from the design objective. Thus, using Fig. 35.18a as an example, a shaft with a diameter of 40.03 mm normally would be considered acceptable and would pass inspections. In the Taguchi approach, however, the part with this diameter represents a deviation from the design objective. Such deviations generally reduce the robustness and performance of products, especially in complex systems.

36.5.1 Robustness

Another aspect of quality is a concept originally suggested by Taguchi that continuously has grown in importance and is referred to as *robustness*. A robust design, process, or system is one that continues to function within acceptable parameters despite variabilities (often unanticipated) in its environment. In other words, its outputs (such as its function and performance) have minimal sensitivity to its input variations (such as environment, load, and power source). In addition, robustness refers to a product or machine performance being insensitive to tolerance changes that should not deteriorate significantly over its intended life.

For example, in a robust design, a part will function sufficiently well even if the loads applied (or their directions) go beyond anticipated values. Likewise, a robust machine or system will undergo minimal deterioration in performance even if it experiences unanticipated variations in environmental conditions, such as temperature, humidity, and vibrations. Also, a robust machine will have no significant reduction in performance over its life, whereas a less robust design will perform less efficiently as time passes.

As a simple illustration of robust design, consider a sheet-metal mounting bracket to be attached to a wall with two bolts (Fig. 36.1a). The positioning of the two mounting holes on the bracket will include some error due to the manufacturing process involved. This error then will prevent the top edge of the bracket from being perfectly horizontal.

A more robust design is shown in Fig. 36.1b, in which the mounting holes have been moved twice as far apart as in the first design. Even though the precision of hole location remains the same and the manufacturing cost also is the same, the variability in the top edge of the bracket (from the horizontal) now has been reduced by one–half. However, if the bracket is subjected to vibration, the bolts may loosen over time. An even more robust design approach would be to use an adhesive to hold the threads in place or to use some type of fastener which would not loosen over time.

36.5.2 Taguchi loss function

The *Taguchi loss function* was introduced because traditional accounting practices had no real way of calculating losses on parts that met design specifications. In the traditional accounting approach, a part is defective and incurs a loss to the company when it exceeds its design tolerances; otherwise, there is no loss to the company.

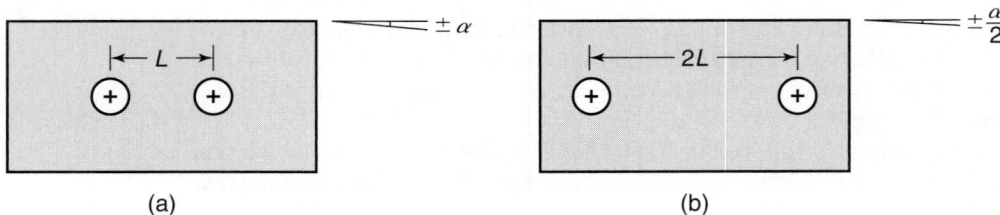

(a)   (b)

FIGURE 36.1   A simple example of robust design. (a) Location of two mounting holes on a sheet-metal bracket where the deviation keeping the top surface of the bracket from being perfectly horizontal is $\pm\alpha$. (b) New location of holes where the deviation (keeping the top surface of the bracket from being perfectly horizontal) is reduced to $\pm\alpha/2$.

The Taguchi loss function is a tool for comparing quality based on minimizing variations. It calculates an increasing loss to the company when the component is further from the design objective. This function is defined as a parabola where one point is the cost of replacement (including shipping, scrapping, and handling costs) at an extreme of the tolerances, while a second point corresponds to zero loss at the design objective.

Mathematically, the loss cost can be written as

$$\text{Loss cost} = k[(Y - T)^2 + \sigma^2] \tag{36.1}$$

where Y is the mean value from manufacturing, T is the target value from design, σ is the standard deviation of parts from manufacturing, and k is a constant, defined as

$$k = \frac{\text{Replacement cost}}{(\text{LSL} - T)^2} \tag{36.2}$$

where LSL is the lower specification limit. When the lower (LSL) and upper (USL) specification limits are the same (that is, the tolerances are balanced), either of the limits can be used in this equation.

EXAMPLE 36.1 Production of polymer tubing

High-quality polymer tubes are being produced for medical applications where the target wall thickness is 2.6 mm, an upper specification limit is 3.2 mm, and a lower specification limit is 2.0 mm (2.6 ± 0.6 mm). If the units are defective, they are replaced at a shipping-included cost of \$10.00. The current process produces parts with a mean of 2.6 mm and a standard deviation of 0.2 mm. The current volume is 10,000 sections of tube per month.

An improvement is being considered for the extruder heating system. This improvement will cut the variation in half, but it costs \$50,000. Determine the Taguchi loss function and the payback period for both cases.

Solution We first identify the quantities involved as follows: USL = 3.2 mm, LSL = 2.0 mm, T = 2.6 mm, σ = 0.2 mm, and Y = 2.6 mm.

The quantity k is given by

$$k = \frac{(\$10.00)}{(3.2 - 2.6)^2} = \$27.28$$

The loss cost is then

$$\text{Loss cost} = (27.78)[(2.6 - 2.6)^2 + 0.2^2] = \$1.11 \text{ per unit}$$

After the improvement, the standard deviation is 0.1 mm; thus, the loss cost is

$$\text{Loss cost} = (27.78)[(2.6 - 2.6)^2 + 0.1^2] = \$0.28 \text{ per unit}$$

The savings are then ($\$1.11 - \$0.28$)(10,000) = \$8300 per month. Hence, the payback period for the investment is \$50,000/(\$8300/month) = 6.02 months.

CASE STUDY 36.1 | Manufacture of Televisions by Sony Corporation

Sony Corporation executives found a confusing situation confronting them in the mid-1980s. Televisions manufactured in Japanese production facilities were better sellers than those produced in a San Diego facility, even though they were produced from identical designs and blueprints. There were no identifications to distinguish the televisions made in Japan from those made in the U.S., so there was no apparent reason for this discrepancy.

However, investigations revealed that the televisions produced in Japan were superior to the American versions; color sharpness was better and hues were more brilliant. Since the televisions were on display in stores, consumers could easily detect and purchase the model that had the best picture.

The difference in picture quality was obvious, but the reasons for this difference were not clear. A further point of confusion was the constant assurance that the San Diego facility had a total quality program in place and that the plant was maintaining quality-control standards so that no defective parts were produced. The Japanese facility did not have a total quality program, but there was an emphasis on reducing variation from part to part.

Further investigations found a typical pattern in an integrated circuit that was critical in affecting color density. The distribution of parts meeting the color-design objective is shown in Fig. 36.2a; the Taguchi loss function for these parts is shown in Fig. 36.2b. In the San Diego facility where the number of defective parts was minimized (zero in this case), a uniform distribution within the specification limits was achieved.

The Japanese facility actually produced parts outside of the design specification, but the standard deviation about the mean was lower. Using the Taguchi loss-function approach (see Example 36.1), the San Diego facility lost about $1.33 per unit, while the Japanese facility lost $0.44 per unit.

Traditional quality viewpoints would find the uniform distribution without defects to be superior to the distribution where a few defects are produced but the majority of parts are closer to the design target values. However, consumers readily can detect which product is superior, and the marketplace proves that minimizing deviations is a worthwhile quality goal.

Source: Courtesy of D.M. Byrne and S. Taguchi.

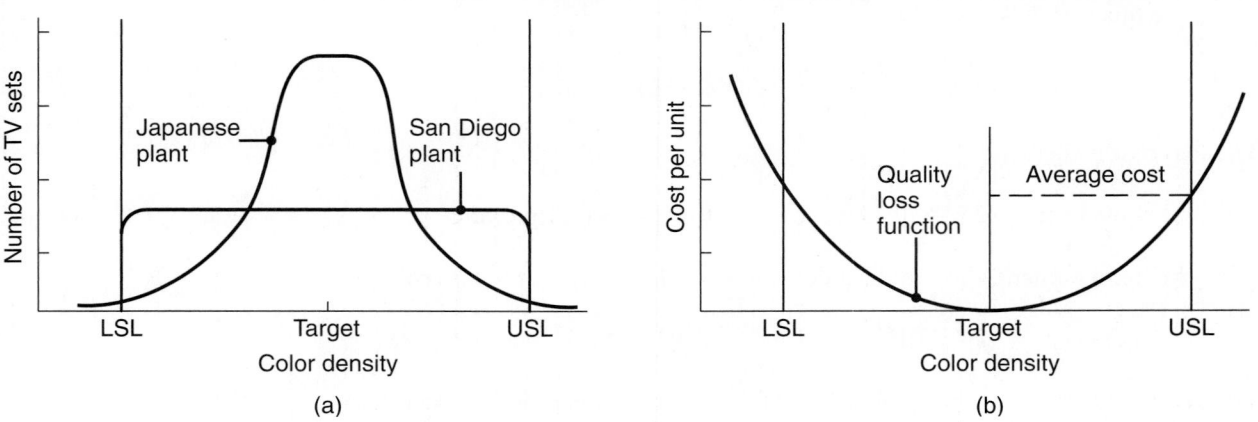

FIGURE 36.2 (a) Objective-function value distribution of color density for television sets. (b) Taguchi loss function showing the average replacement cost per unit to correct quality problems. *Source:* After G. Taguchi.

EXAMPLE 36.2 Increasing quality without increasing the cost of a product

A manufacturer of clay tiles noticed that excessive scrap was being produced because of temperature variations in the kiln used to fire the tiles, thus adversely affecting the company's profits. The first solution which the manufacturer considered was purchasing new kilns with better temperature controls.

However, this solution would require a major capital investment. A study then was undertaken to determine whether modifications could be made in the composition of the clay so that it would be less sensitive to temperature fluctuations during firing.

Based on factorial experiment design in which the factors involved in a process and their interactions are studied simultaneously, it was found that increasing the lime content of the clay made the tiles less sensitive to temperature variations during firing. This modification (which was also the low-cost alternative) was implemented, reducing scrap substantially, and improving tile quality.

36.6 | The ISO and QS Standards

With increasing international trade, global manufacturing, and price-sensitive competition have come a wide choice of industrial and consumer products. Customers increasingly are demanding *high-quality products* and services at low prices and are looking for suppliers that can respond to this demand consistently and reliably.

In turn, this trend has created the need for international conformity and consensus regarding the establishment of methods for quality control, reliability, and safety of a product. In addition to these considerations, equally important concerns regarding the environment and quality of life are being addressed with new international standards. In this section, we describe the standards that are relevant to product quality and environmental issues.

36.6.1 The ISO 9000 standard

First published in 1987 and then revised in 1994, the ISO 9000 standard (*Quality Management and Quality Assurance Standards*) is a deliberately generic series of quality-system management standards. The ISO 9000 standard permanently has influenced the manner in which manufacturing companies conduct business in world trade and has become the world standard for quality.

The ISO 9000 series includes the following standards:

ISO 9001—*Quality systems: Model for quality assurance in design/development, production, installation, and servicing.*

ISO 9002—*Quality systems: Model for quality assurance in production and installation.*

ISO 9003—*Quality systems: Model for quality assurance in final inspection and test.*

ISO 9004—*Quality management and quality system elements: Guidelines.*

Companies voluntarily register for these standards and are issued certificates. Registration may be sought generally for ISO 9001 or 9002, and some companies have registration up to ISO 9003. The 9004 standard is simply a guideline and not a model or a basis for registration. For certification, a company's plants are visited and audited by accredited and independent third-party teams to certify that the standard's 20 key elements are in place and are functioning properly.

Depending on the extent to which a company fails to meet the requirements of the standard, registration may or may not be recommended at that time. The audit team does not advise or consult with the company on how to fix discrepancies but merely describes the nature of the noncompliance. Periodic audits are required to maintain certification. The certification process can take from six months to a year or more and can cost tens of thousands of dollars. The cost depends on the company's size, number of plants, and product line.

The ISO 9000 standard is not a product certification but a **quality process certification**. Companies establish their own criteria and practices for quality. However, the documented quality system must be in compliance with the ISO 9000 standard. Thus, a company cannot write into the system any criterion that opposes the intent of the standard.

Registration symbolizes a company's commitment to conform to consistent practices, as specified by the company's own quality system (such as quality in design, development, production, installation, and servicing), including proper documentation of such practices. In this way, customers (including government agencies) are assured that the supplier of the product or service (which may or may not be within the same country) is following specified practices. In fact, manufacturing companies are themselves assured of such practice regarding their own suppliers who have ISO 9000 registration; thus, suppliers also must be registered.

36.6.2 The QS 9000 standard

Jointly developed by Chrysler, Ford, and General Motors, the QS 9000 standard first was published in August of 1994. Prior to the development of QS 9000, each of the "Big Three" automotive companies had its own standard for quality system requirements.

Tier I suppliers have been required to obtain third-party registration to QS 9000 before the dates established by each of the Big Three companies. Very often, QS 9000 has been described as an "ISO 9000 chassis with a lot of extras." This is a good description, noting that all of the ISO 9000 clauses serve as the foundation of QS 9000. However, the little "extras" are substantial.

The February 1995 edition of QS 9000 has three sections. Section I contains all 20 of the ISO 9001 clauses, but almost every clause has additional requirements for QS 9000. Section II has three sections: Production Part Approval Process, Continuous Improvement, and Manufacturing Capabilities. Section III is entitled Customer-Specific Requirements, and contains separate sections for Chrysler, General Motors, Ford, and Truck Manufacturers, respectively. Existing QS 9000 registrations are being upgraded continuously to comply with new editions of QS 9000.

36.6.3 The ISO 14000 standard

ISO 14000 is a family of standards first published in September of 1996 and pertaining to international **Environmental Management Systems** (EMS). It concerns the way an organization's activities affect the environment throughout the life of its products (see also Section I.6 in the General Introduction). These activities (a) may be internal or external to the organization, (b) range from production to ultimate disposal of the product after its useful life, and (c) include effects on the environment, such as pollution, waste generation and disposal, noise, depletion of natural resources, and energy use.

A rapidly increasing number of companies in many countries (with Japan leading) have been obtaining certification for this standard. The ISO 14000 family of standards has several sections: Guidelines for Environmental Auditing, Environmental Assessment, Environmental Labels and Declarations, and Environmental Management. ISO 14001: *Environmental Management System Requirements* consists of sections on General Requirements, Environmental Policy, Planning, Implementation and Operation, Checking and Corrective Action, and Management Review.

36.7 | **Statistical Methods of Quality Control**

Because of the numerous variables involved in manufacturing processes and operations, the implementation of *statistical techniques* is essential. The following are some of the commonly observed variables in manufacturing:

- Cutting tools, dies, and molds are subject to wear. Thus, part dimensions and surface characteristics vary over time.
- Machinery performs differently depending on its quality, age, condition, and maintenance. Thus, older machines tend to vibrate, are difficult to adjust, and do not maintain tolerances as well as new machines.
- Metalworking fluids perform differently as they degrade. Thus, tool and die life, surface finish of the workpiece, and forces and energy requirements are affected.
- Environmental conditions (such as temperature, humidity, and air quality in the plant) may change from one hour to the next, affecting machines, workspaces, and employees.
- Different shipments of raw materials to a plant may have significantly different dimensions, properties, and surface characteristics.
- Operator skill and attention may vary during the day, from machine to machine or from operator to operator.

In this list, those events that occur randomly—that is, without any particular trend or pattern—are called **chance variations** or **special cause.** Those events that can be traced to specific causes are called **assignable variations** or **common cause.**

The existence of *variability* in production operations has been recognized for centuries, but it was Eli Whitney (1765–1825) who first grasped its full significance when he found that interchangeable parts were indispensable to the mass production of firearms. Modern statistical concepts relevant to manufacturing engineering first were developed in the early 1900s, notably through the work of W.A. Shewhart (1891–1967).

36.7.1 Statistical quality control

To understand **statistical quality control** (SQC), let's first review the terms that are used commonly.

- **Sample size:** The number of parts to be inspected in a sample. The properties of the parts in the sample are studied to gain information about the whole population.
- **Random sampling:** Taking a sample from a population or lot in which each item has an equal chance of being included in the sample. Thus, when taking samples from a large bin, the inspector should not take only those that happen to be within reach.
- **Population:** The total number of individual parts of the same design from which samples are taken; also called the **universe.**

- **Lot size:** A subset of population. A lot or several lots can be considered subsets of the population and may be considered as representative of the population.

The sample is inspected for several characteristics and features (such as tolerances, surface finish, and defects) using the instruments and techniques described in Chapter 35 and in Sections 36.10 and 36.11. These characteristics fall into two categories: those that can be measured quantitatively (method of variables) and those that are measured qualitatively (method of attributes).

1. The **method of variables** is the *quantitative measurement* of the part's characteristics, such as dimensions, tolerances, surface finish, and physical or mechanical properties. The measurements are made for each of the units in the group under consideration, and the results are compared against specifications.

2. The **method of attributes** involves observing the presence or absence of *qualitative characteristics* (such as external or internal defects in machined, formed, or welded parts, or dents in sheet-metal products) in each of the units in the group under consideration. Sample size for attributes-type data generally is larger than for variables-type data.

During the inspection process, the results of the measurement will vary. For example, assume that you are measuring the diameter of turned shafts as they are produced on a lathe using a micrometer (Fig. 35.2). You soon note that their diameters vary, even though (ideally) you want all of the shafts to be exactly the same size. Let's now turn to the consideration of statistical quality-control techniques, which allow us to evaluate these variations and set limits for the acceptance of parts.

If you list the measured diameters of the turned shafts in a given population, you will note that one or more parts have the smallest diameter and one or more have the largest diameter. The rest of the turned shafts have diameters that lie between these two extremes.

When these diameters are grouped and plotted, the plot consists of a *histogram* (bar graph) representing the number of parts in each diameter group (Fig. 36.3a). The bars show a **distribution** (also called a *spread* or *dispersion*) of the shaft-diameter measurements. The *bell-shaped curve* in Fig. 36.3a is called a **frequency distribution** and is the frequency with which parts of each diameter size are being produced.

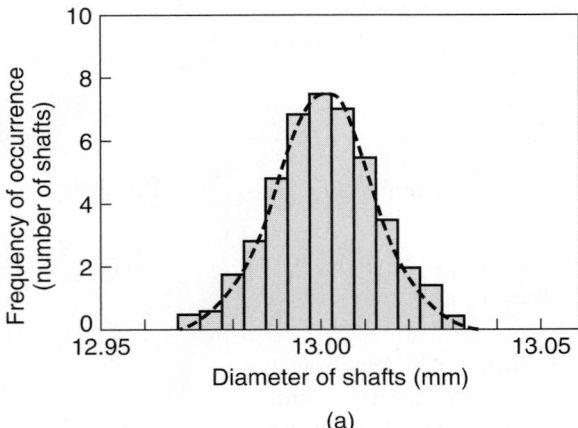

(a)

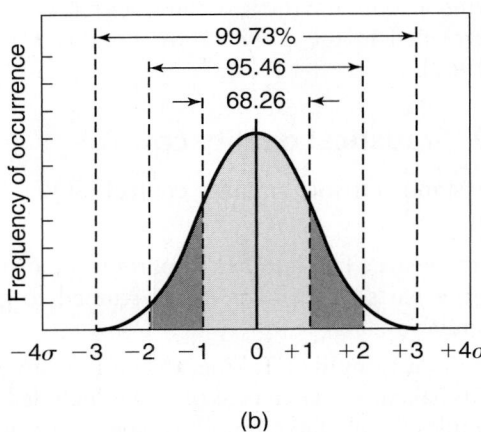

(b)

FIGURE 36.3   (a) A histogram of the number of shafts measured and their respective diameters. This type of curve is called a frequency distribution. (b) A normal distribution curve indicating areas within each range of standard deviation. *Note:* The greater the range, the higher is the percentage of parts that fall within it.

Data from manufacturing processes often fit curves represented by a mathematically derived **normal distribution curve** (Fig. 36.3b). This type of curve also is called *Gaussian*, after K.F. Gauss (1777–1855), who developed it on the basis of probability. The bell-shaped normal distribution curve fitted to the data in Fig. 36.3a has two important features.

First, it shows that most part diameters tend to cluster around an *average value* (**arithmetic mean**). This average usually is designated as $\bar{x}$ and is calculated from the expression

$$\bar{x} = \frac{(x_1 + x_2 + x_3 + \cdots + x_n)}{n} \tag{36.3}$$

where the numerator is the sum of all measured values (shaft diameters) and n is the number of measurements (number of shafts).

The second feature of this curve is its width, indicating the **dispersion** of the diameters measured; the wider the curve, the greater is the dispersion. The difference between the largest value and the smallest value is called the **range**, R:

$$R = x_{\max} - x_{\min} \tag{36.4}$$

Dispersion is estimated by the **standard deviation**, σ, and is given by the expression

$$\sigma = \frac{\sqrt{(x_1 - \bar{x})^2 + (x_2 - \bar{x})^2 + \cdots + (x_n - \bar{x})^2}}{n - 1} \tag{36.5}$$

where x_i is the measured value for each part.

Note from the numerator in Eq. (36.5) that, as the curve widens, the standard deviation becomes greater. Also note that σ has the unit of linear dimension. In comparing Eqs. (36.4) and (36.5), it can be seen that the range, R, is a simpler and more convenient measure of dispersion.

Since we know the number of turned parts that fall within each group, we can calculate the percentage of the total population represented by each group. Thus, Fig. 36.3b shows that, in the measurement of shaft diameters,

99.73% fall within the range of $\pm 3\sigma$
95.46% within $\pm 2\sigma$
68.26% within $\pm 1\sigma$

It can be seen that only 0.27% fall outside the $\pm 3\sigma$ range. This means that there are 2700 defective parts per million produced. In modern manufacturing, this is an unacceptable rate in view of the observation that at this level of defects no modern computer would function reliably. However, note that these quantities only are valid for distributions that are normal (as shown in Fig. 36.3) and are not skewed.

36.7.2 Six sigma

A quality standard in increasingly wider practice is the **six sigma** requirement. First implemented by Motorola in the 1980s, this concept refers to the range of the distribution from manufacturing. Six sigma means that there are only 3.4 defective parts per million parts on each side of the distribution (or a total of 6.8 parts per million). Because of the nature of statistical distributions, this tacitly forces improvements in process capability to reduce variability. The six sigma concept also has been extended to business and office processes and operations.

36.8 | Statistical Process Control

If the number of parts that do not meet set standards (defective parts) increases during a production run, we must be able to determine the cause (incoming materials, machine controls, degradation of metalworking fluids, operator boredom, or other factors) and take appropriate action. Although this statement at first appears to be self-evident, it was only in the early 1950s that a systematic statistical approach was developed to guide operators in manufacturing plants.

This approach advises the operator to take certain measures and actions and tells the operator when to take them in order to avoid producing further defective parts. Known as **statistical process control** (SPC), this technique consists of various elements:

- Control charts and control limits
- Capabilities of the particular manufacturing process
- Characteristics of the machinery involved

36.8.1 $\bar{x}$ and R charts (Shewhart control charts)

The frequency distribution curve in Fig. 36.3b shows a range of shaft diameters being produced that may fall beyond the design tolerance range. The same bell-shaped curve is shown in Fig. 36.4, which now includes the specified tolerances for the diameter of the turned shafts.

Control charts graphically represent the variations of a process over a period of time. They consist of data plotted during production. Typically, there are two plots. The quantity $\bar{x}$ (Fig. 36.5a) is the average for each subset of samples taken and inspected, say, a subset consisting of 5 parts. A sample size of between 2 and 10 parts is sufficiently accurate (although more parts are better) provided that sample size is held constant throughout the inspection.

The frequency of sampling depends on the nature of the process. Some processes may require continual sampling, whereas others may require only one sample per day. Quality-control analysts are best qualified to determine this frequency for a particular operation. Since the measurements in Fig. 36.5a are made consecutively, the abscissa of the control charts also represents time.

The solid horizontal line in this figure is the **average of averages** (**grand average**), denoted as $\bar{\bar{x}}$ and represents the population mean. The upper and lower horizontal broken lines in these control charts indicate the **control limits** for the process. The

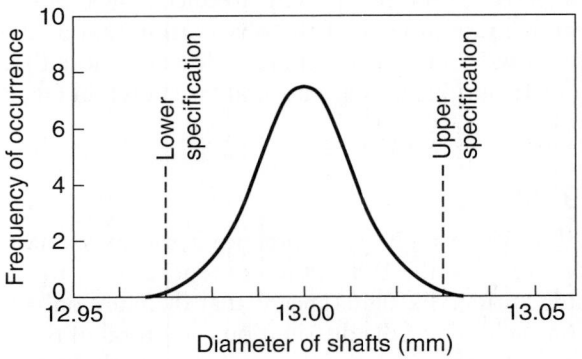

FIGURE 36.4   Frequency distribution curve showing lower and upper specification limits.

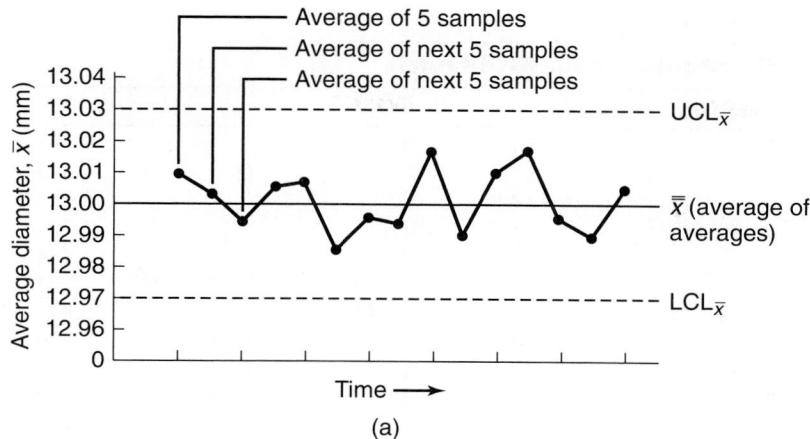

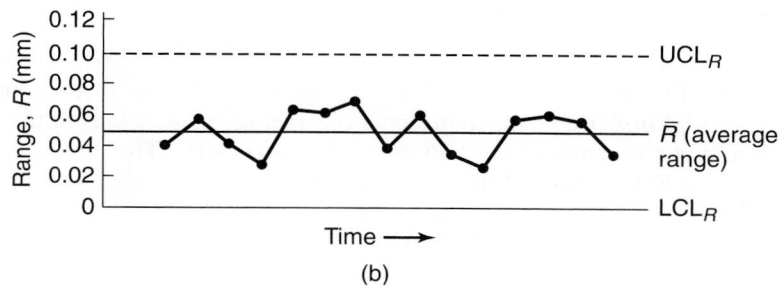

FIGURE 36.5   Control charts used in statistical quality control. The process shown is in good statistical control because all points fall within the lower and upper control limits. In this illustration, the sample size is 5, and the number of samples is 15.

control limits are set on these charts according to statistical-control formulas designed to keep actual production within acceptable variation levels. One common approach is to make sure that all parts are within three standard deviations of the mean ($\pm 3\sigma$).

The standard deviation also can be expressed as a function of range. Thus, for $\bar{x}$, we have

$$\text{Upper control limit (UCL}_{\bar{x}}) = \bar{x} + 3\sigma = \bar{\bar{x}} + A_2\bar{R} \qquad (36.6)$$

and

$$\text{Lower control limit (LCL}_{\bar{x}}) = \bar{x} - 3\sigma = \bar{\bar{x}} - A_2\bar{R} \qquad (36.7)$$

where A_2 is obtained from Table 36.3 and $\bar{R}$ is the average of R values. The quantities $\bar{\bar{x}}$ and $\bar{R}$ are estimated from the measurements taken.

These control limits are calculated on the basis of the past production capability of the equipment itself; they are not associated with either design tolerance specifications or dimensions. They indicate the limits within which a certain percentage of measured values normally are expected to fall because of the inherent variations of the process itself and upon which the limits are based.

The major goal of statistical process control is to improve the manufacturing process with the aid of control charts so as to eliminate assignable causes. The control chart continually indicates progress in this area.

TABLE 36.3

Constants for Control Charts				
Sample size	A_2	D_4	D_3	d_2
2	1.880	3.267	0	1.128
3	1.023	2.575	0	1.693
4	0.729	2.282	0	2.059
5	0.577	2.115	0	2.326
6	0.483	2.004	0	2.534
7	0.419	1.924	0.078	2.704
8	0.373	1.864	0.136	2.847
9	0.337	1.816	0.184	2.970
10	0.308	1.777	0.223	3.078
12	0.266	1.716	0.284	3.258
15	0.223	1.652	0.348	3.472
20	0.180	1.586	0.414	3.735

The second control chart (Fig. 36.5b) shows the range, R, in each subset of samples. The solid horizontal line represents the average of R values in the lot, denoted as $\overline{R}$ and is a measure of the variability of the samples. The upper and lower control limits for R are obtained from the equations

$$\text{UCL}_R = D_4\overline{R} \qquad (36.8)$$

and

$$\text{LCL}_R = D_3\overline{R} \qquad (36.9)$$

where the constants D_4 and D_3 take on the values given in Table 36.3. This table also includes the constant d_2, which is used to estimate the standard deviation of the process distribution shown in Fig. 36.4 from the equation

$$\sigma = \frac{\overline{R}}{d_2} \qquad (36.10)$$

When the curve of a control chart is like the one shown in Fig. 36.5a, we say that the process is *in good statistical control*. In other words:

- There is no discernible trend in the pattern of the curve.
- The points (measured values) are random with time.
- The points do not exceed the control limits.

However, we can see that in curves such as those in Figs. 36.6a, b, and c there are certain *trends*. For example, note in the middle of curve (a) that the diameter of the shafts is increasing with time. The reason for this increase may be a change in one of the process variables, such as wear of the cutting tool.

If the trend is consistently towards large diameters, as in curve (b), with diameters hovering around the upper control limit, it could mean that the tool settings on the lathe may be incorrect, and as a result, the parts being turned are consistently too large. Curve (c) shows two distinct trends that may be caused by factors like a change in the properties of the incoming material or a change in the performance of the cutting fluid (for example, its degradation). These situations place the process *out of control*. Warning limits to this effect are sometimes set at $\pm 2\sigma$.

Analyzing patterns and trends in control charts requires considerable experience so that one may identify the specific cause(s) of an out-of-control situation. Such

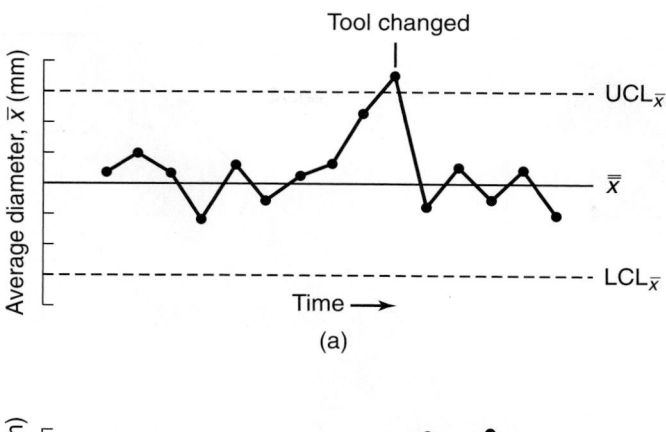

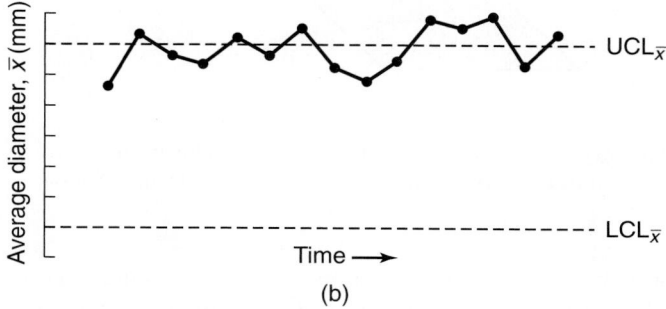

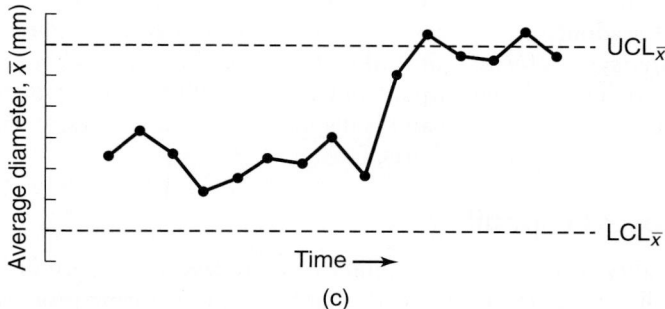

FIGURE 36.6 Control charts. (a) Process begins to become out of control because of such factors as tool wear (*drift*). The tool is changed and the process is then in statistical control. (b) Process parameters are not set properly; thus, all parts are around the upper control limit (*shift in mean*). (c) Process becomes out of control because of factors such as a change in the properties of the incoming material (*shift in mean*).

causes may be those variables listed at the beginning of Section 36.7. Overcontrol of the manufacturing process (that is, setting upper and lower control limits too close to each other, hence a smaller standard deviation range) is a further cause for out-of-control situations. This is the reason why control limits are calculated on the basis of *process variability* rather than on potentially inapplicable criteria.

It is evident that operator training is critical for successful implementation of SPC on the shop floor. Once process control guidelines are established and in the interest of efficiency of operation, operators also should have some responsibility for

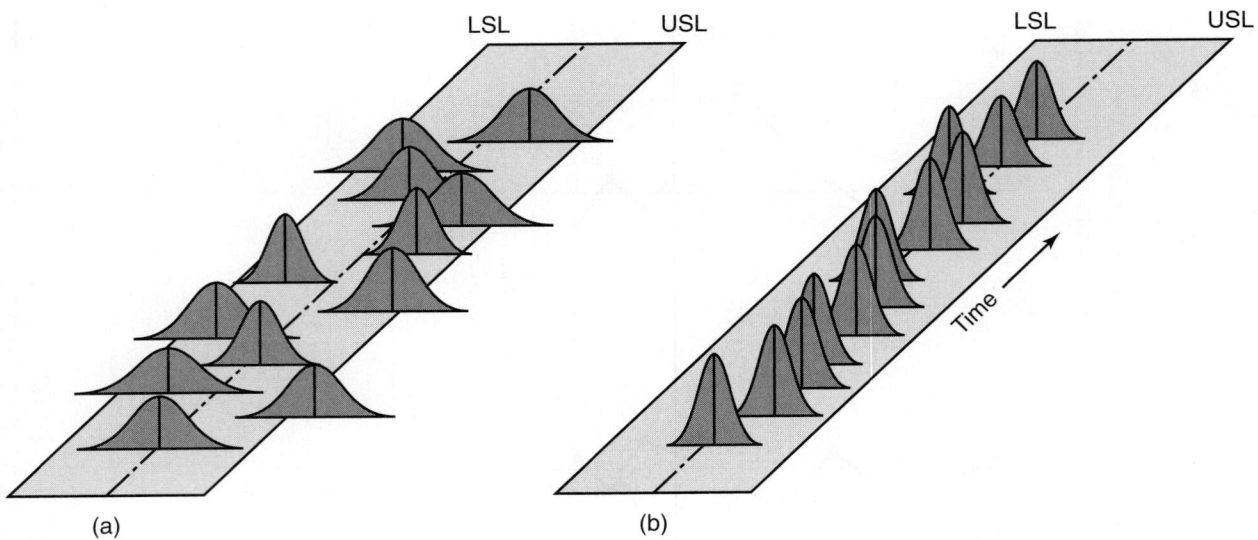

FIGURE 36.7   Illustration of processes that are (a) unstable or out of control and (b) stable or in control. Note in sketch (b) that all distributions have lower standard deviations and have means closer to the desired value. *Source:* After K. Crow.

making adjustments in production processes that are beginning to become out of control. The capabilities of individual operators also should be taken into account so that they are not overloaded with data input and the proper interpretation of this data.

This task now is easier due to the availability of a variety of software. For example, digital readouts on electronic measuring devices now are directly integrated into a computer system for real-time SPC. Figure 35.2 shows such a multifunctional computer system in which the output from a digital caliper or micrometer is analyzed by a microprocessor in real time and is displayed in several ways, such as frequency distribution curves and control charts.

36.8.2 Process capability

Process capability is defined as the ability of a process to produce defect-free products in controlled production. It tells us that the manufacturing process can produce parts consistently and repeatedly within specific limits of precision (Fig. 36.7). Various indicators (indices) are used to determine process capability, describing the relationship between the variability of a process and the spread of lower and upper specifications limits. Since a manufacturing process typically involves materials, machinery, and operators, each factor can be analyzed individually to identify a problem when process capabilities do not meet specification limits. Among various factors to be considered are variations in machine performance, operator skills, and incoming raw materials.

EXAMPLE 36.3 Calculation of control limits and standard deviation

The data given in Table 36.4 show length measurements (in.) taken on a machined workpiece. The sample size is 5, and the number of samples is 10; thus, the total number of parts measured is 50. The quantity $\bar{x}$ is the average of five measurements in each sample.

TABLE 36.4

Sample number	x_1	x_2	x_3	x_4	x_5	$\bar{x}$	R
1	4.46	4.40	4.44	4.46	4.43	4.438	0.06
2	4.45	4.43	4.47	4.39	4.40	4.428	0.08
3	4.38	4.48	4.42	4.42	4.35	4.410	0.13
4	4.42	4.44	4.53	4.49	4.35	4.446	0.18
5	4.42	4.45	4.43	4.44	4.41	4.430	0.04
6	4.44	4.45	4.44	4.39	4.40	4.424	0.06
7	4.39	4.41	4.42	4.46	4.47	4.430	0.08
8	4.45	4.41	4.43	4.41	4.50	4.440	0.09
9	4.44	4.46	4.30	4.38	4.49	4.414	0.19
10	4.42	4.43	4.37	4.47	4.49	4.436	0.12

Solution   We first calculate the average of averages, $\bar{\bar{x}}$,

$$\bar{\bar{x}} = \frac{44.296}{10} = 4.430 \text{ in.}$$

and the average of R values,

$$\bar{R} = \frac{1.03}{10} = 0.103 \text{ in.}$$

Since the sample size is 5, we determine the following constants from Table 36.3: $A_2 = 0.577$, $D_4 = 2.115$, and $D_3 = 0$. The control limits now can be calculated using Eqs. (36.4) through (36.7). Thus, for averages,

$$\text{UCL}_{\bar{x}} = 4.430 + (0.577)(0.103) = 4.489 \text{ in.}$$

$$\text{LCL}_{\bar{x}} = 4.430 - (0.577)(0.103) = 4.371 \text{ in.}$$

And, for ranges,

$$\text{UCL}_R = (2.115)(0.103) = 0.218 \text{ in.}$$

$$\text{LCL}_R = (0)(0.103) = 0 \text{ in.}$$

We now can estimate the standard deviation for the process population of individuals using Eq. (36.10) and for a value of $d_2 = 2.326$. Thus,

$$\sigma = \frac{0.103}{2.326} = 0.044 \text{ in.}$$

CASE STUDY 36.2 | **Dimensional Control of Plastic Parts in Saturn Automobiles**

A typical Saturn automobile has some 38 different injection-molded interior plastic parts (polycarbonate, polypropylene, and ABS), such as door panels, air-inlet ducts, consoles, and trim. All of these parts must conform to tight dimensional tolerances so that they fit and snap properly during assembly without unsightly gaps or buckles.

However, dimensions of these plastic parts change with temperature and humidity, and because of their flexibility, they also tend to bend and curl. For this reason, measurement and inspection of plastic parts (including the use of coordinate measuring machines (CMMs); see Section 35.5.1) can be difficult. Although traditional gages also are used for monitoring process parameters in making these parts, a superior inspection system has been developed whereby SPC feedback is received from a direct computer-controlled CMM so that they are molded properly.

The system compensates for the flexibility of the parts, allows automatic measurement of various part features, and makes measurements of the mold at periodic intervals. The data are analyzed on a regular basis, and when necessary, corrective actions are taken and changes are made in materials, processing, or mold design, so that the parts being molded will maintain good dimensional stability.

Source: Courtesy of Saturn Corp. and *Manufacturing Engineering.*

36.8.3 Acceptance Sampling and Control

Acceptance sampling consists of taking only a few random samples from a lot, and inspecting them to judge whether the entire lot is acceptable or whether it should be rejected or reworked. Developed in the 1920s and used extensively during World War II for military hardware (MIL STD 105), this statistical technique is used widely and has become valuable. Acceptance sampling is useful particularly for inspecting high-production-rate parts where 100% inspection would be too costly. However, there are certain critical devices, such as pacemakers, prosthetic devices, and components of the space shuttle, which must be inspected 100%.

A variety of acceptance sampling plans have been prepared for both military and national standards based on an acceptable, predetermined, and limiting percentage of nonconforming parts in the sample. If this percentage is exceeded, the entire lot is rejected, or it is reworked if economically feasible. Note that the actual number of samples (but not the percentages of the lot that are in the sample) can be significant in acceptance sampling.

The greater the number of samples taken from a lot, the greater is the chance that the sample will contain nonconforming parts, and the lower is the probability of the lot's acceptance. **Probability** is defined as the relative occurrence of an event. The probability of acceptance is obtained from various operating characteristics curves, one example of which is shown in Fig. 36.8.

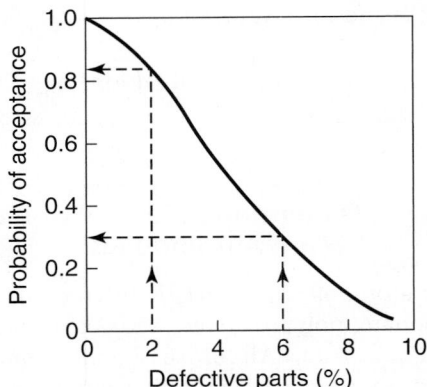

FIGURE 36.8 A typical operating-characteristics curve used in acceptance sampling. The higher the percentage of defective parts, the lower is the probability of acceptance by the consumer.

The **acceptance quality level** (AQL) commonly is defined as the level at which there is a 95% acceptance probability for the lot. This percentage indicates to the manufacturer that 5% of the parts in the lot may be rejected by the consumer (**producer's risk**). Likewise, the consumer knows that 95% of the parts are acceptable (**consumer's risk**).

The manufacturer can salvage those lots that do not meet the desired quality standards through a secondary rectifying inspection. In this method, a 100% inspection is made of the rejected lot, and the defective parts are removed. This time-consuming and costly process is an incentive for the manufacturer to better control the production process.

Acceptance sampling requires less time and fewer inspections than do other sampling methods. Consequently, inspection of the parts can be more detailed. Automated inspection techniques have been developed so that 100% inspection of all parts is possible and inspection also can be economical.

36.9 | Reliability of Products and Processes

All products eventually fail in some manner or other: automobile tires become worn and the treads become smooth, electric motors burn out, water heaters begin to leak, dies and cutting tools wear out, and machinery stops functioning properly. **Product reliability** may be defined as the probability that a product will perform, while in normal use by the customer and without failure, its intended function in a given environment and for a specified period of time.

The more critical the application of a particular product, the higher its reliability must be. Thus, the reliability of an aircraft jet engine, a medical instrument, or an elevator cable must be much higher than that of a kitchen faucet or a mechanical pencil. From the topics described in this chapter, it can be seen that, as the quality of each component of a product increases, so too does the reliability of the whole product.

For an ordinary steel chain, the reliability of each link in the chain is critical. Similarly, the reliability of each gear in a gear train for a machine or an automobile is critical as well. This condition is known as *series reliability*. On the other hand, for a steel cable consisting of many individual wires, the reliability of each individual wire is not as critical because the cable consists of many wires. This condition is known as *parallel reliability*. The parallel reliability concept is important in the design of backup systems, which permit a product to continue functioning in the event that one of its components fails. Electrical or hydraulic systems in an aircraft, for example, typically are backed by mechanical systems. Such systems are called *redundant systems*.

Predicting reliability is an important science and involves complex mathematical relationships and calculations. The importance of predicting the reliability of the critical components of civilian or military aircraft is obvious. The reliability of an automated and computer-controlled high-speed production line with all of its complex mechanical and electronic components is also important, as its failure can result in major economic losses to the manufacturer.

Process reliability may be defined as the capability of a particular manufacturing process to operate predictably and smoothly over a period of time. Thus, it is implicit that there will be no deterioration in performance, which otherwise would require downtime on machines, interruption of production, and result in economic loss.

36.10 | Nondestructive Testing

Nondestructive testing (NDT) is carried out in such a manner that product integrity and surface texture remain unchanged. These techniques generally require considerable operator skill, and interpreting test results accurately may be difficult because the observations can be subjective. However, the use of computer graphics and other enhancement techniques have reduced significantly the likelihood of human error. The present systems have various capabilities for data acquisition and qualitative and quantitative inspection and analysis.

Listed here are the basic principles of major nondestructive testing techniques.

Liquid penetrants. In this technique, fluids are applied to the surfaces of the part and allowed to penetrate into cracks, seams, and pores (Fig. 36.9). By capillary action, the penetrant can seep into cracks as small as 0.1 μm (4 μin.) in width. Two common types of liquids used for this test are (a) *fluorescent penetrants* with various sensitivities, which fluoresce under ultraviolet light, and (b) *visible penetrants* using dyes (usually red in color) which appear as bright outlines on the workpiece surface.

This method can be used to detect a variety of surface defects. The equipment is simple and easy to use, can be portable, and is less costly to operate than that of other methods. However, this method only can detect defects that are open to the surface or are external.

Magnetic-particle inspection. This technique consists of placing fine ferromagnetic particles on the surface of the part. The particles can be applied either dry or in a liquid carrier, such as water or oil. When magnetized with a magnetic field, a discontinuity (defect) on the surface causes the particles to gather visibly around the defect (Fig. 36.10).

Thus, the defect becomes a magnet due to flux leakages where magnetic-field lines are interrupted by the defect. This in turn creates a small-scale *N*–*S* pole at either side of the defect as field lines exit the surface. The particles generally take the shape and size of the defect. Subsurface defects also can be detected by this method, provided they are not deep. The ferromagnetic particles may be colored with pigments for better visibility on metal surfaces.

The magnetic fields can be generated either with direct current or alternating current using yokes, bars, and coils. Subsurface defects can be detected best with

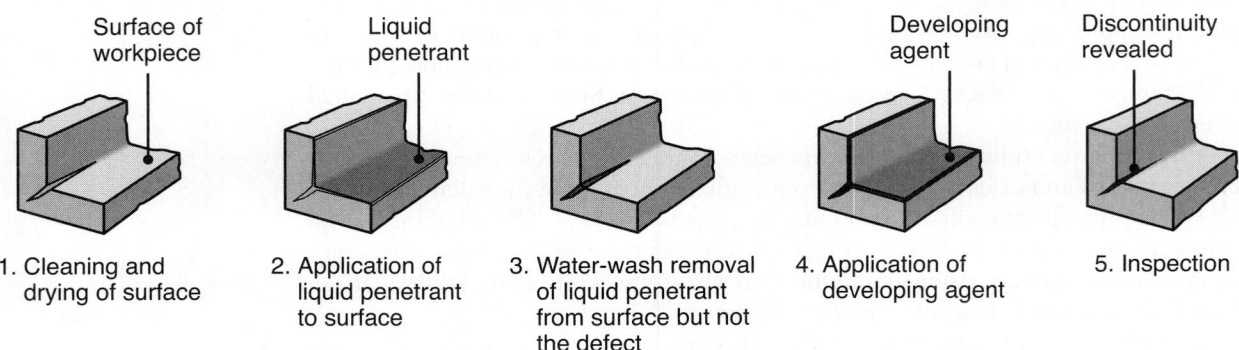

1. Cleaning and drying of surface
2. Application of liquid penetrant to surface
3. Water-wash removal of liquid penetrant from surface but not the defect
4. Application of developing agent
5. Inspection

FIGURE 36.9 Sequence of operations for liquid-penetrant inspection to detect the presence of cracks and other flaws in a workpiece. *Source:* ASM International.

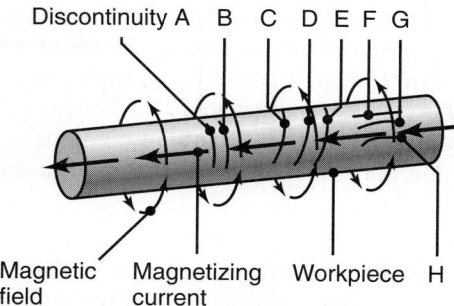

Discontinuity A B C D E F G

Magnetic   Magnetizing   Workpiece   H
field   current

FIGURE 36.10   Schematic illustration of magnetic-particle inspection of a part with a defect in it. Cracks that are in a direction parallel to the magnetic field, (such as in A) would not be detected, whereas the others shown would. Cracks F, G, and H are the easiest to detect. *Source*: ASM International.

direct current. The magnetic-particle method also can be used on ferromagnetic materials only, but parts have to be demagnetized and cleaned after inspection. The equipment may be portable, or it may be stationary.

Ultrasonic inspection.   In this technique, an ultrasonic beam travels through the part. An internal defect (such as a crack) interrupts the beam and reflects back a portion of the ultrasonic energy. The amplitude of the energy reflected and the time required for its return indicate the presence and location of any flaws in the workpiece.

The ultrasonic waves are generated by transducers (called *search units* or *probes*), available in various types and shapes. Transducers operate on the principle of *piezoelectricity* (see Section 3.7) using materials such as quartz, lithium sulfate, or various ceramics. Most inspections are carried out at a frequency range of 1 to 25 MHz. Couplants are used to transmit the ultrasonic waves from the transducer to the test piece; typical couplants are water, oil, glycerin, and grease.

The ultrasonic-inspection method has high penetrating power and sensitivity. It can be used from various directions to inspect flaws in large parts, such as railroad wheels, pressure vessels, and die blocks. This method requires experienced personnel to properly conduct the inspection and to correctly interpret the results.

Acoustic methods.   The *acoustic-emission technique* detects signals (high-frequency stress waves) generated by the workpiece itself during plastic deformation, crack initiation and propagation, phase transformation, and abrupt reorientation of grain boundaries. Bubble formation during boiling of a liquid and friction and wear of sliding interfaces are other sources of acoustic signals (see also Section 21.5.4).

Acoustic-emission inspection typically is performed by elastically stressing the part or structure, such as bending a beam, applying torque to a shaft, or internally pressurizing a vessel. Sensors typically consisting of piezoelectric ceramic elements detect acoustic emissions. This method is effective particularly for continuous surveillance of load-bearing structures.

The **acoustic-impact technique** consists of tapping the surface of an object, listening to, and analyzing the signals to detect discontinuities and flaws. The principle is basically the same as when one taps walls, desktops, or countertops in various locations with a finger or a hammer and listens to the sound emitted. Vitrified grinding wheels (Section 26.2) are tested in a similar manner (*ring test*) to detect cracks in the wheel that may not be visible to the naked eye. The acoustic-impact technique is easy to perform and can be instrumented and automated. However, the results depend on the geometry and mass of the part, so a reference standard is necessary for identifying flaws.

Radiography.   Radiography uses x-ray inspection to detect such internal flaws as cracks and porosity. The principle involves differences in density within a part. For example, the metal surrounding a defect typically is denser and, hence, shows up as lighter than the flaws on an x-ray film. This is similar to the way bones and teeth show up lighter than the rest of the body in x-ray films.

The radiation source is typically an x-ray tube, and a visible, permanent image is made on a film or radiographic paper (Fig. 36.11a). Fluoroscopes also are used to produce x-ray images very quickly, and it is a real-time radiography technique that shows events as they are occurring. Radiography requires expensive equipment, proper interpretation of results, and can be a radiation hazard.

- **Digital radiography** is when the film is replaced by a linear array of detectors (Fig. 36.11b). The x-ray beam is collimated into a fan beam (compare Fig. 36.11a and b), and the workpiece is moved vertically. The detectors digitally sample the radiation, and the data are stored in computer memory. The monitor then displays the data as a two-dimensional image of the workpiece.

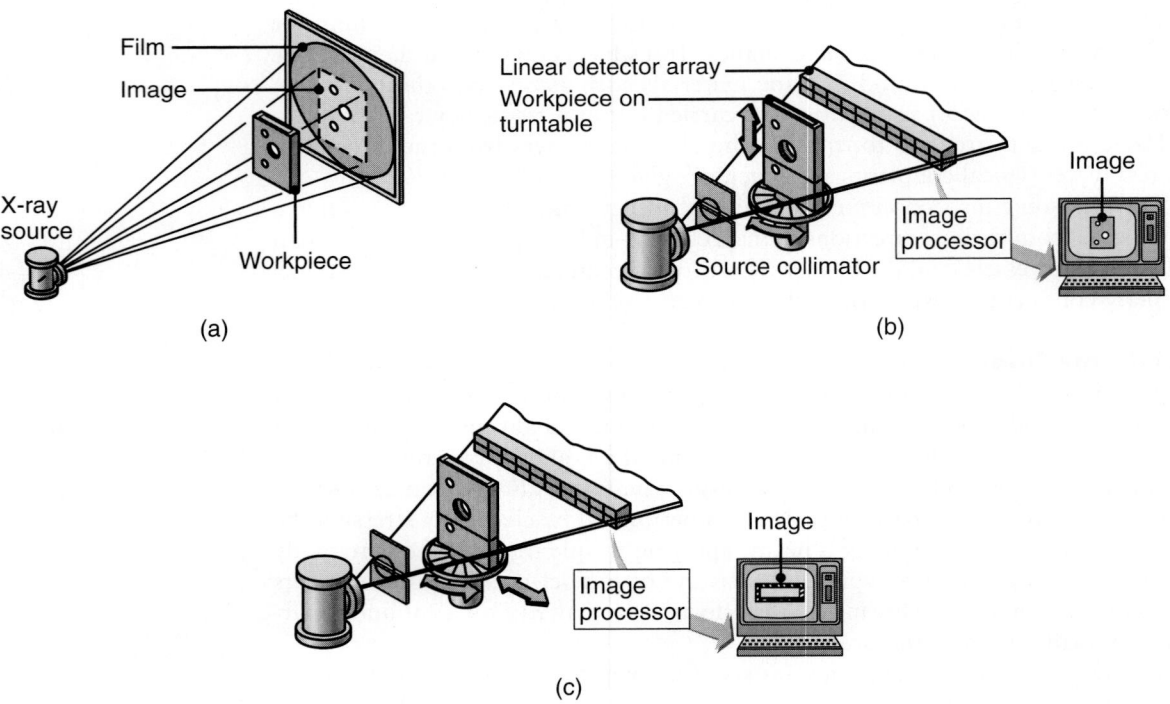

FIGURE 36.11   Three methods of radiographic inspection: (a) conventional radiography, (b) digital radiography, and (c) computed tomography. *Source:* ASM International.

- **Computed tomography** is based on the same system as described for digital radiography, except that the workpiece is rotated along a vertical axis as it is being moved vertically (Fig. 36.11c) and the monitor produces x-ray images of thin cross-sections of the workpiece. The translation and rotation of the workpiece provide several angles from which to precisely view the object.

- **Computer-assisted tomography (catscan)** also is based on the same principle and is used widely in medical practice and diagnosis.

Eddy-current inspection. This method is based on the principle of electromagnetic induction. The part is placed in or adjacent to an electric coil through which alternating current (exciting current) flows at frequencies ranging from 60 Hz to 6 MHz. This current causes eddy currents to flow in the part. Defects in the part impede and change the direction of eddy currents (Fig. 36.12) and cause changes in the electromagnetic field. These changes affect the exciting coil (inspection coil)—the voltage of which is monitored to determine the presence of flaws.

The inspection coils can be made in various sizes and shapes to suit the geometry of the part being inspected. Parts must be conductive electrically and flaw depths detected usually are limited to 13 mm (0.5 in.). The technique requires the use of a standard reference sample to set the sensitivity of the tester.

Thermal inspection. Thermal inspection involves using contact- or noncontact-type heat-sensing devices that detect temperature changes. Defects in the workpiece (such as cracks, debonded regions in laminated structures, and poor joints) cause a change in temperature distribution.

In **thermographic inspection**, materials (such as heat-sensitive paints and papers, liquid crystals, and other coatings) are applied to the workpiece surface. Any changes in their color or appearance indicate defects. The most common method of noncontact-thermographic inspection uses infrared detectors (usually infrared scanning microscopes and cameras), which have a high response time and sensitivies of 1°C (2°F). **Thermometric inspection** utilizes devices (such as thermocouples, radiometers, pyrometers) and sometimes meltable materials, such as wax-like crayons.

Holography. The holography technique creates a three-dimensional image of the part utilizing an optical system (Fig. 36.13). Generally used on simple shapes and highly polished surfaces, this technique records the image on a photographic film.

Its use has been extended to **holographic interferometry** for the inspection of parts with various shapes and surface features. In response to double- and multiple-

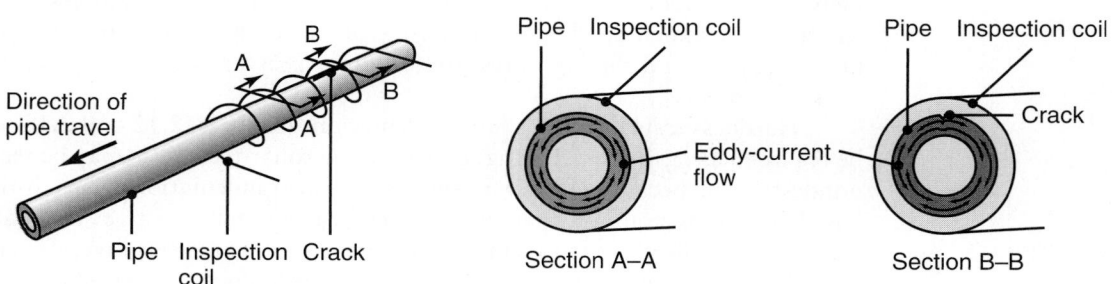

FIGURE 36.12 Changes in eddy-current flow caused by a defect in a workpiece. *Source:* ASM International.

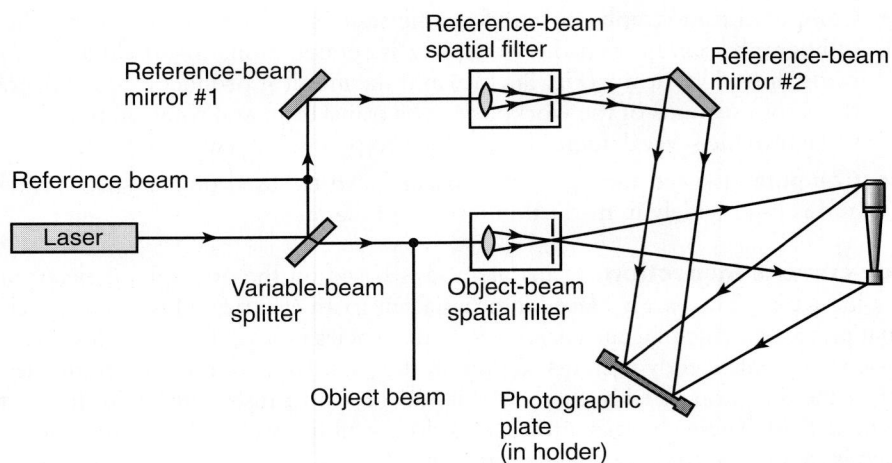

FIGURE 36.13   Schematic illustration of the basic optical system used in holography elements in radiography for detecting flaws in workpieces. *Source:* ASM International.

exposure techniques while the part is being subjected to external forces or time-dependent variations, changes in the images reveal defects in the part.

- In **acoustic holography**, information on internal defects is obtained directly from the image of the interior of the part. In *liquid-surface acoustical holography*, the workpiece and two ultrasonic transducers (one for the object beam and the other for the reference beam) are immersed in a water-filled tank. The holographic image then is obtained from the ripples in the tank.

- In **scanning acoustical holography**, only one transducer is used, and the hologram is produced by electronic-phase detection. In addition to being more sensitive, the equipment usually is portable and can accommodate very large workpieces by using a water column instead of a tank.

36.11 | Destructive Testing

As the name suggests, the part tested using *destructive testing* methods no longer maintains its integrity, original shape, or surface characteristics. The mechanical test methods described in Chapter 2 are all destructive, in that a sample or specimen has to be removed from the product in order to test it. Examples of other destructive tests include the speed testing of grinding wheels to determine their bursting speed and the high-pressure testing of pressure vessels to determine their bursting pressure.

Hardness tests that leave large indentations (see Figs. 2.12 and 2.13) also may be regarded as destructive testing. However, microhardness tests may be regarded as nondestructive because of the very small permanent indentations. This distinction is based on the assumption that the material is not notch-sensitive. Generally, most glasses, highly heat-treated metals, and ceramics are notch-sensitive. Consequently, a small indentation produced by the indenter can reduce their strength and toughness significantly.

36.12 | Automated Inspection

Traditionally, individual parts and assemblies of parts have been manufactured in batches, sent to inspection in quality-control rooms (**post-process inspection**) and, if approved, placed into inventory. If the parts do not pass the quality inspection, they are either scrapped or kept and used on the basis of having a certain acceptable deviation from the standard.

In contrast, *automated inspection* uses a variety of sensor systems that monitor the relevant parameters *during* the manufacturing process (called **on-line inspection**). Using these measurements, the process automatically corrects itself to produce acceptable parts. Thus, further inspection of the part at another location in the plant is unnecessary. Parts also may be inspected immediately after they are produced (called **in-process inspection**).

The development of accurate sensors and advanced computer-control systems has enabled the integration of automated inspection into manufacturing operations (Chapters 37 and 38). Such a system ensures that no part is moved from one manufacturing process to another (for example, a turning operation followed by cylindrical grinding), unless the part is made correctly and meets the standards in the first operation.

Automated inspection is flexible and responsive to product design changes. Furthermore, because of automated equipment, less operator skill is required, productivity is increased, and parts have higher quality, reliability, and dimensional accuracy.

Sensors for automated inspection. Continued advances in **sensor technology** (to be described in Section 37.7) have made on-line or real-time monitoring of manufacturing processes feasible. Directly or indirectly and with the use of various *probes*, sensors can detect dimensions, surface roughness, temperature, force, power, vibration, tool wear, and the presence of external or internal defects.

Sensors may operate on the principles of strain gages, inductance, capacitance, ultrasonics, acoustics, pneumatics, infrared radiation, optics, lasers, and various electronic gages. Sensors may be *tactile* (touching) or *nontactile*. They are linked to microprocessors and computers for graphic data display (see also *programmable logic controllers*, Section 37.2.6). This capability allows the rapid on-line adjustment of any processing parameter, thus resulting in the production of parts that consistently are within specified standards of dimensional tolerance and quality. For example, such systems already are standard equipment on many metal-cutting machine tools and grinding machines described in Part IV of this book.

SUMMARY

- Quality must be built into products. Quality assurance concerns various aspects of production, such as design, manufacturing, and assembly, and especially inspection at each step of production for conformance to specifications.

- Statistical quality control and process control are indispensable in modern manufacturing. They are important particularly in the production of interchangeable parts and in the reduction of manufacturing costs.

- Although all quality-control approaches have their limits of applicability in the production of items, implementation of total quality management, the ISO and QSO 9000 standard, and the ISO 14000 standard are among the most significant developments in quality control in manufacturing.

- Several nondestructive and destructive testing techniques (each of which has its own applications, advantages, and limitations) are available for inspection of completed parts and products.
- The traditional approach of inspecting the part or product after it is manufactured has been replaced largely by on-line and 100% inspection of all parts and products being manufactured.

KEY TERMS

Acceptance quality level	ISO standards	Range
Acceptance sampling	Juran methods	Reliability
Assignable variations	Kaizen	Return on quality
Automated inspection	Lot size	Robustness
Chance variations	Lower control limit	Sample size
Common cause	Method of attributes	Sensors
Consumer's risk	Method of variables	Shewhart control charts
Continuous improvement	Nondestructive testing	Six sigma
Control charts	Normal distribution curve	Special cause
Control limits	Population	Specification limits
Defect prevention	Probability	Standard deviation
Deming methods	Process capability	Statistical process control
Destructive testing	Process reliability	Statistical quality control
Dispersion	Product reliability	Statistics
Distribution	Producer's risk	Taguchi loss function
Environmental management systems	QS standards	Taguchi methods
Experimental design	Quality	Total quality control
Factorial design	Quality assurance	Total quality management
Frequency distribution	Quality circle	Upper control limit
Grand average	Random sampling	Variability

BIBLIOGRAPHY

Aft, L.S., *Fundamentals of Industrial Quality Control*, 3rd ed., Addison-Wesley, 1998.

Anderson, D.M., *Design for Manufacturability, Optimizing Cost, Quality, and Time-to-Market*, CIM Press, 1991.

ASM Handbook, Vol. 17: *Nondestructive Evaluation and Quality Control*, ASM International, 1989.

Bentley, J.P., *Introduction to Reliability and Quality Control*, 2nd ed., Addison-Wesley, 1999.

Besterfield, D.H., *Quality Control*, 5th ed., Prentice Hall, 1997.

Bothe, D.R., *Measuring Process Capability: Techniques and Calculations for Quality and Manufacturing Engineers*, McGraw-Hill, 1997.

Bray, D.E., and Stanley, R.K., *Nondestructive Evaluation: A Tool in Design, Manufacturing, and Service*, Rev. ed., CRC Press, 1997.

Breyfogle, F.W., III, *Statistical Methods for Testing, Development, and Manufacturing*, Wiley Interscience, 1992.

Cartz, L., *Nondestructive Testing*, ASM International, 1995.

Clements, R.B., *The Handbook of Statistical Methods in Manufacturing*, Prentice Hall, 1991.

Connor, G., *Six Sigma and other Continuous Improvement Tools for the Small Shop*, Society of Manufacturing Engineers, 2001.

Deming, W.E., *Out of the Crisis*, MIT Press, 1986.

DeVor, R.E., Chang T., and Sutherland, J.W., *Statistical Quality Design and Control*, Macmillan, 1992.

Diamond, W.J., *Practical Experimental Designs for Engineers and Scientists* 2nd ed., Van Nostrand Reinhold, 1989.

Doty, L.A., *SPC for Short Run Manufacturing*, Hanser Gardner, 1997.

———, *Statistical Process Control*, 2nd ed., Industrial Press, 1996.

Elsayed, E.A., *Reliability Engineering*, Addison-Wesley, 1996.

Feigenbaum, A.V., *Total Quality Control*, 3rd ed., Revised, McGraw-Hill, 1991.

Fowlkes, W.Y., and Creveling, C.M., *Engineering Methods for Robust Product Design: Using Taguchi Methods in Technology and Product Development*, Addison-Wesley, 1995.

Grant, E.L., and Leavenworth, R.S., *Statistical Quality Control*, McGraw-Hill, 1997.

Imai, M., *Gemba Kaizen: A Commonsense, Low-Cost Approach to Management*, McGraw-Hill, 1997.

Ireson, W.G., Coombs, C.F., Jr., and Moss, R.Y., *Handbook of Reliability Engineering and Management*, 2nd ed., McGraw-Hill, 1996.

Juran, J.M., and Godfrey, A.B. (eds.), *Juran's Quality Handbook*, 5th ed., McGraw-Hill, 1999.

Kales, P., *Reliability: For Technology, Engineering, and Management*, Prentice Hall, 1997.

Kane, V.E., *Defect Prevention: Use of Simple Statistical Tools*, Marcel Dekker, 1989.

Kear, F.W., *Statistical Process Control in Manufacturing Practice*, Marcel Dekker, 1997.

Kolarik, W.J., *Creating Quality*, McGraw-Hill, 1995.

Lamprecht, J.L., *Implementing the ISO 9000 Series*, Marcel Dekker, 1993.

Levinson, W.A., and Tumbelty, F., *SPC Essentials and Productivity Improvement: A Manufacturing Approach*, American Society for Quality Control, 1996.

Lewis, E.E., *Introduction to Reliability Engineering*, 2nd ed., Wiley, 1995.

Menon, H.G., *TQM in New Product Manufacturing*, McGraw-Hill, 1992.

Monden, Y., *Toyota Production System*, Industrial Engineering and Management Press, 1993.

Montgomery, D.C., *Introduction to Statistical Quality Control*, 3rd ed., Wiley, 1996.

Oakland, J.S., *Statistical Process Control*, 3rd ed., Butterworth-Heinemann, 1996.

Ott, E.R., and Schilling, E.G., *Process Quality Control*, 2nd ed., McGraw-Hill, 1990.

Park, S., *Robust Design and Analysis for Quality Engineering*, Chapman & Hall, 1997.

Quensenberry, C.P., *SPC Methods for Quality Improvement*, Wiley, 1997.

Raj, B, Jaykumar, T., and Thavasimuthu, M. (eds.); *Practical Non-Destructive Testing*, 2nd ed., ASM International, 2002.

Ross, J.E., *Total Quality Management—Text, Cases and Readings*, St. Lucie Press, 1993.

Ross, P.J., *Taguchi Techniques for Quality Engineering*, 2nd ed., McGraw-Hill, 1996.

Rothery, B., *ISO 14000 and ISO 9000*, Gower Publishing Co., 1995.

Smith, G.M., *Statistical Process Control and Quality Improvement*, American Society for Quality Control, 1997.

Taguchi, G., *Introduction to Quality Engineering*, UNIPUB/Kraus International, 1986.

———, *Taguchi on Robust Technology Development: Bringing Quality Engineering Upstream*, ASME Press, 1993.

———, ElSayed E.A., and Hsiang, T., *Quality Engineering in Production Systems*, McGraw-Hill, 1989.

Taylor, W., *Optimization and Variation in Reduction in Quality*, McGraw-Hill, 1991.

Tool and Manufacturing Engineers Handbook, 4th ed., Vol. 4, *Assembly, Testing, and Quality Control*, Society of Manufacturing Engineers, 1986.

Tool and Manufacturing Engineers Handbook, 4th ed., Vol. 7, *Continuous Improvement*, Society of Manufacturing Engineers, 1993.

Tozawa, B., and Bodek, N., *The Idea Generator: Quick and Easy Kaizen*, PCS Press, 2001.

REVIEW QUESTIONS

36.1 Explain why major efforts continually are being made to build quality into products.

36.2 What are chance variations?

36.3 Define the terms: sample size, random sampling, population, and lot size.

36.4 Explain the difference between method of variables and method of attributes.

36.5 Define standard deviation. Why is it important in manufacturing?

36.6 Describe what is meant by statistical process control.

36.7 Why are control charts made? How are they used?

36.8 What do control limits indicate?

36.9 Define process capability. How is it used?

36.10 What is acceptance sampling? Why was it developed?

36.11 What is the difference between series and parallel reliability?

36.12 What is a Taguchi loss function? What is its significance?

36.13 What is meant by six sigma quality?

36.14 What is the difference between (a) probability and reliability and (b) robustness and reliability?

36.15 Make a list of destructive techniques and comment.

QUALITATIVE PROBLEMS

36.16 What is the consequence of setting lower and upper specifications closer to the peak of the curve in Fig. 36.4?

36.17 Identify several factors that can cause a process to become out of control.

36.18 Describe situations in which the need for destructive testing techniques is unavoidable.

36.19 Which of the nondestructive inspection techniques are suitable for nonmetallic materials? Why?

36.20 Give examples of products where 100% sampling is not possible or feasible.

36.21 What are the advantages of automated inspection? Why has it become an important part of manufacturing engineering?

36.22 Why is reliability important in manufacturing engineering? Give several examples.

36.23 Give examples of the acoustic-impact inspection technique other than those given in the chapter.

36.24 Explain why *GO* and *NOTGO* gages (see Section 35.4.4) are incompatible with the Taguchi philosophy.

36.25 Describe your thoughts regarding Table 36.1.

36.26 Give examples of robust design in addition to that in Fig. 36.1.

QUANTITATIVE PROBLEMS

36.27 Beverage can manufacturers try to achieve failure rates of less than one can in ten thousand. If this corresponds to n-sigma quality, find n.

36.28 Assume that in Example 36.3, the number of samples was 8 instead of 10. Using the top half of the data in Table 36.4 recalculate the control limits and the standard deviation. Compare your observations with the results obtained by using 10 samples.

36.29 Calculate the control limits for averages and ranges for the following: (a) number of samples = 6, (b) $\overline{\overline{x}} = 65$, and (c) $R = 5$.

36.30 Calculate the control limits for the following: (a) number of samples = 5, (b) $\overline{x} = 36.5$, and (c) $\mathrm{UCL}_R = 5.75$.

36.31 In an inspection with sample size 12 and a sample number of 50, it was found that the average range was 12 and the average-of-averages was 75. Calculate the control limits for averages and for ranges.

36.32 Determine the control limits for the data shown in the table at the bottom of the first column.

36.33 The average-of-averages of a number of samples of size 8 was determined to be 124. The average range was 17.82 and the standard deviation was 4. The following measurements were taken in a sample: 120, 132, 124, 130, 118, 132, 135, 121, and 127. Is the process in control?

36.34 The trend in the electronics and computer-chip industries is to make products where it is stated that the quality is approaching six sigma (6σ). What would be the reject rate per million parts?

36.35 A manufacturer is ring-rolling ball-bearing races (see Fig. 13.15). The inner surface has a surface roughness specification of $0.10 \pm 0.06 \ \mu$m. Measurements taken from the rolled rings indicate a mean roughness of $0.112 \ \mu$m with a standard deviation of $0.02 \ \mu$m. 50,000 rings per month are manufactured, and the cost of discarding a defective ring is $5.00. It is known that, by changing lubricants to a special emulsion, the mean roughness essentially could be made equal to the design specification. What additional cost per month can be justified for the lubricant?

36.36 For the data of Problem 36.35, assume that the lubricant change can cause the manufacturing process to achieve a roughness of $0.10 \pm 0.01 \ \mu$m. What additional cost per month for the lubricant can be justified? What if the lubricant did not add any new cost?

x_1	x_2	x_3	x_4
0.57	0.61	0.50	0.55
0.59	0.55	0.60	0.58
0.55	0.50	0.55	0.51
0.54	0.57	0.50	0.50
0.58	0.58	0.60	0.56
0.60	0.61	0.55	0.61

SYNTHESIS, DESIGN, AND PROJECTS

36.37 Which aspects of the quality-control concepts of Deming, Taguchi, and Juran would be difficult to implement in a typical manufacturing facility? Why?

36.38 Should products be designed and built for a certain expected life? Explain.

36.39 Survey the available technical literature, contact various associations, and prepare a comprehensive table concerning the life expectancy of various consumer products.

36.40 Is there a relationship between design specifications on blueprints and limits in control charts?

36.41 Would it be desirable to incorporate nondestructive inspection techniques in various metalworking machinery? Give a specific example, make a sketch of such a machine, and explain its features.

36.42 Name several material and process variables that can influence product quality in metal (a) casting, (b) forming, and (c) machining.

36.43 Identify the nondestructive techniques that are capable of detecting internal flaws and those that detect external flaws only.

36.44 Explain the difference between in-process and post-process inspection of manufactured parts. What trends are there in such inspections?

36.45 Assume that you are in charge of manufacturing operations for a company that has not yet adopted statistical process-control techniques. Describe how you would go about developing a plan to do so, including providing for the training of your personnel.

36.46 Many components in products have minimal effect on part robustness and quality. For example, the hinges in the glove compartment of an automobile do not really impact the owner's satisfaction, and the glove compartment is opened so infrequently that a robust design is easy to achieve. Would you advocate using Taguchi methods (like the loss function) on this type of component? Explain.

Manufacturing in a Competitive Environment

In an increasingly global and competitive marketplace for consumer and industrial goods, advances in production processes, machinery, tooling, and operations have been driven mainly by the goals we described in the General Introduction. A review of these goals will provide the proper context for studying and appreciating the various topics and examples presented in the final part of this book. These goals may be summarized as follows:

- Products must fully meet **design and service requirements** and specifications and standards.
- **Quality** must be built into the product at each stage of design and manufacturing.
- The most **economical methods** of manufacturing must be explored and implemented.
- Manufacturing processes and operations must have sufficient **flexibility** to respond rapidly to constantly changing global market demands.
- Manufacturing must strive continually for higher levels of quality and **productivity**.

In the following chapters, we will describe the significance and importance of *viewing manufacturing as a large system* in which all of its parts and activities must be integrated. Although numerical control of machine tools in the early 1950s was a key factor in setting the stage for modern manufacturing, much of the progress in manufacturing activities stems from our ability to view these activities and operations as a large system with complex interactions among all of its components. In implementing a *systems approach* to manufacturing, we can *integrate* and *optimize* various functions and activities that had been separate and distinct entities for a long time.

We begin with the description of the concept of automation in Chapter 37. Its implementation in terms of key developments in numerical control (and later in computer numerical control) then are discussed. We then describe further advances in automation and controls that involve such major topics as adaptive control, industrial robots, sensor technology, and material handling and movement. The chapter ends with assembly systems and how they are implemented in modern production.

Manufacturing systems and how their individual components and operations are integrated are described in Chapter 38, along with the critical role of computers and various enabling technologies as an aid in such activities as design, engineering, manufacturing, and process planning. The important concept of group technology also is discussed. Computer-integrated manufacturing, with its various features such as cellular manufacturing, flexible manufacturing systems, just-in-time production, and lean manufacturing then are described in Chapter 39, including the new concept of holonic manufacturing. This leads to the optimal production of the desired products, the importance and significance of communication networks, and the use of artificial intelligence. We will show that each of these technologies has an impact on flexibility of operation, productivity, product quality, and manufacturing costs.

The purpose of the final chapter of this book is to recognize the importance of the numerous factors that significantly affect competitive manufacturing in a global marketplace. As described in Chapter 40, these factors involve product design, quality, and product life cycle; selection of materials and processes and their substitution in economical production; process capabilities; and costs involved, including those of machinery, tooling, and labor. This chapter thus will help us respond to the goals outlined at the beginning of this introduction.

CHAPTER

37

Automation of Manufacturing Processes

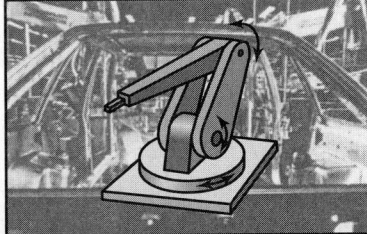

This chapter describes the art and science of automation in manufacturing processes and operations. The ability to reliably and economically produce parts with high accuracy and production rates is aided greatly by the use of automation. The topics considered in this chapter include:

- The use of hard automation for very large production runs.
- Numerical control of machines for improved productivity and increased flexibility in manufacturing.
- Control strategies including open-loop and closed-loop and the adaptive control of processes.
- The use of industrial robots in various phases of manufacturing operations.
- The impact on automation on product design and process economics.

37.1 | Introduction

Until the early 1950s, most manufacturing operations in a typical plant were carried out on traditional machinery, such as lathes, milling machines, drill presses, and various equipment for forming, shaping, and joining materials. However, such equipment generally lacked flexibility, and it required considerable skilled labor to produce parts with acceptable dimensions and characteristics. Each time a different product was to be manufactured, the machinery had to be retooled, fixtures had to be prepared, and the movement of materials among various machines had to be rearranged. The development of new products and of parts with complex shapes required numerous trial-and-error attempts by the operator to set the proper processing parameters on the machine. Furthermore, because of human involvement, making parts that were exactly alike was difficult, time-consuming, and costly.

These circumstances meant that processing methods generally were inefficient and that labor costs were a significant portion of the overall production cost. The necessity for reducing the labor share of product cost became increasingly apparent, as did the need to improve the efficiency and flexibility of manufacturing operations, particularly because of increased domestic and global competition.

Productivity also became a major concern. Generally defined as "output per employee per hour," productivity basically measures operating efficiency. Thus, an efficient operation makes optimum use of all resources, such as materials, energy, capital, labor, machinery, and up-to-date technology. With rapid advances in the science and technology of manufacturing, the efficiency of manufacturing operations began to improve, and as a result, the percentage of total cost represented by labor began to decline.

In improving productivity, the important elements have been *mechanization, automation,* and *control* of manufacturing equipment and systems. **Mechanization** controls a machine or process with the use of various mechanical, hydraulic, pneumatic, or electrical devices. It reached its peak by the 1940s. In spite of its benefits, in mechanized operations the worker is still involved in a particular process directly and must check each step of a machine's performance. For example, if (a) a cutting tool breaks during machining, (b) parts are overheated during heat treatment, (c) surface finish begins to deteriorate during grinding, or (d) dimensional tolerances become too large in sheet-metal forming, the operator must intervene and change one or more of the relevant process parameters and machine settings, which requires considerable experience.

The next step in improving the efficiency of manufacturing operations was **automation.** The word automation was coined in the mid-1940s by the U.S. automobile industry to indicate *automatic handling and processing* of parts in and among production machines. Although there is no precise definition, automation generally means the methodology and system of operating a machine or process by highly *automatic* means and comes from the Greek word *automatos*—meaning self-acting.

There are various types and levels (orders) of automation. The order begins with simple hand tools and continues on to manual machines, power tools, single-cycle automatic machines, multiple-cycle machines, transfer lines, numerically controlled machines, and, ultimately, the implementation of expert systems and artificial intelligence, as described in Chapter 39. Rapid advances in automation and enabling technologies have been made possible largely through developments in **control systems,** with the help of increasingly powerful and sophisticated computers and software.

This chapter follows the outline shown in Fig. 37.1. First, it reviews the history and principles of automation and how it has helped the integration of various key operations and activities in a manufacturing plant. Then, it introduces the concept

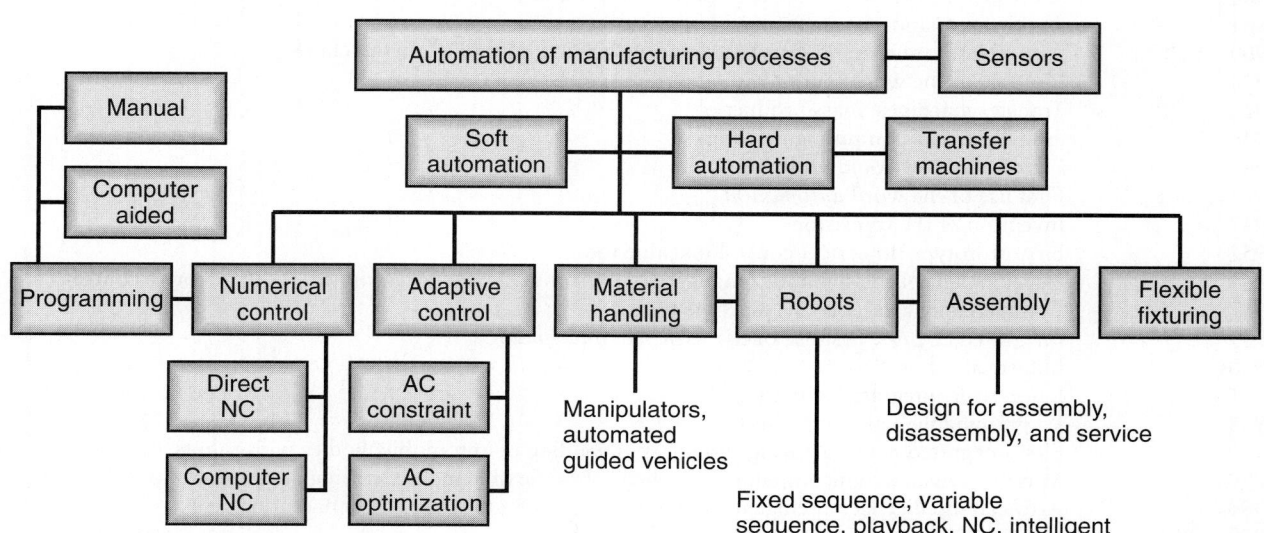

FIGURE 37.1   Outline of topics covered in Chapter 37.

of the control of machines and systems through *numerical control* and *adaptive control* techniques. We also describe how the important activity of material handling and movements have been developed into various systems, particulary ones including the use of *industrial robots* to improve handling efficiency.

The subject of *sensor technology* then is described; this is a topic that is an essential element in the control and optimization of machinery, processes, and systems. We also cover significant developments in *flexible fixturing* and *assembly operations*. These methods enable us to take full advantage of advanced manufacturing technologies, particularly flexible manufacturing systems. This chapter also includes a discussion of the guidelines for *designs for assembly, disassembly, and service* with specific recommendations to improve the efficiency of these operations. The final topic describes the *economic* aspects of the processes and operations covered in this chapter.

37.2 | Automation

Although there can be variations in definition, *automation* generally is defined as the process of enabling machines to follow a predetermined sequence of operations with little or no human labor and using specialized equipment and devices that perform and control manufacturing processes (Table 37.1). Full automation is achieved through various devices, sensors, actuators, techniques, and equipment that are

TABLE 37.1

Development in the History of the Automation of Manufacturing Processes	
Date	Development
1500–1600	Water power for metalworking; rolling mills for coinage strips
1600–1700	Hand lathe for wood; mechanical calculator
1700–1800	Boring, turning, and screw-cutting lathe; drill press
1800–1900	Copying lathe, turret lathe, universal milling machine; advanced mechanical calculators
1808	Sheet-metal cards with punched holes for automatic control of weaving patterns in looms
1863	Automatic piano player (Pianola)
1900–1920	Geared lathe; automatic screw machine; automatic bottle-making machine
1920	First use of the word *robot*
1920–1940	Transfer machines; mass production
1940	First electronic computing machine
1943	First digital electronic computer
1945	First use of the word *automation*
1947	Invention of the transistor
1952	First prototype numerical-control machine tool
1954	Development of the symbolic language APT (Automatically Programmed Tool); adaptive control
1957	Commercially available NC machine tools
1959	Integrated circuits; first use of the term *group technology*
1960s	Industrial robots
1965	Large-scale integrated circuits
1968	Programmable logic controllers
1970	First integrated manufacturing system; spot welding of automobile bodies with robots
1970s	Microprocessors; minicomputer-controlled robot; flexible manufacturing systems; group technology
1980s	Artificial intelligence; intelligent robots; smart sensors; untended manufacturing cells
1990–2000	Integrated manufacturing systems; intelligent and sensor-based machines; telecommunications and global manufacturing networks; fuzzy-logic devices; artificial neural networks; Internet tools

capable of (a) monitoring all aspects of the manufacturing operation, (b) making decisions concerning the changes that should be made in the operation, and (c) controlling all aspects of it.

Automation is an *evolutionary* rather than a revolutionary concept. In manufacturing plants, it has been implemented especially in the following basic areas of activity:

- **Manufacturing processes:** Machining, forging, cold extrusion, casting, powder metallurgy, and grinding operations are major examples of processes that have been automated extensively.

- **Material handling and movement:** Materials and parts in various stages of completion are moved throughout a plant by computer-controlled equipment with little or no human guidance.

- **Inspection:** Parts are inspected automatically for dimensional accuracy, surface finish, quality, and various specific characteristics while being made (*in-process inspection*).

- **Assembly:** Individually manufactured parts and components are assembled automatically into subassemblies and assemblies to form a product.

- **Packaging:** Products are packaged automatically for shipment.

37.2.1 Evolution of automation

Some metalworking processes were developed as early as 4000 B.C., as shown in Table I.2 in the General Introduction. However, it was not until the beginning of the Industrial Revolution in the 1750s (also referred to as the First Industrial Revolution) that automation began to be introduced in the production of goods. Machine tools (such as turret lathes, automatic screw machines, and automatic bottle-making equipment) began to be developed in the late 1890s. Mass-production techniques and transfer machines were developed in the 1920s. These machines had *fixed* automatic mechanisms and were designed to produce *specific* products best represented by the automobile industry which produced passenger cars at a high production rate and low cost.

The major breakthrough in automation began with numerical control (NC) of machine tools. Since this historic development, rapid progress has been made in automating most aspects of manufacturing. These developments involve the introduction of computers into automation, computerized numerical control (CNC), adaptive control (AC), industrial robots, computer-aided design, engineering, and manufacturing (CAD/CAE/CAM), and computer-integrated manufacturing (CIM) systems.

Manufacturing involves various levels of automation, depending on the processes used, the product desired, and production volumes. Manufacturing systems in order of increasing automation (see Fig. 37.2) include the following classifications:

- **Job shops:** These facilities use general-purpose machines and machining centers with high levels of human labor involvement.

- **Stand-alone NC production:** This uses **numerically controlled machines** (Section 37.3) but with significant operator/machine interaction.

- **Manufacturing cells:** These use a designed cluster of machines with integrated computer control and flexible material handling—often with industrial robots (Section 37.6).

- **Flexible manufacturing systems:** Described in Section 39.3, these use computer control of all aspects of manufacturing, the simultaneous incorporation of a number of manufacturing cells, and automated material-handling systems.

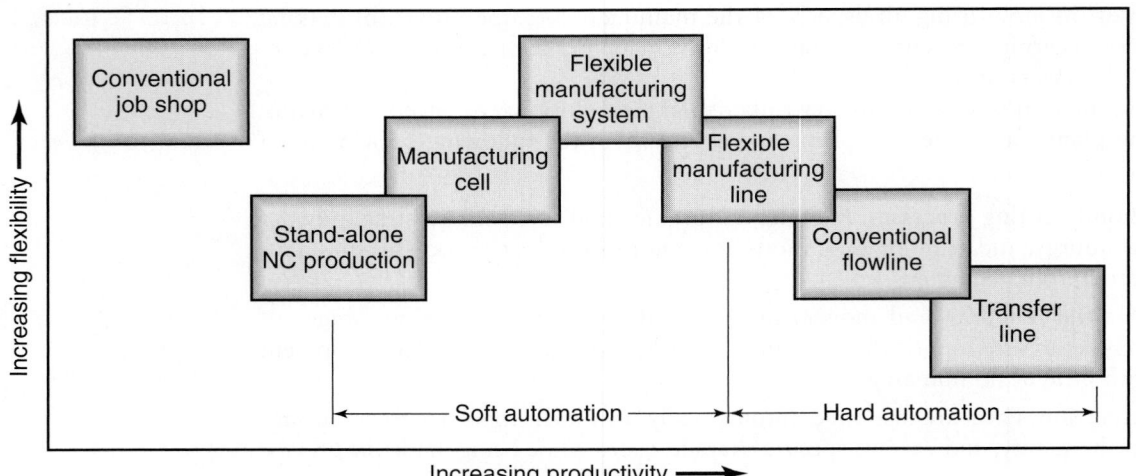

FIGURE 37.2   Flexibility and productivity of various manufacturing systems. Note the overlap between the systems; it is due to the various levels of automation and computer control that are possible in each group. See also Chapter 39 for details.

- **Flexible manufacturing lines**: Organize computer-controlled machinery in production lines instead of cells. Part transfer is through hard automation, product flow is more limited than in flexible manufacturing systems, but the throughput is larger for higher production quantities.
- **Flowlines** and **transfer lines**: Consist of organized groupings of machinery with automated material handling between machines. The manufacturing line is designed with limited or no flexibility, since the goal is to produce a single part.

37.2.2 Implementation of automation

Automation has the following primary goals:

- **Integrate** various aspects of manufacturing operations so as to improve product quality and uniformity, minimize cycle times and effort, and reduce labor costs.
- **Improve productivity** by reducing manufacturing costs through better control of production. Parts are loaded, fed, and unloaded on machines more efficiently, machines are used more effectively, and production is organized more efficiently.
- **Improve quality** by using more repeatable processes.
- **Reduce human involvement**, boredom, and thus the possibility of human error.
- **Reduce workpiece damage** caused by the manual handling of parts.
- **Raise the level of safety** for personnel, especially under hazardous working conditions.
- **Economize on floor space** in the plant by arranging machines, material handling and movement, and auxiliary equipment more efficiently.

Automation and production quantity.   Production quantity is crucial in determining the type of machinery and the level of automation required to produce parts economically. *Total production quantity* is defined as the total number of parts to be made. This volume can be produced in individual batches of various *lot sizes*. Lot size greatly influences the economics of production. *Production rate* is defined as

TABLE 37.2

Approximate Annual Production Quantity		
Type of production	Number produced	Typical products
Experimental or prototype	1–10	All products
Piece or small-batch	10–5000	Aircraft, missiles, special machinery, dies, jewelry, and orthopedic implants
Batch or high-volume	5000–100,000	Trucks, agricultural machinery, jet engines, diesel engines, computer components, and sporting goods
Mass production	100,000 and over	Automobiles, appliances, fasteners, and food and beverage containers

the number of parts produced per unit time, such as per day, per month, or per year. The approximate and generally accepted ranges of production volume are shown in Table 37.2 for some typical applications. As expected, experimental or prototype products represent the lowest volume. (See also Chapter 20.)

Job shops typically produce small quantities per year (Fig. 37.2) using various standard general-purpose machine tools (*stand-alone machines*) or *machining centers* (see Chapter 25). The operations performed typically have high part variety—meaning that different parts can be produced in a short time without extensive changes in tooling and in operations. On the other hand, machinery in job shops generally requires skilled labor to operate and production quantities and rates are low. As a result, production cost per part is high (Fig. 37.3). When parts involve a large labor component, the production is called *labor intensive*.

- **Piece-part production** usually involves very small quantities and is suitable for job shops. The majority of piece-part production involves lot sizes of 50 or less.

- **Small-batch production** quantities typically range from 10 to 100, and the equipment used consists of general-purpose machines and machining centers with various computer controls.

- **Batch production** usually involves lot sizes between 100 and 5000. It utilizes machinery similar to that used for small-batch production but with specially designed fixtures for higher productivity.

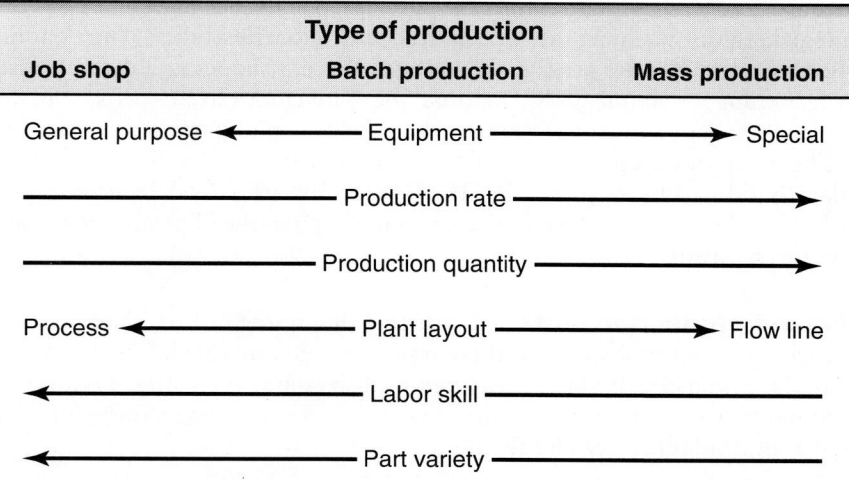

FIGURE 37.3   General characteristics of three types of production methods: job shop, batch, and mass production.

- **Mass production** involves quantities often over 100,000. It requires special-purpose machinery (called *dedicated machines*) and automated equipment for transferring materials and parts in progress. Although the machinery, equipment, and specialized tooling are expensive, both the labor skills required and the labor costs are relatively low. These production systems are organized for a specific type of product; hence, they lack flexibility.

37.2.3 Applications of automation

Automation can be applied to the manufacturing of all types of goods, from raw materials to finished products, and in all types of production, from job shops to large manufacturing facilities. The decision to automate a new or existing production facility requires taking into account the following additional considerations:

- Type of product manufactured
- Production quantity and rate of production required
- Particular phase of the manufacturing operation to be automated
- Level of skill in the available workforce
- Any reliability and maintenance problems that may be associated with automated systems
- Economics of the process

Because automation generally involves high initial equipment cost and requires a knowledge of operation and maintenance principles, a decision about the implementation of even low levels of automation must involve a careful study of the true needs of an organization. In some situations, **selective automation** rather than total automation of a facility is desirable. As we will see in the rest of Part IX of this book, there are several important and complex issues involved in the decision making about the appropriate level of automation.

37.2.4 Hard automation

In *hard automation* or **fixed-position automation**, the machines are designed to produce a standard product, such as a gear, a shaft, or an engine block. Although product size and processing parameters (such as speed, feed, and depth of cut) can be changed, these machines are specialized and lack flexibility. They cannot be modified to accommodate products that have different shapes and dimensions. (See *group technology*, Section 38.8). Because these machines are expensive to design and build, their economical use requires the production of parts in very large quantities. The machines used in hard-automation applications usually are built on the **modular (building-block) principle**. They generally are called *transfer machines* (see below) and consist of two major components: powerhead production units and transfer mechanisms.

Powerhead production units.   Consisting of a frame or bed, electric drive motors, gearboxes, and tool spindles, these units are self-contained. Their components are available commercially in various standard sizes and capacities. Because of this inherent modularity, they easily can be regrouped for producing a different part and thus have some adaptability and flexibility.

Transfer machines.   Typically consisting of two or more powerhead units, these machines can be arranged on the shop floor in linear, circular, or U-shaped patterns.

The weight and shape of the workpieces influence the arrangement selected, which is important for continuity of operation in the event of tool failure or machine breakdown in one or more of the units. *Buffer storage* features are incorporated in these machines to permit continued operation in such an event.

Transfer mechanisms and transfer lines. Transfer mechanisms are used to move the workpiece from one station to another in the machine or from one machine to another to enable various operations to be performed on the part. Workpieces are transferred by methods, including (a) rails along which the parts (which usually are placed on pallets) are pushed or pulled by various mechanisms (Fig. 37.4a), (b) rotary indexing tables (Fig. 37.4b), and (c) overhead conveyors.

The transfer of parts from station to station usually is controlled by sensors and other devices. Tools on transfer machines easily can be changed using tool-holders with quick-change features, and the machines can be equipped with various automatic gaging and inspection systems. These systems are utilized between operations to ensure that the dimensions of a part produced in one station are within acceptable tolerances before that part is transferred to the next station. Transfer machines also are used extensively in automated assembly, as described in Section 37.9.

The **transfer lines** or **flow lines** in a very large system for producing cylinder heads for engine blocks consisting of a number of transfer machines are shown in Fig. 37.5. This system is capable of producing 100 cylinder heads per hour. Note the various machining operations performed: milling, drilling, reaming, boring, tapping, honing, washing, and gaging.

37.2.5 Soft automation

We have seen that hard automation generally involves mass-production machines that lack flexibility. In *soft automation* (also called **flexible** or **programmable automation**), greater flexibility is achieved through the use of computer control of the machine and of its functions; thus, it can can produce parts having complex shapes. Soft automation is an important development, because the machine can be reprogrammed easily and readily to produce a part that has a shape or dimensions different from the one produced just prior to it. Further advances in flexible automation include the extensive use of modern computers leading to the development of **flexible manufacturing systems** (Section 39.3) with high levels of efficiency and productivity.

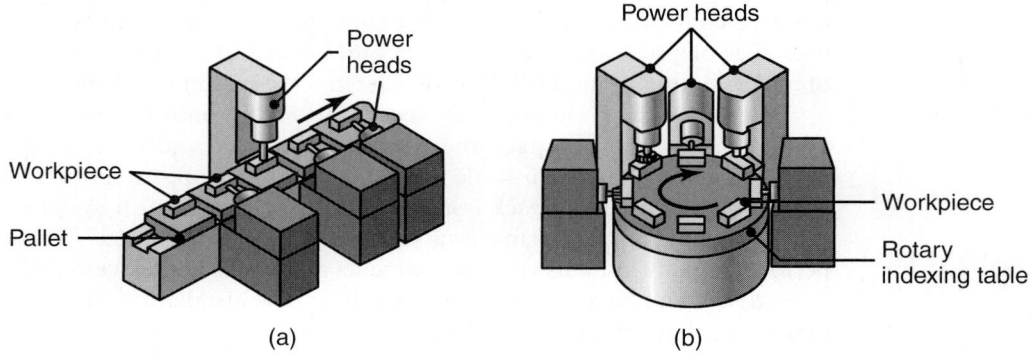

(a)   (b)

FIGURE 37.4 Two types of transfer mechanisms: (a) straight rails and (b) circular or rotary patterns.

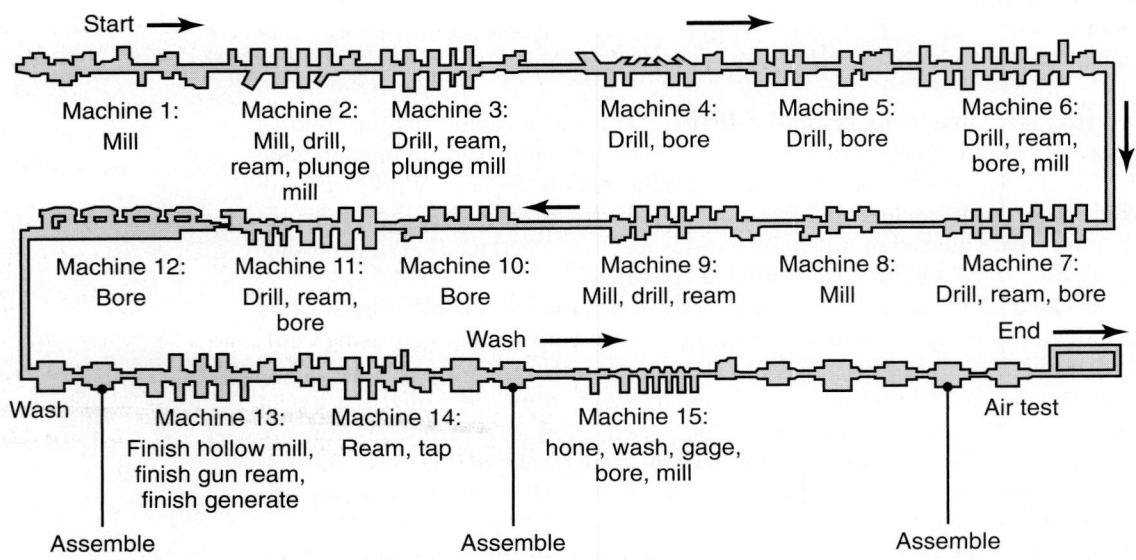

FIGURE 37.5   A large transfer line for producing engine blocks and cylinder heads. *Source:* Courtesy of Ford Motor Company.

37.2.6 Programmable logic controllers

The control of a manufacturing process in the proper sequence (especially one involving groups of machines and material-handling equipment) traditionally has been performed by timers, switches, relays, counters, and similar hardwired devices that are based on mechanical, electromechanical, and pneumatic principles. Beginning in 1968, *programmable logic controllers* (PLC) were introduced to replace these hardwired devices.

The programmable logic controller has been defined by the National Electrical Manufacturers Association (NEMA) as "a digitally operating electronic apparatus which uses a programmable memory for the internal storage of instructions for implementing specific functions such as logic, sequencing, timing, counting, and arithmetic to control, through digital or analog input/output modules, various types of machines or processes." A digital computer, which is used to control the functions of a programmable controller, is considered to be within this scope.

Because PLCs eliminate the need for relay control panels, can be reprogrammed, and take less space, they have been adopted widely in manufacturing systems and operations. Their basic functions are (a) on–off, (b) motion, (c) sequential operations, and (d) feedback control. PLCs also are used in system control with high-speed digital-processing and communication capabilities. These controllers perform reliably in industrial environments and improve the overall efficiency of the operation. PLCs have become less popular in new installations because of advances in numerical-control machines, but they still represent a very large installation base. Microcomputers are used extensively because they are less expensive than PLCs and are easier to program and to network.

37.2.7 Total productive maintenance

The management and maintenance of a wide variety of machines, equipment, and systems are among the important aspects affecting the productivity of a manufacturing organization. The concepts of *total productive maintenance* (TPM) and *total productive equipment management* (TPEM) include continued analysis of such factors as (a) equipment breakdown and equipment problems; (b) monitoring and improving equipment productivity; (c) implementation of preventive and predictive maintenance; (d) reduction of setup time, idle time, and cycle time; (e) full utilization of machinery and equipment and the improvement of their effectiveness; and (f) reduction of product defects. Teamwork is an important component of this activity and involves the full cooperation of the machine operators, maintenance personnel, engineers, and the management of the organization. (See also **kaizen**, Section 36.1.)

37.3 | Numerical Control

Numerical control (NC) is a method of controlling the movements of machine components by directly inserting coded instructions in the form of numbers and letters into the system. The system automatically interprets these data and converts them to output signals. These signals, in turn, control various machine components, such as turning spindles on and off, changing tools, moving the workpiece or the tools along specific paths, and turning cutting fluids on and off.

In order to appreciate the importance of the numerical control of machines, let's briefly review how a process such as machining traditionally has been carried out. After studying the working drawings of a part, the operator sets up the appropriate process parameters (such as cutting speed, feed, depth of cut, cutting fluid, and so on), determines the sequence of the machining operations to be performed, clamps the workpiece in a workholding device, and proceeds with the machining of the part.

Depending on part shape and on the dimensional accuracy specified, this approach usually requires skilled operators. The machining procedure may depend on the particular operator, and because of the possibilities of human error, even the parts produced by the same operator may not be identical. Therefore, part quality will depend on the particular operator or even on the same operator on a particular day of the week or hour of the day. Because of the need for improved product quality and reduced manufacturing costs, such variability in performance and its effects on product quality are no longer acceptable. This situation can be eliminated by the numerical control of the machining operation.

The importance of numerical control can be illustrated further by the following example. Assume that several holes are to be drilled on a part in the positions shown in Fig. 37.6.

In the traditional manual method of machining this part, the operator positions the drill bit with respect to the workpiece using reference points given by any of the three methods shown in the figure. The operator then proceeds to drill the holes. Let's first assume that 100 parts, all having exactly the same shape and dimensional accuracy, are to be drilled. Obviously, this operation is going to be tedious, because the operator has to go through the same motions repeatedly. Moreover, the probability is high that, for various reasons, some of the parts machined will be different from others.

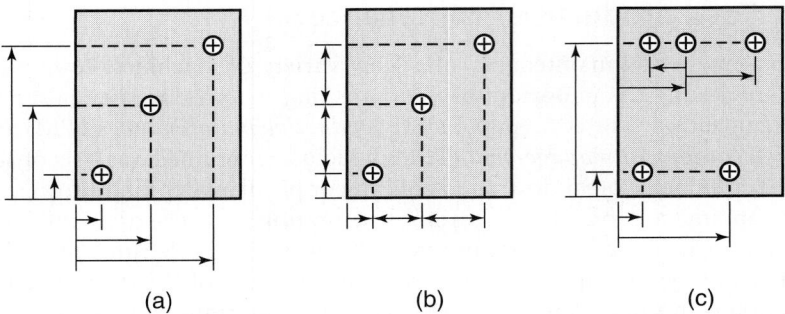

(a)   (b)   (c)

FIGURE 37.6   Positions of drilled holes in a workpiece. Three methods of measurements are shown: (a) absolute dimensioning referenced from one point at the lower left of the part; (b) incremental dimensioning made sequentially from one hole to another; and (c) mixed dimensioning—a combination of both methods.

Let's now assume that during this production run the order for these parts is changed, and ten of the parts now require holes in different positions. The machinist now has to reposition the work table; this operation will be time consuming and is subject to error. Such operations can be performed easily by numerical-control machines that are capable of producing parts repeatedly and accurately and of handling different parts by simply loading different part programs.

In operations under numerical control, data concerning all aspects of the machining operation (such as tool locations, speeds, feeds, and cutting fluids) are stored in hard disks. On the basis of input information, relays and other devices (*hard-wired controls*) can be actuated to obtain a desired machine setup. Complex operations (such as turning a part having various contours or die sinking in a milling machine) are carried out easily. NC machines are used extensively in small- and medium-quantity production (typically 500 parts or less) of a wide variety of parts both in small shops and in large manufacturing facilities.

EXAMPLE 37.1 Historical origin of numerical control

The basic concept behind numerical control apparently was implemented in the early 1800s when punched holes in sheet-metal cards were used to automatically control the movements of weaving machines. Needles were activated by sensing of the presence or absence of a hole in the card. This invention was followed by automatic piano players (*Pianola*), in which the keys were activated by air flowing through holes punched in a perforated roll of paper.

The principle of numerically controlling the movements of machine tools was first conceived in the 1940s by J. Parsons (1913 –) in his attempt to machine complex helicopter blades. The first prototype NC machine was built in 1952 at the Massachusetts Institute of Technology. It was a vertical-spindle, two-axis copy milling machine retrofitted with servomotors, and the machining operations performed consisted of end milling and face milling on a thick aluminum plate.

The numerical data to be punched into the paper tapes were generated by a digital computer, another invention which was being developed at the same time at MIT. In the experiments, parts were machined successfully, accurately, and repeatedly without operator intervention. On the basis of this success, the machine-tool industry began designing, building, and marketing NC machine tools. Later, these machines were equipped with computer-numerical controls (CNC) yielding greater flexibility, accuracy, versatility, and ease of operation. The latest developments are *machining centers*, the principles of which were described in Chapter 25.

37.3.1 Computer numerical control

In the next step in the development of numerical control, the control hardware (mounted on the NC machine) was converted to local computer control by software. Two types of computerized systems were developed: direct numerical control and computer numerical control.

- In **direct numerical control** (DNC), several machines are controlled directly—step by step—by a central mainframe computer. In this system, the operator has access to the central computer through a remote terminal. With DNC, the status of all machines in a manufacturing facility can be monitored and assessed from a central computer. However, DNC has a crucial disadvantage: If the computer shuts down, all of the machines become inoperative.

  A more recent definition of DNC is **distributed numerical control**, in which a central computer serves as the control system over a number of individual CNC machines having onboard microcomputers. This system provides large memory and computational facilities and offers flexibility while overcoming the disadvantage of direct numerical control.

- **Computer numerical control** (CNC) is a system in which a control microcomputer is an integral part of a machine (*onboard computer*). The machine operator can program onboard computers, modify the programs directly, prepare programs for different parts, and store the programs. CNC systems are used widely today because of the availability of (a) small computers with large memory, (b) low-cost programmable controllers and microprocessors, and (c) program-editing capabilities.

37.3.2 Principles of NC machines

The basic elements and operation of a typical NC machine are shown in Fig. 37.7. The functional elements in numerical control and the components involved are described here.

- **Data input:** The numerical information is read and stored in computer memory.
- **Data processing:** The programs are read into the machine control unit for processing.
- **Data output:** This information is translated into commands (typically pulsed commands) to the servomotor (Fig. 37.8). The servomotor then moves the work table (on which the workpiece is mounted) to specific positions through linear or rotary movements by means of stepping motors, leadscrews, and other similar devices.

Types of control circuits.   An NC machine can be controlled through two types of circuits: open-loop and closed-loop. In the **open-loop** system (Fig. 37.8a), the signals are sent to the servomotor by the controller, but the movements and

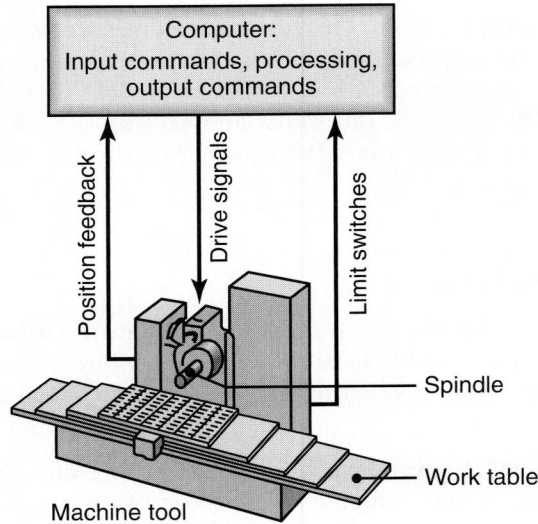

FIGURE 37.7   Schematic illustration of the major components for position control on a numerical-control machine tool.

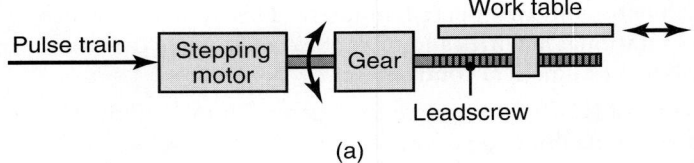

(a)

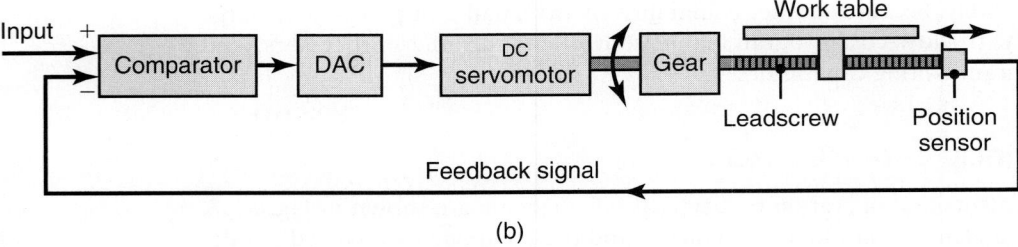

(b)

FIGURE 37.8   Schematic illustration of the components of (a) an open-loop and (b) a closed-loop control system for a numerical-control machine. DAC means "digital-to-analog converter."

final positions of the work table are not checked for accuracy. The **closed-loop** system (Fig. 37.8b) is equipped with various transducers, sensors, and counters that accurately measure the position of the work table. Through **feedback control**, the position of the work table is compared against the signal, and the table movements terminate when the proper coordinates are reached.

Position measurement in NC machines is carried out through indirect or direct methods (Fig. 37.9). In *indirect measuring systems*, *rotary encoders* or *resolvers* convert rotary movement to translation movement. However, backlash (the play between two adjacent mating gear teeth) can affect measurement accuracy significantly. Position feedback mechanisms utilize various sensors that are based mainly on

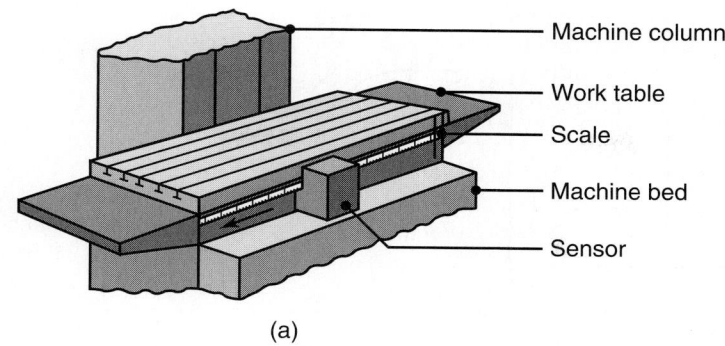

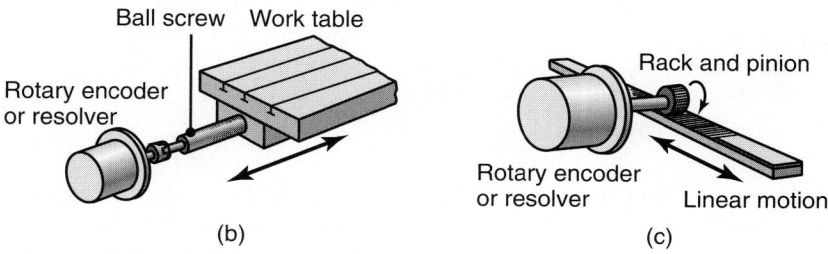

FIGURE 37.9 (a) Direct measurement of the linear displacement of a machine-tool work table. (b) and (c) Indirect measurement methods.

magnetic and photoelectric principles. In *direct measuring systems*, a sensing device reads a graduated scale on the machine table or slide for linear movement (Fig. 37.9c). This system is the more accurate because the scale is built into the machine and backlash in the mechanisms is not significant.

37.3.3 Types of control systems

There are two basic types of control systems in numerical control: point-to-point and contouring.

1. In a **point-to-point system** (also called *positioning system*) each axis of the machine is driven separately by leadscrews and at different velocities, depending on the type of operation. The machine moves initially at maximum velocity in order to reduce nonproductive time but decelerates as the tool approaches its numerically defined position. Thus, in an operation such as drilling a hole, the positioning and drilling take place sequentially (Fig. 37.10a).

   After the hole is drilled, the tool retracts upward and moves rapidly to another position, and the operation is repeated. The path followed from one position to another is important in only one respect: It must be chosen to minimize the time of travel for better efficiency. Point-to-point systems are used mainly in drilling, punching, and straight milling operations.

2. In a **contouring system** (also known as a *continuous path system*), both the positioning and the operations are performed along controlled paths but at different velocities. Because the tool acts as it travels along a prescribed path (Fig. 37.10b), accurate control and synchronization of velocities and movements are important. The contouring system typically is used on lathes, milling machines, grinders, welding machinery, and machining centers.

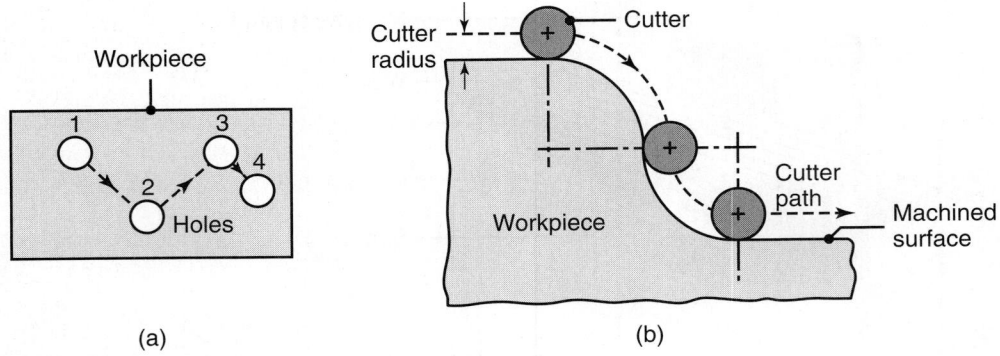

FIGURE 37.10 Movement of tools in numerical-control machining. (a) Point-to-point, in which the drill bit drills a hole at position 1, is retracted and moved to position 2 and so on. (b) Continuous path by a milling cutter. Note that the cutter path is compensated for by the cutter radius. This path also can be compensated for cutter wear.

Interpolation. Movement along the path (*interpolation*) occurs incrementally by one of several basic methods (Fig. 37.11). Examples of actual paths in drilling, boring, and milling operations are shown in Fig. 37.12. In all interpolations, the path controlled is that of the *center of rotation* of the tool. Compensation for different types of tools, for different diameters of tools, or for tool wear during machining can be made in the NC program.

- In **linear interpolation**, the tool moves in a straight line from start to end (Fig. 37.11a) along two or three axes. Theoretically, all types of profiles can be produced by this method by making the increments between the points small (Fig. 37.11b). However, a large amount of data has to be processed in order to do so.

- In **circular interpolation** (Fig. 37.11c), the inputs required for the path are the coordinates of the end points, the coordinates of the center of the circle and its radius, and the direction of the tool along the arc.

- In **parabolic interpolation** and **cubic interpolation**, the path is approximated by curves using higher-order mathematical equations. This method is effective in 5-axis machines and is useful in die sinking operations for the sheet-forming of automotive bodies. These interpolations also are used for the movements of industrial robots, as to discussed in Section 37.6.

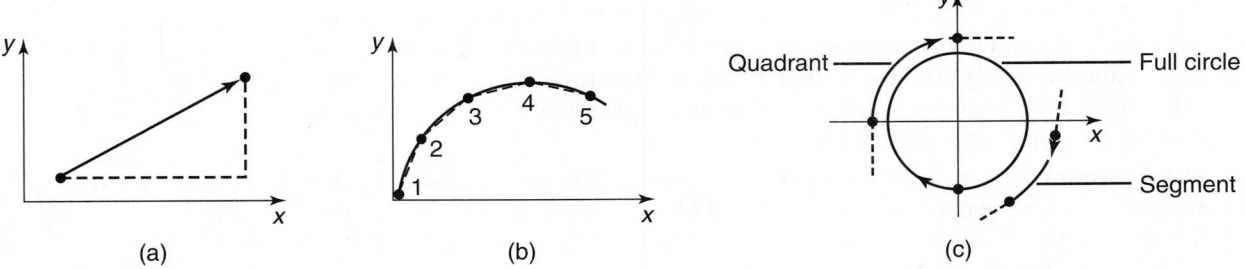

FIGURE 37.11 Types of interpolation in numerical control: (a) linear, (b) continuous path approximated by incremental straight lines, and (c) circular.

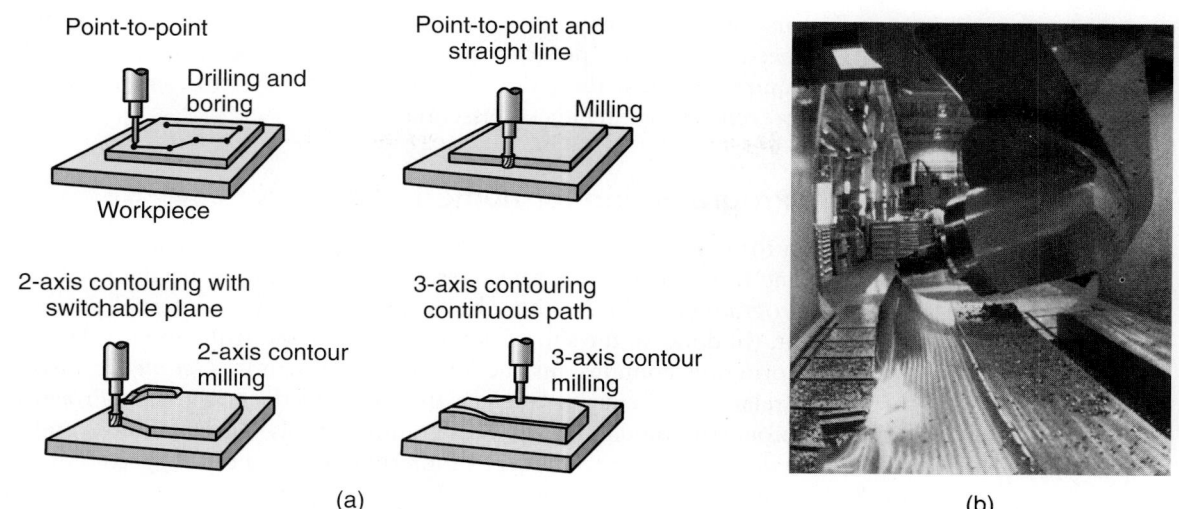

Point-to-point
Drilling and
boring
Workpiece

Point-to-point and
straight line
Milling

2-axis contouring with
switchable plane
2-axis contour
milling

3-axis contouring
continuous path
3-axis contour
milling

(a)

(b)

FIGURE 37.12   (a) Schematic illustration of drilling, boring, and milling with various paths. (b) Machining a sculptured surface on a 5-axis numerical-control machine. *Source:* Courtesy of The Ingersoll Milling Machine Co.

37.3.4 Accuracy in numerical control

Positioning accuracy in numerical-control machines is defined by how accurately the machine can be positioned to a certain coordinate system. *Repeat accuracy* is defined as the closeness of the agreement of the repeated position movements under the same operating conditions of the machine. *Resolution* (also called *sensitivity*) is the smallest increment of motion of the machine components.

The *stiffness* of the machine tool and the *backlash* in gear drives and leadscrews are important for achieving dimensional accuracy. Backlash in modern machines is eliminated by using pre-loaded ball screws. Also, rapid response to command signals requires that friction in machine slideways and inertia be minimized; the latter can be achieved by reducing the mass of moving components of the machine, such as using lightweight materials, inlcuding ceramics.

37.3.5 Advantages and limitations of numerical control

Numerical control has the following advantages over conventional methods of machine control:

- Flexibility of operation is improved, as well as the ability to produce complex shapes with good dimensional accuracy, good repeatability, reduced scrap loss, high production rates, productivity, and product quality.

- Tooling costs are reduced, because templates and other fixtures are not required.

- Machine adjustments are easy to make.

- More operations can be performed with each setup, and the lead time for setup and machining required is less when compared to conventional methods.

- Programs can be prepared rapidly, and they can be recalled at any time by utilizing microprocessors.

- Faster prototype production is possible.

- Operator skill required is less than that for a qualified machinist, and the operator has more time to attend to other tasks in the work area.

The major limitations of NC are the relatively high initial cost of the equipment, the need and cost for programming and computer time, and the special maintenance required. Because these machines are complex systems, breakdowns can be costly, so preventive maintenance is essential.

37.3.6 Programming for numerical control

A program for numerical control consists of a sequence of directions that causes an NC machine to carry out a certain operation, machining being the most common process. *Programming for NC* may be (a) performed by an internal programming department, (b) done on the shop floor, or (c) purchased from an outside source.

The program contains instructions and commands. *Geometric instructions* pertain to relative movements between the tool and the workpiece. *Processing instructions* concern spindle speeds, feeds, cutting tools, cutting fluids, and so on. *Travel instructions* pertain to the type of interpolation and to the speed of movement of the tool or the work table. *Switching instructions* concern the on–off position for coolant supplies, direction or lack of spindle rotation, tool changes, workpiece feeding, clamping, and so on.

Manual part programming consists first of calculating the dimensional relationships of the tool, workpiece, and work table on the basis of the engineering drawings of the part (including CAD), the manufacturing operations to be performed, and their sequence. A program sheet then is prepared, detailing the necessary information to carry out the particular operation. The part program then is prepared on the basis of this information. However, the work involved is tedious, time-consuming, and uneconomical. Consequently, manual programming is used mostly in simple point-to-point applications. With the wider availability of manufacturing software, manual part programming is becoming less common.

Computer-aided part programming involves special symbolic *programming languages* that determine the coordinate points of corners, edges, and surfaces of the part. A **programming language** is a means of communicating with the computer; it involves the use of symbolic characters. The programmer describes in this language the component to be processed, and the computer converts that description to commands for the NC machine. Several languages are available commercially.

Complex parts are machined using graphics-based, computer-aided machining programs. A tool path is created in a largely graphic environment that is similar to a CAD program. The machine code (**G-Code**) is created automatically by the program. This code is valuable to communicating machining instructions to the CNC hardware, but it is difficult to edit and troubleshoot without software interpreters. When minor difficulties are encountered, such as (a) using different end-mill diameters than originally programmed or (b) changing the cutting speed to avoid chatter, the machine operator needs to modify the program, which is difficult using G-Code.

In *shop-floor programming*, CNC programming software is used directly on the machine-tool controller. This allows higher-level geometry and processing information to be sent to the CNC controller instead of G-code. G-code is then developed by the dedicated computer under control of the machine operator. The advantage is that any changes that are made to the machining program are sent back to the programming group, are stored as a shop-proven design iteration, and can be re-used or standardized across parts. Before production begins, the programs must be verified, either by viewing a simulation of the process on a monitor or by first making the part from an inexpensive material, such as aluminum, wood, wax, or plastic.

37.4 | Adaptive Control

Adaptive control (AC) is basically a dynamic-feedback system in which the operating parameters automatically adapt themselves to conform to new circumstances. Human reactions to occurrences in everyday life already contain dynamic-feedback control. For example, driving a car on a smooth road is relatively easy, and we need to make few (if any) adjustments. However, on a rough road, we may have to steer to avoid potholes by visually and continuously observing the condition of the road. Also, our body feels the car's rough movements and vibrations; we then react by changing the direction and/or the speed of the car to minimize the effects of the rough road and to increase the comfort of the ride.

An **adaptive controller** continuously checks road conditions, adapts an appropriate desired braking profile (for example, an antilock brake system and traction control), and then uses feedback to implement it. Adaptive control is a logical extension of computer numerical-control systems. As described in Section 37.3, the part programmer sets the processing parameters based on the existing knowledge of the workpiece material and various data on the particular manufacturing process. In CNC machines, these parameters are held constant during a particular process cycle. In AC, on the other hand, the system is capable of automatic adjustments *during* processing through closed-loop feedback control (Fig. 37.13). Several adaptive-control systems are available commercially for a variety of applications.

Adaptive control in manufacturing.   The application of AC in manufacturing is important particularly in situations where workpiece dimensions and quality are not uniform, such as in a poor casting or an improperly heat-treated part. The major purposes of adaptive control in manufacturing are to (a) optimize production rate, (b) optimize product quality, and (c) minimize production cost.

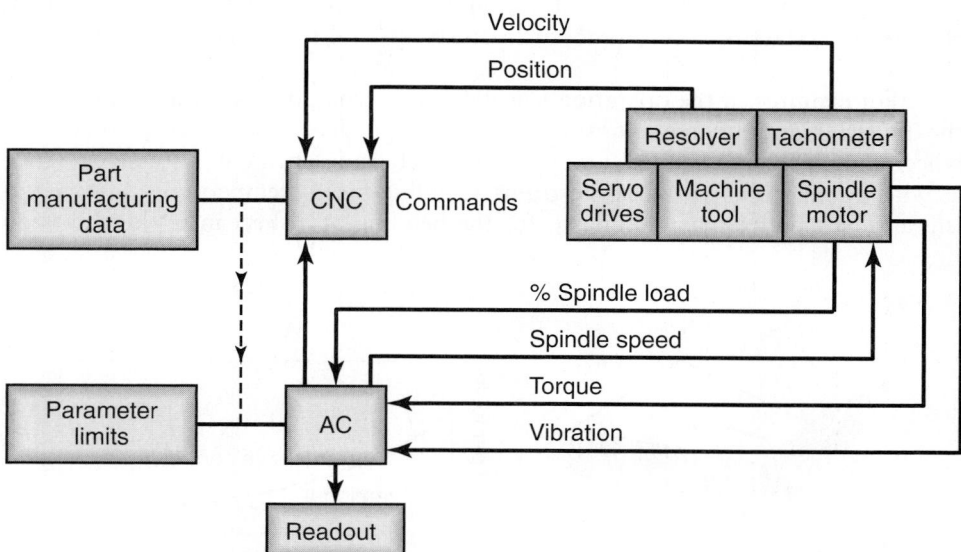

FIGURE 37.13   Schematic illustration of the application of adaptive control (AC) for a turning operation. The system monitors such parameters as cutting force, torque, and vibrations. If these parameters are excessive, it modifies process variables (such as feed and depth of cut) to bring them back to acceptable levels.

As an example of adaptive control, let's consider a machining operation such as turning on a lathe. The adaptive control system senses parameters *in real time*, such as cutting forces, spindle torque, temperature, tool-wear rate, tool condition, and surface finish of the workpiece. The system converts this information into commands that then modify the process parameters on the machine tool to hold them constant, hold them within certain limits, or to optimize the machining operation.

Those systems that place a constraint on a process variable (such as force, torque, or temperature) are called *adaptive control constraint* (ACC) systems. Thus, if the thrust force and the cutting force (and hence the torque) increase excessively (such as because of the presence of a hard region in a cast workpiece), the system changes the speed or the feed in order to lower the cutting force to an acceptable level (Fig. 37.14). Without adaptive control or the direct intervention of the operator (as is the case in traditional machining operations), high cutting forces may cause tool failure or may cause the workpiece to deflect or distort excessively. As a result, workpiece dimensional accuracy and surface finish begin to deteriorate. Those systems that optimize an operation are called *adaptive control optimization* (ACO) systems. Optimization may involve maximizing the material-removal rate between tool changes or improving surface finish.

Response time must be short for AC to be effective, particularly in high-speed machining operations (see Section 25.5). Assume, for example, that a turning operation is being performed on a lathe at a spindle speed of 1000 rpm, and the tool suddenly breaks, thus adversely affecting the surface finish and dimensional accuracy of the part. In order for the AC system to be effective, the sensing system must respond within a very short time, otherwise the damage to the workpiece will be extensive.

For adaptive control to be effective in manufacturing operations, quantitative relationships must be established and stored in the computer software as mathematical models. For example, if the tool-wear rate in a machining operation is excessive, the computer must be able to decide how much of a change in speed or feed is necessary (and whether to increase it or decrease it) in order to reduce the wear rate to an acceptable level. The system also should be able to compensate for dimensional changes in the workpiece due to such causes as tool wear and temperature rise (Fig. 37.15).

For example, if the operation is grinding, the computer software must reflect the desired quantitative relationships among independent process variables (such as wheel and work speeds, feed, and type of wheel) and dependent parameters (such as wheel wear, dulling of abrasive grains, grinding forces, temperature, surface finish, and part deflections). Similarly, for the bending of a sheet in a V-die, data on

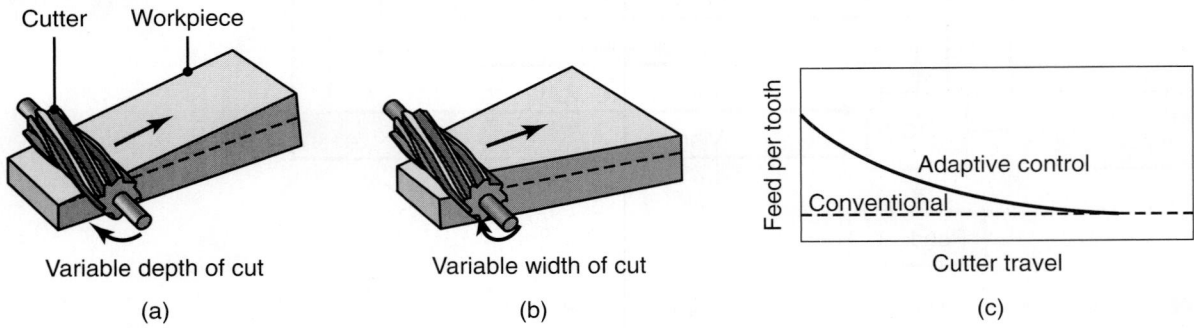

FIGURE 37.14   An example of adaptive control in milling. As the depth of cut (a) or the width of cut (b) increases, the cutting forces and the torque increase. The system senses this increase and automatically reduces the feed (c) to avoid excessive forces or tool breakage in order to maintain cutting efficiency. *Source:* After Y. Koren.

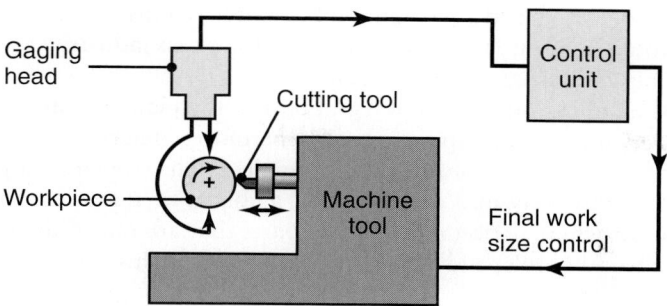

FIGURE 37.15   In-process inspection of workpiece diameter in a turning operation. The system automatically adjusts the radial position of the cutting tool in order to produce the correct diameter.

the dependence of springback on punch travel and on other material and process variables must be stored in the computer memory.

Compared to the other parameters involved in manufacturing operations, cutting forces and spindle torque in machining have been found to be the easiest to monitor by adaptive control. Various solid-state power controls are available commercially, in which power is displayed or interfaced with data-acquisition systems. Coupled with CNC, adaptive control is a powerful tool in optimizing manufacturing operations.

37.5 | Material Handling and Movement

During a typical manufacturing operation, raw materials and parts in progress are moved from storage to machines, from machine to machine, from inspection to assembly and inventory, and finally to shipment. For example, (a) workpieces are loaded on machines, such as a forging mounted on a milling-machine bed for further processing, (b) sheet metal is fed into a press for stamping, (c) parts are removed from one machine and loaded onto another (such as a machined forging to be subsequently ground for better surface finish and dimensional accuracy), and (d) finished parts are assembled into a finished product.

Similarly, tools, molds, dies, and various other equipment and fixtures also are moved around in manufacturing plants. Cutting tools are mounted on lathes, dies are placed in presses or hammers, grinding wheels are mounted on spindles, and parts are mounted on special fixtures for dimensional measurement and inspection.

These materials must be moved either manually or by some mechanical means, and it takes time to transport them from one location to another. **Material handling** is defined as the functions and systems associated with the transportation, storage, and control of materials and parts in the total manufacturing cycle of a product. The total time required for actual manufacturing depends on part size and shape and on the set of operations to be performed. It also should be noted that idle time and the time required for transporting materials can constitute the majority of the time consumed.

Plant layout is an important aspect of the orderly flow of materials and components throughout the manufacturing cycle. The time and distances required for moving raw materials and parts should be minimized, and storage areas and service centers should be organized accordingly. For parts requiring multiple operations, equipment should be grouped around the operator or an industrial robot or robots (see *cellular manufacturing*, Section 39.2).

Material handling must be an integral part of the planning, implementing, and controlling of manufacturing operations. Furthermore, material handling should be repeatable and predictable. Consider, for example, what happens if a part or workpiece is loaded improperly into a forging die, a chuck, or the collet on a lathe. The consequences of such an action may well be parts that are out of dimensional tolerance, broken dies and tools, and improperly made parts. This action also can present safety hazards.

Methods of material handling.   Several factors have to be considered in selecting a suitable material-handling method for a particular manufacturing operation:

1. Shape, weight, and characteristics of the parts.
2. Types and distances of movements, and the position and orientation of the parts during movement and at their final destination.
3. Conditions of the path along which the parts are to be transported.
4. Degree of automation, the level of control desired, and integration with other systems and equipment.
5. Operator skill required.
6. Economic considerations.

For small-batch manufacturing operations, raw materials and parts can be handled and transported by hand, but this method is generally costly. Moreover, because it involves humans, this practice can be unpredictable and unreliable. It even can be unsafe for the operator because of the weight and shape of the parts to be moved and environmental factors (such as heat and smoke) in foundries and forging plants. In automated manufacturing plants, computer-controlled material and parts flow are implemented commonly. These changes have resulted in improved repeatability and, of course, in lower labor costs.

Equipment.   Various types of equipment can be used to move materials, such as conveyors, rollers, self-powered monorails, carts, forklift trucks, and various mechanical, electrical, magnetic, pneumatic, and hydraulic devices and manipulators. *Manipulators* can be designed to be controlled directly by the operator, or they can be automated for repeated operations (such as the loading and unloading of parts from machine tools, presses, and furnaces). Manipulators are capable of gripping and moving heavy parts and of orienting them as required between the manufacturing and assembly operations. Machines often are used in a sequence, so workpieces are transferred directly from machine to machine. Machinery combinations having the capability of conveying parts without the use of additional material-handling apparatus are called *integral transfer devices*.

Flexible material handling and movement with real-time control has become an integral part of modern manufacturing. Industrial robots, specially designed pallets, and **automated guided vehicles** (AGVs) are used extensively in flexible manufacturing systems to move parts and to orient them as required (Fig. 37.16).

Automated guided vehicles (which are the latest development in material movement in plants) operate automatically along pathways with in-floor wiring (or tapes for optical scanning) without operator intervention. This transport system has high flexibility and is capable of random delivery to different workstations. It optimizes the movement of materials and parts in cases of congestion around workstations, machine breakdown (downtime), or the failure of one section of the system.

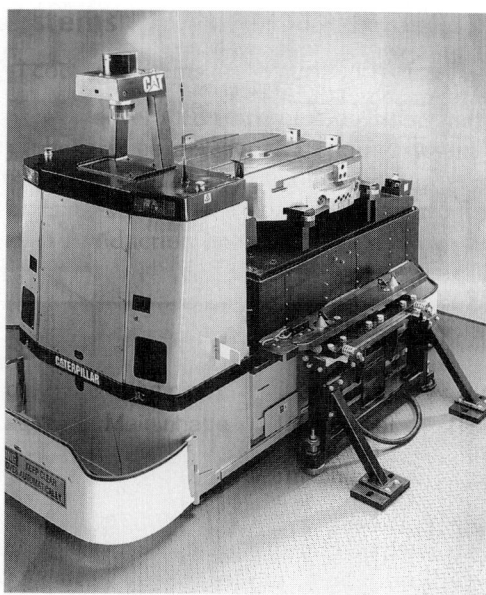

FIGURE 37.16 A self-guided vehicle (Caterpillar Model SGC-M) carrying a machining pallet. The vehicle is aligned next to a stand on the floor. Instead of following a wire or stripe path on the factory floor, this vehicle calculates its own path and automatically corrects for any deviations. *Source*: Courtesy of Caterpillar Industrial, Inc.

The movements of AGVs are planned so that they interface with **automated storage/retrieval systems** (AS/RS) in order to utilize warehouse space efficiently and to reduce labor costs. Nowadays, however, these systems are considered undesirable because of the current focus on minimal inventory and on *just-in-time* production methods. (See Section 39.5.)

Coding systems that have been developed to locate and identify parts throughout the manufacturing system and to transfer them to their appropriate stations are outlined here:

- **Bar coding** is the most widely used system and the least costly.
- **Magnetic strips** constitute the second most common coding system.
- **RF** (radio frequency) **tags** are popular. Although expensive, they do not need the clear line of sight needed by the previous two systems, they have a long range, and they are rewritable.
- Other identification systems are based on **acoustic waves, optical character recognition**, and **machine vision**.

37.6 | Industrial Robots

The word **robot** was coined in 1920 by the Czech author K. Čapek in his play *R.U.R.* (Rossum's Universal Robots). It is derived from the Czech word *robota*, meaning worker. An **industrial robot** has been described by the International Organization for Standardization (ISO) as "a machine formed by a mechanism including several degrees of freedom, often having the appearance of one or several arms ending in a wrist capable of holding a tool, a workpiece, or an inspection device." In particular, its control unit must use a memorizing device, and it sometimes may use sensing or adaptation appliances to take into account environment and circumstances. These multipurpose machines generally are designed to carry out a repetitive function and can be adapted to other operations.

The first industrial robots were introduced in the early 1960s. Computer-controlled robots were commercialized in the early 1970s, and the first robot controlled by a microcomputer appeared in 1974. Industrial robots first were used in hazardous operations (such as for the handling of toxic and radioactive materials) and the loading and unloading of hot workpieces from furnaces and in foundries. Simple, rule-of-thumb applications for robots are the "three D's" (*dull, dirty,* and *dangerous,* including *demeaning* but necessary tasks) and the three H's (*hot, heavy,* and *hazardous*).

Industrial robots have become important components in manufacturing processes and systems. They have helped improve productivity and product quality, and have reduced labor costs significantly. Some modern robots are *anthropomorphic,* meaning that they resemble humans in shape and in movement. These complex mechanisms are made possible by powerful computer processors and fast motors that can maintain a robot's balance and accurate movement control.

37.6.1 Robot components

To appreciate the functions of robot components and their capabilities, we simultaneously might observe the flexibility and capability of diverse movements of our own arm, wrist, hand, and fingers in reaching for and grabbing an object from a shelf, in using a hand tool, or in operating a car or a machine. Described next are the basic components of an industrial robot (Fig. 37.17).

Manipulator.   Also called **arm and wrist,** the *manipulator* is a mechanical unit that provides motions (trajectories) similar to those of a human arm and hand. The end of the wrist can reach a point in space having a specific set of coordinates and in a specific orientation. Most robots have six rotational joints. Seven degrees of freedom or *redundant* robots for special applications also are available.

End effector.   The end of the wrist in a robot is equipped with an *end effector,* which also is called *end-of-arm tooling.* Depending on the type of operation, conventional end effectors may be equipped with any of the following (Fig. 37.18):

* Grippers, hooks, scoops, electromagnets, vacuum cups, and adhesive fingers for material handling
* Spray guns for painting
* Attachments for spot and arc welding and for arc cutting
* Power tools (such as drills, nut drivers, and burrs)
* Measuring instruments

End effectors generally are custom-made to meet specific handling requirements. Mechanical grippers are used the most commonly and are equipped with two or more fingers. *Compliant end effectors* are used to handle fragile materials or to facilitate assembly. These end effectors can use elastic mechanisms to limit the force which can be applied to the workpiece, or they can be designed with a desired stiffness.

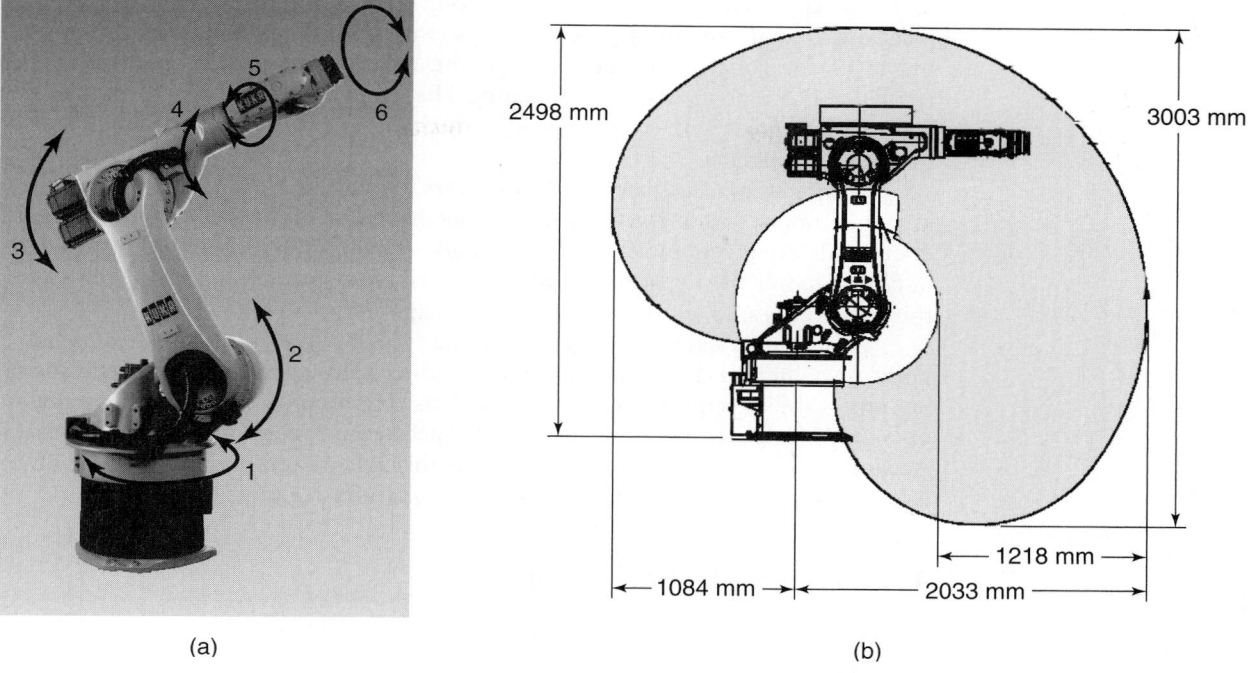

(a)   (b)

FIGURE 37.17   (a) Schematic illustration of a 6-axis KR-30 KUKA robot. The payload at the wrist is 30 kg and repeatability is ±0.15 mm (±0.006 in.). The robot has mechanical brakes on all of its axes, which are coupled directly. (b) The work envelope of the robot, as viewed from the side. *Source:* Courtesy of KUKA Robotics Corp.

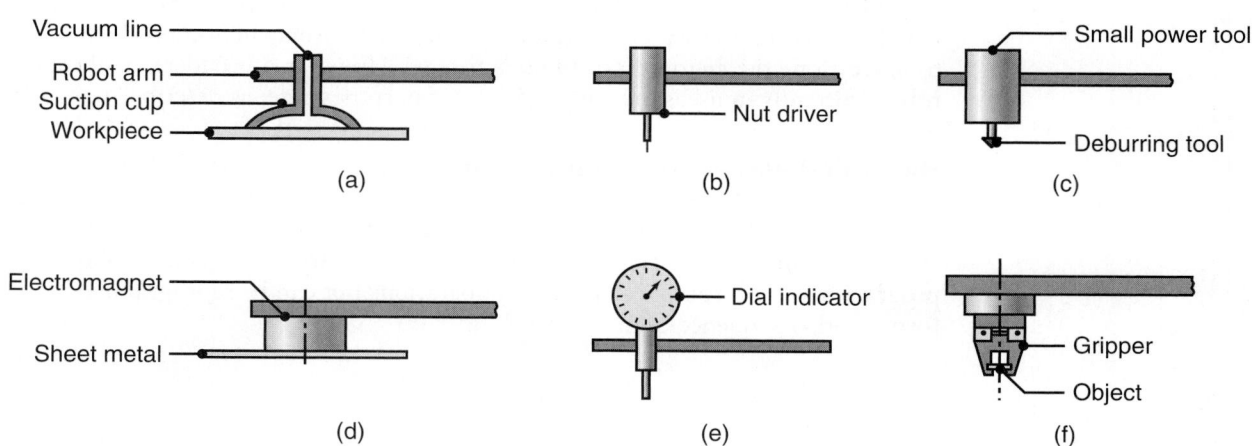

FIGURE 37.18   Types of devices and tools attached to end effectors to perform a variety of operations.

The selection of an appropriate end effector for a specific application depends on such factors as the payload, environment, reliability, and cost.

Power supply.   Each motion of the manipulator (linear and rotational) is controlled and regulated by independent actuators that use an electrical, a pneumatic, or a hydraulic power supply. Each source of energy and each type of motor has its own characteristics, advantages, and limitations.

Control system.   Also known as the *controller*, the control system is the communications and information-processing system that gives commands for the movements of the robat. It is the brain of the robot and stores data to initiate and terminate movements of the manipulator. The control system is also the *nerves* of the robot; it interfaces with computers and other equipment, such as manufacturing cells or assembly systems.

Feedback devices (such as transducers) are an important part of the control system. Robots with a fixed set of motions have *open-loop control*. In this system, commands are given and the robot arm goes through its motions. Unlike feedback in *closed-loop systems*, accuracy of the movements is not monitored. Consequently, this system does not have a self-correcting capability.

As in numerical control machines, the types of control in industrial robots are point-to-point and continuous-path. Depending on the particular task, the positioning repeatability required may be as small as 0.050 mm (0.002 in.), as in assembly operations for electronic printed circuitry. Specialized robots can reach such accuracy. Accuracy and repeatability vary greatly with payload and with position within the work envelope and, as such, are difficult to quantify for most robots.

37.6.2 Classification of robots

Robots may be classified by basic type, as follows (Fig. 37.19):

   a. Cartesian or rectilinear
   b. Cylindrical
   c. Spherical or polar
   d. Articulated, revolute, jointed, or anthropomorphic

Robots may be attached permanently to the floor of a manufacturing plant, they may move along overhead rails (*gantry robots*), or they may be equipped with wheels to move along the factory floor (*mobile robots*). However, a broader classification of robots currently in use is most helpful for our purposes here, as described next.

Fixed- and variable-sequence robots.   The fixed-sequence robot (also called a *pick-and-place robot*) is programmed for a specific sequence of operations. Its movements are from point to point, and the cycle is repeated continuously. These robots are simple and relatively inexpensive. The variable-sequence robot can be programmed for a specific sequence of operations but can be reprogrammed to perform another sequence of operations.

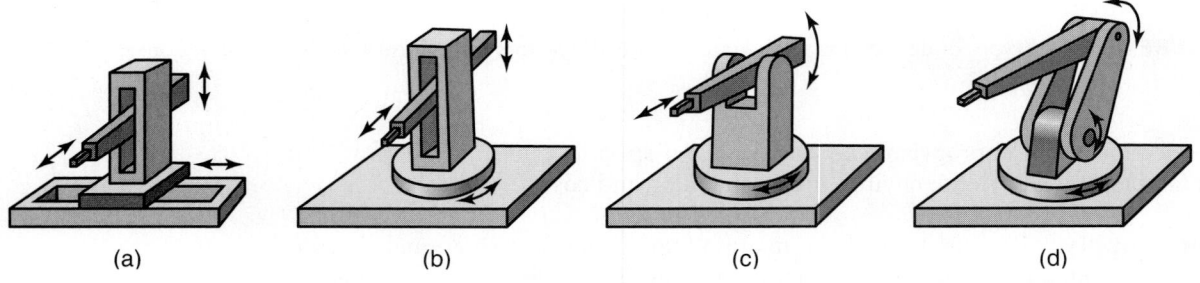

| (a) | (b) | (c) | (d) |

FIGURE 37.19   Four types of industrial robots: (a) cartesian (rectilinear), (b) cylindrical, (c) spherical (polar), and (d) articulated (revolute, jointed, or anthropomorphic).

Playback robot.   An operator leads or walks the playback robot and its end effector through the desired path; in other words, the operator teaches the robot by showing it what to do. The robot records the path and sequence of motions and can repeat them continually without any further action or guidance by the operator. Another type is the **teach pendant**, which utilizes hand-held button boxes that are connected to the control panel; they are used to control and guide the robot and its tooling through the work to be performed. These movements then are registered in the memory of the controller and are reenacted automatically by the robot whenever needed.

Numerically controlled robot.   The numerically controlled robot is programmed and operated much like a numerically controlled machine. The robot is servocontrolled by digital data, and its sequence of movements can be changed with relative ease. As in NC machines, there are two basic types of controls: point-to-point and continuous-path.

Point-to-point robots are easy to program and have a higher load-carrying capacity and a larger **work envelope**, which is the maximum extent or reach of the robot hand or working tool in all directions (Fig. 37.20). Continuous-path robots have greater accuracy than point-to-point robots, but they have lower load-carrying capacity. Advanced robots have a complex system of path control, enabling high-speed movements with great accuracy.

Intelligent (Sensory) robot.   The intelligent robot is capable of performing some of the functions and tasks carried out by humans. It is equipped with a variety of sensors with *visual (computer vision)* and *tactile* capabilities (see Section 37.7). Much like humans, the robot observes and evaluates the immediate environment and its proximity to other objects, especially machinery, by perception and pattern recognition. It then makes appropriate decisions for the next movement and proceeds accordingly. Because its operation is very complex, powerful computers are required to control this type of robot.

Developments in intelligent robots include:

- Behaving more like humans, performing tasks such as moving among a variety of machines and equipment on the shop floor and avoiding collisions.

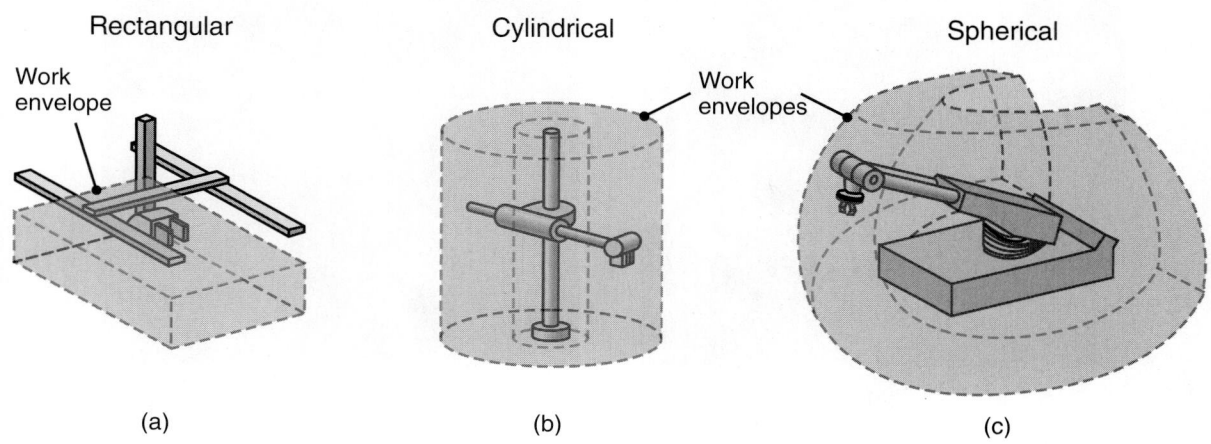

FIGURE 37.20   Work envelopes for three types of robots. The choice depends on the particular application. (See also Fig. 37.17b).

- Recognizing, selecting, and properly gripping the correct raw material or workpiece.
- Transporting the part to a machine for further processing or inspection.
- Assembling the components into subassemblies or a final product.

37.6.3 Applications and selection of robots

Major applications of industrial robots include the following:

- Material handling consists of the loading, unloading, and transferring of workpieces in manufacturing facilities. These operations can be performed reliably and repeatedly with robots, thereby improving quality and reducing scrap losses. Some examples are: (a) casting and molding operations in which molten metal, raw materials, lubricants, and parts in various stages of completion are handled without operator interference; (b) heat-treating operations in which parts are loaded and unloaded from furnaces and quench baths; (c) forming operations in which parts are loaded and unloaded from presses and various other types of metalworking machinery.
- Spot welding unitizes automobile and truck bodies, producing welds of good quality (Fig. 37.21a). Robots also perform other similar operations, such as arc welding, arc cutting, and riveting.
- Operations such as deburring, grinding, and polishing can be done by using appropriate tools attached to the end effectors.
- Applying adhesives and sealants, such as in the automobile frame shown in Fig. 37.21b.
- Spray painting (particularly of complex shapes) and cleaning operations are frequent applications because the motions for one piece are repeated accurately for the next piece.
- Automated assembly (Fig. 37.22).
- Inspection and gaging at speeds much higher than those that can be achieved by humans.

(a)   (b)

FIGURE 37.21   Examples of industrial robot applications. (a) Spot welding automobile bodies with industrial robots. (b) Sealing joints of an automobile body with an industrial robot. *Source:* Courtesy of Cincinnati Milacron, Inc.

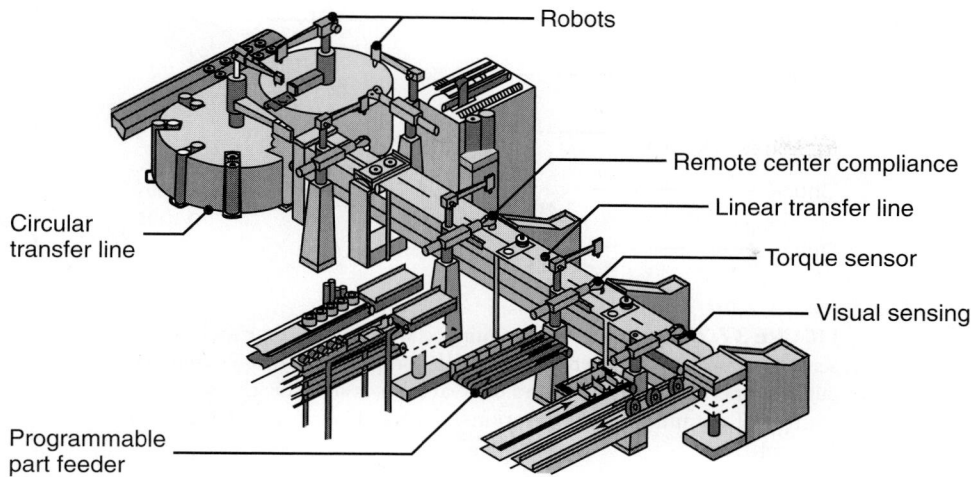

FIGURE 37.22 Automated assembly operations using industrial robots and circular and linear transfer lines.

Robot selection. Factors that influence the selection of robots in manufacturing are as follows:

- Load-carrying capacity
- Speed of movement
- Reliability
- Repeatability
- Arm configuration
- Degrees of freedom
- The control system
- Program memory
- Work envelope (see Fig. 37.17b)

Economics. In addition to the technical factors, cost and benefit considerations also are significant aspects of robot selection and their use. The increasing availability and reliability and the reduced costs of sophisticated, intelligent robots are having a major economic impact on manufacturing operations. Such robots gradually are displacing human labor.

Robot safety. Depending on the size of the robot's work envelope, speed, and proximity to humans, safety considerations in a robot environment are important, particularly for programmers and maintenance personnel who are in direct physical interaction with robots. In addition, the movement of the robot with respect to other machinery requires a high level of reliability in order to avoid collisions and damage to equipment. Its material-handling activities require the proper securing of raw materials and parts in the robot gripper at various stages in the production line.

37.7 | Sensor Technology

A *sensor* is a device which produces a signal in response to its detecting or measuring a property, such as position, force, torque, pressure, temperature, humidity, speed,

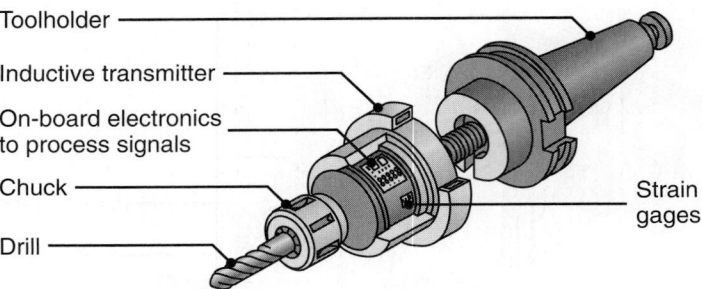

FIGURE 37.23   A toolholder equipped with thrust-force and torque sensors (smart toolholder), capable of continuously monitoring the cutting operation. Such toolholders are necessary for the adaptive control of manufacturing operations. *Source:* Courtesy of Cincinnati Milacron, Inc.

acceleration, or vibration. Traditionally, sensors (such as actuators and switches) have been used to set limits on the performance of machines. Common examples are (a) stops on machine tools to restrict work table movements, (b) pressure and temperature gages with automatic shut-off features, and (c) governors on engines to prevent excessive speed of operation. Sensor technology has become an important aspect of manufacturing processes and systems. It is essential for proper data acquisition and for the monitoring, communication, and computer control of machines and systems (Fig. 37.23).

Because they convert one quantity to another, sensors often are referred to as *transducers. Analog sensors* produce a signal, such as voltage, that is proportional to the measured quantity. *Digital sensors* have numeric or digital outputs that can be transferred to computers directly. **Analog-to-digital converters** (ADC) are available for interfacing analog sensors with computers.

37.7.1 Sensor classification

Sensors that are of interest in manufacturing may be classified generally as follows.

- **Mechanical** sensors measure such quantities as position, shape, velocity, force, torque, pressure, vibration, strain, and mass.
- **Electrical** sensors measure voltage, current, charge, and conductivity.
- **Magnetic** sensors measure magnetic field, flux, and permeability.
- **Thermal** sensors measure temperature, flux, conductivity, and specific heat.
- Other types are *acoustic, ultrasonic, chemical, optical, radiation, laser, and fiber-optic.*

Depending on its application, a sensor may consist of metallic, nonmetallic, organic, or inorganic materials, as well as fluids, gases, plasmas, or semiconductors. Using the special characteristics of these materials, sensors convert the quantity or property measured to analog or digital output. The operation of an ordinary mercury thermometer, for example, is based on the difference between the thermal expansion of mercury and that of glass.

Similarly, a machine part, a physical obstruction, or barrier in a space can be detected by breaking the beam of light when sensed by a photoelectric cell. A *proximity sensor* (which senses and measures the distance between it and an object or a moving member of a machine) can be based on acoustics, magnetism, capacitance, or optics. Other actuators contact the object and take appropriate action (usually by electromechanical means). Sensors are essential to the control of intelligent robots, and are being developed with capabilities that resemble those of humans (*smart sensors*, see the following).

Tactile sensing. Tactile sensing is the continuous sensing of variable contact forces, commonly by an array of sensors. Such a system is capable of performing within an arbitrary three-dimensional space. Fragile parts (such as glass bottles and electronic devices) can be handled by robots with *compliant (smart) end effectors* (Fig. 37.24). These effectors can sense the force applied to the object being handled using piezoelectric devices, strain gages, magnetic induction, ultrasonics, and optical systems of fiber optics and light-emitting diodes.

The force sensed is monitored and controlled through closed-loop feedback devices. Compliant grippers having force feedback and sensory perception are complicated and require powerful computers. *Anthropomorphic end effectors* are designed to simulate the human hand and fingers and to have the capability of sensing touch, force, and movement. The ideal tactile sensor must also sense *slip*—a capability of human fingers and hand that we tend to take for granted.

Visual sensing. *(machine vision, computer vision).* In visual sensing, cameras optically sense the presence and shape of the object (Fig. 37.25). A microprocessor then processes the image (usually in less than one second), the image is measured, and the measurements are digitized (**image recognition**). There are two basic systems of machine vision: linear array and matrix array.

FIGURE 37.24 A robot gripper with tactile sensors. In spite of their capabilities, tactile sensors are used less frequently because of their high cost and their low durability in industrial environments. *Source:* Courtesy of Lord Corporation.

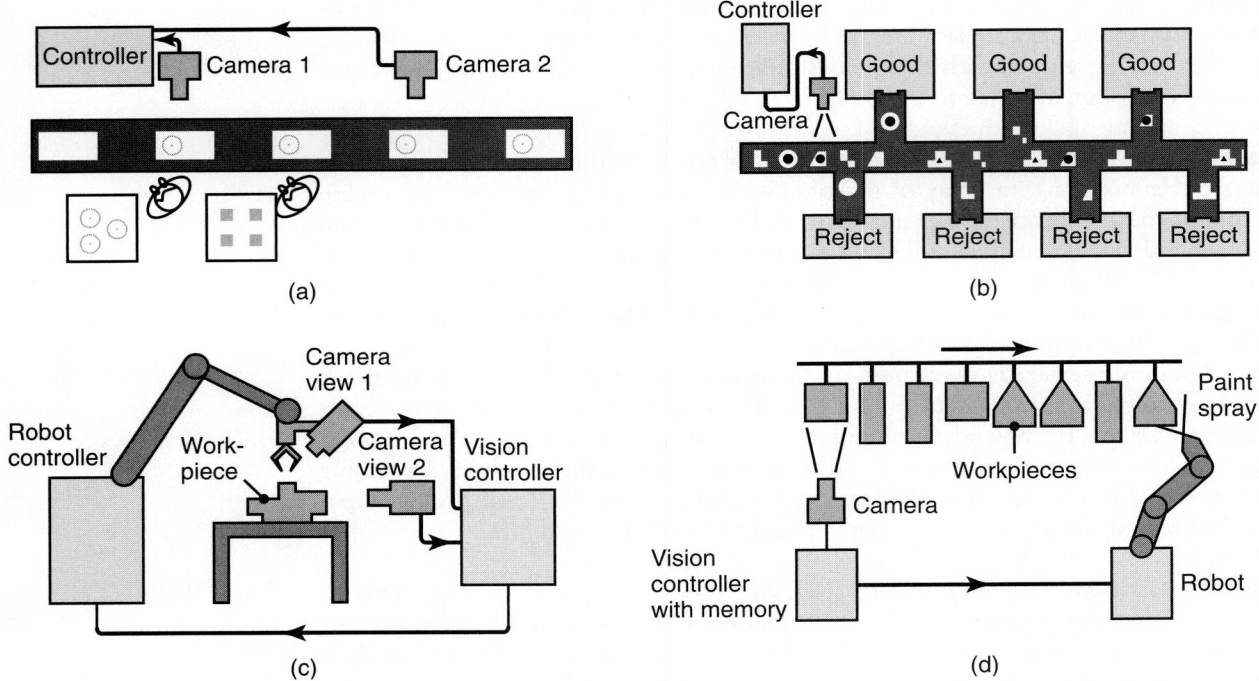

FIGURE 37.25   Examples of machine-vision applications. (a) In-line inspection of parts. (b) Identification of parts with various shapes and inspection and rejection of defective parts. (c) Use of cameras to provide positional input to a robot relative to the workpiece. (d) Painting parts having different shapes by means of input from a camera. The system's memory allows the robot to identify the particular shape to be painted and to proceed with the correct movements of a paint spray attached to the end effector.

In *linear array*, only one dimension is sensed, such as the presence of an object or some feature on its surface. *Matrix arrays* sense two or even three dimensions and are capable of detecting, for example, a properly inserted component in a printed circuit or a properly made solder joint (assembly verification). When used in automated inspection systems, these sensors also can detect cracks and flaws.

Machine vision is suitable particularly for inaccessible parts, in hostile manufacturing environments, for measuring a large number of small features, and in situations where physical contact with the part may cause damage. Applications of machine vision include (a) on-line, real-time inspection in sheet-metal stamping lines and (b) sensors for machine tools that can sense tool offset and tool breakage, verify part placement and fixturing, and monitor surface finish.

Machine vision is capable of in-line identification and inspection of parts and of rejecting defective ones. Several applications of machine vision in manufacturing are shown in Fig. 37.25. With visual sensing capabilities, end effectors are able to pick up parts and grip them in the proper orientation and location.

The selection of a sensor for a particular application depends on such factors as (a) the particular quantity to be measured or sensed, (b) the sensor's interaction with other components in the system, (c) expected service life, (d) required level of sophistication, (e) difficulties associated with its use, (f) power source, and (g) cost. A further consideration in sensor selection is the environment in which the sensors are to be used. *Robust sensors* have been developed to withstand extremes of temperature, shock and vibration, humidity, corrosion, dust and various contaminants, fluids, electromagnetic radiation, and other interferences.

Smart sensors.   These sensors have the capability to perform a logic function, to conduct two-way communication, and to make decisions and take appropriate actions. The necessary input and the knowledge required to make a decision can be built into a smart sensor. For example, a computer chip with sensors can be programmed to turn a machine tool off when a cutting tool fails. Likewise, a smart sensor can stop a mobile robot or a robot arm from accidentally coming in contact with an object or people by sensing quantities such as distance, heat, and noise.

37.7.2 Sensor fusion

Sensor fusion basically involves the integration of multiple sensors in such a manner that the individual data from each of the sensors (such as force, vibration, temperature, and dimensions) are combined to provide a higher level of information and reliability. A common application of sensor fusion occurs when someone drinks from a cup of hot coffee. Although we take such an everyday event for granted, it readily can be seen that this process involves data input from the person's eyes, lips, tongue, and hands. Through our basic senses of sight, hearing, smell, taste, and touch, there is real-time monitoring of relative movements, positions, and temperatures. Thus, if the coffee is too hot, the hand movement of the cup toward the lip is controlled and adjusted accordingly.

The earliest applications of sensor fusion were in robot movement control, missile flight tracking, and similar military applications, primarily because these activities involve movements that mimic human behavior. Another example of sensor fusion is a machining operation in which a set of different but integrated sensors monitors (a) the dimensions and surface finish of the workpiece, (b) tool forces, vibrations, and wear, (c) the temperatures in various regions of the tool–workpiece system, and (d) the spindle power.

An important aspect in sensor fusion is *sensor validation*: The failure of one particular sensor is detected so that the control system maintains high reliability. For this application, the receiving of redundant data from different sensors is essential. It can be seen that the receiving, integrating, and processing of all data from various sensors can be a complex problem.

With advances in sensor size, quality, and technology and continued developments in computer-control systems, artificial intelligence, expert systems, and artificial neural networks (all described in Chapter 39), sensor fusion has become practical and available at relatively low cost.

EXAMPLE 37.2 Special applications of sensors

Three special applications of sensors are described in this example.

1. Fiber-optic sensors are being developed for gas-turbine engines. The size of the head of a pin, these sensors will be installed in critical locations and will monitor the conditions inside the engine, such as temperature, pressure, and flow of gases. Continuous monitoring of the signals from these sensors will help detect possible engine problems and also provide the necessary data for improving the efficiency of the engines. *Source:* Courtesy of Prime Research.

2. It has been determined that, in the United States, more than a quarter of the passenger-vehicle tires are underinflated—a condition that can lead to tire separation and blowouts. Tire-pressure remote sensors have been developed and installed in some automobiles. Underinflated tires can become hot due

to excessive flexing and internal friction. Based on this observation, a special rubber has been developed that changes its color from black to red when its temperature rises above 77°C (170°F). Thus, a small strip of this material can be built into the sidewalls of tires, which would be visible as a red ring when the tire gets too hot. *Source: After F. Kelley and B. Rosenbaum.*

3. Electronic sensors have been developed to distinguish basic tastes, such as the sourness of milk and the bitterness of medications. Coffee, however, has subtle taste variations that are difficult to detect by these sensors, including sensing characteristics such as acidity, sourness, bitterness, and the caffeine level. Every year, $70 billion worth of coffee is sold around the world, and the quality of coffee beans traditionally have been checked and ensured by human tasters.

   A new sensor is "an electronic tongue," developed jointly by the University of Pennsylvania and Embrapa Research Agency in Brazil. The sensor consists of ten gold electrodes covered with thin films of an electrically conducting polymer. It is immersed in liquid coffee, and various molecules in it are absorbed by these films, changing their electrical properties. It is claimed that the sensor is 1000 times more sensitive and the cost is about half of others available.

   Source: After L.H. Mattoso and Technology Review.

37.8 | Flexible Fixturing

In describing *workholding devices* for the manufacturing operations throughout this book, the words clamp, jig, and fixture often were used interchangeably and sometimes in pairs, such as in jigs and fixtures. **Clamps** are simple multifunctional workholding devices, and **jigs** have various reference surfaces and points for accurate alignment of parts or tools. **Fixtures** generally are designed for specific purposes. Other workholding devices also include *chucks, collets,* and *mandrels,* many of which usually are operated manually. Workholding devices also are designed and operated at various levels of mechanization and automation (such as power chucks) and are driven by mechanical, hydraulic, or electrical means.

Workholding devices generally have specific ranges of capacity. For example, (a) a particular collet can accommodate bars only within a certain range of diameters; (b) four-jaw chucks can accommodate square or prismatic workpieces having certain dimensions; and (c) various other devices and fixtures are designed and made for specific workpiece shapes and dimensions and for specific tasks, called *dedicated fixtures*. If the part has curved surfaces, it is possible to shape the contacting surfaces of the jaws themselves by machining them (*machinable jaws*) to conform to the workpiece surfaces.

The emergence of flexible manufacturing systems has necessitated the design and use of workholding devices and fixtures which have *built-in flexibility*. There are several methods of *flexible fixturing* based on different principles that also are called **intelligent fixturing systems**. These devices are capable of quickly accommodating a range of part shapes and dimensions—without the necessity of extensive changes, adjustments, or requiring operator intervention—both of which would affect productivity adversely.

Modular fixturing. Modular fixturing often is used for small or moderate lot sizes, especially when the cost of dedicated fixtures and the time required to produce them are difficult to justify. Complex workpieces can be located within machines through fixtures produced quickly from standard components and can be disassembled when a production run is completed. Modular fixtures usually are based on tooling plates or blocks configured with grid holes or T-slots upon which a fixture is constructed.

A number of other standard components (such as locating pins, adjustable stops, workpiece supports, V-blocks, clamps and springs) can be mounted onto the base plate or block to quickly produce a fixture. By computer-aided fixture planning for specific applications, such fixtures can be assembled and modified using industrial robots. As compared with dedicated fixturing, modular fixturing has been shown to be low in cost, have a shorter lead time, easier to repair damaged components, and possess more intrinsic flexibility of application.

Bed-of-nails device. This fixture consists of a series of air-actuated pins that conform to the shape of the external surfaces of the part. Each pin moves as necessary to conform to the shape at its point of contact with the part. The pins are then mechanically locked against the part. The device is compact, has high stiffness, and is reconfigurable.

Adjustable-force clamping. A schematic illustration of such a system is shown in Fig. 37.26. The strain gage mounted on the clamp senses the magnitude of the clamping force. The system then adjusts this force to keep the workpiece securely clamped to the workpiece for the particular application. It also can prevent excessive clamping forces that otherwise may damage the workpiece surface, particularly if it is soft or slender.

Phase-change materials. There are two methods that can hold irregular-shaped or curved workpieces in a medium other than hard tooling. In the first and older method, a *low-melting-point metal* is used as the clamping medium. For example, an irregular-shaped workpiece is dipped into molten lead and allowed to set (like the wooden stick in a popsicle—a process similar to *insert molding*). After setting, the solidified lead block is clamped in a simple fixture. The possibly adverse effect of such materials as lead on the workpiece to be clamped (*liquid–metal embrittlement*) must be considered.

In the second method, the supporting medium is a *magnetorheological* (MR) or *electrorheological* (ER) *fluid*. In the MR application, the particles are ferromagnetic or paramagnetic, μm in size as well as using nanoparticles, and in a nonmagnetic

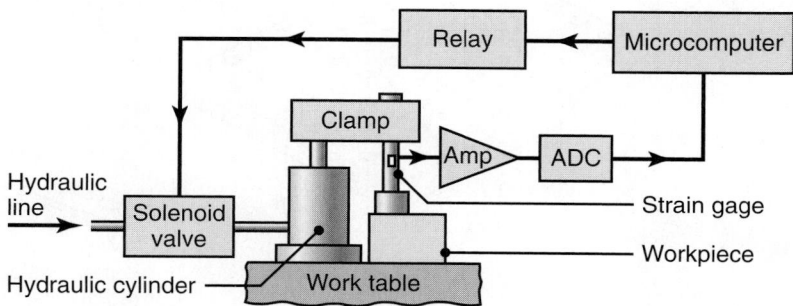

FIGURE 37.26 Schematic illustration of an adjustable-force clamping system. The clamping force is sensed by the strain gage, and the system automatically adjusts this force. *Source:* After P. K. Wright.

fluid. Surfactants are added to alleviate the particles from settling. After immersing the workpiece in the fluid, an external magnetic field is applied, whereby the particles are polarized and the behavior of the fluid changes from a liquid to that of a solid. Afterwards, the workpiece is retrieved by removing the external magnetic field. The process is suitable particularly for nonferrous metal parts. In the ER application, the fluid is a suspension of fine dielectric particles in a liquid of low dielectric constant. After the application of an electrical field, the liquid becomes a solid.

CASE STUDY 37.1 | Development of a Modular Fixture

Figure 37.27 shows a round cast-iron housing that requires a fixture for milling the housing flat, counterboring the center hole, drilling the four corner holes, and machining the circular mounting area. These operations are to be performed in moderate lot sizes on a CNC milling machine; thus, a fixture which accurately and consistently locates the workpiece is required. The lot size is not large enough to justify the design and fabrication of dedicated fixtures; hence, a modular fixture is constructed from the components illustrated in Fig. 37.28.

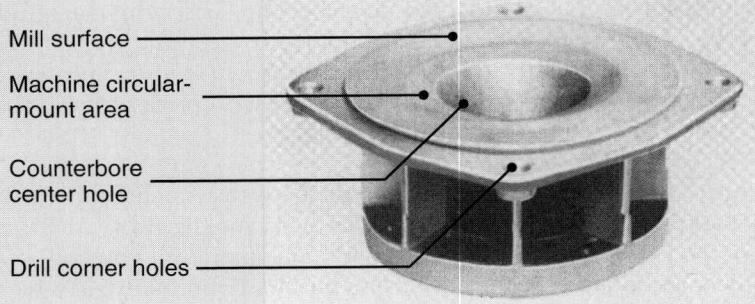

Mill surface

Machine circular-mount area

Counterbore center hole

Drill corner holes

FIGURE 37.27   Cast-iron housing and the machining operations required.

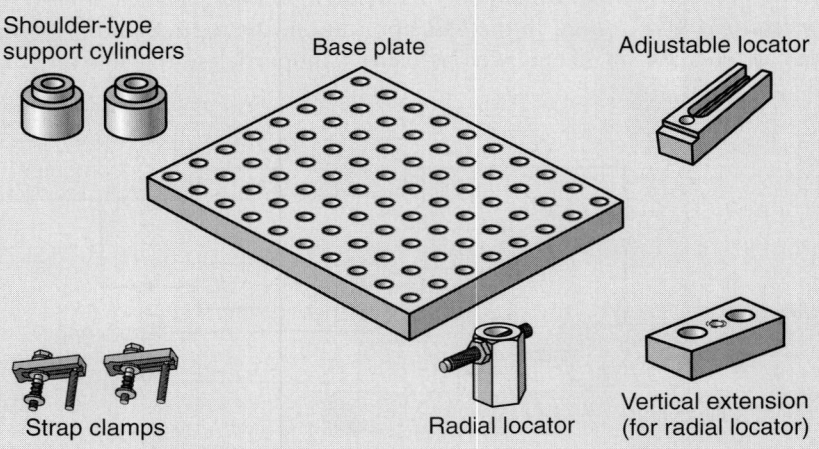

Shoulder-type support cylinders

Base plate

Adjustable locator

Strap clamps

Radial locator

Vertical extension (for radial locator)

FIGURE 37.28   Modular components used to construct the fixture for CNC machining of the cast-iron housing depicted in Fig. 37.27.

The first step in designing such a fixture is to select a tooling plate. Tooling plates with T-slots or gridded holes are the most common alternatives for modular fixtures. For this case, a rectangular plate with gridded holes is selected, which has a sufficiently large surface area to accommodate both the workpiece and the fixturing elements.

The workpiece must not be clamped on the surface that is to be machined; however, there is a lower flange which is suitable for clamping. Clamping generally is accomplished by exerting a force with a clamp against a locating button. For reasons of stability, this round workpiece should be located on the lower flange at three points and spaced approximately 120° apart. The first locating elements selected are shoulder-type support cylinders. The flange rests on the shoulders, and the cylinders support the workpiece on its diameter and also raise the workpiece above the tooling plate. This method has the advantage of eliminating the adverse effects of chips (from machining) on the tooling plate which could interfere with the orientation of the workpiece.

The support cylinders are mounted onto the tooling plate using locating screws. An adjustable extension support then is mounted to the tooling plate in a position where the correct diameter is set between the three locators. The extension support is positioned to contact only the workpiece's bottom edge, and sufficient space is allowed so that the locator will not bind the housing on the locating diameter.

The workpiece must be oriented consistently in order to maintain the required dimensional tolerances on the corner holes. An adjustable stop is utilized to locate the workpiece. This stop uses a threaded locating button which is adjusted to orient the workpiece. An extension support is used so that the stop is vertically located directly below the surface to be machined.

Strap clamps then are used to hold the workpiece in place. The springs and washers on the clamping stud permit the strap clamp to rise automatically when loosened, so the clamp will not fall when the workpiece is removed. The modular fixture with the workpiece in place is shown in Fig. 37.29. Note that planar, concentric, and radial locations all have been defined accurately by the fixture.

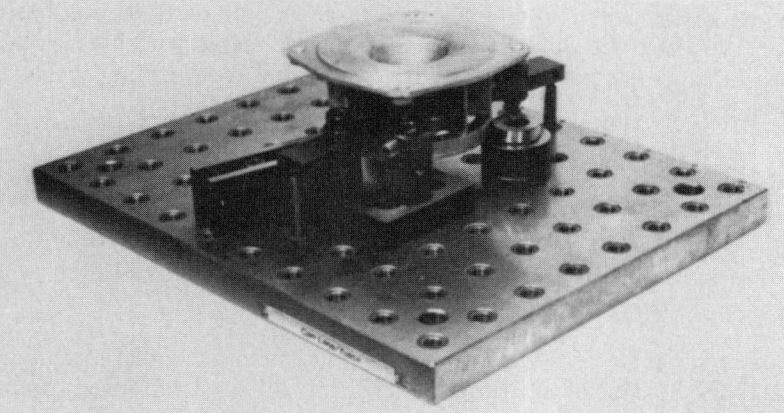

FIGURE 37.29 Completed modular fixture with cast-iron housing in place, as would be assembled for use in a machining center or CNC milling machine. *Source:* Courtesy of Carr Lane Manufacturing Company.

37.9 | Assembly Systems

The individual parts and components produced by various manufacturing processes are **assembled** into finished products by various methods. Some products are simple and have only two or three components to assemble. Operations that can be performed with such relative ease are an ordinary pencil with an eraser, a frying pan with a handle, or an aluminum beverage can. However, most products consist of numerous parts, and their assembly requires considerable planning.

Traditionally, assembly has involved much manual work and, thus, has contributed significantly to product cost. The total assembly operation usually is broken into individual assembly operations (*subassemblies*) with an operator assigned to carry out each step. Assembly costs are typically 25 to 50% of the total cost of manufacturing, with the percentage of workers involved in assembly operations ranging from 20 to 60%. In the electronics industries, typically some 40 to 60% of total wages are paid to assembly workers.

As production costs and quantities of products to be assembled increase, the necessity for automated assembly become obvious. Beginning with the hand assembly of muskets in the late 1700s and the early 1800s—with *interchangeable parts*—assembly methods have been improved upon vastly over the years. The first application of large-scale efficient assembly was the assembly of flywheel magnetos for the Model T Ford. This experience eventually led to mass production of the whole automobile. The choice of an assembly method and system depends on the required production rate, the total quantity to be produced, the product's life cycle, labor availability, and cost. **Automated assembly** effectively can reduce the overall product cost.

As we have seen, parts are manufactured within certain dimensional tolerance ranges. Taking ball bearings as an example, we know that although they all have the same nominal dimensions, some balls in a lot will be smaller than others by a very small amount. Likewise, some bearing races will be smaller than others in the lot. There are two methods of assembly for such high-volume products: random assembly and selective assembly.

In **random assembly**, the components are put together by selecting them randomly from the lots produced. In **selective assembly**, the balls and races are segregated by groups of sizes (from smallest to largest). The parts then are selected to mate properly. Thus, the smallest diameter balls are mated with inner races having the largest outside diameter and, likewise, with outer races having the smallest inside diameters.

Methods and systems of assembly.   There are three basic methods of assembly: manual, high-speed automatic, and robotic. These methods can be used individually or, as is the case on most applications in practice, in combination. An analysis of the product design must first be made (Fig. 37.30) with regard to the appropriate and economical method of assembly.

1. **Manual assembly** uses relatively simple tools and generally is economical for small lots. Because of the dexterity of the human hand and fingers and their capability for feedback through various senses, workers can manually assemble even complex parts without much difficulty. In spite of the use of sophisticated mechanisms, robots, and computer controls, the aligning and placing of a simple square peg into a square hole involving small clearances can be difficult in automated assembly—yet the human hand is capable of doing this simple operation with relative ease.

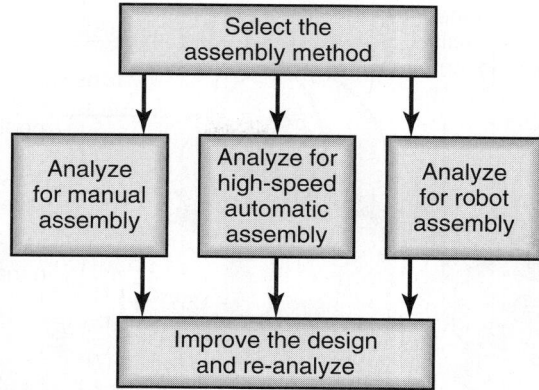

FIGURE 37.30   Stages in the design-for-assembly analysis. *Source: After G. Boothroyd and P. Dewhurst.*

2. **High-speed automated assembly** utilizes *transfer mechanisms* designed specially for assembly. Two examples of such assembly are shown in Fig. 37.31, in which individual assembly is carried out on products that are indexed for proper positioning during assembly.

3. In **robotic assembly**, one or two general-purpose robots operate at a single workstation (Fig. 37.32), or the robots operate at a multistation assembly system.

There are three basic types of assembly systems: synchronous, nonsynchronous, and continuous.

Synchronous systems.   Also called *indexing*, individual parts and components are supplied and assembled at a constant rate at fixed individual stations. The rate of movement is based on the station that takes the longest time to complete its portion of the assembly. This system is used primarily for high-volume, high-speed assembly of small products. Transfer systems move the partial assemblies from workstation to workstation by various mechanical means. Two typical transfer systems (*rotary indexing* and *in-line indexing*) are shown in Fig. 37.32. These systems can operate in either a fully automatic mode or a semiautomatic mode. However, a breakdown of one station will shut down the whole assembly operation.

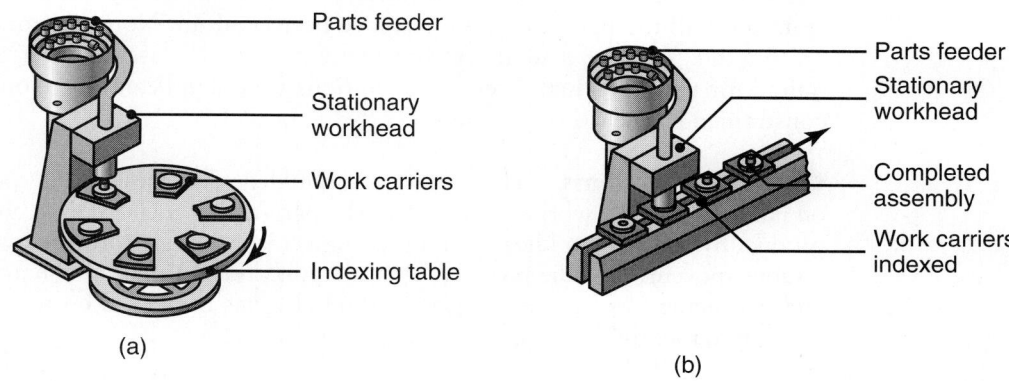

FIGURE 37.31   Transfer systems for automated assembly: (a) rotary indexing machine and (b) in-line indexing machine. *Source: After G. Boothroyd.*

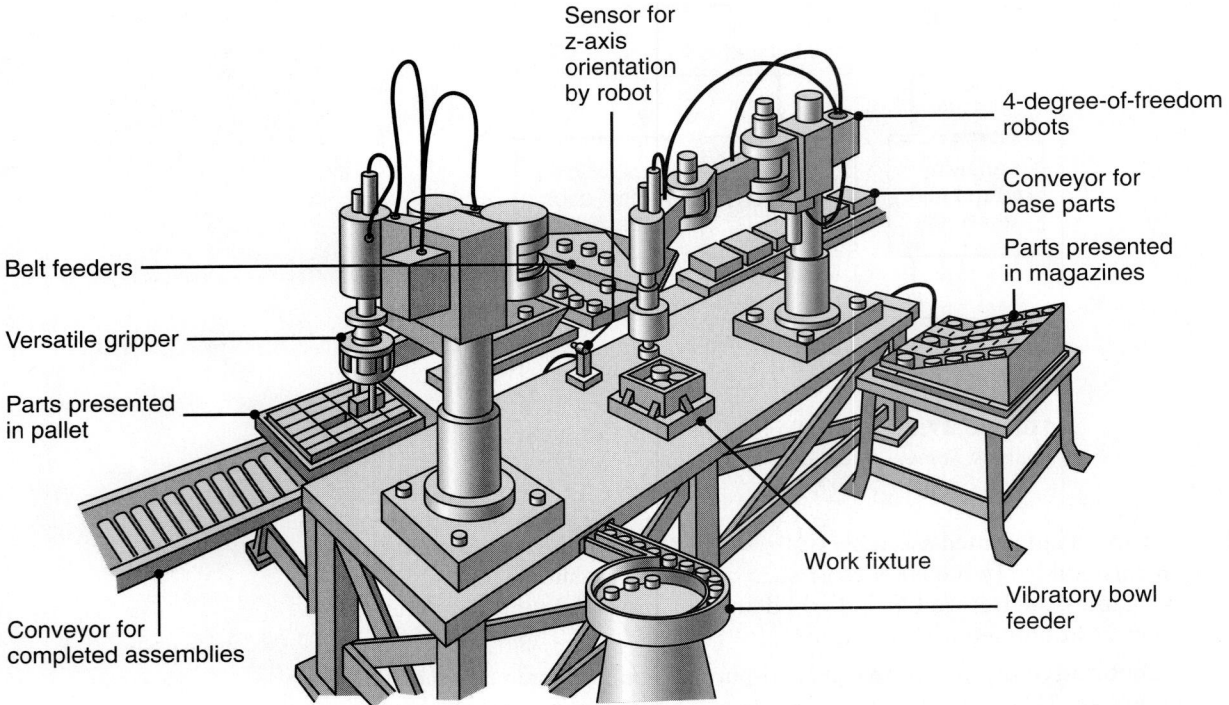

FIGURE 37.32 A two-arm robot assembly station. *Source: Product Design for Assembly*, 1989 edition, by G. Boothroyd and P. Dewhurst. Reproduced with permission.

The part feeders supply the individual parts to be assembled and place them on other components, which are secured on work carriers or fixtures. The feeders move the individual parts, by vibratory or other means, through delivery chutes and ensure their proper orientation by various ingenious means (Fig. 37.33). Orienting parts properly and avoiding jamming are essential in all automated assembly operations.

Nonsynchronous systems. Each station operates independently, and any imbalance is accommodated in storage (*buffer*) between stations. The station continues operating until the next buffer is full or the previous buffer is empty. Furthermore, if one station becomes inoperative for some reason, the assembly line continues to operate until all the parts in the buffer have been used up. Nonsynchronous systems are suitable for large assemblies with many parts to be assembled. If the times required for the individual assembly operations vary significantly, the output will be constrained by the slowest station.

Continuous systems. The product is assembled while moving at a constant speed on pallets or similar workpiece carriers. The parts to be assembled are brought to the product by various workheads, and their movements are synchronized with the continuous movement of the product. Typical applications of this system are in bottling and packaging plants, although the method also has been used on mass-production lines for automobiles and appliances.

Flexible assembly systems. Assembly systems generally are set up for a specific product line. However, they can be modified for increased flexibility in order to assemble product lines that have a variety of product models. Such *flexible assembly*

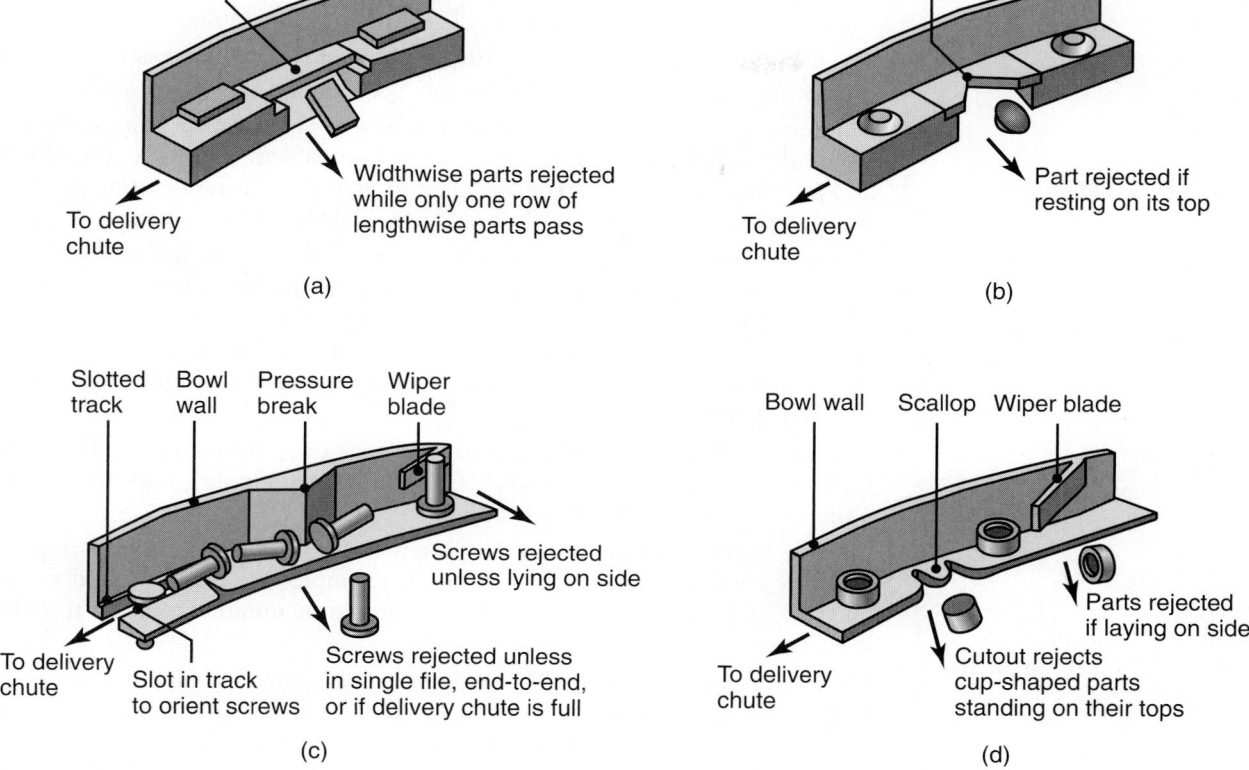

FIGURE 37.33   Examples of guides to ensure that parts are properly oriented for automated assembly. *Source:* After G. Boothroyd.

systems (FAS) utilize computer controls, interchangeable and programmable work-heads and feeding devices, coded pallets, and automated guiding devices. For example, the General Motors plant for the Saturn subcompact automobile is designed with a flexible assembly system. This system is capable of assembling up to a dozen different transmission and engine combinations and power steering and air conditioning units.

37.10   Design Considerations for Fixturing, Assembly, Disassembly, and Servicing

As in many aspects of manufacturing processes and systems, design is an integral part of the topics described in this chapter as well. These topics require special design considerations, as described here.

37.10.1 Design for fixturing

The proper design, construction, and operation of flexible workholding devices and fixtures are essential to the operation and efficiency of advanced manufacturing systems. The major design issues involved are described here.

- Workholding devices must position the workpiece automatically and accurately. They must maintain its location precisely and with sufficient clamping force to withstand the particular manufacturing operation. Fixtures also should be able to accommodate parts repeatedly in the same position.

- The fixtures must have sufficient stiffness to resist (without excessive distortion) the normal and shear stresses developed at the workpiece–fixture interfaces.

- The presence of loose chips and other debris between the locating surfaces of the workpiece and the fixture can be a serious problem. Chips from machining and grinding operations are most likely to be present where cutting fluids are used, as they tend to stick to the wet surfaces due to surface–tension forces.

- A flexible fixture should accommodate parts to be made by different processes and ones with dimensions and surface features that vary from part to part. These considerations are even more important when the workpiece (a) is fragile or made of a brittle material, (b) is made of a relatively soft and flexible material (such as plastic or rubber parts) or (c) has a relatively soft coating on its contacting surfaces.

- Clamps and fixtures should have low profiles to avoid collision with cutting tools. Collision avoidance also is an important factor in programming tool paths in machining operations.

- Flexible fixturing must also meet special requirements in manufacturing cells and flexible manufacturing systems. For example, the time required to load and unload parts on modern machinery should be minimal in order to reduce cycle times.

- Workpieces should be designed to allow locating and clamping within the fixture. Flanges, flats, or other locating surfaces should be incorporated in the design to simplify fixture design and to aid in part transfer into machinery.

37.10.2 Design for assembly, disassembly, and servicing

Design for assembly.   Although the functions of a product and its design for manufacture have been matters of major interest for some time, *design for assembly* (DFA) has attracted special attention (particularly design for automated assembly) because of the need to reduce assembly costs. In **manual assembly** a major advantage is that humans easily can pick the correct similar or different parts from bulk (such as from a nearby bin), and the human senses can guide the hands for proper assembly. However, in **high-speed automated assembly**, automatic handling generally requires that parts be separated from the bulk, conveyed by hoppers or vibratory feeders (see Fig. 37.31), and assembled in their proper locations and orientations.

General guidelines for design for assembly are summarized below:

1. Reduce the number and variety of parts in a product. Simplify the product design and incorporate multiple functions into a single part. Use common parts as much as possible. Consider subassemblies which would serve as modules.

2. Parts should have a high degree of symmetry (such as round or square) or a high degree of asymmetry (such as oval or rectangular) so that they cannot be installed incorrectly or do not require locating, aligning, or adjusting. Design parts for easy insertion.

3. Designs should allow parts to be assembled without any obstructions. There should be a direct line of sight. Assemblies should not be turned over for insertion of parts.

4. Consider methods such as snap fits and avoid the need for fasteners such as bolts, nuts, and screws. If used, fastener variety should be minimized, and they should be spaced and located so that tools can be used without obstruction.

5. Part designs should consider such factors as size, shape, weight, flexibility, abrasiveness, and possible entanglement with other parts.

6. Parts should be inserted from a single direction, preferably vertically (from above) to take advantage of gravity. Assembly from two or more directions can be difficult.

7. Products should be designed or existing products redesigned so that there are no physical obstructions to the free movement of the parts during assembly. For example, sharp external and internal corners should be replaced with chamfers, tapers, or radii.

8. Color codes should be used on those parts that may appear to be similar but are different.

Robotic assembly. Design guidelines for robotic assembly include the following additional considerations:

- Parts should be designed so that they can be gripped and manipulated by the same gripper (end effector) of the robot. Such a design avoids the need for different grippers. Parts should be made available to the gripper in the proper orientation.

- Assembly that involves threaded fasteners (bolts, nuts, and screws) may be difficult for robots. One exception is the use of self-threading screws for sheet metal, plastics, and wooden parts. Also, note that robots easily can handle snap fits, rivets, welds, and adhesives.

The development of compliant end effectors and dexterous manipulators has made robotic assembly even more attractive.

Design for disassembly. The manner and ease with which a product may be taken apart for maintenance or replacement of its parts is another important consideration in product design. Consider, for example, the difficulties one has in removing certain components from under the hood of some automobiles. Similar difficulties exist in the disassembly of several other products. The general approach to design for disassembly requires the consideration of factors that are similar to those outlined above for assembly. Analysis of computer or physical models of products and their components with regard to disassembly generally indicate any potential problems, such as obstructions, size of passageways, lack of line of sight, and the difficulty of firmly gripping and guiding objects.

An important aspect of design for disassembly is how, after its life cycle, a product is to be taken apart for *recycling*, especially the more valuable components. For example, note that depending on (a) their design and location, (b) the type of tools used, and (c) whether manual or power tools are used, rivets may take longer to remove than screws or snap fits and that a bonded layer of valuable material on a component would be very difficult (if not impossible) to remove for recycling or reuse. Obviously, the longer it takes to take components apart, the higher is the cost of doing so. It then is possible that this cost becomes prohibitive. Consequently, the time required for disassembly has been studied and measured. Although it depends on the manner in which it is done, some examples are: cutting wire at 0.25 seconds; disconnecting wire at 1.5 seconds; snap fits and clips at 1 to 3 seconds; and screws and bolts at 0.15 to 0.6 seconds per revolution.

Design for servicing.   Design for assembly and disassembly includes taking into account the ease with which a product can be serviced and, if necessary, repaired. Design for servicing essentially is based on the concept that the elements that are most likely to need servicing be at the outer layers of the product. In this way, individual parts are easier to reach and service without the need to remove various other parts in order to do so.

Evaluating assembly efficiency.   Significant effort has been directed towards the development of analytical and computer-based tools to estimate the efficiency of assembly operations. These tools provide a basis for comparisons of designs and the objective selection of design attributes that make assembly easier.

To evaluate assembly efficiency, each component of an assembly is evaluated with respect to its features that can affect assembly and a baseline estimated time required to incorporate the part into the assembly. (This also can be measured for existing products.) The assembly efficiency, η, then is given by

$$\eta = \frac{Nt}{t_{\text{tot}}} \tag{37.1}$$

where N is the number of parts, t_{tot} is the total assembly time and t is the ideal assembly time for a small part that presents no difficulties in handling, orientation or assembly, and commonly is taken as three seconds. Using Eq. (37.1), competing designs can be evaluated with respect to design-for-assembly. It has been noted that products that are in need of redesign to facilitate assembly usually have assembly efficiencies around 5 to 10%, while well-designed parts have assembly efficiencies around 25%. It should be noted that assembly efficiencies near 100% are unlikely to be achieved in practice.

37.11 | Economic Considerations

As will be described in greater detail in Chapter 40 and as we have seen throughout many chapters in this book, there are numerous considerations involved in determining the overall economics of production operations. Because all production systems essentially are a combination of machines and people, important factors influencing the final decisions include the type and cost of machinery, equipment, and tooling; cost of its operation; the skill level and amount of labor required; and production quantity. We have seen that lot size and production rate greatly influence the economics of production.

Small quantities per year can be manufactured in job shops. However, the type of machinery in job shops generally requires skilled labor to operate; furthermore, production quantity and rate are low. As a result, cost per part can be high. At the other extreme, there is the production of very large quantities of parts using conventional flow lines and transfer lines and involving special-purpose machinery, equipment, specialized tooling, and computer-control systems. Although all of these components constitute major investments, both the level of labor skill required and the labor costs are relatively low because of the high level of automation implemented. However, these production systems are organized for a specific type of product and, hence, lack flexibility.

Because most manufacturing operations are between these two extremes, an appropriate decision has to be made regarding the desirable level of automation to be implemented. In many situations, selective automation rather than total automation of a facility has been found to be cost-effective. Generally, the higher the level of skill available in the workforce, the lower is the need for automation, provided that higher labor costs are justified and that there are sufficient qualified workers available. Conversely, if a manufacturing facility already has been automated, the skill level required is relatively lower.

In addition, some products have a large labor component; thus, their production is labor intensive. This is especially the case with products requiring extensive assembly. Examples of labor-intensive products are: aircraft, software, bicycles, pianos, furniture, toys, shoes, textiles, and garments. This is a major reason why so many household as well as high-tech products are now made or assembled in countries where labor costs are low, such as in China, India, Mexico, and Pacific-rim countries (see Table I.4). However, in the United States, for example, improved product design and the availability, reliability, reduced cost, and use of industrial robots are having a major economic impact on manufacturing operations, significantly reducing the need for labor.

SUMMARY

- Automation has been implemented in manufacturing processes, material handling, inspection, assembly, and packaging at increasing rates. There are several levels of automation, ranging from simple automation of machines to untended manufacturing cells.

- True automation began with the numerical control of machines, which has the capability of flexibility of operation, lower cost, and ease of making different parts with less operator skill. Production quantity and rate are important factors in determining the economic levels of automation.

- Manufacturing operations are optimized further (both in quality and in cost) by adaptive control techniques, which continuously monitor the operation and make necessary adjustments in the processing parameters.

- Major advances have been made in material handling, particularly with the implementation of industrial robots and automated guided vehicles.

- Sensors are essential in the implementation of these modern technologies; a wide variety of sensors based on various principles have been developed and installed.

- Other advances include flexible fixturing and automated assembly techniques that reduce the need for worker intervention and that lower manufacturing costs. The effective and economic implementation of these techniques requires that design for assembly, disassembly, and servicing be recognized as an important factor in the total design and manufacturing process.

- The efficient and economic implementation of these techniques requires that design for assembly, disassembly, and servicing be recognized as an important factor in the total design and manufacturing process.

KEY TERMS

Adaptive control	Closed-loop control	Contouring
Assembly	Compliant end effectors	Control systems
Automated guided vehicle	Computer numerical control	Dedicated machines
Automation	Computer vision	End effector
Buffer	Continuous path	Feedback

Flexible assembly systems
Flexible fixturing
Hard automation
Hard-wired controls
Industrial robot
Intelligent robot
Interpolation
Machine vision
Manipulators
Material handling
Mechanization
Numerical control

Open-loop control
Part programming
Positioning
Powerhead production units
Productivity
Programmable controller
Programming language
Random assembly
Repeat accuracy
Resolution
Robot
Selective assembly

Selective automation
Sensor fusion
Sensors
Smart sensors
Soft automation
Stand-alone machines
Tactile sensing
Total productive maintenance
Transfer lines
Visual sensing
Work envelope

BIBLIOGRAPHY

Amic, P.J., *Computer Numerical Control Programming*, Prentice Hall, 1996.

Asfahl, C.R., *Robots and Manufacturing Automation*, 2nd ed., Wiley, 1992.

Astrom, K.J. and Wittenmark, B., *Adaptive Control*, 2nd ed., Addison-Wesley, 1994.

Bolhouse, V., *Fundamentals of Machine Vision*, Robotic Industries Association, 1997.

Boothroyd, G., *Assembly Automation and Product Design*, Marcel Dekker, 1989.

Boothroyd, G.; Dewhurst, P. and Knight, W., *Product Design for Manufacture and Assembly*, 2nd ed., Marcel Dekker, 2001.

Brooks, R.R. and Iyengar, S., *Multi-Sensor Fusion: Fundamentals and Applications with Software*, Prentice Hall, 1997.

Burke, M., *Handbook of Machine Vision Engineering*, Chapman & Hall, 1999.

Busch-Vishniac, I., *Electromechanical Sensors and Actuators*, Springer, 1999.

Chow, W., *Assembly Line Design: Methodology and Applications*, Marcel Dekker, 1990.

Coiffet, P. and Chizouze, M., *Introduction to Robot Technology*, McGraw-Hill, 1993.

Fraden, J., *Handbook of Modern Sensors: Physics, Designs, and Applications*, 2nd ed., Springer-Verlag, 1996.

Galbiati, L.J., *Machine Vision and Digital Image Processing Fundamentals*, Prentice Hall, 1997.

Gibbs, D. and Crandell, T., *CNC: An Introduction to Machining and Part Programming*, Industrial Press, 1991.

Ioannu, P.A., *Robust Adaptive Control*, Prentice Hall, 1995.

Jain, R. (ed.), *Machine Vision*, Prentice Hall, 1995.

Lynch, M., *Computer Numerical Control for Machining*, McGraw-Hill, 1992.

Molloy, O.; Warman, E.A. and Tilley, S., *Design for Manufacturing and Assembly: Concepts, Architectures and Implementation*, Kluwer, 1998.

Myler, H.R., *Fundamentals of Machine Vision*, Society of Photo-optical Instrumentation Engineers, 1998.

Nanfara, F.; Uccello, T. and Murphy, D., *The CNC Workbook: An Introduction to Computer Numerical Controls*, Addison-Wesley, 1995.

Nof, S.Y. (ed.), *Handbook of Industrial Robotics*, Wiley, 1999.

Nof, S.Y.; Wilhelm, W.E. and Warnecke, H.-J., *Industrial Assembly*, Chapman & Hall, 1998.

Rampersad, H.K. *Integral and Simultaneous Design for Robotic Assembly*, Wiley, 1995.

Rehg, J.A., *Introduction to Robotics in CIM Systems*, 5th ed., Prentice Hall, 2002.

Sandler, B.-Z., *Robotics: Designing the Mechanisms for Automated Machinery*, 2nd ed., Prentice Hall, 1999.

Seames, W., *Computer Numerical Control: Concepts and Programming*, 3rd ed., Delmar, 1995.

Soloman, S., *Sensors Handbook*, McGraw-Hill, 1997.

Stenerson, J. and Curran, K.S., *Computer Numerical Control: Operation and Programming*, 2nd ed, Prentice Hall, 2000.

Tool and Manufacturing Engineers Handbook, 4th ed., Vol. 4, *Assembly, Testing, and Quality Control*, Society of Manufacturing Engineers, 1986.

_____ Vol. 9, *Material and Part Handling in Manufacturing*, Society of Manufacturing Engineers, 1998.

Valentino, J.V. and Goldenberg, J., *Introduction to Computer Numerical Control*, 2nd ed., Regents/Prentice Hall, 1999.

Williams, D.J., *Manufacturing Systems*, 2nd ed., Chapman & Hall, 1994.

Zuech, N., *Understanding and Applying Machine Vision*, 2nd ed., Marcel Dekker, 1999.

REVIEW QUESTIONS

37.1 Describe the differences between mechanization and automation. Give several specific examples for each.

37.2 Why is automation generally regarded as evolutionary rather than revolutionary?

37.3 Are there activities in manufacturing operations that cannot be automated? Explain.

37.4 Explain the difference between hard and soft automation. Why are they so called ?

37.5 Describe the principle of numerical control of machines. What factors led to the need for and development of numerical control?

37.6 Describe open-loop and closed-loop control circuits.

37.7 Describe the principle and purposes of adaptive control.

37.8 What factors have led to the development of automated guided vehicles? Do they have any disadvantages? Explain your answers.

37.9 Describe the features of an industrial robot. Why are these features necessary?

37.10 Discuss the principles of various types of sensors.

37.11 Describe the concept of design for assembly. Why has it become an important factor in manufacturing?

37.12 Is it possible to have partial automation in assembly? Explain.

37.13 What are the advantages of flexible fixturing?

37.14 How are robots programmed to follow a certain path?

QUALITATIVE PROBLEMS

37.15 Giving specific examples, discuss your observations concerning Fig. 37.2.

37.16 What are the relative advantages and limitations of the two arrangements for power heads shown in Fig. 37.4?

37.17 Discuss methods of on-line gaging of workpiece diameters in turning operations other than that shown in Fig. 37.15.

37.18 Are drilling and punching the only applications for the point-to-point system shown in Fig. 37.10a? Explain.

37.19 What determines the number of robots in an automated assembly line such as that shown in Figs. 37.22 and 37.32?

37.20 Describe situations in which the shape and size of the work envelope of a robot (Fig. 37.20) can be critical.

37.21 Explain the functions of each of the components of an industrial robot.

37.22 Explain the difference between an automated guided vehicle and a self-guided vehicle.

37.23 Give specific examples for which (a) an open-loop and (b) a closed-loop control system would be desirable.

37.24 Explain why sensors have become so essential in the development of automated manufacturing systems.

37.25 Why is there a need for flexible fixturing for holding workpieces? Are there any disadvantages? Explain.

37.26 Table 37.2 shows a few examples of typical products for each category. Add several other examples to this list.

37.27 Describe applications of machine vision for specific parts that are similar to the examples shown in Fig. 37.25.

37.28 Sketch the workspace (envelope) of each of the robots in Fig. 37.19.

SYNTHESIS, DESIGN, AND PROJECTS

37.29 Give an example of a metal-forming operation that is suitable for adaptive control.

37.30 List and discuss the factors that should be considered in choosing a suitable material-handling system for a particular manufacturing facility.

37.31 Describe possible applications for industrial robots not discussed in this chapter.

37.32 Design two different systems of mechanical grippers for widely different applications.

37.33 Give some applications for the systems shown in Fig. 37.25a and c.

37.34 For a system similar to that shown in Fig. 37.26, design a flexible fixturing setup for a lathe chuck.

37.35 Think of other part shapes that can be guided such as those shown in Fig. 37.33.

37.36 Give examples of products that are suitable for the three types of production shown in Fig. 37.3.

37.37 Give examples where tactile sensors would not be suitable. Explain why.

37.38 Describe situations where machine vision cannot be applied properly and reliably. Explain why.

37.39 Choose one machine each from Parts II through IV and design a system in which sensor fusion can be used effectively.

37.40 Why should the level of automation in a manufacturing facility depend on production quantity and production rate?

37.41 Think of a product and design a transfer line for it that is similar to that shown in Fig. 37.5. Specify the types and the number of machines required.

37.42 Describe your thoughts on the usefulness and applications of modular fixturing consisting of various individual clamps, pins, supports, and attachments mounted on a base plate.

37.43 Inspect several household products and describe the manner in which they have been assembled. Comment on any product design changes you would make so that assembly, disassembly, and servicing are simpler and faster.

37.44 Inspect Table 37.1 on the history of automation and describe your thoughts as to what new developments possibly could be added to the bottom of the list in the near future.

37.45 Design a robot gripper that will pick up and place the following: (a) eggs—without breaking them; (b) an object made of very soft rubber; (c) a metal ball with a very smooth and polished surface; (d) a newspaper; and (e) tableware, such as knives, spoons, and forks.

37.46 Review the specifications of various numerical-control machines and make a list of typical numbers for their (a) positioning accuracy, (b) repeat accuracy, and (c) resolution. Comment on your observations.

37.47 Obtain an old toaster and disassemble it. Explain how you would go about assembling it by automated assembly.

37.48 Perform a literature search and determine how a bowling pinsetter operates. Explain how the pins are never placed upside down in a bowling lane.

Computer-Aided Manufacturing

This chapter describes how computers impact product design and manufacturing processes. It shows how computer software has aided engineers throughout the following tasks:

- Computer-aided design by assisting in the graphical descriptions of parts.
- The use of computers in the direct control of manufacturing processes and computer-aided manufacturing.
- How computers can simulate manufacturing processes and systems.
- Group technology approaches to allow the rapid recovery of previous design and manufacturing experience and to apply information to new situations in a straightforward manner.

38.1 | Introduction

The importance of product quality was emphasized in Chapter 36, along with the necessity for the commitment of a company's management to total quality control. Recall the statements that *quality must be built into the product*, that *high quality does not necessarily mean higher cost*, and that *marketing poor-quality products indeed can be very costly to the manufacturer*. It has been shown that high quality is far more attainable and less expensive if design and manufacturing activities are integrated properly rather than if they were treated as separate entities.

Integration can be performed successfully and effectively through *computer-aided design, engineering, manufacturing, process planning*, and the *simulation of processes and systems*, as will be described throughout this chapter. The widespread availability of high-speed processors and highly developed software has allowed computers to proliferate into all areas of manufacturing. Computers are used to perform a wide variety of tasks ranging from individual part drafting, to modeling manufacturing processes and systems, and to performing database management.

38.2 | Manufacturing Systems

The word *system* is derived from the Greek word *systema*, meaning to combine. System has come to mean an arrangement of physical entities with identifiable and quantifiable interacting parameters. As we have seen throughout various chapters, manufacturing entails a large number of interdependent activities consisting of distinct entities; thus, it can be treated as a system.

Manufacturing is a complex system, because it consists of many diverse physical as well as human elements—some of which are difficult to predict and control. These difficulties include such factors as the supply and cost of raw materials, national and global market changes, the impact of constantly developing technologies, and human behavior and performance. Ideally, a manufacturing system should be represented by mathematical and physical models which show the nature and extent of the interdependence of all of the variables involved. In this way, the effects of a change or a disturbance that occurs anywhere in the system can be analyzed, and necessary adjustments can be made.

For example, the supply of a particular raw material may be reduced significantly due to, say, global demands, wars, strikes, or geopolitics. Because the raw-material cost now will rise, alternative materials may have to be considered and selected. This selection must be made after careful consideration of several factors because such a change may have adverse effects on product quality, production rate, and manufacturing costs. For example, the material selected may not be as easy to form, machine, or weld, or the product integrity may suffer during its processing.

In a constantly changing marketplace, the demand for a product may fluctuate randomly and rapidly for a variety of reasons. As examples, note the downsizing of automobiles during the 1980s in response to fuel shortages and, in contrast, the recent popularity of sport-utility vehicles and current interest in gas-electric hybrids. The system thus must be able to produce the modified product on a relatively short lead time and minimize large expenditures in machinery and tooling.

Such a complex system can be difficult to analyze and model because of a lack of comprehensive or reliable data on all of the variables involved. Furthermore, it is not easy to correctly predict and to control some of these variables. For example: (a) machine-tool characteristics, their performance, and their response to random external disturbances cannot be precisely modeled, (b) raw-material costs are difficult to predict accurately, and (c) human behavior and performance are difficult to model. In spite of the difficulties, much progress has been made in modeling and simulating manufacturing systems.

38.3 | Computer-Integrated Manufacturing

The various levels of automation in manufacturing operations (described in Chapter 37) have been extended further by including *information processing functions* and by utilizing an extensive network of interactive computers. This has led to *computer-integrated manufacturing* (CIM), which describes the computerized integration of all aspects of product design, process planning, production, and distribution, as well as the management and operation of the whole manufacturing organization.

Computer-integrated manufacturing is a *methodology* rather than an assemblage of machines, equipment, and computers. The effectiveness of CIM critically depends on the use of a *large-scale integrated communications system* involving computers, machines, and their controls (described in Section 39.7). Because CIM ideally should involve the total operation of an organization, it must have an extensive database concerning the technical and business aspects of the operation. Consequently, if planned all at once, CIM can be prohibitively expensive, particularly for small- and medium-sized companies.

Implementation of CIM in existing plants may begin with modules in various phases of the total operation. For new manufacturing plants, on the other hand, comprehensive and long-range strategic planning covering all phases of the operation is essential in order to fully benefit from CIM. Such planning must take into account the following considerations:

- The mission, goals, and culture of the organization.
- Availability of financial, technical, and human resources.
- Existing as well as emerging technologies in the areas of the products to be manufactured.
- The level of integration required.

Subsystems of CIM.   Computer-integrated manufacturing systems consist of subsystems that are integrated into a whole. These subsystems consist of the following (see Fig. 38.1):

- Business planning and support
- Product design

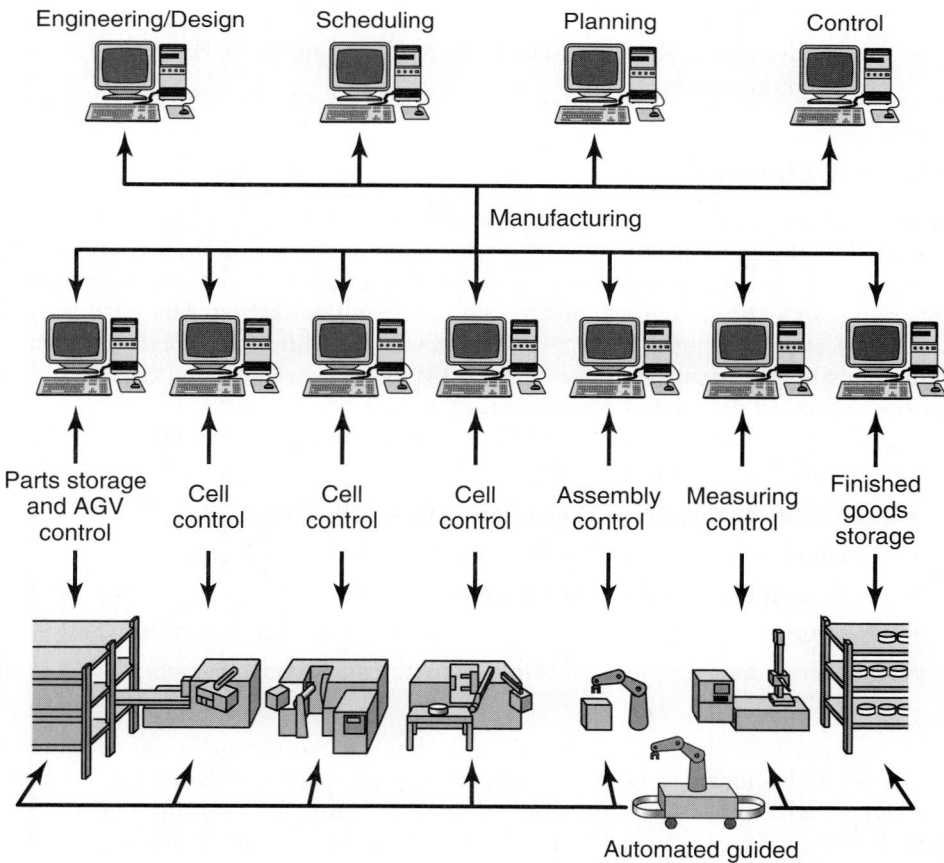

FIGURE 38.1   A schematic illustration of a computer-integrated manufacturing system. The manufacturing cells and their controls shown at the lower left are described in Section 39.2. *Source*: After U. Rembold.

- Manufacturing process planning
- Process automation and control
- Production monitoring systems

The subsystems are designed, developed, and implemented in such a manner that the output of one subsystem serves as the input of another subsystem. Organizationally, these subsystems usually are divided into two functions:

- **Business planning functions:** These include activities such as forecasting, scheduling, material-requirements planning, invoicing, and accounting.
- **Business execution functions:** Includes production and process control, material handling, testing, and inspection of the system.

The major benefits of CIM are:

- Emphasis on product quality and uniformity, as implemented through better process control.
- Efficient use of materials, machinery, and personnel and major reduction of work-in-progress inventory, all of which improve productivity and lower product cost.
- Total control of the production, schedules, and management of the entire manufacturing operation.
- Responsiveness to shorter product life cycles, changing market demands, and global competition.

38.3.1 Database

An efficient computer-integrated manufacturing system requires a single *database* to be shared by the entire manufacturing organization. Databases consist of up-to-date, detailed, and accurate information relating to products, designs, machines, processes, materials, production, finances, purchasing, sales, and marketing. This vast array of data is stored in computer memory and recalled or modified as necessary either by individuals in the organization or by the CIM system itself, while it is controlling various aspects of design and production.

A database generally consists of the following items, some of which are classified as technical and others as nontechnical:

- **Product data:** Part shape, dimensions, and specifications
- **Data management attributes:** Revision level, and part number
- **Production data:** Manufacturing processes used
- **Operational data:** Scheduling, lot sizes, and assembly requirements
- **Resources data:** Capital, machines, equipment, tooling, personnel, and their capabilities

Databases are built by individuals and using various sensors in the production machinery and equipment. Data are collected automatically by a **data-acquisition system** (DAS), which can report the number of parts being produced per unit of time, their dimensional accuracy, surface finish, weight, and other characteristics at specified rates of sampling.

The components of DAS include microprocessors, transducers, and analog-to-digital converters (ADCs). Data-acquisition systems also are capable of analyzing the data and transferring them to other computers for purposes such as statistical analysis, data presentation, and the forecasting of product demand.

Several factors are important in the use and implementation of databases:

1. They should be timely, accurate, easily accessible, easily shared, and user friendly.

2. Because it is used for various purposes and by many people, the database must be flexible and responsive to the needs of different users.

3. CIM systems can be accessed by designers, manufacturing engineers, process planners, financial officers, and the management of the company by using appropriate *access codes*. Of course, companies must protect data against tampering or unauthorized use.

4. If there are problems with data accuracy, the correct data should be recovered and restored.

38.4 | Computer-Aided Design and Engineering

Computer-aided design (CAD) involves the use of computers to create design drawings and product models. (See Fig. I.2 in the General Introduction.) Computer-aided design is usually associated with **interactive computer graphics**, known as a **CAD system**. Computer-aided design systems are powerful tools and are used in the design and geometric modeling of components and products.

Drawings are generated at **workstations**, and the design is displayed continuously on the monitor in different colors for its various components. The designer easily can conceptualize the part to be designed on the graphics screen and can consider alternative designs or modify a particular design to meet specific design requirements quickly. Using powerful software such as CATIA (after *c*omputer-*a*ided *t*hree-dimensional *i*nteractive *a*pplications), the design can be subjected to engineering analysis and can identify potential problems, such as an excessive load, deflection, or interference at mating surfaces during assembly. Information (such as a list of materials, specifications, and manufacturing instructions) also is stored in the CAD database. Using this information, the designer can analyze the manufacturing economics of alternative designs.

Computer-aided engineering (CAE) allows several applications to share the information in the database. These applications include (a) finite-element analysis of stresses, strains, deflections, and temperature distribution in structures and load-bearing members, (b) the generation, storage, and retrieval of NC data, and (c) the design of integrated circuits and various electronic devices.

38.4.1 Exchange specifications

Because of the availability of a variety of CAD systems with different characteristics supplied by different vendors, effective communication and exchange of data between these systems is essential. **Drawing exchange format** (DFX) was developed for use with *Autodesk* and basically has become a standard because of the long-term success of this software package. DFX is limited to transferring geometry information only. Similarly, *stereolithography* (STL) formats are used to export three-dimensional geometries—initially only to rapid prototyping systems (Chapter 20)—but recently, it has become a format for data exchange between different CAD systems.

The necessity for a single, neutral format for better compatibility and for the transfer of more information than geometry alone is currently filled mainly by **initial graphics exchange specification** (IGES). This is used for translation in two directions (in and out of a system) and also is used widely for translation of three-dimensional line and surface data. Because IGES is evolving, there are many variations of IGES in existence.

Another useful format is a solid-model-based standard called the **Product Data Exchange Specification** (PDES), which is based on the Standard for the Exchange of Product Model Data (STEP) developed by the International Standards Organization. PDES allows information on shape, design, manufacturing, quality assurance, testing, maintenance, etc., to be transferred between CAD systems.

38.4.2 Elements of CAD systems

The design process in a CAD system consists of the four stages described in this section.

Geometric modeling.   In geometric modeling, a physical object or any of its parts is described mathematically or analytically. The designer first constructs a geometric model by giving commands that create or modify lines, surfaces, solids, dimensions, and text. Together, these present an accurate and complete two- or three-dimensional representation of the object. The results of these commands are displayed and can be moved around on the screen, and any section desired can be magnified to view details. These data are stored in the database contained in computer memory.

The models in a CAD system can be presented in three different ways.

1. In **line representation** (**wire-frame**, Fig. 38.2), all of the edges of the model are visible as solid lines. However, this image can be ambiguous, particularly for complex shapes. Therefore, various colors generally are used for different parts of the object to make it easier to visualize.

   The three types of wire-frame representations are two, two and one-half, and three dimensional. A two-dimensional image shows the profile of the object. A two and one-half dimensional image can be obtained by translational sweep, that is, by moving the two-dimensional object along the z axis. For round objects, a two and one-half dimensional model can be generated by simply rotating a two-dimensional model around its axis.

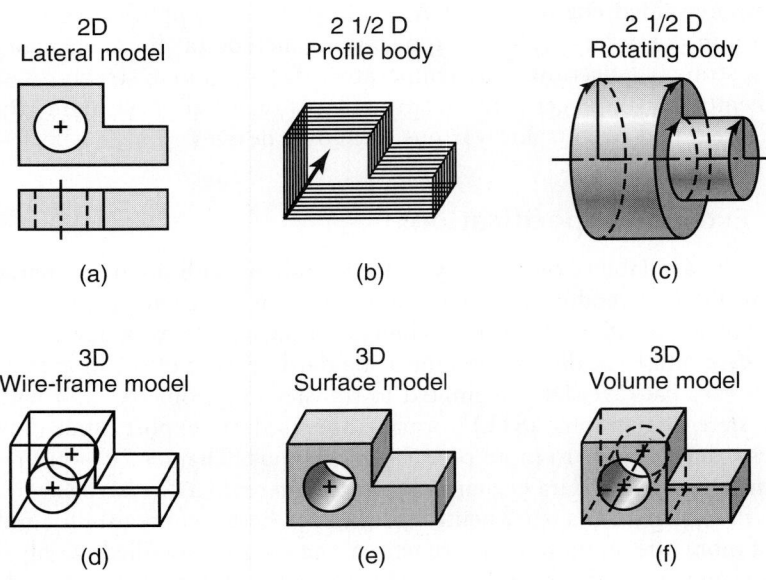

FIGURE 38.2   Various types of modeling for CAD.

2. In the **surface model**, all visible surfaces are shown in the model. Surface models define surface features and edges of objects. Modern CAD programs use Bezier curves, B-splines, or nonuniform rational B-splines (NURBS) for surface modeling. Each of these uses control points to define a polynomial curve or surface. A Bezier curve passes through the first and last vertex and uses the other control points to generate a blended curve.

The drawback to Bezier curves is that the modification of one control point will affect the entire curve. B-splines are a blended piecewise polynomial curve, where modification of a control point only affects the curve in the area of the modification. Examples of two-dimensional Bezier curves and B-splines are shown in Fig. 38.3. A *NURBS* (*non uniform rational B-spline*) is a special kind of B-spline where each control point has a weight associated with it.

3. In the **solid model**, all surfaces are shown, but the data describe the interior volume. Solid models can be constructed from "swept volumes" (Fig. 38.2b and c) or by the techniques shown in Fig. 38.4. In **boundary representation** (B-rep), surfaces are combined to develop a solid model (Fig. 38.4a). In **constructive solid geometry** (CSG), simple shapes such as spheres, cubes, blocks, cylinders, and cones (called **primitives of solids**) are combined to develop a solid model (Fig. 38.4b).

Programs are available whereby the user selects any combination of these primitives and their sizes and combines them into the desired solid model. Although solid models have certain advantages (such as ease of design analysis and ease of preparation for manufacturing the part), they require more computer memory and processing time than the wire-frame and surface models shown in Fig. 38.2.

A special kind of solid model is a **parametric model**, where a part is stored not only in terms of a B-rep or CSG definition but is derived from the dimensions and constraints that define the features (Fig. 38.5). Whenever a change is made, the part is recreated from these definitions. This allows simple and straightforward updates and changes to be made to the models.

The **octree representation** of a solid object is shown in Fig. 38.6; it is a three-dimensional analog to pixels on a television screen. Just as any area can be broken down into quadrants, any volume can be broken down into octants, which are then identified as solid, void, or partially filled. Partially filled *voxels* (from *vo*lume pi*xels*) are broken into smaller octants and are reclassified. With increasing resolution, exceptional part detail is achieved. This process may appear to be somewhat cumbersome,

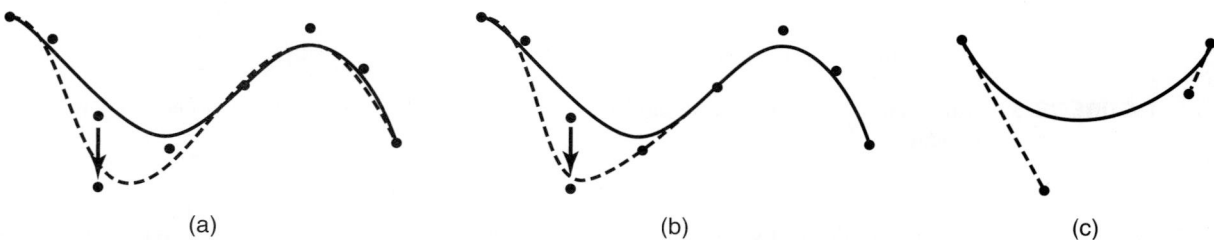

(a)   (b)   (c)

FIGURE 38.3   Types of splines. (a) A Bezier curve passes through the first and last control point but generates a curve from the other points. Changing a control point modifies the entire curve. (b) A B-spline is constructed piecewise so that changing a vertex affects the curve only in the vicinity of the changed control point. (c) A third-order (cubic) piecewise Bezier curve constructed through two adjacent control points with two other control points defining the curve slope at the endpoints. A third-order piecewise Bezier curve is continuous, but its slope may be discontinuous.

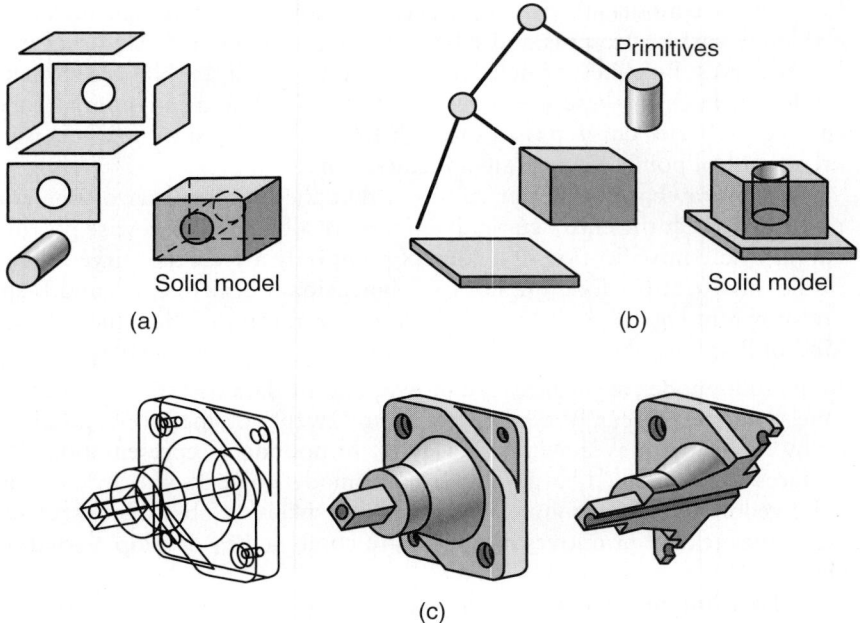

FIGURE 38.4 (a) Boundary representation of solids showing the enclosing surfaces of the solid model and the generated solid model. (b) A solid model represented as compositions of solid primitives. (c) Three representations of the same part by CAD. *Source*: After P. Ranky.

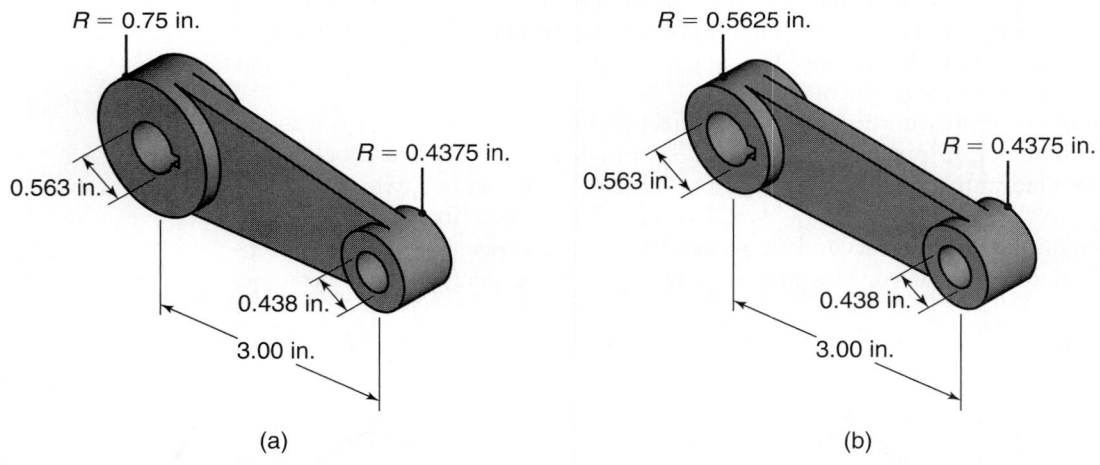

FIGURE 38.5 An example of parametric design. Dimensions of part features can be modified easily to quickly obtain an updated solid model.

but it allows for accurate description of complex surfaces. It is used particularly in bio-medical applications, such as for modeling bone geometries.

A **skeleton** (Fig. 38.7) commonly is used for kinematic analysis of parts or assemblies. A skeleton is the family of lines, planes, and curves that describe a part without the detail of surface models. Conceptually, a skeleton can be constructed by fitting the largest circle (or sphere for three-dimensional objects) within the geometry.

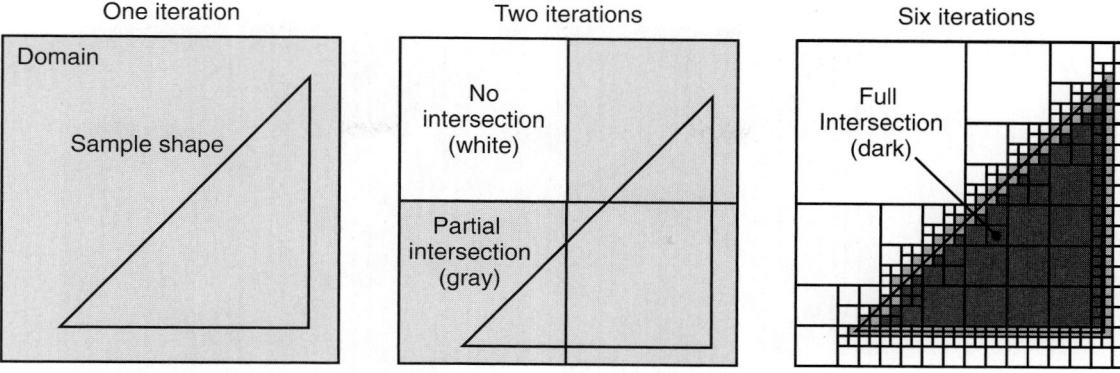

FIGURE 38.6   The octree representation of a solid object. Any volume can be broken down into octants, which are then identified as solid, void, or partially filled. Shown is a two-dimensional version (or quadtree) for the representation of shapes in a plane.

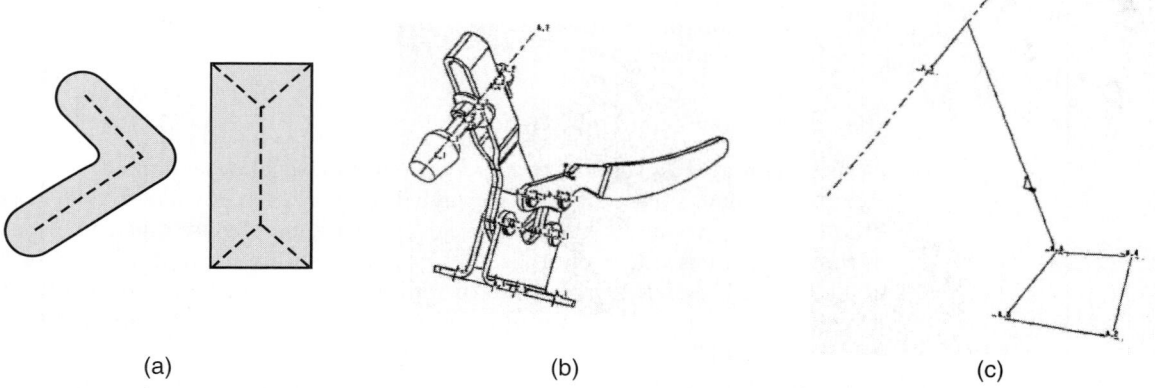

(a)   (b)   (c)

FIGURE 38.7   (a) Illustration of the skeleton data structure for solid objects. The skeleton is the dashed line in the object interior. (b) A skeleton model used for the kinematic analysis of a clamp. *Source:* S.D. Lockhart and C.M. Johnson, *Engineering Design Communication*, Prentice Hall, 2000.

The skeleton is the set of points that connect the centers of the circles or spheres. A current area of research involves using skeleton models instead of conventional surface or solid models.

CASE STUDY 38.1 | **CAD Model Development for Automotive Components**

CAD models are used extensively for a great variety of tasks. In the automotive industry, for example, it is especially important to have a detailed CAD model of a particular component in the product database in order to ensure that all of those who will be working on it have all of the data they need to perform their tasks. Special care is taken to build very precise CAD models of the automobile components that passengers will see and interact with on a regular basis. Examples of such components are outer-body panels, handles, seats, and the instrument panel (Fig. 38.8). The quality of the *visible* (Class I) surfaces has a major impact on overall vehicle quality and customer perception of the look and feel of the automobile.

FIGURE 38.8   Every vehicle component, from body panels to knobs on the instrument panel, has a solid model associated with it. *Source:* Courtesy of Ford Motor Company.

Two-dimensional concept sketches

Stylists with a background and experience in industrial design and/or art first develop two-dimensional concepts through a series of sketches. These sketches most frequently are drawn by hand, although software may be used instead, especially if the stylist starts with a photograph or a scanned drawing that needs to be modified. Concept sketches provide an overall feel for the aesthetics of the object and frequently are very detailed and show texture, color, and where individual surfaces on a vehicle should meet.

Most often, stylists are given a set of packaging constraints, such as (a) how the component should be assembled with other components, (b) what the size of the component should be, and (c) what the size and shape of any structure lying behind the visible surfaces should be. The time involved in producing a series of such concept sketches for an individual component or a set of components typically ranges from a few days to several weeks.

Three-dimensional surface model

As a concept is being reviewed and refined, several very accurate surface models of the component are constructed. To start the surface model, a computer-controlled optical scanner scans a conceptualized clay model, producing a *cloud-of-points* organized along the scan lines. Depending on the size of the component, a point cloud may consist of hundreds of thousands to millions of points. The point cloud is read into point-processing software (such as Paraform[1] or ICEM Surf[2]) to further organize the points and filter out noise. Scanning can take anywhere from several hours to a day to be completed. If, however, a digital three-dimensional clay model of the component already is available, it is converted into a point cloud organized into scan lines without the need for physical scanning and point-cloud post-processing.

---

[1]FreeForm is a product of SensAble Technologies, Inc.
[2]Paraform is a product of Paraform, Inc.

Next, the scan lines from the point cloud are used to construct mathematical surfaces using software such as ICEM Surf[3] and Alias/Wavefront StudioTools.[4] To construct the surfaces, first freeform NURBS curves that interpolate or approximate the scan lines are constructed. A NURBS surface patch then is fit through the curves. An individual surface patch models a small region of a single component's face. Several patches are constructed and joined smoothly at common edges to form the entire face.

Faces join each other at common edges to model an entire component. A great deal of experience is required to determine how to divide a face into a collection of patches that can be fit with the simplest low-order surfaces possible and still meet smoothly at common boundaries. A surfacing specialist performs this task in conjunction with the stylist to ensure that the surfaces are of high quality and that they capture the stylist's intent. A single component may take as long as a week to model.

Surface models are passed along to the various downstream departments to be used for tooling design, feasibility checks, analysis, and for the design of *nonvisible* (Class II) surfaces. As the design evolves, the dimensional tolerances on the surfaces are tightened gradually.

When designing outer-body panels, a major milestone is finalizing what is called *first flange and fillet*, shown in Fig. 38.9, in which the edges of the body panels are turned under or hemmed (see Fig. 16.23c) to provide a flange to connect to the inner panel (see Fig. 16.25). The shape of the flange and fillet affects the overall aesthetics and shape of the body panel; hence, their careful design is important. After the shape of the fillet and flange has been decided, the inner-panel designs can be completed.

Surface verification

Once the final surface model is completed, it must be verified and evaluated for surface quality and aesthetics. NC tool paths are generated automatically from the surfaces, either from within the surfacing software or through specialized machining software. The tool paths then are used to CNC machine the surfaces in clay. If the component is small, instead of machining a clay model, an STL file can be generated and the mock-up can be built on a rapid prototyping machine (see Chapter 20). It takes anywhere from several hours to several days to machine a component and perform any hand finishing that may be required.

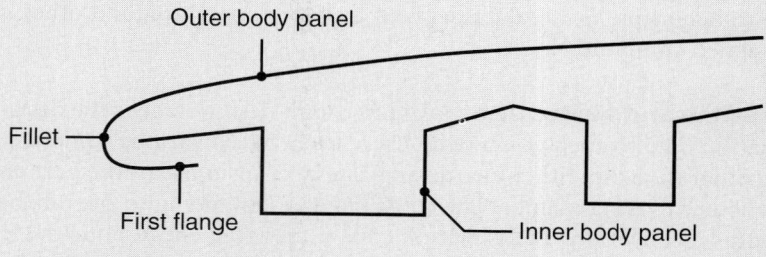

FIGURE 38.9   First flange and fillet.

[3]ICEM Surf is a product of ICEM Technologies, Inc.
[4]StudioTools is a product of Alias/Wavefront.

Clay models can be coated with a thin layer of latex material and painted to make them look more realistic and to help evaluate surface quality. The coating may be modified to improve surface smoothness between patches or to change the way light reflects off the surface of the model. All changes made to the clay must be translated back to the digital surface model, either through rescanning and refitting the surface patches or by tweaking the shape of the existing patches. At this stage of surface verification, only very minor changes to within ± 0.5 mm (0.04 in.) can be made.

Solid-model construction

After the surface model is finalized, it is used to develop a solid model. For sheet-metal components (such as body panels), body specialists offset the surfaces to form a solid. For other components (such as instrument panels, door handles, or wheels), they add manufacturing features to the surface model (such as flanges, bosses, and ribs). They collaborate with the manufacturers to determine what features must be added and where they should be placed to ensure that the component can be fabricated from the desired materials and at the target cost.

In the process of making a solid model, it may be discovered that the surface model must be modified because of (a) changes in packaging constraints, (b) the component failing to meet minimal manufacturing requirements, or (c) surfaces not matching up properly at common edges with the appropriate smoothness. These changes are communicated back to the surface modeler and the stylist so that the surfaces can be modified and re-verified. Finally, the solid model is entered into a product database where it is then available to suppliers and engineers for further analysis and manufacturing.

Source: Courtesy of A. Marsan and P. Stewart, Ford Motor Company.

Design analysis and optimization.   After the geometric features of a particular design have been determined, the design is subjected to an engineering analysis. This phase may consist of analyzing (for example) stresses, strains, deflections, vibrations, heat transfer, temperature distribution, or dimensional tolerances. Various sophisticated software packages are available, each having the capabilities to compute these quantities accurately and rapidly.

Because of the relative ease with which such analyses can now be done, designers increasingly are willing to analyze a design more thoroughly before it moves on to production. Experiments and measurements in the field nonetheless may be necessary to determine the actual effects of loads, temperature, and other variables on the designed components.

Design review and evaluation.   An important design stage is the design review and evaluation used to check for any interference between various components. This is done in order to avoid difficulties during assembly or in use of the part and to determine whether moving members (such as linkages) are going to operate as intended. Software is available with animation capabilities to identify potential problems with moving members and other dynamic situations. During the design review and evaluation stage, the part is dimensioned and toleranced precisely to the full degree required for manufacturing it.

Documentation and drafting.   After the preceding stages have been completed, the design is reproduced by automated drafting machines for documentation and reference. At this stage, detailed and working drawings also are developed and printed. The CAD system also is capable of developing and drafting sectional views of the part, scaling the drawings, and performing transformations in order to present various views of the part.

Database.   Many components either are standard components that are mass produced according to a given design specification (such as bolts or gears) or are identical to parts used in previous designs. Modern CAD systems thus have a built-in database-management system that allows designers to locate, view, and adopt parts from a stock part library. These parts can be modeled parametrically to allow cost-effective updating of the geometry. Some databases are available commercially with extensive parts libraries; many vendors make their part libraries available on the world wide web.

38.5 | Computer-Aided Manufacturing

Computer-aided manufacturing (CAM) involves the use of computers to assist in all phases of manufacturing a product. It encompasses many of the technologies described in Chapter 37 and in this chapter. Because of the joint benefits, computer-aided design and computer-aided manufacturing often are combined into **CAD/CAM systems**. This combination allows the transfer of information from the design stage into the stage of planning for manufacture without the need to reenter the data on part geometry manually. The database developed during CAD is stored and processed further by CAM into the necessary data and instructions for operating and controlling production machinery, material-handling equipment, and automated testing and inspection for product quality. CAD/CAM systems also are capable of coding and classifying parts into groups that have similar shapes using alphanumeric coding. (See *group technology*, Section 38.8.)

An important feature of CAD/CAM in machining operations is its capability to describe the *tool path*. (See Figs. 23.11, 23.12, 24.2, 25.9, 26.12, and 26.20.) The instructions (programs) are computer generated, and they can be modified by the programmer to optimize the tool path. The engineer or technician then can display and check the tool path for possible tool collisions with clamps, fixtures, or other interferences visually.

By standardizing product development and reducing design effort, tryout, and prototype work, CAD/CAM has made possible significantly reduced manufacturing costs and improved productivity. For example, the two-engine Boeing 777 passenger airplane was designed completely by computer (**paperless design**) with 2000 workstations linked to eight computers. The plane was constructed directly from the CAD/CAM software developed (an enhanced CATIA system), and no prototypes or mockups were built—as were required for previous models. The cost for this development was on the order of $6 billion.

Some typical applications of CAD/CAM are as follows:

- Programming for numerical control and industrial robots.
- Design of dies and molds for casting, for example, in which shrinkage allowances are preprogrammed.
- Dies for metalworking operations, such as complex dies for sheet forming and progressive dies for stamping.
- Design of tooling and fixtures and EDM electrodes.

- Quality control and inspection, such as coordinate-measuring machines programmed on a CAD/CAM workstation.
- Process planning and scheduling.
- Plant layout.

38.6 | Computer-Aided Process Planning

Process planning is concerned with selecting methods of production: tooling, fixtures, machinery, sequence of operations, and assembly. All of these diverse activities must be planned, which traditionally has been done by process planners. The sequence of processes and operations to be performed, the machines to be used, the standard time for each operation, and similar information all are documented on a **routing sheet** (Fig. 38.10).

When done manually, this task is highly labor-intensive and time-consuming and relies heavily on the experience of the process planner. Modern practice in routing sheets is to store the relevant data in computers and affix a *bar code* (or other identification) to the part. The data can then be reviewed at a dedicated monitor.

ROUTING SHEET		
CUSTOMER'S NAME: Midwest Valve Co.		PART NAME: Valve body
QUANTITY: 15		PART NO.: 302
Operation No.	Description of operation	Machine
10	Inspect forging, check hardness	Rockwell tester
20	Rough machine flanges	Lathe No. 5
30	Finish machine flanges	Lathe No. 5
40	Bore and counter bore hole	Boring mill No. 1
50	Turn internal grooves	Boring mill No. 1
60	Drill and tap holes	Drill press No. 2
70	Grind flange end faces	Grinder No. 2
80	Grind bore	Internal grinder No. 1
90	Clean	Vapor degreaser
100	Inspect	Ultrasonic tester

FIGURE 38.10   An example of a simple routing sheet. These *operation sheets* may include additional information on materials, tooling, estimated time for each operation, processing parameters (such as cutting speeds and feeds), and other information. The routing sheet travels with the part from operation to operation. The current practice is to store all relevant data in computers and to affix to the part a bar code that serves as a key to the database of parts information.

Computer-aided process planning (CAPP) accomplishes this complex task by viewing the total operation as an integrated system, so that the individual steps involved in making each part are coordinated with others and are performed efficiently and reliably. Although CAPP requires extensive software and good coordination with CAD/CAM (as well as other aspects of integrated manufacturing systems discussed in the rest of this chapter), it is a powerful tool for efficiently planning and scheduling manufacturing operations. CAPP is effective particularly in small-volume, high-variety parts production.

38.6.1 Elements of CAPP systems

There are two types of computer-aided process-planning systems: *variant* and *generative*.

Variant system.   Also called the **derivative system**, these computer files contain a standard process plan for the part to be manufactured. A search for a standard plan (based on its shape and its manufacturing characteristics) is made in the database by a code number for the part. The plan is retrieved, displayed for review, and printed as a routing sheet.

The variant-process plan includes information such as the types of tools and machines to be used, the sequence of manufacturing operations to be performed, the speeds, the feeds, and the time required for each sequence. Minor modifications of an existing process plan (which usually are necessary) also can be made. In the variant system, if the standard plan for a particular part is not in the computer files, a plan that is close to it (with a similar code number and an existing routing sheet) is retrieved. If a routing sheet does not exist, one is made for the new part and stored in computer memory.

Generative system.   A process plan is automatically generated on the basis of the same logical procedures that would be followed by a traditional process planner in making that particular part. However, the generative system is complex because it must contain comprehensive and detailed information of the part shape and dimensions; process capabilities; selection of manufacturing methods, machinery, and tools; and the sequence of operations to be performed. These capabilities of computers (known as **expert systems**) are described in Section 39.8.

The generative system is capable of creating a new plan instead of having to use and modify an existing plan (as the variant system must). Although generally used less commonly than the variant system, this system has such advantages as (a) flexibility and consistency for the process planning for new parts and (b) higher overall planning quality because of the capability of the decision logic in the system to optimize the planning and to utilize up-to-date manufacturing technology.

The process-planning capabilities of computers also can be integrated into the planning and control of production systems. These activities are a subsystem of computer-integrated manufacturing, as described in Section 38.3. Several functions can be performed, such as **capacity planning** for plants to meet production schedules, control of inventory, purchasing, and production scheduling.

38.6.2 Material-requirements planning and manufacturing resource planning

Computer-based systems for managing inventories and delivery schedules of raw materials and tools are called *material-requirements planning* (MRP). This activity (also regarded as a method of *inventory control*) involves the keeping of complete records of inventories of materials, supplies, parts in various stages of production (called work in progress or WIP), orders, purchasing, and scheduling.

Several files of data usually are involved in a master production schedule. These files pertain to the raw materials required (*bill of materials*), product structure levels (individual items that compose a product, such as components, subassemblies, and assemblies), and scheduling.

Manufacturing resource planning (MRP-II) controls all aspects of manufacturing planning through feedback. Although the system is complex, it is capable of final production scheduling, of monitoring actual results in terms of performance and output, and of comparing those results against the master production schedule.

38.6.3 Enterprise resource planning

Beginning in the 1990s, *enterprise resource planning* (ERP) became an important trend. It is basically an extension of MRP-II, and although there are variations, it also is a method for effective planning and control of all the resources needed in a business enterprise (i.e., companies) to take orders for products, produce them, ship them to the customer, and service them. ERP thus attempts to coordinate, optimize, and dynamically integrate all information sources and the widely diverse technical and financial activities in a manufacturing organization.

Effective implementation of ERP is a difficult and challenging task because of the following factors:

- The difficulties encountered in timely, effective, and reliable communication among all parties involved, especially in a global business enterprise; thus, teamwork is essential.

- The need for changing and evolving business practices in an age where information systems and e-commerce have become highly relevant to the success of business organizations.

- Meeting extensive and specific hardware and software requirements for ERP.

38.7 Computer Simulation of Manufacturing Processes and Systems

With the increasing sophistication of computer hardware and software, *computer simulation* of manufacturing processes and systems has advanced rapidly. Simulation takes two basic forms:

- It is a model of a specific operation intended to determine the viability of a process or to optimize or improve its performance.

- It models multiple processes and their interactions to help process planners and plant designers in the layout of machinery and facilities.

Individual processes have been modeled using various mathematical schemes (See Fig. 10.17). Finite-element analysis has been applied increasingly in software packages (**process simulation**) that are available commercially and are inexpensive. Typical problems addressed are *process viability* (such as the formability of sheet metal in a certain die) and *process optimization* (such as material flow in forging a given die to identify potential defects, or mold design in casting to eliminate hot spots, promote uniform cooling, and minimize defects).

Simulation of an entire manufacturing system involving multiple processes and equipment helps plant engineers to organize machinery and to identify critical machinery elements. In addition, such models can assist manufacturing engineers with scheduling and routing (by *discrete event simulation*). Commercially available software packages often are used for these simulations, but dedicated software programs written for a specific company also can be developed.

EXAMPLE 38.1 **Simulation of plant-scaled manufacturing**

The proliferation of low-cost, high-performance computer systems and the development of advanced software have allowed the simulation of plant-scaled manufacturing systems, and have led to the optimization of manufacturing and assembly operations.

As an example, Envision software, produced by the Delmia Corporation, allows the simulation of manufacturing processes in three dimensions (including human manikins) to identify safety hazards, manufacturing problems, or bottlenecks; improve machining accuracy; or optimize tooling organization. Since simulation can be performed prior to building an assembly line, it significantly can reduce development time and cost. Figure 38.11 shows a simulation of a robotic welding line in an automotive plant where a collision between a robot and a vehicle has been detected. The robot program can be modified to prevent such collisions before the line is put into operation. While this is a powerful demonstration of the utility of system simulation, the more common application is to optimize the sequence of operations and organization of machinery to reduce manufacturing costs.

Software also has the capability of conducting ergonomic analyses of various operations and machinery setups and, thus, can identify bottlenecks in

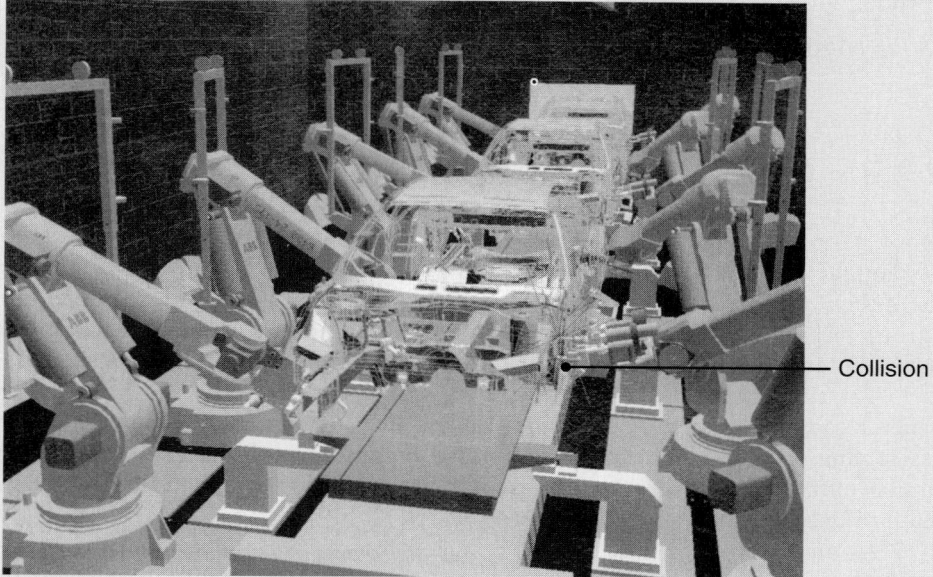

— Collision

FIGURE 38.11   Simulation of a robotic welding station. A collision has been detected that production engineers can rectify before building the assembly line, thus reducing development time and cost.

the movement of parts, equipment, or personnel. The bottleneck then can be relieved by the process planner by adjusting the automated or manual procedures at these locations. Using such techniques, a Daimler-Chrysler facility in Rastatt, Germany, was able to balance its production lines so that each worker is busy an average of 85 to 95% of the time.

Another application of systems simulation is to plan manufacturing operations to optimize production and to plan for just-in-time production (Section 39.5). For example, if a car manufacturer needs to produce 1000 vehicles in a given period, production can be optimized by certain strategies (such as distributing the number of vehicles with sunroofs throughout the day, or grouping vehicles by color), so the number of paint changes in paint booths are minimized. Software (such as that produced by ILOG Corporation) can plan and schedule plant operations far enough in advance to order material as it is needed, thus eliminating stockpiled inventory.

The Nissan Corporation used the ILOG software to expand production at its Sunderland, England, facility. The plant already was producing 7000 vehicles per week when Nissan decided to increase production by 3000 vehicles per week. The conventional solution would have been to construct an additional assembly line at great capital investment. However, it was noted that the existing scheduling software (based on plant-manager input) did not take into account variables such as part availability and tool changeover; the plant was adhering to its schedule only three percent of the time.

The ILOG software provides optimization of the manufacturing sequence and schedule and enables cars to cross over from assembly lines to paint lines and tooling lines (as needed) to increase throughput. The result was that the Sunderland plant met the vehicle scheduling targets 85% of the time by eliminating much of the time-consuming task of rescheduling vehicles while they wait in storage buffers between the main sections of the plant. This approach led to increased production, and the cost of the software was recovered in three days of production.

Source: Adapted from *Mechanical Engineering*, March 2001.

38.8 | Group Technology

Many parts in products have certain similarities in their shape and in their method of manufacture. *Group technology* (GT) is a concept that seeks to take advantage of the *design and processing similarities* among the parts to be produced. The term "group technology" first was used in 1959, but not until the use of interactive computers became widespread in the 1970s did this technology develop significantly.

The similarity in the characteristics of similar parts (Fig. 38.12) suggests that benefits can be obtained by *classifying* and *coding* these parts into families. By disassembling each product into its individual components and then identifying the similar parts, one company found that 90% of the 3000 parts made by a company fell into only five major families of parts.

As an example, a pump can be broken into such basic components as the motor, housing, shaft, flanges, and seals. In spite of the variety of pumps manufactured, each of these components is basically the same in terms of its design and manufacturing characteristics. Consequently, all shafts can be placed in one *family* of shafts and so on. Furthermore, questions also have been raised as to why a particular product should have so many different sizes of fasteners.

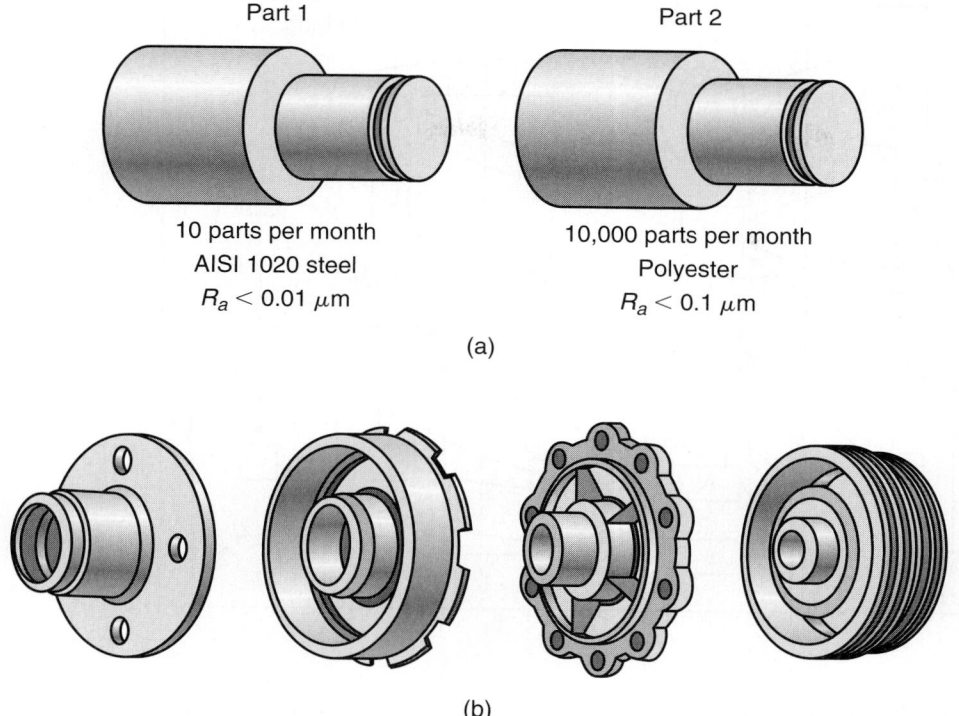

FIGURE 38.12   Grouping parts according to their (a) geometric similarities and (b) manufacturing attributes.

Group technology becomes especially attractive because of the ever-greater variety of products available to consumers, which are often produced in batches. Since nearly 75% of manufacturing today is batch production, improving the efficiency of batch production becomes important.

The traditional product flow in batch manufacturing is shown in Fig. 38.13a. Note that machines of the same type are arranged in groups—that is, groups of lathes, milling machines, drill presses, and grinders. In such a layout (called **functional layout**), there usually is considerable random movement, as shown by the arrows indicating movement of materials and parts. Such an arrangement is not efficient, because it wastes time and effort. A more efficient product flow line to take advantage of group technology is the **group layout** (Fig. 38.13b). (See also *cellular manufacturing*, Section 39.2.)

38.8.1 Advantages of group technology

The major advantages of group technology include the following:

- It makes possible the standardization of part design and the minimization of design duplication. New part designs can be developed using similar (yet previously used) designs, and in this way, a significant amount of time and effort can be saved. The product designer quickly can determine whether data on a similar part already exists in the computer files.

- Data that reflect the experience of the designer and the manufacturing process planner are stored in the database. Thus, a new and less experienced engineer quickly can benefit from that experience by retrieving any of the previous designs and process plans.

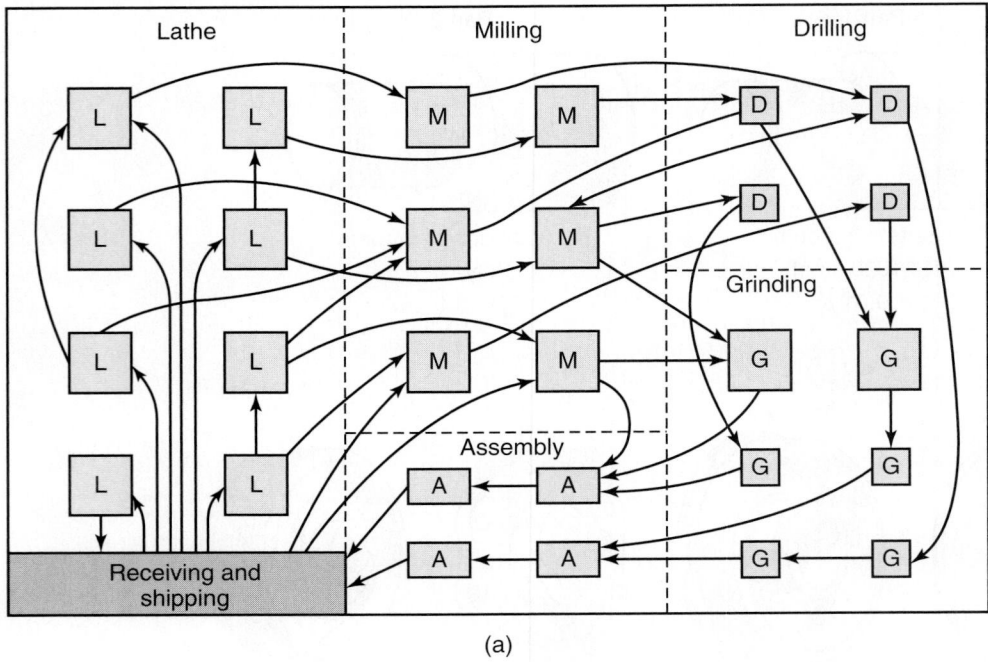

(a)

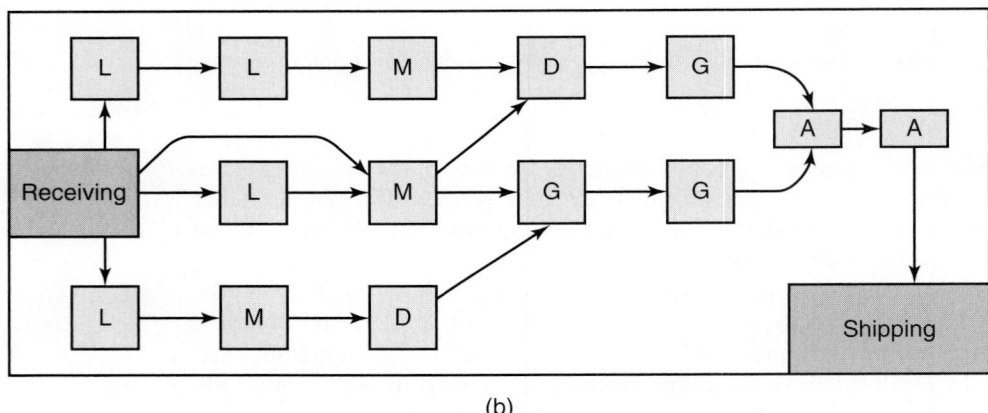

(b)

FIGURE 38.13   (a) Functional layout of machine tools in a traditional plant. Arrows indicate the flow of materials and parts in various stages of completion. (b) Group-technology (cellular) layout. *Legend:* L = lathe, M = milling machine, D = drilling machine, G = grinding machine, A = assembly. *Source:* After M.P. Groover.

- Manufacturing costs can be estimated more easily, and the relevant statistics on materials, processes, number of parts produced, and other factors can be obtained more easily.

- Process plans are standardized and scheduled more efficiently, orders are grouped for more efficient production, and machine utilization is improved. Setup times are reduced, and parts are produced more efficiently and with better and more consistent product quality. Similar tools, fixtures, and machinery are shared in the production of a family of parts. Programming for NC is automated more fully.

- With the implementation of CAD/CAM, cellular manufacturing, and CIM, group technology is capable of greatly improving the productivity and reducing the costs in batch production—approaching the benefits of mass production. Depending on the level of implementation, potential savings in each of the various design and manufacturing phases can range from 5 to 75%.

38.8.2 Classification and coding of parts

In group technology, parts are identified and grouped into families by **classification and coding (C/C) systems**. This process is a critical and complex first step and is done according to the part's design attributes and manufacturing attributes. (See Fig. 38.12.)

Design attributes.   These pertain to similarities in geometric features and consist of the following:

- External and internal shapes and dimensions
- Aspect ratios (such as length-to-width or length-to-diameter)
- Dimensional tolerances
- Surface finish
- Part functions

Manufacturing attributes.   This system uses the similarities in the methods and sequence of the manufacturing operations performed on the part. As we have seen, selection of a manufacturing process (or processes) depends on many factors, among which are the shape, the dimensions, and other geometric features of the part. Consequently, manufacturing and design attributes are interrelated. The manufacturing attributes of a part consist of the following:

- Primary processes used
- Secondary and finishing processes used
- Dimensional tolerances and surface finish
- Sequence of operations performed
- Tools, dies, fixtures, and machinery used
- Production quantity and production rate

Coding can be time consuming, and considerable experience is required. The coding can be done simply by viewing the shapes of the parts in a generic way and then classifying the parts accordingly (such as parts having rotational symmetry, parts having rectilinear shape, and parts having large surface-to-thickness ratios). The parts being reviewed and classified should be representatives of the company's product lines. A more thorough method is to review all of the data and drawings concerning the design *and* manufacture of all of the parts.

Parts also may be classified by studying their production flow during the manufacturing cycle—an approach called **production flow analysis** (PFA). Recall from Section 38.6 that routing sheets clearly show process plans and the sequence of operations to be performed. One drawback to PFA is that a particular routing sheet does not necessarily indicate that the total manufacturing operation is optimized.

38.8.3 Coding

The code for parts can be based on a company's own system of coding, or it can be based on one of several classification and coding systems available commercially.

Whether it was developed in-house or it was purchased, the system must be compatible with the company's other systems (such as NC machinery and CAPP systems). The code structure for part families typically consists of numbers, letters, or a combination of the two. Each specific component of a product is assigned a code. This code may pertain to design attributes only (generally less than 12 digits) or to manufacturing attributes only (although most advanced systems include both, using as many as 30 digits).

The three basic *levels of coding* vary in degree of complexity.

Hierarchical coding.   In this code (also called **monocode**), the interpretation of each succeeding digit depends on the value of the preceding digit. Each symbol amplifies the information contained in the preceding digit; therefore, a digit in the code cannot be interpreted alone. The advantage of this system is that a short code can contain a large amount of information. However, this method is difficult to apply in a computerized system.

Polycodes.   Each digit in this code (also known as **chain type**) has its own interpretation, which does not depend on the preceding digit. This structure tends to be relatively long, but it allows the identification of specific part attributes and is well suited to computer implementation.

Decision-tree coding.   This system (also called **hybrid codes**) is the most advanced, and it combines both design and manufacturing attributes (Fig. 38.14).

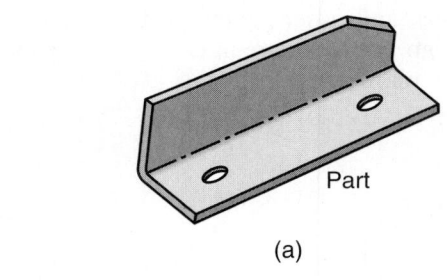

(a)

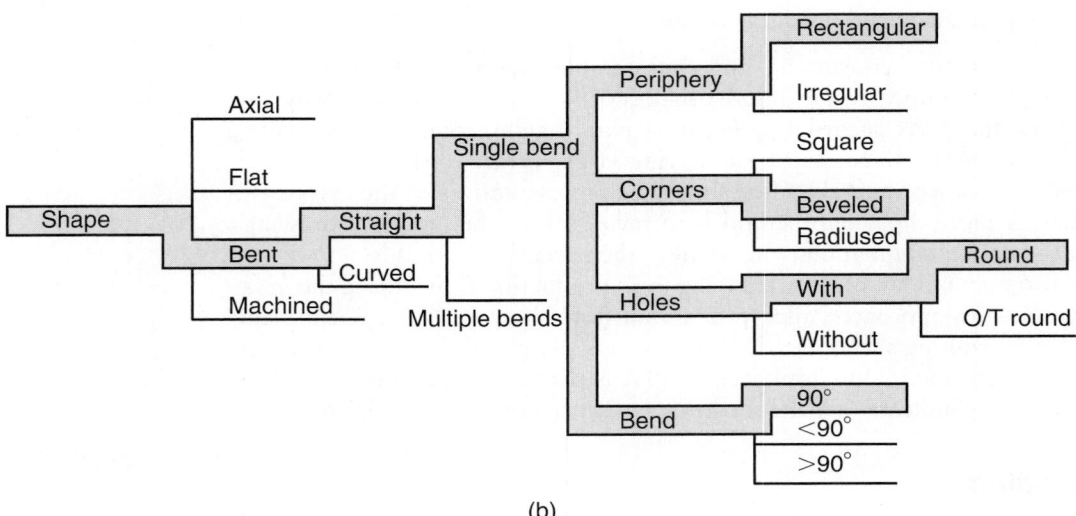

(b)

FIGURE 38.14   Decision-tree classification for a sheet-metal bracket. *Source*: After G. W. Millar.

38.8.4 Coding systems

Three major industrial coding systems are described here.

1. The **Opitz system** was developed in the 1960s in Germany by H. Opitz (1905–1977), and was the first comprehensive coding system presented. The basic code consists of nine digits (12345 6789) representing design and manufacturing data (Fig. 38.15). Four additional codes (ABCD) may be used to identify the type and sequence of production operations.

		Form Code					Supplementary Code			
		1st digit Part class	2nd digit Main shape	3rd digit Rotational surface machining	4th digit Plane surface machining	5th digit Auxiliary holes, gear teeth, forming	Digit 1	2	3	4
0	Rotational parts	$\frac{L}{D} \leqq 0.5$	External shape, external shape elements	Internal shape, internal shape elements	Plane surface machining	Auxiliary holes	Dimension	Material	Original shape of raw material	Accuracy
1		$0.5 < \frac{L}{D} < 3$								
2		$\frac{L}{D} \geqq 3$				Gear teeth				
3		$\frac{L}{D} \leqq 2$ with deviation	Main shape	Rotational machining, internal and external shape elements		Auxiliary holes, gear teeth, and forming				
4		$\frac{L}{D} > 2$ with deviation								
5		Special								
6	Nonrotational parts	$\frac{A}{B} \leqq 3, \frac{A}{C} \geqq 4$ Flat parts	Main shape	Principal bores	Plane surface machining	Auxiliary holes, gear teeth, and forming				
7		$\frac{A}{B} > 3$ Long parts	Main shape							
8		$\frac{A}{B} \leqq 3, \frac{A}{C} \geqq 4$ Cubic parts	Main shape							
9		Special								

FIGURE 38.15 Classification and coding system according to Opitz, consisting of a form code of 5 digits and a supplementary code of 4 digits.

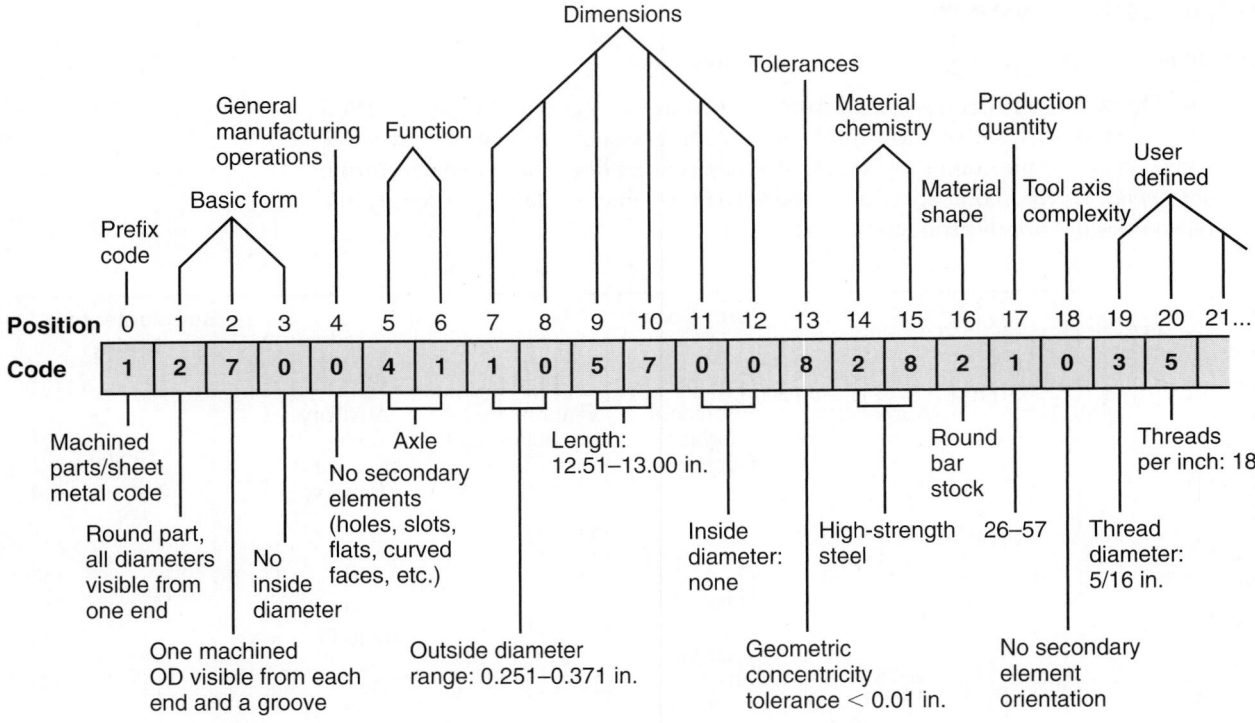

FIGURE 38.16   Typical multiClass code for a machined part. *Source*: Courtesy of Organization for Industrial Research.

Digit	Items	(Rotational component)
1	Parts name	General classification
2		Detail classification
3	Materials	General classification
4		Detail classification
5	Major dimensions	Length
6		Diameter
7	Primary shapes and ratio of major dimensions	
8	Shape details and kinds of processes / External surface	External surface and outer primary shape
9		Concentric screw-threaded parts
10		Functional cut-off parts
11		Extraordinary shaped parts
12		Forming
13		Cylindrical surface
14	Internal surface	Internal primary shape
15		Internal curved surface
16		Internal flat and cylindrical surface
17	End surface	
18	Nonconcentric holes	Regularly located holes
19		Special holes
20	Noncutting process	
21	Accuracy	

FIGURE 38.17   The structure of a KK-3 system for rotational components. *Source*: Courtesy of Japan Society for the Promotion of Machine Industry.

This system has two drawbacks: (a) it is possible to have different codes for parts that have similar manufacturing attributes, and (b) a number of parts with different shapes can have the same code.

2. The **multiClass system** was developed to help automate and standardize several design, production, and management functions and involves up to 30 digits (Fig 38.16). It is used interactively with a computer that asks the user a number of questions. On the basis of the answers given, the computer automatically assigns a code number to the part.

3. The **KK-3 system** is a general-purpose system for parts that are to be machined or ground. It uses a 21-digit decimal system. This code is much greater in length than the two previous systems described, but it classifies dimensions and dimensional ratios, such as the length-to-diameter of the part. The structure of a KK-3 system for rotational components is shown in Fig. 38.17.

SUMMARY

- Integrated manufacturing systems are implemented to various degrees to optimize operations, improve product quality, and reduce production costs.

- Computer-integrated manufacturing processes have become the most important means of improving productivity, responding to rapidly changing market demands, and improving the control of both manufacturing and the management functions of an organization.

- Advanced software and powerful computers allow the description of part geometry in several different formats, including wireframe, octree, surface models, solid models, skeletons, and boundary representations.

- Computers also are used to simulate manufacturing operations and systems, as well as to aid in the selection of manufacturing processes.

- Group technology is a powerful approach that allows the rapid recovery of previous designs and manufacturing experiences by encoding a part based on its geometric features or manufacturing concerns. A number of group-technology coding systems are available.

KEY TERMS

Classification and coding systems	Design attributes	Octree representation
Coding	Enterprise resource planning	Paperless design
Computer-aided design and engineering	Exchange specifications	Process planning
Computer-aided manufacturing	Functional layout	Production flow analysis
Computer-aided process planning	Group layout	Routing sheet
Computer-integrated manufacturing	Group technology	Solid model
Computer simulation	Manufacturing attributes	Surface model
Data-acquisition system	Manufacturing resource planning	Wire frame
Database	Material-requirements planning	
	Modeling	

BIBLIOGRAPHY

Amirouche, F.M.L., *Principles of Computer-Aided Design and Manufacturing 2nd ed.*, Prentice Hall, 2003.

Badiru, A.B., *Expert Systems Applications in Engineering and Manufacturing*, Prentice Hall, 1992.

Burbidge, J.L., *Production Flow Analysis for Planning Group Technology*, Oxford, 1997.

Chang, T.-C., Wysk, R.A., and Wang, H.P., *Computer-Aided Manufacturing*, 2nd ed., Prentice Hall, 1997.

Chorafas, D.N., *Expert Systems in Manufacturing*, Van Nostrand Reinhold, 1992.

Corbett, J., Dooner, M., Meleka, J., and Pym, C., *Design for Manufacture: Strategies, Principles and Techniques*, Addison-Wesley, 1991.

Famili, A., Nau, D.S., and Kim, S.H., *Artificial Intelligence Applications in Manufacturing*, American Association for Artificial Intelligence, 1992.

Foston, A.L., Smith, C.L., and Au, T., *Fundamentals of Computer-Integrated Manufacturing*, Prentice Hall, 1991.

Gu, P. and Norrie, D.H., *Intelligent Manufacturing Planning*, Chapman & Hall, 1995.

Higgins, P., Roy, L.R., and Tierney, L., *Manufacturing Planning and Control: Beyond MRP II*, Chapman & Hall, 1996.

Kusiak, A., *Concurrent Engineering: Automation, Tools, and Techniques*, Wiley, 1993.

_____, *Intelligent Design and Manufacturing*, Wiley, 1992.

_____, *Intelligent Manufacturing Systems*, Prentice Hall, 1990.

McMahon, C. and Browne, J., *CADCAM—From Principles to Practice*, Addison-Wesley, 1993.

Vajpayee, S.K., *Principles of Computer-Integrated Manufacturing*, Prentice Hall, 1998.

Williams, D.J., *Manufacturing Systems*, 2nd ed., Chapman & Hall, 1994.

Wu, J.-K., *Neural Networks and Simulation Methods*, Marcel Dekker, 1994.

REVIEW QUESTIONS

38.1   In what ways have computers had an impact on manufacturing?

38.2   Discuss the benefits of computer-integrated manufacturing operations.

38.3   What is a database? Why is it necessary? Why should the management of a company have access to databases?

38.4   What are the differences between the terms "computer-aided" and "computer-integrated"?

38.5   Explain how a CAD system operates.

38.6   What are the advantages of CAD systems over traditional methods of design? Are there any limitations?

38.7   Describe the purposes of process planning. How are computers used in such planning?

38.8   Explain the features of two types of CAPP systems.

38.9   Describe the features of a routing sheet. Why is it necessary?

38.10   What is group technology? Why was it developed? Explain its advantages.

38.11   What is a NURB?

38.12   Describe what is meant by the term "manufacturing" system.

38.13   What does classification and coding mean in group technology?

QUALITATIVE PROBLEMS

38.14   Describe your observations regarding Fig. 38.1.

38.15   Give examples of primitives of solids other than those shown in Fig. 38.4a and b.

38.16   Describe your understanding of the octree representation in Fig. 38.6.

38.17   Explain the logic behind the arrangements shown in Fig. 38.13b.

38.18   Describe your observations regarding Fig. 38.2.

38.19   What are the advantages of hierarchical coding?

38.20   Referring to Fig. 38.3, what are the advantages of a third-order piecewise Bezier curve over a B-spline or a conventional Bezier curve?

38.21   Describe situations that would require the design change at its larger end of the part in Fig. 38.5.

SYNTHESIS, DESIGN, AND PROJECTS

38.22 Review various manufactured parts described in this book and group them in a manner similar to those shown in Fig. 38.12.

38.23 Think of a product and make a decision-tree chart similar to that shown in Fig. 38.14.

38.24 Outline the benefits of a manufacturing system, giving specific examples.

38.25 How would you describe the principle of computer-aided manufacturing to an older worker in a manufacturing facility who is not familiar with computers?

38.26 Think of a simple product and make a routing sheet, similar to that shown in Fig. 38.10. If the same part is given to another person, what is the likelihood that the routing sheet developed will be the same? Explain.

38.27 Review Fig. 38.10, then suggest a routing sheet for one of the following: (a) an automotive connecting rod, (b) a compressor blade, (c) a glass bottle, (d) an injection-molding die, or (e) a bevel gear.

This chapter shows how computer systems and communications networks impact product development and manufacturing through the integration of all of its activities. Specifically, we describe:

• The principles of manned and untended manufacturing cells and their features.
• How cells are integrated into flexible manufacturing systems.
• The new concept of holonic manufacturing and its applications.
• The principles of just-in-time manufacturing and lean manufacturing and their benefits.
• Importance and features of communication systems in manufacturing.
• How artificial intelligence and expert systems are applied to manufacturing.

39.1 | Introduction

On several occasions, we have described the implementation and benefits of automation and computer control at various stages of manufacturing operations. This chapter focuses on the **computer integration** of **manufacturing activities**. Integration means that manufacturing processes, operations, and their management are treated as a system.

A major advantage of such an approach is that machines, tooling, and manufacturing operations now acquire a built-in flexibility, called **flexible manufacturing systems**. As a result, the system is capable of rapidly responding to changes in product types and fluctuating demands, as well as ensuring *on-time delivery* of products to the customer. Failure of on-time delivery in a highly competitive global environment can upset management plans and production schedules severely and, consequently, can have major adverse effects on a company's operations.

This chapter describes the key elements that enable the execution of the functions necessary for such a flexible system. We begin with *cellular manufacturing*, which is the basic unit of flexibility in the production of goods. We show that manufacturing cells can be broadened into *flexible manufacturing systems*, with major implications for the production capabilities of an operation. *Holonic manufacturing* is then described, which is a new concept of how manufacturing units can be organized for higher efficiency of operation.

The important concept of *just-in-time production* is described, in which parts are produced "just in time" to be made into subassemblies, assemblies, and final products. This method eliminates the need for inventories (which can be a financial burden to a company) as well as significantly saving on space and storage facilities. Because of the necessity and extensive use of computer controls, hardware, and software in all the activities outlined above, the planning and effective implementation of *communication networks* is a critical component of the overall operation.

This chapter concludes with a review of *artificial intelligence*, which consists of expert systems, natural language processing, machine vision, artificial neural networks, and fuzzy logic. We describe how these developments impact manufacturing activities and their future.

39.2 | Cellular Manufacturing

The concept of group technology (which we described in Section 38.8) can be implemented effectively in *cellular manufacturing*. A **manufacturing cell** is a small unit, consisting of one to several workstations. A workstation usually contains either one machine (called a *single-machine cell*) or several machines (called a *group-machine cell*) with each machine performing a different operation on the part. The machines can be modified, retooled, and regrouped for different product lines within the same family of parts.

Cellular manufacturing primarily has been utilized in machining and sheet-metal forming operations. The machine tools commonly used in the cells are lathes, milling machines, drills, grinders, and electrical-discharge machines. For sheet forming, the equipment typically consists of shearing, punching, bending, and other forming machines. This equipment also may include special-purpose machines and CNC machines. Automated inspection and testing equipment also are generally a part of this cell.

The capabilities of cellular manufacturing typically involve the following operations:

- Loading and unloading raw materials and workpieces at workstations.
- Changing tools at workstations.
- Transferring workpieces and tooling between workstations.
- Scheduling and controlling the total operation in the cell.

In attended (*manned*) machining cells, materials can be moved and transferred manually by the operator (unless the parts are too heavy or the movements are too hazardous) or by an industrial robot located centrally in the cell.

Flexible manufacturing cells (FMC).   Manufacturing cells can be made flexible by using machining centers, CNC machines, and industrial robots or other mechanized systems for handling materials and work in progress. An example of a *manned, flexible manufacturing cell* (FMC) for machining operations is shown in Fig. 39.1. This is a manned (tended) cell, consisting of machine tools such as lathes, milling machines, a grinder, and an inspection and measuring station (such as a coordinate-measuring machine). This cell can be manned by a single operator, or two operators if there are more machine tools in the cell. Flexible manufacturing cells also can be designed and operated with a *central robot*. Cell design and operation are more exacting, as the machines, the robots, the end effectors, and the control systems must function properly.

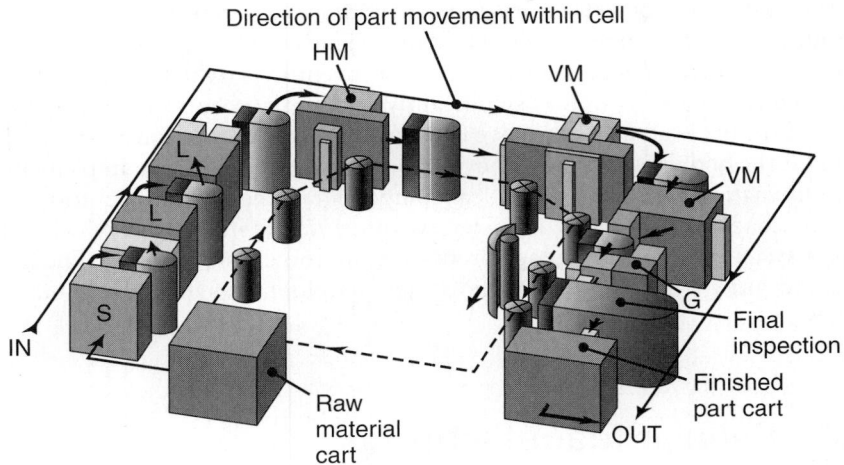

FIGURE 39.1   Schematic illustration of a manned, flexible manufacturing cell showing various machine tools and an inspection station. S, saw; L, lathe; HM, horizontal milling machine; VM, vertical milling machine; G, grinder; x, worker positions. *Source:* After J T. Black.

Cell design.   Because of the unique features of manufacturing cells, their design and placement in traditional plants requires the reorganization of the plant and the rearrangement of existing product flow lines. The machines may be arranged along a line, in a U-shape, an L-shape, or a loop. Selecting the best machine and material-handling equipment arrangement also involves taking into account such factors as the production rate, the type of product, and its shape, size, and weight.

The likelihood of significant change in demand for part families also should be considered during cell design to ensure that the machines and equipment involved have the proper flexibility and capacity. The cost of flexible cells can be high, but this is outweighed by increased productivity, flexibility, and controllability.

EXAMPLE 39.1 Manufacturing cells in a small machine shop

The following is an actual example of the application of the manufacturing-cell concept in a small shop. Company A has only 10 employees, 11 milling machines, and 11 machining centers. These machines are set up in cells (milling cells and turning cells). The machines in the cells are arranged to allow an operator to machine the part in the most efficient and precise manner. Each cell allows the operator to monitor the performance of the machines in the cell.

Over 1200 different product lots have been produced over the years with quantities ranging from one part to as many as 35,000 parts of the same design. The parts are inspected as they are produced. Each employee in the shop is involved in the programming and the running of the machines and in the in-process inspection of parts.

39.3 | Flexible Manufacturing Systems

A *flexible manufacturing system* (FMS) integrates all of the major elements of manufacturing into a highly automated system (Fig 39.2). First utilized in the late 1960s, an FMS consists of a number of manufacturing cells, each containing an industrial robot (serving several CNC machines) and an automated material-handling system, and all interfaced with a central computer. Different computer instructions can be downloaded for each successive part passing through a particular workstation. The system can handle a *variety of part configurations* and produce them *in any order*. A general view of an FMS installation in a plant is shown in Fig. 39.3.

This highly automated system is capable of optimizing each step of the total operation. These steps may involve (a) one or more processes and operations, such as machining, grinding, cutting, forming, powder metallurgy, heat treating, and finishing, (b) handling of raw materials, (c) measurement and inspection, and (d) assembly. The most common applications of FMS to date have been in machining and assembly operations.

FMS can be regarded as a system that combines the benefits of two other systems: (1) the highly productive but inflexible transfer lines and (2) job-shop production, which can produce large product variety on stand-alone machines but is inefficient. The relative characteristics of transfer lines and FMS are shown in Table 39.1. Note that in FMS, the time required for changeover to a different part is very short. The quick response to product and market-demand variations is a major attribute of FMS.

Compared to conventional manufacturing systems, the major benefits of FMS are the following:

- Parts can be produced randomly, in batch sizes as small as one, and at a lower unit cost.

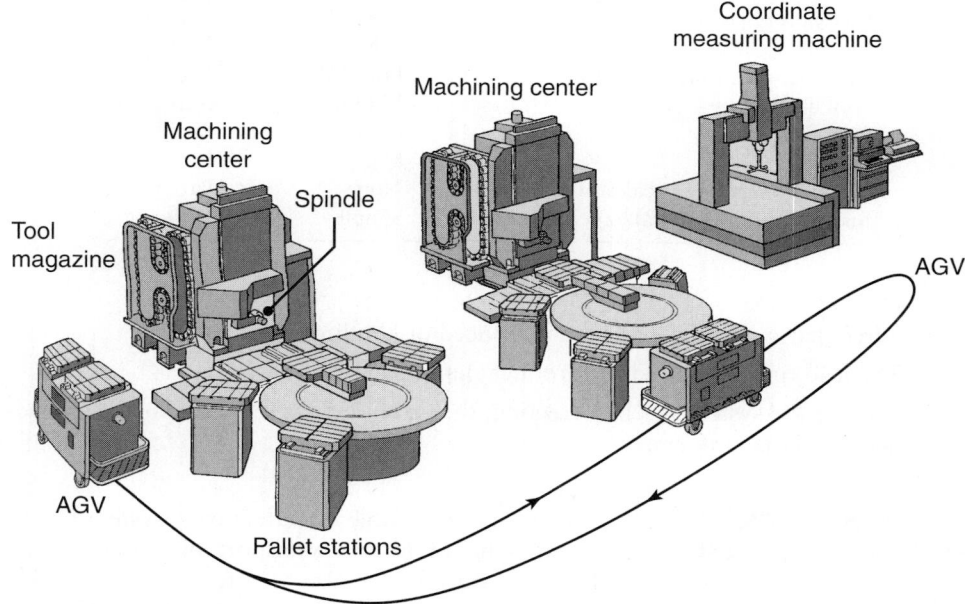

FIGURE 39.2 A schematic illustration of a flexible manufacturing system showing machining centers, a measuring and inspection station, and automated guided vehicles. *Source*: After J T. Black.

FIGURE 39.3   A general view of a flexible manufacturing system in a plant showing several machining centers and automated guided vehicles moving along the white line in the aisle. *Source:* Courtesy of Cincinnati Milacron, Inc.

TABLE 39.1

Comparison of General Characteristics of Transfer Lines and Flexible Manufacturing Systems		
Characteristic	Transfer line	FMS
Part variety	Few	Infinite
Lot size	>100	1–50
Part-changing time	Long	Very short
Tool change	Manual	Automatic
Adaptive control	Difficult	Available
Inventory	High	Low
Production during breakdown	None	Partial
Justification for capital expenditure	Simple	Difficult

- Direct labor and inventories are reduced or eliminated.
- The lead times required for product changes are shorter.
- Because the system is self-correcting, the production is more reliable and product quality is uniform.

Elements of FMS.   The basic elements of a flexible manufacturing system are (a) workstations and cells, (b) automated handling and transport of materials and parts, and (c) control systems. The workstations are arranged to yield the greatest efficiency in production with an orderly flow of materials and parts in progress through the system.

The kinds of machines in workstations depend on the type of production. For example, for machining operations, they usually consist of a variety of three- to five-axis machining centers, CNC lathes, milling machines, drill presses, and grinders. Also included are various other equipment, such as for automated inspection (including coordinate-measuring machines), assembly, and cleaning. Other types of operations suitable for FMS include sheet-metal forming, punching and shearing, and forging. They may incorporate furnaces, various machines, trimming presses, heat-treating facilities, and cleaning equipment.

Because of the flexibility of FMS, material-handling systems are very important. This system is controlled by a central computer and performed by automated guided vehicles, conveyors, and various transfer mechanisms. This system is capable of transporting raw materials, blanks, and parts in various stages of completion to any machine (in random order) and at any time. *Prismatic parts* usually are moved on specially designed **pallets**. Parts having *rotational symmetry* (such as those in turning operations) usually are moved by robots and various mechanical devices.

Scheduling.   Because FMS involves a major capital investment, efficient machine utilization is essential. Machines must not stand idle. Consequently, proper scheduling and process planning are crucial. Scheduling for FMS is *dynamic*, unlike that in job shops where a relatively rigid schedule is followed to perform a set of operations. The scheduling system in FMS specifies the types of operations to be performed on each part and identifies the machines or manufacturing cells where these operations are to take place. Dynamic scheduling is capable of responding to quick changes in product type; hence, it is responsive to real-time decisions.

Because of the flexibility in FMS, no setup time is wasted in switching between manufacturing operations. However, the characteristics, performance, and reliability of each unit in the system must be monitored to ensure that parts moving from one workstation are of acceptable quality and dimensional accuracy before they move on to the next workstation.

Economic justification of FMS.   FMS installations are very capital-intensive—costing millions of dollars. Consequently, a thorough cost-benefit analysis must be conducted before any final decision is made. This analysis should include such factors as the cost of capital, energy, materials, and labor; the expected markets for the products to be manufactured; and any anticipated fluctuations in market demand and product type. An additional consideration is the time and effort required for installing and debugging the system.

As can be seen in Fig. 37.2, the most effective FMS applications are in medium-quantity batch production. When a variety of parts is to be produced, FMS is suitable for production quantities typically of 15,000 to 35,000 aggregate parts per year. For individual parts with the same configuration, production may reach 100,000 parts per year. In contrast, high-volume, low-variety parts production is best obtained from transfer machines (dedicated equipment). Finally, low-volume, high-variety parts production can be done best on conventional standard machinery (with or without numerical control) or by machining centers.

EXAMPLE 39.2 Flexible manufacturing systems in large and small companies

Because of the advantages of FMS technology, many manufacturers have long considered implementing a large-scale system in their facilities. However, after detailed review and on the basis of the experience of other companies, most manufacturers have decided on some smaller, simpler, modular, and less expensive system that is more cost effective. These systems include flexible manufacturing cells (the cost of which would be on the order of a few hundred thousand dollars), stand-alone machining centers, and various CNC machine tools that are easier to control than an FMS.

There is a general feeling that, when FMS became an established alternative, the expectations were high. In some cases, extensive computerization has led to much confusion and inefficiency in company operations. Particularly for smaller companies, important considerations include not only the fact that large capital investment and major hardware and software acquisitions are necessary but also that the efficient operation of a large FMS requires the extensive training of personnel.

In contrast to the experience of small companies, there are several examples of the successful and economically viable implementation of an FMS in a large company. The results of a survey of 20 such operating systems in the United States have indicated improvements gained over prior methods. Some systems now are capable of economically producing lot sizes of one part. In spite of the high cost, the system has paid for itself in a number of companies.

39.4 | Holonic Manufacturing

Holonic manufacturing is a new concept describing a unique organization of manufacturing units. The word holonic is from the Greek *holos* (meaning whole) and the suffix *on* (meaning a part of). Thus, each component in a holonic manufacturing system (at the same time) is an independent entity (or *whole*) and a subservient *part* of a hierarchical organization. We describe this system here because of its potential beneficial impact on computer-integrated manufacturing operations.

Holonic organizational systems have been studied since the 1960s, and there are a number of examples in biological systems. Three fundamental observations of these systems can be stated as:

1. Complex systems will evolve from simple systems much more rapidly if there are stable intermediate forms than if there are none. Also, stable and complex systems require a hierarchical system for evolution.

2. Holons simultaneously are self-contained wholes to their subordinated parts and the dependent parts of other systems. Holons are autonomous and self-reliant units, which have a degree of independence and can handle contingencies without asking higher levels in the hierarchical system for instructions. At the same time, holons are subject to control from multiple sources of higher system levels.

3. A holarchy consists of (a) autonomous wholes in charge of their parts and (b) dependent parts controlled by higher levels of a hierarchy and (c) are coordinated according to their local environment.

In biological systems, hierarchies have the characteristics of stability in the face of disturbances, optimum use of available resources, and a high level of flexibility when their environment changes.

A *manufacturing holon* is an autonomous and cooperative building block of a manufacturing system for the production, storage, and transfer of objects or of information. It consists of a control part and an optional physical-processing part. For example, a holon can be a combination of a CNC milling machine and an operator interacting via a suitable interface. A holon also can consist of other holons which provide the necessary processing, information, and human interfaces to the outside world, such as a group of manufacturing cells. Holarchies can be created and dissolved dynamically, depending on the current needs of the particular manufacturing process.

A holonic-systems view of the manufacturing operation is one of creating a working manufacturing environment from the bottom up. Maximum flexibility can be achieved by providing intelligence within holons to both (a) support all production and control functions required to complete production tasks and (b) manage the underlying equipment and systems. The manufacturing system can dynamically reconfigure into operational hierarchies to optimally produce the desired products with holons or elements being added or removed as needed.

Holarchical manufacturing systems rely on fast and effective communication between holons, as opposed to traditional hierarchical control where individual processing power is essential. A large number of specific arrangements and software algorithms have been proposed for holarchical systems. A detailed description of these is beyond the scope of this book. However, the general sequence of events can be outlined as follows:

1. A factory consists of a number of resource holons, available as separate entities in a resource pool. For example, available holons may consist of a (a) CNC milling machine and operator, (b) CNC grinder and operator, and (c) CNC lathe and operator.

2. Upon receipt of an order or directive from higher levels in the factory hierarchical structure, an order holon is formed and begins communicating and negotiating with the available resource holons.

3. The negotiations lead to a self-organized grouping of resource holons, which are assigned based on product requirements, resource holon availability, and customer requirements. For example, a given product may require a CNC lathe, CNC grinder, and automated inspection station to organize it into a production holon.

4. In case of breakdown, lack of machine availability, or changing customer requirements, other holons from the resource pool can be added or subtracted as needed, allowing a reorganization of the production holon. Production bottlenecks can be identified and eliminated through communication and negotiation between the holons in the resource pool.

Step 4 has been referred to as *plug and play*, a term borrowed from the computer industry where hardware components seamlessly integrate into a system.

39.5 | Just-in-Time Production

The *just-in-time production* (JIT) concept was originated in the United States decades ago but was first implemented on a large scale in 1953 at the Toyota Motor Company in Japan to eliminate waste of materials, machines, capital, manpower, and inventory throughout the manufacturing system. The JIT concept has the following goals:

- Receive supplies just in time to be used.
- Produce parts just in time to be made into subassemblies.
- Produce subassemblies just in time to be assembled into finished products.
- Produce and deliver finished products just in time to be sold.

In traditional manufacturing, the parts are made in batches, placed in inventory, and used whenever necessary. This approach is known as a **push system**, meaning that parts are made according to a schedule and are placed in inventory to be used whenever they are needed. In contrast, just-in-time is a **pull system**, meaning that parts are produced to order and the production is matched with demand for the final assembly of products.

There are no stockpiles—the ideal production quantity being one (this also is called *zero inventory, stockless production,* or *demand scheduling*). Moreover, parts are inspected as they are manufactured and are used within a short period of time. In this way, a worker maintains continuous production control, immediately identifying defective parts and reducing process variation to produce quality products.

Implementation of the JIT concept requires that all aspects of manufacturing operations be monitored and reviewed so that all of those operations and resources that do not add value are eliminated. This approach emphasizes (a) pride and dedication in producing high-quality products, (b) the elimination of idle resources, and (c) teamwork among workers, engineers, and management to quickly solve any problems that arise during production or assembly.

The ability to detect production problems as the parts are being made has been likened to the level of water (representing inventory levels) in a lake covering a bed of boulders (representing production problems). When the water level is high (high inventories associated with push production), the boulders are not exposed. By contrast, when the level is low (low inventories associated with pull production), the boulders are exposed and can be identified and removed. This analogy indicates that high-inventory levels can mask quality and production problems with parts that already are made and stockpiled.

The JIT concept requires the timely delivery of all supplies and parts from outside sources and from other divisions of a company; thus, it significantly reduces or eliminates in-plant inventory. Suppliers are expected to deliver—often on a daily basis—pre-inspected goods as they are needed for production and assembly. This approach requires reliable suppliers, close cooperation and trust between the company and its vendors, and a reliable system of transportation. Also important for smoother operation is the reduction of the number of suppliers. In one example, an Apple Computer plant reduced the number of suppliers from 300 to 70.

Advantages of JIT.   Summarized here are the major advantages of just-in-time production:

- Low inventory-carrying costs.
- Fast detection of defects in the production or the delivery of supplies and, hence, low scrap loss.
- Reduced inspection and reworking of parts.
- High-quality products made at low cost.

Although there can be significant variations in performance, implementation of just-in-time production has resulted in reductions of 20 to 40% in product cost, 60 to 80% in inventory, up to 90% in rejection rates, 90% in lead times, and 50% in scrap, rework, and warranty costs. Increases of 30 to 50% in direct-labor productivity and of 60% in indirect-labor productivity also have been attained.

Kanban.   The implementation of JIT in Japan involved *kanban*, meaning visible record. These records originally consisted of two types of cards (called kanbans—now replaced by bar-coded plastic tags and other devices):

- The *production card*, which authorizes the production of one container or cart of identical, specified parts at a workstation.

- The *conveyance* or *move card*, which authorizes the transfer of one container or cart of parts from that particular workstation to the workstation where the parts will be used.

The cards contain information on the type of part, the location where issued, the part number, and the number of items in the container. The number of containers in circulation at any time is controlled completely and can be scheduled as desired for maximum production efficiency.

EXAMPLE 39.3 Applications of JIT in the U.S. automotive industry

In one application of JIT, automotive seats are made in the supplier's plant just two hours before they are needed at the assembly plant, which is 120 km (75 miles) away. The seats are unloaded at the assembly plant and transferred quickly in the proper sequence of style and color (each seat being marked beforehand for a specific automobile moving down an assembly line). Each seat arrives on the assembly line just in time for installation.

Because the system operates efficiently, the major U.S. automakers prefer to have suppliers of various components to relocate plants closer to the assembly lines. However, this is a costly and capital-intensive undertaking on the part of suppliers. An alternative is to have a dependable delivery system of transportation by truck and/or rail for reliable and timely delivery of supplies.

39.6 | Lean Manufacturing

In a modern manufacturing environment, companies must be responsive to the needs of the customers and their specific requirements and to fluctuating global market demands. At the same time, the manufacturing enterprise must be conducted with a minimum amount of wasted resources to ensure competitiveness. This realization has lead to *lean production* or *lean manufacturing* strategies.

Lean manufacturing is a systematic approach to *identifying and eliminating waste* (that is, non-value-added activities) in every area of manufacturing—through continuous improvement and by emphasizing product flow in a pull system. When applied on a large scale, lean manufacturing generally is referred to as *agile manufacturing*. Lean production requires that a manufacturer examine all of its activities from the viewpoint of the customer and optimize processes to maximize added value. This viewpoint is critically important, because it helps identify whether or not an activity:

- Clearly adds value
- Adds no value but cannot be avoided
- Adds no value but can be avoided

The focus of lean production is on the entire process flow and not just the improvement of one or more individual operations. Typical wastes to be considered and either reduced or eliminated in lean manufacturing include the following:

- Using just-in-time production approaches to eliminate inventory, because inventory represents cost, leads to defects, and reduces responsiveness to changing market demands.

- The elimination of waiting time, which may be caused by unbalanced work loads, unplanned maintenance, and quality problems. Therefore, maximize the efficiency of workers at all times.

- The elimination of unnecessary processes and steps, because they represent costs.

- Minimizing or eliminating product transportation, because it represents an activity that adds no value. This waste can be eliminated (such as by forming machining cells) or minimized (such as with better plant layout).

- Performing time and motion studies to identify inefficient workers or unnecessary product motions.

- The elimination of part defects.

39.7 | Communications Networks in Manufacturing

In order to maintain a high level of coordination and efficiency of operation in integrated manufacturing, an extensive, high-speed, and interactive *communications network* is essential. The *local area network* (LAN) is a hardware-and-software system in which logically related groups of machines and equipment communicate with each other. A local area network links these groups to each other, bringing different phases of manufacturing into a unified operation.

A local area network can be very large and complex (i.e., linking hundreds or even thousands of machines and devices located in several buildings). Various network layouts (Fig. 39.4) of fiber optics or copper cables are used over distances ranging from a few meters to as much as 32 km (20 mi). For larger distances, *wide area networks* (WAN) are used.

Different types of networks can be linked or integrated through "gateways" and "bridges." Access control to the network is important; otherwise, collisions can

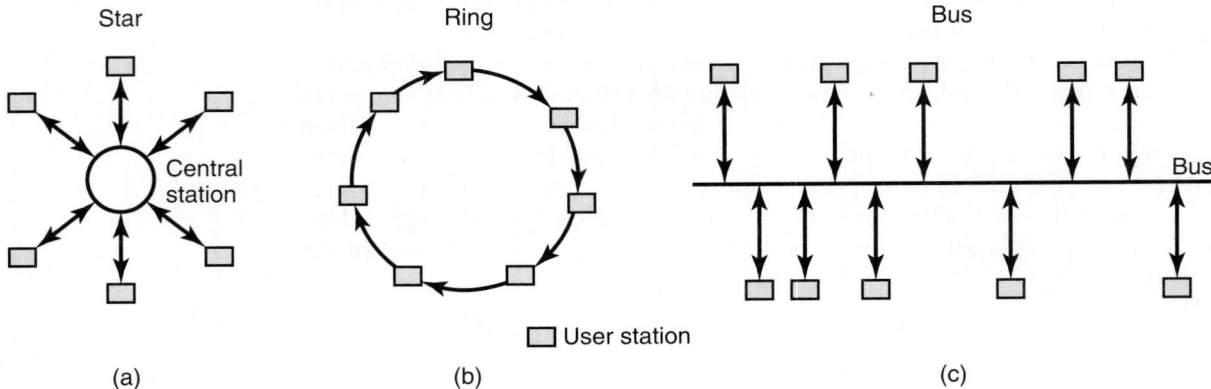

FIGURE 39.4   Three basic types of topology for a local area network (LAN). (a) The *star* topology is suitable for situations that are not subject to frequent configuration changes. All messages pass through a central station. Telephone systems in office buildings usually have this type of topology. (b) In the *ring* topology, all individual user stations are connected in a continuous ring. The message is forwarded from one station to the next until it reaches its assigned destination. Although the wiring is relatively simple, the failure of one station shuts down the entire network. (c) In the *bus* topology, all stations have independent access to the bus. This system is reliable and is easier than the other two to service. Because its arrangement is similar to the layout of the machines in the factory, its installation is relatively easy, and it can be rearranged when the machines are rearranged.

occur when several workstations are transmitting simultaneously. Continuous scanning of the transmitting medium is essential.

In the 1970s, a *carrier-sense multiple access with collision detection* (CSMA/CD) *system* was developed and implemented in *Ethernet*, which has become the industry standard. Other access-control methods are the *token ring* and *token bus*—in which a token (special message) is passed from device to device. Only the device that has the token is allowed to transmit, while all of the other devices only receive.

Conventional LANs require the routing of wires (often through masonry walls or other permanent structures) and require computers or machinery to remain stationary. *Wireless local area networks* (WLAN) allow equipment such as mobile test stands or data-collection devices (i.e., bar code readers) to easily maintain a network connection. A communication standard (IEEE 802.11) currently defines frequencies and specifications of signals, and two radio frequency and one infrared methods for WLANs. Although wireless networks are slower than those that are hard wired, their flexibility makes them desirable, especially for situations where slow tasks (such as machine monitoring) are the main application.

Personal area networks (PAN) are used for electronic devices (such as cellular telephones and personal data assistants) but are not as widespread for manufacturing applications. PANs are based on communications standards (such as Bluetooth, IrDA, and HomeRF) and are designed to allow data and voice communication over short distances. For example, a short-range Bluetooth device will allow communication over a 10 m (32 ft) distance. PANs are undergoing major changes, and communications standards are continually being refined.

Communications standards.   Typically, one manufacturing cell is built with machines and equipment purchased from one vendor, another cell with machines purchased from another vendor, and a third purchased from yet another vendor. As a result, a variety of programmable devices are involved and are driven by several computers and microprocessors purchased at various times from different vendors and having various capacities and levels of sophistication.

Each cell's computers have their own specifications and proprietary standards, and they cannot communicate beyond the cell with others unless equipped with custom-built interfaces. This situation created *islands of automation*; in some cases, up to 50% of the cost of automation was related to overcoming difficulties in the communications between individual manufacturing cells and other parts of the organization.

The existence of automated cells that could function independently from each other (without a common base for information transfer) led to the need for a *communications standard* to improve communications and the efficiency of computer-integrated manufacturing. The first step toward standardization began in 1980. After considerable effort and on the basis of existing national and international standards, a set of communications standards known as *manufacturing automation protocol* (MAP) was developed.

The International Organization for Standardization (ISO)/Open System Interconnect (OSI) reference model is accepted worldwide. The ISO/OSI model has a hierarchical structure, in which communication between two users is divided into seven layers (Fig. 39.5). Each layer has a special task:

- Mechanical and electronic means of data transmission
- Error detection and correction
- Correct transmission of the message
- Control of the dialog between users

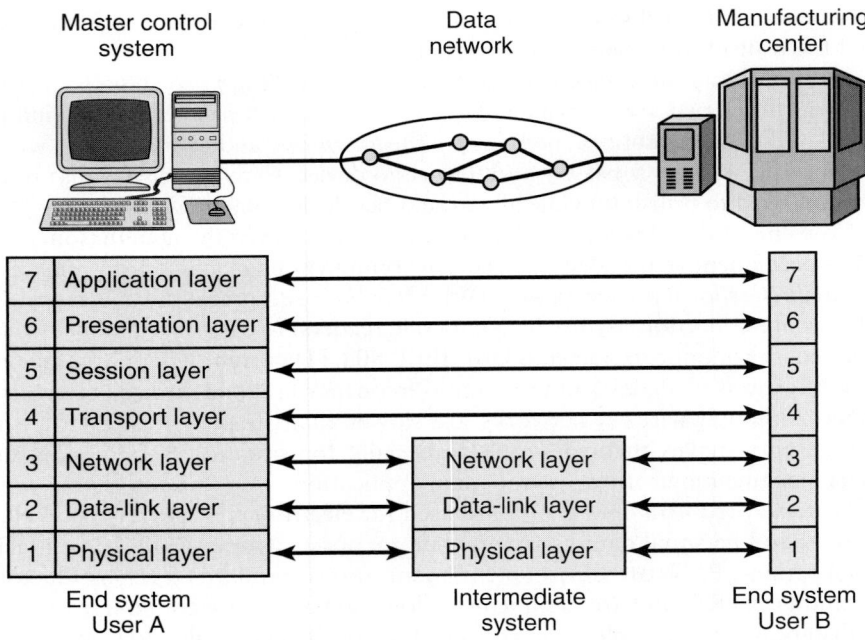

FIGURE 39.5   The ISO/OSI reference model for open communication. *Source:* After U. Rembold.

- Translation of the message into a common syntax
- Verification that the data transferred has been understood

The operation of this system is complex. Basically, each standard-sized chunk of message or data from User A to be transmitted to User B moves sequentially through the successive layers at A's end from Layer 7 to Layer 1. More information is added to the original message as it travels through each layer. The complete *packet* is transmitted through the physical communications medium to User B and then moves through the layers (from 1 to 7) at B's end. The transmission takes place through coaxial cable, fiber-optic cable, microwaves, and similar devices.

Communication protocols have been extended to office automation as well with the development of **technical and office protocol** (TOP), which is based on the ISO/OSI reference model. In this way, total communication (MAP/TOP) is established among the factory floor and offices at all levels of an organization. A common practice is the use of **Internet tools** (hardware, software, and protocols) within a company to link all departments and functions into a self-contained and fully compatible **Intranet**. Several tools for implementing this link are available commercially; they are inexpensive and are easy to install, integrate, and use.

39.8 | Artificial Intelligence

Artificial intelligence (AI) is that part of computer science concerned with systems that exhibit some characteristics usually associated with intelligence in human behavior (such as learning, reasoning, problem solving, and understanding language). The goal of AI is to *simulate* such human behaviors on the computer. The art of bringing relevant principles and tools of AI to bear on difficult application problems is known as **knowledge engineering**.

Artificial intelligence has a major effect on the design, the automation, and the overall economics of manufacturing operations—in large part because of advances in computer memory expansion (VLSI chip design) and decreasing costs. Artificial intelligence packages costing as much as a few thousand dollars have been developed, many of which can be run on personal computers. Thus, AI has become accessible to office desks and shop floors.

Expert systems. An *expert system* (ES, also called a **knowledge-based system**) generally is defined as an intelligent computer program that has the capability to solve difficult real-life problems by the use of **knowledge-based** and **inference** procedures (Fig. 39.6). The goal of an expert system is the capability to conduct an intellectually demanding task in the way that a human expert would.

The field of knowledge required to perform this task is called the **domain** of the expert system. Expert systems utilize a knowledge base containing facts, data, definitions, and assumptions. They also have the capacity for a **heuristic** approach (that is, making good judgments on the basis of discovery and revelation and making high-probability guesses—just as a human expert would).

The knowledge base is expressed in computer codes (usually in the form of **if-then rules**) and can generate a series of questions. The mechanism for using these rules to solve problems is called an **inference engine**. Expert systems also can communicate with other computer software packages.

To construct expert systems for solving the complex design and manufacturing problems encountered, one needs (a) a great deal of knowledge and (b) a mechanism for manipulating this knowledge to create solutions. Because of the difficulty involved in accurately modeling the many years of experience of an expert (or a team of experts)

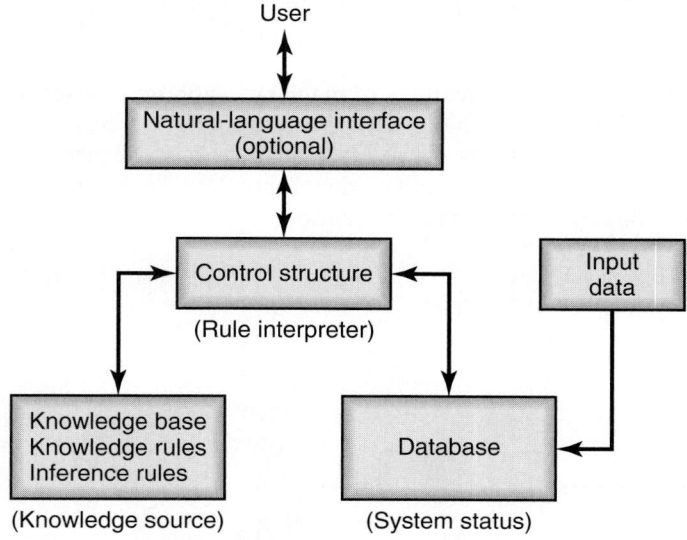

FIGURE 39.6 Basic structure of an expert system. The knowledge base consists of knowledge rules (general information about the problem) and inference rules (the way conclusions are reached). The results may be communicated to the user through the natural-language interface.

and the complex inductive reasoning and decision-making capabilities of humans (including the capacity to learn from mistakes), developing knowledge-based systems requires considerable time and effort.

Expert systems operate on a real-time basis, and their short reaction times provide rapid responses to problems. The programming languages most commonly used for these applications are C++, LISP, and PROLOG (other languages also can be used). An important development is expert-system software **shells** or **environments** (also called **framework systems**). These software packages are essentially expert-system outlines that allow a person to write specific applications to suit special needs. Writing these programs requires considerable experience and time.

Several expert systems have been developed and used since the early 1970s and utilize computers with various capacities and for such specialized applications as the following:

- Problem diagnosis in various types of machines and equipment, and determination of corrective actions.
- Modeling and simulation of production facilities.
- Computer-aided design, process planning, and production scheduling.
- Management of a company's manufacturing strategy.

Natural-language processing.   Traditionally, obtaining information from a database in the computer memory has required the utilization of computer programmers to translate questions in natural language into "queries" in some machine language. Natural-language interfaces with database systems are in various stages of development. These systems allow a user to obtain information by entering English or other language commands in the form of simple, typed questions.

Software shells are available, and they are used in such applications as the scheduling of material flow in manufacturing and the analyzing of information in databases. Significant progress continually is being made on computer software that will have speech synthesis and recognition (**voice recognition**) capabilities in order to eliminate the need to type commands on keyboards.

Machine vision.   The basic features of machine vision were described in Section 37.7. Computers and software, implementing artificial intelligence, are combined with cameras and other optical sensors. These machines then perform such operations as inspecting, identifying, sorting parts, and guiding robots (*intelligent robots*)—operations that otherwise would require human intervention.

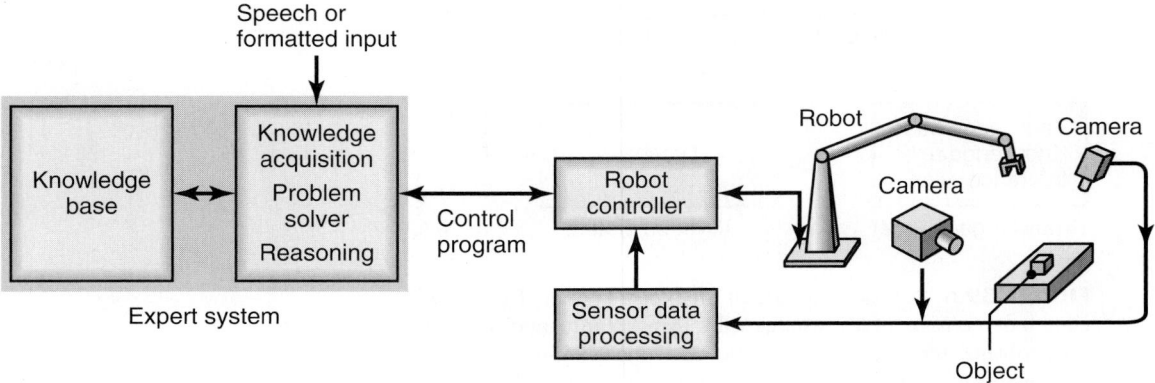

FIGURE 39.7   Expert system as applied to an industrial robot guided by machine vision.

Artificial neural networks. Although computers are much faster than the human brain at sequential tasks, humans are much better at pattern-based tasks that can be performed with parallel processing, such as recognizing features (in faces and voices, even under noisy conditions), assessing situations quickly, and adjusting to new and dynamic conditions. These advantages also are due partly to the ability of humans to use several senses (sight, hearing, smell, taste, and touch) simultaneously (*data fusion*) and in real time. The branch of AI called *artificial neural networks* (ANN) attempts to gain some of these capabilities through computer imitation of the way that data is processed by the human brain.

The human brain has about 100 billion linked **neurons** (cells that are the fundamental functional units of nerve tissue) and more than a thousand times that many connections. Each neuron performs only one simple task: It receives input signals from a fixed set of neurons, and when those input signals are related in a certain way (specific to that particular neuron), it generates an electrochemical output signal which goes to a fixed set of neurons. It now is believed that human learning is accomplished by changes in the strengths of these signal connections between neurons.

Artificial neural networks are used in such applications as noise reduction (in telephones), speech recognition, and process control. For example, they can be used for predicting the surface finish of a workpiece obtained by end milling on the basis of input parameters such as cutting force, torque, acoustic emission, and spindle acceleration. Although still controversial, the opinion of many is that true artificial intelligence will evolve only through advances in ANN.

Fuzzy logic. An element of AI having important applications in control systems and pattern recognition is *fuzzy logic* (also called *fuzzy models*). Introduced in 1965 and based on the observation that people can make good decisions on the basis of imprecise and nonnumeric information, fuzzy models are mathematical means of representing vagueness and imprecise information (hence the term "fuzzy").

These models have the ability to recognize, represent, manipulate, interpret, and utilize data and information that are vague or lack precision. These methods deal with reasoning and decision making at a level higher than neural networks. Typical linguistic examples are the following: few, very, more or less, small, medium, extremely, and almost all.

Fuzzy technologies and devices have been developed (and successfully applied) in areas such as robotics and motion control, image processing and machine vision, machine learning, and the design of intelligent systems. Some applications are in (a) the automatic transmission of automobiles; (b) a washing machine that automatically adjusts the washing cycle for load size, fabric type, and amount of dirt; and (c) a helicopter that obeys vocal commands to go forward, up, left, right, hover, and land.

39.9 | Economic Considerations

The economic considerations in implementing the various computer-integrated activities described in this chapter are crucial in view of the complexities and the high costs involved. Flexible manufacturing system installations are very capital intensive. Consequently, a thorough cost-benefit analysis must be conducted before a final decision is made. This analysis should include such factors as:

- Cost of capital, energy, materials, and labor.
- Expected markets for the products to be produced.

- Anticipated fluctuations in market demand and product type.
- Time and effort required for installing and debugging the system.

Typically, an FMS system can take two to five years to install and at least six months to debug. Although FMS requires few (if any) machine operators, the personnel in charge of the total operation must be trained and highly skilled. These personnel include manufacturing engineers, computer programmers, and maintenance engineers. The most effective FMS applications have been in medium-volume batch production. When a variety of parts is to be produced, FMS is suitable for production volumes of 15,000–35,000 aggregate parts per year. For individual parts of the same configuration, production may reach 100,000 units per year. In contrast, high-volume, low-variety parts production best is obtained from transfer machines (dedicated equipment). Low-volume, high-variety parts production best can be done on conventional standard machinery (with or without NC) or by machining centers.

SUMMARY

- Integrated manufacturing systems are implemented to various degrees to optimize operations, improve product quality, and reduce costs.
- Computer-integrated manufacturing systems have become the most important means of improving productivity, responding to changing market demands, and better controlling manufacturing and management functions. With extensive use of computers and the rapid developments in sophisticated software, product designs, their analysis, and their simulation are now very detailed and thorough.
- Artificial intelligence continues to create new opportunities in all aspects of manufacturing science, engineering, and technology.
- Economic considerations in the design and implementation of computer-integrated manufacturing systems, especially flexible manufacturing systems, are particularly crucial because of the major capital expenditures required.

KEY TERMS

Artificial intelligence
Artificial neural networks
Cellular manufacturing
Communications network
Communications standard
Computer-integrated manufacturing systems
Environments
Ethernet
Expert systems
Flexible manufacturing
Framework systems

Functional layout
Fuzzy logic
Holonic manufacturing
If-then rules
Inference engine
Internet tools
Just-in-time production
Kanban
Knowledge engineering
Knowledge-based system
Lean manufacturing
Local area network

Machine vision
Manufacturing cell
Natural-language processing
Pallet
Pull system
Push system
Technical and office protocol
Waste
Wireless networks
Zero inventory

BIBLIOGRAPHY

Allen, J.; Robinson, C.; and Stewart, D., *Lean Manufacturing: A Plant Floor Guide,* Society of Manufacturing Engineers, 2001.

Askin, R.G. and Standridge, C.R. *Modeling and Analysis of Manufacturing Systems,* Wiley, 1993.

Badiru, A.B., *Expert Systems Applications in Engineering and Manufacturing,* Prentice Hall, 1992.

Baudin, M., *Lean Assembly,* Productivity Press, 2002.

Bedworth, D.D.; Henderson, M.R.; and Wolf, P.M., *Computer-Integrated Design and Manufacturing,* McGraw-Hill, 1991.

Biekert, R., *CIM Technology—Fundamentals and Applications,* Goodheart-Willcox, 1998.

Black, J.T., *The Design of the Factory with a Future,* McGraw-Hill, 1981.

———, and Hunter, S.L., *Lean Manufacturing Systems and Cell Design,* Society of Manufacturing Engineers, 2003.

Bodek, N., *Kaikaku: The Power and Magic of Lean,* PRC Press, 2004.

Burbidge, J.L., *Production Flow Analysis for Planning Group Technology,* Oxford, 1997.

Chang, T.-C.; Wysk, R.A.; and Wang, H.P., *Computer-Aided Manufacturing,* 2nd ed., Prentice Hall, 1997.

Cheng, T.C.E. and Podolsky, S., *Just-in-Time Manufacturing: An Introduction,* 2nd ed., Chapman & Hall, 1996.

Chorafas, D.N., *Expert Systems in Manufacturing,* Van Nostrand Reinhold, 1992.

Connor, G., *Lean Manufacturing for the Small Shop,* Society of Manufacturing Engineers, 2001.

Famili, A.; Nau, D.S.; and Kim, S.H., *Artificial Intelligence Applications in Manufacturing,* American Association for Artificial Intelligence, 1992.

Foston, A.L.; Smith, C.L.; and Au, T., *Fundamentals of Computer-Integrated Manufacturing,* Prentice Hall, 1991.

Goldratt, E.M., *Theory of Constraints,* North River Press, 1990.

Hales, H.L.; Anderson, B.; and Fillmore, W., *Planning Manufacturing Cell Textbook,* Society of Manufacturing Engineers, 2002.

Hannam, R., *CIM: From Concept to Realisation,* Addison-Wesley, 1998.

Harmon, R.L., *Reinventing the Factory II: Managing the World Class Factory,* Free Press, 1992.

Hitomi, K., *Manufacturing Systems Engineering,* 2nd ed., Taylor & Francis, 1996.

Irani, S.A. (ed.), *Handbook of Cellular Manufacturing Systems,* Wiley, 1999.

Jordan, J. and Michel, F., *Lean Company: Making the Right Choices,* Society of Manufacturing Engineers, 2001.

Kasabov, N.K., *Foundations of Neural Networks, Fuzzy Systems, and Knowledge Engineering,* The MIT Press, 1996.

Krishnamoorty, C.S. and Rajeev, S., *Artificial Intelligence and Expert Systems for Engineers,* CRC Press, 1996.

Kusiak, A., *Concurrent Engineering: Automation, Tools, and Techniques,* Wiley, 1993.

———, *Intelligent Design and Manufacturing,* Wiley, 1992.

———, *Intelligent Manufacturing Systems,* Prentice Hall, 1990.

Leondes, C.T. (ed.), *Fuzzy Logic and Expert Systems Applications,* Academic Press, 1998.

Liebowitz, J. (ed.), *The Handbook of Applied Expert Systems,* CRC Press, 1997.

Louis, R.S., *Integrating Kanban with MRP II: Automating a Pull System for Enhanced JIT Inventory Management,* Productivity Press, 1997.

Luggen, W.W., *Flexible Manufacturing Cells and Systems,* Prentice Hall, 1991.

Majima, I., *The Shift to JIT: How People Make the Difference,* Productivity Press, 1992.

Maus, R. and Keyes J. (eds.), *Handbook of Expert Systems in Manufacturing,* McGraw-Hill, 1991.

Mitchell, F.H., Jr., *CIM Systems: An Introduction to Computer-Integrated Manufacturing,* Prentice Hall, 1991.

Monden, Y., *Toyota Production System: An Integrated Approach to Just-in-Time,* 3rd ed., Institute of Industrial Engineers, 1998.

Popovic, D. and Bhatkar, V. *Methods and Tools for Applied Artificial Intelligence.* Marcel Dekker, 1994.

Rehg, J.A., *Computer-Integrated Manufacturing,* Prentice Hall, 1994.

———, *Introduction to Robotics in CIM Systems,* 3rd ed., Prentice Hall, 1997.

Rembold, U.; Nnaji, B.O.; and Storr, A., *Computer Integrated Manufacturing and Engineering,* Addison-Wesley, 1993.

Sandras, W.W., *Just-in-Time: Making It Happen,* Wiley, 1997.

Shingo, S., *A Study of the Toyota Production System,* Productivity Press, 1989.

Singh, N., *Systems Approach to Computer-Integrated Design and Manufacturing,* Wiley, 1995.

Singh, N. and Rajamani, D., *Cellular Manufacturing Systems: Design, Planning and Control,* Chapman & Hall, 1996.

Standard, C. and Davis, S., *Running Today's Factory,* Society of Manufacturing Engineers and Hanser-Gardner, 1999.

Tool and Manufacturing Engineers Handbook, 4th ed., Vol. 5: *Manufacturing Management,* Society of Manufacturing Engineers, 1988.

——— Vol. 7. *Continuous Improvement.* Society of Manufacturing Engineers, 1993.

Vajpayee, S.K., *Principles of Computer-Integrated Manufacturing,* Prentice Hall, 1998.

Williams, D.J., *Manufacturing Systems,* 2nd ed., Chapman & Hall, 1994.

Wu, J.-K., *Neural Networks and Simulation Methods,* Marcel Dekker, 1994.

REVIEW QUESTIONS

39.1 In what ways have computers had an impact on manufacturing?

39.2 What advantages are there in viewing manufacturing as a system? What are the components of a manufacturing system?

39.3 Discuss the benefits of computer-integrated manufacturing operations.

39.4 What is a manufacturing cell? Why was it developed?

39.5 Describe the principle of flexible manufacturing systems. Why do they require major capital investment?

39.6 Why is a flexible manufacturing system capable of producing a wide range of lot sizes?

39.7 What are the benefits of just-in-time production? Why is it called a pull system?

39.8 Explain the function of a local area network.

39.9 What are the advantages of a communications standard?

39.10 What are the differences between ring and star networks?

39.11 What is meant by the term "holonic manufacturing"?

39.12 What is an FMC, and what is an FMS? What are the differences?

39.13 What is Kanban?

39.14 What is lean manufacturing? Why is just-in-time required in lean manufacturing?

QUALITATIVE PROBLEMS

39.15 Would machining centers be suitable for just-in-time production? Explain.

39.16 Give an example of a push system and of a pull system. Indicate the fundamental difference between the two methods.

39.17 Is there a minimum to the number of machines in a manufacturing cell? Explain.

39.18 Are robots always a component of a FMC? Explain.

39.19 Describe the elements of artificial intelligence. Is machine vision a part of it? Explain.

39.20 Are there any disadvantages to zero inventory? Explain.

39.21 Give examples in manufacturing processes and operations in which artificial intelligence could be effective.

39.22 Describe your opinions concerning the voice-recognition capabilities of future machines and controls.

39.23 Evaluate a process from a lean-production perspective. For example, closely observe the following and identify, eliminate (when possible), or optimize the steps that produce waste when (a) preparing breakfast for a group of eight, (b) washing clothes or cars, (c) using internet browsing software, and (d) studying for an exam, writing a report, or writing a term paper.

SYNTHESIS, DESIGN, AND PROJECTS

39.24 Think of a product line for a commonly used household item and design a manufacturing cell for making it. Describe the features of the machines and equipment involved.

39.25 What types of (a) products and (b) production machines would not be suitable for FMC? What design or manufacturing features make them unsuitable? Explain with examples.

39.26 Surveys have indicated that 95% of all the different parts made in the United States are produced in lots of 50 or less. Comment on this observation and describe your thoughts regarding the implementation of the technologies outlined in Chapters 37 through 39.

39.27 Can a factory ever be completely untended? Explain.

39.28 Assume that you own a manufacturing company, and that you are aware that you have not taken full advantage of the technological advances in manufacturing. However, now you would like to do so, and you now have the necessary capital. Describe how you would go about analyzing your company's needs and how you would plan to implement these technologies. Consider technical as well as human aspects.

39.29 How would you describe the benefits of FMS to an older worker in a manufacturing facility whose experience has been running only simple machine tools?

39.30 Artificial neural networks are useful particularly where problems are ill defined and the data is vague.

Give examples in manufacturing where ANN can be useful.

39.31 Consider a routing sheet for a product, as discussed in Section 38.6. If a holonic manufacturing system is to automatically generate the manufacturing sequence:

a. What is the likelihood that it will be the same as one produced manually?

b. What is the likelihood that a holonic system will produce the same sequence every time? Explain.

39.32 It has been suggested by some that artificial intelligence systems ultimately will be able to replace the human brain. Do you agree? Explain.

Product Design and Process Selection in a Competitive Environment

40

Manufacturing high-quality products at the lowest possible cost requires an understanding of the often complex relationships among many factors. This chapter describes the interrelated factors in product design, development, and manufacturing, including:

• Consideration of a product's entire life cycle and the effects on product design.

• The importance of material and process selection on design.

• The economic significance of manufacturing operations with a description of the important factors in costs associated with a product.

• How value analysis can help optimize manufacturing and minimize product cost.

40.1 | Introduction

In view of the extensive variety of materials and manufacturing processes available today, the task of producing a high-quality product by selecting the best materials and the best processes while minimizing production costs is a major challenge as well as an opportunity. The cost of a product often determines its marketability and customer acceptance in the global marketplace. Meeting this challenge requires not only a thorough knowledge of the characteristics of materials and processes but also innovative and creative approaches to product design, manufacturing, and assembly technologies.

This chapter begins with important considerations in **product design**, which involve numerous factors regarding not only the design itself but also the ease of manufacturing that product and the time required to produce it. There are many opportunities to simplify designs, reduce the number of components in the product, and reduce the size and the dimensions of the components, especially to save on costly materials.

Product quality and **life expectancy** are discussed, outlining the relevant parameters involved and including the concept of return on quality. Increasingly important are the **life-cycle assessment** and **life-cycle engineering** of products, services, and systems, particularly regarding their potentially adverse impact on the environment. The major emphasis in *sustainable manufacturing* is to reduce or eliminate any adverse effects of manufacturing both on the environment and on society in general while still allowing a company to be profitable.

Although the *selection of materials* for products traditionally requires much experience, there now are several databases and expert systems available that facilitate the selection process to meet several specific requirements. Also, in reviewing the materials used in existing products (from simple hand tools to automobiles and aircraft), there are numerous opportunities for the *substitution of materials* for better performance and, especially, for cost savings.

In the production phase, it is imperative that the *capabilities of manufacturing processes* be assessed properly as a basis and an essential guide to the ultimate selection of an appropriate process or series of processes. As described throughout this book and depending on the product design and materials specified, there usually is more than one method of manufacturing a product, its components, and its subassemblies. An improper selection can have a major effect not only on product quality but on its cost as well.

Although the economics of individual manufacturing processes have been described at the end of individual chapters, this chapter takes a broader view and summarizes the important overall manufacturing cost factors. We will investigate cost-reduction methods, including **value analysis**—a powerful tool that can be used to evaluate the cost of each manufacturing step relative to its contribution to the value of the product.

The term **world class** now is used frequently to describe a certain quality and level of manufacturing activities and to signify the fact that products must meet international standards and be acceptable worldwide. It also should be recognized that being world class (like quality) is not a fixed target for a manufacturing company to reach; it is a moving target, rising to higher and higher levels as time passes.

40.2 | Product Design

We have highlighted throughout various chapters those aspects that are relevant to *design for manufacture and assembly* (DFMA) as well as to competitive manufacturing. Various guidelines for materials and manufacturing are given in the references listed in Table 40.1.

Major advances are being made continually in designs for manufacture and assembly and for which software packages now are readily available. Although their use requires considerable training and knowledge, these advances greatly help designers develop products that require fewer components, reduce assembly time, reduce time for manufacture, and, consequently, reduce total production cost.

Product design considerations.   In addition to the design guidelines we frequently have given for individual manufacturing processes, there are general product design considerations. (See also *robust design*, Section 36.5.1.) Designers often must question if they have addressed considerations such as:

- Have all alternative product designs been investigated thoroughly?
- Can the design be simplified and the number of its components minimized without adversely affecting its intended functions and performance?
- Can unnecessary features of the product or some of its components be eliminated or combined with other features?
- Are some of the components commercially available?
- Have modular design and building-block concepts been considered for a family of similar products and for servicing and repair, upgrading, and installing options?
- Can the design be made smaller and lighter?

TABLE 40.1

References to Various Topics in This Book

Design Considerations in Processing

Metal casting	Section 12.2
Forging	Section 14.6
Sheet-metal forming	Section 16.13
Powder metallurgy	Section 17.6
Ceramics shaping	Section 18.5
Polymers processing	Section 19.15
Machining	Various sections in Chapters 23 and 24
Abrasive processes	Section 26.5
Advanced machining	Various sections in Chapter 27
Joining processes	Various sections in Chapters 30 through 32

Material Properties

Tables 2.1, 2.2, 2.3, and Figs. 2.4, 2.6, 2.8, 2.14, 2.15, 2.16, and 2.28
Tables 3.1, 3.2, 3.3, and Figs. 3.1 and 3.2
Tables 5.2, 5.4 and 5.6
Tables 6.2 through 6.10
Tables 7.1 through 7.3
Tables 8.2 and 8.3
Table 9.2 and Figs. 9.3, 9.5, and 9.7
Table 10.1
Tables 12.3, 12.4, and 12.5, and Fig. 12.4
Tables 16.3 and 16.4
Tables 17.3, 17.4, 17.5, and Fig. 17.10
Table 20.2
Tables 22.1, 22.2, 22.5 and
Figs. 22.1 and 22.9
Table 26.1
Table 32.3

Table 18.1
Tables 19.1 and 19.2
Table 20.1
Tables 23.1, 23.6, 23.8, and 23.10
Tables 26.3 and 26.4
Table 27.1
Tables 28.1 and 28.2
Table 29.1
Table VI.1
Table 30.1
Table 32.4
Table 34.1
Tables 40.2 and 40.3

Manufacturing Characteristics of Materials

Table I.3
Table 5.6
Table 10.1
Tables 12.2 and 12.6
Table 14.3
Tables 16.2 and 16.3 and Fig. 16.34
Section 21.7
Fig. 22.2
Section 30.9.2

Dimensional Tolerances and Surface Finish

Table 11.2
Table 23.1 and Figs. 23.13 and 23.14
Fig. 27.4

General Costs and Economics

Table I.4
Table 6.1
Section 12.6
Section 14.9
Section 16.15
Section 17.8 and Table 17.6
Section 19.16
Fig. 24.34
Section 25.8
Section 26.9 and Fig. 26.34
Section 27.10
Section 31.8
Section 32.7
Tables 40.2, 40.5, and 40.9

Capabilities of Manufacturing Processes

Tables 11.1 and 11.2
Table III.1
Table 14.1
Table 16.1
Section 17.7 and Fig. 17.14

- Are the specified dimensional tolerances and surface finish unnecessarily tight, thereby significantly increasing product cost?
- Can those dimensional tolerances and surface finish be relaxed without any adverse effects?
- Will the product be difficult or excessively time consuming to assemble and disassemble for maintenance, servicing, or recycling of some or all of its components?
- Have subassemblies been considered?
- Is the use of fasteners (including their quantity and variety) minimized?

As some examples of the foregoing considerations, observe how (a) the sizes of products such as cameras, computers, calculators, electronic equipment, and cell phones have been reduced; (b) the repair of products (including small and large appliances and automobiles) often is done by replacing subassemblies and modules, and (c) fewer fasteners are used in products and more assemblies are made using snap-on fasteners.

Product design and quantity of materials.   With high production rates possible because of modern equipment and reduced labor rates due to automation, the cost of materials can become a significant portion of a product's total cost. Although the material cost cannot be reduced below the market level, reductions can be made in the *quantity of materials* used in the components to be produced. The wide use of techniques (such as finite-element analysis, minimum-weight design, design optimization, and computer-aided design and manufacturing) have facilitated design analysis, material selection, material usage, and overall optimization greatly.

Typically, reductions in the quantity of materials used can be achieved by reducing the component's volume. This approach may include the selection of materials having high strength-to-weight or stiffness-to-weight ratios. Obviously, higher ratios can be obtained by improving and optimizing the product design and by selecting different cross-sections—such as those having a high moment of inertia (i.e., I-beams and channels) or by using tubular or hollow components instead of solid bars.

However, implementing such design changes and minimizing the amount of materials used can lead to thin cross-sections and present significant problems in manufacturing, such as:

- Casting or molding thin cross-sections can present difficulties in die and mold filling and in maintaining specified dimensional accuracy and surface finish.
- Forging thin sections requires high forces, due to friction and rapid chilling, on thin sections in hot forging.
- Impact extrusion of thin-walled parts can be difficult, especially when requiring high dimensional accuracy and symmetry.
- Formability in sheet-metal forming may be reduced as sheet thickness decreases, but this also can lead to buckling of the part under the high compressive stresses developed in the plane of the sheet during forming.
- Machining and grinding thin workpieces can present difficulties, such as part distortion, poor dimensional accuracy, and chatter. Consequently, advanced machining processes have to be considered where the forces on the workpiece are minimal or nil.
- Welding thin sheets and slender structures can cause significant product distortion due to thermal gradients developed during the welding cycle.

Conversely, making parts having thick cross-sections can have their own adverse effects. Examples are:

- In processes such as casting and injection molding, the production rate can be slower because of the increased cycle time required to allow sufficient time for the part to cool in the mold before removing it.
- Unless controlled, porosity can develop in thicker sections of castings.
- The bendability of sheet metals decreases as their thickness increases.
- In powder metallurgy, there typically can be significant variations in density and properties throughout parts with varying thicknesses.
- Welding thick sections can present problems in the depth and quality of the welded joint.

- Strain-hardening materials will not achieve the same degree of cold work; therefore, parts with thick walls may be weaker than their thinner-walled counterparts.
- In die-cast parts, thinner sections will have a higher strength per unit thickness (because of the smaller grain size developed) than will thicker sections.
- Processing polymeric parts requires increased cycle times as their thickness or volume increases because of the longer time required for the parts to sufficiently cool to be removed from the molds.

EXAMPLE 40.1 An application of design for manufacturing and assembly

Numerous examples can be given regarding the benefits of applying design for manufacturing and assembly principles and techniques in the early stages of product concept and development. These principles also can be applied to modifying existing designs and in selecting appropriate production methods. In one example, the redesign of the pilot's instrument panel for a military helicopter designed and built by McDonnell-Douglas was considered.

The components of the panel consisted of sheet metal, extrusions, and rivets. Using DFMA software and analyzing this panel, it was estimated that the redesign would lead to the following changes: (a) number of parts from 74 to 9; (b) panel weight from 3.00 kg to 2.74 kg; (c) fabrication time from 305 hrs to 20 hrs; (d) assembly time from 149 hrs to 8 hours; and (e) total production time from 697 hrs to 181 hrs. It also was estimated that, as a result of this redesign, cost savings would be 74%. Based on these results, other components of the instrument panel were subjected to similar analysis.

40.3 | Product Quality and Life Expectancy

Product quality and the techniques involved in quality assurance and control were described in detail in Chapter 36. The word *quality* is difficult to define precisely, partly because it includes not only well-defined technical considerations but also human (hence subjective) opinion. However, a high-quality product generally is considered to have the following characteristics:

- It satisfies the needs and expectations of the customer, including cost.
- It is compatible with the customer's working environment.
- It functions reliably and safely over its intended life.
- The aesthetics are pleasing.
- Installation, maintenance, and future improvements are easy to perform and at low cost.
- It is available in the quantities desired when needed.

In view of the global economy and competition, a major priority in product quality is the concept of *continuous improvement*—as exemplified by the Japanese term **Kaizen**, meaning never-ending improvement. Still, the level of quality that a manufacturer chooses for its products depends on the market for which the products are intended. Low-cost, low-quality products have their own market niche, just as there is a market for high-quality, expensive products, such as a high-precision machine tool, a Rolls-Royce automobile, a private airplane or yacht, or sporting equipment.

We have seen that (through *concurrent engineering*, Section I.2) design and manufacturing engineers have the responsibility and freedom to select and specify materials for the particular products to be made. Quality considerations always are a part of such tasks. Let's take a simple case of selecting materials for a common screwdriver. Based on its functions, one can specify stem materials that have high yield and tensile strengths, torsional stiffness, hardness, and resistance to wear and corrosion. As a result, this screwdriver will perform better and last longer than one made of materials with inferior properties.

On the other hand, materials with better properties generally may be somewhat more expensive due partly to their being more difficult and exacting to produce than others. The same considerations are relevant to the screwdriver handles as well—typical material choices being plastic or wood. Each of these two materials has its own advantages and limitations. Consequently, the manufacturing operations and systems involved in a particular product must be reviewed thoroughly to keep product costs low, which can be done based on the numerous technical and economic considerations discussed throughout this book. Unfortunately, design engineers often are faced with this dilemma, which succinctly has been stated as "Good, fast, or cheap; pick any two."

Return on quality.   Because of the significant costs that can be incurred in considering quality, the concept of **return on quality** (ROQ) can be important. The basic components are that

- Because of its major influence on customer satisfaction, quality must be viewed as an investment.
- There has to be a certain limit to how much should be spent on quality improvements.
- Expenditure specifically should be made toward quality improvement.
- An incremental improvement in quality versus the additional costs involved must be reviewed carefully.

As we have pointed out, quality is fairly subjective. Clearly, parts or products that do not serve their intended purpose detract from quality, and there is a clearly defined cost that can be calculated for such defects.

Contrary to a general perception, high-quality products do not necessarily cost more. For example, in industries making computer chips and computer hardware, the ROQ is minimized at a value approaching zero defects (see *six sigma*; Section 36.7.2), while in others (such as ordinary water faucets or door hinges), the cost to eliminate the final few defects can be unnecessarily high. Customer satisfaction is a qualitative factor and is difficult to include in calculations, but satisfaction is increased and the customers are more likely to be retained when there are no defects in products. It has been estimated that the relative costs involved in identifying and repairing defects in products grow by orders of magnitude, also called the *rule of ten*, as shown in Table 40.2.

Life expectancy of products.   Numerous surveys have indicated that the *average life expectancies* of products are as shown in Table 40.3. The life expectancies of products can vary significantly, depending on the materials and processes employed (hence the quality) and on many other factors—especially the frequency and quality of maintenance required. Numerous examples can be given where a choice has to be made between different processes and materials to manufacture a product. For example, consider the following: (a) sheet metal versus cast-iron frying pans, (b) carbon-steel versus stainless-steel exhaust systems for automobiles, (c) wood versus metal handles for hammers, (d) plastic versus metal outdoor furniture, and (e) aluminum versus reinforced-plastic ladders.

TABLE 40.2

Relative Cost of Repair at Various Stages of Product Development and Sale	
Stage	Relative cost of repair
When the part is made	1
Sub-assembly of part	10
Final assembly of product	100
Product is at dealership	1000
Product is at the customer	10,000

TABLE 40.3

Average Life Expectancy for Various Products. See also Table 36.1.	
Product	Years
Car battery	4
Central air-conditioning unit	15
Clothes dryer (gas)	13
Clothes washer	13
Dishwasher	10
Furnace (gas)	18
Hair dryer	5
Kitchen disposal	10
Machinery	30
Manufacturing cell	15
Passenger car	8
Personal computer	4
Refrigerator	17
Vacuum cleaner	10
Water heater (electric)	14
Water heater (gas)	12

40.4 Life-Cycle Assessment and Engineering; Sustainable Manufacturing

Life cycle (also called *cradle-to-grave*) can be defined as consecutive and interlinked stages of a product or service system, including:

- Extraction of natural resources
- Processing of the raw materials
- Manufacturing of products
- Transportation and distribution of the product to the customer
- Use, maintenance, and reuse of the product
- Disposal of the product or recovery and recycling of its components

Note that all of these factors basically are applicable to any product. Each type of product has its own metallic and nonmetallic materials that have been processed into individual parts and assembled, and each product has its own life cycle. Note also that some products are disposable and others are reusable. In addition to materials, other important considerations are the use of fluids (such as lubricants, coolants, toxic solutions, and those used in heat treating and plating processes), which can have serious environmental effects. Government regulations in many countries now have moved towards the direction of making manufacturers become life-cycle accountable.

Life cycle assessment (LCA). According to the ISO 14000 standard (see Section 36.6) life cycle assessment is defined as a systematic set of procedures for compiling and examining the inputs and outputs of materials and energy and the associated environmental impacts or burdens directly attributable to the functioning of a product, process, or service system throughout its entire life cycle.

Life cycle engineering (LCE). Concerned with environmental factors as in life cycle assessment, life cycle engineering deals in greater depth with design, optimization, and various technical considerations regarding each component of a product or process life cycle. A major aim of life cycle engineering is to consider reusing and recycling the components of a product from the earliest stage of discussing and considering the product design (also called *green design* or *green engineering*).

Although life cycle analysis and engineering are comprehensive and powerful tools, their implementation can be costly, challenging, and time consuming. This is largely because of uncertainties in the input data (regarding materials, processes, long-term effects, costs, etc.) and the time required to gather reliable data to properly assess the often complex interrelationships among various components of the whole system. Various software is being developed to expedite these analyses, particularly in the chemical and process industries because of the higher potential for environmental damage in their operations.

Sustainable manufacturing. In recent years, we all have become increasingly aware that, ultimately, there are limited natural resources on this earth, clearly necessitating the need to conserve materials and energy. The term *sustainable manufacturing* now is used to indicate and emphasize the need for conserving resources, particularly through proper maintenance and reuse. While maintaining a company's profitability, this is to be done for the purposes of (a) increasing the life cycle of products, (b) eliminating damage to the environment, and (c) ensuring our collective social well-being, especially for future generations.

EXAMPLE 40.2 Sustainable manufacturing in the production of Nike athletic shoes

Among numerous examples from industry, the production of Nike shoes clearly has indicated the benefits of sustainable manufacturing. The athletic shoes are assembled using adhesives. Up to around 1990, the adhesives used contained petroleum-based solvents, which pose health hazards to humans and contribute to petrochemical smog. The company worked with chemical suppliers to successfully develop water-based adhesive technology, which now is used for the majority of shoe-assembly operations. As a result, solvent use in all manufacturing processes in the subcontracted facilities in Asia has been reduced by 67% since 1995. Also, in 1997, 834,000 gallons of hazardous solvents were replaced with 1290 tons of water-based adhesives.

Regarding another component of the shoe, the rubber outsoles are made by a process which results in significant amounts of extra rubber around the periphery of the sole (called flashing, which is similar to the flash shown in Fig. 14.5d or Fig. 19.19c). With about 40 factories using thousands of molds and producing over a million outsoles a day, the flashing constitutes the largest chunk of waste in the manufacturing process for the shoes. In order to reduce this waste, the company developed a technology which grinds the flashing into 500μm rubber powder, which is then added back into the rubber mixture needed to make the outsole. Waste reduction was reduced by 40%, but also, it was found that the mixed rubber had better abrasion resistance, durability, and overall performance than the highest premium rubber!

40.5 | Material Selection for Products

In selecting materials for a product, it is essential to have a clear understanding of the *functional requirements* for each of its individual components. Although the general criteria for selecting materials were described in Section I.4 of the General Introduction, this section will discuss them in further detail.

Mechanical, physical, and chemical properties.   As we described in Chapter 2, mechanical properties include strength, toughness, ductility, stiffness, hardness, and resistance to fatigue, creep, and impact. Physical properties include density, melting point, specific heat, thermal and electrical conductivity, thermal expansion, and magnetic properties. Chemical properties that are of primary concern in manufacturing are susceptibility to oxidation and corrosion and to the various surface-treatment processes described in Chapter 34.

Material selection now is easier and faster because of the availability of extensive computer databases that provide greater accessibility and accuracy. However, in order to facilitate the selection of materials and other parameters that we will describe later, *expert-system software* (**smart databases**; see also Section 39.8) has been developed. With the proper input of product design and functional requirements, these systems are capable of identifying appropriate materials for a particular application, just as an expert or a team of experts would.

The following considerations are significant in the material selection for various products:

- Do the materials selected have the appropriate manufacturing characteristics?
- Can some of the materials be replaced by others that are less expensive?
- Do the materials considered have properties that unnecessarily exceed minimum requirements and specifications?
- Are the raw materials (stock) specified available in standard shapes, dimensions, tolerances, and surface characteristics?
- Is the material supply reliable? Can the material be delivered within the required quantities in the required time frame? Are there likely to be significant price increases or fluctuations?
- Does the material present any environmental hazards and concerns?

Shapes of commercially available materials.   After selecting the necessary materials, we need to know the shapes and the sizes in which these materials are available

TABLE 40.4

Commercially Available Shapes of Materials	
Material	Available as
Aluminum	B, F, I, P, S, T, W
Ceramics	B, p, s, T
Copper and brass	B, f, I, P, s, T, W
Elastomers	b, P, T
Glass	B, P, s, T, W
Graphite	B, P, s, T, W
Magnesium	B, I, P, S, T, w
Plastics	B, f, P, T, w
Precious metals	B, F, I, P, t, W
Steels and stainless steels	B, I, P, S, T, W
Zinc	F, I, P, W

Note: B = bar and rod; F = foil; I = ingots; P = plate and sheet; S = structural shapes; T = tubing; W = wire.
Lowercase letters indicate limited availability.
Most of the metals also are available in powder form, including pre-alloyed powders.

commercially (Table 40.4). Materials generally are available in various forms, such as castings, extrusions, forgings, bars, plates, sheets, foils, rods, wires, and metal powders.

Purchasing materials in shapes that require the least additional processing is an important consideration. However, such characteristics as the surface quality, the dimensional tolerances, and the straightness also must be taken into account. Obviously, the better and the more consistent these characteristics are, the less additional processing is required.

For example, if we want to produce simple shafts having good dimensional accuracy, roundness, straightness, and surface finish, we could purchase round bars that are turned and centerless-ground to the dimensions specified. Unless our facilities are capable of producing round bars economically, it generally is cheaper to purchase them.

On the other hand, if we need to make a stepped shaft (one having different diameters along its length), we could purchase a round bar (having a diameter at least equal to the largest diameter of the final stepped shaft) and turn it on a lathe or process it by some other means in order to reduce its diameter. If the stock has wide dimensional tolerances, is warped, or is out of round, we must order a larger size to ensure proper dimensional control of the final shaft.

As we have seen, each manufacturing step produces parts having specific shapes, surface finishes, and dimensional accuracies. Examples include:

- Hot-rolled or hot-drawn products have a rougher surface finish and greater dimensional tolerances than cold-rolled or cold-drawn products.

- Castings generally have less dimensional accuracy and poorer surface finish than parts made by cold extrusion or powder metallurgy.

- Round bars turned on a lathe generally have a rougher surface finish than bars that are ground on cylindrical grinding machines.

- The wall thickness of welded tubing is generally more uniform than that of seamless tubing (produced by the Mannesmann process; see Fig. 13.18).

- Extrusions have smaller cross-sectional tolerances than parts made by roll-forming sheet metal.

Manufacturing characteristics of materials.   As we have described throughout this book, manufacturing characteristics of materials typically include castability, workability, formability, machinability, weldability, and hardenability by heat treatment. Because raw materials have to be formed, shaped, machined, ground, fabricated, or heat treated into individual components having specific shapes and dimensions, their manufacturing characteristics are crucial to the proper selection of these materials.

Recall also that the quality of the raw material can influence its manufacturing properties greatly. The following are some examples:

- A rod or bar with a longitudinal seam (lap) will develop cracks during simple upsetting and heading operations.
- Round bars with internal defects and hard inclusions will crack during seamless-tube production.
- Porous castings will produce a poor surface finish when machined.
- Blanks that are heat treated nonuniformly and bars that are not stress-relieved will distort during subsequent processing.
- Incoming stock that has variations in composition and microstructure cannot be heat treated or machined consistently and uniformly.
- Sheet-metal stock having variations in its cold-worked conditions will exhibit springback during bending and other forming operations because of differences in yield stress.
- If prelubricated sheet-metal blanks are supplied with nonuniform lubricant distribution and thickness, their formability, surface finish, and overall quality in stamping operations will be affected adversely.

Reliability of material supply.   In Section I.4 we pointed out geopolitical factors that can affect the supply of strategic materials. Other factors such as strikes, shortages, and the reluctance of suppliers to produce materials in a particular shape, quality, or quantity (because there is no economic incentive) also affect reliability of supply. Even though the availability of materials throughout the country as a whole may not be a significant concern, it can be a problem for a certain business because of the particular location of a manufacturing plant.

Recycling considerations.   We repeatedly have emphasized the importance of recycling in material selection. It is essential to remember that recycling requires that individual components in a product must be taken apart. Also, if much effort and time has to be expended in doing so, recycling may become prohibitively expensive. Here we summarize some general guidelines to facilitate recycling.

- Reduce the number of parts and types of materials in products.
- Use a modular design to facilitate disassembly.
- For plastic parts, use single types of polymers (as much as possible).
- Mark plastic parts for ease of identification (as is done with plastic food containers and bottles).
- Avoid using coatings, paints, and plating; instead, use molded-in colors in plastic parts.
- Avoid using adhesives, rivets, and other permanent joining methods in assembly; instead, use fasteners—especially snap-in fasteners.

As an example of this type of approach to recycling, one manufacturer of laser-jet printers has reduced the number of parts in the cartridge by 32% and the number of plastic resins by 55%.

Cost of materials and processing.   Because of its processing history, the unit cost of a raw material depends not only on the material itself but also on its shape, size, and condition. For example, because more operations are involved in the production of thin wire than in that of round rod, the unit cost of the thin wire is much higher. Similarly, powder metals generally are more expensive than bulk metals. Furthermore, the cost of materials typically decreases as the quantity purchased increases. Likewise, certain segments of industry (such as automotive companies) purchase materials in very large quantities; the larger the quantity, the lower is the cost per unit weight (bulk discount).

The cost of materials may be determined in terms of cost per unit weight or cost per unit volume. Table 40.5 shows the cost per unit volume relative to that for carbon steel. The benefit of cost per volume can be seen by the following simple example. In the design of a steel cantilevered rectangular beam supporting a certain load at its end, a maximum deflection is specified. Using equations developed in mechanics of solids and assuming that the weight of the beam can be neglected, we can determine an appropriate cross-section of the beam. Since all dimensions are known now, the volume of the beam can be calculated, and if we know the cost of the material per unit volume, we easily can calculate the cost of the beam. If the cost is per unit weight, we first have to calculate its weight and then determine the cost.

The cost of a particular material is subject to fluctuations caused by factors as simple as supply and demand or as complex as geopolitics. If a product is no longer cost competitive, alternative and less costly materials can be selected. For example, the copper shortage in the 1940s led the U.S. government to mint pennies from zinc-plated steel. Similarly, when the price of copper increased substantially during the 1960s, the electrical wiring being installed in homes (for a time) was made of aluminum. It should be noted, however, that this substitution led to the redesign of switches and outlets to avoid excessive heating at the junctions.

When scrap is produced during manufacturing, as in sheet-metal fabricating, forging, and machining, the value of the scrap is deducted from the material's cost in order to obtain the net material cost. The scrap produced in selected manufacturing processes is given in Table 40.6. Note that, in machining, scrap can be very high,

TABLE 40.5

Approximate Cost per Unit Volume for Wrought Metals and Polymers Relative to Cost of Carbon Steel			
Gold	60,000	Aluminum alloys	2–3
Silver	600	High-strength low-alloy steels	1.4
Molybdenum alloys	200–250	Gray cast iron	1.2
Nickel	35	Carbon steel	1
Titanium alloys	20–40	Nylons, acetals, and silicon	1.1–2
Copper alloys	5–6	Rubber	0.2–1
Stainless steels	2–9	Other plastics and elastomers*	02–2
Magnesium alloys	2–4		

* As molding compounds
Note: Costs vary significantly with quantity purchased, supply and demand, size and shape, and various other factors.

TABLE 40.6

Approximate Ranges of Scrap Produced in Various Manufacturing Processes			
Process	Scrap (%)	Process	Scrap (%)
Machining	10–60	Permanent-mold casting	10
Hot forging	20–25	Powder metallurgy	<5
Sheet-metal forming	10–25	Rolling	<1
Hot extrusion	15		

whereas rolling, ring rolling, and powder metallurgy (all of which are net- or near-net-shape processes) produce the least scrap. As expected, the value of the scrap depends on the type of metal and on the demand for it; typically, it is between 10 and 40% of the original cost of the material.

EXAMPLE 40.3 Effect of workpiece hardness on cost in drilling

Gear blanks forged from 8617 alloy steel and having a hardness range of from 149 HB to 156 HB required the drilling of a hole 75 mm (3 in.) in diameter in the hub. The blanks were drilled with a standard helix drill. After only 10 pieces, however, the drill began to gall, temperatures increased excessively, the drill became dull, and the drilled holes had developed a rough internal surface finish.

In order to improve machinability and reduce galling, the hardness of the gear blanks was increased to a range from 217 to 241 HB by heating them to 840°C (1540°F) and then quenching them in oil. When blanks at this hardness level were drilled, galling was reduced, the surface finish was improved, the drill life increased to 50 pieces, and the cost of drilling was reduced by 80%.

Source: ASM International.

40.6 | Material Substitution

There is hardly a product on the global market today for which the *substitution of materials* has not played a major role in helping companies maintain their competitive positions. Automobile and aircraft manufacturing are typical examples of major industries in which the substitution of materials is an ongoing activity. A similar trend is evident in sporting goods and numerous other products.

Although new products continually appear on the market, the majority of the design and manufacturing effort is concerned with improving existing products. There are several reasons for substituting materials in existing products:

1. Reduce the costs of materials and processing.

2. Improve manufacturing, assembly, and installation and allow conversion to automated assembly.

3. Improve the performance of the product (such as by reducing its weight and by improving resistance to wear, fatigue, and corrosion).

4. Increase stiffness-to-weight and strength-to-weight ratios.

5. Reduce the need for maintenance and repair.

6. Reduce vulnerability to the unreliability of the supply of materials.

7. Improve compliance with legislation and regulations prohibiting the use of certain materials for environmental reasons, as well as to respond to other social concerns.

8. Improve robustness to reduce performance variations or environmental sensitivity of the product.

Substitution of materials in the automobile industry. The automobile is a good example of the effective substitution of materials in order to achieve one or more of the objectives outlined previously. Some examples of material substitution are:

- Certain components of the metal body replaced with plastic or reinforced-plastic parts.

- Metal bumpers, gears, pumps, fuel tanks, housings, covers, clamps, and various other components replaced with plastics or composites.

- Metallic engine components replaced with ceramic and composites parts.

- All-metal driveshafts replaced with composite-material driveshafts.

- Cast-iron engine blocks changed to cast-aluminum, forged crankshafts to cast crankshafts, and forged connecting rods to cast, powder-metallurgy, or composite-material connecting rods.

- Leather seats in automobiles in some luxury cars (including Mercedes) can now be replaced (offered as an option) with synthetic materials in response to concerns raised by advocacy groups.

Because the automobile industry is a major consumer of both metallic and nonmetallic materials, there is constant competition among suppliers, particularly in the steel, aluminum, and plastics industries. Industry engineers and management continually are investigating the relative advantages and limitations of these principal materials in their applications, recycling and other environmental considerations, and relative costs and benefits (in particular).

Substitution of materials in the aircraft industry. In the aircraft and aerospace industries:

- Conventional aluminum alloys (2000 and 7000 series) are being replaced with aluminum-lithium alloys and titanium alloys (because of their higher strength-to-weight ratios).

- Forged parts are being replaced with powder-metallurgy parts that are manufactured with better control of impurities and microstructure; the powder-metallurgy parts also require less machining and produce less scrap of expensive materials.

- Advanced composite materials and honeycomb structures are replacing traditional aluminum airframe components (Fig. 40.1), and metal-matrix composites are replacing some of the aluminum and titanium in structural components.

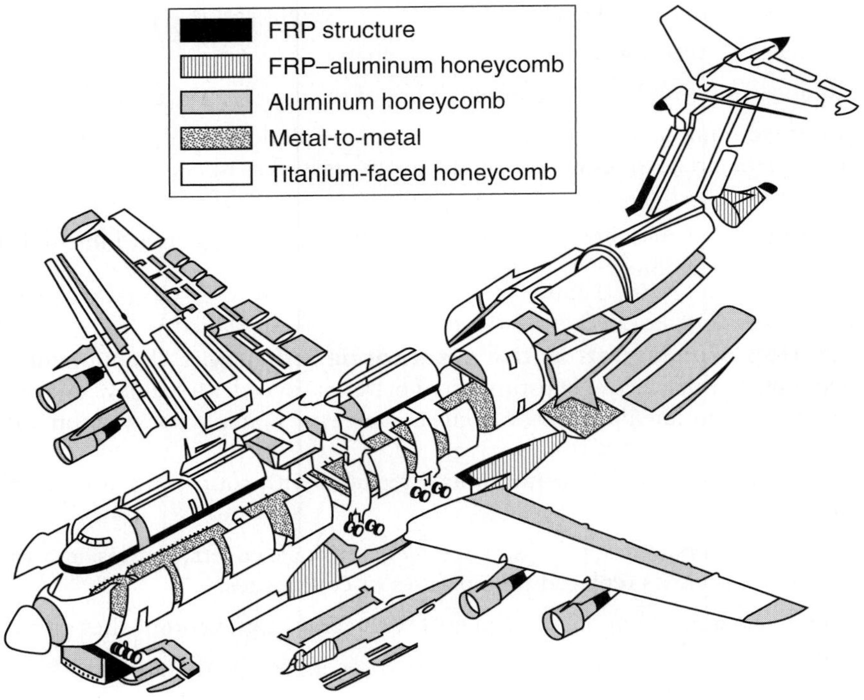

FIGURE 40.1 Advanced materials in the Lockheed C-5A transport aircraft. (*Note:* FRP is fiber-reinforced plastic.)

EXAMPLE 40.4 Material changes between C-5A and C-5B military cargo aircraft

Table 40.7 shows the changes made in materials for various components of the two aircraft and the reasons for these changes.

Source: After H.B. Allison, Lockheed-Georgia.

TABLE 40.7

Material Changes from C-5A to C-5B Military Cargo Aircraft

Item	C-5A Material	C-5B Material	Reason for change
Wing panels	7075-T6511	7175-T73511	Durability
Main frame			
Forgings	7075-F	7049-01	Stress-corrosion resistance
Machined frames	7075-T6	7049-T73	
Frame straps	7075-T6 plate	7050-T7651 plate	
Fuselage skin	7079-T6	7475-T61	Material availability
Fuselage under-floor end fittings	7075-T6 forging	7049-T73 forging	Stress-corrosion resistance
Wing/pylon attach fitting	4340 alloy steel	PH13-8Mo	Corrosion prevention
Aft ramp lock hooks	D6-AC	PH13-8Mo	Corrosion prevention
Hydraulic lines	AM350 stainless steel	21-6-9 stainless steel	Improved field repair
Fuselage failsafe straps	6AI-4V titanium	7475-T61 aluminum	Titanium strap debonding

40.7 | Manufacturing Process Capabilities

We have seen that each manufacturing process has its particular advantages and limitations. For example, casting and powder injection molding generally produce more complex shapes than can forging. However, by subsequent machining and finishing operations, forgings can be made into complex shapes and with high dimensional accuracy and surface finish. Furthermore, forgings have strength and toughness that generally is superior to that of castings and powder-metallurgy products.

The shape of a product may be such that it best can be fabricated from several individual components and then joined with fasteners or with such techniques as brazing, welding, and adhesive bonding. For another product, manufacturing it in one piece may be more economical because of the significant assembly costs otherwise involved.

Other factors that must be considered in process selection are the minimum section size and dimensions that can be produced satisfactorily (Fig. 40.2). For example, very thin sections can be produced by cold rolling but would be difficult or impossible using processes such as sand casting, forging, or powder metallurgy.

Dimensional tolerances and surface finish. The dimensional tolerances and surface finish produced are important particularly in subsequent assembly operations (because of possible difficulties in fitting the parts together for assembly), and in the proper operation of machines and instruments (because their performance can affect tolerances and finish). The dimensional tolerance and surface finish typically obtained by various manufacturing processes are illustrated qualitatively in Fig. 40.3.

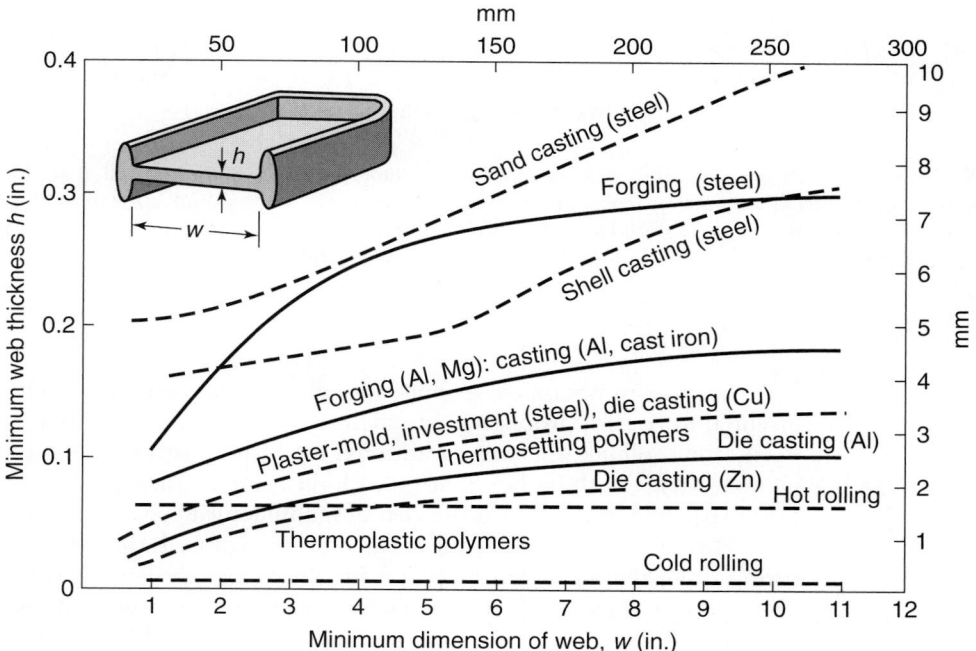

FIGURE 40.2 Manufacturing process capabilities for minimum part dimensions. *Source:* After J.A. Schey.

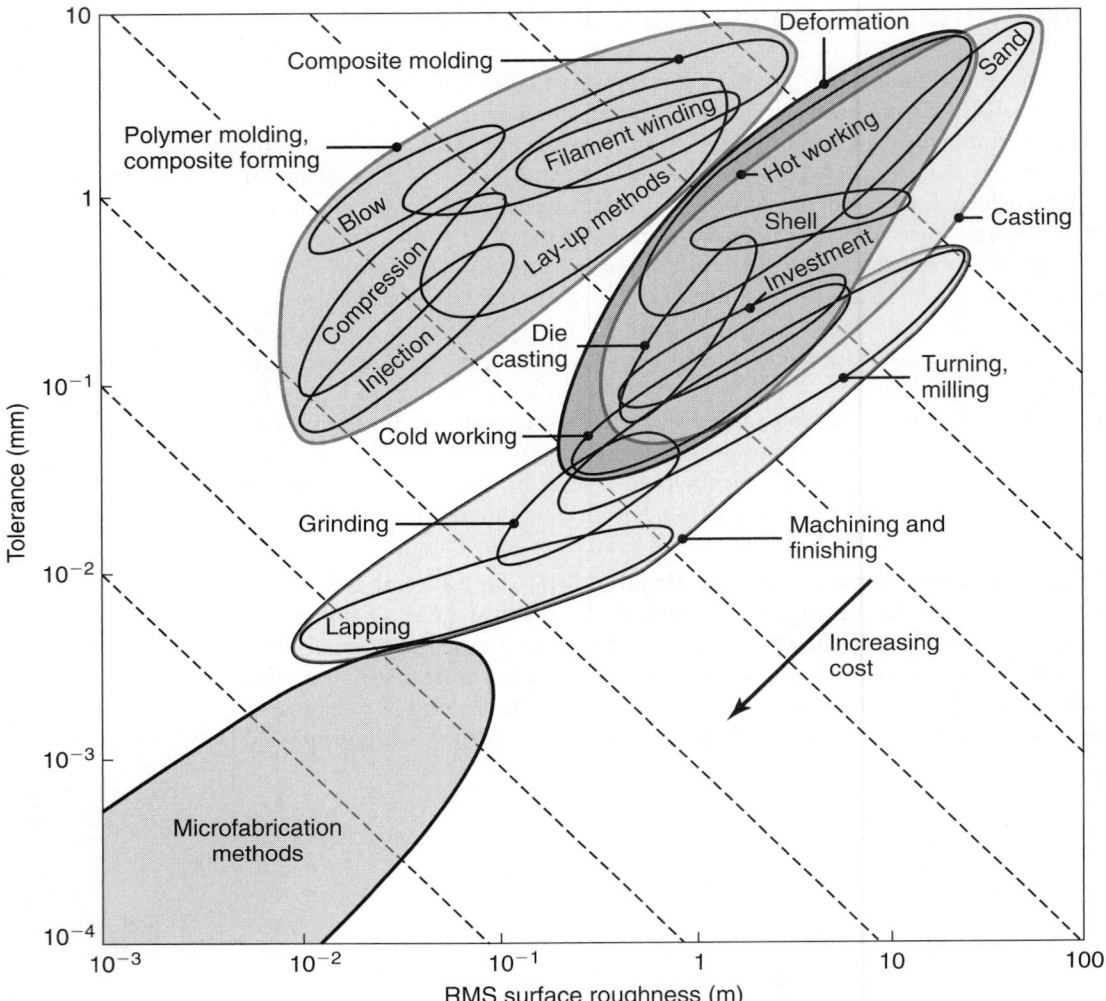

FIGURE 40.3   A plot of achievable tolerance versus surface roughness for assorted manufacturing operations. The dashed lines indicate cost factors where an increase in precision corresponding to the separation of two neighboring lines gives an increase in cost for a given process (or a factor of two). *Source:* M.F. Ashby, *Materials Selection in Design*. Butterworth-Heineman, 1999.

In order to obtain closer dimensional tolerances and better surface finish, it is essential to control processing parameters and use higher-quality machinery, equipment, and tooling; otherwise, additional (and thus costly) finishing operations may be necessary. On the other hand, the closer the tolerance and the finer the surface finish specified, the higher is the cost of manufacturing, as shown in Fig. 40.4.

Furthermore, the finer the surface finish required, the longer is the manufacturing time, the more processes are involved, and the higher is the product cost (Fig. 40.5). For example, in the machining of aircraft structural members made of titanium alloys, it has been observed that as much as 60% of the cost of machining the part may be expended in the final machining pass in order to maintain proper tolerances and surface finish. Unless otherwise required (with appropriate technical and economic justification), parts should be made with as rough a surface

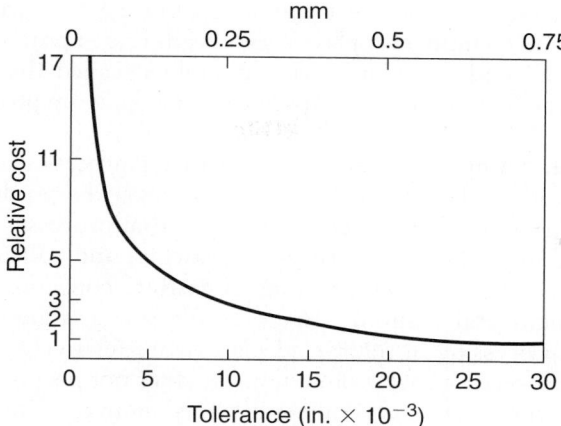

FIGURE 40.4   Dependence of manufacturing cost on dimensional tolerances.

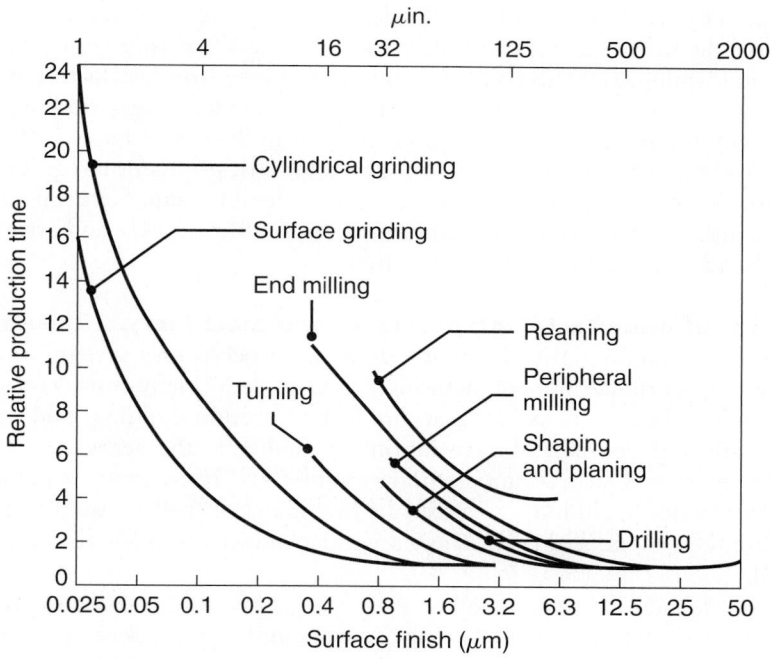

FIGURE 40.5   Relative production time as a function of surface finish produced by various manufacturing processes. (See also Fig. 26.34.)

finish and as wide a dimensional tolerance as functionally and aesthetically will be acceptable.

Production quantity.   Depending on the type of product, the production quantity (*lot size*) varies widely. For example, bearings, bolts, spark plugs, ammunition, plastic containers, tires, automobiles, and lawn mowers are produced in very large quantities.

In contrast, jet engines, diesel engines, locomotives, and ships are produced in limited quantities. Production quantity also plays a significant role in process and equipment selection. In fact, an entire manufacturing discipline (called the *economic order quantity*) is devoted to determining mathematically the optimum production quantity.

Production rate.   A major factor in manufacturing process selection is the production rate. This is defined as the number of pieces to be produced per unit of time (such as per hour, per day, or per year). Recall that processes such as die casting, powder metallurgy, deep drawing, wire drawing, and roll forming are high production-rate operations. By contrast, sand casting, conventional and electrochemical machining, metal spinning, superplastic forming, adhesive and diffusion bonding, and the processing of reinforced plastics are relatively slow production-rate operations. However, a low production rate does not necessarily mean that a particular manufacturing process is inherently uneconomical. The production rate obviously can be increased by using multiple equipment and highly automated machines.

Lead time.   Lead time generally is defined as the length of time between the receipt of an order for a product and its delivery to the customer at the specified time. The selection of a manufacturing process and operation is influenced greatly by the time required to begin production. Thus, depending on the die-shape complexity, its size, and the die material, the lead time for such processes as forging, extrusion, die casting, roll forming, and sheet-metal forming can range from weeks to months.

In contrast, material-removal processes such as machining, grinding, and advanced machining techniques have significant built-in flexibility, because they utilize machines and tooling that easily can be adapted to most production requirements in a very short time. Recall that machining centers, flexible manufacturing cells, and flexible manufacturing systems are capable of responding quickly and effectively to product changes and to production quantities.

Robustness of manufacturing processes and machinery.   Robustness was described in Section 36.5.1 in terms of a design, a process, or a system. In order to appreciate its importance in manufacturing processes, let's briefly consider a situation where a simple plastic gear is being produced by injection molding, and that there are significant and unpredictable variations in quality as the gears are being produced. There are several well understood variables and parameters in the injection molding of plastics, including the effects of raw-material (pellets) quality, temperatures within the system, and time. These are independent variables, thus they can be controlled.

However, there are certain other variables (called *noise*) that largely are beyond the control of the operator. Among these are ambient-temperature and humidity variations in the plant during the day, dust in the air entering the plant from an open door (thus possibly contaminating the pellets being fed into the hoppers of the injection-molding machine), and the variability in the performance of different operators during different shifts. Obviously these variables are difficult or impossible to precisely control.

In order to obtain sustained good product quality, it first is necessary to understand the effects (if any) of each element of noise in the operation. For example: (a) Why and how does the ambient temperature affect the quality and surface characteristics of the molded gears? (b) Why and how is the dust coating on a pellet affecting its performance in the molding machine? (c) How different are the performances of different operators during different shifts and why? (d) Are there inherent variations in machine performance during the day and, if so, why?

It then will be possible to establish new operating parameters so that variations in, say, ambient temperature and plant environment do not affect gear quality adversely. Note that these considerations are valid equally for any other production process, although some are less sensitive to noise (such as bulk-deformation processes) than others (such as microelectronics manufacturing). Because this is a complex subject, it is beyond the scope of this book to further discuss it.

40.8 | Process Selection

Based on the preceding discussions and considerations, we now proceed with the topic of *selection of manufacturing processes and machinery*, and how this selection process relates to various factors (see also Section I.5). From the discussions in various chapters in this book, recall that process selection is related intimately to the characteristics of the materials to be processed (Table 40.8). Note also that most manufacturing processes and operations now are automated and computer controlled with the purpose of optimizing all aspects of the total operation, increasing product reliability and product quality, and reducing costs (see Section 40.9).

The major factors involved in process selection are discussed next.

Characteristics and properties of the workpiece materials.   Some materials can be processed at room temperature, whereas others require elevated temperatures (and thus the need for furnaces, appropriate tooling, and various controls). Some materials are easy to work because they are soft and ductile. Others (being hard, brittle, and abrasive) require special processing techniques and tool and die materials.

Materials have different manufacturing characteristics, such as castability, forgeability, workability, machinability, and weldability. Few materials have favorable characteristics in all of these relevant categories. For example, a material that is castable or forgeable may later present difficulties in the machining, grinding, or finishing operations that may be required for acceptable surface finish, dimensional accuracy, and quality.

Materials also have different responses to the rate of deformation (strain-rate sensitivity) to which they are subjected. Thus, the speed at which a machine operates will affect the quality of the product, including the development of external and internal defects. For example, impact extrusion or drop forging may not be appropriate for materials with high strain-rate sensitivity, whereas they will perform well in a hydraulic press or direct extrusion.

Geometric features of the part.   Features such as part shape, size, thickness, dimensional tolerances, and surface-finish requirements greatly influence the selection of a process or processes, as we have described throughout this chapter and other chapters in the book.

Production rate and quantity.   These requirements dictate process selection by way of the productivity of a process, machine, or system, as we have described throughout this book.

Process selection considerations.   We may summarize those factors involved in process selection with the following questions.

- Does each component of the product have to be manufactured in the plant? Are some of the parts commercially available as standard items?

TABLE 40.8

General Characteristics of Manufacturing Processes for Various Metals and Alloys

	Carbon steels	Alloy steels	Stainless steels	Tool and die steels	Aluminum alloys	Magnesium alloys	Copper alloys	Nickel alloys	Titanium alloys	Refractory alloys
Casting										
Sand	A	A	A	B	A	A	A	A	B	A
Plaster	—	—	—	—	A	A	A	—	—	—
Ceramic	A	A	A	A	B	B	A	A	B	A
Investment	A	A	A	—	A	B	A	A	A	A
Permanent	B	B	—	—	A	A	A	—	—	—
Die	—	—	—	—	A	A	A	—	—	—
Forging										
Hot	A	A	A	A	A	A	A	A	A	A
Extrusion										
Hot	A	A	A	B	A	A	A	A	A	A
Cold	A	B	A	—	A	—	A	B	—	—
Impact	—	—	—	—	A	A	A	—	—	—
Rolling	A	A	A	—	A	A	A	A	A	B
Powder metals	A	A	A	A	A	A	A	A	A	A
Sheet-metal forming	A	A	A	—	A	A	A	A	A	A
Machining	A	A	A	—	A	A	A	B	A	B
Chemical	A	B	A	B	A	A	A	B	B	B
ECM	—	A	B	A	—	—	B	A	A	A
EDM	—	B	B	A	B	—	B	B	B	A
Grinding	A	A	A	A	A	A	A	A	A	A
Welding	A	A	A	—	A	A	A	A	A	A

Note: A-Generally processed by this method; B-Can be processed by this method, but may present some difficulties; — Usually not processed by this method. Product quality and productivity depend greatly on the techniques and equipment used, operator skill, and proper control of processing variables.

- Is the tooling required available in the plant? Can it be purchased as a standard item?

- Can group technology be implemented for parts with similar geometric and manufacturing attributes?

- Have all alternative manufacturing processes been investigated?

- Are the methods selected economical for the type of material, the part shape to be produced, and the required production rate?

- Can the requirements for dimensional tolerances, surface finish, and product quality be met consistently, or can they be relaxed?

- Can the part be produced to final dimensions without requiring additional processing?

- Are all processing parameters optimized?

- Is scrap produced, and if so, is it minimized? What is the value of the scrap?

- Have all the automation and computer-control possibilities been explored for all phases of the total manufacturing cycle?

- Are in-line, automated inspection techniques and quality control being implemented properly?

EXAMPLE 40.5 Process selection in making a single part

Assume that you are asked to make the simple part shown in Fig. 40.6a. First, we should consider and determine the part's function, the types of load and environment to which it is to be subjected, the dimensional tolerances and surface finish specified, and the quantity and production rate required. For the sake of discussion, let's assume that the part is round, that it is 125 mm (5 in.) long, and that its large and small diameters are 38 mm and 25 mm (1.5 in. and 1.0 in.), respectively. Further assume because of functional requirements (such as strength, stiffness, hardness, and resistance to wear and elevated temperatures) that this part should be made of metal.

   Which manufacturing process would you choose and how would you organize the production facilities to manufacture a cost-competitive, high-quality product? Recall that, as much as possible, parts should be produced at or near their final shape (net- or near-net-shape manufacturing), using an approach that largely eliminates much secondary processing (such as machining, grinding, and other finishing operations) and thus reduces the total manufacturing time and cost. Because it is relatively simple, this part can be manufactured by such methods as

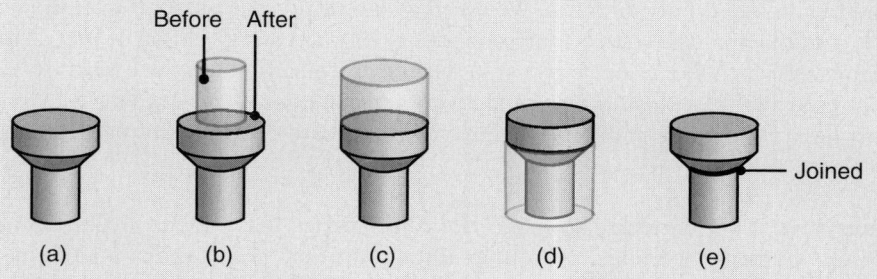

FIGURE 40.6   Various methods of making a simple part: (a) casting or powder metallurgy, (b) forging or upsetting, (c) extrusion, (d) machining, and (e) joining two pieces.

(a) casting or powder metallurgy, (b) forging or upsetting, (c) extrusion, (d) machining, or (e) joining together two separate pieces.

For net-shape production, the two suitable processes are *casting* and *powder metallurgy*—each process having its own characteristics, need for specific tooling and labor skill, and costs. This part also can be made by cold, warm, or hot *forming*. One method is the upsetting (heading) of a round bar, 25 mm (1 in.) in diameter, in a suitable die to form the larger end. Another possibility is the partial direct *extrusion* of a 38-mm (1.5-in.) diameter bar and reduce its diameter to 25 mm. Note that each of these processes produces little or no material waste.

This part also can be made by *machining* a 38-mm diameter bar stock to obtain the 25-mm diameter section. However, machining requires much longer time than forming processes, and some material inevitably will be wasted as metal chips. On the other hand, machining involves no special tooling (unlike net-shape processes which generally require special dies), and this operation can be carried out easily on a common lathe or a CNC lathe. Note also that this part can be made in two separate pieces, and then *joined* by welding, brazing, or adhesive bonding.

Because of the different operations required in producing the raw materials, costs depend not only on the type of material (ingot, powder, drawn rod, extrusion) but also on its size and shape. Thus, per unit weight, (a) square bars are more expensive than round bars, (b) cold-rolled plate is more expensive than hot-rolled plate or sheet, (c) hot-rolled bars are much less expensive than metal powders of the same type.

It appears that (if only a few parts are needed) machining this part is the most economical method. If the production quantity and rate requirements are high, producing this part by a heading operation or by cold extrusion (a form of closed-die forging) would be the appropriate choice. If the top and bottom portions of the part must be made of different materials, then joining them would be the most appropriate choice.

EXAMPLE 40.6 Manufacturing a sheet-metal part by different methods

Let's form a simple, dish-shaped part from sheet metal. Such a part can be formed by placing a round, flat piece of sheet metal between a pair of male and female dies, then closing the dies by applying a vertical force in a press (Fig. 40.7a). Parts can be formed at high production rates by this method, generally known as *stamping* or *pressworking*.

Assume now that the size of the part is very large, say, 2 m (80 in.) in diameter, and the lot size is only 50 parts. We now have to reconsider the total operation and ask a number of questions. Is it economical to manufacture a set of dies 2 m in diameter (which would be very costly) when the production quantity is so low? Are presses available with sufficient capacity to accommodate such large dies? Does the part have to be made in one piece? Are there alternative methods of manufacturing this part?

Solution This part can be made by *welding* together smaller pieces of metal formed by other methods. Note that large municipal water tanks and ships are made by this method. Would a part manufactured by welding be acceptable for its intended purpose in the environment in which it will be used? Will it have the required properties and the correct shape after welding, or will it require additional processing?

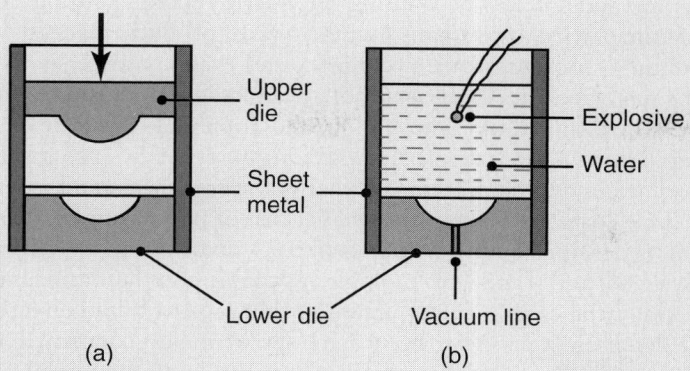

FIGURE 40.7 Two methods of making a dish-shaped sheet-metal part: (a) pressworking using a male and female die, (b) explosive forming using one die only.

This part also can be made by *explosive forming*, as shown in Fig. 40.7b. It will be noted, however, that the deformation of the material in explosive forming takes place at a very high rate. Consequently, a series of questions have to be asked:

a. Is the material capable of undergoing deformation at high rates without fracture or any detrimental effect on the final properties of the formed part?

b. Can the dimensional tolerances and surface finish be held within acceptable limits?

c. Is the life of the die sufficiently long under the high transient pressures generated in explosive forming?

d. Can this operation be performed in a manufacturing plant within the city limits or should it be carried out in open country?

e. Is the production rate sufficiently high for this type of product?

f. Although explosive forming has the advantage of requiring only one die, is the operation economical?

From this brief discussion, it can be seen that for each part or component of a product similar approaches are required to arrive at a conclusion as to which process is both the most suitable for specific requirements and the most economical.

40.9 │ Manufacturing Costs and Cost Reduction

In order for a product to be marketed successfully, its cost must be competitive with that of similar products, particularly in the global marketplace. The total cost of a product consists of several categories, such as material cost, tooling cost, fixed costs, variable costs, direct-labor costs, and indirect-labor costs. As a general guide to the costs involved, note that we have included sections on the *economics* of the following individual groups of manufacturing processes and operations: Part II (casting); Part III (rolling, forging, extrusion, drawing, sheet-metal working, powder metallurgy, ceramics, polymer processing); Part IV (machining, abrasive processing, advanced machining); and Part VI (welding and various joining processes).

Various methods of cost accounting are used by manufacturing organizations. Depending on the particular company and the type of products made, the methodologies of accounting procedures can be complex and even controversial. Furthermore, because of the many technical and operational factors involved, calculating individual cost factors can be difficult, time consuming, and not always accurate or reliable.

Costing systems (also called *cost justification*) typically include the following considerations: (a) intangible benefits of quality improvements and inventory reduction, (b) life-cycle costs, (c) machine usage, (d) cost of purchasing machinery as compared to that of leasing them, (e) financial risks involved in implementing highly automated systems, and (f) new technologies available and their impact on products.

Additionally, the costs to a manufacturer that are attributed directly to *product liability* and defense claims have been a matter of major concern. Every modern product now has a built-in added cost to cover possible product liability claims. For example, it is estimated that (a) liability suits against car manufacturers in the United States add about $500 to the indirect cost of an automobile, and (b) 20% of the price we pay for a ladder is attributed to potential product liability costs. We outline here the various major *cost factors* in manufacturing.

Materials costs.   Some cost data are given in various tables throughout this book. See also Table 40.1.

Tooling costs.   These are the costs involved in making the tools, dies, molds, patterns, and special jigs and fixtures required for manufacturing a product. High tooling costs may be justified in high-volume production of a single item. The expected life of tools and dies and their obsolescence (because of product changes) also are important considerations.

The tooling cost is influenced greatly by the production process selected. For example, (a) the tooling cost for die casting is higher than that of sand casting; (b) the tooling cost in machining or grinding is much lower than that of powder metallurgy, forging, or extrusion; (c) carbide tools are more expensive than high-speed steel tools, but their life is longer; (d) if a part is to be manufactured by spinning, the tooling cost for conventional spinning is much lower than that of shear spinning; and (e) tooling for rubber-forming processes is less expensive than that of the die sets (male and female) used for the deep drawing and stamping of sheet metals.

Fixed costs.   These costs include electric power, fuel, taxes on real estate, rent, insurance, and capital (including depreciation and interest). The company must meet these costs regardless of whether or not it has made a particular product. Consequently, fixed costs are not sensitive to production volume.

Capital costs.   These costs represent the investment in buildings, land, machinery, tooling, and equipment. As can be seen in Table 40.9, the cost of machines and systems can vary widely. In view of the generally high equipment costs, particularly those involving transfer lines and flexible manufacturing cells and systems, high production rates and production quantities are essential to justify such large expenditures and to maintain product costs at or below the all-important competitive level. Lower unit costs (cost-per-piece) can be achieved by continuous production involving around-the-clock operation—but only as long as demand warrants it. Equipment maintenance also is essential to ensure high productivity. Any breakdown of machinery leading to *downtime* can be very costly—by as much as thousands of dollars per hour.

Direct-labor costs.   The direct-labor cost is for the labor directly involved in manufacturing (*productive labor*). This cost includes all labor from the time raw materials

TABLE 40.9

Relative Costs for Machinery and Equipment (Costs vary greatly, depending on size, capacity, options, and level of automation and computer controls. See also the sections on economics in various chapters.)	
Automatic screw machine	M-H
Boring mill, horizontal	M-H
Broaching	M-H
Deep drawing	M-H
Die casting	M-H
Drilling	L-M
Electrical-discharge machining	L-M
Electron-beam welding	M-H
Extruder, polymer	L-M
Extrusion press	M-H
Flexible manufacturing cell and system	H-VH
Forging	M-H
Fused deposition modeling	L
Gas tungsten-arc welding	L
Gear shaping	L-H
Grinding	L-H
Headers	L-M
Honing, lapping	L-M
Injection molding	M-H
Laser-beam welding	M-H
Lathes	L-M
Machining center	L-M
Mechanical press	L-M
Milling	L-M
Powder-injection molding	M-H
Powder metallurgy	L-M
Powder metallurgy, HIP	M-H
Resistance spot welding	L-M
Ring rolling	M-H
Robots	L-M
Roll forming	L-M
Rubber forming	L-M
Sand casting	L-M
Spinning	L-M
Stereolithography	L-M
Stamping	L-M
Stretch forming	M-H
Transfer lines	H-VH
Ultrasonic welding	L-M

Note: L, low; M, medium; H, high; VH, very high.

first are handled to the time when the product is finished. This period generally is referred to as *floor-to-floor time*. For example, a machine operator picks up a round bar from a bin, machines it into the shape of a threaded rod, and places it into another bin. The direct-labor cost is calculated by multiplying the labor rate (hourly wage, including benefits) by the time that the worker spends producing the part.

The time required for producing a particular part depends not only on its size, shape, and dimensional accuracy but also on the workpiece material. For example, the cutting speeds for high-temperature alloys are lower than those for aluminum or plain-carbon steels. Consequently, the cost of machining aerospace materials is much higher than that for machining more common alloys, such as those of aluminum and steel.

Labor costs in manufacturing and assembly vary greatly from country to country (see Table I.4 in the General Introduction). It is not surprising that many of the products one purchases today—from clothing to high-tech toys and products—are either made or assembled in countries where labor costs are low. On the other hand, firms located in countries with high labor rates tend to emphasize high value-added manufacturing tasks or high automation levels, so the labor component of the cost is reduced significantly. For example, labor cost associated with some products manufactured in the United States accounts for only 5% of total product cost.

For labor-intensive industries, many manufacturers have attempted to move manufacturing activities to countries with a lower labor rate. However, while this has sometimes been financially attractive, often the cost savings anticipated are not realized. Some of the hidden costs associated with outsourcing are the following:

- International shipping is far more complicated and time-consuming than domestic shipping. For example, it takes roughly four to six weeks for a container ship to bring product from China to the United States or Europe, and this time is increasing as homeland security issues continue to unfold.

- Because of the lengthy shipping times, just-in-time manufacturing approaches and their associated cost savings cannot be realized (see Section 39.5). Also, because of the long shipping times, schedules are rigid, design modifications cannot be made easily, and companies readily cannot address changes in the market or in demand. Thus, companies that outsource tend to lose agility and have difficulty following lean-manufacturing approaches.

- Legal systems are not as established in countries with lower labor rates. Procedures that are common in the United States or Europe, such as accounting audits, protection of patented designs and intellectual property, and conflict resolution are more difficult to enforce or obtain in other countries.

- Payments often are expected based on units completed, product defect rates can be significant, and thus, parts may require reworking upon receipt.

- There are numerous hidden costs, such as increased paperwork and documentation, lower productivity from existing employees because of lower morale, and difficulties in communication.

Outsourcing has been an issue of major attention, notably in the popular media. Undoubtedly, it will continue to be a significant concern. Low labor costs (such as those in China, India, Mexico, and Pacific Rim countries) can be countered effectively in the West and in Japan by improving productivity, reducing the labor component of products, and further improving the efficiency of manufacturing operations. The economic justification for outsourcing is complex and controversial, and it is beyond the scope of this book. Some of the texts in the Bibliography of this chapter can be helpful in gaining a proper perspective of this topic.

Indirect-labor costs.   These costs are generated in the servicing of the total manufacturing operation, consisting of such activities as supervision, repair, maintenance, quality control, engineering, research, and sales; it also includes the cost of office staff. Because they do not contribute directly to the production of finished parts, or they are not chargeable to a specific product, these costs are referred to as **overhead** (**burden rate**) and are charged proportionally to all products. The personnel involved in these activities are categorized as **nonproductive labor.**

Manufacturing costs and production quantity. One of the most significant factors in manufacturing costs is production quantity. Obviously, large production quantity requires high production rates. High production rates, in turn, require the use of mass-production techniques that involve special machinery (*dedicated machinery*) and employ proportionally less direct labor as well as that of plants that operate on two or three shifts. At the other extreme, smaller production quantity usually means a larger direct-labor involvement.

Small batch production usually is done on general-purpose machines, such as lathes, milling machines, and hydraulic presses. The equipment is versatile, and parts with different shapes and sizes can be produced by appropriate changes in the tooling. However, direct-labor costs are high because these machines usually are operated by skilled labor.

For larger quantities (*medium-batch production*), these same general-purpose machines can be equipped with various jigs and fixtures or can be computer-controlled. To further reduce labor costs, machining centers and flexible manufacturing systems are important alternatives. Generally, for quantities of 100,000 or more, the machines are designed for specific purposes, and they perform a variety of specific operations with very little direct labor.

Cost reduction. Cost reduction requires a study on how the costs described previously are interrelated, using *relative costs* as an important parameter. As we have seen, the unit cost of a product can vary widely. For example, some parts may be made from expensive materials but require very little processing, such as for minted gold coins. Consequently, the cost of materials relative to that of direct labor is high.

By contrast, some products may require several complex and expensive production steps to process relatively inexpensive materials, such as for carbon steels. For example, an electric motor is made of relatively inexpensive materials, yet several different manufacturing operations are involved in the making of the housing, rotor, bearings, brushes, and various other components. Unless highly automated, assembly operations for such products can become a significant portion of the overall cost.

A typical approximate breakdown of costs in manufacturing today is as follows:

Design	5%
Material	50%
Direct Labor	15%
Overhead	30%

In the 1960s, labor accounted for as much as 40% of the production cost. Today, it can be as low as 5%, depending on the type of product and level of automation. In the breakdown above, note the very small contribution of the design phase, yet the design phase generally has the largest influence on the *quality* and *success* of a product in the marketplace. Cost reductions can be achieved by a thorough analysis of all the costs incurred in each phase in the manufacture of a product. The methods employed are described in detail in some of the references in the Bibliography at the end of this chapter.

We have emphasized the opportunities for cost reduction in various chapters throughout this book. Among these are the following:

- Simplifying part design and the number of subassemblies required.
- Specifying broader dimensional tolerances and allowing rougher surface finish.
- Using less expensive materials.
- Investigating alternative methods of manufacturing.
- Using more efficient machines and equipment.

The introduction of more automated systems and of up-to-date technology in a manufacturing facility is an obvious means of reducing certain types of costs. However, this approach must be undertaken with due care and only after a thorough **cost-benefit analysis**, which requires reliable data input and consideration of the technical as well as the human factors involved. Advanced technologies (which can be very costly) should be implemented only after a complete analysis of the more obvious cost factors—known as **return on investment** (ROI).

You have undoubtedly noted that (over a period of time) the prices of some products (such as calculators, computers, and digital watches) have decreased greatly, while the prices of other products (such as housing and books) have gone up. These trends are partly a result of the inevitable impact of computer-controlled machinery and operations on all aspects of product design and manufacturing. They also result from changes in the various component costs over time, including those due to labor, machinery, domestic and international competition, and worldwide economic trends (such as supply and demand, exchange rates, and tariffs).

40.9.1 Value analysis

Manufacturing adds *value* to materials as they become discrete products and are marketed. Because this value is added in individual stages during the creation of the product, the utilization of value analysis (also called *value engineering, value control*, and *value management*) is important. *Value analysis* is a system that evaluates each step in design, materials, processes, and operations in order to manufacture a product that performs all of its intended functions and does so at the lowest possible cost. We have described the important considerations in each step of value analysis throughout this chapter.

A monetary value is established for each of two product attributes: (a) **use value**, reflecting the functions of the product and (b) **esteem value** or **prestige value**, reflecting the attractiveness of the product that makes its ownership desirable. The *value of a product* then is defined as

$$\text{Value} = \frac{\text{Product function and performance}}{\text{Product cost}} \tag{40.1}$$

Thus, the goal of value analysis is to obtain maximum performance per unit cost. Value analysis generally consists of the following six phases:

1. *Information phase*: to gather data and determine costs.
2. *Analysis phase*: to define functions and identify problem areas as well as opportunities.
3. *Creativity phase*: to seek ideas to respond to problems and opportunities without judging the value of each of these ideas.
4. *Evaluation phase*: to select the ideas to be developed and to identify the costs involved.
5. *Implementation phase*: to present facts, costs, and values to the company management; to develop a plan, and to motivate positive action, all in order to obtain a commitment of the resources necessary to accomplish the task.
6. *Review phase*: in which the overall value-analysis process is reviewed and necessary adjustments are made.

Value analysis is an important and all-encompassing interdisciplinary activity. Usually it is coordinated by a value engineer and is conducted jointly by designers, engineers, quality control, purchasing, marketing personnel, and managers. In order for value analysis to be effective, it must have the full support of a company's top management. Implementation of value analysis in manufacturing results in such benefits as (a) significant cost reduction, (b) reduced lead times, (c) better product quality and performance, (d) reduced time for manufacturing the product, and (e) reduced product weight and size.

An example of product weight reduction is the development of the antilock braking system (ABS) for automotive applications. In 1989, the typical weight of a Bosch brand ABS was 6.2 kg (13.6 lb). In 2001, its weight was 1.8 kg (4 lb.), representing a reduction of 70%, which also helped reduce the weight of the automobile. Note that (since weight is related to the product's volume and considering its function) reducing the size indicates that the ratio of surface area to volume increases. Consequently, characteristics (such as surface finish and surface integrity) become important with respect to their possible effects on product performance.

CASE STUDY 40.1 | Concurrent Engineering for Intravenous Solution Containers

Baxter Healthcare manufactures over one million intravenous (IV) solution containers every day in the United States, providing critical therapies within the health care industry. Recognizing that patients' lives depend upon the safe delivery of medical solutions, the introduction of new or improved products is highly regulated by internal company standards and external government agencies.

A well-defined product-development process provides the framework to consistently meet the regulated quality, reliability, and manufacturing design requirements. More importantly, a concurrent engineering environment catalyzes the development process to minimize development cost and time to market.

In the 1990s, the company focused development efforts on a new set of materials for flexible-IV containers. The container system being developed was expected to be more environmentally friendly, compatible with a larger variety of new critical-care drugs, exceed all quality requirements, and remain cost effective. Some of the key design issues shown in Table 40.10 would allow the product to be safe for patients and maintain economic viability.

The Matrixed Team

Given the extensive requirements, Baxter formed a multifunctional team of over 25 individuals. As the requirements and goals indicated, marketing, manufacturing, and development team members worked side by side with specialists in materials science, regulatory affairs, clinical affairs, toxicology, chemical stability, and sterility assurance. The product team members each recognized their responsibility for the success or failure of the product design effort.

The Active Team

All team members contributed to the master requirements definition during the product conceptualization phase. All team members led and communicated the test and development activities within their respective fields to the team at large during the development phase. All disciplines offered and accepted peer-review criticism of product or process designs at major development steps. All team members assured that the product designs, quality assessment methods, and fabrication techniques were transferred efficiently to the designated plants during the implementation phase. The product team avoided costly design iterations and revisions and minimized duplicative efforts by maintaining a matrixed, active team throughout the process.

TABLE 40.10

Design Issues for Intravenous Solution Containers

Container product requirements	Container processing requirements
Provide physical and chemical shelf-life of up to several years without compromising the solution	Implement the new product with existing equipment and plant personnel
Allow for the addition and mixing of various drug solutions in the hospital or alternate site pharmacy	Allow product manufacturing at rates up to one million per day without additional production cells Maintain fabrication throughput at over 60 containers per minute per machine Maintain printing and filling speeds over 120 per minute
Provide a surface for printing fade, flake, and smear-proof labeling	
Provide a surface for adherence of various pressure-sensitive adhesive labels at room, refrigerated, and elevated temperatures	
Withstand pressurization when an infusion device (cuff) is placed on the container for controlled solution delivery to the patient	Withstand steam sterilization temperatures, pressures, and times
Maintain compatibility with available handling and administration-set devices	Allow packing in cartons that provide the fullest density of pallet loading Withstand air and ground shipping and through climate extremes in areas as diverse as Arizona and Alaska

The Virtual Team

Many team members were located at manufacturing plants or development centers throughout northern Illinois and the greater United States. By necessity, team meetings leveraged teleconference and videoconference capabilities, which also minimized travel expense. Electronic-mail access provided the rapid and broad communication of impending issues and resolutions. A company-wide intranet site was developed to share documents and design prints, which assured that all locations were utilizing the most recent project advancements. The team customized emerging communication technologies to enhance the concurrent development of new materials, processing, and fabrication technologies.

Finished Goods

The engineering effort of the matrixed, active, and virtual teams resulted in a product introduction within three years. New materials and components were developed which met all product and process requirements and ultimately satisfied patient needs. The new product was transparent, and patient-care professionals continued to use their existing training and well-practiced techniques. Disposal of the product was streamlined; the new materials now could be disposed of safely or recycled. Reliability and user satisfaction remained high. The new materials were compatible with an extended array of solution format drugs packaged in IV containers. Indeed, team success will be measured by the number of additional products introduced in the following years.

Source: Courtesy of K. Anderson, Baxter Healthcare Corporation, and S. Petronis, Zimmer Inc. (formerly of Baxter Healthcare Corporation).

SUMMARY

- Regardless of how well a product meets design specifications and quality standards, it must also meet economic criteria in order to be competitive in the domestic and global marketplace. Guidelines have been established for designing products for economic production.

- Important considerations in product design and manufacturing include manufacturing characteristics of materials, product life expectancy, and life cycle engineering.

- Substitution of materials, modification of product design, and relaxing of dimensional tolerance and surface finish requirements are important methods of cost reduction.

- The total cost of a product includes several elements, such as cost of materials, tooling, capital, labor, and overhead. Materials costs can be reduced through careful selection without compromising design and service requirements, functions, specifications, and standards for good product quality.

- Labor costs generally are becoming an increasingly smaller percentage of production costs, but to counteract lower wages in other countries, they can be reduced further through highly automated and computer-controlled manufacturing operations.

KEY TERMS

Burden rate

Capital costs

Cost-benefit analysis

Cost justification

Cost reduction

Dedicated machines

Direct labor

Downtime

Economic order quantity

Fixed costs

Floor-to-floor time

Indirect labor

Lead time

Nonproductive labor

Outsourcing

Overhead

Process capabilities

Production rate

Production quantity

Relative costs

Return on investment

Scrap

Smart databases

Value

Value analysis

BIBLIOGRAPHY

Anderson, D.M., *Design for Manufacturability, Optimizing Cost, Quality, and Time-to-Market*, 2nd ed., CIM Press, 2001.

Ashby, M.F., *Materials Selection in Mechanical Design*, 2nd. ed., Pergamon, 1999.

ASM Handbook, Vol. 20: *Materials Selection and Design*, ASM International, 1997.

Baxter, M., *Product Design: A Practical Guide to Systematic Methods of New Product Development*, Chapman & Hall, 1995.

Benhabib, B., *Manufacturing: Design, Production, Automation, and Integration*, Marcel Dekker, 2003.

Billatos, S. and Basaly, N., *Green Technology and Design for the Environment*, Taylor & Francis, 1997.

Boothroyd, G., Dewhurst, P., and Knight, W., *Product Design for Manufacture and Assembly*, 2nd ed., Marcel Dekker, 2001.

Bralla, J.G., *Design for Manufacturability Handbook*, McGraw-Hill, 1999.

Brown, J., *Value Engineering: A Blueprint*, Industrial Press, 1992.

Cattanach, R.E. (ed.), *The Handbook of Environmentally Conscious Manufacturing*. McGraw-Hill, 1994.

Corbett, J., Donner, M., Maleka, J., and Pym, C., *Design for; Manufacture: Strategies, Principles and Techniques*, Addison-Wesley, 1991.

Cosumano, M.S., *Thinking Beyond Lean*, The Free Press, 1998.

Creese, R.C., Adithan, M., and Pabla, B.S., *Estimating and Costing for the Metal Manufacturing Industries*, Marcel Dekker, 1992.

Demaid, A. and DeWit, J.H.W. (eds.), *Case Studies in Manufacturing With Advanced Materials*, North-Holland, 1995.

Dettmer, W.H., *Breaking the Constraints to World-Class Performance*, ASQ Quality Press, 1998.

Dieter, G.E., *Engineering Design: A Materials and Processing Approach*, 3rd ed., McGraw-Hill, 1999.

Dieter, G.E.; Kuhn, H.A.; and Semiatin, S.L., *Handbook of Workability and Process Design*, ASM International, 2003.

Erhorn, C. and Stark, J., *Competing by Design: Creating Value and Market Advantage in New Product Development*, Wiley, 1995.

Farag, M.M., *Materials Selection for Engineering Design*, Prentice Hall, 1997.

Fleischer, M. and Liker, J.K., *Concurrent Engineering Effectiveness: Integrating Product Development Across Organizations*, Hanser Gardner, 1997.

Ghosh, A., Miyamoto, Y., Reimanis, I., and Lannutti, J.J., *Functionally Graded Materials: Manufacture, Properties, and Applications*, American Ceramic Society, 1997.

Graedel, T.E. and Allenby, B.R., *Design for Environment*, Prentice Hall, 1997.

Halevi, G., *Restructuring the Manufacturing Process: Applying the Matrix Method*, St. Lucie Press, 1998.

Hartley, J.R., and Okamoto, S., *Concurrent Engineering: Shortening Lead Times, Raising Quality, and Lowering Costs*, Productivity Press, 1998.

Kaufman, J.J, and Sato, Y., *Value Analysis Tear-Down: A New Process for Product Development*, Industrial Press and Society of Manufacturing Engineers, 2004.

Koenig, D.T., *Manufacturing Engineering*, 2nd ed., Taylor and Francis, 1994.

Kusiak, A. (ed.), *Concurrent Engineering: Automation, Tools and Techniques*, Wiley, 1993.

Lesco, J., *Industrial Design: Guide to Materials and Manufacturing*, Van Nostrand Reinhold, 1998.

Magrab, E.B., *Integrated Product and Process Design and Development: The Product Realization Process*, CRC Press, 1997.

Mahoney, R.M., *High-Mix, Low-Volume Manufacturing*, Wiley, 1992.

Mangonon, P.C., *The Principles of Materials Selection for Design*, Prentice Hall, 1999.

Mather, H., *Competitive Manufacturing*, 2nd ed., CRC Press, 1998.

Meyers, F.E., *Motion and Time Study for Lean Manufacturing*, Prentice Hall, 1998.

Miller, J.G., *Benchmarking Global Manufacturing*, Irwin, 1990.

Nevins, J.L. and Whitney, D.E., *Concurrent Design of Products and Processes*, McGraw-Hill, 1989.

Oleson, J.D., *Pathways to Agility: Mass Customization in Action*, Wiley, 1998.

Otto, K.N. and Wood, K.L., *Product Design: Techniques in Reverse Engineering and New Product Development*, Prentice Hall, 2001.

Ostwald, P.R., *Engineering Cost Estimating*, 3rd ed., Prentice Hall, 1992.

Paashuis, V., *The Organization of Integrated Product Development*, Springer, 1997.

Park, R.J., *Value Engineering: A Plan for Invention*, St. Lucie Press, 1999.

Park, S., *Robust Design and Analysis for Quality Engineering*, Chapman & Hall, 1997.

Prasad, B., *Concurrent Engineering Fundamentals*, 2 vols., Prentice Hall, 1995.

Pugh, S., Clausing, D., and Andrade, R. (eds.), *Creating Innovative Products Using Total Design*, Addison-Wesley, 1996.

Pugh, S., *Total Design: Integrated Methods for Successful Product Engineering*, Addison-Wesley, 1991.

Ranky, P.G., *An Introduction to Concurrent/Simultaneous Engineering: Methods, Tools, and Case Studies*, CIMWare USA, 1997.

Rhyder, R.F., *Manufacturing Process Design and Optimization*, Marcel Dekker, 1997.

Roosenburg, N.F.M. and Eekels, J., *Product Design: Fundamentals and Methods*, Wiley, 1995.

Shetty, D., *Design for Product Success*, Society of Manufacturing Engineers, 2002.

Shina, S.G. (ed.), *Successful Implementation of Concurrent Engineering Products and Processes*, Wiley, 1997.

Sims, E.R., Jr., *Precision Manufacturing Costing*, Marcel Dekker, 1995.

Stoll, H.W., *Product Design Methods and Practices*, Marcel Dekker, 1999.

Swift, K.G. and Booker, J.D., *Process Selection: From Design to Manufacture*, Wiley, 1997.

Tool and Manufacturing Engineers Handbook, 4th ed., Vol 5: *Manufacturing Engineering Management*, Society of Manufacturing Engineers, 1988.

_____, Vol. 6: *Design for Manufacturability*, Society of Manufacturing Engineers, 1992.

_____, Vol. 7: *Continuous Improvement*, Society of Manufacturing Engineers, 1993.

Ulrich, K.T. and Eppinger, S.D., *Product Design and Development*, McGraw-Hill, 1995.

Vollman, T.E., *Manufacturing Planning and Control Systems*, Irwin, 1997.

Walker, J.M. (ed.), *Handbook of Manufacturing Engineering*, Marcel Dekker, 1996.

Wang, B., *Integrated Product, Process, and Enterprise Design*, Chapman & Hall, 1997.

Wenzel, H., Hauschild, M., and Alting, L., *Environmental Assessment of Products*, Vol. 1, Chapman & Hall, 1997.

Wenzel, H. and Hauschild, M., *Environmental Assessment of Products*, Vol. 2, Chapman & Hall, 1997.

Winchell, W. (ed.), *Realistic Cost Estimating for Manufacturing*, 2nd ed., Society of Manufacturing Engineers, 1989.

Woeppel, M., *The Manufacturer's Guide to Implementing the Theory of Constraints*, St. Lucie Press, 1999.

Womack, J.P. and Jones, D.T., *Lean Thinking*, Simon & Schuster, 1996.

Wroblewski, A.J. and Vanka, S., *MaterialTool: A Selection Guide of Materials and Processes for Designers*, Prentice Hall, 1997.

REVIEW QUESTIONS

40.1 Describe the major considerations involved in selecting materials for products.

40.2 Why is a knowledge of commercially available shapes of materials important? Give five different examples.

40.3 What is meant by "manufacturing properties" of materials? Give three examples demonstrating the importance of this information.

40.4 Why is material substitution an important aspect of manufacturing engineering? Give five examples from your own experience or observations.

40.5 What factors are involved in the selection of manufacturing processes? Explain why they are important.

40.6 What is meant by manufacturing process capabilities? Select four different and specific manufacturing processes and describe their capabilities.

40.7 Is production quantity significant in process selection? Explain your answer.

40.8 Describe the various costs involved in manufacturing.

40.9 Explain the difference between direct-labor cost and indirect-labor cost.

40.10 List the advantages and disadvantages of outsourcing manufacturing activities to countries with low labor costs.

40.11 Describe your understanding of the following terms: (a) life expectancy, (b) life cycle engineering, and (c) sustainable manufacturing.

40.12 How is value defined? Explain the reasons behind it.

40.13 What factors are involved in cost justification?

40.14 What is the meaning and significance of the term "return on investment"? Explain.

QUALITATIVE PROBLEMS

40.15 Explain why the value of the scrap produced in a manufacturing process depends on the type of material and processes involved.

40.16 Comment on the magnitude and range of scrap shown in Table 40.6, and the reasons for it

40.17 Describe your observations concerning the information given in Table 40.5, and the reasons for it.

40.18 Other than the size of the machine, what factors are involved in the range of prices in each machine category shown in Table 40.9?

40.19 Explain how the high cost of some of the machinery listed in Table 40.9 can be justified.

40.20 Based on the topics covered in this book, explain the reasons for the relative positions of the curves shown in Fig. 40.2.

40.21 What factors are involved in the shape of the curve shown in Fig. 40.4? Explain.

40.22 Is it always desirable to purchase stock that is close to the final dimensions of a part to be manufactured? Explain why and give some examples.

40.23 What course of action would you take if the supply of a raw material selected for a product line becomes unreliable?

40.24 Describe the potential problems involved in reducing the quantity of materials in products. Give some examples.

40.25 Explain the reasons why there is a strong desire in industry to practice near-net-shape manufacturing.

40.26 Estimate the position of the following processes in Fig. 40.5: (a) centerless grinding, (b) electrochemical machining, (c) chemical milling, and (d) extrusion.

40.27 In Section 40.9, there is a breakdown of costs in today's manufacturing environment maintaining that design costs contribute only 5% to the total cost. Explain why this figure is reasonable.

SYNTHESIS, DESIGN, AND PROJECTS

40.28 As you can see, Table 40.8 on manufacturing processes includes only metals and their alloys. Based on the information given in this book and other sources, prepare a similar table for nonmetallic materials including ceramics, plastics, reinforced plastics, and both metal-matrix and ceramic-matrix composite materials.

40.29 Review Fig. I.4 in the General Introduction and present your thoughts concerning the two flow charts. Would you want to make any modifications, and if so, what would they be?

40.30 Over the years, numerous consumer products have become obsolete (or nearly so—such as rotary-dial telephones, analog radio tuners, turntables, and vacuum tubes), while many new products have entered the market. Make a comprehensive list of obsolete products and one of new products. Comment on the reasons for the changes you observe. Discuss how different manufacturing methods and systems have evolved in order to produce the new products.

40.31 Describe how you could reduce each of the costs involved in the manufacturing of products.

40.32 Make suggestions as to how to reduce the dependence of production time on surface finish (shown in Fig. 40.5).

40.33 There is a period between the time that an employee is hired and the time that the employee finishes with training during which the employee is paid and receives benefits but produces nothing. Where should such costs be placed among the categories given in this chapter?

40.34 Select three different products and make a survey of the changes in their prices over the past ten years. Discuss the possible reasons for the changes.

40.35 Figure 2.1a shows the shape of a typical tensile-test specimen having a round cross-section. Assuming that the starting material (stock) is a round rod and only one specimen is needed, discuss the processes and the machinery by which the specimen can be made, including their relative advantages and limitations. Describe how the process you selected can be changed for economical production as the number of specimens required increases.

40.36 Present your thoughts concerning the replacement of aluminum beverage cans with steel ones.

40.37 Table 40.4 listed several materials and their commercially available shapes. By contacting suppliers of the following materials, extend that list to include: (a) titanium, (b) superalloys, (c) lead, (d) tungsten, and (e) amorphous metals.

40.38 Select three different products commonly found in homes. State your opinions on (a) what materials were used in each product, (b) why they were chosen, (c) how the products were manufactured, and (d) why those particular processes were used.

40.39 Inspect the components under the hood of your automobile. Identify several parts that have been produced to net-shape or near-net-shape condition. Comment on the design and production aspects of these parts and on how the manufacturer achieved the near-net-shape condition.

40.40 Comment on the differences (if any) between the designs, the materials, and the processing and assembly methods used for making products (such as hand tools and ladders) for professional use and those for consumer use.

40.41 The capabilities of some machining processes are shown in Fig. 23.1. Inspect the various part shapes produced and suggest alternative manufacturing processes. Comment on the properties of materials that would influence your suggestions.

40.42 If the dimensions of the parts in Problem 40.38 were (a) two times larger or (b) five times larger, how different would your answer be? Explain your responses.

40.43 The cross-section of a jet engine is shown in Fig. 6.1. On the basis of the topics covered in this book, select any three individual components of such an engine and describe the materials and processes that you would use in making them in quantities of, say, 1000. Remember that these parts must be manufactured at minimum cost yet maintain their quality, integrity, and reliability.

40.44 Discuss the tradeoffs involved in selecting between the two materials for each of the applications listed here:

  a. steel or plastic paper clips

  b. forged or cast crankshafts

  c. forged or powder-metallurgy connecting rods

  d. plastic or sheet-metal light-switch plates

  e. metal or plastic intake manifolds

  f. sheet-metal or cast hubcaps

  g. steel or copper nails

  h. wood or metal handle for hammers

  i. steel or aluminum lawn chairs

Also, discuss the typical conditions to which these products are subjected in their normal use.

40.45 Discuss the manufacturing process (or processes) suitable for making the products listed in Problem 40.44. Explain whether they would require additional operations (such as coating, plating, heat treating, and finishing). If so, make recommendations and give the reasons for them.

40.46 Inspect some products around your home and describe how you would go about taking them completely apart quickly and recycling their components. Comment on their design regarding the ease with which they can be disassembled.

40.47 What products do you know of that would be very difficult to disassemble for recycling purposes?

40.48 Junction boxes for electrical wiring are available either in galvanized sheet metal or in injection-molded plastic (in colors such as white or blue). Considering all of their various features, describe your thoughts on how you would go about deciding which one to purchase, and explain why.

40.49 Discuss the factors that influence the choice between each of the following pairs of processes:

a. sand casting versus die casting of a fractional electric-motor housing

b. machining versus forming of a large gear

c. forging versus powder-metallurgy production of a gear

d. casting versus stamping a sheet-metal frying pan

e. making outdoor furniture from aluminum tubing versus cast iron

f. welding versus casting of machine-tool structures

g. thread rolling versus machining a bolt for a high-strength application

h. thermoforming a plastic versus molding a thermoset to make a fan blade for a household fan

40.50 through 40.58 Review the products illustrated below and describe your thoughts on (a) the materials that could be used, your own selection, and your reasons for it, (b) manufacturing processes and why you would select them, and (c) based on your review, any design changes that you would like to recommend.

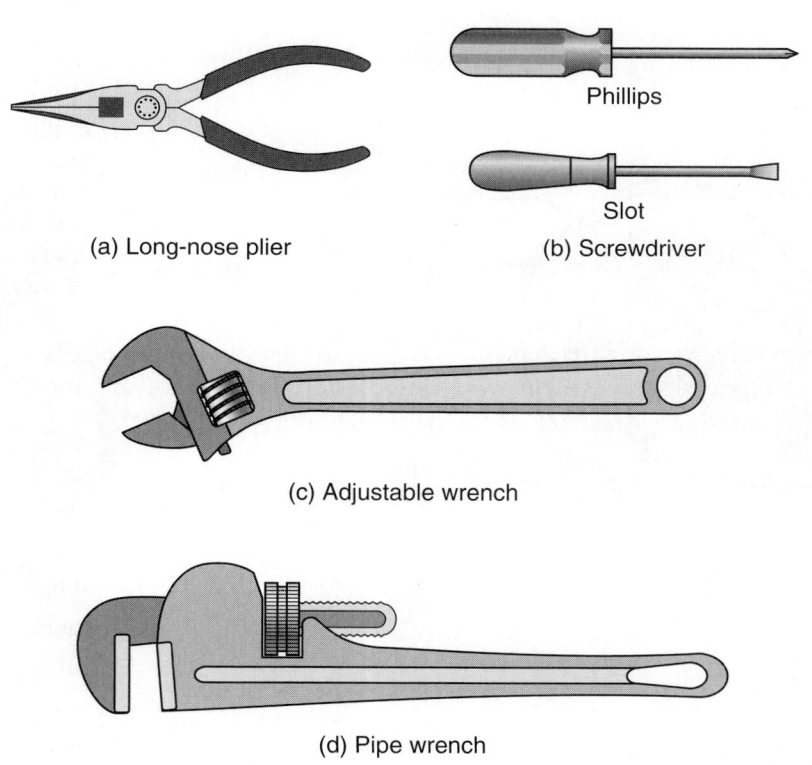

(a) Long-nose plier

(b) Screwdriver

Phillips

Slot

(c) Adjustable wrench

(d) Pipe wrench

FIGURE P40.50

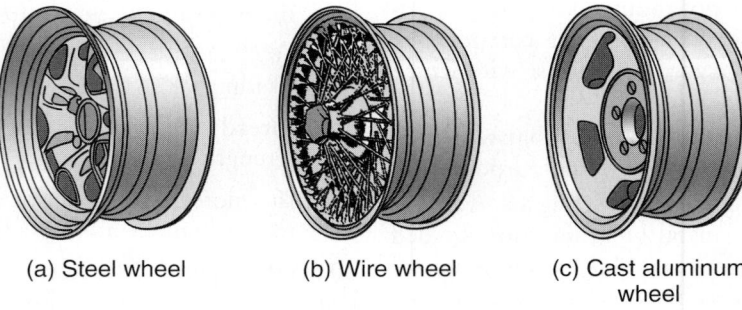

(a) Steel wheel   (b) Wire wheel   (c) Cast aluminum wheel

FIGURE P40.51

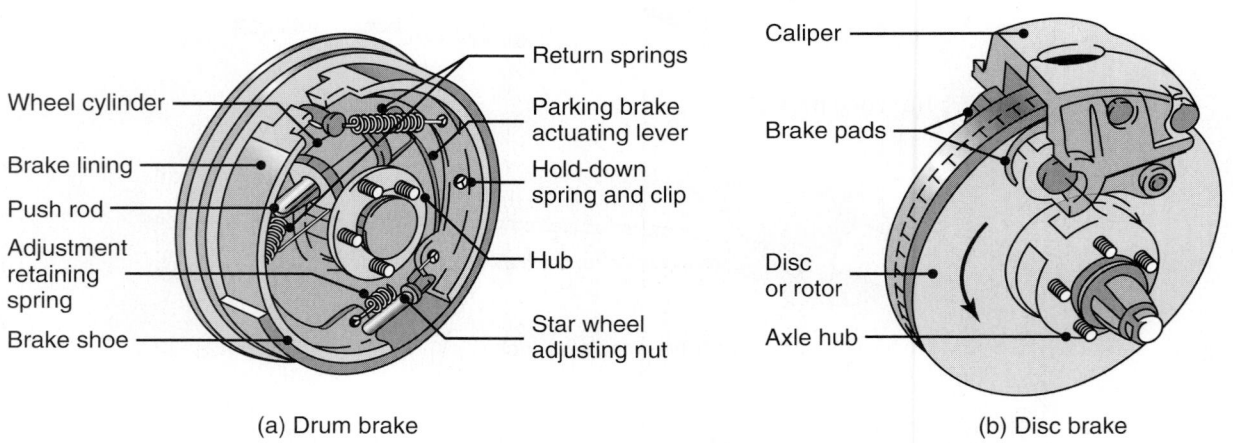

Wheel cylinder —   Return springs

  Parking brake actuating lever

Brake lining —

Push rod —   Hold-down spring and clip

Adjustment retaining spring

Brake shoe —   Hub

  Star wheel adjusting nut

Caliper —

Brake pads —

Disc or rotor

Axle hub —

(a) Drum brake   (b) Disc brake

FIGURE P40.52

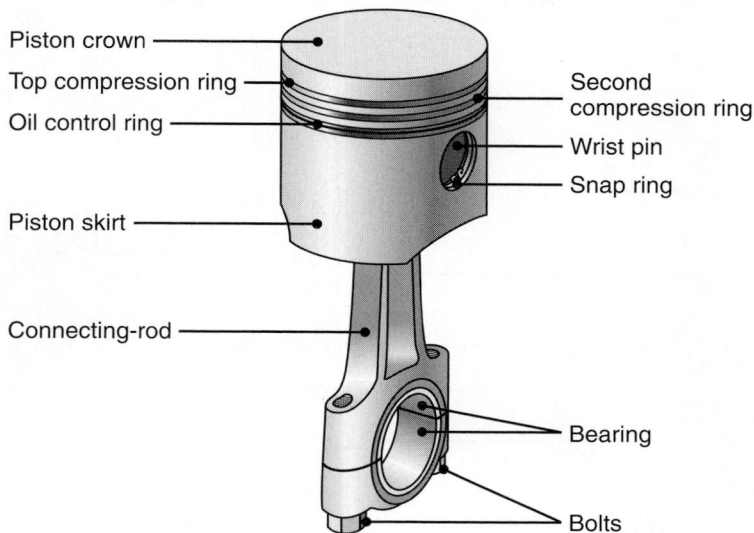

Piston crown
Top compression ring
Oil control ring
Piston skirt
Connecting-rod
Second compression ring
Wrist pin
Snap ring
Bearing
Bolts

FIGURE P40.53

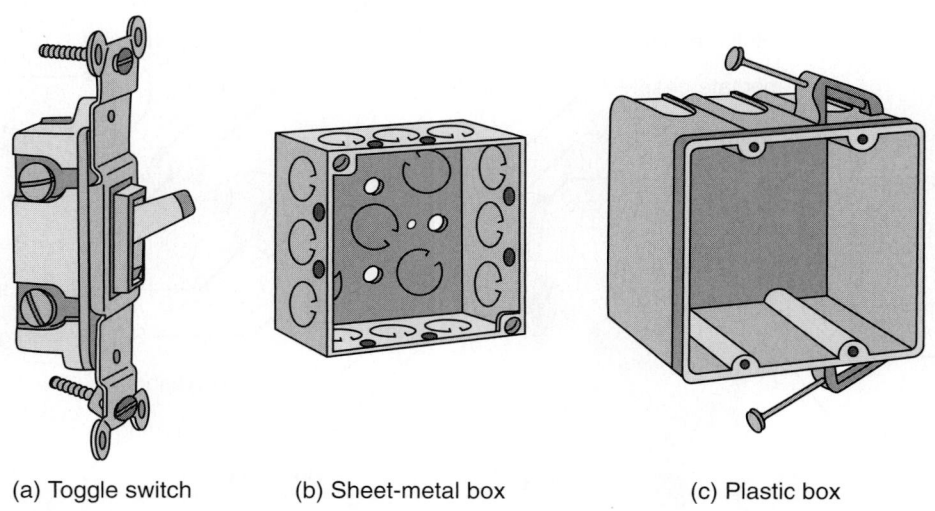

(a) Toggle switch   (b) Sheet-metal box   (c) Plastic box

FIGURE P40.54

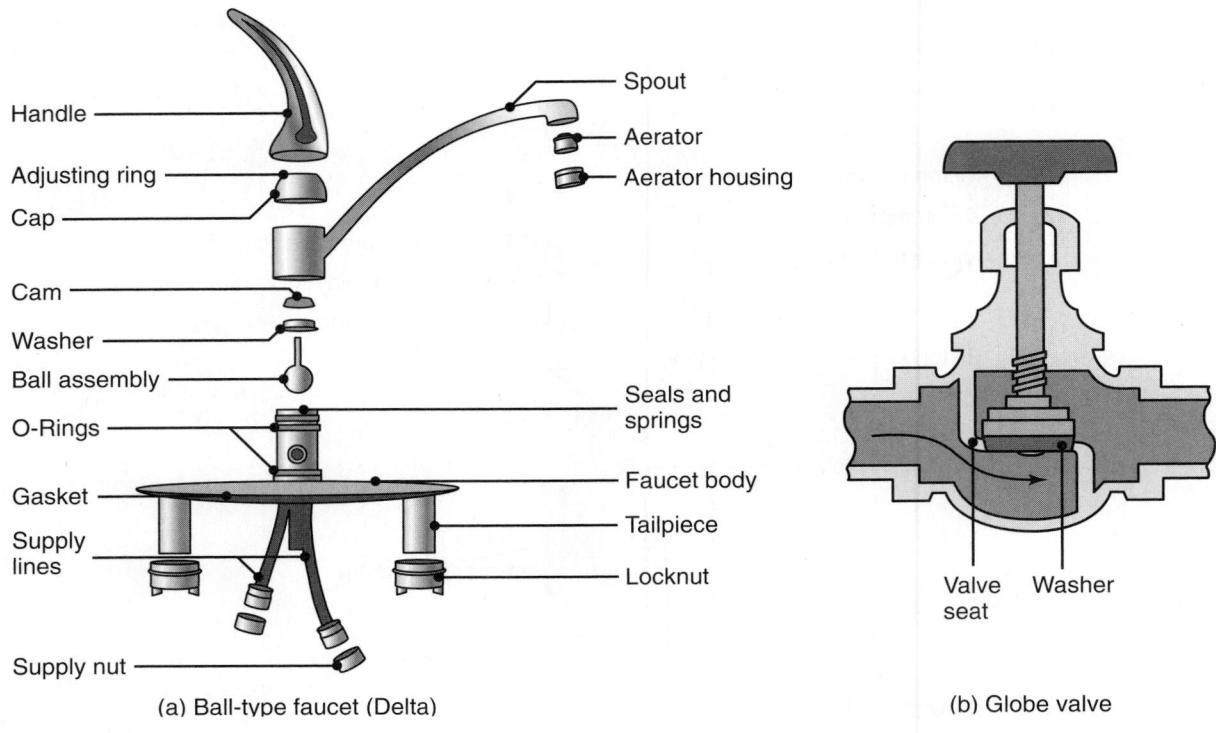

Handle

Adjusting ring

Cap

Cam

Washer

Ball assembly

O-Rings

Gasket

Supply lines

Supply nut

Spout

Aerator

Aerator housing

Seals and springs

Faucet body

Tailpiece

Locknut

(a) Ball-type faucet (Delta)

Valve seat

Washer

(b) Globe valve

FIGURE P40.55

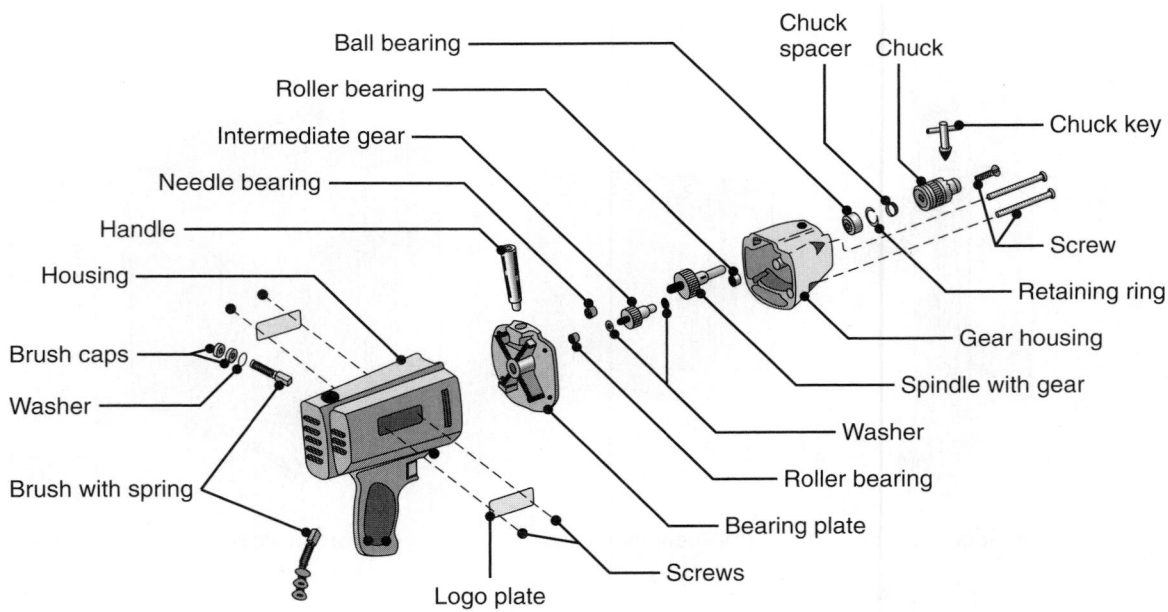

Ball bearing

Roller bearing

Intermediate gear

Needle bearing

Handle

Housing

Brush caps

Washer

Brush with spring

Logo plate

Chuck spacer

Chuck

Chuck key

Screw

Retaining ring

Gear housing

Spindle with gear

Washer

Roller bearing

Bearing plate

Screws

FIGURE P40.56

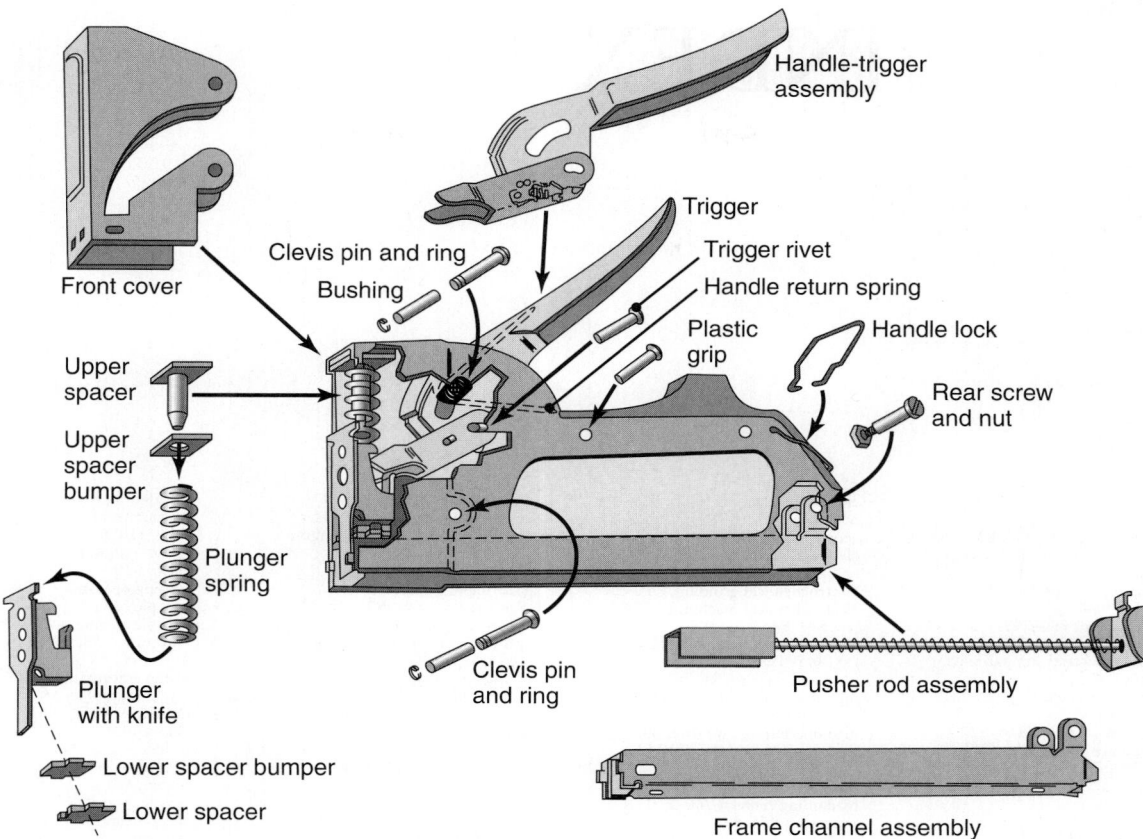

Handle-trigger assembly

Trigger

Trigger rivet

Handle return spring

Clevis pin and ring

Bushing

Plastic grip

Handle lock

Rear screw and nut

Front cover

Upper spacer

Upper spacer bumper

Plunger spring

Plunger with knife

Lower spacer bumper

Lower spacer

Clevis pin and ring

Pusher rod assembly

Frame channel assembly

FIGURE P40.57

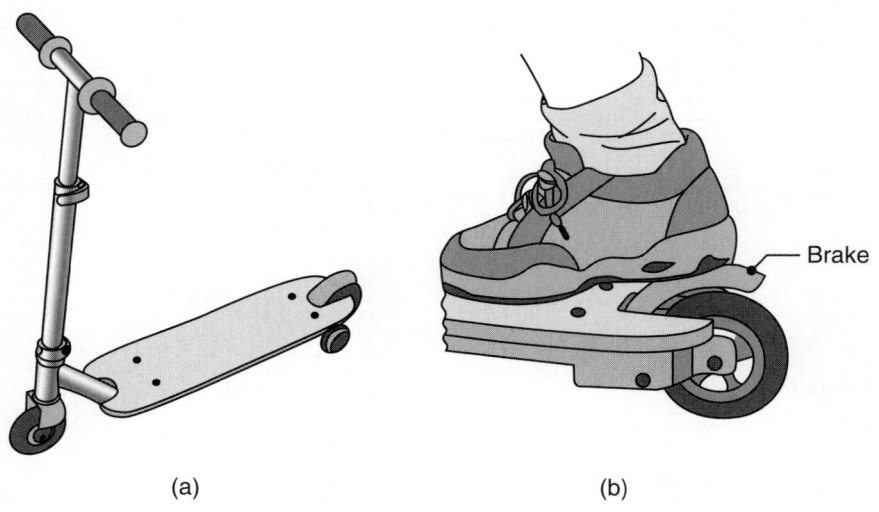

Brake

(a)

(b)

FIGURE P40.58

INDEX

REFERENCES TO VARIOUS TOPICS IN THIS BOOK

Design Considerations in Processing

Metal casting	Section 12.2
Forging	Section 14.6
Sheet-metal forming	Section 16.13
Powder metallurgy	Section 17.6
Ceramics shaping	Section 18.5
Polymers processing	Section 19.15
Machining	Various sections in Chapters 23 and 24
Abrasive processes	Section 26.5
Advanced machining	Various sections in Chapter 27
Joining processes	Various sections in Chapters 30 through 32

Material Properties

Tables 2.1, 2.2, and 2.3, and Figs. 2.4, 2.6, 2.8, 2.14, 2.15, 2.16, and 2.28
Tables 3.1, 3.2, and 3.3, and Figs. 3.1 and 3.2
Tables 5.2, 5.4, and 5.6
Tables 6.2 through 6.10
Tables 7.1 through 7.3
Tables 8.2 and 8.3
Table 9.2 and Figs. 9.3, 9.5, and 9.7
Table 10.1
Table 11.3
Tables 12.3, 12.4, and 12.5, and Fig. 12.4
Tables 16.3 and 16.4
Tables 17.3, 17.4, and 17.5, and Fig. 17.10
Table 20.2
Tables 22.1, 22.2, and 22.5, and Figs. 22.1 and 22.9
Table 26.1
Table 32.3

Manufacturing Characteristics of Materials

Table I.3
Table 5.6
Table 6.4
Table 8.3
Table 10.1
Tables 12.2 and 12.6
Table 14.3
Tables 16.2 and 16.3, and Fig. 16.34
Section 21.7
Fig. 22.2
Section 30.9.2

Capabilities of Manufacturing Processes

Tables 11.1 and 11.2
Table III.1
Table 14.1
Table 16.1
Section 17.7 and Fig. 17.14
Table 18.1
Tables 19.1 and 19.2
Table 20.1
Tables 23.1, 23.6, 23.8, and 23.10
Tables 26.3 and 26.4
Table 27.1
Tables 28.1 and 28.2
Table 29.1
Table VI.1
Table 30.1
Table 32.4
Table 34.1
Tables 40.2 and 40.3

Dimensional Tolerances and Surface Finish

Table 11.2
Table 23.1 and Figs. 23.13 and 23.14
Fig. 27.4

General Costs and Economics

Table I.4
Table 6.1
Section 12.6
Section 14.9
Section 16.15
Section 17.8 and Table 17.6
Section 19.16
Fig. 24.34
Section 25.8
Section 26.9 and Fig. 26.34
Section 27.10
Section 31.8
Section 32.7
Tables 40.2, 40.5, and 40.9